Win the complete library of RSMeans Online data!

ACCURATE   EFFICIENT   CUSTOMIZABLE

RSMeans Online
FROM THE GORDIAN GROUP

D1709874

Register your book below to receive
your **free** quarterly updates.
Plus you will be entered into a
quarterly drawing to win the **complete**
RSMeans Online library of 2016 data!

Be sure to keep up-to-date in 2016!

Fill out the card below for RSMeans' free quarterly updates,
as well as a chance to win the complete RSMeans Online library.
Please provide your name, address, and email below and return
this card by mail, or register online:

info.thegordiangroup.com/2016updates.html

Name _____

Email _____ Title _____

Company _____

Street _____

City/Town _____ State/Prov. _____ Zip/Postal Code _____

RSMeans
FROM THE GORDIAN GROUP

Win the complete library of RSMeans Online data!

ACCURATE   EFFICIENT   CUSTOMIZABLE

RSMeansOnline
FROM THE GORDIAN GROUP®

Register your book below to receive your **free** quarterly updates.
Plus you will be entered into a quarterly drawing to win the **complete** RSMeans Online library of 2016 data!

Commercial Renovation Cost Data

RSMeans
FROM THE GORDIAN GROUP®

2016
37th annual edition

Bob Mewis, CCP, Senior Editor

Engineering Director
Bob Mewis, CCP (*1, 2, 4, 13, 14, 31, 32, 33, 34, 35, 41, 44, 46*)

Contributing Editors
Christopher Babbitt
Adrian C. Charest, PE
Cheryl Elsmore
Wafaa Hamitou
Joseph Kelble
Charles Kibbee
Robert J. Kuchta (*8*)
Michael Landry
Thomas Lane (*6, 7*)
Genevieve Medeiros
Elisa Mello
Chris Morris (*26, 27, 28, 48*)

Melville J. Mossman, PE (*21, 22, 23*)
Marilyn Phelan, AIA (*9, 10, 11, 12*)
Stephen C. Plotner (*3, 5*)
Stephen Rosenberg
Kevin Souza
Keegan Spraker
Tim Tonello
David Yazbek

Vice President Data & Engineering
Chris Anderson

Product Manager
Andrea Sillah

Production Manager
Debbie Panarelli

Production
Sharon Larsen
Jonathan Forgit
Sheryl Rose
Mary Lou Geary

Technical Support
Gary L. Hoitt
Kathryn S. Rodriguez

Cover Design
Blaire Gaddis

Numbers in italics are the divisional responsibilities for each editor. Please contact the designated editor directly with any questions.

RSMeans
Construction Publishers & Consultants
1099 Hingham Street, Suite 201
Rockland, MA 02370
United States of America
1-877-742-4719
www.RSMeans.com

Copyright 2015 by RSMeans
All rights reserved.
Cover photo ©
Audrius Priveda/iStock/Thinkstock

Printed in the United States of America
ISSN 2161-4628
ISBN 978-1-943215-02-7

0046 $219.99 per copy (in United States)
Price is subject to change without prior notice.

Related RSMeans Products and Services

This data set is aimed primarily at commercial renovation projects. The information contained within has provisions for estimating small jobs with small quantities of work and large jobs that encompass major renovations. As with all RSMeans cost data, with reasonable exercise of judgment, this data can be used for estimating jobs with a wide range of costs.

The engineers at RSMeans suggest the following products and services as companion information resources to *RSMeans Commercial Renovation Cost Data*:

Construction Cost Data
Facilities Construction Cost Data 2016
Facilities Maintenance & Repair Cost Data 2016

Reference Books
Estimating Building Costs
RSMeans Estimating Handbook
Green Building: Project Planning & Estimating
How to Estimate with RSMeans Data
Plan Reading & Material Takeoff
Project Scheduling & Management for Construction
Universal Design for Style, Comfort & Safety

Seminars and In-House Training
Facilities Construction Estimating
RSMeans Online® Training
Practical Project Management for Construction Professionals
Scheduling with MSProject for Construction Professionals
Mechanical & Electrical Estimating

RSMeans Online Store
Visit RSMeans at www.RSMeans.com for the most reliable and current resources available on the market. Learn more about our more than 20 data sets available in Online, Book, eBook, and CD formats. Our library of reference books is also available, along with professional development seminars aimed at improving cost estimating, project management, administration, and facilities management skills.

RSMeans Electronic Data
Receive the most up-to-date cost information with RSMeans Online. This web-based service is quick, intuitive, and easy to use, giving you instant access to RSMeans' comprehensive database. Learn more at: **www.RSMeans.com/Online**.

RSMeans Custom Solutions
Building owners, facility managers, building product manufacturers, attorneys, and even insurance firms across the public and private sectors have engaged in RSMeans' custom solutions to solve their estimating needs including:

Custom Cost Engineering Solutions
Knowing the lifetime maintenance cost of your building, how much it will cost to build a specific building type in different locations across the country and globe, and if the estimate from which you are basing a major decision is accurate are all imperative for building owners and their facility managers.

Market & Custom Analytics
Taking a construction product to market requires accurate market research in order to make critical business development decisions. Once the product goes to market, one way to differentiate your product from the competition is to make any cost saving claims you have to offer.

Third-Party Legal Resources
Natural disasters can lead to reconstruction and repairs just as new developments and expansions can lead to issues of eminent domain. In these cases where one party must pay, it is plausible for legal disputes over costs and estimates to arise.

Construction Costs for Software Applications
More than 25 unit price and assemblies cost databases are available through a number of leading estimating and facilities management software partners. For more information, see "Other RSMeans Products and Services" at the back of this publication.

RSMeans data is also available to federal, state, and local government agencies as multi-year, multi-seat licenses.

For procurement construction cost data such as Job Order Contracting and Change Order Management please refer to The Gordian Group's procurement solutions.

For information on our current partners, call 1-877-742-4719.

Table of Contents

Foreword

Who We Are

Since 1942, RSMeans has delivered construction cost estimating information and consulting throughout North America. In 2014, RSMeans was acquired by The Gordian Group, combining two industry-leading construction cost databases. Through the RSMeans line of products and services, The Gordian Group provides innovative construction cost estimating data to organizations pursuing efficient and effective construction planning, estimating, procurement, and information solutions.

Our Offerings

With RSMeans' construction cost estimating data, contractors, architects, engineers, facility owners, and managers can utilize the most up-to-date data available in many formats to meet their specific planning and estimating needs.

When you purchase information from RSMeans, you are, in effect, hiring the services of a full-time staff of construction and engineering professionals.

Our thoroughly-experienced and highly-qualified staff works daily to collect, analyze, and disseminate comprehensive cost information to meet your needs.

These staff members have years of practical construction experience and engineering training prior to joining the firm. As a result, you can count on them not only for accurate cost figures, but also for additional background reference information that will help you create a realistic estimate.

The RSMeans organization is equipped to help you solve construction problems through its variety of data solutions.

Access our comprehensive database electronically with RSMeans Online. Quick, intuitive, easy to use, and updated continuously throughout the year, RSMeans Online is the most accurate and up-to-date cost estimating data.

This up-to-date, accurate, and localized data can be leveraged for a broad range of applications, such as Custom Cost Engineering solutions, Market and Custom Analytics, and Third-Party Legal Resources.

To ensure you are getting the most from your cost estimating data, we also offer a myriad of reference guides, training, and professional seminars (learn more at **www.RSMeans.com/Learn**).

In short, RSMeans can provide you with the tools and expertise for developing accurate and dependable construction estimates and budgets in a variety of ways.

Our Commitment

Today at RSMeans, we do more than talk about the quality of our data and the usefulness of the information. We stand behind all of our data—from historical cost indexes to construction materials and techniques—to craft current costs and predict future trends.

If you have any questions about our products or services, please call us toll-free at 1-877-742-4719. You can also visit our website at: **www.RSMeans.com**.

How the Cost Data Is Built: An Overview

A Powerful Construction Tool

You now have one of the most powerful construction tools available today. A successful project is built on the foundation of an accurate and dependable estimate. This tool will enable you to construct such an estimate.

For the casual user the information is designed to be:

- quickly and easily understood so you can get right to your estimate.
- filled with valuable information so you can understand the necessary factors that go into a cost estimate.

For the professional user, the information is designed to be:

- a handy reference that can be quickly referred to for key costs.
- a comprehensive, fully reliable source of current construction costs and productivity rates so you'll be prepared to estimate any project.
- a source for preliminary project costs, product selections, and alternate materials and methods.

To meet all of these requirements, we have organized the information into the following clearly defined sections.

Estimating with RSMeans Unit Price Cost Data*

Please refer to these steps for guidance on completing an estimate using RSMeans unit price cost data.

How to Use the Information: The Details

This section contains an in-depth explanation of how the information is arranged and how you can use it to determine a reliable construction cost estimate. It includes how we develop our cost figures and how to prepare your estimate.

Unit Prices*

All cost data has been divided into 50 divisions according to the MasterFormat system of classification and numbering.

Assemblies*

The cost data in this section has been organized in an "Assemblies" format. These assemblies are the functional elements of a building and are arranged according to the 7 elements of the UNIFORMAT II classification system. For a complete explanation of a typical "Assembly", see "How RSMeans Assemblies Data Works."

Residential Models*

Model buildings for four classes of construction—economy, average, custom, and luxury—are developed and shown with complete costs per square foot.

Commercial/Industrial/ Institutional Models*

This section contains complete costs for 77 typical model buildings expressed as costs per square foot.

Green Commercial/Industrial/ Institutional Models*

This section contains complete costs for 25 green model buildings expressed as costs per square foot.

References*

This section includes information on Equipment Rental Costs, Crew Listings, Historical Cost Indexes, City Cost Indexes, Location Factors, Reference Tables, and Change Orders, as well as a listing of abbreviations.

- **Equipment Rental Costs:** Included are the average costs to rent and operate hundreds of pieces of construction equipment.
- **Crew Listings:** This section lists all the crews referenced in the cost data. A crew is composed of more than one trade classification and/or the addition of power equipment to any trade classification. Power equipment is included in the cost of the crew. Costs are shown both with bare labor rates and with the installing contractor's overhead and profit added. For each, the total crew cost per eight-hour day and the composite cost per labor-hour are listed.

- **Historical Cost Indexes:** These indexes provide you with data to adjust construction costs over time.
- **City Cost Indexes:** All costs in this data set are U.S. national averages. Costs vary by region. You can adjust for this by CSI Division to over 700 locations throughout the U.S. and Canada by using this data.
- **Location Factors:** You can adjust total project costs to over 900 locations throughout the U.S. and Canada by using the weighted number, which applies across all divisions.
- **Reference Tables:** At the beginning of selected major classifications in the Unit Prices are reference numbers indicators. These numbers refer you to related information in the Reference Section. In this section, you'll find reference tables, explanations, and estimating information that support how we develop the unit price data, technical data, and estimating procedures.
- **Change Orders:** This section includes information on the factors that influence the pricing of change orders.
- **Abbreviations:** A listing of abbreviations used throughout this information, along with the terms they represent, is included.

Index (printed versions only)

A comprehensive listing of all terms and subjects will help you quickly find what you need when you are not sure where it occurs in MasterFormat.

Conclusion

This information is designed to be as comprehensive and easy to use as possible.

The Construction Specifications Institute (CSI) and Construction Specifications Canada (CSC) have produced the 2014 edition of MasterFormat®, a system of titles and numbers used extensively to organize construction information.

All unit prices in the RSMeans cost data are now arranged in the 50-division MasterFormat® 2014 system.

---

* Not all information is available in all data sets

Note: The material prices in RSMeans cost data are "contractor's prices." They are the prices that contractors can expect to pay at the lumberyards, suppliers'/distributors' warehouses, etc. Small orders of specialty items would be higher than the costs shown, while very large orders, such as truckload lots, would be less. The variation would depend on the size, timing, and negotiating power of the contractor. The labor costs are primarily for new construction or major renovation rather than repairs or minor alterations. With reasonable exercise of judgment, the figures can be used for any building work.

Estimating with RSMeans Unit Prices

Following these steps will allow you to complete an accurate estimate using RSMeans Unit Price cost data.

1. Scope Out the Project

- Think through the project and identify the CSI Divisions needed in your estimate.
- Identify the individual work tasks that will need to be covered in your estimate.
- The Unit Price data has been divided into 50 Divisions according to the CSI MasterFormat® 2014.
- In printed versions, the Unit Price Section Table of Contents on page 1 may also be helpful when scoping out your project.
- Experienced estimators find it helpful to begin with Division 2 and continue through completion. Division 1 can be estimated after the full project scope is known.

2. Quantify

- Determine the number of units required for each work task that you identified.
- Experienced estimators include an allowance for waste in their quantities. (Waste is not included in RSMeans Unit Price line items unless otherwise stated.)

3. Price the Quantities

- Use the search tools available to locate individual Unit Price line items for your estimate.
- Reference Numbers indicated within a Unit Price section refer to additional information that you may find useful.
- The crew indicates who is performing the work for that task. Crew codes are expanded in the Crew Listings in the Reference Section to include all trades and equipment that comprise the crew.
- The Daily Output is the amount of work the crew is expected to do in one day.
- The Labor-Hours value is the amount of time it will take for the crew to install one unit of work.
- The abbreviated Unit designation indicates the unit of measure upon which the crew, productivity, and prices are based.
- Bare Costs are shown for materials, labor, and equipment needed to complete the Unit Price line item. Bare costs do not include waste, project overhead, payroll insurance, payroll taxes, main office overhead, or profit.
- The Total Incl O&P cost is the billing rate or invoice amount of the installing contractor or subcontractor who performs the work for the Unit Price line item.

4. Multiply

- Multiply the total number of units needed for your project by the Total Incl O&P cost for each Unit Price line item.
- Be careful that your take off unit of measure matches the unit of measure in the Unit column.
- The price you calculate is an estimate for a completed item of work.
- Keep scoping individual tasks, determining the number of units required for those tasks, matching each task with individual Unit Price line items, and multiplying quantities by Total Incl O&P costs.
- An estimate completed in this manner is priced as if a subcontractor, or set of subcontractors, is performing the work. The estimate does not yet include Project Overhead or Estimate Summary components such as general contractor markups on subcontracted work, general contractor office overhead and profit, contingency, and location factors.

5. Project Overhead

- Include project overhead items from Division 1-General Requirements.
- These items are needed to make the job run. They are typically, but not always, provided by the general contractor. Items include, but are not limited to, field personnel, insurance, performance bond, permits, testing, temporary utilities, field office and storage facilities, temporary scaffolding and platforms, equipment mobilization and demobilization, temporary roads and sidewalks, winter protection, temporary barricades and fencing, temporary security, temporary signs, field engineering and layout, final cleaning, and commissioning.
- Each item should be quantified and matched to individual Unit Price line items in Division 1, then priced and added to your estimate.
- An alternate method of estimating project overhead costs is to apply a percentage of the total project cost, usually 5% to 15% with an average of 10% (see General Conditions).
- Include other project related expenses in your estimate such as:
  - Rented equipment not itemized in the Crew Listings
  - Rubbish handling throughout the project (see section 02 41 19.19)

6. Estimate Summary

- Include sales tax as required by laws of your state or county.
- Include the general contractor's markup on self-performed work, usually 5% to 15% with an average of 10%.
- Include the general contractor's markup on subcontracted work, usually 5% to 15% with an average of 10%.
- Include the general contractor's main office overhead and profit:
  - RSMeans gives general guidelines on the general contractor's main office overhead (see section 01 31 13.60 and Reference Number R013113-50).
  - RSMeans gives no guidance on the general contractor's profit.
  - Markups will depend on the size of the general contractor's operations, projected annual revenue, the level of risk, and the level of competition in the local area and for this project in particular.
- Include a contingency, usually 3% to 5%, if appropriate.
- Adjust your estimate to the project's location by using the City Cost Indexes or the Location Factors in the Reference Section:
  - Look at the rules in "How to Use the City Cost Indexes" to see how to apply the Indexes for your location.
  - When the proper Index or Factor has been identified for the project's location, convert it to a multiplier by dividing it by 100, then multiply that multiplier by your estimated total cost. The original estimated total cost will now be adjusted up or down from the national average to a total that is appropriate for your location.

Editors' Note:
We urge you to spend time reading and
understanding the supporting material.
An accurate estimate requires experience,
knowledge, and careful calculation. The more
you know about how we at RSMeans developed
the data, the more accurate your estimate will
be. In addition, it is important to take into
consideration the reference material such as
Equipment Listings, Crew Listings, City Cost
Indexes, Location Factors, and Reference
Tables.

How to Use the Cost Data: The Details

What's Behind the Numbers? The Development of Cost Data

The staff at RSMeans continually monitors developments in the construction industry in order to ensure reliable, thorough, and up-to-date cost information. While overall construction costs may vary relative to general economic conditions, price fluctuations within the industry are dependent upon many factors. Individual price variations may, in fact, be opposite to overall economic trends. Therefore, costs are constantly tracked, and complete updates are performed yearly. Also, new items are frequently added in response to changes in materials and methods.

Costs—$ (U.S.)

All costs represent U.S. national averages and are given in U.S. dollars. The RSMeans City Cost Indexes (CCI) can be used to adjust costs to a particular location. The CCI for Canada can be used to adjust U.S. national averages to local costs in Canadian dollars. No exchange rate conversion is necessary because it has already been factored in.

[G] The processes or products identified by the green symbol in our publications have been determined to be environmentally responsible and/or resource-efficient solely by the RSMeans engineering staff. The inclusion of the green symbol does not represent compliance with any specific industry association or standard.

Material Costs

The RSMeans staff contacts manufacturers, dealers, distributors, and contractors all across the U.S. and Canada to determine national average material costs. If you have access to current material costs for your specific location, you may wish to make adjustments to reflect differences from the national average. Included within material costs are fasteners for a normal installation. RSMeans engineers use manufacturers' recommendations, written specifications, and/or standard construction practices for size and spacing of fasteners. Adjustments to material costs may be required for your specific application or location. The manufacturer's warranty is assumed. Extended warranties are not included in the material costs. **Material costs do not include sales tax.**

Labor Costs

Labor costs are based upon a mathematical average of trade-specific wages in 30 major U.S. cities. The type of wage (union, open shop, or residential) is identified on the inside back cover of printed publications or is selected by the estimator when using the electronic products. Markups for the wages can also be found on the inside back cover of printed publications and/or under the labor references found in the electronic products.

- If wage rates in your area vary from those used, or if rate increases are expected within a given year, labor costs should be adjusted accordingly.

Labor costs reflect productivity based on actual working conditions. In addition to actual installation, these figures include time spent during a normal weekday on tasks, such as material receiving and handling, mobilization at site, site movement, breaks, and cleanup.

Productivity data is developed over an extended period so as not to be influenced by abnormal variations and reflects a typical average.

Equipment Costs

Equipment costs include not only rental, but also operating costs for equipment under normal use. The operating costs include parts and labor for routine servicing, such as repair and replacement of pumps, filters, and worn lines. Normal operating expendables, such as fuel, lubricants, tires, and electricity (where applicable), are also included. Extraordinary operating expendables with highly variable wear patterns, such as diamond bits and blades, are excluded. These costs are included under materials. Equipment rental rates are obtained from industry sources throughout North America—contractors, suppliers, dealers, manufacturers, and distributors.

Rental rates can also be treated as reimbursement costs for contractor-owned equipment. Owned equipment costs include depreciation, loan payments, interest, taxes, insurance, storage, and major repairs.

Equipment costs do not include operators' wages.

Equipment Cost/Day—The cost of equipment required for each crew is included in the Crew Listings in the Reference Section (small tools that are considered essential everyday tools are not listed out separately). The Crew Listings itemize specialized tools and heavy equipment along with labor trades. The daily cost of itemized equipment included in a crew is based on dividing the weekly bare rental rate by 5 (number of working days per week), then adding the hourly operating cost times 8 (the number of hours per day). This Equipment Cost/Day is shown in the last column of the Equipment Rental Costs in the Reference Section.

Mobilization/Demobilization—The cost to move construction equipment from an equipment yard or rental company to the job site and back again is not included in equipment costs. Mobilization (to the site) and demobilization (from the site) costs can be found in the Unit Price Section. If a piece of equipment is already at the job site, it is not appropriate to utilize mob./demob. costs again in an estimate.

Overhead and Profit

Total Cost including O&P for the installing contractor is shown in the last column of the Unit Price and/or Assemblies. This figure is the sum of the bare material cost plus 10% for profit, the bare labor cost plus total overhead and profit, and the bare equipment cost plus 10% for profit. Details for the calculation of overhead and profit on labor are shown on the inside back cover of the printed product and in the Reference Section of the electronic product.

General Conditions

Cost data in this data set is presented in two ways: Bare Costs and Total Cost including O&P (Overhead and Profit). General Conditions, or General Requirements, of the contract should also be added to the Total Cost including O&P when applicable. Costs for General Conditions are listed in Division 1 of the Unit Price Section and in the Reference Section.

General Conditions for the installing contractor may range from 0% to 10% of the Total Cost including O&P. For the general or prime contractor, costs for General Conditions may range from 5% to 15% of the Total Cost including O&P, with a figure of 10% as the most typical allowance. If applicable, the Assemblies and Models sections use costs that include the installing contractor's overhead and profit (O&P).

Factors Affecting Costs

Costs can vary depending upon a number of variables. Here's a listing of some factors that affect costs and points to consider.

Quality—The prices for materials and the workmanship upon which productivity is based represent sound construction work. They are also in line with industry standard and manufacturer specifications and are frequently used by federal, state, and local governments.

Overtime—We have made no allowance for overtime. If you anticipate premium time or work beyond normal working hours, be sure to make an appropriate adjustment to your labor costs.

Productivity—The productivity, daily output, and labor-hour figures for each line item are based on working an eight-hour day in daylight hours in moderate temperatures. For work that extends beyond normal work hours or is performed under adverse conditions, productivity may decrease.

Size of Project—The size, scope of work, and type of construction project will have a significant impact on cost. Economies of scale can reduce costs for large projects. Unit costs can often run higher for small projects.

Location—Material prices are for metropolitan areas. However, in dense urban areas, traffic and site storage limitations may increase costs. Beyond a 20-mile radius of metropolitan areas, extra trucking or transportation charges may also increase the material costs slightly. On the other hand, lower wage rates may be in effect. Be sure to consider both of these factors when preparing an estimate, particularly if the job site is located in a central city or remote rural location. In addition, highly specialized subcontract items may require travel and per-diem expenses for mechanics.

Other Factors—

- season of year
- contractor management
- weather conditions
- local union restrictions
- building code requirements
- availability of:
  - adequate energy
  - skilled labor
  - building materials
- owner's special requirements/restrictions
- safety requirements
- environmental considerations
- access

Unpredictable Factors—General business conditions influence "in-place" costs of all items. Substitute materials and construction methods may have to be employed. These may affect the installed cost and/or life cycle costs. Such factors may be difficult to evaluate and cannot necessarily be predicted on the basis of the job's location in a particular section of the country. Thus, where these factors apply, you may find significant but unavoidable cost variations for which you will have to apply a measure of judgment to your estimate.

Rounding of Costs

In printed publications only, all unit prices in excess of $5.00 have been rounded to make them easier to use and still maintain adequate precision of the results.

How Subcontracted Items Affect Costs

A considerable portion of all large construction jobs is usually subcontracted. In fact, the percentage done by subcontractors is constantly increasing and may run over 90%. Since the workers employed by these companies do nothing else but install their particular product, they soon become experts in that line. The result is, installation by these firms is accomplished so efficiently that the total in-place cost, even adding the general contractor's overhead and profit, is no more, and often less, than if the principal contractor had handled the installation. Companies that deal with construction specialties are anxious to have their product perform well and, consequently, the installation will be the best possible.

Contingencies

The allowance for contingencies generally provides for unforeseen construction difficulties. On alterations or repair jobs, 20% is not too much. If drawings are final and only field contingencies are being considered, 2% or 3% is probably sufficient, and often nothing needs to be added. Contractually, changes in plans will be covered by extras. The contractor should consider inflationary price trends and possible material shortages during the course of the job. These escalation factors are dependent upon both economic conditions and the anticipated time between the estimate and actual construction. If drawings are not complete or approved, or a budget cost is wanted, it is wise to add 5% to 10%. Contingencies, then, are a matter of judgment.

Important Estimating Considerations

The productivity, or daily output, of each craftsman or crew assumes a well-managed job where tradesmen with the proper tools and equipment, along with the appropriate construction materials, are present. Included are daily set-up and cleanup time, break time, and plan layout time. Unless otherwise indicated, time for material movement on site (for items that can be transported by hand) of up to 200' into the building and to the first or second floor is also included. If material has to be transported by other means, over greater distances, or to higher floors, an additional allowance should be considered by the estimator.

While horizontal movement is typically a sole function of distances, vertical transport introduces other variables that can significantly impact productivity. In an occupied building, the use of elevators (assuming access, size, and required protective measures are acceptable) must be understood at the time of the estimate. For new construction, hoist wait and cycle times can easily be 15 minutes and may result in scheduled access extending beyond the normal work day. Finally, all vertical transport will impose strict weight limits likely to preclude the use of any motorized material handling.

The productivity, or daily output, also assumes installation that meets manufacturer/ designer/standard specifications. A time allowance for quality control checks, minor adjustments, and any task required to ensure the proper function or operation is also included. For items that require connections to services, time is included for positioning, leveling, securing the unit, and for making all the necessary connections (and start up where applicable), ensuring a complete installation. Estimating of the services themselves (electrical, plumbing, water, steam, hydraulics, dust collection, etc.) is separate.

In some cases, the estimator must consider the use of a crane and an appropriate crew for the installation of large or heavy items.

For those situations where a crane is not included in the assigned crew and as part of the line item cost, equipment rental costs, mobilization and demobilization costs, and operator and support personnel costs must be considered.

Labor-Hours

The labor-hours expressed in this publication are derived by dividing the total daily labor-hours for the crew by the daily output. Based on average installation time and the assumptions listed above, the labor-hours include: direct labor, indirect labor, and nonproductive time. A typical day for a craftsman might include, but is not limited to:

- Direct Work
  - ☐ Measuring and layout
  - ☐ Preparing materials
  - ☐ Actual installation
  - ☐ Quality assurance/quality control
- Indirect Work
  - ☐ Reading plans or specifications
  - ☐ Preparing space
  - ☐ Receiving materials
  - ☐ Material movement
  - ☐ Giving or receiving instruction
  - ☐ Miscellaneous
- Non-Work
  - ☐ Chatting
  - ☐ Personal issues
  - ☐ Breaks
  - ☐ Interruptions (i.e., sickness, weather, material or equipment shortages, etc.)

If any of the items for a typical day do not apply to the particular work or project situation, the estimator should make any necessary adjustments.

Final Checklist

Estimating can be a straightforward process provided you remember the basics. Here's a checklist of some of the steps you should remember to complete before finalizing your estimate.

Did you remember to:

- factor in the City Cost Index for your locale?
- take into consideration which items have been marked up and by how much?
- mark up the entire estimate sufficiently for your purposes?
- read the background information on techniques and technical matters that could impact your project time span and cost?
- include all components of your project in the final estimate?
- double check your figures for accuracy?
- call RSMeans if you have any questions about your estimate or the data you've used? Remember, RSMeans stands behind its information. If you have any questions about your estimate, about the costs you've used from our data, or even about the technical aspects of the job that may affect your estimate, feel free to call the RSMeans editors at 1-877-742-4719.

Free Quarterly Updates

Stay up-to-date throughout the year with RSMeans' free cost data updates four times a year. Sign up online to make sure you have access to the newest data. Every quarter we provide the city cost adjustment factors for hundreds of cities and key materials.

Register at: **http://info.TheGordianGroup.com/ 2016Updates.html**.

Using Minimum Labor/Equipment Charges for Small Quantities

Estimating small construction or repair tasks often creates situations in which the quantity of work to be performed is very small. When this occurs, the labor and/or equipment costs to perform the work may be too low to allow for the crew to get to the job, receive instructions, find materials, get set up, perform the work, clean up, and get to the next job. In these situations, the estimator should compare the developed labor and/or equipment costs for performing the work (e.g.,, quantity × labor and/or equipment costs) with the *minimum labor/equipment charge* within the Unit Price section of the data set. (These minimum labor/equipment charge line items appear only in *Facilities Construction Cost Data* and *Commercial Renovation Cost Data*.)

If the labor and/or equipment costs developed by the estimator are LOWER THAN the *minimum labor/equipment charge* listed at the bottom of specific sections of Unit Price costs, the estimator should adjust the developed costs upward to the *minimum labor/equipment charge*. The proper use of a *minimum labor/equipment charge* results in having enough money in the estimate to cover the contractor's higher cost of performing a very small amount of work during a partial workday.

A *minimum labor/equipment charge* should be used only when the task being estimated is the only task the crew will perform at the job site that day. **If, however, the crew will be able to perform other tasks at the job site that day, the use of a minimum labor/equipment charge is not appropriate.**

08 52 Wood Windows

08 52 10 – Plain Wood Windows

08 52 10.40 Casement Window		Crew	Daily Output	Labor-Hours	Unit	Material	2016 Bare Costs Labor	2016 Bare Costs Equipment	Total	Total Incl O&P
0010	**CASEMENT WINDOW**, including frame, screen and grilles									
0100	2'-0" x 3'-0" H, dbl. insulated glass	G	1 Carp	10	.800	Ea.	273	39		312	365
0150	Low E glass	G		10	.800		259	39		298	350
0200	2'-0" x 4'-6" high, double insulated glass	G		9	.889		370	43		413	475
0250	Low E glass	G		9	.889		370	43		413	475
9000	Minimum labor/equipment charge		▼	3	2.667	Job		129		129	211

Example:

Establish the bid price to install two casement windows. Assume installation of 2'×4'-6" metal clad windows with insulating glass [Unit Price line number 08 52 10.40 0250], and that this is the only task this crew will perform at the job site that day.

Solution:

Step One – Develop the Bare Labor Cost for this task:
Bare Labor Cost = 2 windows @ $43/each = $86.00

Step Two – Evaluate the *minimum labor/equipment charge* for this Unit Price section against the developed Bare Labor Cost for this task:
Minimum labor/equipment charge = $129.00 (compare with $86.00)

Step Three – Choose to adjust the developed labor cost upward to the *minimum labor/equipment charge*.

Step Four – Develop the bid price for this task (including O&P):
Add together the marked-up Bare Material Cost for this task and the marked-up *minimum labor/equipment charge* for this Unit Price section.

2 × ($370.00 + 10%) + ($129.00 + 63.9%)

= 2 × ($370.00 + $37.00) + ($129.00 + $82.43)

= 2 × ($407.00) + $211.43

= $814.00 + $211.43

= $1,025.43

ANSWER: $1,025.43 is the correct bid price to use. This sum takes into consideration the material cost (with 10% for profit) for these two windows, plus the *minimum labor/equipment charge* (with O&P included) for this section of the Unit Price.

Unit Price Section

Table of Contents

1

Table of Contents (cont.)

Table of Contents (cont.)

How RSMeans Unit Price Data Works

All RSMeans unit price data is organized in the same way.

It is important to understand the structure, so that you can find information easily and use it correctly.

03 30 Cast-In-Place Concrete

03 30 53 – Miscellaneous Cast-In-Place Concrete

03 30 53.40 Concrete In Place		Crew	Daily Output	Labor-Hours	Unit	Material	2016 Bare Costs Labor	Equipment	Total	Total Incl O&P
0010	**CONCRETE IN PLACE**	R033105-10								
0020	Including forms (4 uses), Grade 60 rebar, concrete (Portland cement	R033105-20								
0050	Type I), placement and finishing unless otherwise indicated	R033105-70								
0300	Beams (3500 psi), 5 kip per L.F., 10' span	R033105-85	C-14A	15.62	12.804	C.Y.	320	620	48.50	988.50	1,400
0350	25' span		"	18.55	10.782		335	520	41	896	1,275
0500	Chimney foundations (5000 psi), over 5 C.Y.		C-14C	32.22	3.476		155	160	.97	315.97	430
0510	(3500 psi), under 5 C.Y.		"	23.71	4.724		182	217	1.32	400.32	555
0700	Columns, square (4000 psi), 12" x 12", up to 1% reinforcing by area		C-14A	11.96	16.722		365	810	63.50	1,238.50	1,800
3540	Equipment pad (3000 psi), 3' x 3' x 6" thick		C-14H	45	1.067	Ea.	47.50	50.50	.70	98.70	135
3550	4' x 4' x 6" thick			30	1.600		70.50	75.50	1.06	147.06	202
3560	5' x 5' x 8" thick			18	2.667		124	126	1.76	251.76	345
3570	6' x 6' x 8" thick			14	3.429		167	162	2.26	331.26	450
3580	8' x 8' x 10" thick			8	6		360	284	3.96	647.96	860
3590	10' x 10' x 12" thick			5	9.600		615	455	6.35	1,076.35	1,425

① Line Numbers

RSMeans **Line Numbers** consist of 12 characters, which identify a unique location in the database for each task. The first 6 or 8 digits conform to the Construction Specifications Institute MasterFormat® 2014. The remainder of the digits are a further breakdown by RSMeans in order to arrange items in understandable groups of similar tasks. Line numbers are consistent across all RSMeans publications, so a line number in any RSMeans product will always refer to the same unit of work.

② Descriptions

RSMeans **Descriptions** are shown in a hierarchical structure to make them readable. In order to read a complete description, read up through the indents to the top of the section. Include everything that is above and to the left that is not contradicted by information below. For instance, the complete description for line 03 30 53.40 3550 is "Concrete in place, including forms (4 uses), Grade 60 rebar, concrete (Portland cement Type 1), placement and finishing unless otherwise indicated; Equipment pad (3000 psi), 4' x 4' x 6" thick."

③ RSMeans Data

When using **RSMeans data**, it is important to read through an entire section to ensure that you use the data that most closely matches your work. Note that sometimes there is additional information shown in the section that may improve your price. There are frequently lines that further describe, add to, or adjust data for specific situations.

④ Reference Information

RSMeans engineers have created **reference** information to assist you in your estimate. **If** there is information that applies to a section, it will be indicated at the start of the section. The Reference Section is located in the back of the data set.

⑤ Crews

Crews include labor and/or equipment necessary to accomplish each task. In this case, Crew C-14H is used.

RSMeans selects a crew to represent the workers and equipment that are typically used for that task. In this case, Crew C-14H consists of one carpenter foreman (outside), two carpenters, one rodman, one laborer, one cement finisher, and one gas engine vibrator. Details of all crews can be found in the reference section.

Crews - Renovation

Crew No. Crew C-14H	Bare Costs		Incl. Subs O&P		Cost Per Labor-Hour	
	Hr.	Daily	Hr.	Daily	Bare Costs	Incl. O&P
1 Carpenter Foreman (outside)	$50.45	$403.60	$82.45	$659.60	$47.32	$76.79
2 Carpenters	48.45	775.20	79.15	1266.40		
1 Rodman (reinf.)	53.00	424.00	85.80	686.40		
1 Laborer	37.90	303.20	61.95	495.60		
1 Cement Finisher	45.65	365.20	72.25	578.00		
1 Gas Engine Vibrator		31.60		34.76	.66	.72
48 L.H., Daily Totals		$2302.80		$3720.76	$47.98	$77.52

6 Daily Output

The **Daily Output** is the amount of work that the crew can do in a normal 8-hour workday, including mobilization, layout, movement of materials, and cleanup. In this case, crew C-14H can install thirty 4' x 4' x 6" thick concrete pads in a day. Daily output is variable and based on many factors, including the size of the job, location, and environmental conditions. RSMeans data represents work done in daylight (or adequate lighting) and temperate conditions.

7 Labor-Hours

The figure in the **Labor-Hours** column is the amount of labor required to perform one unit of work–in this case the amount of labor required to construct one 4' x 4' equipment pad. This figure is calculated by dividing the number of hours of labor in the crew by the daily output (48 labor hours divided by 30 pads = 1.6 hours of labor per pad). Multiply 1.600 times 60 to see the value in minutes: 60 x 1.6 = 96 minutes. Note: the labor-hour figure is not dependent on the crew size. A change in crew size will result in a corresponding change in daily output, but the labor-hours per unit of work will not change.

8 Unit of Measure

All RSMeans unit cost data includes the typical **Unit of Measure** used for estimating that item. For concrete-in-place the typical unit is cubic yards (C.Y.) or each (Ea.). For installing broadloom carpet it is square yard, and for gypsum board it is square foot. The estimator needs to take special care that the unit in the data matches the unit in the take-off. Unit conversions may be found in the Reference Section.

9 Bare Costs

Bare Costs are the costs of materials, labor, and equipment that the installing contractor pays. They represent the cost, in U.S. dollars, for one unit of work. They do not include any markups for profit or labor burden.

10 Bare Total

The **Total column** represents the total bare cost for the installing contractor, in U.S. dollars. In this case, the sum of $70.50 for material + $75.50 for labor + $1.06 for equipment is $147.06.

11 Total Incl O&P

The **Total Incl O&P column** is the total cost, including overhead and profit, that the installing contractor will charge the customer. This represents the cost of materials plus 10% profit, the cost of labor plus labor burden and 10% profit, and the cost of equipment plus 10% profit. It does not include the general contractor's overhead and profit. Note: See the inside back cover of the printed product or the reference section of the electronic product for details of how RSMeans calculates labor burden.

National Average

The data in RSMeans print publications represents a "national average" cost. This data should be modified to the project location using the **City Cost Indexes** or **Location Factors** tables found in the Reference Section. Use the location factors to adjust estimate totals if the project covers multiple trades. Use the city cost indexes (CCI) for single trade projects or projects where a more detailed analysis is required. All figures in the two tables are derived from the same research. The last row of data in the CCI, the weighted average, is the same as the numbers reported for each location in the location factor table.

How RSMeans Unit Price Data Works (Continued)

Project Name: Pre-Engineered Steel Building				Architect: As Shown				
Location:	Anywhere, USA						01/01/16	R+R
Line Number	Description	Qty	Unit	Material	Labor	Equipment	SubContract	Estimate Total
03 30 53.40 3940	Strip footing, 12" x 24", reinforced	34	C.Y.	$4,896.00	$3,638.00	$22.10	$0.00	
03 30 53.40 3950	Strip footing, 12" x 36", reinforced	15	C.Y.	$2,070.00	$1,290.00	$7.80	$0.00	
03 11 13.65 3000	Concrete slab edge forms	500	L.F.	$165.00	$1,220.00	$0.00	$0.00	
03 22 11.10 0200	Welded wire fabric reinforcing	150	C.S.F.	$2,632.50	$4,125.00	$0.00	$0.00	
03 31 13.35 0300	Ready mix concrete, 4000 psi for slab on grade	278	C.Y.	$31,414.00	$0.00	$0.00	$0.00	
03 31 13.70 4300	Place, strike off & consolidate concrete slab	278	C.Y.	$0.00	$4,795.50	$161.24	$0.00	
03 35 13.30 0250	Machine float & trowel concrete slab	15,000	S.F.	$0.00	$9,000.00	$450.00	$0.00	
03 15 16.20 0140	Cut control joints in concrete slab	950	L.F.	$47.50	$389.50	$85.50	$0.00	
03 39 23.13 0300	Sprayed concrete curing membrane	150	C.S.F.	$1,905.00	$960.00	$0.00	$0.00	
Division 03	**Subtotal**			**$43,130.00**	**$25,418.00**	**$726.64**	**$0.00**	**$69,274.64**
08 36 13.10 2650	Manual 10' x 10' steel sectional overhead door	8	Ea.	$9,800.00	$3,440.00	$0.00	$0.00	
08 36 13.10 2860	Insulation and steel back panel for OH door	800	S.F.	$3,800.00	$0.00	$0.00	$0.00	
Division 08	**Subtotal**			**$13,600.00**	**$3,440.00**	**$0.00**	**$0.00**	**$17,040.00**
13 34 19.50 1100	Pre-Engineered Steel Building, 100' x 150' x 24'	15,000	SF Flr.	$0.00	$0.00	$0.00	$360,000.00	
13 34 19.50 6050	Framing for PESB door opening, 3' x 7'	4	Opng.	$0.00	$0.00	$0.00	$2,320.00	
13 34 19.50 6100	Framing for PESB door opening, 10' x 10'	8	Opng.	$0.00	$0.00	$0.00	$9,400.00	
13 34 19.50 6200	Framing for PESB window opening, 4' x 3'	6	Opng.	$0.00	$0.00	$0.00	$3,390.00	
13 34 19.50 5750	PESB door, 3' x 7', single leaf	4	Opng.	$2,380.00	$680.00	$0.00	$0.00	
13 34 19.50 7750	PESB sliding window, 4' x 3' with screen	6	Opng.	$2,430.00	$582.00	$67.80	$0.00	
13 34 19.50 6550	PESB gutter, eave type, 26 ga., painted	300	L.F.	$2,040.00	$798.00	$0.00	$0.00	
13 34 19.50 8650	PESB roof vent, 12" wide x 10' long	15	Ea.	$540.00	$3,195.00	$0.00	$0.00	
13 34 19.50 6900	PESB insulation, vinyl faced, 4" thick	27,400	S.F.	$12,330.00	$9,316.00	$0.00	$0.00	
Division 13	**Subtotal**			**$19,720.00**	**$14,571.00**	**$67.80**	**$375,110.00**	**$409,468.80**
	Subtotal			$76,450.00	$43,429.00	$794.44	$375,110.00	$495,783.44
Division 01	**General Requirements @ 7%**			5,351.50	3,040.03	55.61	26,257.70	
	Estimate Subtotal			$81,801.50	$46,469.03	$850.05	$401,367.70	$495,783.44
	Sales Tax @ 5%			4,090.08		42.50	10,034.19	
	Subtotal A			85,891.58	46,469.03	892.55	411,401.89	
	GC O & P			8,589.16	28,810.80	89.26	41,140.19	
	Subtotal B			94,480.73	75,279.83	981.81	452,542.08	$623,284.45
	Contingency @ 5%							31,164.22
	Subtotal C							$654,448.67
	Bond @ $12/1000 +10% O&P							8,638.72
	Subtotal D							$663,087.40
	Location Adjustment Factor				102.30			15,251.01
	Grand Total							$678,338.41

This estimate is based on an interactive spreadsheet. A copy of this spreadsheet is located on the RSMeans website at **www.RSMeans.com/2016extras.** You are free to download it and adjust it to your methodology.

6

Sample Estimate

This sample demonstrates the elements of an estimate, including a tally of the RSMeans data lines and a summary of the markups on a contractor's work to arrive at a total cost to the owner. The RSMeans Location Factor is added at the bottom of the estimate to adjust the cost of the work to a specific location.

---

1 Work Performed

The body of the estimate shows the RSMeans data selected, including the line number, a brief description of each item, its take-off unit and quantity, and the bare costs of materials, labor, and equipment. This estimate also includes a column titled "SubContract." This data is taken from the RSMeans column "Total Incl O&P" and represents the total that a subcontractor would charge a general contractor for the work, including the sub's markup for overhead and profit.

2 Division 1, General Requirements

This is the first division numerically but the last division estimated. Division 1 includes project-wide needs provided by the general contractor. These requirements vary by project but may include temporary facilities and utilities, security, testing, project cleanup, etc. For small projects a percentage can be used, typically between 5% and 15% of project cost. For large projects the costs may be itemized and priced individually.

3 Sales Tax

If the work is subject to state or local sales taxes, the amount must be added to the estimate. Sales tax may be added to material costs, equipment costs, and subcontracted work. In this case, sales tax was added in all three categories. It was assumed that approximately half the subcontracted work would be material cost, so the tax was applied to 50% of the subcontract total.

4 GC O&P

This entry represents the general contractor's markup on material, labor, equipment, and subcontractor costs. RSMeans' standard markup on materials, equipment, and subcontracted work is 10%. In this estimate, the markup on the labor performed by the GC's workers uses "Skilled Workers Average" shown in Column F on the table "Installing Contractor's Overhead & Profit," which can be found on the inside-back cover of the printed product or in the Reference Section of the electronic product.

5 Contingency

A factor for contingency may be added to any estimate to represent the cost of unknowns that may occur between the time that the estimate is performed and the time the project is constructed. The amount of the allowance will depend on the stage of design at which the estimate is done and the contractor's assessment of the risk involved. Refer to section 01 21 16.50 for contingency allowances.

6 Bonds

Bond costs should be added to the estimate. The figures here represent a typical performance bond, ensuring the owner that if the general contractor does not complete the obligations in the construction contract the bonding company will pay the cost for completion of the work.

7 Location Adjustment

RSMeans published data is based on national average costs. If necessary, adjust the total cost of the project using a location factor from the "Location Factor" table or the "City Cost Index" table. Use location factors if the work is general, covering multiple trades. If the work is by a single trade (e.g.,, masonry) use the more specific data found in the "City Cost Indexes."

Estimating Tips
01 20 00 Price and Payment Procedures
- Allowances that should be added to estimates to cover contingencies and job conditions that are not included in the national average material and labor costs are shown in section 01 21.
- When estimating historic preservation projects (depending on the condition of the existing structure and the owner's requirements), a 15%–20% contingency or allowance is recommended, regardless of the stage of the drawings.

01 30 00 Administrative Requirements
- Before determining a final cost estimate, it is a good practice to review all the items listed in Subdivisions 01 31 and 01 32 to make final adjustments for items that may need customizing to specific job conditions.
- Requirements for initial and periodic submittals can represent a significant cost to the General Requirements of a job. Thoroughly check the submittal specifications when estimating a project to determine any costs that should be included.

01 40 00 Quality Requirements
- All projects will require some degree of Quality Control. This cost is not included in the unit cost of construction listed in each division. Depending upon the terms of the contract, the various costs of inspection and testing can be the responsibility of either the owner or the contractor. Be sure to include the required costs in your estimate.

01 50 00 Temporary Facilities and Controls
- Barricades, access roads, safety nets, scaffolding, security, and many more requirements for the execution of a safe project are elements of direct cost. These costs can easily be overlooked when preparing an estimate. When looking through the major classifications of this subdivision, determine which items apply to each division in your estimate.
- Construction Equipment Rental Costs can be found in the Reference Section in section 01 54 33. Operators' wages are not included in equipment rental costs.
- Equipment mobilization and demobilization costs are not included in equipment rental costs and must be considered separately.

- The cost of small tools provided by the installing contractor for his workers is covered in the "Overhead" column on the "Installing Contractor's Overhead and Profit" table that lists labor trades, base rates and markups and, therefore, is included in the "Total Incl. O&P" cost of any Unit Price line item.

01 70 00 Execution and Closeout Requirements
- When preparing an estimate, thoroughly read the specifications to determine the requirements for Contract Closeout. Final cleaning, record documentation, operation and maintenance data, warranties and bonds, and spare parts and maintenance materials can all be elements of cost for the completion of a contract. Do not overlook these in your estimate.

Reference Numbers
Reference numbers are shown at the beginning of some major classifications. These numbers refer to related items in the Reference Section. The reference information may be an estimating procedure, an alternate pricing method, or technical information.

Note: Not all subdivisions listed here necessarily appear. ■

Did you know?
RSMeans Online gives you the same access to RSMeans' data with 24/7 access:
- Quickly locate costs in the searchable database.
- Build cost lists, estimates, and reports in minutes.
- Adjust costs to any location in the U.S. and Canada with the click of a button.

Start your free trial today at **www.RSMeansOnline.com**

01 11 Summary of Work

01 11 31 – Professional Consultants

01 11 31.10 Architectural Fees		Crew	Daily Output	Labor-Hours	Unit	Material	2016 Bare Costs Labor	Equipment	Total	Total Incl O&P
0010	**ARCHITECTURAL FEES** R011110-10									
0020	For new construction									
0060	Minimum				Project				4.90%	4.90%
0090	Maximum								16%	16%
0100	For alteration work, to $500,000, add to new construction fee								50%	50%
0150	Over $500,000, add to new construction fee								25%	25%

01 11 31.20 Construction Management Fees										
0010	**CONSTRUCTION MANAGEMENT FEES**									
0060	For work to $100,000				Project				10%	10%
0070	To $250,000								9%	9%
0090	To $1,000,000								6%	6%
0100	To $5,000,000								5%	5%
0110	To $10,000,000								4%	4%

01 11 31.30 Engineering Fees										
0010	**ENGINEERING FEES** R011110-30									
0020	Educational planning consultant, minimum				Project				.50%	.50%
0100	Maximum				"				2.50%	2.50%
0400	Elevator & conveying systems, minimum				Contrct				2.50%	2.50%
0500	Maximum								5%	5%
1000	Mechanical (plumbing & HVAC), minimum								4.10%	4.10%
1100	Maximum								10.10%	10.10%
1200	Structural, minimum				Project				1%	1%
1300	Maximum				"				2.50%	2.50%

01 11 31.75 Renderings										
0010	**RENDERINGS** Color, matted, 20" x 30", eye level,									
0020	1 building, minimum				Ea.	2,050			2,050	2,275
0050	Average					2,875			2,875	3,175
0100	Maximum					4,625			4,625	5,100
1000	5 buildings, minimum					4,125			4,125	4,525
1100	Maximum					8,250			8,250	9,075

01 21 Allowances

01 21 16 – Contingency Allowances

01 21 16.50 Contingencies

01 21 16.50 Contingencies										
0010	**CONTINGENCIES**, Add to estimate									
0020	Conceptual stage				Project				20%	20%
0050	Schematic stage								15%	15%
0100	Preliminary working drawing stage (Design Dev.)								10%	10%
0150	Final working drawing stage								3%	3%

01 21 53 – Factors Allowance

01 21 53.50 Factors

01 21 53.50 Factors										
0010	**FACTORS** Cost adjustments R012153-10									
0100	Add to construction costs for particular job requirements									
0500	Cut & patch to match existing construction, add, minimum				Costs	2%	3%			
0550	Maximum					5%	9%			
0800	Dust protection, add, minimum					1%	2%			
0850	Maximum					4%	11%			
1100	Equipment usage curtailment, add, minimum					1%	1%			
1150	Maximum					3%	10%			
1400	Material handling & storage limitation, add, minimum					1%	1%			

01 21 Allowances

01 21 53 – Factors Allowance

01 21 53.50 Factors

		Crew	Daily Output	Labor-Hours	Unit	Material	2016 Bare Costs Labor	Equipment	Total	Total Incl O&P
1450	Maximum				Costs	6%	7%			
1700	Protection of existing work, add, minimum					2%	2%			
1750	Maximum					5%	7%			
2000	Shift work requirements, add, minimum						5%			
2050	Maximum						30%			
2300	Temporary shoring and bracing, add, minimum					2%	5%			
2350	Maximum					5%	12%			
2400	Work inside prisons and high security areas, add, minimum						30%			
2450	Maximum						50%			

01 21 55 – Job Conditions Allowance

01 21 55.50 Job Conditions

		Crew	Daily Output	Labor-Hours	Unit	Material	2016 Bare Costs Labor	Equipment	Total	Total Incl O&P
0010	**JOB CONDITIONS** Modifications to applicable									
0020	cost summaries									
0100	Economic conditions, favorable, deduct				Project				2%	2%
0200	Unfavorable, add								5%	5%
0300	Hoisting conditions, favorable, deduct								2%	2%
0400	Unfavorable, add								5%	5%
0500	General Contractor management, experienced, deduct								2%	2%
0600	Inexperienced, add								10%	10%
0700	Labor availability, surplus, deduct								1%	1%
0800	Shortage, add								10%	10%
0900	Material storage area, available, deduct								1%	1%
1000	Not available, add								2%	2%
1100	Subcontractor availability, surplus, deduct								5%	5%
1200	Shortage, add								12%	12%
1300	Work space, available, deduct								2%	2%
1400	Not available, add								5%	5%

01 21 63 – Taxes

01 21 63.10 Taxes

			Crew	Daily Output	Labor-Hours	Unit	Material	2016 Bare Costs Labor	Equipment	Total	Total Incl O&P
0010	**TAXES**	R012909-80									
0020	Sales tax, State, average					%	5.08%				
0050	Maximum	R012909-85					7.50%				
0200	Social Security, on first $118,500 of wages							7.65%			
0300	Unemployment, combined Federal and State, minimum							.60%			
0350	Average							9.60%			
0400	Maximum							12%			

01 31 Project Management and Coordination

01 31 13 – Project Coordination

01 31 13.20 Field Personnel

		Crew	Daily Output	Labor-Hours	Unit	Material	2016 Bare Costs Labor	Equipment	Total	Total Incl O&P
0010	**FIELD PERSONNEL**									
0020	Clerk, average				Week		460		460	745
0160	General purpose laborer, average						1,525		1,525	2,475
0180	Project manager, minimum						2,025		2,025	3,300
0200	Average						2,325		2,325	3,775
0220	Maximum						2,675		2,675	4,325
0240	Superintendent, minimum						1,975		1,975	3,200
0260	Average						2,175		2,175	3,525
0280	Maximum						2,475		2,475	4,000

01 31 Project Management and Coordination

01 31 13 – Project Coordination

01 31 13.30 Insurance

01 31 13.30 Insurance		Crew	Daily Output	Labor-Hours	Unit	Material	2016 Bare Costs Labor	Equipment	Total	Total Incl O&P
0010	**INSURANCE** R013113-40									
0020	Builders risk, standard, minimum				Job				.24%	.24%
0050	Maximum R013113-60								.64%	.64%
0200	All-risk type, minimum								.25%	.25%
0250	Maximum				↓				.62%	.62%
0400	Contractor's equipment floater, minimum				Value				.50%	.50%
0450	Maximum				"				1.50%	1.50%
0600	Public liability, average				Job				2.02%	2.02%
0810	Workers' compensation & employer's liability									
2000	Range of 35 trades in 50 states, excl. wrecking & clerical, min.				Payroll	1.42%				
2100	Average					12.61%				
2200	Maximum				↓	107.12%				

01 31 13.40 Main Office Expense

01 31 13.40 Main Office Expense		Crew	Daily Output	Labor-Hours	Unit	Material	Labor	Equipment	Total	Total Incl O&P
0010	**MAIN OFFICE EXPENSE** Average for General Contractors									
0020	As a percentage of their annual volume									
0030	Annual volume to $300,000, minimum				% Vol.				20%	
0040	Maximum								30%	
0060	To $500,000, minimum								17%	
0070	Maximum								22%	
0080	To $1,000,000, minimum								16%	
0090	Maximum								19%	
0110	To $3,000,000, minimum								14%	
0120	Maximum								16%	
0130	To $5,000,000, minimum								8%	
0140	Maximum				↓				10%	

01 31 13.50 General Contractor's Mark-Up

01 31 13.50 General Contractor's Mark-Up		Crew	Daily Output	Labor-Hours	Unit	Material	Labor	Equipment	Total	Total Incl O&P
0010	**GENERAL CONTRACTOR'S MARK-UP** on Change Orders									
0200	Extra work, by subcontractors, add				%				10%	10%
0250	By General Contractor, add								15%	15%
0400	Omitted work, by subcontractors, deduct all but								5%	5%
0450	By General Contractor, deduct all but								7.50%	7.50%
0600	Overtime work, by subcontractors, add								15%	15%
0650	By General Contractor, add				↓				10%	10%

01 31 13.60 General Contractor's Main Office Overhead

01 31 13.60 General Contractor's Main Office Overhead		Crew	Daily Output	Labor-Hours	Unit	Material	Labor	Equipment	Total	Total Incl O&P
0010	**GENERAL CONTRACTOR'S MAIN OFFICE OVERHEAD** R013113-50									
0020	As percent of direct costs, minimum				%				5%	
0050	Average								13%	
0100	Maximum				↓				30%	

01 31 13.80 Overhead and Profit

01 31 13.80 Overhead and Profit		Crew	Daily Output	Labor-Hours	Unit	Material	Labor	Equipment	Total	Total Incl O&P
0010	**OVERHEAD & PROFIT** Allowance to add to items in this R013113-50									
0020	data set that do not include Subs O&P, average				%				25%	
0100	Allowance to add to items in this data set that									
0110	do include Subs O&P, minimum				%				5%	5%
0150	Average								10%	10%
0200	Maximum								15%	15%
0290	Typical, by size of project, under $50,000								40%	
0310	$50,000 to $100,000								35%	
0320	$100,000 to $500,000								25%	
0330	$500,000 to $1,000,000				↓				20%	

01 31 Project Management and Coordination

01 31 13 – Project Coordination

01 31 13.90 Performance Bond	Crew	Daily Output	Labor-Hours	Unit	Material	2016 Bare Costs Labor	Equipment	Total	Total Incl O&P
0010 **PERFORMANCE BOND**									
0020 For buildings, minimum				Job				.60%	.60%
0100 Maximum				"				2.50%	2.50%

01 41 Regulatory Requirements

01 41 26 – Permit Requirements

01 41 26.50 Permits

0010 **PERMITS**									
0020 Rule of thumb, most cities, minimum				Job				.50%	.50%
0100 Maximum				"				2%	2%

01 45 Quality Control

01 45 23 – Testing and Inspecting Services

01 45 23.50 Testing

0010 **TESTING** and Inspecting Services									
1800 Compressive test, cylinder, delivered to lab, ASTM C 39				Ea.				12	13
1900 Picked up by lab, minimum								14	15
1950 Average								18	20
2000 Maximum								27	30
2200 Compressive strength, cores (not incl. drilling), ASTM C 42								36	40
2300 Patching core holes				▼				22	24
4730 Soil testing									
4735 Soil density, nuclear method, ASTM D2922				Ea.				35	38.50
4740 Sand cone method, ASTM D1556								27	30
4750 Moisture content, ASTM D 2216								9	10
4780 Permeability test, double ring infiltrometer								500	550
4800 Permeability, var. or constant head, undist., ASTM D 2434								227	250
4850 Recompacted								250	275
4900 Proctor compaction, 4" standard mold, ASTM D 698								123	135
4950 6" modified mold								68	75
5100 Shear tests, triaxial, minimum								410	450
5150 Maximum								545	600
5550 Technician for inspection, per day, earthwork								320	350
5650 Bolting								400	440
5750 Roofing								480	530
5790 Welding				▼				480	530
5820 Non-destructive metal testing, dye penetrant				Day				310	340
5840 Magnetic particle								310	340
5860 Radiography								450	495
5880 Ultrasonic				▼				310	340
6000 Welding certification, minimum				Ea.				91	100
6100 Maximum				"				250	275
7000 Underground storage tank									
7500 Volumetric tightness test ,<=12,000 gal.				Ea.				435	480
7510 <=30,000 gal.				"				615	675
7600 Vadose zone (soil gas) sampling, 10-40 samples, min.				Day				1,375	1,500
7610 Maximum				"				2,275	2,500
7700 Ground water monitoring incl. drilling 3 wells, min.				Total				4,550	5,000
7710 Maximum				"				6,375	7,000

For customer support on your Commercial Renovation Cost Data, call 877.742.4719.

13

01 45 Quality Control

01 45 23 – Testing and Inspecting Services

01 45 23.50 Testing	Crew	Daily Output	Labor-Hours	Unit	Material	2016 Bare Costs Labor	Equipment	Total	Total Incl O&P
8000	X-ray concrete slabs				Ea.				182	200

01 51 Temporary Utilities

01 51 13 – Temporary Electricity

01 51 13.80 Temporary Utilities

		Crew	Daily Output	Labor-Hours	Unit	Material	Labor	Equipment	Total	Total Incl O&P
0010	**TEMPORARY UTILITIES**									
0350	Lighting, lamps, wiring, outlets, 40,000 S.F. building, 8 strings	1 Elec	34	.235	CSF Flr	5	12.95		17.95	25.50
0360	16 strings	"	17	.471		10	26		36	51
0400	Power for temp lighting only, 6.6 KWH, per month								.92	1.01
0430	11.8 KWH, per month								1.65	1.82
0450	23.6 KWH, per month								3.30	3.63
0600	Power for job duration incl. elevator, etc., minimum								47	51.50
0650	Maximum								110	121

01 52 Construction Facilities

01 52 13 – Field Offices and Sheds

01 52 13.20 Office and Storage Space

		Crew	Daily Output	Labor-Hours	Unit	Material	Labor	Equipment	Total	Total Incl O&P
0010	**OFFICE AND STORAGE SPACE**									
0020	Office trailer, furnished, no hookups, 20' x 8', buy	2 Skwk	1	16	Ea.	10,700	800		11,500	13,100
0250	Rent per month					193			193	213
0300	32' x 8', buy	2 Skwk	.70	22.857		16,000	1,150		17,150	19,500
0350	Rent per month					246			246	270
0400	50' x 10', buy	2 Skwk	.60	26.667		24,600	1,325		25,925	29,300
0450	Rent per month					350			350	385
0500	50' x 12', buy	2 Skwk	.50	32		30,100	1,600		31,700	35,700
0550	Rent per month					440			440	485
0700	For air conditioning, rent per month, add					48.50			48.50	53.50
0800	For delivery, add per mile				Mile	11			11	12.10
0890	Delivery each way				Ea.	1,525			1,525	1,700
1200	Storage boxes, 20' x 8', buy	2 Skwk	1.80	8.889		2,975	445		3,420	4,000
1250	Rent per month					78			78	85.50
1300	40' x 8', buy	2 Skwk	1.40	11.429		3,825	570		4,395	5,125
1350	Rent per month					108			108	118

01 54 Construction Aids

01 54 16 – Temporary Hoists

01 54 16.50 Weekly Forklift Crew

		Crew	Daily Output	Labor-Hours	Unit	Material	Labor	Equipment	Total	Total Incl O&P
0010	**WEEKLY FORKLIFT CREW**									
0100	All-terrain forklift, 45' lift, 35' reach, 9000 lb. capacity	A-3P	.20	40	Week		1,975	2,675	4,650	6,050

01 54 19 – Temporary Cranes

01 54 19.50 Daily Crane Crews

		Crew	Daily Output	Labor-Hours	Unit	Material	Labor	Equipment	Total	Total Incl O&P
0010	**DAILY CRANE CREWS** for small jobs, portal to portal									
0100	12-ton truck-mounted hydraulic crane	A-3H	1	8	Day		420	875	1,295	1,625
0200	25-ton	A-3I	1	8			420	1,025	1,445	1,775
0300	40-ton	A-3J	1	8			420	1,250	1,670	2,025
0400	55-ton	A-3K	1	16			785	1,650	2,435	3,050
0500	80-ton	A-3L	1	16			785	2,375	3,160	3,850

01 54 Construction Aids

01 54 19 – Temporary Cranes

01 54 19.50 Daily Crane Crews		Crew	Daily Output	Labor-Hours	Unit	Material	2016 Bare Costs Labor	Equipment	Total	Total Incl O&P
0600	100-ton	A-3M	1	16	Day		785	2,375	3,160	3,875
0900	If crane is needed on a Saturday, Sunday or Holiday									
0910	At time-and-a-half, add				Day		50%			
0920	At double time, add				"		100%			

01 54 19.60 Monthly Tower Crane Crew										
0010	**MONTHLY TOWER CRANE CREW**, excludes concrete footing									
0100	Static tower crane, 130' high, 106' jib, 6200 lb. capacity	A-3N	.05	176	Month		9,200	25,200	34,400	42,300

01 54 23 – Temporary Scaffolding and Platforms

01 54 23.60 Pump Staging		Crew	Daily Output	Labor-Hours	Unit	Material	Labor	Equipment	Total	Total Incl O&P
0010	**PUMP STAGING**, Aluminum R015423-20									
0200	24' long pole section, buy				Ea.	420			420	460
0300	18' long pole section, buy					325			325	355
0400	12' long pole section, buy					219			219	241
0500	6' long pole section, buy					115			115	127
0600	6' long splice joint section, buy					85.50			85.50	94
0700	Pump jack, buy					174			174	192
0900	Foldable brace, buy					68.50			68.50	75.50
1000	Workbench/back safety rail support, buy					92			92	101
1100	Scaffolding planks/workbench, 14" wide x 24' long, buy					765			765	840
1200	Plank end safety rail, buy					350			350	385
1250	Safety net, 22' long, buy					420			420	460
1300	System in place, 50' working height, per use based on 50 uses	2 Carp	84.80	.189	C.S.F.	6.95	9.15		16.10	22.50
1400	100 uses		84.80	.189		3.49	9.15		12.64	18.80
1500	150 uses		84.80	.189		2.34	9.15		11.49	17.50

01 54 23.70 Scaffolding		Crew	Daily Output	Labor-Hours	Unit	Material	Labor	Equipment	Total	Total Incl O&P
0010	**SCAFFOLDING** R015423-10									
0015	Steel tube, regular, no plank, labor only to erect & dismantle									
0090	Building exterior, wall face, 1 to 5 stories, 6'-4" x 5' frames	3 Carp	8	3	C.S.F.		145		145	237
0200	6 to 12 stories	4 Carp	8	4			194		194	315
0301	13 to 20 stories	5 Clab	8	5			190		190	310
0460	Building interior, wall face area, up to 16' high	3 Carp	12	2			97		97	158
0560	16' to 40' high		10	2.400			116		116	190
0800	Building interior floor area, up to 30' high		150	.160	C.C.F.		7.75		7.75	12.65
0900	Over 30' high	4 Carp	160	.200	"		9.70		9.70	15.85
0906	Complete system for face of walls, no plank, material only rent/mo				C.S.F.	35.50			35.50	39
0908	Interior spaces, no plank, material only rent/mo				C.C.F.	4.30			4.30	4.73
0910	Steel tubular, heavy duty shoring, buy									
0920	Frames 5' high 2' wide				Ea.	81.50			81.50	90
0925	5' high 4' wide					93			93	102
0930	6' high 2' wide					93.50			93.50	103
0935	6' high 4' wide					109			109	120
0940	Accessories									
0945	Cross braces				Ea.	15.50			15.50	17.05
0950	U-head, 8" x 8"					19.10			19.10	21
0955	J-head, 4" x 8"					13.90			13.90	15.30
0960	Base plate, 8" x 8"					15.50			15.50	17.05
0965	Leveling jack					33.50			33.50	36.50
1000	Steel tubular, regular, buy									
1100	Frames 3' high 5' wide				Ea.	89			89	97.50
1150	5' high 5' wide					104			104	114
1200	6'-4" high 5' wide					147			147	162
1350	7'-6" high 6' wide					158			158	173

For customer support on your Commercial Renovation Cost Data, call 877.742.4719.

15

01 54 23 – Temporary Scaffolding and Platforms

01 54 23.70 Scaffolding		Crew	Daily Output	Labor-Hours	Unit	Material	2016 Bare Costs Labor	Equipment	Total	Total Incl O&P
1500	Accessories cross braces				Ea.	18.10			18.10	19.90
1550	Guardrail post					16.40			16.40	18.05
1600	Guardrail 7' section					8.70			8.70	9.60
1650	Screw jacks & plates					21.50			21.50	24
1700	Sidearm brackets					30.50			30.50	33.50
1750	8" casters					31			31	34
1800	Plank 2" x 10" x 16'-0"					59.50			59.50	65.50
1900	Stairway section					286			286	315
1910	Stairway starter bar					32			32	35
1920	Stairway inside handrail					58.50			58.50	64
1930	Stairway outside handrail					86.50			86.50	95
1940	Walk-thru frame guardrail				▼	41.50			41.50	46
2000	Steel tubular, regular, rent/mo.									
2100	Frames 3' high 5' wide				Ea.	5			5	5.50
2150	5' high 5' wide					5			5	5.50
2200	6'-4" high 5' wide					5.35			5.35	5.90
2250	7'-6" high 6' wide					10			10	11
2500	Accessories, cross braces					1			1	1.10
2550	Guardrail post					1			1	1.10
2600	Guardrail 7' section					1			1	1.10
2650	Screw jacks & plates					2			2	2.20
2700	Sidearm brackets					2			2	2.20
2750	8" casters					8			8	8.80
2800	Outrigger for rolling tower					3			3	3.30
2850	Plank 2" x 10" x 16'-0"					10			10	11
2900	Stairway section					35			35	38.50
2940	Walk-thru frame guardrail				▼	2.50			2.50	2.75
3000	Steel tubular, heavy duty shoring, rent/mo.									
3250	5' high 2' & 4' wide				Ea.	8.50			8.50	9.35
3300	6' high 2' & 4' wide					8.50			8.50	9.35
3500	Accessories, cross braces					1			1	1.10
3600	U - head, 8" x 8"					2.50			2.50	2.75
3650	J - head, 4" x 8"					2.50			2.50	2.75
3700	Base plate, 8" x 8"					1			1	1.10
3750	Leveling jack					2.50			2.50	2.75
5700	Planks, 2" x 10" x 16'-0", labor only to erect & remove to 50' H	3 Carp	72	.333			16.15		16.15	26.50
5800	Over 50' high	4 Carp	80	.400	▼		19.40		19.40	31.50

01 54 23.75 Scaffolding Specialties

		Crew	Daily Output	Labor-Hours	Unit	Material	2016 Bare Costs Labor	Equipment	Total	Total Incl O&P
0010	**SCAFFOLDING SPECIALTIES**									
1200	Sidewalk bridge, heavy duty steel posts & beams, including									
1210	parapet protection & waterproofing (material cost is rent/month)									
1220	8' to 10' wide, 2 posts	3 Carp	15	1.600	L.F.	43.50	77.50		121	175
1230	3 posts	"	10	2.400	"	67.50	116		183.50	265
1500	Sidewalk bridge using tubular steel scaffold frames including									
1510	planking (material cost is rent/month)	3 Carp	45	.533	L.F.	8.55	26		34.55	51.50
1600	For 2 uses per month, deduct from all above					50%				
1700	For 1 use every 2 months, add to all above					100%				
1900	Catwalks, 20" wide, no guardrails, 7' span, buy				Ea.	166			166	183
2000	10' span, buy					233			233	256
3720	Putlog, standard, 8' span, with hangers, buy					66.50			66.50	73
3730	Rent per month					20			20	22
3750	12' span, buy				▼	99			99	109

01 54 23 – Temporary Scaffolding and Platforms

01 54 23.75 Scaffolding Specialties	Crew	Daily Output	Labor-Hours	Unit	Material	2016 Bare Costs Labor	Equipment	Total	Total Incl O&P
3755	Rent per month				Ea.	25			25	27.50
3760	Trussed type, 16' span, buy					229			229	252
3770	Rent per month					30			30	33
3790	22' span, buy					274			274	300
3795	Rent per month					40			40	44
4000	7 step					815			815	900
4100	Rolling towers, buy, 5' wide, 7' long, 10' high					1,300			1,300	1,450
4200	For 5' high added sections, to buy, add				↓	244			244	269
4300	Complete incl. wheels, railings, outriggers,									
4350	21' high, to buy				Ea.	2,225			2,225	2,425
4400	Rent/month = 5% of purchase cost				"	200			200	220
5000	Motorized work platform, mast climber									
5050	Base unit, 50' W, less than 100' tall, rent/mo				Ea.	2,875			2,875	3,175
5100	Less than 200' tall, rent/mo					3,500			3,500	3,850
5150	Less than 300' tall, rent/mo					4,375			4,375	4,800
5200	Less than 400' tall, rent/mo				↓	5,125			5,125	5,625
5250	Set up and demob, per unit, less than 100' tall	B-68F	16.60	1.446	C.S.F.		73	20	93	140
5300	Less than 200' tall		25	.960			48.50	13.35	61.85	93
5350	Less than 300' tall		30	.800			40.50	11.15	51.65	78
5400	Less than 400' tall		33	.727	↓		37	10.10	47.10	70.50
5500	Mobilization (price includes freight in and out) per unit	↓			Ea.				1,100	1,225

01 54 23.80 Staging Aids

		Crew	Daily Output	Labor-Hours	Unit	Material	2016 Bare Costs Labor	Equipment	Total	Total Incl O&P
0010	**STAGING AIDS** and fall protection equipment									
0100	Sidewall staging bracket, tubular, buy				Ea.	51			51	56
0110	Cost each per day, based on 250 days use				Day	.20			.20	.22
0200	Guard post, buy				Ea.	52			52	57
0210	Cost each per day, based on 250 days use				Day	.21			.21	.23
0300	End guard chains, buy per pair				Pair	38			38	41.50
0310	Cost per set per day, based on 250 days use				Day	.20			.20	.22
1010	Cost each per day, based on 250 days use				"	.04			.04	.04
1100	Wood bracket, buy				Ea.	21			21	23.50
1110	Cost each per day, based on 250 days use				Day	.08			.08	.09
2010	Cost per pair per day, based on 250 days use				"	.49			.49	.53
2100	Steel siderail jack, buy per pair				Pair	97.50			97.50	107
2110	Cost per pair per day, based on 250 days use				Day	.39			.39	.43
3010	Cost each per day, based on 250 days use				"	.24			.24	.26
3100	Aluminum scaffolding plank, 20" wide x 24' long, buy				Ea.	790			790	870
3110	Cost each per day, based on 250 days use				Day	3.16			3.16	3.48
4000	Nylon full body harness, lanyard and rope grab				Ea.	175			175	193
4010	Cost each per day, based on 250 days use				Day	.70			.70	.77
4100	Rope for safety line, 5/8" x 100' nylon, buy				Ea.	56			56	61.50
4110	Cost each per day, based on 250 days use				Day	.22			.22	.25
4200	Permanent U-Bolt roof anchor, buy				Ea.	35.50			35.50	39.50
4300	Temporary (one use) roof ridge anchor, buy				"	32			32	35.50
5000	Installation (setup and removal) of staging aids									
5010	Sidewall staging bracket	2 Carp	64	.250	Ea.		12.10		12.10	19.80
5020	Guard post with 2 wood rails	"	64	.250			12.10		12.10	19.80
5030	End guard chains, set	1 Carp	64	.125			6.05		6.05	9.90
5100	Roof shingling bracket		96	.083			4.04		4.04	6.60
5200	Ladder jack	↓	64	.125	↓		6.05		6.05	9.90
5300	Wood plank, 2" x 10" x 16'	2 Carp	80	.200			9.70		9.70	15.85
5310	Aluminum scaffold plank, 20" x 24'	"	40	.400	↓		19.40		19.40	31.50

For customer support on your Commercial Renovation Cost Data, call 877.742.4719.

17

01 54 Construction Aids

01 54 23 – Temporary Scaffolding and Platforms

01 54 23.80 Staging Aids	Crew	Daily Output	Labor-Hours	Unit	Material	2016 Bare Costs Labor	2016 Bare Costs Equipment	Total	Total Incl O&P
5410	Safety rope	1 Carp	40	.200	Ea.		9.70		9.70	15.85
5420	Permanent U-Bolt roof anchor (install only)	2 Carp	40	.400			19.40		19.40	31.50
5430	Temporary roof ridge anchor (install only)	1 Carp	64	.125			6.05		6.05	9.90

01 54 26 – Temporary Swing Staging

01 54 26.50 Swing Staging

		Crew	Daily Output	Labor-Hours	Unit	Material	2016 Bare Costs Labor	2016 Bare Costs Equipment	Total	Total Incl O&P
0010	**SWING STAGING**, 500 lb. cap., 2' wide to 24' long, hand operated									
0020	steel cable type, with 60' cables, buy				Ea.	5,275			5,275	5,800
0030	Rent per month				"	530			530	580
0600	Lightweight (not for masons) 24' long for 150' height,									
0610	manual type, buy				Ea.	11,000			11,000	12,100
0620	Rent per month					1,100			1,100	1,200
0700	Powered, electric or air, to 150' high, buy					27,700			27,700	30,500
0710	Rent per month					1,950			1,950	2,125
0780	To 300' high, buy					28,100			28,100	30,900
0800	Rent per month					1,975			1,975	2,175
1000	Bosun's chair or work basket 3' x 3.5', to 300' high, electric, buy					11,200			11,200	12,400
1010	Rent per month					785			785	865
2200	Move swing staging (setup and remove)	E-4	2	16	Move		860	73.50	933.50	1,600

01 54 36 – Equipment Mobilization

01 54 36.50 Mobilization

		Crew	Daily Output	Labor-Hours	Unit	Material	2016 Bare Costs Labor	2016 Bare Costs Equipment	Total	Total Incl O&P
0010	**MOBILIZATION** (Use line item again for demobilization)	R015436-50								
0015	Up to 25 mi. haul dist. (50 mi. RT for mob/demob crew)									
1200	Small equipment, placed in rear of, or towed by pickup truck	A-3A	4	2	Ea.		98.50	39	137.50	199
1300	Equipment hauled on 3-ton capacity towed trailer	A-3Q	2.67	3			147	67.50	214.50	310
1400	20-ton capacity	B-34U	2	8			370	236	606	850
1500	40-ton capacity	B-34N	2	8			375	375	750	1,025
1600	50-ton capacity	B-34V	1	24			1,150	1,050	2,200	3,000
1700	Crane, truck-mounted, up to 75 ton (driver only)	1 Eqhv	4	2			105		105	165
1800	Over 75 ton (with chase vehicle)	A-3E	2.50	6.400			305	62.50	367.50	555
2400	Crane, large lattice boom, requiring assembly	B-34W	.50	144			6,575	7,525	14,100	18,800
2500	For each additional 5 miles haul distance, add						10%	10%		
3000	For large pieces of equipment, allow for assembly/knockdown									
3100	For mob/demob of micro-tunneling equip, see Section 33 05 23.19									
3200	For mob/demob of pile driving equip, see Section 31 62 19.10									

01 55 Vehicular Access and Parking

01 55 23 – Temporary Roads

01 55 23.50 Roads and Sidewalks

		Crew	Daily Output	Labor-Hours	Unit	Material	2016 Bare Costs Labor	2016 Bare Costs Equipment	Total	Total Incl O&P
0010	**ROADS AND SIDEWALKS** Temporary									
0050	Roads, gravel fill, no surfacing, 4" gravel depth	B-14	715	.067	S.Y.	3.70	2.69	.51	6.90	9
0100	8" gravel depth	"	615	.078	"	7.40	3.13	.59	11.12	13.90

01 56 Temporary Barriers and Enclosures

01 56 13 – Temporary Air Barriers

01 56 13.60 Tarpaulins

	01 56 13.60 Tarpaulins	Crew	Daily Output	Labor-Hours	Unit	Material	2016 Bare Costs Labor	Equipment	Total	Total Incl O&P
0010	**TARPAULINS**									
0020	Cotton duck, 10 oz. to 13.13 oz. per S.Y., 6' x 8'				S.F.	.88			.88	.97
0050	30' x 30'					.64			.64	.70
0100	Polyvinyl coated nylon, 14 oz. to 18 oz., minimum					1.42			1.42	1.56
0150	Maximum					1.43			1.43	1.57
0200	Reinforced polyethylene 3 mils thick, white					.04			.04	.04
0300	4 mils thick, white, clear or black					.11			.11	.12
0400	5.5 mils thick, clear					.19			.19	.21
0500	White, fire retardant					.56			.56	.62
0600	12 mils, oil resistant, fire retardant					.43			.43	.47
0700	8.5 mils, black					.23			.23	.25
0720	Steel reinforced polyethylene, 20 mils thick					.52			.52	.57
0730	Polyester reinforced w/integral fastening system, 11 mils thick					.22			.22	.24
0740	Mylar polyester, non-reinforced, 7 mils thick					1.17			1.17	1.29

01 56 13.90 Winter Protection

	01 56 13.90 Winter Protection	Crew	Daily Output	Labor-Hours	Unit	Material	2016 Bare Costs Labor	Equipment	Total	Total Incl O&P
0010	**WINTER PROTECTION**									
0100	Framing to close openings	2 Clab	500	.032	S.F.	.43	1.21		1.64	2.45
0200	Tarpaulins hung over scaffolding, 8 uses, not incl. scaffolding		1500	.011		.24	.40		.64	.92
0250	Tarpaulin polyester reinf. w/integral fastening system, 11 mils thick		1600	.010		.22	.38		.60	.86
0300	Prefab fiberglass panels, steel frame, 8 uses		1200	.013		2.40	.51		2.91	3.47

01 56 16 – Temporary Dust Barriers

01 56 16.10 Dust Barriers, Temporary

	01 56 16.10 Dust Barriers, Temporary	Crew	Daily Output	Labor-Hours	Unit	Material	2016 Bare Costs Labor	Equipment	Total	Total Incl O&P
0010	**DUST BARRIERS, TEMPORARY**									
0020	Spring loaded telescoping pole & head, to 12', erect and dismantle	1 Clab	240	.033	Ea.		1.26		1.26	2.06
0025	Cost per day (based upon 250 days)				Day	.26			.26	.29
0030	To 21', erect and dismantle	1 Clab	240	.033	Ea.		1.26		1.26	2.06
0035	Cost per day (based upon 250 days)				Day	.54			.54	.60
0040	Accessories, caution tape reel, erect and dismantle	1 Clab	480	.017	Ea.		.63		.63	1.03
0045	Cost per day (based upon 250 days)				Day	.36			.36	.40
0060	Foam rail and connector, erect and dismantle	1 Clab	240	.033	Ea.		1.26		1.26	2.06
0065	Cost per day (based upon 250 days)				Day	.12			.12	.13
0070	Caution tape	1 Clab	384	.021	C.L.F.	3.13	.79		3.92	4.73
0080	Zipper, standard duty		60	.133	Ea.	8	5.05		13.05	17.05
0090	Heavy duty		48	.167	"	9.75	6.30		16.05	21
0100	Polyethylene sheet, 4 mil		37	.216	Sq.	2.55	8.20		10.75	16.20
0110	6 mil		37	.216	"	3.70	8.20		11.90	17.45
1000	Dust partition, 6 mil polyethylene, 1" x 3" frame	2 Carp	2000	.008	S.F.	.30	.39		.69	.95
1080	2" x 4" frame	"	2000	.008	"	.33	.39		.72	.99

01 56 23 – Temporary Barricades

01 56 23.10 Barricades

	01 56 23.10 Barricades	Crew	Daily Output	Labor-Hours	Unit	Material	2016 Bare Costs Labor	Equipment	Total	Total Incl O&P
0010	**BARRICADES**									
0020	5' high, 3 rail @ 2" x 8", fixed	2 Carp	20	.800	L.F.	5.70	39		44.70	70
0150	Movable	"	30	.533	"	4.74	26		30.74	47
0300	Stock units, 6' high, 8' wide, plain, buy				Ea.	390			390	430
0350	With reflective tape, buy				"	405			405	445
0400	Break-a-way 3" PVC pipe barricade									
0410	with 3 ea. 1' x 4' reflectorized panels, buy				Ea.	106			106	117
0500	Barricades, plastic, 8 x 24" wide, foldable					66			66	72.50
0800	Traffic cones, PVC, 18" high					13.25			13.25	14.60
0850	28" high					20.50			20.50	22.50
0900	Barrels, 55 gal., with flasher	1 Clab	96	.083		147	3.16		150.16	166

For customer support on your Commercial Renovation Cost Data, call 877.742.4719.

19

01 56 Temporary Barriers and Enclosures

01 56 23 – Temporary Barricades

01 56 23.10 Barricades		Crew	Daily Output	Labor-Hours	Unit	Material	2016 Bare Costs Labor	Equipment	Total	Total Incl O&P
1000	Guardrail, wooden, 3' high, 1" x 6", on 2" x 4" posts	2 Carp	200	.080	L.F.	1.22	3.88		5.10	7.70
1100	2" x 6", on 4" x 4" posts	"	165	.097		2.41	4.70		7.11	10.35
1200	Portable metal with base pads, buy					15.55			15.55	17.10
1250	Typical installation, assume 10 reuses	2 Carp	600	.027	↓	2.55	1.29		3.84	4.92
1300	Barricade tape, polyethylene, 7 mil, 3" wide x 500' long roll				Ea.	25			25	27.50
3000	Detour signs, set up and remove									
3010	Reflective aluminum, MUTCD, 24" x 24", post mounted	1 Clab	20	.400	Ea.	2.41	15.15		17.56	27.50
4000	Roof edge portable barrier stands and warning flags, 50 uses	1 Rohe	9100	.001	L.F.	.07	.03		.10	.12
4010	100 uses	"	9100	.001	"	.03	.03		.06	.09

01 56 26 – Temporary Fencing

01 56 26.50 Temporary Fencing

		Crew	Daily Output	Labor-Hours	Unit	Material	2016 Bare Costs Labor	Equipment	Total	Total Incl O&P
0010	**TEMPORARY FENCING**									
0020	Chain link, 11 ga., 4' high	2 Clab	400	.040	L.F.	1.59	1.52		3.11	4.23
0100	6' high		300	.053		3.75	2.02		5.77	7.45
0200	Rented chain link, 6' high, to 1000' (up to 12 mo.)		400	.040		2.89	1.52		4.41	5.65
0250	Over 1000' (up to 12 mo.)	↓	300	.053		2.89	2.02		4.91	6.50
0350	Plywood, painted, 2" x 4" frame, 4' high	A-4	135	.178		6.25	8.15		14.40	20
0400	4" x 4" frame, 8' high	"	110	.218		12.20	10		22.20	29.50
0500	Wire mesh on 4" x 4" posts, 4' high	2 Carp	100	.160		9.95	7.75		17.70	23.50
0550	8' high	"	80	.200	↓	15.10	9.70		24.80	32.50

01 56 29 – Temporary Protective Walkways

01 56 29.50 Protection

		Crew	Daily Output	Labor-Hours	Unit	Material	2016 Bare Costs Labor	Equipment	Total	Total Incl O&P
0010	**PROTECTION**									
0020	Stair tread, 2" x 12" planks, 1 use	1 Carp	75	.107	Tread	5.05	5.15		10.20	14
0100	Exterior plywood, 1/2" thick, 1 use		65	.123		2.04	5.95		7.99	12
0200	3/4" thick, 1 use		60	.133	↓	2.73	6.45		9.18	13.55
2200	Sidewalks, 2" x 12" planks, 2 uses		350	.023	S.F.	.84	1.11		1.95	2.73
2300	Exterior plywood, 2 uses, 1/2" thick		750	.011		.34	.52		.86	1.21
2400	5/8" thick		650	.012		.38	.60		.98	1.39
2500	3/4" thick	↓	600	.013	↓	.46	.65		1.11	1.56

01 56 32 – Temporary Security

01 56 32.50 Watchman

		Crew	Daily Output	Labor-Hours	Unit	Material	2016 Bare Costs Labor	Equipment	Total	Total Incl O&P
0010	**WATCHMAN**									
0020	Service, monthly basis, uniformed person, minimum				Hr.				25	27.50
0100	Maximum								45.50	50
0200	Person and command dog, minimum								31	34
0300	Maximum				↓				54.50	60
0500	Sentry dog, leased, with job patrol (yard dog), 1 dog				Week				290	320
0600	2 dogs				"				390	430
0800	Purchase, trained sentry dog, minimum				Ea.				1,375	1,500
0900	Maximum				"				2,725	3,000

01 58 Project Identification

01 58 13 – Temporary Project Signage

01 58 13.50 Signs	Crew	Daily Output	Labor-Hours	Unit	Material	2016 Bare Costs Labor	Equipment	Total	Total Incl O&P
0010 **SIGNS**									
0020 High intensity reflectorized, no posts, buy				Ea.	25			25	27.50

01 71 Examination and Preparation

01 71 23 – Field Engineering

01 71 23.13 Construction Layout

	Crew	Daily Output	Labor-Hours	Unit	Material	2016 Bare Costs Labor	Equipment	Total	Total Incl O&P
0010 **CONSTRUCTION LAYOUT**									
1100 Crew for layout of building, trenching or pipe laying, 2 person crew	A-6	1	16	Day		775	43.50	818.50	1,300
1200 3 person crew	A-7	1	24			1,275	43.50	1,318.50	2,075
1400 Crew for roadway layout, 4 person crew	A-8	1	32			1,650	44	1,694	2,675

01 74 Cleaning and Waste Management

01 74 13 – Progress Cleaning

01 74 13.20 Cleaning Up

	Crew	Daily Output	Labor-Hours	Unit	Material	2016 Bare Costs Labor	Equipment	Total	Total Incl O&P
0010 **CLEANING UP**									
0020 After job completion, allow, minimum				Job				.30%	.30%
0040 Maximum				"				1%	1%
0052 Cleanup of floor area, continuous, per day, during const.	A-5	16	1.125	M.S.F.	2.23	43	3.88	49.11	77
0100 Final by GC at end of job	"	11.50	1.565	"	2.36	60	5.40	67.76	107

01 76 Protecting Installed Construction

01 76 13 – Temporary Protection of Installed Construction

01 76 13.20 Temporary Protection

	Crew	Daily Output	Labor-Hours	Unit	Material	2016 Bare Costs Labor	Equipment	Total	Total Incl O&P
0010 **TEMPORARY PROTECTION**									
0020 Flooring, 1/8" tempered hardboard, taped seams	2 Carp	1500	.011	S.F.	.42	.52		.94	1.30
0030 Peel away carpet protection	1 Clab	3200	.003	"	.13	.09		.22	.29

01 91 Commissioning

01 91 13 – General Commissioning Requirements

01 91 13.50 Building Commissioning

	Crew	Daily Output	Labor-Hours	Unit	Material	2016 Bare Costs Labor	Equipment	Total	Total Incl O&P
0010 **BUILDING COMMISSIONING**									
0100 Systems operation and verification during turnover				%				.25%	.25%
0150 Including all systems subcontractors								.50%	.50%
0200 Systems design assistance, operation, verification and training								.50%	.50%
0250 Including all systems subcontractors								1%	1%

01 93 09 – Facility Equipment

01 93 09.50 Moving Equipment	Crew	Daily Output	Labor-Hours	Unit	Material	2016 Bare Costs Labor	2016 Bare Costs Equipment	Total	Total Incl O&P
0010 **MOVING EQUIPMENT**, Remove and reset, 100' distance,									
0020 No obstructions, no assembly or leveling unless noted									
0100 Annealing furnace, 24' overall	B-67	4	4	Ea.		201	78.50	279.50	400
0200 Annealing oven, small		14	1.143			57.50	22.50	80	115
0240 Very large		1	16			800	315	1,115	1,600
0400 Band saw, small		12	1.333			67	26	93	134
0440 Large		8	2			100	39	139	201
0500 Blue print copy machine		7	2.286			115	45	160	230
0600 Bonding mill, 6"		7	2.286			115	45	160	230
0620 12"		6	2.667			134	52.50	186.50	268
0640 18"		4	4			201	78.50	279.50	400
0660 24"		2	8			400	157	557	800
0700 Boring machine (jig)	B-68	7	3.429			173	45	218	320
0800 Bridgeport mill, standard	B-67	14	1.143			57.50	22.50	80	115
1000 Calibrator, 6 unit	"	14	1.143			57.50	22.50	80	115
1100 Comparitor, bench top	2 Clab	14	1.143			43.50		43.50	71
1140 Floor mounted	B-67	7	2.286			115	45	160	230
1200 Computer, desk top	2 Clab	25	.640			24.50		24.50	39.50
1300 Copy machine	"	25	.640			24.50		24.50	39.50
1500 Deflasher	B-67	14	1.143			57.50	22.50	80	115
1600 Degreaser, small		14	1.143			57.50	22.50	80	115
1640 Large 24' overall		1	16			800	315	1,115	1,600
1700 Desk with chair	2 Clab	25	.640			24.50		24.50	39.50
1725 Desk only	"	22	.727			27.50		27.50	45
1750 Chair only	1 Clab	300	.027			1.01		1.01	1.65
1800 Dial press	B-67	7	2.286			115	45	160	230
1900 Drafting table	2 Clab	14	1.143			43.50		43.50	71
2000 Drill press, bench top	"	14	1.143			43.50		43.50	71
2040 Floor mounted	B-67	14	1.143			57.50	22.50	80	115
2080 Industrial radial	"	7	2.286			115	45	160	230
2100 Dust collector, portable	2 Clab	25	.640			24.50		24.50	39.50
2140 Stationary, small	B-67	7	2.286			115	45	160	230
2180 Stationary, large	"	2	8			400	157	557	800
2300 Electric discharge machine	B-68	7	3.429			173	45	218	320
2400 Environmental chamber walls, including assembly	4 Clab	18	1.778	L.F.		67.50		67.50	110
2600 File cabinet	2 Clab	25	.640	Ea.		24.50		24.50	39.50
2800 Grinder/sander, pedestal mount	B-67	14	1.143			57.50	22.50	80	115
3000 Hack saw, power	2 Clab	24	.667			25.50		25.50	41.50
3100 Hydraulic press	B-67	14	1.143			57.50	22.50	80	115
3500 Laminar flow tables	"	14	1.143			57.50	22.50	80	115
3600 Lathe, bench	2 Clab	14	1.143			43.50		43.50	71
3640 6"	B-67	14	1.143			57.50	22.50	80	115
3680 10"		13	1.231			61.50	24	85.50	124
3720 12"		12	1.333			67	26	93	134
4000 Milling machine		8	2			100	39	139	201
4100 Molding press, 25 ton		5	3.200			160	62.50	222.50	320
4140 60 ton		4	4			201	78.50	279.50	400
4180 100 ton		2	8			400	157	557	800
4220 150 ton		1.50	10.667			535	209	744	1,075
4260 200 ton		1	16			800	315	1,115	1,600
4300 300 ton		.75	21.333			1,075	420	1,495	2,125
4700 Oil pot stand		14	1.143			57.50	22.50	80	115
5000 Press, 10 ton		14	1.143			57.50	22.50	80	115

01 93 Facility Maintenance

01 93 09 – Facility Equipment

01 93 09.50 Moving Equipment	Crew	Daily Output	Labor- Hours	Unit	Material	2016 Bare Costs Labor	Equipment	Total	Total Incl O&P
5040	15 ton	B-67	12	1.333	Ea.		67	26	93	134
5080	20 ton		10	1.600			80	31.50	111.50	161
5120	30 ton		8	2			100	39	139	201
5160	45 ton		6	2.667			134	52.50	186.50	268
5200	60 ton		4	4			201	78.50	279.50	400
5240	75 ton		2.50	6.400			320	125	445	645
5280	100 ton		2	8			400	157	557	800
5500	Raised floor, including assembly	2 Carp	250	.064	S.F.		3.10		3.10	5.05
5600	Rolling mill, 6"	B-67	7	2.286	Ea.		115	45	160	230
5640	9"		6	2.667			134	52.50	186.50	268
5680	12"		4	4			201	78.50	279.50	400
5720	13"		3.50	4.571			229	89.50	318.50	460
5760	18"		2	8			400	157	557	800
5800	25"		1	16			800	315	1,115	1,600
6000	Sander, floor stand		14	1.143			57.50	22.50	80	115
6100	Screw machine		7	2.286			115	45	160	230
6200	Shaper, 16"		14	1.143			57.50	22.50	80	115
6300	Shear, power assist		4	4			201	78.50	279.50	400
6400	Slitter, 6"		14	1.143			57.50	22.50	80	115
6440	8"		13	1.231			61.50	24	85.50	124
6480	10"		12	1.333			67	26	93	134
6520	12"		11	1.455			73	28.50	101.50	147
6560	16"		10	1.600			80	31.50	111.50	161
6600	20"		8	2			100	39	139	201
6640	24"		6	2.667			134	52.50	186.50	268
6800	Snag and tap machine		7	2.286			115	45	160	230
6900	Solder machine (auto)		7	2.286			115	45	160	230
7000	Storage cabinet metal, small	2 Clab	36	.444			16.85		16.85	27.50
7040	Large		25	.640			24.50		24.50	39.50
7100	Storage rack open, small		14	1.143			43.50		43.50	71
7140	Large		7	2.286			86.50		86.50	142
7200	Surface bench, small	B-67	14	1.143			57.50	22.50	80	115
7240	Large	"	5	3.200			160	62.50	222.50	320
7300	Surface grinder, large wet	B-68	5	4.800			242	62.50	304.50	450
7500	Time check machine	2 Clab	14	1.143			43.50		43.50	71
8000	Welder, 30 kVA (bench)		14	1.143			43.50		43.50	71
8100	Work bench with chair		25	.640			24.50		24.50	39.50
8110	For haul up to 200 feet, add				%		25%			

For customer support on your Commercial Renovation Cost Data, call 877.742.4719.

23

Division Notes

	CREW	DAILY OUTPUT	LABOR-HOURS	UNIT	BARE COSTS				TOTAL INCL O&P
					MAT.	LABOR	EQUIP.	TOTAL	

Estimating Tips
02 30 00 Subsurface Investigation
In preparing estimates on structures involving earthwork or foundations, all information concerning soil characteristics should be obtained. Look particularly for hazardous waste, evidence of prior dumping of debris, and previous stream beds.

02 40 00 Demolition and Structure Moving
The costs shown for selective demolition do not include rubbish handling or disposal. These items should be estimated separately using RSMeans data or other sources.
- Historic preservation often requires that the contractor remove materials from the existing structure, rehab them, and replace them. The estimator must be aware of any related measures and precautions that must be taken when doing

selective demolition and cutting and patching. Requirements may include special handling and storage, as well as security.
- In addition to Subdivision 02 41 00, you can find selective demolition items in each division. Example: Roofing demolition is in Division 7.
- Absent of any other specific reference, an approximate demolish-in-place cost can be obtained by halving the new-install labor cost. To remove for reuse, allow the entire new-install labor figure.

02 40 00 Building Deconstruction
This section provides costs for the careful dismantling and recycling of most of low-rise building materials.

02 50 00 Containment of Hazardous Waste
This section addresses on-site hazardous waste disposal costs.

02 80 00 Hazardous Material Disposal/ Remediation
This subdivision includes information on hazardous waste handling, asbestos remediation, lead remediation, and mold remediation. See reference R028213-20 and R028319-60 for further guidance in using these unit price lines.

02 90 00 Monitoring Chemical Sampling, Testing Analysis
This section provides costs for on-site sampling and testing hazardous waste.

Reference Numbers
Reference numbers are shown at the beginning of some major classifications. These numbers refer to related items in the Reference Section. The reference information may be an estimating procedure, an alternate pricing method, or technical information.

Note: Not all subdivisions listed here necessarily appear. ■

Did you know?
RSMeans Online gives you the same access to RSMeans' data with 24/7 access:
- Quickly locate costs in the searchable database.
- Build cost lists, estimates, and reports in minutes.
- Adjust costs to any location in the U.S. and Canada with the click of a button.

Start your free trial today at **www.RSMeansOnline.com**

RSMeans Online
FROM THE GORDIAN GROUP

02 21 Surveys

02 21 13 – Site Surveys

02 21 13.09 Topographical Surveys	Crew	Daily Output	Labor-Hours	Unit	Material	2016 Bare Costs Labor	Equipment	Total	Total Incl O&P
0010 **TOPOGRAPHICAL SURVEYS**									
0020 Topographical surveying, conventional, minimum	A-7	3.30	7.273	Acre	20	385	13.25	418.25	650
0100 Maximum	A-8	.60	53.333	"	60	2,725	73	2,858	4,525

02 21 13.13 Boundary and Survey Markers

	Crew	Daily Output	Labor-Hours	Unit	Material	2016 Bare Costs Labor	Equipment	Total	Total Incl O&P
0010 **BOUNDARY AND SURVEY MARKERS**									
0300 Lot location and lines, large quantities, minimum	A-7	2	12	Acre	35	635	22	692	1,100
0320 Average	"	1.25	19.200		55	1,025	35	1,115	1,725
0400 Small quantities, maximum	A-8	1	32	↓	75	1,650	44	1,769	2,750

02 32 Geotechnical Investigations

02 32 13 – Subsurface Drilling and Sampling

02 32 13.10 Boring and Exploratory Drilling

	Crew	Daily Output	Labor-Hours	Unit	Material	2016 Bare Costs Labor	Equipment	Total	Total Incl O&P
0010 **BORING AND EXPLORATORY DRILLING**									
0020 Borings, initial field stake out & determination of elevations	A-6	1	16	Day		775	43.50	818.50	1,300
0100 Drawings showing boring details				Total		335		335	425
0200 Report and recommendations from P.E.						775		775	970
0300 Mobilization and demobilization	B-55	4	6	↓		236	273	509	685
0350 For over 100 miles, per added mile		450	.053	Mile		2.09	2.43	4.52	6.05
0600 Auger holes in earth, no samples, 2-1/2" diameter		78.60	.305	L.F.		12	13.90	25.90	35
0650 4" diameter		67.50	.356			13.95	16.20	30.15	40.50
0800 Cased borings in earth, with samples, 2-1/2" diameter		55.50	.432		15	16.95	19.70	51.65	65.50
0850 4" diameter	↓	32.60	.736		18	29	33.50	80.50	104
1000 Drilling in rock, "BX" core, no sampling	B-56	34.90	.458			19.95	45.50	65.45	82
1050 With casing & sampling		31.70	.505		15	22	50	87	108
1200 "NX" core, no sampling		25.92	.617			27	61.50	88.50	111
1250 With casing and sampling	↓	25	.640	↓	16	28	63.50	107.50	132
1400 Borings, earth, drill rig and crew with truck mounted auger	B-55	1	24	Day		940	1,100	2,040	2,725
1450 Rock using crawler type drill	B-56	1	16	"		695	1,600	2,295	2,875
1500 For inner city borings add, minimum								10%	10%
1510 Maximum								20%	20%

02 41 Demolition

02 41 13 – Selective Site Demolition

02 41 13.15 Hydrodemolition

	Crew	Daily Output	Labor-Hours	Unit	Material	2016 Bare Costs Labor	Equipment	Total	Total Incl O&P
0010 **HYDRODEMOLITION**									
0015 Hydrodemolition, concrete pavement									
0120 20,000 PSI, Crew to include loader/vacuum truck as required									
0130 2" depth	B-5	1000	.056	S.F.		2.35	1.43	3.78	5.35
0410 4" depth		800	.070			2.94	1.78	4.72	6.70
0420 6" depth	↓	600	.093	↓		3.92	2.38	6.30	8.95

02 41 13.17 Demolish, Remove Pavement and Curb

	Crew	Daily Output	Labor-Hours	Unit	Material	2016 Bare Costs Labor	Equipment	Total	Total Incl O&P
0010 **DEMOLISH, REMOVE PAVEMENT AND CURB** R024119-10									
5010 Pavement removal, bituminous roads, up to 3" thick	B-38	690	.058	S.Y.		2.50	1.83	4.33	6.05
5050 4" to 6" thick		420	.095			4.11	3.01	7.12	9.90
5100 Bituminous driveways		640	.063			2.70	1.98	4.68	6.50
5200 Concrete to 6" thick, hydraulic hammer, mesh reinforced		255	.157			6.75	4.96	11.71	16.35
5300 Rod reinforced		200	.200	↓		8.65	6.30	14.95	21
5400 Concrete, 7" to 24" thick, plain		33	1.212	C.Y.		52.50	38.50	91	127
5500 Reinforced	↓	24	1.667	"		72	52.50	124.50	174

26

02 41 Demolition

02 41 13 – Selective Site Demolition

02 41 13.17 Demolish, Remove Pavement and Curb

		Crew	Daily Output	Labor-Hours	Unit	Material	2016 Bare Costs Labor	Equipment	Total	Total Incl O&P
5590	Minimum labor/equipment charge	B-38	6	6.667	Job		288	211	499	695
5600	With hand held air equipment, bituminous, to 6" thick	B-39	1900	.025	S.F.		1.01	.12	1.13	1.78
5700	Concrete to 6" thick, no reinforcing		1600	.030			1.20	.14	1.34	2.11
5800	Mesh reinforced		1400	.034			1.38	.16	1.54	2.41
5900	Rod reinforced		765	.063			2.52	.30	2.82	4.42
5990	Minimum labor/equipment charge	B-38	6	6.667	Job		288	211	499	695
6000	Curbs, concrete, plain	B-6	360	.067	L.F.		2.78	1.02	3.80	5.60
6100	Reinforced		275	.087			3.63	1.33	4.96	7.30
6200	Granite		360	.067			2.78	1.02	3.80	5.60
6300	Bituminous		528	.045			1.89	.69	2.58	3.82
6390	Minimum labor/equipment charge		6	4	Job		167	61	228	335

02 41 13.23 Utility Line Removal

		Crew	Daily Output	Labor-Hours	Unit	Material	Labor	Equipment	Total	Total Incl O&P
0010	**UTILITY LINE REMOVAL**									
0015	No hauling, abandon catch basin or manhole	B-6	7	3.429	Ea.		143	52	195	289
0020	Remove existing catch basin or manhole, masonry		4	6			250	91.50	341.50	505
0030	Catch basin or manhole frames and covers, stored		13	1.846			77	28	105	155
0040	Remove and reset		7	3.429			143	52	195	289
0900	Hydrants, fire, remove only	B-21A	5	8			380	96	476	705
0950	Remove and reset	"	2	20			945	240	1,185	1,775
0990	Minimum labor/equipment charge	2 Plum	2	8	Job		475		475	740
2900	Pipe removal, sewer/water, no excavation, 12" diameter	B-6	175	.137	L.F.		5.70	2.09	7.79	11.50
2930	15"-18" diameter	B-12Z	150	.160			6.85	10.55	17.40	22.50
2960	21"-24" diameter		120	.200			8.55	13.20	21.75	28.50
3000	27"-36" diameter		90	.267			11.40	17.60	29	37.50
3200	Steel, welded connections, 4" diameter	B-6	160	.150			6.25	2.28	8.53	12.60
3300	10" diameter		80	.300			12.50	4.57	17.07	25
3390	Minimum labor/equipment charge		3	8	Job		335	122	457	675

02 41 13.30 Minor Site Demolition

		Crew	Daily Output	Labor-Hours	Unit	Material	Labor	Equipment	Total	Total Incl O&P
0010	**MINOR SITE DEMOLITION** R024119-10									
0100	Roadside delineators, remove only	B-80	175	.183	Ea.		7.70	4.16	11.86	17.05
0110	Remove and reset	"	100	.320	"		13.50	7.30	20.80	30
0400	Minimum labor/equipment charge	B-6	4	6	Job		250	91.50	341.50	505
0800	Guiderail, corrugated steel, remove only	B-80A	100	.240	L.F.		9.10	3.07	12.17	18.25
0850	Remove and reset	"	40	.600	"		22.50	7.70	30.20	45.50
0860	Guide posts, remove only	B-80B	120	.267	Ea.		10.85	2.07	12.92	19.85
0870	Remove and reset	B-55	50	.480	"		18.85	22	40.85	54.50
0890	Minimum labor/equipment charge	2 Clab	4	4	Job		152		152	248
1000	Masonry walls, block, solid	B-5	1800	.031	C.F.		1.31	.79	2.10	2.98
1200	Brick, solid		900	.062			2.61	1.58	4.19	5.95
1400	Stone, with mortar		900	.062			2.61	1.58	4.19	5.95
1500	Dry set		1500	.037			1.57	.95	2.52	3.58
1600	Median barrier, precast concrete, remove and store	B-3	430	.112	L.F.		4.71	6	10.71	14.20
1610	Remove and reset	"	390	.123	"		5.20	6.60	11.80	15.65
1650	Minimum labor/equipment charge	A-1	4	2	Job		76	20.50	96.50	147
4000	Sidewalk removal, bituminous, 2" thick	B-6	350	.069	S.Y.		2.86	1.04	3.90	5.75
4010	2-1/2" thick		325	.074			3.08	1.12	4.20	6.20
4050	Brick, set in mortar		185	.130			5.40	1.98	7.38	10.85
4100	Concrete, plain, 4"		160	.150			6.25	2.28	8.53	12.60
4110	Plain, 5"		140	.171			7.15	2.61	9.76	14.40
4120	Plain, 6"		120	.200			8.35	3.05	11.40	16.80
4200	Mesh reinforced, concrete, 4"		150	.160			6.65	2.44	9.09	13.45
4210	5" thick		131	.183			7.65	2.79	10.44	15.35

For customer support on your Commercial Renovation Cost Data, call 877.742.4719.

27

02 41 Demolition

02 41 13 – Selective Site Demolition

02 41 13.30 Minor Site Demolition

		Crew	Daily Output	Labor-Hours	Unit	Material	2016 Bare Costs Labor	2016 Bare Costs Equipment	Total	Total Incl O&P
4220	6" thick	B-6	112	.214	S.Y.		8.95	3.26	12.21	18
4290	Minimum labor/equipment charge	B-39	12	4	Job		160	19.05	179.05	282

02 41 13.33 Railtrack Removal

		Crew	Daily Output	Labor-Hours	Unit	Material	2016 Bare Costs Labor	2016 Bare Costs Equipment	Total	Total Incl O&P
0010	**RAILTRACK REMOVAL**									
3500	Railroad track removal, ties and track	B-13	330	.170	L.F.		7.05	2.27	9.32	13.85
3600	Ballast	B-14	500	.096	C.Y.		3.85	.73	4.58	7.05
3700	Remove and re-install, ties & track using new bolts & spikes		50	.960	L.F.		38.50	7.30	45.80	70.50
3800	Turnouts using new bolts and spikes		1	48	Ea.		1,925	365	2,290	3,525
3890	Minimum labor/equipment charge		5	9.600	Job		385	73	458	705

02 41 13.60 Selective Demolition Fencing

		Crew	Daily Output	Labor-Hours	Unit	Material	2016 Bare Costs Labor	2016 Bare Costs Equipment	Total	Total Incl O&P
0010	**SELECTIVE DEMOLITION FENCING** R024119-10									
1600	Fencing, barbed wire, 3 strand	2 Clab	430	.037	L.F.		1.41		1.41	2.31
1650	5 strand	"	280	.057			2.17		2.17	3.54
1700	Chain link, posts & fabric, 8' to 10' high, remove only	B-6	445	.054			2.25	.82	3.07	4.53

02 41 16 – Structure Demolition

02 41 16.13 Building Demolition

		Crew	Daily Output	Labor-Hours	Unit	Material	2016 Bare Costs Labor	2016 Bare Costs Equipment	Total	Total Incl O&P
0010	**BUILDING DEMOLITION** Large urban projects, incl. 20 mi. haul R024119-10									
0011	No foundation or dump fees, C.F. is vol. of building standing									
0020	Steel	B-8	21500	.003	C.F.		.13	.15	.28	.38
0050	Concrete		15300	.004			.18	.22	.40	.53
0080	Masonry		20100	.003			.14	.17	.31	.40
0100	Mixture of types		20100	.003			.14	.17	.31	.40
0500	Small bldgs, or single bldgs, no salvage included, steel	B-3	14800	.003			.14	.17	.31	.41
0600	Concrete		11300	.004			.18	.23	.41	.54
0650	Masonry		14800	.003			.14	.17	.31	.41
0700	Wood		14800	.003			.14	.17	.31	.41
0750	For buildings with no interior walls, deduct								30%	30%
1000	Demoliton single family house, one story, wood 1600 S.F.	B-3	1	48	Ea.		2,025	2,575	4,600	6,100
1020	3200 S.F.		.50	96			4,050	5,150	9,200	12,200
1200	Demoliton two family house, two story, wood 2400 S.F.		.67	71.964			3,025	3,875	6,900	9,150
1220	4200 S.F.		.38	128			5,400	6,875	12,275	16,300
1300	Demoliton three family house, three story, wood 3200 S.F.		.50	96			4,050	5,150	9,200	12,200
1320	5400 S.F.		.30	160			6,750	8,600	15,350	20,400
5000	For buildings with no interior walls, deduct								30%	30%

02 41 16.17 Building Demolition Footings and Foundations

		Crew	Daily Output	Labor-Hours	Unit	Material	2016 Bare Costs Labor	2016 Bare Costs Equipment	Total	Total Incl O&P
0010	**BUILDING DEMOLITION FOOTINGS AND FOUNDATIONS** R024119-10									
0200	Floors, concrete slab on grade,									
0240	4" thick, plain concrete	B-13L	5000	.003	S.F.		.17	.41	.58	.71
0280	Reinforced, wire mesh		4000	.004			.21	.51	.72	.89
0300	Rods		4500	.004			.19	.46	.65	.79
0400	6" thick, plain concrete		4000	.004			.21	.51	.72	.89
0420	Reinforced, wire mesh		3200	.005			.26	.64	.90	1.12
0440	Rods		3600	.004			.23	.57	.80	1
1000	Footings, concrete, 1' thick, 2' wide		300	.053	L.F.		2.79	6.85	9.64	11.90
1080	1'-6" thick, 2' wide		250	.064			3.34	8.20	11.54	14.30
1120	3' wide		200	.080			4.18	10.25	14.43	17.90
1140	2' thick, 3' wide		175	.091			4.78	11.70	16.48	20.50
1200	Average reinforcing, add								10%	10%
1220	Heavy reinforcing, add								20%	20%
2000	Walls, block, 4" thick	B-13L	8000	.002	S.F.		.10	.26	.36	.45
2040	6" thick		6000	.003			.14	.34	.48	.60

02 41 Demolition

02 41 16 – Structure Demolition

02 41 16.17 Building Demolition Footings and Foundations

		Crew	Daily Output	Labor-Hours	Unit	Material	2016 Bare Costs Labor	2016 Bare Costs Equipment	Total	Total Incl O&P
2080	8" thick	B-13L	4000	.004	S.F.		.21	.51	.72	.89
2100	12" thick	↓	3000	.005			.28	.68	.96	1.19
2200	For horizontal reinforcing, add								10%	10%
2220	For vertical reinforcing, add								20%	20%
2400	Concrete, plain concrete, 6" thick	B-13L	4000	.004			.21	.51	.72	.89
2420	8" thick		3500	.005			.24	.59	.83	1.02
2440	10" thick		3000	.005			.28	.68	.96	1.19
2500	12" thick	↓	2500	.006			.33	.82	1.15	1.43
2600	For average reinforcing, add								10%	10%
2620	For heavy reinforcing, add				↓				20%	20%
9000	Minimum labor/equipment charge	A-1	2	4	Job		152	40.50	192.50	293

02 41 16.19 Minor Building Deconstruction

			Crew	Daily Output	Labor-Hours	Unit	Material	2016 Bare Costs Labor	2016 Bare Costs Equipment	Total	Total Incl O&P
0010	**MINOR BUILDING DECONSTRUCTION**	R024119-10									
0011	For salvage, avg house										
0225	2 story, pre 1970 house, 1400 S.F., labor for all salv mat, min	G	6 Clab	25	1.920	SF Flr.		73		73	119
0230	Maximum	G	"	15	3.200	"		121		121	198
0235	Salvage of carpet, tackless	G	2 Clab	2400	.007	S.F.		.25		.25	.41
0240	Wood floors, incl denailing and packaging	G	3 Clab	270	.089	"		3.37		3.37	5.50
0260	Wood doors and trim, standard	G	1 Clab	16	.500	Ea.		18.95		18.95	31
0280	Base or cove mouldings, incl denailing and packaging	G		500	.016	L.F.		.61		.61	.99
0320	Closet shelving and trim, incl denailing and packaging	G		18	.444	Set		16.85		16.85	27.50
0340	Kitchen cabinets, uppers and lowers, prefab type	G	2 Clab	24	.667	L.F.		25.50		25.50	41.50
0360	Kitchen cabinets, uppers and lowers, built-in	G		12	1.333	"		50.50		50.50	82.50
0370	Bath fixt, incl toilet, tub, vanity, and med cabinet	G	↓	3	5.333	Set		202		202	330

02 41 19 – Selective Demolition

02 41 19.13 Selective Building Demolition

0010	**SELECTIVE BUILDING DEMOLITION**
0020	Costs related to selective demolition of specific building components
0025	are included under Common Work Results (XX 05)
0030	in the component's appropriate division.

02 41 19.16 Selective Demolition, Cutout

			Crew	Daily Output	Labor-Hours	Unit	Material	2016 Bare Costs Labor	2016 Bare Costs Equipment	Total	Total Incl O&P
0010	**SELECTIVE DEMOLITION, CUTOUT**	R024119-10									
0020	Concrete, elev. slab, light reinforcement, under 6 C.F.		B-9	65	.615	C.F.		23.50	3.52	27.02	42.50
0050	Light reinforcing, over 6 C.F.			75	.533	"		20.50	3.05	23.55	37
0200	Slab on grade to 6" thick, not reinforced, under 8 S.F.			85	.471	S.F.		18	2.69	20.69	32.50
0250	8 – 16 S.F.		↓	175	.229	"		8.75	1.31	10.06	15.75
0255	For over 16 S.F. see Line 02 41 16.17 0400										
0600	Walls, not reinforced, under 6 C.F.		B-9	60	.667	C.F.		25.50	3.81	29.31	45.50
0650	6 – 12 C.F.		"	80	.500	"		19.15	2.86	22.01	34.50
0655	For over 12 C.F. see Line 02 41 16.17 2500										
1000	Concrete, elevated slab, bar reinforced, under 6 C.F.		B-9	45	.889	C.F.		34	5.10	39.10	61
1050	Bar reinforced, over 6 C.F.			50	.800	"		30.50	4.58	35.08	55
1200	Slab on grade to 6" thick, bar reinforced, under 8 S.F.			75	.533	S.F.		20.50	3.05	23.55	37
1250	8 – 16 S.F.		↓	150	.267	"		10.20	1.53	11.73	18.40
1255	For over 16 S.F. see Line 02 41 16.17 0440										
1400	Walls, bar reinforced, under 6 C.F.		B-9	50	.800	C.F.		30.50	4.58	35.08	55
1450	6 – 12 C.F.		"	70	.571	"		22	3.27	25.27	39.50
1455	For over 12 C.F. see Lines 02 41 16.17 2500 and 2600										
2000	Brick, to 4 S.F. opening, not including toothing										
2040	4" thick		B-9	30	1.333	Ea.		51	7.65	58.65	92
2060	8" thick			18	2.222			85	12.70	97.70	153
2080	12" thick		↓	10	4	↓		153	23	176	275

For customer support on your Commercial Renovation Cost Data, call 877.742.4719.

29

02 41 19 – Selective Demolition

02 41 19.16 Selective Demolition, Cutout

		Crew	Daily Output	Labor-Hours	Unit	Material	2016 Bare Costs Labor	Equipment	Total	Total Incl O&P
2400	Concrete block, to 4 S.F. opening, 2" thick	B-9	35	1.143	Ea.		44	6.55	50.55	78.50
2420	4" thick		30	1.333			51	7.65	58.65	92
2440	8" thick		27	1.481			56.50	8.45	64.95	102
2460	12" thick		24	1.667			64	9.55	73.55	115
2600	Gypsum block, to 4 S.F. opening, 2" thick		80	.500			19.15	2.86	22.01	34.50
2620	4" thick		70	.571			22	3.27	25.27	39.50
2640	8" thick		55	.727			28	4.16	32.16	50
2800	Terra cotta, to 4 S.F. opening, 4" thick		70	.571			22	3.27	25.27	39.50
2840	8" thick		65	.615			23.50	3.52	27.02	42.50
2880	12" thick		50	.800			30.50	4.58	35.08	55
4000	For toothing masonry, see Section 04 01 20.50									
6000	Walls, interior, not including re-framing,									
6010	openings to 5 S.F.									
6100	Drywall to 5/8" thick	1 Clab	24	.333	Ea.		12.65		12.65	20.50
6200	Paneling to 3/4" thick		20	.400			15.15		15.15	25
6300	Plaster, on gypsum lath		20	.400			15.15		15.15	25
6340	On wire lath		14	.571			21.50		21.50	35.50
7000	Wood frame, not including re-framing, openings to 5 S.F.									
7200	Floors, sheathing and flooring to 2" thick	1 Clab	5	1.600	Ea.		60.50		60.50	99
7310	Roofs, sheathing to 1" thick, not including roofing		6	1.333			50.50		50.50	82.50
7410	Walls, sheathing to 1" thick, not including siding		7	1.143			43.50		43.50	71
8500	Minimum labor/equipment charge		4	2	Job		76		76	124

02 41 19.18 Selective Demolition, Disposal Only

		Crew	Daily Output	Labor-Hours	Unit	Material	2016 Bare Costs Labor	Equipment	Total	Total Incl O&P
0010	**SELECTIVE DEMOLITION, DISPOSAL ONLY**	R024119-10								
0015	Urban bldg w/salvage value allowed									
0020	Including loading and 5 mile haul to dump									
0200	Steel frame	B-3	430	.112	C.Y.		4.71	6	10.71	14.20
0300	Concrete frame		365	.132			5.55	7.05	12.60	16.70
0400	Masonry construction		445	.108			4.55	5.80	10.35	13.70
0500	Wood frame		247	.194			8.20	10.45	18.65	25

02 41 19.19 Selective Demolition

		Crew	Daily Output	Labor-Hours	Unit	Material	2016 Bare Costs Labor	Equipment	Total	Total Incl O&P
0010	**SELECTIVE DEMOLITION**, Rubbish Handling	R024119-10								
0020	The following are to be added to the demolition prices									
0050	The following are components for a complete chute system									
0100	Top chute circular steel, 4' long, 18" diameter	B-1C	15	1.600	Ea.	266	61.50	30	357.50	425
0102	23" diameter		15	1.600		288	61.50	30	379.50	450
0104	27" diameter		15	1.600		310	61.50	30	401.50	475
0106	30" diameter		15	1.600		330	61.50	30	421.50	500
0108	33" diameter		15	1.600		355	61.50	30	446.50	525
0110	36" diameter		15	1.600		375	61.50	30	466.50	550
0112	Regular chute, 18" diameter		15	1.600		199	61.50	30	290.50	355
0114	23" diameter		15	1.600		222	61.50	30	313.50	380
0116	27" diameter		15	1.600		243	61.50	30	334.50	400
0118	30" diameter		15	1.600		266	61.50	30	357.50	425
0120	33" diameter		15	1.600		288	61.50	30	379.50	450
0122	36" diameter		15	1.600		310	61.50	30	401.50	475
0124	Control door chute, 18" diameter		15	1.600		375	61.50	30	466.50	550
0126	23" diameter		15	1.600		400	61.50	30	491.50	575
0128	27" diameter		15	1.600		420	61.50	30	511.50	600
0130	30" diameter		15	1.600		445	61.50	30	536.50	625
0132	33" diameter		15	1.600		465	61.50	30	556.50	645
0134	36" diameter		15	1.600		490	61.50	30	581.50	670

02 41 19.19 Selective Demolition	Crew	Daily Output	Labor-Hours	Unit	Material	2016 Bare Costs Labor	Equipment	Total	Total Incl O&P
0136	Chute liners, 14 ga., 18-30" diameter	B-1C	15	1.600	Ea.	223	61.50	30	314.50	380
0138	33-36" diameter		15	1.600		280	61.50	30	371.50	445
0140	17% thinner chute, 30" diameter		15	1.600		222	61.50	30	313.50	380
0142	33% thinner chute, 30" diameter		15	1.600		167	61.50	30	258.50	315
0144	Top chute cover	1 Clab	24	.333		153	12.65		165.65	189
0146	Door chute cover	"	24	.333		153	12.65		165.65	189
0148	Top chute trough	2 Clab	12	1.333		510	50.50		560.50	645
0150	Bolt down frame & counter weights, 250 lb.	B-1	4	6		4,250	231		4,481	5,025
0152	500 lb.		4	6		6,200	231		6,431	7,200
0154	750 lb.		4	6		9,100	231		9,331	10,400
0156	1000 lb.		2.67	8.989		10,200	345		10,545	11,800
0158	1500 lb.		2.67	8.989		13,000	345		13,345	14,900
0160	Chute warning light system, 5 stories	B-1C	4	6		8,975	231	112	9,318	10,400
0162	10 stories	"	2	12		14,300	465	224	14,989	16,700
0164	Dust control device for dumpsters	1 Clab	8	1		102	38		140	174
0166	Install or replace breakaway cord		8	1		27	38		65	91.50
0168	Install or replace warning sign		16	.500		11.45	18.95		30.40	43.50
0600	Dumpster, weekly rental, 1 dump/week, 6 C.Y. capacity (2 Tons)				Week	415			415	455
0700	10 C.Y. capacity (3 Tons)					480			480	530
0725	20 C.Y. capacity (5 Tons)					565			565	625
0800	30 C.Y. capacity (7 Tons)					730			730	800
0840	40 C.Y. capacity (10 Tons)					775			775	850
0900	Alternate pricing for dumpsters									
0910	Delivery, average for all sizes				Ea.	75			75	82.50
0920	Haul, average for all sizes					235			235	259
0930	Rent per day, average for all sizes					20			20	22
0940	Rent per month, average for all sizes					80			80	88
0950	Disposal fee per ton, average for all sizes				Ton	88			88	97
2000	Load, haul, dump and return, 0 – 50' haul, hand carried	2 Clab	24	.667	C.Y.		25.50		25.50	41.50
2005	Wheeled		37	.432			16.40		16.40	27
2040	0 – 100' haul, hand carried		16.50	.970			37		37	60
2045	Wheeled		25	.640			24.50		24.50	39.50
2080	Haul and return, add per each extra 100' haul, hand carried		35.50	.451			17.10		17.10	28
2085	Wheeled		54	.296			11.25		11.25	18.35
2120	For travel in elevators, up to 10 floors, add		140	.114			4.33		4.33	7.10
2130	0 – 50' haul, incl. up to 5 riser stair, hand carried		23	.696			26.50		26.50	43
2135	Wheeled		35	.457			17.35		17.35	28.50
2140	6 – 10 riser stairs, hand carried		22	.727			27.50		27.50	45
2145	Wheeled		34	.471			17.85		17.85	29
2150	11 – 20 riser stairs, hand carried		20	.800			30.50		30.50	49.50
2155	Wheeled		31	.516			19.55		19.55	32
2160	21 – 40 riser stairs, hand carried		16	1			38		38	62
2165	Wheeled		24	.667			25.50		25.50	41.50
2170	0 – 100' haul, incl. 5 riser stair, hand carried		15	1.067			40.50		40.50	66
2175	Wheeled		23	.696			26.50		26.50	43
2180	6 – 10 riser stair, hand carried		14	1.143			43.50		43.50	71
2185	Wheeled		21	.762			29		29	47
2190	11 – 20 riser stair, hand carried		12	1.333			50.50		50.50	82.50
2195	Wheeled		18	.889			33.50		33.50	55
2200	21 – 40 riser stair, hand carried		8	2			76		76	124
2205	Wheeled		12	1.333			50.50		50.50	82.50
2210	Haul and return, add per each extra 100' haul, hand carried		35.50	.451			17.10		17.10	28
2215	Wheeled		54	.296			11.25		11.25	18.35

R024119-20

02 41 Demolition

02 41 19 – Selective Demolition

02 41 19.19 Selective Demolition

		Crew	Daily Output	Labor-Hours	Unit	Material	2016 Bare Costs Labor	2016 Bare Costs Equipment	Total	Total Incl O&P
2220	For each additional flight of stairs, up to 5 risers, add	2 Clab	550	.029	Flight		1.10		1.10	1.80
2225	6 – 10 risers, add		275	.058			2.21		2.21	3.60
2230	11 – 20 risers, add		138	.116			4.39		4.39	7.20
2235	21 – 40 risers, add		69	.232			8.80		8.80	14.35
3000	Loading & trucking, including 2 mile haul, chute loaded	B-16	45	.711	C.Y.		28.50	15.35	43.85	63
3040	Hand loading truck, 50' haul	"	48	.667			26.50	14.40	40.90	59
3080	Machine loading truck	B-17	120	.267			11.20	6.50	17.70	25.50
3120	Wheeled 50' and ramp dump loaded	2 Clab	24	.667			25.50		25.50	41.50
5000	Haul, per mile, up to 8 C.Y. truck	B-34B	1165	.007			.30	.59	.89	1.13
5100	Over 8 C.Y. truck	"	1550	.005			.22	.45	.67	.85

02 41 19.20 Selective Demolition, Dump Charges

			Crew	Daily Output	Labor-Hours	Unit	Material	2016 Bare Costs Labor	2016 Bare Costs Equipment	Total	Total Incl O&P
0010	**SELECTIVE DEMOLITION, DUMP CHARGES**	R024119-10									
0020	Dump charges, typical urban city, tipping fees only										
0100	Building construction materials					Ton	74			74	81
0200	Trees, brush, lumber						63			63	69.50
0300	Rubbish only						63			63	69.50
0500	Reclamation station, usual charge						74			74	81

02 41 19.21 Selective Demolition, Gutting

			Crew	Daily Output	Labor-Hours	Unit	Material	2016 Bare Costs Labor	2016 Bare Costs Equipment	Total	Total Incl O&P
0010	**SELECTIVE DEMOLITION, GUTTING**	R024119-10									
0020	Building interior, including disposal, dumpster fees not included										
0500	Residential building										
0560	Minimum		B-16	400	.080	SF Flr.		3.18	1.73	4.91	7.05
0580	Maximum		"	360	.089	"		3.53	1.92	5.45	7.85
0900	Commercial building										
1000	Minimum		B-16	350	.091	SF Flr.		3.63	1.97	5.60	8.05
1020	Maximum			250	.128	"		5.10	2.76	7.86	11.35
3000	Minimum labor/equipment charge			4	8	Job		320	173	493	705

02 41 19.25 Selective Demolition, Saw Cutting

			Crew	Daily Output	Labor-Hours	Unit	Material	2016 Bare Costs Labor	2016 Bare Costs Equipment	Total	Total Incl O&P
0010	**SELECTIVE DEMOLITION, SAW CUTTING**	R024119-10									
0015	Asphalt, up to 3" deep		B-89	1050	.015	L.F.	.12	.69	.47	1.28	1.76
0020	Each additional inch of depth		"	1800	.009		.04	.40	.27	.71	.99
1200	Masonry walls, hydraulic saw, brick, per inch of depth		B-89B	300	.053		.04	2.43	2.84	5.31	7.05
1220	Block walls, solid, per inch of depth		"	250	.064		.04	2.92	3.41	6.37	8.45
2000	Brick or masonry w/hand held saw, per inch of depth		A-1	125	.064		.03	2.43	.65	3.11	4.72
5000	Wood sheathing to 1" thick, on walls		1 Carp	200	.040			1.94		1.94	3.17
5020	On roof		"	250	.032			1.55		1.55	2.53
9000	Minimum labor/equipment charge		A-1	2	4	Job		152	40.50	192.50	293

02 41 19.27 Selective Demolition, Torch Cutting

			Crew	Daily Output	Labor-Hours	Unit	Material	2016 Bare Costs Labor	2016 Bare Costs Equipment	Total	Total Incl O&P
0010	**SELECTIVE DEMOLITION, TORCH CUTTING**	R024119-10									
0020	Steel, 1" thick plate		E-25	333	.024	L.F.	.84	1.33	.03	2.20	3.30
0040	1" diameter bar		"	600	.013	Ea.	.14	.74	.02	.90	1.47
1000	Oxygen lance cutting, reinforced concrete walls										
1040	12" to 16" thick walls		1 Clab	10	.800	L.F.		30.50		30.50	49.50
1080	24" thick walls		"	6	1.333	"		50.50		50.50	82.50
1090	Minimum labor/equipment charge		E-25	2	4	Job		221	5.70	226.70	395
1100	See Section 05 05 21.10										

02 42 Removal and Salvage of Construction Materials

02 42 10 – Building Deconstruction

02 42 10.10 Estimated Salvage Value or Savings		Crew	Daily Output	Labor-Hours	Unit	Material	2016 Bare Costs Labor	2016 Bare Costs Equipment	Total	Total Incl O&P
0010	**ESTIMATED SALVAGE VALUE OR SAVINGS**									
0015	Excludes material handling, packaging, container costs and									
0020	transportation for salvage or disposal									
0050	All Items in Section 02 42 10.10 are credit deducts and not costs									
0100	Copper Wire Salvage Value	G				Lb.			1.60	1.60
0110	Disposal Savings	G							.04	.04
0200	Copper Pipe Salvage Value	G							2.50	2.50
0210	Disposal Savings	G							.05	.05
0300	Steel Pipe Salvage Value	G							.06	.06
0310	Disposal Savings	G							.03	.03
0400	Cast Iron Pipe Salvage Value	G							.03	.03
0410	Disposal Savings	G							.01	.01
0500	Steel Doors or Windows Salvage Value	G							.06	.06
0510	Aluminum	G							.55	.55
0520	Disposal Savings	G							.03	.03
0600	Aluminum Siding Salvage Value	G							.49	.49
0630	Disposal Savings	G							.03	.03
0640	Wood Siding (no lead or asbestos)	G				C.Y.			12	12
0800	Clean Concrete Disposal Savings	G				Ton			62	62
0850	Asphalt Shingles Disposal Savings	G				"			60	60
1000	Wood wall framing clean salvage value	G				M.B.F.			55	55
1010	Painted	G							44	44
1020	Floor framing	G							55	55
1030	Painted	G							44	44
1050	Roof framing	G							55	55
1060	Painted	G							44	44
1100	Wood beams salvage value	G							55	55
1200	Wood framing and beams disposal savings	G				Ton			66	66
1220	Wood sheathing and sub-base flooring	G							72.50	72.50
1230	Wood wall paneling (1/4 inch thick)	G							66	66
1300	Wood panel 3/4-1 inch thick low salvage value	G				S.F.			.55	.55
1350	High salvage value	G				"			2.20	2.20
1400	Disposal savings	G				Ton			66	66
1500	Flooring tongue and groove 25/32 inch thick low salvage value	G				S.F.			.55	.55
1530	High salvage value	G				"			1.10	1.10
1560	Disposal savings	G				Ton			66	66
1600	Drywall or sheet rock salvage value	G							22	22
1650	Disposal savings	G							66	66

02 42 10.20 Deconstruction of Building Components

		Crew	Daily Output	Labor-Hours	Unit	Material	Labor	Equipment	Total	Total Incl O&P
0010	**DECONSTRUCTION OF BUILDING COMPONENTS**									
0012	Buildings one or two stories only									
0015	Excludes material handling, packaging, container costs and									
0020	transportation for salvage or disposal									
0050	Deconstruction of Plumbing Fixtures									
0100	Wall hung or countertop lavatory	G	2 Clab	16	1	Ea.		38		38	62
0110	Single or double compartment kitchen sink	G		14	1.143			43.50		43.50	71
0120	Wall hung urinal	G		14	1.143			43.50		43.50	71
0130	Floor mounted	G		8	2			76		76	124
0140	Floor mounted water closet	G		16	1			38		38	62
0150	Wall hung	G		14	1.143			43.50		43.50	71
0160	Water fountain, free standing	G		16	1			38		38	62
0170	Wall hung or deck mounted	G		12	1.333			50.50		50.50	82.50

For customer support on your Commercial Renovation Cost Data, call 877.742.4719.

33

02 42 Removal and Salvage of Construction Materials

02 42 10 – Building Deconstruction

02 42 10.20 Deconstruction of Building Components

		Crew	Daily Output	Labor-Hours	Unit	Material	2016 Bare Costs Labor	Equipment	Total	Total Incl O&P
0180	Bathtub, steel or fiberglass	G 2 Clab	10	1.600	Ea.		60.50		60.50	99
0190	Cast iron	G	8	2			76		76	124
0200	Shower, single	G	6	2.667			101		101	165
0210	Group	G	7	2.286			86.50		86.50	142
0300	Deconstruction of Electrical Fixtures									
0310	Surface mounted incandescent fixtures	G 2 Clab	48	.333	Ea.		12.65		12.65	20.50
0320	Fluorescent, 2 lamp	G	32	.500			18.95		18.95	31
0330	4 lamp	G	24	.667			25.50		25.50	41.50
0340	Strip Fluorescent, 1 lamp	G	40	.400			15.15		15.15	25
0350	2 lamp	G	32	.500			18.95		18.95	31
0400	Recessed drop-in fluorescent fixture, 2 lamp	G	27	.593			22.50		22.50	36.50
0410	4 lamp	G	18	.889			33.50		33.50	55
0500	Deconstruction of appliances									
0510	Cooking stoves	G 2 Clab	26	.615	Ea.		23.50		23.50	38
0520	Dishwashers	G "	26	.615	"		23.50		23.50	38
0600	Deconstruction of millwork and trim									
0610	Cabinets, wood	G 2 Carp	40	.400	L.F.		19.40		19.40	31.50
0620	Countertops	G	100	.160	"		7.75		7.75	12.65
0630	Wall paneling, 1 inch thick	G	500	.032	S.F.		1.55		1.55	2.53
0640	Ceiling trim	G	500	.032	L.F.		1.55		1.55	2.53
0650	Wainscoting	G	500	.032	S.F.		1.55		1.55	2.53
0660	Base, 3/4" to 1" thick	G	600	.027	L.F.		1.29		1.29	2.11
0700	Deconstruction of doors and windows									
0710	Doors, wrap, interior, wood, single, no closers	G 2 Carp	21	.762	Ea.	4.75	37		41.75	66
0720	Double	G	13	1.231		9.50	59.50		69	108
0730	Solid core, single, exterior or interior	G	10	1.600		4.75	77.50		82.25	132
0740	Double	G	8	2		9.50	97		106.50	168
0810	Windows, wrap, wood, single									
0812	with no casement or cladding	G 2 Carp	21	.762	Ea.	4.75	37		41.75	66
0820	with casement and/or cladding	G "	18	.889	"	4.75	43		47.75	76
0900	Deconstruction of interior finishes									
0910	Drywall for recycling	G 2 Clab	1775	.009	S.F.		.34		.34	.56
0920	Plaster wall, first floor	G	1775	.009			.34		.34	.56
0930	Second floor	G	1330	.012			.46		.46	.75
1000	Deconstruction of roofing and accessories									
1010	Built-up roofs	G 2 Clab	570	.028	S.F.		1.06		1.06	1.74
1020	Gutters, fascia and rakes	G "	1140	.014	L.F.		.53		.53	.87
2000	Deconstruction of wood components									
2010	Roof sheeting	G 2 Clab	570	.028	S.F.		1.06		1.06	1.74
2020	Main roof framing	G	760	.021	L.F.		.80		.80	1.30
2030	Porch roof framing	G	445	.036			1.36		1.36	2.23
2040	Beams 4" x 8"	G B-1	375	.064			2.47		2.47	4.03
2050	4" x 10"	G	300	.080			3.09		3.09	5.05
2055	4" x 12"	G	250	.096			3.70		3.70	6.05
2060	6" x 8"	G	250	.096			3.70		3.70	6.05
2065	6" x 10"	G	200	.120			4.63		4.63	7.55
2070	6" x 12"	G	170	.141			5.45		5.45	8.90
2075	8" x 12"	G	126	.190			7.35		7.35	12
2080	10" x 12"	G	100	.240			9.25		9.25	15.15
2100	Ceiling joists	G 2 Clab	800	.020			.76		.76	1.24
2150	Wall framing, interior	G	1230	.013			.49		.49	.81
2160	Sub-floor	G	2000	.008	S.F.		.30		.30	.50
2170	Floor joists	G	2000	.008	L.F.		.30		.30	.50

For customer support on your Commercial Renovation Cost Data, call 877.742.4719.

02 42 Removal and Salvage of Construction Materials

02 42 10 – Building Deconstruction

02 42 10.20 Deconstruction of Building Components

		Crew	Daily Output	Labor-Hours	Unit	Material	2016 Bare Costs Labor	2016 Bare Costs Equipment	Total	Total Incl O&P
2200	Wood siding (no lead or asbestos)	G 2 Clab	1300	.012	S.F.		.47		.47	.76
2300	Wall framing, exterior	G	1600	.010	L.F.		.38		.38	.62
2400	Stair risers	G	53	.302	Ea.		11.45		11.45	18.70
2500	Posts	G	800	.020	L.F.		.76		.76	1.24
3000	Deconstruction of exterior brick walls									
3010	Exterior brick walls, first floor	G 2 Clab	200	.080	S.F.		3.03		3.03	4.96
3020	Second floor	G	64	.250	"		9.50		9.50	15.50
3030	Brick chimney	G	100	.160	C.F.		6.05		6.05	9.90
4000	Deconstruction of concrete									
4010	Slab on grade, 4" thick, plain concrete	G B-9	500	.080	S.F.		3.06	.46	3.52	5.50
4020	Wire mesh reinforced	G	470	.085			3.26	.49	3.75	5.90
4030	Rod reinforced	G	400	.100			3.83	.57	4.40	6.90
4110	Foundation wall, 6" thick, plain concrete	G	160	.250			9.60	1.43	11.03	17.20
4120	8" thick	G	140	.286			10.95	1.63	12.58	19.70
4130	10" thick	G	120	.333			12.75	1.91	14.66	23
9000	Deconstruction process, support equipment as needed									
9010	Daily use, portal to portal, 12-ton truck-mounted hydraulic crane crew	G A-3H	1	8	Day		420	875	1,295	1,625
9020	Daily use, skid steer and operator	G A-3C	1	8			395	315	710	965
9030	Daily use, backhoe 48 H.P., operator and labor	G "	1	8			395	315	710	965

02 42 10.30 Deconstruction Material Handling

		Crew	Daily Output	Labor-Hours	Unit	Material	2016 Bare Costs Labor	2016 Bare Costs Equipment	Total	Total Incl O&P
0010	**DECONSTRUCTION MATERIAL HANDLING**									
0012	Buildings one or two stories only									
0100	Clean and stack brick on pallet	G 2 Clab	1200	.013	Ea.		.51		.51	.83
0200	Haul 50' and load rough lumber up to 2" x 8" size	G	2000	.008	"		.30		.30	.50
0210	Lumber larger than 2" x 8"	G	3200	.005	B.F.		.19		.19	.31
0300	Finish wood for recycling stack and wrap per pallet	G	8	2	Ea.	38	76		114	166
0350	Light fixtures		6	2.667		68.50	101		169.50	240
0375	Windows		6	2.667		64.50	101		165.50	236
0400	Miscellaneous materials		8	2		19	76		95	145
1000	See Section 02 41 19.19 for bulk material handling									

02 43 Structure Moving

02 43 13 – Structure Relocation

02 43 13.13 Building Relocation

		Crew	Daily Output	Labor-Hours	Unit	Material	2016 Bare Costs Labor	2016 Bare Costs Equipment	Total	Total Incl O&P
0010	**BUILDING RELOCATION**									
0011	One day move, up to 24' wide									
0020	Reset on existing foundation				Total				11,500	11,500
0040	Wood or steel frame bldg., based on ground floor area	G B-4	185	.259	S.F.		10.15	2.78	12.93	19.60
0060	Masonry bldg., based on ground floor area	G "	137	.350			13.70	3.75	17.45	26.50
0200	For 24' to 42' wide, add								15%	15%

For customer support on your Commercial Renovation Cost Data, call 877.742.4719.

35

02 58 Snow Control

02 58 13 – Snow Fencing

02 58 13.10 Snow Fencing System	Crew	Daily Output	Labor-Hours	Unit	Material	2016 Bare Costs Labor	Equipment	Total	Total Incl O&P
0010 **SNOW FENCING SYSTEM**									
7001 Snow fence on steel posts 10' O.C., 4' high	B-1	500	.048	L.F.	.45	1.85		2.30	3.53

02 65 Underground Storage Tank Removal

02 65 10 – Underground Tank and Contaminated Soil Removal

02 65 10.30 Removal of Underground Storage Tanks

		Crew	Daily Output	Labor-Hours	Unit	Material	2016 Bare Costs Labor	Equipment	Total	Total Incl O&P
0010	**REMOVAL OF UNDERGROUND STORAGE TANKS** R026510-20									
0011	Petroleum storage tanks, non-leaking									
0100	Excavate & load onto trailer									
0110	3000 gal. to 5000 gal. tank	G B-14	4	12	Ea.		480	91.50	571.50	880
0120	6000 gal. to 8000 gal. tank	G B-3A	3	13.333			540	345	885	1,250
0130	9000 gal. to 12000 gal. tank	"	2	20			810	515	1,325	1,900
0190	Known leaking tank, add				%				100%	100%
0200	Remove sludge, water and remaining product from tank bottom									
0201	of tank with vacuum truck									
0300	3000 gal. to 5000 gal. tank	G A-13	5	1.600	Ea.		78.50	152	230.50	291
0310	6000 gal. to 8000 gal. tank	G	4	2			98.50	190	288.50	365
0320	9000 gal. to 12000 gal. tank	G	3	2.667			131	253	384	485
0390	Dispose of sludge off-site, average				Gal.				6.25	6.80
0400	Insert inert solid CO_2 "dry ice" into tank									
0401	For cleaning/transporting tanks (1.5 lb./100 gal. cap)	G 1 Clab	500	.016	Lb.	1.17	.61		1.78	2.28
0403	Insert solid carbon dioxide, 1.5 lb./100 gal.	G "	400	.020	"	1.17	.76		1.93	2.53
0503	Disconnect and remove piping	G 1 Plum	160	.050	L.F.		2.96		2.96	4.62
0603	Transfer liquids, 10% of volume	G "	1600	.005	Gal.		.30		.30	.46
0703	Cut accessway into underground storage tank	G 1 Clab	5.33	1.501	Ea.		57		57	93
0813	Remove sludge, wash and wipe tank, 500 gal.	G 1 Plum	8	1			59		59	92.50
0823	3,000 gal.	G	6.67	1.199			71		71	111
0833	5,000 gal.	G	6.15	1.301			77		77	120
0843	8,000 gal.	G	5.33	1.501			89		89	139
0853	10,000 gal.	G	4.57	1.751			104		104	162
0863	12,000 gal.	G	4.21	1.900			112		112	175
1020	Haul tank to certified salvage dump, 100 miles round trip									
1023	3000 gal. to 5000 gal. tank				Ea.				760	830
1026	6000 gal. to 8000 gal. tank								880	960
1029	9,000 gal. to 12,000 gal. tank								1,050	1,150
1100	Disposal of contaminated soil to landfill									
1110	Minimum				C.Y.				145	160
1111	Maximum				"				400	440
1120	Disposal of contaminated soil to									
1121	bituminous concrete batch plant									
1130	Minimum				C.Y.				80	88
1131	Maximum				"				115	125
1203	Excavate, pull, & load tank, backfill hole, 8,000 gal. +	G B-12C	.50	32	Ea.		1,450	2,350	3,800	4,900
1213	Haul tank to certified dump, 100 miles rt, 8,000 gal. +	G B-34K	1	8			345	950	1,295	1,600
1223	Excavate, pull, & load tank, backfill hole, 500 gal.	G B-11C	1	16			710	365	1,075	1,550
1233	Excavate, pull, & load tank, backfill hole, 3,000 – 5,000 gal.	G B-11M	.50	32			1,425	790	2,215	3,150
1243	Haul tank to certified dump, 100 miles rt, 500 gal.	G B-34L	1	8			395	248	643	895
1253	Haul tank to certified dump, 100 miles rt, 3,000 – 5,000 gal.	G B-34M	1	8			395	305	700	960
2010	Decontamination of soil on site incl poly tarp on top/bottom									
2011	Soil containment berm and chemical treatment									
2020	Minimum	G B-11C	100	.160	C.Y.	7.80	7.10	3.65	18.55	24

02 65 Underground Storage Tank Removal

02 65 10 – Underground Tank and Contaminated Soil Removal

02 65 10.30 Removal of Underground Storage Tanks		Crew	Daily Output	Labor-Hours	Unit	Material	2016 Bare Costs Labor	Equipment	Total	Total Incl O&P
2021	Maximum	G B-11C	100	.160	C.Y.	10.10	7.10	3.65	20.85	26.50
2050	Disposal of decontaminated soil, minimum								135	150
2055	Maximum			↓					400	440

02 81 Transportation and Disposal of Hazardous Materials

02 81 20 – Hazardous Waste Handling

02 81 20.10 Hazardous Waste Cleanup/Pickup/Disposal

		Crew	Daily Output	Labor-Hours	Unit	Material	Labor	Equipment	Total	Total Incl O&P
0010	**HAZARDOUS WASTE CLEANUP/PICKUP/DISPOSAL**									
0100	For contractor rental equipment, i.e., Dozer,									
0110	Front end loader, Dump truck, etc., see 01 54 33 Reference Section									
1000	Solid pickup									
1100	55 gal. drums				Ea.				240	265
1120	Bulk material, minimum				Ton				190	210
1130	Maximum				"				595	655
1200	Transportation to disposal site									
1220	Truckload = 80 drums or 25 C.Y. or 18 tons									
1260	Minimum				Mile				3.95	4.45
1270	Maximum				"				7.25	7.35
3000	Liquid pickup, vacuum truck, stainless steel tank									
3100	Minimum charge, 4 hours									
3110	1 compartment, 2200 gallon				Hr.				140	155
3120	2 compartment, 5000 gallon				"				200	225
3400	Transportation in 6900 gallon bulk truck				Mile				7.95	8.75
3410	In teflon lined truck				"				10.20	11.25
5000	Heavy sludge or dry vacuumable material				Hr.				140	160
6000	Dumpsite disposal charge, minimum				Ton				140	155
6020	Maximum				"				415	455

02 82 Asbestos Remediation

02 82 13 – Asbestos Abatement

02 82 13.39 Asbestos Remediation Plans and Methods

		Crew	Daily Output	Labor-Hours	Unit	Material	Labor	Equipment	Total	Total Incl O&P
0010	**ASBESTOS REMEDIATION PLANS AND METHODS**									
0100	Building Survey-Commercial Building				Ea.				2,200	2,400
0200	Asbestos Abatement Remediation Plan				"				1,350	1,475

02 82 13.41 Asbestos Abatement Equipment

		Crew	Daily Output	Labor-Hours	Unit	Material	Labor	Equipment	Total	Total Incl O&P
0010	**ASBESTOS ABATEMENT EQUIPMENT** R028213-20									
0011	Equipment and supplies, buy									
0200	Air filtration device, 2000 CFM				Ea.	975			975	1,075
0250	Large volume air sampling pump, minimum					355			355	390
0260	Maximum					340			340	375
0300	Airless sprayer unit, 2 gun					3,275			3,275	3,600
0350	Light stand, 500 watt				↓	51			51	56
0400	Personal respirators									
0410	Negative pressure, 1/2 face, dual operation, min.				Ea.	27			27	29.50
0420	Maximum					30			30	33
0450	P.A.P.R., full face, minimum					127			127	140
0460	Maximum					169			169	185
0470	Supplied air, full face, incl. air line, minimum					172			172	189
0480	Maximum				↓	410			410	450

For customer support on your Commercial Renovation Cost Data, call 877.742.4719.

37

02 82 13 – Asbestos Abatement

02 82 13.41 Asbestos Abatement Equipment

		Crew	Daily Output	Labor-Hours	Unit	Material	2016 Bare Costs Labor	Equipment	Total	Total Incl O&P
0500	Personnel sampling pump				Ea.	230			230	254
1500	Power panel, 20 unit, incl. GFI					615			615	675
1600	Shower unit, including pump and filters					1,300			1,300	1,450
1700	Supplied air system (type C)					3,475			3,475	3,825
1750	Vacuum cleaner, HEPA, 16 gal., stainless steel, wet/dry					440			440	485
1760	55 gallon					570			570	630
1800	Vacuum loader, 9 – 18 ton/hr.					96,000			96,000	106,000
1900	Water atomizer unit, including 55 gal. drum					286			286	315
2000	Worker protection, whole body, foot, head cover & gloves, plastic					9			9	9.90
2500	Respirator, single use					25.50			25.50	28
2550	Cartridge for respirator					5.85			5.85	6.45
2570	Glove bag, 7 mil, 50" x 64"					9.35			9.35	10.30
2580	10 mil, 44" x 60"					5.90			5.90	6.45
3000	HEPA vacuum for work area, minimum					310			310	345
3050	Maximum					830			830	915
6000	Disposable polyethylene bags, 6 mil, 3 C.F.					.84			.84	.92
6300	Disposable fiber drums, 3 C.F.					18.55			18.55	20.50
6400	Pressure sensitive caution labels, 3" x 5"					3.54			3.54	3.89
6450	11" x 17"					7.50			7.50	8.25
6500	Negative air machine, 1800 CFM					865			865	950

02 82 13.42 Preparation of Asbestos Containment Area

		Crew	Daily Output	Labor-Hours	Unit	Material	2016 Bare Costs Labor	Equipment	Total	Total Incl O&P
0010	**PREPARATION OF ASBESTOS CONTAINMENT AREA**									
0100	Pre-cleaning, HEPA vacuum and wet wipe, flat surfaces	A-9	12000	.005	S.F.	.02	.28		.30	.48
0200	Protect carpeted area, 2 layers 6 mil poly on 3/4" plywood	"	1000	.064		2	3.42		5.42	7.70
0300	Separation barrier, 2" x 4" @ 16", 1/2" plywood ea. side, 8' high	2 Carp	400	.040		3.40	1.94		5.34	6.90
0310	12' high		320	.050		3.40	2.42		5.82	7.70
0320	16' high		200	.080		2.30	3.88		6.18	8.90
0400	Personnel decontam. chamber, 2" x 4" @ 16", 3/4" ply ea. side		280	.057		4.40	2.77		7.17	9.35
0450	Waste decontam. chamber, 2" x 4" studs @ 16", 3/4" ply ea. side		360	.044		4.40	2.15		6.55	8.35
0500	Cover surfaces with polyethylene sheeting									
0501	Including glue and tape									
0550	Floors, each layer, 6 mil	A-9	8000	.008	S.F.	.04	.43		.47	.73
0551	4 mil		9000	.007		.03	.38		.41	.64
0560	Walls, each layer, 6 mil		6000	.011		.04	.57		.61	.96
0561	4 mil		7000	.009		.03	.49		.52	.81
0570	For heights above 12', add						20%			
0575	For heights above 20', add						30%			
0580	For fire retardant poly, add					100%				
0590	For large open areas, deduct					10%	20%			
0600	Seal floor penetrations with foam firestop to 36 sq. in.	2 Carp	200	.080	Ea.	9.50	3.88		13.38	16.85
0610	36 sq. in. to 72 sq. in.		125	.128		19.05	6.20		25.25	31
0615	72 sq. in. to 144 sq. in.		80	.200		38	9.70		47.70	58
0620	Wall penetrations, to 36 square inches		180	.089		9.50	4.31		13.81	17.55
0630	36 sq. in. to 72 sq. in.		100	.160		19.05	7.75		26.80	33.50
0640	72 sq. in. to 144 sq. in.		60	.267		38	12.90		50.90	63
0800	Caulk seams with latex	1 Carp	230	.035	L.F.	.17	1.69		1.86	2.94
0900	Set up neg. air machine, 1-2k CFM/25 M.C.F. volume	1 Asbe	4.30	1.860	Ea.		99.50		99.50	160
0950	Set up and remove portable shower unit	2 Asbe	4	4	"		214		214	345

02 82 13.43 Bulk Asbestos Removal

		Crew	Daily Output	Labor-Hours	Unit	Material	2016 Bare Costs Labor	2016 Bare Costs Equipment	Total	Total Incl O&P
0010	**BULK ASBESTOS REMOVAL**									
0020	Includes disposable tools and 2 suits and 1 respirator filter/day/worker									
0100	Beams, W 10 x 19	A-9	235	.272	L.F.	.81	14.55		15.36	24.50
0110	W 12 x 22		210	.305		.91	16.30		17.21	27
0120	W 14 x 26		180	.356		1.06	19		20.06	31.50
0130	W 16 x 31		160	.400		1.19	21.50		22.69	36
0140	W 18 x 40		140	.457		1.36	24.50		25.86	41
0150	W 24 x 55		110	.582		1.74	31		32.74	52
0160	W 30 x 108		85	.753		2.25	40.50		42.75	67
0170	W 36 x 150		72	.889		2.65	47.50		50.15	79.50
0200	Boiler insulation		480	.133	S.F.	.47	7.15		7.62	11.95
0210	With metal lath, add				%				50%	50%
0300	Boiler breeching or flue insulation	A-9	520	.123	S.F.	.37	6.60		6.97	10.95
0310	For active boiler, add				%				100%	100%
0400	Duct or AHU insulation	A-10B	440	.073	S.F.	.22	3.89		4.11	6.50
0500	Duct vibration isolation joints, up to 24 sq. in. duct	A-9	56	1.143	Ea.	3.41	61		64.41	102
0520	25 sq. in. to 48 sq. in. duct		48	1.333		3.98	71.50		75.48	118
0530	49 sq. in. to 76 sq. in. duct		40	1.600		4.77	85.50		90.27	142
0600	Pipe insulation, air cell type, up to 4" diameter pipe		900	.071	L.F.	.21	3.80		4.01	6.35
0610	4" to 8" diameter pipe		800	.080		.24	4.28		4.52	7.10
0620	10" to 12" diameter pipe		700	.091		.27	4.89		5.16	8.15
0630	14" to 16" diameter pipe		550	.116		.35	6.20		6.55	10.40
0650	Over 16" diameter pipe		650	.098	S.F.	.29	5.25		5.54	8.75
0700	With glove bag up to 3" diameter pipe		200	.320	L.F.	9.30	17.10		26.40	38
1000	Pipe fitting insulation up to 4" diameter pipe		320	.200	Ea.	.60	10.70		11.30	17.80
1100	6" to 8" diameter pipe		304	.211		.63	11.25		11.88	18.75
1110	10" to 12" diameter pipe		192	.333		.99	17.80		18.79	29.50
1120	14" to 16" diameter pipe		128	.500		1.49	26.50		27.99	44.50
1130	Over 16" diameter pipe		176	.364	S.F.	1.08	19.45		20.53	32
1200	With glove bag, up to 8" diameter pipe		75	.853	L.F.	6.40	45.50		51.90	80.50
2000	Scrape foam fireproofing from flat surface		2400	.027	S.F.	.08	1.43		1.51	2.38
2100	Irregular surfaces		1200	.053		.16	2.85		3.01	4.75
3000	Remove cementitious material from flat surface		1800	.036		.11	1.90		2.01	3.17
3100	Irregular surface		1000	.064		.14	3.42		3.56	5.65
4000	Scrape acoustical coating/fireproofing, from ceiling		3200	.020		.06	1.07		1.13	1.79
5000	Remove VAT and mastic from floor by hand		2400	.027		.08	1.43		1.51	2.38
5100	By machine	A-11	4800	.013		.04	.71	.01	.76	1.19
5150	For 2 layers, add				%				50%	50%
6000	Remove contaminated soil from crawl space by hand	A-9	400	.160	C.F.	.48	8.55		9.03	14.25
6100	With large production vacuum loader	A-12	700	.091	"	.27	4.89	1.08	6.24	9.35
7000	Radiator backing, not including radiator removal	A-9	1200	.053	S.F.	.16	2.85		3.01	4.75
8000	Cement-asbestos transite board and cement wall board	2 Asbe	1000	.016		.15	.85		1	1.53
8100	Transite shingle siding	A-10B	750	.043		.22	2.28		2.50	3.91
8200	Shingle roofing	"	2000	.016		.08	.86		.94	1.47
8250	Built-up, no gravel, non-friable	B-2	1400	.029		.08	1.09		1.17	1.88
8260	Bituminous flashing	1 Rofc	300	.027		.08	1.11		1.19	2.10
8300	Asbestos millboard, flat board and VAT contaminated plywood	2 Asbe	1000	.016		.08	.85		.93	1.46
9000	For type B (supplied air) respirator equipment, add				%				10%	10%

02 82 13.44 Demolition In Asbestos Contaminated Area

		Crew	Daily Output	Labor-Hours	Unit	Material	2016 Bare Costs Labor	2016 Bare Costs Equipment	Total	Total Incl O&P
0010	**DEMOLITION IN ASBESTOS CONTAMINATED AREA**									
0200	Ceiling, including suspension system, plaster and lath	A-9	2100	.030	S.F.	.09	1.63		1.72	2.72
0210	Finished plaster, leaving wire lath		585	.109		.33	5.85		6.18	9.75

For customer support on your Commercial Renovation Cost Data, call 877.742.4719.

39

02 82 Asbestos Remediation

02 82 13 – Asbestos Abatement

02 82 13.44 Demolition In Asbestos Contaminated Area	Crew	Daily Output	Labor-Hours	Unit	Material	2016 Bare Costs Labor	Equipment	Total	Total Incl O&P
0220	Suspended acoustical tile	A-9	3500	.018	S.F.	.05	.98		1.03	1.63
0230	Concealed tile grid system		3000	.021		.06	1.14		1.20	1.90
0240	Metal pan grid system		1500	.043		.13	2.28		2.41	3.80
0250	Gypsum board		2500	.026		.08	1.37		1.45	2.28
0260	Lighting fixtures up to 2' x 4'		72	.889	Ea.	2.65	47.50		50.15	79.50
0400	Partitions, non load bearing									
0410	Plaster, lath, and studs	A-9	690	.093	S.F.	.91	4.96		5.87	8.95
0450	Gypsum board and studs	"	1390	.046	"	.14	2.46		2.60	4.10
9000	For type B (supplied air) respirator equipment, add				%				10%	10%

02 82 13.45 OSHA Testing

		Crew	Daily Output	Labor-Hours	Unit	Material	2016 Bare Costs Labor	Equipment	Total	Total Incl O&P
0010	**OSHA TESTING**									
0100	Certified technician, minimum				Day				200	220
0110	Maximum								300	330
0120	Industrial hygienist, minimum								250	250
0130	Maximum								400	440
0200	Asbestos sampling and PCM analysis, NIOSH 7400, minimum	1 Asbe	8	1	Ea.	14	53.50		67.50	101
0210	Maximum		4	2		22	107		129	196
1000	Cleaned area samples		8	1		200	53.50		253.50	305
1100	PCM air sample analysis, NIOSH 7400, minimum		8	1		14	53.50		67.50	101
1110	Maximum		4	2		2.20	107		109.20	174
1200	TEM air sample analysis, NIOSH 7402, minimum								80	106
1210	Maximum								360	450

02 82 13.46 Decontamination of Asbestos Containment Area

		Crew	Daily Output	Labor-Hours	Unit	Material	2016 Bare Costs Labor	Equipment	Total	Total Incl O&P
0010	**DECONTAMINATION OF ASBESTOS CONTAINMENT AREA**									
0100	Spray exposed substrate with surfactant (bridging)									
0200	Flat surfaces	A-9	6000	.011	S.F.	.40	.57		.97	1.36
0250	Irregular surfaces		4000	.016	"	.40	.86		1.26	1.81
0300	Pipes, beams, and columns		2000	.032	L.F.	.60	1.71		2.31	3.41
1000	Spray encapsulate polyethylene sheeting		8000	.008	S.F.	.40	.43		.83	1.13
1100	Roll down polyethylene sheeting		8000	.008	"		.43		.43	.69
1500	Bag polyethylene sheeting		400	.160	Ea.	1	8.55		9.55	14.85
2000	Fine clean exposed substrate, with nylon brush		2400	.027	S.F.		1.43		1.43	2.29
2500	Wet wipe substrate		4800	.013			.71		.71	1.14
2600	Vacuum surfaces, fine brush		6400	.010			.53		.53	.86
3000	Structural demolition									
3100	Wood stud walls	A-9	2800	.023	S.F.		1.22		1.22	1.96
3500	Window manifolds, not incl. window replacement		4200	.015			.81		.81	1.31
3600	Plywood carpet protection		2000	.032			1.71		1.71	2.75
4000	Remove custom decontamination facility	A-10A	8	3	Ea.	15	161		176	275
4100	Remove portable decontamination facility	3 Asbe	12	2	"	13.35	107		120.35	187
5000	HEPA vacuum, shampoo carpeting	A-9	4800	.013	S.F.	.10	.71		.81	1.25
9000	Final cleaning of protected surfaces	A-10A	8000	.003	"		.16		.16	.26

02 82 13.47 Asbestos Waste Pkg., Handling, and Disp.

		Crew	Daily Output	Labor-Hours	Unit	Material	2016 Bare Costs Labor	Equipment	Total	Total Incl O&P
0010	**ASBESTOS WASTE PACKAGING, HANDLING, AND DISPOSAL**									
0100	Collect and bag bulk material, 3 C.F. bags, by hand	A-9	400	.160	Ea.	.84	8.55		9.39	14.65
0200	Large production vacuum loader	A-12	880	.073		1	3.89	.86	5.75	8.30
1000	Double bag and decontaminate	A-9	960	.067		.84	3.56		4.40	6.60
2000	Containerize bagged material in drums, per 3 C.F. drum	"	800	.080		18.55	4.28		22.83	27.50
3000	Cart bags 50' to dumpster	2 Asbe	400	.040			2.14		2.14	3.43
5000	Disposal charges, not including haul, minimum				C.Y.				61	67
5020	Maximum				"				355	395
9000	For type B (supplied air) respirator equipment, add				%				10%	10%

02 82 Asbestos Remediation

02 82 13 – Asbestos Abatement

02 82 13.48 Asbestos Encapsulation With Sealants

	02 82 13.48 Asbestos Encapsulation With Sealants	Crew	Daily Output	Labor-Hours	Unit	Material	2016 Bare Costs Labor	Equipment	Total	Total Incl O&P
0010	**ASBESTOS ENCAPSULATION WITH SEALANTS**									
0100	Ceilings and walls, minimum	A-9	21000	.003	S.F.	.40	.16		.56	.70
0110	Maximum		10600	.006		.50	.32		.82	1.07
0200	Columns and beams, minimum		13300	.005		.40	.26		.66	.85
0210	Maximum		5325	.012		.60	.64		1.24	1.69
0300	Pipes to 12" diameter including minor repairs, minimum		800	.080	L.F.	.50	4.28		4.78	7.40
0310	Maximum		400	.160	"	1.20	8.55		9.75	15.05

02 83 Lead Remediation

02 83 19 – Lead-Based Paint Remediation

02 83 19.21 Lead Paint Remediation Plans and Methods

		Crew	Daily Output	Labor-Hours	Unit	Material	2016 Bare Costs Labor	Equipment	Total	Total Incl O&P
0010	**LEAD PAINT REMEDIATION PLANS AND METHODS**									
0100	Building Survey-Commercial Building				Ea.				2,050	2,250
0200	Lead Abatement Remediation Plan								1,225	1,350
0300	Lead Paint Testing, AAS Analysis								51	56
0400	Lead Paint Testing, X-Ray Fluorescence								51	56

02 83 19.22 Preparation of Lead Containment Area

		Crew	Daily Output	Labor-Hours	Unit	Material	2016 Bare Costs Labor	Equipment	Total	Total Incl O&P
0010	**PREPARATION OF LEAD CONTAINMENT AREA**									
0020	Lead abatement work area, test kit, per swab	1 Skwk	16	.500	Ea.	3.12	25		28.12	44
0025	For dust barriers see Section 01 56 16.10									
0050	Caution sign	1 Skwk	48	.167	Ea.	8.35	8.30		16.65	22.50
0100	Pre-cleaning, HEPA vacuum and wet wipe, floor and wall surfaces	3 Skwk	5000	.005	S.F.	.01	.24		.25	.41
0105	Ceiling, 6' - 11' high		4100	.006		.08	.29		.37	.56
0108	12' - 15' high		3550	.007		.07	.34		.41	.63
0115	Over 15' high		3000	.008		.09	.40		.49	.75
0500	Cover surfaces with polyethylene sheeting									
0550	Floors, each layer, 6 mil	3 Skwk	8000	.003	S.F.	.04	.15		.19	.28
0560	Walls, each layer, 6 mil	"	6000	.004	"	.04	.20		.24	.36
0570	For heights above 12', add						20%			
2400	Post abatement cleaning of protective sheeting, HEPA vacuum & wet wipe	3 Skwk	5000	.005	S.F.	.01	.24		.25	.41
2450	Doff, bag and seal protective sheeting		12000	.002		.01	.10		.11	.17
2500	Post abatement cleaning, HEPA vacuum & wet wipe		5000	.005		.01	.24		.25	.41

02 83 19.23 Encapsulation of Lead-Based Paint

		Crew	Daily Output	Labor-Hours	Unit	Material	2016 Bare Costs Labor	Equipment	Total	Total Incl O&P
0010	**ENCAPSULATION OF LEAD-BASED PAINT**									
0020	Interior, brushwork, trim, under 6"	1 Pord	240	.033	L.F.	2.30	1.34		3.64	4.69
0030	6" to 12" wide		180	.044		3	1.79		4.79	6.20
0040	Balustrades		300	.027		2	1.08		3.08	3.93
0050	Pipe to 4" diameter		500	.016		1.20	.65		1.85	2.36
0060	To 8" diameter		375	.021		1.50	.86		2.36	3.03
0070	To 12" diameter		250	.032		2.20	1.29		3.49	4.50
0080	To 16" diameter		170	.047		3.30	1.90		5.20	6.70
0090	Cabinets, ornate design		200	.040	S.F.	3	1.61		4.61	5.90
0100	Simple design		250	.032	"	2.30	1.29		3.59	4.61
0110	Doors, 3' x 7', both sides, incl. frame & trim									
0120	Flush	1 Pord	6	1.333	Ea.	28	54		82	118
0130	French, 10 – 15 lite		3	2.667		6	108		114	180
0140	Panel		4	2		35	80.50		115.50	169
0150	Louvered		2.75	2.909		32	117		149	224
0160	Windows, per interior side, per 15 S.F.									
0170	1 to 6 lite	1 Pord	14	.571	Ea.	20	23		43	59

For customer support on your Commercial Renovation Cost Data, call 877.742.4719.

41

02 83 Lead Remediation

02 83 19 – Lead-Based Paint Remediation

02 83 19.23 Encapsulation of Lead-Based Paint		Crew	Daily Output	Labor-Hours	Unit	Material	2016 Bare Costs Labor	2016 Bare Costs Equipment	Total	Total Incl O&P
0180	7 to 10 lite	1 Pord	7.50	1.067	Ea.	22	43		65	93
0190	12 lite		5.75	1.391		30	56		86	123
0200	Radiators		8	1	▼	70	40.50		110.50	142
0210	Grilles, vents		275	.029	S.F.	2	1.17		3.17	4.09
0220	Walls, roller, drywall or plaster		1000	.008		.60	.32		.92	1.18
0230	With spunbonded reinforcing fabric		720	.011		.70	.45		1.15	1.49
0240	Wood		800	.010		.70	.40		1.10	1.42
0250	Ceilings, roller, drywall or plaster		900	.009		.70	.36		1.06	1.35
0260	Wood		700	.011	▼	.80	.46		1.26	1.62
0270	Exterior, brushwork, gutters and downspouts		300	.027	L.F.	1.90	1.08		2.98	3.82
0280	Columns		400	.020	S.F.	1.40	.81		2.21	2.84
0290	Spray, siding	▼	600	.013	"	1	.54		1.54	1.96
0300	Miscellaneous									
0310	Electrical conduit, brushwork, to 2" diameter	1 Pord	500	.016	L.F.	1.20	.65		1.85	2.36
0320	Brick, block or concrete, spray		500	.016	S.F.	1.20	.65		1.85	2.36
0330	Steel, flat surfaces and tanks to 12"		500	.016		1.20	.65		1.85	2.36
0340	Beams, brushwork		400	.020		1.40	.81		2.21	2.84
0350	Trusses	▼	400	.020	▼	1.40	.81		2.21	2.84

02 83 19.26 Removal of Lead-Based Paint

02 83 19.26 Removal of Lead-Based Paint		Crew	Daily Output	Labor-Hours	Unit	Material	2016 Bare Costs Labor	2016 Bare Costs Equipment	Total	Total Incl O&P
0010	**REMOVAL OF LEAD-BASED PAINT** R028319-60									
0011	By chemicals, per application									
0050	Baseboard, to 6" wide	1 Pord	64	.125	L.F.	.80	5.05		5.85	9
0070	To 12" wide		32	.250	"	1.47	10.10		11.57	17.80
0200	Balustrades, one side		28	.286	S.F.	1.50	11.55		13.05	20
1400	Cabinets, simple design		32	.250		1.37	10.10		11.47	17.70
1420	Ornate design		25	.320		1.60	12.90		14.50	23
1600	Cornice, simple design		60	.133		1.60	5.40		7	10.40
1620	Ornate design		20	.400		5.50	16.15		21.65	32
2800	Doors, one side, flush		84	.095		1.73	3.84		5.57	8.10
2820	Two panel		80	.100		1.37	4.04		5.41	8
2840	Four panel		45	.178	▼	1.43	7.15		8.58	13.10
2880	For trim, one side, add		64	.125	L.F.	.77	5.05		5.82	8.95
3000	Fence, picket, one side		30	.267	S.F.	1.37	10.75		12.12	18.80
3200	Grilles, one side, simple design		30	.267		1.37	10.75		12.12	18.80
3220	Ornate design		25	.320	▼	1.47	12.90		14.37	22.50
3240	Handrails		90	.089	L.F.	1.37	3.59		4.96	7.25
4400	Pipes, to 4" diameter		90	.089		1.73	3.59		5.32	7.65
4420	To 8" diameter		50	.160		3.47	6.45		9.92	14.20
4440	To 12" diameter		36	.222		5.20	8.95		14.15	20
4460	To 16" diameter		20	.400	▼	6.95	16.15		23.10	33.50
4500	For hangers, add		40	.200	Ea.	2.50	8.05		10.55	15.70
4800	Siding		90	.089	S.F.	1.30	3.59		4.89	7.20
5000	Trusses, open		55	.145	SF Face	2	5.85		7.85	11.65
6200	Windows, one side only, double hung, 1/1 light, 24" x 48" high		4	2	Ea.	24	80.50		104.50	157
6220	30" x 60" high		3	2.667		32	108		140	208
6240	36" x 72" high		2.50	3.200		38	129		167	250
6280	40" x 80" high		2	4		47	161		208	310
6400	Colonial window, 6/6 light, 24" x 48" high		2	4		47	161		208	310
6420	30" x 60" high		1.50	5.333		62	215		277	415
6440	36" x 72" high		1	8		93	325		418	620
6480	40" x 80" high		1	8		93	325		418	620
6600	8/8 light, 24" x 48" high	▼	2	4	▼	47	161		208	310

02 83 Lead Remediation

02 83 19 – Lead-Based Paint Remediation

02 83 19.26 Removal of Lead-Based Paint

		Crew	Daily Output	Labor-Hours	Unit	Material	2016 Bare Costs Labor	Equipment	Total	Total Incl O&P
6620	40" x 80" high	1 Pord	1	8	Ea.	93	325		418	620
6800	12/12 light, 24" x 48" high		1	8		93	325		418	620
6820	40" x 80" high	↓	.75	10.667	↓	125	430		555	830
6840	Window frame & trim items, included in pricing above									
7000	Hand scraping and HEPA vacuum, less than 4 S.F.	1 Pord	8	1	Ea.	.84	40.50		41.34	66
8000	Collect and bag bulk material, 3 C.F. bags, by hand		30	.267	"	.84	10.75		11.59	18.20
9000	Minimum labor/equipment charge	↓	3	2.667	Job		108		108	173

02 85 Mold Remediation

02 85 16 – Mold Remediation Preparation and Containment

02 85 16.40 Mold Remediation Plans and Methods

		Crew	Daily Output	Labor-Hours	Unit	Material	2016 Bare Costs Labor	Equipment	Total	Total Incl O&P
0010	**MOLD REMEDIATION PLANS AND METHODS**									
0020	Initial inspection, areas to 2500 S.F.				Total				268	295
0030	Areas to 5000 S.F.								440	485
0032	Areas to 10000 S.F.								440	485
0040	Testing, air sample each								126	139
0050	Swab sample								115	127
0060	Tape sample								125	140
0070	Post remediation air test								126	139
0080	Mold abatement plan, area to 2500 S.F.								1,225	1,325
0090	Areas to 5000 S.F.								1,625	1,775
0095	Areas to 10000 S.F.								2,550	2,800
0100	Packup & removal of contents, average 3 bedroom home, excl storage								8,150	8,975
0110	Average 5 bedroom home, excl storage				▼				15,300	16,800
0600	For demolition in mold contaminated areas, see Section 02 85 33.50									
0610	For personal protection equipment, see Section 02 82 13.41									

02 85 16.50 Preparation of Mold Containment Area

		Crew	Daily Output	Labor-Hours	Unit	Material	2016 Bare Costs Labor	Equipment	Total	Total Incl O&P
0010	**PREPARATION OF MOLD CONTAINMENT AREA**									
0100	Pre-cleaning, HEPA vacuum and wet wipe, flat surfaces	A-9	12000	.005	S.F.	.02	.28		.30	.48
0300	Separation barrier, 2" x 4" @ 16", 1/2" plywood ea. side, 8' high	2 Carp	400	.040		3.40	1.94		5.34	6.90
0310	12' high		320	.050		3.40	2.42		5.82	7.70
0320	16' high		200	.080		2.30	3.88		6.18	8.90
0400	Personnel decontam. chamber, 2" x 4" @ 16", 3/4" ply ea. side		280	.057		4.40	2.77		7.17	9.35
0450	Waste decontam. chamber, 2" x 4" studs @ 16", 3/4" ply each side	↓	360	.044	↓	4.40	2.15		6.55	8.35
0500	Cover surfaces with polyethylene sheeting									
0501	Including glue and tape									
0550	Floors, each layer, 6 mil	A-9	8000	.008	S.F.	.04	.43		.47	.73
0551	4 mil		9000	.007		.03	.38		.41	.64
0560	Walls, each layer, 6 mil		6000	.011		.04	.57		.61	.96
0561	4 mil	↓	7000	.009	↓	.03	.49		.52	.81
0570	For heights above 12', add						20%			
0575	For heights above 20', add						30%			
0580	For fire retardant poly, add					100%				
0590	For large open areas, deduct					10%	20%			
0600	Seal floor penetrations with foam firestop to 36 sq. in.	2 Carp	200	.080	Ea.	9.50	3.88		13.38	16.85
0610	36 sq. in. to 72 sq. in.		125	.128		19.05	6.20		25.25	31
0615	72 sq. in. to 144 sq. in.		80	.200		38	9.70		47.70	58
0620	Wall penetrations, to 36 square inches		180	.089		9.50	4.31		13.81	17.55
0630	36 sq. in. to 72 sq. in.		100	.160		19.05	7.75		26.80	33.50
0640	72 sq. in. to 144 sq. in.	↓	60	.267	↓	38	12.90		50.90	63
0800	Caulk seams with latex caulk	1 Carp	230	.035	L.F.	.17	1.69		1.86	2.94

For customer support on your Commercial Renovation Cost Data, call 877.742.4719.

43

02 85 Mold Remediation

02 85 16 – Mold Remediation Preparation and Containment

02 85 16.50 Preparation of Mold Containment Area	Crew	Daily Output	Labor-Hours	Unit	Material	2016 Bare Costs Labor	Equipment	Total	Total Incl O&P
0900	Set up neg. air machine, 1-2k CFM/25 M.C.F. volume	1 Asbe	4.30	1.860	Ea.		99.50		99.50	160

02 85 33 – Removal and Disposal of Materials with Mold

02 85 33.50 Demolition in Mold Contaminated Area

	02 85 33.50	Crew	Daily Output	Labor-Hours	Unit	Material	Labor	Equipment	Total	Total Incl O&P
0010	**DEMOLITION IN MOLD CONTAMINATED AREA**									
0200	Ceiling, including suspension system, plaster and lath	A-9	2100	.030	S.F.	.09	1.63		1.72	2.72
0210	Finished plaster, leaving wire lath		585	.109		.33	5.85		6.18	9.75
0220	Suspended acoustical tile		3500	.018		.05	.98		1.03	1.63
0230	Concealed tile grid system		3000	.021		.06	1.14		1.20	1.90
0240	Metal pan grid system		1500	.043		.13	2.28		2.41	3.80
0250	Gypsum board		2500	.026		.08	1.37		1.45	2.28
0255	Plywood		2500	.026		.08	1.37		1.45	2.28
0260	Lighting fixtures up to 2' x 4'		72	.889	Ea.	2.65	47.50		50.15	79.50
0400	Partitions, non load bearing									
0410	Plaster, lath, and studs	A-9	690	.093	S.F.	.91	4.96		5.87	8.95
0450	Gypsum board and studs		1390	.046		.14	2.46		2.60	4.10
0465	Carpet & pad		1390	.046		.14	2.46		2.60	4.10
0600	Pipe insulation, air cell type, up to 4" diameter pipe		900	.071	L.F.	.21	3.80		4.01	6.35
0610	4" to 8" diameter pipe		800	.080		.24	4.28		4.52	7.10
0620	10" to 12" diameter pipe		700	.091		.27	4.89		5.16	8.15
0630	14" to 16" diameter pipe		550	.116		.35	6.20		6.55	10.40
0650	Over 16" diameter pipe		650	.098	S.F.	.29	5.25		5.54	8.75
9000	For type B (supplied air) respirator equipment, add				%				10%	10%

Estimating Tips

General

- Carefully check all the plans and specifications. Concrete often appears on drawings other than structural drawings, including mechanical and electrical drawings for equipment pads. The cost of cutting and patching is often difficult to estimate. See Subdivision 03 81 for Concrete Cutting, Subdivision 02 41 19.16 for Cutout Demolition, Subdivision 03 05 05.10 for Concrete Demolition, and Subdivision 02 41 19.19 for Rubbish Handling (handling, loading and hauling of debris).

- Always obtain concrete prices from suppliers near the job site. A volume discount can often be negotiated, depending upon competition in the area. Remember to add for waste, particularly for slabs and footings on grade.

03 10 00 Concrete Forming and Accessories

- A primary cost for concrete construction is forming. Most jobs today are constructed with prefabricated forms. The selection of the forms best suited for the job and the total square feet of forms required for efficient concrete forming and placing are key elements in estimating concrete construction. Enough forms must be available for erection to make efficient use of the concrete placing equipment and crew.

- Concrete accessories for forming and placing depend upon the systems used. Study the plans and specifications to ensure that all special accessory requirements have been included in the cost estimate, such as anchor bolts, inserts, and hangers.

- Included within costs for forms-in-place are all necessary bracing and shoring.

03 20 00 Concrete Reinforcing

- Ascertain that the reinforcing steel supplier has included all accessories, cutting, bending, and an allowance for lapping, splicing, and waste. A good rule of thumb is 10% for lapping, splicing, and waste. Also, 10% waste should be allowed for welded wire fabric.

- The unit price items in the subdivisions for Reinforcing In Place, Glass Fiber Reinforcing, and Welded Wire Fabric include the labor to install accessories such as beam and slab bolsters, high chairs, and bar ties and tie wire. The material cost for these accessories is not included; they may be obtained from the Accessories Subdivisions.

03 30 00 Cast-In-Place Concrete

- When estimating structural concrete, pay particular attention to requirements for concrete additives, curing methods, and surface treatments. Special consideration for climate, hot or cold, must be included in your estimate. Be sure to include requirements for concrete placing equipment, and concrete finishing.

- For accurate concrete estimating, the estimator must consider each of the following major components individually: forms, reinforcing steel, ready-mix concrete, placement of the concrete, and finishing of the top surface. For faster estimating, Subdivision 03 30 53.40 for Concrete-In-Place can be used; here, various items of concrete work are presented that include the costs of all five major components (unless specifically stated otherwise).

03 40 00 Precast Concrete
03 50 00 Cast Decks and Underlayment

- The cost of hauling precast concrete structural members is often an important factor. For this reason, it is important to get a quote from the nearest supplier. It may become economically feasible to set up precasting beds on the site if the hauling costs are prohibitive.

Reference Numbers

Reference numbers are shown at the beginning of some major classifications. These numbers refer to related items in the Reference Section. The reference information may be an estimating procedure, an alternate pricing method, or technical information.

Note: Not all subdivisions listed here necessarily appear. ■

Did you know?

RSMeans Online gives you the same access to RSMeans' data with 24/7 access:

- Quickly locate costs in the searchable database.
- Build cost lists, estimates, and reports in minutes.
- Adjust costs to any location in the U.S. and Canada with the click of a button.

Start your free trial today at **www.RSMeansOnline.com**

03 01 Maintenance of Concrete

03 01 30 – Maintenance of Cast-In-Place Concrete

03 01 30.62 Concrete Patching

		Crew	Daily Output	Labor-Hours	Unit	Material	2016 Bare Costs Labor	Equipment	Total	Total Incl O&P
0010	**CONCRETE PATCHING**									
0100	Floors, 1/4" thick, small areas, regular grout	1 Cefi	170	.047	S.F.	1.49	2.15		3.64	5.05
0150	Epoxy grout	"	100	.080	"	8.25	3.65		11.90	14.85
2000	Walls, including chipping, cleaning and epoxy grout									
2100	1/4" deep	1 Cefi	65	.123	S.F.	7.25	5.60		12.85	16.85
2150	1/2" deep		50	.160		14.45	7.30		21.75	27.50
2200	3/4" deep		40	.200		21.50	9.15		30.65	38.50
9000	Minimum labor/equipment charge		4.50	1.778	Job		81		81	128

03 01 30.71 Concrete Crack Repair

		Crew	Daily Output	Labor-Hours	Unit	Material	2016 Bare Costs Labor	Equipment	Total	Total Incl O&P
0010	**CONCRETE CRACK REPAIR**									
1000	Structural repair of concrete cracks by epoxy injection (ACI RAP-1)									
1001	suitable for horizontal, vertical and overhead repairs									
1010	Clean/grind concrete surface(s) free of contaminants	1 Cefi	400	.020	L.F.		.91		.91	1.45
1015	Rout crack with v-notch crack chaser, if needed	C-32	600	.027		.03	1.11	.15	1.29	2
1020	Blow out crack with oil-free dry compressed air (1 pass)	C-28	3000	.003			.12	.01	.13	.20
1030	Install surface-mounted entry ports (spacing = concrete depth)	1 Cefi	400	.020	Ea.	1.42	.91		2.33	3.01
1040	Cap crack at surface with epoxy gel (per side/face)		400	.020	L.F.	.35	.91		1.26	1.83
1050	Snap off ports, grind off epoxy cap residue after injection		200	.040	"		1.83		1.83	2.89
1100	Manual injection with 2-part epoxy cartridge, excludes prep									
1110	Up to 1/32" (0.03125") wide x 4" deep	1 Cefi	160	.050	L.F.	.17	2.28		2.45	3.80
1120	6" deep		120	.067		.26	3.04		3.30	5.10
1130	8" deep		107	.075		.35	3.41		3.76	5.80
1140	10" deep		100	.080		.43	3.65		4.08	6.30
1150	12" deep		96	.083		.52	3.80		4.32	6.55
1210	Up to 1/16" (0.0625") wide x 4" deep		160	.050		.35	2.28		2.63	3.99
1220	6" deep		120	.067		.52	3.04		3.56	5.40
1230	8" deep		107	.075		.69	3.41		4.10	6.15
1240	10" deep		100	.080		.87	3.65		4.52	6.75
1250	12" deep		96	.083		1.04	3.80		4.84	7.15
1310	Up to 3/32" (0.09375") wide x 4" deep		160	.050		.52	2.28		2.80	4.18
1320	6" deep		120	.067		.78	3.04		3.82	5.70
1330	8" deep		107	.075		1.04	3.41		4.45	6.55
1340	10" deep		100	.080		1.30	3.65		4.95	7.25
1350	12" deep		96	.083		1.56	3.80		5.36	7.70
1410	Up to 1/8" (0.125") wide x 4" deep		160	.050		.69	2.28		2.97	4.37
1420	6" deep		120	.067		1.04	3.04		4.08	5.95
1430	8" deep		107	.075		1.39	3.41		4.80	6.95
1440	10" deep		100	.080		1.74	3.65		5.39	7.70
1450	12" deep		96	.083		2.08	3.80		5.88	8.30
1500	Pneumatic injection with 2-part bulk epoxy, excludes prep									
1510	Up to 5/32" (0.15625") wide x 4" deep	C-31	240	.033	L.F.	.25	1.52	1.55	3.32	4.39
1520	6" deep		180	.044		.37	2.03	2.07	4.47	5.90
1530	8" deep		160	.050		.49	2.28	2.33	5.10	6.70
1540	10" deep		150	.053		.62	2.43	2.49	5.54	7.25
1550	12" deep		144	.056		.74	2.54	2.59	5.87	7.70
1610	Up to 3/16" (0.1875") wide x 4" deep		240	.033		.30	1.52	1.55	3.37	4.45
1620	6" deep		180	.044		.44	2.03	2.07	4.54	6
1630	8" deep		160	.050		.59	2.28	2.33	5.20	6.80
1640	10" deep		150	.053		.74	2.43	2.49	5.66	7.40
1650	12" deep		144	.056		.89	2.54	2.59	6.02	7.85
1710	Up to 1/4" (0.1875") wide x 4" deep		240	.033		.40	1.52	1.55	3.47	4.55
1720	6" deep		180	.044		.59	2.03	2.07	4.69	6.15

46

For customer support on your Commercial Renovation Cost Data, call 877.742.4719.

03 01 30 – Maintenance of Cast-In-Place Concrete

03 01 30.71 Concrete Crack Repair		Crew	Daily Output	Labor-Hours	Unit	Material	2016 Bare Costs			Total	Total Incl O&P
							Labor	Equipment			
1730	8" deep	C-31	160	.050	L.F.	.79	2.28	2.33		5.40	7.05
1740	10" deep		150	.053		.99	2.43	2.49		5.91	7.70
1750	12" deep	↓	144	.056	↓	1.19	2.54	2.59		6.32	8.15
2000	Non-structural filling of concrete cracks by gravity-fed resin (ACI RAP-2)										
2001	suitable for individual cracks in stable horizontal surfaces only										
2010	Clean/grind concrete surface(s) free of contaminants	1 Cefi	400	.020	L.F.		.91			.91	1.45
2020	Rout crack with v-notch crack chaser, if needed	C-32	600	.027		.03	1.11	.15		1.29	2
2030	Blow out crack with oil-free dry compressed air (1 pass)	C-28	3000	.003			.12	.01		.13	.20
2040	Cap crack with epoxy gel at underside of elevated slabs, if needed	1 Cefi	400	.020		.35	.91			1.26	1.83
2050	Insert backer rod into crack, if needed		400	.020		.02	.91			.93	1.48
2060	Partially fill crack with fine dry sand, if needed		800	.010		.02	.46			.48	.74
2070	Apply two beads of sealant alongside crack to form reservoir, if needed	↓	400	.020	↓	.25	.91			1.16	1.72
2100	Manual filling with squeeze bottle of 2-part epoxy resin										
2110	Full depth crack up to 1/16" (0.0625") wide x 4" deep	1 Cefi	300	.027	L.F.	.09	1.22			1.31	2.03
2120	6" deep		240	.033		.13	1.52			1.65	2.56
2130	8" deep		200	.040		.18	1.83			2.01	3.08
2140	10" deep		185	.043		.22	1.97			2.19	3.36
2150	12" deep		175	.046		.26	2.09			2.35	3.59
2210	Full depth crack up to 1/8" (0.125") wide x 4" deep		270	.030		.18	1.35			1.53	2.33
2220	6" deep		215	.037		.26	1.70			1.96	2.98
2230	8" deep		180	.044		.35	2.03			2.38	3.60
2240	10" deep		165	.048		.44	2.21			2.65	3.98
2250	12" deep		155	.052		.53	2.36			2.89	4.31
2310	Partial depth crack up to 3/16" (0.1875") wide x 1" deep		300	.027		.07	1.22			1.29	2
2410	Up to 1/4" (0.250") wide x 1" deep		270	.030		.09	1.35			1.44	2.24
2510	Up to 5/16" (0.3125") wide x 1" deep		250	.032		.11	1.46			1.57	2.43
2610	Up to 3/8" (0.375") wide x 1" deep	↓	240	.033	↓	.13	1.52			1.65	2.56

03 01 30.72 Concrete Surface Repairs

		Crew	Daily Output	Labor-Hours	Unit	Material	Labor	Equipment		Total	Total Incl O&P
0010	**CONCRETE SURFACE REPAIRS**										
9000	Surface repair by Methacrylate flood coat (ACI RAP-13)										
9010	Suitable for healing and sealing horizontal surfaces only										
9020	Large cracks must previously have been repaired or filled										
9100	Shotblast entire surface to remove contaminants	A-1A	4000	.002	S.F.		.10	.05		.15	.22
9200	Blow off dust and debris with oil-free dry compressed air	C-28	16000	.001			.02			.02	.04
9300	Flood coat surface w/Methacrylate, distribute w/broom/squeegee, no prep.	3 Cefi	8000	.003		.98	.14			1.12	1.30
9400	Lightly broadcast even coat of dry silica sand while sealer coat is tacky	1 Cefi	8000	.001	↓	.01	.05			.06	.08

03 05 Common Work Results for Concrete

03 05 05 – Selective Demolition for Concrete

03 05 05.10 Selective Demolition, Concrete

		Crew	Daily Output	Labor-Hours	Unit	Material	Labor	Equipment		Total	Total Incl O&P
0010	**SELECTIVE DEMOLITION, CONCRETE** R024119-10										
0012	Excludes saw cutting, torch cutting, loading or hauling										
0050	Break into small pieces, reinf. less than 1% of cross-sectional area	B-9	24	1.667	C.Y.		64	9.55		73.55	115
0060	Reinforcing 1% to 2% of cross-sectional area		16	2.500			96	14.30		110.30	173
0070	Reinforcing more than 2% of cross-sectional area	↓	8	5	↓		192	28.50		220.50	345
0150	Remove whole pieces, up to 2 tons per piece	E-18	36	1.111	Ea.		59	27		86	132
0160	2 – 5 tons per piece		30	1.333			71	32		103	159
0170	5 – 10 tons per piece		24	1.667			88.50	40.50		129	198
0180	10 – 15 tons per piece	↓	18	2.222			118	53.50		171.50	263
0250	Precast unit embedded in masonry, up to 1 C.F.	D-1	16	1			42			42	68.50
0260	1 – 2 C.F.	↓	12	1.333			56			56	91.50

For customer support on your Commercial Renovation Cost Data, call 877.742.4719.

47

03 05 Common Work Results for Concrete

03 05 05 – Selective Demolition for Concrete

03 05 05.10 Selective Demolition, Concrete	Crew	Daily Output	Labor-Hours	Unit	Material	2016 Bare Costs Labor	2016 Bare Costs Equipment	Total	Total Incl O&P
0270	2 – 5 C.F.	D-1	10	1.600	Ea.		67.50		67.50	110
0280	5 – 10 C.F.	↓	8	2	↓		84.50		84.50	137
0990	For hydrodemolition see Section 02 41 13.15									
1910	Minimum labor/equipment charge	B-9	2	20	Job		765	114	879	1,375

03 05 13 – Basic Concrete Materials

03 05 13.20 Concrete Admixtures and Surface Treatments

		Crew	Daily Output	Labor-Hours	Unit	Material	2016 Bare Costs Labor	2016 Bare Costs Equipment	Total	Total Incl O&P
0010	**CONCRETE ADMIXTURES AND SURFACE TREATMENTS**									
0040	Abrasives, aluminum oxide, over 20 tons				Lb.	1.88			1.88	2.07
0050	1 to 20 tons					2.02			2.02	2.22
0070	Under 1 ton					2.10			2.10	2.31
0100	Silicon carbide, black, over 20 tons					2.87			2.87	3.16
0110	1 to 20 tons					3.04			3.04	3.35
0120	Under 1 ton				↓	3.17			3.17	3.49
0200	Air entraining agent, .7 to 1.5 oz. per bag, 55 gallon drum				Gal.	14			14	15.40
0220	5 gallon pail					19.25			19.25	21
0300	Bonding agent, acrylic latex, 250 S.F. per gallon, 5 gallon pail					23			23	25.50
0320	Epoxy resin, 80 S.F. per gallon, 4 gallon case				↓	62			62	68.50
0400	Calcium chloride, 50 lb. bags, TL lots				Ton	810			810	890
0420	Less than truckload lots				Bag	18.60			18.60	20.50
0500	Carbon black, liquid, 2 to 8 lb. per bag of cement				Lb.	9.10			9.10	10
0600	Colored pigments, integral, 2 to 10 lb. per bag of cement, subtle colors					2.43			2.43	2.67
0610	Standard colors					3.28			3.28	3.61
0620	Premium colors				↓	5.10			5.10	5.60
0920	Dustproofing compound, 250 S.F./gal., 5 gallon pail				Gal.	6.95			6.95	7.65
1010	Epoxy based, 125 S.F./gal., 5 gallon pail				"	56.50			56.50	62
1100	Hardeners, metallic, 55 lb. bags, natural (grey)				Lb.	.76			.76	.84
1200	Colors					1.45			1.45	1.60
1300	Non-metallic, 55 lb. bags, natural grey					.42			.42	.46
1320	Colors				↓	.97			.97	1.07
1550	Release agent, for tilt slabs, 5 gallon pail				Gal.	17.90			17.90	19.65
1570	For forms, 5 gallon pail					11.45			11.45	12.60
1590	Concrete release agent for forms, 100% biodegradable, zero VOC, 5 gal pail [G]					21			21	23
1595	55 gallon drum [G]					16.80			16.80	18.50
1600	Sealer, hardener and dustproofer, epoxy-based, 125 S.F./gal., 5 gallon unit					56.50			56.50	62
1620	3 gallon unit					62			62	68
1630	Sealer, solvent-based, 250 S.F./gal., 55 gallon drum					21			21	23
1640	5 gallon pail					31.50			31.50	35
1650	Sealer, water based, 350 S.F., 55 gallon drum					20.50			20.50	22.50
1660	5 gallon pail					23			23	25.50
1900	Set retarder, 100 S.F./gal., 1 gallon pail				↓	6.25			6.25	6.90
2000	Waterproofing, integral 1 lb. per bag of cement				Lb.	2.12			2.12	2.33
2100	Powdered metallic, 40 lbs. per 100 S.F., standard colors					2.52			2.52	2.77
2120	Premium colors				↓	3.53			3.53	3.88
3000	For integral colored pigments, 2500 psi (5 bag mix)									
3100	Standard colors, 1.8 lbs. per bag, add				C.Y.	22			22	24
3200	9.4 lbs. per bag, add					114			114	126
3400	Premium colors, 1.8 lbs. per bag, add					29.50			29.50	32.50
3500	7.5 lbs. per bag, add					123			123	135
3700	Ultra premium colors, 1.8 lbs. per bag, add					46			46	50.50
3800	7.5 lbs. per bag, add				↓	192			192	211

03 05 Common Work Results for Concrete

03 05 13 – Basic Concrete Materials

03 05 13.85 Winter Protection		Crew	Daily Output	Labor-Hours	Unit	Material	2016 Bare Costs Labor	Equipment	Total	Total Incl O&P
0010	**WINTER PROTECTION**									
0012	For heated ready mix, add				C.Y.	4.45			4.45	4.90
0100	Temporary heat to protect concrete, 24 hours	2 Clab	50	.320	M.S.F.	200	12.15		212.15	239
0200	Temporary shelter for slab on grade, wood frame/polyethylene sheeting									
0201	Build or remove, light framing for short spans	2 Carp	10	1.600	M.S.F.	292	77.50		369.50	445
0210	Large framing for long spans	"	3	5.333	"	380	258		638	840
0710	Electrically, heated pads, 15 watts/S.F., 20 uses				S.F.	.55			.55	.61

03 11 Concrete Forming

03 11 13 – Structural Cast-In-Place Concrete Forming

03 11 13.20 Forms In Place, Beams and Girders

			Crew	Daily Output	Labor-Hours	Unit	Material	2016 Bare Costs Labor	Equipment	Total	Total Incl O&P
0010	**FORMS IN PLACE, BEAMS AND GIRDERS**	R031113-40									
0500	Exterior spandrel, job-built plywood, 12" wide, 1 use	R031113-60	C-2	225	.213	SFCA	3.16	10.05		13.21	19.85
0650	4 use			310	.155		1.03	7.30		8.33	13.05
1000	18" wide, 1 use			250	.192		2.85	9.05		11.90	17.90
1150	4 use			315	.152		.94	7.15		8.09	12.75
1500	24" wide, 1 use			265	.181		2.62	8.50		11.12	16.80
1650	4 use			325	.148		.85	6.95		7.80	12.30
2000	Interior beam, job-built plywood, 12" wide, 1 use			300	.160		3.62	7.50		11.12	16.30
2150	4 use			377	.127		1.18	6		7.18	11.10
2500	24" wide, 1 use			320	.150		2.68	7.05		9.73	14.45
2650	4 use			395	.122		.86	5.70		6.56	10.30
3000	Encasing steel beam, hung, job-built plywood, 1 use			325	.148		3.33	6.95		10.28	15
3150	4 use			430	.112		1.08	5.25		6.33	9.80
9000	Minimum labor/equipment charge		2 Carp	2	8	Job		390		390	635

03 11 13.25 Forms In Place, Columns

			Crew	Daily Output	Labor-Hours	Unit	Material	2016 Bare Costs Labor	Equipment	Total	Total Incl O&P
0010	**FORMS IN PLACE, COLUMNS**	R031113-40									
0500	Round fiberglass, 4 use per mo., rent, 12" diameter		C-1	160	.200	L.F.	9.85	9.15		19	26
0550	16" diameter	R031113-60		150	.213		11.75	9.75		21.50	29
0600	18" diameter			140	.229		13.15	10.45		23.60	31.50
0650	24" diameter			135	.237		16.35	10.85		27.20	35.50
0700	28" diameter			130	.246		18.25	11.30		29.55	38.50
0800	30" diameter			125	.256		19.05	11.75		30.80	40
0850	36" diameter			120	.267		25.50	12.20		37.70	48
1500	Round fiber tube, recycled paper, 1 use, 8" diameter	G		155	.206		1.82	9.45		11.27	17.45
1550	10" diameter	G		155	.206		2.46	9.45		11.91	18.15
1600	12" diameter	G		150	.213		2.84	9.75		12.59	19.05
1650	14" diameter	G		145	.221		4.02	10.10		14.12	21
1700	16" diameter	G		140	.229		4.97	10.45		15.42	22.50
1720	18" diameter	G		140	.229		5.80	10.45		16.25	23.50
1750	20" diameter	G		135	.237		7.25	10.85		18.10	26
1800	24" diameter	G		130	.246		9.10	11.30		20.40	28.50
1850	30" diameter	G		125	.256		13.95	11.75		25.70	34.50
1900	36" diameter	G		115	.278		16.55	12.75		29.30	39
1950	42" diameter	G		100	.320		42	14.65		56.65	70
2000	48" diameter	G		85	.376		49.50	17.25		66.75	82.50
2200	For seamless type, add						15%				
3000	Round, steel, 4 use per mo., rent, regular duty, 14" diameter	G	C-1	145	.221	L.F.	18.45	10.10		28.55	37
3050	16" diameter	G		125	.256		18.80	11.75		30.55	39.50
3100	Heavy duty, 20" diameter	G		105	.305		20	13.95		33.95	45.50

For customer support on your Commercial Renovation Cost Data, call 877.742.4719.

49

03 11 13 – Structural Cast-In-Place Concrete Forming

03 11 13.25 Forms In Place, Columns

		Crew	Daily Output	Labor-Hours	Unit	Material	2016 Bare Costs Labor	Equipment	Total	Total Incl O&P
3150	24" diameter	G	C-1	85	.376	L.F.	22	17.25		39.25	52
3200	30" diameter	G		70	.457		25.50	21		46.50	62
3250	36" diameter	G		60	.533		27	24.50		51.50	70
3300	48" diameter	G		50	.640		40.50	29.50		70	92.50
3350	60" diameter	G		45	.711		49.50	32.50		82	108
4500	For second and succeeding months, deduct						50%				
5000	Job-built plywood, 8" x 8" columns, 1 use		C-1	165	.194	SFCA	2.71	8.90		11.61	17.50
5050	2 use			195	.164		1.55	7.50		9.05	14
5100	3 use			210	.152		1.08	7		8.08	12.60
5150	4 use			215	.149		.89	6.80		7.69	12.15
5500	12" x 12" columns, 1 use			180	.178		2.59	8.15		10.74	16.15
5550	2 use			210	.152		1.42	7		8.42	12.95
5600	3 use			220	.145		1.04	6.65		7.69	12.05
5650	4 use			225	.142		.84	6.50		7.34	11.60
6000	16" x 16" columns, 1 use			185	.173		2.58	7.90		10.48	15.80
6050	2 use			215	.149		1.38	6.80		8.18	12.65
6100	3 use			230	.139		1.04	6.35		7.39	11.55
6150	4 use			235	.136		.84	6.25		7.09	11.15
6500	24" x 24" columns, 1 use			190	.168		2.91	7.70		10.61	15.80
6550	2 use			216	.148		1.60	6.80		8.40	12.85
6600	3 use			230	.139		1.16	6.35		7.51	11.70
6650	4 use			238	.134		.95	6.15		7.10	11.10
7000	36" x 36" columns, 1 use			200	.160		2.46	7.35		9.81	14.70
7050	2 use			230	.139		1.39	6.35		7.74	11.95
7100	3 use			245	.131		.98	6		6.98	10.90
7150	4 use			250	.128		.80	5.85		6.65	10.50
7400	Steel framed plywood, based on 50 uses of purchased										
7420	forms, and 4 uses of bracing lumber										
7500	8" x 8" column		C-1	340	.094	SFCA	2.17	4.31		6.48	9.45
7550	10" x 10"			350	.091		1.89	4.19		6.08	8.95
7600	12" x 12"			370	.086		1.60	3.96		5.56	8.20
7650	16" x 16"			400	.080		1.25	3.66		4.91	7.35
7700	20" x 20"			420	.076		1.10	3.49		4.59	6.90
7750	24" x 24"			440	.073		.80	3.33		4.13	6.35
7755	30" x 30"			440	.073		1.02	3.33		4.35	6.55
7760	36" x 36"			460	.070		.89	3.19		4.08	6.20
9000	Minimum labor/equipment charge		2 Carp	2	8	Job		390		390	635

03 11 13.35 Forms In Place, Elevated Slabs

			Crew	Daily Output	Labor-Hours	Unit	Material	2016 Bare Costs Labor	Equipment	Total	Total Incl O&P
0010	**FORMS IN PLACE, ELEVATED SLABS**	R031113-40									
1000	Flat plate, job-built plywood, to 15' high, 1 use		C-2	470	.102	S.F.	3.76	4.80		8.56	12
1150	4 use	R031113-60		560	.086		1.22	4.03		5.25	7.95
1500	15' to 20' high ceilings, 4 use			495	.097		1.34	4.56		5.90	8.90
2000	Flat slab, drop panels, job-built plywood, to 15' high, 1 use			449	.107		4.31	5.05		9.36	12.95
2150	4 use			544	.088		1.40	4.15		5.55	8.35
2250	15' to 20' high ceilings, 4 use			480	.100		2.52	4.70		7.22	10.45
3500	Floor slab, with 1-way joist pans, 1 use			415	.116		6.45	5.45		11.90	16
3650	4 use			500	.096		3.52	4.51		8.03	11.30
4500	With 2-way waffle domes, 1 use			405	.119		6.60	5.55		12.15	16.35
4550	4 use			470	.102		3.68	4.80		8.48	11.90
5000	Box out for slab openings, over 16" deep, 1 use			190	.253	SFCA	4.50	11.90		16.40	24.50
5050	2 use			240	.200	"	2.48	9.40		11.88	18.05
5500	Shallow slab box outs, to 10 S.F.			42	1.143	Ea.	11.20	54		65.20	100

03 11 Concrete Forming

03 11 13 – Structural Cast-In-Place Concrete Forming

03 11 13.35 Forms In Place, Elevated Slabs

		Crew	Daily Output	Labor-Hours	Unit	Material	2016 Bare Costs Labor	Equipment	Total	Total Incl O&P
5550	Over 10 S.F. (use perimeter)	C-2	600	.080	L.F.	1.50	3.76		5.26	7.80
6000	Bulkhead forms for slab, with keyway, 1 use, 2 piece		500	.096		2.10	4.51		6.61	9.70
6100	3 piece (see also edge forms)		460	.104		2.28	4.91		7.19	10.50
6200	Slab bulkhead form, 4-1/2" high, exp metal, w/keyway & stakes [G]	C-1	1200	.027		.91	1.22		2.13	3
6210	5-1/2" high [G]		1100	.029		1.15	1.33		2.48	3.45
6215	7-1/2" high [G]		960	.033		1.26	1.53		2.79	3.88
6220	9-1/2" high [G]		840	.038		1.38	1.75		3.13	4.37
6500	Curb forms, wood, 6" to 12" high, on elevated slabs, 1 use		180	.178	SFCA	1.62	8.15		9.77	15.10
6550	2 use		205	.156		.89	7.15		8.04	12.70
6600	3 use		220	.145		.65	6.65		7.30	11.60
6650	4 use		225	.142		.53	6.50		7.03	11.25
7000	Edge forms to 6" high, on elevated slab, 4 use		500	.064	L.F.	.20	2.93		3.13	5
7070	7" to 12" high, 1 use		162	.198	SFCA	1.23	9.05		10.28	16.15
7080	2 use		198	.162		.68	7.40		8.08	12.85
7090	3 use		222	.144		.49	6.60		7.09	11.35
7101	4 use		350	.091		.20	4.19		4.39	7.05
7500	Depressed area forms to 12" high, 4 use		300	.107	L.F.	1	4.89		5.89	9.10
7550	12" to 24" high, 4 use		175	.183		1.36	8.40		9.76	15.20
8000	Perimeter deck and rail for elevated slabs, straight		90	.356		12.80	16.30		29.10	40.50
8050	Curved		65	.492		17.60	22.50		40.10	56.50
8500	Void forms, round plastic, 8" high x 3" diameter [G]		450	.071	Ea.	1.54	3.26		4.80	7
8550	4" diameter [G]		425	.075		2.29	3.45		5.74	8.15
8600	6" diameter [G]		400	.080		3.77	3.66		7.43	10.15
8650	8" diameter [G]		375	.085		6.75	3.91		10.66	13.85
9000	Minimum labor/equipment charge	2 Carp	2	8	Job		390		390	635

03 11 13.40 Forms In Place, Equipment Foundations

		Crew	Daily Output	Labor-Hours	Unit	Material	2016 Bare Costs Labor	Equipment	Total	Total Incl O&P
0010	**FORMS IN PLACE, EQUIPMENT FOUNDATIONS** R031113-40									
0020	1 use	C-2	160	.300	SFCA	3.35	14.10		17.45	26.50
0050	2 use R031113-60		190	.253		1.84	11.90		13.74	21.50
0100	3 use		200	.240		1.34	11.30		12.64	19.90
0150	4 use		205	.234		1.09	11		12.09	19.20
9000	Minimum labor/equipment charge	1 Carp	3	2.667	Job		129		129	211

03 11 13.45 Forms In Place, Footings

		Crew	Daily Output	Labor-Hours	Unit	Material	2016 Bare Costs Labor	Equipment	Total	Total Incl O&P
0010	**FORMS IN PLACE, FOOTINGS** R031113-40									
0020	Continuous wall, plywood, 1 use	C-1	375	.085	SFCA	6.80	3.91		10.71	13.90
0150	4 use R031113-60	"	485	.066	"	2.21	3.02		5.23	7.40
1500	Keyway, 4 use, tapered wood, 2" x 4"	1 Carp	530	.015	L.F.	.21	.73		.94	1.42
1550	2" x 6"	"	500	.016	"	.30	.78		1.08	1.60
3000	Pile cap, square or rectangular, job-built plywood, 1 use	C-1	290	.110	SFCA	2.85	5.05		7.90	11.40
3150	4 use		383	.084		.92	3.83		4.75	7.25
5000	Spread footings, job-built lumber, 1 use		305	.105		2.15	4.81		6.96	10.20
5150	4 use		414	.077		.70	3.54		4.24	6.55
9000	Minimum labor/equipment charge	1 Carp	3	2.667	Job		129		129	211

03 11 13.47 Forms In Place, Gas Station Forms

		Crew	Daily Output	Labor-Hours	Unit	Material	2016 Bare Costs Labor	Equipment	Total	Total Incl O&P
0010	**FORMS IN PLACE, GAS STATION FORMS**									
0050	Curb fascia, with template, 12 ga. steel, left in place, 9" high [G]	1 Carp	50	.160	L.F.	13.90	7.75		21.65	28
1000	Sign or light bases, 18" diameter, 9" high [G]		9	.889	Ea.	87.50	43		130.50	167
1050	30" diameter, 13" high [G]		8	1	"	139	48.50		187.50	232
1990	Minimum labor/equipment charge		2	4	Job		194		194	315
2000	Island forms, 10' long, 9" high, 3'- 6" wide [G]	C-1	10	3.200	Ea.	390	147		537	665
2050	4' wide [G]		9	3.556		400	163		563	705
2500	20' long, 9" high, 4' wide [G]		6	5.333		645	244		889	1,100

For customer support on your Commercial Renovation Cost Data, call 877.742.4719.

51

03 11 Concrete Forming

03 11 13 – Structural Cast-In-Place Concrete Forming

03 11 13.47 Forms In Place, Gas Station Forms		Crew	Daily Output	Labor-Hours	Unit	Material	2016 Bare Costs Labor	Equipment	Total	Total Incl O&P
2550	5' wide	G	C-1	5	6.400	Ea.	670	293		963	1,225
9000	Minimum labor/equipment charge		1 Carp	3	2.667	Job		129		129	211

03 11 13.50 Forms In Place, Grade Beam

0010	**FORMS IN PLACE, GRADE BEAM**	R031113-40									
0020	Job-built plywood, 1 use		C-2	530	.091	SFCA	3.06	4.26		7.32	10.30
0150	4 use	R031113-60	"	605	.079	"	1	3.73		4.73	7.20
9000	Minimum labor/equipment charge		2 Carp	2	8	Job		390		390	635

03 11 13.65 Forms In Place, Slab On Grade

0010	**FORMS IN PLACE, SLAB ON GRADE**	R031113-40									
1000	Bulkhead forms w/keyway, wood, 6" high, 1 use	R031113-60	C-1	510	.063	L.F.	.97	2.87		3.84	5.75
1400	Bulkhead form for slab, 4-1/2" high, exp metal, incl keyway & stakes	G		1200	.027		.91	1.22		2.13	3
1410	5-1/2" high	G		1100	.029		1.15	1.33		2.48	3.45
1420	7-1/2" high	G		960	.033		1.26	1.53		2.79	3.88
1430	9-1/2" high	G		840	.038	↓	1.38	1.75		3.13	4.37
2000	Curb forms, wood, 6" to 12" high, on grade, 1 use			215	.149	SFCA	2.64	6.80		9.44	14.05
2150	4 use			275	.116	"	.86	5.35		6.21	9.65
3000	Edge forms, wood, 4 use, on grade, to 6" high			600	.053	L.F.	.33	2.44		2.77	4.36
3050	7" to 12" high			435	.074	SFCA	.83	3.37		4.20	6.40
3060	Over 12"			350	.091	"	.88	4.19		5.07	7.80
3500	For depressed slabs, 4 use, to 12" high			300	.107	L.F.	.72	4.89		5.61	8.80
3550	To 24" high			175	.183		.95	8.40		9.35	14.75
4000	For slab blockouts, to 12" high, 1 use			200	.160		.78	7.35		8.13	12.85
4050	To 24" high, 1 use			120	.267		.99	12.20		13.19	21
4100	Plastic (extruded), to 6" high, multiple use, on grade		↓	800	.040	↓	8.20	1.83		10.03	12
5000	Screed, 24 ga. metal key joint, see Section 03 15 16.30										
5020	Wood, incl. wood stakes, 1" x 3"		C-1	900	.036	L.F.	.78	1.63		2.41	3.51
5050	2" x 4"			900	.036	"	.78	1.63		2.41	3.52
6000	Trench forms in floor, wood, 1 use			160	.200	SFCA	1.79	9.15		10.94	16.90
6150	4 use			185	.173	"	.58	7.90		8.48	13.60
8760	Void form, corrugated fiberboard, 4" x 12", 4' long	G		3000	.011	S.F.	3.33	.49		3.82	4.47
8770	6" x 12", 4' long		↓	3000	.011		4.10	.49		4.59	5.30
8780	1/4" thick hardboard protective cover for void form		2 Carp	1500	.011		.63	.52		1.15	1.53
9000	Minimum labor/equipment charge		1 Carp	2	4	Job		194		194	315

03 11 13.85 Forms In Place, Walls

0010	**FORMS IN PLACE, WALLS**	R031113-10									
0100	Box out for wall openings, to 16" thick, to 10 S.F.		C-2	24	2	Ea.	25.50	94		119.50	183
0150	Over 10 S.F. (use perimeter)	R031113-40	"	280	.171	L.F.	2.20	8.05		10.25	15.55
0250	Brick shelf, 4" w, add to wall forms, use wall area above shelf										
0260	1 use	R031113-60	C-2	240	.200	SFCA	2.37	9.40		11.77	17.95
0350	4 use			300	.160	"	.95	7.50		8.45	13.35
0500	Bulkhead, wood with keyway, 1 use, 2 piece		↓	265	.181	L.F.	2.07	8.50		10.57	16.20
0600	Bulkhead forms with keyway, 1 piece expanded metal, 8" wall	G	C-1	1000	.032		1.26	1.47		2.73	3.79
0610	10" wall	G		800	.040		1.38	1.83		3.21	4.51
0620	12" wall	G	↓	525	.061	↓	1.66	2.79		4.45	6.40
0700	Buttress, to 8' high, 1 use		C-2	350	.137	SFCA	4.46	6.45		10.91	15.45
0850	4 use			480	.100	"	1.47	4.70		6.17	9.30
1000	Corbel or haunch, to 12" wide, add to wall forms, 1 use			150	.320	L.F.	2.31	15.05		17.36	27
1150	4 use			180	.267	"	.75	12.55		13.30	21.50
2000	Wall, job-built plywood, to 8' high, 1 use			370	.130	SFCA	2.68	6.10		8.78	12.90
2050	2 use			435	.110		1.70	5.20		6.90	10.35
2100	3 use			495	.097		1.24	4.56		5.80	8.80
2150	4 use		↓	505	.095	↓	1	4.47		5.47	8.40

03 11 Concrete Forming

03 11 13 – Structural Cast-In-Place Concrete Forming

03 11 13.85 Forms In Place, Walls

		Crew	Daily Output	Labor-Hours	Unit	Material	2016 Bare Costs Labor	Equipment	Total	Total Incl O&P
2400	Over 8' to 16' high, 1 use	C-2	280	.171	SFCA	2.96	8.05		11.01	16.40
2450	2 use		345	.139		1.29	6.55		7.84	12.10
2500	3 use		375	.128		.92	6		6.92	10.85
2550	4 use		395	.122		.76	5.70		6.46	10.20
2700	Over 16' high, 1 use		235	.204		2.62	9.60		12.22	18.60
2750	2 use		290	.166		1.44	7.80		9.24	14.30
2800	3 use		315	.152		1.05	7.15		8.20	12.85
2850	4 use		330	.145		.85	6.85		7.70	12.10
4000	Radial, smooth curved, job-built plywood, 1 use		245	.196		2.50	9.20		11.70	17.80
4150	4 use		335	.143		.81	6.75		7.56	11.90
4200	Below grade, job-built plywood, 1 use		225	.213		2.62	10.05		12.67	19.30
4210	2 use		225	.213		1.44	10.05		11.49	18
4220	3 use		225	.213		1.20	10.05		11.25	17.70
4230	4 use		225	.213		.85	10.05		10.90	17.35
4600	Retaining wall, battered, job-built plyw'd, to 8' high, 1 use		300	.160		1.97	7.50		9.47	14.45
4750	4 use		390	.123		.64	5.80		6.44	10.15
4900	Over 8' to 16' high, 1 use		240	.200		2.14	9.40		11.54	17.70
5050	4 use		320	.150		.70	7.05		7.75	12.25
7500	Lintel or sill forms, 1 use	1 Carp	30	.267		3.13	12.90		16.03	24.50
7560	4 use	"	37	.216		1.02	10.50		11.52	18.20
7800	Modular prefabricated plywood, based on 20 uses of purchased									
7820	forms, and 4 uses of bracing lumber									
7860	To 8' high	C-2	800	.060	SFCA	.90	2.82		3.72	5.60
8060	Over 8' to 16' high		600	.080		.95	3.76		4.71	7.20
8600	Pilasters, 1 use		270	.178		3.11	8.35		11.46	17.05
8660	4 use		385	.125		1.01	5.85		6.86	10.70
9010	Steel framed plywood, based on 50 uses of purchased									
9020	forms, and 4 uses of bracing lumber									
9060	To 8' high	C-2	600	.080	SFCA	.72	3.76		4.48	6.95
9260	Over 8' to 16' high		450	.107		.72	5		5.72	9
9460	Over 16' to 20' high		400	.120		.72	5.65		6.37	10
9475	For elevated walls, add						10%			
9480	For battered walls, 1 side battered, add					10%	10%			
9485	For battered walls, 2 sides battered, add					15%	15%			
9900	Minimum labor/equipment charge	2 Carp	2	8	Job		390		390	635

03 11 16 – Architectural Cast-in-Place Concrete Forming

03 11 16.13 Concrete Form Liners

		Crew	Daily Output	Labor-Hours	Unit	Material	2016 Bare Costs Labor	Equipment	Total	Total Incl O&P
0010	**CONCRETE FORM LINERS**									
5750	Liners for forms (add to wall forms), ABS plastic									
5800	Aged wood, 4" wide, 1 use	1 Carp	256	.031	SFCA	3.47	1.51		4.98	6.30
5820	2 use		256	.031		1.91	1.51		3.42	4.57
5830	3 use		256	.031		1.39	1.51		2.90	4
5840	4 use		256	.031		1.13	1.51		2.64	3.71
5900	Fractured rope rib, 1 use		192	.042		5.05	2.02		7.07	8.85
5925	2 use		192	.042		2.77	2.02		4.79	6.35
5950	3 use		192	.042		2.01	2.02		4.03	5.50
6000	4 use		192	.042		1.63	2.02		3.65	5.10
6100	Ribbed, 3/4" deep x 1-1/2" O.C., 1 use		224	.036		5.05	1.73		6.78	8.40
6125	2 use		224	.036		2.77	1.73		4.50	5.85
6150	3 use		224	.036		2.01	1.73		3.74	5.05
6200	4 use		224	.036		1.63	1.73		3.36	4.63
6300	Rustic brick pattern, 1 use		224	.036		3.47	1.73		5.20	6.65

For customer support on your Commercial Renovation Cost Data, call 877.742.4719.

53

03 11 Concrete Forming

03 11 16 – Architectural Cast-in-Place Concrete Forming

03 11 16.13 Concrete Form Liners		Crew	Daily Output	Labor-Hours	Unit	Material	2016 Bare Costs Labor	Equipment	Total	Total Incl O&P
6325	2 use	1 Carp	224	.036	SFCA	1.91	1.73		3.64	4.93
6350	3 use		224	.036		1.39	1.73		3.12	4.36
6400	4 use		224	.036		1.13	1.73		2.86	4.07
6500	3/8" striated, random, 1 use		224	.036		3.47	1.73		5.20	6.65
6525	2 use		224	.036		1.91	1.73		3.64	4.93
6550	3 use		224	.036		1.39	1.73		3.12	4.36
6600	4 use	▼	224	.036	▼	1.13	1.73		2.86	4.07

03 15 Concrete Accessories

03 15 05 – Concrete Forming Accessories

03 15 05.70 Shores

			Crew	Daily Output	Labor-Hours	Unit	Material	2016 Bare Costs Labor	Equipment	Total	Total Incl O&P
0010	**SHORES**										
0020	Erect and strip, by hand, horizontal members										
0500	Aluminum joists and stringers	G	2 Carp	60	.267	Ea.		12.90		12.90	21
0600	Steel, adjustable beams	G		45	.356			17.25		17.25	28
0700	Wood joists			50	.320			15.50		15.50	25.50
0800	Wood stringers			30	.533			26		26	42
1000	Vertical members to 10' high	G		55	.291			14.10		14.10	23
1050	To 13' high	G		50	.320			15.50		15.50	25.50
1100	To 16' high	G		45	.356	▼		17.25		17.25	28
1500	Reshoring	G		1400	.011	S.F.	.61	.55		1.16	1.57
1600	Flying truss system	G	C-17D	9600	.009	SFCA		.44	.08	.52	.80
1760	Horizontal, aluminum joists, 6-1/4" high x 5' to 21' span, buy	G				L.F.	16.25			16.25	17.85
1770	Beams, 7-1/4" high x 4' to 30' span	G				"	19			19	21
1810	Horizontal, steel beam, W8x10, 7' span, buy	G				Ea.	61			61	67.50
1830	10' span	G					71			71	78
1920	15' span	G					122			122	135
1940	20' span	G				▼	172			172	189
1970	Steel stringer, W8x10, 4' to 16' span, buy	G				L.F.	7.10			7.10	7.85
3000	Rent for job duration, aluminum joist @ 2' O.C., per mo.	G				SF Flr.	.41			.41	.45
3050	Steel W8x10	G					.18			.18	.20
3060	Steel adjustable	G				▼	.18			.18	.20
3500	#1 post shore, steel, 5'-7" to 9'-6" high, 10000# cap., buy	G				Ea.	155			155	171
3550	#2 post shore, 7'-3" to 12'-10" high, 7800# capacity	G					180			180	198
3600	#3 post shore, 8'-10" to 16'-1" high, 3800# capacity	G				▼	196			196	216
5010	Frame shoring systems, steel, 12000#/leg, buy										
5040	Frame, 2' wide x 6' high	G				Ea.	93.50			93.50	103
5250	X-brace	G					15.50			15.50	17.05
5550	Base plate	G					15.50			15.50	17.05
5600	Screw jack	G					33.50			33.50	36.50
5650	U-head, 8" x 8"	G				▼	19.10			19.10	21

03 15 05.85 Stair Tread Inserts

			Crew	Daily Output	Labor-Hours	Unit	Material	2016 Bare Costs Labor	Equipment	Total	Total Incl O&P
0010	**STAIR TREAD INSERTS**										
0105	Cast nosing insert, abrasive surface, pre-drilled, includes screws										
0110	Aluminum, 3" wide x 3' long		1 Cefi	32	.250	Ea.	57.50	11.40		68.90	81
0120	4' long			31	.258		73	11.80		84.80	99
0130	5' long		▼	30	.267	▼	90.50	12.15		102.65	119
0135	Extruded nosing insert, black abrasive strips, continuous anchor										
0140	Aluminum, 3" wide x 3' long		1 Cefi	64	.125	Ea.	37	5.70		42.70	50
0150	4' long			60	.133		57.50	6.10		63.60	72.50
0160	5' long		▼	56	.143		64.50	6.50		71	81

03 15 Concrete Accessories

03 15 05 – Concrete Forming Accessories

03 15 05.85 Stair Tread Inserts

		Crew	Daily Output	Labor-Hours	Unit	Material	2016 Bare Costs Labor	2016 Bare Costs Equipment	Total	Total Incl O&P
0165	Extruded nosing insert, black abrasive strips, pre-drilled, incl. screws									
0170	Aluminum, 3" wide x 3' long	1 Cefi	32	.250	Ea.	51	11.40		62.40	74
0180	4' long		31	.258		66	11.80		77.80	91.50
0190	5' long		30	.267		77.50	12.15		89.65	105
9000	Minimum labor/equipment charge	1 Carp	4	2	Job		97		97	158

03 15 13 – Waterstops

03 15 13.50 Waterstops

		Crew	Daily Output	Labor-Hours	Unit	Material	2016 Bare Costs Labor	2016 Bare Costs Equipment	Total	Total Incl O&P
0010	**WATERSTOPS**, PVC and Rubber									
0020	PVC, ribbed 3/16" thick, 4" wide	1 Carp	155	.052	L.F.	1.32	2.50		3.82	5.55
0050	6" wide		145	.055		2.29	2.67		4.96	6.90
0500	With center bulb, 6" wide, 3/16" thick		135	.059		2.02	2.87		4.89	6.90
0550	3/8" thick		130	.062		3.79	2.98		6.77	9.05
0600	9" wide x 3/8" thick		125	.064		6.15	3.10		9.25	11.80

03 15 16 – Concrete Construction Joints

03 15 16.20 Control Joints, Saw Cut

		Crew	Daily Output	Labor-Hours	Unit	Material	2016 Bare Costs Labor	2016 Bare Costs Equipment	Total	Total Incl O&P
0010	**CONTROL JOINTS, SAW CUT**									
0100	Sawcut control joints in green concrete									
0120	1" depth	C-27	2000	.008	L.F.	.04	.37	.08	.49	.71
0140	1-1/2" depth		1800	.009		.05	.41	.09	.55	.80
0160	2" depth		1600	.010		.07	.46	.11	.64	.92
0180	Sawcut joint reservoir in cured concrete									
0182	3/8" wide x 3/4" deep, with single saw blade	C-27	1000	.016	L.F.	.05	.73	.17	.95	1.41
0184	1/2" wide x 1" deep, with double saw blades		900	.018		.10	.81	.19	1.10	1.60
0186	3/4" wide x 1-1/2" deep, with double saw blades		800	.020		.21	.91	.21	1.33	1.91
0190	Water blast joint to wash away laitance, 2 passes	C-29	2500	.003			.12	.03	.15	.23
0200	Air blast joint to blow out debris and air dry, 2 passes	C-28	2000	.004			.18	.01	.19	.30
0300	For backer rod, see Section 07 91 23.10									
0340	For joint sealant, see Section 03 15 16.30 or 07 92 13.20									
0900	For replacement of joint sealant, see Section 07 01 90.81									

03 15 16.30 Expansion Joints

			Crew	Daily Output	Labor-Hours	Unit	Material	2016 Bare Costs Labor	2016 Bare Costs Equipment	Total	Total Incl O&P
0010	**EXPANSION JOINTS**										
0020	Keyed, cold, 24 ga., incl. stakes, 3-1/2" high	G	1 Carp	200	.040	L.F.	.86	1.94		2.80	4.12
0050	4-1/2" high	G		200	.040		.91	1.94		2.85	4.17
0100	5-1/2" high	G		195	.041		1.15	1.99		3.14	4.52
0150	7-1/2" high	G		190	.042		1.26	2.04		3.30	4.72
0160	9-1/2" high	G		185	.043		1.38	2.10		3.48	4.94
0300	Poured asphalt, plain, 1/2" x 1"		1 Clab	450	.018		.78	.67		1.45	1.96
0350	1" x 2"			400	.020		3.12	.76		3.88	4.67
0500	Neoprene, liquid, cold applied, 1/2" x 1"			450	.018		2.28	.67		2.95	3.61
0550	1" x 2"			400	.020		9.10	.76		9.86	11.30
0700	Polyurethane, poured, 2 part, 1/2" x 1"			400	.020		1.38	.76		2.14	2.76
0750	1" x 2"			350	.023		5.50	.87		6.37	7.45
0900	Rubberized asphalt, hot or cold applied, 1/2" x 1"			450	.018		.38	.67		1.05	1.52
0950	1" x 2"			400	.020		1.51	.76		2.27	2.90
1100	Hot applied, fuel resistant, 1/2" x 1"			450	.018		.57	.67		1.24	1.73
1150	1" x 2"			400	.020		2.27	.76		3.03	3.73
2000	Premolded, bituminous fiber, 1/2" x 6"		1 Carp	375	.021		.40	1.03		1.43	2.13
2050	1" x 12"			300	.027		1.97	1.29		3.26	4.28
2140	Concrete expansion joint, recycled paper and fiber, 1/2" x 6"	G		390	.021		.43	.99		1.42	2.09
2150	1/2" x 12"	G		360	.022		.85	1.08		1.93	2.70
2250	Cork with resin binder, 1/2" x 6"			375	.021		1.16	1.03		2.19	2.97

For customer support on your Commercial Renovation Cost Data, call 877.742.4719.

55

03 15 16 – Concrete Construction Joints

03 15 16.30 Expansion Joints		Crew	Daily Output	Labor-Hours	Unit	Material	2016 Bare Costs Labor	Equipment	Total	Total Incl O&P
2300	1" x 12"	1 Carp	300	.027	L.F.	3.20	1.29		4.49	5.65
2500	Neoprene sponge, closed cell, 1/2" x 6"		375	.021		2.41	1.03		3.44	4.34
2550	1" x 12"		300	.027		8.40	1.29		9.69	11.35
2750	Polyethylene foam, 1/2" x 6"		375	.021		.64	1.03		1.67	2.39
2800	1" x 12"		300	.027		2.31	1.29		3.60	4.65
3000	Polyethylene backer rod, 3/8" diameter		460	.017		.03	.84		.87	1.42
3050	3/4" diameter		460	.017		.06	.84		.90	1.45
3100	1" diameter		460	.017		.12	.84		.96	1.51
3500	Polyurethane foam, with polybutylene, 1/2" x 1/2"		475	.017		1.14	.82		1.96	2.58
3550	1" x 1"		450	.018		2.88	.86		3.74	4.58
3750	Polyurethane foam, regular, closed cell, 1/2" x 6"		375	.021		.84	1.03		1.87	2.61
3800	1" x 12"		300	.027		3	1.29		4.29	5.40
4000	Polyvinyl chloride foam, closed cell, 1/2" x 6"		375	.021		2.28	1.03		3.31	4.20
4050	1" x 12"		300	.027		7.85	1.29		9.14	10.75
4250	Rubber, gray sponge, 1/2" x 6"		375	.021		1.92	1.03		2.95	3.80
4300	1" x 12"		300	.027		6.90	1.29		8.19	9.70
4400	Redwood heartwood, 1" x 4"		400	.020		1.16	.97		2.13	2.86
4450	1" x 6"		375	.021		1.77	1.03		2.80	3.64
5000	For installation in walls, add						75%			
5250	For installation in boxouts, add						25%			

03 15 19 – Cast-In Concrete Anchors

03 15 19.05 Anchor Bolt Accessories

		Crew	Daily Output	Labor-Hours	Unit	Material	2016 Bare Costs Labor	Equipment	Total	Total Incl O&P
0010	**ANCHOR BOLT ACCESSORIES**									
0015	For anchor bolts set in fresh concrete, see Section 03 15 19.10									
8150	Anchor bolt sleeve, plastic, 1" diam. bolts	1 Carp	60	.133	Ea.	12.95	6.45		19.40	25
8500	1-1/2" diameter		28	.286		19.20	13.85		33.05	43.50
8600	2" diameter		24	.333		19.45	16.15		35.60	48
8650	3" diameter		20	.400		35.50	19.40		54.90	70.50

03 15 19.10 Anchor Bolts

			Crew	Daily Output	Labor-Hours	Unit	Material	2016 Bare Costs Labor	Equipment	Total	Total Incl O&P
0010	**ANCHOR BOLTS**										
0015	Made from recycled materials										
0025	Single bolts installed in fresh concrete, no templates										
0030	Hooked w/nut and washer, 1/2" diameter, 8" long	G	1 Carp	132	.061	Ea.	1.32	2.94		4.26	6.25
0040	12" long	G		131	.061		1.47	2.96		4.43	6.45
0050	5/8" diameter, 8" long	G		129	.062		3.09	3		6.09	8.30
0060	12" long	G		127	.063		3.81	3.05		6.86	9.20
0070	3/4" diameter, 8" long	G		127	.063		3.81	3.05		6.86	9.20
0080	12" long	G		125	.064		4.76	3.10		7.86	10.30
0090	2-bolt pattern, including job-built 2-hole template, per set										
0100	J-type, incl. hex nut & washer, 1/2" diameter x 6" long	G	1 Carp	21	.381	Set	5.55	18.45		24	36
0110	12" long	G		21	.381		6.15	18.45		24.60	37
0120	18" long	G		21	.381		7	18.45		25.45	37.50
0130	3/4" diameter x 8" long	G		20	.400		10.80	19.40		30.20	43.50
0140	12" long	G		20	.400		12.70	19.40		32.10	45.50
0150	18" long	G		20	.400		15.55	19.40		34.95	48.50
0160	1" diameter x 12" long	G		19	.421		22	20.50		42.50	57.50
0170	18" long	G		19	.421		25.50	20.50		46	61.50
0180	24" long	G		19	.421		30.50	20.50		51	67
0190	36" long	G		18	.444		40.50	21.50		62	79.50
0200	1-1/2" diameter x 18" long	G		17	.471		42	23		65	83.50
0210	24" long	G		16	.500		49	24		73	93.50
0300	L-type, incl. hex nut & washer, 3/4" diameter x 12" long	G		20	.400		13.75	19.40		33.15	46.50

56

For customer support on your Commercial Renovation Cost Data, call 877.742.4719.

03 15 Concrete Accessories

03 15 19 – Cast-In Concrete Anchors

03 15 19.10 Anchor Bolts

		Crew	Daily Output	Labor-Hours	Unit	Material	2016 Bare Costs Labor	Equipment	Total	Total Incl O&P
0310	18" long	G 1 Carp	20	.400	Set	16.70	19.40		36.10	50
0320	24" long	G	20	.400		19.65	19.40		39.05	53
0330	30" long	G	20	.400		24	19.40		43.40	58
0340	36" long	G	20	.400		27	19.40		46.40	61.50
0350	1" diameter x 12" long	G	19	.421		21	20.50		41.50	56.50
0360	18" long	G	19	.421		25.50	20.50		46	61.50
0370	24" long	G	19	.421		30.50	20.50		51	67
0380	30" long	G	19	.421		35.50	20.50		56	72.50
0390	36" long	G	18	.444		40	21.50		61.50	79
0400	42" long	G	18	.444		48	21.50		69.50	88
0410	48" long	G	18	.444		53.50	21.50		75	94
0420	1-1/4" diameter x 18" long	G	18	.444		30	21.50		51.50	68
0430	24" long	G	18	.444		35	21.50		56.50	73.50
0440	30" long	G	17	.471		40	23		63	81.50
0450	36" long	G	17	.471		45	23		68	87
1000	4-bolt pattern, including job-built 4-hole template, per set									
1100	J-type, incl. hex nut & washer, 1/2" diameter x 6" long	G 1 Carp	19	.421	Set	7.90	20.50		28.40	42
1110	12" long	G	19	.421		9.05	20.50		29.55	43.50
1120	18" long	G	18	.444		10.85	21.50		32.35	47
1130	3/4" diameter x 8" long	G	17	.471		18.40	23		41.40	58
1140	12" long	G	17	.471		22	23		45	62
1150	18" long	G	17	.471		28	23		51	68
1160	1" diameter x 12" long	G	16	.500		40.50	24		64.50	84
1170	18" long	G	15	.533		48	26		74	95
1180	24" long	G	15	.533		58	26		84	106
1190	36" long	G	15	.533		78	26		104	128
1200	1-1/2" diameter x 18" long	G	13	.615		80.50	30		110.50	137
1210	24" long	G	12	.667		95	32.50		127.50	158
1300	L-type, incl. hex nut & washer, 3/4" diameter x 12" long	G	17	.471		24.50	23		47.50	64
1310	18" long	G	17	.471		30	23		53	70.50
1320	24" long	G	17	.471		36	23		59	77.50
1330	30" long	G	16	.500		45	24		69	89
1340	36" long	G	16	.500		51	24		75	95.50
1350	1" diameter x 12" long	G	16	.500		38.50	24		62.50	82
1360	18" long	G	15	.533		47.50	26		73.50	94
1370	24" long	G	15	.533		58	26		84	106
1380	30" long	G	15	.533		67.50	26		93.50	117
1390	36" long	G	15	.533		77	26		103	127
1400	42" long	G	14	.571		93	27.50		120.50	147
1410	48" long	G	14	.571		104	27.50		131.50	159
1420	1-1/4" diameter x 18" long	G	14	.571		57	27.50		84.50	108
1430	24" long	G	14	.571		67	27.50		94.50	119
1440	30" long	G	13	.615		77	30		107	133
1450	36" long	G	13	.615		87	30		117	144

03 15 19.45 Machinery Anchors

		Crew	Daily Output	Labor-Hours	Unit	Material	2016 Bare Costs Labor	Equipment	Total	Total Incl O&P
0010	**MACHINERY ANCHORS**, heavy duty, incl. sleeve, floating base nut,									
0020	lower stud & coupling nut, fiber plug, connecting stud, washer & nut.									
0030	For flush mounted embedment in poured concrete heavy equip. pads.									
0200	Stud & bolt, 1/2" diameter	G E-16	40	.400	Ea.	52.50	21.50	3.68	77.68	100
0300	5/8" diameter	G	35	.457		61	25	4.21	90.21	115
0500	3/4" diameter	G	30	.533		72.50	29	4.91	106.41	136
0600	7/8" diameter	G	25	.640		82.50	34.50	5.90	122.90	158

For customer support on your Commercial Renovation Cost Data, call 877.742.4719.

57

03 15 Concrete Accessories

03 15 19 – Cast-In Concrete Anchors

03 15 19.45 Machinery Anchors		Crew	Daily Output	Labor-Hours	Unit	Material	2016 Bare Costs Labor	Equipment	Total	Total Incl O&P
0800	1" diameter	G E-16	20	.800	Ea.	88.50	43.50	7.35	139.35	182
0900	1-1/4" diameter	G	15	1.067		118	58	9.80	185.80	243

03 21 Reinforcement Bars

03 21 11 – Plain Steel Reinforcement Bars

03 21 11.60 Reinforcing In Place

		Crew	Daily Output	Labor-Hours	Unit	Material	2016 Bare Costs Labor	Equipment	Total	Total Incl O&P
0010	**REINFORCING IN PLACE**, 50-60 ton lots, A615 Grade 60 R032110-10									
0020	Includes labor, but not material cost, to install accessories									
0030	Made from recycled materials									
0102	Beams & Girders, #3 to #7	G 4 Rodm	3200	.010	Lb.	.48	.53		1.01	1.39
0152	#8 to #18	G	5400	.006		.48	.31		.79	1.04
0202	Columns, #3 to #7	G	3000	.011		.48	.57		1.05	1.45
0252	#8 to #18	G	4600	.007		.48	.37		.85	1.13
0402	Elevated slabs, #4 to #7	G	5800	.006		.48	.29		.77	1
0502	Footings, #4 to #7	G	4200	.008		.48	.40		.88	1.18
0552	#8 to #18	G	7200	.004		.48	.24		.72	.91
0602	Slab on grade, #3 to #7	G	4200	.008		.48	.40		.88	1.18
0702	Walls, #3 to #7	G	6000	.005		.48	.28		.76	.99
0752	#8 to #18	G	8000	.004		.48	.21		.69	.87
0900	For other than 50 – 60 ton lots									
1000	Under 10 ton job, #3 to #7, add					25%	10%			
1010	#8 to #18, add					20%	10%			
1050	10 – 50 ton job, #3 to #7, add					10%				
1060	#8 to #18, add					5%				
1100	60 – 100 ton job, #3 to #7, deduct					5%				
1110	#8 to #18, deduct					10%				
1150	Over 100 ton job, #3 to #7, deduct					10%				
1160	#8 to #18, deduct					15%				
2000	Unloading & sorting, add to above	C-5	100	.560	Ton		29	7.50	36.50	55.50
2200	Crane cost for handling, 90 picks/day, up to 1.5 Tons/bundle, add to above		135	.415			21.50	5.55	27.05	41
2210	1.0 Ton/bundle		92	.609			32	8.15	40.15	60
2220	0.5 Ton/bundle		35	1.600			83.50	21.50	105	158
2400	Dowels, 2 feet long, deformed, #3	G 2 Rodm	520	.031	Ea.	.40	1.63		2.03	3.08
2410	#4	G	480	.033		.71	1.77		2.48	3.64
2420	#5	G	435	.037		1.10	1.95		3.05	4.37
2430	#6	G	360	.044		1.59	2.36		3.95	5.55
2450	Longer and heavier dowels, add	G	725	.022	Lb.	.53	1.17		1.70	2.47
2500	Smooth dowels, 12" long, 1/4" or 3/8" diameter	G	140	.114	Ea.	.72	6.05		6.77	10.60
2520	5/8" diameter	G	125	.128		1.26	6.80		8.06	12.40
2530	3/4" diameter	G	110	.145		1.57	7.70		9.27	14.20
2600	Dowel sleeves for CIP concrete, 2-part system									
2610	Sleeve base, plastic, for 5/8" smooth dowel sleeve, fasten to edge form	1 Rodm	200	.040	Ea.	.53	2.12		2.65	4.01
2615	Sleeve, plastic, 12" long, for 5/8" smooth dowel, snap onto base		400	.020		1.33	1.06		2.39	3.18
2620	Sleeve base, for 3/4" smooth dowel sleeve		175	.046		.53	2.42		2.95	4.50
2625	Sleeve, 12" long, for 3/4" smooth dowel		350	.023		1.61	1.21		2.82	3.73
2630	Sleeve base, for 1" smooth dowel sleeve		150	.053		.68	2.83		3.51	5.35
2635	Sleeve, 12" long, for 1" smooth dowel		300	.027		1.42	1.41		2.83	3.85
2700	Dowel caps, visual warning only, plastic, #3 to #8	2 Rodm	800	.020		.30	1.06		1.36	2.05
2720	#8 to #18		750	.021		.74	1.13		1.87	2.65
2750	Impalement protective, plastic, #4 to #9		800	.020		1.66	1.06		2.72	3.55
9000	Minimum labor/equipment charge	1 Rodm	4	2	Job		106		106	172

03 21 Reinforcement Bars

03 21 21 – Composite Reinforcement Bars

03 21 21.11 Glass Fiber-Reinforced Polymer Reinf. Bars	Crew	Daily Output	Labor-Hours	Unit	Material	2016 Bare Costs Labor	Equipment	Total	Total Incl O&P
0010 **GLASS FIBER-REINFORCED POLYMER REINFORCEMENT BARS**									
0020 Includes labor, but not material cost, to install accessories									
0050 #2 bar, .043 lb./L.F.	4 Rodm	9500	.003	L.F.	.40	.18		.58	.73
0100 #3 bar, .092 lb./L.F.		9300	.003		.49	.18		.67	.84
0150 #4 bar, .160 lb./L.F.		9100	.004		.71	.19		.90	1.08
0200 #5 bar, .258 lb./L.F.		8700	.004		1.09	.20		1.29	1.52
0250 #6 bar, .372 lb./L.F.		8300	.004		1.41	.20		1.61	1.88
0300 #7 bar, .497 lb./L.F.		7900	.004		1.84	.21		2.05	2.37
0350 #8 bar, .620 lb./L.F.		7400	.004		2.38	.23		2.61	2.99
0400 #9 bar, .800 lb./L.F.		6800	.005		3.10	.25		3.35	3.81
0450 #10 bar, 1.08 lb./L.F.	▼	5800	.006	▼	3.75	.29		4.04	4.60
0500 For Bends, add per bend				Ea.	1.48			1.48	1.63

03 22 Fabric and Grid Reinforcing

03 22 11 – Plain Welded Wire Fabric Reinforcing

03 22 11.10 Plain Welded Wire Fabric

		Crew	Daily Output	Labor-Hours	Unit	Material	2016 Bare Costs Labor	Equipment	Total	Total Incl O&P
0010	**PLAIN WELDED WIRE FABRIC** ASTM A185 R032205-30									
0020	Includes labor, but not material cost, to install accessories									
0030	Made from recycled materials									
0050	Sheets									
0100	6 x 6 - W1.4 x W1.4 (10 x 10) 21 lb. per C.S.F.	G	2 Rodm	35	.457	C.S.F.	13.50	24		37.50	54
0200	6 x 6 - W2.1 x W2.1 (8 x 8) 30 lb. per C.S.F.	G		31	.516		17.55	27.50		45.05	64
0300	6 x 6 - W2.9 x W2.9 (6 x 6) 42 lb. per C.S.F.	G		29	.552		23	29		52	72.50
0400	6 x 6 - W4 x W4 (4 x 4) 58 lb. per C.S.F.	G		27	.593		31.50	31.50		63	85.50
0500	4 x 4 - W1.4 x W1.4 (10 x 10) 31 lb. per C.S.F.	G		31	.516		20	27.50		47.50	66.50
0600	4 x 4 - W2.1 x W2.1 (8 x 8) 44 lb. per C.S.F.	G		29	.552		25	29		54	75
0650	4 x 4 - W2.9 x W2.9 (6 x 6) 61 lb. per C.S.F.	G		27	.593		40.50	31.50		72	95.50
0700	4 x 4 - W4 x W4 (4 x 4) 85 lb. per C.S.F.	G	▼	25	.640	▼	50.50	34		84.50	111
0750	Rolls									
0800	2 x 2 - #14 galv., 21 lb./C.S.F., beam & column wrap	G	2 Rodm	6.50	2.462	C.S.F.	41.50	130		171.50	257
0900	2 x 2 - #12 galv. for gunite reinforcing	G	"	6.50	2.462	"	62.50	130		192.50	280
9000	Minimum labor/equipment charge		1 Rodm	4	2	Job		106		106	172

03 22 13 – Galvanized Welded Wire Fabric Reinforcing

03 22 13.10 Galvanized Welded Wire Fabric

		Unit	Material	Labor		Total	Total Incl O&P
0010	**GALVANIZED WELDED WIRE FABRIC**						
0100	Add to plain welded wire pricing for galvanized welded wire	Lb.	.23			.23	.25

03 22 16 – Epoxy-Coated Welded Wire Fabric Reinforcing

03 22 16.10 Epoxy-Coated Welded Wire Fabric

		Unit	Material	Labor		Total	Total Incl O&P
0010	**EPOXY-COATED WELDED WIRE FABRIC**						
0100	Add to plain welded wire pricing for epoxy-coated welded wire	Lb.	.22			.22	.24

For customer support on your Commercial Renovation Cost Data, call 877.742.4719.

59

03 24 Fibrous Reinforcing

03 24 05 – Reinforcing Fibers

03 24 05.30 Synthetic Fibers

		Crew	Daily Output	Labor-Hours	Unit	Material	2016 Bare Costs Labor	Equipment	Total	Total Incl O&P
0010	**SYNTHETIC FIBERS**									
0100	Synthetic fibers, add to concrete				Lb.	4.40			4.40	4.84
0110	1-1/2 lb. per C.Y.				C.Y.	6.80			6.80	7.50

03 24 05.70 Steel Fibers

		Crew	Daily Output	Labor-Hours	Unit	Material	2016 Bare Costs Labor	Equipment	Total	Total Incl O&P
0010	**STEEL FIBERS**									
0140	ASTM A850, Type V, continuously deformed, 1-1/2" long x 0.045" diam.									
0150	Add to price of ready mix concrete	G			Lb.	1.13			1.13	1.24
0205	Alternate pricing, dosing at 5 lb. per C.Y., add to price of RMC	G			C.Y.	5.65			5.65	6.20
0210	10 lb. per C.Y.	G				11.30			11.30	12.45
0215	15 lb. per C.Y.	G				16.95			16.95	18.65
0220	20 lb. per C.Y.	G				22.50			22.50	25
0225	25 lb. per C.Y.	G				28.50			28.50	31
0230	30 lb. per C.Y.	G				34			34	37.50
0235	35 lb. per C.Y.	G				39.50			39.50	43.50
0240	40 lb. per C.Y.	G				45			45	49.50
0250	50 lb. per C.Y.	G				56.50			56.50	62
0275	75 lb. per C.Y.	G				85			85	93
0300	100 lb. per C.Y.	G				113			113	124

03 30 Cast-In-Place Concrete

03 30 53 – Miscellaneous Cast-In-Place Concrete

03 30 53.40 Concrete In Place

		Crew	Daily Output	Labor-Hours	Unit	Material	2016 Bare Costs Labor	Equipment	Total	Total Incl O&P
0010	**CONCRETE IN PLACE**	R033105-10								
0020	Including forms (4 uses), Grade 60 rebar, concrete (Portland cement	R033105-20								
0050	Type I), placement and finishing unless otherwise indicated	R033105-70								
0300	Beams (3500 psi), 5 kip per L.F., 10' span	R033105-85	C-14A	15.62	12.804	C.Y.	320	620	48.50	988.50	1,400
0350	25' span		"	18.55	10.782		335	520	41	896	1,275
0500	Chimney foundations (5000 psi), over 5 C.Y.		C-14C	32.22	3.476		155	160	.97	315.97	430
0510	(3500 psi), under 5 C.Y.		"	23.71	4.724		182	217	1.32	400.32	555
0700	Columns, square (4000 psi), 12" x 12", up to 1% reinforcing by area		C-14A	11.96	16.722		365	810	63.50	1,238.50	1,800
0720	Up to 2% reinforcing by area			10.13	19.743		565	955	75	1,595	2,250
0740	Up to 3% reinforcing by area			9.03	22.148		835	1,075	84	1,994	2,775
0800	16" x 16", up to 1% reinforcing by area			16.22	12.330		290	595	47	932	1,350
0820	Up to 2% reinforcing by area			12.57	15.911		480	770	60.50	1,310.50	1,850
0840	Up to 3% reinforcing by area			10.25	19.512		735	945	74	1,754	2,400
0900	24" x 24", up to 1% reinforcing by area			23.66	8.453		245	410	32	687	970
0920	Up to 2% reinforcing by area			17.71	11.293		425	545	43	1,013	1,400
0940	Up to 3% reinforcing by area			14.15	14.134		670	685	53.50	1,408.50	1,925
1000	36" x 36", up to 1% reinforcing by area			33.69	5.936		216	287	22.50	525.50	735
1020	Up to 2% reinforcing by area			23.32	8.576		375	415	32.50	822.50	1,125
1040	Up to 3% reinforcing by area			17.82	11.223		625	545	42.50	1,212.50	1,625
1100	Columns, round (4000 psi), tied, 12" diameter, up to 1% reinforcing by area			20.97	9.537		320	460	36	816	1,150
1120	Up to 2% reinforcing by area			15.27	13.098		520	635	50	1,205	1,650
1140	Up to 3% reinforcing by area			12.11	16.515		780	800	63	1,643	2,225
1200	16" diameter, up to 1% reinforcing by area			31.49	6.351		295	305	24	624	850
1220	Up to 2% reinforcing by area			19.12	10.460		495	505	40	1,040	1,425
1240	Up to 3% reinforcing by area			13.77	14.524		735	705	55	1,495	2,025
1300	20" diameter, up to 1% reinforcing by area			41.04	4.873		289	236	18.50	543.50	725
1320	Up to 2% reinforcing by area			24.05	8.316		470	405	31.50	906.50	1,200
1340	Up to 3% reinforcing by area			17.01	11.758		730	570	44.50	1,344.50	1,775
1400	24" diameter, up to 1% reinforcing by area			51.85	3.857		269	187	14.65	470.65	615

03 30 53 – Miscellaneous Cast-In-Place Concrete

03 30 53.40 Concrete In Place	Crew	Daily Output	Labor-Hours	Unit	Material	2016 Bare Costs Labor	Equipment	Total	Total Incl O&P
1420 Up to 2% reinforcing by area	C-14A	27.06	7.391	C.Y.	470	360	28	858	1,125
1440 Up to 3% reinforcing by area		18.29	10.935		715	530	41.50	1,286.50	1,700
1500 36" diameter, up to 1% reinforcing by area		75.04	2.665		269	129	10.15	408.15	515
1520 Up to 2% reinforcing by area		37.49	5.335		445	258	20.50	723.50	935
1540 Up to 3% reinforcing by area		22.84	8.757		695	425	33.50	1,153.50	1,475
1900 Elevated slab (4000 psi), flat slab with drops, 125 psf Sup. Load, 20' span	C-14B	38.45	5.410		266	261	19.75	546.75	740
1950 30' span		50.99	4.079		280	197	14.90	491.90	645
2100 Flat plate, 125 psf Sup. Load, 15' span		30.24	6.878		244	330	25	599	835
2150 25' span		49.60	4.194		253	203	15.30	471.30	625
2300 Waffle const., 30" domes, 125 psf Sup. Load, 20' span		37.07	5.611		269	271	20.50	560.50	760
2350 30' span		44.07	4.720		250	228	17.25	495.25	665
2500 One way joists, 30" pans, 125 psf Sup. Load, 15' span		27.38	7.597		330	365	27.50	722.50	985
2550 25' span		31.15	6.677		305	325	24.50	654.50	885
2700 One way beam & slab, 125 psf Sup. Load, 15' span		20.59	10.102		264	490	37	791	1,125
2750 25' span		28.36	7.334		249	355	27	631	880
2900 Two way beam & slab, 125 psf Sup. Load, 15' span		24.04	8.652		254	420	31.50	705.50	995
2950 25' span		35.87	5.799		221	280	21	522	720
3100 Elevated slabs, flat plate, including finish, not									
3110 including forms or reinforcing									
3150 Regular concrete (4000 psi), 4" slab	C-8	2613	.021	S.F.	1.52	.91	.28	2.71	3.44
3200 6" slab		2585	.022		2.21	.92	.28	3.41	4.21
3250 2-1/2" thick floor fill		2685	.021		1	.88	.27	2.15	2.82
3300 Lightweight, 110# per C.F., 2-1/2" thick floor fill		2585	.022		1.48	.92	.28	2.68	3.41
3400 Cellular concrete, 1-5/8" fill, under 5000 S.F.		2000	.028		1	1.18	.36	2.54	3.42
3450 Over 10,000 S.F.		2200	.025		.96	1.08	.33	2.37	3.15
3500 Add per floor for 3 to 6 stories high		31800	.002			.07	.02	.09	.15
3520 For 7 to 20 stories high		21200	.003			.11	.03	.14	.22
3540 Equipment pad (3000 psi), 3' x 3' x 6" thick	C-14H	45	1.067	Ea.	47.50	50.50	.70	98.70	135
3550 4' x 4' x 6" thick		30	1.600		70.50	75.50	1.06	147.06	202
3560 5' x 5' x 8" thick		18	2.667		124	126	1.76	251.76	345
3570 6' x 6' x 8" thick		14	3.429		167	162	2.26	331.26	450
3580 8' x 8' x 10" thick		8	6		360	284	3.96	647.96	860
3590 10' x 10' x 12" thick		5	9.600		615	455	6.35	1,076.35	1,425
3800 Footings (3000 psi), spread under 1 C.Y.	C-14C	28	4	C.Y.	171	184	1.12	356.12	490
3825 1 C.Y. to 5 C.Y.		43	2.605		207	120	.73	327.73	425
3850 Over 5 C.Y.		75	1.493		190	68.50	.42	258.92	320
3900 Footings, strip (3000 psi), 18" x 9", unreinforced	C-14L	40	2.400		131	108	.79	239.79	320
3920 18" x 9", reinforced	C-14C	35	3.200		154	147	.90	301.90	410
3925 20" x 10", unreinforced	C-14L	45	2.133		127	95.50	.70	223.20	297
3930 20" x 10", reinforced	C-14C	40	2.800		146	129	.78	275.78	370
3935 24" x 12", unreinforced	C-14L	55	1.745		125	78.50	.58	204.08	267
3940 24" x 12", reinforced	C-14C	48	2.333		144	107	.65	251.65	335
3945 36" x 12", unreinforced	C-14L	70	1.371		121	61.50	.45	182.95	233
3950 36" x 12", reinforced	C-14C	60	1.867		138	86	.52	224.52	293
4000 Foundation mat (3000 psi), under 10 C.Y.		38.67	2.896		209	133	.81	342.81	450
4050 Over 20 C.Y.		56.40	1.986		184	91.50	.56	276.06	350
4200 Wall, free-standing (3000 psi), 8" thick, 8' high	C-14D	45.83	4.364		166	210	16.60	392.60	540
4250 14' high		27.26	7.337		196	350	28	574	820
4260 12" thick, 8' high		64.32	3.109		150	149	11.80	310.80	420
4270 14' high		40.01	4.999		160	240	19	419	585
4300 15" thick, 8' high		80.02	2.499		145	120	9.50	274.50	365
4350 12' high		51.26	3.902		145	187	14.85	346.85	480
4500 18' high		48.85	4.094		163	197	15.55	375.55	515

For customer support on your Commercial Renovation Cost Data, call 877.742.4719.

61

03 30 53.40 Concrete In Place		Crew	Daily Output	Labor-Hours	Unit	Material	2016 Bare Costs Labor	Equipment	Total	Total Incl O&P
4520	Handicap access ramp (4000 psi), railing both sides, 3' wide	C-14H	14.58	3.292	L.F.	315	156	2.17	473.17	605
4525	5' wide		12.22	3.928		325	186	2.59	513.59	665
4530	With 6" curb and rails both sides, 3' wide		8.55	5.614		325	266	3.71	594.71	795
4535	5' wide		7.31	6.566		330	310	4.33	644.33	875
4650	Slab on grade (3500 psi), not including finish, 4" thick	C-14E	60.75	1.449	C.Y.	129	68.50	.52	198.02	254
4700	6" thick	"	92	.957	"	124	45	.34	169.34	210
4701	Thickened slab edge (3500 psi), for slab on grade poured									
4702	monolithically with slab; depth is in addition to slab thickness;									
4703	formed vertical outside edge, earthen bottom and inside slope									
4705	8" deep x 8" wide bottom, unreinforced	C-14L	2190	.044	L.F.	3.57	1.97	.01	5.55	7.15
4710	8" x 8", reinforced	C-14C	1670	.067		5.85	3.09	.02	8.96	11.45
4715	12" deep x 12" wide bottom, unreinforced	C-14L	1800	.053		7.30	2.39	.02	9.71	11.90
4720	12" x 12", reinforced	C-14C	1310	.086		11.40	3.94	.02	15.36	19
4725	16" deep x 16" wide bottom, unreinforced	C-14L	1440	.067		12.30	2.99	.02	15.31	18.45
4730	16" x 16", reinforced	C-14C	1120	.100		17.25	4.60	.03	21.88	26.50
4735	20" deep x 20" wide bottom, unreinforced	C-14L	1150	.083		18.70	3.75	.03	22.48	26.50
4740	20" x 20", reinforced	C-14C	920	.122		25	5.60	.03	30.63	36.50
4745	24" deep x 24" wide bottom, unreinforced	C-14L	930	.103		26.50	4.63	.03	31.16	36.50
4750	24" x 24", reinforced	C-14C	740	.151		34.50	6.95	.04	41.49	49.50
4751	Slab on grade (3500 psi), incl. troweled finish, not incl. forms									
4760	or reinforcing, over 10,000 S.F., 4" thick	C-14F	3425	.021	S.F.	1.43	.91	.01	2.35	3.03
4820	6" thick		3350	.021		2.08	.93	.01	3.02	3.79
4840	8" thick		3184	.023		2.85	.98	.01	3.84	4.71
4900	12" thick		2734	.026		4.28	1.14	.01	5.43	6.55
4950	15" thick		2505	.029		5.35	1.24	.01	6.60	7.90
5000	Slab on grade (3000 psi), incl. broom finish, not incl. forms									
5001	or reinforcing, 4" thick	C-14G	2873	.019	S.F.	1.39	.83	.01	2.23	2.87
5010	6" thick		2590	.022		2.18	.92	.01	3.11	3.89
5020	8" thick		2320	.024		2.84	1.03	.01	3.88	4.79
5200	Lift slab in place above the foundation, incl. forms, reinforcing,									
5210	concrete (4000 psi) and columns, over 20,000 S.F. per floor	C-14B	2113	.098	S.F.	7.05	4.75	.36	12.16	15.90
5250	10,000 S.F. to 20,000 S.F. per floor		1650	.126		7.70	6.10	.46	14.26	18.85
5300	Under 10,000 S.F. per floor		1500	.139		8.35	6.70	.51	15.56	20.50
5500	Lightweight, ready mix, including screed finish only,									
5510	not including forms or reinforcing									
5550	1:4 (2500 psi) for structural roof decks	C-14B	260	.800	C.Y.	166	38.50	2.92	207.42	249
5600	1:6 (3000 psi) for ground slab with radiant heat	C-14F	92	.783		168	34	.34	202.34	239
5650	1:3:2 (2000 psi) with sand aggregate, roof deck	C-14B	260	.800		164	38.50	2.92	205.42	247
5700	Ground slab (2000 psi)	C-14F	107	.673		164	29	.30	193.30	228
5900	Pile caps (3000 psi), incl. forms and reinf., sq. or rect., under 10 C.Y.	C-14C	54.14	2.069		176	95	.58	271.58	350
5950	Over 10 C.Y.		75	1.493		164	68.50	.42	232.92	292
6000	Triangular or hexagonal, under 10 C.Y.		53	2.113		129	97.50	.59	227.09	300
6050	Over 10 C.Y.		85	1.318		145	60.50	.37	205.87	258
6200	Retaining walls (3000 psi), gravity, 4' high see Section 32 32	C-14D	66.20	3.021		145	145	11.50	301.50	410
6250	10' high		125	1.600		139	77	6.10	222.10	285
6300	Cantilever, level backfill loading, 8' high		70	2.857		156	137	10.85	303.85	410
6350	16' high		91	2.198		151	106	8.35	265.35	345
6800	Stairs (3500 psi), not including safety treads, free standing, 3'-6" wide	C-14H	83	.578	LF Nose	5.70	27.50	.38	33.58	51
6850	Cast on ground		125	.384	"	4.73	18.15	.25	23.13	35
7000	Stair landings, free standing		200	.240	S.F.	4.61	11.35	.16	16.12	23.50
7050	Cast on ground		475	.101	"	3.61	4.78	.07	8.46	11.80
9000	Minimum labor/equipment charge	2 Carp	1	16	Job		775		775	1,275

03 31 Structural Concrete

03 31 13 – Heavyweight Structural Concrete

03 31 13.25 Concrete, Hand Mix	Crew	Daily Output	Labor-Hours	Unit	Material	2016 Bare Costs Labor	Equipment	Total	Total Incl O&P
0010 **CONCRETE, HAND MIX** for small quantities or remote areas									
0050 Includes bulk local aggregate, bulk sand, bagged Portland									
0060 cement (Type I) and water, using gas powered cement mixer									
0125 2500 psi	C-30	135	.059	C.F.	3.50	2.25	1.28	7.03	8.95
0130 3000 psi		135	.059		3.75	2.25	1.28	7.28	9.20
0135 3500 psi		135	.059		3.89	2.25	1.28	7.42	9.35
0140 4000 psi		135	.059		4.06	2.25	1.28	7.59	9.55
0145 4500 psi		135	.059		4.26	2.25	1.28	7.79	9.75
0150 5000 psi		135	.059		4.53	2.25	1.28	8.06	10.05
0300 Using pre-bagged dry mix and wheelbarrow (80-lb. bag = 0.6 C.F.)									
0340 4000 psi	1 Clab	48	.167	C.F.	6.50	6.30		12.80	17.50

03 31 13.30 Concrete, Volumetric Site-Mixed

03 31 13.30 Concrete, Volumetric Site-Mixed	Crew	Daily Output	Labor-Hours	Unit	Material	Labor	Equipment	Total	Total Incl O&P
0010 **CONCRETE, VOLUMETRIC SITE-MIXED**									
0015 Mixed on-site in volumetric truck									
0020 Includes local aggregate, sand, Portland cement (Type I) and water									
0025 Excludes all additives and treatments									
0100 3000 psi, 1 C.Y. mixed and discharged				C.Y.	193			193	213
0110 2 C.Y.					147			147	162
0120 3 C.Y.					131			131	144
0130 4 C.Y.					118			118	130
0140 5 C.Y.					110			110	121
0200 For truck holding/waiting time past first 2 on-site hours, add				Hr.	88			88	97
0210 For trip charge beyond first 20 miles, each way, add				Mile	3.50			3.50	3.85
0220 For each additional increase of 500 psi, add				Ea.	4.17			4.17	4.59

03 31 13.35 Heavyweight Concrete, Ready Mix

03 31 13.35 Heavyweight Concrete, Ready Mix	Crew	Daily Output	Labor-Hours	Unit	Material	Labor	Equipment	Total	Total Incl O&P
0010 **HEAVYWEIGHT CONCRETE, READY MIX**, delivered									
0012 Includes local aggregate, sand, Portland cement (Type I) and water									
0015 Excludes all additives and treatments R033105-20									
0020 2000 psi				C.Y.	102			102	113
0100 2500 psi					105			105	115
0150 3000 psi					107			107	118
0200 3500 psi					110			110	121
0300 4000 psi					113			113	124
0350 4500 psi					116			116	127
0400 5000 psi					119			119	131
0411 6000 psi					122			122	134
0412 8000 psi					129			129	141
0413 10,000 psi					135			135	149
0414 12,000 psi					142			142	156
1000 For high early strength (Portland cement Type III), add					10%				
1300 For winter concrete (hot water), add					4.45			4.45	4.90
1410 For mid-range water reducer, add					3.33			3.33	3.66
1420 For high-range water reducer/superplasticizer, add					6			6	6.60
1430 For retarder, add					3.23			3.23	3.55
1440 For non-Chloride accelerator, add					5.70			5.70	6.25
1450 For Chloride accelerator, per 1%, add					3			3	3.30
1460 For fiber reinforcing, synthetic (1 lb./C.Y.), add					6.70			6.70	7.35
1500 For Saturday delivery, add					7.50			7.50	8.25
1510 For truck holding/waiting time past 1st hour per load, add				Hr.	96.50			96.50	106
1520 For short load (less than 4 C.Y.), add per load				Ea.	74.50			74.50	82
2000 For all lightweight aggregate, add				C.Y.	45%				

For customer support on your Commercial Renovation Cost Data, call 877.742.4719.

63

03 31 13.70 Placing Concrete	Crew	Daily Output	Labor-Hours	Unit	Material	2016 Bare Costs Labor	2016 Bare Costs Equipment	Total	Total Incl O&P
0010 PLACING CONCRETE R033105-70									
0020 Includes labor and equipment to place, level (strike off) and consolidate									
0050 Beams, elevated, small beams, pumped	C-20	60	1.067	C.Y.		43.50	13.20	56.70	85
0100 With crane and bucket	C-7	45	1.600			66	27	93	137
0200 Large beams, pumped	C-20	90	.711			29	8.80	37.80	56.50
0250 With crane and bucket	C-7	65	1.108			46	18.60	64.60	94.50
0400 Columns, square or round, 12" thick, pumped	C-20	60	1.067			43.50	13.20	56.70	85
0450 With crane and bucket	C-7	40	1.800			74.50	30	104.50	154
0600 18" thick, pumped	C-20	90	.711			29	8.80	37.80	56.50
0650 With crane and bucket	C-7	55	1.309			54	22	76	112
0800 24" thick, pumped	C-20	92	.696			28.50	8.60	37.10	55.50
0850 With crane and bucket	C-7	70	1.029			42.50	17.30	59.80	87.50
1000 36" thick, pumped	C-20	140	.457			18.65	5.65	24.30	36
1050 With crane and bucket	C-7	100	.720			30	12.10	42.10	61.50
1400 Elevated slabs, less than 6" thick, pumped	C-20	140	.457			18.65	5.65	24.30	36
1450 With crane and bucket	C-7	95	.758			31.50	12.75	44.25	64.50
1500 6" to 10" thick, pumped	C-20	160	.400			16.30	4.94	21.24	32
1550 With crane and bucket	C-7	110	.655			27	11	38	55.50
1600 Slabs over 10" thick, pumped	C-20	180	.356			14.50	4.39	18.89	28.50
1650 With crane and bucket	C-7	130	.554			23	9.30	32.30	47.50
1900 Footings, continuous, shallow, direct chute	C-6	120	.400			15.80	.53	16.33	26
1950 Pumped	C-20	150	.427			17.40	5.25	22.65	34
2000 With crane and bucket	C-7	90	.800			33	13.45	46.45	68.50
2100 Footings, continuous, deep, direct chute	C-6	140	.343			13.55	.45	14	22.50
2150 Pumped	C-20	160	.400			16.30	4.94	21.24	32
2200 With crane and bucket	C-7	110	.655			27	11	38	55.50
2400 Footings, spread, under 1 C.Y., direct chute	C-6	55	.873			34.50	1.15	35.65	57.50
2450 Pumped	C-20	65	.985			40	12.15	52.15	78.50
2500 With crane and bucket	C-7	45	1.600			66	27	93	137
2600 Over 5 C.Y., direct chute	C-6	120	.400			15.80	.53	16.33	26
2650 Pumped	C-20	150	.427			17.40	5.25	22.65	34
2700 With crane and bucket	C-7	100	.720			30	12.10	42.10	61.50
2900 Foundation mats, over 20 C.Y., direct chute	C-6	350	.137			5.40	.18	5.58	9
2950 Pumped	C-20	400	.160			6.50	1.98	8.48	12.70
3000 With crane and bucket	C-7	300	.240			9.95	4.03	13.98	20.50
3200 Grade beams, direct chute	C-6	150	.320			12.65	.42	13.07	21
3250 Pumped	C-20	180	.356			14.50	4.39	18.89	28.50
3300 With crane and bucket	C-7	120	.600			25	10.10	35.10	51
3500 High rise, for more than 5 stories, pumped, add per story	C-20	2100	.030			1.24	.38	1.62	2.42
3510 With crane and bucket, add per story	C-7	2100	.034			1.42	.58	2	2.92
3700 Pile caps, under 5 C.Y., direct chute	C-6	90	.533			21	.70	21.70	35.50
3750 Pumped	C-20	110	.582			23.50	7.20	30.70	46.50
3800 With crane and bucket	C-7	80	.900			37	15.10	52.10	76.50
3850 Pile cap, 5 C.Y. to 10 C.Y., direct chute	C-6	175	.274			10.85	.36	11.21	18
3900 Pumped	C-20	200	.320			13.05	3.96	17.01	25.50
3950 With crane and bucket	C-7	150	.480			19.85	8.05	27.90	41
4000 Over 10 C.Y., direct chute	C-6	215	.223			8.85	.29	9.14	14.65
4050 Pumped	C-20	240	.267			10.85	3.30	14.15	21
4100 With crane and bucket	C-7	185	.389			16.10	6.55	22.65	33
4300 Slab on grade, up to 6" thick, direct chute	C-6	110	.436			17.25	.58	17.83	28.50
4350 Pumped	C-20	130	.492			20	6.10	26.10	39
4400 With crane and bucket	C-7	110	.655			27	11	38	55.50

03 31 Structural Concrete

03 31 13 – Heavyweight Structural Concrete

03 31 13.70 Placing Concrete

		Crew	Daily Output	Labor-Hours	Unit	Material	2016 Bare Costs Labor	Equipment	Total	Total Incl O&P
4600	Over 6" thick, direct chute	C-6	165	.291	C.Y.		11.50	.38	11.88	19.10
4650	Pumped	C-20	185	.346			14.10	4.28	18.38	27.50
4700	With crane and bucket	C-7	145	.497			20.50	8.35	28.85	42
4900	Walls, 8" thick, direct chute	C-6	90	.533			21	.70	21.70	35.50
4950	Pumped	C-20	100	.640			26	7.90	33.90	51
5000	With crane and bucket	C-7	80	.900			37	15.10	52.10	76.50
5050	12" thick, direct chute	C-6	100	.480			18.95	.63	19.58	31.50
5100	Pumped	C-20	110	.582			23.50	7.20	30.70	46.50
5200	With crane and bucket	C-7	90	.800			33	13.45	46.45	68.50
5300	15" thick, direct chute	C-6	105	.457			18.05	.60	18.65	30
5350	Pumped	C-20	120	.533			21.50	6.60	28.10	42.50
5400	With crane and bucket	C-7	95	.758			31.50	12.75	44.25	64.50
5600	Wheeled concrete dumping, add to placing costs above									
5610	Walking cart, 50' haul, add	C-18	32	.281	C.Y.		10.70	1.88	12.58	19.55
5620	150' haul, add		24	.375			14.30	2.50	16.80	26.50
5700	250' haul, add		18	.500			19.05	3.34	22.39	34.50
5800	Riding cart, 50' haul, add	C-19	80	.113			4.29	1.25	5.54	8.35
5810	150' haul, add		60	.150			5.70	1.66	7.36	11.20
5900	250' haul, add		45	.200			7.60	2.22	9.82	14.90
6000	Concrete in-fill for pan-type metal stairs and landings. Manual placement									
6010	includes up to 50' horizontal haul from point of concrete discharge.									
6100	Stair pan treads, 2" deep									
6110	Flights in 1st floor level up/down from discharge point	C-8A	3200	.015	S.F.		.61		.61	.99
6120	2nd floor level		2500	.019			.78		.78	1.27
6130	3rd floor level		2000	.024			.98		.98	1.58
6140	4th floor level		1800	.027			1.09		1.09	1.76
6200	Intermediate stair landings, pan-type 4" deep									
6210	Flights in 1st floor level up/down from discharge point	C-8A	2000	.024	S.F.		.98		.98	1.58
6220	2nd floor level		1500	.032			1.31		1.31	2.11
6230	3rd floor level		1200	.040			1.63		1.63	2.64
6240	4th floor level		1000	.048			1.96		1.96	3.16
9000	Minimum labor/equipment charge	C-6	2	24	Job		950	31.50	981.50	1,575

03 35 Concrete Finishing

03 35 13 – High-Tolerance Concrete Floor Finishing

03 35 13.30 Finishing Floors, High Tolerance

		Crew	Daily Output	Labor-Hours	Unit	Material	2016 Bare Costs Labor	Equipment	Total	Total Incl O&P
0010	**FINISHING FLOORS, HIGH TOLERANCE**									
0012	Finishing of fresh concrete flatwork requires that concrete									
0013	first be placed, struck off & consolidated									
0015	Basic finishing for various unspecified flatwork									
0100	Bull float only	C-10	4000	.006	S.F.		.26		.26	.41
0125	Bull float & manual float		2000	.012			.52		.52	.83
0150	Bull float, manual float, & broom finish, w/edging & joints		1850	.013			.56		.56	.89
0200	Bull float, manual float & manual steel trowel		1265	.019			.82		.82	1.31
0210	For specified Random Access Floors in ACI Classes 1, 2, 3 and 4 to achieve									
0215	Composite Overall Floor Flatness and Levelness values up to FF35/FL25									
0250	Bull float, machine float & machine trowel (walk-behind)	C-10C	1715	.014	S.F.		.60	.03	.63	.99
0300	Power screed, bull float, machine float & trowel (walk-behind)	C-10D	2400	.010			.43	.05	.48	.74
0350	Power screed, bull float, machine float & trowel (ride-on)	C-10E	4000	.006			.26	.06	.32	.48
0352	For specified Random Access Floors in ACI Classes 5, 6, 7 and 8 to achieve									
0354	Composite Overall Floor Flatness and Levelness values up to FF50/FL50									

For customer support on your Commercial Renovation Cost Data, call 877.742.4719.

65

03 35 Concrete Finishing

03 35 13 – High-Tolerance Concrete Floor Finishing

03 35 13.30 Finishing Floors, High Tolerance

		Crew	Daily Output	Labor-Hours	Unit	Material	2016 Bare Costs Labor	2016 Bare Costs Equipment	Total	Total Incl O&P
0356	Add for two-dimensional restraightening after power float	C-10	6000	.004	S.F.		.17		.17	.28
0358	For specified Random or Defined Access Floors in ACI Class 9 to achieve									
0360	Composite Overall Floor Flatness and Levelness values up to FF100/FL100									
0362	Add for two-dimensional restraightening after bull float & power float	C-10	3000	.008	S.F.		.34		.34	.55
0364	For specified Superflat Defined Access Floors in ACI Class 9 to achieve									
0366	Minimum Floor Flatness and Levelness values of FF100/FL100									
0368	Add for 2-dim'l restraightening after bull float, power float, power trowel	C-10	2000	.012	S.F.		.52		.52	.83
9100	Minimum labor/equipment charge	"	2	12	Job		515		515	825

03 35 29 – Tooled Concrete Finishing

03 35 29.60 Finishing Walls

		Crew	Daily Output	Labor-Hours	Unit	Material	2016 Bare Costs Labor	2016 Bare Costs Equipment	Total	Total Incl O&P
0010	**FINISHING WALLS**									
0020	Break ties and patch voids	1 Cefi	540	.015	S.F.	.04	.68		.72	1.12
0050	Burlap rub with grout	"	450	.018		.04	.81		.85	1.33
0300	Bush hammer, green concrete	B-39	1000	.048			1.93	.23	2.16	3.38
0350	Cured concrete	"	650	.074			2.96	.35	3.31	5.20
0500	Acid etch	1 Cefi	575	.014		.14	.64		.78	1.17
0700	Sandblast, light penetration	E-11	1100	.029		.52	1.23	.21	1.96	2.91
0750	Heavy penetration	"	375	.085		1.04	3.62	.62	5.28	8.05
0850	Grind form fins flush	1 Clab	700	.011	L.F.		.43		.43	.71
9000	Minimum labor/equipment charge	C-10	2	12	Job		515		515	825

03 35 33 – Stamped Concrete Finishing

03 35 33.50 Slab Texture Stamping

		Crew	Daily Output	Labor-Hours	Unit	Material	2016 Bare Costs Labor	2016 Bare Costs Equipment	Total	Total Incl O&P
0010	**SLAB TEXTURE STAMPING**									
0050	Stamping requires that concrete first be placed, struck off, consolidated,									
0060	bull floated and free of bleed water. Decorative stamping tasks include:									
0100	Step 1 - first application of dry shake colored hardener	1 Cefi	6400	.001	S.F.	.44	.06		.50	.58
0110	Step 2 - bull float		6400	.001			.06		.06	.09
0130	Step 3 - second application of dry shake colored hardener		6400	.001		.22	.06		.28	.33
0140	Step 4 - bull float, manual float & steel trowel	3 Cefi	1280	.019			.86		.86	1.35
0150	Step 5 - application of dry shake colored release agent	1 Cefi	6400	.001		.09	.06		.15	.19
0160	Step 6 - place, tamp & remove mats	3 Cefi	2400	.010		.88	.46		1.34	1.69
0170	Step 7 - touch up edges, mat joints & simulated grout lines	1 Cefi	1280	.006			.29		.29	.45
0300	Alternate stamping estimating method includes all tasks above	4 Cefi	800	.040		1.63	1.83		3.46	4.69
0400	Step 8 - pressure wash @ 3000 psi after 24 hours	1 Cefi	1600	.005			.23		.23	.36
0500	Step 9 - roll 2 coats cure/seal compound when dry	"	800	.010		.63	.46		1.09	1.41

03 35 43 – Polished Concrete Finishing

03 35 43.10 Polished Concrete Floors

		Crew	Daily Output	Labor-Hours	Unit	Material	2016 Bare Costs Labor	2016 Bare Costs Equipment	Total	Total Incl O&P
0010	**POLISHED CONCRETE FLOORS**									
0015	Processing of cured concrete to include grinding, honing,									
0020	and polishing of interior floors with 22" segmented diamond									
0025	planetary floor grinder (2 passes in different directions per grit)									
0100	Removal of pre-existing coatings, dry, with carbide discs using									
0105	dry vacuum pick-up system, final hand sweeping									
0110	Glue, adhesive or tar	J-4	1.60	15	M.S.F.	20.50	645	139	804.50	1,200
0120	Paint, epoxy, 1 coat		3.60	6.667		20.50	287	61.50	369	550
0130	2 coats		1.80	13.333		20.50	575	123	718.50	1,075
0200	Grinding and edging, wet, including wet vac pick-up and auto									
0205	scrubbing between grit changes									
0210	40-grit diamond/metal matrix	J-4A	1.60	20	M.S.F.	32.50	835	289	1,156.50	1,700
0220	80-grit diamond/metal matrix		2	16		32.50	670	231	933.50	1,375
0230	120-grit diamond/metal matrix		2.40	13.333		32.50	555	193	780.50	1,150

03 35 Concrete Finishing

03 35 43 – Polished Concrete Finishing

03 35 43.10 Polished Concrete Floors	Crew	Daily Output	Labor-Hours	Unit	Material	2016 Bare Costs Labor	Equipment	Total	Total Incl O&P
0240	200-grit diamond/metal matrix	J-4A	2.80	11.429	M.S.F.	32.50	475	165	672.50	985
0300	Spray on dye or stain (1 coat)	1 Cefi	16	.500		220	23		243	278
0400	Spray on densifier/hardener (2 coats)	"	8	1		350	45.50		395.50	460
0410	Auto scrubbing after 2nd coat, when dry	J-4B	16	.500	↓		18.95	15.05	34	47.50
0500	Honing and edging, wet, including wet vac pick-up and auto									
0505	scrubbing between grit changes									
0510	100-grit diamond/resin matrix	J-4A	2.80	11.429	M.S.F.	32.50	475	165	672.50	985
0520	200-grit diamond/resin matrix	"	2.80	11.429	"	32.50	475	165	672.50	985
0530	Dry, including dry vacuum pick-up system, final hand sweeping									
0540	400-grit diamond/resin matrix	J-4A	2.80	11.429	M.S.F.	32.50	475	165	672.50	985
0600	Polishing and edging, dry, including dry vac pick-up and hand									
0605	sweeping between grit changes									
0610	800-grit diamond/resin matrix	J-4A	2.80	11.429	M.S.F.	32.50	475	165	672.50	985
0620	1500-grit diamond/resin matrix		2.80	11.429		32.50	475	165	672.50	985
0630	3000-grit diamond/resin matrix	↓	2.80	11.429		32.50	475	165	672.50	985
0700	Auto scrubbing after final polishing step	J-4B	16	.500	↓		18.95	15.05	34	47.50

03 37 Specialty Placed Concrete

03 37 13 – Shotcrete

03 37 13.30 Gunite (Dry-Mix)

		Crew	Daily Output	Labor-Hours	Unit	Material	2016 Bare Costs Labor	Equipment	Total	Total Incl O&P
0010	**GUNITE (DRY-MIX)**									
0020	Typical in place, 1" layers, no mesh included	C-16	2000	.028	S.F.	.37	1.18	.20	1.75	2.54
0300	Typical in place, including mesh, 2" thick, flat surfaces		1000	.056		1.37	2.37	.39	4.13	5.75
0350	Curved surfaces		500	.112		1.37	4.74	.79	6.90	9.95
0500	4" thick, flat surfaces		750	.075		2.11	3.16	.52	5.79	8
0550	Curved surfaces	↓	350	.160		2.11	6.75	1.12	9.98	14.45
0900	Prepare old walls, no scaffolding, good condition	C-10	1000	.024			1.03		1.03	1.65
0950	Poor condition	"	275	.087			3.76		3.76	6
1100	For high finish requirement or close tolerance, add						50%			
1150	Very high				↓		110%			
9000	Minimum labor/equipment charge	C-10	1	24	Job		1,025		1,025	1,650

03 37 13.60 Shotcrete (Wet-Mix)

		Crew	Daily Output	Labor-Hours	Unit	Material	2016 Bare Costs Labor	Equipment	Total	Total Incl O&P
0010	**SHOTCRETE (WET-MIX)**									
0020	Wet mix, placed @ up to 12 C.Y. per hour, 3000 psi	C-8C	80	.600	C.Y.	120	25	5.75	150.75	179
0100	Up to 35 C.Y. per hour	C-8E	240	.200	"	108	8.30	2.18	118.48	135
1010	Fiber reinforced, 1" thick	C-8C	1740	.028	S.F.	.88	1.15	.26	2.29	3.12
1020	2" thick		900	.053		1.76	2.23	.51	4.50	6.10
1030	3" thick		825	.058		2.63	2.43	.56	5.62	7.45
1040	4" thick	↓	750	.064	↓	3.51	2.67	.61	6.79	8.85

03 39 Concrete Curing

03 39 13 – Water Concrete Curing

03 39 13.50 Water Curing	Crew	Daily Output	Labor-Hours	Unit	Material	2016 Bare Costs Labor	Equipment	Total	Total Incl O&P
0010 **WATER CURING**									
0015 With burlap, 4 uses assumed, 7.5 oz.	2 Clab	55	.291	C.S.F.	13.95	11.05		25	33.50
0100 10 oz.	"	55	.291	"	25	11.05		36.05	45.50
0400 Curing blankets, 1" to 2" thick, buy				S.F.	.26			.26	.29
9000 Minimum labor/equipment charge	1 Clab	5	1.600	Job		60.50		60.50	99

03 39 23 – Membrane Concrete Curing

03 39 23.13 Chemical Compound Membrane Concrete Curing

	Crew	Daily Output	Labor-Hours	Unit	Material	Labor	Equipment	Total	Total Incl O&P
0010 **CHEMICAL COMPOUND MEMBRANE CONCRETE CURING**									
0300 Sprayed membrane curing compound	2 Clab	95	.168	C.S.F.	12.70	6.40		19.10	24.50
0700 Curing compound, solvent based, 400 S.F./gal., 55 gallon lots				Gal.	21			21	23
0720 5 gallon lots					31.50			31.50	35
0800 Curing compound, water based, 250 S.F./gal., 55 gallon lots					20.50			20.50	22.50
0820 5 gallon lots					23			23	25.50

03 39 23.23 Sheet Membrane Concrete Curing

	Crew	Daily Output	Labor-Hours	Unit	Material	Labor	Equipment	Total	Total Incl O&P
0010 **SHEET MEMBRANE CONCRETE CURING**									
0200 Curing blanket, burlap/poly, 2-ply	2 Clab	70	.229	C.S.F.	29	8.65		37.65	45.50

03 41 Precast Structural Concrete

03 41 13 – Precast Concrete Hollow Core Planks

03 41 13.50 Precast Slab Planks

		Crew	Daily Output	Labor-Hours	Unit	Material	Labor	Equipment	Total	Total Incl O&P
0010 **PRECAST SLAB PLANKS**	R034105-30									
0020 Prestressed roof/floor members, grouted, solid, 4" thick		C-11	2400	.030	S.F.	6.15	1.58	.78	8.51	10.35
0050 6" thick			2800	.026		6.90	1.35	.67	8.92	10.65
0100 Hollow, 8" thick			3200	.023		7.70	1.18	.58	9.46	11.15
0150 10" thick			3600	.020		8	1.05	.52	9.57	11.20
0200 12" thick			4000	.018		8.30	.95	.47	9.72	11.25

03 41 23 – Precast Concrete Stairs

03 41 23.50 Precast Stairs

	Crew	Daily Output	Labor-Hours	Unit	Material	Labor	Equipment	Total	Total Incl O&P
0010 **PRECAST STAIRS**									
0020 Precast concrete treads on steel stringers, 3' wide	C-12	75	.640	Riser	135	30.50	8.80	174.30	207
0300 Front entrance, 5' wide with 48" platform, 2 risers		16	3	Flight	485	143	41	669	810
0350 5 risers		12	4		765	191	55	1,011	1,200
0500 6' wide, 2 risers		15	3.200		535	153	44	732	885
0550 5 risers		11	4.364		845	208	60	1,113	1,325
1200 Basement entrance stairwell, 6 steps, incl. steel bulkhead door	B-51	22	2.182		1,450	85	11.30	1,546.30	1,725
1250 14 steps	"	11	4.364		2,400	170	22.50	2,592.50	2,925

03 41 33 – Precast Structural Pretensioned Concrete

03 41 33.10 Precast Beams

		Crew	Daily Output	Labor-Hours	Unit	Material	Labor	Equipment	Total	Total Incl O&P
0010 **PRECAST BEAMS**	R034105-30									
0011 L-shaped, 20' span, 12" x 20"		C-11	32	2.250	Ea.	3,150	118	58.50	3,326.50	3,750
0060 18" x 36"			24	3		4,325	158	78	4,561	5,100
0100 24" x 44"			22	3.273		5,175	172	85	5,432	6,100
0150 30' span, 12" x 36"			24	3		5,825	158	78	6,061	6,775
0200 18" x 44"			20	3.600		7,100	189	93.50	7,382.50	8,250
0250 24" x 52"			16	4.500		8,525	236	117	8,878	9,925
0400 40' span, 12" x 52"			20	3.600		9,050	189	93.50	9,332.50	10,400
0450 18" x 52"			16	4.500		10,100	236	117	10,453	11,600
0500 24" x 52"			12	6		11,400	315	156	11,871	13,200
1200 Rectangular, 20' span, 12" x 20"			32	2.250		3,050	118	58.50	3,226.50	3,650

03 41 Precast Structural Concrete

03 41 33 – Precast Structural Pretensioned Concrete

03 41 33.10 Precast Beams

		Crew	Daily Output	Labor-Hours	Unit	Material	2016 Bare Costs Labor	2016 Bare Costs Equipment	Total	Total Incl O&P
1250	18" x 36"	C-11	24	3	Ea.	3,800	158	78	4,036	4,525
1300	24" x 44"		22	3.273		4,525	172	85	4,782	5,375
1400	30' span, 12" x 36"		24	3		5,100	158	78	5,336	5,975
1450	18" x 44"		20	3.600		6,150	189	93.50	6,432.50	7,200
1500	24" x 52"		16	4.500		7,425	236	117	7,778	8,725
1600	40' span, 12" x 52"		20	3.600		7,575	189	93.50	7,857.50	8,775
1650	18" x 52"		16	4.500		8,625	236	117	8,978	10,000
1700	24" x 52"		12	6		9,900	315	156	10,371	11,600
2000	"T" shaped, 20' span, 12" x 20"		32	2.250		3,600	118	58.50	3,776.50	4,225
2050	18" x 36"		24	3		4,850	158	78	5,086	5,675
2100	24" x 44"		22	3.273		5,800	172	85	6,057	6,775
2200	30' span, 12" x 36"		24	3		6,650	158	78	6,886	7,650
2250	18" x 44"		20	3.600		8,050	189	93.50	8,332.50	9,275
2300	24" x 52"		16	4.500		9,625	236	117	9,978	11,100
2500	40' span, 12" x 52"		20	3.600		10,900	189	93.50	11,182.50	12,400
2550	18" x 52"		16	4.500		11,600	236	117	11,953	13,300
2600	24" x 52"		12	6		12,800	315	156	13,271	14,800

03 41 33.15 Precast Columns

			Crew	Daily Output	Labor-Hours	Unit	Material	2016 Bare Costs Labor	2016 Bare Costs Equipment	Total	Total Incl O&P
0010	**PRECAST COLUMNS**	R034105-30									
0020	Rectangular to 12' high, 16" x 16"		C-11	120	.600	L.F.	180	31.50	15.55	227.05	270
0050	24" x 24"		"	96	.750	"	243	39.50	19.45	301.95	355

03 45 Precast Architectural Concrete

03 45 13 – Faced Architectural Precast Concrete

03 45 13.50 Precast Wall Panels

		Crew	Daily Output	Labor-Hours	Unit	Material	2016 Bare Costs Labor	2016 Bare Costs Equipment	Total	Total Incl O&P
0010	**PRECAST WALL PANELS**									
2200	Fiberglass reinforced cement with urethane core									
2210	R20, 8' x 8', 5" plain finish	E-2	750	.075	S.F.	21.50	3.91	2.03	27.44	33
2220	Exposed aggregate or brick finish	"	600	.093	"	32.50	4.88	2.53	39.91	47

03 48 Precast Concrete Specialties

03 48 43 – Precast Concrete Trim

03 48 43.40 Precast Lintels

		Crew	Daily Output	Labor-Hours	Unit	Material	2016 Bare Costs Labor	2016 Bare Costs Equipment	Total	Total Incl O&P
0010	**PRECAST LINTELS**, smooth gray, prestressed, stock units only									
0800	4" wide, 8" high, x 4' long	D-10	28	1.143	Ea.	25	53	17.10	95.10	131
0850	8' long		24	1.333		63	61.50	19.95	144.45	191
1000	6" wide, 8" high, x 4' long		26	1.231		37	57	18.45	112.45	153
1050	10' long		22	1.455		92.50	67	22	181.50	234
1200	8" wide, 8" high, x 4' long		24	1.333		43.50	61.50	19.95	124.95	170
1250	12' long		20	1.600		141	74	24	239	300
1275	For custom sizes, types, colors, or finishes of precast lintels, add					150%				

03 48 43.90 Precast Window Sills

		Crew	Daily Output	Labor-Hours	Unit	Material	2016 Bare Costs Labor	2016 Bare Costs Equipment	Total	Total Incl O&P
0010	**PRECAST WINDOW SILLS**									
0600	Precast concrete, 4" tapers to 3", 9" wide	D-1	70	.229	L.F.	14.65	9.65		24.30	32
0650	11" wide	"	60	.267	"	24	11.25		35.25	44.50

69

For customer support on your Commercial Renovation Cost Data, call 877.742.4719.

03 51 Cast Roof Decks

03 51 13 – Cementitious Wood Fiber Decks

03 51 13.50 Cementitious/Wood Fiber Planks		Crew	Daily Output	Labor-Hours	Unit	Material	2016 Bare Costs Labor	Equipment	Total	Total Incl O&P
0010	**CEMENTITIOUS/WOOD FIBER PLANKS**									
0050	Plank, beveled edge, 1" thick	2 Carp	1000	.016	S.F.	2.63	.78		3.41	4.16
0100	1-1/2" thick		975	.016		3.41	.80		4.21	5.05
0150	T & G, 2" thick		950	.017		2.89	.82		3.71	4.51
0200	2-1/2" thick		925	.017		3.15	.84		3.99	4.84
1000	Bulb tee, sub-purlin and grout, 6' span, add	E-1	5000	.005		2.07	.25	.03	2.35	2.74
1100	8' span	"	4200	.006		2.07	.30	.04	2.41	2.83

03 52 Lightweight Concrete Roof Insulation

03 52 16 – Lightweight Insulating Concrete

03 52 16.13 Lightweight Cellular Insulating Concrete

		Crew	Daily Output	Labor-Hours	Unit	Material	2016 Bare Costs Labor	Equipment	Total	Total Incl O&P
0010	**LIGHTWEIGHT CELLULAR INSULATING CONCRETE** R035216-10									
0020	Portland cement and foaming agent G	C-8	50	1.120	C.Y.	123	47.50	14.55	185.05	227

03 52 16.16 Lightweight Aggregate Insulating Concrete

		Crew	Daily Output	Labor-Hours	Unit	Material	2016 Bare Costs Labor	Equipment	Total	Total Incl O&P
0010	**LIGHTWEIGHT AGGREGATE INSULATING CONCRETE** R035216-10									
0100	Poured vermiculite or perlite, field mix,									
0110	1:6 field mix G	C-8	50	1.120	C.Y.	263	47.50	14.55	325.05	380
0200	Ready mix, 1:6 mix, roof fill, 2" thick G		10000	.006	S.F.	1.46	.24	.07	1.77	2.07
0250	3" thick G		7700	.007	"	2.19	.31	.09	2.59	3

03 53 Concrete Topping

03 53 16 – Iron-Aggregate Concrete Topping

03 53 16.50 Floor Topping

		Crew	Daily Output	Labor-Hours	Unit	Material	2016 Bare Costs Labor	Equipment	Total	Total Incl O&P
0010	**FLOOR TOPPING**									
0400	Integral topping/finish, on fresh concrete, using 1:1:2 mix, 3/16" thick	C-10B	1000	.040	S.F.	.11	1.64	.27	2.02	3.06
0450	1/2" thick		950	.042		.29	1.73	.29	2.31	3.42
0500	3/4" thick		850	.047		.44	1.93	.32	2.69	3.94
0600	1" thick		750	.053		.58	2.19	.37	3.14	4.56
0800	Granolithic topping, on fresh or cured concrete, 1:1:1-1/2 mix, 1/2" thick		590	.068		.32	2.78	.46	3.56	5.35
0820	3/4" thick		580	.069		.49	2.83	.47	3.79	5.60
0850	1" thick		575	.070		.65	2.85	.48	3.98	5.85
0950	2" thick		500	.080		1.30	3.28	.55	5.13	7.35

03 54 Cast Underlayment

03 54 13 – Gypsum Cement Underlayment

03 54 13.50 Poured Gypsum Underlayment

		Crew	Daily Output	Labor-Hours	Unit	Material	2016 Bare Costs Labor	Equipment	Total	Total Incl O&P
0010	**POURED GYPSUM UNDERLAYMENT**									
0400	Underlayment, gypsum based, self-leveling 2500 psi, pumped, 1/2" thick	C-8	24000	.002	S.F.	.36	.10	.03	.49	.59
0500	3/4" thick		20000	.003		.54	.12	.04	.70	.83
0600	1" thick		16000	.004		.72	.15	.05	.92	1.09
1400	Hand placed, 1/2" thick	C-18	450	.020		.36	.76	.13	1.25	1.80
1500	3/4" thick	"	300	.030		.54	1.14	.20	1.88	2.69

03 54 16 – Hydraulic Cement Underlayment

03 54 16.50 Cement Underlayment

		Crew	Daily Output	Labor-Hours	Unit	Material	2016 Bare Costs Labor	Equipment	Total	Total Incl O&P
0010	**CEMENT UNDERLAYMENT**									
2510	Underlayment, P.C. based, self-leveling, 4100 psi, pumped, 1/4" thick	C-8	20000	.003	S.F.	1.58	.12	.04	1.74	1.96
2520	1/2" thick		19000	.003		3.15	.12	.04	3.31	3.71

03 54 Cast Underlayment

03 54 16 – Hydraulic Cement Underlayment

03 54 16.50 Cement Underlayment		Crew	Daily Output	Labor-Hours	Unit	Material	2016 Bare Costs Labor	Equipment	Total	Total Incl O&P
2530	3/4" thick	C-8	18000	.003	S.F.	4.73	.13	.04	4.90	5.45
2540	1" thick		17000	.003		6.30	.14	.04	6.48	7.20
2550	1-1/2" thick	↓	15000	.004		9.45	.16	.05	9.66	10.70
2560	Hand placed, 1/2" thick	C-18	450	.020		3.15	.76	.13	4.04	4.87
2610	Topping, P.C. based, self-leveling, 6100 psi, pumped, 1/4" thick	C-8	20000	.003		2.37	.12	.04	2.53	2.84
2620	1/2" thick		19000	.003		4.75	.12	.04	4.91	5.45
2630	3/4" thick		18000	.003		7.10	.13	.04	7.27	8.10
2660	1" thick		17000	.003		9.50	.14	.04	9.68	10.70
2670	1-1/2" thick	↓	15000	.004		14.25	.16	.05	14.46	15.95
2680	Hand placed, 1/2" thick	C-18	450	.020	↓	4.75	.76	.13	5.64	6.60

03 63 Epoxy Grouting

03 63 05 – Grouting of Dowels and Fasteners

03 63 05.10 Epoxy Only

		Crew	Daily Output	Labor-Hours	Unit	Material	2016 Bare Costs Labor	Equipment	Total	Total Incl O&P
0010	**EPOXY ONLY**									
1500	Chemical anchoring, epoxy cartridge, excludes layout, drilling, fastener									
1530	For fastener 3/4" diam. x 6" embedment	2 Skwk	72	.222	Ea.	4.84	11.10		15.94	23.50
1535	1" diam. x 8" embedment		66	.242		7.25	12.10		19.35	27.50
1540	1-1/4" diam. x 10" embedment		60	.267		14.50	13.30		27.80	37.50
1545	1-3/4" diam. x 12" embedment		54	.296		24	14.80		38.80	50.50
1550	14" embedment		48	.333		29	16.65		45.65	59
1555	2" diam. x 12" embedment		42	.381		38.50	19		57.50	73.50
1560	18" embedment	↓	32	.500	↓	48.50	25		73.50	93.50

03 81 Concrete Cutting

03 81 13 – Flat Concrete Sawing

03 81 13.50 Concrete Floor/Slab Cutting

		Crew	Daily Output	Labor-Hours	Unit	Material	2016 Bare Costs Labor	Equipment	Total	Total Incl O&P
0010	**CONCRETE FLOOR/SLAB CUTTING**									
0050	Includes blade cost, layout and set-up time									
0300	Saw cut concrete slabs, plain, up to 3" deep	B-89	1060	.015	L.F.	.13	.69	.47	1.29	1.75
0320	Each additional inch of depth		3180	.005		.04	.23	.16	.43	.59
0400	Mesh reinforced, up to 3" deep		980	.016		.14	.74	.50	1.38	1.89
0420	Each additional inch of depth		2940	.005		.05	.25	.17	.47	.63
0500	Rod reinforced, up to 3" deep		800	.020		.18	.91	.62	1.71	2.32
0520	Each additional inch of depth		2400	.007	↓	.06	.30	.21	.57	.77
0590	Minimum labor/equipment charge	↓	2	8	Job		365	247	612	850

03 81 13.75 Concrete Saw Blades

		Crew	Daily Output	Labor-Hours	Unit	Material	2016 Bare Costs Labor	Equipment	Total	Total Incl O&P
0010	**CONCRETE SAW BLADES**									
3000	Blades for saw cutting, included in cutting line items									
3020	Diamond, 12" diameter				Ea.	248			248	273
3040	18" diameter					420			420	465
3080	24" diameter					775			775	850
3120	30" diameter					1,000			1,000	1,100
3160	36" diameter					1,350			1,350	1,475
3200	42" diameter				↓	2,700			2,700	2,975

For customer support on your Commercial Renovation Cost Data, call 877.742.4719.

71

03 81 Concrete Cutting

03 81 16 – Track Mounted Concrete Wall Sawing

03 81 16.50 Concrete Wall Cutting	Crew	Daily Output	Labor-Hours	Unit	Material	2016 Bare Costs Labor	Equipment	Total	Total Incl O&P
0010 **CONCRETE WALL CUTTING**									
0750 Includes blade cost, layout and set-up time									
0800 Concrete walls, hydraulic saw, plain, per inch of depth	B-89B	250	.064	L.F.	.04	2.92	3.41	6.37	8.45
0820 Rod reinforcing, per inch of depth		150	.107	"	.06	4.86	5.70	10.62	14.05
0890 Minimum labor/equipment charge	↓	2	8	Job		365	425	790	1,050

03 82 Concrete Boring

03 82 13 – Concrete Core Drilling

03 82 13.10 Core Drilling

	Crew	Daily Output	Labor-Hours	Unit	Material	2016 Bare Costs Labor	Equipment	Total	Total Incl O&P
0010 **CORE DRILLING**									
0015 Includes bit cost, layout and set-up time									
0020 Reinforced concrete slab, up to 6" thick									
0100 1" diameter core	B-89A	17	.941	Ea.	.18	41.50	6.90	48.58	75
0150 For each additional inch of slab thickness in same hole, add		1440	.011		.03	.49	.08	.60	.91
0200 2" diameter core		16.50	.970		.28	42.50	7.10	49.88	77
0250 For each additional inch of slab thickness in same hole, add		1080	.015		.05	.65	.11	.81	1.23
0300 3" diameter core		16	1		.38	44	7.30	51.68	80
0350 For each additional inch of slab thickness in same hole, add		720	.022		.06	.98	.16	1.20	1.84
0500 4" diameter core		15	1.067		.50	47	7.80	55.30	85
0550 For each additional inch of slab thickness in same hole, add		480	.033		.08	1.46	.24	1.78	2.74
0700 6" diameter core		14	1.143		.76	50	8.35	59.11	91.50
0750 For each additional inch of slab thickness in same hole, add		360	.044		.13	1.95	.32	2.40	3.67
0900 8" diameter core		13	1.231		1.04	54	9	64.04	99
0950 For each additional inch of slab thickness in same hole, add		288	.056		.17	2.44	.41	3.02	4.61
1100 10" diameter core		12	1.333		1.48	58.50	9.75	69.73	107
1150 For each additional inch of slab thickness in same hole, add		240	.067		.25	2.93	.49	3.67	5.55
1300 12" diameter core		11	1.455		1.77	64	10.65	76.42	118
1350 For each additional inch of slab thickness in same hole, add		206	.078		.29	3.41	.57	4.27	6.50
1500 14" diameter core		10	1.600		2.07	70	11.70	83.77	129
1550 For each additional inch of slab thickness in same hole, add		180	.089		.34	3.90	.65	4.89	7.45
1700 18" diameter core		9	1.778		2.78	78	13	93.78	144
1750 For each additional inch of slab thickness in same hole, add		144	.111		.46	4.88	.81	6.15	9.35
1754 24" diameter core		8	2		3.97	88	14.60	106.57	163
1756 For each additional inch of slab thickness in same hole, add	↓	120	.133		.66	5.85	.97	7.48	11.30
1760 For horizontal holes, add to above				↓		20%	20%		
1770 Prestressed hollow core plank, 8" thick									
1780 1" diameter core	B-89A	17.50	.914	Ea.	.24	40	6.70	46.94	73
1790 For each additional inch of plank thickness in same hole, add		3840	.004		.03	.18	.03	.24	.36
1794 2" diameter core		17.25	.928		.38	40.50	6.80	47.68	74
1796 For each additional inch of plank thickness in same hole, add		2880	.006		.05	.24	.04	.33	.49
1800 3" diameter core		17	.941		.51	41.50	6.90	48.91	75
1810 For each additional inch of plank thickness in same hole, add		1920	.008		.06	.37	.06	.49	.73
1820 4" diameter core		16.50	.970		.66	42.50	7.10	50.26	77.50
1830 For each additional inch of plank thickness in same hole, add		1280	.013		.08	.55	.09	.72	1.08
1840 6" diameter core		15.50	1.032		1.02	45.50	7.55	54.07	83
1850 For each additional inch of plank thickness in same hole, add		960	.017		.13	.73	.12	.98	1.46
1860 8" diameter core		15	1.067		1.39	47	7.80	56.19	86
1870 For each additional inch of plank thickness in same hole, add		768	.021		.17	.91	.15	1.23	1.85
1880 10" diameter core		14	1.143		1.97	50	8.35	60.32	93
1890 For each additional inch of plank thickness in same hole, add	↓	640	.025	↓	.25	1.10	.18	1.53	2.26

03 82 Concrete Boring

03 82 13 – Concrete Core Drilling

03 82 13.10 Core Drilling

		Crew	Daily Output	Labor-Hours	Unit	Material	2016 Bare Costs Labor	Equipment	Total	Total Incl O&P
1900	12" diameter core	B-89A	13.50	1.185	Ea.	2.35	52	8.65	63	96.50
1910	For each additional inch of plank thickness in same hole, add		548	.029		.29	1.28	.21	1.78	2.63
1999	Drilling, core, minimum labor/equipment charge		2	8	Job		350	58.50	408.50	635
3000	Bits for core drilling, included in drilling line items									
3010	Diamond, premium, 1" diameter				Ea.	71.50			71.50	78.50
3020	2" diameter					114			114	125
3030	3" diameter					154			154	169
3040	4" diameter					199			199	219
3060	6" diameter					305			305	335
3080	8" diameter					415			415	460
3110	10" diameter					590			590	650
3120	12" diameter					705			705	775
3140	14" diameter					825			825	910
3180	18" diameter					1,125			1,125	1,225
3240	24" diameter					1,575			1,575	1,750

03 82 16 – Concrete Drilling

03 82 16.10 Concrete Impact Drilling

		Crew	Daily Output	Labor-Hours	Unit	Material	2016 Bare Costs Labor	Equipment	Total	Total Incl O&P
0010	**CONCRETE IMPACT DRILLING**									
0020	Includes bit cost, layout and set-up time, no anchors									
0050	Up to 4" deep in concrete/brick floors/walls									
0100	Holes, 1/4" diameter	1 Carp	75	.107	Ea.	.07	5.15		5.22	8.55
0150	For each additional inch of depth in same hole, add		430	.019		.02	.90		.92	1.49
0200	3/8" diameter		63	.127		.06	6.15		6.21	10.10
0250	For each additional inch of depth in same hole, add		340	.024		.01	1.14		1.15	1.88
0300	1/2" diameter		50	.160		.06	7.75		7.81	12.70
0350	For each additional inch of depth in same hole, add		250	.032		.02	1.55		1.57	2.55
0400	5/8" diameter		48	.167		.10	8.10		8.20	13.30
0450	For each additional inch of depth in same hole, add		240	.033		.03	1.61		1.64	2.67
0500	3/4" diameter		45	.178		.13	8.60		8.73	14.20
0550	For each additional inch of depth in same hole, add		220	.036		.03	1.76		1.79	2.92
0600	7/8" diameter		43	.186		.18	9		9.18	14.95
0650	For each additional inch of depth in same hole, add		210	.038		.04	1.85		1.89	3.07
0700	1" diameter		40	.200		.16	9.70		9.86	16
0750	For each additional inch of depth in same hole, add		190	.042		.04	2.04		2.08	3.37
0800	1-1/4" diameter		38	.211		.27	10.20		10.47	16.95
0850	For each additional inch of depth in same hole, add		180	.044		.07	2.15		2.22	3.59
0900	1-1/2" diameter		35	.229		.41	11.05		11.46	18.55
0950	For each additional inch of depth in same hole, add		165	.048		.10	2.35		2.45	3.95
1000	For ceiling installations, add						40%			

Division Notes

	CREW	DAILY OUTPUT	LABOR-HOURS	UNIT	BARE COSTS				TOTAL INCL O&P
					MAT.	LABOR	EQUIP.	TOTAL	

Estimating Tips

04 05 00 Common Work Results for Masonry

- The terms mortar and grout are often used interchangeably, and incorrectly. Mortar is used to bed masonry units, seal the entry of air and moisture, provide architectural appearance, and allow for size variations in the units. Grout is used primarily in reinforced masonry construction and is used to bond the masonry to the reinforcing steel. Common mortar types are M(2500 psi), S(1800 psi), N(750 psi), and O(350 psi), and conform to ASTM C270. Grout is either fine or coarse and conforms to ASTM C476, and in-place strengths generally exceed 2500 psi. Mortar and grout are different components of masonry construction and are placed by entirely different methods. An estimator should be aware of their unique uses and costs.

- Mortar is included in all assembled masonry line items. The mortar cost, part of the assembled masonry material cost, includes all ingredients, all labor, and all equipment required. Please see reference number R040513-10.

- Waste, specifically the loss/ droppings of mortar and the breakage of brick and block, is included in all unit cost lines that include mortar and masonry units

in this division. A factor of 25% is added for mortar and 3% for brick and concrete masonry units.

- Scaffolding or staging is not included in any of the Division 4 costs. Refer to Subdivision 01 54 23 for scaffolding and staging costs.

04 20 00 Unit Masonry

- The most common types of unit masonry are brick and concrete masonry. The major classifications of brick are building brick (ASTM C62), facing brick (ASTM C216), glazed brick, fire brick, and pavers. Many varieties of texture and appearance can exist within these classifications, and the estimator would be wise to check local custom and availability within the project area. For repair and remodeling jobs, matching the existing brick may be the most important criteria.

- Brick and concrete block are priced by the piece and then converted into a price per square foot of wall. Openings less than two square feet are generally ignored by the estimator because any savings in units used is offset by the cutting and trimming required.

- It is often difficult and expensive to find and purchase small lots of historic brick. Costs can vary widely. Many design issues affect costs, selection of mortar mix, and repairs or replacement of masonry

materials. Cleaning techniques must be reflected in the estimate.

- All masonry walls, whether interior or exterior, require bracing. The cost of bracing walls during construction should be included by the estimator, and this bracing must remain in place until permanent bracing is complete. Permanent bracing of masonry walls is accomplished by masonry itself, in the form of pilasters or abutting wall corners, or by anchoring the walls to the structural frame. Accessories in the form of anchors, anchor slots, and ties are used, but their supply and installation can be by different trades. For instance, anchor slots on spandrel beams and columns are supplied and welded in place by the steel fabricator, but the ties from the slots into the masonry are installed by the bricklayer. Regardless of the installation method, the estimator must be certain that these accessories are accounted for in pricing.

Reference Numbers

Reference numbers are shown at the beginning of some major classifications. These numbers refer to related items in the Reference Section. The reference information may be an estimating procedure, an alternate pricing method, or technical information.

Note: Not all subdivisions listed here necessarily appear. ■

04 01 Maintenance of Masonry

04 01 20 – Maintenance of Unit Masonry

04 01 20.20 Pointing Masonry

04 01 20.20 Pointing Masonry		Crew	Daily Output	Labor-Hours	Unit	Material	2016 Bare Costs Labor	Equipment	Total	Total Incl O&P
0010	**POINTING MASONRY**									
0300	Cut and repoint brick, hard mortar, running bond	1 Bric	80	.100	S.F.	.58	4.63		5.21	8.20
0320	Common bond		77	.104		.58	4.81		5.39	8.45
0360	Flemish bond		70	.114		.61	5.30		5.91	9.25
0400	English bond		65	.123		.61	5.70		6.31	9.90
0600	Soft old mortar, running bond		100	.080		.58	3.70		4.28	6.65
0620	Common bond		96	.083		.58	3.85		4.43	6.90
0640	Flemish bond		90	.089		.61	4.11		4.72	7.35
0680	English bond		82	.098		.61	4.51		5.12	8
0700	Stonework, hard mortar		140	.057	L.F.	.77	2.64		3.41	5.15
0720	Soft old mortar		160	.050	"	.77	2.31		3.08	4.61
1000	Repoint, mask and grout method, running bond		95	.084	S.F.	.77	3.89		4.66	7.20
1020	Common bond		90	.089		.77	4.11		4.88	7.55
1040	Flemish bond		86	.093		.81	4.30		5.11	7.90
1060	English bond		77	.104		.81	4.81		5.62	8.70
2000	Scrub coat, sand grout on walls, thin mix, brushed		120	.067		3.22	3.08		6.30	8.55
2020	Troweled		98	.082		4.49	3.78		8.27	11.10
9000	Minimum labor/equipment charge		3	2.667	Job		123		123	201

04 01 20.30 Pointing CMU

04 01 20.30 Pointing CMU		Crew	Daily Output	Labor-Hours	Unit	Material	Labor	Equipment	Total	Total Incl O&P
0010	**POINTING CMU**									
0300	Cut and repoint block, hard mortar, running bond	1 Bric	190	.042	S.F.	.24	1.95		2.19	3.43
0310	Stacked bond		200	.040		.24	1.85		2.09	3.27
0600	Soft old mortar, running bond		230	.035		.24	1.61		1.85	2.88
0610	Stacked bond		245	.033		.24	1.51		1.75	2.72

04 01 20.41 Unit Masonry Stabilization

04 01 20.41 Unit Masonry Stabilization		Crew	Daily Output	Labor-Hours	Unit	Material	Labor	Equipment	Total	Total Incl O&P
0010	**UNIT MASONRY STABILIZATION**									
0100	Structural repointing method									
0110	Cut/grind mortar joint	1 Bric	240	.033	L.F.		1.54		1.54	2.51
0120	Clean and mask joint		2500	.003		.11	.15		.26	.36
0130	Epoxy paste and 1/4" FRP rod		240	.033		1.77	1.54		3.31	4.45
0132	3/8" FRP rod		160	.050		2.50	2.31		4.81	6.50
0140	Remove masking		14400	.001			.03		.03	.04
0300	Structural fabric method									
0310	Primer	1 Bric	600	.013	S.F.	.93	.62		1.55	2.02
0320	Apply filling/leveling paste		720	.011		.73	.51		1.24	1.64
0330	Epoxy, glass fiber fabric		720	.011		8.35	.51		8.86	10.05
0340	Carbon fiber fabric		720	.011		20	.51		20.51	23

04 01 20.50 Toothing Masonry

04 01 20.50 Toothing Masonry		Crew	Daily Output	Labor-Hours	Unit	Material	Labor	Equipment	Total	Total Incl O&P
0010	**TOOTHING MASONRY**									
0500	Brickwork, soft old mortar	1 Clab	40	.200	V.L.F.		7.60		7.60	12.40
0520	Hard mortar		30	.267			10.10		10.10	16.50
0700	Blockwork, soft old mortar		70	.114			4.33		4.33	7.10
0720	Hard mortar		50	.160			6.05		6.05	9.90
9000	Minimum labor/equipment charge		4	2	Job		76		76	124

04 01 30 – Unit Masonry Cleaning

04 01 30.20 Cleaning Masonry

04 01 30.20 Cleaning Masonry		Crew	Daily Output	Labor-Hours	Unit	Material	Labor	Equipment	Total	Total Incl O&P
0010	**CLEANING MASONRY**									
0200	By chemical, brush and rinse, new work, light construction dust	D-1	1000	.016	S.F.	.06	.67		.73	1.16
0220	Medium construction dust		800	.020		.09	.84		.93	1.47
0240	Heavy construction dust, drips or stains		600	.027		.12	1.12		1.24	1.96
0260	Low pressure wash and rinse, light restoration, light soil		800	.020		.14	.84		.98	1.52

For customer support on your Commercial Renovation Cost Data, call 877.742.4719.

04 01 30 – Unit Masonry Cleaning

04 01 30.20 Cleaning Masonry		Crew	Daily Output	Labor-Hours	Unit	Material	2016 Bare Costs Labor	Equipment	Total	Total Incl O&P
0270	Average soil, biological staining	D-1	400	.040	S.F.	.20	1.69		1.89	2.96
0280	Heavy soil, biological and mineral staining, paint		330	.048		.27	2.04		2.31	3.62
0300	High pressure wash and rinse, heavy restoration, light soil		600	.027		.12	1.12		1.24	1.96
0310	Average soil, biological staining		400	.040		.18	1.69		1.87	2.94
0320	Heavy soil, biological and mineral staining, paint	↓	250	.064		.24	2.70		2.94	4.65
0400	High pressure wash, water only, light soil	C-29	500	.016			.61	.14	.75	1.15
0420	Average soil, biological staining		375	.021			.81	.19	1	1.53
0440	Heavy soil, biological and mineral staining, paint		250	.032			1.21	.29	1.50	2.30
0800	High pressure water and chemical, light soil		450	.018		.16	.67	.16	.99	1.46
0820	Average soil, biological staining		300	.027		.24	1.01	.24	1.49	2.18
0840	Heavy soil, biological and mineral staining, paint	↓	200	.040		.33	1.52	.36	2.21	3.23
1200	Sandblast, wet system, light soil	J-6	1750	.018		.35	.77	.13	1.25	1.76
1220	Average soil, biological staining		1100	.029		.52	1.22	.21	1.95	2.76
1240	Heavy soil, biological and mineral staining, paint		700	.046		.70	1.92	.33	2.95	4.22
1400	Dry system, light soil		2500	.013		.35	.54	.09	.98	1.34
1420	Average soil, biological staining		1750	.018		.52	.77	.13	1.42	1.95
1440	Heavy soil, biological and mineral staining, paint	↓	1000	.032		.70	1.34	.23	2.27	3.19
1800	For walnut shells, add					.69			.69	.76
1820	For corn chips, add					.69			.69	.76
2000	Steam cleaning, light soil	A-1H	750	.011			.40	.10	.50	.77
2020	Average soil, biological staining		625	.013			.49	.12	.61	.93
2040	Heavy soil, biological and mineral staining	↓	375	.021			.81	.21	1.02	1.55
4000	Add for masking doors and windows	1 Clab	800	.010	↓	.07	.38		.45	.70
4200	Add for pedestrian protection				Job				10%	10%
9000	Minimum labor/equipment charge	D-4	2	16	"		685	67.50	752.50	1,175

04 01 30.60 Brick Washing

0010	BRICK WASHING									
0012	Acid cleanser, smooth brick surface	1 Bric	560	.014	S.F.	.05	.66		.71	1.13
0050	Rough brick		400	.020		.06	.93		.99	1.58
0060	Stone, acid wash	↓	600	.013	↓	.08	.62		.70	1.08
1000	Muriatic acid, price per gallon in 5 gallon lots				Gal.	9.45			9.45	10.35

04 05 05 – Selective Demolition for Masonry

04 05 05.10 Selective Demolition

0010	SELECTIVE DEMOLITION	R024119-10								
0200	Bond beams, 8" block with #4 bar	2 Clab	32	.500	L.F.		18.95		18.95	31
0300	Concrete block walls, unreinforced, 2" thick		1200	.013	S.F.		.51		.51	.83
0310	4" thick		1150	.014			.53		.53	.86
0320	6" thick		1100	.015			.55		.55	.90
0330	8" thick		1050	.015			.58		.58	.94
0340	10" thick		1000	.016			.61		.61	.99
0360	12" thick		950	.017			.64		.64	1.04
0380	Reinforced alternate courses, 2" thick		1130	.014			.54		.54	.88
0390	4" thick		1080	.015			.56		.56	.92
0400	6" thick		1035	.015			.59		.59	.96
0410	8" thick		990	.016			.61		.61	1
0420	10" thick		940	.017			.65		.65	1.05
0430	12" thick		890	.018			.68		.68	1.11
0440	Reinforced alternate courses & vertically 48" OC, 4" thick		900	.018			.67		.67	1.10
0450	6" thick	↓	850	.019	↓		.71		.71	1.17

For customer support on your Commercial Renovation Cost Data, call 877.742.4719.

77

04 05 Common Work Results for Masonry

04 05 05 – Selective Demolition for Masonry

04 05 05.10 Selective Demolition	Crew	Daily Output	Labor-Hours	Unit	Material	2016 Bare Costs Labor	Equipment	Total	Total Incl O&P
0460	8" thick	2 Clab	800	.020	S.F.		.76		.76	1.24
0480	10" thick		750	.021			.81		.81	1.32
0490	12" thick		700	.023			.87		.87	1.42
1000	Chimney, 16" x 16", soft old mortar	1 Clab	55	.145	C.F.		5.50		5.50	9
1020	Hard mortar		40	.200			7.60		7.60	12.40
1030	16" x 20", soft old mortar		55	.145			5.50		5.50	9
1040	Hard mortar		40	.200			7.60		7.60	12.40
1050	16" x 24", soft old mortar		55	.145			5.50		5.50	9
1060	Hard mortar		40	.200			7.60		7.60	12.40
1080	20" x 20", soft old mortar		55	.145			5.50		5.50	9
1100	Hard mortar		40	.200			7.60		7.60	12.40
1110	20" x 24", soft old mortar		55	.145			5.50		5.50	9
1120	Hard mortar		40	.200			7.60		7.60	12.40
1140	20" x 32", soft old mortar		55	.145			5.50		5.50	9
1160	Hard mortar		40	.200			7.60		7.60	12.40
1200	48" x 48", soft old mortar		55	.145			5.50		5.50	9
1220	Hard mortar		40	.200			7.60		7.60	12.40
1250	Metal, high temp steel jacket, 24" diameter	E-2	130	.431	V.L.F.		22.50	11.70	34.20	51.50
1260	60" diameter	"	60	.933			49	25.50	74.50	112
1280	Flue lining, up to 12" x 12"	1 Clab	200	.040			1.52		1.52	2.48
1282	Up to 24" x 24"		150	.053			2.02		2.02	3.30
2000	Columns, 8" x 8", soft old mortar		48	.167			6.30		6.30	10.35
2020	Hard mortar		40	.200			7.60		7.60	12.40
2060	16" x 16", soft old mortar		16	.500			18.95		18.95	31
2100	Hard mortar		14	.571			21.50		21.50	35.50
2140	24" x 24", soft old mortar		8	1			38		38	62
2160	Hard mortar		6	1.333			50.50		50.50	82.50
2200	36" x 36", soft old mortar		4	2			76		76	124
2220	Hard mortar		3	2.667			101		101	165
2230	Alternate pricing method, soft old mortar		30	.267	C.F.		10.10		10.10	16.50
2240	Hard mortar		23	.348	"		13.20		13.20	21.50
3000	Copings, precast or masonry, to 8" wide									
3020	Soft old mortar	1 Clab	180	.044	L.F.		1.68		1.68	2.75
3040	Hard mortar	"	160	.050	"		1.90		1.90	3.10
3100	To 12" wide									
3120	Soft old mortar	1 Clab	160	.050	L.F.		1.90		1.90	3.10
3140	Hard mortar	"	140	.057	"		2.17		2.17	3.54
4000	Fireplace, brick, 30" x 24" opening									
4020	Soft old mortar	1 Clab	2	4	Ea.		152		152	248
4040	Hard mortar		1.25	6.400			243		243	395
4100	Stone, soft old mortar		1.50	5.333			202		202	330
4120	Hard mortar		1	8			305		305	495
5000	Veneers, brick, soft old mortar		140	.057	S.F.		2.17		2.17	3.54
5020	Hard mortar		125	.064			2.43		2.43	3.96
5050	Glass block, up to 4" thick		500	.016			.61		.61	.99
5100	Granite and marble, 2" thick		180	.044			1.68		1.68	2.75
5120	4" thick		170	.047			1.78		1.78	2.92
5140	Stone, 4" thick		180	.044			1.68		1.68	2.75
5160	8" thick		175	.046			1.73		1.73	2.83
5400	Alternate pricing method, stone, 4" thick		60	.133	C.F.		5.05		5.05	8.25
5420	8" thick		85	.094	"		3.57		3.57	5.85
9000	Minimum labor/equipment charge		2	4	Job		152		152	248

For customer support on your Commercial Renovation Cost Data, call 877.742.4719.

04 05 Common Work Results for Masonry

04 05 13 – Masonry Mortaring

04 05 13.10 Cement	Crew	Daily Output	Labor-Hours	Unit	Material	2016 Bare Costs Labor	Equipment	Total	Total Incl O&P
0010 **CEMENT**									
0100 Masonry, 70 lb. bag, T.L. lots				Bag	12			12	13.20
0150 L.T.L. lots					12.75			12.75	14
0200 White, 70 lb. bag, T.L. lots					15.70			15.70	17.30
0250 L.T.L. lots				↓	16.55			16.55	18.20

04 05 13.20 Lime									
0010 **LIME**									
0020 Masons, hydrated, 50 lb. bag, T.L. lots				Bag	10.35			10.35	11.40
0050 L.T.L. lots					11.40			11.40	12.55
0200 Finish, double hydrated, 50 lb. bag, T.L. lots					9.20			9.20	10.15
0250 L.T.L. lots				↓	10.15			10.15	11.15

04 05 13.23 Surface Bonding Masonry Mortaring									
0010 **SURFACE BONDING MASONRY MORTARING**									
0020 Gray or white colors, not incl. block work	1 Bric	540	.015	S.F.	.14	.69		.83	1.26

04 05 13.30 Mortar									
0010 **MORTAR**									
0020 With masonry cement									
0100 Type M, 1:1:6 mix	1 Brhe	143	.056	C.F.	5.50	2.13		7.63	9.50
0200 Type N, 1:3 mix		143	.056		5.50	2.13		7.63	9.50
0300 Type O, 1:3 mix		143	.056		3.96	2.13		6.09	7.80
0400 Type PM, 1:1:6 mix, 2500 psi		143	.056		6	2.13		8.13	10.05
0500 Type S, 1/2:1:4 mix	↓	143	.056	↓	5.25	2.13		7.38	9.25
2000 With Portland cement and lime									
2100 Type M, 1:1/4:3 mix	1 Brhe	143	.056	C.F.	9.40	2.13		11.53	13.80
2200 Type N, 1:1:6 mix, 750 psi		143	.056		7.40	2.13		9.53	11.60
2300 Type O, 1:2:9 mix (Pointing Mortar)		143	.056		8.10	2.13		10.23	12.35
2400 Type PL, 1:1/2:4 mix, 2500 psi		143	.056		6	2.13		8.13	10.05
2600 Type S, 1:1/2:4 mix, 1800 psi	↓	143	.056		8.95	2.13		11.08	13.30
2650 Pre-mixed, type S or N					5.30			5.30	5.85
2700 Mortar for glass block	1 Brhe	143	.056	↓	11.30	2.13		13.43	15.85
2900 Mortar for fire brick, dry mix, 10 lb. pail				Ea.	27			27	29.50

04 05 13.91 Masonry Restoration Mortaring									
0010 **MASONRY RESTORATION MORTARING**									
0020 Masonry restoration mix				Lb.	.25			.25	.28
0050 White				"	.25			.25	.28

04 05 13.93 Mortar Pigments									
0010 **MORTAR PIGMENTS**, 50 lb. bags (2 bags per M bricks)									
0020 Color admixture, range 2 to 10 lb. per bag of cement, light colors				Lb.	5.55			5.55	6.10
0050 Medium colors					7.40			7.40	8.15
0100 Dark colors				↓	15.50			15.50	17.05

04 05 13.95 Sand									
0010 **SAND**, screened and washed at pit									
0020 For mortar, per ton				Ton	21.50			21.50	23.50
0050 With 10 mile haul					35.50			35.50	39
0100 With 30 mile haul				↓	68.50			68.50	75.50
0200 Screened and washed, at the pit				C.Y.	29.50			29.50	32.50
0250 With 10 mile haul					49			49	54
0300 With 30 mile haul				↓	95			95	105

For customer support on your Commercial Renovation Cost Data, call 877.742.4719.

79

04 05 13 – Masonry Mortaring

04 05 13.98 Mortar Admixtures	Crew	Daily Output	Labor-Hours	Unit	Material	2016 Bare Costs Labor	Equipment	Total	Total Incl O&P
0010 **MORTAR ADMIXTURES**									
0020 Waterproofing admixture, per quart (1 qt. to 2 bags of masonry cement)				Qt.	3.22			3.22	3.54

04 05 16 – Masonry Grouting

04 05 16.30 Grouting

		Crew	Daily Output	Labor-Hours	Unit	Material	2016 Bare Costs Labor	Equipment	Total	Total Incl O&P
0010	**GROUTING**									
0011	Bond beams & lintels, 8" deep, 6" thick, 0.15 C.F. per L.F.	D-4	1480	.022	L.F.	.69	.93	.09	1.71	2.36
0020	8" thick, 0.2 C.F. per L.F.		1400	.023		1.11	.98	.10	2.19	2.91
0050	10" thick, 0.25 C.F. per L.F.		1200	.027		1.16	1.14	.11	2.41	3.24
0060	12" thick, 0.3 C.F. per L.F.		1040	.031		1.39	1.32	.13	2.84	3.80
0200	Concrete block cores, solid, 4" thk., by hand, 0.067 C.F./S.F. of wall	D-8	1100	.036	S.F.	.31	1.56		1.87	2.88
0210	6" thick, pumped, 0.175 C.F. per S.F.	D-4	720	.044		.81	1.90	.19	2.90	4.17
0250	8" thick, pumped, 0.258 C.F. per S.F.		680	.047		1.19	2.02	.20	3.41	4.79
0300	10" thick, pumped, 0.340 C.F. per S.F.		660	.048		1.57	2.08	.20	3.85	5.30
0350	12" thick, pumped, 0.422 C.F. per S.F.		640	.050		1.95	2.14	.21	4.30	5.85
0500	Cavity walls, 2" space, pumped, 0.167 C.F./S.F. of wall		1700	.019		.77	.81	.08	1.66	2.24
0550	3" space, 0.250 C.F./S.F.		1200	.027		1.16	1.14	.11	2.41	3.24
0600	4" space, 0.333 C.F. per S.F.		1150	.028		1.54	1.19	.12	2.85	3.75
0700	6" space, 0.500 C.F. per S.F.		800	.040		2.31	1.71	.17	4.19	5.50
0800	Door frames, 3' x 7' opening, 2.5 C.F. per opening		60	.533	Opng.	11.55	23	2.25	36.80	52
0850	6' x 7' opening, 3.5 C.F. per opening		45	.711	"	16.20	30.50	3	49.70	70
2000	Grout, C476, for bond beams, lintels and CMU cores		350	.091	C.F.	4.63	3.92	.39	8.94	11.85
9000	Minimum labor/equipment charge	1 Bric	2	4	Job		185		185	300

04 05 19 – Masonry Anchorage and Reinforcing

04 05 19.05 Anchor Bolts

		Crew	Daily Output	Labor-Hours	Unit	Material	2016 Bare Costs Labor	Equipment	Total	Total Incl O&P
0010	**ANCHOR BOLTS**									
0015	Installed in fresh grout in CMU bond beams or filled cores, no templates									
0020	Hooked, with nut and washer, 1/2" diam., 8" long	1 Bric	132	.061	Ea.	1.32	2.80		4.12	6
0030	12" long		131	.061		1.47	2.82		4.29	6.20
0040	5/8" diameter, 8" long		129	.062		3.09	2.87		5.96	8.05
0050	12" long		127	.063		3.81	2.91		6.72	8.95
0060	3/4" diameter, 8" long		127	.063		3.81	2.91		6.72	8.95
0070	12" long		125	.064		4.76	2.96		7.72	10.05

04 05 19.16 Masonry Anchors

		Crew	Daily Output	Labor-Hours	Unit	Material	2016 Bare Costs Labor	Equipment	Total	Total Incl O&P
0010	**MASONRY ANCHORS**									
0020	For brick veneer, galv., corrugated, 7/8" x 7", 22 Ga.	1 Bric	10.50	.762	C	15.15	35		50.15	74
0100	24 Ga.		10.50	.762		9.90	35		44.90	68.50
0150	16 Ga.		10.50	.762		29	35		64	89
0200	Buck anchors, galv., corrugated, 16 ga., 2" bend, 8" x 2"		10.50	.762		56.50	35		91.50	120
0250	8" x 3"		10.50	.762		58.50	35		93.50	122
0300	Adjustable, rectangular, 4-1/8" wide									
0350	Anchor and tie, 3/16" wire, mill galv.									
0400	2-3/4" eye, 3-1/4" tie	1 Bric	1.05	7.619	M	460	350		810	1,075
0500	4-3/4" tie		1.05	7.619		495	350		845	1,125
0520	5-1/2" tie		1.05	7.619		535	350		885	1,175
0550	4-3/4" eye, 3-1/4" tie		1.05	7.619		505	350		855	1,125
0570	4-3/4" tie		1.05	7.619		550	350		900	1,175
0580	5-1/2" tie		1.05	7.619		595	350		945	1,225
0660	Cavity wall, Z-type, galvanized, 6" long, 1/8" diam.		10.50	.762	C	25	35		60	85
0670	3/16" diameter		10.50	.762		30.50	35		65.50	91
0680	1/4" diameter		10.50	.762		39	35		74	101
0850	8" long, 3/16" diameter		10.50	.762		26.50	35		61.50	86.50

04 05 Common Work Results for Masonry

04 05 19 – Masonry Anchorage and Reinforcing

04 05 19.16 Masonry Anchors		Crew	Daily Output	Labor-Hours	Unit	Material	2016 Bare Costs Labor	Equipment	Total	Total Incl O&P
0855	1/4" diameter	1 Bric	10.50	.762	C	48	35		83	111
1000	Rectangular type, galvanized, 1/4" diameter, 2" x 6"		10.50	.762		73	35		108	138
1050	4" x 6"		10.50	.762		88	35		123	155
1100	3/16" diameter, 2" x 6"		10.50	.762		45.50	35		80.50	108
1150	4" x 6"		10.50	.762		52	35		87	115
1200	Mesh wall tie, 1/2" mesh, hot dip galvanized									
1400	16 ga., 12" long, 3" wide	1 Bric	9	.889	C	87.50	41		128.50	164
1420	6" wide		9	.889		128	41		169	207
1440	12" wide		8.50	.941		204	43.50		247.50	295
1500	Rigid partition anchors, plain, 8" long, 1" x 1/8"		10.50	.762		243	35		278	325
1550	1" x 1/4"		10.50	.762		285	35		320	375
1580	1-1/2" x 1/8"		10.50	.762		267	35		302	350
1600	1-1/2" x 1/4"		10.50	.762		335	35		370	425
1650	2" x 1/8"		10.50	.762		315	35		350	405
1700	2" x 1/4"		10.50	.762		415	35		450	515
2000	Column flange ties, wire, galvanized									
2300	3/16" diameter, up to 3" wide	1 Bric	10.50	.762	C	89.50	35		124.50	156
2350	To 5" wide		10.50	.762		97.50	35		132.50	165
2400	To 7" wide		10.50	.762		105	35		140	174
2600	To 9" wide		10.50	.762		112	35		147	182
2650	1/4" diameter, up to 3" wide		10.50	.762		112	35		147	182
2700	To 5" wide		10.50	.762		122	35		157	192
2800	To 7" wide		10.50	.762		137	35		172	208
2850	To 9" wide		10.50	.762		147	35		182	220
2900	For hot dip galvanized, add					35%				
4000	Channel slots, 1-3/8" x 1/2" x 8"									
4100	12 ga., plain	1 Bric	10.50	.762	C	220	35		255	300
4150	16 ga., galvanized	"	10.50	.762	"	150	35		185	223
4200	Channel slot anchors									
4300	16 ga., galvanized, 1-1/4" x 3-1/2"				C	59			59	65
4350	1-1/4" x 5-1/2"					69.50			69.50	76
4400	1-1/4" x 7-1/2"					78.50			78.50	86.50
4500	1/8" plain, 1-1/4" x 3-1/2"					147			147	162
4550	1-1/4" x 5-1/2"					157			157	172
4600	1-1/4" x 7-1/2"					171			171	188
4700	For corrugation, add					79			79	87
4750	For hot dip galvanized, add					35%				
5000	Dowels									
5100	Plain, 1/4" diameter, 3" long				C	57			57	62.50
5150	4" long					63			63	69.50
5200	6" long					77.50			77.50	85
5300	3/8" diameter, 3" long					75.50			75.50	83
5350	4" long					93.50			93.50	103
5400	6" long					108			108	119
5500	1/2" diameter, 3" long					110			110	121
5550	4" long					130			130	143
5600	6" long					169			169	186
5700	5/8" diameter, 3" long					151			151	166
5750	4" long					185			185	204
5800	6" long					254			254	280
6000	3/4" diameter, 3" long					187			187	206
6100	4" long					237			237	261
6150	6" long					340			340	375

For customer support on your Commercial Renovation Cost Data, call 877.742.4719.

81

04 05 19 – Masonry Anchorage and Reinforcing

04 05 19.16 Masonry Anchors	Crew	Daily Output	Labor-Hours	Unit	Material	2016 Bare Costs Labor	Equipment	Total	Total Incl O&P
6300	For hot dip galvanized, add				C	35%				

04 05 19.26 Masonry Reinforcing Bars

04 05 19.16 Masonry Anchors		Crew	Daily Output	Labor-Hours	Unit	Material	Labor	Equipment	Total	Total Incl O&P
0010	**MASONRY REINFORCING BARS**									
0015	Steel bars A615, placed horiz., #3 & #4 bars	1 Bric	450	.018	Lb.	.48	.82		1.30	1.87
0020	#5 & #6 bars		800	.010		.48	.46		.94	1.28
0050	Placed vertical, #3 & #4 bars		350	.023		.48	1.06		1.54	2.25
0060	#5 & #6 bars		650	.012		.48	.57		1.05	1.46
0200	Joint reinforcing, regular truss, to 6" wide, mill std galvanized		30	.267	C.L.F.	22.50	12.35		34.85	45
0250	12" wide		20	.400		26.50	18.50		45	59
0400	Cavity truss with drip section, to 6" wide		30	.267		23	12.35		35.35	45.50
0450	12" wide		20	.400		26.50	18.50		45	59.50
0500	Joint reinforcing, ladder type, mill std galvanized									
0600	9 ga. sides, 9 ga. ties, 4" wall	1 Bric	30	.267	C.L.F.	21	12.35		33.35	43
0650	6" wall		30	.267		18.60	12.35		30.95	40.50
0700	8" wall		25	.320		19.90	14.80		34.70	46
0750	10" wall		20	.400		22	18.50		40.50	54.50
0800	12" wall		20	.400		23	18.50		41.50	55.50
1000	Truss type									
1100	9 ga. sides, 9 ga. ties, 4" wall	1 Bric	30	.267	C.L.F.	22	12.35		34.35	44
1150	6" wall		30	.267		22.50	12.35		34.85	44.50
1200	8" wall		25	.320		26.50	14.80		41.30	53.50
1250	10" wall		20	.400		22	18.50		40.50	54
1300	12" wall		20	.400		22.50	18.50		41	55
1500	3/16" sides, 9 ga. ties, 4" wall		30	.267		25.50	12.35		37.85	48
1550	6" wall		30	.267		32	12.35		44.35	55
1600	8" wall		25	.320		33.50	14.80		48.30	60.50
1650	10" wall		20	.400		34	18.50		52.50	67.50
1700	12" wall		20	.400		34	18.50		52.50	67.50
2000	3/16" sides, 3/16" ties, 4" wall		30	.267		36	12.35		48.35	59.50
2050	6" wall		30	.267		37	12.35		49.35	60.50
2100	8" wall		25	.320		38	14.80		52.80	66
2150	10" wall		20	.400		40	18.50		58.50	74
2200	12" wall		20	.400		42.50	18.50		61	76.50
2500	Cavity truss type, galvanized									
2600	9 ga. sides, 9 ga. ties, 4" wall	1 Bric	25	.320	C.L.F.	38	14.80		52.80	65.50
2650	6" wall		25	.320		35.50	14.80		50.30	63
2700	8" wall		20	.400		39.50	18.50		58	73.50
2750	10" wall		15	.533		41.50	24.50		66	86
2800	12" wall		15	.533		38	24.50		62.50	82
3000	3/16" sides, 9 ga. ties, 4" wall		25	.320		36	14.80		50.80	63.50
3050	6" wall		25	.320		33	14.80		47.80	60.50
3100	8" wall		20	.400		47.50	18.50		66	82.50
3150	10" wall		15	.533		33.50	24.50		58	77
3200	12" wall		15	.533		38	24.50		62.50	81.50
3500	For hot dip galvanizing, add				Ton	460			460	505

04 05 23 – Masonry Accessories

04 05 23.13 Masonry Control and Expansion Joints

04 05 23.13 Masonry Control and Expansion Joints		Crew	Daily Output	Labor-Hours	Unit	Material	Labor	Equipment	Total	Total Incl O&P
0010	**MASONRY CONTROL AND EXPANSION JOINTS**									
0020	Rubber, for double wythe 8" minimum wall (Brick/CMU)	1 Bric	400	.020	L.F.	2.08	.93		3.01	3.80
0025	"T" shaped		320	.025		1.25	1.16		2.41	3.26
0030	Cross-shaped for CMU units		280	.029		1.47	1.32		2.79	3.77
0050	PVC, for double wythe 8" minimum wall (Brick/CMU)		400	.020		1.48	.93		2.41	3.14

04 05 Common Work Results for Masonry

04 05 23 – Masonry Accessories

04 05 23.13 Masonry Control and Expansion Joints

	04 05 23.13 Masonry Control and Expansion Joints	Crew	Daily Output	Labor-Hours	Unit	Material	2016 Bare Costs Labor	Equipment	Total	Total Incl O&P
0120	"T" shaped	1 Bric	320	.025	L.F.	.78	1.16		1.94	2.74
0160	Cross-shaped for CMU units	↓	280	.029	↓	.97	1.32		2.29	3.22

04 05 23.19 Masonry Cavity Drainage, Weepholes, and Vents

		Crew	Daily Output	Labor-Hours	Unit	Material	2016 Bare Costs Labor	Equipment	Total	Total Incl O&P
0010	**MASONRY CAVITY DRAINAGE, WEEPHOLES, AND VENTS**									
0020	Extruded aluminum, 4" deep, 2-3/8" x 8-1/8"	1 Bric	30	.267	Ea.	35	12.35		47.35	58
0050	5" x 8-1/8"		25	.320		45.50	14.80		60.30	74
0100	2-1/4" x 25"		25	.320		79.50	14.80		94.30	112
0150	5" x 16-1/2"		22	.364		65.50	16.80		82.30	99.50
0175	5" x 24"		22	.364		85.50	16.80		102.30	122
0200	6" x 16-1/2"		22	.364		92.50	16.80		109.30	130
0250	7-3/4" x 16-1/2"		20	.400		79	18.50		97.50	117
0400	For baked enamel finish, add					35%				
0500	For cast aluminum, painted, add					60%				
1000	Stainless steel ventilators, 6" x 6"	1 Bric	25	.320		210	14.80		224.80	256
1050	8" x 8"		24	.333		232	15.40		247.40	281
1100	12" x 12"		23	.348		266	16.10		282.10	320
1150	12" x 6"		24	.333		254	15.40		269.40	305
1200	Foundation block vent, galv., 1-1/4" thk, 8" high, 16" long, no damper	↓	30	.267		16.20	12.35		28.55	38
1250	For damper, add				↓	3.11			3.11	3.42

04 05 23.95 Wall Plugs

		Crew	Daily Output	Labor-Hours	Unit	Material	2016 Bare Costs Labor	Equipment	Total	Total Incl O&P
0010	**WALL PLUGS** (for nailing to brickwork)									
0020	25 ga., galvanized, plain	1 Bric	10.50	.762	C	25.50	35		60.50	85.50
0050	Wood filled	"	10.50	.762	"	64.50	35		99.50	129

04 21 Clay Unit Masonry

04 21 13 – Brick Masonry

04 21 13.13 Brick Veneer Masonry

		Crew	Daily Output	Labor-Hours	Unit	Material	2016 Bare Costs Labor	Equipment	Total	Total Incl O&P
0010	**BRICK VENEER MASONRY**, T.L. lots, excl. scaff., grout & reinforcing R042110-20									
0015	Material costs incl. 3% brick and 25% mortar waste									
2000	Standard, sel. common, 4" x 2-2/3" x 8", (6.75/S.F.) R042110-50	D-8	230	.174	S.F.	4.34	7.45		11.79	16.90
2020	Red, 4" x 2-2/3" x 8", running bond		220	.182		3.93	7.80		11.73	17
2050	Full header every 6th course (7.88/S.F.)		185	.216		4.58	9.30		13.88	20
2100	English, full header every 2nd course (10.13/S.F.)		140	.286		5.85	12.25		18.10	26.50
2150	Flemish, alternate header every course (9.00/S.F.)		150	.267		5.20	11.45		16.65	24.50
2200	Flemish, alt. header every 6th course (7.13/S.F.)		205	.195		4.15	8.40		12.55	18.20
2250	Full headers throughout (13.50/S.F.)		105	.381		7.80	16.35		24.15	35
2300	Rowlock course (13.50/S.F.)		100	.400		7.80	17.20		25	36.50
2350	Rowlock stretcher (4.50/S.F.)		310	.129		2.64	5.55		8.19	11.90
2400	Soldier course (6.75/S.F.)		200	.200		3.93	8.60		12.53	18.30
2450	Sailor course (4.50/S.F.)		290	.138		2.64	5.90		8.54	12.55
2600	Buff or gray face, running bond (6.75/S.F.)		220	.182		4.15	7.80		11.95	17.25
2700	Glazed face brick, running bond		210	.190		12.20	8.20		20.40	26.50
2750	Full header every 6th course (7.88/S.F.)		170	.235		14.20	10.10		24.30	32
3000	Jumbo, 6" x 4" x 12" running bond (3.00/S.F.)		435	.092		4.85	3.95		8.80	11.80
3050	Norman, 4" x 2-2/3" x 12" running bond (4.5/S.F.)		320	.125		6.05	5.35		11.40	15.40
3100	Norwegian, 4" x 3-1/5" x 12" (3.75/S.F.)		375	.107		5.35	4.58		9.93	13.35
3150	Economy, 4" x 4" x 8" (4.50/S.F.)		310	.129		4.24	5.55		9.79	13.65
3200	Engineer, 4" x 3-1/5" x 8" (5.63/S.F.)		260	.154		3.76	6.60		10.36	14.90
3250	Roman, 4" x 2" x 12" (6.00/S.F.)		250	.160		7.20	6.85		14.05	19.10
3300	SCR, 6" x 2-2/3" x 12" (4.50/S.F.)	↓	310	.129	↓	6.20	5.55		11.75	15.85

For customer support on your Commercial Renovation Cost Data, call 877.742.4719.

83

04 21 Clay Unit Masonry

04 21 13 – Brick Masonry

04 21 13.13 Brick Veneer Masonry	Crew	Daily Output	Labor-Hours	Unit	Material	2016 Bare Costs Labor	Equipment	Total	Total Incl O&P
3350	Utility, 4" x 4" x 12" (3.00/S.F.)	D-8	360	.111	S.F.	4.78	4.77		9.55	13
3360	For less than truck load lots, add				M	15%				
3400	For cavity wall construction, add						15%			
3450	For stacked bond, add						10%			
3500	For interior veneer construction, add						15%			
3550	For curved walls, add						30%			
9000	Minimum labor/equipment charge	D-1	2	8	Job		335		335	550

04 21 13.14 Thin Brick Veneer

04 21 13.14 Thin Brick Veneer	Crew	Daily Output	Labor-Hours	Unit	Material	2016 Bare Costs Labor	Equipment	Total	Total Incl O&P
0010	**THIN BRICK VENEER**									
0015	Material costs incl. 3% brick and 25% mortar waste									
0020	On & incl. metal panel support sys, modular, 2-2/3" x 5/8" x 8", red	D-7	92	.174	S.F.	9.35	6.90		16.25	21
0100	Closure, 4" x 5/8" x 8"		110	.145		9.10	5.75		14.85	19.10
0110	Norman, 2-2/3" x 5/8" x 12"		110	.145		9.15	5.75		14.90	19.15
0120	Utility, 4" x 5/8" x 12"		125	.128		8.85	5.10		13.95	17.75
0130	Emperor, 4" x 3/4" x 16"		175	.091		10	3.63		13.63	16.70
0140	Super emperor, 8" x 3/4" x 16"		195	.082		10.30	3.25		13.55	16.50
0150	For L shaped corners with 4" return, add				L.F.	9.25			9.25	10.20
0200	On masonry/plaster back-up, modular, 2-2/3" x 5/8" x 8", red	D-7	137	.117	S.F.	4.36	4.63		8.99	12.10
0210	Closure, 4" x 5/8" x 8"		165	.097		4.12	3.84		7.96	10.60
0220	Norman, 2-2/3" x 5/8" x 12"		165	.097		4.16	3.84		8	10.65
0230	Utility, 4" x 5/8" x 12"		185	.086		3.88	3.43		7.31	9.65
0240	Emperor, 4" x 3/4" x 16"		260	.062		5	2.44		7.44	9.35
0250	Super emperor, 8" x 3/4" x 16"		285	.056		5.35	2.23		7.58	9.35
0260	For L shaped corners with 4" return, add				L.F.	9.25			9.25	10.20
0270	For embedment into pre-cast concrete panels, add				S.F.	14.40			14.40	15.85

04 21 13.15 Chimney

04 21 13.15 Chimney	Crew	Daily Output	Labor-Hours	Unit	Material	2016 Bare Costs Labor	Equipment	Total	Total Incl O&P
0010	**CHIMNEY**, excludes foundation, scaffolding, grout and reinforcing									
0100	Brick, 16" x 16", 8" flue	D-1	18.20	.879	V.L.F.	25	37		62	88
0150	16" x 20" with one 8" x 12" flue		16	1		39	42		81	112
0200	16" x 24" with two 8" x 8" flues		14	1.143		57	48		105	142
0250	20" x 20" with one 12" x 12" flue		13.70	1.168		46.50	49		95.50	131
0300	20" x 24" with two 8" x 12" flues		12	1.333		65	56		121	163
0350	20" x 32" with two 12" x 12" flues		10	1.600		81.50	67.50		149	200

04 21 13.18 Columns

04 21 13.18 Columns	Crew	Daily Output	Labor-Hours	Unit	Material	2016 Bare Costs Labor	Equipment	Total	Total Incl O&P
0010	**COLUMNS**, solid, excludes scaffolding, grout and reinforcing									
0050	Brick, 8" x 8", 9 brick per V.L.F.	D-1	56	.286	V.L.F.	5	12.05		17.05	25
0100	12" x 8", 13.5 brick per V.L.F.		37	.432		7.50	18.20		25.70	38
0200	12" x 12", 20 brick per V.L.F.		25	.640		11.10	27		38.10	56.50
0300	16" x 12", 27 brick per V.L.F.		19	.842		15	35.50		50.50	74
0400	16" x 16", 36 brick per V.L.F.		14	1.143		20	48		68	101
0500	20" x 16", 45 brick per V.L.F.		11	1.455		25	61.50		86.50	127
0600	20" x 20", 56 brick per V.L.F.		9	1.778		31	75		106	156
0700	24" x 20", 68 brick per V.L.F.		7	2.286		38	96.50		134.50	199
0800	24" x 24", 81 brick per V.L.F.		6	2.667		45	112		157	233
1000	36" x 36", 182 brick per V.L.F.		3	5.333		101	225		326	475
9000	Minimum labor/equipment charge		2	8	Job		335		335	550

04 21 13.30 Oversized Brick

04 21 13.30 Oversized Brick	Crew	Daily Output	Labor-Hours	Unit	Material	2016 Bare Costs Labor	Equipment	Total	Total Incl O&P
0010	**OVERSIZED BRICK**, excludes scaffolding, grout and reinforcing									
0100	Veneer, 4" x 2.25" x 16"	D-8	387	.103	S.F.	5.30	4.44		9.74	13.05
0102	8" x 2.25" x 16", multicell		265	.151		16.65	6.50		23.15	29
0105	4" x 2.75" x 16"		412	.097		5.50	4.17		9.67	12.90
0107	8" x 2.75" x 16", multicell		295	.136		16.70	5.80		22.50	28

84

04 21 13 – Brick Masonry

04 21 13.30 Oversized Brick

		Crew	Daily Output	Labor-Hours	Unit	Material	2016 Bare Costs Labor	2016 Bare Costs Equipment	Total	Total Incl O&P
0110	4" x 4" x 16"	D-8	460	.087	S.F.	3.61	3.73		7.34	10.05
0120	4" x 8" x 16"		533	.075		4.25	3.22		7.47	9.90
0122	4" x 8" x 16" multicell		327	.122		15.65	5.25		20.90	26
0125	Loadbearing, 6" x 4" x 16", grouted and reinforced		387	.103		10.70	4.44		15.14	18.95
0130	8" x 4" x 16", grouted and reinforced		327	.122		11.75	5.25		17	21.50
0132	10" x 4" x 16", grouted and reinforced		327	.122		23.50	5.25		28.75	34.50
0135	6" x 8" x 16", grouted and reinforced		440	.091		13.90	3.90		17.80	21.50
0140	8" x 8" x 16", grouted and reinforced		400	.100		14.85	4.30		19.15	23.50
0145	Curtainwall/reinforced veneer, 6" x 4" x 16"		387	.103		15.25	4.44		19.69	24
0150	8" x 4" x 16"		327	.122		18.55	5.25		23.80	29
0152	10" x 4" x 16"		327	.122		26	5.25		31.25	37.50
0155	6" x 8" x 16"		440	.091		18.95	3.90		22.85	27.50
0160	8" x 8" x 16"		400	.100		26.50	4.30		30.80	36
0200	For 1 to 3 slots in face, add					15%				
0210	For 4 to 7 slots in face, add					25%				
0220	For bond beams, add					20%				
0230	For bullnose shapes, add					20%				
0240	For open end knockout, add					10%				
0250	For white or gray color group, add					10%				
0260	For 135 degree corner, add					250%				

04 21 13.35 Common Building Brick

		Crew	Daily Output	Labor-Hours	Unit	Material	2016 Bare Costs Labor	2016 Bare Costs Equipment	Total	Total Incl O&P
0010	**COMMON BUILDING BRICK**, C62, TL lots, material only R042110-20									
0020	Standard				M	495			495	545
0050	Select				"	520			520	570

04 21 13.40 Structural Brick

		Crew	Daily Output	Labor-Hours	Unit	Material	2016 Bare Costs Labor	2016 Bare Costs Equipment	Total	Total Incl O&P
0010	**STRUCTURAL BRICK** C652, Grade SW, incl. mortar, scaffolding not incl.									
0100	Standard unit, 4-5/8" x 2-3/4" x 9-5/8"	D-8	245	.163	S.F.	4.21	7		11.21	16.05
0120	Bond beam		225	.178		4.21	7.65		11.86	17.10
0140	V cut bond beam		225	.178		4.21	7.65		11.86	17.10
0160	Stretcher quoin, 5-5/8" x 2-3/4" x 9-5/8"		245	.163		7.80	7		14.80	20
0180	Corner quoin		245	.163		7.80	7		14.80	20
0200	Corner, 45 deg, 4-5/8" x 2-3/4" x 10-7/16"		235	.170		7.80	7.30		15.10	20.50

04 21 13.45 Face Brick

		Crew	Daily Output	Labor-Hours	Unit	Material	2016 Bare Costs Labor	2016 Bare Costs Equipment	Total	Total Incl O&P
0010	**FACE BRICK** Material Only, C216, TL lots R042110-20									
0300	Standard modular, 4" x 2-2/3" x 8"				M	460			460	505
0450	Economy, 4" x 4" x 8"					805			805	885
0510	Economy, 4" x 4" x 12"					1,275			1,275	1,400
0550	Jumbo, 6" x 4" x 12"					1,450			1,450	1,600
0610	Jumbo, 8" x 4" x 12"					1,450			1,450	1,600
0650	Norwegian, 4" x 3-1/5" x 12"					1,275			1,275	1,400
0710	Norwegian, 6" x 3-1/5" x 12"					1,575			1,575	1,725
0850	Standard glazed, plain colors, 4" x 2-2/3" x 8"					1,650			1,650	1,800
1000	Deep trim shades, 4" x 2-2/3" x 8"					2,100			2,100	2,325
1080	Jumbo utility, 4" x 4" x 12"					1,425			1,425	1,575
1120	4" x 8" x 8"					1,825			1,825	2,025
1140	4" x 8" x 16"					5,400			5,400	5,950
1260	Engineer, 4" x 3-1/5" x 8"					540			540	595
1350	King, 4" x 2-3/4" x 10"					500			500	550
1770	Standard modular, double glazed, 4" x 2-2/3" x 8"					2,475			2,475	2,725
1850	Jumbo, colored glazed ceramic, 6" x 4" x 12"					2,625			2,625	2,875
2050	Jumbo utility, glazed, 4" x 4" x 12"					4,925			4,925	5,400
2100	4" x 8" x 8"					5,800			5,800	6,375

For customer support on your Commercial Renovation Cost Data, call 877.742.4719.

85

04 21 13 – Brick Masonry

04 21 13.45 Face Brick		Crew	Daily Output	Labor-Hours	Unit	Material	2016 Bare Costs Labor	Equipment	Total	Total Incl O&P
2150	4" x 16" x 8"				M	6,775			6,775	7,450
2170	For less than truck load lots, add					15			15	16.50
2180	For buff or gray brick, add					16			16	17.60
3050	Used brick					405			405	445
3150	Add for brick to match existing work, minimum					5%				
3200	Maximum					50%				

04 21 26 – Glazed Structural Clay Tile Masonry

04 21 26.10 Structural Facing Tile		Crew	Daily Output	Labor-Hours	Unit	Material	2016 Bare Costs Labor	Equipment	Total	Total Incl O&P
0010	**STRUCTURAL FACING TILE**, std. colors, excl. scaffolding, grout, reinforcing									
0020	6T series, 5-1/3" x 12", 2.3 pieces per S.F., glazed 1 side, 2" thick	D-8	225	.178	S.F.	8.55	7.65		16.20	22
0100	4" thick		220	.182		11.65	7.80		19.45	25.50
0150	Glazed 2 sides		195	.205		15.50	8.80		24.30	31.50
0250	6" thick		210	.190		17.45	8.20		25.65	32.50
0300	Glazed 2 sides		185	.216		20.50	9.30		29.80	37.50
0400	8" thick		180	.222		23	9.55		32.55	41
0500	Special shapes, group 1		400	.100	Ea.	7.40	4.30		11.70	15.15
0550	Group 2		375	.107		12	4.58		16.58	20.50
0600	Group 3		350	.114		15.45	4.91		20.36	25
0650	Group 4		325	.123		32	5.30		37.30	43.50
0700	Group 5		300	.133		38	5.75		43.75	51.50
0750	Group 6		275	.145		52	6.25		58.25	67
1000	Fire rated, 4" thick, 1 hr. rating		210	.190	S.F.	16.60	8.20		24.80	31.50
1050	2 hr. rating		170	.235		25	10.10		35.10	44
1300	Acoustic, 4" thick		210	.190		34	8.20		42.20	51
1310	6" thick		170	.235		35	10.10		45.10	55
1320	8" thick		145	.276		37	11.85		48.85	60.50
1350	8T acoustic, 4" thick		265	.151		17.55	6.50		24.05	30
1360	6" thick		256	.156		25	6.70		31.70	38.50
1370	8" thick		208	.192		26.50	8.25		34.75	43
2000	8W series, 8" x 16", 1.125 pieces per S.F.									
2050	2" thick, glazed 1 side	D-8	360	.111	S.F.	10	4.77		14.77	18.80
2100	4" thick, glazed 1 side		345	.116		13.70	4.98		18.68	23
2150	Glazed 2 sides		325	.123		16.30	5.30		21.60	26.50
2200	6" thick, glazed 1 side		330	.121		23.50	5.20		28.70	34.50
2210	Glazed 2 sides		250	.160		27	6.85		33.85	40.50
2250	8" thick, glazed 1 side		310	.129		24.50	5.55		30.05	36
2260	Glazed 2 sides		220	.182		33.50	7.80		41.30	49
2500	Special shapes, group 1		300	.133	Ea.	14.30	5.75		20.05	25
2550	Group 2		280	.143		19.10	6.15		25.25	31
2600	Group 3		260	.154		20.50	6.60		27.10	33.50
2650	Group 4		250	.160		42.50	6.85		49.35	58
2700	Group 5		240	.167		38	7.15		45.15	53.50
2750	Group 6		230	.174		82	7.45		89.45	103
2900	Fire rated, 2" thick, glazed 1 side		360	.111	S.F.	13.15	4.77		17.92	22.50
3000	4" thick, glazed 1 side		345	.116		13.20	4.98		18.18	22.50
3020	Gazed 2 sides		345	.116		20.50	4.98		25.48	30.50
3040	6" thick, glazed 1 side		330	.121		24.50	5.20		29.70	35.50
3045	Glazed 2 sides		250	.160		21.50	6.85		28.35	34.50
3047	8" thick, glazed 1 side		310	.129		29	5.55		34.55	41
3048	Glazed 2 sides		220	.182		30.50	7.80		38.30	46
3050	Special shapes, group 1, fire rated		300	.133		18.25	5.75		24	29.50
3060	Group 2		280	.143		24.50	6.15		30.65	37

04 21 Clay Unit Masonry

04 21 26 – Glazed Structural Clay Tile Masonry

04 21 26.10 Structural Facing Tile

		Crew	Daily Output	Labor-Hours	Unit	Material	2016 Bare Costs Labor	Equipment	Total	Total Incl O&P
3070	Group 3	D-8	260	.154	S.F.	25.50	6.60		32.10	39
3080	Group 4		250	.160		47.50	6.85		54.35	63
3090	Group 5		240	.167		53	7.15		60.15	70
3100	Acoustic, 4" thick		345	.116		18.55	4.98		23.53	28.50
3120	4W series, 8" x 8", 2.25 pieces per S.F.									
3125	2" thick, glazed 1 side	D-8	360	.111	S.F.	9.90	4.77		14.67	18.65
3130	4" thick, glazed 1 side		345	.116		11.65	4.98		16.63	21
3135	Glazed 2 sides		325	.123		16.05	5.30		21.35	26.50
3140	6" thick, glazed 1 side		330	.121		16.60	5.20		21.80	26.50
3145	Glazed 2 sides		250	.160		20	6.85		26.85	33
3150	8" thick, glazed 1 side		310	.129		24	5.55		29.55	35.50
3155	Special shapes, group 1		300	.133	Ea.	7.75	5.75		13.50	17.85
3160	Group 2		280	.143	"	8.95	6.15		15.10	19.85
3200	For designer colors, add					25%				
3300	For epoxy mortar joints, add				S.F.	1.74			1.74	1.91
9000	Minimum labor/equipment charge	D-1	2	8	Job		335		335	550

04 21 29 – Terra Cotta Masonry

04 21 29.10 Terra Cotta Masonry Components

		Crew	Daily Output	Labor-Hours	Unit	Material	2016 Bare Costs Labor	Equipment	Total	Total Incl O&P
0010	**TERRA COTTA MASONRY COMPONENTS**									
0020	Coping, split type, not glazed, 9" wide	D-1	90	.178	L.F.	14.75	7.50		22.25	28.50
0100	13" wide		80	.200		20.50	8.45		28.95	36
0200	Coping, split type, glazed, 9" wide		90	.178		12.60	7.50		20.10	26
0250	13" wide		80	.200		16.80	8.45		25.25	32
0500	Partition or back-up blocks, scored, in C.L. lots									
0700	Non-load bearing 12" x 12", 3" thick, special order	D-8	550	.073	S.F.	17.70	3.12		20.82	24.50
0750	4" thick, standard		500	.080		5.80	3.44		9.24	12
0800	6" thick		450	.089		7.80	3.82		11.62	14.75
0850	8" thick		400	.100		9.85	4.30		14.15	17.80
1000	Load bearing, 12" x 12", 4" thick, in walls		500	.080		6.20	3.44		9.64	12.40
1050	In floors		750	.053		6.20	2.29		8.49	10.55
1200	6" thick, in walls		450	.089		8.30	3.82		12.12	15.35
1250	In floors		675	.059		8.30	2.55		10.85	13.30
1400	8" thick, in walls		400	.100		10.40	4.30		14.70	18.40
1450	In floors		575	.070		10.40	2.99		13.39	16.25
1600	10" thick, in walls, special order		350	.114		25	4.91		29.91	36
1650	In floors, special order		500	.080		25	3.44		28.44	33.50
1800	12" thick, in walls, special order		300	.133		25	5.75		30.75	37
1850	In floors, special order		450	.089		25	3.82		28.82	33.50
2000	For reinforcing with steel rods, add to above					15%	5%			
2025	Placed vertical, #3 & #4 bars	1 Bric	350	.023	Lb.	.48	1.06		1.54	2.25
2100	For smooth tile instead of scored, add				S.F.	3.99			3.99	4.39
2200	For L.C.L. quantities, add				"	10%	10%			
9000	Minimum labor/equipment charge	D-1	2	8	Job		335		335	550

04 21 29.20 Terra Cotta Tile

		Crew	Daily Output	Labor-Hours	Unit	Material	2016 Bare Costs Labor	Equipment	Total	Total Incl O&P
0010	**TERRA COTTA TILE**, on walls, dry set, 1/2" thick									
0100	Square, hexagonal or lattice shapes, unglazed	1 Tilf	135	.059	S.F.	4.73	2.64		7.37	9.35
0300	Glazed, plain colors		130	.062		7.25	2.74		9.99	12.25
0400	Intense colors		125	.064		8.50	2.85		11.35	13.85

For customer support on your Commercial Renovation Cost Data, call 877.742.4719.

87

04 22 10 – Concrete Masonry Units

04 22 10.11 Autoclave Aerated Concrete Block		Crew	Daily Output	Labor-Hours	Unit	Material	2016 Bare Costs Labor	Equipment	Total	Total Incl O&P
0010	**AUTOCLAVE AERATED CONCRETE BLOCK**, excl. scaffolding, grout & reinforcing									
0050	Solid, 4" x 8" x 24", incl. mortar	G	D-8	600	.067	S.F.	1.55	2.86		4.41	6.35
0060	6" x 8" x 24"	G		600	.067		2.34	2.86		5.20	7.25
0070	8" x 8" x 24"	G		575	.070		3.11	2.99		6.10	8.30
0080	10" x 8" x 24"	G		575	.070		3.80	2.99		6.79	9.05
0090	12" x 8" x 24"	G		550	.073		4.66	3.12		7.78	10.25

04 22 10.14 Concrete Block, Back-Up		Crew	Daily Output	Labor-Hours	Unit	Material	Labor	Equipment	Total	Total Incl O&P
0010	**CONCRETE BLOCK, BACK-UP**, C90, 2000 psi									
0020	Normal weight, 8" x 16" units, tooled joint 1 side									
0050	Not-reinforced, 2000 psi, 2" thick	D-8	475	.084	S.F.	1.53	3.62		5.15	7.60
0200	4" thick		460	.087		1.84	3.73		5.57	8.10
0300	6" thick		440	.091		2.40	3.90		6.30	9
0350	8" thick		400	.100		2.56	4.30		6.86	9.80
0400	10" thick		330	.121		3.03	5.20		8.23	11.80
0450	12" thick	D-9	310	.155		4.17	6.50		10.67	15.20
1000	Reinforced, alternate courses, 4" thick	D-8	450	.089		2	3.82		5.82	8.40
1100	6" thick		430	.093		2.56	4		6.56	9.30
1150	8" thick		395	.101		2.76	4.35		7.11	10.15
1200	10" thick		320	.125		3.19	5.35		8.54	12.25
1250	12" thick	D-9	300	.160		4.34	6.75		11.09	15.75
9000	Minimum labor/equipment charge	D-1	2	8	Job		335		335	550

04 22 10.16 Concrete Block, Bond Beam		Crew	Daily Output	Labor-Hours	Unit	Material	Labor	Equipment	Total	Total Incl O&P
0010	**CONCRETE BLOCK, BOND BEAM**, C90, 2000 psi									
0020	Not including grout or reinforcing									
0125	Regular block, 6" thick	D-8	584	.068	L.F.	2.73	2.94		5.67	7.80
0130	8" high, 8" thick	"	565	.071		2.82	3.04		5.86	8.05
0150	12" thick	D-9	510	.094		3.83	3.97		7.80	10.65
0525	Lightweight, 6" thick	D-8	592	.068		2.85	2.90		5.75	7.85
0530	8" high, 8" thick	"	575	.070		3.43	2.99		6.42	8.65
0550	12" thick	D-9	520	.092		4.63	3.89		8.52	11.45
2000	Including grout and 2 #5 bars									
2100	Regular block, 8" high, 8" thick	D-8	300	.133	L.F.	4.93	5.75		10.68	14.70
2150	12" thick	D-9	250	.192		6.55	8.10		14.65	20.50
2500	Lightweight, 8" high, 8" thick	D-8	305	.131		5.55	5.65		11.20	15.25
2550	12" thick	D-9	255	.188		7.35	7.95		15.30	21
9000	Minimum labor/equipment charge	D-1	2	8	Job		335		335	550

04 22 10.19 Concrete Block, Insulation Inserts		Crew	Daily Output	Labor-Hours	Unit	Material	Labor	Equipment	Total	Total Incl O&P
0010	**CONCRETE BLOCK, INSULATION INSERTS**									
0100	Styrofoam, plant installed, add to block prices									
0200	8" x 16" units, 6" thick				S.F.	1.20			1.20	1.32
0250	8" thick					1.35			1.35	1.49
0300	10" thick					1.40			1.40	1.54
0350	12" thick					1.55			1.55	1.71
0500	8" x 8" units, 8" thick					1.20			1.20	1.32
0550	12" thick					1.40			1.40	1.54

04 22 10.23 Concrete Block, Decorative		Crew	Daily Output	Labor-Hours	Unit	Material	Labor	Equipment	Total	Total Incl O&P
0010	**CONCRETE BLOCK, DECORATIVE**, C90, 2000 psi									
0020	Embossed, simulated brick face									
0100	8" x 16" units, 4" thick	D-8	400	.100	S.F.	2.80	4.30		7.10	10.10
0200	8" thick		340	.118		3.06	5.05		8.11	11.55
0250	12" thick		300	.133		5.15	5.75		10.90	15
0400	Embossed both sides									

04 22 Concrete Unit Masonry

04 22 10 – Concrete Masonry Units

04 22 10.23 Concrete Block, Decorative	Crew	Daily Output	Labor-Hours	Unit	Material	2016 Bare Costs Labor	Equipment	Total	Total Incl O&P
0500	8" thick	D-8	300	.133	S.F.	4.13	5.75		9.88	13.85
0550	12" thick	"	275	.145	"	5.40	6.25		11.65	16.10
1000	Fluted high strength									
1100	8" x 16" x 4" thick, flutes 1 side,	D-8	345	.116	S.F.	3.98	4.98		8.96	12.50
1150	Flutes 2 sides		335	.119		4.57	5.15		9.72	13.40
1200	8" thick		300	.133		5.35	5.75		11.10	15.20
1250	For special colors, add					.64			.64	.71
1400	Deep grooved, smooth face									
1450	8" x 16" x 4" thick	D-8	345	.116	S.F.	2.56	4.98		7.54	10.90
1500	8" thick	"	300	.133	"	4.05	5.75		9.80	13.75
2000	Formblock, incl. inserts & reinforcing									
2100	8" x 16" x 8" thick	D-8	345	.116	S.F.	3.57	4.98		8.55	12.05
2150	12" thick	"	310	.129	"	4.73	5.55		10.28	14.20
2500	Ground face									
2600	8" x 16" x 4" thick	D-8	345	.116	S.F.	3.88	4.98		8.86	12.35
2650	6" thick		325	.123		4.30	5.30		9.60	13.35
2700	8" thick		300	.133		4.74	5.75		10.49	14.50
2725	10" thick		280	.143		5.55	6.15		11.70	16.10
2750	12" thick	D-9	265	.181		6.10	7.65		13.75	19.10
2900	For special colors, add, minimum					15%				
2950	For special colors, add, maximum					45%				
4000	Slump block									
4100	4" face height x 16" x 4" thick	D-1	165	.097	S.F.	4.14	4.09		8.23	11.20
4150	6" thick		160	.100		6.25	4.21		10.46	13.70
4200	8" thick		155	.103		6	4.35		10.35	13.70
4250	10" thick		140	.114		11.55	4.82		16.37	20.50
4300	12" thick		130	.123		13.10	5.20		18.30	23
4400	6" face height x 16" x 6" thick		155	.103		5.80	4.35		10.15	13.45
4450	8" thick		150	.107		8.75	4.49		13.24	16.95
4500	10" thick		130	.123		13.40	5.20		18.60	23
4550	12" thick		120	.133		14.05	5.60		19.65	24.50
5000	Split rib profile units, 1" deep ribs, 8 ribs									
5100	8" x 16" x 4" thick	D-8	345	.116	S.F.	3.98	4.98		8.96	12.50
5150	6" thick		325	.123		4.50	5.30		9.80	13.55
5200	8" thick		300	.133		5.10	5.75		10.85	14.90
5225	10" thick		300	.133		5.60	5.75		11.35	15.45
5250	12" thick	D-9	275	.175		5.95	7.35		13.30	18.50
5400	For special deeper colors, 4" thick, add					1.29			1.29	1.42
5450	12" thick, add					1.33			1.33	1.47
5600	For white, 4" thick, add					1.29			1.29	1.42
5650	6" thick, add					1.31			1.31	1.44
5700	8" thick, add					1.34			1.34	1.48
5750	12" thick, add					1.39			1.39	1.53
6000	Split face									
6100	8" x 16" x 4" thick	D-8	350	.114	S.F.	3.59	4.91		8.50	11.95
6150	6" thick		325	.123		4.09	5.30		9.39	13.10
6200	8" thick		300	.133		4.60	5.75		10.35	14.35
6225	10" thick		275	.145		5.20	6.25		11.45	15.90
6250	12" thick	D-9	270	.178		5.55	7.50		13.05	18.30
6300	For scored, add					.38			.38	.42
6400	For special deeper colors, 4" thick, add					.62			.62	.68
6450	6" thick, add					.75			.75	.82
6500	8" thick, add					.77			.77	.85

For customer support on your Commercial Renovation Cost Data, call 877.742.4719.

89

04 22 Concrete Unit Masonry

04 22 10 – Concrete Masonry Units

04 22 10.23 Concrete Block, Decorative

		Crew	Daily Output	Labor-Hours	Unit	Material	2016 Bare Costs Labor	Equipment	Total	Total Incl O&P
6550	12" thick, add				S.F.	.79			.79	.87
6650	For white, 4" thick, add					1.28			1.28	1.40
6700	6" thick, add					1.29			1.29	1.42
6750	8" thick, add					1.30			1.30	1.43
6800	12" thick, add					1.33			1.33	1.47
7000	Scored ground face, 2 to 5 scores									
7100	8" x 16" x 4" thick	D-8	340	.118	S.F.	7.85	5.05		12.90	16.80
7150	6" thick		310	.129		8.70	5.55		14.25	18.55
7200	8" thick		290	.138		9.80	5.90		15.70	20.50
7250	12" thick	D-9	265	.181		13.15	7.65		20.80	27
8000	Hexagonal face profile units, 8" x 16" units									
8100	4" thick, hollow	D-8	340	.118	S.F.	3.66	5.05		8.71	12.25
8200	Solid		340	.118		4.69	5.05		9.74	13.35
8300	6" thick, hollow		310	.129		3.96	5.55		9.51	13.35
8350	8" thick, hollow		290	.138		4.53	5.90		10.43	14.65
8500	For stacked bond, add						26%			
8550	For high rise construction, add per story	D-8	67.80	.590	M.S.F.		25.50		25.50	41
8600	For scored block, add					10%				
8650	For honed or ground face, per face, add				Ea.	1.18			1.18	1.29
8700	For honed or ground end, per end, add				"	1.18			1.18	1.29
8750	For bullnose block, add					10%				
8800	For special color, add					13%				
9000	Minimum labor/equipment charge	D-1	2	8	Job		335		335	550

04 22 10.24 Concrete Block, Exterior

		Crew	Daily Output	Labor-Hours	Unit	Material	2016 Bare Costs Labor	Equipment	Total	Total Incl O&P
0010	**CONCRETE BLOCK, EXTERIOR**, C90, 2000 psi									
0020	Reinforced alt courses, tooled joints 2 sides									
0100	Normal weight, 8" x 16" x 6" thick	D-8	395	.101	S.F.	2.35	4.35		6.70	9.70
0200	8" thick		360	.111		3.62	4.77		8.39	11.75
0250	10" thick		290	.138		4.19	5.90		10.09	14.25
0300	12" thick	D-9	250	.192		4.89	8.10		12.99	18.55
9000	Minimum labor/equipment charge	D-1	2	8	Job		335		335	550

04 22 10.26 Concrete Block Foundation Wall

		Crew	Daily Output	Labor-Hours	Unit	Material	2016 Bare Costs Labor	Equipment	Total	Total Incl O&P
0010	**CONCRETE BLOCK FOUNDATION WALL**, C90/C145									
0050	Normal-weight, cut joints, horiz joint reinf, no vert reinf.									
0200	Hollow, 8" x 16" x 6" thick	D-8	455	.088	S.F.	3.02	3.78		6.80	9.45
0250	8" thick		425	.094		3.24	4.04		7.28	10.15
0300	10" thick		350	.114		3.70	4.91		8.61	12.05
0350	12" thick	D-9	300	.160		4.88	6.75		11.63	16.30
0500	Solid, 8" x 16" block, 6" thick	D-8	440	.091		3.08	3.90		6.98	9.75
0550	8" thick	"	415	.096		4.29	4.14		8.43	11.45
0600	12" thick	D-9	350	.137		6.15	5.80		11.95	16.15
1000	Reinforced, #4 vert @ 48"									
1100	Hollow, 8" x 16" block, 4" thick	D-8	455	.088	S.F.	2.99	3.78		6.77	9.45
1125	6" thick		445	.090		3.92	3.86		7.78	10.60
1150	8" thick		415	.096		4.53	4.14		8.67	11.75
1200	10" thick		340	.118		5.35	5.05		10.40	14.10
1250	12" thick	D-9	290	.166		6.90	6.95		13.85	18.95
1500	Solid, 8" x 16" block, 6" thick	D-8	430	.093		3.10	4		7.10	9.90
1600	8" thick	"	405	.099		4.29	4.24		8.53	11.60
1650	12" thick	D-9	340	.141		6.15	5.95		12.10	16.45
9000	Minimum labor/equipment charge	D-1	2	8	Job		335		335	550

04 22 Concrete Unit Masonry

04 22 10 – Concrete Masonry Units

04 22 10.30 Concrete Block, Interlocking	Crew	Daily Output	Labor-Hours	Unit	Material	2016 Bare Costs Labor	Equipment	Total	Total Incl O&P
0010 **CONCRETE BLOCK, INTERLOCKING**									
0100 Not including grout or reinforcing									
0200 8" x 16" units, 2,000 psi, 8" thick	D-1	245	.065	S.F.	3.03	2.75		5.78	7.80
0300 12" thick	"	220	.073		4.49	3.06		7.55	9.95
0400 Including grout & reinforcing, 8" thick	D-4	245	.131		8.05	5.60	.55	14.20	18.50
0450 12" thick		220	.145		9.70	6.25	.61	16.56	21.50
0500 16" thick		185	.173		12.15	7.40	.73	20.28	26
9000 Minimum labor/equipment charge	D-1	2	8	Job		335		335	550

04 22 10.32 Concrete Block, Lintels	Crew	Daily Output	Labor-Hours	Unit	Material	2016 Bare Costs Labor	Equipment	Total	Total Incl O&P
0010 **CONCRETE BLOCK, LINTELS**, C90, normal weight									
0100 Including grout and horizontal reinforcing									
0200 8" x 8" x 8", 1 #4 bar	D-4	300	.107	L.F.	3.87	4.57	.45	8.89	12.15
0250 2 #4 bars		295	.108		4.09	4.65	.46	9.20	12.50
0400 8" x 16" x 8", 1 #4 bar		275	.116		3.90	4.99	.49	9.38	12.90
0450 2 #4 bars		270	.119		4.12	5.10	.50	9.72	13.30
1000 12" x 8" x 8", 1 #4 bar		275	.116		5.30	4.99	.49	10.78	14.45
1100 2 #4 bars		270	.119		5.50	5.10	.50	11.10	14.80
1150 2 #5 bars		270	.119		5.75	5.10	.50	11.35	15.10
1200 2 #6 bars		265	.121		6.05	5.15	.51	11.71	15.55
1500 12" x 16" x 8", 1 #4 bar		250	.128		5.95	5.50	.54	11.99	16
1600 2 #3 bars		245	.131		6	5.60	.55	12.15	16.25
1650 2 #4 bars		245	.131		6.15	5.60	.55	12.30	16.45
1700 2 #5 bars		240	.133		6.40	5.70	.56	12.66	16.90
9000 Minimum labor/equipment charge	D-1	2	8	Job		335		335	550

04 22 10.33 Lintel Block	Crew	Daily Output	Labor-Hours	Unit	Material	2016 Bare Costs Labor	Equipment	Total	Total Incl O&P
0010 **LINTEL BLOCK**									
3481 Lintel block 6" x 8" x 8"	D-1	300	.053	Ea.	1.32	2.25		3.57	5.10
3501 6" x 16" x 8"		275	.058		2.07	2.45		4.52	6.25
3521 8" x 8" x 8"		275	.058		1.19	2.45		3.64	5.30
3561 8" x 16" x 8"		250	.064		2.09	2.70		4.79	6.70

04 22 10.34 Concrete Block, Partitions	Crew	Daily Output	Labor-Hours	Unit	Material	2016 Bare Costs Labor	Equipment	Total	Total Incl O&P
0010 **CONCRETE BLOCK, PARTITIONS**, excludes scaffolding									
1000 Lightweight block, tooled joints, 2 sides, hollow									
1100 Not reinforced, 8" x 16" x 4" thick	D-8	440	.091	S.F.	1.85	3.90		5.75	8.40
1150 6" thick		410	.098		2.63	4.19		6.82	9.70
1200 8" thick		385	.104		3.20	4.46		7.66	10.75
1250 10" thick		370	.108		3.88	4.64		8.52	11.80
1300 12" thick	D-9	350	.137		4.09	5.80		9.89	13.90
1500 Reinforced alternate courses, 4" thick	D-8	435	.092		2.01	3.95		5.96	8.65
1600 6" thick		405	.099		2.76	4.24		7	9.95
1650 8" thick		380	.105		3.35	4.52		7.87	11.05
1700 10" thick		365	.110		4.05	4.71		8.76	12.10
1750 12" thick	D-9	345	.139		4.26	5.85		10.11	14.25
4000 Regular block, tooled joints, 2 sides, hollow									
4100 Not reinforced, 8" x 16" x 4" thick	D-8	430	.093	S.F.	1.74	4		5.74	8.40
4150 6" thick		400	.100		2.30	4.30		6.60	9.55
4200 8" thick		375	.107		2.47	4.58		7.05	10.15
4250 10" thick		360	.111		2.93	4.77		7.70	11
4300 12" thick	D-9	340	.141		4.08	5.95		10.03	14.20
4500 Reinforced alternate courses, 8" x 16" x 4" thick	D-8	425	.094		1.91	4.04		5.95	8.70
4550 6" thick		395	.101		2.46	4.35		6.81	9.80

04 22 Concrete Unit Masonry

04 22 10 – Concrete Masonry Units

04 22 10.34 Concrete Block, Partitions

		Crew	Daily Output	Labor-Hours	Unit	Material	2016 Bare Costs Labor	Equipment	Total	Total Incl O&P
4600	8" thick	D-8	370	.108	S.F.	2.67	4.64		7.31	10.50
4650	10" thick		355	.113		4.10	4.84		8.94	12.40
4700	12" thick	D-9	335	.143		4.25	6.05		10.30	14.45
9000	Minimum labor/equipment charge	D-1	2	8	Job		335		335	550

04 22 10.38 Concrete Brick

		Crew	Daily Output	Labor-Hours	Unit	Material	2016 Bare Costs Labor	Equipment	Total	Total Incl O&P
0010	**CONCRETE BRICK**, C55, grade N, type 1									
0100	Regular, 4 x 2-1/4 x 8	D-8	660	.061	Ea.	.55	2.60		3.15	4.85
0125	Rusticated, 4 x 2-1/4 x 8		660	.061		.62	2.60		3.22	4.92
0150	Frog, 4 x 2-1/4 x 8		660	.061		.59	2.60		3.19	4.89
0200	Double, 4 x 4-7/8 x 8		535	.075		.96	3.21		4.17	6.30

04 22 10.44 Glazed Concrete Block

		Crew	Daily Output	Labor-Hours	Unit	Material	2016 Bare Costs Labor	Equipment	Total	Total Incl O&P
0010	**GLAZED CONCRETE BLOCK** C744									
0100	Single face, 8" x 16" units, 2" thick	D-8	360	.111	S.F.	9.85	4.77		14.62	18.60
0200	4" thick		345	.116		10.15	4.98		15.13	19.25
0250	6" thick		330	.121		10.95	5.20		16.15	20.50
0300	8" thick		310	.129		11.95	5.55		17.50	22
0350	10" thick		295	.136		13.40	5.80		19.20	24.50
0400	12" thick	D-9	280	.171		14.30	7.20		21.50	27.50
0700	Double face, 8" x 16" units, 4" thick	D-8	340	.118		13.85	5.05		18.90	23.50
0750	6" thick		320	.125		17.40	5.35		22.75	28
0800	8" thick		300	.133		18.25	5.75		24	29.50
1000	Jambs, bullnose or square, single face, 8" x 16", 2" thick		315	.127	Ea.	18.80	5.45		24.25	29.50
1050	4" thick		285	.140	"	19.70	6.05		25.75	31.50
1200	Caps, bullnose or square, 8" x 16", 2" thick		420	.095	L.F.	17.45	4.09		21.54	26
1250	4" thick		380	.105	"	18.50	4.52		23.02	28
1256	Corner, bullnose or square, 2" thick		280	.143	Ea.	20.50	6.15		26.65	32.50
1258	4" thick		270	.148		21.50	6.35		27.85	34
1260	6" thick		260	.154		27.50	6.60		34.10	41
1270	8" thick		250	.160		33.50	6.85		40.35	48
1280	10" thick		240	.167		34	7.15		41.15	49
1290	12" thick		230	.174		36	7.45		43.45	51.50
1500	Cove base, 8" x 16", 2" thick		315	.127	L.F.	8.60	5.45		14.05	18.30
1550	4" thick		285	.140		8.70	6.05		14.75	19.35
1600	6" thick		265	.151		9.30	6.50		15.80	21
1650	8" thick		245	.163		9.75	7		16.75	22
9000	Minimum labor/equipment charge	D-1	2	8	Job		335		335	550

04 23 Glass Unit Masonry

04 23 13 – Vertical Glass Unit Masonry

04 23 13.10 Glass Block

		Crew	Daily Output	Labor-Hours	Unit	Material	2016 Bare Costs Labor	Equipment	Total	Total Incl O&P
0010	**GLASS BLOCK**									
0100	Plain, 4" thick, under 1,000 S.F., 6" x 6"	D-8	115	.348	S.F.	27.50	14.95		42.45	55
0150	8" x 8"		160	.250		16	10.75		26.75	35
0160	end block		160	.250		60.50	10.75		71.25	84
0170	90 deg corner		160	.250		56	10.75		66.75	79
0180	45 deg corner		160	.250		46	10.75		56.75	68
0200	12" x 12"		175	.229		24	9.80		33.80	42.50
0210	4" x 8"		160	.250		30	10.75		40.75	50.50
0220	6" x 8"		160	.250		21.50	10.75		32.25	41
0300	1,000 to 5,000 S.F., 6" x 6"		135	.296		27	12.75		39.75	50.50

04 23 Glass Unit Masonry

04 23 13 – Vertical Glass Unit Masonry

04 23 13.10 Glass Block		Crew	Daily Output	Labor-Hours	Unit	Material	2016 Bare Costs Labor	Equipment	Total	Total Incl O&P
0350	8" x 8"	D-8	190	.211	S.F.	15.65	9.05		24.70	32
0400	12" x 12"		215	.186		23.50	8		31.50	39
0410	4" x 8"		215	.186		29.50	8		37.50	45.50
0420	6" x 8"		215	.186		21	8		29	36
0500	Over 5,000 S.F., 6" x 6"	↓	145	.276	↓	26.50	11.85		38.35	48.50
0700	For solar reflective blocks, add					100%				
1000	Thinline, plain, 3-1/8" thick, under 1,000 S.F., 6" x 6"	D-8	115	.348	S.F.	22.50	14.95		37.45	49
1050	8" x 8"		160	.250		12.55	10.75		23.30	31.50
1250	8" x 8"		215	.186		12	8		20	26
1400	For cleaning block after installation (both sides), add	↓	1000	.040	↓	.16	1.72		1.88	2.98
9000	Minimum labor/equipment charge	D-1	2	8	Job		335		335	550

04 27 Multiple-Wythe Unit Masonry

04 27 10 – Multiple-Wythe Masonry

04 27 10.10 Cornices

		Crew	Daily Output	Labor-Hours	Unit	Material	2016 Bare Costs Labor	Equipment	Total	Total Incl O&P
0010	**CORNICES**									
0110	Face bricks, 12 brick/S.F.	D-1	30	.533	SF Face	5.85	22.50		28.35	43
0150	15 brick/S.F.		23	.696	"	7	29.50		36.50	55
9000	Minimum labor/equipment charge	↓	1.50	10.667	Job		450		450	730

04 27 10.30 Brick Walls

		Crew	Daily Output	Labor-Hours	Unit	Material	2016 Bare Costs Labor	Equipment	Total	Total Incl O&P
0010	**BRICK WALLS**, including mortar, excludes scaffolding									
0020	Estimating by number of brick									
0140	Face brick, 4" thick wall, 6.75 brick/S.F.	D-8	1.45	27.586	M	570	1,175		1,745	2,550
0150	Common brick, 4" thick wall, 6.75 brick/S.F.		1.60	25		610	1,075		1,685	2,425
0204	8" thick, 13.50 bricks per S.F.		1.80	22.222		630	955		1,585	2,250
0250	12" thick, 20.25 bricks per S.F.		1.90	21.053		635	905		1,540	2,175
0304	16" thick, 27.00 bricks per S.F.		2	20		645	860		1,505	2,100
0500	Reinforced, face brick, 4" thick wall, 6.75 brick/S.F.		1.40	28.571		595	1,225		1,820	2,650
0520	Common brick, 4" thick wall, 6.75 brick/S.F.		1.55	25.806		635	1,100		1,735	2,500
0550	8" thick, 13.50 bricks per S.F.		1.75	22.857		655	980		1,635	2,325
0600	12" thick, 20.25 bricks per S.F.		1.85	21.622		660	930		1,590	2,225
0650	16" thick, 27.00 bricks per S.F.		1.95	20.513	↓	670	880		1,550	2,150
0790	Alternate method of figuring by square foot									
0800	Face brick, 4" thick wall, 6.75 brick/S.F.	D-8	215	.186	S.F.	3.86	8		11.86	17.25
0850	Common brick, 4" thick wall, 6.75 brick/S.F. R042110-50		240	.167		4.12	7.15		11.27	16.20
0900	8" thick, 13.50 bricks per S.F.		135	.296		8.55	12.75		21.30	30
1000	12" thick, 20.25 bricks per S.F.		95	.421		12.85	18.10		30.95	43.50
1050	16" thick, 27.00 bricks per S.F.		75	.533		17.40	23		40.40	56.50
1200	Reinforced, face brick, 4" thick wall, 6.75 brick/S.F.		210	.190		4.02	8.20		12.22	17.70
1220	Common brick, 4" thick wall, 6.75 brick/S.F.		235	.170		4.28	7.30		11.58	16.60
1250	8" thick, 13.50 bricks per S.F.		130	.308		8.85	13.20		22.05	31.50
1300	12" thick, 20.25 bricks per S.F.		90	.444		13.35	19.10		32.45	45.50
1350	16" thick, 27.00 bricks per S.F.	↓	70	.571	↓	18.05	24.50		42.55	60
9000	Minimum labor/equipment charge	D-1	2	8	Job		335		335	550

04 27 10.40 Steps

		Crew	Daily Output	Labor-Hours	Unit	Material	2016 Bare Costs Labor	Equipment	Total	Total Incl O&P
0010	**STEPS**									
0012	Entry steps, select common brick	D-1	.30	53.333	M	520	2,250		2,770	4,225

For customer support on your Commercial Renovation Cost Data, call 877.742.4719.

93

04 41 Dry-Placed Stone

04 41 10 – Dry Placed Stone

04 41 10.10 Rough Stone Wall

		Crew	Daily Output	Labor-Hours	Unit	Material	2016 Bare Costs Labor	2016 Bare Costs Equipment	Total	Total Incl O&P
0011	**ROUGH STONE WALL**, Dry									
0012	Dry laid (no mortar), under 18" thick	G D-1	60	.267	C.F.	12.65	11.25		23.90	32
0100	Random fieldstone, under 18" thick	G D-12	60	.533		12.65	22.50		35.15	51
0150	Over 18" thick	G "	63	.508		15.15	21.50		36.65	51.50
0500	Field stone veneer	G D-8	120	.333	S.F.	12.60	14.30		26.90	37.50
0510	Valley stone veneer	G	120	.333		12.60	14.30		26.90	37.50
0520	River stone veneer	G	120	.333		12.60	14.30		26.90	37.50
0600	Rubble stone walls, in mortar bed, up to 18" thick	G D-11	75	.320	C.F.	15.25	14.15		29.40	40
9000	Minimum labor/equipment charge	D-1	2	8	Job		335		335	550

04 43 Stone Masonry

04 43 10 – Masonry with Natural and Processed Stone

04 43 10.05 Ashlar Veneer

		Crew	Daily Output	Labor-Hours	Unit	Material	2016 Bare Costs Labor	2016 Bare Costs Equipment	Total	Total Incl O&P
0011	**ASHLAR VENEER** 4" + or - thk, random or random rectangular									
0150	Sawn face, split joints, low priced stone	D-8	140	.286	S.F.	11.95	12.25		24.20	33
0200	Medium priced stone		130	.308		13.55	13.20		26.75	36.50
0300	High priced stone		120	.333		18	14.30		32.30	43.50
0600	Seam face, split joints, medium price stone		125	.320		19	13.75		32.75	43.50
0700	High price stone		120	.333		18.75	14.30		33.05	44
1000	Split or rock face, split joints, medium price stone		125	.320		11.40	13.75		25.15	35
1100	High price stone		120	.333		17.25	14.30		31.55	42.50

04 43 10.10 Bluestone

		Crew	Daily Output	Labor-Hours	Unit	Material	2016 Bare Costs Labor	2016 Bare Costs Equipment	Total	Total Incl O&P
0010	**BLUESTONE**, cut to size									
0500	Sills, natural cleft, 10" wide to 6' long, 1-1/2" thick	D-11	70	.343	L.F.	13.35	15.15		28.50	39
0550	2" thick		63	.381		14.60	16.85		31.45	43.50
0600	Smooth finish, 1-1/2" thick		70	.343		12	15.15		27.15	37.50
0650	2" thick		63	.381		13	16.85		29.85	42
0800	Thermal finish, 1-1/2" thick		70	.343		12	15.15		27.15	37.50
0850	2" thick		63	.381		13	16.85		29.85	42
1000	Stair treads, natural cleft, 12" wide, 6' long, 1-1/2" thick	D-10	115	.278		12	12.85	4.17	29.02	39
1050	2" thick		105	.305		13	14.10	4.57	31.67	42
1100	Smooth finish, 1-1/2" thick		115	.278		12	12.85	4.17	29.02	39
1150	2" thick		105	.305		13	14.10	4.57	31.67	42
1300	Thermal finish, 1-1/2" thick		115	.278		12	12.85	4.17	29.02	39
1350	2" thick		105	.305		13	14.10	4.57	31.67	42
9000	Minimum labor/equipment charge	D-1	2.50	6.400	Job		270		270	440

04 43 10.45 Granite

		Crew	Daily Output	Labor-Hours	Unit	Material	2016 Bare Costs Labor	2016 Bare Costs Equipment	Total	Total Incl O&P
0010	**GRANITE**, cut to size									
0050	Veneer, polished face, 3/4" to 1-1/2" thick									
0150	Low price, gray, light gray, etc.	D-10	130	.246	S.F.	26.50	11.35	3.69	41.54	52
0180	Medium price, pink, brown, etc.		130	.246		29.50	11.35	3.69	44.54	55
0220	High price, red, black, etc.		130	.246		42	11.35	3.69	57.04	69
0300	1-1/2" to 2-1/2" thick, veneer									
0350	Low price, gray, light gray, etc.	D-10	130	.246	S.F.	28.50	11.35	3.69	43.54	54
0500	Medium price, pink, brown, etc.	"	130	.246	"	33.50	11.35	3.69	48.54	59.50
2500	Steps, copings, etc., finished on more than one surface									
2550	Low price, gray, light gray, etc.	D-10	50	.640	C.F.	91	29.50	9.60	130.10	158
2575	Medium price, pink, brown, etc.		50	.640		118	29.50	9.60	157.10	188
2600	High price, red, black, etc.		50	.640		146	29.50	9.60	185.10	218
9000	Minimum labor/equipment charge	D-1	2	8	Job		335		335	550

04 43 10 – Masonry with Natural and Processed Stone

04 43 10.55 Limestone

		Crew	Daily Output	Labor-Hours	Unit	Material	2016 Bare Costs Labor	2016 Bare Costs Equipment	Total	Total Incl O&P
0010	**LIMESTONE**, cut to size									
0020	Veneer facing panels									
0500	Texture finish, light stick, 4-1/2" thick, 5' x 12'	D-4	300	.107	S.F.	20.50	4.57	.45	25.52	30.50
0750	5" thick, 5' x 14' panels	D-10	275	.116		21.50	5.35	1.74	28.59	34.50
1000	Sugarcube finish, 2" thick, 3' x 5' panels		275	.116		29	5.35	1.74	36.09	42.50
1050	3" thick, 4' x 9' panels		275	.116		25.50	5.35	1.74	32.59	39
1200	4" thick, 5' x 11' panels		275	.116		31.50	5.35	1.74	38.59	45
1400	Sugarcube, textured finish, 4-1/2" thick, 5' x 12'		275	.116		32.50	5.35	1.74	39.59	46.50
1450	5" thick, 5' x 14' panels		275	.116	↓	34	5.35	1.74	41.09	47.50
2000	Coping, sugarcube finish, top & 2 sides		30	1.067	C.F.	68.50	49.50	16	134	173
2100	Sills, lintels, jambs, trim, stops, sugarcube finish, simple		20	1.600		68.50	74	24	166.50	221
2150	Detailed		20	1.600	↓	68.50	74	24	166.50	221
2300	Steps, extra hard, 14" wide, 6" rise	↓	50	.640	L.F.	25	29.50	9.60	64.10	85.50
3000	Quoins, plain finish, 6" x 12" x 12"	D-12	25	1.280	Ea.	41	54.50		95.50	135
3050	6" x 16" x 24"	"	25	1.280	"	55	54.50		109.50	150
9000	Minimum labor/equipment charge	D-1	2	8	Job		335		335	550

04 43 10.60 Marble

		Crew	Daily Output	Labor-Hours	Unit	Material	2016 Bare Costs Labor	2016 Bare Costs Equipment	Total	Total Incl O&P
0011	**MARBLE**, ashlar, split face, 4" + or - thick, random									
0040	Lengths 1' to 4' & heights 2" to 7-1/2", average	D-8	175	.229	S.F.	18.30	9.80		28.10	36
0100	Base, polished, 3/4" or 7/8" thick, polished, 6" high	D-10	65	.492	L.F.	12	22.50	7.35	41.85	58
0300	Carvings or bas-relief, from templates, simple design		80	.400	S.F.	149	18.50	6	173.50	201
0350	Intricate design	↓	80	.400	"	345	18.50	6	369.50	415
1000	Facing, polished finish, cut to size, 3/4" to 7/8" thick									
1050	Carrara or equal	D-10	130	.246	S.F.	22.50	11.35	3.69	37.54	47.50
1100	Arabescato or equal		130	.246		39.50	11.35	3.69	54.54	66
1300	1-1/4" thick, Botticino Classico or equal		125	.256		24	11.80	3.83	39.63	50
1350	Statuarietto or equal		125	.256		41.50	11.80	3.83	57.13	69
1500	2" thick, Crema Marfil or equal		120	.267		48.50	12.30	3.99	64.79	78
1550	Cafe Pinta or equal	↓	120	.267	↓	70	12.30	3.99	86.29	101
1700	Rubbed finish, cut to size, 4" thick									
1740	Average	D-10	100	.320	S.F.	41	14.80	4.79	60.59	75
1780	Maximum	"	100	.320	"	71.50	14.80	4.79	91.09	108
2200	Window sills, 6" x 3/4" thick	D-1	85	.188	L.F.	11.10	7.95		19.05	25
2500	Flooring, polished tiles, 12" x 12" x 3/8" thick									
2510	Thin set, Giallo Solare or equal	D-11	90	.267	S.F.	17.50	11.80		29.30	38.50
2600	Sky Blue or equal		90	.267		16	11.80		27.80	37
2700	Mortar bed, Giallo Solare or equal		65	.369		17.50	16.30		33.80	46
2740	Sky Blue or equal	↓	65	.369		16	16.30		32.30	44
2780	Travertine, 3/8" thick, Sierra or equal	D-10	130	.246		9.25	11.35	3.69	24.29	32.50
2790	Silver or equal	"	130	.246		25.50	11.35	3.69	40.54	50.50
2800	Patio tile, non-slip, 1/2" thick, flame finish	D-11	75	.320	↓	11.25	14.15		25.40	35.50
2900	Shower or toilet partitions, 7/8" thick partitions									
3050	3/4" or 1-1/4" thick stiles, polished 2 sides, average	D-11	75	.320	S.F.	46.50	14.15		60.65	74
3201	Soffits, add to above prices				"	20%	100%			
3210	Stairs, risers, 7/8" thick x 6" high	D-10	115	.278	L.F.	15.55	12.85	4.17	32.57	42.50
3360	Treads, 12" wide x 1-1/4" thick	"	115	.278	"	44	12.85	4.17	61.02	74
3500	Thresholds, 3' long, 7/8" thick, 4" to 5" wide, plain	D-12	24	1.333	Ea.	35.50	57		92.50	132
3550	Beveled		24	1.333	"	71.50	57		128.50	171
3700	Window stools, polished, 7/8" thick, 5" wide	↓	85	.376	L.F.	21.50	16.05		37.55	50
9000	Minimum labor/equipment charge	D-1	2	8	Job		335		335	550

For customer support on your Commercial Renovation Cost Data, call 877.742.4719.

95

04 43 10 – Masonry with Natural and Processed Stone

04 43 10.75 Sandstone or Brownstone	Crew	Daily Output	Labor-Hours	Unit	Material	2016 Bare Costs Labor	Equipment	Total	Total Incl O&P
0011 **SANDSTONE OR BROWNSTONE**									
0100 Sawed face veneer, 2-1/2" thick, to 2' x 4' panels	D-10	130	.246	S.F.	19.35	11.35	3.69	34.39	44
0150 4" thick, to 3'-6" x 8' panels		100	.320		19.35	14.80	4.79	38.94	51
0300 Split face, random sizes		100	.320		14.15	14.80	4.79	33.74	45
9000 Minimum labor/equipment charge	D-1	2.50	6.400	Job		270		270	440

04 43 10.80 Slate

	Crew	Daily Output	Labor-Hours	Unit	Material	Labor	Equipment	Total	Total Incl O&P
0010 **SLATE**									
0040 Pennsylvania - blue gray to black									
0050 Vermont - unfading green, mottled green & purple, gray & purple									
0100 Virginia - blue black									
3100 Stair landings, 1" thick, black, clear	D-1	65	.246	S.F.	21	10.35		31.35	40.50
3200 Ribbon	"	65	.246	"	23.50	10.35		33.85	42.50
3500 Stair treads, sand finish, 1" thick x 12" wide									
3550 Under 3 L.F.	D-10	85	.376	L.F.	23.50	17.40	5.65	46.55	59.50
3600 3 L.F. to 6 L.F.	"	120	.267	"	25	12.30	3.99	41.29	52
3700 Ribbon, sand finish, 1" thick x 12" wide									
3750 To 6 L.F.	D-10	120	.267	L.F.	21	12.30	3.99	37.29	48
4000 Stools or sills, sand finish, 1" thick, 6" wide	D-12	160	.200		12.20	8.55		20.75	27.50
4100 Honed finish		160	.200		11.65	8.55		20.20	26.50
4200 10" wide		90	.356		18.80	15.15		33.95	45
4250 Honed finish		90	.356		17.50	15.15		32.65	44
4400 2" thick, 6" wide		140	.229		19.60	9.75		29.35	37.50
4450 Honed finish		140	.229		18.65	9.75		28.40	36.50
4600 10" wide		90	.356		30.50	15.15		45.65	58.50
4650 Honed finish		90	.356		29	15.15		44.15	56.50
4800 For lengths over 3', add					25%				
9000 Minimum labor/equipment charge	D-1	2.50	6.400	Job		270		270	440

04 43 10.85 Window Sill

	Crew	Daily Output	Labor-Hours	Unit	Material	Labor	Equipment	Total	Total Incl O&P
0010 **WINDOW SILL**									
0020 Bluestone, thermal top, 10" wide, 1-1/2" thick	D-1	85	.188	S.F.	10	7.95		17.95	24
0050 2" thick		75	.213	"	10	9		19	25.50
0100 Cut stone, 5" x 8" plain		48	.333	L.F.	12.55	14.05		26.60	37
0200 Face brick on edge, brick, 8" wide		80	.200		5.25	8.45		13.70	19.50
0400 Marble, 9" wide, 1" thick		85	.188		8.95	7.95		16.90	23
0900 Slate, colored, unfading, honed, 12" wide, 1" thick		85	.188		8.50	7.95		16.45	22.50
0950 2" thick		70	.229		8.50	9.65		18.15	25
9000 Minimum labor/equipment charge	1 Bric	2	4	Job		185		185	300

04 54 Refractory Brick Masonry

04 54 10 – Refractory Brick Work

04 54 10.10 Fire Brick

	Crew	Daily Output	Labor-Hours	Unit	Material	Labor	Equipment	Total	Total Incl O&P
0010 **FIRE BRICK**									
0012 Low duty, 2000°F, 9" x 2-1/2" x 4-1/2"	D-1	.60	26.667	M	1,775	1,125		2,900	3,775
0050 High duty, 3000°F	"	.60	26.667	"	2,650	1,125		3,775	4,750

04 54 10.20 Fire Clay

	Crew	Daily Output	Labor-Hours	Unit	Material	Labor	Equipment	Total	Total Incl O&P
0010 **FIRE CLAY**									
0020 Gray, high duty, 100 lb. bag				Bag	35			35	38.50
0050 100 lb. drum, premixed (400 brick per drum)				Drum	41			41	45

04 57 Masonry Fireplaces

04 57 10 – Brick or Stone Fireplaces

04 57 10.10 Fireplace

		Crew	Daily Output	Labor-Hours	Unit	Material	2016 Bare Costs Labor	Equipment	Total	Total Incl O&P
0010	**FIREPLACE**									
0100	Brick fireplace, not incl. foundations or chimneys									
0110	30" x 29" opening, incl. chamber, plain brickwork	D-1	.40	40	Ea.	570	1,675		2,245	3,375
0200	Fireplace box only (110 brick)	"	2	8	"	162	335		497	730
0300	For elaborate brickwork and details, add					35%	35%			
0400	For hearth, brick & stone, add	D-1	2	8	Ea.	212	335		547	785
0410	For steel, damper, cleanouts, add		4	4		18	169		187	294
0600	Plain brickwork, incl. metal circulator		.50	32		900	1,350		2,250	3,200
0800	Face brick only, standard size, 8" x 2-2/3" x 4"		.30	53.333	M	600	2,250		2,850	4,300
0900	Stone fireplace, fieldstone, add				SF Face	8.50			8.50	9.35
1000	Cut stone, add				"	8			8	8.80
9000	Minimum labor/equipment charge	D-1	2	8	Job		335		335	550

04 71 Manufactured Brick Masonry

04 71 10 – Simulated or Manufactured Brick

04 71 10.10 Simulated Brick

		Crew	Daily Output	Labor-Hours	Unit	Material	2016 Bare Costs Labor	Equipment	Total	Total Incl O&P
0010	**SIMULATED BRICK**									
0020	Aluminum, baked on colors	1 Carp	200	.040	S.F.	4.43	1.94		6.37	8.05
0050	Fiberglass panels		200	.040		8.05	1.94		9.99	12
0100	Urethane pieces cemented in mastic		150	.053		8.60	2.58		11.18	13.65
0150	Vinyl siding panels		200	.040		10.50	1.94		12.44	14.70

04 72 Cast Stone Masonry

04 72 10 – Cast Stone Masonry Features

04 72 10.10 Coping

		Crew	Daily Output	Labor-Hours	Unit	Material	2016 Bare Costs Labor	Equipment	Total	Total Incl O&P
0010	**COPING**, stock units									
0050	Precast concrete, 10" wide, 4" tapers to 3-1/2", 8" wall	D-1	75	.213	L.F.	17.45	9		26.45	34
0100	12" wide, 3-1/2" tapers to 3", 10" wall		70	.229		18.80	9.65		28.45	36
0110	14" wide, 4" tapers to 3-1/2", 12" wall		65	.246		21.50	10.35		31.85	40.50
0150	16" wide, 4" tapers to 3-1/2", 14" wall		60	.267		23	11.25		34.25	44
0300	Limestone for 12" wall, 4" thick		90	.178		14.70	7.50		22.20	28.50
0350	6" thick		80	.200		22	8.45		30.45	38
0500	Marble, to 4" thick, no wash, 9" wide		90	.178		12.85	7.50		20.35	26.50
0550	12" wide		80	.200		18	8.45		26.45	33.50
0700	Terra cotta, 9" wide		90	.178		6.45	7.50		13.95	19.30
0750	12" wide		80	.200		8.80	8.45		17.25	23.50
0800	Aluminum, for 12" wall		80	.200		8.90	8.45		17.35	23.50
9000	Minimum labor/equipment charge		2	8	Job		335		335	550

04 72 20 – Cultured Stone Veneer

04 72 20.10 Cultured Stone Veneer Components

		Crew	Daily Output	Labor-Hours	Unit	Material	2016 Bare Costs Labor	Equipment	Total	Total Incl O&P
0010	**CULTURED STONE VENEER COMPONENTS**									
0110	On wood frame and sheathing substrate, random sized cobbles, corner stones	D-8	70	.571	V.L.F.	10.25	24.50		34.75	51.50
0120	Field stones		140	.286	S.F.	7.45	12.25		19.70	28
0130	Random sized flats, corner stones		70	.571	V.L.F.	10.55	24.50		35.05	51.50
0140	Field stones		140	.286	S.F.	8.75	12.25		21	29.50
0150	Horizontal lined ledgestones, corner stones		75	.533	V.L.F.	10.25	23		33.25	49
0160	Field stones		150	.267	S.F.	7.45	11.45		18.90	27
0170	Random shaped flats, corner stones		65	.615	V.L.F.	10.25	26.50		36.75	54.50
0180	Field stones		150	.267	S.F.	7.45	11.45		18.90	27

For customer support on your Commercial Renovation Cost Data, call 877.742.4719.

97

04 72 Cast Stone Masonry

04 72 20 – Cultured Stone Veneer

04 72 20.10 Cultured Stone Veneer Components	Crew	Daily Output	Labor-Hours	Unit	Material	2016 Bare Costs Labor	Equipment	Total	Total Incl O&P
0190	Random shaped/textured face, corner stones	D-8	65	.615	V.L.F.	10.25	26.50		36.75	54.50
0200	Field stones		130	.308	S.F.	7.45	13.20		20.65	29.50
0210	Random shaped river rock, corner stones		65	.615	V.L.F.	10.25	26.50		36.75	54.50
0220	Field stones		130	.308	S.F.	7.45	13.20		20.65	29.50
0240	On concrete or CMU substrate, random sized cobbles, corner stones		70	.571	V.L.F.	9.60	24.50		34.10	50.50
0250	Field stones		140	.286	S.F.	7.10	12.25		19.35	28
0260	Random sized flats, corner stones		70	.571	V.L.F.	9.90	24.50		34.40	51
0270	Field stones		140	.286	S.F.	8.45	12.25		20.70	29.50
0280	Horizontal lined ledgestones, corner stones		75	.533	V.L.F.	9.60	23		32.60	48
0290	Field stones		150	.267	S.F.	7.10	11.45		18.55	26.50
0300	Random shaped flats, corner stones		70	.571	V.L.F.	9.60	24.50		34.10	50.50
0310	Field stones		140	.286	S.F.	7.10	12.25		19.35	28
0320	Random shaped/textured face, corner stones		65	.615	V.L.F.	9.60	26.50		36.10	53.50
0330	Field stones		130	.308	S.F.	7.10	13.20		20.30	29.50
0340	Random shaped river rock, corner stones		65	.615	V.L.F.	9.60	26.50		36.10	53.50
0350	Field stones	▼	130	.308	S.F.	7.10	13.20		20.30	29.50
0360	Cultured stone veneer, #15 felt weather resistant barrier	1 Clab	3700	.002	Sq.	5.30	.08		5.38	6
0370	Expanded metal lath, diamond, 2.5 lb./S.Y., galvanized	1 Lath	85	.094	S.Y.	2.82	4.45		7.27	10.10
0390	Water table or window sill, 18" long	1 Bric	80	.100	Ea.	9.80	4.63		14.43	18.35

04 73 Manufactured Stone Masonry

04 73 20 – Simulated or Manufactured Stone

04 73 20.10 Simulated Stone

		Crew	Daily Output	Labor-Hours	Unit	Material	2016 Bare Costs Labor	Equipment	Total	Total Incl O&P
0010	**SIMULATED STONE**									
0100	Insulated fiberglass panels, 5/8" ply backer	L-4	200	.120	S.F.	12.05	5.45		17.50	22

Estimating Tips
05 05 00 Common Work Results for Metals

- Nuts, bolts, washers, connection angles, and plates can add a significant amount to both the tonnage of a structural steel job and the estimated cost. As a rule of thumb, add 10% to the total weight to account for these accessories.

- Type 2 steel construction, commonly referred to as "simple construction," consists generally of field-bolted connections with lateral bracing supplied by other elements of the building, such as masonry walls or x-bracing. The estimator should be aware, however, that shop connections may be accomplished by welding or bolting. The method may be particular to the fabrication shop and may have an impact on the estimated cost.

05 10 00 Structural Steel

- Steel items can be obtained from two sources: a fabrication shop or a metals service center. Fabrication shops can fabricate items under more controlled conditions than can crews in the field. They are also more efficient and can produce items more economically. Metal service centers serve as a source of long mill shapes to both fabrication shops and contractors.

- Most line items in this structural steel subdivision, and most items in 05 50 00 Metal Fabrications, are indicated as being shop fabricated. The bare material cost for these shop fabricated items is the "Invoice Cost" from the shop and includes the mill base price of steel plus mill extras, transportation to the shop, shop drawings and detailing where warranted, shop fabrication and handling, sandblasting and a shop coat of primer paint, all necessary structural bolts, and delivery to the job site. The bare labor cost

and bare equipment cost for these shop fabricated items is for field installation or erection.

- Line items in Subdivision 05 12 23.40 Lightweight Framing, and other items scattered in Division 5, are indicated as being field fabricated. The bare material cost for these field fabricated items is the "Invoice Cost" from the metals service center and includes the mill base price of steel plus mill extras, transportation to the metals service center, material handling, and delivery of long lengths of mill shapes to the job site. Material costs for structural bolts and welding rods should be added to the estimate. The bare labor cost and bare equipment cost for these items is for both field fabrication and field installation or erection, and include time for cutting, welding and drilling in the fabricated metal items. Drilling into concrete and fasteners to fasten field fabricated items to other work are not included and should be added to the estimate.

05 20 00 Steel Joist Framing

- In any given project the total weight of open web steel joists is determined by the loads to be supported and the design. However, economies can be realized in minimizing the amount of labor used to place the joists. This is done by maximizing the joist spacing, and therefore minimizing the number of joists required to be installed on the job. Certain spacings and locations may be required by the design, but in other cases maximizing the spacing and keeping it as uniform as possible will keep the costs down.

05 30 00 Steel Decking

- The takeoff and estimating of metal deck involves more than simply the area of the floor or roof and the type of deck specified or shown on the drawings. Many different sizes and types of openings may

exist. Small openings for individual pipes or conduits may be drilled after the floor/roof is installed, but larger openings may require special deck lengths as well as reinforcing or structural support. The estimator should determine who will be supplying this reinforcing. Additionally, some deck terminations are part of the deck package, such as screed angles and pour stops, and others will be part of the steel contract, such as angles attached to structural members and cast-in-place angles and plates. The estimator must ensure that all pieces are accounted for in the complete estimate.

05 50 00 Metal Fabrications

- The most economical steel stairs are those that use common materials, standard details, and most importantly, a uniform and relatively simple method of field assembly. Commonly available A36/A992 channels and plates are very good choices for the main stringers of the stairs, as are angles and tees for the carrier members. Risers and treads are usually made by specialty shops, and it is most economical to use a typical detail in as many places as possible. The stairs should be pre-assembled and shipped directly to the site. The field connections should be simple and straightforward to be accomplished efficiently, and with minimum equipment and labor.

Reference Numbers

Reference numbers are shown at the beginning of some major classifications. These numbers refer to related items in the Reference Section. The reference information may be an estimating procedure, an alternate pricing method, or technical information.

Note: Not all subdivisions listed here necessarily appear. ■

05 05 Common Work Results for Metals

05 05 05 – Selective Demolition for Metals

05 05 05.10 Selective Demolition, Metals

05 05 05.10 Selective Demolition, Metals	Crew	Daily Output	Labor-Hours	Unit	Material	2016 Bare Costs Labor	Equipment	Total	Total Incl O&P
0010 **SELECTIVE DEMOLITION, METALS** R024119-10									
0015 Excludes shores, bracing, cutting, loading, hauling, dumping									
0020 Remove nuts only up to 3/4" diameter	1 Sswk	480	.017	Ea.		.89		.89	1.56
0030 7/8" to 1-1/4" diameter		240	.033			1.77		1.77	3.13
0040 1-3/8" to 2" diameter		160	.050			2.66		2.66	4.69
0060 Unbolt and remove structural bolts up to 3/4" diameter		240	.033			1.77		1.77	3.13
0070 7/8" to 2" diameter		160	.050			2.66		2.66	4.69
0140 Light weight framing members, remove whole or cut up, up to 20 lb.		240	.033			1.77		1.77	3.13
0150 21 – 40 lb.	2 Sswk	210	.076			4.05		4.05	7.15
0160 41 – 80 lb.	3 Sswk	180	.133			7.10		7.10	12.50
0170 81 – 120 lb.	4 Sswk	150	.213			11.35		11.35	20
0230 Structural members, remove whole or cut up, up to 500 lb.	E-19	48	.500			26.50	20	46.50	67
0240 1/4 – 2 tons	E-18	36	1.111			59	27	86	132
0250 2 – 5 tons	E-24	30	1.067			56	25	81	124
0260 5 – 10 tons	E-20	24	2.667			140	49.50	189.50	295
0270 10 – 15 tons	E-2	18	3.111			163	84.50	247.50	370
0340 Fabricated item, remove whole or cut up, up to 20 lb.	1 Sswk	96	.083			4.43		4.43	7.80
0350 21 – 40 lb.	2 Sswk	84	.190			10.15		10.15	17.90
0360 41 – 80 lb.	3 Sswk	72	.333			17.75		17.75	31.50
0370 81 – 120 lb.	4 Sswk	60	.533			28.50		28.50	50
0380 121 – 500 lb.	E-19	48	.500			26.50	20	46.50	67
0390 501 – 1000 lb.	"	36	.667			35	27	62	89.50
0500 Steel roof decking, uncovered, bare	B-2	5000	.008	S.F.		.31		.31	.50
2950 Minimum labor/equipment charge	E-19	2	12	Job		630	485	1,115	1,600

05 05 19 – Post-Installed Concrete Anchors

05 05 19.10 Chemical Anchors

	Crew	Daily Output	Labor-Hours	Unit	Material	Labor	Equipment	Total	Total Incl O&P
0010 **CHEMICAL ANCHORS**									
0020 Includes layout & drilling									
1430 Chemical anchor, w/rod & epoxy cartridge, 3/4" diam. x 9-1/2" long	B-89A	27	.593	Ea.	9.25	26	4.33	39.58	57.50
1435 1" diameter x 11-3/4" long		24	.667		16.95	29.50	4.87	51.32	71.50
1440 1-1/4" diameter x 14" long		21	.762		34	33.50	5.55	73.05	98
1445 1-3/4" diameter x 15" long		20	.800		63.50	35	5.85	104.35	133
1450 18" long		17	.941		76.50	41.50	6.90	124.90	159
1455 2" diameter x 18" long		16	1		101	44	7.30	152.30	191
1460 24" long		15	1.067		132	47	7.80	186.80	230

05 05 19.20 Expansion Anchors

		Crew	Daily Output	Labor-Hours	Unit	Material	Labor	Equipment	Total	Total Incl O&P
0010 **EXPANSION ANCHORS**										
0100 Anchors for concrete, brick or stone, no layout and drilling										
0200 Expansion shields, zinc, 1/4" diameter, 1-5/16" long, single	G	1 Carp	90	.089	Ea.	.44	4.31		4.75	7.55
0300 1-3/8" long, double	G		85	.094		.55	4.56		5.11	8.05
0400 3/8" diameter, 1-1/2" long, single	G		85	.094		.66	4.56		5.22	8.20
0500 2" long, double	G		80	.100		1.20	4.85		6.05	9.20
0600 1/2" diameter, 2-1/16" long, single	G		80	.100		1.21	4.85		6.06	9.25
0700 2-1/2" long, double	G		75	.107		1.94	5.15		7.09	10.60
0800 5/8" diameter, 2-5/8" long, single	G		75	.107		2.01	5.15		7.16	10.65
0900 2-3/4" long, double	G		70	.114		2.74	5.55		8.29	12.05
1000 3/4" diameter, 2-3/4" long, single	G		70	.114		3.18	5.55		8.73	12.55
1100 3-15/16" long, double	G		65	.123		5.10	5.95		11.05	15.35
2100 Hollow wall anchors for gypsum wall board, plaster or tile										
2300 1/8" diameter, short	G	1 Carp	160	.050	Ea.	.25	2.42		2.67	4.24
2400 Long	G		150	.053		.25	2.58		2.83	4.50
2500 3/16" diameter, short	G		150	.053		.46	2.58		3.04	4.73

05 05 19 – Post-Installed Concrete Anchors

05 05 19.20 Expansion Anchors

05 05 19.20 Expansion Anchors		Crew	Daily Output	Labor-Hours	Unit	Material	2016 Bare Costs Labor	Equipment	Total	Total Incl O&P
2600	Long	G	1 Carp	140	.057	Ea.	.64	2.77		3.41	5.20
2700	1/4" diameter, short	G		140	.057		.64	2.77		3.41	5.20
2800	Long	G		130	.062		.70	2.98		3.68	5.65
3000	Toggle bolts, bright steel, 1/8" diameter, 2" long	G		85	.094		.21	4.56		4.77	7.70
3100	4" long	G		80	.100		.27	4.85		5.12	8.20
3200	3/16" diameter, 3" long	G		80	.100		.28	4.85		5.13	8.20
3300	6" long	G		75	.107		.39	5.15		5.54	8.90
3400	1/4" diameter, 3" long	G		75	.107		.38	5.15		5.53	8.85
3500	6" long	G		70	.114		.55	5.55		6.10	9.65
3600	3/8" diameter, 3" long	G		70	.114		.93	5.55		6.48	10.05
3700	6" long	G		60	.133		1.43	6.45		7.88	12.10
3800	1/2" diameter, 4" long	G		60	.133		2	6.45		8.45	12.75
3900	6" long	G		50	.160		2.59	7.75		10.34	15.50
4000	Nailing anchors										
4100	Nylon nailing anchor, 1/4" diameter, 1" long		1 Carp	3.20	2.500	C	22	121		143	222
4200	1-1/2" long			2.80	2.857		25.50	138		163.50	254
4300	2" long			2.40	3.333		30	162		192	297
4400	Metal nailing anchor, 1/4" diameter, 1" long	G		3.20	2.500		18.60	121		139.60	219
4500	1-1/2" long	G		2.80	2.857		24.50	138		162.50	253
4600	2" long	G		2.40	3.333		29	162		191	296
8000	Wedge anchors, not including layout or drilling										
8050	Carbon steel, 1/4" diameter, 1-3/4" long	G	1 Carp	150	.053	Ea.	.43	2.58		3.01	4.69
8100	3-1/4" long	G		140	.057		.56	2.77		3.33	5.15
8150	3/8" diameter, 2-1/4" long	G		145	.055		.52	2.67		3.19	4.94
8200	5" long	G		140	.057		.91	2.77		3.68	5.50
8250	1/2" diameter, 2-3/4" long	G		140	.057		1.02	2.77		3.79	5.65
8300	7" long	G		125	.064		1.75	3.10		4.85	6.95
8350	5/8" diameter, 3-1/2" long	G		130	.062		1.92	2.98		4.90	7
8400	8-1/2" long	G		115	.070		4.08	3.37		7.45	10
8450	3/4" diameter, 4-1/4" long	G		115	.070		3.01	3.37		6.38	8.80
8500	10" long	G		95	.084		6.85	4.08		10.93	14.15
8550	1" diameter, 6" long	G		100	.080		9.55	3.88		13.43	16.85
8575	9" long	G		85	.094		12.40	4.56		16.96	21
8600	12" long	G		75	.107		13.35	5.15		18.50	23
8650	1-1/4" diameter, 9" long	G		70	.114		25	5.55		30.55	37
8700	12" long	G		60	.133		32.50	6.45		38.95	46
8750	For type 303 stainless steel, add						350%				
8800	For type 316 stainless steel, add						450%				
8950	Self-drilling concrete screw, hex washer head, 3/16" diam. x 1-3/4" long	G	1 Carp	300	.027	Ea.	.21	1.29		1.50	2.34
8960	2-1/4" long	G		250	.032		.24	1.55		1.79	2.79
8970	Phillips flat head, 3/16" diam. x 1-3/4" long	G		300	.027		.21	1.29		1.50	2.34
8980	2-1/4" long	G		250	.032		.24	1.55		1.79	2.79
9000	Minimum labor/equipment charge			4	2	Job		97		97	158

05 05 21 – Fastening Methods for Metal

05 05 21.10 Cutting Steel

		Crew	Daily Output	Labor-Hours	Unit	Material	2016 Bare Costs Labor	Equipment	Total	Total Incl O&P
0010	**CUTTING STEEL**										
0020	Hand burning, incl. preparation, torch cutting & grinding, no staging										
0050	Steel to 1/4" thick	E-25	400	.020	L.F.	.19	1.10	.03	1.32	2.18
0100	1/2" thick		320	.025		.35	1.38	.04	1.77	2.86
0150	3/4" thick		260	.031		.58	1.70	.04	2.32	3.69
0200	1" thick		200	.040		.84	2.21	.06	3.11	4.87
9000	Minimum labor/equipment charge		2	4	Job		221	5.70	226.70	395

For customer support on your Commercial Renovation Cost Data, call 877.742.4719.

101

05 05 21 – Fastening Methods for Metal

05 05 21.15 Drilling Steel	Crew	Daily Output	Labor-Hours	Unit	Material	2016 Bare Costs Labor	Equipment	Total	Total Incl O&P
0010 **DRILLING STEEL**									
1910 Drilling & layout for steel, up to 1/4" deep, no anchor									
1920 Holes, 1/4" diameter	1 Sswk	112	.071	Ea.	.10	3.80		3.90	6.80
1925 For each additional 1/4" depth, add		336	.024		.10	1.27		1.37	2.34
1930 3/8" diameter		104	.077		.09	4.09		4.18	7.30
1935 For each additional 1/4" depth, add		312	.026		.09	1.36		1.45	2.51
1940 1/2" diameter		96	.083		.11	4.43		4.54	7.95
1945 For each additional 1/4" depth, add		288	.028		.11	1.48		1.59	2.74
1950 5/8" diameter		88	.091		.16	4.84		5	8.75
1955 For each additional 1/4" depth, add		264	.030		.16	1.61		1.77	3.02
1960 3/4" diameter		80	.100		.20	5.30		5.50	9.60
1965 For each additional 1/4" depth, add		240	.033		.20	1.77		1.97	3.35
1970 7/8" diameter		72	.111		.26	5.90		6.16	10.75
1975 For each additional 1/4" depth, add		216	.037		.26	1.97		2.23	3.77
1980 1" diameter		64	.125		.23	6.65		6.88	12
1985 For each additional 1/4" depth, add		192	.042		.23	2.22		2.45	4.16
1990 For drilling up, add						40%			
2000 Minimum labor/equipment charge	1 Carp	4	2	Job		97		97	158

05 05 21.90 Welding Steel

	Crew	Daily Output	Labor-Hours	Unit	Material	2016 Bare Costs Labor	Equipment	Total	Total Incl O&P
0010 **WELDING STEEL**, Structural R050521-20									
0020 Field welding, 1/8" E6011, cost per welder, no operating engineer	E-14	8	1	Hr.	4.46	55	18.40	77.86	122
0200 With 1/2 operating engineer	E-13	8	1.500		4.46	80	18.40	102.86	161
0300 With 1 operating engineer	E-12	8	2		4.46	104	18.40	126.86	200
0500 With no operating engineer, 2# weld rod per ton	E-14	8	1	Ton	4.46	55	18.40	77.86	122
0600 8# E6011 per ton	"	2	4		17.85	221	73.50	312.35	490
0800 With one operating engineer per welder, 2# E6011 per ton	E-12	8	2		4.46	104	18.40	126.86	200
0900 8# E6011 per ton	"	2	8		17.85	415	73.50	506.35	800
1200 Continuous fillet, down welding									
1300 Single pass, 1/8" thick, 0.1#/L.F.	E-14	150	.053	L.F.	.22	2.94	.98	4.14	6.55
1400 3/16" thick, 0.2#/L.F.		75	.107		.45	5.90	1.96	8.31	13.05
1500 1/4" thick, 0.3#/L.F.		50	.160		.67	8.85	2.94	12.46	19.60
1610 5/16" thick, 0.4#/L.F.		38	.211		.89	11.60	3.87	16.36	25.50
1800 3 passes, 3/8" thick, 0.5#/L.F.		30	.267		1.12	14.70	4.91	20.73	32.50
2010 4 passes, 1/2" thick, 0.7#/L.F.		22	.364		1.56	20	6.70	28.26	44.50
2200 5 to 6 passes, 3/4" thick, 1.3#/L.F.		12	.667		2.90	37	12.25	52.15	81.50
2400 8 to 11 passes, 1" thick, 2.4#/L.F.		6	1.333		5.35	73.50	24.50	103.35	163
2600 For vertical joint welding, add						20%			
2700 Overhead joint welding, add						300%			
2900 For semi-automatic welding, obstructed joints, deduct						5%			
3000 Exposed joints, deduct						15%			
4000 Cleaning and welding plates, bars, or rods									
4010 to existing beams, columns, or trusses	E-14	12	.667	L.F.	1.12	37	12.25	50.37	79.50
9000 Minimum labor/equipment charge	"	4	2	Job		110	37	147	236

05 05 23 – Metal Fastenings

05 05 23.30 Lag Screws

		Crew	Daily Output	Labor-Hours	Unit	Material	2016 Bare Costs Labor	Equipment	Total	Total Incl O&P
0010 **LAG SCREWS**										
0020 Steel, 1/4" diameter, 2" long	G	1 Carp	200	.040	Ea.	.09	1.94		2.03	3.27
0200 1/2" diameter, 3" long	G		130	.062		.68	2.98		3.66	5.60
0300 5/8" diameter, 3" long	G		120	.067		1.22	3.23		4.45	6.65

05 05 Common Work Results for Metals

05 05 23 – Metal Fastenings

05 05 23.50 Powder Actuated Tools and Fasteners		Crew	Daily Output	Labor-Hours	Unit	Material	2016 Bare Costs Labor	Equipment	Total	Total Incl O&P
0010	**POWDER ACTUATED TOOLS & FASTENERS**									
0020	Stud driver, .22 caliber, single shot				Ea.	149			149	164
0100	.27 caliber, semi automatic, strip				"	450			450	495
0300	Powder load, single shot, .22 cal, power level 2, brown				C	5.50			5.50	6.05
0400	Strip, .27 cal, power level 4, red					7.90			7.90	8.70
0600	Drive pin, .300 x 3/4" long	G	1 Carp	4.80	1.667		4.23	81		85.23	137
0700	.300 x 3" long with washer	G	"	4	2		13	97		110	172

05 05 23.70 Structural Blind Bolts

		Crew	Daily Output	Labor-Hours	Unit	Material	Labor	Equipment	Total	Total Incl O&P
0010	**STRUCTURAL BLIND BOLTS**									
0100	1/4" diameter x 1/4" grip	G	1 Sswk	240	.033	Ea.	1.44	1.77		3.21	4.71
0150	1/2" grip	G		216	.037		1.55	1.97		3.52	5.20
0200	3/8" diameter x 1/2" grip	G		232	.034		2.04	1.83		3.87	5.50
0250	3/4" grip	G		208	.038		2.14	2.05		4.19	5.95
0300	1/2" diameter x 1/2" grip	G		224	.036		4.66	1.90		6.56	8.50
0350	3/4" grip	G		200	.040		6.50	2.13		8.63	10.90
0400	5/8" diameter x 3/4" grip	G		216	.037		9.60	1.97		11.57	14.10
0450	1" grip	G		192	.042		11.05	2.22		13.27	16.05

05 05 23.87 Weld Studs

		Crew	Daily Output	Labor-Hours	Unit	Material	Labor	Equipment	Total	Total Incl O&P
0010	**WELD STUDS**									
0020	1/4" diameter, 2-11/16" long	G	E-10	1120	.014	Ea.	.35	.77	.41	1.53	2.21
0100	4-1/8" long	G		1080	.015		.33	.80	.42	1.55	2.24
0200	3/8" diameter, 4-1/8" long	G		1080	.015		.38	.80	.42	1.60	2.30
0300	6-1/8" long	G		1040	.015		.49	.83	.44	1.76	2.49
9000	Minimum labor/equipment charge		1 Sswk	2	4	Job		213		213	375

05 12 Structural Steel Framing

05 12 23 – Structural Steel for Buildings

05 12 23.05 Canopy Framing

		Crew	Daily Output	Labor-Hours	Unit	Material	Labor	Equipment	Total	Total Incl O&P
0010	**CANOPY FRAMING**									
0020	6" and 8" members, shop fabricated	G	E-4	3000	.011	Lb.	1.59	.57	.05	2.21	2.81
9000	Minimum labor/equipment charge		1 Sswk	1	8	Job		425		425	750

05 12 23.10 Ceiling Supports

		Crew	Daily Output	Labor-Hours	Unit	Material	Labor	Equipment	Total	Total Incl O&P
0010	**CEILING SUPPORTS**									
1000	Entrance door/folding partition supports, shop fabricated	G	E-4	60	.533	L.F.	26.50	28.50	2.45	57.45	82
1100	Linear accelerator door supports	G		14	2.286		121	123	10.50	254.50	360
1200	Lintels or shelf angles, hung, exterior hot dipped galv.	G		267	.120		18.10	6.45	.55	25.10	32
1250	Two coats primer paint instead of galv.	G		267	.120		15.65	6.45	.55	22.65	29
1400	Monitor support, ceiling hung, expansion bolted	G		4	8	Ea.	420	430	37	887	1,250
1450	Hung from pre-set inserts	G		6	5.333		450	286	24.50	760.50	1,025
1600	Motor supports for overhead doors	G		4	8		214	430	37	681	1,025
1700	Partition support for heavy folding partitions, without pocket	G		24	1.333	L.F.	60.50	71.50	6.15	138.15	199
1750	Supports at pocket only	G		12	2.667		121	143	12.25	276.25	400
2000	Rolling grilles & fire door supports	G		34	.941		51.50	50.50	4.33	106.33	151
2100	Spider-leg light supports, expansion bolted to ceiling slab	G		8	4	Ea.	172	215	18.40	405.40	590
2150	Hung from pre-set inserts	G		12	2.667	"	186	143	12.25	341.25	470
2400	Toilet partition support	G		36	.889	L.F.	60.50	47.50	4.09	112.09	155
2500	X-ray travel gantry support	G		12	2.667	"	207	143	12.25	362.25	495

For customer support on your Commercial Renovation Cost Data, call 877.742.4719.

103

05 12 23 – Structural Steel for Buildings

05 12 23.17 Columns, Structural

		Crew	Daily Output	Labor-Hours	Unit	Material	2016 Bare Costs Labor	2016 Bare Costs Equipment	Total	Total Incl O&P
0010	**COLUMNS, STRUCTURAL** R051223-10									
0015	Made from recycled materials									
0020	Shop fab'd for 100-ton, 1-2 story project, bolted connections									
0800	Steel, concrete filled, extra strong pipe, 3-1/2" diameter	E-2	660	.085	L.F.	43.50	4.44	2.30	50.24	58
0830	4" diameter		780	.072		48.50	3.76	1.95	54.21	62
0890	5" diameter		1020	.055		58	2.87	1.49	62.36	70.50
0930	6" diameter		1200	.047		77	2.44	1.27	80.71	90
0940	8" diameter		1100	.051		77	2.66	1.38	81.04	90.50
1500	Steel pipe, extra strong, no concrete, 3" to 5" diameter	G	16000	.004	Lb.	1.33	.18	.10	1.61	1.87
3300	Structural tubing, square, A500GrB, 4" to 6" square, light section	G	11270	.005	"	1.33	.26	.13	1.72	2.06
8090	For projects 75 to 99 tons, add				%	10%				
8092	50 to 74 tons, add					20%				
8094	25 to 49 tons, add					30%	10%			
8096	10 to 24 tons, add					50%	25%			
8098	2 to 9 tons, add					75%	50%			
8099	Less than 2 tons, add					100%	100%			
9000	Minimum labor/equipment charge	1 Sswk	1	8	Job		425		425	750

05 12 23.20 Curb Edging

		Crew	Daily Output	Labor-Hours	Unit	Material	2016 Bare Costs Labor	2016 Bare Costs Equipment	Total	Total Incl O&P
0010	**CURB EDGING**									
0020	Steel angle w/anchors, shop fabricated, on forms, 1" x 1", 0.8#/L.F.	G E-4	350	.091	L.F.	1.69	4.91	.42	7.02	10.95
0300	4" x 4" angles, 8.2#/L.F.	G	275	.116	"	13.95	6.25	.54	20.74	27
9000	Minimum labor/equipment charge		4	8	Job		430	37	467	800

05 12 23.40 Lightweight Framing

		Crew	Daily Output	Labor-Hours	Unit	Material	2016 Bare Costs Labor	2016 Bare Costs Equipment	Total	Total Incl O&P
0010	**LIGHTWEIGHT FRAMING**									
0015	Made from recycled materials									
0200	For load-bearing steel studs see Section 05 41 13.30									
0400	Angle framing, field fabricated, 4" and larger	G E-3	440	.055	Lb.	.77	2.94	.33	4.04	6.40
0450	Less than 4" angles	G	265	.091		.80	4.88	.56	6.24	10.10
0600	Channel framing, field fabricated, 8" and larger	G	500	.048		.80	2.59	.29	3.68	5.75
0650	Less than 8" channels	G	335	.072		.80	3.86	.44	5.10	8.15
1000	Continuous slotted channel framing system, shop fab, simple framing	G 2 Sswk	2400	.007		4.11	.35		4.46	5.15
1200	Complex framing	G "	1600	.010		4.64	.53		5.17	6.05
1250	Plate & bar stock for reinforcing beams and trusses	G				1.46			1.46	1.60
1300	Cross bracing, rods, shop fabricated, 3/4" diameter	G E-3	700	.034		1.59	1.85	.21	3.65	5.25
1310	7/8" diameter	G	850	.028		1.59	1.52	.17	3.28	4.62
1320	1" diameter	G	1000	.024		1.59	1.29	.15	3.03	4.19
1330	Angle, 5" x 5" x 3/8"	G	2800	.009		1.59	.46	.05	2.10	2.62
1350	Hanging lintels, shop fabricated	G	850	.028		1.59	1.52	.17	3.28	4.62
1380	Roof frames, shop fabricated, 3'-0" square, 5' span	G E-2	4200	.013		1.59	.70	.36	2.65	3.35
1400	Tie rod, not upset, 1-1/2" to 4" diameter, with turnbuckle	G 2 Sswk	800	.020		1.72	1.06		2.78	3.77
1420	No turnbuckle	G	700	.023		1.66	1.22		2.88	3.97
1500	Upset, 1-3/4" to 4" diameter, with turnbuckle	G	800	.020		1.72	1.06		2.78	3.77
1520	No turnbuckle	G	700	.023		1.66	1.22		2.88	3.97
9000	Minimum labor/equipment charge		2	8	Job		425		425	750

05 12 23.45 Lintels

		Crew	Daily Output	Labor-Hours	Unit	Material	2016 Bare Costs Labor	2016 Bare Costs Equipment	Total	Total Incl O&P
0010	**LINTELS**									
0015	Made from recycled materials									
0020	Plain steel angles, shop fabricated, under 500 lb.	G 1 Bric	550	.015	Lb.	1.02	.67		1.69	2.21
0100	500 to 1000 lb.	G	640	.013		.99	.58		1.57	2.03
0200	1,000 to 2,000 lb.	G	640	.013		.97	.58		1.55	2
0300	2,000 to 4,000 lb.	G	640	.013		.94	.58		1.52	1.97

05 12 Structural Steel Framing

05 12 23 – Structural Steel for Buildings

05 12 23.45 Lintels		Crew	Daily Output	Labor-Hours	Unit	Material	2016 Bare Costs Labor	Equipment	Total	Total Incl O&P
0500	For built-up angles and plates, add to above	G			Lb.	1.33			1.33	1.46
0700	For engineering, add to above					.13			.13	.15
0900	For galvanizing, add to above, under 500 lb.					.30			.30	.33
0950	500 to 2,000 lb.					.28			.28	.30
1000	Over 2,000 lb.					.25			.25	.28
2000	Steel angles, 3-1/2" x 3", 1/4" thick, 2'-6" long	G	1 Bric	47	.170	Ea.	14.30	7.85		22.15	28.50
2100	4'-6" long	G		26	.308		26	14.25		40.25	51.50
2500	3-1/2" x 3-1/2" x 5/16", 5'-0" long	G		18	.444		38	20.50		58.50	75.50
2600	4" x 3-1/2", 1/4" thick, 5'-0" long	G		21	.381		33	17.60		50.60	64.50
2700	9'-0" long	G		12	.667		59	31		90	115
2800	4" x 3-1/2" x 5/16", 7'-0" long	G		12	.667		57	31		88	113
2900	5" x 3-1/2" x 5/16", 10'-0" long	G		8	1		92	46.50		138.50	177
9000	Minimum labor/equipment charge			4	2	Job		92.50		92.50	151

05 12 23.60 Pipe Support Framing

		Crew	Daily Output	Labor-Hours	Unit	Material	2016 Bare Costs Labor	Equipment	Total	Total Incl O&P
0010	**PIPE SUPPORT FRAMING**									
0020	Under 10#/L.F., shop fabricated	G	E-4	3900	.008	Lb.	1.78	.44	.04	2.26	2.77
0200	10.1 to 15#/L.F.	G		4300	.007		1.75	.40	.03	2.18	2.66
0400	15.1 to 20#/L.F.	G		4800	.007		1.72	.36	.03	2.11	2.55
0600	Over 20#/L.F.	G		5400	.006		1.70	.32	.03	2.05	2.46

05 12 23.65 Plates

		Crew	Daily Output	Labor-Hours	Unit	Material	2016 Bare Costs Labor	Equipment	Total	Total Incl O&P
0010	**PLATES**									
0015	Made from recycled materials									
0020	For connections & stiffener plates, shop fabricated									
0050	1/8" thick (5.1 lb./S.F.)	G				S.F.	6.75			6.75	7.45
0100	1/4" thick (10.2 lb./S.F.)	G					13.50			13.50	14.85
0300	3/8" thick (15.3 lb./S.F.)	G					20.50			20.50	22.50
0400	1/2" thick (20.4 lb./S.F.)	G					27			27	29.50
0450	3/4" thick (30.6 lb./S.F.)	G					40.50			40.50	44.50
0500	1" thick (40.8 lb./S.F.)	G					54			54	59.50
2000	Steel plate, warehouse prices, no shop fabrication									
2100	1/4" thick (10.2 lb./S.F.)	G				S.F.	6.85			6.85	7.55

05 12 23.75 Structural Steel Members

		Crew	Daily Output	Labor-Hours	Unit	Material	2016 Bare Costs Labor	Equipment	Total	Total Incl O&P
0010	**STRUCTURAL STEEL MEMBERS** R051223-10									
0015	Made from recycled materials									
0020	Shop fab'd for 100-ton, 1-2 story project, bolted connections									
0102	Beam or girder, W 6 x 9 R051223-15	G	E-2	600	.093	L.F.	13.10	4.88	2.53	20.51	25.50
0302	W 8 x 10	G		600	.093		14.60	4.88	2.53	22.01	27
0502	x 31	G		550	.102		45	5.35	2.76	53.11	61.50
0702	W 10 x 22	G		600	.093		32	4.88	2.53	39.41	46.50
0902	x 49	G		550	.102		71.50	5.35	2.76	79.61	90.50
1102	W 12 x 16	G		880	.064		23.50	3.33	1.73	28.56	33
1302	x 22	G		880	.064		32	3.33	1.73	37.06	43
1502	x 26	G		880	.064		38	3.33	1.73	43.06	49
1902	W 14 x 26	G		990	.057		38	2.96	1.54	42.50	48.50
2102	x 30	G		900	.062		43.50	3.26	1.69	48.45	55.50
2702	W 16 x 26	G		1000	.056		38	2.93	1.52	42.45	48
2902	x 31	G		900	.062		45	3.26	1.69	49.95	57
8490	For projects 75 to 99 tons, add						10%				
8492	50 to 74 tons, add						20%				
8494	25 to 49 tons, add						30%	10%			
8496	10 to 24 tons, add						50%	25%			
8498	2 to 9 tons, add						75%	50%			

For customer support on your Commercial Renovation Cost Data, call 877.742.4719.

105

05 12 23 – Structural Steel for Buildings

05 12 23.75 Structural Steel Members		Crew	Daily Output	Labor-Hours	Unit	Material	2016 Bare Costs Labor	Equipment	Total	Total Incl O&P
8499	Less than 2 tons, add				L.F.	100%	100%			
9000	Minimum labor/equipment charge	E-2	2	28	Job		1,475	760	2,235	3,350

05 12 23.77 Structural Steel Projects

		Crew	Daily Output	Labor-Hours	Unit	Material	Labor	Equipment	Total	Total Incl O&P
0010	**STRUCTURAL STEEL PROJECTS**									
0015	Made from recycled materials									
0020	Shop fab'd for 100-ton, 1-2 story project, bolted connections									
0700	Offices, hospitals, etc., steel bearing, 1 to 2 stories R050521-20 G	E-5	10.30	7.767	Ton	2,650	410	162	3,222	3,825
1300	Industrial bldgs., 1 story, beams & girders, steel bearing G		12.90	6.202		2,650	325	129	3,104	3,625
1400	Masonry bearing R051223-10 G		10	8		2,650	420	167	3,237	3,850
1600	1 story with roof trusses, steel bearing G		10.60	7.547		3,125	400	157	3,682	4,325
1700	Masonry bearing R051223-15 G		8.30	9.639		3,125	510	201	3,836	4,550
5390	For projects 75 to 99 tons, add					10%				
5392	50 to 74 tons, add					20%				
5394	25 to 49 tons, add					30%	10%			
5396	10 to 24 tons, add					50%	25%			
5398	2 to 9 tons, add					75%	50%			
5399	Less than 2 tons, add					100%	100%			

05 12 23.78 Structural Steel Secondary Members

		Crew	Daily Output	Labor-Hours	Unit	Material	Labor	Equipment	Total	Total Incl O&P
0010	**STRUCTURAL STEEL SECONDARY MEMBERS**									
0015	Made from recycled materials									
0020	Shop fabricated for 20-ton girt/purlin framing package, materials only									
0100	Girts/purlins, C/Z-shapes, includes clips and bolts									
0110	6" x 2-1/2" x 2-1/2", 16 ga., 3.0 lb./L.F.				L.F.	3.58			3.58	3.94
0115	14 ga., 3.5 lb./L.F.					4.17			4.17	4.59
0120	8" x 2-3/4" x 2-3/4", 16 ga., 3.4 lb./L.F.					4.05			4.05	4.46
0125	14 ga., 4.1 lb./L.F.					4.89			4.89	5.40
0130	12 ga., 5.6 lb./L.F.					6.70			6.70	7.35
0135	10" x 3-1/2" x 3-1/2", 14 ga., 4.7 lb./L.F.					5.60			5.60	6.15
0140	12 ga., 6.7 lb./L.F.					8			8	8.80
0145	12" x 3-1/2" x 3-1/2", 14 ga., 5.3 lb./L.F.					6.30			6.30	6.95
0150	12 ga., 7.4 lb./L.F.					8.80			8.80	9.70
0200	Eave struts, C-shape, includes clips and bolts									
0210	6" x 4" x 3", 16 ga., 3.1 lb./L.F.				L.F.	3.70			3.70	4.07
0215	14 ga., 3.9 lb./L.F.					4.65			4.65	5.10
0220	8" x 4" x 3", 16 ga., 3.5 lb./L.F.					4.17			4.17	4.59
0225	14 ga., 4.4 lb./L.F.					5.25			5.25	5.75
0230	12 ga., 6.2 lb./L.F.					7.40			7.40	8.15
0235	10" x 5" x 3", 14 ga., 5.2 lb./L.F.					6.20			6.20	6.80
0240	12 ga., 7.3 lb./L.F.					8.70			8.70	9.60
0245	12" x 5" x 4", 14 ga., 6.0 lb./L.F.					7.15			7.15	7.85
0250	12 ga., 8.4 lb./L.F.					10			10	11
0300	Rake/base angle, excludes concrete drilling and expansion anchors									
0310	2" x 2", 14 ga., 1.0 lb./L.F.	2 Sswk	640	.025	L.F.	1.19	1.33		2.52	3.66
0315	3" x 2", 14 ga., 1.3 lb./L.F.		535	.030		1.55	1.59		3.14	4.52
0320	3" x 3", 14 ga., 1.6 lb./L.F.		500	.032		1.91	1.70		3.61	5.10
0325	4" x 3", 14 ga., 1.8 lb./L.F.		480	.033		2.15	1.77		3.92	5.50
0600	Installation of secondary members, erection only									
0610	Girts, purlins, eave struts, 16 ga., 6" deep	E-18	100	.400	Ea.		21.50	9.65	31.15	47.50
0615	8" deep		80	.500			26.50	12.10	38.60	59.50
0620	14 ga., 6" deep		80	.500			26.50	12.10	38.60	59.50
0625	8" deep		65	.615			32.50	14.90	47.40	73
0630	10" deep		55	.727			38.50	17.60	56.10	86.50

05 12 Structural Steel Framing

05 12 23 – Structural Steel for Buildings

05 12 23.78 Structural Steel Secondary Members		Crew	Daily Output	Labor-Hours	Unit	Material	2016 Bare Costs Labor	Equipment	Total	Total Incl O&P
0635	12" deep	E-18	50	.800	Ea.		42.50	19.35	61.85	95
0640	12 ga., 8" deep		50	.800			42.50	19.35	61.85	95
0645	10" deep		45	.889			47.50	21.50	69	105
0650	12" deep	↓	40	1	↓		53	24	77	119
0900	For less than 20-ton job lots									
0905	For 15 to 19 tons, add				%	10%				
0910	For 10 to 14 tons, add					25%				
0915	For 5 to 9 tons, add					50%	50%	50%		
0920	For 1 to 4 tons, add					75%	75%	75%		
0925	For less than 1 ton, add				↓	100%	100%	100%		
0990	Minimum labor/equipment charge	E-18	1	40	Job		2,125	965	3,090	4,750

05 15 Wire Rope Assemblies

05 15 16 – Steel Wire Rope Assemblies

05 15 16.70 Temporary Cable Safety Railing

		Crew	Daily Output	Labor-Hours	Unit	Material	2016 Bare Costs Labor	Equipment	Total	Total Incl O&P
0010	**TEMPORARY CABLE SAFETY RAILING**, Each 100' strand incl.									
0020	2 eyebolts, 1 turnbuckle, 100' cable, 2 thimbles, 6 clips									
0025	Made from recycled materials									
0100	One strand using 1/4" cable & accessories	G	2 Sswk	4	4	C.L.F.	194	213		407	590
0200	1/2" cable & accessories	G	"	2	8	"	425	425		850	1,225

05 21 Steel Joist Framing

05 21 13 – Deep Longspan Steel Joist Framing

05 21 13.50 Deep Longspan Joists

		Crew	Daily Output	Labor-Hours	Unit	Material	2016 Bare Costs Labor	Equipment	Total	Total Incl O&P
0010	**DEEP LONGSPAN JOISTS**									
3010	DLH series, 40-ton job lots, bolted cross bridging, shop primer									
3015	Made from recycled materials									
3040	Spans to 144' (shipped in 2 pieces)	G	E-7	13	6.154	Ton	1,925	325	140	2,390	2,850
3500	For less than 40-ton job lots									
3502	For 30 to 39 tons, add					%	10%			
3504	20 to 29 tons, add						20%			
3506	10 to 19 tons, add						30%			
3507	5 to 9 tons, add						50%	25%		
3508	1 to 4 tons, add						75%	50%		
3509	Less than 1 ton, add					↓	100%	100%		
4010	SLH series, 40-ton job lots, bolted cross bridging, shop primer									
4040	Spans to 200' (shipped in 3 pieces)	G	E-7	13	6.154	Ton	1,975	325	140	2,440	2,900
6100	For less than 40-ton job lots									
6102	For 30 to 39 tons, add					%	10%			
6104	20 to 29 tons, add						20%			
6106	10 to 19 tons, add						30%			
6107	5 to 9 tons, add						50%	25%		
6108	1 to 4 tons, add						75%	50%		
6109	Less than 1 ton, add					↓	100%	100%		

05 21 Steel Joist Framing

05 21 16 – Longspan Steel Joist Framing

05 21 16.50 Longspan Joists

		Crew	Daily Output	Labor-Hours	Unit	Material	2016 Bare Costs Labor	Equipment	Total	Total Incl O&P
0010	**LONGSPAN JOISTS**									
2000	LH series, 40-ton job lots, bolted cross bridging, shop primer									
2015	Made from recycled materials									
2040	Longspan joists, LH series, up to 96'	G E-7	13	6.154	Ton	1,825	325	140	2,290	2,750
2600	For less than 40-ton job lots									
2602	For 30 to 39 tons, add				%	10%				
2604	20 to 29 tons, add					20%				
2606	10 to 19 tons, add					30%				
2607	5 to 9 tons, add					50%	25%			
2608	1 to 4 tons, add					75%	50%			
2609	Less than 1 ton, add					100%	100%			

05 21 19 – Open Web Steel Joist Framing

05 21 19.10 Open Web Joists

		Crew	Daily Output	Labor-Hours	Unit	Material	2016 Bare Costs Labor	Equipment	Total	Total Incl O&P
0010	**OPEN WEB JOISTS**									
0015	Made from recycled materials									
0050	K series, 40-ton lots, horiz. bridging, spans to 30', shop primer	G E-7	12	6.667	Ton	1,650	350	151	2,151	2,600
0440	K series, 30' to 50' spans	G "	17	4.706	"	1,625	248	107	1,980	2,350
0800	For less than 40-ton job lots									
0802	For 30 to 39 tons, add				%	10%				
0804	20 to 29 tons, add					20%				
0806	10 to 19 tons, add					30%				
0807	5 to 9 tons, add					50%	25%			
0808	1 to 4 tons, add					75%	50%			
0809	Less than 1 ton, add					100%	100%			
1010	CS series, 40-ton job lots, horizontal bridging, shop primer									
1040	Spans to 30'	G E-7	12	6.667	Ton	1,700	350	151	2,201	2,650
1500	For less than 40-ton job lots									
1502	For 30 to 39 tons, add				%	10%				
1504	20 to 29 tons, add					20%				
1506	10 to 19 tons, add					30%				
1507	5 to 9 tons, add					50%	25%			
1508	1 to 4 tons, add					75%	50%			
1509	Less than 1 ton, add					100%	100%			
6400	Individual steel bearing plate, 6" x 6" x 1/4" with J-hook	G 1 Bric	160	.050	Ea.	7.95	2.31		10.26	12.50
9000	Minimum labor and equipment, bar joists	F-6	1	40	Job		1,800	660	2,460	3,650

05 31 Steel Decking

05 31 13 – Steel Floor Decking

05 31 13.50 Floor Decking

			Crew	Daily Output	Labor-Hours	Unit	Material	2016 Bare Costs Labor	Equipment	Total	Total Incl O&P
0010	**FLOOR DECKING**	R053100-10									
0015	Made from recycled materials										
5100	Non-cellular composite decking, galvanized, 1-1/2" deep, 16 ga.	G	E-4	3500	.009	S.F.	3.48	.49	.04	4.01	4.75
5120	18 ga.	G		3650	.009		2.81	.47	.04	3.32	3.96
5140	20 ga.	G		3800	.008		2.24	.45	.04	2.73	3.30
5200	2" deep, 22 ga.	G		3860	.008		1.95	.45	.04	2.44	2.98
5300	20 ga.	G		3600	.009		2.15	.48	.04	2.67	3.26
5400	18 ga.	G		3380	.009		2.75	.51	.04	3.30	3.98
5500	16 ga.	G		3200	.010		3.44	.54	.05	4.03	4.78
5700	3" deep, 22 ga.	G		3200	.010		2.13	.54	.05	2.72	3.34
5800	20 ga.	G		3000	.011		2.37	.57	.05	2.99	3.66

05 31 Steel Decking

05 31 13 – Steel Floor Decking

05 31 13.50 Floor Decking		Crew	Daily Output	Labor-Hours	Unit	Material	2016 Bare Costs Labor	Equipment	Total	Total Incl O&P
5900	18 ga.	G	E-4	2850	.011	S.F.	2.93	.60	.05	3.58	4.34
6000	16 ga.	G	↓	2700	.012	↓	3.91	.64	.05	4.60	5.50
9000	Minimum labor/equipment charge		1 Sswk	1	8	Job		425		425	750

05 31 23 – Steel Roof Decking

05 31 23.50 Roof Decking

			Crew	Daily Output	Labor-Hours	Unit	Material	Labor	Equipment	Total	Total Incl O&P
0010	**ROOF DECKING**										
0015	Made from recycled materials										
2100	Open type, 1-1/2" deep, Type B, wide rib, galv., 22 ga., under 50 sq.	G	E-4	4500	.007	S.F.	2.07	.38	.03	2.48	2.99
2400	Over 500 squares	G	"	5100	.006	"	1.49	.34	.03	1.86	2.26

05 31 33 – Steel Form Decking

05 31 33.50 Form Decking

			Crew	Daily Output	Labor-Hours	Unit	Material	Labor	Equipment	Total	Total Incl O&P
0010	**FORM DECKING**										
0015	Made from recycled materials										
6100	Slab form, steel, 28 ga., 9/16" deep, Type UFS, uncoated	G	E-4	4000	.008	S.F.	1.56	.43	.04	2.03	2.52
6200	Galvanized	G	"	4000	.008	"	1.38	.43	.04	1.85	2.32

05 35 Raceway Decking Assemblies

05 35 13 – Steel Cellular Decking

05 35 13.50 Cellular Decking

			Crew	Daily Output	Labor-Hours	Unit	Material	Labor	Equipment	Total	Total Incl O&P
0010	**CELLULAR DECKING**										
0015	Made from recycled materials										
0200	Cellular units, galv, 1-1/2" deep, Type BC, 20-20 ga., over 15 squares	G	E-4	1460	.022	S.F.	8.15	1.18	.10	9.43	11.20
0400	3" deep, Type NC, galvanized, 20-20 ga.	G		1375	.023		9	1.25	.11	10.36	12.20
1000	4-1/2" deep, Type JC, galvanized, 18-20 ga.	G	↓	1100	.029	↓	12.50	1.56	.13	14.19	16.65
1500	For acoustical deck, add						15%				
1900	For multi-story or congested site, add							50%			

05 41 Structural Metal Stud Framing

05 41 13 – Load-Bearing Metal Stud Framing

05 41 13.05 Bracing

			Crew	Daily Output	Labor-Hours	Unit	Material	Labor	Equipment	Total	Total Incl O&P
0010	**BRACING**, shear wall X-bracing, per 10' x 10' bay, one face										
0015	Made of recycled materials										
0120	Metal strap, 20 ga. x 4" wide	G	2 Carp	18	.889	Ea.	18.90	43		61.90	91.50
0130	6" wide	G		18	.889		31.50	43		74.50	105
0160	18 ga. x 4" wide	G		16	1		33	48.50		81.50	115
0170	6" wide	G	↓	16	1	↓	48.50	48.50		97	133
0410	Continuous strap bracing, per horizontal row on both faces										
0420	Metal strap, 20 ga. x 2" wide, studs 12" O.C.	G	1 Carp	7	1.143	C.L.F.	57	55.50		112.50	154
0430	16" O.C.	G		8	1		57	48.50		105.50	142
0440	24" O.C.	G		10	.800		57	39		96	127
0450	18 ga. x 2" wide, studs 12" O.C.	G		6	1.333		79	64.50		143.50	193
0460	16" O.C.	G		7	1.143		79	55.50		134.50	178
0470	24" O.C.	G	↓	8	1	↓	79	48.50		127.50	166

05 41 13.10 Bridging

			Crew	Daily Output	Labor-Hours	Unit	Material	Labor	Equipment	Total	Total Incl O&P
0010	**BRIDGING**, solid between studs w/1-1/4" leg track, per stud bay										
0015	Made from recycled materials										
0200	Studs 12" O.C., 18 ga. x 2-1/2" wide	G	1 Carp	125	.064	Ea.	.95	3.10		4.05	6.10
0210	3-5/8" wide	G	↓	120	.067	↓	1.16	3.23		4.39	6.55

For customer support on your Commercial Renovation Cost Data, call 877.742.4719.

109

05 41 13 – Load-Bearing Metal Stud Framing

05 41 13.10 Bridging

			Crew	Daily Output	Labor-Hours	Unit	Material	2016 Bare Costs Labor	Equipment	Total	Total Incl O&P
0220	4" wide	G	1 Carp	120	.067	Ea.	1.22	3.23		4.45	6.65
0230	6" wide	G		115	.070		1.60	3.37		4.97	7.25
0240	8" wide	G		110	.073		1.97	3.52		5.49	7.90
0300	16 ga. x 2-1/2" wide	G		115	.070		1.21	3.37		4.58	6.85
0310	3-5/8" wide	G		110	.073		1.47	3.52		4.99	7.35
0320	4" wide	G		110	.073		1.57	3.52		5.09	7.50
0330	6" wide	G		105	.076		2.01	3.69		5.70	8.25
0340	8" wide	G		100	.080		2.51	3.88		6.39	9.10
1200	Studs 16" O.C., 18 ga. x 2-1/2" wide	G		125	.064		1.22	3.10		4.32	6.40
1210	3-5/8" wide	G		120	.067		1.49	3.23		4.72	6.95
1220	4" wide	G		120	.067		1.57	3.23		4.80	7
1230	6" wide	G		115	.070		2.05	3.37		5.42	7.75
1240	8" wide	G		110	.073		2.52	3.52		6.04	8.55
1300	16 ga. x 2-1/2" wide	G		115	.070		1.55	3.37		4.92	7.20
1310	3-5/8" wide	G		110	.073		1.88	3.52		5.40	7.80
1320	4" wide	G		110	.073		2.01	3.52		5.53	7.95
1330	6" wide	G		105	.076		2.57	3.69		6.26	8.90
1340	8" wide	G		100	.080		3.22	3.88		7.10	9.90
2200	Studs 24" O.C., 18 ga. x 2-1/2" wide	G		125	.064		1.77	3.10		4.87	7
2210	3-5/8" wide	G		120	.067		2.15	3.23		5.38	7.65
2220	4" wide	G		120	.067		2.27	3.23		5.50	7.80
2230	6" wide	G		115	.070		2.96	3.37		6.33	8.75
2240	8" wide	G		110	.073		3.65	3.52		7.17	9.75
2300	16 ga. x 2-1/2" wide	G		115	.070		2.24	3.37		5.61	7.95
2310	3-5/8" wide	G		110	.073		2.72	3.52		6.24	8.75
2320	4" wide	G		110	.073		2.91	3.52		6.43	8.95
2330	6" wide	G		105	.076		3.72	3.69		7.41	10.15
2340	8" wide	G		100	.080		4.65	3.88		8.53	11.45
3000	Continuous bridging, per row										
3100	16 ga. x 1-1/2" channel thru studs 12" O.C.	G	1 Carp	6	1.333	C.L.F.	51.50	64.50		116	163
3110	16" O.C.	G		7	1.143		51.50	55.50		107	148
3120	24" O.C.	G		8.80	.909		51.50	44		95.50	129
4100	2" x 2" angle x 18 ga., studs 12" O.C.	G		7	1.143		80.50	55.50		136	179
4110	16" O.C.	G		9	.889		80.50	43		123.50	159
4120	24" O.C.	G		12	.667		80.50	32.50		113	142
4200	16 ga., studs 12" O.C.	G		5	1.600		101	77.50		178.50	238
4210	16" O.C.	G		7	1.143		101	55.50		156.50	202
4220	24" O.C.	G		10	.800		101	39		140	175

05 41 13.25 Framing, Boxed Headers/Beams

			Crew	Daily Output	Labor-Hours	Unit	Material	2016 Bare Costs Labor	Equipment	Total	Total Incl O&P
0010	**FRAMING, BOXED HEADERS/BEAMS**										
0015	Made from recycled materials										
0200	Double, 18 ga. x 6" deep	G	2 Carp	220	.073	L.F.	5.50	3.52		9.02	11.80
0210	8" deep	G		210	.076		6.05	3.69		9.74	12.70
0220	10" deep	G		200	.080		7.40	3.88		11.28	14.50
0230	12" deep	G		190	.084		8.05	4.08		12.13	15.50
0300	16 ga. x 8" deep	G		180	.089		7	4.31		11.31	14.75
0310	10" deep	G		170	.094		8.45	4.56		13.01	16.70
0320	12" deep	G		160	.100		9.15	4.85		14	18
0400	14 ga. x 10" deep	G		140	.114		9.75	5.55		15.30	19.75
0410	12" deep	G		130	.123		10.65	5.95		16.60	21.50
1210	Triple, 18 ga. x 8" deep	G		170	.094		8.75	4.56		13.31	17.10
1220	10" deep	G		165	.097		10.60	4.70		15.30	19.35

05 41 Structural Metal Stud Framing

05 41 13 – Load-Bearing Metal Stud Framing

05 41 13.25 Framing, Boxed Headers/Beams		Crew	Daily Output	Labor-Hours	Unit	Material	2016 Bare Costs Labor	Equipment	Total	Total Incl O&P
1230	12" deep	G 2 Carp	160	.100	L.F.	11.60	4.85		16.45	20.50
1300	16 ga. x 8" deep	G	145	.110		10.15	5.35		15.50	19.90
1310	10" deep	G	140	.114		12.15	5.55		17.70	22.50
1320	12" deep	G	135	.119		13.25	5.75		19	24
1400	14 ga. x 10" deep	G	115	.139		13.25	6.75		20	25.50
1410	12" deep	G	110	.145		14.65	7.05		21.70	27.50

05 41 13.30 Framing, Stud Walls		Crew	Daily Output	Labor-Hours	Unit	Material	2016 Bare Costs Labor	Equipment	Total	Total Incl O&P
0010	**FRAMING, STUD WALLS** w/top & bottom track, no openings,									
0020	Headers, beams, bridging or bracing									
0025	Made from recycled materials									
4100	8' high walls, 18 ga. x 2-1/2" wide, studs 12" O.C.	G 2 Carp	54	.296	L.F.	9.10	14.35		23.45	33.50
4110	16" O.C.	G	77	.208		7.25	10.05		17.30	24.50
4120	24" O.C.	G	107	.150		5.45	7.25		12.70	17.85
4130	3-5/8" wide, studs 12" O.C.	G	53	.302		10.85	14.65		25.50	36
4140	16" O.C.	G	76	.211		8.70	10.20		18.90	26
4150	24" O.C.	G	105	.152		6.55	7.40		13.95	19.25
4160	4" wide, studs 12" O.C.	G	52	.308		11.30	14.90		26.20	37
4170	16" O.C.	G	74	.216		9.05	10.50		19.55	27
4180	24" O.C.	G	103	.155		6.80	7.55		14.35	19.80
4190	6" wide, studs 12" O.C.	G	51	.314		14.40	15.20		29.60	41
4200	16" O.C.	G	73	.219		11.55	10.60		22.15	30
4210	24" O.C.	G	101	.158		8.70	7.70		16.40	22
4220	8" wide, studs 12" O.C.	G	50	.320		17.45	15.50		32.95	44.50
4230	16" O.C.	G	72	.222		14	10.75		24.75	33
4240	24" O.C.	G	100	.160		10.60	7.75		18.35	24.50
4300	16 ga. x 2-1/2" wide, studs 12" O.C.	G	47	.340		10.85	16.50		27.35	39
4310	16" O.C.	G	68	.235		8.60	11.40		20	28
4320	24" O.C.	G	94	.170		6.35	8.25		14.60	20.50
4330	3-5/8" wide, studs 12" O.C.	G	46	.348		12.95	16.85		29.80	41.50
4340	16" O.C.	G	66	.242		10.25	11.75		22	30.50
4350	24" O.C.	G	92	.174		7.55	8.45		16	22
4360	4" wide, studs 12" O.C.	G	45	.356		13.55	17.25		30.80	43
4370	16" O.C.	G	65	.246		10.75	11.95		22.70	31.50
4380	24" O.C.	G	90	.178		7.95	8.60		16.55	23
4390	6" wide, studs 12" O.C.	G	44	.364		17.05	17.60		34.65	48
4400	16" O.C.	G	64	.250		13.55	12.10		25.65	34.50
4410	24" O.C.	G	88	.182		10.05	8.80		18.85	25.50
4420	8" wide, studs 12" O.C.	G	43	.372		21	18.05		39.05	52.50
4430	16" O.C.	G	63	.254		16.60	12.30		28.90	38.50
4440	24" O.C.	G	86	.186		12.30	9		21.30	28.50
5100	10' high walls, 18 ga. x 2-1/2" wide, studs 12" O.C.	G	54	.296		10.90	14.35		25.25	35.50
5110	16" O.C.	G	77	.208		8.65	10.05		18.70	26
5120	24" O.C.	G	107	.150		6.35	7.25		13.60	18.85
5130	3-5/8" wide, studs 12" O.C.	G	53	.302		13	14.65		27.65	38.50
5140	16" O.C.	G	76	.211		10.30	10.20		20.50	28
5150	24" O.C.	G	105	.152		7.60	7.40		15	20.50
5160	4" wide, studs 12" O.C.	G	52	.308		13.55	14.90		28.45	39.50
5170	16" O.C.	G	74	.216		10.75	10.50		21.25	29
5180	24" O.C.	G	103	.155		7.95	7.55		15.50	21
5190	6" wide, studs 12" O.C.	G	51	.314		17.25	15.20		32.45	44
5200	16" O.C.	G	73	.219		13.70	10.60		24.30	32.50
5210	24" O.C.	G	101	.158		10.15	7.70		17.85	23.50

For customer support on your Commercial Renovation Cost Data, call 877.742.4719.

111

05 41 13.30 Framing, Stud Walls		Crew	Daily Output	Labor-Hours	Unit	Material	2016 Bare Costs Labor	2016 Bare Costs Equipment	Total	Total Incl O&P
5220	8" wide, studs 12" O.C.	G 2 Carp	50	.320	L.F.	21	15.50		36.50	48.50
5230	16" O.C.	G	72	.222		16.55	10.75		27.30	36
5240	24" O.C.	G	100	.160		12.30	7.75		20.05	26
5300	16 ga. x 2-1/2" wide, studs 12" O.C.	G	47	.340		13.10	16.50		29.60	41.50
5310	16" O.C.	G	68	.235		10.30	11.40		21.70	30
5320	24" O.C.	G	94	.170		7.45	8.25		15.70	21.50
5330	3-5/8" wide, studs 12" O.C.	G	46	.348		15.60	16.85		32.45	44.50
5340	16" O.C.	G	66	.242		12.25	11.75		24	32.50
5350	24" O.C.	G	92	.174		8.90	8.45		17.35	23.50
5360	4" wide, studs 12" O.C.	G	45	.356		16.35	17.25		33.60	46
5370	16" O.C.	G	65	.246		12.85	11.95		24.80	33.50
5380	24" O.C.	G	90	.178		9.35	8.60		17.95	24.50
5390	6" wide, studs 12" O.C.	G	44	.364		20.50	17.60		38.10	51.50
5400	16" O.C.	G	64	.250		16.15	12.10		28.25	37.50
5410	24" O.C.	G	88	.182		11.80	8.80		20.60	27.50
5420	8" wide, studs 12" O.C.	G	43	.372		25	18.05		43.05	57
5430	16" O.C.	G	63	.254		19.80	12.30		32.10	42
5440	24" O.C.	G	86	.186		14.45	9		23.45	30.50
6190	12' high walls, 18 ga. x 6" wide, studs 12" O.C.	G	41	.390		20	18.90		38.90	53
6200	16" O.C.	G	58	.276		15.80	13.35		29.15	39.50
6210	24" O.C.	G	81	.198		11.55	9.55		21.10	28.50
6220	8" wide, studs 12" O.C.	G	40	.400		24.50	19.40		43.90	58
6230	16" O.C.	G	57	.281		19.15	13.60		32.75	43
6240	24" O.C.	G	80	.200		14	9.70		23.70	31.50
6390	16 ga. x 6" wide, studs 12" O.C.	G	35	.457		24	22		46	62.50
6400	16" O.C.	G	51	.314		18.80	15.20		34	45.50
6410	24" O.C.	G	70	.229		13.55	11.05		24.60	33
6420	8" wide, studs 12" O.C.	G	34	.471		29.50	23		52.50	70
6430	16" O.C.	G	50	.320		23	15.50		38.50	51
6440	24" O.C.	G	69	.232		16.60	11.25		27.85	36.50
6530	14 ga. x 3-5/8" wide, studs 12" O.C.	G	34	.471		22.50	23		45.50	62.50
6540	16" O.C.	G	48	.333		17.75	16.15		33.90	46
6550	24" O.C.	G	65	.246		12.75	11.95		24.70	33.50
6560	4" wide, studs 12" O.C.	G	33	.485		24	23.50		47.50	65
6570	16" O.C.	G	47	.340		18.75	16.50		35.25	47.50
6580	24" O.C.	G	64	.250		13.50	12.10		25.60	34.50
6730	12 ga. x 3-5/8" wide, studs 12" O.C.	G	31	.516		31.50	25		56.50	75.50
6740	16" O.C.	G	43	.372		24.50	18.05		42.55	56
6750	24" O.C.	G	59	.271		17.15	13.15		30.30	40.50
6760	4" wide, studs 12" O.C.	G	30	.533		33.50	26		59.50	79
6770	16" O.C.	G	42	.381		26	18.45		44.45	58.50
6780	24" O.C.	G	58	.276		18.30	13.35		31.65	42
7390	16' high walls, 16 ga. x 6" wide, studs 12" O.C.	G	33	.485		31	23.50		54.50	72.50
7400	16" O.C.	G	48	.333		24	16.15		40.15	53
7410	24" O.C.	G	67	.239		17.05	11.55		28.60	37.50
7420	8" wide, studs 12" O.C.	G	32	.500		38	24		62	81.50
7430	16" O.C.	G	47	.340		29.50	16.50		46	59.50
7440	24" O.C.	G	66	.242		21	11.75		32.75	42
7560	14 ga. x 4" wide, studs 12" O.C.	G	31	.516		31	25		56	75
7570	16" O.C.	G	45	.356		24	17.25		41.25	54.50
7580	24" O.C.	G	61	.262		17	12.70		29.70	39.50
7590	6" wide, studs 12" O.C.	G	30	.533		39	26		65	85
7600	16" O.C.	G	44	.364		30	17.60		47.60	62

05 41 Structural Metal Stud Framing

05 41 13 – Load-Bearing Metal Stud Framing

05 41 13.30 Framing, Stud Walls		Crew	Daily Output	Labor-Hours	Unit	Material	2016 Bare Costs Labor	Equipment	Total	Total Incl O&P
7610	24" O.C.	G 2 Carp	60	.267	L.F.	21.50	12.90		34.40	44.50
7760	12 ga. x 4" wide, studs 12" O.C.	G	29	.552		44	26.50		70.50	91.50
7770	16" O.C.	G	40	.400		33.50	19.40		52.90	68.50
7780	24" O.C.	G	55	.291		23.50	14.10		37.60	48.50
7790	6" wide, studs 12" O.C.	G	28	.571		55.50	27.50		83	106
7800	16" O.C.	G	39	.410		42.50	19.90		62.40	79
7810	24" O.C.	G	54	.296		29.50	14.35		43.85	56
8590	20' high walls, 14 ga. x 6" wide, studs 12" O.C.	G	29	.552		48	26.50		74.50	96
8600	16" O.C.	G	42	.381		37	18.45		55.45	70.50
8610	24" O.C.	G	57	.281		26	13.60		39.60	50.50
8620	8" wide, studs 12" O.C.	G	28	.571		52	27.50		79.50	102
8630	16" O.C.	G	41	.390		40	18.90		58.90	75
8640	24" O.C.	G	56	.286		28.50	13.85		42.35	53.50
8790	12 ga. x 6" wide, studs 12" O.C.	G	27	.593		68	28.50		96.50	122
8800	16" O.C.	G	37	.432		52	21		73	91.50
8810	24" O.C.	G	51	.314		36	15.20		51.20	64.50
8820	8" wide, studs 12" O.C.	G	26	.615		82.50	30		112.50	140
8830	16" O.C.	G	36	.444		63	21.50		84.50	105
8840	24" O.C.	G	50	.320		43.50	15.50		59	73.50
9000	Minimum labor/equipment charge		4	4	Job		194		194	315

05 42 Cold-Formed Metal Joist Framing

05 42 13 – Cold-Formed Metal Floor Joist Framing

05 42 13.05 Bracing

		Crew	Daily Output	Labor-Hours	Unit	Material	Labor	Equipment	Total	Total Incl O&P
0010	**BRACING**, continuous, per row, top & bottom									
0015	Made from recycled materials									
0120	Flat strap, 20 ga. x 2" wide, joists at 12" O.C.	G 1 Carp	4.67	1.713	C.L.F.	60	83		143	202
0130	16" O.C.	G	5.33	1.501		57.50	72.50		130	183
0140	24" O.C.	G	6.66	1.201		55.50	58		113.50	156
0150	18 ga. x 2" wide, joists at 12" O.C.	G	4	2		78.50	97		175.50	245
0160	16" O.C.	G	4.67	1.713		77	83		160	221
0170	24" O.C.	G	5.33	1.501		75.50	72.50		148	202

05 42 13.10 Bridging

		Crew	Daily Output	Labor-Hours	Unit	Material	Labor	Equipment	Total	Total Incl O&P
0010	**BRIDGING**, solid between joists w/1-1/4" leg track, per joist bay									
0015	Made from recycled materials									
0230	Joists 12" O.C., 18 ga. track x 6" wide	G 1 Carp	80	.100	Ea.	1.60	4.85		6.45	9.65
0240	8" wide	G	75	.107		1.97	5.15		7.12	10.60
0250	10" wide	G	70	.114		2.46	5.55		8.01	11.75
0260	12" wide	G	65	.123		2.78	5.95		8.73	12.80
0330	16 ga. track x 6" wide	G	70	.114		2.01	5.55		7.56	11.25
0340	8" wide	G	65	.123		2.51	5.95		8.46	12.50
0350	10" wide	G	60	.133		3.10	6.45		9.55	13.95
0360	12" wide	G	55	.145		3.58	7.05		10.63	15.45
0440	14 ga. track x 8" wide	G	60	.133		3.13	6.45		9.58	14
0450	10" wide	G	55	.145		3.89	7.05		10.94	15.80
0460	12" wide	G	50	.160		4.49	7.75		12.24	17.60
0550	12 ga. track x 10" wide	G	45	.178		5.70	8.60		14.30	20.50
0560	12" wide	G	40	.200		5.80	9.70		15.50	22.50
1230	16" O.C., 18 ga. track x 6" wide	G	80	.100		2.05	4.85		6.90	10.15
1240	8" wide	G	75	.107		2.52	5.15		7.67	11.25
1250	10" wide	G	70	.114		3.15	5.55		8.70	12.50

For customer support on your Commercial Renovation Cost Data, call 877.742.4719.

113

05 42 Cold-Formed Metal Joist Framing

05 42 13 – Cold-Formed Metal Floor Joist Framing

05 42 13.10 Bridging

		Crew	Daily Output	Labor-Hours	Unit	Material	2016 Bare Costs Labor	Equipment	Total	Total Incl O&P
1260	12" wide	1 Carp G	65	.123	Ea.	3.56	5.95		9.51	13.65
1330	16 ga. track x 6" wide	G	70	.114		2.57	5.55		8.12	11.90
1340	8" wide	G	65	.123		3.22	5.95		9.17	13.30
1350	10" wide	G	60	.133		3.98	6.45		10.43	14.90
1360	12" wide	G	55	.145		4.59	7.05		11.64	16.55
1440	14 ga. track x 8" wide	G	60	.133		4.01	6.45		10.46	14.95
1450	10" wide	G	55	.145		4.98	7.05		12.03	17
1460	12" wide	G	50	.160		5.75	7.75		13.50	19
1550	12 ga. track x 10" wide	G	45	.178		7.30	8.60		15.90	22
1560	12" wide	G	40	.200		7.45	9.70		17.15	24
2230	24" O.C., 18 ga. track x 6" wide	G	80	.100		2.96	4.85		7.81	11.15
2240	8" wide	G	75	.107		3.65	5.15		8.80	12.45
2250	10" wide	G	70	.114		4.56	5.55		10.11	14.05
2260	12" wide	G	65	.123		5.15	5.95		11.10	15.40
2330	16 ga. track x 6" wide	G	70	.114		3.72	5.55		9.27	13.15
2340	8" wide	G	65	.123		4.65	5.95		10.60	14.85
2350	10" wide	G	60	.133		5.75	6.45		12.20	16.90
2360	12" wide	G	55	.145		6.65	7.05		13.70	18.80
2440	14 ga. track x 8" wide	G	60	.133		5.80	6.45		12.25	16.95
2450	10" wide	G	55	.145		7.20	7.05		14.25	19.45
2460	12" wide	G	50	.160		8.35	7.75		16.10	22
2550	12 ga. track x 10" wide	G	45	.178		10.55	8.60		19.15	25.50
2560	12" wide	G	40	.200		10.80	9.70		20.50	27.50

05 42 13.25 Framing, Band Joist

0010	**FRAMING, BAND JOIST** (track) fastened to bearing wall									
0015	Made from recycled materials									
0220	18 ga. track x 6" deep	2 Carp G	1000	.016	L.F.	1.30	.78		2.08	2.70
0230	8" deep	G	920	.017		1.61	.84		2.45	3.15
0240	10" deep	G	860	.019		2.01	.90		2.91	3.68
0320	16 ga. track x 6" deep	G	900	.018		1.64	.86		2.50	3.21
0330	8" deep	G	840	.019		2.05	.92		2.97	3.76
0340	10" deep	G	780	.021		2.53	.99		3.52	4.40
0350	12" deep	G	740	.022		2.92	1.05		3.97	4.92
0430	14 ga. track x 8" deep	G	750	.021		2.55	1.03		3.58	4.50
0440	10" deep	G	720	.022		3.17	1.08		4.25	5.25
0450	12" deep	G	700	.023		3.66	1.11		4.77	5.85
0540	12 ga. track x 10" deep	G	670	.024		4.64	1.16		5.80	7
0550	12" deep	G	650	.025		4.75	1.19		5.94	7.15

05 42 13.30 Framing, Boxed Headers/Beams

0010	**FRAMING, BOXED HEADERS/BEAMS**									
0015	Made from recycled materials									
0200	Double, 18 ga. x 6" deep	2 Carp G	220	.073	L.F.	5.50	3.52		9.02	11.80
0210	8" deep	G	210	.076		6.05	3.69		9.74	12.70
0220	10" deep	G	200	.080		7.40	3.88		11.28	14.50
0230	12" deep	G	190	.084		8.05	4.08		12.13	15.50
0300	16 ga. x 8" deep	G	180	.089		7	4.31		11.31	14.75
0310	10" deep	G	170	.094		8.45	4.56		13.01	16.70
0320	12" deep	G	160	.100		9.15	4.85		14	18
0400	14 ga. x 10" deep	G	140	.114		9.75	5.55		15.30	19.75
0410	12" deep	G	130	.123		10.65	5.95		16.60	21.50
0500	12 ga. x 10" deep	G	110	.145		12.85	7.05		19.90	25.50
0510	12" deep	G	100	.160		14.15	7.75		21.90	28.50

05 42 Cold-Formed Metal Joist Framing

05 42 13 – Cold-Formed Metal Floor Joist Framing

05 42 13.30 Framing, Boxed Headers/Beams

05 42 13.30 Framing, Boxed Headers/Beams		Crew	Daily Output	Labor-Hours	Unit	Material	2016 Bare Costs Labor	Equipment	Total	Total Incl O&P
1210	Triple, 18 ga. x 8" deep [G]	2 Carp	170	.094	L.F.	8.75	4.56		13.31	17.10
1220	10" deep [G]		165	.097		10.60	4.70		15.30	19.35
1230	12" deep [G]		160	.100		11.60	4.85		16.45	20.50
1300	16 ga. x 8" deep [G]		145	.110		10.15	5.35		15.50	19.90
1310	10" deep [G]		140	.114		12.15	5.55		17.70	22.50
1320	12" deep [G]		135	.119		13.25	5.75		19	24
1400	14 ga. x 10" deep [G]		115	.139		14.10	6.75		20.85	26.50
1410	12" deep [G]		110	.145		15.50	7.05		22.55	28.50
1500	12 ga. x 10" deep [G]		90	.178		18.75	8.60		27.35	34.50
1510	12" deep [G]		85	.188		21	9.10		30.10	38

05 42 13.40 Framing, Joists

05 42 13.40 Framing, Joists		Crew	Daily Output	Labor-Hours	Unit	Material	2016 Bare Costs Labor	Equipment	Total	Total Incl O&P
0010	**FRAMING, JOISTS**, no band joists (track), web stiffeners, headers,									
0020	Beams, bridging or bracing									
0025	Made from recycled materials									
0030	Joists (2" flange) and fasteners, materials only									
0220	18 ga. x 6" deep [G]				L.F.	1.69			1.69	1.86
0230	8" deep [G]					1.98			1.98	2.18
0240	10" deep [G]					2.34			2.34	2.58
0320	16 ga. x 6" deep [G]					2.07			2.07	2.28
0330	8" deep [G]					2.47			2.47	2.71
0340	10" deep [G]					2.89			2.89	3.18
0350	12" deep [G]					3.28			3.28	3.60
0430	14 ga. x 8" deep [G]					3.10			3.10	3.41
0440	10" deep [G]					3.57			3.57	3.93
0450	12" deep [G]					4.06			4.06	4.47
0540	12 ga. x 10" deep [G]					5.20			5.20	5.70
0550	12" deep [G]					5.90			5.90	6.50
1010	Installation of joists to band joists, beams & headers, labor only									
1220	18 ga. x 6" deep	2 Carp	110	.145	Ea.		7.05		7.05	11.50
1230	8" deep		90	.178			8.60		8.60	14.05
1240	10" deep		80	.200			9.70		9.70	15.85
1320	16 ga. x 6" deep		95	.168			8.15		8.15	13.35
1330	8" deep		70	.229			11.05		11.05	18.10
1340	10" deep		60	.267			12.90		12.90	21
1350	12" deep		55	.291			14.10		14.10	23
1430	14 ga. x 8" deep		65	.246			11.95		11.95	19.50
1440	10" deep		45	.356			17.25		17.25	28
1450	12" deep		35	.457			22		22	36
1540	12 ga. x 10" deep		40	.400			19.40		19.40	31.50
1550	12" deep		30	.533			26		26	42
9000	Minimum labor/equipment charge		4	4	Job		194		194	315

05 42 13.45 Framing, Web Stiffeners

05 42 13.45 Framing, Web Stiffeners		Crew	Daily Output	Labor-Hours	Unit	Material	2016 Bare Costs Labor	Equipment	Total	Total Incl O&P
0010	**FRAMING, WEB STIFFENERS** at joist bearing, fabricated from									
0020	Stud piece (1-5/8" flange) to stiffen joist (2" flange)									
0025	Made from recycled materials									
2120	For 6" deep joist, with 18 ga. x 2-1/2" stud [G]	1 Carp	120	.067	Ea.	.91	3.23		4.14	6.30
2130	3-5/8" stud [G]		110	.073		1.08	3.52		4.60	6.95
2140	4" stud [G]		105	.076		1.12	3.69		4.81	7.30
2150	6" stud [G]		100	.080		1.42	3.88		5.30	7.90
2160	8" stud [G]		95	.084		1.71	4.08		5.79	8.55
2220	8" deep joist, with 2-1/2" stud [G]		120	.067		1.22	3.23		4.45	6.65
2230	3-5/8" stud [G]		110	.073		1.45	3.52		4.97	7.35

For customer support on your Commercial Renovation Cost Data, call 877.742.4719.

115

05 42 Cold-Formed Metal Joist Framing

05 42 13 – Cold-Formed Metal Floor Joist Framing

05 42 13.45 Framing, Web Stiffeners		Crew	Daily Output	Labor-Hours	Unit	Material	2016 Bare Costs Labor	Equipment	Total	Total Incl O&P
2240	4" stud	1 Carp	105	.076	Ea.	1.50	3.69		5.19	7.70
2250	6" stud		100	.080		1.90	3.88		5.78	8.45
2260	8" stud		95	.084		2.29	4.08		6.37	9.15
2320	10" deep joist, with 2-1/2" stud		110	.073		1.51	3.52		5.03	7.40
2330	3-5/8" stud		100	.080		1.79	3.88		5.67	8.30
2340	4" stud		95	.084		1.86	4.08		5.94	8.70
2350	6" stud		90	.089		2.36	4.31		6.67	9.65
2360	8" stud		85	.094		2.84	4.56		7.40	10.55
2420	12" deep joist, with 2-1/2" stud		110	.073		1.82	3.52		5.34	7.75
2430	3-5/8" stud		100	.080		2.16	3.88		6.04	8.75
2440	4" stud		95	.084		2.24	4.08		6.32	9.10
2450	6" stud		90	.089		2.84	4.31		7.15	10.15
2460	8" stud		85	.094		3.42	4.56		7.98	11.20
3130	For 6" deep joist, with 16 ga. x 3-5/8" stud		100	.080		1.34	3.88		5.22	7.80
3140	4" stud		95	.084		1.40	4.08		5.48	8.20
3150	6" stud		90	.089		1.75	4.31		6.06	9
3160	8" stud		85	.094		2.14	4.56		6.70	9.80
3230	8" deep joist, with 3-5/8" stud		100	.080		1.80	3.88		5.68	8.35
3240	4" stud		95	.084		1.88	4.08		5.96	8.70
3250	6" stud		90	.089		2.35	4.31		6.66	9.65
3260	8" stud		85	.094		2.87	4.56		7.43	10.60
3330	10" deep joist, with 3-5/8" stud		85	.094		2.22	4.56		6.78	9.90
3340	4" stud		80	.100		2.32	4.85		7.17	10.45
3350	6" stud		75	.107		2.91	5.15		8.06	11.65
3360	8" stud		70	.114		3.55	5.55		9.10	12.95
3430	12" deep joist, with 3-5/8" stud		85	.094		2.68	4.56		7.24	10.40
3440	4" stud		80	.100		2.80	4.85		7.65	11
3450	6" stud		75	.107		3.50	5.15		8.65	12.30
3460	8" stud		70	.114		4.28	5.55		9.83	13.75
4230	For 8" deep joist, with 14 ga. x 3-5/8" stud		90	.089		2.22	4.31		6.53	9.50
4240	4" stud		85	.094		2.35	4.56		6.91	10.05
4250	6" stud		80	.100		2.95	4.85		7.80	11.15
4260	8" stud		75	.107		3.15	5.15		8.30	11.90
4330	10" deep joist, with 3-5/8" stud		75	.107		2.76	5.15		7.91	11.50
4340	4" stud		70	.114		2.91	5.55		8.46	12.25
4350	6" stud		65	.123		3.65	5.95		9.60	13.75
4360	8" stud		60	.133		3.90	6.45		10.35	14.85
4430	12" deep joist, with 3-5/8" stud		75	.107		3.32	5.15		8.47	12.10
4440	4" stud		70	.114		3.50	5.55		9.05	12.90
4450	6" stud		65	.123		4.40	5.95		10.35	14.60
4460	8" stud		60	.133		4.70	6.45		11.15	15.70
5330	For 10" deep joist, with 12 ga. x 3-5/8" stud		65	.123		3.97	5.95		9.92	14.10
5340	4" stud		60	.133		4.23	6.45		10.68	15.20
5350	6" stud		55	.145		5.35	7.05		12.40	17.40
5360	8" stud		50	.160		6.45	7.75		14.20	19.75
5430	12" deep joist, with 3-5/8" stud		65	.123		4.78	5.95		10.73	15
5440	4" stud		60	.133		5.10	6.45		11.55	16.15
5450	6" stud		55	.145		6.45	7.05		13.50	18.60
5460	8" stud		50	.160		7.80	7.75		15.55	21

05 42 Cold-Formed Metal Joist Framing

05 42 23 – Cold-Formed Metal Roof Joist Framing

05 42 23.05 Framing, Bracing

		Crew	Daily Output	Labor-Hours	Unit	Material	2016 Bare Costs Labor	Equipment	Total	Total Incl O&P
0010	**FRAMING, BRACING**									
0015	Made from recycled materials									
0020	Continuous bracing, per row									
0100	16 ga. x 1-1/2" channel thru rafters/trusses @ 16" O.C.	G 1 Carp	4.50	1.778	C.L.F.	51.50	86		137.50	198
0120	24" O.C.	G	6	1.333		51.50	64.50		116	163
0300	2" x 2" angle x 18 ga., rafters/trusses @ 16" O.C.	G	6	1.333		80.50	64.50		145	195
0320	24" O.C.	G	8	1		80.50	48.50		129	168
0400	16 ga., rafters/trusses @ 16" O.C.	G	4.50	1.778		101	86		187	252
0420	24" O.C.	G	6.50	1.231		101	59.50		160.50	209

05 42 23.10 Framing, Bridging

		Crew	Daily Output	Labor-Hours	Unit	Material	2016 Bare Costs Labor	Equipment	Total	Total Incl O&P
0010	**FRAMING, BRIDGING**									
0015	Made from recycled materials									
0020	Solid, between rafters w/1-1/4" leg track, per rafter bay									
1200	Rafters 16" O.C., 18 ga. x 4" deep	G 1 Carp	60	.133	Ea.	1.57	6.45		8.02	12.25
1210	6" deep	G	57	.140		2.05	6.80		8.85	13.35
1220	8" deep	G	55	.145		2.52	7.05		9.57	14.30
1230	10" deep	G	52	.154		3.15	7.45		10.60	15.65
1240	12" deep	G	50	.160		3.56	7.75		11.31	16.55
2200	24" O.C., 18 ga. x 4" deep	G	60	.133		2.27	6.45		8.72	13.05
2210	6" deep	G	57	.140		2.96	6.80		9.76	14.35
2220	8" deep	G	55	.145		3.65	7.05		10.70	15.50
2230	10" deep	G	52	.154		4.56	7.45		12.01	17.20
2240	12" deep	G	50	.160		5.15	7.75		12.90	18.30

05 42 23.50 Framing, Parapets

		Crew	Daily Output	Labor-Hours	Unit	Material	2016 Bare Costs Labor	Equipment	Total	Total Incl O&P
0010	**FRAMING, PARAPETS**									
0015	Made from recycled materials									
0100	3' high installed on 1st story, 18 ga. x 4" wide studs, 12" O.C.	G 2 Carp	100	.160	L.F.	5.70	7.75		13.45	18.90
0110	16" O.C.	G	150	.107		4.85	5.15		10	13.80
0120	24" O.C.	G	200	.080		4.01	3.88		7.89	10.75
0200	6" wide studs, 12" O.C.	G	100	.160		7.30	7.75		15.05	20.50
0210	16" O.C.	G	150	.107		6.25	5.15		11.40	15.30
0220	24" O.C.	G	200	.080		5.15	3.88		9.03	12.05
1100	Installed on 2nd story, 18 ga. x 4" wide studs, 12" O.C.	G	95	.168		5.70	8.15		13.85	19.60
1110	16" O.C.	G	145	.110		4.85	5.35		10.20	14.10
1120	24" O.C.	G	190	.084		4.01	4.08		8.09	11.05
1200	6" wide studs, 12" O.C.	G	95	.168		7.30	8.15		15.45	21.50
1210	16" O.C.	G	145	.110		6.25	5.35		11.60	15.60
1220	24" O.C.	G	190	.084		5.15	4.08		9.23	12.35
2100	Installed on gable, 18 ga. x 4" wide studs, 12" O.C.	G	85	.188		5.70	9.10		14.80	21
2110	16" O.C.	G	130	.123		4.85	5.95		10.80	15.10
2120	24" O.C.	G	170	.094		4.01	4.56		8.57	11.85
2200	6" wide studs, 12" O.C.	G	85	.188		7.30	9.10		16.40	23
2210	16" O.C.	G	130	.123		6.25	5.95		12.20	16.60
2220	24" O.C.	G	170	.094		5.15	4.56		9.71	13.15

05 42 23.60 Framing, Roof Rafters

		Crew	Daily Output	Labor-Hours	Unit	Material	2016 Bare Costs Labor	Equipment	Total	Total Incl O&P
0010	**FRAMING, ROOF RAFTERS**									
0015	Made from recycled materials									
0100	Boxed ridge beam, double, 18 ga. x 6" deep	G 2 Carp	160	.100	L.F.	5.50	4.85		10.35	13.95
0110	8" deep	G	150	.107		6.05	5.15		11.20	15.10
0120	10" deep	G	140	.114		7.40	5.55		12.95	17.20
0130	12" deep	G	130	.123		8.05	5.95		14	18.60

For customer support on your Commercial Renovation Cost Data, call 877.742.4719.

117

05 42 Cold-Formed Metal Joist Framing

05 42 23 – Cold-Formed Metal Roof Joist Framing

05 42 23.60 Framing, Roof Rafters

		Crew	Daily Output	Labor-Hours	Unit	Material	2016 Bare Costs Labor	Equipment	Total	Total Incl O&P
0200	16 ga. x 6" deep	G 2 Carp	150	.107	L.F.	6.20	5.15		11.35	15.30
0210	8" deep	G	140	.114		7	5.55		12.55	16.75
0220	10" deep	G	130	.123		8.45	5.95		14.40	19
0230	12" deep	G	120	.133		9.15	6.45		15.60	20.50
1100	Rafters, 2" flange, material only, 18 ga. x 6" deep	G				1.69			1.69	1.86
1110	8" deep	G				1.98			1.98	2.18
1120	10" deep	G				2.34			2.34	2.58
1130	12" deep	G				2.70			2.70	2.97
1200	16 ga. x 6" deep	G				2.07			2.07	2.28
1210	8" deep	G				2.47			2.47	2.71
1220	10" deep	G				2.89			2.89	3.18
1230	12" deep	G				3.28			3.28	3.60
2100	Installation only, ordinary rafter to 4:12 pitch, 18 ga. x 6" deep	2 Carp	35	.457	Ea.		22		22	36
2110	8" deep		30	.533			26		26	42
2120	10" deep		25	.640			31		31	50.50
2130	12" deep		20	.800			39		39	63.50
2200	16 ga. x 6" deep		30	.533			26		26	42
2210	8" deep		25	.640			31		31	50.50
2220	10" deep		20	.800			39		39	63.50
2230	12" deep		15	1.067			51.50		51.50	84.50
8100	Add to labor, ordinary rafters on steep roofs						25%			
8110	Dormers & complex roofs						50%			
8200	Hip & valley rafters to 4:12 pitch						25%			
8210	Steep roofs						50%			
8220	Dormers & complex roofs						75%			
8300	Hip & valley jack rafters to 4:12 pitch						50%			
8310	Steep roofs						75%			
8320	Dormers & complex roofs						100%			
9000	Minimum labor/equipment charge	2 Carp	4	4	Job		194		194	315

05 42 23.70 Framing, Soffits and Canopies

		Crew	Daily Output	Labor-Hours	Unit	Material	2016 Bare Costs Labor	Equipment	Total	Total Incl O&P
0010	**FRAMING, SOFFITS & CANOPIES**									
0015	Made from recycled materials									
0130	Continuous ledger track @ wall, studs @ 16" O.C., 18 ga. x 4" wide	G 2 Carp	535	.030	L.F.	1.05	1.45		2.50	3.52
0140	6" wide	G	500	.032		1.36	1.55		2.91	4.03
0150	8" wide	G	465	.034		1.68	1.67		3.35	4.57
0160	10" wide	G	430	.037		2.10	1.80		3.90	5.25
0230	Studs @ 24" O.C., 18 ga. x 4" wide	G	800	.020		1	.97		1.97	2.68
0240	6" wide	G	750	.021		1.30	1.03		2.33	3.12
0250	8" wide	G	700	.023		1.61	1.11		2.72	3.58
0260	10" wide	G	650	.025		2.01	1.19		3.20	4.16
1000	Horizontal soffit and canopy members, material only									
1030	1-5/8" flange studs, 18 ga. x 4" deep	G			L.F.	1.34			1.34	1.48
1040	6" deep	G				1.70			1.70	1.87
1050	8" deep	G				2.05			2.05	2.26
1140	2" flange joists, 18 ga. x 6" deep	G				1.93			1.93	2.13
1150	8" deep	G				2.27			2.27	2.49
1160	10" deep	G				2.68			2.68	2.94
4030	Installation only, 18 ga., 1-5/8" flange x 4" deep	2 Carp	130	.123	Ea.		5.95		5.95	9.75
4040	6" deep		110	.145			7.05		7.05	11.50
4050	8" deep		90	.178			8.60		8.60	14.05
4140	2" flange, 18 ga. x 6" deep		110	.145			7.05		7.05	11.50
4150	8" deep		90	.178			8.60		8.60	14.05

05 42 Cold-Formed Metal Joist Framing

05 42 23 – Cold-Formed Metal Roof Joist Framing

05 42 23.70 Framing, Soffits and Canopies

		Crew	Daily Output	Labor-Hours	Unit	Material	2016 Bare Costs Labor	Equipment	Total	Total Incl O&P
4160	10" deep	2 Carp	80	.200	Ea.		9.70		9.70	15.85
6010	Clips to attach fascia to rafter tails, 2" x 2" x 18 ga. angle ⒢	1 Carp	120	.067		.95	3.23		4.18	6.35
6020	16 ga. angle ⒢	"	100	.080	▼	1.20	3.88		5.08	7.65
9000	Minimum labor/equipment charge	2 Carp	4	4	Job		194		194	315

05 44 Cold-Formed Metal Trusses

05 44 13 – Cold-Formed Metal Roof Trusses

05 44 13.60 Framing, Roof Trusses

			Crew	Daily Output	Labor-Hours	Unit	Material	2016 Bare Costs Labor	Equipment	Total	Total Incl O&P
0010	**FRAMING, ROOF TRUSSES**										
0015	Made from recycled materials										
0020	Fabrication of trusses on ground, Fink (W) or King Post, to 4:12 pitch										
0120	18 ga. x 4" chords, 16' span	⒢	2 Carp	12	1.333	Ea.	62.50	64.50		127	175
0130	20' span	⒢		11	1.455		78.50	70.50		149	201
0140	24' span	⒢		11	1.455		94	70.50		164.50	218
0150	28' span	⒢		10	1.600		110	77.50		187.50	248
0160	32' span	⒢		10	1.600		125	77.50		202.50	265
0250	6" chords, 28' span	⒢		9	1.778		139	86		225	294
0260	32' span	⒢		9	1.778		159	86		245	315
0270	36' span	⒢		8	2		179	97		276	355
0280	40' span	⒢		8	2		199	97		296	375
1120	5:12 to 8:12 pitch, 18 ga. x 4" chords, 16' span	⒢		10	1.600		71.50	77.50		149	206
1130	20' span	⒢		9	1.778		89.50	86		175.50	240
1140	24' span	⒢		9	1.778		108	86		194	259
1150	28' span	⒢		8	2		125	97		222	296
1160	32' span	⒢		8	2		143	97		240	315
1250	6" chords, 28' span	⒢		7	2.286		159	111		270	355
1260	32' span	⒢		7	2.286		182	111		293	380
1270	36' span	⒢		6	2.667		204	129		333	435
1280	40' span	⒢		6	2.667		227	129		356	460
2120	9:12 to 12:12 pitch, 18 ga. x 4" chords, 16' span	⒢		8	2		89.50	97		186.50	257
2130	20' span	⒢		7	2.286		112	111		223	305
2140	24' span	⒢		7	2.286		134	111		245	330
2150	28' span	⒢		6	2.667		157	129		286	385
2160	32' span	⒢		6	2.667		179	129		308	410
2250	6" chords, 28' span	⒢		5	3.200		199	155		354	470
2260	32' span	⒢		5	3.200		227	155		382	505
2270	36' span	⒢		4	4		256	194		450	595
2280	40' span	⒢		4	4	▼	284	194		478	625
4900	Minimum labor/equipment charge		▼	4	4	Job		194		194	315
5120	Erection only of roof trusses, to 4:12 pitch, 16' span		F-6	48	.833	Ea.		37.50	13.70	51.20	76
5130	20' span			46	.870			39	14.30	53.30	79.50
5140	24' span			44	.909			41	14.95	55.95	83
5150	28' span			42	.952			43	15.70	58.70	87
5160	32' span			40	1			45	16.45	61.45	91
5170	36' span			38	1.053			47.50	17.35	64.85	96
5180	40' span			36	1.111			50	18.30	68.30	101
5220	5:12 to 8:12 pitch, 16' span			42	.952			43	15.70	58.70	87
5230	20' span			40	1			45	16.45	61.45	91
5240	24' span			38	1.053			47.50	17.35	64.85	96
5250	28' span			36	1.111			50	18.30	68.30	101
5260	32' span			34	1.176			53	19.35	72.35	108

For customer support on your Commercial Renovation Cost Data, call 877.742.4719.

119

05 44 Cold-Formed Metal Trusses

05 44 13 – Cold-Formed Metal Roof Trusses

05 44 13.60 Framing, Roof Trusses

		Crew	Daily Output	Labor-Hours	Unit	Material	2016 Bare Costs Labor	Equipment	Total	Total Incl O&P
5270	36' span	F-6	32	1.250	Ea.		56	20.50	76.50	114
5280	40' span		30	1.333			60	22	82	122
5320	9:12 to 12:12 pitch, 16' span		36	1.111			50	18.30	68.30	101
5330	20' span		34	1.176			53	19.35	72.35	108
5340	24' span		32	1.250			56	20.50	76.50	114
5350	28' span		30	1.333			60	22	82	122
5360	32' span		28	1.429			64.50	23.50	88	130
5370	36' span		26	1.538			69	25.50	94.50	140
5380	40' span		24	1.667	▼		75	27.50	102.50	152
9000	Minimum labor/equipment charge	▼	2	20	Job		900	330	1,230	1,800

05 51 Metal Stairs

05 51 13 – Metal Pan Stairs

05 51 13.50 Pan Stairs

			Crew	Daily Output	Labor-Hours	Unit	Material	2016 Bare Costs Labor	Equipment	Total	Total Incl O&P
0010	**PAN STAIRS**, shop fabricated, steel stringers										
0015	Made from recycled materials										
0200	Cement fill metal pan, picket rail, 3'-6" wide	G	E-4	35	.914	Riser	500	49	4.21	553.21	640
0300	4'-0" wide	G		30	1.067		560	57.50	4.91	622.41	720
0350	Wall rail, both sides, 3'-6" wide	G		53	.604	▼	380	32.50	2.78	415.28	480
1500	Landing, steel pan, conventional	G	▼	160	.200	S.F.	66	10.75	.92	77.67	93

05 51 16 – Metal Floor Plate Stairs

05 51 16.50 Floor Plate Stairs

			Crew	Daily Output	Labor-Hours	Unit	Material	2016 Bare Costs Labor	Equipment	Total	Total Incl O&P
0010	**FLOOR PLATE STAIRS**, shop fabricated, steel stringers										
0015	Made from recycled materials										
0400	Cast iron tread and pipe rail, 3'-6" wide	G	E-4	35	.914	Riser	535	49	4.21	588.21	675
0450	4'-0" wide	G		30	1.067		560	57.50	4.91	622.41	720
0475	5'-0" wide	G		28	1.143		615	61.50	5.25	681.75	795
0500	Checkered plate tread, industrial, 3'-6" wide	G		28	1.143		330	61.50	5.25	396.75	475
0550	Circular, for tanks, 3'-0" wide	G	▼	33	.970		370	52	4.46	426.46	500
0600	For isolated stairs, add							100%			
0800	Custom steel stairs, 3'-6" wide, economy	G	E-4	35	.914		500	49	4.21	553.21	640
0810	Medium priced	G		30	1.067		660	57.50	4.91	722.41	835
0900	Deluxe	G	▼	20	1.600		825	86	7.35	918.35	1,075
1100	For 4' wide stairs, add							5%	5%		
1300	For 5' wide stairs, add					▼	10%	10%			

05 51 19 – Metal Grating Stairs

05 51 19.50 Grating Stairs

			Crew	Daily Output	Labor-Hours	Unit	Material	2016 Bare Costs Labor	Equipment	Total	Total Incl O&P
0010	**GRATING STAIRS**, shop fabricated, steel stringers, safety nosing on treads										
0015	Made from recycled materials										
0020	Grating tread and pipe railing, 3'-6" wide	G	E-4	35	.914	Riser	330	49	4.21	383.21	450
0100	4'-0" wide	G		30	1.067	"	430	57.50	4.91	492.41	580
9000	Minimum labor/equipment charge		▼	2	16	Job		860	73.50	933.50	1,600

05 51 23 – Metal Fire Escapes

05 51 23.25 Fire Escapes

			Crew	Daily Output	Labor-Hours	Unit	Material	2016 Bare Costs Labor	Equipment	Total	Total Incl O&P
0010	**FIRE ESCAPES**, shop fabricated										
0200	2' wide balcony, 1" x 1/4" bars 1-1/2" O.C., with railing	G	2 Sswk	10	1.600	L.F.	58.50	85		143.50	215
0400	1st story cantilevered stair, standard, with railing	G		.50	32	Ea.	2,425	1,700		4,125	5,675
0500	Cable counterweighted, with railing	G		.40	40	"	2,250	2,125		4,375	6,225
0700	36" x 40" platform & fixed stair, with railing	G	▼	.40	40	Flight	1,075	2,125		3,200	4,925

05 51 Metal Stairs

05 51 23 – Metal Fire Escapes

05 51 23.25 Fire Escapes		Crew	Daily Output	Labor-Hours	Unit	Material	2016 Bare Costs Labor	Equipment	Total	Total Incl O&P
0900	For 3'-6" wide escapes, add to above					100%	150%			

05 51 23.50 Fire Escape Stairs

0010	**FIRE ESCAPE STAIRS**, portable									
0100	Portable ladder				Ea.	122			122	135

05 51 33 – Metal Ladders

05 51 33.13 Vertical Metal Ladders

			Crew	Daily Output	Labor-Hours	Unit	Material	Labor	Equipment	Total	Total Incl O&P
0010	**VERTICAL METAL LADDERS**, shop fabricated										
0015	Made from recycled materials										
0020	Steel, 20" wide, bolted to concrete, with cage	G	E-4	50	.640	V.L.F.	65.50	34.50	2.94	102.94	136
0100	Without cage	G		85	.376		36.50	20	1.73	58.23	77.50
0300	Aluminum, bolted to concrete, with cage	G		50	.640		120	34.50	2.94	157.44	196
0400	Without cage	G		85	.376		51	20	1.73	72.73	93.50
9000	Minimum labor/equipment charge			2	16	Job		860	73.50	933.50	1,600

05 51 33.16 Inclined Metal Ladders

			Crew	Daily Output	Labor-Hours	Unit	Material	Labor	Equipment	Total	Total Incl O&P
0010	**INCLINED METAL LADDERS**, shop fabricated										
0015	Made from recycled materials										
3900	Industrial ships ladder, steel, 24" W, grating treads, 2 line pipe rail	G	E-4	30	1.067	Riser	193	57.50	4.91	255.41	320
4000	Aluminum	G	"	30	1.067	"	270	57.50	4.91	332.41	405

05 51 33.23 Alternating Tread Ladders

		Crew	Daily Output	Labor-Hours	Unit	Material	Labor	Equipment	Total	Total Incl O&P
0010	**ALTERNATING TREAD LADDERS**, shop fabricated									
0015	Made from recycled materials									
0800	Alternating tread ladders, 68-degree angle of incline									
0810	8 foot vertical rise, steel, 149 lb., standard paint color	B-68G	3	5.333	Ea.	2,300	284	111	2,695	3,150
0820	Non-standard paint color		3	5.333		2,650	284	111	3,045	3,550
0830	Galvanized		3	5.333		2,675	284	111	3,070	3,575
0840	Stainless		3	5.333		3,875	284	111	4,270	4,900
0850	Aluminum, 87 lb.		3	5.333		2,825	284	111	3,220	3,750
1010	10 foot vertical rise, steel, 181 lb., standard paint color		2.75	5.818		2,775	310	121	3,206	3,725
1020	Non-standard paint color		2.75	5.818		3,175	310	121	3,606	4,175
1030	Galvanized		2.75	5.818		3,225	310	121	3,656	4,225
1040	Stainless		2.75	5.818		4,675	310	121	5,106	5,825
1050	Aluminum, 103 lb.		2.75	5.818		3,425	310	121	3,856	4,425
1210	12 foot vertical rise, steel, 245 lb., standard paint color		2.50	6.400		3,250	340	134	3,724	4,325
1220	Non-standard paint color		2.50	6.400		3,700	340	134	4,174	4,825
1230	Galvanized		2.50	6.400		3,800	340	134	4,274	4,925
1240	Stainless		2.50	6.400		5,450	340	134	5,924	6,750
1250	Aluminum, 103 lb.		2.50	6.400		4,000	340	134	4,474	5,150
1410	14 foot vertical rise, steel, 281 lb., standard paint color		2.25	7.111		3,725	380	148	4,253	4,925
1420	Non-standard paint color		2.25	7.111		4,225	380	148	4,753	5,475
1430	Galvanized		2.25	7.111		4,350	380	148	4,878	5,625
1440	Stainless		2.25	7.111		6,250	380	148	6,778	7,700
1450	Aluminum, 136 lb.		2.25	7.111		4,575	380	148	5,103	5,850
1610	16 foot vertical rise, steel, 317 lb., standard paint color		2	8		4,225	425	167	4,817	5,575
1620	Non-standard paint color		2	8		4,725	425	167	5,317	6,125
1630	Galvanized		2	8		4,925	425	167	5,517	6,325
1640	Stainless		2	8		7,050	425	167	7,642	8,675
1650	Aluminum, 153 lb.		2	8		5,150	425	167	5,742	6,600

For customer support on your Commercial Renovation Cost Data, call 877.742.4719.

121

05 52 Metal Railings

05 52 13 - Pipe and Tube Railings

05 52 13.50 Railings, Pipe		Crew	Daily Output	Labor-Hours	Unit	Material	2016 Bare Costs Labor	Equipment	Total	Total Incl O&P
0010	**RAILINGS, PIPE**, shop fab'd, 3'-6" high, posts @ 5' O.C.									
0015	Made from recycled materials									
0020	Aluminum, 2 rail, satin finish, 1-1/4" diameter	G	E-4	160	.200	L.F.	50.50	10.75	.92	62.17	75.50
0030	Clear anodized	G		160	.200		62	10.75	.92	73.67	88
0040	Dark anodized	G		160	.200		69	10.75	.92	80.67	96
0080	1-1/2" diameter, satin finish	G		160	.200		59	10.75	.92	70.67	85
0090	Clear anodized	G		160	.200		66.50	10.75	.92	78.17	93
0100	Dark anodized	G		160	.200		73	10.75	.92	84.67	100
0140	Aluminum, 3 rail, 1-1/4" diam., satin finish	G		137	.234		65	12.55	1.07	78.62	94.50
0150	Clear anodized	G		137	.234		81.50	12.55	1.07	95.12	113
0160	Dark anodized	G		137	.234		90	12.55	1.07	103.62	122
0200	1-1/2" diameter, satin finish	G		137	.234		77.50	12.55	1.07	91.12	109
0210	Clear anodized	G		137	.234		88.50	12.55	1.07	102.12	121
0220	Dark anodized	G		137	.234		96.50	12.55	1.07	110.12	129
0500	Steel, 2 rail, on stairs, primed, 1-1/4" diameter	G		160	.200		25	10.75	.92	36.67	47.50
0520	1-1/2" diameter	G		160	.200		27	10.75	.92	38.67	50
0540	Galvanized, 1-1/4" diameter	G		160	.200		34	10.75	.92	45.67	57.50
0560	1-1/2" diameter	G		160	.200		38.50	10.75	.92	50.17	62
0580	Steel, 3 rail, primed, 1-1/4" diameter	G		137	.234		37.50	12.55	1.07	51.12	64
0600	1-1/2" diameter	G		137	.234		39	12.55	1.07	52.62	66
0620	Galvanized, 1-1/4" diameter	G		137	.234		52.50	12.55	1.07	66.12	80.50
0640	1-1/2" diameter	G		137	.234		60.50	12.55	1.07	74.12	89.50
0700	Stainless steel, 2 rail, 1-1/4" diam. #4 finish	G		137	.234		110	12.55	1.07	123.62	144
0720	High polish	G		137	.234		178	12.55	1.07	191.62	219
0740	Mirror polish	G		137	.234		223	12.55	1.07	236.62	268
0760	Stainless steel, 3 rail, 1-1/2" diam., #4 finish	G		120	.267		166	14.30	1.23	181.53	210
0770	High polish	G		120	.267		275	14.30	1.23	290.53	330
0780	Mirror finish	G		120	.267		335	14.30	1.23	350.53	395
0900	Wall rail, alum. pipe, 1-1/4" diam., satin finish	G		213	.150		24.50	8.05	.69	33.24	41.50
0905	Clear anodized	G		213	.150		30	8.05	.69	38.74	48
0910	Dark anodized	G		213	.150		35.50	8.05	.69	44.24	54
0915	1-1/2" diameter, satin finish	G		213	.150		27	8.05	.69	35.74	45
0920	Clear anodized	G		213	.150		34	8.05	.69	42.74	52
0925	Dark anodized	G		213	.150		42	8.05	.69	50.74	61
0930	Steel pipe, 1-1/4" diameter, primed	G		213	.150		15	8.05	.69	23.74	31.50
0935	Galvanized	G		213	.150		22	8.05	.69	30.74	39
0940	1-1/2" diameter	G		176	.182		15.50	9.75	.84	26.09	35
0945	Galvanized	G		213	.150		22	8.05	.69	30.74	39
0955	Stainless steel pipe, 1-1/2" diam., #4 finish	G		107	.299		88	16.05	1.38	105.43	127
0960	High polish	G		107	.299		179	16.05	1.38	196.43	227
0965	Mirror polish	G		107	.299		212	16.05	1.38	229.43	263
2000	2-line pipe rail (1-1/2" T&B) with 1/2" pickets @ 4-1/2" O.C.,										
2005	attached handrail on brackets										
2010	42" high aluminum, satin finish, straight & level	G	E-4	120	.267	L.F.	257	14.30	1.23	272.53	310
2050	42" high steel, primed, straight & level	G	"	120	.267		127	14.30	1.23	142.53	167
4000	For curved and level rails, add						10%	10%			
4100	For sloped rails for stairs, add						30%	30%			
9000	Minimum labor/equipment charge		1 Sswk	2	4	Job		213		213	375

05 52 Metal Railings

05 52 16 – Industrial Railings

05 52 16.50 Railings, Industrial		Crew	Daily Output	Labor-Hours	Unit	Material	2016 Bare Costs Labor	Equipment	Total	Total Incl O&P
0010	**RAILINGS, INDUSTRIAL**, shop fab'd, 3'-6" high, posts @ 5' O.C.	G								
0020	2 rail, 3'-6" high, 1-1/2" pipe	G	E-4	255	.125	L.F.	27	6.75	.58	34.33	42.50
0200	For 4" high kick plate, 10 ga., add	G					5.65			5.65	6.20
0500	For curved level rails, add						10%	10%			
0550	For sloped rails for stairs, add						30%	30%			
9000	Minimum labor/equipment charge		1 Sswk	2	4	Job		213		213	375

05 55 Metal Stair Treads and Nosings

05 55 13 – Metal Stair Treads

05 55 13.50 Stair Treads

05 55 13.50 Stair Treads		Crew	Daily Output	Labor-Hours	Unit	Material	2016 Bare Costs Labor	Equipment	Total	Total Incl O&P
0010	**STAIR TREADS**, stringers and bolts not included										
3000	Diamond plate treads, steel, 1/8" thick										
3005	Open riser, black enamel										
3010	9" deep x 36" long	G	2 Sswk	48	.333	Ea.	91.50	17.75		109.25	132
3020	42" long	G		48	.333		96.50	17.75		114.25	138
3030	48" long	G		48	.333		101	17.75		118.75	143
3040	11" deep x 36" long	G		44	.364		96.50	19.35		115.85	140
3050	42" long	G		44	.364		102	19.35		121.35	146
3060	48" long	G		44	.364		108	19.35		127.35	153
3110	Galvanized, 9" deep x 36" long	G		48	.333		144	17.75		161.75	191
3120	42" long	G		48	.333		154	17.75		171.75	201
3130	48" long	G		48	.333		163	17.75		180.75	211
3140	11" deep x 36" long	G		44	.364		149	19.35		168.35	198
3150	42" long	G		44	.364		163	19.35		182.35	213
3160	48" long	G		44	.364		169	19.35		188.35	220
3200	Closed riser, black enamel										
3210	12" deep x 36" long	G	2 Sswk	40	.400	Ea.	110	21.50		131.50	159
3220	42" long	G		40	.400		119	21.50		140.50	169
3230	48" long	G		40	.400		126	21.50		147.50	176
3240	Galvanized, 12" deep x 36" long	G		40	.400		173	21.50		194.50	229
3250	42" long	G		40	.400		189	21.50		210.50	246
3260	48" long	G		40	.400		198	21.50		219.50	256
4000	Bar grating treads										
4005	Steel, 1-1/4" x 3/16" bars, anti-skid nosing, black enamel										
4010	8-5/8" deep x 30" long	G	2 Sswk	48	.333	Ea.	54	17.75		71.75	91
4020	36" long	G		48	.333		63.50	17.75		81.25	101
4030	48" long	G		48	.333		97.50	17.75		115.25	139
4040	10-15/16" deep x 36" long	G		44	.364		70	19.35		89.35	111
4050	48" long	G		44	.364		100	19.35		119.35	144
4060	Galvanized, 8-5/8" deep x 30" long	G		48	.333		62	17.75		79.75	100
4070	36" long	G		48	.333		73.50	17.75		91.25	113
4080	48" long	G		48	.333		108	17.75		125.75	151
4090	10-15/16" deep x 36" long	G		44	.364		86.50	19.35		105.85	129
4100	48" long	G		44	.364		112	19.35		131.35	157
4200	Aluminum, 1-1/4" x 3/16" bars, serrated, with nosing										
4210	7-5/8" deep x 18" long	G	2 Sswk	52	.308	Ea.	50	16.35		66.35	84
4220	24" long	G		52	.308		59	16.35		75.35	94
4230	30" long	G		52	.308		68.50	16.35		84.85	104
4240	36" long	G		52	.308		177	16.35		193.35	224
4250	8-13/16" deep x 18" long	G		48	.333		67	17.75		84.75	106

For customer support on your Commercial Renovation Cost Data, call 877.742.4719.

123

05 55 13 – Metal Stair Treads

05 55 13.50 Stair Treads		Crew	Daily Output	Labor-Hours	Unit	Material	2016 Bare Costs Labor	Equipment	Total	Total Incl O&P
4260	24" long	G 2 Sswk	48	.333	Ea.	99.50	17.75		117.25	141
4270	30" long	G	48	.333		115	17.75		132.75	158
4280	36" long	G	48	.333		194	17.75		211.75	246
4290	10" deep x 18" long	G	44	.364		130	19.35		149.35	177
4300	30" long	G	44	.364		173	19.35		192.35	224
4310	36" long	G	44	.364		210	19.35		229.35	265
5000	Channel grating treads									
5005	Steel, 14 ga., 2-1/2" thick, galvanized									
5010	9" deep x 36" long	G 2 Sswk	48	.333	Ea.	103	17.75		120.75	145
5020	48" long	G	48	.333	"	139	17.75		156.75	184
9000	Minimum labor/equipment charge		4	4	Job		213		213	375

05 55 19 – Metal Stair Tread Covers

05 55 19.50 Stair Tread Covers for Renovation

		Crew	Daily Output	Labor-Hours	Unit	Material	2016 Bare Costs Labor	Equipment	Total	Total Incl O&P
0010	**STAIR TREAD COVERS FOR RENOVATION**									
0205	Extruded tread cover with nosing, pre-drilled, includes screws									
0210	Aluminum with black abrasive strips, 9" wide x 3' long	1 Carp	24	.333	Ea.	110	16.15		126.15	148
0220	4' long		22	.364		145	17.60		162.60	188
0230	5' long		20	.400		180	19.40		199.40	230
0240	11" wide x 3' long		24	.333		142	16.15		158.15	183
0250	4' long		22	.364		182	17.60		199.60	230
0260	5' long		20	.400		235	19.40		254.40	290
0305	Black abrasive strips with yellow front strips									
0310	Aluminum, 9" wide x 3' long	1 Carp	24	.333	Ea.	118	16.15		134.15	157
0320	4' long		22	.364		157	17.60		174.60	202
0330	5' long		20	.400		200	19.40		219.40	252
0340	11" wide x 3' long		24	.333		152	16.15		168.15	194
0350	4' long		22	.364		196	17.60		213.60	245
0360	5' long		20	.400		253	19.40		272.40	310
0405	Black abrasive strips with photoluminescent front strips									
0410	Aluminum, 9" wide x 3' long	1 Carp	24	.333	Ea.	141	16.15		157.15	182
0420	4' long		22	.364		175	17.60		192.60	222
0430	5' long		20	.400		219	19.40		238.40	273
0440	11" wide x 3' long		24	.333		156	16.15		172.15	198
0450	4' long		22	.364		208	17.60		225.60	258
0460	5' long		20	.400		260	19.40		279.40	320
9000	Minimum labor/equipment charge		4	2	Job		97		97	158

05 56 Metal Castings

05 56 13 – Metal Construction Castings

05 56 13.50 Construction Castings

		Crew	Daily Output	Labor-Hours	Unit	Material	2016 Bare Costs Labor	Equipment	Total	Total Incl O&P
0010	**CONSTRUCTION CASTINGS**									
0020	Manhole covers and frames, see Section 33 44 13.13									
0100	Column bases, cast iron, 16" x 16", approx. 65 lb.	G E-4	46	.696	Ea.	142	37.50	3.20	182.70	226
0200	32" x 32", approx. 256 lb.	G	23	1.391	"	520	74.50	6.40	600.90	715
0600	Miscellaneous C.I. castings, light sections, less than 150 lb.	G	3200	.010	Lb.	8.95	.54	.05	9.54	10.85
1300	Special low volume items	G	3200	.010	"	11.25	.54	.05	11.84	13.35

05 58 Formed Metal Fabrications

05 58 13 – Column Covers

05 58 13.05 Column Covers		Crew	Daily Output	Labor-Hours	Unit	Material	2016 Bare Costs Labor	Equipment	Total	Total Incl O&P
0010	**COLUMN COVERS**									
0015	Made from recycled materials									
0020	Excludes structural steel, light ga. metal framing, misc. metals, sealants									
0100	Round covers, 2 halves with 2 vertical joints for backer rod and sealant									
0110	Up to 12' high, no horizontal joints									
0120	12" diameter, 0.125" aluminum, anodized/painted finish	G	2 Sswk	32	.500	V.L.F.	31	26.50		57.50	81
0130	Type 304 stainless steel, 16 gauge, #4 brushed finish	G		32	.500		46	26.50		72.50	97.50
0140	Type 316 stainless steel, 16 gauge, #4 brushed finish	G		32	.500		54.50	26.50		81	107
0150	18" diameter, aluminum	G		32	.500		46.50	26.50		73	98
0160	Type 304 stainless steel	G		32	.500		69	26.50		95.50	123
0170	Type 316 stainless steel	G		32	.500		82	26.50		108.50	138
0180	24" diameter, aluminum	G		32	.500		62	26.50		88.50	115
0190	Type 304 stainless steel	G		32	.500		91.50	26.50		118	148
0200	Type 316 stainless steel	G		32	.500		109	26.50		135.50	167
0210	30" diameter, aluminum	G		30	.533		77.50	28.50		106	135
0220	Type 304 stainless steel	G		30	.533		115	28.50		143.50	176
0230	Type 316 stainless steel	G		30	.533		137	28.50		165.50	201
0240	36" diameter, aluminum	G		30	.533		93	28.50		121.50	152
0250	Type 304 stainless steel	G		30	.533		138	28.50		166.50	201
0260	Type 316 stainless steel	G		30	.533		164	28.50		192.50	231
0400	Up to 24' high, 2 stacked sections with 1 horizontal joint									
0410	18" diameter, aluminum	G	2 Sswk	28	.571	V.L.F.	48.50	30.50		79	107
0450	Type 304 stainless steel	G		28	.571		72	30.50		102.50	133
0460	Type 316 stainless steel	G		28	.571		86	30.50		116.50	149
0470	24" diameter, aluminum	G		28	.571		65	30.50		95.50	125
0480	Type 304 stainless steel	G		28	.571		96.50	30.50		127	160
0490	Type 316 stainless steel	G		28	.571		115	30.50		145.50	180
0500	30" diameter, aluminum	G		24	.667		81	35.50		116.50	152
0510	Type 304 stainless steel	G		24	.667		120	35.50		155.50	195
0520	Type 316 stainless steel	G		24	.667		144	35.50		179.50	221
0530	36" diameter, aluminum	G		24	.667		97.50	35.50		133	170
0540	Type 304 stainless steel	G		24	.667		144	35.50		179.50	222
0550	Type 316 stainless steel	G		24	.667		172	35.50		207.50	253

05 58 23 – Formed Metal Guards

05 58 23.90 Window Guards

		Crew	Daily Output	Labor-Hours	Unit	Material	2016 Bare Costs Labor	Equipment	Total	Total Incl O&P
0010	**WINDOW GUARDS,** shop fabricated									
0015	Expanded metal, steel angle frame, permanent	G	E-4	350	.091	S.F.	23	4.91	.42	28.33	34
0025	Steel bars, 1/2" x 1/2", spaced 5" O.C.	G	"	290	.110	"	15.90	5.95	.51	22.36	28.50
0030	Hinge mounted, add	G				Opng.	46			46	50.50
0040	Removable type, add	G				"	29			29	32
0050	For galvanized guards, add					S.F.	35%				
0070	For pivoted or projected type, add						105%	40%			
0100	Mild steel, stock units, economy	G	E-4	405	.079		6.25	4.24	.36	10.85	14.80
0200	Deluxe	G		405	.079		12.70	4.24	.36	17.30	22
0400	Woven wire, stock units, 3/8" channel frame, 3' x 5' opening	G		40	.800	Opng.	169	43	3.68	215.68	266
0500	4' x 6' opening	G		38	.842		270	45	3.87	318.87	380
0800	Basket guards for above, add	G					233			233	256
1000	Swinging guards for above, add	G					79.50			79.50	87.50
9000	Minimum labor/equipment charge		1 Sswk	2	4	Job		213		213	375

For customer support on your Commercial Renovation Cost Data, call 877.742.4719.

125

05 58 Formed Metal Fabrications

05 58 25 – Formed Lamp Posts

05 58 25.40 Lamp Posts		Crew	Daily Output	Labor-Hours	Unit	Material	2016 Bare Costs Labor	Equipment	Total	Total Incl O&P
0010	**LAMP POSTS**									
0020	Aluminum, 7' high, stock units, post only	G	1 Carp	16	.500	Ea.	85	24		109	133
0100	Mild steel, plain	G		16	.500	"	73.50	24		97.50	121
9000	Minimum labor/equipment charge			4	2	Job		97		97	158

05 71 Decorative Metal Stairs

05 71 13 – Fabricated Metal Spiral Stairs

05 71 13.50 Spiral Stairs

05 71 13.50 Spiral Stairs		Crew	Daily Output	Labor-Hours	Unit	Material	2016 Bare Costs Labor	Equipment	Total	Total Incl O&P
0010	**SPIRAL STAIRS**										
1805	Shop fabricated, custom ordered										
1810	Aluminum, 5'-0" diameter, plain units	G	E-4	45	.711	Riser	530	38	3.27	571.27	655
1820	Fancy units	G		45	.711		925	38	3.27	966.27	1,100
1900	Cast iron, 4'-0" diameter, plain units	G		45	.711		665	38	3.27	706.27	800
1920	Fancy units	G		25	1.280		1,150	68.50	5.90	1,224.40	1,400
2000	Steel, industrial checkered plate, 4' diameter	G		45	.711		440	38	3.27	481.27	555
2200	6' diameter	G		40	.800		620	43	3.68	666.68	760
3100	Spiral stair kits, 12 stacking risers to fit exact floor height										
3110	Steel, flat metal treads, primed, 3'-6" diameter	G	2 Carp	1.60	10	Flight	1,300	485		1,785	2,225
3120	4'-0" diameter	G		1.45	11.034		1,475	535		2,010	2,475
3130	4'-6" diameter	G		1.35	11.852		1,625	575		2,200	2,750
3140	5'-0" diameter			1.25	12.800		1,775	620		2,395	3,000
3210	Galvanized, 3'-6" diameter	G		1.60	10		1,800	485		2,285	2,800
3220	4'-0" diameter	G		1.45	11.034		2,150	535		2,685	3,225
3230	4'-6" diameter	G		1.35	11.852		2,350	575		2,925	3,550
3240	5'-0" diameter	G		1.25	12.800		2,575	620		3,195	3,875
3310	Checkered plate tread, primed, 3'-6" diameter			1.45	11.034		1,550	535		2,085	2,575
3320	4'-0" diameter			1.35	11.852		1,725	575		2,300	2,850
3330	4'-6" diameter			1.25	12.800		1,900	620		2,520	3,100
3340	5'-0" diameter			1.15	13.913		2,075	675		2,750	3,375
3410	Galvanized, 3'-6" diameter	G		1.45	11.034		2,150	535		2,685	3,250
3420	4'-0" diameter			1.35	11.852		2,500	575		3,075	3,675
3430	4'-6" diameter			1.25	12.800		2,700	620		3,320	4,000
3440	5'-0" diameter	G		1.15	13.913		2,900	675		3,575	4,300
3510	Red oak covers on flat metal treads, 3'-6" diameter			1.35	11.852		2,350	575		2,925	3,550
3520	4'-0" diameter			1.25	12.800		2,800	620		3,420	4,100
3530	4'-6" diameter			1.15	13.913		3,025	675		3,700	4,425
3540	5'-0" diameter			1.05	15.238		3,250	740		3,990	4,775

05 73 Decorative Metal Railings

05 73 16 – Wire Rope Decorative Metal Railings

05 73 16.10 Cable Railings

05 73 16.10 Cable Railings		Crew	Daily Output	Labor-Hours	Unit	Material	2016 Bare Costs Labor	Equipment	Total	Total Incl O&P
0010	**CABLE RAILINGS**, with 316 stainless steel 1 x 19 cable, 3/16" diameter										
0015	Made from recycled materials										
0100	1-3/4" diameter stainless steel posts x 42" high, cables 4" OC	G	2 Sswk	25	.640	L.F.	43	34		77	108

05 73 Decorative Metal Railings

05 73 23 – Ornamental Railings

05 73 23.50 Railings, Ornamental		Crew	Daily Output	Labor-Hours	Unit	Material	2016 Bare Costs Labor	Equipment	Total	Total Incl O&P
0010	**RAILINGS, ORNAMENTAL**, 3'-6" high, posts @ 6' O.C.									
0020	Bronze or stainless, hand forged, plain	G	2 Sswk	24	.667	L.F.	260	35.50		295.50	350
0100	Fancy	G		18	.889		520	47.50		567.50	655
0200	Aluminum, panelized, plain	G		24	.667		12.10	35.50		47.60	76
0300	Fancy	G		18	.889		26	47.50		73.50	112
0400	Wrought iron, hand forged, plain	G		24	.667		81.50	35.50		117	153
0500	Fancy	G		18	.889		231	47.50		278.50	340
0550	Steel, panelized, plain	G		24	.667		18.65	35.50		54.15	83
0560	Fancy	G		18	.889		28	47.50		75.50	115
0600	Composite metal/wood/glass, plain			18	.889		143	47.50		190.50	241
0700	Fancy			12	1.333		286	71		357	440
9000	Minimum labor/equipment charge		1 Sswk	2	4	Job		213		213	375

05 75 Decorative Formed Metal

05 75 13 – Columns

05 75 13.10 Aluminum Columns

			Crew	Daily Output	Labor-Hours	Unit	Material	2016 Bare Costs Labor	Equipment	Total	Total Incl O&P
0010	**ALUMINUM COLUMNS**										
0015	Made from recycled materials										
0020	Aluminum, extruded, stock units, no cap or base, 6" diameter	G	E-4	240	.133	L.F.	11.85	7.15	.61	19.61	26.50
0100	8" diameter	G	"	170	.188	"	15.05	10.10	.87	26.02	35.50
0460	Caps, ornamental, plain	G				Set	238			238	262
0470	Fancy	G				"	1,250			1,250	1,375
0500	For square columns, add to column prices above					L.F.	50%				

For customer support on your Commercial Renovation Cost Data, call 877.742.4719.

127

Division Notes

	CREW	DAILY OUTPUT	LABOR-HOURS	UNIT	BARE COSTS				TOTAL INCL O&P
					MAT.	LABOR	EQUIP.	TOTAL	

Estimating Tips
06 05 00 Common Work Results for Wood, Plastics, and Composites

- Common to any wood-framed structure are the accessory connector items such as screws, nails, adhesives, hangers, connector plates, straps, angles, and hold-downs. For typical wood-framed buildings, such as residential projects, the aggregate total for these items can be significant, especially in areas where seismic loading is a concern. For floor and wall framing, the material cost is based on 10 to 25 lbs. of accessory connectors per MBF. Hold-downs, hangers, and other connectors should be taken off by the piece.

  Included with material costs are fasteners for a normal installation. RSMeans engineers use manufacturer's recommendations, written specifications, and/or standard construction practice for size and spacing of fasteners. Prices for various fasteners are shown for informational purposes only. Adjustments should be made if unusual fastening conditions exist.

06 10 00 Carpentry
- Lumber is a traded commodity and therefore sensitive to supply and demand in the marketplace. Even in "budgetary" estimating of wood-framed projects, it is advisable to call local suppliers for the latest market pricing.

- Common quantity units for wood-framed projects are "thousand board feet" (MBF). A board foot is a volume of wood, 1" x 1' x 1', or 144 cubic inches. Board-foot quantities are generally calculated using nominal material dimensions—dressed sizes are ignored. Board foot per lineal foot of any stick of lumber can be calculated by dividing the nominal cross-sectional area by 12. As an example, 2,000 lineal feet of 2 x 12 equates to 4 MBF by dividing the nominal area, 2 x 12, by 12, which equals 2, and multiplying by 2,000 to give 4,000 board feet. This simple rule applies to all nominal dimensioned lumber.

- Waste is an issue of concern at the quantity takeoff for any area of construction. Framing lumber is sold in even foot lengths, i.e., 8', 10', 12', 14', 16' and, depending on spans, wall heights, and the grade of lumber, waste is inevitable. A rule of thumb for lumber waste is 5%–10% depending on material quality and the complexity of the framing.

- Wood in various forms and shapes is used in many projects, even where the main structural framing is steel, concrete, or masonry. Plywood as a back-up partition material and 2x boards used as blocking and cant strips around roof edges are two common examples. The estimator should ensure that the costs of all wood materials are included in the final estimate.

06 20 00 Finish Carpentry
- It is necessary to consider the grade of workmanship when estimating labor costs for erecting millwork and interior finish. In practice, there are three grades: premium, custom, and economy. The RSMeans daily output for base and case moldings is in the range of 200 to 250 L.F. per carpenter per day. This is appropriate for most average custom-grade projects. For premium projects, an adjustment to productivity of 25%–50% should be made, depending on the complexity of the job.

Reference Numbers
Reference numbers are shown at the beginning of some major classifications. These numbers refer to related items in the Reference Section. The reference information may be an estimating procedure, an alternate pricing method, or technical information.

Note: Not all subdivisions listed here necessarily appear. ∎

06 05 05 – Selective Demolition for Wood, Plastics, and Composites

06 05 05.10 Selective Demolition Wood Framing	Crew	Daily Output	Labor-Hours	Unit	Material	2016 Bare Costs Labor	Equipment	Total	Total Incl O&P
0010 **SELECTIVE DEMOLITION WOOD FRAMING** R024119-10									
0100 Timber connector, nailed, small	1 Clab	96	.083	Ea.		3.16		3.16	5.15
0110 Medium		60	.133			5.05		5.05	8.25
0120 Large		48	.167			6.30		6.30	10.35
0130 Bolted, small		48	.167			6.30		6.30	10.35
0140 Medium		32	.250			9.50		9.50	15.50
0150 Large		24	.333			12.65		12.65	20.50
2958 Beams, 2" x 6"	2 Clab	1100	.015	L.F.		.55		.55	.90
2960 2" x 8"		825	.019			.73		.73	1.20
2965 2" x 10"		665	.024			.91		.91	1.49
2970 2" x 12"		550	.029			1.10		1.10	1.80
2972 2" x 14"		470	.034			1.29		1.29	2.11
2975 4" x 8"	B-1	413	.058			2.24		2.24	3.66
2980 4" x 10"		330	.073			2.81		2.81	4.58
2985 4" x 12"		275	.087			3.37		3.37	5.50
3000 6" x 8"		275	.087			3.37		3.37	5.50
3040 6" x 10"		220	.109			4.21		4.21	6.90
3080 6" x 12"		185	.130			5		5	8.20
3120 8" x 12"		140	.171			6.60		6.60	10.80
3160 10" x 12"		110	.218			8.40		8.40	13.75
3162 Alternate pricing method		1.10	21.818	M.B.F.		840		840	1,375
3170 Blocking, in 16" OC wall framing, 2" x 4"	1 Clab	600	.013	L.F.		.51		.51	.83
3172 2" x 6"		400	.020			.76		.76	1.24
3174 In 24" OC wall framing, 2" x 4"		600	.013			.51		.51	.83
3176 2" x 6"		400	.020			.76		.76	1.24
3178 Alt method, wood blocking removal from wood framing		.40	20	M.B.F.		760		760	1,250
3179 Wood blocking removal from steel framing		.36	22.222	"		840		840	1,375
3180 Bracing, let in, 1" x 3", studs 16" OC		1050	.008	L.F.		.29		.29	.47
3181 Studs 24" OC		1080	.007			.28		.28	.46
3182 1" x 4", studs 16" OC		1050	.008			.29		.29	.47
3183 Studs 24" OC		1080	.007			.28		.28	.46
3184 1" x 6", studs 16" OC		1050	.008			.29		.29	.47
3185 Studs 24" OC		1080	.007			.28		.28	.46
3186 2" x 3", studs 16" OC		800	.010			.38		.38	.62
3187 Studs 24" OC		830	.010			.37		.37	.60
3188 2" x 4", studs 16" OC		800	.010			.38		.38	.62
3189 Studs 24" OC		830	.010			.37		.37	.60
3190 2" x 6", studs 16" OC		800	.010			.38		.38	.62
3191 Studs 24" OC		830	.010			.37		.37	.60
3192 2" x 8", studs 16" OC		800	.010			.38		.38	.62
3193 Studs 24" OC		830	.010			.37		.37	.60
3194 "T" shaped metal bracing, studs at 16" OC		1060	.008			.29		.29	.47
3195 Studs at 24" OC		1200	.007			.25		.25	.41
3196 Metal straps, studs at 16" OC		1200	.007			.25		.25	.41
3197 Studs at 24" OC		1240	.006			.24		.24	.40
3200 Columns, round, 8' to 14' tall		40	.200	Ea.		7.60		7.60	12.40
3202 Dimensional lumber sizes	2 Clab	1.10	14.545	M.B.F.		550		550	900
3250 Blocking, between joists	1 Clab	320	.025	Ea.		.95		.95	1.55
3252 Bridging, metal strap, between joists		320	.025	Pr.		.95		.95	1.55
3254 Wood, between joists		320	.025	"		.95		.95	1.55
3260 Door buck, studs, header & access., 8' high 2" x 4" wall, 3' wide		32	.250	Ea.		9.50		9.50	15.50
3261 4' wide		32	.250			9.50		9.50	15.50
3262 5' wide		32	.250			9.50		9.50	15.50

For customer support on your Commercial Renovation Cost Data, call 877.742.4719.

06 05 05 – Selective Demolition for Wood, Plastics, and Composites

06 05 05.10 Selective Demolition Wood Framing	Crew	Daily Output	Labor-Hours	Unit	Material	2016 Bare Costs Labor	Equipment	Total	Total Incl O&P
3263	6' wide	1 Clab	32	.250	Ea.		9.50		9.50	15.50
3264	8' wide		30	.267			10.10		10.10	16.50
3265	10' wide		30	.267			10.10		10.10	16.50
3266	12' wide		30	.267			10.10		10.10	16.50
3267	2" x 6" wall, 3' wide		32	.250			9.50		9.50	15.50
3268	4' wide		32	.250			9.50		9.50	15.50
3269	5' wide		32	.250			9.50		9.50	15.50
3270	6' wide		32	.250			9.50		9.50	15.50
3271	8' wide		30	.267			10.10		10.10	16.50
3272	10' wide		30	.267			10.10		10.10	16.50
3273	12' wide		30	.267			10.10		10.10	16.50
3274	Window buck, studs, header & access, 8' high 2" x 4" wall, 2' wide		24	.333			12.65		12.65	20.50
3275	3' wide		24	.333			12.65		12.65	20.50
3276	4' wide		24	.333			12.65		12.65	20.50
3277	5' wide		24	.333			12.65		12.65	20.50
3278	6' wide		24	.333			12.65		12.65	20.50
3279	7' wide		24	.333			12.65		12.65	20.50
3280	8' wide		22	.364			13.80		13.80	22.50
3281	10' wide		22	.364			13.80		13.80	22.50
3282	12' wide		22	.364			13.80		13.80	22.50
3283	2" x 6" wall, 2' wide		24	.333			12.65		12.65	20.50
3284	3' wide		24	.333			12.65		12.65	20.50
3285	4' wide		24	.333			12.65		12.65	20.50
3286	5' wide		24	.333			12.65		12.65	20.50
3287	6' wide		24	.333			12.65		12.65	20.50
3288	7' wide		24	.333			12.65		12.65	20.50
3289	8' wide		22	.364			13.80		13.80	22.50
3290	10' wide		22	.364			13.80		13.80	22.50
3291	12' wide		22	.364			13.80		13.80	22.50
3360	Deck or porch decking		825	.010	L.F.		.37		.37	.60
3400	Fascia boards, 1" x 6"		500	.016			.61		.61	.99
3440	1" x 8"		450	.018			.67		.67	1.10
3480	1" x 10"		400	.020			.76		.76	1.24
3490	2" x 6"		450	.018			.67		.67	1.10
3500	2" x 8"		400	.020			.76		.76	1.24
3510	2" x 10"		350	.023			.87		.87	1.42
3610	Furring, on wood walls or ceiling		4000	.002	S.F.		.08		.08	.12
3620	On masonry or concrete walls or ceiling		1200	.007	"		.25		.25	.41
3800	Headers over openings, 2 @ 2" x 6"		110	.073	L.F.		2.76		2.76	4.51
3840	2 @ 2" x 8"		100	.080			3.03		3.03	4.96
3880	2 @ 2" x 10"		90	.089			3.37		3.37	5.50
3885	Alternate pricing method		.26	30.651	M.B.F.		1,150		1,150	1,900
3920	Joists, 1" x 4"		1250	.006	L.F.		.24		.24	.40
3930	1" x 6"		1135	.007			.27		.27	.44
3940	1" x 8"		1000	.008			.30		.30	.50
3950	1" x 10"		895	.009			.34		.34	.55
3960	1" x 12"		765	.010			.40		.40	.65
4200	2" x 4"	2 Clab	1000	.016			.61		.61	.99
4230	2" x 6"		970	.016			.63		.63	1.02
4240	2" x 8"		940	.017			.65		.65	1.05
4250	2" x 10"		910	.018			.67		.67	1.09
4280	2" x 12"		880	.018			.69		.69	1.13
4281	2" x 14"		850	.019			.71		.71	1.17

For customer support on your Commercial Renovation Cost Data, call 877.742.4719.

131

06 05 05.10 Selective Demolition Wood Framing	Crew	Daily Output	Labor-Hours	Unit	Material	2016 Bare Costs Labor	Equipment	Total	Total Incl O&P
4282 Composite joists, 9-1/2"	2 Clab	960	.017	L.F.		.63		.63	1.03
4283 11-7/8"		930	.017			.65		.65	1.07
4284 14"		897	.018			.68		.68	1.11
4285 16"		865	.019			.70		.70	1.15
4290 Wood joists, alternate pricing method		1.50	10.667	M.B.F.		405		405	660
4500 Open web joist, 12" deep		500	.032	L.F.		1.21		1.21	1.98
4505 14" deep		475	.034			1.28		1.28	2.09
4510 16" deep		450	.036			1.35		1.35	2.20
4520 18" deep		425	.038			1.43		1.43	2.33
4530 24" deep		400	.040			1.52		1.52	2.48
4550 Ledger strips, 1" x 2"	1 Clab	1200	.007			.25		.25	.41
4560 1" x 3"		1200	.007			.25		.25	.41
4570 1" x 4"		1200	.007			.25		.25	.41
4580 2" x 2"		1100	.007			.28		.28	.45
4590 2" x 4"		1000	.008			.30		.30	.50
4600 2" x 6"		1000	.008			.30		.30	.50
4601 2" x 8 or 2" x 10"		800	.010			.38		.38	.62
4602 4" x 6"		600	.013			.51		.51	.83
4604 4" x 8"		450	.018			.67		.67	1.10
5400 Posts, 4" x 4"	2 Clab	800	.020			.76		.76	1.24
5405 4" x 6"		550	.029			1.10		1.10	1.80
5410 4" x 8"		440	.036			1.38		1.38	2.25
5425 4" x 10"		390	.041			1.56		1.56	2.54
5430 4" x 12"		350	.046			1.73		1.73	2.83
5440 6" x 6"		400	.040			1.52		1.52	2.48
5445 6" x 8"		350	.046			1.73		1.73	2.83
5450 6" x 10"		320	.050			1.90		1.90	3.10
5455 6" x 12"		290	.055			2.09		2.09	3.42
5480 8" x 8"		300	.053			2.02		2.02	3.30
5500 10" x 10"		240	.067			2.53		2.53	4.13
5660 Tongue and groove floor planks		2	8	M.B.F.		305		305	495
5682 Rafters, ordinary, 16" OC, 2" x 4"		880	.018	S.F.		.69		.69	1.13
5683 2" x 6"		840	.019			.72		.72	1.18
5684 2" x 8"		820	.020			.74		.74	1.21
5685 2" x 10"		820	.020			.74		.74	1.21
5686 2" x 12"		810	.020			.75		.75	1.22
5687 24" OC, 2" x 4"		1170	.014			.52		.52	.85
5688 2" x 6"		1117	.014			.54		.54	.89
5689 2" x 8"		1091	.015			.56		.56	.91
5690 2" x 10"		1091	.015			.56		.56	.91
5691 2" x 12"		1077	.015			.56		.56	.92
5795 Rafters, ordinary, 2" x 4" (alternate method)		862	.019	L.F.		.70		.70	1.15
5800 2" x 6" (alternate method)		850	.019			.71		.71	1.17
5840 2" x 8" (alternate method)		837	.019			.72		.72	1.18
5855 2" x 10" (alternate method)		825	.019			.73		.73	1.20
5865 2" x 12" (alternate method)		812	.020			.75		.75	1.22
5870 Sill plate, 2" x 4"	1 Clab	1170	.007			.26		.26	.42
5871 2" x 6"		780	.010			.39		.39	.64
5872 2" x 8"		586	.014			.52		.52	.85
5873 Alternate pricing method		.78	10.256	M.B.F.		390		390	635
5885 Ridge board, 1" x 4"	2 Clab	900	.018	L.F.		.67		.67	1.10
5886 1" x 6"		875	.018			.69		.69	1.13
5887 1" x 8"		850	.019			.71		.71	1.17

06 05 Common Work Results for Wood, Plastics, and Composites

06 05 05 - Selective Demolition for Wood, Plastics, and Composites

06 05 05.10 Selective Demolition Wood Framing	Crew	Daily Output	Labor-Hours	Unit	Material	2016 Bare Costs Labor	Equipment	Total	Total Incl O&P
5888	1" x 10"	2 Clab	825	.019	L.F.		.73		.73	1.20
5889	1" x 12"		800	.020			.76		.76	1.24
5890	2" x 4"		900	.018			.67		.67	1.10
5892	2" x 6"		875	.018			.69		.69	1.13
5894	2" x 8"		850	.019			.71		.71	1.17
5896	2" x 10"		825	.019			.73		.73	1.20
5898	2" x 12"		800	.020			.76		.76	1.24
5900	Hip & valley rafters, 2" x 6"		500	.032			1.21		1.21	1.98
5940	2" x 8"		420	.038			1.44		1.44	2.36
6050	Rafter tie, 1" x 4"		1250	.013			.49		.49	.79
6052	1" x 6"		1135	.014			.53		.53	.87
6054	2" x 4"		1000	.016			.61		.61	.99
6056	2" x 6"		970	.016			.63		.63	1.02
6070	Sleepers, on concrete, 1" x 2"	1 Clab	4700	.002			.06		.06	.11
6075	1" x 3"		4000	.002			.08		.08	.12
6080	2" x 4"		3000	.003			.10		.10	.17
6085	2" x 6"		2600	.003			.12		.12	.19
6086	Sheathing from roof, 5/16"	2 Clab	1600	.010	S.F.		.38		.38	.62
6088	3/8"		1525	.010			.40		.40	.65
6090	1/2"		1400	.011			.43		.43	.71
6092	5/8"		1300	.012			.47		.47	.76
6094	3/4"		1200	.013			.51		.51	.83
6096	Board sheathing from roof		1400	.011			.43		.43	.71
6100	Sheathing, from walls, 1/4"		1200	.013			.51		.51	.83
6110	5/16"		1175	.014			.52		.52	.84
6120	3/8"		1150	.014			.53		.53	.86
6130	1/2"		1125	.014			.54		.54	.88
6140	5/8"		1100	.015			.55		.55	.90
6150	3/4"		1075	.015			.56		.56	.92
6152	Board sheathing from walls		1500	.011			.40		.40	.66
6158	Subfloor/roof deck, with boards		2200	.007			.28		.28	.45
6159	Subfloor/roof deck, with tongue & groove boards		2000	.008			.30		.30	.50
6160	Plywood, 1/2" thick		768	.021			.79		.79	1.29
6162	5/8" thick		760	.021			.80		.80	1.30
6164	3/4" thick		750	.021			.81		.81	1.32
6165	1-1/8" thick		720	.022			.84		.84	1.38
6166	Underlayment, particle board, 3/8" thick	1 Clab	780	.010			.39		.39	.64
6168	1/2" thick		768	.010			.39		.39	.65
6170	5/8" thick		760	.011			.40		.40	.65
6172	3/4" thick		750	.011			.40		.40	.66
6200	Stairs and stringers, straight run	2 Clab	40	.400	Riser		15.15		15.15	25
6240	With platforms, winders or curves	"	26	.615	"		23.50		23.50	38
6300	Components, tread	1 Clab	110	.073	Ea.		2.76		2.76	4.51
6320	Riser		80	.100	"		3.79		3.79	6.20
6390	Stringer, 2" x 10"		260	.031	L.F.		1.17		1.17	1.91
6400	2" x 12"		260	.031			1.17		1.17	1.91
6410	3" x 10"		250	.032			1.21		1.21	1.98
6420	3" x 12"		250	.032			1.21		1.21	1.98
6590	Wood studs, 2" x 3"	2 Clab	3076	.005			.20		.20	.32
6600	2" x 4"		2000	.008			.30		.30	.50
6640	2" x 6"		1600	.010			.38		.38	.62
6720	Wall framing, including studs, plates and blocking, 2" x 4"	1 Clab	600	.013	S.F.		.51		.51	.83
6740	2" x 6"		480	.017	"		.63		.63	1.03

For customer support on your Commercial Renovation Cost Data, call 877.742.4719.

133

06 05 05.10 Selective Demolition Wood Framing

		Crew	Daily Output	Labor-Hours	Unit	Material	2016 Bare Costs Labor	Equipment	Total	Total Incl O&P
6750	Headers, 2" x 4"	1 Clab	1125	.007	L.F.		.27		.27	.44
6755	2" x 6"		1125	.007			.27		.27	.44
6760	2" x 8"		1050	.008			.29		.29	.47
6765	2" x 10"		1050	.008			.29		.29	.47
6770	2" x 12"		1000	.008			.30		.30	.50
6780	4" x 10"		525	.015			.58		.58	.94
6785	4" x 12"		500	.016			.61		.61	.99
6790	6" x 8"		560	.014			.54		.54	.89
6795	6" x 10"		525	.015			.58		.58	.94
6797	6" x 12"		500	.016			.61		.61	.99
7000	Trusses									
7050	12' span	2 Clab	74	.216	Ea.		8.20		8.20	13.40
7150	24' span	F-3	66	.606			30	10	40	59.50
7200	26' span		64	.625			31	10.30	41.30	61.50
7250	28' span		62	.645			32	10.60	42.60	63
7300	30' span		58	.690			34	11.35	45.35	67.50
7350	32' span		56	.714			35	11.75	46.75	70
7400	34' span		54	.741			36.50	12.20	48.70	72.50
7450	36' span		52	.769			38	12.65	50.65	75.50
8000	Soffit, T & G wood	1 Clab	520	.015	S.F.		.58		.58	.95
8010	Hardboard, vinyl or aluminum	"	640	.013			.47		.47	.77
8030	Plywood	2 Carp	315	.051			2.46		2.46	4.02
9000	Minimum labor/equipment charge	1 Clab	4	2	Job		76		76	124
9500	See Section 02 41 19.19 for rubbish handling									

06 05 05.20 Selective Demolition Millwork and Trim

		Crew	Daily Output	Labor-Hours	Unit	Material	2016 Bare Costs Labor	Equipment	Total	Total Incl O&P
0010	**SELECTIVE DEMOLITION MILLWORK AND TRIM** R024119-10									
1000	Cabinets, wood, base cabinets, per L.F.	2 Clab	80	.200	L.F.		7.60		7.60	12.40
1020	Wall cabinets, per L.F.	"	80	.200	"		7.60		7.60	12.40
1060	Remove and reset, base cabinets	2 Carp	18	.889	Ea.		43		43	70.50
1070	Wall cabinets		20	.800			39		39	63.50
1072	Oven cabinet, 7' high		11	1.455			70.50		70.50	115
1074	Cabinet door, up to 2' high	1 Clab	66	.121			4.59		4.59	7.50
1076	2' - 4' high	"	46	.174			6.60		6.60	10.75
1100	Steel, painted, base cabinets	2 Clab	60	.267	L.F.		10.10		10.10	16.50
1120	Wall cabinets		60	.267	"		10.10		10.10	16.50
1200	Casework, large area		320	.050	S.F.		1.90		1.90	3.10
1220	Selective		200	.080	"		3.03		3.03	4.96
1500	Counter top, straight runs		200	.080	L.F.		3.03		3.03	4.96
1510	L, U or C shapes		120	.133			5.05		5.05	8.25
1550	Remove and reset, straight runs	2 Carp	50	.320			15.50		15.50	25.50
1560	L, U or C shape	"	40	.400			19.40		19.40	31.50
2000	Paneling, 4' x 8' sheets	2 Clab	2000	.008	S.F.		.30		.30	.50
2100	Boards, 1" x 4"		700	.023			.87		.87	1.42
2120	1" x 6"		750	.021			.81		.81	1.32
2140	1" x 8"		800	.020			.76		.76	1.24
3000	Trim, baseboard, to 6" wide		1200	.013	L.F.		.51		.51	.83
3040	Greater than 6" and up to 12" wide		1000	.016			.61		.61	.99
3080	Remove and reset, minimum	2 Carp	400	.040			1.94		1.94	3.17
3090	Maximum	"	300	.053			2.58		2.58	4.22
3100	Ceiling trim	2 Clab	1000	.016			.61		.61	.99
3120	Chair rail		1200	.013			.51		.51	.83
3140	Railings with balusters		240	.067			2.53		2.53	4.13

06 05 Common Work Results for Wood, Plastics, and Composites

06 05 05 – Selective Demolition for Wood, Plastics, and Composites

06 05 05.20 Selective Demolition Millwork and Trim	Crew	Daily Output	Labor-Hours	Unit	Material	2016 Bare Costs Labor	2016 Bare Costs Equipment	Total	Total Incl O&P
3160	Wainscoting	2 Clab	700	.023	S.F.		.87		.87	1.42
9000	Minimum labor/equipment charge	1 Clab	4	2	Job		76		76	124

06 05 23 – Wood, Plastic, and Composite Fastenings

06 05 23.10 Nails

		Crew	Daily Output	Labor-Hours	Unit	Material	2016 Bare Costs Labor	2016 Bare Costs Equipment	Total	Total Incl O&P
0010	**NAILS**, material only, based upon 50# box purchase									
0020	Copper nails, plain				Lb.	10.10			10.10	11.15
0400	Stainless steel, plain					9.05			9.05	10
0500	Box, 3d to 20d, bright					1.40			1.40	1.54
0520	Galvanized					1.70			1.70	1.87
0600	Common, 3d to 60d, plain					1.37			1.37	1.51
0700	Galvanized					2			2	2.20
0800	Aluminum					10.90			10.90	11.95
1000	Annular or spiral thread, 4d to 60d, plain					2.58			2.58	2.84
1200	Galvanized					3.10			3.10	3.41
1400	Drywall nails, plain					1.40			1.40	1.54
1600	Galvanized					2.10			2.10	2.31
1800	Finish nails, 4d to 10d, plain					1.20			1.20	1.32
2000	Galvanized					1.83			1.83	2.01
2100	Aluminum					5.80			5.80	6.40
2300	Flooring nails, hardened steel, 2d to 10d, plain					3.33			3.33	3.66
2400	Galvanized					4			4	4.40
2500	Gypsum lath nails, 1-1/8", 13 ga. flathead, blued					2.05			2.05	2.26
2600	Masonry nails, hardened steel, 3/4" to 3" long, plain					1.70			1.70	1.87
2700	Galvanized					3.25			3.25	3.58
2900	Roofing nails, threaded, galvanized					1.75			1.75	1.93
3100	Aluminum					4.80			4.80	5.30
3300	Compressed lead head, threaded, galvanized					2.45			2.45	2.70
3600	Siding nails, plain shank, galvanized					2			2	2.20
3800	Aluminum					5.95			5.95	6.55
5000	Add to prices above for cement coating					.12			.12	.13
5200	Zinc or tin plating					.15			.15	.17
5500	Vinyl coated sinkers, 8d to 16d					2.40			2.40	2.64

06 05 23.50 Wood Screws

		Crew	Daily Output	Labor-Hours	Unit	Material	2016 Bare Costs Labor	2016 Bare Costs Equipment	Total	Total Incl O&P
0010	**WOOD SCREWS**									
0020	#8 x 1" long, steel				C	2.75			2.75	3.03
0100	Brass					12.25			12.25	13.50
0200	#8, 2" long, steel					4.25			4.25	4.68
0300	Brass					22.50			22.50	24.50
0400	#10, 1" long, steel					3.45			3.45	3.80
0500	Brass					15.60			15.60	17.15
0600	#10, 2" long, steel					5.25			5.25	5.80
0700	Brass					27			27	29.50
0800	#10, 3" long, steel					8.90			8.90	9.80
1000	#12, 2" long, steel					7.35			7.35	8.10
1100	Brass					35			35	38
1500	#12, 3" long, steel					11.20			11.20	12.30
2000	#12, 4" long, steel					19.20			19.20	21

06 05 23.60 Timber Connectors

		Crew	Daily Output	Labor-Hours	Unit	Material	2016 Bare Costs Labor	2016 Bare Costs Equipment	Total	Total Incl O&P
0010	**TIMBER CONNECTORS**									
0020	Add up cost of each part for total cost of connection									
0100	Connector plates, steel, with bolts, straight	2 Carp	75	.213	Ea.	30.50	10.35		40.85	50.50
0110	Tee, 7 ga.		50	.320		35	15.50		50.50	64

For customer support on your Commercial Renovation Cost Data, call 877.742.4719.

135

06 05 23 – Wood, Plastic, and Composite Fastenings

06 05 23.60 Timber Connectors		Crew	Daily Output	Labor-Hours	Unit	Material	2016 Bare Costs Labor	Equipment	Total	Total Incl O&P
0120	T-Strap, 14 ga., 12" x 8" x 2"	2 Carp	50	.320	Ea.	35	15.50		50.50	64
0150	Anchor plates, 7 ga., 9" x 7"		75	.213		30.50	10.35		40.85	50.50
0200	Bolts, machine, sq. hd. with nut & washer, 1/2" diameter, 4" long	1 Carp	140	.057		.75	2.77		3.52	5.35
0300	7-1/2" long		130	.062		1.37	2.98		4.35	6.40
0500	3/4" diameter, 7-1/2" long		130	.062		3.20	2.98		6.18	8.40
0610	Machine bolts, w/nut, washer, 3/4" diam., 15" L, HD's & beam hangers		95	.084		5.95	4.08		10.03	13.20
0800	Drilling bolt holes in timber, 1/2" diameter		450	.018	Inch		.86		.86	1.41
0900	1" diameter		350	.023	"		1.11		1.11	1.81
1100	Framing anchor, angle, 3" x 3" x 1-1/2", 12 ga		175	.046	Ea.	2.56	2.21		4.77	6.45
1150	Framing anchors, 18 ga., 4-1/2" x 2-3/4"		175	.046		2.56	2.21		4.77	6.45
1160	Framing anchors, 18 ga., 4-1/2" x 3"		175	.046		2.56	2.21		4.77	6.45
1170	Clip anchors plates, 18 ga., 12" x 1-1/8"		175	.046		2.56	2.21		4.77	6.45
1250	Holdowns, 3 ga. base, 10 ga. body		8	1		25	48.50		73.50	107
1260	Holdowns, 7 ga. 11-1/16" x 3-1/4"		8	1		25	48.50		73.50	107
1270	Holdowns, 7 ga. 14-3/8" x 3-1/8"		8	1		25	48.50		73.50	107
1275	Holdowns, 12 ga. 8" x 2-1/2"		8	1		25	48.50		73.50	107
1300	Joist and beam hangers, 18 ga. galv., for 2" x 4" joist		175	.046		.72	2.21		2.93	4.41
1400	2" x 6" to 2" x 10" joist		165	.048		1.36	2.35		3.71	5.35
1600	16 ga. galv., 3" x 6" to 3" x 10" joist		160	.050		2.92	2.42		5.34	7.15
1700	3" x 10" to 3" x 14" joist		160	.050		4.88	2.42		7.30	9.30
1800	4" x 6" to 4" x 10" joist		155	.052		3.04	2.50		5.54	7.40
1900	4" x 10" to 4" x 14" joist		155	.052		4.96	2.50		7.46	9.55
2000	Two-2" x 6" to two-2" x 10" joists		150	.053		4.16	2.58		6.74	8.80
2100	Two-2" x 10" to two-2" x 14" joists		150	.053		4.64	2.58		7.22	9.30
2300	3/16" thick, 6" x 8" joist		145	.055		65	2.67		67.67	76
2400	6" x 10" joist		140	.057		67.50	2.77		70.27	79
2500	6" x 12" joist		135	.059		70.50	2.87		73.37	82
2700	1/4" thick, 6" x 14" joist		130	.062		73	2.98		75.98	85.50
2900	Plywood clips, extruded aluminum H clip, for 3/4" panels					.22			.22	.24
3000	Galvanized 18 ga. back-up clip					.20			.20	.22
3200	Post framing, 16 ga. galv. for 4" x 4" base, 2 piece	1 Carp	130	.062		16.35	2.98		19.33	23
3300	Cap		130	.062		22.50	2.98		25.48	30
3500	Rafter anchors, 18 ga. galv., 1-1/2" wide, 5-1/4" long		145	.055		.48	2.67		3.15	4.90
3600	10-3/4" long		145	.055		1.44	2.67		4.11	5.95
3800	Shear plates, 2-5/8" diameter		120	.067		2.44	3.23		5.67	8
3900	4" diameter		115	.070		5.70	3.37		9.07	11.80
4000	Sill anchors, embedded in concrete or block, 25-1/2" long		115	.070		12.70	3.37		16.07	19.50
4100	Spike grids, 3" x 6"		120	.067		.94	3.23		4.17	6.35
4400	Split rings, 2-1/2" diameter		120	.067		2.02	3.23		5.25	7.50
4500	4" diameter		110	.073		2.98	3.52		6.50	9.05
4550	Tie plate, 20 ga., 7" x 3 1/8"		110	.073		2.98	3.52		6.50	9.05
4560	Tie plate, 20 ga., 5" x 4 1/8"		110	.073		2.98	3.52		6.50	9.05
4575	Twist straps, 18 ga., 12" x 1 1/4"		110	.073		2.98	3.52		6.50	9.05
4580	Twist straps, 18 ga., 16" x 1 1/4"		110	.073		2.98	3.52		6.50	9.05
4600	Strap ties, 20 ga., 2-1/16" wide, 12 13/16" long		180	.044		.93	2.15		3.08	4.54
4700	Strap ties, 16 ga., 1-3/8" wide, 12" long		180	.044		.93	2.15		3.08	4.54
4800	21-5/8" x 1-1/4"		160	.050		2.94	2.42		5.36	7.20
5000	Toothed rings, 2-5/8" or 4" diameter		90	.089		1.79	4.31		6.10	9
5200	Truss plates, nailed, 20 ga., up to 32' span		17	.471	Truss	12.85	23		35.85	51.50
5400	Washers, 2" x 2" x 1/8"				Ea.	.38			.38	.42
5500	3" x 3" x 3/16"				"	1.04			1.04	1.14
9000	Minimum labor/equipment charge	1 Carp	4	2	Job		97		97	158

06 05 Common Work Results for Wood, Plastics, and Composites

06 05 23 – Wood, Plastic, and Composite Fastenings

06 05 23.80 Metal Bracing		Crew	Daily Output	Labor-Hours	Unit	Material	2016 Bare Costs Labor	Equipment	Total	Total Incl O&P
0010	**METAL BRACING**									
0302	Let-in, "T" shaped, 22 ga. galv. steel, studs at 16" O.C.	1 Carp	580	.014	L.F.	.83	.67		1.50	2
0402	Studs at 24" O.C.		600	.013		.83	.65		1.48	1.97
0502	Steel straps, 16 ga. galv. steel, studs at 16" O.C.		600	.013		1.07	.65		1.72	2.24
0602	Studs at 24" O.C.	↓	620	.013	↓	1.07	.63		1.70	2.20

06 11 Wood Framing

06 11 10 – Framing with Dimensional, Engineered or Composite Lumber

06 11 10.01 Forest Stewardship Council Certification

0010	**FOREST STEWARDSHIP COUNCIL CERTIFICATION**									
0020	For Forest Stewardship Council (FSC) cert dimension lumber, add G					65%				

06 11 10.02 Blocking

06 11 10.02 Blocking		Crew	Daily Output	Labor-Hours	Unit	Material	Labor	Equipment	Total	Total Incl O&P
0010	**BLOCKING**									
2600	Miscellaneous, to wood construction									
2620	2" x 4"	1 Carp	.17	47.059	M.B.F.	595	2,275		2,870	4,375
2625	Pneumatic nailed		.21	38.095		600	1,850		2,450	3,675
2660	2" x 8"		.27	29.630		635	1,425		2,060	3,050
2665	Pneumatic nailed	↓	.33	24.242	↓	640	1,175		1,815	2,625
2720	To steel construction									
2740	2" x 4"	1 Carp	.14	57.143	M.B.F.	595	2,775		3,370	5,175
2780	2" x 8"		.21	38.095	"	635	1,850		2,485	3,725
9000	Minimum labor/equipment charge	↓	4	2	Job		97		97	158

06 11 10.04 Wood Bracing

06 11 10.04 Wood Bracing		Crew	Daily Output	Labor-Hours	Unit	Material	Labor	Equipment	Total	Total Incl O&P
0010	**WOOD BRACING**									
0012	Let-in, with 1" x 6" boards, studs @ 16" O.C.	1 Carp	150	.053	L.F.	.71	2.58	·	3.29	5
0202	Studs @ 24" O.C.	"	230	.035	"	.71	1.69		2.40	3.53

06 11 10.06 Bridging

06 11 10.06 Bridging		Crew	Daily Output	Labor-Hours	Unit	Material	Labor	Equipment	Total	Total Incl O&P
0010	**BRIDGING**									
0012	Wood, for joists 16" O.C., 1" x 3"	1 Carp	130	.062	Pr.	.64	2.98		3.62	5.60
0017	Pneumatic nailed		170	.047		.69	2.28		2.97	4.48
0102	2" x 3" bridging		130	.062		.66	2.98		3.64	5.60
0107	Pneumatic nailed		170	.047		.66	2.28		2.94	4.45
0302	Steel, galvanized, 18 ga., for 2" x 10" joists at 12" O.C.		130	.062		.94	2.98		3.92	5.90
0352	16" O.C.		135	.059		.94	2.87		3.81	5.75
0402	24" O.C.		140	.057		1.94	2.77		4.71	6.65
0602	For 2" x 14" joists at 16" O.C.		130	.062		1.43	2.98		4.41	6.45
0902	Compression type, 16" O.C., 2" x 8" joists		200	.040		.94	1.94		2.88	4.21
1002	2" x 12" joists	↓	200	.040	↓	.94	1.94		2.88	4.21

06 11 10.10 Beam and Girder Framing

06 11 10.10 Beam and Girder Framing		Crew	Daily Output	Labor-Hours	Unit	Material	Labor	Equipment	Total	Total Incl O&P
0010	**BEAM AND GIRDER FRAMING** R061110-30									
1000	Single, 2" x 6"	2 Carp	700	.023	L.F.	.61	1.11		1.72	2.48
1005	Pneumatic nailed		812	.020		.62	.95		1.57	2.24
1020	2" x 8"		650	.025		.85	1.19		2.04	2.88
1025	Pneumatic nailed		754	.021		.85	1.03		1.88	2.62
1040	2" x 10"		600	.027		1.34	1.29		2.63	3.58
1045	Pneumatic nailed		696	.023		1.34	1.11		2.45	3.30
1060	2" x 12"		550	.029		1.70	1.41		3.11	4.17
1065	Pneumatic nailed		638	.025		1.71	1.22		2.93	3.87
1080	2" x 14"	↓	500	.032	↓	2.15	1.55		3.70	4.89

For customer support on your Commercial Renovation Cost Data, call 877.742.4719.

137

06 11 Wood Framing

06 11 10 - Framing with Dimensional, Engineered or Composite Lumber

06 11 10.10 Beam and Girder Framing

		Crew	Daily Output	Labor-Hours	Unit	Material	2016 Bare Costs Labor	Equipment	Total	Total Incl O&P
1085	Pneumatic nailed	2 Carp	580	.028	L.F.	2.16	1.34		3.50	4.55
1100	3" x 8"		550	.029		2.72	1.41		4.13	5.30
1120	3" x 10"		500	.032		3.43	1.55		4.98	6.30
1140	3" x 12"		450	.036		4.11	1.72		5.83	7.35
1160	3" x 14"		400	.040		4.80	1.94		6.74	8.45
1170	4" x 6"	F-3	1100	.036		2.90	1.79	.60	5.29	6.75
1180	4" x 8"		1000	.040		2.96	1.97	.66	5.59	7.15
1200	4" x 10"		950	.042		4.70	2.07	.69	7.46	9.25
1220	4" x 12"		900	.044		5.20	2.19	.73	8.12	10.05
1240	4" x 14"		850	.047		6.05	2.32	.77	9.14	11.30
1250	6" x 8"		525	.076		7.50	3.75	1.25	12.50	15.75
1260	6" x 10"		500	.080		6.55	3.94	1.32	11.81	15.05
1290	8" x 12"		300	.133		15.70	6.55	2.19	24.44	30.50
2000	Double, 2" x 6"	2 Carp	625	.026		1.22	1.24		2.46	3.37
2005	Pneumatic nailed		725	.022		1.23	1.07		2.30	3.10
2020	2" x 8"		575	.028		1.70	1.35		3.05	4.07
2025	Pneumatic nailed		667	.024		1.71	1.16		2.87	3.78
2040	2" x 10"		550	.029		2.67	1.41		4.08	5.25
2045	Pneumatic nailed		638	.025		2.69	1.22		3.91	4.95
2060	2" x 12"		525	.030		3.40	1.48		4.88	6.15
2065	Pneumatic nailed		610	.026		3.42	1.27		4.69	5.85
2080	2" x 14"		475	.034		4.30	1.63		5.93	7.40
2085	Pneumatic nailed		551	.029		4.32	1.41		5.73	7.05
3000	Triple, 2" x 6"		550	.029		1.83	1.41		3.24	4.32
3005	Pneumatic nailed		638	.025		1.85	1.22		3.07	4.02
3020	2" x 8"		525	.030		2.54	1.48		4.02	5.20
3025	Pneumatic nailed		609	.026		2.56	1.27		3.83	4.90
3040	2" x 10"		500	.032		4.01	1.55		5.56	6.95
3045	Pneumatic nailed		580	.028		4.03	1.34		5.37	6.60
3060	2" x 12"		475	.034		5.10	1.63		6.73	8.25
3065	Pneumatic nailed		551	.029		5.15	1.41		6.56	7.95
3080	2" x 14"		450	.036		7.15	1.72		8.87	10.65
3085	Pneumatic nailed		522	.031		6.50	1.49		7.99	9.55
9000	Minimum labor/equipment charge	1 Carp	2	4	Job		194		194	315

06 11 10.12 Ceiling Framing

		Crew	Daily Output	Labor-Hours	Unit	Material	2016 Bare Costs Labor	Equipment	Total	Total Incl O&P
0010	**CEILING FRAMING**									
6000	Suspended, 2" x 3"	2 Carp	1000	.016	L.F.	.38	.78		1.16	1.68
6050	2" x 4"		900	.018		.40	.86		1.26	1.85
6100	2" x 6"		800	.020		.61	.97		1.58	2.25
6150	2" x 8"		650	.025		.85	1.19		2.04	2.88
9000	Minimum labor/equipment charge	1 Carp	4	2	Job		97		97	158

06 11 10.14 Posts and Columns

		Crew	Daily Output	Labor-Hours	Unit	Material	2016 Bare Costs Labor	Equipment	Total	Total Incl O&P
0010	**POSTS AND COLUMNS**									
0100	4" x 4"	2 Carp	390	.041	L.F.	1.73	1.99		3.72	5.15
0150	4" x 6"		275	.058		2.90	2.82		5.72	7.80
0200	4" x 8"		220	.073		2.96	3.52		6.48	9
0250	6" x 6"		215	.074		5.05	3.61		8.66	11.50
0300	6" x 8"		175	.091		7.50	4.43		11.93	15.50
0350	6" x 10"		150	.107		6.55	5.15		11.70	15.65
9000	Minimum labor/equipment charge	1 Carp	2	4	Job		194		194	315

06 11 Wood Framing

06 11 10 – Framing with Dimensional, Engineered or Composite Lumber

06 11 10.18 Joist Framing

		Crew	Daily Output	Labor-Hours	Unit	Material	2016 Bare Costs Labor	Equipment	Total	Total Incl O&P
0010	**JOIST FRAMING** R061110-30									
2000	Joists, 2" x 4"	2 Carp	1250	.013	L.F.	.40	.62		1.02	1.45
2005	Pneumatic nailed		1438	.011		.40	.54		.94	1.32
2100	2" x 6"		1250	.013		.61	.62		1.23	1.68
2105	Pneumatic nailed		1438	.011		.62	.54		1.16	1.56
2150	2" x 8"		1100	.015		.85	.70		1.55	2.08
2155	Pneumatic nailed		1265	.013		.85	.61		1.46	1.94
2200	2" x 10"		900	.018		1.34	.86		2.20	2.88
2205	Pneumatic nailed		1035	.015		1.34	.75		2.09	2.70
2250	2" x 12"		875	.018		1.70	.89		2.59	3.32
2255	Pneumatic nailed		1006	.016		1.71	.77		2.48	3.14
2300	2" x 14"		770	.021		2.15	1.01		3.16	4
2305	Pneumatic nailed		886	.018		2.16	.88		3.04	3.80
2350	3" x 6"		925	.017		1.89	.84		2.73	3.45
2400	3" x 10"		780	.021		3.43	.99		4.42	5.40
2450	3" x 12"		600	.027		4.11	1.29		5.40	6.65
2500	4" x 6"		800	.020		2.90	.97		3.87	4.77
2550	4" x 10"		600	.027		4.70	1.29		5.99	7.25
2600	4" x 12"		450	.036		5.20	1.72		6.92	8.50
2605	Sister joist, 2" x 6"		800	.020		.61	.97		1.58	2.25
2606	Pneumatic nailed		960	.017		.62	.81		1.43	2
2610	2" x 8"		640	.025		.85	1.21		2.06	2.91
2611	Pneumatic nailed		768	.021		.85	1.01		1.86	2.59
2615	2" x 10"		535	.030		1.34	1.45		2.79	3.84
2616	Pneumatic nailed		642	.025		1.34	1.21		2.55	3.45
2620	2" x 12"		455	.035		1.70	1.70		3.40	4.65
2625	Pneumatic nailed		546	.029		1.71	1.42		3.13	4.20
3000	Composite wood joist 9-1/2" deep		.90	17.778	M.L.F.	1,850	860		2,710	3,450
3010	11-1/2" deep		.88	18.182		2,050	880		2,930	3,700
3020	14" deep		.82	19.512		2,125	945		3,070	3,900
3030	16" deep		.78	20.513		3,575	995		4,570	5,575
4000	Open web joist 12" deep		.88	18.182		3,325	880		4,205	5,100
4002	Per linear foot		880	.018	L.F.	3.32	.88		4.20	5.10
4004	Treated, per linear foot		880	.018	"	4.21	.88		5.09	6.05
4010	14" deep		.82	19.512	M.L.F.	3,625	945		4,570	5,550
4012	Per linear foot		820	.020	L.F.	3.63	.95		4.58	5.55
4014	Treated, per linear foot		820	.020	"	4.66	.95		5.61	6.70
4020	16" deep		.78	20.513	M.L.F.	3,800	995		4,795	5,800
4022	Per linear foot		780	.021	L.F.	3.79	.99		4.78	5.80
4024	Treated, per linear foot		780	.021	"	4.97	.99		5.96	7.05
4030	18" deep		.74	21.622	M.L.F.	3,850	1,050		4,900	5,950
4032	Per linear foot		740	.022	L.F.	3.85	1.05		4.90	5.95
4034	Treated, per linear foot		740	.022	"	5.20	1.05		6.25	7.40
6000	Composite rim joist, 1-1/4" x 9-1/2"		.90	17.778	M.L.F.	1,975	860		2,835	3,575
6010	1-1/4" x 11-1/2"		.88	18.182		2,275	880		3,155	3,950
6020	1-1/4" x 14-1/2"		.82	19.512		2,750	945		3,695	4,575
6030	1-1/4" x 16-1/2"		.78	20.513		3,175	995		4,170	5,100
9000	Minimum labor/equipment charge	1 Carp	4	2	Job		97		97	158

06 11 10.24 Miscellaneous Framing

		Crew	Daily Output	Labor-Hours	Unit	Material	2016 Bare Costs Labor	Equipment	Total	Total Incl O&P
0010	**MISCELLANEOUS FRAMING**									
2000	Firestops, 2" x 4"	2 Carp	780	.021	L.F.	.40	.99		1.39	2.06
2005	Pneumatic nailed		952	.017		.40	.81		1.21	1.77

For customer support on your Commercial Renovation Cost Data, call 877.742.4719.

139

06 11 10.24 Miscellaneous Framing

		Crew	Daily Output	Labor-Hours	Unit	Material	2016 Bare Costs Labor	Equipment	Total	Total Incl O&P
2100	2" x 6"	2 Carp	600	.027	L.F.	.61	1.29		1.90	2.78
2105	Pneumatic nailed		732	.022		.62	1.06		1.68	2.41
5000	Nailers, treated, wood construction, 2" x 4"		800	.020		.52	.97		1.49	2.16
5005	Pneumatic nailed		960	.017		.53	.81		1.34	1.90
5100	2" x 6"		750	.021		.77	1.03		1.80	2.54
5105	Pneumatic nailed		900	.018		.78	.86		1.64	2.27
5120	2" x 8"		700	.023		1.06	1.11		2.17	2.97
5125	Pneumatic nailed		840	.019		1.06	.92		1.98	2.68
5200	Steel construction, 2" x 4"		750	.021		.52	1.03		1.55	2.27
5220	2" x 6"		700	.023		.77	1.11		1.88	2.66
5240	2" x 8"		650	.025		1.06	1.19		2.25	3.11
7000	Rough bucks, treated, for doors or windows, 2" x 6"		400	.040		.77	1.94		2.71	4.02
7005	Pneumatic nailed		480	.033		.78	1.61		2.39	3.50
7100	2" x 8"		380	.042		1.06	2.04		3.10	4.49
7105	Pneumatic nailed		456	.035		1.06	1.70		2.76	3.95
8000	Stair stringers, 2" x 10"		130	.123		1.34	5.95		7.29	11.20
8100	2" x 12"		130	.123		1.70	5.95		7.65	11.60
8150	3" x 10"		125	.128		3.43	6.20		9.63	13.90
8200	3" x 12"		125	.128		4.11	6.20		10.31	14.65
8870	Laminated structural lumber, 1-1/4" x 11-1/2"		130	.123		2.26	5.95		8.21	12.25
8880	1-1/4" x 14-1/2"		130	.123		2.72	5.95		8.67	12.75
9000	Minimum labor/equipment charge	1 Carp	4	2	Job		97		97	158

06 11 10.26 Partitions

		Crew	Daily Output	Labor-Hours	Unit	Material	2016 Bare Costs Labor	Equipment	Total	Total Incl O&P
0010	**PARTITIONS**									
0020	Single bottom and double top plate, no waste, std. & better lumber									
0180	2" x 4" studs, 8' high, studs 12" O.C.	2 Carp	80	.200	L.F.	4.37	9.70		14.07	20.50
0185	12" O.C., pneumatic nailed		96	.167		4.40	8.10		12.50	18.05
0200	16" O.C.		100	.160		3.58	7.75		11.33	16.60
0205	16" O.C., pneumatic nailed		120	.133		3.60	6.45		10.05	14.50
0300	24" O.C.		125	.128		2.78	6.20		8.98	13.20
0305	24" O.C., pneumatic nailed		150	.107		2.80	5.15		7.95	11.55
0380	10' high, studs 12" O.C.		80	.200		5.15	9.70		14.85	21.50
0385	12" O.C., pneumatic nailed		96	.167		5.20	8.10		13.30	18.90
0400	16" O.C.		100	.160		4.17	7.75		11.92	17.25
0405	16" O.C., pneumatic nailed		120	.133		4.20	6.45		10.65	15.15
0500	24" O.C.		125	.128		3.18	6.20		9.38	13.65
0505	24" O.C., pneumatic nailed		150	.107		3.20	5.15		8.35	11.95
0580	12' high, studs 12" O.C.		65	.246		5.95	11.95		17.90	26
0585	12" O.C., pneumatic nailed		78	.205		6	9.95		15.95	23
0600	16" O.C.		80	.200		4.77	9.70		14.47	21
0605	16" O.C., pneumatic nailed		96	.167		4.80	8.10		12.90	18.50
0700	24" O.C.		100	.160		3.58	7.75		11.33	16.60
0705	24" O.C., pneumatic nailed		120	.133		3.60	6.45		10.05	14.50
0780	2" x 6" studs, 8' high, studs 12" O.C.		70	.229		6.70	11.05		17.75	25.50
0785	12" O.C., pneumatic nailed		84	.190		6.75	9.25		16	22.50
0800	16" O.C.		90	.178		5.50	8.60		14.10	20
0805	16" O.C., pneumatic nailed		108	.148		5.55	7.20		12.75	17.85
0900	24" O.C.		115	.139		4.28	6.75		11.03	15.70
0905	24" O.C., pneumatic nailed		138	.116		4.31	5.60		9.91	13.95
0980	10' high, studs 12" O.C.		70	.229		7.95	11.05		19	27
0985	12" O.C., pneumatic nailed		84	.190		8	9.25		17.25	24
1000	16" O.C.		90	.178		6.40	8.60		15	21

For customer support on your Commercial Renovation Cost Data, call 877.742.4719.

06 11 Wood Framing

06 11 10 – Framing with Dimensional, Engineered or Composite Lumber

06 11 10.26 Partitions

		Crew	Daily Output	Labor-Hours	Unit	Material	2016 Bare Costs Labor	Equipment	Total	Total Incl O&P
1005	16" O.C., pneumatic nailed	2 Carp	108	.148	L.F.	6.45	7.20		13.65	18.85
1100	24" O.C.		115	.139		4.89	6.75		11.64	16.40
1105	24" O.C., pneumatic nailed		138	.116		4.92	5.60		10.52	14.60
1180	12' high, studs 12" O.C.		55	.291		9.15	14.10		23.25	33
1185	12" O.C., pneumatic nailed		66	.242		9.25	11.75		21	29.50
1200	16" O.C.		70	.229		7.35	11.05		18.40	26
1205	16" O.C., pneumatic nailed		84	.190		7.40	9.25		16.65	23
1300	24" O.C.		90	.178		5.50	8.60		14.10	20
1305	24" O.C., pneumatic nailed		108	.148		5.55	7.20		12.75	17.85
1400	For horizontal blocking, 2" x 4", add		600	.027		.40	1.29		1.69	2.55
1500	2" x 6", add		600	.027		.61	1.29		1.90	2.78
1600	For openings, add	▼	250	.064	▼		3.10		3.10	5.05
1702	Headers for above openings, material only, add				B.F.	.70			.70	.77
9000	Minimum labor/equipment charge	1 Carp	4	2	Job		97		97	158

06 11 10.28 Porch or Deck Framing

		Crew	Daily Output	Labor-Hours	Unit	Material	2016 Bare Costs Labor	Equipment	Total	Total Incl O&P
0010	**PORCH OR DECK FRAMING**									
0100	Treated lumber, posts or columns, 4" x 4"	2 Carp	390	.041	L.F.	1.35	1.99		3.34	4.73
0110	4" x 6"		275	.058		2.02	2.82		4.84	6.80
0120	4" x 8"		220	.073		3.87	3.52		7.39	10
0130	Girder, single, 4" x 4"		675	.024		1.35	1.15		2.50	3.36
0140	4" x 6"		600	.027		2.02	1.29		3.31	4.33
0150	4" x 8"		525	.030		3.87	1.48		5.35	6.65
0160	Double, 2" x 4"		625	.026		1.07	1.24		2.31	3.21
0170	2" x 6"		600	.027		1.59	1.29		2.88	3.85
0180	2" x 8"		575	.028		2.17	1.35		3.52	4.58
0190	2" x 10"		550	.029		2.76	1.41		4.17	5.35
0200	2" x 12"		525	.030		3.87	1.48		5.35	6.65
0210	Triple, 2" x 4"		575	.028		1.61	1.35		2.96	3.97
0220	2" x 6"		550	.029		2.38	1.41		3.79	4.92
0230	2" x 8"		525	.030		3.25	1.48		4.73	6
0240	2" x 10"		500	.032		4.13	1.55		5.68	7.10
0250	2" x 12"		475	.034		5.80	1.63		7.43	9.05
0260	Ledger, bolted 4' O.C., 2" x 4"		400	.040		.69	1.94		2.63	3.92
0270	2" x 6"		395	.041		.93	1.96		2.89	4.24
0280	2" x 8"		390	.041		1.21	1.99		3.20	4.58
0290	2" x 10"		385	.042		1.50	2.01		3.51	4.94
0300	2" x 12"		380	.042		2.05	2.04		4.09	5.60
0310	Joists, 2" x 4"		1250	.013		.53	.62		1.15	1.60
0320	2" x 6"		1250	.013		.79	.62		1.41	1.88
0330	2" x 8"		1100	.015		1.08	.70		1.78	2.33
0340	2" x 10"		900	.018		1.37	.86		2.23	2.92
0350	2" x 12"	▼	875	.018		1.64	.89		2.53	3.26
0360	Railings and trim , 1" x 4"	1 Carp	300	.027		.46	1.29		1.75	2.62
0370	2" x 2"		300	.027		.33	1.29		1.62	2.48
0380	2" x 4"		300	.027		.53	1.29		1.82	2.69
0390	2" x 6"		300	.027	▼	.78	1.29		2.07	2.97
0400	Decking, 1" x 4"		275	.029	S.F.	2.16	1.41		3.57	4.68
0410	2" x 4"		300	.027		1.78	1.29		3.07	4.06
0420	2" x 6"		320	.025		1.68	1.21		2.89	3.83
0430	5/4" x 6"		320	.025	▼	2.25	1.21		3.46	4.46
0440	Balusters, square, 2" x 2"	2 Carp	660	.024	L.F.	.34	1.17		1.51	2.29
0450	Turned, 2" x 2"	▼	420	.038	▼	.45	1.85		2.30	3.51

For customer support on your Commercial Renovation Cost Data, call 877.742.4719.

141

06 11 10.28 Porch or Deck Framing

		Crew	Daily Output	Labor-Hours	Unit	Material	2016 Bare Costs Labor	Equipment	Total	Total Incl O&P
0460	Stair stringer, 2" x 10"	2 Carp	130	.123	L.F.	1.37	5.95		7.32	11.25
0470	2" x 12"		130	.123		1.64	5.95		7.59	11.55
0480	Stair treads, 1" x 4"		140	.114		2.17	5.55		7.72	11.45
0490	2" x 4"		140	.114		.53	5.55		6.08	9.65
0500	2" x 6"		160	.100		.80	4.85		5.65	8.80
0510	5/4" x 6"		160	.100		1.05	4.85		5.90	9.05
0520	Turned handrail post, 4" x 4"		64	.250	Ea.	31	12.10		43.10	54
0530	Lattice panel, 4' x 8', 1/2"		1600	.010	S.F.	.81	.48		1.29	1.68
0535	3/4"		1600	.010	"	1.21	.48		1.69	2.12
0540	Cedar, posts or columns, 4" x 4"		390	.041	L.F.	3.96	1.99		5.95	7.60
0550	4" x 6"		275	.058		7.70	2.82		10.52	13.05
0560	4" x 8"		220	.073		10.45	3.52		13.97	17.20
0800	Decking, 1" x 4"		550	.029		1.91	1.41		3.32	4.40
0810	2" x 4"		600	.027		3.86	1.29		5.15	6.35
0820	2" x 6"		640	.025		6.95	1.21		8.16	9.65
0830	5/4" x 6"		640	.025		4.45	1.21		5.66	6.85
0840	Railings and trim, 1" x 4"		600	.027		1.91	1.29		3.20	4.21
0860	2" x 4"		600	.027		3.86	1.29		5.15	6.35
0870	2" x 6"		600	.027		6.95	1.29		8.24	9.75
0920	Stair treads, 1" x 4"		140	.114		1.91	5.55		7.46	11.15
0930	2" x 4"		140	.114		3.86	5.55		9.41	13.30
0940	2" x 6"		160	.100		6.95	4.85		11.80	15.55
0950	5/4" x 6"		160	.100		4.45	4.85		9.30	12.80
0980	Redwood, posts or columns, 4" x 4"		390	.041		6.45	1.99		8.44	10.35
0990	4" x 6"		275	.058		12.60	2.82		15.42	18.45
1000	4" x 8"		220	.073		23.50	3.52		27.02	32
1240	Decking, 1" x 4"	1 Carp	275	.029	S.F.	3.97	1.41		5.38	6.65
1260	2" x 6"		340	.024		7.50	1.14		8.64	10.10
1270	5/4" x 6"		320	.025		4.83	1.21		6.04	7.30
1280	Railings and trim, 1" x 4"	2 Carp	600	.027	L.F.	1.17	1.29		2.46	3.40
1310	2" x 6"		600	.027		7.50	1.29		8.79	10.35
1420	Alternative decking, wood/plastic composite, 5/4" x 6" [G]		640	.025		3.14	1.21		4.35	5.45
1440	1" x 4" square edge fir		550	.029		2.18	1.41		3.59	4.70
1450	1" x 4" tongue and groove fir		450	.036		1.46	1.72		3.18	4.41
1460	1" x 4" mahogany		550	.029		2.01	1.41		3.42	4.51
1462	5/4" x 6" PVC		550	.029		3.71	1.41		5.12	6.40
1465	Framing, porch or deck, alt deck fastening, screws, add	1 Carp	240	.033	S.F.		1.61		1.61	2.64
1470	Accessories, joist hangers, 2" x 4"		160	.050	Ea.	.72	2.42		3.14	4.75
1480	2" x 6" through 2" x 12"		150	.053		1.36	2.58		3.94	5.70
1530	Post footing, incl excav, backfill, tube form & concrete, 4' deep, 8" dia	F-7	12	2.667		12.65	115		127.65	202
1540	10" diameter		11	2.909		18.40	126		144.40	226
1550	12" diameter		10	3.200		24	138		162	253

06 11 10.30 Roof Framing

		Crew	Daily Output	Labor-Hours	Unit	Material	2016 Bare Costs Labor	Equipment	Total	Total Incl O&P
0010	**ROOF FRAMING** R061110-30									
1900	Rough fascia, 2" x 6"	2 Carp	250	.064	L.F.	.61	3.10		3.71	5.70
2000	2" x 8"		225	.071		.85	3.45		4.30	6.60
2100	2" x 10"		180	.089		1.34	4.31		5.65	8.50
5000	Rafters, to 4 in 12 pitch, 2" x 6", ordinary		1000	.016		.61	.78		1.39	1.94
5060	2" x 8", ordinary		950	.017		.85	.82		1.67	2.26
5120	2" x 10", ordinary		630	.025		1.34	1.23		2.57	3.48
5180	2" x 12", ordinary		575	.028		1.70	1.35		3.05	4.07
5250	Composite rafter, 9-1/2" deep		575	.028		1.85	1.35		3.20	4.24

06 11 10 – Framing with Dimensional, Engineered or Composite Lumber

06 11 10.30 Roof Framing

		Crew	Daily Output	Labor-Hours	Unit	Material	2016 Bare Costs Labor	Equipment	Total	Total Incl O&P
5260	11-1/2" deep	2 Carp	575	.028	L.F.	2.04	1.35		3.39	4.45
5300	Hip and valley rafters, 2" x 6", ordinary		760	.021		.61	1.02		1.63	2.34
5360	2" x 8", ordinary		720	.022		.85	1.08		1.93	2.69
5420	2" x 10", ordinary		570	.028		1.34	1.36		2.70	3.69
5480	2" x 12", ordinary		525	.030		1.70	1.48		3.18	4.28
5540	Hip and valley jacks, 2" x 6", ordinary		600	.027		.61	1.29		1.90	2.78
5600	2" x 8", ordinary		490	.033		.85	1.58		2.43	3.51
5660	2" x 10", ordinary		450	.036		1.34	1.72		3.06	4.28
5720	2" x 12", ordinary	▼	375	.043	▼	1.70	2.07		3.77	5.25
5761	For slopes steeper than 4 in 12, add						30%			
5780	Rafter tie, 1" x 4", #3	2 Carp	800	.020	L.F.	.46	.97		1.43	2.09
5790	2" x 4", #3		800	.020		.40	.97		1.37	2.02
5800	Ridge board, #2 or better, 1" x 6"		600	.027		.71	1.29		2	2.89
5820	1" x 8"		550	.029		1.18	1.41		2.59	3.60
5840	1" x 10"		500	.032		1.53	1.55		3.08	4.22
5860	2" x 6"		500	.032		.61	1.55		2.16	3.20
5880	2" x 8"		450	.036		.85	1.72		2.57	3.74
5900	2" x 10"		400	.040		1.34	1.94		3.28	4.64
5920	Roof cants, split, 4" x 4"		650	.025		1.73	1.19		2.92	3.86
5940	6" x 6"		600	.027		5.05	1.29		6.34	7.70
5960	Roof curbs, untreated, 2" x 6"		520	.031		.61	1.49		2.10	3.11
5980	2" x 12"		400	.040		1.70	1.94		3.64	5.05
6000	Sister rafters, 2" x 6"		800	.020		.61	.97		1.58	2.25
6020	2" x 8"		640	.025		.85	1.21		2.06	2.91
6040	2" x 10"		535	.030		1.34	1.45		2.79	3.84
6060	2" x 12"	▼	455	.035	▼	1.70	1.70		3.40	4.65
9000	Minimum labor/equipment charge	1 Carp	4	2	Job		97		97	158

06 11 10.32 Sill and Ledger Framing

		Crew	Daily Output	Labor-Hours	Unit	Material	2016 Bare Costs Labor	Equipment	Total	Total Incl O&P
0010	**SILL AND LEDGER FRAMING**									
0020	Extruded polystyrene sill sealer, 5-1/2" wide	1 Carp	1600	.005	L.F.	.14	.24		.38	.55
2002	Ledgers, nailed, 2" x 4"	2 Carp	755	.021		.40	1.03		1.43	2.12
2052	2" x 6"		600	.027		.61	1.29		1.90	2.78
2102	Bolted, not including bolts, 3" x 6"		325	.049		1.88	2.39		4.27	5.95
2152	3" x 12"		233	.069		4.08	3.33		7.41	9.95
2602	Mud sills, redwood, construction grade, 2" x 4"		895	.018		2.26	.87		3.13	3.91
2622	2" x 6"		780	.021		3.40	.99		4.39	5.35
4002	Sills, 2" x 4"		600	.027		.39	1.29		1.68	2.54
4052	2" x 6"		550	.029		.60	1.41		2.01	2.96
4082	2" x 8"		500	.032		.83	1.55		2.38	3.45
4101	2" x 10"		450	.036		1.32	1.72		3.04	4.26
4121	2" x 12"		400	.040		1.68	1.94		3.62	5
4202	Treated, 2" x 4"		550	.029		.52	1.41		1.93	2.87
4222	2" x 6"		500	.032		.76	1.55		2.31	3.37
4242	2" x 8"		450	.036		1.04	1.72		2.76	3.96
4261	2" x 10"		400	.040		1.33	1.94		3.27	4.63
4281	2" x 12"		350	.046		1.88	2.21		4.09	5.70
4402	4" x 4"		450	.036		1.31	1.72		3.03	4.25
4422	4" x 6"		350	.046		1.96	2.21		4.17	5.80
4462	4" x 8"		300	.053		3.79	2.58		6.37	8.40
4481	4" x 10"	▼	260	.062	▼	4.73	2.98		7.71	10.05
9000	Minimum labor/equipment charge	1 Carp	4	2	Job		97		97	158

For customer support on your Commercial Renovation Cost Data, call 877.742.4719.

143

06 11 Wood Framing

06 11 10 - Framing with Dimensional, Engineered or Composite Lumber

06 11 10.34 Sleepers

		Crew	Daily Output	Labor-Hours	Unit	Material	2016 Bare Costs Labor	Equipment	Total	Total Incl O&P
0010	**SLEEPERS**									
0100	On concrete, treated, 1" x 2"	2 Carp	2350	.007	L.F.	.27	.33		.60	.84
0150	1" x 3"		2000	.008		.44	.39		.83	1.12
0200	2" x 4"		1500	.011		.56	.52		1.08	1.46
0250	2" x 6"	↓	1300	.012	↓	.86	.60		1.46	1.91
9000	Minimum labor/equipment charge	1 Carp	4	2	Job		97		97	158

06 11 10.36 Soffit and Canopy Framing

		Crew	Daily Output	Labor-Hours	Unit	Material	2016 Bare Costs Labor	Equipment	Total	Total Incl O&P
0010	**SOFFIT AND CANOPY FRAMING**									
1002	Canopy or soffit framing , 1" x 4"	2 Carp	900	.018	L.F.	.46	.86		1.32	1.92
1021	1" x 6"		850	.019		.71	.91		1.62	2.27
1042	1" x 8"		750	.021		1.18	1.03		2.21	2.99
1102	2" x 4"		620	.026		.40	1.25		1.65	2.48
1121	2" x 6"		560	.029		.61	1.38		1.99	2.93
1142	2" x 8"		500	.032		.85	1.55		2.40	3.46
1202	3" x 4"		500	.032		1.14	1.55		2.69	3.79
1221	3" x 6"		400	.040		1.89	1.94		3.83	5.25
1242	3" x 10"	↓	300	.053	↓	3.43	2.58		6.01	8
9000	Minimum labor/equipment charge	1 Carp	4	2	Job		97		97	158

06 11 10.38 Treated Lumber Framing Material

		Crew	Daily Output	Labor-Hours	Unit	Material	2016 Bare Costs Labor	Equipment	Total	Total Incl O&P
0010	**TREATED LUMBER FRAMING MATERIAL**									
0100	2" x 4"				M.B.F.	775			775	850
0110	2" x 6"					765			765	840
0120	2" x 8"					780			780	860
0130	2" x 10"					795			795	875
0140	2" x 12"					940			940	1,025
0200	4" x 4"					980			980	1,075
0210	4" x 6"					980			980	1,075
0220	4" x 8"				↓	1,425			1,425	1,550

06 11 10.40 Wall Framing

		Crew	Daily Output	Labor-Hours	Unit	Material	2016 Bare Costs Labor	Equipment	Total	Total Incl O&P
0010	**WALL FRAMING** R061110-30									
0100	Door buck, studs, header, access, 8' H, 2" x 4" wall, 3' W	1 Carp	32	.250	Ea.	16.65	12.10		28.75	38
0110	4' wide		32	.250		17.90	12.10		30	39.50
0120	5' wide		32	.250		21.50	12.10		33.60	43.50
0130	6' wide		32	.250		23	12.10		35.10	45.50
0140	8' wide		30	.267		34.50	12.90		47.40	59
0150	10' wide		30	.267		47.50	12.90		60.40	73
0160	12' wide		30	.267		65	12.90		77.90	92.50
0170	2" x 6" wall, 3' wide		32	.250		23.50	12.10		35.60	46
0180	4' wide		32	.250		24.50	12.10		36.60	47
0190	5' wide		32	.250		28.50	12.10		40.60	51.50
0200	6' wide		32	.250		30	12.10		42.10	53
0210	8' wide		30	.267		41.50	12.90		54.40	66.50
0220	10' wide		30	.267		54	12.90		66.90	80.50
0230	12' wide		30	.267		72	12.90		84.90	100
0240	Window buck, studs, header & access, 8' high 2" x 4" wall, 2' wide		24	.333		17.70	16.15		33.85	46
0250	3' wide		24	.333		20.50	16.15		36.65	49
0260	4' wide		24	.333		22.50	16.15		38.65	51.50
0270	5' wide		24	.333		26.50	16.15		42.65	55.50
0280	6' wide		24	.333		29	16.15		45.15	58.50
0290	7' wide		24	.333		38.50	16.15		54.65	69
0300	8' wide	↓	22	.364	↓	43	17.60		60.60	76

144

06 11 Wood Framing

06 11 10 – Framing with Dimensional, Engineered or Composite Lumber

06 11 10.40 Wall Framing	Crew	Daily Output	Labor-Hours	Unit	Material	2016 Bare Costs Labor	Equipment	Total	Total Incl O&P
0310	10' wide	1 Carp	22	.364	Ea.	57	17.60		74.60	91.50
0320	12' wide		22	.364		77	17.60		94.60	114
0330	2" x 6" wall, 2' wide		24	.333		26	16.15		42.15	55.50
0340	3' wide		24	.333		29	16.15		45.15	58.50
0350	4' wide		24	.333		32	16.15		48.15	61.50
0360	5' wide		24	.333		36	16.15		52.15	66
0370	6' wide		24	.333		39.50	16.15		55.65	69.50
0380	7' wide		24	.333		50	16.15		66.15	81.50
0390	8' wide		22	.364		54.50	17.60		72.10	88.50
0400	10' wide		22	.364		69	17.60		86.60	105
0410	12' wide	▼	22	.364	▼	90	17.60		107.60	128
2002	Headers over openings, 2" x 6"	2 Carp	360	.044	L.F.	.61	2.15		2.76	4.19
2007	2" x 6", pneumatic nailed		432	.037		.62	1.79		2.41	3.61
2052	2" x 8"		340	.047		.85	2.28		3.13	4.65
2057	2" x 8", pneumatic nailed		408	.039		.85	1.90		2.75	4.04
2101	2" x 10"		320	.050		1.34	2.42		3.76	5.45
2106	2" x 10", pneumatic nailed		384	.042		1.34	2.02		3.36	4.78
2152	2" x 12"		300	.053		1.70	2.58		4.28	6.10
2157	2" x 12", pneumatic nailed		360	.044		1.71	2.15		3.86	5.40
2180	4" x 8"		260	.062		2.96	2.98		5.94	8.15
2185	4" x 8", pneumatic nailed		312	.051		2.97	2.48		5.45	7.35
2191	4" x 10"		240	.067		4.70	3.23		7.93	10.45
2196	4" x 10", pneumatic nailed		288	.056		4.72	2.69		7.41	9.60
2202	4" x 12"		190	.084		5.20	4.08		9.28	12.35
2207	4" x 12", pneumatic nailed		228	.070		5.20	3.40		8.60	11.30
2241	6" x 10"		165	.097		6.55	4.70		11.25	14.90
2246	6" x 10", pneumatic nailed		198	.081		6.60	3.92		10.52	13.65
2251	6" x 12"		140	.114		8.25	5.55		13.80	18.10
2256	6" x 12", pneumatic nailed		168	.095		8.25	4.61		12.86	16.65
5002	Plates, untreated, 2" x 3"		850	.019		.38	.91		1.29	1.90
5007	2" x 3", pneumatic nailed		1020	.016		.38	.76		1.14	1.66
5022	2" x 4"		800	.020		.40	.97		1.37	2.02
5027	2" x 4", pneumatic nailed		960	.017		.40	.81		1.21	1.76
5041	2" x 6"		750	.021		.61	1.03		1.64	2.36
5046	2" x 6", pneumatic nailed		900	.018		.62	.86		1.48	2.09
5122	Studs, 8' high wall, 2" x 3"		1200	.013		.38	.65		1.03	1.47
5127	2" x 3", pneumatic nailed		1440	.011		.38	.54		.92	1.30
5142	2" x 4"		1100	.015		.39	.70		1.09	1.58
5147	2" x 4", pneumatic nailed		1320	.012		.40	.59		.99	1.40
5162	2" x 6"		1000	.016		.61	.78		1.39	1.94
5167	2" x 6", pneumatic nailed		1200	.013		.62	.65		1.27	1.74
5182	3" x 4"		800	.020		1.14	.97		2.11	2.84
5187	3" x 4", pneumatic nailed	▼	960	.017	▼	1.15	.81		1.96	2.58
8200	For 12' high walls, deduct						5%			
8220	For stub wall, 6' high, add						20%			
8240	3' high, add						40%			
8250	For second story & above, add						5%			
8300	For dormer & gable, add						15%			
9000	Minimum labor/equipment charge	1 Carp	4	2	Job		97		97	158

For customer support on your Commercial Renovation Cost Data, call 877.742.4719.

145

06 11 Wood Framing

06 11 10 – Framing with Dimensional, Engineered or Composite Lumber

06 11 10.42 Furring

		Crew	Daily Output	Labor-Hours	Unit	Material	2016 Bare Costs Labor	Equipment	Total	Total Incl O&P
0010	**FURRING**									
0012	Wood strips, 1" x 2", on walls, on wood	1 Carp	550	.015	L.F.	.24	.70		.94	1.42
0015	On wood, pneumatic nailed		710	.011		.24	.55		.79	1.16
0300	On masonry		495	.016		.26	.78		1.04	1.57
0400	On concrete		260	.031		.26	1.49		1.75	2.73
0600	1" x 3", on walls, on wood		550	.015		.39	.70		1.09	1.58
0605	On wood, pneumatic nailed		710	.011		.39	.55		.94	1.32
0700	On masonry		495	.016		.42	.78		1.20	1.74
0800	On concrete		260	.031		.42	1.49		1.91	2.90
0850	On ceilings, on wood		350	.023		.39	1.11		1.50	2.24
0855	On wood, pneumatic nailed		450	.018		.39	.86		1.25	1.84
0900	On masonry		320	.025		.42	1.21		1.63	2.44
0950	On concrete		210	.038		.42	1.85		2.27	3.48
9000	Minimum labor/equipment charge		4	2	Job		97		97	158

06 11 10.44 Grounds

		Crew	Daily Output	Labor-Hours	Unit	Material	2016 Bare Costs Labor	Equipment	Total	Total Incl O&P
0010	**GROUNDS**									
0020	For casework, 1" x 2" wood strips, on wood	1 Carp	330	.024	L.F.	.24	1.17		1.41	2.19
0100	On masonry		285	.028		.26	1.36		1.62	2.51
0200	On concrete		250	.032		.26	1.55		1.81	2.82
0400	For plaster, 3/4" deep, on wood		450	.018		.24	.86		1.10	1.68
0500	On masonry		225	.036		.26	1.72		1.98	3.10
0600	On concrete		175	.046		.26	2.21		2.47	3.91
0700	On metal lath		200	.040		.26	1.94		2.20	3.46
9000	Minimum labor/equipment charge		4	2	Job		97		97	158

06 12 Structural Panels

06 12 10 – Structural Insulated Panels

06 12 10.10 OSB Faced Panels

			Crew	Daily Output	Labor-Hours	Unit	Material	2016 Bare Costs Labor	Equipment	Total	Total Incl O&P
0010	**OSB FACED PANELS**										
0100	Structural insul. panels, 7/16" OSB both faces, EPS insul, 3-5/8" T	G	F-3	2075	.019	S.F.	3.65	.95	.32	4.92	5.90
0110	5-5/8" thick	G		1725	.023		4.10	1.14	.38	5.62	6.80
0120	7-3/8" thick	G		1425	.028		4.45	1.38	.46	6.29	7.65
0130	9-3/8" thick	G		1125	.036		4.75	1.75	.59	7.09	8.75
0140	7/16" OSB one face, EPS insul, 3-5/8" thick	G		2175	.018		3.75	.91	.30	4.96	5.95
0150	5-5/8" thick	G		1825	.022		4.35	1.08	.36	5.79	6.95
0160	7-3/8" thick	G		1525	.026		4.85	1.29	.43	6.57	7.90
0170	9-3/8" thick	G		1225	.033		5.35	1.61	.54	7.50	9.10
0190	7/16" OSB - 1/2" GWB faces, EPS insul, 3-5/8" T	G		2075	.019		3.35	.95	.32	4.62	5.60
0200	5-5/8" thick	G		1725	.023		4	1.14	.38	5.52	6.65
0210	7-3/8" thick	G		1425	.028		4.50	1.38	.46	6.34	7.70
0220	9-3/8" thick	G		1125	.036		5.10	1.75	.59	7.44	9.10
0240	7/16" OSB - 1/2" MRGWB faces, EPS insul, 3-5/8" T	G		2075	.019		3.45	.95	.32	4.72	5.70
0250	5-5/8" thick	G		1725	.023		4.10	1.14	.38	5.62	6.80
0260	7-3/8" thick	G		1425	.028		4.55	1.38	.46	6.39	7.75
0270	9-3/8" thick	G		1125	.036		5.20	1.75	.59	7.54	9.20
0300	For 1/2" GWB added to OSB skin, add	G					1.35			1.35	1.49
0310	For 1/2" MRGWB added to OSB skin, add	G					1.35			1.35	1.49
0320	For one T1-11 skin, add to OSB-OSB	G					1.95			1.95	2.15
0330	For one 19/32" CDX skin, add to OSB-OSB	G					1.50			1.50	1.65
0500	Structural insulated panel, 7/16" OSB both sides, straw core										

06 12 Structural Panels

06 12 10 – Structural Insulated Panels

06 12 10.10 OSB Faced Panels

		Crew	Daily Output	Labor-Hours	Unit	Material	2016 Bare Costs Labor	Equipment	Total	Total Incl O&P
0510	4-3/8" T, walls (w/sill, splines, plates) G	F-6	2400	.017	S.F.	7.55	.75	.27	8.57	9.80
0520	Floors (w/splines) G		2400	.017		7.55	.75	.27	8.57	9.80
0530	Roof (w/splines) G		2400	.017		7.55	.75	.27	8.57	9.80
0550	7-7/8" T, walls (w/sill, splines, plates) G		2400	.017		11.40	.75	.27	12.42	14.05
0560	Floors (w/splines) G		2400	.017		11.40	.75	.27	12.42	14.05
0570	Roof (w/splines) G		2400	.017		11.40	.75	.27	12.42	14.05

06 12 19 – Composite Shearwall Panels

06 12 19.10 Steel and Wood Composite Shearwall Panels

		Crew	Daily Output	Labor-Hours	Unit	Material	2016 Bare Costs Labor	Equipment	Total	Total Incl O&P
0010	**STEEL & WOOD COMPOSITE SHEARWALL PANELS**									
0020	Anchor bolts, 36" long (must be placed in wet concrete)	1 Carp	150	.053	Ea.	34	2.58		36.58	41.50
0030	On concrete, 2 x 4 & 2 x 6 walls, 7' - 10' high, 360 lb. shear, 12" wide	2 Carp	8	2		460	97		557	665
0040	715 lb. shear, 15" wide		8	2		510	97		607	720
0050	1860 lb. shear, 18" wide		8	2		525	97		622	740
0060	2780 lb. shear, 21" wide		8	2		560	97		657	775
0070	3790 lb. shear, 24" wide		8	2		640	97		737	860
0080	2 x 6 walls, 11' to 13' high, 1180 lb. shear, 18" wide		6	2.667		645	129		774	920
0090	1555 lb. shear, 21" wide		6	2.667		720	129		849	1,000
0100	2280 lb. shear, 24" wide		6	2.667		805	129		934	1,100
0110	For installing above on wood floor frame, add									
0120	Coupler nuts, threaded rods, bolts, shear transfer plate kit	1 Carp	16	.500	Ea.	65	24		89	111
0130	Framing anchors, angle (2 required)	"	96	.083	"	2.56	4.04		6.60	9.40
0140	For blocking see Section 06 11 10.02									
0150	For installing above, first floor to second floor, wood floor frame, add									
0160	Add stack option to first floor wall panel				Ea.	71.50			71.50	78.50
0170	Threaded rods, bolts, shear transfer plate kit	1 Carp	16	.500		75	24		99	122
0180	Framing anchors, angle (2 required)	"	96	.083		2.56	4.04		6.60	9.40
0190	For blocking see section 06 11 10.02									
0200	For installing stacked panels, balloon framing									
0210	Add stack option to first floor wall panel				Ea.	71.50			71.50	78.50
0220	Threaded rods, bolts kit	1 Carp	16	.500	"	45	24		69	89

06 13 Heavy Timber Construction

06 13 23 – Heavy Timber Framing

06 13 23.10 Heavy Framing

		Crew	Daily Output	Labor-Hours	Unit	Material	2016 Bare Costs Labor	Equipment	Total	Total Incl O&P
0010	**HEAVY FRAMING**									
0020	Beams, single 6" x 10"	2 Carp	1.10	14.545	M.B.F.	1,525	705		2,230	2,825
0100	Single 8" x 16"		1.20	13.333	"	2,075	645		2,720	3,350
0202	Built from 2" lumber, multiple 2" x 14"		900	.018	B.F.	.91	.86		1.77	2.41
0212	Built from 3" lumber, multiple 3" x 6"		700	.023		1.25	1.11		2.36	3.19
0222	Multiple 3" x 8"		800	.020		1.35	.97		2.32	3.07
0232	Multiple 3" x 10"		900	.018		1.36	.86		2.22	2.91
0242	Multiple 3" x 12"		1000	.016		1.36	.78		2.14	2.77
0252	Built from 4" lumber, multiple 4" x 6"		800	.020		1.44	.97		2.41	3.16
0262	Multiple 4" x 8"		900	.018		1.10	.86		1.96	2.62
0272	Multiple 4" x 10"		1000	.016		1.40	.78		2.18	2.81
0282	Multiple 4" x 12"		1100	.015		1.29	.70		1.99	2.57
0292	Columns, structural grade, 1500f, 4" x 4"		450	.036	L.F.	1.66	1.72		3.38	4.64
0302	6" x 6"		225	.071		4.27	3.45		7.72	10.35
0402	8" x 8"		240	.067		6.90	3.23		10.13	12.85
0502	10" x 10"		90	.178		12.45	8.60		21.05	28

For customer support on your Commercial Renovation Cost Data, call 877.742.4719.

147

06 13 Heavy Timber Construction

06 13 23 – Heavy Timber Framing

06 13 23.10 Heavy Framing	Crew	Daily Output	Labor-Hours	Unit	Material	2016 Bare Costs Labor	Equipment	Total	Total Incl O&P
0602	12" x 12"	2 Carp	70	.229	L.F.	17.50	11.05		28.55	37.50
0802	Floor planks, 2" thick, T & G, 2" x 6"		1050	.015	B.F.	1.59	.74		2.33	2.96
0902	2" x 10"		1100	.015		1.60	.70		2.30	2.91
1102	3" thick, 3" x 6"		1050	.015		1.59	.74		2.33	2.96
1202	3" x 10"		1100	.015		1.59	.70		2.29	2.90
1402	Girders, structural grade, 12" x 12"		800	.020		1.46	.97		2.43	3.18
1502	10" x 16"		1000	.016		2.50	.78		3.28	4.02
2302	Roof purlins, 4" thick, structural grade		1050	.015		1.10	.74		1.84	2.42
9000	Minimum labor/equipment charge	1 Carp	2	4	Job		194		194	315

06 15 Wood Decking

06 15 16 – Wood Roof Decking

06 15 16.10 Solid Wood Roof Decking

		Crew	Daily Output	Labor-Hours	Unit	Material	2016 Bare Costs Labor	Equipment	Total	Total Incl O&P
0010	**SOLID WOOD ROOF DECKING**									
0350	Cedar planks, 2" thick	2 Carp	350	.046	S.F.	6.45	2.21		8.66	10.70
0400	3" thick		320	.050		9.70	2.42		12.12	14.60
0500	4" thick		250	.064		12.90	3.10		16	19.25
0550	6" thick		200	.080		19.35	3.88		23.23	28
0650	Douglas fir, 2" thick		350	.046		2.73	2.21		4.94	6.60
0700	3" thick		320	.050		4.10	2.42		6.52	8.45
0800	4" thick		250	.064		5.45	3.10		8.55	11.05
0850	6" thick		200	.080		8.20	3.88		12.08	15.35
0950	Hemlock, 2" thick		350	.046		2.78	2.21		4.99	6.70
1000	3" thick		320	.050		4.17	2.42		6.59	8.55
1100	4" thick		250	.064		5.55	3.10		8.65	11.15
1150	6" thick		200	.080		8.35	3.88		12.23	15.50
1250	Western white spruce, 2" thick		350	.046		1.77	2.21		3.98	5.55
1300	3" thick		320	.050		2.66	2.42		5.08	6.90
1400	4" thick		250	.064		3.55	3.10		6.65	8.95
1450	6" thick		200	.080		5.30	3.88		9.18	12.20
9000	Minimum labor/equipment charge	1 Carp	2	4	Job		194		194	315

06 15 23 – Laminated Wood Decking

06 15 23.10 Laminated Roof Deck

		Crew	Daily Output	Labor-Hours	Unit	Material	2016 Bare Costs Labor	Equipment	Total	Total Incl O&P
0010	**LAMINATED ROOF DECK**									
0020	Pine or hemlock, 3" thick	2 Carp	425	.038	S.F.	6.25	1.82		8.07	9.90
0100	4" thick		325	.049		8.15	2.39		10.54	12.85
0300	Cedar, 3" thick		425	.038		7.10	1.82		8.92	10.80
0400	4" thick		325	.049		9.50	2.39		11.89	14.35
0600	Fir, 3" thick		425	.038		5.50	1.82		7.32	9.05
0700	4" thick		325	.049		7.50	2.39		9.89	12.15
9000	Minimum labor/equipment charge	1 Carp	3	2.667	Job		129		129	211

06 16 Sheathing

06 16 13 – Insulating Sheathing

06 16 13.10 Insulating Sheathing		Crew	Daily Output	Labor-Hours	Unit	Material	2016 Bare Costs Labor	Equipment	Total	Total Incl O&P
0010	**INSULATING SHEATHING**									
0020	Expanded polystyrene, 1#/C.F. density, 3/4" thick R2.89	G	2 Carp	1400	.011	S.F.	.35	.55		.90	1.29
0030	1" thick R3.85	G		1300	.012		.42	.60		1.02	1.43
0040	2" thick R7.69	G		1200	.013		.68	.65		1.33	1.81
0050	Extruded polystyrene, 15 PSI compressive strength, 1" thick, R5	G		1300	.012		.76	.60		1.36	1.80
0060	2" thick, R10	G		1200	.013		.94	.65		1.59	2.09
0070	Polyisocyanurate, 2#/C.F. density, 3/4" thick	G		1400	.011		.62	.55		1.17	1.58
0080	1" thick	G		1300	.012		.63	.60		1.23	1.66
0090	1-1/2" thick	G		1250	.013		.80	.62		1.42	1.89
0100	2" thick	G		1200	.013		.98	.65		1.63	2.14

06 16 23 – Subflooring

06 16 23.10 Subfloor

06 16 23.10 Subfloor		Crew	Daily Output	Labor-Hours	Unit	Material	2016 Bare Costs Labor	Equipment	Total	Total Incl O&P
0010	**SUBFLOOR** R061636-20									
0011	Plywood, CDX, 1/2" thick	2 Carp	1500	.011	SF Flr.	.68	.52		1.20	1.59
0015	Pneumatic nailed		1860	.009		.68	.42		1.10	1.43
0102	5/8" thick		1350	.012		.76	.57		1.33	1.78
0107	Pneumatic nailed		1674	.010		.76	.46		1.22	1.60
0202	3/4" thick		1250	.013		.91	.62		1.53	2.01
0207	Pneumatic nailed		1550	.010		.91	.50		1.41	1.82
0302	1-1/8" thick, 2-4-1 including underlayment		1050	.015		2.03	.74		2.77	3.44
0440	With boards, 1" x 6", S4S, laid regular		900	.018		1.54	.86		2.40	3.10
0452	1" x 8", laid regular		1000	.016		1.88	.78		2.66	3.34
0462	Laid diagonal		850	.019		1.88	.91		2.79	3.56
0502	1" x 10", laid regular		1100	.015		1.92	.70		2.62	3.26
0602	Laid diagonal		900	.018		1.92	.86		2.78	3.52
8990	Subfloor adhesive, 3/8" bead	1 Carp	2300	.003	L.F.	.12	.17		.29	.41
9000	Minimum labor/equipment charge	"	4	2	Job		97		97	158

06 16 26 – Underlayment

06 16 26.10 Wood Product Underlayment

06 16 26.10 Wood Product Underlayment		Crew	Daily Output	Labor-Hours	Unit	Material	2016 Bare Costs Labor	Equipment	Total	Total Incl O&P
0010	**WOOD PRODUCT UNDERLAYMENT** R061636-20									
0015	Plywood, underlayment grade, 1/4" thick	2 Carp	1500	.011	S.F.	.89	.52		1.41	1.82
0018	Pneumatic nailed		1860	.009		.89	.42		1.31	1.66
0030	3/8" thick		1500	.011		.99	.52		1.51	1.93
0070	Pneumatic nailed		1860	.009		.99	.42		1.41	1.77
0102	1/2" thick		1450	.011		1.17	.53		1.70	2.16
0107	Pneumatic nailed		1798	.009		1.17	.43		1.60	1.99
0202	5/8" thick		1400	.011		1.29	.55		1.84	2.32
0207	Pneumatic nailed		1736	.009		1.29	.45		1.74	2.15
0302	3/4" thick		1300	.012		1.48	.60		2.08	2.60
0306	Pneumatic nailed		1612	.010		1.48	.48		1.96	2.42
0502	Particle board, 3/8" thick		1500	.011		.41	.52		.93	1.29
0507	Pneumatic nailed		1860	.009		.41	.42		.83	1.13
0602	1/2" thick		1450	.011		.45	.53		.98	1.37
0607	Pneumatic nailed		1798	.009		.45	.43		.88	1.20
0802	5/8" thick		1400	.011		.54	.55		1.09	1.49
0807	Pneumatic nailed		1736	.009		.54	.45		.99	1.32
0902	3/4" thick		1300	.012		.68	.60		1.28	1.72
0907	Pneumatic nailed		1612	.010		.68	.48		1.16	1.54
1100	Hardboard, underlayment grade, 4' x 4', .215" thick	G	1500	.011		.63	.52		1.15	1.53
9000	Minimum labor/equipment charge	1 Carp	4	2	Job		97		97	158

For customer support on your Commercial Renovation Cost Data, call 877.742.4719.

149

06 16 Sheathing

06 16 33 – Wood Board Sheathing

06 16 33.10 Board Sheathing	Crew	Daily Output	Labor-Hours	Unit	Material	2016 Bare Costs Labor	Equipment	Total	Total Incl O&P
0009	**BOARD SHEATHING**									
0010	Roof, 1" x 6" boards, laid horizontal	2 Carp	725	.022	S.F.	1.54	1.07		2.61	3.44
0020	On steep roof		520	.031		1.54	1.49		3.03	4.13
0040	On dormers, hips, & valleys		480	.033		1.54	1.61		3.15	4.33
0050	Laid diagonal		650	.025		1.54	1.19		2.73	3.64
0070	1" x 8" boards, laid horizontal		875	.018		1.88	.89		2.77	3.52
0080	On steep roof		635	.025		1.88	1.22		3.10	4.06
0090	On dormers, hips, & valleys		580	.028		1.92	1.34		3.26	4.29
0100	Laid diagonal		725	.022		1.88	1.07		2.95	3.82
0110	Skip sheathing, 1" x 4", 7" OC	1 Carp	1200	.007		.58	.32		.90	1.16
0120	1" x 6", 9" OC		1450	.006		.71	.27		.98	1.22
0180	Tongue and groove sheathing/decking, 1" x 6"		1000	.008		1.82	.39		2.21	2.63
0190	2" x 6"		1000	.008		3.81	.39		4.20	4.82
0200	Walls, 1" x 6" boards, laid regular	2 Carp	650	.025		1.54	1.19		2.73	3.64
0210	Laid diagonal		585	.027		1.54	1.33		2.87	3.85
0220	1" x 8" boards, laid regular		765	.021		1.88	1.01		2.89	3.73
0230	Laid diagonal		650	.025		1.88	1.19		3.07	4.02

06 16 36 – Wood Panel Product Sheathing

06 16 36.10 Sheathing

06 16 36.10 Sheathing		Crew	Daily Output	Labor-Hours	Unit	Material	2016 Bare Costs Labor	Equipment	Total	Total Incl O&P
0010	**SHEATHING**	R061110-30									
0012	Plywood on roofs, CDX										
0032	5/16" thick		2 Carp	1600	.010	S.F.	.62	.48		1.10	1.48
0037	Pneumatic nailed	R061636-20		1952	.008		.62	.40		1.02	1.34
0052	3/8" thick			1525	.010		.62	.51		1.13	1.51
0057	Pneumatic nailed			1860	.009		.62	.42		1.04	1.36
0102	1/2" thick			1400	.011		.68	.55		1.23	1.65
0103	Pneumatic nailed			1708	.009		.68	.45		1.13	1.49
0202	5/8" thick			1300	.012		.76	.60		1.36	1.81
0207	Pneumatic nailed			1586	.010		.76	.49		1.25	1.64
0302	3/4" thick			1200	.013		.91	.65		1.56	2.06
0307	Pneumatic nailed			1464	.011		.91	.53		1.44	1.87
0502	Plywood on walls, with exterior CDX, 3/8" thick			1200	.013		.62	.65		1.27	1.74
0507	Pneumatic nailed			1488	.011		.62	.52		1.14	1.53
0602	1/2" thick			1125	.014		.68	.69		1.37	1.88
0607	Pneumatic nailed			1395	.011		.68	.56		1.24	1.66
0702	5/8" thick			1050	.015		.76	.74		1.50	2.05
0707	Pneumatic nailed			1302	.012		.76	.60		1.36	1.81
0802	3/4" thick			975	.016		.91	.80		1.71	2.30
0807	Pneumatic nailed			1209	.013		.91	.64		1.55	2.05
1000	For shear wall construction, add							20%			
1200	For structural 1 exterior plywood, add					S.F.	10%				
3000	Wood fiber, regular, no vapor barrier, 1/2" thick		2 Carp	1200	.013		.64	.65		1.29	1.76
3100	5/8" thick			1200	.013		.76	.65		1.41	1.90
3300	No vapor barrier, in colors, 1/2" thick			1200	.013		.77	.65		1.42	1.91
3400	5/8" thick			1200	.013		.81	.65		1.46	1.95
3600	With vapor barrier one side, white, 1/2" thick			1200	.013		.63	.65		1.28	1.75
3700	Vapor barrier 2 sides, 1/2" thick			1200	.013		.86	.65		1.51	2.01
3800	Asphalt impregnated, 25/32" thick			1200	.013		.33	.65		.98	1.42
3850	Intermediate, 1/2" thick			1200	.013		.25	.65		.90	1.34
4500	Oriented strand board, on roof, 7/16" thick	G		1460	.011		.50	.53		1.03	1.42
4505	Pneumatic nailed	G		1780	.009		.50	.44		.94	1.26
4550	1/2" thick	G		1400	.011		.50	.55		1.05	1.45

150

06 16 Sheathing

06 16 36 – Wood Panel Product Sheathing

06 16 36.10 Sheathing		Crew	Daily Output	Labor-Hours	Unit	Material	2016 Bare Costs Labor	Equipment	Total	Total Incl O&P
4555	Pneumatic nailed	G 2 Carp	1736	.009	S.F.	.50	.45		.95	1.28
4600	5/8" thick	G	1300	.012		.69	.60		1.29	1.73
4605	Pneumatic nailed	G	1586	.010		.69	.49		1.18	1.56
4610	On walls, 7/16" thick	G	1200	.013		.50	.65		1.15	1.61
4615	Pneumatic nailed	G	1488	.011		.50	.52		1.02	1.40
4620	1/2" thick	G	1195	.013		.50	.65		1.15	1.61
4625	Pneumatic nailed	G	1325	.012		.50	.59		1.09	1.51
4630	5/8" thick	G	1050	.015		.69	.74		1.43	1.97
4635	Pneumatic nailed	G	1302	.012		.69	.60		1.29	1.73
4700	Oriented strand board, factory laminated W.R. barrier, on roof, 1/2" thick	G	1400	.011		.78	.55		1.33	1.76
4705	Pneumatic nailed	G	1736	.009		.78	.45		1.23	1.59
4720	5/8" thick	G	1300	.012		.93	.60		1.53	1.99
4725	Pneumatic nailed	G	1586	.010		.93	.49		1.42	1.82
4730	5/8" thick, T&G	G	1150	.014		.93	.67		1.60	2.12
4735	Pneumatic nailed, T&G	G	1400	.011		.93	.55		1.48	1.92
4740	On walls, 7/16" thick	G	1200	.013		.66	.65		1.31	1.79
4745	Pneumatic nailed	G	1488	.011		.66	.52		1.18	1.58
4750	1/2" thick	G	1195	.013		.78	.65		1.43	1.92
4755	Pneumatic nailed	G	1325	.012		.78	.59		1.37	1.82
4800	Joint sealant tape, 3-1/2"		7600	.002	L.F.	.31	.10		.41	.51
4810	Joint sealant tape, 6"		7600	.002	"	.44	.10		.54	.65
9000	Minimum labor/equipment charge	1 Carp	2	4	Job		194		194	315

06 16 43 – Gypsum Sheathing

06 16 43.10 Gypsum Sheathing

		Crew	Daily Output	Labor-Hours	Unit	Material	Labor	Equipment	Total	Total Incl O&P
0010	**GYPSUM SHEATHING**									
0020	Gypsum, weatherproof, 1/2" thick	2 Carp	1125	.014	S.F.	.46	.69		1.15	1.64
0040	With embedded glass mats	"	1100	.015	"	.70	.70		1.40	1.92

06 17 Shop-Fabricated Structural Wood

06 17 33 – Wood I-Joists

06 17 33.10 Wood and Composite I-Joists

		Crew	Daily Output	Labor-Hours	Unit	Material	Labor	Equipment	Total	Total Incl O&P
0010	**WOOD AND COMPOSITE I-JOISTS**									
0100	Plywood webs, incl. bridging & blocking, panels 24" O.C.									
1200	15' to 24' span, 50 psf live load	F-5	2400	.013	SF Flr.	2.09	.65		2.74	3.37
1300	55 psf live load		2250	.014		2.30	.70		3	3.68
1400	24' to 30' span, 45 psf live load		2600	.012		2.40	.60		3	3.62
1500	55 psf live load		2400	.013		4.05	.65		4.70	5.55

06 17 53 – Shop-Fabricated Wood Trusses

06 17 53.10 Roof Trusses

		Crew	Daily Output	Labor-Hours	Unit	Material	Labor	Equipment	Total	Total Incl O&P
0010	**ROOF TRUSSES**									
0100	Fink (W) or King post type, 2'-0" O.C.									
0200	Metal plate connected, 4 in 12 slope									
0210	24' to 29' span	F-3	3000	.013	SF Flr.	1.70	.66	.22	2.58	3.17
0300	30' to 43' span		3000	.013		2.14	.66	.22	3.02	3.65
0400	44' to 60' span		3000	.013		2.25	.66	.22	3.13	3.77
0700	Glued and nailed, add					50%				

For customer support on your Commercial Renovation Cost Data, call 877.742.4719.

151

06 18 Glued-Laminated Construction

06 18 13 – Glued-Laminated Beams

06 18 13.20 Laminated Framing	Crew	Daily Output	Labor-Hours	Unit	Material	2016 Bare Costs Labor	Equipment	Total	Total Incl O&P
0010 **LAMINATED FRAMING**									
0020 30 lb., short term live load, 15 lb. dead load									
0200 Straight roof beams, 20' clear span, beams 8' O.C.	F-3	2560	.016	SF Flr.	2.08	.77	.26	3.11	3.82
0300 Beams 16' O.C.		3200	.013		1.52	.62	.21	2.35	2.90
0500 40' clear span, beams 8' O.C.		3200	.013		3.95	.62	.21	4.78	5.60
0600 Beams 16' O.C.		3840	.010		3.25	.51	.17	3.93	4.60
0800 60' clear span, beams 8' O.C.	F-4	2880	.017		6.75	.81	.39	7.95	9.20
0900 Beams 16' O.C.	"	3840	.013		5.05	.61	.29	5.95	6.85
1100 Tudor arches, 30' to 40' clear span, frames 8' O.C.	F-3	1680	.024		8.80	1.17	.39	10.36	12.05
1200 Frames 16' O.C.	"	2240	.018		6.90	.88	.29	8.07	9.35
1400 50' to 60' clear span, frames 8' O.C.	F-4	2200	.022		9.50	1.06	.51	11.07	12.75
1500 Frames 16' O.C.		2640	.018		8.10	.89	.43	9.42	10.80
1700 Radial arches, 60' clear span, frames 8' O.C.		1920	.025		8.85	1.22	.59	10.66	12.35
1800 Frames 16' O.C.		2880	.017		7	.81	.39	8.20	9.45
2000 100' clear span, frames 8' O.C.		1600	.030		9.15	1.46	.70	11.31	13.20
2100 Frames 16' O.C.		2400	.020		8.05	.97	.47	9.49	10.95
2300 120' clear span, frames 8' O.C.		1440	.033		12.15	1.62	.78	14.55	16.85
2400 Frames 16' O.C.		1920	.025		11.10	1.22	.59	12.91	14.80
2600 Bowstring trusses, 20' O.C., 40' clear span	F-3	2400	.017		5.50	.82	.27	6.59	7.70
2700 60' clear span	F-4	3600	.013		4.95	.65	.31	5.91	6.85
2800 100' clear span		4000	.012		7	.58	.28	7.86	8.95
2900 120' clear span		3600	.013		7.45	.65	.31	8.41	9.60
3000 For less than 1000 B.F., add					20%				
3100 For premium appearance, add to S.F. prices					5%				
3300 For industrial type, deduct					15%				
3500 For stain and varnish, add					5%				
3900 For 3/4" laminations, add to straight					25%				
4100 Add to curved					15%				
4300 Alternate pricing method: (use nominal footage of									
4310 components). Straight beams, camber less than 6"	F-3	3.50	11.429	M.B.F.	2,925	560	188	3,673	4,325
4400 Columns, including hardware		2	20		3,125	985	330	4,440	5,400
4600 Curved members, radius over 32'		2.50	16		3,200	785	263	4,248	5,100
4700 Radius 10' to 32'		3	13.333		3,175	655	219	4,049	4,825
4900 For complicated shapes, add maximum					100%				
5100 For pressure treating, add to straight					35%				
5200 Add to curved					45%				
6000 Laminated veneer members, southern pine or western species									
6050 1-3/4" wide x 5-1/2" deep	2 Carp	480	.033	L.F.	3.44	1.61		5.05	6.40
6100 9-1/2" deep		480	.033		4.74	1.61		6.35	7.85
6150 14" deep		450	.036		7.85	1.72		9.57	11.45
6200 18" deep		450	.036		10.65	1.72		12.37	14.55
6300 Parallel strand members, southern pine or western species									
6350 1-3/4" wide x 9-1/4" deep	2 Carp	480	.033	L.F.	5.10	1.61		6.71	8.25
6400 11-1/4" deep		450	.036		5.75	1.72		7.47	9.15
6450 14" deep		400	.040		7.85	1.94		9.79	11.80
6500 3-1/2" wide x 9-1/4" deep		480	.033		16.30	1.61		17.91	20.50
6550 11-1/4" deep		450	.036		20.50	1.72		22.22	25.50
6600 14" deep		400	.040		24	1.94		25.94	29.50
6650 7" wide x 9-1/4" deep		450	.036		34.50	1.72		36.22	41
6700 11-1/4" deep		420	.038		43.50	1.85		45.35	50.50
6750 14" deep		400	.040		52	1.94		53.94	60
9000 Minimum labor/equipment charge	F-3	2.50	16	Job		785	263	1,048	1,575

152

06 22 Millwork

06 22 13 – Standard Pattern Wood Trim

06 22 13.15 Moldings, Base	Crew	Daily Output	Labor-Hours	Unit	Material	2016 Bare Costs Labor	Equipment	Total	Total Incl O&P
0010 MOLDINGS, BASE									
5100 Classic profile, 5/8" x 5-1/2", finger jointed and primed	1 Carp	250	.032	L.F.	1.49	1.55		3.04	4.17
5105 Poplar		240	.033		1.57	1.61		3.18	4.36
5110 Red oak		220	.036		2.41	1.76		4.17	5.55
5115 Maple		220	.036		3.64	1.76		5.40	6.90
5120 Cherry		220	.036		4.32	1.76		6.08	7.65
5125 3/4 x 7-1/2", finger jointed and primed		250	.032		1.89	1.55		3.44	4.61
5130 Poplar		240	.033		2.51	1.61		4.12	5.40
5135 Red oak		220	.036		3.50	1.76		5.26	6.75
5140 Maple		220	.036		4.84	1.76		6.60	8.20
5145 Cherry		220	.036		5.75	1.76		7.51	9.25
5150 Modern profile, 5/8" x 3-1/2", finger jointed and primed		250	.032		.92	1.55		2.47	3.54
5155 Poplar		240	.033		1	1.61		2.61	3.74
5160 Red oak		220	.036		1.65	1.76		3.41	4.69
5165 Maple		220	.036		2.55	1.76		4.31	5.70
5170 Cherry		220	.036		2.56	1.76		4.32	5.70
5175 Ogee profile, 7/16" x 3", finger jointed and primed		250	.032		.63	1.55		2.18	3.22
5180 Poplar		240	.033		.66	1.61		2.27	3.36
5185 Red oak		220	.036		.93	1.76		2.69	3.90
5200 9/16" x 3-1/2", finger jointed and primed		250	.032		.66	1.55		2.21	3.25
5205 Pine		240	.033		1.13	1.61		2.74	3.88
5210 Red oak		220	.036		2.38	1.76		4.14	5.50
5215 9/16" x 4-1/2", red oak		220	.036		3.56	1.76		5.32	6.80
5220 5/8" x 3-1/2", finger jointed and primed		250	.032		.91	1.55		2.46	3.53
5225 Poplar		240	.033		1	1.61		2.61	3.74
5230 Red oak		220	.036		1.65	1.76		3.41	4.69
5235 Maple		220	.036		2.55	1.76		4.31	5.70
5240 Cherry		220	.036		2.75	1.76		4.51	5.90
5245 5/8" x 4", finger jointed and primed		250	.032		1.17	1.55		2.72	3.81
5250 Poplar		240	.033		1.29	1.61		2.90	4.06
5255 Red oak		220	.036		1.85	1.76		3.61	4.91
5260 Maple		220	.036		2.85	1.76		4.61	6
5265 Cherry		220	.036		2.99	1.76		4.75	6.15
5270 Rectangular profile, oak, 3/8" x 1-1/4"		260	.031		1.20	1.49		2.69	3.76
5275 1/2" x 2-1/2"		255	.031		1.94	1.52		3.46	4.61
5280 1/2" x 3-1/2"		250	.032		2.36	1.55		3.91	5.10
5285 1" x 6"		240	.033		4.21	1.61		5.82	7.25
5290 1" x 8"		240	.033		5.80	1.61		7.41	9.05
5295 Pine, 3/8" x 1-3/4"		260	.031		.46	1.49		1.95	2.94
5300 7/16" x 2-1/2"		255	.031		.74	1.52		2.26	3.29
5305 1" x 6"		240	.033		.83	1.61		2.44	3.55
5310 1" x 8"		240	.033		1.06	1.61		2.67	3.80
5315 Shoe, 1/2" x 3/4", primed		260	.031		.50	1.49		1.99	2.99
5320 Pine		240	.033		.30	1.61		1.91	2.97
5325 Poplar		240	.033		.39	1.61		2	3.07
5330 Red oak		220	.036		.53	1.76		2.29	3.46
5335 Maple		220	.036		.70	1.76		2.46	3.65
5340 Cherry		220	.036		.80	1.76		2.56	3.76
5345 11/16" x 1-1/2", pine		240	.033		.70	1.61		2.31	3.41
5350 Caps, 11/16" x 1-3/8", pine		240	.033		.55	1.61		2.16	3.24
5355 3/4" x 1-3/4", finger jointed and primed		260	.031		.71	1.49		2.20	3.22
5360 Poplar		240	.033		1.01	1.61		2.62	3.75
5365 Red oak		220	.036		1.21	1.76		2.97	4.21

06 22 13 – Standard Pattern Wood Trim

06 22 13.15 Moldings, Base		Crew	Daily Output	Labor-Hours	Unit	Material	2016 Bare Costs Labor	Equipment	Total	Total Incl O&P
5370	Maple	1 Carp	220	.036	L.F.	1.58	1.76		3.34	4.62
5375	Cherry		220	.036		3.20	1.76		4.96	6.40
5380	Combination base & shoe, 9/16" x 3-1/2" & 1/2" x 3/4", pine		125	.064		1.43	3.10		4.53	6.60
5385	Three piece oak, 6" high		80	.100		5.95	4.85		10.80	14.45
5390	Including 3/4" x 1" base shoe		70	.114		6.10	5.55		11.65	15.80
5395	Flooring cant strip, 3/4" x 3/4", pre-finished pine		260	.031		.51	1.49		2	3
5400	For pre-finished, stain and clear coat, add					.53			.53	.58
5405	Clear coat only, add					.45			.45	.50
9000	Minimum labor/equipment charge	1 Carp	4	2	Job		97		97	158

06 22 13.30 Moldings, Casings

06 22 13.30 Moldings, Casings		Crew	Daily Output	Labor-Hours	Unit	Material	2016 Bare Costs Labor	Equipment	Total	Total Incl O&P
0010	**MOLDINGS, CASINGS**									
0085	Apron, 9/16" x 2-1/2", pine	1 Carp	250	.032	L.F.	1.63	1.55		3.18	4.32
0090	5/8" x 2-1/2", pine		250	.032		1.63	1.55		3.18	4.32
0110	5/8" x 3-1/2", pine		220	.036		1.86	1.76		3.62	4.92
0300	Band, 11/16" x 1-1/8", pine		270	.030		.75	1.44		2.19	3.17
0310	11/16" x 1-1/2", finger jointed and primed		270	.030		.76	1.44		2.20	3.18
0320	Pine		270	.030		.96	1.44		2.40	3.40
0330	11/16" x 1-3/4", finger jointed and primed		270	.030		.95	1.44		2.39	3.39
0350	Pine		270	.030		1.16	1.44		2.60	3.62
0355	Beaded, 3/4" x 3-1/2", finger jointed and primed		220	.036		1.11	1.76		2.87	4.10
0360	Poplar		220	.036		1.11	1.76		2.87	4.10
0365	Red oak		220	.036		1.65	1.76		3.41	4.69
0370	Maple		220	.036		2.55	1.76		4.31	5.70
0375	Cherry		220	.036		3.09	1.76		4.85	6.30
0380	3/4" x 4", finger jointed and primed		220	.036		1.17	1.76		2.93	4.16
0385	Poplar		220	.036		1.45	1.76		3.21	4.47
0390	Red oak		220	.036		2.05	1.76		3.81	5.15
0395	Maple		220	.036		2.65	1.76		4.41	5.80
0400	Cherry		220	.036		3.33	1.76		5.09	6.55
0405	3/4" x 5-1/2", finger jointed and primed		200	.040		1.17	1.94		3.11	4.45
0410	Poplar		200	.040		1.93	1.94		3.87	5.30
0415	Red oak		200	.040		2.84	1.94		4.78	6.30
0420	Maple		200	.040		3.73	1.94		5.67	7.25
0425	Cherry		200	.040		4.31	1.94		6.25	7.90
0430	Classic profile, 3/4" x 2-3/4", finger jointed and primed		250	.032		.80	1.55		2.35	3.41
0435	Poplar		250	.032		1.02	1.55		2.57	3.65
0440	Red oak		250	.032		1.48	1.55		3.03	4.16
0445	Maple		250	.032		2.09	1.55		3.64	4.83
0450	Cherry		250	.032		2.39	1.55		3.94	5.15
0455	Fluted, 3/4" x 3-1/2", poplar		220	.036		1.12	1.76		2.88	4.11
0460	Red oak		220	.036		1.66	1.76		3.42	4.70
0465	Maple		220	.036		2.53	1.76		4.29	5.65
0470	Cherry		220	.036		3.09	1.76		4.85	6.30
0475	3/4" x 4", poplar		220	.036		1.44	1.76		3.20	4.46
0480	Red oak		220	.036		2.03	1.76		3.79	5.10
0485	Maple		220	.036		2.65	1.76		4.41	5.80
0490	Cherry		220	.036		3.33	1.76		5.09	6.55
0495	3/4" x 5-1/2", poplar		200	.040		1.93	1.94		3.87	5.30
0500	Red oak		200	.040		2.86	1.94		4.80	6.30
0505	Maple		200	.040		3.73	1.94		5.67	7.25
0510	Cherry		200	.040		4.32	1.94		6.26	7.90
0515	3/4" x 7-1/2", poplar		190	.042		1.22	2.04		3.26	4.67

154

For customer support on your Commercial Renovation Cost Data, call 877.742.4719.

06 22 13 – Standard Pattern Wood Trim

06 22 13.30 Moldings, Casings

		Crew	Daily Output	Labor-Hours	Unit	Material	2016 Bare Costs Labor	Equipment	Total	Total Incl O&P
0520	Red oak	1 Carp	190	.042	L.F.	3.92	2.04		5.96	7.65
0525	Maple		190	.042		5.60	2.04		7.64	9.50
0530	Cherry		190	.042		6.70	2.04		8.74	10.70
0535	3/4" x 9-1/2", poplar		180	.044		4.13	2.15		6.28	8.05
0540	Red oak		180	.044		6.55	2.15		8.70	10.70
0545	Maple		180	.044		8.95	2.15		11.10	13.30
0550	Cherry		180	.044		9.75	2.15		11.90	14.20
0555	Modern profile, 9/16" x 2-1/4", poplar		250	.032		.59	1.55		2.14	3.18
0560	Red oak		250	.032		.79	1.55		2.34	3.40
0565	11/16" x 2-1/2", finger jointed & primed		250	.032		.84	1.55		2.39	3.45
0570	Pine		250	.032		1.28	1.55		2.83	3.94
0575	3/4" x 2-1/2", poplar		250	.032		.90	1.55		2.45	3.52
0580	Red oak		250	.032		1.21	1.55		2.76	3.86
0585	Maple		250	.032		1.81	1.55		3.36	4.52
0590	Cherry		250	.032		2.30	1.55		3.85	5.05
0595	Mullion, 5/16" x 2", pine		270	.030		.86	1.44		2.30	3.29
0600	9/16" x 2-1/2", fingerjointed and primed		250	.032		.98	1.55		2.53	3.61
0605	Pine		250	.032		1.28	1.55		2.83	3.94
0610	Red oak		250	.032		2.86	1.55		4.41	5.65
0615	1-1/16" x 3-3/4", red oak		220	.036		6.95	1.76		8.71	10.50
0620	Ogee, 7/16" x 2-1/2", poplar		250	.032		.58	1.55		2.13	3.17
0625	Red oak		250	.032		.79	1.55		2.34	3.40
0630	9/16" x 2-1/4", finger jointed and primed		250	.032		.51	1.55		2.06	3.09
0635	Poplar		250	.032		.59	1.55		2.14	3.18
0640	Red oak		250	.032		.78	1.55		2.33	3.39
0645	11/16" x 2-1/2", finger jointed and primed		250	.032		.72	1.55		2.27	3.32
0700	Pine		250	.032		1.41	1.55		2.96	4.08
0701	Red oak		250	.032		2.70	1.55		4.25	5.50
0730	11/16" x 3-1/2", finger jointed and primed		220	.036		1.11	1.76		2.87	4.10
0750	Pine		220	.036		1.72	1.76		3.48	4.77
0755	3/4" x 2-1/2", finger jointed and primed		250	.032		.93	1.55		2.48	3.55
0760	Poplar		250	.032		.89	1.55		2.44	3.51
0765	Red oak		250	.032		1.20	1.55		2.75	3.85
0770	Maple		250	.032		1.80	1.55		3.35	4.51
0775	Cherry		250	.032		2.27	1.55		3.82	5
0780	3/4" x 3-1/2", finger jointed and primed		220	.036		.92	1.76		2.68	3.89
0785	Poplar		220	.036		1.12	1.76		2.88	4.11
0790	Red oak		220	.036		1.66	1.76		3.42	4.70
0795	Maple		220	.036		2.56	1.76		4.32	5.70
0800	Cherry		220	.036		3.11	1.76		4.87	6.30
4700	Square profile, 1" x 1", teak		215	.037		2.26	1.80		4.06	5.45
4800	Rectangular profile, 1" x 3", teak		200	.040		6.20	1.94		8.14	10
9000	Minimum labor/equipment charge		4	2	Job		97		97	158

06 22 13.35 Moldings, Ceilings

		Crew	Daily Output	Labor-Hours	Unit	Material	2016 Bare Costs Labor	Equipment	Total	Total Incl O&P
0010	**MOLDINGS, CEILINGS**									
0600	Bed, 9/16" x 1-3/4", pine	1 Carp	270	.030	L.F.	1.06	1.44		2.50	3.51
0650	9/16" x 2", pine		270	.030		1.14	1.44		2.58	3.60
0710	9/16" x 1-3/4", oak		270	.030		1.86	1.44		3.30	4.39
1200	Cornice, 9/16" x 1-3/4", pine		270	.030		.96	1.44		2.40	3.40
1300	9/16" x 2-1/4", pine		265	.030		1.27	1.46		2.73	3.78
1350	Cove, 1/2" x 2-1/4", poplar		265	.030		1.04	1.46		2.50	3.53
1360	Red oak		265	.030		1.37	1.46		2.83	3.89

06 22 13.35 Moldings, Ceilings		Crew	Daily Output	Labor-Hours	Unit	Material	2016 Bare Costs Labor	Equipment	Total	Total Incl O&P
1370	Hard maple	1 Carp	265	.030	L.F.	1.86	1.46		3.32	4.43
1380	Cherry		265	.030		2.28	1.46		3.74	4.89
2400	9/16" x 1-3/4", pine		270	.030		.97	1.44		2.41	3.41
2500	11/16" x 2-3/4", pine		265	.030		1.80	1.46		3.26	4.37
2510	Crown, 5/8" x 5/8", poplar		300	.027		.44	1.29		1.73	2.59
2520	Red oak		300	.027		.52	1.29		1.81	2.68
2530	Hard maple		300	.027		.70	1.29		1.99	2.88
2540	Cherry		300	.027		.74	1.29		2.03	2.92
2600	9/16" x 3-5/8", pine		250	.032		2.02	1.55		3.57	4.75
2700	11/16" x 4-1/4", pine		250	.032		2.90	1.55		4.45	5.70
2705	Oak		250	.032		6.40	1.55		7.95	9.60
2710	3/4" x 1-3/4", poplar		270	.030		.71	1.44		2.15	3.13
2720	Red oak		270	.030		1.05	1.44		2.49	3.50
2730	Hard maple		270	.030		1.39	1.44		2.83	3.88
2740	Cherry		270	.030		1.71	1.44		3.15	4.23
2750	3/4" x 2", poplar		270	.030		.94	1.44		2.38	3.38
2760	Red oak		270	.030		1.26	1.44		2.70	3.73
2770	Hard maple		270	.030		1.68	1.44		3.12	4.20
2780	Cherry		270	.030		1.88	1.44		3.32	4.42
2790	3/4" x 2-3/4", poplar		265	.030		1.02	1.46		2.48	3.51
2800	Red oak		265	.030		1.57	1.46		3.03	4.11
2810	Hard maple		265	.030		2.37	1.46		3.83	4.99
2820	Cherry		265	.030		2.40	1.46		3.86	5.05
2830	3/4" x 3-1/2", poplar		250	.032		1.30	1.55		2.85	3.96
2840	Red oak		250	.032		1.96	1.55		3.51	4.68
2850	Hard maple		250	.032		2.56	1.55		4.11	5.35
2860	Cherry		250	.032		3	1.55		4.55	5.85
2870	FJP poplar		250	.032		.92	1.55		2.47	3.54
2880	3/4" x 5", poplar		245	.033		1.93	1.58		3.51	4.70
2890	Red oak		245	.033		2.85	1.58		4.43	5.70
2900	Hard maple		245	.033		3.74	1.58		5.32	6.70
2910	Cherry		245	.033		4.32	1.58		5.90	7.35
2920	FJP poplar		245	.033		1.32	1.58		2.90	4.03
2930	3/4" x 6-1/4", poplar		240	.033		2.29	1.61		3.90	5.15
2940	Red oak		240	.033		3.46	1.61		5.07	6.45
2950	Hard maple		240	.033		4.52	1.61		6.13	7.60
2960	Cherry		240	.033		5.35	1.61		6.96	8.55
2970	7/8" x 8-3/4", poplar		220	.036		3.56	1.76		5.32	6.80
2980	Red oak		220	.036		4.97	1.76		6.73	8.35
2990	Hard maple		220	.036		6.75	1.76		8.51	10.35
3000	Cherry		220	.036		7.60	1.76		9.36	11.30
3010	1" x 7-1/4", poplar		220	.036		3.77	1.76		5.53	7.05
3020	Red oak		220	.036		5.85	1.76		7.61	9.35
3030	Hard maple		220	.036		8.15	1.76		9.91	11.85
3040	Cherry		220	.036		9.50	1.76		11.26	13.35
3050	1-1/16" x 4-1/4", poplar		250	.032		2.29	1.55		3.84	5.05
3060	Red oak		250	.032		2.69	1.55		4.24	5.50
3070	Hard maple		250	.032		3.56	1.55		5.11	6.45
3080	Cherry		250	.032		5.10	1.55		6.65	8.15
3090	Dentil crown, 3/4" x 5", poplar		250	.032		1.93	1.55		3.48	4.65
3100	Red oak		250	.032		2.85	1.55		4.40	5.65
3110	Hard maple		250	.032		3.74	1.55		5.29	6.65
3120	Cherry		250	.032		4.32	1.55		5.87	7.30

156

06 22 13 – Standard Pattern Wood Trim

06 22 13.35 Moldings, Ceilings		Crew	Daily Output	Labor-Hours	Unit	Material	2016 Bare Costs Labor	Equipment	Total	Total Incl O&P
3130	Dentil piece for above, 1/2" x 1/2", poplar	1 Carp	300	.027	L.F.	2.73	1.29		4.02	5.10
3140	Red oak		300	.027		3.14	1.29		4.43	5.55
3150	Hard maple		300	.027		3.57	1.29		4.86	6.05
3160	Cherry		300	.027		3.89	1.29		5.18	6.40
9000	Minimum labor/equipment charge		4	2	Job		97		97	158

06 22 13.40 Moldings, Exterior		Crew	Daily Output	Labor-Hours	Unit	Material	2016 Bare Costs Labor	Equipment	Total	Total Incl O&P
0010	**MOLDINGS, EXTERIOR**									
0100	Band board, cedar, rough sawn, 1" x 2"	1 Carp	300	.027	L.F.	.55	1.29		1.84	2.71
0110	1" x 3"		300	.027		.81	1.29		2.10	3
0120	1" x 4"		250	.032		1.08	1.55		2.63	3.72
0130	1" x 6"		250	.032		1.63	1.55		3.18	4.32
0140	1" x 8"		225	.036		2.17	1.72		3.89	5.20
0150	1" x 10"		225	.036		2.70	1.72		4.42	5.80
0160	1" x 12"		200	.040		3.24	1.94		5.18	6.75
0240	STK, 1" x 2"		300	.027		.53	1.29		1.82	2.70
0250	1" x 3"		300	.027		.59	1.29		1.88	2.75
0260	1" x 4"		250	.032		.71	1.55		2.26	3.31
0270	1" x 6"		250	.032		1.13	1.55		2.68	3.77
0280	1" x 8"		225	.036		1.63	1.72		3.35	4.60
0290	1" x 10"		225	.036		2.12	1.72		3.84	5.15
0300	1" x 12"		200	.040		3.62	1.94		5.56	7.15
0310	Pine, #2, 1" x 2"		300	.027		.25	1.29		1.54	2.39
0320	1" x 3"		300	.027		.40	1.29		1.69	2.55
0330	1" x 4"		250	.032		.49	1.55		2.04	3.07
0340	1" x 6"		250	.032		.73	1.55		2.28	3.34
0350	1" x 8"		225	.036		1.20	1.72		2.92	4.13
0360	1" x 10"		225	.036		1.57	1.72		3.29	4.53
0370	1" x 12"		200	.040		1.92	1.94		3.86	5.30
0380	D & better, 1" x 2"		300	.027		.43	1.29		1.72	2.59
0390	1" x 3"		300	.027		.66	1.29		1.95	2.84
0400	1" x 4"		250	.032		.89	1.55		2.44	3.51
0410	1" x 6"		250	.032		1.12	1.55		2.67	3.76
0420	1" x 8"		225	.036		1.62	1.72		3.34	4.59
0430	1" x 10"		225	.036		2.03	1.72		3.75	5.05
0440	1" x 12"		200	.040		2.63	1.94		4.57	6.05
0450	Redwood, clear all heart, 1" x 2"		300	.027		.63	1.29		1.92	2.81
0460	1" x 3"		300	.027		.94	1.29		2.23	3.15
0470	1" x 4"		250	.032		1.19	1.55		2.74	3.84
0480	1" x 6"		252	.032		1.80	1.54		3.34	4.49
0490	1" x 8"		225	.036		2.36	1.72		4.08	5.40
0500	1" x 10"		225	.036		3.84	1.72		5.56	7.05
0510	1" x 12"		200	.040		4.75	1.94		6.69	8.40
0530	Corner board, cedar, rough sawn, 1" x 2"		225	.036		.55	1.72		2.27	3.41
0540	1" x 3"		225	.036		.81	1.72		2.53	3.70
0550	1" x 4"		200	.040		1.08	1.94		3.02	4.36
0560	1" x 6"		200	.040		1.63	1.94		3.57	4.96
0570	1" x 8"		200	.040		2.17	1.94		4.11	5.55
0580	1" x 10"		175	.046		2.70	2.21		4.91	6.60
0590	1" x 12"		175	.046		3.24	2.21		5.45	7.20
0670	STK, 1" x 2"		225	.036		.53	1.72		2.25	3.40
0680	1" x 3"		225	.036		.59	1.72		2.31	3.45
0690	1" x 4"		200	.040		.69	1.94		2.63	3.93

For customer support on your Commercial Renovation Cost Data, call 877.742.4719.

157

06 22 13.40 Moldings, Exterior	Crew	Daily Output	Labor-Hours	Unit	Material	2016 Bare Costs Labor	Equipment	Total	Total Incl O&P
0700	1" x 6"	1 Carp	200	.040	L.F.	1.13	1.94		3.07	4.41
0710	1" x 8"		200	.040		1.63	1.94		3.57	4.96
0720	1" x 10"		175	.046		2.12	2.21		4.33	5.95
0730	1" x 12"		175	.046		3.62	2.21		5.83	7.60
0740	Pine, #2, 1" x 2"		225	.036		.25	1.72		1.97	3.09
0750	1" x 3"		225	.036		.40	1.72		2.12	3.25
0760	1" x 4"		200	.040		.49	1.94		2.43	3.71
0770	1" x 6"		200	.040		.73	1.94		2.67	3.98
0780	1" x 8"		200	.040		1.20	1.94		3.14	4.49
0790	1" x 10"		175	.046		1.57	2.21		3.78	5.35
0800	1" x 12"		175	.046		1.92	2.21		4.13	5.75
0810	D & better, 1" x 2"		225	.036		.43	1.72		2.15	3.29
0820	1" x 3"		225	.036		.66	1.72		2.38	3.54
0830	1" x 4"		200	.040		.89	1.94		2.83	4.15
0840	1" x 6"		200	.040		1.12	1.94		3.06	4.40
0850	1" x 8"		200	.040		1.62	1.94		3.56	4.95
0860	1" x 10"		175	.046		2.03	2.21		4.24	5.85
0870	1" x 12"		175	.046		2.63	2.21		4.84	6.50
0880	Redwood, clear all heart, 1" x 2"		225	.036		.63	1.72		2.35	3.51
0890	1" x 3"		225	.036		.94	1.72		2.66	3.85
0900	1" x 4"		200	.040		1.19	1.94		3.13	4.48
0910	1" x 6"		200	.040		1.80	1.94		3.74	5.15
0920	1" x 8"		200	.040		2.36	1.94		4.30	5.75
0930	1" x 10"		175	.046		3.84	2.21		6.05	7.85
0940	1" x 12"		175	.046		4.75	2.21		6.96	8.85
0950	Cornice board, cedar, rough sawn, 1" x 2"		330	.024		.55	1.17		1.72	2.52
0960	1" x 3"		290	.028		.81	1.34		2.15	3.07
0970	1" x 4"		250	.032		1.08	1.55		2.63	3.72
0980	1" x 6"		250	.032		1.63	1.55		3.18	4.32
0990	1" x 8"		200	.040		2.17	1.94		4.11	5.55
1000	1" x 10"		180	.044		2.70	2.15		4.85	6.50
1010	1" x 12"		180	.044		3.24	2.15		5.39	7.10
1020	STK, 1" x 2"		330	.024		.53	1.17		1.70	2.51
1030	1" x 3"		290	.028		.59	1.34		1.93	2.82
1040	1" x 4"		250	.032		.71	1.55		2.26	3.31
1050	1" x 6"		250	.032		1.13	1.55		2.68	3.77
1060	1" x 8"		200	.040		1.63	1.94		3.57	4.96
1070	1" x 10"		180	.044		2.12	2.15		4.27	5.85
1080	1" x 12"		180	.044		3.62	2.15		5.77	7.50
1500	Pine, #2, 1" x 2"		330	.024		.25	1.17		1.42	2.20
1510	1" x 3"		290	.028		.27	1.34		1.61	2.48
1600	1" x 4"		250	.032		.49	1.55		2.04	3.07
1700	1" x 6"		250	.032		.73	1.55		2.28	3.34
1800	1" x 8"		200	.040		1.20	1.94		3.14	4.49
1900	1" x 10"		180	.044		1.57	2.15		3.72	5.25
2000	1" x 12"		180	.044		1.92	2.15		4.07	5.65
2020	D & better, 1" x 2"		330	.024		.43	1.17		1.60	2.40
2030	1" x 3"		290	.028		.66	1.34		2	2.91
2040	1" x 4"		250	.032		.89	1.55		2.44	3.51
2050	1" x 6"		250	.032		1.12	1.55		2.67	3.76
2060	1" x 8"		200	.040		1.62	1.94		3.56	4.95
2070	1" x 10"		180	.044		2.03	2.15		4.18	5.75
2080	1" x 12"		180	.044		2.63	2.15		4.78	6.40

06 22 Millwork

06 22 13 – Standard Pattern Wood Trim

06 22 13.40 Moldings, Exterior		Crew	Daily Output	Labor-Hours	Unit	Material	2016 Bare Costs Labor	Equipment	Total	Total Incl O&P
2090	Redwood, clear all heart, 1" x 2"	1 Carp	330	.024	L.F.	.63	1.17		1.80	2.62
2100	1" x 3"		290	.028		.94	1.34		2.28	3.22
2110	1" x 4"		250	.032		1.19	1.55		2.74	3.84
2120	1" x 6"		250	.032		1.80	1.55		3.35	4.51
2130	1" x 8"		200	.040		2.36	1.94		4.30	5.75
2140	1" x 10"		180	.044		3.84	2.15		5.99	7.75
2150	1" x 12"		180	.044		4.75	2.15		6.90	8.75
2160	3 piece, 1" x 2", 1" x 4", 1" x 6", rough sawn cedar		80	.100		3.27	4.85		8.12	11.50
2180	STK cedar		80	.100		2.37	4.85		7.22	10.50
2200	#2 pine		80	.100		1.47	4.85		6.32	9.50
2210	D & better pine		80	.100		2.44	4.85		7.29	10.60
2220	Clear all heart redwood		80	.100		3.62	4.85		8.47	11.90
2230	1" x 8", 1" x 10", 1" x 12", rough sawn cedar		65	.123		8.10	5.95		14.05	18.65
2240	STK cedar		65	.123		7.35	5.95		13.30	17.85
2300	#2 pine		65	.123		4.66	5.95		10.61	14.90
2320	D & better pine		65	.123		6.25	5.95		12.20	16.65
2330	Clear all heart redwood		65	.123		10.90	5.95		16.85	22
2340	Door/window casing, cedar, rough sawn, 1" x 2"		275	.029		.55	1.41		1.96	2.90
2350	1" x 3"		275	.029		.81	1.41		2.22	3.19
2360	1" x 4"		250	.032		1.08	1.55		2.63	3.72
2370	1" x 6"		250	.032		1.63	1.55		3.18	4.32
2380	1" x 8"		230	.035		2.17	1.69		3.86	5.15
2390	1" x 10"		230	.035		2.70	1.69		4.39	5.70
2395	1" x 12"		210	.038		3.24	1.85		5.09	6.60
2410	STK, 1" x 2"		275	.029		.53	1.41		1.94	2.89
2420	1" x 3"		275	.029		.59	1.41		2	2.94
2430	1" x 4"		250	.032		.71	1.55		2.26	3.31
2440	1" x 6"		250	.032		1.13	1.55		2.68	3.77
2450	1" x 8"		230	.035		1.63	1.69		3.32	4.54
2460	1" x 10"		230	.035		2.12	1.69		3.81	5.10
2470	1" x 12"		210	.038		3.62	1.85		5.47	7
2550	Pine, #2, 1" x 2"		275	.029		.25	1.41		1.66	2.58
2560	1" x 3"		275	.029		.40	1.41		1.81	2.74
2570	1" x 4"		250	.032		.49	1.55		2.04	3.07
2580	1" x 6"		250	.032		.73	1.55		2.28	3.34
2590	1" x 8"		230	.035		1.20	1.69		2.89	4.07
2600	1" x 10"		230	.035		1.57	1.69		3.26	4.47
2610	1" x 12"		210	.038		1.92	1.85		3.77	5.15
2620	Pine, D & better, 1" x 2"		275	.029		.43	1.41		1.84	2.78
2630	1" x 3"		275	.029		.66	1.41		2.07	3.03
2640	1" x 4"		250	.032		.89	1.55		2.44	3.51
2650	1" x 6"		250	.032		1.12	1.55		2.67	3.76
2660	1" x 8"		230	.035		1.62	1.69		3.31	4.53
2670	1" x 10"		230	.035		2.03	1.69		3.72	4.98
2680	1" x 12"		210	.038		2.63	1.85		4.48	5.90
2690	Redwood, clear all heart, 1" x 2"		275	.029		.63	1.41		2.04	3
2695	1" x 3"		275	.029		.94	1.41		2.35	3.34
2710	1" x 4"		250	.032		1.19	1.55		2.74	3.84
2715	1" x 6"		250	.032		1.80	1.55		3.35	4.51
2730	1" x 8"		230	.035		2.36	1.69		4.05	5.35
2740	1" x 10"		230	.035		3.84	1.69		5.53	6.95
2750	1" x 12"		210	.038		4.75	1.85		6.60	8.25
3500	Bellyband, pine, 11/16" x 4-1/4"		250	.032		3.23	1.55		4.78	6.10

For customer support on your Commercial Renovation Cost Data, call 877.742.4719.

159

06 22 Millwork

06 22 13 – Standard Pattern Wood Trim

06 22 13.40 Moldings, Exterior		Crew	Daily Output	Labor-Hours	Unit	Material	2016 Bare Costs Labor	Equipment	Total	Total Incl O&P
3610	Brickmold, pine, 1-1/4" x 2"	1 Carp	200	.040	L.F.	2.43	1.94		4.37	5.85
3620	FJP, 1-1/4" x 2"		200	.040		1.09	1.94		3.03	4.37
5100	Fascia, cedar, rough sawn, 1" x 2"		275	.029		.55	1.41		1.96	2.90
5110	1" x 3"		275	.029		.81	1.41		2.22	3.19
5120	1" x 4"		250	.032		1.08	1.55		2.63	3.72
5200	1" x 6"		250	.032		1.63	1.55		3.18	4.32
5300	1" x 8"		230	.035		2.17	1.69		3.86	5.15
5310	1" x 10"		230	.035		2.70	1.69		4.39	5.70
5320	1" x 12"		210	.038		3.24	1.85		5.09	6.60
5400	2" x 4"		220	.036		1.07	1.76		2.83	4.05
5500	2" x 6"		220	.036		1.60	1.76		3.36	4.64
5600	2" x 8"		200	.040		2.14	1.94		4.08	5.50
5700	2" x 10"		180	.044		2.66	2.15		4.81	6.45
5800	2" x 12"		170	.047		6.35	2.28		8.63	10.65
6120	STK, 1" x 2"		275	.029		.53	1.41		1.94	2.89
6130	1" x 3"		275	.029		.59	1.41		2	2.94
6140	1" x 4"		250	.032		.71	1.55		2.26	3.31
6150	1" x 6"		250	.032		1.13	1.55		2.68	3.77
6160	1" x 8"		230	.035		1.63	1.69		3.32	4.54
6170	1" x 10"		230	.035		2.12	1.69		3.81	5.10
6180	1" x 12"		210	.038		3.62	1.85		5.47	7
6185	2" x 2"		260	.031		.65	1.49		2.14	3.16
6190	Pine, #2, 1" x 2"		275	.029		.25	1.41		1.66	2.58
6200	1" x 3"		275	.029		.40	1.41		1.81	2.74
6210	1" x 4"		250	.032		.49	1.55		2.04	3.07
6220	1" x 6"		250	.032		.73	1.55		2.28	3.34
6230	1" x 8"		230	.035		1.20	1.69		2.89	4.07
6240	1" x 10"		230	.035		1.57	1.69		3.26	4.47
6250	1" x 12"		210	.038		1.92	1.85		3.77	5.15
6260	D & better, 1" x 2"		275	.029		.43	1.41		1.84	2.78
6270	1" x 3"		275	.029		.66	1.41		2.07	3.03
6280	1" x 4"		250	.032		.89	1.55		2.44	3.51
6290	1" x 6"		250	.032		1.12	1.55		2.67	3.76
6300	1" x 8"		230	.035		1.62	1.69		3.31	4.53
6310	1" x 10"		230	.035		2.03	1.69		3.72	4.98
6312	1" x 12"		210	.038		2.63	1.85		4.48	5.90
6330	Southern yellow, 1-1/4" x 5"		240	.033		2.43	1.61		4.04	5.30
6340	1-1/4" x 6"		240	.033		2.52	1.61		4.13	5.40
6350	1-1/4" x 8"		215	.037		3.43	1.80		5.23	6.70
6360	1-1/4" x 12"		190	.042		5.05	2.04		7.09	8.90
6370	Redwood, clear all heart, 1" x 2"		275	.029		.63	1.41		2.04	3
6380	1" x 3"		275	.029		1.19	1.41		2.60	3.61
6390	1" x 4"		250	.032		1.19	1.55		2.74	3.84
6400	1" x 6"		250	.032		1.80	1.55		3.35	4.51
6410	1" x 8"		230	.035		2.36	1.69		4.05	5.35
6420	1" x 10"		230	.035		3.84	1.69		5.53	6.95
6430	1" x 12"		210	.038		4.75	1.85		6.60	8.25
6440	1-1/4" x 5"		240	.033		1.88	1.61		3.49	4.71
6450	1-1/4" x 6"		240	.033		2.25	1.61		3.86	5.10
6460	1-1/4" x 8"		215	.037		3.53	1.80		5.33	6.85
6470	1-1/4" x 12"		190	.042		7.10	2.04		9.14	11.20
6580	Frieze, cedar, rough sawn, 1" x 2"		275	.029		.55	1.41		1.96	2.90
6590	1" x 3"		275	.029		.81	1.41		2.22	3.19

06 22 Millwork

06 22 13 – Standard Pattern Wood Trim

06 22 13.40 Moldings, Exterior		Crew	Daily Output	Labor-Hours	Unit	Material	2016 Bare Costs Labor	Equipment	Total	Total Incl O&P
6600	1" x 4"	1 Carp	250	.032	L.F.	1.08	1.55		2.63	3.72
6610	1" x 6"		250	.032		1.63	1.55		3.18	4.32
6620	1" x 8"		250	.032		2.17	1.55		3.72	4.92
6630	1" x 10"		225	.036		2.70	1.72		4.42	5.80
6640	1" x 12"		200	.040		3.21	1.94		5.15	6.70
6650	STK, 1" x 2"		275	.029		.53	1.41		1.94	2.89
6660	1" x 3"		275	.029		.59	1.41		2	2.94
6670	1" x 4"		250	.032		.71	1.55		2.26	3.31
6680	1" x 6"		250	.032		1.13	1.55		2.68	3.77
6690	1" x 8"		250	.032		1.63	1.55		3.18	4.32
6700	1" x 10"		225	.036		2.12	1.72		3.84	5.15
6710	1" x 12"		200	.040		3.62	1.94		5.56	7.15
6790	Pine, #2, 1" x 2"		275	.029		.25	1.41		1.66	2.58
6800	1" x 3"		275	.029		.40	1.41		1.81	2.74
6810	1" x 4"		250	.032		.49	1.55		2.04	3.07
6820	1" x 6"		250	.032		.73	1.55		2.28	3.34
6830	1" x 8"		250	.032		1.20	1.55		2.75	3.85
6840	1" x 10"		225	.036		1.57	1.72		3.29	4.53
6850	1" x 12"		200	.040		1.92	1.94		3.86	5.30
6860	D & better, 1" x 2"		275	.029		.43	1.41		1.84	2.78
6870	1" x 3"		275	.029		.66	1.41		2.07	3.03
6880	1" x 4"		250	.032		.89	1.55		2.44	3.51
6890	1" x 6"		250	.032		1.12	1.55		2.67	3.76
6900	1" x 8"		250	.032		1.62	1.55		3.17	4.31
6910	1" x 10"		225	.036		2.03	1.72		3.75	5.05
6920	1" x 12"		200	.040		2.63	1.94		4.57	6.05
6930	Redwood, clear all heart, 1" x 2"		275	.029		.63	1.41		2.04	3
6940	1" x 3"		275	.029		.94	1.41		2.35	3.34
6950	1" x 4"		250	.032		1.19	1.55		2.74	3.84
6960	1" x 6"		250	.032		1.80	1.55		3.35	4.51
6970	1" x 8"		250	.032		2.36	1.55		3.91	5.15
6980	1" x 10"		225	.036		3.84	1.72		5.56	7.05
6990	1" x 12"		200	.040		4.75	1.94		6.69	8.40
7000	Grounds, 1" x 1", cedar, rough sawn		300	.027		.28	1.29		1.57	2.42
7010	STK		300	.027		.33	1.29		1.62	2.47
7020	Pine, #2		300	.027		.16	1.29		1.45	2.28
7030	D & better		300	.027		.27	1.29		1.56	2.40
7050	Redwood		300	.027		.39	1.29		1.68	2.53
7060	Rake/verge board, cedar, rough sawn, 1" x 2"		225	.036		.55	1.72		2.27	3.41
7070	1" x 3"		225	.036		.81	1.72		2.53	3.70
7080	1" x 4"		200	.040		1.08	1.94		3.02	4.36
7090	1" x 6"		200	.040		1.63	1.94		3.57	4.96
7100	1" x 8"		190	.042		2.17	2.04		4.21	5.70
7110	1" x 10"		190	.042		2.70	2.04		4.74	6.30
7120	1" x 12"		180	.044		3.24	2.15		5.39	7.10
7130	STK, 1" x 2"		225	.036		.53	1.72		2.25	3.40
7140	1" x 3"		225	.036		.59	1.72		2.31	3.45
7150	1" x 4"		200	.040		.71	1.94		2.65	3.95
7160	1" x 6"		200	.040		1.13	1.94		3.07	4.41
7170	1" x 8"		190	.042		1.63	2.04		3.67	5.10
7180	1" x 10"		190	.042		2.12	2.04		4.16	5.65
7190	1" x 12"		180	.044		3.62	2.15		5.77	7.50
7200	Pine, #2, 1" x 2"		225	.036		.25	1.72		1.97	3.09

For customer support on your Commercial Renovation Cost Data, call 877.742.4719.

161

06 22 13 – Standard Pattern Wood Trim

06 22 13.40 Moldings, Exterior	Crew	Daily Output	Labor-Hours	Unit	Material	2016 Bare Costs Labor	Equipment	Total	Total Incl O&P
7210	1" x 3"	1 Carp	225	.036	L.F.	.40	1.72		2.12	3.25
7220	1" x 4"		200	.040		.49	1.94		2.43	3.71
7230	1" x 6"		200	.040		.73	1.94		2.67	3.98
7240	1" x 8"		190	.042		1.20	2.04		3.24	4.65
7250	1" x 10"		190	.042		1.57	2.04		3.61	5.05
7260	1" x 12"		180	.044		1.92	2.15		4.07	5.65
7340	D & better, 1" x 2"		225	.036		.43	1.72		2.15	3.29
7350	1" x 3"		225	.036		.66	1.72		2.38	3.54
7360	1" x 4"		200	.040		.89	1.94		2.83	4.15
7370	1" x 6"		200	.040		1.12	1.94		3.06	4.40
7380	1" x 8"		190	.042		1.62	2.04		3.66	5.10
7390	1" x 10"		190	.042		2.03	2.04		4.07	5.55
7400	1" x 12"		180	.044		2.63	2.15		4.78	6.40
7410	Redwood, clear all heart, 1" x 2"		225	.036		.63	1.72		2.35	3.51
7420	1" x 3"		225	.036		.94	1.72		2.66	3.85
7430	1" x 4"		200	.040		1.19	1.94		3.13	4.48
7440	1" x 6"		200	.040		1.80	1.94		3.74	5.15
7450	1" x 8"		190	.042		2.36	2.04		4.40	5.95
7460	1" x 10"		190	.042		3.84	2.04		5.88	7.55
7470	1" x 12"		180	.044		4.75	2.15		6.90	8.75
7480	2" x 4"		200	.040		2.29	1.94		4.23	5.70
7490	2" x 6"		182	.044		3.43	2.13		5.56	7.25
7500	2" x 8"		165	.048		4.56	2.35		6.91	8.85
7630	Soffit, cedar, rough sawn, 1" x 2"	2 Carp	440	.036		.55	1.76		2.31	3.48
7640	1" x 3"		440	.036		.81	1.76		2.57	3.77
7650	1" x 4"		420	.038		1.08	1.85		2.93	4.21
7660	1" x 6"		420	.038		1.63	1.85		3.48	4.81
7670	1" x 8"		420	.038		2.17	1.85		4.02	5.40
7680	1" x 10"		400	.040		2.70	1.94		4.64	6.15
7690	1" x 12"		400	.040		3.24	1.94		5.18	6.75
7700	STK, 1" x 2"		440	.036		.53	1.76		2.29	3.47
7710	1" x 3"		440	.036		.59	1.76		2.35	3.52
7720	1" x 4"		420	.038		.71	1.85		2.56	3.80
7730	1" x 6"		420	.038		1.13	1.85		2.98	4.26
7740	1" x 8"		420	.038		1.63	1.85		3.48	4.81
7750	1" x 10"		400	.040		2.12	1.94		4.06	5.50
7760	1" x 12"		400	.040		3.62	1.94		5.56	7.15
7770	Pine, #2, 1" x 2"		440	.036		.25	1.76		2.01	3.16
7780	1" x 3"		440	.036		.40	1.76		2.16	3.32
7790	1" x 4"		420	.038		.49	1.85		2.34	3.56
7800	1" x 6"		420	.038		.73	1.85		2.58	3.83
7810	1" x 8"		420	.038		1.20	1.85		3.05	4.34
7820	1" x 10"		400	.040		1.57	1.94		3.51	4.89
7830	1" x 12"		400	.040		1.92	1.94		3.86	5.30
7840	D & better, 1" x 2"		440	.036		.43	1.76		2.19	3.36
7850	1" x 3"		440	.036		.66	1.76		2.42	3.61
7860	1" x 4"		420	.038		.89	1.85		2.74	4
7870	1" x 6"		420	.038		1.12	1.85		2.97	4.25
7880	1" x 8"		420	.038		1.62	1.85		3.47	4.80
7890	1" x 10"		400	.040		2.03	1.94		3.97	5.40
7900	1" x 12"		400	.040		2.63	1.94		4.57	6.05
7910	Redwood, clear all heart, 1" x 2"		440	.036		.63	1.76		2.39	3.58
7920	1" x 3"		440	.036		.94	1.76		2.70	3.92

06 22 13 – Standard Pattern Wood Trim

06 22 13.40 Moldings, Exterior

		Crew	Daily Output	Labor-Hours	Unit	Material	2016 Bare Costs Labor	Equipment	Total	Total Incl O&P
7930	1" x 4"	2 Carp	420	.038	L.F.	1.19	1.85		3.04	4.33
7940	1" x 6"		420	.038		1.80	1.85		3.65	5
7950	1" x 8"		420	.038		2.36	1.85		4.21	5.60
7960	1" x 10"		400	.040		3.84	1.94		5.78	7.40
7970	1" x 12"		400	.040		4.75	1.94		6.69	8.40
8050	Trim, crown molding, pine, 11/16" x 4-1/4"	1 Carp	250	.032		4.54	1.55		6.09	7.50
8060	Back band, 11/16" x 1-1/16"		250	.032		.99	1.55		2.54	3.62
8070	Insect screen frame stock, 1-1/16" x 1-3/4"		395	.020		2.39	.98		3.37	4.23
8080	Dentils, 2-1/2" x 2-1/2" x 4", 6" O.C.		30	.267		1.23	12.90		14.13	22.50
8100	Fluted, 5-1/2"		165	.048		5.05	2.35		7.40	9.40
8110	Stucco bead, 1-3/8" x 1-5/8"		250	.032		2.50	1.55		4.05	5.30
9000	Minimum labor/equipment charge		4	2	Job		97		97	158

06 22 13.45 Moldings, Trim

		Crew	Daily Output	Labor-Hours	Unit	Material	2016 Bare Costs Labor	Equipment	Total	Total Incl O&P
0010	**MOLDINGS, TRIM**									
0200	Astragal, stock pine, 11/16" x 1-3/4"	1 Carp	255	.031	L.F.	1.42	1.52		2.94	4.04
0250	1-5/16" x 2-3/16"		240	.033		2.10	1.61		3.71	4.95
0800	Chair rail, stock pine, 5/8" x 2-1/2"		270	.030		1.58	1.44		3.02	4.09
0900	5/8" x 3-1/2"		240	.033		2.40	1.61		4.01	5.30
1000	Closet pole, stock pine, 1-1/8" diameter		200	.040		1.16	1.94		3.10	4.45
1100	Fir, 1-5/8" diameter		200	.040		2.20	1.94		4.14	5.60
3300	Half round, stock pine, 1/4" x 1/2"		270	.030		.24	1.44		1.68	2.61
3350	1/2" x 1"		255	.031		.73	1.52		2.25	3.28
3400	Handrail, fir, single piece, stock, hardware not included									
3450	1-1/2" x 1-3/4"	1 Carp	80	.100	L.F.	2.56	4.85		7.41	10.70
3470	Pine, 1-1/2" x 1-3/4"		80	.100		2.40	4.85		7.25	10.55
3500	1-1/2" x 2-1/2"		76	.105		2.46	5.10		7.56	11.05
3600	Lattice, stock pine, 1/4" x 1-1/8"		270	.030		.35	1.44		1.79	2.74
3700	1/4" x 1-3/4"		250	.032		.92	1.55		2.47	3.54
3800	Miscellaneous, custom, pine, 1" x 1"		270	.030		.43	1.44		1.87	2.83
3850	1" x 2"		265	.030		.87	1.46		2.33	3.35
3900	1" x 3"		240	.033		1.30	1.61		2.91	4.07
4100	Birch or oak, nominal 1" x 1"		240	.033		.39	1.61		2	3.07
4200	Nominal 1" x 3"		215	.037		1.18	1.80		2.98	4.25
4400	Walnut, nominal 1" x 1"		215	.037		.62	1.80		2.42	3.63
4500	Nominal 1" x 3"		200	.040		1.86	1.94		3.80	5.20
4700	Teak, nominal 1" x 1"		215	.037		2.64	1.80		4.44	5.85
4800	Nominal 1" x 3"		200	.040		7.90	1.94		9.84	11.85
4900	Quarter round, stock pine, 1/4" x 1/4"		275	.029		.24	1.41		1.65	2.56
4950	3/4" x 3/4"		255	.031		.50	1.52		2.02	3.03
5600	Wainscot moldings, 1-1/8" x 9/16", 2' high, minimum		76	.105	S.F.	11.65	5.10		16.75	21
5700	Maximum		65	.123	"	15.80	5.95		21.75	27
9000	Minimum labor/equipment charge		4	2	Job		97		97	158

06 22 13.50 Moldings, Window and Door

		Crew	Daily Output	Labor-Hours	Unit	Material	2016 Bare Costs Labor	Equipment	Total	Total Incl O&P
0010	**MOLDINGS, WINDOW AND DOOR**									
2800	Door moldings, stock, decorative, 1-1/8" wide, plain	1 Carp	17	.471	Set	47.50	23		70.50	90
2900	Detailed		17	.471	"	91.50	23		114.50	139
2960	Clear pine door jamb, no stops, 11/16" x 4-9/16"		240	.033	L.F.	5.20	1.61		6.81	8.35
3150	Door trim set, 1 head and 2 sides, pine, 2-1/2 wide		12	.667	Opng.	24	32.50		56.50	79
3170	3-1/2" wide		11	.727	"	29	35		64	89.50
3250	Glass beads, stock pine, 3/8" x 1/2"		275	.029	L.F.	.33	1.41		1.74	2.66
3270	3/8" x 7/8"		270	.030		.42	1.44		1.86	2.81
4850	Parting bead, stock pine, 3/8" x 3/4"		275	.029		.44	1.41		1.85	2.78

For customer support on your Commercial Renovation Cost Data, call 877.742.4719.

163

06 22 Millwork

06 22 13 – Standard Pattern Wood Trim

06 22 13.50 Moldings, Window and Door

		Crew	Daily Output	Labor-Hours	Unit	Material	2016 Bare Costs Labor	Equipment	Total	Total Incl O&P
4870	1/2" x 3/4"	1 Carp	255	.031	L.F.	.42	1.52		1.94	2.94
5000	Stool caps, stock pine, 11/16" x 3-1/2"		200	.040		2.07	1.94		4.01	5.45
5100	1-1/16" x 3-1/4"		150	.053		3.24	2.58		5.82	7.80
5300	Threshold, oak, 3' long, inside, 5/8" x 3-5/8"		32	.250	Ea.	6.50	12.10		18.60	27
5400	Outside, 1-1/2" x 7-5/8"		16	.500	"	45	24		69	89
5900	Window trim sets, including casings, header, stops,									
5910	stool and apron, 2-1/2" wide, FJP	1 Carp	13	.615	Opng.	31.50	30		61.50	83
5950	Pine		10	.800		37	39		76	105
6000	Oak		6	1.333		64.50	64.50		129	177
9000	Minimum labor/equipment charge		4	2	Job		97		97	158

06 22 13.60 Moldings, Soffits

		Crew	Daily Output	Labor-Hours	Unit	Material	2016 Bare Costs Labor	Equipment	Total	Total Incl O&P
0010	**MOLDINGS, SOFFITS** R061110-30									
0200	Soffits, pine, 1" x 4"	2 Carp	420	.038	L.F.	.46	1.85		2.31	3.53
0210	1" x 6"		420	.038		.71	1.85		2.56	3.80
0220	1" x 8"		420	.038		1.17	1.85		3.02	4.31
0230	1" x 10"		400	.040		1.53	1.94		3.47	4.85
0240	1" x 12"		400	.040		1.88	1.94		3.82	5.25
0250	STK cedar, 1" x 4"		420	.038		.68	1.85		2.53	3.77
0260	1" x 6"		420	.038		1.10	1.85		2.95	4.23
0270	1" x 8"		420	.038		1.60	1.85		3.45	4.78
0280	1" x 10"		400	.040		2.08	1.94		4.02	5.45
0290	1" x 12"		400	.040		3.58	1.94		5.52	7.10
1000	Exterior AC plywood, 1/4" thick		400	.040	S.F.	.95	1.94		2.89	4.22
1050	3/8" thick		400	.040		.99	1.94		2.93	4.26
1100	1/2" thick		400	.040		1.17	1.94		3.11	4.46
1150	Polyvinyl chloride, white, solid	1 Carp	230	.035		2.10	1.69		3.79	5.05
1160	Perforated	"	230	.035		2.10	1.69		3.79	5.05
1170	Accessories, "J" channel 5/8"	2 Carp	700	.023	L.F.	.49	1.11		1.60	2.35
9000	Minimum labor/equipment charge	"	5	3.200	Job		155		155	253

06 25 Prefinished Paneling

06 25 13 – Prefinished Hardboard Paneling

06 25 13.10 Paneling, Hardboard

			Crew	Daily Output	Labor-Hours	Unit	Material	2016 Bare Costs Labor	Equipment	Total	Total Incl O&P
0010	**PANELING, HARDBOARD**										
0050	Not incl. furring or trim, hardboard, tempered, 1/8" thick	G	2 Carp	500	.032	S.F.	.42	1.55		1.97	2.99
0100	1/4" thick	G		500	.032		.65	1.55		2.20	3.25
0300	Tempered pegboard, 1/8" thick	G		500	.032		.42	1.55		1.97	2.99
0400	1/4" thick	G		500	.032		.68	1.55		2.23	3.28
0600	Untempered hardboard, natural finish, 1/8" thick	G		500	.032		.43	1.55		1.98	3
0700	1/4" thick	G		500	.032		.55	1.55		2.10	3.14
0900	Untempered pegboard, 1/8" thick	G		500	.032		.46	1.55		2.01	3.04
1000	1/4" thick	G		500	.032		.48	1.55		2.03	3.06
1200	Plastic faced hardboard, 1/8" thick	G		500	.032		.66	1.55		2.21	3.26
1300	1/4" thick	G		500	.032		.89	1.55		2.44	3.51
1500	Plastic faced pegboard, 1/8" thick	G		500	.032		.67	1.55		2.22	3.27
1600	1/4" thick	G		500	.032		.85	1.55		2.40	3.47
1800	Wood grained, plain or grooved, 1/8" thick	G		500	.032		.69	1.55		2.24	3.29
1900	1/4" thick	G		425	.038		1.40	1.82		3.22	4.52
2100	Moldings, wood grained MDF			500	.032	L.F.	.41	1.55		1.96	2.98
2200	Pine			425	.038	"	1.40	1.82		3.22	4.52
9000	Minimum labor/equipment charge		1 Carp	2	4	Job		194		194	315

06 25 Prefinished Paneling

06 25 16 – Prefinished Plywood Paneling

06 25 16.10 Paneling, Plywood	Crew	Daily Output	Labor-Hours	Unit	Material	2016 Bare Costs Labor	Equipment	Total	Total Incl O&P
0010 **PANELING, PLYWOOD** R061636-20									
2400 Plywood, prefinished, 1/4" thick, 4' x 8' sheets									
2410 with vertical grooves. Birch faced, economy	2 Carp	500	.032	S.F.	1.45	1.55		3	4.13
2420 Average		420	.038		1.20	1.85		3.05	4.34
2430 Custom		350	.046		1.25	2.21		3.46	5
2600 Mahogany, African		400	.040		2.80	1.94		4.74	6.25
2700 Philippine (Lauan)		500	.032		.65	1.55		2.20	3.25
2900 Oak		500	.032		1.40	1.55		2.95	4.07
3000 Cherry		400	.040		2.10	1.94		4.04	5.50
3200 Rosewood		320	.050		2.95	2.42		5.37	7.20
3400 Teak		400	.040		2.95	1.94		4.89	6.40
3600 Chestnut		375	.043		4.85	2.07		6.92	8.75
3800 Pecan		400	.040		2.55	1.94		4.49	6
3900 Walnut, average		500	.032		2.40	1.55		3.95	5.15
3950 Custom		400	.040		5.25	1.94		7.19	8.95
4000 Plywood, prefinished, 3/4" thick, stock grades, economy		320	.050		1.60	2.42		4.02	5.70
4100 Average		224	.071		4.70	3.46		8.16	10.80
4300 Architectural grade, custom		224	.071		5.20	3.46		8.66	11.35
4400 Luxury		160	.100		5.20	4.85		10.05	13.60
4600 Plywood, "A" face, birch, VC, 1/2" thick, natural		450	.036		2.05	1.72		3.77	5.05
4700 Select		450	.036		2.15	1.72		3.87	5.20
4900 Veneer core, 3/4" thick, natural		320	.050		2.24	2.42		4.66	6.40
5000 Select		320	.050		2.44	2.42		4.86	6.65
5200 Lumber core, 3/4" thick, natural		320	.050		3.05	2.42		5.47	7.30
5500 Plywood, knotty pine, 1/4" thick, A2 grade		450	.036		1.70	1.72		3.42	4.68
5600 A3 grade		450	.036		2.15	1.72		3.87	5.20
5800 3/4" thick, veneer core, A2 grade		320	.050		2.20	2.42		4.62	6.40
5900 A3 grade		320	.050		2.45	2.42		4.87	6.65
6100 Aromatic cedar, 1/4" thick, plywood		400	.040		2.25	1.94		4.19	5.65
6200 1/4" thick, particle board		400	.040		1.15	1.94		3.09	4.44
9000 Minimum labor/equipment charge	1 Carp	2	4	Job		194		194	315

06 25 26 – Panel System

06 25 26.10 Panel Systems

	Crew	Daily Output	Labor-Hours	Unit	Material	2016 Bare Costs Labor	Equipment	Total	Total Incl O&P
0010 **PANEL SYSTEMS**									
0100 Raised panel, eng. wood core w/wood veneer, std., paint grade	2 Carp	300	.053	S.F.	11.20	2.58		13.78	16.50
0110 Oak veneer		300	.053		25.50	2.58		28.08	32.50
0120 Maple veneer		300	.053		32.50	2.58		35.08	40
0130 Cherry veneer		300	.053		37	2.58		39.58	44.50
0300 Class I fire rated, paint grade		300	.053		13	2.58		15.58	18.50
0310 Oak veneer		300	.053		30	2.58		32.58	37
0320 Maple veneer		300	.053		40.50	2.58		43.08	48.50
0330 Cherry veneer		300	.053		49	2.58		51.58	57.50
0510 Beadboard, 5/8" MDF, standard, primed		300	.053		8.45	2.58		11.03	13.50
0520 Oak veneer, unfinished		300	.053		13.40	2.58		15.98	18.95
0530 Maple veneer, unfinished		300	.053		14.60	2.58		17.18	20.50
0610 Rustic paneling, 5/8" MDF, standard, maple veneer, unfinished		300	.053		18.20	2.58		20.78	24

06 26 Board Paneling

06 26 13 – Profile Board Paneling

06 26 13.10 Paneling, Boards	Crew	Daily Output	Labor-Hours	Unit	Material	2016 Bare Costs Labor	Equipment	Total	Total Incl O&P
0010 **PANELING, BOARDS**									
6400 Wood board paneling, 3/4" thick, knotty pine	2 Carp	300	.053	S.F.	1.95	2.58		4.53	6.35
6500 Rough sawn cedar		300	.053		3.20	2.58		5.78	7.75
6700 Redwood, clear, 1" x 4" boards		300	.053		4.96	2.58		7.54	9.65
6900 Aromatic cedar, closet lining, boards	↓	275	.058	↓	2.40	2.82		5.22	7.25
9000 Minimum labor/equipment charge	1 Carp	2	4	Job		194		194	315

06 43 Wood Stairs and Railings

06 43 13 – Wood Stairs

06 43 13.20 Prefabricated Wood Stairs

		Crew	Daily Output	Labor-Hours	Unit	Material	2016 Bare Costs Labor	Equipment	Total	Total Incl O&P
0010	**PREFABRICATED WOOD STAIRS**									
0100	Box stairs, prefabricated, 3'-0" wide									
0110	Oak treads, up to 14 risers	2 Carp	39	.410	Riser	92.50	19.90		112.40	135
0600	With pine treads for carpet, up to 14 risers	"	39	.410	"	59.50	19.90		79.40	98
1100	For 4' wide stairs, add				Flight	25%				
1550	Stairs, prefabricated stair handrail with balusters	1 Carp	30	.267	L.F.	78.50	12.90		91.40	107
1700	Basement stairs, prefabricated, pine treads									
1710	Pine risers, 3' wide, up to 14 risers	2 Carp	52	.308	Riser	59.50	14.90		74.40	90
4000	Residential, wood, oak treads, prefabricated		1.50	10.667	Flight	1,200	515		1,715	2,175
4200	Built in place	↓	.44	36.364	"	2,175	1,750		3,925	5,250
4400	Spiral, oak, 4'-6" diameter, unfinished, prefabricated,									
4500	incl. railing, 9' high	2 Carp	1.50	10.667	Flight	3,425	515		3,940	4,625
9000	Minimum labor/equipment charge	"	3	5.333	Job		258		258	420

06 43 13.40 Wood Stair Parts

		Crew	Daily Output	Labor-Hours	Unit	Material	2016 Bare Costs Labor	Equipment	Total	Total Incl O&P
0010	**WOOD STAIR PARTS**									
0020	Pin top balusters, 1-1/4", oak, 34"	1 Carp	96	.083	Ea.	5.10	4.04		9.14	12.20
0030	38"		96	.083		5.85	4.04		9.89	13.05
0040	42"		96	.083		6.35	4.04		10.39	13.60
0050	Poplar, 34"		96	.083		3.05	4.04		7.09	9.95
0060	38"		96	.083		3.88	4.04		7.92	10.85
0070	42"		96	.083		8.85	4.04		12.89	16.30
0080	Maple, 34"		96	.083		4.90	4.04		8.94	12
0090	38"		96	.083		5.60	4.04		9.64	12.80
0100	42"		96	.083		6.50	4.04		10.54	13.75
0130	Primed, 34"		96	.083		3	4.04		7.04	9.90
0140	38"		96	.083		3.62	4.04		7.66	10.60
0150	42"		96	.083		4.42	4.04		8.46	11.45
0180	Box top balusters, 1-1/4", oak, 34"		60	.133		8.95	6.45		15.40	20.50
0190	38"		60	.133		9.95	6.45		16.40	21.50
0200	42"		60	.133		10.95	6.45		17.40	22.50
0210	Poplar, 34"		60	.133		6.25	6.45		12.70	17.45
0220	38"		60	.133		6.95	6.45		13.40	18.20
0230	42"		60	.133		7.50	6.45		13.95	18.80
0240	Maple, 34"		60	.133		8.25	6.45		14.70	19.65
0250	38"		60	.133		9	6.45		15.45	20.50
0260	42"		60	.133		10	6.45		16.45	21.50
0290	Primed, 34"		60	.133		7	6.45		13.45	18.25
0300	38"		60	.133		8	6.45		14.45	19.35
0310	42"		60	.133	↓	8.35	6.45		14.80	19.75
0340	Square balusters, cut from lineal stock, pine, 1-1/16" x 1-1/16"		180	.044	L.F.	1.50	2.15		3.65	5.15
0350	1-5/16" x 1-5/16"		180	.044		2.10	2.15		4.25	5.85

06 43 Wood Stairs and Railings

06 43 13 – Wood Stairs

06 43 13.40 Wood Stair Parts		Crew	Daily Output	Labor-Hours	Unit	Material	2016 Bare Costs Labor	Equipment	Total	Total Incl O&P
0360	1-5/8" x 1-5/8"	1 Carp	180	.044	L.F.	3.30	2.15		5.45	7.15
0370	Turned newel, oak, 3-1/2" square, 48" high		8	1	Ea.	92	48.50		140.50	180
0380	62" high		8	1		92	48.50		140.50	180
0390	Poplar, 3-1/2" square, 48" high		8	1		54	48.50		102.50	139
0400	62" high		8	1		68	48.50		116.50	154
0410	Maple, 3-1/2" square, 48" high		8	1		72	48.50		120.50	158
0420	62" high		8	1		92	48.50		140.50	180
0430	Square newel, oak, 3-1/2" square, 48" high		8	1		54	48.50		102.50	139
0440	58" high		8	1		68	48.50		116.50	154
0450	Poplar, 3-1/2" square, 48" high		8	1		35	48.50		83.50	118
0460	58" high		8	1		42	48.50		90.50	125
0470	Maple, 3" square, 48" high		8	1		52	48.50		100.50	136
0480	58" high		8	1		64	48.50		112.50	150
0490	Railings, oak, economy		96	.083	L.F.	8.65	4.04		12.69	16.10
0500	Average		96	.083		13	4.04		17.04	21
0510	Custom		96	.083		16.50	4.04		20.54	25
0520	Maple, economy		96	.083		11	4.04		15.04	18.70
0530	Average		96	.083		13.50	4.04		17.54	21.50
0540	Custom		96	.083		16.95	4.04		20.99	25.50
0550	Oak, for bending rail, economy		48	.167		23.50	8.10		31.60	39
0560	Average		48	.167		26	8.10		34.10	41.50
0570	Custom		48	.167		29.50	8.10		37.60	45
0580	Maple, for bending rail, economy		48	.167		32	8.10		40.10	48
0590	Average		48	.167		32	8.10		40.10	48
0600	Custom		48	.167		32	8.10		40.10	48
0610	Risers, oak, 3/4" x 8", 36" long		80	.100	Ea.	13	4.85		17.85	22
0620	42" long		70	.114		15.15	5.55		20.70	25.50
0630	48" long		63	.127		17.30	6.15		23.45	29
0640	54" long		56	.143		19.50	6.90		26.40	33
0650	60" long		50	.160		21.50	7.75		29.25	36.50
0660	72" long		42	.190		26	9.25		35.25	43.50
0670	Poplar, 3/4" x 8", 36" long		80	.100		12.50	4.85		17.35	21.50
0680	42" long		71	.113		14.55	5.45		20	25
0690	48" long		63	.127		16.65	6.15		22.80	28.50
0700	54" long		56	.143		18.70	6.90		25.60	32
0710	60" long		50	.160		21	7.75		28.75	35.50
0720	72" long		42	.190		25	9.25		34.25	42.50
0730	Pine, 1" x 8", 36" long		80	.100		3.52	4.85		8.37	11.75
0740	42" long		70	.114		4.11	5.55		9.66	13.55
0750	48" long		63	.127		4.69	6.15		10.84	15.20
0760	54" long		56	.143		5.30	6.90		12.20	17.10
0770	60" long		50	.160		5.85	7.75		13.60	19.10
0780	72" long		42	.190		7.05	9.25		16.30	23
0790	Treads, oak, no returns, 1-1/32" x 11-1/2" x 36" long		32	.250		27	12.10		39.10	49.50
0800	42" long		32	.250		31.50	12.10		43.60	54.50
0810	48" long		32	.250		36	12.10		48.10	59.50
0820	54" long		32	.250		40.50	12.10		52.60	64.50
0830	60" long		32	.250		45	12.10		57.10	69.50
0840	72" long		32	.250		54	12.10		66.10	79.50
0850	Mitred return one end, 1-1/32" x 11-1/2" x 36" long		24	.333		36	16.15		52.15	66
0860	42" long		24	.333		42	16.15		58.15	72.50
0870	48" long		24	.333		48	16.15		64.15	79.50
0880	54" long		24	.333		54	16.15		70.15	86

For customer support on your Commercial Renovation Cost Data, call 877.742.4719.

167

06 43 Wood Stairs and Railings

06 43 13 – Wood Stairs

06 43 13.40 Wood Stair Parts

		Crew	Daily Output	Labor-Hours	Unit	Material	2016 Bare Costs Labor	Equipment	Total	Total Incl O&P
0890	60" long	1 Carp	24	.333	Ea.	60	16.15		76.15	92.50
0900	72" long		24	.333		72	16.15		88.15	106
0910	Mitred return two ends, 1-1/32" x 11-1/2" x 36" long		12	.667		46	32.50		78.50	104
0920	42" long		12	.667		53.50	32.50		86	112
0930	48" long		12	.667		61.50	32.50		94	121
0940	54" long		12	.667		69	32.50		101.50	129
0950	60" long		12	.667		76.50	32.50		109	138
0960	72" long		12	.667		92	32.50		124.50	154
0970	Starting step, oak, 48", bullnose		8	1		172	48.50		220.50	268
0980	Double end bullnose		8	1		254	48.50		302.50	360
1030	Skirt board, pine, 1" x 10"		55	.145	L.F.	1.53	7.05		8.58	13.20
1040	1" x 12"		52	.154	"	1.88	7.45		9.33	14.25
1050	Oak landing tread, 1-1/16" thick		54	.148	S.F.	9	7.20		16.20	21.50
1060	Oak cove molding		96	.083	L.F.	1	4.04		5.04	7.70
1070	Oak stringer molding		96	.083	"	4	4.04		8.04	11
1090	Rail bolt, 5/16" x 3-1/2"		48	.167	Ea.	2.75	8.10		10.85	16.25
1100	5/16" x 4-1/2"		48	.167		2.75	8.10		10.85	16.25
1120	Newel post anchor		16	.500		13	24		37	54
1130	Tapered plug, 1/2"		240	.033		1	1.61		2.61	3.74
1140	1"		240	.033		.99	1.61		2.60	3.73
9000	Minimum labor/equipment charge		3	2.667	Job		129		129	211

06 43 16 – Wood Railings

06 43 16.10 Wood Handrails and Railings

		Crew	Daily Output	Labor-Hours	Unit	Material	2016 Bare Costs Labor	Equipment	Total	Total Incl O&P
0010	**WOOD HANDRAILS AND RAILINGS**									
0020	Custom design, architectural grade, hardwood, plain	1 Carp	38	.211	L.F.	12.05	10.20		22.25	30
0100	Shaped		30	.267		62.50	12.90		75.40	90
0300	Stock interior railing with spindles 4" O.C., 4' long		40	.200		38.50	9.70		48.20	58.50
0400	8' long		48	.167		38.50	8.10		46.60	55.50
9000	Minimum labor/equipment charge		3	2.667	Job		129		129	211

06 44 Ornamental Woodwork

06 44 19 – Wood Grilles

06 44 19.10 Grilles

		Crew	Daily Output	Labor-Hours	Unit	Material	2016 Bare Costs Labor	Equipment	Total	Total Incl O&P
0010	**GRILLES** and panels, hardwood, sanded									
0020	2' x 4' to 4' x 8', custom designs, unfinished, economy	1 Carp	38	.211	S.F.	62	10.20		72.20	84.50
0050	Average		30	.267		70	12.90		82.90	98
0100	Custom		19	.421		72	20.50		92.50	113
9000	Minimum labor/equipment charge		2	4	Job		194		194	315

06 44 33 – Wood Mantels

06 44 33.10 Fireplace Mantels

		Crew	Daily Output	Labor-Hours	Unit	Material	2016 Bare Costs Labor	Equipment	Total	Total Incl O&P
0010	**FIREPLACE MANTELS**									
0015	6" molding, 6' x 3'-6" opening, plain, paint grade	1 Carp	5	1.600	Opng.	440	77.50		517.50	605
0100	Ornate, oak		5	1.600		590	77.50		667.50	775
0300	Prefabricated pine, colonial type, stock, deluxe		2	4		1,500	194		1,694	1,975
0400	Economy		3	2.667		690	129		819	970
9000	Minimum labor/equipment charge		3	2.667	Job		129		129	211

06 44 Ornamental Woodwork

06 44 33 – Wood Mantels

06 44 33.20 Fireplace Mantel Beam	Crew	Daily Output	Labor-Hours	Unit	Material	2016 Bare Costs Labor	Equipment	Total	Total Incl O&P
0010 **FIREPLACE MANTEL BEAM**									
0020 Rough texture wood, 4" x 8"	1 Carp	36	.222	L.F.	8.30	10.75		19.05	27
0100 4" x 10"		35	.229	"	10.80	11.05		21.85	30
0300 Laminated hardwood, 2-1/4" x 10-1/2" wide, 6' long		5	1.600	Ea.	125	77.50		202.50	265
0400 8' long		5	1.600	"	150	77.50		227.50	292
0600 Brackets for above, rough sawn		12	.667	Pr.	12	32.50		44.50	66
0700 Laminated		12	.667	"	16	32.50		48.50	70.50
9000 Minimum labor/equipment charge		4	2	Job		97		97	158

06 44 39 – Wood Posts and Columns

06 44 39.10 Decorative Beams

	Crew	Daily Output	Labor-Hours	Unit	Material	2016 Bare Costs Labor	Equipment	Total	Total Incl O&P
0010 **DECORATIVE BEAMS**									
0020 Rough sawn cedar, non-load bearing, 4" x 4"	2 Carp	180	.089	L.F.	1.25	4.31		5.56	8.45
0100 4" x 6"		170	.094		1.80	4.56		6.36	9.45
0200 4" x 8"		160	.100		2.40	4.85		7.25	10.55
0300 4" x 10"		150	.107		3.65	5.15		8.80	12.45
0400 4" x 12"		140	.114		4.61	5.55		10.16	14.10
0500 8" x 8"		130	.123		4.80	5.95		10.75	15.05
9000 Minimum labor/equipment charge	1 Carp	3	2.667	Job		129		129	211

06 44 39.20 Columns

	Crew	Daily Output	Labor-Hours	Unit	Material	2016 Bare Costs Labor	Equipment	Total	Total Incl O&P
0010 **COLUMNS**									
0050 Aluminum, round colonial, 6" diameter	2 Carp	80	.200	V.L.F.	19	9.70		28.70	37
0100 8" diameter		62.25	.257		22	12.45		34.45	44.50
0200 10" diameter		55	.291		22.50	14.10		36.60	48
0250 Fir, stock units, hollow round, 6" diameter		80	.200		29.50	9.70		39.20	48.50
0300 8" diameter		80	.200		35.50	9.70		45.20	55
0350 10" diameter		70	.229		44.50	11.05		55.55	67
0360 12" diameter		65	.246		54.50	11.95		66.45	79.50
0400 Solid turned, to 8' high, 3-1/2" diameter		80	.200		9.60	9.70		19.30	26.50
0500 4-1/2" diameter		75	.213		11.90	10.35		22.25	30
0600 5-1/2" diameter		70	.229		16	11.05		27.05	35.50
0800 Square columns, built-up, 5" x 5"		65	.246		14.20	11.95		26.15	35
0900 Solid, 3-1/2" x 3-1/2"		130	.123		9.60	5.95		15.55	20.50
1600 Hemlock, tapered, T & G, 12" diam., 10' high		100	.160		39.50	7.75		47.25	56
1700 16' high		65	.246		73	11.95		84.95	99.50
1900 14" diameter, 10' high		100	.160		113	7.75		120.75	138
2000 18' high		65	.246		103	11.95		114.95	133
2200 18" diameter, 12' high		65	.246		165	11.95		176.95	202
2300 20' high		50	.320		118	15.50		133.50	156
2500 20" diameter, 14' high		40	.400		180	19.40		199.40	230
2600 20' high		35	.457		170	22		192	223
2800 For flat pilasters, deduct					33%				
3000 For splitting into halves, add				Ea.	106			106	117
4000 Rough sawn cedar posts, 4" x 4"	2 Carp	250	.064	V.L.F.	3.95	3.10		7.05	9.40
4100 4" x 6"		235	.068		6.80	3.30		10.10	12.90
4200 6" x 6"		220	.073		9.90	3.52		13.42	16.65
4300 8" x 8"		200	.080		19.20	3.88		23.08	27.50
9000 Minimum labor/equipment charge	1 Carp	3	2.667	Job		129		129	211

For customer support on your Commercial Renovation Cost Data, call 877.742.4719.

169

06 48 Wood Frames

06 48 13 – Exterior Wood Door Frames

06 48 13.10 Exterior Wood Door Frames and Accessories	Crew	Daily Output	Labor-Hours	Unit	Material	2016 Bare Costs Labor	Equipment	Total	Total Incl O&P
0010	**EXTERIOR WOOD DOOR FRAMES AND ACCESSORIES**									
0400	Exterior frame, incl. ext. trim, pine, 5/4 x 4-9/16" deep	2 Carp	375	.043	L.F.	6.60	2.07		8.67	10.70
0420	5-3/16" deep		375	.043		7.90	2.07		9.97	12.05
0440	6-9/16" deep		375	.043		8.80	2.07		10.87	13.05
0600	Oak, 5/4 x 4-9/16" deep		350	.046		19.75	2.21		21.96	25
0620	5-3/16" deep		350	.046		21.50	2.21		23.71	27.50
0640	6-9/16" deep		350	.046		19.50	2.21		21.71	25
1000	Sills, 8/4 x 8" deep, oak, no horns		100	.160		6.65	7.75		14.40	19.95
1020	2" horns		100	.160		20.50	7.75		28.25	35
1040	3" horns		100	.160		20.50	7.75		28.25	35
1100	8/4 x 10" deep, oak, no horns		90	.178		6.40	8.60		15	21
1120	2" horns		90	.178		26.50	8.60		35.10	43
1140	3" horns		90	.178		26.50	8.60		35.10	43
2000	Wood frame & trim, ext, colonial, 3' opng, fluted pilasters, flat head		22	.727	Ea.	505	35		540	615
2010	Dentil head		21	.762		580	37		617	695
2020	Ram's head		20	.800		695	39		734	825
2100	5'-4" opening, in-swing, fluted pilasters, flat head		17	.941		440	45.50		485.50	560
2120	Ram's head		15	1.067		1,400	51.50		1,451.50	1,600
2140	Out swing, fluted pilasters, flat head		17	.941		520	45.50		565.50	645
2160	Ram's head		15	1.067		1,475	51.50		1,526.50	1,700
2400	6'-0" opening, in-swing, fluted pilasters, flat head		16	1		520	48.50		568.50	650
2420	Ram's head		10	1.600		1,475	77.50		1,552.50	1,750
2460	Out-swing, fluted pilasters, flat head		16	1		520	48.50		568.50	650
2480	Ram's head		10	1.600		1,475	77.50		1,552.50	1,750
2600	For two sidelights, flat head, add		30	.533	Opng.	226	26		252	291
2620	Ram's head, add		20	.800	"	840	39		879	990
2700	Custom birch frame, 3'-0" opening		16	1	Ea.	240	48.50		288.50	345
2750	6'-0" opening		16	1		360	48.50		408.50	475
2900	Exterior, modern, plain trim, 3' opng., in-swing, FJP		26	.615		46.50	30		76.50	99.50
2920	Fir		24	.667		55	32.50		87.50	113
2940	Oak		22	.727		63	35		98	127

06 48 16 – Interior Wood Door Frames

06 48 16.10 Interior Wood Door Jamb and Frames

		Crew	Daily Output	Labor-Hours	Unit	Material	2016 Bare Costs Labor	Equipment	Total	Total Incl O&P
0010	**INTERIOR WOOD DOOR JAMB AND FRAMES**									
3000	Interior frame, pine, 11/16" x 3-5/8" deep	2 Carp	375	.043	L.F.	4.44	2.07		6.51	8.25
3020	4-9/16" deep		375	.043		4.92	2.07		6.99	8.80
3200	Oak, 11/16" x 3-5/8" deep		350	.046		9.85	2.21		12.06	14.40
3220	4-9/16" deep		350	.046		9.95	2.21		12.16	14.55
3240	5-3/16" deep		350	.046		13.90	2.21		16.11	18.85
3400	Walnut, 11/16" x 3-5/8" deep		350	.046		9	2.21		11.21	13.50
3420	4-9/16" deep		350	.046		9.45	2.21		11.66	14
3440	5-3/16" deep		350	.046		9.55	2.21		11.76	14.10
3600	Pocket door frame		16	1	Ea.	81	48.50		129.50	168
3800	Threshold, oak, 5/8" x 3-5/8" deep		200	.080	L.F.	3.64	3.88		7.52	10.35
3820	4-5/8" deep		190	.084		4.35	4.08		8.43	11.45
3840	5-5/8" deep		180	.089		6.75	4.31		11.06	14.50
9000	Minimum labor/equipment charge	1 Carp	4	2	Job		97		97	158

06 49 19.10 Shutters, Exterior	Crew	Daily Output	Labor-Hours	Unit	Material	2016 Bare Costs Labor	Equipment	Total	Total Incl O&P
0010 **SHUTTERS, EXTERIOR**									
0012 Aluminum, louvered, 1'-4" wide, 3'-0" long	1 Carp	10	.800	Pr.	200	39		239	284
0200 4'-0" long		10	.800		240	39		279	330
0300 5'-4" long		10	.800		280	39		319	375
0400 6'-8" long		9	.889		355	43		398	460
1000 Pine, louvered, primed, each 1'-2" wide, 3'-3" long		10	.800		220	39		259	305
1100 4'-7" long		10	.800		270	39		309	360
1250 Each 1'-4" wide, 3'-0" long		10	.800		220	39		259	305
1350 5'-3" long		10	.800		330	39		369	425
1500 Each 1'-6" wide, 3'-3" long		10	.800		244	39		283	330
1600 4'-7" long		10	.800		325	39		364	420
1620 Cedar, louvered, 1'-2" wide, 5'-7" long		10	.800		315	39		354	410
1630 Each 1'-4" wide, 2'-2" long		10	.800		175	39		214	257
1640 3'-0" long		10	.800		217	39		256	305
1650 3'-3" long		10	.800		228	39		267	315
1660 3'-11" long		10	.800		265	39		304	355
1670 4'-3" long		10	.800		275	39		314	370
1680 5'-3" long		10	.800		290	39		329	385
1690 5'-11" long		10	.800		350	39		389	450
1700 Door blinds, 6'-9" long, each 1'-3" wide		9	.889		380	43		423	485
1710 1'-6" wide		9	.889		435	43		478	550
1720 Cedar, solid raised panel, each 1'-4" wide, 3'-3" long		10	.800		315	39		354	410
1730 3'-11" long		10	.800		360	39		399	460
1740 4'-3" long		10	.800		365	39		404	465
1750 4'-7" long		10	.800		390	39		429	495
1760 4'-11" long		10	.800		415	39		454	525
1770 5'-11" long		10	.800		520	39		559	635
1800 Door blinds, 6'-9" long, each 1'-3" wide		9	.889		535	43		578	660
1900 1'-6" wide		9	.889		630	43		673	765
2500 Polystyrene, solid raised panel, each 1'-4" wide, 3'-3" long		10	.800		72	39		111	143
2600 3'-11" long		10	.800		93	39		132	166
2700 4'-7" long		10	.800		105	39		144	180
2800 5'-3" long		10	.800		120	39		159	196
2900 6'-8" long		9	.889		152	43		195	238
4500 Polystyrene, louvered, each 1'-2" wide, 3'-3" long		10	.800		36	39		75	103
4600 4'-7" long		10	.800		46	39		85	114
4750 5'-3" long		10	.800		59	39		98	129
4850 6'-8" long		9	.889		68	43		111	146
6000 Vinyl, louvered, each 1'-2" x 4'-7" long		10	.800		64	39		103	134
6200 Each 1'-4" x 6'-8" long		9	.889	▼	76	43		119	154
9000 Minimum labor/equipment charge	▼	4	2	Job		97		97	158

For customer support on your Commercial Renovation Cost Data, call 877.742.4719.

171

06 51 13 – Plastic Lumber

06 51 13.10 Recycled Plastic Lumber

		Crew	Daily Output	Labor-Hours	Unit	Material	2016 Bare Costs Labor	Equipment	Total	Total Incl O&P
0010	**RECYCLED PLASTIC LUMBER**									
4000	Sheeting, recycled plastic, black or white, 4' x 8' x 1/8"	G	2 Carp	1100	.015	S.F.	1.26	.70		1.96	2.54
4010	4' x 8' x 3/16"	G		1100	.015		1.90	.70		2.60	3.24
4020	4' x 8' x 1/4"	G		950	.017		2.25	.82		3.07	3.81
4030	4' x 8' x 3/8"	G		950	.017		3.80	.82		4.62	5.50
4040	4' x 8' x 1/2"	G		900	.018		5	.86		5.86	6.90
4050	4' x 8' x 5/8"	G		900	.018		7.50	.86		8.36	9.65
4060	4' x 8' x 3/4"	G		850	.019		9.40	.91		10.31	11.85
4070	Add for colors	G				Ea.	5%				
8500	100% recycled plastic, var colors, NLB, 2" x 2"	G				L.F.	1.91			1.91	2.10
8510	2" x 4"	G					3.95			3.95	4.35
8520	2" x 6"	G					6.20			6.20	6.80
8530	2" x 8"	G					8.50			8.50	9.35
8540	2" x 10"	G					12.40			12.40	13.60
8550	5/4" x 4"	G					4.45			4.45	4.90
8560	5/4" x 6"	G					5.60			5.60	6.15
8570	1" x 6"	G					3.06			3.06	3.37
8580	1/2" x 8"	G					3.45			3.45	3.80
8590	2" x 10" T & G	G					12.20			12.20	13.40
8600	3" x 10" T & G	G					17.15			17.15	18.85
8610	Add for premium colors	G					20%				

06 51 13.12 Structural Plastic Lumber

		Crew	Daily Output	Labor-Hours	Unit	Material	2016 Bare Costs Labor	Equipment	Total	Total Incl O&P
0010	**STRUCTURAL PLASTIC LUMBER**									
1320	Plastic lumber, posts or columns, 4" x 4"	2 Carp	390	.041	L.F.	9.25	1.99		11.24	13.40
1325	4" x 6"		275	.058		13.25	2.82		16.07	19.20
1330	4" x 8"		220	.073		19.35	3.52		22.87	27.50
1340	Girder, single, 4" x 4"		675	.024		9.25	1.15		10.40	12.05
1345	4" x 6"		600	.027		13.25	1.29		14.54	16.70
1350	4" x 8"		525	.030		19.35	1.48		20.83	24
1352	Double, 2" x 4"		625	.026		8.55	1.24		9.79	11.45
1354	2" x 6"		600	.027		12.85	1.29		14.14	16.25
1356	2" x 8"		575	.028		16.90	1.35		18.25	21
1358	2" x 10"		550	.029		20.50	1.41		21.91	25
1360	2" x 12"		525	.030		25	1.48		26.48	30
1362	Triple, 2" x 4"		575	.028		12.80	1.35		14.15	16.30
1364	2" x 6"		550	.029		19.30	1.41		20.71	23.50
1366	2" x 8"		525	.030		25.50	1.48		26.98	30.50
1368	2" x 10"		500	.032		31	1.55		32.55	36.50
1370	2" x 12"		475	.034		37.50	1.63		39.13	43.50
1372	Ledger, bolted 4' O.C., 2" x 4"		400	.040		4.42	1.94		6.36	8.05
1374	2" x 6"		550	.029		6.50	1.41		7.91	9.45
1376	2" x 8"		390	.041		8.55	1.99		10.54	12.70
1378	2" x 10"		385	.042		10.40	2.01		12.41	14.75
1380	2" x 12"		380	.042		12.55	2.04		14.59	17.20
1382	Joists, 2" x 4"		1250	.013		4.27	.62		4.89	5.70
1384	2" x 6"		1250	.013		6.45	.62		7.07	8.05
1386	2" x 8"		1100	.015		8.45	.70		9.15	10.45
1388	2" x 10"		500	.032		10.40	1.55		11.95	14
1390	2" x 12"		875	.018		12.45	.89		13.34	15.15
1392	Railings and trim , 5/4" x 4"	1 Carp	300	.027		4.75	1.29		6.04	7.35
1394	2" x 2"		300	.027		2.40	1.29		3.69	4.75
1396	2" x 4"		300	.027		4.25	1.29		5.54	6.80

06 51 Structural Plastic Shapes and Plates

06 51 13 – Plastic Lumber

06 51 13.12 Structural Plastic Lumber		Crew	Daily Output	Labor-Hours	Unit	Material	2016 Bare Costs Labor	Equipment	Total	Total Incl O&P
1398	2" x 6"	1 Carp	300	.027	L.F.	6.40	1.29		7.69	9.15

06 52 Plastic Structural Assemblies

06 52 10 – Fiberglass Structural Assemblies

06 52 10.30 Fiberglass Grating

		Crew	Daily Output	Labor-Hours	Unit	Material	2016 Bare Costs Labor	Equipment	Total	Total Incl O&P
0010	**FIBERGLASS GRATING**									
0100	Molded, green (for mod. corrosive environment)									
0140	1" x 4" mesh, 1" thick	2 Sswk	400	.040	S.F.	8.50	2.13		10.63	13.10
0180	1-1/2" square mesh, 1" thick		400	.040		18.70	2.13		20.83	24.50
0220	1-1/4" thick		400	.040		9.35	2.13		11.48	14
0260	1-1/2" thick		400	.040		23	2.13		25.13	29
0300	2" square mesh, 2" thick		320	.050		26	2.66		28.66	33.50
1000	Orange (for highly corrosive environment)									
1040	1" x 4" mesh, 1" thick	2 Sswk	400	.040	S.F.	16.50	2.13		18.63	22
1080	1-1/2" square mesh, 1" thick		400	.040		19.55	2.13		21.68	25.50
1120	1-1/4" thick		400	.040		20	2.13		22.13	26
1160	1-1/2" thick		400	.040		20.50	2.13		22.63	26.50
1200	2" square mesh, 2" thick		320	.050		20.50	2.66		23.16	27
3000	Pultruded, green (for mod. corrosive environment)									
3040	1" O.C. bar spacing, 1" thick	2 Sswk	400	.040	S.F.	14.85	2.13		16.98	20
3080	1-1/2" thick		320	.050		15.30	2.66		17.96	21.50
3120	1-1/2" O.C. bar spacing, 1" thick		400	.040		13.70	2.13		15.83	18.80
3160	1-1/2" thick		400	.040		18.05	2.13		20.18	23.50
4000	Grating support legs, fixed height, no base				Ea.	46.50			46.50	51.50
4040	With base					41			41	45
4080	Adjustable to 60"					61.50			61.50	67.50

06 52 10.40 Fiberglass Floor Grating

		Crew	Daily Output	Labor-Hours	Unit	Material	2016 Bare Costs Labor	Equipment	Total	Total Incl O&P
0010	**FIBERGLASS FLOOR GRATING**									
0100	Reinforced polyester, fire retardant, 1" x 4" grid, 1" thick	E-4	510	.063	S.F.	14	3.37	.29	17.66	21.50
0200	1-1/2" x 6" mesh, 1-1/2" thick		500	.064		16.50	3.44	.29	20.23	24.50
0300	With grit surface, 1-1/2" x 6" grid, 1-1/2" thick		500	.064		16.80	3.44	.29	20.53	25

06 63 Plastic Railings

06 63 10 – Plastic (PVC) Railings

06 63 10.10 Plastic Railings

		Crew	Daily Output	Labor-Hours	Unit	Material	2016 Bare Costs Labor	Equipment	Total	Total Incl O&P
0010	**PLASTIC RAILINGS**									
0100	Horizontal PVC handrail with balusters, 3-1/2" wide, 36" high	1 Carp	96	.083	L.F.	25.50	4.04		29.54	34.50
0150	42" high		96	.083		28.50	4.04		32.54	38
0200	Angled PVC handrail with balusters, 3-1/2" wide, 36" high		72	.111		29	5.40		34.40	41
0250	42" high		72	.111		32.50	5.40		37.90	44.50
0300	Post sleeve for 4 x 4 post		96	.083		13.10	4.04		17.14	21
0400	Post cap for 4 x 4 post, flat profile		48	.167	Ea.	13.20	8.10		21.30	27.50
0450	Newel post style profile		48	.167		24	8.10		32.10	39.50
0500	Raised corbeled profile		48	.167		36.50	8.10		44.60	53
0550	Post base trim for 4 x 4 post		96	.083		18.80	4.04		22.84	27

For customer support on your Commercial Renovation Cost Data, call 877.742.4719.

173

06 65 10.10 PVC Trim, Exterior		Crew	Daily Output	Labor-Hours	Unit	Material	2016 Bare Costs Labor	Equipment	Total	Total Incl O&P
0010	**PVC TRIM, EXTERIOR**									
0100	Cornerboards, 5/4" x 6" x 6"	1 Carp	240	.033	L.F.	8	1.61		9.61	11.45
0110	Door/window casing, 1" x 4"		200	.040		1.46	1.94		3.40	4.78
0120	1" x 6"		200	.040		2.20	1.94		4.14	5.60
0130	1" x 8"		195	.041		2.93	1.99		4.92	6.45
0140	1" x 10"		195	.041		3.81	1.99		5.80	7.45
0150	1" x 12"		190	.042		4.36	2.04		6.40	8.15
0160	5/4" x 4"		195	.041		2.19	1.99		4.18	5.65
0170	5/4" x 6"		195	.041		3.15	1.99		5.14	6.70
0180	5/4" x 8"		190	.042		4.19	2.04		6.23	7.95
0190	5/4" x 10"		190	.042		5.10	2.04		7.14	8.95
0200	5/4" x 12"		185	.043		5.60	2.10		7.70	9.55
0210	Fascia, 1" x 4"		250	.032		1.46	1.55		3.01	4.14
0220	1" x 6"		250	.032		2.20	1.55		3.75	4.95
0230	1" x 8"		225	.036		2.93	1.72		4.65	6.05
0240	1" x 10"		225	.036		3.81	1.72		5.53	7
0250	1" x 12"		200	.040		4.36	1.94		6.30	7.95
0260	5/4" x 4"		240	.033		2.19	1.61		3.80	5.05
0270	5/4" x 6"		240	.033		3.15	1.61		4.76	6.10
0280	5/4" x 8"		215	.037		4.19	1.80		5.99	7.55
0290	5/4" x 10"		215	.037		5.10	1.80		6.90	8.55
0300	5/4" x 12"		190	.042		5.60	2.04		7.64	9.50
0310	Frieze, 1" x 4"		250	.032		1.46	1.55		3.01	4.14
0320	1" x 6"		250	.032		2.20	1.55		3.75	4.95
0330	1" x 8"		225	.036		2.93	1.72		4.65	6.05
0340	1" x 10"		225	.036		3.81	1.72		5.53	7
0350	1" x 12"		200	.040		4.36	1.94		6.30	7.95
0360	5/4" x 4"		240	.033		2.19	1.61		3.80	5.05
0370	5/4" x 6"		240	.033		3.15	1.61		4.76	6.10
0380	5/4" x 8"		215	.037		4.19	1.80		5.99	7.55
0390	5/4" x 10"		215	.037		5.10	1.80		6.90	8.55
0400	5/4" x 12"		190	.042		5.60	2.04		7.64	9.50
0410	Rake, 1" x 4"		200	.040		1.46	1.94		3.40	4.78
0420	1" x 6"		200	.040		2.20	1.94		4.14	5.60
0430	1" x 8"		190	.042		2.93	2.04		4.97	6.55
0440	1" x 10"		190	.042		3.81	2.04		5.85	7.50
0450	1" x 12"		180	.044		4.36	2.15		6.51	8.30
0460	5/4" x 4"		195	.041		2.19	1.99		4.18	5.65
0470	5/4" x 6"		195	.041		3.15	1.99		5.14	6.70
0480	5/4" x 8"		185	.043		4.19	2.10		6.29	8.05
0490	5/4" x 10"		185	.043		5.10	2.10		7.20	9
0500	5/4" x 12"		175	.046		5.60	2.21		7.81	9.75
0510	Rake trim, 1" x 4"		225	.036		1.46	1.72		3.18	4.42
0520	1" x 6"		225	.036		2.20	1.72		3.92	5.25
0560	5/4" x 4"		220	.036		2.19	1.76		3.95	5.30
0570	5/4" x 6"		220	.036		3.15	1.76		4.91	6.35
0610	Soffit, 1" x 4"	2 Carp	420	.038		1.46	1.85		3.31	4.63
0620	1" x 6"		420	.038		2.20	1.85		4.05	5.45
0630	1" x 8"		420	.038		2.93	1.85		4.78	6.25
0640	1" x 10"		400	.040		3.81	1.94		5.75	7.35
0650	1" x 12"		400	.040		4.36	1.94		6.30	7.95
0660	5/4" x 4"		410	.039		2.19	1.89		4.08	5.50
0670	5/4" x 6"		410	.039		3.15	1.89		5.04	6.55

06 65 Plastic Trim

06 65 10 – PVC Trim

06 65 10.10 PVC Trim, Exterior

		Crew	Daily Output	Labor-Hours	Unit	Material	2016 Bare Costs Labor	Equipment	Total	Total Incl O&P
0680	5/4" x 8"	2 Carp	410	.039	L.F.	4.19	1.89		6.08	7.70
0690	5/4" x 10"		390	.041		5.10	1.99		7.09	8.85
0700	5/4" x 12"		390	.041		5.60	1.99		7.59	9.40

06 80 Composite Fabrications

06 80 10 – Composite Decking

06 80 10.10 Woodgrained Composite Decking

		Crew	Daily Output	Labor-Hours	Unit	Material	2016 Bare Costs Labor	Equipment	Total	Total Incl O&P
0010	**WOODGRAINED COMPOSITE DECKING**									
0100	Woodgrained composite decking, 1" x 6"	2 Carp	640	.025	L.F.	3.65	1.21		4.86	6
0110	Grooved edge		660	.024		3.79	1.17		4.96	6.10
0120	2" x 6"		640	.025		3.78	1.21		4.99	6.15
0130	Encased, 1" x 6"		640	.025		3.78	1.21		4.99	6.15
0140	Grooved edge		660	.024		3.92	1.17		5.09	6.25
0150	2" x 6"		640	.025		5.10	1.21		6.31	7.65

06 81 Composite Railings

06 81 10 – Encased Railings

06 81 10.10 Encased Composite Railings

		Crew	Daily Output	Labor-Hours	Unit	Material	2016 Bare Costs Labor	Equipment	Total	Total Incl O&P
0010	**ENCASED COMPOSITE RAILINGS**									
0100	Encased composite railing, 6' long, 36" high, incl. balusters	1 Carp	16	.500	Ea.	161	24		185	217
0110	42" high, incl. balusters		16	.500		181	24		205	239
0120	8' long, 36" high, incl. balusters		12	.667		161	32.50		193.50	230
0130	42" high, incl. balusters		12	.667		181	32.50		213.50	252
0140	Accessories, post sleeve, 4" x 4", 39" long		32	.250		23.50	12.10		35.60	46
0150	96" long		24	.333		89	16.15		105.15	125
0160	6" x 6", 39" long		32	.250		48	12.10		60.10	73
0170	96" long		24	.333		141	16.15		157.15	182
0180	Accessories, post skirt, 4" x 4"		96	.083		4.25	4.04		8.29	11.30
0190	6" x 6"		96	.083		5	4.04		9.04	12.10
0200	Post cap, 4" x 4", flat		48	.167		6.30	8.10		14.40	20
0210	Pyramid		48	.167		6.30	8.10		14.40	20
0220	Post cap, 6" x 6", flat		48	.167		10.20	8.10		18.30	24.50
0230	Pyramid		48	.167		10.20	8.10		18.30	24.50

For customer support on your Commercial Renovation Cost Data, call 877.742.4719.

175

Division Notes

		CREW	DAILY OUTPUT	LABOR-HOURS	UNIT	BARE COSTS				TOTAL INCL O&P
						MAT.	LABOR	EQUIP.	TOTAL	

Estimating Tips

07 10 00 Dampproofing and Waterproofing

- Be sure of the job specifications before pricing this subdivision. The difference in cost between waterproofing and dampproofing can be great. Waterproofing will hold back standing water. Dampproofing prevents the transmission of water vapor. Also included in this section are vapor retarding membranes.

07 20 00 Thermal Protection

- Insulation and fireproofing products are measured by area, thickness, volume or R-value. Specifications may give only what the specific R-value should be in a certain situation. The estimator may need to choose the type of insulation to meet that R-value.

07 30 00 Steep Slope Roofing
07 40 00 Roofing and Siding Panels

- Many roofing and siding products are bought and sold by the square. One square is equal to an area that measures 100 square feet.

  This simple change in unit of measure could create a large error if the estimator is not observant. Accessories necessary for a complete installation must be figured into any calculations for both material and labor.

07 50 00 Membrane Roofing
07 60 00 Flashing and Sheet Metal
07 70 00 Roofing and Wall Specialties and Accessories

- The items in these subdivisions compose a roofing system. No one component completes the installation, and all must be estimated. Built-up or single-ply membrane roofing systems are made up of many products and installation trades. Wood blocking at roof perimeters or penetrations, parapet coverings, reglets, roof drains, gutters, downspouts, sheet metal flashing, skylights, smoke vents, and roof hatches all need to be considered along with the roofing material. Several different installation trades will need to work together on the roofing system. Inherent difficulties in the scheduling and coordination of various trades must be accounted for when estimating labor costs.

07 90 00 Joint Protection

- To complete the weather-tight shell, the sealants and caulkings must be estimated. Where different materials meet—at expansion joints, at flashing penetrations, and at hundreds of other locations throughout a construction project—caulking and sealants provide another line of defense against water penetration. Often, an entire system is based on the proper location and placement of caulking or sealants. The detailed drawings that are included as part of a set of architectural plans show typical locations for these materials. When caulking or sealants are shown at typical locations, this means the estimator must include them for all the locations where this detail is applicable. Be careful to keep different types of sealants separate, and remember to consider backer rods and primers if necessary.

Reference Numbers

Reference numbers are shown at the beginning of some major classifications. These numbers refer to related items in the Reference Section. The reference information may be an estimating procedure, an alternate pricing method, or technical information.

Note: Not all subdivisions listed here necessarily appear. ■

07 01 50 – Maintenance of Membrane Roofing

07 01 50.10 Roof Coatings		Crew	Daily Output	Labor-Hours	Unit	Material	2016 Bare Costs Labor	Equipment	Total	Total Incl O&P
0010	**ROOF COATINGS**									
0012	Asphalt, brush grade, material only				Gal.	8.95			8.95	9.80
0200	Asphalt base, fibered aluminum coating	G				8.25			8.25	9.10
0300	Asphalt primer, 5 gallon				↓	7.50			7.50	8.25
0600	Coal tar pitch, 200 lb. barrels				Ton	1,250			1,250	1,375
0700	Tar roof cement, 5 gal. lots				Gal.	14.15			14.15	15.55
0800	Glass fibered roof & patching cement, 5 gallon				"	8.25			8.25	9.10
0900	Reinforcing glass membrane, 450 S.F./roll				Ea.	59.50			59.50	65.50
1000	Neoprene roof coating, 5 gal., 2 gal./sq.				Gal.	33.50			33.50	36.50
1100	Roof patch & flashing cement, 5 gallon					8.75			8.75	9.60
1200	Roof resaturant, glass fibered, 3 gal./sq.					8.85			8.85	9.75
1600	Reflective roof coating, white, elastomeric, approx. 50 S.F. per gal.	G			↓	17.20			17.20	18.90

07 01 90 – Maintenance of Joint Protection

07 01 90.81 Joint Sealant Replacement

		Crew	Daily Output	Labor-Hours	Unit	Material	Labor	Equipment	Total	Total Incl O&P
0010	**JOINT SEALANT REPLACEMENT**									
0050	Control joints in concrete floors/slabs									
0100	Option 1 for joints with hard dry sealant									
0110	Step 1: Sawcut to remove 95% of old sealant									
0112	1/4" wide x 1/2" deep, with single saw blade	C-27	4800	.003	L.F.	.02	.15	.04	.21	.30
0114	3/8" wide x 3/4" deep, with single saw blade		4000	.004		.03	.18	.04	.25	.37
0116	1/2" wide x 1" deep, with double saw blades		3600	.004		.06	.20	.05	.31	.43
0118	3/4" wide x 1-1/2" deep, with double saw blades	↓	3200	.005		.12	.23	.05	.40	.55
0120	Step 2: Water blast joint faces and edges	C-29	2500	.003			.12	.03	.15	.23
0130	Step 3: Air blast joint faces and edges	C-28	2000	.004			.18	.01	.19	.30
0140	Step 4: Sand blast joint faces and edges	E-11	2000	.016			.68	.12	.80	1.29
0150	Step 5: Air blast joint faces and edges	C-28	2000	.004	↓		.18	.01	.19	.30
0200	Option 2 for joints with soft pliable sealant									
0210	Step 1: Plow joint with rectangular blade	B-62	2600	.009	L.F.		.38	.07	.45	.69
0220	Step 2: Sawcut to re-face joint faces									
0222	1/4" wide x 1/2" deep, with single saw blade	C-27	2400	.007	L.F.	.02	.30	.07	.39	.58
0224	3/8" wide x 3/4" deep, with single saw blade		2000	.008		.04	.37	.08	.49	.71
0226	1/2" wide x 1" deep, with double saw blades		1800	.009		.08	.41	.09	.58	.82
0228	3/4" wide x 1-1/2" deep, with double saw blades	↓	1600	.010		.16	.46	.11	.73	1.01
0230	Step 3: Water blast joint faces and edges	C-29	2500	.003			.12	.03	.15	.23
0240	Step 4: Air blast joint faces and edges	C-28	2000	.004			.18	.01	.19	.30
0250	Step 5: Sand blast joint faces and edges	E-11	2000	.016			.68	.12	.80	1.29
0260	Step 6: Air blast joint faces and edges	C-28	2000	.004	↓		.18	.01	.19	.30
0290	For saw cutting new control joints, see Section 03 15 16.20									
8910	For backer rod, see Section 07 91 23.10									
8920	For joint sealant, see Section 03 15 16.30 or 07 92 13.20									

07 05 Common Work Results for Thermal and Moisture Protection

07 05 05 – Selective Demolition for Thermal and Moisture Protection

07 05 05.10 Selective Demo., Thermal and Moist. Protection	Crew	Daily Output	Labor-Hours	Unit	Material	2016 Bare Costs Labor	2016 Bare Costs Equipment	Total	Total Incl O&P
0010	**SELECTIVE DEMO., THERMAL AND MOISTURE PROTECTION**								
0020	Caulking/sealant, to 1" x 1" joint R024119-10	1 Clab	600	.013	L.F.		.51		.51	.83
0120	Downspouts, including hangers		350	.023	"		.87		.87	1.42
0220	Flashing, sheet metal		290	.028	S.F.		1.05		1.05	1.71
0420	Gutters, aluminum or wood, edge hung		240	.033	L.F.		1.26		1.26	2.06
0520	Built-in		100	.080	"		3.03		3.03	4.96
0620	Insulation, air/vapor barrier		3500	.002	S.F.		.09		.09	.14
0670	Batts or blankets		1400	.006	C.F.		.22		.22	.35
0720	Foamed or sprayed in place	2 Clab	1000	.016	B.F.		.61		.61	.99
0770	Loose fitting	1 Clab	3000	.003	C.F.		.10		.10	.17
0870	Rigid board		3450	.002	B.F.		.09		.09	.14
1120	Roll roofing, cold adhesive		12	.667	Sq.		25.50		25.50	41.50
1170	Roof accessories, adjustable metal chimney flashing		9	.889	Ea.		33.50		33.50	55
1325	Plumbing vent flashing		32	.250	"		9.50		9.50	15.50
1375	Ridge vent strip, aluminum		310	.026	L.F.		.98		.98	1.60
1620	Skylight to 10 S.F.		8	1	Ea.		38		38	62
2120	Roof edge, aluminum soffit and fascia		570	.014	L.F.		.53		.53	.87
2170	Concrete coping, up to 12" wide	2 Clab	160	.100			3.79		3.79	6.20
2220	Drip edge	1 Clab	1000	.008			.30		.30	.50
2270	Gravel stop		950	.008			.32		.32	.52
2370	Sheet metal coping, up to 12" wide		240	.033			1.26		1.26	2.06
2470	Roof insulation board, over 2" thick	B-2	7800	.005	B.F.		.20		.20	.32
2520	Up to 2" thick	"	3900	.010	S.F.		.39		.39	.64
2620	Roof ventilation, louvered gable vent	1 Clab	16	.500	Ea.		18.95		18.95	31
2670	Remove, roof hatch	G-3	15	2.133			102		102	162
2675	Rafter vents	1 Clab	960	.008			.32		.32	.52
2720	Soffit vent and/or fascia vent		575	.014	L.F.		.53		.53	.86
2775	Soffit vent strip, aluminum, 3" to 4" wide		160	.050			1.90		1.90	3.10
2820	Roofing accessories, shingle moulding, to 1" x 4"		1600	.005			.19		.19	.31
2870	Cant strip	B-2	2000	.020			.77		.77	1.25
2920	Concrete block walkway	1 Clab	230	.035			1.32		1.32	2.15
3070	Roofing, felt paper, 15#		70	.114	Sq.		4.33		4.33	7.10
3125	#30 felt		30	.267	"		10.10		10.10	16.50
3170	Asphalt shingles, 1 layer	B-2	3500	.011	S.F.		.44		.44	.72
3180	2 layers		1750	.023	"		.88		.88	1.43
3370	Modified bitumen		26	1.538	Sq.		59		59	96.50
3420	Built-up, no gravel, 3 ply		25	1.600			61.50		61.50	100
3470	4 ply		21	1.905			73		73	119
3620	5 ply		1600	.025	S.F.		.96		.96	1.57
3720	5 ply, with gravel		890	.045			1.72		1.72	2.81
3725	Loose gravel removal		5000	.008			.31		.31	.50
3730	Embedded gravel removal		2000	.020			.77		.77	1.25
3870	Fiberglass sheet		1200	.033			1.28		1.28	2.09
4120	Slate shingles		1900	.021			.81		.81	1.32
4170	Ridge shingles, clay or slate		2000	.020	L.F.		.77		.77	1.25
4320	Single ply membrane, attached at seams		52	.769	Sq.		29.50		29.50	48
4370	Ballasted		75	.533			20.50		20.50	33.50
4420	Fully adhered		39	1.026	"		39.50		39.50	64
4550	Roof hatch, 2'-6" x 3'-0"	1 Clab	10	.800	Ea.		30.50		30.50	49.50
4670	Wood shingles	B-2	2200	.018	S.F.		.70		.70	1.14
4820	Sheet metal roofing	"	2150	.019			.71		.71	1.16
4970	Siding, horizontal wood clapboards	1 Clab	380	.021			.80		.80	1.30
5025	Exterior insulation finish system	"	120	.067			2.53		2.53	4.13

07 05 Common Work Results for Thermal and Moisture Protection

07 05 05 – Selective Demolition for Thermal and Moisture Protection

07 05 05.10 Selective Demo., Thermal and Moist. Protection	Crew	Daily Output	Labor-Hours	Unit	Material	2016 Bare Costs Labor	Equipment	Total	Total Incl O&P
5070	Tempered hardboard, remove and reset	1 Carp	380	.021	S.F.		1.02		1.02	1.67
5120	Tempered hardboard sheet siding	"	375	.021	↓		1.03		1.03	1.69
5170	Metal, corner strips	1 Clab	850	.009	L.F.		.36		.36	.58
5225	Horizontal strips		444	.018	S.F.		.68		.68	1.12
5320	Vertical strips		400	.020			.76		.76	1.24
5520	Wood shingles		350	.023			.87		.87	1.42
5620	Stucco siding		360	.022			.84		.84	1.38
5670	Textured plywood		725	.011			.42		.42	.68
5720	Vinyl siding		510	.016	↓		.59		.59	.97
5770	Corner strips		900	.009	L.F.		.34		.34	.55
5870	Wood, boards, vertical	↓	400	.020	S.F.		.76		.76	1.24
5920	Waterproofing, protection/drain board	2 Clab	3900	.004	B.F.		.16		.16	.25
5970	Over 1/2" thick		1750	.009	S.F.		.35		.35	.57
6020	To 1/2" thick	↓	2000	.008	"		.30		.30	.50
9000	Minimum labor/equipment charge	1 Clab	2	4	Job		152		152	248

07 11 Dampproofing

07 11 13 – Bituminous Dampproofing

07 11 13.10 Bituminous Asphalt Coating

		Crew	Daily Output	Labor-Hours	Unit	Material	Labor	Equipment	Total	Total Incl O&P
0010	**BITUMINOUS ASPHALT COATING**									
0030	Brushed on, below grade, 1 coat	1 Rofc	665	.012	S.F.	.22	.50		.72	1.16
0100	2 coat		500	.016		.45	.67		1.12	1.69
0300	Sprayed on, below grade, 1 coat		830	.010		.22	.40		.62	.98
0400	2 coat	↓	500	.016	↓	.44	.67		1.11	1.68
0500	Asphalt coating, with fibers				Gal.	8.25			8.25	9.10
0600	Troweled on, asphalt with fibers, 1/16" thick	1 Rofc	500	.016	S.F.	.36	.67		1.03	1.59
0700	1/8" thick		400	.020		.63	.83		1.46	2.21
1000	1/2" thick		350	.023	↓	2.06	.95		3.01	3.99
9000	Minimum labor/equipment charge	↓	3	2.667	Job		111		111	201

07 11 16 – Cementitious Dampproofing

07 11 16.20 Cementitious Parging

		Crew	Daily Output	Labor-Hours	Unit	Material	Labor	Equipment	Total	Total Incl O&P
0010	**CEMENTITIOUS PARGING**									
0020	Portland cement, 2 coats, 1/2" thick	D-1	250	.064	S.F.	.35	2.70		3.05	4.77
0100	Waterproofed Portland cement, 1/2" thick, 2 coats	"	250	.064	"	3.95	2.70		6.65	8.75

07 12 Built-up Bituminous Waterproofing

07 12 13 – Built-Up Asphalt Waterproofing

07 12 13.20 Membrane Waterproofing

		Crew	Daily Output	Labor-Hours	Unit	Material	Labor	Equipment	Total	Total Incl O&P
0010	**MEMBRANE WATERPROOFING**									
0012	On slabs, 1 ply, felt, mopped	G-1	3000	.019	S.F.	.42	.73	.19	1.34	1.98
0015	On walls, 1 ply, felt, mopped		3000	.019		.42	.73	.19	1.34	1.98
0100	On slabs, 1 ply, glass fiber fabric, mopped		2100	.027		.47	1.04	.28	1.79	2.70
0105	On walls, 1 ply, glass fiber fabric, mopped		2100	.027		.47	1.04	.28	1.79	2.70
0300	On slabs, 2 ply, felt, mopped		2500	.022		.83	.87	.23	1.93	2.74
0305	On walls, 2 ply, felt, mopped		2500	.022		.83	.87	.23	1.93	2.74
0400	On slabs, 2 ply, glass fiber fabric, mopped		1650	.034		1.04	1.32	.35	2.71	3.92
0405	On walls, 2 ply, glass fiber fabric, mopped		1650	.034		1.04	1.32	.35	2.71	3.92
0600	On slabs, 3 ply, felt, mopped		2100	.027		1.25	1.04	.28	2.57	3.55
0605	On walls, 3 ply, felt, mopped	↓	2100	.027		1.25	1.04	.28	2.57	3.55

07 12 Built-up Bituminous Waterproofing

07 12 13 – Built-Up Asphalt Waterproofing

07 12 13.20 Membrane Waterproofing	Crew	Daily Output	Labor-Hours	Unit	Material	2016 Bare Costs Labor	Equipment	Total	Total Incl O&P
0700	On slabs, 3 ply, glass fiber fabric, mopped	G-1	1550	.036	S.F.	1.41	1.41	.37	3.19	4.51
0705	On walls, 3 ply, glass fiber fabric, mopped		1550	.036		1.41	1.41	.37	3.19	4.51
0710	Asphaltic hardboard protection board, 1/8" thick	2 Rofc	500	.032		.65	1.33		1.98	3.13
0715	1/4" thick		450	.036		1.56	1.48		3.04	4.40
1000	EPS membrane protection board, 1/4"		3500	.005		.34	.19		.53	.71
1050	3/8" thick		3500	.005		.37	.19		.56	.74
1060	1/2" thick		3500	.005		.40	.19		.59	.78
1070	Fiberglass fabric, black, 20/10 mesh		116	.138	Sq.	15.85	5.75		21.60	28
9000	Minimum labor/equipment charge		2	8	Job		335		335	600

07 13 Sheet Waterproofing

07 13 53 – Elastomeric Sheet Waterproofing

07 13 53.10 Elastomeric Sheet Waterproofing and Access.

		Crew	Daily Output	Labor-Hours	Unit	Material	2016 Bare Costs Labor	Equipment	Total	Total Incl O&P
0010	**ELASTOMERIC SHEET WATERPROOFING AND ACCESS.**									
0090	EPDM, plain, 45 mils thick	2 Rofc	580	.028	S.F.	1.44	1.15		2.59	3.67
0100	60 mils thick		570	.028		1.50	1.17		2.67	3.76
0300	Nylon reinforced sheets, 45 mils thick		580	.028		1.56	1.15		2.71	3.80
0400	60 mils thick		570	.028		1.66	1.17		2.83	3.93
0600	Vulcanizing splicing tape for above, 2" wide				C.L.F.	61			61	67
0700	4" wide				"	122			122	134
0900	Adhesive, bonding, 60 S.F. per gal.				Gal.	28			28	31
1000	Splicing, 75 S.F. per gal.				"	40			40	44
1200	Neoprene sheets, plain, 45 mils thick	2 Rofc	580	.028	S.F.	1.82	1.15		2.97	4.08
1300	60 mils thick		570	.028		2.08	1.17		3.25	4.40
1500	Nylon reinforced, 45 mils thick		580	.028		2.07	1.15		3.22	4.36
1600	60 mils thick		570	.028		2.67	1.17		3.84	5.05
1800	120 mils thick		500	.032		5.50	1.33		6.83	8.45
1900	Adhesive, splicing, 150 S.F. per gal. per coat				Gal.	40			40	44
2100	Fiberglass reinforced, fluid applied, 1/8" thick	2 Rofc	500	.032	S.F.	1.76	1.33		3.09	4.35
2200	Polyethylene and rubberized asphalt sheets, 60 mils thick		550	.029		.86	1.21		2.07	3.14
2400	Polyvinyl chloride sheets, plain, 10 mils thick		580	.028		.15	1.15		1.30	2.25
2500	20 mils thick		570	.028		.20	1.17		1.37	2.33
2700	30 mils thick		560	.029		.25	1.19		1.44	2.43
3000	Adhesives, trowel grade, 40-100 S.F. per gal.				Gal.	24			24	26.50
3100	Brush grade, 100-250 S.F. per gal.				"	21.50			21.50	24
3300	Bitumen modified polyurethane, fluid applied, 55 mils thick	2 Rofc	665	.024	S.F.	.92	1		1.92	2.82
9000	Minimum labor/equipment charge	"	2	8	Job		335		335	600

07 16 Cementitious and Reactive Waterproofing

07 16 16 – Crystalline Waterproofing

07 16 16.20 Cementitious Waterproofing

		Crew	Daily Output	Labor-Hours	Unit	Material	2016 Bare Costs Labor	Equipment	Total	Total Incl O&P
0010	**CEMENTITIOUS WATERPROOFING**									
0020	1/8" application, sprayed on	G-2A	1000	.024	S.F.	.74	.89	.71	2.34	3.14
0050	4 coat cementitious metallic slurry	1 Cefi	1.20	6.667	C.S.F.	35	305		340	520

For customer support on your Commercial Renovation Cost Data, call 877.742.4719.

181

07 19 Water Repellents

07 19 19 – Silicone Water Repellents

07 19 19.10 Silicone Based Water Repellents		Crew	Daily Output	Labor-Hours	Unit	Material	2016 Bare Costs Labor	Equipment	Total	Total Incl O&P
0010	**SILICONE BASED WATER REPELLENTS**									
0020	Water base liquid, roller applied	2 Rofc	7000	.002	S.F.	.53	.10		.63	.75
0200	Silicone or stearate, sprayed on CMU, 1 coat	1 Rofc	4000	.002		.37	.08		.45	.55
0300	2 coats		3000	.003		.73	.11		.84	1.01
9000	Minimum labor/equipment charge		3	2.667	Job		111		111	201

07 21 Thermal Insulation

07 21 13 – Board Insulation

07 21 13.10 Rigid Insulation

07 21 13.10 Rigid Insulation			Crew	Daily Output	Labor-Hours	Unit	Material	2016 Bare Costs Labor	Equipment	Total	Total Incl O&P
0010	**RIGID INSULATION**, for walls										
0040	Fiberglass, 1.5#/C.F., unfaced, 1" thick, R4.1	G	1 Carp	1000	.008	S.F.	.29	.39		.68	.95
0060	1-1/2" thick, R6.2	G		1000	.008		.38	.39		.77	1.05
0080	2" thick, R8.3	G		1000	.008		.49	.39		.88	1.17
0120	3" thick, R12.4	G		800	.010		.56	.48		1.04	1.41
0370	3#/C.F., unfaced, 1" thick, R4.3	G		1000	.008		.52	.39		.91	1.20
0390	1-1/2" thick, R6.5	G		1000	.008		.78	.39		1.17	1.49
0400	2" thick, R8.7	G		890	.009		1.05	.44		1.49	1.87
0420	2-1/2" thick, R10.9	G		800	.010		1.10	.48		1.58	2
0440	3" thick, R13	G		800	.010		1.59	.48		2.07	2.54
0520	Foil faced, 1" thick, R4.3	G		1000	.008		.90	.39		1.29	1.62
0540	1-1/2" thick, R6.5	G		1000	.008		1.35	.39		1.74	2.12
0560	2" thick, R8.7	G		890	.009		1.69	.44		2.13	2.57
0580	2-1/2" thick, R10.9	G		800	.010		1.98	.48		2.46	2.97
0600	3" thick, R13	G		800	.010		2.18	.48		2.66	3.19
1600	Isocyanurate, 4' x 8' sheet, foil faced, both sides										
1610	1/2" thick	G	1 Carp	800	.010	S.F.	.31	.48		.79	1.13
1620	5/8" thick	G		800	.010		.33	.48		.81	1.15
1630	3/4" thick	G		800	.010		.38	.48		.86	1.21
1640	1" thick	G		800	.010		.51	.48		.99	1.35
1650	1-1/2" thick	G		730	.011		.59	.53		1.12	1.52
1660	2" thick	G		730	.011		.75	.53		1.28	1.70
1670	3" thick	G		730	.011		1.85	.53		2.38	2.91
1680	4" thick	G		730	.011		2.10	.53		2.63	3.18
1700	Perlite, 1" thick, R2.77	G		800	.010		.44	.48		.92	1.27
1750	2" thick, R5.55	G		730	.011		.77	.53		1.30	1.72
1900	Extruded polystyrene, 25 PSI compressive strength, 1" thick, R5	G		800	.010		.56	.48		1.04	1.41
1940	2" thick R10	G		730	.011		1.13	.53		1.66	2.11
1960	3" thick, R15	G		730	.011		1.58	.53		2.11	2.61
2100	Expanded polystyrene, 1" thick, R3.85	G		800	.010		.26	.48		.74	1.08
2120	2" thick, R7.69	G		730	.011		.52	.53		1.05	1.44
2140	3" thick, R11.49	G		730	.011		.78	.53		1.31	1.73
9000	Minimum labor/equipment charge			4	2	Job		97		97	158

07 21 13.13 Foam Board Insulation

07 21 13.13 Foam Board Insulation			Crew	Daily Output	Labor-Hours	Unit	Material	2016 Bare Costs Labor	Equipment	Total	Total Incl O&P
0010	**FOAM BOARD INSULATION**										
0600	Polystyrene, expanded, 1" thick, R4	G	1 Carp	680	.012	S.F.	.26	.57		.83	1.22
0700	2" thick, R8	G		675	.012	"	.52	.57		1.09	1.51
9000	Minimum labor/equipment charge			4	2	Job		97		97	158

07 21 Thermal Insulation

07 21 16 – Blanket Insulation

07 21 16.10 Blanket Insulation for Floors/Ceilings		Crew	Daily Output	Labor-Hours	Unit	Material	2016 Bare Costs Labor	Equipment	Total	Total Incl O&P
0010	**BLANKET INSULATION FOR FLOORS/CEILINGS**									
0020	Including spring type wire fasteners									
2000	Fiberglass, blankets or batts, paper or foil backing									
2100	3-1/2" thick, R13	G	1 Carp	700	.011	S.F.	.40	.55		.95	1.34
2150	6-1/4" thick, R19	G		600	.013		.51	.65		1.16	1.62
2210	9-1/2" thick, R30	G		500	.016		.73	.78		1.51	2.07
2220	12" thick, R38	G		475	.017		1.05	.82		1.87	2.49
3000	Unfaced, 3-1/2" thick, R13	G		600	.013		.33	.65		.98	1.42
3010	6-1/4" thick, R19	G		500	.016		.39	.78		1.17	1.70
3020	9-1/2" thick, R30	G		450	.018		.60	.86		1.46	2.07
3030	12" thick, R38	G		425	.019		.75	.91		1.66	2.32
9000	Minimum labor/equipment charge			4	2	Job		97		97	158

07 21 16.20 Blanket Insulation for Walls

		Crew	Daily Output	Labor-Hours	Unit	Material	2016 Bare Costs Labor	Equipment	Total	Total Incl O&P
0010	**BLANKET INSULATION FOR WALLS**									
0020	Kraft faced fiberglass, 3-1/2" thick, R11, 15" wide	G	1 Carp	1350	.006	S.F.	.31	.29		.60	.81
0030	23" wide	G		1600	.005		.31	.24		.55	.74
0060	R13, 11" wide	G		1150	.007		.33	.34		.67	.91
0080	15" wide	G		1350	.006		.33	.29		.62	.83
0100	23" wide	G		1600	.005		.33	.24		.57	.76
0110	R15, 11" wide	G		1150	.007		.49	.34		.83	1.09
0120	15" wide	G		1350	.006		.49	.29		.78	1.01
0130	23" wide	G		1600	.005		.49	.24		.73	.94
0140	6" thick, R19, 11" wide	G		1150	.007		.44	.34		.78	1.03
0160	15" wide	G		1350	.006		.44	.29		.73	.95
0180	23" wide	G		1600	.005		.44	.24		.68	.88
0182	R21, 11" wide	G		1150	.007		.64	.34		.98	1.25
0184	15" wide	G		1350	.006		.64	.29		.93	1.17
0186	23" wide	G		1600	.005		.64	.24		.88	1.10
0188	9" thick, R30, 11" wide	G		985	.008		.73	.39		1.12	1.44
0200	15" wide	G		1150	.007		.73	.34		1.07	1.35
0220	23" wide	G		1350	.006		.73	.29		1.02	1.27
0230	12" thick, R38, 11" wide	G		985	.008		1.05	.39		1.44	1.80
0240	15" wide	G		1150	.007		1.05	.34		1.39	1.71
0260	23" wide	G		1350	.006		1.05	.29		1.34	1.63
0410	Foil faced fiberglass, 3-1/2" thick, R13, 11" wide	G		1150	.007		.47	.34		.81	1.07
0420	15" wide	G		1350	.006		.47	.29		.76	.99
0440	23" wide	G		1600	.005		.47	.24		.71	.92
0442	R15, 11" wide	G		1150	.007		.48	.34		.82	1.08
0444	15" wide	G		1350	.006		.48	.29		.77	1
0446	23" wide	G		1600	.005		.48	.24		.72	.93
0448	6" thick, R19, 11" wide	G		1150	.007		.62	.34		.96	1.23
0460	15" wide	G		1350	.006		.62	.29		.91	1.15
0480	23" wide	G		1600	.005		.62	.24		.86	1.08
0482	R21, 11" wide	G		1150	.007		.64	.34		.98	1.25
0484	15" wide	G		1350	.006		.64	.29		.93	1.17
0486	23" wide	G		1600	.005		.64	.24		.88	1.10
0488	9" thick, R30, 11" wide	G		985	.008		.94	.39		1.33	1.67
0500	15" wide	G		1150	.007		.94	.34		1.28	1.58
0550	23" wide	G		1350	.006		.94	.29		1.23	1.50
0560	12" thick, R38, 11" wide	G		985	.008		1.15	.39		1.54	1.91
0570	15" wide	G		1150	.007		1.15	.34		1.49	1.82
0580	23" wide	G		1350	.006		1.15	.29		1.44	1.74

For customer support on your Commercial Renovation Cost Data, call 877.742.4719.

183

07 21 Thermal Insulation

07 21 16 – Blanket Insulation

07 21 16.20 Blanket Insulation for Walls

		Crew	Daily Output	Labor-Hours	Unit	Material	2016 Bare Costs Labor	Equipment	Total	Total Incl O&P
0620	Unfaced fiberglass, 3-1/2" thick, R13, 11" wide	G 1 Carp	1150	.007	S.F.	.33	.34		.67	.91
0820	15" wide	G	1350	.006		.33	.29		.62	.83
0830	23" wide	G	1600	.005		.33	.24		.57	.76
0832	R15, 11" wide	G	1150	.007		.46	.34		.80	1.06
0836	23" wide	G	1600	.005		.46	.24		.70	.91
0838	6" thick, R19, 11" wide	G	1150	.007		.39	.34		.73	.98
0860	15" wide	G	1150	.007		.39	.34		.73	.98
0880	23" wide	G	1350	.006		.39	.29		.68	.90
0882	R21, 11" wide	G	1150	.007		.50	.34		.84	1.10
0886	15" wide	G	1350	.006		.50	.29		.79	1.02
0888	23" wide	G	1600	.005		.50	.24		.74	.95
0890	9" thick, R30, 11" wide	G	985	.008		.60	.39		.99	1.30
0900	15" wide	G	1150	.007		.60	.34		.94	1.21
0920	23" wide	G	1350	.006		.60	.29		.89	1.13
0930	12" thick, R38, 11" wide	G	985	.008		.75	.39		1.14	1.47
0940	15" wide	G	1000	.008		.75	.39		1.14	1.46
0960	23" wide	G	1150	.007		.75	.34		1.09	1.38
1300	Wall or ceiling insulation, mineral wool batts									
1320	3-1/2" thick, R15	G 1 Carp	1600	.005	S.F.	.67	.24		.91	1.14
1340	5-1/2" thick, R23	G	1600	.005		1.05	.24		1.29	1.56
1380	7-1/4" thick, R30	G	1350	.006		1.39	.29		1.68	2
1700	Non-rigid insul, recycled blue cotton fiber, unfaced batts, R13, 16" wide	G	1600	.005		1.01	.24		1.25	1.51
1710	R19, 16" wide	G	1600	.005		1.38	.24		1.62	1.92
1850	Friction fit wire insulation supports, 16" O.C.		960	.008	Ea.	.07	.40		.47	.74
9000	Minimum labor/equipment charge		4	2	Job		97		97	158

07 21 19 – Foamed In Place Insulation

07 21 19.10 Masonry Foamed In Place Insulation

		Crew	Daily Output	Labor-Hours	Unit	Material	2016 Bare Costs Labor	Equipment	Total	Total Incl O&P
0010	**MASONRY FOAMED IN PLACE INSULATION**									
0100	Amino-plast foam, injected into block core, 6" block	G G-2A	6000	.004	Ea.	.16	.15	.12	.43	.57
0110	8" block	G	5000	.005		.20	.18	.14	.52	.69
0120	10" block	G	4000	.006		.25	.22	.18	.65	.86
0130	12" block	G	3000	.008		.33	.30	.24	.87	1.14
0140	Injected into cavity wall	G	13000	.002	B.F.	.06	.07	.05	.18	.24
0150	Preparation, drill holes into mortar joint every 4 VLF, 5/8" dia	1 Clab	960	.008	Ea.		.32		.32	.52
0160	7/8" dia		680	.012			.45		.45	.73
0170	Patch drilled holes, 5/8" diameter		1800	.004		.04	.17		.21	.32
0180	7/8" diameter		1200	.007		.05	.25		.30	.47
9000	Minimum labor/equipment charge	G-2A	2	12	Job		445	355	800	1,175
9010	Minimum labor/equipment charge	1 Clab	4	2	"		76		76	124

07 21 23 – Loose-Fill Insulation

07 21 23.10 Poured Loose-Fill Insulation

		Crew	Daily Output	Labor-Hours	Unit	Material	2016 Bare Costs Labor	Equipment	Total	Total Incl O&P
0010	**POURED LOOSE-FILL INSULATION**									
0020	Cellulose fiber, R3.8 per inch	G 1 Carp	200	.040	C.F.	.69	1.94		2.63	3.93
0021	4" thick	G	1000	.008	S.F.	.16	.39		.55	.81
0022	6" thick	G	800	.010	"	.28	.48		.76	1.10
0080	Fiberglass wool, R4 per inch	G	200	.040	C.F.	.55	1.94		2.49	3.77
0081	4" thick	G	600	.013	S.F.	.19	.65		.84	1.27
0082	6" thick	G	400	.020	"	.26	.97		1.23	1.87
0100	Mineral wool, R3 per inch	G	200	.040	C.F.	.45	1.94		2.39	3.67
0101	4" thick	G	600	.013	S.F.	.15	.65		.80	1.22
0102	6" thick	G	400	.020	"	.23	.97		1.20	1.83
0300	Polystyrene, R4 per inch	G	200	.040	C.F.	1.60	1.94		3.54	4.93

07 21 Thermal Insulation

07 21 23 – Loose-Fill Insulation

07 21 23.10 Poured Loose-Fill Insulation

		Crew	Daily Output	Labor-Hours	Unit	Material	2016 Bare Costs Labor	Equipment	Total	Total Incl O&P
0301	4" thick	G 1 Carp	600	.013	S.F.	.53	.65		1.18	1.64
0302	6" thick	G	400	.020	"	.80	.97		1.77	2.46
0400	Perlite, R2.78 per inch	G	200	.040	C.F.	5.25	1.94		7.19	8.95
0401	4" thick	G	1000	.008	S.F.	1.75	.39		2.14	2.55
0402	6" thick	G	800	.010	"	2.63	.48		3.11	3.68
9000	Minimum labor/equipment charge		4	2	Job		97		97	158

07 21 23.20 Masonry Loose-Fill Insulation

		Crew	Daily Output	Labor-Hours	Unit	Material	2016 Bare Costs Labor	Equipment	Total	Total Incl O&P
0010	**MASONRY LOOSE-FILL INSULATION**, vermiculite or perlite									
0100	In cores of concrete block, 4" thick wall, .115 C.F./S.F.	G D-1	4800	.003	S.F.	.60	.14		.74	.89
0200	6" thick wall, .175 C.F./S.F.	G	3000	.005		.92	.22		1.14	1.38
0300	8" thick wall, .258 C.F./S.F.	G	2400	.007		1.35	.28		1.63	1.95
0400	10" thick wall, .340 C.F./S.F.	G	1850	.009		1.79	.36		2.15	2.55
0500	12" thick wall, .422 C.F./S.F.	G	1200	.013		2.22	.56		2.78	3.35
0600	Poured cavity wall, vermiculite or perlite, water repellent	G	250	.064	C.F.	5.25	2.70		7.95	10.20
0700	Foamed in place, urethane in 2-5/8" cavity	G G-2A	1035	.023	S.F.	1.38	.86	.69	2.93	3.78
0800	For each 1" added thickness, add	G "	2372	.010	"	.53	.37	.30	1.20	1.56

07 21 26 – Blown Insulation

07 21 26.10 Blown Insulation

		Crew	Daily Output	Labor-Hours	Unit	Material	2016 Bare Costs Labor	Equipment	Total	Total Incl O&P
0010	**BLOWN INSULATION** Ceilings, with open access									
0020	Cellulose, 3-1/2" thick, R13	G G-4	5000	.005	S.F.	.24	.19	.08	.51	.65
0030	5-3/16" thick, R19	G	3800	.006		.35	.24	.11	.70	.91
0050	6-1/2" thick, R22	G	3000	.008		.45	.31	.14	.90	1.15
0100	8-11/16" thick, R30	G	2600	.009		.61	.36	.16	1.13	1.42
0120	10-7/8" thick, R38	G	1800	.013		.78	.51	.23	1.52	1.94
1000	Fiberglass, 5.5" thick, R11	G	3800	.006		.19	.24	.11	.54	.72
1050	6" thick, R12	G	3000	.008		.26	.31	.14	.71	.94
1100	8.8" thick, R19	G	2200	.011		.33	.42	.19	.94	1.26
1200	10" thick, R22	G	1800	.013		.38	.51	.23	1.12	1.51
1300	11.5" thick, R26	G	1500	.016		.46	.62	.27	1.35	1.81
1350	13" thick, R30	G	1400	.017		.53	.66	.29	1.48	1.98
1450	16" thick, R38	G	1145	.021		.68	.81	.36	1.85	2.45
1500	20" thick, R49	G	920	.026		.89	1.01	.45	2.35	3.11
9000	Minimum labor/equipment charge		4	6	Job		231	103	334	495

07 21 27 – Reflective Insulation

07 21 27.10 Reflective Insulation Options

		Crew	Daily Output	Labor-Hours	Unit	Material	2016 Bare Costs Labor	Equipment	Total	Total Incl O&P
0010	**REFLECTIVE INSULATION OPTIONS**									
0020	Aluminum foil on reinforced scrim	G 1 Carp	19	.421	C.S.F.	14.90	20.50		35.40	50
0100	Reinforced with woven polyolefin	G	19	.421		22	20.50		42.50	57.50
0500	With single bubble air space, R8.8	G	15	.533		26	26		52	70.50
0600	With double bubble air space, R9.8	G	15	.533		31	26		57	76
9000	Minimum labor/equipment charge		4	2	Job		97		97	158

07 21 29 – Sprayed Insulation

07 21 29.10 Sprayed-On Insulation

		Crew	Daily Output	Labor-Hours	Unit	Material	2016 Bare Costs Labor	Equipment	Total	Total Incl O&P
0010	**SPRAYED-ON INSULATION**									
0020	Fibrous/cementitious, finished wall, 1" thick, R3.7	G G-2	2050	.012	S.F.	.34	.47	.07	.88	1.20
0100	Attic, 5.2" thick, R19	G	1550	.015	"	.42	.62	.09	1.13	1.56
0200	Fiberglass, R4 per inch, vertical	G	1600	.015	B.F.	.18	.60	.08	.86	1.26
0210	Horizontal	G	1200	.020	"	.18	.81	.11	1.10	1.62
0300	Closed cell, spray polyurethane foam, 2 pounds per cubic foot density									
0310	1" thick	G G-2A	6000	.004	S.F.	.53	.15	.12	.80	.97
0320	2" thick	G	3000	.008		1.05	.30	.24	1.59	1.94

For customer support on your Commercial Renovation Cost Data, call 877.742.4719.

185

07 21 Thermal Insulation

07 21 29 – Sprayed Insulation

07 21 29.10 Sprayed-On Insulation		Crew	Daily Output	Labor-Hours	Unit	Material	2016 Bare Costs Labor	Equipment	Total	Total Incl O&P
0330	3" thick	G	G-2A	2000	.012	S.F.	1.58	.44	.36	2.38	2.90
0335	3-1/2" thick	G		1715	.014		1.84	.52	.42	2.78	3.38
0340	4" thick	G		1500	.016		2.10	.59	.48	3.17	3.86
0350	5" thick	G		1200	.020		2.63	.74	.59	3.96	4.83
0355	5-1/2" thick	G		1090	.022		2.89	.81	.65	4.35	5.30
0360	6" thick	G		1000	.024		3.15	.89	.71	4.75	5.80
9000	Minimum labor/equipment charge		G-2	2	12	Job		485	67.50	552.50	850

07 22 Roof and Deck Insulation

07 22 16 – Roof Board Insulation

07 22 16.10 Roof Deck Insulation

			Crew	Daily Output	Labor-Hours	Unit	Material	2016 Bare Costs Labor	Equipment	Total	Total Incl O&P
0010	**ROOF DECK INSULATION**, fastening excluded										
0016	Asphaltic cover board, fiberglass lined, 1/8" thick		1 Rofc	1400	.006	S.F.	.47	.24		.71	.95
0018	1/4" thick			1400	.006		.94	.24		1.18	1.46
0020	Fiberboard low density, 1/2" thick R1.39	G		1300	.006		.33	.26		.59	.82
0030	1" thick R2.78	G		1040	.008		.55	.32		.87	1.19
0080	1-1/2" thick R4.17	G		1040	.008		.84	.32		1.16	1.50
0100	2" thick R5.56	G		1040	.008		1.10	.32		1.42	1.79
0110	Fiberboard high density, 1/2" thick R1.3	G		1300	.006		.30	.26		.56	.79
0120	1" thick R2.5	G		1040	.008		.56	.32		.88	1.20
0130	1-1/2" thick R3.8	G		1040	.008		.86	.32		1.18	1.53
0200	Fiberglass, 3/4" thick R2.78	G		1300	.006		.61	.26		.87	1.13
0400	15/16" thick R3.70	G		1300	.006		.81	.26		1.07	1.35
0460	1-1/16" thick R4.17	G		1300	.006		1.01	.26		1.27	1.57
0600	1-5/16" thick R5.26	G		1300	.006		1.37	.26		1.63	1.97
0650	2-1/16" thick R8.33	G		1040	.008		1.45	.32		1.77	2.18
0700	2-7/16" thick R10	G		1040	.008		1.68	.32		2	2.43
0800	Gypsum cover board, fiberglass mat facer, 1/4" thick			1400	.006		.47	.24		.71	.95
0810	1/2" thick			1300	.006		.57	.26		.83	1.09
0820	5/8" thick			1200	.007		.61	.28		.89	1.17
0830	Primed fiberglass mat facer, 1/4" thick			1400	.006		.48	.24		.72	.96
0840	1/2" thick			1300	.006		.57	.26		.83	1.09
0850	5/8" thick			1200	.007		.60	.28		.88	1.16
1650	Perlite, 1/2" thick R1.32	G		1365	.006		.26	.24		.50	.73
1655	3/4" thick R2.08	G		1040	.008		.37	.32		.69	.99
1660	1" thick R2.78	G		1040	.008		.53	.32		.85	1.16
1670	1-1/2" thick R4.17	G		1040	.008		.78	.32		1.10	1.44
1680	2" thick R5.56	G		910	.009		1.06	.37		1.43	1.83
1685	2-1/2" thick R6.67	G		910	.009		1.40	.37		1.77	2.20
1690	Tapered for drainage	G		1040	.008	B.F.	1.05	.32		1.37	1.74
1700	Polyisocyanurate, 2#/C.F. density, 3/4" thick	G		1950	.004	S.F.	.46	.17		.63	.82
1705	1" thick	G		1820	.004		.47	.18		.65	.85
1715	1-1/2" thick	G		1625	.005		.64	.21		.85	1.07
1725	2" thick	G		1430	.006		.82	.23		1.05	1.32
1735	2-1/2" thick	G		1365	.006		1.10	.24		1.34	1.65
1745	3" thick	G		1300	.006		1.22	.26		1.48	1.80
1755	3-1/2" thick	G		1300	.006		1.90	.26		2.16	2.55
1765	Tapered for drainage	G		1820	.004	B.F.	.59	.18		.77	.98
1900	Extruded Polystyrene										
1910	15 PSI compressive strength, 1" thick, R5	G	1 Rofc	1950	.004	S.F.	.60	.17		.77	.97
1920	2" thick, R10	G		1625	.005		.78	.21		.99	1.23

07 22 Roof and Deck Insulation

07 22 16 – Roof Board Insulation

07 22 16.10 Roof Deck Insulation		Crew	Daily Output	Labor-Hours	Unit	Material	2016 Bare Costs Labor	Equipment	Total	Total Incl O&P
1930	3" thick, R15	G	1 Rofc	1300	.006	S.F.	1.56	.26		1.82	2.18
1932	4" thick, R20	G		1300	.006	↓	2.10	.26		2.36	2.77
1934	Tapered for drainage	G		1950	.004	B.F.	.62	.17		.79	.99
1940	25 PSI compressive strength, 1" thick, R5	G		1950	.004	S.F.	.69	.17		.86	1.07
1942	2" thick, R10	G		1625	.005		1.31	.21		1.52	1.81
1944	3" thick, R15	G		1300	.006		2	.26		2.26	2.66
1946	4" thick, R20	G		1300	.006	↓	2.76	.26		3.02	3.50
1948	Tapered for drainage	G		1950	.004	B.F.	.61	.17		.78	.98
1950	40 PSI compressive strength, 1" thick, R5	G		1950	.004	S.F.	.53	.17		.70	.89
1952	2" thick, R10	G		1625	.005		1.01	.21		1.22	1.48
1954	3" thick, R15	G		1300	.006		1.46	.26		1.72	2.06
1956	4" thick, R20	G		1300	.006	↓	1.91	.26		2.17	2.56
1958	Tapered for drainage	G		1820	.004	B.F.	.76	.18		.94	1.17
1960	60 PSI compressive strength, 1" thick, R5	G		1885	.004	S.F.	.74	.18		.92	1.13
1962	2" thick, R10	G		1560	.005		1.41	.21		1.62	1.94
1964	3" thick, R15	G		1270	.006		2.29	.26		2.55	2.99
1966	4" thick, R20	G		1235	.006	↓	2.85	.27		3.12	3.62
1968	Tapered for drainage	G		1820	.004	B.F.	.96	.18		1.14	1.39
2010	Expanded polystyrene, 1#/C.F. density, 3/4" thick, R2.89	G		1950	.004	S.F.	.20	.17		.37	.52
2020	1" thick, R3.85	G		1950	.004		.26	.17		.43	.60
2100	2" thick, R7.69	G		1625	.005		.52	.21		.73	.94
2110	3" thick, R11.49	G		1625	.005		.78	.21		.99	1.23
2120	4" thick, R15.38	G		1625	.005		1.04	.21		1.25	1.51
2130	5" thick, R19.23	G		1495	.005		1.30	.22		1.52	1.83
2140	6" thick, R23.26	G		1495	.005	↓	1.56	.22		1.78	2.12
2150	Tapered for drainage	G	↓	1950	.004	B.F.	.53	.17		.70	.89
2400	Composites with 2" EPS										
2410	1" fiberboard	G	1 Rofc	1325	.006	S.F.	1.40	.25		1.65	1.99
2420	7/16" oriented strand board	G		1040	.008		1.13	.32		1.45	1.82
2430	1/2" plywood	G		1040	.008		1.39	.32		1.71	2.11
2440	1" perlite	G	↓	1040	.008	↓	1.14	.32		1.46	1.83
2450	Composites with 1-1/2" polyisocyanurate										
2460	1" fiberboard	G	1 Rofc	1040	.008	S.F.	1.19	.32		1.51	1.89
2470	1" perlite	G		1105	.007		1.06	.30		1.36	1.71
2480	7/16" oriented strand board	G		1040	.008	↓	.93	.32		1.25	1.60
3000	Fastening alternatives, coated screws, 2" long			3744	.002	Ea.	.06	.09		.15	.23
3010	4" long			3120	.003		.11	.11		.22	.31
3020	6" long			2675	.003		.18	.12		.30	.43
3030	8" long			2340	.003		.27	.14		.41	.56
3040	10" long			1872	.004		.45	.18		.63	.82
3050	Pre-drill and drive wedge spike, 2-1/2"			1248	.006		.39	.27		.66	.91
3060	3-1/2"			1101	.007		.52	.30		.82	1.12
3070	4-1/2"			936	.009		.64	.36		1	1.34
3075	3" galvanized deck plates		↓	7488	.001	↓	.08	.04		.12	.17
3080	Spot mop asphalt		G-1	295	.190	Sq.	5.55	7.40	1.96	14.91	21.50
3090	Full mop asphalt		"	192	.292		11.10	11.40	3.01	25.51	36
3110	Low-rise polyurethane adhesive, from 5 gallon kit, 12" OC beads		1 Rofc	45	.178		33	7.40		40.40	50
3120	6" OC beads			32	.250		66	10.45		76.45	91.50
3130	4" OC beads			30	.267	↓	99	11.10		110.10	129
9000	Minimum labor/equipment charge		↓	3.25	2.462	Job		103		103	185

For customer support on your Commercial Renovation Cost Data, call 877.742.4719.

187

07 24 Exterior Insulation and Finish Systems

07 24 13 – Polymer-Based Exterior Insulation and Finish System

07 24 13.10 Exterior Insulation and Finish Systems		Crew	Daily Output	Labor-Hours	Unit	Material	2016 Bare Costs Labor	Equipment	Total	Total Incl O&P
0010	**EXTERIOR INSULATION AND FINISH SYSTEMS**									
0095	Field applied, 1" EPS insulation	G	J-1	390	.103	S.F.	1.93	4.32	.36	6.61	9.40
0100	With 1/2" cement board sheathing	G		268	.149		2.71	6.30	.52	9.53	13.60
0105	2" EPS insulation	G		390	.103		2.19	4.32	.36	6.87	9.70
0110	With 1/2" cement board sheathing	G		268	.149		2.97	6.30	.52	9.79	13.90
0115	3" EPS insulation	G		390	.103		2.45	4.32	.36	7.13	10
0120	With 1/2" cement board sheathing	G		268	.149		3.23	6.30	.52	10.05	14.20
0125	4" EPS insulation	G		390	.103		2.71	4.32	.36	7.39	10.30
0130	With 1/2" cement board sheathing	G		268	.149		4.28	6.30	.52	11.10	15.35
0140	Premium finish add			1265	.032		.35	1.33	.11	1.79	2.64
0150	Heavy duty reinforcement add			914	.044		.85	1.84	.15	2.84	4.05
0160	2.5#/S.Y. metal lath substrate add		1 Lath	75	.107	S.Y.	2.82	5.05		7.87	11.05
0170	3.4#/S.Y. metal lath substrate add		"	75	.107	"	4.18	5.05		9.23	12.55
0180	Color or texture change		J-1	1265	.032	S.F.	.80	1.33	.11	2.24	3.13
0190	With substrate leveling base coat		1 Plas	530	.015		.80	.68		1.48	1.96
0210	With substrate sealing base coat		1 Pord	1224	.007		.12	.26		.38	.55
0370	V groove shape in panel face					L.F.	.65			.65	.72
0380	U groove shape in panel face					"	.85			.85	.94
0433	Crack repair, acrylic rubber, fluid applied, 20 mils thick		1 Plas	350	.023	S.F.	1.50	1.03		2.53	3.29
0437	50 mils thick, reinforced		"	200	.040	"	2.70	1.80		4.50	5.85
0440	For higher than one story, add							25%			

07 25 Weather Barriers

07 25 10 – Weather Barriers or Wraps

07 25 10.10 Weather Barriers

		Crew	Daily Output	Labor-Hours	Unit	Material	2016 Bare Costs Labor	Equipment	Total	Total Incl O&P
0010	**WEATHER BARRIERS**									
0400	Asphalt felt paper, 15#	1 Carp	37	.216	Sq.	5.30	10.50		15.80	23
0401	Per square foot	"	3700	.002	S.F.	.05	.10		.15	.23
0450	Housewrap, exterior, spun bonded polypropylene									
0470	Small roll	1 Carp	3800	.002	S.F.	.15	.10		.25	.34
0480	Large roll	"	4000	.002	"	.14	.10		.24	.31
2100	Asphalt felt roof deck vapor barrier, class 1 metal decks	1 Rofc	37	.216	Sq.	21.50	9		30.50	40.50
2200	For all other decks	"	37	.216		16.40	9		25.40	34.50
2800	Asphalt felt, 50% recycled content, 15 lb., 4 sq. per roll	1 Carp	36	.222		5.50	10.75		16.25	23.50
2810	30 lb., 2 sq. per roll	"	36	.222		10.85	10.75		21.60	29.50
3000	Building wrap, spunbonded polyethylene	2 Carp	8000	.002	S.F.	.16	.10		.26	.34
9960	Minimum labor/equipment charge	1 Carp	2	4	Job		194		194	315

07 26 Vapor Retarders

07 26 10 – Above-Grade Vapor Retarders

07 26 10.10 Vapor Retarders

			Crew	Daily Output	Labor-Hours	Unit	Material	2016 Bare Costs Labor	Equipment	Total	Total Incl O&P
0010	**VAPOR RETARDERS**										
0020	Aluminum and kraft laminated, foil 1 side	G	1 Carp	37	.216	Sq.	11.95	10.50		22.45	30.50
0100	Foil 2 sides	G		37	.216		14	10.50		24.50	32.50
0600	Polyethylene vapor barrier, standard, 2 mil	G		37	.216		1.56	10.50		12.06	18.80
0700	4 mil	G		37	.216		2.55	10.50		13.05	19.90
0900	6 mil	G		37	.216		3.70	10.50		14.20	21
1200	10 mil	G		37	.216		7.75	10.50		18.25	25.50
1300	Clear reinforced, fire retardant, 8 mil	G		37	.216		10.60	10.50		21.10	29

07 26 Vapor Retarders

07 26 10 – Above-Grade Vapor Retarders

07 26 10.10 Vapor Retarders

	07 26 10.10 Vapor Retarders	Crew	Daily Output	Labor-Hours	Unit	Material	2016 Bare Costs Labor	Equipment	Total	Total Incl O&P
1350	Cross laminated type, 3 mil **G**	1 Carp	37	.216	Sq.	7.70	10.50		18.20	25.50
1400	4 mil **G**		37	.216		8.10	10.50		18.60	26
1800	Reinf. waterproof, 2 mil polyethylene backing, 1 side		37	.216		6.10	10.50		16.60	24
1900	2 sides		37	.216		8	10.50		18.50	26
2400	Waterproofed kraft with sisal or fiberglass fibers		37	.216		13	10.50		23.50	31.50
9950	Minimum labor/equipment charge		4	2	Job		97		97	158

07 27 Air Barriers

07 27 13 – Modified Bituminous Sheet Air Barriers

07 27 13.10 Modified Bituminous Sheet Air Barrier

		Crew	Daily Output	Labor-Hours	Unit	Material	2016 Bare Costs Labor	Equipment	Total	Total Incl O&P
0010	**MODIFIED BITUMINOUS SHEET AIR BARRIER**									
0100	SBS Modified sheet laminated to polyethylene sheet, 40 mils, 4" wide	1 Carp	1200	.007	L.F.	.37	.32		.69	.94
0120	6" wide		1100	.007		.51	.35		.86	1.14
0140	9" wide		1000	.008		.71	.39		1.10	1.41
0160	12" wide		900	.009		.91	.43		1.34	1.70
0180	18" wide	2 Carp	1700	.009	S.F.	.85	.46		1.31	1.67
0200	36" wide	"	1800	.009		.83	.43		1.26	1.61
0220	Adhesive for above	1 Carp	1400	.006		.35	.28		.63	.83

07 27 26 – Fluid-Applied Membrane Air Barriers

07 27 26.10 Fluid Applied Membrane Air Barrier

		Crew	Daily Output	Labor-Hours	Unit	Material	2016 Bare Costs Labor	Equipment	Total	Total Incl O&P
0010	**FLUID APPLIED MEMBRANE AIR BARRIER**									
0100	Spray applied vapor barrier, 25 S.F./gallon	1 Pord	1375	.006	S.F.	.01	.23		.24	.40

07 31 Shingles and Shakes

07 31 13 – Asphalt Shingles

07 31 13.10 Asphalt Roof Shingles

		Crew	Daily Output	Labor-Hours	Unit	Material	2016 Bare Costs Labor	Equipment	Total	Total Incl O&P
0010	**ASPHALT ROOF SHINGLES**									
0100	Standard strip shingles									
0150	Inorganic, class A, 25 year	1 Rofc	5.50	1.455	Sq.	81	60.50		141.50	198
0155	Pneumatic nailed		7	1.143		81	47.50		128.50	175
0200	30 year		5	1.600		96	66.50		162.50	226
0205	Pneumatic nailed		6.25	1.280		96	53.50		149.50	203
0250	Standard laminated multi-layered shingles									
0300	Class A, 240-260 lb./square	1 Rofc	4.50	1.778	Sq.	109	74		183	254
0305	Pneumatic nailed		5.63	1.422		109	59.50		168.50	227
0350	Class A, 250-270 lb./square		4	2		109	83.50		192.50	271
0355	Pneumatic nailed		5	1.600		109	66.50		175.50	240
0400	Premium, laminated multi-layered shingles									
0450	Class A, 260-300 lb./square	1 Rofc	3.50	2.286	Sq.	158	95.50		253.50	345
0455	Pneumatic nailed		4.37	1.831		158	76.50		234.50	310
0500	Class A, 300-385 lb./square		3	2.667		241	111		352	465
0505	Pneumatic nailed		3.75	2.133		241	89		330	425
0800	#15 felt underlayment		64	.125		5.30	5.20		10.50	15.25
0825	#30 felt underlayment		58	.138		10.30	5.75		16.05	22
0850	Self adhering polyethylene and rubberized asphalt underlayment		22	.364		75	15.15		90.15	110
0900	Ridge shingles		330	.024	L.F.	2.20	1.01		3.21	4.24
0905	Pneumatic nailed		412.50	.019	"	2.20	.81		3.01	3.88
1000	For steep roofs (7 to 12 pitch or greater), add						50%			
9000	Minimum labor/equipment charge	1 Rofc	3	2.667	Job		111		111	201

For customer support on your Commercial Renovation Cost Data, call 877.742.4719.

189

07 31 Shingles and Shakes

07 31 16 – Metal Shingles

07 31 16.10 Aluminum Shingles		Crew	Daily Output	Labor-Hours	Unit	Material	2016 Bare Costs Labor	Equipment	Total	Total Incl O&P
0010	**ALUMINUM SHINGLES**									
0020	Mill finish, .019 thick	1 Carp	5	1.600	Sq.	221	77.50		298.50	370
0100	.020" thick	"	5	1.600		247	77.50		324.50	400
0300	For colors, add					21			21	23
0600	Ridge cap, .024" thick	1 Carp	170	.047	L.F.	3.68	2.28		5.96	7.75
0700	End wall flashing, .024" thick		170	.047		2.15	2.28		4.43	6.10
0900	Valley section, .024" thick		170	.047		3.65	2.28		5.93	7.75
1000	Starter strip, .024" thick		400	.020		1.73	.97		2.70	3.48
1200	Side wall flashing, .024" thick		170	.047		2.10	2.28		4.38	6.05
1500	Gable flashing, .024" thick		400	.020		1.70	.97		2.67	3.45
9000	Minimum labor/equipment charge		3	2.667	Job		129		129	211

07 31 16.20 Steel Shingles

0010	**STEEL SHINGLES**									
0012	Galvanized, 26 ga.	1 Rots	2.20	3.636	Sq.	345	152		497	655
0200	24 ga.	"	2.20	3.636		335	152		487	645
0300	For colored galvanized shingles, add					56			56	61.50
9000	Minimum labor/equipment charge	1 Rots	3	2.667	Job		111		111	201

07 31 26 – Slate Shingles

07 31 26.10 Slate Roof Shingles

0010	**SLATE ROOF SHINGLES**									
0100	Buckingham Virginia black, 3/16" - 1/4" thick [G]	1 Rots	1.75	4.571	Sq.	540	191		731	940
0200	1/4" thick [G]		1.75	4.571		540	191		731	940
0900	Pennsylvania black, Bangor, #1 clear [G]		1.75	4.571		490	191		681	885
1200	Vermont, unfading, green, mottled green [G]		1.75	4.571		495	191		686	890
1300	Semi-weathering green & gray [G]		1.75	4.571		355	191		546	735
1400	Purple [G]		1.75	4.571		435	191		626	820
1500	Black or gray [G]		1.75	4.571		475	191		666	870
2500	Slate roof repair, extensive replacement		1	8		600	335		935	1,250
2600	Repair individual pieces, scattered		19	.421	Ea.	7	17.60		24.60	39
2700	Ridge shingles, slate		200	.040	L.F.	10	1.67		11.67	14
9000	Minimum labor/equipment charge		3	2.667	Job		111		111	201

07 31 29 – Wood Shingles and Shakes

07 31 29.13 Wood Shingles

0010	**WOOD SHINGLES** R061110-30									
0012	16" No. 1 red cedar shingles, 5" exposure, on roof	1 Carp	2.50	3.200	Sq.	281	155		436	565
0015	Pneumatic nailed		3.25	2.462		281	119		400	505
0200	7-1/2" exposure, on walls		2.05	3.902		187	189		376	515
0205	Pneumatic nailed		2.67	2.996		187	145		332	445
0300	18" No. 1 red cedar perfections, 5-1/2" exposure, on roof		2.75	2.909		248	141		389	505
0305	Pneumatic nailed		3.57	2.241		248	109		357	450
0500	7-1/2" exposure, on walls		2.25	3.556		182	172		354	480
0505	Pneumatic nailed		2.92	2.740		182	133		315	420
0600	Resquared and rebutted, 5-1/2" exposure, on roof		3	2.667		282	129		411	520
0605	Pneumatic nailed		3.90	2.051		282	99.50		381.50	470
0900	7-1/2" exposure, on walls		2.45	3.265		207	158		365	485
0905	Pneumatic nailed		3.18	2.516		207	122		329	425
1000	Add to above for fire retardant shingles					56			56	61.50
1060	Preformed ridge shingles	1 Carp	400	.020	L.F.	5	.97		5.97	7.10
2000	White cedar shingles, 16" long, extras, 5" exposure, on roof		2.40	3.333	Sq.	193	162		355	475
2005	Pneumatic nailed		3.12	2.564		193	124		317	415
2050	5" exposure on walls		2	4		193	194		387	525

07 31 Shingles and Shakes

07 31 29 – Wood Shingles and Shakes

07 31 29.13 Wood Shingles

		Crew	Daily Output	Labor-Hours	Unit	Material	2016 Bare Costs Labor	2016 Bare Costs Equipment	Total	Total Incl O&P
2055	Pneumatic nailed	1 Carp	2.60	3.077	Sq.	193	149		342	455
2100	7-1/2" exposure, on walls		2	4		138	194		332	465
2105	Pneumatic nailed		2.60	3.077		138	149		287	395
2150	"B" grade, 5" exposure on walls		2	4		165	194		359	495
2155	Pneumatic nailed		2.60	3.077		165	149		314	425
2300	For 15# organic felt underlayment on roof, 1 layer, add		64	.125		5.30	6.05		11.35	15.75
2400	2 layers, add		32	.250		10.60	12.10		22.70	31.50
2600	For steep roofs (7/12 pitch or greater), add to above						50%			
2700	Panelized systems, No.1 cedar shingles on 5/16" CDX plywood									
2800	On walls, 8' strips, 7" or 14" exposure	2 Carp	700	.023	S.F.	6	1.11		7.11	8.40
3500	On roofs, 8' strips, 7" or 14" exposure	1 Carp	3	2.667	Sq.	600	129		729	870
3505	Pneumatic nailed		4	2	"	600	97		697	820
9000	Minimum labor/equipment charge		3	2.667	Job		129		129	211

07 31 29.16 Wood Shakes

		Crew	Daily Output	Labor-Hours	Unit	Material	2016 Bare Costs Labor	2016 Bare Costs Equipment	Total	Total Incl O&P
0010	**WOOD SHAKES**									
1100	Hand-split red cedar shakes, 1/2" thick x 24" long, 10" exp. on roof	1 Carp	2.50	3.200	Sq.	287	155		442	570
1105	Pneumatic nailed		3.25	2.462		287	119		406	510
1110	3/4" thick x 24" long, 10" exp. on roof		2.25	3.556		287	172		459	595
1115	Pneumatic nailed		2.92	2.740		287	133		420	530
1200	1/2" thick, 18" long, 8-1/2" exp. on roof		2	4		179	194		373	510
1205	Pneumatic nailed		2.60	3.077		179	149		328	440
1210	3/4" thick x 18" long, 8 1/2" exp. on roof		1.80	4.444		179	215		394	545
1215	Pneumatic nailed		2.34	3.419		179	166		345	470
1255	10" exp. on walls		2	4		173	194		367	505
1260	10" exposure on walls, pneumatic nailed		2.60	3.077		173	149		322	435
1700	Add to above for fire retardant shakes, 24" long					56			56	61.50
1800	18" long					56			56	61.50
1810	Ridge shakes	1 Carp	350	.023	L.F.	5.50	1.11		6.61	7.85

07 32 Roof Tiles

07 32 13 – Clay Roof Tiles

07 32 13.10 Clay Tiles

		Crew	Daily Output	Labor-Hours	Unit	Material	2016 Bare Costs Labor	2016 Bare Costs Equipment	Total	Total Incl O&P
0010	**CLAY TILES**, including accessories									
0300	Flat shingle, interlocking, 15", 166 pcs/sq, fireflashed blend	3 Rots	6	4	Sq.	465	167		632	810
0500	Terra cotta red		6	4		510	167		677	860
0600	Roman pan and top, 18", 102 pcs/sq, fireflashed blend		5.50	4.364		500	182		682	880
0640	Terra cotta red	1 Rots	2.40	3.333		555	139		694	860
1100	Barrel mission tile, 18", 166 pcs/sq, fireflashed blend	3 Rots	5.50	4.364		410	182		592	780
1140	Terra cotta red		5.50	4.364		410	182		592	780
1700	Scalloped edge flat shingle, 14", 145 pcs/sq, fireflashed blend		6	4		1,125	167		1,292	1,550
1800	Terra cotta red		6	4		1,025	167		1,192	1,425
3010	#15 felt underlayment	1 Rofc	64	.125		5.30	5.20		10.50	15.25
3020	#30 felt underlayment		58	.138		10.30	5.75		16.05	22
3040	Polyethylene and rubberized asph. underlayment		22	.364		75	15.15		90.15	110
9010	Minimum labor/equipment charge	1 Rots	2	4	Job		167		167	300

For customer support on your Commercial Renovation Cost Data, call 877.742.4719.

191

07 32 Roof Tiles

07 32 16 – Concrete Roof Tiles

07 32 16.10 Concrete Tiles	Crew	Daily Output	Labor-Hours	Unit	Material	2016 Bare Costs Labor	Equipment	Total	Total Incl O&P
0010 **CONCRETE TILES**									
0020 Corrugated, 13" x 16-1/2", 90 per sq., 950 lb. per sq.									
0050 Earthtone colors, nailed to wood deck	1 Rots	1.35	5.926	Sq.	103	247		350	560
0150 Blues		1.35	5.926		103	247		350	560
0200 Greens		1.35	5.926		103	247		350	560
0250 Premium colors		1.35	5.926		103	247		350	560
0500 Shakes, 13" x 16-1/2", 90 per sq., 950 lb. per sq.									
0600 All colors, nailed to wood deck	1 Rots	1.50	5.333	Sq.	124	223		347	535
1500 Accessory pieces, ridge & hip, 10" x 16-1/2", 8 lb. each	"	120	.067	Ea.	3.70	2.78		6.48	9.05
1700 Rake, 6-1/2" x 16-3/4", 9 lb. each					3.70			3.70	4.07
1800 Mansard hip, 10" x 16-1/2", 9.2 lb. each					3.70			3.70	4.07
1900 Hip starter, 10" x 16-1/2", 10.5 lb. each					10.40			10.40	11.45
2000 3 or 4 way apex, 10" each side, 11.5 lb. each					12			12	13.20
9000 Minimum labor/equipment charge	1 Rots	3	2.667	Job		111		111	201

07 32 19 – Metal Roof Tiles

07 32 19.10 Metal Roof Tiles

	Crew	Daily Output	Labor-Hours	Unit	Material	2016 Bare Costs Labor	Equipment	Total	Total Incl O&P
0010 **METAL ROOF TILES**									
0020 Accessories included, .032" thick aluminum, mission tile	1 Carp	2.50	3.200	Sq.	825	155		980	1,175
0200 Spanish tiles		3	2.667	"	540	129		669	805
9000 Minimum labor/equipment charge		3	2.667	Job		129		129	211

07 33 Natural Roof Coverings

07 33 63 – Vegetated Roofing

07 33 63.10 Green Roof Systems

		Crew	Daily Output	Labor-Hours	Unit	Material	2016 Bare Costs Labor	Equipment	Total	Total Incl O&P
0010 **GREEN ROOF SYSTEMS**										
0020 Soil mixture for green roof 30% sand, 55% gravel, 15% soil										
0100 Hoist and spread soil mixture 4 inch depth up to five stories tall roof	G	B-13B	4000	.014	S.F.	.25	.58	.28	1.11	1.52
0150 6 inch depth	G		2667	.021		.37	.87	.42	1.66	2.28
0200 8 inch depth	G		2000	.028		.50	1.16	.56	2.22	3.04
0250 10 inch depth	G		1600	.035		.62	1.45	.70	2.77	3.79
0300 12 inch depth	G		1335	.042		.74	1.74	.84	3.32	4.56
0310 Alt. man-made soil mix, hoist & spread, 4" deep up to 5 stories tall roof	G		4000	.014		1.90	.58	.28	2.76	3.34
0350 Mobilization 55 ton crane to site	G	1 Eqhv	3.60	2.222	Ea.		116		116	184
0355 Hoisting cost to five stories per day (Avg. 28 picks per day)	G	B-13B	1	56	Day		2,325	1,125	3,450	4,975
0360 Mobilization or demobilization, 100 ton crane to site driver & escort	G	A-3E	2.50	6.400	Ea.		305	62.50	367.50	555
0365 Hoisting cost six to ten stories per day (Avg. 21 picks per day)	G	B-13C	1	56	Day		2,325	1,675	4,000	5,575
0370 Hoist and spread soil mixture 4 inch depth six to ten stories tall roof	G		4000	.014	S.F.	.25	.58	.42	1.25	1.67
0375 6 inch depth	G		2667	.021		.37	.87	.63	1.87	2.51
0380 8 inch depth	G		2000	.028		.50	1.16	.83	2.49	3.34
0385 10 inch depth	G		1600	.035		.62	1.45	1.04	3.11	4.17
0390 12 inch depth	G		1335	.042		.74	1.74	1.25	3.73	5
0400 Green roof edging treated lumber 4" x 4", no hoisting included	G	2 Carp	400	.040	L.F.	1.35	1.94		3.29	4.65
0410 4" x 6"	G		400	.040		2.02	1.94		3.96	5.40
0420 4" x 8"	G		360	.044		3.87	2.15		6.02	7.75
0430 4" x 6" double stacked	G		300	.053		4.04	2.58		6.62	8.65
0500 Green roof edging redwood lumber 4" x 4", no hoisting included	G		400	.040		6.45	1.94		8.39	10.25
0510 4" x 6"	G		400	.040		12.60	1.94		14.54	17
0520 4" x 8"	G		360	.044		23.50	2.15		25.65	29.50
0530 4" x 6" double stacked	G		300	.053		25	2.58		27.58	31.50
0550 Components, not including membrane or insulation:										

07 33 Natural Roof Coverings

07 33 63 – Vegetated Roofing

07 33 63.10 Green Roof Systems		Crew	Daily Output	Labor-Hours	Unit	Material	2016 Bare Costs Labor	Equipment	Total	Total Incl O&P
0560	Fluid applied rubber membrane, reinforced, 215 mil thick	G	G-5	350	.114	S.F.	.30	4.34	.55	5.19	8.80
0570	Root barrier	G	2 Rofc	775	.021		.68	.86		1.54	2.30
0580	Moisture retention barrier and reservoir	G	"	900	.018		2.59	.74		3.33	4.19
0600	Planting sedum, light soil, potted, 2-1/4" diameter, two per S.F.	G	1 Clab	420	.019		5.65	.72		6.37	7.40
0610	one per S.F.	G	"	840	.010		2.82	.36		3.18	3.69
0630	Planting sedum mat per S.F. including shipping (4000 S.F. min)	G	4 Clab	4000	.008		6.75	.30		7.05	7.95
0640	Installation sedum mat system (no soil required) per S.F. (4000 S.F. min)	G	"	4000	.008		9.40	.30		9.70	10.85
0645	Note: pricing of sedum mats shipped in full truck loads (4000-5000 S.F.)										

07 41 Roof Panels

07 41 13 – Metal Roof Panels

07 41 13.10 Aluminum Roof Panels

		Crew	Daily Output	Labor-Hours	Unit	Material	2016 Bare Costs Labor	Equipment	Total	Total Incl O&P
0010	**ALUMINUM ROOF PANELS**									
0020	Corrugated or ribbed, .0155" thick, natural	G-3	1200	.027	S.F.	1	1.27		2.27	3.13
0300	Painted		1200	.027		1.45	1.27		2.72	3.63
0400	Corrugated, .018" thick, on steel frame, natural finish		1200	.027		1.25	1.27		2.52	3.41
0600	Painted		1200	.027		1.55	1.27		2.82	3.74
0700	Corrugated, on steel frame, natural, .024" thick		1200	.027		1.80	1.27		3.07	4.01
0800	Painted		1200	.027		2.20	1.27		3.47	4.45
0900	.032" thick, natural		1200	.027		2.57	1.27		3.84	4.86
1200	Painted		1200	.027		3.18	1.27		4.45	5.55
9000	Minimum labor/equipment charge	1 Rofc	3	2.667	Job		111		111	201

07 41 13.20 Steel Roofing Panels

		Crew	Daily Output	Labor-Hours	Unit	Material	2016 Bare Costs Labor	Equipment	Total	Total Incl O&P
0010	**STEEL ROOFING PANELS**									
0012	Corrugated or ribbed, on steel framing, 30 ga. galv	G-3	1100	.029	S.F.	1.64	1.38		3.02	4.02
0100	28 ga.		1050	.030		1.69	1.45		3.14	4.18
0300	26 ga.		1000	.032		1.95	1.52		3.47	4.59
0400	24 ga.		950	.034		2.90	1.60		4.50	5.75
0600	Colored, 28 ga.		1050	.030		1.72	1.45		3.17	4.21
0700	26 ga.		1000	.032		2.04	1.52		3.56	4.68
0710	Flat profile, 1-3/4" standing seams, 10" wide, standard finish, 26 ga.		1000	.032		3.75	1.52		5.27	6.55
0715	24 ga.		950	.034		4.35	1.60		5.95	7.35
0720	22 ga.		900	.036		5.40	1.69		7.09	8.65
0725	Zinc aluminum alloy finish, 26 ga.		1000	.032		3	1.52		4.52	5.75
0730	24 ga.		950	.034		3.55	1.60		5.15	6.45
0735	22 ga.		900	.036		4.05	1.69		5.74	7.15
0740	12" wide, standard finish, 26 ga.		1000	.032		3.75	1.52		5.27	6.55
0745	24 ga.		950	.034		4.90	1.60		6.50	7.95
0750	Zinc aluminum alloy finish, 26 ga.		1000	.032		4.26	1.52		5.78	7.15
0755	24 ga.		950	.034		3.50	1.60		5.10	6.40
0840	Flat profile, 1" x 3/8" batten, 12" wide, standard finish, 26 ga.		1000	.032		3.30	1.52		4.82	6.05
0845	24 ga.		950	.034		3.90	1.60		5.50	6.85
0850	22 ga.		900	.036		4.65	1.69		6.34	7.80
0855	Zinc aluminum alloy finish, 26 ga.		1000	.032		3.20	1.52		4.72	5.95
0860	24 ga.		950	.034		3.60	1.60		5.20	6.50
0865	22 ga.		900	.036		4.15	1.69		5.84	7.30
0870	16-1/2" wide, standard finish, 24 ga.		950	.034		3.85	1.60		5.45	6.80
0875	22 ga.		900	.036		4.30	1.69		5.99	7.45
0880	Zinc aluminum alloy finish, 24 ga.		950	.034		3.35	1.60		4.95	6.25
0885	22 ga.		900	.036		3.75	1.69		5.44	6.85
0890	Flat profile, 2" x 2" batten, 12" wide, standard finish, 26 ga.		1000	.032		3.85	1.52		5.37	6.70

For customer support on your Commercial Renovation Cost Data, call 877.742.4719.

193

07 41 Roof Panels

07 41 13 – Metal Roof Panels

07 41 13.20 Steel Roofing Panels		Crew	Daily Output	Labor-Hours	Unit	Material	2016 Bare Costs Labor	Equipment	Total	Total Incl O&P
0895	24 ga.	G-3	950	.034	S.F.	4.55	1.60		6.15	7.55
0900	22 ga.		900	.036		5.55	1.69		7.24	8.80
0905	Zinc aluminum alloy finish, 26 ga.		1000	.032		3.55	1.52		5.07	6.35
0910	24 ga.		950	.034		4.05	1.60		5.65	7
0915	22 ga.		900	.036		4.70	1.69		6.39	7.85
0920	16-1/2" wide, standard finish, 24 ga.		950	.034		4.20	1.60		5.80	7.20
0925	22 ga.		900	.036		4.90	1.69		6.59	8.10
0930	Zinc aluminum alloy finish, 24 ga.		950	.034		3.80	1.60		5.40	6.75
0935	22 ga.	↓	900	.036	↓	4.30	1.69		5.99	7.45
9000	Minimum labor/equipment charge	1 Rofc	2	4	Job		167		167	300

07 41 33 – Plastic Roof Panels

07 41 33.10 Fiberglass Panels

		Crew	Daily Output	Labor-Hours	Unit	Material	2016 Bare Costs Labor	Equipment	Total	Total Incl O&P
0010	**FIBERGLASS PANELS**									
0012	Corrugated panels, roofing, 8 oz. per S.F.	G-3	1000	.032	S.F.	2.15	1.52		3.67	4.81
0100	12 oz. per S.F.		1000	.032		4.20	1.52		5.72	7.05
0300	Corrugated siding, 6 oz. per S.F.		880	.036		1.90	1.73		3.63	4.86
0400	8 oz. per S.F.		880	.036		2.15	1.73		3.88	5.15
0500	Fire retardant		880	.036		3.80	1.73		5.53	6.95
0600	12 oz. siding, textured		880	.036		3.75	1.73		5.48	6.90
0700	Fire retardant		880	.036		4.35	1.73		6.08	7.55
0900	Flat panels, 6 oz. per S.F., clear or colors		880	.036		2.40	1.73		4.13	5.40
1100	Fire retardant, class A		880	.036		3.45	1.73		5.18	6.55
1300	8 oz. per S.F., clear or colors	↓	880	.036	↓	2.50	1.73		4.23	5.50
9000	Minimum labor/equipment charge	1 Rofc	2	4	Job		167		167	300

07 42 Wall Panels

07 42 13 – Metal Wall Panels

07 42 13.10 Mansard Panels

		Crew	Daily Output	Labor-Hours	Unit	Material	2016 Bare Costs Labor	Equipment	Total	Total Incl O&P
0010	**MANSARD PANELS**									
0600	Aluminum, stock units, straight surfaces	1 Shee	115	.070	S.F.	4.30	3.98	•	8.28	11.05
0700	Concave or convex surfaces, add		75	.107	"	2.20	6.10		8.30	12.05
0800	For framing, to 5' high, add		115	.070	L.F.	3.75	3.98		7.73	10.45
0900	Soffits, to 1' wide		125	.064	S.F.	2.50	3.66		6.16	8.55
9000	Minimum labor/equipment charge	↓	2.50	3.200	Job		183		183	289

07 42 13.20 Aluminum Siding Panels

		Crew	Daily Output	Labor-Hours	Unit	Material	2016 Bare Costs Labor	Equipment	Total	Total Incl O&P
0010	**ALUMINUM SIDING PANELS**									
0012	Corrugated, on steel framing, .019" thick, natural finish	G-3	775	.041	S.F.	1.63	1.96		3.59	4.93
0100	Painted		775	.041		1.77	1.96		3.73	5.10
0400	Farm type, .021" thick on steel frame, natural		775	.041		1.65	1.96		3.61	4.96
0600	Painted		775	.041		1.76	1.96		3.72	5.10
0700	Industrial type, corrugated, on steel, .024" thick, mill		775	.041		2.30	1.96		4.26	5.65
0900	Painted		775	.041		2.47	1.96		4.43	5.85
1000	.032" thick, mill		775	.041		2.50	1.96		4.46	5.90
1200	Painted		775	.041		3	1.96		4.96	6.45
1300	V-Beam, on steel frame, .032" thick, mill		775	.041		2.83	1.96		4.79	6.25
1500	Painted		775	.041		3.15	1.96		5.11	6.60
1600	.040" thick, mill		775	.041		3.40	1.96		5.36	6.90
1800	Painted	↓	775	.041		3.95	1.96		5.91	7.50
3800	Horizontal, colored clapboard, 8" wide, plain	2 Carp	515	.031		2.47	1.51		3.98	5.20
3810	Insulated	↓	515	.031	↓	2.85	1.51		4.36	5.60

07 42 Wall Panels

07 42 13 – Metal Wall Panels

07 42 13.20 Aluminum Siding Panels		Crew	Daily Output	Labor-Hours	Unit	Material	2016 Bare Costs Labor	Equipment	Total	Total Incl O&P
3830	8" embossed, painted	2 Carp	515	.031	S.F.	2.85	1.51		4.36	5.60
3840	Insulated		515	.031		3.05	1.51		4.56	5.80
3860	12" painted, smooth		600	.027		2.65	1.29		3.94	5.05
3870	Insulated		600	.027		2.85	1.29		4.14	5.25
3890	12" embossed, painted		600	.027		2.84	1.29		4.13	5.25
3900	Insulated		515	.031		2.72	1.51		4.23	5.45
4000	Vertical board & batten, colored, non-insulated		515	.031		2.05	1.51		3.56	4.72
4200	For simulated wood design, add					.15			.15	.17
4300	Corners for above, outside	2 Carp	515	.031	V.L.F.	3.60	1.51		5.11	6.40
4500	Inside corners	"	515	.031	"	1.72	1.51		3.23	4.35
9000	Minimum labor/equipment charge	1 Carp	3	2.667	Job		129		129	211

07 42 13.30 Steel Siding

		Crew	Daily Output	Labor-Hours	Unit	Material	2016 Bare Costs Labor	Equipment	Total	Total Incl O&P
0010	**STEEL SIDING**									
0020	Beveled, vinyl coated, 8" wide	1 Carp	265	.030	S.F.	1.85	1.46		3.31	4.43
0050	10" wide	"	275	.029		1.90	1.41		3.31	4.39
0080	Galv, corrugated or ribbed, on steel frame, 30 ga.	G-3	800	.040		1.20	1.90		3.10	4.37
0100	28 ga.		795	.040		1.30	1.92		3.22	4.50
0300	26 ga.		790	.041		1.80	1.93		3.73	5.05
0400	24 ga.		785	.041		2	1.94		3.94	5.30
0600	22 ga.		770	.042		2.25	1.98		4.23	5.65
0700	Colored, corrugated/ribbed, on steel frame, 10 yr. finish, 28 ga.		800	.040		1.95	1.90		3.85	5.20
0900	26 ga.		795	.040		2.12	1.92		4.04	5.40
1000	24 ga.		790	.041		2.38	1.93		4.31	5.70
1020	20 ga.		785	.041		3.02	1.94		4.96	6.40
9000	Minimum labor/equipment charge	1 Carp	3	2.667	Job		129		129	211

07 46 Siding

07 46 23 – Wood Siding

07 46 23.10 Wood Board Siding

		Crew	Daily Output	Labor-Hours	Unit	Material	2016 Bare Costs Labor	Equipment	Total	Total Incl O&P
0010	**WOOD BOARD SIDING**									
3200	Wood, cedar bevel, A grade, 1/2" x 6"	1 Carp	295	.027	S.F.	4.25	1.31		5.56	6.85
3300	1/2" x 8"		330	.024		6.50	1.17		7.67	9.05
3500	3/4" x 10", clear grade		375	.021		6.75	1.03		7.78	9.15
3600	"B" grade		375	.021		3.79	1.03		4.82	5.85
3800	Cedar, rough sawn, 1" x 4", A grade, natural		220	.036		6.55	1.76		8.31	10.15
3900	Stained		220	.036		6.70	1.76		8.46	10.25
4100	1" x 12", board & batten, #3 & Btr., natural		420	.019		4.48	.92		5.40	6.45
4200	Stained		420	.019		4.81	.92		5.73	6.80
4400	1" x 8" channel siding, #3 & Btr., natural		330	.024		4.57	1.17		5.74	6.90
4500	Stained		330	.024		4.92	1.17		6.09	7.30
4700	Redwood, clear, beveled, vertical grain, 1/2" x 4"		220	.036		4.15	1.76		5.91	7.45
4750	1/2" x 6"		295	.027		4.79	1.31		6.10	7.40
4800	1/2" x 8"		330	.024		5.20	1.17		6.37	7.60
5000	3/4" x 10"		375	.021		4.90	1.03		5.93	7.10
5200	Channel siding, 1" x 10", B grade		375	.021		4.50	1.03		5.53	6.65
5250	Redwood, T&G boards, B grade, 1" x 4"		220	.036		3.25	1.76		5.01	6.45
5270	1" x 8"		330	.024		5.20	1.17		6.37	7.60
5400	White pine, rough sawn, 1" x 8", natural		330	.024		2.19	1.17		3.36	4.33
5500	Stained		330	.024		2.19	1.17		3.36	4.33
9000	Minimum labor/equipment charge		2	4	Job		194		194	315

For customer support on your Commercial Renovation Cost Data, call 877.742.4719.

195

07 46 29 – Plywood Siding

07 46 29.10 Plywood Siding Options	Crew	Daily Output	Labor-Hours	Unit	Material	2016 Bare Costs Labor	Equipment	Total	Total Incl O&P
0010 **PLYWOOD SIDING OPTIONS**									
0900 Plywood, medium density overlaid, 3/8" thick	2 Carp	750	.021	S.F.	1.29	1.03		2.32	3.11
1000 1/2" thick		700	.023		1.50	1.11		2.61	3.46
1100 3/4" thick		650	.025		1.99	1.19		3.18	4.14
1600 Texture 1-11, cedar, 5/8" thick, natural		675	.024		2.70	1.15		3.85	4.85
1700 Factory stained		675	.024		2.80	1.15		3.95	4.96
1900 Texture 1-11, fir, 5/8" thick, natural		675	.024		1.46	1.15		2.61	3.49
2000 Factory stained		675	.024		1.80	1.15		2.95	3.86
2050 Texture 1-11, S.Y.P., 5/8" thick, natural		675	.024		1.37	1.15		2.52	3.39
2100 Factory stained		675	.024		1.44	1.15		2.59	3.46
2200 Rough sawn cedar, 3/8" thick, natural		675	.024		1.25	1.15		2.40	3.26
2300 Factory stained		675	.024		1.50	1.15		2.65	3.53
2500 Rough sawn fir, 3/8" thick, natural		675	.024		.90	1.15		2.05	2.87
2600 Factory stained		675	.024		1.04	1.15		2.19	3.02
2800 Redwood, textured siding, 5/8" thick		675	.024		1.92	1.15		3.07	3.99
3000 Polyvinyl chloride coated, 3/8" thick	↓	750	.021	↓	1.11	1.03		2.14	2.91
9000 Minimum labor/equipment charge	1 Carp	2	4	Job		194		194	315

07 46 33 – Plastic Siding

07 46 33.10 Vinyl Siding

	Crew	Daily Output	Labor-Hours	Unit	Material	2016 Bare Costs Labor	Equipment	Total	Total Incl O&P
0010 **VINYL SIDING**									
3995 Clapboard profile, woodgrain texture, .048 thick, double 4	2 Carp	495	.032	S.F.	1.06	1.57		2.63	3.73
4000 Double 5		550	.029		1.06	1.41		2.47	3.47
4005 Single 8		495	.032		1.35	1.57		2.92	4.05
4010 Single 10		550	.029		1.62	1.41		3.03	4.09
4015 .044 thick, double 4		495	.032		1.04	1.57		2.61	3.71
4020 Double 5		550	.029		1.04	1.41		2.45	3.45
4025 .042 thick, double 4		495	.032		1.04	1.57		2.61	3.71
4030 Double 5		550	.029		1.04	1.41		2.45	3.45
4035 Cross sawn texture, .040 thick, double 4		495	.032		.72	1.57		2.29	3.36
4040 Double 5		550	.029		.72	1.41		2.13	3.10
4045 Smooth texture, .042 thick, double 4		495	.032		.79	1.57		2.36	3.43
4050 Double 5		550	.029		.79	1.41		2.20	3.17
4055 Single 8		495	.032		.79	1.57		2.36	3.43
4060 Cedar texture, .044 thick, double 4		495	.032		1.12	1.57		2.69	3.80
4065 Double 6		600	.027		1.12	1.29		2.41	3.35
4070 Dutch lap profile, woodgrain texture, .048 thick, double 5		550	.029		1.06	1.41		2.47	3.47
4075 .044 thick, double 4.5		525	.030		1.06	1.48		2.54	3.58
4080 .042 thick, double 4.5		525	.030		.90	1.48		2.38	3.41
4085 .040 thick, double 4.5		525	.030		.72	1.48		2.20	3.21
4100 Shake profile, 10" wide		400	.040		3.61	1.94		5.55	7.15
4105 Vertical pattern, .046 thick, double 5		550	.029		1.56	1.41		2.97	4.02
4110 .044 thick, triple 3		550	.029		1.74	1.41		3.15	4.22
4115 .040 thick, triple 4		550	.029		1.61	1.41		3.02	4.08
4120 .040 thick, triple 2.66		550	.029		1.74	1.41		3.15	4.22
4125 Insulation, fan folded extruded polystyrene, 1/4"		2000	.008		.29	.39		.68	.95
4130 3/8"		2000	.008	↓	.32	.39		.71	.99
4135 Accessories, J channel, 5/8" pocket		700	.023	L.F.	.50	1.11		1.61	2.36
4140 3/4" pocket		695	.023		.53	1.12		1.65	2.40
4145 1-1/4" pocket		680	.024		.84	1.14		1.98	2.78
4150 Flexible, 3/4" pocket		600	.027		2.33	1.29		3.62	4.67
4155 Under sill finish trim		500	.032		.54	1.55		2.09	3.12
4160 Vinyl starter strip		700	.023	↓	.65	1.11		1.76	2.52

07 46 Siding

07 46 33 – Plastic Siding

07 46 33.10 Vinyl Siding

		Crew	Daily Output	Labor-Hours	Unit	Material	2016 Bare Costs Labor	Equipment	Total	Total Incl O&P
4165	Aluminum starter strip	2 Carp	700	.023	L.F.	.29	1.11		1.40	2.13
4170	Window casing, 2-1/2" wide, 3/4" pocket		510	.031		1.68	1.52		3.20	4.33
4175	Outside corner, woodgrain finish, 4" face, 3/4" pocket		700	.023		2.08	1.11		3.19	4.10
4180	5/8" pocket		700	.023		2.08	1.11		3.19	4.10
4185	Smooth finish, 4" face, 3/4" pocket		700	.023		2.08	1.11		3.19	4.10
4190	7/8" pocket		690	.023		1.96	1.12		3.08	4
4195	1-1/4" pocket		700	.023		1.38	1.11		2.49	3.33
4200	Soffit and fascia, 1' overhang, solid		120	.133		4.65	6.45		11.10	15.65
4205	Vented		120	.133		4.65	6.45		11.10	15.65
4207	18" overhang, solid		110	.145		5.45	7.05		12.50	17.50
4208	Vented		110	.145		5.45	7.05		12.50	17.50
4210	2' overhang, solid		100	.160		6.20	7.75		13.95	19.50
4215	Vented		100	.160		6.20	7.75		13.95	19.50
4217	3' overhang, solid		100	.160		7.75	7.75		15.50	21
4218	Vented		100	.160		7.75	7.75		15.50	21
4220	Colors for siding and soffits, add				S.F.	.14			.14	.15
4225	Colors for accessories and trim, add				L.F.	.30			.30	.33
9000	Minimum labor/equipment charge	1 Carp	3	2.667	Job		129		129	211

07 46 33.20 Polypropylene Siding

		Crew	Daily Output	Labor-Hours	Unit	Material	2016 Bare Costs Labor	Equipment	Total	Total Incl O&P
0010	**POLYPROPYLENE SIDING**									
4090	Shingle profile, random grooves, double 7	2 Carp	400	.040	S.F.	3.11	1.94		5.05	6.60
4092	Cornerpost for above	1 Carp	365	.022	L.F.	12.65	1.06		13.71	15.65
4095	Triple 5	2 Carp	400	.040	S.F.	3.11	1.94		5.05	6.60
4097	Cornerpost for above	1 Carp	365	.022	L.F.	11.80	1.06		12.86	14.75
5000	Staggered butt, double 7"	2 Carp	400	.040	S.F.	3.61	1.94		5.55	7.15
5002	Cornerpost for above	1 Carp	365	.022	L.F.	12.75	1.06		13.81	15.80
5010	Half round, double 6-1/4"	2 Carp	360	.044	S.F.	3.61	2.15		5.76	7.50
5020	Shake profile, staggered butt, double 9"	"	510	.031	"	3.61	1.52		5.13	6.45
5022	Cornerpost for above	1 Carp	365	.022	L.F.	9.60	1.06		10.66	12.30
5030	Straight butt, double 7"	2 Carp	400	.040	S.F.	3.61	1.94		5.55	7.15
5032	Cornerpost for above	1 Carp	365	.022	L.F.	12.60	1.06		13.66	15.65
6000	Accessories, J channel, 5/8" pocket	2 Carp	700	.023		.50	1.11		1.61	2.36
6010	3/4" pocket		695	.023		.53	1.12		1.65	2.40
6020	1-1/4" pocket		680	.024		.84	1.14		1.98	2.78
6030	Aluminum starter strip		700	.023		.29	1.11		1.40	2.13

07 46 46 – Fiber Cement Siding

07 46 46.10 Fiber Cement Siding

		Crew	Daily Output	Labor-Hours	Unit	Material	2016 Bare Costs Labor	Equipment	Total	Total Incl O&P
0010	**FIBER CEMENT SIDING**									
0020	Lap siding, 5/16" thick, 6" wide, 4-3/4" exposure, smooth texture	2 Carp	415	.039	S.F.	1.29	1.87		3.16	4.46
0025	Woodgrain texture		415	.039		1.29	1.87		3.16	4.46
0030	7-1/2" wide, 6-1/4" exposure, smooth texture		425	.038		1.59	1.82		3.41	4.73
0035	Woodgrain texture		425	.038		1.59	1.82		3.41	4.73
0040	8" wide, 6-3/4" exposure, smooth texture		425	.038		1.19	1.82		3.01	4.29
0045	Rough sawn texture		425	.038		1.19	1.82		3.01	4.29
0050	9-1/2" wide, 8-1/4" exposure, smooth texture		440	.036		1.26	1.76		3.02	4.27
0055	Woodgrain texture		440	.036		1.26	1.76		3.02	4.27
0060	12" wide, 10-3/8" exposure, smooth texture		455	.035		2.01	1.70		3.71	4.99
0065	Woodgrain texture		455	.035		2.01	1.70		3.71	4.99
0070	Panel siding, 5/16" thick, smooth texture		750	.021		1.31	1.03		2.34	3.13
0075	Stucco texture		750	.021		1.31	1.03		2.34	3.13
0080	Grooved woodgrain texture		750	.021		1.31	1.03		2.34	3.13
0085	V - grooved woodgrain texture		750	.021		1.31	1.03		2.34	3.13

07 46 Siding

07 46 46 – Fiber Cement Siding

07 46 46.10 Fiber Cement Siding	Crew	Daily Output	Labor-Hours	Unit	Material	2016 Bare Costs Labor	Equipment	Total	Total Incl O&P
0088	Shingle siding, 48" x 15-1/4" panels, 7" exposure	2 Carp	700	.023	S.F.	4.12	1.11		5.23	6.35
0090	Wood starter strip	▼	400	.040	L.F.	.41	1.94		2.35	3.62

07 46 73 – Soffit

07 46 73.10 Soffit Options

		Crew	Daily Output	Labor-Hours	Unit	Material	Labor	Equipment	Total	Total Incl O&P
0010	**SOFFIT OPTIONS**									
0012	Aluminum, residential, .020" thick	1 Carp	210	.038	S.F.	2	1.85		3.85	5.20
0100	Baked enamel on steel, 16 or 18 ga.		105	.076		5.90	3.69		9.59	12.55
0300	Polyvinyl chloride, white, solid		230	.035		2.10	1.69		3.79	5.05
0400	Perforated	▼	230	.035		2.10	1.69		3.79	5.05
0500	For colors, add				▼	.15			.15	.17
9000	Minimum labor/equipment charge	1 Carp	3	2.667	Job		129		129	211

07 51 Built-Up Bituminous Roofing

07 51 13 – Built-Up Asphalt Roofing

07 51 13.10 Built-Up Roofing Components

		Crew	Daily Output	Labor-Hours	Unit	Material	Labor	Equipment	Total	Total Incl O&P
0010	**BUILT-UP ROOFING COMPONENTS**									
0012	Asphalt saturated felt, #30, 2 square per roll	1 Rofc	58	.138	Sq.	10.30	5.75		16.05	22
0200	#15, 4 sq. per roll, plain or perforated, not mopped		58	.138		5.30	5.75		11.05	16.25
0300	Roll roofing, smooth, #65		15	.533		10.10	22		32.10	51
0500	#90		12	.667		44	28		72	98
0520	Mineralized		12	.667		35	28		63	88.50
0540	D.C. (double coverage), 19" selvage edge	▼	10	.800	▼	54.50	33.50		88	120
0580	Adhesive (lap cement)				Gal.	8.85			8.85	9.75
0800	Steep, flat or dead level asphalt, 10 ton lots, packaged				Ton	925			925	1,025
9000	Minimum labor/equipment charge	1 Rofc	4	2	Job		83.50		83.50	151

07 51 13.13 Cold-Applied Built-Up Asphalt Roofing

		Crew	Daily Output	Labor-Hours	Unit	Material	Labor	Equipment	Total	Total Incl O&P
0010	**COLD-APPLIED BUILT-UP ASPHALT ROOFING**									
0020	3 ply system, installation only (components listed below)	G-5	50	.800	Sq.		30.50	3.86	34.36	59.50
0100	Spunbond poly. fabric, 1.35 oz./S.Y., 36" W, 10.8 sq./roll				Ea.	138			138	152
0500	Base & finish coat, 3 gal./sq., 5 gal./can				Gal.	7.85			7.85	8.65
0600	Coating, ceramic granules, 1/2 sq./bag				Ea.	23			23	25.50
0700	Aluminum, 2 gal./sq.				Gal.	13.50			13.50	14.85
0800	Emulsion, fibered or non-fibered, 4 gal./sq.				"	6.50			6.50	7.15

07 51 13.20 Built-Up Roofing Systems

		Crew	Daily Output	Labor-Hours	Unit	Material	Labor	Equipment	Total	Total Incl O&P
0010	**BUILT-UP ROOFING SYSTEMS**									
0120	Asphalt flood coat with gravel/slag surfacing, not including									
0140	Insulation, flashing or wood nailers									
0200	Asphalt base sheet, 3 plies #15 asphalt felt, mopped	G-1	22	2.545	Sq.	99.50	99.50	26.50	225.50	315
0350	On nailable decks		21	2.667		103	104	27.50	234.50	330
0500	4 plies #15 asphalt felt, mopped		20	2.800		137	109	29	275	380
0550	On nailable decks		19	2.947		121	115	30.50	266.50	375
0700	Coated glass base sheet, 2 plies glass (type IV), mopped		22	2.545		110	99.50	26.50	236	330
0850	3 plies glass, mopped		20	2.800		134	109	29	272	375
0950	On nailable decks		19	2.947		126	115	30.50	271.50	380
1100	4 plies glass fiber felt (type IV), mopped		20	2.800		166	109	29	304	410
1150	On nailable decks		19	2.947		150	115	30.50	295.50	405
1200	Coated & saturated base sheet, 3 plies #15 asph. felt, mopped		20	2.800		111	109	29	249	350
1250	On nailable decks		19	2.947		104	115	30.50	249.50	355
1300	4 plies #15 asphalt felt, mopped	▼	22	2.545	▼	130	99.50	26.50	256	350
2000	Asphalt flood coat, smooth surface									

07 51 Built-Up Bituminous Roofing

07 51 13 – Built-Up Asphalt Roofing

07 51 13.20 Built-Up Roofing Systems

		Crew	Daily Output	Labor-Hours	Unit	Material	2016 Bare Costs Labor	Equipment	Total	Total Incl O&P
2200	Asphalt base sheet & 3 plies #15 asphalt felt, mopped	G-1	24	2.333	Sq.	106	91	24	221	305
2400	On nailable decks		23	2.435		98	95	25	218	305
2600	4 plies #15 asphalt felt, mopped		24	2.333		124	91	24	239	325
2700	On nailable decks	↓	23	2.435	↓	116	95	25	236	325
2900	Coated glass fiber base sheet, mopped, and 2 plies of									
2910	glass fiber felt (type IV)	G-1	25	2.240	Sq.	106	87.50	23	216.50	300
3100	On nailable decks		24	2.333		100	91	24	215	300
3200	3 plies, mopped		23	2.435		129	95	25	249	340
3300	On nailable decks		22	2.545		121	99.50	26.50	247	340
3800	4 plies glass fiber felt (type IV), mopped		23	2.435		153	95	25	273	365
3900	On nailable decks		22	2.545		145	99.50	26.50	271	370
4000	Coated & saturated base sheet, 3 plies #15 asph. felt, mopped		24	2.333		107	91	24	222	310
4200	On nailable decks		23	2.435		99	95	25	219	310
4300	4 plies #15 organic felt, mopped	↓	22	2.545	↓	125	99.50	26.50	251	345
4500	Coal tar pitch with gravel/slag surfacing									
4600	4 plies #15 tarred felt, mopped	G-1	21	2.667	Sq.	200	104	27.50	331.50	440
4800	3 plies glass fiber felt (type IV), mopped	"	19	2.947	"	165	115	30.50	310.50	425
5000	Coated glass fiber base sheet, and 2 plies of									
5010	glass fiber felt, (type IV), mopped	G-1	19	2.947	Sq.	169	115	30.50	314.50	430
5300	On nailable decks		18	3.111		148	121	32	301	415
5600	4 plies glass fiber felt (type IV), mopped		21	2.667		229	104	27.50	360.50	470
5800	On nailable decks	↓	20	2.800	↓	207	109	29	345	455

07 51 13.30 Cants

		Crew	Daily Output	Labor-Hours	Unit	Material	2016 Bare Costs Labor	Equipment	Total	Total Incl O&P
0010	**CANTS**									
0012	Lumber, treated, 4" x 4" cut diagonally	1 Rofc	325	.025	L.F.	1.80	1.03		2.83	3.83
0300	Mineral or fiber, trapezoidal, 1" x 4" x 48"		325	.025		.30	1.03		1.33	2.18
0400	1-1/2" x 5-5/8" x 48"		325	.025	↓	.48	1.03		1.51	2.38
9000	Minimum labor/equipment charge	↓	4	2	Job		83.50		83.50	151

07 51 13.40 Felts

		Crew	Daily Output	Labor-Hours	Unit	Material	2016 Bare Costs Labor	Equipment	Total	Total Incl O&P
0010	**FELTS**									
0012	Glass fibered roofing felt, #15, not mopped	1 Rofc	58	.138	Sq.	10.85	5.75		16.60	22.50
0300	Base sheet, #80, channel vented		58	.138		42.50	5.75		48.25	57.50
0400	#70, coated		58	.138		17.80	5.75		23.55	30
0500	Cap, #87, mineral surfaced		58	.138		81	5.75		86.75	99.50
0600	Flashing membrane, #65		16	.500		10.10	21		31.10	48.50
0800	Coal tar fibered, #15, no mopping		58	.138		18	5.75		23.75	30
0900	Asphalt felt, #15, 4 sq. per roll, no mopping		58	.138		5.30	5.75		11.05	16.25
1100	#30, 2 sq. per roll		58	.138		10.30	5.75		16.05	22
1200	Double coated, #33		58	.138		11	5.75		16.75	22.50
1400	#40, base sheet		58	.138		10.95	5.75		16.70	22.50
1450	Coated and saturated		58	.138		12	5.75		17.75	23.50
1500	Tarred felt, organic, #15, 4 sq. rolls		58	.138		13.40	5.75		19.15	25
1550	#30, 2 sq. roll	↓	58	.138		26.50	5.75		32.25	39.50
1700	Add for mopping above felts, per ply, asphalt, 24 lb. per sq.	G-1	192	.292		11.10	11.40	3.01	25.51	36
1800	Coal tar mopping, 30 lb. per sq.		186	.301		18.85	11.75	3.11	33.71	45
1900	Flood coat, with asphalt, 60 lb. per sq.		60	.933		28	36.50	9.65	74.15	107
2000	With coal tar, 75 lb. per sq.	↓	56	1	↓	47	39	10.35	96.35	134
9000	Minimum labor/equipment charge	1 Rofc	4	2	Job		83.50		83.50	151

07 51 13.50 Walkways for Built-Up Roofs

		Crew	Daily Output	Labor-Hours	Unit	Material	2016 Bare Costs Labor	Equipment	Total	Total Incl O&P
0010	**WALKWAYS FOR BUILT-UP ROOFS**									
0020	Asphalt impregnated, 3' x 6' x 1/2" thick	1 Rofc	400	.020	S.F.	1.79	.83		2.62	3.47
0100	3' x 3' x 3/4" thick	↓	400	.020	"	5.15	.83		5.98	7.15

For customer support on your Commercial Renovation Cost Data, call 877.742.4719.

199

07 51 Built-Up Bituminous Roofing

07 51 13 – Built-Up Asphalt Roofing

07 51 13.50 Walkways for Built-Up Roofs		Crew	Daily Output	Labor-Hours	Unit	Material	2016 Bare Costs Labor	2016 Bare Costs Equipment	Total	Total Incl O&P
0600	100% recycled rubber, 3' x 4' x 3/8"	G	1 Rofc	400	.020	L.F.	6.45	.83		7.28	8.60
0610	3' x 4' x 1/2"	G		400	.020		6.80	.83		7.63	9
0620	3' x 4' x 3/4"	G		400	.020		8.40	.83		9.23	10.75
9000	Minimum labor/equipment charge			2.75	2.909	Job		121		121	219

07 52 Modified Bituminous Membrane Roofing

07 52 13 – Atactic-Polypropylene-Modified Bituminous Membrane Roofing

07 52 13.10 APP Modified Bituminous Membrane

		Crew	Daily Output	Labor-Hours	Unit	Material	Labor	Equipment	Total	Total Incl O&P
0010	**APP MODIFIED BITUMINOUS MEMBRANE** R075213-30									
0020	Base sheet, #15 glass fiber felt, nailed to deck	1 Rofc	58	.138	Sq.	11.85	5.75		17.60	23.50
0030	Spot mopped to deck	G-1	295	.190		16.40	7.40	1.96	25.76	33.50
0040	Fully mopped to deck	"	192	.292		22	11.40	3.01	36.41	48
0050	#15 organic felt, nailed to deck	1 Rofc	58	.138		6.30	5.75		12.05	17.30
0060	Spot mopped to deck	G-1	295	.190		10.85	7.40	1.96	20.21	27.50
0070	Fully mopped to deck	"	192	.292		16.40	11.40	3.01	30.81	42
2100	APP mod., smooth surf. cap sheet, poly. reinf., torched, 160 mils	G-5	2100	.019	S.F.	.74	.72	.09	1.55	2.21
2150	170 mils		2100	.019		.75	.72	.09	1.56	2.23
2200	Granule surface cap sheet, poly. reinf., torched, 180 mils		2000	.020		.93	.76	.10	1.79	2.50
2250	Smooth surface flashing, torched, 160 mils		1260	.032		.74	1.21	.15	2.10	3.15
2300	170 mils		1260	.032		.75	1.21	.15	2.11	3.17
2350	Granule surface flashing, torched, 180 mils		1260	.032		.93	1.21	.15	2.29	3.36
2400	Fibrated aluminum coating	1 Rofc	3800	.002		.08	.09		.17	.25
2450	Seam heat welding	"	205	.039	L.F.	.08	1.63		1.71	3.03

07 52 16 – Styrene-Butadiene-Styrene Modified Bituminous Membrane Roofing

07 52 16.10 SBS Modified Bituminous Membrane

		Crew	Daily Output	Labor-Hours	Unit	Material	Labor	Equipment	Total	Total Incl O&P
0010	**SBS MODIFIED BITUMINOUS MEMBRANE**									
0080	Mod bit rfng, SBS mod, gran surf cap sheet, poly reinf									
1150	For reflective granules, add				S.F.	.71	1.02	.28	2.01	2.93
1600	Smooth surface cap sheet, mopped, 145 mils	G-1	2100	.027		.80	1.04	.28	2.12	3.06
1620	Lightweight base sheet, fiberglass reinforced, 35 to 47 mil		2100	.027		.27	1.04	.28	1.59	2.48
1625	Heavyweight base/ply sheet, reinforced, 87 to 120 mil thick		2100	.027		.88	1.04	.28	2.20	3.15
1650	Granulated walkpad, 180 – 220 mils	1 Rofc	400	.020		1.77	.83		2.60	3.45
1700	Smooth surface flashing, 145 mils	G-1	1260	.044		.80	1.73	.46	2.99	4.51
1800	150 mils		1260	.044		.50	1.73	.46	2.69	4.18
1900	Granular surface flashing, 150 mils		1260	.044		.70	1.73	.46	2.89	4.40
2000	160 mils		1260	.044		.72	1.73	.46	2.91	4.42
2010	Elastomeric asphalt primer	1 Rofc	2600	.003		.16	.13		.29	.41
2015	Roofing asphalt, 30 lb. per square	G-1	19000	.003		.14	.12	.03	.29	.39
2020	Cold process adhesive, 20 – 30 mils thick	1 Rofc	750	.011		.23	.44		.67	1.06
2025	Self adhering vapor retarder, 30 to 45 mils thick	G-5	2150	.019		1	.71	.09	1.80	2.47
2050	Seam heat welding	1 Rofc	205	.039	L.F.	.08	1.63		1.71	3.03

07 53 Elastomeric Membrane Roofing

07 53 16 – Chlorosulfonate-Polyethylene Roofing

07 53 16.10 Chlorosulfonated Polyethylene Roofing

	Crew	Daily Output	Labor-Hours	Unit	Material	2016 Bare Costs Labor	2016 Bare Costs Equipment	Total	Total Incl O&P
0010 **CHLOROSULFONATED POLYETHYLENE ROOFING**									
0800 Chlorosulfonated polyethylene (CSPE)									
0900 45 mils, heat welded seams, plate attachment	G-5	35	1.143	Sq.	251	43.50	5.50	300	360
1100 Heat welded seams, plate attachment and ballasted		26	1.538		260	58.50	7.45	325.95	400
1200 60 mils, heat welded seams, plate attachment		35	1.143		325	43.50	5.50	374	445
1300 Heat welded seams, plate attachment and ballasted		26	1.538		335	58.50	7.45	400.95	485

07 53 23 – Ethylene-Propylene-Diene-Monomer Roofing

07 53 23.20 Ethylene-Propylene-Diene-Monomer Roofing

	Crew	Daily Output	Labor-Hours	Unit	Material	2016 Bare Costs Labor	2016 Bare Costs Equipment	Total	Total Incl O&P
0010 **ETHYLENE-PROPYLENE-DIENE-MONOMER ROOFING (EPDM)**									
3500 Ethylene-propylene-diene-monomer (EPDM), 45 mils, 0.28 psf									
3600 Loose-laid & ballasted with stone (10 psf)	G-5	51	.784	Sq.	83.50	30	3.79	117.29	150
3700 Mechanically attached		35	1.143		78.50	43.50	5.50	127.50	171
3800 Fully adhered with adhesive		26	1.538		111	58.50	7.45	176.95	235
4500 60 mils, 0.40 psf									
4600 Loose-laid & ballasted with stone (10 psf)	G-5	51	.784	Sq.	95	30	3.79	128.79	163
4700 Mechanically attached		35	1.143		89	43.50	5.50	138	183
4800 Fully adhered with adhesive		26	1.538		122	58.50	7.45	187.95	247
4810 45 mil, .28 psf, membrane only					50			50	55
4820 60 mil, .40 psf, membrane only					59.50			59.50	65.50
4850 Seam tape for membrane, 3" x 100' roll				Ea.	43.50			43.50	48
4900 Batten strips, 10' sections					3.43			3.43	3.77
4910 Cover tape for batten strips, 6" x 100' roll					183			183	202
4930 Plate anchors				M	78			78	86
4970 Adhesive for fully adhered systems, 60 S.F./gal.				Gal.	20			20	22

07 53 29 – Polyisobutylene Roofing

07 53 29.10 Polyisobutylene Roofing

	Crew	Daily Output	Labor-Hours	Unit	Material	2016 Bare Costs Labor	2016 Bare Costs Equipment	Total	Total Incl O&P
0010 **POLYISOBUTYLENE ROOFING**									
7500 Polyisobutylene (PIB), 100 mils, 0.57 psf									
7600 Loose-laid & ballasted with stone/gravel (10 psf)	G-5	51	.784	Sq.	200	30	3.79	233.79	278
7700 Partially adhered with adhesive		35	1.143		240	43.50	5.50	289	350
7800 Hot asphalt attachment		35	1.143		230	43.50	5.50	279	340
7900 Fully adhered with contact cement		26	1.538		250	58.50	7.45	315.95	390

07 54 Thermoplastic Membrane Roofing

07 54 19 – Polyvinyl-Chloride Roofing

07 54 19.10 Polyvinyl-Chloride Roofing (PVC)

	Crew	Daily Output	Labor-Hours	Unit	Material	2016 Bare Costs Labor	2016 Bare Costs Equipment	Total	Total Incl O&P
0010 **POLYVINYL-CHLORIDE ROOFING (PVC)**									
8200 Heat welded seams									
8700 Reinforced, 48 mils, 0.33 psf									
8750 Loose-laid & ballasted with stone/gravel (12 psf)	G-5	51	.784	Sq.	117	30	3.79	150.79	187
8800 Mechanically attached		35	1.143		112	43.50	5.50	161	208
8850 Fully adhered with adhesive		26	1.538		154	58.50	7.45	219.95	283
8860 Reinforced, 60 mils, .40 psf									
8870 Loose-laid & ballasted with stone/gravel (12 psf)	G-5	51	.784	Sq.	119	30	3.79	152.79	188
8880 Mechanically attached		35	1.143		113	43.50	5.50	162	209
8890 Fully adhered with adhesive		26	1.538		156	58.50	7.45	221.95	284

For customer support on your Commercial Renovation Cost Data, call 877.742.4719.

201

07 54 Thermoplastic Membrane Roofing

07 54 23 – Thermoplastic-Polyolefin Roofing

07 54 23.10 Thermoplastic Polyolefin Roofing (T.P.O)	Crew	Daily Output	Labor-Hours	Unit	Material	2016 Bare Costs Labor	Equipment	Total	Total Incl O&P
0010 **THERMOPLASTIC POLYOLEFIN ROOFING (T.P.O.)**									
0100 45 mil, loose laid & ballasted with stone (1/2 ton/sq.)	G-5	51	.784	Sq.	87	30	3.79	120.79	153
0120 Fully adhered		25	1.600		78	60.50	7.75	146.25	205
0140 Mechanically attached		34	1.176		77.50	44.50	5.70	127.70	172
0160 Self adhered		35	1.143		84	43.50	5.50	133	177
0180 60 mil membrane, heat welded seams, ballasted		50	.800		100	30.50	3.86	134.36	169
0200 Fully adhered		25	1.600		91.50	60.50	7.75	159.75	219
0220 Mechanically attached		34	1.176		94.50	44.50	5.70	144.70	191
0240 Self adhered		35	1.143		103	43.50	5.50	152	199

07 54 30 – Ketone Ethylene Ester Roofing

07 54 30.10 Ketone Ethylene Ester Roofing	Crew	Daily Output	Labor-Hours	Unit	Material	2016 Bare Costs Labor	Equipment	Total	Total Incl O&P
0010 **KETONE ETHYLENE ESTER ROOFING**									
0100 Ketone ethylene ester roofing, 50 mil, fully adhered	G-5	26	1.538	Sq.	190	58.50	7.45	255.95	320
0120 Mechanically attached		35	1.143		123	43.50	5.50	172	220
0140 Ballasted with stone		51	.784		129	30	3.79	162.79	200
0160 50 mil, fleece backed, adhered w/hot asphalt	G-1	26	2.154		152	84	22.50	258.50	345
0180 Accessories, pipe boot	1 Rofc	32	.250	Ea.	24	10.45		34.45	45.50
0200 Pre-formed corners		32	.250	"	8.50	10.45		18.95	28
0220 Ketone clad metal, including up to 4 bends		330	.024	S.F.	3.88	1.01		4.89	6.10
0240 Walkway pad	2 Rofc	800	.020	"	4.08	.83		4.91	6
0260 Stripping material	1 Rofc	310	.026	L.F.	.94	1.08		2.02	2.97

07 55 Protected Membrane Roofing

07 55 10 – Protected Membrane Roofing Components

07 55 10.10 Protected Membrane Roofing Components	Crew	Daily Output	Labor-Hours	Unit	Material	2016 Bare Costs Labor	Equipment	Total	Total Incl O&P
0010 **PROTECTED MEMBRANE ROOFING COMPONENTS**									
0100 Choose roofing membrane from 07 50									
0120 Then choose roof deck insulation from 07 22									
0130 Filter fabric	2 Rofc	10000	.002	S.F.	.10	.07		.17	.23
0140 Ballast, 3/8" - 1/2" in place	G-1	36	1.556	Ton	18.20	60.50	16.05	94.75	148
0150 3/4" - 1-1/2" in place	"	36	1.556	"	18.20	60.50	16.05	94.75	148
0200 2" concrete blocks, natural	1 Clab	115	.070	S.F.	3.37	2.64		6.01	8
0210 Colors	"	115	.070	"	3.73	2.64		6.37	8.40

07 56 Fluid-Applied Roofing

07 56 10 – Fluid-Applied Roofing Elastomers

07 56 10.10 Elastomeric Roofing	Crew	Daily Output	Labor-Hours	Unit	Material	2016 Bare Costs Labor	Equipment	Total	Total Incl O&P
0010 **ELASTOMERIC ROOFING**									
0020 Acrylic, 44% solids, 2 coats, on corrugated metal	2 Rofc	2400	.007	S.F.	.56	.28		.84	1.11
0025 On smooth metal		3000	.005		.44	.22		.66	.89
0030 On foam or modified bitumen		1500	.011		.89	.44		1.33	1.78
0035 On concrete		1500	.011		.89	.44		1.33	1.78
0040 On tar and gravel		1500	.011		.89	.44		1.33	1.78
0045 36% solids, 2 coats, on corrugated metal		2400	.007		.56	.28		.84	1.11
0050 On smooth metal		3000	.005		.44	.22		.66	.89
0055 On foam or modified bitumen		1500	.011		.89	.44		1.33	1.78
0060 On concrete		1500	.011		.89	.44		1.33	1.78
0065 On tar and gravel		1500	.011		.89	.44		1.33	1.78

07 56 Fluid-Applied Roofing

07 56 10 – Fluid-Applied Roofing Elastomers

07 56 10.10 Elastomeric Roofing	Crew	Daily Output	Labor-Hours	Unit	Material	2016 Bare Costs Labor	Equipment	Total	Total Incl O&P
0070	Primer if required, 2 coats on corrugated metal	2 Rofc	2400	.007	S.F.	.51	.28		.79	1.06
0075	On smooth metal		3000	.005		.51	.22		.73	.96
0080	On foam or modified bitumen		1500	.011		.74	.44		1.18	1.61
0085	On concrete		1500	.011		.55	.44		.99	1.41
0090	On tar & gravel/rolled roof		1500	.011		.74	.44		1.18	1.61
0110	Acrylic rubber, fluid applied, 20 mils thick	G-5	2000	.020		2.20	.76	.10	3.06	3.90
0120	50 mils, reinforced		1200	.033		3.25	1.27	.16	4.68	6.05
0130	For walking surface, add		900	.044		1.20	1.69	.21	3.10	4.60
0300	Neoprene, fluid applied, 20 mil thick, not reinforced	G-1	1135	.049		1.34	1.93	.51	3.78	5.50
0600	Non woven polyester, reinforced		960	.058		1.47	2.28	.60	4.35	6.40
0700	5 coat neoprene deck, 60 mil thick, under 10,000 S.F.		325	.172		4.48	6.75	1.78	13.01	19.05
0900	Over 10,000 S.F.		625	.090		4.48	3.50	.93	8.91	12.25
9000	Minimum labor/equipment charge	1 Rofc	2	4	Job		167		167	300

07 57 Coated Foamed Roofing

07 57 13 – Sprayed Polyurethane Foam Roofing

07 57 13.10 Sprayed Polyurethane Foam Roofing (S.P.F.)	Crew	Daily Output	Labor-Hours	Unit	Material	2016 Bare Costs Labor	Equipment	Total	Total Incl O&P
0010	**SPRAYED POLYURETHANE FOAM ROOFING (S.P.F.)**									
0100	Primer for metal substrate (when required)	G-2A	3000	.008	S.F.	.48	.30	.24	1.02	1.31
0200	Primer for non-metal substrate (when required)		3000	.008		.19	.30	.24	.73	.99
0300	Closed cell spray, polyurethane foam, 3 lb. per C.F. density, 1", R6.7		15000	.002		.66	.06	.05	.77	.88
0400	2", R13.4		13125	.002		1.32	.07	.05	1.44	1.63
0500	3", R18.6		11485	.002		1.98	.08	.06	2.12	2.39
0550	4", R24.8		10080	.002		2.64	.09	.07	2.80	3.14
0700	Spray-on silicone coating		2500	.010		1.21	.36	.29	1.86	2.27
0800	Warranty 5-20 year manufacturer's								.15	.15
0900	Warranty 20 year, no dollar limit								.20	.20

07 58 Roll Roofing

07 58 10 – Asphalt Roll Roofing

07 58 10.10 Roll Roofing	Crew	Daily Output	Labor-Hours	Unit	Material	2016 Bare Costs Labor	Equipment	Total	Total Incl O&P
0010	**ROLL ROOFING**									
0100	Asphalt, mineral surface									
0200	1 ply #15 organic felt, 1 ply mineral surfaced									
0300	Selvage roofing, lap 19", nailed & mopped	G-1	27	2.074	Sq.	77	81	21.50	179.50	254
0400	3 plies glass fiber felt (type IV), 1 ply mineral surfaced									
0500	Selvage roofing, lapped 19", mopped	G-1	25	2.240	Sq.	132	87.50	23	242.50	330
0600	Coated glass fiber base sheet, 2 plies of glass fiber									
0700	Felt (type IV), 1 ply mineral surfaced selvage									
0800	Roofing, lapped 19", mopped	G-1	25	2.240	Sq.	139	87.50	23	249.50	335
0900	On nailable decks	"	24	2.333	"	128	91	24	243	330
1000	3 plies glass fiber felt (type III), 1 ply mineral surfaced									
1100	Selvage roofing, lapped 19", mopped	G-1	25	2.240	Sq.	132	87.50	23	242.50	330

203

For customer support on your Commercial Renovation Cost Data, call 877.742.4719.

07 61 Sheet Metal Roofing

07 61 13 – Standing Seam Sheet Metal Roofing

07 61 13.10 Standing Seam Sheet Metal Roofing, Field Fab.		Crew	Daily Output	Labor-Hours	Unit	Material	2016 Bare Costs Labor	Equipment	Total	Total Incl O&P
0010	STANDING SEAM SHEET METAL ROOFING, FIELD FABRICATED									
0400	Copper standing seam roofing, over 10 squares, 16 oz., 125 lb. per sq.	1 Shee	1.30	6.154	Sq.	985	350		1,335	1,625
0600	18 oz., 140 lb. per sq.		1.20	6.667		1,100	380		1,480	1,800
0700	20 oz., 150 lb. per sq.		1.10	7.273		1,300	415		1,715	2,075
1200	For abnormal conditions or small areas, add					25%	100%			
1300	For lead-coated copper, add					25%				
1400	Lead standing seam roofing, 5 lb. per S.F.	1 Shee	1.30	6.154		1,300	350		1,650	1,975
1500	Zinc standing seam roofing, .020" thick		1.20	6.667		1,225	380		1,605	1,950
1510	.027" thick		1.15	6.957		1,525	400		1,925	2,325
1520	.032" thick		1.10	7.273		1,775	415		2,190	2,600
1530	.040" thick		1.05	7.619		2,200	435		2,635	3,125
9000	Minimum labor/equipment charge		2	4	Job		229		229	360

07 61 16 – Batten Seam Sheet Metal Roofing

07 61 16.10 Batten Seam Sheet Metal Roofing, Field Fabricated

		Crew	Daily Output	Labor-Hours	Unit	Material	Labor	Equipment	Total	Total Incl O&P
0010	BATTEN SEAM SHEET METAL ROOFING, FIELD FABRICATED									
0012	Copper batten seam roofing, over 10 sq., 16 oz., 130 lb. per sq.	1 Shee	1.10	7.273	Sq.	1,250	415		1,665	2,025
0020	Lead batten seam roofing, 5 lb. per S.F.		1.20	6.667		1,350	380		1,730	2,100
0100	Zinc/copper alloy batten seam roofing, .020 thick		1.20	6.667		1,275	380		1,655	2,000
0200	Copper roofing, batten seam, over 10 sq., 18 oz., 145 lb. per sq.		1	8		1,400	460		1,860	2,250
0300	20 oz., 160 lb. per sq.		1	8		1,625	460		2,085	2,525
0500	Stainless steel batten seam roofing, type 304, 28 ga.		1.20	6.667		620	380		1,000	1,275
0600	26 ga.		1.15	6.957		660	400		1,060	1,350
0800	Zinc, copper alloy roofing, batten seam, .027" thick		1.15	6.957		1,600	400		2,000	2,400
0900	.032" thick		1.10	7.273		1,850	415		2,265	2,675
1000	.040" thick		1.05	7.619		2,300	435		2,735	3,225
9000	Minimum labor/equipment charge		2	4	Job		229		229	360

07 61 19 – Flat Seam Sheet Metal Roofing

07 61 19.10 Flat Seam Sheet Metal Roofing, Field Fabricated

		Crew	Daily Output	Labor-Hours	Unit	Material	Labor	Equipment	Total	Total Incl O&P
0010	FLAT SEAM SHEET METAL ROOFING, FIELD FABRICATED									
0900	Copper flat seam roofing, over 10 squares, 16 oz., 115 lb./sq.	1 Shee	1.20	6.667	Sq.	915	380		1,295	1,600
0950	18 oz., 130 lb./sq.		1.15	6.957		1,025	400		1,425	1,750
1000	20 oz., 145 lb./sq.		1.10	7.273		1,200	415		1,615	1,975
1008	Zinc flat seam roofing, .020" thick		1.20	6.667		1,100	380		1,480	1,800
1010	.027" thick		1.15	6.957		1,375	400		1,775	2,125
1020	.032" thick		1.12	7.143		1,575	410		1,985	2,375
1030	.040" thick		1.05	7.619		1,975	435		2,410	2,875
1100	Lead flat seam roofing, 5 lb. per S.F.		1.30	6.154		1,150	350		1,500	1,825
9000	Minimum labor/equipment charge		2.75	2.909	Job		167		167	263

07 62 Sheet Metal Flashing and Trim

07 62 10 – Sheet Metal Trim

07 62 10.10 Sheet Metal Cladding

		Crew	Daily Output	Labor-Hours	Unit	Material	Labor	Equipment	Total	Total Incl O&P
0010	SHEET METAL CLADDING									
0100	Aluminum, up to 6 bends, .032" thick, window casing	1 Carp	180	.044	S.F.	1.65	2.15		3.80	5.35
0200	Window sill		72	.111	L.F.	1.65	5.40		7.05	10.60
0300	Door casing		180	.044	S.F.	1.65	2.15		3.80	5.35
0400	Fascia		250	.032		1.65	1.55		3.20	4.35
0500	Rake trim		225	.036		1.65	1.72		3.37	4.63
0700	.024" thick, window casing		180	.044		1.45	2.15		3.60	5.10
0800	Window sill		72	.111	L.F.	1.45	5.40		6.85	10.40

07 62 Sheet Metal Flashing and Trim

07 62 10 – Sheet Metal Trim

07 62 10.10 Sheet Metal Cladding	Crew	Daily Output	Labor-Hours	Unit	Material	2016 Bare Costs Labor	Equipment	Total	Total Incl O&P
0900	Door casing	1 Carp	180	.044	S.F.	1.45	2.15		3.60	5.10
1000	Fascia		250	.032		1.45	1.55		3	4.13
1100	Rake trim		225	.036		1.45	1.72		3.17	4.41
1200	Vinyl coated aluminum, up to 6 bends, window casing		180	.044	↓	1.54	2.15		3.69	5.20
1300	Window sill		72	.111	L.F.	1.54	5.40		6.94	10.50
1400	Door casing		180	.044	S.F.	1.54	2.15		3.69	5.20
1500	Fascia		250	.032		1.54	1.55		3.09	4.22
1600	Rake trim	↓	225	.036	↓	1.54	1.72		3.26	4.50

07 65 Flexible Flashing

07 65 10 – Sheet Metal Flashing

07 65 10.10 Sheet Metal Flashing and Counter Flashing

		Crew	Daily Output	Labor-Hours	Unit	Material	2016 Bare Costs Labor	Equipment	Total	Total Incl O&P
0010	**SHEET METAL FLASHING AND COUNTER FLASHING**									
0011	Including up to 4 bends									
0020	Aluminum, mill finish, .013" thick	1 Rofc	145	.055	S.F.	.83	2.30		3.13	5.05
0030	.016" thick		145	.055		.95	2.30		3.25	5.20
0060	.019" thick		145	.055		1.25	2.30		3.55	5.55
0100	.032" thick		145	.055		1.35	2.30		3.65	5.65
0200	.040" thick		145	.055		2.31	2.30		4.61	6.70
0300	.050" thick		145	.055	↓	2.64	2.30		4.94	7.05
0325	Mill finish 5" x 7" step flashing, .016" thick		1920	.004	Ea.	.15	.17		.32	.48
0350	Mill finish 12" x 12" step flashing, .016" thick	↓	1600	.005	"	.55	.21		.76	.99
0400	Painted finish, add				S.F.	.32			.32	.35
1000	Mastic-coated 2 sides, .005" thick	1 Rofc	330	.024		1.86	1.01		2.87	3.87
1100	.016" thick		330	.024		2.05	1.01		3.06	4.08
1600	Copper, 16 oz., sheets, under 1000 lb.		115	.070		7.85	2.90		10.75	13.90
1700	Over 4000 lb.		155	.052		7.85	2.15		10	12.55
1900	20 oz. sheets, under 1000 lb.		110	.073		10.30	3.03		13.33	16.80
2000	Over 4000 lb.		145	.055		9.80	2.30		12.10	14.90
2200	24 oz. sheets, under 1000 lb.		105	.076		14.60	3.18		17.78	22
2300	Over 4000 lb.		135	.059		13.85	2.47		16.32	19.70
2500	32 oz. sheets, under 1000 lb.		100	.080		18.50	3.34		21.84	26.50
2600	Over 4000 lb.		130	.062	↓	17.60	2.57		20.17	24
2700	W shape for valleys, 16 oz., 24" wide		100	.080	L.F.	16.20	3.34		19.54	24
5800	Lead, 2.5 lb. per S.F., up to 12" wide		135	.059	S.F.	5.40	2.47		7.87	10.40
5900	Over 12" wide		135	.059		3.99	2.47		6.46	8.85
8900	Stainless steel sheets, 32 ga.,		155	.052		3.25	2.15		5.40	7.45
9000	28 ga.		155	.052		4.20	2.15		6.35	8.50
9100	26 ga.		155	.052		4.45	2.15		6.60	8.80
9200	24 ga.		155	.052	↓	4.95	2.15		7.10	9.35
9290	For mechanically keyed flashing, add					40%				
9320	Steel sheets, galvanized, 20 ga.	1 Rofc	130	.062	S.F.	1.25	2.57		3.82	6
9322	22 ga.		135	.059		1.21	2.47		3.68	5.80
9324	24 ga.		140	.057		.92	2.38		3.30	5.30
9326	26 ga.		148	.054		.80	2.25		3.05	4.95
9328	28 ga.		155	.052		.69	2.15		2.84	4.64
9340	30 ga.		160	.050		.58	2.09		2.67	4.40
9400	Terne coated stainless steel, .015" thick, 28 ga		155	.052		8.05	2.15		10.20	12.75
9500	.018" thick, 26 ga		155	.052		9	2.15		11.15	13.80
9600	Zinc and copper alloy (brass), .020" thick		155	.052		10.20	2.15		12.35	15.10
9700	.027" thick	↓	155	.052	↓	12	2.15		14.15	17.10

For customer support on your Commercial Renovation Cost Data, call 877.742.4719.

205

07 65 Flexible Flashing

07 65 10 – Sheet Metal Flashing

07 65 10.10 Sheet Metal Flashing and Counter Flashing

		Crew	Daily Output	Labor-Hours	Unit	Material	2016 Bare Costs Labor	Equipment	Total	Total Incl O&P
9800	.032" thick	1 Rofc	155	.052	S.F.	15.40	2.15		17.55	21
9900	.040" thick		155	.052	↓	20	2.15		22.15	26
9950	Minimum labor/equipment charge	↓	3	2.667	Job		111		111	201

07 65 12 – Fabric and Mastic Flashings

07 65 12.10 Fabric and Mastic Flashing and Counter Flashing

		Crew	Daily Output	Labor-Hours	Unit	Material	2016 Bare Costs Labor	Equipment	Total	Total Incl O&P
0010	**FABRIC AND MASTIC FLASHING AND COUNTER FLASHING**									
1300	Asphalt flashing cement, 5 gallon				Gal.	9.65			9.65	10.60
4900	Fabric, asphalt-saturated cotton, specification grade	1 Rofc	35	.229	S.Y.	3.05	9.55		12.60	20.50
5000	Utility grade		35	.229		1.45	9.55		11	18.80
5300	Close-mesh fabric, saturated, 17 oz. per S.Y.		35	.229		2.20	9.55		11.75	19.60
5500	Fiberglass, resin-coated		35	.229	↓	1	9.55		10.55	18.30
8500	Shower pan, bituminous membrane, 7 oz.	↓	155	.052	S.F.	1.72	2.15		3.87	5.75

07 65 13 – Laminated Sheet Flashing

07 65 13.10 Laminated Sheet Flashing

		Crew	Daily Output	Labor-Hours	Unit	Material	2016 Bare Costs Labor	Equipment	Total	Total Incl O&P
0010	**LAMINATED SHEET FLASHING**, Including up to 4 bends									
0500	Aluminum, fabric-backed 2 sides, mill finish, .004" thick	1 Rofc	330	.024	S.F.	1.60	1.01		2.61	3.58
0700	.005" thick		330	.024		1.85	1.01		2.86	3.86
0750	Mastic-backed, self adhesive		460	.017		3.60	.73		4.33	5.25
0800	Mastic-coated 2 sides, .004" thick		330	.024		1.60	1.01		2.61	3.58
2800	Copper, paperbacked 1 side, 2 oz.		330	.024		1.84	1.01		2.85	3.84
2900	3 oz.		330	.024		2.20	1.01		3.21	4.24
3100	Paperbacked 2 sides, 2 oz.		330	.024		2.05	1.01		3.06	4.08
3150	3 oz.		330	.024		2.30	1.01		3.31	4.35
3200	5 oz.		330	.024		3.35	1.01		4.36	5.50
3250	7 oz.		330	.024		6.70	1.01		7.71	9.15
3400	Mastic-backed 2 sides, copper, 2 oz.		330	.024		1.95	1.01		2.96	3.97
3500	3 oz.		330	.024		2.50	1.01		3.51	4.57
3700	5 oz.		330	.024		3.80	1.01		4.81	6
3800	Fabric-backed 2 sides, copper, 2 oz.		330	.024		2.10	1.01		3.11	4.13
4000	3 oz.		330	.024		2.80	1.01		3.81	4.90
4100	5 oz.		330	.024		3.90	1.01		4.91	6.10
4300	Copper-clad stainless steel, .015" thick, under 500 lb.		115	.070		6.40	2.90		9.30	12.30
4400	Over 2000 lb.		155	.052		6.70	2.15		8.85	11.25
4600	.018" thick, under 500 lb.		100	.080		7.85	3.34		11.19	14.65
4700	Over 2000 lb.		145	.055		7.50	2.30		9.80	12.40
8550	Shower pan, 3 ply copper and fabric, 3 oz.		155	.052		4	2.15		6.15	8.30
8600	7 oz.		155	.052		4.20	2.15		6.35	8.50
9300	Stainless steel, paperbacked 2 sides, .005" thick	↓	330	.024	↓	3.88	1.01		4.89	6.10

07 65 19 – Plastic Sheet Flashing

07 65 19.10 Plastic Sheet Flashing and Counter Flashing

		Crew	Daily Output	Labor-Hours	Unit	Material	2016 Bare Costs Labor	Equipment	Total	Total Incl O&P
0010	**PLASTIC SHEET FLASHING AND COUNTER FLASHING**									
7300	Polyvinyl chloride, black, 10 mil	1 Rofc	285	.028	S.F.	.21	1.17		1.38	2.34
7400	20 mil		285	.028		.26	1.17		1.43	2.40
7600	30 mil		285	.028		.34	1.17		1.51	2.48
7700	60 mil		285	.028		.82	1.17		1.99	3.01
7900	Black or white for exposed roofs, 60 mil	↓	285	.028	↓	1.80	1.17		2.97	4.09
8060	PVC tape, 5" x 45 mils, for joint covers, 100 L.F./roll				Ea.	172			172	189
8850	Polyvinyl chloride, 30 mil	1 Rofc	160	.050	S.F.	1.45	2.09		3.54	5.35

07 65 Flexible Flashing

07 65 23 – Rubber Sheet Flashing

07 65 23.10 Rubber Sheet Flashing and Counterflashing	Crew	Daily Output	Labor-Hours	Unit	Material	2016 Bare Costs Labor	Equipment	Total	Total Incl O&P
0010 **RUBBER SHEET FLASHING AND COUNTERFLASHING**									
4810 EPDM 90 mils, 1" diameter pipe flashing	1 Rofc	32	.250	Ea.	19.35	10.45		29.80	40.50
4820 2" diameter		30	.267		19.30	11.10		30.40	41
4830 3" diameter		28	.286		20.50	11.90		32.40	44.50
4840 4" diameter		24	.333		24	13.90		37.90	51
4850 6" diameter		22	.364		23.50	15.15		38.65	53.50
8100 Rubber, butyl, 1/32" thick		285	.028	S.F.	1.96	1.17		3.13	4.27
8200 1/16" thick		285	.028		2.65	1.17		3.82	5.05
8300 Neoprene, cured, 1/16" thick		285	.028		2.55	1.17		3.72	4.92
8400 1/8" thick		285	.028		6.05	1.17		7.22	8.75

07 65 26 – Self-Adhering Sheet Flashing

07 65 26.10 Self-Adhering Sheet or Roll Flashing

	Crew	Daily Output	Labor-Hours	Unit	Material	Labor	Equipment	Total	Total Incl O&P
0010 **SELF-ADHERING SHEET OR ROLL FLASHING**									
0020 Self-adhered flashing, 25 mil cross laminated HDPE, 4" wide	1 Rofc	960	.008	L.F.	.20	.35		.55	.85
0040 6" wide		896	.009		.30	.37		.67	.99
0060 9" wide		832	.010		.44	.40		.84	1.21
0080 12" wide		768	.010		.59	.43		1.02	1.43

07 71 Roof Specialties

07 71 16 – Manufactured Counterflashing Systems

07 71 16.10 Roof Drain Boot

	Crew	Daily Output	Labor-Hours	Unit	Material	Labor	Equipment	Total	Total Incl O&P
0010 **ROOF DRAIN BOOT**									
0100 Cast iron, 4" diameter	1 Shee	125	.064	L.F.	50	3.66		53.66	61.50
0300 4" x 3"		125	.064		57.50	3.66		61.16	69.50
0400 5" x 4"		125	.064		117	3.66		120.66	135

07 71 19 – Manufactured Gravel Stops and Fasciae

07 71 19.10 Gravel Stop

	Crew	Daily Output	Labor-Hours	Unit	Material	Labor	Equipment	Total	Total Incl O&P
0010 **GRAVEL STOP**									
0020 Aluminum, .050" thick, 4" face height, mill finish	1 Shee	145	.055	L.F.	6.40	3.16		9.56	12.05
0080 Duranodic finish		145	.055		7.35	3.16		10.51	13.10
0100 Painted		145	.055		7.45	3.16		10.61	13.20
0300 6" face height		135	.059		6.60	3.39		9.99	12.60
0350 Duranodic finish		135	.059		7.75	3.39		11.14	13.90
0400 Painted		135	.059		8.70	3.39		12.09	14.90
0600 8" face height		125	.064		7.55	3.66		11.21	14.10
0650 Duranodic finish		125	.064		9.10	3.66		12.76	15.80
0700 Painted		125	.064		9.55	3.66		13.21	16.30
0900 12" face height, .080" thick, 2 piece		100	.080		10.65	4.58		15.23	18.95
0950 Duranodic finish		100	.080		10.80	4.58		15.38	19.15
1000 Painted		100	.080		13.20	4.58		17.78	22
1200 Copper, 16 oz., 3" face height		145	.055		24	3.16		27.16	31.50
1300 6" face height		135	.059		33	3.39		36.39	42
1350 Galv steel, 24 ga., 4" leg, plain, with continuous cleat, 4" face		145	.055		6.25	3.16		9.41	11.90
1360 6" face height		145	.055		6.50	3.16		9.66	12.15
1500 Polyvinyl chloride, 6" face height		135	.059		5.70	3.39		9.09	11.60
1600 9" face height		125	.064		6.60	3.66		10.26	13.05
1800 Stainless steel, 24 ga., 6" face height		135	.059		15.30	3.39		18.69	22
1900 12" face height		100	.080		22.50	4.58		27.08	32.50
2100 20 ga., 6" face height		135	.059		19	3.39		22.39	26.50

For customer support on your Commercial Renovation Cost Data, call 877.742.4719.

207

07 71 Roof Specialties

07 71 19 – Manufactured Gravel Stops and Fasciae

07 71 19.10 Gravel Stop

		Crew	Daily Output	Labor-Hours	Unit	Material	2016 Bare Costs Labor	Equipment	Total	Total Incl O&P
2200	12" face height	1 Shee	100	.080	L.F.	27	4.58		31.58	37
9000	Minimum labor/equipment charge	↓	3.50	2.286	Job		131		131	207

07 71 19.30 Fascia

		Crew	Daily Output	Labor-Hours	Unit	Material	2016 Bare Costs Labor	Equipment	Total	Total Incl O&P
0010	**FASCIA**									
0100	Aluminum, reverse board and batten, .032" thick, colored, no furring incl	1 Shee	145	.055	S.F.	6.75	3.16		9.91	12.45
0200	Residential type, aluminum	1 Carp	200	.040	L.F.	1.95	1.94		3.89	5.30
0220	Vinyl	"	200	.040	"	2.10	1.94		4.04	5.50
0300	Steel, galv and enameled, stock, no furring, long panels	1 Shee	145	.055	S.F.	5.25	3.16		8.41	10.80
0600	Short panels		115	.070	"	5.85	3.98		9.83	12.75
9000	Minimum labor/equipment charge	↓	4	2	Job		115		115	181

07 71 23 – Manufactured Gutters and Downspouts

07 71 23.10 Downspouts

		Crew	Daily Output	Labor-Hours	Unit	Material	2016 Bare Costs Labor	Equipment	Total	Total Incl O&P
0010	**DOWNSPOUTS**									
0020	Aluminum, embossed, .020" thick, 2" x 3"	1 Shee	190	.042	L.F.	1.05	2.41		3.46	4.96
0100	Enameled		190	.042		1.39	2.41		3.80	5.35
0300	.024" thick, 2' x 3"		180	.044		2.05	2.54		4.59	6.30
0400	3" x 4"		140	.057		2.37	3.27		5.64	7.75
0600	Round, corrugated aluminum, 3" diameter, .020" thick		190	.042		2	2.41		4.41	6
0700	4" diameter, .025" thick		140	.057	↓	2.95	3.27		6.22	8.40
0900	Wire strainer, round, 2" diameter		155	.052	Ea.	1.84	2.95		4.79	6.70
1000	4" diameter		155	.052		2.40	2.95		5.35	7.30
1200	Rectangular, perforated, 2" x 3"		145	.055		2.33	3.16		5.49	7.55
1300	3" x 4"		145	.055	↓	3.33	3.16		6.49	8.65
1500	Copper, round, 16 oz., stock, 2" diameter		190	.042	L.F.	7.60	2.41		10.01	12.15
1600	3" diameter		190	.042		8	2.41		10.41	12.60
1800	4" diameter		145	.055		9.70	3.16		12.86	15.65
1900	5" diameter		130	.062		15.75	3.52		19.27	23
2100	Rectangular, corrugated copper, stock, 2" x 3"		190	.042		8.75	2.41		11.16	13.40
2200	3" x 4"		145	.055		10.40	3.16		13.56	16.45
2400	Rectangular, plain copper, stock, 2" x 3"		190	.042		11.05	2.41		13.46	15.95
2500	3" x 4"		145	.055	↓	13.95	3.16		17.11	20.50
2700	Wire strainers, rectangular, 2" x 3"		145	.055	Ea.	17.05	3.16		20.21	23.50
2800	3" x 4"		145	.055		17.80	3.16		20.96	24.50
3000	Round, 2" diameter		145	.055		6.40	3.16		9.56	12.05
3100	3" diameter		145	.055		7.10	3.16		10.26	12.80
3300	4" diameter		145	.055		12.10	3.16		15.26	18.30
3400	5" diameter		115	.070	↓	22.50	3.98		26.48	31
3600	Lead-coated copper, round, stock, 2" diameter		190	.042	L.F.	22	2.41		24.41	28.50
3700	3" diameter		190	.042		22.50	2.41		24.91	28.50
3900	4" diameter		145	.055		23.50	3.16		26.66	31
4000	5" diameter, corrugated		130	.062		23	3.52		26.52	30.50
4200	6" diameter, corrugated		105	.076		30	4.36		34.36	40
4300	Rectangular, corrugated, stock, 2" x 3"		190	.042		15.70	2.41		18.11	21
4500	Plain, stock, 2" x 3"		190	.042		24	2.41		26.41	30.50
4600	3" x 4"		145	.055		33	3.16		36.16	41.50
4800	Steel, galvanized, round, corrugated, 2" or 3" diameter, 28 ga.		190	.042		2.10	2.41		4.51	6.10
4900	4" diameter, 28 ga.		145	.055		2.52	3.16		5.68	7.75
5100	5" diameter, 26 ga.		130	.062		3.73	3.52		7.25	9.65
5400	6" diameter, 28 ga.		105	.076		4.15	4.36		8.51	11.45
5500	26 ga.		105	.076		3.50	4.36		7.86	10.75
5700	Rectangular, corrugated, 28 ga., 2" x 3"		190	.042		1.94	2.41		4.35	5.95
5800	3" x 4"	↓	145	.055	↓	1.82	3.16		4.98	7

07 71 Roof Specialties

07 71 23 - Manufactured Gutters and Downspouts

07 71 23.10 Downspouts

		Crew	Daily Output	Labor-Hours	Unit	Material	2016 Bare Costs Labor	2016 Bare Costs Equipment	Total	Total Incl O&P
6000	Rectangular, plain, 28 ga., galvanized, 2" x 3"	1 Shee	190	.042	L.F.	3.65	2.41		6.06	7.80
6100	3" x 4"		145	.055		4.06	3.16		7.22	9.45
6300	Epoxy painted, 24 ga., corrugated, 2" x 3"		190	.042		2.33	2.41		4.74	6.35
6400	3" x 4"		145	.055	▼	2.83	3.16		5.99	8.10
6600	Wire strainers, rectangular, 2" x 3"		145	.055	Ea.	17.05	3.16		20.21	23.50
6700	3" x 4"		145	.055		17.80	3.16		20.96	24.50
6900	Round strainers, 2" or 3" diameter		145	.055		3.86	3.16		7.02	9.25
7000	4" diameter		145	.055	▼	5.80	3.16		8.96	11.35
9000	Minimum labor/equipment charge	▼	4	2	Job		115		115	181

07 71 23.20 Downspout Elbows

		Crew	Daily Output	Labor-Hours	Unit	Material	2016 Bare Costs Labor	2016 Bare Costs Equipment	Total	Total Incl O&P
0010	**DOWNSPOUT ELBOWS**									
0020	Aluminum, embossed, 2" x 3", .020" thick	1 Shee	100	.080	Ea.	.95	4.58		5.53	8.30
0100	Enameled		100	.080		1.75	4.58		6.33	9.20
0200	Embossed, 3" x 4", .025" thick		100	.080		4.51	4.58		9.09	12.20
0300	Enameled		100	.080		3.75	4.58		8.33	11.40
0400	Embossed, corrugated, 3" diameter, .020" thick		100	.080		2.74	4.58		7.32	10.25
0500	4" diameter, .025" thick		100	.080		5.80	4.58		10.38	13.60
0600	Copper, 16 oz., 2" diameter		100	.080		9.45	4.58		14.03	17.65
0700	3" diameter		100	.080		9	4.58		13.58	17.15
0800	4" diameter		100	.080		13.90	4.58		18.48	22.50
1000	Rectangular, 2" x 3" corrugated		100	.080		9.35	4.58		13.93	17.55
1100	3" x 4" corrugated		100	.080		13.25	4.58		17.83	22
1300	Vinyl, 2-1/2" diameter, 45 or 75 degree bend		100	.080	▼	3.88	4.58		8.46	11.50
1400	Tee Y junction		75	.107	▼	12.90	6.10		19	24
9000	Minimum labor/equipment charge	▼	4	2	Job		115		115	181

07 71 23.30 Gutters

		Crew	Daily Output	Labor-Hours	Unit	Material	2016 Bare Costs Labor	2016 Bare Costs Equipment	Total	Total Incl O&P
0010	**GUTTERS**									
0012	Aluminum, stock units, 5" K type, .027" thick, plain	1 Shee	125	.064	L.F.	2.81	3.66		6.47	8.90
0100	Enameled		125	.064		2.79	3.66		6.45	8.85
0300	5" K type, .032" thick, plain		125	.064		3.47	3.66		7.13	9.60
0400	Enameled		125	.064		3.59	3.66		7.25	9.75
0700	Copper, half round, 16 oz., stock units, 4" wide		125	.064		9.50	3.66		13.16	16.25
0900	5" wide		125	.064		7.25	3.66		10.91	13.75
1000	6" wide		118	.068		11.85	3.88		15.73	19.15
1200	K type, 16 oz., stock, 5" wide		125	.064		8.05	3.66		11.71	14.65
1300	6" wide		125	.064		8.50	3.66		12.16	15.15
1500	Lead coated copper, 16 oz., half round, stock, 4" wide		125	.064		14.70	3.66		18.36	22
1600	6" wide		118	.068		22	3.88		25.88	30
1800	K type, stock, 5" wide		125	.064		18.35	3.66		22.01	26
1900	6" wide		125	.064		19.65	3.66		23.31	27.50
2100	Copper clad stainless steel, K type, 5" wide		125	.064		7.50	3.66		11.16	14.05
2200	6" wide		125	.064		9.45	3.66		13.11	16.20
2400	Steel, galv, half round or box, 28 ga., 5" wide, plain		125	.064		2.10	3.66		5.76	8.10
2500	Enameled		125	.064		2.18	3.66		5.84	8.20
2700	26 ga., stock, 5" wide		125	.064		2.40	3.66		6.06	8.45
2800	6" wide	▼	125	.064		2.78	3.66		6.44	8.85
3000	Vinyl, O.G., 4" wide	1 Carp	115	.070		1.30	3.37		4.67	6.95
3100	5" wide		115	.070		1.60	3.37		4.97	7.25
3200	4" half round, stock units	▼	115	.070	▼	1.35	3.37		4.72	7
3250	Joint connectors				Ea.	2.95			2.95	3.25
3300	Wood, clear treated cedar, fir or hemlock, 3" x 4"	1 Carp	100	.080	L.F.	10	3.88		13.88	17.35
3400	4" x 5"	"	100	.080	"	16.75	3.88		20.63	25

For customer support on your Commercial Renovation Cost Data, call 877.742.4719.

209

07 71 23 – Manufactured Gutters and Downspouts

07 71 23.30 Gutters		Crew	Daily Output	Labor-Hours	Unit	Material	2016 Bare Costs Labor	Equipment	Total	Total Incl O&P
5000	Accessories, end cap, K type, aluminum 5"	1 Shee	625	.013	Ea.	.70	.73		1.43	1.93
5010	6"		625	.013		1.50	.73		2.23	2.81
5020	Copper, 5"		625	.013		3.22	.73		3.95	4.70
5030	6"		625	.013		3.54	.73		4.27	5.05
5040	Lead coated copper, 5"		625	.013		12.50	.73		13.23	14.90
5050	6"		625	.013		13.35	.73		14.08	15.85
5060	Copper clad stainless steel, 5"		625	.013		3	.73		3.73	4.46
5070	6"		625	.013		3.60	.73		4.33	5.10
5080	Galvanized steel, 5"		625	.013		1.28	.73		2.01	2.57
5090	6"		625	.013		2.18	.73		2.91	3.56
5100	Vinyl, 4"	1 Carp	625	.013		13.15	.62		13.77	15.50
5110	5"	"	625	.013		13.15	.62		13.77	15.50
5120	Half round, copper, 4"	1 Shee	625	.013		4.53	.73		5.26	6.15
5130	5"		625	.013		4.53	.73		5.26	6.15
5140	6"		625	.013		7.65	.73		8.38	9.60
5150	Lead coated copper, 5"		625	.013		14.15	.73		14.88	16.70
5160	6"		625	.013		21	.73		21.73	24.50
5170	Copper clad stainless steel, 5"		625	.013		4.50	.73		5.23	6.10
5180	6"		625	.013		4.50	.73		5.23	6.10
5190	Galvanized steel, 5"		625	.013		2.25	.73		2.98	3.64
5200	6"		625	.013		2.80	.73		3.53	4.24
5210	Outlet, aluminum, 2" x 3"		420	.019		.62	1.09		1.71	2.40
5220	3" x 4"		420	.019		1.05	1.09		2.14	2.88
5230	2-3/8" round		420	.019		.56	1.09		1.65	2.34
5240	Copper, 2" x 3"		420	.019		6.50	1.09		7.59	8.85
5250	3" x 4"		420	.019		7.85	1.09		8.94	10.35
5260	2-3/8" round		420	.019		4.55	1.09		5.64	6.70
5270	Lead coated copper, 2" x 3"		420	.019		25.50	1.09		26.59	29.50
5280	3" x 4"		420	.019		28.50	1.09		29.59	33
5290	2-3/8" round		420	.019		25.50	1.09		26.59	29.50
5300	Copper clad stainless steel, 2" x 3"		420	.019		6.50	1.09		7.59	8.85
5310	3" x 4"		420	.019		7.85	1.09		8.94	10.35
5320	2-3/8" round		420	.019		4.55	1.09		5.64	6.70
5330	Galvanized steel, 2" x 3"		420	.019		3.18	1.09		4.27	5.20
5340	3" x 4"		420	.019		4.90	1.09		5.99	7.10
5350	2-3/8" round		420	.019		3.98	1.09		5.07	6.10
5360	K type mitres, aluminum		65	.123		3	7.05		10.05	14.40
5370	Copper		65	.123		17.40	7.05		24.45	30
5380	Lead coated copper		65	.123		54.50	7.05		61.55	71
5390	Copper clad stainless steel		65	.123		27.50	7.05		34.55	41.50
5400	Galvanized steel		65	.123		22.50	7.05		29.55	36
5420	Half round mitres, copper		65	.123		63.50	7.05		70.55	81
5430	Lead coated copper		65	.123		89.50	7.05		96.55	110
5440	Copper clad stainless steel		65	.123		56	7.05		63.05	73
5450	Galvanized steel		65	.123		25.50	7.05		32.55	39
5460	Vinyl mitres and outlets		65	.123		9.75	7.05		16.80	22
5470	Sealant		940	.009	L.F.	.01	.49		.50	.78
5480	Soldering		96	.083	"	.20	4.77		4.97	7.75
9000	Minimum labor/equipment charge		3.75	2.133	Job		122		122	193

210

For customer support on your Commercial Renovation Cost Data, call 877.742.4719.

07 71 Roof Specialties

07 71 23 – Manufactured Gutters and Downspouts

07 71 23.35 Gutter Guard

07 71 23.35 Gutter Guard	Crew	Daily Output	Labor-Hours	Unit	Material	2016 Bare Costs Labor	2016 Bare Costs Equipment	Total	Total Incl O&P
0010 **GUTTER GUARD**									
0020 6" wide strip, aluminum mesh	1 Carp	500	.016	L.F.	2.50	.78		3.28	4.02
0100 Vinyl mesh		500	.016	"	2.70	.78		3.48	4.24
9000 Minimum labor/equipment charge		4	2	Job		97		97	158

07 71 26 – Reglets

07 71 26.10 Reglets and Accessories

07 71 26.10 Reglets and Accessories	Crew	Daily Output	Labor-Hours	Unit	Material	2016 Bare Costs Labor	2016 Bare Costs Equipment	Total	Total Incl O&P
0010 **REGLETS AND ACCESSORIES**									
0020 Reglet, aluminum, .025" thick, in parapet	1 Carp	225	.036	L.F.	1.98	1.72		3.70	4.99
0300 16 oz. copper		225	.036		6.50	1.72		8.22	9.95
0400 Galvanized steel, 24 ga.		225	.036		1.36	1.72		3.08	4.31
0600 Stainless steel, .020" thick		225	.036		3.80	1.72		5.52	7
0900 Counter flashing for above, 12" wide, .032" aluminum	1 Shee	150	.053		2.20	3.05		5.25	7.25
1200 16 oz. copper		150	.053		6.25	3.05		9.30	11.70
1300 Galvanized steel, 26 ga.		150	.053		1.30	3.05		4.35	6.25
1500 Stainless steel, .020" thick		150	.053		6.30	3.05		9.35	11.75
9000 Minimum labor/equipment charge	1 Carp	3	2.667	Job		129		129	211

07 71 29 – Manufactured Roof Expansion Joints

07 71 29.10 Expansion Joints

07 71 29.10 Expansion Joints	Crew	Daily Output	Labor-Hours	Unit	Material	2016 Bare Costs Labor	2016 Bare Costs Equipment	Total	Total Incl O&P
0010 **EXPANSION JOINTS**									
0300 Butyl or neoprene center with foam insulation, metal flanges									
0400 Aluminum, .032" thick for openings to 2-1/2"	1 Rofc	165	.048	L.F.	11.65	2.02		13.67	16.45
0600 For joint openings to 3-1/2"		165	.048		11.65	2.02		13.67	16.45
0610 For joint openings to 5"		165	.048		13.50	2.02		15.52	18.50
0620 For joint openings to 8"		165	.048		16.30	2.02		18.32	21.50
0700 Copper, 16 oz. for openings to 2-1/2"		165	.048		18.40	2.02		20.42	23.50
0900 For joint openings to 3-1/2"		165	.048		19.25	2.02		21.27	24.50
0910 For joint openings to 5"		165	.048		21	2.02		23.02	26.50
0920 For joint openings to 8"		165	.048		24	2.02		26.02	30
1000 Galvanized steel, 26 ga. for openings to 2-1/2"		165	.048		9.80	2.02		11.82	14.45
1200 For joint openings to 3-1/2"		165	.048		9.80	2.02		11.82	14.45
1210 For joint openings to 5"		165	.048		11.30	2.02		13.32	16.10
1220 For joint openings to 8"		165	.048		14.80	2.02		16.82	19.95
1300 Lead-coated copper, 16 oz. for openings to 2-1/2"		165	.048		33.50	2.02		35.52	40.50
1500 For joint openings to 3-1/2"		165	.048		33.50	2.02		35.52	40.50
1600 Stainless steel, .018", for openings to 2-1/2"		165	.048		13.20	2.02		15.22	18.15
1800 For joint openings to 3-1/2"		165	.048		13.20	2.02		15.22	18.15
1810 For joint openings to 5"		165	.048		14.50	2.02		16.52	19.60
1820 For joint openings to 8"		165	.048		21	2.02		23.02	26.50
1900 Neoprene, double-seal type with thick center, 4-1/2" wide		125	.064		14.75	2.67		17.42	21
1950 Polyethylene bellows, with galv steel flat flanges		100	.080		6.40	3.34		9.74	13.05
1960 With galvanized angle flanges		100	.080		6.60	3.34		9.94	13.25
2000 Roof joint with extruded aluminum cover, 2"	1 Shee	115	.070		28	3.98		31.98	37.50
2100 Roof joint, plastic curbs, foam center, standard	1 Rofc	100	.080		13.20	3.34		16.54	20.50
2200 Large	"	100	.080		17.20	3.34		20.54	25
2500 Roof to wall joint with extruded aluminum cover	1 Shee	115	.070		28	3.98		31.98	37.50
2700 Wall joint, closed cell foam on PVC cover, 9" wide	1 Rofc	125	.064		5.40	2.67		8.07	10.75
2800 12" wide	"	115	.070		6.50	2.90		9.40	12.40
9000 Minimum labor/equipment charge	1 Shee	3	2.667	Job		153		153	241

For customer support on your Commercial Renovation Cost Data, call 877.742.4719.

211

07 71 Roof Specialties

07 71 43 – Drip Edge

07 71 43.10 Drip Edge, Rake Edge, Ice Belts	Crew	Daily Output	Labor-Hours	Unit	Material	2016 Bare Costs Labor	Equipment	Total	Total Incl O&P
0010 **DRIP EDGE, RAKE EDGE, ICE BELTS**									
0020 Aluminum, .016" thick, 5" wide, mill finish	1 Carp	400	.020	L.F.	.57	.97		1.54	2.21
0100 White finish		400	.020		.63	.97		1.60	2.27
0200 8" wide, mill finish		400	.020		1.45	.97		2.42	3.18
0300 Ice belt, 28" wide, mill finish		100	.080		7.65	3.88		11.53	14.80
0310 Vented, mill finish		400	.020		1.99	.97		2.96	3.77
0320 Painted finish		400	.020		2.24	.97		3.21	4.04
0400 Galvanized, 5" wide		400	.020		.59	.97		1.56	2.23
0500 8" wide, mill finish		400	.020		.82	.97		1.79	2.48
0510 Rake edge, aluminum, 1-1/2" x 1-1/2"		400	.020		.33	.97		1.30	1.94
0520 3-1/2" x 1-1/2"		400	.020		.45	.97		1.42	2.08
9000 Minimum labor/equipment charge		4	2	Job		97		97	158

07 72 Roof Accessories

07 72 23 – Relief Vents

07 72 23.10 Roof Vents

	Crew	Daily Output	Labor-Hours	Unit	Material	2016 Bare Costs Labor	Equipment	Total	Total Incl O&P
0010 **ROOF VENTS**									
0020 Mushroom shape, for built-up roofs, aluminum	1 Rofc	30	.267	Ea.	65	11.10		76.10	91.50
0100 PVC, 6" high		30	.267	"	21	11.10		32.10	43.50
9000 Minimum labor/equipment charge		2.75	2.909	Job		121		121	219

07 72 23.20 Vents

	Crew	Daily Output	Labor-Hours	Unit	Material	2016 Bare Costs Labor	Equipment	Total	Total Incl O&P
0010 **VENTS**									
0100 Soffit or eave, aluminum, mill finish, strips, 2-1/2" wide	1 Carp	200	.040	L.F.	.46	1.94		2.40	3.68
0200 3" wide		200	.040		.48	1.94		2.42	3.70
0300 Enamel finish, 3" wide		200	.040		.50	1.94		2.44	3.72
0400 Mill finish, rectangular, 4" x 16"		72	.111	Ea.	1.52	5.40		6.92	10.45
0500 8" x 16"		72	.111		2.33	5.40		7.73	11.35
2420 Roof ventilator	Q-9	16	1		54	51.50		105.50	141
2500 Vent, roof vent	1 Rofc	24	.333		27	13.90		40.90	54.50

07 72 26 – Ridge Vents

07 72 26.10 Ridge Vents and Accessories

	Crew	Daily Output	Labor-Hours	Unit	Material	2016 Bare Costs Labor	Equipment	Total	Total Incl O&P
0010 **RIDGE VENTS AND ACCESSORIES**									
0100 Aluminum strips, mill finish	1 Rofc	160	.050	L.F.	2.86	2.09		4.95	6.90
0150 Painted finish		160	.050	"	4.05	2.09		6.14	8.20
0200 Connectors		48	.167	Ea.	4.75	6.95		11.70	17.80
0300 End caps		48	.167	"	2.05	6.95		9	14.80
0400 Galvanized strips		160	.050	L.F.	3.60	2.09		5.69	7.70
0430 Molded polyethylene, shingles not included		160	.050	"	2.75	2.09		4.84	6.80
0440 End plugs		48	.167	Ea.	2.05	6.95		9	14.80
0450 Flexible roll, shingles not included		160	.050	L.F.	2.34	2.09		4.43	6.35
2300 Ridge vent strip, mill finish	1 Shee	155	.052	"	3.88	2.95		6.83	8.95

07 72 33 – Roof Hatches

07 72 33.10 Roof Hatch Options

	Crew	Daily Output	Labor-Hours	Unit	Material	2016 Bare Costs Labor	Equipment	Total	Total Incl O&P
0010 **ROOF HATCH OPTIONS**									
0500 2'-6" x 3', aluminum curb and cover	G-3	10	3.200	Ea.	980	152		1,132	1,325
0520 Galvanized steel curb and aluminum cover		10	3.200		635	152		787	945
0540 Galvanized steel curb and cover		10	3.200		705	152		857	1,025
0600 2'-6" x 4'-6", aluminum curb and cover		9	3.556		1,275	169		1,444	1,675
0800 Galvanized steel curb and aluminum cover		9	3.556		885	169		1,054	1,250

07 72 Roof Accessories

07 72 33 – Roof Hatches

07 72 33.10 Roof Hatch Options

07 72 33.10	Roof Hatch Options	Crew	Daily Output	Labor-Hours	Unit	Material	2016 Bare Costs Labor	Equipment	Total	Total Incl O&P
0900	Galvanized steel curb and cover	G-3	9	3.556	Ea.	965	169		1,134	1,325
1100	4' x 4' aluminum curb and cover		8	4		1,650	190		1,840	2,125
1120	Galvanized steel curb and aluminum cover		8	4		1,700	190		1,890	2,175
1140	Galvanized steel curb and cover		8	4		1,175	190		1,365	1,600
1200	2'-6" x 8'-0", aluminum curb and cover		6.60	4.848		1,900	231		2,131	2,450
1400	Galvanized steel curb and aluminum cover		6.60	4.848		1,800	231		2,031	2,350
1500	Galvanized steel curb and cover	↓	6.60	4.848		1,450	231		1,681	1,950
1800	For plexiglass panels, 2'-6" x 3'-0", add to above				↓	450			450	495
9000	Minimum labor/equipment charge	2 Carp	2	8	Job		390		390	635

07 72 36 – Smoke Vents

07 72 36.10 Smoke Hatches

0010	SMOKE HATCHES									
0200	For 3'-0" long, add to roof hatches from Section 07 72 33.10				Ea.	25%	5%			
0250	For 4'-0" long, add to roof hatches from Section 07 72 33.10					20%	5%			
0300	For 8'-0" long, add to roof hatches from Section 07 72 33.10				↓	10%	5%			

07 72 36.20 Smoke Vent Options

0010	SMOKE VENT OPTIONS									
0100	4' x 4' aluminum cover and frame	G-3	13	2.462	Ea.	2,050	117		2,167	2,425
0200	Galvanized steel cover and frame		13	2.462		1,550	117		1,667	1,875
0300	4' x 8' aluminum cover and frame		8	4		2,650	190		2,840	3,225
0400	Galvanized steel cover and frame	↓	8	4	↓	2,200	190		2,390	2,725
9000	Minimum labor/equipment charge	2 Carp	2	8	Job		390		390	635

07 72 53 – Snow Guards

07 72 53.10 Snow Guard Options

0010	SNOW GUARD OPTIONS									
0100	Slate & asphalt shingle roofs, fastened with nails	1 Rofc	160	.050	Ea.	12.55	2.09		14.64	17.55
0200	Standing seam metal roofs, fastened with set screws		48	.167		16.95	6.95		23.90	31
0300	Surface mount for metal roofs, fastened with solder		48	.167	↓	6.05	6.95		13	19.20
0400	Double rail pipe type, including pipe	↓	130	.062	L.F.	34	2.57		36.57	42

07 72 73 – Pitch Pockets

07 72 73.10 Pitch Pockets, Variable Sizes

0010	PITCH POCKETS, VARIABLE SIZES									
0100	Adjustable, 4" to 7", welded corners, 4" deep	1 Rofc	48	.167	Ea.	16.45	6.95		23.40	30.50
0200	Side extenders, 6"	"	240	.033	"	2.75	1.39		4.14	5.55

07 72 80 – Vents

07 72 80.30 Vent Options

0010	VENT OPTIONS									
0020	Plastic, for insulated decks, 1 per M.S.F.	1 Rofc	40	.200	Ea.	21	8.35		29.35	38
0100	Heavy duty		20	.400		38	16.70		54.70	72
0300	Aluminum	↓	30	.267		23	11.10		34.10	45.50
0800	Polystyrene baffles, 12" wide for 16" O.C. rafter spacing	1 Carp	90	.089	↓	.44	4.31		4.75	7.55
0900	For 24" O.C. rafter spacing		110	.073	↓	.78	3.52		4.30	6.60
9000	Minimum labor/equipment charge	↓	3	2.667	Job		129		129	211

For customer support on your Commercial Renovation Cost Data, call 877.742.4719.

213

07 76 Roof Pavers

07 76 16 – Roof Decking Pavers

07 76 16.10 Roof Pavers and Supports	Crew	Daily Output	Labor-Hours	Unit	Material	2016 Bare Costs Labor	Equipment	Total	Total Incl O&P
0010 **ROOF PAVERS AND SUPPORTS**									
1000 Roof decking pavers, concrete blocks, 2" thick, natural	1 Clab	115	.070	S.F.	3.37	2.64		6.01	8
1100 Colors		115	.070	"	3.73	2.64		6.37	8.40
1200 Support pedestal, bottom cap		960	.008	Ea.	3	.32		3.32	3.82
1300 Top cap		960	.008		4.80	.32		5.12	5.80
1400 Leveling shims, 1/16"		1920	.004		1.20	.16		1.36	1.58
1500 1/8"		1920	.004		1.20	.16		1.36	1.58
1600 Buffer pad		960	.008		2.50	.32		2.82	3.27
1700 PVC legs (4" SDR 35)		2880	.003	Inch	.12	.11		.23	.31
2000 Alternate pricing method, system in place		101	.079	S.F.	7	3		10	12.60

07 81 Applied Fireproofing

07 81 16 – Cementitious Fireproofing

07 81 16.10 Sprayed Cementitious Fireproofing

	Crew	Daily Output	Labor-Hours	Unit	Material	2016 Bare Costs Labor	Equipment	Total	Total Incl O&P
0010 **SPRAYED CEMENTITIOUS FIREPROOFING**									
0050 Not including canvas protection, normal density									
0100 Per 1" thick, on flat plate steel	G-2	3000	.008	S.F.	.53	.32	.05	.90	1.16
0200 Flat decking		2400	.010		.53	.40	.06	.99	1.30
0400 Beams		1500	.016		.53	.64	.09	1.26	1.73
0500 Corrugated or fluted decks		1250	.019		.80	.77	.11	1.68	2.24
0700 Columns, 1-1/8" thick		1100	.022		.60	.88	.12	1.60	2.21
0800 2-3/16" thick		700	.034		1.35	1.38	.19	2.92	3.92
0900 For canvas protection, add		5000	.005		.10	.19	.03	.32	.45
1000 Not including canvas protection, high density									
1100 Per 1" thick, on flat plate steel	G-2	3000	.008	S.F.	2.13	.32	.05	2.50	2.91
1110 On flat decking		2400	.010		2.13	.40	.06	2.59	3.05
1120 On beams		1500	.016		2.13	.64	.09	2.86	3.48
1130 Corrugated or fluted decks		1250	.019		2.13	.77	.11	3.01	3.70
1140 Columns, 1-1/8" thick		1100	.022		2.39	.88	.12	3.39	4.18
1150 2-3/16" thick		1100	.022		4.79	.88	.12	5.79	6.80
1170 For canvas protection, add		5000	.005		.10	.19	.03	.32	.45
1200 Not including canvas protection, retrofitting									
1210 Per 1" thick, on flat plate steel	G-2	1500	.016	S.F.	.46	.64	.09	1.19	1.64
1220 On flat decking		1200	.020		.46	.81	.11	1.38	1.92
1230 On beams		750	.032		.46	1.29	.18	1.93	2.77
1240 Corrugated or fluted decks		625	.038		.69	1.55	.22	2.46	3.48
1250 Columns, 1-1/8" thick		550	.044		.51	1.76	.25	2.52	3.67
1260 2-3/16" thick		500	.048		1.03	1.93	.27	3.23	4.54
1400 Accessories, preliminary spattered texture coat		4500	.005		.04	.21	.03	.28	.42
1410 Bonding agent	1 Plas	1000	.008		.10	.36		.46	.68
9000 Minimum labor/equipment charge	G-2	3	8	Job		320	45	365	570

07 84 Firestopping

07 84 13 – Penetration Firestopping

07 84 13.10 Firestopping		Crew	Daily Output	Labor-Hours	Unit	Material	2016 Bare Costs Labor	Equipment	Total	Total Incl O&P
0010	**FIRESTOPPING**	R078413-30								
0100	Metallic piping, non insulated									
0110	Through walls, 2" diameter	1 Carp	16	.500	Ea.	3.60	24		27.60	43.50
0120	4" diameter		14	.571		6.30	27.50		33.80	52
0130	6" diameter		12	.667		9.25	32.50		41.75	63
0140	12" diameter		10	.800		18.45	39		57.45	84
0150	Through floors, 2" diameter		32	.250		1.80	12.10		13.90	22
0160	4" diameter		28	.286		3.15	13.85		17	26
0170	6" diameter		24	.333		4.60	16.15		20.75	31.50
0180	12" diameter		20	.400		9.25	19.40		28.65	41.50
0190	Metallic piping, insulated									
0200	Through walls, 2" diameter	1 Carp	16	.500	Ea.	6.30	24		30.30	46.50
0210	4" diameter		14	.571		9.25	27.50		36.75	55
0220	6" diameter		12	.667		12.40	32.50		44.90	66.50
0230	12" diameter		10	.800		21	39		60	87
0240	Through floors, 2" diameter		32	.250		3.15	12.10		15.25	23.50
0250	4" diameter		28	.286		4.60	13.85		18.45	27.50
0260	6" diameter		24	.333		6.20	16.15		22.35	33.50
0270	12" diameter		20	.400		10.60	19.40		30	43
0280	Non metallic piping, non insulated									
0290	Through walls, 2" diameter	1 Carp	12	.667	Ea.	45.50	32.50		78	103
0300	4" diameter		10	.800		87	39		126	159
0310	6" diameter		8	1		170	48.50		218.50	266
0330	Through floors, 2" diameter		16	.500		30	24		54	72.50
0340	4" diameter		6	1.333		55.50	64.50		120	167
0350	6" diameter		6	1.333		103	64.50		167.50	219
0370	Ductwork, insulated & non insulated, round									
0380	Through walls, 6" diameter	1 Carp	12	.667	Ea.	12.40	32.50		44.90	66.50
0390	12" diameter		10	.800		21	39		60	87
0400	18" diameter		8	1		25.50	48.50		74	107
0410	Through floors, 6" diameter		16	.500		6.20	24		30.20	46.50
0420	12" diameter		14	.571		10.60	27.50		38.10	56.50
0430	18" diameter		12	.667		12.85	32.50		45.35	67
0440	Ductwork, insulated & non insulated, rectangular									
0450	With stiffener/closure angle, through walls, 6" x 12"	1 Carp	8	1	Ea.	16.20	48.50		64.70	97
0460	12" x 24"		6	1.333		32.50	64.50		97	142
0470	24" x 48"		4	2		65	97		162	230
0480	With stiffener/closure angle, through floors, 6" x 12"		10	.800		8.10	39		47.10	72.50
0490	12" x 24"		8	1		16.20	48.50		64.70	97
0500	24" x 48"		6	1.333		32.50	64.50		97	142
0510	Multi trade openings									
0520	Through walls, 6" x 12"	1 Carp	2	4	Ea.	45	194		239	365
0530	12" x 24"	"	1	8		147	390		537	795
0540	24" x 48"	2 Carp	1	16		525	775		1,300	1,850
0550	48" x 96"	"	.75	21.333		1,975	1,025		3,000	3,875
0560	Through floors, 6" x 12"	1 Carp	2	4		49.50	194		243.50	370
0570	12" x 24"	"	1	8		68	390		458	710
0580	24" x 48"	2 Carp	.75	21.333		138	1,025		1,163	1,850
0590	48" x 96"	"	.50	32		390	1,550		1,940	2,950
0600	Structural penetrations, through walls									
0610	Steel beams, W8 x 10	1 Carp	8	1	Ea.	68.50	48.50		117	154
0620	W12 x 14		6	1.333		99	64.50		163.50	215
0630	W21 x 44		5	1.600		138	77.50		215.50	279

07 84 Firestopping

07 84 13 – Penetration Firestopping

07 84 13.10 Firestopping		Crew	Daily Output	Labor-Hours	Unit	Material	2016 Bare Costs Labor	Equipment	Total	Total Incl O&P
0640	W36 x 135	1 Carp	3	2.667	Ea.	219	129		348	450
0650	Bar joists, 18" deep		6	1.333		40.50	64.50		105	151
0660	24" deep		6	1.333		54	64.50		118.50	165
0670	36" deep		5	1.600		81	77.50		158.50	216
0680	48" deep	↓	4	2	↓	108	97		205	276
0690	Construction joints, floor slab at exterior wall									
0700	Precast, brick, block or drywall exterior									
0710	2" wide joint	1 Carp	125	.064	L.F.	9.50	3.10		12.60	15.50
0720	4" wide joint	"	75	.107	"	14.85	5.15		20	25
0730	Metal panel, glass or curtain wall exterior									
0740	2" wide joint	1 Carp	40	.200	L.F.	9.50	9.70		19.20	26.50
0750	4" wide joint	"	25	.320	"	14.85	15.50		30.35	42
0760	Floor slab to drywall partition									
0770	Flat joint	1 Carp	100	.080	L.F.	13.20	3.88		17.08	21
0780	Fluted joint		50	.160		15.05	7.75		22.80	29
0790	Etched fluted joint	↓	75	.107	↓	18.50	5.15		23.65	29
0800	Floor slab to concrete/masonry partition									
0810	Flat joint	1 Carp	75	.107	L.F.	13.20	5.15		18.35	23
0820	Fluted joint	"	50	.160	"	15.05	7.75		22.80	29
0830	Concrete/CMU wall joints									
0840	1" wide	1 Carp	100	.080	L.F.	18.85	3.88		22.73	27
0850	2" wide		75	.107		23.50	5.15		28.65	34
0860	4" wide	↓	50	.160		33.50	7.75		41.25	49
0870	Concrete/CMU floor joints									
0880	1" wide	1 Carp	200	.040	L.F.	18.85	1.94		20.79	23.50
0890	2" wide		150	.053		23.50	2.58		26.08	29.50
0900	4" wide	↓	100	.080	↓	33.50	3.88		37.38	43

07 91 Preformed Joint Seals

07 91 13 – Compression Seals

07 91 13.10 Compression Seals

		Crew	Daily Output	Labor-Hours	Unit	Material	2016 Bare Costs Labor	Equipment	Total	Total Incl O&P
0010	**COMPRESSION SEALS**									
4900	O-ring type cord, 1/4"	1 Bric	472	.017	L.F.	.43	.78		1.21	1.75
4910	1/2"		440	.018		1.20	.84		2.04	2.69
4920	3/4"		424	.019		2.40	.87		3.27	4.06
4930	1"		408	.020		4.08	.91		4.99	5.95
4940	1-1/4"		384	.021		7.40	.96		8.36	9.70
4950	1-1/2"		368	.022		9.20	1.01		10.21	11.75
4960	1-3/4"		352	.023		18.55	1.05		19.60	22
4970	2"	↓	344	.023		22	1.08		23.08	26.50

07 91 16 – Joint Gaskets

07 91 16.10 Joint Gaskets

		Crew	Daily Output	Labor-Hours	Unit	Material	2016 Bare Costs Labor	Equipment	Total	Total Incl O&P
0010	**JOINT GASKETS**									
4400	Joint gaskets, neoprene, closed cell w/adh, 1/8" x 3/8"	1 Bric	240	.033	L.F.	.32	1.54		1.86	2.86
4500	1/4" x 3/4"		215	.037		.62	1.72		2.34	3.48
4700	1/2" x 1"		200	.040		1.45	1.85		3.30	4.61
4800	3/4" x 1-1/2"	↓	165	.048	↓	1.60	2.24		3.84	5.40

For customer support on your Commercial Renovation Cost Data, call 877.742.4719.

07 91 Preformed Joint Seals

07 91 23 – Backer Rods

07 91 23.10 Backer Rods	Crew	Daily Output	Labor-Hours	Unit	Material	2016 Bare Costs Labor	Equipment	Total	Total Incl O&P
0010 **BACKER RODS**									
0032 Backer rod, polyethylene, 1/4" diameter	1 Bric	460	.017	L.F.	.02	.80		.82	1.34
0052 1/2" diameter		460	.017		.04	.80		.84	1.35
0072 3/4" diameter		460	.017		.06	.80		.86	1.38
0092 1" diameter	▼	460	.017	▼	.12	.80		.92	1.44

07 91 26 – Joint Fillers

07 91 26.10 Joint Fillers	Crew	Daily Output	Labor-Hours	Unit	Material	2016 Bare Costs Labor	Equipment	Total	Total Incl O&P
0010 **JOINT FILLERS**									
4360 Butyl rubber filler, 1/4" x 1/4"	1 Bric	290	.028	L.F.	.22	1.28		1.50	2.33
4365 1/2" x 1/2"		250	.032		.90	1.48		2.38	3.40
4370 1/2" x 3/4"		210	.038		1.35	1.76		3.11	4.35
4375 3/4" x 3/4"		230	.035		2.02	1.61		3.63	4.84
4380 1" x 1"	▼	180	.044	▼	2.70	2.06		4.76	6.30
4390 For coloring, add					12%				
4980 Polyethylene joint backing, 1/4" x 2"	1 Bric	2.08	3.846	C.L.F.	12.90	178		190.90	305
4990 1/4" x 6"		1.28	6.250	"	28.50	289		317.50	500
5600 Silicone, room temp vulcanizing foam seal, 1/4" x 1/2"		1312	.006	L.F.	.46	.28		.74	.96
5610 1/2" x 1/2"		656	.012		.91	.56		1.47	1.92
5620 1/2" x 3/4"		442	.018		1.37	.84		2.21	2.86
5630 3/4" x 3/4"		328	.024		2.05	1.13		3.18	4.09
5640 1/8" x 1"		1312	.006		.46	.28		.74	.96
5650 1/8" x 3"		442	.018		1.37	.84		2.21	2.86
5670 1/4" x 3"		295	.027		2.73	1.25		3.98	5.05
5680 1/4" x 6"		148	.054		5.45	2.50		7.95	10.05
5690 1/2" x 6"		82	.098		10.90	4.51		15.41	19.35
5700 1/2" x 9"		52.50	.152		16.40	7.05		23.45	29.50
5710 1/2" x 12"	▼	33	.242	▼	22	11.20		33.20	42.50

07 92 Joint Sealants

07 92 13 – Elastomeric Joint Sealants

07 92 13.10 Masonry Joint Sealants	Crew	Daily Output	Labor-Hours	Unit	Material	2016 Bare Costs Labor	Equipment	Total	Total Incl O&P
0010 **MASONRY JOINT SEALANTS**, 1/2" x 1/2" joint									
0050 Re-caulk only, oil base	1 Bric	225	.036	L.F.	.52	1.64		2.16	3.25
0100 Acrylic latex		205	.039		.23	1.80		2.03	3.20
0200 Polyurethane		200	.040		4.43	1.85		6.28	7.90
0300 Silicone		195	.041		4.42	1.90		6.32	7.95
1000 Cut out and re-caulk, oil base		145	.055		.52	2.55		3.07	4.72
1050 Acrylic latex		130	.062		.23	2.85		3.08	4.89
1100 Polyurethane		125	.064		4.43	2.96		7.39	9.70
1150 Silicone		120	.067	▼	4.42	3.08		7.50	9.85
9000 Minimum labor/equipment charge	▼	4	2	Job		92.50		92.50	151

07 92 13.20 Caulking and Sealant Options	Crew	Daily Output	Labor-Hours	Unit	Material	2016 Bare Costs Labor	Equipment	Total	Total Incl O&P
0010 **CAULKING AND SEALANT OPTIONS**									
0050 Latex acrylic based, bulk				Gal.	27			27	29.50
0055 Bulk in place 1/4" x 1/4" bead	1 Bric	300	.027	L.F.	.08	1.23		1.31	2.10
0060 1/4" x 3/8"		294	.027		.14	1.26		1.40	2.20
0065 1/4" x 1/2"		288	.028		.19	1.28		1.47	2.30
0075 3/8" x 3/8"		284	.028		.21	1.30		1.51	2.35
0080 3/8" x 1/2"		280	.029		.28	1.32		1.60	2.46
0085 3/8" x 5/8"	▼	276	.029	▼	.35	1.34		1.69	2.57

For customer support on your Commercial Renovation Cost Data, call 877.742.4719.

217

07 92 Joint Sealants

07 92 13 – Elastomeric Joint Sealants

07 92 13.20 Caulking and Sealant Options	Crew	Daily Output	Labor-Hours	Unit	Material	2016 Bare Costs Labor	Equipment	Total	Total Incl O&P
0095	3/8" x 3/4"	1 Bric	272	.029	L.F.	.42	1.36		1.78	2.67
0100	1/2" x 1/2"		275	.029		.38	1.35		1.73	2.60
0105	1/2" x 5/8"		269	.030		.47	1.38		1.85	2.76
0110	1/2" x 3/4"		263	.030		.56	1.41		1.97	2.91
0115	1/2" x 7/8"		256	.031		.66	1.45		2.11	3.07
0120	1/2" x 1"		250	.032		.75	1.48		2.23	3.24
0125	3/4" x 3/4"		244	.033		.85	1.52		2.37	3.40
0130	3/4" x 1"		225	.036		1.13	1.64		2.77	3.92
0135	1" x 1"		200	.040		1.50	1.85		3.35	4.66
0190	Cartridges				Gal.	32.50			32.50	36
0200	11 fl. oz. cartridge				Ea.	2.80			2.80	3.08
0500	1/4" x 1/2"	1 Bric	288	.028	L.F.	.23	1.28		1.51	2.34
0600	1/2" x 1/2"		275	.029		.46	1.35		1.81	2.69
0800	3/4" x 3/4"		244	.033		1.03	1.52		2.55	3.60
0900	3/4" x 1"		225	.036		1.37	1.64		3.01	4.19
1000	1" x 1"		200	.040		1.72	1.85		3.57	4.90
1400	Butyl based, bulk				Gal.	35			35	38.50
1500	Cartridges				"	38.50			38.50	42
1700	1/4" x 1/2", 154 L.F./gal.	1 Bric	288	.028	L.F.	.23	1.28		1.51	2.34
1800	1/2" x 1/2", 77 L.F./gal.	"	275	.029	"	.45	1.35		1.80	2.69
2300	Polysulfide compounds, 1 component, bulk				Gal.	82			82	90
2400	Cartridges				"	128			128	141
2600	1 or 2 component, in place, 1/4" x 1/4", 308 L.F./gal.	1 Bric	300	.027	L.F.	.27	1.23		1.50	2.30
2700	1/2" x 1/4", 154 L.F./gal.		288	.028		.53	1.28		1.81	2.68
2900	3/4" x 3/8", 68 L.F./gal.		272	.029		1.21	1.36		2.57	3.54
3000	1" x 1/2", 38 L.F./gal.		250	.032		2.16	1.48		3.64	4.78
3200	Polyurethane, 1 or 2 component				Gal.	53			53	58.50
3300	Cartridges				"	65			65	71.50
3500	Bulk, in place, 1/4" x 1/4"	1 Bric	300	.027	L.F.	.17	1.23		1.40	2.20
3655	1/2" x 1/4"		288	.028		.34	1.28		1.62	2.47
3800	3/4" x 3/8"		272	.029		.78	1.36		2.14	3.07
3900	1" x 1/2"		250	.032		1.38	1.48		2.86	3.93
4100	Silicone rubber, bulk				Gal.	54			54	59.50
4200	Cartridges				"	53			53	58.50
4300	Bulk in place, 1/4" x 1/2", 154 L.F./gal.	1 Bric	235	.034	L.F.	.35	1.57		1.92	2.95

07 92 16 – Rigid Joint Sealants

07 92 16.10 Rigid Joint Sealants

		Crew	Daily Output	Labor-Hours	Unit	Material	Labor	Equipment	Total	Total Incl O&P
0010	**RIGID JOINT SEALANTS**									
5802	Tapes, sealant, PVC foam adhesive, 1/16" x 1/4"				L.F.	.10			.10	.11
5902	1/16" x 1/2"					.10			.10	.11
5952	1/16" x 1"					.16			.16	.18
6002	1/8" x 1/2"					.10			.10	.10

07 92 19 – Acoustical Joint Sealants

07 92 19.10 Acoustical Sealant

		Crew	Daily Output	Labor-Hours	Unit	Material	Labor	Equipment	Total	Total Incl O&P
0010	**ACOUSTICAL SEALANT**									
0020	Acoustical sealant, elastomeric, cartridges				Ea.	8.50			8.50	9.35
0025	In place, 1/4" x 1/4"	1 Bric	300	.027	L.F.	.35	1.23		1.58	2.39
0030	1/4" x 1/2"		288	.028		.69	1.28		1.97	2.85
0035	1/2" x 1/2"		275	.029		1.39	1.35		2.74	3.71
0040	1/2" x 3/4"		263	.030		2.08	1.41		3.49	4.58
0045	3/4" x 3/4"		244	.033		3.12	1.52		4.64	5.90
0050	1" x 1"		200	.040		5.55	1.85		7.40	9.10

218

07 95 Expansion Control

07 95 13 – Expansion Joint Cover Assemblies

07 95 13.50 Expansion Joint Assemblies	Crew	Daily Output	Labor-Hours	Unit	Material	2016 Bare Costs Labor	Equipment	Total	Total Incl O&P
0010 **EXPANSION JOINT ASSEMBLIES**									
0200 Floor cover assemblies, 1" space, aluminum	1 Sswk	38	.211	L.F.	18.65	11.20		29.85	40.50
0300 Bronze		38	.211		59.50	11.20		70.70	85.50
0500 2" space, aluminum		38	.211		18.15	11.20		29.35	40
0600 Bronze		38	.211		57.50	11.20		68.70	83.50
0800 Wall and ceiling assemblies, 1" space, aluminum		38	.211		16.50	11.20		27.70	38
0900 Bronze		38	.211		51	11.20		62.20	76
1100 2" space, aluminum		38	.211		16.85	11.20		28.05	38.50
1200 Bronze		38	.211		51.50	11.20		62.70	77
1400 Floor to wall assemblies, 1" space, aluminum		38	.211		19.50	11.20		30.70	41.50
1500 Bronze or stainless		38	.211		62	11.20		73.20	88
1700 Gym floor angle covers, aluminum, 3" x 3" angle		46	.174		19.50	9.25		28.75	38
1800 3" x 4" angle		46	.174		23	9.25		32.25	42
2000 Roof closures, aluminum, flat roof, low profile, 1" space		57	.140		24	7.45		31.45	39.50
2100 High profile		57	.140		26	7.45		33.45	41.50
2300 Roof to wall, low profile, 1" space		57	.140		23	7.45		30.45	38.50
2400 High profile		57	.140		26	7.45		33.45	41.50
9000 Minimum labor/equipment charge		2	4	Job		213		213	375

For customer support on your Commercial Renovation Cost Data, call 877.742.4719.

219

Division Notes

		CREW	DAILY OUTPUT	LABOR-HOURS	UNIT	BARE COSTS				TOTAL INCL O&P
	I					MAT.	LABOR	EQUIP.	TOTAL	

Estimating Tips

08 10 00 Doors and Frames

All exterior doors should be addressed for their energy conservation (insulation and seals).

- Most metal doors and frames look alike, but there may be significant differences among them. When estimating these items, be sure to choose the line item that most closely compares to the specification or door schedule requirements regarding:
  - □ type of metal
  - □ metal gauge
  - □ door core material
  - □ fire rating
  - □ finish
- Wood and plastic doors vary considerably in price. The primary determinant is the veneer material. Lauan, birch, and oak are the most common veneers. Other variables include the following:
  - □ hollow or solid core
  - □ fire rating
  - □ flush or raised panel
  - □ finish
- Door pricing includes bore for cylindrical lockset and mortise for hinges.

08 30 00 Specialty Doors and Frames

- There are many varieties of special doors, and they are usually priced per each. Add frames, hardware, or operators required for a complete installation.

08 40 00 Entrances, Storefronts, and Curtain Walls

- Glazed curtain walls consist of the metal tube framing and the glazing material. The cost data in this subdivision is presented for the metal tube framing alone or the composite wall. If your estimate requires a detailed takeoff of the framing, be sure to add the glazing cost and any tints.

08 50 00 Windows

- Most metal windows are delivered preglazed. However, some metal windows are priced without glass. Refer to 08 80 00 Glazing for glass pricing. The grade C indicates commercial grade windows, usually ASTM C-35.
- All wood windows and vinyl are priced preglazed. The glazing is insulating glass. Add the cost of screens and grills if required, and not already included.

08 70 00 Hardware

- Hardware costs add considerably to the cost of a door. The most efficient method to determine the hardware requirements for a project is to review the door and hardware schedule together. One type of door may have different hardware, depending on the door usage.
- Door hinges are priced by the pair, with most doors requiring 1-1/2 pairs per door. The hinge prices do not include installation labor, because it is included in door installation.

Hinges are classified according to the frequency of use, base material, and finish.

08 80 00 Glazing

- Different openings require different types of glass. The most common types are:
  - □ float
  - □ tempered
  - □ insulating
  - □ impact-resistant
  - □ ballistic-resistant
- Most exterior windows are glazed with insulating glass. Entrance doors and window walls, where the glass is less than 18" from the floor, are generally glazed with tempered glass. Interior windows and some residential windows are glazed with float glass.
- Coastal communities require the use of impact-resistant glass, dependant on wind speed.
- The insulation or 'u' value is a strong consideration, along with solar heat gain, to determine total energy efficiency.

Reference Numbers

Reference numbers are shown at the beginning of some major classifications. These numbers refer to related items in the Reference Section. The reference information may be an estimating procedure, an alternate pricing method, or technical information.

Note: Not all subdivisions listed here necessarily appear. ■

08 01 Operation and Maintenance of Openings

08 01 11 – Operation and Maintenance of Metal Doors and Frames

08 01 11.10 Door and Window Maintenance		Crew	Daily Output	Labor-Hours	Unit	Material	2016 Bare Costs Labor	Equipment	Total	Total Incl O&P
0010	**DOOR & WINDOW MAINTENANCE**									
1130	Install door	2 Carp	13.07	1.224	Ea.		59.50		59.50	97
1140	Remove deadbolt	1 Carp	13.85	.578			28		28	45.50
1150	Remove panic bar		7.69	1.040			50.50		50.50	82.50
1420	Remove door, plane to fit, rail, reinstall door	↓	9.60	.833	↓		40.50		40.50	66

08 01 14 – Operation and Maintenance of Wood Doors

08 01 14.15 Door and Window Maintenance

		Crew	Daily Output	Labor-Hours	Unit	Material	Labor	Equipment	Total	Total Incl O&P
0010	**DOOR & WINDOW MAINTENANCE**									
0050	Rehang single door	1 Carp	32	.250	Ea.		12.10		12.10	19.80
0100	Rehang double door		16	.500	Pr.		24		24	39.50
0350	Install window sash cord	↓	16	.500	Ea.	4.40	24		28.40	44.50

08 01 53 – Operation and Maintenance of Plastic Windows

08 01 53.81 Solid Vinyl Replacement Windows

			Crew	Daily Output	Labor-Hours	Unit	Material	Labor	Equipment	Total	Total Incl O&P
0010	**SOLID VINYL REPLACEMENT WINDOWS**	R085313-20									
0020	Double hung, insulated glass, up to 83 united inches	G	2 Carp	8	2	Ea.	345	97		442	540
0040	84 to 93	G		8	2		385	97		482	585
0060	94 to 101	G		6	2.667		385	129		514	635
0080	102 to 111	G		6	2.667		405	129		534	655
0100	112 to 120	G		6	2.667	↓	440	129		569	695
0120	For each united inch over 120, add	G		800	.020	Inch	5.60	.97		6.57	7.75
0140	Casement windows, one operating sash, 42 to 60 united inches	G		8	2	Ea.	240	97		337	420
0160	61 to 70	G		8	2		270	97		367	455
0180	71 to 80	G		8	2		290	97		387	480
0200	81 to 96	G		8	2		310	97		407	500
0220	Two operating sash, 58 to 78 united inches	G		8	2		475	97		572	685
0240	79 to 88	G		8	2		505	97		602	715
0260	89 to 98	G		8	2		550	97		647	765
0280	99 to 108	G		6	2.667		575	129		704	845
0300	109 to 121	G		6	2.667		615	129		744	885
0320	Two operating, one fixed sash, 73 to 108 united inches	G		8	2		745	97		842	980
0340	109 to 118	G		8	2		785	97		882	1,025
0360	119 to 128	G		6	2.667		805	129		934	1,100
0380	129 to 138	G		6	2.667		870	129		999	1,175
0400	139 to 156	G		6	2.667		945	129		1,074	1,250
0420	Four operating sash, 98 to 118 united inches	G		8	2		1,075	97		1,172	1,325
0440	119 to 128	G		8	2		1,150	97		1,247	1,400
0460	129 to 138	G		6	2.667		1,200	129		1,329	1,525
0480	139 to 148	G		6	2.667		1,275	129		1,404	1,600
0500	149 to 168	G		6	2.667		1,350	129		1,479	1,675
0520	169 to 178	G		6	2.667		1,450	129		1,579	1,800
0560	Fixed picture window, up to 63 united inches	G		8	2		180	97		277	355
0580	64 to 83	G		8	2		206	97		303	385
0600	84 to 101	G		8	2	↓	257	97		354	440
0620	For each united inch over 101, add	G	↓	900	.018	Inch	3.10	.86		3.96	4.82
0800	Cellulose fiber insulation, poured into sash balance cavity	G	1 Carp	36	.222	C.F.	.69	10.75		11.44	18.35
0820	Silicone caulking at perimeter	G	"	800	.010	L.F.	.17	.48		.65	.98
2000	Impact resistant replacement windows										
2005	Laminated glass, 120 mph rating, measure in united inches										
2010	Installation labor does not cover any rework of the window opening										
2020	Double hung, insulated glass, up to 101 united inches		2 Carp	8	2	Ea.	480	97		577	685
2025	For each united inch over 101, add			80	.200	Inch	3.25	9.70		12.95	19.45
2100	Casement windows, impact resistant, up to 60 united inches		↓	8	2	Ea.	450	97		547	655

For customer support on your Commercial Renovation Cost Data, call 877.742.4719.

08 01 Operation and Maintenance of Openings

08 01 53 – Operation and Maintenance of Plastic Windows

08 01 53.81 Solid Vinyl Replacement Windows

	08 01 53.81 Solid Vinyl Replacement Windows	Crew	Daily Output	Labor-Hours	Unit	Material	2016 Bare Costs Labor	Equipment	Total	Total Incl O&P
2120	61 to 70	2 Carp	8	2	Ea.	480	97		577	685
2130	71 to 80		8	2		500	97		597	710
2140	81 to 100		8	2		515	97		612	725
2150	For each united inch over 100, add		80	.200	Inch	4.55	9.70		14.25	21
2200	Awning windows, impact resistant, up to 60 united inches		8	2	Ea.	470	97		567	675
2220	61 to 70		8	2		480	97		577	685
2230	71 to 80		8	2		500	97		597	710
2240	For each united inch over 80, add		80	.200	Inch	4.55	9.70		14.25	21
2300	Picture windows, impact resistant, up to 63 united inches		8	2	Ea.	345	97		442	540
2320	63 to 83		8	2		370	97		467	565
2330	84 to 101		8	2		415	97		512	615
2340	For each united inch over 101, add		80	.200	Inch	3.25	9.70		12.95	19.45

08 05 Common Work Results for Openings

08 05 05 – Selective Demolition for Openings

08 05 05.10 Selective Demolition Doors

		Crew	Daily Output	Labor-Hours	Unit	Material	2016 Bare Costs Labor	Equipment	Total	Total Incl O&P
0010	**SELECTIVE DEMOLITION DOORS** R024119-10									
0200	Doors, exterior, 1-3/4" thick, single, 3' x 7' high	1 Clab	16	.500	Ea.		18.95		18.95	31
0202	Doors, exterior, 3' - 6' wide x 7' high		12	.667			25.50		25.50	41.50
0210	3' x 8' high		10	.800			30.50		30.50	49.50
0215	Double, 3' x 8' high		6	1.333			50.50		50.50	82.50
0220	Double, 6' x 7' high		12	.667			25.50		25.50	41.50
0500	Interior, 1-3/8" thick, single, 3' x 7' high		20	.400			15.15		15.15	25
0520	Double, 6' x 7' high		16	.500			18.95		18.95	31
0700	Bi-folding, 3' x 6'-8" high		20	.400			15.15		15.15	25
0720	6' x 6'-8" high		18	.444			16.85		16.85	27.50
0900	Bi-passing, 3' x 6'-8" high		16	.500			18.95		18.95	31
0940	6' x 6'-8" high		14	.571			21.50		21.50	35.50
1500	Remove and reset, hollow core	1 Carp	8	1			48.50		48.50	79
1520	Solid		6	1.333			64.50		64.50	106
2000	Frames, including trim, metal		8	1			48.50		48.50	79
2200	Wood	2 Carp	32	.500			24		24	39.50
2201	Alternate pricing method	1 Carp	200	.040	L.F.		1.94		1.94	3.17
2205	Remove door hardware	"	45.70	.175	Ea.		8.50		8.50	13.85
3000	Special doors, counter doors	2 Carp	6	2.667			129		129	211
3100	Double acting		10	1.600			77.50		77.50	127
3200	Floor door (trap type), or access type		8	2			97		97	158
3300	Glass, sliding, including frames		12	1.333			64.50		64.50	106
3400	Overhead, commercial, 12' x 12' high		4	4			194		194	315
3440	up to 20' x 16' high		3	5.333			258		258	420
3445	up to 35' x 30' high		1	16			775		775	1,275
3500	Residential, 9' x 7' high		8	2			97		97	158
3540	16' x 7' high		7	2.286			111		111	181
3600	Remove and reset, small		4	4			194		194	315
3620	Large		2.50	6.400			310		310	505
3700	Roll-up grille		5	3.200			155		155	253
3800	Revolving door		2	8			390		390	635
3900	Storefront swing door		3	5.333			258		258	420
3902	Cafe/bar swing door	2 Clab	8	2			76		76	124
4000	Residential lockset, exterior	1 Carp	28	.286			13.85		13.85	22.50
4224	Pocket door		8	1			48.50		48.50	79

For customer support on your Commercial Renovation Cost Data, call 877.742.4719.

223

08 05 Common Work Results for Openings

08 05 05 – Selective Demolition for Openings

08 05 05.10 Selective Demolition Doors

		Crew	Daily Output	Labor-Hours	Unit	Material	2016 Bare Costs Labor	2016 Bare Costs Equipment	Total	Total Incl O&P
4300	Remove and reset deadbolt	1 Carp	8	1	Ea.		48.50		48.50	79
5585	Remove panic device	1 Clab	10	.800			30.50		30.50	49.50
5590	Remove mail slot		45	.178			6.75		6.75	11
5595	Remove peep hole		45	.178			6.75		6.75	11
5604	Remove 9' x 7' garage door	2 Carp	9.50	1.684			81.50		81.50	133
5624	16' x 7' garage door		8	2			97		97	158
5626	9' x 7' swing up garage door		7	2.286			111		111	181
5628	16' x 7' swing up garage door		7	2.286			111		111	181
5630	Garage door demolition, remove garage door track	1 Carp	6	1.333			64.50		64.50	106
5644	Remove overhead door opener	1 Clab	5	1.600			60.50		60.50	99
5774	Remove heavy gage sectional door, 20 ga., 8' x 8'	2 Carp	7	2.286			111		111	181
5784	10' x 10'		6	2.667			129		129	211
5794	12' x 12'		5	3.200			155		155	253
5805	14' x 14'		4	4			194		194	315
6384	Remove French door unit	1 Clab	8	1			38		38	62
7100	Remove double swing pneumatic doors, openers and sensors	2 Skwk	.50	32	Opng.		1,600		1,600	2,575
7570	Remove shock absorbing door	2 Sswk	1.90	8.421	"		450		450	790
9000	Minimum labor/equipment charge	1 Carp	4	2	Job		97		97	158

08 05 05.20 Selective Demolition of Windows

		Crew	Daily Output	Labor-Hours	Unit	Material	2016 Bare Costs Labor	2016 Bare Costs Equipment	Total	Total Incl O&P
0010	**SELECTIVE DEMOLITION OF WINDOWS** R024119-10									
0200	Aluminum, including trim, to 12 S.F.	1 Clab	16	.500	Ea.		18.95		18.95	31
0240	To 25 S.F.		11	.727			27.50		27.50	45
0280	To 50 S.F.		5	1.600			60.50		60.50	99
0320	Storm windows/screens, to 12 S.F.		27	.296			11.25		11.25	18.35
0360	To 25 S.F.		21	.381			14.45		14.45	23.50
0400	To 50 S.F.		16	.500			18.95		18.95	31
0600	Glass, up to 10 S.F. per window		200	.040	S.F.		1.52		1.52	2.48
0620	Over 10 S.F. per window		150	.053	"		2.02		2.02	3.30
1000	Steel, including trim, to 12 S.F.		13	.615	Ea.		23.50		23.50	38
1020	To 25 S.F.		9	.889			33.50		33.50	55
1040	To 50 S.F.		4	2			76		76	124
2000	Wood, including trim, to 12 S.F.		22	.364			13.80		13.80	22.50
2020	To 25 S.F.		18	.444			16.85		16.85	27.50
2060	To 50 S.F.		13	.615			23.50		23.50	38
2065	To 180 S.F.		8	1			38		38	62
4400	Remove skylight, prefabricated glass block with metal frame	G-3	180	.178	S.F.		8.45		8.45	13.55
4410	Remove skylight, plstc domes, flush/curb mtd	"	395	.081	"		3.85		3.85	6.15
5020	Remove and reset window, up to a 2' x 2' window	1 Carp	6	1.333	Ea.		64.50		64.50	106
5040	Up to a 3' x 3' window		4	2			97		97	158
5080	Up to a 4' x 5' window		2	4			194		194	315
6000	Screening only	1 Clab	4000	.002	S.F.		.08		.08	.12
9000	Minimum labor/equipment charge	"	4	2	Job		76		76	124

08 11 Metal Doors and Frames

08 11 16 – Aluminum Doors and Frames

08 11 16.10 Entrance Doors	Crew	Daily Output	Labor-Hours	Unit	Material	2016 Bare Costs Labor	Equipment	Total	Total Incl O&P
0010 **ENTRANCE DOORS** and frame, aluminum, narrow stile									
0011 Including standard hardware, clear finish, no glass									
0012 Top and bottom offset pivots, 1/4" beveled glass stops, threshold									
0013 Dead bolt lock with inside thumb screw, standard push pull									
0020 3'-0" x 7'-0" opening	2 Sswk	2	8	Ea.	935	425		1,360	1,775
0025 Anodizing aluminum entr. door & frame, add					104			104	114
0030 3'-6" x 7'-0" opening	2 Sswk	2	8		880	425		1,305	1,725
0100 3'-0" x 10'-0" opening, 3' high transom		1.80	8.889		1,300	475		1,775	2,275
0200 3'-6" x 10'-0" opening, 3' high transom		1.80	8.889		1,375	475		1,850	2,325
0280 5'-0" x 7'-0" opening		2	8		1,500	425		1,925	2,400
0300 6'-0" x 7'-0" opening		1.30	12.308	↓	1,275	655		1,930	2,550
0301 6'-0" x 7'-0" opening		1.30	12.308	Pr.	1,275	655		1,930	2,550
0400 6'-0" x 10'-0" opening, 3' high transom		1.10	14.545	"	1,700	775		2,475	3,250
0520 3'-0" x 7'-0" opening, wide stile		2	8	Ea.	1,000	425		1,425	1,850
0540 3'-6" x 7'-0" opening		2	8		1,250	425		1,675	2,125
0560 5'-0" x 7'-0" opening		2	8	↓	1,625	425		2,050	2,525
0580 6'-0" x 7'-0" opening		1.30	12.308	Pr.	1,675	655		2,330	2,975
0600 7'-0" x 7'-0" opening	↓	1	16	"	1,800	850		2,650	3,475
1200 For non-standard size, add				Leaf	80%				
1250 For installation of non-standard size, add						20%			
1300 Light bronze finish, add				Leaf	36%				
1400 Dark bronze finish, add					25%				
1500 For black finish, add					40%				
1600 Concealed panic device, add				↓	955			955	1,050
1700 Electric striker release, add				Opng.	286			286	315
1800 Floor check, add				Leaf	660			660	730
1900 Concealed closer, add				"	540			540	595
9000 Minimum labor/equipment charge	2 Carp	4	4	Job		194		194	315

08 11 63 – Metal Screen and Storm Doors and Frames

08 11 63.23 Aluminum Screen and Storm Doors and Frames

	Crew	Daily Output	Labor-Hours	Unit	Material	2016 Bare Costs Labor	Equipment	Total	Total Incl O&P
0010 **ALUMINUM SCREEN AND STORM DOORS AND FRAMES**									
0020 Combination storm and screen									
0420 Clear anodic coating, 2'-8" wide	2 Carp	14	1.143	Ea.	207	55.50		262.50	320
0440 3'-0" wide	"	14	1.143	"	180	55.50		235.50	288
0500 For 7' door height, add					8%				
1020 Mill finish, 2'-8" wide	2 Carp	14	1.143	Ea.	240	55.50		295.50	355
1040 3'-0" wide	"	14	1.143	↓	263	55.50		318.50	380
1100 For 7'-0" door, add					8%				
1520 White painted, 2'-8" wide	2 Carp	14	1.143		288	55.50		343.50	405
1540 3'-0" wide	"	14	1.143		310	55.50		365.50	430
1600 For 7'-0" door, add				↓	8%				
2000 Wood door & screen, see Section 08 14 33.20									
9000 Minimum labor/equipment charge	1 Carp	4	2	Job		97		97	158

For customer support on your Commercial Renovation Cost Data, call 877.742.4719.

225

08 12 13.13 Standard Hollow Metal Frames		Crew	Daily Output	Labor-Hours	Unit	Material	2016 Bare Costs Labor	Equipment	Total	Total Incl O&P
0010	**STANDARD HOLLOW METAL FRAMES**									
0020	16 ga., up to 5-3/4" jamb depth									
0025	3'-0" x 6'-8" single	G	2 Carp	16	1	Ea.	148	48.50		196.50	242
0028	3'-6" wide, single	G		16	1		160	48.50		208.50	255
0030	4'-0" wide, single	G		16	1		159	48.50		207.50	253
0040	6'-0" wide, double	G		14	1.143		204	55.50		259.50	315
0045	8'-0" wide, double	G		14	1.143		213	55.50		268.50	325
0100	3'-0" x 7'-0" single	G		16	1		155	48.50		203.50	250
0110	3'-6" wide, single			16	1		165	48.50		213.50	261
0112	4'-0" wide, single			16	1		165	48.50		213.50	261
0140	6'-0" wide, double			14	1.143		194	55.50		249.50	305
0145	8'-0" wide, double			14	1.143		229	55.50		284.50	345
1000	16 ga., up to 4-7/8" deep, 3'-0" x 7'-0" single	G		16	1		166	48.50		214.50	262
1140	6'-0" wide, double	G		14	1.143		188	55.50		243.50	298
1200	16 ga., 8-3/4" deep, 3'-0" x 7'-0" single	G		16	1		199	48.50		247.50	298
1240	6'-0" wide, double	G		14	1.143		234	55.50		289.50	350
2800	14 ga., up to 3-7/8" deep, 3'-0" x 7'-0" single	G		16	1		181	48.50		229.50	278
2840	6'-0" wide, double	G		14	1.143		217	55.50		272.50	330
3000	14 ga., up to 5-3/4" deep, 3'-0" x 6'-8" single	G		16	1		155	48.50		203.50	250
3002	3'-6" wide, single	G		16	1		204	48.50		252.50	305
3005	4'-0" wide, single	G		16	1		210	48.50		258.50	310
3600	up to 5-3/4" jamb depth, 4'-0" x 7'-0" single			15	1.067		185	51.50		236.50	288
3620	6'-0" wide, double			12	1.333		235	64.50		299.50	365
3640	8'-0" wide, double			12	1.333		246	64.50		310.50	375
3700	8'-0" high, 4'-0" wide, single			15	1.067		235	51.50		286.50	345
3740	8'-0" wide, double			12	1.333		290	64.50		354.50	425
4000	6-3/4" deep, 4'-0" x 7'-0" single	G		15	1.067		223	51.50		274.50	330
4020	6'-0" wide, double	G		12	1.333		278	64.50		342.50	410
4040	8'-0", wide double	G		12	1.333		287	64.50		351.50	420
4100	8'-0" high, 4'-0" wide, single	G		15	1.067		276	51.50		327.50	390
4140	8'-0" wide, double	G		12	1.333		320	64.50		384.50	460
4400	8-3/4" deep, 4'-0" x 7'-0", single			15	1.067		252	51.50		303.50	360
4440	8'-0" wide, double			12	1.333		315	64.50		379.50	450
4500	4'-0" x 8'-0", single			15	1.067		283	51.50		334.50	395
4540	8'-0" wide, double			12	1.333		350	64.50		414.50	490
4900	For welded frames, add						63			63	69.50
5380	Steel frames, KD, 14 ga., "B" label, to 5-3/4" throat, to 3'-0" x 7'-0"		2 Carp	15	1.067		221	51.50		272.50	330
5400	14 ga., "B" label, up to 5-3/4" deep, 4'-0" x 7'-0" single	G		15	1.067		211	51.50		262.50	320
5440	8'-0" wide, double	G		12	1.333		274	64.50		338.50	405
5800	6-3/4" deep, 7'-0" high, 4'-0" wide, single	G		15	1.067		218	51.50		269.50	325
5840	8'-0" wide, double	G		12	1.333		390	64.50		454.50	530
6200	8-3/4" deep, 4'-0" x 7'-0" single	G		15	1.067		296	51.50		347.50	410
6240	8'-0" wide, double	G		12	1.333		380	64.50		444.50	525
6300	For "A" label use same price as "B" label										
6400	For baked enamel finish, add						30%	15%			
6500	For galvanizing, add						20%				
6600	For hospital stop, add					Ea.	295			295	325
6620	For hospital stop, stainless steel add					"	380			380	420
7900	Transom lite frames, fixed, add		2 Carp	155	.103	S.F.	54	5		59	67.50
8000	Movable, add		"	130	.123	"	68	5.95		73.95	85
9000	Minimum labor/equipment charge		1 Carp	4	2	Job		97		97	158

08 12 13 – Hollow Metal Frames

08 12 13.15 Borrowed Lites	Crew	Daily Output	Labor-Hours	Unit	Material	2016 Bare Costs Labor	Equipment	Total	Total Incl O&P
0010	**BORROWED LITES**								
0100	Hollow metal section 20 ga., 3-1/2" with glass stop	2 Carp	100	.160	L.F.	8.15	7.75		15.90	21.50
0110	3-3/4" with glass stop		100	.160		8.30	7.75		16.05	22
0120	4" with glass stop		100	.160		8.45	7.75		16.20	22
0130	4-5/8" with glass stop		100	.160		8.75	7.75		16.50	22.50
0140	4-7/8" with glass stop		100	.160		8.90	7.75		16.65	22.50
0150	5" with glass stop		100	.160		9	7.75		16.75	22.50
0160	5-3/8" with glass stop		100	.160		9.25	7.75		17	23
0300	Hollow metal section 18 ga., 3-1/2" with glass stop		80	.200		9	9.70		18.70	26
0310	3-3/4" with glass stop		80	.200		9.20	9.70		18.90	26
0320	4" with glass stop		80	.200		9.30	9.70		19	26
0330	4-5/8" with glass stop		80	.200		9.75	9.70		19.45	26.50
0340	4-7/8" with glass stop		80	.200		10	9.70		19.70	27
0350	5" with glass stop		80	.200		10.10	9.70		19.80	27
0360	5-3/8" with glass stop		80	.200		10.35	9.70		20.05	27.50
0370	6-5/8" with glass stop		80	.200		11.25	9.70		20.95	28.50
0380	6-7/8" with glass stop		80	.200		11.40	9.70		21.10	28.50
0390	7-1/4" with glass stop		80	.200		11.90	9.70		21.60	29
0500	Mullion section 18 ga., 3-1/2" with double stop		50	.320		21	15.50		36.50	48.50
0510	3-3/4" with double stop		50	.320		21.50	15.50		37	49
0530	4-5/8" with double stop		50	.320		23	15.50		38.50	50.50
0540	4-7/8" with double stop		50	.320		23.50	15.50		39	51
0550	5" with double stop		50	.320		23.50	15.50		39	51
0560	5-3/8" with double stop		50	.320		24	15.50		39.50	52
0570	6-5/8" with double stop		50	.320		26	15.50		41.50	54.50
0580	6-7/8" with double stop		50	.320		26.50	15.50		42	55
0590	7-1/4" with double stop		50	.320		27	15.50		42.50	55.50
1000	Assembled frame 2'-0" x 2'-0" 16 ga., 3-3/4" with glass stop		10	1.600	Ea.	180	77.50		257.50	325
1010	3'-0" x 2'-0"		10	1.600		180	77.50		257.50	325
1020	4'-0" x 2'-0"		10	1.600		180	77.50		257.50	325
1030	5'-0" x 2'-0"		10	1.600		180	77.50		257.50	325
1040	6'-0" x 2'-0"		8	2		270	97		367	455
1050	7'-0" x 2'-0"		8	2		270	97		367	455
1060	8'-0" x 2'-0"		8	2		270	97		367	455
1100	2'-0" x 7'-0"		8	2		270	97		367	455
1110	3'-0" x 7'-0"		8	2		270	97		367	455
1120	4'-0" x 7'-0"		8	2		270	97		367	455
1130	5'-0" x 7'-0"		8	2		270	97		367	455
1140	6'-0" x 7'-0"		6	2.667		295	129		424	535
1150	7'-0" x 7'-0"		6	2.667		295	129		424	535
1160	8'-0" x 7'-0"		6	2.667		295	129		424	535
1200	Assembled frame 2'-0" x 2'-0" 16 ga., 4-7/8" with glass stop		10	1.600		180	77.50		257.50	325
1210	3'-0" x 2'-0"		10	1.600		180	77.50		257.50	325
1220	4'-0" x 2'-0"		8	2		180	97		277	355
1230	5'-0" x 2'-0"		8	2		180	97		277	355
1240	6'-0" x 2'-0"		8	2		270	97		367	455
1250	7'-0" x 2'-0"		8	2		270	97		367	455
1260	8'-0" x 2'-0"		8	2		270	97		367	455
1300	2'-0" x 7'-0"		8	2		270	97		367	455
1310	3'-0" x 7'-0"		8	2		270	97		367	455
1320	4'-0" x 7'-0"		8	2		270	97		367	455
1330	5'-0" x 7'-0"		6	2.667		270	129		399	510

For customer support on your Commercial Renovation Cost Data, call 877.742.4719.

227

08 12 Metal Frames

08 12 13 – Hollow Metal Frames

08 12 13.15 Borrowed Lites		Crew	Daily Output	Labor-Hours	Unit	Material	2016 Bare Costs Labor	2016 Bare Costs Equipment	Total	Total Incl O&P
1340	6'-0" x 7'-0"	2 Carp	6	2.667	Ea.	295	129		424	535
1350	7'-0" x 7'-0"		6	2.667		270	129		399	510
1360	8'-0" x 7'-0"		6	2.667		295	129		424	535
1400	Assembled frame 2'-0" x 3'-0" 16 ga., 6-1/8" with glass stop		6	2.667		180	129		309	410
1410	3'-0" x 3'-0"		8	2		180	97		277	355
1420	4'-0" x 3'-0"		8	2		180	97		277	355
1430	5'-0" x 3'-0"		6	2.667		180	129		309	410
1440	6'-0" x 3'-0"		6	2.667		270	129		399	510
1450	7'-0" x 3'-0"		6	2.667		270	129		399	510
1460	8'-0" x 3'-0"	↓	10	1.600	↓	270	77.50		347.50	425

08 12 13.25 Channel Metal Frames

			Crew	Daily Output	Labor-Hours	Unit	Material	2016 Bare Costs Labor	2016 Bare Costs Equipment	Total	Total Incl O&P
0010	**CHANNEL METAL FRAMES**										
0020	Steel channels with anchors and bar stops										
0100	6" channel @ 8.2#/L.F., 3' x 7' door, weighs 150#	G	E-4	13	2.462	Ea.	239	132	11.30	382.30	505
0200	8" channel @ 11.5#/L.F., 6' x 8' door, weighs 275#	G		9	3.556		435	191	16.35	642.35	835
0300	8' x 12' door, weighs 400#	G		6.50	4.923		635	264	22.50	921.50	1,200
0400	10" channel @ 15.3#/L.F., 10' x 10' door, weighs 500#	G		6	5.333		795	286	24.50	1,105.50	1,400
0500	12' x 12' door, weighs 600#	G		5.50	5.818		955	310	27	1,292	1,625
0600	12" channel @ 20.7#/L.F., 12' x 12' door, weighs 825#	G		4.50	7.111		1,300	380	32.50	1,712.50	2,150
0700	12' x 16' door, weighs 1000#	G	↓	4	8		1,600	430	37	2,067	2,550
0800	For frames without bar stops, light sections, deduct						15%				
0900	Heavy sections, deduct				↓		10%				
9000	Minimum labor/equipment charge		E-4	4	8	Job		430	37	467	800

08 13 Metal Doors

08 13 13 – Hollow Metal Doors

08 13 13.13 Standard Hollow Metal Doors

			Crew	Daily Output	Labor-Hours	Unit	Material	2016 Bare Costs Labor	2016 Bare Costs Equipment	Total	Total Incl O&P
0010	**STANDARD HOLLOW METAL DOORS** R081313-20										
0015	Flush, full panel, hollow core										
0017	When noted doors are prepared but do not include glass or louvers										
0020	1-3/8" thick, 20 ga., 2'-0" x 6'-8"	G	2 Carp	20	.800	Ea.	335	39		374	435
0040	2'-8" x 6'-8"	G		18	.889		350	43		393	455
0060	3'-0" x 6'-8"	G		17	.941		350	45.50		395.50	460
0100	3'-0" x 7'-0"		↓	17	.941		360	45.50		405.50	475
0120	For vision lite, add						96			96	106
0140	For narrow lite, add						104			104	114
0320	Half glass, 20 ga., 2'-0" x 6'-8"	G	2 Carp	20	.800		490	39		529	605
0340	2'-8" x 6'-8"	G		18	.889		515	43		558	635
0360	3'-0" x 6'-8"	G		17	.941		510	45.50		555.50	635
0400	3'-0" x 7'-0"			17	.941		630	45.50		675.50	765
0410	1-3/8" thick, 18 ga., 2'-0" x 6'-8"			20	.800		400	39		439	505
0420	3'-0" x 6'-8"			17	.941		405	45.50		450.50	520
0425	3'-0" x 7'-0"		↓	17	.941		425	45.50		470.50	540
0450	For vision lite, add						96			96	106
0452	For narrow lite, add						104			104	114
0460	Half glass, 18 ga., 2'-0" x 6'-8"	G	2 Carp	20	.800		555	39		594	680
0465	2'-8" x 6'-8"	G		18	.889		575	43		618	705
0470	3'-0" x 6'-8"	G		17	.941		565	45.50		610.50	695
0475	3'-0" x 7'-0"	G		17	.941		575	45.50		620.50	710
0500	Hollow core, 1-3/4" thick, full panel, 20 ga., 2'-8" x 6'-8"	G		18	.889		420	43		463	530
0520	3'-0" x 6'-8"	G	↓	17	.941		420	45.50		465.50	535

08 13 13 – Hollow Metal Doors

08 13 13.13 Standard Hollow Metal Doors		Crew	Daily Output	Labor-Hours	Unit	Material	2016 Bare Costs Labor	Equipment	Total	Total Incl O&P
0640	3'-0" x 7'-0"	G 2 Carp	17	.941	Ea.	440	45.50		485.50	555
0680	4'-0" x 7'-0"	G	15	1.067		630	51.50		681.50	780
0700	4'-0" x 8'-0"	G	13	1.231		735	59.50		794.50	910
1000	18 ga., 2'-8" x 6'-8"	G	17	.941		490	45.50		535.50	615
1020	3'-0" x 6'-8"	G	16	1		475	48.50		523.50	605
1120	3'-0" x 7'-0"	G	17	.941		515	45.50		560.50	645
1150	3'-6" x 7'-0"	1 Carp	14	.571		570	27.50		597.50	675
1180	4'-0" x 7'-0"	G 2 Carp	14	1.143		630	55.50		685.50	785
1200	4'-0" x 8'-0"	G "	17	.941		735	45.50		780.50	885
1212	For vision lite, add					96			96	106
1214	For narrow lite, add					104			104	114
1230	Half glass, 20 ga., 2'-8" x 6'-8"	G 2 Carp	20	.800		575	39		614	700
1240	3'-0" x 6'-8"	G	18	.889		575	43		618	705
1260	3'-0" x 7'-0"	G	18	.889		595	43		638	720
1280	Embossed panel, 1-3/4" thick, poly core, 20 ga., 3'-0" x 7'-0"	G	18	.889		415	43		458	525
1290	Half glass, 1-3/4" thick, poly core, 20 ga., 3'-0" x 7'-0"	G	18	.889		610	43		653	740
1320	18 ga., 2'-8" x 6'-8"	G	18	.889		640	43		683	775
1340	3'-0" x 6'-8"	G	17	.941		635	45.50		680.50	770
1360	3'-0" x 7'-0"	G	17	.941		645	45.50		690.50	785
1380	4'-0" x 7'-0"	G	15	1.067		785	51.50		836.50	950
1400	4'-0" x 8'-0"	G	14	1.143		890	55.50		945.50	1,075
1500	Flush full panel, 16 ga., steel hollow core									
1520	2'-0" x 6'-8"	G 2 Carp	20	.800	Ea.	555	39		594	675
1530	2'-8" x 6'-8"	G	20	.800		555	39		594	680
1540	3'-0" x 6'-8"	G	20	.800		550	39		589	670
1560	2'-8" x 7'-0"	G	18	.889		575	43		618	700
1570	3'-0" x 7'-0"	G	18	.889		560	43		603	685
1580	3'-6" x 7'-0"	G	18	.889		655	43		698	790
1590	4'-0" x 7'-0"	G	18	.889		725	43		768	865
1600	2'-8" x 8'-0"	G	18	.889		700	43		743	840
1620	3'-0" x 8'-0"	G	18	.889		720	43		763	860
1630	3'-6" x 8'-0"	G	18	.889		800	43		843	950
1640	4'-0" x 8'-0"	G	18	.889		850	43		893	1,000
1650	1-13/16", 14 ga., 2'-8" x 7'-0"	G	10	1.600		1,125	77.50		1,202.50	1,375
1670	3'-0" x 7'-0"	G	10	1.600		1,075	77.50		1,152.50	1,325
1690	3'-6" x 7'-0"	G	10	1.600		1,200	77.50		1,277.50	1,425
1700	4'-0" x 7'-0"	G	10	1.600		1,250	77.50		1,327.50	1,500
1720	Insulated, 1-3/4" thick, full panel, 18 ga., 3'-0" x 6'-8"	G	15	1.067		475	51.50		526.50	610
1740	2'-8" x 7'-0"	G	16	1		505	48.50		553.50	635
1760	3'-0" x 7'-0"	G	15	1.067		490	51.50		541.50	625
1800	4'-0" x 8'-0"	G	13	1.231		740	59.50		799.50	910
1820	Half glass, 18 ga., 3'-0" x 6'-8"	G	16	1		635	48.50		683.50	775
1840	2'-8" x 7'-0"	G	17	.941		660	45.50		705.50	800
1860	3'-0" x 7'-0"	G	16	1		690	48.50		738.50	835
1900	4'-0" x 8'-0"	G	14	1.143		670	55.50		725.50	825
2000	For vision lite, add					96			96	106
2010	For narrow lite, add					104			104	114
8100	For bottom louver, add					285			285	315
8110	For baked enamel finish, add					30%	15%			
8120	For galvanizing, add					20%				
8190	Commercial door, flush steel, fire door see section 08 13 13 15									
8270	For dutch door with shelf, add to standard door				Ea.	300			300	330
9000	Minimum labor/equipment charge	1 Carp	4	2	Job		97		97	158

For customer support on your Commercial Renovation Cost Data, call 877.742.4719.

229

08 13 Metal Doors

08 13 13 – Hollow Metal Doors

08 13 13.15 Metal Fire Doors

		Crew	Daily Output	Labor-Hours	Unit	Material	2016 Bare Costs Labor	2016 Bare Costs Equipment	Total	Total Incl O&P
0010	**METAL FIRE DOORS** R081313-20									
0015	Steel, flush, "B" label, 90 minute									
0020	Full panel, 20 ga., 2'-0" x 6'-8"	2 Carp	20	.800	Ea.	420	39		459	525
0040	2'-8" x 6'-8"		18	.889		435	43		478	550
0060	3'-0" x 6'-8"		17	.941		435	45.50		480.50	555
0080	3'-0" x 7'-0"		17	.941		455	45.50		500.50	575
0140	18 ga., 3'-0" x 6'-8"		16	1		495	48.50		543.50	625
0160	2'-8" x 7'-0"		17	.941		520	45.50		565.50	645
0180	3'-0" x 7'-0"		16	1		505	48.50		553.50	635
0200	4'-0" x 7'-0"	↓	15	1.067	↓	650	51.50		701.50	800
0220	For "A" label, 3 hour, 18 ga., use same price as "B" label									
0240	For vision lite, add				Ea.	156			156	172
0300	Full panel, 16 ga., 2'-0" x 6'-8"	2 Carp	20	.800		545	39		584	665
0310	2'-8" x 6'-8"		18	.889		545	43		588	670
0320	3'-0" x 6'-8"		17	.941		540	45.50		585.50	670
0350	2'-8" x 7'-0"		17	.941		565	45.50		610.50	695
0360	3'-0" x 7'-0"		16	1		545	48.50		593.50	680
0370	4'-0" x 7'-0"		15	1.067		705	51.50		756.50	865
0520	Flush, "B" label 90 min., egress core, 20 ga., 2'-0" x 6'-8"		18	.889		655	43		698	790
0540	2'-8" x 6'-8"		17	.941		665	45.50		710.50	805
0560	3'-0" x 6'-8"		16	1		665	48.50		713.50	810
0580	3'-0" x 7'-0"		16	1		685	48.50		733.50	830
0640	Flush, "A" label 3 hour, egress core, 18 ga., 3'-0" x 6'-8"		15	1.067		720	51.50		771.50	880
0660	2'-8" x 7'-0"		16	1		750	48.50		798.50	900
0680	3'-0" x 7'-0"	↓	15	1.067		740	51.50		791.50	900
0700	4'-0" x 7'-0"		14	1.143	↓	885	55.50		940.50	1,050
9000	Minimum labor/equipment charge	1 Carp	4	2	Job		97		97	158

08 13 13.20 Residential Steel Doors

			Crew	Daily Output	Labor-Hours	Unit	Material	2016 Bare Costs Labor	2016 Bare Costs Equipment	Total	Total Incl O&P
0010	**RESIDENTIAL STEEL DOORS**										
0020	Prehung, insulated, exterior										
0030	Embossed, full panel, 2'-8" x 6'-8"	G	2 Carp	17	.941	Ea.	315	45.50		360.50	420
0040	3'-0" x 6'-8"	G		15	1.067		273	51.50		324.50	385
0060	3'-0" x 7'-0"	G		15	1.067		345	51.50		396.50	465
0070	5'-4" x 6'-8", double	G		8	2		630	97		727	855
0220	Half glass, 2'-8" x 6'-8"	G		17	.941		320	45.50		365.50	425
0240	3'-0" x 6'-8"	G		16	1		320	48.50		368.50	430
0260	3'-0" x 7'-0"	G		16	1		370	48.50		418.50	485
0270	5'-4" x 6'-8", double	G		8	2		650	97		747	875
1320	Flush face, full panel, 2'-8" x 6'-8"	G		16	1		264	48.50		312.50	370
1340	3'-0" x 6'-8"	G		15	1.067		264	51.50		315.50	375
1360	3'-0" x 7'-0"	G		15	1.067		296	51.50		347.50	410
1380	5'-4" x 6'-8", double	G		8	2		560	97		657	775
1420	Half glass, 2'-8" x 6'-8"	G		17	.941		325	45.50		370.50	430
1440	3'-0" x 6'-8"	G		16	1		325	48.50		373.50	435
1460	3'-0" x 7'-0"	G		16	1		400	48.50		448.50	520
1480	5'-4" x 6'-8", double	G	↓	8	2		650	97		747	875
1500	Sidelight, full lite, 1'-0" x 6'-8" with grille	G					252			252	277
1510	1'-0" x 6'-8", low e	G					274			274	300
1520	1'-0" x 6'-8", half lite	G					280			280	310
1530	1'-0" x 6'-8", half lite, low e	G					284			284	310
2300	Interior, residential, closet, bi-fold, 2'-0" x 6'-8"	G	2 Carp	16	1		161	48.50		209.50	256
2330	3'-0" wide	G	↓	16	1	↓	200	48.50		248.50	299

For customer support on your Commercial Renovation Cost Data, call 877.742.4719.

08 13 Metal Doors

08 13 13 – Hollow Metal Doors

08 13 13.20 Residential Steel Doors		Crew	Daily Output	Labor-Hours	Unit	Material	2016 Bare Costs Labor	Equipment	Total	Total Incl O&P
2360	4'-0" wide	G 2 Carp	15	1.067	Ea.	260	51.50		311.50	370
2400	5'-0" wide	G	14	1.143		315	55.50		370.50	440
2420	6'-0" wide	G	13	1.231		335	59.50		394.50	470
9000	Minimum labor/equipment charge	1 Carp	4	2	Job		97		97	158

08 13 13.25 Doors Hollow Metal

		Crew	Daily Output	Labor-Hours	Unit	Material	2016 Bare Costs Labor	Equipment	Total	Total Incl O&P
0010	**DOORS HOLLOW METAL**									
0500	Exterior, commercial, flush, 20 ga., 1-3/4" x 7'-0" x 2'-6" wide	G 2 Carp	15	1.067	Ea.	395	51.50		446.50	515
0530	2'-8" wide	G	15	1.067		435	51.50		486.50	565
0560	3'-0" wide	G	14	1.143		435	55.50		490.50	570
0590	3'-6" wide	G	14	1.143		445	55.50		500.50	580
1000	18 ga., 1-3/4" x 7'-0" x 2'-6" wide	G	15	1.067		495	51.50		546.50	630
1030	2'-8" wide	G	15	1.067		505	51.50		556.50	640
1060	3'-0" wide	G	14	1.143		490	55.50		545.50	630
1500	16 ga., 1-3/4 x 7'-0" x 2'-6" wide	G	15	1.067		570	51.50		621.50	710
1530	2'-8" wide	G	15	1.067		570	51.50		621.50	715
1560	3'-0" wide	G	14	1.143		565	55.50		620.50	710
1590	3'-6" wide	G	14	1.143		650	55.50		705.50	805
2900	Fire door, "A" label, 18 gauge, 1-3/4" x 2'-6" x 7'-0"	G	15	1.067		650	51.50		701.50	800
2930	2'-8" wide	G	15	1.067		670	51.50		721.50	820
2960	3'-0" wide	G	14	1.143		640	55.50		695.50	795
2990	3'-6" wide	G	14	1.143		735	55.50		790.50	900
3100	"B" label, 2'-6" wide	G	15	1.067		590	51.50		641.50	730
3130	2'-8" wide	G	15	1.067		590	51.50		641.50	735
3160	3'-0" wide	G	14	1.143		585	55.50		640.50	735

08 13 16 – Aluminum Doors

08 13 16.10 Commercial Aluminum Doors

		Crew	Daily Output	Labor-Hours	Unit	Material	2016 Bare Costs Labor	Equipment	Total	Total Incl O&P
0010	**COMMERCIAL ALUMINUM DOORS**, flush, no glazing									
5000	Flush panel doors, pair of 2'-6" x 7'-0"	2 Sswk	2	8	Pr.	1,500	425		1,925	2,400
5050	3'-0" x 7'-0", single		2.50	6.400	Ea.	840	340		1,180	1,525
5100	Pair of 3'-0" x 7'-0"		2	8	Pr.	1,600	425		2,025	2,500
5150	3'-6" x 7'-0", single		2.50	6.400	Ea.	1,000	340		1,340	1,725

08 14 Wood Doors

08 14 13 – Carved Wood Doors

08 14 13.10 Types of Wood Doors, Carved

		Crew	Daily Output	Labor-Hours	Unit	Material	2016 Bare Costs Labor	Equipment	Total	Total Incl O&P
0010	**TYPES OF WOOD DOORS, CARVED**									
3000	Solid wood, 1-3/4" thick stile and rail									
3020	Mahogany, 3'-0" x 7'-0", six panel	2 Carp	14	1.143	Ea.	1,275	55.50		1,330.50	1,500
3030	With two lites		10	1.600		1,950	77.50		2,027.50	2,275
3040	3'-6" x 8'-0", six panel		10	1.600		1,550	77.50		1,627.50	1,825
3050	With two lites		8	2		2,600	97		2,697	3,000
3100	Pine, 3'-0" x 7'-0", six panel		14	1.143		570	55.50		625.50	715
3110	With two lites		10	1.600		825	77.50		902.50	1,025
3120	3'-6" x 8'-0", six panel		10	1.600		950	77.50		1,027.50	1,175
3130	With two lites		8	2		1,825	97		1,922	2,175
3200	Red oak, 3'-0" x 7'-0", six panel		14	1.143		1,850	55.50		1,905.50	2,125
3210	With two lites		10	1.600		2,450	77.50		2,527.50	2,825
3220	3'-6" x 8'-0", six panel		10	1.600		2,800	77.50		2,877.50	3,200
3230	With two lites		8	2		3,400	97		3,497	3,900
4000	Hand carved door, mahogany									

For customer support on your Commercial Renovation Cost Data, call 877.742.4719.

231

08 14 Wood Doors

08 14 13 – Carved Wood Doors

08 14 13.10 Types of Wood Doors, Carved	Crew	Daily Output	Labor-Hours	Unit	Material	2016 Bare Costs Labor	Equipment	Total	Total Incl O&P
4020	3'-0" x 7'-0", simple design	2 Carp	14	1.143	Ea.	1,750	55.50		1,805.50	2,025
4030	Intricate design		11	1.455		3,700	70.50		3,770.50	4,200
4040	3'-6" x 8'-0", simple design		10	1.600		3,000	77.50		3,077.50	3,425
4050	Intricate design	↓	8	2		3,700	97		3,797	4,225
4400	For custom finish, add					475			475	525
4600	Side light, mahogany, 7'-0" x 1'-6" wide, 4 lites	2 Carp	18	.889		1,100	43		1,143	1,275
4610	6 lites		14	1.143		2,625	55.50		2,680.50	3,000
4620	8'-0" x 1'-6" wide, 4 lites		14	1.143		1,800	55.50		1,855.50	2,075
4630	6 lites		10	1.600		2,100	77.50		2,177.50	2,425
4640	Side light, oak, 7'-0" x 1'-6" wide, 4 lites		18	.889		1,200	43		1,243	1,400
4650	6 lites		14	1.143		2,100	55.50		2,155.50	2,400
4660	8'-0" x 1'-6" wide, 4 lites		14	1.143		1,100	55.50		1,155.50	1,300
4670	6 lites	↓	10	1.600	↓	2,100	77.50		2,177.50	2,425

08 14 16 – Flush Wood Doors

08 14 16.09 Smooth Wood Doors

		Crew	Daily Output	Labor-Hours	Unit	Material	2016 Bare Costs Labor	Equipment	Total	Total Incl O&P
0010	**SMOOTH WOOD DOORS**									
0015	Flush, interior, hollow core									
0025	Lauan face, 1-3/8", 3'-0" x 6'-8"	2 Carp	17	.941	Ea.	54.50	45.50		100	135
0030	4'-0" x 6'-8"		16	1		126	48.50		174.50	218
0080	1-3/4", 2'-0" x 6'-8"	↓	17	.941		48.50	45.50		94	128
0085	3'-0" x 6'-8"	1 Carp	16	.500		54.50	24		78.50	99.50
0108	3'-0" x 7'-0"	2 Carp	16	1	↓	145	48.50		193.50	239
0112	Pair of 3'-0" x 7'-0"		9	1.778	Pr.	125	86		211	278
0140	Birch face, 1-3/8", 2'-6" x 6'-8"		17	.941	Ea.	85	45.50		130.50	168
0180	3'-0" x 6'-8"		17	.941		95.50	45.50		141	180
0200	4'-0" x 6'-8"		16	1		158	48.50		206.50	253
0202	1-3/4", 2'-0" x 6'-8"		17	.941		57	45.50		102.50	138
0204	2'-4" x 7'-0"		16	1		121	48.50		169.50	212
0206	2'-6" x 7'-0"		16	1		125	48.50		173.50	217
0210	3'-0" x 7'-0"		16	1	↓	138	48.50		186.50	231
0214	Pair of 3'-0" x 7'-0"		9	1.778	Pr.	266	86		352	435
0220	Oak face, 1-3/8", 2'-0" x 6'-8"		17	.941	Ea.	104	45.50		149.50	189
0280	3'-0" x 6'-8"		17	.941		115	45.50		160.50	202
0300	4'-0" x 6'-8"		16	1		139	48.50		187.50	232
0305	1-3/4", 2'-6" x 6'-8"		17	.941		110	45.50		155.50	196
0310	3'-0" x 7'-0"		16	1		211	48.50		259.50	310
0320	Walnut face, 1-3/8", 2'-0" x 6'-8"		17	.941		182	45.50		227.50	276
0340	2'-6" x 6'-8"		17	.941		189	45.50		234.50	283
0380	3'-0" x 6'-8"		17	.941		197	45.50		242.50	291
0400	4'-0" x 6'-8"	↓	16	1		216	48.50		264.50	315
0430	For 7'-0" high, add					26			26	28.50
0440	For 8'-0" high, add					36			36	39.50
0480	For prefinishing, clear, add					46			46	50.50
0500	For prefinishing, stain, add					57			57	62.50
1320	M.D. overlay on hardboard, 1-3/8", 2'-0" x 6'-8"	2 Carp	17	.941		115	45.50		160.50	202
1340	2'-6" x 6'-8"		17	.941		115	45.50		160.50	202
1380	3'-0" x 6'-8"		17	.941		126	45.50		171.50	214
1400	4'-0" x 6'-8"	↓	16	1		178	48.50		226.50	275
1420	For 7'-0" high, add					15.75			15.75	17.35
1440	For 8'-0" high, add					31.50			31.50	34.50
1720	H.P. plastic laminate, 1-3/8", 2'-0" x 6'-8"	2 Carp	16	1		260	48.50		308.50	365
1740	2'-6" x 6'-8"	↓	16	1	Ea.	260	48.50		308.50	365

08 14 Wood Doors

08 14 16 – Flush Wood Doors

08 14 16.09 Smooth Wood Doors		Crew	Daily Output	Labor-Hours	Unit	Material	2016 Bare Costs Labor	Equipment	Total	Total Incl O&P
1780	3'-0" x 6'-8"	2 Carp	15	1.067	Ea.	290	51.50		341.50	405
1800	4'-0" x 6'-8"	↓	14	1.143		385	55.50		440.50	510
1820	For 7'-0" high, add					15.75			15.75	17.35
1840	For 8'-0" high, add					31.50			31.50	34.50
2020	Particle core, lauan face, 1-3/8", 2'-6" x 6'-8"	2 Carp	15	1.067		92	51.50		143.50	186
2040	3'-0" x 6'-8"		14	1.143		94	55.50		149.50	194
2080	3'-0" x 7'-0"		13	1.231		101	59.50		160.50	209
2085	4'-0" x 7'-0"		12	1.333		123	64.50		187.50	241
2110	1-3/4", 3'-0" x 7'-0"		13	1.231		145	59.50		204.50	258
2120	Birch face, 1-3/8", 2'-6" x 6'-8"		15	1.067		103	51.50		154.50	198
2140	3'-0" x 6'-8"		14	1.143		113	55.50		168.50	215
2180	3'-0" x 7'-0"		13	1.231		123	59.50		182.50	233
2200	4'-0" x 7'-0"		12	1.333		139	64.50		203.50	259
2205	1-3/4", 3'-0" x 7'-0"		13	1.231		123	59.50		182.50	233
2220	Oak face, 1-3/8", 2'-6" x 6'-8"		15	1.067		116	51.50		167.50	213
2240	3'-0" x 6'-8"		14	1.143		128	55.50		183.50	232
2280	3'-0" x 7'-0"		13	1.231		133	59.50		192.50	244
2300	4'-0" x 7'-0"		12	1.333		156	64.50		220.50	278
2305	1-3/4" 3'-0" x 7'-0"		13	1.231		190	59.50		249.50	305
2320	Walnut face, 1-3/8", 2'-0" x 6'-8"		15	1.067		123	51.50		174.50	220
2340	2'-6" x 6'-8"		14	1.143		139	55.50		194.50	244
2380	3'-0" x 6'-8"		13	1.231		156	59.50		215.50	270
2400	4'-0" x 6'-8"	↓	12	1.333		206	64.50		270.50	335
2440	For 8'-0" high, add					41			41	45
2460	For 8'-0" high walnut, add					36			36	39.50
2720	For prefinishing, clear, add					36			36	39.50
2740	For prefinishing, stain, add					53			53	58.50
3320	M.D. overlay on hardboard, 1-3/8", 2'-6" x 6'-8"	2 Carp	14	1.143		106	55.50		161.50	208
3340	3'-0" x 6'-8"		13	1.231		115	59.50		174.50	225
3380	3'-0" x 7'-0"		12	1.333		117	64.50		181.50	235
3400	4'-0" x 7'-0"	↓	10	1.600		155	77.50		232.50	298
3440	For 8'-0" height, add					37			37	40.50
3460	For solid wood core, add					42			42	46
3720	H.P. plastic laminate, 1-3/8", 2'-6" x 6'-8"	2 Carp	13	1.231		160	59.50		219.50	274
3740	3'-0" x 6'-8"		12	1.333		185	64.50		249.50	310
3780	3'-0" x 7'-0"		11	1.455		190	70.50		260.50	325
3800	4'-0" x 7'-0"	↓	8	2		225	97		322	405
3840	For 8'-0" height, add					37			37	40.50
3860	For solid wood core, add					42			42	46
4000	Exterior, flush, solid core, birch, 1-3/4" x 2'-6" x 7'-0"	2 Carp	15	1.067		173	51.50		224.50	275
4020	2'-8" wide		15	1.067		159	51.50		210.50	260
4040	3'-0" wide	↓	14	1.143		209	55.50		264.50	320
4045	3'-0" x 8'-0"	1 Carp	8	1		430	48.50		478.50	555
4100	Oak faced 1-3/4" x 2'-6" x 7'-0"	2 Carp	15	1.067		215	51.50		266.50	320
4120	2'-8" wide		15	1.067		225	51.50		276.50	335
4140	3'-0" wide	↓	14	1.143		230	55.50		285.50	345
4160	Walnut faced, 1-3/4" x 3'-0" x 6'-8"	1 Carp	17	.471		295	23		318	365
4180	3'-6" wide	"	17	.471		365	23		388	440
4200	Walnut faced, 1-3/4" x 2'-6" x 7'-0"	2 Carp	15	1.067		310	51.50		361.50	425
4220	2'-8" wide		15	1.067		315	51.50		366.50	430
4240	3'-0" wide	↓	14	1.143		320	55.50		375.50	440
4300	For 6'-8" high door, deduct from 7'-0" door					16.80			16.80	18.50
5000	Wood doors, for vision lite, add					96			96	106

08 14 Wood Doors

08 14 16 – Flush Wood Doors

08 14 16.09 Smooth Wood Doors	Crew	Daily Output	Labor-Hours	Unit	Material	2016 Bare Costs Labor	Equipment	Total	Total Incl O&P
5010 Wood doors, for narrow lite, add				Ea.	104			104	114
5015 Wood doors, for bottom (or top) louver, add				▼	285			285	315
9000 Minimum labor/equipment charge	1 Carp	4	2	Job		97		97	158

08 14 16.10 Wood Doors Decorator

	Crew	Daily Output	Labor-Hours	Unit	Material	2016 Bare Costs Labor	Equipment	Total	Total Incl O&P
0010 **WOOD DOORS DECORATOR**									
1800 Exterior, flush, solid wood core, birch 1-3/4" x 2'-6" x 7'-0"	2 Carp	15	1.067	Ea.	320	51.50		371.50	435
1820 2'-8" wide		15	1.067		325	51.50		376.50	445
1840 3'-0" wide		14	1.143		335	55.50	·	390.50	460
1900 Oak faced, 1-3/4" x 2'-6" x 7'-0"		15	1.067		320	51.50		371.50	435
1920 2'-8" wide		15	1.067		330	51.50		381.50	450
1940 3'-0" wide		14	1.143		340	55.50		395.50	465
2100 Walnut faced, 1-3/4" x 2'-6"x 7'-0"		15	1.067		385	51.50		436.50	510
2120 2'-8" wide		15	1.067		400	51.50		451.50	525
2140 3'-0" wide	▼	14	1.143	▼	430	55.50		485.50	565

08 14 16.20 Wood Fire Doors

	Crew	Daily Output	Labor-Hours	Unit	Material	2016 Bare Costs Labor	Equipment	Total	Total Incl O&P
0010 **WOOD FIRE DOORS**									
0020 Particle core, 7 face plys, "B" label,									
0040 1 hour, birch face, 1-3/4" x 2'-6" x 6'-8"	2 Carp	14	1.143	Ea.	420	55.50		475.50	550
0080 3'-0" x 6'-8"		13	1.231		400	59.50		459.50	540
0090 3'-0" x 7'-0"		12	1.333		435	64.50		499.50	585
0100 4'-0" x 7'-0"		12	1.333		540	64.50		604.50	700
0140 Oak face, 2'-6" x 6'-8"		14	1.143		450	55.50		505.50	585
0180 3'-0" x 6'-8"		13	1.231		455	59.50		514.50	600
0190 3'-0" x 7'-0"		12	1.333		460	64.50		524.50	610
0200 4'-0" x 7'-0"		12	1.333		590	64.50		654.50	755
0240 Walnut face, 2'-6" x 6'-8"		14	1.143		470	55.50		525.50	605
0280 3'-0" x 6'-8"		13	1.231		490	59.50		549.50	640
0290 3'-0" x 7'-0"		12	1.333		510	64.50		574.50	665
0300 4'-0" x 7'-0"		12	1.333		630	64.50		694.50	800
0440 M.D. overlay on hardboard, 2'-6" x 6'-8"		15	1.067		315	51.50		366.50	430
0480 3'-0" x 6'-8"		14	1.143		375	55.50		430.50	505
0490 3'-0" x 7'-0"		13	1.231		395	59.50		454.50	535
0500 4'-0" x 7'-0"		12	1.333		415	64.50		479.50	560
0740 90 minutes, birch face, 1-3/4" x 2'-6" x 6'-8"		14	1.143		325	55.50		380.50	445
0780 3'-0" x 6'-8"		13	1.231		320	59.50		379.50	450
0790 3'-0" x 7'-0"		12	1.333		395	64.50		459.50	535
0800 4'-0" x 7'-0"		12	1.333		545	64.50		609.50	700
0840 Oak face, 2'-6" x 6'-8"		14	1.143		430	55.50		485.50	565
0880 3'-0" x 6'-8"		13	1.231		440	59.50		499.50	585
0890 3'-0" x 7'-0"		12	1.333		455	64.50		519.50	605
0900 4'-0" x 7'-0"		12	1.333		590	64.50		654.50	755
0940 Walnut face, 2'-6" x 6'-8"		14	1.143		400	55.50		455.50	530
0980 3'-0" x 6'-8"		13	1.231		410	59.50		469.50	550
0990 3'-0" x 7'-0"		12	1.333		470	64.50		534.50	620
1000 4'-0" x 7'-0"		12	1.333		620	64.50		684.50	785
1140 M.D. overlay on hardboard, 2'-6" x 6'-8"		15	1.067		355	51.50		406.50	475
1180 3'-0" x 6'-8"		14	1.143		375	55.50		430.50	505
1190 3'-0" x 7'-0"		13	1.231		405	59.50		464.50	545
1200 4'-0" x 7'-0"	▼	12	1.333		455	64.50		519.50	605
1240 For 8'-0" height, add					75			75	82.50
1260 For 8'-0" height walnut, add				▼	90			90	99
2200 Custom architectural "B" label, flush, 1-3/4" thick, birch,									

08 14 Wood Doors

08 14 16 – Flush Wood Doors

08 14 16.20 Wood Fire Doors	Crew	Daily Output	Labor-Hours	Unit	Material	2016 Bare Costs Labor	Equipment	Total	Total Incl O&P
2210	Solid core									
2220	2'-6" x 7'-0"	2 Carp	15	1.067	Ea.	305	51.50		356.50	420
2260	3'-0" x 7'-0"		14	1.143		315	55.50		370.50	435
2300	4'-0" x 7'-0"		13	1.231		400	59.50		459.50	540
2420	4'-0" x 8'-0"		11	1.455		430	70.50		500.50	590
2480	For oak veneer, add					50%				
2500	For walnut veneer, add					75%				
9000	Minimum labor/equipment charge	1 Carp	4	2	Job		97		97	158

08 14 33 – Stile and Rail Wood Doors

08 14 33.10 Wood Doors Paneled

08 14 33.10 Wood Doors Paneled		Crew	Daily Output	Labor-Hours	Unit	Material	2016 Bare Costs Labor	Equipment	Total	Total Incl O&P
0010	**WOOD DOORS PANELED**									
0020	Interior, six panel, hollow core, 1-3/8" thick									
0040	Molded hardboard, 2'-0" x 6'-8"	2 Carp	17	.941	Ea.	62	45.50		107.50	143
0060	2'-6" x 6'-8"		17	.941		64	45.50		109.50	145
0070	2'-8" x 6'-8"		17	.941		67	45.50		112.50	148
0080	3'-0" x 6'-8"		17	.941		72	45.50		117.50	154
0140	Embossed print, molded hardboard, 2'-0" x 6'-8"		17	.941		64	45.50		109.50	145
0160	2'-6" x 6'-8"		17	.941		64	45.50		109.50	145
0180	3'-0" x 6'-8"		17	.941		72	45.50		117.50	154
0540	Six panel, solid, 1-3/8" thick, pine, 2'-0" x 6'-8"		15	1.067		155	51.50		206.50	256
0560	2'-6" x 6'-8"		14	1.143		170	55.50		225.50	278
0580	3'-0" x 6'-8"		13	1.231		145	59.50		204.50	258
1020	Two panel, bored rail, solid, 1-3/8" thick, pine, 1'-6" x 6'-8"		16	1		270	48.50		318.50	375
1040	2'-0" x 6'-8"		15	1.067		355	51.50		406.50	475
1060	2'-6" x 6'-8"		14	1.143		400	55.50		455.50	530
1340	Two panel, solid, 1-3/8" thick, fir, 2'-0" x 6'-8"		15	1.067		160	51.50		211.50	261
1360	2'-6" x 6'-8"		14	1.143		210	55.50		265.50	320
1380	3'-0" x 6'-8"		13	1.231		415	59.50		474.50	555
1740	Five panel, solid, 1-3/8" thick, fir, 2'-0" x 6'-8"		15	1.067		280	51.50		331.50	395
1760	2'-6" x 6'-8"		14	1.143		420	55.50		475.50	550
1780	3'-0" x 6'-8"		13	1.231		420	59.50		479.50	560
4190	Exterior, Knotty pine, paneled, 1-3/4", 3'-0" x 6'-8"		16	1		780	48.50		828.50	940
4195	Double 1-3/4", 3'-0" x 6'-8"		16	1		1,550	48.50		1,598.50	1,800
4200	Ash, paneled, 1-3/4", 3'-0" x 6'-8"		16	1		890	48.50		938.50	1,050
4205	Double 1-3/4", 3'-0" x 6'-8"		16	1		1,775	48.50		1,823.50	2,025
4210	Cherry, paneled, 1-3/4", 3'-0" x 6'-8"		16	1		890	48.50		938.50	1,050
4215	Double 1-3/4", 3'-0" x 6'-8"		16	1		1,775	48.50		1,823.50	2,025
4230	Ash, paneled, 1-3/4", 3'-0" x 8'-0"		16	1		950	48.50		998.50	1,125
4235	Double 1-3/4", 3'-0" x 8'-0"		16	1		1,900	48.50		1,948.50	2,175
4240	Hard Maple, paneled, 1-3/4", 3'-0" x 8'-0"		16	1		950	48.50		998.50	1,125
4245	Double 1-3/4", 3'-0" x 8'-0"		16	1		1,900	48.50		1,948.50	2,175
4250	Cherry, paneled, 1-3/4", 3'-0" x 8'-0"		16	1		1,000	48.50		1,048.50	1,175
4255	Double 1-3/4", 3'-0" x 8'-0"		16	1		2,000	48.50		2,048.50	2,275
9000	Minimum labor/equipment charge	1 Carp	4	2	Job		97		97	158

08 14 33.20 Wood Doors Residential

08 14 33.20 Wood Doors Residential		Crew	Daily Output	Labor-Hours	Unit	Material	2016 Bare Costs Labor	Equipment	Total	Total Incl O&P
0010	**WOOD DOORS RESIDENTIAL**									
0200	Exterior, combination storm & screen, pine									
0260	2'-8" wide	2 Carp	10	1.600	Ea.	300	77.50		377.50	455
0280	3'-0" wide		9	1.778		310	86		396	480
0300	7'-1" x 3'-0" wide		9	1.778		365	86		451	540
0400	Full lite, 6'-9" x 2'-6" wide		11	1.455		320	70.50		390.50	465
0420	2'-8" wide		10	1.600		320	77.50		397.50	475

For customer support on your Commercial Renovation Cost Data, call 877.742.4719.

235

08 14 Wood Doors

08 14 33 – Stile and Rail Wood Doors

08 14 33.20 Wood Doors Residential	Crew	Daily Output	Labor-Hours	Unit	Material	2016 Bare Costs Labor	Equipment	Total	Total Incl O&P
0440	3'-0" wide	2 Carp	9	1.778	Ea.	325	86		411	495
0500	7'-1" x 3'-0" wide		9	1.778		355	86		441	530
0700	Dutch door, pine, 1-3/4" x 2'-8" x 6'-8", 6 panel		12	1.333		825	64.50		889.50	1,025
0720	Half glass		10	1.600		900	77.50		977.50	1,125
0800	3'-0" wide, 6 panel		12	1.333		790	64.50		854.50	970
0820	Half glass		10	1.600		975	77.50		1,052.50	1,200
1000	Entrance door, colonial, 1-3/4" x 6'-8" x 2'-8" wide		16	1		550	48.50		598.50	685
1020	6 panel pine, 3'-0" wide		15	1.067		445	51.50		496.50	575
1100	8 panel pine, 2'-8" wide		16	1		600	48.50		648.50	740
1120	3'-0" wide		15	1.067		570	51.50		621.50	710
1200	For tempered safety glass lites (min of 2), add					79			79	87
1300	Flush, birch, solid core, 1-3/4" x 6'-8" x 2'-8" wide	2 Carp	16	1		113	48.50		161.50	204
1320	3'-0" wide		15	1.067		119	51.50		170.50	215
1350	7'-0" x 2'-8" wide		16	1		123	48.50		171.50	214
1360	3'-0" wide		15	1.067		143	51.50		194.50	242
1380	For tempered safety glass lites, add					105			105	116
1420	6'-8" x 3'-0" wide, fir	2 Carp	16	1		470	48.50		518.50	595
1720	Carved mahogany 3'-0" x 6'-8"		15	1.067		1,375	51.50		1,426.50	1,600
1800	Lauan, solid core, 1-3/4" x 7'-0" x 2'-4" wide		16	1		129	48.50		177.50	221
1810	2'-6" wide		15	1.067		134	51.50		185.50	232
1820	2'-8" wide		9	1.778		139	86		225	294
1830	3'-0" wide		16	1		152	48.50		200.50	247
1840	3'-4" wide		16	1		242	48.50		290.50	345
1850	Pair of 3'-0" wide		15	1.067	Pr.	305	51.50		356.50	420
2700	Interior, closet, bi-fold, w/hardware, no frame or trim incl.									
2720	Flush, birch, 2'-6" x 6'-8"	2 Carp	13	1.231	Ea.	67.50	59.50		127	172
2740	3'-0" wide		13	1.231		71.50	59.50		131	176
2760	4'-0" wide		12	1.333		109	64.50		173.50	226
2780	5'-0" wide		11	1.455		105	70.50		175.50	231
2800	6'-0" wide		10	1.600		128	77.50		205.50	268
2820	Flush, hardboard, primed, 6'-8" x 2'-6" wide		13	1.231		67.50	59.50		127	172
2840	3'-0" wide		13	1.231		75	59.50		134.50	180
2860	4'-0" wide		12	1.333		150	64.50		214.50	271
2880	5'-0" wide		11	1.455		158	70.50		228.50	288
2900	6'-0" wide		10	1.600		159	77.50		236.50	300
3000	Raised panel pine, 6'-6" or 6'-8" x 2'-6" wide		13	1.231		205	59.50		264.50	325
3020	3'-0" wide		13	1.231		282	59.50		341.50	410
3040	4'-0" wide		12	1.333		305	64.50		369.50	440
3060	5'-0" wide		11	1.455		365	70.50		435.50	515
3080	6'-0" wide		10	1.600		400	77.50		477.50	565
3200	Louvered, pine, 6'-6" or 6'-8" x 2'-6" wide		13	1.231		144	59.50		203.50	256
3220	3'-0" wide		13	1.231		210	59.50		269.50	330
3240	4'-0" wide		12	1.333		235	64.50		299.50	365
3260	5'-0" wide		11	1.455		262	70.50		332.50	405
3280	6'-0" wide		10	1.600		289	77.50		366.50	445
4400	Bi-passing closet, incl. hardware and frame, no trim incl.									
4420	Flush, lauan, 6'-8" x 4'-0" wide	2 Carp	12	1.333	Opng.	176	64.50		240.50	300
4440	5'-0" wide		11	1.455		187	70.50		257.50	320
4460	6'-0" wide		10	1.600		209	77.50		286.50	355
4600	Flush, birch, 6'-8" x 4'-0" wide		12	1.333		242	64.50		306.50	370
4620	5'-0" wide		11	1.455		212	70.50		282.50	350
4640	6'-0" wide		10	1.600		283	77.50		360.50	435
4800	Louvered, pine, 6'-8" x 4'-0" wide		12	1.333		485	64.50		549.50	635

08 14 Wood Doors

08 14 33 – Stile and Rail Wood Doors

08 14 33.20 Wood Doors Residential	Crew	Daily Output	Labor-Hours	Unit	Material	2016 Bare Costs Labor	Equipment	Total	Total Incl O&P
4820	5'-0" wide	2 Carp	11	1.455	Opng.	525	70.50		595.50	695
4840	6'-0" wide		10	1.600		625	77.50		702.50	815
5000	Paneled, pine, 6'-8" x 4'-0" wide		12	1.333		470	64.50		534.50	625
5020	5'-0" wide		11	1.455		545	70.50		615.50	715
5040	6'-0" wide		10	1.600		730	77.50		807.50	930
5042	8'-0" wide	▼	12	1.333	▼	940	64.50		1,004.50	1,125
6100	Folding accordion, closet, including track and frame									
6120	Vinyl, 2 layer, stock	2 Carp	10	1.600	Ea.	66	77.50		143.50	200
6200	Rigid PVC	"	10	1.600	"	55.50	77.50		133	188
7310	Passage doors, flush, no frame included									
7320	Hardboard, hollow core, 1-3/8" x 6'-8" x 1'-6" wide	2 Carp	18	.889	Ea.	40.50	43		83.50	115
7330	2'-0" wide		18	.889		43	43		86	118
7340	2'-6" wide		18	.889		47.50	43		90.50	123
7350	2'-8" wide		18	.889		49	43		92	125
7360	3'-0" wide		17	.941		52	45.50		97.50	132
7420	Lauan, hollow core, 1-3/8" x 6'-8" x 1'-6" wide		18	.889		35.50	43		78.50	110
7440	2'-0" wide		18	.889		35	43		78	109
7450	2'-4" wide		18	.889		38.50	43		81.50	113
7460	2'-6" wide		18	.889		38.50	43		81.50	113
7480	2'-8" wide		18	.889		40	43		83	115
7500	3'-0" wide		17	.941		42.50	45.50		88	121
7700	Birch, hollow core, 1-3/8" x 6'-8" x 1'-6" wide		18	.889		41	43		84	116
7720	2'-0" wide		18	.889		42.50	43		85.50	117
7740	2'-6" wide		18	.889		50	43		93	126
7760	2'-8" wide		18	.889		50.50	43		93.50	126
7780	3'-0" wide		17	.941		52	45.50		97.50	132
8000	Pine louvered, 1-3/8" x 6'-8" x 1'-6" wide		19	.842		120	41		161	199
8020	2'-0" wide		18	.889		131	43		174	215
8040	2'-6" wide		18	.889		150	43		193	236
8060	2'-8" wide		18	.889		161	43		204	248
8080	3'-0" wide		17	.941		175	45.50		220.50	267
8300	Pine paneled, 1-3/8" x 6'-8" x 1'-6" wide		19	.842		128	41		169	208
8320	2'-0" wide		18	.889		153	43		196	239
8330	2'-4" wide		18	.889		182	43		225	271
8340	2'-6" wide		18	.889		182	43		225	271
8360	2'-8" wide		18	.889		185	43		228	275
8380	3'-0" wide		17	.941		204	45.50		249.50	299
8490	3'-0" wide		17	.941		276	45.50		321.50	380
8804	Pocket door, 6 panel pine, 2'-6" x 6'-8" with frame		10.50	1.524		345	74		419	500
8814	2'-8" x 6'-8"		10.50	1.524		350	74		424	505
8824	3'-0" x 6'-8"		10.50	1.524		355	74		429	510
9000	Passage doors, flush, no frame, birch, solid core, 1-3/8" x 2'-4" x 7'-0"		16	1		118	48.50		166.50	209
9020	2'-8" wide		16	1		125	48.50		173.50	217
9040	3'-0" wide		16	1		135	48.50		183.50	228
9060	3'-4" wide		15	1.067	▼	156	51.50		207.50	257
9080	Pair of 3'-0" wide		9	1.778	Pr.	267	86		353	435
9100	Lauan, solid core, 1-3/8" x 7'-0" x 2'-4" wide		16	1	Ea.	137	48.50		185.50	229
9120	2'-8" wide		16	1		146	48.50		194.50	240
9140	3'-0" wide		16	1		155	48.50		203.50	249
9160	3'-4" wide		15	1.067	▼	163	51.50		214.50	264
9180	Pair of 3'-0" wide		9	1.778	Pr.	281	86		367	450
9200	Hardboard, solid core, 1-3/8" x 7'-0" x 2'-4" wide		16	1	Ea.	169	48.50		217.50	265
9220	2'-8" wide	▼	16	1	▼	175	48.50		223.50	272

For customer support on your Commercial Renovation Cost Data, call 877.742.4719.

237

08 14 Wood Doors

08 14 33 – Stile and Rail Wood Doors

08 14 33.20 Wood Doors Residential		Crew	Daily Output	Labor-Hours	Unit	Material	2016 Bare Costs Labor	Equipment	Total	Total Incl O&P
9240	3'-0" wide	2 Carp	16	1	Ea.	180	48.50		228.50	277
9260	3'-4" wide		15	1.067		199	51.50		250.50	305
9900	Minimum labor/equipment charge	1 Carp	4	2	Job		97		97	158

08 14 35 – Torrified Doors

08 14 35.10 Torrified Exterior Doors

		Crew	Daily Output	Labor-Hours	Unit	Material	2016 Bare Costs Labor	Equipment	Total	Total Incl O&P
0010	**TORRIFIED EXTERIOR DOORS**									
0020	Wood doors made from torrified wood, exterior									
0030	All doors require a finish be applied, all glass is insulated									
0040	All doors require pilot holes for all fasteners									
0100	6 panel, paint grade poplar, 1-3/4" x 3'-0"x 6'-8"	2 Carp	12	1.333	Ea.	1,075	64.50		1,139.50	1,275
0120	Half glass 3'-0"x 6'-8"	"	12	1.333		1,175	64.50		1,239.50	1,400
0200	Side lite, full glass, 1-3/4" x 1'-2" x 6'-8"					905			905	995
0220	Side lite, half glass, 1-3/4" x 1'-2" x 6'-8"					905			905	995
0300	Raised face, 2 panel, paint grade poplar, 1-3/4" x 3'-0"x 7'-0"	2 Carp	12	1.333		1,275	64.50		1,339.50	1,500
0320	Side lite, raised face, half glass, 1-3/4" x 1'-2" x 7'-0"					1,050			1,050	1,175
0500	6 panel, Fir, 1-3/4" x 3'-0"x 6'-8"	2 Carp	12	1.333		1,550	64.50		1,614.50	1,800
0520	Half glass 3'-0"x 6'-8"	"	12	1.333		1,650	64.50		1,714.50	1,925
0600	Side lite, full glass, 1-3/4" x 1'-2" x 6'-8"					1,150			1,150	1,275
0620	Side lite, half glass, 1-3/4" x 1'-2" x 6'-8"					1,450			1,450	1,600
0700	6 panel, Mahogany, 1-3/4" x 3'-0"x 6'-8"	2 Carp	12	1.333		1,625	64.50		1,689.50	1,875
0800	Side lite, full glass, 1-3/4" x 1'-2" x 6'-8"					1,225			1,225	1,350
0820	Side lite, half glass, 1-3/4" x 1'-2" x 6'-8"					1,200			1,200	1,325

08 14 40 – Interior Cafe Doors

08 14 40.10 Cafe Style Doors

		Crew	Daily Output	Labor-Hours	Unit	Material	2016 Bare Costs Labor	Equipment	Total	Total Incl O&P
0010	**CAFE STYLE DOORS**									
6520	Interior cafe doors, 2'-6" opening, stock, panel pine	2 Carp	16	1	Ea.	218	48.50		266.50	320
6540	3'-0" opening	"	16	1	"	240	48.50		288.50	345
6550	Louvered pine									
6560	2'-6" opening	2 Carp	16	1	Ea.	180	48.50		228.50	277
8000	3'-0" opening		16	1		193	48.50		241.50	291
8010	2'-6" opening, hardwood		16	1		264	48.50		312.50	370
8020	3'-0" opening		16	1		281	48.50		329.50	390
9000	Minimum labor/equipment charge	1 Carp	4	2	Job		97		97	158

08 16 Composite Doors

08 16 13 – Fiberglass Doors

08 16 13.10 Entrance Doors, Fibrous Glass

			Crew	Daily Output	Labor-Hours	Unit	Material	2016 Bare Costs Labor	Equipment	Total	Total Incl O&P
0010	**ENTRANCE DOORS, FIBROUS GLASS**										
0020	Exterior, fiberglass, door, 2'-8" wide x 6'-8" high	G	2 Carp	15	1.067	Ea.	270	51.50		321.50	380
0040	3'-0" wide x 6'-8" high	G		15	1.067		270	51.50		321.50	380
0060	3'-0" wide x 7'-0" high	G		15	1.067		460	51.50		511.50	590
0080	3'-0" wide x 6'-8" high, with two lites	G		15	1.067		315	51.50		366.50	430
0100	3'-0" wide x 8'-0" high, with two lites	G		15	1.067		525	51.50		576.50	665
0110	Half glass, 3'-0" wide x 6'-8" high	G		15	1.067		435	51.50		486.50	565
0120	3'-0" wide x 6'-8" high, low e	G		15	1.067		465	51.50		516.50	595
0130	3'-0" wide x 8'-0" high	G		15	1.067		585	51.50		636.50	730
0140	3'-0" wide x 8'-0" high, low e	G		15	1.067		655	51.50		706.50	805
0150	Side lights, 1'-0" wide x 6'-8" high	G					266			266	293
0160	1'-0" wide x 6'-8" high, low e	G					281			281	310
0180	1'-0" wide x 6'-8" high, full glass	G					310			310	340

08 16 Composite Doors

08 16 13 – Fiberglass Doors

08 16 13.10 Entrance Doors, Fibrous Glass	Crew	Daily Output	Labor-Hours	Unit	Material	2016 Bare Costs Labor	Equipment	Total	Total Incl O&P
0190	1'-0" wide x 6'-8" high, low e	G			Ea.	340			340	370

08 16 14 – French Doors

08 16 14.10 Exterior Doors With Glass Lites

		Crew	Daily Output	Labor-Hours	Unit	Material	2016 Bare Costs Labor	Equipment	Total	Total Incl O&P
0010	**EXTERIOR DOORS WITH GLASS LITES**									
0020	French, Fir, 1-3/4", 3'-0" wide x 6'-8" high	2 Carp	12	1.333	Ea.	600	64.50		664.50	765
0025	Double		12	1.333		1,200	64.50		1,264.50	1,425
0030	Maple, 1-3/4", 3'-0" wide x 6'-8" high		12	1.333		675	64.50		739.50	850
0035	Double		12	1.333		1,350	64.50		1,414.50	1,575
0040	Cherry, 1-3/4", 3'-0" wide x 6'-8" high		12	1.333		790	64.50		854.50	970
0045	Double		12	1.333		1,575	64.50		1,639.50	1,825
0100	Mahogany, 1-3/4", 3'-0" wide x 8'-0" high		10	1.600		800	77.50		877.50	1,000
0105	Double		10	1.600		1,600	77.50		1,677.50	1,875
0110	Fir, 1-3/4", 3'-0" wide x 8'-0" high		10	1.600		1,200	77.50		1,277.50	1,450
0115	Double		10	1.600		2,400	77.50		2,477.50	2,775
0120	Oak, 1-3/4", 3'-0" wide x 8'-0" high		10	1.600		1,825	77.50		1,902.50	2,125
0125	Double		10	1.600		3,650	77.50		3,727.50	4,125

08 17 Integrated Door Opening Assemblies

08 17 13 – Integrated Metal Door Opening Assemblies

08 17 13.10 Hollow Metal Doors and Frames

		Crew	Daily Output	Labor-Hours	Unit	Material	2016 Bare Costs Labor	Equipment	Total	Total Incl O&P
0010	**HOLLOW METAL DOORS AND FRAMES**									
0100	Prehung, flush, 18 ga., 1-3/4" x 6'-8" x 2'-8" wide	1 Carp	16	.500	Ea.	665	24		689	775
0120	3'-0" wide		15	.533		625	26		651	725
0130	3'-6" wide		13	.615		715	30		745	835
0140	4'-0" wide		10	.800		775	39		814	915
0300	Double, 1-3/4" x 3'-0" x 6'-8"		5	1.600		700	77.50		777.50	895
0350	3'-0" x 8'-0"		4	2		1,550	97		1,647	1,850

08 17 13.20 Stainless Steel Doors and Frames

			Crew	Daily Output	Labor-Hours	Unit	Material	2016 Bare Costs Labor	Equipment	Total	Total Incl O&P
0010	**STAINLESS STEEL DOORS AND FRAMES**										
0020	Stainless Steel(304) prehung 24 g 2'-6" x 6'-8" door w/16 g frame	G	2 Carp	6	2.667	Ea.	1,800	129		1,929	2,175
0025	2'-8" x 6'-8"	G		6	2.667		1,825	129		1,954	2,225
0030	3'-0" x 6'-8"	G		6	2.667		1,825	129		1,954	2,200
0040	3'-0" x 7'-0"	G		6	2.667		1,850	129		1,979	2,225
0050	4'-0" x 7'-0"	G		5	3.200		2,400	155		2,555	2,900
0100	Stainless Steel(316) prehung 24 g 2'-6" x 6'-8" door w/16 g frame	G		6	2.667		2,700	129		2,829	3,175
0110	2'-8" x 6'-8"	G		6	2.667		2,725	129		2,854	3,175
0120	3'-0" x 6'-8"	G		6	2.667		2,500	129		2,629	2,950
0150	3'-0" x 7'-0"	G		6	2.667		2,675	129		2,804	3,150
0160	4'-0" x 7'-0"	G		6	2.667		3,300	129		3,429	3,825
0300	Stainless Steel(304) prehung 18 g 2'-6" x 6'-8" door w/16 g frame	G		6	2.667		3,075	129		3,204	3,575
0310	2'-8" x 6'-8"	G		6	2.667		3,100	129		3,229	3,600
0320	3'-0" x 6'-8"	G		6	2.667		3,150	129		3,279	3,675
0350	3'-0" x 7'-0"	G		6	2.667		3,200	129		3,329	3,725
0360	4'-0" x 7'-0"	G		5	3.200		3,400	155		3,555	3,975
0500	Stainless steel, prehung door, foam core, 14 ga, 3'-0" x 7'-0"	G		5	3.200		3,400	155		3,555	4,000
0600	Stainless steel, prehung double door, foam core, 14 ga, 3'-0" x 7'-0"	G		4	4		6,700	194		6,894	7,700

08 17 23 – Integrated Wood Door Opening Assemblies

08 17 23.10 Pre-Hung Doors		Crew	Daily Output	Labor-Hours	Unit	Material	2016 Bare Costs Labor	Equipment	Total	Total Incl O&P
0010	**PRE-HUNG DOORS**									
0300	Exterior, wood, comb. storm & screen, 6'-9" x 2'-6" wide	2 Carp	15	1.067	Ea.	296	51.50		347.50	410
0320	2'-8" wide		15	1.067		296	51.50		347.50	410
0340	3'-0" wide		15	1.067		305	51.50		356.50	420
0360	For 7'-0" high door, add					30			30	33
1600	Entrance door, flush, birch, solid core									
1620	4-5/8" solid jamb, 1-3/4" x 6'-8" x 2'-8" wide	2 Carp	16	1	Ea.	289	48.50		337.50	400
1640	3'-0" wide		16	1		375	48.50		423.50	495
1642	5-5/8" jamb		16	1		325	48.50		373.50	440
1680	For 7'-0" high door, add					25			25	27.50
2000	Entrance door, colonial, 6 panel pine									
2020	4-5/8" solid jamb, 1-3/4" x 6'-8" x 2'-8" wide	2 Carp	16	1	Ea.	640	48.50		688.50	785
2040	3'-0" wide	"	16	1		675	48.50		723.50	820
2060	For 7'-0" high door, add					54			54	59
2200	For 5-5/8" solid jamb, add					41.50			41.50	46
2230	French style, exterior, 1 lite, 1-3/4" x 3'-0" x 6'-8"	1 Carp	14	.571		590	27.50		617.50	695
2235	9 lites	"	14	.571		660	27.50		687.50	770
2245	15 lites	2 Carp	14	1.143		640	55.50		695.50	795
2250	Double, 15 lites, 2'-0" x 6'-8", 4'-0" opening		7	2.286	Pr.	1,200	111		1,311	1,500
2260	2'-6" x 6'-8", 5'-0" opening		7	2.286		1,325	111		1,436	1,625
2280	3'-0" x 6'-8", 6'-0" opening		7	2.286		1,350	111		1,461	1,650
2430	3'-0" x 7'-0", 15 lites		14	1.143	Ea.	950	55.50		1,005.50	1,150
2432	Two 3'-0" x 7'-0"		7	2.286	Pr.	1,975	111		2,086	2,325
2435	3'-0" x 8'-0"		14	1.143	Ea.	1,000	55.50		1,055.50	1,200
2437	Two, 3'-0" x 8'-0"		7	2.286	Pr.	2,075	111		2,186	2,475
4000	Interior, passage door, 4-5/8" solid jamb									
4350	Paneled, primed, hollow core, 2'-8" wide	2 Carp	17	.941	Ea.	116	45.50		161.50	203
4360	3'-0" wide		17	.941		180	45.50		225.50	273
4400	Lauan, flush, solid core, 1-3/8" x 6'-8" x 2'-6" wide		17	.941		186	45.50		231.50	280
4420	2'-8" wide		17	.941		186	45.50		231.50	280
4440	3'-0" wide		16	1		201	48.50		249.50	300
4600	Hollow core, 1-3/8" x 6'-8" x 2'-6" wide		17	.941		125	45.50		170.50	213
4620	2'-8" wide		17	.941		125	45.50		170.50	212
4640	3'-0" wide		16	1		140	48.50		188.50	233
4700	For 7'-0" high door, add					35.50			35.50	39
5000	Birch, flush, solid core, 1-3/8" x 6'-8" x 2'-6" wide	2 Carp	17	.941		275	45.50		320.50	380
5020	2'-8" wide		17	.941		201	45.50		246.50	296
5040	3'-0" wide		16	1		300	48.50		348.50	410
5200	Hollow core, 1-3/8" x 6'-8" x 2'-6" wide		17	.941		221	45.50		266.50	320
5220	2'-8" wide		17	.941		241	45.50		286.50	340
5240	3'-0" wide		16	1		242	48.50		290.50	345
5280	For 7'-0" high door, add					30.50			30.50	33.50
5500	Hardboard paneled, 1-3/8" x 6'-8" x 2'-6" wide	2 Carp	17	.941		145	45.50		190.50	234
5520	2'-8" wide		17	.941		153	45.50		198.50	243
5540	3'-0" wide		16	1		150	48.50		198.50	244
6000	Pine paneled, 1-3/8" x 6'-8" x 2'-6" wide		17	.941		255	45.50		300.50	355
6020	2'-8" wide		17	.941		274	45.50		319.50	375
6040	3'-0" wide		16	1		282	48.50		330.50	390
8200	Birch, flush, solid core, 1-3/4" x 6'-8" x 2'-4" wide	1 Carp	17	.471		229	23		252	290
8220	2'-6" wide		17	.471		227	23		250	288
8240	2'-8" wide		17	.471		221	23		244	281
8260	3'-0" wide		16	.500		229	24		253	292

08 17 Integrated Door Opening Assemblies

08 17 23 – Integrated Wood Door Opening Assemblies

08 17 23.10 Pre-Hung Doors

08 17 23.10 Pre-Hung Doors		Crew	Daily Output	Labor-Hours	Unit	Material	2016 Bare Costs Labor	Equipment	Total	Total Incl O&P
8280	3'-6" wide	1 Carp	15	.533	Ea.	360	26		386	435
8500	Pocket door frame with lauan, flush, hollow core, 1-3/8" x 3'-0" x 6'-8"		17	.471		223	23		246	284
9000	Minimum labor/equipment charge		4	2	Job		97		97	158

08 31 Access Doors and Panels

08 31 13 – Access Doors and Frames

08 31 13.10 Types of Framed Access Doors

08 31 13.10 Types of Framed Access Doors		Crew	Daily Output	Labor-Hours	Unit	Material	2016 Bare Costs Labor	Equipment	Total	Total Incl O&P
0010	**TYPES OF FRAMED ACCESS DOORS**									
1000	Fire rated door with lock									
1100	Metal, 12" x 12"	1 Carp	10	.800	Ea.	165	39		204	246
1150	18" x 18"		9	.889		220	43		263	315
1200	24" x 24"		9	.889		325	43		368	430
1250	24" x 36"		8	1		335	48.50		383.50	445
1300	24" x 48"		8	1		430	48.50		478.50	555
1350	36" x 36"		7.50	1.067		495	51.50		546.50	630
1400	48" x 48"		7.50	1.067		635	51.50		686.50	780
1600	Stainless steel, 12" x 12"		10	.800		282	39		321	375
1650	18" x 18"		9	.889		410	43		453	520
1700	24" x 24"		9	.889		500	43		543	620
1750	24" x 36"		8	1		640	48.50		688.50	785
2000	Flush door for finishing									
2100	Metal 8" x 8"	1 Carp	10	.800	Ea.	38	39		77	106
2150	12" x 12"	"	10	.800	"	45	39		84	113
3000	Recessed door for acoustic tile									
3100	Metal, 12" x 12"	1 Carp	4.50	1.778	Ea.	82	86		168	231
3150	12" x 24"		4.50	1.778		100	86		186	251
3200	24" x 24"		4	2		130	97		227	300
3250	24" x 36"		4	2		180	97		277	355
4000	Recessed door for drywall									
4100	Metal 12" x 12"	1 Carp	6	1.333	Ea.	80	64.50		144.50	194
4150	12" x 24"		5.50	1.455		114	70.50		184.50	240
4200	24" x 36"		5	1.600		182	77.50		259.50	325
6000	Standard door									
6100	Metal, 8" x 8"	1 Carp	10	.800	Ea.	45	39		84	113
6150	12" x 12"		10	.800		50	39		89	119
6200	18" x 18"		9	.889		70	43		113	148
6250	24" x 24"		9	.889		85	43		128	164
6300	24" x 36"		8	1		125	48.50		173.50	217
6350	36" x 36"		8	1		145	48.50		193.50	239
6500	Stainless steel, 8" x 8"		10	.800		85	39		124	157
6550	12" x 12"		10	.800		110	39		149	185
6600	18" x 18"		9	.889		205	43		248	297
6650	24" x 24"		9	.889		265	43		308	365
9000	Minimum labor/equipment charge		4	2	Job		97		97	158

08 31 13.20 Bulkhead/Cellar Doors

08 31 13.20 Bulkhead/Cellar Doors		Crew	Daily Output	Labor-Hours	Unit	Material	2016 Bare Costs Labor	Equipment	Total	Total Incl O&P
0010	**BULKHEAD/CELLAR DOORS**									
0020	Steel, not incl. sides, 44" x 62"	1 Carp	5.50	1.455	Ea.	600	70.50		670.50	775
0100	52" x 73"		5.10	1.569		850	76		926	1,050
0500	With sides and foundation plates, 57" x 45" x 24"		4.70	1.702		910	82.50		992.50	1,125
0600	42" x 49" x 51"		4.30	1.860		925	90		1,015	1,175
9000	Minimum labor/equipment charge		2	4	Job		194		194	315

For customer support on your Commercial Renovation Cost Data, call 877.742.4719.

241

08 31 Access Doors and Panels

08 31 13 – Access Doors and Frames

08 31 13.30 Commercial Floor Doors	Crew	Daily Output	Labor-Hours	Unit	Material	2016 Bare Costs Labor	Equipment	Total	Total Incl O&P
0010 **COMMERCIAL FLOOR DOORS**									
0020 Aluminum tile, steel frame, one leaf, 2' x 2' opng.	2 Sswk	3.50	4.571	Opng.	880	243		1,123	1,400
0050 3'-6" x 3'-6" opening		3.50	4.571		1,600	243		1,843	2,175
0500 Double leaf, 4' x 4' opening		3	5.333		1,750	284		2,034	2,425
0550 5' x 5' opening		3	5.333	▼	3,100	284		3,384	3,900
9000 Minimum labor/equipment charge	▼	2	8	Job		425		425	750

08 31 13.35 Industrial Floor Doors

	Crew	Daily Output	Labor-Hours	Unit	Material	2016 Bare Costs Labor	Equipment	Total	Total Incl O&P
0010 **INDUSTRIAL FLOOR DOORS**									
0020 Steel 300 psf L.L., single leaf, 2' x 2', 175#	2 Sswk	6	2.667	Opng.	760	142		902	1,075
0050 3' x 3' opening, 300#		5.50	2.909		1,075	155		1,230	1,450
0300 Double leaf, 4' x 4' opening, 455#		5	3.200		2,275	170		2,445	2,800
0350 5' x 5' opening, 645#		4.50	3.556		3,300	189		3,489	3,950
1000 Aluminum, 300 psf L.L., single leaf, 2' x 2', 60#		6	2.667		800	142		942	1,125
1050 3' x 3' opening, 100#		5.50	2.909		1,300	155		1,455	1,700
1500 Double leaf, 4' x 4' opening, 160#		5	3.200		2,100	170		2,270	2,600
1550 5' x 5' opening, 235#		4.50	3.556	▼	2,800	189		2,989	3,400
9000 Minimum labor/equipment charge	▼	2	8	Job		425		425	750

08 32 Sliding Glass Doors

08 32 13 – Sliding Aluminum-Framed Glass Doors

08 32 13.10 Sliding Aluminum Doors

	Crew	Daily Output	Labor-Hours	Unit	Material	2016 Bare Costs Labor	Equipment	Total	Total Incl O&P
0010 **SLIDING ALUMINUM DOORS**									
0350 Aluminum, 5/8" tempered insulated glass, 6' wide									
0400 Premium	2 Carp	4	4	Ea.	1,625	194		1,819	2,100
0450 Economy		4	4		820	194		1,014	1,225
0500 8' wide, premium		3	5.333		1,675	258		1,933	2,250
0550 Economy		3	5.333		1,450	258		1,708	2,000
0600 12' wide, premium		2.50	6.400		3,000	310		3,310	3,800
0650 Economy		2.50	6.400		1,550	310		1,860	2,200
4000 Aluminum, baked on enamel, temp glass, 6'-8" x 10'-0" wide		4	4		1,100	194		1,294	1,525
4020 Insulating glass, 6'-8" x 6'-0" wide		4	4		955	194		1,149	1,375
4040 8'-0" wide		3	5.333		1,125	258		1,383	1,650
4060 10'-0" wide		2	8		1,375	390		1,765	2,125
4080 Anodized, temp glass, 6'-8" x 6'-0" wide		4	4		455	194		649	815
4100 8'-0" wide		3	5.333		575	258		833	1,050
4120 10'-0" wide	▼	2	8	▼	645	390		1,035	1,350
5000 Aluminum sliding glass door system									
5010 Sliding door 4' wide opening single side	2 Carp	2	8	Ea.	6,000	390		6,390	7,225
5015 8' wide opening single side		2	8		9,000	390		9,390	10,500
5020 Telescoping glass door system, 4' wide opening biparting		2	8		5,000	390		5,390	6,125
5025 8' wide opening biparting		2	8		6,000	390		6,390	7,225
5030 Folding glass door, 4' wide opening biparting		2	8		8,000	390		8,390	9,425
5035 8' wide opening biparting		2	8		10,000	390		10,390	11,600
5040 ICU-CCU sliding telescoping glass door, 4' x 7', single side opening		2	8		3,000	390		3,390	3,925
5045 8' x 7', single side opening	▼	2	8		4,625	390		5,015	5,725
7000 Electric swing door operator and control, single door w/sensors	1 Carp	4	2		2,800	97		2,897	3,225
7005 Double door w/sensors		2	4		5,500	194		5,694	6,375
7010 Electric folding door operator and control, single door w/sensors		4	2		6,500	97		6,597	7,300
7015 Bi-folding door		4	2		7,800	97		7,897	8,725
7020 Electric swing door operator and control, single door	▼	4	2	▼	3,000	97		3,097	3,450

08 32 Sliding Glass Doors

08 32 19 – Sliding Wood-Framed Glass Doors

08 32 19.10 Sliding Wood Doors

		Crew	Daily Output	Labor-Hours	Unit	Material	2016 Bare Costs Labor	2016 Bare Costs Equipment	Total	Total Incl O&P
0010	**SLIDING WOOD DOORS**									
0020	Wood, tempered insul. glass, 6' wide, premium	2 Carp	4	4	Ea.	1,450	194		1,644	1,925
0100	Economy		4	4		1,175	194		1,369	1,600
0150	8' wide, wood, premium		3	5.333		1,850	258		2,108	2,450
0200	Economy		3	5.333		1,500	258		1,758	2,075
0235	10' wide, wood, premium		2.50	6.400		2,600	310		2,910	3,350
0240	Economy		2.50	6.400		2,200	310		2,510	2,925
0250	12' wide, wood, premium		2.50	6.400		3,000	310		3,310	3,800
0300	Economy		2.50	6.400		2,450	310		2,760	3,200
9000	Minimum labor/equipment charge	1 Carp	2	4	Job		194		194	315

08 32 19.15 Sliding Glass Vinyl-Clad Wood Doors

			Crew	Daily Output	Labor-Hours	Unit	Material	2016 Bare Costs Labor	2016 Bare Costs Equipment	Total	Total Incl O&P
0010	**SLIDING GLASS VINYL-CLAD WOOD DOORS**										
0020	Glass, sliding vinyl clad, insul. glass, 6'-0" x 6'-8"	G	2 Carp	4	4	Opng.	1,500	194		1,694	2,000
0025	6'-0" x 6'-10" high	G		4	4		1,650	194		1,844	2,150
0030	6'-0" x 8'-0" high	G		4	4		1,975	194		2,169	2,500
0050	5'-0" x 6'-8" high	G		4	4		1,500	194		1,694	1,975
0100	8'-0" x 6'-10" high	G		4	4		2,025	194		2,219	2,550
0150	8'-0" x 8'-0" high	G		4	4		2,225	194		2,419	2,775
0500	4 leaf, 9'-0" x 6'-10" high	G		3	5.333		3,250	258		3,508	4,000
0550	9'-0" x 8'-0" high	G		3	5.333		3,775	258		4,033	4,575
0600	12'-0" x 6'-10" high	G		3	5.333		3,875	258		4,133	4,700
0650	12'-0" x 8'-0" high	G		3	5.333		3,875	258		4,133	4,700
9000	Minimum labor/equipment charge		1 Carp	4	2	Job		97		97	158

08 33 Coiling Doors and Grilles

08 33 13 – Coiling Counter Doors

08 33 13.10 Counter Doors, Coiling Type

		Crew	Daily Output	Labor-Hours	Unit	Material	2016 Bare Costs Labor	2016 Bare Costs Equipment	Total	Total Incl O&P
0010	**COUNTER DOORS, COILING TYPE**									
0020	Manual, incl. frame and hardware, galv. stl., 4' roll-up, 6' long	2 Carp	2	8	Opng.	1,225	390		1,615	1,975
0300	Galvanized steel, UL label		1.80	8.889		1,300	430		1,730	2,125
0600	Stainless steel, 4' high roll-up, 6' long		2	8		2,150	390		2,540	2,975
0700	10' long		1.80	8.889		2,600	430		3,030	3,550
2000	Aluminum, 4' high, 4' long		2.20	7.273		1,550	350		1,900	2,275
2020	6' long		2	8		1,850	390		2,240	2,650
2040	8' long		1.90	8.421		2,100	410		2,510	2,975
2060	10' long		1.80	8.889		2,150	430		2,580	3,050
2080	14' long		1.40	11.429		2,750	555		3,305	3,925
2100	6' high, 4' long		2	8		1,900	390		2,290	2,700
2120	6' long		1.60	10		1,700	485		2,185	2,675
2140	10' long		1.40	11.429		2,250	555		2,805	3,375
9000	Minimum labor/equipment charge	1 Carp	2	4	Job		194		194	315

08 33 16 – Coiling Counter Grilles

08 33 16.10 Coiling Grilles

		Crew	Daily Output	Labor-Hours	Unit	Material	2016 Bare Costs Labor	2016 Bare Costs Equipment	Total	Total Incl O&P
0010	**COILING GRILLES**									
2020	Aluminum, manual operated, mill finish	2 Sswk	82	.195	S.F.	29	10.40		39.40	50
2040	Bronze anodized		82	.195	"	45	10.40		55.40	68
2060	Steel, manual operated, 10' x 10' high		1	16	Opng.	2,600	850		3,450	4,350
2080	15' x 8' high		.80	20	"	2,900	1,075		3,975	5,075
3000	For safety edge bottom bar, electric, add				L.F.	49			49	54
8000	For motor operation, add	2 Sswk	5	3.200	Opng.	1,250	170		1,420	1,675

For customer support on your Commercial Renovation Cost Data, call 877.742.4719.

243

08 33 Coiling Doors and Grilles

08 33 16 – Coiling Counter Grilles

08 33 16.10 Coiling Grilles	Crew	Daily Output	Labor-Hours	Unit	Material	2016 Bare Costs Labor	Equipment	Total	Total Incl O&P
9000	Minimum labor/equipment charge	2 Sswk	1	16	Job		850		850	1,500

08 33 23 – Overhead Coiling Doors

08 33 23.10 Coiling Service Doors

08 33 23.10 Coiling Service Doors	Crew	Daily Output	Labor-Hours	Unit	Material	Labor	Equipment	Total	Total Incl O&P
0010	**COILING SERVICE DOORS** Steel, manual, 20 ga., incl. hardware									
0050	8' x 8' high	2 Sswk	1.60	10	Ea.	1,175	530		1,705	2,250
0130	12' x 12' high, standard		1.20	13.333		2,200	710		2,910	3,675
0160	10' x 20' high, standard		.50	32		2,675	1,700		4,375	5,950
2000	Class A fire doors, manual, 20 ga., 8' x 8' high		1.40	11.429		1,625	610		2,235	2,875
2100	10' x 10' high		1.10	14.545		2,225	775		3,000	3,825
2200	20' x 10' high		.80	20		4,550	1,075		5,625	6,875
2300	12' x 12' high		1	16		3,500	850		4,350	5,350
2400	20' x 12' high		.80	20		4,925	1,075		6,000	7,300
2500	14' x 14' high		.60	26.667		3,825	1,425		5,250	6,725
2600	20' x 16' high		.50	32		6,025	1,700		7,725	9,625
2700	10' x 20' high	↓	.40	40	↓	4,750	2,125		6,875	8,975
3000	For 18 ga. doors, add				S.F.	1.75			1.75	1.93
3300	For enamel finish, add				"	2.10			2.10	2.31
3600	For safety edge bottom bar, pneumatic, add				L.F.	23			23	25.50
4000	For weatherstripping, extruded rubber, jambs, add					15.20			15.20	16.70
4100	Hood, add					8.55			8.55	9.40
4200	Sill, add				↓	5.85			5.85	6.45
4500	Motor operators, to 14' x 14' opening	2 Sswk	5	3.200	Ea.	1,175	170		1,345	1,600
4700	For fire door, additional fusible link, add				"	30			30	33
9000	Minimum labor/equipment charge	2 Sswk	1	16	Job		850		850	1,500

08 34 Special Function Doors

08 34 13 – Cold Storage Doors

08 34 13.10 Doors for Cold Area Storage

08 34 13.10 Doors for Cold Area Storage	Crew	Daily Output	Labor-Hours	Unit	Material	Labor	Equipment	Total	Total Incl O&P
0010	**DOORS FOR COLD AREA STORAGE**									
0020	Single, 20 ga. galvanized steel									
0300	Horizontal sliding, 5' x 7', manual operation, 3.5" thick	2 Carp	2	8	Ea.	3,325	390		3,715	4,275
0400	4" thick		2	8		3,300	390		3,690	4,275
0500	6" thick		2	8		3,500	390		3,890	4,475
0800	5' x 7', power operation, 2" thick		1.90	8.421		5,700	410		6,110	6,950
0900	4" thick		1.90	8.421		5,800	410		6,210	7,050
1000	6" thick		1.90	8.421		6,600	410		7,010	7,925
1300	9' x 10', manual operation, 2" insulation		1.70	9.412		4,550	455		5,005	5,750
1400	4" insulation		1.70	9.412		4,775	455		5,230	6,000
1500	6" insulation		1.70	9.412		5,700	455		6,155	7,000
1800	Power operation, 2" insulation		1.60	10		7,825	485		8,310	9,425
1900	4" insulation		1.60	10		8,000	485		8,485	9,600
2000	6" insulation		1.70	9.412	↓	9,000	455		9,455	10,600
2300	For stainless steel face, add					25%				
3000	Hinged, lightweight, 3' x 7'-0", 2" thick	2 Carp	2	8	Ea.	1,450	390		1,840	2,200
3050	4" thick		1.90	8.421		1,800	410		2,210	2,650
3300	Polymer doors, 3' x 7'-0"		1.90	8.421		1,375	410		1,785	2,175
3350	6" thick		1.40	11.429		2,375	555		2,930	3,525
3600	Stainless steel, 3' x 7'-0", 4" thick		1.90	8.421		1,775	410		2,185	2,625
3650	6" thick		1.40	11.429		3,000	555		3,555	4,200
3900	Painted, 3' x 7'-0", 4" thick	↓	1.90	8.421	↓	1,300	410		1,710	2,100

08 34 13 – Cold Storage Doors

08 34 13.10 Doors for Cold Area Storage	Crew	Daily Output	Labor-Hours	Unit	Material	2016 Bare Costs Labor	Equipment	Total	Total Incl O&P
3950 6" thick	2 Carp	1.40	11.429	Ea.	2,375	555		2,930	3,525
5000 Bi-parting, electric operated									
5010 6' x 8' opening, galv. faces, 4" thick for cooler	2 Carp	.80	20	Opng.	7,350	970		8,320	9,675
5050 For freezer, 4" thick		.80	20		8,100	970		9,070	10,500
5300 For door buck framing and door protection, add		2.50	6.400		655	310		965	1,225
6000 Galvanized batten door, galvanized hinges, 4' x 7'		2	8		1,875	390		2,265	2,675
6050 6' x 8'		1.80	8.889		2,525	430		2,955	3,475
6500 Fire door, 3 hr., 6' x 8', single slide		.80	20		8,400	970		9,370	10,800
6550 Double, bi-parting		.70	22.857		12,400	1,100		13,500	15,400
9000 Minimum labor/equipment charge	1 Carp	2	4	Job		194		194	315

08 34 36 – Darkroom Doors

08 34 36.10 Various Types of Darkroom Doors

	Crew	Daily Output	Labor-Hours	Unit	Material	Labor	Equipment	Total	Total Incl O&P
0010 **VARIOUS TYPES OF DARKROOM DOORS**									
0015 Revolving, standard, 2 way, 36" diameter	2 Carp	3.10	5.161	Opng.	2,900	250		3,150	3,575
0020 41" diameter		3.10	5.161		3,000	250		3,250	3,725
0050 3 way, 51" diameter		1.40	11.429		3,825	555		4,380	5,100
1000 4 way, 49" diameter		1.40	11.429		3,950	555		4,505	5,250
2000 Hinged safety, 2 way, 41" diameter		2.30	6.957		3,750	335		4,085	4,650
2500 3 way, 51" diameter		1.40	11.429		4,050	555		4,605	5,375
3000 Pop out safety, 2 way, 41" diameter		3.10	5.161		4,025	250		4,275	4,825
4000 3 way, 51" diameter		1.40	11.429		4,875	555		5,430	6,275
5000 Wheelchair-type, pop out, 51" diameter		1.40	11.429		5,500	555		6,055	6,950
5020 72" diameter		.90	17.778		8,950	860		9,810	11,300

08 34 53 – Security Doors and Frames

08 34 53.20 Steel Door

	Crew	Daily Output	Labor-Hours	Unit	Material	Labor	Equipment	Total	Total Incl O&P
0010 **STEEL DOOR** flush with ballistic core and welded frame both 14 ga.									
0120 UL 752 Level 3, 1-3/4", 3'-0" x 7'-0"	2 Carp	1.50	10.667	Opng.	2,525	515		3,040	3,625
0125 1-3/4", 3'-6" x 7'-0"		1.50	10.667		2,750	515		3,265	3,875
0130 1-3/4", 4'-0" x 7'-0"		1.20	13.333		2,950	645		3,595	4,300
0150 UL 752 Level 8, 1-3/4", 3'-0" x 7'-0"		1.50	10.667		12,800	515		13,315	14,900
0155 1-3/4", 3'-6" x 7'-0"		1.50	10.667		13,900	515		14,415	16,100
0160 1-3/4", 4'-0" x 7'-0"		1.20	13.333		14,100	645		14,745	16,600
1000 Safe room sliding door and hardware, 1-3/4", 3'-0" x 7'-0" UL 752 Level 3		.50	32		23,500	1,550		25,050	28,400
1050 Safe room swinging door and hardware, 1-3/4", 3'-0" x 7'-0" UL 752 Level 3		.50	32		28,100	1,550		29,650	33,400

08 34 53.30 Wood Ballistic Doors

	Crew	Daily Output	Labor-Hours	Unit	Material	Labor	Equipment	Total	Total Incl O&P
0010 **WOOD BALLISTIC DOORS** with frames and hardware									
0050 Wood, 1-3/4", 3'-0" x 7'-0" UL 752 Level 3	2 Carp	1.50	10.667	Opng.	2,500	515		3,015	3,600

08 34 56 – Security Gates

08 34 56.10 Gates

	Crew	Daily Output	Labor-Hours	Unit	Material	Labor	Equipment	Total	Total Incl O&P
0010 **GATES**									
0015 Driveway gates include mounting hardware									
0500 Wood, security gate, driveway, dual, 10' wide	H-4	.80	25	Opng.	3,900	1,125		5,025	6,100
0505 12' wide		.80	25		4,300	1,125		5,425	6,525
0510 15' wide		.80	25		5,100	1,125		6,225	7,400
0600 Steel, security gate, driveway, single, 10' wide		.80	25		2,150	1,125		3,275	4,150
0605 12' wide		.80	25		2,225	1,125		3,350	4,225
0620 Steel, security gate, driveway, dual, 12' wide		.80	25		2,050	1,125		3,175	4,075
0625 14' wide		.80	25		2,200	1,125		3,325	4,225
0630 16' wide		.80	25		2,525	1,125		3,650	4,575
0700 Aluminum, security gate, driveway, dual, 10' wide		.80	25		3,000	1,125		4,125	5,100
0705 12' wide		.80	25		3,200	1,125		4,325	5,325

For customer support on your Commercial Renovation Cost Data, call 877.742.4719.

245

08 34 Special Function Doors

08 34 56 – Security Gates

08 34 56.10 Gates

	08 34 56.10 Gates	Crew	Daily Output	Labor-Hours	Unit	Material	2016 Bare Costs Labor	Equipment	Total	Total Incl O&P
0710	16' wide	H-4	.80	25	Opng.	3,700	1,125		4,825	5,875
1000	Security gate, driveway, opener 12 VDC				Ea.	800			800	880
1010	Wireless					1,800			1,800	1,975
1020	Security gate, driveway, opener 24 VDC					1,300			1,300	1,425
1030	Wireless					1,450			1,450	1,600
1040	Security gate, driveway, opener 12 VDC, solar panel 10 watt	1 Elec	2	4		300	220		520	670
1050	20 watt	"	2	4		450	220		670	835

08 34 73 – Sound Control Door Assemblies

08 34 73.10 Acoustical Doors

	08 34 73.10 Acoustical Doors	Crew	Daily Output	Labor-Hours	Unit	Material	2016 Bare Costs Labor	Equipment	Total	Total Incl O&P
0010	**ACOUSTICAL DOORS**									
0020	Including framed seals, 3' x 7', wood, 40 STC rating	2 Carp	1.50	10.667	Ea.	1,400	515		1,915	2,400
0100	Steel, 41 STC rating		1.50	10.667		3,300	515		3,815	4,475
0200	45 STC rating		1.50	10.667		3,725	515		4,240	4,950
0300	48 STC rating		1.50	10.667		4,350	515		4,865	5,625
0400	52 STC rating		1.50	10.667		5,000	515		5,515	6,350
9000	Minimum labor/equipment charge	1 Carp	4	2	Job		97		97	158

08 36 Panel Doors

08 36 13 – Sectional Doors

08 36 13.10 Overhead Commercial Doors

	08 36 13.10 Overhead Commercial Doors	Crew	Daily Output	Labor-Hours	Unit	Material	2016 Bare Costs Labor	Equipment	Total	Total Incl O&P
0010	**OVERHEAD COMMERCIAL DOORS**									
1000	Stock, sectional, heavy duty, wood, 1-3/4" thick, 8' x 8' high	2 Carp	2	8	Ea.	1,150	390		1,540	1,875
1200	12' x 12' high		1.50	10.667		2,225	515		2,740	3,300
1300	Chain hoist, 14' x 14' high		1.30	12.308		3,375	595		3,970	4,675
1600	20' x 16' high		.65	24.615		5,650	1,200		6,850	8,150
2100	For medium duty custom door, deduct					5%	5%			
2150	For medium duty stock doors, deduct					10%	5%			
2300	Fiberglass and aluminum, heavy duty, sectional, 12' x 12' high	2 Carp	1.50	10.667	Ea.	3,050	515		3,565	4,200
2450	Chain hoist, 20' x 20' high	"	.50	32		7,225	1,550		8,775	10,500
2900	For electric trolley operator, 1/3 H.P., to 12' x 12', add	1 Carp	2	4		1,050	194		1,244	1,500
2950	Over 12' x 12', 1/2 H.P., add		1	8		1,125	390		1,515	1,875
2980	Overhead, for row of clear lites add		1	8		110	390		500	755
9000	Minimum labor/equipment charge	2 Carp	1.50	10.667	Job		515		515	845

08 36 13.20 Residential Garage Doors

	08 36 13.20 Residential Garage Doors	Crew	Daily Output	Labor-Hours	Unit	Material	2016 Bare Costs Labor	Equipment	Total	Total Incl O&P
0010	**RESIDENTIAL GARAGE DOORS**									
0050	Hinged, wood, custom, double door, 9' x 7'	2 Carp	4	4	Ea.	815	194		1,009	1,225
0070	16' x 7'		3	5.333		1,225	258		1,483	1,775
0200	Overhead, sectional, incl. hardware, fiberglass, 9' x 7', standard		5	3.200		960	155		1,115	1,300
0220	Deluxe		5	3.200		1,150	155		1,305	1,500
0300	16' x 7', standard		6	2.667		1,575	129		1,704	1,925
0320	Deluxe		6	2.667		2,150	129		2,279	2,550
0500	Hardboard, 9' x 7', standard		8	2		685	97		782	915
0520	Deluxe		8	2		815	97		912	1,050
0600	16' x 7', standard		6	2.667		1,225	129		1,354	1,550
0620	Deluxe		6	2.667		1,425	129		1,554	1,775
0700	Metal, 9' x 7', standard		8	2		835	97		932	1,075
0720	Deluxe		6	2.667		960	129		1,089	1,250
0800	16' x 7', standard		6	2.667		1,025	129		1,154	1,325
0820	Deluxe		5	3.200		1,400	155		1,555	1,800
0900	Wood, 9' x 7', standard		8	2		970	97		1,067	1,225

08 36 Panel Doors

08 36 13 – Sectional Doors

08 36 13.20 Residential Garage Doors		Crew	Daily Output	Labor-Hours	Unit	Material	2016 Bare Costs Labor	Equipment	Total	Total Incl O&P
0920	Deluxe	2 Carp	8	2	Ea.	2,150	97		2,247	2,500
1000	16' x 7', standard		6	2.667		1,625	129		1,754	2,000
1020	Deluxe		6	2.667		3,025	129		3,154	3,525
1800	Door hardware, sectional	1 Carp	4	2		350	97		447	545
1810	Door tracks only		4	2		163	97		260	340
1820	One side only		7	1.143		120	55.50		175.50	223
3000	Swing-up, including hardware, fiberglass, 9' x 7', standard	2 Carp	8	2		1,000	97		1,097	1,250
3020	Deluxe		8	2		1,100	97		1,197	1,350
3100	16' x 7', standard		6	2.667		1,250	129		1,379	1,575
3120	Deluxe		6	2.667		1,600	129		1,729	1,950
3200	Hardboard, 9' x 7', standard		8	2		550	97		647	765
3220	Deluxe		8	2		650	97		747	875
3300	16' x 7', standard		6	2.667		670	129		799	945
3320	Deluxe		6	2.667		850	129		979	1,150
3400	Metal, 9' x 7', standard		8	2		600	97		697	820
3420	Deluxe		8	2		975	97		1,072	1,225
3500	16' x 7', standard		6	2.667		800	129		929	1,100
3520	Deluxe		6	2.667		1,100	129		1,229	1,400
3600	Wood, 9' x 7', standard		8	2		700	97		797	930
3620	Deluxe		8	2		1,125	97		1,222	1,400
3700	16' x 7', standard		6	2.667		900	129		1,029	1,200
3720	Deluxe		6	2.667		2,100	129		2,229	2,500
3900	Door hardware only, swing up	1 Carp	4	2		148	97		245	320
3920	One side only		7	1.143		90	55.50		145.50	190
4000	For electric operator, economy, add		8	1		425	48.50		473.50	550
4100	Deluxe, including remote control		8	1		610	48.50		658.50	755
4500	For transmitter/receiver control, add to operator				Total	110			110	121
4600	Transmitters, additional				"	60			60	66
6000	Replace section, on sectional door, fiberglass, 9' x 7'	1 Carp	4	2	Ea.	640	97		737	860
6020	16' x 7'		3.50	2.286		765	111		876	1,025
6200	Hardboard, 9' x 7'		4	2		157	97		254	330
6220	16' x 7'		3.50	2.286		230	111		341	435
6300	Metal, 9' x 7'		4	2		200	97		297	380
6320	16' x 7'		3.50	2.286		320	111		431	530
6500	Wood, 9' x 7'		4	2		120	97		217	290
6520	16' x 7'		3.50	2.286		240	111		351	445
7010	Garage doors, row of lites					120			120	132
9000	Minimum labor/equipment charge	1 Carp	2.50	3.200	Job		155		155	253

08 36 19 – Multi-Leaf Vertical Lift Doors

08 36 19.10 Sectional Vertical Lift Doors

		Crew	Daily Output	Labor-Hours	Unit	Material	2016 Bare Costs Labor	Equipment	Total	Total Incl O&P
0010	**SECTIONAL VERTICAL LIFT DOORS**									
0020	Motorized, 14 ga. steel, incl. frame and control panel									
0050	16' x 16' high	L-10	.50	48	Ea.	21,500	2,575	1,325	25,400	29,500
0100	10' x 20' high		1.30	18.462		35,000	990	505	36,495	40,700
0120	15' x 20' high		1.30	18.462		42,800	990	505	44,295	49,300
0140	20' x 20' high		1	24		50,000	1,275	660	51,935	58,000
0160	25' x 20' high		1	24		56,500	1,275	660	58,435	65,000
0170	32' x 24' high		.75	32		48,700	1,725	880	51,305	57,500
0180	20' x 25' high		1	24		58,000	1,275	660	59,935	66,500
0200	25' x 25' high		.70	34.286		70,500	1,825	940	73,265	81,500
0220	25' x 30' high		.70	34.286		75,500	1,825	940	78,265	87,000
0240	30' x 30' high		.70	34.286		86,000	1,825	940	88,765	99,000

247

08 36 Panel Doors

08 36 19 – Multi-Leaf Vertical Lift Doors

08 36 19.10 Sectional Vertical Lift Doors	Crew	Daily Output	Labor-Hours	Unit	Material	2016 Bare Costs Labor	Equipment	Total	Total Incl O&P
0260	35' x 30' high	L-10	.70	34.286	Ea.	92,500	1,825	940	95,265	106,000

08 38 Traffic Doors

08 38 19 – Rigid Traffic Doors

08 38 19.20 Double Acting Swing Doors

		Crew	Daily Output	Labor-Hours	Unit	Material	Labor	Equipment	Total	Total Incl O&P
0010	**DOUBLE ACTING SWING DOORS**									
0020	Including frame, closer, hardware and vision panel									
1000	Polymer, 7'-0" high, 4'-0" wide	2 Carp	4.20	3.810	Pr.	2,100	185		2,285	2,600
1025	6'-0" wide		4	4		2,200	194		2,394	2,750
1050	6'-8" wide		4	4		2,400	194		2,594	2,975
2000	3/4" thick, stainless steel									
2010	Stainless steel, 7' high opening, 4' wide	2 Carp	4	4	Pr.	2,600	194		2,794	3,175
2050	7' wide		3.80	4.211	"	2,800	204		3,004	3,400
9000	Minimum labor/equipment charge		2	8	Job		390		390	635

08 38 19.30 Shock Absorbing Doors

		Crew	Daily Output	Labor-Hours	Unit	Material	Labor	Equipment	Total	Total Incl O&P
0010	**SHOCK ABSORBING DOORS**									
0020	Rigid, no frame, 1-1/2" thick, 5' x 7'	2 Sswk	1.90	8.421	Opng.	1,525	450		1,975	2,475
0100	8' x 8'		1.80	8.889		2,000	475		2,475	3,025
0500	Flexible, no frame, insulated, .16" thick, economy, 5' x 7'		2	8		1,750	425		2,175	2,675
0600	Deluxe		1.90	8.421		2,625	450		3,075	3,675
1000	8' x 8' opening, economy		2	8		2,750	425		3,175	3,775
1100	Deluxe		1.90	8.421		3,500	450		3,950	4,650
9000	Minimum labor/equipment charge		2	8	Job		425		425	750

08 41 Entrances and Storefronts

08 41 13 – Aluminum-Framed Entrances and Storefronts

08 41 13.20 Tube Framing

		Crew	Daily Output	Labor-Hours	Unit	Material	Labor	Equipment	Total	Total Incl O&P
0010	**TUBE FRAMING**, For window walls and store fronts, aluminum stock									
0050	Plain tube frame, mill finish, 1-3/4" x 1-3/4"	2 Glaz	103	.155	L.F.	10.10	7.25		17.35	23
0150	1-3/4" x 4"		98	.163		13.55	7.60		21.15	27.50
0200	1-3/4" x 4-1/2"		95	.168		16.30	7.85		24.15	30.50
0250	2" x 6"		89	.180		24	8.40		32.40	40
0350	4" x 4"		87	.184		27.50	8.60		36.10	44
0400	4-1/2" x 4-1/2"		85	.188		29	8.80		37.80	46
0450	Glass bead		240	.067		3.15	3.11		6.26	8.50
1000	Flush tube frame, mill finish, 1/4" glass, 1-3/4" x 4", open header		80	.200		13.55	9.35		22.90	30
1050	Open sill		82	.195		11.05	9.10		20.15	27
1100	Closed back header		83	.193		19.40	9		28.40	36
1150	Closed back sill		85	.188		18.55	8.80		27.35	35
1160	Tube frmg., spandrel cover both sides, alum 1" wide	1 Sswk	85	.094	S.F.	99	5		104	118
1170	Tube frmg., spandrel cover both sides, alum 2" wide	"	85	.094	"	39	5		44	52
1200	Vertical mullion, one piece	2 Glaz	75	.213	L.F.	20.50	9.95		30.45	38.50
1250	Two piece		73	.219		21.50	10.25		31.75	40
1300	90° or 180° vertical corner post		75	.213		33	9.95		42.95	52
1400	1-3/4" x 4-1/2", open header		80	.200		16.30	9.35		25.65	33
1450	Open sill		82	.195		13.80	9.10		22.90	30
1500	Closed back header		83	.193		19.65	9		28.65	36
1550	Closed back sill		85	.188		19.35	8.80		28.15	36
1600	Vertical mullion, one piece		75	.213		21.50	9.95		31.45	40

08 41 13 – Aluminum-Framed Entrances and Storefronts

08 41 13.20 Tube Framing		Crew	Daily Output	Labor-Hours	Unit	Material	2016 Bare Costs Labor	Equipment	Total	Total Incl O&P
1650	Two piece	2 Glaz	73	.219	L.F.	23	10.25		33.25	42
1700	90° or 180° vertical corner post		75	.213		23.50	9.95		33.45	41.50
2000	Flush tube frame, mil fin.,ins. glass w/thml brk, 2" x 4-1/2", open header		75	.213		16.70	9.95		26.65	34.50
2050	Open sill		77	.208		14.05	9.70		23.75	31
2100	Closed back header		78	.205		15.85	9.60		25.45	33
2150	Closed back sill		80	.200		15.20	9.35		24.55	32
2200	Vertical mullion, one piece		70	.229		17.65	10.65		28.30	37
2250	Two piece		68	.235		19.10	11		30.10	39
2300	90° or 180° vertical corner post		70	.229		18.15	10.65		28.80	37.50
5000	Flush tube frame, mill fin., thermal brk., 2-1/4" x 4-1/2", open header		74	.216		17.55	10.10		27.65	35.50
5050	Open sill		75	.213		15.45	9.95		25.40	33
5100	Vertical mullion, one piece		69	.232		18.20	10.85		29.05	37.50
5150	Two piece		67	.239		22	11.15		33.15	42
5200	90° or 180° vertical corner post		69	.232		19.60	10.85		30.45	39
5295	Flush tube frame, mill finish, 4-1/2" x 4-1/2"		85	.188		29	8.80		37.80	46
5300	Plain tube frame, mill finish, 2" x 3" jamb		115	.139		11.10	6.50		17.60	23
5310	2" x 4"		115	.139		16.90	6.50		23.40	29
5320	3" x 5-1/2"		107	.150		27.50	7		34.50	41.50
5330	Head, 2" x 3"		200	.080		20.50	3.74		24.24	28.50
5340	Mullion, 2" x 3"		200	.080		21	3.74		24.74	29
5350	2" x 4"		115	.139		24.50	6.50		31	37.50
5360	3" x 5-1/2"		107	.150		32	7		39	47
5370	4" corner mullion		90	.178		30	8.30		38.30	46.50
5380	Horizontal, 2" x 3"		115	.139		24	6.50		30.50	37
5390	3" x 5-1/2"		107	.150		30	7		37	44.50
5430	Sill section, 1/8" x 6"		133	.120		37	5.60		42.60	49.50
5440	1/8" x 7"		133	.120		44	5.60		49.60	57.50
5450	1/8" x 8-1/2"		133	.120		51.50	5.60		57.10	66
5460	Column covers, aluminum, 1/8" x 26"		57	.281		67	13.10		80.10	95
5470	1/8" x 34"		53	.302		86	14.10		100.10	118
5480	1/8" x 38"		53	.302		95.50	14.10		109.60	128
5700	Vertical mullions, clear finish, 1/4" thick glass		528	.030		12.85	1.42		14.27	16.45
5720	3/8" thick		445	.036		12.85	1.68		14.53	16.85
5740	1/2" thick		400	.040		13.30	1.87		15.17	17.70
5760	3/4" thick		344	.047		13.75	2.17		15.92	18.65
5780	1" thick		304	.053		14	2.46		16.46	19.40
6980	Door stop (snap in)		380	.042		3.45	1.97		5.42	7
7000	For joints, 90°, clip type, add				Ea.	26.50			26.50	29
7050	Screw spline joint, add					24.50			24.50	26.50
7100	For joint other than 90°, add					50.50			50.50	56
8000	For bronze anodized aluminum, add					15%				
8020	For black finish, add					30%				
8050	For stainless steel materials, add					350%				
8100	For monumental grade, add					53%				
8150	For steel stiffener, add	2 Glaz	200	.080	L.F.	11.60	3.74		15.34	18.80
8200	For 2 to 5 stories, add per story				Story		8%			
9000	Minimum labor/equipment charge	2 Glaz	2	8	Job		375		375	605

08 41 19 – Stainless-Steel-Framed Entrances and Storefronts

08 41 19.10 Stainless-Steel and Glass Entrance Unit

0010	**STAINLESS-STEEL AND GLASS ENTRANCE UNIT**, narrow stiles									
0020	3' x 7' opening, including hardware, minimum	2 Sswk	1.60	10	Opng.	6,900	530		7,430	8,550
0050	Average		1.40	11.429		7,375	610		7,985	9,200

For customer support on your Commercial Renovation Cost Data, call 877.742.4719.

249

08 41 Entrances and Storefronts

08 41 19 – Stainless-Steel-Framed Entrances and Storefronts

08 41 19.10 Stainless-Steel and Glass Entrance Unit	Crew	Daily Output	Labor-Hours	Unit	Material	2016 Bare Costs Labor	Equipment	Total	Total Incl O&P
0100 Maximum	2 Sswk	1.20	13.333	Opng.	7,950	710		8,660	10,000
1000 For solid bronze entrance units, statuary finish, add					64%				
1100 Without statuary finish, add				↓	45%				
2000 Balanced doors, 3' x 7', economy	2 Sswk	.90	17.778	Ea.	9,375	945		10,320	12,000
2100 Premium		.70	22.857	"	15,700	1,225		16,925	19,500
9000 Minimum labor/equipment charge	↓	2	8	Job		425		425	750

08 41 26 – All-Glass Entrances and Storefronts

08 41 26.10 Window Walls Aluminum, Stock

		Crew	Daily Output	Labor-Hours	Unit	Material	Labor	Equipment	Total	Total Incl O&P
0010	**WINDOW WALLS ALUMINUM, STOCK**, including glazing									
0020	Minimum	H-2	160	.150	S.F.	47.50	6.55		54.05	63
0050	Average		140	.171		65.50	7.50		73	84
0100	Maximum	↓	110	.218	↓	176	9.55		185.55	209
0500	For translucent sandwich wall systems, see Section 07 41 33.10									
0850	Cost of the above walls depends on material,									
0860	finish, repetition, and size of units.									
0870	The larger the opening, the lower the S.F. cost									

08 42 Entrances

08 42 26 – All-Glass Entrances

08 42 26.10 Swinging Glass Doors

		Crew	Daily Output	Labor-Hours	Unit	Material	Labor	Equipment	Total	Total Incl O&P
0010	**SWINGING GLASS DOORS**									
0020	Including hardware, 1/2" thick, tempered, 3' x 7' opening	2 Glaz	2	8	Opng.	2,275	375		2,650	3,100
0100	6' x 7' opening		1.40	11.429	"	4,450	535		4,985	5,775
9000	Minimum labor/equipment charge	↓	2	8	Job		375		375	605

08 42 36 – Balanced Door Entrances

08 42 36.10 Balanced Entrance Doors

		Crew	Daily Output	Labor-Hours	Unit	Material	Labor	Equipment	Total	Total Incl O&P
0010	**BALANCED ENTRANCE DOORS**									
0020	Hardware & frame, alum. & glass, 3' x 7', econ.	2 Sswk	.90	17.778	Ea.	6,700	945		7,645	9,050
0150	Premium		.70	22.857	"	8,025	1,225		9,250	11,000
9000	Minimum labor/equipment charge	↓	1	16	Job		850		850	1,500

08 43 Storefronts

08 43 13 – Aluminum-Framed Storefronts

08 43 13.10 Aluminum-Framed Entrance Doors and Frames

		Crew	Daily Output	Labor-Hours	Unit	Material	Labor	Equipment	Total	Total Incl O&P
0010	**ALUMINUM-FRAMED ENTRANCE DOORS AND FRAMES**									
0015	Standard hardware and glass stops but no glass									
0020	Entrance door, 3' x 7' opening, clear anodized finish	2 Sswk	7	2.286	Opng.	565	122		687	835
0040	Bronze finish		7	2.286		570	122		692	840
0060	Black finish		7	2.286		615	122		737	890
0200	3'-6" x 7'-0", mill finish		7	2.286		660	122		782	940
0220	Bronze finish		7	2.286		680	122		802	960
0240	Black finish		7	2.286		780	122		902	1,075
0500	6' x 7' opening, clear finish		6	2.667		880	142		1,022	1,225
0520	Bronze finish		6	2.667		955	142		1,097	1,300
0600	Door frame for above doors 3'-0" x 7'-0", mill finish		6	2.667		415	142		557	705
0620	Bronze finish		6	2.667		480	142		622	780
0640	Black finish		6	2.667		525	142		667	825
0700	3'-6" x 7'-0", mill finish	↓	6	2.667	↓	350	142		492	635

08 43 Storefronts

08 43 13 – Aluminum-Framed Storefronts

08 43 13.10 Aluminum-Framed Entrance Doors and Frames		Crew	Daily Output	Labor-Hours	Unit	Material	2016 Bare Costs Labor	Equipment	Total	Total Incl O&P
0720	Bronze finish	2 Sswk	6	2.667	Opng.	350	142		492	635
0740	Black finish		6	2.667		350	142		492	635
0800	6'-0" x 7'-0", mill finish		6	2.667		350	142		492	635
0820	Bronze finish		6	2.667		355	142		497	640
0840	Black finish		6	2.667		385	142		527	670
1000	With 3' high transom above, 3' x 7' opening, clear finish		5.50	2.909		515	155		670	840
1050	Bronze finish		5.50	2.909		530	155		685	855
1100	Black finish		5.50	2.909		555	155		710	885
1300	3'-6" x 7'-0" opening, clear finish		5.50	2.909		365	155		520	675
1320	Bronze finish		5.50	2.909		385	155		540	695
1340	Black finish		5.50	2.909		390	155		545	705
1500	6' x 7' opening, clear finish		5.50	2.909		620	155		775	960
1550	Bronze finish		5.50	2.909		640	155		795	980
1600	Black finish	▼	5.50	2.909	▼	705	155		860	1,050
8000	For 8' high doors, add				Ea.	200			200	220
9000	Minimum labor/equipment charge	2 Sswk	4	4	Job		213		213	375

08 43 13.20 Storefront Systems

		Crew	Daily Output	Labor-Hours	Unit	Material	2016 Bare Costs Labor	Equipment	Total	Total Incl O&P
0010	**STOREFRONT SYSTEMS**, aluminum frame clear 3/8" plate glass									
0020	incl. 3' x 7' door with hardware (400 sq. ft. max. wall)									
0500	Wall height to 12' high, commercial grade	2 Glaz	150	.107	S.F.	23.50	4.98		28.48	33.50
0600	Institutional grade		130	.123		28.50	5.75		34.25	41
0700	Monumental grade		115	.139		41	6.50		47.50	55.50
1000	6' x 7' door with hardware, commercial grade		135	.119		30.50	5.55		36.05	42.50
1100	Institutional grade		115	.139		29	6.50		35.50	42.50
1200	Monumental grade	▼	100	.160		55.50	7.45		62.95	73
1500	For bronze anodized finish, add					15%				
1600	For black anodized finish, add					36%				
1700	For stainless steel framing, add to monumental				▼	78%				
9000	Minimum labor/equipment charge	2 Glaz	1	16	Job		745		745	1,200

08 43 29 – Sliding Storefronts

08 43 29.10 Sliding Panels

		Crew	Daily Output	Labor-Hours	Unit	Material	2016 Bare Costs Labor	Equipment	Total	Total Incl O&P
0010	**SLIDING PANELS**									
0020	Mall fronts, aluminum & glass, 15' x 9' high	2 Glaz	1.30	12.308	Opng.	3,550	575		4,125	4,825
0100	24' x 9' high		.70	22.857		5,025	1,075		6,100	7,250
0200	48' x 9' high, with fixed panels	▼	.90	17.778		9,275	830		10,105	11,600
0500	For bronze finish, add				▼	17%				
9000	Minimum labor/equipment charge	2 Glaz	1	16	Job		745		745	1,200

08 45 Translucent Wall and Roof Assemblies

08 45 10 – Translucent Roof Assemblies

08 45 10.10 Skyroofs

		Crew	Daily Output	Labor-Hours	Unit	Material	2016 Bare Costs Labor	Equipment	Total	Total Incl O&P
0010	**SKYROOFS**									
1200	Skylights, circular, clear, double glazed acrylic									
1230	30" diameter	2 Carp	3	5.333	Ea.	3,000	258		3,258	3,725
1250	60" diameter		3	5.333		4,000	258		4,258	4,825
1290	96" diameter	▼	2	8	▼	5,000	390		5,390	6,125
1300	Skylight Barrel Vault, clear, double glazed, acrylic									
1330	3'-0" X 12'-0"	G-3	3	10.667	Ea.	5,000	510		5,510	6,300
1350	4'-0" X 12'-0"		3	10.667		5,500	510		6,010	6,850
1390	5'-0" X 12'-0"	▼	2	16	▼	6,000	760		6,760	7,825

For customer support on your Commercial Renovation Cost Data, call 877.742.4719.

251

08 45 Translucent Wall and Roof Assemblies

08 45 10 – Translucent Roof Assemblies

08 45 10.10 Skyroofs		Crew	Daily Output	Labor-Hours	Unit	Material	2016 Bare Costs Labor	Equipment	Total	Total Incl O&P
1400	Skylight Pyramid, aluminum frame, clear low-E laminated glass									
1410	The glass is installed in the frame except where noted									
1430	Square, 3' x 3'	G-3	3	10.667	Ea.	6,000	510		6,510	7,400
1440	4' x 4'		3	10.667		7,000	510		7,510	8,500
1450	5' x 5', glass must be field installed		3	10.667		8,000	510		8,510	9,600
1460	6' x 6', glass must be field installed		2	16		10,000	760		10,760	12,200
1550	Install pre-cut laminated glass in aluminum frame on a flat roof	2 Glaz	55	.291	SF Surf		13.60		13.60	22
1560	Install pre-cut laminated glass in aluminum frame on a sloped roof	"	40	.400	"		18.70		18.70	30.50
9000	Minimum labor/equipment charge	2 Carp	8	2	Job		97		97	158

08 51 Metal Windows

08 51 13 – Aluminum Windows

08 51 13.10 Aluminum Sash

08 51 13.10 Aluminum Sash		Crew	Daily Output	Labor-Hours	Unit	Material	2016 Bare Costs Labor	Equipment	Total	Total Incl O&P
0010	**ALUMINUM SASH**									
0020	Stock, grade C, glaze & trim not incl., casement	2 Sswk	200	.080	S.F.	39.50	4.26		43.76	50.50
0050	Double hung		200	.080		39.50	4.26		43.76	51
0100	Fixed casement		200	.080		17.35	4.26		21.61	26.50
0150	Picture window		200	.080		18.50	4.26		22.76	28
0200	Projected window		200	.080		35.50	4.26		39.76	46.50
0250	Single hung		200	.080		16.55	4.26		20.81	25.50
0300	Sliding		200	.080		21.50	4.26		25.76	31
1000	Mullions for above, tubular		240	.067	L.F.	6.15	3.55		9.70	13.05
9000	Minimum labor/equipment charge	1 Sswk	2	4	Job		213		213	375

08 51 13.20 Aluminum Windows

08 51 13.20 Aluminum Windows		Crew	Daily Output	Labor-Hours	Unit	Material	2016 Bare Costs Labor	Equipment	Total	Total Incl O&P
0010	**ALUMINUM WINDOWS**, incl. frame and glazing, commercial grade									
1000	Stock units, casement, 3'-1" x 3'-2" opening	2 Sswk	10	1.600	Ea.	375	85		460	560
1050	Add for storms					120			120	132
1600	Projected, with screen, 3'-1" x 3'-2" opening	2 Sswk	10	1.600		355	85		440	540
1700	Add for storms					117			117	129
2000	4'-5" x 5'-3" opening	2 Sswk	8	2		400	106		506	630
2100	Add for storms					126			126	139
2500	Enamel finish windows, 3'-1" x 3'-2"	2 Sswk	10	1.600		360	85		445	545
2600	4'-5" x 5'-3"		8	2		405	106		511	635
3000	Single hung, 2' x 3' opening, enameled, standard glazed		10	1.600		208	85		293	380
3100	Insulating glass		10	1.600		252	85		337	425
3300	2'-8" x 6'-8" opening, standard glazed		8	2		365	106		471	590
3400	Insulating glass		8	2		475	106		581	710
3700	3'-4" x 5'-0" opening, standard glazed		9	1.778		300	94.50		394.50	495
3800	Insulating glass		9	1.778		335	94.50		429.50	530
4000	Sliding aluminum, 3' x 2' opening, standard glazed		10	1.600		217	85		302	390
4100	Insulating glass		10	1.600		232	85		317	405
4300	5' x 3' opening, standard glazed		9	1.778		330	94.50		424.50	530
4400	Insulating glass		9	1.778		385	94.50		479.50	590
4600	8' x 4' opening, standard glazed		6	2.667		350	142		492	635
4700	Insulating glass		6	2.667		565	142		707	870
5000	9' x 5' opening, standard glazed		4	4		530	213		743	960
5100	Insulating glass		4	4		850	213		1,063	1,300
5500	Sliding, with thermal barrier and screen, 6' x 4', 2 track		8	2		725	106		831	985
5700	4 track		8	2		910	106		1,016	1,200
6000	For above units with bronze finish, add					15%				
6200	For installation in concrete openings, add					8%				

08 51 Metal Windows

08 51 13 – Aluminum Windows

08 51 13.20 Aluminum Windows

		Crew	Daily Output	Labor-Hours	Unit	Material	2016 Bare Costs Labor	Equipment	Total	Total Incl O&P
7000	Double hung, insulating glass, 2'-0" x 2'-0"	2 Sswk	11	1.455	Ea.	117	77.50		194.50	266
7020	2'-0" x 2'-6"		11	1.455		129	77.50		206.50	279
7040	3'-0" x 1'-6"		10	1.600		270	85		355	445
7060	3'-0" x 2'-0"		10	1.600		315	85		400	495
7080	3'-0" x 2'-6"		10	1.600		380	85		465	570
7100	3'-0" x 3'-0"		10	1.600		425	85		510	615
7120	3'-0" x 3'-6"		10	1.600		445	85		530	640
7140	3'-0" x 4'-0"		10	1.600		480	85		565	680
7160	3'-0" x 5'-0"		10	1.600		530	85		615	730
7180	3'-0" x 6'-0"		9	1.778		560	94.50		654.50	780

08 51 13.30 Impact Resistant Aluminum Windows

		Crew	Daily Output	Labor-Hours	Unit	Material	2016 Bare Costs Labor	Equipment	Total	Total Incl O&P
0010	**IMPACT RESISTANT ALUMINUM WINDOWS**, incl. frame and glazing									
0100	Single hung, impact resistant, 2'-8" x 5'-0"	2 Carp	9	1.778	Ea.	1,250	86		1,336	1,525
0120	3'-0" x 5'-0"		9	1.778		1,300	86		1,386	1,575
0130	4'-0" x 5'-0"		9	1.778		1,300	86		1,386	1,575
0250	Horizontal slider, impact resistant, 5'-5" x 5'-2"		9	1.778		1,550	86		1,636	1,850

08 51 23 – Steel Windows

08 51 23.10 Steel Sash

		Crew	Daily Output	Labor-Hours	Unit	Material	2016 Bare Costs Labor	Equipment	Total	Total Incl O&P
0010	**STEEL SASH** Custom units, glazing and trim not included									
0100	Casement, 100% vented	2 Sswk	200	.080	S.F.	67	4.26		71.26	81
0200	50% vented		200	.080		55	4.26		59.26	68
0300	Fixed		200	.080		29.50	4.26		33.76	40
1000	Projected, commercial, 40% vented		200	.080		51.50	4.26		55.76	64
1100	Intermediate, 50% vented		200	.080		58.50	4.26		62.76	72
1500	Industrial, horizontally pivoted		200	.080		53.50	4.26		57.76	66.50
1600	Fixed		200	.080		30.50	4.26		34.76	41
2000	Industrial security sash, 50% vented		200	.080		58	4.26		62.26	71.50
2100	Fixed		200	.080		47	4.26		51.26	59
2500	Picture window		200	.080		30.50	4.26		34.76	41
3000	Double hung		200	.080		60	4.26		64.26	73.50
5000	Mullions for above, open interior face		240	.067	L.F.	10.55	3.55		14.10	17.85
5100	With interior cover		240	.067	"	17.40	3.55		20.95	25.50
5200	Single glazing for above, add	2 Glaz	200	.080	S.F.	6.90	3.74		10.64	13.65
6000	Double glazing for above, add		200	.080		13.10	3.74		16.84	20.50
6100	Triple glazing for above, add		85	.188		12.50	8.80		21.30	28
9000	Minimum labor/equipment charge	1 Sswk	2	4	Job		213		213	375

08 51 23.20 Steel Windows

		Crew	Daily Output	Labor-Hours	Unit	Material	2016 Bare Costs Labor	Equipment	Total	Total Incl O&P
0010	**STEEL WINDOWS** Stock, including frame, trim and insul. glass R085123-10									
1000	Custom units, double hung, 2'-8" x 4'-6" opening	2 Sswk	12	1.333	Ea.	705	71		776	905
1100	2'-4" x 3'-9" opening		12	1.333		585	71		656	765
1500	Commercial projected, 3'-9" x 5'-5" opening		10	1.600		1,225	85		1,310	1,500
1600	6'-9" x 4'-1" opening		7	2.286		1,625	122		1,747	2,000
2000	Intermediate projected, 2'-9" x 4'-1" opening		12	1.333		685	71		756	880
2100	4'-1" x 5'-5" opening		10	1.600		1,400	85		1,485	1,700
9000	Minimum labor/equipment charge	1 Sswk	3	2.667	Job		142		142	250

08 51 66 – Metal Window Screens

08 51 66.10 Screens

		Crew	Daily Output	Labor-Hours	Unit	Material	2016 Bare Costs Labor	Equipment	Total	Total Incl O&P
0010	**SCREENS**									
0020	For metal sash, aluminum or bronze mesh, flat screen	2 Sswk	1200	.013	S.F.	4.30	.71		5.01	6
0500	Wicket screen, inside window		1000	.016		6.65	.85		7.50	8.80
0800	Security screen, aluminum frame with stainless steel cloth		1200	.013		24	.71		24.71	27.50

For customer support on your Commercial Renovation Cost Data, call 877.742.4719.

253

08 51 Metal Windows

08 51 66 – Metal Window Screens

08 51 66.10 Screens		Crew	Daily Output	Labor-Hours	Unit	Material	2016 Bare Costs Labor	Equipment	Total	Total Incl O&P
0900	Steel grate, painted, on steel frame	2 Sswk	1600	.010	S.F.	13.10	.53		13.63	15.35
1000	Screens for solar louvers	↓	160	.100	↓	24.50	5.30		29.80	36.50
4000	See Section 05 58 23.90									

08 52 Wood Windows

08 52 10 – Plain Wood Windows

08 52 10.20 Awning Window

			Crew	Daily Output	Labor-Hours	Unit	Material	Labor	Equipment	Total	Total Incl O&P
0010	**AWNING WINDOW**, Including frame, screens and grilles										
0100	34" x 22", insulated glass		1 Carp	10	.800	Ea.	271	39		310	360
0200	Low E glass			10	.800		272	39		311	365
0300	40" x 28", insulated glass			9	.889		315	43		358	415
0400	Low E Glass			9	.889		340	43		383	445
0500	48" x 36", insulated glass			8	1		470	48.50		518.50	595
0600	Low E glass		↓	8	1		495	48.50		543.50	620
4000	Impact windows, minimum, add						60%				
4010	Impact windows, maximum, add					↓	160%				
9000	Minimum labor/equipment charge		1 Carp	4	2	Job		97		97	158

08 52 10.30 Wood Windows

			Crew	Daily Output	Labor-Hours	Unit	Material	Labor	Equipment	Total	Total Incl O&P
0010	**WOOD WINDOWS**, double hung										
0020	Including frame, double insulated glass, screens and grilles										
0040	Double hung, 2'-2" x 3'-4" high		2 Carp	15	1.067	Ea.	212	51.50		263.50	320
0060	2'-2" x 4'-4"			14	1.143		231	55.50		286.50	345
0080	2'-6" x 3'-4"			13	1.231		221	59.50		280.50	340
0100	2'-6" x 4'-0"			12	1.333		230	64.50		294.50	360
0120	2'-6" x 4'-8"			12	1.333		253	64.50		317.50	385
0140	2'-10" x 3'-4"			10	1.600		226	77.50		303.50	375
0160	2'-10" x 4'-0"			10	1.600		248	77.50		325.50	400
0180	3'-7" x 3'-4"			9	1.778		256	86		342	425
0200	3'-7" x 5'-4"			9	1.778		286	86		372	455
0220	3'-10" x 5'-4"		↓	8	2		515	97		612	725
3800	Triple glazing for above, add					↓	25%				

08 52 10.40 Casement Window

			Crew	Daily Output	Labor-Hours	Unit	Material	Labor	Equipment	Total	Total Incl O&P
0010	**CASEMENT WINDOW**, including frame, screen and grilles										
0100	2'-0" x 3'-0" H, dbl. insulated glass	G	1 Carp	10	.800	Ea.	273	39		312	365
0150	Low E glass	G		10	.800		259	39		298	350
0200	2'-0" x 4'-6" high, double insulated glass	G		9	.889		370	43		413	475
0250	Low E glass	G		9	.889		370	43		413	475
0260	Casement 4'-2" x 4'-2" double insulated glass	G		11	.727		890	35		925	1,050
0270	4'-0" x 4'-0" Low E glass	G		11	.727		535	35		570	650
0290	6'-4" x 5'-7" Low E glass	G		9	.889		1,125	43		1,168	1,325
0300	2'-4" x 6'-0" high, double insulated glass	G		8	1		440	48.50		488.50	565
0350	Low E glass	G		8	1		440	48.50		488.50	565
0522	Vinyl clad, premium, double insulated glass, 2'-0" x 3'-0"	G		10	.800		271	39		310	360
0524	2'-0" x 4'-0"	G		9	.889		315	43		358	420
0525	2'-0" x 5'-0"	G		8	1		360	48.50		408.50	480
0528	2'-0" x 6'-0"	G		8	1		380	48.50		428.50	500
0600	3'-0" x 5'-0"	G		8	1		665	48.50		713.50	810
0700	4'-0" x 3'-0"	G		8	1		730	48.50		778.50	885
0710	4'-0" x 4'-0"	G		8	1		625	48.50		673.50	765
0720	4'-8" x 4'-0"	G		8	1		690	48.50		738.50	840

08 52 Wood Windows

08 52 10 – Plain Wood Windows

08 52 10.40 Casement Window		Crew	Daily Output	Labor-Hours	Unit	Material	2016 Bare Costs Labor	Equipment	Total	Total Incl O&P
0730	4'-8" x 5'-0"	**G** 1 Carp	6	1.333	Ea.	790	64.50		854.50	970
0740	4'-8" x 6'-0"	**G**	6	1.333		880	64.50		944.50	1,075
0750	6'-0" x 4'-0"	**G**	6	1.333		805	64.50		869.50	990
0800	6'-0" x 5'-0"	**G**	6	1.333		895	64.50		959.50	1,100
0900	5'-6" x 5'-6"	**G** 2 Carp	15	1.067		1,425	51.50		1,476.50	1,650
8190	For installation, add per leaf						15%			
8200	For multiple leaf units, deduct for stationary sash									
8220	2' high				Ea.	23.50			23.50	25.50
8240	4'-6" high					26.50			26.50	29
8260	6' high					35			35	38.50
8300	Impact windows, minimum, add					60%				
8310	Impact windows, maximum, add					160%				
9000	Minimum labor/equipment charge	1 Carp	3	2.667	Job		129		129	211

08 52 10.50 Double Hung

		Crew	Daily Output	Labor-Hours	Unit	Material	Labor	Equipment	Total	Total Incl O&P
0010	**DOUBLE HUNG**, Including frame, screens and grilles									
0100	2'-0" x 3'-0" high, low E insul. glass	**G** 1 Carp	10	.800	Ea.	210	39		249	295
0200	3'-0" x 4'-0" high, double insulated glass	**G**	9	.889		279	43		322	375
0300	4'-0" x 4'-6" high, low E insulated glass	**G**	8	1		320	48.50		368.50	435
8000	Impact windows, minimum, add					60%				
8010	Impact windows, maximum, add					160%				
9000	Minimum labor/equipment charge	1 Carp	3	2.667	Job		129		129	211

08 52 10.55 Picture Window

		Crew	Daily Output	Labor-Hours	Unit	Material	Labor	Equipment	Total	Total Incl O&P
0010	**PICTURE WINDOW**, Including frame and grilles									
0100	3'-6" x 4'-0" high, dbl. insulated glass	2 Carp	12	1.333	Ea.	420	64.50		484.50	570
0150	Low E glass		12	1.333		435	64.50		499.50	580
0200	4'-0" x 4'-6" high, double insulated glass		11	1.455		550	70.50		620.50	720
0250	Low E glass		11	1.455		530	70.50		600.50	700
0300	5'-0" x 4'-0" high, double insulated glass		11	1.455		580	70.50		650.50	755
0350	Low E glass		11	1.455		605	70.50		675.50	780
0400	6'-0" x 4'-6" high, double insulated glass		10	1.600		640	77.50		717.50	825
0450	Low E glass		10	1.600		640	77.50		717.50	830

08 52 10.65 Wood Sash

		Crew	Daily Output	Labor-Hours	Unit	Material	Labor	Equipment	Total	Total Incl O&P
0010	**WOOD SASH**, Including glazing but not trim									
0050	Custom, 5'-0" x 4'-0", 1" dbl. glazed, 3/16" thick lites	2 Carp	3.20	5	Ea.	230	242		472	650
0100	1/4" thick lites		5	3.200		245	155		400	525
0200	1" thick, triple glazed		5	3.200		415	155		570	710
0300	7'-0" x 4'-6" high, 1" double glazed, 3/16" thick lites		4.30	3.721		420	180		600	755
0400	1/4" thick lites		4.30	3.721		475	180		655	815
0500	1" thick, triple glazed		4.30	3.721		540	180		720	890
0600	8'-6" x 5'-0" high, 1" double glazed, 3/16" thick lites		3.50	4.571		565	221		786	985
0700	1/4" thick lites		3.50	4.571		620	221		841	1,050
0800	1" thick, triple glazed		3.50	4.571		625	221		846	1,050
0900	Window frames only, based on perimeter length				L.F.	4.02			4.02	4.42
1200	Window sill, stock, per lineal foot					8.50			8.50	9.35
1250	Casing, stock					3.30			3.30	3.63

08 52 10.70 Sliding Windows

		Crew	Daily Output	Labor-Hours	Unit	Material	Labor	Equipment	Total	Total Incl O&P
0010	**SLIDING WINDOWS**									
0100	3'-0" x 3'-0" high, double insulated	**G** 1 Carp	10	.800	Ea.	284	39		323	375
0120	Low E glass	**G**	10	.800		310	39		349	405
0200	4'-0" x 3'-6" high, double insulated	**G**	9	.889		355	43		398	460
0220	Low E glass	**G**	9	.889		360	43		403	470
0300	6'-0" x 5'-0" high, double insulated	**G**	8	1		490	48.50		538.50	615

For customer support on your Commercial Renovation Cost Data, call 877.742.4719.

255

08 52 Wood Windows

08 52 10 – Plain Wood Windows

08 52 10.70 Sliding Windows		Crew	Daily Output	Labor-Hours	Unit	Material	2016 Bare Costs Labor	Equipment	Total	Total Incl O&P
0320	Low E glass	G 1 Carp	8	1	Ea.	530	48.50		578.50	660
9000	Minimum labor/equipment charge	↓	3	2.667	Job		129		129	211

08 52 13 – Metal-Clad Wood Windows

08 52 13.10 Awning Windows, Metal-Clad

		Crew	Daily Output	Labor-Hours	Unit	Material	Labor	Equipment	Total	Total Incl O&P
0010	**AWNING WINDOWS, METAL-CLAD**									
2000	Metal clad, awning deluxe, double insulated glass, 34" x 22"	1 Carp	9	.889	Ea.	249	43		292	345
2050	36" x 25"		9	.889		276	43		319	375
2100	40" x 22"		9	.889		292	43		335	390
2150	40" x 30"		9	.889		340	43		383	445
2200	48" x 28"		8	1		350	48.50		398.50	465
2250	60" x 36"	↓	8	1	↓	375	48.50		423.50	490

08 52 13.20 Casement Windows, Metal-Clad

		Crew	Daily Output	Labor-Hours	Unit	Material	Labor	Equipment	Total	Total Incl O&P
0010	**CASEMENT WINDOWS, METAL-CLAD**									
0100	Metal clad, deluxe, dbl. insul. glass, 2'-0" x 3'-0" high	G 1 Carp	10	.800	Ea.	285	39		324	380
0120	2'-0" x 4'-0" high	G	9	.889		320	43		363	420
0130	2'-0" x 5'-0" high	G	8	1		330	48.50		378.50	445
0140	2'-0" x 6'-0" high	G ↓	8	1		370	48.50		418.50	485
0300	Metal clad, casement, bldrs mdl, 6'-0" x 4'-0", dbl. insltd gls, 3 panels	2 Carp	10	1.600		1,200	77.50		1,277.50	1,450
0310	9'-0" x 4'-0" , 4 panels		8	2		1,550	97		1,647	1,850
0320	10'-0" x 5'-0", 5 panels		7	2.286		2,100	111		2,211	2,500
0330	12'-0" x 6'-0", 6 panels	↓	6	2.667	↓	2,700	129		2,829	3,150

08 52 13.30 Double-Hung Windows, Metal-clad

		Crew	Daily Output	Labor-Hours	Unit	Material	Labor	Equipment	Total	Total Incl O&P
0010	**DOUBLE-HUNG WINDOWS, METAL-CLAD**									
0100	Metal clad, deluxe, dbl. insul. glass, 2'-6" x 3'-0" high	G 1 Carp	10	.800	Ea.	280	39		319	375
0120	3'-0" x 3'-6" high	G	10	.800		320	39		359	415
0140	3'-0" x 4'-0" high	G	9	.889		335	43		378	440
0160	3'-0" x 4'-6" high	G	9	.889		350	43		393	455
0180	3'-0" x 5'-0" high	G	8	1		380	48.50		428.50	495
0200	3'-6" x 6'-0" high	G ↓	8	1	↓	455	48.50		503.50	585

08 52 13.35 Picture and Sliding Windows Metal-Clad

		Crew	Daily Output	Labor-Hours	Unit	Material	Labor	Equipment	Total	Total Incl O&P
0010	**PICTURE AND SLIDING WINDOWS METAL-CLAD**									
2000	Metal clad, dlx picture, dbl. insul. glass, 4'-0" x 4'-0" high	2 Carp	12	1.333	Ea.	380	64.50		444.50	525
2100	4'-0" x 6'-0" high		11	1.455		560	70.50		630.50	730
2200	5'-0" x 6'-0" high		10	1.600		620	77.50		697.50	805
2300	6'-0" x 6'-0" high	↓	10	1.600		710	77.50		787.50	905
2400	Metal clad, dlx sliding, double insulated glass, 3'-0" x 3'-0" high	G 1 Carp	10	.800		330	39		369	425
2420	4'-0" x 3' 6" high	G	9	.889		400	43		443	510
2440	5'-0" x 4'-0" high	G	9	.889		480	43		523	595
2460	6'-0" x 5'-0" high	G ↓	8	1	↓	730	48.50		778.50	880
9000	Minimum labor/equipment charge	2 Carp	2.75	5.818	Job		282		282	460

08 52 13.40 Bow and Bay Windows, Metal-Clad

		Crew	Daily Output	Labor-Hours	Unit	Material	Labor	Equipment	Total	Total Incl O&P
0010	**BOW AND BAY WINDOWS, METAL-CLAD**									
0100	Metal clad, deluxe, dbl. insul. glass, 8'-0" x 5'-0" high, 4 panels	2 Carp	10	1.600	Ea.	1,700	77.50		1,777.50	2,000
0120	10'-0" x 5'-0" high, 5 panels		8	2		1,825	97		1,922	2,150
0140	10'-0" x 6'-0" high, 5 panels		7	2.286		2,150	111		2,261	2,550
0160	12'-0" x 6'-0" high, 6 panels		6	2.667		2,925	129		3,054	3,425
0400	Double hung, bldrs. model, bay, 8' x 4' high, dbl. insulated glass		10	1.600		1,350	77.50		1,427.50	1,600
0440	Low E glass		10	1.600		1,450	77.50		1,527.50	1,725
0460	9'-0" x 5'-0" high, double insulated glass		6	2.667		1,450	129		1,579	1,800
0480	Low E glass		6	2.667		1,525	129		1,654	1,875
0500	Metal clad, deluxe, dbl. insul. glass, 7'-0" x 4'-0" high		10	1.600		1,300	77.50		1,377.50	1,550

08 52 Wood Windows

08 52 13 – Metal-Clad Wood Windows

08 52 13.40 Bow and Bay Windows, Metal-Clad		Crew	Daily Output	Labor-Hours	Unit	Material	2016 Bare Costs Labor	Equipment	Total	Total Incl O&P
0520	8'-0" x 4'-0" high	2 Carp	8	2	Ea.	1,350	97		1,447	1,625
0540	8'-0" x 5'-0" high		7	2.286		1,375	111		1,486	1,700
0560	9'-0" x 5'-0" high	↓	6	2.667	↓	1,475	129		1,604	1,825

08 52 16 – Plastic-Clad Wood Windows

08 52 16.10 Bow Window

		Crew	Daily Output	Labor-Hours	Unit	Material	2016 Bare Costs Labor	Equipment	Total	Total Incl O&P
0010	**BOW WINDOW** Including frames, screens, and grilles									
0020	End panels operable									
1000	Bow type, casement, wood, bldrs. mdl., 8' x 5' dbl. insltd glass, 4 panel	2 Carp	10	1.600	Ea.	1,500	77.50		1,577.50	1,775
1050	Low E glass		10	1.600		1,325	77.50		1,402.50	1,575
1100	10'-0" x 5'-0", double insulated glass, 6 panels		6	2.667		1,350	129		1,479	1,700
1200	Low E glass, 6 panels		6	2.667		1,450	129		1,579	1,800
1300	Vinyl clad, bldrs. model, double insulated glass, 6'-0" x 4'-0", 3 panel		10	1.600		1,025	77.50		1,102.50	1,250
1340	9'-0" x 4'-0", 4 panel		8	2		1,350	97		1,447	1,650
1380	10'-0" x 6'-0", 5 panels		7	2.286		2,250	111		2,361	2,650
1420	12'-0" x 6'-0", 6 panels		6	2.667		2,925	129		3,054	3,425
2000	Bay window, 8' x 5', dbl. insul glass		10	1.600		1,925	77.50		2,002.50	2,225
2050	Low E glass		10	1.600		2,325	77.50		2,402.50	2,675
2100	12'-0" x 6'-0", double insulated glass, 6 panels		6	2.667		2,400	129		2,529	2,825
2200	Low E glass		6	2.667		3,175	129		3,304	3,700
2300	Vinyl clad, premium, double insulated glass, 8'-0" x 5'-0"		10	1.600		1,750	77.50		1,827.50	2,050
2340	10'-0" x 5'-0"		8	2		2,300	97		2,397	2,675
2380	10'-0" x 6'-0"		7	2.286		2,625	111		2,736	3,050
2420	12'-0" x 6'-0"		6	2.667		3,200	129		3,329	3,725
3300	Vinyl clad, premium, double insulated glass, 7'-0" x 4'-6"		10	1.600		1,400	77.50		1,477.50	1,650
3340	8'-0" x 4'-6"		8	2		1,425	97		1,522	1,700
3380	8'-0" x 5'-0"		7	2.286		1,500	111		1,611	1,825
3420	9'-0" x 5'-0"		6	2.667	↓	1,525	129		1,654	1,875
9000	Minimum labor/equipment charge	↓	2.50	6.400	Job		310		310	505

08 52 16.15 Awning Window Vinyl-Clad

		Crew	Daily Output	Labor-Hours	Unit	Material	2016 Bare Costs Labor	Equipment	Total	Total Incl O&P
0010	**AWNING WINDOW VINYL-CLAD** Including frames, screens, and grilles									
0240	Vinyl clad, 34" x 22"	1 Carp	10	.800	Ea.	267	39		306	355
0280	36" x 28"		9	.889		305	43		348	405
0300	36" x 36"		9	.889		345	43		388	445
0340	40" x 22"		10	.800		291	39		330	385
0360	48" x 28"		8	1		370	48.50		418.50	485
0380	60" x 36"	↓	8	1	↓	500	48.50		548.50	630

08 52 16.20 Half Round, Vinyl Clad

		Crew	Daily Output	Labor-Hours	Unit	Material	2016 Bare Costs Labor	Equipment	Total	Total Incl O&P
0010	**HALF ROUND, VINYL CLAD**, double insulated glass, incl. grille									
0800	14" height x 24" base	2 Carp	9	1.778	Ea.	415	86		501	595
1040	15" height x 25" base		8	2		385	97		482	580
1060	16" height x 28" base		7	2.286		405	111		516	625
1080	17" height x 29" base	↓	7	2.286		420	111		531	640
2000	19" height x 33" base	1 Carp	6	1.333		465	64.50		529.50	615
2100	20" height x 35" base		6	1.333		465	64.50		529.50	615
2200	21" height x 37" base	↓	6	1.333		480	64.50		544.50	635
2250	23" height x 41" base	2 Carp	6	2.667		520	129		649	780
2300	26" height x 48" base		6	2.667		535	129		664	795
2350	30" height x 56" base	↓	6	2.667		615	129		744	885
3000	36" height x 67" base	1 Carp	4	2		1,050	97		1,147	1,300
3040	38" height x 71" base	2 Carp	5	3.200		975	155		1,130	1,325
3050	40" height x 75" base	"	5	3.200		1,275	155		1,430	1,650
5000	Elliptical, 71" x 16"	1 Carp	11	.727	↓	975	35		1,010	1,125

For customer support on your Commercial Renovation Cost Data, call 877.742.4719.

257

08 52 Wood Windows

08 52 16 – Plastic-Clad Wood Windows

08 52 16.20 Half Round, Vinyl Clad		Crew	Daily Output	Labor-Hours	Unit	Material	2016 Bare Costs Labor	2016 Bare Costs Equipment	Total	Total Incl O&P
5100	95" x 21"	1 Carp	10	.800	Ea.	1,375	39		1,414	1,600

08 52 16.35 Double-Hung Window

			Crew	Daily Output	Labor-Hours	Unit	Material	Labor	Equipment	Total	Total Incl O&P
0010	**DOUBLE-HUNG WINDOW** Including frames, screens, and grilles										
0300	Vinyl clad, premium, double insulated glass, 2'-6" x 3'-0"	G	1 Carp	10	.800	Ea.	310	39		349	405
0305	2'-6" x 4'-0"	G		10	.800		360	39		399	460
0400	3'-0" x 3'-6"	G		10	.800		335	39		374	435
0500	3'-0" x 4'-0"	G		9	.889		395	43		438	505
0600	3'-0" x 4'-6"	G		9	.889		410	43		453	520
0700	3'-0" x 5'-0"	G		8	1		445	48.50		493.50	565
0790	3'-4" x 5'-0"	G		8	1		455	48.50		503.50	580
0800	3'-6" x 6'-0"	G		8	1		490	48.50		538.50	620
0820	4'-0" x 5'-0"	G		7	1.143		560	55.50		615.50	705
0830	4'-0" x 6'-0"	G		7	1.143		700	55.50		755.50	855

08 52 16.40 Transom Windows

		Crew	Daily Output	Labor-Hours	Unit	Material	Labor	Equipment	Total	Total Incl O&P
0010	**TRANSOM WINDOWS**									
0050	Vinyl clad, premium, double insulated glass, 32" x 8"	1 Carp	16	.500	Ea.	188	24		212	246
0100	36" x 8"		16	.500		199	24		223	259
0110	36" x 12"		16	.500		212	24		236	273
2000	Custom sizes, up to 350 sq. in.		12	.667		256	32.50		288.50	335
2100	351 to 750 sq. in.		12	.667		325	32.50		357.50	410
2200	751 to 1150 sq. in.		11	.727		400	35		435	500
2300	1151 to 1450 sq. in.		11	.727		460	35		495	570
2400	1451 to 1850 sq. in.	2 Carp	12	1.333		530	64.50		594.50	690
2500	1851 to 2250 sq. in.		12	1.333		605	64.50		669.50	770
2600	2251 to 2650 sq. in.		11	1.455		680	70.50		750.50	865
2700	2651 to 3050 sq. in.		11	1.455		685	70.50		755.50	870
2800	3051 to 3450 sq. in.		11	1.455		755	70.50		825.50	945
2900	3451 to 3850 sq. in.		10	1.600		810	77.50		887.50	1,025
3000	3851 to 4250 sq. in.		10	1.600		870	77.50		947.50	1,075
3100	4251 to 4650 sq. in.		10	1.600		890	77.50		967.50	1,100
3200	4651 to 5050 sq. in.		10	1.600		935	77.50		1,012.50	1,150
3300	5051 to 5450 sq. in.		9	1.778		1,000	86		1,086	1,250
3400	5451 to 5850 sq. in.		9	1.778		975	86		1,061	1,225
3600	6251 to 6650 sq. in.		8	2		1,175	97		1,272	1,425
3700	6651 to 7050 sq. in.		8	2		1,250	97		1,347	1,525

08 52 16.45 Trapezoid Windows

		Crew	Daily Output	Labor-Hours	Unit	Material	Labor	Equipment	Total	Total Incl O&P
0010	**TRAPEZOID WINDOWS**									
0900	20" base x 44" leg x 53" leg	2 Carp	13	1.231	Ea.	390	59.50		449.50	530
1000	24" base x 90" leg x 102" leg		8	2		670	97		767	900
3000	36" base x 40" leg x 22" leg		12	1.333		430	64.50		494.50	575
3010	36" base x 44" leg x 25" leg		13	1.231		450	59.50		509.50	595
3050	36" base x 26" leg x 48" leg		9	1.778		455	86		541	640
3100	36" base x 42" legs, 50" peak		9	1.778		545	86		631	740
3200	36" base x 60" leg x 81" leg		11	1.455		710	70.50		780.50	900
4320	44" base x 23" leg x 56" leg		11	1.455		560	70.50		630.50	735
4350	44" base x 59" leg x 92" leg		10	1.600		840	77.50		917.50	1,050
4500	46" base x 15" leg x 46" leg		8	2		425	97		522	625
4550	46" base x 16" leg x 48" leg		8	2		450	97		547	655
4600	46" base x 50" leg x 80" leg		7	2.286		680	111		791	930
6600	66" base x 12" leg x 42" leg		8	2		575	97		672	795
6650	66" base x 12" legs, 28" peak		9	1.778		455	86		541	640
6700	68" base x 23" legs, 31" peak		8	2		575	97		672	795

For customer support on your Commercial Renovation Cost Data, call 877.742.4719.

08 52 Wood Windows

08 52 16 – Plastic-Clad Wood Windows

08 52 16.70 Vinyl Clad, Premium, Dbl. Insulated Glass	Crew	Daily Output	Labor-Hours	Unit	Material	2016 Bare Costs Labor	2016 Bare Costs Equipment	Total	Total Incl O&P
0010 **VINYL CLAD, PREMIUM, DBL. INSULATED GLASS**									
1000 Sliding, 3'-0" x 3'-0" ⒢	1 Carp	10	.800	Ea.	605	39		644	730
1050 4'-0" x 3'-6" ⒢		9	.889		685	43		728	825
1100 5'-0" x 4'-0" ⒢		9	.889		900	43		943	1,050
1150 6'-0" x 5'-0" ⒢	↓	8	1	↓	1,125	48.50		1,173.50	1,325

08 52 50 – Window Accessories

08 52 50.10 Window Grille or Muntin

	Crew	Daily Output	Labor-Hours	Unit	Material	Labor	Equipment	Total	Total Incl O&P
0010 **WINDOW GRILLE OR MUNTIN**, snap in type									
0020 Standard pattern interior grilles									
2000 Wood, awning window, glass size 28" x 16" high	1 Carp	30	.267	Ea.	28	12.90		40.90	52
2060 44" x 24" high		32	.250		40	12.10		52.10	64
2100 Casement, glass size, 20" x 36" high		30	.267		32	12.90		44.90	56
2180 20" x 56" high		32	.250	↓	43	12.10		55.10	67.50
2200 Double hung, glass size, 16" x 24" high		24	.333	Set	51	16.15		67.15	83
2280 32" x 32" high		34	.235	"	131	11.40		142.40	163
2500 Picture, glass size, 48" x 48" high		30	.267	Ea.	120	12.90		132.90	153
2580 60" x 68" high		28	.286	"	183	13.85		196.85	225
2600 Sliding, glass size, 14" x 36" high		24	.333	Set	35.50	16.15		51.65	65.50
2680 36" x 36" high		22	.364	"	43.50	17.60		61.10	76.50
9000 Minimum labor/equipment charge	↓	5	1.600	Job		77.50		77.50	127

08 52 66 – Wood Window Screens

08 52 66.10 Wood Screens

	Crew	Daily Output	Labor-Hours	Unit	Material	Labor	Equipment	Total	Total Incl O&P
0010 **WOOD SCREENS**									
0020 Over 3 S.F., 3/4" frames	2 Carp	375	.043	S.F.	4.89	2.07		6.96	8.80
0100 1-1/8" frames		375	.043		8.20	2.07		10.27	12.40
0200 Rescreen wood frame	↓	500	.032	↓	1.61	1.55		3.16	4.30
9000 Minimum labor/equipment charge	1 Carp	4	2	Job		97		97	158

08 52 69 – Wood Storm Windows

08 52 69.10 Storm Windows

	Crew	Daily Output	Labor-Hours	Unit	Material	Labor	Equipment	Total	Total Incl O&P
0010 **STORM WINDOWS**, aluminum residential									
0300 Basement, mill finish, incl. fiberglass screen									
0320 1'-10" x 1'-0" high ⒢	2 Carp	30	.533	Ea.	36	26		62	81.50
0340 2'-9" x 1'-6" high ⒢		30	.533		39	26		65	85
0360 3'-4" x 2'-0" high ⒢	↓	30	.533	↓	46	26		72	92.50
1600 Double-hung, combination, storm & screen									
1700 Custom, clear anodic coating, 2'-0" x 3'-5" high	2 Carp	30	.533	Ea.	97	26		123	149
1720 2'-6" x 5'-0" high		28	.571		117	27.50		144.50	174
1740 4'-0" x 6'-0" high		25	.640		235	31		266	310
1800 White painted, 2'-0" x 3'-5" high		30	.533		112	26		138	165
1820 2'-6" x 5'-0" high		28	.571		173	27.50		200.50	235
1840 4'-0" x 6'-0" high		25	.640		286	31		317	365
2000 Clear anodic coating, 2'-0" x 3'-5" high ⒢		30	.533		97	26		123	149
2020 2'-6" x 5'-0" high ⒢		28	.571		121	27.50		148.50	178
2040 4'-0" x 6'-0" high ⒢		25	.640		133	31		164	197
2400 White painted, 2'-0" x 3'-5" high ⒢		30	.533		92	26		118	143
2420 2'-6" x 5'-0" high ⒢		28	.571		97	27.50		124.50	152
2440 4'-0" x 6'-0" high ⒢		25	.640		112	31		143	174
2600 Mill finish, 2'-0" x 3'-5" high ⒢		30	.533		87	26		113	138
2620 2'-6" x 5'-0" high ⒢		28	.571		92	27.50		119.50	146
2640 4'-0" x 6'-8" high ⒢	↓	25	.640	↓	112	31		143	174

08 52 Wood Windows

08 52 69 – Wood Storm Windows

08 52 69.10 Storm Windows		Crew	Daily Output	Labor-Hours	Unit	Material	2016 Bare Costs Labor	Equipment	Total	Total Incl O&P
4000	Picture window, storm, 1 lite, white or bronze finish									
4020	4'-6" x 4'-6" high	2 Carp	25	.640	Ea.	133	31		164	197
4040	5'-8" x 4'-6" high		20	.800		150	39		189	229
4400	Mill finish, 4'-6" x 4'-6" high		25	.640		133	31		164	197
4420	5'-8" x 4'-6" high		20	.800		154	39		193	233
4600	3 lite, white or bronze finish									
4620	4'-6" x 4'-6" high	2 Carp	25	.640	Ea.	163	31		194	230
4640	5'-8" x 4'-6" high		20	.800		173	39		212	254
4800	Mill finish, 4'-6" x 4'-6" high		25	.640		154	31		185	220
4820	5'-8" x 4'-6" high		20	.800		163	39		202	243
6000	Sliding window, storm, 2 lite, white or bronze finish									
6020	3'-4" x 2'-7" high	2 Carp	28	.571	Ea.	135	27.50		162.50	194
6040	4'-4" x 3'-3" high		25	.640		150	31		181	216
6060	5'-4" x 6'-0" high		20	.800		240	39		279	330
8000	PVC framed									
8100	Double-hung, combination, storm & screen									
8120	2'-6" x 3'-5"	2 Carp	15	1.067	Ea.	135	51.50		186.50	234
8140	4' x 6'	"	12.50	1.280	"	171	62		233	289
8600	Single lite picture storm									
8620	4'-6" x 4'-6"	2 Carp	12.50	1.280	Ea.	171	62		233	289
8640	5'-8" x 4'-6"	"	10	1.600	"	187	77.50		264.50	335
9000	Interior storm window									
9100	Storm window interior glass	1 Glaz	107	.075	S.F.	6.20	3.49		9.69	12.50
9410	Minimum labor/equipment charge	1 Carp	4	2	Job		97		97	158

08 53 Plastic Windows

08 53 13 – Vinyl Windows

08 53 13.20 Vinyl Single Hung Windows

08 53 13.20 Vinyl Single Hung Windows			Crew	Daily Output	Labor-Hours	Unit	Material	2016 Bare Costs Labor	Equipment	Total	Total Incl O&P
0010	**VINYL SINGLE HUNG WINDOWS**, insulated glass										
0020	Grids, low e, J fin, extension jambs										
0130	25" x 41"	G	2 Carp	20	.800	Ea.	200	39		239	284
0140	25" x 49"	G		18	.889		210	43		253	300
0150	25" x 57"	G		17	.941		220	45.50		265.50	315
0160	25" x 65"	G		16	1		240	48.50		288.50	345
0170	29" x 41"	G		18	.889		205	43		248	297
0180	29" x 53"	G		18	.889		220	43		263	315
0190	29" x 57"	G		17	.941		230	45.50		275.50	330
0200	29" x 65"	G		16	1		240	48.50		288.50	345
0210	33" x 41"	G		20	.800		215	39		254	300
0220	33" x 53"	G		18	.889		230	43		273	325
0230	33" x 57"	G		17	.941		240	45.50		285.50	340
0240	33" x 65"	G		16	1		250	48.50		298.50	355
0250	37" x 41"	G		20	.800		242	39		281	330
0260	37" x 53"	G		18	.889		255	43		298	350
0270	37" x 57"	G		17	.941		265	45.50		310.50	365
0280	37" x 65"	G		16	1		290	48.50		338.50	400

08 53 13.30 Vinyl Double Hung Windows

08 53 13.30 Vinyl Double Hung Windows			Crew	Daily Output	Labor-Hours	Unit	Material	2016 Bare Costs Labor	Equipment	Total	Total Incl O&P
0010	**VINYL DOUBLE HUNG WINDOWS**, insulated glass										
0100	Grids, low E, J fin, ext. jambs, 21" x 53"	G	2 Carp	18	.889	Ea.	215	43		258	310
0102	21" x 37"	G		18	.889		223	43		266	315
0104	21" x 41"	G		18	.889		232	43		275	325

08 53 Plastic Windows

08 53 13 – Vinyl Windows

08 53 13.30 Vinyl Double Hung Windows		Crew	Daily Output	Labor-Hours	Unit	Material	2016 Bare Costs Labor	Equipment	Total	Total Incl O&P
0106	21" x 49"	G 2 Carp	18	.889	Ea.	247	43		290	345
0110	21" x 57"	G	17	.941		263	45.50		308.50	365
0120	21" x 65"	G	16	1		279	48.50		327.50	385
0128	25" x 37"	G	20	.800		234	39		273	320
0130	25" x 41"	G	20	.800		242	39		281	330
0140	25" x 49"	G	18	.889		258	43		301	355
0145	25" x 53"	G	18	.889		265	43		308	365
0150	25" x 57"	G	17	.941		274	45.50		319.50	375
0160	25" x 65"	G	16	1		290	48.50		338.50	400
0162	25" x 69"	G	16	1		297	48.50		345.50	405
0164	25" x 77"	G	16	1		325	48.50		373.50	435
0168	29" x 37"	G	18	.889		242	43		285	335
0170	29" x 41"	G	18	.889		250	43		293	345
0172	29" x 49"	G	18	.889		267	43		310	365
0180	29" x 53"	G	18	.889		271	43		314	370
0190	29" x 57"	G	17	.941		280	45.50		325.50	385
0200	29" x 65"	G	16	1		300	48.50		348.50	410
0202	29" x 69"	G	16	1		310	48.50		358.50	420
0205	29" x 77"	G	16	1		335	48.50		383.50	445
0208	33" x 37"	G	20	.800		255	39		294	345
0210	33" x 41"	G	20	.800		263	39		302	355
0215	33" x 49"	G	20	.800		280	39		319	375
0220	33" x 53"	G	18	.889		285	43		328	385
0230	33" x 57"	G	17	.941		297	45.50		342.50	400
0240	33" x 65"	G	16	1		315	48.50		363.50	425
0242	33" x 69"	G	16	1		330	48.50		378.50	440
0246	33" x 77"	G	16	1		350	48.50		398.50	465
0250	37" x 41"	G	20	.800		289	39		328	385
0255	37" x 49"	G	20	.800		282	39		321	375
0260	37" x 53"	G	18	.889		315	43		358	415
0270	37" x 57"	G	17	.941		325	45.50		370.50	435
0280	37" x 65"	G	16	1		350	48.50		398.50	465
0282	37" x 69"	G	16	1		360	48.50		408.50	475
0286	37" x 77"	G	16	1		380	48.50		428.50	495
0300	Solid vinyl, average quality, double insulated glass, 2'-0" x 3'-0"	G 1 Carp	10	.800		291	39		330	385
0310	3'-0" x 4'-0"	G	9	.889		209	43		252	300
0330	Premium, double insulated glass, 2'-6" x 3'-0"	G	10	.800		271	39		310	360
0340	3'-0" x 3'-6"	G	9	.889		305	43		348	405
0350	3'-0" x 4'-0"	G	9	.889		325	43		368	425
0360	3'-0" x 4'-6"	G	9	.889		335	43		378	440
0370	3'-0" x 5'-0"	G	8	1		360	48.50		408.50	475
0380	3'-6" x 6'-0"	G	8	1		390	48.50		438.50	505

08 53 13.40 Vinyl Casement Windows

		Crew	Daily Output	Labor-Hours	Unit	Material	2016 Bare Costs Labor	Equipment	Total	Total Incl O&P
0010	**VINYL CASEMENT WINDOWS**, insulated glass									
0015	Grids, low E, J fin, extension jambs, screens									
0100	One lite, 21" x 41"	G 2 Carp	20	.800	Ea.	295	39		334	390
0110	21" x 47"	G	20	.800		320	39		359	420
0120	21" x 53"	G	20	.800		345	39		384	445
0128	24" x 35"	G	19	.842		283	41		324	375
0130	24" x 41"	G	19	.842		310	41		351	405
0140	24" x 47"	G	19	.842		335	41		376	430
0150	24" x 53"	G	19	.842		360	41		401	460

For customer support on your Commercial Renovation Cost Data, call 877.742.4719.

261

08 53 13.40 Vinyl Casement Windows

			Crew	Daily Output	Labor-Hours	Unit	Material	2016 Bare Costs Labor	Equipment	Total	Total Incl O&P
0158	28" x 35"	G	2 Carp	19	.842	Ea.	300	41		341	395
0160	28" x 41"	G		19	.842		330	41		371	430
0170	28" x 47"	G		19	.842		350	41		391	450
0180	28" x 53"	G		19	.842		385	41		426	490
0184	28" x 59"	G		19	.842		395	41		436	500
0188	Two lites, 33" x 35"	G		18	.889		470	43		513	585
0190	33" x 41"	G		18	.889		500	43		543	620
0200	33" x 47"	G		18	.889		540	43		583	660
0210	33" x 53"	G		18	.889		575	43		618	700
0212	33" x 59"	G		18	.889		610	43		653	740
0215	33" x 72"	G		18	.889		635	43		678	765
0220	41" x 41"	G		18	.889		545	43		588	670
0230	41" x 47"	G		18	.889		585	43		628	710
0240	41" x 53"	G		17	.941		620	45.50		665.50	755
0242	41" x 59"	G		17	.941		650	45.50		695.50	790
0246	41" x 72"	G		17	.941		685	45.50		730.50	830
0250	47" x 41"	G		17	.941		555	45.50		600.50	685
0260	47" x 47"	G		17	.941		590	45.50		635.50	725
0270	47" x 53"	G		17	.941		625	45.50		670.50	765
0272	47" x 59"	G		17	.941		680	45.50		725.50	825
0280	56" x 41"	G		15	1.067		595	51.50		646.50	740
0290	56" x 47"	G		15	1.067		625	51.50		676.50	775
0300	56" x 53"	G		15	1.067		680	51.50		731.50	835
0302	56" x 59"	G		15	1.067		710	51.50		761.50	865
0310	56" x 72"	G		15	1.067		770	51.50		821.50	935
0340	Solid vinyl, premium, double insulated glass, 2'-0" x 3'-0" high	G	1 Carp	10	.800		270	39		309	360
0360	2'-0" x 4'-0" high	G		9	.889		299	43		342	400
0380	2'-0" x 5'-0" high	G		8	1		335	48.50		383.50	450

08 53 13.50 Vinyl Picture Windows

			Crew	Daily Output	Labor-Hours	Unit	Material	2016 Bare Costs Labor	Equipment	Total	Total Incl O&P
0010	**VINYL PICTURE WINDOWS**, insulated glass										
0120	Grids, low e, J fin, ext. jambs, 47" x 35"		2 Carp	12	1.333	Ea.	300	64.50		364.50	435
0130	47" x 41"			12	1.333		390	64.50		454.50	535
0140	47" x 47"			12	1.333		345	64.50		409.50	480
0150	47" x 53"			11	1.455		370	70.50		440.50	520
0160	71" x 35"			11	1.455		390	70.50		460.50	545
0170	71" x 41"			11	1.455		410	70.50		480.50	565
0180	71" x 47"			11	1.455		440	70.50		510.50	600

08 53 13.60 Vinyl Half Round Windows

			Crew	Daily Output	Labor-Hours	Unit	Material	2016 Bare Costs Labor	Equipment	Total	Total Incl O&P
0010	**VINYL HALF ROUND WINDOWS**, Including grille, j fin, low E, ext. jambs										
0100	10" height x 20" base		2 Carp	8	2	Ea.	475	97		572	680
0110	15" height x 30" base			8	2		445	97		542	650
0120	17" height x 34" base			7	2.286		325	111		436	540
0130	19" height x 38" base			7	2.286		355	111		466	570
0140	19" height x 33" base			7	2.286		555	111		666	795
0150	24" height x 48" base		1 Carp	6	1.333		370	64.50		434.50	510
0160	25" height x 50" base		"	6	1.333		755	64.50		819.50	935
0170	30" height x 60" base		2 Carp	6	2.667		665	129		794	940

08 54 Composite Windows

08 54 13 – Fiberglass Windows

08 54 13.10 Fiberglass Single Hung Windows

		Crew	Daily Output	Labor-Hours	Unit	Material	2016 Bare Costs Labor	Equipment	Total	Total Incl O&P
0010	**FIBERGLASS SINGLE HUNG WINDOWS**									
0100	Grids, low E, 18" x 24" G	2 Carp	18	.889	Ea.	335	43		378	440
0110	18" x 40" G		17	.941		340	45.50		385.50	450
0130	24" x 40" G		20	.800		360	39		399	460
0230	36" x 36" G		17	.941		370	45.50		415.50	480
0250	36" x 48" G		20	.800		405	39		444	510
0260	36" x 60" G		18	.889		445	43		488	560
0280	36" x 72" G		16	1		470	48.50		518.50	595
0290	48" x 40" G		16	1		470	48.50		518.50	595

08 54 13.30 Fiberglass Slider Windows

		Crew	Daily Output	Labor-Hours	Unit	Material	2016 Bare Costs Labor	Equipment	Total	Total Incl O&P
0010	**FIBERGLASS SLIDER WINDOWS**									
0100	Grids, low E, 36" x 24" G	2 Carp	20	.800	Ea.	340	39		379	440
0110	36" x 36" G	"	20	.800	"	390	39		429	495

08 54 13.50 Fiberglass Bay Windows

		Crew	Daily Output	Labor-Hours	Unit	Material	2016 Bare Costs Labor	Equipment	Total	Total Incl O&P
0010	**FIBERGLASS BAY WINDOWS**									
0150	48" x 36" G	2 Carp	11	1.455	Ea.	1,100	70.50		1,170.50	1,325

08 56 Special Function Windows

08 56 46 – Radio-Frequency-Interference Shielding Windows

08 56 46.10 Radio-Frequency-Interference Mesh

		Crew	Daily Output	Labor-Hours	Unit	Material	2016 Bare Costs Labor	Equipment	Total	Total Incl O&P
0010	**RADIO-FREQUENCY-INTERFERENCE MESH**									
0100	16 mesh copper 0.011" wire				S.F.	4.94			4.94	5.45
0150	22 mesh copper 0.015" wire					6.10			6.10	6.70
0200	100 mesh copper 0.022" wire					8.75			8.75	9.60
0250	100 mesh stainless steel 0.0012" wire					14.80			14.80	16.30

08 56 63 – Detention Windows

08 56 63.13 Visitor Cubicle Windows

		Crew	Daily Output	Labor-Hours	Unit	Material	2016 Bare Costs Labor	Equipment	Total	Total Incl O&P
0010	**VISITOR CUBICLE WINDOWS**									
4000	Visitor cubicle, vision panel, no intercom	E-4	2	16	Ea.	3,200	860	73.50	4,133.50	5,125

08 61 Roof Windows

08 61 13 – Metal Roof Windows

08 61 13.10 Roof Windows

		Crew	Daily Output	Labor-Hours	Unit	Material	2016 Bare Costs Labor	Equipment	Total	Total Incl O&P
0010	**ROOF WINDOWS**, fixed high perf tmpd glazing, metallic framed									
0020	46" x 21-1/2", Flashed for shingled roof	1 Carp	8	1	Ea.	281	48.50		329.50	390
0100	46" x 28"		8	1		305	48.50		353.50	415
0125	57" x 44"		6	1.333		375	64.50		439.50	520
0130	72" x 28"		7	1.143		375	55.50		430.50	505
0150	Fixed, laminated tempered glazing, 46" x 21-1/2"		8	1		465	48.50		513.50	590
0175	46" x 28"		8	1		515	48.50		563.50	645
0200	57" x 44"		6	1.333		485	64.50		549.50	635
0500	Vented flashing set for shingled roof, 46" x 21-1/2"		7	1.143		465	55.50		520.50	600
0525	46" x 28"		6	1.333		515	64.50		579.50	670
0550	57" x 44"		5	1.600		645	77.50		722.50	835
0560	72" x 28"		5	1.600		645	77.50		722.50	835
0575	Flashing set for low pitched roof, 46" x 21-1/2"		7	1.143		535	55.50		590.50	680
0600	46" x 28"		7	1.143		585	55.50		640.50	735
0625	57" x 44"		5	1.600		730	77.50		807.50	925

For customer support on your Commercial Renovation Cost Data, call 877.742.4719.

263

08 61 Roof Windows

08 61 13 – Metal Roof Windows

08 61 13.10 Roof Windows		Crew	Daily Output	Labor-Hours	Unit	Material	2016 Bare Costs Labor	Equipment	Total	Total Incl O&P
0650	Flashing set for curb 46" x 21-1/2"	1 Carp	7	1.143	Ea.	620	55.50		675.50	770
0675	46" x 28"		7	1.143		670	55.50		725.50	830
0700	57" x 44"		5	1.600		825	77.50		902.50	1,025

08 62 Unit Skylights

08 62 13 – Domed Unit Skylights

08 62 13.10 Domed Skylights

			Crew	Daily Output	Labor-Hours	Unit	Material	2016 Bare Costs Labor	Equipment	Total	Total Incl O&P
0010	**DOMED SKYLIGHTS**										
0020	Skylight, fixed dome type, 22" x 22"	G	G-3	12	2.667	Ea.	216	127		343	440
0030	22" x 46"	G		10	3.200		269	152		421	540
0040	30" x 30"	G		12	2.667		286	127		413	520
0050	30" x 46"	G		10	3.200		380	152		532	665
0110	Fixed, double glazed, 22" x 27"	G		12	2.667		275	127		402	510
0120	22" x 46"	G		10	3.200		293	152		445	565
0130	44" x 46"	G		10	3.200		430	152		582	720
0210	Operable, double glazed, 22" x 27"	G		12	2.667		405	127		532	650
0220	22" x 46"	G		10	3.200		465	152		617	760
0230	44" x 46"	G		10	3.200		905	152		1,057	1,250
9000	Minimum labor/equipment charge			2	16	Job		760		760	1,225

08 62 13.20 Skylights

			Crew	Daily Output	Labor-Hours	Unit	Material	2016 Bare Costs Labor	Equipment	Total	Total Incl O&P
0010	**SKYLIGHTS**, flush or curb mounted										
2120	Ventilating insulated plexiglass dome with										
2130	curb mounting, 36" x 36"	G	G-3	12	2.667	Ea.	480	127		607	735
2150	52" x 52"	G		12	2.667		670	127		797	940
2160	28" x 52"	G		10	3.200		490	152		642	785
2170	36" x 52"	G		10	3.200		545	152		697	840
2180	For electric opening system, add	G					315			315	345
2210	Operating skylight, with thermopane glass, 24" x 48"	G	G-3	10	3.200		585	152		737	890
2220	32" x 48"	G		9	3.556		615	169		784	945
2300	Insulated safety glass with aluminum frame	G		160	.200	S.F.	94	9.50		103.50	118
4000	Skylight, solar tube kit, incl dome, flashing, diffuser, 1 pipe, 10" dia.	G	1 Carp	2	4	Ea.	242	194		436	580
4010	14" diam.	G		2	4		335	194		529	685
4020	21" diam.	G		2	4		420	194		614	775
4030	Accessories for, 1' long x 9" dia pipe	G		24	.333		50	16.15		66.15	81.50
4040	2' long x 9" dia pipe	G		24	.333		43	16.15		59.15	74
4050	4' long x 9" dia pipe	G		20	.400		75	19.40		94.40	114
4060	1' long x 13" dia pipe	G		24	.333		70	16.15		86.15	104
4070	2' long x 13" dia pipe	G		24	.333		55	16.15		71.15	87
4080	4' long x 13" dia pipe	G		20	.400		102	19.40		121.40	144
4090	6.5" turret ext for 21" dia pipe	G		16	.500		110	24		134	161
4100	12' long x 21" dia flexible pipe	G		12	.667		88	32.50		120.50	150
4110	45 degree elbow, 10"	G		16	.500		152	24		176	207
4120	14"	G		16	.500		82	24		106	130
4130	Interior decorative ring, 9"	G		20	.400		40	19.40		59.40	75.50
4140	13"	G		20	.400		60	19.40		79.40	97.50

08 63 Metal-Framed Skylights

08 63 13 – Domed Metal-Framed Skylights

08 63 13.20 Skylight Rigid Metal-Framed	Crew	Daily Output	Labor-Hours	Unit	Material	2016 Bare Costs Labor	Equipment	Total	Total Incl O&P
0010 **SKYLIGHT RIGID METAL-FRAMED** skylight framing is aluminum									
0050 Fixed acrylic double domes, curb mount, 25-1/2" x 25-1/2"	G-3	10	3.200	Ea.	210	152		362	475
0060 25-1/2" x 33-1/2"		10	3.200		242	152		394	510
0070 25-1/2" x 49-1/2"		6	5.333		279	254		533	710
0080 33-1/2" x 33-1/2"		8	4		305	190		495	640
0090 37-1/2" x 25-1/2"		8	4		250	190		440	580
0100 37-1/2" x 37-1/2"		8	4		245	190		435	575
0110 37-1/2" x 49-1/2"		6	5.333		420	254		674	865
0120 49-1/2" x 33-1/2"		6	5.333		375	254		629	815
0130 49-1/2" x 49-1/2"		6	5.333		470	254		724	920
1000 Fixed tempered glass, curb mount, 17-1/2" x 33-1/2"		10	3.200		170	152		322	430
1020 17-1/2" x 49-1/2"		6	5.333		190	254		444	615
1030 25-1/2" x 25-1/2"		10	3.200		170	152		322	430
1040 25-1/2" x 33-1/2"		10	3.200		200	152		352	465
1050 25-1/2" x 37-1/2"		8	4		210	190		400	535
1060 25-1/2" x 49-1/2"		6	5.333		222	254		476	650
1070 25-1/2" x 73-1/2"		6	5.333		375	254		629	815
1080 33-1/2" x 33-1/2"		8	4		235	190		425	565
2000 Manual vent tempered glass & screen, curb, 25-1/2" x 25-1/2"		10	3.200		410	152		562	695
2020 25-1/2" x 37-1/2"		10	3.200		465	152		617	755
2030 25-1/2" x 49-1/2"		8	4		505	190		695	860
2040 33-1/2" x 33-1/2"		8	4		535	190		725	895
2050 33-1/2" x 49-1/2"		6	5.333		675	254		929	1,150
2060 37-1/2" x 37-1/2"		6	5.333		615	254		869	1,075
2070 49-1/2" x 49-1/2"		6	5.333		790	254		1,044	1,275
3000 Electric vent tempered glass, curb mount, 25-1/2" x 25-1/2"		10	3.200		960	152		1,112	1,300
3020 25-1/2" x 37-1/2"		10	3.200		1,050	152		1,202	1,400
3030 25-1/2" x 49-1/2"		8	4		1,125	190		1,315	1,525
3040 33-1/2" x 33-1/2"		8	4		1,125	190		1,315	1,550
3050 33-1/2" x 49-1/2"		6	5.333		1,225	254		1,479	1,725
3060 37-1/2" x 37-1/2"		6	5.333		1,200	254		1,454	1,700
3070 49-1/2" x 49-1/2"		6	5.333		1,325	254		1,579	1,850

08 71 Door Hardware

08 71 13 – Automatic Door Operators

08 71 13.10 Automatic Openers Commercial

	Crew	Daily Output	Labor-Hours	Unit	Material	2016 Bare Costs Labor	Equipment	Total	Total Incl O&P
0010 **AUTOMATIC OPENERS COMMERCIAL**									
0020 Pneumatic, incl opener, motion sens, control box, tubing, compressor									
0050 For single swing door, per opening	2 Skwk	.80	20	Ea.	4,550	1,000		5,550	6,625
0100 Pair, per opening		.50	32	Opng.	7,375	1,600		8,975	10,700
1000 For single sliding door, per opening		.60	26.667		4,975	1,325		6,300	7,625
1300 Bi-parting pair		.50	32		7,475	1,600		9,075	10,800
1420 Electronic door opener incl motion sens, 12 V control box, motor									
1450 For single swing door, per opening	2 Skwk	.80	20	Opng.	3,675	1,000		4,675	5,675
1500 Pair, per opening		.50	32		6,800	1,600		8,400	10,100
1600 For single sliding door, per opening		.60	26.667		4,500	1,325		5,825	7,100
1700 Bi-parting pair		.50	32		5,475	1,600		7,075	8,600
1750 Handicap actuator buttons, 2, including 12 V DC wiring, add	1 Carp	1.50	5.333	Pr.	470	258		728	935
2000 Electric Panic Button for door	2 Skwk	.50	32	Opng.	203	1,600		1,803	2,800

For customer support on your Commercial Renovation Cost Data, call 877.742.4719.

265

08 71 Door Hardware

08 71 20 – Hardware

08 71 20.15 Hardware	Crew	Daily Output	Labor-Hours	Unit	Material	2016 Bare Costs Labor	Equipment	Total	Total Incl O&P
0010 **HARDWARE**									
0020 Average percentage for hardware, total job cost									
0500 Total hardware for building, average distribution				Job	85%	15%			
1000 Door hardware, apartment, interior	1 Carp	4	2	Door	465	97		562	670
1300 Average, door hardware, motel/hotel interior, with access card		4	2		600	97		697	820
1500 Hospital bedroom, average quality		4	2		650	97		747	875
2000 High quality		3	2.667		750	129		879	1,025
2250 School, single exterior, incl. lever, incl. panic device		3	2.667		1,325	129		1,454	1,650
2500 Single interior, regular use, lever included		3	2.667		645	129		774	920
2600 Average, door hdwe. set, school, classroom, ANSI F88, incl. lever		3	2.667		925	129		1,054	1,225
2850 Stairway, single interior		3	2.667		605	129		734	875
3100 Double exterior, with panic device		2	4	Pr.	2,450	194		2,644	3,025
6020 Add for fire alarm door holder, electro-magnetic	1 Elec	4	2	Ea.	103	110		213	283

08 71 20.30 Door Closers

08 71 20.30	Crew	Daily Output	Labor-Hours	Unit	Material	2016 Bare Costs Labor	Equipment	Total	Total Incl O&P
0010 **DOOR CLOSERS** Adjustable backcheck, multiple mounting									
0015 and rack and pinion									
0020 Standard Regular Arm	1 Carp	6	1.333	Ea.	192	64.50		256.50	315
0040 Hold open arm		6	1.333		200	64.50		264.50	325
0100 Fusible link		6.50	1.231		165	59.50		224.50	280
1500 Concealed closers, normal use, head, pivot hung, interior		5.50	1.455		360	70.50		430.50	515
1510 Exterior		5.50	1.455		420	70.50		490.50	575
1520 Overhead concealed, all sizes, regular arm		5.50	1.455		200	70.50		270.50	335
1525 Concealed arm		5	1.600		315	77.50		392.50	470
1530 Concealed in door, all sizes, regular arm		5.50	1.455		325	70.50		395.50	475
1535 Concealed arm		5	1.600		255	77.50		332.50	410
1560 Floor concealed, all sizes, single acting		2.20	3.636		500	176		676	840
1565 Double acting		2.20	3.636		475	176		651	815
1570 Interior, floor, offset pivot, single acting		3.50	2.286		830	111		941	1,100
1590 Exterior		3.50	2.286		905	111		1,016	1,175
1610 Hold open arm		6	1.333		430	64.50		494.50	580
1620 Double acting, standard arm		6	1.333		780	64.50		844.50	965
1630 Hold open arm		6	1.333		785	64.50		849.50	965
1640 Floor, center hung, single acting, bottom arm		6	1.333		420	64.50		484.50	565
1650 Double acting		6	1.333		475	64.50		539.50	630
1660 Offset hung, single acting, bottom arm		6	1.333		565	64.50		629.50	725
2000 Backcheck and adjustable power, hinge face mount									
5000 For cast aluminum cylinder, deduct				Ea.	35			35	38.50
5040 For delayed action, add					46			46	50.50
5080 For fusible link arm, add					35			35	38.50
5120 For shock absorbing arm, add					50			50	55
5160 For spring power adjustment, add					40			40	44
6000 Closer-holder, hinge face mount, all sizes, exposed arm	1 Carp	6.50	1.231		205	59.50		264.50	325
6500 Electro magnetic closer/holder									
6510 Single point, no detector	1 Carp	4	2	Ea.	520	97		617	730
6515 Including detector		4	2		685	97		782	910
6520 Multi-point, no detector		4	2		910	97		1,007	1,150
6524 Including detector		4	2		1,350	97		1,447	1,650
6550 Electric automatic operators									
6555 Operator	1 Carp	4	2	Ea.	2,025	97		2,122	2,375
6570 Wall plate actuator		4	2		222	97		319	405
7000 Electronic closer-holder, hinge facemount, concealed arm		5	1.600		420	77.50		497.50	585
7400 With built-in detector		5	1.600		605	77.50		682.50	790

08 71 Door Hardware

08 71 20 – Hardware

08 71 20.30 Door Closers	Crew	Daily Output	Labor-Hours	Unit	Material	2016 Bare Costs Labor	Equipment	Total	Total Incl O&P
9000	Minimum labor/equipment charge	1 Carp	4	2	Job		97		97	158

08 71 20.36 Panic Devices

	08 71 20.36 Panic Devices	Crew	Daily Output	Labor-Hours	Unit	Material	Labor	Equipment	Total	Total Incl O&P
0010	**PANIC DEVICES** R087110-10									
0015	Touch bars various styles									
0040	Single door exit only, rim device, wide stile									
0050	Economy US28	1 Carp	5	1.600	Ea.	231	77.50		308.50	380
0060	Standard duty US28		5	1.600		665	77.50		742.50	860
0065	US26D		5	1.600		705	77.50		782.50	900
0070	US10		5	1.600		725	77.50		802.50	925
0075	US3		5	1.600		795	77.50		872.50	1,000
0080	Night Latch, Economy US28		5	1.600		315	77.50		392.50	470
0085	Standard duty, Night Latch US28		5	1.600		970	77.50		1,047.50	1,200
0090	US26D		5	1.600		1,075	77.50		1,152.50	1,300
0095	US10		5	1.600		1,075	77.50		1,152.50	1,325
0100	US3		5	1.600		1,175	77.50		1,252.50	1,425
0500	Single door exit only, rim device, narrow stile									
0540	Economy US28	1 Carp	5	1.600	Ea.	465	77.50		542.50	635
0550	Standard duty US28		5	1.600		830	77.50		907.50	1,025
0565	US26D		5	1.600		900	77.50		977.50	1,125
0570	US10		5	1.600		745	77.50		822.50	945
0575	US3		5	1.600		950	77.50		1,027.50	1,175
0580	Economy with night latch US28		5	1.600		320	77.50		397.50	480
0585	Standard duty US28		5	1.600		870	77.50		947.50	1,075
0590	US26D		5	1.600		1,050	77.50		1,127.50	1,275
0595	US10		5	1.600		870	77.50		947.50	1,075
0600	US32D		5	1.600		1,200	77.50		1,277.50	1,450
1040	Single door exit only, surface vertical rod									
1050	Surface Vertical Rod, Economy US28	1 Carp	4	2	Ea.	525	97		622	740
1060	Surface Vertical Rod, Standard duty US28		4	2		965	97		1,062	1,200
1065	US26D		4	2		1,100	97		1,197	1,375
1070	US10		4	2		1,025	97		1,122	1,275
1075	US3		4	2		1,200	97		1,297	1,475
1080	US32D		4	2		1,200	97		1,297	1,475
1500	Single door exit, rim, narrow stile, economy, surface vertical rod									
1540	Surface Vertical Rod, Narrow, Economy US28	1 Carp	4	2	Ea.	420	97		517	625
2040	Single door exit only, concealed vertical rod									
2042	These devices will require separate and extra door preparation									
2060	Concealed Rod, standard duty US28	1 Carp	4	2	Ea.	1,000	97		1,097	1,250
2065	US26D		4	2		1,125	97		1,222	1,400
2070	US10		4	2		1,275	97		1,372	1,550
2560	Concealed Rod, narrow, standard duty US28		4	2		990	97		1,087	1,250
2565	US26D		4	2		1,075	97		1,172	1,350
2567	US26		4	2		1,225	97		1,322	1,500
2570	US10		4	2		1,025	97		1,122	1,275
2575	US3		4	2		1,225	97		1,322	1,500
3040	Single door exit only, mortise device, wide stile									
3042	These devices must be combined with a mortise lock									
3060	Mortise, standard duty US28	1 Carp	4	2	Ea.	925	97		1,022	1,175
3065	US26D		4	2		465	97		562	675
3067	US26		4	2		1,150	97		1,247	1,425
3070	US10		4	2		1,000	97		1,097	1,250
3075	US3		4	2		1,025	97		1,122	1,275

08 71 Door Hardware

08 71 20 – Hardware

08 71 20.36 Panic Devices

		Crew	Daily Output	Labor-Hours	Unit	Material	2016 Bare Costs Labor	Equipment	Total	Total Incl O&P
3080	US32D	1 Carp	4	2	Ea.	1,025	97		1,122	1,275
3500	Single door exit only, mortise device, narrow stile									
3510	These devices must be combined with a mortise lock									
3540	Mortise, economy US28	1 Carp	4	2	Ea.	385	97		482	580

08 71 20.40 Lockset

		Crew	Daily Output	Labor-Hours	Unit	Material	2016 Bare Costs Labor	Equipment	Total	Total Incl O&P
0010	**LOCKSET**, Standard duty									
0020	Non-keyed, passage, w/sect. trim	1 Carp	12	.667	Ea.	70	32.50		102.50	130
0100	Privacy		12	.667		75	32.50		107.50	136
0400	Keyed, single cylinder function		10	.800		108	39		147	182
0420	Hotel (see also Section 08 71 20.15)		8	1		200	48.50		248.50	299
0500	Lever handled, keyed, single cylinder function		10	.800		128	39		167	205
1000	Heavy duty with sectional trim, non-keyed, passages		12	.667		125	32.50		157.50	191
1100	Privacy		12	.667		155	32.50		187.50	224
1400	Keyed, single cylinder function		10	.800		185	39		224	268
1420	Hotel		8	1		500	48.50		548.50	630
1600	Communicating		10	.800		300	39		339	395
1690	For re-core cylinder, add					50			50	55
3800	Cipher lockset w/key pad (security item)	1 Carp	13	.615		940	30		970	1,075
3900	Cipher lockset with dial for swinging doors (security item)		13	.615		1,800	30		1,830	2,050
3920	with dial for swinging doors & drill resistant plate (security item)		12	.667		2,200	32.50		2,232.50	2,475
3950	Cipher lockset with dial for safe/vault door (security item)		12	.667		1,500	32.50		1,532.50	1,700
3980	Keyless, pushbutton type									
4000	Residential/light commercial, deadbolt, standard	1 Carp	9	.889	Ea.	140	43		183	225
4010	Heavy duty		9	.889		230	43		273	325
4020	Industrial, heavy duty, with deadbolt		9	.889		380	43		423	485
4030	Key override		9	.889		390	43		433	500
4040	Lever activated handle		9	.889		415	43		458	525
4050	Key override		9	.889		440	43		483	555
4060	Double sided pushbutton type		8	1		750	48.50		798.50	905
4070	Key override		8	1		790	48.50		838.50	950
9000	Minimum labor/equipment charge		6	1.333	Job		64.50		64.50	106

08 71 20.41 Dead Locks

		Crew	Daily Output	Labor-Hours	Unit	Material	2016 Bare Costs Labor	Equipment	Total	Total Incl O&P
0010	**DEAD LOCKS**									
0011	Mortise heavy duty outside key (security item)	1 Carp	9	.889	Ea.	175	43		218	264
0020	Double cylinder		9	.889		175	43		218	264
0100	Medium duty, outside key		10	.800		110	39		149	185
0110	Double cylinder		10	.800		120	39		159	196
1000	Tubular, standard duty, outside key		10	.800		45	39		84	113
1010	Double cylinder		10	.800		60	39		99	130
1200	Night latch, outside key		10	.800		45	39		84	113
1203	Deadlock night latch		7.70	1.039		45	50.50		95.50	132
1420	Deadbolt lock, single cylinder		10	.800		48.50	39		87.50	117
1440	Double cylinder		10	.800		63	39		102	133

08 71 20.42 Mortise Locksets

		Crew	Daily Output	Labor-Hours	Unit	Material	2016 Bare Costs Labor	Equipment	Total	Total Incl O&P
0010	**MORTISE LOCKSETS**, Comm., wrought knobs & full escutcheon trim									
0015	Assumes mortise is cut									
0020	Non-keyed, passage, grade 3	1 Carp	9	.889	Ea.	149	43		192	235
0030	Grade 1		8	1		405	48.50		453.50	525
0040	Privacy set Grade 3		9	.889		169	43		212	257
0050	Grade 1		8	1		455	48.50		503.50	580
0100	Keyed, office/entrance/apartment, Grade 2		8	1		197	48.50		245.50	296
0110	Grade 1		7	1.143		515	55.50		570.50	660

08 71 Door Hardware

08 71 20 – Hardware

08 71 20.42 Mortise Locksets

		Crew	Daily Output	Labor-Hours	Unit	Material	2016 Bare Costs Labor	Equipment	Total	Total Incl O&P
0120	Single cylinder, typical, Grade 3	1 Carp	8	1	Ea.	189	48.50		237.50	287
0130	Grade 1		7	1.143		500	55.50		555.50	640
0200	Hotel, room, Grade 3		7	1.143		190	55.50		245.50	300
0210	Grade 1 (see also Section 08 71 20.15)		6	1.333		510	64.50		574.50	665
0300	Double cylinder, Grade 3		8	1		225	48.50		273.50	325
0310	Grade 1		7	1.143		515	55.50		570.50	660
1000	Wrought knobs and sectional trim, non-keyed, passage, Grade 3		10	.800		130	39		169	207
1010	Grade 1		9	.889		405	43		448	515
1040	Privacy, Grade 3		10	.800		145	39		184	224
1050	Grade 1		9	.889		455	43		498	570
1100	Keyed, entrance, office/apartment, Grade 3		9	.889		220	43		263	315
1103	Install lockset		6.92	1.156		220	56		276	335
1110	Grade 1		8	1		520	48.50		568.50	650
1120	Single cylinder, Grade 3		9	.889		225	43		268	320
1130	Grade 1		8	1		500	48.50		548.50	630
2000	Cast knobs and full escutcheon trim									
2010	Non-keyed, passage, Grade 3	1 Carp	9	.889	Ea.	275	43		318	375
2020	Grade 1		8	1		380	48.50		428.50	500
2040	Privacy, Grade 3		9	.889		320	43		363	420
2050	Grade 1		8	1		440	48.50		488.50	565
2120	Keyed, single cylinder, Grade 3		8	1		330	48.50		378.50	445
2123	Mortise lock		6.15	1.301		330	63		393	470
2130	Grade 1		7	1.143		525	55.50		580.50	670
3000	Cast knob and sectional trim, non-keyed, passage, Grade 3		10	.800		210	39		249	295
3010	Grade 1		10	.800		380	39		419	485
3040	Privacy, Grade 3		10	.800		225	39		264	310
3050	Grade 1		10	.800		440	39		479	550
3100	Keyed, office/entrance/apartment, Grade 3		9	.889		255	43		298	350
3110	Grade 1		9	.889		580	43		623	710
3120	Single cylinder, Grade 3		9	.889		260	43		303	355
3130	Grade 1		9	.889		525	43		568	650
3190	For re-core cylinder, add					75			75	82.50

08 71 20.45 Peepholes

		Crew	Daily Output	Labor-Hours	Unit	Material	2016 Bare Costs Labor	Equipment	Total	Total Incl O&P
0010	**PEEPHOLES**									
2010	Peephole	1 Carp	32	.250	Ea.	15.80	12.10		27.90	37
2020	Peephole, wide view	"	32	.250	"	16.50	12.10		28.60	38

08 71 20.50 Door Stops

		Crew	Daily Output	Labor-Hours	Unit	Material	2016 Bare Costs Labor	Equipment	Total	Total Incl O&P
0010	**DOOR STOPS**									
0020	Holder & bumper, floor or wall	1 Carp	32	.250	Ea.	35	12.10		47.10	59
1300	Wall bumper, 4" diameter, with rubber pad, aluminum		32	.250		12.30	12.10		24.40	33.50
1600	Door bumper, floor type, aluminum		32	.250		8.70	12.10		20.80	29.50
1620	Brass		32	.250		10	12.10		22.10	31
1630	Bronze		32	.250		18.35	12.10		30.45	40
1900	Plunger type, door mounted		32	.250		29.50	12.10		41.60	52.50
2500	Holder, floor type, aluminum		32	.250		35	12.10		47.10	58.50
2520	Wall type, aluminum		32	.250		35.50	12.10		47.60	59
2530	Overhead type, bronze		32	.250		105	12.10		117.10	135
2540	Plunger type, aluminum		32	.250		29.50	12.10		41.60	52.50
2560	Brass		32	.250		45	12.10		57.10	69
3000	Electro-magnetic, wall mounted, US3		3	2.667		263	129		392	500
3020	Floor mounted, US3		3	2.667		315	129		444	560
4000	Doorstop, ceiling mounted		3	2.667		20	129		149	233

For customer support on your Commercial Renovation Cost Data, call 877.742.4719.

269

08 71 20.50 Door Stops

		Crew	Daily Output	Labor-Hours	Unit	Material	2016 Bare Costs Labor	Equipment	Total	Total Incl O&P
4030	Doorstop, header	1 Carp	3	2.667	Ea.	45	129		174	261
9000	Minimum labor/equipment charge	▼	6	1.333	Job		64.50		64.50	106

08 71 20.55 Push-Pull Plates

		Crew	Daily Output	Labor-Hours	Unit	Material	2016 Bare Costs Labor	Equipment	Total	Total Incl O&P
0010	**PUSH-PULL PLATES**									
0090	Push plate, 0.050 thick, 3" x 12", aluminum	1 Carp	12	.667	Ea.	6.25	32.50		38.75	60
0100	4" x 16"		12	.667		12.90	32.50		45.40	67
0110	6" x 16"		12	.667		8.50	32.50		41	62.50
0120	8" x 16"		12	.667		9.90	32.50		42.40	64
0200	Push plate, 0.050 thick, 3" x 12", brass		12	.667		14	32.50		46.50	68.50
0210	4" x 16"		12	.667		17.50	32.50		50	72.50
0220	6" x 16"		12	.667		27	32.50		59.50	82.50
0230	8" x 16"		12	.667		35	32.50		67.50	91.50
0250	Push plate, 0.050 thick, 3" x 12", satin brass		12	.667		14.25	32.50		46.75	68.50
0260	4" x 16"		12	.667		17.70	32.50		50.20	72.50
0270	6" x 16"		12	.667		27	32.50		59.50	82.50
0280	8" x 16"		12	.667		35	32.50		67.50	91.50
0490	Push plate, 0.050 thick, 3" x 12", bronze		12	.667		17	32.50		49.50	71.50
0500	4" x 16"		12	.667		25.50	32.50		58	81
0510	6" x 16"		12	.667		32	32.50		64.50	88
0520	8" x 16"		12	.667		38.50	32.50		71	95.50
0600	Push plate, antimicrobial copper alloy finish, 3.5" x 15"		13	.615		20.50	30		50.50	71.50
0610	4" x 16"		13	.615		24.50	30		54.50	75
0620	6" x 16"		13	.615		27	30		57	78
0630	6" x 20"		13	.615		27	30		57	78
0740	Push plate, 0.050 thick, 3" x 12", stainless steel		12	.667		12	32.50		44.50	66
0750	4" x 16"		12	.667		27.50	32.50		60	83.50
0760	6" x 16"		12	.667		18.65	32.50		51.15	73.50
0780	8" x 16"		12	.667		23.50	32.50		56	79
0790	Push plate, 0.050 thick, 3" x 12", satin stainless steel		12	.667		7	32.50		39.50	60.50
0810	4" x 16"		12	.667		8.50	32.50		41	62.50
0820	6" x 16"		12	.667		11.70	32.50		44.20	66
0830	8" x 16"		12	.667		16	32.50		48.50	70.50
0980	Pull plate, 0.050 thick, 3" x 12", aluminum		12	.667		25.50	32.50		58	81
1000	4" x 16"		12	.667		25.50	32.50		58	81
1050	Pull plate, 0.050 thick, 3" x 12", brass		12	.667		39.50	32.50		72	96.50
1060	4" x 16"		12	.667		34.50	32.50		67	91
1080	Pull plate, 0.050 thick, 3" x 12", bronze		12	.667		50	32.50		82.50	108
1100	4" x 16"		12	.667		56.50	32.50		89	115
1180	Pull plate, 0.050 thick, 3" x 12", stainless steel		12	.667		45.50	32.50		78	103
1200	4" x 16"		12	.667		59	32.50		91.50	118
1250	Pull plate, 0.050 thick, 3" x 12", chrome		12	.667		44.50	32.50		77	102
1270	4" x 16"		12	.667		46.50	32.50		79	104
1500	Pull handle and push bar, aluminum		11	.727		122	35		157	192
2000	Bronze		10	.800	▼	164	39		203	244
9800	Minimum labor/equipment charge	▼	5	1.600	Job		77.50		77.50	127

08 71 20.60 Entrance Locks

		Crew	Daily Output	Labor-Hours	Unit	Material	2016 Bare Costs Labor	Equipment	Total	Total Incl O&P
0010	**ENTRANCE LOCKS**									
0015	Cylinder, grip handle deadlocking latch	1 Carp	9	.889	Ea.	175	43		218	264
0020	Deadbolt		8	1		175	48.50		223.50	272
0100	Push and pull plate, dead bolt		8	1		225	48.50		273.50	325
0900	For handicapped lever, add				▼	150			150	165

08 71 Door Hardware

08 71 20 – Hardware

08 71 20.65 Thresholds		Crew	Daily Output	Labor-Hours	Unit	Material	2016 Bare Costs Labor	Equipment	Total	Total Incl O&P
0010	**THRESHOLDS**									
0011	Threshold 3' long saddles aluminum	1 Carp	48	.167	L.F.	9.70	8.10		17.80	24
0100	Aluminum, 8" wide, 1/2" thick		12	.667	Ea.	49	32.50		81.50	107
0500	Bronze		60	.133	L.F.	43	6.45		49.45	58
0600	Bronze, panic threshold, 5" wide, 1/2" thick		12	.667	Ea.	158	32.50		190.50	226
0700	Rubber, 1/2" thick, 5-1/2" wide		20	.400		40	19.40		59.40	75.50
0800	2-3/4" wide		20	.400		45.50	19.40		64.90	81.50
1950	ADA Compliant Thresholds									
2000	Threshold, wood oak 3-1/2" wide x 24" long	1 Carp	12	.667	Ea.	11	32.50		43.50	65
2010	3-1/2" wide x 36" long		12	.667		15	32.50		47.50	69.50
2020	3-1/2" wide x 48" long		12	.667		20	32.50		52.50	75
2030	4-1/2" wide x 24" long		12	.667		13	32.50		45.50	67.50
2040	4-1/2" wide x 36" long		12	.667		19	32.50		51.50	74
2050	4-1/2" wide x 48" long		12	.667		25	32.50		57.50	80.50
2060	6-1/2" wide x 24" long		12	.667		18	32.50		50.50	73
2070	6-1/2" wide x 36" long		12	.667		27	32.50		59.50	82.50
2080	6-1/2" wide x 48" long		12	.667		36	32.50		68.50	92.50
2090	Threshold, wood cherry 3-1/2" wide x 24" long		12	.667		15	32.50		47.50	69.50
2100	3-1/2" wide x 36" long		12	.667		22	32.50		54.50	77
2110	3-1/2" wide x 48" long		12	.667		29	32.50		61.50	85
2120	4-1/2" wide x 24" long		12	.667		19	32.50		51.50	74
2130	4-1/2" wide x 36" long		12	.667		28	32.50		60.50	84
2140	4-1/2" wide x 48" long		12	.667		37	32.50		69.50	93.50
2150	6-1/2" wide x 24" long		12	.667		27	32.50		59.50	82.50
2160	6-1/2" wide x 36" long		12	.667		40	32.50		72.50	97
2170	6-1/2" wide x 48" long		12	.667		53	32.50		85.50	112
2180	Threshold, wood walnut 3-1/2" wide x 24" long		12	.667		18	32.50		50.50	73
2190	3-1/2" wide x 36" long		12	.667		27	32.50		59.50	82.50
2200	3-1/2" wide x 48" long		12	.667		36	32.50		68.50	92.50
2210	4-1/2" wide x 24" long		12	.667		26	32.50		58.50	81.50
2220	4-1/2" wide x 36" long		12	.667		38	32.50		70.50	95
2230	4-1/2" wide x 48" long		12	.667		50	32.50		82.50	108
2240	6-1/2" wide x 24" long		12	.667		43	32.50		75.50	101
2250	6-1/2" wide x 36" long		12	.667		63	32.50		95.50	123
2260	6-1/2" wide x 48" long		12	.667		84	32.50		116.50	146
2300	Threshold, aluminum 4" wide x 36" long		12	.667		33	32.50		65.50	89.50
2310	4" wide x 48" long		12	.667		41	32.50		73.50	98
2320	4" wide x 72" long		12	.667		66	32.50		98.50	126
2330	5" wide x 36" long		12	.667		45	32.50		77.50	103
2340	5" wide x 48" long		12	.667		56	32.50		88.50	115
2350	5" wide x 72" long		12	.667		89	32.50		121.50	151
2360	6" wide x 36" long		12	.667		53	32.50		85.50	112
2370	6" wide x 48" long		12	.667		68	32.50		100.50	128
2380	6" wide x 72" long		12	.667		106	32.50		138.50	170
2390	7" wide x 36" long		12	.667		69	32.50		101.50	129
2400	7" wide x 48" long		12	.667		90	32.50		122.50	152
2410	7" wide x 72" long		12	.667		137	32.50		169.50	204
2500	Threshold, ramp, aluminum or rubber 24" x 24"		12	.667		190	32.50		222.50	262
9000	Minimum labor/equipment charge		4	2	Job		97		97	158

For customer support on your Commercial Renovation Cost Data, call 877.742.4719.

271

08 71 Door Hardware

08 71 20 – Hardware

08 71 20.75 Door Hardware Accessories	Crew	Daily Output	Labor-Hours	Unit	Material	2016 Bare Costs Labor	Equipment	Total	Total Incl O&P
0010 **DOOR HARDWARE ACCESSORIES**									
0050 Door closing coordinator, 36" (for paired openings up to 56")	1 Carp	8	1	Ea.	105	48.50		153.50	195
0060 48" (for paired openings up to 84")		8	1		113	48.50		161.50	203
0070 56" (for paired openings up to 96")		8	1		124	48.50		172.50	215
1000 Knockers, brass, standard		16	.500		49	24		73	93.50
1100 Deluxe		10	.800		121	39		160	197
4500 Rubber door silencers		540	.015		.45	.72		1.17	1.67
9000 Minimum labor/equipment charge		6	1.333	Job		64.50		64.50	106

08 71 20.80 Hasps

	Crew	Daily Output	Labor-Hours	Unit	Material	Labor	Equipment	Total	Total Incl O&P
0010 **HASPS**, steel assembly									
0015 3"	1 Carp	26	.308	Ea.	4.90	14.90		19.80	30
0020 4-1/2"		13	.615		6.60	30		36.60	56
0040 6"		12.50	.640		9.35	31		40.35	61

08 71 20.90 Hinges

	Crew	Daily Output	Labor-Hours	Unit	Material	Labor	Equipment	Total	Total Incl O&P
0010 **HINGES**									
0012 Full mortise, avg. freq., steel base, USP, 4-1/2" x 4-1/2"				Pr.	37			37	40.50
0100 5" x 5", USP					57.50			57.50	63
0200 6" x 6", USP					119			119	131
0400 Brass base, 4-1/2" x 4-1/2", US10					60			60	66
0500 5" x 5", US10					86			86	94.50
0600 6" x 6", US10					162			162	178
0800 Stainless steel base, 4-1/2" x 4-1/2", US32					76.50			76.50	84
0900 For non removable pin, add (security item)				Ea.	5.25			5.25	5.80
0910 For floating pin, driven tips, add					3.30			3.30	3.63
0930 For hospital type tip on pin, add					13.95			13.95	15.35
0940 For steeple type tip on pin, add					18.95			18.95	21
0950 Full mortise, high frequency, steel base, 3-1/2" x 3-1/2", US26D				Pr.	30.50			30.50	33.50
1000 4-1/2" x 4-1/2", USP					66			66	73
1100 5" x 5", USP					52			52	57.50
1200 6" x 6", USP					138			138	152
1400 Brass base, 3-1/2" x 3-1/2", US4					51			51	56
1430 4-1/2" x 4-1/2", US10					75.50			75.50	83
1500 5" x 5", US10					126			126	138
1600 6" x 6", US10					169			169	186
1800 Stainless steel base, 4-1/2" x 4-1/2", US32					103			103	114
1810 5" x 4-1/2", US32					137			137	151
1930 For hospital type tip on pin, add				Ea.	14.60			14.60	16.05
1950 Full mortise, low frequency, steel base, 3-1/2" x 3-1/2", US26D				Pr.	25			25	27.50
2000 4-1/2" x 4-1/2", USP					23			23	25
2100 5" x 5", USP					48			48	53
2200 6" x 6", USP					93.50			93.50	103
2300 4-1/2" x 4-1/2", US3					17.35			17.35	19.05
2310 5" x 5", US3					42			42	46
2400 Brass bass, 4-1/2" x 4-1/2", US10					54.50			54.50	59.50
2500 5" x 5", US10					78			78	86
2800 Stainless steel base, 4-1/2" x 4-1/2", US32					76			76	84
8000 Install hinge	1 Carp	34	.235			11.40		11.40	18.60

272

08 71 Door Hardware

08 71 20 – Hardware

08 71 20.91 Special Hinges

08 71 20.91 Special Hinges	Crew	Daily Output	Labor-Hours	Unit	Material	Labor	2016 Bare Costs Equipment	Total	Total Incl O&P
0010 **SPECIAL HINGES**									
0015 Paumelle, high frequency									
0020 Steel base, 6" x 4-1/2", US10				Pr.	145			145	160
0100 Brass base, 5" x 4-1/2", US10				Ea.	248			248	273
0200 Paumelle, average frequency, steel base, 4-1/2" x 3-1/2", US10				Pr.	98.50			98.50	108
0400 Olive knuckle, low frequency, brass base, 6" x 4-1/2", US10				Ea.	144			144	159
1000 Electric hinge with concealed conductor, average frequency									
1010 Steel base, 4-1/2" x 4-1/2", US26D				Pr.	310			310	340
1100 Bronze base, 4-1/2" x 4-1/2", US26D				"	335			335	365
1200 Electric hinge with concealed conductor, high frequency									
1210 Steel base, 4-1/2" x 4-1/2", US26D				Pr.	262			262	288
1600 Double weight, 800 lb., steel base, removable pin, 5" x 6", USP					420			420	460
1700 Steel base-welded pin, 5" x 6", USP					166			166	183
1800 Triple weight, 2000 lb., steel base, welded pin, 5" x 6", USP					545			545	600
2000 Pivot reinf., high frequency, steel base, 7-3/4" door plate, USP					149			149	164
2200 Bronze base, 7-3/4" door plate, US10					232			232	255
3000 Swing clear, full mortise, full or half surface, high frequency,									
3010 Steel base, 5" high, USP				Pr.	143			143	157
3200 Swing clear, full mortise, average frequency									
3210 Steel base, 4-1/2" high, USP				Pr.	128			128	141
4000 Wide throw, average frequency, steel base, 4-1/2" x 6", USP					94.50			94.50	104
4200 High frequency, steel base, 4-1/2" x 6", USP					112			112	124
4440 US26D					89.50			89.50	98
4600 Spring hinge, single acting, 6" flange, steel				Ea.	50			50	55
4700 Brass					94.50			94.50	104
4900 Double acting, 6" flange, steel					80			80	88
4950 Brass					133			133	146
5000 T-strap, galvanized, 4"				Pr.	17.25			17.25	19
5010 6"					29.50			29.50	32.50
5020 8"					52.50			52.50	58
8000 Continuous hinges									
8010 Steel, piano, 2" x 72"	1 Carp	20	.400	Ea.	22	19.40		41.40	55.50
8020 Brass, piano, 1-1/16" x 30"		30	.267		8	12.90		20.90	30
8030 Acrylic, piano, 1-3/4" x 12"		40	.200		15	9.70		24.70	32.50
8040 Aluminum, door, standard duty, 7'		3	2.667		135	129		264	360
8050 Heavy duty, 7'		3	2.667		142	129		271	365
8060 8'		3	2.667		160	129		289	385
8070 Steel, door, heavy duty, 7'		3	2.667		200	129		329	430
8080 8'		3	2.667		250	129		379	485
8090 Stainless steel, door, heavy duty, 7'		3	2.667		260	129		389	495
8100 8'		3	2.667		280	129		409	520
9000 Continuous hinge, steel, full mortise, heavy duty, 96 inch		2	4		630	194		824	1,000
9200 Continuous geared hinge, aluminum, full mortise, standard duty, 83 inch		3	2.667		130	129		259	355
9250 Continuous geared hinge, aluminum, full mortise, heavy duty, 83 inch		3	2.667		215	129		344	450

08 71 20.95 Kick Plates

	Crew	Daily Output	Labor-Hours	Unit	Material	Labor	Equipment	Total	Total Incl O&P
0010 **KICK PLATES**									
0020 Stainless steel, .050, 16 ga., 8" x 28", US32	1 Carp	15	.533	Ea.	38	26		64	84
0030 8" x 30"		15	.533		41	26		67	87
0040 8" x 34"		15	.533		46	26		72	92.50
0050 10" x 28"		15	.533		76	26		102	126
0060 10" x 30"		15	.533		82	26		108	132
0070 10" x 34"		15	.533		92	26		118	143

For customer support on your Commercial Renovation Cost Data, call 877.742.4719.

273

08 71 20.95 Kick Plates		Crew	Daily Output	Labor-Hours	Unit	Material	2016 Bare Costs Labor	Equipment	Total	Total Incl O&P
0080	Mop/Kick, 4" x 28"	1 Carp	15	.533	Ea.	34	26		60	79.50
0090	4" x 30"		15	.533		36	26		62	81.50
0100	4" x 34"		15	.533		41	26		67	87
0110	6" x 28"		15	.533		43	26		69	89.50
0120	6" x 30"		15	.533		47	26		73	93.50
0130	6" x 34"		15	.533		53	26		79	101
0500	Bronze, .050", 8" x 28"		15	.533		67.50	26		93.50	116
0510	8" x 30"		15	.533		65	26		91	114
0520	8" x 34"		15	.533		73	26		99	123
0530	10" x 28"		15	.533		75	26		101	125
0540	10" x 30"		15	.533		80	26		106	130
0550	10" x 34"		15	.533		91	26		117	142
0560	Mop/Kick, 4" x 28"		15	.533		33	26		59	78.50
0570	4" x 30"		15	.533		36	26		62	81.50
0580	4" x 34"		15	.533		37	26		63	82.50
0590	6" x 28"		15	.533		46	26		72	92.50
0600	6" x 30"		15	.533		52	26		78	99
0610	6" x 34"		15	.533		56	26		82	104
1000	Acrylic, .125", 8" x 26"		15	.533		28	26		54	73
1010	8" x 36"		15	.533		38	26		64	84
1020	8" x 42"		15	.533		45	26		71	91.50
1030	10" x 26"		15	.533		35	26		61	80.50
1040	10" x 36"		15	.533		48	26		74	95
1050	10" x 42"		15	.533		68.50	26		94.50	118
1060	Mop/Kick, 4" x 26"		15	.533		17	26		43	60.50
1070	4" x 36"		15	.533		24	26		50	68.50
1080	4" x 42"		15	.533		27	26		53	71.50
1090	6" x 26"		15	.533		23	26		49	67.50
1100	6" x 36"		15	.533		34	26		60	79.50
1110	6" x 42"		15	.533		39	26		65	85
1220	Brass, .050", 8" x 26"		15	.533		56	26		82	104
1230	8" x 36"		15	.533		75	26		101	125
1240	8" x 42"		15	.533		86	26		112	137
1250	10" x 26"		15	.533		71	26		97	120
1260	10" x 36"		15	.533		91	26		117	142
1270	10" x 42"		15	.533		105	26		131	158
1320	Mop/Kick, 4" x 26"		15	.533		28	26		54	73
1330	4" x 36"		15	.533		39	26		65	85
1340	4" x 42"		15	.533		44	26		70	90.50
1350	6" x 26"		15	.533		38	26		64	84
1360	6" x 36"		15	.533		48	26		74	95
1370	6" x 42"		15	.533		55	26		81	103
1800	Aluminum, .050", 8" x 26"		15	.533		35	26		61	80.50
1810	8" x 36"		15	.533		40	26		66	86
1820	8" x 42"		15	.533		47	26		73	93.50
1830	10" x 26"		15	.533		36	26		62	81.50
1840	10" x 36"		15	.533		50	26		76	97
1850	10" x 42"		15	.533		59	26		85	107
1860	Mop/Kick, 4" x 26"		15	.533		15	26		41	58.50
1870	4" x 36"		15	.533		20	26		46	64
1880	4" x 42"		15	.533		24	26		50	68.50
1890	6" x 26"		15	.533		22	26		48	66
1900	6" x 36"		15	.533		30	26		56	75

08 71 Door Hardware

08 71 20 – Hardware

08 71 20.95 Kick Plates	Crew	Daily Output	Labor-Hours	Unit	Material	2016 Bare Costs Labor	Equipment	Total	Total Incl O&P
1910	6" x 42"	1 Carp	15	.533	Ea.	35	26		61	80.50
9000	Minimum labor/equipment charge	↓	6	1.333	Job		64.50		64.50	106

08 71 21 – Astragals

08 71 21.10 Exterior Mouldings, Astragals

		Crew	Daily Output	Labor-Hours	Unit	Material	2016 Bare Costs Labor	Equipment	Total	Total Incl O&P
0010	**EXTERIOR MOULDINGS, ASTRAGALS**									
0400	One piece, overlapping cadmium plated steel, flat, 3/16" x 2"	1 Carp	90	.089	L.F.	4	4.31		8.31	11.45
0600	Prime coated steel, flat, 1/8" x 3"		90	.089		5.75	4.31		10.06	13.40
0800	Stainless steel, flat, 3/32" x 1-5/8"		90	.089		16	4.31		20.31	24.50
1000	Aluminum, flat, 1/8" x 2"		90	.089		4.10	4.31		8.41	11.55
1200	Nail on, "T" extrusion		120	.067		1.90	3.23		5.13	7.40
1300	Vinyl bulb insert		105	.076		2.50	3.69		6.19	8.80
1600	Screw on, "T" extrusion		90	.089		3.75	4.31		8.06	11.20
1700	Vinyl insert		75	.107		4.50	5.15		9.65	13.40
2000	"L" extrusion, neoprene bulbs		75	.107		4.10	5.15		9.25	12.95
2100	Neoprene sponge insert		75	.107		6.90	5.15		12.05	16.05
2200	Magnetic		75	.107		10.60	5.15		15.75	20
2400	Spring hinged security seal, with cam		75	.107		6.80	5.15		11.95	15.95
2600	Spring loaded locking bolt, vinyl insert		45	.178		9.20	8.60		17.80	24
2800	Neoprene sponge strip, "Z" shaped, aluminum		60	.133		8.30	6.45		14.75	19.70
2900	Solid neoprene strip, nail on aluminum strip	↓	90	.089	↓	4.05	4.31		8.36	11.50
3000	One piece stile protection									
3020	Neoprene fabric loop, nail on aluminum strips	1 Carp	60	.133	L.F.	1.10	6.45		7.55	11.75
3110	Flush mounted aluminum extrusion, 1/2" x 1-1/4"		60	.133		6.90	6.45		13.35	18.15
3140	3/4" x 1-3/8"		60	.133		4.10	6.45		10.55	15.05
3160	1-1/8" x 1-3/4"		60	.133		4.70	6.45		11.15	15.70
3300	Mortise, 9/16" x 3/4"		60	.133		4.10	6.45		10.55	15.05
3320	13/16" x 1-3/8"		60	.133		4.30	6.45		10.75	15.30
3600	Spring bronze strip, nail on type		105	.076		1.85	3.69		5.54	8.10
3620	Screw on, with retainer		75	.107		2.70	5.15		7.85	11.40
3800	Flexible stainless steel housing, pile insert, 1/2" door		105	.076		7.25	3.69		10.94	14.05
3820	3/4" door		105	.076		8.10	3.69		11.79	14.95
4000	Extruded aluminum retainer, flush mount, pile insert		105	.076		2.25	3.69		5.94	8.55
4080	Mortise, felt insert		90	.089		4.55	4.31		8.86	12.05
4160	Mortise with spring, pile insert		90	.089		3.40	4.31		7.71	10.80
4400	Rigid vinyl retainer, mortise, pile insert		105	.076		2.70	3.69		6.39	9
4600	Wool pile filler strip, aluminum backing	↓	105	.076	↓	2.70	3.69		6.39	9
5000	Two piece overlapping astragal, extruded aluminum retainer									
5010	Pile insert	1 Carp	60	.133	L.F.	3.35	6.45		9.80	14.25
5020	Vinyl bulb insert		60	.133		1.85	6.45		8.30	12.60
5040	Vinyl flap insert		60	.133		3.65	6.45		10.10	14.55
5060	Solid neoprene flap insert		60	.133		6.55	6.45		13	17.75
5080	Hypalon rubber flap insert		60	.133		6.65	6.45		13.10	17.85
5090	Snap on cover, pile insert		60	.133		9.45	6.45		15.90	21
5400	Magnetic aluminum, surface mounted		60	.133		23	6.45		29.45	35.50
5500	Interlocking aluminum, 5/8" x 1" neoprene bulb insert		45	.178		5.70	8.60		14.30	20.50
5600	Adjustable aluminum, 9/16" x 21/32", pile insert		45	.178		17.55	8.60		26.15	33.50
5800	Magnetic, adjustable, 9/16" x 21/32"	↓	45	.178	↓	23	8.60		31.60	39
6000	Two piece stile protection									
6010	Cloth backed rubber loop, 1" gap, nail on aluminum strips	1 Carp	45	.178	L.F.	4.35	8.60		12.95	18.85
6040	Screw on aluminum strips		45	.178		6.55	8.60		15.15	21.50
6100	1-1/2" gap, screw on aluminum extrusion		45	.178		5.85	8.60		14.45	20.50
6240	Vinyl fabric loop, slotted aluminum extrusion, 1" gap	↓	45	.178		2.20	8.60		10.80	16.45

For customer support on your Commercial Renovation Cost Data, call 877.742.4719.

275

08 71 Door Hardware

08 71 21 – Astragals

08 71 21.10 Exterior Mouldings, Astragals	Crew	Daily Output	Labor-Hours	Unit	Material	2016 Bare Costs Labor	Equipment	Total	Total Incl O&P
6300	1-1/4" gap	1 Carp	45	.178	L.F.	6.20	8.60		14.80	21

08 71 25 – Weatherstripping

08 71 25.10 Mechanical Seals, Weatherstripping

	08 71 25.10 Mechanical Seals, Weatherstripping	Crew	Daily Output	Labor-Hours	Unit	Material	2016 Bare Costs Labor	Equipment	Total	Total Incl O&P
0010	**MECHANICAL SEALS, WEATHERSTRIPPING**									
1000	Doors, wood frame, interlocking, for 3' x 7' door, zinc	1 Carp	3	2.667	Opng.	44	129		173	260
1100	Bronze		3	2.667		56	129		185	273
1300	6' x 7' opening, zinc		2	4		54	194		248	375
1400	Bronze		2	4		65	194		259	385
1700	Wood frame, spring type, bronze									
1800	3' x 7' door	1 Carp	7.60	1.053	Opng.	23.50	51		74.50	109
1900	6' x 7' door	"	7	1.143	"	31	55.50		86.50	125
2200	Metal frame, spring type, bronze									
2300	3' x 7' door	1 Carp	3	2.667	Opng.	47.50	129		176.50	263
2400	6' x 7' door	"	2.50	3.200	"	52	155		207	310
2500	For stainless steel, spring type, add					133%				
2700	Metal frame, extruded sections, 3' x 7' door, aluminum	1 Carp	3	2.667	Opng.	28	129		157	242
2800	Bronze		3	2.667		82	129		211	300
3100	6' x 7' door, aluminum		1.50	5.333		35	258		293	460
3200	Bronze		1.50	5.333		137	258		395	570
3500	Threshold weatherstripping									
3650	Door sweep, flush mounted, aluminum	1 Carp	25	.320	Ea.	19	15.50		34.50	46.50
3700	Vinyl		25	.320		18	15.50		33.50	45.50
5000	Garage door bottom weatherstrip, 12' aluminum, clear		14	.571		25	27.50		52.50	72.50
5010	Bronze		14	.571		90	27.50		117.50	144
5050	Bottom protection, Rubber		14	.571		37	27.50		64.50	85.50
5100	Threshold		14	.571		73	27.50		100.50	126
9000	Minimum labor/equipment charge		3	2.667	Job		129		129	211

08 74 Access Control Hardware

08 74 13 – Card Key Access Control Hardware

08 74 13.50 Card Key Access

	08 74 13.50 Card Key Access	Crew	Daily Output	Labor-Hours	Unit	Material	2016 Bare Costs Labor	Equipment	Total	Total Incl O&P
0010	**CARD KEY ACCESS**									
0020	Computerized system, processor, proximity reader and cards									
0030	Does not include door hardware, lockset or wiring									
0040	Card key system for 1 door				Ea.	1,275			1,275	1,400
0060	Card key system for 2 doors					2,175			2,175	2,375
0080	Card key system for 4 doors					2,675			2,675	2,950
0100	Processor for card key access system					900			900	990
0160	Magnetic lock for electric access, 600 pound holding force					200			200	220
0170	Magnetic lock for electric access, 1200 pound holding force					200			200	220
0200	Proximity card reader					130			130	143

08 74 19 – Biometric Identity Access Control Hardware

08 74 19.50 Biometric Identity Access

	08 74 19.50 Biometric Identity Access	Crew	Daily Output	Labor-Hours	Unit	Material	2016 Bare Costs Labor	Equipment	Total	Total Incl O&P
0010	**BIOMETRIC IDENTITY ACCESS**									
0220	Hand geometry scanner, mem of 512 users, excl striker/power	1 Elec	3	2.667	Ea.	2,100	147		2,247	2,525
0230	Memory upgrade for, adds 9,700 user profiles		8	1		300	55		355	415
0240	Adds 32,500 user profiles		8	1		600	55		655	745
0250	Prison type, memory of 256 users, excl striker/power		3	2.667		2,600	147		2,747	3,075
0260	Memory upgrade for, adds 3,300 user profiles		8	1		250	55		305	360
0270	Adds 9,700 user profiles		8	1		460	55		515	590

08 74 Access Control Hardware

08 74 19 – Biometric Identity Access Control Hardware

08 74 19.50 Biometric Identity Access		Crew	Daily Output	Labor-Hours	Unit	Material	2016 Bare Costs Labor	Equipment	Total	Total Incl O&P
0280	Adds 27,900 user profiles	1 Elec	8	1	Ea.	610	55		665	755
0290	All weather, mem of 512 users, excl striker/power		3	2.667		3,900	147		4,047	4,525
0300	Facial & fingerprint scanner, combination unit, excl striker/power		3	2.667		4,300	147		4,447	4,950
0310	Access for, for initial setup, excl striker/power		3	2.667		1,100	147		1,247	1,425

08 75 Window Hardware

08 75 10 – Window Handles and Latches

08 75 10.10 Handles and Latches

		Crew	Daily Output	Labor-Hours	Unit	Material	2016 Bare Costs Labor	Equipment	Total	Total Incl O&P
0010	**HANDLES AND LATCHES**									
1000	Handles, surface mounted, aluminum	1 Carp	24	.333	Ea.	4.56	16.15		20.71	31.50
1020	Brass		24	.333		5.85	16.15		22	33
1040	Chrome		24	.333		7.30	16.15		23.45	34.50
1500	Recessed, aluminum		12	.667		2.64	32.50		35.14	56
1520	Brass		12	.667		4.14	32.50		36.64	57.50
1540	Chrome		12	.667		3.67	32.50		36.17	57
2000	Latches, aluminum		20	.400		3.48	19.40		22.88	35.50
2020	Brass		20	.400		4.65	19.40		24.05	36.50
2040	Chrome		20	.400		3.39	19.40		22.79	35
9000	Minimum labor/equipment charge		6	1.333	Job		64.50		64.50	106

08 75 30 – Weatherstripping

08 75 30.10 Mechanical Weather Seals

		Crew	Daily Output	Labor-Hours	Unit	Material	2016 Bare Costs Labor	Equipment	Total	Total Incl O&P
0010	**MECHANICAL WEATHER SEALS**, Window, double hung, 3' x 5'									
0020	Zinc	1 Carp	7.20	1.111	Opng.	20	54		74	110
0100	Bronze		7.20	1.111		40	54		94	132
0200	Vinyl V strip		7	1.143		9.50	55.50		65	101
0500	As above but heavy duty, zinc		4.60	1.739		20	84.50		104.50	160
0600	Bronze		4.60	1.739		70	84.50		154.50	215
9000	Minimum labor/equipment charge	1 Clab	4.60	1.739	Job		66		66	108

08 79 Hardware Accessories

08 79 20 – Door Accessories

08 79 20.10 Door Hardware Accessories

		Crew	Daily Output	Labor-Hours	Unit	Material	2016 Bare Costs Labor	Equipment	Total	Total Incl O&P
0010	**DOOR HARDWARE ACCESSORIES**									
0140	Door bolt, surface, 4"	1 Carp	32	.250	Ea.	13	12.10		25.10	34
0160	Door latch	"	12	.667	"	8.55	32.50		41.05	62.50
0200	Sliding closet door									
0220	Track and hanger, single	1 Carp	10	.800	Ea.	65	39		104	135
0240	Double		8	1		80	48.50		128.50	167
0260	Door guide, single		48	.167		30	8.10		38.10	46
0280	Double		48	.167		40	8.10		48.10	57
0600	Deadbolt and lock cover plate, brass or stainless steel		30	.267		28	12.90		40.90	52
0620	Hole cover plate, brass or chrome		35	.229		8	11.05		19.05	27
2240	Mortise lockset, passage, lever handle		9	.889		160	43		203	247
4000	Security chain, standard		18	.444		10	21.50		31.50	46
4100	Deluxe		18	.444		45	21.50		66.50	84.50

For customer support on your Commercial Renovation Cost Data, call 877.742.4719.

277

08 81 Glass Glazing

08 81 10 – Float Glass

08 81 10.10 Various Types and Thickness of Float Glass

		Crew	Daily Output	Labor-Hours	Unit	Material	2016 Bare Costs Labor	Equipment	Total	Total Incl O&P
0010	**VARIOUS TYPES AND THICKNESS OF FLOAT GLASS**									
0020	3/16" Plain	2 Glaz	130	.123	S.F.	5.10	5.75		10.85	14.90
0200	Tempered, clear		130	.123		7	5.75		12.75	17
0300	Tinted		130	.123		8.10	5.75		13.85	18.20
0600	1/4" thick, clear, plain		120	.133		6.40	6.25		12.65	17.15
0700	Tinted		120	.133		9.10	6.25		15.35	20
0800	Tempered, clear		120	.133		9	6.25		15.25	20
0900	Tinted		120	.133		11.35	6.25		17.60	22.50
1600	3/8" thick, clear, plain		75	.213		10.40	9.95		20.35	27.50
1700	Tinted		75	.213		15.70	9.95		25.65	33.50
1800	Tempered, clear		75	.213		16.75	9.95		26.70	34.50
1900	Tinted		75	.213		19.10	9.95		29.05	37
2200	1/2" thick, clear, plain		55	.291		17.65	13.60		31.25	41.50
2300	Tinted		55	.291		27.50	13.60		41.10	52
2400	Tempered, clear		55	.291		25	13.60		38.60	49.50
2500	Tinted		55	.291		26	13.60		39.60	51
2800	5/8" thick, clear, plain		45	.356		27.50	16.60		44.10	57
2900	Tempered, clear		45	.356		31.50	16.60		48.10	61.50
3200	3/4" thick, clear, plain		35	.457		35.50	21.50		57	73.50
3300	Tempered, clear		35	.457		41	21.50		62.50	80
3600	1" thick, clear, plain	↓	30	.533		59	25		84	105
8900	For low emissivity coating for 3/16" & 1/4" only, add to above					18%				
9000	Minimum labor/equipment charge	1 Glaz	2	4	Job		187		187	305

08 81 17 – Fire Glass

08 81 17.10 Fire Resistant Glass

		Crew	Daily Output	Labor-Hours	Unit	Material	2016 Bare Costs Labor	Equipment	Total	Total Incl O&P
0010	**FIRE RESISTANT GLASS**									
0020	Fire Glass Minimum	2 Glaz	40	.400	S.F.	37	18.70		55.70	71
0030	Mid Range		40	.400		80	18.70		98.70	119
0050	High End	↓	40	.400	↓	355	18.70		373.70	420

08 81 20 – Vision Panels

08 81 20.10 Full Vision

		Crew	Daily Output	Labor-Hours	Unit	Material	2016 Bare Costs Labor	Equipment	Total	Total Incl O&P
0010	**FULL VISION**, window system with 3/4" glass mullions									
0020	Up to 10' high	H-2	130	.185	S.F.	64	8.10		72.10	83.50
0100	10' to 20' high, minimum		110	.218		68.50	9.55		78.05	90.50
0150	Average		100	.240		73.50	10.50		84	98
0200	Maximum	↓	80	.300	↓	82.50	13.15		95.65	113
9000	Minimum labor/equipment charge	1 Glaz	2	4	Job		187		187	305

08 81 25 – Glazing Variables

08 81 25.10 Applications of Glazing

		Crew	Daily Output	Labor-Hours	Unit	Material	2016 Bare Costs Labor	Equipment	Total	Total Incl O&P
0010	**APPLICATIONS OF GLAZING**	R088110-10								
0600	For glass replacement, add				S.F.		100%			
0700	For gasket settings, add				L.F.	5.75			5.75	6.35
0900	For sloped glazing, add				S.F.		26%			
2000	Fabrication, polished edges, 1/4" thick				Inch	.55			.55	.61
2100	1/2" thick					1.30			1.30	1.43
2500	Mitered edges, 1/4" thick					1.30			1.30	1.43
2600	1/2" thick				↓	2.15			2.15	2.37

08 81 Glass Glazing

08 81 30 – Insulating Glass

08 81 30.10 Reduce Heat Transfer Glass

		Crew	Daily Output	Labor-Hours	Unit	Material	2016 Bare Costs Labor	Equipment	Total	Total Incl O&P
0010	**REDUCE HEAT TRANSFER GLASS**									
0015	2 lites 1/8" float, 1/2" thk under 15 S.F.									
0020	Clear [G] R088110-10	2 Glaz	95	.168	S.F.	10.30	7.85		18.15	24
0100	Tinted [G]		95	.168		14.25	7.85		22.10	28.50
0200	2 lites 3/16" float, for 5/8" thk unit, 15 to 30 S.F., clear [G]		90	.178		13.70	8.30		22	28.50
0400	1" thk, dbl. glazed, 1/4" float, 30 to 70 S.F., clear [G]		75	.213		17.15	9.95		27.10	35
0500	Tinted [G]		75	.213		23.50	9.95		33.45	42
2000	Both lites, light & heat reflective [G]		85	.188		31.50	8.80		40.30	49
2500	Heat reflective, film inside, 1" thick unit, clear [G]		85	.188		27.50	8.80		36.30	45
2600	Tinted [G]		85	.188		28.50	8.80		37.30	46
3000	Film on weatherside, clear, 1/2" thick unit [G]		95	.168		19.60	7.85		27.45	34.50
3100	5/8" thick unit [G]		90	.178		20	8.30		28.30	35.50
3200	1" thick unit [G]		85	.188		27	8.80		35.80	44.50
3350	Clear heat reflective film on inside [G]	1 Glaz	50	.160		12.90	7.45		20.35	26.50
3360	Heat reflective film on inside, trinted [G]		25	.320		13.70	14.95		28.65	39
3370	Heat reflective film on the inside, metalized [G]		20	.400		14	18.70		32.70	46
5000	Spectrally selective film, on ext, blocks solar gain/allows 70% of light [G]	2 Glaz	95	.168		14.25	7.85		22.10	28.50
9000	Minimum labor/equipment charge	1 Glaz	2	4	Job		187		187	305

08 81 40 – Plate Glass

08 81 40.10 Plate Glass

		Crew	Daily Output	Labor-Hours	Unit	Material	2016 Bare Costs Labor	Equipment	Total	Total Incl O&P
0010	**PLATE GLASS** Twin ground, polished,									
0020	3/16" thick	2 Glaz	100	.160	S.F.	5.35	7.45		12.80	18
0100	1/4" thick		94	.170		7.35	7.95		15.30	21
0200	3/8" thick		60	.267		12.60	12.45		25.05	34
0300	1/2" thick		40	.400		24.50	18.70		43.20	57.50

08 81 55 – Window Glass

08 81 55.10 Sheet Glass

		Crew	Daily Output	Labor-Hours	Unit	Material	2016 Bare Costs Labor	Equipment	Total	Total Incl O&P
0010	**SHEET GLASS** (window), clear float, stops, putty bed									
0015	1/8" thick, clear float	2 Glaz	480	.033	S.F.	3.71	1.56		5.27	6.60
0500	3/16" thick, clear		480	.033		6	1.56		7.56	9.10
0600	Tinted		480	.033		7.45	1.56		9.01	10.70
0700	Tempered		480	.033		9.20	1.56		10.76	12.60
9000	Minimum labor/equipment charge		5	3.200	Job		149		149	242

08 81 65 – Wire Glass

08 81 65.10 Glass Reinforced With Wire

		Crew	Daily Output	Labor-Hours	Unit	Material	2016 Bare Costs Labor	Equipment	Total	Total Incl O&P
0010	**GLASS REINFORCED WITH WIRE**									
0012	1/4" thick rough obscure	2 Glaz	135	.119	S.F.	23.50	5.55		29.05	35
1000	Polished wire, 1/4" thick, diamond, clear		135	.119		28	5.55		33.55	39.50
1500	Pinstripe, obscure		135	.119		41	5.55		46.55	54

For customer support on your Commercial Renovation Cost Data, call 877.742.4719.

279

08 83 Mirrors

08 83 13 – Mirrored Glass Glazing

08 83 13.10 Mirrors

		Crew	Daily Output	Labor-Hours	Unit	Material	2016 Bare Costs Labor	Equipment	Total	Total Incl O&P
0010	**MIRRORS**, No frames, wall type, 1/4" plate glass, polished edge									
0100	Up to 5 S.F.	2 Glaz	125	.128	S.F.	9.50	6		15.50	20
0200	Over 5 S.F.		160	.100		9.25	4.67		13.92	17.75
0500	Door type, 1/4" plate glass, up to 12 S.F.		160	.100		8.80	4.67		13.47	17.25
1000	Float glass, up to 10 S.F., 1/8" thick		160	.100		5.90	4.67		10.57	14.05
1100	3/16" thick		150	.107		7.30	4.98		12.28	16.10
1500	12" x 12" wall tiles, square edge, clear		195	.082		2.43	3.83		6.26	8.85
1600	Veined		195	.082		6.20	3.83		10.03	13.05
2000	1/4" thick, stock sizes, one way transparent		125	.128		19.70	6		25.70	31
2010	Bathroom, unframed, laminated		160	.100		14	4.67		18.67	23
2500	Tempered		160	.100		17.80	4.67		22.47	27

08 83 13.15 Reflective Glass

			Crew	Daily Output	Labor-Hours	Unit	Material	2016 Bare Costs Labor	Equipment	Total	Total Incl O&P
0010	**REFLECTIVE GLASS**										
0100	1/4" float with fused metallic oxide fixed	G	2 Glaz	115	.139	S.F.	16.95	6.50		23.45	29
0500	1/4" float glass with reflective applied coating	G	"	115	.139	"	13.70	6.50		20.20	25.50

08 84 Plastic Glazing

08 84 10 – Plexiglass Glazing

08 84 10.10 Plexiglass Acrylic

		Crew	Daily Output	Labor-Hours	Unit	Material	2016 Bare Costs Labor	Equipment	Total	Total Incl O&P
0010	**PLEXIGLASS ACRYLIC**, clear, masked,									
0020	1/8" thick, cut sheets	2 Glaz	170	.094	S.F.	12.20	4.40		16.60	20.50
0200	Full sheets		195	.082		5.15	3.83		8.98	11.85
0500	1/4" thick, cut sheets		165	.097		14.30	4.53		18.83	23
0600	Full sheets		185	.086		9.20	4.04		13.24	16.65
0900	3/8" thick, cut sheets		155	.103		20	4.82		24.82	30
1000	Full sheets		180	.089		15.30	4.15		19.45	23.50
1300	1/2" thick, cut sheets		135	.119		28.50	5.55		34.05	40.50
1400	Full sheets		150	.107		20	4.98		24.98	30
1700	3/4" thick, cut sheets		115	.139		70.50	6.50		77	88
1800	Full sheets		130	.123		40.50	5.75		46.25	54
2100	1" thick, cut sheets		105	.152		79	7.10		86.10	98.50
2200	Full sheets		125	.128		49	6		55	63.50
3000	Colored, 1/8" thick, cut sheets		170	.094		18.35	4.40		22.75	27
3200	Full sheets		195	.082		11.20	3.83		15.03	18.50
3500	1/4" thick, cut sheets		165	.097		20.50	4.53		25.03	30
3600	Full sheets		185	.086		14.30	4.04		18.34	22.50
4000	Mirrors, untinted, cut sheets, 1/8" thick		185	.086		12.25	4.04		16.29	20
4200	1/4" thick		180	.089		16.30	4.15		20.45	24.50

08 84 20 – Polycarbonate

08 84 20.10 Thermoplastic

		Crew	Daily Output	Labor-Hours	Unit	Material	2016 Bare Costs Labor	Equipment	Total	Total Incl O&P
0010	**THERMOPLASTIC**, clear, masked, cut sheets									
0020	1/8" thick	2 Glaz	170	.094	S.F.	14.30	4.40		18.70	23
0500	3/16" thick		165	.097		17.35	4.53		21.88	26.50
1000	1/4" thick		155	.103		17.35	4.82		22.17	27
1500	3/8" thick		150	.107		27	4.98		31.98	37.50
9000	Minimum labor/equipment charge	1 Glaz	2	4	Job		187		187	305

08 85 Glazing Accessories

08 85 10 - Miscellaneous Glazing Accessories

08 85 10.10 Glazing Gaskets	Crew	Daily Output	Labor-Hours	Unit	Material	2016 Bare Costs Labor	Equipment	Total	Total Incl O&P
0010 **GLAZING GASKETS**, Neoprene for glass tongued mullion									
0015 1/4" glass	2 Glaz	200	.080	L.F.	.77	3.74		4.51	6.90
0020 3/8"		200	.080		.96	3.74		4.70	7.10
0040 1/2"		200	.080		1.09	3.74		4.83	7.25
0060 3/4"		180	.089		1.59	4.15		5.74	8.50
0080 1"		180	.089		1.29	4.15		5.44	8.15
1000 Glazing compound, wood	1 Glaz	58	.138		1.10	6.45		7.55	11.65
1005 Glazing compound, per window, up to 30 L.F.					1.10			1.10	1.21
1006 Glazing compound, per window, up to 30 L.F.				Ea.	33			33	36.50
1020 Metal	1 Glaz	58	.138	L.F.	1.10	6.45		7.55	11.65

08 85 20 - Structural Glazing Sealants

08 85 20.10 Structural Glazing Adhesives

	Crew	Daily Output	Labor-Hours	Unit	Material	2016 Bare Costs Labor	Equipment	Total	Total Incl O&P
0010 **STRUCTURAL GLAZING ADHESIVES**									
0050 Structural glazing adhesive, 1/4" x 1/4" joint seal	1 Glaz	200	.040	L.F.	14.50	1.87		16.37	19

08 87 Glazing Surface Films

08 87 13 - Solar Control Films

08 87 13.10 Solar Films On Glass

	Crew	Daily Output	Labor-Hours	Unit	Material	2016 Bare Costs Labor	Equipment	Total	Total Incl O&P
0010 **SOLAR FILMS ON GLASS** (glass not included)									
1000 Bronze, 20% VLT	H-2	180	.133	S.F.	.80	5.85		6.65	10.40
1020 50% VLT		180	.133		1.20	5.85		7.05	10.80
1050 Neutral, 20% VLT		180	.133		1.20	5.85		7.05	10.80
1100 Silver, 15% VlT		180	.133		.85	5.85		6.70	10.45
1120 35% VLT		180	.133		3.21	5.85		9.06	13.05
1150 68% VLT		180	.133		3.74	5.85		9.59	13.60
3000 One way mirror with night vision 5% in and 15% out VLT		180	.133		1.60	5.85		7.45	11.25

08 87 16 - Glass Safety Films

08 87 16.10 Safety Films On Glass

	Crew	Daily Output	Labor-Hours	Unit	Material	2016 Bare Costs Labor	Equipment	Total	Total Incl O&P
0010 **SAFETY FILMS ON GLASS** (glass not included)									
0015 Safety film helps hold glass together when broken									
0050 Safety film, clear, 2 mil	H-2	180	.133	S.F.	1.24	5.85		7.09	10.85
0100 Safety film, clear, 4 mil		180	.133		1.71	5.85		7.56	11.40
0110 Safety film, tinted, 4 mil		180	.133		1.34	5.85		7.19	10.95
0150 Safety film, black tint, 8 mil		180	.133		2.04	5.85		7.89	11.75

08 87 53 - Security Films On Glass

08 87 53.10 Security Film Adhered On Glass

	Crew	Daily Output	Labor-Hours	Unit	Material	2016 Bare Costs Labor	Equipment	Total	Total Incl O&P
0010 **SECURITY FILM ADHERED ON GLASS** (glass not included)									
0020 Security film, clear, 7 mil	1 Glaz	200	.040	S.F.	1.03	1.87		2.90	4.16
0030 8 mil		200	.040		1.86	1.87		3.73	5.10
0040 9 mil		200	.040		2.07	1.87		3.94	5.30
0050 10 mil		200	.040		2.24	1.87		4.11	5.50
0060 12 mil		200	.040		1.96	1.87		3.83	5.20
0075 15 mil		200	.040		2.38	1.87		4.25	5.65
0100 Security film, sealed with structural adhesive, 7 mil		180	.044		5.55	2.08		7.63	9.45
0110 8 mil		180	.044		6.50	2.08		8.58	10.50
0140 14 mil		180	.044		7.10	2.08		9.18	11.15

For customer support on your Commercial Renovation Cost Data, call 877.742.4719.

281

08 88 Special Function Glazing

08 88 52 – Prefabricated Glass Block Windows

08 88 52.10 Prefabricated Glass Block Windows	Crew	Daily Output	Labor-Hours	Unit	Material	2016 Bare Costs Labor	Equipment	Total	Total Incl O&P
0010 **PREFABRICATED GLASS BLOCK WINDOWS**									
0015 Includes frame and silicone seal									
0020 Glass Block Window 16" x 8"	2 Glaz	6	2.667	Ea.	185	125		310	405
0025 16" x 16"		6	2.667		223	125		348	445
0030 16" x 24"		6	2.667		265	125		390	495
0050 16" x 48"		4	4		380	187		567	725
0100 Glass Block Window 24" x 8"		6	2.667		211	125		336	435
0110 24" x 16"		6	2.667		265	125		390	495
0120 24" x 24"		6	2.667		315	125		440	545
0150 24" x 48"		6	2.667		470	125		595	715
0200 Glass Block Window 32" x 8"		6	2.667		211	125		336	435
0210 32" x 16"		6	2.667		300	125		425	530
0220 32" x 24"		4	4		365	187		552	705
0250 32" x 48"		4	4		555	187		742	915
0300 Glass Block Window 40" x 8"		6	2.667		271	125		396	500
0310 40" x 16"		6	2.667		340	125		465	575
0320 40" x 24"		6	2.667		410	125		535	655
0350 40" x 48"		6	2.667		645	125		770	910
0400 Glass Block Window 48" x 8"		6	2.667		300	125		425	530
0410 48" x 16"		6	2.667		380	125		505	615
0420 48" x 24"		6	2.667		470	125		595	715
0450 48" x 48"		6	2.667		730	125		855	1,000
0500 Glass Block Window 56" x 8"		6	2.667		330	125		455	565
0510 56" x 16"		4	4		415	187		602	765
0520 56" x 24"		4	4		520	187		707	875
0550 56" x 48"		4	4		825	187		1,012	1,200

08 88 52.20 Prefabricated Glass Block Windows Hurricane Resistant

08 88 52.20 Prefabricated Glass Block Windows Hurricane Resistant	Crew	Daily Output	Labor-Hours	Unit	Material	2016 Bare Costs Labor	Equipment	Total	Total Incl O&P
0010 **PREFABRICATED GLASS BLOCK WINDOWS HURRICANE RESISTANT**									
0015 Includes frame and silicone seal									
3020 Glass Block Window 16" x 16"	2 Glaz	6	2.667	Ea.	500	125		625	750
3030 16" x 24"		6	2.667		580	125		705	840
3040 16" x 32"		4	4		640	187		827	1,000
3050 16" x 40"		4	4		690	187		877	1,075
3060 16" x 48"		4	4		700	187		887	1,075
3070 16" x 56"		4	4		880	187		1,067	1,275
3080 Glass Block Window 24" x 24"		6	2.667		640	125		765	905
3090 24" x 32"		4	4		700	187		887	1,075
3100 24" x 40"		4	4		880	187		1,067	1,275
3110 24" x 48"		4	4		910	187		1,097	1,300
3120 24" x 56"		4	4		950	187		1,137	1,350
3130 Glass Block Window 32" x 32"		4	4		800	187		987	1,175
3140 32" x 40"		4	4		900	187		1,087	1,300
3150 32" x 48"		4	4		1,050	187		1,237	1,450
3160 32" x 56"		4	4		1,200	187		1,387	1,625
3200 Glass Block Window 40" x 40"		4	4		1,025	187		1,212	1,425
3210 40" x 48"		3	5.333		1,250	249		1,499	1,775
3220 40" x 56"		3	5.333		1,400	249		1,649	1,950
3260 Glass Block Window 48" x 48"		3	5.333		1,450	249		1,699	2,000
3270 48" x 56"		3	5.333		1,625	249		1,874	2,175
3310 Glass Block Window 56" x 56"		3	5.333		1,775	249		2,024	2,350

08 88 Special Function Glazing

08 88 56 – Ballistics-Resistant Glazing

08 88 56.10 Laminated Glass

		Crew	Daily Output	Labor-Hours	Unit	Material	2016 Bare Costs Labor	2016 Bare Costs Equipment	Total	Total Incl O&P
0010	**LAMINATED GLASS**									
0020	Clear float .03" vinyl 1/4"	2 Glaz	90	.178	S.F.	12.50	8.30		20.80	27
0100	3/8" thick		78	.205		22.50	9.60		32.10	40
0200	.06" vinyl, 1/2" thick		65	.246		25.50	11.50		37	46.50
1000	5/8" thick		90	.178		29.50	8.30		37.80	46
2000	Bullet-resisting, 1-3/16" thick, to 15 S.F.		16	1		113	46.50		159.50	201
2100	Over 15 S.F.		16	1		126	46.50		172.50	215
2200	2" thick, to 15 S.F.		10	1.600		140	74.50		214.50	275
2300	Over 15 S.F.		10	1.600		120	74.50		194.50	253
2500	2-1/4" thick, to 15 S.F.		12	1.333		185	62.50		247.50	305
2600	Over 15 S.F.		12	1.333		175	62.50		237.50	294
2700	Level 2 (.357 magnum), NIJ and UL		12	1.333		83	62.50		145.50	193
2750	Level 3A (.44 magnum) NIJ, UL 3		12	1.333		88	62.50		150.50	198
2800	Level 4 (AK-47) NIJ, UL 7 & 8		12	1.333		122	62.50		184.50	235
2850	Level 5 (M-16) UL		12	1.333		125	62.50		187.50	239
2900	Level 3 (7.62 Armor Piercing) NIJ, UL 4 & 5		12	1.333		147	62.50		209.50	263

08 91 Louvers

08 91 16 – Operable Wall Louvers

08 91 16.10 Movable Blade Louvers

		Crew	Daily Output	Labor-Hours	Unit	Material	2016 Bare Costs Labor	2016 Bare Costs Equipment	Total	Total Incl O&P
0010	**MOVABLE BLADE LOUVERS**									
0100	PVC, commercial grade, 12" x 12"	1 Shee	20	.400	Ea.	82	23		105	126
0110	16" x 16"		14	.571		96	32.50		128.50	158
0120	18" x 18"		14	.571		99	32.50		131.50	161
0130	24" x 24"		14	.571		140	32.50		172.50	206
0140	30" x 30"		12	.667		168	38		206	245
0150	36" x 36"		12	.667		197	38		235	277
0300	Stainless steel, commercial grade, 12" x 12"		20	.400		219	23		242	277
0310	16" x 16"		14	.571		265	32.50		297.50	345
0320	18" x 18"		14	.571		282	32.50		314.50	360
0330	20" x 20"		14	.571		325	32.50		357.50	405
0340	24" x 24"		14	.571		385	32.50		417.50	475
0350	30" x 30"		12	.667		430	38		468	535
0360	36" x 36"		12	.667		520	38		558	635

08 91 19 – Fixed Louvers

08 91 19.10 Aluminum Louvers

		Crew	Daily Output	Labor-Hours	Unit	Material	2016 Bare Costs Labor	2016 Bare Costs Equipment	Total	Total Incl O&P
0010	**ALUMINUM LOUVERS**									
0020	Aluminum with screen, residential, 8" x 8"	1 Carp	38	.211	Ea.	20	10.20		30.20	38.50
0100	12" x 12"		38	.211		16	10.20		26.20	34.50
0200	12" x 18"		35	.229		20	11.05		31.05	40
0250	14" x 24"		30	.267		29	12.90		41.90	53
0300	18" x 24"		27	.296		32	14.35		46.35	58.50
0500	24" x 30"		24	.333		60	16.15		76.15	92.50
0700	Triangle, adjustable, small		20	.400		58	19.40		77.40	95.50
0800	Large		15	.533		79.50	26		105.50	129
2100	Midget, aluminum, 3/4" deep, 1" diameter		85	.094		.77	4.56		5.33	8.30
2150	3" diameter		60	.133		2.48	6.45		8.93	13.30
2200	4" diameter		50	.160		4.95	7.75		12.70	18.10
2250	6" diameter		30	.267		4.10	12.90		17	25.50
3000	PVC, commercial grade, 12" x 12"	1 Shee	20	.400		97	23		120	143

08 91 19 – Fixed Louvers

08 91 19.10 Aluminum Louvers

		Crew	Daily Output	Labor-Hours	Unit	Material	2016 Bare Costs Labor	Equipment	Total	Total Incl O&P
3010	12" x 18"	1 Shee	20	.400	Ea.	109	23		132	156
3020	12" x 24"		20	.400		146	23		169	197
3030	14" x 24"		20	.400		152	23		175	203
3100	Aluminum, commercial grade, 12" x 12"		20	.400		171	23		194	224
3110	24" x 24"		16	.500		249	28.50		277.50	320
3120	24" x 36"		16	.500		300	28.50		328.50	375
3130	36" x 24"		16	.500		310	28.50		338.50	385
3140	36" x 36"		14	.571		390	32.50		422.50	480
3150	36" x 48"		10	.800		400	46		446	515
3160	48" x 36"		10	.800		390	46		436	505
3170	48" x 48"		10	.800		435	46		481	550
3180	60" x 48"		8	1		485	57.50		542.50	625
3190	60" x 60"		8	1		600	57.50		657.50	750

08 91 19.20 Steel Louvers

		Crew	Daily Output	Labor-Hours	Unit	Material	2016 Bare Costs Labor	Equipment	Total	Total Incl O&P
0010	**STEEL LOUVERS**									
3300	Galvanized steel, fixed blades, commercial grade, 18" x 18"	1 Shee	20	.400	Ea.	197	23		220	253
3310	24" x 24"		20	.400		232	23		255	291
3320	24" x 36"		16	.500		300	28.50		328.50	375
3330	36" x 24"		16	.500		298	28.50		326.50	375
3340	36" x 36"		14	.571		380	32.50		412.50	470
3350	36" x 48"		10	.800		385	46		431	500
3360	48" x 36"		10	.800		385	46		431	500
3370	48" x 48"		10	.800		425	46		471	540
3380	60" x 48"		10	.800		490	46		536	615
3390	60" x 60"		10	.800		590	46		636	720

08 91 26 – Door Louvers

08 91 26.10 Steel Louvers, 18 Gauge, Fixed Blade

		Crew	Daily Output	Labor-Hours	Unit	Material	2016 Bare Costs Labor	Equipment	Total	Total Incl O&P
0010	**STEEL LOUVERS, 18 GAUGE, FIXED BLADE**									
0050	12" x 12", with enamel or powder coat	1 Carp	20	.400	Ea.	82	19.40		101.40	122
0055	18" x 12"		20	.400		87.50	19.40		106.90	128
0060	18" x 18"		20	.400		106	19.40		125.40	149
0065	24" x 12"		20	.400		108	19.40		127.40	151
0070	24" x 18"		20	.400		117	19.40		136.40	161
0075	24" x 24"		20	.400		144	19.40		163.40	190
0100	12" x 12", galvanized		20	.400		74	19.40		93.40	113
0105	18" x 12"		20	.400		87.50	19.40		106.90	128
0115	24" x 12"		20	.400		104	19.40		123.40	146
0125	24" x 24"		20	.400		144	19.40		163.40	190
3000	Fire rated									
3010	12" x 12"	1 Carp	10	.800	Ea.	201	39		240	285
3050	18" x 18"		10	.800		205	39		244	289
3100	24" x 12"		10	.800		237	39		276	325
3120	24" x 18"		10	.800		294	39		333	390
3130	24" x 24"		10	.800		287	39		326	380

08 95 Vents

08 95 13 – Soffit Vents

08 95 13.10 Wall Louvers

		Crew	Daily Output	Labor-Hours	Unit	Material	2016 Bare Costs Labor	Equipment	Total	Total Incl O&P
0010	**WALL LOUVERS**									
2330	Soffit vent, continuous, 3" wide, aluminum, mill finish	1 Carp	200	.040	L.F.	.76	1.94		2.70	4.01
2340	Baked enamel finish		200	.040	"	6.05	1.94		7.99	9.80
2400	Under eaves vent, aluminum, mill finish, 16" x 4"		48	.167	Ea.	1.93	8.10		10.03	15.30
2500	16" x 8"		48	.167	"	2.23	8.10		10.33	15.65

08 95 16 – Wall Vents

08 95 16.10 Louvers

		Crew	Daily Output	Labor-Hours	Unit	Material	2016 Bare Costs Labor	Equipment	Total	Total Incl O&P
0010	**LOUVERS**									
0020	Redwood, 2'-0" diameter, full circle	1 Carp	16	.500	Ea.	190	24		214	249
0100	Half circle		16	.500		175	24		199	233
0200	Octagonal		16	.500		142	24		166	196
0300	Triangular, 5/12 pitch, 5'-0" at base		16	.500		200	24		224	260
1000	Rectangular, 1'-4" x 1'-3"		16	.500		19.50	24		43.50	61
1100	Rectangular, 1'-4" x 1'-8"		16	.500		28	24		52	70.50
1200	1'-4" x 2'-2"		15	.533		31	26		57	76.50
1300	1'-9" x 2'-2"		15	.533		37.50	26		63.50	83.50
1400	2'-3" x 2'-2"		14	.571		49.50	27.50		77	99.50
1700	2'-4" x 2'-11"		13	.615		49.50	30		79.50	103
2000	Aluminum, 12" x 16"		25	.320		20	15.50		35.50	47.50
2010	16" x 20"		25	.320		27.50	15.50		43	56
2020	24" x 30"		25	.320		59	15.50		74.50	90.50
2100	6' triangle		12	.667		168	32.50		200.50	238
3100	Round, 2'-2" diameter		16	.500		124	24		148	176
7000	Vinyl gable vent, 8" x 8"		38	.211		14	10.20		24.20	32
7020	12" x 12"		38	.211		28	10.20		38.20	47.50
7080	12" x 18"		35	.229		37	11.05		48.05	58.50
7200	18" x 24"		30	.267		45	12.90		57.90	70.50
9000	Minimum labor/equipment charge		3.50	2.286	Job		111		111	181

For customer support on your Commercial Renovation Cost Data, call 877.742.4719.

285

Division Notes

	CREW	DAILY OUTPUT	LABOR-HOURS	UNIT	BARE COSTS				TOTAL INCL O&P
					MAT.	LABOR	EQUIP.	TOTAL	

Estimating Tips

General

- Room Finish Schedule: A complete set of plans should contain a room finish schedule. If one is not available, it would be well worth the time and effort to obtain one.

09 20 00 Plaster and Gypsum Board

- Lath is estimated by the square yard plus a 5% allowance for waste. Furring, channels, and accessories are measured by the linear foot. An extra foot should be allowed for each accessory miter or stop.
- Plaster is also estimated by the square yard. Deductions for openings vary by preference, from zero deduction to 50% of all openings over 2 feet in width. The estimator should allow one extra square foot for each linear foot of horizontal interior or exterior angle located below the ceiling level. Also, double the areas of small radius work.
- Drywall accessories, studs, track, and acoustical caulking are all measured by the linear foot. Drywall taping is figured by the square foot. Gypsum wallboard is estimated by the square foot. No material deductions should be made for door or window openings under 32 S.F.

09 60 00 Flooring

- Tile and terrazzo areas are taken off on a square foot basis. Trim and base materials are measured by the linear foot. Accent tiles are listed per each. Two basic methods of installation are used. Mud set is approximately 30% more expensive than thin set. In terrazzo work, be sure to include the linear footage of embedded decorative strips, grounds, machine rubbing, and power cleanup.
- Wood flooring is available in strip, parquet, or block configuration. The latter two types are set in adhesives with quantities estimated by the square foot. The laying pattern will influence labor costs and material waste. In addition to the material and labor for laying wood floors, the estimator must make allowances for sanding and finishing these areas, unless the flooring is prefinished.
- Sheet flooring is measured by the square yard. Roll widths vary, so consideration should be given to use the most economical width, as waste must be figured into the total quantity. Consider also the installation methods available, direct glue down or stretched.

09 70 00 Wall Finishes

- Wall coverings are estimated by the square foot. The area to be covered is measured, length by height of wall above baseboards, to calculate the square footage of each wall. This figure is divided by the number of square feet in the single roll which is being used. Deduct, in full, the areas of openings such as doors and windows. Where a pattern match is required allow 25%–30% waste.

09 80 00 Acoustic Treatment

- Acoustical systems fall into several categories. The takeoff of these materials should be by the square foot of area with a 5% allowance for waste. Do not forget about scaffolding, if applicable, when estimating these systems.

09 90 00 Painting and Coating

- A major portion of the work in painting involves surface preparation. Be sure to include cleaning, sanding, filling, and masking costs in the estimate.
- Protection of adjacent surfaces is not included in painting costs. When considering the method of paint application, an important factor is the amount of protection and masking required. These must be estimated separately and may be the determining factor in choosing the method of application.

Reference Numbers

Reference numbers are shown at the beginning of some major classifications. These numbers refer to related items in the Reference Section. The reference information may be an estimating procedure, an alternate pricing method, or technical information.

Note: Not all subdivisions listed here necessarily appear. ■

Did you know?

RSMeans Online gives you the same access to RSMeans' data with 24/7 access:

- Quickly locate costs in the searchable database.
- Build cost lists, estimates, and reports in minutes.
- Adjust costs to any location in the U.S. and Canada with the click of a button.

Start your free trial today at **www.RSMeansOnline.com**

RSMeans Online
FROM THE G◉RDIAN GROUP®

09 01 Maintenance of Finishes

09 01 70 – Maintenance of Wall Finishes

09 01 70.10 Gypsum Wallboard Repairs

	Crew	Daily Output	Labor-Hours	Unit	Material	2016 Bare Costs Labor	Equipment	Total	Total Incl O&P
0010	**GYPSUM WALLBOARD REPAIRS**								
0100	Fill and sand, pin/nail holes	1 Carp	960	.008	Ea.		.40		.40	.66
0110	Screw head pops		480	.017			.81		.81	1.32
0120	Dents, up to 2" square		48	.167		.01	8.10		8.11	13.20
0130	2" to 4" square		24	.333		.03	16.15		16.18	26.50
0140	Cut square, patch, sand and finish, holes, up to 2" square		12	.667		.03	32.50		32.53	53
0150	2" to 4" square		11	.727		.09	35		35.09	57.50
0160	4" to 8" square		10	.800		.24	39		39.24	64
0170	8" to 12" square		8	1		.48	48.50		48.98	79.50
0180	12" to 32" square		6	1.333		1.58	64.50		66.08	108
0210	16" by 48"		5	1.600		2.71	77.50		80.21	130
0220	32" by 48"		4	2		4.43	97		101.43	163
0230	48" square		3.50	2.286		6.25	111		117.25	188
0240	60" square		3.20	2.500		9.75	121		130.75	209
0500	Skim coat surface with joint compound		1600	.005	S.F.	.03	.24		.27	.44
0510	Prepare, retape and refinish joints		60	.133	L.F.	.64	6.45		7.09	11.25
9000	Minimum labor/equipment charge		2	4	Job		194		194	315

09 05 Common Work Results for Finishes

09 05 05 – Selective Demolition for Finishes

09 05 05.10 Selective Demolition, Ceilings

	Crew	Daily Output	Labor-Hours	Unit	Material	2016 Bare Costs Labor	Equipment	Total	Total Incl O&P
0010	**SELECTIVE DEMOLITION, CEILINGS** R024119-10								
0200	Ceiling, drywall, furred and nailed or screwed	2 Clab	800	.020	S.F.		.76		.76	1.24
0220	On metal frame		760	.021			.80		.80	1.30
0240	On suspension system, including system		720	.022			.84		.84	1.38
1000	Plaster, lime and horse hair, on wood lath, incl. lath		700	.023			.87		.87	1.42
1020	On metal lath		570	.028			1.06		1.06	1.74
1100	Gypsum, on gypsum lath		720	.022			.84		.84	1.38
1120	On metal lath		500	.032			1.21		1.21	1.98
1200	Suspended ceiling, mineral fiber, 2' x 2' or 2' x 4'		1500	.011			.40		.40	.66
1250	On suspension system, incl. system		1200	.013			.51		.51	.83
1500	Tile, wood fiber, 12" x 12", glued		900	.018			.67		.67	1.10
1540	Stapled		1500	.011			.40		.40	.66
1580	On suspension system, incl. system		760	.021			.80		.80	1.30
2000	Wood, tongue and groove, 1" x 4"		1000	.016			.61		.61	.99
2040	1" x 8"		1100	.015			.55		.55	.90
2400	Plywood or wood fiberboard, 4' x 8' sheets		1200	.013			.51		.51	.83
9000	Minimum labor/equipment charge	1 Clab	2	4	Job		152		152	248

09 05 05.20 Selective Demolition, Flooring

	Crew	Daily Output	Labor-Hours	Unit	Material	2016 Bare Costs Labor	Equipment	Total	Total Incl O&P
0010	**SELECTIVE DEMOLITION, FLOORING** R024119-10								
0200	Brick with mortar	2 Clab	475	.034	S.F.		1.28		1.28	2.09
0400	Carpet, bonded, including surface scraping		2000	.008			.30		.30	.50
0480	Tackless		9000	.002			.07		.07	.11
0550	Carpet tile, releasable adhesive		5000	.003			.12		.12	.20
0560	Permanent adhesive		1850	.009			.33		.33	.54
0600	Composition, acrylic or epoxy		400	.040			1.52		1.52	2.48
0700	Concrete, scarify skin	A-1A	225	.036			1.77	.96	2.73	3.93
0800	Resilient, sheet goods	2 Clab	1400	.011			.43		.43	.71
0820	For gym floors	"	900	.018			.67		.67	1.10
0850	Vinyl or rubber cove base	1 Clab	1000	.008	L.F.		.30		.30	.50
0860	Vinyl or rubber cove base, molded corner	"	1000	.008	Ea.		.30		.30	.50

09 05 Common Work Results for Finishes

09 05 05 – Selective Demolition for Finishes

09 05 05.20 Selective Demolition, Flooring

		Crew	Daily Output	Labor-Hours	Unit	Material	2016 Bare Costs Labor	Equipment	Total	Total Incl O&P
0870	For glued and caulked installation, add to labor						50%			
0900	Vinyl composition tile, 12" x 12"	2 Clab	1000	.016	S.F.		.61		.61	.99
2000	Tile, ceramic, thin set		675	.024			.90		.90	1.47
2020	Mud set		625	.026			.97		.97	1.59
2200	Marble, slate, thin set		675	.024			.90		.90	1.47
2220	Mud set		625	.026			.97		.97	1.59
2600	Terrazzo, thin set		450	.036			1.35		1.35	2.20
2620	Mud set		425	.038			1.43		1.43	2.33
2640	Terrazzo, cast in place	▼	300	.053			2.02		2.02	3.30
3000	Wood, block, on end	1 Carp	400	.020			.97		.97	1.58
3200	Parquet		450	.018			.86		.86	1.41
3400	Strip flooring, interior, 2-1/4" x 25/32" thick		325	.025			1.19		1.19	1.95
3500	Exterior, porch flooring, 1" x 4"		220	.036			1.76		1.76	2.88
3800	Subfloor, tongue and groove, 1" x 6"		325	.025			1.19		1.19	1.95
3820	1" x 8"		430	.019			.90		.90	1.47
3840	1" x 10"		520	.015			.75		.75	1.22
4000	Plywood, nailed		600	.013			.65		.65	1.06
4100	Glued and nailed		400	.020			.97		.97	1.58
4200	Hardboard, 1/4" thick	▼	760	.011			.51		.51	.83
8000	Remove flooring, bead blast, simple floor plan	A-1A	1000	.008			.40	.22	.62	.89
8100	Complex floor plan		400	.020			1	.54	1.54	2.21
8150	Mastic only	▼	1500	.005	▼		.27	.14	.41	.59
9000	Minimum labor/equipment charge	1 Clab	4	2	Job		76		76	124

09 05 05.30 Selective Demolition, Walls and Partitions

		Crew	Daily Output	Labor-Hours	Unit	Material	2016 Bare Costs Labor	Equipment	Total	Total Incl O&P
0010	**SELECTIVE DEMOLITION, WALLS AND PARTITIONS** R024119-10									
0020	Walls, concrete, reinforced	B-39	120	.400	C.F.		16.05	1.90	17.95	28
0025	Plain	"	160	.300			12.05	1.43	13.48	21
0100	Brick, 4" to 12" thick	B-9	220	.182	▼		6.95	1.04	7.99	12.55
0200	Concrete block, 4" thick		1150	.035	S.F.		1.33	.20	1.53	2.40
0280	8" thick		1050	.038			1.46	.22	1.68	2.63
0300	Exterior stucco 1" thick over mesh	▼	3200	.013			.48	.07	.55	.86
1000	Drywall, nailed or screwed	1 Clab	1000	.008			.30		.30	.50
1010	2 layers		400	.020			.76		.76	1.24
1020	Glued and nailed		900	.009			.34		.34	.55
1500	Fiberboard, nailed		900	.009			.34		.34	.55
1520	Glued and nailed		800	.010			.38		.38	.62
1568	Plenum barrier, sheet lead		300	.027			1.01		1.01	1.65
1600	Glass block		65	.123			4.66		4.66	7.60
2000	Movable walls, metal, 5' high		300	.027			1.01		1.01	1.65
2020	8' high	▼	400	.020			.76		.76	1.24
2200	Metal or wood studs, finish 2 sides, fiberboard	B-1	520	.046			1.78		1.78	2.91
2250	Lath and plaster		260	.092			3.56		3.56	5.80
2300	Plasterboard (drywall)		520	.046			1.78		1.78	2.91
2350	Plywood	▼	450	.053			2.06		2.06	3.36
2800	Paneling, 4' x 8' sheets	1 Clab	475	.017			.64		.64	1.04
3000	Plaster, lime and horsehair, on wood lath		400	.020			.76		.76	1.24
3020	On metal lath		335	.024			.91		.91	1.48
3400	Gypsum or perlite, on gypsum lath		410	.020			.74		.74	1.21
3420	On metal lath		300	.027	▼		1.01		1.01	1.65
3450	Plaster, interior gypsum, acoustic, or cement		60	.133	S.Y.		5.05		5.05	8.25
3500	Stucco, on masonry		145	.055			2.09		2.09	3.42
3510	Commercial 3-coat	▼	80	.100	▼		3.79		3.79	6.20

For customer support on your Commercial Renovation Cost Data, call 877.742.4719.

289

09 05 Common Work Results for Finishes

09 05 05 – Selective Demolition for Finishes

09 05 05.30 Selective Demolition, Walls and Partitions

	09 05 05.30 Selective Demolition, Walls and Partitions	Crew	Daily Output	Labor-Hours	Unit	Material	2016 Bare Costs Labor	Equipment	Total	Total Incl O&P
3520	Interior stucco	1 Clab	25	.320	S.Y.		12.15		12.15	19.80
3753	Remove damaged glass block	B-1	134.62	.178	S.F.		6.90		6.90	11.25
3760	Tile, ceramic, on walls, thin set	1 Clab	300	.027			1.01		1.01	1.65
3765	Mud set		250	.032	↓		1.21		1.21	1.98
3800	Toilet partitions, slate or marble		5	1.600	Ea.		60.50		60.50	99
3820	Metal or plastic	↓	8	1	"		38		38	62
5000	Wallcovering, vinyl	1 Pape	700	.011	S.F.		.46		.46	.75
5010	With release agent		1500	.005			.22		.22	.35
5025	Wallpaper, 2 layers or less, by hand		250	.032			1.30		1.30	2.09
5035	3 layers or more		165	.048			1.97		1.97	3.16
5040	Designer	↓	480	.017	↓		.68		.68	1.09
9000	Minimum labor/equipment charge	1 Clab	4	2	Job		76		76	124

09 05 71 – Acoustic Underlayment

09 05 71.10 Acoustical Underlayment

		Crew	Daily Output	Labor-Hours	Unit	Material	2016 Bare Costs Labor	Equipment	Total	Total Incl O&P
0010	**ACOUSTICAL UNDERLAYMENT**									
4000	Nylon matting 0.4" thick, with carbon black spinerette									
4010	plus polyester fabric, on floor	D-7	1600	.010	S.F.	1.36	.40		1.76	2.13
4200	Fiberglass reinf. backer board underlayment, 7/16" thick, on floor	"	1500	.011	"	2.77	.42		3.19	3.72

09 21 Plaster and Gypsum Board Assemblies

09 21 16 – Gypsum Board Assemblies

09 21 16.23 Gypsum Board Shaft Wall Assemblies

		Crew	Daily Output	Labor-Hours	Unit	Material	2016 Bare Costs Labor	Equipment	Total	Total Incl O&P
0010	**GYPSUM BOARD SHAFT WALL ASSEMBLIES**									
0020	Cavity type on 25 ga. J-track & C-H studs, 24" O.C.									
0030	1" thick coreboard wall liner on shaft side									
0040	2-hour assembly with double layer									
0060	5/8" f.r. gyp bd on rm side, 2-1/2" J-track & C-H studs	2 Carp	220	.073	S.F.	2.87	3.52		6.39	8.90
0065	4" J-track & C-H studs		220	.073		3.02	3.52		6.54	9.05
0070	6" J-track & C-H studs	↓	220	.073	↓	3.23	3.52		6.75	9.30
0100	3-hour assembly with triple layer									
0300	5/8" f.r. gyp bd on rm side, 2-1/2" J track & C-H studs	2 Carp	180	.089	S.F.	2.53	4.31		6.84	9.85
0305	4" J track & C-H studs		180	.089		2.68	4.31		6.99	10
0310	6" J track & C-H studs	↓	180	.089	↓	2.89	4.31		7.20	10.20
0400	4-hour assembly, 1" coreboard, 5/8" fire rated gypsum board									
0600	and 3/4" galv. metal furring channels, 24" O.C., with									
0700	Dbl. layer 5/8" f.r. gyp bd on rm side, 2-1/2" trk. & C-H studs	2 Carp	110	.145	S.F.	3.23	7.05		10.28	15.05
0705	4" J track & C-H studs		110	.145		3.02	7.05		10.07	14.80
0710	6" J track & C-H studs	↓	110	.145		3.23	7.05		10.28	15.05
0900	For taping & finishing, add per side	1 Carp	1050	.008	↓	.05	.37		.42	.65
1000	For insulation, see Section 07 21									
5200	For work over 8' high, add	2 Carp	3060	.005	S.F.		.25		.25	.41
5300	For distribution cost over 3 stories high, add per story	"	6100	.003	"		.13		.13	.21

09 21 16.33 Partition Wall

		Crew	Daily Output	Labor-Hours	Unit	Material	2016 Bare Costs Labor	Equipment	Total	Total Incl O&P
0010	**PARTITION WALL** Stud wall, 8' to 12' high									
0050	1/2", interior, gypsum board, std, tape & finish 2 sides									
0500	Installed on and incl., 2" x 4" wood studs, 16" O.C.	2 Carp	310	.052	S.F.	1.19	2.50		3.69	5.40
1000	Metal studs, NLB, 25 ga., 16" O.C., 3-5/8" wide		350	.046		1.12	2.21		3.33	4.86
1200	6" wide		330	.048		1.25	2.35		3.60	5.20
1400	Water resistant, on 2" x 4" wood studs, 16" O.C.		310	.052		1.35	2.50		3.85	5.55
1600	Metal studs, NLB, 25 ga., 16" O.C., 3-5/8" wide	↓	350	.046	↓	1.28	2.21		3.49	5.05

09 21 Plaster and Gypsum Board Assemblies

09 21 16 – Gypsum Board Assemblies

09 21 16.33 Partition Wall	Crew	Daily Output	Labor-Hours	Unit	Material	2016 Bare Costs Labor	Equipment	Total	Total Incl O&P
1800	6" wide	2 Carp	330	.048	S.F.	1.41	2.35		3.76	5.40
2000	Fire res., 2 layers, 1-1/2 hr., on 2" x 4" wood studs, 16" O.C.		210	.076		1.97	3.69		5.66	8.20
2200	Metal studs, NLB, 25 ga., 16" O.C., 3-5/8" wide		250	.064		1.90	3.10		5	7.15
2400	6" wide		230	.070		2.03	3.37		5.40	7.75
2600	Fire & water res., 2 layers, 1-1/2 hr., 2" x 4" studs, 16" O.C.		210	.076		1.97	3.69		5.66	8.20
2800	Metal studs, NLB, 25 ga., 16" O.C., 3-5/8" wide		250	.064		1.90	3.10		5	7.15
3000	6" wide		230	.070		2.03	3.37		5.40	7.75
3200	5/8", interior, gypsum board, standard, tape & finish 2 sides									
3400	Installed on and including 2" x 4" wood studs, 16" O.C.	2 Carp	300	.053	S.F.	1.21	2.58		3.79	5.55
3600	24" O.C.		330	.048		1.12	2.35		3.47	5.05
3800	Metal studs, NLB, 25 ga., 16" O.C., 3-5/8" wide		340	.047		1.14	2.28		3.42	4.98
4000	6" wide		320	.050		1.27	2.42		3.69	5.35
4200	24" O.C., 3-5/8" wide		360	.044		1.05	2.15		3.20	4.67
4400	6" wide		340	.047		1.14	2.28		3.42	4.97
4800	Water resistant, on 2" x 4" wood studs, 16" O.C.		300	.053		1.39	2.58		3.97	5.75
5000	24" O.C.		330	.048		1.30	2.35		3.65	5.25
5200	Metal studs, NLB, 25 ga. 16" O.C., 3-5/8" wide		340	.047		1.32	2.28		3.60	5.20
5400	6" wide		320	.050		1.45	2.42		3.87	5.55
5600	24" O.C., 3-5/8" wide		360	.044		1.23	2.15		3.38	4.87
5800	6" wide		340	.047		1.32	2.28		3.60	5.15
6000	Fire resistant, 2 layers, 2 hr., on 2" x 4" wood studs, 16" O.C.		205	.078		1.84	3.78		5.62	8.20
6200	24" O.C.		235	.068		1.84	3.30		5.14	7.40
6400	Metal studs, NLB, 25 ga., 16" O.C., 3-5/8" wide		245	.065		1.90	3.16		5.06	7.25
6600	6" wide		225	.071		1.99	3.45		5.44	7.85
6800	24" O.C., 3-5/8" wide		265	.060		1.77	2.93		4.70	6.70
7000	6" wide		245	.065		1.86	3.16		5.02	7.20
7200	Fire & water resistant, 2 layers, 2 hr., 2" x 4" studs, 16" O.C.		205	.078		1.93	3.78		5.71	8.35
7400	24" O.C.		235	.068		1.84	3.30		5.14	7.40
7600	Metal studs, NLB, 25 ga., 16" O.C., 3-5/8" wide		245	.065		1.86	3.16		5.02	7.20
7800	6" wide		225	.071		1.99	3.45		5.44	7.85
8000	24" O.C., 3-5/8" wide		265	.060		1.77	2.93		4.70	6.70
8200	6" wide		245	.065		1.86	3.16		5.02	7.20
8600	1/2" blueboard, mesh tape both sides									
8620	Installed on and including 2" x 4" wood studs, 16" O.C.	2 Carp	300	.053	S.F.	1.21	2.58		3.79	5.55
8640	Metal studs, NLB, 25 ga., 16" O.C., 3-5/8" wide		340	.047		1.14	2.28		3.42	4.98
8660	6" wide		320	.050		1.27	2.42		3.69	5.35
8800	Hospital security partition, 5/8" fiber reinf. high abuse gyp. bd.									
8810	Mtl. studs, NLB, 20 ga., 16" O.C., 3-5/8" wide, w/sec. mesh, gyp. bd.	2 Carp	208	.077	S.F.	4.20	3.73		7.93	10.70
9000	Exterior, 1/2" gypsum sheathing, 1/2" gypsum finished, interior,									
9100	including foil faced insulation, metal studs, 20 ga.									
9200	16" O.C., 3-5/8" wide	2 Carp	290	.055	S.F.	1.77	2.67		4.44	6.30
9400	6" wide		270	.059		1.95	2.87		4.82	6.85
9600	Partitions, for work over 8' high, add		1530	.010			.51		.51	.83

09 22 03 – Fastening Methods for Finishes

09 22 03.20 Drilling Plaster/Drywall	Crew	Daily Output	Labor-Hours	Unit	Material	2016 Bare Costs Labor	Equipment	Total	Total Incl O&P
0010 **DRILLING PLASTER/DRYWALL**									
1100 Drilling & layout for drywall/plaster walls, up to 1" deep, no anchor									
1200 Holes, 1/4" diameter	1 Carp	150	.053	Ea.	.01	2.58		2.59	4.23
1300 3/8" diameter		140	.057		.01	2.77		2.78	4.53
1400 1/2" diameter		130	.062		.01	2.98		2.99	4.88
1500 3/4" diameter		120	.067		.02	3.23		3.25	5.30
1600 1" diameter		110	.073		.02	3.52		3.54	5.75
1700 1-1/4" diameter		100	.080		.03	3.88		3.91	6.40
1800 1-1/2" diameter		90	.089		.05	4.31		4.36	7.10
1900 For ceiling installations, add						40%			

09 22 13 – Metal Furring

09 22 13.13 Metal Channel Furring

	Crew	Daily Output	Labor-Hours	Unit	Material	2016 Bare Costs Labor	Equipment	Total	Total Incl O&P
0010 **METAL CHANNEL FURRING**									
0030 Beams and columns, 7/8" channels, galvanized, 12" O.C.	1 Lath	155	.052	S.F.	.42	2.44		2.86	4.32
0050 16" O.C.		170	.047		.35	2.23		2.58	3.89
0070 24" O.C.		185	.043		.23	2.05		2.28	3.47
0100 Ceilings, on steel, 7/8" channels, galvanized, 12" O.C.		210	.038		.38	1.80		2.18	3.26
0300 16" O.C.		290	.028		.35	1.31		1.66	2.44
0400 24" O.C.		420	.019		.23	.90		1.13	1.67
0600 1-5/8" channels, galvanized, 12" O.C.		190	.042		.51	1.99		2.50	3.71
0700 16" O.C.		260	.031		.46	1.46		1.92	2.80
0900 24" O.C.		390	.021		.31	.97		1.28	1.87
0930 7/8" channels with sound isolation clips, 12" O.C.		120	.067		1.76	3.15		4.91	6.90
0940 16" O.C.		100	.080		1.32	3.78		5.10	7.40
0950 24" O.C.		165	.048		.88	2.29		3.17	4.58
0960 1-5/8" channels, galvanized, 12" O.C.		110	.073		1.89	3.44		5.33	7.50
0970 16" O.C.		100	.080		1.42	3.78		5.20	7.50
0980 24" O.C.		155	.052		.94	2.44		3.38	4.89
1000 Walls, 7/8" channels, galvanized, 12" O.C.		235	.034		.38	1.61		1.99	2.96
1200 16" O.C.		265	.030		.35	1.43		1.78	2.63
1300 24" O.C.		350	.023		.23	1.08		1.31	1.95
1500 1-5/8" channels, galvanized, 12" O.C.		210	.038		.51	1.80		2.31	3.41
1600 16" O.C.		240	.033		.46	1.58		2.04	2.99
1800 24" O.C.		305	.026		.31	1.24		1.55	2.30
1920 7/8" channels with sound isolation clips, 12" O.C.		125	.064		1.76	3.03		4.79	6.70
1940 16" O.C.		100	.080		1.32	3.78		5.10	7.40
1950 24" O.C.		150	.053		.88	2.52		3.40	4.95
1960 1-5/8" channels, galvanized, 12" O.C.		115	.070		1.89	3.29		5.18	7.30
1970 16" O.C.		95	.084		1.42	3.98		5.40	7.85
1980 24" O.C.		140	.057		.94	2.70		3.64	5.30
9000 Minimum labor/equipment charge		4	2	Job		94.50		94.50	149

09 22 16 – Non-Structural Metal Framing

09 22 16.13 Non-Structural Metal Stud Framing

	Crew	Daily Output	Labor-Hours	Unit	Material	2016 Bare Costs Labor	Equipment	Total	Total Incl O&P
0010 **NON-STRUCTURAL METAL STUD FRAMING**									
1600 Non-load bearing, galv., 8' high, 25 ga. 1-5/8" wide, 16" O.C.	1 Carp	619	.013	S.F.	.26	.63		.89	1.31
1610 24" O.C.		950	.008		.20	.41		.61	.89
1620 2-1/2" wide, 16" O.C.		613	.013		.33	.63		.96	1.39
1630 24" O.C.		938	.009		.24	.41		.65	.95
1640 3-5/8" wide, 16" O.C.		600	.013		.39	.65		1.04	1.49
1650 24" O.C.		925	.009		.29	.42		.71	1
1660 4" wide, 16" O.C.		594	.013		.44	.65		1.09	1.55
1670 24" O.C.		925	.009		.33	.42		.75	1.04

09 22 16.13 Non-Structural Metal Stud Framing	Crew	Daily Output	Labor-Hours	Unit	Material	2016 Bare Costs Labor	Equipment	Total	Total Incl O&P
1680	6" wide, 16" O.C.	1 Carp	588	.014	S.F.	.52	.66		1.18	1.65
1690	24" O.C.		906	.009		.39	.43		.82	1.13
1700	20 ga. studs, 1-5/8" wide, 16" O.C.		494	.016		.33	.78		1.11	1.65
1710	24" O.C.		763	.010		.25	.51		.76	1.10
1720	2-1/2" wide, 16" O.C.		488	.016		.42	.79		1.21	1.77
1730	24" O.C.		750	.011		.32	.52		.84	1.19
1740	3-5/8" wide, 16" O.C.		481	.017		.49	.81		1.30	1.85
1750	24" O.C.		738	.011		.36	.53		.89	1.26
1760	4" wide, 16" O.C.		475	.017		.58	.82		1.40	1.97
1770	24" O.C.		738	.011		.43	.53		.96	1.34
1780	6" wide, 16" O.C.		469	.017		.67	.83		1.50	2.09
1790	24" O.C.		725	.011		.51	.53		1.04	1.43
2000	Non-load bearing, galv., 10' high, 25 ga. 1-5/8" wide, 16" O.C.		495	.016		.25	.78		1.03	1.55
2100	24" O.C.		760	.011		.18	.51		.69	1.03
2200	2-1/2" wide, 16" O.C.		490	.016		.31	.79		1.10	1.63
2250	24" O.C.		750	.011		.23	.52		.75	1.09
2300	3-5/8" wide, 16" O.C.		480	.017		.37	.81		1.18	1.73
2350	24" O.C.		740	.011		.27	.52		.79	1.16
2400	4" wide, 16" O.C.		475	.017		.41	.82		1.23	1.78
2450	24" O.C.		740	.011		.30	.52		.82	1.19
2500	6" wide, 16" O.C.		470	.017		.49	.82		1.31	1.89
2550	24" O.C.		725	.011		.36	.53		.89	1.27
2600	20 ga. studs, 1-5/8" wide, 16" O.C.		395	.020		.32	.98		1.30	1.95
2650	24" O.C.		610	.013		.23	.64		.87	1.30
2700	2-1/2" wide, 16" O.C.		390	.021		.40	.99		1.39	2.06
2750	24" O.C.		600	.013		.30	.65		.95	1.39
2800	3-5/8" wide, 16" OC		385	.021		.46	1.01		1.47	2.15
2850	24" O.C.		590	.014		.34	.66		1	1.44
2900	4" wide, 16" O.C.		380	.021		.55	1.02		1.57	2.27
2950	24" O.C.		590	.014		.40	.66		1.06	1.51
3000	6" wide, 16" O.C.		375	.021		.64	1.03		1.67	2.39
3050	24" O.C.		580	.014		.47	.67		1.14	1.61
3060	Non-load bearing, galv., 12' high, 25 ga. 1-5/8" wide, 16" O.C.		413	.019		.24	.94		1.18	1.79
3070	24" O.C.		633	.013		.17	.61		.78	1.19
3080	2-1/2" wide, 16" O.C.		408	.020		.29	.95		1.24	1.87
3090	24" O.C.		625	.013		.22	.62		.84	1.25
3100	3-5/8" wide, 16" O.C.		400	.020		.35	.97		1.32	1.97
3110	24" O.C.		617	.013		.26	.63		.89	1.31
3120	4" wide, 16" O.C.		396	.020		.39	.98		1.37	2.03
3130	24" O.C.		617	.013		.29	.63		.92	1.35
3140	6" wide, 16" O.C.		392	.020		.47	.99		1.46	2.14
3150	24" O.C.		604	.013		.34	.64		.98	1.43
3160	20 ga. studs, 1-5/8" wide, 16" O.C.		329	.024		.30	1.18		1.48	2.25
3170	24" O.C.		508	.016		.22	.76		.98	1.49
3180	2-1/2" wide, 16" O.C.		325	.025		.38	1.19		1.57	2.37
3190	24" O.C.		500	.016		.28	.78		1.06	1.58
3200	3-5/8" wide, 16" O.C.		321	.025		.44	1.21		1.65	2.45
3210	24" O.C.		492	.016		.32	.79		1.11	1.64
3220	4" wide, 16" O.C.		317	.025		.52	1.22		1.74	2.58
3230	24" O.C.		492	.016		.38	.79		1.17	1.71
3240	6" wide, 16" O.C.		313	.026		.61	1.24		1.85	2.69
3250	24" O.C.		483	.017		.44	.80		1.24	1.80
5000	For load bearing studs, see Section 05 41 13.30									

09 22 Supports for Plaster and Gypsum Board

09 22 16 – Non-Structural Metal Framing

09 22 16.13 Non-Structural Metal Stud Framing	Crew	Daily Output	Labor-Hours	Unit	Material	2016 Bare Costs Labor	Equipment	Total	Total Incl O&P
9000	Minimum labor/equipment charge	1 Carp	4	2	Job		97		97	158

09 22 26 – Suspension Systems

09 22 26.13 Ceiling Suspension Systems

		Crew	Daily Output	Labor-Hours	Unit	Material	Labor	Equipment	Total	Total Incl O&P
0010	**CEILING SUSPENSION SYSTEMS** for gypsum board or plaster									
8000	Suspended ceilings, including carriers									
8200	1-1/2" carriers, 24" O.C. with:									
8300	7/8" channels, 16" O.C.	1 Lath	275	.029	S.F.	.57	1.38		1.95	2.79
8320	24" O.C.		310	.026		.45	1.22		1.67	2.41
8400	1-5/8" channels, 16" O.C.		205	.039		.68	1.85		2.53	3.66
8420	24" O.C.		250	.032		.53	1.51		2.04	2.97
8600	2" carriers, 24" O.C. with:									
8700	7/8" channels, 16" O.C.	1 Lath	250	.032	S.F.	.61	1.51		2.12	3.06
8720	24" O.C.		285	.028		.49	1.33		1.82	2.63
8800	1-5/8" channels, 16" O.C.		190	.042		.72	1.99		2.71	3.93
8820	24" O.C.		225	.036		.57	1.68		2.25	3.28

09 22 36 – Lath

09 22 36.13 Gypsum Lath

		Crew	Daily Output	Labor-Hours	Unit	Material	Labor	Equipment	Total	Total Incl O&P
0010	**GYPSUM LATH** R092000-50									
0020	Plain or perforated, nailed, 3/8" thick	1 Lath	85	.094	S.Y.	3.06	4.45		7.51	10.35
0100	1/2" thick		80	.100		2.43	4.73		7.16	10.10
0300	Clipped to steel studs, 3/8" thick		75	.107		3.06	5.05		8.11	11.30
0400	1/2" thick		70	.114		2.43	5.40		7.83	11.15
1500	For ceiling installations, add		216	.037			1.75		1.75	2.76
1600	For columns and beams, add		170	.047			2.23		2.23	3.51
9000	Minimum labor/equipment charge		4.25	1.882	Job		89		89	140

09 22 36.23 Metal Lath

		Crew	Daily Output	Labor-Hours	Unit	Material	Labor	Equipment	Total	Total Incl O&P
0010	**METAL LATH** R092000-50									
0020	Diamond, expanded, 2.5 lb. per S.Y., painted				S.Y.	4.09			4.09	4.50
0100	Galvanized					2.82			2.82	3.10
0300	3.4 lb. per S.Y., painted					4.19			4.19	4.61
0400	Galvanized					4.18			4.18	4.60
0600	For 15# asphalt sheathing paper, add					.48			.48	.52
0900	Flat rib, 1/8" high, 2.75 lb., painted					3.48			3.48	3.83
1000	Foil backed					3.67			3.67	4.04
1200	3.4 lb. per S.Y., painted					4.32			4.32	4.75
1300	Galvanized					4.45			4.45	4.90
1500	For 15# asphalt sheathing paper, add					.48			.48	.52
1800	High rib, 3/8" high, 3.4 lb. per S.Y., painted					4.52			4.52	4.97
1900	Galvanized					3.84			3.84	4.22
2400	3/4" high, painted, .60 lb. per S.F.				S.F.	.65			.65	.72
2500	.75 lb. per S.F.				"	1.40			1.40	1.54
2800	Stucco mesh, painted, 3.6 lb.				S.Y.	3.85			3.85	4.24
3000	K-lath, perforated, absorbent paper, regular					4.42			4.42	4.86
3100	Heavy duty					5.20			5.20	5.75
3300	Waterproof, heavy duty, grade B backing					5.10			5.10	5.60
3400	Fire resistant backing					5.65			5.65	6.20
3600	2.5 lb. diamond painted, on wood framing, on walls	1 Lath	85	.094		4.09	4.45		8.54	11.50
3700	On ceilings		75	.107		4.09	5.05		9.14	12.45
3900	3.4 lb. diamond painted, on wood framing, on walls		80	.100		4.32	4.73		9.05	12.20
4000	On ceilings		70	.114		4.32	5.40		9.72	13.25
4200	3.4 lb. diamond painted, wired to steel framing		75	.107		4.32	5.05		9.37	12.70

09 22 Supports for Plaster and Gypsum Board

09 22 36 – Lath

09 22 36.23 Metal Lath

	Crew	Daily Output	Labor-Hours	Unit	Material	2016 Bare Costs Labor	2016 Bare Costs Equipment	Total	Total Incl O&P
4300	On ceilings	1 Lath	60	.133	S.Y.	4.32	6.30		10.62	14.70
4600	Cornices, wired to steel		35	.229		4.32	10.80		15.12	22
4800	Screwed to steel studs, 2.5 lb.		80	.100		4.09	4.73		8.82	11.95
4900	3.4 lb.		75	.107		4.19	5.05		9.24	12.55
5100	Rib lath, painted, wired to steel, on walls, 2.5 lb.		75	.107		3.48	5.05		8.53	11.80
5200	3.4 lb.		70	.114		4.52	5.40		9.92	13.45
5400	4.0 lb.		65	.123		5.65	5.80		11.45	15.40
5500	For self-furring lath, add					.11			.11	.12
5700	Suspended ceiling system, incl. 3.4 lb. diamond lath, painted	1 Lath	15	.533		4.31	25		29.31	44.50
5800	Galvanized	"	15	.533		4.32	25		29.32	45
6000	Hollow metal stud partitions, 3.4 lb. painted lath both sides									
6010	Non-load bearing, 25 ga., w/rib lath, 2-1/2" studs, 12" O.C.	1 Lath	20.30	.394	S.Y.	12.55	18.65		31.20	43.50
6300	16" O.C.		21.10	.379		11.80	17.95		29.75	41.50
6350	24" O.C.		22.70	.352		11.10	16.65		27.75	38.50
6400	3-5/8" studs, 16" O.C.		19.50	.410		12.35	19.40		31.75	44
6600	24" O.C.		20.40	.392		11.50	18.55		30.05	41.50
6700	4" studs, 16" O.C.		20.40	.392		12.75	18.55		31.30	43
6900	24" O.C.		21.60	.370		11.75	17.50		29.25	40.50
7000	6" studs, 16" O.C.		19.50	.410		13.45	19.40		32.85	45.50
7100	24" O.C.		21.10	.379		12.30	17.95		30.25	42
7200	L.B. partitions, 16 ga., w/rib lath, 2-1/2" studs, 16" O.C.		20	.400		13.25	18.90		32.15	44.50
7300	3-5/8" studs, 16 ga.		19.70	.406		15	19.20		34.20	47
7500	4" studs, 16 ga.		19.50	.410		15.55	19.40		34.95	47.50
7600	6" studs, 16 ga.		18.70	.428		18.55	20		38.55	52.50
9000	Minimum labor/equipment charge		4.25	1.882	Job		89		89	140

09 23 Gypsum Plastering

09 23 13 – Acoustical Gypsum Plastering

09 23 13.10 Perlite or Vermiculite Plaster

		Crew	Daily Output	Labor-Hours	Unit	Material	2016 Bare Costs Labor	2016 Bare Costs Equipment	Total	Total Incl O&P
0010	**PERLITE OR VERMICULITE PLASTER** R092000-50									
0020	In 100 lb. bags, under 200 bags				Bag	17.55			17.55	19.35
0300	2 coats, no lath included, on walls	J-1	92	.435	S.Y.	5.85	18.35	1.53	25.73	37
0400	On ceilings		79	.506		5.85	21.50	1.78	29.13	42.50
0900	3 coats, no lath included, on walls		74	.541		6.40	23	1.90	31.30	45.50
1000	On ceilings		63	.635		6.40	27	2.23	35.63	52
1700	For irregular or curved surfaces, add to above						30%			
1800	For columns and beams, add to above						50%			
1900	For soffits, add to ceiling prices						40%			
9000	Minimum labor/equipment charge	1 Plas	1	8	Job		360		360	575

09 23 20 – Gypsum Plaster

09 23 20.10 Gypsum Plaster On Walls and Ceilings

		Crew	Daily Output	Labor-Hours	Unit	Material	2016 Bare Costs Labor	2016 Bare Costs Equipment	Total	Total Incl O&P
0010	**GYPSUM PLASTER ON WALLS AND CEILINGS** R092000-50									
0020	80# bag, less than 1 ton				Bag	15.95			15.95	17.55
0300	2 coats, no lath included, on walls	J-1	105	.381	S.Y.	3.70	16.05	1.34	21.09	31
0400	On ceilings		92	.435		3.70	18.35	1.53	23.58	35
0900	3 coats, no lath included, on walls		87	.460		5.35	19.40	1.61	26.36	38.50
1000	On ceilings		78	.513		5.35	21.50	1.80	28.65	42.50
1600	For irregular or curved surfaces, add						30%			
1800	For columns & beams, add						50%			
9000	Minimum labor/equipment charge	1 Plas	1	8	Job		360		360	575

295

For customer support on your Commercial Renovation Cost Data, call 877.742.4719.

09 24 Cement Plastering

09 24 23 – Cement Stucco

09 24 23.40 Stucco		Crew	Daily Output	Labor-Hours	Unit	Material	2016 Bare Costs Labor	Equipment	Total	Total Incl O&P
0010	**STUCCO** R092000-50									
0015	3 coats 1" thick, float finish, with mesh, on wood frame	J-2	63	.762	S.Y.	6.40	33	2.22	41.62	61.50
0100	On masonry construction, no mesh incl.	J-1	67	.597		2.55	25	2.10	29.65	45
0300	For trowel finish, add	1 Plas	170	.047			2.11		2.11	3.37
0400	For 3/4" thick, on masonry, deduct	J-1	880	.045		.63	1.92	.16	2.71	3.93
0600	For coloring, add		685	.058		.40	2.46	.20	3.06	4.60
0700	For special texture, add		200	.200		1.41	8.45	.70	10.56	15.75
0900	For soffits, add	J-2	155	.310		2.18	13.30	.90	16.38	24.50
1000	Exterior stucco, with bonding agent, 3 coats, on walls, no mesh incl.	J-1	200	.200		3.65	8.45	.70	12.80	18.25
1200	Ceilings		180	.222		3.65	9.35	.78	13.78	19.85
1300	Beams		80	.500		3.65	21	1.76	26.41	39.50
1500	Columns		100	.400		3.65	16.85	1.40	21.90	32.50
1550	Minimum labor/equipment charge	1 Plas	1	8	Job		360		360	575
1600	Mesh, painted, nailed to wood, 1.8 lb.	1 Lath	60	.133	S.Y.	6.50	6.30		12.80	17.10
1800	3.6 lb.		55	.145		3.85	6.90		10.75	15.10
1900	Wired to steel, painted, 1.8 lb.		53	.151		6.50	7.15		13.65	18.40
2100	3.6 lb.		50	.160		3.85	7.55		11.40	16.20
9000	Minimum labor/equipment charge		4	2	Job		94.50		94.50	149

09 25 Other Plastering

09 25 23 – Lime Based Plastering

09 25 23.10 Venetian Plaster

09 25 23.10 Venetian Plaster		Crew	Daily Output	Labor-Hours	Unit	Material	2016 Bare Costs Labor	Equipment	Total	Total Incl O&P
0010	**VENETIAN PLASTER**									
0100	Walls, 1 coat primer, roller applied	1 Plas	950	.008	S.F.	.16	.38		.54	.78
0200	Plaster, 3 coats, incl. sanding	2 Plas	700	.023		.46	1.03		1.49	2.14
0210	For pigment, light colors add per S.F. plaster					.02			.02	.02
0220	For pigment, dark colors add per S.F. plaster					.05			.05	.06
0300	For sealer/wax coat incl. burnishing, add	1 Plas	300	.027		.41	1.20		1.61	2.36

09 26 Veneer Plastering

09 26 13 – Gypsum Veneer Plastering

09 26 13.20 Blueboard

09 26 13.20 Blueboard		Crew	Daily Output	Labor-Hours	Unit	Material	2016 Bare Costs Labor	Equipment	Total	Total Incl O&P
0010	**BLUEBOARD** For use with thin coat									
0100	plaster application see Section 09 26 13.80									
1000	3/8" thick, on walls or ceilings, standard, no finish included	2 Carp	1900	.008	S.F.	.35	.41		.76	1.06
1100	With thin coat plaster finish		875	.018		.46	.89		1.35	1.96
1400	On beams, columns, or soffits, standard, no finish included		675	.024		.40	1.15		1.55	2.32
1450	With thin coat plaster finish		475	.034		.51	1.63		2.14	3.24
3000	1/2" thick, on walls or ceilings, standard, no finish included		1900	.008		.34	.41		.75	1.04
3100	With thin coat plaster finish		875	.018		.45	.89		1.34	1.95
3300	Fire resistant, no finish included		1900	.008		.34	.41		.75	1.04
3400	With thin coat plaster finish		875	.018		.45	.89		1.34	1.95
3450	On beams, columns, or soffits, standard, no finish included		675	.024		.39	1.15		1.54	2.31
3500	With thin coat plaster finish		475	.034		.50	1.63		2.13	3.22
3700	Fire resistant, no finish included		675	.024		.39	1.15		1.54	2.31
3800	With thin coat plaster finish		475	.034		.50	1.63		2.13	3.22
5000	5/8" thick, on walls or ceilings, fire resistant, no finish included		1900	.008		.34	.41		.75	1.04
5100	With thin coat plaster finish		875	.018		.45	.89		1.34	1.95
5500	On beams, columns, or soffits, no finish included		675	.024		.39	1.15		1.54	2.31

09 26 Veneer Plastering

09 26 13 – Gypsum Veneer Plastering

09 26 13.20 Blueboard		Crew	Daily Output	Labor-Hours	Unit	Material	2016 Bare Costs Labor	Equipment	Total	Total Incl O&P
5600	With thin coat plaster finish	2 Carp	475	.034	S.F.	.50	1.63		2.13	3.22
6000	For high ceilings, over 8' high, add		3060	.005			.25		.25	.41
6500	For over 3 stories high, add per story		6100	.003			.13		.13	.21
9000	Minimum labor/equipment charge	1 Carp	2	4	Job		194		194	315

09 26 13.80 Thin Coat Plaster										
0010	THIN COAT PLASTER	R092000-50								
0012	1 coat veneer, not incl. lath	J-1	3600	.011	S.F.	.11	.47	.04	.62	.91
1000	In 50 lb. bags				Bag	15.05			15.05	16.55

09 28 Backing Boards and Underlayments

09 28 13 – Cementitious Backing Boards

09 28 13.10 Cementitious Backerboard

		Crew	Daily Output	Labor-Hours	Unit	Material	2016 Bare Costs Labor	Equipment	Total	Total Incl O&P
0010	CEMENTITIOUS BACKERBOARD									
0070	Cementitious backerboard, on floor, 3' x 4' x 1/2" sheets	2 Carp	525	.030	S.F.	.84	1.48		2.32	3.33
0080	3' x 5' x 1/2" sheets		525	.030		.81	1.48		2.29	3.30
0090	3' x 6' x 1/2" sheets		525	.030		.79	1.48		2.27	3.27
0100	3' x 4' x 5/8" sheets		525	.030		1.01	1.48		2.49	3.52
0110	3' x 5' x 5/8" sheets		525	.030		1.01	1.48		2.49	3.52
0120	3' x 6' x 5/8" sheets		525	.030		1	1.48		2.48	3.50
0150	On wall, 3' x 4' x 1/2" sheets		350	.046		.84	2.21		3.05	4.54
0160	3' x 5' x 1/2" sheets		350	.046		.81	2.21		3.02	4.51
0170	3' x 6' x 1/2" sheets		350	.046		.79	2.21		3	4.48
0180	3' x 4' x 5/8" sheets		350	.046		1.01	2.21		3.22	4.73
0190	3' x 5' x 5/8" sheets		350	.046		1.01	2.21		3.22	4.73
0200	3' x 6' x 5/8" sheets		350	.046		1	2.21		3.21	4.71
0250	On counter, 3' x 4' x 1/2" sheets		180	.089		.84	4.31		5.15	7.95
0260	3' x 5' x 1/2" sheets		180	.089		.81	4.31		5.12	7.95
0270	3' x 6' x 1/2" sheets		180	.089		.79	4.31		5.10	7.90
0300	3' x 4' x 5/8" sheets		180	.089		1.01	4.31		5.32	8.15
0310	3' x 5' x 5/8" sheets		180	.089		1.01	4.31		5.32	8.15
0320	3' x 6' x 5/8" sheets		180	.089		1	4.31		5.31	8.15

09 29 Gypsum Board

09 29 10 – Gypsum Board Panels

09 29 10.30 Gypsum Board

		Crew	Daily Output	Labor-Hours	Unit	Material	2016 Bare Costs Labor	Equipment	Total	Total Incl O&P
0010	GYPSUM BOARD on walls & ceilings	R092910-10								
0100	Nailed or screwed to studs unless otherwise noted									
0110	1/4" thick, on walls or ceilings, standard, no finish included	2 Carp	1330	.012	S.F.	.36	.58		.94	1.35
0115	1/4" thick, on walls or ceilings, flexible, no finish included		1050	.015		.51	.74		1.25	1.77
0117	1/4" thick, on columns or soffits, flexible, no finish included		1050	.015		.51	.74		1.25	1.77
0130	1/4" thick, standard, no finish included, less than 800 S.F.		510	.031		.36	1.52		1.88	2.88
0150	3/8" thick, on walls, standard, no finish included		2000	.008		.35	.39		.74	1.02
0200	On ceilings, standard, no finish included		1800	.009		.35	.43		.78	1.09
0250	On beams, columns, or soffits, no finish included		675	.024		.35	1.15		1.50	2.27
0300	1/2" thick, on walls, standard, no finish included		2000	.008		.33	.39		.72	.99
0350	Taped and finished (level 4 finish)		965	.017		.38	.80		1.18	1.73
0390	With compound skim coat (level 5 finish)		775	.021		.43	1		1.43	2.10
0400	Fire resistant, no finish included		2000	.008		.36	.39		.75	1.03
0450	Taped and finished (level 4 finish)		965	.017		.41	.80		1.21	1.76

297

For customer support on your Commercial Renovation Cost Data, call 877.742.4719.

09 29 10 – Gypsum Board Panels

09 29 10.30 Gypsum Board	Crew	Daily Output	Labor-Hours	Unit	Material	2016 Bare Costs Labor	Equipment	Total	Total Incl O&P
0490	With compound skim coat (level 5 finish)	2 Carp	775	.021	S.F.	.46	1		1.46	2.14
0500	Water resistant, no finish included		2000	.008		.41	.39		.80	1.08
0550	Taped and finished (level 4 finish)		965	.017		.46	.80		1.26	1.81
0590	With compound skim coat (level 5 finish)		775	.021		.51	1		1.51	2.19
0600	Prefinished, vinyl, clipped to studs		900	.018		.49	.86		1.35	1.95
0700	Mold resistant, no finish included		2000	.008		.43	.39		.82	1.10
0710	Taped and finished (level 4 finish)		965	.017		.48	.80		1.28	1.84
0720	With compound skim coat (level 5 finish)		775	.021		.53	1		1.53	2.21
1000	On ceilings, standard, no finish included		1800	.009		.33	.43		.76	1.06
1050	Taped and finished (level 4 finish)		765	.021		.38	1.01		1.39	2.08
1090	With compound skim coat (level 5 finish)		610	.026		.43	1.27		1.70	2.55
1100	Fire resistant, no finish included		1800	.009		.36	.43		.79	1.10
1150	Taped and finished (level 4 finish)		765	.021		.41	1.01		1.42	2.11
1195	With compound skim coat (level 5 finish)		610	.026		.46	1.27		1.73	2.59
1200	Water resistant, no finish included		1800	.009		.41	.43		.84	1.15
1250	Taped and finished (level 4 finish)		765	.021		.46	1.01		1.47	2.16
1290	With compound skim coat (level 5 finish)		610	.026		.51	1.27		1.78	2.64
1310	Mold resistant, no finish included		1800	.009		.43	.43		.86	1.17
1320	Taped and finished (level 4 finish)		765	.021		.48	1.01		1.49	2.19
1330	With compound skim coat (level 5 finish)		610	.026		.53	1.27		1.80	2.66
1350	Sag resistant, no finish included		1600	.010		.35	.48		.83	1.18
1360	Taped and finished (level 4 finish)		765	.021		.40	1.01		1.41	2.10
1370	With compound skim coat (level 5 finish)		610	.026		.45	1.27		1.72	2.58
1500	On beams, columns, or soffits, standard, no finish included		675	.024		.38	1.15		1.53	2.30
1550	Taped and finished (level 4 finish)		540	.030		.38	1.44		1.82	2.77
1590	With compound skim coat (level 5 finish)		475	.034		.43	1.63		2.06	3.14
1600	Fire resistant, no finish included		675	.024		.36	1.15		1.51	2.28
1650	Taped and finished (level 4 finish)		540	.030		.41	1.44		1.85	2.80
1690	With compound skim coat (level 5 finish)		475	.034		.46	1.63		2.09	3.18
1700	Water resistant, no finish included		675	.024		.47	1.15		1.62	2.40
1750	Taped and finished (level 4 finish)		540	.030		.46	1.44		1.90	2.85
1790	With compound skim coat (level 5 finish)		475	.034		.51	1.63		2.14	3.23
1800	Mold resistant, no finish included		675	.024		.49	1.15		1.64	2.42
1810	Taped and finished (level 4 finish)		540	.030		.48	1.44		1.92	2.88
1820	With compound skim coat (level 5 finish)		475	.034		.53	1.63		2.16	3.25
1850	Sag resistant, no finish included		675	.024		.40	1.15		1.55	2.32
1860	Taped and finished (level 4 finish)		540	.030		.40	1.44		1.84	2.79
1870	With compound skim coat (level 5 finish)		475	.034		.45	1.63		2.08	3.17
2000	5/8" thick, on walls, standard, no finish included		2000	.008		.34	.39		.73	1
2050	Taped and finished (level 4 finish)		965	.017		.39	.80		1.19	1.74
2090	With compound skim coat (level 5 finish)		775	.021		.44	1		1.44	2.11
2100	Fire resistant, no finish included		2000	.008		.35	.39		.74	1.02
2150	Taped and finished (level 4 finish)		965	.017		.40	.80		1.20	1.75
2195	With compound skim coat (level 5 finish)		775	.021		.45	1		1.45	2.13
2200	Water resistant, no finish included		2000	.008		.43	.39		.82	1.10
2250	Taped and finished (level 4 finish)		965	.017		.48	.80		1.28	1.84
2290	With compound skim coat (level 5 finish)		775	.021		.53	1		1.53	2.21
2300	Prefinished, vinyl, clipped to studs		900	.018		.76	.86		1.62	2.25
2510	Mold resistant, no finish included		2000	.008		.47	.39		.86	1.15
2520	Taped and finished (level 4 finish)		965	.017		.52	.80		1.32	1.88
2530	With compound skim coat (level 5 finish)		775	.021		.57	1		1.57	2.26
3000	On ceilings, standard, no finish included		1800	.009		.34	.43		.77	1.07
3050	Taped and finished (level 4 finish)		765	.021		.39	1.01		1.40	2.09

09 29 10.30 Gypsum Board		Crew	Daily Output	Labor-Hours	Unit	Material	2016 Bare Costs Labor	Equipment	Total	Total Incl O&P
3090	With compound skim coat (level 5 finish)	2 Carp	615	.026	S.F.	.44	1.26		1.70	2.54
3100	Fire resistant, no finish included		1800	.009		.35	.43		.78	1.09
3150	Taped and finished (level 4 finish)		765	.021		.40	1.01		1.41	2.10
3190	With compound skim coat (level 5 finish)		615	.026		.45	1.26		1.71	2.56
3200	Water resistant, no finish included		1800	.009		.43	.43		.86	1.17
3250	Taped and finished (level 4 finish)		765	.021		.48	1.01		1.49	2.19
3290	With compound skim coat (level 5 finish)		615	.026		.53	1.26		1.79	2.64
3300	Mold resistant, no finish included		1800	.009		.47	.43		.90	1.22
3310	Taped and finished (level 4 finish)		765	.021		.52	1.01		1.53	2.23
3320	With compound skim coat (level 5 finish)		615	.026		.57	1.26		1.83	2.69
3500	On beams, columns, or soffits, no finish included		675	.024		.39	1.15		1.54	2.31
3550	Taped and finished (level 4 finish)		475	.034		.45	1.63		2.08	3.16
3590	With compound skim coat (level 5 finish)		380	.042		.51	2.04		2.55	3.89
3600	Fire resistant, no finish included		675	.024		.40	1.15		1.55	2.32
3650	Taped and finished (level 4 finish)		475	.034		.46	1.63		2.09	3.17
3690	With compound skim coat (level 5 finish)		380	.042		.45	2.04		2.49	3.83
3700	Water resistant, no finish included		675	.024		.49	1.15		1.64	2.42
3750	Taped and finished (level 4 finish)		475	.034		.53	1.63		2.16	3.25
3790	With compound skim coat (level 5 finish)		380	.042		.55	2.04		2.59	3.93
3800	Mold resistant, no finish included		675	.024		.54	1.15		1.69	2.47
3810	Taped and finished (level 4 finish)		475	.034		.57	1.63		2.20	3.30
3820	With compound skim coat (level 5 finish)		380	.042		.59	2.04		2.63	3.98
4000	Fireproofing, beams or columns, 2 layers, 1/2" thick, incl finish		330	.048		.81	2.35		3.16	4.74
4010	Mold resistant		330	.048		.95	2.35		3.30	4.89
4050	5/8" thick		300	.053		.79	2.58		3.37	5.10
4060	Mold resistant		300	.053		1.03	2.58		3.61	5.35
4100	3 layers, 1/2" thick		225	.071		1.22	3.45		4.67	7
4110	Mold resistant		225	.071		1.43	3.45		4.88	7.25
4150	5/8" thick		210	.076		1.19	3.69		4.88	7.35
4160	Mold resistant		210	.076		1.55	3.69		5.24	7.75
5050	For 1" thick coreboard on columns	▼	480	.033		.75	1.61		2.36	3.47
5100	For foil-backed board, add					.15			.15	.17
5200	For work over 8' high, add	2 Carp	3060	.005			.25		.25	.41
5270	For textured spray, add	2 Lath	1600	.010		.04	.47		.51	.79
5300	For distribution cost over 3 stories high, add per story	2 Carp	6100	.003	▼		.13		.13	.21
5350	For finishing inner corners, add		950	.017	L.F.	.10	.82		.92	1.44
5355	For finishing outer corners, add	▼	1250	.013		.23	.62		.85	1.27
5500	For acoustical sealant, add per bead	1 Carp	500	.016	▼	.04	.78		.82	1.32
5550	Sealant, 1 quart tube				Ea.	7.05			7.05	7.80
6000	Gypsum sound dampening panels									
6010	1/2" thick on walls, multi-layer, light weight, no finish included	2 Carp	1500	.011	S.F.	2.02	.52		2.54	3.06
6015	Taped and finished (level 4 finish)		725	.022		2.07	1.07		3.14	4.02
6020	With compound skim coat (level 5 finish)		580	.028		2.12	1.34		3.46	4.51
6025	5/8" thick on walls, for wood studs, no finish included		1500	.011		2.23	.52		2.75	3.29
6030	Taped and finished (level 4 finish)		725	.022		2.28	1.07		3.35	4.26
6035	With compound skim coat (level 5 finish)		580	.028		2.33	1.34		3.67	4.74
6040	For metal stud, no finish included		1500	.011		2.18	.52		2.70	3.24
6045	Taped and finished (level 4 finish)		725	.022		2.23	1.07		3.30	4.20
6050	With compound skim coat (level 5 finish)		580	.028		2.28	1.34		3.62	4.69
6055	Abuse resist, no finish included		1500	.011		3.95	.52		4.47	5.20
6060	Taped and finished (level 4 finish)		725	.022		4	1.07		5.07	6.15
6065	With compound skim coat (level 5 finish)		580	.028		4.05	1.34		5.39	6.65
6070	Shear rated, no finish included	▼	1500	.011	▼	4.53	.52		5.05	5.80

09 29 10.30 Gypsum Board	Crew	Daily Output	Labor-Hours	Unit	Material	2016 Bare Costs Labor	Equipment	Total	Total Incl O&P
6075	Taped and finished (level 4 finish)	2 Carp	725	.022	S.F.	4.58	1.07		5.65	6.80
6080	With compound skim coat (level 5 finish)		580	.028		4.63	1.34		5.97	7.30
6085	For SCIF applications, no finish included		1500	.011		4.96	.52		5.48	6.30
6090	Taped and finished (level 4 finish)		725	.022		5	1.07		6.07	7.25
6095	With compound skim coat (level 5 finish)		580	.028		5.05	1.34		6.39	7.75
6100	1-3/8" thick on walls, THX Certified, no finish included		1500	.011		8.40	.52		8.92	10.10
6105	Taped and finished (level 4 finish)		725	.022		8.45	1.07		9.52	11.05
6110	With compound skim coat (level 5 finish)		580	.028		8.50	1.34		9.84	11.55
6115	5/8" thick on walls, score & snap installation, no finish included		2000	.008		1.79	.39		2.18	2.60
6120	Taped and finished (level 4 finish)		965	.017		1.84	.80		2.64	3.33
6125	With compound skim coat (level 5 finish)		775	.021		1.89	1		2.89	3.71
7020	5/8" thick on ceilings, for wood joists, no finish included		1200	.013		2.23	.65		2.88	3.51
7025	Taped and finished (level 4 finish)		510	.031		2.28	1.52		3.80	4.99
7030	With compound skim coat (level 5 finish)		410	.039		2.33	1.89		4.22	5.65
7035	For metal joists, no finish included		1200	.013		2.18	.65		2.83	3.46
7040	Taped and finished (level 4 finish)		510	.031		2.23	1.52		3.75	4.93
7045	With compound skim coat (level 5 finish)		410	.039		2.28	1.89		4.17	5.60
7050	Abuse resist, no finish included		1200	.013		3.95	.65		4.60	5.40
7055	Taped and finished (level 4 finish)		510	.031		4	1.52		5.52	6.90
7060	With compound skim coat (level 5 finish)		410	.039		4.05	1.89		5.94	7.55
7065	Shear rated, no finish included		1200	.013		4.53	.65		5.18	6.05
7070	Taped and finished (level 4 finish)		510	.031		4.58	1.52		6.10	7.55
7075	With compound skim coat (level 5 finish)		410	.039		4.63	1.89		6.52	8.20
7080	For SCIF applications, no finish included		1200	.013		4.96	.65		5.61	6.50
7085	Taped and finished (level 4 finish)		510	.031		5	1.52		6.52	8
7090	With compound skim coat (level 5 finish)		410	.039		5.05	1.89		6.94	8.65
8010	5/8" thick on ceilings, score & snap installation, no finish included		1600	.010		1.79	.48		2.27	2.76
8015	Taped and finished (level 4 finish)		680	.024		1.84	1.14		2.98	3.88
8020	With compound skim coat (level 5 finish)		545	.029		1.89	1.42		3.31	4.40
9000	Minimum labor/equipment charge	1 Carp	2	4	Job		194		194	315

09 29 10.50 High Abuse Gypsum Board

		Crew	Daily Output	Labor-Hours	Unit	Material	2016 Bare Costs Labor	Equipment	Total	Total Incl O&P
0010	**HIGH ABUSE GYPSUM BOARD**, fiber reinforced, nailed or									
0100	screwed to studs unless otherwise noted									
0110	1/2" thick, on walls, no finish included	2 Carp	1800	.009	S.F.	.72	.43		1.15	1.49
0120	Taped and finished (level 4 finish)		870	.018		.77	.89		1.66	2.30
0130	With compound skim coat (level 5 finish)		700	.023		.82	1.11		1.93	2.71
0150	On ceilings, no finish included		1620	.010		.72	.48		1.20	1.57
0160	Taped and finished (level 4 finish)		690	.023		.77	1.12		1.89	2.68
0170	With compound skim coat (level 5 finish)		550	.029		.82	1.41		2.23	3.20
0210	5/8" thick, on walls, no finish included		1800	.009		.82	.43		1.25	1.60
0220	Taped and finished (level 4 finish)		870	.018		.87	.89		1.76	2.41
0230	With compound skim coat (level 5 finish)		700	.023		.92	1.11		2.03	2.82
0250	On ceilings, no finish included		1620	.010		.82	.48		1.30	1.68
0260	Taped and finished (level 4 finish)		690	.023		.87	1.12		1.99	2.79
0270	With compound skim coat (level 5 finish)		550	.029		.92	1.41		2.33	3.31
0272	5/8" thick, on roof sheathing for 1-hour rating		1620	.010		.82	.48		1.30	1.68
0310	5/8" thick, on walls, very high impact, no finish included		1800	.009		1	.43		1.43	1.80
0320	Taped and finished (level 4 finish)		870	.018		1.05	.89		1.94	2.61
0330	With compound skim coat (level 5 finish)		700	.023		1.10	1.11		2.21	3.02
0350	On ceilings, no finish included		1620	.010		1	.48		1.48	1.88
0360	Taped and finished (level 4 finish)		690	.023		1.05	1.12		2.17	2.99
0370	With compound skim coat (level 5 finish)		550	.029		1.10	1.41		2.51	3.51

09 29 Gypsum Board

09 29 10 – Gypsum Board Panels

09 29 10.50 High Abuse Gypsum Board

		Crew	Daily Output	Labor-Hours	Unit	Material	2016 Bare Costs Labor	Equipment	Total	Total Incl O&P
0400	High abuse, gypsum core, paper face									
0410	1/2" thick, on walls, no finish included	2 Carp	1800	.009	S.F.	.63	.43		1.06	1.39
0420	Taped and finished (level 4 finish)		870	.018		.68	.89		1.57	2.21
0430	With compound skim coat (level 5 finish)		700	.023		.73	1.11		1.84	2.61
0450	On ceilings, no finish included		1620	.010		.63	.48		1.11	1.47
0460	Taped and finished (level 4 finish)		690	.023		.68	1.12		1.80	2.59
0470	With compound skim coat (level 5 finish)		550	.029		.73	1.41		2.14	3.10
0510	5/8" thick, on walls, no finish included		1800	.009		.67	.43		1.10	1.44
0520	Taped and finished (level 4 finish)		870	.018		.72	.89		1.61	2.25
0530	With compound skim coat (level 5 finish)		700	.023		.77	1.11		1.88	2.66
0550	On ceilings, no finish included		1620	.010		.67	.48		1.15	1.52
0560	Taped and finished (level 4 finish)		690	.023		.72	1.12		1.84	2.63
0570	With compound skim coat (level 5 finish)		550	.029		.77	1.41		2.18	3.15
1000	For high ceilings, over 8' high, add		2750	.006			.28		.28	.46
1010	For distribution cost over 3 stories high, add per story		5500	.003			.14		.14	.23

09 29 15 – Gypsum Board Accessories

09 29 15.10 Accessories, Gypsum Board

		Crew	Daily Output	Labor-Hours	Unit	Material	2016 Bare Costs Labor	Equipment	Total	Total Incl O&P
0010	**ACCESSORIES, GYPSUM BOARD**									
0020	Casing bead, galvanized steel	1 Carp	2.90	2.759	C.L.F.	24	134		158	245
0100	Vinyl		3	2.667		22.50	129		151.50	236
0300	Corner bead, galvanized steel, 1" x 1"		4	2		15.45	97		112.45	175
0400	1-1/4" x 1-1/4"		3.50	2.286		16.60	111		127.60	199
0600	Vinyl		4	2		20.50	97		117.50	181
0900	Furring channel, galv. steel, 7/8" deep, standard		2.60	3.077		33.50	149		182.50	281
1000	Resilient		2.55	3.137		25	152		177	276
1100	J trim, galvanized steel, 1/2" wide		3	2.667		22	129		151	235
1120	5/8" wide		2.95	2.712		32	131		163	250
1500	Z stud, galvanized steel, 1-1/2" wide		2.60	3.077		39.50	149		188.50	288
9000	Minimum labor/equipment charge		3	2.667	Job		129		129	211

09 30 Tiling

09 30 13 – Ceramic Tiling

09 30 13.10 Ceramic Tile

		Crew	Daily Output	Labor-Hours	Unit	Material	2016 Bare Costs Labor	Equipment	Total	Total Incl O&P
0010	**CERAMIC TILE**									
0600	Cove base, 4-1/4" x 4-1/4" high, mud set	D-7	91	.176	L.F.	4.19	6.95		11.14	15.60
0700	Thin set		128	.125		4.07	4.96		9.03	12.30
0900	6" x 4-1/4" high, mud set		100	.160		4.37	6.35		10.72	14.80
1000	Thin set		137	.117		4.25	4.63		8.88	11.95
1200	Sanitary cove base, 6" x 4-1/4" high, mud set		93	.172		4.67	6.80		11.47	15.90
1300	Thin set		124	.129		4.55	5.10		9.65	13.05
1500	6" x 6" high, mud set		84	.190		5.45	7.55		13	17.90
1600	Thin set		117	.137		5.30	5.40		10.70	14.40
2400	Bullnose trim, 4-1/4" x 4-1/4", mud set		82	.195		4.15	7.75		11.90	16.75
2500	Thin set		128	.125		4.07	4.96		9.03	12.25
2700	2" x 6" bullnose trim, mud set		84	.190		4.38	7.55		11.93	16.70
2800	Thin set		124	.129		4.32	5.10		9.42	12.80
3000	Floors, natural clay, random or uniform, thin set, color group 1		183	.087	S.F.	4.15	3.47		7.62	10
3100	Color group 2		183	.087		5.75	3.47		9.22	11.80
3300	Porcelain type, 1 color, color group 2, 1" x 1"		183	.087		5.20	3.47		8.67	11.15
3310	2" x 2" or 2" x 1", thin set		190	.084		6.40	3.34		9.74	12.30

09 30 Tiling

09 30 13 – Ceramic Tiling

09 30 13.10 Ceramic Tile		Crew	Daily Output	Labor-Hours	Unit	Material	2016 Bare Costs Labor	Equipment	Total	Total Incl O&P
3350	For random blend, 2 colors, add				S.F.	1			1	1.10
3360	4 colors, add					1.50			1.50	1.65
4300	Specialty tile, 4-1/4" x 4-1/4" x 1/2", decorator finish	D-7	183	.087		10.40	3.47		13.87	16.90
4500	Add for epoxy grout, 1/16" joint, 1" x 1" tile		800	.020		.67	.79		1.46	1.99
4600	2" x 2" tile		820	.020		.62	.77		1.39	1.90
4610	Add for epoxy grout, 1/8" joint, 8" x 8" x 3/8" tile, add		900	.018		1.39	.71		2.10	2.64
4800	Pregrouted sheets, walls, 4-1/4" x 4-1/4", 6" x 4-1/4"									
4810	and 8-1/2" x 4-1/4", 4 S.F. sheets, silicone grout	D-7	240	.067	S.F.	5.40	2.64		8.04	10.10
5100	Floors, unglazed, 2 S.F. sheets,									
5110	urethane adhesive	D-7	180	.089	S.F.	5.10	3.52		8.62	11.15
5400	Walls, interior, thin set, 4-1/4" x 4-1/4" tile		190	.084		2.49	3.34		5.83	8
5500	6" x 4-1/4" tile		190	.084		3.01	3.34		6.35	8.55
5700	8-1/2" x 4-1/4" tile		190	.084		5.25	3.34		8.59	11.05
5800	6" x 6" tile		175	.091		3.46	3.63		7.09	9.50
5810	8" x 8" tile		170	.094		4.75	3.73		8.48	11.10
5820	12" x 12" tile		160	.100		4.55	3.97		8.52	11.25
5830	16" x 16" tile		150	.107		5.05	4.23		9.28	12.25
6000	Decorated wall tile, 4-1/4" x 4-1/4", color group 1		270	.059		3.53	2.35		5.88	7.60
6100	Color group 4		180	.089		52	3.52		55.52	62.50
6600	Crystalline glazed, 4-1/4" x 4-1/4", mud set, plain		100	.160		4.52	6.35		10.87	14.95
6700	4-1/4" x 4-1/4", scored tile		100	.160		6	6.35		12.35	16.60
6900	6" x 6" plain		93	.172		6.80	6.80		13.60	18.25
7000	For epoxy grout, 1/16" joints, 4-1/4" tile, add		800	.020		.41	.79		1.20	1.70
7200	For tile set in dry mortar, add		1735	.009			.37		.37	.58
7300	For tile set in Portland cement mortar, add		290	.055		.16	2.19		2.35	3.63
9300	Ceramic tiles, recycled glass, standard colors, 2" x 2" thru 6" x 6" G		190	.084		21.50	3.34		24.84	29
9310	6" x 6" G		175	.091		21.50	3.63		25.13	29.50
9320	8" x 8" G		170	.094		23	3.73		26.73	31.50
9330	12" x 12" G		160	.100		23	3.97		26.97	32
9340	Earthtones, 2" x 2" to 4" x 8" G		190	.084		25.50	3.34		28.84	33.50
9350	6" x 6" G		175	.091		25.50	3.63		29.13	33.50
9360	8" x 8" G		170	.094		26.50	3.73		30.23	35
9370	12" x 12" G		160	.100		26.50	3.97		30.47	35.50
9380	Deep colors, 2" x 2" to 4" x 8" G		190	.084		30	3.34		33.34	38.50
9390	6" x 6" G		175	.091		30	3.63		33.63	38.50
9400	8" x 8" G		170	.094		31.50	3.73		35.23	40.50
9410	12" x 12" G		160	.100		31.50	3.97		35.47	41
9500	Minimum labor/equipment charge		3.25	4.923	Job		195		195	310

09 30 13.20 Ceramic Tile Repairs

		Crew	Daily Output	Labor-Hours	Unit	Material	Labor	Equipment	Total	Total Incl O&P
0010	**CERAMIC TILE REPAIRS**									
1000	Grout removal, carbide tipped, rotary grinder	1 Clab	240	.033	L.F.		1.26		1.26	2.06

09 30 13.45 Ceramic Tile Accessories

		Crew	Daily Output	Labor-Hours	Unit	Material	Labor	Equipment	Total	Total Incl O&P
0010	**CERAMIC TILE ACCESSORIES**									
0100	Spacers, 1/8"				C	1.98			1.98	2.18
1310	Sealer for natural stone tile, installed	1 Tilf	650	.012	S.F.	.05	.55		.60	.92

09 30 16 – Quarry Tiling

09 30 16.10 Quarry Tile

		Crew	Daily Output	Labor-Hours	Unit	Material	Labor	Equipment	Total	Total Incl O&P
0010	**QUARRY TILE**									
0100	Base, cove or sanitary, mud set, to 5" high, 1/2" thick	D-7	110	.145	L.F.	5.85	5.75		11.60	15.55
0300	Bullnose trim, red, mud set, 6" x 6" x 1/2" thick		120	.133		4.70	5.30		10	13.50
0400	4" x 4" x 1/2" thick		110	.145		4.83	5.75		10.58	14.40
0600	4" x 8" x 1/2" thick, using 8" as edge		130	.123		4.79	4.88		9.67	12.95

09 30 Tiling

09 30 16 – Quarry Tiling

09 30 16.10 Quarry Tile

09 30 16.10	Quarry Tile	Crew	Daily Output	Labor-Hours	Unit	Material	2016 Bare Costs Labor	Equipment	Total	Total Incl O&P
0700	Floors, mud set, 1,000 S.F. lots, red, 4" x 4" x 1/2" thick	D-7	120	.133	S.F.	8.05	5.30		13.35	17.20
0900	6" x 6" x 1/2" thick		140	.114		7.95	4.53		12.48	15.90
1000	4" x 8" x 1/2" thick	↓	130	.123		5	4.88		9.88	13.20
1300	For waxed coating, add					.76			.76	.84
1500	For non-standard colors, add					.46			.46	.51
1600	For abrasive surface, add					.52			.52	.57
1800	Brown tile, imported, 6" x 6" x 3/4"	D-7	120	.133		7.40	5.30		12.70	16.50
1900	8" x 8" x 1"		110	.145		8.70	5.75		14.45	18.65
2100	For thin set mortar application, deduct		700	.023			.91		.91	1.43
2700	Stair tread, 6" x 6" x 3/4", plain		50	.320		7.40	12.70		20.10	28
2800	Abrasive		47	.340		6.80	13.50		20.30	29
3000	Wainscot, 6" x 6" x 1/2", thin set, red		105	.152		5.55	6.05		11.60	15.65
3100	Non-standard colors		105	.152	↓	5.50	6.05		11.55	15.60
3300	Window sill, 6" wide, 3/4" thick		90	.178	L.F.	5.45	7.05		12.50	17.10
3400	Corners		80	.200	Ea.	6.10	7.95		14.05	19.25
9000	Minimum labor/equipment charge	↓	3.25	4.923	Job		195		195	310

09 30 23 – Glass Mosaic Tiling

09 30 23.10 Glass Mosaics

09 30 23.10	Glass Mosaics	Crew	Daily Output	Labor-Hours	Unit	Material	Labor	Equipment	Total	Total Incl O&P
0010	**GLASS MOSAICS** 3/4" tile on 12" sheets, standard grout									
1020	1" tile on 12" sheets, opalescent finish	D-7	73	.219	S.F.	17.50	8.70		26.20	33
1040	1" x 2" tile on 12" sheet, blend		73	.219		19.05	8.70		27.75	34.50
1060	2" tile on 12" sheet, blend		73	.219		16.50	8.70		25.20	32
1080	5/8" x random tile, linear, on 12" sheet, blend		73	.219		25	8.70		33.70	41
1600	Dots on 12" sheet		73	.219		25	8.70		33.70	41
1700	For glass mosaic tiles set in dry mortar, add		290	.055		.45	2.19		2.64	3.95
1720	For glass mosaic tile set in Portland cement mortar, add		290	.055	↓	.01	2.19		2.20	3.46
1730	For polyblend sanded tile grout	↓	96.15	.166	Lb.	2.19	6.60		8.79	12.80

09 30 29 – Metal Tiling

09 30 29.10 Metal Tile

09 30 29.10	Metal Tile	Crew	Daily Output	Labor-Hours	Unit	Material	Labor	Equipment	Total	Total Incl O&P
0010	**METAL TILE** 4' x 4' sheet, 24 ga., tile pattern, nailed									
0200	Stainless steel	2 Carp	512	.031	S.F.	28	1.51		29.51	33.50
0400	Aluminized steel	"	512	.031	"	15.10	1.51		16.61	19.05
9000	Minimum labor/equipment charge	1 Carp	4	2	Job		97		97	158

09 30 95 – Tile & Stone Setting Materials and Specialties

09 30 95.10 Moisture Resistant, Anti-Fracture Membrane

09 30 95.10	Moisture Resistant, Anti-Fracture Membrane	Crew	Daily Output	Labor-Hours	Unit	Material	Labor	Equipment	Total	Total Incl O&P
0010	**MOISTURE RESISTANT, ANTI-FRACTURE MEMBRANE**									
0200	Elastomeric membrane, 1/16" thick	D-7	275	.058	S.F.	1.07	2.31		3.38	4.82

09 34 Waterproofing-Membrane Tiling

09 34 13 – Waterproofing-Membrane Ceramic Tiling

09 34 13.10 Ceramic Tile Waterproofing Membrane

09 34 13.10	Ceramic Tile Waterproofing Membrane	Crew	Daily Output	Labor-Hours	Unit	Material	Labor	Equipment	Total	Total Incl O&P
0010	**CERAMIC TILE WATERPROOFING MEMBRANE**									
0020	On floors, including thinset									
0030	Fleece laminated polyethylene grid, 1/8" thick	D-7	250	.064	S.F.	2.28	2.54		4.82	6.50
0040	5/16" thick	"	250	.064	"	2.60	2.54		5.14	6.85
0050	On walls, including thinset									
0060	Fleece laminated polyethylene sheet, 8 mil thick	D-7	480	.033	S.F.	2.28	1.32		3.60	4.58
0070	Accessories, including thinset									
0080	Joint and corner sheet, 4 mils thick, 5" wide	1 Tilf	240	.033	L.F.	1.35	1.48		2.83	3.82

For customer support on your Commercial Renovation Cost Data, call 877.742.4719.

303

09 34 Waterproofing-Membrane Tiling

09 34 13 – Waterproofing-Membrane Ceramic Tiling

09 34 13.10 Ceramic Tile Waterproofing Membrane	Crew	Daily Output	Labor-Hours	Unit	Material	2016 Bare Costs Labor	Equipment	Total	Total Incl O&P
0090	7-1/4" wide	1 Tilf	180	.044	L.F.	1.71	1.98		3.69	5
0100	10" wide		120	.067	▼	2.08	2.97		5.05	6.95
0110	Pre-formed corners, inside		32	.250	Ea.	6.95	11.15		18.10	25
0120	Outside		32	.250		7.65	11.15		18.80	26
0130	2" flanged floor drain with 6" stainless steel grate		16	.500	▼	370	22.50		392.50	445
0140	EPS, sloped shower floor		480	.017	S.F.	4.97	.74		5.71	6.60
0150	Curb	▼	32	.250	L.F.	14.05	11.15		25.20	33

09 51 Acoustical Ceilings

09 51 23 – Acoustical Tile Ceilings

09 51 23.10 Suspended Acoustic Ceiling Tiles

		Crew	Daily Output	Labor-Hours	Unit	Material	2016 Bare Costs Labor	Equipment	Total	Total Incl O&P
0010	**SUSPENDED ACOUSTIC CEILING TILES**, not including									
0100	suspension system									
0300	Fiberglass boards, film faced, 2' x 2' or 2' x 4', 5/8" thick	1 Carp	625	.013	S.F.	1.25	.62		1.87	2.39
0400	3/4" thick		600	.013		2.63	.65		3.28	3.95
0500	3" thick, thermal, R11		450	.018		2.62	.86		3.48	4.29
0600	Glass cloth faced fiberglass, 3/4" thick		500	.016		2.72	.78		3.50	4.26
0700	1" thick		485	.016		3.31	.80		4.11	4.95
0820	1-1/2" thick, nubby face		475	.017		2.60	.82		3.42	4.19
1110	Mineral fiber tile, lay-in, 2' x 2' or 2' x 4', 5/8" thick, fine texture		625	.013		.96	.62		1.58	2.07
1115	Rough textured		625	.013		.86	.62		1.48	1.96
1125	3/4" thick, fine textured		600	.013		1.99	.65		2.64	3.25
1130	Rough textured		600	.013		1.60	.65		2.25	2.82
1135	Fissured		600	.013		1.85	.65		2.50	3.10
1150	Tegular, 5/8" thick, fine textured		470	.017		1.04	.82		1.86	2.49
1155	Rough textured		470	.017		1.23	.82		2.05	2.70
1165	3/4" thick, fine textured		450	.018		2.08	.86		2.94	3.70
1170	Rough textured		450	.018		1.41	.86		2.27	2.96
1175	Fissured	▼	450	.018		2.02	.86		2.88	3.63
1185	For plastic film face, add					.79			.79	.87
1190	For fire rating, add					.45			.45	.50
1300	Metal panel, lay-in, 2' x 2', sq. edge	1 Carp	500	.016		9.55	.78		10.33	11.75
1350	Tegular edge		500	.016		13.20	.78		13.98	15.75
1400	2' x 4', sq. edge		500	.016		12.90	.78		13.68	15.45
1450	Tegular edge		500	.016		13.20	.78		13.98	15.75
1500	Perforated alum. clip-in, 2' x 2'		500	.016		13.45	.78		14.23	16
1550	2' x 4'		500	.016		10.85	.78		11.63	13.20
1600	Solid alum. planks, 3-1/4"x12', open reveal		500	.016		2.35	.78		3.13	3.86
1650	Closed reveal		500	.016		3	.78		3.78	4.57
1700	7-1/4" x 12', open reveal		500	.016		4	.78		4.78	5.65
1750	Closed reveal		500	.016		5.10	.78		5.88	6.85
1775	Metal, open cell, 2' x 2', 6" cell		500	.016		8.25	.78		9.03	10.35
1800	8" cell		500	.016		9.15	.78		9.93	11.30
1825	2' x 4', 6" cell		500	.016		5.35	.78		6.13	7.15
1850	8" cell		500	.016		5.35	.78		6.13	7.15
1870	Translucent lay-in panels, 2' x 2'		500	.016		23	.78		23.78	26.50
1890	2' x 6'		500	.016		17.20	.78		17.98	20
3720	Mineral fiber, 24" x 24" or 48", reveal edge, painted, 5/8" thick		600	.013		1.19	.65		1.84	2.37
3740	3/4" thick		575	.014		1.62	.67		2.29	2.88
5020	66 – 78% recycled content, 3/4" thick Ⓖ		600	.013		2.03	.65		2.68	3.29
5040	Mylar, 42% recycled content, 3/4" thick Ⓖ	▼	600	.013	▼	4.77	.65		5.42	6.30

09 51 Acoustical Ceilings

09 51 23 – Acoustical Tile Ceilings

09 51 23.10 Suspended Acoustic Ceiling Tiles

	09 51 23.10 Suspended Acoustic Ceiling Tiles	Crew	Daily Output	Labor-Hours	Unit	Material	2016 Bare Costs Labor	Equipment	Total	Total Incl O&P
6000	Remove and replace ceiling tiles, min fiber, 2' x 2' or 2' x 4', 5/8"thk.	1 Carp	335	.024	S.F.	.96	1.16		2.12	2.95
9000	Minimum labor/equipment charge		4	2	Job		97		97	158

09 51 23.30 Suspended Ceilings, Complete

	09 51 23.30 Suspended Ceilings, Complete	Crew	Daily Output	Labor-Hours	Unit	Material	2016 Bare Costs Labor	Equipment	Total	Total Incl O&P
0010	**SUSPENDED CEILINGS, COMPLETE**, including standard									
0100	suspension system but not incl. 1-1/2" carrier channels									
0600	Fiberglass ceiling board, 2' x 4' x 5/8", plain faced	1 Carp	500	.016	S.F.	2.02	.78		2.80	3.49
0700	Offices, 2' x 4' x 3/4"		380	.021		3.40	1.02		4.42	5.40
0800	Mineral fiber, on 15/16" T bar susp. 2' x 2' x 3/4" lay-in board		345	.023		2.98	1.12		4.10	5.10
0810	2' x 4' x 5/8" tile		380	.021		2.43	1.02		3.45	4.34
0820	Tegular, 2' x 2' x 5/8" tile on 9/16" grid		250	.032		2.45	1.55		4	5.25
0830	2' x 4' x 3/4" tile		275	.029		2.59	1.41		4	5.15
0900	Luminous panels, prismatic, acrylic		255	.031		3.41	1.52		4.93	6.25
1200	Metal pan with acoustic pad, steel		75	.107		5.15	5.15		10.30	14.10
1300	Painted aluminum		75	.107		3.23	5.15		8.38	12
1500	Aluminum, degreased finish		75	.107		5.25	5.15		10.40	14.20
1600	Stainless steel		75	.107		9.75	5.15		14.90	19.20
1800	Tile, Z bar suspension, 5/8" mineral fiber tile		150	.053		2.32	2.58		4.90	6.75
1900	3/4" mineral fiber tile		150	.053		2.35	2.58		4.93	6.80
2402	For strip lighting, see Section 26 51 13.50									
2500	For rooms under 500 S.F., add				S.F.		25%			
9000	Minimum labor/equipment charge	1 Carp	2	4	Job		194		194	315

09 51 53 – Direct-Applied Acoustical Ceilings

09 51 53.10 Ceiling Tile

	09 51 53.10 Ceiling Tile	Crew	Daily Output	Labor-Hours	Unit	Material	2016 Bare Costs Labor	Equipment	Total	Total Incl O&P
0010	**CEILING TILE**, stapled or cemented									
0100	12" x 12" or 12" x 24", not including furring									
0600	Mineral fiber, vinyl coated, 5/8" thick	1 Carp	300	.027	S.F.	2.23	1.29		3.52	4.56
0700	3/4" thick		300	.027		2.93	1.29		4.22	5.35
0900	Fire rated, 3/4" thick, plain faced		300	.027		1.31	1.29		2.60	3.55
1000	Plastic coated face		300	.027		1.93	1.29		3.22	4.23
1200	Aluminum faced, 5/8" thick, plain		300	.027		1.74	1.29		3.03	4.02
3000	Wood fiber tile, 1/2" thick		400	.020		1.15	.97		2.12	2.85
3100	3/4" thick		400	.020		1.23	.97		2.20	2.93
3300	For flameproofing, add					.10			.10	.11
3400	For sculptured 3 dimensional, add					.31			.31	.34
3900	For ceiling primer, add					.13			.13	.14
4000	For ceiling cement, add					.40			.40	.44
9000	Minimum labor/equipment charge	1 Carp	4	2	Job		97		97	158

09 53 Acoustical Ceiling Suspension Assemblies

09 53 23 – Metal Acoustical Ceiling Suspension Assemblies

09 53 23.30 Ceiling Suspension Systems

	09 53 23.30 Ceiling Suspension Systems	Crew	Daily Output	Labor-Hours	Unit	Material	2016 Bare Costs Labor	Equipment	Total	Total Incl O&P
0010	**CEILING SUSPENSION SYSTEMS** for boards and tile									
0050	Class A suspension system, 15/16" T bar, 2' x 4' grid	1 Carp	800	.010	S.F.	.77	.48		1.25	1.63
0300	2' x 2' grid		650	.012		.99	.60		1.59	2.06
0310	25% recycled steel, 2' x 4' grid ⬛G		800	.010		.80	.48		1.28	1.67
0320	2' x 2' grid ⬛G		650	.012		1	.60		1.60	2.07
0350	For 9/16" grid, add					.16			.16	.18
0360	For fire rated grid, add					.09			.09	.10
0370	For colored grid, add					.21			.21	.23
0400	Concealed Z bar suspension system, 12" module	1 Carp	520	.015		.86	.75		1.61	2.17

For customer support on your Commercial Renovation Cost Data, call 877.742.4719.

305

09 53 23 – Metal Acoustical Ceiling Suspension Assemblies

09 53 23.30 Ceiling Suspension Systems	Crew	Daily Output	Labor-Hours	Unit	Material	2016 Bare Costs Labor	Equipment	Total	Total Incl O&P
0600	1-1/2" carrier channels, 4' O.C., add	1 Carp	470	.017	S.F.	.12	.82		.94	1.48
0700	Carrier channels for ceilings with									
0900	recessed lighting fixtures, add	1 Carp	460	.017	S.F.	.22	.84		1.06	1.62
3000	Seismic ceiling bracing, IBC Site Class D, Occupancy Category II									
3050	For ceilings less than 2500 S.F.									
3060	Seismic clips at attached walls	1 Carp	180	.044	Ea.	1.06	2.15		3.21	4.68
3100	For ceilings greater than 2500 S.F., add									
3120	Seismic clips, joints at cross tees	1 Carp	120	.067	Ea.	3.02	3.23		6.25	8.60
3140	At cross tees and mains, mains field cut	"	60	.133	"	3.02	6.45		9.47	13.85
3200	Compression posts, telescopic, attached to structure above									
3210	To 30" high	1 Carp	26	.308	Ea.	36.50	14.90		51.40	65
3220	30" to 48" high		25.50	.314		40	15.20		55.20	69
3230	48" to 84" high		25	.320		48.50	15.50		64	79
3240	84" to 102" high		24.50	.327		55.50	15.80		71.30	87
3250	102" to 120" high		24	.333		79.50	16.15		95.65	114
3260	120" to 144" high		24	.333		88	16.15		104.15	124
3300	Stabilizer bars									
3310	12" long	1 Carp	240	.033	Ea.	1.02	1.61		2.63	3.76
3320	24" long		235	.034		.97	1.65		2.62	3.76
3330	36" long		230	.035		.93	1.69		2.62	3.78
3340	48" long		220	.036		.77	1.76		2.53	3.73
3400	Wire support for light fixtures, per L.F. height to structure above									
3410	Less than 10 lb.	1 Carp	400	.020	L.F.	.28	.97		1.25	1.89
3420	10 lb. to 56 lb.	"	240	.033	"	.57	1.61		2.18	3.26
3500	Retrofit existing suspended ceiling grid to current code									
3510	Less than 2500 S.F. (using clips @ perimeter)	1 Carp	1455	.006	S.F.	.07	.27		.34	.51
3520	Greater than 2500 S.F. (using compression posts and clips)		550	.015		.41	.70		1.11	1.60
4000	Remove and replace ceiling grid, class A, 15/16" T bar, 2' x 2' grid		435	.018		.99	.89		1.88	2.55
4010	2' x 4' grid		535	.015		.77	.72		1.49	2.02

09 54 Specialty Ceilings

09 54 26 – Suspended Wood Ceilings

09 54 26.10 Wood Ceilings

0010	**WOOD CEILINGS**									
1000	4" - 6" wood slats on heavy duty 15/16" T-bar grid	2 Carp	250	.064	S.F.	25	3.10		28.10	32

09 54 33 – Decorative Panel Ceilings

09 54 33.20 Metal Panel Ceilings

0010	**METAL PANEL CEILINGS**									
0020	Lay-in or screwed to furring, not including grid									
0100	Tin ceilings, 2' x 2' or 2' x 4', bare steel finish	2 Carp	300	.053	S.F.	2.51	2.58		5.09	7
0120	Painted white finish		300	.053	"	3.69	2.58		6.27	8.30
0140	Copper, chrome or brass finish		300	.053	L.F.	6.50	2.58		9.08	11.35
0200	Cornice molding, 2-1/2" to 3-1/2" wide, 4' long, bare steel finish		200	.080	S.F.	2.30	3.88		6.18	8.90
0220	Painted white finish		200	.080		2.84	3.88		6.72	9.45
0240	Copper, chrome or brass finish		200	.080		4.01	3.88		7.89	10.75
0320	5" to 6-1/2" wide, 4' long, bare steel finish		150	.107		3.23	5.15		8.38	12
0340	Painted white finish		150	.107		4.09	5.15		9.24	12.95
0360	Copper, chrome or brass finish		150	.107		6.50	5.15		11.65	15.60
0420	Flat molding, 3-1/2" to 5" wide, 4' long, bare steel finish		250	.064		3.71	3.10		6.81	9.15
0440	Painted white finish		250	.064		3.96	3.10		7.06	9.40

09 54 Specialty Ceilings

09 54 33 – Decorative Panel Ceilings

09 54 33.20 Metal Panel Ceilings	Crew	Daily Output	Labor-Hours	Unit	Material	2016 Bare Costs Labor	Equipment	Total	Total Incl O&P
0460 Copper, chrome or brass finish	2 Carp	250	.064	S.F.	7.50	3.10		10.60	13.30

09 61 Flooring Treatment

09 61 19 – Concrete Floor Staining

09 61 19.40 Floors, Interior

		Crew	Daily Output	Labor-Hours	Unit	Material	Labor	Equipment	Total	Total Incl O&P
0010	**FLOORS, INTERIOR**									
0300	Acid stain and sealer									
0310	Stain, one coat	1 Pord	650	.012	S.F.	.12	.50		.62	.94
0320	Two coats		570	.014		.25	.57		.82	1.18
0330	Acrylic sealer, one coat		2600	.003		.23	.12		.35	.46
0340	Two coats		1400	.006		.47	.23		.70	.88

09 62 Specialty Flooring

09 62 19 – Laminate Flooring

09 62 19.10 Floating Floor

		Crew	Daily Output	Labor-Hours	Unit	Material	Labor	Equipment	Total	Total Incl O&P
0010	**FLOATING FLOOR**									
8300	Floating floor, laminate, wood pattern strip, complete	1 Clab	133	.060	S.F.	4.50	2.28		6.78	8.70
8310	Components, T & G wood composite strips					4.02			4.02	4.42
8320	Film					.17			.17	.18
8330	Foam					.26			.26	.29
8340	Adhesive					.71			.71	.79
8350	Installation kit					.19			.19	.21
8360	Trim, 2" wide x 3' long				L.F.	4.30			4.30	4.73
8370	Reducer moulding				"	5.70			5.70	6.25

09 62 23 – Bamboo Flooring

09 62 23.10 Flooring, Bamboo

			Crew	Daily Output	Labor-Hours	Unit	Material	Labor	Equipment	Total	Total Incl O&P
0010	**FLOORING, BAMBOO**										
8600	Flooring, wood, bamboo strips, unfinished, 5/8" x 4" x 3'	G	1 Carp	255	.031	S.F.	4.60	1.52		6.12	7.55
8610	5/8" x 4" x 4'	G		275	.029		4.77	1.41		6.18	7.55
8620	5/8" x 4" x 6'	G		295	.027		5.25	1.31		6.56	7.90
8630	Finished, 5/8" x 4" x 3'	G		255	.031		5.05	1.52		6.57	8.05
8640	5/8" x 4" x 4'	G		275	.029		5.30	1.41		6.71	8.15
8650	5/8" x 4" x 6'	G		295	.027		4.94	1.31		6.25	7.60
8660	Stair treads, unfinished, 1-1/16" x 11-1/2" x 4'	G		18	.444	Ea.	44	21.50		65.50	83.50
8670	Finished, 1-1/16" x 11-1/2" x 4'	G		18	.444		82	21.50		103.50	126
8680	Stair risers, unfinished, 5/8" x 7-1/2" x 4'	G		18	.444		16.30	21.50		37.80	53
8690	Finished, 5/8" x 7-1/2" x 4'	G		18	.444		31	21.50		52.50	69
8700	Stair nosing, unfinished, 6' long	G		16	.500		36	24		60	79
8710	Finished, 6' long	G		16	.500		42	24		66	85.50

09 62 29 – Cork Flooring

09 62 29.10 Cork Tile Flooring

			Crew	Daily Output	Labor-Hours	Unit	Material	Labor	Equipment	Total	Total Incl O&P
0010	**CORK TILE FLOORING**										
2200	Cork tile, standard finish, 1/8" thick		1 Tilf	315	.025	S.F.	6.95	1.13		8.08	9.45
2250	3/16" thick	G		315	.025		5.55	1.13		6.68	7.95
2300	5/16" thick	G		315	.025		8.35	1.13		9.48	11
2350	1/2" thick	G		315	.025		10.95	1.13		12.08	13.80
2500	Urethane finish, 1/8" thick	G		315	.025		8.10	1.13		9.23	10.70
2550	3/16" thick	G		315	.025		8.25	1.13		9.38	10.85

For customer support on your Commercial Renovation Cost Data, call 877.742.4719.

307

09 62 Specialty Flooring

09 62 29 – Cork Flooring

09 62 29.10 Cork Tile Flooring		Crew	Daily Output	Labor-Hours	Unit	Material	2016 Bare Costs Labor	Equipment	Total	Total Incl O&P
2600	5/16" thick	G 1 Tilf	315	.025	S.F.	8.85	1.13		9.98	11.55
2650	1/2" thick	G ↓	315	.025	↓	12.15	1.13		13.28	15.15

09 63 Masonry Flooring

09 63 13 – Brick Flooring

09 63 13.10 Miscellaneous Brick Flooring

		Crew	Daily Output	Labor-Hours	Unit	Material	Labor	Equipment	Total	Total Incl O&P
0010	**MISCELLANEOUS BRICK FLOORING**									
0020	Acid-proof shales, red, 8" x 3-3/4" x 1-1/4" thick	D-7	.43	37.209	M	705	1,475		2,180	3,100
0050	2-1/4" thick	D-1	.40	40		965	1,675		2,640	3,800
0200	Acid-proof clay brick, 8" x 3-3/4" x 2-1/4" thick	G	.40	40	↓	925	1,675		2,600	3,775
0250	9" x 4-1/2" x 3"	G ↓	95	.168	S.F.	4.23	7.10		11.33	16.20
0260	Cast ceramic, pressed, 4" x 8" x 1/2", unglazed	D-7	100	.160		6.35	6.35		12.70	16.95
0270	Glazed		100	.160		8.45	6.35		14.80	19.30
0280	Hand molded flooring, 4" x 8" x 3/4", unglazed		95	.168		8.35	6.70		15.05	19.75
0290	Glazed		95	.168		10.50	6.70		17.20	22
0300	8" hexagonal, 3/4" thick, unglazed		85	.188		9.20	7.45		16.65	22
0310	Glazed	↓	85	.188		16.60	7.45		24.05	30
0400	Heavy duty industrial, cement mortar bed, 2" thick, not incl. brick	D-1	80	.200		1.06	8.45		9.51	14.85
0450	Acid-proof joints, 1/4" wide	"	65	.246		1.46	10.35		11.81	18.45
0500	Pavers, 8" x 4", 1" to 1-1/4" thick, red	D-7	95	.168		3.69	6.70		10.39	14.60
0510	Ironspot	"	95	.168		5.20	6.70		11.90	16.30
0540	1-3/8" to 1-3/4" thick, red	D-1	95	.168		3.56	7.10		10.66	15.45
0560	Ironspot		95	.168		5.15	7.10		12.25	17.25
0580	2-1/4" thick, red		90	.178		3.62	7.50		11.12	16.20
0590	Ironspot		90	.178		5.60	7.50		13.10	18.40
0700	Paver, adobe brick, 6" x 12", 1/2" joint	G ↓	42	.381		1.32	16.05		17.37	27.50
0710	Mexican red, 12" x 12"	G 1 Tilf	48	.167		1.72	7.40		9.12	13.60
0720	Saltillo, 12" x 12"	G "	48	.167	↓	1.49	7.40		8.89	13.35
0800	For sidewalks and patios with pavers, see Section 32 14 16.10									
0870	For epoxy joints, add	D-1	600	.027	S.F.	2.81	1.12		3.93	4.92
0880	For Furan underlayment, add	"	600	.027		2.33	1.12		3.45	4.39
0890	For waxed surface, steam cleaned, add	A-1H	1000	.008	↓	.20	.30	.08	.58	.81
9000	Minimum labor/equipment charge	1 Bric	2	4	Job		185		185	300

09 63 40 – Stone Flooring

09 63 40.10 Marble

		Crew	Daily Output	Labor-Hours	Unit	Material	Labor	Equipment	Total	Total Incl O&P
0010	**MARBLE**									
0020	Thin gauge tile, 12" x 6", 3/8", white Carara	D-7	60	.267	S.F.	14.95	10.55		25.50	33
0100	Travertine		60	.267		9.95	10.55		20.50	27.50
0200	12" x 12" x 3/8", thin set, floors		60	.267		10.30	10.55		20.85	28
0300	On walls		52	.308	↓	9.85	12.20		22.05	30
1000	Marble threshold, 4" wide x 36" long x 5/8" thick, white		60	.267	Ea.	10.30	10.55		20.85	28
9000	Minimum labor/equipment charge	↓	3	5.333	Job		211		211	335

09 63 40.20 Slate Tile

		Crew	Daily Output	Labor-Hours	Unit	Material	Labor	Equipment	Total	Total Incl O&P
0010	**SLATE TILE**									
0020	Vermont, 6" x 6" x 1/4" thick, thin set	D-7	180	.089	S.F.	7.50	3.52		11.02	13.80
9000	Minimum labor/equipment charge	"	3	5.333	Job		211		211	335

09 64 Wood Flooring

09 64 16 – Wood Block Flooring

09 64 16.10 End Grain Block Flooring

	Crew	Daily Output	Labor-Hours	Unit	Material	2016 Bare Costs Labor	Equipment	Total	Total Incl O&P
0010 **END GRAIN BLOCK FLOORING**									
0020 End grain flooring, coated, 2" thick	1 Carp	295	.027	S.F.	3.57	1.31		4.88	6.10
0400 Natural finish, 1" thick, fir		125	.064		3.69	3.10		6.79	9.10
0600 1-1/2" thick, pine		125	.064		3.62	3.10		6.72	9.05
0700 2" thick, pine		125	.064		4.49	3.10		.7.59	10
9000 Minimum labor/equipment charge		2	4	Job		194		194	315

09 64 23 – Wood Parquet Flooring

09 64 23.10 Wood Parquet

	Crew	Daily Output	Labor-Hours	Unit	Material	2016 Bare Costs Labor	Equipment	Total	Total Incl O&P
0010 **WOOD PARQUET** flooring									
5200 Parquetry, 5/16" thk, no finish, oak, plain pattern	1 Carp	160	.050	S.F.	5.40	2.42		7.82	9.90
5300 Intricate pattern		100	.080		9.65	3.88		13.53	17
5500 Teak, plain pattern		160	.050		6.30	2.42		8.72	10.90
5600 Intricate pattern		100	.080		10.70	3.88		14.58	18.10
5650 13/16" thick, select grade oak, plain pattern		160	.050		9.75	2.42		12.17	14.70
5700 Intricate pattern		100	.080		17.25	3.88		21.13	25.50
5800 Custom parquetry, including finish, plain pattern		100	.080		17.05	3.88		20.93	25
5900 Intricate pattern		50	.160		25	7.75		32.75	40
6700 Parquetry, prefinished white oak, 5/16" thick, plain pattern		160	.050		7.90	2.42		10.32	12.65
6800 Intricate pattern		100	.080		8.45	3.88		12.33	15.60
7000 Walnut or teak, parquetry, plain pattern		160	.050		8.40	2.42		10.82	13.20
7100 Intricate pattern		100	.080		13.50	3.88		17.38	21
7200 Acrylic wood parquet blocks, 12" x 12" x 5/16",									
7210 Irradiated, set in epoxy	1 Carp	160	.050	S.F.	10.20	2.42		12.62	15.15

09 64 29 – Wood Strip and Plank Flooring

09 64 29.10 Wood

	Crew	Daily Output	Labor-Hours	Unit	Material	2016 Bare Costs Labor	Equipment	Total	Total Incl O&P
0010 **WOOD** R061110-30									
0020 Fir, vertical grain, 1" x 4", not incl. finish, grade B & better	1 Carp	255	.031	S.F.	2.79	1.52		4.31	5.55
0100 C grade & better		255	.031		2.63	1.52		4.15	5.35
0300 Flat grain, 1" x 4", not incl. finish, B & better		255	.031		3.19	1.52		4.71	6
0400 C & better		255	.031		3.07	1.52		4.59	5.85
4000 Maple, strip, 25/32" x 2-1/4", not incl. finish, select		170	.047		4.90	2.28		7.18	9.10
4100 #2 & better		170	.047		4.42	2.28		6.70	8.60
4300 33/32" x 3-1/4", not incl. finish, #1 grade		170	.047		4.56	2.28		6.84	8.70
4400 #2 & better		170	.047		4.06	2.28		6.34	8.20
4600 Oak, white or red, 25/32" x 2-1/4", not incl. finish									
4700 #1 common	1 Carp	170	.047	S.F.	3.33	2.28		5.61	7.40
4900 Select quartered, 2-1/4" wide		170	.047		4.19	2.28		6.47	8.35
5000 Clear		170	.047		4.10	2.28		6.38	8.25
6100 Prefinished, white oak, prime grade, 2-1/4" wide		170	.047		4.83	2.28		7.11	9
6200 3-1/4" wide		185	.043		5.25	2.10		7.35	9.15
6400 Ranch plank		145	.055		7.15	2.67		9.82	12.20
6500 Hardwood blocks, 9" x 9", 25/32" thick		160	.050		6	2.42		8.42	10.55
7400 Yellow pine, 3/4" x 3-1/8", T & G, C & better, not incl. finish		200	.040		1.53	1.94		3.47	4.85
7500 Refinish wood floor, sand, 2 coats poly, wax, soft wood	1 Clab	400	.020		.90	.76		1.66	2.23
7600 Hard wood		130	.062		1.34	2.33		3.67	5.30
7800 Sanding and finishing, 2 coats polyurethane		295	.027		.90	1.03		1.93	2.67
7900 Subfloor and underlayment, see Section 06 16									
8015 Transition molding, 2 1/4" wide, 5' long	1 Carp	19.20	.417	Ea.	10.85	20		30.85	45
9000 Minimum labor/equipment charge	"	2	4	Job		194		194	315

09 65 Resilient Flooring

09 65 10 – Resilient Tile Underlayment

09 65 10.10 Latex Underlayment		Crew	Daily Output	Labor-Hours	Unit	Material	2016 Bare Costs Labor	Equipment	Total	Total Incl O&P
0010	**LATEX UNDERLAYMENT**									
3600	Latex underlayment, 1/8" thk., cementitious for resilient flooring	1 Tilf	160	.050	S.F.	1.22	2.23		3.45	4.85
4000	Liquid, fortified				Gal.	32			32	35

09 65 13 – Resilient Base and Accessories

09 65 13.13 Resilient Base

		Crew	Daily Output	Labor-Hours	Unit	Material	Labor	Equipment	Total	Total Incl O&P
0010	**RESILIENT BASE**									
0690	1/8" vinyl base, 2 1/2" H, straight or cove, standard colors	1 Tilf	315	.025	L.F.	.68	1.13		1.81	2.53
0700	4" high		315	.025		1.21	1.13		2.34	3.11
0710	6" high		315	.025		1.38	1.13		2.51	3.30
0720	Corners, 2 1/2" high		315	.025	Ea.	2.14	1.13		3.27	4.13
0730	4" high		315	.025		2.35	1.13		3.48	4.37
0740	6" high		315	.025		2.65	1.13		3.78	4.70
0800	1/8" rubber base, 2 1/2" H, straight or cove, standard colors		315	.025	L.F.	1.11	1.13		2.24	3
1100	4" high		315	.025		1.14	1.13		2.27	3.03
1110	6" high		315	.025		1.91	1.13		3.04	3.89
1150	Corners, 2 1/2" high		315	.025	Ea.	2.25	1.13		3.38	4.26
1153	4" high		315	.025		2.30	1.13		3.43	4.31
1155	6" high		315	.025		2.86	1.13		3.99	4.93
1450	For premium color/finish add					50%				
1500	Millwork profile	1 Tilf	315	.025	L.F.	6.15	1.13		7.28	8.55

09 65 13.23 Resilient Stair Treads and Risers

		Crew	Daily Output	Labor-Hours	Unit	Material	Labor	Equipment	Total	Total Incl O&P
0010	**RESILIENT STAIR TREADS AND RISERS**									
0300	Rubber, molded tread, 12" wide, 5/16" thick, black	1 Tilf	115	.070	L.F.	15	3.10		18.10	21.50
0400	Colors		115	.070		15.35	3.10		18.45	21.50
0600	1/4" thick, black		115	.070		13.35	3.10		16.45	19.60
0700	Colors		115	.070		14.95	3.10		18.05	21.50
0900	Grip strip safety tread, colors, 5/16" thick		115	.070		20	3.10		23.10	27
1000	3/16" thick		120	.067		14.75	2.97		17.72	21
1200	Landings, smooth sheet rubber, 1/8" thick		120	.067	S.F.	8.15	2.97		11.12	13.65
1300	3/16" thick		120	.067	"	8.95	2.97		11.92	14.55
1500	Nosings, 3" wide, 3/16" thick, black		140	.057	L.F.	4.67	2.54		7.21	9.15
1600	Colors		140	.057		5.30	2.54		7.84	9.80
1800	Risers, 7" high, 1/8" thick, flat		250	.032		8.10	1.42		9.52	11.15
1900	Coved		250	.032		9.35	1.42		10.77	12.55
2100	Vinyl, molded tread, 12" wide, colors, 1/8" thick		115	.070		5.35	3.10		8.45	10.75
2200	1/4" thick		115	.070		6.70	3.10		9.80	12.30
2300	Landing material, 1/8" thick		200	.040	S.F.	6.05	1.78		7.83	9.45
2400	Riser, 7" high, 1/8" thick, coved		175	.046	L.F.	2.85	2.03		4.88	6.35
2500	Tread and riser combined, 1/8" thick		80	.100	"	10.25	4.45		14.70	18.30
9000	Minimum labor/equipment charge		3	2.667	Job		119		119	187

09 65 16 – Resilient Sheet Flooring

09 65 16.10 Rubber and Vinyl Sheet Flooring

			Crew	Daily Output	Labor-Hours	Unit	Material	Labor	Equipment	Total	Total Incl O&P
0010	**RUBBER AND VINYL SHEET FLOORING**										
5500	Linoleum, sheet goods	G	1 Tilf	360	.022	S.F.	3.59	.99		4.58	5.50
5900	Rubber, sheet goods, 36" wide, 1/8" thick			120	.067		7.10	2.97		10.07	12.50
5950	3/16" thick			100	.080		10.45	3.56		14.01	17.10
6000	1/4" thick			90	.089		12.30	3.96		16.26	19.80
8000	Vinyl sheet goods, backed, .065" thick, plain pattern/colors			250	.032		3.69	1.42		5.11	6.30
8050	Intricate pattern/colors			200	.040		4.34	1.78		6.12	7.60
8100	.080" thick, plain pattern/colors			230	.035		4.12	1.55		5.67	6.95
8150	Intricate pattern/colors			200	.040		6.45	1.78		8.23	9.90

09 65 Resilient Flooring

09 65 16 – Resilient Sheet Flooring

09 65 16.10 Rubber and Vinyl Sheet Flooring

09 65 16.10 Rubber and Vinyl Sheet Flooring	Crew	Daily Output	Labor-Hours	Unit	Material	2016 Bare Costs Labor	Equipment	Total	Total Incl O&P
8200	.125" thick, plain pattern/colors	1 Tilf	230	.035	S.F.	3.87	1.55		5.42	6.70
8250	Intricate pattern/colors		200	.040		7.50	1.78		9.28	11.05
8400	For welding seams, add		100	.080	L.F.	.23	3.56		3.79	5.85
8450	For integral cove base, add		175	.046	"	.83	2.03		2.86	4.12
8700	Adhesive cement, 1 gallon per 200 to 300 S.F.				Gal.	29.50			29.50	32
8800	Asphalt primer, 1 gallon per 300 S.F.					14			14	15.40
8900	Emulsion, 1 gallon per 140 S.F.					18.85			18.85	20.50

09 65 19 – Resilient Tile Flooring

09 65 19.19 Vinyl Composition Tile Flooring

09 65 19.19 Vinyl Composition Tile Flooring	Crew	Daily Output	Labor-Hours	Unit	Material	2016 Bare Costs Labor	Equipment	Total	Total Incl O&P
0010	**VINYL COMPOSITION TILE FLOORING**									
7000	Vinyl composition tile, 12" x 12", 1/16" thick	1 Tilf	500	.016	S.F.	1.19	.71		1.90	2.43
7050	Embossed		500	.016		2.21	.71		2.92	3.55
7100	Marbleized		500	.016		2.21	.71		2.92	3.55
7150	Solid		500	.016		2.85	.71		3.56	4.26
7200	3/32" thick, embossed		500	.016		1.43	.71		2.14	2.69
7250	Marbleized		500	.016		2.54	.71		3.25	3.91
7300	Solid		500	.016		2.36	.71		3.07	3.72
7350	1/8" thick, marbleized		500	.016		2.44	.71		3.15	3.80
7400	Solid		500	.016		1.55	.71		2.26	2.83
7450	Conductive		500	.016		6.05	.71		6.76	7.75

09 65 19.23 Vinyl Tile Flooring

09 65 19.23 Vinyl Tile Flooring	Crew	Daily Output	Labor-Hours	Unit	Material	2016 Bare Costs Labor	Equipment	Total	Total Incl O&P
0010	**VINYL TILE FLOORING**									
7500	Vinyl tile, 12" x 12", 3/32" thick, standard colors/patterns	1 Tilf	500	.016	S.F.	3.71	.71		4.42	5.20
7550	1/8" thick, standard colors/patterns		500	.016		5.20	.71		5.91	6.80
7600	1/8" thick, premium colors/patterns		500	.016		7	.71		7.71	8.80
7650	Solid colors		500	.016		3.25	.71		3.96	4.70
7700	Marbleized or Travertine pattern		500	.016		5.90	.71		6.61	7.60
7750	Florentine pattern		500	.016		6.30	.71		7.01	8.05
7800	Premium colors/patterns		500	.016		6.25	.71		6.96	7.95
9500	Minimum labor/equipment charge		4	2	Job		89		89	140

09 65 19.33 Rubber Tile Flooring

09 65 19.33 Rubber Tile Flooring	Crew	Daily Output	Labor-Hours	Unit	Material	2016 Bare Costs Labor	Equipment	Total	Total Incl O&P
0010	**RUBBER TILE FLOORING**									
6050	Rubber tile, marbleized colors, 12" x 12", 1/8" thick	1 Tilf	400	.020	S.F.	5.75	.89		6.64	7.70
6100	3/16" thick		400	.020		9.05	.89		9.94	11.35
6300	Special tile, plain colors, 1/8" thick		400	.020		7.95	.89		8.84	10.15
6350	3/16" thick		400	.020		9.05	.89		9.94	11.35

09 65 33 – Conductive Resilient Flooring

09 65 33.10 Conductive Rubber and Vinyl Flooring

09 65 33.10 Conductive Rubber and Vinyl Flooring	Crew	Daily Output	Labor-Hours	Unit	Material	2016 Bare Costs Labor	Equipment	Total	Total Incl O&P
0010	**CONDUCTIVE RUBBER AND VINYL FLOORING**									
1700	Conductive flooring, rubber tile, 1/8" thick	1 Tilf	315	.025	S.F.	6.95	1.13		8.08	9.45
1800	Homogeneous vinyl tile, 1/8" thick	"	315	.025	"	6.70	1.13		7.83	9.20

09 65 66 – Resilient Athletic Flooring

09 65 66.10 Resilient Athletic Flooring

09 65 66.10 Resilient Athletic Flooring	Crew	Daily Output	Labor-Hours	Unit	Material	2016 Bare Costs Labor	Equipment	Total	Total Incl O&P
0010	**RESILIENT ATHLETIC FLOORING**									
1000	Recycled rubber rolled goods, for weight rooms, 3/8" thk.	1 Tilf	315	.025	S.F.	2.97	1.13		4.10	5.05
2000	Vinyl sheet flooring, 1/4" thk.	"	315	.025	"	4.39	1.13		5.52	6.60

09 66 Terrazzo Flooring

09 66 13 – Portland Cement Terrazzo Flooring

09 66 13.10 Portland Cement Terrazzo	Crew	Daily Output	Labor-Hours	Unit	Material	2016 Bare Costs Labor	Equipment	Total	Total Incl O&P
0010 **PORTLAND CEMENT TERRAZZO**, cast-in-place									
0020 Cove base, 6" high, 16 ga. zinc	1 Mstz	20	.400	L.F.	3.30	17.85		21.15	31.50
0100 Curb, 6" high and 6" wide		6	1.333		5.65	59.50		65.15	100
0300 Divider strip for floors, 14 ga., 1-1/4" deep, zinc		375	.021		1.54	.95		2.49	3.19
0400 Brass		375	.021		2.61	.95		3.56	4.37
0600 Heavy top strip 1/4" thick, 1-1/4" deep, zinc		300	.027		2.15	1.19		3.34	4.25
1200 For thin set floors, 16 ga., 1/2" x 1/2", zinc	▼	350	.023	▼	1.19	1.02		2.21	2.92
1500 Floor, bonded to concrete, 1-3/4" thick, gray cement	J-3	75	.213	S.F.	3.35	8.75	4.13	16.23	22
1600 White cement, mud set		75	.213		3.73	8.75	4.13	16.61	22.50
1800 Not bonded, 3" total thickness, gray cement		70	.229		4.17	9.35	4.43	17.95	24.50
1900 White cement, mud set	▼	70	.229	▼	4.86	9.35	4.43	18.64	25
9000 Minimum labor/equipment charge	1 Mstz	1	8	Job		355		355	565

09 66 16 – Terrazzo Floor Tile

09 66 16.10 Tile or Terrazzo Base

	Crew	Daily Output	Labor-Hours	Unit	Material	2016 Bare Costs Labor	Equipment	Total	Total Incl O&P
0010 **TILE OR TERRAZZO BASE**									
0020 Scratch coat only	1 Mstz	150	.053	S.F.	.43	2.38		2.81	4.22
0500 Scratch and brown coat only		75	.107	"	.82	4.76		5.58	8.40
9000 Minimum labor/equipment charge	▼	1	8	Job		355		355	565

09 66 16.13 Portland Cement Terrazzo Floor Tile

	Crew	Daily Output	Labor-Hours	Unit	Material	2016 Bare Costs Labor	Equipment	Total	Total Incl O&P
0010 **PORTLAND CEMENT TERRAZZO FLOOR TILE**									
1200 Floor tiles, non-slip, 1" thick, 12" x 12"	D-1	60	.267	S.F.	20.50	11.25		31.75	41
1300 1-1/4" thick, 12" x 12"		60	.267		20.50	11.25		31.75	41.50
1500 16" x 16"		50	.320		22.50	13.50		36	46.50
1600 1-1/2" thick, 16" x 16"	▼	45	.356		20.50	15		35.50	47
1800 For Venetian terrazzo, add					6.15			6.15	6.80
1900 For white cement, add				▼	.58			.58	.64

09 66 16.30 Terrazzo, Precast

	Crew	Daily Output	Labor-Hours	Unit	Material	2016 Bare Costs Labor	Equipment	Total	Total Incl O&P
0010 **TERRAZZO, PRECAST**									
0020 Base, 6" high, straight	1 Mstz	70	.114	L.F.	12.10	5.10		17.20	21.50
0100 Cove		60	.133		13.20	5.95		19.15	24
0300 8" high, straight		60	.133		11.80	5.95		17.75	22.50
0400 Cove	▼	50	.160		17.35	7.15		24.50	30.50
0600 For white cement, add					.46			.46	.51
0700 For 16 ga. zinc toe strip, add					1.81			1.81	1.99
0900 Curbs, 4" x 4" high	1 Mstz	40	.200		33.50	8.90		42.40	50.50
1000 8" x 8" high	"	30	.267		38.50	11.90		50.40	61.50
2400 Stair treads, 1-1/2" thick, non-slip, three line pattern	2 Mstz	70	.229		41	10.20		51.20	61
2500 Nosing and two lines		70	.229		41	10.20		51.20	61
2700 2" thick treads, straight		60	.267		46.50	11.90		58.40	70.50
2800 Curved		50	.320		61	14.25		75.25	89.50
3000 Stair risers, 1" thick, to 6" high, straight sections		60	.267		11.10	11.90		23	31
3100 Cove		50	.320		15.80	14.25		30.05	40
3300 Curved, 1" thick, to 6" high, vertical		48	.333		21.50	14.85		36.35	47
3400 Cove		38	.421		41	18.80		59.80	74.50
3600 Stair tread and riser, single piece, straight, smooth surface		60	.267		52.50	11.90		64.40	77
3700 Non skid surface		40	.400		68	17.85		85.85	103
3900 Curved tread and riser, smooth surface		40	.400		74	17.85		91.85	109
4000 Non skid surface		32	.500		93	22.50		115.50	137
4200 Stair stringers, notched, 1" thick		25	.640		30	28.50		58.50	78
4300 2" thick	▼	22	.727		35.50	32.50		68	90
4500 Stair landings, structural, non-slip, 1-1/2" thick		85	.188	S.F.	33	8.40		41.40	49.50
4600 3" thick	▼	75	.213	▼	46.50	9.50		56	66.50

09 66 Terrazzo Flooring

09 66 16 – Terrazzo Floor Tile

09 66 16.30 Terrazzo, Precast	Crew	Daily Output	Labor-Hours	Unit	Material	2016 Bare Costs Labor	Equipment	Total	Total Incl O&P
4800	Wainscot, 12" x 12" x 1" tiles	1 Mstz	12	.667	S.F.	7.10	29.50		36.60	55
4900	16" x 16" x 1-1/2" tiles	"	8	1	▼	14.25	44.50		58.75	86
9500	Minimum labor/equipment charge	1 Tilf	2	4	Job		178		178	281

09 66 23 – Resinous Matrix Terrazzo Flooring

09 66 23.16 Epoxy-Resin Terrazzo Flooring

		Crew	Daily Output	Labor-Hours	Unit	Material	Labor	Equipment	Total	Total Incl O&P
0010	**EPOXY-RESIN TERRAZZO FLOORING**									
1800	Epoxy terrazzo, 1/4" thick, chemical resistant, granite chips	J-3	200	.080	S.F.	6	3.28	1.55	10.83	13.50
1900	Recycled porcelain	"	150	.107	"	9.10	4.37	2.07	15.54	19.20

09 67 Fluid-Applied Flooring

09 67 13 – Elastomeric Liquid Flooring

09 67 13.13 Elastomeric Liquid Flooring

		Crew	Daily Output	Labor-Hours	Unit	Material	Labor	Equipment	Total	Total Incl O&P
0010	**ELASTOMERIC LIQUID FLOORING**									
0020	Cementitious acrylic, 1/4" thick	C-6	520	.092	S.F.	1.66	3.65	.12	5.43	7.90
0200	Methyl methachrylate, 1/4" thick	C-8A	3000	.016	"	6.50	.65		7.15	8.20

09 67 26 – Quartz Flooring

09 67 26.26 Quartz Flooring

		Crew	Daily Output	Labor-Hours	Unit	Material	Labor	Equipment	Total	Total Incl O&P
0010	**QUARTZ FLOORING**									
0600	Epoxy, with colored quartz chips, broadcast, 3/8" thick	C-6	675	.071	S.F.	2.81	2.81	.09	5.71	7.75
0700	1/2" thick		490	.098		4.05	3.87	.13	8.05	10.90
0900	Troweled, minimum		560	.086		3.57	3.39	.11	7.07	9.55
1000	Maximum	▼	480	.100	▼	5.40	3.95	.13	9.48	12.45
1200	Heavy duty epoxy topping, 1/4" thick,									
1300	500 to 1,000 S.F.	C-6	420	.114	S.F.	5.60	4.52	.15	10.27	13.65
1500	1,000 to 2,000 S.F.		450	.107		5.10	4.22	.14	9.46	12.60
1600	Over 10,000 S.F.	▼	480	.100	▼	4.70	3.95	.13	8.78	11.70

09 68 Carpeting

09 68 05 – Carpet Accessories

09 68 05.11 Flooring Transition Strip

		Crew	Daily Output	Labor-Hours	Unit	Material	Labor	Equipment	Total	Total Incl O&P
0010	**FLOORING TRANSITION STRIP**									
0107	Clamp down brass divider, 12' strip, vinyl to carpet	1 Tilf	31.25	.256	Ea.	13.95	11.40		25.35	33.50
0117	Vinyl to hard surface	"	31.25	.256	"	13.95	11.40		25.35	33.50

09 68 10 – Carpet Pad

09 68 10.10 Commercial Grade Carpet Pad

		Crew	Daily Output	Labor-Hours	Unit	Material	Labor	Equipment	Total	Total Incl O&P
0010	**COMMERCIAL GRADE CARPET PAD**									
9000	Sponge rubber pad, 20 oz./sq. yd.	1 Tilf	150	.053	S.Y.	4.60	2.37		6.97	8.80
9100	40 to 62 oz./sq. yd.		150	.053		8.75	2.37		11.12	13.40
9200	Felt pad, 20 oz./sq. yd.		150	.053		5.60	2.37		7.97	9.95
9300	32 to 56 oz./sq. yd.		150	.053		9.30	2.37		11.67	14
9400	Bonded urethane pad, 2.7 density		150	.053		5.95	2.37		8.32	10.30
9500	13.0 Density		150	.053		8	2.37		10.37	12.55
9600	Prime urethane pad, 2.7 density		150	.053		3.25	2.37		5.62	7.30
9700	13.0 density	▼	150	.053	▼	4.97	2.37		7.34	9.20

313

For customer support on your Commercial Renovation Cost Data, call 877.742.4719.

09 68 Carpeting

09 68 13 – Tile Carpeting

09 68 13.10 Carpet Tile

		Crew	Daily Output	Labor-Hours	Unit	Material	2016 Bare Costs Labor	Equipment	Total	Total Incl O&P
0010	**CARPET TILE**									
0100	Tufted nylon, 18" x 18", hard back, 20 oz.	1 Tilf	80	.100	S.Y.	24	4.45		28.45	33.50
0110	26 oz.		80	.100		29	4.45		33.45	39
0200	Cushion back, 20 oz.		80	.100		28	4.45		32.45	37.50
0210	26 oz.		80	.100		43	4.45		47.45	54
1100	Tufted, 24" x 24", hard back, 24 oz. nylon		80	.100		30.50	4.45		34.95	40.50
1180	35 oz.		80	.100		35.50	4.45		39.95	46
5060	42 oz.	↓	80	.100	↓	46	4.45		50.45	58

09 68 16 – Sheet Carpeting

09 68 16.10 Sheet Carpet

		Crew	Daily Output	Labor-Hours	Unit	Material	2016 Bare Costs Labor	Equipment	Total	Total Incl O&P
0010	**SHEET CARPET**									
0700	Nylon, level loop, 26 oz., light to medium traffic	1 Tilf	75	.107	S.Y.	21.50	4.75		26.25	31
0720	28 oz., light to medium traffic		75	.107		29.50	4.75		34.25	40
0900	32 oz., medium traffic		75	.107		39.50	4.75		44.25	51
1100	40 oz., medium to heavy traffic		75	.107		45	4.75		49.75	57
2920	Nylon plush, 30 oz., medium traffic		75	.107		31	4.75		35.75	41.50
3000	36 oz., medium traffic		75	.107		35	4.75		39.75	46
3100	42 oz., medium to heavy traffic		70	.114		45	5.10		50.10	57.50
3200	46 oz., medium to heavy traffic		70	.114		50.50	5.10		55.60	63.50
3300	54 oz., heavy traffic		70	.114		55	5.10		60.10	68.50
4110	Wool, level loop, 40 oz., medium traffic		70	.114		106	5.10		111.10	125
4500	50 oz., medium to heavy traffic		70	.114		100	5.10		105.10	118
4700	Patterned, 32 oz., medium to heavy traffic		70	.114		91	5.10		96.10	108
4900	48 oz., heavy traffic	↓	70	.114	↓	100	5.10		105.10	118
5000	For less than full roll (approx. 1500 S.F.), add					25%				
5100	For small rooms, less than 12' wide, add						25%			
5200	For large open areas (no cuts), deduct						25%			
5600	For bound carpet baseboard, add	1 Tilf	300	.027	L.F.	3	1.19		4.19	5.15
5610	For stairs, not incl. price of carpet, add	"	30	.267	Riser		11.85		11.85	18.70
5620	For borders and patterns, add to labor						18%			
8950	For tackless, stretched installation, add padding from 09 68 10.10 to above									
9850	For brand-named specific fiber, add				S.Y.	25%				
9900	Carpet cleaning machine, rent				Day				38	42
9910	Minimum labor/equipment charge	1 Tilf	3	2.667	Job		119		119	187

09 68 20 – Athletic Carpet

09 68 20.10 Indoor Athletic Carpet

		Crew	Daily Output	Labor-Hours	Unit	Material	2016 Bare Costs Labor	Equipment	Total	Total Incl O&P
0010	**INDOOR ATHLETIC CARPET**									
3700	Polyethylene, in rolls, no base incl., landscape surfaces	1 Tilf	275	.029	S.F.	4.16	1.29		5.45	6.60
3800	Nylon action surface, 1/8" thick		275	.029		3.76	1.29		5.05	6.20
3900	1/4" thick		275	.029		5.45	1.29		6.74	8
4000	3/8" thick		275	.029		6.80	1.29		8.09	9.55
5500	Polyvinyl chloride, sheet goods for gyms, 1/4" thick		80	.100		6.50	4.45		10.95	14.15
5600	3/8" thick	↓	60	.133	↓	9.55	5.95		15.50	19.85

09 69 Access Flooring

09 69 13 – Rigid-Grid Access Flooring

09 69 13.10 Access Floors	Crew	Daily Output	Labor-Hours	Unit	Material	2016 Bare Costs Labor	Equipment	Total	Total Incl O&P
0010 **ACCESS FLOORS**									
0015 Access floor pkg. including panel, pedestal, & stringers 1500 lbs load									
0100 Package pricing, conc. fill panels, no fin., 6" ht.	4 Carp	750	.043	S.F.	15.85	2.07		17.92	21
0105 12" ht.		750	.043		16.40	2.07		18.47	21.50
0110 18" ht.		750	.043		16.05	2.07		18.12	21
0115 24" ht.		750	.043		17.05	2.07		19.12	22
0120 Package pricing, steel panels, no fin., 6" ht.		750	.043		12.55	2.07		14.62	17.20
0125 12" ht.		750	.043		15.20	2.07		17.27	20
0130 18" ht.		750	.043		15.25	2.07		17.32	20
0135 24" ht.		750	.043		15.35	2.07		17.42	20.50
0140 Package pricing, wood core panels, no fin., 6" ht.		750	.043		11.50	2.07		13.57	16.05
0145 12" ht.		750	.043		11.50	2.07		13.57	16.05
0150 18" ht.		750	.043		11.65	2.07		13.72	16.20
0155 24" ht.		750	.043		11.75	2.07		13.82	16.35
0160 Pkg. pricing, conc. fill pnls, no stringers, no fin., 6" ht, 1250 lbs load		700	.046		10.20	2.21		12.41	14.85
0165 12" ht.		700	.046		10	2.21		12.21	14.60
0170 18" ht.		700	.046		10.55	2.21		12.76	15.20
0175 24" ht.		700	.046		11.10	2.21		13.31	15.80
0250 Panels, 2' x 2' conc. fill, no fin.	2 Carp	500	.032		20.50	1.55		22.05	25
0255 With 1/8" high pressure laminate		500	.032		61	1.55		62.55	70
0260 Metal panels with 1/8" high pressure laminate		500	.032		72.50	1.55		74.05	82
0265 Wood core panels with 1/8" high pressure laminate		500	.032		47.50	1.55		49.05	55
0400 Aluminum panels, no fin.		500	.032		33	1.55		34.55	38.50
0600 For carpet covering, add					8.80			8.80	9.70
0700 For vinyl floor covering, add					9.20			9.20	10.15
0900 For high pressure laminate covering, add					7.50			7.50	8.25
0910 For snap on stringer system, add	2 Carp	1000	.016		1.56	.78		2.34	2.99
1000 Machine cutouts after initial installation	1 Carp	50	.160	Ea.	20	7.75		27.75	34.50
1050 Pedestals, 6" to 12"	2 Carp	85	.188		8.40	9.10		17.50	24
1100 Air conditioning grilles, 4" x 12"	1 Carp	17	.471		71	23		94	116
1150 4" x 18"	"	14	.571		97.50	27.50		125	152
1200 Approach ramps, steel	2 Carp	60	.267	S.F.	25	12.90		37.90	49
1300 Aluminum	"	40	.400	"	34	19.40		53.40	69
1500 Handrail, 2 rail, aluminum	1 Carp	15	.533	L.F.	109	26		135	161

09 72 Wall Coverings

09 72 19 – Textile Wall Coverings

09 72 19.10 Textile Wall Covering

	Crew	Daily Output	Labor-Hours	Unit	Material	Labor	Equipment	Total	Total Incl O&P
0010 **TEXTILE WALL COVERING**, including sizing; add 10-30% waste @ takeoff									
0020 Silk	1 Pape	640	.013	S.F.	4.23	.51		4.74	5.45
0030 Cotton		640	.013		6.75	.51		7.26	8.25
0040 Linen		640	.013		1.89	.51		2.40	2.90
0050 Blend		640	.013		3.03	.51		3.54	4.15

09 72 20 – Natural Fiber Wall Covering

09 72 20.10 Natural Fiber Wall Covering

	Crew	Daily Output	Labor-Hours	Unit	Material	Labor	Equipment	Total	Total Incl O&P
0010 **NATURAL FIBER WALL COVERING**, including sizing; add 10-30% waste @ takeoff									
0015 Bamboo	1 Pape	640	.013	S.F.	2.31	.51		2.82	3.36
0030 Burlap		640	.013		2.07	.51		2.58	3.10
0045 Jute		640	.013		1.32	.51		1.83	2.27
0060 Sisal		640	.013		1.49	.51		2	2.46

09 72 23 – Wallpapering

09 72 23.10 Wallpaper	Crew	Daily Output	Labor-Hours	Unit	Material	2016 Bare Costs Labor	Equipment	Total	Total Incl O&P
0010 **WALLPAPER** including sizing; add 10-30 percent waste @ takeoff									
0050 Aluminum foil	1 Pape	275	.029	S.F.	1.04	1.18		2.22	3.04
0100 Copper sheets, .025" thick, vinyl backing		240	.033		5.55	1.35		6.90	8.25
0300 Phenolic backing		240	.033		7.20	1.35		8.55	10.10
0600 Cork tiles, light or dark, 12" x 12" x 3/16"		240	.033		4.66	1.35		6.01	7.30
0700 5/16" thick		235	.034		3.13	1.38		4.51	5.65
0900 1/4" basketweave		240	.033		3.50	1.35		4.85	6
1000 1/2" natural, non-directional pattern		240	.033		6.55	1.35		7.90	9.35
1100 3/4" natural, non-directional pattern		240	.033		11.95	1.35		13.30	15.30
1200 Granular surface, 12" x 36", 1/2" thick		385	.021		1.31	.84		2.15	2.80
1300 1" thick		370	.022		1.68	.88		2.56	3.26
1500 Polyurethane coated, 12" x 12" x 3/16" thick		240	.033		4.08	1.35		5.43	6.65
1600 5/16" thick		235	.034		6.60	1.38		7.98	9.45
1800 Cork wallpaper, paperbacked, natural		480	.017		2	.68		2.68	3.29
1900 Colors		480	.017		2.87	.68		3.55	4.25
2100 Flexible wood veneer, 1/32" thick, plain woods		100	.080		2.44	3.25		5.69	7.90
2200 Exotic woods	▼	95	.084	▼	3.69	3.42		7.11	9.55
2400 Gypsum-based, fabric-backed, fire resistant									
2500 for masonry walls, 21 oz./S.Y.	1 Pape	800	.010	S.F.	.77	.41		1.18	1.50
2600 Average		720	.011		1.27	.45		1.72	2.12
2700 Small quantities	▼	640	.013		.70	.51		1.21	1.59
2750 Acrylic, modified, semi-rigid PVC, .028" thick	2 Carp	330	.048		1.28	2.35		3.63	5.25
2800 .040" thick	"	320	.050		1.85	2.42		4.27	6
3000 Vinyl wall covering, fabric-backed, lightweight, type 1 (12-15 oz./S.Y.)	1 Pape	640	.013		.97	.51		1.48	1.89
3300 Medium weight, type 2 (20-24 oz./S.Y.)		480	.017		.90	.68		1.58	2.08
3400 Heavy weight, type 3 (28 oz./S.Y.)	▼	435	.018		1.37	.75		2.12	2.71
3600 Adhesive, 5 gal. lots (18 S.Y./gal.)				Gal.	14			14	15.40
3700 Wallpaper, average workmanship, solid pattern, low cost paper	1 Pape	640	.013	S.F.	.59	.51		1.10	1.47
3900 Basic patterns (matching required), avg. cost paper		535	.015		1.13	.61		1.74	2.22
4000 Paper at $85 per double roll, quality workmanship		435	.018		1.72	.75		2.47	3.09
4190 Grass cloth, natural fabric G		400	.020		1.86	.81		2.67	3.36
4200 Grass cloths with lining paper G		400	.020		.96	.81		1.77	2.37
4300 Premium texture/color G		350	.023		2.85	.93		3.78	4.63
5990 Wallpaper removal, 1 layer		800	.010		.03	.41		.44	.68
6000 Wallpaper removal, 3 layer	▼	400	.020	▼	.08	.81		.89	1.40
9000 Minimum labor/equipment charge	▼	2	4	Job		162		162	261

09 77 Special Wall Surfacing

09 77 30 – Fiberglass Reinforced Panels

09 77 30.10 Fiberglass Reinforced Plastic Panels

	Crew	Daily Output	Labor-Hours	Unit	Material	2016 Bare Costs Labor	Equipment	Total	Total Incl O&P
0010 **FIBERGLASS REINFORCED PLASTIC PANELS**, .090" thick									
0020 On walls, adhesive mounted, embossed surface	2 Carp	640	.025	S.F.	1.11	1.21		2.32	3.20
0030 Smooth surface		640	.025		1.37	1.21		2.58	3.49
0040 Fire rated, embossed surface		640	.025		1.99	1.21		3.20	4.17
0050 Nylon rivet mounted, on drywall, embossed surface		480	.033		1.11	1.61		2.72	3.86
0060 Smooth surface		480	.033		1.37	1.61		2.98	4.15
0070 Fire rated, embossed surface		480	.033		1.99	1.61		3.60	4.83
0080 On masonry, embossed surface		320	.050		1.11	2.42		3.53	5.20
0090 Smooth surface		320	.050		1.37	2.42		3.79	5.45
0100 Fire rated, embossed surface		320	.050		1.99	2.42		4.41	6.15

09 77 Special Wall Surfacing

09 77 30 – Fiberglass Reinforced Panels

09 77 30.10 Fiberglass Reinforced Plastic Panels	Crew	Daily Output	Labor-Hours	Unit	Material	2016 Bare Costs Labor	Equipment	Total	Total Incl O&P
0110	Nylon rivet and adhesive mounted, on drywall, embossed surface	2 Carp	240	.067	S.F.	1.26	3.23		4.49	6.70
0120	Smooth surface		240	.067		1.26	3.23		4.49	6.70
0130	Fire rated, embossed surface		240	.067		2.19	3.23		5.42	7.70
0140	On masonry, embossed surface		190	.084		1.26	4.08		5.34	8.05
0150	Smooth surface		190	.084		1.26	4.08		5.34	8.05
0160	Fire rated, embossed surface	▼	190	.084	▼	2.19	4.08		6.27	9.05
0170	For moldings, add	1 Carp	250	.032	L.F.	.26	1.55		1.81	2.82
0180	On ceilings, for lay in grid system, embossed surface		400	.020	S.F.	1.11	.97		2.08	2.80
0190	Smooth surface		400	.020		1.37	.97		2.34	3.09
0200	Fire rated, embossed surface	▼	400	.020	▼	1.99	.97		2.96	3.77

09 77 43 – Panel Systems

09 77 43.20 Slatwall Panels and Accessories

09 77 43.20 Slatwall Panels and Accessories	Crew	Daily Output	Labor-Hours	Unit	Material	2016 Bare Costs Labor	Equipment	Total	Total Incl O&P
0010	**SLATWALL PANELS AND ACCESSORIES**									
0100	Slatwall panel, 4' x 8' x 3/4" T, MDF, paint grade	1 Carp	500	.016	S.F.	1.42	.78		2.20	2.83
0110	Melamine finish		500	.016		1.92	.78		2.70	3.38
0120	High pressure plastic laminate finish		500	.016		3.74	.78		4.52	5.40
0125	Wood veneer	▼	500	.016		3.55	.78		4.33	5.20
0130	Aluminum channel inserts, add				▼	2.59			2.59	2.85
0200	Accessories, corner forms, 8' L				L.F.	6			6	6.60
0210	T-connector, 8' L					8.60			8.60	9.45
0220	J-mold, 8' L					1.25			1.25	1.38
0230	Edge cap, 8' L					1.46			1.46	1.61
0240	Finish end cap, 8' L				▼	3.37			3.37	3.71
0300	Display hook, metal, 4" L				Ea.	.43			.43	.47
0310	6" L					.48			.48	.53
0320	8" L					.52			.52	.57
0330	10" L					.59			.59	.65
0340	12" L					.64			.64	.70
0350	Acrylic, 4" L					.93			.93	1.02
0360	6" L					1.07			1.07	1.18
0370	8" L					1.13			1.13	1.24
0380	10" L					1.28			1.28	1.41
0400	Waterfall hanger, metal, 12" - 16"					3.84			3.84	4.22
0410	Acrylic					11.50			11.50	12.65
0500	Shelf bracket, metal, 8"					1.79			1.79	1.97
0510	10"					1.97			1.97	2.17
0520	12"					2.17			2.17	2.39
0530	14"					2.39			2.39	2.63
0540	16"					2.81			2.81	3.09
0550	Acrylic, 8"					3.81			3.81	4.19
0560	10"					4.06			4.06	4.47
0570	12"					4.29			4.29	4.72
0580	14"					4.52			4.52	4.97
0600	Shelf, acrylic, 12" x 16" x 1/4"					20			20	22
0610	12" x 24" x 1/4"				▼	24.50			24.50	27

09 81 Acoustic Insulation

09 81 16 – Acoustic Blanket Insulation

09 81 16.10 Sound Attenuation Blanket		Crew	Daily Output	Labor-Hours	Unit	Material	2016 Bare Costs Labor	Equipment	Total	Total Incl O&P
0010	**SOUND ATTENUATION BLANKET**									
0020	Blanket, 1" thick	1 Carp	925	.009	S.F.	.25	.42		.67	.96
0500	1-1/2" thick		920	.009		.25	.42		.67	.97
1000	2" thick		915	.009		.36	.42		.78	1.09
1500	3" thick		910	.009		.52	.43		.95	1.27
2000	Wall hung, STC 18 – 21, 1" thick, 4' x 20'	2 Carp	22	.727	Ea.	390	35		425	490
2010	10' x 20'	"	19	.842		975	41		1,016	1,150
2020	Wall hung, STC 27 – 28, 3" thick, 4' x 20'	3 Carp	12	2		560	97		657	775
2030	10' x 20'	"	9	2.667		1,400	129		1,529	1,725
3400	Urethane plastic foam, open cell, on wall, 2" thick	2 Carp	2050	.008	S.F.	3.11	.38		3.49	4.04
3500	3" thick		1550	.010		4.13	.50		4.63	5.35
3600	4" thick		1050	.015		5.80	.74		6.54	7.55
3700	On ceiling, 2" thick		1700	.009		3.10	.46		3.56	4.15
3800	3" thick		1300	.012		4.13	.60		4.73	5.50
3900	4" thick		900	.018		5.80	.86		6.66	7.75
9000	Minimum labor/equipment charge	1 Carp	5	1.600	Job		77.50		77.50	127

09 84 Acoustic Room Components

09 84 13 – Fixed Sound-Absorptive Panels

09 84 13.10 Fixed Panels

09 84 13.10 Fixed Panels		Crew	Daily Output	Labor-Hours	Unit	Material	2016 Bare Costs Labor	Equipment	Total	Total Incl O&P
0010	**FIXED PANELS** Perforated steel facing, painted with									
0100	Fiberglass or mineral filler, no backs, 2-1/4" thick, modular									
0200	space units, ceiling or wall hung, white or colored	1 Carp	100	.080	S.F.	8.90	3.88		12.78	16.15
0300	Fiberboard sound deadening panels, 1/2" thick	"	600	.013	"	.33	.65		.98	1.42
0500	Fiberglass panels, 4' x 8' x 1" thick, with									
0600	glass cloth face for walls, cemented	1 Carp	155	.052	S.F.	8.55	2.50		11.05	13.50
0700	1-1/2" thick, dacron covered, inner aluminum frame,									
0710	wall mounted	1 Carp	300	.027	S.F.	8.90	1.29		10.19	11.90
0900	Mineral fiberboard panels, fabric covered, 30" x 108",									
1000	3/4" thick, concealed spline, wall mounted	1 Carp	150	.053	S.F.	6.40	2.58		8.98	11.25
9000	Minimum labor/equipment charge	"	4	2	Job		97		97	158

09 91 Painting

09 91 03 – Paint Restoration

09 91 03.20 Sanding

09 91 03.20 Sanding		Crew	Daily Output	Labor-Hours	Unit	Material	2016 Bare Costs Labor	Equipment	Total	Total Incl O&P
0010	**SANDING** and puttying interior trim, compared to									
0100	Painting 1 coat, on quality work				L.F.		100%			
0300	Medium work						50%			
0400	Industrial grade						25%			
0500	Surface protection, placement and removal									
0510	Basic drop cloths	1 Pord	6400	.001	S.F.		.05		.05	.08
0520	Masking with paper		800	.010		.07	.40		.47	.73
0530	Volume cover up (using plastic sheathing or building paper)		16000	.001			.02		.02	.03

09 91 03.30 Exterior Surface Preparation

09 91 03.30 Exterior Surface Preparation		Crew	Daily Output	Labor-Hours	Unit	Material	2016 Bare Costs Labor	Equipment	Total	Total Incl O&P
0010	**EXTERIOR SURFACE PREPARATION**									
0015	Doors, per side, not incl. frames or trim									
0020	Scrape & sand									
0030	Wood, flush	1 Pord	616	.013	S.F.		.52		.52	.84
0040	Wood, detail		496	.016			.65		.65	1.05

318

09 91 03.30 Exterior Surface Preparation	Crew	Daily Output	Labor-Hours	Unit	Material	2016 Bare Costs Labor	Equipment	Total	Total Incl O&P
0050	Wood, louvered	1 Pord	280	.029	S.F.		1.15		1.15	1.85
0060	Wood, overhead	↓	616	.013	↓		.52		.52	.84
0070	Wire brush									
0080	Metal, flush	1 Pord	640	.013	S.F.		.50		.50	.81
0090	Metal, detail		520	.015			.62		.62	1
0100	Metal, louvered		360	.022			.90		.90	1.44
0110	Metal or fibr., overhead		640	.013			.50		.50	.81
0120	Metal, roll up		560	.014			.58		.58	.93
0130	Metal, bulkhead	↓	640	.013	↓		.50		.50	.81
0140	Power wash, based on 2500 lb. operating pressure									
0150	Metal, flush	A-1H	2240	.004	S.F.		.14	.03	.17	.26
0160	Metal, detail		2120	.004			.14	.04	.18	.27
0170	Metal, louvered		2000	.004			.15	.04	.19	.29
0180	Metal or fibr., overhead		2400	.003			.13	.03	.16	.25
0190	Metal, roll up		2400	.003			.13	.03	.16	.25
0200	Metal, bulkhead	↓	2200	.004	↓		.14	.04	.18	.27
0400	Windows, per side, not incl. trim									
0410	Scrape & sand									
0420	Wood, 1-2 lite	1 Pord	320	.025	S.F.		1.01		1.01	1.62
0430	Wood, 3-6 lite		280	.029			1.15		1.15	1.85
0440	Wood, 7-10 lite		240	.033			1.34		1.34	2.16
0450	Wood, 12 lite		200	.040			1.61		1.61	2.59
0460	Wood, Bay/Bow	↓	320	.025	↓		1.01		1.01	1.62
0470	Wire brush									
0480	Metal, 1-2 lite	1 Pord	480	.017	S.F.		.67		.67	1.08
0490	Metal, 3-6 lite		400	.020			.81		.81	1.30
0500	Metal, Bay/Bow	↓	480	.017	↓		.67		.67	1.08
0510	Power wash, based on 2500 lb. operating pressure									
0520	1-2 lite	A-1H	4400	.002	S.F.		.07	.02	.09	.13
0530	3-6 lite		4320	.002			.07	.02	.09	.13
0540	7-10 lite		4240	.002			.07	.02	.09	.14
0550	12 lite		4160	.002			.07	.02	.09	.14
0560	Bay/Bow	↓	4400	.002	↓		.07	.02	.09	.13
0600	Siding, scrape and sand, light=10-30%, med.=30-70%									
0610	Heavy=70-100% of surface to sand									
0650	Texture 1-11, light	1 Pord	480	.017	S.F.		.67		.67	1.08
0660	Med.		440	.018			.73		.73	1.18
0670	Heavy		360	.022			.90		.90	1.44
0680	Wood shingles, shakes, light		440	.018			.73		.73	1.18
0690	Med.		360	.022			.90		.90	1.44
0700	Heavy		280	.029			1.15		1.15	1.85
0710	Clapboard, light		520	.015			.62		.62	1
0720	Med.		480	.017			.67		.67	1.08
0730	Heavy	↓	400	.020	↓		.81		.81	1.30
0740	Wire brush									
0750	Aluminum, light	1 Pord	600	.013	S.F.		.54		.54	.86
0760	Med.		520	.015			.62		.62	1
0770	Heavy	↓	440	.018	↓		.73		.73	1.18
0780	Pressure wash, based on 2500 lb. operating pressure									
0790	Stucco	A-1H	3080	.003	S.F.		.10	.03	.13	.19
0800	Aluminum or vinyl		3200	.003			.09	.02	.11	.18
0810	Siding, masonry, brick & block	↓	2400	.003	↓		.13	.03	.16	.25
1300	Miscellaneous, wire brush									

For customer support on your Commercial Renovation Cost Data, call 877.742.4719.

319

09 91 Painting

09 91 03 – Paint Restoration

09 91 03.30 Exterior Surface Preparation

		Crew	Daily Output	Labor-Hours	Unit	Material	2016 Bare Costs Labor	Equipment	Total	Total Incl O&P
1310	Metal, pedestrian gate	1 Pord	100	.080	S.F.		3.23		3.23	5.20
1320	Aluminum chain link, both sides		250	.032			1.29		1.29	2.08
1400	Existing galvanized surface, clean and prime, prep for painting	↓	380	.021	↓	.12	.85		.97	1.51
8000	For chemical washing, see Section 04 01 30									
8010	For steam cleaning, see Section 04 01 30.20									

09 91 03.40 Interior Surface Preparation

		Crew	Daily Output	Labor-Hours	Unit	Material	2016 Bare Costs Labor	Equipment	Total	Total Incl O&P
0010	**INTERIOR SURFACE PREPARATION**									
0020	Doors, per side, not incl. frames or trim									
0030	Scrape & sand									
0040	Wood, flush	1 Pord	616	.013	S.F.		.52		.52	.84
0050	Wood, detail		496	.016			.65		.65	1.05
0060	Wood, louvered	↓	280	.029			1.15		1.15	1.85
0070	Wire brush									
0080	Metal, flush	1 Pord	640	.013	S.F.		.50		.50	.81
0090	Metal, detail		520	.015			.62		.62	1
0100	Metal, louvered	↓	360	.022	↓		.90		.90	1.44
0110	Hand wash									
0120	Wood, flush	1 Pord	2160	.004	S.F.		.15		.15	.24
0130	Wood, detailed		2000	.004			.16		.16	.26
0140	Wood, louvered		1360	.006			.24		.24	.38
0150	Metal, flush		2160	.004			.15		.15	.24
0160	Metal, detail		2000	.004			.16		.16	.26
0170	Metal, louvered	↓	1360	.006	↓		.24		.24	.38
0400	Windows, per side, not incl. trim									
0410	Scrape & sand									
0420	Wood, 1-2 lite	1 Pord	360	.022	S.F.		.90		.90	1.44
0430	Wood, 3-6 lite		320	.025			1.01		1.01	1.62
0440	Wood, 7-10 lite		280	.029			1.15		1.15	1.85
0450	Wood, 12 lite		240	.033			1.34		1.34	2.16
0460	Wood, Bay/Bow	↓	360	.022	↓		.90		.90	1.44
0470	Wire brush									
0480	Metal, 1-2 lite	1 Pord	520	.015	S.F.		.62		.62	1
0490	Metal, 3-6 lite		440	.018			.73		.73	1.18
0500	Metal, Bay/Bow	↓	520	.015	↓		.62		.62	1
0600	Walls, sanding, light=10-30%, medium - 30-70%,									
0610	heavy=70-100% of surface to sand									
0650	Walls, sand									
0660	Gypsum board or plaster, light	1 Pord	3077	.003	S.F.		.10		.10	.17
0670	Gypsum board or plaster, medium		2160	.004			.15		.15	.24
0680	Gypsum board or plaster, heavy		923	.009			.35		.35	.56
0690	Wood, T&G, light		2400	.003			.13		.13	.22
0700	Wood, T&G, med.		1600	.005			.20		.20	.32
0710	Wood, T&G, heavy	↓	800	.010	↓		.40		.40	.65
0720	Walls, wash									
0730	Gypsum board or plaster	1 Pord	3200	.003	S.F.		.10		.10	.16
0740	Wood, T&G		3200	.003	↓		.10		.10	.16
0750	Masonry, brick & block, smooth		2800	.003			.12		.12	.19
0760	Masonry, brick & block, coarse	↓	2000	.004	↓		.16		.16	.26
8000	For chemical washing, see Section 04 01 30									
8010	For steam cleaning, see Section 04 01 30.20									
9010	Minimum labor/equipment charge	1 Pord	3	2.667	Job		108		108	173

09 91 Painting

09 91 03 – Paint Restoration

09 91 03.41 Scrape After Fire Damage

		Crew	Daily Output	Labor-Hours	Unit	Material	2016 Bare Costs Labor	Equipment	Total	Total Incl O&P
0010	**SCRAPE AFTER FIRE DAMAGE**									
0050	Boards, 1" x 4"	1 Pord	336	.024	L.F.		.96		.96	1.54
0060	1" x 6"		260	.031			1.24		1.24	2
0070	1" x 8"		207	.039			1.56		1.56	2.51
0080	1" x 10"		174	.046			1.86		1.86	2.98
0500	Framing, 2" x 4"		265	.030			1.22		1.22	1.96
0510	2" x 6"		221	.036			1.46		1.46	2.35
0520	2" x 8"		190	.042			1.70		1.70	2.73
0530	2" x 10"		165	.048			1.96		1.96	3.14
0540	2" x 12"		144	.056			2.24		2.24	3.60
1000	Heavy framing, 3" x 4"		226	.035			1.43		1.43	2.30
1010	4" x 4"		210	.038			1.54		1.54	2.47
1020	4" x 6"		191	.042			1.69		1.69	2.72
1030	4" x 8"		165	.048			1.96		1.96	3.14
1040	4" x 10"		144	.056			2.24		2.24	3.60
1060	4" x 12"		131	.061			2.46		2.46	3.96
2900	For sealing, light damage		825	.010	S.F.	.14	.39		.53	.78
2920	Heavy damage		460	.017	"	.31	.70		1.01	1.47
3000	For sandblasting, see Section 04 01 30.20									
9000	Minimum labor/equipment charge	1 Pord	3	2.667	Job		108		108	173

09 91 13 – Exterior Painting

09 91 13.30 Fences

		Crew	Daily Output	Labor-Hours	Unit	Material	2016 Bare Costs Labor	Equipment	Total	Total Incl O&P
0010	**FENCES**									
0100	Chain link or wire metal, one side, water base									
0110	Roll & brush, first coat	1 Pord	960	.008	S.F.	.08	.34		.42	.62
0120	Second coat		1280	.006		.07	.25		.32	.49
0130	Spray, first coat		2275	.004		.08	.14		.22	.31
0140	Second coat		2600	.003		.08	.12		.20	.28
0150	Picket, water base									
0160	Roll & brush, first coat	1 Pord	865	.009	S.F.	.08	.37		.45	.69
0170	Second coat		1050	.008		.08	.31		.39	.58
0180	Spray, first coat		2275	.004		.08	.14		.22	.32
0190	Second coat		2600	.003		.08	.12		.20	.29
0200	Stockade, water base									
0210	Roll & brush, first coat	1 Pord	1040	.008	S.F.	.08	.31		.39	.59
0220	Second coat		1200	.007		.08	.27		.35	.52
0230	Spray, first coat		2275	.004		.08	.14		.22	.32
0240	Second coat		2600	.003		.08	.12		.20	.29
9000	Minimum labor/equipment charge		2	4	Job		161		161	259

09 91 13.42 Miscellaneous, Exterior

		Crew	Daily Output	Labor-Hours	Unit	Material	2016 Bare Costs Labor	Equipment	Total	Total Incl O&P
0010	**MISCELLANEOUS, EXTERIOR**									
0100	Railing, ext., decorative wood, incl. cap & baluster									
0110	Newels & spindles @ 12" O.C.									
0120	Brushwork, stain, sand, seal & varnish									
0130	First coat	1 Pord	90	.089	L.F.	.85	3.59		4.44	6.70
0140	Second coat	"	120	.067	"	.85	2.69		3.54	5.25
0150	Rough sawn wood, 42" high, 2" x 2" verticals, 6" O.C.									
0160	Brushwork, stain, each coat	1 Pord	90	.089	L.F.	.26	3.59		3.85	6.05
0170	Wrought iron, 1" rail, 1/2" sq. verticals									
0180	Brushwork, zinc chromate, 60" high, bars 6" O.C.									
0190	Primer	1 Pord	130	.062	L.F.	.88	2.48		3.36	4.96

For customer support on your Commercial Renovation Cost Data, call 877.742.4719.

321

09 91 13 – Exterior Painting

09 91 13.42 Miscellaneous, Exterior

	Crew	Daily Output	Labor-Hours	Unit	Material	2016 Bare Costs Labor	Equipment	Total	Total Incl O&P
0200 Finish coat	1 Pord	130	.062	L.F.	1.13	2.48		3.61	5.25
0210 Additional coat	↓	190	.042	↓	1.32	1.70		3.02	4.18
0220 Shutters or blinds, single panel, 2' x 4', paint all sides									
0230 Brushwork, primer	1 Pord	20	.400	Ea.	.67	16.15		16.82	26.50
0240 Finish coat, exterior latex		20	.400		.62	16.15		16.77	26.50
0250 Primer & 1 coat, exterior latex		13	.615		1.14	25		26.14	41.50
0260 Spray, primer		35	.229		.98	9.20		10.18	15.90
0270 Finish coat, exterior latex		35	.229		1.31	9.20		10.51	16.25
0280 Primer & 1 coat, exterior latex	↓	20	.400	↓	1.06	16.15		17.21	27
0290 For louvered shutters, add				S.F.	10%				
0300 Stair stringers, exterior, metal									
0310 Roll & brush, zinc chromate, to 14", each coat	1 Pord	320	.025	L.F.	.38	1.01		1.39	2.03
0320 Rough sawn wood, 4" x 12"									
0330 Roll & brush, exterior latex, each coat	1 Pord	215	.037	L.F.	.09	1.50		1.59	2.51
0340 Trellis/lattice, 2" x 2" @ 3" O.C. with 2" x 8" supports									
0350 Spray, latex, per side, each coat	1 Pord	475	.017	S.F.	.09	.68		.77	1.19
0450 Decking, ext., sealer, alkyd, brushwork, sealer coat		1140	.007		.10	.28		.38	.57
0460 1st coat		1140	.007		.10	.28		.38	.57
0470 2nd coat		1300	.006		.07	.25		.32	.48
0500 Paint, alkyd, brushwork, primer coat		1140	.007		.12	.28		.40	.59
0510 1st coat		1140	.007		.13	.28		.41	.60
0520 2nd coat		1300	.006		.09	.25		.34	.50
0600 Sand paint, alkyd, brushwork, 1 coat		150	.053	↓	.13	2.15		2.28	3.61
9000 Minimum labor/equipment charge	↓	2	4	Job		161		161	259

09 91 13.60 Siding Exterior

	Crew	Daily Output	Labor-Hours	Unit	Material	2016 Bare Costs Labor	Equipment	Total	Total Incl O&P
0010 **SIDING EXTERIOR**, Alkyd (oil base)									
0450 Steel siding, oil base, paint 1 coat, brushwork	2 Pord	2015	.008	S.F.	.10	.32		.42	.62
0500 Spray		4550	.004		.15	.14		.29	.40
0800 Paint 2 coats, brushwork		1300	.012		.20	.50		.70	1.02
1000 Spray		2750	.006		.17	.23		.40	.57
1200 Stucco, rough, oil base, paint 2 coats, brushwork		1300	.012		.20	.50		.70	1.02
1400 Roller		1625	.010		.21	.40		.61	.87
1600 Spray		2925	.005		.22	.22		.44	.60
1800 Texture 1-11 or clapboard, oil base, primer coat, brushwork		1300	.012		.15	.50		.65	.97
2000 Spray		4550	.004		.15	.14		.29	.40
2400 Paint 2 coats, brushwork		810	.020		.30	.80		1.10	1.60
2600 Spray		2600	.006		.33	.25		.58	.76
3400 Stain 2 coats, brushwork		950	.017		.17	.68		.85	1.28
4000 Spray		3050	.005		.19	.21		.40	.55
4200 Wood shingles, oil base primer coat, brushwork		1300	.012		.14	.50		.64	.95
4400 Spray		3900	.004		.13	.17		.30	.41
5000 Paint 2 coats, brushwork		810	.020		.25	.80		1.05	1.55
5200 Spray		2275	.007		.23	.28		.51	.72
6500 Stain 2 coats, brushwork		950	.017		.17	.68		.85	1.28
7000 Spray	↓	2660	.006		.24	.24		.48	.65
8000 For latex paint, deduct					10%				
8100 For work over 12' H, from pipe scaffolding, add						15%			
8200 For work over 12' H, from extension ladder, add						25%			
8300 For work over 12' H, from swing staging, add				↓		35%			
9000 Minimum labor/equipment charge	1 Pord	2	4	Job		161		161	259

322

For customer support on your Commercial Renovation Cost Data, call 877.742.4719.

09 91 13.62 Siding, Misc.		Crew	Daily Output	Labor-Hours	Unit	Material	2016 Bare Costs Labor	Equipment	Total	Total Incl O&P
0010	**SIDING, MISC.**, latex paint									
0100	Aluminum siding									
0110	Brushwork, primer	2 Pord	2275	.007	S.F.	.06	.28		.34	.53
0120	Finish coat, exterior latex		2275	.007		.06	.28		.34	.52
0130	Primer & 1 coat exterior latex		1300	.012		.13	.50		.63	.94
0140	Primer & 2 coats exterior latex		975	.016		.19	.66		.85	1.27
0150	Mineral fiber shingles									
0160	Brushwork, primer	2 Pord	1495	.011	S.F.	.15	.43		.58	.86
0170	Finish coat, industrial enamel		1495	.011		.17	.43		.60	.87
0180	Primer & 1 coat enamel		810	.020		.32	.80		1.12	1.63
0190	Primer & 2 coats enamel		540	.030		.48	1.20		1.68	2.45
0200	Roll, primer		1625	.010		.17	.40		.57	.83
0210	Finish coat, industrial enamel		1625	.010		.18	.40		.58	.84
0220	Primer & 1 coat enamel		975	.016		.35	.66		1.01	1.44
0230	Primer & 2 coats enamel		650	.025		.53	.99		1.52	2.18
0240	Spray, primer		3900	.004		.13	.17		.30	.41
0250	Finish coat, industrial enamel		3900	.004		.15	.17		.32	.43
0260	Primer & 1 coat enamel		2275	.007		.28	.28		.56	.77
0270	Primer & 2 coats enamel		1625	.010		.43	.40		.83	1.11
0280	Waterproof sealer, first coat		4485	.004		.10	.14		.24	.34
0290	Second coat		5235	.003		.09	.12		.21	.30
0300	Rough wood incl. shingles, shakes or rough sawn siding									
0310	Brushwork, primer	2 Pord	1280	.013	S.F.	.14	.50		.64	.96
0320	Finish coat, exterior latex		1280	.013		.10	.50		.60	.92
0330	Primer & 1 coat exterior latex		960	.017		.24	.67		.91	1.34
0340	Primer & 2 coats exterior latex		700	.023		.34	.92		1.26	1.86
0350	Roll, primer		2925	.005		.18	.22		.40	.55
0360	Finish coat, exterior latex		2925	.005		.12	.22		.34	.49
0370	Primer & 1 coat exterior latex		1790	.009		.30	.36		.66	.92
0380	Primer & 2 coats exterior latex		1300	.012		.43	.50		.93	1.27
0390	Spray, primer		3900	.004		.15	.17		.32	.44
0400	Finish coat, exterior latex		3900	.004		.10	.17		.27	.38
0410	Primer & 1 coat exterior latex		2600	.006		.25	.25		.50	.67
0420	Primer & 2 coats exterior latex		2080	.008		.34	.31		.65	.88
0430	Waterproof sealer, first coat		4485	.004		.18	.14		.32	.42
0440	Second coat		4485	.004		.10	.14		.24	.34
0450	Smooth wood incl. butt, T&G, beveled, drop or B&B siding									
0460	Brushwork, primer	2 Pord	2325	.007	S.F.	.10	.28		.38	.56
0470	Finish coat, exterior latex		1280	.013		.10	.50		.60	.92
0480	Primer & 1 coat exterior latex		800	.020		.20	.81		1.01	1.52
0490	Primer & 2 coats exterior latex		630	.025		.30	1.02		1.32	1.98
0500	Roll, primer		2275	.007		.11	.28		.39	.58
0510	Finish coat, exterior latex		2275	.007		.11	.28		.39	.58
0520	Primer & 1 coat exterior latex		1300	.012		.22	.50		.72	1.04
0530	Primer & 2 coats exterior latex		975	.016		.33	.66		.99	1.42
0540	Spray, primer		4550	.004		.08	.14		.22	.32
0550	Finish coat, exterior latex		4550	.004		.10	.14		.24	.34
0560	Primer & 1 coat exterior latex		2600	.006		.18	.25		.43	.60
0570	Primer & 2 coats exterior latex		1950	.008		.28	.33		.61	.83
0580	Waterproof sealer, first coat		5230	.003		.10	.12		.22	.31
0590	Second coat		5980	.003		.10	.11		.21	.28
0600	For oil base paint, add					10%				

For customer support on your Commercial Renovation Cost Data, call 877.742.4719.

323

09 91 13.62 Siding, Misc.

	Crew	Daily Output	Labor-Hours	Unit	Material	2016 Bare Costs Labor	2016 Bare Costs Equipment	Total	Total Incl O&P
9000	Minimum labor/equipment charge	1 Pord	2	4	Job		161		161	259

09 91 13.70 Doors and Windows, Exterior

		Crew	Daily Output	Labor-Hours	Unit	Material	Labor	Equipment	Total	Total Incl O&P
0010	**DOORS AND WINDOWS, EXTERIOR**									
0100	Door frames & trim, only									
0110	Brushwork, primer	1 Pord	512	.016	L.F.	.06	.63		.69	1.08
0120	Finish coat, exterior latex		512	.016		.08	.63		.71	1.09
0130	Primer & 1 coat, exterior latex		300	.027		.14	1.08		1.22	1.88
0135	2 coats, exterior latex, both sides		15	.533	Ea.	6.85	21.50		28.35	42
0140	Primer & 2 coats, exterior latex		265	.030	L.F.	.22	1.22		1.44	2.20
0150	Doors, flush, both sides, incl. frame & trim									
0160	Roll & brush, primer	1 Pord	10	.800	Ea.	4.65	32.50		37.15	57
0170	Finish coat, exterior latex		10	.800		5.85	32.50		38.35	58.50
0180	Primer & 1 coat, exterior latex		7	1.143		10.50	46		56.50	85.50
0190	Primer & 2 coats, exterior latex		5	1.600		16.35	64.50		80.85	122
0200	Brushwork, stain, sealer & 2 coats polyurethane		4	2		30	80.50		110.50	163
0210	Doors, French, both sides, 10-15 lite, incl. frame & trim									
0220	Brushwork, primer	1 Pord	6	1.333	Ea.	2.32	54		56.32	89
0230	Finish coat, exterior latex		6	1.333		2.93	54		56.93	89.50
0240	Primer & 1 coat, exterior latex		3	2.667		5.25	108		113.25	179
0250	Primer & 2 coats, exterior latex		2	4		8	161		169	268
0260	Brushwork, stain, sealer & 2 coats polyurethane		2.50	3.200		10.70	129		139.70	220
0270	Doors, louvered, both sides, incl. frame & trim									
0280	Brushwork, primer	1 Pord	7	1.143	Ea.	4.65	46		50.65	79
0290	Finish coat, exterior latex		7	1.143		5.85	46		51.85	80.50
0300	Primer & 1 coat, exterior latex		4	2		10.50	80.50		91	142
0310	Primer & 2 coats, exterior latex		3	2.667		16	108		124	191
0320	Brushwork, stain, sealer & 2 coats polyurethane		4.50	1.778		30	71.50		101.50	148
0330	Doors, panel, both sides, incl. frame & trim									
0340	Roll & brush, primer	1 Pord	6	1.333	Ea.	4.65	54		58.65	91.50
0350	Finish coat, exterior latex		6	1.333		5.85	54		59.85	93
0360	Primer & 1 coat, exterior latex		3	2.667		10.50	108		118.50	185
0370	Primer & 2 coats, exterior latex		2.50	3.200		16	129		145	226
0380	Brushwork, stain, sealer & 2 coats polyurethane		3	2.667		30	108		138	206
0400	Windows, per ext. side, based on 15 S.F.									
0410	1 to 6 lite									
0420	Brushwork, primer	1 Pord	13	.615	Ea.	.92	25		25.92	41
0430	Finish coat, exterior latex		13	.615		1.15	25		26.15	41.50
0440	Primer & 1 coat, exterior latex		8	1		2.07	40.50		42.57	67.50
0450	Primer & 2 coats, exterior latex		6	1.333		3.16	54		57.16	90
0460	Stain, sealer & 1 coat varnish		7	1.143		4.23	46		50.23	78.50
0470	7 to 10 lite									
0480	Brushwork, primer	1 Pord	11	.727	Ea.	.92	29.50		30.42	48
0490	Finish coat, exterior latex		11	.727		1.15	29.50		30.65	48.50
0500	Primer & 1 coat, exterior latex		7	1.143		2.07	46		48.07	76.50
0510	Primer & 2 coats, exterior latex		5	1.600		3.16	64.50		67.66	107
0520	Stain, sealer & 1 coat varnish		6	1.333		4.23	54		58.23	91
0530	12 lite									
0540	Brushwork, primer	1 Pord	10	.800	Ea.	.92	32.50		33.42	53
0550	Finish coat, exterior latex		10	.800		1.15	32.50		33.65	53.50
0560	Primer & 1 coat, exterior latex		6	1.333		2.07	54		56.07	89
0570	Primer & 2 coats, exterior latex		5	1.600		3.16	64.50		67.66	107
0580	Stain, sealer & 1 coat varnish		6	1.333		4.33	54		58.33	91.50

09 91 Painting

09 91 13 – Exterior Painting

09 91 13.70 Doors and Windows, Exterior	Crew	Daily Output	Labor-Hours	Unit	Material	2016 Bare Costs Labor	2016 Bare Costs Equipment	Total	Total Incl O&P
0590	For oil base paint, add				Ea.	10%				
9000	Minimum labor/equipment charge	1 Pord	2	4	Job		161		161	259

09 91 13.80 Trim, Exterior

		Crew	Daily Output	Labor-Hours	Unit	Material	Labor	Equipment	Total	Total Incl O&P
0010	**TRIM, EXTERIOR**									
0100	Door frames & trim (see Doors, interior or exterior)									
0110	Fascia, latex paint, one coat coverage									
0120	1" x 4", brushwork	1 Pord	640	.013	L.F.	.02	.50		.52	.84
0130	Roll		1280	.006		.03	.25		.28	.44
0140	Spray		2080	.004		.02	.16		.18	.27
0150	1" x 6" to 1" x 10", brushwork		640	.013		.08	.50		.58	.90
0160	Roll		1230	.007		.09	.26		.35	.51
0170	Spray		2100	.004		.07	.15		.22	.32
0180	1" x 12", brushwork		640	.013		.08	.50		.58	.90
0190	Roll		1050	.008		.09	.31		.40	.58
0200	Spray		2200	.004		.07	.15		.22	.31
0210	Gutters & downspouts, metal, zinc chromate paint									
0220	Brushwork, gutters, 5", first coat	1 Pord	640	.013	L.F.	.40	.50		.90	1.25
0230	Second coat		960	.008		.38	.34		.72	.95
0240	Third coat		1280	.006		.30	.25		.55	.75
0250	Downspouts, 4", first coat		640	.013		.40	.50		.90	1.25
0260	Second coat		960	.008		.38	.34		.72	.95
0270	Third coat		1280	.006		.30	.25		.55	.75
0280	Gutters & downspouts, wood									
0290	Brushwork, gutters, 5", primer	1 Pord	640	.013	L.F.	.06	.50		.56	.88
0300	Finish coat, exterior latex		640	.013		.07	.50		.57	.89
0310	Primer & 1 coat exterior latex		400	.020		.14	.81		.95	1.45
0320	Primer & 2 coats exterior latex		325	.025		.22	.99		1.21	1.84
0330	Downspouts, 4", primer		640	.013		.06	.50		.56	.88
0340	Finish coat, exterior latex		640	.013		.07	.50		.57	.89
0350	Primer & 1 coat exterior latex		400	.020		.14	.81		.95	1.45
0360	Primer & 2 coats exterior latex		325	.025		.11	.99		1.10	1.72
0370	Molding, exterior, up to 14" wide									
0380	Brushwork, primer	1 Pord	640	.013	L.F.	.07	.50		.57	.89
0390	Finish coat, exterior latex		640	.013		.08	.50		.58	.90
0400	Primer & 1 coat exterior latex		400	.020		.17	.81		.98	1.48
0410	Primer & 2 coats exterior latex		315	.025		.17	1.02		1.19	1.83
0420	Stain & fill		1050	.008		.10	.31		.41	.60
0430	Shellac		1850	.004		.14	.17		.31	.44
0440	Varnish		1275	.006		.10	.25		.35	.52
9000	Minimum labor/equipment charge		2	4	Job		161		161	259

09 91 13.90 Walls, Masonry (CMU), Exterior

		Crew	Daily Output	Labor-Hours	Unit	Material	Labor	Equipment	Total	Total Incl O&P
0010	**WALLS, MASONRY (CMU), EXTERIOR**									
0360	Concrete masonry units (CMU), smooth surface									
0370	Brushwork, latex, first coat	1 Pord	640	.013	S.F.	.07	.50		.57	.89
0380	Second coat		960	.008		.06	.34		.40	.60
0390	Waterproof sealer, first coat		736	.011		.26	.44		.70	.98
0400	Second coat		1104	.007		.26	.29		.55	.75
0410	Roll, latex, paint, first coat		1465	.005		.09	.22		.31	.44
0420	Second coat		1790	.004		.06	.18		.24	.36
0430	Waterproof sealer, first coat		1680	.005		.26	.19		.45	.59
0440	Second coat		2060	.004		.26	.16		.42	.53
0450	Spray, latex, paint, first coat		1950	.004		.07	.17		.24	.34

For customer support on your Commercial Renovation Cost Data, call 877.742.4719.

325

09 91 Painting

09 91 13 – Exterior Painting

09 91 13.90 Walls, Masonry (CMU), Exterior

		Crew	Daily Output	Labor-Hours	Unit	Material	2016 Bare Costs Labor	2016 Bare Costs Equipment	Total	Total Incl O&P
0460	Second coat	1 Pord	2600	.003	S.F.	.05	.12		.17	.26
0470	Waterproof sealer, first coat		2245	.004		.26	.14		.40	.51
0480	Second coat		2990	.003		.26	.11		.37	.45
0490	Concrete masonry unit (CMU), porous									
0500	Brushwork, latex, first coat	1 Pord	640	.013	S.F.	.14	.50		.64	.97
0510	Second coat		960	.008		.07	.34		.41	.62
0520	Waterproof sealer, first coat		736	.011		.26	.44		.70	.98
0530	Second coat		1104	.007		.26	.29		.55	.75
0540	Roll latex, first coat		1465	.005		.11	.22		.33	.47
0550	Second coat		1790	.004		.07	.18		.25	.37
0560	Waterproof sealer, first coat		1680	.005		.26	.19		.45	.59
0570	Second coat		2060	.004		.26	.16		.42	.53
0580	Spray latex, first coat		1950	.004		.08	.17		.25	.36
0590	Second coat		2600	.003		.05	.12		.17	.26
0600	Waterproof sealer, first coat		2245	.004		.26	.14		.40	.51
0610	Second coat		2990	.003		.26	.11		.37	.45
9000	Minimum labor/equipment charge		2	4	Job		161		161	259

09 91 23 – Interior Painting

09 91 23.20 Cabinets and Casework

		Crew	Daily Output	Labor-Hours	Unit	Material	2016 Bare Costs Labor	2016 Bare Costs Equipment	Total	Total Incl O&P
0010	**CABINETS AND CASEWORK**									
1000	Primer coat, oil base, brushwork	1 Pord	650	.012	S.F.	.06	.50		.56	.87
2000	Paint, oil base, brushwork, 1 coat		650	.012		.11	.50		.61	.92
3000	Stain, brushwork, wipe off		650	.012		.09	.50		.59	.90
4000	Shellac, 1 coat, brushwork		650	.012		.12	.50		.62	.93
4500	Varnish, 3 coats, brushwork, sand after 1st coat		325	.025		.26	.99		1.25	1.88
5000	For latex paint, deduct					10%				
6300	Strip, prep and refinish wood furniture									
6310	Remove paint using chemicals, wood furniture	1 Pord	28	.286	S.F.	1.60	11.55		13.15	20.50
6320	Prep for painting, sanding		75	.107		.24	4.30		4.54	7.15
6350	Stain and wipe, brushwork		600	.013		.09	.54		.63	.96
6355	Spray applied		900	.009		.08	.36		.44	.67
6360	Sealer or varnish, brushwork		1080	.007		.09	.30		.39	.57
6365	Spray applied		2100	.004		.09	.15		.24	.34
6370	Paint, primer, brushwork		720	.011		.06	.45		.51	.79
6375	Spray applied		2100	.004		.06	.15		.21	.32
6380	Finish coat, brushwork		810	.010		.11	.40		.51	.76
6385	Spray applied		2100	.004		.10	.15		.25	.36
9010	Minimum labor/equipment charge		2	4	Job		161		161	259

09 91 23.33 Doors and Windows, Interior Alkyd (Oil Base)

		Crew	Daily Output	Labor-Hours	Unit	Material	2016 Bare Costs Labor	2016 Bare Costs Equipment	Total	Total Incl O&P
0010	**DOORS AND WINDOWS, INTERIOR ALKYD (OIL BASE)**									
0500	Flush door & frame, 3' x 7', oil, primer, brushwork	1 Pord	10	.800	Ea.	3.70	32.50		36.20	56
1000	Paint, 1 coat		10	.800		4.16	32.50		36.66	56.50
1400	Stain, brushwork, wipe off		18	.444		1.83	17.95		19.78	31
1600	Shellac, 1 coat, brushwork		25	.320		2.47	12.90		15.37	23.50
1800	Varnish, 3 coats, brushwork, sand after 1st coat		9	.889		5.40	36		41.40	63.50
2000	Panel door & frame, 3' x 7', oil, primer, brushwork		6	1.333		2.41	54		56.41	89
2200	Paint, 1 coat		6	1.333		4.16	54		58.16	91
2600	Stain, brushwork, panel door, 3' x 7', not incl. frame		16	.500		1.83	20		21.83	34.50
2800	Shellac, 1 coat, brushwork		22	.364		2.47	14.65		17.12	26
3000	Varnish, 3 coats, brushwork, sand after 1st coat		7.50	1.067		5.40	43		48.40	75
4400	Windows, including frame and trim, per side									
4600	Colonial type, 6/6 lites, 2' x 3', oil, primer, brushwork	1 Pord	14	.571	Ea.	.38	23		23.38	37.50

09 91 23 – Interior Painting

09 91 23.33 Doors and Windows, Interior Alkyd (Oil Base)		Crew	Daily Output	Labor-Hours	Unit	Material	2016 Bare Costs Labor	Equipment	Total	Total Incl O&P
5800	Paint, 1 coat	1 Pord	14	.571	Ea.	.66	23		23.66	37.50
6200	3' x 5' opening, 6/6 lites, primer coat, brushwork		12	.667		.95	27		27.95	44
6400	Paint, 1 coat		12	.667		1.64	27		28.64	45
6800	4' x 8' opening, 6/6 lites, primer coat, brushwork		8	1		2.03	40.50		42.53	67
7000	Paint, 1 coat		8	1		3.50	40.50		44	69
8000	Single lite type, 2' x 3', oil base, primer coat, brushwork		33	.242		.38	9.80		10.18	16.10
8200	Paint, 1 coat		33	.242		.66	9.80		10.46	16.40
8600	3' x 5' opening, primer coat, brushwork		20	.400		.95	16.15		17.10	27
8800	Paint, 1 coat		20	.400		1.64	16.15		17.79	28
9200	4' x 8' opening, primer coat, brushwork		14	.571		2.03	23		25.03	39
9400	Paint, 1 coat		14	.571		3.50	23		26.50	41
9900	Minimum labor/equipment charge		2	4	Job		161		161	259

09 91 23.35 Doors and Windows, Interior Latex

		Crew	Daily Output	Labor-Hours	Unit	Material	2016 Bare Costs Labor	Equipment	Total	Total Incl O&P
0010	**DOORS & WINDOWS, INTERIOR LATEX**									
0100	Doors, flush, both sides, incl. frame & trim									
0110	Roll & brush, primer	1 Pord	10	.800	Ea.	4.15	32.50		36.65	56.50
0120	Finish coat, latex		10	.800		5.20	32.50		37.70	58
0130	Primer & 1 coat latex		7	1.143		9.35	46		55.35	84.50
0140	Primer & 2 coats latex		5	1.600		14.25	64.50		78.75	120
0160	Spray, both sides, primer		20	.400		4.36	16.15		20.51	31
0170	Finish coat, latex		20	.400		5.45	16.15		21.60	32
0180	Primer & 1 coat latex		11	.727		9.90	29.50		39.40	58
0190	Primer & 2 coats latex		8	1		15.10	40.50		55.60	81.50
0200	Doors, French, both sides, 10-15 lite, incl. frame & trim									
0210	Roll & brush, primer	1 Pord	6	1.333	Ea.	2.07	54		56.07	89
0220	Finish coat, latex		6	1.333		2.61	54		56.61	89.50
0230	Primer & 1 coat latex		3	2.667		4.68	108		112.68	178
0240	Primer & 2 coats latex		2	4		7.15	161		168.15	267
0260	Doors, louvered, both sides, incl. frame & trim									
0270	Roll & brush, primer	1 Pord	7	1.143	Ea.	4.15	46		50.15	78.50
0280	Finish coat, latex		7	1.143		5.20	46		51.20	80
0290	Primer & 1 coat, latex		4	2		9.10	80.50		89.60	140
0300	Primer & 2 coats, latex		3	2.667		14.55	108		122.55	189
0320	Spray, both sides, primer		20	.400		4.36	16.15		20.51	31
0330	Finish coat, latex		20	.400		5.45	16.15		21.60	32
0340	Primer & 1 coat, latex		11	.727		9.90	29.50		39.40	58
0350	Primer & 2 coats, latex		8	1		15.45	40.50		55.95	82
0360	Doors, panel, both sides, incl. frame & trim									
0370	Roll & brush, primer	1 Pord	6	1.333	Ea.	4.36	54		58.36	91.50
0380	Finish coat, latex		6	1.333		5.20	54		59.20	92.50
0390	Primer & 1 coat, latex		3	2.667		9.35	108		117.35	183
0400	Primer & 2 coats, latex		2.50	3.200		14.55	129		143.55	224
0420	Spray, both sides, primer		10	.800		4.36	32.50		36.86	57
0430	Finish coat, latex		10	.800		5.45	32.50		37.95	58
0440	Primer & 1 coat, latex		5	1.600		9.90	64.50		74.40	115
0450	Primer & 2 coats, latex		4	2		15.45	80.50		95.95	147
0460	Windows, per interior side, based on 15 S.F.									
0470	1 to 6 lite									
0480	Brushwork, primer	1 Pord	13	.615	Ea.	.82	25		25.82	41
0490	Finish coat, enamel		13	.615		1.03	25		26.03	41
0500	Primer & 1 coat enamel		8	1		1.85	40.50		42.35	67
0510	Primer & 2 coats enamel		6	1.333		2.88	54		56.88	89.50

For customer support on your Commercial Renovation Cost Data, call 877.742.4719.

327

09 91 Painting

09 91 23 – Interior Painting

09 91 23.35 Doors and Windows, Interior Latex

		Crew	Daily Output	Labor-Hours	Unit	Material	2016 Bare Costs Labor	Equipment	Total	Total Incl O&P
0530	7 to 10 lite									
0540	Brushwork, primer	1 Pord	11	.727	Ea.	.82	29.50		30.32	48
0550	Finish coat, enamel		11	.727		1.03	29.50		30.53	48
0560	Primer & 1 coat enamel		7	1.143		1.85	46		47.85	76
0570	Primer & 2 coats enamel		5	1.600		2.88	64.50		67.38	107
0590	12 lite									
0600	Brushwork, primer	1 Pord	10	.800	Ea.	.82	32.50		33.32	53
0610	Finish coat, enamel		10	.800		1.03	32.50		33.53	53
0620	Primer & 1 coat enamel		6	1.333		1.85	54		55.85	88.50
0630	Primer & 2 coats enamel		5	1.600		2.88	64.50		67.38	107
0650	For oil base paint, add					10%				
9000	Minimum labor/equipment charge	1 Pord	2	4	Job		161		161	259

09 91 23.39 Doors and Windows, Interior Latex, Zero Voc

			Crew	Daily Output	Labor-Hours	Unit	Material	2016 Bare Costs Labor	Equipment	Total	Total Incl O&P
0010	**DOORS & WINDOWS, INTERIOR LATEX, ZERO VOC**										
0100	Doors flush, both sides, incl. frame & trim										
0110	Roll & brush, primer	G	1 Pord	10	.800	Ea.	5.10	32.50		37.60	57.50
0120	Finish coat, latex	G		10	.800		5.55	32.50		38.05	58
0130	Primer & 1 coat latex	G		7	1.143		10.65	46		56.65	85.50
0140	Primer & 2 coats latex	G		5	1.600		15.90	64.50		80.40	121
0160	Spray, both sides, primer	G		20	.400		5.40	16.15		21.55	32
0170	Finish coat, latex	G		20	.400		5.80	16.15		21.95	32.50
0180	Primer & 1 coat latex	G		11	.727		11.25	29.50		40.75	59.50
0190	Primer & 2 coats latex	G		8	1		16.80	40.50		57.30	83.50
0200	Doors, French, both sides, 10-15 lite, incl. frame & trim										
0210	Roll & brush, primer	G	1 Pord	6	1.333	Ea.	2.56	54		56.56	89.50
0220	Finish coat, latex	G		6	1.333		2.77	54		56.77	89.50
0230	Primer & 1 coat latex	G		3	2.667		5.35	108		113.35	179
0240	Primer & 2 coats latex	G		2	4		7.95	161		168.95	268
0360	Doors, panel, both sides, incl. frame & trim										
0370	Roll & brush, primer	G	1 Pord	6	1.333	Ea.	5.40	54		59.40	92.50
0380	Finish coat, latex	G		6	1.333		5.55	54		59.55	92.50
0390	Primer & 1 coat, latex	G		3	2.667		10.65	108		118.65	185
0400	Primer & 2 coats, latex	G		2.50	3.200		16.20	129		145.20	226
0420	Spray, both sides, primer	G		10	.800		5.40	32.50		37.90	58
0430	Finish coat, latex	G		10	.800		5.80	32.50		38.30	58.50
0440	Primer & 1 coat, latex	G		5	1.600		11.25	64.50		75.75	116
0450	Primer & 2 coats, latex	G		4	2		17.15	80.50		97.65	149
0460	Windows, per interior side, based on 15 S.F.										
0470	1 to 6 lite										
0480	Brushwork, primer	G	1 Pord	13	.615	Ea.	1.01	25		26.01	41
0490	Finish coat, enamel	G		13	.615		1.09	25		26.09	41
0500	Primer & 1 coat enamel	G		8	1		2.10	40.50		42.60	67.50
0510	Primer & 2 coats enamel	G		6	1.333		3.20	54		57.20	90
9000	Minimum labor/equipment charge			2	4	Job		161		161	259

09 91 23.40 Floors, Interior

		Crew	Daily Output	Labor-Hours	Unit	Material	2016 Bare Costs Labor	Equipment	Total	Total Incl O&P
0010	**FLOORS, INTERIOR**									
0100	Concrete paint, latex									
0110	Brushwork									
0120	1st coat	1 Pord	975	.008	S.F.	.14	.33		.47	.68
0130	2nd coat		1150	.007		.09	.28		.37	.55
0140	3rd coat		1300	.006		.07	.25		.32	.48
0150	Roll									

09 91 Painting

09 91 23 – Interior Painting

09 91 23.40 Floors, Interior		Crew	Daily Output	Labor-Hours	Unit	Material	2016 Bare Costs Labor	Equipment	Total	Total Incl O&P
0160	1st coat	1 Pord	2600	.003	S.F.	.19	.12		.31	.41
0170	2nd coat		3250	.002		.11	.10		.21	.28
0180	3rd coat	↓	3900	.002	↓	.08	.08		.16	.22
0190	Spray									
0200	1st coat	1 Pord	2600	.003	S.F.	.16	.12		.28	.38
0210	2nd coat		3250	.002		.09	.10		.19	.26
0220	3rd coat	↓	3900	.002	↓	.07	.08		.15	.21

09 91 23.44 Anti-Slip Floor Treatments

		Crew	Daily Output	Labor-Hours	Unit	Material	2016 Bare Costs Labor	Equipment	Total	Total Incl O&P
0010	**ANTI-SLIP FLOOR TREATMENTS**									
1000	Walking surface treatment, ADA compliant, mop on and rinse									
1100	For tile, terrazzo, stone or smooth concrete	1 Pord	4000	.002	S.F.	.32	.08		.40	.49
1110	For marble		4000	.002		.21	.08		.29	.37
1120	For wood		4000	.002		.20	.08		.28	.35
1130	For baths and showers	↓	500	.016		.37	.65		1.02	1.45
2000	Granular additive for paint or sealer, add to paint cost				↓	.02			.02	.02

09 91 23.52 Miscellaneous, Interior

		Crew	Daily Output	Labor-Hours	Unit	Material	2016 Bare Costs Labor	Equipment	Total	Total Incl O&P
0010	**MISCELLANEOUS, INTERIOR**									
2400	Floors, conc./wood, oil base, primer/sealer coat, brushwork	2 Pord	1950	.008	S.F.	.10	.33		.43	.63
2450	Roller		5200	.003		.10	.12		.22	.31
2600	Spray		6000	.003		.10	.11		.21	.28
2650	Paint 1 coat, brushwork		1950	.008		.10	.33		.43	.64
2800	Roller		5200	.003		.10	.12		.22	.31
2850	Spray		6000	.003		.11	.11		.22	.29
3000	Stain, wood floor, brushwork, 1 coat		4550	.004		.09	.14		.23	.33
3200	Roller		5200	.003		.09	.12		.21	.30
3250	Spray		6000	.003		.09	.11		.20	.27
3400	Varnish, wood floor, brushwork		4550	.004		.09	.14		.23	.32
3450	Roller		5200	.003		.09	.12		.21	.30
3600	Spray	↓	6000	.003		.10	.11		.21	.28
3800	Grilles, per side, oil base, primer coat, brushwork	1 Pord	520	.015		.13	.62		.75	1.14
3850	Spray		1140	.007		.13	.28		.41	.61
3920	Paint 2 coats, brushwork		325	.025		.43	.99		1.42	2.07
3940	Spray	↓	650	.012	↓	.49	.50		.99	1.34
4600	Miscellaneous surfaces, metallic paint, spray applied									
4610	Water based, non-tintable, warm silver	1 Pord	1140	.007	S.F.	.78	.28		1.06	1.31
4620	Rusted iron		1140	.007		1	.28		1.28	1.56
4630	Low VOC, tintable	↓	1140	.007	↓	.35	.28		.63	.84
5000	Pipe, 1" - 4" diameter, primer or sealer coat, oil base, brushwork	2 Pord	1250	.013	L.F.	.10	.52		.62	.94
5100	Spray		2165	.007		.09	.30		.39	.58
5350	Paint 2 coats, brushwork		775	.021		.19	.83		1.02	1.55
5400	Spray		1240	.013		.22	.52		.74	1.08
6300	13" - 16" diameter, primer or sealer coat, brushwork		310	.052		.40	2.08		2.48	3.79
6450	Spray		540	.030		.44	1.20		1.64	2.41
6500	Paint 2 coats, brushwork		195	.082		.78	3.31		4.09	6.15
6550	Spray	↓	310	.052	↓	.86	2.08		2.94	4.30
7000	Trim, wood, incl. puttying, under 6" wide									
7200	Primer coat, oil base, brushwork	1 Pord	650	.012	L.F.	.03	.50		.53	.83
7250	Paint, 1 coat, brushwork		650	.012		.05	.50		.55	.86
7450	3 coats		325	.025		.16	.99		1.15	1.77
7500	Over 6" wide, primer coat, brushwork		650	.012		.06	.50		.56	.87
7550	Paint, 1 coat, brushwork		650	.012		.11	.50		.61	.92
7650	3 coats	↓	325	.025	↓	.32	.99		1.31	1.95

For customer support on your Commercial Renovation Cost Data, call 877.742.4719.

329

09 91 Painting

09 91 23 – Interior Painting

09 91 23.52 Miscellaneous, Interior

		Crew	Daily Output	Labor-Hours	Unit	Material	2016 Bare Costs Labor	Equipment	Total	Total Incl O&P
8000	Cornice, simple design, primer coat, oil base, brushwork	1 Pord	650	.012	S.F.	.06	.50		.56	.87
8250	Paint, 1 coat		650	.012		.11	.50		.61	.92
8350	Ornate design, primer coat		350	.023		.06	.92		.98	1.55
8400	Paint, 1 coat		350	.023		.11	.92		1.03	1.60
8600	Balustrades, primer coat, oil base, brushwork		520	.015		.06	.62		.68	1.07
8900	Trusses and wood frames, primer coat, oil base, brushwork		800	.010		.06	.40		.46	.72
8950	Spray		1200	.007		.07	.27		.34	.50
9220	Paint 2 coats, brushwork		500	.016		.21	.65		.86	1.27
9240	Spray		600	.013		.24	.54		.78	1.12
9260	Stain, brushwork, wipe off		600	.013		.09	.54		.63	.96
9280	Varnish, 3 coats, brushwork		275	.029		.26	1.17		1.43	2.17
9350	For latex paint, deduct					10%				
9900	Minimum labor/equipment charge	1 Pord	2	4	Job		161		161	259

09 91 23.62 Electrostatic Painting

		Crew	Daily Output	Labor-Hours	Unit	Material	2016 Bare Costs Labor	Equipment	Total	Total Incl O&P
0010	**ELECTROSTATIC PAINTING**									
0100	In shop									
0200	Flat surfaces (lockers, casework, elevator doors, etc.)									
0300	One coat	1 Pord	200	.040	S.F.	.57	1.61		2.18	3.21
0400	Two coats	"	120	.067	"	.82	2.69		3.51	5.25
0500	Irregular surfaces (furniture, door frames, etc.)									
0600	One coat	1 Pord	150	.053	S.F.	.57	2.15		2.72	4.08
0700	Two coats	"	100	.080	"	.82	3.23		4.05	6.10
0800	On site									
0900	Flat surfaces (lockers, casework, elevator doors, etc.)									
1000	One coat	1 Pord	150	.053	S.F.	.57	2.15		2.72	4.08
1100	Two coats	"	100	.080	"	.82	3.23		4.05	6.10
1200	Irregular surfaces (furniture, door frames, etc.)									
1300	One coat	1 Pord	115	.070	S.F.	.57	2.81		3.38	5.15
1400	Two coats		70	.114		.82	4.61		5.43	8.30
2000	Anti-microbial coating, hospital application		150	.053		.06	2.15		2.21	3.53

09 91 23.72 Walls and Ceilings, Interior

		Crew	Daily Output	Labor-Hours	Unit	Material	2016 Bare Costs Labor	Equipment	Total	Total Incl O&P
0010	**WALLS AND CEILINGS, INTERIOR**									
0100	Concrete, drywall or plaster, latex, primer or sealer coat									
0200	Smooth finish, brushwork	1 Pord	1150	.007	S.F.	.06	.28		.34	.52
0240	Roller		1350	.006		.06	.24		.30	.45
0280	Spray		2750	.003		.05	.12		.17	.25
0300	Sand finish, brushwork		975	.008		.06	.33		.39	.60
0340	Roller		1150	.007		.06	.28		.34	.52
0380	Spray		2275	.004		.05	.14		.19	.29
0800	Paint 2 coats, smooth finish, brushwork		680	.012		.14	.47		.61	.91
0840	Roller		800	.010		.14	.40		.54	.80
0880	Spray		1625	.005		.13	.20		.33	.46
0900	Sand finish, brushwork		605	.013		.14	.53		.67	1.01
0940	Roller		1020	.008		.14	.32		.46	.66
0980	Spray		1700	.005		.13	.19		.32	.45
1600	Glaze coating, 2 coats, spray, clear		1200	.007		.49	.27		.76	.97
1640	Multicolor		1200	.007		.97	.27		1.24	1.50
1660	Painting walls, complete, including surface prep, primer &									
1670	2 coats finish, on drywall or plaster, with roller	1 Pord	325	.025	S.F.	.21	.99		1.20	1.83
1700	For oil base paint, add					10%				
1800	For ceiling installations, add						25%			
2000	Masonry or concrete block, primer/sealer, latex paint									

09 91 Painting

09 91 23.72 Walls and Ceilings, Interior

		Crew	Daily Output	Labor-Hours	Unit	Material	2016 Bare Costs Labor	Equipment	Total	Total Incl O&P
2100	Primer, smooth finish, brushwork	1 Pord	1000	.008	S.F.	.11	.32		.43	.64
2110	Roller		1150	.007		.11	.28		.39	.57
2180	Spray		2400	.003		.10	.13		.23	.33
2200	Sand finish, brushwork		850	.009		.11	.38		.49	.73
2210	Roller		975	.008		.11	.33		.44	.65
2280	Spray		2050	.004		.10	.16		.26	.36
2800	Primer plus one finish coat, smooth brush		525	.015		.30	.61		.91	1.32
2810	Roller		615	.013		.19	.53		.72	1.05
2880	Spray		1200	.007		.17	.27		.44	.61
2900	Sand finish, brushwork		450	.018		.19	.72		.91	1.36
2910	Roller		515	.016		.19	.63		.82	1.22
2980	Spray		1025	.008		.17	.31		.48	.69
3600	Glaze coating, 3 coats, spray, clear		900	.009		.70	.36		1.06	1.35
3620	Multicolor		900	.009		1.14	.36		1.50	1.83
4000	Block filler, 1 coat, brushwork		425	.019		.12	.76		.88	1.36
4100	Silicone, water repellent, 2 coats, spray	▼	2000	.004		.37	.16		.53	.67
4120	For oil base paint, add					10%				
8200	For work 8' - 15' H, add						10%			
8300	For work over 15' H, add						20%			
8400	For light textured surfaces, add						10%			
8410	Heavy textured, add				▼		25%			
9900	Minimum labor/equipment charge	1 Pord	2	4	Job		161		161	259

09 91 23.74 Walls and Ceilings, Interior, Zero VOC Latex

			Crew	Daily Output	Labor-Hours	Unit	Material	2016 Bare Costs Labor	Equipment	Total	Total Incl O&P
0010	**WALLS AND CEILINGS, INTERIOR, ZERO VOC LATEX**										
0100	Concrete, dry wall or plaster, latex, primer or sealer coat										
0200	Smooth finish, brushwork	G	1 Pord	1150	.007	S.F.	.07	.28		.35	.52
0240	Roller	G		1350	.006		.06	.24		.30	.45
0280	Spray	G		2750	.003		.05	.12		.17	.24
0300	Sand finish, brushwork	G		975	.008		.06	.33		.39	.60
0340	Roller	G		1150	.007		.07	.28		.35	.53
0380	Spray	G		2275	.004		.05	.14		.19	.29
0800	Paint 2 coats, smooth finish, brushwork	G		680	.012		.15	.47		.62	.93
0840	Roller	G		800	.010		.16	.40		.56	.82
0880	Spray	G		1625	.005		.14	.20		.34	.47
0900	Sand finish, brushwork	G		605	.013		.15	.53		.68	1.02
0940	Roller	G		1020	.008		.16	.32		.48	.68
0980	Spray	G	▼	1700	.005		.14	.19		.33	.46
1800	For ceiling installations, add	G						25%			
8200	For work 8' - 15' H, add							10%			
8300	For work over 15' H, add					▼		20%			
9900	Minimum labor/equipment charge		1 Pord	2	4	Job		161		161	259

09 91 23.75 Dry Fall Painting

		Crew	Daily Output	Labor-Hours	Unit	Material	2016 Bare Costs Labor	Equipment	Total	Total Incl O&P
0010	**DRY FALL PAINTING**									
0100	Sprayed on walls, gypsum board or plaster									
0220	One coat	1 Pord	2600	.003	S.F.	.07	.12		.19	.28
0250	Two coats		1560	.005		.15	.21		.36	.49
0280	Concrete or textured plaster, one coat		1560	.005		.07	.21		.28	.41
0310	Two coats		1300	.006		.15	.25		.40	.56
0340	Concrete block, one coat		1560	.005		.07	.21		.28	.41
0370	Two coats		1300	.006		.15	.25		.40	.56
0400	Wood, one coat		877	.009		.07	.37		.44	.67
0430	Two coats	▼	650	.012	▼	.15	.50		.65	.96

For customer support on your Commercial Renovation Cost Data, call 877.742.4719.

331

09 91 Painting

09 91 23 – Interior Painting

09 91 23.75 Dry Fall Painting	Crew	Daily Output	Labor-Hours	Unit	Material	2016 Bare Costs Labor	Equipment	Total	Total Incl O&P
0440	On ceilings, gypsum board or plaster									
0470	One coat	1 Pord	1560	.005	S.F.	.07	.21		.28	.41
0500	Two coats		1300	.006		.15	.25		.40	.56
0530	Concrete or textured plaster, one coat		1560	.005		.07	.21		.28	.41
0560	Two coats		1300	.006		.15	.25		.40	.56
0570	Structural steel, bar joists or metal deck, one coat		1560	.005		.07	.21		.28	.41
0580	Two coats		1040	.008		.15	.31		.46	.66
9900	Minimum labor/equipment charge		2	4	Job		161		161	259

09 93 Staining and Transparent Finishing

09 93 23 – Interior Staining and Finishing

09 93 23.10 Varnish

		Crew	Daily Output	Labor-Hours	Unit	Material	Labor	Equipment	Total	Total Incl O&P
0010	**VARNISH**									
0012	1 coat + sealer, on wood trim, brush, no sanding included	1 Pord	400	.020	S.F.	.07	.81		.88	1.38
0020	1 coat + sealer, on wood trim, brush, no sanding included, no VOC		400	.020		.21	.81		1.02	1.53
0100	Hardwood floors, 2 coats, no sanding included, roller		1890	.004		.15	.17		.32	.44
9000	Minimum labor/equipment charge		4	2	Job		80.50		80.50	130

09 96 High-Performance Coatings

09 96 23 – Graffiti-Resistant Coatings

09 96 23.10 Graffiti Resistant Treatments

		Crew	Daily Output	Labor-Hours	Unit	Material	Labor	Equipment	Total	Total Incl O&P
0010	**GRAFFITI RESISTANT TREATMENTS**, sprayed on walls									
0100	Non-sacrificial, permanent non-stick coating, clear, on metals	1 Pord	2000	.004	S.F.	2.06	.16		2.22	2.53
0200	Concrete		2000	.004		2.35	.16		2.51	2.84
0300	Concrete block		2000	.004		3.03	.16		3.19	3.60
0400	Brick		2000	.004		3.44	.16		3.60	4.04
0500	Stone		2000	.004		3.44	.16		3.60	4.04
0600	Unpainted wood		2000	.004		3.97	.16		4.13	4.63
2000	Semi-permanent cross linking polymer primer, on metals		2000	.004		.70	.16		.86	1.03
2100	Concrete		2000	.004		.84	.16		1	1.18
2200	Concrete block		2000	.004		1.05	.16		1.21	1.41
2300	Brick		2000	.004		.84	.16		1	1.18
2400	Stone		2000	.004		.84	.16		1	1.18
2500	Unpainted wood		2000	.004		1.16	.16		1.32	1.54
3000	Top coat, on metals		2000	.004		.55	.16		.71	.86
3100	Concrete		2000	.004		.62	.16		.78	.95
3200	Concrete block		2000	.004		.87	.16		1.03	1.22
3300	Brick		2000	.004		.73	.16		.89	1.06
3400	Stone		2000	.004		.73	.16		.89	1.06
3500	Unpainted wood		2000	.004		.87	.16		1.03	1.22
5000	Sacrificial, water based, on metal		2000	.004		.32	.16		.48	.62
5100	Concrete		2000	.004		.32	.16		.48	.62
5200	Concrete block		2000	.004		.32	.16		.48	.62
5300	Brick		2000	.004		.32	.16		.48	.62
5400	Stone		2000	.004		.32	.16		.48	.62
5500	Unpainted wood		2000	.004		.32	.16		.48	.62
8000	Cleaner for use after treatment									
8100	Towels or wipes, per package of 30				Ea.	.63			.63	.70
8200	Aerosol spray, 24 oz. can				"	18			18	19.80

09 96 High-Performance Coatings

09 96 46 – Intumescent Painting

09 96 46.10 Coatings, Intumescent	Crew	Daily Output	Labor-Hours	Unit	Material	2016 Bare Costs Labor	Equipment	Total	Total Incl O&P
0010 **COATINGS, INTUMESCENT**, spray applied									
0100 On exterior structural steel, 0.25" d.f.t.	1 Pord	475	.017	S.F.	.41	.68		1.09	1.54
0150 0.51" d.f.t.		350	.023		.41	.92		1.33	1.93
0200 0.98" d.f.t.		280	.029		.41	1.15		1.56	2.30
0300 On interior structural steel, 0.108" d.f.t.		300	.027		.41	1.08		1.49	2.18
0350 0.310" d.f.t.		150	.053		.41	2.15		2.56	3.91
0400 0.670" d.f.t.		100	.080		.41	3.23		3.64	5.65

09 96 53 – Elastomeric Coatings

09 96 53.10 Coatings, Elastomeric

	Crew	Daily Output	Labor-Hours	Unit	Material	Labor	Equipment	Total	Total Incl O&P
0010 **COATINGS, ELASTOMERIC**									
0020 High build, water proof, one coat system									
0100 Concrete, brush	1 Pord	650	.012	S.F.	.26	.50		.76	1.09

09 96 56 – Epoxy Coatings

09 96 56.20 Wall Coatings

	Crew	Daily Output	Labor-Hours	Unit	Material	Labor	Equipment	Total	Total Incl O&P
0010 **WALL COATINGS**									
0100 Acrylic glazed coatings, matte	1 Pord	525	.015	S.F.	.31	.61		.92	1.33
0200 Gloss		305	.026		.65	1.06		1.71	2.42
0300 Epoxy coatings, solvent based		525	.015		.40	.61		1.01	1.43
0400 Water based		170	.047		.27	1.90		2.17	3.35
2400 Sprayed perlite or vermiculite, 1/16" thick, solvent based		2935	.003		.27	.11		.38	.48
2500 Water based		640	.013		.74	.50		1.24	1.62
2700 Vinyl plastic wall coating, solvent based		735	.011		.33	.44		.77	1.07
2800 Water based		240	.033		.82	1.34		2.16	3.06
3000 Urethane on smooth surface, 2 coats, solvent based		1135	.007		.31	.28		.59	.80
3100 Water based		665	.012		.59	.49		1.08	1.43

For customer support on your Commercial Renovation Cost Data, call 877.742.4719.

333

Division Notes

	CREW	DAILY OUTPUT	LABOR-HOURS	UNIT	BARE COSTS				TOTAL INCL O&P
					MAT.	LABOR	EQUIP.	TOTAL	

Estimating Tips
General

- The items in this division are usually priced per square foot or each.

- Many items in Division 10 require some type of support system or special anchors that are not usually furnished with the item. The required anchors must be added to the estimate in the appropriate division.

- Some items in Division 10, such as lockers, may require assembly before installation. Verify the amount of assembly required. Assembly can often exceed installation time.

10 20 00 Interior Specialties

- Support angles and blocking are not included in the installation of toilet compartments, shower/dressing compartments, or cubicles. Appropriate line items from Divisions 5 or 6 may need to be added to support the installations.

- Toilet partitions are priced by the stall. A stall consists of a side wall, pilaster, and door with hardware. Toilet tissue holders and grab bars are extra.

- The required acoustical rating of a folding partition can have a significant impact on costs. Verify the sound transmission coefficient rating of the panel priced to the specification requirements.

- Grab bar installation does not include supplemental blocking or backing to support the required load. When grab bars are installed at an existing facility, provisions must be made to attach the grab bars to solid structure.

Reference Numbers

Reference numbers are shown at the beginning of some major classifications. These numbers refer to related items in the Reference Section. The reference information may be an estimating procedure, an alternate pricing method, or technical information.

Note: Not all subdivisions listed here necessarily appear. ∎

Did you know?

RSMeans Online gives you the same access to RSMeans' data with 24/7 access:

- Quickly locate costs in the searchable database.

- Build cost lists, estimates, and reports in minutes.

- Adjust costs to any location in the U.S. and Canada with the click of a button.

Start your free trial today at **www.RSMeansOnline.com**

RSMeans Online
FROM THE G⊕RDIAN GROUP®

10 11 13 – Chalkboards

10 11 13.13 Fixed Chalkboards

		Crew	Daily Output	Labor-Hours	Unit	Material	2016 Bare Costs Labor	Equipment	Total	Total Incl O&P
0010	FIXED CHALKBOARDS Porcelain enamel steel									
3900	Wall hung									
4000	Aluminum frame and chalktrough									
4300	3' x 5'	2 Carp	15	1.067	Ea.	325	51.50		376.50	440
4600	4' x 12'	"	13	1.231	"	605	59.50		664.50	770
4700	Wood frame and chalktrough									
4800	3' x 4'	2 Carp	16	1	Ea.	199	48.50		247.50	298
5300	4' x 8'	"	13	1.231	"	320	59.50		379.50	450
5400	Liquid chalk, white porcelain enamel, wall hung									
5420	Deluxe units, aluminum trim and chalktrough									
5450	4' x 4'	2 Carp	16	1	Ea.	269	48.50		317.50	375
5550	4' x 12'	"	12	1.333	"	545	64.50		609.50	705
5700	Wood trim and chalktrough									
5900	4' x 4'	2 Carp	16	1	Ea.	715	48.50		763.50	865
6200	4' x 8'		14	1.143	"	970	55.50		1,025.50	1,175
9000	Minimum labor/equipment charge		3	5.333	Job		258		258	420

10 11 13.43 Portable Chalkboards

		Crew	Daily Output	Labor-Hours	Unit	Material	2016 Bare Costs Labor	Equipment	Total	Total Incl O&P
0010	PORTABLE CHALKBOARDS									
0100	Freestanding, reversible									
0120	Economy, wood frame, 4' x 6'									
0140	Chalkboard both sides				Ea.	600			600	660
0160	Chalkboard one side, cork other side				"	570			570	625
0200	Standard, lightweight satin finished aluminum, 4' x 6'									
0220	Chalkboard both sides				Ea.	660			660	725
0240	Chalkboard one side, cork other side				"	665			665	730
0300	Deluxe, heavy duty extruded aluminum, 4' x 6'									
0320	Chalkboard both sides				Ea.	1,075			1,075	1,175
0340	Chalkboard one side, cork other side				"	985			985	1,075

10 11 16 – Markerboards

10 11 16.53 Electronic Markerboards

		Crew	Daily Output	Labor-Hours	Unit	Material	2016 Bare Costs Labor	Equipment	Total	Total Incl O&P
0010	ELECTRONIC MARKERBOARDS									
0100	Wall hung or free standing, 3' x 4' to 4' x 6'	2 Carp	8	2	S.F.	87.50	97		184.50	254
0150	5' x 6' to 4' x 8'		8	2	"	64.50	97		161.50	229
0500	Interactive projection module for existing whiteboards		8	2	Ea.	1,300	97		1,397	1,575

10 11 23 – Tackboards

10 11 23.10 Fixed Tackboards

		Crew	Daily Output	Labor-Hours	Unit	Material	2016 Bare Costs Labor	Equipment	Total	Total Incl O&P
0010	FIXED TACKBOARDS									
2120	Prefabricated, 1/4" cork, 3' x 5' with aluminum frame	2 Carp	16	1	Ea.	135	48.50		183.50	227
2140	Wood frame	"	16	1	"	159	48.50		207.50	254
2300	Glass enclosed cabinets, alum., cork panel, hinged doors									
2600	4' x 7', 3 door	2 Carp	10	1.600	Ea.	1,775	77.50		1,852.50	2,075
9000	Minimum labor/equipment charge	"	4	4	Job		194		194	315

10 11 23.20 Control Boards

		Crew	Daily Output	Labor-Hours	Unit	Material	2016 Bare Costs Labor	Equipment	Total	Total Incl O&P
0010	CONTROL BOARDS									
1000	Hospital patient display board, 4-color custom design									
1010	Porcelain steel dry erase board, 36" x 24"	2 Carp	7.50	2.133	Ea.	238	103		341	430

10 13 Directories

10 13 10 – Building Directories

10 13 10.10 Directory Boards		Crew	Daily Output	Labor-Hours	Unit	Material	2016 Bare Costs Labor	Equipment	Total	Total Incl O&P
0010	**DIRECTORY BOARDS**									
0050	Plastic, glass covered, 30" x 20"	2 Carp	3	5.333	Ea.	198	258		456	635
0100	36" x 48"		2	8		845	390		1,235	1,575
0900	Outdoor, weatherproof, black plastic, 36" x 24"		2	8		760	390		1,150	1,475
1000	36" x 36"	↓	1.50	10.667	↓	880	515		1,395	1,825
9000	Minimum labor/equipment charge	1 Carp	1	8	Job		390		390	635

10 14 Signage

10 14 19 – Dimensional Letter Signage

10 14 19.10 Exterior Signs

		Crew	Daily Output	Labor-Hours	Unit	Material	Labor	Equipment	Total	Total Incl O&P
0010	**EXTERIOR SIGNS**									
3900	Plaques, custom, 20" x 30", for up to 450 letters, cast aluminum	2 Carp	4	4	Ea.	1,850	194		2,044	2,350
4000	Cast bronze		4	4		1,775	194		1,969	2,275
4200	30" x 36", up to 900 letters, cast aluminum		3	5.333		2,675	258		2,933	3,350
4300	Cast bronze	↓	3	5.333		4,100	258		4,358	4,925
5100	Exit signs, 24 ga. alum., 14" x 12" surface mounted	1 Carp	30	.267		50	12.90		62.90	76
5200	10" x 7"	"	20	.400		27.50	19.40		46.90	61.50
6400	Replacement sign faces, 6" or 8"	1 Clab	50	.160	↓	66	6.05		72.05	82.50
9000	Minimum labor/equipment charge	1 Carp	4	2	Job		97		97	158

10 14 23 – Panel Signage

10 14 23.13 Engraved Panel Signage

		Crew	Daily Output	Labor-Hours	Unit	Material	Labor	Equipment	Total	Total Incl O&P
0010	**ENGRAVED PANEL SIGNAGE**, interior									
1010	Flexible door sign, adhesive back, w/Braille, 5/8" letters, 4" x 4"	1 Clab	32	.250	Ea.	33	9.50		42.50	52
1050	6" x 6"		32	.250		48.50	9.50		58	69
1100	8" x 2"		32	.250		33	9.50		42.50	52
1150	8" x 4"		32	.250		43.50	9.50		53	63
1200	8" x 8"		32	.250		53	9.50		62.50	74
1250	12" x 2"		32	.250		36	9.50		45.50	55
1300	12" x 6"		32	.250		39	9.50		48.50	58.50
1350	12" x 12"		32	.250		150	9.50		159.50	181
1500	Graphic symbols, 2" x 2"		32	.250		12	9.50		21.50	28.50
1550	6" x 6"		32	.250		31	9.50		40.50	49.50
1600	8" x 8"	↓	32	.250		39	9.50		48.50	58
2010	Corridor, stock acrylic, 2-sided, with mounting bracket, 2" x 8"	1 Carp	24	.333		24.50	16.15		40.65	53.50
2020	2" x 10"		24	.333		35.50	16.15		51.65	65.50
2050	3" x 8"		24	.333		28.50	16.15		44.65	58
2060	3" x 10"		24	.333		38	16.15		54.15	68
2070	3" x 12"		24	.333		37	16.15		53.15	67
2100	4" x 8"		24	.333		21	16.15		37.15	49.50
2110	4" x 10"		24	.333		38	16.15		54.15	68.50
2120	4" x 12"	↓	24	.333	↓	53.50	16.15		69.65	85

10 14 53 – Traffic Signage

10 14 53.20 Traffic Signs

		Crew	Daily Output	Labor-Hours	Unit	Material	Labor	Equipment	Total	Total Incl O&P
0010	**TRAFFIC SIGNS**									
0012	Stock, 24" x 24", no posts, .080" alum. reflectorized	B-80	70	.457	Ea.	86.50	19.30	10.40	116.20	138
0100	High intensity		70	.457		99.50	19.30	10.40	129.20	152
0300	30" x 30", reflectorized		70	.457		125	19.30	10.40	154.70	179
0400	High intensity		70	.457		135	19.30	10.40	164.70	190
0600	Guide and directional signs, 12" x 18", reflectorized		70	.457		37.50	19.30	10.40	67.20	84
0700	High intensity	↓	70	.457	↓	52.50	19.30	10.40	82.20	100

10 14 Signage

10 14 53 – Traffic Signage

10 14 53.20 Traffic Signs

		Crew	Daily Output	Labor-Hours	Unit	Material	2016 Bare Costs Labor	2016 Bare Costs Equipment	Total	Total Incl O&P
0900	18" x 24", stock signs, reflectorized	B-80	70	.457	Ea.	47.50	19.30	10.40	77.20	95
1000	High intensity		70	.457		52.50	19.30	10.40	82.20	100
1200	24" x 24", stock signs, reflectorized		70	.457		57.50	19.30	10.40	87.20	106
1300	High intensity		70	.457		63.50	19.30	10.40	93.20	112
1500	Add to above for steel posts, galvanized, 10'-0" upright, bolted		200	.160		32.50	6.75	3.64	42.89	51
1600	12'-0" upright, bolted		140	.229	▼	39	9.65	5.20	53.85	64.50
1800	Highway road signs, aluminum, over 20 S.F., reflectorized		350	.091	S.F.	34	3.86	2.08	39.94	46
2000	High intensity		350	.091		34	3.86	2.08	39.94	46
2200	Highway, suspended over road, 80 S.F. min., reflectorized		165	.194		32.50	8.20	4.41	45.11	54
2300	High intensity	▼	165	.194	▼	31	8.20	4.41	43.61	52.50

10 17 Telephone Specialties

10 17 16 – Telephone Enclosures

10 17 16.10 Commercial Telephone Enclosures

		Crew	Daily Output	Labor-Hours	Unit	Material	Labor	Equipment	Total	Total Incl O&P
0010	**COMMERCIAL TELEPHONE ENCLOSURES**									
0300	Shelf type, wall hung, recessed	2 Carp	5	3.200	Ea.	745	155		900	1,075
0400	Surface mount		5	3.200	"	1,625	155		1,780	2,025
9000	Minimum labor/equipment charge	▼	4	4	Job		194		194	315

10 21 Compartments and Cubicles

10 21 13 – Toilet Compartments

10 21 13.13 Metal Toilet Compartments

		Crew	Daily Output	Labor-Hours	Unit	Material	Labor	Equipment	Total	Total Incl O&P
0010	**METAL TOILET COMPARTMENTS**									
0110	Cubicles, ceiling hung									
0200	Powder coated steel	2 Carp	4	4	Ea.	525	194		719	890
0500	Stainless steel	"	4	4		1,100	194		1,294	1,550
0600	For handicap units, incl. 52" grab bars, add				▼	450			450	495
0900	Floor and ceiling anchored									
1000	Powder coated steel	2 Carp	5	3.200	Ea.	570	155		725	885
1300	Stainless steel	"	5	3.200		1,225	155		1,380	1,600
1400	For handicap units, incl. 52" grab bars, add				▼	370			370	405
1610	Floor anchored									
1700	Powder coated steel	2 Carp	7	2.286	Ea.	600	111		711	840
2000	Stainless steel	"	7	2.286		1,425	111		1,536	1,750
2100	For handicap units, incl. 52" grab bars, add					335			335	370
2200	For juvenile units, deduct				▼	41.50			41.50	45.50
2450	Floor anchored, headrail braced									
2500	Powder coated steel	2 Carp	6	2.667	Ea.	380	129		509	630
2804	Stainless steel	"	6	2.667		1,025	129		1,154	1,325
2900	For handicap units, incl. 52" grab bars, add					370			370	410
3000	Wall hung partitions, powder coated steel	2 Carp	7	2.286		635	111		746	880
3300	Stainless steel	"	7	2.286		1,650	111		1,761	2,000
3400	For handicap units, incl. 52" grab bars, add				▼	370			370	410
4000	Screens, entrance, floor mounted, 58" high, 48" wide									
4200	Powder coated steel	2 Carp	15	1.067	Ea.	242	51.50		293.50	350
4500	Stainless steel	"	15	1.067	"	910	51.50		961.50	1,075
4650	Urinal screen, 18" wide									
4704	Powder coated steel	2 Carp	6.15	2.602	Ea.	209	126		335	435
5004	Stainless steel	"	6.15	2.602	"	535	126		661	795

10 21 13 – Toilet Compartments

10 21 13.13 Metal Toilet Compartments	Crew	Daily Output	Labor-Hours	Unit	Material	2016 Bare Costs Labor	Equipment	Total	Total Incl O&P
5100 Floor mounted, head rail braced									
5300 Powder coated steel	2 Carp	8	2	Ea.	230	97		327	410
5600 Stainless steel	"	8	2	"	580	97		677	800
5750 Pilaster, flush									
5800 Powder coated steel	2 Carp	10	1.600	Ea.	270	77.50		347.50	425
6100 Stainless steel		10	1.600		610	77.50		687.50	795
6300 Post braced, powder coated steel		10	1.600		170	77.50		247.50	315
6600 Stainless steel		10	1.600		450	77.50		527.50	620
6700 Wall hung, bracket supported									
6800 Powder coated steel	2 Carp	10	1.600	Ea.	170	77.50		247.50	315
7100 Stainless steel		10	1.600		290	77.50		367.50	445
7400 Flange supported, powder coated steel		10	1.600		114	77.50		191.50	252
7700 Stainless steel		10	1.600		315	77.50		392.50	475
7800 Wedge type, powder coated steel		10	1.600		134	77.50		211.50	275
8100 Stainless steel		10	1.600		575	77.50		652.50	760
9000 Minimum labor/equipment charge	1 Carp	2.50	3.200	Job		155		155	253

10 21 13.14 Metal Toilet Compartment Components

	Crew	Daily Output	Labor-Hours	Unit	Material	2016 Bare Costs Labor	Equipment	Total	Total Incl O&P
0010 **METAL TOILET COMPARTMENT COMPONENTS**									
0100 Pilasters									
0110 Overhead braced, powder coated steel, 7" wide x 82" high	2 Carp	22.20	.721	Ea.	75	35		110	140
0120 Stainless steel		22.20	.721		135	35		170	205
0130 Floor braced, powder coated steel, 7" wide x 70" high		23.30	.687		128	33.50		161.50	196
0140 Stainless steel		23.30	.687		253	33.50		286.50	335
0150 Ceiling hung, powder coated steel, 7" wide x 83" high		13.30	1.203		137	58.50		195.50	246
0160 Stainless steel		13.30	1.203		283	58.50		341.50	405
0170 Wall hung, powder coated steel, 3" wide x 58" high		18.90	.847		133	41		174	213
0180 Stainless steel		18.90	.847		202	41		243	289
0200 Panels									
0210 Powder coated steel, 31" wide x 58" high	2 Carp	18.90	.847	Ea.	138	41		179	218
0220 Stainless steel		18.90	.847		360	41		401	460
0230 Powder coated steel, 53" wide x 58" high		18.90	.847		173	41		214	257
0240 Stainless steel		18.90	.847		470	41		511	580
0250 Powder coated steel, 63" wide x 58" high		18.90	.847		209	41		250	297
0260 Stainless steel		18.90	.847		540	41		581	660
0300 Doors									
0310 Powder coated steel, 24" wide x 58" high	2 Carp	14.10	1.135	Ea.	145	55		200	249
0320 Stainless steel		14.10	1.135		305	55		360	425
0330 Powder coated steel, 26" wide x 58" high		14.10	1.135		147	55		202	252
0340 Stainless steel		14.10	1.135		315	55		370	435
0350 Powder coated steel, 28" wide x 58" high		14.10	1.135		166	55		221	273
0360 Stainless steel		14.10	1.135		350	55		405	475
0370 Powder coated steel, 36" wide x 58" high		14.10	1.135		180	55		235	288
0380 Stainless steel		14.10	1.135		390	55		445	520
0400 Headrails									
0410 For powder coated steel, 62" long	2 Carp	65	.246	Ea.	22	11.95		33.95	44
0420 Stainless steel		65	.246		22	11.95		33.95	44
0430 For powder coated steel, 84" long		50	.320		32	15.50		47.50	60.50
0440 Stainless steel		50	.320		31.50	15.50		47	60.50
0450 For powder coated steel, 120" long		30	.533		44	26		70	90.50
0460 Stainless steel		30	.533		44	26		70	90.50
9000 Minimum labor/equipment charge	1 Carp	4	2	Job		97		97	158

For customer support on your Commercial Renovation Cost Data, call 877.742.4719.

339

10 21 13.16 Plastic-Laminate-Clad Toilet Compartments	Crew	Daily Output	Labor-Hours	Unit	Material	2016 Bare Costs Labor	Equipment	Total	Total Incl O&P
0010 **PLASTIC-LAMINATE-CLAD TOILET COMPARTMENTS**									
0110 Cubicles, ceiling hung									
0300 Plastic laminate on particle board	2 Carp	4	4	Ea.	525	194		719	895
0600 For handicap units, incl. 52" grab bars, add				"	450			450	495
0900 Floor and ceiling anchored									
1100 Plastic laminate on particle board	2 Carp	5	3.200	Ea.	820	155		975	1,150
1400 For handicap units, incl. 52" grab bars, add				"	370			370	405
1610 Floor mounted									
1800 Plastic laminate on particle board	2 Carp	7	2.286	Ea.	535	111		646	765
2450 Floor mounted, headrail braced									
2600 Plastic laminate on particle board	2 Carp	6	2.667	Ea.	765	129		894	1,050
3400 For handicap units, incl. 52" grab bars, add					370			370	410
4300 Entrance screen, floor mtd., plas. lam., 58" high, 48" wide	2 Carp	15	1.067		630	51.50		681.50	775
4800 Urinal screen, 18" wide, ceiling braced, plastic laminate		8	2		204	97		301	380
5400 Floor mounted, headrail braced		8	2		204	97		301	380
5900 Pilaster, flush, plastic laminate		10	1.600		520	77.50		597.50	700
6400 Post braced, plastic laminate		10	1.600		315	77.50		392.50	475
6700 Wall hung, bracket supported									
6900 Plastic laminate on particle board	2 Carp	10	1.600	Ea.	94.50	77.50		172	231
7450 Flange supported									
7500 Plastic laminate on particle board	2 Carp	10	1.600	Ea.	240	77.50		317.50	390
9000 Minimum labor/equipment charge	1 Carp	2.50	3.200	Job		155		155	253

10 21 13.17 Plastic-Lam. Clad Toilet Compart. Components

	Crew	Daily Output	Labor-Hours	Unit	Material	2016 Bare Costs Labor	Equipment	Total	Total Incl O&P
0010 **PLASTIC-LAMINATE CLAD TOILET COMPARTMENT COMPONENTS**									
0100 Pilasters									
0110 Overhead braced, 7" wide x 82" high	2 Carp	22.20	.721	Ea.	101	35		136	168
0130 Floor anchored, 7" wide x 70" high		23.30	.687		101	33.50		134.50	166
0150 Ceiling hung, 7" wide x 83" high		13.30	1.203		105	58.50		163.50	211
0180 Wall hung, 3" wide x 58" high		18.90	.847		93	41		134	169
0200 Panels									
0210 31" wide x 58" high	2 Carp	18.90	.847	Ea.	150	41		191	232
0230 51" wide x 58" high		18.90	.847		205	41		246	293
0250 63" wide x 58" high		18.90	.847		239	41		280	330
0300 Doors									
0310 24" wide x 58" high	2 Carp	14.10	1.135	Ea.	148	55		203	253
0330 26" wide x 58" high		14.10	1.135		153	55		208	258
0350 28" wide x 58" high		14.10	1.135		158	55		213	264
0370 36" wide x 58" high		14.10	1.135		188	55		243	297
0400 Headrails									
0410 62" long	2 Carp	65	.246	Ea.	22.50	11.95		34.45	44.50
0430 84" long		60	.267		31	12.90		43.90	55
0450 120" long		30	.533		43	26		69	89.50
9000 Minimum labor/equipment charge	1 Carp	4	2	Job		97		97	158

10 21 13.19 Plastic Toilet Compartments

	Crew	Daily Output	Labor-Hours	Unit	Material	2016 Bare Costs Labor	Equipment	Total	Total Incl O&P
0010 **PLASTIC TOILET COMPARTMENTS**									
0110 Cubicles, ceiling hung									
0250 Phenolic	2 Carp	4	4	Ea.	890	194		1,084	1,300
0600 For handicap units, incl. 52" grab bars, add				"	450			450	495
0900 Floor and ceiling anchored									
1050 Phenolic	2 Carp	5	3.200	Ea.	840	155		995	1,175
1400 For handicap units, incl. 52" grab bars, add				"	370			370	405
1610 Floor mounted									

10 21 Compartments and Cubicles

10 21 13 – Toilet Compartments

10 21 13.19 Plastic Toilet Compartments		Crew	Daily Output	Labor-Hours	Unit	Material	2016 Bare Costs Labor	Equipment	Total	Total Incl O&P
1750	Phenolic	2 Carp	7	2.286	Ea.	750	111		861	1,000
2100	For handicap units, incl. 52" grab bars, add					335			335	370
2200	For juvenile units, deduct					41.50			41.50	45.50
2450	Floor mounted, headrail braced									
2550	Phenolic	2 Carp	6	2.667	Ea.	750	129		879	1,025
9000	Minimum labor/equipment charge	1 Carp	2.50	3.200	Job		155		155	253

10 21 13.20 Plastic Toilet Compartment Components

		Crew	Daily Output	Labor-Hours	Unit	Material	2016 Bare Costs Labor	Equipment	Total	Total Incl O&P
0010	**PLASTIC TOILET COMPARTMENT COMPONENTS**									
0100	Pilasters									
0110	Overhead braced, polymer plastic, 7" wide x 82" high	2 Carp	22.20	.721	Ea.	129	35		164	199
0120	Phenolic		22.20	.721		152	35		187	225
0130	Floor braced, polymer plastic, 7" wide x 70" high		23.30	.687		178	33.50		211.50	251
0140	Phenolic		23.30	.687		144	33.50		177.50	213
0150	Ceiling hung, polymer plastic, 7" wide x 83" high		13.30	1.203		171	58.50		229.50	283
0160	Phenolic		13.30	1.203		163	58.50		221.50	274
0180	Wall hung, phenolic, 3" wide x 58" high		18.90	.847		98.50	41		139.50	176
0200	Panels									
0203	Polymer plastic, 18" high x 55" high	2 Carp	18.90	.847	Ea.	274	41		315	365
0206	Phenolic, 18" wide x 58" high		18.90	.847		244	41		285	335
0210	Polymer plastic, 31" high x 55" high		18.90	.847		320	41		361	420
0220	Phenolic, 31" wide x 58" high		18.90	.847		274	41		315	365
0223	Polymer plastic, 48" high x 55" high		18.90	.847		495	41		536	610
0226	Phenolic, 48" wide x 58" high		18.90	.847		465	41		506	575
0230	Polymer plastic, 51" wide x 55" high		18.90	.847		435	41		476	545
0240	Phenolic, 51" wide x 58" high		18.90	.847		415	41		456	520
0250	Polymer plastic, 63" wide x 55" high		18.90	.847		590	41		631	710
0260	Phenolic, 63" wide x 58" high		18.90	.847		440	41		481	550
0300	Doors									
0310	Polymer plastic, 24" wide x 55" high	2 Carp	14.10	1.135	Ea.	217	55		272	330
0320	Phenolic, 24" wide x 58" high		14.10	1.135		305	55		360	425
0330	Polymer plastic, 26" high x 55" high		14.10	1.135		232	55		287	345
0340	Phenolic, 26" wide x 58" high		14.10	1.135		325	55		380	445
0350	Polymer plastic, 28" wide x 55" high		14.10	1.135		254	55		309	370
0360	Phenolic, 28" wide x 58" high		14.10	1.135		340	55		395	465
0370	Polymer plastic, 36" wide x 55" high		14.10	1.135		300	55		355	420
0380	Phenolic, 36" wide x 58" high		14.10	1.135		425	55		480	555
0400	Headrails									
0410	For polymer plastic, 62" long	2 Carp	65	.246	Ea.	23	11.95		34.95	45
0420	Phenolic		65	.246		22.50	11.95		34.45	44.50
0430	For polymer plastic, 84" long		50	.320		32.50	15.50		48	61.50
0440	Phenolic		50	.320		32	15.50		47.50	61
0450	For polymer plastic, 120" long		30	.533		44.50	26		70.50	91
0460	Phenolic		30	.533		44	26		70	90.50
9000	Minimum labor/equipment charge	1 Carp	4	2	Job		97		97	158

10 21 13.40 Stone Toilet Compartments

		Crew	Daily Output	Labor-Hours	Unit	Material	2016 Bare Costs Labor	Equipment	Total	Total Incl O&P
0010	**STONE TOILET COMPARTMENTS**									
0100	Cubicles, ceiling hung, marble	2 Marb	2	8	Ea.	1,800	365		2,165	2,575
0600	For handicap units, incl. 52" grab bars, add					450			450	495
0800	Floor & ceiling anchored, marble	2 Marb	2.50	6.400		1,975	294		2,269	2,650
1400	For handicap units, incl. 52" grab bars, add					370			370	405
1600	Floor mounted, marble	2 Marb	3	5.333		1,225	245		1,470	1,750
2400	Floor mounted, headrail braced, marble	"	3	5.333		1,150	245		1,395	1,675

10 21 Compartments and Cubicles

10 21 13 – Toilet Compartments

10 21 13.40 Stone Toilet Compartments	Crew	Daily Output	Labor-Hours	Unit	Material	2016 Bare Costs Labor	Equipment	Total	Total Incl O&P
2900	For handicap units, incl. 52" grab bars, add				Ea.	370			370	410
4100	Entrance screen, floor mounted marble, 58" high, 48" wide	2 Marb	9	1.778		795	81.50		876.50	1,000
4600	Urinal screen, 18" wide, ceiling braced, marble	D-1	6	2.667	↓	755	112		867	1,025
5100	Floor mounted, head rail braced									
5200	Marble	D-1	6	2.667	Ea.	645	112		757	895
5700	Pilaster, flush, marble		9	1.778		840	75		915	1,050
6200	Post braced, marble	↓	9	1.778	↓	825	75		900	1,025
9000	Minimum labor/equipment charge	1 Carp	2.50	3.200	Job		155		155	253

10 21 23 – Cubicle Curtains and Track

10 21 23.16 Cubicle Track and Hardware

		Crew	Daily Output	Labor-Hours	Unit	Material	Labor	Equipment	Total	Total Incl O&P
0010	**CUBICLE TRACK AND HARDWARE**									
0020	Curtain track, box channel, ceiling mounted	1 Carp	135	.059	L.F.	6.25	2.87		9.12	11.55
0100	Suspended		100	.080		8.30	3.88		12.18	15.45
0550	Polyester, antimicrobial, 9' ceiling height		425	.019		19.65	.91		20.56	23
0560	8' ceiling height	↓	425	.019	↓	17.50	.91		18.41	20.50

10 22 Partitions

10 22 16 – Folding Gates

10 22 16.10 Security Gates

		Crew	Daily Output	Labor-Hours	Unit	Material	Labor	Equipment	Total	Total Incl O&P
0010	**SECURITY GATES** for roll up type, see Section 08 33 13.10									
0300	Scissors type folding gate, ptd. steel, single, 6-1/2' high, 5-1/2' wide	2 Sswk	4	4	Opng.	229	213		442	625
0350	6-1/2' wide		4	4		248	213		461	645
0400	7-1/2' wide		4	4		241	213		454	640
0600	Double gate, 8' high, 8' wide		2.50	6.400		395	340		735	1,025
0650	10' wide		2.50	6.400		430	340		770	1,075
0700	12' wide		2	8		625	425		1,050	1,425
0750	14' wide		2	8		635	425		1,060	1,450
0900	Door gate, folding steel, 4' wide, 61" high		4	4		141	213		354	530
1000	71" high		4	4		169	213		382	560
1200	81" high		4	4		195	213		408	590
1300	Window gates, 2' to 4' wide, 31" high		4	4		83	213		296	465
1500	55" high		3.75	4.267		122	227		349	535
1600	79" high	↓	3.50	4.571	↓	144	243		387	590

10 22 19 – Demountable Partitions

10 22 19.43 Demountable Composite Partitions

		Crew	Daily Output	Labor-Hours	Unit	Material	Labor	Equipment	Total	Total Incl O&P
0010	**DEMOUNTABLE COMPOSITE PARTITIONS**, add for doors									
0100	Do not deduct door openings from total L.F.									
0900	Demountable gypsum system on 2" to 2-1/2"									
1000	Steel studs, 9' high, 3" to 3-3/4" thick									
1200	Vinyl clad gypsum	2 Carp	48	.333	L.F.	60	16.15		76.15	92.50
1300	Fabric clad gypsum		44	.364		150	17.60		167.60	194
1500	Steel clad gypsum	↓	40	.400	↓	167	19.40		186.40	216
1600	1.75 system, aluminum framing, vinyl clad hardboard,									
1800	Paper honeycomb core panel, 1-3/4" to 2-1/2" thick									
1900	9' high	2 Carp	48	.333	L.F.	101	16.15		117.15	138
2100	7' high		60	.267		90.50	12.90		103.40	121
2200	5' high	↓	80	.200		76.50	9.70		86.20	100
2250	Unitized gypsum system									
2300	Unitized panel, 9' high, 2" to 2-1/2" thick									
2350	Vinyl clad gypsum	2 Carp	48	.333	L.F.	130	16.15		146.15	170

10 22 Partitions

10 22 19 – Demountable Partitions

10 22 19.43 Demountable Composite Partitions	Crew	Daily Output	Labor-Hours	Unit	Material	2016 Bare Costs Labor	Equipment	Total	Total Incl O&P
2400	Fabric clad gypsum	2 Carp	44	.364	L.F.	214	17.60		231.60	264
2500	Unitized mineral fiber system									
2510	Unitized panel, 9' high, 2-1/4" thick, aluminum frame									
2550	Vinyl clad mineral fiber	2 Carp	48	.333	L.F.	129	16.15		145.15	169
2600	Fabric clad mineral fiber	"	44	.364	"	193	17.60		210.60	241
2800	Movable steel walls, modular system									
2900	Unitized panels, 9' high, 48" wide									
3100	Baked enamel, pre-finished	2 Carp	60	.267	L.F.	146	12.90		158.90	182
3200	Fabric clad steel	"	56	.286	"	212	13.85		225.85	256
5500	For acoustical partitions, add, unsealed				S.F.	2.36			2.36	2.60
5550	Sealed				"	11			11	12.10
5700	For doors, see Sections 08 16									
5800	For door hardware, see Section 08 71									
6100	In-plant modular office system, w/prehung hollow core door									
6200	3" thick polystyrene core panels									
6250	12' x 12', 2 wall	2 Clab	3.80	4.211	Ea.	4,125	160		4,285	4,775
6300	4 wall		1.90	8.421		6,275	320		6,595	7,425
6350	16' x 16', 2 wall		3.60	4.444		6,250	168		6,418	7,150
6400	4 wall		1.80	8.889		8,475	335		8,810	9,875
9000	Minimum labor/equipment charge	2 Carp	3	5.333	Job		258		258	420

10 22 23 – Portable Partitions, Screens, and Panels

10 22 23.13 Wall Screens

10 22 23.13 Wall Screens	Crew	Daily Output	Labor-Hours	Unit	Material	2016 Bare Costs Labor	Equipment	Total	Total Incl O&P
0010	**WALL SCREENS**, divider panels, free standing, fiber core									
0020	Fabric face straight									
0100	3'-0" long, 4'-0" high	2 Carp	100	.160	L.F.	123	7.75		130.75	148
0200	5'-0" high		90	.178		103	8.60		111.60	127
0500	6'-0" high		75	.213		104	10.35		114.35	132
0900	5'-0" long, 4'-0" high		175	.091		67.50	4.43		71.93	82
1000	5'-0" high		150	.107		75.50	5.15		80.65	91.50
1500	6"-0" high		125	.128		90.50	6.20		96.70	110
1600	6'-0" long, 5'-0" high		162	.099		75.50	4.79		80.29	91
3200	Economical panels, fabric face, 4'-0" long, 5'-0" high		132	.121		50	5.85		55.85	64.50
3250	6'-0" high		112	.143		55.50	6.90		62.40	72.50
3300	5'-0" long, 5'-0" high		150	.107		54.50	5.15		59.65	68.50
3350	6'-0" high		125	.128		50	6.20		56.20	65
3450	Acoustical panels, 60 to 90 NRC, 3'-0" long, 5'-0" high		90	.178		76.50	8.60		85.10	98
3550	6'-0" high		75	.213		89.50	10.35		99.85	115
3600	5'-0" long, 5'-0" high		150	.107		61	5.15		66.15	75.50
3650	6'-0" high		125	.128		69	6.20		75.20	85.50
3700	6'-0" long, 5'-0" high		162	.099		53	4.79		57.79	66.50
3750	6'-0" high		138	.116		82	5.60		87.60	99
3800	Economy acoustical panels, 40 N.R.C., 4'-0" long, 5'-0" high		132	.121		50	5.85		55.85	64.50
3850	6'-0" high		112	.143		55.50	6.90		62.40	72.50
3900	5'-0" long, 6'-0" high		125	.128		50	6.20		56.20	65
3950	6'-0" long, 5'-0" high		162	.099		47	4.79		51.79	60
9000	Minimum labor/equipment charge		3	5.333	Job		258		258	420

10 22 33 – Accordion Folding Partitions

10 22 33.10 Partitions, Accordion Folding

10 22 33.10 Partitions, Accordion Folding	Crew	Daily Output	Labor-Hours	Unit	Material	2016 Bare Costs Labor	Equipment	Total	Total Incl O&P
0010	**PARTITIONS, ACCORDION FOLDING**									
0100	Vinyl covered, over 150 S.F., frame not included									
0300	Residential, 1.25 lb. per S.F., 8' maximum height	2 Carp	300	.053	S.F.	25	2.58		27.58	31.50
0400	Commercial, 1.75 lb. per S.F., 8' maximum height		225	.071		28.50	3.45		31.95	37

For customer support on your Commercial Renovation Cost Data, call 877.742.4719.

343

10 22 Partitions

10 22 33 – Accordion Folding Partitions

10 22 33.10 Partitions, Accordion Folding	Crew	Daily Output	Labor-Hours	Unit	Material	2016 Bare Costs Labor	Equipment	Total	Total Incl O&P
0900	Acoustical, 3 lb. per S.F., 17' maximum height	2 Carp	100	.160	S.F.	31.50	7.75		39.25	47.50
1200	5 lb. per S.F., 20' maximum height		95	.168		44	8.15		52.15	62
1500	Vinyl clad wood or steel, electric operation, 5.0 psf		160	.100		63	4.85		67.85	77
1900	Wood, non-acoustic, birch or mahogany, to 10' high		300	.053		34	2.58		36.58	41
9000	Minimum labor/equipment charge		4	4	Job		194		194	315

10 22 39 – Folding Panel Partitions

10 22 39.10 Partitions, Folding Panel	Crew	Daily Output	Labor-Hours	Unit	Material	2016 Bare Costs Labor	Equipment	Total	Total Incl O&P
0010	**PARTITIONS, FOLDING PANEL**, acoustic, wood									
0100	Vinyl faced, to 18' high, 6 psf, economy trim	2 Carp	60	.267	S.F.	57	12.90		69.90	83.50
0150	Standard trim		45	.356		68	17.25		85.25	103
0200	Premium trim		30	.533		87.50	26		113.50	139
0400	Plastic laminate or hardwood finish, standard trim		60	.267		58.50	12.90		71.40	85.50
0500	Premium trim		30	.533		62.50	26		88.50	111
0600	Wood, low acoustical type, 4.5 psf, to 14' high		50	.320		42.50	15.50		58	72.50
9000	Minimum labor/equipment charge		4	4	Job		194		194	315

10 26 Wall and Door Protection

10 26 13 – Corner Guards

10 26 13.10 Metal Corner Guards	Crew	Daily Output	Labor-Hours	Unit	Material	2016 Bare Costs Labor	Equipment	Total	Total Incl O&P
0010	**METAL CORNER GUARDS**									
0020	Steel angle w/anchors, 1" x 1" x 1/4", 1.5#/L.F.	2 Carp	160	.100	L.F.	7.45	4.85		12.30	16.10
0100	2" x 2" x 1/4" angles, 3.2#/L.F.	"	150	.107	"	10.70	5.15		15.85	20
1000	Wall protection, stainless steel, 16 ga, 48" x 36" tall, screwed to studs	2 Skwk	500	.032	S.F.	8.25	1.60		9.85	11.70
1050	Wall end guard, stainless steel, 16 ga, 36" tall, screwed to studs	1 Skwk	30	.267	Ea.	25	13.30		38.30	49
9000	Minimum labor/equipment charge	1 Carp	2	4	Job		194		194	315

10 28 Toilet, Bath, and Laundry Accessories

10 28 13 – Toilet Accessories

10 28 13.13 Commercial Toilet Accessories	Crew	Daily Output	Labor-Hours	Unit	Material	2016 Bare Costs Labor	Equipment	Total	Total Incl O&P
0010	**COMMERCIAL TOILET ACCESSORIES**									
0200	Curtain rod, stainless steel, 5' long, 1" diameter	1 Carp	13	.615	Ea.	28	30		58	79.50
0300	1-1/4" diameter		13	.615		30.50	30		60.50	82.50
0350	Chrome, 1" diameter		13	.615		32.50	30		62.50	84
0360	For vinyl curtain, add		1950	.004	S.F.	.97	.20		1.17	1.39
0400	Diaper changing station, horizontal, wall mounted, plastic		10	.800	Ea.	245	39		284	335
0420	Vertical		10	.800		223	39		262	310
0430	Oval shaped		10	.800		223	39		262	310
0440	Recessed, with stainless steel flange		6	1.333		510	64.50		574.50	665
0500	Dispenser units, combined soap & towel dispensers,									
0510	Mirror and shelf, flush mounted	1 Carp	10	.800	Ea.	310	39		349	410
0600	Towel dispenser and waste receptacle,									
0610	18 gallon capacity	1 Carp	10	.800	Ea.	310	39		349	405
0800	Grab bar, straight, 1-1/4" diameter, stainless steel, 18" long		24	.333		30.50	16.15		46.65	60
0900	24" long		23	.348		30.50	16.85		47.35	61
1000	30" long		22	.364		33	17.60		50.60	65.50
1100	36" long		20	.400		37	19.40		56.40	72
1105	42" long		20	.400		39	19.40		58.40	74.50
1120	Corner, 36" long		20	.400		90	19.40		109.40	131
1200	1-1/2" diameter, 24" long		23	.348		32.50	16.85		49.35	63.50

10 28 13.13 Commercial Toilet Accessories		Crew	Daily Output	Labor-Hours	Unit	Material	2016 Bare Costs Labor	2016 Bare Costs Equipment	Total	Total Incl O&P
1300	36" long	1 Carp	20	.400	Ea.	35.50	19.40		54.90	70.50
1310	42" long		18	.444		39	21.50		60.50	77.50
1500	Tub bar, 1-1/4" diameter, 24" x 36"		14	.571		100	27.50		127.50	155
1600	Plus vertical arm		12	.667		90.50	32.50		123	153
1900	End tub bar, 1" diameter, 90° angle, 16" x 32"		12	.667		101	32.50		133.50	164
2010	Tub/shower/toilet, 2-wall, 36" x 24"		12	.667		93.50	32.50		126	156
2110	Antimicrobial copper alloy finish, straight, 18" long		24	.333		70	16.15		86.15	104
2120	24" long		23	.348		77	16.85		93.85	112
2130	36" long		20	.400		91.50	19.40		110.90	133
2140	48" long		19	.421		105	20.50		125.50	150
2300	Hand dryer, surface mounted, electric, 115 volt, 20 amp		4	2		480	97		577	685
2400	230 volt, 10 amp		4	2		680	97		777	905
2450	Hand dryer, touch free, 1400 watt, 81,000 rpm		4	2		1,025	97		1,122	1,275
2600	Hat and coat strip, stainless steel, 4 hook, 36" long		24	.333		68	16.15		84.15	102
2700	6 hook, 60" long		20	.400		124	19.40		143.40	169
3000	Mirror, with stainless steel 3/4" square frame, 18" x 24"		20	.400		49.50	19.40		68.90	86
3100	36" x 24"		15	.533		110	26		136	163
3200	48" x 24"		10	.800		152	39		191	232
3300	72" x 24"		6	1.333		305	64.50		369.50	440
3500	With 5" stainless steel shelf, 18" x 24"		20	.400		198	19.40		217.40	250
3600	36" x 24"		15	.533		243	26		269	310
3700	48" x 24"		10	.800		257	39		296	345
3800	72" x 24"		6	1.333		259	64.50		323.50	390
4100	Mop holder strip, stainless steel, 5 holders, 48" long		20	.400		78	19.40		97.40	118
4200	Napkin/tampon dispenser, recessed		15	.533		630	26		656	735
4220	Semi-recessed		6.50	1.231		310	59.50		369.50	440
4250	Napkin receptacle, recessed		6.50	1.231		177	59.50		236.50	293
4300	Robe hook, single, regular		96	.083		19.40	4.04		23.44	28
4400	Heavy duty, concealed mounting		56	.143		22.50	6.90		29.40	36.50
4600	Soap dispenser, chrome, surface mounted, liquid		20	.400		50	19.40		69.40	86.50
5000	Recessed stainless steel, liquid		10	.800		166	39		205	246
5600	Shelf, stainless steel, 5" wide, 18 ga., 24" long		24	.333		85	16.15		101.15	120
5700	48" long		16	.500		157	24		181	212
5800	8" wide shelf, 18 ga., 24" long		22	.364		72.50	17.60		90.10	109
5900	48" long		14	.571		127	27.50		154.50	185
6000	Toilet seat cover dispenser, stainless steel, recessed		20	.400		181	19.40		200.40	231
6050	Surface mounted		15	.533		35	26		61	81
6100	Toilet tissue dispenser, surface mounted, SS, single roll		30	.267		18.95	12.90		31.85	42
6200	Double roll		24	.333		24.50	16.15		40.65	53.50
6240	Plastic, twin/jumbo dbl. roll		24	.333		28.50	16.15		44.65	57.50
6290	Toilet seat	1 Plum	40	.200		31.50	11.85		43.35	53
6400	Towel bar, stainless steel, 18" long	1 Carp	23	.348		41.50	16.85		58.35	73
6500	30" long	"	21	.381		66	18.45		84.45	103
6610	Antimicrobial copper alloy finish, 3/4" round, straight, w/o mounting				L.F.	11.20			11.20	12.30
6620	Antimicrobial copper alloy finish, 1" round, straight, w/o mounting				"	13.45			13.45	14.75
6630	24" long, including mounting	1 Carp	23	.348	Ea.	95	16.85		111.85	133
6700	Towel dispenser, stainless steel, surface mounted		16	.500		47.50	24		71.50	91.50
6800	Flush mounted, recessed		10	.800		282	39		321	375
6900	Plastic, touchless, battery operated		16	.500		89	24		113	137
7000	Towel holder, hotel type, 2 guest size		20	.400		53	19.40		72.40	90
7200	Towel shelf, stainless steel, 24" long, 8" wide		20	.400		64	19.40		83.40	102
7400	Tumbler holder, for tumbler only		30	.267		18.75	12.90		31.65	41.50
7410	Tumbler holder, recessed		20	.400		8	19.40		27.40	40.50

For customer support on your Commercial Renovation Cost Data, call 877.742.4719.

345

10 28 Toilet, Bath, and Laundry Accessories

10 28 13 – Toilet Accessories

10 28 13.13 Commercial Toilet Accessories	Crew	Daily Output	Labor-Hours	Unit	Material	2016 Bare Costs Labor	Equipment	Total	Total Incl O&P
7500	Soap, tumbler & toothbrush	1 Carp	30	.267	Ea.	19.60	12.90		32.50	42.50
7510	Tumbler & toothbrush holder		20	.400		13.50	19.40		32.90	46.50
7700	Wall urn ash receiver, surface mount, 11" long		12	.667		95	32.50		127.50	157
8000	Waste receptacles, stainless steel, with top, 13 gallon		10	.800		315	39		354	410
8100	36 gallon		8	1		410	48.50		458.50	535
9000	Minimum labor/equipment charge		5	1.600	Job		77.50		77.50	127

10 28 16 – Bath Accessories

10 28 16.20 Medicine Cabinets

10 28 16.20 Medicine Cabinets	Crew	Daily Output	Labor-Hours	Unit	Material	2016 Bare Costs Labor	Equipment	Total	Total Incl O&P
0010	**MEDICINE CABINETS**									
0020	With mirror, sst frame, 16" x 22", unlighted	1 Carp	14	.571	Ea.	99	27.50		126.50	154
0100	Wood frame		14	.571		140	27.50		167.50	199
0300	Sliding mirror doors, 20" x 16" x 4-3/4", unlighted		7	1.143		133	55.50		188.50	237
0400	24" x 19" x 8-1/2", lighted		5	1.600		184	77.50		261.50	330
0600	Triple door, 30" x 32", unlighted, plywood body		7	1.143		325	55.50		380.50	450
0700	Steel body		7	1.143		375	55.50		430.50	500
0900	Oak door, wood body, beveled mirror, single door		7	1.143		199	55.50		254.50	310
1000	Double door		6	1.333		415	64.50		479.50	560
1200	Hotel cabinets, stainless, with lower shelf, unlighted		10	.800		200	39		239	283
1300	Lighted		5	1.600		305	77.50		382.50	460
9000	Minimum labor/equipment charge		4	2	Job		97		97	158

10 28 19 – Tub and Shower Enclosures

10 28 19.10 Partitions, Shower

10 28 19.10 Partitions, Shower	Crew	Daily Output	Labor-Hours	Unit	Material	2016 Bare Costs Labor	Equipment	Total	Total Incl O&P
0010	**PARTITIONS, SHOWER** floor mounted, no plumbing									
0400	Cabinet, one piece, fiberglass, 32" x 32"	2 Carp	5	3.200	Ea.	535	155		690	845
0420	36" x 36"		5	3.200		625	155		780	940
0440	36" x 48"		5	3.200		1,375	155		1,530	1,750
0460	Acrylic, 32" x 32"		5	3.200		340	155		495	630
0480	36" x 36"		5	3.200		1,050	155		1,205	1,400
0500	36" x 48"		5	3.200		1,400	155		1,555	1,800
0520	Shower door for above, clear plastic, 24" wide	1 Carp	8	1		188	48.50		236.50	286
0540	28" wide		8	1		225	48.50		273.50	325
0560	Tempered glass, 24" wide		8	1		220	48.50		268.50	320
0580	28" wide		8	1		252	48.50		300.50	355
4100	Shower doors, economy plastic, 24" wide	1 Shee	9	.889		144	51		195	239
4200	Tempered glass door, economy		8	1		258	57.50		315.50	375
4700	Deluxe, tempered glass, chrome on brass frame, 42" to 44"		8	1		415	57.50		472.50	545
4800	39" to 48" wide		1	8		675	460		1,135	1,475
9990	Minimum labor/equipment charge		2.50	3.200	Job		183		183	289

10 31 Manufactured Fireplaces

10 31 13 – Manufactured Fireplace Chimneys

10 31 13.10 Fireplace Chimneys

10 31 13.10 Fireplace Chimneys	Crew	Daily Output	Labor-Hours	Unit	Material	2016 Bare Costs Labor	Equipment	Total	Total Incl O&P
0010	**FIREPLACE CHIMNEYS**									
0500	Chimney dbl. wall, all stainless, over 8'-6", 7" diam., add to fireplace	1 Carp	33	.242	V.L.F.	84	11.75		95.75	112
0600	10" diameter, add to fireplace		32	.250		112	12.10		124.10	144
0700	12" diameter, add to fireplace		31	.258		167	12.50		179.50	204
0800	14" diameter, add to fireplace		30	.267		227	12.90		239.90	271
1000	Simulated brick chimney top, 4' high, 16" x 16"		10	.800	Ea.	435	39		474	545
1100	24" x 24"		7	1.143	"	555	55.50		610.50	700

10 31 Manufactured Fireplaces

10 31 13 – Manufactured Fireplace Chimneys

10 31 13.20 Chimney Accessories

	10 31 13.20 Chimney Accessories	Crew	Daily Output	Labor-Hours	Unit	Material	2016 Bare Costs Labor	Equipment	Total	Total Incl O&P
0010	**CHIMNEY ACCESSORIES**									
0020	Chimney screens, galv., 13" x 13" flue	1 Bric	8	1	Ea.	58.50	46.50		105	140
0050	24" x 24" flue		5	1.600		124	74		198	257
0200	Stainless steel, 13" x 13" flue		8	1		97.50	46.50		144	183
0250	20" x 20" flue		5	1.600		152	74		226	287
2400	Squirrel and bird screens, galvanized, 8" x 8" flue		16	.500		54	23		77	97
2450	13" x 13" flue		12	.667		55	31		86	111
9000	Minimum labor/equipment charge		3.50	2.286	Job		106		106	172

10 31 16 – Manufactured Fireplace Forms

10 31 16.10 Fireplace Forms

	10 31 16.10 Fireplace Forms	Crew	Daily Output	Labor-Hours	Unit	Material	2016 Bare Costs Labor	Equipment	Total	Total Incl O&P
0010	**FIREPLACE FORMS**									
1800	Fireplace forms, no accessories, 32" opening	1 Bric	3	2.667	Ea.	705	123		828	975
1900	36" opening		2.50	3.200		895	148		1,043	1,225
2000	40" opening		2	4		1,200	185		1,385	1,600
2100	78" opening		1.50	5.333		1,725	247		1,972	2,300

10 31 23 – Prefabricated Fireplaces

10 31 23.10 Fireplace, Prefabricated

	10 31 23.10 Fireplace, Prefabricated	Crew	Daily Output	Labor-Hours	Unit	Material	2016 Bare Costs Labor	Equipment	Total	Total Incl O&P
0010	**FIREPLACE, PREFABRICATED**, free standing or wall hung									
0100	With hood & screen, painted	1 Carp	1.30	6.154	Ea.	1,475	298		1,773	2,100
0150	Average		1	8		1,700	390		2,090	2,500
0200	Stainless steel		.90	8.889		3,050	430		3,480	4,050
1500	Simulated logs, gas fired, 40,000 BTU, 2' long, manual safety pilot		7	1.143	Set	440	55.50		495.50	575
1600	Adjustable flame remote pilot		6	1.333		1,200	64.50		1,264.50	1,425
1700	Electric, 1,500 BTU, 1'-6" long, incandescent flame		7	1.143		205	55.50		260.50	315
1800	1,500 BTU, LED flame		6	1.333		320	64.50		384.50	455
2000	Fireplace, built-in, 36" hearth, radiant		1.30	6.154	Ea.	680	298		978	1,225
2100	Recirculating, small fan		1	8		900	390		1,290	1,625
2150	Large fan		.90	8.889		2,000	430		2,430	2,900
2200	42" hearth, radiant		1.20	6.667		915	325		1,240	1,525
2300	Recirculating, small fan		.90	8.889		1,200	430		1,630	2,025
2350	Large fan		.80	10		1,350	485		1,835	2,275
2400	48" hearth, radiant		1.10	7.273		2,175	350		2,525	2,950
2500	Recirculating, small fan		.80	10		2,475	485		2,960	3,525
2550	Large fan		.70	11.429		2,475	555		3,030	3,625
3000	See through, including doors		.80	10		2,200	485		2,685	3,200
3200	Corner (2 wall)		1	8		3,425	390		3,815	4,375

10 32 Fireplace Specialties

10 32 13 – Fireplace Dampers

10 32 13.10 Dampers

	10 32 13.10 Dampers	Crew	Daily Output	Labor-Hours	Unit	Material	2016 Bare Costs Labor	Equipment	Total	Total Incl O&P
0010	**DAMPERS**									
0800	Damper, rotary control, steel, 30" opening	1 Bric	6	1.333	Ea.	119	61.50		180.50	231
0850	Cast iron, 30" opening		6	1.333		125	61.50		186.50	237
0880	36" opening		6	1.333		127	61.50		188.50	240
0900	48" opening		6	1.333		167	61.50		228.50	284
0920	60" opening		6	1.333		355	61.50		416.50	490
1000	84" opening, special order		5	1.600		910	74		984	1,125
1050	96" opening, special order		4	2		925	92.50		1,017.50	1,175
1200	Steel plate, poker control, 60" opening		8	1		320	46.50		366.50	430

10 32 Fireplace Specialties

10 32 13 – Fireplace Dampers

10 32 13.10 Dampers	Crew	Daily Output	Labor-Hours	Unit	Material	2016 Bare Costs Labor	Equipment	Total	Total Incl O&P
1250	84" opening, special order	1 Bric	5	1.600	Ea.	585	74		659	765
1400	"Universal" type, chain operated, 32" x 20" opening		8	1		250	46.50		296.50	350
1450	48" x 24" opening		5	1.600		375	74		449	530

10 32 23 – Fireplace Doors

10 32 23.10 Doors

		Crew	Daily Output	Labor-Hours	Unit	Material	Labor	Equipment	Total	Total Incl O&P
0010	**DOORS**									
0400	Cleanout doors and frames, cast iron, 8" x 8"	1 Bric	12	.667	Ea.	42.50	31		73.50	96.50
0450	12" x 12"		10	.800		97	37		134	167
0500	18" x 24"		8	1		150	46.50		196.50	241
0550	Cast iron frame, steel door, 24" x 30"		5	1.600		315	74		389	465
1600	Dutch Oven door and frame, cast iron, 12" x 15" opening		13	.615		131	28.50		159.50	192
1650	Copper plated, 12" x 15" opening		13	.615		257	28.50		285.50	330

10 35 Stoves

10 35 13 – Heating Stoves

10 35 13.10 Woodburning Stoves

		Crew	Daily Output	Labor-Hours	Unit	Material	Labor	Equipment	Total	Total Incl O&P
0010	**WOODBURNING STOVES**									
0015	Cast iron, less than 1500 SF	2 Carp	1.30	12.308	Ea.	1,300	595		1,895	2,400
0020	1500 to 2000 S.F.		1	16		2,050	775		2,825	3,525
0030	greater than 2,000 SF		.80	20		2,800	970		3,770	4,650
0050	For gas log lighter, add					46.50			46.50	51

10 43 Emergency Aid Specialties

10 43 13 – Defibrillator Cabinets

10 43 13.05 Defibrillator Cabinets

		Crew	Daily Output	Labor-Hours	Unit	Material	Labor	Equipment	Total	Total Incl O&P
0010	**DEFIBRILLATOR CABINETS**, not equipped, stainless steel									
0050	Defibrillator cabinet, stainless steel with strobe & alarm 12" x 27"	1 Carp	10	.800	Ea.	385	39		424	490
0100	Automatic External Defibrillator	"	30	.267	"	1,375	12.90		1,387.90	1,550

10 44 Fire Protection Specialties

10 44 13 – Fire Protection Cabinets

10 44 13.53 Fire Equipment Cabinets

		Crew	Daily Output	Labor-Hours	Unit	Material	Labor	Equipment	Total	Total Incl O&P
0010	**FIRE EQUIPMENT CABINETS**, not equipped, 20 ga. steel box									
0040	Recessed, D.S. glass in door, box size given									
1000	Portable extinguisher, single, 8" x 12" x 27", alum. door & frame	Q-12	8	2	Ea.	160	104		264	340
1100	Steel door and frame	"	8	2	"	122	104		226	296
3000	Hose rack assy., 1-1/2" valve & 100' hose, 24" x 40" x 5-1/2"									
3200	Steel door and frame	Q-12	6	2.667	Ea.	258	138		396	500
4000	Hose rack assy., 2-1/2" x 1-1/2" valve, 100' hose, 24" x 40" x 8"									
4200	Steel door and frame	Q-12	6	2.667	Ea.	265	138		403	510
5000	Hose rack assy., 2-1/2" x 1-1/2" valve, 100' hose									
5010	and extinguisher, 30" x 40" x 8"									
5200	Steel door and frame	Q-12	5	3.200	Ea.	277	166		443	565

10 44 Fire Protection Specialties

10 44 16 – Fire Extinguishers

10 44 16.13 Portable Fire Extinguishers	Crew	Daily Output	Labor-Hours	Unit	Material	2016 Bare Costs Labor	Equipment	Total	Total Incl O&P
0010 **PORTABLE FIRE EXTINGUISHERS**									
0140 CO$_2$, with hose and "H" horn, 10 lb.				Ea.	288			288	315
1000 Dry chemical, pressurized									
1040 Standard type, portable, painted, 2-1/2 lb.				Ea.	38.50			38.50	42.50
1080 10 lb.					82			82	90
1100 20 lb.					135			135	149
1120 30 lb.					435			435	480
2000 ABC all purpose type, portable, 2-1/2 lb.					21			21	23.50
2080 9-1/2 lb.					44			44	48
3500 Halotron 1, 2-1/2 lb.					127			127	140
3600 5 lb.					199			199	219
3700 11 lb.					395			395	435

10 51 Lockers

10 51 13 – Metal Lockers

10 51 13.10 Lockers

	Crew	Daily Output	Labor-Hours	Unit	Material	2016 Bare Costs Labor	Equipment	Total	Total Incl O&P
0011 **LOCKERS** steel, baked enamel, pre-assembled									
0110 Single tier box locker, 12" x 15" x 72"	1 Shee	20	.400	Ea.	231	23		254	290
0120 18" x 15" x 72"		20	.400		245	23		268	305
0130 12" x 18" x 72"		20	.400		235	23		258	294
0140 18" x 18" x 72"		20	.400		285	23		308	350
0410 Double tier, 12" x 15" x 36"		30	.267		233	15.25		248.25	280
0420 18" x 15" x 36"		30	.267		235	15.25		250.25	283
0430 12" x 18" x 36"		30	.267		265	15.25		280.25	315
0440 18" x 18" x 36"		30	.267		225	15.25		240.25	272
0500 Two person, 18" x 15" x 72"		20	.400		297	23		320	360
0510 18" x 18" x 72"		20	.400		330	23		353	395
0520 Duplex, 15" x 15" x 72"		20	.400		335	23		358	405
0530 15" x 21" x 72"		20	.400		370	23		393	445
1100 Wire meshed wardrobe, floor mtd., open front varsity type		7.50	1.067		277	61		338	400
2400 16-person locker unit with clothing rack									
2500 72" wide x 15" deep x 72" high	1 Shee	15	.533	Ea.	495	30.50		525.50	590
2550 18" deep		15	.533	"	650	30.50		680.50	765
3250 Rack w/24 wire mesh baskets		1.50	5.333	Set	410	305		715	930
3260 30 baskets		1.25	6.400		360	365		725	975
3270 36 baskets		.95	8.421		470	480		950	1,275
3280 42 baskets		.80	10		455	575		1,030	1,400
3600 For hanger rods, add				Ea.	2			2	2.20
3650 For number plate kit, 100 plates #1 - #100, add	1 Shee	4	2		82	115		197	271
3700 For locker base, closed front panel		90	.089		7.20	5.10		12.30	15.95
3710 End panel, bolted		36	.222		9.30	12.70		22	30
3800 For sloping top, 12" wide		24	.333		32.50	19.10		51.60	66
3810 15" wide		24	.333		35.50	19.10		54.60	69
3820 18" wide		24	.333		35.50	19.10		54.60	69
3850 Sloping top end panel, 12" deep		72	.111		12.20	6.35		18.55	23.50
3860 15" deep		72	.111		12.20	6.35		18.55	23.50
3870 18" deep		72	.111		12.20	6.35		18.55	23.50
3900 For finish end panels, steel, 60" high, 15" deep		12	.667		36	38		74	99.50
3910 72" high, 12" deep		12	.667		29	38		67	92
3920 18" deep		12	.667		43.50	38		81.50	108

For customer support on your Commercial Renovation Cost Data, call 877.742.4719.

349

10 51 Lockers

10 51 13 – Metal Lockers

10 51 13.10 Lockers	Crew	Daily Output	Labor-Hours	Unit	Material	2016 Bare Costs Labor	Equipment	Total	Total Incl O&P
5000	For "ready to assemble" lockers,								
5010	Add to labor						75%		
5020	Deduct from material					20%			
6000	Heavy duty for detention facility, tamper proof, 14 ga. welded steel, solid								
6100	24" W x 24" D x 74" H, single tier	1 Shee	18	.444	Ea.	525	25.50		550.50	615
6110	Double tier		18	.444		515	25.50		540.50	605
6120	Triple tier		18	.444		530	25.50		555.50	625
9000	Minimum labor/equipment charge		2.50	3.200	Job		183		183	289

10 55 Postal Specialties

10 55 23 – Mail Boxes

10 55 23.10 Commercial Mail Boxes

		Crew	Daily Output	Labor-Hours	Unit	Material	Labor	Equipment	Total	Total Incl O&P
0010	**COMMERCIAL MAIL BOXES**									
0020	Horiz., key lock, 5"H x 6"W x 15"D, alum., rear load	1 Carp	34	.235	Ea.	42.50	11.40		53.90	65.50
0100	Front loading		34	.235		42.50	11.40		53.90	65.50
0200	Double, 5"H x 12"W x 15"D, rear loading		26	.308		60.50	14.90		75.40	91
0300	Front loading		26	.308		69	14.90		83.90	100
0500	Quadruple, 10"H x 12"W x 15"D, rear loading		20	.400		101	19.40		120.40	143
0600	Front loading		20	.400		92.50	19.40		111.90	134
1600	Vault type, horizontal, for apartments, 4" x 5"		34	.235		26.50	11.40		37.90	47.50
1700	Alphabetical directories, 120 names		10	.800		125	39		164	202
1900	Letter slot, residential		20	.400		80	19.40		99.40	120
2000	Post office type		8	1		119	48.50		167.50	210
9000	Minimum labor/equipment charge		5	1.600	Job		77.50		77.50	127

10 56 Storage Assemblies

10 56 13 – Metal Storage Shelving

10 56 13.10 Shelving

		Crew	Daily Output	Labor-Hours	Unit	Material	Labor	Equipment	Total	Total Incl O&P
0010	**SHELVING**									
0020	Metal, industrial, cross-braced, 3' wide, 12" deep	1 Sswk	175	.046	SF Shlf	7.20	2.43		9.63	12.20
0100	24" deep		330	.024		5.20	1.29		6.49	7.95
2200	Wide span, 1600 lb. capacity per shelf, 6' wide, 24" deep		380	.021		7.65	1.12		8.77	10.40
2400	36" deep		440	.018		6.75	.97		7.72	9.10
9000	Minimum labor/equipment charge	1 Carp	4	2	Job		97		97	158

10 57 Wardrobe and Closet Specialties

10 57 23 – Closet and Utility Shelving

10 57 23.19 Wood Closet and Utility Shelving

		Crew	Daily Output	Labor-Hours	Unit	Material	Labor	Equipment	Total	Total Incl O&P
0010	**WOOD CLOSET AND UTILITY SHELVING**									
0020	Pine, clear grade, no edge band, 1" x 8"	1 Carp	115	.070	L.F.	3.47	3.37		6.84	9.30
0100	1" x 10"		110	.073		4.32	3.52		7.84	10.50
0200	1" x 12"		105	.076		5.20	3.69		8.89	11.80
0600	Plywood, 3/4" thick with lumber edge, 12" wide		75	.107		1.91	5.15		7.06	10.55
0700	24" wide		70	.114		3.39	5.55		8.94	12.80
0900	Bookcase, clear grade pine, shelves 12" O.C., 8" deep, per S.F. shelf		70	.114	S.F.	11.30	5.55		16.85	21.50
1000	12" deep shelves		65	.123	"	16.95	5.95		22.90	28.50
1200	Adjustable closet rod and shelf, 12" wide, 3' long		20	.400	Ea.	11.60	19.40		31	44.50

10 57 Wardrobe and Closet Specialties

10 57 23 – Closet and Utility Shelving

10 57 23.19 Wood Closet and Utility Shelving	Crew	Daily Output	Labor-Hours	Unit	Material	2016 Bare Costs Labor	Equipment	Total	Total Incl O&P
1300	8' long	1 Carp	15	.533	Ea.	21	26		47	65.50
1500	Prefinished shelves with supports, stock, 8" wide		75	.107	L.F.	4.89	5.15		10.04	13.85
1600	10" wide		70	.114	"	5.70	5.55		11.25	15.30
9000	Minimum labor/equipment charge		4	2	Job		97		97	158

10 73 Protective Covers

10 73 13 – Awnings

10 73 13.10 Awnings, Fabric

		Crew	Daily Output	Labor-Hours	Unit	Material	2016 Bare Costs Labor	Equipment	Total	Total Incl O&P
0010	**AWNINGS, FABRIC**									
0020	Including acrylic canvas and frame, standard design									
0100	Door and window, slope, 3' high, 4' wide	1 Carp	4.50	1.778	Ea.	755	86		841	970
0110	6' wide		3.50	2.286		970	111		1,081	1,250
0120	8' wide		3	2.667		1,200	129		1,329	1,500
0200	Quarter round convex, 4' wide		3	2.667		1,175	129		1,304	1,500
0210	6' wide		2.25	3.556		1,525	172		1,697	1,950
0220	8' wide		1.80	4.444		1,875	215		2,090	2,400
0300	Dome, 4' wide		7.50	1.067		455	51.50		506.50	585
0310	6' wide		3.50	2.286		1,025	111		1,136	1,300
0320	8' wide		2	4		1,800	194		1,994	2,325
0350	Elongated dome, 4' wide		1.33	6.015		1,700	291		1,991	2,350
0360	6' wide		1.11	7.207		2,025	350		2,375	2,800
0370	8' wide		1	8		2,375	390		2,765	3,250
1000	Entry or walkway, peak, 12' long, 4' wide	2 Carp	.90	17.778		5,475	860		6,335	7,425
1010	6' wide		.60	26.667		8,450	1,300		9,750	11,400
1020	8' wide		.40	40		11,700	1,950		13,650	16,000
1100	Radius with dome end, 4' wide		1.10	14.545		4,150	705		4,855	5,725
1110	6' wide		.70	22.857		6,675	1,100		7,775	9,150
1120	8' wide		.50	32		9,500	1,550		11,050	12,900
2000	Retractable lateral arm awning, manual									
2010	To 12' wide, 8'-6" projection	2 Carp	1.70	9.412	Ea.	1,250	455		1,705	2,100
2020	To 14' wide, 8'-6" projection		1.10	14.545		1,450	705		2,155	2,750
2030	To 19' wide, 8'-6" projection		.85	18.824		1,950	910		2,860	3,650
2040	To 24' wide, 8'-6" projection		.67	23.881		2,475	1,150		3,625	4,625
2050	Motor for above, add	1 Carp	2.67	3		1,050	145		1,195	1,400
3000	Patio/deck canopy with frame									
3010	12' wide, 12' projection	2 Carp	2	8	Ea.	1,750	390		2,140	2,550
3020	16' wide, 14' projection	"	1.20	13.333		2,725	645		3,370	4,050
9000	For fire retardant canvas, add					7%				
9010	For lettering or graphics, add					35%				
9020	For painted or coated acrylic canvas, deduct					8%				
9030	For translucent or opaque vinyl canvas, add					10%				
9040	For 6 or more units, deduct					20%	15%			

10 73 16 – Canopies

10 73 16.10 Canopies, Residential

		Crew	Daily Output	Labor-Hours	Unit	Material	2016 Bare Costs Labor	Equipment	Total	Total Incl O&P
0010	**CANOPIES, RESIDENTIAL** Prefabricated									
0500	Carport, free standing, baked enamel, alum., .032", 40 psf									
0520	16' x 8', 4 posts	2 Carp	3	5.333	Ea.	4,725	258		4,983	5,625
0600	20' x 10', 6 posts		2	8		4,825	390		5,215	5,950
0605	30' x 10', 8 posts		2	8		7,250	390		7,640	8,600
1000	Door canopies, extruded alum., .032", 42" projection, 4' wide	1 Carp	8	1		203	48.50		251.50	300

10 73 Protective Covers

10 73 16 – Canopies

10 73 16.10 Canopies, Residential

		Crew	Daily Output	Labor-Hours	Unit	Material	2016 Bare Costs Labor	Equipment	Total	Total Incl O&P
1020	6' wide	1 Carp	6	1.333	Ea.	287	64.50		351.50	420
1040	8' wide	2 Carp	9	1.778		380	86		466	555
1060	10' wide		7	2.286		405	111		516	625
1080	12' wide		5	3.200		635	155		790	955
1200	54" projection, 4' wide	1 Carp	8	1		251	48.50		299.50	355
1220	6' wide	"	6	1.333		294	64.50		358.50	430
1240	8' wide	2 Carp	9	1.778		355	86		441	530
1260	10' wide		7	2.286		530	111		641	765
1280	12' wide		5	3.200		575	155		730	885
1300	Painted, add					20%				
1310	Bronze anodized, add					50%				
3000	Window awnings, aluminum, window 3' high, 4' wide	1 Carp	10	.800		130	39		169	207
3020	6' wide	"	8	1		185	48.50		233.50	283
3040	9' wide	2 Carp	9	1.778		310	86		396	485
3060	12' wide	"	5	3.200		310	155		465	600
3100	Window, 4' high, 4' wide	1 Carp	10	.800		164	39		203	244
3120	6' wide	"	8	1		236	48.50		284.50	340
3140	9' wide	2 Carp	9	1.778		385	86		471	565
3160	12' wide	"	5	3.200		385	155		540	680
3200	Window, 6' high, 4' wide	1 Carp	10	.800		295	39		334	390
3220	6' wide	"	8	1		340	48.50		388.50	455
3240	9' wide	2 Carp	9	1.778		935	86		1,021	1,175
3260	12' wide	"	5	3.200		1,275	155		1,430	1,650
3400	Roll-up aluminum, 2'-6" wide	1 Carp	14	.571		190	27.50		217.50	254
3420	3' wide		12	.667		195	32.50		227.50	267
3440	4' wide		10	.800		244	39		283	335
3460	6' wide		8	1		290	48.50		338.50	400
3480	9' wide	2 Carp	9	1.778		400	86		486	580
3500	12' wide	"	5	3.200		535	155		690	845
3600	Window awnings, canvas, 24" drop, 3' wide	1 Carp	30	.267	L.F.	50	12.90		62.90	76
3620	4' wide		40	.200		43	9.70		52.70	63.50
3700	30" drop, 3' wide		30	.267		51.50	12.90		64.40	78
3720	4' wide		40	.200		44.50	9.70		54.20	65
3740	5' wide		45	.178		40	8.60		48.60	58
3760	6' wide		48	.167		37	8.10		45.10	53.50
3780	8' wide		48	.167		34.50	8.10		42.60	51
3800	10' wide		50	.160		42	7.75		49.75	58.50
3900	Repair canvas, minimum		8	1	Ea.		48.50		48.50	79
3920	Maximum		4	2	"		97		97	158
9000	Minimum labor/equipment charge		3	2.667	Job		129		129	211

10 73 16.20 Metal Canopies

		Crew	Daily Output	Labor-Hours	Unit	Material	2016 Bare Costs Labor	Equipment	Total	Total Incl O&P
0010	**METAL CANOPIES**									
0020	Wall hung, .032", aluminum, prefinished, 8' x 10'	K-2	1.30	18.462	Ea.	2,350	925	236	3,511	4,450
0300	8' x 20'		1.10	21.818		3,850	1,100	279	5,229	6,400
1000	12' x 20'		1	24		5,750	1,200	305	7,255	8,750
2300	Aluminum entrance canopies, flat soffit, .032"									
2500	3'-6" x 4'-0", clear anodized	2 Carp	4	4	Ea.	935	194		1,129	1,350
9000	Minimum labor/equipment charge	"	2	8	Job		390		390	635

10 74 Manufactured Exterior Specialties

10 74 23 – Cupolas

10 74 23.10 Wood Cupolas

	10 74 23.10 Wood Cupolas	Crew	Daily Output	Labor-Hours	Unit	Material	2016 Bare Costs Labor	Equipment	Total	Total Incl O&P
0010	**WOOD CUPOLAS**									
0020	Stock units, pine, painted, 18" sq., 28" high, alum. roof	1 Carp	4.10	1.951	Ea.	184	94.50		278.50	355
0100	Copper roof		3.80	2.105		255	102		357	450
0300	23" square, 33" high, aluminum roof		3.70	2.162		385	105		490	590
0400	Copper roof		3.30	2.424		505	117		622	750
0600	30" square, 37" high, aluminum roof		3.70	2.162		560	105		665	785
0700	Copper roof		3.30	2.424		670	117		787	925
0900	Hexagonal, 31" wide, 46" high, copper roof		4	2		930	97		1,027	1,175
1000	36" wide, 50" high, copper roof		3.50	2.286		1,525	111		1,636	1,850
1200	For deluxe stock units, add to above					25%				
1400	For custom built units, add to above					50%	50%			
9000	Minimum labor/equipment charge	1 Carp	2.75	2.909	Job		141		141	230

10 74 29 – Steeples

10 74 29.10 Prefabricated Steeples

	10 74 29.10 Prefabricated Steeples	Crew	Daily Output	Labor-Hours	Unit	Material	2016 Bare Costs Labor	Equipment	Total	Total Incl O&P
0010	**PREFABRICATED STEEPLES**									
4000	Steeples, translucent fiberglass, 30" square, 15' high	F-3	2	20	Ea.	8,525	985	330	9,840	11,400
4150	25' high	"	1.80	22.222		9,675	1,100	365	11,140	12,800
4600	Aluminum, baked finish, 16" square, 14' high					5,825			5,825	6,425
4640	35' high, 8' base					36,700			36,700	40,400
4680	152' high, custom					600,500			600,500	660,500

10 74 33 – Weathervanes

10 74 33.10 Residential Weathervanes

	10 74 33.10 Residential Weathervanes	Crew	Daily Output	Labor-Hours	Unit	Material	2016 Bare Costs Labor	Equipment	Total	Total Incl O&P
0010	**RESIDENTIAL WEATHERVANES**									
0020	Residential types, 18" to 24"	1 Carp	8	1	Ea.	96	48.50		144.50	185
0100	24" to 48"		2	4	"	1,700	194		1,894	2,200
9000	Minimum labor/equipment charge		4	2	Job		97		97	158

10 74 46 – Window Wells

10 74 46.10 Area Window Wells

	10 74 46.10 Area Window Wells	Crew	Daily Output	Labor-Hours	Unit	Material	2016 Bare Costs Labor	Equipment	Total	Total Incl O&P
0010	**AREA WINDOW WELLS**, Galvanized steel									
0020	20 ga., 3'-2" wide, 1' deep	1 Sswk	29	.276	Ea.	17.25	14.70		31.95	45
0100	2' deep		23	.348		31.50	18.50		50	67
0300	16 ga., 3'-2" wide, 1' deep		29	.276		23	14.70		37.70	51.50
0400	3' deep		23	.348		48	18.50		66.50	85
0600	Welded grating for above, 15 lb., painted		45	.178		91.50	9.45		100.95	118
0700	Galvanized		45	.178		124	9.45		133.45	154
0900	Translucent plastic cap for above		60	.133		20.50	7.10		27.60	35

10 86 Security Mirrors and Domes

10 86 10 – Security Mirrors

10 86 10.10 Exterior Traffic Control Mirrors

	10 86 10.10 Exterior Traffic Control Mirrors	Crew	Daily Output	Labor-Hours	Unit	Material	2016 Bare Costs Labor	Equipment	Total	Total Incl O&P
0010	**EXTERIOR TRAFFIC CONTROL MIRRORS**									
0100	Convex, stainless steel, 20 ga., 26" diameter	1 Carp	12	.667	Ea.	221	32.50		253.50	296

10 86 20 – Security Domes

10 86 20.10 Domes

	10 86 20.10 Domes	Crew	Daily Output	Labor-Hours	Unit	Material	2016 Bare Costs Labor	Equipment	Total	Total Incl O&P
0010	**DOMES** for security cameras (CCTV)									
0100	Ceiling mounted, 10" diameter	1 Carp	30	.267	Ea.	12.55	12.90		25.45	35
0110	12" diameter	"	30	.267	"	9	12.90		21.90	31

For customer support on your Commercial Renovation Cost Data, call 877.742.4719.

353

Division Notes

	CREW	DAILY OUTPUT	LABOR-HOURS	UNIT	BARE COSTS				TOTAL INCL O&P
					MAT.	LABOR	EQUIP.	TOTAL	

Estimating Tips

General

- The items in this division are usually priced per square foot or each. Many of these items are purchased by the owner for installation by the contractor. Check the specifications for responsibilities and include time for receiving, storage, installation, and mechanical and electrical hookups in the appropriate divisions.

- Many items in Division 11 require some type of support system that is not usually furnished with the item. Examples of these systems include blocking for the attachment of casework and support angles for ceiling-hung projection screens. The required blocking or supports must be added to the estimate in the appropriate division.

- Some items in Division 11 may require assembly or electrical hookups. Verify the amount of assembly required or the need for a hard electrical connection and add the appropriate costs.

Reference Numbers

Reference numbers are shown at the beginning of some major classifications. These numbers refer to related items in the Reference Section. The reference information may be an estimating procedure, an alternate pricing method, or technical information.

Note: Not all subdivisions listed here necessarily appear. ■

11 11 Vehicle Service Equipment

11 11 13 – Compressed-Air Vehicle Service Equipment

11 11 13.10 Compressed Air Equipment	Crew	Daily Output	Labor-Hours	Unit	Material	2016 Bare Costs Labor	Equipment	Total	Total Incl O&P
0010 **COMPRESSED AIR EQUIPMENT**									
0030 Compressors, electric, 1-1/2 H.P., standard controls	L-4	1.50	16	Ea.	470	725		1,195	1,700
0550 Dual controls		1.50	16		765	725		1,490	2,025
0600 5 H.P., 115/230 volt, standard controls		1	24		1,850	1,075		2,925	3,800
0650 Dual controls		1	24		2,825	1,075		3,900	4,875

11 11 19 – Vehicle Lubrication Equipment

11 11 19.10 Lubrication Equipment

	Crew	Daily Output	Labor-Hours	Unit	Material	2016 Bare Costs Labor	Equipment	Total	Total Incl O&P
0010 **LUBRICATION EQUIPMENT**									
3000 Lube equipment, 3 reel type, with pumps, not including piping	L-4	.50	48	Set	8,800	2,175		10,975	13,200
3100 Hose reel, including hose, oil/lube, 1000 PSI	2 Sswk	2	8	Ea.	410	425		835	1,200
3200 Grease, 5000 PSI		2	8		500	425		925	1,300
3300 Air, 50 feet, 160 PSI		2	8		750	425		1,175	1,575
3350 25 feet, 160 PSI		2	8		450	425		875	1,250

11 11 33 – Vehicle Spray Painting Equipment

11 11 33.10 Spray Painting Equipment

	Crew	Daily Output	Labor-Hours	Unit	Material	2016 Bare Costs Labor	Equipment	Total	Total Incl O&P
0010 **SPRAY PAINTING EQUIPMENT**									
4000 Spray painting booth, 26' long, complete	L-4	.40	60	Ea.	10,400	2,725		13,125	15,900

11 12 Parking Control Equipment

11 12 13 – Parking Key and Card Control Units

11 12 13.10 Parking Control Units

	Crew	Daily Output	Labor-Hours	Unit	Material	2016 Bare Costs Labor	Equipment	Total	Total Incl O&P
0010 **PARKING CONTROL UNITS**									
5100 Card reader	1 Elec	2	4	Ea.	1,450	220		1,670	1,950
5120 Proximity with customer display	2 Elec	1	16		5,925	880		6,805	7,900
6000 Parking control software, basic functionality	1 Elec	.50	16		22,600	880		23,480	26,300
6020 Multi-function	"	.20	40		92,500	2,200		94,700	105,000

11 12 16 – Parking Ticket Dispensers

11 12 16.10 Ticket Dispensers

	Crew	Daily Output	Labor-Hours	Unit	Material	2016 Bare Costs Labor	Equipment	Total	Total Incl O&P
0010 **TICKET DISPENSERS**									
5900 Ticket spitter with time/date stamp, standard	2 Elec	2	8	Ea.	6,800	440		7,240	8,150
5920 Mag stripe encoding	"	2	8	"	18,800	440		19,240	21,400

11 12 26 – Parking Fee Collection Equipment

11 12 26.13 Parking Fee Coin Collection Equipment

	Crew	Daily Output	Labor-Hours	Unit	Material	2016 Bare Costs Labor	Equipment	Total	Total Incl O&P
0010 **PARKING FEE COIN COLLECTION EQUIPMENT**									
5200 Cashier booth, average	B-22	1	30	Ea.	10,100	1,325	211	11,636	13,500
5300 Collector station, pay on foot	2 Elec	.20	80		111,000	4,400		115,400	129,000
5320 Credit card only	"	.50	32		20,500	1,775		22,275	25,200

11 12 26.23 Fee Equipment

	Crew	Daily Output	Labor-Hours	Unit	Material	2016 Bare Costs Labor	Equipment	Total	Total Incl O&P
0010 **FEE EQUIPMENT**									
5600 Fee computer	1 Elec	1.50	5.333	Ea.	12,000	294		12,294	13,700

11 12 33 – Parking Gates

11 12 33.13 Lift Arm Parking Gates

	Crew	Daily Output	Labor-Hours	Unit	Material	2016 Bare Costs Labor	Equipment	Total	Total Incl O&P
0010 **LIFT ARM PARKING GATES**									
5000 Barrier gate with programmable controller	2 Elec	3	5.333	Ea.	3,425	294		3,719	4,225
5020 Industrial		3	5.333		5,200	294		5,494	6,150
5050 Non-programmable, with reader and 12' arm		3	5.333		1,900	294		2,194	2,525
5500 Exit verifier		1	16		19,200	880		20,080	22,500
5700 Full sign, 4" letters	1 Elec	2	4		1,225	220		1,445	1,700

11 12 Parking Control Equipment

11 12 33 – Parking Gates

11 12 33.13 Lift Arm Parking Gates	Crew	Daily Output	Labor-Hours	Unit	Material	2016 Bare Costs Labor	2016 Bare Costs Equipment	Total	Total Incl O&P
5800	Inductive loop	2 Elec	4	4	Ea.	207	220		427	570
5950	Vehicle detector, microprocessor based	1 Elec	3	2.667		435	147		582	705
7100	Traffic spike unit, flush mount, spring loaded, 72" L	B-89	4	4		1,275	182	123	1,580	1,825
7200	Surface mount, 72" L	2 Skwk	10	1.600	▼	2,350	80		2,430	2,700

11 13 Loading Dock Equipment

11 13 13 – Loading Dock Bumpers

11 13 13.10 Dock Bumpers

		Crew	Daily Output	Labor-Hours	Unit	Material	2016 Bare Costs Labor	2016 Bare Costs Equipment	Total	Total Incl O&P
0010	**DOCK BUMPERS** Bolts not included									
0012	2" x 6" to 4" x 8", average	1 Carp	300	.027	B.F.	.70	1.29		1.99	2.88
0050	Bumpers, lam. rubber blocks 4-1/2" thick, 10" high, 14" long		26	.308	Ea.	60	14.90		74.90	90.50
0200	24" long		22	.364		95	17.60		112.60	134
0300	36" long		17	.471		131	23		154	182
0500	12" high, 14" long	▼	25	.320	▼	74.50	15.50		90	107

11 13 16 – Loading Dock Seals and Shelters

11 13 16.10 Dock Seals and Shelters

		Crew	Daily Output	Labor-Hours	Unit	Material	2016 Bare Costs Labor	2016 Bare Costs Equipment	Total	Total Incl O&P
0010	**DOCK SEALS AND SHELTERS**									
6200	Shelters, fabric, for truck or train, scissor arms, minimum	1 Carp	1	8	Ea.	1,875	390		2,265	2,675
6300	Maximum	"	.50	16	"	2,500	775		3,275	4,025

11 13 19 – Stationary Loading Dock Equipment

11 13 19.10 Dock Equipment

		Crew	Daily Output	Labor-Hours	Unit	Material	2016 Bare Costs Labor	2016 Bare Costs Equipment	Total	Total Incl O&P
0010	**DOCK EQUIPMENT**									
2200	Dock boards, heavy duty, 60" x 60", aluminum, 5,000 lb. capacity				Ea.	1,350			1,350	1,500
4200	Platform lifter, 6' x 6', portable, 3,000 lb. capacity				"	9,250			9,250	10,200
6000	Dock leveler, 15 ton capacity									
6100	I beam construction, mechanical, 6' x 8'	E-16	.50	32	Ea.	4,700	1,725	294	6,719	8,525
6150	Hydraulic		.50	32		5,525	1,725	294	7,544	9,450
6200	Formed beam deck construction, mechanical, 6' x 8'		.50	32		3,475	1,725	294	5,494	7,175
6250	Hydraulic		.50	32		4,750	1,725	294	6,769	8,600
6300	Edge of dock leveler, mechanical, 15 ton capacity	▼	2	8	▼	1,025	435	73.50	1,533.50	2,000
7000	22.5 ton capacity									
7100	Vertical storing dock leveler, hydraulic, 6' x 6'	E-16	.40	40	Ea.	7,425	2,175	370	9,970	12,400

11 21 Retail and Service Equipment

11 21 13 – Cash Registers and Checking Equipment

11 21 13.10 Checkout Counter

		Crew	Daily Output	Labor-Hours	Unit	Material	2016 Bare Costs Labor	2016 Bare Costs Equipment	Total	Total Incl O&P
0010	**CHECKOUT COUNTER**									
0020	Supermarket conveyor, single belt	2 Clab	10	1.600	Ea.	3,175	60.50		3,235.50	3,600
0100	Double belt, power take-away		9	1.778		4,575	67.50		4,642.50	5,125
0400	Double belt, power take-away, incl. side scanning		7	2.286		5,375	86.50		5,461.50	6,050
0800	Warehouse or bulk type	▼	6	2.667	▼	6,350	101		6,451	7,150
1000	Scanning system, 2 lanes, w/registers, scan gun & memory				System	16,600			16,600	18,300
1100	10 lanes, single processor, full scan, with scales				"	158,000			158,000	173,500
2000	Register, restaurant, minimum				Ea.	740			740	815
2100	Maximum					3,125			3,125	3,425
2150	Store, minimum					740			740	815
2200	Maximum				▼	3,125			3,125	3,425

11 21 Retail and Service Equipment

11 21 33 – Checkroom Equipment

11 21 33.10 Clothes Check Equipment	Crew	Daily Output	Labor-Hours	Unit	Material	2016 Bare Costs Labor	Equipment	Total	Total Incl O&P
0010 **CLOTHES CHECK EQUIPMENT**									
0030 Clothes check rack, free standing, st. stl., 2-tier, 90 bag capacity				Ea.	1,575			1,575	1,725
0050 Wall mounted, 45 bag capacity	L-2	8	2		830	84.50		914.50	1,050
0100 Garment checking bag, green mesh fabric, 21" H x 17" W with 4.5" hook					25.50			25.50	28

11 21 53 – Barber and Beauty Shop Equipment

11 21 53.10 Barber Equipment

11 21 53.10 Barber Equipment	Crew	Daily Output	Labor-Hours	Unit	Material	2016 Bare Costs Labor	Equipment	Total	Total Incl O&P
0010 **BARBER EQUIPMENT**									
0020 Chair, hydraulic, movable, minimum	1 Carp	24	.333	Ea.	625	16.15		641.15	715
0050 Maximum	"	16	.500		3,975	24		3,999	4,425
0200 Wall hung styling station with mirrors, minimum	L-2	8	2		510	84.50		594.50	705
0300 Maximum	"	4	4		2,650	169		2,819	3,200
0500 Sink, hair washing basin, rough plumbing not incl.	1 Plum	8	1		520	59		579	670
1000 Sterilizer, liquid solution for tools					169			169	186
1100 Total equipment, rule of thumb, per chair, minimum	L-8	1	20		2,100	1,000		3,100	3,925
1150 Maximum	"	1	20		5,900	1,000		6,900	8,100

11 21 73 – Commercial Laundry and Dry Cleaning Equipment

11 21 73.13 Dry Cleaning Equipment

11 21 73.13 Dry Cleaning Equipment	Crew	Daily Output	Labor-Hours	Unit	Material	2016 Bare Costs Labor	Equipment	Total	Total Incl O&P
0010 **DRY CLEANING EQUIPMENT**									
2000 Dry cleaners, electric, 20 lb. capacity, not incl. rough-in	L-1	.20	80	Ea.	34,100	4,575		38,675	44,600
2050 25 lb. capacity		.17	94.118		49,400	5,375		54,775	63,000
2100 30 lb. capacity		.15	106		52,000	6,100		58,100	67,000
2150 60 lb. capacity		.09	177		79,000	10,200		89,200	103,000

11 21 73.16 Drying and Conditioning Equipment

11 21 73.16 Drying and Conditioning Equipment	Crew	Daily Output	Labor-Hours	Unit	Material	2016 Bare Costs Labor	Equipment	Total	Total Incl O&P
0010 **DRYING AND CONDITIONING EQUIPMENT**									
0100 Dryers, not including rough-in									
1500 Industrial, 30 lb. capacity	1 Plum	2	4	Ea.	3,225	237		3,462	3,925
1600 50 lb. capacity	"	1.70	4.706		3,650	279		3,929	4,450
4700 Lint collector, ductwork not included, 8,000 to 10,000 CFM	Q-10	.30	80		9,350	4,275		13,625	17,100

11 21 73.19 Finishing Equipment

11 21 73.19 Finishing Equipment	Crew	Daily Output	Labor-Hours	Unit	Material	2016 Bare Costs Labor	Equipment	Total	Total Incl O&P
0010 **FINISHING EQUIPMENT**									
3500 Folders, blankets & sheets, minimum	1 Elec	.17	47.059	Ea.	33,700	2,600		36,300	41,000
3700 King size with automatic stacker		.10	80		61,000	4,400		65,400	74,000
3800 For conveyor delivery, add		.45	17.778		14,900	980		15,880	17,900

11 21 73.23 Commercial Ironing Equipment

11 21 73.23 Commercial Ironing Equipment	Crew	Daily Output	Labor-Hours	Unit	Material	2016 Bare Costs Labor	Equipment	Total	Total Incl O&P
0010 **COMMERCIAL IRONING EQUIPMENT**									
4500 Ironers, institutional, 110", single roll	1 Elec	.20	40	Ea.	32,600	2,200		34,800	39,300

11 21 73.26 Commercial Washers and Extractors

11 21 73.26 Commercial Washers and Extractors	Crew	Daily Output	Labor-Hours	Unit	Material	2016 Bare Costs Labor	Equipment	Total	Total Incl O&P
0010 **COMMERCIAL WASHERS AND EXTRACTORS**, not including rough-in									
6000 Combination washer/extractor, 20 lb. capacity	L-6	1.50	8	Ea.	6,025	465		6,490	7,350
6100 30 lb. capacity		.80	15		9,375	865		10,240	11,700
6200 50 lb. capacity		.68	17.647		11,000	1,025		12,025	13,700
6300 75 lb. capacity		.30	40		20,600	2,325		22,925	26,300
6350 125 lb. capacity		.16	75		27,700	4,325		32,025	37,300

11 21 73.33 Coin-Operated Laundry Equipment

11 21 73.33 Coin-Operated Laundry Equipment	Crew	Daily Output	Labor-Hours	Unit	Material	2016 Bare Costs Labor	Equipment	Total	Total Incl O&P
0010 **COIN-OPERATED LAUNDRY EQUIPMENT**									
0990 Dryer, gas fired									
1000 Commercial, 30 lb. capacity, coin operated, single	1 Plum	3	2.667	Ea.	3,350	158		3,508	3,950
1100 Double stacked	"	2	4	"	7,500	237		7,737	8,625
5290 Clothes washer									
5300 Commercial, coin operated, average	1 Plum	3	2.667	Ea.	1,250	158		1,408	1,625

11 21 Retail and Service Equipment

11 21 83 – Photo Processing Equipment

11 21 83.13 Darkroom Equipment

		Crew	Daily Output	Labor-Hours	Unit	Material	2016 Bare Costs Labor	2016 Bare Costs Equipment	Total	Total Incl O&P
0010	**DARKROOM EQUIPMENT**									
0020	Developing sink, 5" deep, 24" x 48"	Q-1	2	8	Ea.	4,150	425		4,575	5,250
0050	48" x 52"		1.70	9.412		4,425	500		4,925	5,650
0200	10" deep, 24" x 48"		1.70	9.412		1,575	500		2,075	2,500
0250	24" x 108"		1.50	10.667		3,725	570		4,295	4,975
0500	Dryers, dehumidified filtered air, 36" x 25" x 68" high	L-7	6	4.667		4,450	216		4,666	5,250
0550	48" x 25" x 68" high		5	5.600		10,400	260		10,660	11,800
2000	Processors, automatic, color print, minimum		4	7		16,000	325		16,325	18,100
2050	Maximum		.60	46.667		25,100	2,175		27,275	31,100
2300	Black and white print, minimum		2	14		12,100	650		12,750	14,400
2350	Maximum		.80	35		62,000	1,625		63,625	70,500
2600	Manual processor, 16" x 20" maximum print size		2	14		9,575	650		10,225	11,600
2650	20" x 24" maximum print size		1	28		9,100	1,300		10,400	12,100
3000	Viewing lights, 20" x 24"		6	4.667		280	216		496	660
3100	20" x 24" with color correction		6	4.667		400	216		616	790
3500	Washers, round, minimum sheet 11" x 14"	Q-1	2	8		3,250	425		3,675	4,250
3550	Maximum sheet 20" x 24"		1	16		3,575	850		4,425	5,250
3800	Square, minimum sheet 20" x 24"		1	16		3,575	850		4,425	5,250
3900	Maximum sheet 50" x 56"		.80	20		4,900	1,075		5,975	7,050
4500	Combination tank sink, tray sink, washers, with									
4510	Dry side tables, average	Q-1	.45	35.556	Ea.	10,800	1,900		12,700	14,800

11 22 Banking Equipment

11 22 13 – Vault Equipment

11 22 13.16 Safes

		Crew	Daily Output	Labor-Hours	Unit	Material	2016 Bare Costs Labor	2016 Bare Costs Equipment	Total	Total Incl O&P
0010	**SAFES**									
0800	Money, "B" label, 9" x 14" x 14"				Ea.	540			540	595
0900	Tool resistive, 24" x 24" x 20"					4,175			4,175	4,600
1050	Tool and torch resistive, 24" x 24" x 20"					7,875			7,875	8,650

11 22 16 – Teller and Service Equipment

11 22 16.13 Teller Equipment Systems

		Crew	Daily Output	Labor-Hours	Unit	Material	2016 Bare Costs Labor	2016 Bare Costs Equipment	Total	Total Incl O&P
0010	**TELLER EQUIPMENT SYSTEMS**									
0020	Alarm system, police	2 Elec	1.60	10	Ea.	4,975	550		5,525	6,325
0400	Bullet resistant teller window, 44" x 60"	1 Glaz	.60	13.333		4,225	625		4,850	5,650
0500	48" x 60"	"	.60	13.333		5,900	625		6,525	7,475
3000	Counters for banks, frontal only	2 Carp	1	16	Station	1,950	775		2,725	3,425
3100	Complete with steel undercounter		.50	32	"	3,625	1,550		5,175	6,500
5400	Partitions, bullet-resistant, 1-3/16" glass, 8' high		10	1.600	L.F.	201	77.50		278.50	350
5600	Pass thru, bullet-res. window, painted steel, 24" x 36"	2 Sswk	1.60	10	Ea.	2,800	530		3,330	4,025

For customer support on your Commercial Renovation Cost Data, call 877.742.4719.

359

11 30 13 – Residential Appliances

11 30 13.15 Cooking Equipment		Crew	Daily Output	Labor-Hours	Unit	Material	2016 Bare Costs Labor	Equipment	Total	Total Incl O&P
0010	**COOKING EQUIPMENT**									
0020	Cooking range, 30" free standing, 1 oven, minimum	2 Clab	10	1.600	Ea.	470	60.50		530.50	615
0050	Maximum	"	4	4		1,900	152		2,052	2,325
0350	Built-in, 30" wide, 1 oven, minimum	1 Elec	6	1.333		900	73.50		973.50	1,100
0400	Maximum	2 Carp	2	8		1,575	390		1,965	2,350
0900	Countertop cooktops, 4 burner, standard, minimum	1 Elec	6	1.333		350	73.50		423.50	500
0950	Maximum		3	2.667		1,700	147		1,847	2,100
1250	Microwave oven, minimum		4	2		119	110		229	300
1300	Maximum		2	4		510	220		730	900
5380	Oven, built in, standard		4	2		900	110		1,010	1,150
5390	Deluxe		2	4		2,525	220		2,745	3,125

11 30 13.16 Refrigeration Equipment

		Crew	Daily Output	Labor-Hours	Unit	Material	Labor	Equipment	Total	Total Incl O&P
0010	**REFRIGERATION EQUIPMENT**									
5450	Refrigerator, no frost, 6 C.F.	2 Clab	15	1.067	Ea.	229	40.50		269.50	320
5500	Refrigerator, no frost, 10 C.F. to 12 C.F., minimum		10	1.600		475	60.50		535.50	620
5600	Maximum		6	2.667		555	101		656	780
6790	Energy-star qualified, 18 C.F., minimum **G**	2 Carp	4	4		650	194		844	1,025
6795	Maximum **G**		2	8		1,250	390		1,640	2,000
6797	21.7 C.F., minimum **G**		4	4		1,125	194		1,319	1,550
6799	Maximum **G**		4	4		2,075	194		2,269	2,600

11 30 13.17 Kitchen Cleaning Equipment

		Crew	Daily Output	Labor-Hours	Unit	Material	Labor	Equipment	Total	Total Incl O&P
0010	**KITCHEN CLEANING EQUIPMENT**									
2750	Dishwasher, built-in, 2 cycles, minimum	L-1	4	4	Ea.	299	229		528	685
2800	Maximum		2	8		450	455		905	1,200
3100	Energy-star qualified, minimum **G**		4	4		490	229		719	895
3110	Maximum **G**		2	8		1,650	455		2,105	2,525

11 30 13.18 Waste Disposal Equipment

		Crew	Daily Output	Labor-Hours	Unit	Material	Labor	Equipment	Total	Total Incl O&P
0010	**WASTE DISPOSAL EQUIPMENT**									
1750	Compactor, residential size, 4 to 1 compaction, minimum	1 Carp	5	1.600	Ea.	690	77.50		767.50	885
1800	Maximum	"	3	2.667		1,150	129		1,279	1,475
3300	Garbage disposal, sink type, minimum	L-1	10	1.600		93	91.50		184.50	244
3350	Maximum	"	10	1.600		243	91.50		334.50	410

11 30 13.19 Kitchen Ventilation Equipment

		Crew	Daily Output	Labor-Hours	Unit	Material	Labor	Equipment	Total	Total Incl O&P
0010	**KITCHEN VENTILATION EQUIPMENT**									
4150	Hood for range, 2 speed, vented, 30" wide, minimum	L-3	5	3.200	Ea.	83	167		250	360
4200	Maximum		3	5.333		855	279		1,134	1,375
4300	42" wide, minimum		5	3.200		166	167		333	450
4330	Custom		5	3.200		1,625	167		1,792	2,075
4350	Maximum		3	5.333		1,975	279		2,254	2,625
4400	Ventless hood, 2 speed, 30" wide, minimum	1 Elec	8	1		99.50	55		154.50	194
4450	Maximum		7	1.143		585	63		648	745
4510	36" wide, minimum		7	1.143		75	63		138	180
4550	Maximum		6	1.333		690	73.50		763.50	875
4580	42" wide, minimum		6	1.333		187	73.50		260.50	320
4600	Maximum		5	1.600		680	88		768	885
4650	For vented 1 speed, deduct from maximum					55			55	60.50

11 30 13.24 Washers

		Crew	Daily Output	Labor-Hours	Unit	Material	Labor	Equipment	Total	Total Incl O&P
0010	**WASHERS**									
5000	Residential, 4 cycle, average	1 Plum	3	2.667	Ea.	890	158		1,048	1,225
6750	Energy star qualified, front loading, minimum **G**		3	2.667		700	158		858	1,025
6760	Maximum **G**		1	8		1,600	475		2,075	2,500

11 30 Residential Equipment

11 30 13 – Residential Appliances

11 30 13.24 Washers

		Crew	Daily Output	Labor-Hours	Unit	Material	2016 Bare Costs Labor	Equipment	Total	Total Incl O&P
6764	Top loading, minimum	G 1 Plum	3	2.667	Ea.	570	158		728	870
6766	Maximum	G ↓	3	2.667	↓	1,400	158		1,558	1,800

11 30 13.25 Dryers

		Crew	Daily Output	Labor-Hours	Unit	Material	2016 Bare Costs Labor	Equipment	Total	Total Incl O&P
0010	**DRYERS**									
0500	Gas fired residential, 16 lb. capacity, average	1 Plum	3	2.667	Ea.	685	158		843	1,000
6770	Electric, front loading, energy-star qualified, minimum	G L-2	3	5.333		475	225		700	895
6780	Maximum	G "	2	8	↓	1,600	340		1,940	2,300

11 30 15 – Miscellaneous Residential Appliances

11 30 15.13 Sump Pumps

		Crew	Daily Output	Labor-Hours	Unit	Material	2016 Bare Costs Labor	Equipment	Total	Total Incl O&P
0010	**SUMP PUMPS**									
6400	Cellar drainer, pedestal, 1/3 H.P., molded PVC base	1 Plum	3	2.667	Ea.	140	158		298	400
6450	Solid brass	"	2	4	"	297	237		534	695
6460	Sump pump, see also Section 22 14 29.16									

11 30 33 – Retractable Stairs

11 30 33.10 Disappearing Stairway

		Crew	Daily Output	Labor-Hours	Unit	Material	2016 Bare Costs Labor	Equipment	Total	Total Incl O&P
0010	**DISAPPEARING STAIRWAY** No trim included									
0020	One piece, yellow pine, 8'-0" ceiling	2 Carp	4	4	Ea.	250	194		444	590
0030	9'-0" ceiling		4	4		284	194		478	625
0040	10'-0" ceiling		3	5.333		254	258		512	700
0050	11'-0" ceiling	↓	3	5.333		285	258		543	735
0060	12'-0" ceiling		3	5.333		360	258		618	815
0100	Custom grade, pine, 8'-6" ceiling, minimum	1 Carp	4	2		195	97		292	370
0150	Average		3.50	2.286		244	111		355	450
0200	Maximum		3	2.667		305	129		434	545
0500	Heavy duty, pivoted, from 7'-7" to 12'-10" floor to floor		3	2.667		1,025	129		1,154	1,325
0600	16'-0" ceiling		2	4		1,600	194		1,794	2,075
0800	Economy folding, pine, 8'-6" ceiling		4	2		176	97		273	350
0900	9'-6" ceiling	↓	4	2		196	97		293	375
1100	Automatic electric, aluminum, floor to floor height, 8' to 9'	2 Carp	1	16		8,700	775		9,475	10,900
1400	11' to 12'		.90	17.778		9,250	860		10,110	11,600
1700	14' to 15'	↓	.70	22.857	↓	9,975	1,100		11,075	12,800
9000	Minimum labor/equipment charge	1 Carp	2	4	Job		194		194	315

11 32 Unit Kitchens

11 32 13 – Metal Unit Kitchens

11 32 13.10 Commercial Unit Kitchens

		Crew	Daily Output	Labor-Hours	Unit	Material	2016 Bare Costs Labor	Equipment	Total	Total Incl O&P
0010	**COMMERCIAL UNIT KITCHENS**									
1500	Combination range, refrigerator and sink, 30" wide, minimum	L-1	2	8	Ea.	1,100	455		1,555	1,925
1550	Maximum	"	1	16	"	1,250	915		2,165	2,800
1640	Combination range, refrigerator, sink, microwave									
1660	Oven and ice maker	L-1	.80	20	Ea.	4,550	1,150		5,700	6,775

For customer support on your Commercial Renovation Cost Data, call 877.742.4719.

361

11 41 Foodservice Storage Equipment

11 41 13 – Refrigerated Food Storage Cases

11 41 13.10 Refrigerated Food Cases	Crew	Daily Output	Labor-Hours	Unit	Material	2016 Bare Costs Labor	Equipment	Total	Total Incl O&P
0010 **REFRIGERATED FOOD CASES**									
0030 Dairy, multi-deck, 12' long	Q-5	3	5.333	Ea.	11,300	291		11,591	12,900
0100 For rear sliding doors, add					1,850			1,850	2,025
0200 Delicatessen case, service deli, 12' long, single deck	Q-5	3.90	4.103		8,400	224		8,624	9,600
0300 Multi-deck, 18 S.F. shelf display		3	5.333		7,650	291		7,941	8,875
0400 Freezer, self-contained, chest-type, 30 C.F.		3.90	4.103		9,125	224		9,349	10,400
0500 Glass door, upright, 78 C.F.		3.30	4.848		12,000	265		12,265	13,600
0600 Frozen food, chest type, 12' long		3.30	4.848		8,225	265		8,490	9,475
0700 Glass door, reach-in, 5 door		3	5.333		14,100	291		14,391	16,000
0800 Island case, 12' long, single deck		3.30	4.848		7,275	265		7,540	8,450
0900 Multi-deck		3	5.333		8,500	291		8,791	9,800
1000 Meat case, 12' long, single deck		3.30	4.848		7,275	265		7,540	8,450
1050 Multi-deck		3.10	5.161		10,400	282		10,682	11,800
1100 Produce, 12' long, single deck		3.30	4.848		6,875	265		7,140	7,975
1200 Multi-deck		3.10	5.161		8,725	282		9,007	10,000

11 41 13.20 Refrigerated Food Storage Equipment

11 41 13.20	Crew	Daily Output	Labor-Hours	Unit	Material	Labor	Equipment	Total	Total Incl O&P
0010 **REFRIGERATED FOOD STORAGE EQUIPMENT**									
2350 Cooler, reach-in, beverage, 6' long	Q-1	6	2.667	Ea.	3,375	142		3,517	3,925
4300 Freezers, reach-in, 44 C.F.		4	4		5,325	213		5,538	6,175
4500 68 C.F.		3	5.333		4,550	284		4,834	5,475
4600 Freezer, pre-fab, 8' x 8' w/refrigeration	2 Carp	.45	35.556		10,600	1,725		12,325	14,400
4620 8' x 12'		.35	45.714		12,100	2,225		14,325	16,900
4640 8' x 16'		.25	64		15,000	3,100		18,100	21,600
4660 8' x 20'		.17	94.118		21,700	4,550		26,250	31,400
4680 Reach-in, 1 compartment	Q-1	4	4		2,725	213		2,938	3,325
4685 Energy star rated **G**	R-18	7.80	3.333		2,800	161		2,961	3,325
4700 2 compartment	Q-1	3	5.333		4,275	284		4,559	5,150
4705 Energy star rated **G**	R-18	6.20	4.194		3,075	203		3,278	3,725
4710 3 compartment	Q-1	3	5.333		5,725	284		6,009	6,750
4715 Energy star rated **G**	R-18	5.60	4.643		4,475	225		4,700	5,275
8320 Refrigerator, reach-in, 1 compartment		7.80	3.333		2,550	161		2,711	3,075
8325 Energy star rated **G**		7.80	3.333		2,800	161		2,961	3,325
8330 2 compartment		6.20	4.194		3,900	203		4,103	4,625
8335 Energy star rated **G**		6.20	4.194		3,075	203		3,278	3,725
8340 3 compartment		5.60	4.643		4,950	225		5,175	5,775
8345 Energy star rated **G**		5.60	4.643		4,475	225		4,700	5,275
8350 Pre-fab, with refrigeration, 8' x 8'	2 Carp	.45	35.556		6,875	1,725		8,600	10,400
8360 8' x 12'		.35	45.714		8,525	2,225		10,750	13,000
8370 8' x 16'		.25	64		12,500	3,100		15,600	18,800
8380 8' x 20'		.17	94.118		15,800	4,550		20,350	24,900
8390 Pass-thru/roll-in, 1 compartment	R-18	7.80	3.333		4,525	161		4,686	5,225
8400 2 compartment		6.24	4.167		6,350	202		6,552	7,300
8410 3 compartment		5.60	4.643		8,675	225		8,900	9,875
8420 Walk-in, alum, door & floor only, no refrig, 6' x 6' x 7'-6"	2 Carp	1.40	11.429		8,725	555		9,280	10,500
8430 10' x 6' x 7'-6"		.55	29.091		12,600	1,400		14,000	16,200
8440 12' x 14' x 7'-6"		.25	64		17,400	3,100		20,500	24,200
8450 12' x 20' x 7'-6"		.17	94.118		18,700	4,550		23,250	28,100
8460 Refrigerated cabinets, mobile					3,925			3,925	4,325
8470 Refrigerator/freezer, reach-in, 1 compartment	R-18	5.60	4.643		5,750	225		5,975	6,675
8480 2 compartment	"	4.80	5.417		7,200	262		7,462	8,325

11 41 Foodservice Storage Equipment

11 41 13 – Refrigerated Food Storage Cases

11 41 13.30 Wine Cellar

		Crew	Daily Output	Labor-Hours	Unit	Material	Labor	2016 Bare Costs Equipment	Total	Total Incl O&P
0010	WINE CELLAR, refrigerated, Redwood interior, carpeted, walk-in type									
0020	6'-8" high, including racks									
0200	80" W x 48" D for 900 bottles	2 Carp	1.50	10.667	Ea.	4,250	515		4,765	5,525
0250	80" W x 72" D for 1300 bottles		1.33	12.030		5,175	585		5,760	6,650
0300	80" W x 94" D for 1900 bottles		1.17	13.675		6,275	665		6,940	7,975
0400	80" W x 124" D for 2500 bottles	↓	1	16	↓	7,375	775		8,150	9,375

11 41 33 – Foodservice Shelving

11 41 33.20 Metal Food Storage Shelving

		Crew	Daily Output	Labor-Hours	Unit	Material	Labor	2016 Bare Costs Equipment	Total	Total Incl O&P
0010	METAL FOOD STORAGE SHELVING									
8600	Stainless steel shelving, louvered 4-tier, 20" x 3'	1 Clab	6	1.333	Ea.	1,400	50.50		1,450.50	1,625
8605	20" x 4'		6	1.333		1,575	50.50		1,625.50	1,825
8610	20" x 6'		6	1.333		1,675	50.50		1,725.50	1,900
8615	24" x 3'		6	1.333		1,975	50.50		2,025.50	2,250
8620	24" x 4'		6	1.333		2,350	50.50		2,400.50	2,675
8625	24" x 6'		6	1.333		3,275	50.50		3,325.50	3,675
8630	Flat 4-tier, 20" x 3'		6	1.333		1,150	50.50		1,200.50	1,350
8635	20" x 4'		6	1.333		1,400	50.50		1,450.50	1,600
8640	20" x 5'		6	1.333		1,600	50.50		1,650.50	1,825
8645	24" x 3'		6	1.333		1,250	50.50		1,300.50	1,450
8650	24" x 4'		6	1.333		2,225	50.50		2,275.50	2,525
8655	24" x 6'		6	1.333		2,625	50.50		2,675.50	2,975
8700	Galvanized shelving, louvered 4-tier, 20" x 3'		6	1.333		760	50.50		810.50	920
8705	20" x 4'		6	1.333		860	50.50		910.50	1,025
8710	20" x 6'		6	1.333		1,000	50.50		1,050.50	1,175
8715	24" x 3'		6	1.333		695	50.50		745.50	850
8720	24" x 4'		6	1.333		920	50.50		970.50	1,075
8725	24" x 6'		6	1.333		1,325	50.50		1,375.50	1,525
8730	Flat 4-tier, 20" x 3'		6	1.333		700	50.50		750.50	855
8735	20" x 4'		6	1.333		620	50.50		670.50	765
8740	20" x 6'		6	1.333		910	50.50		960.50	1,075
8745	24" x 3'		6	1.333		690	50.50		740.50	840
8750	24" x 4'		6	1.333		715	50.50		765.50	875
8755	24" x 6'		6	1.333		820	50.50		870.50	990
8760	Stainless steel dunnage rack, 24" x 3'		8	1		395	38		433	495
8765	24" x 4'		8	1		430	38		468	535
8770	Galvanized dunnage rack, 24" x 3'		8	1		169	38		207	248
8775	24" x 4'	↓	8	1	↓	187	38		225	267

11 42 Food Preparation Equipment

11 42 10 – Commercial Food Preparation Equipment

11 42 10.10 Choppers, Mixers and Misc. Equipment

		Crew	Daily Output	Labor-Hours	Unit	Material	Labor	2016 Bare Costs Equipment	Total	Total Incl O&P
0010	CHOPPERS, MIXERS AND MISC. EQUIPMENT									
1700	Choppers, 5 pounds	R-18	7	3.714	Ea.	2,200	180		2,380	2,700
1720	16 pounds		5	5.200		2,325	252		2,577	2,950
1740	35 to 40 pounds	↓	4	6.500		3,550	315		3,865	4,375
1840	Coffee brewer, 5 burners	1 Plum	3	2.667		1,325	158		1,483	1,725
1850	Coffee urn, twin 6 gallon urns		2	4		2,450	237		2,687	3,050
1860	Single, 3 gallon	↓	3	2.667		1,800	158		1,958	2,225
3000	Fast food equipment, total package, minimum	6 Skwk	.08	600		203,500	29,900		233,400	272,500
3100	Maximum	"	.07	685		277,500	34,200		311,700	360,500

For customer support on your Commercial Renovation Cost Data, call 877.742.4719.

363

11 42 Food Preparation Equipment

11 42 10 – Commercial Food Preparation Equipment

11 42 10.10 Choppers, Mixers and Misc. Equipment		Crew	Daily Output	Labor-Hours	Unit	Material	2016 Bare Costs Labor	2016 Bare Costs Equipment	Total	Total Incl O&P
3800	Food mixers, bench type, 20 quarts	L-7	7	4	Ea.	2,875	186		3,061	3,475
3850	40 quarts		5.40	5.185		6,450	241		6,691	7,500
3900	60 quarts		5	5.600		11,700	260		11,960	13,300
4040	80 quarts		3.90	7.179		12,200	335		12,535	14,000
4100	Floor type, 20 quarts		15	1.867		3,225	86.50		3,311.50	3,700
4120	60 quarts		14	2		10,200	93		10,293	11,400
4140	80 quarts		12	2.333		16,200	108		16,308	18,000
4160	140 quarts	▼	8.60	3.256		27,300	151		27,451	30,200
6700	Peelers, small	R-18	8	3.250		2,000	157		2,157	2,450
6720	Large	"	6	4.333		4,850	210		5,060	5,650
6800	Pulper/extractor, close coupled, 5 HP	1 Plum	1.90	4.211		3,575	249		3,824	4,350
8580	Slicer with table	R-18	9	2.889	▼	4,675	140		4,815	5,375

11 43 Food Delivery Carts and Conveyors

11 43 13 – Food Delivery Carts

11 43 13.10 Mobile Carts, Racks and Trays

		Crew	Daily Output	Labor-Hours	Unit	Material	Labor	Equipment	Total	Total Incl O&P
0010	**MOBILE CARTS, RACKS AND TRAYS**									
1650	Cabinet, heated, 1 compartment, reach-in	R-18	5.60	4.643	Ea.	3,725	225		3,950	4,450
1655	Pass-thru roll-in		5.60	4.643		3,850	225		4,075	4,575
1660	2 compartment, reach-in	▼	4.80	5.417		9,050	262		9,312	10,400
1670	Mobile					3,600			3,600	3,975
2000	Hospital food cart, hot and cold service, 20 tray capacity					15,300			15,300	16,900
6850	Mobile rack w/pan slide					1,400			1,400	1,550
9180	Tray and silver dispenser, mobile	1 Clab	16	.500	▼	925	18.95		943.95	1,050

11 44 Food Cooking Equipment

11 44 13 – Commercial Ranges

11 44 13.10 Cooking Equipment

		Crew	Daily Output	Labor-Hours	Unit	Material	Labor	Equipment	Total	Total Incl O&P
0010	**COOKING EQUIPMENT**									
0020	Bake oven, gas, one section	Q-1	8	2	Ea.	5,525	107		5,632	6,250
0300	Two sections		7	2.286		9,150	122		9,272	10,300
0600	Three sections	▼	6	2.667		10,700	142		10,842	12,000
0900	Electric convection, single deck	L-7	4	7		6,225	325		6,550	7,375
1300	Broiler, without oven, standard	Q-1	8	2		3,575	107		3,682	4,100
1550	Infrared	L-7	4	7		7,450	325		7,775	8,725
4750	Fryer, with twin baskets, modular model	Q-1	7	2.286		1,250	122		1,372	1,575
5000	Floor model, on 6" legs	"	5	3.200		2,400	171		2,571	2,925
5100	Extra single basket, large					100			100	110
5170	Energy star rated, 50 lb. capacity G	R-18	4	6.500		4,625	315		4,940	5,550
5175	85 lb. capacity G	"	4	6.500		10,000	315		10,315	11,500
5300	Griddle, SS, 24" plate, w/4" legs, elec, 208 V, 3 phase, 3' long	Q-1	7	2.286		1,400	122		1,522	1,750
5550	4' long	"	6	2.667		2,250	142		2,392	2,700
6200	Iced tea brewer	1 Plum	3.44	2.326		755	138		893	1,050
6350	Kettle, w/steam jacket, tilting, w/positive lock, SS, 20 gallons	L-7	7	4		8,700	186		8,886	9,875
6600	60 gallons	"	6	4.667		11,800	216		12,016	13,400
6900	Range, restaurant type, 6 burners and 1 standard oven, 36" wide	Q-1	7	2.286		2,850	122		2,972	3,325
6950	Convection		7	2.286		4,200	122		4,322	4,825
7150	2 standard ovens, 24" griddle, 60" wide		6	2.667		4,750	142		4,892	5,450
7200	1 standard, 1 convection oven	▼	6	2.667	▼	9,200	142		9,342	10,300

11 44 Food Cooking Equipment

11 44 13 - Commercial Ranges

11 44 13.10 Cooking Equipment	Crew	Daily Output	Labor-Hours	Unit	Material	2016 Bare Costs Labor	Equipment	Total	Total Incl O&P
7450	Heavy duty, single 34" standard oven, open top	Q-1	5	3.200	Ea.	5,600	171		5,771	6,425
7500	Convection oven		5	3.200		6,325	171		6,496	7,225
7700	Griddle top		6	2.667		3,025	142		3,167	3,575
7750	Convection oven		6	2.667		7,350	142		7,492	8,325
7760	Induction cooker, electric	L-7	7	4		1,825	186		2,011	2,325
8850	Steamer, electric 27 KW		7	4		11,200	186		11,386	12,600
9100	Electric, 10 KW or gas 100,000 BTU		5	5.600		6,325	260		6,585	7,375
9150	Toaster, conveyor type, 16-22 slices per minute					1,175			1,175	1,300
9160	Pop-up, 2 slot					620			620	685
9200	For deluxe models of above equipment, add					75%				
9400	Rule of thumb: Equipment cost based									
9410	on kitchen work area									
9420	Office buildings, minimum	L-7	77	.364	S.F.	90.50	16.85		107.35	127
9450	Maximum		58	.483		153	22.50		175.50	205
9550	Public eating facilities, minimum		77	.364		119	16.85		135.85	159
9600	Maximum		46	.609		193	28		221	258
9750	Hospitals, minimum		58	.483		122	22.50		144.50	171
9800	Maximum		39	.718		225	33.50		258.50	300

11 46 Food Dispensing Equipment

11 46 16 - Service Line Equipment

11 46 16.10 Commercial Food Dispensing Equipment	Crew	Daily Output	Labor-Hours	Unit	Material	2016 Bare Costs Labor	Equipment	Total	Total Incl O&P
0010	**COMMERCIAL FOOD DISPENSING EQUIPMENT**									
1050	Butter pat dispenser	1 Clab	13	.615	Ea.	870	23.50		893.50	1,000
1100	Bread dispenser, counter top		13	.615		890	23.50		913.50	1,025
1900	Cup and glass dispenser, drop in		4	2		1,075	76		1,151	1,325
1920	Disposable cup, drop in		16	.500		550	18.95		568.95	635
2650	Dish dispenser, drop in, 12"		11	.727		2,000	27.50		2,027.50	2,250
2660	Mobile		10	.800		3,050	30.50		3,080.50	3,400
3300	Food warmer, counter, 1.2 KW					700			700	770
3550	1.6 KW					2,200			2,200	2,425
3600	Well, hot food, built-in, rectangular, 12" x 20"	R-30	10	2.600		805	116		921	1,075
3610	Circular, 7 qt.		10	2.600		415	116		531	645
3620	Refrigerated, 2 compartments		10	2.600		3,050	116		3,166	3,550
3630	3 compartments		9	2.889		3,775	129		3,904	4,350
3640	4 compartments		8	3.250		4,350	145		4,495	5,025
4720	Frost cold plate		9	2.889		19,000	129		19,129	21,100
5700	Hot chocolate dispenser	1 Plum	4	2		1,175	118		1,293	1,475
5750	Ice dispenser 567 pound	Q-1	6	2.667		5,375	142		5,517	6,150
6250	Jet spray dispenser	R-18	4.50	5.778		2,700	279		2,979	3,400
6300	Juice dispenser, concentrate	"	4.50	5.778		1,900	279		2,179	2,500
6690	Milk dispenser, bulk, 2 flavor	R-30	8	3.250		1,750	145		1,895	2,150
6695	3 flavor	"	8	3.250		2,375	145		2,520	2,825
8800	Serving counter, straight	1 Carp	40	.200	L.F.	910	9.70		919.70	1,025
8820	Curved section	"	30	.267	"	1,175	12.90		1,187.90	1,325
8825	Solid surface, see Section 12 36 61.16									
8860	Sneeze guard with lights, 60" L	1 Clab	16	.500	Ea.	365	18.95		383.95	430
8900	Sneeze guard, stainless steel and glass, single sided									
8910	Portable, 48" W				Ea.	315			315	345
8920	Portable, 72" W					335			335	370
8930	Adjustable, 36" W	1 Carp	24	.333		250	16.15		266.15	300

For customer support on your Commercial Renovation Cost Data, call 877.742.4719.

365

11 46 Food Dispensing Equipment

11 46 16 – Service Line Equipment

11 46 16.10 Commercial Food Dispensing Equipment	Crew	Daily Output	Labor-Hours	Unit	Material	2016 Bare Costs Labor	2016 Bare Costs Equipment	Total	Total Incl O&P
8940	Adjustable, 48" W	1 Carp	20	.400	Ea.	315	19.40		334.40	380
9100	Soft serve ice cream machine, medium	R-18	11	2.364		9,200	114		9,314	10,300
9110	Large	"	9	2.889	↓	21,800	140		21,940	24,100

11 46 83 – Ice Machines

11 46 83.10 Commercial Ice Equipment

		Crew	Daily Output	Labor-Hours	Unit	Material	Labor	Equipment	Total	Total Incl O&P
0010	**COMMERCIAL ICE EQUIPMENT**									
5800	Ice cube maker, 50 pounds per day	Q-1	6	2.667	Ea.	1,625	142		1,767	2,000
5810	65 pounds per day, energy star rated		6	2.667		1,775	142		1,917	2,175
5900	250 pounds per day		1.20	13.333		2,700	710		3,410	4,075
5950	300 pounds per day, remote condensing		1.20	13.333		2,300	710		3,010	3,625
6050	500 pounds per day		4	4		2,950	213		3,163	3,550
6060	With bin		1.20	13.333		3,600	710		4,310	5,075
6070	Modular, with bin and condenser		1.20	13.333		3,875	710		4,585	5,375
6090	1000 pounds per day, with bin		1	16		5,125	850		5,975	6,975
6100	Ice flakers, 300 pounds per day		1.60	10		3,175	535		3,710	4,300
6120	600 pounds per day		.95	16.842		4,475	895		5,370	6,325
6130	1000 pounds per day		.75	21.333		4,575	1,125		5,700	6,825
6140	2000 pounds per day	↓	.65	24.615		21,600	1,300		22,900	25,800
6160	Ice storage bin, 500 pound capacity	Q-5	1	16		1,000	875		1,875	2,475
6180	1000 pound	"	.56	28.571	↓	2,750	1,550		4,300	5,450

11 48 Foodservice Cleaning and Disposal Equipment

11 48 13 – Commercial Dishwashers

11 48 13.10 Dishwashers

		Crew	Daily Output	Labor-Hours	Unit	Material	Labor	Equipment	Total	Total Incl O&P
0010	**DISHWASHERS**									
2700	Dishwasher, commercial, rack type									
2720	10 to 12 racks per hour	Q-1	3.20	5	Ea.	3,525	266		3,791	4,300
2730	Energy star rated, 35 to 40 racks/hour G		1.30	12.308		4,700	655		5,355	6,175
2740	50 to 60 racks/hour G	↓	1.30	12.308		11,600	655		12,255	13,800
2800	Automatic, 190 to 230 racks per hour	L-6	.35	34.286		14,000	1,975		15,975	18,500
2820	235 to 275 racks per hour		.25	48		31,500	2,775		34,275	38,900
2840	8,750 to 12,500 dishes per hour	↓	.10	120	↓	55,500	6,950		62,450	72,000
2950	Dishwasher hood, canopy type	L-3A	10	1.200	L.F.	1,025	63.50		1,088.50	1,225
2960	Pant leg type	"	2.50	4.800	Ea.	8,700	253		8,953	9,975
5200	Garbage disposal 1.5 HP, 100 GPH	L-1	4.80	3.333		2,000	191		2,191	2,500
5210	3 HP, 120 GPH		4.60	3.478		2,525	199		2,724	3,100
5220	5 HP, 250 GPH	↓	4.50	3.556	↓	3,125	203		3,328	3,750
6750	Pot sink, 3 compartment	1 Plum	7.25	1.103	L.F.	930	65.50		995.50	1,125
6760	Pot washer, low temp wash/rinse		1.60	5	Ea.	5,075	296		5,371	6,025
6770	High pressure wash, high temperature rinse	↓	1.20	6.667		36,600	395		36,995	40,800
9170	Trash compactor, small, up to 125 lb. compacted weight	L-4	4	6		23,600	272		23,872	26,400
9175	Large, up to 175 lb. compacted weight	"	3	8	↓	28,100	360		28,460	31,500

11 52 Audio-Visual Equipment

11 52 13 - Projection Screens

11 52 13.10 Projection Screens, Wall or Ceiling Hung

11 52 13.10 Projection Screens, Wall or Ceiling Hung	Crew	Daily Output	Labor-Hours	Unit	Material	2016 Bare Costs Labor	Equipment	Total	Total Incl O&P
0010 **PROJECTION SCREENS, WALL OR CEILING HUNG**, matte white									
0100 Manually operated, economy	2 Carp	500	.032	S.F.	6.45	1.55		8	9.60
0300 Intermediate		450	.036		7.85	1.72		9.57	11.45
0400 Deluxe		400	.040		9.75	1.94		11.69	13.90
9000 Minimum labor/equipment charge		3	5.333	Job		258		258	420

11 52 16 - Projectors

11 52 16.10 Movie Equipment

	Crew	Daily Output	Labor-Hours	Unit	Material	2016 Bare Costs Labor	Equipment	Total	Total Incl O&P
0011 **MOVIE EQUIPMENT**									
0020 Changeover, minimum				Ea.	485			485	535
3000 Projection screens, rigid, in wall, acrylic, 1/4" thick	2 Glaz	195	.082	S.F.	43.50	3.83		47.33	54
3100 1/2" thick	"	130	.123	"	50.50	5.75		56.25	65
3700 Sound systems, incl. amplifier, mono, minimum	1 Elec	.90	8.889	Ea.	3,450	490		3,940	4,550
3800 Dolby/Super Sound, maximum		.40	20		18,800	1,100		19,900	22,400
4100 Dual system, 2 channel, front surround, minimum		.70	11.429		4,825	630		5,455	6,275
4200 Dolby/Super Sound, 4 channel, maximum		.40	20		17,200	1,100		18,300	20,600
5700 Seating, painted steel, upholstered, minimum	2 Carp	35	.457		131	22		153	180
5800 Maximum	"	28	.571		450	27.50		477.50	535

11 53 Laboratory Equipment

11 53 33 - Emergency Safety Appliances

11 53 33.13 Emergency Equipment

	Crew	Daily Output	Labor-Hours	Unit	Material	2016 Bare Costs Labor	Equipment	Total	Total Incl O&P
0010 **EMERGENCY EQUIPMENT**									
1400 Safety equipment, eye wash, hand held				Ea.	405			405	445
1450 Deluge shower				"	775			775	850

11 53 43 - Service Fittings and Accessories

11 53 43.13 Fittings

	Crew	Daily Output	Labor-Hours	Unit	Material	2016 Bare Costs Labor	Equipment	Total	Total Incl O&P
0010 **FITTINGS**									
1600 Sink, one piece plastic, flask wash, hose, free standing	1 Plum	1.60	5	Ea.	1,950	296		2,246	2,600
1610 Epoxy resin sink, 25" x 16" x 10"	"	2	4	"	221	237		458	615
1630 Steel table, open underneath				L.F.	510			510	560
1640 Doors underneath				"	800			800	880
1900 Utility cabinet, stainless steel				Ea.	1,500			1,500	1,625
1950 Utility table, acid resistant top with drawers	2 Carp	30	.533	L.F.	165	26		191	223
8000 Alternate pricing method: as percent of lab furniture									
8050 Installation, not incl. plumbing & duct work				% Furn.				22%	22%
8100 Plumbing, final connections, simple system								10%	10%
8110 Moderately complex system								15%	15%
8120 Complex system								20%	20%
8150 Electrical, simple system								10%	10%
8160 Moderately complex system								20%	20%
8170 Complex system								35%	35%

11 53 53 - Biological Safety Cabinets

11 53 53.10 Pharmacy Cabinets

	Crew	Daily Output	Labor-Hours	Unit	Material	2016 Bare Costs Labor	Equipment	Total	Total Incl O&P
0010 **PHARMACY CABINETS**, vertical flow									
0100 Class II, type B2, 6' L	2 Carp	1.50	10.667	Ea.	13,300	515		13,815	15,500

For customer support on your Commercial Renovation Cost Data, call 877.742.4719.

367

11 57 Vocational Shop Equipment

11 57 10 – Shop Equipment

11 57 10.10 Vocational School Shop Equipment	Crew	Daily Output	Labor-Hours	Unit	Material	2016 Bare Costs Labor	2016 Bare Costs Equipment	Total	Total Incl O&P
0010 **VOCATIONAL SCHOOL SHOP EQUIPMENT**									
0020 Benches, work, wood, average	2 Carp	5	3.200	Ea.	665	155		820	990
0100 Metal, average		5	3.200		575	155		730	885
0400 Combination belt & disc sander, 6"		4	4		1,725	194		1,919	2,225
0700 Drill press, floor mounted, 12", 1/2 H.P.		4	4		425	194		619	780
0800 Dust collector, not incl. ductwork, 6" diameter	1 Shee	1.10	7.273		4,950	415		5,365	6,100
0810 Dust collector bag, 20" diameter	"	5	1.600		465	91.50		556.50	655
1000 Grinders, double wheel, 1/2 H.P.	2 Carp	5	3.200		225	155		380	500
1300 Jointer, 4", 3/4 H.P.		4	4		1,425	194		1,619	1,875
1600 Kilns, 16 C.F., to 2000°		4	4		1,500	194		1,694	1,975
1900 Lathe, woodworking, 10", 1/2 H.P.		4	4		580	194		774	955
2200 Planer, 13" x 6"		4	4		1,075	194		1,269	1,525
2500 Potter's wheel, motorized		4	4		1,250	194		1,444	1,700
2800 Saws, band, 14", 3/4 H.P.		4	4		960	194		1,154	1,375
3100 Metal cutting band saw, 14"		4	4		2,600	194		2,794	3,175
3400 Radial arm saw, 10", 2 H.P.		4	4		1,375	194		1,569	1,850
3700 Scroll saw, 24"		4	4		600	194		794	975
4000 Table saw, 10", 3 H.P.		4	4		2,550	194		2,744	3,150
4300 Welder AC arc, 30 amp capacity		4	4		3,075	194		3,269	3,700

11 61 Broadcast, Theater, and Stage Equipment

11 61 23 – Folding and Portable Stages

11 61 23.10 Portable Stages

	Crew	Daily Output	Labor-Hours	Unit	Material	Labor	Equipment	Total	Total Incl O&P
0010 **PORTABLE STAGES**									
5000 Stages, portable with steps, folding legs, stock, 8" high				SF Stg.	33.50			33.50	37
5100 16" high					52			52	57
5200 32" high					53.50			53.50	59
5300 40" high					60			60	66

11 61 33 – Rigging Systems and Controls

11 61 33.10 Controls

	Crew	Daily Output	Labor-Hours	Unit	Material	Labor	Equipment	Total	Total Incl O&P
0010 **CONTROLS**									
0050 Control boards with dimmers and breakers, minimum	1 Elec	1	8	Ea.	12,700	440		13,140	14,600
0150 Maximum	"	.20	40	"	130,500	2,200		132,700	147,000

11 61 43 – Stage Curtains

11 61 43.10 Curtains

	Crew	Daily Output	Labor-Hours	Unit	Material	Labor	Equipment	Total	Total Incl O&P
0010 **CURTAINS**									
0500 Curtain track, straight, light duty	2 Carp	20	.800	L.F.	28.50	39		67.50	94.50
0700 Curved sections		12	1.333	"	183	64.50		247.50	305
1000 Curtains, velour, medium weight		600	.027	S.F.	8.25	1.29		9.54	11.15

11 66 Athletic Equipment

11 66 13 – Exercise Equipment

11 66 13.10 Physical Training Equipment

	11 66 13.10 Physical Training Equipment	Crew	Daily Output	Labor-Hours	Unit	Material	2016 Bare Costs Labor	Equipment	Total	Total Incl O&P
0010	**PHYSICAL TRAINING EQUIPMENT**									
0020	Abdominal rack, 2 board capacity				Ea.	490			490	535
0050	Abdominal board, upholstered					720			720	790
0200	Bicycle trainer, minimum					500			500	550
0300	Deluxe, electric					4,700			4,700	5,175
0400	Barbell set, chrome plated steel, 25 lb.					289			289	320
0420	100 lb.					580			580	635
0450	200 lb.					800			800	880
0500	Weight plates, cast iron, per lb.				Lb.	4			4	4.40
0520	Storage rack, 10 station				Ea.	995			995	1,100
0600	Circuit training apparatus, 12 machines minimum	2 Clab	1.25	12.800	Set	30,300	485		30,785	34,100
0700	Average		1	16		37,800	605		38,405	42,600
0800	Maximum		.75	21.333		45,300	810		46,110	51,000
0820	Dumbbell set, cast iron, with rack and 5 pair					485			485	535
0900	Squat racks	2 Clab	5	3.200	Ea.	925	121		1,046	1,225
4150	Exercise equipment, bicycle trainer					760			760	835
4180	Chinning bar, adjustable, wall mounted	1 Carp	5	1.600		212	77.50		289.50	360
4200	Exercise ladder, 16' x 1'-7", suspended	L-2	3	5.333		1,400	225		1,625	1,900
4210	High bar, floor plate attached	1 Carp	4	2		2,000	97		2,097	2,350
4240	Parallel bars, adjustable		4	2		1,650	97		1,747	1,950
4270	Uneven parallel bars, adjustable		4	2		3,425	97		3,522	3,925
4280	Wall mounted, adjustable	L-2	1.50	10.667	Set	910	450		1,360	1,750
4300	Rope, ceiling mounted, 18' long	1 Carp	3.66	2.186	Ea.	196	106		302	390
4330	Side horse, vaulting		5	1.600		1,375	77.50		1,452.50	1,625
4360	Treadmill, motorized, deluxe, training type		5	1.600		3,900	77.50		3,977.50	4,400
4390	Weight lifting multi-station, minimum	2 Clab	1	16		350	605		955	1,375

11 66 23 – Gymnasium Equipment

11 66 23.13 Basketball Equipment

	11 66 23.13 Basketball Equipment	Crew	Daily Output	Labor-Hours	Unit	Material	Labor	Equipment	Total	Total Incl O&P
0010	**BASKETBALL EQUIPMENT**									
1000	Backstops, wall mtd., 6' extended, fixed, minimum	L-2	1	16	Ea.	1,500	675		2,175	2,750
1100	Maximum		1	16		1,800	675		2,475	3,075
1200	Swing up, minimum		1	16		1,600	675		2,275	2,850
1250	Maximum		1	16		2,925	675		3,600	4,325
1300	Portable, manual, heavy duty, spring operated		1.90	8.421		12,800	355		13,155	14,700
1400	Ceiling suspended, stationary, minimum		.78	20.513		4,200	865		5,065	6,050
1450	Fold up, with accessories, maximum		.40	40		6,175	1,700		7,875	9,575
1600	For electrically operated, add	1 Elec	1	8		2,400	440		2,840	3,300
5800	Wall pads, 1-1/2" thick, standard (not fire rated)	2 Carp	640	.025	S.F.	4.77	1.21		5.98	7.25

11 66 23.19 Boxing Ring

	11 66 23.19 Boxing Ring	Crew	Daily Output	Labor-Hours	Unit	Material	Labor	Equipment	Total	Total Incl O&P
0010	**BOXING RING**									
4100	Elevated, 22' x 22'	L-4	.10	240	Ea.	6,750	10,900		17,650	25,100
4110	For cellular plastic foam padding, add		.10	240		1,125	10,900		12,025	19,000
4120	Floor level, including posts and ropes only, 20' x 20'		.80	30		4,650	1,350		6,000	7,325
4130	Canvas, 30' x 30'		5	4.800		1,350	217		1,567	1,850

11 66 23.47 Gym Mats

	11 66 23.47 Gym Mats	Crew	Daily Output	Labor-Hours	Unit	Material	Labor	Equipment	Total	Total Incl O&P
0010	**GYM MATS**									
5500	2" thick, naugahyde covered				S.F.	4.68			4.68	5.15
5600	Vinyl/nylon covered					8.30			8.30	9.10
6000	Wrestling mats, 1" thick, heavy duty					4.69			4.69	5.15

11 66 Athletic Equipment

11 66 43 – Interior Scoreboards

11 66 43.10 Scoreboards	Crew	Daily Output	Labor-Hours	Unit	Material	2016 Bare Costs Labor	Equipment	Total	Total Incl O&P
0010 **SCOREBOARDS**									
7000 Baseball, minimum	R-3	1.30	15.385	Ea.	3,600	840	108	4,548	5,375
7200 Maximum		.05	400		20,300	21,900	2,800	45,000	59,500
7300 Football, minimum		.86	23.256		5,300	1,275	163	6,738	7,975
7400 Maximum		.20	100		18,300	5,475	700	24,475	29,400
7500 Basketball (one side), minimum		2.07	9.662		2,500	530	68	3,098	3,650
7600 Maximum		.30	66.667		7,475	3,650	470	11,595	14,400
7700 Hockey-basketball (four sides), minimum		.25	80		8,925	4,375	560	13,860	17,200
7800 Maximum	▼	.15	133	▼	15,800	7,300	935	24,035	29,600

11 66 53 – Gymnasium Dividers

11 66 53.10 Divider Curtains

	Crew	Daily Output	Labor-Hours	Unit	Material	2016 Bare Costs Labor	Equipment	Total	Total Incl O&P
0010 **DIVIDER CURTAINS**									
4500 Gym divider curtain, mesh top, vinyl bottom, manual	L-4	500	.048	S.F.	9.50	2.17		11.67	14
4700 Electric roll up	L-7	400	.070	"	12.65	3.25		15.90	19.15

11 71 Medical Sterilizing Equipment

11 71 10 – Medical Sterilizers & Distillers

11 71 10.10 Sterilizers and Distillers

	Crew	Daily Output	Labor-Hours	Unit	Material	2016 Bare Costs Labor	Equipment	Total	Total Incl O&P
0010 **STERILIZERS AND DISTILLERS**									
3010 Portable, top loading, 105 – 135 degree C, 3 to 30 psi, 50 L chamber				Ea.	9,400			9,400	10,300
3020 Stainless steel basket, 10.7" diam. x 11.8" H					209			209	230
3025 Stainless steel pail, 10.7" diam. x 10.7" H					293			293	320
3050 85 L chamber					16,300			16,300	17,900
3060 Stainless steel basket, 15.3" diam. x 11.5" H					375			375	410
3065 Stainless steel pail, 15.3" diam. x 11" H				▼	595			595	655

11 76 Operating Room Equipment

11 76 10 – Operating Room Equipment

11 76 10.10 Surgical Equipment

	Crew	Daily Output	Labor-Hours	Unit	Material	2016 Bare Costs Labor	Equipment	Total	Total Incl O&P
0010 **SURGICAL EQUIPMENT**									
6550 Major surgery table, minimum	1 Sswk	.50	16	Ea.	26,300	850		27,150	30,400
6570 Maximum		.50	16		28,300	850		29,150	32,700
6600 Hydraulic, hand-held control, general surgery		.60	13.333		30,400	710		31,110	34,700
6650 Stationary, universal	▼	.50	16		39,200	850		40,050	44,600
6800 Surgical lights, major operating room, dual head, minimum	2 Elec	1	16		4,250	880		5,130	6,050
6850 Maximum		1	16		30,500	880		31,380	35,000
6900 Ceiling mount articulation, single arm	▼	1	16	▼	3,800	880		4,680	5,550

11 81 Facility Maintenance Equipment

11 81 19 – Vacuum Cleaning Systems

11 81 19.10 Vacuum Cleaning	Crew	Daily Output	Labor-Hours	Unit	Material	2016 Bare Costs Labor	Equipment	Total	Total Incl O&P
0010 **VACUUM CLEANING**									
0020 Central, 3 inlet, residential	1 Skwk	.90	8.889	Total	1,100	445		1,545	1,925
0200 Commercial		.70	11.429		1,275	570		1,845	2,325
0400 5 inlet system, residential		.50	16		1,550	800		2,350	3,000
0600 7 inlet system, commercial	↓	.40	20		1,750	1,000		2,750	3,550
4010 Rule of thumb: First 1200 S.F., installed								1,425	1,575
4020 For each additional S.F., add				S.F.				.26	.26

11 82 Facility Solid Waste Handling Equipment

11 82 26 – Facility Waste Compactors

11 82 26.10 Compactors

	Crew	Daily Output	Labor-Hours	Unit	Material	2016 Bare Costs Labor	Equipment	Total	Total Incl O&P
0010 **COMPACTORS**									
0020 Compactors, 115 volt, 250#/hr., chute fed	L-4	1	24	Ea.	12,200	1,075		13,275	15,200
0100 Hand fed		2.40	10		15,400	455		15,855	17,600
1000 Heavy duty industrial compactor, 0.5 C.Y. capacity		1	24		10,300	1,075		11,375	13,200
1050 1.0 C.Y. capacity	↓	1	24		15,800	1,075		16,875	19,200
1400 For handling hazardous waste materials, 55 gallon drum packer, std.					20,100			20,100	22,100
1410 55 gallon drum packer w/HEPA filter					25,100			25,100	27,600
1420 55 gallon drum packer w/charcoal & HEPA filter					33,400			33,400	36,800
1430 All of the above made explosion proof, add					1,475			1,475	1,625

11 82 39 – Medical Waste Disposal Systems

11 82 39.10 Off-Site Disposal

	Crew	Daily Output	Labor-Hours	Unit	Material	2016 Bare Costs Labor	Equipment	Total	Total Incl O&P
0010 **OFF-SITE DISPOSAL**									
0100 Medical waste disposal, Red Bag system, pick up & treat, 200 lb. per week				Week	182			182	201
0110 Per month				Month	700			700	770
0150 Red bags, 7-10 gal., 1.2 mil, pkg of 500				Ea.	67			67	74
0200 15 gal., package of 250					64			64	70
0250 33 gal., package of 250					66.50			66.50	73
0300 45 gal., package of 100					56.50			56.50	62

11 82 39.20 Disposal Carts

	Crew	Daily Output	Labor-Hours	Unit	Material	2016 Bare Costs Labor	Equipment	Total	Total Incl O&P
0010 **DISPOSAL CARTS**									
2010 Medical waste disposal cart, HDPE, w/lid, 28 gal. capacity				Ea.	267			267	294
2020 96 gal. capacity					310			310	340
2030 150 gal. capacity, low profile					480			480	530
2040 200 gal. capacity					930			930	1,025

11 82 39.30 Medical Waste Sanitizers

	Crew	Daily Output	Labor-Hours	Unit	Material	2016 Bare Costs Labor	Equipment	Total	Total Incl O&P
0010 **MEDICAL WASTE SANITIZERS**									
2010 Small, hand loaded, 1.5 C.Y., 225 lb. capacity				Ea.	79,500			79,500	87,500
2020 Medium, cart loaded, 6.25 C.Y., 938 lb. capacity					106,500			106,500	117,000
2030 Large, cart loaded, 15 C.Y., 2250 lb. capacity					133,000			133,000	146,000
3010 Cart, aluminum, 75 lb. capacity					2,200			2,200	2,425
3020 95 lb. capacity					2,400			2,400	2,625
4010 Stainless steel, 173 lb. capacity					3,125			3,125	3,450
4020 232 lb. capacity					3,500			3,500	3,850
4030 Cart lift, hydraulic scissor type					6,225			6,225	6,825
4040 Portable aluminum ramp					1,750			1,750	1,925
4050 Fold-down steel tracks					1,425			1,425	1,575
4060 Pull-out drawer, small					6,700			6,700	7,375
4070 Medium					9,675			9,675	10,600
4080 Large					13,600			13,600	15,000

For customer support on your Commercial Renovation Cost Data, call 877.742.4719.

371

11 82 Facility Solid Waste Handling Equipment

11 82 39 – Medical Waste Disposal Systems

11 82 39.30 Medical Waste Sanitizers	Crew	Daily Output	Labor-Hours	Unit	Material	2016 Bare Costs Labor	Equipment	Total	Total Incl O&P
5000	Medical waste treatment, sanitize, on-site									
5010	Less than 15,000 lb. per month				Lb.	.20			.20	.22
5020	Over 15,000 lb. per month				"	.16			.16	.18

11 91 Religious Equipment

11 91 13 – Baptisteries

11 91 13.10 Baptistry

		Crew	Daily Output	Labor-Hours	Unit	Material	Labor	Equipment	Total	Total Incl O&P
0010	**BAPTISTRY**									
0150	Fiberglass, 3'-6" deep, x 13'-7" long,									
0160	steps at both ends, incl. plumbing, minimum	L-8	1	20	Ea.	5,600	1,000		6,600	7,800
0200	Maximum	"	.70	28.571		9,425	1,450		10,875	12,700
0250	Add for filter, heater and lights					1,850			1,850	2,050

11 91 23 – Sanctuary Equipment

11 91 23.10 Sanctuary Furnishings

		Crew	Daily Output	Labor-Hours	Unit	Material	Labor	Equipment	Total	Total Incl O&P
0010	**SANCTUARY FURNISHINGS**									
0020	Altar, wood, custom design, plain	1 Carp	1.40	5.714	Ea.	2,600	277		2,877	3,325
0050	Deluxe		.20	40	"	12,500	1,950		14,450	17,000
5000	Wall cross, aluminum, extruded, 2" x 2" section		34	.235	L.F.	222	11.40		233.40	263
5150	4" x 4" section		29	.276		320	13.35		333.35	370
5300	Bronze, extruded, 1" x 2" section		31	.258		440	12.50		452.50	500
5350	2-1/2" x 2-1/2" section		34	.235		665	11.40		676.40	750

11 97 Security Equipment

11 97 30 – Security Drawers

11 97 30.10 Pass Through Drawer

		Crew	Daily Output	Labor-Hours	Unit	Material	Labor	Equipment	Total	Total Incl O&P
0010	**PASS THROUGH DRAWER**									
0100	Pass-thru drawer for personal items, 18" x 15" x 24"	1 Skwk	2	4	Ea.	2,775	200		2,975	3,375
0110	Including speakers	"	1.50	5.333	"	3,125	266		3,391	3,875

11 98 Detention Equipment

11 98 30 – Detention Cell Equipment

11 98 30.10 Cell Equipment

		Crew	Daily Output	Labor-Hours	Unit	Material	Labor	Equipment	Total	Total Incl O&P
0010	**CELL EQUIPMENT**									
3000	Toilet apparatus including wash basin, average	L-8	1.50	13.333	Ea.	3,450	675		4,125	4,875

For customer support on your Commercial Renovation Cost Data, call 877.742.4719.

Estimating Tips
General

- The items in this division are usually priced per square foot or each. Most of these items are purchased by the owner and installed by the contractor. Do not assume the items in Division 12 will be purchased and installed by the contractor. Check the specifications for responsibilities and include receiving, storage, installation, and mechanical and electrical hookups in the appropriate divisions.

- Some items in this division require some type of support system that is not usually furnished with the item. Examples of these systems include blocking for the attachment of casework and heavy drapery rods. The required blocking must be added to the estimate in the appropriate division.

Reference Numbers

Reference numbers are shown at the beginning of some major classifications. These numbers refer to related items in the Reference Section. The reference information may be an estimating procedure, an alternate pricing method, or technical information.

Note: Not all subdivisions listed here necessarily appear. ■

12 21 Window Blinds

12 21 13 – Horizontal Louver Blinds

12 21 13.13 Metal Horizontal Louver Blinds	Crew	Daily Output	Labor-Hours	Unit	Material	2016 Bare Costs Labor	Equipment	Total	Total Incl O&P
0010 **METAL HORIZONTAL LOUVER BLINDS**									
0020 Horizontal, 1" aluminum slats, solid color, stock	1 Carp	590	.014	S.F.	4.69	.66		5.35	6.20

12 21 13.33 Vinyl Horizontal Louver Blinds	Crew	Daily Output	Labor-Hours	Unit	Material	2016 Bare Costs Labor	Equipment	Total	Total Incl O&P
0010 **VINYL HORIZONTAL LOUVER BLINDS**									
0100 2" composite, 48" wide, 48" high	1 Carp	30	.267	Ea.	91	12.90		103.90	121
0120 72" high		29	.276		123	13.35		136.35	157
0140 96" high		28	.286		167	13.85		180.85	207
0200 60" wide, 60" high		27	.296		114	14.35		128.35	150
0220 72" high		25	.320		131	15.50		146.50	170
0240 96" high		24	.333		196	16.15		212.15	243
0300 72" wide, 72" high		25	.320		181	15.50		196.50	225
0320 96" high		23	.348		250	16.85		266.85	305
0400 96" wide, 96" high		20	.400		325	19.40		344.40	385
1000 2" faux wood, 48" wide, 48" high		30	.267		60	12.90		72.90	87
1020 72" high		29	.276		78	13.35		91.35	108
1040 96" high		28	.286		93.50	13.85		107.35	126
1300 72" wide, 72" high		25	.320		141	15.50		156.50	182
1320 96" high		23	.348		212	16.85		228.85	261
1400 96" wide, 96" high		20	.400		223	19.40		242.40	277

12 23 Interior Shutters

12 23 10 – Wood Interior Shutters

12 23 10.10 Wood Interior Shutters	Crew	Daily Output	Labor-Hours	Unit	Material	2016 Bare Costs Labor	Equipment	Total	Total Incl O&P
0010 **WOOD INTERIOR SHUTTERS**, louvered									
0200 Two panel, 27" wide, 36" high	1 Carp	5	1.600	Set	158	77.50		235.50	300
0300 33" wide, 36" high		5	1.600		204	77.50		281.50	350
0500 47" wide, 36" high		5	1.600		273	77.50		350.50	425
1000 Four panel, 27" wide, 36" high		5	1.600		158	77.50		235.50	300
1100 33" wide, 36" high		5	1.600		202	77.50		279.50	350
1300 47" wide, 36" high		5	1.600		270	77.50		347.50	425
1400 Plantation shutters, 16" x 48"		5	1.600	Ea.	182	77.50		259.50	325
1450 16" x 96"		4	2		299	97		396	490
1460 36" x 96"		3	2.667		660	129		789	935

12 23 10.13 Wood Panels	Crew	Daily Output	Labor-Hours	Unit	Material	2016 Bare Costs Labor	Equipment	Total	Total Incl O&P
0010 **WOOD PANELS**									
3000 Wood folding panels with movable louvers, 7" x 20" each	1 Carp	17	.471	Pr.	82.50	23		105.50	128
4000 Fixed louver type, stock units, 8" x 20" each		17	.471		94	23		117	141
4450 18" x 40" each		17	.471		134	23		157	186

12 24 Window Shades

12 24 13 – Roller Window Shades

12 24 13.10 Shades

		Crew	Daily Output	Labor-Hours	Unit	Material	2016 Bare Costs Labor	Equipment	Total	Total Incl O&P
0010	**SHADES**									
0020	Basswood, roll-up, stain finish, 3/8" slats	1 Carp	300	.027	S.F.	17.15	1.29		18.44	21
0030	Double layered, heat reflective		685	.012		10.35	.57		10.92	12.25
0200	7/8" slats		300	.027		17	1.29		18.29	21
0900	Mylar, single layer, non-heat reflective		685	.012		3.20	.57		3.77	4.44
0910	Mylar, single layer, heat reflective		685	.012		2.52	.57		3.09	3.69
1000	Double layered, heat reflective		685	.012		6.10	.57		6.67	7.60
1100	Triple layered, heat reflective		685	.012		7.75	.57		8.32	9.45
2000	Polyester, room darkening, with continuous cord, GEI									
2010	36" x 72"	1 Carp	38	.211	Ea.	139	10.20		149.20	170
2020	48" x 72"		28	.286		148	13.85		161.85	186
2030	60" x 72"		23	.348		201	16.85		217.85	249
2040	72" x 72"		19	.421		211	20.50		231.50	266
5000	Thermal, roll up, R4		44	.182	S.F.	12.40	8.80		21.20	28
5030	R10.7		44	.182	"	14.30	8.80		23.10	30
5050	Magnetic clips, set of 20				Set	26.50			26.50	29.50
6011	Solar screening, fiberglass [G]	1 Carp	85	.094	S.F.	6.90	4.56		11.46	15.05

12 32 Manufactured Wood Casework

12 32 16 – Manufactured Plastic-Laminate-Clad Casework

12 32 16.20 Plastic Laminate Casework Doors

0010	**PLASTIC LAMINATE CASEWORK DOORS**									
1000	For casework frames, see Section 12 32 23.15									
1100	For casework hardware, see Section 12 32 23.35									

12 32 23 – Hardwood Casework

12 32 23.10 Manufactured Wood Casework, Stock Units

		Crew	Daily Output	Labor-Hours	Unit	Material	Labor	Equipment	Total	Total Incl O&P
0010	**MANUFACTURED WOOD CASEWORK, STOCK UNITS**									
0300	Built-in drawer units, pine, 18" deep, 32" high, unfinished									
0400	Minimum	2 Carp	53	.302	L.F.	123	14.65		137.65	160
0500	Maximum	"	40	.400	"	145	19.40		164.40	192
0700	Kitchen base cabinets, hardwood, not incl. counter tops,									
0710	24" deep, 35" high, prefinished									
0800	One top drawer, one door below, 12" wide	2 Carp	24.80	.645	Ea.	271	31.50		302.50	350
0820	15" wide		24	.667		282	32.50		314.50	365
0840	18" wide		23.30	.687		305	33.50		338.50	395
0860	21" wide		22.70	.705		315	34		349	405
0880	24" wide		22.30	.717		370	35		405	465
1000	Four drawers, 12" wide		24.80	.645		285	31.50		316.50	365
1020	15" wide		24	.667		289	32.50		321.50	375
1040	18" wide		23.30	.687		320	33.50		353.50	405
1060	24" wide		22.30	.717		350	35		385	440
1200	Two top drawers, two doors below, 27" wide		22	.727		400	35		435	500
1220	30" wide		21.40	.748		440	36		476	540
1240	33" wide		20.90	.766		455	37		492	560
1260	36" wide		20.30	.788		470	38		508	585
1280	42" wide		19.80	.808		495	39		534	610
1300	48" wide		18.90	.847		535	41		576	650
1500	Range or sink base, two doors below, 30" wide		21.40	.748		365	36		401	460
1520	33" wide		20.90	.766		390	37		427	490
1540	36" wide		20.30	.788		405	38		443	515

For customer support on your Commercial Renovation Cost Data, call 877.742.4719.

375

12 32 23 – Hardwood Casework

12 32 23.10 Manufactured Wood Casework, Stock Units		Crew	Daily Output	Labor-Hours	Unit	Material	2016 Bare Costs Labor	Equipment	Total	Total Incl O&P
1560	42" wide	2 Carp	19.80	.808	Ea.	430	39		469	535
1580	48" wide		18.90	.847		450	41		491	560
1800	For sink front units, deduct					161			161	177
2000	Corner base cabinets, 36" wide, standard	2 Carp	18	.889		670	43		713	810
2100	Lazy Susan with revolving door	"	16.50	.970		855	47		902	1,025
4000	Kitchen wall cabinets, hardwood, 12" deep with two doors									
4050	12" high, 30" wide	2 Carp	24.80	.645	Ea.	245	31.50		276.50	320
4100	36" wide		24	.667		291	32.50		323.50	375
4400	15" high, 30" wide		24	.667		249	32.50		281.50	325
4420	33" wide		23.30	.687		310	33.50		343.50	395
4440	36" wide		22.70	.705		299	34		333	385
4450	42" wide		22.70	.705		335	34		369	425
4700	24" high, 30" wide		23.30	.687		335	33.50		368.50	420
4720	36" wide		22.70	.705		365	34		399	460
4740	42" wide		22.30	.717		415	35		450	515
5000	30" high, one door, 12" wide		22	.727		234	35		269	315
5020	15" wide		21.40	.748		246	36		282	330
5040	18" wide		20.90	.766		270	37		307	360
5060	24" wide		20.30	.788		315	38		353	415
5300	Two doors, 27" wide		19.80	.808		350	39		389	450
5320	30" wide		19.30	.829		370	40		410	470
5340	36" wide		18.80	.851		420	41		461	530
5360	42" wide		18.50	.865		455	42		497	570
5380	48" wide		18.40	.870		510	42		552	635
6000	Corner wall, 30" high, 24" wide		18	.889		360	43		403	465
6050	30" wide		17.20	.930		380	45		425	495
6100	36" wide		16.50	.970		435	47		482	555
6500	Revolving Lazy Susan		15.20	1.053		490	51		541	625
7000	Broom cabinet, 84" high, 24" deep, 18" wide		10	1.600		670	77.50		747.50	865
7500	Oven cabinets, 84" high, 24" deep, 27" wide		8	2		1,025	97		1,122	1,275
7750	Valance board trim		396	.040	L.F.	16.25	1.96		18.21	21
7780	Toe kick trim	1 Carp	256	.031	"	3.10	1.51		4.61	5.90
7790	Base cabinet corner filler		16	.500	Ea.	43.50	24		67.50	87
7800	Cabinet filler, 3" x 24"		20	.400		17.10	19.40		36.50	50.50
7810	3" x 30"		20	.400		21.50	19.40		40.90	55
7820	3" x 42"		18	.444		30	21.50		51.50	68
7830	3" x 80"		16	.500		57	24		81	103
7850	Cabinet panel		50	.160	S.F.	9.30	7.75		17.05	23
9000	For deluxe models of all cabinets, add					40%				
9500	For custom built in place, add					25%	10%			
9558	Rule of thumb, kitchen cabinets not including									
9560	appliances & counter top, minimum	2 Carp	30	.533	L.F.	182	26		208	242
9600	Maximum	"	25	.640	"	405	31		436	495
9700	Minimum labor/equipment charge	1 Carp	3	2.667	Job		129		129	211

12 32 23.30 Manufactured Wood Casework Vanities

		Crew	Daily Output	Labor-Hours	Unit	Material	2016 Bare Costs Labor	Equipment	Total	Total Incl O&P
0010	**MANUFACTURED WOOD CASEWORK VANITIES**									
8000	Vanity bases, 2 doors, 30" high, 21" deep, 24" wide	2 Carp	20	.800	Ea.	315	39		354	410
8050	30" wide		16	1		375	48.50		423.50	495
8100	36" wide		13.33	1.200		365	58		423	495
8150	48" wide		11.43	1.400		480	68		548	635
9000	For deluxe models of all vanities, add to above					40%				
9500	For custom built in place, add to above					25%	10%			

12 32 Manufactured Wood Casework

12 32 23 – Hardwood Casework

12 32 23.35 Manufactured Wood Casework Hardware	Crew	Daily Output	Labor-Hours	Unit	Material	2016 Bare Costs Labor	Equipment	Total	Total Incl O&P
0010 **MANUFACTURED WOOD CASEWORK HARDWARE**									
1000 Catches, minimum	1 Carp	235	.034	Ea.	1.30	1.65		2.95	4.12
1040 Maximum	"	80	.100	"	7.75	4.85		12.60	16.45
2000 Door/drawer pulls, handles									
2200 Handles and pulls, projecting, metal, minimum	1 Carp	48	.167	Ea.	5.20	8.10		13.30	18.90
2240 Maximum		36	.222		10.65	10.75		21.40	29.50
2300 Wood, minimum		48	.167		5.30	8.10		13.40	19.05
2340 Maximum		36	.222		9.80	10.75		20.55	28.50
2400 Drawer pulls, antimicrobial copper alloy finish		50	.160		19	7.75		26.75	33.50
2600 Flush, metal, minimum		48	.167		5.30	8.10		13.40	19.05
2640 Maximum		36	.222		9.90	10.75		20.65	28.50
2900 Drawer knobs, antimicrobial copper alloy finish		50	.160		12.55	7.75		20.30	26.50
3000 Drawer tracks/glides, minimum		48	.167	Pr.	8.95	8.10		17.05	23
3040 Maximum		24	.333		26	16.15		42.15	55
4000 Cabinet hinges, minimum		160	.050		3.10	2.42		5.52	7.35
4040 Maximum		68	.118		12.40	5.70		18.10	23
7000 Appliance pulls, antimicrobial copper alloy finish		50	.160	L.F.	68	7.75		75.75	87.50

12 34 Manufactured Plastic Casework

12 34 16 – Manufactured Solid-Plastic Casework

12 34 16.10 Outdoor Casework

	Crew	Daily Output	Labor-Hours	Unit	Material	2016 Bare Costs Labor	Equipment	Total	Total Incl O&P
0010 **OUTDOOR CASEWORK**									
0020 Cabinet, base, sink/range, 36"	2 Carp	20.30	.788	Ea.	1,625	38		1,663	1,875
0100 Base, 36"		20.30	.788		1,500	38		1,538	1,725
0200 Filler strip, 1" x 30"		158	.101		43	4.91		47.91	55
0210 Filler strip, 2" x 30"		158	.101		43	4.91		47.91	55

12 35 Specialty Casework

12 35 53 – Laboratory Casework

12 35 53.13 Metal Laboratory Casework

	Crew	Daily Output	Labor-Hours	Unit	Material	2016 Bare Costs Labor	Equipment	Total	Total Incl O&P
0010 **METAL LABORATORY CASEWORK**									
0020 Cabinets, base, door units, metal	2 Carp	18	.889	L.F.	238	43		281	330
0300 Drawer units		18	.889		530	43		573	650
0700 Tall storage cabinets, open, 7' high		20	.800		510	39		549	625
0900 With glazed doors		20	.800		765	39		804	905
1300 Wall cabinets, metal, 12-1/2" deep, open		20	.800		184	39		223	267
1500 With doors		20	.800		365	39		404	465

12 35 70 – Healthcare Casework

12 35 70.13 Hospital Casework

	Crew	Daily Output	Labor-Hours	Unit	Material	2016 Bare Costs Labor	Equipment	Total	Total Incl O&P
0010 **HOSPITAL CASEWORK**									
3000 Hospital cabinets, stainless steel with glass door(s), lockable									
3010 One door, 24" W x 18" D x 60" H	2 Clab	18	.889	Ea.	2,525	33.50		2,558.50	2,825
3020 Two doors, 36" W x 18" D x 60" H		15	1.067		3,100	40.50		3,140.50	3,475
3030 36" W x 24" D x 67" H		15	1.067		3,675	40.50		3,715.50	4,125
3040 48" W x 24" D x 66" H		12	1.333		4,200	50.50		4,250.50	4,700
3050 48" W x 24" D x 72" H		12	1.333		5,100	50.50		5,150.50	5,675
3060 60" W x 24" D x 72" H		9	1.778		6,050	67.50		6,117.50	6,750

For customer support on your Commercial Renovation Cost Data, call 877.742.4719.

377

12 36 Countertops

12 36 16 – Metal Countertops

12 36 16.10 Stainless Steel Countertops		Crew	Daily Output	Labor-Hours	Unit	Material	2016 Bare Costs Labor	Equipment	Total	Total Incl O&P
0010	**STAINLESS STEEL COUNTERTOPS**									
3200	Stainless steel, custom	1 Carp	24	.333	S.F.	155	16.15		171.15	198

12 36 19 – Wood Countertops

12 36 19.10 Maple Countertops

		Crew	Daily Output	Labor-Hours	Unit	Material	2016 Bare Costs Labor	Equipment	Total	Total Incl O&P
0010	**MAPLE COUNTERTOPS**									
2900	Solid, laminated, 1-1/2" thick, no splash	1 Carp	28	.286	L.F.	76.50	13.85		90.35	107
3000	With square splash		28	.286	"	91	13.85		104.85	123
3400	Recessed cutting block with trim, 16" x 20" x 1"	↓	8	1	Ea.	93	48.50		141.50	182

12 36 23 – Plastic Countertops

12 36 23.13 Plastic-Laminate-Clad Countertops

		Crew	Daily Output	Labor-Hours	Unit	Material	2016 Bare Costs Labor	Equipment	Total	Total Incl O&P
0010	**PLASTIC-LAMINATE-CLAD COUNTERTOPS**									
0020	Stock, 24" wide w/backsplash, minimum	1 Carp	30	.267	L.F.	17.25	12.90		30.15	40
0100	Maximum		25	.320		35	15.50		50.50	64
0300	Custom plastic, 7/8" thick, aluminum molding, no splash		30	.267		30.50	12.90		43.40	54.50
0400	Cove splash		30	.267		29.50	12.90		42.40	53.50
0600	1-1/4" thick, no splash		28	.286		36	13.85		49.85	62
0700	Square splash		28	.286		43	13.85		56.85	70
0900	Square edge, plastic face, 7/8" thick, no splash		30	.267		33.50	12.90		46.40	58
1000	With splash	↓	30	.267		40	12.90		52.90	65
1200	For stainless channel edge, 7/8" thick, add					3.17			3.17	3.49
1300	1-1/4" thick, add					3.78			3.78	4.16
1500	For solid color suede finish, add			↓		4.14			4.14	4.55
1700	For end splash, add				Ea.	18.65			18.65	20.50
1900	For cut outs, standard, add, minimum	1 Carp	32	.250		12.40	12.10		24.50	33.50
2000	Maximum		8	1	↓	6.20	48.50		54.70	86
2100	Postformed, including backsplash and front edge		30	.267	L.F.	10.35	12.90		23.25	32.50
2110	Mitred, add		12	.667	Ea.		32.50		32.50	53
2200	Built-in place, 25" wide, plastic laminate		25	.320	L.F.	42	15.50		57.50	71.50
9000	Minimum labor/equipment charge	↓	3.75	2.133	Job		103		103	169

12 36 33 – Tile Countertops

12 36 33.10 Ceramic Tile Countertops

		Crew	Daily Output	Labor-Hours	Unit	Material	2016 Bare Costs Labor	Equipment	Total	Total Incl O&P
0010	**CERAMIC TILE COUNTERTOPS**									
2300	Ceramic tile mosaic	1 Carp	25	.320	L.F.	34	15.50		49.50	63

12 36 40 – Stone Countertops

12 36 40.10 Natural Stone Countertops

		Crew	Daily Output	Labor-Hours	Unit	Material	2016 Bare Costs Labor	Equipment	Total	Total Incl O&P
0010	**NATURAL STONE COUNTERTOPS**									
2500	Marble, stock, with splash, 1/2" thick, minimum	1 Bric	17	.471	L.F.	42.50	22		64.50	82
2700	3/4" thick, maximum	"	13	.615	"	107	28.50		135.50	164

12 36 53 – Laboratory Countertops

12 36 53.10 Laboratory Countertops and Sinks

		Crew	Daily Output	Labor-Hours	Unit	Material	2016 Bare Costs Labor	Equipment	Total	Total Incl O&P
0010	**LABORATORY COUNTERTOPS AND SINKS**									
0020	Countertops, epoxy resin, not incl. base cabinets, acid-proof, minimum	2 Carp	82	.195	S.F.	42.50	9.45		51.95	62.50
0030	Maximum		70	.229		52	11.05		63.05	75.50
0040	Stainless steel	↓	82	.195	↓	150	9.45		159.45	180

12 36 Countertops

12 36 61 – Simulated Stone Countertops

12 36 61.16 Solid Surface Countertops		Crew	Daily Output	Labor-Hours	Unit	Material	2016 Bare Costs Labor	Equipment	Total	Total Incl O&P
0010	**SOLID SURFACE COUNTERTOPS**, Acrylic polymer									
0020	Pricing for orders of 100 L.F. or greater									
0100	25" wide, solid colors	2 Carp	28	.571	L.F.	54.50	27.50		82	105
0200	Patterned colors		28	.571		69	27.50		96.50	121
0300	Premium patterned colors		28	.571		86.50	27.50		114	140
0400	With silicone attached 4" backsplash, solid colors		27	.593		60	28.50		88.50	113
0500	Patterned colors		27	.593		76	28.50		104.50	131
0600	Premium patterned colors		27	.593		94.50	28.50		123	151
0700	With hard seam attached 4" backsplash, solid colors		23	.696		60	33.50		93.50	121
0800	Patterned colors		23	.696		76	33.50		109.50	139
0900	Premium patterned colors		23	.696		94.50	33.50		128	159
1000	Pricing for order of 51 – 99 L.F.									
1100	25" wide, solid colors	2 Carp	24	.667	L.F.	63	32.50		95.50	122
1200	Patterned colors		24	.667		79.50	32.50		112	141
1300	Premium patterned colors		24	.667		99.50	32.50		132	162
1400	With silicone attached 4" backsplash, solid colors		23	.696		69	33.50		102.50	131
1500	Patterned colors		23	.696		87.50	33.50		121	151
1600	Premium patterned colors		23	.696		109	33.50		142.50	175
1700	With hard seam attached 4" backsplash, solid colors		20	.800		69	39		108	140
1800	Patterned colors		20	.800		87.50	39		126.50	160
1900	Premium patterned colors		20	.800		109	39		148	184
2000	Pricing for order of 1 – 50 L.F.									
2100	25" wide, solid colors	2 Carp	20	.800	L.F.	73.50	39		112.50	145
2200	Patterned colors		20	.800		93.50	39		132.50	167
2300	Premium patterned colors		20	.800		117	39		156	193
2400	With silicone attached 4" backsplash, solid colors		19	.842		81	41		122	156
2500	Patterned colors		19	.842		102	41		143	180
2600	Premium patterned colors		19	.842		128	41		169	207
2700	With hard seam attached 4" backsplash, solid colors		15	1.067		81	51.50		132.50	174
2800	Patterned colors		15	1.067		102	51.50		153.50	198
2900	Premium patterned colors		15	1.067		128	51.50		179.50	225
3000	Sinks, pricing for order of 100 or greater units									
3100	Single bowl, hard seamed, solid colors, 13" x 17"	1 Carp	3	2.667	Ea.	370	129		499	615
3200	10" x 15"		7	1.143		170	55.50		225.50	278
3300	Cutouts for sinks		8	1			48.50		48.50	79
3400	Sinks, pricing for order of 51 – 99 units									
3500	Single bowl, hard seamed, solid colors, 13" x 17"	1 Carp	2.55	3.137	Ea.	425	152		577	715
3600	10" x 15"		6	1.333		196	64.50		260.50	320
3700	Cutouts for sinks		7	1.143			55.50		55.50	90.50
3800	Sinks, pricing for order of 1 – 50 units									
3900	Single bowl, hard seamed, solid colors, 13" x 17"	1 Carp	2	4	Ea.	500	194		694	865
4000	10" x 15"		4.55	1.758		230	85		315	390
4100	Cutouts for sinks		5.25	1.524			74		74	121
4200	Cooktop cutouts, pricing for 100 or greater units		4	2		27	97		124	188
4300	51 – 99 units		3.40	2.353		31.50	114		145.50	221
4400	1 – 50 units		3	2.667		36.50	129		165.50	252

12 36 61.17 Solid Surface Vanity Tops

		Crew	Daily Output	Labor-Hours	Unit	Material	2016 Bare Costs Labor	Equipment	Total	Total Incl O&P
0010	**SOLID SURFACE VANITY TOPS**									
0015	Solid surface, center bowl, 17" x 19"	1 Carp	12	.667	Ea.	190	32.50		222.50	262
0020	19" x 25"		12	.667		194	32.50		226.50	267
0030	19" x 31"		12	.667		227	32.50		259.50	305
0040	19" x 37"		12	.667		264	32.50		296.50	345

For customer support on your Commercial Renovation Cost Data, call 877.742.4719.

379

12 36 Countertops

12 36 61 – Simulated Stone Countertops

12 36 61.17 Solid Surface Vanity Tops		Crew	Daily Output	Labor-Hours	Unit	Material	2016 Bare Costs Labor	Equipment	Total	Total Incl O&P
0050	22" x 25"	1 Carp	10	.800	Ea.	345	39		384	445
0060	22" x 31"		10	.800		405	39		444	510
0070	22" x 37"		10	.800		470	39		509	585
0080	22" x 43"		10	.800		535	39		574	655
0090	22" x 49"		10	.800		595	39		634	720
0110	22" x 55"		8	1		675	48.50		723.50	825
0120	22" x 61"		8	1		770	48.50		818.50	930
0220	Double bowl, 22" x 61"		8	1		870	48.50		918.50	1,050
0230	Double bowl, 22" x 73"		8	1		950	48.50		998.50	1,125
0240	For aggregate colors, add					35%				
0250	For faucets and fittings, see Section 22 41 39.10									

12 36 61.19 Quartz Agglomerate Countertops

		Crew	Daily Output	Labor-Hours	Unit	Material	2016 Bare Costs Labor	Equipment	Total	Total Incl O&P
0010	**QUARTZ AGGLOMERATE COUNTERTOPS**									
0100	25" wide, 4" backsplash, color group A, minimum	2 Carp	15	1.067	L.F.	65	51.50		116.50	156
0110	Maximum		15	1.067		91	51.50		142.50	185
0120	Color group B, minimum		15	1.067		67	51.50		118.50	158
0130	Maximum		15	1.067		95	51.50		146.50	190
0140	Color group C, minimum		15	1.067		78	51.50		129.50	171
0150	Maximum		15	1.067		107	51.50		158.50	203
0160	Color group D, minimum		15	1.067		85	51.50		136.50	178
0170	Maximum		15	1.067		115	51.50		166.50	212

12 48 Rugs and Mats

12 48 13 – Entrance Floor Mats and Frames

12 48 13.13 Entrance Floor Mats

			Crew	Daily Output	Labor-Hours	Unit	Material	2016 Bare Costs Labor	Equipment	Total	Total Incl O&P
0010	**ENTRANCE FLOOR MATS**										
2000	Recycled rubber tire tile, 12" x 12" x 3/8" thick	G	1 Clab	125	.064	S.F.	10.05	2.43		12.48	15
2510	Natural cocoa fiber, 1/2" thick	G		125	.064		8.30	2.43		10.73	13.10
2520	3/4" thick	G		125	.064		8.15	2.43		10.58	12.90
2530	1" thick	G		125	.064		11.15	2.43		13.58	16.25
3000	Hospital tacky mats, package of 30 with frame					Ea.	59			59	65
3010	4 packages of 30					"	85			85	93.50

12 51 Office Furniture

12 51 16 – Case Goods

12 51 16.13 Metal Case Goods

			Crew	Daily Output	Labor-Hours	Unit	Material	2016 Bare Costs Labor	Equipment	Total	Total Incl O&P
0010	**METAL CASE GOODS**										
0020	Desks, 29" high, double pedestal, 30" x 60", metal, minimum					Ea.	645			645	710
0030	Maximum					"	960			960	1,050

12 54 Hospitality Furniture

12 54 13 – Hotel and Motel Furniture

12 54 13.10 Hotel Furniture

		Crew	Daily Output	Labor-Hours	Unit	Material	2016 Bare Costs Labor	Equipment	Total	Total Incl O&P
0010	**HOTEL FURNITURE**									
0020	Standard quality set, minimum				Room	2,525			2,525	2,775
0200	Maximum				"	8,600			8,600	9,475

12 54 16 – Restaurant Furniture

12 54 16.20 Furniture, Restaurant

		Crew	Daily Output	Labor-Hours	Unit	Material	2016 Bare Costs Labor	Equipment	Total	Total Incl O&P
0010	**FURNITURE, RESTAURANT**									
0020	Bars, built-in, front bar	1 Carp	5	1.600	L.F.	285	77.50		362.50	440
0200	Back bar		5	1.600	"	208	77.50		285.50	355
0500	Booth unit, molded plastic, stub wall and 2 seats, minimum		2	4	Set	380	194		574	735
0600	Maximum		1.50	5.333	"	1,425	258		1,683	1,975
0800	Booth seat, upholstered, foursome, single (end) minimum		5	1.600	Ea.	780	77.50		857.50	985
0900	Maximum		4	2		1,000	97		1,097	1,250
1000	Foursome, double, minimum		4	2		1,150	97		1,247	1,425
1100	Maximum		3	2.667		1,775	129		1,904	2,150
1300	Circle booth, upholstered, 1/4 circle, minimum		3	2.667		1,350	129		1,479	1,675
1400	Maximum		2	4		1,700	194		1,894	2,200
1500	3/4 circle, minimum		1.50	5.333		6,450	258		6,708	7,525
1600	Maximum		1	8		6,850	390		7,240	8,150

12 55 Detention Furniture

12 55 13 – Detention Bunks

12 55 13.13 Cots

		Crew	Daily Output	Labor-Hours	Unit	Material	2016 Bare Costs Labor	Equipment	Total	Total Incl O&P
0010	**COTS**									
2500	Bolted, single, painted steel	E-4	20	1.600	Ea.	340	86	7.35	433.35	535
2700	Stainless steel	"	20	1.600	"	985	86	7.35	1,078.35	1,225

12 56 Institutional Furniture

12 56 43 – Dormitory Furniture

12 56 43.10 Dormitory Furnishings

		Crew	Daily Output	Labor-Hours	Unit	Material	2016 Bare Costs Labor	Equipment	Total	Total Incl O&P
0010	**DORMITORY FURNISHINGS**									
0300	Bunkable bed, twin, minimum				Ea.	395			395	430
0320	Maximum				"	585			585	645

12 56 51 – Library Furniture

12 56 51.10 Library Furnishings

		Crew	Daily Output	Labor-Hours	Unit	Material	2016 Bare Costs Labor	Equipment	Total	Total Incl O&P
0010	**LIBRARY FURNISHINGS**									
1710	Carrels, hardwood, 36" x 24", minimum	1 Carp	5	1.600	Ea.	750	77.50		827.50	950
1720	Maximum		4	2		2,025	97		2,122	2,400
1730	Metal, minimum		5	1.600		263	77.50		340.50	415
1740	Maximum		4	2		460	97		557	665
6010	Bookshelf, metal, 90" high, 10" shelf, double face		11.50	.696	L.F.	215	33.50		248.50	291
6020	Single face		12	.667	"	122	32.50		154.50	187
6050	For 8" shelving, subtract from above					10%				
6070	For 42" high with countertop, subtract from above					20%				

For customer support on your Commercial Renovation Cost Data, call 877.742.4719.

381

12 56 Institutional Furniture

12 56 70 – Healthcare Furniture

12 56 70.10 Furniture, Hospital		Crew	Daily Output	Labor-Hours	Unit	Material	2016 Bare Costs Labor	Equipment	Total	Total Incl O&P
0010	**FURNITURE, HOSPITAL**									
0020	Beds, manual, minimum				Ea.	785			785	865
0100	Maximum				"	2,450			2,450	2,700
1100	Patient wall systems, not incl. plumbing, minimum				Room	1,425			1,425	1,575
1200	Maximum				"	1,925			1,925	2,125

12 63 Stadium and Arena Seating

12 63 13 – Stadium and Arena Bench Seating

12 63 13.13 Bleachers

		Crew	Daily Output	Labor-Hours	Unit	Material	2016 Bare Costs Labor	Equipment	Total	Total Incl O&P
0010	**BLEACHERS**									
3000	Telescoping, manual to 15 tier, minimum	F-5	65	.492	Seat	96	24		120	146
3100	Maximum		60	.533		144	26		170	202
3300	16 to 20 tier, minimum		60	.533		231	26		257	297
3400	Maximum		55	.582		288	28.50		316.50	360
3600	21 to 30 tier, minimum		50	.640		240	31.50		271.50	315
3700	Maximum		40	.800		315	39		354	415
3900	For integral power operation, add, minimum	2 Elec	300	.053		48	2.94		50.94	57.50
4000	Maximum	"	250	.064		77	3.53		80.53	90
5000	Benches, folding, in wall, 14' table, 2 benches	L-4	2	12	Set	815	545		1,360	1,775

12 67 Pews and Benches

12 67 13 – Pews

12 67 13.13 Sanctuary Pews

		Crew	Daily Output	Labor-Hours	Unit	Material	2016 Bare Costs Labor	Equipment	Total	Total Incl O&P
0010	**SANCTUARY PEWS**									
1500	Bench type, hardwood, minimum	1 Carp	20	.400	L.F.	101	19.40		120.40	143
1550	Maximum	"	15	.533		176	26		202	236
1570	For kneeler, add					22.50			22.50	24.50

12 93 Interior Public Space Furnishings

12 93 23 – Trash and Litter Receptacles

12 93 23.10 Trash Receptacles

			Crew	Daily Output	Labor-Hours	Unit	Material	2016 Bare Costs Labor	Equipment	Total	Total Incl O&P
0010	**TRASH RECEPTACLES**										
0500	Recycled plastic, var colors, round, 32 gal., 28" x 38" H	G	2 Clab	5	3.200	Ea.	660	121		781	925
0510	32 gal., 31" x 32" H	G		5	3.200		760	121		881	1,025
9110	Plastic, with dome lid, 32 gal. capacity			35	.457		61	17.35		78.35	96
9120	Recycled plastic slats, plastic dome lid, 32 gal. capacity			35	.457		300	17.35		317.35	365

Estimating Tips
General
- The items and systems in this division are usually estimated, purchased, supplied, and installed as a unit by one or more subcontractors. The estimator must ensure that all parties are operating from the same set of specifications and assumptions, and that all necessary items are estimated and will be provided. Many times the complex items and systems are covered, but the more common ones, such as excavation or a crane, are overlooked for the very reason that everyone assumes nobody could miss them. The estimator should be the central focus and be able to ensure that all systems are complete.
- Another area where problems can develop in this division is at the interface between systems. The estimator must ensure, for instance, that anchor bolts, nuts, and washers are estimated and included for the air-supported structures and pre-engineered buildings to be bolted to their foundations. Utility supply is a common area where essential items or pieces of equipment can be missed or overlooked, because each subcontractor may feel it is another's responsibility. The estimator should also be aware of certain items which may be supplied as part of a package but installed by others, and ensure that the installing contractor's estimate includes the cost of installation. Conversely, the estimator must also ensure that items are not costed by two different subcontractors, resulting in an inflated overall estimate.

13 30 00 Special Structures
- The foundations and floor slab, as well as rough mechanical and electrical, should be estimated, as this work is required for the assembly and erection of the structure. Generally, as noted in the data set, the pre-engineered building comes as a shell. Pricing is based on the size and structural design parameters stated in the reference section. Additional features, such as windows and doors with their related structural framing, must also be included by the estimator. Here again, the estimator must have a clear understanding of the scope of each portion of the work and all the necessary interfaces.

Reference Numbers
Reference numbers are shown at the beginning of some major classifications. These numbers refer to related items in the Reference Section. The reference information may be an estimating procedure, an alternate pricing method, or technical information.

Note: Not all subdivisions listed here necessarily appear. ■

Did you know?
RSMeans Online gives you the same access to RSMeans' data with 24/7 access:
- Quickly locate costs in the searchable database.
- Build cost lists, estimates, and reports in minutes.
- Adjust costs to any location in the U.S. and Canada with the click of a button.

Start your free trial today at **www.RSMeansOnline.com**

RSMeans Online
FROM THE GORDIAN GROUP

13 05 05 – Selective Demolition for Special Construction

13 05 05.10 Selective Demolition, Air Supported Structures	Crew	Daily Output	Labor-Hours	Unit	Material	2016 Bare Costs Labor	Equipment	Total	Total Incl O&P
0010	**SELECTIVE DEMOLITION, AIR SUPPORTED STRUCTURES**								
0020	Tank covers, scrim, dbl. layer, vinyl poly w/hdwe., blower & controls								
0050	Round and rectangular	B-2	9000	.004	S.F.		.17		.17	.28
0100	Warehouse structures								
0120	Poly/vinyl fabric, 28 oz., incl. tension cables & inflation system	4 Clab	9000	.004	SF Flr.		.13		.13	.22
0150	Reinforced vinyl, 12 oz., 3000 S.F.	"	5000	.006			.24		.24	.40
0200	12,000 to 24,000 S.F.	8 Clab	20000	.003			.12		.12	.20
0250	Tedlar vinyl fabric, 28 oz. w/liner, to 3000 S.F.	4 Clab	5000	.006			.24		.24	.40
0300	12,000 to 24,000 S.F.	8 Clab	20000	.003			.12		.12	.20
0350	Greenhouse/shelter, woven polyethylene with liner								
0400	3000 S.F.	4 Clab	5000	.006	SF Flr.		.24		.24	.40
0450	12,000 to 24,000 S.F.	8 Clab	20000	.003			.12		.12	.20
0500	Tennis/gymnasium, poly/vinyl fabric, 28 oz., incl. thermal liner	4 Clab	9000	.004			.13		.13	.22
0600	Stadium/convention center, teflon coated fiberglass, incl. thermal liner	9 Clab	40000	.002			.07		.07	.11
0700	Doors, air lock, 15' long, 10' x 10'	2 Carp	1.50	10.667	Ea.		515		515	845
0720	15' x 15'		.80	20			970		970	1,575
0750	Revolving personnel door, 6' diam. x 6'-6" high		1.50	10.667			515		515	845

13 05 05.20 Selective Demolition, Garden Houses

		Crew	Daily Output	Labor-Hours	Unit	Material	Labor	Equipment	Total	Total Incl O&P
0010	**SELECTIVE DEMOLITION, GARDEN HOUSES**									
0020	Prefab, wood, excl foundation, average	2 Clab	400	.040	SF Flr.		1.52		1.52	2.48

13 05 05.25 Selective Demolition, Geodesic Domes

		Crew	Daily Output	Labor-Hours	Unit	Material	Labor	Equipment	Total	Total Incl O&P
0010	**SELECTIVE DEMOLITION, GEODESIC DOMES**									
0050	Shell only, interlocking plywood panels, 30' diameter	F-5	3.20	10	Ea.		490		490	800
0060	34' diameter		2.30	13.913			680		680	1,125
0070	39' diameter		2	16			785		785	1,275
0080	45' diameter	F-3	2.20	18.182			895	299	1,194	1,775
0090	55' diameter		2	20			985	330	1,315	1,950
0100	60' diameter		2	20			985	330	1,315	1,950
0110	65' diameter		1.60	25			1,225	410	1,635	2,450

13 05 05.30 Selective Demolition, Greenhouses

		Crew	Daily Output	Labor-Hours	Unit	Material	Labor	Equipment	Total	Total Incl O&P
0010	**SELECTIVE DEMOLITION, GREENHOUSES**									
0020	Resi-type, free standing, excl. foundations, 9' long x 8' wide	2 Clab	160	.100	SF Flr.		3.79		3.79	6.20
0030	9' long x 11' wide		170	.094			3.57		3.57	5.85
0040	9' long x 14' wide		220	.073			2.76		2.76	4.51
0050	9' long x 17' wide		320	.050			1.90		1.90	3.10
0060	Lean-to type, 4' wide		64	.250			9.50		9.50	15.50
0070	7' wide		120	.133			5.05		5.05	8.25
0080	Geodesic hemisphere, 1/8" plexiglass glazing, 8' diam.		4	4	Ea.		152		152	248
0090	24' diam.		.80	20			760		760	1,250
0100	48' diam.		.40	40			1,525		1,525	2,475

13 05 05.35 Selective Demolition, Hangars

		Crew	Daily Output	Labor-Hours	Unit	Material	Labor	Equipment	Total	Total Incl O&P
0010	**SELECTIVE DEMOLITION, HANGARS**									
0020	T type hangars, prefab, steel , galv roof & walls, incl doors, excl fndtn	E-2	2550	.022	SF Flr.		1.15	.60	1.75	2.63
0030	Circular type, prefab, steel frame, plastic skin, incl foundation, 80' diam	"	.50	112	Total		5,850	3,050	8,900	13,500

13 05 05.45 Selective Demolition, Lightning Protection

		Crew	Daily Output	Labor-Hours	Unit	Material	Labor	Equipment	Total	Total Incl O&P
0010	**SELECTIVE DEMOLITION, LIGHTNING PROTECTION**									
0020	Air terminal & base, copper, 3/8" diam. x 10", to 75' h	1 Clab	16	.500	Ea.		18.95		18.95	31
0030	1/2" diam. x 12", over 75' h		16	.500			18.95		18.95	31
0050	Aluminum, 1/2" diam. x 12", to 75' h		16	.500			18.95		18.95	31
0060	5/8" diam. x 12", over 75' h		16	.500			18.95		18.95	31
0070	Cable, copper, 220 lb. per thousand feet, to 75' high		640	.013	L.F.		.47		.47	.77

R024119-10 (reference, row 0050)

384

For customer support on your Commercial Renovation Cost Data, call 877.742.4719.

13 05 Common Work Results for Special Construction

13 05 05 – Selective Demolition for Special Construction

13 05 05.45 Selective Demolition, Lightning Protection	Crew	Daily Output	Labor-Hours	Unit	Material	2016 Bare Costs Labor	Equipment	Total	Total Incl O&P
0080	375 lb. per thousand feet, over 75' high	1 Clab	460	.017	L.F.		.66		.66	1.08
0090	Aluminum, 101 lb. per thousand feet, to 75' high		560	.014			.54		.54	.89
0100	199 lb. per thousand feet, over 75' high		480	.017			.63		.63	1.03
0110	Arrester, 175 V AC, to ground		16	.500	Ea.		18.95		18.95	31
0120	650 V AC, to ground		13	.615	"		23.50		23.50	38

13 05 05.50 Selective Demolition, Pre-Engineered Steel Buildings

		Crew	Daily Output	Labor-Hours	Unit	Material	Labor	Equipment	Total	Total Incl O&P
0010	**SELECTIVE DEMOLITION, PRE-ENGINEERED STEEL BUILDINGS**									
0500	Pre-engd. steel bldgs., rigid frame, clear span & multi post, excl. salvage									
0550	3,500 to 7,500 S.F.	L-10	1000	.024	SF Flr.		1.29	.66	1.95	2.91
0600	7,501 to 12,500 S.F.		1500	.016			.86	.44	1.30	1.94
0650	12,501 S.F. or greater		1650	.015			.78	.40	1.18	1.77
0700	Pre-engd. steel building components									
0710	Entrance canopy, including frame 4' x 4'	E-24	8	4	Ea.		211	93.50	304.50	465
0720	4' x 8'	"	7	4.571			241	107	348	535
0730	HM doors, self framing, single leaf	2 Skwk	8	2			100		100	162
0740	Double leaf		5	3.200			160		160	259
0760	Gutter, eave type		600	.027	L.F.		1.33		1.33	2.16
0770	Sash, single slide, double slide or fixed		24	.667	Ea.		33.50		33.50	54
0780	Skylight, fiberglass, to 30 S.F.		16	1			50		50	81
0785	Roof vents, circular, 12" to 24" diameter		12	1.333			66.50		66.50	108
0790	Continuous, 10' long		8	2			100		100	162
0900	Shelters, aluminum frame									
0910	Acrylic glazing, 3' x 9' x 8' high	2 Skwk	2	8	Ea.		400		400	645
0920	9' x 12' x 8' high	"	1.50	10.667	"		530		530	860

13 05 05.60 Selective Demolition, Silos

		Crew	Daily Output	Labor-Hours	Unit	Material	Labor	Equipment	Total	Total Incl O&P
0010	**SELECTIVE DEMOLITION, SILOS**									
0020	Conc stave, indstrl, conical/sloping bott, excl fndtn, 12' diam., 35' h	E-24	.18	177	Ea.		9,375	4,150	13,525	20,700
0030	16' diam., 45' h		.12	266			14,000	6,250	20,250	31,100
0040	25' diam., 75' h		.08	400			21,100	9,350	30,450	46,500
0050	Steel, factory fabricated, 30,000 gal. cap, painted or epoxy lined	L-5	2	28			1,500	375	1,875	3,000

13 05 05.65 Selective Demolition, Sound Control

		Crew	Daily Output	Labor-Hours	Unit	Material	Labor	Equipment	Total	Total Incl O&P
0010	**SELECTIVE DEMOLITION, SOUND CONTROL**									
0120	Acoustical enclosure, 4" thick walls & ceiling panels, 8 lb./S.F.	3 Carp	144	.167	SF Surf		8.10		8.10	13.20
0130	10.5 lb./S.F.		128	.188			9.10		9.10	14.85
0140	Reverb chamber, parallel walls, 4" thick		120	.200			9.70		9.70	15.85
0150	Skewed walls, parallel roof, 4" thick		110	.218			10.55		10.55	17.25
0160	Skewed walls/roof, 4" layer/air space		96	.250			12.10		12.10	19.80
0170	Sound-absorbing panels, painted metal, 2'-6" x 8', under 1,000 S.F.		430	.056			2.70		2.70	4.42
0180	Over 1,000 S.F.		480	.050			2.42		2.42	3.96
0190	Flexible transparent curtain, clear	3 Shee	430	.056			3.20		3.20	5.05
0192	50% clear, 50% foam		430	.056			3.20		3.20	5.05
0194	25% clear, 75% foam		430	.056			3.20		3.20	5.05
0196	100% foam		430	.056			3.20		3.20	5.05
0200	Audio-masking sys., incl. speakers, amplfr., signal gnrtr.									
0205	Ceiling mounted, 5,000 S.F.	2 Elec	4800	.003	S.F.		.18		.18	.28
0210	10,000 S.F.		5600	.003			.16		.16	.24
0220	Plenum mounted, 5,000 S.F.		7600	.002			.12		.12	.18
0230	10,000 S.F.		8800	.002			.10		.10	.16

13 05 05.70 Selective Demolition, Special Purpose Rooms

		Crew	Daily Output	Labor-Hours	Unit	Material	Labor	Equipment	Total	Total Incl O&P
0010	**SELECTIVE DEMOLITION, SPECIAL PURPOSE ROOMS**									
0100	Audiometric rooms, under 500 S.F. surface	4 Carp	200	.160	SF Surf		7.75		7.75	12.65
0110	Over 500 S.F. surface	"	240	.133	"		6.45		6.45	10.55

For customer support on your Commercial Renovation Cost Data, call 877.742.4719.

385

13 05 05 – Selective Demolition for Special Construction

13 05 05.70 Selective Demolition, Special Purpose Rooms		Crew	Daily Output	Labor-Hours	Unit	Material	2016 Bare Costs Labor	Equipment	Total	Total Incl O&P
0200	Clean rooms, 12' x 12' soft wall, class 100	1 Carp	.30	26.667	Ea.		1,300		1,300	2,100
0210	Class 1000		.30	26.667			1,300		1,300	2,100
0220	Class 10,000		.35	22.857			1,100		1,100	1,800
0230	Class 100,000		.35	22.857			1,100		1,100	1,800
0300	Darkrooms, shell complete, 8' high	2 Carp	220	.073	SF Flr.		3.52		3.52	5.75
0310	12' high		110	.145	"		7.05		7.05	11.50
0350	Darkrooms doors, mini-cylindrical, revolving		4	4	Ea.		194		194	315
0400	Music room, practice modular		140	.114	SF Surf		5.55		5.55	9.05
0500	Refrigeration structures and finishes									
0510	Wall finish, 2 coat Portland cement plaster, 1/2" thick	1 Clab	200	.040	S.F.		1.52		1.52	2.48
0520	Fiberglass panels, 1/8" thick		400	.020			.76		.76	1.24
0530	Ceiling finish, polystyrene plastic, 1" to 2" thick		500	.016			.61		.61	.99
0540	4" thick		450	.018			.67		.67	1.10
0550	Refrigerator, prefab aluminum walk-in, 7'-6" high, 6' x 6' OD	2 Carp	100	.160	SF Flr.		7.75		7.75	12.65
0560	10' x 10' OD		160	.100			4.85		4.85	7.90
0570	Over 150 S.F.		200	.080			3.88		3.88	6.35
0600	Sauna, prefabricated, including heater & controls, 7' high, to 30 S.F.		120	.133			6.45		6.45	10.55
0610	To 40 S.F.		140	.114			5.55		5.55	9.05
0620	To 60 S.F.		175	.091			4.43		4.43	7.25
0630	To 100 S.F.		220	.073			3.52		3.52	5.75
0640	To 130 S.F.		250	.064			3.10		3.10	5.05
0650	Steam bath, heater, timer, head, single, to 140 C.F.	1 Plum	2.20	3.636	Ea.		215		215	335
0660	To 300 C.F.		2.20	3.636			215		215	335
0670	Steam bath, comm. size, w/blow-down assembly, to 800 C.F.		1.80	4.444			263		263	410
0680	To 2500 C.F.		1.60	5			296		296	460
0690	Steam bath, comm. size, multiple, for motels, apts, 500 C.F., 2 baths		2	4			237		237	370
0700	1,000 C.F., 4 baths		1.40	5.714			340		340	525

13 05 05.75 Selective Demolition, Storage Tanks

		Crew	Daily Output	Labor-Hours	Unit	Material	Labor	Equipment	Total	Total Incl O&P
0010	**SELECTIVE DEMOLITION, STORAGE TANKS**									
0500	Steel tank, single wall, above ground, not incl. fdn., pumps or piping									
0510	Single wall, 275 gallon	Q-1	3	5.333	Ea.		284		284	445
0520	550 thru 2,000 gallon	B-34P	2	12			615	335	950	1,350
0530	5,000 thru 10,000 gallon	B-34Q	2	12			620	640	1,260	1,675
0540	15,000 thru 30,000 gallon	B-34S	2	16			865	1,775	2,640	3,325
0600	Steel tank, double wall, above ground not incl. fdn., pumps & piping									
0620	500 thru 2,000 gallon	B-34P	2	12	Ea.		615	335	950	1,350

13 05 05.85 Selective Demolition, Swimming Pool Equip

		Crew	Daily Output	Labor-Hours	Unit	Material	Labor	Equipment	Total	Total Incl O&P
0010	**SELECTIVE DEMOLITION, SWIMMING POOL EQUIP**									
0020	Diving stand, stainless steel, 3 meter	2 Clab	3	5.333	Ea.		202		202	330
0030	1 meter		5	3.200			121		121	198
0040	Diving board, 16' long, aluminum		5.40	2.963			112		112	184
0050	Fiberglass		5.40	2.963			112		112	184
0070	Ladders, heavy duty, stainless steel, 2 tread		14	1.143			43.50		43.50	71
0080	4 tread		12	1.333			50.50		50.50	82.50
0090	Lifeguard chair, stainless steel, fixed		5	3.200			121		121	198
0100	Slide, tubular, fiberglass, aluminum handrails & ladder, 5', straight		4	4			152		152	248
0110	8', curved		6	2.667			101		101	165
0120	10', curved		3	5.333			202		202	330
0130	12' straight, with platform		2.50	6.400			243		243	395
0140	Removable access ramp, stainless steel		4	4			152		152	248
0150	Removable stairs, stainless steel, collapsible		4	4			152		152	248

13 05 Common Work Results for Special Construction

13 05 05 – Selective Demolition for Special Construction

13 05 05.90 Selective Demolition, Tension Structures	Crew	Daily Output	Labor-Hours	Unit	Material	2016 Bare Costs Labor	Equipment	Total	Total Incl O&P
0010 **SELECTIVE DEMOLITION, TENSION STRUCTURES**									
0020 Steel/alum. frame, fabric shell, 60' clear span, 6,000 S.F.	B-41	2000	.022	SF Flr.		.86	.15	1.01	1.58
0030 12,000 S.F.		2200	.020			.79	.14	.93	1.43
0040 80' clear span, 20,800 S.F.		2440	.018			.71	.12	.83	1.29
0050 100' clear span, 10,000 S.F.	L-5	4350	.013			.69	.17	.86	1.38
0060 26,000 S.F.		4600	.012			.65	.16	.81	1.31
0070 36,000 S.F.		5000	.011			.60	.15	.75	1.20

13 05 05.95 Selective Demo, X-Ray/Radio Freq Protection

	Crew	Daily Output	Labor-Hours	Unit	Material	Labor	Equipment	Total	Total Incl O&P
0010 **SELECTIVE DEMO, X-RAY/RADIO FREQ PROTECTION**									
0020 Shielding lead, lined door frame, excl. hdwe., 1/16" thick	1 Clab	4.80	1.667	Ea.		63		63	103
0030 Lead sheets, 1/16" thick	2 Clab	270	.059	S.F.		2.25		2.25	3.67
0040 1/8" thick		240	.067			2.53		2.53	4.13
0050 Lead shielding, 1/4" thick		270	.059			2.25		2.25	3.67
0060 1/2" thick		240	.067			2.53		2.53	4.13
0070 Lead glass, 1/4" thick, 2.0 mm LE, 12" x 16"	2 Glaz	16	1	Ea.		46.50		46.50	75.50
0080 24" x 36"		8	2			93.50		93.50	151
0090 36" x 60"		4	4			187		187	305
0100 Lead glass window frame, with 1/16" lead & voice passage, 36" x 60"		4	4			187		187	305
0110 Lead glass window frame, 24" x 36"		8	2			93.50		93.50	151
0120 Lead gypsum board, 5/8" thick with 1/16" lead	2 Clab	320	.050	S.F.		1.90		1.90	3.10
0130 1/8" lead		280	.057			2.17		2.17	3.54
0140 1/32" lead		400	.040			1.52		1.52	2.48
0150 Butt joints, 1/8" lead or thicker, 2" x 7' long batten strip		480	.033	Ea.		1.26		1.26	2.06
0160 X-ray protection, average radiography room, up to 300 S.F., 1/16" lead, min		.50	32	Total		1,225		1,225	1,975
0170 Maximum		.30	53.333			2,025		2,025	3,300
0180 Deep therapy X-ray room, 250 kV cap, up to 300 S.F., 1/4" lead, min		.20	80			3,025		3,025	4,950
0190 Maximum		.12	133			5,050		5,050	8,250
0880 Radio frequency shielding, prefab or screen-type copper or steel, minimum		360	.044	SF Surf		1.68		1.68	2.75
0890 Average		310	.052			1.96		1.96	3.20
0895 Maximum		290	.055			2.09		2.09	3.42

13 12 Fountains

13 12 13 – Exterior Fountains

13 12 13.10 Outdoor Fountains

	Crew	Daily Output	Labor-Hours	Unit	Material	Labor	Equipment	Total	Total Incl O&P
0010 **OUTDOOR FOUNTAINS**									
0100 Outdoor fountain, 48" high with bowl and figures	2 Clab	2	8	Ea.	570	305		875	1,125
0200 Commercial, concrete or cast stone, 40-60" H, simple		2	8		760	305		1,065	1,325
0220 Average		2	8		1,300	305		1,605	1,925
0240 Ornate		2	8		2,475	305		2,780	3,225
0260 Metal, 72" high		2	8		965	305		1,270	1,550
0280 90" High		2	8		1,250	305		1,555	1,900
0300 120" High		2	8		5,000	305		5,305	6,000
0320 Resin or fiberglass, 40-60" H, wall type		2	8		475	305		780	1,025
0340 Waterfall type		2	8		1,200	305		1,505	1,800

13 12 23 – Interior Fountains

13 12 23.10 Indoor Fountains

	Crew	Daily Output	Labor-Hours	Unit	Material	Labor	Equipment	Total	Total Incl O&P
0010 **INDOOR FOUNTAINS**									
0100 Commercial, floor type, resin or fiberglass, lighted, cascade type	2 Clab	2	8	Ea.	292	305		597	815
0120 Tiered type		2	8		287	305		592	810
0140 Waterfall type		2	8		330	305		635	860

For customer support on your Commercial Renovation Cost Data, call 877.742.4719.

387

13 17 Tubs and Pools

13 17 13 – Hot Tubs

13 17 13.10 Redwood Hot Tub System	Crew	Daily Output	Labor-Hours	Unit	Material	2016 Bare Costs Labor	Equipment	Total	Total Incl O&P
0010 **REDWOOD HOT TUB SYSTEM**									
7050 4' diameter x 4' deep	Q-1	1	16	Ea.	3,225	850		4,075	4,875
7150 6' diameter x 4' deep		.80	20		4,950	1,075		6,025	7,075
7200 8' diameter x 4' deep		.80	20		7,250	1,075		8,325	9,625

13 17 33 – Whirlpool Tubs

13 17 33.10 Whirlpool Bath	Crew	Daily Output	Labor-Hours	Unit	Material	2016 Bare Costs Labor	Equipment	Total	Total Incl O&P
0010 **WHIRLPOOL BATH**									
6000 Whirlpool, bath with vented overflow, molded fiberglass									
6100 66" x 36" x 24"	Q-1	1	16	Ea.	3,475	850		4,325	5,150
6400 72" x 36" x 21"		1	16		1,800	850		2,650	3,325
6500 60" x 34" x 21"		1	16		1,800	850		2,650	3,300
6600 72" x 42" x 23"		1	16		2,175	850		3,025	3,725
6710 For color, add					10%				
6711 For designer colors and trim, add					25%				

13 21 Controlled Environment Rooms

13 21 26 – Cold Storage Rooms

13 21 26.50 Refrigeration	Crew	Daily Output	Labor-Hours	Unit	Material	2016 Bare Costs Labor	Equipment	Total	Total Incl O&P
0010 **REFRIGERATION**									
0020 Curbs, 12" high, 4" thick, concrete	2 Carp	58	.276	L.F.	5.15	13.35		18.50	27.50
6300 Rule of thumb for complete units, w/o doors & refrigeration, cooler		146	.110	SF Flr.	149	5.30		154.30	173
6400 Freezer		109.60	.146	"	176	7.05		183.05	206

13 24 Special Activity Rooms

13 24 16 – Saunas

13 24 16.50 Saunas and Heaters	Crew	Daily Output	Labor-Hours	Unit	Material	2016 Bare Costs Labor	Equipment	Total	Total Incl O&P
0010 **SAUNAS AND HEATERS**									
0020 Prefabricated, incl. heater & controls, 7' high, 6' x 4', C/C	L-7	2.20	12.727	Ea.	5,325	590		5,915	6,800
1700 Door only, cedar, 2' x 6', with 1' x 4' tempered insulated glass window	2 Carp	3.40	4.706		780	228		1,008	1,225
1800 Prehung, incl. jambs, pulls & hardware	"	12	1.333		760	64.50		824.50	945
2500 Heaters only (incl. above), wall mounted, to 200 C.F.					730			730	805
4480 For additional equipment, see Section 11 66 13.10									

13 24 26 – Steam Baths

13 24 26.50 Steam Baths and Components	Crew	Daily Output	Labor-Hours	Unit	Material	2016 Bare Costs Labor	Equipment	Total	Total Incl O&P
0010 **STEAM BATHS AND COMPONENTS**									
0020 Heater, timer & head, single, to 140 C.F.	1 Plum	1.20	6.667	Ea.	2,200	395		2,595	3,025
0500 To 300 C.F.	"	1.10	7.273		2,425	430		2,855	3,350
2000 Multiple, motels, apts., 2 baths, w/blow-down assm., 500 C.F.	Q-1	1.30	12.308		6,325	655		6,980	8,000
2500 4 baths	"	.70	22.857		10,200	1,225		11,425	13,100

13 28 Athletic and Recreational Special Construction

13 28 33 – Athletic and Recreational Court Walls

13 28 33.50 Sport Court

		Crew	Daily Output	Labor-Hours	Unit	Material	2016 Bare Costs Labor	Equipment	Total	Total Incl O&P
0010	**SPORT COURT**									
0020	Floors, No. 2 & better maple, 25/32" thick				SF Flr.				6.05	6.65
0300	Squash, regulation court in existing building, minimum				Court	36,800			36,800	40,400
0400	Maximum				"	41,000			41,000	45,000
0450	Rule of thumb for components:									
0470	Walls	3 Carp	.15	160	Court	11,000	7,750		18,750	24,800
0500	Floor	"	.25	96		8,725	4,650		13,375	17,200
0550	Lighting	2 Elec	.60	26.667	↓	2,100	1,475		3,575	4,575

13 34 Fabricated Engineered Structures

13 34 13 – Glazed Structures

13 34 13.13 Greenhouses

		Crew	Daily Output	Labor-Hours	Unit	Material	2016 Bare Costs Labor	Equipment	Total	Total Incl O&P
0010	**GREENHOUSES,** Shell only, stock units, not incl. 2' stub walls,									
0020	foundation, floors, heat or compartments									
0300	Residential type, free standing, 8'-6" long x 7'-6" wide	2 Carp	59	.271	SF Flr.	25	13.15		38.15	49
0400	10'-6" wide		85	.188		41	9.10		50.10	60
0600	13'-6" wide		108	.148		43.50	7.20		50.70	59.50
0700	17'-0" wide		160	.100		47	4.85		51.85	59.50
0900	Lean-to type, 3'-10" wide		34	.471		43.50	23		66.50	85
1000	6'-10" wide		58	.276		29	13.35		42.35	54
1050	8'-0" wide	↓	60	.267	↓	63	12.90		75.90	90
1100	Wall mounted to existing window, 3' x 3'	1 Carp	4	2	Ea.	1,600	97		1,697	1,900
1120	4' x 5'	"	3	2.667		1,950	129		2,079	2,350
3900	Cooling, 1200 CFM exhaust fan, add					310			310	340
4000	7850 CFM					1,050			1,050	1,150
4200	For heaters, 10 MBH, add					215			215	237
4300	60 MBH, add					780			780	855
4500	For benches, 2' x 8', add					160			160	176
4600	4' x 10', add				↓	195			195	214
4800	For ventilation & humidity control w/ 4 integrated outlets, add				Total	240			240	264
4900	For environmental controls and automation, 8 outputs, 9 stages, add				"	765			765	840
5100	For humidification equipment, add				Ea.	299			299	330
5200	For vinyl shading, add				S.F.	.50	·		.50	.55

13 34 13.19 Swimming Pool Enclosures

		Crew	Daily Output	Labor-Hours	Unit	Material	2016 Bare Costs Labor	Equipment	Total	Total Incl O&P
0010	**SWIMMING POOL ENCLOSURES** Translucent, free standing									
0020	not including foundations, heat or light									
0200	Economy	2 Carp	200	.080	SF Hor.	33.50	3.88		37.38	43.50
0600	Deluxe	"	70	.229	"	93	11.05		104.05	120

13 34 23 – Fabricated Structures

13 34 23.16 Fabricated Control Booths

		Crew	Daily Output	Labor-Hours	Unit	Material	2016 Bare Costs Labor	Equipment	Total	Total Incl O&P
0010	**FABRICATED CONTROL BOOTHS**									
0100	Guard House, prefab conc. w/bullet resistant doors & windows, roof & wiring									
0110	8' x 8', Level III	L-10	1	24	Ea.	51,000	1,275	660	52,935	59,000
0120	8' x 8', Level IV	"	1	24	"	58,000	1,275	660	59,935	67,000

13 34 23.45 Kiosks

		Crew	Daily Output	Labor-Hours	Unit	Material	2016 Bare Costs Labor	Equipment	Total	Total Incl O&P
0010	**KIOSKS**									
0020	Round, advertising type, 5' diameter, 7' high, aluminum wall, illuminated				Ea.	25,100			25,100	27,600
0100	Aluminum wall, non-illuminated					24,000			24,000	26,500
0500	Rectangular, 5' x 9', 7'-6" high, aluminum wall, illuminated					27,300			27,300	30,000
0600	Aluminum wall, non-illuminated				↓	25,700			25,700	28,200

13 34 Fabricated Engineered Structures

13 34 63 – Natural Fiber Construction

13 34 63.50 Straw Bale Construction		Crew	Daily Output	Labor-Hours	Unit	Material	2016 Bare Costs Labor	Equipment	Total	Total Incl O&P
0010	**STRAW BALE CONSTRUCTION**									
2020	Straw bales in walls w/modified post and beam frame	[G] 2 Carp	320	.050	S.F.	6.55	2.42		8.97	11.15

13 42 Building Modules

13 42 63 – Detention Cell Modules

13 42 63.16 Steel Detention Cell Modules

		Crew	Daily Output	Labor-Hours	Unit	Material	Labor	Equipment	Total	Total Incl O&P
0010	**STEEL DETENTION CELL MODULES**									
2000	Cells, prefab., 5' to 6' wide, 7' to 8' high, 7' to 8' deep,									
2010	bar front, cot, not incl. plumbing	E-4	1.50	21.333	Ea.	9,800	1,150	98	11,048	12,900

13 48 Sound, Vibration, and Seismic Control

13 48 13 – Manufactured Sound and Vibration Control Components

13 48 13.50 Audio Masking

		Crew	Daily Output	Labor-Hours	Unit	Material	Labor	Equipment	Total	Total Incl O&P
0010	**AUDIO MASKING**, acoustical enclosure, 4" thick wall and ceiling									
0020	8# per S.F., up to 12' span	3 Carp	72	.333	SF Surf	33.50	16.15		49.65	63.50
0300	Better quality panels, 10.5# per S.F.		64	.375		38	18.15		56.15	71.50
0400	Reverb-chamber, 4" thick, parallel walls		60	.400		47.50	19.40		66.90	83.50

13 49 Radiation Protection

13 49 13 – Integrated X-Ray Shielding Assemblies

13 49 13.50 Lead Sheets

		Crew	Daily Output	Labor-Hours	Unit	Material	Labor	Equipment	Total	Total Incl O&P
0010	**LEAD SHEETS**									
0300	Lead sheets, 1/16" thick	2 Lath	135	.119	S.F.	10.50	5.60		16.10	20.50
0400	1/8" thick		120	.133		31.50	6.30		37.80	44.50
0500	Lead shielding, 1/4" thick		135	.119		41.50	5.60		47.10	55
0550	1/2" thick		120	.133		73.50	6.30		79.80	91
1200	X-ray protection, average radiography or fluoroscopy									
1210	room, up to 300 S.F. floor, 1/16" lead, economy	2 Lath	.25	64	Total	10,100	3,025		13,125	15,900
1500	7'-0" walls, deluxe	"	.15	106	"	12,200	5,050		17,250	21,400
1600	Deep therapy X-ray room, 250 kV capacity,									
1800	up to 300 S.F. floor, 1/4" lead, economy	2 Lath	.08	200	Total	28,300	9,450		37,750	46,000
1900	7'-0" walls, deluxe		.06	266	"	34,900	12,600		47,500	58,500
1999	Minimum labor/equipment charge		4.50	3.556	Job		168		168	265

13 49 21 – Lead Glazing

13 49 21.50 Lead Glazing

		Crew	Daily Output	Labor-Hours	Unit	Material	Labor	Equipment	Total	Total Incl O&P
0010	**LEAD GLAZING**									
0600	Lead glass, 1/4" thick, 2.0 mm LE, 12" x 16"	2 Glaz	13	1.231	Ea.	380	57.50		437.50	510
0700	24" x 36"	"	8	2	"	1,325	93.50		1,418.50	1,600
2000	X-ray viewing panels, clear lead plastic									
2010	7 mm thick, 0.3 mm LE, 2.3 lb./S.F.	H-3	139	.115	S.F.	248	4.76		252.76	281
2020	12 mm thick, 0.5 mm LE, 3.9 lb./S.F.		82	.195		365	8.05		373.05	415
2030	18 mm thick, 0.8 mm LE, 5.9 lb./S.F.		54	.296		420	12.25		432.25	480
2040	22 mm thick, 1.0 mm LE, 7.2 lb./S.F.		44	.364		545	15.05		560.05	625
2050	35 mm thick, 1.5 mm LE, 11.5 lb./S.F.		28	.571		835	23.50		858.50	960
2060	46 mm thick, 2.0 mm LE, 15.0 lb./S.F.		21	.762		1,100	31.50		1,131.50	1,250
2090	For panels 12 S.F. to 48 S.F., add crating charge				Ea.				50	50

13 49 Radiation Protection

13 49 23 – Integrated RFI/EMI Shielding Assemblies

13 49 23.50 Modular Shielding Partitions	Crew	Daily Output	Labor-Hours	Unit	Material	2016 Bare Costs Labor	Equipment	Total	Total Incl O&P
0010 **MODULAR SHIELDING PARTITIONS**									
4000 X-ray barriers, modular, panels mounted within framework for									
4002 attaching to floor, wall or ceiling, upper portion is clear lead									
4005 plastic window panels 48"H, lower portion is opaque leaded									
4008 steel panels 36"H, structural supports not incl.									
4010 1-section barrier, 36"W x 84"H overall									
4020 0.5 mm LE panels	H-3	6.40	2.500	Ea.	8,375	103		8,478	9,400
4030 0.8 mm LE panels		6.40	2.500		9,050	103		9,153	10,100
4040 1.0 mm LE panels		5.33	3.002		10,600	124		10,724	11,800
4050 1.5 mm LE panels		5.33	3.002		14,100	124		14,224	15,700
4060 2-section barrier, 72"W x 84"H overall									
4070 0.5 mm LE panels	H-3	4	4	Ea.	12,100	165		12,265	13,700
4080 0.8 mm LE panels		4	4		13,500	165		13,665	15,100
4090 1.0 mm LE panels		3.56	4.494		16,500	186		16,686	18,500
5000 1.5 mm LE panels		3.20	5		23,500	207		23,707	26,200
5010 3-section barrier, 108"W x 84"H overall									
5020 0.5 mm LE panels	H-3	3.20	5	Ea.	18,200	207		18,407	20,300
5030 0.8 mm LE panels		3.20	5		20,200	207		20,407	22,500
5040 1.0 mm LE panels		2.67	5.993		24,800	248		25,048	27,600
5050 1.5 mm LE panels		2.46	6.504		35,200	269		35,469	39,200
7000 X-ray barriers, mobile, mounted within framework w/casters on									
7005 bottom, clear lead plastic window panels on upper portion,									
7010 opaque on lower, 30"W x 75"H overall, incl. framework									
7020 24"H upper w/0.5 mm LE, 48"H lower w/0.8 mm LE	1 Carp	16	.500	Ea.	4,025	24		4,049	4,475
7030 48"W x 75"H overall, incl. framework									
7040 36"H upper w/0.5 mm LE, 36"H lower w/0.8 mm LE	1 Carp	16	.500	Ea.	6,350	24		6,374	7,050
7050 36"H upper w/1.0 mm LE, 36"H lower w/1.5 mm LE	"	16	.500	"	7,525	24		7,549	8,325
7060 72"W x 75"H overall, incl. framework									
7070 36"H upper w/0.5 mm LE, 36"H lower w/0.8 mm LE	1 Carp	16	.500	Ea.	7,525	24		7,549	8,325
7080 36"H upper w/1.0 mm LE, 36"H lower w/1.5 mm LE	"	16	.500	"	9,425	24		9,449	10,400

For customer support on your Commercial Renovation Cost Data, call 877.742.4719.

391

Division Notes

		CREW	DAILY OUTPUT	LABOR-HOURS	UNIT	BARE COSTS				TOTAL INCL O&P
						MAT.	LABOR	EQUIP.	TOTAL	

Estimating Tips
General
- Many products in Division 14 will require some type of support or blocking for installation not included with the item itself. Examples are supports for conveyors or tube systems, attachment points for lifts, and footings for hoists or cranes. Add these supports in the appropriate division.

14 10 00 Dumbwaiters
14 20 00 Elevators
- Dumbwaiters and elevators are estimated and purchased in a method similar to buying a car. The manufacturer has a base unit with standard features. Added to this base unit price will be whatever options the owner or specifications require. Increased load capacity, additional vertical travel, additional stops, higher speed, and cab finish options are items to be considered. When developing an estimate for dumbwaiters and elevators, remember that some items needed by the installers may have to be included as part of the general contract.

Examples are:
- □ shaftway
- □ rail support brackets
- □ machine room
- □ electrical supply
- □ sill angles
- □ electrical connections
- □ pits
- □ roof penthouses
- □ pit ladders

Check the job specifications and drawings before pricing.

- Installation of elevators and handicapped lifts in historic structures can require significant additional costs. The associated structural requirements may involve cutting into and repairing finishes, moldings, flooring, etc. The estimator must account for these special conditions.

14 30 00 Escalators and Moving Walks
- Escalators and moving walks are specialty items installed by specialty contractors. There are numerous options associated with these items. For specific options, contact a manufacturer or contractor. In a method similar to estimating dumbwaiters and elevators, you should verify the extent of general contract work and add items as necessary.

14 40 00 Lifts
14 90 00 Other Conveying Equipment
- Products such as correspondence lifts, chutes, and pneumatic tube systems, as well as other items specified in this subdivision, may require trained installers. The general contractor might not have any choice as to who will perform the installation, or when it will be performed. Long lead times are often required for these products, making early decisions in scheduling necessary.

Reference Numbers
Reference numbers are shown at the beginning of some major classifications. These numbers refer to related items in the Reference Section. The reference information may be an estimating procedure, an alternate pricing method, or technical information.

Note: Not all subdivisions listed here necessarily appear. ∎

14 11 Manual Dumbwaiters

14 11 10 – Hand Operated Dumbwaiters

14 11 10.20 Manual Dumbwaiters	Crew	Daily Output	Labor-Hours	Unit	Material	2016 Bare Costs Labor	Equipment	Total	Total Incl O&P
0010 **MANUAL DUMBWAITERS**									
0020 2 stop, hand powered, up to 75 lb. capacity	2 Elev	.75	21.333	Ea.	3,100	1,675		4,775	6,000
0100 76 lb. capacity and up		.50	32	"	6,900	2,525		9,425	11,500
0300 For each additional stop, add		.75	21.333	Stop	1,125	1,675		2,800	3,850

14 12 Electric Dumbwaiters

14 12 10 – Dumbwaiters

14 12 10.10 Electric Dumbwaiters

	Crew	Daily Output	Labor-Hours	Unit	Material	2016 Bare Costs Labor	Equipment	Total	Total Incl O&P
0010 **ELECTRIC DUMBWAITERS**									
0020 2 stop, up to 75 lb. capacity	2 Elev	.13	123	Ea.	7,625	9,725		17,350	23,400
0100 76 lb. capacity and up		.11	145	"	23,000	11,500		34,500	43,000
0600 For each additional stop, add		.54	29.630	Stop	3,400	2,350		5,750	7,375

14 21 Electric Traction Elevators

14 21 13 – Electric Traction Freight Elevators

14 21 13.10 Electric Traction Freight Elevators and Options

		Crew	Daily Output	Labor-Hours	Unit	Material	2016 Bare Costs Labor	Equipment	Total	Total Incl O&P
0010 **ELECTRIC TRACTION FREIGHT ELEVATORS AND OPTIONS**	R142000-10									
0425 Electric freight, base unit, 4000 lb., 200 fpm, 4 stop, std. fin.		2 Elev	.05	320	Ea.	115,000	25,300		140,300	165,500
0450 For 5000 lb. capacity, add	R142000-40					6,000			6,000	6,600
0500 For 6000 lb. capacity, add						14,700			14,700	16,100
0525 For 7000 lb. capacity, add						18,300			18,300	20,200
0550 For 8000 lb. capacity, add						22,900			22,900	25,200
0575 For 10000 lb. capacity, add						31,400			31,400	34,600
0600 For 12000 lb. capacity, add						38,200			38,200	42,000
0625 For 16000 lb. capacity, add						46,300			46,300	51,000
0650 For 20000 lb. capacity, add						51,500			51,500	56,500
0675 For increased speed, 250 fpm, add						17,100			17,100	18,800
0700 300 fpm, geared electric, add						20,700			20,700	22,800
0725 350 fpm, geared electric, add						25,200			25,200	27,700
0750 400 fpm, geared electric, add						29,600			29,600	32,600
0775 500 fpm, gearless electric, add						36,900			36,900	40,600
0800 600 fpm, gearless electric, add						44,100			44,100	48,500
0825 700 fpm, gearless electric, add						52,500			52,500	58,000
0850 800 fpm, gearless electric, add						60,500			60,500	66,500
0875 For class "B" loading, add						4,850			4,850	5,325
0900 For class "C-1" loading, add						7,100			7,100	7,800
0925 For class "C-2" loading, add						8,150			8,150	8,975
0950 For class "C-3" loading, add						10,900			10,900	12,000
0975 For travel over 40 V.L.F., add		2 Elev	7.25	2.207	V.L.F.	655	175		830	995
1000 For number of stops over 4, add		"	.27	59.259	Stop	4,400	4,675		9,075	12,100

14 21 23 – Electric Traction Passenger Elevators

14 21 23.10 Electric Traction Passenger Elevators and Options

	Crew	Daily Output	Labor-Hours	Unit	Material	2016 Bare Costs Labor	Equipment	Total	Total Incl O&P
0010 **ELECTRIC TRACTION PASSENGER ELEVATORS AND OPTIONS**									
1625 Electric pass., base unit, 2000 lb., 200 fpm, 4 stop, std. fin.	2 Elev	.05	320	Ea.	94,000	25,300		119,300	142,500
1650 For 2500 lb. capacity, add					3,950			3,950	4,350
1675 For 3000 lb. capacity, add					4,475			4,475	4,925
1700 For 3500 lb. capacity, add					5,925			5,925	6,500
1725 For 4000 lb. capacity, add					6,975			6,975	7,675
1750 For 4500 lb. capacity, add					9,550			9,550	10,500

For customer support on your Commercial Renovation Cost Data, call 877.742.4719.

14 21 Electric Traction Elevators

14 21 23 – Electric Traction Passenger Elevators

14 21 23.10 Electric Traction Passenger Elevators and Options	Crew	Daily Output	Labor-Hours	Unit	Material	2016 Bare Costs Labor	Equipment	Total	Total Incl O&P
1775 For 5000 lb. capacity, add				Ea.	11,700			11,700	12,800
1800 For increased speed, 250 fpm, geared electric, add					4,875			4,875	5,350
1825 300 fpm, geared electric, add					6,725			6,725	7,375
1850 350 fpm, geared electric, add					8,950			8,950	9,850
1875 400 fpm, geared electric, add					11,700			11,700	12,900
1900 500 fpm, gearless electric, add					29,400			29,400	32,300
1925 600 fpm, gearless electric, add					44,700			44,700	49,200
1950 700 fpm, gearless electric, add					52,500			52,500	58,000
1975 800 fpm, gearless electric, add					58,000			58,000	64,000
2000 For travel over 40 V.L.F., add	2 Elev	7.25	2.207	V.L.F.	715	175		890	1,050
2025 For number of stops over 4, add		.27	59.259	Stop	3,150	4,675		7,825	10,700
2400 Electric hospital, base unit, 4000 lb., 200 fpm, 4 stop, std fin.		.05	320	Ea.	84,000	25,300		109,300	131,500
2425 For 4500 lb. capacity, add					5,775			5,775	6,375
2450 For 5000 lb. capacity, add					7,575			7,575	8,350
2475 For increased speed, 250 fpm, geared electric, add					5,025			5,025	5,525
2500 300 fpm, geared electric, add					7,800			7,800	8,575
2525 350 fpm, geared electric, add					9,000			9,000	9,900
2550 400 fpm, geared electric, add					11,900			11,900	13,100
2575 500 fpm, gearless electric, add					34,200			34,200	37,700
2600 600 fpm, gearless electric, add					50,000			50,000	55,000
2625 700 fpm, gearless electric, add					55,500			55,500	61,000
2650 800 fpm, gearless electric, add					62,000			62,000	68,500
2675 For travel over 40 V.L.F., add	2 Elev	7.25	2.207	V.L.F.	440	175		615	755
2700 For number of stops over 4, add	"	.27	59.259	Stop	4,550	4,675		9,225	12,200

14 21 33 – Electric Traction Residential Elevators

14 21 33.20 Residential Elevators

0010 **RESIDENTIAL ELEVATORS**									
7000 Residential, cab type, 1 floor, 2 stop, economy model	2 Elev	.20	80	Ea.	11,400	6,325		17,725	22,400
7100 Custom model		.10	160		19,300	12,700		32,000	40,700
7200 2 floor, 3 stop, economy model		.12	133		17,000	10,500		27,500	35,000
7300 Custom model		.06	266		27,700	21,100		48,800	63,000

14 24 Hydraulic Elevators

14 24 13 – Hydraulic Freight Elevators

14 24 13.10 Hydraulic Freight Elevators and Options

0010 **HYDRAULIC FREIGHT ELEVATORS AND OPTIONS**									
1025 Hydraulic freight, base unit, 2000 lb., 50 fpm, 2 stop, std. fin.	2 Elev	.10	160	Ea.	81,000	12,700		93,700	109,000
1050 For 2500 lb. capacity, add					3,975			3,975	4,375
1075 For 3000 lb. capacity, add					5,350			5,350	5,875
1100 For 3500 lb. capacity, add					8,600			8,600	9,475
1125 For 4000 lb. capacity, add					9,925			9,925	10,900
1150 For 4500 lb. capacity, add					14,500			14,500	15,900
1175 For 5000 lb. capacity, add					15,500			15,500	17,100
1200 For 6000 lb. capacity, add					16,100			16,100	17,700
1225 For 7000 lb. capacity, add					24,300			24,300	26,700
1250 For 8000 lb. capacity, add					31,400			31,400	34,500
1275 For 10000 lb. capacity, add					38,400			38,400	42,300
1300 For 12000 lb. capacity, add					49,800			49,800	55,000
1325 For 16000 lb. capacity, add					71,000			71,000	78,000
1350 For 20000 lb. capacity, add					82,000			82,000	90,000

For customer support on your Commercial Renovation Cost Data, call 877.742.4719.

395

14 24 Hydraulic Elevators

14 24 13 – Hydraulic Freight Elevators

14 24 13.10 Hydraulic Freight Elevators and Options	Crew	Daily Output	Labor-Hours	Unit	Material	2016 Bare Costs Labor	Equipment	Total	Total Incl O&P
1375	For increased speed, 100 fpm, add				Ea.	1,575			1,575	1,725
1400	125 fpm, add					3,475			3,475	3,825
1425	150 fpm, add					4,775			4,775	5,250
1450	175 fpm, add					7,275			7,275	8,000
1475	For class "B" loading, add					4,575			4,575	5,025
1500	For class "C-1" loading, add					7,050			7,050	7,750
1525	For class "C-2" loading, add					8,125			8,125	8,925
1550	For class "C-3" loading, add					11,200			11,200	12,300
1575	For travel over 20 V.L.F., add	2 Elev	7.25	2.207	V.L.F.	870	175		1,045	1,225
1600	For number of stops over 2, add	"	.27	59.259	Stop	2,200	4,675		6,875	9,625

14 24 23 – Hydraulic Passenger Elevators

14 24 23.10 Hydraulic Passenger Elevators and Options

		Crew	Daily Output	Labor-Hours	Unit	Material	Labor	Equipment	Total	Total Incl O&P
0010	**HYDRAULIC PASSENGER ELEVATORS AND OPTIONS**									
2050	Hyd. pass., base unit, 1500 lb., 100 fpm, 2 stop, std. fin.	2 Elev	.10	160	Ea.	38,600	12,700		51,300	62,000
2075	For 2000 lb. capacity, add					830			830	915
2100	For 2500 lb. capacity, add					2,875			2,875	3,175
2125	For 3000 lb. capacity, add					4,100			4,100	4,500
2150	For 3500 lb. capacity, add					7,225			7,225	7,950
2175	For 4000 lb. capacity, add					8,675			8,675	9,525
2200	For 4500 lb. capacity, add					11,400			11,400	12,500
2225	For 5000 lb. capacity, add					16,100			16,100	17,700
2250	For increased speed, 125 fpm, add					2,025			2,025	2,225
2275	150 fpm, add					2,625			2,625	2,900
2300	175 fpm, add					5,100			5,100	5,625
2325	200 fpm, add					9,750			9,750	10,700
2350	For travel over 12 V.L.F., add	2 Elev	7.25	2.207	V.L.F.	710	175		885	1,050
2375	For number of stops over 2, add		.27	59.259	Stop	985	4,675		5,660	8,300
2725	Hydraulic hospital, base unit, 4000 lb., 100 fpm, 2 stop, std. fin.		.10	160	Ea.	62,500	12,700		75,200	88,000
2775	For 4500 lb. capacity, add					7,025			7,025	7,725
2800	For 5000 lb. capacity, add					10,300			10,300	11,300
2825	For increased speed, 125 fpm, add					2,500			2,500	2,750
2850	150 fpm, add					3,600			3,600	3,975
2875	175 fpm, add					6,050			6,050	6,650
2900	200 fpm, add					8,875			8,875	9,775
2925	For travel over 12 V.L.F., add	2 Elev	7.25	2.207	V.L.F.	535	175		710	860
2950	For number of stops over 2, add	"	.27	59.259	Stop	4,350	4,675		9,025	12,000

14 27 Custom Elevator Cabs and Doors

14 27 13 – Custom Elevator Cab Finishes

14 27 13.10 Cab Finishes

		Crew	Daily Output	Labor-Hours	Unit	Material	Labor	Equipment	Total	Total Incl O&P
0010	**CAB FINISHES**									
3325	Passenger elevator cab finishes (based on 3500 lb. cab size)									
3350	Acrylic panel ceiling				Ea.	755			755	830
3375	Aluminum eggcrate ceiling					865			865	950
3400	Stainless steel doors					3,950			3,950	4,350
3425	Carpet flooring					610			610	670
3450	Epoxy flooring					465			465	510
3475	Quarry tile flooring					875			875	960
3500	Slate flooring					1,600			1,600	1,750
3525	Textured rubber flooring					650			650	715

14 27 Custom Elevator Cabs and Doors

14 27 13 - Custom Elevator Cab Finishes

14 27 13.10 Cab Finishes	Crew	Daily Output	Labor-Hours	Unit	Material	2016 Bare Costs Labor	Equipment	Total	Total Incl O&P
3550	Stainless steel walls				Ea.	4,100			4,100	4,500
3575	Stainless steel returns at door				▼	1,175			1,175	1,275
4450	Hospital elevator cab finishes (based on 3500 lb. cab size)									
4475	Aluminum eggcrate ceiling				Ea.	900			900	990
4500	Stainless steel doors					3,950			3,950	4,350
4525	Epoxy flooring					465			465	510
4550	Quarry tile flooring					875			875	960
4575	Textured rubber flooring					650			650	715
4600	Stainless steel walls					4,550			4,550	5,000
4625	Stainless steel returns at door				▼	960			960	1,050

14 28 Elevator Equipment and Controls

14 28 10 - Elevator Equipment and Control Options

14 28 10.10 Elevator Controls and Doors

		Crew	Daily Output	Labor-Hours	Unit	Material	2016 Bare Costs Labor	Equipment	Total	Total Incl O&P
0010	**ELEVATOR CONTROLS AND DOORS**									
2975	Passenger elevator options									
3000	2 car group automatic controls	2 Elev	.66	24.242	Ea.	4,675	1,925		6,600	8,100
3025	3 car group automatic controls		.44	36.364		8,900	2,875		11,775	14,200
3050	4 car group automatic controls		.33	48.485		17,800	3,825		21,625	25,500
3075	5 car group automatic controls		.26	61.538		32,100	4,875		36,975	42,800
3100	6 car group automatic controls		.22	72.727		64,500	5,750		70,250	79,500
3125	Intercom service		3	5.333		965	420		1,385	1,700
3150	Duplex car selective collective		.66	24.242		8,175	1,925		10,100	11,900
3175	Center opening 1 speed doors		2	8		1,950	635		2,585	3,125
3200	Center opening 2 speed doors		2	8		2,750	635		3,385	4,000
3225	Rear opening doors (opposite front)		2	8		4,225	635		4,860	5,625
3250	Side opening 2 speed doors		2	8		4,350	635		4,985	5,750
3275	Automatic emergency power switching		.66	24.242		1,200	1,925		3,125	4,250
3300	Manual emergency power switching	▼	8	2		515	158		673	810
3625	Hall finishes, stainless steel doors					1,425			1,425	1,575
3650	Stainless steel frames					1,450			1,450	1,600
3675	12 month maintenance contract								3,600	3,950
3700	Signal devices, hall lanterns	2 Elev	8	2		520	158		678	815
3725	Position indicators, up to 3		9.40	1.702		475	135		610	735
3750	Position indicators, per each over 3	▼	32	.500		420	39.50		459.50	520
3775	High speed heavy duty door opener					3,250			3,250	3,575
3800	Variable voltage, O.H. gearless machine, min.	2 Elev	.16	100		34,000	7,900		41,900	49,600
3815	Maximum		.07	228		81,000	18,100		99,100	117,000
3825	Basement installed geared machine		.33	48.485	▼	49,200	3,825		53,025	60,000
3850	Freight elevator options									
3875	Doors, bi-parting	2 Elev	.66	24.242	Ea.	7,550	1,925		9,475	11,300
3900	Power operated door and gate	"	.66	24.242		24,300	1,925		26,225	29,700
3925	Finishes, steel plate floor					1,825			1,825	2,000
3950	14 ga. 1/4" x 4' steel plate walls					2,000			2,000	2,200
3975	12 month maintenance contract								3,600	3,950
4000	Signal devices, hall lanterns	2 Elev	8	2		505	158		663	800
4025	Position indicators, up to 3		9.40	1.702		490	135		625	745
4050	Position indicators, per each over 3		32	.500		425	39.50		464.50	530
4075	Variable voltage basement installed geared machine	▼	.66	24.242	▼	20,000	1,925		21,925	25,000
4100	Hospital elevator options									
4125	2 car group automatic controls	2 Elev	.66	24.242	Ea.	4,850	1,925		6,775	8,300

For customer support on your Commercial Renovation Cost Data, call 877.742.4719.

397

14 28 10 – Elevator Equipment and Control Options

14 28 10.10 Elevator Controls and Doors		Crew	Daily Output	Labor-Hours	Unit	Material	2016 Bare Costs Labor	Equipment	Total	Total Incl O&P
4150	3 car group automatic controls	2 Elev	.44	36.364	Ea.	9,275	2,875		12,150	14,600
4175	4 car group automatic controls		.33	48.485		12,400	3,825		16,225	19,600
4200	5 car group automatic controls		.26	61.538		33,500	4,875		38,375	44,400
4225	6 car group automatic controls		.22	72.727		67,500	5,750		73,250	83,500
4250	Intercom service		3	5.333		1,000	420		1,420	1,750
4275	Duplex car selective collective		.66	24.242		8,100	1,925		10,025	11,900
4300	Center opening 1 speed doors		2	8		1,950	635		2,585	3,125
4325	Center opening 2 speed doors		2	8		2,575	635		3,210	3,825
4350	Rear opening doors (opposite front)		2	8		4,250	635		4,885	5,650
4375	Side opening 2 speed doors		2	8		6,350	635		6,985	7,950
4400	Automatic emergency power switching		.66	24.242		1,175	1,925		3,100	4,225
4425	Manual emergency power switching		8	2		505	158		663	800
4675	Hall finishes, stainless steel doors					1,600			1,600	1,750
4700	Stainless steel frames					1,475			1,475	1,625
4725	12 month maintenance contract								3,600	3,950
4750	Signal devices, hall lanterns	2 Elev	8	2		490	158		648	780
4775	Position indicators, up to 3		9.40	1.702		475	135		610	730
4800	Position indicators, per each over 3		32	.500		415	39.50		454.50	520
4825	High speed heavy duty door opener					3,250			3,250	3,575
4850	Variable voltage, O.H. gearless machine, min.	2 Elev	.16	100		51,500	7,900		59,400	68,500
4865	Maximum		.07	228		79,500	18,100		97,600	115,500
4875	Basement installed geared machine		.33	48.485		20,300	3,825		24,125	28,300
5000	Drilling for piston, casing included, 18" diameter	B-48	80	.700	V.L.F.	62.50	30	36.50	129	158

14 42 13 – Inclined Wheelchair Lifts

14 42 13.10 Inclined Wheelchair Lifts and Stairclimbers

		Crew	Daily Output	Labor-Hours	Unit	Material	2016 Bare Costs Labor	Equipment	Total	Total Incl O&P
0010	**INCLINED WHEELCHAIR LIFTS AND STAIRCLIMBERS**									
7700	Stair climber (chair lift), single seat, minimum	2 Elev	1	16	Ea.	5,325	1,275		6,600	7,825
7800	Maximum		.20	80		7,350	6,325		13,675	17,800
8700	Stair lift, minimum		1	16		14,500	1,275		15,775	17,900
8900	Maximum		.20	80		22,900	6,325		29,225	35,000

14 42 16 – Vertical Wheelchair Lifts

14 42 16.10 Wheelchair Lifts

		Crew	Daily Output	Labor-Hours	Unit	Material	2016 Bare Costs Labor	Equipment	Total	Total Incl O&P
0010	**WHEELCHAIR LIFTS**									
8000	Wheelchair lift, minimum	2 Elev	1	16	Ea.	7,325	1,275		8,600	10,000
8500	Maximum	"	.50	32	"	17,300	2,525		19,825	22,900

14 45 10 – Hydraulic Vehicle Lifts

14 45 10.10 Hydraulic Lifts

		Crew	Daily Output	Labor-Hours	Unit	Material	2016 Bare Costs Labor	Equipment	Total	Total Incl O&P
0010	**HYDRAULIC LIFTS**									
2200	Single post, 8000 lb. capacity	L-4	.40	60	Ea.	5,500	2,725		8,225	10,500
2810	Double post, 6000 lb. capacity		2.67	8.989		7,750	405		8,155	9,175
2815	9000 lb. capacity		2.29	10.480		18,400	475		18,875	21,000
2820	15,000 lb. capacity		2	12		20,800	545		21,345	23,800
2822	Four post, 26,000 lb. capacity		1.80	13.333		13,000	605		13,605	15,300
2825	30,000 lb. capacity		1.60	15		46,000	680		46,680	51,500
2830	Ramp style, 4 post, 25,000 lb. capacity		2	12		18,000	545		18,545	20,700

14 45 Vehicle Lifts

14 45 10 – Hydraulic Vehicle Lifts

14 45 10.10 Hydraulic Lifts		Crew	Daily Output	Labor-Hours	Unit	Material	2016 Bare Costs Labor	Equipment	Total	Total Incl O&P
2835	35,000 lb. capacity	L-4	1	24	Ea.	84,000	1,075		85,075	94,500
2840	50,000 lb. capacity		1	24		94,000	1,075		95,075	105,000
2845	75,000 lb. capacity	↓	1	24		109,000	1,075		110,075	122,000
2850	For drive thru tracks, add, minimum					1,050			1,050	1,150
2855	Maximum					1,800			1,800	1,975
2860	Ramp extensions, 3' (set of 2)					950			950	1,050
2865	Rolling jack platform					3,300			3,300	3,625
2870	Electric/hydraulic jacking beam					8,825			8,825	9,725
2880	Scissor lift, portable, 6000 lb. capacity				↓	8,675			8,675	9,525

14 91 Facility Chutes

14 91 33 – Laundry and Linen Chutes

14 91 33.10 Chutes

		Crew	Daily Output	Labor-Hours	Unit	Material	Labor	Equipment	Total	Total Incl O&P
0011	**CHUTES**, linen, trash or refuse									
0050	Aluminized steel, 16 ga., 18" diameter	2 Shee	3.50	4.571	Floor	1,700	262		1,962	2,300
0100	24" diameter		3.20	5		1,775	286		2,061	2,400
0400	Galvanized steel, 16 ga., 18" diameter		3.50	4.571		1,000	262		1,262	1,525
0500	24" diameter		3.20	5		1,125	286		1,411	1,700
0800	Stainless steel, 18" diameter		3.50	4.571		3,000	262		3,262	3,725
0900	24" diameter	↓	3.20	5	↓	3,150	286		3,436	3,900

14 91 82 – Trash Chutes

14 91 82.10 Trash Chutes and Accessories

		Crew	Daily Output	Labor-Hours	Unit	Material	Labor	Equipment	Total	Total Incl O&P
0010	**TRASH CHUTES AND ACCESSORIES**									
9000	Minimum labor/equipment charge	1 Shee	1	8	Job		460		460	725

14 92 Pneumatic Tube Systems

14 92 10 – Conventional, Automatic and Computer Controlled Pneumatic Tube Systems

14 92 10.10 Pneumatic Tube Systems

		Crew	Daily Output	Labor-Hours	Unit	Material	Labor	Equipment	Total	Total Incl O&P
0010	**PNEUMATIC TUBE SYSTEMS**									
0020	100' long, single tube, 2 stations, stock									
0100	3" diameter	2 Stpi	.12	133	Total	3,375	8,100		11,475	16,300
0300	4" diameter	"	.09	177	"	4,275	10,800		15,075	21,500
0400	Twin tube, two stations or more, conventional system									
0700	3" round	2 Stpi	46	.348	L.F.	38	21		59	75
1050	Add for blower		2	8	System	5,200	485		5,685	6,475
1110	Plus for each round station, add		7.50	2.133	Ea.	1,350	129		1,479	1,700
1200	Alternate pricing method: base cost, economy model		.75	21.333	Total	5,750	1,300		7,050	8,350
1300	Custom model		.25	64	"	11,500	3,875		15,375	18,800
1500	Plus total system length, add, for economy model		93.40	.171	L.F.	8.25	10.40		18.65	25.50
1600	For custom model	↓	37.60	.426	"	25	26		51	68

For customer support on your Commercial Renovation Cost Data, call 877.742.4719.

399

Division Notes

	CREW	DAILY OUTPUT	LABOR-HOURS	UNIT	BARE COSTS				TOTAL INCL O&P
					MAT.	LABOR	EQUIP.	TOTAL	

Estimating Tips

Pipe for fire protection and all uses is located in Subdivisions 21 11 13 and 22 11 13.

The labor adjustment factors listed in Subdivision 22 01 02.20 also apply to Division 21.

Many, but not all, areas in the U.S. require backflow protection in the fire system. It is advisable to check local building codes for specific requirements.

For your reference, the following is a list of the most applicable Fire Codes and Standards which may be purchased from the NFPA, 1 Batterymarch Park, Quincy, MA 02169-7471.

- NFPA 1: Uniform Fire Code
- NFPA 10: Portable Fire Extinguishers
- NFPA 11: Low-, Medium-, and High-Expansion Foam
- NFPA 12: Carbon Dioxide Extinguishing Systems (Also companion 12A)
- NFPA 13: Installation of Sprinkler Systems (Also companion 13D, 13E, and 13R)
- NFPA 14: Installation of Standpipe and Hose Systems
- NFPA 15: Water Spray Fixed Systems for Fire Protection
- NFPA 16: Installation of Foam-Water Sprinkler and Foam-Water Spray Systems
- NFPA 17: Dry Chemical Extinguishing Systems (Also companion 17A)
- NFPA 18: Wetting Agents
- NFPA 20: Installation of Stationary Pumps for Fire Protection
- NFPA 22: Water Tanks for Private Fire Protection
- NFPA 24: Installation of Private Fire Service Mains and their Appurtenances
- NFPA 25: Inspection, Testing and Maintenance of Water-Based Fire Protection

Reference Numbers

Reference numbers are shown at the beginning of some major classifications. These numbers refer to related items in the Reference Section. The reference information may be an estimating procedure, an alternate pricing method, or technical information.

Note: Not all subdivisions listed here necessarily appear. ■

21 05 Common Work Results for Fire Suppression

21 05 23 – General-Duty Valves for Water-Based Fire-Suppression Piping

21 05 23.50 General-Duty Valves	Crew	Daily Output	Labor-Hours	Unit	Material	2016 Bare Costs Labor	Equipment	Total	Total Incl O&P
0010 **GENERAL-DUTY VALVES**, for water-based fire suppression									
6500 Check, swing, C.I. body, brass fittings, auto. ball drip									
6520 4" size	Q-12	3	5.333	Ea.	370	277		647	840
9990 Minimum labor/equipment charge	1 Spri	3	2.667	Job		154		154	241

21 11 Facility Fire-Suppression Water-Service Piping

21 11 16 – Facility Fire Hydrants

21 11 16.50 Fire Hydrants for Buildings

	Crew	Daily Output	Labor-Hours	Unit	Material	Labor	Equipment	Total	Total Incl O&P
0010 **FIRE HYDRANTS FOR BUILDINGS**									
3750 Hydrants, wall, w/caps, single, flush, polished brass									
3800 2-1/2" x 2-1/2"	Q-12	5	3.200	Ea.	224	166		390	505
4350 Double, projecting, polished brass									
4400 2-1/2" x 2-1/2" x 4"	Q-12	5	3.200	Ea.	267	166		433	555

21 11 19 – Fire-Department Connections

21 11 19.50 Connections for the Fire-Department

	Crew	Daily Output	Labor-Hours	Unit	Material	Labor	Equipment	Total	Total Incl O&P
0010 **CONNECTIONS FOR THE FIRE-DEPARTMENT**									
7140 Standpipe connections, wall, w/plugs & chains									
7280 Double, flush, polished brass									
7300 2-1/2" x 2-1/2" x 4"	Q-12	5	3.200	Ea.	545	166		711	860
9990 Minimum labor/equipment charge	1 Spri	4	2	Job		115		115	180

21 12 Fire-Suppression Standpipes

21 12 13 – Fire-Suppression Hoses and Nozzles

21 12 13.50 Fire Hoses and Nozzles

	Crew	Daily Output	Labor-Hours	Unit	Material	Labor	Equipment	Total	Total Incl O&P
0010 **FIRE HOSES AND NOZZLES**									
0200 Adapters, rough brass, straight hose threads									
0220 One piece, female to male, rocker lugs									
0240 1" x 1"				Ea.	50.50			50.50	56
0320 2" x 2"					42			42	46
0380 2-1/2" x 2-1/2"					19			19	21
2200 Hose, less couplings									
2260 Synthetic jacket, lined, 300 lb. test, 1-1/2" diameter	Q-12	2600	.006	L.F.	3.40	.32		3.72	4.24
2280 2-1/2" diameter		2200	.007		5.85	.38		6.23	7.05
2360 High strength, 500 lb. test, 1-1/2" diameter		2600	.006		3.51	.37		3.83	4.36
2380 2-1/2" diameter		2200	.007		6.20	.38		6.58	7.45
5600 Nozzles, brass									
5620 Adjustable fog, 3/4" booster line				Ea.	119			119	131
5640 1-1/2" leader line					80			80	88
5660 2-1/2" direct connection					164			164	180
5680 2-1/2" playpipe nozzle					220			220	242
5780 For chrome plated, add					8%				
9900 Minimum labor/equipment charge	1 Plum	2	4	Job		237		237	370

21 12 19 – Fire-Suppression Hose Racks

21 12 19.50 Fire Hose Racks

	Crew	Daily Output	Labor-Hours	Unit	Material	Labor	Equipment	Total	Total Incl O&P
0010 **FIRE HOSE RACKS**									
2600 Hose rack, swinging, for 1-1/2" diameter hose,									
2620 Enameled steel, 50' & 75' lengths of hose	Q-12	20	.800	Ea.	64	41.50		105.50	136

21 12 Fire-Suppression Standpipes

21 12 23 – Fire-Suppression Hose Valves

21 12 23.70 Fire Hose Valves

		Crew	Daily Output	Labor-Hours	Unit	Material	2016 Bare Costs Labor	2016 Bare Costs Equipment	Total	Total Incl O&P
0010	**FIRE HOSE VALVES**									
3000	Gate, hose, wheel handle, N.R.S., rough brass, 1-1/2"	1 Spri	12	.667	Ea.	141	38.50		179.50	215
3040	2-1/2", 300 lb.	"	7	1.143		199	66		265	320
3080	For polished brass, add					40%				
3090	For polished chrome, add					50%				

21 13 Fire-Suppression Sprinkler Systems

21 13 13 – Wet-Pipe Sprinkler Systems

21 13 13.50 Wet-Pipe Sprinkler System Components

		Crew	Daily Output	Labor-Hours	Unit	Material	2016 Bare Costs Labor	2016 Bare Costs Equipment	Total	Total Incl O&P
0010	**WET-PIPE SPRINKLER SYSTEM COMPONENTS**									
1220	Water motor, complete with gong	1 Spri	4	2	Ea.	435	115		550	660
1800	Firecycle system, controls, includes panel,									
1820	batteries, solenoid valves and pressure switches	Q-13	1	32	Ea.	22,000	1,750		23,750	27,000
1860	Detector	1 Spri	16	.500	"	705	29		734	820
1900	Flexible sprinkler head connectors									
1910	Braided stainless steel hose with mounting bracket									
1920	1/2" and 3/4" outlet size									
1940	40" length	1 Spri	30	.267	Ea.	74	15.35		89.35	106
1960	60" length	"	22	.364	"	85	21		106	127
1982	May replace hard-pipe armovers									
1984	For wet, pre-action, deluge or dry pipe systems									
2600	Sprinkler heads, not including supply piping									
3700	Standard spray, pendent or upright, brass, 135°F to 286°F									
3740	1/2" NPT, 1/2" orifice	1 Spri	16	.500	Ea.	10.40	29		39.40	56.50
3790	For riser & feeder piping, see Section 22 11 13.00									

21 13 16 – Dry-Pipe Sprinkler Systems

21 13 16.50 Dry-Pipe Sprinkler System Components

		Crew	Daily Output	Labor-Hours	Unit	Material	2016 Bare Costs Labor	2016 Bare Costs Equipment	Total	Total Incl O&P
0010	**DRY-PIPE SPRINKLER SYSTEM COMPONENTS**									
0600	Accelerator	1 Spri	8	1	Ea.	795	57.50		852.50	965
0800	Air compressor for dry pipe system, automatic, complete									
0820	30 gal. system capacity, 3/4 HP	1 Spri	1.30	6.154	Ea.	810	355		1,165	1,450
2600	Sprinkler heads, not including supply piping									
2640	Dry, pendent, 1/2" orifice, 3/4" or 1" NPT									
2660	1/2" to 6" length	1 Spri	14	.571	Ea.	132	33		165	197
2700	15-1/4" to 18" length	"	14	.571	"	154	33		187	221
6330	Valves and components									
8200	Dry pipe valve, incl. trim and gauges, 3" size	Q-12	2	8	Ea.	2,700	415		3,115	3,625
8220	4" size	"	1	16	"	3,000	830		3,830	4,600

21 13 39 – Foam-Water Systems

21 13 39.50 Foam-Water System Components

		Crew	Daily Output	Labor-Hours	Unit	Material	2016 Bare Costs Labor	2016 Bare Costs Equipment	Total	Total Incl O&P
0010	**FOAM-WATER SYSTEM COMPONENTS**									
2600	Sprinkler heads, not including supply piping									
3600	Foam-water, pendent or upright, 1/2" NPT	1 Spri	12	.667	Ea.	216	38.50		254.50	298

For customer support on your Commercial Renovation Cost Data, call 877.742.4719.

403

Division Notes

	CREW	DAILY OUTPUT	LABOR-HOURS	UNIT	BARE COSTS				TOTAL INCL O&P
					MAT.	LABOR	EQUIP.	TOTAL	

Estimating Tips

22 10 00 Plumbing Piping and Pumps

This subdivision is primarily basic pipe and related materials. The pipe may be used by any of the mechanical disciplines, i.e., plumbing, fire protection, heating, and air conditioning.

Note: CPVC plastic piping approved for fire protection is located in 21 11 13.

- The labor adjustment factors listed in Subdivision 22 01 02.20 apply throughout Divisions 21, 22, and 23. CAUTION: the correct percentage may vary for the same items. For example, the percentage add for the basic pipe installation should be based on the maximum height that the craftsman must install for that particular section. If the pipe is to be located 14' above the floor but it is suspended on threaded rod from beams, the bottom flange of which is 18' high (4' rods), then the height is actually 18' and the add is 20%. The pipe coverer, however, does not have to go above the 14', and so the add should be 10%.

- Most pipe is priced first as straight pipe with a joint (coupling, weld, etc.) every 10' and a hanger usually every 10'. There are exceptions with hanger spacing such as for cast iron pipe (5') and plastic pipe (3 per 10'). Following each type of pipe there are several lines listing sizes and the amount to be subtracted to delete couplings and hangers. This is for pipe that is to be buried or supported together on trapeze hangers. The reason that the couplings are deleted is that these runs are usually long, and frequently longer lengths of pipe are used. By deleting the couplings, the estimator is expected to look up and add back the correct reduced number of couplings.

- When preparing an estimate, it may be necessary to approximate the fittings. Fittings usually run between 25% and 50% of the cost of the pipe. The lower percentage is for simpler runs, and the higher number is for complex areas, such as mechanical rooms.

- For historic restoration projects, the systems must be as invisible as possible, and pathways must be sought for pipes, conduit, and ductwork. While installations in accessible spaces (such as basements and attics) are relatively straightforward to estimate, labor costs may be more difficult to determine when delivery systems must be concealed.

22 40 00 Plumbing Fixtures

- Plumbing fixture costs usually require two lines: the fixture itself and its "rough-in, supply, and waste."

- In the Assemblies Section (Plumbing D2010) for the desired fixture, the System Components Group at the center of the page shows the fixture on the first line. The rest of the list (fittings, pipe, tubing, etc.) will total up to what we refer to in the Unit Price section as "Rough-in, supply, waste, and vent." Note that for most fixtures we allow a nominal 5' of tubing to reach from the fixture to a main or riser.

- Remember that gas- and oil-fired units need venting.

Reference Numbers

Reference numbers are shown at the beginning of some major classifications. These numbers refer to related items in the Reference Section. The reference information may be an estimating procedure, an alternate pricing method, or technical information.

Note: Not all subdivisions listed here necessarily appear. ∎

22 01 02.20 Labor Adjustment Factors		Crew	Daily Output	Labor-Hours	Unit	Material	2016 Bare Costs Labor	Equipment	Total	Total Incl O&P
0010	**LABOR ADJUSTMENT FACTORS** (For Div. 21, 22 and 23) R220102-20									
0100	Labor factors, The below are reasonable suggestions, however									
0110	each project must be evaluated for its own peculiarities, and									
0120	the adjustments be increased or decreased depending on the									
0130	severity of the special conditions.									
1000	Add to labor for elevated installation (Above floor level)									
1080	10' to 14.5' high						10%			
1100	15' to 19.5' high						20%			
1120	20' to 24.5' high						25%			
1140	25' to 29.5' high						35%			
1160	30' to 34.5' high						40%			
1180	35' to 39.5' high						50%			
1200	40' and higher						55%			
2000	Add to labor for crawl space									
2100	3' high						40%			
2140	4' high						30%			
3000	Add to labor for multi-story building									
3010	For new construction (No elevator available)									
3100	Add for floors 3 thru 10						5%			
3110	Add for floors 11 thru 15						10%			
3120	Add for floors 16 thru 20						15%			
3130	Add for floors 21 thru 30						20%			
3140	Add for floors 31 and up						30%			
3170	For existing structure (Elevator available)									
3180	Add for work on floor 3 and above						2%			
4000	Add to labor for working in existing occupied buildings									
4100	Hospital						35%			
4140	Office building						25%			
4180	School						20%			
4220	Factory or warehouse						15%			
4260	Multi dwelling						15%			
5000	Add to labor, miscellaneous									
5100	Cramped shaft						35%			
5140	Congested area						15%			
5180	Excessive heat or cold						30%			
9000	Labor factors, The above are reasonable suggestions, however									
9010	each project should be evaluated for its own peculiarities.									
9100	Other factors to be considered are:									
9140	Movement of material and equipment through finished areas									
9180	Equipment room									
9220	Attic space									
9260	No service road									
9300	Poor unloading/storage area									
9340	Congested site area/heavy traffic									

22 05 Common Work Results for Plumbing

22 05 05 – Selective Demolition for Plumbing

22 05 05.10 Plumbing Demolition

		Crew	Daily Output	Labor-Hours	Unit	Material	2016 Bare Costs Labor	Equipment	Total	Total Incl O&P
0010	**PLUMBING DEMOLITION** R220105-10									
1020	Fixtures, including 10' piping									
1100	Bathtubs, cast iron R024119-10	1 Plum	4	2	Ea.		118		118	185
1120	Fiberglass		6	1.333			79		79	123
1140	Steel		5	1.600			94.50		94.50	148
1200	Lavatory, wall hung		10	.800			47.50		47.50	74
1220	Counter top		8	1			59		59	92.50
1300	Sink, single compartment		8	1			59		59	92.50
1320	Double compartment		7	1.143			67.50		67.50	105
1400	Water closet, floor mounted		8	1			59		59	92.50
1420	Wall mounted		7	1.143			67.50		67.50	105
1500	Urinal, floor mounted		4	2			118		118	185
1520	Wall mounted		7	1.143			67.50		67.50	105
1600	Water fountains, free standing		8	1			59		59	92.50
1620	Wall or deck mounted		6	1.333			79		79	123
2000	Piping, metal, up thru 1-1/2" diameter		200	.040	L.F.		2.37		2.37	3.69
2050	2" thru 3-1/2" diameter		150	.053			3.16		3.16	4.92
2100	4" thru 6" diameter	2 Plum	100	.160			9.45		9.45	14.75
2150	8" thru 14" diameter	.''	60	.267			15.80		15.80	24.50
2153	16" thru 20" diameter	Q-18	70	.343			19.40	.83	20.23	31.50
2155	24" thru 26" diameter		55	.436			24.50	1.06	25.56	39.50
2156	30" thru 36" diameter		40	.600			34	1.45	35.45	54.50
2160	Plastic pipe with fittings, up thru 1-1/2" diameter	1 Plum	250	.032			1.89		1.89	2.95
2162	2" thru 3" diameter	''	200	.040			2.37		2.37	3.69
2164	4" thru 6" diameter	Q-1	200	.080			4.26		4.26	6.65
2166	8" thru 14" diameter		150	.107			5.70		5.70	8.85
2168	16" diameter		100	.160			8.50		8.50	13.30
2250	Water heater, 40 gal.	1 Plum	6	1.333	Ea.		79		79	123
6000	Remove and reset fixtures, easy access		6	1.333			79		79	123
6100	Difficult access		4	2			118		118	185
9000	Minimum labor/equipment charge		2	4	Job		237		237	370

22 05 23 – General-Duty Valves for Plumbing Piping

22 05 23.20 Valves, Bronze

		Crew	Daily Output	Labor-Hours	Unit	Material	2016 Bare Costs Labor	Equipment	Total	Total Incl O&P
0010	**VALVES, BRONZE**									
1020	Angle, 150 lb., rising stem, threaded									
1070	3/4"	1 Plum	20	.400	Ea.	203	23.50		226.50	261
1080	1"		19	.421		293	25		318	365
1100	1-1/2"		13	.615		495	36.50		531.50	595
1102	Soldered same price as threaded									
1110	2"	1 Plum	11	.727	Ea.	795	43		838	940
1300	Ball									
1398	Threaded, 150 psi									
1460	3/4"	1 Plum	20	.400	Ea.	23.50	23.50		47	62.50
1522	Solder the same price as threaded									
1750	Check, swing, class 150, regrinding disc, threaded									
1850	1/2"	1 Plum	24	.333	Ea.	79.50	19.75		99.25	119
1860	3/4"		20	.400		105	23.50		128.50	153
1870	1"		19	.421		152	25		177	206
1890	1-1/2"		13	.615		255	36.50		291.50	340
1900	2"		11	.727		375	43		418	480
2850	Gate, N.R.S., soldered, 125 psi									
2920	1/2"	1 Plum	24	.333	Ea.	56.50	19.75		76.25	93

For customer support on your Commercial Renovation Cost Data, call 877.742.4719.

407

22 05 23 – General-Duty Valves for Plumbing Piping

22 05 23.20 Valves, Bronze

		Crew	Daily Output	Labor-Hours	Unit	Material	2016 Bare Costs Labor	Equipment	Total	Total Incl O&P
2940	3/4"	1 Plum	20	.400	Ea.	66	23.50		89.50	110
2950	1"		19	.421		81.50	25		106.50	129
2970	1-1/2"		13	.615		152	36.50		188.50	224
2980	2"		11	.727		198	43		241	284
4850	Globe, class 150, rising stem, threaded									
4950	1/2"	1 Plum	24	.333	Ea.	101	19.75		120.75	142
4960	3/4"	"	20	.400	"	103	23.50		126.50	151
5600	Relief, pressure & temperature, self-closing, ASME, threaded									
5640	3/4"	1 Plum	28	.286	Ea.	206	16.90		222.90	254
5650	1"		24	.333		330	19.75		349.75	395
5660	1-1/4"		20	.400		655	23.50		678.50	755
5670	1-1/2"		18	.444		1,300	26.50		1,326.50	1,475
5680	2"		16	.500		1,325	29.50		1,354.50	1,500
6400	Pressure, water, ASME, threaded									
6440	3/4"	1 Plum	28	.286	Ea.	181	16.90		197.90	226
6450	1"		24	.333		278	19.75		297.75	335
6470	1-1/2"		18	.444		610	26.50		636.50	710
6900	Reducing, water pressure									
6920	300 psi to 25-75 psi, threaded or sweat									
6940	1/2"	1 Plum	24	.333	Ea.	425	19.75		444.75	495
6950	3/4"		20	.400		435	23.50		458.50	510
6960	1"		19	.421		670	25		695	780
8350	Tempering, water, sweat connections									
8400	1/2"	1 Plum	24	.333	Ea.	102	19.75		121.75	143
8440	3/4"	"	20	.400	"	136	23.50		159.50	187
8650	Threaded connections									
8700	1/2"	1 Plum	24	.333	Ea.	151	19.75		170.75	197
8740	3/4"	"	20	.400	"	825	23.50		848.50	945
8800	Water heater water & gas safety shut off									
8810	Protection against a leaking water heater									
8814	Shut off valve	1 Plum	16	.500	Ea.	171	29.50		200.50	234
8818	Water heater dam		32	.250		25	14.80		39.80	50.50
8822	Gas control wiring harness		32	.250		21.50	14.80		36.30	47
8830	Whole house flood safety shut off									
8834	Connections									
8838	3/4" NPT	1 Plum	12	.667	Ea.	960	39.50		999.50	1,100
8842	1" NPT		11	.727		985	43		1,028	1,150
8846	1-1/4" NPT		10	.800		1,025	47.50		1,072.50	1,200
9000	Minimum labor/equipment charge		4	2	Job		118		118	185

22 05 23.60 Valves, Plastic

		Crew	Daily Output	Labor-Hours	Unit	Material	2016 Bare Costs Labor	Equipment	Total	Total Incl O&P
0010	**VALVES, PLASTIC**									
1150	Ball, PVC, socket or threaded, true union									
1230	1/2"	1 Plum	26	.308	Ea.	42.50	18.20		60.70	75.50
1240	3/4"		25	.320		52	18.95		70.95	86.50
1250	1"		23	.348		62	20.50		82.50	100
1260	1-1/4"		21	.381		88	22.50		110.50	132
1270	1-1/2"		20	.400		98.50	23.50		122	145
1280	2"		17	.471		133	28		161	191
1650	CPVC, socket or threaded, single union									
1700	1/2"	1 Plum	26	.308	Ea.	61	18.20		79.20	96
1720	3/4"		25	.320		77	18.95		95.95	114
1730	1"		23	.348		92.50	20.50		113	134

22 05 23 – General-Duty Valves for Plumbing Piping

22 05 23.60 Valves, Plastic		Crew	Daily Output	Labor-Hours	Unit	Material	2016 Bare Costs Labor	2016 Bare Costs Equipment	Total	Total Incl O&P
1750	1-1/4"	1 Plum	21	.381	Ea.	147	22.50		169.50	197
1760	1-1/2"		20	.400		147	23.50		170.50	199
1770	2"	↓	17	.471	↓	203	28		231	267
3150	Ball check, PVC, socket or threaded									
3290	2"	1 Plum	17	.471	Ea.	154	28		182	213
9000	Minimum labor/equipment charge	"	3.50	2.286	Job		135		135	211

22 05 76 – Facility Drainage Piping Cleanouts

22 05 76.10 Cleanouts

		Crew	Daily Output	Labor-Hours	Unit	Material	2016 Bare Costs Labor	2016 Bare Costs Equipment	Total	Total Incl O&P
0010	**CLEANOUTS**									
0060	Floor type									
0080	Round or square, scoriated nickel bronze top									
0100	2" pipe size	1 Plum	10	.800	Ea.	197	47.50		244.50	291
0120	3" pipe size		8	1		295	59		354	420
0140	4" pipe size	↓	6	1.333	↓	295	79		374	450
0980	Round top, recessed for terrazzo									
1000	2" pipe size	1 Plum	9	.889	Ea.	197	52.50		249.50	299
1080	3" pipe size		6	1.333		295	79		374	450
1100	4" pipe size	↓	4	2		295	118		413	510
1120	5" pipe size	Q-1	6	2.667	↓	375	142		517	635
9000	Minimum labor/equipment charge	1 Plum	3	2.667	Job		158		158	246

22 05 76.20 Cleanout Tees

		Crew	Daily Output	Labor-Hours	Unit	Material	2016 Bare Costs Labor	2016 Bare Costs Equipment	Total	Total Incl O&P
0010	**CLEANOUT TEES**									
0100	Cast iron, B&S, with countersunk plug									
0220	3" pipe size	1 Plum	3.60	2.222	Ea.	297	132		429	530
0240	4" pipe size	"	3.30	2.424	"	370	144		514	630
0500	For round smooth access cover, same price									
4000	Plastic, tees and adapters. Add plugs									
4010	ABS, DWV									
4020	Cleanout tee, 1-1/2" pipe size	1 Plum	15	.533	Ea.	8.65	31.50		40.15	58.50
9000	Minimum labor/equipment charge	"	2.75	2.909	Job		172		172	269

22 07 Plumbing Insulation

22 07 16 – Plumbing Equipment Insulation

22 07 16.10 Insulation for Plumbing Equipment

			Crew	Daily Output	Labor-Hours	Unit	Material	2016 Bare Costs Labor	2016 Bare Costs Equipment	Total	Total Incl O&P
0010	**INSULATION FOR PLUMBING EQUIPMENT**										
2900	Domestic water heater wrap kit										
2920	1-1/2" with vinyl jacket, 20 to 60 gal.	G	1 Plum	8	1	Ea.	15.10	59		74.10	109
2925	50 to 80 gallons	G	"	8	1	"	23.50	59		82.50	119

22 07 19 – Plumbing Piping Insulation

22 07 19.10 Piping Insulation

			Crew	Daily Output	Labor-Hours	Unit	Material	2016 Bare Costs Labor	2016 Bare Costs Equipment	Total	Total Incl O&P
0010	**PIPING INSULATION**										
0100	Rule of thumb, as a percentage of total mechanical costs					Job				10%	10%
0110	Insulation req'd. is based on the surface size/area to be covered										
0230	Insulated protectors (ADA)										
0235	For exposed piping under sinks or lavatories	♿									
0240	Vinyl coated foam, velcro tabs										
0245	P Trap, 1-1/4" or 1-1/2"		1 Plum	32	.250	Ea.	16.85	14.80		31.65	41.50
0260	Valve and supply cover										
0265	1/2", 3/8", and 7/16" pipe size	♿	1 Plum	32	.250	Ea.	16.45	14.80		31.25	41
0280	Tailpiece offset (wheelchair)										

For customer support on your Commercial Renovation Cost Data, call 877.742.4719.

409

22 07 19 - Plumbing Piping Insulation

22 07 19.10 Piping Insulation		Crew	Daily Output	Labor-Hours	Unit	Material	2016 Bare Costs Labor	Equipment	Total	Total Incl O&P
0285	1-1/4" pipe size	1 Plum	32	.250	Ea.	13.65	14.80		28.45	38
0600	Pipe covering (price copper tube one size less than IPS)									
6120	2" wall, 1/2" iron pipe size **G**	Q-14	145	.110	L.F.	6.20	5.30		11.50	15.30
6210	4" iron pipe size **G**	"	125	.128	"	11.25	6.15		17.40	22.50
6600	Fiberglass, with all service jacket									
6840	1" wall, 1/2" iron pipe size **G**	Q-14	240	.067	L.F.	.83	3.20		4.03	6.05
6860	3/4" iron pipe size **G**		230	.070		.90	3.34		4.24	6.35
6870	1" iron pipe size **G**		220	.073		.97	3.49		4.46	6.65
6880	1-1/4" iron pipe size **G**		210	.076		1.05	3.66		4.71	7.05
6890	1-1/2" iron pipe size **G**		210	.076		1.13	3.66		4.79	7.15
6900	2" iron pipe size **G**		200	.080		1.22	3.84		5.06	7.50
7080	1-1/2" wall, 1/2" iron pipe size **G**		230	.070		1.57	3.34		4.91	7.10
7140	2" iron pipe size **G**		190	.084		2.11	4.05		6.16	8.80
7160	3" iron pipe size **G**	↓	170	.094	↓	2.37	4.52		6.89	9.85
7879	Rubber tubing, flexible closed cell foam									
8100	1/2" wall, 1/4" iron pipe size **G**	1 Asbe	90	.089	L.F.	.54	4.75		5.29	8.20
8120	3/8" iron pipe size **G**		90	.089		.60	4.75		5.35	8.25
8130	1/2" iron pipe size **G**		89	.090		.67	4.80		5.47	8.45
8140	3/4" iron pipe size **G**		89	.090		.75	4.80		5.55	8.55
8150	1" iron pipe size **G**		88	.091		.82	4.85		5.67	8.70
8170	1-1/2" iron pipe size **G**		87	.092		1.15	4.91		6.06	9.15
8180	2" iron pipe size **G**		86	.093		1.47	4.97		6.44	9.60
8200	3" iron pipe size **G**		85	.094		2.06	5.05		7.11	10.30
8220	4" iron pipe size **G**		80	.100		3.06	5.35		8.41	11.95
8300	3/4" wall, 1/4" iron pipe size **G**		90	.089		.85	4.75		5.60	8.55
8330	1/2" iron pipe size **G**		89	.090		1.10	4.80		5.90	8.90
8340	3/4" iron pipe size **G**		89	.090		1.35	4.80		6.15	9.20
8350	1" iron pipe size **G**		88	.091		1.54	4.85		6.39	9.50
8360	1-1/4" iron pipe size **G**		87	.092		2.05	4.91		6.96	10.15
8370	1-1/2" iron pipe size **G**		87	.092		2.32	4.91		7.23	10.45
8380	2" iron pipe size **G**		86	.093		2.68	4.97		7.65	10.95
8444	1" wall, 1/2" iron pipe size **G**		86	.093		2.05	4.97		7.02	10.25
8445	3/4" iron pipe size **G**		84	.095		2.48	5.10		7.58	10.90
8446	1" iron pipe size **G**		84	.095		2.89	5.10		7.99	11.35
8447	1-1/4" iron pipe size **G**		82	.098		3.23	5.20		8.43	11.90
8448	1-1/2" iron pipe size **G**		82	.098		3.76	5.20		8.96	12.50
8449	2" iron pipe size **G**		80	.100		4.95	5.35		10.30	14.05
8450	2-1/2" iron pipe size **G**	↓	80	.100	↓	6.45	5.35		11.80	15.70
8456	Rubber insulation tape, 1/8" x 2" x 30' **G**				Ea.	21			21	23
9600	Minimum labor/equipment charge	1 Plum	4	2	Job		118		118	185

22 11 13 – Facility Water Distribution Piping

22 11 13.14 Pipe, Brass

	Crew	Daily Output	Labor-Hours	Unit	Material	2016 Bare Costs Labor	Equipment	Total	Total Incl O&P
0010 **PIPE, BRASS**, Plain end									
0900 Field threaded, coupling & clevis hanger assembly 10' O.C.									
0920 Regular weight									
1120 1/2" diameter	1 Plum	48	.167	L.F.	6.90	9.85		16.75	23
1140 3/4" diameter		46	.174		9.55	10.30		19.85	26.50
1160 1" diameter		43	.186		13.65	11		24.65	32
1180 1-1/4" diameter	Q-1	72	.222		20.50	11.85		32.35	41
1200 1-1/2" diameter		65	.246		24.50	13.10		37.60	47
1220 2" diameter		53	.302		33.50	16.10		49.60	62
9000 Minimum labor/equipment charge	1 Plum	4	2	Job		118		118	185

22 11 13.16 Pipe Fittings, Brass

	Crew	Daily Output	Labor-Hours	Unit	Material	2016 Bare Costs Labor	Equipment	Total	Total Incl O&P
0010 **PIPE FITTINGS, BRASS**, Rough bronze, threaded.									
1000 Standard wt., 90° Elbow									
1100 1/2"	1 Plum	12	.667	Ea.	18	39.50		57.50	81.50
1120 3/4"		11	.727		24	43		67	93.50
1140 1"		10	.800		39	47.50		86.50	117
1160 1-1/4"	Q-1	17	.941		63.50	50		113.50	148
1180 1-1/2"	"	16	1		78.50	53.50		132	169
1500 45° Elbow, 1/8"	1 Plum	13	.615		22	36.50		58.50	81.50
1580 1/2"		12	.667		22	39.50		61.50	86
1600 3/4"		11	.727		31.50	43		74.50	102
1620 1"		10	.800		53.50	47.50		101	133
1640 1-1/4"	Q-1	17	.941		85	50		135	172
1660 1-1/2"	"	16	1		107	53.50		160.50	201
2000 Tee, 1/8"	1 Plum	9	.889		21	52.50		73.50	106
2080 1/2"		8	1		21	59		80	116
2100 3/4"		7	1.143		30	67.50		97.50	138
2120 1"		6	1.333		54.50	79		133.50	183
2140 1-1/4"	Q-1	10	1.600		93.50	85.50		179	236
2160 1-1/2"	"	9	1.778		105	94.50		199.50	264
2500 Coupling, 1/8"	1 Plum	26	.308		15.05	18.20		33.25	45
2580 1/2"		15	.533		11.85	31.50		43.35	62
2600 3/4"		14	.571		21	34		55	76
2620 1"		13	.615		36	36.50		72.50	96.50
2640 1-1/4"	Q-1	22	.727		60	39		99	127
2660 1-1/2"		20	.800		78.50	42.50		121	153
2680 2"		18	.889		130	47.50		177.50	216
9000 Minimum labor/equipment charge	1 Plum	4	2	Job		118		118	185

22 11 13.23 Pipe/Tube, Copper

	Crew	Daily Output	Labor-Hours	Unit	Material	2016 Bare Costs Labor	Equipment	Total	Total Incl O&P
0010 **PIPE/TUBE, COPPER**, Solder joints									
1000 Type K tubing, couplings & clevis hanger assemblies 10' O.C.									
1100 1/4" diameter	1 Plum	84	.095	L.F.	3.83	5.65		9.48	13
1200 1" diameter		66	.121		10.60	7.20		17.80	23
1260 2" diameter		40	.200		24.50	11.85		36.35	45.50
2000 Type L tubing, couplings & clevis hanger assemblies 10' O.C.									
2100 1/4" diameter	1 Plum	88	.091	L.F.	2.80	5.40		8.20	11.50
2120 3/8" diameter		84	.095		3.50	5.65		9.15	12.65
2140 1/2" diameter		81	.099		3.68	5.85		9.53	13.15
2160 5/8" diameter		79	.101		5.30	6		11.30	15.20
2180 3/4" diameter		76	.105		5.25	6.25		11.50	15.45
2200 1" diameter		68	.118		7.40	6.95		14.35	19
2220 1-1/4" diameter		58	.138		10.05	8.15		18.20	24

For customer support on your Commercial Renovation Cost Data, call 877.742.4719.

411

22 11 13.23 Pipe/Tube, Copper

		Crew	Daily Output	Labor-Hours	Unit	Material	2016 Bare Costs Labor	Equipment	Total	Total Incl O&P
2240	1-1/2" diameter	1 Plum	52	.154	L.F.	12.60	9.10		21.70	28
2260	2" diameter	↓	42	.190		18.40	11.30		29.70	37.50
2280	2-1/2" diameter	Q-1	62	.258		28	13.75		41.75	52
2300	3" diameter		56	.286		37	15.20		52.20	64
2320	3-1/2" diameter		43	.372		50.50	19.85		70.35	86.50
2340	4" diameter		39	.410		57.50	22		79.50	97.50
2360	5" diameter	↓	34	.471		106	25		131	156
2380	6" diameter	Q-2	40	.600		146	33		179	212
2410	For other than full hard temper, add				↓	21%				
2590	For silver solder, add						15%			
3000	Type M tubing, couplings & clevis hanger assemblies 10' O.C.									
3140	1/2" diameter	1 Plum	84	.095	L.F.	3.21	5.65		8.86	12.35
3180	3/4" diameter		78	.103		4.39	6.05		10.44	14.25
3200	1" diameter	↓	70	.114	↓	6.60	6.75		13.35	17.80
4000	Type DWV tubing, couplings & clevis hanger assemblies 10' O.C.									
4100	1-1/4" diameter	1 Plum	60	.133	L.F.	9.20	7.90		17.10	22.50
4120	1-1/2" diameter		54	.148		11.30	8.75		20.05	26
4140	2" diameter	↓	44	.182		14.75	10.75		25.50	33
4160	3" diameter	Q-1	58	.276		26.50	14.70		41.20	52
4180	4" diameter		40	.400		42.50	21.50		64	79.50
4200	5" diameter	↓	36	.444		103	23.50		126.50	151
4220	6" diameter	Q-2	42	.571	↓	148	31.50		179.50	211
9000	Minimum labor/equipment charge	1 Plum	4	2	Job		118		118	185

22 11 13.25 Pipe/Tube Fittings, Copper

		Crew	Daily Output	Labor-Hours	Unit	Material	2016 Bare Costs Labor	Equipment	Total	Total Incl O&P
0010	**PIPE/TUBE FITTINGS, COPPER**, Wrought unless otherwise noted									
0020	For silver solder, add						15%			
0040	Solder joints, copper x copper									
0070	90° elbow, 1/4"	1 Plum	22	.364	Ea.	3.28	21.50		24.78	37
0100	1/2"		20	.400		1.13	23.50		24.63	38
0120	3/4"		19	.421		2.53	25		27.53	42
0130	1"		16	.500		6.20	29.50		35.70	53
0250	45° elbow, 1/4"		22	.364		5.85	21.50		27.35	40
0270	3/8"		22	.364		5	21.50		26.50	39
0280	1/2"		20	.400		2.06	23.50		25.56	39.50
0290	5/8"		19	.421		8.75	25		33.75	48.50
0300	3/4"		19	.421		3.52	25		28.52	43
0310	1"		16	.500		8.85	29.50		38.35	55.50
0320	1-1/4"		15	.533		11.85	31.50		43.35	62
0330	1-1/2"		13	.615		14.55	36.50		51.05	73
0340	2"	↓	11	.727		24.50	43		67.50	93.50
0350	2-1/2"	Q-1	13	1.231		47	65.50		112.50	154
0360	3"		13	1.231		66	65.50		131.50	175
0370	3-1/2"		10	1.600		113	85.50		198.50	257
0380	4"		9	1.778		136	94.50		230.50	297
0390	5"	↓	6	2.667		490	142		632	760
0400	6"	Q-2	9	2.667		775	147		922	1,075
0450	Tee, 1/4"	1 Plum	14	.571		6.55	34		40.55	59.50
0470	3/8"		14	.571		5.25	34		39.25	58.50
0480	1/2"		13	.615		1.92	36.50		38.42	59
0490	5/8"		12	.667		12.05	39.50		51.55	75
0500	3/4"		12	.667		4.64	39.50		44.14	66.50
0510	1"	↓	10	.800		13.95	47.50		61.45	89.50

22 11 13 – Facility Water Distribution Piping

22 11 13.25 Pipe/Tube Fittings, Copper		Crew	Daily Output	Labor-Hours	Unit	Material	2016 Bare Costs Labor	Equipment	Total	Total Incl O&P
0520	1-1/4"	1 Plum	9	.889	Ea.	19.25	52.50		71.75	103
0530	1-1/2"		8	1		29	59		88	125
0540	2"	↓	7	1.143		46	67.50		113.50	156
0550	2-1/2"	Q-1	8	2		85	107		192	260
0560	3"		7	2.286		122	122		244	325
0570	3-1/2"		6	2.667		355	142		497	610
0580	4"		5	3.200		286	171		457	580
0590	5"	↓	4	4		875	213		1,088	1,300
0600	6"	Q-2	6	4		1,200	221		1,421	1,675
0612	Tee, reducing on the outlet, 1/4"	1 Plum	15	.533		10.75	31.50		42.25	61
0613	3/8"		15	.533		10.40	31.50		41.90	60.50
0614	1/2"		14	.571		9.65	34		43.65	63
0615	5/8"		13	.615		19.40	36.50		55.90	78.50
0616	3/4"		12	.667		6.45	39.50		45.95	68.50
0617	1"		11	.727		21	43		64	90
0618	1-1/4"		10	.800		24	47.50		71.50	101
0619	1-1/2"		9	.889		25	52.50		77.50	110
0620	2"	↓	8	1		40	59		99	137
0621	2-1/2"	Q-1	9	1.778		99	94.50		193.50	257
0622	3"		8	2		103	107		210	279
0623	4"		6	2.667		220	142		362	465
0624	5"	↓	5	3.200		1,125	171		1,296	1,525
0625	6"	Q-2	7	3.429		1,550	189		1,739	2,000
0626	8"	"	6	4		5,950	221		6,171	6,900
0630	Tee, reducing on the run, 1/4"	1 Plum	15	.533		13.90	31.50		45.40	64.50
0631	3/8"		15	.533		19.60	31.50		51.10	70.50
0632	1/2"		14	.571		13.25	34		47.25	67
0633	5/8"		13	.615		19.55	36.50		56.05	78.50
0634	3/4"		12	.667		14.30	39.50		53.80	77
0635	1"		11	.727		16.95	43		59.95	85.50
0636	1-1/4"		10	.800		27.50	47.50		75	104
0637	1-1/2"		9	.889		47	52.50		99.50	134
0638	2"	↓	8	1		60	59		119	159
0639	2-1/2"	Q-1	9	1.778		128	94.50		222.50	289
0640	3"		8	2		187	107		294	370
0641	4"		6	2.667		380	142		522	640
0642	5"	↓	5	3.200		1,075	171		1,246	1,450
0643	6"	Q-2	7	3.429		1,625	189		1,814	2,100
0644	8"	"	6	4		5,950	221		6,171	6,900
0650	Coupling, 1/4"	1 Plum	24	.333		.77	19.75		20.52	32
0670	3/8"		24	.333		1	19.75		20.75	32
0680	1/2"		22	.364		.85	21.50		22.35	34.50
0690	5/8"		21	.381		2.50	22.50		25	38
0700	3/4"		21	.381		1.72	22.50		24.22	37
0710	1"		18	.444		3.39	26.50		29.89	44.50
0715	1-1/4"		17	.471		5.85	28		33.85	50
0716	1-1/2"		15	.533		7.85	31.50		39.35	57.50
0718	2"	↓	13	.615		13.15	36.50		49.65	71.50
0721	2-1/2"	Q-1	15	1.067		25.50	57		82.50	117
0722	3"		13	1.231		36	65.50		101.50	142
0724	3-1/2"		8	2		69	107		176	242
0726	4"		7	2.286		73	122		195	271
0728	5"	↓	6	2.667		169	142		311	405

For customer support on your Commercial Renovation Cost Data, call 877.742.4719.

413

22 11 13.25 Pipe/Tube Fittings, Copper

		Crew	Daily Output	Labor-Hours	Unit	Material	2016 Bare Costs Labor	Equipment	Total	Total Incl O&P
0731	6"	Q-2	8	3	Ea.	279	166		445	565
2000	DWV, solder joints, copper x copper									
2030	90° Elbow, 1-1/4"	1 Plum	13	.615	Ea.	12.05	36.50		48.55	70.50
2050	1-1/2"		12	.667		16	39.50		55.50	79
2070	2"		10	.800		25.50	47.50		73	102
2090	3"	Q-1	10	1.600		62	85.50		147.50	201
2100	4"	"	9	1.778		300	94.50		394.50	480
2250	Tee, Sanitary, 1-1/4"	1 Plum	9	.889		21	52.50		73.50	105
2270	1-1/2"		8	1		26.50	59		85.50	122
2290	2"		7	1.143		36	67.50		103.50	145
2310	3"	Q-1	7	2.286		138	122		260	340
2330	4"	"	6	2.667		330	142		472	585
2400	Coupling, 1-1/4"	1 Plum	14	.571		5	34		39	58
2420	1-1/2"		13	.615		6.20	36.50		42.70	64
2440	2"		11	.727		8.65	43		51.65	76.50
2460	3"	Q-1	11	1.455		20	77.50		97.50	143
2480	4"	"	10	1.600		44	85.50		129.50	182
9000	Minimum labor/equipment charge	1 Plum	4	2	Job		118		118	185

22 11 13.44 Pipe, Steel

		Crew	Daily Output	Labor-Hours	Unit	Material	2016 Bare Costs Labor	Equipment	Total	Total Incl O&P
0010	**PIPE, STEEL** R221113-50									
0020	All pipe sizes are to Spec. A-53 unless noted otherwise									
0050	Schedule 40, threaded, with couplings, and clevis hanger									
0060	assemblies sized for covering, 10' O.C.									
0540	Black, 1/4" diameter	1 Plum	66	.121	L.F.	4.96	7.20		12.16	16.65
0560	1/2" diameter		63	.127		2.90	7.50		10.40	14.90
0570	3/4" diameter		61	.131		3.35	7.75		11.10	15.80
0580	1" diameter		53	.151		4.30	8.95		13.25	18.70
0590	1-1/4" diameter	Q-1	89	.180		5.20	9.60		14.80	20.50
0600	1-1/2" diameter		80	.200		5.90	10.65		16.55	23
0610	2" diameter		64	.250		7.45	13.30		20.75	29
0620	2-1/2" diameter		50	.320		11.55	17.05		28.60	39
0630	3" diameter		43	.372		14.60	19.85		34.45	47
0640	3-1/2" diameter		40	.400		20.50	21.50		42	55.50
0650	4" diameter		36	.444		22	23.50		45.50	61.50
1280	All pipe sizes are to Spec. A-53 unless noted otherwise									
1281	Schedule 40, threaded, with couplings and clevis hanger									
1282	assemblies sized for covering, 10' O.C.									
1290	Galvanized, 1/4" diameter	1 Plum	66	.121	L.F.	6.70	7.20		13.90	18.55
1310	1/2" diameter		63	.127		3.18	7.50		10.68	15.20
1320	3/4" diameter		61	.131		3.60	7.75		11.35	16.05
1330	1" diameter		53	.151		4.83	8.95		13.78	19.25
1350	1-1/2" diameter	Q-1	80	.200		6.60	10.65		17.25	24
1360	2" diameter		64	.250		8.45	13.30		21.75	30.50
1370	2-1/2" diameter		50	.320		13.40	17.05		30.45	41
1380	3" diameter		43	.372		16.85	19.85		36.70	49.50
1400	4" diameter		36	.444		25	23.50		48.50	64
2000	Welded, sch. 40, on yoke & roll hanger assy's, sized for covering, 10' O.C.									
2040	Black, 1" diameter	Q-15	93	.172	L.F.	4.14	9.15	.62	13.91	19.55
2070	2" diameter		61	.262		6.65	14	.95	21.60	30.50
2090	3" diameter		43	.372		12.55	19.85	1.35	33.75	46.50
2110	4" diameter		37	.432		17.50	23	1.57	42.07	57
2120	5" diameter		32	.500		37	26.50	1.82	65.32	84

22 11 13.44 Pipe, Steel

		Crew	Daily Output	Labor-Hours	Unit	Material	2016 Bare Costs Labor	2016 Bare Costs Equipment	Total	Total Incl O&P
2130	6" diameter	Q-16	36	.667	L.F.	46	37	1.61	84.61	110
9990	Minimum labor/equipment charge	1 Plum	3	2.667	Job		158		158	246

22 11 13.45 Pipe Fittings, Steel, Threaded

		Crew	Daily Output	Labor-Hours	Unit	Material	2016 Bare Costs Labor	2016 Bare Costs Equipment	Total	Total Incl O&P
0010	**PIPE FITTINGS, STEEL, THREADED** R221113-50									
0020	Cast Iron									
0040	Standard weight, black									
0060	90° Elbow, straight									
0070	1/4"	1 Plum	16	.500	Ea.	10.15	29.50		39.65	57
0080	3/8"		16	.500		14.65	29.50		44.15	62
0090	1/2"		15	.533		6.40	31.50		37.90	56
0100	3/4"		14	.571		6.70	34		40.70	60
0110	1"		13	.615		7.90	36.50		44.40	65.50
0130	1-1/2"	Q-1	20	.800		15.55	42.50		58.05	83.50
0140	2"		18	.889		24.50	47.50		72	101
0150	2-1/2"		14	1.143		58.50	61		119.50	159
0160	3"		10	1.600		95.50	85.50		181	238
0180	4"		6	2.667		177	142		319	415
0500	Tee, straight									
0510	1/4"	1 Plum	10	.800	Ea.	15.85	47.50		63.35	91.50
0520	3/8"		10	.800		15.40	47.50		62.90	91
0540	3/4"		9	.889		11.70	52.50		64.20	95
0550	1"		8	1		10.40	59		69.40	104
0570	1-1/2"	Q-1	13	1.231		24.50	65.50		90	129
0580	2"		11	1.455		34.50	77.50		112	159
0590	2-1/2"		9	1.778		89	94.50		183.50	246
0600	3"		6	2.667		137	142		279	370
0620	4"		4	4		276	213		489	635
0700	Standard weight, galvanized cast iron									
0720	90° Elbow, straight									
0730	1/4"	1 Plum	16	.500	Ea.	16.95	29.50		46.45	64.50
0740	3/8"		16	.500		16.95	29.50		46.45	64.50
0750	1/2"		15	.533		19.35	31.50		50.85	70.50
0760	3/4"		14	.571		19	34		53	73.50
0770	1"		13	.615		22	36.50		58.50	81
0790	1-1/2"	Q-1	20	.800		47	42.50		89.50	118
0800	2"		18	.889		69.50	47.50		117	150
0810	2-1/2"		14	1.143		142	61		203	251
0820	3"		10	1.600		216	85.50		301.50	370
0840	4"		6	2.667		395	142		537	655
1100	Tee, straight									
1110	1/4"	1 Plum	10	.800	Ea.	17.80	47.50		65.30	93.50
1120	3/8"		10	.800		20.50	47.50		68	96.50
1140	3/4"		9	.889		27.50	52.50		80	112
1150	1"		8	1		29.50	59		88.50	125
1170	1-1/2"	Q-1	13	1.231		68.50	65.50		134	178
1180	2"		11	1.455		85.50	77.50		163	215
1190	2-1/2"		9	1.778		174	94.50		268.50	340
1200	3"		6	2.667		390	142		532	650
1220	4"		4	4		550	213		763	935
5000	Malleable iron, 150 lb.									
5020	Black									
5040	90° elbow, straight									

22 11 Facility Water Distribution

22 11 13 – Facility Water Distribution Piping

22 11 13.45 Pipe Fittings, Steel, Threaded

		Crew	Daily Output	Labor-Hours	Unit	Material	2016 Bare Costs Labor	Equipment	Total	Total Incl O&P
5060	1/4"	1 Plum	16	.500	Ea.	4.98	29.50		34.48	51.50
5070	3/8"		16	.500		4.98	29.50		34.48	51.50
5090	3/4"		14	.571		4.17	34		38.17	57
5100	1"		13	.615		7.25	36.50		43.75	65
5120	1-1/2"	Q-1	20	.800		15.70	42.50		58.20	84
5130	2"		18	.889		28	47.50		75.50	105
5140	2-1/2"		14	1.143		60.50	61		121.50	162
5150	3"		10	1.600		88.50	85.50		174	230
5170	4"		6	2.667		190	142		332	430
5450	Tee, straight									
5470	1/4"	1 Plum	10	.800	Ea.	7.25	47.50		54.75	82
5480	3/8"		10	.800		7.25	47.50		54.75	82
5500	3/4"		9	.889		6.65	52.50		59.15	89.50
5510	1"		8	1		11.35	59		70.35	105
5520	1-1/4"	Q-1	14	1.143		18.40	61		79.40	115
5530	1-1/2"		13	1.231		23	65.50		88.50	127
5540	2"		11	1.455		39	77.50		116.50	164
5550	2-1/2"		9	1.778		84	94.50		178.50	241
5560	3"		6	2.667		124	142		266	355
5570	3-1/2"		5	3.200		287	171		458	580
5580	4"		4	4		299	213		512	660
5650	Coupling									
5670	1/4"	1 Plum	19	.421	Ea.	6.20	25		31.20	46
5680	3/8"		19	.421		6.20	25		31.20	46
5690	1/2"		19	.421		4.78	25		29.78	44.50
5700	3/4"		18	.444		5.60	26.50		32.10	47
5710	1"		15	.533		8.40	31.50		39.90	58
5720	1-1/4"	Q-1	26	.615		10.85	33		43.85	63
5730	1-1/2"		24	.667		14.65	35.50		50.15	71.50
5740	2"		21	.762		21.50	40.50		62	87.50
5750	2-1/2"		18	.889		60	47.50		107.50	140
5760	3"		14	1.143		81	61		142	184
5780	4"		10	1.600		163	85.50		248.50	315
6000	For galvanized elbows, tees, and couplings, add					20%				
9990	Minimum labor/equipment charge	1 Plum	4	2	Job		118		118	185

22 11 13.47 Pipe Fittings, Steel

		Crew	Daily Output	Labor-Hours	Unit	Material	2016 Bare Costs Labor	Equipment	Total	Total Incl O&P
0010	**PIPE FITTINGS, STEEL**, Flanged, Welded & Special									
3000	Weld joint, butt, carbon steel, standard weight									
3040	90° elbow, long radius									
3050	1/2" pipe size	Q-15	16	1	Ea.	45	53.50	3.63	102.13	136
3060	3/4" pipe size		16	1		45	53.50	3.63	102.13	136
3070	1" pipe size		16	1		21	53.50	3.63	78.13	110
3080	1-1/4" pipe size		14	1.143		21	61	4.15	86.15	123
3090	1-1/2" pipe size		13	1.231		21	65.50	4.47	90.97	130
3100	2" pipe size		10	1.600		22.50	85.50	5.80	113.80	164
3110	2-1/2" pipe size		8	2		27.50	107	7.25	141.75	205
3120	3" pipe size		7	2.286		29	122	8.30	159.30	231
3130	4" pipe size		5	3.200		48	171	11.60	230.60	330
3136	5" pipe size		4	4		103	213	14.50	330.50	460
3140	6" pipe size	Q-16	5	4.800		106	265	11.60	382.60	545
3350	Tee, straight									
3360	1/2" pipe size	Q-15	10	1.600	Ea.	110	85.50	5.80	201.30	260

22 11 Facility Water Distribution

22 11 13 – Facility Water Distribution Piping

22 11 13.47 Pipe Fittings, Steel

		Crew	Daily Output	Labor-Hours	Unit	Material	2016 Bare Costs Labor	Equipment	Total	Total Incl O&P
3370	3/4" pipe size	Q-15	10	1.600	Ea.	110	85.50	5.80	201.30	260
3380	1" pipe size		10	1.600		54.50	85.50	5.80	145.80	199
3390	1-1/4" pipe size		9	1.778		68.50	94.50	6.45	169.45	231
3400	1-1/2" pipe size		8	2		68.50	107	7.25	182.75	250
3410	2" pipe size		6	2.667		54.50	142	9.70	206.20	292
3420	2-1/2" pipe size		5	3.200		75.50	171	11.60	258.10	360
3430	3" pipe size		4	4		84	213	14.50	311.50	440
3440	4" pipe size		3	5.333		118	284	19.35	421.35	595
3446	5" pipe size	▼	2.50	6.400		195	340	23	558	770
3450	6" pipe size	Q-16	3	8	▼	203	440	19.35	662.35	935
9990	Minimum labor/equipment charge	Q-15	3	5.333	Job		284	19.35	303.35	465

22 11 13.48 Pipe, Fittings and Valves, Steel, Grooved-Joint

		Crew	Daily Output	Labor-Hours	Unit	Material	2016 Bare Costs Labor	Equipment	Total	Total Incl O&P
0010	**PIPE, FITTINGS AND VALVES, STEEL, GROOVED-JOINT**									
0012	Fittings are ductile iron. Steel fittings noted.									
0020	Pipe includes coupling & clevis type hanger assemblies, 10' O.C.									
1000	Schedule 40, black									
1040	3/4" diameter	1 Plum	71	.113	L.F.	5	6.65		11.65	15.90
1050	1" diameter		63	.127		4.81	7.50		12.31	17
1060	1-1/4" diameter		58	.138		5.95	8.15		14.10	19.30
1070	1-1/2" diameter		51	.157		6.60	9.30		15.90	22
1080	2" diameter	▼	40	.200		7.65	11.85		19.50	27
1090	2-1/2" diameter	Q-1	57	.281		11.60	14.95		26.55	36.50
1100	3" diameter		50	.320		14	17.05		31.05	42
1110	4" diameter		45	.356		19.55	18.95		38.50	51
1120	5" diameter	▼	37	.432	▼	40.50	23		63.50	80.50
3990	Fittings: coupling material required at joints not incl. in fitting price.									
3994	Add 1 selected coupling, material only, per joint for installed price.									
4000	Elbow, 90° or 45°, painted									
4030	3/4" diameter	1 Plum	50	.160	Ea.	63.50	9.45		72.95	85
4040	1" diameter		50	.160		34	9.45		43.45	52.50
4050	1-1/4" diameter		40	.200		34	11.85		45.85	56
4060	1-1/2" diameter		33	.242		34	14.35		48.35	60
4070	2" diameter	▼	25	.320		34	18.95		52.95	67
4080	2-1/2" diameter	Q-1	40	.400		34	21.50		55.50	70.50
4090	3" diameter		33	.485		60	26		86	107
4100	4" diameter		25	.640		65	34		99	125
4110	5" diameter	▼	20	.800		155	42.50		197.50	238
4250	For galvanized elbows, add				▼	26%				
4690	Tee, painted									
4700	3/4" diameter	1 Plum	38	.211	Ea.	68.50	12.45		80.95	94.50
4740	1" diameter		33	.242		53	14.35		67.35	80.50
4750	1-1/4" diameter		27	.296		53	17.55		70.55	85.50
4760	1-1/2" diameter		22	.364		53	21.50		74.50	91.50
4770	2" diameter	▼	17	.471		53	28		81	102
4780	2-1/2" diameter	Q-1	27	.593		53	31.50		84.50	107
4790	3" diameter		22	.727		72	39		111	140
4800	4" diameter		17	.941		110	50		160	199
4810	5" diameter	▼	13	1.231		256	65.50		321.50	385
4900	For galvanized tees, add					24%				
4933	20" diameter	Q-2	16	1.500	▼	565	83		648	750
4940	Flexible, standard, painted									
4950	3/4" diameter	1 Plum	100	.080	Ea.	19	4.74		23.74	28.50

22 11 13.48 Pipe, Fittings and Valves, Steel, Grooved-Joint	Crew	Daily Output	Labor-Hours	Unit	Material	2016 Bare Costs Labor	Equipment	Total	Total Incl O&P
4960	1" diameter	1 Plum	100	.080	Ea.	19	4.74		23.74	28.50
4970	1-1/4" diameter		80	.100		25	5.90		30.90	37
4980	1-1/2" diameter		67	.119		27	7.05		34.05	40.50
4990	2" diameter		50	.160		29	9.45		38.45	47
5000	2-1/2" diameter	Q-1	80	.200		33.50	10.65		44.15	53.50
5010	3" diameter		67	.239		37.50	12.70		50.20	61
5020	3-1/2" diameter		57	.281		53.50	14.95		68.45	82
5030	4" diameter		50	.320		54	17.05		71.05	85.50
5040	5" diameter		40	.400		81	21.50		102.50	123
5050	6" diameter	Q-2	50	.480		96	26.50		122.50	148
5200	For galvanized couplings, add					33%				
7790	Valves: coupling material required at joints not incl. in valve price.									
7794	Add 1 selected coupling, material only, per joint for installed price.									
9990	Minimum labor/equipment charge	1 Plum	4	2	Job		118		118	185

22 11 13.74 Pipe, Plastic

		Crew	Daily Output	Labor-Hours	Unit	Material	2016 Bare Costs Labor	Equipment	Total	Total Incl O&P
0010	**PIPE, PLASTIC**									
1800	PVC, couplings 10' O.C., clevis hanger assemblies, 3 per 10'									
1820	Schedule 40									
1860	1/2" diameter	1 Plum	54	.148	L.F.	4.96	8.75		13.71	19.10
1870	3/4" diameter		51	.157		5.30	9.30		14.60	-20.50
1880	1" diameter		46	.174		5.95	10.30		16.25	22.50
1890	1-1/4" diameter		42	.190		6.70	11.30		18	25
1900	1-1/2" diameter		36	.222		7	13.15		20.15	28
1910	2" diameter	Q-1	59	.271		8.25	14.45		22.70	31.50
1920	2-1/2" diameter		56	.286		10.65	15.20		25.85	35.50
1930	3" diameter		53	.302		13	16.10		29.10	39.50
1940	4" diameter		48	.333		16.60	17.75		34.35	46
1950	5" diameter		43	.372		27.50	19.85		47.35	61
1960	6" diameter		39	.410		28.50	22		50.50	65.50
4100	DWV type, schedule 40, couplings 10' O.C., clevis hanger assy's, 3 per 10'									
4210	ABS, schedule 40, foam core type									
4212	Plain end black									
4214	1-1/2" diameter	1 Plum	39	.205	L.F.	5.30	12.15		17.45	25
4216	2" diameter	Q-1	62	.258		5.70	13.75		19.45	28
4218	3" diameter		56	.286		8.15	15.20		23.35	32.50
4220	4" diameter		51	.314		10.45	16.70		27.15	37.50
4222	6" diameter		42	.381		18.80	20.50		39.30	52
4240	To delete coupling & hangers, subtract									
4244	1-1/2" diam. to 6" diam.					43%	48%			
4400	PVC									
4410	1-1/4" diameter	1 Plum	42	.190	L.F.	5.50	11.30		16.80	23.50
4460	2" diameter	Q-1	59	.271		5.70	14.45		20.15	29
4470	3" diameter		53	.302		8.20	16.10		24.30	34
4480	4" diameter		48	.333		10.25	17.75		28	39
4490	6" diameter		39	.410		16.90	22		38.90	52.50
5300	CPVC, socket joint, couplings 10' O.C., clevis hanger assemblies, 3 per 10'									
5302	Schedule 40									
5304	1/2" diameter	1 Plum	54	.148	L.F.	6.05	8.75		14.80	20.50
5305	3/4" diameter		51	.157		6.95	9.30		16.25	22
5306	1" diameter		46	.174		8.35	10.30		18.65	25.50
5307	1-1/4" diameter		42	.190		10	11.30		21.30	28.50
5308	1-1/2" diameter		36	.222		11.20	13.15		24.35	33

For customer support on your Commercial Renovation Cost Data, call 877.742.4719.

22 11 13.74 Pipe, Plastic

		Crew	Daily Output	Labor-Hours	Unit	Material	2016 Bare Costs Labor	Equipment	Total	Total Incl O&P
5309	2" diameter	Q-1	59	.271	L.F.	13.30	14.45		27.75	37
5310	2-1/2" diameter		56	.286		20.50	15.20		35.70	46.50
5311	3" diameter	↓	53	.302	↓	24.50	16.10		40.60	52
5360	CPVC, threaded, couplings 10' O.C., clevis hanger assemblies, 3 per 10'									
5380	Schedule 40									
5460	1/2" diameter	1 Plum	54	.148	L.F.	6.90	8.75		15.65	21.50
5470	3/4" diameter		51	.157		8.35	9.30		17.65	23.50
5480	1" diameter		46	.174		9.85	10.30		20.15	27
5490	1-1/4" diameter		42	.190		11.15	11.30		22.45	30
5500	1-1/2" diameter	↓	36	.222		12.20	13.15		25.35	34
5510	2" diameter	Q-1	59	.271		14.55	14.45		29	38.50
5520	2-1/2" diameter		56	.286		22	15.20		37.20	47.50
5530	3" diameter	↓	53	.302	↓	26.50	16.10		42.60	54
9900	Minimum labor/equipment charge	1 Plum	4	2	Job		118		118	185

22 11 13.76 Pipe Fittings, Plastic

		Crew	Daily Output	Labor-Hours	Unit	Material	2016 Bare Costs Labor	Equipment	Total	Total Incl O&P
0010	**PIPE FITTINGS, PLASTIC**									
2700	PVC (white), schedule 40, socket joints									
2760	90° elbow, 1/2"	1 Plum	33.30	.240	Ea.	.50	14.20		14.70	22.50
2810	2"	Q-1	36.40	.440	"	3.03	23.50		26.53	40
4500	DWV, ABS, non pressure, socket joints									
4540	1/4 Bend, 1-1/4"	1 Plum	20.20	.396	Ea.	3.17	23.50		26.67	40
4560	1-1/2"	"	18.20	.440		2.36	26		28.36	43
4570	2"	Q-1	33.10	.483	↓	3.75	26		29.75	44
4650	1/8 Bend, same as 1/4 Bend									
4800	Tee, sanitary									
4820	1-1/4"	1 Plum	13.50	.593	Ea.	4.21	35		39.21	59
4830	1-1/2"	"	12.10	.661		3.60	39		42.60	65
4840	2"	Q-1	20	.800	↓	5.55	42.50		48.05	72.50
5000	DWV, PVC, schedule 40, socket joints									
5040	1/4 bend, 1-1/4"	1 Plum	20.20	.396	Ea.	6.40	23.50		29.90	43.50
5060	1-1/2"	"	18.20	.440		1.82	26		27.82	42.50
5070	2"	Q-1	33.10	.483		2.87	26		28.87	43
5080	3"		20.80	.769		8.45	41		49.45	73.50
5090	4"		16.50	.970		16.60	51.50		68.10	99
5100	6"	↓	10.10	1.584		56.50	84.50		141	194
5105	8"	Q-2	9.30	2.581		75.50	143		218.50	305
5110	1/4 bend, long sweep, 1-1/2"	1 Plum	18.20	.440		4.21	26		30.21	45
5112	2"	Q-1	33.10	.483		4.70	26		30.70	45
5114	3"		20.80	.769		10.75	41		51.75	76
5116	4"	↓	16.50	.970		20.50	51.50		72	103
5150	1/8 bend, 1-1/4"	1 Plum	20.20	.396		4.32	23.50		27.82	41.50
5215	8"	Q-2	9.30	2.581		165	143		308	405
5250	Tee, sanitary 1-1/4"	1 Plum	13.50	.593		6.65	35		41.65	62
5254	1-1/2"	"	12.10	.661		3.17	39		42.17	64.50
5255	2"	Q-1	20	.800		4.67	42.50		47.17	71.50
5256	3"		13.90	1.151		12.25	61.50		73.75	109
5257	4"		11	1.455		22.50	77.50		100	146
5259	6"	↓	6.70	2.388		91	127		218	298
5261	8"	Q-2	6.20	3.871		199	214		413	555
5264	2" x 1-1/2"	Q-1	22	.727		4.13	39		43.13	65
5266	3" x 1-1/2"		15.50	1.032		8.95	55		63.95	95.50
5268	4" x 3"	↓	12.10	1.322	↓	26.50	70.50		97	139

22 11 13 – Facility Water Distribution Piping

22 11 13.76 Pipe Fittings, Plastic		Crew	Daily Output	Labor-Hours	Unit	Material	2016 Bare Costs Labor	Equipment	Total	Total Incl O&P
5271	6" x 4"	Q-1	6.90	2.319	Ea.	88	124		212	290
5314	Combination Y & 1/8 bend, 1-1/2"	1 Plum	12.10	.661		7.70	39		46.70	69.50
5315	2"	Q-1	20	.800		8.20	42.50		50.70	75.50
5317	3"		13.90	1.151		21	61.50		82.50	119
5318	4"	↓	11	1.455	↓	42	77.50		119.50	167
5324	Combination Y & 1/8 bend, reducing									
5325	2" x 2" x 1-1/2"	Q-1	22	.727	Ea.	10.80	39		49.80	72.50
5327	3" x 3" x 1-1/2"		15.50	1.032		19.25	55		74.25	107
5328	3" x 3" x 2"		15.30	1.046		14.40	55.50		69.90	103
5329	4" x 4" x 2"	↓	12.20	1.311		22	70		92	133
5331	Wye, 1-1/4"	1 Plum	13.50	.593		8.50	35		43.50	64
5332	1-1/2"	"	12.10	.661		5.80	39		44.80	67.50
5333	2"	Q-1	20	.800		5.70	42.50		48.20	73
5334	3"		13.90	1.151		15.35	61.50		76.85	112
5335	4"		11	1.455		28	77.50		105.50	152
5336	6"	↓	6.70	2.388		81.50	127		208.50	288
5337	8"	Q-2	6.20	3.871		144	214		358	495
5341	2" x 1-1/2"	Q-1	22	.727		7	39		46	68
5342	3" x 1-1/2"		15.50	1.032		10.40	55		65.40	97
5343	4" x 3"		12.10	1.322		22.50	70.50		93	135
5344	6" x 4"	↓	6.90	2.319		62	124		186	261
5345	8" x 6"	Q-2	6.40	3.750		135	207		342	475
5347	Double wye, 1-1/2"	1 Plum	9.10	.879		13	52		65	95.50
5348	2"	Q-1	16.60	.964		14.50	51.50		66	96
5349	3"		10.40	1.538		30.50	82		112.50	162
5350	4"	↓	8.25	1.939	↓	62	103		165	229
5353	Double wye, reducing									
5354	2" x 2" x 1-1/2" x 1-1/2"	Q-1	16.80	.952	Ea.	13.25	50.50		63.75	93.50
5355	3" x 3" x 2" x 2"		10.60	1.509		22.50	80.50		103	150
5356	4" x 4" x 3" x 3"		8.45	1.893		49	101		150	211
5357	6" x 6" x 4" x 4"	↓	7.25	2.207		170	118		288	370
5374	Coupling, 1-1/4"	1 Plum	20.20	.396		4	23.50		27.50	41
5376	1-1/2"	"	18.20	.440		.85	26		26.85	41.50
5378	2"	Q-1	33.10	.483		1.17	26		27.17	41.50
5380	3"		20.80	.769		4.05	41		45.05	68.50
5390	4"		16.50	.970		6.95	51.50		58.45	88
5410	Reducer bushing, 2" x 1-1/4"		36.50	.438		3.35	23.50		26.85	40
5412	3" x 1-1/2"		27.30	.586		7.20	31		38.20	56.50
5414	4" x 2"		18.20	.879		12.55	47		59.55	87
5416	6" x 4"	↓	11.10	1.441		32	77		109	156
5418	8" x 6"	Q-2	10.20	2.353	↓	63	130		193	273
5500	CPVC, Schedule 80, threaded joints									
5540	90° Elbow, 1/4"	1 Plum	32	.250	Ea.	12.75	14.80		27.55	37
5560	1/2"		30.30	.264		7.40	15.65		23.05	32.50
5570	3/4"		26	.308		11.05	18.20		29.25	40.50
5580	1"		22.70	.352		15.55	21		36.55	49.50
5590	1-1/4"		20.20	.396		30	23.50		53.50	69.50
5600	1-1/2"	↓	18.20	.440		32.50	26		58.50	76
5610	2"	Q-1	33.10	.483		43	26		69	87.50
5620	2-1/2"		24.20	.661		134	35		169	202
5630	3"	↓	20.80	.769		144	41		185	222
5730	Coupling, 1/4"	1 Plum	32	.250		16.25	14.80		31.05	41
5732	1/2"	↓	30.30	.264	↓	13.35	15.65		29	39

For customer support on your Commercial Renovation Cost Data, call 877.742.4719.

22 11 Facility Water Distribution

22 11 13 – Facility Water Distribution Piping

22 11 13.76 Pipe Fittings, Plastic		Crew	Daily Output	Labor-Hours	Unit	Material	2016 Bare Costs Labor	Equipment	Total	Total Incl O&P
5734	3/4"	1 Plum	26	.308	Ea.	21	18.20		39.20	51.50
5736	1"		22.70	.352		24.50	21		45.50	59.50
5738	1-1/4"		20.20	.396		26	23.50		49.50	65
5740	1-1/2"	↓	18.20	.440		28	26		54	71
5742	2"	Q-1	33.10	.483		33	26		59	76
5744	2-1/2"		24.20	.661		59	35		94	120
5746	3"	↓	20.80	.769	↓	69	41		110	140
5900	CPVC, Schedule 80, socket joints									
5904	90° Elbow, 1/4"	1 Plum	32	.250	Ea.	12.15	14.80		26.95	36.50
5906	1/2"		30.30	.264		4.76	15.65		20.41	30
5908	3/4"		26	.308		6.10	18.20		24.30	35
5910	1"		22.70	.352		9.65	21		30.65	43
5912	1-1/4"		20.20	.396		21	23.50		44.50	59.50
5914	1-1/2"	↓	18.20	.440		23.50	26		49.50	66
5916	2"	Q-1	33.10	.483		28	26		54	71
5918	2-1/2"		24.20	.661		64.50	35		99.50	126
5920	3"	↓	20.80	.769		73	41		114	145
5930	45° Elbow, 1/4"	1 Plum	32	.250		18.05	14.80		32.85	43
5932	1/2"		30.30	.264		5.80	15.65		21.45	31
5934	3/4"		26	.308		8.35	18.20		26.55	37.50
5936	1"		22.70	.352		13.35	21		34.35	47
5938	1-1/4"		20.20	.396		26.50	23.50		50	65.50
5940	1-1/2"	↓	18.20	.440		27	26		53	70
5942	2"	Q-1	33.10	.483		30.50	26		56.50	73.50
5944	2-1/2"		24.20	.661		62	35		97	123
5946	3"	↓	20.80	.769		79.50	41		120.50	152
5990	Coupling, 1/4"	1 Plum	32	.250		12.90	14.80		27.70	37
5992	1/2"		30.30	.264		5.05	15.65		20.70	30
5994	3/4"		26	.308		7.05	18.20		25.25	36.50
5996	1"		22.70	.352		9.45	21		30.45	43
5998	1-1/4"		20.20	.396		14.15	23.50		37.65	52
6000	1-1/2"	↓	18.20	.440		17.85	26		43.85	60
6002	2"	Q-1	33.10	.483		20.50	26		46.50	63
6004	2-1/2"		24.20	.661		46	35		81	106
6006	3"	↓	20.80	.769	↓	50	41		91	119
9990	Minimum labor/equipment charge	1 Plum	4	2	Job		118		118	185

22 11 19 – Domestic Water Piping Specialties

22 11 19.10 Flexible Connectors

		Crew	Daily Output	Labor-Hours	Unit	Material	2016 Bare Costs Labor	Equipment	Total	Total Incl O&P
0010	**FLEXIBLE CONNECTORS**, Corrugated, 7/8" O.D., 1/2" I.D.									
0050	Gas, seamless brass, steel fittings									
0200	12" long	1 Plum	36	.222	Ea.	6.20	13.15		19.35	27.50
0220	18" long		36	.222		7.70	13.15		20.85	29
0240	24" long		34	.235		9.15	13.95		23.10	31.50
0260	30" long		34	.235	↓	9.85	13.95		23.80	32.50
9000	Minimum labor/equipment charge	↓	4	2	Job		118		118	185

22 11 19.18 Mixing Valve

		Crew	Daily Output	Labor-Hours	Unit	Material	2016 Bare Costs Labor	Equipment	Total	Total Incl O&P
0010	**MIXING VALVE**, Automatic, water tempering.									
0040	1/2" size	1 Stpi	19	.421	Ea.	555	25.50		580.50	650
0050	3/4" size		18	.444	"	555	27		582	650
9000	Minimum labor/equipment charge	↓	5	1.600	Job		97		97	151

For customer support on your Commercial Renovation Cost Data, call 877.742.4719.

421

22 11 Facility Water Distribution

22 11 19 – Domestic Water Piping Specialties

22 11 19.38 Water Supply Meters	Crew	Daily Output	Labor-Hours	Unit	Material	2016 Bare Costs Labor	Equipment	Total	Total Incl O&P
0010 **WATER SUPPLY METERS**									
2000 Domestic/commercial, bronze									
2020 Threaded									
2060 5/8" diameter, to 20 GPM	1 Plum	16	.500	Ea.	50	29.50		79.50	101
2080 3/4" diameter, to 30 GPM		14	.571		91	34		125	153
2100 1" diameter, to 50 GPM	↓	12	.667	↓	138	39.50		177.50	214
2300 Threaded/flanged									
2340 1-1/2" diameter, to 100 GPM	1 Plum	8	1	Ea.	340	59		399	465
9000 Minimum labor/equipment charge	"	3.25	2.462	Job		146		146	227

22 11 19.42 Backflow Preventers

	Crew	Daily Output	Labor-Hours	Unit	Material	2016 Bare Costs Labor	Equipment	Total	Total Incl O&P
0010 **BACKFLOW PREVENTERS**, Includes valves									
0020 and four test cocks, corrosion resistant, automatic operation									
1000 Double check principle									
1010 Threaded, with ball valves									
1020 3/4" pipe size	1 Plum	16	.500	Ea.	205	29.50		234.50	271
1030 1" pipe size		14	.571		234	34		268	310
1040 1-1/2" pipe size		10	.800		500	47.50		547.50	625
1050 2" pipe size	↓	7	1.143	↓	580	67.50		647.50	745
1080 Threaded, with gate valves									
1100 3/4" pipe size	1 Plum	16	.500	Ea.	960	29.50		989.50	1,100
1120 1" pipe size		14	.571		970	34		1,004	1,125
1140 1-1/2" pipe size		10	.800		1,250	47.50		1,297.50	1,450
1160 2" pipe size	↓	7	1.143	↓	1,525	67.50		1,592.50	1,775
4000 Reduced pressure principle									
4100 Threaded, bronze, valves are ball									
4120 3/4" pipe size	1 Plum	16	.500	Ea.	450	29.50		479.50	540
4140 1" pipe size		14	.571		485	34		519	590
4150 1-1/4" pipe size		12	.667		890	39.50		929.50	1,050
4160 1-1/2" pipe size		10	.800		975	47.50		1,022.50	1,150
4180 2" pipe size	↓	7	1.143	↓	1,100	67.50		1,167.50	1,300
5000 Flanged, bronze, valves are OS&Y									
5060 2-1/2" pipe size	Q-1	5	3.200	Ea.	4,950	171		5,121	5,725
5080 3" pipe size		4.50	3.556		5,625	189		5,814	6,500
5100 4" pipe size	↓	3	5.333		7,050	284		7,334	8,200
5120 6" pipe size	Q-2	3	8	↓	10,200	440		10,640	12,000
5600 Flanged, iron, valves are OS&Y									
5660 2-1/2" pipe size	Q-1	5	3.200	Ea.	2,525	171		2,696	3,050
5680 3" pipe size		4.50	3.556		2,650	189		2,839	3,225
5700 4" pipe size	↓	3	5.333		3,350	284		3,634	4,125
5720 6" pipe size	Q-2	3	8		4,825	440		5,265	6,025
5800 Rebuild 4" diameter reduced pressure	1 Plum	4	2		355	118		473	575
5810 6" diameter		2.66	3.008		375	178		553	695
5820 8" diameter		2	4		465	237		702	885
5830 10" diameter	↓	1.60	5	↓	545	296		841	1,050
9010 Minimum labor/equipment charge		2	4	Job		237		237	370

22 11 19.50 Vacuum Breakers

	Crew	Daily Output	Labor-Hours	Unit	Material	2016 Bare Costs Labor	Equipment	Total	Total Incl O&P
0010 **VACUUM BREAKERS**									
0013 See also backflow preventers Section 22 11 19.42									
1000 Anti-siphon continuous pressure type									
1010 Max. 150 PSI - 210°F									
1020 Bronze body									
1030 1/2" size	1 Stpi	24	.333	Ea.	161	20		181	209

422

22 11 Facility Water Distribution

22 11 19 – Domestic Water Piping Specialties

22 11 19.50 Vacuum Breakers

		Crew	Daily Output	Labor-Hours	Unit	Material	2016 Bare Costs Labor	Equipment	Total	Total Incl O&P
1040	3/4" size	1 Stpi	20	.400	Ea.	161	24.50		185.50	215
1050	1" size		19	.421		166	25.50		191.50	223
1060	1-1/4" size		15	.533		325	32.50		357.50	410
1070	1-1/2" size		13	.615		405	37.50		442.50	505
1080	2" size		11	.727		415	44		459	525
1200	Max. 125 PSI with atmospheric vent									
1210	Brass, in-line construction									
1220	1/4" size	1 Stpi	24	.333	Ea.	126	20		146	171
1230	3/8" size	"	24	.333		126	20		146	171
1260	For polished chrome finish, add					13%				
2000	Anti-siphon, non-continuous pressure type									
2010	Hot or cold water 125 PSI - 210°F									
2020	Bronze body									
2030	1/4" size	1 Stpi	24	.333	Ea.	69	20		89	108
2040	3/8" size		24	.333		69	20		89	108
2050	1/2" size		24	.333		78	20		98	117
2060	3/4" size		20	.400		93	24.50		117.50	140
2070	1" size		19	.421		144	25.50		169.50	198
2080	1-1/4" size		15	.533		252	32.50		284.50	330
2090	1-1/2" size		13	.615		296	37.50		333.50	385
2100	2" size		11	.727		460	44		504	575
2110	2-1/2" size		8	1		1,325	60.50		1,385.50	1,550
2120	3" size		6	1.333		1,750	81		1,831	2,050
2150	For polished chrome finish, add					50%				
3000	Air gap fitting									
3020	1/2" NPT size	1 Plum	19	.421	Ea.	54	25		79	98
3030	1" NPT size	"	15	.533		62	31.50		93.50	117
3040	2" NPT size	Q-1	21	.762		121	40.50		161.50	197
3050	3" NPT size		14	1.143		239	61		300	360
3060	4" NPT size		10	1.600		239	85.50		324.50	395

22 11 19.54 Water Hammer Arresters/Shock Absorbers

		Crew	Daily Output	Labor-Hours	Unit	Material	2016 Bare Costs Labor	Equipment	Total	Total Incl O&P
0010	**WATER HAMMER ARRESTERS/SHOCK ABSORBERS**									
0490	Copper									
0500	3/4" male I.P.S. For 1 to 11 fixtures	1 Plum	12	.667	Ea.	28	39.50		67.50	92.50
0600	1" male I.P.S. For 12 to 32 fixtures		8	1		47.50	59		106.50	145
0700	1-1/4" male I.P.S. For 33 to 60 fixtures		8	1		48	59		107	146
0800	1-1/2" male I.P.S. For 61 to 113 fixtures		8	1		69	59		128	169
0900	2" male I.P.S. For 114 to 154 fixtures		8	1		101	59		160	204
1000	2-1/2" male I.P.S. For 155 to 330 fixtures		4	2		305	118		423	520
9000	Minimum labor/equipment charge		3.50	2.286	Job		135		135	211

22 11 19.64 Hydrants

		Crew	Daily Output	Labor-Hours	Unit	Material	2016 Bare Costs Labor	Equipment	Total	Total Incl O&P
0010	**HYDRANTS**									
0050	Wall type, moderate climate, bronze, encased									
0200	3/4" IPS connection	1 Plum	16	.500	Ea.	800	29.50		829.50	925
1000	Non-freeze, bronze, exposed									
1100	3/4" IPS connection, 4" to 9" thick wall	1 Plum	14	.571	Ea.	515	34		549	620
1120	10" to 14" thick wall	"	12	.667		560	39.50		599.50	675
1280	For anti-siphon type, add					136			136	149
9000	Minimum labor/equipment charge	1 Plum	3	2.667	Job		158		158	246

For customer support on your Commercial Renovation Cost Data, call 877.742.4719.

423

22 13 Facility Sanitary Sewerage

22 13 16 – Sanitary Waste and Vent Piping

22 13 16.20 Pipe, Cast Iron

		Crew	Daily Output	Labor-Hours	Unit	Material	2016 Bare Costs Labor	Equipment	Total	Total Incl O&P
0010	**PIPE, CAST IRON**, Soil, on clevis hanger assemblies, 5' O.C. R221113-50									
0020	Single hub, service wt., lead & oakum joints 10' O.C.									
2120	2" diameter	Q-1	63	.254	L.F.	11.70	13.55		25.25	34
2140	3" diameter		60	.267		15.45	14.20		29.65	39
2160	4" diameter	↓	55	.291		19.65	15.50		35.15	45.50
2180	5" diameter	Q-2	76	.316		26.50	17.45		43.95	56
2200	6" diameter	"	73	.329		32.50	18.15		50.65	64
2220	8" diameter	Q-3	59	.542	↓	48.50	30.50		79	101
4000	No hub, couplings 10' O.C.									
4100	1-1/2" diameter	Q-1	71	.225	L.F.	11.50	12		23.50	31.50
4120	2" diameter		67	.239		12.05	12.70		24.75	33
4140	3" diameter		64	.250		15.45	13.30		28.75	38
4160	4" diameter	↓	58	.276		19.70	14.70		34.40	44.50
4180	5" diameter	Q-2	83	.289	↓	27	16		43	54.50
9000	Minimum labor/equipment charge	1 Plum	4	2	Job		118		118	185

22 13 16.30 Pipe Fittings, Cast Iron

		Crew	Daily Output	Labor-Hours	Unit	Material	2016 Bare Costs Labor	Equipment	Total	Total Incl O&P
0010	**PIPE FITTINGS, CAST IRON**, Soil R221113-50									
0040	Hub and spigot, service weight, lead & oakum joints									
0080	1/4 bend, 2"	Q-1	16	1	Ea.	20.50	53.50		74	106
0120	3"		14	1.143		27.50	61		88.50	125
0140	4"	↓	13	1.231		43	65.50		108.50	150
0160	5"	Q-2	18	1.333		60	73.50		133.50	181
0180	6"	"	17	1.412		75	78		153	204
0200	8"	Q-3	11	2.909		225	164		389	505
0340	1/8 bend, 2"	Q-1	16	1		14.65	53.50		68.15	99
0350	3"		14	1.143		23	61		84	120
0360	4"	↓	13	1.231		33.50	65.50		99	139
0380	5"	Q-2	18	1.333		47.50	73.50		121	167
0400	6"	"	17	1.412		57	78		135	185
0420	8"	Q-3	11	2.909		170	164		334	445
0500	Sanitary tee, 2"	Q-1	10	1.600		28.50	85.50		114	165
0540	3"		9	1.778		46.50	94.50		141	199
0620	4"	↓	8	2		57	107		164	229
0700	5"	Q-2	12	2		113	111		224	297
0800	6"	"	11	2.182		128	121		249	330
0880	8"	Q-3	7	4.571	↓	340	258		598	775
5990	No hub									
6000	Cplg. & labor required at joints not incl. in fitting									
6010	price. Add 1 coupling per joint for installed price									
6020	1/4 Bend, 1-1/2"				Ea.	10.55			10.55	11.65
6060	2"					11.55			11.55	12.70
6080	3"					16.10			16.10	17.70
6120	4"					24			24	26
6184	1/4 Bend, long sweep, 1-1/2"					27			27	29.50
6186	2"					25			25	27.50
6188	3"					30.50			30.50	33.50
6189	4"					48.50			48.50	53.50
6190	5"					92.50			92.50	102
6191	6"					107			107	118
6192	8"					291			291	320
6193	10"					585			585	645
6200	1/8 Bend, 1-1/2"					8.90			8.90	9.80

22 13 Facility Sanitary Sewerage

22 13 16 – Sanitary Waste and Vent Piping

22 13 16.30 Pipe Fittings, Cast Iron

		Crew	Daily Output	Labor-Hours	Unit	Material	2016 Bare Costs Labor	Equipment	Total	Total Incl O&P
6210	2"				Ea.	9.95			9.95	10.95
6212	3"					13.30			13.30	14.65
6214	4"					17.45			17.45	19.15
6380	Sanitary tee, tapped, 1-1/2"					21			21	23
6382	2" x 1-1/2"					18.60			18.60	20.50
6384	2"					19.95			19.95	22
6386	3" x 2"					29.50			29.50	32.50
6388	3"					51			51	56.50
6390	4" x 1-1/2"					26.50			26.50	29
6392	4" x 2"					30			30	33
6393	4"					30			30	33
6394	6" x 1-1/2"					68.50			68.50	75.50
6396	6" x 2"					70			70	77
6459	Sanitary tee, 1-1/2"					14.85			14.85	16.30
6460	2"					15.90			15.90	17.50
6470	3"					19.60			19.60	21.50
6472	4"					37			37	41
8000	Coupling, standard (by CISPI Mfrs.)									
8020	1-1/2"	Q-1	48	.333	Ea.	13.95	17.75		31.70	43
8040	2"		44	.364		15.25	19.35		34.60	47
8080	3"		38	.421		17.10	22.50		39.60	54
8120	4"		33	.485		19.90	26		45.90	62.50
8300	Coupling, cast iron clamp & neoprene gasket (by MG)									
8310	1-1/2"	Q-1	48	.333	Ea.	8.25	17.75		26	36.50
8320	2"		44	.364		9.30	19.35		28.65	40
8330	3"		38	.421		11.10	22.50		33.60	47
8340	4"		33	.485		15.30	26		41.30	57.50
8600	Coupling, Stainless steel, heavy duty									
8620	1-1/2"	Q-1	48	.333	Ea.	5.90	17.75		23.65	34
8630	2"		44	.364		6.10	19.35		25.45	36.50
8640	2" x 1-1/2"		44	.364		10.15	19.35		29.50	41
8650	3"		38	.421		6.65	22.50		29.15	42.50
8660	4"		33	.485		7.50	26		33.50	49
9000	Minimum labor/equipment charge	1 Plum	4	2	Job		118		118	185

22 13 16.60 Traps

		Crew	Daily Output	Labor-Hours	Unit	Material	2016 Bare Costs Labor	Equipment	Total	Total Incl O&P
0010	**TRAPS**									
0030	Cast iron, service weight									
0050	Running P trap, without vent									
1100	2"	Q-1	16	1	Ea.	148	53.50		201.50	246
1150	4"	"	13	1.231		148	65.50		213.50	265
1160	6"	Q-2	17	1.412		655	78		733	840
3000	P trap, B&S, 2" pipe size	Q-1	16	1		35	53.50		88.50	122
3040	3" pipe size	"	14	1.143		52.50	61		113.50	153
4700	Copper, drainage, drum trap									
4840	3" x 6" swivel, 1-1/2" pipe size	1 Plum	16	.500	Ea.	160	29.50		189.50	222
5100	P trap, standard pattern									
5200	1-1/4" pipe size	1 Plum	18	.444	Ea.	78.50	26.50		105	127
5240	1-1/2" pipe size		17	.471		72	28		100	123
5260	2" pipe size		15	.533		111	31.50		142.50	171
5280	3" pipe size		11	.727		281	43		324	375
9000	Minimum labor/equipment charge		3	2.667	Job		158		158	246

For customer support on your Commercial Renovation Cost Data, call 877.742.4719.

425

22 13 Facility Sanitary Sewerage

22 13 16 – Sanitary Waste and Vent Piping

22 13 16.80 Vent Flashing and Caps

		Crew	Daily Output	Labor-Hours	Unit	Material	2016 Bare Costs Labor	Equipment	Total	Total Incl O&P
0010	**VENT FLASHING AND CAPS**									
0120	Vent caps									
0140	Cast iron									
0180	2-1/2" - 3-5/8" pipe	1 Plum	21	.381	Ea.	45	22.50		67.50	84.50
0190	4" - 4-1/8" pipe	"	19	.421	"	55	25		80	99.50
0900	Vent flashing									
1000	Aluminum with lead ring									
1050	3" pipe	1 Plum	17	.471	Ea.	5.50	28		33.50	49.50
1060	4" pipe	"	16	.500	"	6.65	29.50		36.15	53.50
1350	Copper with neoprene ring									
1440	2" pipe	1 Plum	18	.444	Ea.	68	26.50		94.50	116
1450	3" pipe		17	.471		82	28		110	134
1460	4" pipe		16	.500		82	29.50		111.50	136
9000	Minimum labor/equipment charge		4	2	Job		118		118	185

22 13 19 – Sanitary Waste Piping Specialties

22 13 19.13 Sanitary Drains

		Crew	Daily Output	Labor-Hours	Unit	Material	2016 Bare Costs Labor	Equipment	Total	Total Incl O&P
0010	**SANITARY DRAINS**									
0400	Deck, auto park, C.I., 13" top									
0440	3", 4", 5", and 6" pipe size	Q-1	8	2	Ea.	1,450	107		1,557	1,750
0480	For galvanized body, add				"	780			780	855
2000	Floor, medium duty, C.I., deep flange, 7" diam. top									
2040	2" and 3" pipe size	Q-1	12	1.333	Ea.	232	71		303	365
2080	For galvanized body, add					96.50			96.50	106
2120	With polished bronze top					315			315	345
2500	Heavy duty, cleanout & trap w/bucket, C.I., 15" top									
2540	2", 3", and 4" pipe size	Q-1	6	2.667	Ea.	6,575	142		6,717	7,450
2560	For galvanized body, add					1,675			1,675	1,850
2580	With polished bronze top					7,300			7,300	8,025

22 13 19.14 Floor Receptors

		Crew	Daily Output	Labor-Hours	Unit	Material	2016 Bare Costs Labor	Equipment	Total	Total Incl O&P
0010	**FLOOR RECEPTORS**, For connection to 2", 3" & 4" diameter pipe									
0200	12-1/2" square top, 25 sq. in. open area	Q-1	10	1.600	Ea.	975	85.50		1,060.50	1,200
0300	For grate with 4" diam. x 3-3/4" high funnel, add					315			315	345
0400	For grate with 6" diameter x 6" high funnel, add					223			223	246
0700	For acid-resisting bucket, add					274			274	300
0900	For stainless steel mesh bucket liner, add					216			216	238
2000	12-5/8" diameter top, 40 sq. in. open area	Q-1	10	1.600		790	85.50		875.50	1,000
2100	For options, add same prices as square top									
3000	8" x 4" rectangular top, 7.5 sq. in. open area	Q-1	14	1.143	Ea.	805	61		866	980
3100	For trap primer connections, add				"	88			88	97
9000	Minimum labor/equipment charge	Q-1	3	5.333	Job		284		284	445

22 13 19.15 Sink Waste Treatment

		Crew	Daily Output	Labor-Hours	Unit	Material	2016 Bare Costs Labor	Equipment	Total	Total Incl O&P
0010	**SINK WASTE TREATMENT**, System for commercial kitchens									
0100	includes clock timer & fittings									
0200	System less chemical, wall mounted cabinet	1 Plum	16	.500	Ea.	580	29.50		609.50	685
2000	Chemical, 1 gallon, add					34.50			34.50	38
2100	6 gallons, add					157			157	172
2200	15 gallons, add					425			425	470
2300	30 gallons, add					800			800	880
2400	55 gallons, add					1,350			1,350	1,500

22 13 Facility Sanitary Sewerage

22 13 23 – Sanitary Waste Interceptors

22 13 23.10 Interceptors	Crew	Daily Output	Labor-Hours	Unit	Material	2016 Bare Costs Labor	Equipment	Total	Total Incl O&P
0010 **INTERCEPTORS**									
0150 Grease, fabricated steel, 4 GPM, 8 lb. fat capacity	1 Plum	4	2	Ea.	1,150	118		1,268	1,450
0200 7 GPM, 14 lb. fat capacity		4	2		1,600	118		1,718	1,950
1000 10 GPM, 20 lb. fat capacity	↓	4	2		1,875	118		1,993	2,250
1120 50 GPM, 100 lb. fat capacity	Q-1	2	8		6,300	425		6,725	7,600
1160 100 GPM, 200 lb. fat capacity	↓	2	8		14,300	425		14,725	16,400
1240 300 GPM, 600 lb. fat capacity	↓	1	16	↓	31,500	850		32,350	36,000
9000 Minimum labor/equipment charge	1 Plum	3	2.667	Job		158		158	246

22 13 29 – Sanitary Sewerage Pumps

22 13 29.14 Sewage Ejector Pumps

	Crew	Daily Output	Labor-Hours	Unit	Material	2016 Bare Costs Labor	Equipment	Total	Total Incl O&P
0010 **SEWAGE EJECTOR PUMPS**, With operating and level controls									
0100 Simplex system incl. tank, cover, pump 15' head									
0500 37 gal. PE tank, 12 GPM, 1/2 HP, 2" discharge	Q-1	3.20	5	Ea.	490	266		756	955
0510 3" discharge		3.10	5.161		535	275		810	1,025
0530 87 GPM, .7 HP, 2" discharge		3.20	5		755	266		1,021	1,250
0540 3" discharge		3.10	5.161		820	275		1,095	1,325
0600 45 gal. coated stl. tank, 12 GPM, 1/2 HP, 2" discharge		3	5.333		880	284		1,164	1,425
0610 3" discharge		2.90	5.517		915	294		1,209	1,450
0630 87 GPM, .7 HP, 2" discharge		3	5.333		1,125	284		1,409	1,700
0640 3" discharge		2.90	5.517		1,200	294		1,494	1,750
0660 134 GPM, 1 HP, 2" discharge		2.80	5.714		1,225	305		1,530	1,825
0680 3" discharge		2.70	5.926		1,275	315		1,590	1,925
0700 70 gal. PE tank, 12 GPM, 1/2 HP, 2" discharge		2.60	6.154		950	330		1,280	1,550
0710 3" discharge		2.40	6.667		1,000	355		1,355	1,650
0730 87 GPM, 0.7 HP, 2" discharge		2.50	6.400		1,225	340		1,565	1,875
0740 3" discharge		2.30	6.957		1,300	370		1,670	2,000
0760 134 GPM, 1 HP, 2" discharge		2.20	7.273		1,325	385		1,710	2,075
0770 3" discharge		2	8	↓	1,425	425		1,850	2,250
9000 Minimum labor/equipment charge	↓	2.50	6.400	Job		340		340	530

22 14 Facility Storm Drainage

22 14 26 – Facility Storm Drains

22 14 26.13 Roof Drains

	Crew	Daily Output	Labor-Hours	Unit	Material	2016 Bare Costs Labor	Equipment	Total	Total Incl O&P
0010 **ROOF DRAINS**									
0140 Cornice, C.I., 45° or 90° outlet									
0200 3" and 4" pipe size	Q-1	12	1.333	Ea.	355	71		426	500
0260 For galvanized body, add					75.50			75.50	83
0280 For polished bronze dome, add				↓	85.50			85.50	94
3860 Roof, flat metal deck, C.I. body, 12" C.I. dome									
3890 3" pipe size	Q-1	14	1.143	Ea.	425	61		486	560
3900 4" pipe size	"	13	1.231	"	425	65.50		490.50	565
4620 Main, all aluminum, 12" low profile dome									
4640 2", 3" and 4" pipe size	Q-1	14	1.143	Ea.	475	61		536	620
9000 Minimum labor/equipment charge	1 Plum	4	2	Job		118		118	185

For customer support on your Commercial Renovation Cost Data, call 877.742.4719.

427

22 14 Facility Storm Drainage

22 14 29 – Sump Pumps

22 14 29.16 Submersible Sump Pumps

		Crew	Daily Output	Labor-Hours	Unit	Material	2016 Bare Costs Labor	Equipment	Total	Total Incl O&P
0010	**SUBMERSIBLE SUMP PUMPS**									
1000	Elevator sump pumps, automatic									
1010	Complete systems, pump, oil detector, controls and alarm									
1020	1-1/2"discharge, does not include the sump pit/tank									
1040	1/3 HP, 115 V	1 Plum	4.40	1.818	Ea.	1,900	108		2,008	2,275
1050	1/2 HP, 115 V		4	2		1,975	118		2,093	2,350
1060	1/2 HP, 230 V		4	2		2,025	118		2,143	2,400
1070	3/4 HP, 115 V		3.60	2.222		2,050	132		2,182	2,475
1080	3/4 HP, 230 V	↓	3.60	2.222	↓	2,100	132		2,232	2,525
1100	Sump pump only									
1110	1/3 HP, 115 V	1 Plum	6.40	1.250	Ea.	174	74		248	305
1120	1/2 HP, 115 V		5.80	1.379		248	81.50		329.50	400
1130	1/2 HP, 230 V		5.80	1.379		293	81.50		374.50	445
1140	3/4 HP, 115 V		5.40	1.481		330	87.50		417.50	500
1150	3/4 HP, 230 V	↓	5.40	1.481	↓	380	87.50		467.50	550
1200	Oil detector, control and alarm only									
1210	115 V	1 Plum	8	1	Ea.	1,725	59		1,784	2,000
1220	230 V	"	8	1	"	1,725	59		1,784	2,000
7000	Sump pump, automatic									
7100	Plastic, 1-1/4" discharge, 1/4 HP	1 Plum	6.40	1.250	Ea.	150	74		224	280
7140	1/3 HP		6	1.333		217	79		296	360
7180	1-1/2" discharge, 1/2 HP		5.20	1.538		305	91		396	475
7500	Cast iron, 1-1/4" discharge, 1/4 HP		6	1.333		211	79		290	355
7540	1/3 HP		6	1.333		248	79		327	395
7560	1/2 HP		5	1.600	↓	300	94.50		394.50	480
9000	Minimum labor/equipment charge	↓	4	2	Job		118		118	185

22 15 General Service Compressed-Air Systems

22 15 19 – General Service Packaged Air Compressors and Receivers

22 15 19.10 Air Compressors

		Crew	Daily Output	Labor-Hours	Unit	Material	2016 Bare Costs Labor	Equipment	Total	Total Incl O&P
0010	**AIR COMPRESSORS**									
5250	Air, reciprocating air cooled, splash lubricated, tank mounted									
5300	Single stage, 1 phase, 140 psi									
5303	1/2 HP, 17 gal. tank	1 Stpi	3	2.667	Ea.	2,275	162		2,437	2,750
5305	3/4 HP, 30 gal. tank		2.60	3.077		2,450	187		2,637	3,000
5307	1 HP, 30 gal. tank	↓	2.20	3.636	↓	2,550	221		2,771	3,150

22 31 Domestic Water Softeners

22 31 13 – Residential Domestic Water Softeners

22 31 13.10 Residential Water Softeners

		Crew	Daily Output	Labor-Hours	Unit	Material	2016 Bare Costs Labor	Equipment	Total	Total Incl O&P
0010	**RESIDENTIAL WATER SOFTENERS**									
7350	Water softener, automatic, to 30 grains per gallon	2 Plum	5	3.200	Ea.	400	189		589	735
7400	To 100 grains per gallon	"	4	4	"	850	237		1,087	1,300

22 33 Electric Domestic Water Heaters

22 33 13 – Instantaneous Electric Domestic Water Heaters

22 33 13.20 Instantaneous Elec. Point-Of-Use Water Heaters

		Crew	Daily Output	Labor-Hours	Unit	Material	2016 Bare Costs Labor	2016 Bare Costs Equipment	Total	Total Incl O&P
0010	**INSTANTANEOUS ELECTRIC POINT-OF-USE WATER HEATERS**									
8965	Point of use, electric, glass lined									
8969	Energy saver									
8970	2.5 gal. single element [G]	1 Plum	2.80	2.857	Ea.	296	169		465	590
8971	4 gal. single element [G]		2.80	2.857		305	169		474	· 600
8974	6 gal. single element [G]		2.50	3.200		315	189		504	640
8975	10 gal. single element [G]		2.50	3.200		395	189		584	730
8976	15 gal. single element [G]		2.40	3.333		445	197		642	800
8977	20 gal. single element [G]		2.40	3.333		490	197		687	850
8978	30 gal. single element [G]		2.30	3.478		580	206		786	960
8979	40 gal. single element [G]		2.20	3.636		960	215		1,175	1,375
8988	Commercial (ASHRAE energy std. 90)									
8989	6 gallon [G]	1 Plum	2.50	3.200	Ea.	835	189		1,024	1,225
8990	10 gallon [G]		2.50	3.200		890	189		1,079	1,275
8991	15 gallon [G]		2.40	3.333		945	197		1,142	1,350
8992	20 gallon [G]		2.40	3.333		1,025	197		1,222	1,425
8993	30 gallon [G]		2.30	3.478		2,150	206		2,356	2,700
8995	Under the sink, copper, w/bracket									
8996	2.5 gallon [G]	1 Plum	4	2	Ea.	485	118		603	720
9000	Minimum labor/equipment charge	"	1.75	4.571	Job		271		271	420

22 33 30 – Residential, Electric Domestic Water Heaters

22 33 30.13 Residential, Small-Capacity Elec. Water Heaters

		Crew	Daily Output	Labor-Hours	Unit	Material	Labor	Equipment	Total	Total Incl O&P
0010	**RESIDENTIAL, SMALL-CAPACITY ELECTRIC DOMESTIC WATER HEATERS**									
1000	Residential, electric, glass lined tank, 5 yr., 10 gal., single element	1 Plum	2.30	3.478	Ea.	395	206		601	755
1060	30 gallon, double element		2.20	3.636		905	215		1,120	1,325
1080	40 gallon, double element		2	4		965	237		1,202	1,450
1100	52 gallon, double element		2	4		1,075	237		1,312	1,575
1120	66 gallon, double element		1.80	4.444		1,475	263		1,738	2,025
1140	80 gallon, double element		1.60	5		1,650	296		1,946	2,250

22 33 33 – Light-Commercial Electric Domestic Water Heaters

22 33 33.10 Commercial Electric Water Heaters

		Crew	Daily Output	Labor-Hours	Unit	Material	Labor	Equipment	Total	Total Incl O&P
0010	**COMMERCIAL ELECTRIC WATER HEATERS**									
4000	Commercial, 100° rise. NOTE: for each size tank, a range of									
4010	heaters between the ones shown are available									
4020	Electric									
4100	5 gal., 3 kW, 12 GPH, 208 volt	1 Plum	2	4	Ea.	4,275	237		4,512	5,100
4160	50 gal., 36 kW, 148 GPH, 208 volt	"	1.80	4.444		10,100	263		10,363	11,500
4480	400 gal., 210 kW, 860 GPH, 480 volt	Q-1	1	16		60,000	850		60,850	67,500

22 34 Fuel-Fired Domestic Water Heaters

22 34 13 – Instantaneous, Tankless, Gas Domestic Water Heaters

22 34 13.10 Instantaneous, Tankless, Gas Water Heaters

		Crew	Daily Output	Labor-Hours	Unit	Material	Labor	Equipment	Total	Total Incl O&P
0010	**INSTANTANEOUS, TANKLESS, GAS WATER HEATERS**									
9410	Natural gas/propane, 3.2 GPM [G]	1 Plum	2	4	Ea.	460	237		697	875
9420	6.4 GPM [G]		1.90	4.211		625	249		874	1,075
9430	8.4 GPM [G]		1.80	4.444		730	263		993	1,200
9440	9.5 GPM [G]		1.60	5		920	296		1,216	1,450

For customer support on your Commercial Renovation Cost Data, call 877.742.4719.

429

22 34 Fuel-Fired Domestic Water Heaters

22 34 30 – Residential Gas Domestic Water Heaters

22 34 30.13 Residential, Atmos, Gas Domestic Wtr Heaters	Crew	Daily Output	Labor-Hours	Unit	Material	2016 Bare Costs Labor	2016 Bare Costs Equipment	Total	Total Incl O&P
0010 **RESIDENTIAL, ATMOSPHERIC, GAS DOMESTIC WATER HEATERS**									
2000 Gas fired, foam lined tank, 10 yr., vent not incl.									
2040 30 gallon	1 Plum	2	4	Ea.	1,625	237		1,862	2,150
2060 40 gallon		1.90	4.211		1,625	249		1,874	2,175
2100 75 gallon	↓	1.50	5.333	↓	2,425	315		2,740	3,175
3000 Tank leak safety, water & gas shut off see 22 05 23.20 8800									

22 34 36 – Commercial Gas Domestic Water Heaters

22 34 36.13 Commercial, Atmos., Gas Domestic Water Htrs.

	Crew	Daily Output	Labor-Hours	Unit	Material	Labor	Equipment	Total	Total Incl O&P
0010 **COMMERCIAL, ATMOSPHERIC, GAS DOMESTIC WATER HEATERS**									
6000 Gas fired, flush jacket, std. controls, vent not incl.									
6040 75 MBH input, 73 GPH	1 Plum	1.40	5.714	Ea.	3,425	340		3,765	4,300
6060 98 MBH input, 95 GPH		1.40	5.714		6,700	340		7,040	7,900
6180 200 MBH input, 192 GPH		.60	13.333		9,175	790		9,965	11,300
6900 For low water cutoff, add		8	1		360	59		419	490
6960 For bronze body hot water circulator, add	↓	4	2	↓	2,050	118		2,168	2,425

22 34 46 – Oil-Fired Domestic Water Heaters

22 34 46.10 Residential Oil-Fired Water Heaters

	Crew	Daily Output	Labor-Hours	Unit	Material	Labor	Equipment	Total	Total Incl O&P
0010 **RESIDENTIAL OIL-FIRED WATER HEATERS**									
3000 Oil fired, glass lined tank, 5 yr., vent not included, 30 gallon	1 Plum	2	4	Ea.	1,175	237		1,412	1,675
3040 50 gallon	"	1.80	4.444	"	1,400	263		1,663	1,925

22 34 46.20 Commercial Oil-Fired Water Heaters

	Crew	Daily Output	Labor-Hours	Unit	Material	Labor	Equipment	Total	Total Incl O&P
0010 **COMMERCIAL OIL-FIRED WATER HEATERS**									
8000 Oil fired, glass lined, UL listed, std. controls, vent not incl.									
8060 140 gal., 140 MBH input, 134 GPH	Q-1	2.13	7.512	Ea.	23,400	400		23,800	26,300
8080 140 gal., 199 MBH input, 191 GPH		2	8		24,200	425		24,625	27,300
8160 140 gal., 540 MBH input, 519 GPH	↓	.96	16.667		33,000	890		33,890	37,700
8280 201 gal., 1250 MBH input, 1200 GPH	Q-2	1.22	19.672	↓	51,500	1,075		52,575	58,000

22 41 Residential Plumbing Fixtures

22 41 13 – Residential Water Closets, Urinals, and Bidets

22 41 13.13 Water Closets

	Crew	Daily Output	Labor-Hours	Unit	Material	Labor	Equipment	Total	Total Incl O&P
0010 **WATER CLOSETS**									
0032 For automatic flush, see Line 22 42 39.10 0972									
0150 Tank type, vitreous china, incl. seat, supply pipe w/stop, 1.6 gpf or noted									
0200 Wall hung									
0400 Two piece, close coupled	Q-1	5.30	3.019	Ea.	525	161		686	830
0960 For rough-in, supply, waste, vent and carrier	"	2.73	5.861	"	945	310		1,255	1,525
0999 Floor mounted									
1100 Two piece, close coupled	Q-1	5.30	3.019	Ea.	230	161		391	505
1102 Economy		5.30	3.019		125	161		286	390
1110 Two piece, close coupled, dual flush		5.30	3.019		320	161		481	600
1140 Two piece, close coupled, 1.28 gpf, ADA G	↓	5.30	3.019	↓	310	161		471	590
1960 For color, add					30%				
1980 For rough-in, supply, waste and vent	Q-1	3.05	5.246	Ea.	325	280		605	795

22 41 Residential Plumbing Fixtures

22 41 16 – Residential Lavatories and Sinks

22 41 16.13 Lavatories	Crew	Daily Output	Labor-Hours	Unit	Material	2016 Bare Costs Labor	Equipment	Total	Total Incl O&P
0010 **LAVATORIES**, With trim, white unless noted otherwise									
0500 Vanity top, porcelain enamel on cast iron									
0600 20" x 18"	Q-1	6.40	2.500	Ea.	315	133		448	555
0640 33" x 19" oval	"	6.40	2.500	"	475	133		608	730
0860 For color, add					25%				
1000 Cultured marble, 19" x 17", single bowl	Q-1	6.40	2.500	Ea.	151	133		284	375
1120 25" x 22", single bowl	"	6.40	2.500	"	190	133		323	415
1580 For color, same price									
1900 Stainless steel, self-rimming, 25" x 22", single bowl, ledge	Q-1	6.40	2.500	Ea.	340	133		473	585
1960 17" x 22", single bowl		6.40	2.500		330	133		463	575
2600 Steel, enameled, 20" x 17", single bowl		5.80	2.759		150	147		297	395
2900 Vitreous china, 20" x 16", single bowl		5.40	2.963		236	158		394	505
2960 20" x 17", single bowl		5.40	2.963		138	158		296	395
3580 Rough-in, supply, waste and vent for all above lavatories	▼	2.30	6.957	▼	218	370		588	820
4000 Wall hung									
4040 Porcelain enamel on cast iron, 16" x 14", single bowl	Q-1	8	2	Ea.	500	107		607	715
4180 20" x 18", single bowl	"	8	2	"	265	107		372	455
4580 For color, add					30%				
6000 Vitreous china, 18" x 15", single bowl with backsplash	Q-1	7	2.286	Ea.	186	122		308	395
6960 Rough-in, supply, waste and vent for above lavatories	"	1.66	9.639	"	435	515		950	1,275
7000 Pedestal type									
7600 Vitreous china, 27" x 21", white	Q-1	6.60	2.424	Ea.	680	129		809	950
7610 27" x 21", colored		6.60	2.424		860	129		989	1,150
7620 27" x 21", premium color		6.60	2.424		975	129		1,104	1,275
7660 26" x 20", white		6.60	2.424		670	129		799	935
7670 26" x 20", colored		6.60	2.424		845	129		974	1,125
7680 26" x 20", premium color		6.60	2.424		960	129		1,089	1,250
7700 24" x 20", white		6.60	2.424		395	129		524	635
7710 24" x 20", colored		6.60	2.424		490	129		619	740
7720 24" x 20", premium color		6.60	2.424		550	129		679	810
7760 21" x 18", white		6.60	2.424		277	129		406	505
7770 21" x 18", colored		6.60	2.424		335	129		464	570
7990 Rough-in, supply, waste and vent for pedestal lavatories	▼	1.66	9.639	▼	435	515		950	1,275
9000 Minimum labor/equipment charge	1 Plum	3	2.667	Job		158		158	246

22 41 16.16 Sinks

	Crew	Daily Output	Labor-Hours	Unit	Material	2016 Bare Costs Labor	Equipment	Total	Total Incl O&P
0010 **SINKS**, With faucets and drain									
2000 Kitchen, counter top style, P.E. on C.I., 24" x 21" single bowl	Q-1	5.60	2.857	Ea.	297	152		449	560
2100 31" x 22" single bowl		5.60	2.857		550	152		702	835
2200 32" x 21" double bowl		4.80	3.333		370	178		548	680
3000 Stainless steel, self rimming, 19" x 18" single bowl		5.60	2.857		605	152		757	900
3100 25" x 22" single bowl		5.60	2.857		675	152		827	975
3200 33" x 22" double bowl		4.80	3.333		975	178		1,153	1,350
3300 43" x 22" double bowl		4.80	3.333		1,125	178		1,303	1,525
4000 Steel, enameled, with ledge, 24" x 21" single bowl		5.60	2.857		525	152		677	815
4100 32" x 21" double bowl	▼	4.80	3.333		520	178		698	845
4960 For color sinks except stainless steel, add					10%				
4980 For rough-in, supply, waste and vent, counter top sinks	Q-1	2.14	7.477		247	400		647	890
5790 For rough-in, supply, waste & vent, sinks	"	1.85	8.649	▼	247	460		707	990

431

For customer support on your Commercial Renovation Cost Data, call 877.742.4719.

22 41 19 – Residential Bathtubs

22 41 19.10 Baths	Crew	Daily Output	Labor-Hours	Unit	Material	2016 Bare Costs Labor	Equipment	Total	Total Incl O&P
0010 **BATHS**									
0100 Tubs, recessed porcelain enamel on cast iron, with trim									
0180 48" x 42"	Q-1	4	4	Ea.	2,675	213		2,888	3,250
0220 72" x 36"	"	3	5.333	"	2,750	284		3,034	3,475
0300 Mat bottom									
0380 5' long	Q-1	4.40	3.636	Ea.	1,150	194		1,344	1,550
0560 Corner 48" x 44"		4.40	3.636		2,675	194		2,869	3,225
2000 Enameled formed steel, 4'-6" long		5.80	2.759		485	147		632	765
4600 Module tub & showerwall surround, molded fiberglass									
4610 5' long x 34" wide x 76" high	Q-1	4	4	Ea.	825	213		1,038	1,250
4750 Handicap with 1-1/2" OD grab bar, antiskid bottom									
4760 60" x 32-3/4" x 72" high	Q-1	4	4		825	213		1,038	1,225
4770 60" x 30" x 71" high with molded seat		3.50	4.571		860	244		1,104	1,325
9600 Rough-in, supply, waste and vent, for all above tubs, add		2.07	7.729		355	410		765	1,025
9900 Minimum labor/equipment charge		3	5.333	Job		284		284	445

22 41 23 – Residential Showers

22 41 23.20 Showers

	Crew	Daily Output	Labor-Hours	Unit	Material	Labor	Equipment	Total	Total Incl O&P
0011 **SHOWERS**, Stall, with drain only									
1520 32" square	Q-1	5	3.200	Ea.	1,175	171		1,346	1,550
1530 36" square		4.80	3.333		2,925	178		3,103	3,475
1540 Terrazzo receptor, 32" square		5	3.200		1,350	171		1,521	1,750
1580 36" corner angle		4.80	3.333		1,725	178		1,903	2,175
3000 Fiberglass, one piece, with 3 walls, 32" x 32" square		5.50	2.909		335	155		490	610
3100 36" x 36" square		5.50	2.909		385	155		540	660
4200 Rough-in, supply, waste and vent for above showers		2.05	7.805		345	415		760	1,025

22 41 23.40 Shower System Components

	Crew	Daily Output	Labor-Hours	Unit	Material	Labor	Equipment	Total	Total Incl O&P
0010 **SHOWER SYSTEM COMPONENTS**									
5500 Head, water economizer, 1.6 GPM [G]	1 Plum	24	.333	Ea.	45	19.75		64.75	80.50

22 41 36 – Residential Laundry Trays

22 41 36.10 Laundry Sinks

	Crew	Daily Output	Labor-Hours	Unit	Material	Labor	Equipment	Total	Total Incl O&P
0010 **LAUNDRY SINKS**, With trim									
0020 Porcelain enamel on cast iron, black iron frame									
0050 24" x 21", single compartment	Q-1	6	2.667	Ea.	590	142		732	865
0100 26" x 21", single compartment	"	6	2.667	"	615	142		757	895
2000 Molded stone, on wall hanger or legs									
2020 22" x 23", single compartment	Q-1	6	2.667	Ea.	179	142		321	420
2100 45" x 21", double compartment	"	5	3.200	"	345	171		516	640
3000 Plastic, on wall hanger or legs									
3020 18" x 23", single compartment	Q-1	6.50	2.462	Ea.	154	131		285	375
3100 20" x 24", single compartment		6.50	2.462		169	131		300	390
3200 36" x 23", double compartment		5.50	2.909		199	155		354	460
3300 40" x 24", double compartment		5.50	2.909		295	155		450	565
5000 Stainless steel, counter top, 22" x 17" single compartment		6	2.667		76.50	142		218.50	305
5200 33" x 22", double compartment		5	3.200		91	171		262	365
9600 Rough-in, supply, waste and vent, for all laundry sinks		2.14	7.477		247	400		647	890
9810 Minimum labor/equipment charge	1 Plum	3	2.667	Job		158		158	246

22 41 39 – Residential Faucets, Supplies and Trim

22 41 39.10 Faucets and Fittings

	Crew	Daily Output	Labor-Hours	Unit	Material	Labor	Equipment	Total	Total Incl O&P
0010 **FAUCETS AND FITTINGS**									
0150 Bath, faucets, diverter spout combination, sweat	1 Plum	8	1	Ea.	86	59		145	187

22 41 Residential Plumbing Fixtures

22 41 39 – Residential Faucets, Supplies and Trim

22 41 39.10 Faucets and Fittings	Crew	Daily Output	Labor-Hours	Unit	Material	2016 Bare Costs Labor	Equipment	Total	Total Incl O&P
0200	For integral stops, IPS unions, add				Ea.	109			109	120
0420	Bath, press-bal mix valve w/diverter, spout, shower head, arm/flange	1 Plum	8	1		173	59		232	283
0500	Drain, central lift, 1-1/2" IPS male		20	.400		49	23.50		72.50	91
0600	Trip lever, 1-1/2" IPS male		20	.400		55	23.50		78.50	97.50
0810	Bidet									
0812	Fitting, over the rim, swivel spray/pop-up drain	1 Plum	8	1	Ea.	212	59		271	325
1000	Kitchen sink faucets, top mount, cast spout		10	.800		77.50	47.50		125	159
1100	For spray, add		24	.333		16.95	19.75		36.70	49.50
1200	Wall type, swing tube spout		10	.800		73.50	47.50		121	155
1300	Single control lever handle									
1310	With pull out spray									
1320	Polished chrome	1 Plum	10	.800	Ea.	200	47.50		247.50	293
2000	Laundry faucets, shelf type, IPS or copper unions		12	.667		61.50	39.50		101	129
2100	Lavatory faucet, centerset, without drain		10	.800		63.50	47.50		111	144
2120	With pop-up drain		6.66	1.201		83	71		154	202
2210	Porcelain cross handles and pop-up drain									
2220	Polished chrome	1 Plum	6.66	1.201	Ea.	198	71		269	330
2230	Polished brass	"	6.66	1.201	"	305	71		376	445
2260	Single lever handle and pop-up drain									
2280	Satin nickel	1 Plum	6.66	1.201	Ea.	283	71		354	420
2290	Polished chrome		6.66	1.201		198	71		269	330
2800	Self-closing, center set		10	.800		148	47.50		195.50	237
2810	Automatic sensor and operator, with faucet head [G]		6.15	1.301		455	77		532	620
4000	Shower by-pass valve with union		18	.444		57	26.50		83.50	104
4100	Shower arm with flange and head		22	.364		19.95	21.50		41.45	55.50
4140	Shower, hand held, pin mount, massage action, chrome		22	.364		76	21.50		97.50	117
4142	Polished brass		22	.364		142	21.50		163.50	190
4144	Shower, hand held, wall mtd, adj. spray, 2 wall mounts, chrome		20	.400		102	23.50		125.50	149
4146	Polished brass		20	.400		192	23.50		215.50	249
4148	Shower, hand held head, bar mounted 24", adj. spray, chrome		20	.400		164	23.50		187.50	217
4150	Polished brass		20	.400		340	23.50		363.50	405
4200	Shower thermostatic mixing valve, concealed, with shower head trim kit		8	1		345	59		404	475
4220	Shower pressure balancing mixing valve									
4230	With shower head, arm, flange and diverter tub spout									
4240	Chrome	1 Plum	6.14	1.303	Ea.	375	77		452	535
4250	Satin nickel		6.14	1.303		505	77		582	675
4260	Polished graphite		6.14	1.303		505	77		582	675
5000	Sillcock, compact, brass, IPS or copper to hose		24	.333		10.20	19.75		29.95	42.50
6000	Stop and waste valves, bronze									
6100	Angle, solder end 1/2"	1 Plum	24	.333	Ea.	12.65	19.75		32.40	45
6110	3/4"		20	.400		14.85	23.50		38.35	53.50
6300	Straightway, solder end 3/8"		24	.333		17.50	19.75		37.25	50.50
6310	1/2"		24	.333		17.50	19.75		37.25	50.50
6410	Straightway, threaded 1/2"		24	.333		21.50	19.75		41.25	54.50
6420	3/4"		20	.400		23.50	23.50		47	62.50
6430	1"		19	.421		27	25		52	69
7800	Water closet, wax gasket		96	.083		1.39	4.93		6.32	9.25
7820	Gasket toilet tank to bowl		32	.250		2.96	14.80		17.76	26.50
7830	Replacement diaphragm washer assy for ballcock valve		12	.667		2.96	39.50		42.46	65
7850	Dual flush valve		12	.667		20	39.50		59.50	83.50
8000	Water supply stops, polished chrome plate									
8200	Angle, 3/8"	1 Plum	24	.333	Ea.	8.60	19.75		28.35	40.50
8300	1/2"		22	.364		8.60	21.50		30.10	43

For customer support on your Commercial Renovation Cost Data, call 877.742.4719.

433

22 41 Residential Plumbing Fixtures

22 41 39 – Residential Faucets, Supplies and Trim

22 41 39.10 Faucets and Fittings	Crew	Daily Output	Labor-Hours	Unit	Material	2016 Bare Costs Labor	Equipment	Total	Total Incl O&P
8400	Straight, 3/8"	1 Plum	26	.308	Ea.	9	18.20		27.20	38.50
8500	1/2"		24	.333		9	19.75		28.75	41
8600	Water closet, angle, w/flex riser, 3/8"		24	.333		29.50	19.75		49.25	63
9000	Minimum labor/equipment charge		4	2	Job		118		118	185

22 42 Commercial Plumbing Fixtures

22 42 13 – Commercial Water Closets, Urinals, and Bidets

22 42 13.13 Water Closets

22 42 13.13 Water Closets	Crew	Daily Output	Labor-Hours	Unit	Material	2016 Bare Costs Labor	Equipment	Total	Total Incl O&P
0010	**WATER CLOSETS**									
3000	Bowl only, with flush valve, seat, 1.6 gpf unless noted									
3100	Wall hung	Q-1	5.80	2.759	Ea.	965	147		1,112	1,275
3200	For rough-in, supply, waste and vent, single WC		2.56	6.250		980	335		1,315	1,600
3360	With floor outlet, 1.28 gpf [G]		5.80	2.759		550	147		697	835
3362	With floor outlet, 1.28 gpf, ADA [G]		5.80	2.759		575	147		722	860
3370	For rough-in, supply, waste and vent, single WC		2.84	5.634		365	300		665	870
3390	Floor mounted children's size, 10-3/4" high									
3392	With automatic flush sensor, 1.6 gpf	Q-1	6.20	2.581	Ea.	625	138		763	900
3396	With automatic flush sensor, 1.28 gpf		6.20	2.581	"	625	138		763	900
9000	Minimum labor/equipment charge		4	4	Job		213		213	330

22 42 13.16 Urinals

22 42 13.16 Urinals	Crew	Daily Output	Labor-Hours	Unit	Material	2016 Bare Costs Labor	Equipment	Total	Total Incl O&P
0010	**URINALS**									
3000	Wall hung, vitreous china, with self-closing valve									
3100	Siphon jet type	Q-1	3	5.333	Ea.	277	284		561	750
3120	Blowout type		3	5.333		465	284		749	955
3140	Water saving .5 gpf [G]		3	5.333		555	284		839	1,050
3300	Rough-in, supply, waste & vent		2.83	5.654		590	300		890	1,125
5000	Stall type, vitreous china, includes valve		2.50	6.400		745	340		1,085	1,350
6980	Rough-in, supply, waste and vent		1.99	8.040		360	430		790	1,075
9000	Minimum labor/equipment charge		4	4	Job		213		213	330

22 42 16 – Commercial Lavatories and Sinks

22 42 16.13 Lavatories

22 42 16.13 Lavatories	Crew	Daily Output	Labor-Hours	Unit	Material	2016 Bare Costs Labor	Equipment	Total	Total Incl O&P
0010	**LAVATORIES**, With trim, white unless noted otherwise									
0020	Commercial lavatories same as residential. See Section 22 41 16									

22 42 16.40 Service Sinks

22 42 16.40 Service Sinks	Crew	Daily Output	Labor-Hours	Unit	Material	2016 Bare Costs Labor	Equipment	Total	Total Incl O&P
0010	**SERVICE SINKS**									
6650	Service, floor, corner, P.E. on C.I., 28" x 28"	Q-1	4.40	3.636	Ea.	1,025	194		1,219	1,425
6790	For rough-in, supply, waste & vent, floor service sinks		1.64	9.756	"	715	520		1,235	1,600
9000	Minimum labor/equipment charge		4	4	Job		213		213	330

22 42 23 – Commercial Showers

22 42 23.30 Group Showers

22 42 23.30 Group Showers	Crew	Daily Output	Labor-Hours	Unit	Material	2016 Bare Costs Labor	Equipment	Total	Total Incl O&P
0010	**GROUP SHOWERS**									
6000	Group, w/pressure balancing valve, rough-in and rigging not included									
6800	Column, 6 heads, no receptors, less partitions	Q-1	3	5.333	Ea.	9,300	284		9,584	10,600
6900	With stainless steel partitions		1	16		12,000	850		12,850	14,500
7600	5 heads, no receptors, less partitions		3	5.333		6,400	284		6,684	7,500
7620	4 heads (1 handicap) no receptors, less partitions		3	5.333		5,700	284		5,984	6,725
7700	With stainless steel partitions		1	16		6,400	850		7,250	8,350
9000	Minimum labor/equipment charge		4	4	Job		213		213	330

22 42 Commercial Plumbing Fixtures

22 42 33 – Wash Fountains

22 42 33.20 Commercial Wash Fountains	Crew	Daily Output	Labor-Hours	Unit	Material	2016 Bare Costs Labor	Equipment	Total	Total Incl O&P
0010 **COMMERCIAL WASH FOUNTAINS**									
1900 Group, foot control									
2000 Precast terrazzo, circular, 36" diam., 5 or 6 persons	Q-2	3	8	Ea.	7,350	440		7,790	8,775
2100 54" diameter for 8 or 10 persons		2.50	9.600		9,150	530		9,680	10,900
2400 Semi-circular, 36" diam. for 3 persons		3	8		6,450	440		6,890	7,775
2500 54" diam. for 4 or 5 persons		2.50	9.600		8,650	530		9,180	10,400
5610 Group, infrared control, barrier free									
5614 Precast terrazzo									
5620 Semi-circular 36" diam. for 3 persons	Q-2	3	8	Ea.	7,325	440		7,765	8,750
5630 46" diam. for 4 persons		2.80	8.571		7,900	475		8,375	9,450
5640 Circular, 54" diam. for 8 persons, button control		2.50	9.600		9,575	530		10,105	11,300
5700 Rough-in, supply, waste and vent for above wash fountains	Q-1	1.82	8.791		375	470		845	1,150
9000 Minimum labor/equipment charge	Q-2	3	8	Job		440		440	690

22 42 39 – Commercial Faucets, Supplies, and Trim

22 42 39.10 Faucets and Fittings

	Crew	Daily Output	Labor-Hours	Unit	Material	2016 Bare Costs Labor	Equipment	Total	Total Incl O&P
0010 **FAUCETS AND FITTINGS**									
0840 Flush valves, with vacuum breaker									
0850 Water closet									
0860 Exposed, rear spud	1 Plum	8	1	Ea.	152	59		211	260
0870 Top spud		8	1		152	59		211	260
0880 Concealed, rear spud		8	1		204	59		263	320
0890 Top spud		8	1		162	59		221	271
0900 Wall hung		8	1		181	59		240	292
0910 Dual flush flushometer		12	.667		217	39.50		256.50	300
0912 Flushometer retrofit kit		18	.444		17.10	26.50		43.60	60
0920 Urinal									
0930 Exposed, stall	1 Plum	8	1	Ea.	152	59		211	260
0940 Wall (washout)		8	1		139	59		198	246
0950 Pedestal, top spud		8	1		142	59		201	249
0960 Concealed, stall		8	1		159	59		218	268
0970 Wall (washout)		8	1		172	59		231	282
0971 Automatic flush sensor and operator for									
0972 urinals or water closets, standard [G]	1 Plum	8	1	Ea.	460	59		519	600
0980 High efficiency water saving									
0984 Water closets, 1.28 gpf [G]	1 Plum	8	1	Ea.	415	59		474	555
0988 Urinals, .5 gpf [G]	"	8	1	"	415	59		474	555
2790 Faucets for lavatories									
2800 Self-closing, center set	1 Plum	10	.800	Ea.	148	47.50		195.50	237
2810 Automatic sensor and operator, with faucet head		6.15	1.301		455	77		532	620
3000 Service sink faucet, cast spout, pail hook, hose end		14	.571		76	34		110	136

22 42 39.30 Carriers and Supports

	Crew	Daily Output	Labor-Hours	Unit	Material	2016 Bare Costs Labor	Equipment	Total	Total Incl O&P
0010 **CARRIERS AND SUPPORTS**, For plumbing fixtures									
3000 Lavatory, concealed arm									
3050 Floor mounted, single									
3100 High back fixture	1 Plum	6	1.333	Ea.	570	79		649	750
3200 Flat slab fixture		6	1.333		455	79		534	625
3220 Paraplegic		6	1.333		590	79		669	775
6980 Water closet, siphon jet									
7000 Horizontal, adjustable, caulk									
7040 Single, 4" pipe size	1 Plum	5.33	1.501	Ea.	810	89		899	1,025
7060 5" pipe size		5.33	1.501		965	89		1,054	1,200

For customer support on your Commercial Renovation Cost Data, call 877.742.4719.

435

22 42 Commercial Plumbing Fixtures

22 42 39 – Commercial Faucets, Supplies, and Trim

22 42 39.30 Carriers and Supports		Crew	Daily Output	Labor-Hours	Unit	Material	2016 Bare Costs Labor	Equipment	Total	Total Incl O&P
7100	Double, 4" pipe size	1 Plum	5	1.600	Ea.	1,650	94.50		1,744.50	1,975
7120	5" pipe size	↓	5	1.600	↓	1,700	94.50		1,794.50	2,025
8200	Water closet, residential									
8220	Vertical centerline, floor mount									
8240	Single, 3" caulk, 2" or 3" vent	1 Plum	6	1.333	Ea.	655	79		734	850
8260	4" caulk, 2" or 4" vent		6	1.333	"	845	79		924	1,050
9990	Minimum labor/equipment charge	↓	3.50	2.286	Job		135		135	211

22 43 Healthcare Plumbing Fixtures

22 43 39 – Healthcare Faucets

22 43 39.10 Faucets and Fittings

		Crew	Daily Output	Labor-Hours	Unit	Material	2016 Bare Costs Labor	Equipment	Total	Total Incl O&P
0010	**FAUCETS AND FITTINGS**									
2850	Medical, bedpan cleanser, with pedal valve,	1 Plum	12	.667	Ea.	780	39.50		819.50	920
2860	With screwdriver stop valve		12	.667		405	39.50		444.50	505
2870	With self-closing spray valve		12	.667		249	39.50		288.50	335
2900	Faucet, gooseneck spout, wrist handles, grid drain		10	.800		191	47.50		238.50	284
2940	Mixing valve, knee action, screwdriver stops	↓	4	2	↓	435	118		553	665

22 45 Emergency Plumbing Fixtures

22 45 13 – Emergency Showers

22 45 13.10 Emergency Showers

		Crew	Daily Output	Labor-Hours	Unit	Material	2016 Bare Costs Labor	Equipment	Total	Total Incl O&P
0010	**EMERGENCY SHOWERS**, Rough-in not included									
5000	Shower, single head, drench, ball valve, pull, freestanding	Q-1	4	4	Ea.	380	213		593	745
5200	Horizontal or vertical supply		4	4		570	213		783	955
6000	Multi-nozzle, eye/face wash combination		4	4		680	213		893	1,075
6400	Multi-nozzle, 12 spray, shower only		4	4		2,050	213		2,263	2,575
6600	For freeze-proof, add		6	2.667	↓	480	142		622	750
9000	Minimum labor/equipment charge	↓	3	5.333	Job		284		284	445

22 45 16 – Eyewash Equipment

22 45 16.10 Eyewash Safety Equipment

		Crew	Daily Output	Labor-Hours	Unit	Material	2016 Bare Costs Labor	Equipment	Total	Total Incl O&P
0010	**EYEWASH SAFETY EQUIPMENT**, Rough-in not included									
1000	Eye wash fountain									
1400	Plastic bowl, pedestal mounted	Q-1	4	4	Ea.	288	213		501	645

22 47 Drinking Fountains and Water Coolers

22 47 13 – Drinking Fountains

22 47 13.10 Drinking Water Fountains

		Crew	Daily Output	Labor-Hours	Unit	Material	2016 Bare Costs Labor	Equipment	Total	Total Incl O&P
0010	**DRINKING WATER FOUNTAINS**, For connection to cold water supply									
1000	Wall mounted, non-recessed									
2700	Stainless steel, single bubbler, no back	1 Plum	4	2	Ea.	1,075	118		1,193	1,350
2740	With back		4	2		575	118		693	820
2780	Dual handle & wheelchair projection type		4	2		720	118		838	975
2820	Dual level for handicapped type		3.20	2.500		1,575	148		1,723	1,950
3980	For rough-in, supply and waste, add	↓	2.21	3.620	↓	180	214		394	535
4000	Wall mounted, semi-recessed									
4200	Poly-marble, single bubbler	1 Plum	4	2	Ea.	990	118		1,108	1,275
4600	Stainless steel, satin finish, single bubbler	"	4	2	"	1,325	118		1,443	1,650

22 47 Drinking Fountains and Water Coolers

22 47 13 – Drinking Fountains

22 47 13.10 Drinking Water Fountains

	22 47 13.10 Drinking Water Fountains	Crew	Daily Output	Labor-Hours	Unit	Material	2016 Bare Costs Labor	Equipment	Total	Total Incl O&P
6000	Wall mounted, fully recessed									
6400	Poly-marble, single bubbler	1 Plum	4	2	Ea.	1,675	118		1,793	2,000
6800	Stainless steel, single bubbler		4	2		1,525	118		1,643	1,850
7580	For rough-in, supply and waste, add	↓	1.83	4.372	↓	180	259		439	605
7600	Floor mounted, pedestal type									
8600	Enameled iron, heavy duty service, 2 bubblers	1 Plum	2	4	Ea.	2,650	237		2,887	3,275
8880	For freeze-proof valve system, add		2	4		705	237		942	1,150
8900	For rough-in, supply and waste, add		1.83	4.372	↓	180	259		439	605
9000	Minimum labor/equipment charge	↓	2	4	Job		237		237	370

22 47 16 – Pressure Water Coolers

22 47 16.10 Electric Water Coolers

	22 47 16.10 Electric Water Coolers	Crew	Daily Output	Labor-Hours	Unit	Material	2016 Bare Costs Labor	Equipment	Total	Total Incl O&P
0010	**ELECTRIC WATER COOLERS**									
0100	Wall mounted, non-recessed									
0140	4 GPH	Q-1	4	4	Ea.	680	213		893	1,075
0160	8 GPH, barrier free, sensor operated		4	4		1,075	213		1,288	1,500
1000	Dual height, 8.2 GPH		3.80	4.211		2,150	224		2,374	2,725
1040	14.3 GPH		3.80	4.211		1,075	224		1,299	1,525
3300	Semi-recessed, 8.1 GPH	↓	4	4	↓	785	213		998	1,200
4600	Floor mounted, flush-to-wall									
4640	4 GPH	1 Plum	3	2.667	Ea.	775	158		933	1,100
4980	For stainless steel cabinet, add					138			138	152
5000	Dual height, 8.2 GPH	1 Plum	2	4	↓	1,200	237		1,437	1,700
9000	Minimum labor/equipment charge	"	2	4	Job		237		237	370

22 66 Chemical-Waste Systems for Lab. and Healthcare Facilities

22 66 53 – Laboratory Chemical-Waste and Vent Piping

22 66 53.30 Glass Pipe

	22 66 53.30 Glass Pipe	Crew	Daily Output	Labor-Hours	Unit	Material	2016 Bare Costs Labor	Equipment	Total	Total Incl O&P
0010	**GLASS PIPE**, Borosilicate, couplings & clevis hanger assemblies, 10' O.C.									
0020	Drainage									
1100	1-1/2" diameter	Q-1	52	.308	L.F.	11.55	16.40		27.95	38
1120	2" diameter		44	.364		14.85	19.35		34.20	46.50
1140	3" diameter		39	.410		19.70	22		41.70	55.50
1160	4" diameter		30	.533		34.50	28.50		63	82.50
1180	6" diameter	↓	26	.615	↓	59.50	33		92.50	116
9000	Minimum labor/equipment charge	1 Plum	4	2	Job		118		118	185

22 66 53.40 Pipe Fittings, Glass

	22 66 53.40 Pipe Fittings, Glass	Crew	Daily Output	Labor-Hours	Unit	Material	2016 Bare Costs Labor	Equipment	Total	Total Incl O&P
0010	**PIPE FITTINGS, GLASS**									
0020	Drainage, beaded ends									
0040	Coupling & labor required at joints not incl. in fitting									
0050	price. Add 1 per joint for installed price									
0070	90° Bend or sweep, 1-1/2"				Ea.	32			32	35
0090	2"					40.50			40.50	44.50
0100	3"					66.50			66.50	73.50
0110	4"					107			107	117
0120	6" (sweep only)					335			335	365
0350	Tee, single sanitary, 1-1/2"					51.50			51.50	57
0370	2"					51.50			51.50	57
0380	3"					77			77	85
0390	4"					138			138	152
0400	6"				↓	370			370	410

22 66 53 – Laboratory Chemical-Waste and Vent Piping

22 66 53.40	Pipe Fittings, Glass	Crew	Daily Output	Labor-Hours	Unit	Material	2016 Bare Costs Labor	Equipment	Total	Total Incl O&P
0500	Coupling, stainless steel, TFE seal ring									
0520	1-1/2"	Q-1	32	.500	Ea.	23.50	26.50		50	67.50
0530	2"		30	.533		29.50	28.50		58	77
0540	3"		25	.640		40	34		74	97
0550	4"		23	.696		68.50	37		105.50	134
0560	6"		20	.800		154	42.50		196.50	236
9000	Minimum labor/equipment charge	1 Plum	4	2	Job		118		118	185

22 66 53.60	Corrosion Resistant Pipe	Crew	Daily Output	Labor-Hours	Unit	Material	2016 Bare Costs Labor	Equipment	Total	Total Incl O&P
0010	**CORROSION RESISTANT PIPE**, No couplings or hangers									
0020	Iron alloy, drain, mechanical joint									
1000	1-1/2" diameter	Q-1	70	.229	L.F.	65.50	12.20		77.70	91
1100	2" diameter		66	.242		67	12.90		79.90	94
1120	3" diameter		60	.267		86.50	14.20		100.70	117
1140	4" diameter		52	.308		111	16.40		127.40	148
2980	Plastic, epoxy, fiberglass filament wound, B&S joint									
3000	2" diameter	Q-1	62	.258	L.F.	12.10	13.75		25.85	35
3100	3" diameter		51	.314		14.15	16.70		30.85	41.50
3120	4" diameter		45	.356		20	18.95		38.95	51.50
3160	8" diameter	Q-2	38	.632		44.50	35		79.50	104
3200	12" diameter	"	28	.857		73	47.50		120.50	154
9800	Minimum labor/equipment charge	1 Plum	4	2	Job		118		118	185

22 66 53.70	Pipe Fittings, Corrosion Resistant	Crew	Daily Output	Labor-Hours	Unit	Material	2016 Bare Costs Labor	Equipment	Total	Total Incl O&P
0010	**PIPE FITTINGS, CORROSION RESISTANT**									
0030	Iron alloy									
0050	Mechanical joint									
0060	1/4 Bend, 1-1/2"	Q-1	12	1.333	Ea.	110	71		181	232
0080	2"		10	1.600		180	85.50		265.50	330
0090	3"		9	1.778		216	94.50		310.50	385
0100	4"		8	2		248	107		355	440
0160	Tee and Y, sanitary, straight									
0170	1-1/2"	Q-1	8	2	Ea.	120	107		227	298
0180	2"		7	2.286		160	122		282	365
0190	3"		6	2.667		248	142		390	495
0200	4"		5	3.200		455	171		626	765
0360	Coupling, 1-1/2"		14	1.143		68	61		129	170
0380	2"		12	1.333		77	71		148	196
0390	3"		11	1.455		81	77.50		158.50	210
0400	4"		10	1.600		92	85.50		177.50	234
3000	Epoxy, filament wound									
3030	Quick-lock joint									
3040	90° Elbow, 2"	Q-1	28	.571	Ea.	98.50	30.50		129	156
3060	3"		16	1		113	53.50		166.50	208
3070	4"		13	1.231		154	65.50		219.50	272
3190	Tee, 2"		19	.842		235	45		280	330
3200	3"		11	1.455		283	77.50		360.50	430
3210	4"		9	1.778		340	94.50		434.50	525
9000	Minimum labor/equipment charge	1 Plum	4	2	Job		118		118	185

Estimating Tips

The labor adjustment factors listed in Subdivision 22 01 02.20 also apply to Division 23.

23 10 00 Facility Fuel Systems

- The prices in this subdivision for above- and below-ground storage tanks do not include foundations or hold-down slabs, unless noted. The estimator should refer to Divisions 3 and 31 for foundation system pricing. In addition to the foundations, required tank accessories, such as tank gauges, leak detection devices, and additional manholes and piping, must be added to the tank prices.

23 50 00 Central Heating Equipment

- When estimating the cost of an HVAC system, check to see who is responsible for providing and installing the temperature control system. It is possible to overlook controls, assuming that they would be included in the electrical estimate.

- When looking up a boiler, be careful on specified capacity. Some manufacturers rate their products on output while others use input.

- Include HVAC insulation for pipe, boiler, and duct (wrap and liner).

- Be careful when looking up mechanical items to get the correct pressure rating and connection type (thread, weld, flange).

23 70 00 Central HVAC Equipment

- Combination heating and cooling units are sized by the air conditioning requirements. (See Reference No. R236000-20 for preliminary sizing guide.)

- A ton of air conditioning is nominally 400 CFM.

- Rectangular duct is taken off by the linear foot for each size, but its cost is usually estimated by the pound. Remember that SMACNA standards now base duct on internal pressure.

- Prefabricated duct is estimated and purchased like pipe: straight sections and fittings.

- Note that cranes or other lifting equipment are not included on any lines in Division 23. For example, if a crane is required to lift a heavy piece of pipe into place high above a gym floor, or to put a rooftop unit on the roof of a four-story building, etc., it must be added. Due to the potential for extreme variation—from nothing additional required to a major crane or helicopter—we feel that including a nominal amount for "lifting contingency" would be useless and detract from the accuracy of the estimate. When using equipment rental cost data from RSMeans, do not forget to include the cost of the operator(s).

Reference Numbers

Reference numbers are shown at the beginning of some major classifications. These numbers refer to related items in the Reference Section. The reference information may be an estimating procedure, an alternate pricing method, or technical information.

Note: Not all subdivisions listed here necessarily appear. ■

Did you know?

RSMeans Online gives you the same access to RSMeans' data with 24/7 access:

- Quickly locate costs in the searchable database.
- Build cost lists, estimates, and reports in minutes.
- Adjust costs to any location in the U.S. and Canada with the click of a button.

Start your free trial today at **www.RSMeansOnline.com**

RSMeans Online
FROM THE GORDIAN GROUP®

23 05 Common Work Results for HVAC

23 05 05 – Selective Demolition for HVAC

23 05 05.10 HVAC Demolition

23 05 05.10 HVAC Demolition		Crew	Daily Output	Labor-Hours	Unit	Material	2016 Bare Costs Labor	2016 Bare Costs Equipment	Total	Total Incl O&P
0010	**HVAC DEMOLITION** R220105-10									
0100	Air conditioner, split unit, 3 ton	Q-5	2	8	Ea.		435		435	680
0150	Package unit, 3 ton R024119-10	Q-6	3	8	"		455		455	705
0298	Boilers									
0300	Electric, up thru 148 kW	Q-19	2	12	Ea.		655		655	1,025
0310	150 thru 518 kW	"	1	24			1,325		1,325	2,050
0320	550 thru 2000 kW	Q-21	.40	80			4,500		4,500	7,000
0330	2070 kW and up	"	.30	106			6,000		6,000	9,350
0340	Gas and/or oil, up thru 150 MBH	Q-7	2.20	14.545			840		840	1,300
0350	160 thru 2000 MBH		.80	40			2,300		2,300	3,600
0360	2100 thru 4500 MBH		.50	64			3,700		3,700	5,775
0370	4600 thru 7000 MBH		.30	106			6,175		6,175	9,600
0380	7100 thru 12,000 MBH		.16	200			11,600		11,600	18,000
0390	12,200 thru 25,000 MBH		.12	266			15,400		15,400	24,000
1000	Ductwork, 4" high, 8" wide	1 Clab	200	.040	L.F.		1.52		1.52	2.48
1020	10" wide		190	.042			1.60		1.60	2.61
1040	14" wide		180	.044			1.68		1.68	2.75
1100	6" high, 8" wide		165	.048			1.84		1.84	3
1120	12" wide		150	.053			2.02		2.02	3.30
1140	18" wide		135	.059			2.25		2.25	3.67
1200	10" high, 12" wide		125	.064			2.43		2.43	3.96
1220	18" wide		115	.070			2.64		2.64	4.31
1240	24" wide		110	.073			2.76		2.76	4.51
1300	12"-14" high, 16"-18" wide		85	.094			3.57		3.57	5.85
1320	24" wide		75	.107			4.04		4.04	6.60
1340	48" wide		71	.113			4.27		4.27	7
1400	18" high, 24" wide		67	.119			4.53		4.53	7.40
1420	36" wide		63	.127			4.81		4.81	7.85
1440	48" wide		59	.136			5.15		5.15	8.40
1500	30" high, 36" wide		56	.143			5.40		5.40	8.85
1520	48" wide		53	.151			5.70		5.70	9.35
1540	72" wide		50	.160			6.05		6.05	9.90
1550	Duct heater, electric strip	1 Elec	8	1	Ea.		55		55	85
1850	Minimum labor/equipment charge	1 Clab	3	2.667	Job		101		101	165
2200	Furnace, electric	Q-20	2	10	Ea.		520		520	820
2300	Gas or oil, under 120 MBH	Q-9	4	4			206		206	325
2340	Over 120 MBH	"	3	5.333			275		275	435
2800	Heat pump, package unit, 3 ton	Q-5	2.40	6.667			365		365	570
2840	Split unit, 3 ton		2	8			435		435	680
3000	Mechanical equipment, light items. Unit is weight, not cooling.		.90	17.778	Ton		970		970	1,525
3600	Heavy items		1.10	14.545	"		795		795	1,250
5090	Remove refrigerant from system	1 Stpi	40	.200	Lb.		12.15		12.15	18.95
9000	Minimum labor/equipment charge	Q-6	3	8	Job		455		455	705

23 05 23 – General-Duty Valves for HVAC Piping

23 05 23.30 Valves, Iron Body

23 05 23.30 Valves, Iron Body		Crew	Daily Output	Labor-Hours	Unit	Material	2016 Bare Costs Labor	2016 Bare Costs Equipment	Total	Total Incl O&P
0010	**VALVES, IRON BODY**									
1020	Butterfly, wafer type, gear actuator, 200 lb.									
1030	2"	1 Plum	14	.571	Ea.	96	34		130	158
1060	4"	Q-1	5	3.200	"	113	171		284	390
1650	Gate, 125 lb., N.R.S.									
2150	Flanged									
2240	2-1/2"	Q-1	5	3.200	Ea.	765	171		936	1,100

23 05 Common Work Results for HVAC

23 05 23 – General-Duty Valves for HVAC Piping

23 05 23.30 Valves, Iron Body

		Crew	Daily Output	Labor-Hours	Unit	Material	2016 Bare Costs Labor	Equipment	Total	Total Incl O&P
2260	3"	Q-1	4.50	3.556	Ea.	855	189		1,044	1,225
2280	4"	↓	3	5.333	↓	1,225	284		1,509	1,800
3550	OS&Y, 125 lb., flanged									
3680	4"	Q-1	3	5.333	Ea.	810	284		1,094	1,325
3690	5"	Q-2	3.40	7.059		1,300	390		1,690	2,025
3700	6"	"	3	8	↓	1,300	440		1,740	2,125
9000	Minimum labor/equipment charge	1 Plum	3	2.667	Job		158		158	246

23 07 HVAC Insulation

23 07 13 – Duct Insulation

23 07 13.10 Duct Thermal Insulation

			Crew	Daily Output	Labor-Hours	Unit	Material	2016 Bare Costs Labor	Equipment	Total	Total Incl O&P
0010	**DUCT THERMAL INSULATION**										
0100	Rule of thumb, as a percentage of total mechanical costs					Job				10%	10%
0110	Insulation req'd. is based on the surface size/area to be covered										
3000	Ductwork										
3020	Blanket type, fiberglass, flexible										
3030	Fire rated for grease and hazardous exhaust ducts										
3060	1-1/2" thick		Q-14	84	.190	S.F.	4.54	9.15		13.69	19.70
3090	Fire rated for plenums										
3100	1/2" x 24" x 25'		Q-14	1.94	8.247	Roll	167	395		562	820
3110	1/2" x 24" x 25'			98	.163	S.F.	3.35	7.85		11.20	16.30
3120	1/2" x 48" x 25'			1.04	15.385	Roll	335	740		1,075	1,550
3126	1/2" x 48" x 25'		↓	104	.154	S.F.	3.35	7.40		10.75	15.55
3140	FSK vapor barrier wrap, .75 lb. density										
3160	1" thick	G	Q-14	350	.046	S.F.	.18	2.20		2.38	3.73
3170	1-1/2" thick	G		320	.050		.22	2.40		2.62	4.10
3180	2" thick	G		300	.053		.26	2.56		2.82	4.41
3190	3" thick	G		260	.062		.36	2.96		3.32	5.15
3200	4" thick	G	↓	242	.066	↓	.51	3.18		3.69	5.65
3210	Vinyl jacket, same as FSK										
9600	Minimum labor/equipment charge		1 Stpi	4	2	Job		121		121	189

23 07 16 – HVAC Equipment Insulation

23 07 16.10 HVAC Equipment Thermal Insulation

		Crew	Daily Output	Labor-Hours	Unit	Material	2016 Bare Costs Labor	Equipment	Total	Total Incl O&P
0010	**HVAC EQUIPMENT THERMAL INSULATION**									
0100	Rule of thumb, as a percentage of total mechanical costs				Job				10%	10%
0110	Insulation req'd. is based on the surface size/area to be covered									
9610	Minimum labor/equipment charge	1 Stpi	4	2	Job		121		121	189

23 09 Instrumentation and Control for HVAC

23 09 53 – Pneumatic and Electric Control System for HVAC

23 09 53.10 Control Components

			Crew	Daily Output	Labor-Hours	Unit	Material	2016 Bare Costs Labor	Equipment	Total	Total Incl O&P
0010	**CONTROL COMPONENTS**										
5000	Thermostats										
5030	Manual		1 Stpi	8	1	Ea.	53	60.50		113.50	153
5040	1 set back, electric, timed	G		8	1		38	60.50		98.50	136
5050	2 set back, electric, timed	G	↓	8	1		190	60.50		250.50	305
5200	24 hour, automatic, clock	G	1 Shee	8	1	↓	165	57.50		222.50	273
6000	Valves, motorized zone										
6100	Sweat connections, 1/2" C x C		1 Stpi	20	.400	Ea.	171	24.50		195.50	226

441

23 09 Instrumentation and Control for HVAC

23 09 53 – Pneumatic and Electric Control System for HVAC

23 09 53.10 Control Components		Crew	Daily Output	Labor-Hours	Unit	Material	2016 Bare Costs Labor	2016 Bare Costs Equipment	Total	Total Incl O&P
6110	3/4" C x C	1 Stpi	20	.400	Ea.	172	24.50		196.50	227
6120	1" C x C	↓	19	.421	↓	223	25.50		248.50	285
9000	Minimum labor/equipment charge	1 Plum	4	2	Job		118		118	185

23 21 Hydronic Piping and Pumps

23 21 20 – Hydronic HVAC Piping Specialties

23 21 20.34 Dielectric Unions

		Crew	Daily Output	Labor-Hours	Unit	Material	Labor	Equipment	Total	Total Incl O&P
0010	**DIELECTRIC UNIONS**, Standard gaskets for water and air									
0020	250 psi maximum pressure									
0280	Female IPT to sweat, straight									
0340	3/4" pipe size	1 Plum	20	.400	Ea.	8.30	23.50		31.80	46
0780	Female IPT to female IPT, straight									
0800	1/2" pipe size	1 Plum	24	.333	Ea.	15	19.75		34.75	47.50
0840	3/4" pipe size		20	.400	"	17.05	23.50		40.55	56
9000	Minimum labor/equipment charge	↓	4	2	Job		118		118	185

23 21 20.46 Expansion Tanks

		Crew	Daily Output	Labor-Hours	Unit	Material	Labor	Equipment	Total	Total Incl O&P
0010	**EXPANSION TANKS**									
2000	Steel, liquid expansion, ASME, painted, 15 gallon capacity	Q-5	17	.941	Ea.	665	51.50		716.50	810
2040	30 gallon capacity		12	1.333	↓	740	73		813	930
2080	60 gallon capacity		8	2		1,050	109		1,159	1,325
2120	100 gallon capacity		6	2.667		1,500	146		1,646	1,900
3000	Steel ASME expansion, rubber diaphragm, 19 gal. cap. accept.		12	1.333		2,525	73		2,598	2,900
3020	31 gallon capacity		8	2		2,800	109		2,909	3,275
3040	61 gallon capacity		6	2.667		3,950	146		4,096	4,575
3080	119 gallon capacity		4	4	↓	4,250	219		4,469	5,025
9000	Minimum labor/equipment charge	↓	4	4	Job		219		219	340

23 21 20.70 Steam Traps

		Crew	Daily Output	Labor-Hours	Unit	Material	Labor	Equipment	Total	Total Incl O&P
0010	**STEAM TRAPS**									
0030	Cast iron body, threaded									
0040	Inverted bucket									
0050	1/2" pipe size	1 Stpi	12	.667	Ea.	142	40.50		182.50	219
0100	1" pipe size		9	.889		380	54		434	505
0120	1-1/4" pipe size	↓	8	1	↓	575	60.50		635.50	730
1000	Float & thermostatic, 15 psi									
1010	3/4" pipe size	1 Stpi	16	.500	Ea.	135	30.50		165.50	196
1020	1" pipe size		15	.533		196	32.50		228.50	266
1040	1-1/2" pipe size		9	.889		285	54		339	400
1060	2" pipe size		6	1.333	↓	525	81		606	700
9000	Minimum labor/equipment charge	↓	4	2	Job		121		121	189

23 21 20.76 Strainers, Y Type, Bronze Body

		Crew	Daily Output	Labor-Hours	Unit	Material	Labor	Equipment	Total	Total Incl O&P
0010	**STRAINERS, Y TYPE, BRONZE BODY**									
0050	Screwed, 125 lb., 1/4" pipe size	1 Stpi	24	.333	Ea.	27	20		47	61
0100	1/2" pipe size		20	.400		33.50	24.50		58	75
0120	3/4" pipe size		19	.421		41.50	25.50		67	85.50
0140	1" pipe size		17	.471		49	28.50		77.50	98.50
0160	1-1/2" pipe size		14	.571		106	34.50		140.50	170
0180	2" pipe size		13	.615		139	37.50		176.50	212
0182	3" pipe size	↓	12	.667		930	40.50		970.50	1,100
0200	300 lb., 2-1/2" pipe size	Q-5	17	.941		660	51.50		711.50	805
0220	3" pipe size		16	1		1,175	54.50		1,229.50	1,350

23 21 Hydronic Piping and Pumps

23 21 20 – Hydronic HVAC Piping Specialties

23 21 20.76 Strainers, Y Type, Bronze Body		Crew	Daily Output	Labor-Hours	Unit	Material	2016 Bare Costs Labor	Equipment	Total	Total Incl O&P
0240	4" pipe size	Q-5	15	1.067	Ea.	2,300	58.50		2,358.50	2,625
1000	Flanged, 150 lb., 1-1/2" pipe size	1 Stpi	11	.727		475	44		519	590
1020	2" pipe size	"	8	1		640	60.50		700.50	800
1030	2-1/2" pipe size	Q-5	5	3.200		950	175		1,125	1,325
1040	3" pipe size		4.50	3.556		1,175	194		1,369	1,575
1060	4" pipe size		3	5.333		1,775	291		2,066	2,400
1100	6" pipe size	Q-6	3	8		3,400	455		3,855	4,425
1106	8" pipe size	"	2.60	9.231		3,725	525		4,250	4,925
1500	For 300 lb. rating, add					40%				
9000	Minimum labor/equipment charge	1 Stpi	3.75	2.133	Job		129		129	202

23 21 23 – Hydronic Pumps

23 21 23.13 In-Line Centrifugal Hydronic Pumps

		Crew	Daily Output	Labor-Hours	Unit	Material	2016 Bare Costs Labor	Equipment	Total	Total Incl O&P
0010	**IN-LINE CENTRIFUGAL HYDRONIC PUMPS**									
0600	Bronze, sweat connections, 1/40 HP, in line									
0640	3/4" size	Q-1	16	1	Ea.	225	53.50		278.50	330
1000	Flange connection, 3/4" to 1-1/2" size									
1040	1/12 HP	Q-1	6	2.667	Ea.	580	142		722	860
1060	1/8 HP		6	2.667		1,000	142		1,142	1,325
1100	1/3 HP		6	2.667		1,125	142		1,267	1,450
1140	2" size, 1/6 HP		5	3.200		1,425	171		1,596	1,850
1180	2-1/2" size, 1/4 HP		5	3.200		1,825	171		1,996	2,300
1220	3" size, 1/4 HP		4	4		1,925	213		2,138	2,425
9000	Minimum labor/equipment charge		3.25	4.923	Job		262		262	410

23 23 Refrigerant Piping

23 23 16 – Refrigerant Piping Specialties

23 23 16.16 Refrigerant Line Sets

		Crew	Daily Output	Labor-Hours	Unit	Material	2016 Bare Costs Labor	Equipment	Total	Total Incl O&P
0010	**REFRIGERANT LINE SETS**									
0100	Copper tube									
0110	1/2" insulation, both tubes									
0120	Combination 1/4" and 1/2" tubes									
0130	10' set	Q-5	42	.381	Ea.	70.50	21		91.50	110
0140	20' set		40	.400		110	22		132	155
0150	30' set		37	.432		148	23.50		171.50	199
0160	40' set		35	.457		193	25		218	252
0170	50' set		32	.500		229	27.50		256.50	295
0180	100' set		22	.727		520	39.50		559.50	635
0300	Combination 1/4" and 3/4" tubes									
0310	10' set	Q-5	40	.400	Ea.	78.50	22		100.50	121
0320	20' set		38	.421		139	23		162	189
0330	30' set		35	.457		208	25		233	268
0340	40' set		33	.485		272	26.50		298.50	340
0350	50' set		30	.533		345	29		374	420
0380	100' set		20	.800		780	43.50		823.50	930
0500	Combination 3/8" & 3/4" tubes									
0510	10' set	Q-5	28	.571	Ea.	90.50	31		121.50	148
0520	20' set		36	.444		141	24.50		165.50	193
0530	30' set		34	.471		191	25.50		216.50	251
0540	40' set		31	.516		252	28		280	320
0550	50' set		28	.571		296	31		327	375

For customer support on your Commercial Renovation Cost Data, call 877.742.4719.

443

23 23 Refrigerant Piping

23 23 16 – Refrigerant Piping Specialties

23 23 16.16 Refrigerant Line Sets		Crew	Daily Output	Labor-Hours	Unit	Material	2016 Bare Costs Labor	Equipment	Total	Total Incl O&P
0580	100' set	Q-5	18	.889	Ea.	875	48.50		923.50	1,025
0700	Combination 3/8" & 1-1/8" tubes									
0710	10' set	Q-5	36	.444	Ea.	158	24.50		182.50	212
0720	20' set		33	.485		265	26.50		291.50	335
0730	30' set		31	.516		355	28		383	435
0740	40' set		28	.571		535	31		566	635
0750	50' set		26	.615		615	33.50		648.50	735
0900	Combination 1/2" & 3/4" tubes									
0910	10' set	Q-5	37	.432	Ea.	95.50	23.50		119	142
0920	20' set		35	.457		169	25		194	225
0930	30' set		33	.485		255	26.50		281.50	320
0940	40' set		30	.533		335	29		364	415
0950	50' set		27	.593		420	32.50		452.50	515
0980	100' set		17	.941		950	51.50		1,001.50	1,125
2100	Combination 1/2" & 1-1/8" tubes									
2110	10' set	Q-5	35	.457	Ea.	163	25		188	218
2120	20' set		31	.516		279	28		307	350
2130	30' set		29	.552		420	30		450	505
2140	40' set		25	.640		565	35		600	675
2150	50' set		14	1.143		625	62.50		687.50	785
2300	For 1" thick insulation add					30%	15%			

23 31 HVAC Ducts and Casings

23 31 13 – Metal Ducts

23 31 13.13 Rectangular Metal Ducts

		Crew	Daily Output	Labor-Hours	Unit	Material	2016 Bare Costs Labor	Equipment	Total	Total Incl O&P
0010	**RECTANGULAR METAL DUCTS**									
0020	Fabricated rectangular, includes fittings, joints, supports,									
0021	allowance for flexible connections and field sketches.									
0030	Does not include "as-built dwgs." or insulation.									
0031	NOTE: Fabrication and installation are combined									
0040	as LABOR cost. Approx. 25% fittings assumed.									
0042	Fabrication/Inst. is to commercial quality standards									
0043	(SMACNA or equiv.) for structure, sealing, leak testing, etc.									
0100	Aluminum, alloy 3003-H14, under 100 lb.	Q-10	75	.320	Lb.	3.20	17.10		20.30	30.50
0110	100 to 500 lb.		80	.300		1.88	16.05		17.93	27.50
0120	500 to 1,000 lb.		95	.253		1.82	13.50		15.32	23.50
0160	Over 5,000 lb.		145	.166		1.77	8.85		10.62	15.90
0500	Galvanized steel, under 200 lb.		235	.102		.57	5.45		6.02	9.25
0520	200 to 500 lb.		245	.098		.56	5.25		5.81	8.85
0540	500 to 1,000 lb.		255	.094		.55	5.05		5.60	8.55
0560	1,000 to 2,000 lb.		265	.091		.54	4.84		5.38	8.25
0580	Over 5,000 lb.		285	.084		.53	4.50		5.03	7.70
1000	Stainless steel, type 304, under 100 lb.		165	.145		4.20	7.75		11.95	16.85
1020	100 to 500 lb.		175	.137		3.85	7.35		11.20	15.80
1030	500 to 1,000 lb.		190	.126		2.68	6.75		9.43	13.60

23 31 HVAC Ducts and Casings

23 31 13 – Metal Ducts

23 31 13.16 Round and Flat-Oval Spiral Ducts

	23 31 13.16 Round and Flat-Oval Spiral Ducts	Crew	Daily Output	Labor-Hours	Unit	Material	2016 Bare Costs Labor	Equipment	Total	Total Incl O&P
0010	**ROUND AND FLAT-OVAL SPIRAL DUCTS**									
0020	Fabricated round and flat oval spiral, includes hangers,									
0021	supports and field sketches.									
5400	Spiral preformed, steel, galv., straight lengths, Max 10" spwg.									
5410	4" diameter, 26 ga.	Q-9	360	.044	L.F.	1.86	2.29		4.15	5.65
5416	5" diameter, 26 ga.		320	.050		1.86	2.58		4.44	6.10
5420	6" diameter, 26 ga.		280	.057		1.86	2.94		4.80	6.70
5425	7" diameter, 26 ga.		240	.067		2.17	3.44		5.61	7.80
5430	8" diameter, 26 ga.		200	.080		2.48	4.12		6.60	9.25
5440	10" diameter, 26 ga.		160	.100		3.10	5.15		8.25	11.55
9990	Minimum labor/equipment charge	1 Shee	3	2.667	Job		153		153	241

23 31 16 – Nonmetal Ducts

23 31 16.13 Fibrous-Glass Ducts

		Crew	Daily Output	Labor-Hours	Unit	Material	2016 Bare Costs Labor	Equipment	Total	Total Incl O&P
0010	**FIBROUS-GLASS DUCTS**									
3490	Rigid fiberglass duct board, foil reinf. kraft facing									
3500	Rectangular, 1" thick, alum. faced, (FRK), std. weight	Q-10	350	.069	SF Surf	.79	3.66		4.45	6.65
9990	Minimum labor/equipment charge	1 Shee	3	2.667	Job		153		153	241

23 31 16.19 PVC Ducts

		Crew	Daily Output	Labor-Hours	Unit	Material	2016 Bare Costs Labor	Equipment	Total	Total Incl O&P
0010	**PVC DUCTS**									
4000	Rigid plastic, corrosive fume resistant PVC									
4020	Straight, 6" diameter	Q-9	220	.073	L.F.	12.30	3.75		16.05	19.40
4040	8" diameter		160	.100		17.80	5.15		22.95	27.50
4060	10" diameter		120	.133		22.50	6.85		29.35	35.50
4070	12" diameter		100	.160		26	8.25		34.25	41.50
4080	14" diameter		70	.229		32.50	11.80		44.30	54.50
4090	16" diameter		62	.258		37.50	13.30		50.80	62.50
4100	18" diameter		58	.276		51.50	14.20		65.70	79.50
4110	20" diameter	Q-10	75	.320		58.50	17.10		75.60	91
4130	24" diameter	"	55	.436		66	23.50		89.50	110
4250	Coupling, 6" diameter	Q-9	88	.182	Ea.	25	9.35		34.35	42.50
4270	8" diameter		78	.205		32.50	10.55		43.05	52.50
4290	10" diameter		70	.229		37.50	11.80		49.30	60
4300	12" diameter		55	.291		41	15		56	68.50
4310	14" diameter		44	.364		53	18.75		71.75	87.50
4320	16" diameter		40	.400		56	20.50		76.50	94
4330	18" diameter		38	.421		72.50	21.50		94	114
4340	20" diameter	Q-10	55	.436		87	23.50		110.50	133
4360	24" diameter	"	45	.533		109	28.50		137.50	165
4470	Elbow, 90°, 6" diameter	Q-9	44	.364		114	18.75		132.75	156
4490	8" diameter		28	.571		128	29.50		157.50	188
4510	10" diameter		18	.889		156	46		202	244
4520	12" diameter		15	1.067		177	55		232	282
4540	16" diameter		10	1.600		267	82.50		349.50	425
4550	18" diameter	Q-10	15	1.600		370	85.50		455.50	540
4560	20" diameter		14	1.714		550	91.50		641.50	750
4580	24" diameter		12	2		675	107		782	915
4750	Elbow 45°, use 90° and deduct					25%				
9990	Minimum labor/equipment charge	1 Shee	3	2.667	Job		153		153	241

For customer support on your Commercial Renovation Cost Data, call 877.742.4719.

445

23 33 13 – Dampers

23 33 13.13 Volume-Control Dampers	Crew	Daily Output	Labor-Hours	Unit	Material	2016 Bare Costs Labor	Equipment	Total	Total Incl O&P
0010 VOLUME-CONTROL DAMPERS									
5990 Multi-blade dampers, opposed blade, 8" x 6"	1 Shee	24	.333	Ea.	21.50	19.10		40.60	54
5994 8" x 8"		22	.364		22.50	21		43.50	57.50
5996 10" x 10"		21	.381		26.50	22		48.50	64
6000 12" x 12"		21	.381		29.50	22		51.50	67
6020 12" x 18"		18	.444		39.50	25.50		65	83.50
6030 14" x 10"		20	.400		29	23		52	67.50
6031 14" x 14"		17	.471		35.50	27		62.50	81.50
6033 16" x 12"		17	.471		35.50	27		62.50	81.50
6035 16" x 16"		16	.500		44	28.50		72.50	93.50
6037 18" x 16"		15	.533		48	30.50		78.50	101
6038 18" x 18"		15	.533		52	30.50		82.50	105
6070 20" x 16"		14	.571		52	32.50		84.50	109
6072 20" x 20"		13	.615		62.50	35		97.50	125
6074 22" x 18"		14	.571		62.50	32.50		95	121
6076 24" x 16"		11	.727		61	41.50		102.50	133
6078 24" x 20"		8	1		72.50	57.50		130	171
6080 24" x 24"		8	1		85	57.50		142.50	184
6110 26" x 26"		6	1.333		95	76.50		171.50	225
6133 30" x 30"	Q-9	6.60	2.424		137	125		262	345
6135 32" x 32"		6.40	2.500		154	129		283	370
6180 48" x 36"		5.60	2.857		253	147		400	510
7500 Variable volume modulating motorized damper, incl. elect. mtr.									
7504 8" x 6"	1 Shee	15	.533	Ea.	118	30.50		148.50	178
7506 10" x 6"		14	.571		118	32.50		150.50	182
7510 10" x 10"		13	.615		122	35		157	190
7520 12" x 12"		12	.667		127	38		165	199
7522 12" x 16"		11	.727		127	41.50		168.50	206
7524 16" x 10"		12	.667		125	38		163	197
7526 16" x 14"		10	.800		130	46		176	216
7528 16" x 18"		9	.889		135	51		186	229
7542 18" x 18"		8	1		138	57.50		195.50	242
7544 20" x 14"		8	1		141	57.50		198.50	246
7546 20" x 18"		7	1.143		147	65.50		212.50	265
7560 24" x 12"		8	1		143	57.50		200.50	248
7562 24" x 18"		7	1.143		154	65.50		219.50	272
7564 24" x 24"		6	1.333		160	76.50		236.50	296
7568 28" x 10"		7	1.143		143	65.50		208.50	260
7590 30" x 14"		5	1.600		151	91.50		242.50	310
7600 30" x 18"		4	2		198	115		313	400
7610 30" x 24"		3.80	2.105		264	121		385	480
7700 For thermostat, add		8	1		31	57.50		88.50	125
8000 Multi-blade dampers, parallel blade									
8100 8" x 8"	1 Shee	24	.333	Ea.	82	19.10		101.10	120
8140 16" x 10"		20	.400		103	23		126	150
8160 18" x 12"		18	.444		113	25.50		138.50	164
8220 28" x 16"		10	.800		150	46		196	238

23 33 13.16 Fire Dampers

	Crew	Daily Output	Labor-Hours	Unit	Material	Labor	Equipment	Total	Total Incl O&P
0010 FIRE DAMPERS									
3000 Fire damper, curtain type, 1-1/2 hr. rated, vertical, 6" x 6"	1 Shee	24	.333	Ea.	24	19.10		43.10	56.50
3020 8" x 6"		22	.364		24	21		45	59.50
3240 16" x 14"		18	.444		44	25.50		69.50	88.50

23 33 Air Duct Accessories

23 33 13 – Dampers

23 33 13.16 Fire Dampers		Crew	Daily Output	Labor-Hours	Unit	Material	2016 Bare Costs Labor	Equipment	Total	Total Incl O&P
3400	24" x 20"	1 Shee	8	1	Ea.	54	57.50		111.50	150

23 33 19 – Duct Silencers

23 33 19.10 Duct Silencers

		Crew	Daily Output	Labor-Hours	Unit	Material	2016 Bare Costs Labor	Equipment	Total	Total Incl O&P
0010	**DUCT SILENCERS**									
9000	Silencers, noise control for air flow, duct				MCFM	57			57	63

23 33 23 – Turning Vanes

23 33 23.13 Air Turning Vanes

		Crew	Daily Output	Labor-Hours	Unit	Material	2016 Bare Costs Labor	Equipment	Total	Total Incl O&P
0010	**AIR TURNING VANES**									
9400	Turning vane components									
9410	Turning vane rail	1 Shee	160	.050	L.F.	.75	2.86		3.61	5.35
9420	Double thick, factory fab. vane		300	.027		1.17	1.53		2.70	3.70
9428	12" high set		170	.047		1.68	2.69		4.37	6.10
9432	14" high set		160	.050		1.96	2.86		4.82	6.70
9900	Minimum labor/equipment charge		4	2	Job		115		115	181

23 33 46 – Flexible Ducts

23 33 46.10 Flexible Air Ducts

		Crew	Daily Output	Labor-Hours	Unit	Material	2016 Bare Costs Labor	Equipment	Total	Total Incl O&P
0010	**FLEXIBLE AIR DUCTS**									
1280	Add to labor for elevated installation									
1282	of prefabricated (purchased) ductwork									
1283	10' to 15' high						10%			
1284	15' to 20' high						20%			
1285	20' to 25' high						25%			
1286	25' to 30' high						35%			
1287	30' to 35' high						40%			
1288	35' to 40' high						50%			
1289	Over 40' high						55%			
1300	Flexible, coated fiberglass fabric on corr. resist. metal helix									
1400	pressure to 12" (WG) UL-181									
1500	Noninsulated, 3" diameter	Q-9	400	.040	L.F.	1.12	2.06		3.18	4.48
1540	5" diameter		320	.050		1.30	2.58		3.88	5.50
1560	6" diameter		280	.057		1.50	2.94		4.44	6.30
1580	7" diameter		240	.067		1.53	3.44		4.97	7.10
1600	8" diameter		200	.080		1.91	4.12		6.03	8.60
1640	10" diameter		160	.100		2.46	5.15		7.61	10.85
1660	12" diameter		120	.133		2.94	6.85		9.79	14.10
1900	Insulated, 1" thick, PE jacket, 3" diameter [G]		380	.042		2.60	2.17		4.77	6.30
1910	4" diameter [G]		340	.047		2.60	2.43		5.03	6.70
1920	5" diameter [G]		300	.053		2.60	2.75		5.35	7.20
1940	6" diameter [G]		260	.062		2.94	3.17		6.11	8.25
1960	7" diameter [G]		220	.073		3.20	3.75		6.95	9.40
1980	8" diameter [G]		180	.089		3.49	4.58		8.07	11.10
2020	10" diameter [G]		140	.114		4.25	5.90		10.15	14
2040	12" diameter [G]		100	.160		4.90	8.25		13.15	18.40
2060	14" diameter [G]		80	.200		6.05	10.30		16.35	23
9990	Minimum labor/equipment charge	1 Shee	3	2.667	Job		153		153	241

23 33 Air Duct Accessories

23 33 53 – Duct Liners

23 33 53.10 Duct Liner Board

23 33 53.10 Duct Liner Board		Crew	Daily Output	Labor-Hours	Unit	Material	2016 Bare Costs Labor	Equipment	Total	Total Incl O&P
0010	**DUCT LINER BOARD**									
3490	Board type, fiberglass liner, 3 lb. density									
3680	No finish									
3700	1" thick	G	Q-14	170	.094	S.F.	.44	4.52		4.96	7.75
3710	1-1/2" thick	G		140	.114		.66	5.50		6.16	9.55
3720	2" thick	G	↓	130	.123	↓	.88	5.90		6.78	10.45
3940	Board type, non-fibrous foam									
3950	Temperature, bacteria and fungi resistant									
3960	1" thick	G	Q-14	150	.107	S.F.	2.38	5.15		7.53	10.85
3970	1-1/2" thick	G		130	.123		3.28	5.90		9.18	13.10
3980	2" thick	G	↓	120	.133		3.98	6.40		10.38	14.70

23 34 HVAC Fans

23 34 13 – Axial HVAC Fans

23 34 13.10 Axial Flow HVAC Fans

		Crew	Daily Output	Labor-Hours	Unit	Material	Labor	Equipment	Total	Total Incl O&P
0010	**AXIAL FLOW HVAC FANS**									
0020	Air conditioning and process air handling									

23 34 14 – Blower HVAC Fans

23 34 14.10 Blower Type HVAC Fans

		Crew	Daily Output	Labor-Hours	Unit	Material	Labor	Equipment	Total	Total Incl O&P
0010	**BLOWER TYPE HVAC FANS**									
2500	Ceiling fan, right angle, extra quiet, 0.10" S.P.									
2540	210 CFM	Q-20	19	1.053	Ea.	370	55		425	490
2580	885 CFM		16	1.250		925	65.50		990.50	1,125
2620	2,960 CFM	↓	11	1.818		1,700	95		1,795	2,025
2640	For wall or roof cap, add	1 Shee	16	.500	↓	315	28.50		343.50	390

23 34 16 – Centrifugal HVAC Fans

23 34 16.10 Centrifugal Type HVAC Fans

		Crew	Daily Output	Labor-Hours	Unit	Material	Labor	Equipment	Total	Total Incl O&P
0010	**CENTRIFUGAL TYPE HVAC FANS**									
0200	In-line centrifugal, supply/exhaust booster									
0220	aluminum wheel/hub, disconnect switch, 1/4" S.P.									
0240	500 CFM, 10" diameter connection	Q-20	3	6.667	Ea.	1,450	350		1,800	2,125
0280	1,520 CFM, 16" diameter connection		2	10		1,500	520		2,020	2,475
0320	3,480 CFM, 20" diameter connection		.80	25		1,950	1,300		3,250	4,200
0326	5,080 CFM, 20" diameter connection	↓	.75	26.667	↓	2,250	1,400		3,650	4,675
5000	Utility set, centrifugal, V belt drive, motor									
5020	1/4" S.P., 1200 CFM, 1/4 HP	Q-20	6	3.333	Ea.	1,750	174		1,924	2,200
5040	1520 CFM, 1/3 HP		5	4		2,300	209		2,509	2,850
5060	1850 CFM, 1/2 HP		4	5		2,275	261		2,536	2,900
5080	2180 CFM, 3/4 HP	↓	3	6.667	↓	2,700	350		3,050	3,500
7000	Roof exhauster, centrifugal, aluminum housing, 12" galvanized									
7020	curb, bird screen, back draft damper, 1/4" S.P.									
7100	Direct drive, 320 CFM, 11" sq. damper	Q-20	7	2.857	Ea.	725	149		874	1,025
7120	600 CFM, 11" sq. damper	"	6	3.333	"	925	174		1,099	1,300
9900	Minimum labor/equipment charge	1 Elec	4	2	Job		110		110	170

23 34 HVAC Fans

23 34 23 – HVAC Power Ventilators

23 34 23.10 HVAC Power Circulators and Ventilators	Crew	Daily Output	Labor-Hours	Unit	Material	2016 Bare Costs Labor	Equipment	Total	Total Incl O&P
0010 **HVAC POWER CIRCULATORS AND VENTILATORS**									
6650 Residential, bath exhaust, grille, back draft damper									
6660 50 CFM	Q-20	24	.833	Ea.	65	43.50		108.50	140
6670 110 CFM		22	.909		100	47.50		147.50	185
6680 Light combination, squirrel cage, 100 watt, 70 CFM		24	.833		114	43.50		157.50	194
6700 Light/heater combination, ceiling mounted									
6710 70 CFM, 1450 watt	Q-20	24	.833	Ea.	165	43.50		208.50	250
6800 Heater combination, recessed, 70 CFM		24	.833		68.50	43.50		112	144
6820 With 2 infrared bulbs		23	.870		107	45.50		152.50	190
6900 Kitchen exhaust, grille, complete, 160 CFM		22	.909		111	47.50		158.50	197
6910 180 CFM		20	1		93.50	52		145.50	185
6920 270 CFM		18	1.111		176	58		234	285
6940 Residential roof jacks and wall caps									
6944 Wall cap with back draft damper									
6946 3" & 4" diam. round duct	1 Shee	11	.727	Ea.	26	41.50		67.50	94
6948 6" diam. round duct	"	11	.727	"	65	41.50		106.50	137
6958 Roof jack with bird screen and back draft damper									
6960 3" & 4" diam. round duct	1 Shee	11	.727	Ea.	26	41.50		67.50	94.50
6962 3-1/4" x 10" rectangular duct	"	10	.800	"	49	46		95	127
6980 Transition									
6982 3-1/4" x 10" to 6" diam. round	1 Shee	20	.400	Ea.	33	23		56	72.50
8020 Attic, roof type									
8030 Aluminum dome, damper & curb									
8080 12" diameter, 1000 CFM (gravity)	1 Elec	10	.800	Ea.	595	44		639	725
8090 16" diameter, 1500 CFM (gravity)		9	.889		720	49		769	865
8100 20" diameter, 2500 CFM (gravity)		8	1		880	55		935	1,050
8110 26" diameter, 4000 CFM (gravity)		7	1.143		1,075	63		1,138	1,275
8120 32" diameter, 6500 CFM (gravity)		6	1.333		1,475	73.50		1,548.50	1,725
8130 38" diameter, 8000 CFM (gravity)		5	1.600		2,175	88		2,263	2,525
8140 50" diameter, 13,000 CFM (gravity)		4	2		3,150	110		3,260	3,650
8160 Plastic, ABS dome									
8180 1050 CFM	1 Elec	14	.571	Ea.	175	31.50		206.50	242
8200 1600 CFM	"	12	.667	"	262	36.50		298.50	345
8240 Attic, wall type, with shutter, one speed									
8250 12" diameter, 1000 CFM	1 Elec	14	.571	Ea.	410	31.50		441.50	500
8260 14" diameter, 1500 CFM		12	.667		445	36.50		481.50	545
8270 16" diameter, 2000 CFM		9	.889		505	49		554	630
8290 Whole house, wall type, with shutter, one speed									
8300 30" diameter, 4800 CFM	1 Elec	7	1.143	Ea.	1,075	63		1,138	1,275
8310 36" diameter, 7000 CFM		6	1.333		1,175	73.50		1,248.50	1,425
8320 42" diameter, 10,000 CFM		5	1.600		1,325	88		1,413	1,575
8330 48" diameter, 16,000 CFM		4	2		1,625	110		1,735	1,975
8340 For two speed, add					98			98	108
8350 Whole house, lay-down type, with shutter, one speed									
8360 30" diameter, 4500 CFM	1 Elec	8	1	Ea.	1,150	55		1,205	1,325
8370 36" diameter, 6500 CFM		7	1.143		1,225	63		1,288	1,450
8380 42" diameter, 9000 CFM		6	1.333		1,350	73.50		1,423.50	1,625
8390 48" diameter, 12,000 CFM		5	1.600		1,550	88		1,638	1,825
8440 For two speed, add					74			74	81
8450 For 12 hour timer switch, add	1 Elec	32	.250		74	13.80		87.80	103

For customer support on your Commercial Renovation Cost Data, call 877.742.4719.

449

23 36 Air Terminal Units

23 36 13 – Constant-Air-Volume Units

23 36 13.10 Constant Volume Mixing Boxes	Crew	Daily Output	Labor-Hours	Unit	Material	2016 Bare Costs Labor	Equipment	Total	Total Incl O&P
0010 **CONSTANT VOLUME MIXING BOXES**									
5180 Mixing box, includes electric or pneumatic motor									
5200 Constant volume, 150 to 270 CFM	Q-9	12	1.333	Ea.	715	68.50		783.50	895
5210 270 to 600 CFM	"	11	1.455	"	735	75		810	930

23 36 16 – Variable-Air-Volume Units

23 36 16.10 Variable Volume Mixing Boxes

	Crew	Daily Output	Labor-Hours	Unit	Material	2016 Bare Costs Labor	Equipment	Total	Total Incl O&P
0010 **VARIABLE VOLUME MIXING BOXES**									
5180 Mixing box, includes electric or pneumatic motor									
5500 VAV Cool only, pneumatic, pressure independent 300 to 600 CFM	Q-9	11	1.455	Ea.	745	75		820	940
5510 500 to 1000 CFM		9	1.778		765	91.50		856.50	985
5520 800 to 1600 CFM		9	1.778		790	91.50		881.50	1,025
5530 1100 to 2000 CFM		8	2		810	103		913	1,050
5540 1500 to 3000 CFM		7	2.286		860	118		978	1,125
5550 2000 to 4000 CFM		6	2.667		870	137		1,007	1,175
5560 For electric, w/thermostat, pressure dependent, add					23			23	25.50

23 37 Air Outlets and Inlets

23 37 13 – Diffusers, Registers, and Grilles

23 37 13.10 Diffusers

	Crew	Daily Output	Labor-Hours	Unit	Material	2016 Bare Costs Labor	Equipment	Total	Total Incl O&P
0010 **DIFFUSERS**, Aluminum, opposed blade damper unless noted									
0100 Ceiling, linear, also for sidewall									
0120 2" wide	1 Shee	32	.250	L.F.	16.80	14.30		31.10	41
0160 4" wide		26	.308		22.50	17.60		40.10	52.50
0180 6" wide		24	.333		27.50	19.10		46.60	60.50
0200 8" wide		22	.364		32.50	21		53.50	68.50
0500 Perforated, 24" x 24" lay-in panel size, 6" x 6"		16	.500	Ea.	157	28.50		185.50	217
0520 8" x 8"		15	.533		165	30.50		195.50	229
0530 9" x 9"		14	.571		167	32.50		199.50	236
0540 10" x 10"		14	.571		168	32.50		200.50	236
0560 12" x 12"		12	.667		175	38		213	252
0590 16" x 16"		11	.727		195	41.50		236.50	281
0600 18" x 18"		10	.800		209	46		255	305
0610 20" x 20"		10	.800		226	46		272	320
0620 24" x 24"		9	.889		248	51		299	355
1000 Rectangular, 1 to 4 way blow, 6" x 6"		16	.500		50	28.50		78.50	100
1010 8" x 8"		15	.533		58	30.50		88.50	112
1014 9" x 9"		15	.533		66	30.50		96.50	121
1016 10" x 10"		15	.533		80	30.50		110.50	136
1020 12" x 6"		15	.533		71.50	30.50		102	127
1040 12" x 9"		14	.571		75.50	32.50		108	135
1060 12" x 12"		12	.667		84.50	38		122.50	153
1070 14" x 6"		13	.615		77.50	35		112.50	141
1074 14" x 14"		12	.667		128	38		166	200
1150 18" x 18"		9	.889		138	51		189	232
1170 24" x 12"		10	.800		163	46		209	253
2000 T bar mounting, 24" x 24" lay-in frame, 6" x 6"		16	.500		72	28.50		100.50	125
2020 8" x 8"		14	.571		72.50	32.50		105	131
2040 12" x 12"		12	.667		88.50	38		126.50	158
2060 16" x 16"		11	.727		119	41.50		160.50	197
2080 18" x 18"		10	.800		133	46		179	219

23 37 Air Outlets and Inlets

23 37 13 – Diffusers, Registers, and Grilles

23 37 13.10 Diffusers		Crew	Daily Output	Labor-Hours	Unit	Material	2016 Bare Costs Labor	2016 Bare Costs Equipment	Total	Total Incl O&P
6000	For steel diffusers instead of aluminum, deduct				Ea.	10%				
9000	Minimum labor/equipment charge	1 Shee	4	2	Job		115		115	181

23 37 13.30 Grilles

	23 37 13.30 Grilles	Crew	Daily Output	Labor-Hours	Unit	Material	Labor	Equipment	Total	Total Incl O&P
0010	**GRILLES**									
0020	Aluminum, unless noted otherwise									
1000	Air return, steel, 6" x 6"	1 Shee	26	.308	Ea.	18.70	17.60		36.30	48.50
1020	10" x 6"		24	.333		18.70	19.10		37.80	50.50
1080	16" x 8"		22	.364		26.50	21		47.50	62
1100	12" x 12"		22	.364		26.50	21		47.50	62
1120	24" x 12"		18	.444		36	25.50		61.50	79.50
1300	48" x 24"		12	.667		128	38		166	200
9000	Minimum labor/equipment charge		4	2	Job		115		115	181

23 37 13.60 Registers

	23 37 13.60 Registers	Crew	Daily Output	Labor-Hours	Unit	Material	Labor	Equipment	Total	Total Incl O&P
0010	**REGISTERS**									
0980	Air supply									
1000	Ceiling/wall, O.B. damper, anodized aluminum									
1010	One or two way deflection, adj. curved face bars									
1120	12" x 12"	1 Shee	18	.444	Ea.	22	25.50		47.50	64
1140	14" x 8"		17	.471		19.10	27		46.10	63.50
1240	20" x 6"		18	.444		19.60	25.50		45.10	61.50
1340	24" x 8"		13	.615		26.50	35		61.50	84.50
1350	24" x 18"		12	.667		49	38		87	114
2700	Above registers in steel instead of aluminum, deduct					10%				
3000	Baseboard, hand adj. damper, enameled steel									
3012	8" x 6"	1 Shee	26	.308	Ea.	5.05	17.60		22.65	33.50
3020	10" x 6"		24	.333		5.50	19.10		24.60	36
3040	12" x 5"		23	.348		6	19.90		25.90	38
3060	12" x 6"		23	.348		6.75	19.90		26.65	39
3080	12" x 8"		22	.364		8.70	21		29.70	42.50
3100	14" x 6"		20	.400		6.50	23		29.50	43
9000	Minimum labor/equipment charge		4	2	Job		115		115	181

23 37 23 – HVAC Gravity Ventilators

23 37 23.10 HVAC Gravity Air Ventilators

	23 37 23.10 HVAC Gravity Air Ventilators	Crew	Daily Output	Labor-Hours	Unit	Material	Labor	Equipment	Total	Total Incl O&P
0010	**HVAC GRAVITY AIR VENTILATORS**, Includes base									
1280	Rotary ventilators, wind driven, galvanized									
1300	4" neck diameter	Q-9	20	.800	Ea.	47.50	41		88.50	118
1400	12" neck diameter		10	1.600		67	82.50		149.50	204
1500	24" neck diameter		8	2		299	103		402	495
1600	For aluminum, add					30%				
9000	Minimum labor/equipment charge	1 Plum	2	4	Job		237		237	370

451

For customer support on your Commercial Renovation Cost Data, call 877.742.4719.

23 38 Ventilation Hoods

23 38 13 – Commercial-Kitchen Hoods

23 38 13.10 Hood and Ventilation Equipment	Crew	Daily Output	Labor-Hours	Unit	Material	2016 Bare Costs Labor	Equipment	Total	Total Incl O&P
0010 **HOOD AND VENTILATION EQUIPMENT**									
2970 Exhaust hood, sst, gutter on all sides, 4' x 4' x 2'	1 Carp	1.80	4.444	Ea.	4,725	215		4,940	5,550
2980 4' x 4' x 7'	"	1.60	5	"	7,525	242		7,767	8,675

23 41 Particulate Air Filtration

23 41 13 – Panel Air Filters

23 41 13.10 Panel Type Air Filters

	Crew	Daily Output	Labor-Hours	Unit	Material	Labor	Equipment	Total	Total Incl O&P
0010 **PANEL TYPE AIR FILTERS**									
2950 Mechanical media filtration units									
3000 High efficiency type, with frame, non-supported G				MCFM	35			35	38.50
3100 Supported type G				"	50			50	55
5500 Throwaway glass or paper media type, 12" x 36" x 1"				Ea.	2.39			2.39	2.63

23 41 16 – Renewable-Media Air Filters

23 41 16.10 Disposable Media Air Filters

	Crew	Daily Output	Labor-Hours	Unit	Material	Labor	Equipment	Total	Total Incl O&P
0010 **DISPOSABLE MEDIA AIR FILTERS**									
5000 Renewable disposable roll				C.S.F.	5.25			5.25	5.75

23 41 19 – Washable Air Filters

23 41 19.10 Permanent Air Filters

	Crew	Daily Output	Labor-Hours	Unit	Material	Labor	Equipment	Total	Total Incl O&P
0010 **PERMANENT AIR FILTERS**									
4500 Permanent washable G				MCFM	20			20	22

23 41 23 – Extended Surface Filters

23 41 23.10 Expanded Surface Filters

	Crew	Daily Output	Labor-Hours	Unit	Material	Labor	Equipment	Total	Total Incl O&P
0010 **EXPANDED SURFACE FILTERS**									
4000 Medium efficiency, extended surface G				MCFM	6			6	6.60

23 42 Gas-Phase Air Filtration

23 42 13 – Activated-Carbon Air Filtration

23 42 13.10 Charcoal Type Air Filtration

	Crew	Daily Output	Labor-Hours	Unit	Material	Labor	Equipment	Total	Total Incl O&P
0010 **CHARCOAL TYPE AIR FILTRATION**									
0050 Activated charcoal type, full flow				MCFM	600			600	660
0060 Full flow, impregnated media 12" deep					225			225	248
0070 HEPA filter & frame for field erection					400			400	440
0080 HEPA filter-diffuser, ceiling install.				↓	300			300	330

23 43 Electronic Air Cleaners

23 43 13 – Washable Electronic Air Cleaners

23 43 13.10 Electronic Air Cleaners

	Crew	Daily Output	Labor-Hours	Unit	Material	Labor	Equipment	Total	Total Incl O&P
0010 **ELECTRONIC AIR CLEANERS**									
2000 Electronic air cleaner, duct mounted									
2150 1000 CFM	1 Shee	4	2	Ea.	420	115		535	645
2200 1200 CFM		3.80	2.105		485	121		606	725
2250 1400 CFM	↓	3.60	2.222	↓	500	127		627	750

23 51 Breechings, Chimneys, and Stacks

23 51 13 – Draft Control Devices

23 51 13.16 Vent Dampers

		Crew	Daily Output	Labor-Hours	Unit	Material	2016 Bare Costs Labor	2016 Bare Costs Equipment	Total	Total Incl O&P
0010	**VENT DAMPERS**									
5000	Vent damper, bi-metal, gas, 3" diameter	Q-9	24	.667	Ea.	51.50	34.50		86	111
5010	4" diameter		24	.667		51.50	34.50		86	111
5020	5" diameter		23	.696		51.50	36		87.50	114
5030	6" diameter		22	.727		51.50	37.50		89	116
5040	7" diameter		21	.762		53.50	39.50		93	121
5050	8" diameter		20	.800		57	41		98	128
9000	Minimum labor/equipment charge	1 Shee	4	2	Job		115		115	181

23 51 13.19 Barometric Dampers

		Crew	Daily Output	Labor-Hours	Unit	Material	2016 Bare Costs Labor	2016 Bare Costs Equipment	Total	Total Incl O&P
0010	**BAROMETRIC DAMPERS**									
1000	Barometric, gas fired system only, 6" size for 5" and 6" pipes	1 Shee	20	.400	Ea.	74.50	23		97.50	118
1040	8" size, for 7" and 8" pipes	"	18	.444	"	103	25.50		128.50	154
2000	All fuel, oil, oil/gas, coal									
2020	10" for 9" and 10" pipes	1 Shee	15	.533	Ea.	173	30.50		203.50	238
3260	For thermal switch for above, add	"	24	.333	"	71.50	19.10		90.60	109

23 51 23 – Gas Vents

23 51 23.10 Gas Chimney Vents

		Crew	Daily Output	Labor-Hours	Unit	Material	2016 Bare Costs Labor	2016 Bare Costs Equipment	Total	Total Incl O&P
0010	**GAS CHIMNEY VENTS**, Prefab metal, U.L. listed									
0020	Gas, double wall, galvanized steel									
0080	3" diameter	Q-9	72	.222	V.L.F.	6.60	11.45		18.05	25.50
0100	4" diameter		68	.235		8.15	12.10		20.25	28
0120	5" diameter		64	.250		9.25	12.90		22.15	30.50
0140	6" diameter		60	.267		11.05	13.75		24.80	33.50
0160	7" diameter		56	.286		21	14.70		35.70	46
0180	8" diameter		52	.308		22.50	15.85		38.35	50
0200	10" diameter		48	.333		43	17.20		60.20	74.50
0220	12" diameter		44	.364		51.50	18.75		70.25	86
0260	16" diameter		40	.400		123	20.50		143.50	168
0300	20" diameter	Q-10	36	.667		179	35.50		214.50	252
0340	24" diameter	"	32	.750		280	40		320	375

23 51 26 – All-Fuel Vent Chimneys

23 51 26.30 All-Fuel Vent Chimneys, Double Wall, St. Stl.

		Crew	Daily Output	Labor-Hours	Unit	Material	2016 Bare Costs Labor	2016 Bare Costs Equipment	Total	Total Incl O&P
0010	**ALL-FUEL VENT CHIMNEYS, DOUBLE WALL, STAINLESS STEEL**									
7780	All fuel, pressure tight, double wall, 4" insulation, U.L. listed, 1400°F.									
7790	304 stainless steel liner, aluminized steel outer jacket									
7800	6" diameter	Q-9	60	.267	V.L.F.	65	13.75		78.75	93
7804	8" diameter		52	.308		75	15.85		90.85	107
7806	10" diameter		48	.333		83.50	17.20		100.70	119
7808	12" diameter		44	.364		95.50	18.75		114.25	135
7810	14" diameter		42	.381		107	19.65		126.65	149
7880	For 316 stainless steel liner, add				L.F.	30%				
8000	All fuel, double wall, stainless steel fittings									
8010	Roof support, 6" diameter	Q-9	30	.533	Ea.	115	27.50		142.50	171
8030	8" diameter		26	.615		134	31.50		165.50	198
8040	10" diameter		24	.667		142	34.50		176.50	210
8050	12" diameter		22	.727		151	37.50		188.50	225
8060	14" diameter		21	.762		158	39.50		197.50	236
8100	Elbow 45°, 6" diameter		30	.533		246	27.50		273.50	315
8140	8" diameter		26	.615		276	31.50		307.50	355
8160	10" diameter		24	.667		315	34.50		349.50	400
8180	12" diameter		22	.727		360	37.50		397.50	455

For customer support on your Commercial Renovation Cost Data, call 877.742.4719.

453

23 51 26 – All-Fuel Vent Chimneys

23 51 26.30 All-Fuel Vent Chimneys, Double Wall, St. Stl.		Crew	Daily Output	Labor-Hours	Unit	Material	2016 Bare Costs Labor	Equipment	Total	Total Incl O&P
8200	14" diameter	Q-9	21	.762	Ea.	400	39.50		439.50	500
8300	Insulated tee, 6" diameter		30	.533		289	27.50		316.50	365
8360	8" diameter		26	.615		320	31.50		351.50	400
8380	10" diameter		24	.667		355	34.50		389.50	445
8400	12" diameter		22	.727		405	37.50		442.50	505
8420	14" diameter		20	.800		470	41		511	585
8500	Boot tee, 6" diameter		28	.571		550	29.50		579.50	650
8520	8" diameter		24	.667		600	34.50		634.50	715
8530	10" diameter		22	.727		700	37.50		737.50	830
8540	12" diameter		20	.800		830	41		871	975
8550	14" diameter		18	.889		945	46		991	1,100
8600	Rain cap with bird screen, 6" diameter		30	.533		296	27.50		323.50	370
8640	8" diameter		26	.615		345	31.50		376.50	425
8660	10" diameter		24	.667		400	34.50		434.50	495
8680	12" diameter		22	.727		460	37.50		497.50	570
8700	14" diameter		21	.762		525	39.50		564.50	635
8800	Flat roof flashing, 6" diameter		30	.533		108	27.50		135.50	163
8840	8" diameter		26	.615		118	31.50		149.50	180
8860	10" diameter		24	.667		127	34.50		161.50	193
8880	12" diameter		22	.727		139	37.50		176.50	212
8900	14" diameter		21	.762		142	39.50		181.50	218

23 52 Heating Boilers

23 52 13 – Electric Boilers

23 52 13.10 Electric Boilers, ASME

		Crew	Daily Output	Labor-Hours	Unit	Material	2016 Bare Costs Labor	Equipment	Total	Total Incl O&P
0010	**ELECTRIC BOILERS, ASME**, Standard controls and trim									
1000	Steam, 6 KW, 20.5 MBH	Q-19	1.20	20	Ea.	4,100	1,100		5,200	6,200
1160	60 KW, 205 MBH		1	24		6,925	1,325		8,250	9,650
1300	296 KW, 1010 MBH		.45	53.333		25,000	2,925		27,925	32,200
1400	592 KW, 2020 MBH	Q-21	.34	94.118		35,400	5,300		40,700	47,200
1540	1110 KW, 3788 MBH	"	.19	168		49,600	9,475		59,075	69,000
2000	Hot water, 7.5 KW, 25.6 MBH	Q-19	1.30	18.462		4,950	1,000		5,950	7,025
2040	30 KW, 102 MBH		1.20	20		5,325	1,100		6,425	7,550
2060	45 KW, 164 MBH		1.20	20		5,425	1,100		6,525	7,650
2070	60 KW, 205 MBH		1.20	20		5,525	1,100		6,625	7,775
2080	75 KW, 256 MBH		1.10	21.818		5,900	1,200		7,100	8,350
2100	90 KW, 307 MBH		1.10	21.818		6,000	1,200		7,200	8,450
9000	Minimum labor/equipment charge	Q-20	1	20	Job		1,050		1,050	1,650

23 52 23 – Cast-Iron Boilers

23 52 23.20 Gas-Fired Boilers

		Crew	Daily Output	Labor-Hours	Unit	Material	2016 Bare Costs Labor	Equipment	Total	Total Incl O&P
0010	**GAS-FIRED BOILERS**, Natural or propane, standard controls, packaged									
1000	Cast iron, with insulated jacket									
2000	Steam, gross output, 81 MBH	Q-7	1.40	22.857	Ea.	2,450	1,325		3,775	4,750
2060	163 MBH		.90	35.556		3,275	2,050		5,325	6,800
2200	440 MBH		.51	62.500		6,350	3,600		9,950	12,600
2320	1,875 MBH		.30	106		25,200	6,175		31,375	37,300
2400	3570 MBH		.18	181		35,800	10,500		46,300	56,000
2540	6,970 MBH		.10	320		98,500	18,500		117,000	137,500
3000	Hot water, gross output, 80 MBH		1.46	21.918		1,975	1,275		3,250	4,150
3020	100 MBH		1.35	23.704		2,550	1,375		3,925	4,950

23 52 Heating Boilers

23 52 23 – Cast-Iron Boilers

23 52 23.20 Gas-Fired Boilers

		Crew	Daily Output	Labor-Hours	Unit	Material	2016 Bare Costs Labor	Equipment	Total	Total Incl O&P
3040	122 MBH	Q-7	1.10	29.091	Ea.	2,675	1,675		4,350	5,550
3060	163 MBH		1	32		3,150	1,850		5,000	6,350
3200	440 MBH		.58	54.983		5,950	3,175		9,125	11,500
3320	2,000 MBH		.26	125		21,400	7,225		28,625	34,800
3540	6,970 MBH	↓	.09	359		131,500	20,800		152,300	177,000
7000	For tankless water heater, add					10%				
7050	For additional zone valves up to 312 MBH, add			↓		194			194	214
9900	Minimum labor/equipment charge	Q-6	1	24	Job		1,350		1,350	2,125

23 52 23.30 Gas/Oil Fired Boilers

		Crew	Daily Output	Labor-Hours	Unit	Material	2016 Bare Costs Labor	Equipment	Total	Total Incl O&P
0010	**GAS/OIL FIRED BOILERS**, Combination with burners and controls, packaged									
1000	Cast iron with insulated jacket									
2000	Steam, gross output, 720 MBH	Q-7	.43	74.074	Ea.	14,700	4,275		18,975	22,900
2140	2,700 MBH		.19	165		28,100	9,575		37,675	45,800
2380	6,970 MBH	↓	.09	372	↓	102,500	21,500		124,000	146,500
2900	Hot water, gross output									
2910	200 MBH	Q-6	.62	39.024	Ea.	10,600	2,200		12,800	15,200
2920	300 MBH		.49	49.080		10,600	2,775		13,375	16,000
2930	400 MBH		.41	57.971		12,400	3,275		15,675	18,800
2940	500 MBH	↓	.36	67.039		13,400	3,800		17,200	20,600
3000	584 MBH	Q-7	.44	72.072		14,800	4,175		18,975	22,800
3060	1,460 MBH		.28	113		36,800	6,525		43,325	50,500
3300	13,500 MBH, 403.3 BHP	↓	.04	727	↓	192,500	42,000		234,500	277,000

23 52 23.40 Oil-Fired Boilers

		Crew	Daily Output	Labor-Hours	Unit	Material	2016 Bare Costs Labor	Equipment	Total	Total Incl O&P
0010	**OIL-FIRED BOILERS**, Standard controls, flame retention burner, packaged									
1000	Cast iron, with insulated flush jacket									
2000	Steam, gross output, 109 MBH	Q-7	1.20	26.667	Ea.	2,325	1,550		3,875	4,950
2060	207 MBH		.90	35.556		3,175	2,050		5,225	6,700
2180	1,084 MBH		.38	85.106		10,900	4,925		15,825	19,700
2200	1,360 MBH		.33	98.160		12,500	5,675		18,175	22,600
2240	2,175 MBH		.24	133		18,100	7,750		25,850	32,100
2340	4,360 MBH		.15	214		42,800	12,400		55,200	66,500
2460	6,970 MBH	↓	.09	363	↓	101,000	21,000		122,000	144,000
3000	Hot water, same price as steam									
4000	For tankless coil in smaller sizes, add				Ea.	15%				

23 52 26 – Steel Boilers

23 52 26.40 Oil-Fired Boilers

		Crew	Daily Output	Labor-Hours	Unit	Material	2016 Bare Costs Labor	Equipment	Total	Total Incl O&P
0010	**OIL-FIRED BOILERS**, Standard controls, flame retention burner									
5000	Steel, with insulated flush jacket									
7000	Hot water, gross output, 103 MBH	Q-6	1.60	15	Ea.	1,825	850		2,675	3,325
7020	122 MBH		1.45	16.506		2,000	935		2,935	3,650
7040	137 MBH		1.36	17.595		2,150	995		3,145	3,900
7060	168 MBH		1.30	18.405		2,250	1,050		3,300	4,100
7080	225 MBH		1.22	19.704		3,250	1,125		4,375	5,325
7100	315 MBH		.96	25.105		7,000	1,425		8,425	9,925
7120	420 MBH		.70	34.483	↓	7,350	1,950		9,300	11,200
9000	Minimum labor/equipment charge	↓	1.75	13.714	Job		775		775	1,200

For customer support on your Commercial Renovation Cost Data, call 877.742.4719.

455

23 52 Heating Boilers

23 52 88 – Burners

23 52 88.10 Replacement Type Burners	Crew	Daily Output	Labor-Hours	Unit	Material	2016 Bare Costs Labor	Equipment	Total	Total Incl O&P
0010 **REPLACEMENT TYPE BURNERS**									
0990 Residential, conversion, gas fired, LP or natural									
1000 Gun type, atmospheric input 50 to 225 MBH	Q-1	2.50	6.400	Ea.	915	340		1,255	1,525
1020 100 to 400 MBH		2	8		1,525	425		1,950	2,350
1040 300 to 1000 MBH	↓	1.70	9.412	↓	4,600	500		5,100	5,850
3000 Flame retention oil fired assembly, input									
3020 .50 to 2.25 GPH	Q-1	2.40	6.667	Ea.	266	355		621	850
3040 2.0 to 5.0 GPH		2	8		335	425		760	1,025
3060 3.0 to 7.0 GPH		1.80	8.889		575	475		1,050	1,375
3080 6.0 to 12.0 GPH	↓	1.60	10	↓	920	535		1,455	1,825

23 54 Furnaces

23 54 13 – Electric-Resistance Furnaces

23 54 13.10 Electric Furnaces

	Crew	Daily Output	Labor-Hours	Unit	Material	2016 Bare Costs Labor	Equipment	Total	Total Incl O&P
0010 **ELECTRIC FURNACES**, Hot air, blowers, std. controls									
0011 not including gas, oil or flue piping									
1000 Electric, UL listed									
1100 34.1 MBH	Q-20	4.40	4.545	Ea.	455	237		692	875

23 54 16 – Fuel-Fired Furnaces

23 54 16.13 Gas-Fired Furnaces

	Crew	Daily Output	Labor-Hours	Unit	Material	2016 Bare Costs Labor	Equipment	Total	Total Incl O&P
0010 **GAS-FIRED FURNACES**									
3000 Gas, AGA certified, upflow, direct drive models									
3020 45 MBH input	Q-9	4	4	Ea.	580	206		786	960
3040 60 MBH input		3.80	4.211		595	217		812	995
3060 75 MBH input		3.60	4.444		660	229		889	1,075
3100 100 MBH input		3.20	5		690	258		948	1,175
3120 125 MBH input		3	5.333		725	275		1,000	1,225
3130 150 MBH input		2.80	5.714		1,375	294		1,669	1,975
3140 200 MBH input	↓	2.60	6.154	↓	2,950	315		3,265	3,750

23 54 16.16 Oil-Fired Furnaces

	Crew	Daily Output	Labor-Hours	Unit	Material	2016 Bare Costs Labor	Equipment	Total	Total Incl O&P
0010 **OIL-FIRED FURNACES**									
6000 Oil, UL listed, atomizing gun type burner									
6020 56 MBH output	Q-9	3.60	4.444	Ea.	1,700	229		1,929	2,225
6040 95 MBH output		3.40	4.706		2,575	242		2,817	3,200
6060 134 MBH output		3.20	5		2,625	258		2,883	3,275
6080 151 MBH output		3	5.333		2,725	275		3,000	3,425
6100 200 MBH input		2.60	6.154	↓	3,000	315		3,315	3,825
9000 Minimum labor/equipment charge	↓	2.75	5.818	Job		300		300	475

23 54 24 – Furnace Components for Cooling

23 54 24.10 Furnace Components and Combinations

	Crew	Daily Output	Labor-Hours	Unit	Material	2016 Bare Costs Labor	Equipment	Total	Total Incl O&P
0010 **FURNACE COMPONENTS AND COMBINATIONS**									
0080 Coils, A.C. evaporator, for gas or oil furnaces									
0090 Add-on, with holding charge									
0100 Upflow									
0120 1-1/2 ton cooling	Q-5	4	4	Ea.	222	219		441	585
0130 2 ton cooling		3.70	4.324		241	236		477	635
0140 3 ton cooling		3.30	4.848		400	265		665	855
0150 4 ton cooling		3	5.333		530	291		821	1,050
0160 5 ton cooling	↓	2.70	5.926	↓	550	325		875	1,100

23 54 Furnaces

23 54 24 – Furnace Components for Cooling

23 54 24.10 Furnace Components and Combinations	Crew	Daily Output	Labor-Hours	Unit	Material	2016 Bare Costs Labor	Equipment	Total	Total Incl O&P
0300	Downflow									
0330	2-1/2 ton cooling	Q-5	3	5.333	Ea.	330	291		621	820
0340	3-1/2 ton cooling		2.60	6.154		505	335		840	1,075
0350	5 ton cooling	↓	2.20	7.273	↓	550	395		945	1,225
0600	Horizontal									
0630	2 ton cooling	Q-5	3.90	4.103	Ea.	269	224		493	645
0640	3 ton cooling		3.50	4.571		380	250		630	810
0650	4 ton cooling		3.20	5		445	273		718	915
0660	5 ton cooling	↓	2.90	5.517	↓	445	300		745	960
2000	Cased evaporator coils for air handlers									
2100	1-1/2 ton cooling	Q-5	4.40	3.636	Ea.	251	199		450	585
2110	2 ton cooling		4.10	3.902		330	213		543	695
2120	2-1/2 ton cooling		3.90	4.103		340	224		564	725
2130	3 ton cooling		3.70	4.324		365	236		601	770
2140	3-1/2 ton cooling		3.50	4.571		455	250		705	895
2150	4 ton cooling		3.20	5		495	273		768	970
2160	5 ton cooling	↓	2.90	5.517	↓	500	300		800	1,025
3010	Air handler, modular									
3100	With cased evaporator cooling coil									
3120	1-1/2 ton cooling	Q-5	3.80	4.211	Ea.	855	230		1,085	1,300
3130	2 ton cooling		3.50	4.571		945	250		1,195	1,450
3140	2-1/2 ton cooling		3.30	4.848		1,000	265		1,265	1,525
3150	3 ton cooling		3.10	5.161		1,150	282		1,432	1,725
3160	3-1/2 ton cooling		2.90	5.517		1,325	300		1,625	1,950
3170	4 ton cooling		2.50	6.400		1,425	350		1,775	2,125
3180	5 ton cooling	↓	2.10	7.619	↓	1,700	415		2,115	2,525
3500	With no cooling coil									
3520	1-1/2 ton coil size	Q-5	12	1.333	Ea.	720	73		793	910
3530	2 ton coil size		10	1.600		720	87.50		807.50	930
3540	2-1/2 ton coil size		10	1.600		730	87.50		817.50	940
3554	3 ton coil size		9	1.778		750	97		847	975
3560	3-1/2 ton coil size		9	1.778		770	97		867	995
3570	4 ton coil size		8.50	1.882		780	103		883	1,025
3580	5 ton coil size	↓	8	2	↓	790	109		899	1,050
4000	With heater									
4120	5 kW, 17.1 MBH	Q-5	16	1	Ea.	455	54.50		509.50	585
4130	7.5 kW, 25.6 MBH		15.60	1.026		1,100	56		1,156	1,300
4140	10 kW, 34.2 MBH		15.20	1.053		1,175	57.50		1,232.50	1,400
4150	12.5 kW, 42.7 MBH	↓	14.80	1.081	↓	1,550	59		1,609	1,800

23 55 Fuel-Fired Heaters

23 55 13 – Fuel-Fired Duct Heaters

23 55 13.16 Gas-Fired Duct Heaters

		Crew	Daily Output	Labor-Hours	Unit	Material	2016 Bare Costs Labor	Equipment	Total	Total Incl O&P
0010	**GAS-FIRED DUCT HEATERS**, Includes burner, controls, stainless steel									
0020	heat exchanger. Gas fired, electric ignition									
1000	Outdoor installation, with power venter									
1120	225 MBH output	Q-5	2.50	6.400	Ea.	5,100	350		5,450	6,175
1160	375 MBH output		1.60	10		8,325	545		8,870	10,000
1180	450 MBH output	↓	1.40	11.429	↓	9,550	625		10,175	11,500

For customer support on your Commercial Renovation Cost Data, call 877.742.4719.

457

23 55 Fuel-Fired Heaters

23 55 33 – Fuel-Fired Unit Heaters

23 55 33.16 Gas-Fired Unit Heaters		Crew	Daily Output	Labor-Hours	Unit	Material	2016 Bare Costs Labor	Equipment	Total	Total Incl O&P
0010	**GAS-FIRED UNIT HEATERS**, Cabinet, grilles, fan, ctrls., burner, no piping									
0022	thermostat, no piping. For flue see Section 23 51 23.10									
1000	Gas fired, floor mounted									
1100	60 MBH output	Q-5	10	1.600	Ea.	925	87.50		1,012.50	1,150
1180	180 MBH output		6	2.667		1,475	146		1,621	1,825
2000	Suspension mounted, propeller fan, 20 MBH output		8.50	1.882		1,150	103		1,253	1,400
2040	60 MBH output		7	2.286		1,750	125		1,875	2,125
2100	130 MBH output		5	3.200		2,250	175		2,425	2,750
2240	320 MBH output	▼	2	8		4,200	435		4,635	5,275
2500	For powered venter and adapter, add					440			440	485
5000	Wall furnace, 17.5 MBH output	Q-5	6	2.667		745	146		891	1,050
5020	24 MBH output		5	3.200		745	175		920	1,100
5040	35 MBH output		4	4	▼	825	219		1,044	1,250
9000	Minimum labor/equipment charge	▼	3.50	4.571	Job		250		250	390

23 56 Solar Energy Heating Equipment

23 56 16 – Packaged Solar Heating Equipment

23 56 16.40 Solar Heating Systems

		Crew	Daily Output	Labor-Hours	Unit	Material	2016 Bare Costs Labor	Equipment	Total	Total Incl O&P
0010	**SOLAR HEATING SYSTEMS**									
0020	System/Package prices, not including connecting									
0030	pipe, insulation, or special heating/plumbing fixtures									
0500	Hot water, standard package, low temperature									
0580	2 collectors, circulator, fittings, 120 gal. tank	[G]	Q-1	.40	40	Ea.	5,075	2,125		7,200	8,925

23 56 19 – Solar Heating Components

23 56 19.50 Solar Heating Ancillary

		Crew	Daily Output	Labor-Hours	Unit	Material	2016 Bare Costs Labor	Equipment	Total	Total Incl O&P
0010	**SOLAR HEATING ANCILLARY**									
2300	Circulators, air									
2310	Blowers									
2550	Booster fan 6" diameter, 120 CFM	[G]	Q-9	16	1	Ea.	36.50	51.50		88	122
2580	8" diameter, 150 CFM	[G]		16	1		41	51.50		92.50	127
2660	Shutter/damper	[G]		12	1.333		56	68.50		124.50	170
2670	Shutter motor	[G]	▼	16	1		139	51.50		190.50	235
2800	Circulators, liquid, 1/25 HP, 5.3 GPM	[G]	Q-1	14	1.143		181	61		242	294
2820	1/20 HP, 17 GPM	[G]	"	12	1.333	▼	263	71		334	400
3000	Collector panels, air with aluminum absorber plate									
3010	Wall or roof mount									
3040	Flat black, plastic glazing									
3080	4' x 8'	[G]	Q-9	6	2.667	Ea.	660	137		797	945
3300	Collector panels, liquid with copper absorber plate									
3320	Black chrome, tempered glass glazing									
3330	Alum. frame, 4' x 8', 5/32" single glazing	[G]	Q-1	9.50	1.684	Ea.	1,000	89.50		1,089.50	1,250
3450	Flat black, alum. frame, 3.5' x 7.5'	[G]		9	1.778		880	94.50		974.50	1,125
3500	4' x 8'	[G]		5.50	2.909		1,050	155		1,205	1,400
3600	Liquid, full wetted, plastic, alum. frame, 4' x 10'	[G]		5	3.200		320	171		491	620
3650	Collector panel mounting, flat roof or ground rack	[G]		7	2.286	▼	245	122		367	460
3670	Roof clamps	[G]	▼	70	.229	Set	2.83	12.20		15.03	22
3700	Roof strap, teflon	[G]	1 Plum	205	.039	L.F.	25	2.31		27.31	31
3900	Differential controller with two sensors									
3930	Thermostat, hard wired	[G]	1 Plum	8	1	Ea.	101	59		160	204
4050	Pool valve system	[G]	▼	2.50	3.200	▼	161	189		350	470

23 56 19.50 Solar Heating Ancillary		Crew	Daily Output	Labor-Hours	Unit	Material	2016 Bare Costs Labor	Equipment	Total	Total Incl O&P
4070	With 12 VAC actuator	1 Plum	2	4	Ea.	325	237		562	725
4150	Sensors									
4220	Freeze prevention	1 Plum	32	.250	Ea.	22.50	14.80		37.30	48
4300	Heat exchanger									
4315	includes coil, blower, circulator									
4316	and controller for DHW and space hot air									
4330	Fluid to air coil, up flow, 45 MBH	Q-1	4	4	Ea.	315	213		528	680
4380	70 MBH		3.50	4.571		360	244		604	775
4400	80 MBH		3	5.333		475	284		759	970
4580	Fluid to fluid package includes two circulating pumps									
4590	expansion tank, check valve, relief valve									
4600	controller, high temperature cutoff and sensors	Q-1	2.50	6.400	Ea.	800	340		1,140	1,400
4650	Heat transfer fluid									
4700	Propylene glycol, inhibited anti-freeze	1 Plum	28	.286	Gal.	15	16.90		31.90	43
4800	Solar storage tanks, knocked down									
5190	6'-3" high, 7' x 7' = 306 C.F./2000 gallons	Q-10	1.20	20	Ea.	14,200	1,075		15,275	17,400
5210	7' x 14' = 613 C.F./4000 gallons		.60	40		21,400	2,125		23,525	26,900
5230	10'-6" x 14' = 919 C.F./6000 gallons		.40	60		24,900	3,200		28,100	32,500
5250	14' x 17'-6" = 1531 C.F./10,000 gallons	Q-11	.30	106		32,100	5,825		37,925	44,500
7000	Solar control valves and vents									
7070	Air eliminator, automatic 3/4" size	1 Plum	32	.250	Ea.	30	14.80		44.80	56
7090	Air vent, automatic, 1/8" fitting		32	.250		17.25	14.80		32.05	42
7100	Manual, 1/8" NPT		32	.250		2.44	14.80		17.24	25.50
7120	Backflow preventer, 1/2" pipe size		16	.500		79	29.50		108.50	133
7130	3/4" pipe size		16	.500		80.50	29.50		110	135
7150	Balancing valve, 3/4" pipe size		20	.400		67.50	23.50		91	112
7180	Draindown valve, 1/2" copper tube		9	.889		212	52.50		264.50	315
7200	Flow control valve, 1/2" pipe size		22	.364		130	21.50		151.50	177
7220	Expansion tank, up to 5 gal.		32	.250		68	14.80		82.80	98
7400	Pressure gauge, 2" dial		32	.250		24	14.80		38.80	49.50
7450	Relief valve, temp. and pressure 3/4" pipe size		30	.267		22.50	15.80		38.30	49
7500	Solenoid valve, normally closed									
7520	Brass, 3/4" NPT, 24V	1 Plum	9	.889	Ea.	128	52.50		180.50	223
7530	1" NPT, 24V		9	.889		207	52.50		259.50	310
7750	Vacuum relief valve, 3/4" pipe size		32	.250		29.50	14.80		44.30	55.50
7800	Thermometers									
7820	Digital temperature monitoring, 4 locations	1 Plum	2.50	3.200	Ea.	137	189		326	445
7900	Upright, 1/2" NPT	"	8	1	"	39.50	59		98.50	136
8250	Water storage tank with heat exchanger and electric element									
8270	66 gal. with 2" x 2 lb. density insulation	1 Plum	1.60	5	Ea.	1,600	296		1,896	2,225
8300	80 gal. with 2" x 2 lb. density insulation	"	1.60	5	"	1,600	296		1,896	2,225
8500	Water storage module, plastic									
8600	Tubular, 12" diameter, 4' high	1 Carp	48	.167	Ea.	89	8.10		97.10	111
8610	12" diameter, 8' high	"	40	.200		143	9.70		152.70	173
8650	Cap, 12" diameter					19.80			19.80	22
9000	Minimum labor/equipment charge	1 Plum	2	4	Job		237		237	370

For customer support on your Commercial Renovation Cost Data, call 877.742.4719.

459

23 61 Refrigerant Compressors

23 61 16 – Reciprocating Refrigerant Compressors

23 61 16.10 Reciprocating Compressors	Crew	Daily Output	Labor-Hours	Unit	Material	2016 Bare Costs Labor	Equipment	Total	Total Incl O&P
0010 **RECIPROCATING COMPRESSORS**									
0990 Refrigeration, recip. hermetic, switches & protective devices									
1000 10 ton	Q-5	1	16	Ea.	12,800	875		13,675	15,500
1100 20 ton	Q-6	.72	33.333		27,200	1,900		29,100	32,900
1400 50 ton	"	.20	120		32,100	6,800		38,900	45,900
1600 130 ton	Q-7	.21	152	↓	40,500	8,800		49,300	58,500

23 62 Packaged Compressor and Condenser Units

23 62 13 – Packaged Air-Cooled Refrigerant Compressor and Condenser Units

23 62 13.10 Packaged Air-Cooled Refrig. Condensing Units

	Crew	Daily Output	Labor-Hours	Unit	Material	2016 Bare Costs Labor	Equipment	Total	Total Incl O&P
0010 **PACKAGED AIR-COOLED REFRIGERANT CONDENSING UNITS**									
0020 Condensing unit									
0030 Air cooled, compressor, standard controls									
0050 1.5 ton	Q-5	2.50	6.400	Ea.	1,275	350		1,625	1,950
0200 2.5 ton		1.70	9.412		1,425	515		1,940	2,375
0300 3 ton		1.30	12.308		1,475	670		2,145	2,675
0400 4 ton		.90	17.778		1,950	970		2,920	3,650
0500 5 ton	↓	.60	26.667	↓	2,325	1,450		3,775	4,850

23 64 Packaged Water Chillers

23 64 19 – Reciprocating Water Chillers

23 64 19.10 Reciprocating Type Water Chillers

	Crew	Daily Output	Labor-Hours	Unit	Material	2016 Bare Costs Labor	Equipment	Total	Total Incl O&P
0010 **RECIPROCATING TYPE WATER CHILLERS**, With standard controls									
0494 Water chillers, integral air cooled condenser									
0980 Water cooled, multiple compressor, semi-hermetic, tower not incl.									
1090 45 ton cooling	Q-7	.29	111	Ea.	22,900	6,425		29,325	35,200
1100 50 ton cooling		.28	113		29,700	6,575		36,275	43,000
1160 100 ton cooling		.18	179		51,500	10,400		61,900	72,500
1200 145 ton cooling	↓	.16	202	↓	60,500	11,700		72,200	84,500

23 64 23 – Scroll Water Chillers

23 64 23.10 Scroll Water Chillers

	Crew	Daily Output	Labor-Hours	Unit	Material	2016 Bare Costs Labor	Equipment	Total	Total Incl O&P
0010 **SCROLL WATER CHILLERS**, With standard controls									
0480 Packaged w/integral air cooled condenser									
0482 10 ton cooling	Q-7	.34	94.118	Ea.	15,200	5,450		20,650	25,200
0490 15 ton cooling		.37	86.486		18,300	5,000		23,300	28,000
0500 20 ton cooling		.34	94.118		21,700	5,450		27,150	32,300
0520 40 ton cooling	↓	.30	108	↓	33,200	6,250		39,450	46,300
0680 Scroll water cooled, single compressor, hermetic, tower not incl.									
0760 10 ton cooling	Q-6	.36	67.039	Ea.	6,625	3,800		10,425	13,200

23 64 26 – Rotary-Screw Water Chillers

23 64 26.10 Rotary-Screw Type Water Chillers

	Crew	Daily Output	Labor-Hours	Unit	Material	2016 Bare Costs Labor	Equipment	Total	Total Incl O&P
0010 **ROTARY-SCREW TYPE WATER CHILLERS**, With standard controls									
0110 Screw, liquid chiller, air cooled, insulated evaporator									
0120 130 ton	Q-7	.14	228	Ea.	93,500	13,200		106,700	123,500
0124 160 ton	"	.13	246	"	115,000	14,200		129,200	148,500

23 64 Packaged Water Chillers

23 64 33 – Direct Expansion Water Chillers

23 64 33.10 Direct Expansion Type Water Chillers	Crew	Daily Output	Labor- Hours	Unit	Material	2016 Bare Costs Labor	Equipment	Total	Total Incl O&P
0010 **DIRECT EXPANSION TYPE WATER CHILLERS**, With standard controls									
8000 Direct expansion, shell and tube type, for built up systems									
8020 1 ton	Q-5	2	8	Ea.	7,475	435		7,910	8,900
8030 5 ton		1.90	8.421		12,400	460		12,860	14,400
8040 10 ton	↓	1.70	9.412	↓	15,400	515		15,915	17,700
9000 Minimum labor/equipment charge	Q-6	1	24	Job		1,350		1,350	2,125

23 65 Cooling Towers

23 65 13 – Forced-Draft Cooling Towers

23 65 13.10 Forced-Draft Type Cooling Towers

	Crew	Daily Output	Labor- Hours	Unit	Material	2016 Bare Costs Labor	Equipment	Total	Total Incl O&P
0010 **FORCED-DRAFT TYPE COOLING TOWERS**, Packaged units									
0070 Galvanized steel									
0080 Induced draft, crossflow									
0100 Vertical, belt drive, 61 tons	Q-6	90	.267	TonAC	215	15.10		230.10	261
0150 100 ton		100	.240		208	13.60		221.60	249
0200 115 ton		109	.220		180	12.45		192.45	218
0250 131 ton		120	.200		217	11.35		228.35	256
0260 162 ton	↓	132	.182	↓	175	10.30		185.30	209
1000 For higher capacities, use multiples									
1500 Induced air, double flow									
1900 Vertical, gear drive, 167 ton	Q-6	126	.190	TonAC	170	10.80		180.80	204
2000 297 ton		129	.186		105	10.55		115.55	131
2100 582 ton		132	.182		58.50	10.30		68.80	80
2200 1016 ton	↓	150	.160	↓	79	9.05		88.05	101
3000 For higher capacities, use multiples									
3500 For pumps and piping, add	Q-6	38	.632	TonAC	108	36		144	174
4000 For absorption systems, add				"	75%	75%			
4100 Cooling water chemical feeder	Q-5	3	5.333	Ea.	380	291		671	870
5000 Fiberglass tower on galvanized steel support structure									
5010 Draw thru									
5100 100 ton	Q-6	1.40	17.143	Ea.	13,900	970		14,870	16,800
5120 120 ton		1.20	20		16,200	1,125		17,325	19,600
5140 140 ton		1	24		17,500	1,350		18,850	21,300
5160 160 ton		.80	30		19,400	1,700		21,100	24,100
5180 180 ton		.65	36.923		22,100	2,100		24,200	27,600
5200 200 ton	↓	.48	50	↓	25,000	2,825		27,825	31,900
5300 For stainless steel support structure, add					30%				
5360 For higher capacities, use multiples of each size									
6000 Stainless steel									
6010 Induced draft, crossflow, horizontal, belt drive									
6100 57 ton	Q-6	1.50	16	Ea.	27,100	905		28,005	31,200
6120 91 ton		.99	24.242		32,700	1,375		34,075	38,200
6140 111 ton		.43	55.814		43,900	3,150		47,050	53,000
6160 126 ton	↓	.22	109	↓	45,600	6,175		51,775	59,500
6170 Induced draft, crossflow, vertical, gear drive									
6172 167 ton	Q-6	.75	32	Ea.	53,000	1,825		54,825	61,500
6174 297 ton		.43	55.814		59,000	3,150		62,150	70,000
6176 582 ton		.23	104		99,500	5,900		105,400	118,500
6180 1016 ton	↓	.15	160	↓	159,000	9,075		168,075	188,500
9000 Minimum labor/equipment charge	↓	1	24	Job		1,350		1,350	2,125

For customer support on your Commercial Renovation Cost Data, call 877.742.4719.

461

23 72 Air-to-Air Energy Recovery Equipment

23 72 13 – Heat-Wheel Air-to-Air Energy-Recovery Equipment

23 72 13.10 Heat-Wheel Air-to-Air Energy Recovery Equip.	Crew	Daily Output	Labor-Hours	Unit	Material	2016 Bare Costs Labor	2016 Bare Costs Equipment	Total	Total Incl O&P
0010 **HEAT-WHEEL AIR-TO-AIR ENERGY RECOVERY EQUIPMENT**									
0100 Air to air									
9000 Minimum labor/equipment charge	1 Shee	4	2	Job		115		115	181

23 73 Indoor Central-Station Air-Handling Units

23 73 39 – Indoor, Direct Gas-Fired Heating and Ventilating Units

23 73 39.10 Make-Up Air Unit

	Crew	Daily Output	Labor-Hours	Unit	Material	2016 Bare Costs Labor	2016 Bare Costs Equipment	Total	Total Incl O&P
0010 **MAKE-UP AIR UNIT**									
0020 Indoor suspension, natural/LP gas, direct fired,									
0032 standard control. For flue see Section 23 51 23.10									
0040 70°F temperature rise, MBH is input									
0100 75 MBH input	Q-6	3.60	6.667	Ea.	5,050	380		5,430	6,150
0220 225 MBH input		2.40	10		6,875	565		7,440	8,425
0300 400 MBH input		1.60	15		16,200	850		17,050	19,100

23 74 Packaged Outdoor HVAC Equipment

23 74 23 – Packaged, Outdoor, Heating-Only Makeup-Air Units

23 74 23.16 Packaged, Ind.-Fired, In/Outdoor

	Crew	Daily Output	Labor-Hours	Unit	Material	2016 Bare Costs Labor	2016 Bare Costs Equipment	Total	Total Incl O&P
0010 **PACKAGED, IND.-FIRED, IN/OUTDOOR**, Heat-Only Makeup-Air Units									
1000 Rooftop unit, natural gas, gravity vent, S.S. exchanger									
1010 70°F temperature rise, MBH is input									
1020 250 MBH	Q-6	4	6	Ea.	15,600	340		15,940	17,700
1040 400 MBH		3.60	6.667		18,100	380		18,480	20,500
1060 550 MBH		3.30	7.273		20,400	410		20,810	23,100
1600 For cleanable filters, add					5%				
1700 For electric modulating gas control, add					10%				
9000 Minimum labor/equipment charge	Q-6	2.75	8.727	Job		495		495	770

23 74 33 – Dedicated Outdoor-Air Units

23 74 33.10 Rooftop Air Conditioners

	Crew	Daily Output	Labor-Hours	Unit	Material	2016 Bare Costs Labor	2016 Bare Costs Equipment	Total	Total Incl O&P
0010 **ROOFTOP AIR CONDITIONERS**, Standard controls, curb, economizer									
1000 Single zone, electric cool, gas heat									
1100 3 ton cooling, 60 MBH heating	Q-5	.70	22.857	Ea.	2,800	1,250		4,050	5,025
1120 4 ton cooling, 95 MBH heating		.61	26.403		3,250	1,450		4,700	5,825
1150 7.5 ton cooling, 170 MBH heating		.50	32.258		5,875	1,750		7,625	9,225
1156 8.5 ton cooling, 170 MBH heating		.46	34.783		7,100	1,900		9,000	10,800
1160 10 ton cooling, 200 MBH heating	Q-6	.67	35.982		9,100	2,050		11,150	13,200
1220 30 ton cooling, 540 MBH heating	Q-7	.47	68.376		35,600	3,950		39,550	45,400
1240 40 ton cooling, 675 MBH heating	"	.35	91.168		43,900	5,275		49,175	56,500
2000 Multizone, electric cool, gas heat, economizer									
2120 20 ton cooling, 360 MBH heating	Q-7	.53	60.038	Ea.	71,000	3,475		74,475	84,000
2180 30 ton cooling, 540 MBH heating		.37	85.562		109,000	4,950		113,950	127,500
2210 50 ton cooling, 540 MBH heating		.23	142		155,000	8,225		163,225	183,500
2220 70 ton cooling, 1500 MBH heating		.16	198		168,000	11,500		179,500	203,000
2240 80 ton cooling, 1500 MBH heating		.14	228		192,000	13,200		205,200	231,500
2280 105 ton cooling, 1500 MBH heating		.11	290		222,000	16,800		238,800	270,000
2400 For hot water heat coil, deduct					5%				
2500 For steam heat coil, deduct					2%				
2600 For electric heat, deduct					3%	5%			
9000 Minimum labor/equipment charge	Q-5	1.50	10.667	Job		585		585	910

23 81 Decentralized Unitary HVAC Equipment

23 81 19 – Self-Contained Air-Conditioners

23 81 19.10 Window Unit Air Conditioners

		Crew	Daily Output	Labor-Hours	Unit	Material	2016 Bare Costs Labor	Equipment	Total	Total Incl O&P
0010	**WINDOW UNIT AIR CONDITIONERS**									
4000	Portable/window, 15 amp, 125 V grounded receptacle required									
4500	10,000 BTUH	1 Carp	6	1.333	Ea.	590	64.50		654.50	755
4520	12,000 BTUH	L-2	8	2	"	740	84.50		824.50	955
9000	Minimum labor/equipment charge	1 Carp	2	4	Job		194		194	315

23 81 19.20 Self-Contained Single Package

		Crew	Daily Output	Labor-Hours	Unit	Material	2016 Bare Costs Labor	Equipment	Total	Total Incl O&P
0010	**SELF-CONTAINED SINGLE PACKAGE**									
0100	Air cooled, for free blow or duct, not incl. remote condenser									
0110	Constant volume									
0200	3 ton cooling	Q-5	1	16	Ea.	3,700	875		4,575	5,450
0210	4 ton cooling	"	.80	20		4,025	1,100		5,125	6,125
0240	10 ton cooling	Q-7	1	32		7,925	1,850		9,775	11,600
0280	30 ton cooling	"	.80	40		25,300	2,300		27,600	31,400
1000	Water cooled for free blow or duct, not including tower									
1010	Constant volume									
1100	3 ton cooling	Q-6	1	24	Ea.	3,575	1,350		4,925	6,050
1120	5 ton cooling	"	1	24		4,650	1,350		6,000	7,225
1140	10 ton cooling	Q-7	.90	35.556		8,825	2,050		10,875	12,900
1180	30 ton cooling	"	.70	45.714		37,900	2,650		40,550	45,800
1240	60 ton cooling	Q-8	.30	106		69,500	6,175	194	75,869	86,500
1300	For hot water or steam heat coils, add					12%	10%			
9000	Minimum labor/equipment charge	Q-5	1	16	Job		875		875	1,375

23 81 23 – Computer-Room Air-Conditioners

23 81 23.10 Computer Room Units

		Crew	Daily Output	Labor-Hours	Unit	Material	2016 Bare Costs Labor	Equipment	Total	Total Incl O&P
0010	**COMPUTER ROOM UNITS**									
1000	Air cooled, includes remote condenser but not									
1020	interconnecting tubing or refrigerant									
1160	6 ton	Q-5	.30	53.333	Ea.	38,500	2,925		41,425	47,000
1240	10 ton		.25	64		40,700	3,500		44,200	50,500
1260	12 ton		.24	66.667		42,100	3,650		45,750	52,000
1300	20 ton	Q-6	.26	92.308		53,500	5,225		58,725	67,000
1320	22 ton	"	.24	100		54,500	5,675		60,175	68,500

23 81 26 – Split-System Air-Conditioners

23 81 26.10 Split Ductless Systems

		Crew	Daily Output	Labor-Hours	Unit	Material	2016 Bare Costs Labor	Equipment	Total	Total Incl O&P
0010	**SPLIT DUCTLESS SYSTEMS**									
0100	Cooling only, single zone									
0110	Wall mount									
0120	3/4 ton cooling	Q-5	2	8	Ea.	1,100	435		1,535	1,900
0130	1 ton cooling		1.80	8.889		1,225	485		1,710	2,100
0140	1-1/2 ton cooling		1.60	10		1,975	545		2,520	3,025
0150	2 ton cooling		1.40	11.429		2,250	625		2,875	3,450
1000	Ceiling mount									
1020	2 ton cooling	Q-5	1.40	11.429	Ea.	2,000	625		2,625	3,175
1030	3 ton cooling	"	1.20	13.333	"	2,550	730		3,280	3,925
3000	Multizone									
3010	Wall mount									
3020	2 @ 3/4 ton cooling	Q-5	1.80	8.889	Ea.	3,375	485		3,860	4,450
5000	Cooling/Heating									
5010	Wall mount									
5110	1 ton cooling	Q-5	1.70	9.412	Ea.	1,300	515		1,815	2,225
5120	1-1/2 ton cooling	"	1.50	10.667	"	1,975	585		2,560	3,075

For customer support on your Commercial Renovation Cost Data, call 877.742.4719.

463

23 81 Decentralized Unitary HVAC Equipment

23 81 26 – Split-System Air-Conditioners

23 81 26.10 Split Ductless Systems

		Crew	Daily Output	Labor-Hours	Unit	Material	2016 Bare Costs Labor	2016 Bare Costs Equipment	Total	Total Incl O&P
7000	Accessories for all split ductless systems									
7010	Add for ambient frost control	Q-5	8	2	Ea.	126	109		235	310
7020	Add for tube/wiring kit, (Line sets)									
7030	15' kit	Q-5	32	.500	Ea.	93	27.50		120.50	145
7036	25' kit		28	.571		186	31		217	253
7040	35' kit		24	.667		196	36.50		232.50	273
7050	50' kit	↓	20	.800	↓	237	43.50		280.50	330

23 81 29 – Variable Refrigerant Flow HVAC Systems

23 81 29.10 Heat Pump, Gas Driven

		Crew	Daily Output	Labor-Hours	Unit	Material	2016 Bare Costs Labor	2016 Bare Costs Equipment	Total	Total Incl O&P
0010	**HEAT PUMP, GAS DRIVEN**, Variable refrigerant volume (VRV) type									
0020	Not including interconnecting tubing or multi-zone controls									
1000	For indoor fan VRV type AHU see 23 82 19.40									
1010	Multi-zone split									
1020	Outdoor unit									
1100	8 tons cooling, for up to 17 zones	Q-5	1.30	12.308	Ea.	29,100	670		29,770	33,100
1110	Isolation rails		2.60	6.154	Pair	1,025	335		1,360	1,650
1160	15 tons cooling, for up to 33 zones		1	16	Ea.	38,600	875		39,475	43,800
1170	Isolation rails	↓	2.60	6.154	Pair	1,200	335		1,535	1,850
2000	Packaged unit									
2020	Outdoor unit									
2200	11 tons cooling	Q-5	1.30	12.308	Ea.	37,200	670		37,870	42,000
2210	Roof curb adapter	"	2.60	6.154		645	335		980	1,225
2220	Thermostat	1 Stpi	1.30	6.154	↓	385	375		760	1,000

23 81 43 – Air-Source Unitary Heat Pumps

23 81 43.10 Air-Source Heat Pumps

		Crew	Daily Output	Labor-Hours	Unit	Material	2016 Bare Costs Labor	2016 Bare Costs Equipment	Total	Total Incl O&P
0010	**AIR-SOURCE HEAT PUMPS**, Not including interconnecting tubing									
1000	Air to air, split system, not including curbs, pads, fan coil and ductwork									
1012	Outside condensing unit only, for fan coil see Section 23 82 19.10									
1020	2 ton cooling, 8.5 MBH heat @ 0°F	Q-5	2	8	Ea.	1,600	435		2,035	2,425
1040	3 ton cooling, 13 MBH heat @ 0°F		1.20	13.333		2,050	730		2,780	3,375
1080	7.5 ton cooling, 33 MBH heat @ 0°F	↓	.45	35.556		4,625	1,950		6,575	8,100
1120	15 ton cooling, 64 MBH heat @ 0°F	Q-6	.50	48		8,450	2,725		11,175	13,600
1130	20 ton cooling, 85 MBH heat @ 0°F		.35	68.571		17,000	3,875		20,875	24,800
1140	25 ton cooling, 119 MBH heat @ 0°F	↓	.25	96	↓	20,400	5,450		25,850	31,000
1500	Single package, not including curbs, pads, or plenums									
1520	2 ton cooling, 6.5 MBH heat @ 0°F	Q-5	1.50	10.667	Ea.	3,100	585		3,685	4,325
1560	3 ton cooling, 10 MBH heat @ 0°F	"	1.20	13.333	"	3,575	730		4,305	5,050
6000	Air to water, single package, excluding storage tank and ductwork									
6010	Includes circulating water pump, air duct connections, digital temperature									
6020	controller with remote tank temp. probe and sensor for storage tank.									
6040	Water heating - air cooling capacity									
6110	35.5 MBH heat water, 2.3 Ton cool air	Q-5	1.60	10	Ea.	17,100	545		17,645	19,700
6120	58 MBH heat water, 3.8 Ton cool air		1.10	14.545		19,500	795		20,295	22,800
6130	76 MBH heat water, 4.9 Ton cool air		.87	18.391		23,300	1,000		24,300	27,200
6140	98 MBH heat water, 6.5 Ton cool air		.62	25.806		30,100	1,400		31,500	35,300
6150	113 MBH heat water, 7.4 Ton cool air		.59	27.119		33,000	1,475		34,475	38,600
6160	142 MBH heat water, 9.2 Ton cool air		.52	30.769		40,000	1,675		41,675	46,600
6170	171 MBH heat water, 11.1 Ton cool air	↓	.49	32.653	↓	47,100	1,775		48,875	55,000

23 81 Decentralized Unitary HVAC Equipment

23 81 46 – Water-Source Unitary Heat Pumps

23 81 46.10 Water Source Heat Pumps	Crew	Daily Output	Labor-Hours	Unit	Material	2016 Bare Costs Labor	Equipment	Total	Total Incl O&P
0010 **WATER SOURCE HEAT PUMPS**, Not incl. connecting tubing or water source									
2000 Water source to air, single package									
2100 1 ton cooling, 13 MBH heat @ 75°F	Q-5	2	8	Ea.	1,775	435		2,210	2,625
2140 2 ton cooling, 19 MBH heat @ 75°F		1.70	9.412		2,200	515		2,715	3,225
2220 5 ton cooling, 29 MBH heat @ 75°F		.90	17.778		3,275	970		4,245	5,125
9000 Minimum labor/equipment charge		1.75	9.143	Job		500		500	780

23 82 Convection Heating and Cooling Units

23 82 16 – Air Coils

23 82 16.10 Flanged Coils

	Crew	Daily Output	Labor-Hours	Unit	Material	2016 Bare Costs Labor	Equipment	Total	Total Incl O&P
0010 **FLANGED COILS**									
1500 Hot water heating, 1 row, 24" x 48"	Q-5	4	4	Ea.	1,675	219		1,894	2,200
9000 Minimum labor/equipment charge	1 Plum	1	8	Job		475		475	740

23 82 16.20 Duct Heaters

	Crew	Daily Output	Labor-Hours	Unit	Material	2016 Bare Costs Labor	Equipment	Total	Total Incl O&P
0010 **DUCT HEATERS**, Electric, 480 V, 3 Ph.									
0020 Finned tubular insert, 500°F									
0100 8" wide x 6" high, 4.0 kW	Q-20	16	1.250	Ea.	765	65.50		830.50	950
0120 12" high, 8.0 kW		15	1.333		1,275	69.50		1,344.50	1,500
0140 18" high, 12.0 kW		14	1.429		1,775	74.50		1,849.50	2,075
0160 24" high, 16.0 kW		13	1.538		2,300	80.50		2,380.50	2,650
0180 30" high, 20.0 kW		12	1.667		2,800	87		2,887	3,200
0300 12" wide x 6" high, 6.7 kW		15	1.333		815	69.50		884.50	1,000
0320 12" high, 13.3 kW		14	1.429		1,325	74.50		1,399.50	1,575
0340 18" high, 20.0 kW		13	1.538		1,850	80.50		1,930.50	2,150
8000 To obtain BTU multiply kW by 3413									
9900 Minimum labor/equipment charge	1 Shee	4	2	Job		115		115	181

23 82 19 – Fan Coil Units

23 82 19.10 Fan Coil Air Conditioning

	Crew	Daily Output	Labor-Hours	Unit	Material	2016 Bare Costs Labor	Equipment	Total	Total Incl O&P
0010 **FAN COIL AIR CONDITIONING**									
0030 Fan coil AC, cabinet mounted, filters and controls									
0100 Chilled water, 1/2 ton cooling	Q-5	8	2	Ea.	460	109		569	675
0120 1 ton cooling		6	2.667		685	146		831	980
0160 2.5 ton cooling		5	3.200		1,450	175		1,625	1,875
0180 3 ton cooling		4	4		1,625	219		1,844	2,125
0262 For hot water coil, add					40%	10%			
0940 Direct expansion, for use w/air cooled condensing unit, 1.5 ton cooling	Q-5	5	3.200	Ea.	515	175		690	845
1000 5 ton cooling	"	3	5.333		1,025	291		1,316	1,575
1040 10 ton cooling	Q-6	2.60	9.231		2,075	525		2,600	3,100
1060 20 ton cooling	"	.70	34.286		4,000	1,950		5,950	7,425
1500 For hot water coil, add					40%	10%			
1512 For condensing unit add see Section 23 62									
3000 Chilled water, horizontal unit, housing, 2 pipe, fan control, no valves									
3100 1/2 ton cooling	Q-5	8	2	Ea.	1,275	109		1,384	1,575
3110 1 ton cooling		6	2.667		1,500	146		1,646	1,875
3120 1.5 ton cooling		5.50	2.909		1,725	159		1,884	2,150
3130 2 ton cooling		5.25	3.048		2,075	166		2,241	2,525
3140 3 ton cooling		4	4		2,475	219		2,694	3,075
3150 3.5 ton cooling		3.80	4.211		2,725	230		2,955	3,350
3160 4 ton cooling		3.80	4.211		2,725	230		2,955	3,350
4000 With electric heat, 2 pipe									

For customer support on your Commercial Renovation Cost Data, call 877.742.4719.

465

23 82 Convection Heating and Cooling Units

23 82 19 – Fan Coil Units

23 82 19.10 Fan Coil Air Conditioning

		Crew	Daily Output	Labor-Hours	Unit	Material	2016 Bare Costs Labor	Equipment	Total	Total Incl O&P
4100	1/2 ton cooling	Q-5	8	2	Ea.	1,500	109		1,609	1,825
4105	3/4 ton cooling		7	2.286		1,675	125		1,800	2,025
4110	1 ton cooling		6	2.667		1,825	146		1,971	2,250
4120	1.5 ton cooling		5.50	2.909		2,025	159		2,184	2,500
4130	2 ton cooling		5.25	3.048		2,575	166		2,741	3,075
4135	2.5 ton cooling		4.80	3.333		3,550	182		3,732	4,175
4140	3 ton cooling		4	4		4,775	219		4,994	5,600
4150	3.5 ton cooling		3.80	4.211		5,700	230		5,930	6,625
4160	4 ton cooling		3.60	4.444		6,575	243		6,818	7,600
9000	Minimum labor/equipment charge		3.75	4.267	Job		233		233	365

23 82 19.40 Fan Coil Air Conditioning

			Crew	Daily Output	Labor-Hours	Unit	Material	2016 Bare Costs Labor	Equipment	Total	Total Incl O&P
0010	**FAN COIL AIR CONDITIONING**, Variable refrigerant volume (VRV) type										
0020	Not including interconnecting tubing or multi-zone controls										
0030	For VRV condensing unit see section 23 81 29.10										
0050	Indoor type, ducted										
0100	Vertical concealed										
0130	1 ton cooling	G	Q-5	2.97	5.387	Ea.	2,475	294		2,769	3,175
0140	1.5 ton cooling	G		2.77	5.776		2,525	315		2,840	3,275
0150	2 ton cooling	G		2.70	5.926		2,725	325		3,050	3,500
0160	2.5 ton cooling	G		2.60	6.154		2,875	335		3,210	3,675
0170	3 ton cooling	G		2.50	6.400		2,925	350		3,275	3,750
0180	3.5 ton cooling	G		2.30	6.957		3,075	380		3,455	3,975
0190	4 ton cooling	G		2.19	7.306		3,075	400		3,475	4,025
0200	4.5 ton cooling	G		2.08	7.692		3,400	420		3,820	4,375
0230	Outside air connection possible										
1100	Ceiling concealed										
1130	.6 ton cooling	G	Q-5	2.60	6.154	Ea.	1,625	335		1,960	2,300
1140	.75 ton cooling	G		2.60	6.154		1,675	335		2,010	2,375
1150	1 ton cooling	G		2.60	6.154		1,775	335		2,110	2,475
1152	Screening door	G	1 Stpi	5.20	1.538		98	93.50		191.50	254
1160	1.5 ton cooling	G	Q-5	2.60	6.154		1,825	335		2,160	2,550
1162	Screening door	G	1 Stpi	5.20	1.538		113	93.50		206.50	271
1170	2 ton cooling	G	Q-5	2.60	6.154		2,150	335		2,485	2,875
1172	Screening door	G	1 Stpi	5.20	1.538		98	93.50		191.50	254
1180	2.5 ton cooling	G	Q-5	2.60	6.154		2,500	335		2,835	3,275
1182	Screening door	G	1 Stpi	5.20	1.538		170	93.50		263.50	335
1190	3 ton cooling	G	Q-5	2.60	6.154		2,650	335		2,985	3,450
1192	Screening door	G	1 Stpi	5.20	1.538		170	93.50		263.50	335
1200	4 ton cooling	G	Q-5	2.30	6.957		2,725	380		3,105	3,575
1202	Screening door	G	1 Stpi	5.20	1.538		170	93.50		263.50	335
1220	6 ton cooling	G	Q-5	2.08	7.692		4,625	420		5,045	5,725
1230	8 ton cooling	G	"	1.89	8.466		5,200	460		5,660	6,450
1260	Outside air connection possible										
4050	Indoor type, duct-free										
4100	Ceiling mounted cassette										
4130	.75 ton cooling	G	Q-5	3.46	4.624	Ea.	1,875	253		2,128	2,475
4140	1 ton cooling	G		2.97	5.387		2,000	294		2,294	2,650
4150	1.5 ton cooling	G		2.77	5.776		2,050	315		2,365	2,750
4160	2 ton cooling	G		2.60	6.154		2,200	335		2,535	2,950
4170	2.5 ton cooling	G		2.50	6.400		2,250	350		2,600	3,025
4180	3 ton cooling	G		2.30	6.957		2,425	380		2,805	3,250
4190	4 ton cooling	G		2.08	7.692		2,700	420		3,120	3,625

23 82 Convection Heating and Cooling Units

23 82 19 – Fan Coil Units

23 82 19.40 Fan Coil Air Conditioning

	Crew	Daily Output	Labor-Hours	Unit	Material	2016 Bare Costs Labor	Equipment	Total	Total Incl O&P
4198	For ceiling cassette decoration panel, add	G	1 Stpi	5.20	1.538	Ea.	294	93.50		387.50	470
4230	Outside air connection possible										
4500	Wall mounted										
4530	.6 ton cooling	G	Q-5	5.20	3.077	Ea.	995	168		1,163	1,350
4540	.75 ton cooling	G		5.20	3.077		1,050	168		1,218	1,400
4550	1 ton cooling	G		4.16	3.846		1,200	210		1,410	1,625
4560	1.5 ton cooling	G		3.77	4.244		1,300	232		1,532	1,800
4570	2 ton cooling	G		3.46	4.624		1,400	253		1,653	1,950
4590	Condensate pump for the above air handlers	G	1 Stpi	6.46	1.238		211	75		286	350
4700	Floor standing unit										
4730	1 ton cooling	G	Q-5	2.60	6.154	Ea.	1,675	335		2,010	2,375
4740	1.5 ton cooling	G		2.60	6.154		1,900	335		2,235	2,625
4750	2 ton cooling	G		2.60	6.154		2,050	335		2,385	2,775
4800	Floor standing unit, concealed										
4830	1 ton cooling	G	Q-5	2.60	6.154	Ea.	1,600	335		1,935	2,300
4840	1.5 ton cooling	G		2.60	6.154		1,825	335		2,160	2,525
4850	2 ton cooling	G		2.60	6.154		1,950	335		2,285	2,675
4880	Outside air connection possible										
8000	Accessories										
8100	Branch divergence pipe fitting										
8110	Capacity under 76 MBH		1 Stpi	2.50	3.200	Ea.	151	194		345	470
8120	76 MBH to 112 MBH			2.50	3.200		178	194		372	500
8130	112 MBH to 234 MBH			2.50	3.200		505	194		699	860
8200	Header pipe fitting										
8210	Max 4 branches										
8220	Capacity under 76 MBH		1 Stpi	1.70	4.706	Ea.	221	286		507	690
8260	Max 8 branches										
8270	76 MBH to 112 MBH		1 Stpi	1.70	4.706	Ea.	450	286		736	940
8280	112 MBH to 234 MBH		"	1.30	6.154	"	720	375		1,095	1,375

23 82 29 – Radiators

23 82 29.10 Hydronic Heating

	Crew	Daily Output	Labor-Hours	Unit	Material	2016 Bare Costs Labor	Equipment	Total	Total Incl O&P
0010	**HYDRONIC HEATING**, Terminal units, not incl. main supply pipe									
1000	Radiation										
1100	Panel, baseboard, C.I., including supports, no covers	Q-5	46	.348	L.F.	38	19		57	71
3000	Radiators, cast iron										
3100	Free standing or wall hung, 6 tube, 25" high	Q-5	96	.167	Section	47.50	9.10		56.60	66
9000	Minimum labor/equipment charge	"	3.20	5	Job		273		273	425

23 82 36 – Finned-Tube Radiation Heaters

23 82 36.10 Finned Tube Radiation

	Crew	Daily Output	Labor-Hours	Unit	Material	2016 Bare Costs Labor	Equipment	Total	Total Incl O&P
0010	**FINNED TUBE RADIATION**, Terminal units, not incl. main supply pipe									
1150	Fin tube, wall hung, 14" slope top cover, with damper										
1200	1-1/4" copper tube, 4-1/4" alum. fin	Q-5	38	.421	L.F.	43	23		66	83.50
1250	1-1/4" steel tube, 4-1/4" steel fin		36	.444		39.50	24.50		64	81.50
1255	2" steel tube, 4-1/4" steel fin		32	.500		43	27.50		70.50	89.50
1310	Baseboard, pkgd, 1/2" copper tube, alum. fin, 7" high		60	.267		11.20	14.55		25.75	35
1320	3/4" copper tube, alum. fin, 7" high		58	.276		7.35	15.05		22.40	31.50
1340	1" copper tube, alum. fin, 8-7/8" high		56	.286		18.85	15.60		34.45	45
1360	1-1/4" copper tube, alum. fin, 8-7/8" high		54	.296		28	16.20		44.20	55.50
1500	Note: fin tube may also require corners, caps, etc.										

23 82 Convection Heating and Cooling Units

23 82 39 – Unit Heaters

23 82 39.16 Propeller Unit Heaters	Crew	Daily Output	Labor-Hours	Unit	Material	2016 Bare Costs Labor	Equipment	Total	Total Incl O&P
0010 **PROPELLER UNIT HEATERS**									
3950 Unit heaters, propeller, 115 V 2 psi steam, 60°F entering air									
4000 Horizontal, 12 MBH	Q-5	12	1.333	Ea.	385	73		458	540
4060 43.9 MBH		8	2		585	109		694	810
4240 286.9 MBH		2	8		1,775	435		2,210	2,625
4260 364 MBH		1.80	8.889		2,200	485		2,685	3,175
4270 404 MBH		1.60	10		2,300	545		2,845	3,375
4300 Vertical diffuser same price									
4310 Vertical flow, 40 MBH	Q-5	11	1.455	Ea.	575	79.50		654.50	760
4326 131.0 MBH	"	4	4		900	219		1,119	1,325
4354 420 MBH, (460 V)	Q-6	1.80	13.333		2,300	755		3,055	3,700
4358 500 MBH, (460 V)		1.71	14.035		3,050	795		3,845	4,600
4362 570 MBH, (460 V)		1.40	17.143		4,175	970		5,145	6,125
4366 620 MBH, (460 V)		1.30	18.462		4,175	1,050		5,225	6,225
4370 960 MBH, (460 V)		1.10	21.818		7,950	1,225		9,175	10,700

23 83 Radiant Heating Units

23 83 33 – Electric Radiant Heaters

23 83 33.10 Electric Heating	Crew	Daily Output	Labor-Hours	Unit	Material	2016 Bare Costs Labor	Equipment	Total	Total Incl O&P
0010 **ELECTRIC HEATING**, not incl. conduit or feed wiring									
1100 Rule of thumb: Baseboard units, including control	1 Elec	4.40	1.818	kW	105	100		205	271
1300 Baseboard heaters, 2' long, 350 watt		8	1	Ea.	26	55		81	114
1400 3' long, 750 watt		8	1		31.50	55		86.50	120
1600 4' long, 1000 watt		6.70	1.194		35.50	66		101.50	141
1800 5' long, 935 watt		5.70	1.404		46.50	77.50		124	171
2000 6' long, 1500 watt		5	1.600		51	88		139	192
2400 8' long, 2000 watt		4	2		56	110		166	232
2600 9' long, 1680 watt		3.60	2.222		92	122		214	290
2800 10' long, 1875 watt		3.30	2.424		180	134		314	405
2950 Wall heaters with fan, 120 to 277 volt									
3170 1000 watt	1 Elec	6	1.333	Ea.	83.50	73.50		157	206
3180 1250 watt		5	1.600		83.50	88		171.50	228
3190 1500 watt		4	2		83.50	110		193.50	262
3230 2500 watt		3.50	2.286		209	126		335	425
3250 4000 watt		2.70	2.963		310	163		473	590
3600 Thermostats, integral		16	.500		26.50	27.50		54	71.50
3800 Line voltage, 1 pole		8	1		12.85	55		67.85	99
3810 2 pole		8	1		18.50	55		73.50	106
9990 Minimum labor/equipment charge		4	2	Job		110		110	170

23 84 Humidity Control Equipment

23 84 13 – Humidifiers

23 84 13.10 Humidifier Units

23 84 13.10 Humidifier Units	Crew	Daily Output	Labor-Hours	Unit	Material	2016 Bare Costs Labor	Equipment	Total	Total Incl O&P
0010 **HUMIDIFIER UNITS**									
0520 Steam, room or duct, filter, regulators, auto. controls, 220 V									
0560 22 lb. per hour	Q-5	5	3.200	Ea.	2,975	175		3,150	3,550
0580 33 lb. per hour		4	4		3,050	219		3,269	3,725
0620 100 lb. per hour		3	5.333		4,475	291		4,766	5,375
9000 Minimum labor/equipment charge		3.50	4.571	Job		250		250	390

23 91 Prefabricated Equipment Supports

23 91 10 – Prefabricated Curbs, Pads and Stands

23 91 10.10 Prefabricated Pads and Stands

23 91 10.10 Prefabricated Pads and Stands	Crew	Daily Output	Labor-Hours	Unit	Material	2016 Bare Costs Labor	Equipment	Total	Total Incl O&P
0010 **PREFABRICATED PADS AND STANDS**									
6000 Pad, fiberglass reinforced concrete with polystyrene foam core									
6050 Condenser, 2" thick, 20" x 38"	1 Shee	8	1	Ea.	28.50	57.50		86	122
6090 24" x 36"	"	12	.667		36.50	38		74.50	100
6280 36" x 36"	Q-9	8	2		52.50	103		155.50	221
6300 36" x 40"		7	2.286		59.50	118		177.50	252
6320 36" x 48"		7	2.286		69.50	118		187.50	263

For customer support on your Commercial Renovation Cost Data, call 877.742.4719.

469

Division Notes

		CREW	DAILY OUTPUT	LABOR-HOURS	UNIT	BARE COSTS				TOTAL INCL O&P
						MAT.	LABOR	EQUIP.	TOTAL	

Estimating Tips
26 05 00 Common Work Results for Electrical

- Conduit should be taken off in three main categories—power distribution, branch power, and branch lighting—so the estimator can concentrate on systems and components, therefore making it easier to ensure all items have been accounted for.

- For cost modifications for elevated conduit installation, add the percentages to labor according to the height of installation, and only to the quantities exceeding the different height levels, not to the total conduit quantities.

- Remember that aluminum wiring of equal ampacity is larger in diameter than copper and may require larger conduit.

- If more than three wires at a time are being pulled, deduct percentages from the labor hours of that grouping of wires.

- When taking off grounding systems, identify separately the type and size of wire, and list each unique type of ground connection.

- The estimator should take the weights of materials into consideration when completing a takeoff. Topics to consider include: How will the materials be supported? What methods of support are available? How high will the support structure have to reach? Will the final support structure be able to withstand the total burden? Is the support material included or separate from the fixture, equipment, and material specified?

- Do not overlook the costs for equipment used in the installation. If scaffolding or highlifts are available in the field, contractors may use them in lieu of the proposed ladders and rolling staging.

26 20 00 Low-Voltage Electrical Transmission

- Supports and concrete pads may be shown on drawings for the larger equipment, or the support system may be only a piece of plywood for the back of a panelboard. In either case, it must be included in the costs.

26 40 00 Electrical and Cathodic Protection

- When taking off cathodic protections systems, identify the type and size of cable, and list each unique type of anode connection.

26 50 00 Lighting

- Fixtures should be taken off room by room, using the fixture schedule, specifications, and the ceiling plan. For large concentrations of lighting fixtures in the same area, deduct the percentages from labor hours.

Reference Numbers

Reference numbers are shown at the beginning of some major classifications. These numbers refer to related items in the Reference Section. The reference information may be an estimating procedure, an alternate pricing method, or technical information.

Note: Not all subdivisions listed here necessarily appear. ∎

26 01 Operation and Maintenance of Electrical Systems

26 01 02 – Labor Adjustment

26 01 02.20 Labor Adjustment Factors	Crew	Daily Output	Labor-Hours	Unit	Material	2016 Bare Costs Labor	Equipment	Total	Total Incl O&P
0010 **LABOR ADJUSTMENT FACTORS** (For Div. 26, 27 and 28)									
0100 Subtract from labor for Economy of Scale for Wire, R260519-90									
0110 4-5 wires						25%			
0120 6-10 wires						30%			
0130 11-15 wires						35%			
0140 over 15 wires						40%			

26 01 50 – Operation and Maintenance of Lighting

26 01 50.81 Luminaire Replacement

		Crew	Daily Output	Labor-Hours	Unit	Material	2016 Bare Costs Labor	Equipment	Total	Total Incl O&P
0010	**LUMINAIRE REPLACEMENT**									
0962	Remove and install new replacement lens, 2' x 2' troffer	1 Elec	16	.500	Ea.	15.90	27.50		43.40	60
0964	2' x 4' troffer		16	.500		14.90	27.50		42.40	59
3200	Remove and replace (reinstall), lighting fixture	↓	4	2	↓		110		110	170

26 05 Common Work Results for Electrical

26 05 05 – Selective Demolition for Electrical

26 05 05.10 Electrical Demolition

			Crew	Daily Output	Labor-Hours	Unit	Material	2016 Bare Costs Labor	Equipment	Total	Total Incl O&P
0010	**ELECTRICAL DEMOLITION**	R260105-30									
0020	Conduit to 15' high, including fittings & hangers										
0100	Rigid galvanized steel, 1/2" to 1" diameter	R024119-10	1 Elec	242	.033	L.F.		1.82		1.82	2.82
0120	1-1/4" to 2"		"	200	.040			2.20		2.20	3.41
0140	2-1/2" to 3-1/2"		2 Elec	302	.053			2.92		2.92	4.51
0200	Electric metallic tubing (EMT), 1/2" to 1"		1 Elec	394	.020			1.12		1.12	1.73
0220	1-1/4" to 1-1/2"			326	.025			1.35		1.35	2.09
0240	2" to 3"		↓	236	.034	↓		1.87		1.87	2.89
0400	Wiremold raceway, including fittings & hangers										
0420	No. 3000		1 Elec	250	.032	L.F.		1.76		1.76	2.73
0440	No. 4000		"	217	.037	"		2.03		2.03	3.14
0500	Channels, steel, including fittings & hangers										
0520	3/4" x 1-1/2"		1 Elec	308	.026	L.F.		1.43		1.43	2.21
0540	1-1/2" x 1-1/2"		↓	269	.030			1.64		1.64	2.53
0560	1-1/2" x 1-7/8"		↓	229	.035	↓		1.92		1.92	2.98
0600	Copper bus duct, indoor, 3 phase										
0610	Including hangers & supports										
0620	225 amp		2 Elec	135	.119	L.F.		6.55		6.55	10.10
0640	400 amp			106	.151			8.30		8.30	12.85
0660	600 amp			86	.186			10.25		10.25	15.85
0680	1000 amp			60	.267			14.70		14.70	22.50
0700	1600 amp			40	.400			22		22	34
0720	3000 amp		↓	10	1.600	↓		88		88	136
0800	Plug-in switches, 600V 3 ph., incl. disconnecting										
0820	wire, conduit terminations, 30 amp		1 Elec	15.50	.516	Ea.		28.50		28.50	44
0840	60 amp			13.90	.576			31.50		31.50	49
0850	100 amp			10.40	.769			42.50		42.50	65.50
0860	200 amp		↓	6.20	1.290			71		71	110
0880	400 amp		2 Elec	5.40	2.963			163		163	252
0900	600 amp			3.40	4.706			259		259	400
0920	800 amp			2.60	6.154			340		340	525
0940	1200 amp			2	8			440		440	680
0960	1600 amp		↓	1.70	9.412			520		520	800
1010	Safety switches, 250 or 600V, incl. disconnection										

26 05 05.10 Electrical Demolition		Crew	Daily Output	Labor-Hours	Unit	Material	2016 Bare Costs Labor	Equipment	Total	Total Incl O&P
1050	of wire & conduit terminations									
1100	30 amp	1 Elec	12.30	.650	Ea.		36		36	55.50
1120	60 amp		8.80	.909			50		50	77.50
1140	100 amp		7.30	1.096			60.50		60.50	93.50
1160	200 amp		5	1.600			88		88	136
1210	Panel boards, incl. removal of all breakers,									
1220	conduit terminations & wire connections									
1230	3 wire, 120/240 V, 100A, to 20 circuits	1 Elec	2.60	3.077	Ea.		170		170	262
1240	200 amps, to 42 circuits	2 Elec	2.60	6.154			340		340	525
1260	4 wire, 120/208 V, 125A, to 20 circuits	1 Elec	2.40	3.333			184		184	284
1270	200 amps, to 42 circuits	2 Elec	2.40	6.667			365		365	570
1300	Transformer, dry type, 1 phase, incl. removal of									
1320	supports, wire & conduit terminations									
1340	1 kVA	1 Elec	7.70	1.039	Ea.		57.50		57.50	88.50
1360	5 kVA	"	4.70	1.702			94		94	145
1420	75 kVA	2 Elec	2.50	6.400			355		355	545
1440	3 phase to 600V, primary									
1460	3 kVA	1 Elec	3.87	2.067	Ea.		114		114	176
1480	15 kVA	2 Elec	3.67	4.360			240		240	370
1490	25 kVA		3.51	4.558			251		251	390
1500	30 kVA		3.42	4.678			258		258	400
1530	112.5 kVA	R-3	2.90	6.897			375	48.50	423.50	640
1560	500 kVA		1.40	14.286			780	100	880	1,325
1570	750 kVA		1.10	18.182			995	128	1,123	1,700
1600	Pull boxes & cabinets, sheet metal, incl. removal									
1620	of supports and conduit terminations									
1640	6" x 6" x 4"	1 Elec	31.10	.257	Ea.		14.15		14.15	22
1660	12" x 12" x 4"		23.30	.343			18.90		18.90	29.50
1720	Junction boxes, 4" sq. & oct.		80	.100			5.50		5.50	8.50
1740	Handy box		107	.075			4.12		4.12	6.35
1760	Switch box		107	.075			4.12		4.12	6.35
1780	Receptacle & switch plates		257	.031			1.72		1.72	2.65
1800	Wire, THW-THWN-THHN, removed from									
1810	in place conduit, to 15' high									
1830	#14	1 Elec	65	.123	C.L.F.		6.80		6.80	10.50
1840	#12		55	.145			8		8	12.40
1850	#10		45.50	.176			9.70		9.70	15
1880	#4	2 Elec	53	.302			16.65		16.65	25.50
1890	#3		50	.320			17.65		17.65	27.50
1910	1/0		33.20	.482			26.50		26.50	41
1920	2/0		29.20	.548			30		30	46.50
1930	3/0		25	.640			35.50		35.50	54.50
1980	400 kcmil		17	.941			52		52	80
1990	500 kcmil		16.20	.988			54.50		54.50	84
2000	Interior fluorescent fixtures, incl. supports									
2010	& whips, to 15' high									
2100	Recessed drop-in 2' x 2', 2 lamp	2 Elec	35	.457	Ea.		25		25	39
2120	2' x 4', 2 lamp		33	.485			26.50		26.50	41.50
2140	2' x 4', 4 lamp		30	.533			29.50		29.50	45.50
2160	4' x 4', 4 lamp		20	.800			44		44	68
2180	Surface mount, acrylic lens & hinged frame									
2200	1' x 4', 2 lamp	2 Elec	44	.364	Ea.		20		20	31
2220	2' x 2', 2 lamp		44	.364			20		20	31

For customer support on your Commercial Renovation Cost Data, call 877.742.4719.

473

26 05 Common Work Results for Electrical

26 05 05 – Selective Demolition for Electrical

26 05 05.10 Electrical Demolition		Crew	Daily Output	Labor-Hours	Unit	Material	2016 Bare Costs Labor	Equipment	Total	Total Incl O&P
2260	2' x 4', 4 lamp	2 Elec	33	.485	Ea.		26.50		26.50	41.50
2280	4' x 4', 4 lamp	▼	23	.696	▼		38.50		38.50	59.50
2300	Strip fixtures, surface mount									
2320	4' long, 1 lamp	2 Elec	53	.302	Ea.		16.65		16.65	25.50
2340	4' long, 2 lamp		50	.320			17.65		17.65	27.50
2360	8' long, 1 lamp		42	.381			21		21	32.50
2380	8' long, 2 lamp	▼	40	.400	▼		22		22	34
2400	Pendant mount, industrial, incl. removal									
2410	of chain or rod hangers, to 15' high									
2420	4' long, 2 lamp	2 Elec	35	.457	Ea.		25		25	39
2440	8' long, 2 lamp	"	27	.593	"		32.50		32.50	50.50
2460	Interior incandescent, surface, ceiling									
2470	or wall mount, to 12' high									
2480	Metal cylinder type, 75 Watt	2 Elec	62	.258	Ea.		14.20		14.20	22
2500	150 Watt	"	62	.258	"		14.20		14.20	22
2520	Metal halide, high bay									
2540	400 Watt	2 Elec	15	1.067	Ea.		59		59	91
2560	1000 Watt		12	1.333			73.50		73.50	114
2580	150 Watt, low bay	▼	20	.800	▼		44		44	68
2600	Exterior fixtures, incandescent, wall mount									
2620	100 Watt	2 Elec	50	.320	Ea.		17.65		17.65	27.50
2640	Quartz, 500 Watt		33	.485			26.50		26.50	41.50
2660	1500 Watt	▼	27	.593	▼		32.50		32.50	50.50
2680	Wall pack, mercury vapor									
2700	175 Watt	2 Elec	25	.640	Ea.		35.50		35.50	54.50
2720	250 Watt	"	25	.640	"		35.50		35.50	54.50
9000	Minimum labor/equipment charge	1 Elec	4	2	Job		110		110	170
9900	Add to labor for higher elevated installation									
9910	15' to 20' high, add						10%			
9920	20' to 25' high, add						20%			
9930	25' to 30' high, add						25%			
9940	30' to 35' high, add						30%			
9950	35' to 40' high, add						35%			
9960	Over 40' high, add						40%			

26 05 19 – Low-Voltage Electrical Power Conductors and Cables

26 05 19.20 Armored Cable

		Crew	Daily Output	Labor-Hours	Unit	Material	2016 Bare Costs Labor	Equipment	Total	Total Incl O&P
0010	**ARMORED CABLE**									
0050	600 volt, copper (BX), #14, 2 conductor, solid	1 Elec	2.40	3.333	C.L.F.	45.50	184		229.50	335
0100	3 conductor, solid		2.20	3.636		73	200		273	390
0150	#12, 2 conductor, solid		2.30	3.478		46	192		238	345
0200	3 conductor, solid		2	4		74.50	220		294.50	420
0250	#10, 2 conductor, solid		2	4		87.50	220		307.50	435
0300	3 conductor, solid		1.60	5		121	276		397	560
0340	#8, 2 conductor, stranded		1.50	5.333		229	294		523	705
0350	3 conductor, stranded	▼	1.30	6.154		229	340		569	775
9010	600 volt, copper (MC) steel clad, #14, 2 wire	R-1A	8.16	1.961		50.50	97.50		148	206
9020	3 wire		7.21	2.219		74.50	110		184.50	252
9030	4 wire		6.69	2.392		103	119		222	297
9040	#12, 2 wire		7.21	2.219		46.50	110		156.50	221
9050	3 wire		6.69	2.392		79	119		198	270
9070	#10, 2 wire		6.04	2.649		108	131		239	320
9080	3 wire		5.65	2.832		142	140		282	375

474

26 05 19 – Low-Voltage Electrical Power Conductors and Cables

26 05 19.20 Armored Cable		Crew	Daily Output	Labor-Hours	Unit	Material	2016 Bare Costs Labor	Equipment	Total	Total Incl O&P
9100	#8, 2 wire, stranded	R-1A	3.94	4.061	C.L.F.	194	201		395	525
9110	3 wire, stranded	↓	3.40	4.706	↓	272	233		505	660
9900	Minimum labor/equipment charge	R-1C	4	8	Job		395	32.50	427.50	650

26 05 19.50 Mineral Insulated Cable										
0010	**MINERAL INSULATED CABLE** 600 volt									
0100	1 conductor, #12	1 Elec	1.60	5	C.L.F.	375	276		651	840
0200	#10		1.60	5		485	276		761	960
0400	#8		1.50	5.333		540	294		834	1,050
0500	#6	↓	1.40	5.714		640	315		955	1,200
0600	#4	2 Elec	2.40	6.667		860	365		1,225	1,525
0800	#2		2.20	7.273		1,225	400		1,625	1,975
0900	#1		2.10	7.619		1,400	420		1,820	2,200
1000	1/0		2	8		1,650	440		2,090	2,475
1100	2/0		1.90	8.421		1,975	465		2,440	2,900
1200	3/0		1.80	8.889		2,350	490		2,840	3,350
1400	4/0	↓	1.60	10		2,725	550		3,275	3,825
1410	250 kcmil	3 Elec	2.40	10		3,075	550		3,625	4,225
1420	350 kcmil	↓	1.95	12.308		3,500	680		4,180	4,900
1430	500 kcmil	↓	1.95	12.308		4,500	680		5,180	6,000
1500	2 conductor, #12	1 Elec	1.40	5.714		785	315		1,100	1,350
1600	#10		1.20	6.667		965	365		1,330	1,625
1800	#8		1.10	7.273		1,100	400		1,500	1,850
2000	#6	↓	1.05	7.619		1,525	420		1,945	2,325
2100	#4	2 Elec	2	8		2,050	440		2,490	2,950
2200	3 conductor, #12	1 Elec	1.20	6.667		980	365		1,345	1,650
2400	#10		1.10	7.273		1,100	400		1,500	1,850
2600	#8		1.05	7.619		1,400	420		1,820	2,175
2800	#6	↓	1	8		1,825	440		2,265	2,700
3000	#4	2 Elec	1.80	8.889		2,300	490		2,790	3,275
3100	4 conductor, #12	1 Elec	1.20	6.667		1,025	365		1,390	1,700
3200	#10		1.10	7.273		1,250	400		1,650	1,975
3400	#8		1	8		1,625	440		2,065	2,450
3600	#6		.90	8.889		2,100	490		2,590	3,075
3620	7 conductor, #12		1.10	7.273		1,325	400		1,725	2,075
3640	#10		1	8	↓	1,725	440		2,165	2,575
3800	Terminations, 600 volt, 1 conductor, #12		8	1	Ea.	16.60	55		71.60	103
5500	2 conductor, #12		6.70	1.194		16.60	66		82.60	120
5600	#10		6.40	1.250		25	69		94	135
5800	#8		6.20	1.290		25	71		96	138
6000	#6		5.70	1.404		25	77.50		102.50	148
6200	#4		5.30	1.509		55.50	83		138.50	190
6400	3 conductor, #12		5.70	1.404		25	77.50		102.50	148
6500	#10		5.50	1.455		25	80		105	152
6600	#8		5.20	1.538		25	85		110	159
6800	#6		4.80	1.667		25	92		117	170
7200	#4		4.60	1.739		55.50	96		151.50	209
7400	4 conductor, #12		4.60	1.739		27.50	96		123.50	179
7500	#10		4.40	1.818		27.50	100		127.50	186
7600	#8		4.20	1.905		27.50	105		132.50	193
8400	#6		4	2		58	110		168	234
8500	7 conductor, #12		3.50	2.286		27.50	126		153.50	226
8600	#10	↓	3	2.667	↓	60	147		207	293

26 05 19.50 Mineral Insulated Cable

		Crew	Daily Output	Labor-Hours	Unit	Material	2016 Bare Costs Labor	Equipment	Total	Total Incl O&P
8800	Crimping tool, plier type				Ea.	62			62	68
9000	Stripping tool					233			233	257
9200	Hand vise					69.50			69.50	76.50
9500	Minimum labor/equipment charge	1 Elec	4	2	Job		110		110	170

26 05 19.55 Non-Metallic Sheathed Cable

		Crew	Daily Output	Labor-Hours	Unit	Material	2016 Bare Costs Labor	Equipment	Total	Total Incl O&P
0010	**NON-METALLIC SHEATHED CABLE** 600 volt									
0100	Copper with ground wire (Romex)									
0150	#14, 2 conductor	1 Elec	2.70	2.963	C.L.F.	21.50	163		184.50	276
0200	3 conductor		2.40	3.333		30.50	184		214.50	320
0220	4 conductor		2.20	3.636		44	200		244	360
0250	#12, 2 conductor		2.50	3.200		33	176		209	310
0300	3 conductor		2.20	3.636		46.50	200		246.50	360
0320	4 conductor		2	4		68.50	220		288.50	415
0350	#10, 2 conductor		2.20	3.636		50.50	200		250.50	365
0400	3 conductor		1.80	4.444		73.50	245		318.50	460
0430	#8, 2 conductor		1.60	5		79.50	276		355.50	510
0450	3 conductor		1.50	5.333		118	294		412	585
0500	#6, 3 conductor		1.40	5.714		192	315		507	695
0550	SE type SER aluminum cable, 3 RHW and									
0600	1 bare neutral, 3 #8 & 1 #8	1 Elec	1.60	5	C.L.F.	69	276		345	500
0650	3 #6 & 1 #6	"	1.40	5.714		78	315		393	570
0700	3 #4 & 1 #6	2 Elec	2.40	6.667		87.50	365		452.50	665
0750	3 #2 & 1 #4		2.20	7.273		129	400		529	760
0800	3 #1/0 & 1 #2		2	8		195	440		635	895
0850	3 #2/0 & 1 #1		1.80	8.889		229	490		719	1,000
0900	3 #4/0 & 1 #2/0		1.60	10		325	550		875	1,200
6500	Service entrance cap for copper SEU									
6700	150 amp	1 Elec	10	.800	Ea.	13.80	44		57.80	83
6800	200 amp		8	1	"	21	55		76	108
9000	Minimum labor/equipment charge		4	2	Job		110		110	170

26 05 19.90 Wire

		Crew	Daily Output	Labor-Hours	Unit	Material	2016 Bare Costs Labor	Equipment	Total	Total Incl O&P
0010	**WIRE** R260533-22									
0020	600 volt, copper type THW, solid, #14	1 Elec	13	.615	C.L.F.	7	34		41	60
0030	#12		11	.727		10.65	40		50.65	74
0040	#10		10	.800		16.70	44		60.70	86.50
0140	#8		8	1		32.50	55		87.50	121
0160	#6		6.50	1.231		55.50	68		123.50	166
0180	#4	2 Elec	10.60	1.509		85	83		168	223
0200	#3		10	1.600		106	88		194	252
0220	#2		9	1.778		133	98		231	297
0240	#1		8	2		175	110		285	365
0260	1/0		6.60	2.424		218	134		352	445
0280	2/0		5.80	2.759		274	152		426	535
0300	3/0		5	3.200		345	176		521	655
0350	4/0		4.40	3.636		440	200		640	790
0400	250 kcmil	3 Elec	6	4		495	220		715	885
0420	300 kcmil		5.70	4.211		560	232		792	975
0450	350 kcmil		5.40	4.444		650	245		895	1,100
0480	400 kcmil		5.10	4.706		745	259		1,004	1,225
0490	500 kcmil		4.80	5		925	276		1,201	1,450
0510	750 kcmil		3.30	7.273		1,925	400		2,325	2,750
0540	600 volt, aluminum type THHN, stranded, #6	1 Elec	8	1		36.50	55		91.50	125

26 05 Common Work Results for Electrical

26 05 19 - Low-Voltage Electrical Power Conductors and Cables

26 05 19.90 Wire

		Crew	Daily Output	Labor-Hours	Unit	Material	2016 Bare Costs Labor	Equipment	Total	Total Incl O&P
0560	#4	2 Elec	13	1.231	C.L.F.	45.50	68		113.50	155
0580	#2		10.60	1.509		61.50	83		144.50	197
0600	#1		9	1.778		90	98		188	250
0620	1/0		8	2		108	110		218	289
0640	2/0		7.20	2.222		128	122		250	330
0680	3/0		6.60	2.424		158	134		292	380
0700	4/0		6.20	2.581		176	142		318	415
0720	250 kcmil	3 Elec	8.70	2.759		215	152		367	470
0740	300 kcmil		8.10	2.963		297	163		460	575
0760	350 kcmil		7.50	3.200		300	176		476	605
0780	400 kcmil		6.90	3.478		355	192		547	685
0800	500 kcmil		6	4		390	220		610	770
0850	600 kcmil		5.70	4.211		495	232		727	900
0880	700 kcmil		5.10	4.706		570	259		829	1,025
9000	Minimum labor/equipment charge	1 Elec	4	2	Job		110		110	170

26 05 26 - Grounding and Bonding for Electrical Systems

26 05 26.80 Grounding

		Crew	Daily Output	Labor-Hours	Unit	Material	2016 Bare Costs Labor	Equipment	Total	Total Incl O&P
0010	**GROUNDING**									
0030	Rod, copper clad, 8' long, 1/2" diameter	1 Elec	5.50	1.455	Ea.	20	80		100	146
0040	5/8" diameter		5.50	1.455		19.40	80		99.40	146
0050	3/4" diameter		5.30	1.509		34	83		117	167
0080	10' long, 1/2" diameter		4.80	1.667		22.50	92		114.50	167
0090	5/8" diameter		4.60	1.739		24.50	96		120.50	175
0100	3/4" diameter		4.40	1.818		39	100		139	198
0130	15' long, 3/4" diameter		4	2		51	110		161	226
0260	Wire ground bare armored, #8-1 conductor		2	4	C.L.F.	69.50	220		289.50	415
0280	#4-1 conductor		1.60	5		134	276		410	570
0390	Bare copper wire, stranded, #8		11	.727		25	40		65	89.50
0400	#6		10	.800		44.50	44		88.50	117
0600	#2	2 Elec	10	1.600		120	88		208	268
0800	3/0		6.60	2.424		264	134		398	500
1000	4/0		5.70	2.807		335	155		490	610
1200	250 kcmil	3 Elec	7.20	3.333		395	184		579	720
1800	Water pipe ground clamps, heavy duty									
2000	Bronze, 1/2" to 1" diameter	1 Elec	8	1	Ea.	24	55		79	111
2100	1-1/4" to 2" diameter		8	1		33	55		88	122
2200	2-1/2" to 3" diameter		6	1.333		44.50	73.50		118	163
2800	Brazed connections, #6 wire		12	.667		16.35	36.50		52.85	75
3000	#2 wire		10	.800		22	44		66	92
3100	3/0 wire		8	1		32.50	55		87.50	121
3200	4/0 wire		7	1.143		37	63		100	139
3400	250 kcmil wire		5	1.600		43.50	88		131.50	184
3600	500 kcmil wire		4	2		53.50	110		163.50	229
9000	Minimum labor/equipment charge		4	2	Job		110		110	170

26 05 33 - Raceway and Boxes for Electrical Systems

26 05 33.13 Conduit

		Crew	Daily Output	Labor-Hours	Unit	Material	2016 Bare Costs Labor	Equipment	Total	Total Incl O&P
0010	**CONDUIT** To 15' high, includes 2 terminations, 2 elbows, R260533-22									
0020	11 beam clamps, and 11 couplings per 100 L.F.									
1161	Field bends, 45° to 90°, 1/2" diameter	1 Elec	53	.151	Ea.		8.30		8.30	12.85
1162	3/4" diameter		47	.170			9.40		9.40	14.50
1163	1" diameter		44	.182			10		10	15.50
1164	1-1/4" diameter		23	.348			19.15		19.15	29.50

26 05 33.13 Conduit		Crew	Daily Output	Labor-Hours	Unit	Material	2016 Bare Costs Labor	Equipment	Total	Total Incl O&P
1165	1-1/2" diameter	1 Elec	21	.381	Ea.		21		21	32.50
1166	2" diameter		16	.500			27.50		27.50	42.50
1991	Field bends, 45° to 90°, 1/2" diameter		44	.182			10		10	15.50
1992	3/4" diameter		40	.200			11		11	17.05
1993	1" diameter		36	.222			12.25		12.25	18.95
1994	1-1/4" diameter		19	.421			23		23	36
1995	1-1/2" diameter		18	.444			24.50		24.50	38
1996	2" diameter		13	.615			34		34	52.50
2500	Steel, intermediate conduit (IMC), 1/2" diameter		100	.080	L.F.	2.20	4.41		6.61	9.20
2530	3/4" diameter		90	.089		2.75	4.90		7.65	10.60
2550	1" diameter		70	.114		3.93	6.30		10.23	14.05
2570	1-1/4" diameter		65	.123		4.94	6.80		11.74	15.95
2600	1-1/2" diameter		60	.133		6.45	7.35		13.80	18.45
2630	2" diameter		50	.160		7.90	8.80		16.70	22.50
2650	2-1/2" diameter		40	.200		14.10	11		25.10	32.50
2670	3" diameter	2 Elec	60	.267		18.55	14.70		33.25	43
2700	3-1/2" diameter		54	.296		23	16.35		39.35	50.50
2730	4" diameter		50	.320		19.40	17.65		37.05	49
2731	Field bends, 45° to 90°, 1/2" diameter	1 Elec	44	.182	Ea.		10		10	15.50
2732	3/4" diameter		40	.200			11		11	17.05
2733	1" diameter		36	.222			12.25		12.25	18.95
2734	1-1/4" diameter		19	.421			23		23	36
2735	1-1/2" diameter		18	.444			24.50		24.50	38
2736	2" diameter		13	.615			34		34	52.50
5000	Electric metallic tubing (EMT), 1/2" diameter		170	.047	L.F.	.70	2.59		3.29	4.78
5020	3/4" diameter		130	.062		1	3.39		4.39	6.35
5040	1" diameter		115	.070		1.67	3.83		5.50	7.80
5060	1-1/4" diameter		100	.080		2.69	4.41		7.10	9.75
5080	1-1/2" diameter		90	.089		3.43	4.90		8.33	11.35
5100	2" diameter		80	.100		4.24	5.50		9.74	13.15
5120	2-1/2" diameter		60	.133		8.70	7.35		16.05	21
5140	3" diameter	2 Elec	100	.160		10.20	8.80		19	25
5160	3-1/2" diameter		90	.178		12.85	9.80		22.65	29.50
5180	4" diameter		80	.200		13.85	11		24.85	32.50
5200	Field bends, 45° to 90°, 1/2" diameter	1 Elec	89	.090	Ea.		4.95		4.95	7.65
5220	3/4" diameter		80	.100			5.50		5.50	8.50
5240	1" diameter		73	.110			6.05		6.05	9.35
5260	1-1/4" diameter		38	.211			11.60		11.60	17.95
5280	1-1/2" diameter		36	.222			12.25		12.25	18.95
5300	2" diameter		26	.308			16.95		16.95	26
5320	Offsets, 1/2" diameter		65	.123			6.80		6.80	10.50
5340	3/4" diameter		62	.129			7.10		7.10	11
5360	1" diameter		53	.151			8.30		8.30	12.85
5380	1-1/4" diameter		30	.267			14.70		14.70	22.50
5400	1-1/2" diameter		28	.286			15.75		15.75	24.50
7600	EMT, "T" fittings with covers, 1/2" diameter, set screw		16	.500		6.70	27.50		34.20	50
9990	Minimum labor/equipment charge		4	2	Job		110		110	170

26 05 33.16 Outlet Boxes

		Crew	Daily Output	Labor-Hours	Unit	Material	2016 Bare Costs Labor	Equipment	Total	Total Incl O&P
0010	**OUTLET BOXES**									
0020	Pressed steel, octagon, 4"	1 Elec	20	.400	Ea.	2.22	22		24.22	36.50
0060	Covers, blank		64	.125		.93	6.90		7.83	11.65
0100	Extension rings		40	.200		3.68	11		14.68	21

478

26 05 33 – Raceway and Boxes for Electrical Systems

26 05 33.16 Outlet Boxes

		Crew	Daily Output	Labor-Hours	Unit	Material	2016 Bare Costs Labor	2016 Bare Costs Equipment	Total	Total Incl O&P
0150	Square, 4"	1 Elec	20	.400	Ea.	2.88	22		24.88	37
0200	Extension rings		40	.200		3.72	11		14.72	21
0250	Covers, blank		64	.125		1	6.90		7.90	11.75
0300	Plaster rings		64	.125		2.04	6.90		8.94	12.90
0652	Switchbox		24	.333		3.55	18.35		21.90	32.50
1100	Concrete, floor, 1 gang		5.30	1.509		89.50	83		172.50	228
9000	Minimum labor/equipment charge		4	2	Job		110		110	170

26 05 33.17 Outlet Boxes, Plastic

		Crew	Daily Output	Labor-Hours	Unit	Material	2016 Bare Costs Labor	2016 Bare Costs Equipment	Total	Total Incl O&P
0010	**OUTLET BOXES, PLASTIC**									
0050	4" diameter, round with 2 mounting nails	1 Elec	25	.320	Ea.	2.39	17.65		20.04	30
0100	Bar hanger mounted		25	.320		5.10	17.65		22.75	33
0200	4", square with 2 mounting nails		25	.320		4.63	17.65		22.28	32.50
0300	Plaster ring		64	.125		1.98	6.90		8.88	12.85
0400	Switch box with 2 mounting nails, 1 gang		30	.267		4.39	14.70		19.09	27.50
0500	2 gang		25	.320		3.33	17.65		20.98	31
0600	3 gang		20	.400		4.19	22		26.19	38.50
0700	Old work box		30	.267		4.85	14.70		19.55	28
9000	Minimum labor/equipment charge		4	2	Job		110		110	170

26 05 33.18 Pull Boxes

		Crew	Daily Output	Labor-Hours	Unit	Material	2016 Bare Costs Labor	2016 Bare Costs Equipment	Total	Total Incl O&P
0010	**PULL BOXES**									
0100	Steel, pull box, NEMA 1, type SC, 6" W x 6" H x 4" D	1 Elec	8	1	Ea.	9.65	55		64.65	95.50
0200	8" W x 8" H x 4" D		8	1		13.20	55		68.20	99.50
0300	10" W x 12" H x 6" D		5.30	1.509		22.50	83		105.50	154
0400	16" W x 20" H x 8" D		4	2		84.50	110		194.50	263
0500	20" W x 24" H x 8" D		3.20	2.500		98	138		236	320
0600	24" W x 36" H x 8" D		2.70	2.963		153	163		316	420
0650	Pull box, hinged , NEMA 1, 6" W x 6" H x 4" D		8	1		11.60	55		66.60	98
0800	12" W x 16" H x 6" D		4.70	1.702		57	94		151	208
1000	20" W x 20" H x 6" D		3.60	2.222		99	122		221	298
1200	20" W x 20" H x 8" D		3.20	2.500		149	138		287	375
1400	24" W x 36" H x 8" D		2.70	2.963		237	163		400	515
1600	24" W x 42" H x 8" D		2	4		350	220		570	725

26 05 33.23 Wireway

		Crew	Daily Output	Labor-Hours	Unit	Material	2016 Bare Costs Labor	2016 Bare Costs Equipment	Total	Total Incl O&P
0010	**WIREWAY** to 15' high									
0100	NEMA 1, Screw cover w/fittings and supports, 2-1/2" x 2-1/2"	1 Elec	45	.178	L.F.	10.80	9.80		20.60	27
0200	4" x 4"	"	40	.200		11.65	11		22.65	30
0400	6" x 6"	2 Elec	60	.267		19.60	14.70		34.30	44
0600	8" x 8"	"	40	.400		28.50	22		50.50	65.50
4475	NEMA 3R, Screw cover w/fittings and supports, 4" x 4"	1 Elec	36	.222		14	12.25		26.25	34.50
4480	6" x 6"	2 Elec	55	.291		18.75	16.05		34.80	45.50
4485	8" x 8"		36	.444		29	24.50		53.50	70
4490	12" x 12"		18	.889		46.50	49		95.50	127

26 05 33.95 Cutting and Drilling

		Crew	Daily Output	Labor-Hours	Unit	Material	2016 Bare Costs Labor	2016 Bare Costs Equipment	Total	Total Incl O&P
0010	**CUTTING AND DRILLING**									
0100	Hole drilling to 10' high, concrete wall									
0110	8" thick, 1/2" pipe size	R-31	12	.667	Ea.	.24	36.50	5.35	42.09	63
0120	3/4" pipe size		12	.667		.24	36.50	5.35	42.09	63
0130	1" pipe size		9.50	.842		.38	46.50	6.75	53.63	80
0140	1-1/4" pipe size		9.50	.842		.38	46.50	6.75	53.63	80
0150	1-1/2" pipe size		9.50	.842		.38	46.50	6.75	53.63	80
0160	2" pipe size		4.40	1.818		.51	100	14.60	115.11	172
0170	2-1/2" pipe size		4.40	1.818		.51	100	14.60	115.11	172

For customer support on your Commercial Renovation Cost Data, call 877.742.4719.

479

26 05 33.95 Cutting and Drilling	Crew	Daily Output	Labor-Hours	Unit	Material	2016 Bare Costs Labor	Equipment	Total	Total Incl O&P
0180	3" pipe size	R-31	4.40	1.818	Ea.	.51	100	14.60	115.11	172
0190	3-1/2" pipe size		3.30	2.424		.66	134	19.45	154.11	229
0200	4" pipe size		3.30	2.424		.66	134	19.45	154.11	229
0500	12" thick, 1/2" pipe size		9.40	.851		.36	47	6.85	54.21	80.50
0520	3/4" pipe size		9.40	.851		.36	47	6.85	54.21	80.50
0540	1" pipe size		7.30	1.096		.57	60.50	8.80	69.87	104
0560	1-1/4" pipe size		7.30	1.096		.57	60.50	8.80	69.87	104
0570	1-1/2" pipe size		7.30	1.096		.57	60.50	8.80	69.87	104
0580	2" pipe size		3.60	2.222		.77	122	17.85	140.62	209
0590	2-1/2" pipe size		3.60	2.222		.77	122	17.85	140.62	209
0600	3" pipe size		3.60	2.222		.77	122	17.85	140.62	209
0610	3-1/2" pipe size		2.80	2.857		1	157	23	181	269
0630	4" pipe size		2.50	3.200		1	176	25.50	202.50	305
0650	16" thick, 1/2" pipe size		7.60	1.053		.48	58	8.45	66.93	99.50
0670	3/4" pipe size		7	1.143		.48	63	9.20	72.68	108
0690	1" pipe size		6	1.333		.76	73.50	10.70	84.96	127
0710	1-1/4" pipe size		5.50	1.455		.76	80	11.70	92.46	138
0730	1-1/2" pipe size		5.50	1.455		.76	80	11.70	92.46	138
0750	2" pipe size		3	2.667		1.03	147	21.50	169.53	252
0770	2-1/2" pipe size		2.70	2.963		1.03	163	24	188.03	279
0790	3" pipe size		2.50	3.200		1.03	176	25.50	202.53	305
0810	3-1/2" pipe size		2.30	3.478		1.33	192	28	221.33	330
0830	4" pipe size		2	4		1.33	220	32	253.33	375
0850	20" thick, 1/2" pipe size		6.40	1.250		.59	69	10.05	79.64	119
0870	3/4" pipe size		6	1.333		.59	73.50	10.70	84.79	126
0890	1" pipe size		5	1.600		.95	88	12.85	101.80	151
0910	1-1/4" pipe size		4.80	1.667		.95	92	13.40	106.35	158
0930	1-1/2" pipe size		4.60	1.739		.95	96	13.95	110.90	164
0950	2" pipe size		2.70	2.963		1.28	163	24	188.28	279
0970	2-1/2" pipe size		2.40	3.333		1.28	184	27	212.28	315
0990	3" pipe size		2.20	3.636		1.28	200	29	230.28	345
1010	3-1/2" pipe size		2	4		1.66	220	32	253.66	375
1030	4" pipe size		1.70	4.706		1.66	259	38	298.66	445
1050	24" thick, 1/2" pipe size		5.50	1.455		.71	80	11.70	92.41	138
1070	3/4" pipe size		5.10	1.569		.71	86.50	12.60	99.81	149
1090	1" pipe size		4.30	1.860		1.14	103	14.95	119.09	177
1110	1-1/4" pipe size		4	2		1.14	110	16.05	127.19	189
1130	1-1/2" pipe size		4	2		1.14	110	16.05	127.19	189
1150	2" pipe size		2.40	3.333		1.54	184	27	212.54	315
1170	2-1/2" pipe size		2.20	3.636		1.54	200	29	230.54	345
1190	3" pipe size		2	4		1.54	220	32	253.54	375
1210	3-1/2" pipe size		1.80	4.444		1.99	245	35.50	282.49	420
1230	4" pipe size		1.50	5.333		1.99	294	43	338.99	505
1500	Brick wall, 8" thick, 1/2" pipe size		18	.444		.24	24.50	3.57	28.31	42
1520	3/4" pipe size		18	.444		.24	24.50	3.57	28.31	42
1540	1" pipe size		13.30	.602		.38	33	4.83	38.21	57
1560	1-1/4" pipe size		13.30	.602		.38	33	4.83	38.21	57
1580	1-1/2" pipe size		13.30	.602		.38	33	4.83	38.21	57
1600	2" pipe size		5.70	1.404		.51	77.50	11.25	89.26	133
1620	2-1/2" pipe size		5.70	1.404		.51	77.50	11.25	89.26	133
1640	3" pipe size		5.70	1.404		.51	77.50	11.25	89.26	133
1660	3-1/2" pipe size		4.40	1.818		.66	100	14.60	115.26	172
1680	4" pipe size		4	2		.66	110	16.05	126.71	188

26 05 33 – Raceway and Boxes for Electrical Systems

26 05 33.95 Cutting and Drilling		Crew	Daily Output	Labor-Hours	Unit	Material	2016 Bare Costs Labor	Equipment	Total	Total Incl O&P
1700	12" thick, 1/2" pipe size	R-31	14.50	.552	Ea.	.36	30.50	4.43	35.29	52.50
1720	3/4" pipe size		14.50	.552		.36	30.50	4.43	35.29	52.50
1740	1" pipe size		11	.727		.57	40	5.85	46.42	69
1760	1-1/4" pipe size		11	.727		.57	40	5.85	46.42	69
1780	1-1/2" pipe size		11	.727		.57	40	5.85	46.42	69
1800	2" pipe size		5	1.600		.77	88	12.85	101.62	151
1820	2-1/2" pipe size		5	1.600		.77	88	12.85	101.62	151
1840	3" pipe size		5	1.600		.77	88	12.85	101.62	151
1860	3-1/2" pipe size		3.80	2.105		1	116	16.90	133.90	199
1880	4" pipe size		3.30	2.424		1	134	19.45	154.45	230
1900	16" thick, 1/2" pipe size		12.30	.650		.48	36	5.20	41.68	62
1920	3/4" pipe size		12.30	.650		.48	36	5.20	41.68	62
1940	1" pipe size		9.30	.860		.76	47.50	6.90	55.16	82
1960	1-1/4" pipe size		9.30	.860		.76	47.50	6.90	55.16	82
1980	1-1/2" pipe size		9.30	.860		.76	47.50	6.90	55.16	82
2000	2" pipe size		4.40	1.818		1.03	100	14.60	115.63	172
2010	2-1/2" pipe size		4.40	1.818		1.03	100	14.60	115.63	172
2030	3" pipe size		4.40	1.818		1.03	100	14.60	115.63	172
2050	3-1/2" pipe size		3.30	2.424		1.33	134	19.45	154.78	230
2070	4" pipe size		3	2.667		1.33	147	21.50	169.83	252
2090	20" thick, 1/2" pipe size		10.70	.748		.59	41	6	47.59	71
2110	3/4" pipe size		10.70	.748		.59	41	6	47.59	71
2130	1" pipe size		8	1		.95	55	8.05	64	95
2150	1-1/4" pipe size		8	1		.95	55	8.05	64	95
2170	1-1/2" pipe size		8	1		.95	55	8.05	64	95
2190	2" pipe size		4	2		1.28	110	16.05	127.33	189
2210	2-1/2" pipe size		4	2		1.28	110	16.05	127.33	189
2230	3" pipe size		4	2		1.28	110	16.05	127.33	189
2250	3-1/2" pipe size		3	2.667		1.66	147	21.50	170.16	252
2270	4" pipe size		2.70	2.963		1.66	163	24	188.66	280
2290	24" thick, 1/2" pipe size		9.40	.851		.71	47	6.85	54.56	81
2310	3/4" pipe size		9.40	.851		.71	47	6.85	54.56	81
2330	1" pipe size		7.10	1.127		1.14	62	9.05	72.19	107
2350	1-1/4" pipe size		7.10	1.127		1.14	62	9.05	72.19	107
2370	1-1/2" pipe size		7.10	1.127		1.14	62	9.05	72.19	107
2390	2" pipe size		3.60	2.222		1.54	122	17.85	141.39	210
2410	2-1/2" pipe size		3.60	2.222		1.54	122	17.85	141.39	210
2430	3" pipe size		3.60	2.222		1.54	122	17.85	141.39	210
2450	3-1/2" pipe size		2.80	2.857		1.99	157	23	181.99	270
2470	4" pipe size	▼	2.50	3.200	▼	1.99	176	25.50	203.49	305
3000	Knockouts to 8' high, metal boxes & enclosures									
3020	With hole saw, 1/2" pipe size	1 Elec	53	.151	Ea.		8.30		8.30	12.85
3040	3/4" pipe size		47	.170			9.40		9.40	14.50
3050	1" pipe size		40	.200			11		11	17.05
3060	1-1/4" pipe size		36	.222			12.25		12.25	18.95
3070	1-1/2" pipe size		32	.250			13.80		13.80	21.50
3080	2" pipe size		27	.296			16.35		16.35	25
3090	2-1/2" pipe size		20	.400			22		22	34
4010	3" pipe size		16	.500			27.50		27.50	42.50
4030	3-1/2" pipe size		13	.615			34		34	52.50
4050	4" pipe size	▼	11	.727	▼		40		40	62

For customer support on your Commercial Renovation Cost Data, call 877.742.4719.

481

26 05 39 – Underfloor Raceways for Electrical Systems

26 05 39.30 Conduit In Concrete Slab

		Crew	Daily Output	Labor-Hours	Unit	Material	2016 Bare Costs Labor	2016 Bare Costs Equipment	Total	Total Incl O&P
0010	**CONDUIT IN CONCRETE SLAB** Including terminations,									
0020	fittings and supports									
3230	PVC, schedule 40, 1/2" diameter	1 Elec	270	.030	L.F.	.48	1.63		2.11	3.05
3250	3/4" diameter		230	.035		.56	1.92		2.48	3.57
3270	1" diameter		200	.040		.76	2.20		2.96	4.25
3300	1-1/4" diameter		170	.047		1	2.59		3.59	5.10
3330	1-1/2" diameter		140	.057		1.21	3.15		4.36	6.20
3350	2" diameter		120	.067		1.52	3.67		5.19	7.35
4350	Rigid galvanized steel, 1/2" diameter		200	.040		2.35	2.20		4.55	6
4400	3/4" diameter		170	.047		2.48	2.59		5.07	6.75
4450	1" diameter		130	.062		3.44	3.39		6.83	9.05
4500	1-1/4" diameter		110	.073		4.78	4.01		8.79	11.45
4600	1-1/2" diameter		100	.080		5.35	4.41		9.76	12.65
4800	2" diameter		90	.089		6.60	4.90		11.50	14.80
9000	Minimum labor/equipment charge		4	2	Job		110		110	170

26 05 39.40 Conduit In Trench

		Crew	Daily Output	Labor-Hours	Unit	Material	2016 Bare Costs Labor	2016 Bare Costs Equipment	Total	Total Incl O&P
0010	**CONDUIT IN TRENCH** Includes terminations and fittings									
0020	Does not include excavation or backfill, see Section 31 23 16									
0200	Rigid galvanized steel, 2" diameter	1 Elec	150	.053	L.F.	6.25	2.94		9.19	11.45
0400	2-1/2" diameter	"	100	.080		12.15	4.41		16.56	20
0600	3" diameter	2 Elec	160	.100		14.15	5.50		19.65	24
0800	3-1/2" diameter		140	.114		18.40	6.30		24.70	30.50
1000	4" diameter		100	.160		20.50	8.80		29.30	36
1200	5" diameter		80	.200		42	11		53	63
1400	6" diameter		60	.267		58.50	14.70		73.20	86.50
9000	Minimum labor/equipment charge	1 Elec	4	2	Job		110		110	170

26 05 80 – Wiring Connections

26 05 80.10 Motor Connections

		Crew	Daily Output	Labor-Hours	Unit	Material	2016 Bare Costs Labor	2016 Bare Costs Equipment	Total	Total Incl O&P
0010	**MOTOR CONNECTIONS**									
0020	Flexible conduit and fittings, 115 volt, 1 phase, up to 1 HP motor	1 Elec	8	1	Ea.	5.55	55		60.55	91
9000	Minimum labor/equipment charge	"	4	2	Job		110		110	170

26 05 90 – Residential Applications

26 05 90.10 Residential Wiring

		Crew	Daily Output	Labor-Hours	Unit	Material	2016 Bare Costs Labor	2016 Bare Costs Equipment	Total	Total Incl O&P
0010	**RESIDENTIAL WIRING**									
0020	20' avg. runs and #14/2 wiring incl. unless otherwise noted									
1000	Service & panel, includes 24' SE-AL cable, service eye, meter,									
1010	Socket, panel board, main bkr., ground rod, 15 or 20 amp									
1020	1-pole circuit breakers, and misc. hardware									
1100	100 amp, with 10 branch breakers	1 Elec	1.19	6.723	Ea.	485	370		855	1,100
1110	With PVC conduit and wire		.92	8.696		525	480		1,005	1,325
1120	With RGS conduit and wire		.73	10.959		685	605		1,290	1,700
1150	150 amp, with 14 branch breakers		1.03	7.767		705	430		1,135	1,425
1170	With PVC conduit and wire		.82	9.756		785	540		1,325	1,700
1180	With RGS conduit and wire		.67	11.940		1,100	660		1,760	2,225
1200	200 amp, with 18 branch breakers	2 Elec	1.80	8.889		1,050	490		1,540	1,900
1220	With PVC conduit and wire		1.46	10.959		1,125	605		1,730	2,175
1230	With RGS conduit and wire		1.24	12.903		1,525	710		2,235	2,775
1800	Lightning surge suppressor	1 Elec	32	.250		50.50	13.80		64.30	77.50
2000	Switch devices									
2100	Single pole, 15 amp, Ivory, with a 1-gang box, cover plate,									
2110	Type NM (Romex) cable	1 Elec	17.10	.468	Ea.	18.70	26		44.70	60.50

482

For customer support on your Commercial Renovation Cost Data, call 877.742.4719.

26 05 90 – Residential Applications

26 05 90.10 Residential Wiring	Crew	Daily Output	Labor-Hours	Unit	Material	2016 Bare Costs Labor	Equipment	Total	Total Incl O&P
2120 Type MC (BX) cable	1 Elec	14.30	.559	Ea.	27.50	31		58.50	77.50
2130 EMT & wire		5.71	1.401		36	77		113	159
2150 3-way, #14/3, type NM cable		14.55	.550		12.70	30.50		43.20	61
2170 Type MC cable		12.31	.650		25	36		61	83
2180 EMT & wire		5	1.600		30	88		118	169
2200 4-way, #14/3, type NM cable		14.55	.550		20.50	30.50		51	69.50
2220 Type MC cable		12.31	.650		32.50	36		68.50	91.50
2230 EMT & wire		5	1.600		37.50	88		125.50	177
2250 S.P., 20 amp, #12/2, type NM cable		13.33	.600		14.75	33		47.75	67
2270 Type MC cable		11.43	.700		21	38.50		59.50	82.50
2280 EMT & wire		4.85	1.649		33.50	91		124.50	178
2290 S.P. rotary dimmer, 600W, no wiring		17	.471		31.50	26		57.50	74.50
2300 S.P. rotary dimmer, 600W, type NM cable		14.55	.550		35.50	30.50		66	86
2320 Type MC cable		12.31	.650		44.50	36		80.50	104
2330 EMT & wire		5	1.600		54.50	88		142.50	196
2350 3-way rotary dimmer, type NM cable		13.33	.600		19.85	33		52.85	73
2370 Type MC cable		11.43	.700		28.50	38.50		67	91
2380 EMT & wire		4.85	1.649		39	91		130	184
2400 Interval timer wall switch, 20 amp, 1-30 min., #12/2									
2410 Type NM cable	1 Elec	14.55	.550	Ea.	57	30.50		87.50	110
2420 Type MC cable		12.31	.650		62	36		98	124
2430 EMT & wire		5	1.600		76	88		164	220
2500 Decorator style									
2510 S.P., 15 amp, type NM cable	1 Elec	17.10	.468	Ea.	22.50	26		48.50	65
2520 Type MC cable		14.30	.559		31.50	31		62.50	82
2530 EMT & wire		5.71	1.401		40	77		117	163
2550 3-way, #14/3, type NM cable		14.55	.550		16.60	30.50		47.10	65.50
2570 Type MC cable		12.31	.650		29	36		65	87.50
2580 EMT & wire		5	1.600		33.50	88		121.50	173
2600 4-way, #14/3, type NM cable		14.55	.550		24.50	30.50		55	73.50
2620 Type MC cable		12.31	.650		36.50	36		72.50	96
2630 EMT & wire		5	1.600		41.50	88		129.50	182
2650 S.P., 20 amp, #12/2, type NM cable		13.33	.600		18.60	33		51.60	71.50
2670 Type MC cable		11.43	.700		25	38.50		63.50	87
2680 EMT & wire		4.85	1.649		37.50	91		128.50	182
2700 S.P., slide dimmer, type NM cable		17.10	.468		24.50	26		50.50	67
2720 Type MC cable		14.30	.559		33	31		64	84
2730 EMT & wire		5.71	1.401		43.50	77		120.50	167
2750 S.P., touch dimmer, type NM cable		17.10	.468		41	26		67	85
2770 Type MC cable		14.30	.559		50	31		81	102
2780 EMT & wire		5.71	1.401		60	77		137	185
2800 3-way touch dimmer, type NM cable		13.33	.600		51.50	33		84.50	108
2820 Type MC cable		11.43	.700		60	38.50		98.50	126
2830 EMT & wire		4.85	1.649		70	91		161	218
3000 Combination devices									
3100 S.P. switch/15 amp recpt., Ivory, 1-gang box, plate									
3110 Type NM cable	1 Elec	11.43	.700	Ea.	24	38.50		62.50	86
3120 Type MC cable		10	.800		33	44		77	104
3130 EMT & wire		4.40	1.818		43	100		143	203
3150 S.P. switch/pilot light, type NM cable		11.43	.700		24.50	38.50		63	86.50
3170 Type MC cable		10	.800		33	44		77	105
3180 EMT & wire		4.43	1.806		43	99.50		142.50	202
3190 2-S.P. switches, 2-#14/2, no wiring		14	.571		13.70	31.50		45.20	63.50

For customer support on your Commercial Renovation Cost Data, call 877.742.4719.

483

26 05 90.10 Residential Wiring		Crew	Daily Output	Labor-Hours	Unit	Material	2016 Bare Costs Labor	Equipment	Total	Total Incl O&P
3200	2-S.P. switches, 2-#14/2, type NM cables	1 Elec	10	.800	Ea.	26.50	44		70.50	97.50
3220	Type MC cable		8.89	.900		40.50	49.50		90	121
3230	EMT & wire		4.10	1.951		45.50	108		153.50	216
3250	3-way switch/15 amp recpt., #14/3, type NM cable		10	.800		31.50	44		75.50	103
3270	Type MC cable		8.89	.900		44	49.50		93.50	125
3280	EMT & wire		4.10	1.951		48.50	108		156.50	220
3300	2-3 way switches, 2-#14/3, type NM cables		8.89	.900		40.50	49.50		90	121
3320	Type MC cable		8	1		61.50	55		116.50	153
3330	EMT & wire		4	2		55.50	110		165.50	231
3350	S.P. switch/20 amp recpt., #12/2, type NM cable		10	.800		31	44		75	103
3370	Type MC cable		8.89	.900		36	49.50		85.50	116
3380	EMT & wire		4.10	1.951		50	108		158	221
3400	Decorator style									
3410	S.P. switch/15 amp recpt., type NM cable	1 Elec	11.43	.700	Ea.	28	38.50		66.50	90.50
3420	Type MC cable		10	.800		36.50	44		80.50	109
3430	EMT & wire		4.40	1.818		47	100		147	207
3450	S.P. switch/pilot light, type NM cable		11.43	.700		28	38.50		66.50	90.50
3470	Type MC cable		10	.800		37	44		81	109
3480	EMT & wire		4.40	1.818		47	100		147	207
3500	2-S.P. switches, 2-#14/2, type NM cables		10	.800		30.50	44		74.50	102
3520	Type MC cable		8.89	.900		44	49.50		93.50	125
3530	EMT & wire		4.10	1.951		49.50	108		157.50	221
3550	3-way/15 amp recpt., #14/3, type NM cable		10	.800		35.50	44		79.50	107
3570	Type MC cable		8.89	.900		48	49.50		97.50	129
3580	EMT & wire		4.10	1.951		52.50	108		160.50	224
3650	2-3 way switches, 2-#14/3, type NM cables		8.89	.900		44.50	49.50		94	126
3670	Type MC cable		8	1		65	55		120	157
3680	EMT & wire		4	2		59.50	110		169.50	236
3700	S.P. switch/20 amp recpt., #12/2, type NM cable		10	.800		35	44		79	107
3720	Type MC cable		8.89	.900		40	49.50		89.50	121
3730	EMT & wire		4.10	1.951		54	108		162	225
4000	Receptacle devices									
4010	Duplex outlet, 15 amp recpt., Ivory, 1-gang box, plate									
4015	Type NM cable	1 Elec	14.55	.550	Ea.	10.30	30.50		40.80	58.50
4020	Type MC cable		12.31	.650		19	36		55	76.50
4030	EMT & wire		5.33	1.501		28	82.50		110.50	159
4050	With #12/2, type NM cable		12.31	.650		12.65	36		48.65	69.50
4070	Type MC cable		10.67	.750		19	41.50		60.50	85
4080	EMT & wire		4.71	1.699		31.50	93.50		125	180
4100	20 amp recpt., #12/2, type NM cable		12.31	.650		19.30	36		55.30	77
4120	Type MC cable		10.67	.750		25.50	41.50		67	92
4130	EMT & wire		4.71	1.699		38	93.50		131.50	187
4140	For GFI see Section 26 05 90.10 line 4300 below									
4150	Decorator style, 15 amp recpt., type NM cable	1 Elec	14.55	.550	Ea.	14.20	30.50		44.70	62.50
4170	Type MC cable		12.31	.650		23	36		59	80.50
4180	EMT & wire		5.33	1.501		31.50	82.50		114	163
4200	With #12/2, type NM cable		12.31	.650		16.55	36		52.55	73.50
4220	Type MC cable		10.67	.750		23	41.50		64.50	89
4230	EMT & wire		4.71	1.699		35.50	93.50		129	184
4250	20 amp recpt., #12/2, type NM cable		12.31	.650		23	36		59	81
4270	Type MC cable		10.67	.750		29.50	41.50		71	96.50
4280	EMT & wire		4.71	1.699		42	93.50		135.50	191
4300	GFI, 15 amp recpt., type NM cable		12.31	.650		23	36		59	80.50

26 05 Common Work Results for Electrical

26 05 90 – Residential Applications

26 05 90.10 Residential Wiring	Crew	Daily Output	Labor-Hours	Unit	Material	2016 Bare Costs Labor	Equipment	Total	Total Incl O&P
4320	Type MC cable	1 Elec	10.67	.750	Ea.	31.50	41.50		73	98.50
4330	EMT & wire		4.71	1.699		40.50	93.50		134	190
4350	GFI with #12/2, type NM cable		10.67	.750		25	41.50		66.50	91.50
4370	Type MC cable		9.20	.870		31.50	48		79.50	109
4380	EMT & wire		4.21	1.900		44	105		149	211
4400	20 amp recpt., #12/2, type NM cable		10.67	.750		51.50	41.50		93	121
4420	Type MC cable		9.20	.870		58	48		106	138
4430	EMT & wire		4.21	1.900		70.50	105		175.50	240
4500	Weather-proof cover for above receptacles, add		32	.250		2.94	13.80		16.74	24.50
4550	Air conditioner outlet, 20 amp-240 volt recpt.									
4560	30' of #12/2, 2 pole circuit breaker									
4570	Type NM cable	1 Elec	10	.800	Ea.	62	44		106	137
4580	Type MC cable		9	.889		70	49		119	153
4590	EMT & wire		4	2		81	110		191	259
4600	Decorator style, type NM cable		10	.800		67	44		111	142
4620	Type MC cable		9	.889		74.50	49		123.50	158
4630	EMT & wire		4	2		85.50	110		195.50	264
4650	Dryer outlet, 30 amp-240 volt recpt., 20' of #10/3									
4660	2 pole circuit breaker									
4670	Type NM cable	1 Elec	6.41	1.248	Ea.	58	69		127	170
4680	Type MC cable		5.71	1.401		64	77		141	190
4690	EMT & wire		3.48	2.299		75	127		202	279
4700	Range outlet, 50 amp-240 volt recpt., 30' of #8/3									
4710	Type NM cable	1 Elec	4.21	1.900	Ea.	85	105		190	256
4720	Type MC cable		4	2		126	110		236	310
4730	EMT & wire		2.96	2.703		104	149		253	345
4750	Central vacuum outlet, Type NM cable		6.40	1.250		53.50	69		122.50	166
4770	Type MC cable		5.71	1.401		66.50	77		143.50	193
4780	EMT & wire		3.48	2.299		81	127		208	285
4800	30 amp-110 volt locking recpt., #10/2 circ. bkr.									
4810	Type NM cable	1 Elec	6.20	1.290	Ea.	61.50	71		132.50	178
4820	Type MC cable		5.40	1.481		78.50	81.50		160	212
4830	EMT & wire		3.20	2.500		91.50	138		229.50	315
4900	Low voltage outlets									
4910	Telephone recpt., 20' of 4/C phone wire	1 Elec	26	.308	Ea.	8.40	16.95		25.35	35
4920	TV recpt., 20' of RG59U coax wire, F type connector	"	16	.500	"	16	27.50		43.50	60
4950	Door bell chime, transformer, 2 buttons, 60' of bellwire									
4970	Economy model	1 Elec	11.50	.696	Ea.	50	38.50		88.50	115
4980	Custom model		11.50	.696		99.50	38.50		138	169
4990	Luxury model, 3 buttons		9.50	.842		281	46.50		327.50	380
6000	Lighting outlets									
6050	Wire only (for fixture), type NM cable	1 Elec	32	.250	Ea.	6.35	13.80		20.15	28.50
6070	Type MC cable		24	.333		12.55	18.35		30.90	42.50
6080	EMT & wire		10	.800		20.50	44		64.50	90.50
6100	Box (4"), and wire (for fixture), type NM cable		25	.320		13.90	17.65		31.55	43
6120	Type MC cable		20	.400		20	22		42	56
6130	EMT & wire		11	.727		28	40		68	93
6200	Fixtures (use with lines 6050 or 6100 above)									
6210	Canopy style, economy grade	1 Elec	40	.200	Ea.	27	11		38	46.50
6220	Custom grade		40	.200		53.50	11		64.50	76
6250	Dining room chandelier, economy grade		19	.421		67.50	23		90.50	110
6270	Luxury grade		15	.533		820	29.50		849.50	950
6310	Kitchen fixture (fluorescent), economy grade		30	.267		71.50	14.70		86.20	102

485

26 05 90.10 Residential Wiring		Crew	Daily Output	Labor-Hours	Unit	Material	2016 Bare Costs Labor	Equipment	Total	Total Incl O&P
6320	Custom grade	1 Elec	25	.320	Ea.	221	17.65		238.65	272
6350	Outdoor, wall mounted, economy grade		30	.267		25.50	14.70		40.20	50.50
6360	Custom grade		30	.267		120	14.70		134.70	155
6370	Luxury grade		25	.320		248	17.65		265.65	300
6410	Outdoor PAR floodlights, 1 lamp, 150 watt		20	.400		27.50	22		49.50	64
6420	2 lamp, 150 watt each		20	.400		45.50	22		67.50	84
6425	Motion sensing, 2 lamp, 150 watt each		20	.400		75.50	22		97.50	117
6430	For infrared security sensor, add		32	.250		132	13.80		145.80	167
6450	Outdoor, quartz-halogen, 300 watt flood		20	.400		33	22		55	70.50
6600	Recessed downlight, round, pre-wired, 50 or 75 watt trim		30	.267		68	14.70		82.70	97
6610	With shower light trim		30	.267		82.50	14.70		97.20	113
6620	With wall washer trim		28	.286		85	15.75		100.75	118
6630	With eye-ball trim		28	.286		84.50	15.75		100.25	118
6700	Porcelain lamp holder		40	.200		2.52	11		13.52	19.80
6710	With pull switch		40	.200		7.55	11		18.55	25.50
6750	Fluorescent strip, 2-20 watt tube, wrap around diffuser, 24"		24	.333		42	18.35		60.35	75
6760	1-34 watt tube, 48"		24	.333		100	18.35		118.35	139
6770	2-34 watt tubes, 48"		20	.400		118	22		140	164
6800	Bathroom heat lamp, 1-250 watt		28	.286		37.50	15.75		53.25	65.50
6810	2-250 watt lamps		28	.286		59.50	15.75		75.25	90
6820	For timer switch, see Section 26 05 90.10 line 2400									
6900	Outdoor post lamp, incl. post, fixture, 35' of #14/2									
6910	Type NM cable	1 Elec	3.50	2.286	Ea.	285	126		411	510
6920	Photo-eye, add		27	.296		27.50	16.35		43.85	55
6950	Clock dial time switch, 24 hr., w/enclosure, type NM cable		11.43	.700		73	38.50		111.50	140
6970	Type MC cable		11	.727		81.50	40		121.50	152
6980	EMT & wire		4.85	1.649		90.50	91		181.50	241
7000	Alarm systems									
7050	Smoke detectors, box, #14/3, type NM cable	1 Elec	14.55	.550	Ea.	33.50	30.50		64	84
7070	Type MC cable		12.31	.650		44.50	36		80.50	105
7080	EMT & wire		5	1.600		49	88		137	190
7090	For relay output to security system, add					10.20			10.20	11.20
8000	Residential equipment									
8050	Disposal hook-up, incl. switch, outlet box, 3' of flex									
8060	20 amp-1 pole circ. bkr., and 25' of #12/2									
8070	Type NM cable	1 Elec	10	.800	Ea.	36.50	44		80.50	108
8080	Type MC cable		8	1		43.50	55		98.50	133
8090	EMT & wire		5	1.600		58.50	88		146.50	200
8100	Trash compactor or dishwasher hook-up, incl. outlet box,									
8110	3' of flex, 15 amp-1 pole circ. bkr., and 25' of #14/2									
8120	Type NM cable	1 Elec	10	.800	Ea.	21	44		65	91.50
8130	Type MC cable		8	1		31	55		86	119
8140	EMT & wire		5	1.600		43.50	88		131.50	184
8150	Hot water sink dispenser hook-up, use line 8100									
8200	Vent/exhaust fan hook-up, type NM cable	1 Elec	32	.250	Ea.	6.35	13.80		20.15	28.50
8220	Type MC cable		24	.333		12.55	18.35		30.90	42.50
8230	EMT & wire		10	.800		20.50	44		64.50	90.50
8250	Bathroom vent fan, 50 CFM (use with above hook-up)									
8260	Economy model	1 Elec	15	.533	Ea.	19.65	29.50		49.15	67
8270	Low noise model		15	.533		42	29.50		71.50	92
8280	Custom model		12	.667		131	36.50		167.50	201
8300	Bathroom or kitchen vent fan, 110 CFM									
8310	Economy model	1 Elec	15	.533	Ea.	67.50	29.50		97	120

26 05 90 – Residential Applications

26 05 90.10 Residential Wiring	Crew	Daily Output	Labor-Hours	Unit	Material	2016 Bare Costs Labor	Equipment	Total	Total Incl O&P
8320	Low noise model	1 Elec	15	.533	Ea.	92	29.50		121.50	147
8350	Paddle fan, variable speed (w/o lights)									
8360	Economy model (AC motor)	1 Elec	10	.800	Ea.	109	44		153	188
8362	With light kit		10	.800		150	44		194	233
8370	Custom model (AC motor)		10	.800		262	44		306	355
8372	With light kit		10	.800		300	44		344	405
8380	Luxury model (DC motor)		8	1		288	55		343	400
8382	With light kit		8	1		330	55		385	445
8390	Remote speed switch for above, add		12	.667		37.50	36.50		74	98
8500	Whole house exhaust fan, ceiling mount, 36", variable speed									
8510	Remote switch, incl. shutters, 20 amp-1 pole circ. bkr.									
8520	30' of #12/2, type NM cable	1 Elec	4	2	Ea.	1,325	110		1,435	1,650
8530	Type MC cable		3.50	2.286		1,350	126		1,476	1,675
8540	EMT & wire		3	2.667		1,350	147		1,497	1,725
8600	Whirlpool tub hook-up, incl. timer switch, outlet box									
8610	3' of flex, 20 amp-1 pole GFI circ. bkr.									
8620	30' of #12/2, type NM cable	1 Elec	5	1.600	Ea.	157	88		245	310
8630	Type MC cable		4.20	1.905		160	105		265	340
8640	EMT & wire		3.40	2.353		172	130		302	390
8650	Hot water heater hook-up, incl. 1-2 pole circ. bkr., box;									
8660	3' of flex, 20' of #10/2, type NM cable	1 Elec	5	1.600	Ea.	31.50	88		119.50	171
8670	Type MC cable		4.20	1.905		42.50	105		147.50	209
8680	EMT & wire		3.40	2.353		47	130		177	252
9000	Heating/air conditioning									
9050	Furnace/boiler hook-up, incl. firestat, local on-off switch									
9060	Emergency switch, and 40' of type NM cable	1 Elec	4	2	Ea.	52	110		162	227
9070	Type MC cable		3.50	2.286		65.50	126		191.50	267
9080	EMT & wire		1.50	5.333		83.50	294		377.50	545
9100	Air conditioner hook-up, incl. local 60 amp disc. switch									
9110	3' sealtite, 40 amp, 2 pole circuit breaker									
9130	40' of #8/2, type NM cable	1 Elec	3.50	2.286	Ea.	144	126		270	355
9140	Type MC cable		3	2.667		204	147		351	450
9150	EMT & wire		1.30	6.154		185	340		525	730
9200	Heat pump hook-up, 1-40 & 1-100 amp 2 pole circ. bkr.									
9210	Local disconnect switch, 3' sealtite									
9220	40' of #8/2 & 30' of #3/2									
9230	Type NM cable	1 Elec	1.30	6.154	Ea.	460	340		800	1,025
9240	Type MC cable		1.08	7.407		510	410		920	1,200
9250	EMT & wire		.94	8.511		520	470		990	1,300
9500	Thermostat hook-up, using low voltage wire									
9520	Heating only, 25' of #18-3	1 Elec	24	.333	Ea.	8	18.35		26.35	37.50
9530	Heating/cooling, 25' of #18-4	"	20	.400	"	10.50	22		32.50	45.50

For customer support on your Commercial Renovation Cost Data, call 877.742.4719.

487

26 22 Low-Voltage Transformers

26 22 13 – Low-Voltage Distribution Transformers

26 22 13.10 Transformer, Dry-Type	Crew	Daily Output	Labor-Hours	Unit	Material	2016 Bare Costs Labor	Equipment	Total	Total Incl O&P
0010 **TRANSFORMER, DRY-TYPE**									
0050 Single phase, 240/480 volt primary, 120/240 volt secondary									
0100 1 kVA	1 Elec	2	4	Ea.	340	220		560	715
0300 2 kVA		1.60	5		540	276		816	1,025
0500 3 kVA		1.40	5.714		670	315		985	1,225
0700 5 kVA		1.20	6.667		920	365		1,285	1,575
0900 7.5 kVA	2 Elec	2.20	7.273		1,275	400		1,675	2,025
1100 10 kVA		1.60	10		1,450	550		2,000	2,450
1300 15 kVA		1.20	13.333		1,975	735		2,710	3,275
2300 3 phase, 480 volt primary, 120/208 volt secondary									
2310 Ventilated, 3 kVA	1 Elec	1	8	Ea.	1,050	440		1,490	1,825
2700 6 kVA		.80	10		1,175	550		1,725	2,150
2900 9 kVA		.70	11.429		1,175	630		1,805	2,275
3100 15 kVA	2 Elec	1.10	14.545		1,500	800		2,300	2,900
3300 30 kVA		.90	17.778		1,475	980		2,455	3,150
3500 45 kVA		.80	20		1,750	1,100		2,850	3,625
9000 Minimum labor/equipment charge	1 Elec	1	8	Job		440		440	680

26 24 Switchboards and Panelboards

26 24 16 – Panelboards

26 24 16.20 Panelboard and Load Center Circuit Breakers

26 24 16.20 Panelboard and Load Center Circuit Breakers	Crew	Daily Output	Labor-Hours	Unit	Material	2016 Bare Costs Labor	Equipment	Total	Total Incl O&P
0010 **PANELBOARD AND LOAD CENTER CIRCUIT BREAKERS**									
0050 Bolt-on, 10,000 amp I.C., 120 volt, 1 pole									
0100 15 to 50 amp	1 Elec	10	.800	Ea.	19.15	44		63.15	89
0200 60 amp		8	1		19.20	55		74.20	106
0300 70 amp		8	1		28	55		83	116
0350 240 volt, 2 pole									
0400 15 to 50 amp	1 Elec	8	1	Ea.	82	55		137	176
0500 60 amp		7.50	1.067		56	59		115	153
0600 80 to 100 amp		5	1.600		107	88		195	254
0700 3 pole, 15 to 60 amp		6.20	1.290		133	71		204	256
0800 70 amp		5	1.600		168	88		256	320
0900 80 to 100 amp		3.60	2.222		191	122		313	400
1000 22,000 amp I.C., 240 volt, 2 pole, 70 – 225 amp		2.70	2.963		600	163		763	910
1100 3 pole, 70 – 225 amp		2.30	3.478		595	192		787	950
1200 14,000 amp I.C., 277 volts, 1 pole, 15 – 30 amp		8	1		51	55		106	141
1300 22,000 amp I.C., 480 volts, 2 pole, 70 – 225 amp		2.70	2.963		535	163		698	840
1400 3 pole, 70 – 225 amp		2.30	3.478		790	192		982	1,175
2060 Plug-in tandem, 120/240 V, 2-15 A, 1 pole		11	.727		54.50	40		94.50	122
2070 1-15 A & 1-20 A		11	.727		54.50	40		94.50	122
2080 2-20 A		11	.727		54.50	40		94.50	122
2082 Arc fault circuit interrupter, 120/240 V, 1-15 A & 1-20 A, 1 pole		11	.727		56.50	40		96.50	124
9000 Minimum labor/equipment charge		3	2.667	Job		147		147	227

26 24 16.30 Panelboards Commercial Applications

26 24 16.30 Panelboards Commercial Applications	Crew	Daily Output	Labor-Hours	Unit	Material	2016 Bare Costs Labor	Equipment	Total	Total Incl O&P
0010 **PANELBOARDS COMMERCIAL APPLICATIONS**									
0050 NQOD, w/20 amp 1 pole bolt-on circuit breakers									
0100 3 wire, 120/240 volts, 100 amp main lugs									
0150 10 circuits	1 Elec	1	8	Ea.	640	440		1,080	1,375
0200 14 circuits		.88	9.091		715	500		1,215	1,550
0250 18 circuits		.75	10.667		780	590		1,370	1,775
0300 20 circuits		.65	12.308		870	680		1,550	2,000

26 24 Switchboards and Panelboards

26 24 16 – Panelboards

26 24 16.30 Panelboards Commercial Applications	Crew	Daily Output	Labor-Hours	Unit	Material	Labor	2016 Bare Costs Equipment	Total	Total Incl O&P
0600	4 wire, 120/208 volts, 100 amp main lugs, 12 circuits	1 Elec	1	8	Ea.	690	440		1,130	1,450
0650	16 circuits		.75	10.667		790	590		1,380	1,775
0700	20 circuits		.65	12.308		915	680		1,595	2,050
0750	24 circuits		.60	13.333		910	735		1,645	2,125
0800	30 circuits		.53	15.094		1,125	830		1,955	2,525
0850	225 amp main lugs, 32 circuits	2 Elec	.90	17.778		1,275	980		2,255	2,925
0900	34 circuits		.84	19.048		1,300	1,050		2,350	3,050
0950	36 circuits		.80	20		1,325	1,100		2,425	3,175
1000	42 circuits		.68	23.529		1,475	1,300		2,775	3,625
1200	NEHB, w/20 amp, 1 pole bolt-on circuit breakers									
1250	4 wire, 277/480 volts, 100 amp main lugs, 12 circuits	1 Elec	.88	9.091	Ea.	1,325	500		1,825	2,225
1300	20 circuits	"	.60	13.333		1,975	735		2,710	3,300
1350	225 amp main lugs, 24 circuits	2 Elec	.90	17.778		2,225	980		3,205	3,975
1400	30 circuits		.80	20		2,675	1,100		3,775	4,650
1450	36 circuits		.72	22.222		3,125	1,225		4,350	5,325
1600	NQOD panel, w/20 amp, 1 pole, circuit breakers									
2000	4 wire, 120/208 volts with main circuit breaker									
2050	100 amp main, 24 circuits	1 Elec	.47	17.021	Ea.	1,275	940		2,215	2,850
2100	30 circuits	"	.40	20		1,425	1,100		2,525	3,275
2200	225 amp main, 32 circuits	2 Elec	.72	22.222		2,400	1,225		3,625	4,525
2250	42 circuits		.56	28.571		2,625	1,575		4,200	5,300
2300	400 amp main, 42 circuits		.48	33.333		3,525	1,825		5,350	6,725
2350	600 amp main, 42 circuits		.40	40		5,200	2,200		7,400	9,125
2400	NEHB, with 20 amp, 1 pole circuit breaker									
2450	4 wire, 277/480 volts with main circuit breaker									
2500	100 amp main, 24 circuits	1 Elec	.42	19.048	Ea.	2,575	1,050		3,625	4,450
2550	30 circuits	"	.38	21.053		3,025	1,150		4,175	5,125
2600	225 amp main, 30 circuits	2 Elec	.72	22.222		3,775	1,225		5,000	6,075
2650	42 circuits	"	.56	28.571		4,650	1,575		6,225	7,550
9000	Minimum labor/equipment charge	1 Elec	1	8	Job		440		440	680

26 27 Low-Voltage Distribution Equipment

26 27 13 – Electricity Metering

26 27 13.10 Meter Centers and Sockets

26 27 13.10 Meter Centers and Sockets	Crew	Daily Output	Labor-Hours	Unit	Material	Labor	2016 Bare Costs Equipment	Total	Total Incl O&P
0010	METER CENTERS AND SOCKETS									
0100	Sockets, single position, 4 terminal, 100 amp	1 Elec	3.20	2.500	Ea.	43	138		181	261
0200	150 amp		2.30	3.478		48	192		240	350
0300	200 amp		1.90	4.211		92	232		324	460
0400	Transformer rated, 20 amp		3.20	2.500		152	138		290	380
0500	Double position, 4 terminal, 100 amp		2.80	2.857		204	157		361	470
0600	150 amp		2.10	3.810		240	210		450	590
0700	200 amp		1.70	4.706		450	259		709	895
1100	Meter centers and sockets, three phase, single pos, 7 terminal, 100 amp		2.80	2.857		164	157		321	425
1200	200 amp		2.10	3.810		295	210		505	650
1400	400 amp		1.70	4.706		720	259		979	1,200
9000	Minimum labor/equipment charge		3	2.667	Job		147		147	227

26 27 Low-Voltage Distribution Equipment

26 27 16 – Electrical Cabinets and Enclosures

26 27 16.10 Cabinets

	Crew	Daily Output	Labor-Hours	Unit	Material	2016 Bare Costs Labor	2016 Bare Costs Equipment	Total	Total Incl O&P
0010 **CABINETS**									
7000 Cabinets, current transformer									
7050 Single door, 24" H x 24" W x 10" D	1 Elec	1.60	5	Ea.	129	276		405	565
7100 30" H x 24" W x 10" D		1.30	6.154		140	340		480	680
7150 36" H x 24" W x 10" D		1.10	7.273		186	400		586	825
7200 30" H x 30" W x 10" D		1	8		250	440		690	955
7250 36" H x 30" W x 10" D		.90	8.889		300	490		790	1,075
7300 36" H x 36" W x 10" D		.80	10		310	550		860	1,200
7500 Double door, 48" H x 36" W x 10" D		.60	13.333		590	735		1,325	1,775
7550 24" H x 24" W x 12" D		1	8		173	440		613	870
9990 Minimum labor/equipment charge		2	4	Job		220		220	340

26 27 23 – Indoor Service Poles

26 27 23.40 Surface Raceway

	Crew	Daily Output	Labor-Hours	Unit	Material	2016 Bare Costs Labor	2016 Bare Costs Equipment	Total	Total Incl O&P
0010 **SURFACE RACEWAY**									
0090 Metal, straight section									
0100 No. 500	1 Elec	100	.080	L.F.	1.04	4.41		5.45	7.95
0110 No. 700		100	.080		1.17	4.41		5.58	8.10
0400 No. 1500, small pancake		90	.089		2.15	4.90		7.05	9.90
0600 No. 2000, base & cover, blank		90	.089		2.19	4.90		7.09	9.95
0800 No. 3000, base & cover, blank		75	.107		4.18	5.90		10.08	13.70
1000 No. 4000, base & cover, blank		65	.123		6.80	6.80		13.60	18
1200 No. 6000, base & cover, blank		50	.160		11.40	8.80		20.20	26
2400 Fittings, elbows, No. 500		40	.200	Ea.	1.90	11		12.90	19.15
2800 Elbow cover, No. 2000		40	.200		3.57	11		14.57	21
2880 Tee, No. 500		42	.190		3.66	10.50		14.16	20.50
2900 No. 2000		27	.296		11.85	16.35		28.20	38
3000 Switch box, No. 500		16	.500		11	27.50		38.50	54.50
3400 Telephone outlet, No. 1500		16	.500		14.10	27.50		41.60	58
3600 Junction box, No. 1500		16	.500		9.65	27.50		37.15	53
3800 Plugmold wired sections, No. 2000									
4000 1 circuit, 6 outlets, 3 ft. long	1 Elec	8	1	Ea.	36	55		91	125
4100 2 circuits, 8 outlets, 6 ft. long	"	5.30	1.509	"	59	83		142	194
9300 Non-metallic, straight section									
9310 7/16" x 7/8", base & cover, blank	1 Elec	160	.050	L.F.	1.72	2.76		4.48	6.15
9320 Base & cover w/adhesive		160	.050		1.42	2.76		4.18	5.80
9340 7/16" x 1-5/16", base & cover, blank		145	.055		1.88	3.04		4.92	6.75
9350 Base & cover w/adhesive		145	.055		2.17	3.04		5.21	7.10
9370 11/16" x 2-1/4", base & cover, blank		130	.062		2.66	3.39		6.05	8.20
9380 Base & cover w/adhesive		130	.062		3.01	3.39		6.40	8.55
9385 1-11/16" x 5-1/4", two compartment base & cover w/screws		80	.100		7.55	5.50		13.05	16.80
9400 Fittings, elbows, 7/16" x 7/8"		50	.160	Ea.	1.88	8.80		10.68	15.70
9410 7/16" x 1-5/16"		45	.178		1.95	9.80		11.75	17.30
9420 11/16" x 2-1/4"		40	.200		2.11	11		13.11	19.35
9425 1-11/16" x 5-1/4"		28	.286		11.05	15.75		26.80	36.50
9430 Tees, 7/16" x 7/8"		35	.229		2.42	12.60		15.02	22
9440 7/16" x 1-5/16"		32	.250		2.49	13.80		16.29	24
9450 11/16" x 2-1/4"		30	.267		2.56	14.70		17.26	25.50
9455 1-11/16" x 5-1/4"		24	.333		17.75	18.35		36.10	48
9460 Cover clip, 7/16" x 7/8"		80	.100		.49	5.50		5.99	9.05
9470 7/16" x 1-5/16"		72	.111		.44	6.10		6.54	9.95
9480 11/16" x 2-1/4"		64	.125		.75	6.90		7.65	11.50
9484 1-11/16" x 5-1/4"		42	.190		2.61	10.50		13.11	19.10

For customer support on your Commercial Renovation Cost Data, call 877.742.4719.

26 27 Low-Voltage Distribution Equipment

26 27 23 – Indoor Service Poles

26 27 23.40 Surface Raceway		Crew	Daily Output	Labor-Hours	Unit	Material	2016 Bare Costs Labor	Equipment	Total	Total Incl O&P
9486	Wire clip, 1-11/16" x 5-1/4"	1 Elec	68	.118	Ea.	.45	6.50		6.95	10.50
9490	Blank end, 7/16" x 7/8"		50	.160		.70	8.80		9.50	14.40
9500	7/16" x 1-5/16"		45	.178		.78	9.80		10.58	16
9510	11/16" x 2-1/4"		40	.200		1.18	11		12.18	18.35
9515	1-11/16" x 5-1/4"		38	.211		5.45	11.60		17.05	24
9520	Round fixture box, 5.5" dia x 1"		25	.320		11.10	17.65		28.75	39.50
9530	Device box, 1 gang		30	.267		5	14.70		19.70	28
9540	2 gang		25	.320	▼	7.30	17.65		24.95	35.50
9990	Minimum labor/equipment charge	▼	5	1.600	Job		88		88	136

26 27 26 – Wiring Devices

26 27 26.20 Wiring Devices Elements

		Crew	Daily Output	Labor-Hours	Unit	Material	Labor	Equipment	Total	Total Incl O&P
0010	**WIRING DEVICES ELEMENTS**									
0200	Toggle switch, quiet type, single pole, 15 amp	1 Elec	40	.200	Ea.	9.50	11		20.50	27.50
0500	20 amp		27	.296		3.35	16.35		19.70	28.50
0550	Rocker, 15 amp		40	.200		3.86	11		14.86	21.50
0560	20 amp		27	.296		7.80	16.35		24.15	33.50
0600	3 way, 15 amp		23	.348		1.90	19.15		21.05	31.50
0850	Rocker, 15 amp		23	.348		4.50	19.15		23.65	34.50
0860	20 amp		18	.444		10.70	24.50		35.20	50
0900	4 way, 15 amp		15	.533		7.45	29.50		36.95	53.50
1030	Rocker, 15 amp		15	.533		17.50	29.50		47	65
1040	20 amp		11	.727		18.05	40		58.05	82
1650	Dimmer switch, 120 volt, incandescent, 600 watt, 1 pole [G]		16	.500		22.50	27.50		50	67.50
2460	Receptacle, duplex, 120 volt, grounded, 15 amp		40	.200		1.26	11		12.26	18.45
2470	20 amp		27	.296		7.95	16.35		24.30	33.50
2480	Ground fault interrupting, 15 amp		27	.296		13.75	16.35		30.10	40
2482	20 amp		27	.296		40	16.35		56.35	69
2490	Dryer, 30 amp		15	.533		4.46	29.50		33.96	50.50
2500	Range, 50 amp		11	.727		12.15	40		52.15	75.50
2600	Wall plates, stainless steel, 1 gang		80	.100		2.56	5.50		8.06	11.30
2800	2 gang		53	.151		4.33	8.30		12.63	17.60
3200	Lampholder, keyless		26	.308		16.85	16.95		33.80	44.50
3400	Pullchain with receptacle		22	.364	▼	20.50	20		40.50	53.50
9000	Minimum labor/equipment charge	▼	4	2	Job		110		110	170

26 27 73 – Door Chimes

26 27 73.10 Doorbell System

		Crew	Daily Output	Labor-Hours	Unit	Material	Labor	Equipment	Total	Total Incl O&P
0010	**DOORBELL SYSTEM**, incl. transformer, button & signal									
1000	Door chimes, 2 notes	1 Elec	16	.500	Ea.	23.50	27.50		51	68
1020	with ambient light		12	.667		113	36.50		149.50	181
1100	Tube type, 3 tube system		12	.667		193	36.50		229.50	269
1180	4 tube system		10	.800		425	44		469	540
1900	For transformer & button, add		5	1.600		11.45	88		99.45	149
3000	For push button only		24	.333	▼	2.26	18.35		20.61	31
9000	Minimum labor/equipment charge	▼	4	2	Job		110		110	170

26 28 Low-Voltage Circuit Protective Devices

26 28 16 – Enclosed Switches and Circuit Breakers

26 28 16.10 Circuit Breakers

		Crew	Daily Output	Labor-Hours	Unit	Material	2016 Bare Costs Labor	Equipment	Total	Total Incl O&P
0010	**CIRCUIT BREAKERS** (in enclosure)									
0100	Enclosed (NEMA 1), 600 volt, 3 pole, 30 amp	1 Elec	3.20	2.500	Ea.	520	138		658	785
0200	60 amp		2.80	2.857		635	157		792	945
0400	100 amp		2.30	3.478		730	192		922	1,100
0500	200 amp		1.50	5.333		1,525	294		1,819	2,125
0600	225 amp		1.50	5.333		1,675	294		1,969	2,300
0700	400 amp	2 Elec	1.60	10		2,875	550		3,425	4,025
9000	Minimum labor/equipment charge	1 Elec	4	2	Job		110		110	170

26 28 16.20 Safety Switches

		Crew	Daily Output	Labor-Hours	Unit	Material	2016 Bare Costs Labor	Equipment	Total	Total Incl O&P
0010	**SAFETY SWITCHES**									
0100	General duty 240 volt, 3 pole NEMA 1, fusible, 30 amp	1 Elec	3.20	2.500	Ea.	62.50	138		200.50	282
0200	60 amp		2.30	3.478		106	192		298	410
0300	100 amp		1.90	4.211		181	232		413	560
0400	200 amp		1.30	6.154		385	340		725	950
0500	400 amp	2 Elec	1.80	8.889		980	490		1,470	1,825
9990	Minimum labor/equipment charge	1 Elec	3	2.667	Job		147		147	227

26 28 16.40 Time Switches

		Crew	Daily Output	Labor-Hours	Unit	Material	2016 Bare Costs Labor	Equipment	Total	Total Incl O&P
0010	**TIME SWITCHES**									
0100	Single pole, single throw, 24 hour dial	1 Elec	4	2	Ea.	145	110		255	330
0200	24 hour dial with reserve power		3.60	2.222		615	122		737	865
0300	Astronomic dial		3.60	2.222		240	122		362	455
0400	Astronomic dial with reserve power		3.30	2.424		665	134		799	935
0500	7 day calendar dial		3.30	2.424		146	134		280	365
0600	7 day calendar dial with reserve power		3.20	2.500		285	138		423	530
0700	Photo cell 2000 watt		8	1		28.50	55		83.50	117
1080	Load management device, 4 loads		2	4		1,150	220		1,370	1,600
1100	8 loads		1	8		2,325	440		2,765	3,225
9000	Minimum labor/equipment charge		3.50	2.286	Job		126		126	195

26 29 Low-Voltage Controllers

26 29 13 – Enclosed Controllers

26 29 13.40 Relays

		Crew	Daily Output	Labor-Hours	Unit	Material	2016 Bare Costs Labor	Equipment	Total	Total Incl O&P
0010	**RELAYS** Enclosed (NEMA 1)									
0050	600 volt AC, 1 pole, 12 amp	1 Elec	5.30	1.509	Ea.	86	83		169	224
0100	2 pole, 12 amp		5	1.600		86	88		174	231
0200	4 pole, 10 amp		4.50	1.778		115	98		213	277

26 29 23 – Variable-Frequency Motor Controllers

26 29 23.10 Variable Frequency Drives/Adj. Frequency Drives

			Crew	Daily Output	Labor-Hours	Unit	Material	2016 Bare Costs Labor	Equipment	Total	Total Incl O&P
0010	**VARIABLE FREQUENCY DRIVES/ADJ. FREQUENCY DRIVES**										
0100	Enclosed (NEMA 1), 460 volt, for 3 HP motor size	G	1 Elec	.80	10	Ea.	1,900	550		2,450	2,950
0110	5 HP motor size	G		.80	10		2,175	550		2,725	3,250
0120	7.5 HP motor size	G		.67	11.940		2,275	660		2,935	3,550
0130	10 HP motor size	G		.67	11.940		2,500	660		3,160	3,775
0140	15 HP motor size	G	2 Elec	.89	17.978		3,225	990		4,215	5,075
0150	20 HP motor size	G		.89	17.978		4,375	990		5,365	6,350
0160	25 HP motor size	G		.67	23.881		5,400	1,325		6,725	7,950
0170	30 HP motor size	G		.67	23.881		5,850	1,325		7,175	8,475
0180	40 HP motor size	G		.67	23.881		6,775	1,325		8,100	9,475
0190	50 HP motor size	G		.53	30.189		9,975	1,675		11,650	13,600
0200	60 HP motor size	G	R-3	.56	35.714		12,100	1,950	251	14,301	16,600

26 29 Low-Voltage Controllers

26 29 23 – Variable-Frequency Motor Controllers

26 29 23.10 Variable Frequency Drives/Adj. Frequency Drives		Crew	Daily Output	Labor-Hours	Unit	Material	2016 Bare Costs Labor	Equipment	Total	Total Incl O&P
0210	75 HP motor size	G R-3	.56	35.714	Ea.	13,900	1,950	251	16,101	18,600
0220	100 HP motor size	G	.50	40		15,600	2,200	281	18,081	20,900
0230	125 HP motor size	G	.50	40		17,000	2,200	281	19,481	22,300
0240	150 HP motor size	G	.50	40		20,500	2,200	281	22,981	26,200
0250	200 HP motor size	G	.42	47.619		26,900	2,600	335	29,835	34,000
1100	Custom-engineered, 460 volt, for 3 HP motor size	G 1 Elec	.56	14.286		3,075	785		3,860	4,600
1110	5 HP motor size	G	.56	14.286		3,075	785		3,860	4,600
1120	7.5 HP motor size	G	.47	17.021		3,225	940		4,165	5,000
1130	10 HP motor size	G	.47	17.021		3,375	940		4,315	5,150
1140	15 HP motor size	G 2 Elec	.62	25.806		4,225	1,425		5,650	6,850
1150	20 HP motor size	G	.62	25.806		4,750	1,425		6,175	7,425
1160	25 HP motor size	G	.47	34.043		5,550	1,875		7,425	9,025
1170	30 HP motor size	G	.47	34.043		6,950	1,875		8,825	10,500
1180	40 HP motor size	G	.47	34.043		8,200	1,875		10,075	11,900
1190	50 HP motor size	G	.37	43.243		9,425	2,375		11,800	14,100
1200	60 HP motor size	G R-3	.39	51.282		14,200	2,800	360	17,360	20,300
1210	75 HP motor size	G	.39	51.282		15,100	2,800	360	18,260	21,300
1220	100 HP motor size	G	.35	57.143		16,500	3,125	400	20,025	23,500
1230	125 HP motor size	G	.35	57.143		17,700	3,125	400	21,225	24,800
1240	150 HP motor size	G	.35	57.143		20,300	3,125	400	23,825	27,600
1250	200 HP motor size	G	.29	68.966		26,500	3,775	485	30,760	35,500
2000	For complex & special design systems to meet specific									
2010	requirements, obtain quote from vendor.									

26 32 Packaged Generator Assemblies

26 32 13 – Engine Generators

26 32 13.16 Gas-Engine-Driven Generator Sets

		Crew	Daily Output	Labor-Hours	Unit	Material	Labor	Equipment	Total	Total Incl O&P
0010	**GAS-ENGINE-DRIVEN GENERATOR SETS**									
0020	Gas or gasoline operated, includes battery,									
0050	charger, & muffler									
0200	3 phase 4 wire, 277/480 volt, 7.5 kW	R-3	.83	24.096	Ea.	8,550	1,325	169	10,044	11,700
0300	11.5 kW		.71	28.169		12,100	1,550	198	13,848	15,900
0400	20 kW		.63	31.746		14,300	1,725	223	16,248	18,600
0500	35 kW		.55	36.364		17,000	2,000	255	19,255	22,100

26 51 Interior Lighting

26 51 13 – Interior Lighting Fixtures, Lamps, and Ballasts

26 51 13.50 Interior Lighting Fixtures

		Crew	Daily Output	Labor-Hours	Unit	Material	Labor	Equipment	Total	Total Incl O&P
0010	**INTERIOR LIGHTING FIXTURES** Including lamps, mounting									
0030	hardware and connections									
0100	Fluorescent, C.W. lamps, troffer, recess mounted in grid, RS									
0130	Grid ceiling mount									
0200	Acrylic lens, 1'W x 4'L, two 40 watt	1 Elec	5.70	1.404	Ea.	50	77.50		127.50	175
0300	2'W x 2'L, two U40 watt		5.70	1.404		54	77.50		131.50	179
0600	2'W x 4'L, four 40 watt		4.70	1.702		60.50	94		154.50	212
0910	Acrylic lens, 1'W x 4'L, two 32 watt T8	G	5.70	1.404		59.50	77.50		137	186
0930	2'W x 2'L, two U32 watt T8	G	5.70	1.404		93	77.50		170.50	222
0940	2'W x 4'L, two 32 watt T8	G	5.30	1.509		77	83		160	214
0950	2'W x 4'L, three 32 watt T8	G	5	1.600		68	88		156	211

For customer support on your Commercial Renovation Cost Data, call 877.742.4719.

493

26 51 13.50 Interior Lighting Fixtures		Crew	Daily Output	Labor-Hours	Unit	Material	2016 Bare Costs Labor	Equipment	Total	Total Incl O&P
0960	2'W x 4'L, four 32 watt T8	G	1 Elec	4.70	1.702	Ea.	72.50	94		166.50	225
1000	Surface mounted, RS										
1030	Acrylic lens with hinged & latched door frame										
1100	1'W x 4'L, two 40 watt		1 Elec	7	1.143	Ea.	64	63		127	168
1200	2'W x 2'L, two U40 watt			7	1.143		68.50	63		131.50	173
1500	2'W x 4'L, four 40 watt			5.30	1.509		81	83		164	218
2100	Strip fixture										
2130	Surface mounted										
2200	4' long, one 40 watt, RS		1 Elec	8.50	.941	Ea.	23	52		75	106
2300	4' long, two 40 watt, RS	"	8	1		41	55		96	131
2600	8' long, one 75 watt, SL		2 Elec	13.40	1.194		50.50	66		116.50	158
2700	8' long, two 75 watt, SL	"	12.40	1.290		61	71		132	177
3000	Strip, pendent mounted, industrial, white porcelain enamel										
3100	4' long, two 40 watt, RS		1 Elec	5.70	1.404	Ea.	52.50	77.50		130	178
3200	4' long, two 60 watt, HO	"	5	1.600		82.50	88		170.50	227
3300	8' long, two 75 watt, SL		2 Elec	8.80	1.818		98	100		198	263
4220	Metal halide, integral ballast, ceiling, recess mounted										
4230	prismatic glass lens, floating door										
4240	2'W x 2'L, 250 watt		1 Elec	3.20	2.500	Ea.	315	138		453	565
4250	2'W x 2'L, 400 watt		2 Elec	5.80	2.759		365	152		517	635
4260	Surface mounted, 2'W x 2'L, 250 watt		1 Elec	2.70	2.963		345	163		508	630
4270	400 watt		2 Elec	4.80	3.333		405	184		589	730
4280	High bay, aluminum reflector,										
4290	Single unit, 400 watt		2 Elec	4.60	3.478	Ea.	410	192		602	745
4300	Single unit, 1000 watt			4	4		590	220		810	990
4310	Twin unit, 400 watt			3.20	5		820	276		1,096	1,325
4320	Low bay, aluminum reflector, 250W DX lamp		1 Elec	3.20	2.500		360	138		498	615
4340	High pressure sodium integral ballast ceiling, recess mounted										
4350	prismatic glass lens, floating door										
4360	2'W x 2'L, 150 watt lamp		1 Elec	3.20	2.500	Ea.	385	138		523	640
4370	2'W x 2'L, 400 watt lamp		2 Elec	5.80	2.759		460	152		612	740
4380	Surface mounted, 2'W x 2'L, 150 watt lamp		1 Elec	2.70	2.963		470	163		633	770
4390	400 watt lamp		2 Elec	4.80	3.333		525	184		709	865
4400	High bay, aluminum reflector,										
4410	Single unit, 400 watt lamp		2 Elec	4.60	3.478	Ea.	380	192		572	710
4430	Single unit, 1000 watt lamp	"	4	4		545	220		765	940
4440	Low bay, aluminum reflector, 150 watt lamp		1 Elec	3.20	2.500		325	138		463	575
4450	Incandescent, high hat can, round alzak reflector, prewired										
4470	100 watt		1 Elec	8	1	Ea.	71.50	55		126.50	164
4480	150 watt			8	1		104	55		159	199
4500	300 watt			6.70	1.194		241	66		307	365
4600	Square glass lens with metal trim, prewired										
4630	100 watt		1 Elec	6.70	1.194	Ea.	55	66		121	163
4700	200 watt			6.70	1.194		97	66		163	209
6010	Vapor tight, incandescent, ceiling mounted, 200 watt			6.20	1.290		82.50	71		153.50	201
6100	Fluorescent, surface mounted, 2 lamps, 4'L, RS, 40 watt			3.20	2.500		115	138		253	340
6850	Vandalproof, surface mounted, fluorescent, two 32 watt T8	G		3.20	2.500		252	138		390	490
6860	Incandescent, one 150 watt			8	1		90	55		145	184
6900	Mirror light, fluorescent, RS, acrylic enclosure, two 40 watt			8	1		128	55		183	225
6910	One 40 watt			8	1		106	55		161	201
6920	One 20 watt			12	.667		81.50	36.50		118	147
7500	Ballast replacement, by weight of ballast, to 15' high										
7520	Indoor fluorescent, less than 2 lb.		1 Elec	10	.800	Ea.	26	44		70	96.50

For customer support on your Commercial Renovation Cost Data, call 877.742.4719.

26 51 Interior Lighting

26 51 13 – Interior Lighting Fixtures, Lamps, and Ballasts

26 51 13.50 Interior Lighting Fixtures		Crew	Daily Output	Labor-Hours	Unit	Material	2016 Bare Costs Labor	Equipment	Total	Total Incl O&P
7540	Two 40W, watt reducer, 2 to 5 lb.	1 Elec	9.40	.851	Ea.	41	47		88	118
7560	Two F96 slimline, over 5 lb.		8	1		77	55		132	170
7580	Vaportite ballast, less than 2 lb.		9.40	.851		26	47		73	101
7600	2 lb. to 5 lb.		8.90	.899		41	49.50		90.50	122
7620	Over 5 lb.		7.60	1.053		77	58		135	174
7630	Electronic ballast for two tubes		8	1		37	55		92	126
7640	Dimmable ballast one lamp ⒢		8	1		106	55		161	202
7650	Dimmable ballast two-lamp ⒢		7.60	1.053		104	58		162	204
9000	Minimum labor/equipment charge		3	2.667	Job		147		147	227

26 51 13.70 Residential Fixtures

		Crew	Daily Output	Labor-Hours	Unit	Material	2016 Bare Costs Labor	Equipment	Total	Total Incl O&P
0010	**RESIDENTIAL FIXTURES**									
0400	Fluorescent, interior, surface, circline, 32 watt & 40 watt	1 Elec	20	.400	Ea.	141	22		163	189
0500	2' x 2', two U-tube 32 watt T8		8	1		165	55		220	266
0700	Shallow under cabinet, two 20 watt		16	.500		71.50	27.50		99	121
0900	Wall mounted, 4'L, two 32 watt T8, with baffle		10	.800		133	44		177	214
2000	Incandescent, exterior lantern, wall mounted, 60 watt		16	.500		60	27.50		87.50	109
2100	Post light, 150W, with 7' post		4	2		270	110		380	465
2500	Lamp holder, weatherproof with 150W PAR		16	.500		32	27.50		59.50	77.50
2550	With reflector and guard		12	.667		56	36.50		92.50	119
2600	Interior pendent, globe with shade, 150 watt		20	.400		201	22		223	255
9000	Minimum labor/equipment charge		4	2	Job		110		110	170

26 52 Emergency Lighting

26 52 13 – Emergency Lighting Equipments

26 52 13.10 Emergency Lighting and Battery Units

		Crew	Daily Output	Labor-Hours	Unit	Material	2016 Bare Costs Labor	Equipment	Total	Total Incl O&P
0010	**EMERGENCY LIGHTING AND BATTERY UNITS**									
0300	Emergency light units, battery operated									
0350	Twin sealed beam light, 25 watt, 6 volt each									
0500	Lead battery operated	1 Elec	4	2	Ea.	162	110		272	350
0700	Nickel cadmium battery operated		4	2	"	475	110		585	690
9000	Minimum labor/equipment charge		4	2	Job		110		110	170

26 53 Exit Signs

26 53 13 – Exit Lighting

26 53 13.10 Exit Lighting Fixtures

		Crew	Daily Output	Labor-Hours	Unit	Material	2016 Bare Costs Labor	Equipment	Total	Total Incl O&P
0010	**EXIT LIGHTING FIXTURES**									
0080	Exit light ceiling or wall mount, incandescent, single face	1 Elec	8	1	Ea.	49	55		104	139
0100	Double face		6.70	1.194	"	43	66		109	150
9000	Minimum labor/equipment charge		4	2	Job		110		110	170

26 55 Special Purpose Lighting

26 55 33 – Hazard Warning Lighting

26 55 33.10 Warning Beacons	Crew	Daily Output	Labor-Hours	Unit	Material	2016 Bare Costs Labor	Equipment	Total	Total Incl O&P
0010 **WARNING BEACONS**									
0015 Surface mount with colored or clear lens									
0100 Rotating beacon									
0110 120V, 40 watt halogen	1 Elec	3.50	2.286	Ea.	101	126		227	305
0120 24V, 20 watt halogen	"	3.50	2.286	"	211	126		337	430
0200 Steady beacon									
0210 120V, 40 watt halogen	1 Elec	3.50	2.286	Ea.	101	126		227	305
0220 24V, 20W		3.50	2.286		101	126		227	305
0230 12V DC, incandescent		3.50	2.286		101	126		227	305
0300 Flashing beacon									
0310 120V, 40 watt halogen	1 Elec	3.50	2.286	Ea.	101	126		227	305
0320 24V, 20 watt halogen		3.50	2.286		101	126		227	305
0410 12V DC with two 6V lantern batteries		7	1.143		101	63		164	209

26 55 59 – Display Lighting

26 55 59.10 Track Lighting

	Crew	Daily Output	Labor-Hours	Unit	Material	Labor	Equipment	Total	Total Incl O&P
0010 **TRACK LIGHTING**									
0080 Track, 1 circuit, 4' section	1 Elec	6.70	1.194	Ea.	44.50	66		110.50	151
0100 8' section	2 Elec	10.60	1.509		69	83		152	205
0300 3 circuits, 4' section	1 Elec	6.70	1.194		92.50	66		158.50	204
0400 8' section	2 Elec	10.60	1.509		109	83		192	249
9000 Minimum labor/equipment charge	1 Elec	3	2.667	Job		147		147	227

26 56 Exterior Lighting

26 56 23 – Area Lighting

26 56 23.10 Exterior Fixtures

	Crew	Daily Output	Labor-Hours	Unit	Material	Labor	Equipment	Total	Total Incl O&P
0010 **EXTERIOR FIXTURES** With lamps									
0200 Wall mounted, incandescent, 100 watt	1 Elec	8	1	Ea.	36	55		91	125
0400 Quartz, 500 watt		5.30	1.509		52	83		135	187
1100 Wall pack, low pressure sodium, 35 watt		4	2		234	110		344	430
1150 55 watt		4	2		278	110		388	475
1160 High pressure sodium, 70 watt		4	2		230	110		340	425
1170 150 watt		4	2		251	110		361	445
1180 Metal Halide, 175 watt		4	2		255	110		365	450
1190 250 watt		4	2		290	110		400	490
1195 400 watt		4	2		345	110		455	550
1250 Induction lamp, 40 watt		4	2		355	110		465	560
1260 80 watt		4	2		600	110		710	830
1278 LED, poly lens, 26 watt		4	2		214	110		324	405
1280 110 watt		4	2		680	110		790	920
1500 LED, glass lens, 13 watt		4	2		214	110		324	405

26 56 33 – Walkway Lighting

26 56 33.10 Walkway Luminaire

	Crew	Daily Output	Labor-Hours	Unit	Material	Labor	Equipment	Total	Total Incl O&P
0010 **WALKWAY LUMINAIRE**									
9000 Minimum labor/equipment charge	1 Elec	3.75	2.133	Job		118		118	182

26 56 Exterior Lighting

26 56 36 – Flood Lighting

26 56 36.20 Floodlights		Crew	Daily Output	Labor-Hours	Unit	Material	2016 Bare Costs Labor	Equipment	Total	Total Incl O&P
0010	**FLOODLIGHTS** with ballast and lamp,									
1290	floor mtd, mount with swivel bracket									
1400	Pole mounted, pole not included									
2250	Low pressure sodium, 55 watt	1 Elec	2.70	2.963	Ea.	440	163		603	735
2270	90 watt	"	2	4	"	600	220		820	1,000

26 61 Lighting Systems and Accessories

26 61 23 – Lamps Applications

26 61 23.10 Lamps			Crew	Daily Output	Labor-Hours	Unit	Material	2016 Bare Costs Labor	Equipment	Total	Total Incl O&P
0010	**LAMPS**										
0080	Fluorescent, rapid start, cool white, 2' long, 20 watt		1 Elec	1	8	C	360	440		800	1,075
0100	4' long, 40 watt			.90	8.889		275	490		765	1,050
0170	4' long, 34 watt energy saver	G		.90	8.889		430	490		920	1,225
0176	2' long, T8, 17 W energy saver	G		1	8		410	440		850	1,125
0178	3' long, T8, 25 W energy saver	G		.90	8.889		440	490		930	1,225
0180	4' long, T8, 32 watt energy saver	G		.90	8.889		305	490		795	1,100
0560	Twin tube compact lamp	G		.90	8.889		355	490		845	1,150
0570	Double twin tube compact lamp	G		.80	10		725	550		1,275	1,650
0600	Mercury vapor, mogul base, deluxe white, 100 watt			.30	26.667		11,400	1,475		12,875	14,900
0800	400 watt			.30	26.667		5,000	1,475		6,475	7,775
1000	Metal halide, mogul base, 175 watt			.30	26.667		2,175	1,475		3,650	4,675
1100	250 watt			.30	26.667		2,450	1,475		3,925	4,975
1350	High pressure sodium, 70 watt			.30	26.667		2,650	1,475		4,125	5,200
1370	150 watt			.30	26.667		2,825	1,475		4,300	5,400
1500	Low pressure sodium, 35 watt			.30	26.667		37,600	1,475		39,075	43,600
1600	90 watt			.30	26.667		18,300	1,475		19,775	22,400
1800	Incandescent, interior, A21, 100 watt			1.60	5		233	276		509	680
1900	A21, 150 watt			1.60	5		166	276		442	610
2300	R30, 75 watt			1.30	6.154		620	340		960	1,200
2500	Exterior, PAR 38, 75 watt			1.30	6.154		2,100	340		2,440	2,825
2600	PAR 38, 150 watt			1.30	6.154		2,200	340		2,540	2,925
9000	Minimum labor/equipment charge			4	2	Job		110		110	170

Division Notes

		CREW	DAILY OUTPUT	LABOR-HOURS	UNIT	BARE COSTS				TOTAL INCL O&P
						MAT.	LABOR	EQUIP.	TOTAL	

Estimating Tips

27 20 00 Data Communications
27 30 00 Voice Communications
27 40 00 Audio-Video Communications

When estimating material costs for special systems, it is always prudent to obtain manufacturers' quotations for equipment prices and special installation requirements which will affect the total costs.

Reference Numbers

Reference numbers are shown at the beginning of some major classifications. These numbers refer to related items in the Reference Section. The reference information may be an estimating procedure, an alternate pricing method, or technical information.

Note: Not all subdivisions listed here necessarily appear. ∎

27 15 Communications Horizontal Cabling

27 15 10 – Special Communications Cabling

27 15 10.23 Sound and Video Cables and Fittings	Crew	Daily Output	Labor-Hours	Unit	Material	2016 Bare Costs Labor	Equipment	Total	Total Incl O&P
0010 **SOUND AND VIDEO CABLES & FITTINGS**									
1250 Nonshielded, #22-2 conductor	1 Elec	10	.800	C.L.F.	14.25	44		58.25	83.50

27 15 13 – Communications Copper Horizontal Cabling

27 15 13.13 Communication Cables and Fittings

	Crew	Daily Output	Labor-Hours	Unit	Material	2016 Bare Costs Labor	Equipment	Total	Total Incl O&P
0010 **COMMUNICATION CABLES AND FITTINGS**									
5000 High performance unshielded twisted pair (UTP)									
7000 Category 5, #24, 4 pair solid, PVC jacket	1 Elec	7	1.143	C.L.F.	11.20	63		74.20	110
7100 4 pair solid, plenum		7	1.143		38.50	63		101.50	140
7200 4 pair stranded, PVC jacket		7	1.143		22.50	63		85.50	122
7210 Category 5e, #24, 4 pair solid, PVC jacket		7	1.143		16.90	63		79.90	116
7212 4 pair solid, plenum		7	1.143		23.50	63		86.50	124
7214 4 pair stranded, PVC jacket		7	1.143		25.50	63		88.50	126
7240 Category 6, #24, 4 pair solid, PVC jacket		7	1.143		15.60	63		78.60	115
7242 4 pair solid, plenum		7	1.143		15	63		78	114
7244 4 pair stranded, PVC jacket		7	1.143		14.60	63		77.60	114
7300 Connector, RJ-45, category 5		80	.100	Ea.	.74	5.50		6.24	9.30
7302 Shielded RJ-45, category 5		72	.111		1.73	6.10		7.83	11.35
7310 Jack, UTP RJ-45, category 3		72	.111		.40	6.10		6.50	9.90
7312 Category 5		65	.123		5.50	6.80		12.30	16.55
7314 Category 5e		65	.123		4.99	6.80		11.79	16
7316 Category 6		65	.123		3.20	6.80		10	14

27 41 Audio-Video Systems

27 41 33 – Master Antenna Television Systems

27 41 33.10 T.V. Systems

	Crew	Daily Output	Labor-Hours	Unit	Material	2016 Bare Costs Labor	Equipment	Total	Total Incl O&P
0010 **T.V. SYSTEMS**, not including rough-in wires, cables & conduits									
5000 Antenna, small	1 Elec	6	1.333	Ea.	48	73.50		121.50	167
5100 Large	"	4	2	"	202	110		312	390

27 51 Distributed Audio-Video Communications Systems

27 51 16 – Public Address Systems

27 51 16.10 Public Address System

	Crew	Daily Output	Labor-Hours	Unit	Material	2016 Bare Costs Labor	Equipment	Total	Total Incl O&P
0010 **PUBLIC ADDRESS SYSTEM**									
0100 Conventional, office	1 Elec	5.33	1.501	Speaker	139	82.50		221.50	281
0200 Industrial		2.70	2.963	"	268	163		431	545
9000 Minimum labor/equipment charge		3.50	2.286	Job		126		126	195

27 51 19 – Sound Masking Systems

27 51 19.10 Sound System

	Crew	Daily Output	Labor-Hours	Unit	Material	2016 Bare Costs Labor	Equipment	Total	Total Incl O&P
0010 **SOUND SYSTEM**, not including rough-in wires, cables & conduits									
2000 Intercom, 30 station capacity, master station	2 Elec	2	8	Ea.	2,300	440		2,740	3,200
2020 10 station capacity	"	4	4		1,250	220		1,470	1,725
3600 House telephone, talking station	1 Elec	1.60	5		485	276		761	955
3800 Press to talk, release to listen	"	5.30	1.509		113	83		196	253
4000 System-on button					67.50			67.50	74
4200 Door release	1 Elec	4	2		121	110		231	305
4400 Combination speaker and microphone		8	1		205	55		260	310
4600 Termination box		3.20	2.500		64.50	138		202.50	284
4800 Amplifier or power supply		5.30	1.509		740	83		823	945

27 51 19 – Sound Masking Systems

27 51 19.10 Sound System		Crew	Daily Output	Labor-Hours	Unit	Material	2016 Bare Costs Labor	Equipment	Total	Total Incl O&P
5000	Vestibule door unit	1 Elec	16	.500	Name	136	27.50		163.50	193
5200	Strip cabinet		27	.296	Ea.	258	16.35		274.35	310
5400	Directory		16	.500	"	121	27.50		148.50	176
9000	Minimum labor/equipment charge		3.50	2.286	Job		126		126	195

For customer support on your Commercial Renovation Cost Data, call 877.742.4719.

501

Division Notes

	CREW	DAILY OUTPUT	LABOR-HOURS	UNIT	BARE COSTS				TOTAL INCL O&P
					MAT.	LABOR	EQUIP.	TOTAL	

Estimating Tips

- When estimating material costs for electronic safety and security systems, it is always prudent to obtain manufacturers' quotations for equipment prices and special installation requirements that affect the total cost.

- Fire alarm systems consist of control panels, annunciator panels, battery with rack, charger, and fire alarm actuating and indicating devices. Some fire alarm systems include speakers, telephone lines, door closer controls, and other components. Be careful not to overlook the costs related to installation for these items. Also be aware of costs for integrated automation instrumentation and terminal devices, control equipment, control wiring, and programming.

- Security equipment includes items such as CCTV, access control, and other detection and identification systems to perform alert and alarm functions. Be sure to consider the costs related to installation for this security equipment, such as for integrated automation instrumentation and terminal devices, control equipment, control wiring, and programming.

Reference Numbers

Reference numbers are shown at the beginning of some major classifications. These numbers refer to related items in the Reference Section. The reference information may be an estimating procedure, an alternate pricing method, or technical information.

Note: Not all subdivisions listed here necessarily appear. ■

Did you know?

RSMeans Online gives you the same access to RSMeans' data with 24/7 access:

- Quickly locate costs in the searchable database.
- Build cost lists, estimates, and reports in minutes.
- Adjust costs to any location in the U.S. and Canada with the click of a button.

Start your free trial today at **www.RSMeansOnline.com**

RSMeans Online
FROM THE GORDIAN GROUP®

28 13 Access Control

28 13 53 – Security Access Detection

28 13 53.13 Security Access Metal Detectors	Crew	Daily Output	Labor-Hours	Unit	Material	2016 Bare Costs Labor	Equipment	Total	Total Incl O&P
0010 SECURITY ACCESS METAL DETECTORS									
0240 Metal detector, hand-held, wand type, unit only				Ea.	129			129	138
0250 Metal detector, walk through portal type, single zone	1 Elec	2	4		4,400	220		4,620	5,200
0260 Multi-zone	"	2	4		5,450	220		5,670	6,350

28 13 53.16 Security Access X-Ray Equipment	Crew	Daily Output	Labor-Hours	Unit	Material	2016 Bare Costs Labor	Equipment	Total	Total Incl O&P
0010 SECURITY ACCESS X-RAY EQUIPMENT									
0290 X-ray machine, desk top, for mail/small packages/letters	1 Elec	4	2	Ea.	3,475	110		3,585	4,000
0300 Conveyor type, incl monitor		2	4		16,000	220		16,220	17,900
0310 Includes additional features		2	4		28,000	220		28,220	31,100
0320 X-ray machine, large unit, for airports, incl monitor	2 Elec	1	16		40,000	880		40,880	45,400
0330 Full console	"	.50	32		68,500	1,775		70,275	78,000

28 13 53.23 Security Access Explosive Detection Equipment	Crew	Daily Output	Labor-Hours	Unit	Material	2016 Bare Costs Labor	Equipment	Total	Total Incl O&P
0010 SECURITY ACCESS EXPLOSIVE DETECTION EQUIPMENT									
0270 Explosives detector, walk through portal type	1 Elec	2	4	Ea.	44,200	220		44,420	48,900
0280 Hand-held, battery operated				"	25,500				28,100

28 16 Intrusion Detection

28 16 16 – Intrusion Detection Systems Infrastructure

28 16 16.50 Intrusion Detection	Crew	Daily Output	Labor-Hours	Unit	Material	2016 Bare Costs Labor	Equipment	Total	Total Incl O&P
0010 INTRUSION DETECTION, not including wires & conduits									
0100 Burglar alarm, battery operated, mechanical trigger	1 Elec	4	2	Ea.	278	110		388	475
0200 Electrical trigger		4	2		330	110		440	535
0400 For outside key control, add		8	1		86	55		141	180
0600 For remote signaling circuitry, add		8	1		137	55		192	235
0800 Card reader, flush type, standard		2.70	2.963		790	163		953	1,125
1000 Multi-code		2.70	2.963		1,200	163		1,363	1,575
1200 Door switches, hinge switch		5.30	1.509		62.50	83		145.50	198
1400 Magnetic switch		5.30	1.509		77	83		160	214
1600 Exit control locks, horn alarm		4	2		232	110		342	425
1800 Flashing light alarm		4	2		263	110		373	460
2000 Indicating panels, 1 channel		2.70	2.963		268	163		431	545
2200 10 channel	2 Elec	3.20	5		895	276		1,171	1,400
2400 20 channel		2	8		2,500	440		2,940	3,400
2600 40 channel		1.14	14.035		4,450	775		5,225	6,100
2800 Ultrasonic motion detector, 12 volt	1 Elec	2.30	3.478		210	192		402	525
3000 Infrared photoelectric detector		4	2		144	110		254	330
3200 Passive infrared detector		4	2		234	110		344	430
3400 Glass break alarm switch		8	1		117	55		172	214
3420 Switchmats, 30" x 5'		5.30	1.509		104	83		187	243
3440 30" x 25'		4	2		203	110		313	395
3460 Police connect panel		4	2		269	110		379	465
3480 Telephone dialer		5.30	1.509		385	83		468	555
3500 Alarm bell		4	2		104	110		214	285
3520 Siren		4	2		146	110		256	330
3540 Microwave detector, 10' to 200'		2	4		580	220		800	975
3560 10' to 350'		2	4		1,750	220		1,970	2,275

28 23 Video Surveillance

28 23 23 – Video Surveillance Systems Infrastructure

28 23 23.50 Video Surveillance Equipments	Crew	Daily Output	Labor-Hours	Unit	Material	2016 Bare Costs Labor	Equipment	Total	Total Incl O&P
0010 **VIDEO SURVEILLANCE EQUIPMENTS**									
0200 Video cameras, wireless, hidden in exit signs, clocks, etc., incl. receiver	1 Elec	3	2.667	Ea.	152	147		299	395
0210 Accessories for video recorder, single camera		3	2.667		183	147		330	430
0220 For multiple cameras		3	2.667		1,750	147		1,897	2,150
0230 Video cameras, wireless, for under vehicle searching, complete	↓	2	4	↓	9,725	220		9,945	11,000

28 31 Fire Detection and Alarm

28 31 23 – Fire Detection and Alarm Annunciation Panels and Fire Stations

28 31 23.50 Alarm Panels and Devices

	Crew	Daily Output	Labor-Hours	Unit	Material	2016 Bare Costs Labor	Equipment	Total	Total Incl O&P
0010 **ALARM PANELS AND DEVICES**, not including wires & conduits									
3594 Fire, alarm control panel									
3600 4 zone	2 Elec	2	8	Ea.	405	440		845	1,125
3800 8 zone		1	16		690	880		1,570	2,125
4000 12 zone	↓	.67	23.988		2,400	1,325		3,725	4,675
4200 Battery and rack	1 Elec	4	2		410	110		520	625
4400 Automatic charger		8	1		585	55		640	725
4600 Signal bell		8	1		96.50	55		151.50	191
4800 Trouble buzzer or manual station		8	1		83.50	55		138.50	177
5600 Strobe and horn		5.30	1.509		152	83		235	296
5800 Fire alarm horn		6.70	1.194		61	66		127	169
6000 Door holder, electro-magnetic		4	2		103	110		213	283
6200 Combination holder and closer		3.20	2.500		123	138		261	350
6600 Drill switch		8	1		370	55		425	495
6800 Master box		2.70	2.963		6,400	163		6,563	7,300
7000 Break glass station		8	1		55.50	55		110.50	146
7800 Remote annunciator, 8 zone lamp	↓	1.80	4.444		209	245		454	610
8000 12 zone lamp	2 Elec	2.60	6.154		335	340		675	895
8200 16 zone lamp	"	2.20	7.273	↓	420	400		820	1,075

28 31 43 – Fire Detection Sensors

28 31 43.50 Fire and Heat Detectors

	Crew	Daily Output	Labor-Hours	Unit	Material	2016 Bare Costs Labor	Equipment	Total	Total Incl O&P
0010 **FIRE & HEAT DETECTORS**									
5000 Detector, rate of rise	1 Elec	8	1	Ea.	51	55		106	141
5100 Fixed temperature	"	8	1	"	51	55		106	141

28 31 46 – Smoke Detection Sensors

28 31 46.50 Smoke Detectors

	Crew	Daily Output	Labor-Hours	Unit	Material	2016 Bare Costs Labor	Equipment	Total	Total Incl O&P
0010 **SMOKE DETECTORS**									
5200 Smoke detector, ceiling type	1 Elec	6.20	1.290	Ea.	110	71		181	231
5400 Duct type	"	3.20	2.500	"	325	138		463	575

For customer support on your Commercial Renovation Cost Data, call 877.742.4719.

505

28 39 Mass Notification Systems

28 39 10 – Notification Systems

28 39 10.10 Mass Notification System	Crew	Daily Output	Labor-Hours	Unit	Material	2016 Bare Costs Labor	Equipment	Total	Total Incl O&P
0010 **MASS NOTIFICATION SYSTEM**									
0100 Wireless command center, 10,000 devices	2 Elec	1.33	12.030	Ea.	3,500	665		4,165	4,875
0200 Option, email notification					1,700			1,700	1,850
0210 Remote device supervision & monitor					2,425			2,425	2,650
0300 Antenna VHF or UHF, for medium range	1 Elec	4	2		137	110		247	320
0310 For high-power transmitter		2	4		690	220		910	1,100
0400 Transmitter, 25 watt		4	2		2,075	110		2,185	2,450
0410 40 watt		2.66	3.008		1,900	166		2,066	2,350
0420 100 watt		1.33	6.015		6,175	330		6,505	7,300
0500 Wireless receiver/control module for speaker		8	1		250	55		305	360
0600 Desktop paging controller, stand alone		4	2		605	110		715	840

Estimating Tips

31 05 00 Common Work Results for Earthwork

- Estimating the actual cost of performing earthwork requires careful consideration of the variables involved. This includes items such as type of soil, whether water will be encountered, dewatering, whether banks need bracing, disposal of excavated earth, and length of haul to fill or spoil sites, etc. If the project has large quantities of cut or fill, consider raising or lowering the site to reduce costs, while paying close attention to the effect on site drainage and utilities.

- If the project has large quantities of fill, creating a borrow pit on the site can significantly lower the costs.

- It is very important to consider what time of year the project is scheduled for completion. Bad weather can create large cost overruns from dewatering, site repair, and lost productivity from cold weather.

Reference Numbers

Reference numbers are shown at the beginning of some major classifications. These numbers refer to related items in the Reference Section. The reference information may be an estimating procedure, an alternate pricing method, or technical information.

Note: Not all subdivisions listed here necessarily appear. ■

31 05 Common Work Results for Earthwork

31 05 13 – Soils for Earthwork

31 05 13.10 Borrow		Crew	Daily Output	Labor-Hours	Unit	Material	2016 Bare Costs Labor	2016 Bare Costs Equipment	Total	Total Incl O&P
0010	**BORROW**									
0020	Spread, 200 H.P. dozer, no compaction, 2 mi. RT haul									
0200	Common borrow	B-15	600	.047	C.Y.	12.40	2.09	4.63	19.12	22
0700	Screened loam		600	.047		27	2.09	4.63	33.72	38
0800	Topsoil, weed free	↓	600	.047		24.50	2.09	4.63	31.22	35.50
0900	For 5 mile haul, add	B-34B	200	.040	↓		1.73	3.45	5.18	6.60

31 05 16 – Aggregates for Earthwork

31 05 16.10 Borrow		Crew	Daily Output	Labor-Hours	Unit	Material	2016 Bare Costs Labor	2016 Bare Costs Equipment	Total	Total Incl O&P
0010	**BORROW**									
0020	Spread, with 200 H.P. dozer, no compaction, 2 mi. RT haul									
0100	Bank run gravel	B-15	600	.047	L.C.Y.	20.50	2.09	4.63	27.22	31
0300	Crushed stone (1.40 tons per CY), 1-1/2"		600	.047		24	2.09	4.63	30.72	35
0320	3/4"		600	.047		24	2.09	4.63	30.72	35
0340	1/2"		600	.047		28	2.09	4.63	34.72	39.50
0360	3/8"		600	.047		29.50	2.09	4.63	36.22	41
0400	Sand, washed, concrete		600	.047		37.50	2.09	4.63	44.22	50
0500	Dead or bank sand		600	.047		20	2.09	4.63	26.72	30.50
0600	Select structural fill		600	.047		22	2.09	4.63	28.72	32.50
0900	For 5 mile haul, add	B-34B	200	.040	↓		1.73	3.45	5.18	6.60

31 06 Schedules for Earthwork

31 06 60 – Schedules for Special Foundations and Load Bearing Elements

31 06 60.14 Piling Special Costs		Crew	Daily Output	Labor-Hours	Unit	Material	2016 Bare Costs Labor	2016 Bare Costs Equipment	Total	Total Incl O&P
0010	**PILING SPECIAL COSTS**									
0011	Piling special costs, pile caps, see Section 03 30 53.40									
0500	Cutoffs, concrete piles, plain	1 Pile	5.50	1.455	Ea.		70.50		70.50	115
0600	With steel thin shell, add		38	.211			10.20		10.20	16.65
0700	Steel pile or "H" piles		19	.421			20.50		20.50	33.50
0800	Wood piles	↓	38	.211	↓		10.20		10.20	16.65
1000	Testing, any type piles, test load is twice the design load									
1050	50 ton design load, 100 ton test				Ea.				14,000	15,500
1100	100 ton design load, 200 ton test								20,000	22,000
1200	200 ton design load, 400 ton test				↓				28,000	31,000

31 06 60.15 Mobilization		Crew	Daily Output	Labor-Hours	Unit	Material	2016 Bare Costs Labor	2016 Bare Costs Equipment	Total	Total Incl O&P
0010	**MOBILIZATION**									
0020	Set up & remove, air compressor, 600 CFM	A-5	3.30	5.455	Ea.		209	18.80	227.80	360
0100	1200 CFM	"	2.20	8.182			315	28	343	540
0200	Crane, with pile leads and pile hammer, 75 ton	B-19	.60	106			5,250	2,850	8,100	11,600
0300	150 ton	"	.36	177	↓		8,775	4,750	13,525	19,400

31 14 Earth Stripping and Stockpiling

31 14 13 – Soil Stripping and Stockpiling

31 14 13.23 Topsoil Stripping and Stockpiling	Crew	Daily Output	Labor-Hours	Unit	Material	2016 Bare Costs Labor	Equipment	Total	Total Incl O&P
0010 **TOPSOIL STRIPPING AND STOCKPILING**									
1500	Loam or topsoil, remove/stockpile on site								
1510	By hand, 6" deep, 50' haul, less than 100 S.Y.	B-1	100	.240	S.Y.		9.25		9.25	15.15
1520	By skid steer, 6" deep, 100' haul, 101-500 S.Y.	B-62	500	.048			2	.35	2.35	3.61
1530	100' haul, 501-900 S.Y.	"	900	.027			1.11	.19	1.30	2
1540	200' haul, 901-1100 S.Y.	B-63	1000	.040			1.61	.17	1.78	2.79
1550	By dozer, 200' haul, 1101-4000 S.Y.	B-10B	4000	.003			.14	.35	.49	.60

31 23 Excavation and Fill

31 23 16 – Excavation

31 23 16.13 Excavating, Trench

		Crew	Daily Output	Labor-Hours	Unit	Material	Labor	Equipment	Total	Total Incl O&P
0010	**EXCAVATING, TRENCH**									
0011	Or continuous footing									
0020	Common earth with no sheeting or dewatering included									
1400	By hand with pick and shovel 2' to 6' deep, light soil	1 Clab	8	1	B.C.Y.		38		38	62
1500	Heavy soil	"	4	2	"		76		76	124
1700	For tamping backfilled trenches, air tamp, add	A-1G	100	.080	E.C.Y.		3.03	.57	3.60	5.60
1900	Vibrating plate, add	B-18	180	.133	"		5.15	.26	5.41	8.70
2100	Trim sides and bottom for concrete pours, common earth		1500	.016	S.F.		.62	.03	.65	1.04
2300	Hardpan		600	.040	"		1.54	.08	1.62	2.61
9000	Minimum labor/equipment charge	1 Clab	4	2	Job		76		76	124

31 23 16.14 Excavating, Utility Trench

		Crew	Daily Output	Labor-Hours	Unit	Material	Labor	Equipment	Total	Total Incl O&P
0010	**EXCAVATING, UTILITY TRENCH**									
0011	Common earth									
0050	Trenching with chain trencher, 12 H.P., operator walking									
0100	4" wide trench, 12" deep	B-53	800	.010	L.F.		.49	.09	.58	.87
0150	18" deep		750	.011			.52	.09	.61	.93
0200	24" deep		700	.011			.56	.10	.66	1
0300	6" wide trench, 12" deep		650	.012			.61	.10	.71	1.08
0350	18" deep		600	.013			.66	.11	.77	1.16
0400	24" deep		550	.015			.72	.12	.84	1.27
0450	36" deep		450	.018			.87	.15	1.02	1.55
0600	8" wide trench, 12" deep		475	.017			.83	.14	.97	1.47
0650	18" deep		400	.020			.98	.17	1.15	1.75
0700	24" deep		350	.023			1.12	.19	1.31	1.99
0750	36" deep		300	.027			1.31	.23	1.54	2.32
0900	Minimum labor/equipment charge		2	4	Job		197	34	231	350
1000	Backfill by hand including compaction, add									
1050	4" wide trench, 12" deep	A-1G	800	.010	L.F.		.38	.07	.45	.70
1100	18" deep		530	.015			.57	.11	.68	1.05
1150	24" deep		400	.020			.76	.14	.90	1.40
1300	6" wide trench, 12" deep		540	.015			.56	.11	.67	1.04
1350	18" deep		405	.020			.75	.14	.89	1.37
1400	24" deep		270	.030			1.12	.21	1.33	2.07
1450	36" deep		180	.044			1.68	.32	2	3.10
1600	8" wide trench, 12" deep		400	.020			.76	.14	.90	1.40
1650	18" deep		265	.030			1.14	.21	1.35	2.11
1700	24" deep		200	.040			1.52	.28	1.80	2.79
1750	36" deep		135	.059			2.25	.42	2.67	4.13
2000	Chain trencher, 40 H.P. operator riding									
2050	6" wide trench and backfill, 12" deep	B-54	1200	.007	L.F.		.33	.28	.61	.83

For customer support on your Commercial Renovation Cost Data, call 877.742.4719.

509

31 23 Excavation and Fill

31 23 16 – Excavation

31 23 16.14 Excavating, Utility Trench

		Crew	Daily Output	Labor-Hours	Unit	Material	2016 Bare Costs Labor	Equipment	Total	Total Incl O&P
2100	18" deep	B-54	1000	.008	L.F.		.39	.33	.72	.99
2150	24" deep		975	.008			.40	.34	.74	1.02
2200	36" deep		900	.009			.44	.37	.81	1.10
2250	48" deep		750	.011			.52	.45	.97	1.32
2300	60" deep		650	.012			.61	.51	1.12	1.53
2400	8" wide trench and backfill, 12" deep		1000	.008			.39	.33	.72	.99
2450	18" deep		950	.008			.41	.35	.76	1.05
2500	24" deep		900	.009			.44	.37	.81	1.10
2550	36" deep		800	.010			.49	.42	.91	1.24
2600	48" deep		650	.012			.61	.51	1.12	1.53
2700	12" wide trench and backfill, 12" deep		975	.008			.40	.34	.74	1.02
2750	18" deep		860	.009			.46	.39	.85	1.15
2800	24" deep		800	.010			.49	.42	.91	1.24
2850	36" deep		725	.011			.54	.46	1	1.37
3000	16" wide trench and backfill, 12" deep		835	.010			.47	.40	.87	1.19
3050	18" deep		750	.011			.52	.45	.97	1.32
3100	24" deep	▼	700	.011	▼		.56	.48	1.04	1.42
3200	Compaction with vibratory plate, add								35%	35%
5100	Hand excavate and trim for pipe bells after trench excavation									
5200	8" pipe	1 Clab	155	.052	L.F.		1.96		1.96	3.20
5300	18" pipe	"	130	.062	"		2.33		2.33	3.81
9000	Minimum labor/equipment charge	A-1G	4	2	Job		76	14.20	90.20	140

31 23 16.16 Structural Excavation for Minor Structures

		Crew	Daily Output	Labor-Hours	Unit	Material	2016 Bare Costs Labor	Equipment	Total	Total Incl O&P
0010	**STRUCTURAL EXCAVATION FOR MINOR STRUCTURES**									
0015	Hand, pits to 6' deep, sandy soil	1 Clab	8	1	B.C.Y.		38		38	62
0100	Heavy soil or clay		4	2			76		76	124
0300	Pits 6' to 12' deep, sandy soil		5	1.600			60.50		60.50	99
0500	Heavy soil or clay		3	2.667			101		101	165
0700	Pits 12' to 18' deep, sandy soil		4	2			76		76	124
0900	Heavy soil or clay		2	4			152		152	248
1100	Hand loading trucks from stock pile, sandy soil		12	.667			25.50		25.50	41.50
1300	Heavy soil or clay	▼	8	1	▼		38		38	62
1500	For wet or muck hand excavation, add to above								50%	50%
6000	Machine excavation, for spread and mat footings, elevator pits,									
6001	and small building foundations									
6030	Common earth, hydraulic backhoe, 1/2 C.Y. bucket	B-12E	55	.291	B.C.Y.		13.10	8.10	21.20	30
6035	3/4 C.Y. bucket	B-12F	90	.178			8	7.25	15.25	21
6040	1 C.Y. bucket	B-12A	108	.148			6.70	7.55	14.25	19
6050	1-1/2 C.Y. bucket	B-12B	144	.111			5	7.15	12.15	15.90
6060	2 C.Y. bucket	B-12C	200	.080			3.61	5.90	9.51	12.25
6070	Sand and gravel, 3/4 C.Y. bucket	B-12F	100	.160			7.20	6.55	13.75	18.75
6080	1 C.Y. bucket	B-12A	120	.133			6	6.75	12.75	17.10
6090	1-1/2 C.Y. bucket	B-12B	160	.100			4.51	6.45	10.96	14.30
6100	2 C.Y. bucket	B-12C	220	.073			3.28	5.35	8.63	11.15
6110	Clay, till, or blasted rock, 3/4 C.Y. bucket	B-12F	80	.200			9	8.20	17.20	23.50
6120	1 C.Y. bucket	B-12A	95	.168			7.60	8.55	16.15	21.50
6130	1-1/2 C.Y. bucket	B-12B	130	.123			5.55	7.90	13.45	17.60
6140	2 C.Y. bucket	B-12C	175	.091			4.12	6.70	10.82	14
6230	Sandy clay & loam, hydraulic backhoe, 1/2 C.Y. bucket	B-12E	60	.267			12	7.45	19.45	27.50
6235	3/4 C.Y. bucket	B-12F	98	.163			7.35	6.65	14	19.15
6240	1 C.Y. bucket	B-12A	116	.138			6.20	7	13.20	17.70
6250	1-1/2 C.Y. bucket	B-12B	156	.103	▼		4.62	6.60	11.22	14.65

31 23 Excavation and Fill

31 23 16 – Excavation

31 23 16.16 Structural Excavation for Minor Structures

	Crew	Daily Output	Labor-Hours	Unit	Material	2016 Bare Costs Labor	2016 Bare Costs Equipment	Total	Total Incl O&P
9000 Minimum labor/equipment charge	1 Clab	4	2	Job		76		76	124

31 23 16.42 Excavating, Bulk Bank Measure

		Crew	Daily Output	Labor-Hours	Unit	Material	Labor	Equipment	Total	Total Incl O&P
0010	**EXCAVATING, BULK BANK MEASURE** R312316-40									
0011	Common earth piled									
0020	For loading onto trucks, add								15%	15%
0200	Excavator, hydraulic, crawler mtd., 1 C.Y. cap. = 100 C.Y./hr. R312316-45	B-12A	800	.020	B.C.Y.		.90	1.02	1.92	2.57
1200	Front end loader, track mtd., 1-1/2 C.Y. cap. = 70 C.Y./hr.	B-10N	560	.021			1	.94	1.94	2.63
1500	Wheel mounted, 3/4 C.Y. cap. = 45 C.Y./hr.	B-10R	360	.033	▼		1.56	.83	2.39	3.40
5000	Excavating, bulk bank measure, sandy clay & loam piled									
5020	For loading onto trucks, add								15%	15%
5100	Excavator, hydraulic, crawler mtd., 1 C.Y. cap. = 120 C.Y./hr.	B-12A	960	.017	B.C.Y.		.75	.85	1.60	2.14
9000	Minimum labor/equipment charge	B-10L	2	6	Job		280	236	516	705

31 23 16.46 Excavating, Bulk, Dozer

		Crew	Daily Output	Labor-Hours	Unit	Material	Labor	Equipment	Total	Total Incl O&P
0010	**EXCAVATING, BULK, DOZER**									
0011	Open site									
2000	80 H.P., 50' haul, sand & gravel	B-10L	460	.026	B.C.Y.		1.22	1.02	2.24	3.08
2200	150' haul, sand & gravel		230	.052			2.44	2.05	4.49	6.15
2400	300' haul, sand & gravel	▼	120	.100			4.67	3.93	8.60	11.75
3000	105 H.P., 50' haul, sand & gravel	B-10W	700	.017			.80	.86	1.66	2.23
3200	150' haul, sand & gravel		310	.039			1.81	1.95	3.76	5.05
3300	300' haul, sand & gravel	▼	140	.086			4	4.31	8.31	11.15
4000	200 H.P., 50' haul, sand & gravel	B-10B	1400	.009			.40	1	1.40	1.74
4200	150' haul, sand & gravel		595	.020			.94	2.35	3.29	4.08
4400	300' haul, sand & gravel	▼	310	.039	▼		1.81	4.51	6.32	7.85

31 23 23 – Fill

31 23 23.13 Backfill

		Crew	Daily Output	Labor-Hours	Unit	Material	Labor	Equipment	Total	Total Incl O&P
0010	**BACKFILL** R312323-30									
0015	By hand, no compaction, light soil	1 Clab	14	.571	L.C.Y.		21.50		21.50	35.50
0100	Heavy soil		11	.727	"		27.50		27.50	45
0300	Compaction in 6" layers, hand tamp, add to above	▼	20.60	.388	E.C.Y.		14.70		14.70	24
0400	Roller compaction operator walking, add	B-10A	100	.120			5.60	1.85	7.45	11
0500	Air tamp, add	B-9D	190	.211			8.05	1.35	9.40	14.70
0600	Vibrating plate, add	A-1D	60	.133			5.05	.60	5.65	8.90
0800	Compaction in 12" layers, hand tamp, add to above	1 Clab	34	.235			8.90		8.90	14.60
0900	Roller compaction operator walking, add	B-10A	150	.080			3.74	1.23	4.97	7.30
1000	Air tamp, add	B-9	285	.140			5.40	.80	6.20	9.70
1100	Vibrating plate, add	A-1E	90	.089	▼		3.37	.52	3.89	6.05
1200	Trench, dozer, no compaction, 60 HP	B-10L	425	.028	L.C.Y.		1.32	1.11	2.43	3.33
1300	Dozer backfilling, bulk, up to 300' haul, no compaction	B-10B	1200	.010	"		.47	1.16	1.63	2.03
1400	Air tamped, add	B-11B	80	.200	E.C.Y.		8.70	3.66	12.36	18.05
1900	Dozer backfilling, trench, up to 300' haul, no compaction	B-10B	900	.013	L.C.Y.		.62	1.55	2.17	2.70
2000	Air tamped, add	B-11B	80	.200	E.C.Y.		8.70	3.66	12.36	18.05
2350	Spreading in 8" layers, small dozer	B-10B	1060	.011	L.C.Y.		.53	1.32	1.85	2.29
2450	Compacting with vibrating plate, 8" lifts	A-1D	73	.110	E.C.Y.		4.15	.50	4.65	7.35

31 23 23.14 Backfill, Structural

		Crew	Daily Output	Labor-Hours	Unit	Material	Labor	Equipment	Total	Total Incl O&P
0010	**BACKFILL, STRUCTURAL**									
0011	Dozer or F.E. loader									
0020	From existing stockpile, no compaction									
1000	55 H.P. wheeled loader, 50' haul, common earth	B-11C	200	.080	L.C.Y.		3.56	1.83	5.39	7.70
2000	80 H.P., 50' haul, sand & gravel	B-10L	1100	.011			.51	.43	.94	1.28
2010	Sandy clay & loam	▼	1070	.011	▼		.52	.44	.96	1.32

For customer support on your Commercial Renovation Cost Data, call 877.742.4719.

511

31 23 23.14 Backfill, Structural

		Crew	Daily Output	Labor-Hours	Unit	Material	2016 Bare Costs Labor	Equipment	Total	Total Incl O&P
2020	Common earth	B-10L	975	.012	L.C.Y.		.57	.48	1.05	1.45
2040	Clay		850	.014			.66	.55	1.21	1.66
2400	300' haul, sand & gravel		370	.032			1.51	1.27	2.78	3.82
2410	Sandy clay & loam		360	.033			1.56	1.31	2.87	3.93
2420	Common earth		330	.036			1.70	1.43	3.13	4.28
2440	Clay		290	.041			1.93	1.63	3.56	4.88
3000	105 H.P., 50' haul, sand & gravel	B-10W	1350	.009			.42	.45	.87	1.15
3010	Sandy clay & loam		1325	.009			.42	.46	.88	1.18
3020	Common earth		1225	.010			.46	.49	.95	1.27
3040	Clay		1100	.011			.51	.55	1.06	1.41
3300	300' haul, sand & gravel		465	.026			1.21	1.30	2.51	3.35
3310	Sandy clay & loam		455	.026			1.23	1.33	2.56	3.43
3320	Common earth		415	.029			1.35	1.45	2.80	3.76
3340	Clay		370	.032			1.51	1.63	3.14	4.21

31 23 23.16 Fill By Borrow and Utility Bedding

		Crew	Daily Output	Labor-Hours	Unit	Material	2016 Bare Costs Labor	Equipment	Total	Total Incl O&P
0010	**FILL BY BORROW AND UTILITY BEDDING**									
0049	Utility bedding, for pipe & conduit, not incl. compaction									
0050	Crushed or screened bank run gravel	B-6	150	.160	L.C.Y.	23	6.65	2.44	32.09	39
0100	Crushed stone 3/4" to 1/2"		150	.160		24	6.65	2.44	33.09	40
0200	Sand, dead or bank		150	.160		20	6.65	2.44	29.09	35.50
0500	Compacting bedding in trench	A-1D	90	.089	E.C.Y.		3.37	.40	3.77	5.95
0600	If material source exceeds 2 miles, add for extra mileage.									
0610	See Section 31 23 23.20 for hauling mileage add.									

31 23 23.17 General Fill

		Crew	Daily Output	Labor-Hours	Unit	Material	2016 Bare Costs Labor	Equipment	Total	Total Incl O&P
0010	**GENERAL FILL**									
0011	Spread dumped material, no compaction									
0020	By dozer	B-10B	1000	.012	L.C.Y.		.56	1.40	1.96	2.44
0100	By hand	1 Clab	12	.667	"		25.50		25.50	41.50
9000	Minimum labor/equipment charge	"	4	2	Job		76		76	124

31 23 23.20 Hauling

		Crew	Daily Output	Labor-Hours	Unit	Material	2016 Bare Costs Labor	Equipment	Total	Total Incl O&P
0010	**HAULING**									
0011	Excavated or borrow, loose cubic yards									
0012	no loading equipment, including hauling, waiting, loading/dumping									
0013	time per cycle (wait, load, travel, unload or dump & return)									
0014	8 C.Y. truck, 15 MPH ave, cycle 0.5 miles, 10 min. wait/Ld./Uld.	B-34A	320	.025	L.C.Y.		1.08	1.30	2.38	3.17
0016	cycle 1 mile		272	.029			1.27	1.53	2.80	3.74
0018	cycle 2 miles		208	.038			1.66	2.01	3.67	4.88
0020	cycle 4 miles		144	.056			2.40	2.90	5.30	7.05
0022	cycle 6 miles		112	.071			3.09	3.73	6.82	9.05
0024	cycle 8 miles		88	.091			3.93	4.74	8.67	11.50
0026	20 MPH ave, cycle 0.5 mile		336	.024			1.03	1.24	2.27	3.03
0028	cycle 1 mile		296	.027			1.17	1.41	2.58	3.43
0030	cycle 2 miles		240	.033			1.44	1.74	3.18	4.23
0032	cycle 4 miles		176	.045			1.96	2.37	4.33	5.75
0034	cycle 6 miles		136	.059			2.54	3.07	5.61	7.45
0036	cycle 8 miles		112	.071			3.09	3.73	6.82	9.05
0044	25 MPH ave, cycle 4 miles		192	.042			1.80	2.17	3.97	5.30
0046	cycle 6 miles		160	.050			2.16	2.61	4.77	6.35
0048	cycle 8 miles		128	.063			2.70	3.26	5.96	7.95
0050	30 MPH ave, cycle 4 miles		216	.037			1.60	1.93	3.53	4.71
0052	cycle 6 miles		176	.045			1.96	2.37	4.33	5.75
0054	cycle 8 miles		144	.056			2.40	2.90	5.30	7.05

31 23 23 – Fill

31 23 23.20 Hauling		Crew	Daily Output	Labor-Hours	Unit	Material	2016 Bare Costs Labor	Equipment	Total	Total Incl O&P
0114	15 MPH ave, cycle 0.5 mile, 15 min. wait/Ld./Uld.	B-34A	224	.036	L.C.Y.		1.54	1.86	3.40	4.53
0116	cycle 1 mile		200	.040			1.73	2.09	3.82	5.10
0118	cycle 2 miles		168	.048			2.06	2.48	4.54	6.05
0120	cycle 4 miles		120	.067			2.88	3.48	6.36	8.45
0122	cycle 6 miles		96	.083			3.60	4.35	7.95	10.60
0124	cycle 8 miles		80	.100			4.32	5.20	9.52	12.70
0126	20 MPH ave, cycle 0.5 mile		232	.034			1.49	1.80	3.29	4.38
0128	cycle 1 mile		208	.038			1.66	2.01	3.67	4.88
0130	cycle 2 miles		184	.043			1.88	2.27	4.15	5.50
0132	cycle 4 miles		144	.056			2.40	2.90	5.30	7.05
0134	cycle 6 miles		112	.071			3.09	3.73	6.82	9.05
0136	cycle 8 miles		96	.083			3.60	4.35	7.95	10.60
0144	25 MPH ave, cycle 4 miles		152	.053			2.27	2.75	5.02	6.70
0146	cycle 6 miles		128	.063			2.70	3.26	5.96	7.95
0148	cycle 8 miles		112	.071			3.09	3.73	6.82	9.05
0150	30 MPH ave, cycle 4 miles		168	.048			2.06	2.48	4.54	6.05
0152	cycle 6 miles		144	.056			2.40	2.90	5.30	7.05
0154	cycle 8 miles		120	.067			2.88	3.48	6.36	8.45
0214	15 MPH ave, cycle 0.5 mile, 20 min wait/Ld./Uld.		176	.045			1.96	2.37	4.33	5.75
0216	cycle 1 mile		160	.050			2.16	2.61	4.77	6.35
0218	cycle 2 miles		136	.059			2.54	3.07	5.61	7.45
0220	cycle 4 miles		104	.077			3.32	4.01	7.33	9.75
0222	cycle 6 miles		88	.091			3.93	4.74	8.67	11.50
0224	cycle 8 miles		72	.111			4.80	5.80	10.60	14.15
0226	20 MPH ave, cycle 0.5 mile		176	.045			1.96	2.37	4.33	5.75
0228	cycle 1 mile		168	.048			2.06	2.48	4.54	6.05
0230	cycle 2 miles		144	.056			2.40	2.90	5.30	7.05
0232	cycle 4 miles		120	.067			2.88	3.48	6.36	8.45
0234	cycle 6 miles		96	.083			3.60	4.35	7.95	10.60
0236	cycle 8 miles		88	.091			3.93	4.74	8.67	11.50
0244	25 MPH ave, cycle 4 miles		128	.063			2.70	3.26	5.96	7.95
0246	cycle 6 miles		112	.071			3.09	3.73	6.82	9.05
0248	cycle 8 miles		96	.083			3.60	4.35	7.95	10.60
0250	30 MPH ave, cycle 4 miles		136	.059			2.54	3.07	5.61	7.45
0252	cycle 6 miles		120	.067			2.88	3.48	6.36	8.45
0254	cycle 8 miles		104	.077			3.32	4.01	7.33	9.75
0314	15 MPH ave, cycle 0.5 mile, 25 min wait/Ld./Uld.		144	.056			2.40	2.90	5.30	7.05
0316	cycle 1 mile		128	.063			2.70	3.26	5.96	7.95
0318	cycle 2 miles		112	.071			3.09	3.73	6.82	9.05
0320	cycle 4 miles		96	.083			3.60	4.35	7.95	10.60
0322	cycle 6 miles		80	.100			4.32	5.20	9.52	12.70
0324	cycle 8 miles		64	.125			5.40	6.50	11.90	15.85
0326	20 MPH ave, cycle 0.5 mile		144	.056			2.40	2.90	5.30	7.05
0328	cycle 1 mile		136	.059			2.54	3.07	5.61	7.45
0330	cycle 2 miles		120	.067			2.88	3.48	6.36	8.45
0332	cycle 4 miles		104	.077			3.32	4.01	7.33	9.75
0334	cycle 6 miles		88	.091			3.93	4.74	8.67	11.50
0336	cycle 8 miles		80	.100			4.32	5.20	9.52	12.70
0344	25 MPH ave, cycle 4 miles		112	.071			3.09	3.73	6.82	9.05
0346	cycle 6 miles		96	.083			3.60	4.35	7.95	10.60
0348	cycle 8 miles		88	.091			3.93	4.74	8.67	11.50
0350	30 MPH ave, cycle 4 miles		112	.071			3.09	3.73	6.82	9.05
0352	cycle 6 miles		104	.077			3.32	4.01	7.33	9.75

For customer support on your Commercial Renovation Cost Data, call 877.742.4719.

513

31 23 23.20 Hauling		Crew	Daily Output	Labor-Hours	Unit	Material	2016 Bare Costs Labor	Equipment	Total	Total Incl O&P
0354	cycle 8 miles	B-34A	96	.083	L.C.Y.		3.60	4.35	7.95	10.60
0414	15 MPH ave, cycle 0.5 mile, 30 min wait/Ld./Uld.		120	.067			2.88	3.48	6.36	8.45
0416	cycle 1 mile		112	.071			3.09	3.73	6.82	9.05
0418	cycle 2 miles		96	.083			3.60	4.35	7.95	10.60
0420	cycle 4 miles		80	.100			4.32	5.20	9.52	12.70
0422	cycle 6 miles		72	.111			4.80	5.80	10.60	14.15
0424	cycle 8 miles		64	.125			5.40	6.50	11.90	15.85
0426	20 MPH ave, cycle 0.5 mile		120	.067			2.88	3.48	6.36	8.45
0428	cycle 1 mile		112	.071			3.09	3.73	6.82	9.05
0430	cycle 2 miles		104	.077			3.32	4.01	7.33	9.75
0432	cycle 4 miles		88	.091			3.93	4.74	8.67	11.50
0434	cycle 6 miles		80	.100			4.32	5.20	9.52	12.70
0436	cycle 8 miles		72	.111			4.80	5.80	10.60	14.15
0444	25 MPH ave, cycle 4 miles		96	.083			3.60	4.35	7.95	10.60
0446	cycle 6 miles		88	.091			3.93	4.74	8.67	11.50
0448	cycle 8 miles		80	.100			4.32	5.20	9.52	12.70
0450	30 MPH ave, cycle 4 miles		96	.083			3.60	4.35	7.95	10.60
0452	cycle 6 miles		88	.091			3.93	4.74	8.67	11.50
0454	cycle 8 miles		80	.100			4.32	5.20	9.52	12.70
0514	15 MPH ave, cycle 0.5 mile, 35 min wait/Ld./Uld.		104	.077			3.32	4.01	7.33	9.75
0516	cycle 1 mile		96	.083			3.60	4.35	7.95	10.60
0518	cycle 2 miles		88	.091			3.93	4.74	8.67	11.50
0520	cycle 4 miles		72	.111			4.80	5.80	10.60	14.15
0522	cycle 6 miles		64	.125			5.40	6.50	11.90	15.85
0524	cycle 8 miles		56	.143			6.15	7.45	13.60	18.15
0526	20 MPH ave, cycle 0.5 mile		104	.077			3.32	4.01	7.33	9.75
0528	cycle 1 mile		96	.083			3.60	4.35	7.95	10.60
0530	cycle 2 miles		96	.083			3.60	4.35	7.95	10.60
0532	cycle 4 miles		80	.100			4.32	5.20	9.52	12.70
0534	cycle 6 miles		72	.111			4.80	5.80	10.60	14.15
0536	cycle 8 miles		64	.125			5.40	6.50	11.90	15.85
0544	25 MPH ave, cycle 4 miles		88	.091			3.93	4.74	8.67	11.50
0546	cycle 6 miles		80	.100			4.32	5.20	9.52	12.70
0548	cycle 8 miles		72	.111			4.80	5.80	10.60	14.15
0550	30 MPH ave, cycle 4 miles		88	.091			3.93	4.74	8.67	11.50
0552	cycle 6 miles		80	.100			4.32	5.20	9.52	12.70
0554	cycle 8 miles		72	.111			4.80	5.80	10.60	14.15
1014	12 C.Y. truck, cycle 0.5 mile, 15 MPH ave, 15 min. wait/Ld./Uld.	B-34B	336	.024			1.03	2.06	3.09	3.92
1016	cycle 1 mile		300	.027			1.15	2.30	3.45	4.38
1018	cycle 2 miles		252	.032			1.37	2.74	4.11	5.25
1020	cycle 4 miles		180	.044			1.92	3.84	5.76	7.30
1022	cycle 6 miles		144	.056			2.40	4.80	7.20	9.15
1024	cycle 8 miles		120	.067			2.88	5.75	8.63	11
1025	cycle 10 miles		96	.083			3.60	7.20	10.80	13.70
1026	20 MPH ave, cycle 0.5 mile		348	.023			.99	1.99	2.98	3.78
1028	cycle 1 mile		312	.026			1.11	2.21	3.32	4.22
1030	cycle 2 miles		276	.029			1.25	2.50	3.75	4.77
1032	cycle 4 miles		216	.037			1.60	3.20	4.80	6.10
1034	cycle 6 miles		168	.048			2.06	4.11	6.17	7.85
1036	cycle 8 miles		144	.056			2.40	4.80	7.20	9.15
1038	cycle 10 miles		120	.067			2.88	5.75	8.63	11
1040	25 MPH ave, cycle 4 miles		228	.035			1.52	3.03	4.55	5.75
1042	cycle 6 miles		192	.042			1.80	3.60	5.40	6.85

514

31 23 23 – Fill

31 23 23.20 Hauling		Crew	Daily Output	Labor-Hours	Unit	Material	2016 Bare Costs Labor	Equipment	Total	Total Incl O&P
1044	cycle 8 miles	B-34B	168	.048	L.C.Y.		2.06	4.11	6.17	7.85
1046	cycle 10 miles		144	.056			2.40	4.80	7.20	9.15
1050	30 MPH ave, cycle 4 miles		252	.032			1.37	2.74	4.11	5.25
1052	cycle 6 miles		216	.037			1.60	3.20	4.80	6.10
1054	cycle 8 miles		180	.044			1.92	3.84	5.76	7.30
1056	cycle 10 miles		156	.051			2.22	4.43	6.65	8.45
1060	35 MPH ave, cycle 4 miles		264	.030			1.31	2.62	3.93	4.99
1062	cycle 6 miles		228	.035			1.52	3.03	4.55	5.75
1064	cycle 8 miles		204	.039			1.69	3.39	5.08	6.45
1066	cycle 10 miles		180	.044			1.92	3.84	5.76	7.30
1068	cycle 20 miles		120	.067			2.88	5.75	8.63	11
1069	cycle 30 miles		84	.095			4.11	8.20	12.31	15.65
1070	cycle 40 miles		72	.111			4.80	9.60	14.40	18.30
1072	40 MPH ave, cycle 6 miles		240	.033			1.44	2.88	4.32	5.50
1074	cycle 8 miles		216	.037			1.60	3.20	4.80	6.10
1076	cycle 10 miles		192	.042			1.80	3.60	5.40	6.85
1078	cycle 20 miles		120	.067			2.88	5.75	8.63	11
1080	cycle 30 miles		96	.083			3.60	7.20	10.80	13.70
1082	cycle 40 miles		72	.111			4.80	9.60	14.40	18.30
1084	cycle 50 miles		60	.133			5.75	11.50	17.25	22
1094	45 MPH ave, cycle 8 miles		216	.037			1.60	3.20	4.80	6.10
1096	cycle 10 miles		204	.039			1.69	3.39	5.08	6.45
1098	cycle 20 miles		132	.061			2.62	5.25	7.87	9.95
1100	cycle 30 miles		108	.074			3.20	6.40	9.60	12.20
1102	cycle 40 miles		84	.095			4.11	8.20	12.31	15.65
1104	cycle 50 miles		72	.111			4.80	9.60	14.40	18.30
1106	50 MPH ave, cycle 10 miles		216	.037			1.60	3.20	4.80	6.10
1108	cycle 20 miles		144	.056			2.40	4.80	7.20	9.15
1110	cycle 30 miles		108	.074			3.20	6.40	9.60	12.20
1112	cycle 40 miles		84	.095			4.11	8.20	12.31	15.65
1114	cycle 50 miles		72	.111			4.80	9.60	14.40	18.30
1214	15 MPH ave, cycle 0.5 mile, 20 min. wait/Ld./Uld.		264	.030			1.31	2.62	3.93	4.99
1216	cycle 1 mile		240	.033			1.44	2.88	4.32	5.50
1218	cycle 2 miles		204	.039			1.69	3.39	5.08	6.45
1220	cycle 4 miles		156	.051			2.22	4.43	6.65	8.45
1222	cycle 6 miles		132	.061			2.62	5.25	7.87	9.95
1224	cycle 8 miles		108	.074			3.20	6.40	9.60	12.20
1225	cycle 10 miles		96	.083			3.60	7.20	10.80	13.70
1226	20 MPH ave, cycle 0.5 mile		264	.030			1.31	2.62	3.93	4.99
1228	cycle 1 mile		252	.032			1.37	2.74	4.11	5.25
1230	cycle 2 miles		216	.037			1.60	3.20	4.80	6.10
1232	cycle 4 miles		180	.044			1.92	3.84	5.76	7.30
1234	cycle 6 miles		144	.056			2.40	4.80	7.20	9.15
1236	cycle 8 miles		132	.061			2.62	5.25	7.87	9.95
1238	cycle 10 miles		108	.074			3.20	6.40	9.60	12.20
1240	25 MPH ave, cycle 4 miles		192	.042			1.80	3.60	5.40	6.85
1242	cycle 6 miles		168	.048			2.06	4.11	6.17	7.85
1244	cycle 8 miles		144	.056			2.40	4.80	7.20	9.15
1246	cycle 10 miles		132	.061			2.62	5.25	7.87	9.95
1250	30 MPH ave, cycle 4 miles		204	.039			1.69	3.39	5.08	6.45
1252	cycle 6 miles		180	.044			1.92	3.84	5.76	7.30
1254	cycle 8 miles		156	.051			2.22	4.43	6.65	8.45
1256	cycle 10 miles		144	.056			2.40	4.80	7.20	9.15

For customer support on your Commercial Renovation Cost Data, call 877.742.4719.

515

31 23 23.20 Hauling	Crew	Daily Output	Labor-Hours	Unit	Material	2016 Bare Costs Labor	Equipment	Total	Total Incl O&P
1260	35 MPH ave, cycle 4 miles	B-34B	216	.037	L.C.Y.		1.60	3.20	4.80	6.10
1262	cycle 6 miles		192	.042			1.80	3.60	5.40	6.85
1264	cycle 8 miles		168	.048			2.06	4.11	6.17	7.85
1266	cycle 10 miles		156	.051			2.22	4.43	6.65	8.45
1268	cycle 20 miles		108	.074			3.20	6.40	9.60	12.20
1269	cycle 30 miles		72	.111			4.80	9.60	14.40	18.30
1270	cycle 40 miles		60	.133			5.75	11.50	17.25	22
1272	40 MPH ave, cycle 6 miles		192	.042			1.80	3.60	5.40	6.85
1274	cycle 8 miles		180	.044			1.92	3.84	5.76	7.30
1276	cycle 10 miles		156	.051			2.22	4.43	6.65	8.45
1278	cycle 20 miles		108	.074			3.20	6.40	9.60	12.20
1280	cycle 30 miles		84	.095			4.11	8.20	12.31	15.65
1282	cycle 40 miles		72	.111			4.80	9.60	14.40	18.30
1284	cycle 50 miles		60	.133			5.75	11.50	17.25	22
1294	45 MPH ave, cycle 8 miles		180	.044			1.92	3.84	5.76	7.30
1296	cycle 10 miles		168	.048			2.06	4.11	6.17	7.85
1298	cycle 20 miles		120	.067			2.88	5.75	8.63	11
1300	cycle 30 miles		96	.083			3.60	7.20	10.80	13.70
1302	cycle 40 miles		72	.111			4.80	9.60	14.40	18.30
1304	cycle 50 miles		60	.133			5.75	11.50	17.25	22
1306	50 MPH ave, cycle 10 miles		180	.044			1.92	3.84	5.76	7.30
1308	cycle 20 miles		132	.061			2.62	5.25	7.87	9.95
1310	cycle 30 miles		96	.083			3.60	7.20	10.80	13.70
1312	cycle 40 miles		84	.095			4.11	8.20	12.31	15.65
1314	cycle 50 miles		72	.111			4.80	9.60	14.40	18.30
1414	15 MPH ave, cycle 0.5 mile, 25 min. wait/Ld./Uld.		204	.039			1.69	3.39	5.08	6.45
1416	cycle 1 mile		192	.042			1.80	3.60	5.40	6.85
1418	cycle 2 miles		168	.048			2.06	4.11	6.17	7.85
1420	cycle 4 miles		132	.061			2.62	5.25	7.87	9.95
1422	cycle 6 miles		120	.067			2.88	5.75	8.63	11
1424	cycle 8 miles		96	.083			3.60	7.20	10.80	13.70
1425	cycle 10 miles		84	.095			4.11	8.20	12.31	15.65
1426	20 MPH ave, cycle 0.5 mile		216	.037			1.60	3.20	4.80	6.10
1428	cycle 1 mile		204	.039			1.69	3.39	5.08	6.45
1430	cycle 2 miles		180	.044			1.92	3.84	5.76	7.30
1432	cycle 4 miles		156	.051			2.22	4.43	6.65	8.45
1434	cycle 6 miles		132	.061			2.62	5.25	7.87	9.95
1436	cycle 8 miles		120	.067			2.88	5.75	8.63	11
1438	cycle 10 miles		96	.083			3.60	7.20	10.80	13.70
1440	25 MPH ave, cycle 4 miles		168	.048			2.06	4.11	6.17	7.85
1442	cycle 6 miles		144	.056			2.40	4.80	7.20	9.15
1444	cycle 8 miles		132	.061			2.62	5.25	7.87	9.95
1446	cycle 10 miles		108	.074			3.20	6.40	9.60	12.20
1450	30 MPH ave, cycle 4 miles		168	.048			2.06	4.11	6.17	7.85
1452	cycle 6 miles		156	.051			2.22	4.43	6.65	8.45
1454	cycle 8 miles		132	.061			2.62	5.25	7.87	9.95
1456	cycle 10 miles		120	.067			2.88	5.75	8.63	11
1460	35 MPH ave, cycle 4 miles		180	.044			1.92	3.84	5.76	7.30
1462	cycle 6 miles		156	.051			2.22	4.43	6.65	8.45
1464	cycle 8 miles		144	.056			2.40	4.80	7.20	9.15
1466	cycle 10 miles		132	.061			2.62	5.25	7.87	9.95
1468	cycle 20 miles		96	.083			3.60	7.20	10.80	13.70
1469	cycle 30 miles		72	.111			4.80	9.60	14.40	18.30

31 23 23 – Fill

31 23 23.20 Hauling		Crew	Daily Output	Labor-Hours	Unit	Material	2016 Bare Costs Labor	2016 Bare Costs Equipment	Total	Total Incl O&P
1470	cycle 40 miles	B-34B	60	.133	L.C.Y.		5.75	11.50	17.25	22
1472	40 MPH ave, cycle 6 miles		168	.048			2.06	4.11	6.17	7.85
1474	cycle 8 miles		156	.051			2.22	4.43	6.65	8.45
1476	cycle 10 miles		144	.056			2.40	4.80	7.20	9.15
1478	cycle 20 miles		96	.083			3.60	7.20	10.80	13.70
1480	cycle 30 miles		84	.095			4.11	8.20	12.31	15.65
1482	cycle 40 miles		60	.133			5.75	11.50	17.25	22
1484	cycle 50 miles		60	.133			5.75	11.50	17.25	22
1494	45 MPH ave, cycle 8 miles		156	.051			2.22	4.43	6.65	8.45
1496	cycle 10 miles		144	.056			2.40	4.80	7.20	9.15
1498	cycle 20 miles		108	.074			3.20	6.40	9.60	12.20
1500	cycle 30 miles		84	.095			4.11	8.20	12.31	15.65
1502	cycle 40 miles		72	.111			4.80	9.60	14.40	18.30
1504	cycle 50 miles		60	.133			5.75	11.50	17.25	22
1506	50 MPH ave, cycle 10 miles		156	.051			2.22	4.43	6.65	8.45
1508	cycle 20 miles		120	.067			2.88	5.75	8.63	11
1510	cycle 30 miles		96	.083			3.60	7.20	10.80	13.70
1512	cycle 40 miles		72	.111			4.80	9.60	14.40	18.30
1514	cycle 50 miles		60	.133			5.75	11.50	17.25	22
1614	15 MPH ave, cycle 0.5 mile, 30 min. wait/Ld./Uld.		180	.044			1.92	3.84	5.76	7.30
1616	cycle 1 mile		168	.048			2.06	4.11	6.17	7.85
1618	cycle 2 miles		144	.056			2.40	4.80	7.20	9.15
1620	cycle 4 miles		120	.067			2.88	5.75	8.63	11
1622	cycle 6 miles		108	.074			3.20	6.40	9.60	12.20
1624	cycle 8 miles		84	.095			4.11	8.20	12.31	15.65
1625	cycle 10 miles		84	.095			4.11	8.20	12.31	15.65
1626	20 MPH ave, cycle 0.5 mile		180	.044			1.92	3.84	5.76	7.30
1628	cycle 1 mile		168	.048			2.06	4.11	6.17	7.85
1630	cycle 2 miles		156	.051			2.22	4.43	6.65	8.45
1632	cycle 4 miles		132	.061			2.62	5.25	7.87	9.95
1634	cycle 6 miles		120	.067			2.88	5.75	8.63	11
1636	cycle 8 miles		108	.074			3.20	6.40	9.60	12.20
1638	cycle 10 miles		96	.083			3.60	7.20	10.80	13.70
1640	25 MPH ave, cycle 4 miles		144	.056			2.40	4.80	7.20	9.15
1642	cycle 6 miles		132	.061			2.62	5.25	7.87	9.95
1644	cycle 8 miles		108	.074			3.20	6.40	9.60	12.20
1646	cycle 10 miles		108	.074			3.20	6.40	9.60	12.20
1650	30 MPH ave, cycle 4 miles		144	.056			2.40	4.80	7.20	9.15
1652	cycle 6 miles		132	.061			2.62	5.25	7.87	9.95
1654	cycle 8 miles		120	.067			2.88	5.75	8.63	11
1656	cycle 10 miles		108	.074			3.20	6.40	9.60	12.20
1660	35 MPH ave, cycle 4 miles		156	.051			2.22	4.43	6.65	8.45
1662	cycle 6 miles		144	.056			2.40	4.80	7.20	9.15
1664	cycle 8 miles		132	.061			2.62	5.25	7.87	9.95
1666	cycle 10 miles		120	.067			2.88	5.75	8.63	11
1668	cycle 20 miles		84	.095			4.11	8.20	12.31	15.65
1669	cycle 30 miles		72	.111			4.80	9.60	14.40	18.30
1670	cycle 40 miles		60	.133			5.75	11.50	17.25	22
1672	40 MPH ave, cycle 6 miles		144	.056			2.40	4.80	7.20	9.15
1674	cycle 8 miles		132	.061			2.62	5.25	7.87	9.95
1676	cycle 10 miles		120	.067			2.88	5.75	8.63	11
1678	cycle 20 miles		96	.083			3.60	7.20	10.80	13.70
1680	cycle 30 miles		72	.111			4.80	9.60	14.40	18.30

31 23 Excavation and Fill

31 23 23 – Fill

31 23 23.20 Hauling		Crew	Daily Output	Labor-Hours	Unit	Material	2016 Bare Costs Labor	2016 Bare Costs Equipment	Total	Total Incl O&P
1682	cycle 40 miles	B-34B	60	.133	L.C.Y.		5.75	11.50	17.25	22
1684	cycle 50 miles		48	.167			7.20	14.40	21.60	27.50
1694	45 MPH ave, cycle 8 miles		144	.056			2.40	4.80	7.20	9.15
1696	cycle 10 miles		132	.061			2.62	5.25	7.87	9.95
1698	cycle 20 miles		96	.083			3.60	7.20	10.80	13.70
1700	cycle 30 miles		84	.095			4.11	8.20	12.31	15.65
1702	cycle 40 miles		60	.133			5.75	11.50	17.25	22
1704	cycle 50 miles		60	.133			5.75	11.50	17.25	22
1706	50 MPH ave, cycle 10 miles		132	.061			2.62	5.25	7.87	9.95
1708	cycle 20 miles		108	.074			3.20	6.40	9.60	12.20
1710	cycle 30 miles		84	.095			4.11	8.20	12.31	15.65
1712	cycle 40 miles		72	.111			4.80	9.60	14.40	18.30
1714	cycle 50 miles		60	.133			5.75	11.50	17.25	22
2000	Hauling, 8 C.Y. truck, small project cost per hour	B-34A	8	1	Hr.		43	52	95	127
2100	12 C.Y. Truck	B-34B	8	1			43	86.50	129.50	165
2150	16.5 C.Y. Truck	B-34C	8	1			43	90.50	133.50	170
2175	18 C.Y. 8 wheel Truck	B-34I	8	1			43	109	152	190
2200	20 C.Y. Truck	B-34D	8	1			43	92.50	135.50	172
2300	Grading at dump, or embankment if required, by dozer	B-10B	1000	.012	L.C.Y.		.56	1.40	1.96	2.44
2310	Spotter at fill or cut, if required	1 Clab	8	1	Hr.		38		38	62
9014	18 C.Y. truck, 8 wheels,15 min. wait/Ld./Uld.,15 MPH, cycle 0.5 mi.	B-34I	504	.016	L.C.Y.		.69	1.72	2.41	3
9016	cycle 1 mile		450	.018			.77	1.93	2.70	3.37
9018	cycle 2 miles		378	.021			.91	2.30	3.21	4
9020	cycle 4 miles		270	.030			1.28	3.22	4.50	5.60
9022	cycle 6 miles		216	.037			1.60	4.03	5.63	7
9024	cycle 8 miles		180	.044			1.92	4.83	6.75	8.40
9025	cycle 10 miles		144	.056			2.40	6.05	8.45	10.50
9026	20 MPH ave, cycle 0.5 mile		522	.015			.66	1.67	2.33	2.90
9028	cycle 1 mile		468	.017			.74	1.86	2.60	3.23
9030	cycle 2 miles		414	.019			.83	2.10	2.93	3.65
9032	cycle 4 miles		324	.025			1.07	2.68	3.75	4.67
9034	cycle 6 miles		252	.032			1.37	3.45	4.82	6
9036	cycle 8 miles		216	.037			1.60	4.03	5.63	7
9038	cycle 10 miles		180	.044			1.92	4.83	6.75	8.40
9040	25 MPH ave, cycle 4 miles		342	.023			1.01	2.54	3.55	4.43
9042	cycle 6 miles		288	.028			1.20	3.02	4.22	5.25
9044	cycle 8 miles		252	.032			1.37	3.45	4.82	6
9046	cycle 10 miles		216	.037			1.60	4.03	5.63	7
9050	30 MPH ave, cycle 4 miles		378	.021			.91	2.30	3.21	4
9052	cycle 6 miles		324	.025			1.07	2.68	3.75	4.67
9054	cycle 8 miles		270	.030			1.28	3.22	4.50	5.60
9056	cycle 10 miles		234	.034			1.48	3.72	5.20	6.45
9060	35 MPH ave, cycle 4 miles		396	.020			.87	2.20	3.07	3.81
9062	cycle 6 miles		342	.023			1.01	2.54	3.55	4.43
9064	cycle 8 miles		288	.028			1.20	3.02	4.22	5.25
9066	cycle 10 miles		270	.030			1.28	3.22	4.50	5.60
9068	cycle 20 miles		162	.049			2.13	5.35	7.48	9.35
9070	cycle 30 miles		126	.063			2.74	6.90	9.64	12
9072	cycle 40 miles		90	.089			3.84	9.65	13.49	16.85
9074	40 MPH ave, cycle 6 miles		360	.022			.96	2.41	3.37	4.21
9076	cycle 8 miles		324	.025			1.07	2.68	3.75	4.67
9078	cycle 10 miles		288	.028			1.20	3.02	4.22	5.25
9080	cycle 20 miles		180	.044			1.92	4.83	6.75	8.40

518

31 23 23.20 Hauling		Crew	Daily Output	Labor-Hours	Unit	Material	2016 Bare Costs Labor	Equipment	Total	Total Incl O&P
9082	cycle 30 miles	B-34I	144	.056	L.C.Y.		2.40	6.05	8.45	10.50
9084	cycle 40 miles		108	.074			3.20	8.05	11.25	14
9086	cycle 50 miles		90	.089			3.84	9.65	13.49	16.85
9094	45 MPH ave, cycle 8 miles		324	.025			1.07	2.68	3.75	4.67
9096	cycle 10 miles		306	.026			1.13	2.84	3.97	4.94
9098	cycle 20 miles		198	.040			1.75	4.39	6.14	7.65
9100	cycle 30 miles		144	.056			2.40	6.05	8.45	10.50
9102	cycle 40 miles		126	.063			2.74	6.90	9.64	12
9104	cycle 50 miles		108	.074			3.20	8.05	11.25	14
9106	50 MPH ave, cycle 10 miles		324	.025			1.07	2.68	3.75	4.67
9108	cycle 20 miles		216	.037			1.60	4.03	5.63	7
9110	cycle 30 miles		162	.049			2.13	5.35	7.48	9.35
9112	cycle 40 miles		126	.063			2.74	6.90	9.64	12
9114	cycle 50 miles		108	.074			3.20	8.05	11.25	14
9214	20 min. wait/Ld./Uld.,15 MPH, cycle 0.5 mi.		396	.020			.87	2.20	3.07	3.81
9216	cycle 1 mile		360	.022			.96	2.41	3.37	4.21
9218	cycle 2 miles		306	.026			1.13	2.84	3.97	4.94
9220	cycle 4 miles		234	.034			1.48	3.72	5.20	6.45
9222	cycle 6 miles		198	.040			1.75	4.39	6.14	7.65
9224	cycle 8 miles		162	.049			2.13	5.35	7.48	9.35
9225	cycle 10 miles		144	.056			2.40	6.05	8.45	10.50
9226	20 MPH ave, cycle 0.5 mile		396	.020			.87	2.20	3.07	3.81
9228	cycle 1 mile		378	.021			.91	2.30	3.21	4
9230	cycle 2 miles		324	.025			1.07	2.68	3.75	4.67
9232	cycle 4 miles		270	.030			1.28	3.22	4.50	5.60
9234	cycle 6 miles		216	.037			1.60	4.03	5.63	7
9236	cycle 8 miles		198	.040			1.75	4.39	6.14	7.65
9238	cycle 10 miles		162	.049			2.13	5.35	7.48	9.35
9240	25 MPH ave, cycle 4 miles		288	.028			1.20	3.02	4.22	5.25
9242	cycle 6 miles		252	.032			1.37	3.45	4.82	6
9244	cycle 8 miles		216	.037			1.60	4.03	5.63	7
9246	cycle 10 miles		198	.040			1.75	4.39	6.14	7.65
9250	30 MPH ave, cycle 4 miles		306	.026			1.13	2.84	3.97	4.94
9252	cycle 6 miles		270	.030			1.28	3.22	4.50	5.60
9254	cycle 8 miles		234	.034			1.48	3.72	5.20	6.45
9256	cycle 10 miles		216	.037			1.60	4.03	5.63	7
9260	35 MPH ave, cycle 4 miles		324	.025			1.07	2.68	3.75	4.67
9262	cycle 6 miles		288	.028			1.20	3.02	4.22	5.25
9264	cycle 8 miles		252	.032			1.37	3.45	4.82	6
9266	cycle 10 miles		234	.034			1.48	3.72	5.20	6.45
9268	cycle 20 miles		162	.049			2.13	5.35	7.48	9.35
9270	cycle 30 miles		108	.074			3.20	8.05	11.25	14
9272	cycle 40 miles		90	.089			3.84	9.65	13.49	16.85
9274	40 MPH ave, cycle 6 miles		288	.028			1.20	3.02	4.22	5.25
9276	cycle 8 miles		270	.030			1.28	3.22	4.50	5.60
9278	cycle 10 miles		234	.034			1.48	3.72	5.20	6.45
9280	cycle 20 miles		162	.049			2.13	5.35	7.48	9.35
9282	cycle 30 miles		126	.063			2.74	6.90	9.64	12
9284	cycle 40 miles		108	.074			3.20	8.05	11.25	14
9286	cycle 50 miles		90	.089			3.84	9.65	13.49	16.85
9294	45 MPH ave, cycle 8 miles		270	.030			1.28	3.22	4.50	5.60
9296	cycle 10 miles		252	.032			1.37	3.45	4.82	6
9298	cycle 20 miles		180	.044			1.92	4.83	6.75	8.40

31 23 23.20 Hauling		Crew	Daily Output	Labor-Hours	Unit	Material	2016 Bare Costs Labor	2016 Bare Costs Equipment	Total	Total Incl O&P
9300	cycle 30 miles	B-34I	144	.056	L.C.Y.		2.40	6.05	8.45	10.50
9302	cycle 40 miles		108	.074			3.20	8.05	11.25	14
9304	cycle 50 miles		90	.089			3.84	9.65	13.49	16.85
9306	50 MPH ave, cycle 10 miles		270	.030			1.28	3.22	4.50	5.60
9308	cycle 20 miles		198	.040			1.75	4.39	6.14	7.65
9310	cycle 30 miles		144	.056			2.40	6.05	8.45	10.50
9312	cycle 40 miles		126	.063			2.74	6.90	9.64	12
9314	cycle 50 miles		108	.074			3.20	8.05	11.25	14
9414	25 min. wait/Ld./Uld.,15 MPH, cycle 0.5 mi.		306	.026			1.13	2.84	3.97	4.94
9416	cycle 1 mile		288	.028			1.20	3.02	4.22	5.25
9418	cycle 2 miles		252	.032			1.37	3.45	4.82	6
9420	cycle 4 miles		198	.040			1.75	4.39	6.14	7.65
9422	cycle 6 miles		180	.044			1.92	4.83	6.75	8.40
9424	cycle 8 miles		144	.056			2.40	6.05	8.45	10.50
9425	cycle 10 miles		126	.063			2.74	6.90	9.64	12
9426	20 MPH ave, cycle 0.5 mile		324	.025			1.07	2.68	3.75	4.67
9428	cycle 1 mile		306	.026			1.13	2.84	3.97	4.94
9430	cycle 2 miles		270	.030			1.28	3.22	4.50	5.60
9432	cycle 4 miles		234	.034			1.48	3.72	5.20	6.45
9434	cycle 6 miles		198	.040			1.75	4.39	6.14	7.65
9436	cycle 8 miles		180	.044			1.92	4.83	6.75	8.40
9438	cycle 10 miles		144	.056			2.40	6.05	8.45	10.50
9440	25 MPH ave, cycle 4 miles		252	.032			1.37	3.45	4.82	6
9442	cycle 6 miles		216	.037			1.60	4.03	5.63	7
9444	cycle 8 miles		198	.040			1.75	4.39	6.14	7.65
9446	cycle 10 miles		180	.044			1.92	4.83	6.75	8.40
9450	30 MPH ave, cycle 4 miles		252	.032			1.37	3.45	4.82	6
9452	cycle 6 miles		234	.034			1.48	3.72	5.20	6.45
9454	cycle 8 miles		198	.040			1.75	4.39	6.14	7.65
9456	cycle 10 miles		180	.044			1.92	4.83	6.75	8.40
9460	35 MPH ave, cycle 4 miles		270	.030			1.28	3.22	4.50	5.60
9462	cycle 6 miles		234	.034			1.48	3.72	5.20	6.45
9464	cycle 8 miles		216	.037			1.60	4.03	5.63	7
9466	cycle 10 miles		198	.040			1.75	4.39	6.14	7.65
9468	cycle 20 miles		144	.056			2.40	6.05	8.45	10.50
9470	cycle 30 miles		108	.074			3.20	8.05	11.25	14
9472	cycle 40 miles		90	.089			3.84	9.65	13.49	16.85
9474	40 MPH ave, cycle 6 miles		252	.032			1.37	3.45	4.82	6
9476	cycle 8 miles		234	.034			1.48	3.72	5.20	6.45
9478	cycle 10 miles		216	.037			1.60	4.03	5.63	7
9480	cycle 20 miles		144	.056			2.40	6.05	8.45	10.50
9482	cycle 30 miles		126	.063			2.74	6.90	9.64	12
9484	cycle 40 miles		90	.089			3.84	9.65	13.49	16.85
9486	cycle 50 miles		90	.089			3.84	9.65	13.49	16.85
9494	45 MPH ave, cycle 8 miles		234	.034			1.48	3.72	5.20	6.45
9496	cycle 10 miles		216	.037			1.60	4.03	5.63	7
9498	cycle 20 miles		162	.049			2.13	5.35	7.48	9.35
9500	cycle 30 miles		126	.063			2.74	6.90	9.64	12
9502	cycle 40 miles		108	.074			3.20	8.05	11.25	14
9504	cycle 50 miles		90	.089			3.84	9.65	13.49	16.85
9506	50 MPH ave, cycle 10 miles		234	.034			1.48	3.72	5.20	6.45
9508	cycle 20 miles		180	.044			1.92	4.83	6.75	8.40
9510	cycle 30 miles		144	.056			2.40	6.05	8.45	10.50

31 23 23.20 Hauling		Crew	Daily Output	Labor-Hours	Unit	Material	2016 Bare Costs Labor	2016 Bare Costs Equipment	Total	Total Incl O&P
9512	cycle 40 miles	B-34I	108	.074	L.C.Y.		3.20	8.05	11.25	14
9514	cycle 50 miles		90	.089			3.84	9.65	13.49	16.85
9614	30 min. wait/Ld./Uld.,15 MPH, cycle 0.5 mi.		270	.030			1.28	3.22	4.50	5.60
9616	cycle 1 mile		252	.032			1.37	3.45	4.82	6
9618	cycle 2 miles		216	.037			1.60	4.03	5.63	7
9620	cycle 4 miles		180	.044			1.92	4.83	6.75	8.40
9622	cycle 6 miles		162	.049			2.13	5.35	7.48	9.35
9624	cycle 8 miles		126	.063			2.74	6.90	9.64	12
9625	cycle 10 miles		126	.063			2.74	6.90	9.64	12
9626	20 MPH ave, cycle 0.5 mile		270	.030			1.28	3.22	4.50	5.60
9628	cycle 1 mile		252	.032			1.37	3.45	4.82	6
9630	cycle 2 miles		234	.034			1.48	3.72	5.20	6.45
9632	cycle 4 miles		198	.040			1.75	4.39	6.14	7.65
9634	cycle 6 miles		180	.044			1.92	4.83	6.75	8.40
9636	cycle 8 miles		162	.049			2.13	5.35	7.48	9.35
9638	cycle 10 miles		144	.056			2.40	6.05	8.45	10.50
9640	25 MPH ave, cycle 4 miles		216	.037			1.60	4.03	5.63	7
9642	cycle 6 miles		198	.040			1.75	4.39	6.14	7.65
9644	cycle 8 miles		180	.044			1.92	4.83	6.75	8.40
9646	cycle 10 miles		162	.049			2.13	5.35	7.48	9.35
9650	30 MPH ave, cycle 4 miles		216	.037			1.60	4.03	5.63	7
9652	cycle 6 miles		198	.040			1.75	4.39	6.14	7.65
9654	cycle 8 miles		180	.044			1.92	4.83	6.75	8.40
9656	cycle 10 miles		162	.049			2.13	5.35	7.48	9.35
9660	35 MPH ave, cycle 4 miles		234	.034			1.48	3.72	5.20	6.45
9662	cycle 6 miles		216	.037			1.60	4.03	5.63	7
9664	cycle 8 miles		198	.040			1.75	4.39	6.14	7.65
9666	cycle 10 miles		180	.044			1.92	4.83	6.75	8.40
9668	cycle 20 miles		126	.063			2.74	6.90	9.64	12
9670	cycle 30 miles		108	.074			3.20	8.05	11.25	14
9672	cycle 40 miles		90	.089			3.84	9.65	13.49	16.85
9674	40 MPH ave, cycle 6 miles		216	.037			1.60	4.03	5.63	7
9676	cycle 8 miles		198	.040			1.75	4.39	6.14	7.65
9678	cycle 10 miles		180	.044			1.92	4.83	6.75	8.40
9680	cycle 20 miles		144	.056			2.40	6.05	8.45	10.50
9682	cycle 30 miles		108	.074			3.20	8.05	11.25	14
9684	cycle 40 miles		90	.089			3.84	9.65	13.49	16.85
9686	cycle 50 miles		72	.111			4.80	12.10	16.90	21
9694	45 MPH ave, cycle 8 miles		216	.037			1.60	4.03	5.63	7
9696	cycle 10 miles		198	.040			1.75	4.39	6.14	7.65
9698	cycle 20 miles		144	.056			2.40	6.05	8.45	10.50
9700	cycle 30 miles		126	.063			2.74	6.90	9.64	12
9702	cycle 40 miles		108	.074			3.20	8.05	11.25	14
9704	cycle 50 miles		90	.089			3.84	9.65	13.49	16.85
9706	50 MPH ave, cycle 10 miles		198	.040			1.75	4.39	6.14	7.65
9708	cycle 20 miles		162	.049			2.13	5.35	7.48	9.35
9710	cycle 30 miles		126	.063			2.74	6.90	9.64	12
9712	cycle 40 miles		108	.074			3.20	8.05	11.25	14
9714	cycle 50 miles		90	.089			3.84	9.65	13.49	16.85

For customer support on your Commercial Renovation Cost Data, call 877.742.4719.

521

31 23 23.23 Compaction

31 23 23.23 Compaction		Crew	Daily Output	Labor-Hours	Unit	Material	2016 Bare Costs Labor	2016 Bare Costs Equipment	Total	Total Incl O&P
0010	**COMPACTION**									
5000	Riding, vibrating roller, 6" lifts, 2 passes	B-10Y	3000	.004	E.C.Y.		.19	.19	.38	.51
5020	3 passes		2300	.005			.24	.25	.49	.66
5040	4 passes		1900	.006			.30	.30	.60	.80
5050	8" lifts, 2 passes		4100	.003			.14	.14	.28	.37
5060	12" lifts, 2 passes		5200	.002			.11	.11	.22	.29
5080	3 passes		3500	.003			.16	.16	.32	.44
5100	4 passes		2600	.005			.22	.22	.44	.58
5600	Sheepsfoot or wobbly wheel roller, 6" lifts, 2 passes	B-10G	2400	.005			.23	.50	.73	.92
5620	3 passes		1735	.007			.32	.70	1.02	1.29
5640	4 passes		1300	.009			.43	.93	1.36	1.71
5680	12" lifts, 2 passes		5200	.002			.11	.23	.34	.43
5700	3 passes		3500	.003			.16	.35	.51	.64
5720	4 passes		2600	.005			.22	.47	.69	.85
7000	Walk behind, vibrating plate 18" wide, 6" lifts, 2 passes	A-1D	200	.040			1.52	.18	1.70	2.68
7020	3 passes		185	.043			1.64	.20	1.84	2.90
7040	4 passes		140	.057			2.17	.26	2.43	3.82
7200	12" lifts, 2 passes, 21" wide	A-1E	560	.014			.54	.08	.62	.98
7220	3 passes		375	.021			.81	.12	.93	1.46
7240	4 passes		280	.029			1.08	.17	1.25	1.95
7500	Vibrating roller 24" wide, 6" lifts, 2 passes	B-10A	420	.029			1.33	.44	1.77	2.61
7520	3 passes		280	.043			2	.66	2.66	3.93
7540	4 passes		210	.057			2.67	.88	3.55	5.25
7600	12" lifts, 2 passes		840	.014			.67	.22	.89	1.31
7620	3 passes		560	.021			1	.33	1.33	1.96
7640	4 passes		420	.029			1.33	.44	1.77	2.61
8000	Rammer tamper, 6" to 11", 4" lifts, 2 passes	A-1F	130	.062			2.33	.38	2.71	4.23
8050	3 passes		97	.082			3.13	.52	3.65	5.65
8100	4 passes		65	.123			4.66	.77	5.43	8.45
8200	8" lifts, 2 passes		260	.031			1.17	.19	1.36	2.12
8250	3 passes		195	.041			1.56	.26	1.82	2.82
8300	4 passes		130	.062			2.33	.38	2.71	4.23
8400	13" to 18", 4" lifts, 2 passes	A-1G	390	.021			.78	.15	.93	1.43
8450	3 passes		290	.028			1.05	.20	1.25	1.93
8500	4 passes		195	.041			1.56	.29	1.85	2.86
8600	8" lifts, 2 passes		780	.010			.39	.07	.46	.72
8650	3 passes		585	.014			.52	.10	.62	.96
8700	4 passes		390	.021			.78	.15	.93	1.43
9000	Water, 3000 gal. truck, 3 mile haul	B-45	1888	.008		1.22	.40	.49	2.11	2.52
9010	6 mile haul		1444	.011		1.22	.52	.64	2.38	2.87
9020	12 mile haul		1000	.016		1.22	.75	.92	2.89	3.55
9030	6000 gal. wagon, 3 mile haul	B-59	2000	.004		1.22	.17	.25	1.64	1.90
9040	6 mile haul	"	1600	.005		1.22	.22	.31	1.75	2.03

31 23 23.24 Compaction, Structural

31 23 23.24 Compaction, Structural		Crew	Daily Output	Labor-Hours	Unit	Material	2016 Bare Costs Labor	2016 Bare Costs Equipment	Total	Total Incl O&P
0010	**COMPACTION, STRUCTURAL** R312323-30									
0020	Steel wheel tandem roller, 5 tons	B-10E	8	1.500	Hr.		70	20.50	90.50	135
0050	Air tamp, 6" to 8" lifts, common fill	B-9	250	.160	E.C.Y.		6.15	.92	7.07	11
0060	Select fill	"	300	.133			5.10	.76	5.86	9.20
0600	Vibratory plate, 8" lifts, common fill	A-1D	200	.040			1.52	.18	1.70	2.68
0700	Select fill	"	216	.037			1.40	.17	1.57	2.47
9000	Minimum labor/equipment charge	1 Clab	4	2	Job		76		76	124

522

For customer support on your Commercial Renovation Cost Data, call 877.742.4719.

31 25 Erosion and Sedimentation Controls

31 25 14 – Stabilization Measures for Erosion and Sedimentation Control

31 25 14.16 Rolled Erosion Control Mats and Blankets

	31 25 14.16 Rolled Erosion Control Mats and Blankets		Crew	Daily Output	Labor-Hours	Unit	Material	2016 Bare Costs Labor	Equipment	Total	Total Incl O&P
0010	**ROLLED EROSION CONTROL MATS AND BLANKETS**										
0020	Jute mesh, 100 S.Y. per roll, 4' wide, stapled	G	B-80A	2400	.010	S.Y.	.99	.38	.13	1.50	1.85
0070	Paper biodegradable mesh	G	B-1	2500	.010		.17	.37		.54	.80
0100	Plastic netting, stapled, 2" x 1" mesh, 20 mil	G	"	2500	.010		.47	.37		.84	1.13
0120	Revegetation mat, webbed	G	2 Clab	1000	.016		1.57	.61		2.18	2.72
0200	Polypropylene mesh, stapled, 6.5 oz./S.Y.	G	B-1	2500	.010		1.65	.37		2.02	2.43
0300	Tobacco netting, or jute mesh #2, stapled	G	"	2500	.010		.21	.37		.58	.84
1000	Silt fence, install and maintain, remove	G	B-62	1300	.018	L.F.	.25	.77	.13	1.15	1.67
1100	Allow 25% per month for maintenance; 6-month max life										
1200	Place and remove hay bales	G	A-2	3	8	Ton	261	315	82.50	658.50	890
1250	Hay bales, staked	G	"	2500	.010	L.F.	3.48	.38	.10	3.96	4.55

31 31 Soil Treatment

31 31 16 – Termite Control

31 31 16.13 Chemical Termite Control

	31 31 16.13 Chemical Termite Control		Crew	Daily Output	Labor-Hours	Unit	Material	2016 Bare Costs Labor	Equipment	Total	Total Incl O&P
0010	**CHEMICAL TERMITE CONTROL**										
0020	Slab and walls, residential		1 Skwk	1200	.007	SF Flr.	.32	.33		.65	.89
0030	SS mesh, no chemicals, avg 1400 S.F. home, min	G		1000	.008		.32	.40		.72	1
0100	Commercial, minimum			2496	.003		.33	.16		.49	.62
0200	Maximum			1645	.005		.50	.24		.74	.93
0390	Minimum labor/equipment charge			4	2	Job		100		100	162
0400	Insecticides for termite control, minimum			14.20	.563	Gal.	68.50	28		96.50	121
0500	Maximum			11	.727	"	117	36.50		153.50	188

31 41 Shoring

31 41 13 – Timber Shoring

31 41 13.10 Building Shoring

	31 41 13.10 Building Shoring	Crew	Daily Output	Labor-Hours	Unit	Material	2016 Bare Costs Labor	Equipment	Total	Total Incl O&P
0010	**BUILDING SHORING**									
0020	Shoring, existing building, with timber, no salvage allowance	B-51	2.20	21.818	M.B.F.	840	850	113	1,803	2,425
1000	On cribbing with 35 ton screw jacks, per box and jack		3.60	13.333	Jack	64.50	520	69	653.50	990
1090	Minimum labor/equipment charge		2	24	Ea.		935	124	1,059	1,650
1100	Masonry openings in walls, see Section 02 41 19.16									

31 41 16 – Sheet Piling

31 41 16.10 Sheet Piling Systems

	31 41 16.10 Sheet Piling Systems		Crew	Daily Output	Labor-Hours	Unit	Material	2016 Bare Costs Labor	Equipment	Total	Total Incl O&P
0010	**SHEET PILING SYSTEMS**										
0020	Sheet piling, 50,000 psi steel, not incl. wales, 22 psf, left in place		B-40	10.81	5.920	Ton	1,575	292	355	2,222	2,600
0100	Drive, extract & salvage	R314116-40		6	10.667	"	515	525	635	1,675	2,125
1200	15' deep excavation, 22 psf, left in place			983	.065	S.F.	18.40	3.21	3.88	25.49	30
1300	Drive, extract & salvage			545	.117	"	5.80	5.80	7	18.60	23.50
2100	Rent steel sheet piling and wales, first month					Ton	310			310	340
2200	Per added month					"	31			31	34
3900	Wood, solid sheeting, incl. wales, braces and spacers,										
3910	drive, extract & salvage, 8' deep excavation		B-31	330	.121	S.F.	1.74	4.90	.65	7.29	10.65
4520	Left in place, 8' deep, 55 S.F./hr.			440	.091	"	3.13	3.67	.49	7.29	10
4990	Minimum labor/equipment charge			2	20	Job		810	107	917	1,450
5000	For treated lumber add cost of treatment to lumber										

For customer support on your Commercial Renovation Cost Data, call 877.742.4719.

523

31 43 Concrete Raising

31 43 13 – Pressure Grouting

31 43 13.13 Concrete Pressure Grouting	Crew	Daily Output	Labor-Hours	Unit	Material	2016 Bare Costs Labor	Equipment	Total	Total Incl O&P
0010 **CONCRETE PRESSURE GROUTING**									
0020 Grouting, pressure, cement & sand, 1:1 mix, minimum	B-61	124	.323	Bag	12.75	13.10	2.87	28.72	38
0100 Maximum		51	.784	"	12.75	32	6.95	51.70	73
0200 Cement and sand, 1:1 mix, minimum		250	.160	C.F.	25.50	6.50	1.42	33.42	40
0300 Maximum		100	.400		38	16.20	3.56	57.76	72.50
0400 Epoxy cement grout, minimum		137	.292		675	11.85	2.60	689.45	760
0500 Maximum		57	.702		675	28.50	6.25	709.75	795
0700 Alternate pricing method: (Add for materials)									
0710 5 person crew and equipment	B-61	1	40	Day		1,625	355	1,980	3,025

31 46 Needle Beams

31 46 13 – Cantilever Needle Beams

31 46 13.10 Needle Beams

	Crew	Daily Output	Labor-Hours	Unit	Material	2016 Bare Costs Labor	Equipment	Total	Total Incl O&P
0010 **NEEDLE BEAMS**									
0011 Incl. wood shoring 10' x 10' opening									
0400 Block, concrete, 8" thick	B-9	7.10	5.634	Ea.	50	216	32	298	445
0420 12" thick		6.70	5.970		60.50	229	34	323.50	480
0800 Brick, 4" thick with 8" backup block		5.70	7.018		60.50	269	40	369.50	550
1000 Brick, solid, 8" thick		6.20	6.452		50	247	37	334	500
1040 12" thick		4.90	8.163		60.50	315	46.50	422	630
1080 16" thick		4.50	8.889		82	340	51	473	700
2000 Add for additional floors of shoring	B-1	6	4		50	154		204	305
9000 Minimum labor/equipment charge	"	2	12	Job		465		465	755

31 48 Underpinning

31 48 13 – Underpinning Piers

31 48 13.10 Underpinning Foundations

	Crew	Daily Output	Labor-Hours	Unit	Material	2016 Bare Costs Labor	Equipment	Total	Total Incl O&P
0010 **UNDERPINNING FOUNDATIONS**									
0011 Including excavation,									
0020 forming, reinforcing, concrete and equipment									
0100 5' to 16' below grade, 100 to 500 C.Y.	B-52	2.30	24.348	C.Y.	297	1,075	260	1,632	2,350
0200 Over 500 C.Y.		2.50	22.400		267	995	239	1,501	2,150
0400 16' to 25' below grade, 100 to 500 C.Y.		2	28		325	1,250	299	1,874	2,700
0500 Over 500 C.Y.		2.10	26.667		310	1,175	285	1,770	2,575
0700 26' to 40' below grade, 100 to 500 C.Y.		1.60	35		355	1,550	375	2,280	3,325
0800 Over 500 C.Y.		1.80	31.111		325	1,375	335	2,035	2,950
0900 For under 50 C.Y., add					10%	40%			

31 62 Driven Piles

31 62 13 – Concrete Piles

31 62 13.23 Prestressed Concrete Piles

31 62 13.23 Prestressed Concrete Piles	Crew	Daily Output	Labor-Hours	Unit	Material	2016 Bare Costs Labor	Equipment	Total	Total Incl O&P
0010 **PRESTRESSED CONCRETE PILES**, 200 piles									
0020 Unless specified otherwise, not incl. pile caps or mobilization									
3100 Precast, prestressed, 40' long, 10" thick, square	B-19	700	.091	V.L.F.	12.95	4.51	2.44	19.90	24
3200 12" thick, square		680	.094		18.70	4.64	2.51	25.85	31
3400 14" thick, square		600	.107		22	5.25	2.84	30.09	35.50

31 62 16 – Steel Piles

31 62 16.13 Steel Piles

	Crew	Daily Output	Labor-Hours	Unit	Material	2016 Bare Costs Labor	Equipment	Total	Total Incl O&P
0010 **STEEL PILES**									
0100 Step tapered, round, concrete filled									
0110 8" tip, 12" butt, 60 ton capacity, 30' depth	B-19	760	.084	V.L.F.	17.55	4.15	2.24	23.94	28.50
0120 60' depth with extension		740	.086		36	4.27	2.31	42.58	49
0130 80' depth with extensions		700	.091		55.50	4.51	2.44	62.45	71
0250 "H" Sections, 50' long, HP8 x 36		640	.100		16.55	4.93	2.67	24.15	29

31 62 19 – Timber Piles

31 62 19.10 Wood Piles

	Crew	Daily Output	Labor-Hours	Unit	Material	2016 Bare Costs Labor	Equipment	Total	Total Incl O&P
0010 **WOOD PILES**									
0011 Friction or end bearing, not including									
0050 mobilization or demobilization									
0100 Untreated piles, up to 30' long, 12" butts, 8" points	B-19	625	.102	V.L.F.	13.50	5.05	2.73	21.28	26
0200 30' to 39' long, 12" butts, 8" points	"	700	.091	"	13.50	4.51	2.44	20.45	25
0800 Treated piles, 12 lb. per C.F.,									
0810 friction or end bearing, ASTM class B									
1000 Up to 30' long, 12" butts, 8" points	B-19	625	.102	V.L.F.	13.10	5.05	2.73	20.88	25.50
1100 30' to 39' long, 12" butts, 8" points		700	.091		14.50	4.51	2.44	21.45	26
2700 Mobilization for 10,000 L.F. pile job, add		3300	.019			.96	.52	1.48	2.12
2800 25,000 L.F. pile job, add		8500	.008			.37	.20	.57	.82

31 62 23 – Composite Piles

31 62 23.13 Concrete-Filled Steel Piles

	Crew	Daily Output	Labor-Hours	Unit	Material	2016 Bare Costs Labor	Equipment	Total	Total Incl O&P
0010 **CONCRETE-FILLED STEEL PILES** no mobilization or demobilization									
2600 Pipe piles, 50' lg. 8" diam., 29 lb. per L.F., no concrete	B-19	500	.128	V.L.F.	19.30	6.30	3.41	29.01	35
2700 Concrete filled		460	.139		21	6.85	3.71	31.56	38
2900 10" diameter, 34 lb. per L.F., no concrete		500	.128		22	6.30	3.41	31.71	38
3000 Concrete filled		450	.142		25	7	3.79	35.79	43
3200 12" diameter, 44 lb. per L.F., no concrete		475	.135		27	6.65	3.59	37.24	44
3300 Concrete filled		415	.154		32	7.60	4.11	43.71	52
3500 14" diameter, 46 lb. per L.F., no concrete		430	.149		29	7.35	3.97	40.32	48
3600 Concrete filled		355	.180		37	8.90	4.80	50.70	60

31 63 Bored Piles

31 63 26 – Drilled Caissons

31 63 26.13 Fixed End Caisson Piles

	Crew	Daily Output	Labor-Hours	Unit	Material	2016 Bare Costs Labor	Equipment	Total	Total Incl O&P
0010 **FIXED END CAISSON PILES**									
0015 Including excavation, concrete, 50 lb. reinforcing									
0020 per C.Y., not incl. mobilization, boulder removal, disposal									
0100 Open style, machine drilled, to 50' deep, in stable ground, no [R316326-60]									
0110 casings or ground water, 18" diam., 0.065 C.Y./L.F.	B-43	200	.240	V.L.F.	8.50	10.10	12.75	31.35	39.50
0200 24" diameter, 0.116 C.Y./L.F.		190	.253	"	15.20	10.60	13.45	39.25	48.50
0210 4' bell diameter , add		20	2.400	Ea.	262	101	128	491	590
0500 48" diameter, 0.465 C.Y./L.F.		100	.480	V.L.F.	61	20	25.50	106.50	128

For customer support on your Commercial Renovation Cost Data, call 877.742.4719.

525

31 63 26 – Drilled Caissons

31 63 26.13 Fixed End Caisson Piles	Crew	Daily Output	Labor-Hours	Unit	Material	2016 Bare Costs			Total	Total Incl O&P
						Labor	Equipment			
0510　　　　9' bell diameter , add	B-43	2	24	Ea.	1,525	1,000	1,275		3,800	4,700
1200　Open style, machine drilled, to 50' deep, in wet ground, pulled										
1220　　casing and pumping										
1400　　　24" diameter, 0.116 C.Y./L.F.	B-48	125	.448	V.L.F.	15.20	19.25	23.50		57.95	73.50
1410　　　　4' bell diameter , add	"	19.80	2.828	Ea.	262	122	148		532	645
1700　　　48" diameter, 0.465 C.Y./L.F.	B-49	55	1.600	V.L.F.	61	72	66.50		199.50	257
1710　　　　9' bell diameter , add	"	3.30	26.667	Ea.	1,525	1,200	1,100		3,825	4,825
2300　Open style, machine drilled, to 50' deep, in soft rocks and										
2320　　medium hard shales										
2500　　　24" diameter, 0.116 C.Y./L.F.	B-49	30	2.933	V.L.F.	15.20	132	122		269.20	365
2510　　　　4' bell diameter, add		10.90	8.073	Ea.	262	365	335		962	1,250
2800　　　48" diameter, 0.465 C.Y./L.F.		10	8.800	V.L.F.	61	395	365		821	1,100
2810　　　　9' bell diameter, add		1.10	80	Ea.	1,150	3,600	3,325		8,075	10,700
3600　For rock excavation, sockets, add, minimum		120	.733	C.F.		33	30.50		63.50	86.50
3650　　　Average		95	.926			42	38.50		80.50	110
3700　　　Maximum		48	1.833			82.50	76.50		159	217
3900　For 50' to 100' deep, add				V.L.F.					7%	7%
4000　　For 100' to 150' deep, add									25%	25%
4100　　For 150' to 200' deep, add									30%	30%
4200　For casings left in place, add				Lb.	1.32				1.32	1.45
4300　For other than 50 lb. reinf. per C.Y., add or deduct				"	1.20				1.20	1.32
4400　For steel "I" beam cores, add	B-49	8.30	10.602	Ton	2,125	480	440		3,045	3,575
4500　Load and haul excess excavation, 2 miles	B-34B	178	.045	L.C.Y.		1.94	3.88		5.82	7.40
4600　For mobilization, 50 mile radius, rig to 36"	B-43	2	24	Ea.		1,000	1,275		2,275	3,025
4650　　　Rig to 84"	B-48	1.75	32			1,375	1,675		3,050	4,050
4700　For low headroom, add									50%	50%
4750　For difficult access, add									25%	25%

31 63 29 – Drilled Concrete Piers and Shafts

31 63 29.13 Uncased Drilled Concrete Piers

	Crew	Daily Output	Labor-Hours	Unit	Material	Labor	Equipment		Total	Total Incl O&P
0010　**UNCASED DRILLED CONCRETE PIERS**										
0020　　Unless specified otherwise, not incl. pile caps or mobilization										
0800　　Cast in place friction pile, 50' long, fluted,										
0810　　　tapered steel, 4000 psi concrete, no reinforcing										
0900　　　　12" diameter, 7 ga.	B-19	600	.107	V.L.F.	29.50	5.25	2.84		37.59	44
1200　　　　18" diameter, 7 ga.	"	480	.133	"	44.50	6.60	3.55		54.65	63
1300　　End bearing, fluted, constant diameter,										
1320　　　4000 psi concrete, no reinforcing										
1340　　　　12" diameter, 7 ga.	B-19	600	.107	V.L.F.	31	5.25	2.84		39.09	45.50
1400　　　　18" diameter, 7 ga.	"	480	.133	"	49.50	6.60	3.55		59.65	69

Estimating Tips

32 01 00 Operations and Maintenance of Exterior Improvements

- Recycling of asphalt pavement is becoming very popular and is an alternative to removal and replacement. It can be a good value engineering proposal if removed pavement can be recycled, either at the project site or at another site that is reasonably close to the project site. Sections on repair of flexible and rigid pavement are included.

32 10 00 Bases, Ballasts, and Paving

- When estimating paving, keep in mind the project schedule. Also note that prices for asphalt and concrete are generally higher in the cold seasons. Lines for pavement markings, including tactile warning systems and fence lines, are included.

32 90 00 Planting

- The timing of planting and guarantee specifications often dictate the costs for establishing tree and shrub growth and a stand of grass or ground cover. Establish the work performance schedule to coincide with the local planting season. Maintenance and growth guarantees can add from 20%–100% to the total landscaping cost and can be contractually cumbersome. The cost to replace trees and shrubs can be as high as 5% of the total cost, depending on the planting zone, soil conditions, and time of year.

Reference Numbers

Reference numbers are shown at the beginning of some major classifications. These numbers refer to related items in the Reference Section. The reference information may be an estimating procedure, an alternate pricing method, or technical information.

Note: Not all subdivisions listed here necessarily appear. ∎

Did you know?

RSMeans Online gives you the same access to RSMeans' data with 24/7 access:

- Quickly locate costs in the searchable database.
- Build cost lists, estimates, and reports in minutes.
- Adjust costs to any location in the U.S. and Canada with the click of a button.

Start your free trial today at **www.RSMeansOnline.com**

RSMeans Online
FROM THE GORDIAN GROUP®

32 01 Operation and Maintenance of Exterior Improvements

32 01 13 – Flexible Paving Surface Treatment

32 01 13.61 Slurry Seal (Latex Modified)

		Crew	Daily Output	Labor-Hours	Unit	Material	2016 Bare Costs Labor	2016 Bare Costs Equipment	Total	Total Incl O&P
0010	**SLURRY SEAL (LATEX MODIFIED)**									
3600	Waterproofing, membrane, tar and fabric, small area	B-63	233	.172	S.Y.	14.50	6.90	.74	22.14	28
3640	Large area		1435	.028		13.30	1.12	.12	14.54	16.60
3680	Preformed rubberized asphalt, small area		100	.400		19.70	16.05	1.73	37.48	49.50
3720	Large area		367	.109		17.95	4.38	.47	22.80	27.50
3780	Rubberized asphalt (latex) seal	B-45	5000	.003		2.89	.15	.18	3.22	3.62
9000	Minimum labor/equipment charge	1 Clab	2	4	Job		152		152	248

32 01 13.62 Asphalt Surface Treatment

		Crew	Daily Output	Labor-Hours	Unit	Material	2016 Bare Costs Labor	2016 Bare Costs Equipment	Total	Total Incl O&P
0010	**ASPHALT SURFACE TREATMENT**									
3000	Pavement overlay, polypropylene									
3040	6 oz. per S.Y., ideal conditions	B-63	10000	.004	S.Y.	1.01	.16	.02	1.19	1.39
3080	Adverse conditions		1000	.040		1.35	1.61	.17	3.13	4.28
3120	4 oz. per S.Y., ideal conditions		10000	.004		.82	.16	.02	1	1.18
3160	Adverse conditions		1000	.040		1.06	1.61	.17	2.84	3.95
3200	Tack coat, emulsion, .05 gal. per S.Y., 1000 S.Y.	B-45	2500	.006		.30	.30	.37	.97	1.21
3240	10,000 S.Y.		10000	.002		.24	.08	.09	.41	.48
3270	.10 gal. per S.Y., 1000 S.Y.		2500	.006		.56	.30	.37	1.23	1.50
3275	10,000 S.Y.		10000	.002		.45	.08	.09	.62	.72
3280	.15 gal. per S.Y., 1000 S.Y.		2500	.006		.83	.30	.37	1.50	1.79
3320	10,000 S.Y.		10000	.002		.66	.08	.09	.83	.95

32 01 13.64 Sand Seal

		Crew	Daily Output	Labor-Hours	Unit	Material	2016 Bare Costs Labor	2016 Bare Costs Equipment	Total	Total Incl O&P
0010	**SAND SEAL**									
2080	Sand sealing, sharp sand, asphalt emulsion, small area	B-91	10000	.006	S.Y.	1.52	.29	.23	2.04	2.40
2120	Roadway or large area	"	18000	.004	"	1.31	.16	.13	1.60	1.84
3000	Sealing random cracks, min 1/2" wide, to 1-1/2", 1,000 L.F.	B-77	2800	.014	L.F.	1.59	.56	.20	2.35	2.88
3040	10,000 L.F.		4000	.010	"	1.10	.39	.14	1.63	2.01
3080	Alternate method, 1,000 L.F.		200	.200	Gal.	41	7.80	2.86	51.66	61
3120	10,000 L.F.		325	.123	"	33	4.81	1.76	39.57	46
3200	Multi-cracks (flooding), 1 coat, small area	B-92	460	.070	S.Y.	2.41	2.67	1.41	6.49	8.55
3240	Large area		2850	.011		2.14	.43	.23	2.80	3.31
3280	2 coat, small area		230	.139		14.95	5.35	2.81	23.11	28.50
3320	Large area		1425	.022		13.60	.86	.45	14.91	16.85
3360	Alternate method, small area		115	.278	Gal.	14.95	10.70	5.65	31.30	40
3400	Large area		715	.045	"	14.05	1.72	.91	16.68	19.25

32 01 13.66 Fog Seal

		Crew	Daily Output	Labor-Hours	Unit	Material	2016 Bare Costs Labor	2016 Bare Costs Equipment	Total	Total Incl O&P
0010	**FOG SEAL**									
0012	Sealcoating, 2 coat coal tar pitch emulsion over 10,000 S.Y.	B-45	5000	.003	S.Y.	.93	.15	.18	1.26	1.46
0030	1000 to 10,000 S.Y.	"	3000	.005		.93	.25	.31	1.49	1.76
0100	Under 1000 S.Y.	B-1	1050	.023		.93	.88		1.81	2.46
0300	Petroleum resistant, over 10,000 S.Y.	B-45	5000	.003		1.34	.15	.18	1.67	1.91
0320	1000 to 10,000 S.Y.	"	3000	.005		1.34	.25	.31	1.90	2.21
0400	Under 1000 S.Y.	B-1	1050	.023		1.34	.88		2.22	2.91
0600	Non-skid pavement renewal, over 10,000 S.Y.	B-45	5000	.003		1.41	.15	.18	1.74	1.99
0620	1000 to 10,000 S.Y.	"	3000	.005		1.41	.25	.31	1.97	2.29
0700	Under 1000 S.Y.	B-1	1050	.023		1.41	.88		2.29	2.99
0800	Prepare and clean surface for above	A-2	8545	.003			.11	.03	.14	.21
1000	Hand seal asphalt curbing	B-1	4420	.005	L.F.	.64	.21		.85	1.04
1900	Asphalt surface treatment, single course, small area									
1901	0.30 gal/S.Y. asphalt material, 20#/S.Y. aggregate	B-91	5000	.013	S.Y.	1.34	.58	.47	2.39	2.91
1910	Roadway or large area		10000	.006		1.23	.29	.23	1.75	2.09
1950	Asphalt surface treatment, dbl. course for small area		3000	.021		3.03	.97	.78	4.78	5.75
1960	Roadway or large area		6000	.011		2.73	.48	.39	3.60	4.21

32 01 13 – Flexible Paving Surface Treatment

32 01 13.66 Fog Seal

		Crew	Daily Output	Labor-Hours	Unit	Material	2016 Bare Costs Labor	Equipment	Total	Total Incl O&P
1980	Asphalt surface treatment, single course, for shoulders	B-91	7500	.009	S.Y.	1.50	.39	.31	2.20	2.61

32 01 13.68 Slurry Seal

		Crew	Daily Output	Labor-Hours	Unit	Material	2016 Bare Costs Labor	Equipment	Total	Total Incl O&P
0010	**SLURRY SEAL**									
0100	Slurry seal, type I, 8 lb. agg./S.Y., 1 coat, small or irregular area	B-90	2800	.023	S.Y.	1.92	.97	.80	3.69	4.55
0150	Roadway or large area		10000	.006		1.92	.27	.22	2.41	2.80
0200	Type II, 12 lb. aggregate/S.Y., 2 coats, small or irregular area		2000	.032		3.84	1.35	1.12	6.31	7.65
0250	Roadway or large area		8000	.008		3.84	.34	.28	4.46	5.10
0300	Type III, 20 lb. aggregate/S.Y., 2 coats, small or irregular area		1800	.036		4.53	1.50	1.24	7.27	8.80
0350	Roadway or large area		6000	.011		4.53	.45	.37	5.35	6.10
0400	Slurry seal, thermoplastic coal-tar, type I, small or irregular area		2400	.027		3.87	1.13	.93	5.93	7.10
0450	Roadway or large area		8000	.008		3.87	.34	.28	4.49	5.10
0500	Type II, small or irregular area		2400	.027		4.94	1.13	.93	7	8.30
0550	Roadway or large area		7800	.008		4.94	.35	.29	5.58	6.35

32 01 16 – Flexible Paving Rehabilitation

32 01 16.71 Cold Milling Asphalt Paving

		Crew	Daily Output	Labor-Hours	Unit	Material	2016 Bare Costs Labor	Equipment	Total	Total Incl O&P
0010	**COLD MILLING ASPHALT PAVING**									
5200	Cold planing & cleaning, 1" to 3" asphalt pavmt., over 25,000 S.Y.	B-71	6000	.009	S.Y.		.41	1.15	1.56	1.92
5280	5,000 S.Y. to 10,000 S.Y.	"	4000	.014	"		.61	1.72	2.33	2.88
5300	Asphalt pavement removal from conc. base, no haul									
5320	Rip, load & sweep 1" to 3"	B-70	8000	.007	S.Y.		.31	.23	.54	.74
5330	3" to 6" deep	"	5000	.011			.49	.37	.86	1.20
5340	Profile grooving, asphalt pavement load & sweep, 1" deep	B-71	12500	.004			.20	.55	.75	.93
5350	3" deep		9000	.006			.27	.76	1.03	1.28
5360	6" deep		5000	.011			.49	1.38	1.87	2.30

32 01 16.73 In Place Cold Reused Asphalt Paving

		Crew	Daily Output	Labor-Hours	Unit	Material	2016 Bare Costs Labor	Equipment	Total	Total Incl O&P
0010	**IN PLACE COLD REUSED ASPHALT PAVING**									
5000	Reclamation, pulverizing and blending with existing base									
5040	Aggregate base, 4" thick pavement, over 15,000 S.Y.	B-73	2400	.027	S.Y.		1.24	2.15	3.39	4.35
5080	5,000 S.Y. to 15,000 S.Y.		2200	.029			1.35	2.35	3.70	4.74
5120	8" thick pavement, over 15,000 S.Y.		2200	.029			1.35	2.35	3.70	4.74
5160	5,000 S.Y. to 15,000 S.Y.		2000	.032			1.48	2.59	4.07	5.20

32 01 16.74 In Place Hot Reused Asphalt Paving

		Crew	Daily Output	Labor-Hours	Unit	Material	2016 Bare Costs Labor	Equipment	Total	Total Incl O&P
0010	**IN PLACE HOT REUSED ASPHALT PAVING**									
5500	Recycle asphalt pavement at site									
5520	Remove, rejuvenate and spread 4" deep	G B-72	2500	.026	S.Y.	4.62	1.15	4.60	10.37	12
5521	6" deep	G "	2000	.032	"	6.80	1.43	5.75	13.98	16.05

32 01 17 – Flexible Paving Repair

32 01 17.10 Repair of Asphalt Pavement Holes

		Crew	Daily Output	Labor-Hours	Unit	Material	2016 Bare Costs Labor	Equipment	Total	Total Incl O&P
0010	**REPAIR OF ASPHALT PAVEMENT HOLES** (cold patch)									
0100	Flexible pavement repair holes, roadway, light traffic, 1 C.F. size	B-37A	24	1	Ea.	9.65	39.50	16	65.15	92
0150	Group of two, 1 C.F. size each		16	1.500	Set	19.30	59	24	102.30	143
0200	Group of three, 1 C.F. size each		12	2	"	29	78.50	32	139.50	195
0300	Medium traffic, 1 C.F. size each	B-37B	24	1.333	Ea.	9.65	52	15.95	77.60	113
0350	Group of two, 1 C.F. size each		16	2	Set	19.30	78	24	121.30	175
0400	Group of three, 1 C.F. size each		12	2.667	"	29	104	32	165	236
0500	Highway/heavy traffic, 1 C.F. size each	B-37C	24	1.333	Ea.	9.65	53	26.50	89.15	126
0550	Group of two, 1 C.F. size each		16	2	Set	19.30	80	39.50	138.80	195
0600	Group of three, 1 C.F. size each		12	2.667	"	29	106	52.50	187.50	263
0700	Add police officer and car for traffic control				Hr.	45.50			45.50	50
1000	Flexible pavement repair holes, parking lot, bag material, 1 C.F. size each	B-37D	18	.889	Ea.	46.50	35.50	8.10	90.10	118
1010	Economy bag material, 1 C.F. each		18	.889	"	42	35.50	8.10	85.60	113

For customer support on your Commercial Renovation Cost Data, call 877.742.4719.

529

32 01 Operation and Maintenance of Exterior Improvements

32 01 17 – Flexible Paving Repair

32 01 17.10 Repair of Asphalt Pavement Holes

	Crew	Daily Output	Labor-Hours	Unit	Material	2016 Bare Costs Labor	Equipment	Total	Total Incl O&P
1100	Group of two, 1 C.F. size each	B-37D	12	1.333	Set	93.50	53	12.15	158.65	203
1110	Group of two, economy bag, 1 C.F. size each		12	1.333		84	53	12.15	149.15	192
1200	Group of three 1 C.F. size each		10	1.600		140	64	14.60	218.60	274
1210	Economy bag, group of three, 1 C.F. size each		10	1.600		126	64	14.60	204.60	259
1300	Flexible pavement repair holes, parking lot, 1 C.F. single hole	A-3A	4	2	Ea.	46.50	98.50	39	184	251
1310	Economy material, 1 C.F. single hole		4	2	"	42	98.50	39	179.50	246
1400	Flexible pavement repair holes, parking lot, 1 C.F. four holes		2	4	Set	187	197	78.50	462.50	600
1410	Economy material, 1 C.F. four holes		2	4	"	168	197	78.50	443.50	580
1500	Flexible pavement repair holes, large parking lot, bulk matl, 1 C.F. size	B-37A	32	.750	Ea.	9.65	29.50	12	51.15	72

32 01 17.20 Repair of Asphalt Pavement Patches

	Crew	Daily Output	Labor-Hours	Unit	Material	2016 Bare Costs Labor	Equipment	Total	Total Incl O&P
0010	**REPAIR OF ASPHALT PAVEMENT PATCHES**									
0100	Flexible pavement patches, roadway, light traffic, sawcut 10-25 S.F.	B-89	12	1.333	Ea.		60.50	41	101.50	142
0150	Sawcut 26-60 S.F.		10	1.600			73	49.50	122.50	171
0200	Sawcut 61-100 S.F.		8	2			91	61.50	152.50	213
0210	Sawcut groups of small size patches		24	.667			30.50	20.50	51	71
0220	Large size patches		16	1			45.50	31	76.50	107
0300	Flexible pavement patches, roadway, light traffic, digout 10-25 S.F.	B-6	16	1.500			62.50	23	85.50	126
0350	Digout 26-60 S.F.		12	2			83.50	30.50	114	168
0400	Digout 61-100 S.F.		8	3			125	45.50	170.50	253
0450	Add 8 C.Y. truck, small project debris haulaway	B-34A	8	1	Hr.		43	52	95	127
0460	Add 12 C.Y. truck, small project debris haulaway	B-34B	8	1			43	86.50	129.50	165
0480	Add flagger for non-intersection medium traffic	1 Clab	8	1			38		38	62
0490	Add flasher truck for intersection medium traffic or heavy traffic	A-2B	8	1			42	31	73	102
0500	Flexible pavement patches, roadway, repave, cold, 15 S.F., 4" D	B-37	8	6	Ea.	46	241	20.50	307.50	465
0510	6" depth		8	6		69.50	241	20.50	331	490
0520	Repave, cold, 20 S.F., 4" depth		8	6		61.50	241	20.50	323	480
0530	6" depth		8	6		93	241	20.50	354.50	515
0540	Repave, cold, 25 S.F., 4" depth		8	6		76.50	241	20.50	338	495
0550	6" depth		8	6		116	241	20.50	377.50	540
0600	Repave, cold, 30 S.F., 4" depth		8	6		92	241	20.50	353.50	515
0610	6" depth		8	6		139	241	20.50	400.50	565
0640	Repave, cold, 40 S.F., 4" depth		8	6		122	241	20.50	383.50	545
0650	6" depth		8	6		185	241	20.50	446.50	615
0680	Repave, cold, 50 S.F., 4" depth		8	6		153	241	20.50	414.50	580
0690	6" depth		8	6		232	241	20.50	493.50	670
0720	Repave, cold, 60 S.F., 4" depth		8	6		184	241	20.50	445.50	615
0730	6" depth		8	6		278	241	20.50	539.50	720
0800	Repave, cold, 70 S.F., 4" depth		7	6.857		214	275	23.50	512.50	705
0810	6" depth		7	6.857		325	275	23.50	623.50	825
0820	Repave, cold, 75 S.F., 4" depth		7	6.857		228	275	23.50	526.50	720
0830	6" depth		7	6.857		345	275	23.50	643.50	850
0900	Repave, cold, 80 S.F., 4" depth		6	8		245	320	27	592	820
0910	6" depth		6	8		370	320	27	717	960
0940	Repave, cold, 90 S.F., 4" depth		6	8		275	320	27	622	855
0950	6" depth		6	8		420	320	27	767	1,000
0980	Repave, cold, 100 S.F., 4" depth		6	8		305	320	27	652	885
0990	6" depth		6	8		465	320	27	812	1,050
1000	Add flasher truck for paving operations in medium or heavy traffic	A-2B	8	1	Hr.		42	31	73	102
1100	Prime coat for repair 15-40 S.F., 25% overspray	B-37A	48	.500	Ea.	7.50	19.65	8	35.15	49
1150	41-60 S.F.		40	.600		15	23.50	9.60	48.10	65.50
1175	61-80 S.F.		32	.750		20.50	29.50	12	62	83.50
1200	81-100 S.F.		32	.750		25	29.50	12	66.50	88.50

32 01 17 – Flexible Paving Repair

32 01 17.20 Repair of Asphalt Pavement Patches

		Crew	Daily Output	Labor-Hours	Unit	Material	2016 Bare Costs Labor	Equipment	Total	Total Incl O&P
1210	Groups of patches w/25% overspray	B-37A	3600	.007	S.F.	.25	.26	.11	.62	.82
1300	Flexible pavement repair patches, street repave, hot, 60 S.F., 4" D	B-37	8	6	Ea.	101	241	20.50	362.50	525
1310	6" depth		8	6		153	241	20.50	414.50	580
1320	Repave, hot, 70 S.F., 4" depth		7	6.857		118	275	23.50	416.50	600
1330	6" depth		7	6.857		179	275	23.50	477.50	670
1340	Repave, hot, 75 S.F., 4" depth		7	6.857		126	275	23.50	424.50	610
1350	6" depth		7	6.857		192	275	23.50	490.50	680
1360	Repave, hot, 80 S.F., 4" depth		6	8		135	320	27	482	700
1370	6" depth		6	8		205	320	27	552	775
1380	Repave, hot, 90 S.F., 4" depth		6	8		152	320	27	499	715
1390	6" depth		6	8		230	320	27	577	805
1400	Repave, hot, 100 S.F., 4" depth		6	8		169	320	27	516	735
1410	6" depth		6	8		256	320	27	603	830
1420	Pave hot groups of patches, 4" depth		900	.053	S.F.	1.68	2.14	.18	4	5.50
1430	6" depth		900	.053	"	2.56	2.14	.18	4.88	6.50
1500	Add 8 C.Y. truck for hot asphalt paving operations	B-34A	1	8	Day		345	415	760	1,025
1550	Add flasher truck for hot paving operations in med/heavy traffic	A-2B	8	1	Hr.		42	31	73	102
2000	Add police officer and car for traffic control				"	45.50			45.50	50

32 01 17.61 Sealing Cracks In Asphalt Paving

		Crew	Daily Output	Labor-Hours	Unit	Material	2016 Bare Costs Labor	Equipment	Total	Total Incl O&P
0010	**SEALING CRACKS IN ASPHALT PAVING**									
0100	Sealing cracks in asphalt paving, 1/8" wide x 1/2" depth, slow set	B-37A	2000	.012	L.F.	.18	.47	.19	.84	1.18
0110	1/4" wide x 1/2" depth		2000	.012		.21	.47	.19	.87	1.21
0130	3/8" wide x 1/2" depth		2000	.012		.24	.47	.19	.90	1.24
0140	1/2" wide x 1/2" depth		1800	.013		.27	.52	.21	1	1.38
0150	3/4" wide x 1/2" depth		1600	.015		.33	.59	.24	1.16	1.58
0160	1" wide x 1/2" depth		1600	.015		.39	.59	.24	1.22	1.65
0165	1/8" wide x 1" depth, rapid set	B-37F	2000	.016		.25	.62	.29	1.16	1.60
0170	1/4" wide x 1" depth		2000	.016		.48	.62	.29	1.39	1.86
0175	3/8" wide x 1" depth		1800	.018		.72	.69	.33	1.74	2.29
0180	1/2" wide x 1" depth		1800	.018		.97	.69	.33	1.99	2.55
0185	5/8" wide x 1" depth		1800	.018		1.21	.69	.33	2.23	2.82
0190	3/4" wide x 1" depth		1600	.020		1.45	.78	.37	2.60	3.26
0195	1" wide x 1" depth		1600	.020		1.93	.78	.37	3.08	3.79
0200	1/8" wide x 2" depth, rapid set		2000	.016		.48	.62	.29	1.39	1.86
0210	1/4" wide x 2" depth		2000	.016		.97	.62	.29	1.88	2.39
0220	3/8" wide x 2" depth		1800	.018		1.45	.69	.33	2.47	3.08
0230	1/2" wide x 2" depth		1800	.018		1.93	.69	.33	2.95	3.61
0240	5/8" wide x 2" depth		1800	.018		2.41	.69	.33	3.43	4.14
0250	3/4" wide x 2" depth		1600	.020		2.89	.78	.37	4.04	4.85
0260	1" wide x 2" depth		1600	.020		3.86	.78	.37	5.01	5.90
0300	Add flagger for non-intersection medium traffic	1 Clab	8	1	Hr.		38		38	62
0400	Add flasher truck for intersection medium traffic or heavy traffic	A-2B	8	1	"		42	31	73	102

32 01 29 – Rigid Paving Repair

32 01 29.61 Partial Depth Patching of Rigid Pavement

		Crew	Daily Output	Labor-Hours	Unit	Material	2016 Bare Costs Labor	Equipment	Total	Total Incl O&P
0010	**PARTIAL DEPTH PATCHING OF RIGID PAVEMENT**									
0100	Rigid pavement repair, roadway, light traffic, 25% pitting 15 S.F.	B-37F	16	2	Ea.	28.50	78	37	143.50	199
0110	Pitting 20 S.F.		16	2		38	78	37	153	210
0120	Pitting 25 S.F.		16	2		47.50	78	37	162.50	220
0130	Pitting 30 S.F.		16	2		57	78	37	172	230
0140	Pitting 40 S.F.		16	2		76	78	37	191	251
0150	Pitting 50 S.F.		12	2.667		95	104	49	248	330
0160	Pitting 60 S.F.		12	2.667		114	104	49	267	350

32 01 29.61 Partial Depth Patching of Rigid Pavement

		Crew	Daily Output	Labor-Hours	Unit	Material	2016 Bare Costs Labor	2016 Bare Costs Equipment	Total	Total Incl O&P
0170	Pitting 70 S.F.	B-37F	12	2.667	Ea.	133	104	49	286	370
0180	Pitting 75 S.F.		12	2.667		143	104	49	296	380
0190	Pitting 80 S.F.		8	4		152	156	73.50	381.50	500
0200	Pitting 90 S.F.		8	4		171	156	73.50	400.50	520
0210	Pitting 100 S.F.		8	4		190	156	73.50	419.50	545
0300	Roadway, light traffic, 50% pitting 15 S.F.		12	2.667		57	104	49	210	286
0310	Pitting 20 S.F.		12	2.667		76	104	49	229	305
0320	Pitting 25 S.F.		12	2.667		95	104	49	248	330
0330	Pitting 30 S.F.		12	2.667		114	104	49	267	350
0340	Pitting 40 S.F.		12	2.667		152	104	49	305	390
0350	Pitting 50 S.F.		8	4		190	156	73.50	419.50	545
0360	Pitting 60 S.F.		8	4		228	156	73.50	457.50	585
0370	Pitting 70 S.F.		8	4		266	156	73.50	495.50	625
0380	Pitting 75 S.F.		8	4		285	156	73.50	514.50	650
0390	Pitting 80 S.F.		6	5.333		305	208	98	611	785
0400	Pitting 90 S.F.		6	5.333		340	208	98	646	825
0410	Pitting 100 S.F.		6	5.333		380	208	98	686	870
1000	Rigid pavement repair, light traffic, surface patch, 2" deep, 15 S.F.		8	4		114	156	73.50	343.50	460
1010	Surface patch, 2" deep, 20 S.F.		8	4		152	156	73.50	381.50	500
1020	Surface patch, 2" deep, 25 S.F.		8	4		190	156	73.50	419.50	545
1030	Surface patch, 2" deep, 30 S.F.		8	4		228	156	73.50	457.50	585
1040	Surface patch, 2" deep, 40 S.F.		8	4		305	156	73.50	534.50	670
1050	Surface patch, 2" deep, 50 S.F.		6	5.333		380	208	98	686	870
1060	Surface patch, 2" deep, 60 S.F.		6	5.333		455	208	98	761	950
1070	Surface patch, 2" deep, 70 S.F.		6	5.333		530	208	98	836	1,025
1080	Surface patch, 2" deep, 75 S.F.		6	5.333		570	208	98	876	1,075
1090	Surface patch, 2" deep, 80 S.F.		5	6.400		610	249	118	977	1,200
1100	Surface patch, 2" deep, 90 S.F.		5	6.400		685	249	118	1,052	1,300
1200	Surface patch, 2" deep, 100 S.F.	↓	5	6.400	↓	760	249	118	1,127	1,375
2000	Add flagger for non-intersection medium traffic	1 Clab	8	1	Hr.		38		38	62
2100	Add flasher truck for intersection medium traffic or heavy traffic	A-2B	8	1			42	31	73	102
2200	Add police officer and car for traffic control				↓	45.50			45.50	50

32 01 29.70 Full Depth Patching of Rigid Pavement

		Crew	Daily Output	Labor-Hours	Unit	Material	2016 Bare Costs Labor	2016 Bare Costs Equipment	Total	Total Incl O&P
0010	**FULL DEPTH PATCHING OF RIGID PAVEMENT**									
0015	Pavement preparation includes sawcut, remove pavement and replace									
0020	6 inches of subbase and one layer of reinforcement									
0030	Trucking and haulaway of debris excluded									
0100	Rigid pavement replace, light traffic, prep, 15 S.F., 6" depth	B-37E	16	3.500	Ea.	19.70	149	64	232.70	330
0110	Replacement preparation 20 S.F., 6" depth		16	3.500		26	149	64	239	340
0115	25 S.F., 6" depth		16	3.500		33	149	64	246	345
0120	30 S.F., 6" depth		14	4		39.50	170	73	282.50	400
0125	35 S.F., 6" depth		14	4		46	170	73	289	405
0130	40 S.F., 6" depth		14	4		52.50	170	73	295.50	410
0135	45 S.F., 6" depth		14	4		59	170	73	302	420
0140	50 S.F., 6" depth		12	4.667		65.50	199	85.50	350	485
0145	55 S.F., 6" depth		12	4.667		72	199	85.50	356.50	495
0150	60 S.F., 6" depth		10	5.600		78.50	238	102	418.50	585
0155	65 S.F., 6" depth		10	5.600		85.50	238	102	425.50	590
0160	70 S.F., 6" depth		10	5.600		92	238	102	432	600
0165	75 S.F., 6" depth		10	5.600		98.50	238	102	438.50	605
0170	80 S.F., 6" depth		8	7		105	298	128	531	735
0175	85 S.F., 6" depth	↓	8	7		112	298	128	538	745

32 01 29 – Rigid Paving Repair

32 01 29.70 Full Depth Patching of Rigid Pavement	Crew	Daily Output	Labor-Hours	Unit	Material	2016 Bare Costs Labor	Equipment	Total	Total Incl O&P
0180	90 S.F., 6" depth	B-37E	8	7	Ea.	118	298	128	544	750
0185	95 S.F., 6" depth		8	7		125	298	128	551	760
0190	100 S.F., 6" depth		8	7		131	298	128	557	765
0200	Pavement preparation for 8 inch rigid paving same as 6 inch									
0290	Pavement preparation for 9 inch rigid paving same as 10 inch									
0300	Rigid pavement replace, light traffic,prep, 15 S.F.,10" depth	B-37E	16	3.500	Ea.	32.50	149	64	245.50	345
0302	Pavement preparation 10 inch includes sawcut, remove pavement and									
0304	replace 6 inches of subbase and two layers of reinforcement									
0306	Trucking and haulaway of debris excluded									
0310	Replacement preparation 20 S.F., 10" depth	B-37E	16	3.500	Ea.	43.50	149	64	256.50	360
0315	25 S.F., 10" depth		16	3.500		54.50	149	64	267.50	370
0320	30 S.F., 10" depth		14	4		65.50	170	73	308.50	425
0325	35 S.F., 10" depth		14	4		50.50	170	73	293.50	410
0330	40 S.F., 10" depth		14	4		87	170	73	330	450
0335	45 S.F., 10" depth		12	4.667		98	199	85.50	382.50	520
0340	50 S.F., 10" depth		12	4.667		109	199	85.50	393.50	535
0345	55 S.F., 10" depth		12	4.667		120	199	85.50	404.50	545
0350	60 S.F., 10" depth		10	5.600		131	238	102	471	640
0355	65 S.F., 10" depth		10	5.600		141	238	102	481	655
0360	70 S.F., 10" depth		10	5.600		152	238	102	492	665
0365	75 S.F., 10" depth		8	7		163	298	128	589	800
0370	80 S.F., 10" depth		8	7		174	298	128	600	815
0375	85 S.F., 10" depth		8	7		185	298	128	611	825
0380	90 S.F., 10" depth		8	7		196	298	128	622	835
0385	95 S.F., 10" depth		8	7		207	298	128	633	850
0390	100 S.F., 10" depth		8	7		218	298	128	644	860
0500	Add 8 C.Y. truck, small project debris haulaway	B-34A	8	1	Hr.		43	52	95	127
0550	Add 12 C.Y. truck, small project debris haulaway	B-34B	8	1			43	86.50	129.50	165
0600	Add flagger for non-intersection medium traffic	1 Clab	8	1			38		38	62
0650	Add flasher truck for intersection medium traffic or heavy traffic	A-2B	8	1			42	31	73	102
0700	Add police officer and car for traffic control					45.50			45.50	50
0990	Concrete will be replaced using quick set mix with aggregate									
1000	Rigid pavement replace, light traffic,repour, 15 S.F.,6" depth	B-37F	18	1.778	Ea.	167	69	32.50	268.50	335
1010	20 S.F., 6" depth		18	1.778		223	69	32.50	324.50	395
1015	25 S.F., 6" depth		18	1.778		279	69	32.50	380.50	455
1020	30 S.F., 6" depth		16	2		335	78	37	450	540
1025	35 S.F., 6" depth		16	2		390	78	37	505	600
1030	40 S.F., 6" depth		16	2		445	78	37	560	660
1035	45 S.F., 6" depth		16	2		500	78	37	615	720
1040	50 S.F., 6" depth		16	2		555	78	37	670	785
1045	55 S.F., 6" depth		12	2.667		615	104	49	768	900
1050	60 S.F., 6" depth		12	2.667		670	104	49	823	960
1055	65 S.F., 6" depth		12	2.667		725	104	49	878	1,025
1060	70 S.F., 6" depth		12	2.667		780	104	49	933	1,075
1065	75 S.F., 6" depth		10	3.200		835	125	59	1,019	1,200
1070	80 S.F., 6" depth		10	3.200		890	125	59	1,074	1,250
1075	85 S.F., 6" depth		8	4		945	156	73.50	1,174.50	1,375
1080	90 S.F., 6" depth		8	4		1,000	156	73.50	1,229.50	1,425
1085	95 S.F., 6" depth		8	4		1,050	156	73.50	1,279.50	1,500
1090	100 S.F., 6" depth		8	4		1,125	156	73.50	1,354.50	1,550
1099	Concrete will be replaced using 4500 PSI concrete ready mix									
1100	Rigid pavement replace, light traffic,repour, 15 S.F.,6" depth	A-2	12	2	Ea.	305	78.50	20.50	404	485
1110	20-50 S.F., 6" depth		12	2		305	78.50	20.50	404	485

32 01 29 – Rigid Paving Repair

32 01 29.70 Full Depth Patching of Rigid Pavement	Crew	Daily Output	Labor-Hours	Unit	Material	2016 Bare Costs Labor	Equipment	Total	Total Incl O&P
1120	55-65 S.F., 6" depth	A-2	10	2.400	Ea.	335	94	25	454	550
1130	70-80 S.F., 6" depth		10	2.400		365	94	25	484	580
1140	85-100 S.F., 6" depth		8	3		430	118	31	579	695
1200	Repour 15-40 S.F.,8" depth		12	2		305	78.50	20.50	404	485
1210	45-50 S.F.,8" depth		12	2		335	78.50	20.50	434	520
1220	55-60 S.F.,8" depth		10	2.400		365	94	25	484	580
1230	65-80 S.F.,8" depth		10	2.400		430	94	25	549	650
1240	85-90 S.F.,8" depth		8	3		460	118	31	609	735
1250	95-100 S.F.,8" depth		8	3		490	118	31	639	765
1300	Repour 15-30 S.F.,10" depth		12	2		305	78.50	20.50	404	485
1310	35-40 S.F.,10" depth		12	2		335	78.50	20.50	434	520
1320	45 S.F.,10" depth		12	2		365	78.50	20.50	464	550
1330	50-65 S.F.,10" depth		10	2.400		430	94	25	549	650
1340	70 S.F.,10" depth		10	2.400		460	94	25	579	690
1350	75-80 S.F.,10" depth		10	2.400		490	94	25	609	720
1360	85-95 S.F.,10" depth		8	3		555	118	31	704	835
1370	100 S.F.,10" depth		8	3		585	118	31	734	870

32 06 Schedules for Exterior Improvements

32 06 10 – Schedules for Bases, Ballasts, and Paving

32 06 10.10 Sidewalks, Driveways and Patios

		Crew	Daily Output	Labor-Hours	Unit	Material	2016 Bare Costs Labor	Equipment	Total	Total Incl O&P
0010	**SIDEWALKS, DRIVEWAYS AND PATIOS** No base									
0020	Asphaltic concrete, 2" thick	B-37	720	.067	S.Y.	7.25	2.67	.23	10.15	12.60
0100	2-1/2" thick	"	660	.073	"	9.20	2.92	.25	12.37	15.10
0110	Bedding for brick or stone, mortar, 1" thick	D-1	300	.053	S.F.	.96	2.25		3.21	4.71
0120	2" thick	"	200	.080		2.40	3.37		5.77	8.15
0130	Sand, 2" thick	B-18	8000	.003		.31	.12	.01	.44	.55
0140	4" thick	"	4000	.006		.63	.23	.01	.87	1.08
0300	Concrete, 3000 psi, CIP, 6 x 6 - W1.4 x W1.4 mesh,									
0310	broomed finish, no base, 4" thick	B-24	600	.040	S.F.	1.79	1.76		3.55	4.80
0350	5" thick		545	.044		2.38	1.94		4.32	5.75
0400	6" thick		510	.047		2.78	2.07		4.85	6.40
0450	For bank run gravel base, 4" thick, add	B-18	2500	.010		.47	.37	.02	.86	1.15
0520	8" thick, add	"	1600	.015		.96	.58	.03	1.57	2.03
0550	Exposed aggregate finish, add to above, minimum	B-24	1875	.013		.12	.56		.68	1.04
0600	Maximum		455	.053		.39	2.32		2.71	4.18
0950	Concrete tree grate, 5' square	B-6	25	.960	Ea.	400	40	14.60	454.60	520
0955	Tree well & cover, concrete, 3' square		25	.960		300	40	14.60	354.60	410
0960	Cast iron tree grate with frame, 2 piece, round, 5' diameter		25	.960		1,125	40	14.60	1,179.60	1,300
0980	Square, 5' side		25	.960		1,125	40	14.60	1,179.60	1,300
1700	Redwood, prefabricated, 4' x 4' sections	2 Carp	316	.051	S.F.	4.82	2.45		7.27	9.30
1750	Redwood planks, 1" thick, on sleepers	"	240	.067	"	4.82	3.23		8.05	10.60
2250	Stone dust, 4" thick	B-62	900	.027	S.Y.	3.56	1.11	.19	4.86	5.90
9000	Minimum labor/equipment charge	D-1	2	8	Job		335		335	550

32 06 10.20 Steps

		Crew	Daily Output	Labor-Hours	Unit	Material	2016 Bare Costs Labor	Equipment	Total	Total Incl O&P
0010	**STEPS**									
0011	Incl. excav., borrow & concrete base as required									
0100	Brick steps	B-24	35	.686	LF Riser	15.75	30		45.75	66.50
0200	Railroad ties	2 Clab	25	.640		3.59	24.50		28.09	43.50
0300	Bluestone treads, 12" x 2" or 12" x 1-1/2"	B-24	30	.800		33.50	35		68.50	94
0490	Minimum labor/equipment charge	D-1	2	8	Job		335		335	550

32 06 Schedules for Exterior Improvements

32 06 10 – Schedules for Bases, Ballasts, and Paving

32 06 10.20 Steps	Crew	Daily Output	Labor-Hours	Unit	Material	2016 Bare Costs		Total	Total Incl O&P
						Labor	Equipment		
0500	Concrete, cast in place, see Section 03 30 53.40									
0600	Precast concrete, see Section 03 41 23.50									

32 11 Base Courses

32 11 23 – Aggregate Base Courses

32 11 23.23 Base Course Drainage Layers

		Crew	Daily Output	Labor-Hours	Unit	Material	Labor	Equipment	Total	Total Incl O&P
0010	**BASE COURSE DRAINAGE LAYERS**									
0011	For roadways and large areas									
0050	Crushed 3/4" stone base, compacted, 3" deep	B-36C	5200	.008	S.Y.	2.35	.36	.80	3.51	4.05
0100	6" deep		5000	.008		4.71	.38	.83	5.92	6.70
0200	9" deep		4600	.009		7.05	.41	.91	8.37	9.45
0300	12" deep		4200	.010		9.45	.45	.99	10.89	12.15
0301	Crushed 1-1/2" stone base, compacted to 4" deep	B-36B	6000	.011		4.04	.48	.80	5.32	6.10
0302	6" deep		5400	.012		6.05	.54	.89	7.48	8.50
0303	8" deep		4500	.014		8.10	.65	1.07	9.82	11.10
0304	12" deep		3800	.017		12.15	.76	1.26	14.17	15.95
0310	Minimum labor/equipment charge	B-36	1	40	Job		1,750	1,675	3,425	4,650
0350	Bank run gravel, spread and compacted									
0370	6" deep	B-32	6000	.005	S.Y.	3.96	.25	.39	4.60	5.20
0390	9" deep		4900	.007		5.95	.31	.48	6.74	7.60
0400	12" deep		4200	.008		7.95	.36	.56	8.87	9.90
6900	For small and irregular areas, add						50%	50%		
6990	Minimum labor/equipment charge	1 Clab	3	2.667	Job		101		101	165
7000	Prepare and roll sub-base, small areas to 2500 S.Y.	B-32A	1500	.016	S.Y.		.75	.94	1.69	2.22
8000	Large areas over 2500 S.Y.	"	3500	.007			.32	.40	.72	.95
8050	For roadways	B-32	4000	.008			.38	.59	.97	1.26
9000	Minimum labor/equipment charge	1 Clab	4	2	Job		76		76	124

32 12 Flexible Paving

32 12 16 – Asphalt Paving

32 12 16.13 Plant-Mix Asphalt Paving

		Crew	Daily Output	Labor-Hours	Unit	Material	Labor	Equipment	Total	Total Incl O&P
0010	**PLANT-MIX ASPHALT PAVING**									
0080	Binder course, 1-1/2" thick	B-25	7725	.011	S.Y.	5.60	.47	.36	6.43	7.30
0120	2" thick		6345	.014		7.45	.58	.44	8.47	9.65
0130	2-1/2" thick		5620	.016		9.30	.65	.50	10.45	11.85
0160	3" thick		4905	.018		11.15	.75	.57	12.47	14.15
0170	3-1/2" thick		4520	.019		13.05	.81	.62	14.48	16.35
0300	Wearing course, 1" thick	B-25B	10575	.009		3.70	.39	.29	4.38	5
0340	1-1/2" thick		7725	.012		6.20	.53	.39	7.12	8.15
0380	2" thick		6345	.015		8.35	.64	.48	9.47	10.75
0420	2-1/2" thick		5480	.018		10.30	.74	.55	11.59	13.10
0460	3" thick		4900	.020		12.25	.83	.62	13.70	15.50
0470	3-1/2" thick		4520	.021		14.40	.90	.67	15.97	18.05
0480	4" thick		4140	.023		16.45	.98	.73	18.16	20.50
0500	Open graded friction course	B-25C	5000	.010		2.16	.41	.49	3.06	3.58
0800	Alternate method of figuring paving costs									
0810	Binder course, 1-1/2" thick	B-25	630	.140	Ton	68.50	5.80	4.44	78.74	90
0811	2" thick		690	.128		68.50	5.30	4.05	77.85	88.50
0812	3" thick		800	.110		68.50	4.58	3.50	76.58	87

For customer support on your Commercial Renovation Cost Data, call 877.742.4719.

535

32 12 Flexible Paving

32 12 16 – Asphalt Paving

32 12 16.13 Plant-Mix Asphalt Paving

		Crew	Daily Output	Labor-Hours	Unit	Material	2016 Bare Costs Labor	Equipment	Total	Total Incl O&P
0813	4" thick	B-25	900	.098	Ton	68.50	4.08	3.11	75.69	85.50
0850	Wearing course, 1" thick	B-25B	575	.167		75	7.10	5.30	87.40	100
0851	1-1/2" thick		630	.152		75	6.45	4.82	86.27	98.50
0852	2" thick		690	.139		75	5.90	4.40	85.30	97
0853	2-1/2" thick		765	.125		75	-5.35	3.97	84.32	95.50
0854	3" thick		800	.120		75	5.10	3.80	83.90	95
1000	Pavement replacement over trench, 2" thick	B-37	90	.533	S.Y.	7.70	21.50	1.81	31.01	45
1050	4" thick		70	.686		15.20	27.50	2.32	45.02	64
1080	6" thick		55	.873		24	35	2.96	61.96	87
3000	Prime coat, emulsion, .30 gal. per S.Y., 1000 S.Y.	B-45	2500	.006		1.80	.30	.37	2.47	2.86
3100	Tack coat, emulsion, .10 gal. per S.Y., 1000 S.Y.	"	2500	.006		.60	.30	.37	1.27	1.54

32 12 16.14 Asphaltic Concrete Paving

		Crew	Daily Output	Labor-Hours	Unit	Material	2016 Bare Costs Labor	Equipment	Total	Total Incl O&P
0011	**ASPHALTIC CONCRETE PAVING**, parking lots & driveways									
0015	No asphalt hauling included									
0020	6" stone base, 2" binder course, 1" topping	B-25C	9000	.005	S.F.	1.83	.23	.27	2.33	2.68
0025	2" binder course, 2" topping		9000	.005		2.28	.23	.27	2.78	3.18
0030	3" binder course, 2" topping		9000	.005		2.70	.23	.27	3.20	3.64
0035	4" binder course, 2" topping		9000	.005		3.11	.23	.27	3.61	4.09
0040	1.5" binder course, 1" topping		9000	.005		1.62	.23	.27	2.12	2.45
0042	3" binder course, 1" topping		9000	.005		2.24	.23	.27	2.74	3.14
0045	3" binder course, 3" topping		9000	.005		3.15	.23	.27	3.65	4.14
0050	4" binder course, 3" topping		9000	.005		3.57	.23	.27	4.07	4.59
0055	4" binder course, 4" topping		9000	.005		4.02	.23	.27	4.52	5.10
0300	Binder course, 1-1/2" thick		35000	.001		.62	.06	.07	.75	.86
0400	2" thick		25000	.002		.81	.08	.10	.99	1.13
0500	3" thick		15000	.003		1.25	.14	.16	1.55	1.77
0600	4" thick		10800	.004		1.63	.19	.23	2.05	2.35
0800	Sand finish course, 3/4" thick		41000	.001		.33	.05	.06	.44	.51
0900	1" thick		34000	.001		.40	.06	.07	.53	.62
1000	Fill pot holes, hot mix, 2" thick	B-16	4200	.008		.85	.30	.16	1.31	1.61
1100	4" thick		3500	.009		1.25	.36	.20	1.81	2.18
1120	6" thick		3100	.010		1.67	.41	.22	2.30	2.76
1140	Cold patch, 2" thick	B-51	3000	.016		.97	.62	.08	1.67	2.17
1160	4" thick		2700	.018		1.85	.69	.09	2.63	3.26
1180	6" thick		1900	.025		2.87	.98	.13	3.98	4.90
3000	Prime coat, emulsion, .30 gal. per S.Y., 1000 S.Y.	B-45	2500	.006	S.Y.	1.80	.30	.37	2.47	2.86
3100	Tack coat, emulsion, .10 gal. per S.Y., 1000 S.Y.	"	2500	.006	"	.60	.30	.37	1.27	1.54

32 13 Rigid Paving

32 13 13 – Concrete Paving

32 13 13.23 Concrete Paving Surface Treatment

		Crew	Daily Output	Labor-Hours	Unit	Material	2016 Bare Costs Labor	Equipment	Total	Total Incl O&P
0010	**CONCRETE PAVING SURFACE TREATMENT**									
0015	Including joints, finishing and curing									
0020	Fixed form, 12' pass, unreinforced, 6" thick	B-26	3000	.029	S.Y.	23	1.25	1.17	25.42	29
0100	8" thick		2750	.032		32	1.36	1.28	34.64	38.50
0400	12" thick		1800	.049		45.50	2.08	1.95	49.53	55.50
0410	Conc. pavement, w/jt.,fnsh.&curing,fix form,24' pass,unreinforced,6"T		6000	.015		22	.62	.59	23.21	25.50
0430	8" thick		5500	.016		30	.68	.64	31.32	35
0460	12" thick		3600	.024		43.50	1.04	.97	45.51	51
0520	Welded wire fabric, sheets for rigid paving 2.33 lb./S.Y.	2 Rodm	389	.041		1.22	2.18		3.40	4.87
0530	Reinforcing steel for rigid paving 12 lb./S.Y.		666	.024		6.05	1.27		7.32	8.70

32 13 Rigid Paving

32 13 13 - Concrete Paving

32 13 13.23 Concrete Paving Surface Treatment

		Crew	Daily Output	Labor-Hours	Unit	Material	2016 Bare Costs Labor	2016 Bare Costs Equipment	Total	Total Incl O&P
0540	Reinforcing steel for rigid paving 18 lb./S.Y.	2 Rodm	444	.036	S.Y.	9.10	1.91		11.01	13.10
0620	Slip form, 12' pass, unreinforced, 6" thick	B-26A	5600	.016		22.50	.67	.66	23.83	26.50
0624	8" thick		5300	.017		30.50	.71	.69	31.90	35.50
0630	12" thick		3470	.025		44	1.08	1.06	46.14	51.50
0632	15" thick		2890	.030		55.50	1.30	1.27	58.07	65
0640	Slip form, 24' pass, unreinforced, 6" thick		11200	.008		22	.33	.33	22.66	25
0644	8" thick		10600	.008		29.50	.35	.35	30.20	33
0650	12" thick		6940	.013		43	.54	.53	44.07	48.50
0652	15" thick		5780	.015		54	.65	.64	55.29	61.50
0700	Finishing, broom finish small areas	2 Cefi	120	.133			6.10		6.10	9.65
1000	Curing, with sprayed membrane by hand	2 Clab	1500	.011		1.14	.40		1.54	1.92

32 14 Unit Paving

32 14 13 - Precast Concrete Unit Paving

32 14 13.18 Precast Concrete Plantable Pavers

		Crew	Daily Output	Labor-Hours	Unit	Material	2016 Bare Costs Labor	2016 Bare Costs Equipment	Total	Total Incl O&P
0010	**PRECAST CONCRETE PLANTABLE PAVERS** (50% grass)									
0300	3/4" crushed stone base for plantable pavers, 6 inch depth	B-62	1000	.024	S.Y.	4.71	1	.17	5.88	7
0400	8 inch depth		900	.027		6.30	1.11	.19	7.60	8.95
0500	10 inch depth		800	.030		7.85	1.25	.22	9.32	10.90
0600	12 inch depth		700	.034		9.45	1.43	.25	11.13	12.95
0700	Hydro seeding plantable pavers	B-81A	20	.800	M.S.F.	10.65	32	22	64.65	87.50
0800	Apply fertilizer and seed to plantable pavers	1 Clab	8	1	"	50.50	38		88.50	118

32 14 16 - Brick Unit Paving

32 14 16.10 Brick Paving

		Crew	Daily Output	Labor-Hours	Unit	Material	2016 Bare Costs Labor	2016 Bare Costs Equipment	Total	Total Incl O&P
0010	**BRICK PAVING**									
0012	4" x 8" x 1-1/2", without joints (4.5 brick/S.F.)	D-1	110	.145	S.F.	2.24	6.15		8.39	12.40
0100	Grouted, 3/8" joint (3.9 brick/S.F.)		90	.178		2.02	7.50		9.52	14.40
0200	4" x 8" x 2-1/4", without joints (4.5 bricks/S.F.)		110	.145		2.33	6.15		8.48	12.50
0300	Grouted, 3/8" joint (3.9 brick/S.F.)		90	.178		2.02	7.50		9.52	14.40
0455	Pervious brick paving, 4" x 8" x 3-1/4", without joints (4.5 bricks/S.F.)		110	.145		3.49	6.15		9.64	13.80
0500	Bedding, asphalt, 3/4" thick	B-25	5130	.017		.69	.71	.55	1.95	2.51
0540	Course washed sand bed, 1" thick	B-18	5000	.005		.35	.19	.01	.55	.70
0580	Mortar, 1" thick	D-1	300	.053		.80	2.25		3.05	4.54
0620	2" thick		200	.080		1.60	3.37		4.97	7.25
1500	Brick on 1" thick sand bed laid flat, 4.5 per S.F.		100	.160		2.85	6.75		9.60	14.10
2000	Brick pavers, laid on edge, 7.2 per S.F.		70	.229		3.93	9.65		13.58	20
2500	For 4" thick concrete bed and joints, add		595	.027		1.26	1.13		2.39	3.23
2800	For steam cleaning, add	A-1H	950	.008		.09	.32	.08	.49	.71
9000	Minimum labor/equipment charge	1 Bric	2	4	Job		185		185	300

32 14 23 - Asphalt Unit Paving

32 14 23.10 Asphalt Blocks

		Crew	Daily Output	Labor-Hours	Unit	Material	2016 Bare Costs Labor	2016 Bare Costs Equipment	Total	Total Incl O&P
0010	**ASPHALT BLOCKS**									
0020	Rectangular, 6" x 12" x 1-1/4", w/bed & neopr. adhesive	D-1	135	.119	S.F.	5.60	4.99		10.59	14.25
0100	3" thick		130	.123		7.80	5.20		13	17.05
0300	Hexagonal tile, 8" wide, 1-1/4" thick		135	.119		5.60	4.99		10.59	14.25
0400	2" thick		130	.123		7.80	5.20		13	17.05
0500	Square, 8" x 8", 1-1/4" thick		135	.119		5.60	4.99		10.59	14.25
0600	2" thick		130	.123		7.80	5.20		13	17.05
9000	Minimum labor/equipment charge	1 Bric	2	4	Job		185		185	300

For customer support on your Commercial Renovation Cost Data, call 877.742.4719.

537

32 14 Unit Paving

32 14 40 – Stone Paving

32 14 40.10 Stone Pavers	Crew	Daily Output	Labor-Hours	Unit	Material	2016 Bare Costs Labor	Equipment	Total	Total Incl O&P
0010 **STONE PAVERS**									
1300 Slate, natural cleft, irregular, 3/4" thick	D-1	92	.174	S.F.	9	7.35		16.35	22
1350 Random rectangular, gauged, 1/2" thick		105	.152		19.50	6.40		25.90	32
1400 Random rectangular, butt joint, gauged, 1/4" thick	↓	150	.107		21	4.49		25.49	30.50
1450 For sand rubbed finish, add					9			9	9.90
1550 Granite blocks, 3-1/2" x 3-1/2" x 3-1/2"	D-1	92	.174		15	7.35		22.35	28.50
1600 4" to 12" long, 3" to 5" wide, 3" to 5" thick	"	98	.163	↓	12.50	6.90		19.40	25

32 16 Curbs, Gutters, Sidewalks, and Driveways

32 16 13 – Curbs and Gutters

32 16 13.13 Cast-in-Place Concrete Curbs and Gutters

	Crew	Daily Output	Labor-Hours	Unit	Material	2016 Bare Costs Labor	Equipment	Total	Total Incl O&P
0010 **CAST-IN-PLACE CONCRETE CURBS AND GUTTERS**									
0290 Forms only, no concrete									
0300 Concrete, wood forms, 6" x 18", straight	C-2	500	.096	L.F.	3.13	4.51		7.64	10.85
0400 6" x 18", radius	"	200	.240		3.25	11.30		14.55	22
0404 Concrete, wood forms, 6" x 18", straight & concrete	C-2A	500	.096		6.15	4.47		10.62	14
0406 6" x 18", radius	"	200	.240		6.25	11.15		17.40	25
0415 Machine formed, 6" x 18", straight	B-69A	2000	.024		3.53	1	.55	5.08	6.10
0416 6" x 18", radius	"	900	.053	↓	3.56	2.23	1.22	7.01	8.85

32 16 13.23 Precast Concrete Curbs and Gutters

	Crew	Daily Output	Labor-Hours	Unit	Material	2016 Bare Costs Labor	Equipment	Total	Total Incl O&P
0010 **PRECAST CONCRETE CURBS AND GUTTERS**									
0550 Precast, 6" x 18", straight	B-29	700	.080	L.F.	9	3.31	1.26	13.57	16.65
0600 6" x 18", radius	"	325	.172	"	10	7.15	2.71	19.86	25.50

32 16 13.33 Asphalt Curbs

	Crew	Daily Output	Labor-Hours	Unit	Material	2016 Bare Costs Labor	Equipment	Total	Total Incl O&P
0010 **ASPHALT CURBS**									
0012 Curbs, asphaltic, machine formed, 8" wide, 6" high, 40 L.F./ton	B-27	1000	.032	L.F.	1.73	1.23	.30	3.26	4.24
0100 8" wide, 8" high, 30 L.F. per ton		900	.036		2.30	1.37	.33	4	5.15
0150 Asphaltic berm, 12" W, 3"-6" H, 35 L.F./ton, before pavement	↓	700	.046		.04	1.76	.42	2.22	3.37
0200 12" W, 1-1/2" to 4" H, 60 L.F. per ton, laid with pavement	B-2	1050	.038	↓	.02	1.46		1.48	2.42

32 16 13.43 Stone Curbs

	Crew	Daily Output	Labor-Hours	Unit	Material	2016 Bare Costs Labor	Equipment	Total	Total Incl O&P
0010 **STONE CURBS**									
1000 Granite, split face, straight, 5" x 16"	D-13	275	.175	L.F.	14	7.90	1.74	23.64	30
1100 6" x 18"	"	250	.192		18.40	8.70	1.92	29.02	36
1300 Radius curbing, 6" x 18", over 10' radius	B-29	260	.215		22.50	8.90	3.39	34.79	43
1400 Corners, 2' radius	"	80	.700	Ea.	75.50	29	11.05	115.55	142
1600 Edging, 4-1/2" x 12", straight	D-13	300	.160	L.F.	7	7.25	1.60	15.85	21
1800 Curb inlets, (guttermouth) straight	B-29	41	1.366	Ea.	168	56.50	21.50	246	300
2000 Indian granite (Belgian block)									
2100 Jumbo, 10-1/2" x 7-1/2" x 4", grey	D-1	150	.107	L.F.	7.90	4.49		12.39	16
2150 Pink		150	.107		8.60	4.49		13.09	16.75
2200 Regular, 9" x 4-1/2" x 4-1/2", grey		160	.100		4.67	4.21		8.88	12
2250 Pink		160	.100		6.15	4.21		10.36	13.60
2300 Cubes, 4" x 4" x 4", grey		175	.091		3.63	3.85		7.48	10.25
2350 Pink		175	.091		3.81	3.85		7.66	10.45
2400 6" x 6" x 6", pink	↓	155	.103	↓	12.60	4.35		16.95	21
2500 Alternate pricing method for Indian granite									
2550 Jumbo, 10-1/2" x 7-1/2" x 4" (30 lb.), grey				Ton	450			450	495
2600 Pink					500			500	550
2650 Regular, 9" x 4-1/2" x 4-1/2" (20 lb.), grey					330			330	365
2700 Pink					430			430	475

538

32 16 Curbs, Gutters, Sidewalks, and Driveways

32 16 13 – Curbs and Gutters

32 16 13.43 Stone Curbs	Crew	Daily Output	Labor-Hours	Unit	Material	2016 Bare Costs Labor	Equipment	Total	Total Incl O&P
2750	Cubes, 4" x 4" x 4" (5 lb.), grey				Ton	440			440	485
2800	Pink					490			490	540
2850	6" x 6" x 6" (25 lb.), pink					490			490	540
2900	For pallets, add				↓	22			22	24

32 17 Paving Specialties

32 17 13 – Parking Bumpers

32 17 13.13 Metal Parking Bumpers

		Crew	Daily Output	Labor-Hours	Unit	Material	Labor	Equipment	Total	Total Incl O&P
0010	**METAL PARKING BUMPERS**									
0015	Bumper rails for garages, 12 ga. rail, 6" wide, with steel									
0020	posts 12'-6" O.C., minimum	E-4	190	.168	L.F.	19.20	9.05	.77	29.02	38
0030	Average		165	.194		24	10.40	.89	35.29	46
0100	Maximum	↓	140	.229	↓	29	12.25	1.05	42.30	54
1300	Pipe bollards, conc. filled/paint, 8' L x 4' D hole, 6" diam.	B-6	20	1.200	Ea.	610	50	18.30	678.30	770
1400	8" diam.		15	1.600		690	66.50	24.50	781	895
1500	12" diam.	↓	12	2		975	83.50	30.50	1,089	1,250
8900	Security bollards, SS, lighted, hyd., incl. controls, group of 3	L-7	.06	509		48,400	23,600		72,000	91,000
8910	Group of 5	"	.04	682	↓	65,000	31,700		96,700	123,000
9000	Minimum labor/equipment charge	E-4	2	16	Job		860	73.50	933.50	1,600

32 17 13.16 Plastic Parking Bumpers

			Crew	Daily Output	Labor-Hours	Unit	Material	Labor	Equipment	Total	Total Incl O&P
0010	**PLASTIC PARKING BUMPERS**										
1600	Bollards, recycled plastic, 5" x 5" x 5' L	G	B-6	20	1.200	Ea.	90	50	18.30	158.30	200
1610	Wheel stops, recycled plastic, yellow, 4" x 6" x 6' L	G		20	1.200		65	50	18.30	133.30	172
1620	Speed bumps, recycled plastic, yellow, 3" H x 10" W x 6' L	G		20	1.200		174	50	18.30	242.30	293
1630	3" H x 10" W x 9' L	G	↓	20	1.200	↓	169	50	18.30	237.30	287

32 17 13.19 Precast Concrete Parking Bumpers

		Crew	Daily Output	Labor-Hours	Unit	Material	Labor	Equipment	Total	Total Incl O&P
0010	**PRECAST CONCRETE PARKING BUMPERS**									
1000	Wheel stops, precast concrete incl. dowels, 6" x 10" x 6'-0"	B-2	120	.333	Ea.	40	12.75		52.75	65
1100	8" x 13" x 6'-0"	"	120	.333		46	12.75		58.75	72
1540	Bollards, precast concrete, free standing, round, 15" Dia x 36" H	B-6	14	1.714		315	71.50	26	412.50	490
1550	16" Dia x 30" H		14	1.714		315	71.50	26	412.50	490
1560	18" Dia x 34" H		14	1.714		410	71.50	26	507.50	595
1570	18" Dia x 72" H		14	1.714		485	71.50	26	582.50	680
1572	26" Dia x 18" H		12	2		370	83.50	30.50	484	575
1574	26" Dia x 30" H		12	2		610	83.50	30.50	724	845
1578	26" Dia x 50" H		12	2		860	83.50	30.50	974	1,125
1580	Square, 22" x 32" H		12	2		525	83.50	30.50	639	750
1585	22" x 41" H		12	2		540	83.50	30.50	654	765
1590	Hexagon, 26" x 41" H	↓	10	2.400	↓	555	100	36.50	691.50	810

32 17 13.26 Wood Parking Bumpers

		Crew	Daily Output	Labor-Hours	Unit	Material	Labor	Equipment	Total	Total Incl O&P
0010	**WOOD PARKING BUMPERS**									
0020	Parking barriers, timber w/saddles, treated type									
0100	4" x 4" for cars	B-2	520	.077	L.F.	2.95	2.95		5.90	8.05
0200	6" x 6" for trucks	"	520	.077	"	6.15	2.95		9.10	11.60

32 17 23 – Pavement Markings

32 17 23.13 Painted Pavement Markings

		Crew	Daily Output	Labor-Hours	Unit	Material	Labor	Equipment	Total	Total Incl O&P
0010	**PAINTED PAVEMENT MARKINGS**									
0020	Acrylic waterborne, white or yellow, 4" wide, less than 3000 L.F.	B-78	20000	.002	L.F.	.16	.09	.03	.28	.35
0200	6" wide, less than 3000 L.F.		11000	.004		.24	.17	.06	.47	.60
0500	8" wide, less than 3000 L.F.		10000	.005		.31	.19	.06	.56	.72

32 17 Paving Specialties

32 17 23 – Pavement Markings

32 17 23.13 Painted Pavement Markings

		Crew	Daily Output	Labor-Hours	Unit	Material	2016 Bare Costs Labor	Equipment	Total	Total Incl O&P
0600	12" wide, less than 3000 L.F.	B-78	4000	.012	L.F.	.47	.47	.15	1.09	1.45
0620	Arrows or gore lines		2300	.021	S.F.	.22	.81	.26	1.29	1.85
0640	Temporary paint, white or yellow, less than 3000 L.F.	↓	15000	.003	L.F.	.08	.12	.04	.24	.33
0660	Removal	1 Clab	300	.027			1.01		1.01	1.65
0680	Temporary tape	2 Clab	1500	.011		.49	.40		.89	1.20
0710	Thermoplastic, white or yellow, 4" wide, less than 6000 L.F.	B-79	15000	.003		.33	.10	.10	.53	.64
0730	6" wide, less than 6000 L.F.		14000	.003		.50	.11	.11	.72	.84
0740	8" wide, less than 6000 L.F.		12000	.003		.66	.13	.12	.91	1.08
0750	12" wide, less than 6000 L.F.		6000	.007	↓	.97	.26	.25	1.48	1.77
0760	Arrows		660	.061	S.F.	.63	2.37	2.26	5.26	7.05
0770	Gore lines		2500	.016		.63	.63	.60	1.86	2.37
0780	Letters	↓	660	.061	↓	.63	2.37	2.26	5.26	7.05

32 17 23.14 Pavement Parking Markings

		Crew	Daily Output	Labor-Hours	Unit	Material	2016 Bare Costs Labor	Equipment	Total	Total Incl O&P
0010	**PAVEMENT PARKING MARKINGS**									
0790	Layout of pavement marking	A-2	25000	.001	L.F.		.04	.01	.05	.07
0800	Lines on pvmt., parking stall, paint, white, 4" wide	B-78B	400	.045	Stall	4.62	1.76	.95	7.33	9
0825	Parking stall, small quantities	2 Pord	80	.200		9.25	8.05		17.30	23
0830	Lines on pvmt., parking stall, thermoplastic, white, 4" wide	B-79	300	.133	↓	14	5.20	4.97	24.17	29.50
1000	Street letters and numbers	B-78B	1600	.011	S.F.	.69	.44	.24	1.37	1.74

32 31 Fences and Gates

32 31 13 – Chain Link Fences and Gates

32 31 13.20 Fence, Chain Link Industrial

		Crew	Daily Output	Labor-Hours	Unit	Material	2016 Bare Costs Labor	Equipment	Total	Total Incl O&P
0010	**FENCE, CHAIN LINK INDUSTRIAL**									
0011	Schedule 40, including concrete									
0020	3 strands barb wire, 2" post @ 10' O.C., set in concrete, 6' H									
0200	9 ga. wire, galv. steel, in concrete	B-80C	240	.100	L.F.	19.55	3.93	1.07	24.55	29
0248	Fence, add for vinyl coated fabric				S.F.	.67			.67	.74
0300	Aluminized steel	B-80C	240	.100	L.F.	20	3.93	1.07	25	29.50
0500	6 ga. wire, galv. steel		240	.100		23	3.93	1.07	28	32.50
0600	Aluminized steel		240	.100		30.50	3.93	1.07	35.50	41
0800	6 ga. wire, 6' high but omit barbed wire, galv. steel		250	.096		19.95	3.77	1.03	24.75	29.50
0900	Aluminized steel, in concrete		250	.096		24	3.77	1.03	28.80	34
0920	8' H, 6 ga. wire, 2-1/2" line post, galv. steel, in concrete		180	.133		31.50	5.25	1.43	38.18	45
0940	Aluminized steel, in concrete		180	.133	↓	38.50	5.25	1.43	45.18	52.50
1100	Add for corner posts, 3" diam., galv. steel, in concrete		40	.600	Ea.	88.50	23.50	6.40	118.40	143
1200	Aluminized steel, in concrete		40	.600		88.50	23.50	6.40	118.40	143
1300	Add for braces, galv. steel		80	.300		35.50	11.80	3.21	50.51	61.50
1350	Aluminized steel		80	.300		46	11.80	3.21	61.01	73
1400	Gate for 6' high fence, 1-5/8" frame, 3' wide, galv. steel		10	2.400		207	94	25.50	326.50	410
1500	Aluminized steel, in concrete	↓	10	2.400	↓	209	94	25.50	328.50	410
2000	5'-0" high fence, 9 ga., no barbed wire, 2" line post, in concrete									
2010	10' O.C., 1-5/8" top rail, in concrete									
2100	Galvanized steel, in concrete	B-80C	300	.080	L.F.	19.20	3.14	.86	23.20	27
2200	Aluminized steel, in concrete		300	.080	"	17.45	3.14	.86	21.45	25
2400	Gate, 4' wide, 5' high, 2" frame, galv. steel, in concrete		10	2.400	Ea.	201	94	25.50	320.50	405
2500	Aluminized steel, in concrete		10	2.400	"	181	94	25.50	300.50	380
3100	Overhead slide gate, chain link, 6' high, to 18' wide, in concrete	↓	38	.632	L.F.	97	25	6.75	128.75	155
3110	Cantilever type, in concrete	B-80	48	.667		129	28	15.15	172.15	204
3120	8' high, in concrete		24	1.333		155	56.50	30.50	242	296
3130	10' high, in concrete	↓	18	1.778	↓	190	75	40.50	305.50	375

540

32 31 Fences and Gates

32 31 13 – Chain Link Fences and Gates

32 31 13.20 Fence, Chain Link Industrial

	Crew	Daily Output	Labor-Hours	Unit	Material	2016 Bare Costs Labor	Equipment	Total	Total Incl O&P
9000 Minimum labor/equipment charge	B-80	2	16	Job		675	365	1,040	1,500

32 31 13.25 Fence, Chain Link Residential

	Crew	Daily Output	Labor-Hours	Unit	Material	2016 Bare Costs Labor	Equipment	Total	Total Incl O&P
0010 **FENCE, CHAIN LINK RESIDENTIAL**									
0011 Schedule 20, 11 ga. wire, 1-5/8" post									
3300 Residential, 11 ga. wire, 1-5/8" line post @ 10' O.C.									
3310 1-3/8" top rail									
3350 4' high, galvanized steel	B-80C	475	.051	L.F.	12.65	1.98	.54	15.17	17.70
3400 Aluminized		475	.051	"	16.05	1.98	.54	18.57	21.50
3600 Gate, 3' wide, 1-3/8" frame, galv. steel		10	2.400	Ea.	102	94	25.50	221.50	293
3700 Aluminized		10	2.400	"	112	94	25.50	231.50	305
3900 3' high, galvanized steel		620	.039	L.F.	94	1.52	.41	95.93	106
4000 Aluminized		620	.039	"	94.50	1.52	.41	96.43	107
4200 Gate, 3' wide, 1-3/8" frame, galv. steel		12	2	Ea.	96.50	78.50	21.50	196.50	258
4300 Aluminized		12	2	"	110	78.50	21.50	210	273
9000 Minimum labor/equipment charge	B-1	2	12	Job		465		465	755

32 31 13.26 Tennis Court Fences and Gates

	Crew	Daily Output	Labor-Hours	Unit	Material	2016 Bare Costs Labor	Equipment	Total	Total Incl O&P
0010 **TENNIS COURT FENCES AND GATES**									
3150 Tennis courts, 11 ga. wire, 1-3/4" mesh, 2-1/2" line posts									
3170 1-5/8" top rail, 3" corner and gate posts									
3190 10' high, galvanized steel	B-80	190	.168	L.F.	26.50	7.10	3.83	37.43	44.50
3210 Vinyl covered 9 ga. wire	"	190	.168	"	32	7.10	3.83	42.93	51

32 31 13.40 Fence, Fabric and Accessories

	Crew	Daily Output	Labor-Hours	Unit	Material	2016 Bare Costs Labor	Equipment	Total	Total Incl O&P
0010 **FENCE, FABRIC & ACCESSORIES**									
8000 Components only, fabric, galvanized, 4' high	B-80C	585	.041	L.F.	3.51	1.61	.44	5.56	6.95
8040 6' high		430	.056	"	5.30	2.19	.60	8.09	10
8200 Posts, line post, 4' high		34	.706	Ea.	13	27.50	7.55	48.05	67.50
8240 6' high		26	.923	"	16.30	36	9.85	62.15	88
8400 Top rails, 1-3/8" O.D.		1000	.024	L.F.	1.48	.94	.26	2.68	3.44
8440 1-5/8" O.D.		1000	.024	"	2.34	.94	.26	3.54	4.38

32 31 19 – Decorative Metal Fences and Gates

32 31 19.10 Decorative Fence

	Crew	Daily Output	Labor-Hours	Unit	Material	2016 Bare Costs Labor	Equipment	Total	Total Incl O&P
0010 **DECORATIVE FENCE**									
5300 Tubular picket, steel, 6' sections, 1-9/16" posts, 4' high	B-80C	300	.080	L.F.	30.50	3.14	.86	34.50	39.50
5400 2" posts, 5' high		240	.100		34.50	3.93	1.07	39.50	45.50
5600 2" posts, 6' high		200	.120		41.50	4.71	1.28	47.49	54.50
5700 Staggered picket 1-9/16" posts, 4' high		300	.080		30.50	3.14	.86	34.50	39.50
5800 2" posts, 5' high		240	.100		34.50	3.93	1.07	39.50	45.50
5900 2" posts, 6' high		200	.120		41.50	4.71	1.28	47.49	54.50
6200 Gates, 4' high, 3' wide	B-1	10	2.400	Ea.	276	92.50		368.50	455
6300 5' high, 3' wide		10	2.400		340	92.50		432.50	525
6400 6' high, 3' wide		10	2.400		405	92.50		497.50	595
6500 4' wide		10	2.400		415	92.50		507.50	605

32 31 23 – Plastic Fences and Gates

32 31 23.20 Fence, Recycled Plastic

		Crew	Daily Output	Labor-Hours	Unit	Material	2016 Bare Costs Labor	Equipment	Total	Total Incl O&P
0010 **FENCE, RECYCLED PLASTIC**										
9015 Fence rail, made from recycled plastic, various colors, 2 rail	G	B-1	150	.160	L.F.	10.20	6.15		16.35	21.50
9018 3 rail	G		150	.160		13	6.15		19.15	24.50
9020 4 rail	G		150	.160		15.75	6.15		21.90	27.50
9030 Fence pole, made from recycled plastic, various colors, 7'	G		96	.250	Ea.	50	9.65		59.65	71
9040 Stockade fence, made from recycled plastic, various colors, 4' high	G	B-80C	160	.150	L.F.	35	5.90	1.60	42.50	50
9050 6' high	G		160	.150	"	37	5.90	1.60	44.50	52.50

For customer support on your Commercial Renovation Cost Data, call 877.742.4719.

541

32 31 Fences and Gates

32 31 23 – Plastic Fences and Gates

32 31 23.20 Fence, Recycled Plastic

			Crew	Daily Output	Labor-Hours	Unit	Material	2016 Bare Costs Labor	Equipment	Total	Total Incl O&P
9060	6' pole	G	B-80C	96	.250	Ea.	39.50	9.80	2.67	51.97	62.50
9070	9' pole	G		96	.250	"	45	9.80	2.67	57.47	68.50
9080	Picket fence, made from recycled plastic, various colors, 3' high	G		160	.150	L.F.	23.50	5.90	1.60	31	37
9090	4' high	G		160	.150	"	34	5.90	1.60	41.50	49
9100	3' high gate	G		8	3	Ea.	23	118	32	173	252
9110	4' high gate	G		8	3		30	118	32	180	260
9120	5' high pole	G		96	.250		37.50	9.80	2.67	49.97	60
9130	6' high pole	G		96	.250		43.50	9.80	2.67	55.97	67
9140	Pole cap only	G					5.35			5.35	5.90
9150	Keeper pins only	G					.42			.42	.46

32 31 26 – Wire Fences and Gates

32 31 26.10 Fences, Misc. Metal

		Crew	Daily Output	Labor-Hours	Unit	Material	Labor	Equipment	Total	Total Incl O&P
0010	**FENCES, MISC. METAL**									
0012	Chicken wire, posts @ 4', 1" mesh, 4' high	B-80C	410	.059	L.F.	3.37	2.30	.63	6.30	8.15
0100	2" mesh, 6' high		350	.069		3.78	2.69	.73	7.20	9.35
0200	Galv. steel, 12 ga., 2" x 4" mesh, posts 5' O.C., 3' high		300	.080		2.75	3.14	.86	6.75	9.05
0300	5' high		300	.080		3.35	3.14	.86	7.35	9.75
0400	14 ga., 1" x 2" mesh, 3' high		300	.080		3.39	3.14	.86	7.39	9.75
0500	5' high		300	.080		4.52	3.14	.86	8.52	11
1000	Kennel fencing, 1-1/2" mesh, 6' long, 3'-6" wide, 6'-2" high	2 Clab	4	4	Ea.	505	152		657	805
1050	12' long		4	4		710	152		862	1,025
1200	Top covers, 1-1/2" mesh, 6' long		15	1.067		135	40.50		175.50	214
1250	12' long		12	1.333		189	50.50		239.50	291
4500	Security fence, prison grade, set in concrete, 12' high	B-80	25	1.280	L.F.	61	54	29	144	186
4600	16' high	"	20	1.600	"	78.50	67.50	36.50	182.50	236

32 31 29 – Wood Fences and Gates

32 31 29.20 Fence, Wood Rail

		Crew	Daily Output	Labor-Hours	Unit	Material	Labor	Equipment	Total	Total Incl O&P
0010	**FENCE, WOOD RAIL**									
0012	Picket, No. 2 cedar, Gothic, 2 rail, 3' high	B-1	160	.150	L.F.	8	5.80		13.80	18.25
0050	Gate, 3'-6" wide	B-80C	9	2.667	Ea.	77	105	28.50	210.50	287
0600	Open rail, rustic, No. 1 cedar, 2 rail, 3' high		160	.150	L.F.	6	5.90	1.60	13.50	17.90
0650	Gate, 3' wide		9	2.667	Ea.	82	105	28.50	215.50	292
1200	Stockade, No. 2 cedar, treated wood rails, 6' high		160	.150	L.F.	9	5.90	1.60	16.50	21
1250	Gate, 3' wide		9	2.667	Ea.	96	105	28.50	229.50	310
3300	Board, shadow box, 1" x 6", treated pine, 6' high		160	.150	L.F.	12.50	5.90	1.60	20	25
3400	No. 1 cedar, 6' high		150	.160		25	6.30	1.71	33.01	39
3900	Basket weave, No. 1 cedar, 6' high		160	.150		34	5.90	1.60	41.50	48.50
3950	Gate, 3'-6" wide	B-1	8	3	Ea.	232	116		348	445
4000	Treated pine, 6' high		150	.160	L.F.	16.15	6.15		22.30	28
4200	Gate, 3'-6" wide		9	2.667	Ea.	176	103		279	360
8000	Posts only, 4' high		40	.600		12	23		35	51
8040	6' high		30	.800		17.60	31		48.60	70
9000	Minimum labor/equipment charge	1 Clab	2	4	Job		152		152	248

32 32 Retaining Walls

32 32 29 – Timber Retaining Walls

32 32 29.10 Landscape Timber Retaining Walls

		Crew	Daily Output	Labor-Hours	Unit	Material	2016 Bare Costs Labor	Equipment	Total	Total Incl O&P
0010	**LANDSCAPE TIMBER RETAINING WALLS**									
0100	Treated timbers, 6" x 6"	1 Clab	265	.030	L.F.	3	1.14		4.14	5.15
0110	6" x 8"	"	200	.040	"	3.75	1.52		5.27	6.60
0120	Drilling holes in timbers for fastening, 1/2"	1 Carp	450	.018	Inch		.86		.86	1.41
0130	5/8"	"	450	.018	"		.86		.86	1.41
0140	Reinforcing rods for fastening, 1/2"	1 Clab	312	.026	L.F.	.35	.97		1.32	1.98
0150	5/8"	"	312	.026	"	.55	.97		1.52	2.20
0160	Reinforcing fabric	2 Clab	2500	.006	S.Y.	1.95	.24		2.19	2.55
0170	Gravel backfill		28	.571	C.Y.	18.50	21.50		40	56
0180	Perforated pipe, 4" diameter with silt sock		1200	.013	L.F.	1.25	.51		1.76	2.21
0190	Galvanized 60d common nails	1 Clab	625	.013	Ea.	.17	.49		.66	.97
0200	20d common nails	"	3800	.002	"	.04	.08		.12	.17

32 32 53 – Stone Retaining Walls

32 32 53.10 Retaining Walls, Stone

		Crew	Daily Output	Labor-Hours	Unit	Material	2016 Bare Costs Labor	Equipment	Total	Total Incl O&P
0010	**RETAINING WALLS, STONE**									
0015	Including excavation, concrete footing and									
0020	stone 3' below grade. Price is exposed face area.									
0200	Decorative random stone, to 6' high, 1'-6" thick, dry set	D-1	35	.457	S.F.	58.50	19.25		77.75	96
0300	Mortar set		40	.400		60.50	16.85		77.35	94
0500	Cut stone, to 6' high, 1'-6" thick, dry set		35	.457		60.50	19.25		79.75	98.50
0600	Mortar set		40	.400		61.50	16.85		78.35	95
0800	Random stone, 6' to 10' high, 2' thick, dry set		45	.356		66.50	15		81.50	97.50
0900	Mortar set		50	.320		69	13.50		82.50	97.50
1100	Cut stone, 6' to 10' high, 2' thick, dry set		45	.356		67	15		82	98
1200	Mortar set		50	.320		69	13.50		82.50	98
9000	Minimum labor/equipment charge		2	8	Job		335		335	550

32 33 Site Furnishings

32 33 33 – Site Manufactured Planters

32 33 33.10 Planters

		Crew	Daily Output	Labor-Hours	Unit	Material	2016 Bare Costs Labor	Equipment	Total	Total Incl O&P
0010	**PLANTERS**									
0012	Concrete, sandblasted, precast, 48" diameter, 24" high	2 Clab	15	1.067	Ea.	585	40.50		625.50	710
0100	Fluted, precast, 7' diameter, 36" high		10	1.600		1,550	60.50		1,610.50	1,825
0300	Fiberglass, circular, 36" diameter, 24" high		15	1.067		655	40.50		695.50	790
0400	60" diameter, 24" high		10	1.600		1,125	60.50		1,185.50	1,325
9000	Minimum labor/equipment charge	1 Clab	2	4	Job		152		152	248

32 33 43 – Site Seating and Tables

32 33 43.13 Site Seating

		Crew	Daily Output	Labor-Hours	Unit	Material	2016 Bare Costs Labor	Equipment	Total	Total Incl O&P
0010	**SITE SEATING**									
0012	Seating, benches, park, precast conc., w/backs, wood rails, 4' long	2 Clab	5	3.200	Ea.	605	121		726	865
0100	8' long		4	4		1,025	152		1,177	1,375
0300	Fiberglass, without back, one piece, 4' long		10	1.600		610	60.50		670.50	775
0400	8' long		7	2.286		815	86.50		901.50	1,025
0500	Steel barstock pedestals w/backs, 2" x 3" wood rails, 4' long		10	1.600		1,150	60.50		1,210.50	1,375
0510	8' long		7	2.286		1,500	86.50		1,586.50	1,800
0515	Powder coated steel, 4" x 4" plastic slats, 6' L		8	2		510	76		586	685
0520	3" x 8" wood plank, 4' long		10	1.600		1,250	60.50		1,310.50	1,500
0530	8' long		7	2.286		1,300	86.50		1,386.50	1,575
0540	Backless, 4" x 4" wood plank, 4' square		10	1.600		1,000	60.50		1,060.50	1,200
0550	8' long		7	2.286		1,100	86.50		1,186.50	1,350

For customer support on your Commercial Renovation Cost Data, call 877.742.4719.

543

32 33 Site Furnishings

32 33 43 – Site Seating and Tables

32 33 43.13 Site Seating	Crew	Daily Output	Labor-Hours	Unit	Material	2016 Bare Costs Labor	Equipment	Total	Total Incl O&P
0560	Powder coated steel, with back and 2 anti-vagrant dividers, 6' long	2 Clab	8	2	Ea.	1,100	76		1,176	1,325
0600	Aluminum pedestals, with backs, aluminum slats, 8' long		8	2		500	76		576	675
0610	15' long		5	3.200		910	121		1,031	1,200
0620	Portable, aluminum slats, 8' long		8	2		480	76		556	655
0630	15' long		5	3.200		570	121		691	830
0800	Cast iron pedestals, back & arms, wood slats, 4' long		8	2		395	76		471	560
0820	8' long		5	3.200		1,200	121		1,321	1,525
0840	Backless, wood slats, 4' long		8	2		610	76		686	800
0860	8' long		5	3.200		1,150	121		1,271	1,475
1700	Steel frame, fir seat, 10' long		10	1.600		375	60.50		435.50	515
2000	Benches, park, with back, galv. stl. frame, 4" x 4" plastic slats, 6' L		7	2.286		500	86.50		586.50	690
9000	Minimum labor/equipment charge		2	8	Job		305		305	495

32 84 Planting Irrigation

32 84 23 – Underground Sprinklers

32 84 23.10 Sprinkler Irrigation System

		Crew	Daily Output	Labor-Hours	Unit	Material	2016 Bare Costs Labor	Equipment	Total	Total Incl O&P
0010	**SPRINKLER IRRIGATION SYSTEM**									
0011	For lawns									
0800	Residential system, custom, 1" supply	B-20	2000	.012	S.F.	.26	.51		.77	1.12
0900	1-1/2" supply	"	1800	.013	"	.49	.57		1.06	1.46

32 91 Planting Preparation

32 91 13 – Soil Preparation

32 91 13.23 Structural Soil Mixing

		Crew	Daily Output	Labor-Hours	Unit	Material	2016 Bare Costs Labor	Equipment	Total	Total Incl O&P
0010	**STRUCTURAL SOIL MIXING**									
0100	Rake topsoil, site material, harley rock rake, ideal	B-6	33	.727	M.S.F.		30.50	11.10	41.60	61
0200	Adverse	"	7	3.429			143	52	195	289
0300	Screened loam, york rake and finish, ideal	B-62	24	1			41.50	7.20	48.70	75
0400	Adverse	"	20	1.200			50	8.65	58.65	90
1000	Remove topsoil & stock pile on site, 75 HP dozer, 6" deep, 50' haul	B-10L	30	.400			18.70	15.70	34.40	47.50
1050	300' haul		6.10	1.967			92	77.50	169.50	232
1100	12" deep, 50' haul		15.50	.774			36	30.50	66.50	91
1150	300' haul		3.10	3.871			181	152	333	455
1200	200 HP dozer, 6" deep, 50' haul	B-10B	125	.096			4.48	11.20	15.68	19.45
1250	300' haul		30.70	.391			18.25	45.50	63.75	79
1300	12" deep, 50' haul		62	.194			9.05	22.50	31.55	39.50
1350	300' haul		15.40	.779			36.50	90.50	127	158
1400	Alternate method, 75 HP dozer, 50' haul	B-10L	860	.014	C.Y.		.65	.55	1.20	1.64
1450	300' haul	"	114	.105			4.92	4.13	9.05	12.40
1500	200 HP dozer, 50' haul	B-10B	2660	.005			.21	.53	.74	.92
1600	300' haul	"	570	.021			.98	2.45	3.43	4.27
1800	Rolling topsoil, hand push roller	1 Clab	3200	.003	S.F.		.09		.09	.15
1850	Tractor drawn roller	B-66	10666	.001	"		.04	.02	.06	.09
2000	Root raking and loading, residential, no boulders	B-6	53.30	.450	M.S.F.		18.75	6.85	25.60	38
2100	With boulders		32	.750			31	11.40	42.40	63
2200	Municipal, no boulders		200	.120			5	1.83	6.83	10.05
2300	With boulders		120	.200			8.35	3.05	11.40	16.80
2400	Large commercial, no boulders	B-10B	400	.030			1.40	3.49	4.89	6.10
2500	With boulders	"	240	.050			2.34	5.80	8.14	10.15

32 91 13 – Soil Preparation

32 91 13.23 Structural Soil Mixing

		Crew	Daily Output	Labor-Hours	Unit	Material	2016 Bare Costs Labor	2016 Bare Costs Equipment	Total	Total Incl O&P
3000	Scarify subsoil, residential, skid steer loader w/scarifiers, 50 HP	B-66	32	.250	M.S.F.		12.30	8.25	20.55	28.50
3050	Municipal, skid steer loader w/scarifiers, 50 HP	"	120	.067			3.28	2.20	5.48	7.60
3100	Large commercial, 75 HP, dozer w/scarifier	B-10L	240	.050			2.34	1.96	4.30	5.90
3500	Screen topsoil from stockpile, vibrating screen, wet material (organic)	B-10P	200	.060	C.Y.		2.80	6	8.80	11.05
3550	Dry material	"	300	.040			1.87	3.99	5.86	7.35
3600	Mixing with conditioners, manure and peat	B-10R	550	.022			1.02	.54	1.56	2.23
3650	Mobilization add for 2 days or less operation	B-34K	3	2.667	Job		115	315	430	535
3800	Spread conditioned topsoil, 6" deep, by hand	B-1	360	.067	S.Y.	5.45	2.57		8.02	10.15
3850	300 HP dozer	B-10M	27	.444	M.S.F.	585	21	70	676	755
4000	Spread soil conditioners, alum. sulfate, 1#/S.Y., hand push spreader	1 Clab	17500	.001	S.Y.	21	.02		21.02	23
4050	Tractor spreader	B-66	700	.011	M.S.F.	2,325	.56	.38	2,325.94	2,575
4100	Fertilizer, 0.2#/S.Y., push spreader	1 Clab	17500	.001	S.Y.	.14	.02		.16	.18
4150	Tractor spreader	B-66	700	.011	M.S.F.	15.55	.56	.38	16.49	18.40
4200	Ground limestone, 1#/S.Y., push spreader	1 Clab	17500	.001	S.Y.	.12	.02		.14	.16
4250	Tractor spreader	B-66	700	.011	M.S.F.	13.35	.56	.38	14.29	15.95
4400	Manure, 18#/S.Y., push spreader	1 Clab	2500	.003	S.Y.	8.20	.12		8.32	9.20
4450	Tractor spreader	B-66	280	.029	M.S.F.	910	1.40	.94	912.34	1,000
4500	Perlite, 1" deep, push spreader	1 Clab	17500	.001	S.Y.	10.40	.02		10.42	11.50
4550	Tractor spreader	B-66	700	.011	M.S.F.	1,150	.56	.38	1,150.94	1,275
4600	Vermiculite, push spreader	1 Clab	17500	.001	S.Y.	8.50	.02		8.52	9.40
4650	Tractor spreader	B-66	700	.011	M.S.F.	945	.56	.38	945.94	1,050
5000	Spread topsoil, skid steer loader and hand dress	B-62	270	.089	C.Y.	24.50	3.70	.64	28.84	33.50
5100	Articulated loader and hand dress	B-100	320	.038		24.50	1.75	3.01	29.26	33
5200	Articulated loader and 75 HP dozer	B-10M	500	.024		24.50	1.12	3.77	29.39	33
5300	Road grader and hand dress	B-11L	1000	.016		24.50	.71	.71	25.92	29
6000	Tilling topsoil, 20 HP tractor, disk harrow, 2" deep	B-66	450	.018	M.S.F.		.87	.59	1.46	2.02
6050	4" deep		360	.022			1.09	.73	1.82	2.54
6100	6" deep		270	.030			1.46	.98	2.44	3.38
6150	26" rototiller, 2" deep	A-1J	1250	.006	S.Y.		.24	.04	.28	.44
6200	4" deep		1000	.008			.30	.05	.35	.55
6250	6" deep		750	.011			.40	.06	.46	.73

32 91 13.26 Planting Beds

		Crew	Daily Output	Labor-Hours	Unit	Material	2016 Bare Costs Labor	2016 Bare Costs Equipment	Total	Total Incl O&P
0010	**PLANTING BEDS**									
0100	Backfill planting pit, by hand, on site topsoil	2 Clab	18	.889	C.Y.		33.50		33.50	55
0200	Prepared planting mix, by hand	"	24	.667			25.50		25.50	41.50
0300	Skid steer loader, on site topsoil	B-62	340	.071			2.94	.51	3.45	5.30
0400	Prepared planting mix	"	410	.059			2.44	.42	2.86	4.40
1000	Excavate planting pit, by hand, sandy soil	2 Clab	16	1			38		38	62
1100	Heavy soil or clay	"	8	2			76		76	124
1200	1/2 C.Y. backhoe, sandy soil	B-11C	150	.107			4.75	2.44	7.19	10.30
1300	Heavy soil or clay	"	115	.139			6.20	3.18	9.38	13.45
2000	Mix planting soil, incl. loam, manure, peat, by hand	2 Clab	60	.267		42.50	10.10		52.60	63.50
2100	Skid steer loader	B-62	150	.160		42.50	6.65	1.16	50.31	59
3000	Pile sod, skid steer loader	"	2800	.009	S.Y.		.36	.06	.42	.65
3100	By hand	2 Clab	400	.040			1.52		1.52	2.48
4000	Remove sod, F.E. loader	B-10S	2000	.006			.28	.19	.47	.66
4100	Sod cutter	B-12K	3200	.005			.23	.31	.54	.70
4200	By hand	2 Clab	240	.067			2.53		2.53	4.13

For customer support on your Commercial Renovation Cost Data, call 877.742.4719.

545

32 91 Planting Preparation

32 91 19 – Landscape Grading

32 91 19.13 Topsoil Placement and Grading		Crew	Daily Output	Labor-Hours	Unit	Material	2016 Bare Costs Labor	2016 Bare Costs Equipment	Total	Total Incl O&P
0010	**TOPSOIL PLACEMENT AND GRADING**									
0700	Furnish and place, truck dumped, screened, 4" deep	B-10S	1300	.009	S.Y.	3.38	.43	.29	4.10	4.72
0800	6" deep		820	.015	"	4.32	.68	.46	5.46	6.35
0810	Minimum labor/equipment charge	↓	2	6	Ea.		280	189	469	655
0900	Fine grading and seeding, incl. lime, fertilizer & seed,									
1000	With equipment	B-14	1000	.048	S.Y.	.45	1.93	.37	2.75	4.03
2000	Minimum labor/equipment charge	1 Clab	4	2	Job		76		76	124

32 92 Turf and Grasses

32 92 19 – Seeding

32 92 19.13 Mechanical Seeding

		Crew	Daily Output	Labor-Hours	Unit	Material	Labor	Equipment	Total	Total Incl O&P
0010	**MECHANICAL SEEDING**									
0020	Mechanical seeding, 215 lb./acre	B-66	1.50	5.333	Acre	560	262	176	998	1,225
0100	44 lb./M.S.Y.	"	2500	.003	S.Y.	.20	.16	.11	.47	.59
0101	44 lb./M.S.Y.	1 Clab	13950	.001	S.F.	.02	.02		.04	.06
0300	Fine grading and seeding incl. lime, fertilizer & seed,									
0310	with equipment	B-14	1000	.048	S.Y.	.45	1.93	.37	2.75	4.03
0400	Fertilizer hand push spreader, 35 lb. per M.S.F.	1 Clab	200	.040	M.S.F.	9.75	1.52		11.27	13.25
0600	Limestone hand push spreader, 50 lb. per M.S.F.		180	.044		5.40	1.68		7.08	8.65
0800	Grass seed hand push spreader, 4.5 lb. per M.S.F.	↓	180	.044	↓	22	1.68		23.68	27.50
1000	Hydro or air seeding for large areas, incl. seed and fertilizer	B-81	8900	.003	S.Y.	.50	.12	.08	.70	.83
1100	With wood fiber mulch added	"	8900	.003	"	2	.12	.08	2.20	2.48
1300	Seed only, over 100 lb., field seed, minimum				Lb.	1.80			1.80	1.98
1400	Maximum					1.70			1.70	1.87
1500	Lawn seed, minimum					1.40			1.40	1.54
1600	Maximum				↓	2.50			2.50	2.75
1800	Aerial operations, seeding only, field seed	B-58	50	.480	Acre	585	20	63	668	745
1900	Lawn seed		50	.480		455	20	63	538	600
2100	Seed and liquid fertilizer, field seed		50	.480		725	20	63	808	900
2200	Lawn seed	↓	50	.480	↓	595	20	63	678	755
9000	Minimum labor/equipment charge	1 Clab	4	2	Job		76		76	124

32 92 23 – Sodding

32 92 23.10 Sodding Systems

		Crew	Daily Output	Labor-Hours	Unit	Material	Labor	Equipment	Total	Total Incl O&P
0010	**SODDING SYSTEMS**									
0020	Sodding, 1" deep, bluegrass sod, on level ground, over 8 M.S.F.	B-63	22	1.818	M.S.F.	244	73	7.85	324.85	395
0200	4 M.S.F.		17	2.353		260	94.50	10.20	364.70	450
0300	1000 S.F.		13.50	2.963		295	119	12.85	426.85	530
0500	Sloped ground, over 8 M.S.F.		6	6.667		244	268	29	541	735
0600	4 M.S.F.		5	8		260	320	34.50	614.50	845
0700	1000 S.F.		4	10		295	400	43.50	738.50	1,025
1000	Bent grass sod, on level ground, over 6 M.S.F.		20	2		253	80.50	8.65	342.15	420
1100	3 M.S.F.		18	2.222		267	89	9.60	365.60	450
1200	Sodding 1000 S.F. or less		14	2.857		290	115	12.35	417.35	520
1500	Sloped ground, over 6 M.S.F.		15	2.667		253	107	11.55	371.55	465
1600	3 M.S.F.		13.50	2.963		267	119	12.85	398.85	500
1700	1000 S.F.	↓	12	3.333	↓	290	134	14.45	438.45	555

32 93 Plants

32 93 13 – Ground Covers

32 93 13.10 Ground Cover Plants		Crew	Daily Output	Labor-Hours	Unit	Material	2016 Bare Costs Labor	2016 Bare Costs Equipment	Total	Total Incl O&P
0010	**GROUND COVER PLANTS**									
0012	Plants, pachysandra, in prepared beds	B-1	15	1.600	C	83.50	61.50		145	193
0200	Vinca minor, 1 yr., bare root, in prepared beds		12	2	"	71.50	77		148.50	205
0600	Stone chips, in 50 lb. bags, Georgia marble		520	.046	Bag	3.63	1.78		5.41	6.90
0700	Onyx gemstone		260	.092		17	3.56		20.56	24.50
0800	Quartz		260	.092	↓	17	3.56		20.56	24.50
0900	Pea gravel, truckload lots	↓	28	.857	Ton	27.50	33		60.50	84

32 93 33 – Shrubs

32 93 33.10 Shrubs and Trees

32 93 33.10 Shrubs and Trees		Crew	Daily Output	Labor-Hours	Unit	Material	2016 Bare Costs Labor	2016 Bare Costs Equipment	Total	Total Incl O&P
0010	**SHRUBS AND TREES**									
0011	Evergreen, in prepared beds, B & B									
0100	Arborvitae pyramidal, 4'-5'	B-17	30	1.067	Ea.	101	45	26	172	213
0150	Globe, 12"-15"	B-1	96	.250		24.50	9.65		34.15	43
0300	Cedar, blue, 8'-10'	B-17	18	1.778		235	74.50	43.50	353	430
0500	Hemlock, Canadian, 2-1/2'-3'	B-1	36	.667		32	25.50		57.50	77
0550	Holly, Savannah, 8' - 10' H		9.68	2.479		262	95.50		357.50	445
0600	Juniper, andorra, 18"-24"		80	.300		37	11.55		48.55	60
0620	Wiltoni, 15"-18"	↓	80	.300		26	11.55		37.55	48
0640	Skyrocket, 4-1/2'-5'	B-17	55	.582		108	24.50	14.25	146.75	174
0660	Blue pfitzer, 2'-2-1/2'	B-1	44	.545		39	21		60	77
0680	Ketleerie, 2-1/2'-3'		50	.480		53	18.50		71.50	88.50
0700	Pine, black, 2-1/2'-3'		50	.480		60.50	18.50		79	97
0720	Mugo, 18"-24"	↓	60	.400		60.50	15.45		75.95	91.50
0740	White, 4'-5'	B-17	75	.427		52	17.95	10.45	80.40	97.50
0800	Spruce, blue, 18"-24"	B-1	60	.400		68	15.45		83.45	100
0840	Norway, 4'-5'	B-17	75	.427		84.50	17.95	10.45	112.90	134
0900	Yew, denisforma, 12"-15"	B-1	60	.400		35.50	15.45		50.95	64.50
1000	Capitata, 18"-24"		30	.800		33.50	31		64.50	87
1100	Hicksi, 2'-2-1/2'	↓	30	.800	↓	79.50	31		110.50	138

32 93 33.20 Shrubs

32 93 33.20 Shrubs		Crew	Daily Output	Labor-Hours	Unit	Material	2016 Bare Costs Labor	2016 Bare Costs Equipment	Total	Total Incl O&P
0010	**SHRUBS**									
0011	Broadleaf Evergreen, planted in prepared beds									
0100	Andromeda, 15"-18", container	B-1	96	.250	Ea.	33	9.65		42.65	52.50
0200	Azalea, 15" - 18", container		96	.250		30	9.65		39.65	49
0300	Barberry, 9"-12", container		130	.185		17.95	7.10		25.05	31.50
0400	Boxwood, 15"-18", B&B		96	.250		43.50	9.65		53.15	64
0500	Euonymus, emerald gaiety, 12" to 15", container		115	.209		24.50	8.05		32.55	39.50
0600	Holly, 15"-18", B & B		96	.250		37	9.65		46.65	56.50
0900	Mount laurel, 18" - 24", B & B		80	.300		72	11.55		83.55	98
1000	Paxistema, 9" - 12" high		130	.185		21.50	7.10		28.60	35.50
1100	Rhododendron, 18"-24", container		48	.500		38.50	19.30		57.80	73.50
1200	Rosemary, 1 gal. container		600	.040		18.05	1.54		19.59	22.50
2000	Deciduous, planted in prepared beds, amelanchier, 2'-3', B & B		57	.421		121	16.25		137.25	161
2100	Azalea, 15"-18", B & B		96	.250		30	9.65		39.65	48.50
2300	Bayberry, 2'-3', B & B		57	.421		24.50	16.25		40.75	53.50
2600	Cotoneaster, 15"-18", B & B	↓	80	.300		27	11.55		38.55	48.50
2800	Dogwood, 3'-4', B & B	B-17	40	.800		32.50	33.50	19.55	85.55	112
2900	Euonymus, alatus compacta, 15" to 18", container	B-1	80	.300		26.50	11.55		38.05	48.50
3200	Forsythia, 2'-3', container	"	60	.400		18.85	15.45		34.30	45.50
3300	Hibiscus, 3'-4', B & B	B-17	75	.427		49	17.95	10.45	77.40	94.50
3400	Honeysuckle, 3'-4', B & B	B-1	60	.400		27	15.45		42.45	54.50
3500	Hydrangea, 2'-3', B & B	"	57	.421		30.50	16.25		46.75	60

For customer support on your Commercial Renovation Cost Data, call 877.742.4719.

547

32 93 Plants

32 93 33 – Shrubs

32 93 33.20 Shrubs		Crew	Daily Output	Labor-Hours	Unit	Material	2016 Bare Costs Labor	2016 Bare Costs Equipment	Total	Total Incl O&P
3600	Lilac, 3'-4', B & B	B-17	40	.800	Ea.	28.50	33.50	19.55	81.55	108
3900	Privet, bare root, 18"-24"	B-1	80	.300		14.80	11.55		26.35	35
4100	Quince, 2'-3', B & B	"	57	.421		28.50	16.25		44.75	58
4200	Russian olive, 3'-4', B & B	B-17	75	.427		29.50	17.95	10.45	57.90	73
4400	Spirea, 3'-4', B & B	B-1	70	.343		21	13.20		34.20	44.50
4500	Viburnum, 3'-4', B & B	B-17	40	.800		25.50	33.50	19.55	78.55	104

32 93 43 – Trees

32 93 43.20 Trees

			Crew	Daily Output	Labor-Hours	Unit	Material	Labor	Equipment	Total	Total Incl O&P
0010	**TREES**										
0011	Deciduous, in prep. beds, balled & burlapped (B&B)										
0100	Ash, 2" caliper	G	B-17	8	4	Ea.	180	168	98	446	575
0200	Beech, 5'-6'	G		50	.640		181	27	15.65	223.65	261
0300	Birch, 6'-8', 3 stems	G		20	1.600		163	67.50	39	269.50	330
0500	Crabapple, 6'-8'	G		20	1.600		134	67.50	39	240.50	300
0600	Dogwood, 4'-5'	G		40	.800		127	33.50	19.55	180.05	216
0700	Eastern redbud, 4'-5'	G		40	.800		145	33.50	19.55	198.05	235
0800	Elm, 8'-10'	G		20	1.600		250	67.50	39	356.50	425
0900	Ginkgo, 6'-7'	G		24	1.333		144	56	32.50	232.50	285
1000	Hawthorn, 8'-10', 1" caliper	G		20	1.600		157	67.50	39	263.50	325
1100	Honeylocust, 10'-12', 1-1/2" caliper	G		10	3.200		200	135	78.50	413.50	525
1300	Larch, 8'	G		32	1		123	42	24.50	189.50	231
1400	Linden, 8'-10', 1" caliper	G		20	1.600		139	67.50	39	245.50	305
1500	Magnolia, 4'-5'	G		20	1.600		98	67.50	39	204.50	260
1600	Maple, red, 8'-10', 1-1/2" caliper	G		10	3.200		196	135	78.50	409.50	520
1700	Mountain ash, 8'-10', 1" caliper	G		16	2		171	84	49	304	380
1800	Oak, 2-1/2"-3" caliper	G		6	5.333		315	224	130	669	850
2100	Planetree, 9'-11', 1-1/4" caliper	G		10	3.200		229	135	78.50	442.50	555
2200	Plum, 6'-8', 1" caliper	G		20	1.600		79.50	67.50	39	186	240
2300	Poplar, 9'-11', 1-1/4" caliper	G		10	3.200		140	135	78.50	353.50	455
2500	Sumac, 2'-3'	G		75	.427		42.50	17.95	10.45	70.90	87.50
2700	Tulip, 5'-6'	G		40	.800		45.50	33.50	19.55	98.55	126
2800	Willow, 6'-8', 1" caliper	G		20	1.600		93	67.50	39	199.50	254
9000	Minimum labor/equipment charge		1 Clab	4	2	Job		76		76	124

32 94 Planting Accessories

32 94 13 – Landscape Edging

32 94 13.20 Edging

		Crew	Daily Output	Labor-Hours	Unit	Material	Labor	Equipment	Total	Total Incl O&P
0010	**EDGING**									
0050	Aluminum alloy, including stakes, 1/8" x 4", mill finish	B-1	390	.062	L.F.	4	2.37		6.37	8.30
0051	Black paint		390	.062		4.64	2.37		7.01	9
0052	Black anodized		390	.062		5.35	2.37		7.72	9.80
0100	Brick, set horizontally, 1-1/2 bricks per L.F.	D-1	370	.043		1.29	1.82		3.11	4.37
0150	Set vertically, 3 bricks per L.F.	"	135	.119		4	4.99		8.99	12.50
0200	Corrugated aluminum, roll, 4" wide	1 Carp	650	.012		2.20	.60		2.80	3.39
0250	6" wide	"	550	.015		2.75	.70		3.45	4.18
0600	Railroad ties, 6" x 8"	2 Carp	170	.094		2.11	4.56		6.67	9.75
0650	7" x 9"		136	.118		2.34	5.70		8.04	11.90
0750	Redwood 2" x 4"		330	.048		2.26	2.35		4.61	6.35
0800	Steel edge strips, incl. stakes, 1/4" x 5"	B-1	390	.062		5	2.37		7.37	9.40
0850	3/16" x 4"	"	390	.062		3.95	2.37		6.32	8.25

32 94 Planting Accessories

32 94 13 – Landscape Edging

32 94 13.20 Edging	Crew	Daily Output	Labor-Hours	Unit	Material	2016 Bare Costs Labor	Equipment	Total	Total Incl O&P
9000	Minimum labor/equipment charge	1 Carp	4	2	Job		97		97	158

32 94 50 – Tree Guying

32 94 50.10 Tree Guying Systems

		Crew	Daily Output	Labor-Hours	Unit	Material	Labor	Equipment	Total	Total Incl O&P
0010	**TREE GUYING SYSTEMS**									
0015	Tree guying including stakes, guy wire and wrap									
0100	Less than 3" caliper, 2 stakes	2 Clab	35	.457	Ea.	13.10	17.35		30.45	43
0200	3" to 4" caliper, 3 stakes	"	21	.762	"	19.55	29		48.55	68.50
1000	Including arrowhead anchor, cable, turnbuckles and wrap									
1100	Less than 3" caliper, 3" anchors	2 Clab	20	.800	Ea.	28	30.50		58.50	80.50
1200	3" to 6" caliper, 4" anchors		15	1.067		34	40.50		74.50	104
1300	6" caliper, 6" anchors		12	1.333		28.50	50.50		79	114
1400	8" caliper, 8" anchors		9	1.778		114	67.50		181.50	235

32 96 Transplanting

32 96 23 – Plant and Bulb Transplanting

32 96 23.23 Planting

		Crew	Daily Output	Labor-Hours	Unit	Material	Labor	Equipment	Total	Total Incl O&P
0010	**PLANTING**									
0012	Moving shrubs on site, 12" ball	B-62	28	.857	Ea.		35.50	6.20	41.70	64.50
0100	24" ball	"	22	1.091	"		45.50	7.90	53.40	82

32 96 23.43 Moving Trees

		Crew	Daily Output	Labor-Hours	Unit	Material	Labor	Equipment	Total	Total Incl O&P
0010	**MOVING TREES**, On site									
0300	Moving trees on site, 36" ball	B-6	3.75	6.400	Ea.		267	97.50	364.50	535
0400	60" ball	"	1	24	"		1,000	365	1,365	2,025

32 96 43 – Tree Transplanting

32 96 43.20 Tree Removal

		Crew	Daily Output	Labor-Hours	Unit	Material	Labor	Equipment	Total	Total Incl O&P
0010	**TREE REMOVAL**									
0100	Dig & lace, shrubs, broadleaf evergreen, 18"-24" high	B-1	55	.436	Ea.		16.85		16.85	27.50
0200	2'-3'	"	35	.686			26.50		26.50	43
0300	3'-4'	B-6	30	.800			33.50	12.20	45.70	67.50
0400	4'-5'	"	20	1.200			50	18.30	68.30	101
1000	Deciduous, 12"-15"	B-1	110	.218			8.40		8.40	13.75
1100	18"-24"		65	.369			14.25		14.25	23.50
1200	2'-3'		55	.436			16.85		16.85	27.50
1300	3'-4'	B-6	50	.480			20	7.30	27.30	40.50
2000	Evergreen, 18"-24"	B-1	55	.436			16.85		16.85	27.50
2100	2'-0" to 2'-6"		50	.480			18.50		18.50	30.50
2200	2'-6" to 3'-0"		35	.686			26.50		26.50	43
2300	3'-0" to 3'-6"		20	1.200			46.50		46.50	75.50
3000	Trees, deciduous, small, 2'-3'		55	.436			16.85		16.85	27.50
3100	3'-4'	B-6	50	.480			20	7.30	27.30	40.50
3200	4'-5'		35	.686			28.50	10.45	38.95	57.50
3300	5'-6'		30	.800			33.50	12.20	45.70	67.50
4000	Shade, 5'-6'		50	.480			20	7.30	27.30	40.50
4100	6'-8'		35	.686			28.50	10.45	38.95	57.50
4200	8'-10'		25	.960			40	14.60	54.60	80.50
4300	2" caliper		12	2			83.50	30.50	114	168

For customer support on your Commercial Renovation Cost Data, call 877.742.4719.

549

32 96 43.20 Tree Removal		Crew	Daily Output	Labor-Hours	Unit	Material	2016 Bare Costs		Total	Total Incl O&P
							Labor	Equipment		
5000	Evergreen, 4'-5'	B-6	35	.686	Ea.		28.50	10.45	38.95	57.50
5100	5'-6'		25	.960			40	14.60	54.60	80.50
5200	6'-7'		19	1.263			52.50	19.25	71.75	106
5300	7'-8'		15	1.600			66.50	24.50	91	135
5400	8'-10'		11	2.182			91	33	124	184

Estimating Tips

33 10 00 Water Utilities

33 30 00 Sanitary Sewerage Utilities

33 40 00 Storm Drainage Utilities

- Never assume that the water, sewer, and drainage lines will go in at the early stages of the project. Consider the site access needs before dividing the site in half with open trenches, loose pipe, and machinery obstructions. Always inspect the site to establish that the site drawings are complete. Check off all existing utilities on your drawings as you locate them. Be especially careful with underground utilities because appurtenances are sometimes buried during regrading or repaving operations. If you find any discrepancies, mark up the site plan for further research. Differing site conditions can be very costly if discovered later in the project.

- See also Section 33 01 00 for restoration of pipe where removal/replacement may be undesirable. Use of new types of piping materials can reduce the overall project cost. Owners/design engineers should consider the installing contractor as a valuable source of current information on utility products and local conditions that could lead to significant cost savings.

Reference Numbers

Reference numbers are shown at the beginning of some major classifications. These numbers refer to related items in the Reference Section. The reference information may be an estimating procedure, an alternate pricing method, or technical information.

Note: Not all subdivisions listed here necessarily appear. ∎

---

33 01 30 – Operation and Maintenance of Sewer Utilities

33 01 30.16 TV Inspection of Sewer Pipelines

		Crew	Daily Output	Labor-Hours	Unit	Material	2016 Bare Costs Labor	Equipment	Total	Total Incl O&P
0010	**TV INSPECTION OF SEWER PIPELINES**									
0100	Pipe internal cleaning & inspection, cleaning, pressure pipe systems									
0120	Pig method, lengths 1000' to 10,000'									
0140	4" diameter thru 24" diameter, minimum				L.F.				3.60	4.14
0160	Maximum				"				18	21
6000	Sewage/sanitary systems									
6100	Power rodder with header & cutters									
6110	Mobilization charge, minimum				Total				695	800
6120	Mobilization charge, maximum				"				9,125	10,600
6140	Cleaning 4"-12" diameter				L.F.				3.39	3.90
6190	14"-24" diameter								3.98	4.59
6240	30" diameter								5.80	6.65
6250	36" diameter								6.75	7.75
6260	48" diameter								7.70	8.85
6270	60" diameter								8.70	9.95
6280	72" diameter								9.65	11.10
9000	Inspection, television camera with video									
9060	up to 500 linear feet				Total				715	820

33 01 30.72 Relining Sewers

		Crew	Daily Output	Labor-Hours	Unit	Material	2016 Bare Costs Labor	Equipment	Total	Total Incl O&P
0010	**RELINING SEWERS**									
0011	With cement incl. bypass & cleaning									
0020	Less than 10,000 L.F., urban, 6" to 10"	C-17E	130	.615	L.F.	9.55	31	.76	41.31	61.50
0050	10" to 12"		125	.640		11.75	32	.79	44.54	66
0070	12" to 16"		115	.696		12.05	35	.86	47.91	70.50
0100	16" to 20"		95	.842		14.15	42.50	1.04	57.69	85
0200	24" to 36"		90	.889		15.25	44.50	1.09	60.84	90.50
0300	48" to 72"		80	1		24.50	50.50	1.23	76.23	109
0500	Rural, 6" to 10"		180	.444		9.55	22.50	.55	32.60	47
0550	10" to 12"		175	.457		11.75	23	.56	35.31	51
0570	12" to 16"		160	.500		12.05	25	.62	37.67	55
0600	16" to 20"		135	.593		14.15	30	.73	44.88	65
0700	24" to 36"		125	.640		15.25	32	.79	48.04	69.50
0800	48" to 72"		100	.800		24.50	40	.98	65.48	92.50
1000	Greater than 10,000 L.F., urban, 6" to 10"		160	.500		9.55	25	.62	35.17	52
1050	10" to 12"		155	.516		11.75	26	.63	38.38	55.50
1070	12" to 16"		140	.571		12.05	28.50	.70	41.25	60.50
1100	16" to 20"		120	.667		14.15	33.50	.82	48.47	71
1200	24" to 36"		115	.696		15.25	35	.86	51.11	74
1300	48" to 72"		95	.842		24.50	42.50	1.04	68.04	96
1500	Rural, 6" to 10"		215	.372		9.55	18.70	.46	28.71	41.50
1550	10" to 12"		210	.381		11.75	19.15	.47	31.37	44.50
1570	12" to 16"		185	.432		12.05	22	.53	34.58	49
1600	16" to 20"		150	.533		14.15	27	.66	41.81	60
1700	24" to 36"		140	.571		15.25	28.50	.70	44.45	64
1800	48" to 72"		120	.667		24.50	33.50	.82	58.82	82
2000	Cured in place pipe, non-pressure, flexible felt resin, 400' runs									
2100	6" diameter				L.F.				25.50	28
2200	8" diameter								26.50	29
2300	10" diameter								29	32
2400	12" diameter								34.50	38
2500	15" diameter								59	65
2600	18" diameter								82	90

33 01 Operation and Maintenance of Utilities

33 01 30 – Operation and Maintenance of Sewer Utilities

33 01 30.72 Relining Sewers	Crew	Daily Output	Labor-Hours	Unit	Material	2016 Bare Costs Labor	Equipment	Total	Total Incl O&P
2700	21" diameter				L.F.				105	115
2800	24" diameter								177	195
2900	30" diameter								195	215
3000	36" diameter								205	225
3100	48" diameter								218	240

33 05 Common Work Results for Utilities

33 05 16 – Utility Structures

33 05 16.13 Precast Concrete Utility Boxes

		Crew	Daily Output	Labor-Hours	Unit	Material	2016 Bare Costs Labor	Equipment	Total	Total Incl O&P
0010	**PRECAST CONCRETE UTILITY BOXES**, 6" thick									
0050	5' x 10' x 6' high, I.D.	B-13	2	28	Ea.	1,775	1,150	375	3,300	4,225
0350	Hand hole, precast concrete, 1-1/2" thick									
0400	1'-0" x 2'-0" x 1'-9", I.D., light duty	B-1	4	6	Ea.	410	231		641	830
0450	4'-6" x 3'-2" x 2'-0", O.D., heavy duty	B-6	3	8	"	1,475	335	122	1,932	2,275

33 05 23 – Trenchless Utility Installation

33 05 23.19 Microtunneling

		Crew	Daily Output	Labor-Hours	Unit	Material	2016 Bare Costs Labor	Equipment	Total	Total Incl O&P
0010	**MICROTUNNELING**									
0011	Not including excavation, backfill, shoring,									
0020	or dewatering, average 50'/day, slurry method									
0100	24" to 48" outside diameter, minimum				L.F.				875	965
0110	Adverse conditions, add				%				50%	50%
1000	Rent microtunneling machine, average monthly lease				Month				97,500	107,000
1010	Operating technician				Day				630	705
1100	Mobilization and demobilization, minimum				Job				41,200	45,900
1110	Maximum				"				445,500	490,500

33 05 23.20 Horizontal Boring

		Crew	Daily Output	Labor-Hours	Unit	Material	2016 Bare Costs Labor	Equipment	Total	Total Incl O&P
0010	**HORIZONTAL BORING**									
0011	Casing only, 100' minimum,									
0020	not incl. jacking pits or dewatering									
0100	Roadwork, 1/2" thick wall, 24" diameter casing	B-42	20	3.200	L.F.	122	137	68.50	327.50	435
0200	36" diameter		16	4		225	172	86	483	625
0300	48" diameter		15	4.267		310	183	91.50	584.50	740
0500	Railroad work, 24" diameter		15	4.267		122	183	91.50	396.50	535
0600	36" diameter		14	4.571		225	196	98	519	675
0700	48" diameter		12	5.333		310	229	114	653	840
0900	For ledge, add								20%	20%
1000	Small diameter boring, 3", sandy soil	B-82	900	.018		22	.77	.09	22.86	25.50
1040	Rocky soil	"	500	.032		22	1.39	.17	23.56	26.50

33 05 26 – Utility Identification

33 05 26.05 Utility Connection

		Crew	Daily Output	Labor-Hours	Unit	Material	2016 Bare Costs Labor	Equipment	Total	Total Incl O&P
0010	**UTILITY CONNECTION**									
0020	Water, sanitary, stormwater, gas, single connection	B-14	1	48	Ea.	2,925	1,925	365	5,215	6,750
0030	Telecommunication	"	3	16	"	390	640	122	1,152	1,625

33 05 26.10 Utility Accessories

		Crew	Daily Output	Labor-Hours	Unit	Material	2016 Bare Costs Labor	Equipment	Total	Total Incl O&P
0010	**UTILITY ACCESSORIES** R312316-40									
0400	Underground tape, detectable, reinforced, alum. foil core, 2"	1 Clab	150	.053	C.L.F.	6	2.02		8.02	9.90
0500	6"		140	.057	"	25.50	2.17		27.67	31.50
9000	Minimum labor/equipment charge		4	2	Job		76		76	124

For customer support on your Commercial Renovation Cost Data, call 877.742.4719.

553

33 11 Water Utility Distribution Piping

33 11 13 – Public Water Utility Distribution Piping

33 11 13.15 Water Supply, Ductile Iron Pipe		Crew	Daily Output	Labor-Hours	Unit	Material	2016 Bare Costs			Total	Total Incl O&P
							Labor	Equipment			
0010	**WATER SUPPLY, DUCTILE IRON PIPE** R331113-80										
0020	Not including excavation or backfill										
2000	Pipe, class 50 water piping, 18' lengths										
2020	Mechanical joint, 4" diameter	B-21A	200	.200	L.F.	30.50	9.45	2.40		42.35	51
3000	Push-on joint, 4" diameter		400	.100		15.50	4.73	1.20		21.43	26
3020	6" diameter	↓	333.33	.120	↓	18.25	5.70	1.44		25.39	30.50
8000	Piping, fittings, mechanical joint, AWWA C110										
8006	90° bend, 4" diameter	B-20A	16	2	Ea.	155	92			247	315
8020	6" diameter		12.80	2.500		229	115			344	435
8200	Wye or tee, 4" diameter		10.67	2.999		375	138			513	635
8220	6" diameter		8.53	3.751		565	173			738	895
8398	45° bend, 4" diameter		16	2		255	92			347	430
8400	6" diameter		12.80	2.500		199	115			314	400
8405	8" diameter		10.67	2.999		291	138			429	540
8450	Decreaser, 6" x 4" diameter		14.22	2.250		207	104			311	395
8460	8" x 6" diameter	↓	11.64	2.749	↓	310	127			437	540
8550	Piping, butterfly valves, cast iron										
8560	4" diameter	B-20	6	4	Ea.	470	170			640	790
9600	Steel sleeve with tap, 4" diameter		3	8		405	340			745	1,000
9620	6" diameter	↓	2	12	↓	470	510			980	1,350

33 12 Water Utility Distribution Equipment

33 12 13 – Water Service Connections

33 12 13.15 Tapping, Crosses and Sleeves

		Crew	Daily Output	Labor-Hours	Unit	Material	Labor	Equipment	Total	Total Incl O&P
0010	**TAPPING, CROSSES AND SLEEVES**									
4000	Drill and tap pressurized main (labor only)									
4100	6" main, 1" to 2" service	Q-1	3	5.333	Ea.		284		284	445
4150	8" main, 1" to 2" service	"	2.75	5.818	"		310		310	485
4500	Tap and insert gate valve									
4600	8" main, 4" branch	B-21	3.20	8.750	Ea.		385	44	429	675
4650	6" branch		2.70	10.370			455	52	507	795
4800	12" main, 6" branch	↓	2.35	11.915	↓		525	59.50	584.50	915

33 21 Water Supply Wells

33 21 13 – Public Water Supply Wells

33 21 13.10 Wells and Accessories

		Crew	Daily Output	Labor-Hours	Unit	Material	Labor	Equipment	Total	Total Incl O&P
0010	**WELLS & ACCESSORIES** R221113-50									
0011	Domestic									
0100	Drilled, 4" to 6" diameter	B-23	120	.333	L.F.		12.75	24	36.75	47
0200	8" diameter	"	95.20	.420	"		16.10	30	46.10	59.50
1400	Remove & reset pump, minimum	B-21	4	7	Ea.		310	35	345	540
1420	Maximum	"	2	14	"		615	70	685	1,075
1500	Pumps, installed in wells to 100' deep, 4" submersible									
1510	1/2 H.P.	Q-1	3.22	4.969	Ea.	730	265		995	1,225
1520	3/4 H.P.		2.66	6.015		855	320		1,175	1,450
1600	1 H.P.	↓	2.29	6.987		905	370		1,275	1,575
1700	1-1/2 H.P.	Q-22	1.60	10		1,800	535	410	2,745	3,250
1800	2 H.P.		1.33	12.030		1,725	640	495	2,860	3,450
1900	3 H.P.	↓	1.14	14.035	↓	2,150	750	580	3,480	4,175

33 21 Water Supply Wells

33 21 13 – Public Water Supply Wells

33 21 13.10 Wells and Accessories	Crew	Daily Output	Labor-Hours	Unit	Material	2016 Bare Costs Labor	Equipment	Total	Total Incl O&P
2000	5 H.P.	Q-22	1.14	14.035	Ea.	2,925	750	580	4,255	5,025
2050	Remove and install motor only, 4 H.P.	↓	1.14	14.035		1,300	750	580	2,630	3,225
5000	Wells to 180 ft. deep, 4" submersible, 1 HP	B-21	1.10	25.455		2,375	1,125	128	3,628	4,600
5500	2 HP		1.10	25.455		3,275	1,125	128	4,528	5,575
6000	3 HP		1	28		4,425	1,225	140	5,790	7,000
7000	5 HP		.90	31.111	↓	6,050	1,375	156	7,581	9,050
9000	Minimum labor/equipment charge	↓	1.80	15.556	Job		685	78	763	1,175

33 31 Sanitary Utility Sewerage Piping

33 31 13 – Public Sanitary Utility Sewerage Piping

33 31 13.15 Sewage Collection, Concrete Pipe

0010	**SEWAGE COLLECTION, CONCRETE PIPE**									
0020	See Section 33 41 13.60 for sewage/drainage collection, concrete pipe									

33 31 13.25 Sewage Collection, Polyvinyl Chloride Pipe

		Crew	Daily Output	Labor-Hours	Unit	Material	Labor	Equipment	Total	Total Incl O&P
0010	**SEWAGE COLLECTION, POLYVINYL CHLORIDE PIPE**									
0020	Not including excavation or backfill									
2000	20' lengths, SDR 35, B&S, 4" diameter	B-20	375	.064	L.F.	1.49	2.72		4.21	6.10
2040	6" diameter		350	.069		3.28	2.92		6.20	8.35
2080	13' lengths, SDR 35, B&S, 8" diameter	↓	335	.072		6.15	3.05		9.20	11.70
2120	10" diameter	B-21	330	.085		11.10	3.73	.43	15.26	18.70
4000	Piping, DWV PVC, no exc./bkfill., 10' L, Sch 40, 4" diameter	B-20	375	.064		3.97	2.72		6.69	8.80
4010	6" diameter		350	.069		8	2.92		10.92	13.55
4020	8" diameter	↓	335	.072	↓	12.50	3.05		15.55	18.70

33 36 Utility Septic Tanks

33 36 13 – Utility Septic Tank and Effluent Wet Wells

33 36 13.13 Concrete Utility Septic Tank

		Crew	Daily Output	Labor-Hours	Unit	Material	Labor	Equipment	Total	Total Incl O&P
0010	**CONCRETE UTILITY SEPTIC TANK**									
0011	Not including excavation or piping									
0015	Septic tanks, precast, 1,000 gallon	B-21	8	3.500	Ea.	1,025	154	17.55	1,196.55	1,400
0020	1,250 gallon		8	3.500		1,475	154	17.55	1,646.55	1,900
0060	1,500 gallon		7	4		1,575	176	20	1,771	2,050
0100	2,000 gallon		5	5.600		2,200	246	28	2,474	2,850
0140	2,500 gallon		5	5.600		2,375	246	28	2,649	3,025
0180	4,000 gallon	↓	4	7		5,550	310	35	5,895	6,650
0220	5,000 gal., 4 piece	B-13	3	18.667		9,575	775	250	10,600	12,000
0300	15,000 gallon, 4 piece	B-13B	1.70	32.941		20,000	1,375	660	22,035	24,900
0400	25,000 gallon, 4 piece		1.10	50.909		38,800	2,100	1,025	41,925	47,200
0500	40,000 gallon, 4 piece	↓	.80	70		49,700	2,900	1,400	54,000	60,500
0520	50,000 gallon, 5 piece	B-13C	.60	93.333		57,000	3,875	2,775	63,650	72,500
0640	75,000 gallon, cast in place	C-14C	.25	448		69,500	20,600	125	90,225	110,000
0660	100,000 gallon	"	.15	746		86,000	34,400	209	120,609	150,500
1150	Leaching field chambers, 13' x 3'-7" x 1'-4", standard	B-13	16	3.500		485	145	47	677	815
1200	Heavy duty, 8' x 4' x 1'-6"		14	4		284	166	53.50	503.50	635
1300	13' x 3'-9" x 1'-6"		12	4.667		1,125	193	62.50	1,380.50	1,600
1350	20' x 4' x 1'-6"		5	11.200		1,175	465	150	1,790	2,200
1420	Leaching pit, precast concrete, 6' dia, 3' deep	B-21	4.70	5.957		885	262	30	1,177	1,425
1600	Leaching pit, 6'-6" diameter, 6' deep		5	5.600		1,025	246	28	1,299	1,550
1620	8' deep	↓	4	7		1,200	310	35	1,545	1,875

For customer support on your Commercial Renovation Cost Data, call 877.742.4719.

555

33 36 Utility Septic Tanks

33 36 13 – Utility Septic Tank and Effluent Wet Wells

33 36 13.13 Concrete Utility Septic Tank

		Crew	Daily Output	Labor-Hours	Unit	Material	2016 Bare Costs Labor	2016 Bare Costs Equipment	Total	Total Incl O&P
1700	8' diameter, H-20 load, 6' deep	B-21	4	7	Ea.	1,500	310	35	1,845	2,200
1720	8' deep		3	9.333		2,500	410	47	2,957	3,475
2000	Velocity reducing pit, precast conc., 6' diameter, 3' deep		4.70	5.957		1,600	262	30	1,892	2,200

33 36 13.19 Polyethylene Utility Septic Tank

		Crew	Daily Output	Labor-Hours	Unit	Material	2016 Bare Costs Labor	2016 Bare Costs Equipment	Total	Total Incl O&P
0010	**POLYETHYLENE UTILITY SEPTIC TANK**									
0015	High density polyethylene, 1,000 gallon	B-21	8	3.500	Ea.	1,225	154	17.55	1,396.55	1,625
0020	1,250 gallon		8	3.500		1,400	154	17.55	1,571.55	1,825
0025	1,500 gallon		7	4		1,550	176	20	1,746	2,000

33 36 19 – Utility Septic Tank Effluent Filter

33 36 19.13 Utility Septic Tank Effluent Tube Filter

		Crew	Daily Output	Labor-Hours	Unit	Material	2016 Bare Costs Labor	2016 Bare Costs Equipment	Total	Total Incl O&P
0010	**UTILITY SEPTIC TANK EFFLUENT TUBE FILTER**									
3000	Effluent filter, 4" diameter	1 Skwk	8	1	Ea.	43.50	50		93.50	129
3020	6" diameter		7	1.143		254	57		311	375
3030	8" diameter		7	1.143		225	57		282	340
3040	8" diameter, very fine		7	1.143		460	57		517	600
3050	10" diameter, very fine		6	1.333		240	66.50		306.50	370
3060	10" diameter		6	1.333		280	66.50		346.50	420
3080	12" diameter		6	1.333		670	66.50		736.50	850
3090	15" diameter		5	1.600		1,150	80		1,230	1,375

33 36 33 – Utility Septic Tank Drainage Field

33 36 33.13 Utility Septic Tank Tile Drainage Field

		Crew	Daily Output	Labor-Hours	Unit	Material	2016 Bare Costs Labor	2016 Bare Costs Equipment	Total	Total Incl O&P
0010	**UTILITY SEPTIC TANK TILE DRAINAGE FIELD**									
0015	Distribution box, concrete, 5 outlets	2 Clab	20	.800	Ea.	82	30.50		112.50	140
0020	7 outlets		16	1		82	38		120	152
0025	9 outlets		8	2		500	76		576	675
0115	Distribution boxes, HDPE, 5 outlets		20	.800		64.50	30.50		95	121
0117	6 outlets		15	1.067		65	40.50		105.50	138
0118	7 outlets		15	1.067		65	40.50		105.50	138
0120	8 outlets		10	1.600		68.50	60.50		129	175
0240	Distribution boxes, Outlet Flow Leveler	1 Clab	50	.160		2.35	6.05		8.40	12.50
0300	Precast concrete, galley, 4' x 4' x 4'	B-21	16	1.750		250	77	8.75	335.75	410
0350	HDPE infiltration chamber 12" H X 15" W	2 Clab	300	.053	L.F.	7.35	2.02		9.37	11.40
0351	12" H X 15" W End Cap	1 Clab	32	.250	Ea.	19.75	9.50		29.25	37
0355	chamber 12" H X 22" W	2 Clab	300	.053	L.F.	6.80	2.02		8.82	10.80
0356	12" H X 22" W End Cap	1 Clab	32	.250	Ea.	17.30	9.50		26.80	34.50
0360	chamber 13" H X 34" W	2 Clab	300	.053	L.F.	15	2.02		17.02	19.80
0361	13" H X 34" W End Cap	1 Clab	32	.250	Ea.	53.50	9.50		63	74
0365	chamber 16" H X 34" W	2 Clab	300	.053	L.F.	10.75	2.02		12.77	15.10
0366	16" H X 34" W End Cap	1 Clab	32	.250	Ea.	8	9.50		17.50	24.50
0370	chamber 8" H X 16" W	2 Clab	300	.053	L.F.	9.65	2.02		11.67	13.90
0371	8" H X 16" W End Cap	1 Clab	32	.250	Ea.	10.80	9.50		20.30	27.50

33 36 50 – Drainage Field Systems

33 36 50.10 Drainage Field Excavation and Fill

		Crew	Daily Output	Labor-Hours	Unit	Material	2016 Bare Costs Labor	2016 Bare Costs Equipment	Total	Total Incl O&P
0010	**DRAINAGE FIELD EXCAVATION AND FILL**									
2200	Septic tank & drainage field excavation with 3/4 C.Y. backhoe	B-12F	145	.110	C.Y.		4.97	4.51	9.48	12.95
2400	4' trench for disposal field, 3/4 C.Y. backhoe	"	335	.048	L.F.		2.15	1.95	4.10	5.60
2600	Gravel fill, run of bank	B-6	150	.160	C.Y.	18.50	6.65	2.44	27.59	34
2800	Crushed stone, 3/4"	"	150	.160	"	38.50	6.65	2.44	47.59	55.50

33 41 Storm Utility Drainage Piping

33 41 13 – Public Storm Utility Drainage Piping

33 41 13.40 Piping, Storm Drainage, Corrugated Metal

	Crew	Daily Output	Labor-Hours	Unit	Material	2016 Bare Costs Labor	2016 Bare Costs Equipment	Total	Total Incl O&P
0010 **PIPING, STORM DRAINAGE, CORRUGATED METAL** R221113-50									
0020 Not including excavation or backfill									
2000 Corrugated metal pipe, galvanized									
2020 Bituminous coated with paved invert, 20' lengths									
2040 8" diameter, 16 ga.	B-14	330	.145	L.F.	8.75	5.85	1.11	15.71	20.50
2080 12" diameter, 16 ga.	"	210	.229	"	11.10	9.15	1.74	21.99	29

33 41 13.60 Sewage/Drainage Collection, Concrete Pipe

	Crew	Daily Output	Labor-Hours	Unit	Material	2016 Bare Costs Labor	2016 Bare Costs Equipment	Total	Total Incl O&P
0010 **SEWAGE/DRAINAGE COLLECTION, CONCRETE PIPE**									
0020 Not including excavation or backfill									
1000 Non-reinforced pipe, extra strength, B&S or T&G joints									
1010 6" diameter	B-14	265.04	.181	L.F.	7.25	7.25	1.38	15.88	21.50
1020 8" diameter		224	.214		8	8.60	1.63	18.23	24.50
1040 12" diameter		200	.240		9.95	9.65	1.83	21.43	28.50
2000 Reinforced culvert, class 3, no gaskets									
2010 12" diameter	B-14	150	.320	L.F.	12.15	12.85	2.44	27.44	37
2020 15" diameter	"	150	.320	"	15.70	12.85	2.44	30.99	41

33 44 Storm Utility Water Drains

33 44 13 – Utility Area Drains

33 44 13.13 Catch Basins

	Crew	Daily Output	Labor-Hours	Unit	Material	2016 Bare Costs Labor	2016 Bare Costs Equipment	Total	Total Incl O&P
0010 **CATCH BASINS**									
0011 Not including footing & excavation									
1600 Frames & grates, C.I., 24" square, 500 lb.	B-6	7.80	3.077	Ea.	340	128	47	515	635
1700 26" D shape, 600 lb.	"	7	3.429	"	505	143	52	700	845
3320 Frames and covers, existing, raised for paving, 2", including									
3340 row of brick, concrete collar, up to 12" wide frame	B-6	18	1.333	Ea.	46	55.50	20.50	122	163

33 44 13.15 Remove and Replace Catch Basin Cover

	Crew	Daily Output	Labor-Hours	Unit	Material	2016 Bare Costs Labor	2016 Bare Costs Equipment	Total	Total Incl O&P
0010 **REMOVE AND REPLACE CATCH BASIN COVER**									
0023 Remove catch basin cover	1 Clab	80	.100	Ea.		3.79		3.79	6.20
0033 Replace catch basin cover	"	80	.100	"		3.79		3.79	6.20

33 49 Storm Drainage Structures

33 49 13 – Storm Drainage Manholes, Frames, and Covers

33 49 13.10 Storm Drainage Manholes, Frames and Covers

	Crew	Daily Output	Labor-Hours	Unit	Material	2016 Bare Costs Labor	2016 Bare Costs Equipment	Total	Total Incl O&P
0010 **STORM DRAINAGE MANHOLES, FRAMES & COVERS**									
0020 Excludes footing, excavation, backfill (See line items for frame & cover)									
0050 Brick, 4' inside diameter, 4' deep	D-1	1	16	Ea.	535	675		1,210	1,700
0100 6' deep		.70	22.857		760	965		1,725	2,400
0150 8' deep		.50	32		980	1,350		2,330	3,275
0200 For depths over 8', add		4	4	V.L.F.	83.50	169		252.50	365
1110 Precast, 4' I.D., 4' deep	B-22	4.10	7.317	Ea.	740	325	51.50	1,116.50	1,400
1120 6' deep		3	10		1,375	445	70	1,890	2,300
1130 8' deep		2	15		1,800	670	105	2,575	3,175
1140 For depths over 8', add		16	1.875	V.L.F.	163	83.50	13.15	259.65	330

For customer support on your Commercial Renovation Cost Data, call 877.742.4719.

557

33 51 Natural-Gas Distribution

33 51 13 – Natural-Gas Piping

33 51 13.10 Piping, Gas Service and Distribution, P.E.	Crew	Daily Output	Labor-Hours	Unit	Material	2016 Bare Costs Labor	2016 Bare Costs Equipment	Total	Total Incl O&P
0010 **PIPING, GAS SERVICE AND DISTRIBUTION, POLYETHYLENE**									
0020 Not including excavation or backfill									
1000 60 psi coils, compression coupling @ 100', 1/2" diameter, SDR 11	B-20A	608	.053	L.F.	.50	2.43		2.93	4.41
1010 1" diameter, SDR 11		544	.059		1.13	2.71		3.84	5.55
1040 1-1/4" diameter, SDR 11		544	.059		1.62	2.71		4.33	6.10
1100 2" diameter, SDR 11		488	.066		2.25	3.02		5.27	7.30
1160 3" diameter, SDR 11	↓	408	.078		4.70	3.61		8.31	10.90
1500 60 psi 40' joints with coupling, 3" diameter, SDR 11	B-21A	408	.098		9.40	4.64	1.17	15.21	18.95
1540 4" diameter, SDR 11		352	.114		13.50	5.40	1.36	20.26	25
1600 6" diameter, SDR 11		328	.122		33.50	5.75	1.46	40.71	48
1640 8" diameter, SDR 11	↓	272	.147	↓	46.50	6.95	1.76	55.21	64
9000 Minimum labor/equipment charge	B-20	2	12	Job		510		510	830

33 52 Liquid Fuel Distribution

33 52 16 – Gasoline Distribution

33 52 16.13 Gasoline Piping

	Crew	Daily Output	Labor-Hours	Unit	Material	2016 Bare Costs Labor	2016 Bare Costs Equipment	Total	Total Incl O&P
0010 **GASOLINE PIPING**									
0020 Primary containment pipe, fiberglass-reinforced									
0030 Plastic pipe 15' & 30' lengths									
0040 2" diameter	Q-6	425	.056	L.F.	6.80	3.20		10	12.50
0050 3" diameter		400	.060		10.60	3.40		14	16.95
0060 4" diameter	↓	375	.064	↓	13.75	3.63		17.38	21
0100 Fittings									
0110 Elbows, 90° & 45°, bell-ends, 2"	Q-6	24	1	Ea.	43.50	56.50		100	136
0120 3" diameter		22	1.091		55	62		117	157
0130 4" diameter		20	1.200		70	68		138	183
0200 Tees, bell ends, 2"		21	1.143		60.50	64.50		125	168
0210 3" diameter		18	1.333		64	75.50		139.50	189
0220 4" diameter		15	1.600		84	90.50		174.50	234
0230 Flanges bell ends, 2"		24	1		33.50	56.50		90	126
0240 3" diameter		22	1.091		39	62		101	140
0250 4" diameter		20	1.200		45	68		113	156
0260 Sleeve couplings, 2"		21	1.143		12.40	64.50		76.90	115
0270 3" diameter		18	1.333		17.80	75.50		93.30	138
0280 4" diameter		15	1.600		23	90.50		113.50	167
0290 Threaded adapters, 2"		21	1.143		17.95	64.50		82.45	121
0300 3" diameter		18	1.333		34	75.50		109.50	156
0310 4" diameter		15	1.600		38	90.50		128.50	183
0320 Reducers, 2"		27	.889		27.50	50.50		78	109
0330 3" diameter		22	1.091		27.50	62		89.50	127
0340 4" diameter	↓	20	1.200	↓	36.50	68		104.50	147
1010 Gas station product line for secondary containment (double wall)									
1100 Fiberglass reinforced plastic pipe 25' lengths									
1120 Pipe, plain end, 3" diameter	Q-6	375	.064	L.F.	26.50	3.63		30.13	34.50
1130 4" diameter		350	.069		32	3.88		35.88	41.50
1140 5" diameter		325	.074		36	4.18		40.18	46
1150 6" diameter	↓	300	.080	↓	39	4.53		43.53	50
1200 Fittings									
1230 Elbows, 90° & 45°, 3" diameter	Q-6	18	1.333	Ea.	136	75.50		211.50	268
1240 4" diameter		16	1.500		167	85		252	315
1250 5" diameter	↓	14	1.714		184	97		281	355

33 52 Liquid Fuel Distribution

33 52 16 – Gasoline Distribution

33 52 16.13 Gasoline Piping

33 52 16.13 Gasoline Piping		Crew	Daily Output	Labor-Hours	Unit	Material	2016 Bare Costs Labor	Equipment	Total	Total Incl O&P
1260	6" diameter	Q-6	12	2	Ea.	204	113		317	400
1270	Tees, 3" diameter		15	1.600		166	90.50		256.50	325
1280	4" diameter		12	2		202	113		315	400
1290	5" diameter		9	2.667		315	151		466	585
1300	6" diameter		6	4		375	227		602	770
1310	Couplings, 3" diameter		18	1.333		55.50	75.50		131	179
1320	4" diameter		16	1.500		120	85		205	264
1330	5" diameter		14	1.714		215	97		312	385
1340	6" diameter		12	2		315	113		428	525
1350	Cross-over nipples, 3" diameter		18	1.333		10.50	75.50		86	130
1360	4" diameter		16	1.500		12.75	85		97.75	147
1370	5" diameter		14	1.714		15.90	97		112.90	169
1380	6" diameter		12	2		19.10	113		132.10	198
1400	Telescoping, reducers, concentric 4" x 3"		18	1.333		48	75.50		123.50	171
1410	5" x 4"		17	1.412		95.50	80		175.50	230
1420	6" x 5"		16	1.500		234	85		319	390

33 71 Electrical Utility Transmission and Distribution

33 71 19 – Electrical Underground Ducts and Manholes

33 71 19.17 Electric and Telephone Underground

		Crew	Daily Output	Labor-Hours	Unit	Material	2016 Bare Costs Labor	Equipment	Total	Total Incl O&P
0010	**ELECTRIC AND TELEPHONE UNDERGROUND**									
0011	Not including excavation									
0200	backfill and cast in place concrete									
4200	Underground duct, banks ready for concrete fill, min. of 7.5"									
4400	between conduits, center to center									
4600	2 @ 2" diameter	2 Elec	240	.067	L.F.	1.29	3.67		4.96	7.10
4800	4 @ 2" diameter		120	.133		2.58	7.35		9.93	14.20
5600	4 @ 4" diameter		80	.200		5.65	11		16.65	23.50
6200	Rigid galvanized steel, 2 @ 2" diameter		180	.089		12.55	4.90		17.45	21.50
6400	4 @ 2" diameter		90	.178		25	9.80		34.80	42.50
7400	4 @ 4" diameter		34	.471		80	26		106	128
9990	Minimum labor/equipment charge	1 Elec	3.50	2.286	Job		126		126	195

For customer support on your Commercial Renovation Cost Data, call 877.742.4719.

559

Division Notes

	CREW	DAILY OUTPUT	LABOR-HOURS	UNIT	BARE COSTS				TOTAL INCL O&P
					MAT.	LABOR	EQUIP.	TOTAL	

Estimating Tips

34 11 00 Rail Tracks

This subdivision includes items that may involve either repair of existing, or construction of new, railroad tracks. Additional preparation work, such as the roadbed earthwork, would be found in Division 31. Additional new construction siding and turnouts are found in Subdivision 34 72. Maintenance of railroads is found under 34 01 23 Operation and Maintenance of Railways.

34 40 00 Traffic Signals

This subdivision includes traffic signal systems. Other traffic control devices such as traffic signs are found in Subdivision 10 14 53 Traffic Signage.

34 70 00 Vehicle Barriers

This subdivision includes security vehicle barriers, guide and guard rails, crash barriers, and delineators. The actual maintenance and construction of concrete and asphalt pavement is found in Division 32.

Reference Numbers

Reference numbers are shown at the beginning of some major classifications. These numbers refer to related items in the Reference Section. The reference information may be an estimating procedure, an alternate pricing method, or technical information.

Note: Not all subdivisions listed here necessarily appear. ■

34 71 13.26 Vehicle Guide Rails	Crew	Daily Output	Labor-Hours	Unit	Material	2016 Bare Costs Labor	Equipment	Total	Total Incl O&P
0010 **VEHICLE GUIDE RAILS**									
0012 Corrugated stl., galv. stl. posts, 6'-3" O.C.	B-80	850	.038	L.F.	24	1.59	.86	26.45	30
0200 End sections, galvanized, flared		50	.640	Ea.	100	27	14.55	141.55	170
0300 Wrap around end		50	.640	"	143	27	14.55	184.55	217
0400 Timber guide rail, 4" x 8" with 6" x 8" wood posts, treated		960	.033	L.F.	12.25	1.41	.76	14.42	16.55

Estimating Tips

Products such as conveyors, material handling cranes and hoists, as well as other items specified in this division, require trained installers. The general contractor may not have any choice as to who will perform the installation or when it will be performed. Long lead times are often required for these products, making early decisions in purchasing and scheduling necessary. The installation of this type of equipment may require the embedment of mounting hardware during construction of floors, structural walls, or interior walls/partitions. Electrical connections will require coordination with the electrical contractor.

Reference Numbers

Reference numbers are shown at the beginning of some major classifications. These numbers refer to related items in the Reference Section. The reference information may be an estimating procedure, an alternate pricing method, or technical information.

Note: Not all subdivisions listed here necessarily appear. ■

41 22 13.10 Crane Rail

		Crew	Daily Output	Labor-Hours	Unit	Material	2016 Bare Costs Labor	Equipment	Total	Total Incl O&P
0010	**CRANE RAIL**									
0020	Box beam bridge, no equipment included	E-4	3400	.009	Lb.	1.33	.51	.04	1.88	2.40
0200	Running track only, 104 lb. per yard		5600	.006	"	.66	.31	.03	1	1.30
0210	Running track only, 104 lb. per yard, 20' piece	↓	160	.200	L.F.	23	10.75	.92	34.67	45.50

41 22 13.13 Bridge Cranes

		Crew	Daily Output	Labor-Hours	Unit	Material	2016 Bare Costs Labor	Equipment	Total	Total Incl O&P
0010	**BRIDGE CRANES**									
0100	1 girder, 20' span, 3 ton	M-3	1	34	Ea.	25,400	2,000	139	27,539	31,300
0125	5 ton		1	34		27,900	2,000	139	30,039	34,000
0150	7.5 ton		1	34		33,100	2,000	139	35,239	39,700
0175	10 ton		.80	42.500		43,900	2,500	173	46,573	52,500
0200	15 ton		.80	42.500		56,500	2,500	173	59,173	66,000
0225	30' span, 3 ton		1	34		26,500	2,000	139	28,639	32,400
0250	5 ton		1	34		29,100	2,000	139	31,239	35,300
0275	7.5 ton		1	34		34,800	2,000	139	36,939	41,600
0300	10 ton		.80	42.500		45,400	2,500	173	48,073	54,000
0325	15 ton	↓	.80	42.500		59,000	2,500	173	61,673	69,000
0350	2 girder, 40' span, 3 ton	M-4	.50	72		43,600	4,200	355	48,155	55,000
0375	5 ton		.50	72		45,600	4,200	355	50,155	57,000
0400	7.5 ton		.50	72		50,000	4,200	355	54,555	62,000
0425	10 ton		.40	90		58,500	5,250	445	64,195	73,000
0450	15 ton		.40	90		80,000	5,250	445	85,695	96,000
0475	25 ton		.30	120		94,000	6,975	595	101,570	115,000
0500	50' span, 3 ton		.50	72		49,700	4,200	355	54,255	61,500
0525	5 ton		.50	72		51,500	4,200	355	56,055	64,000
0550	7.5 ton		.50	72		55,500	4,200	355	60,055	68,000
0575	10 ton		.40	90		64,000	5,250	445	69,695	78,500
0600	15 ton		.40	90		83,500	5,250	445	89,195	100,500
0625	25 ton	↓	.30	120	↓	98,500	6,975	595	106,070	120,000

41 22 13.19 Jib Cranes

		Crew	Daily Output	Labor-Hours	Unit	Material	2016 Bare Costs Labor	Equipment	Total	Total Incl O&P
0010	**JIB CRANES**									
0020	Jib crane, wall cantilever, 500 lb. capacity, 8' span	2 Mill	1	16	Ea.	1,425	820		2,245	2,850
0040	12' span		1	16		1,550	820		2,370	3,000
0060	16' span		1	16		1,725	820		2,545	3,175
0080	20' span		1	16		2,200	820		3,020	3,675
0100	1000 lb. capacity, 8' span		1	16		1,525	820		2,345	2,950
0120	12' span		1	16		1,650	820		2,470	3,100
0130	16' span		1	16		2,200	820		3,020	3,700
0150	20' span	↓	1	16	↓	2,575	820		3,395	4,100

Assemblies Section

Table of Contents

Table of Contents

How RSMeans Assemblies Data Works

Assemblies estimating provides a fast and reasonably accurate way to develop construction costs. An assembly is the grouping of individual work items, with appropriate quantities, to provide a cost for a major construction component in a convenient unit of measure.

An assemblies estimate is often used during early stages of design development to compare the cost impact of various design alternatives on total building cost.

Assemblies estimates are also used as an efficient tool to verify construction estimates.

Assemblies estimates do not require a completed design or detailed drawings. Instead, they are based on the general size of the structure and other known parameters of the project. The degree of accuracy of an assemblies estimate is generally within +/- 15%.

Most RSMeans assemblies consist of three major elements: a graphic, the system components, and the cost data itself. The **Graphic** is a visual representation showing

1 Unique 12-character Identifier

RSMeans assemblies are identified by a **unique 12-character identifier**. The assemblies are numbered using UNIFORMAT II, ASTM Standard E1557. The first 5 characters represent this system to Level 3. The last 7 characters represent further breakdown by RSMeans in order to arrange items in understandable groups of similar tasks. Line numbers are consistent across all RSMeans publications, so a line number in any RSMeans assemblies data set will always refer to the same work.

2 Narrative Descriptions

RSMeans assemblies descriptions appear in two formats: narrative and table. **Narrative descriptions** are shown in a hierarchical structure to make them readable. In order to read a complete description, read up through the indents to the top of the section. Include everything that is above and to the left that is not contradicted by information below.

A10 Foundations

A1010 Standard Foundations

This data illustrates and describes a spread footing system including concrete, forms, reinforcing and anchor bolts. Lines within System Components give the unit price and total price on a cost each basis for this system. Prices for spread footing systems are on Line Items A1010 230 1200 thru 2300. Both material quantities and labor costs have been adjusted for the system listed.

Factors: To adjust for job conditions other than normal working situations use Lines A1010 230 2900 thru 4000.

Example: You are to install the system in an existing occupied building. Access to the site and protection to the building are mandatory. Go to Lines A1010 230 3600 and 3800 and apply these percentages to the appropriate MAT. and INST. costs.

System Components	QUANTITY	UNIT	COST EACH MAT.	COST EACH INST.	COST EACH TOTAL
Interior column footing, 3' square 1' thick, 2000 psi concrete, Including forms, reinforcing, and anchor bolts.					
Concrete, 2000 psi	.330	C.Y.	37.29		37.29
Placing concrete	.330	C.Y.		18.90	18.90
Forms, footing, 4 uses	12.000	SFCA	9.24	69.60	78.84
Reinforcing	11.000	Lb.	5.83	7.15	12.98
Anchor bolts, 3/4" diameter	2.000	Ea.	2.92	9.12	12.04
TOTAL			55.28	104.77	160.05

A1010 230	Spread Footing	COST EACH MAT.	COST EACH INST.	COST EACH TOTAL
1200	3' square x 1' thick, 2000 psi concrete	55.50	104	159.50
1300	3000 psi concrete	57	104	161
1400	4000 psi concrete	59	104	163
1600	For alternate footing systems:			
1700	4' square x 1' thick, 2000 psi concrete	92.50	149	241.50
1800	3000 psi concrete	95.50	149	244.50
1900	4000 psi concrete	99	149	248
2100	5' square x 1'-3" thick, 2000 psi concrete	173	245	418
2200	3000 psi concrete	179	245	424
2300	4000 psi concrete	186	245	431
2600				
2700				
2900	Cut & patch to match existing construction, add, minimum	2%	3%	
3000	Maximum	5%	9%	
3100	Dust protection, add, minimum	1%	2%	
3200	Maximum	4%	11%	
3300	Equipment usage curtailment, add, minimum	1%	1%	
3400	Maximum	3%	10%	
3500	Material handling & storage limitation, add, minimum	1%	1%	
3600	Maximum	6%	7%	
3700	Protection of existing work, add, minimum	2%	2%	
3800	Maximum	5%	7%	
3900	Shift work requirements, add, minimum		5%	
4000	Maximum		30%	

For supplemental customizable square foot estimating forms, visit: **www.RSMeans.com/2016extras**

the typical appearance of the assembly in question, frequently accompanied by additional explanatory technical information describing the class of items. The **System Components** is a listing of the individual tasks that make up the assembly, including the quantity and unit of measure for each item, along with the cost of material and installation.

The **Assemblies Data** below lists prices for other similar systems with dimensional and/or size variations.

All RSMeans assemblies costs represent the cost for the installing contractor. An allowance for profit has been added to all material, labor, and equipment rental costs. A markup for labor burdens, including workers'

compensation, fixed overhead, and business overhead, is included with installation costs.

The information in RSMeans cost data represents a "national average" cost. This data should be modified to the project location using the **City Cost Indexes** or **Location Factors** tables found in the Reference Section.

A10 Foundations

A1010   Standard Foundations

This data illustrates and describes a spread footing system including concrete, forms, reinforcing and anchor bolts. Lines within System Components give the unit price and total price on a cost each basis for this system. Prices for spread footing systems are on Line Items A1010 230 1200 thru 2300. Both material quantities and labor costs have been adjusted for the system listed.

Factors: To adjust for job conditions other than normal working situations use Lines A1010 230 2900 thru 4000.

Example: You are to install the system in an existing occupied building. Access to the site and protection to the building are mandatory. Go to Lines A1010 230 3600 and 3800 and apply these percentages to the appropriate MAT. and INST. costs.

③ Unit of Measure

All RSMeans assemblies data includes a typical **Unit of Measure** used for estimating that item. For instance, for continuous footings or foundation walls the unit is linear feet (L.F.). For spread footings the unit is each (Ea.). The estimator needs to take special care that the unit in the data matches the unit in the takeoff. Abbreviations and unit conversions can be found in the Reference Section.

④ System Components

System components are listed separately to detail what is included in the development of the total system price.

System Components ④	QUANTITY	UNIT	COST EACH MAT.	COST EACH INST.	COST EACH ③ TOTAL
Interior column footing, 3' square 1' thick, 2000 psi concrete, Including forms, reinforcing, and anchor bolts.					
Concrete, 2000 psi	.330	C.Y.	37.29		37.29
Placing concrete	.330	C.Y.		18.90	18.90
Forms, footing, 4 uses	12.000	SFCA	9.24	69.60	78.84
Reinforcing	11.000	Lb.	5.83	7.15	12.98
Anchor bolts, 3/4" diameter	2.000	Ea.	2.92	9.12	12.04
TOTAL			55.28	104.77	160.05

A1010 230	Spread Footing	COST EACH MAT.	COST EACH INST.	COST EACH TOTAL
1200	3' square x 1' thick, 2000 psi concrete	55.50	104	159.50
1300	3000 psi concrete	57	104	161
1400	4000 psi concrete	59	104	163
1600	For alternate footing systems:			
1700	4' square x 1' thick, 2000 psi concrete	92.50	149	241.50
1800	3000 psi concrete	95.50	149	244.50
1900	4000 psi concrete	99	149	248
2100	5' square x 1'-3" thick, 2000 psi concrete	173	245	418
2200	3000 psi concrete	179	245	424
2300	4000 psi concrete	186	245	431
2600				
2700				
2900	Cut & patch to match existing construction, add, minimum	2%	3%	
3000	Maximum	5%	9%	
3100	Dust protection, add, minimum	1%	2%	
3200	Maximum	4%	11%	
3300	Equipment usage curtailment, add, minimum	1%	1%	
3400	Maximum	3%	10%	
3500	Material handling & storage limitation, add, minimum	1%	1%	
3600	Maximum	6%	7%	
3700	Protection of existing work, add, minimum	2%	2%	
3800	Maximum	5%	7%	
3900	Shift work requirements, add, minimum		5%	
4000	Maximum		30%	

How RSMeans Assemblies Data Works (Continued)

Sample Estimate

This sample demonstrates the elements of an estimate, including a tally of the RSMeans data lines. Published assemblies costs include all markups for labor burden and profit for the installing contractor. This estimate adds a summary of the markups applied by a general contractor on the installing contractor's work. These figures represent the total cost to the owner. The RSMeans location factor is applied at the bottom of the estimate to adjust the cost of the work to a specific location.

Project Name:	Interior Fit-out, ABC Office			
Location:	**Anywhere, USA**	**Date: 1/1/2016**		**R+R**
Assembly Number	**Description**	**Qty.**	**Unit**	**Subtotal**
C1010 114 1200 ❶	Wood partition, 2 x 4 @ 16" OC w/5/8" FR gypsum board	560.000	S.F.	$2,872.80
C1020 114 1800	Metal door & frame, flush hollow core, 3'-0" x 7'-0"	2.000	Ea.	$2,520.00
C3010 230 0080	Painting, brushwork, primer & 2 coats	1,120.000	S.F.	$1,422.40
C3020 410 0140	Carpet, tufted, nylon, roll goods, 12' wide, 26 oz	240.000	S.F.	$825.60
C3030 210 6000	Acoustic ceilings, 24" x 48" tile, tee grid suspension	200.000	S.F.	$1,180.00
D5020 125 0560	Receptacles incl plate, box, conduit, wire, 20 A duplex	8.000	Ea.	$2,232.00
D5020 125 0720	Light switch incl plate, box, conduit, wire, 20 A single pole	2.000	Ea.	$548.00
D5020 210 0560	Fluorescent fixtures, recess mounted, 20 per 1000 SF	200.000	S.F.	$2,236.00
	Assembly Subtotal			**$13,836.80**
	Sales Tax @ ❷	5 %		$ 345.92
	General Requirements @ ❸	7 %		$ 968.58
	Subtotal A			**$15,151.30**
	GC Overhead @ ❹	5 %		$ 757.56
	Subtotal B			**$15,908.86**
	GC Profit @ ❺	5 %		$ 795.44
	Subtotal C			**$16,704.30**
	Adjusted by Location Factor ❻	118.1		$ 19,727.78
	Architects Fee @ ❼	8 %		$ 1,578.22
	Contingency @ ❽	15 %		$ 2,959.17
	Project Total Cost			**$ 24,265.17**

This estimate is based on an interactive spreadsheet.
A copy of this spreadsheet is located on the RSMeans website at
www.RSMeans.com/2016extras.
You are free to download it and adjust it to your methodology.

① Work Performed
The body of the estimate shows the RSMeans data selected, including line numbers, a brief description of each item, its takeoff quantity and unit, and the total installed cost, including the installing contractor's overhead and profit.

② Sales Tax
If the work is subject to state or local sales taxes, the amount must be added to the estimate. In a conceptual estimate it can be assumed that one half of the total represents material costs. Therefore, apply the sales tax rate to 50% of the assembly subtotal.

③ General Requirements
This item covers project-wide needs provided by the general contractor. These items vary by project but may include temporary facilities and utilities, security, testing, project cleanup, etc. In assemblies estimates a percentage is used, typically between 5% and 15% of project cost.

④ General Contractor Overhead
This entry represents the general contractor's markup on all work to cover project administration costs.

⑤ General Contractor Profit
This entry represents the GC's profit on all work performed. The value included here can vary widely by project and is influenced by the GC's perception of the project's financial risk and market conditions.

⑥ Location Factor
RSMeans published data is based on national average costs. If necessary, adjust the total cost of the project using a location factor from the "Location Factor" table or the "City Cost Indexes" table found in the Reference Section. Use location factors if the work is general, covering the work of multiple trades. If the work is by a single trade (e.g., masonry) use the more specific data found in the City Cost Indexes.

To adjust costs by location factors, multiply the base cost by the factor and divide by 100.

⑦ Architect's Fee
If appropriate, add the design cost to the project estimate. These fees vary based on project complexity and size. Typical design and engineering fees can be found in the reference section.

⑧ Contingency
A factor for contingency may be added to any estimate to represent the cost of unknowns that may occur between the time that the estimate is performed and the time the project is constructed. The amount of the allowance will depend on the stage of design at which the estimate is done, and the contractor's assessment of the risk involved.

A1010 Standard Foundations

This data illustrates and describes a strip footing system including concrete, forms, reinforcing, keyway and dowels. Lines within System Components give the unit price and total price per linear foot for this system. Prices for strip footing systems are on Line Items A1010 120 1400 thru 2500. Both material quantities and labor costs have been adjusted for the system listed.

Factors: To adjust for job conditions other than normal working situations use Lines A1010 120 2900 thru 4000.

Example: You are to install this footing, and due to a lack of accessibility, only hand tools can be used. Material handling is also a problem. Go to Lines A1010 120 3400 and 3600 and apply these percentages to the appropriate MAT. and INST. costs.

System Components	QUANTITY	UNIT	COST PER L.F.		
			MAT.	INST.	TOTAL
Strip footing, 2'-0" wide x 1'-0" thick, 2000 psi concrete including forms Reinforcing, keyway, and dowels.					
Concrete, 2000 psi	.074	C.Y.	8.36		8.36
Placing concrete	.074	C.Y.		1.93	1.93
Forms, footing, 4 uses	2.000	S.F.	4.88	9.88	14.76
Reinforcing	3.170	Lb.	1.68	2.06	3.74
Keyway, 2" x 4", 4 uses	1.000	L.F.	.23	1.19	1.42
Dowels, #4 bars, 2' long, 24" O.C.	.500	Ea.	.39	1.43	1.82
TOTAL			15.54	16.49	32.03

A1010 120	Strip Footing	COST PER L.F.		
		MAT.	INST.	TOTAL
1400	2'-0" wide x 1' thick, 2000 psi concrete	15.55	16.50	32.05
1500	3000 psi concrete	15.90	16.50	32.40
1600	4000 psi concrete	16.35	16.50	32.85
1800	For alternate footing systems:			
1900	2'-6" wide x 1' thick, 2000 psi concrete	18.15	17.50	35.65
2000	3000 psi concrete	18.60	17.50	36.10
2100	4000 psi concrete	19.15	17.50	36.65
2300	3'-0" wide x 1' thick, 2000 psi concrete	20.50	18.50	39
2400	3000 psi concrete	21	18.50	39.50
2500	4000 psi concrete	21.50	18.50	40
2700				
2800				
2900	Cut & patch to match existing construction, add, minimum	2%	3%	
3000	Maximum	5%	9%	
3100	Dust protection, add, minimum	1%	2%	
3200	Maximum	4%	11%	
3300	Equipment usage curtailment, add, minimum	1%	1%	
3400	Maximum	3%	10%	
3500	Material handling & storage limitation, add, minimum	1%	1%	
3600	Maximum	6%	7%	
3700	Protection of existing work, add, minimum	2%	2%	
3800	Maximum	5%	7%	
3900	Shift work requirements, add, minimum		5%	
4000	Maximum		30%	

A1010 Standard Foundations

This data illustrates and describes a spread footing system including concrete, forms, reinforcing and anchor bolts. Lines within System Components give the unit price and total price on a cost each basis for this system. Prices for spread footing systems are on Line Items A1010 230 1200 thru 2300. Both material quantities and labor costs have been adjusted for the system listed.

Factors: To adjust for job conditions other than normal working situations use Lines A1010 230 2900 thru 4000.

Example: You are to install the system in an existing occupied building. Access to the site and protection to the building are mandatory. Go to Lines A1010 230 3600 and 3800 and apply these percentages to the appropriate MAT. and INST. costs.

System Components	QUANTITY	UNIT	COST EACH MAT.	COST EACH INST.	COST EACH TOTAL
Interior column footing, 3' square 1' thick, 2000 psi concrete, Including forms, reinforcing, and anchor bolts.					
Concrete, 2000 psi	.330	C.Y.	37.29		37.29
Placing concrete	.330	C.Y.		18.90	18.90
Forms, footing, 4 uses	12.000	SFCA	9.24	69.60	78.84
Reinforcing	11.000	Lb.	5.83	7.15	12.98
Anchor bolts, 3/4" diameter	2.000	Ea.	2.92	9.12	12.04
TOTAL			55.28	104.77	160.05

A1010 230	Spread Footing	MAT.	INST.	TOTAL
1200	3' square x 1' thick, 2000 psi concrete	55.50	104	159.50
1300	3000 psi concrete	57	104	161
1400	4000 psi concrete	59	104	163
1600	For alternate footing systems:			
1700	4' square x 1' thick, 2000 psi concrete	92.50	149	241.50
1800	3000 psi concrete	95.50	149	244.50
1900	4000 psi concrete	99	149	248
2100	5' square x 1'-3" thick, 2000 psi concrete	173	245	418
2200	3000 psi concrete	179	245	424
2300	4000 psi concrete	186	245	431
2600				
2700				
2900	Cut & patch to match existing construction, add, minimum	2%	3%	
3000	Maximum	5%	9%	
3100	Dust protection, add, minimum	1%	2%	
3200	Maximum	4%	11%	
3300	Equipment usage curtailment, add, minimum	1%	1%	
3400	Maximum	3%	10%	
3500	Material handling & storage limitation, add, minimum	1%	1%	
3600	Maximum	6%	7%	
3700	Protection of existing work, add, minimum	2%	2%	
3800	Maximum	5%	7%	
3900	Shift work requirements, add, minimum		5%	
4000	Maximum		30%	

A1010 Standard Foundations

General: Footing drains can be placed either inside or outside of foundation walls depending upon the source of water to be intercepted. If the source of subsurface water is principally from grade or a subsurface stream above the bottom of the footing, outside drains should be used. For high water tables, use inside drains or both inside and outside.

The effectiveness of underdrains depends on good waterproofing. This must be carefully installed and protected during construction.

Costs below include the labor and materials for the pipe and 6" of crushed stone around pipe. Excavation and backfill are not included.

System Components	QUANTITY	UNIT	COST PER L.F.		
			MAT.	INST.	TOTAL
Foundation underdrain, outside, PVC 4" diameter.					
PVC pipe 4" diam. S.D.R. 35	1.000	L.F.	1.64	4.44	6.08
Crushed stone 3/4" to 1/2"	.070	C.Y.	1.86	.94	2.80
TOTAL			3.50	5.38	8.88

A1010 310	Foundation Underdrain	COST PER L.F.		
		MAT.	INST.	TOTAL
1000	Foundation underdrain, outside only, PVC, 4" diameter	3.50	5.40	8.90
1100	6" diameter	6	5.95	11.95
1400	Perforated HDPE, 6" diameter	4.13	2.31	6.44
1450	8" diameter	6.40	2.89	9.29
1500	12" diameter	11.30	7.05	18.35
1600	Corrugated metal, 16 ga. asphalt coated, 6" diameter	9.55	5.60	15.15
1650	8" diameter	12.30	6	18.30
1700	10" diameter	15.20	7.75	22.95
3000	Outside and inside, PVC, 4" diameter	7	10.80	17.80
3100	6" diameter	12	11.95	23.95
3400	Perforated HDPE, 6" diameter	8.25	4.62	12.87
3450	8" diameter	12.75	5.80	18.55
3500	12" diameter	22.50	14.05	36.55
3600	Corrugated metal, 16 ga., asphalt coated, 6" diameter	19.05	11.20	30.25
3650	8" diameter	24.50	11.95	36.45
3700	10" diameter	30.50	15.45	45.95
4700	Cut & patch to match existing construction, add, minimum	2%	3%	
4800	Maximum	5%	9%	
4900	Dust protection, add, minimum	1%	2%	
5000	Maximum	4%	11%	
5100	Equipment usage curtailment, add, minimum	1%	1%	
5200	Maximum	3%	10%	
5300	Material handling & storage limitation, add, minimum	1%	1%	
5400	Maximum	6%	7%	
5500	Protection of existing work, add, minimum	2%	2%	
5600	Maximum	5%	7%	
5700	Shift work requirements, add, minimum		5%	
5800	Maximum		30%	
5900	Temporary shoring and bracing, add, minimum	2%	5%	
6000	Maximum	5%	12%	

A1030   Slab on Grade

This data illustrates and describes a slab on grade system including slab, bank run gravel, bulkhead forms, placing concrete, welded wire fabric, vapor barrier, steel trowel finish and curing paper. Lines within System Components give the unit price and total price per square foot for this system. Prices for slab on grade systems are on Line Items A1030 110 1400 thru 2600. Both material quantities and labor costs have been adjusted for the system listed.

Factors: To adjust for job conditions other than normal working situations use Lines A1030 110 2900 thru 4000.

Example: You are to install the system at a site where protection of the existing building is required. Go to Line A1030 110 3800 and apply these percentages to the appropriate MAT. and INST. costs.

System Components			COST PER S.F.		
	QUANTITY	UNIT	MAT.	INST.	TOTAL
Ground slab, 4″ thick, 3000 psi concrete, 4″ granular base, vapor barrier					
Welded wire fabric, screed and steel trowel finish.					
Concrete, 4″ thick, 3000 psi concrete	.012	C.Y.	1.42		1.42
Bank run gravel, 4″ deep	.074	C.Y.	.27	.10	.37
Polyethylene vapor barrier, 10 mil.	.011	C.S.F.	.09	.19	.28
Bulkhead forms, expansion material	.100	L.F.	.04	.40	.44
Welded wire fabric, 6 x 6 - #10/10	.011	C.S.F.	.16	.43	.59
Place concrete	.012	C.Y.		.35	.35
Screed & steel trowel finish	1.000	S.F.		.99	.99
TOTAL			1.98	2.46	4.44

A1030 110	Interior Slab on Grade	COST PER S.F.		
		MAT.	INST.	TOTAL
1400	4″ thick slab, 3000 PSI concrete, 4″ deep bank run gravel	1.98	2.46	4.44
1500	6″ deep bank run gravel	2.19	2.46	4.65
1600	12″ deep bank run gravel	2.67	2.49	5.16
1700				
1800				
1900				
2000	For alternate slab systems:			
2100	5″ thick slab, 3000 psi concrete, 6″ deep bank run gravel	2.54	2.54	5.08
2200	12″ deep bank run gravel	3.02	2.57	5.59
2300				
2400				
2500	6″ thick slab, 3000 psi concrete, 6″ deep bank run gravel	3.01	2.65	5.66
2600	12″ deep bank run gravel	3.49	2.68	6.17
2700				
2900	Cut & patch to match existing construction, add, minimum	2%	3%	
3000	Maximum	5%	9%	
3100	Dust protection, add, minimum	1%	2%	
3200	Maximum	4%	11%	
3300	Equipment usage curtailment, add, minimum	1%	1%	
3400	Maximum	3%	10%	
3500	Material handling & storage limitation, add, minimum	1%	1%	
3600	Maximum	6%	7%	
3700	Protection of existing work, add, minimum	2%	2%	
3800	Maximum	5%	7%	
3900	Shift work requirements, add, minimum		5%	
4000	Maximum		30%	

A2010 Basement Excavation

This data illustrates and describes continuous footing and trench excavation systems including a wheel mounted backhoe, operator, equipment rental, fuel, oil, mobilization, no hauling or backfill. Lines within System Components give the unit price and total price per linear foot for this system. Prices for alternate continuous footing and trench excavation systems are on Line Items A2010 130 1200 thru 2700. Both material quantities and labor costs have been adjusted for the system listed.

Factors: To adjust for job conditions other than normal working situations use Lines A2010 130 2900 thru 4000.

Example: You are to install the system and supply dust protection. Go to Line A2010 130 3000 and apply this percentage to the appropriate TOTAL costs.

System Components	QUANTITY	UNIT	COST PER L.F.		
			EQUIP.	LABOR	TOTAL
Continuous footing or trench excav. w/ 3/4 C.Y. wheel mntd backhoe Including operator, equipment rental, fuel, oil and mobilization. Trench Is 4' wide at bottom, 3' deep w/sides sloped 1 to 2 and 200' long. Costs Are based on a production rate of 240 C.Y./day. No hauling/backfill incl.					
Equipment operator	4.000	Hr.		322.50	322.50
3/4 C.Y. backhoe, wheel mounted	.500	Day	98.45		98.45
Operating expense (fuel, oil)	4.000	Hr.	102.52		102.52
Mobilization	1.000	Ea.	1,150	1,850	3,000
TOTAL			1,350.97	2,172.50	3,523.47
COST PER L.F.		L.F.	6.75	10.86	17.61

A2010 130	Excavation, Footings or Trench	COST PER L.F.		
		EQUIP.	LABOR	TOTAL
1200	For alternate trench sizes, 4' bottom with sloped sides:			
1300	2' deep, 50' long	7.55	18.25	25.80
1400	100' long	3.96	9.15	13.11
1500	300' long	4.48	7.25	11.73
1600	4' deep, 50' long	7.95	18.25	26.20
1700	100' long	4.35	9.15	13.50
1800	300' long	2.28	4.26	6.54
1900	6' deep, 50' long	8.55	18.25	26.80
2000	100' long	4.56	10.35	14.91
2100	300' long	3.55	6.25	9.80
2200	8' deep, 50' long	9.20	18.25	27.45
2300	100' long	6.60	12.35	18.95
2400	300' long	4.89	6.25	11.14
2500	10' deep, 50' long	11.25	21.50	32.75
2600	100' long	8.40	15.20	23.60
2700	300' long	6.50	10.95	17.45
2800				
2900	Dust protection, add, minimum		2%	2%
3000	Maximum		11%	11%
3100	Equipment usage curtailment, add, minimum		1%	1%
3200	Maximum		10%	10%
3300	Material handling & storage limitation, add, minimum		1%	1%
3400	Maximum		7%	7%
3500	Protection of existing work, add, minimum		2%	2%
3600	Maximum		7%	7%
3700	Shift work requirements, add, minimum		5%	5%
3800	Maximum		30%	30%
3900	Temporary shoring and bracing, add, minimum		5%	5%
4000	Maximum		12%	12%

A2010 Basement Excavation

This data illustrates and describes foundation excavation systems including a backhoe-loader, operator, equipment rental, fuel, oil, mobilization, hauling material, no backfilling. Lines within System Components give the unit price and total price per cubic yard for this system. Prices for foundation excavation systems are on Line Items A2010 140 1600 thru 2200. Both material quantities and labor costs have been adjusted for the system listed.

Factors: To adjust for job conditions other than normal working situations use Lines A2010 140 2900 thru 4000.

Example: You are to install the system with the use of temporary shoring and bracing. Go to Line A2010 140 3900 and apply these percentages to the appropriate TOTAL costs.

System Components	QUANTITY	UNIT	COST PER C.Y. EQUIP.	COST PER C.Y. LABOR	COST PER C.Y. TOTAL
Foundation excav w/ 3/4 C.Y. backhoe-loader, incl. operator, equip Rental, fuel, oil, and mobilization. Hauling of excavated material is Included. Prices based on one day production of 360 C.Y. in medium soil Without backfilling.					
Equipment operator	8.000	Hr.		645	645
Backhoe-loader, 3/4 C.Y.	1.000	Day	196.90		196.90
Operating expense (fuel, oil)	8.000	Hr.	205.04		205.04
Hauling, 12 C.Y. trucks, 1 mile round trip	360.000	L.C.Y.	1,087.20	795.60	1,882.80
Mobilization	1.000	Ea.	260	590	850
TOTAL			1,749.14	2,030.60	3,779.74
COST PER C.Y.		C.Y.	4.85	5.62	10.50

A2010 140	Excavation, Foundation	COST PER C.Y. EQUIP.	COST PER C.Y. LABOR	COST PER C.Y. TOTAL
1600	100 C.Y.	6.20	11.75	17.95
1700	200 C.Y.	5.45	7	12.45
1800	300 C.Y.	5.05	6	11.05
1850	360 C.Y.	4.86	5.63	10.49
1900	400 C.Y.	4.77	5.45	10.22
2000	500 C.Y.	4.66	5.20	9.86
2100	600 C.Y.	4.55	4.98	9.53
2200	700 C.Y.	4.49	4.80	9.29
2300				
2400				
2900	Dust protection, add, minimum		2%	2%
3000	Maximum		11%	11%
3100	Equipment usage curtailment, add, minimum		1%	1%
3200	Maximum		10%	10%
3300	Material handling & storage limitation, add, minimum		1%	1%
3400	Maximum		7%	7%
3500	Protection of existing work, add, minimum		2%	2%
3600	Maximum		7%	7%
3700	Shift work requirements, add, minimum		5%	5%
3800	Maximum		30%	30%
3900	Temporary shoring and bracing, add, minimum		5%	5%
4000	Maximum		12%	12%

579

A2020 Basement Walls

This data illustrates and describes a concrete wall system including concrete, placing concrete, forms, reinforcing, insulation, waterproofing and anchor bolts. Lines within System Components give the unit price and total price per linear foot for this system. Prices for concrete wall systems are on Line Items A2020 120 1400 thru 2600. Both material quantities and labor costs have been adjusted for the system listed.

Factors: To adjust for job conditions other than normal working situations use Lines A2020 120 2900 thru 4000.

Example: You are to install this wall system where delivery of material is difficult. Go to Line A2020 120 3600 and apply these percentages to the appropriate MAT. and INST. costs.

System Components			COST PER L.F.		
	QUANTITY	UNIT	MAT.	INST.	TOTAL
Cast in place concrete foundation wall, 8″ thick, 3′ high, 2500 psi					
Concrete including forms, reinforcing, waterproofing, and anchor bolts.					
Concrete, 2500 psi, 8″ thick, 3′ high	.070	C.Y.	8.05		8.05
Forms, wall, 4 uses	6.000	S.F.	6.60	43.80	50.40
Reinforcing	6.000	Lb.	3.18	2.76	5.94
Placing concrete	.070	C.Y.		3.59	3.59
Waterproofing	3.000	S.F.	1.47	3.60	5.07
Rigid insulaton, 1″ polystyrene	3.000	S.F.	.87	2.79	3.66
Anchor bolts, 1/2″ diameter, 4′ O.C.	.250	Ea.	.41	1.15	1.56
TOTAL			20.58	57.69	78.27

A2020 120	Concrete Wall	COST PER L.F.		
		MAT.	INST.	TOTAL
1400	8″ thick, 2500 psi concrete, 3′ high	20.50	57.50	78
1500	4′ high	28	77	105
1600	6′ high	42	114	156
1700	8′ high	55.50	153	208.50
1800	3500 psi concrete, 4′ high	28.50	77	105.50
1900	6′ high	43	114	157
2000	8′ high	57	153	210
2100	12″ thick, 2500 psi concrete, 4′ high	36	81.50	117.50
2200	6′ high	52	120	172
2300	8′ high	69.50	160	229.50
2400	3500 psi concrete, 4′ high	37	81.50	118.50
2500	8′ high	71	160	231
2600	10′ high	88	199	287
2700				
2900	Cut & patch to match existing construction, add, minimum	2%	3%	
3000	Maximum	5%	9%	
3100	Dust protection, add, minimum	1%	2%	
3200	Maximum	4%	11%	
3300	Equipment usage curtailment, add, minimum	1%	1%	
3400	Maximum	3%	10%	
3500	Material handling & storage limitation, add, minimum	1%	1%	
3600	Maximum	6%	7%	
3700	Protection of existing work, add, minimum	2%	2%	
3800	Maximum	5%	7%	
3900	Shift work requirements, add, minimum		5%	
4000	Maximum		30%	

A20 Basement Construction

A2020 Basement Walls

This data illustrates and describes a concrete block wall system including concrete block, masonry reinforcing, parging, waterproofing insulation and anchor bolts. Lines within System Components give the unit price and total price per linear foot for this system. Prices for concrete block wall systems are on Line Items A2020 140 1200 thru 2600. Both material quantities and labor costs have been adjusted for the system listed.

Factors: To adjust for job conditions other than normal working situations use Lines A2020 140 2900 thru 4000.

Example: You are to install the system to match an existing foundation wall. Go to Line A2020 140 3000 and apply these percentages to the appropriate MAT. and INST. costs.

System Components	QUANTITY	UNIT	COST PER L.F.		
			MAT.	INST.	TOTAL
Concrete block, 8″ thick, masonry reinforcing, parged and					
Waterproofed, insulation and anchor bolts, wall 2′-8″ high.					
Concrete block, 8″ x 8″ x 16″	2.670	S.F.	9.51	17.62	27.13
Masonry reinforcing	2.000	L.F.	.56	.40	.96
Parging	.193	S.Y.	.78	2.75	3.53
Waterproofing	2.670	S.F.	1.31	3.20	4.51
Insulation, 1″ rigid polystyrene	2.670	S.F.	.77	2.48	3.25
Anchor bolts, 1/2″ diameter, 4′ O.C.	.250	Ea.	.41	1.15	1.56
TOTAL			13.34	27.60	40.94

A2020 140	Concrete Block Wall	COST PER L.F.		
		MAT.	INST.	TOTAL
1200	8″ thick block, 2′-8″ high	13.35	27.50	40.85
1300	4′ high	19.80	40.50	60.30
1400	6′ high	29.50	60.50	90
1500	8′ high	39	80.50	119.50
1600	Grouted solid, 4′ high	25	54.50	79.50
1700	6′ high	37	81.50	118.50
1800	8′ high	49.50	108	157.50
2100	12″ thick block, 4′ high	26.50	58.50	85
2200	6′ high	39.50	87	126.50
2300	8′ high	53	115	168
2400	Grouted solid, 4′ high	35.50	73	108.50
2500	6′ high	52.50	109	161.50
2600	8′ high	70	145	215
2700				
2900	Cut & patch to match existing construction, add, minimum	2%	3%	
3000	Maximum	5%	9%	
3100	Dust protection, add, minimum	1%	2%	
3200	Maximum	4%	11%	
3300	Equipment usage curtailment, add, minimum	1%	1%	
3400	Maximum	3%	10%	
3500	Material handling & storage limitation, add, minimum	1%	1%	
3600	Maximum	6%	7%	
3700	Protection of existing work, add, minimum	2%	2%	
3800	Maximum	5%	7%	
3900	Shift work requirements, add, minimum		5%	
4000	Maximum		30%	

B1010 Floor Construction

This data illustrates and describes a reinforced concrete slab system including concrete, placing concrete, formwork, reinforcing, steel trowel finish, V.A. floor tile and acoustical spray ceiling finish. Lines within System Components give the unit price per square foot for this system. Prices for reinforced concrete slab systems are on Line Items B1010 225 1400 thru 1800. Both material quantities and labor costs have been adjusted for the system listed.

Factors: To adjust for job conditions other than normal working situations use Lines B1010 225 2700 thru 4000.

Example: You are to install the system to match an existing floor system. Go to Line B1010 225 2800 and apply these percentages to the appropriate MAT. and INST. costs.

System Components			COST PER S.F.		
	QUANTITY	UNIT	MAT.	INST.	TOTAL
Flat slab system, reinforced concrete with vinyl tile floor, sprayed Acoustical ceiling finish, not including columns.					
Concrete, 4000 psi, 6" thick	.020	C.Y.	2.48		2.48
Placing concrete	.020	C.Y.		.72	.72
Formwork, 4 use	1.000	S.F.	1.34	6.60	7.94
Edge form	.120	S.F.	.07	1.28	1.35
Reinforcing steel	3.000	Lb.	1.59	1.95	3.54
Steel trowel finish	1.000	S.F.		1.31	1.31
Vinyl floor tile, 1/8" thick, premium colors/patterns	1.000	S.F.	7.70	1.12	8.82
Acoustical spray ceiling finish, solvent based	1.000	S.F.	.30	.18	.48
TOTAL			13.48	13.16	26.64

B1010 225	Floor - Ceiling, Concrete Slab		COST PER S.F.		
			MAT.	INST.	TOTAL
1400	Concrete, 4000 psi, 6" thick		13.50	13.15	26.65
1500	7" thick		13.95	13.70	27.65
1600	8" thick		14.65	14.40	29.05
1700	9" thick		15.25	15	30.25
1800	10" thick		15.95	15.70	31.65
1900					
2000					
2100					
2200					
2300					
2400					
2500					
2700	Cut & patch to match existing construction, add, minimum		2%	3%	
2800	Maximum		5%	9%	
2900	Dust protection, add, minimum		1%	2%	
3000	Maximum		4%	11%	
3100	Equipment usage curtailment, add, minimum		1%	1%	
3200	Maximum		3%	10%	
3300	Material handling & storage limitation, add, minimum		1%	1%	
3400	Maximum		6%	7%	
3500	Protection of existing work, add, minimum		2%	2%	
3600	Maximum		5%	7%	
3700	Shift work requirements, add, minimum			5%	
3800	Maximum			30%	
3900	Temporary shoring and bracing, add, minimum		2%	5%	
4000	Maximum		5%	12%	

This data illustrates and describes a hollow core prestressed concrete panel system including hollow core slab, grout, carpet, carpet padding, and sprayed ceiling. Lines within System Components give the unit price and total price per square foot for this system. Prices for hollow core prestressed concrete panel systems are on Line Items B1010 232 1200 thru 1600. Both material and labor costs have been adjusted for the system listed.

Factors: To adjust for job conditions other than normal working situations use Lines B1010 232 2700 thru 3800.

Example: You are to install the system where dust control is a major concern. Go to Line B1010 232 2800 and apply these percentages to the appropriate MAT. and INST. costs.

System Components	QUANTITY	UNIT	COST PER S.F.		
			MAT.	INST.	TOTAL
Precast hollow core plank with carpeted floors, padding					
And sprayed textured ceiling.					
Hollow core plank, 4" thick with grout topping	1.000	S.F.	6.75	3.58	10.33
Nylon carpet, 26 oz. medium traffic	.110	S.Y.	4.79	.83	5.62
Carpet padding, 20 oz./sq. yd.	.110	S.Y.	.56	.41	.97
Sprayed texture ceiling	1.000	S.F.	.04	.75	.79
TOTAL			12.14	5.57	17.71

B1010 232	Floor - Ceiling, Conc. Panel	COST PER S.F.		
		MAT.	INST.	TOTAL
1200	Hollow core concrete plank, with grout, 4" thick	12.15	5.55	17.70
1300	6" thick	13	5.05	18.05
1400	8" thick	13.85	4.67	18.52
1500	10" thick	14.20	4.37	18.57
1600	12" thick	14.50	4.13	18.63
1700				
1800				
1900				
2000				
2100				
2200				
2300				
2400				
2500				
2700	Dust protection, add, minimum	1%	2%	
2800	Maximum	4%	11%	
2900	Equipment usage curtailment, add, minimum	1%	1%	
3000	Maximum	3%	10%	
3100	Material handling & storage limitation, add, minimum	1%	1%	
3200	Maximum	6%	7%	
3300	Protection of existing work, add, minimum	2%	2%	
3400	Maximum	5%	7%	
3500	Shift work requirements, add, minimum		5%	
3600	Maximum		30%	
3700	Temporary shoring and bracing, add, minimum	2%	5%	
3800	Maximum	5%	12%	

585

B10 Superstructure

B1010 Floor Construction

This data illustrates and describes a hollow core plank system including precast hollow core plank, concrete topping, V.A. tile, concealed suspension system and ceiling tile. Lines within System Components give the unit price and total price per square foot for this system. Prices for hollow core plank systems are on Line Items B1010 233 1400 thru 1800. Both material quantities and labor costs have been adjusted for the system listed.

Factors: To adjust for job conditions other than normal working situations use Lines B1010 233 2900 thru 4000.

Example: You are to install the system where equipment usage is a problem. Go to Line B1010 233 3200 and apply these percentages to the appropriate MAT. and INST. costs.

System Components			COST PER S.F.		
	QUANTITY	UNIT	MAT.	INST.	TOTAL
Precast hollow core plank with 2″ topping, vinyl composition floor tile					
And suspended acoustical tile ceiling.					
Hollow core concrete plank, 4″ thick	1.000	S.F.	6.75	3.58	10.33
Finishing floors, granolithic topping, 1:1:1-1/2 mix, 2″ thick	1.000	S.F.	1.43	5.90	7.33
Vinyl composition tile, .08″ thick	1.000	S.F.	1.57	1.12	2.69
Concealed Z bar suspension system, 12″ module	1.000	S.F.	.95	1.22	2.17
Ceiling tile, mineral fiber, 3/4″ thick	1.000	S.F.	3.22	2.11	5.33
TOTAL			13.92	13.93	27.85

B1010 233	Floor - Ceiling, Conc. Plank	COST PER S.F.		
		MAT.	INST.	TOTAL
1400	Hollow core concrete plank, 4″ thick	13.90	13.90	27.80
1500	6″ thick	14.75	13.45	28.20
1600	8″ thick	15.60	13.05	28.65
1700	10″ thick	15.95	12.70	28.65
1800	12″ thick	16.25	12.50	28.75
1900				
2000				
2100				
2200				
2300				
2400				
2500				
2600				
2700				
2900	Dust protection, add, minimum	1%	2%	
3000	Maximum	4%	11%	
3100	Equipment usage curtailment, add, minimum	1%	1%	
3200	Maximum	3%	10%	
3300	Material handling & storage limitation, add, minimum	1%	1%	
3400	Maximum	6%	7%	
3500	Protection of existing work, add, minimum	2%	2%	
3600	Maximum	5%	7%	
3700	Shift work requirements, add, minimum		5%	
3800	Maximum		30%	
3900	Temporary shoring and bracing, add, minimum	2%	5%	
4000	Maximum	5%	12%	

B1010 Floor Construction

This data illustrates and describes a structural steel w/metal decking and concrete system including steel beams, steel decking, shear studs, concrete, placing concrete, edge form, steel trowel finish, curing, wire fabric, fireproofing, beams and decking, tile floor, and suspended ceiling. Lines within System Components give the unit price and total price per square foot for this system. Prices for alternate structural steel w/metal decking and concrete systems are on Line Items B1010 243 1900 thru 2400. Both material quantities and labor costs have been adjusted for the system listed.

Factors: To adjust for job conditions other than normal working situations use Lines B1010 243 2900 thru 4000.

Example: You are to install the system where material handling and storage are a problem. Go to Line B1010 243 3400 and apply these percentages to the appropriate MAT. and INST. costs.

System Components	QUANTITY	UNIT	COST PER S.F. MAT.	COST PER S.F. INST.	COST PER S.F. TOTAL
Composite structural beams, 20 ga. 3″ deep steel decking, shear studs, 3000 psi, concrete, placing concrete, edge forms, steel trowel finish, Welded wire fabric fireproofing, tile floor and suspended ceiling.					
Steel deck, 20 gage, 3″ deep, galvanized	1.000	S.F.	2.60	1.06	3.66
Structural steel framing	.004	Ton	11.70	2.83	14.53
Shear studs, 3/4″	.250	Ea.	.14	.49	.63
Concrete, 3000 psi, 5″ thick	.015	C.Y.	1.77		1.77
Place concrete	.015	C.Y.		.54	.54
Edge form	.140	L.F.	.03	.67	.70
Steel trowel finish	1.000	S.F.		1.31	1.31
Welded wire fabric 6 x 6 - #10/10	.011	C.S.F.	.16	.43	.59
Fireproofing, sprayed	1.880	S.F.	1.11	2.15	3.26
Vinyl composition floor tile	1.000	S.F.	1.31	1.12	2.43
Suspended acoustical ceiling	1.000	S.F.	3.74	1.67	5.41
TOTAL			22.56	12.27	34.83

B1010 243	Floor - Ceiling, Struc. Steel	COST PER S.F. MAT.	COST PER S.F. INST.	COST PER S.F. TOTAL
1900	Composite deck, galvanized, 3″ deep, 22 ga.	22.50	12.20	34.70
1950	20 ga.	22.50	12.25	34.75
2000	18 ga.	23	12.35	35.35
2100	16 ga.	24.50	12.40	36.90
2200	Composite deck, galvanized, 2″ deep, 22 ga.	22	12.05	34.05
2300	20 ga.	22.50	12.10	34.60
2400	16 ga.	23.50	12.20	35.70
2500				
2900	Dust protection, add, minimum	1%	2%	
3000	Maximum	4%	11%	
3100	Equipment usage curtailment, add, minimum	1%	1%	
3200	Maximum	3%	10%	
3300	Material handling & storage limitation, add, minimum	1%	1%	
3400	Maximum	6%	7%	
3500	Protection of existing work, add, minimum	2%	2%	
3600	Maximum	5%	7%	
3700	Shift work requirements, add, minimum		5%	
3800	Maximum		30%	
3900	Temporary shoring and bracing, add, minimum	2%	5%	
4000	Maximum	5%	12%	

B10 Superstructure

B1010   Floor Construction

This data illustrates and describes an open web joist and steel slab-form system including open web steel joist, slab form, concrete, placing concrete, wire fabric, steel trowel finish, tile floor and plasterboard. Lines within System Components give the unit price and total price per square foot for this system. Prices for open web joists and steel slab-form systems are on Line Items B1010 245 1800 thru 2200. Both material quantities and labor costs have been adjusted for the systems listed.

Factors: To adjust for job conditions other than normal working situations use Lines B1010 245 2900 thru 4000.

Example: You are to install the system in a congested commercial area and most work will be done at night. Go to Line B1010 245 3800 and apply this percentage to the appropriate INST. cost.

System Components	QUANTITY	UNIT	COST PER S.F.		
			MAT.	INST.	TOTAL
Open web steel joists, 24″ O.C., slab form, 3000 psi concrete, welded					
Wire fabric, steel trowel finish, floor tile, 5/8″ drywall ceiling.					
Open web steel joists, 12″ deep, 5.2#/L.F., 24″ O.C.	2.630	Lb.	2.58	1.23	3.81
Slab form, 28 gage, 9/16″ deep, galvanized	1.000	S.F.	1.52	.80	2.32
Concrete, 3000 psi, 2-1/2″ thick	.008	C.Y.	.94		.94
Placing concrete	.008	C.Y.		.29	.29
Welded wire fabric 6 x 6 - #10/10	.011	C.S.F.	.16	.43	.59
Steel trowel finish	1.000	S.F.		1.31	1.31
Vinyl composition floor tile	1.000	S.F.	1.31	1.12	2.43
Ceiling furring, 3/4″ channels, 24″ O.C.	1.000	S.F.	.25	1.42	1.67
Gypsum drywall, 5/8″ thick, finished	1.000	S.F.	.43	1.66	2.09
Paint ceiling	1.000	S.F.	.21	.68	.89
TOTAL			7.40	8.94	16.34

B1010 245	Floor - Ceiling, Steel Joists	COST PER S.F.		
		MAT.	INST.	TOTAL
1800	Open web joists, 12″ deep, 5.2#/L.F.	7.40	8.95	16.35
1900	16″ deep, 6.6#/L.F.	8.05	9.25	17.30
2000	20″ deep, 8.4# L.F.	8.95	9.65	18.60
2100	24″ deep, 11.5# L.F.	10.45	10.40	20.85
2200	26″ deep, 12.8# L.F.	11.10	10.75	21.85
2300				
2400				
2500				
2600				
2700				
2900	Dust protection, add, minimum	1%	2%	
3000	Maximum	4%	11%	
3100	Equipment usage curtailment, add, minimum	1%	1%	
3200	Maximum	3%	10%	
3300	Material handling & storage limitation, add, minimum	1%	1%	
3400	Maximum	6%	7%	
3500	Protection of existing work, add, minimum	2%	2%	
3600	Maximum	5%	7%	
3700	Shift work requirements, add, minimum		5%	
3800	Maximum		30%	
3900	Temporary shoring and bracing, add, minimum	2%	5%	
4000	Maximum	5%	12%	

B1010 Floor Construction

Joist Hangers

Existing Girder

Fiberglass Insulation,
Kraft Faced, 9" Thick

Steel Cross Bridging

12" Open Web Wood Joists, 2'-0" o.c.

3/4" CDX Plywood Sub-floor

Existing Foundation

2" x 8" Sill

Anchor Bolts, 1/2" x 8"

Open Web Wood Joists

This data illustrates and describes floor framing systems including sills, joist hangers, joists, bridging, subfloor and insulation. Lines within systems components give the unit price and total price per square foot for this system. Prices for alternate joist types are on line items B1010 262 0100 through 0700. Both material quantities and labor costs have been adjusted for the system listed.

Factors: To adjust for job conditions other than normal working situations use lines B1010 262 2700 through 4000.

Example: You are to install the system where delivery of material is difficult. Go to line B1010 262 3400 and apply these percentages to the appropriate MAT. and INST. costs.

System Components	QUANTITY	UNIT	COST PER S.F.		
			MAT.	INST.	TOTAL
Open web wood joists, sill, butt to girder, 5/8" subfloor, insul.					
Anchor bolt, hook type, with nut and washer, 1/2" x 8" long	.250	Ea.	.04	.11	.15
Joist & beam hanger, 18 ga. galvanized	.750	Ea.	.11	.29	.40
Bridging, steel, compression, for 2" x 12" joists	.010	C.Pr.	.01	.03	.04
Framing, open web wood joists, 12" deep	.001	M.L.F.	3.65	1.45	5.10
Framing, sills, 2" x 8"	.001	M.B.F.	.69	1.90	2.59
Sub-floor, plywood, CDX, 3/4" thick	1.000	S.F.	.84	.94	1.78
Insulation, fiberglass, kraft face, 9" thick x 23" wide	1.000	S.F.	.80	.47	1.27
TOTAL			6.14	5.19	11.33

B1010 262	Open Web Wood Joists	COST PER S.F.		
		MAT.	INST.	TOTAL
0100	12" open web joists, sill, butt to girder, 5/8" sub-floor	6.15	5.20	11.35
0200	14" joists	6.50	5.30	11.80
0300	16" joists	6.65	5.35	12
0400	18" joists	6.75	5.45	12.20
0450	Wood struc." I " joists, 24" O.C., to 24' span, 50 PSF LL, butt to girder	4.79	4.81	9.60
0500	55 PSF LL	5.05	4.88	9.93
0600	to 30' span, 45 PSF LL	5.15	4.72	9.87
0700	55 PSF LL	6.95	4.81	11.76
2700	Cut & patch to match existing construction, add, minimum	2%	3%	
2800	Maximum	5%	9%	
2900	Dust protection, add, minimum	1%	2%	
3000	Maximum	4%	11%	
3100	Equipment usage curtailment, add, minimum	1%	1%	
3200	Maximum	3%	10%	
3300	Material handling & storage limitation, add, minimum	1%	1%	
3400	Maximum	6%	7%	
3500	Protection of existing work, add, minimum	2%	2%	
3600	Maximum	5%	7%	
3700	Shift work requirements, add, minimum		5%	
3800	Maximum		30%	
3900	Temporary shoring and bracing, add, minimum	2%	5%	
4000	Maximum	5%	12%	

589

For customer support on your **Commercial Renovation Cost Data**, call 877.742.4719.

B1010 Floor Construction

This data illustrates and describes a wood joist floor system including wood joist, oak floor, sub-floor, bridging, sand and finish floor, furring, plasterboard, taped, finished and painted ceiling. Lines within System Components give the unit price and total price per square foot for this system. Prices for wood joist floor systems are on Line Items B1010 263 1700 thru 2300. Both material quantities and labor costs have been adjusted for the system listed.

Factors: To adjust for job conditions other than normal working situations use Lines B1010 263 2700 thru 4000.

Example: You are to install the system during off peak hours, 6 P.M. to 2 A.M. Go to Line B1010 263 3800 and apply this percentage to the appropriate INST. costs.

System Components			COST PER S.F.		
	QUANTITY	UNIT	MAT.	INST.	TOTAL
Wood joists, 2″ x 8″, 16″ O.C.,oak floor (sanded & finished),1/2″ sub					
Floor, 1″x 3″ bridging, furring, 5/8″ drywall, taped, finished and painted.					
Wood joists, 2″ x 8″, 16″ O.C.	1.000	L.F.	.93	1.15	2.08
Subfloor plywood CDX 1/2″	1.000	SF Flr.	.75	.84	1.59
Bridging 1″ x 3″	.150	Pr.	.11	.73	.84
Oak flooring, No. 1 common	1.000	S.F.	4.51	3.72	8.23
Sand and finish floor	1.000	S.F.	.99	1.68	2.67
Wood furring, 1″ x 3″, 16″ O.C.	1.000	L.F.	.43	1.81	2.24
Gypsum drywall, 5/8″ thick	1.000	S.F.	.37	.70	1.07
Tape and finishing	1.000	S.F.	.05	.63	.68
Paint ceiling	1.000	S.F.	.21	.68	.89
TOTAL			8.35	11.94	20.29

B1010 263	Floor-Ceiling, Wood Joists	COST PER S.F.		
		MAT.	INST.	TOTAL
1700	16″ on center, 2″ x 6″ joists	8.10	11.80	19.90
1750	2″ x 8″	8.35	11.95	20.30
1800	2″ x 10″	8.90	12.20	21.10
1900	2″ x 12″	9.30	12.25	21.55
2000	2″ x 14″	9.80	12.45	22.25
2100	12″ on center, 2″ x 10″ joists	9.25	12.55	21.80
2200	2″ x 12″	9.75	12.60	22.35
2300	2″ x 14″	10.35	12.85	23.20
2400				
2500				
2700	Cut & patch to match existing construction, add, minimum	2%	3%	
2800	Maximum	5%	9%	
2900	Dust protection, add, minimum	1%	2%	
3000	Maximum	4%	11%	
3100	Equipment usage curtailment, add, minimum	1%	1%	
3200	Maximum	3%	10%	
3300	Material handling & storage limitation, add, minimum	1%	1%	
3400	Maximum	6%	7%	
3500	Protection of existing work, add, minimum	2%	2%	
3600	Maximum	5%	7%	
3700	Shift work requirements, add, minimum		5%	
3800	Maximum		30%	
3900	Temporary shoring and bracing, add, minimum	2%	5%	
4000	Maximum	5%	12%	

This data illustrates and describes a flat roof bar joist system including asphalt and gravel roof, lightweight concrete, metal decking, open web joists, and suspended acoustic ceiling. Lines within System Components give the unit price and total price per square foot for this system. Prices for flat roof bar joist systems are on Line Items B1020 106 1200 thru 1600. Both material quantities and labor costs have been adjusted for the system listed.

Factors: To adjust for job conditions other than normal working situations use Lines B1020 106 2700 thru 4000.

Example: You are to install the system where material handling and storage present some problem. Go to Line B1020 106 3300 and apply these percentages to the appropriate MAT. and INST. costs.

System Components	QUANTITY	UNIT	COST PER S.F.		
			MAT.	INST.	TOTAL
Three ply asphalt and gravel roof on lightweight concrete, metal decking on Web joist 4' O.C., with suspended acoustic ceiling.					
Open web joist, 10" deep, 4' O.C.	1.250	Lb.	1.23	.59	1.82
Metal decking, 1-1/2" deep, 22 Ga., galvanized	1.000	S.F.	2.28	.71	2.99
Gravel roofing, 3 ply asphalt	.010	C.S.F.	1.43	2.08	3.51
Perlite roof fill, 3" thick	1.000	S.F.	2.41	.59	3
Suspended ceiling, 3/4" mineral fiber on "Z" bar suspension	1.000	S.F.	2.59	4.22	6.81
TOTAL			9.94	8.19	18.13

B1020 106	Steel Joist Roof & Ceiling	COST PER S.F.		
		MAT.	INST.	TOTAL
1200	Open web bar joist, 10" deep, 5#/L.F.	9.95	8.20	18.15
1300	16" deep, 5#/L.F.	10.35	8.40	18.75
1400	18" deep, 6.6#/L.F.	10.65	8.55	19.20
1500	20" deep, 9.6#/L.F.	11.05	8.70	19.75
1600	24" deep, 11.52#/L.F.	11.55	8.95	20.50
1700				
1800				
1900				
2000				
2100				
2200				
2300				
2400				
2500				
2700	Cut & patch to match existing construction, add, minimum	2%	3%	
2800	Maximum	5%	9%	
2900	Dust protection, add, minimum	1%	2%	
3000	Maximum	4%	11%	
3100	Equipment usage curtailment, add, minimum	1%	1%	
3200	Maximum	3%	10%	
3300	Material handling & storage limitation, add, minimum	1%	1%	
3400	Maximum	6%	7%	
3500	Protection of existing work, add, minimum	2%	2%	
3600	Maximum	5%	7%	
3700	Shift work requirements, add, minimum		5%	
3800	Maximum		30%	
3900	Temporary shoring and bracing, add, minimum	2%	5%	
4000	Maximum	5%	12%	

591

B1020 Roof Construction

This data illustrates and describes a wood frame roof system including rafters, ceiling joists, sheathing, building paper, asphalt shingles, roof trim, furring, insulation, plaster and paint. Lines within System Components give the unit price and total price per square foot for this system. Prices for wood frame roof systems are on Line Items B1020 202 1800 thru 2700. Both material quantities and labor costs have been adjusted for the system listed.

Factors: To adjust for job conditions other than normal working situations use Lines B1020 202 3300 thru 4000.

Example: You are to install the system while protecting existing work. Go to Line B1020 202 3800 and apply these percentages to the appropriate MAT. and INST. costs.

System Components	QUANTITY	UNIT	COST PER S.F.		
			MAT.	INST.	TOTAL
Wood frame roof system, 4 in 12 pitch, including rafters, sheathing, Shingles, insulation, drywall, thin coat plaster, and painting.					
Rafters, 2" x 6", 16" O.C., 4 in 12 pitch	1.080	L.F.	.72	1.37	2.09
Ceiling joists, 2" x 6", 16 O.C.	1.000	L.F.	.67	1.01	1.68
Sheathing, 1/2" CDX	1.080	S.F.	.81	.97	1.78
Building paper, 15# felt	.011	C.S.F.	.06	.19	.25
Asphalt, standard strip shingles, inorganic, class A, 30 year	.011	C.S.F.	1.17	1.32	2.49
Roof trim	.100	L.F.	.18	.28	.46
Furring, 1" x 3", 16" O.C.	1.000	L.F.	.43	1.81	2.24
Fiberglass insulation, 6" batts	1.000	S.F.	.68	.47	1.15
Gypsum board, 1/2" thick	1.000	S.F.	.36	.70	1.06
Thin coat plaster	1.000	S.F.	.12	.79	.91
Paint, roller, 2 coats	1.000	S.F.	.21	.68	.89
TOTAL			5.41	9.59	15

B1020 202	Wood Frame Roof & Ceiling	COST PER S.F.		
		MAT.	INST.	TOTAL
1800	Rafters 16" O.C., 2" x 6"	5.40	9.60	15
1900	2" x 8"	5.70	9.65	15.35
2000	2" x 10"	6.30	10.40	16.70
2100	2" x 12"	6.70	10.60	17.30
2200	Rafters 24" O.C., 2" x 6"	5.05	9	14.05
2300	2" x 8"	5.25	9.05	14.30
2400	2" x 10"	5.70	9.60	15.30
2500	2" x 12"	6	9.75	15.75
2600	Roof pitch, 6 in 12, add	3%	10%	
2700	8 in 12, add	5%	12%	
2800				
2900				
3000				
3100				
3300	Cut & patch to match existing construction, add, minimum	2%	3%	
3400	Maximum	5%	9%	
3500	Material handling & storage limitation, add, minimum	1%	1%	
3600	Maximum	6%	7%	
3700	Protection of existing work, add, minimum	2%	2%	
3800	Maximum	5%	7%	
3900	Shift work requirements, add, minimum		5%	
4000	Maximum		30%	

B1020 Roof Construction

5/8" Plywood Sheathing

Truss Anchors

Roof Trusses at 2'-0" o.c.

9" Fiberglass Insulation

1" x 8" Fascia Board

Wood Fabricated Roof Truss System

This data illustrates and describes a sloped roof truss framing system including trusses, truss anchors, sheathing, fascia and insulation. Lines B1020 210 1500 through 1800 are for a flat roof framing system including flat roof trusses, truss anchors, sheathing, fascia and insulation. Lines within system components give the unit price and total price per square foot for this system. Prices for alternate sizes and systems are given in line items B1020 210 0100 through 1800. Both material quantities and labor costs have been adjusted for the system listed.

Factors: To adjust for job conditions other than normal working situations use lines B1020 210 2700 through 4000.

Example: You are to install the system where crane placement and movement will be difficult. Go to line B1020 210 3200 and apply these percentages to the appropriate MAT. and INST. costs.

System Components	QUANTITY	UNIT	COST PER S.F.		
			MAT.	INST.	TOTAL
Roof truss, 2' O.C., 4/12 slope, 12' span, 5/8 sheath., fascia, insul.					
Timber connectors, rafter anchors, galv., 1 1/2" x 5 1/4"	.084	Ea.	.04	.37	.41
Plywood sheathing, CDX, 5/8" thick	1.054	S.F.	.89	1.02	1.91
Truss, 2' O.C., metal plate connectors, 12' span	.042	Ea.	1.91	1.95	3.86
Molding, fascia trim, 1" x 8"	.167	L.F.	.30	.46	.76
Insulation, fiberglass, kraft face, 9" thick x 23" wide	1.000	S.F.	.80	.47	1.27
TOTAL			3.94	4.27	8.21

B1020 210	Roof Truss, Wood	COST PER S.F.		
		MAT.	INST.	TOTAL
0100	Roof truss, 2'O.C., 4/12p, 1'ovhg, 12'span, 5/8"sheath, fascia, insul.	3.94	4.27	8.21
0200	20' span	3.85	3.61	7.46
0300	24' span	3.65	3.36	7.01
0400	26' span	5.20	4.49	9.69
0500	28' span	3.70	3.30	7
0600	30' span	3.72	3.25	6.97
0700	32' span	3.78	3.20	6.98
0800	34' span	3.72	3.17	6.89
0890	60' span	4.83	3.22	8.05
0900	8/12 pitch, 20' span	4.17	3.89	8.06
1000	24' span	4.25	3.63	7.88
1100	26' span	4.14	3.55	7.69
1200	28' span	4.30	3.55	7.85
1300	32' span	4.30	3.48	7.78
1400	36' span	4.47	3.40	7.87
1500	Flat roof frame, fab.strl.joists, 2' O.C., 15'-24' span, 50 PSF LL	4.15	2.90	7.05
1600	55 PSF LL	4.39	2.97	7.36
1700	24'-30'span, 45 PSF LL	4.46	2.76	7.22
1800	55 PSF LL	6.30	2.85	9.15
1900				
2700	Cut & patch to match existing construction, add, minimum	2%	3%	
2800	Maximum	5%	9%	
2900	Dust protection, add, minimum	1%	2%	
3000	Maximum	4%	11%	
3100	Equipment usage curtailment, add, minimum	1%	1%	
3200	Maximum	3%	10%	
3300	Material handling & storage limitation, add, minimum	1%	1%	
3400	Maximum	6%	7%	

B1020 Roof Construction

B1020 210	Roof Truss, Wood	COST PER S.F.		
		MAT.	INST.	TOTAL
3500	Protection of existing work, add, minimum	2%	2%	
3600	Maximum	5%	7%	
3700	Shift work requirements, add, minimum		5%	
3800	Maximum		30%	
3900	Temporary shoring and bracing, add, minimum	2%	5%	
4000	Maximum	5%	12%	

This data illustrates and describes a masonry concrete block system including concrete block wall, pointed, reinforcing, waterproofing, gypsum plaster on gypsum lath, and furring. Lines within System Components give the unit price and total price per square foot for this system. Prices for masonry concrete block wall systems are on Line Items B2010 108 1400 thru 2200. Both material quantities and labor costs have been adjusted for the system listed.

Factors: To adjust for job conditions other than normal working situations use Lines B2010 108 2700 thru 4000.

Example: You are to install the system and match existing construction at several locations. Go to Line B2010 108 2800 and apply these percentages to the appropriate MAT. and INST. costs.

System Components	QUANTITY	UNIT	COST PER S.F.		
			MAT.	INST.	TOTAL
Concrete block, 8″ thick, reinforced every 2 courses, waterproofing gypsum					
Plaster over gypsum lath on 1″ x 3″ furring, interior painting & baseboard.					
Concrete block, 8″ x 8″ x 16″, reinforced	1.000	S.F.	4.98	6.75	11.73
Silicone waterproofing, 2 coats	1.000	S.F.	.81	.20	1.01
Bituminous coating, 1/16″ thick	1.000	S.F.	.39	1.20	1.59
On concrete	1.000	L.F.	.46	2.44	2.90
Gypsum lath, 3/8″ thick	.110	S.Y.	.37	.77	1.14
Gypsum plaster, 2 coats	.110	S.Y.	.45	2.97	3.42
Painting, 2 coats	1.000	S.F.	.14	.53	.67
Baseboard wood, 9/16″ x 2-5/8″	.100	L.F.	.12	.26	.38
Paint baseboard, primer + 1 coat enamel	.100	L.F.	.01	.07	.08
TOTAL			7.73	15.19	22.92

B2010 108	Masonry Wall, Concrete Block	COST PER S.F.		
		MAT.	INST.	TOTAL
1400	8″ thick block, regular	7.75	15.20	22.95
1500	Fluted 2 sides	8.65	17.75	26.40
1600	Deep grooved	7.20	17.75	24.95
1700	Slump block	9.35	15.55	24.90
1800	Split rib	8.35	17.75	26.10
1900				
2000	12″ thick block, regular	7.25	18.15	25.40
2100	Slump block	17.15	16.90	34.05
2200	Split rib	9.30	20	29.30
2300				
2700	Cut & patch to match existing construction, add, minimum	2%	3%	
2800	Maximum	5%	9%	
2900	Dust protection, add, minimum	1%	2%	
3000	Maximum	4%	11%	
3100	Equipment usage curtailment, add, minimum	1%	1%	
3200	Maximum	3%	10%	
3300	Material handling & storage limitation, add, minimum	1%	1%	
3400	Maximum	6%	7%	
3500	Protection of existing work, add, minimum	2%	2%	
3600	Maximum	5%	7%	
3700	Shift work requirements, add, minimum		5%	
3800	Maximum		30%	
3900	Temporary shoring and bracing, add, minimum	2%	5%	
4000	Maximum	5%	12%	

595

For customer support on your Commercial Renovation Cost Data, call 877.742.4719.

This data illustrates and describes a masonry wall, brick-stone system including brick, concrete block, durawall, insulation, plasterboard, taped and finished, furring, baseboard and painting interior. Lines within System Components give the unit price and total price per square foot for this system. Prices for masonry wall, brick-stone systems are on Line Item B2010 129 1400 thru 2500. Both material quantities and labor costs have been adjusted for the system listed.

Factors: To adjust for job conditions other than normal working situations use Lines B2010 129 3100 thru 4200.

Example: You are to install the system without damaging the existing work. Go to Line B2010 129 3900 and apply these percentages to the appropriate MAT. and INST. costs.

System Components	QUANTITY	UNIT	COST PER S.F.		
			MAT.	INST.	TOTAL
Face brick, 4″thick, concrete block back-up, reinforce every second course, 3/4″insulation, furring, 1/2″drywall, taped, finish, and painted, baseboard					
Face brick, 4″ brick	1.000	S.F.	4.32	12.70	17.02
Concrete back-up block, reinforced 8″ thick	1.000	S.F.	3.04	7.10	10.14
3/4″ rigid polystyrene insulation	1.000	S.F.	.62	.79	1.41
Furring, 1″ x 3″, wood, 16″ O.C.	1.000	L.F.	.46	1.28	1.74
Drywall, 1/2″ thick	1.000	S.F.	.36	.63	.99
Taping & finishing	1.000	S.F.	.05	.63	.68
Painting, 2 coats	1.000	S.F.	.21	.68	.89
Baseboard, wood, 9/16″ x 2-5/8″	.100	L.F.	.12	.26	.38
Paint baseboard, primer + 1 coat enamel	.100	L.F.	.01	.07	.08
TOTAL			9.19	24.14	33.33

B2010 131	Masonry Wall, Brick - Stone	COST PER S.F.		
		MAT.	INST.	TOTAL
1400	Face brick, standard, red, 4″ x 2-2/3″ x 8″ (6.75/S.F.)	9.20	24	33.20
1500	Norman, 4″ x 2-2/3″ x 12″ (4.5 per S.F.)	11.50	20	31.50
1600	Roman, 4″ x 2″ x 12″ (6.0 per S.F.)	12.75	22.50	35.25
1700	Engineer, 4″ x 3-1/5″ x 8″ (5.63 per S.F.)	9	22	31
1800	S.C.R., 6″ x 2-2/3″ x 12″ (4.5 per S.F.)	11.70	20.50	32.20
1900	Jumbo, 6″ x 4″ x 12″ (3.0 per S.F.)	10.20	17.90	28.10
2000	Norwegian, 6″ x 3-1/5″ x 12″ (3.75 per S.F.)	10.75	18.90	29.65
2100				
2200				
2300	Stone, veneer, fieldstone, 6″ thick	11.80	20.50	32.30
2400	Marble, 2″ thick	58.50	36	94.50
2500	Limestone, 2″ thick	37	22	59
3100	Cut & patch to match existing construction, add, minimum	2%	3%	
3200	Maximum	5%	9%	
3300	Dust protection, add, minimum	1%	2%	
3400	Maximum	4%	11%	
3500	Equipment usage curtailment, add, minimum	1%	1%	
3600	Maximum	3%	10%	
3700	Material handling & storage limitation, add, minimum	1%	1%	
3800	Maximum	6%	7%	
3900	Protection of existing work, add, minimum	2%	2%	
4000	Maximum	5%	7%	
4100	Shift work requirements, add, minimum		5%	
4200	Maximum		30%	

B2010 Exterior Walls

This data illustrates and describes a parapet wall system including wall, coping, flashing and cant strip. Lines within System Components give the unit price and cost per lineal foot for this system. Prices for parapet wall systems are on Lines B2010 142 1300 thru 2400. Both material quantities and labor costs have been adjusted for the system listed.

Factors: To adjust for job conditions other than normal working situations, use Lines B2010 142 2900 thru 4000.

Example: You are to install the system without damaging the adjacent property. Go to Line B2010 142 3600 and apply these percentages to the appropriate MAT. and INST. costs.

System Components	QUANTITY	UNIT	COST PER L.F.		
			MAT.	INST.	TOTAL
Concrete block parapet, incl. reinf., coping, flashing & cant strip, 2' high					
8" concrete block	2.000	S.F.	9.96	13.50	23.46
Masonry reinforcing	1.000	L.F.	.28	.20	.48
Coping, precast	1.000	L.F.	20.50	15.65	36.15
Roof cant	1.000	L.F.	.50	1.85	2.35
16 oz.	1.000	L.F.	7.15	2.81	9.96
Cap flashing	1.000	L.F.	8.65	5.25	13.90
TOTAL			47.04	39.26	86.30

B2010 142	Parapet Wall	COST PER L.F.		
		MAT.	INST.	TOTAL
1300	Concrete block, 8" thick	47	39.50	86.50
1400	12" thick	57	51	108
1500	Split rib block, 8" thick	48.50	44.50	93
1600	12" thick	55	52.50	107.50
1700	Brick, common brick, 8" wall	55.50	66.50	122
1800	12" wall	65	84.50	149.50
1900	4" brick, 4" backup block	50	64	114
2000	8" backup block	51.50	66	117.50
2100	Stucco on masonry	47.50	44	91.50
2200	On wood frame	21	36.50	57.50
2300	Wood, T 1-11 siding	33	44	77
2400	Boards, 1" x 6" cedar	39.50	48.50	88
2500				
2600				
2900	Dust protection, add, minimum	1%	2%	
3000	Maximum	4%	11%	
3100	Equipment usage curtailment, add, minimum	1%	1%	
3200	Maximum	3%	10%	
3300	Material handling & storage limitation, add, minimum	1%	1%	
3400	Maximum	6%	7%	
3500	Protection of existing work, add, minimum	2%	2%	
3600	Maximum	5%	7%	
3700	Shift work requirements, add, minimum		5%	
3800	Maximum		30%	
3900	Temporary shoring and bracing, add, minimum	2%	5%	
4000	Maximum	5%	12%	

597

This data illustrates and describes a masonry cleaning and restoration system, including staging, cleaning, repointing. Lines within System Components give the unit price and cost per square foot for this system. Prices for systems are on Lines B2010 143 1300 thru 2500. Both material quantities and labor costs have been adjusted for the system listed.

Factors: To adjust for conditions other than normal working situations, use Lines B2010 143 2900 thru 4200.

Example: You are to clean a wall and be concerned about dust control. Go to line B2010 143 3100 and apply these percentages to the appropriate MAT. and INST. costs.

System Components	QUANTITY	UNIT	COST PER S.F.		
			MAT.	INST.	TOTAL
Repoint existing building, brick, common bond, high pressure cleaning, **Water only, soft old mortar.**					
Scaffolding, steel tubular, bldg ext wall face, labor only, 1 to 5 stories	.010	C.S.F.		2.37	2.37
Cleaning masonry, high pressure wash, water only, avg soil, bio staining	1.100	S.F.		1.68	1.68
Repoint, brick common bond	1.000	S.F.	.64	6.25	6.89
TOTAL			.64	10.30	10.94

B2010 143	Masonry Restoration - Cleaning	COST PER S.F.		
		MAT.	INST.	TOTAL
1300	Brick, common bond	.64	10.30	10.94
1400	Flemish bond	.67	10.75	11.42
1500	English bond	.67	11.40	12.07
1700	Stone, 2' x 2' blocks	.85	7.85	8.70
1800	2' x 4' blocks	.64	6.90	7.54
2000	Add to above prices for alternate cleaning systems:			
2100	Chemical brush and wash	.05	.69	.74
2200	High pressure chemical and water	.08	.58	.66
2300	Sandblasting, wet system	.43	1.64	2.07
2400	Dry system	.40	.97	1.37
2500	Steam cleaning		.47	.47
2600				
2900	Cut & patch to match existing construction, add, minimum	2%	3%	
3000	Maximum	5%	9%	
3100	Dust protection, add, minimum	1%	2%	
3200	Maximum	4%	11%	
3300	Equipment usage curtailment, add, minimum	1%	1%	
3400	Maximum	3%	10%	
3500	Material handling & storage limitation, add, minimum	1%	1%	
3600	Maximum	6%	7%	
3900	Protection of existing work, add, minimum	2%	2%	
4000	Maximum	5%	7%	
4100	Shift work requirements, add, minimum		5%	
4200	Maximum		30%	

B2010 Exterior Walls

This data illustrates and describes a wood frame exterior wall system including wood studs, sheathing, felt, insulation, plasterboard, taped and finished, baseboard and painted interior. Lines within System Components give the unit price and total price per square foot for this system. Prices for wood frame exterior wall systems are on Line Items B2010 150 1600 thru 2700. Both material quantities and labor costs have been adjusted for the system listed.

Factors: To adjust for job conditions other than normal working situations use Lines B2010 150 3100 thru 4200.

Example: You are to install the system with need for complete temporary bracing. Go to Line B2010 150 4200 and apply these percentages to the appropriate MAT. and INST. costs.

System Components	QUANTITY	UNIT	COST PER S.F. MAT.	COST PER S.F. INST.	COST PER S.F. TOTAL
Wood stud wall, cedar shingle siding, building paper, plywood sheathing, Insulation, 5/8" drywall, taped, finished and painted, baseboard.					
2" x 4" wood studs, 16" O.C.	.100	L.F.	.46	1.27	1.73
1/2" CDX sheathing	1.000	S.F.	.75	.90	1.65
18" No. 1 red cedar shingles, 7-1/2" exposure	.008	C.S.F.	1.61	2.25	3.86
15# felt paper	.010	C.S.F.	.06	.17	.23
3-1/2" fiberglass insulation	1.000	S.F.	.52	.47	.99
5/8" drywall	1.000	S.F.	.37	.63	1
Drywall finishing adder	1.000	S.F.	.05	.63	.68
Baseboard trim, stock pine, 9/16" x 3-1/2", painted	.100	L.F.	.12	.26	.38
Paint baseboard, primer + 1 coat enamel	.100	L.F.	.01	.07	.08
Paint, 2 coats, interior	1.000	S.F.	.21	.68	.89
TOTAL			4.16	7.33	11.49

B2010 150	Wood Frame Exterior Wall	COST PER S.F. MAT.	COST PER S.F. INST.	COST PER S.F. TOTAL
1600	Cedar shingle siding, painted	4.16	7.35	11.51
1700	Aluminum siding, horizontal clapboard	5.25	7.55	12.80
1800	Cedar bevel siding, 1/2" x 6", vertical , painted	7.25	7.25	14.50
1900	Redwood siding 1" x 4" to 1" x 6" vertical, T & G	7.10	7.95	15.05
2000	Board and batten	7.50	6.60	14.10
2100	Ship lap siding	7.55	7	14.55
2200	Plywood, grooved (T1-11) fir	4.16	6.95	11.11
2300	Redwood	5.50	6.95	12.45
2400	Southern yellow pine	4.06	6.95	11.01
2500	Masonry on stud wall, stucco, wire and plaster	3.33	11.05	14.38
2600	Stone veneer	9.50	14.25	23.75
2700	Brick veneer, brick $275 per M	6.85	17.80	24.65
3100	Cut & patch to match existing construction, add, minimum	2%	3%	
3200	Maximum	5%	9%	
3300	Dust protection, add, minimum	1%	2%	
3400	Maximum	4%	11%	
3500	Material handling & storage limitation, add, minimum	1%	1%	
3600	Maximum	6%	7%	
3700	Protection of existing work, add, minimum	2%	2%	
3800	Maximum	5%	7%	
3900	Shift work requirements, add, minimum		5%	
4000	Maximum		30%	
4100	Temporary shoring and bracing, add, minimum	2%	5%	
4200	Maximum	5%	12%	

B2010 Exterior Walls

B2010 190	Selective Price Sheet	COST PER S.F.		
		MAT.	INST.	TOTAL
0100	Exterior surface, masonry, concrete block, standard 4" thick	1.92	6.50	8.42
0200	6" thick	2.53	7	9.53
0300	8" thick	2.71	7.45	10.16
0400	12" thick	4.48	9.70	14.18
0500	Split rib, 4" thick	4.38	8.10	12.48
0600	8" thick	5.60	9.30	14.90
0700	Brick running bond, standard size, 6.75/S.F.	4.32	12.70	17.02
0800	Buff, 6.75/S.F.	4.56	12.70	17.26
0900	Stucco, on frame	.78	6	6.78
1000	On masonry	.31	4.70	5.01
1100	Metal, aluminum, horizontal, plain	2.72	2.46	5.18
1200	Insulated	3.14	2.46	5.60
1300	Vertical, plain	2.26	2.46	4.72
1400	Insulated	2.36	2.57	4.93
1500	Wood, beveled siding, "A" grade cedar, 1/2" x 6"	4.68	2.15	6.83
1600	1/2" x 8"	7.15	1.92	9.07
1700	Shingles, 16" #1 red, 7-1/2" exposure	2.06	3.10	5.16
1800	18" perfections, 7-1/2" exposure	2.28	2.58	4.86
1900	Handsplit, 10" exposure	3.15	2.53	5.68
2000	White cedar, 7-1/2" exposure	1.51	3.15	4.66
2100	Vertical, board & batten, redwood	4.57	2.88	7.45
2200	White pine	2.41	1.92	4.33
2300	T. & G. boards, redwood, 1" x 4"	3.58	2.88	6.46
2400	1' x 8"	5.70	1.92	7.62
2500				
2600	Interior surface, drywall, taped & finished, standard, 1/2"	.42	1.31	1.73
2700	5/8" thick	.43	1.31	1.74
2800	Fire resistant, 1/2" thick	.45	1.31	1.76
2900	5/8" thick	.44	1.31	1.75
3000	Moisture resistant, 1/2" thick	.50	1.31	1.81
3100	5/8" thick	.53	1.31	1.84
3200	Core board, 1" thick	.83	2.64	3.47
3300	Plaster, gypsum, 2 coats	.45	2.99	3.44
3400	3 coats	.65	3.64	4.29
3500	Perlite or vermiculite, 2 coats	.72	3.41	4.13
3600	3 coats	.78	4.28	5.06
3700	Gypsum lath, standard, 3/8" thick	.37	.78	1.15
3800	1/2" thick	.30	.83	1.13
3900	Fire resistant, 3/8" thick	.30	.94	1.24
4000	1/2" thick	.36	1.02	1.38
4100	Metal lath, diamond, 2.5 lb.	.50	.78	1.28
4200	Rib, 3.4 lb.	.55	.94	1.49
4300	Framing metal studs including top and bottom			
4400	Runners, walls 10' high			
4500	24" O.C., non load bearing 20 gauge, 2-1/2" wide	.33	1.06	1.39
4600	3-5/8" wide	.37	1.07	1.44
4700	4" wide	.44	1.07	1.51
4800	6" wide	.52	1.09	1.61
4900	Load bearing 18 gauge, 2-1/2" wide	.70	1.19	1.89
5000	3-5/8" wide	.84	1.21	2.05
5100	4" wide	.94	1.54	2.48
5200	6" wide	1.12	1.26	2.38
5300	16" O.C., non load bearing 20 gauge, 2-1/2" wide	.44	1.62	2.06
5400	3-5/8" wide	.51	1.64	2.15
5500	4" wide	.60	1.67	2.27
5600	6" wide	.70	1.69	2.39
5700	Load bearing 18 gauge, 2-1/2" wide	.95	1.65	2.60
5800	3-5/8" wide	1.14	1.67	2.81

B20 Exterior Enclosure

B2010 Exterior Walls

B2010 190	Selective Price Sheet	COST PER S.F.		
		MAT.	INST.	TOTAL
5900	4" wide	1.18	1.71	2.89
6000	6" wide	1.51	1.74	3.25
6100	Framing wood studs incl. double top plate and			
6200	Single bottom plate, walls 10' high			
6300	24" O.C., 2" x 4"	.35	1.02	1.37
6400	2" x 6"	.54	1.10	1.64
6500	16" O.C., 2" x 4"	.46	1.27	1.73
6600	2" x 6"	.71	1.41	2.12
6700	Sheathing, boards, 1" x 6"	1.69	1.95	3.64
6800	1" x 8"	2.07	1.66	3.73
6900	Plywood, 3/8" thick	.68	1.06	1.74
7000	1/2" thick	.75	1.13	1.88
7100	5/8" thick	.84	1.21	2.05
7200	3/4" thick	1	1.30	2.30
7300	Wood fiber, 5/8" thick	.84	1.06	1.90
7400	Gypsum weatherproof, 1/2" thick	.51	1.13	1.64
7500	Insulation, fiberglass batts, 3-1/2" thick, R13	.44	.90	1.34
7600	6" thick, R19	.56	1.06	1.62
7700	Poured 4" thick, fiberglass wool, R4/inch	.60	3.17	3.77
7800	Mineral wool, R3/inch	.50	3.17	3.67
7900	Polystyrene, R4/inch	1.76	3.17	4.93
8000	Perlite or vermiculite, R2.7/inch	5.80	3.17	8.97
8100	Rigid, fiberglass, R4.3/inch, 1" thick	.57	.63	1.20
8200	R8.7/inch, 2" thick	1.16	.71	1.87

B2020 Exterior Windows

B2020 108	Vinyl Clad Windows

	MATERIAL	TYPE	GLAZING	SIZE	DETAIL	COST PER UNIT		
						MAT.	INST.	TOTAL
1000	Vinyl clad	casement	insul. glass	1'-4" x 4'-0"		425	146	571
1010				2'-0" x 3'-0"		355	146	501
1020				2'-0" x 4'-0"		410	153	563
1030				2'-0" x 5'-0"		460	162	622
1040				2'-0" x 6'-0"		480	162	642
1050				2'-4" x 4'-0"		460	162	622
1060				2'-6" x 5'-0"		565	157	722
1070				3'-0" x 5'-0"		790	162	952
1080				4'-0" x 3'-0"		865	162	1,027
1090				4'-0" x 4'-0"		745	162	907
1100				4'-8" x 4'-0"		820	162	982
1110				4'-8" x 5'-0"		925	189	1,114
1120				4'-8" x 6'-0"		1,025	189	1,214
1130				6'-0" x 4'-0"		945	189	1,134
1140				6'-0" x 5'-0"		1,050	189	1,239
3000		double-hung	insul. glass	2'-0" x 4'-0"		440	140	580
3050				2'-0" x 5'-0"		500	140	640
3100				2'-4" x 4'-0"		455	146	601
3150				2'-4" x 4'-8"		455	146	601
3200				2'-4" x 6'-0"		515	146	661
3250				2'-6" x 4'-0"		455	146	601
3300				2'-8" x 4'-0"		680	189	869
3350				2'-8" x 5'-0"		535	162	697
3375				2'-8" x 6'-0"		545	162	707
3400				3'-0 x 3'-6"		430	146	576
3450				3'-0 x 4'-0"		495	153	648
3500				3'-0 x 4'-8"		530	153	683
3550				3'-0 x 5'-0"		545	162	707
3600				3'-0 x 6'-0"		490	162	652
3700				4'-0 x 5'-0"		675	173	848
3800				4'-0 x 6'-0"		825	173	998

For customer support on your Commercial Renovation Cost Data, call 877.742.4719.

B2020 Exterior Windows

This data illustrates and describes an aluminum window system including double hung aluminum window, exterior and interior trim, hardware and insulating glass. Lines within System Components give the unit price and total price on a cost each basis for this system. Prices for aluminum window systems are on Line Items B2020 110 1000 thru 2400. Both material quantities and labor costs have been adjusted for the system listed.

Factors: To adjust for job conditions other than normal working situations use Lines B2020 110 3100 thru 4000.

Example: You are to install the system and cut and patch to match existing construction. Go to Line B2020 110 3200 and apply these percentages to the appropriate MAT. and INST. costs.

System Components				COST EACH		
	QUANTITY	UNIT	MAT.	INST.	TOTAL
Double hung aluminum window, 2'-4" x 2'-6" exterior and interior trim, Hardware glazed with insulating glass.						
Double hung aluminum window, 2'-4" x 2' 6"	1.000	Ea.	253.61	43.73	297.34
Interior Trim	1.000	Set	34.50	48.50	83
Hardware	1.000	Set	5	26.50	31.50
TOTAL			293.11	118.73	411.84

B2020 110	Windows - Aluminum	COST EACH		
		MAT.	INST.	TOTAL
1000	Aluminum, double hung, 2'-4" x 2'-6"	293	119	412
1100	2'-8" x 4'-6"	570	180	750
1200	3'- 4" x 5'-6"	875	270	1,145
1300	Casement, 3'-6" x 2'-4"	390	136	526
1400	4'-6" x 2'-4"	495	169	664
1500	5'-6" x 2'-4"	625	229	854
1600	Projected window, 2'-1" x 3'-0"	283	122	405
1700	3'-5" x 3'-0"	445	167	612
1800	4'-0" x 4'-0"	700	253	953
1900	Horizontal sliding 3'-0" x 3'-0"	251	143	394
2000	3'-6" x 4'-0"	375	195	570
2100	5'-0" x 6'-0"	780	360	1,140
2200	Picture window, 3'-8" x 3'-1"	271	160	431
2300	4'-4" x 4'-5"	440	233	673
2400	5'-8" x 4'-9"	625	335	960
2500				
2600				
2700				
2800				
2900				
3100	Cut & patch to match existing construction, add, minimum	2%	3%	
3200	Maximum	5%	9%	
3300	Dust protection, add, minimum	1%	2%	
3400	Maximum	4%	11%	
3500	Material handling & storage limitation, add, minimum	1%	1%	
3600	Maximum	6%	7%	
3700	Protection of existing work, add, minimum	2%	2%	
3800	Maximum	5%	7%	
3900	Shift work requirements, add, minimum		5%	
4000	Maximum		30%	

B2020 Exterior Windows

This data illustrates and describes a wood window system including wood picture window, exterior and interior trim, hardware and insulating glass. Lines within System Components give the unit price and total price on a cost each basis for this system. Prices for wood window systems are on Line Items B2020 112 1200 thru 2900. Both material quantities and labor costs have been adjusted for the system listed.

Factors: To adjust for job conditions other than normal working situations use Lines B2020 112 3100 thru 4000.

Example: You are to install the above system where dust control is vital. Go to Line B2020 112 3400 and apply these percentages to the appropriate MAT. and INST. costs.

System Components	QUANTITY	UNIT	COST EACH MAT.	COST EACH INST.	COST EACH TOTAL
Wood picture window 4'-0" x 4'-6", exterior and interior trim, hardware, Glazed with insulating glass.					
4'-0" x 4'-6" wood picture window with insulating glass	1.000	Ea.	585	115	700
Oak interior trim	1.000	Set	71	106	177
Hardware	1.000	Set	5	26.50	31.50
TOTAL			661	247.50	908.50

B2020 112	Windows - Wood	COST EACH MAT.	COST EACH INST.	COST EACH TOTAL
1200	Picture window, 4'-0" x 4'-6"	660	248	908
1300	5'-0" x 4'-0"	740	248	988
1400	6'-0" x 4'-6"	780	260	1,040
1500	Bow, bay window, 8'-0" x 5'-0", standard	2,625	260	2,885
1600	Deluxe	1,950	260	2,210
1700	Bow, bay window, 12'-0" x 6'-0" standard	3,300	345	3,645
1800	Deluxe	2,700	345	3,045
1900	Double hung, 3'-0" x 4'-0"	365	161	526
2000	4'-0" x 4'-6"	455	212	667
2100	Casement 2'-0" x 3'-0"	325	139	464
2200	2 leaf, 4'-0" x 4'-0"	855	231	1,086
2300	3 leaf, 6'-0" x 6'-0"	1,525	370	1,895
2400	Awning, 2'-10" x 1'-10"	340	139	479
2500	3'-6" x 2'-4"	420	161	581
2600	4'-0" x 3'-0"	615	212	827
2700	Horizontal sliding 3'-0" x 2'-0"	380	139	519
2800	4'-0" x 3'-6"	445	161	606
2900	6'-0" x 5'-0"	655	212	867
3100	Cut & patch to match existing construction, add, minimum	2%	3%	
3200	Maximum	5%	9%	
3300	Dust protection, add, minimum	1%	2%	
3400	Maximum	4%	11%	
3500	Material handling & storage limitation, add, minimum	1%	1%	
3600	Maximum	6%	7%	
3700	Protection of existing work, add, minimum	2%	2%	
3800	Maximum	5%	7%	
3900	Shift work requirements, add, minimum		5%	
4000	Maximum		30%	

This data illustrates and describes storm window and door systems based on a cost each price. Prices for storm window and door systems are on Line Items B2020 116 0400 thru 2400. Both material quantities and labor costs have been adjusted for the system listed.

Factors: To adjust for job conditions other than normal working situations use Lines B2020 116 3100 thru 4000 and apply these percentages to the appropriate MAT and INST. costs.

Example: You are to install the system and protect all existing construction. Go to Line B2020 116 3800 and apply these percentages to the appropriate MAT. and INST. costs.

B2020 116	Storm Windows & Doors	COST EACH		
		MAT.	INST.	TOTAL
0350	Storm Window and door systems, single glazed			
0400	Window, custom aluminum anodized, 2'-0" x 3'-5"	107	42	149
0500	2'-6" x 5'-0"	129	45	174
0600	4'-0" x 6'-0"	259	50.50	309.50
0700	White painted aluminum, 2'-0" x 3'-5"	123	42	165
0800	2'-6" x 5'-0"	190	45	235
0900	4'-0" x 6'-0"	315	50.50	365.50
1000	Average quality aluminum, anodized, 2'-0" x 3'-5"	107	42	149
1100	2'-6" x 5'-0"	133	45	178
1200	4'-0" x 6'-0"	146	50.50	196.50
1300	White painted aluminum, 2'-0" x 3'-5"	101	42	143
1400	2'-6" x 5'-0"	107	45	152
1500	4'-0" x 6'-0"	123	50.50	173.50
1600	Mill finish, 2'-0" x 3'-5"	95.50	42	137.50
1700	2'-6" x 5'-0"	101	45	146
1800	4'-0" x 6'-0"	123	50.50	173.50
1900				
2000	Door, aluminum anodized 3'-0" x 6'-8"	197	90.50	287.50
2100	White painted aluminum, 3'-0" x 6'-8"	340	90.50	430.50
2200	Mill finish, 3'-0" x 6'-8"	289	90.50	379.50
2300				
2400	Wood, storm and screen, painted	340	141	481
2500				
2600				
2700				
2800				
3100	Cut & patch to match existing construction, add, minimum	2%	3%	
3200	Maximum	5%	9%	
3300	Dust protection, add, minimum	1%	2%	
3400	Maximum	4%	11%	
3500	Material handling & storage limitation, add, minimum	1%	1%	
3600	Maximum	6%	7%	
3700	Protection of existing work, add, minimum	2%	2%	
3800	Maximum	5%	7%	
3900	Shift work requirements, add, minimum		5%	
4000	Maximum		30%	

B2020 Exterior Windows

This data illustrates and describes a window wall system including aluminum tube framing, caulking, and glass. Lines within System Components give the unit price and total price per square foot for this system. Prices for window wall systems are on Line Items B2020 124 1300 thru 2600. Both material quantities and labor costs have been adjusted for the system listed.

Factors: To adjust for job conditions other than normal working situations use Lines B2020 124 3100 thru 4000.

Example: You are to install the system with need for complete dust protection. Go to Line B2020 124 3400 and apply these percentages to the appropriate MAT. and INST. costs.

System Components	QUANTITY	UNIT	COST PER S.F.		
			MAT.	INST.	TOTAL
Window wall, including aluminum header, sill, mullions, caulking And glass					
Header, mill finish, 1-3/4" x 4-1/2" deep	.167	L.F.	3.59	2.44	6.03
Sill, mill finish, 1-3/4" x 4-1/2" deep	.167	L.F.	3.59	2.38	5.97
Vertical mullion, 1-3/4" x 4-1/2" deep, 6' O.C.	.191	L.F.	9.13	5.94	15.07
Caulking	.381	L.F.	.10	.80	.90
Glass, 1/4" plate	1.000	S.F.	7.05	10.10	17.15
TOTAL			23.46	21.66	45.12

B2020 124	Aluminum Frame, Window Wall	COST PER S.F.		
		MAT.	INST.	TOTAL
1300	Mill finish, 1-3/4" x 4-1/2" deep, 1/4" plate glass	23.50	21.50	45
1400	2" x 4-1/2" deep, insulating glass	24.50	25	49.50
1500	Thermo-break, 2-1/4" x 4-1/2" deep, insulating glass	26	25.50	51.50
1600	Bronze finish, 1-3/4" x 4-1/2" deep, 1/4" plate glass	26.50	23.50	50
1700	2" x 4-1/2" deep, insulating glass	27	27	54
1800	Thermo-break, 2-1/4" x 4-1/2" deep, insulating glass	28.50	27.50	56
2000	Black finish, 1-3/4" x 4-1/2" deep, 1/4" plate glass	28	24.50	52.50
2100	2" x 4-1/2" deep, insulating glass	28	28	56
2200	Thermo-break, 2-1/4" x 4-1/2" deep, insulating glass	30	28.50	58.50
2300				
2400	Stainless steel, 1-3/4" x 4-1/2" deep, 1/4" plate glass	35.50	29.50	65
2500	2" x 4-1/2" deep, insulating glass	34.50	33.50	68
2600	Thermo-break, 2-1/4" x 4-1/2" deep, insulating glass	37	34.50	71.50
2700				
3100	Cut & patch to match existing construction, add, minimum	2%	3%	
3200	Maximum	5%	9%	
3300	Dust protection, add, minimum	1%	2%	
3400	Maximum	4%	11%	
3500	Material handling, & storage limitation, add, minimum	1%	1%	
3600	Maximum	6%	7%	
3700	Protection of existing work, add, minimum	2%	2%	
3800	Maximum	5%	7%	
3900	Shift work requirements, add, minimum		5%	
4000	Maximum		30%	

B2030   Exterior Doors

This data illustrates and describes a commercial metal door system, including a single aluminum and glass door, narrow stiles, jamb, hardware weatherstripping, panic hardware and closer. Lines within System Components give the unit price and total price on a cost each basis for this system. Prices for commercial metal door systems are on Line Items B2030 125 1200 thru 2500. Both material quantities and labor costs have been adjusted for the system listed.

Factors: To adjust for job conditions other than normal working situations, use Lines B2030 125 3100 thru 4000.

Example: You are to install the system and cut and patch to match existing construction. Go to Line B2030 125 3200 and apply these percentages to the appropriate MAT. and INST. costs.

System Components	QUANTITY	UNIT	COST EACH		
			MAT.	INST.	TOTAL
Single aluminum and glass door, 3'-0"x7'-0", with narrow stiles, ext. jamb.					
Weatherstripping, 1/2" tempered insul. glass, panic hardware, and closer.					
Aluminum door, 3'-0" x 7'-0" x 1-3/4", narrow stiles	1.000	Ea.	1,025	750	1,775
Tempered insulating glass, 1/2" thick	20.000	S.F.	550	440	990
Panic hardware	1.000	Set	505	106	611
Surface mount regular arm	1.000	Ea.	174	97.50	271.50
TOTAL			2,254	1,393.50	3,647.50

B2030 125	Doors, Metal - Commercial	COST EACH		
		MAT.	INST.	TOTAL
1200	Single aluminum and glass, 3'-0" x 7'-0"	2,250	1,400	3,650
1300	With transom, 3'-0" x 10'-0"	2,900	1,650	4,550
1400	Anodized aluminum and glass, 3'-0" x 7'-0"	2,625	1,650	4,275
1500	With transom, 3'-0" x 10'-0"	3,400	1,950	5,350
1600	Steel, deluxe, hollow metal 3'-0" x 7'-0"	1,700	535	2,235
1700	With transom 3'-0" x 10'-0"	2,225	610	2,835
1800	Fire door, "A" label, 3'-0" x 7'-0"	1,825	535	2,360
1900	Double, aluminum and glass, 6'-0" x 7'-0"	3,650	2,375	6,025
2000	With transom, 6'-0" x 10'-0"	4,575	2,950	7,525
2100	Anodized aluminum and glass 6'-0" x 7'-0"	4,150	2,775	6,925
2200	With transom, 6'-0" x 10'-0"	5,225	3,450	8,675
2300	Steel, deluxe, hollow metal, 6'-0" x 7'-0"	3,050	935	3,985
2400	With transom, 6'-0" x 10'-0"	4,125	1,075	5,200
2500	Fire door, "A" label, 6'-0" x 7'-0"	3,225	925	4,150
2800				
2900				
3100	Cut & patch to match existing construction, add, minimum	2%	3%	
3200	Maximum	5%	9%	
3300	Dust protection, add, minimum	1%	2%	
3400	Maximum	4%	11%	
3500	Material handling & storage limitation, add, minimum	1%	1%	
3600	Maximum	6%	7%	
3700	Protection of existing work, add, minimum	2%	2%	
3800	Maximum	5%	7%	
3900	Shift work requirements, add, minimum		5%	
4000	Maximum		30%	

B2030 Exterior Doors

This data illustrates and describes sliding door systems including a sliding door, frame, interior and exterior trim with exterior staining. Lines within System Components give the unit price and total price on a cost each basis for this system. Prices for sliding door systems are on Line Items B2030 150 1000 thru 2400. Both material quantities and labor costs have been adjusted for the system listed.

Factors: To adjust for job conditions other than normal working situations use Lines B2030 150 2700 thru 4000.

Example: You are to install the system with temporary shoring and bracing. Go to Line B2030 150 3900 and apply these percentages to the appropriate MAT. and INST. costs.

System Components	QUANTITY	UNIT	COST EACH		
			MAT.	INST.	TOTAL
Sliding wood door, 6'-0" x 6'-8", with wood frame, interior and exterior Trim and exterior staining.					
Sliding wood door, standard, 6'-0" x 6'-8", insulated glass	1.000	Ea.	1,275	315	1,590
Interior & exterior trim	1.000	Set	31	50.60	81.60
Stain door	1.000	Ea.	2.46	30.50	32.96
Stain trim	1.000	Ea.	2.80	13	15.80
TOTAL			1,311.26	409.10	1,720.36

B2030 150	Doors, Sliding - Patio	COST EACH		
		MAT.	INST.	TOTAL
1000	Wood, standard, 6'-0" x 6'-8", insulated glass	1,300	410	1,710
1100	8'-0" x 6'-8"	1,700	520	2,220
1200	12'-0" x 6'-8"	2,750	635	3,385
1300	Vinyl coated, 6'-0" x 6'-8"	1,625	410	2,035
1400	8'-0" x 6'-8"	2,075	520	2,595
1500	12'-0" x 6'-8"	3,350	635	3,985
1700	Aluminum, standard, 6'-0" x 6'-8", insulated glass	960	440	1,400
1800	8'-0" x 6'-8"	1,650	560	2,210
1900	12'-0" x 6'-8"	1,775	670	2,445
2000	Anodized, 6'-0" x 6'-8"	1,825	440	2,265
2100	8'-0" x 6'-8"	1,900	560	2,460
2200	12'-0" x 6'-8"	3,375	670	4,045
2300				
2400	Deduct for single glazing	86		86
2500				
2700	Cut & patch to match existing construction, add, minimum	2%	3%	
2800	Maximum	5%	9%	
2900	Dust protection, add, minimum	1%	2%	
3000	Maximum	4%	11%	
3100	Equipment usage curtailment, add, minimum	1%	1%	
3200	Maximum	3%	10%	
3300	Material handling & storage limitation, add, minimum	1%	1%	
3400	Maximum	6%	7%	
3500	Protection of existing, work, add, minimum	2%	2%	
3600	Maximum	5%	7%	
3700	Shift work requirements, add, minimum		5%	
3800	Maximum		30%	
3900	Temporary shoring and bracing, add, minimum	2%	5%	
4000	Maximum	5%	12%	

B2030   Exterior Doors

This data illustrates and describes residential door systems including a door, frame, trim, hardware, weatherstripping, stained and finished. Lines within System Components give the unit price and total price on a cost each basis for this system. Prices for residential door systems are on Line Items B2030 240 1500 thru 2200. Both material quantities and labor costs have been adjusted for the system listed.

Factors: To adjust for job conditions other than normal working situations use Lines B2030 240 3100 thru 4000.

Example: You are to install the system with a material handling and storage limitation. Go to Line B2030 240 3600 and apply these percentages to the appropriate MAT. and INST. costs.

System Components	QUANTITY	UNIT	COST EACH MAT.	COST EACH INST.	COST EACH TOTAL
Single solid wood colonial door, 3' x 6'-8" with wood frame and trim,					
Stained and finished, including hardware and weatherstripping.					
Solid wood colonial door, fir 3' x 6'-8" x 1-3/4", hinges	1.000	Ea.	490	84.50	574.50
Hinges, ball bearing	1.000	Ea.	40.50		40.50
Exterior frame, with trim	1.000	Set	164.05	57.46	221.51
Interior trim	1.000	Set	26	53	79
Average quality	1.000	Set	55	45	100
Sill, oak 8" deep	3.500	Ea.	78.75	44.28	123.03
Weatherstripping	1.000	Set	25.50	83.50	109
Stained and finished	1.000	Ea.	5.25	65	70.25
TOTAL			885.05	432.74	1,317.79

B2030 240	Doors, Residential - Exterior	COST EACH MAT.	COST EACH INST.	COST EACH TOTAL
1400	Single doors, 3' x 6'-8"			
1500	Solid colonial door, fir	885	435	1,320
1600	Hollow metal exterior door, plain	855	405	1,260
1700	Solid core wood door, plain	525	415	940
1800				
1900	Double doors, 6' x 6'-8"			
2000	Solid colonial double doors, fir	1,475	575	2,050
2100	Hollow metal exterior doors, plain	1,425	540	1,965
2200	Hollow core wood doors, plain	755	560	1,315
2300				
2800				
2900				
3100	Cut & patch to match existing construction, add, minimum	2%	3%	
3200	Maximum	5%	9%	
3300	Dust protection, add, minimum	1%	2%	
3400	Maximum	4%	11%	
3500	Material handling & storage limitation, add, minimum	1%	1%	
3600	Maximum	6%	7%	
3700	Protection of existing work, add, minimum	2%	2%	
3800	Maximum	5%	7%	
3900	Shift work requirements, add, minimum		5%	
4000	Maximum		30%	

B2030 Exterior Doors

This data illustrates and describes overhead door systems including an overhead door, track, hardware, trim, and electric door opener. Lines within System Components give the unit price and total price on a cost each basis for this system. Prices for overhead door systems are on Line Items B2030 410 1200 thru 2300. Both material quantities and labor costs have been adjusted for the system listed.

Factors: To adjust for job conditions other than normal working situations use Lines B2030 410 3100 thru 4000.

Example: You are to install the system and match existing construction. Go to Line B2030 410 3200 and apply these percentages to the appropriate MAT. and INST. costs.

System Components			COST EACH		
	QUANTITY	UNIT	MAT.	INST.	TOTAL
Wood overhead door, commercial sectional, including track, door, hardware					
And trim, electrically operated.					
Commercial, heavy duty wood door, 8' x 8' x 1-3/4" thick	1.000	Ea.	1,250	635	1,885
Frame, 2 x 8, pressure treated	1.000	Set	27.84	79.92	107.76
Wood trim	1.000	Set	29.76	63.36	93.12
Painting, two coats	1.000	Ea.	21.60	207.36	228.96
Electric trolley operator	1.000	Ea.	1,175	315	1,490
TOTAL			2,504.20	1,300.64	3,804.84

B2030 410	Doors, Overhead	COST EACH		
		MAT.	INST.	TOTAL
1200	Commercial, wood, 1-3/4" thick, 8' x 8'	2,500	1,300	3,800
1300	12' x 12'	3,750	1,825	5,575
1400	14' x 14'	5,050	2,150	7,200
1500	Fiberglass & aluminum 12' x 12'	4,650	1,850	6,500
1600	20' x 20'	9,450	4,725	14,175
1700	Residential, wood, 9' x 7'	1,150	490	1,640
1800	16' x 7'	1,900	730	2,630
1900	Hardboard faced, 9' x 7'	825	490	1,315
2000	16' x 7'	1,450	730	2,180
2100	Fiberglass & aluminum, 9' x 7'	1,125	585	1,710
2200	16' x 7'	1,825	730	2,555
2300	For residential electric opener, add	675	79	754
2400				
2500				
2600				
2700				
2800				
2900				
3100	Cut & patch to match existing construction, add, minumum	2%	3%	
3200	Maximum	5%	9%	
3300	Dust protection, add, minimum	1%	2%	
3400	Maximum	4%	11%	
3500	Material handling & storage limitation, add, minimum	1%	1%	
3600	Maximum	6%	7%	
3700	Protection of existing work, add, minimum	2%	2%	
3800	Maximum	5%	7%	
3900	Shift work requirements, add, minimum		5%	
4000	Maximum		30%	

B2030 Exterior Doors

B2030 810	Selective Price Sheet	COST EACH		
		MAT.	INST.	TOTAL
0100	Door closer, rack and pinion	211	106	317
0200	Backcheck and adjustable power	211	106	317
0300	Regular, hinge face mount, all sizes, regular arm	211	106	317
0400	Hold open arm	220	106	326
0500	Top jamb mount, all sizes, regular arm	211	106	317
0600	Hold open arm	220	106	326
0700	Stop face mount, all sizes, regular arm	211	106	317
0800	Hold open arm	220	106	326
0900	Fusible link, hinge face mount, all sizes, regular arm	250	106	356
1000	Hold open arm	259	106	365
1100	Top jamb mount, all sizes, regular arm	250	106	356
1200	Hold open arm	259	106	365
1300	Stop face mount, all sizes, regular arm	250	106	356
1400	Hold open arm	259	106	365
1500				
1600				
1700	Door stops			
1800				
1900	Holder & bumper, floor or wall	39	19.80	58.80
2000	Wall bumper	13.50	19.80	33.30
2100	Floor bumper	9.55	19.80	29.35
2200	Plunger type, door mounted	32.50	19.80	52.30
2300	Hinges, full mortise, material only, per pair			
2400	Low frequency, 4-1/2" x 4-1/2", steel base, USP	25		25
2500	Brass base, US10	59.50		59.50
2600	Stainless steel base, US32	84		84
2700	Average frequency, 4-1/2" x 4-1/2", steel base, USP	40.50		40.50
2800	Brass base, US10	66		66
2900	Stainless steel base, US32	84		84
3000	High frequency, 4-1/2" x 4-1/2", steel base, USP	73		73
3100	Brass base, US10	83		83
3200	Stainless steel base, US32	114		114
3300	Kick plate			
3400				
3500	6" high, for 3'-0" door, aluminum	32.50	42	74.50
3600	Bronze	61.50	42	103.50
3700	Panic device			
3800				
3900	For rim locks, single door, exit	505	106	611
4000	Outside key and pull	630	127	757
4100	Bar and vertical rod, exit only	955	127	1,082
4200	Outside key and pull	1,075	158	1,233
4300	Lockset			
4400				
4500	Heavy duty, cylindrical, passage doors	138	53	191
4600	Classroom	550	79	629
4700	Bedroom, bathroom, and inner office doors	171	53	224
4800	Apartment, office, and corridor doors	330	63.50	393.50
4900	Standard duty, cylindrical, exit doors	118	63.50	181.50
5100	Passage doors	82.50	53	135.50
5200	Public restroom, classroom, & office doors	220	79	299
5300	Deadlock, mortise, heavy duty	193	70.50	263.50
5400	Double cylinder	193	70.50	263.50
5500	Entrance lock, cylinder, deadlocking latch	193	70.50	263.50
5600	Deadbolt	193	79	272
5700	Commercial, mortise, wrought knob, keyed, minimum	208	79	287
5800	Maximum	550	90.50	640.50
5900	Cast knob, keyed, minimum	286	70.50	356.50

611

B2030 Exterior Doors

B2030 810	Selective Price Sheet	COST EACH		
		MAT.	INST.	TOTAL
6000	Maximum	580	70.50	650.50
6100	Push-pull			
6200				
6300	Aluminum	14.20	53	67.20
6400	Bronze	28	53	81
6500	Door pull, designer style, minimum	28	53	81
6600	Maximum	268	115	383
6700	Threshold			
6800				
6900	3'-0" long door saddles, aluminum, minimum	10.65	13.20	23.85
7000	Maximum	53.50	53	106.50
7100	Bronze, minimum	47.50	10.55	58.05
7200	Maximum	173	53	226
7300	Rubber, 1/2" thick, 5-1/2" wide	44	31.50	75.50
7400	2-3/4" wide	50	31.50	81.50
7500	Weatherstripping, per set			
7600				
7700	Doors, wood frame, interlocking for 3' x 7' door, zinc	48.50	211	259.50
7800	Bronze	61.50	211	272.50
7900	Wood frame, spring type for 3' x 7' door, bronze	25.50	83.50	109
8000	Metal frame, spring type for 3' x 7' door, bronze	52	211	263
8100	For stainless steel, spring type, add	133%		
8200				
8300	Metal frame, extruded sections, 3' x 7' door, aluminum	31	211	242
8400	Bronze	90	211	301

B3010 Roof Coverings

B3010 160	Selective Price Sheet	COST PER S.F.		
		MAT.	INST.	TOTAL
0100	Roofing, built-up, asphalt roll roof, 3 ply organic/mineral surface	.85	1.70	2.55
0200	3 plies glass fiber felt type iv, 1 ply mineral surface	1.45	1.84	3.29
0300	Cold applied, 3 ply		.59	.59
0400	Coal tar pitch, 4 ply tarred felt	2.20	2.19	4.39
0500	Mopped, 3 ply glass fiber	1.81	2.42	4.23
0600	4 ply organic felt	2.20	2.19	4.39
0700	Elastomeric, hypalon, neoprene unreinforced	1.48	4.03	5.51
0800	Polyester reinforced	1.62	4.77	6.39
0900	Neoprene, 5 coats 60 mils	4.93	14.10	19.03
1000	Over 10,000 S.F.	4.93	7.30	12.23
1600				
1700	Asphalt, strip, 210-235#/sq.	.89	1.09	1.98
1800	235-240#/sq.	1.06	1.20	2.26
1900	Class A laminated	1.20	1.34	2.54
2000	Class C laminated	1.20	1.51	2.71
2100	Slate, buckingham, 3/16" thick	5.95	3.45	9.40
2200	Black, 1/4" thick	5.95	3.45	9.40
2300	Wood, shingles, 16" no. 1, 5" exp.	3.10	2.53	5.63
2400	Red cedar, 18" perfections	2.73	2.30	5.03
2500	Shakes, 24", 10" exposure	3.15	2.53	5.68
2600	18", 8-1/2" exposure	1.97	3.15	5.12
2700	Insulation, ceiling batts, fiberglass, 3-1/2" thick, R13	.36	.47	.83
2800	6" thick, R19	.48	.47	.95
2900	9" thick, R30	.80	.55	1.35
3000	12" thick, R38	1.16	.55	1.71
3100	Mineral fiber, 3-1/2" thick, R13	.74	.40	1.14
3200	6" thick, R19	1.16	.40	1.56
3300	Roof deck, fiberboard, 1" thick, R2.78	.73	.77	1.50
3400	Mineral fiber, 2" thick, R5.26	1.33	.77	2.10
3500	Perlite boards, 3/4" thick, R2.08	.53	.77	1.30
3600	2" thick, R5.26	1.29	.85	2.14
3700	Polystyrene extruded, 15 PSI, 1" thick, R5.26	.78	.50	1.28
3800	2" thick, R10	.98	.56	1.54
3900	40 PSI, 1" thick, R5	.70	.50	1.20
4000	Tapered for drainage	.96	.52	1.48
4300	Ceiling, plaster, gypsum, 2 coats	.45	3.41	3.86
4400	3 coats	.65	4.05	4.70
4500	Perlite or vermiculite, 2 coats	.72	3.99	4.71
4600	3 coats	.78	4.99	5.77
4700	Gypsum lath, plain 3/8" thick	.37	.78	1.15
4800	1/2" thick	.30	.83	1.13
4900	Firestop, 3/8" thick	.30	.94	1.24
5000	1/2" thick	.36	1.02	1.38
5100	Metal lath, rib, 2.75 lb.	.43	.88	1.31
5200	3.40 lb.	.55	.94	1.49
5300	Diamond, 2.50 lb.	.50	.88	1.38
5400	3.40 lb.	.53	1.10	1.63
5500	Drywall, taped and finished standard, 1/2" thick	.42	1.66	2.08
5600	5/8" thick	.43	1.66	2.09
5700	Fire resistant, 1/2" thick	.45	1.66	2.11
5800	5/8" thick	.44	1.66	2.10
5900	Water resist., 1/2" thick	.50	1.66	2.16
6000	5/8" thick	.53	1.66	2.19
6100	Finish, instead of taping			
6200	For thin coat plaster, add	.12	.79	.91
6300	Finish textured spray, add	.04	.75	.79
6500	Tile, stapled glued, mineral fiber plastic coated, 5/8" thick	2.45	2.11	4.56
6600	3/4" thick	3.22	2.11	5.33

613

B30 Roofing

B3010 Roof Coverings

B3010 160	Selective Price Sheet	COST PER S.F.		
		MAT.	INST.	TOTAL
6700	Wood fiber, 1/2" thick	1.27	1.58	2.85
6800	3/4" thick	1.35	1.58	2.93
6900	Suspended, fiberglass film faced, 5/8" thick	1.38	1.01	2.39
7000	3" thick	2.88	1.41	4.29
7100	Mineral fiber 5/8" thick, standard face	.75	.94	1.69
7200	Aluminum faced	4.05	1.06	5.11
7500	Ceiling suspension systems, for tile, "T" bar, class "A", 2' x 4' grid	.84	.79	1.63
7600	2' x 2' grid	1.09	.97	2.06
7700	Concealed "Z" bar, 12" module	.95	1.22	2.17
7800				
7900	Plaster/drywall, 3/4" channels, steel furring, 16" O.C.	.38	2.06	2.44
8000	24" O.C.	.25	1.42	1.67
8100	1-1/2" channels, 16" O.C.	.51	2.29	2.80
8200	24" O.C.	.34	1.53	1.87
8300	Ceiling framing, 2" x 4" studs, 16" O.C.	.33	1.06	1.39
8400	24" O.C.	.22	.71	.93

For customer support on your Commercial Renovation Cost Data, call 877.742.4719.

B3020 Roof Openings

This data illustrates and describes a roof hatch system. Lines within System Components give the unit price and total cost each for this system. Prices for systems are on Lines B3020 230 1400 thru 2400. Both material quantities and labor costs have been adjusted for the system listed.

Factors: To adjust for job conditions other than normal working situtations use Lines B3020 230 2900 thru 4000.

Example: You are to install the system and cut and patch to match existing construction. Use line B3020 230 3000 and apply these percentages to the appropriate MAT. and INST. costs.

System Components	QUANTITY	UNIT	COST EACH		
			MAT.	INST.	TOTAL
Roof hatch, 2'-6" x 3'-0", aluminum, curb and cover included.					
Through steel construction.					
Roof hatch, 2'-6" x 3'-0", aluminum	1.000	Ea.	1,075	244	1,319
Cutout decking	11.000	L.F.	4.29	27.17	31.46
Frame opening	44.000	L.F.	38.28	405.24	443.52
Flashing, 16 oz. copper	16.000	S.F.	138.40	84	222.40
Cant strip, 4 x 4, mineral fiber	12.000	L.F.	6	22.20	28.20
TOTAL			1,261.97	782.61	2,044.58

B3020 230	Roof Hatches, Skylights	COST EACH		
		MAT.	INST.	TOTAL
1400	Roof hatch, aluminum, curb and cover, 2'-6" x 3'-0"	1,250	785	2,035
1500	2'-6" x 4'-6"	1,625	955	2,580
1600	2'-6" x 8'-0"	2,400	1,375	3,775
1700	Skylight, plexiglass dome, with curb mounting, 30" x 32"	715	745	1,460
1800	30" x 45"	775	925	1,700
1900	40" x 45"	920	1,250	2,170
2000	Smoke hatch, unlabeled, 2'-6" x 3'-0"	1,525	845	2,370
2100	2'-6" x 4'-6"	1,975	1,025	3,000
2200	2'-6" x 8'-0"	2,600	1,425	4,025
2300				
2400	For steel ladder, add per vertical linear foot	40	37.50	77.50
2450				
2600				
2700				
2900	Cut & patch to match existing construction, add, minimum	2%	3%	
3000	Maximum	5%	9%	
3100	Dust protection, add, minimum	1%	2%	
3200	Maximum	4%	11%	
3300	Equipment usage curtailment, add, minimum	1%	1%	
3400	Maximum	3%	10%	
3500	Material handling & storage limitation, add, minimum	1%	1%	
3600	Maximum	6%	7%	
3700	Protection of existing work, add, minimum	2%	2%	
3800	Maximum	5%	7%	
3900	Shift work requirements, add, minimum		5%	
4000	Maximum		30%	

615

B3020 Roof Openings

B3020 240	Selective Price Sheet	COST PER UNIT		
		MAT.	INST.	TOTAL
0100	Downspouts per L.F., aluminum, enameled .024" thick, 2" x 3"	2.26	4.02	6.28
0200	3" x 4"	2.61	5.15	7.76
0300	Round .025" thick, 3" diam.	2.20	3.80	6
0400	4" diam.	3.25	5.15	8.40
0500	Copper, round 16 oz. stock, 2" diam.	8.35	3.80	12.15
0600	3" diam.	8.80	3.80	12.60
0700	4" diam.	10.65	4.98	15.63
0800	5" diam.	17.30	5.55	22.85
0900	Rectangular, 2" x 3"	12.15	3.80	15.95
1000	3" x 4"	15.35	4.98	20.33
1100	Lead coated copper, round, 2" diam.	24.50	3.80	28.30
1200	3" diam.	24.50	3.80	28.30
1300	4" diam.	26	4.98	30.98
1400	5" diam.	25	5.55	30.55
1500	Rectangular, 2" x 3"	26.50	3.80	30.30
1600	3" x 4"	36.50	4.98	41.48
1700	Steel galvanized, round 28 gauge, 3" diam.	2.31	3.80	6.11
1800	4" diam.	2.77	4.98	7.75
1900	5" diam.	4.10	5.55	9.65
2000	6" diam.	4.57	6.90	11.47
2100	Rectangular, 2" x 3"	4.02	3.80	7.82
2200	3" x 4"	4.47	4.98	9.45
2300	Elbows, aluminum, round, 3" diam.	3.01	7.25	10.26
2400	4" diam.	6.35	7.25	13.60
2500	Rectangular, 2" x 3"	1.05	7.25	8.30
2600	3" x 4"	4.96	7.25	12.21
2700	Copper, round 16 oz., 2" diam.	10.40	7.25	17.65
2800	3" diam.	9.90	7.25	17.15
2900	4" diam.	15.30	7.25	22.55
3100	Rectangular, 2" x 3"	10.30	7.25	17.55
3200	3" x 4"	14.55	7.25	21.80
3300	Drip edge per L.F., aluminum, 5" wide	.63	1.58	2.21
3400	8" wide	1.60	1.58	3.18
3500	28" wide	8.45	6.35	14.80
3600				
3700	Steel galvanized, 5" wide	.65	1.58	2.23
3800	8" wide	.90	1.58	2.48
3900				
4000				
4100				
4200				
4300	Flashing 12" wide per S.F., aluminum, mill finish, .013" thick	.91	4.15	5.06
4400	.019" thick	1.38	4.15	5.53
4500	.040" thick	2.54	4.15	6.69
4600	.050" thick	2.90	4.15	7.05
4700	Copper, mill finish, 16 oz.	8.65	5.25	13.90
4800	20 oz.	11.35	5.45	16.80
4900	24 oz.	16.05	5.75	21.80
5000	32 oz.	20.50	6	26.50
5100	Lead, 2.5 lb./S.F., 12" wide	5.95	4.46	10.41
5200	Over 12" wide	4.39	4.46	8.85
5900	Polyvinyl chloride, black, .010" thick	.23	2.11	2.34
6000	.020" thick	.29	2.11	2.40
6100	.030" thick	.37	2.11	2.48
6200	.056" thick	.90	2.11	3.01
6300	Steel, galvanized, 20 gauge	1.38	4.63	6.01
6400	30 gauge	.64	3.76	4.40
6500	Stainless, 32 gauge, .010" thick	3.58	3.88	7.46

B3020 Roof Openings

B3020 240	Selective Price Sheet	COST PER UNIT		
		MAT.	INST.	TOTAL
6600	28 gauge, .015" thick	4.62	3.88	8.50
6700	26 gauge, .018" thick	4.90	3.88	8.78
6800	24 gauge, .025" thick	5.45	3.88	9.33
6900	Gutters per L.F., aluminum, 5" box, .027" thick	3.19	6	9.19
7000	.032" thick	3.81	5.80	9.61
7100	Copper, half round, 4" wide	10.45	5.80	16.25
7200	6" wide	13	6.15	19.15
7300	Steel, 26 gauge galvanized, 5" wide	2.64	5.80	8.44
7400	6" wide	3.06	5.80	8.86
7500	Wood, treated hem-fir, 3" x 4"	11	6.35	17.35
7600	4" x 5"	18.45	6.35	24.80
7700	Reglet per L.F., aluminum, .025" thick	2.18	2.81	4.99
7800	Copper, 10 oz.	7.15	2.81	9.96
7900	Steel, galvanized, 24 gauge	1.50	2.81	4.31
8000	Stainless, .020" thick	4.18	2.81	6.99
8100	Counter flashing 12" wide per L.F., aluminum, .032" thick	2.42	4.82	7.24
8200	Aluminum, .025" thick	2.42	4.82	7.24
8300	Steel, galvanized, 24 gauge	1.43	4.82	6.25
8400	Stainless, .020" thick	6.95	4.82	11.77

C1010 Partitions

This data illustrates and describes a concrete block wall system including concrete block, horizontal reinforcing alternate courses, mortar, and tooled joints both sides. Lines within System Components give the unit price and total price per square foot for this system. Prices for concrete block wall systems are on Line Items C1010 106 0800 thru 2100. Both material quantities and labor costs have been adjusted for the system listed.

Factors: To adjust for job conditions other than normal working situations use Lines C1010 106 2900 thru 4000.

Example: You are to install the system and protect all existing work. Go to Line C1010 106 3600 and apply these percentages to the appropriate MAT. and INST. costs.

System Components	QUANTITY	UNIT	COST PER S.F. MAT.	COST PER S.F. INST.	COST PER S.F. TOTAL
Concrete block partition, including horizontal reinforcing every second					
Course, mortar, tooled joints both sides.					
8" x 16" concrete block, normal weight, 4" thick	1.000	S.F.	1.92	6.50	8.42
Horizontal reinforcing every second course	.750	L.F.	.21	.15	.36
TOTAL			2.13	6.65	8.78

C1010 106	Partitions, Concrete Block	COST PER S.F. MAT.	COST PER S.F. INST.	COST PER S.F. TOTAL
0800	8" x 16" concrete block, normal weight, 4" thick	2.13	6.65	8.78
0900	6" thick	2.74	7.15	9.89
1000	8" thick	2.92	7.60	10.52
1100	10" thick	3.44	7.90	11.34
1200	12" thick	4.69	9.85	14.54
1300	8" x 16" concrete block, lightweight, 4" thick	2.25	6.50	8.75
1400	6" thick	3.10	6.95	10.05
1500	8" thick	3.73	7.40	11.13
1600	10" thick	4.48	7.70	12.18
1700	12" thick	4.71	9.55	14.26
1800	8" x 16" glazed concrete block, 4" thick	11.35	8.25	19.60
1900	8" thick	13.30	9.15	22.45
2000	Structural facing tile, 6T series, glazed 2 sides, 4" thick	17.25	14.50	31.75
2100	6" thick	22.50	15.25	37.75
2200				
2300				
2400				
2500				
2600				
2900	Cut & patch to match existing construction, add, minimum	2%	3%	
3000	Maximum	5%	9%	
3100	Dust protection, add, minimum	1%	2%	
3200	Maximum	4%	11%	
3300	Material handling & storage limitation, add, minimum	1%	1%	
3400	Maximum	6%	7%	
3500	Protection of existing work, add, minimum	2%	2%	
3600	Maximum	5%	7%	
3700	Shift work requirements, add, minimum		5%	
3800	Maximum		30%	
3900	Temporary shoring and bracing, add, minimum	2%	5%	
4000	Maximum	5%	12%	

C1010 Partitions

This data illustrates and describes a wood stud partition system including wood studs with plates, gypsum plasterboard – taped and finished, insulation, baseboard and painting. Lines within System Components give the unit price and total price per square foot for this system. Prices for wood stud partition systems are on Line Items C1010 132 1300 thru 2700. Both material quantities and labor costs have been adjusted for the system listed.

Factors: To adjust for job conditions other than normal working situations use Lines C1010 132 2900 thru 4000.

Example: You are to install the system where material handling and storage present a serious problem. Go to Line C1010 132 3400 and apply these percentages to the appropriate MAT. and INST. costs.

System Components	QUANTITY	UNIT	COST PER S.F. MAT.	COST PER S.F. INST.	COST PER S.F. TOTAL
Wood stud wall, 2″x4″, 16″ O.C., dbl. top plate, sngl bot. plate, 5/8″ dwl.					
taped, finished and painted on 2 faces, insulation, baseboard, wall 8′ high.					
Wood studs, 2″ x 4″, 16″ O.C., 8′ high	1.000	S.F.	.49	1.58	2.07
Gypsum drywall, 5/8″ thick	2.000	S.F.	.74	1.26	2
Taping and finishing	2.000	S.F.	.10	1.26	1.36
Insulation, 3-1/2″ fiberglass batts	1.000	S.F.	.52	.47	.99
Baseboard	.200	L.F.	.31	.66	.97
Paint baseboard, primer + 2 coats	.200	L.F.	.05	.38	.43
Painting, roller, 2 coats	2.000	S.F.	.42	1.36	1.78
TOTAL			2.63	6.97	9.60

C1010 132	Partitions, Wood Stud	COST PER S.F. MAT.	COST PER S.F. INST.	COST PER S.F. TOTAL
1300	2″ x 3″ studs, 8′ high, 16″ O.C.	2.60	6.90	9.50
1400	24″ O.C.	2.50	6.60	9.10
1500	10′ high, 16″ O.C.	2.58	6.60	9.18
1600	24″ O.C.	2.47	6.35	8.82
1650	2″ x 4″ studs, 8′ high, 16″ O.C.	2.63	6.95	9.58
1700	24″ O.C.	2.52	6.65	9.17
1800	10′ high, 16″ O.C.	2.60	6.65	9.25
1900	24″ O.C.	2.49	6.40	8.89
2000	12′ high, 16″ O.C.	2.56	6.65	9.21
2100	24″ O.C.	2.45	6.40	8.85
2200	2″ x 6″ studs, 8′ high, 16″ O.C.	2.90	7.15	10.05
2300	24″ O.C.	2.73	6.75	9.48
2400	10′ high, 16″ O.C.	2.85	6.80	9.65
2500	24″ O.C.	2.68	6.50	9.18
2600	12′ high, 16″ O.C.	2.78	6.85	9.63
2700	24″ O.C.	2.62	6.50	9.12
2900	Cut & patch to match existing construction, add, minimum	2%	3%	
3000	Maximum	5%	9%	
3100	Dust protection, add, minimum	1%	2%	
3200	Maximum	4%	11%	
3300	Material handling & storage limitation, add, minimum	1%	1%	
3400	Maximum	6%	7%	
3500	Protection of existing work, add, minimum	2%	2%	
3600	Maximum	5%	7%	
3700	Shift work requirements, add, minimum		5%	
3800	Maximum		30%	
3900	Temporary shoring and bracing, add, minimum	2%	5%	
4000	Maximum	5%	12%	

621

This data illustrates and describes a non-load bearing metal stud partition system including metal studs with runners, gypsum plasterboard, taped and finished, insulation, baseboard and painting. Lines within System Components give the unit price and total price per square foot for this system. Prices for non-load bearing metal stud partition systems are on Line Items C1010 134 1300 thru 2300. Both material quantities and labor costs have been adjusted for the system listed.

Factors: To adjust for job conditions other than normal working situations use Lines C1010 134 2900 thru 4000.

Example: You are to install the system and cut and patch to match existing construction. Go to Line C1010 134 3000 and apply these percentages to the appropriate MAT. and INST. costs.

System Components	QUANTITY	UNIT	COST PER S.F.		
			MAT.	INST.	TOTAL
Non-load bearing metal studs, including top & bottom runners, 5/8″ drywall,					
Taped, finished and painted 2 faces, insulation, painted baseboard.					
Metal studs, 25 ga., 3-5/8″ wide, 24″ O.C.	1.000	S.F.	.30	.86	1.16
Gypsum drywall, 5/8″ thick	2.000	S.F.	.74	1.26	2
Taping & finishing	2.000	S.F.	.10	1.26	1.36
Insulation, 3-1/2″ fiberglass batts	1.000	S.F.	.52	.47	.99
Baseboard	.200	L.F.	.25	.53	.78
Paint baseboard, primer + 2 coats	.200	L.F.	.04	.31	.35
Painting, roller work, 2 coats	2.000	S.F.	.42	1.36	1.78
TOTAL			2.37	6.05	8.42

C1010 134	Partitions, Metal Stud, NLB	COST PER S.F.		
		MAT.	INST.	TOTAL
1300	Non-load bearing, 25 ga., 24″ O.C., 2-1/2″ wide	2.32	6.05	8.37
1350	3-5/8″ wide	2.37	6.05	8.42
1400	6″ wide	2.47	6.05	8.52
1500	16″ O.C., 2-1/2″ wide	2.38	6.25	8.63
1600	3-5/8″ wide	2.45	6.25	8.70
1700	6″ wide	2.57	6.30	8.87
1800	20 ga., 24″ O.C., 2-1/2″ wide	2.40	6.25	8.65
1900	3-5/8″ wide	2.51	6.25	8.76
2000	6″ wide	2.59	6.30	8.89
2100	16″ O.C., 2-1/2″ wide	2.48	6.50	8.98
2200	3-5/8″ wide	2.62	6.55	9.17
2300	6″ wide	2.72	6.55	9.27
2400				
2500				
2600				
2700				
2900	Cut & patch to match existing construction, add, minimum	2%	3%	
3000	Maximum	5%	9%	
3100	Dust protection, add, minimum	1%	2%	
3200	Maximum	4%	11%	
3300	Material handling & storage limitation, add, minimum	1%	1%	
3400	Maximum	6%	7%	
3500	Protection of existing work, add, minimum	2%	2%	
3600	Maximum	5%	7%	
3700	Shift work requirements, add, minimum		5%	
3800	Maximum		30%	
3900	Temporary shoring and bracing, add, minimum	2%	5%	
4000	Maximum	5%	12%	

C1010 Partitions

This data illustrates and describes a drywall system including gypsum plasterboard, taped and finished, metal studs with runners, insulation, baseboard and painting. Lines within System Components give the unit price and total price per square foot for this system. Prices for drywall systems are on Line Items C1010 136 1200 thru 1900. Both material quantities and labor costs have been adjusted for the system listed.

Factors: To adjust for job conditions other than normal working situations use Lines C1010 136 2900 thru 4000.

Example: You are to install the system and control dust in the work area. Go to Line C1010 136 3100 and apply these percentages to the appropriate MAT. and INST. costs.

System Components	QUANTITY	UNIT	COST PER S.F. MAT.	COST PER S.F. INST.	COST PER S.F. TOTAL
Gypsum drywall, taped, finished and painted 2 faces, galvanized metal studs					
Including top & bottom runners, insulation, painted baseboard, wall 10'high					
Gypsum drywall, 5/8" thick, standard	2.000	S.F.	.74	1.26	2
Taping and finishing	2.000	S.F.	.10	1.26	1.36
Metal studs, 20 ga., 3-5/8" wide, 24" O.C.	1.000	S.F.	.37	1.07	1.44
Insulation, 3-1/2" fiberglass batts	1.000	S.F.	.52	.47	.99
Baseboard	.200	L.F.	.25	.53	.78
Paint baseboard, primer + 2 coats	.200	L.F.	.03	.13	.16
Painting, roller 2 coats	2.000	S.F.	.30	1.30	1.60
TOTAL			2.31	6.02	8.33

C1010 136	Partitions, Drywall	COST PER S.F. MAT.	COST PER S.F. INST.	COST PER S.F. TOTAL
1200	Gypsum drywall, 5/8" thick, standard	2.31	6	8.31
1300	Fire resistant	2.28	5.80	8.08
1400	Water resistant	2.44	5.80	8.24
1500	1/2" thick, standard	2.22	5.80	8.02
1600	Fire resistant	2.30	5.80	8.10
1700	Water resistant	2.40	5.80	8.20
1800	3/8" thick, vinyl faced, standard	2.58	7.35	9.93
1900	5/8" thick, vinyl faced, fire resistant	3.18	7.35	10.53
2000				
2100				
2200				
2300				
2400				
2500				
2600				
2700				
2900	Cut & patch to match existing construction, add, minimum	2%	3%	
3000	Maximum	5%	9%	
3100	Dust protection, add, minimum	1%	2%	
3200	Maximum	4%	11%	
3300	Material handling & storage limitation, add, minimum	1%	1%	
3400	Maximum	6%	7%	
3500	Protection of existing work, add, minimum	2%	2%	
3600	Maximum	5%	7%	
3700	Shift work requirements, add, minimum		5%	
3800	Maximum		30%	
3900	Temporary shoring and bracing, add, minimum	2%	5%	
4000	Maximum	5%	12%	

623

C1010 Partitions

This data illustrates and describes a load bearing metal stud wall system including metal studs, sheetrock–taped and finished, insulation, baseboard and painting. Lines within System Components give the unit price and total price per square foot for this system. Prices for load bearing metal stud wall systems are on Line Items C1010 138 1500 thru 2500. Both material quantities and labor costs have been adjusted for the system listed.

Factors: To adjust for job conditions other than normal working situations use Lines C1010 138 2900 thru 4000.

Example: You are to install the system using temporary shoring and bracing. Go to Line C1010 138 3900 and apply these percentages to the appropriate MAT. and INST. costs.

System Components	QUANTITY	UNIT	COST PER S.F.		
			MAT.	INST.	TOTAL
Load bearing, 18 ga., 3-5/8″, galvanized metal studs, 24″ O.C.,including					
Top and bottom runners, 1/2″ drywall, taped, finished and painted 2					
Faces, 3″ insulation, and painted baseboard, wall 10′ high.					
Metal studs, 24″ O.C., 18 ga., 3-5/8″ wide, galvanized	1.000	S.F.	.84	1.21	2.05
Gypsum drywall 1/2″ thick	2.000	S.F.	.72	1.26	1.98
Taping and finishing	2.000	S.F.	.10	1.26	1.36
Insulation, 3-1/2″ fiberglass batts	1.000	S.F.	.52	.47	.99
Baseboard	.200	L.F.	.25	.53	.78
Paint baseboard, primer + 2 coats	.200	L.F.	.04	.31	.35
Painting, roller 2 coats	2.000	S.F.	.42	1.36	1.78
TOTAL			2.89	6.40	9.29

C1010 138	Partitions, Metal Stud, LB	COST PER S.F.		
		MAT.	INST.	TOTAL
1500	Load bearing, 18 ga., 24″ O.C., 2-1/2″ wide	2.75	6.40	9.15
1550	3-5/8″ wide	2.89	6.40	9.29
1600	6″ wide	3.17	6.45	9.62
1700	16″ O.C. 2-1/2″ wide	2.93	6.65	9.58
1800	3-5/8″ wide	3.09	6.70	9.79
1900	6″ wide	3.44	6.75	10.19
2000	16 ga., 24″ O.C., 2-1/2″ wide	2.87	6.55	9.42
2100	3-5/8″ wide	3.03	6.55	9.58
2200	6″ wide	3.35	6.65	10
2300	16″ O.C., 2-1/2″ wide	3.08	6.85	9.93
2400	3-5/8″ wide	3.28	6.90	10.18
2500	6″ wide	3.67	7	10.67
2600				
2700				
2900	Cut & patch to match existing construction, add, minimum	2%	3%	
3000	Maximum	5%	9%	
3100	Dust protection, add, minumum	1%	2%	
3200	Maximum	4%	11%	
3300	Material handling & storage limitation, add, minimum	1%	1%	
3400	Maximum	6%	7%	
3500	Protection of existing work, add, minimum	2%	2%	
3600	Maximum	5%	7%	
3700	Shift work requirements, add, minimum		5%	
3800	Maximum		30%	
3900	Temporary shoring and bracing, add, minimum	2%	5%	
4000	Maximum	5%	12%	

For customer support on your Commercial Renovation Cost Data, call 877.742.4719.

This data illustrates and describes a plaster and lath system including gypsum plaster, gypsum lath, wood studs with plates, insulation, baseboard and painting. Lines within System Components give the unit price and price per square foot for this system. Prices for plaster and lath systems are on Line Items C1010 148 1400 thru 2500. Both material quantities and labor costs have been adjusted for the system listed.

Factors: To adjust for job conditions other than normal working situations use Lines C1010 148 2900 thru 4000.

Example: You are to install the system during evening hours only. Go to Line C1010 148 3800 and apply this percentage to the appropriate INST. costs.

System Components	QUANTITY	UNIT	COST PER S.F.		
			MAT.	INST.	TOTAL
Gypsum plaster, 2 coats, over 3/8" lath, 2 faces, 2" x 4" wood Stud partition, 24"O.C. incl. double top plate, single bottom plate, 3-1/2" Insulation, baseboard and painting.					
Gypsum plaster, 2 coats	.220	S.Y.	.90	5.93	6.83
Lath, gypsum, 3/8" thick	.220	S.Y.	.74	1.54	2.28
Wood studs, 2" x 4", 24" O.C.	1.000	S.F.	.35	1.02	1.37
Insulation, 3-1/2" fiberglass batts	1.000	S.F.	.52	.47	.99
Baseboard, 9/16" x 3-1/2"	.200	L.F.	.25	.53	.78
Paint baseboard, primer + 2 coats	.200	L.F.	.03	.13	.16
Paint, 2 coats	2.000	S.F.	.42	1.36	1.78
TOTAL			3.21	10.98	14.19

C1010 148	Partitions, Plaster & Lath	COST PER S.F.		
		MAT.	INST.	TOTAL
1400	Gypsum plaster, 2 coats	3.21	10.95	14.16
1500	3 coats	3.60	12.25	15.85
1600	Perlite plaster, 2 coats	3.73	11.80	15.53
1700	3 coats	3.86	13.55	17.41
1800				
1900				
2000	For alternate lath systems:			
2100	Gypsum lath, 1/2" thick	3.58	11.90	15.48
2200	Foil back, 3/8" thick	3.82	12	15.82
2300	1/2" thick	3.95	12.10	16.05
2400	Metal lath, 2.5 Lb. diamond	3.98	11.80	15.78
2500	3.4 Lb. diamond	4.04	12	16.04
2600				
2700				
2900	Cut & patch to match existing construction, add, minimum	2%	3%	
3000	Maximum	5%	9%	
3100	Dust protection, add, minimum	1%	2%	
3200	Maximum	4%	11%	
3300	Material handling & storage limitation, add, minimum	1%	1%	
3400	Maximum	6%	7%	
3500	Protection of existing work, add, minimum	2%	2%	
3600	Maximum	5%	7%	
3700	Shift work requirements, add, minimum		5%	
3800	Maximum		30%	
3900	Temporary shoring and bracing, add, minimum	2%	5%	
4000	Maximum	5%	12%	

C1010 170	Selective Price Sheet	COST PER S.F.		
		MAT.	INST.	TOTAL
0100	Studs			
0200				
0300	24" O.C. metal, 10' high wall, including			
0400	Top and bottom runners			
0500	Non load bearing, galvanized 25 Ga., 1-5/8" wide	.20	.83	1.03
0600	2-1/2" wide	.25	.84	1.09
0700	3-5/8" wide	.30	.86	1.16
0800	4" wide	.33	.86	1.19
0900	6" wide	.40	.87	1.27
1000				
1100	Galvanized 20 Ga., 2-1/2" wide	.33	1.06	1.39
1200	3-5/8" wide	.37	1.07	1.44
1300	4" wide	.44	1.07	1.51
1400	6" wide	.52	1.09	1.61
1500	Load bearing, painted 18 Ga., 2-1/2" wide	.70	1.19	1.89
1600	3-5/8" wide	.84	1.21	2.05
1700	4" wide	.75	1.23	1.98
1800	6" wide	1.12	1.26	2.38
1900	Galvanized 18 Ga., 2-1/2" wide	.70	1.19	1.89
2000	3-5/8" wide	.84	1.21	2.05
2100	4" wide	.75	1.23	1.98
2200	6" wide	1.12	1.26	2.38
2300	Galvanized 16 Ga., 2-1/2" wide	.82	1.35	2.17
2400	3-5/8" wide	.98	1.38	2.36
2500	4" wide	1.03	1.41	2.44
2600	6" wide	1.30	1.44	2.74
2700				
2800				
2900	24" O.C. wood, 10' high wall, including			
3000	Double top plate and shoe			
3100	2" x 4"	.41	1.20	1.61
3200	2" x 6"	.63	1.28	1.91
3300				
3400				
3500	Furring, 24" O.C. 10' high wall, metal, 3/4" channels	.25	1.42	1.67
3600	1-1/2" channels	.34	1.53	1.87
3700	Wood, on wood, 1" x 2" strips	.14	.58	.72
3800	1" x 3" strips	.22	.58	.80
3900	On masonry, 1" x 2" strips	.15	.64	.79
4000	1" x 3" strips	.23	.64	.87
4100	On concrete, 1" x 2" strips	.15	1.22	1.37
4200	1" x 3" strips	.23	1.22	1.45
4300	Studs			
4400				
4500	16" O.C., metal, 10' high wall, including			
4600	Top and bottom runners			
4700	Non load bearing, galvanized 25 Ga., 1-5/8" wide	.27	1.28	1.55
4800	2-1/2" wide	.34	1.29	1.63
4900	3-5/8" wide	.41	1.32	1.73
5000	4" wide	.45	1.33	1.78
5100	6" wide	.54	1.35	1.89
5200				
5300	Galvanized 20 Ga., 2-1/2" wide	.44	1.62	2.06
5400	3-5/8" wide	.51	1.64	2.15
5500	4" wide	.60	1.67	2.27
5600	6" wide	.70	1.69	2.39
5700	Load bearing, painted 18 Ga., 2-1/2" wide	.95	1.65	2.60
5800	3-5/8" wide	1.14	1.67	2.81

C1010 Partitions

C1010 170	Selective Price Sheet	COST PER S.F.		
		MAT.	INST.	TOTAL
5900	4" wide	1.18	1.71	2.89
6000	6" wide	1.51	1.74	3.25
6100	Galvanized 18 Ga., 2-1/2" wide	.95	1.65	2.60
6200	3-5/8" wide	1.14	1.67	2.81
6300	4" wide	1.18	1.71	2.89
6400	6" wide	1.51	1.74	3.25
6500	Galvanized 16 Ga., 2-1/2" wide	1.13	1.86	2.99
6600	3-5/8" wide	1.35	1.92	3.27
6700	4" wide	1.41	1.95	3.36
6800	6" wide	1.78	1.98	3.76
6900				
7000				
7100	16" O.C., wood, 10' high wall, including			
7200	Double top plate and shoe			
7300	2" x 4"	.50	1.38	1.88
7400	2" x 6"	.62	1.24	1.86
7500				
7600				
7700	Furring, 16" O.C. 10' high wall, metal, 3/4" channels	.38	2.06	2.44
7800	1-1/2" channels	.51	2.29	2.80
7900	Wood, on wood, 1" x 2" strips	.20	.86	1.06
8000	1" x 3" strips	.32	.86	1.18
8100	On masonry, 1" x 2" strips	.22	.96	1.18
8200	1" x 3" strips	.35	.96	1.31
8300	On concrete, 1" x 2" strips	.22	1.83	2.05
8400	1" x 3" strips	.35	1.83	2.18

627

C1010 Partitions

C1010 180	Selective Price Sheet, Drywall & Plaster	COST PER S.F.		
		MAT.	INST.	TOTAL
0100	Lath, gypsum perforated			
0200				
0300	Lath, gypsum, perforated, regular, 3/8" thick	.37	.78	1.15
0400	1/2" thick	.30	.94	1.24
0500	Fire resistant, 3/8" thick	.30	.94	1.24
0600	1/2" thick	.36	1.02	1.38
0700	Foil back, 3/8" thick	.42	.88	1.30
0800	1/2" thick	.48	.94	1.42
0900				
1000	Metal lath			
1100	Metal lath, diamond, painted, 2.5 lb.	.50	.78	1.28
1200	3.4 lb.	.53	.88	1.41
1300	Rib painted, 2.75 lb	.43	.88	1.31
1400	3.40 lb	.55	.94	1.49
1500				
1600				
1700	Plaster, gypsum, 2 coats	.45	2.99	3.44
1800	3 coats	.65	3.64	4.29
1900	Perlite/vermiculite, 2 coats	.72	3.41	4.13
2000	3 coats	.78	4.28	5.06
2100	Bondcrete, 1 coat	.45	1.58	2.03
2200				
2500				
2600				
2700	Drywall, standard, 3/8" thick, no finish included	.39	.63	1.02
2800	1/2" thick, no finish included	.36	.63	.99
2900	Taped and finished	.42	1.31	1.73
3000	5/8" thick, no finish included	.37	.63	1
3100	Taped and finished	.43	1.31	1.74
3200	Fire resistant, 1/2" thick, no finish included	.40	.63	1.03
3300	Taped and finished	.45	1.31	1.76
3400	5/8" thick, no finish included	.39	.63	1.02
3500	Taped and finished	.44	1.31	1.75
3600	Water resistant, 1/2" thick, no finish included	.45	.63	1.08
3700	Taped and finished	.50	1.31	1.81
3800	5/8" thick, no finish included	.47	.63	1.10
3900	Taped and finished	.53	1.31	1.84
4000	Finish, instead of taping			
4100	For thin coat plaster, add	.12	.79	.91
4200	Finish, textured spray, add	.04	.75	.79

C1010 Partitions

This data illustrates and describes a movable office partition system including demountable office partitions of various styles and sizes based on a cost per square foot basis. Prices for movable office partition systems are on Line Items C1010 210 0300 thru 3600.

C1010 210	Partitions, Movable Office	COST PER S.F.		
		MAT.	INST.	TOTAL
0250	Office partition, demountable, no deduction for door opening, add for doors			
0300	Air wall, cork finish, semi acoustic, 1-5/8" thick, minimum	42.50	3.90	46.40
0400	Maximum	47	6.65	53.65
0500	Acoustic, 2" thick, minimum	40	4.15	44.15
0600	Maximum	61.50	5.65	67.15
0700				
0800				
0900	In-plant modular office system, w/prehung steel door			
1000	3" thick honeycomb core panels			
1100	12' x 12', 2 wall	9.05	.52	9.57
1200	4 wall	20.50	1.56	22.06
1300				
1400				
1500	Gypsum, demountable, 3" to 3-3/4" thick x 9' high, vinyl clad	7.35	2.94	10.29
1600	Fabric clad	18.30	3.22	21.52
1700	1.75 system, vinyl clad hardboard, paper honeycomb core panel			
1800	1-3/4" to 2-1/2" thick x 9' high	12.30	2.94	15.24
1900	Unitized gypsum panel, 2" to 2-1/2" thick x 9' high, vinyl clad	15.85	2.94	18.79
2000	Fabric clad	26	3.22	29.22
2100				
2200				
2300	Unitized mineral fiber panel system, 2-1/4" thick x 9' high			
2400	Vinyl clad mineral fiber	15.75	2.94	18.69
2500	Fabric clad mineral fiber	23.50	3.22	26.72
2600				
2700	Movable steel walls, modular system			
2800	Unitized panels, 48" wide x 9' high			
2900	Baked enamel, pre-finished	17.85	2.33	20.18
3000	Fabric clad	26	2.50	28.50
3100	For acoustical partitions, add, unsealed	2.60		2.60
3200	Sealed	12.10		12.10
3300				
3400				
3500	Note: For door prices, see Div. 08			
3600	For door hardware prices, see Div. 08			
3700				
3800				
3900				
4000				

C1020 Interior Doors

This data illustrates and describes flush interior door systems including hollow core door, jamb, header and trim with hardware. Lines within System Components give the unit price and total price on a cost each basis for this system. Prices for flush interior door systems are on Line items C1020 106 1000 thru 2400. Both material quantities and labor costs have been adjusted for the system listed.

Factors: To adjust for job conditions other than normal working situations use Lines C1020 106 2700 thru 3800.

Example: You are to install the system in an area where dust protection is vital. Go to Line C1020 106 3000 and apply these percentages to the appropriate MAT. and INST. costs.

System Components			QUANTITY	UNIT	COST EACH		
					MAT.	INST.	TOTAL
Single hollow core door, include jamb, header, trim and hardware, painted.							
Hollow core Lauan, 1-3/8" thick, 2'-0" x 6'-8", painted			1.000	Ea.	38.50	70.50	109
Paint door and frame, 1 coat			1.000	Ea.	2.46	30.50	32.96
Wood jamb, 4-9/16" deep			1.000	Set	86.40	54.08	140.48
Trim, casing			1.000	Set	49.60	80.96	130.56
Hardware, hinges			1.000	Set	25		25
Hardware, lockset			1.000	Set	33	39.50	72.50
		TOTAL			234.96	275.54	510.50

C1020 106		Doors, Interior Flush, Wood	COST EACH		
			MAT.	INST.	TOTAL
1000		Lauan (Mahogany) hollow core, 1-3/8" x 2'-0" x 6'-8"	235	276	511
1100		2'-6" x 6'-8"	243	280	523
1200		2'-8" x 6'-8"	248	283	531
1300		3'-0" x 6'-8"	255	293	548
1500		Birch, hollow core, 1-3/8" x 2'-0" x 6'-8"	243	276	519
1600		2'-6" x 6'-8"	256	280	536
1700		2'-8" x 6'-8"	259	283	542
1800		3'-0" x 6'-8"	266	293	559
1900		Solid core, pre-hung, 1-3/8" x 2'-6" x 6'-8"	505	284	789
2000		2'-8" x 6'-8"	425	287	712
2100		3'-0" x 6'-8"	540	298	838
2200					
2400		For metal frame instead of wood, add	50%	20%	
2600					
2700		Cut & patch to match existing construction, add, minimum	2%	3%	
2800		Maximum	5%	9%	
2900		Dust protection, add, minimum	1%	2%	
3000		Maximum	4%	11%	
3100		Equipment usage curtailment, add, minimum	1%	1%	
3200		Maximum	3%	10%	
3300		Material handling & storage limitation, add, minimum	1%	1%	
3400		Maximum	6%	7%	
3500		Protection of existing work, add, minimum	2%	2%	
3600		Maximum	5%	7%	
3700		Shift work requirements, add, minimum		5%	
3800		Maximum		30%	
3900					

C1020 Interior Doors

This data illustrates and describes interior, solid and louvered door systems including a pine panel door, wood jambs, header, and trim with hardware. Lines within System Components give the unit price and total price on a cost each basis for this system. Prices for interior, solid and louvered systems are on Line Items C1020 108 1100 thru 2400. Both material quantities and labor costs have been adjusted for the system listed.

Factors: To adjust for job conditions other than normal working situations use Lines C1020 108 2900 thru 4000.

Example: You are to install the system during night hours only. Go to Line C1020 108 4000 and apply these percentages to the appropriate INST. costs.

System Components	QUANTITY	UNIT	COST EACH MAT.	COST EACH INST.	COST EACH TOTAL
Single interior door, including jamb, header, trim and hardware, painted.					
Solid pine panel door, 1-3/8" thick, 2'-0" x 6'-8"	1.000	Ea.	168	70.50	238.50
Paint door and frame, 1 coat	1.000	Ea.	2.46	30.50	32.96
Wooden jamb, 4-5/8" deep	1.000	Set	86.40	54.08	140.48
Trim, casing	1.000	Set	49.60	80.96	130.56
Hardware, hinges	1.000	Set	25		25
Hardware, lockset	1.000	Set	33	39.50	72.50
TOTAL			364.46	275.54	640

C1020 108	Doors, Interior Solid & Louvered	COST EACH MAT.	COST EACH INST.	COST EACH TOTAL
1000				
1100	Solid pine, painted raised panel, 1-3/8" x 2'-0" x 6'-8"	365	276	641
1200	2'-6" x 6'-8"	400	280	680
1300	2'-8" x 6'-8"	410	283	693
1400	3'-0" x 6'-8"	430	293	723
1500				
1600	Louvered pine, painted 1'-6" x 6'-8"	330	272	602
1700	2'-0" x 6'-8"	345	280	625
1800	2'-6" x 6'-8"	370	283	653
1900	3'-0" x 6'-8"	400	293	693
2200	For prehung door, deduct	5%	30%	
2300				
2400	For metal frame instead of wood, add	50%	20%	
2500				
2900	Cut & patch to match existing construction, add, minimum	2%	3%	
3000	Maximum	5%	9%	
3100	Dust protection, add, minimum	1%	2%	
3200	Maximum	4%	11%	
3300	Equipment usage curtailment, add, minimum	1%	1%	
3400	Maximum	3%	10%	
3500	Material handling & storage limitation, add, minimum	1%	1%	
3600	Maximum	6%	7%	
3700	Protection of existing work, add, minimum	2%	2%	
3800	Maximum	5%	7%	
3900	Shift work requirements, add, minimum		5%	
4000	Maximum		30%	

C1020 Interior Doors

This data illustrates and describes interior metal door systems including a metal door, metal frame and hardware. Lines within System Components give the unit price and total price on a cost each for this system. Prices for interior metal door systems are on Line Items C1020 110 1000 thru 2000. Both material quantities and labor costs have been adjusted for the system listed.

Factors: To adjust for job conditions other than normal working situations use Lines C1020 110 2900 thru 4000.

Example: You are to install the system while protecting existing construction. Go to Line C1020 110 3700 and apply these percentages to the appropriate MAT. and INST. costs.

System Components	QUANTITY	UNIT	COST EACH MAT.	COST EACH INST.	COST EACH TOTAL
Single metal door, including frame and hardware.					
Hollow metal door, 1-3/8" thick, 2'-6" x 6'-8", painted	1.000	Ea.	545	84.50	629.50
Metal frame, 6-1/4" deep	1.000	Set	186	79	265
Paint door and frame, 1 coat	1.000	Ea.	2.46	30.50	32.96
Hardware, hinges	1.000	Set	94.50		94.50
Hardware, passage lockset	1.000	Set	33	39.50	72.50
TOTAL			860.96	233.50	1,094.46

C1020 110	Doors, Interior Flush, Metal	COST EACH MAT.	COST EACH INST.	COST EACH TOTAL
1000	Hollow metal doors, 1-3/8" thick, 2'-6" x 6'-8"	860	234	1,094
1100	1-3/8" thick, 2'-8" x 6'-8"	870	234	1,104
1200	3'-0" x 7'-0"	855	245	1,100
1300				
1400	Interior fire door, 1-3/8" thick, 2'-6" x 6'-8"	1,075	234	1,309
1500	2'-8" x 6'-8"	1,100	234	1,334
1600	3'-0" x 7'-0"	1,050	245	1,295
1700				
1800	Add to fire doors:			
1900	Baked enamel finish	30%	15%	
2000	Galvanizing	20%		
2200				
2300				
2400				
2900	Cut & patch to match existing construction, add, minimum	2%	3%	
3000	Maximum	5%	9%	
3100	Dust protection, add, minimum	1%	2%	
3200	Maximum	4%	11%	
3300	Equipment usage curtailment, add, minimum	1%	1%	
3400	Maximum	3%	10%	
3500	Material handling & storage limitation, add, minimum	1%	1%	
3600	Maximum	6%	7%	
3700	Protection of existing work, add, minimum	2%	2%	
3800	Maximum	5%	7%	
3900	Shift work requirements, add, minimum		5%	
4000	Maximum		30%	

This data illustrates and describes an interior closet door system including an interior closet door, painted, with trim and hardware. Prices for interior closet door systems are on Line Items C1020 112 0500 thru 2200. Both material quantities and labor costs have been adjusted for the system listed.

Factors: To adjust for job conditions other than normal working situations use Lines C1020 112 2900 thru 4000.

Example: You are to install the system and match the existing construction. Go to Line C1020 112 2900 and apply these percentages to the appropriate MAT. and INST. costs.

C1020 112	Doors, Closet	COST PER SET		
		MAT.	INST.	TOTAL
0350	Interior closet door painted, including frame, trim and hardware, prehung.			
0400	Bi-fold doors			
0500	Pine paneled, 3'-0" x 6'-8"	420	228	648
0600	6'-0" x 6'-8"	575	305	880
0700	Birch, hollow core, 3'-0" x 6'-8"	206	248	454
0800	6'-0" x 6'-8"	305	330	635
0900	Lauan, hollow core, 3'-0" x 6'-8"	190	228	418
1000	6'-0" x 6'-8"	275	305	580
1100	Louvered pine, 3'-0" x 6'-8"	345	228	573
1200	6'-0" x 6'-8"	455	305	760
1300				
1400	Sliding, bi-passing closet doors			
1500	Pine paneled, 4'-0" x 6'-8"	640	243	883
1600	6'-0" x 6'-8"	940	305	1,245
1700	Birch, hollow core, 4'-0" x 6'-8"	385	243	628
1800	6'-0" x 6'-8"	445	305	750
1900	Lauan, hollow core, 4'-0" x 6'-8"	310	243	553
2000	6'-0" x 6'-8"	365	305	670
2100	Louvered pine, 4'-0" x 6'-8"	650	243	893
2200	6'-0" x 6'-8"	825	305	1,130
2300				
2400				
2500				
2600				
2700				
2800				
2900	Cut & patch to match existing construction, add, minimum	2%	3%	
3000	Maximum	5%	9%	
3100	Dust protection, add, minimum	1%	2%	
3200	Maximum	4%	11%	
3300	Equipment usage curtailment, add, minimum	1%	1%	
3400	Maximum	3%	10%	
3500	Material handling & storage limitation, add, minimum	1%	1%	
3600	Maximum	6%	7%	
3700	Protection of existing work, add, minimum	2%	2%	
3800	Maximum	5%	7%	
3900	Shift work requirements, add, minimum		5%	
4000	Maximum		30%	

633

C20 Stairs

C2010 Stair Construction

This data illustrates and describes a stair system based on a cost per flight price. Prices for various stair systems are on Line Items C2010 130 0700 thru 3200. Both material quantities and labor costs have been adjusted for the system listed.

Factors: To adjust for job conditions other than normal working situations use Lines C2010 130 3500 thru 4200.

Example: You are to install the system during evenings only. Go to Line C2010 130 4200 and apply this percentage to the appropriate MAT. and INST. costs.

System Components	QUANTITY	UNIT	COST PER FLIGHT		
			MAT.	INST.	TOTAL
Below are various stair systems based on cost per flight of stairs, no side Walls. Stairs are 4'-0" wide, railings are included unless otherwise noted.					

C2010 130	Stairs	COST PER FLIGHT		
		MAT.	INST.	TOTAL
0700	Concrete, cast in place, no nosings, no railings, 12 risers	300	2,150	2,450
0800	24 risers	605	4,325	4,930
0900	Add for 1 intermediate landing	162	595	757
1000	Concrete, cast in place, with nosings, no railings, 12 risers	1,275	2,375	3,650
1100	24 risers	2,525	4,775	7,300
1200	Add for 1 intermediate landing	242	615	857
1300	Steel, grating tread, safety nosing, 12 risers	4,750	1,200	5,950
1400	24 risers	9,500	2,400	11,900
1500	Add for intermediate landing	1,175	320	1,495
1600	Steel, cement fill pan tread, 12 risers	7,250	1,200	8,450
1700	24 risers	14,500	2,400	16,900
1800	Add for intermediate landing	1,175	320	1,495
1900	Spiral, industrial, 4' - 6" diameter, 12 risers	5,825	855	6,680
2000	24 risers	11,600	1,700	13,300
2100	Wood, box stairs, oak treads, 12 risers	2,375	715	3,090
2200	24 risers	4,750	1,425	6,175
2300	Add for 1 intermediate landing	208	162	370
2400	Wood, basement stairs, no risers, 12 steps	785	294	1,079
2500	24 steps	1,575	590	2,165
2600	Add for 1 intermediate landing	45.50	31	76.50
2700	Wood, open, rough sawn cedar, 12 steps	2,700	580	3,280
2800	24 steps	5,400	1,150	6,550
2900	Add for 1 intermediate landing	24.50	31.50	56
3000	Wood, residential, oak treads, 12 risers	2,625	3,175	5,800
3100	24 risers	5,225	6,325	11,550
3200	Add for 1 intermediate landing	185	146	331
3500	Dust protection, add, minimum	1%	2%	
3600	Maximum	4%	11%	
3700	Material handling & storage limitation, add, minimum	1%	1%	
3800	Maximum	6%	7%	
3900	Protection of existing work, add, minimum	2%	2%	
4000	Maximum	5%	7%	
4100	Shift work requirements, add, minimum		5%	
4200	Maximum		30%	

C30 Interior Finishes

C3010 Wall Finishes

C3010 210	Selective Price Sheet	COST PER S.F.		
		MAT.	INST.	TOTAL
0100	Painting, on plaster or drywall, brushwork, primer and 1 ct.	.15	.76	.91
0200	Primer and 2 ct.	.22	1.21	1.43
0300	Rollerwork, primer and 1 ct.	.15	.65	.80
0400	Primer and 2 ct.	.22	1.03	1.25
0500	Woodwork incl. puttying, brushwork, primer and 1 ct.	.14	1.15	1.29
0600	Primer and 2 ct.	.21	1.53	1.74
0700	Wood trim to 6" wide, enamel, primer and 1 ct.	.14	.65	.79
0800	Primer and 2 ct.	.22	.82	1.04
0900	Cabinets and casework, enamel, primer and 1 ct.	.14	1.30	1.44
1000	Primer and 2 ct.	.22	1.60	1.82
1100	On masonry or concrete, latex, brushwork, primer and 1 ct.	.31	1.08	1.39
1200	Primer and 2 ct.	.40	1.55	1.95
1300	For block filler, add	.23	1.33	1.56
1400				
1500	Varnish, wood trim, sealer 1 ct., sanding, puttying, quality work	.14	2.34	2.48
1600	Medium work	.11	1.82	1.93
1700	Without sanding	.17	.27	.44
1800				
1900	Wall coverings, wall paper, at low price per double roll, avg workmanship	.65	.82	1.47
2000	At medium price per double roll, average workmanship	1.24	.98	2.22
2100	At high price per double roll, quality workmanship	1.89	1.20	3.09
2200				
2300	Grass cloths with lining paper	1.06	1.31	2.37
2400	Premium texture/color	3.14	1.49	4.63
2500	Vinyl, fabric backed, light weight	1.07	.82	1.89
2600	Medium weight	.99	1.09	2.08
2700	Heavy weight	1.51	1.20	2.71
2800				
2900	Cork tiles, 12" x 12", 3/16" thick	5.15	2.17	7.32
3000	5/16" thick	3.44	2.22	5.66
3100	Granular surface, 12" x 36", 1/2" thick	1.44	1.36	2.80
3200	1" thick	1.85	1.41	3.26
3300	Aluminum foil	1.14	1.90	3.04
3400				
3500	Tile, ceramic, adhesive set, 4-1/4" x 4-1/4"	2.74	5.25	7.99
3600	6" x 6"	3.80	5.70	9.50
3700	Decorated, 4-1/4" x 4-1/4", minimum	3.88	3.71	7.59
3800	Color group 4	57	5.55	62.55
3900	For epoxy grout, add	.45	1.25	1.70
4000	Pregrouted sheets	5.95	4.17	10.12
4100	Glass mosaics, 3/4" tile on 12" sheets, minimum	18.90	13.70	32.60
4200	Color group 8	45.50	15.65	61.15
4300	Metal, tile pattern, 4' x 4' sheet, 24 ga., nailed			
4400	Stainless steel	31	2.47	33.47
4500	Aluminized steel	16.60	2.47	19.07
4600				
4700	Brick, interior veneer, 4" face brick, running bond, minimum	4.24	13	17.24
4800	Maximum	4.24	13	17.24
4900	Simulated, urethane pieces, set in mastic	9.45	4.22	13.67
5000	Fiberglass panels	8.85	3.17	12.02
5100	Wall coating, on drywall, thin coat, plain	.12	.79	.91
5200	Stipple	.12	.79	.91
5300	Textured spray	.04	.75	.79
5400				
5500	Paneling not incl. furring or trim, hardboard, tempered, 1/8" thick	.46	2.53	2.99
5600	1/4" thick	.72	2.53	3.25
5700	Plastic faced, 1/8" thick	.73	2.53	3.26
5800	1/4" thick	.98	2.53	3.51

635

C3010 Wall Finishes

C3010 210	Selective Price Sheet	COST PER S.F.		
		MAT.	INST.	TOTAL
5900	Woodgrained, 1/4" thick, minimum	.76	2.53	3.29
6000	Maximum	1.54	2.98	4.52
6100	Plywood, 4' x 8' shts. 1/4" thick, prefin., birch faced, min.	1.60	2.53	4.13
6200	Maximum	1.38	3.62	5
6300	Walnut, minimum	2.64	2.53	5.17
6400	Maximum	5.80	3.17	8.97
6500	Mahogany, African	3.08	3.17	6.25
6600	Philippine	.72	2.53	3.25
6700	Chestnut	5.35	3.38	8.73
6800	Pecan	2.81	3.17	5.98
6900	Rosewood	3.25	3.96	7.21
7000	Teak	3.25	3.17	6.42
7100	Aromatic cedar, plywood	2.48	3.17	5.65
7200	Particle board	1.27	3.17	4.44
7300	Wood board, 3/4" thick, knotty pine	2.15	4.22	6.37
7400	Rough sawn cedar	3.52	4.22	7.74
7500	Redwood, clear	5.45	4.22	9.67
7600	Aromatic cedar	2.64	4.60	7.24

For customer support on your Commercial Renovation Cost Data, call 877.742.4719.

C30 Interior Finishes

C3020 Floor Finishes

C3020 430	Selective Price Sheet	COST PER S.F.		
		MAT.	INST.	TOTAL
0100	Flooring, carpet, acrylic, 26 oz. light traffic	2.61	.83	3.44
0200	35 oz. heavy traffic	5.50	.83	6.33
0300	Nylon anti-static, 15 oz. light traffic	2.22	.83	3.05
0400	22 oz. medium traffic	4.27	.83	5.10
0500	26 oz. heavy traffic	6.15	.89	7.04
0600	28 oz. heavy traffic	6.70	.89	7.59
0700	Tile, foamed back, needle punch	4.75	.99	5.74
0800	Tufted loop	3.62	.99	4.61
0900	Wool, 36 oz. medium traffic	11.10	.89	11.99
1000	42 oz. heavy traffic	12.20	.89	13.09
1100	Composition, epoxy, with colored chips, minimum	3.09	4.67	7.76
1200	Maximum	4.46	6.45	10.91
1300	Trowelled, minimum	3.93	5.60	9.53
1400	Maximum	5.90	6.55	12.45
1500	Terrazzo, 1/4" thick, chemical resistant, minimum	6.60	6.90	13.50
1600	Maximum	10.05	9.15	19.20
1700	Resilient, asphalt tile, 1/8" thick	1.31	1.40	2.71
1800	Conductive flooring, rubber, 1/8" thick	7.65	1.78	9.43
1900	Cork tile 1/8" thick, standard finish	7.65	1.78	9.43
2000	Urethane finish	8.90	1.78	10.68
2100	PVC sheet goods for gyms, 1/4" thick	7.15	7	14.15
2200	3/8" thick	10.50	9.35	19.85
2300	Vinyl composition 12" x 12" tile, plain, 1/16" thick	1.31	1.12	2.43
2400	1/8" thick	1.71	1.12	2.83
2500	Vinyl tile, 12" x 12" x 1/8" thick, minimum	7.70	1.12	8.82
2600	Premium colors/pattern	6.85	1.12	7.97
2700	Vinyl sheet goods, backed, .080" thick, plain pattern/colors	4.53	2.44	6.97
2900	Slate, random rectangular, 1/4" thick	23	7.30	30.30
3000	1/2" thick	21.50	10.45	31.95
3100	Natural cleft, irregular, 3/4" thick	9.90	11.90	21.80
3200	For sand rubbed finish, add	9.90		9.90
3300	Terrazzo, cast in place, bonded 1-3/4" thick, gray cement	3.69	18.35	22.04
3400	White cement	4.10	18.35	22.45
3500	Not bonded 3" thick, gray cement	4.59	19.65	24.24
3600	White cement	5.35	19.65	25
3700	Precast, 12" x 12" x 1" thick	22.50	18.30	40.80
3800	1-1/4" thick	23	18.30	41.30
3900	16" x 16" x 1-1/4" thick	24.50	22	46.50
4000	1-1/2" thick	22.50	24.50	47
4100	Marble travertine, standard, 12" x 12" x 3/4" thick	10.95	16.70	27.65
4200				
4300	Tile, ceramic, natural clay, thin set	4.56	5.45	10.01
4400	Porcelain, thin set	7.05	5.25	12.30
4500	Specialty, decorator finish	11.45	5.45	16.90
4600				
4700	Quarry, red, mud set, 4" x 4" x 1/2" thick	8.85	8.35	17.20
4800	6" x 6" x 1/2" thick	8.75	7.15	15.90
4900	Brown, imported, 6" x 6" x 3/4" thick	8.15	8.35	16.50
5000	8" x 8" x 1" thick	9.55	9.10	18.65
5100	Slate, Vermont, thin set, 6" x 6" x 1/4" thick	8.25	5.55	13.80
5200				
5300	Wood, maple strip, 25/32" x 2-1/4", finished, select	5.55	3.99	9.54
5400	2nd and better	5.05	3.99	9.04
5500	Oak, 25/32" x 2-1/4" finished, clear	3.83	3.99	7.82
5600	No. 1 common	4.68	3.99	8.67
5700	Parquet, standard, 5/16", finished, minimum	6.10	4.23	10.33
5800	Maximum	10.80	6.60	17.40
5900	Custom, finished, plain pattern	18.80	6.35	25.15

637

C3020 Floor Finishes

C3020 430	Selective Price Sheet	COST PER S.F.		
		MAT.	INST.	TOTAL
6000	Intricate pattern	27.50	12.65	40.15
6100	Prefinished, oak, 2-1/4" wide	5.30	3.72	9.02
6200	Ranch plank	7.85	4.37	12.22
6300	Sleepers on concrete, treated, 24" O.C., 1" x 2"	.23	.42	.65
6400	1" x 3"	.41	.53	.94
6500	2" x 4"	.55	.74	1.29
6600	2" x 6"	.83	.85	1.68
6700	Refinish old floors, soft wood	.99	1.24	2.23
6800	Hard wood	1.47	3.81	5.28
6900	Subflooring, plywood, CDX, 1/2" thick	.75	.84	1.59
7000	5/8" thick	.84	.94	1.78
7100	3/4" thick	1	1.01	2.01
7200				
7300	1" x 10" boards, S4S, laid regular	2.11	1.15	3.26
7400	Laid diagonal	2.11	1.41	3.52
7500	1" x 8" boards, S4S, laid regular	2.07	1.27	3.34
7600	Laid diagonal	2.07	1.49	3.56
7700	Underlayment, plywood, underlayment grade, 3/8" thick	1.09	.84	1.93
7800	1/2" thick	1.29	.87	2.16
7900	5/8" thick	1.42	.90	2.32
8000	3/4" thick	1.63	.97	2.60
8100	Particle board, 3/8" thick	.45	.84	1.29
8200	1/2" thick	.50	.87	1.37
8300	5/8" thick	.59	.90	1.49
8400	3/4" thick	.75	.97	1.72

C3030 Ceiling Finishes

This data illustrates suspended acoustical board systems including acoustic ceiling board, hangers, and T bar suspension. Lines within System Components give the unit price and total price per square foot for this system. Prices for suspended acoustical board systems are on Line Items C3030 210 1000 thru 2400. Both material quantities and labor costs have been adjusted for the system listed.

Factors: To adjust for job conditions other than normal working situations use Lines C3030 210 2900 thru 4000.

Example: You are to install the system and protect existing construction. Go to Line C3030 210 3800 and apply these percentages to the appropriate MAT. and INST. costs.

System Components	QUANTITY	UNIT	COST PER S.F.		
			MAT.	INST.	TOTAL
Suspended acoustical ceiling board installed on exposed grid system.					
Fiberglass boards, film faced, 2' x 4', 5/8" thick	1.000	S.F.	1.38	1.01	2.39
Hangers, #12 wire	1.000	S.F.	.03	.61	.64
T bar suspension system, 2' x 4' grid	1.000	S.F.	.84	.79	1.63
TOTAL			2.25	2.41	4.66

C3030 210	Ceiling, Suspended Acoustical	COST PER S.F.		
		MAT.	INST.	TOTAL
1000	2' x 4' grid, fiberglass board, film faced, 5/8" thick	2.25	2.41	4.66
1100	Mineral fiber board, aluminum faced	4.92	2.37	7.29
1200	Standard faced	1.62	2.34	3.96
1300	Plastic faced	3.63	2.98	6.61
1400	Fiberglass, film faced, 3" thick, R11	3.75	2.81	6.56
1500	Grass cloth faced, 3/4" thick	3.86	2.67	6.53
1600	1" thick	4.51	2.71	7.22
1700	1-1/2" thick, nubby face	3.73	2.73	6.46
2200				
2300				
2400	Add for 2' x 2' grid system	.22	.19	.41
2500				
2600				
2700				
2900	Cut & patch to match existing construction, add, minimum	2%	3%	
3000	Maximum	5%	9%	
3100	Dust protection, add, minimum	1%	2%	
3200	Maximum	4%	11%	
3300	Equipment usage curtailment, add, minimum	1%	1%	
3400	Maximum	3%	10%	
3500	Material handling & storage limitation, add, minimum	1%	1%	
3600	Maximum	6%	7%	
3700	Protection of existing work, add, minimum	2%	2%	
3800	Maximum	5%	7%	
3900	Shift work requirements, add, minimum		5%	
4000	Maximum		30%	

C3030 Ceiling Finishes

This data illustrates and describes suspended gypsum board systems including gypsum board, metal furring, taping, finished and painted. Lines within System Components give the unit price and total price per square foot for this system. Prices for suspended gypsum board systems are on Line Items C3030 220 1400 thru 1700. Both material quantities and labor costs have been adjusted for the system listed.

Factors: To adjust for job conditions other than normal working situations use Lines C3030 220 2900 thru 4000.

Example: You are to install the system and control dust in the working area. Go to Line C3030 220 3200 and apply these percentages to the appropriate MAT. and INST. costs.

System Components	QUANTITY	UNIT	COST PER S.F.		
			MAT.	INST.	TOTAL
Suspended ceiling gypsum board, 4' x 8' x 5/8" thick,					
On metal furring, taped, finished and painted.					
Gypsum drywall, 4' x 8', 5/8" thick, screwed	1.000	S.F.	.37	.70	1.07
Main runners, 1-1/2" C.R.C., 4' O.C.	.500	S.F.	.17	.77	.94
25 ga., channels, 2' O.C.	1.000	S.F.	.25	1.42	1.67
Taped and finished	1.000	S.F.	.05	.63	.68
Paint, 2 coats, roller work	1.000	S.F.	.21	.68	.89
TOTAL			1.05	4.20	5.25

C3030 220	Ceilings, Suspended Gypsum Board	COST PER S.F.		
		MAT.	INST.	TOTAL
1400	4' x 8' x 5/8", gypsum drywall, 2 coats paint	1.05	4.20	5.25
1500	Thin coat plaster, 2 coats paint	.91	3.68	4.59
1600	Spray-on sand finish, no paint	.83	3.64	4.47
1700	12" x 12" x 3/4" acoustical wood fiber tile	2.14	4.47	6.61
1800				
1900				
2000				
2100				
2200				
2300				
2400				
2500				
2600				
2700				
2900	Cut & patch to match existing construction, add, minimum	2%	3%	
3000	Maximum	5%	9%	
3100	Dust protection, add, minimum	1%	2%	
3200	Maximum	4%	11%	
3300	Equipment usage curtailment, add, minimum	1%	1%	
3400	Maximum	3%	10%	
3500	Material handling & storage limitation, add, minimum	1%	1%	
3600	Maximum	6%	7%	
3700	Protection of existing work, add, minimum	2%	2%	
3800	Maximum	5%	7%	
3900	Shift work requirements, add, minimum		5%	
4000	Maximum		30%	

C3030 Ceiling Finishes

This data illustrates and describes suspended plaster and lath systems including gypsum plaster, lath, furring and runners with ceiling painted. Lines within System Components give the unit price and total price per square foot for this system. Prices for suspended plaster and lath systems are on Line Items C3030 230 1200 thru 2300. Both material quantities and labor costs have been adjusted for the system listed.

Factors: To adjust for job conditions other than normal working situations use Lines C3030 230 2900 thru 4000.

Example: You are to install the system to match existing construction. Go to Line C3030 230 2900 and apply these percentages to the appropriate MAT. and INST. costs.

System Components			COST PER S.F.		
	QUANTITY	UNIT	MAT.	INST.	TOTAL
Gypsum plaster, 3 coats, on 3.4# rib lath, on 3/4″ C.R.C. furring on 1-1/2″ main runners, ceiling painted.					
Gypsum plaster, 3 coats	.110	S.Y.	.64	4.02	4.66
3.4# rib lath	.110	S.Y.	.55	.94	1.49
Main runners, 1-1/2″ C.R.C. 24″ O.C.	.333	S.F.	.34	1.53	1.87
Furring 3/4″ C.R.C. 16″ O.C.	1.000	S.F.	.38	2.06	2.44
Painting, 2 coats, roller work	1.000	S.F.	.21	.68	.89
TOTAL			2.12	9.23	11.35

C3030 230	Ceiling, Suspended Plaster	COST PER S.F.		
		MAT.	INST.	TOTAL
1200	Gypsum plaster, 3 coats, on 3.4# rib lath	2.12	9.20	11.32
1300	On 2.5# diamond lath	2.07	9.15	11.22
1400	On 3/8″ gypsum lath	1.94	9.15	11.09
1500	2 coats, on 3.4# rib lath	1.86	9.20	11.06
1600	On 2.5# diamond lath	1.88	8.55	10.43
1700	On 3/8″ gypsum lath	1.75	8.55	10.30
1800	Perlite plaster, 3 coats, on 3.4# rib lath	2.26	10.15	12.41
1900	On 2.5# diamond lath	2.21	10.05	12.26
2000	On 3/8″ gypsum lath	2.08	10.05	12.13
2100	2 coats, on 3.4# rib lath	2.19	9.15	11.34
2200	On 2.5# diamond lath	2.14	9.10	11.24
2300	On 3/8″ gypsum lath	2.01	9.10	11.11
2400				
2500				
2600				
2700				
2900	Cut & patch to match existing construction, add, minimum	2%	3%	
3000	Maximum	5%	9%	
3100	Dust protection, add, minimum	1%	2%	
3200	Maximum	4%	11%	
3300	Equipment usage curtailment, add, minimum	1%	1%	
3400	Maximum	3%	10%	
3500	Material handling & storage limitation, add, minimum	1%	1%	
3600	Maximum	6%	7%	
3700	Protection of existing work, add, minimum	2%	2%	
3800	Maximum	5%	7%	
3900	Shift work requirements, add, minimum		5%	
4000	Maximum		30%	

C3030 Ceiling Finishes

C3030 250	Selective Price Sheet	COST PER S.F.		
		MAT.	INST.	TOTAL
0100	Ceiling, plaster, gypsum, 2 coats	.45	3.41	3.86
0200	3 coats	.65	4.05	4.70
0300	Perlite or vermiculite, 2 coats	.72	3.99	4.71
0400	3 coats	.78	4.99	5.77
0500	Gypsum lath, plain, 3/8" thick	.37	.78	1.15
0600	1/2" thick	.30	.83	1.13
0700	Firestop, 3/8" thick	.30	.94	1.24
0800	1/2" thick	.36	1.02	1.38
0900	Metal lath, rib, 2.75 lb.	.43	.88	1.31
1000	3.40 lb.	.55	.94	1.49
1100	Diamond, 2.50 lb.	.50	.88	1.38
1200	3.40 lb.	.53	1.10	1.63
1500	Drywall, standard, 1/2" thick, no finish included	.36	.70	1.06
1600	Taped and finished	.42	1.66	2.08
1700	5/8" thick, no finish included	.37	.70	1.07
1800	Taped and finished	.43	1.66	2.09
1900	Fire resistant, 1/2" thick, no finish included	.40	.70	1.10
2000	Taped and finished	.45	1.66	2.11
2100	5/8" thick, no finish included	.39	.70	1.09
2200	Taped and finished	.44	1.66	2.10
2300	Water resistant, 1/2" thick, no finish included	.45	.70	1.15
2400	Taped and finished	.50	1.66	2.16
2500	5/8" thick, no finish included	.47	.70	1.17
2600	Taped and finished	.53	1.66	2.19
2700	Finish, instead of taping			
2800	For thin coat plaster, add	.12	.79	.91
2900	Finish, textured spray, add	.04	.75	.79
3000				
3100				
3200				
3300	Tile, stapled or glued, mineral fiber plastic coated, 5/8" thick	2.45	2.11	4.56
3400	3/4" thick	3.22	2.11	5.33
3500	Wood fiber, 1/2" thick	1.27	1.58	2.85
3600	3/4" thick	1.35	1.58	2.93
3700	Suspended, fiberglass, film faced, 5/8" thick	1.38	1.01	2.39
3800	3" thick	2.88	1.41	4.29
3900	Mineral fiber, 5/8" thick, standard	.75	.94	1.69
4000	Aluminum	4.05	1.06	5.11
4100	Wood fiber, reveal edge, 1" thick			
4200	3" thick			
4300	Framing, metal furring, 3/4" channels, 12" O.C.	.42	2.84	3.26
4400	16" O.C.	.38	2.06	2.44
4500	24" O.C.	.25	1.42	1.67
4600				
4700	1-1/2" channels, 12" O.C.	.57	3.14	3.71
4800	16" O.C.	.51	2.29	2.80
4900	24" O.C.	.34	1.53	1.87
5000				
5100				
5200				
5300	Ceiling suspension systems, for tile,			
5400	Concealed "Z" bar, 12" module	.95	1.22	2.17
5500	Class A, "T" bar 2'-0" x 4'-0" grid	.84	.79	1.63
5600	2'-0" x 2'-0" grid	1.09	.97	2.06
5700	Carrier channels for lighting fixtures, add	.24	1.38	1.62
5800				
5900				
6000				

C30 Interior Finishes

C3030 Ceiling Finishes

C3030 250	Selective Price Sheet	COST PER S.F.		
		MAT.	INST.	TOTAL
6100	Wood, furring 1" x 3", on wood, 12" O.C.	.43	1.81	2.24
6200	16" O.C..	.32	1.36	1.68
6300	24" O.C.	.22	.91	1.13
6400				
6500	On concrete, 12" O.C.	.46	3.02	3.48
6600	16" O.C.	.35	2.27	2.62
6700	24" O.C.	.23	1.51	1.74
6800				
6900	Joists, 2" x 4", 12" O.C.	.57	1.30	1.87
7000	16" O.C.	.48	1.09	1.57
7100	24" O.C.	.39	.88	1.27
7200	32" O.C.	.29	.67	.96
7300	2" x 6", 12" O.C.	.88	1.32	2.20
7400	16" O.C.	.73	1.09	1.82
7500	24" O.C.	.58	.87	1.45
7600	32" O.C.	.43	.65	1.08

This data illustrates and describes oil hydraulic elevator systems. Prices for various oil hydraulic elevator systems are on Line Items D1010 130 0400 thru 2000. Both material quantities and labor costs have been adjusted for the system listed.

Factors: To adjust for job conditions other than normal working situations use Lines D1010 130 3100 thru 4000.

Example: You are to install the system with limited equipment usage. Go to Line D1010 130 3400 and apply this percentage to the appropriate TOTAL costs.

D1010 130	Oil Hydraulic Elevators	COST EACH		
		MAT.	INST.	TOTAL
0310	Oil-hydraulic elevator systems			
0320	Including piston and piston shaft			
0330				
0400	1500 Lb. passenger, 2 floors	42,500	19,500	62,000
0500	3 floors	53,000	30,100	83,100
0600	4 floors	83,500	42,400	125,900
0700	5 floors	93,500	53,000	146,500
0800	6 floors	103,000	63,000	166,000
0900				
1000	2500 Lb. passenger, 2 floors	45,700	19,500	65,200
1100	3 floors	56,500	30,100	86,600
1200	4 floors	82,500	42,400	124,900
1300	5 floors	92,500	53,000	145,500
1400	6 floors	106,000	63,000	169,000
1500				
1600	4000 Lb. passenger, 2 floors	52,000	19,500	71,500
1700	3 floors	65,000	30,100	95,100
1800	4 floors	93,000	42,400	135,400
1900	5 floors	103,000	53,000	156,000
2000	6 floors	112,500	63,000	175,500
2100				
2200				
2300				
2400				
2500				
2600				
2700				
2800				
2900				
3000				
3100	Dust protection, add, minimum	1%	2%	
3200	Maximum	4%	11%	
3300	Equipment usage curtailment, add, minimum	1%	1%	
3400	Maximum	3%	10%	
3500	Material handling & storage limitation, add, minimum	1%	1%	
3600	Maximum	6%	7%	
3700	Protection of existing work, add, minimum	2%	2%	
3800	Maximum	5%	7%	
3900	Shift work requirements, add, minimum		5%	
4000	Maximum		30%	

D2010 Plumbing Fixtures

This data illustrates and describes a women's public restroom system including a water closet, lavatory, accessories, and service piping. Lines within System Components give the unit price and total price on a cost each basis for this system. Prices for women's public restroom systems are on Line Items D2010 956 1300 thru 1700. Both material quantities and labor costs have been adjusted for the system listed.

Factors: To adjust for job conditions other than normal working situations use Lines D2010 956 2900 thru 4000.

Example: You are to install the system and protect surrounding area from dust. Go to Line D2010 956 3100 and apply these percentages to the MAT. and INST. costs.

System Components	QUANTITY	UNIT	COST EACH MAT.	COST EACH INST.	COST EACH TOTAL
Public women's restroom incl. water closet, lavatory, accessories and Necessary service piping to install this system in one wall.					
Water closet, wall mounted, one piece	1.000	Ea.	1,050	229	1,279
Rough-in waste and vent for water closet	1.000	Set	1,075	520	1,595
Lavatory, 20" x 18" P.E. cast iron with accessories	1.000	Ea.	291	166	457
Rough-in waste and vent for lavatory	1.000	Set	480	800	1,280
Toilet partition, painted metal between walls, floor mounted	1.000	Ea.	420	211	631
For handicap unit, add	1.000	Ea.	410		410
Grab bar, 36" long	1.000	Ea.	81	63	144
Mirror, 18" x 24", with stainless steel shelf	1.000	Ea.	218	31.50	249.50
Napkin/tampon dispenser, recessed	1.000	Ea.	695	42	737
Soap Dispenser, chrome, surface mounted, liquid	1.000	Ea.	55	31.50	86.50
Toilet tissue dispenser, surface mounted, stainless steel	1.000	Ea.	21	21	42
Towel dispenser, surface mounted, stainless steel	1.000	Ea.	52	39.50	91.50
TOTAL			4,848	2,154.50	7,002.50

D2010 956	Plumbing - Public Restroom	COST EACH MAT.	COST EACH INST.	COST EACH TOTAL
1200				
1300	Public women's restroom, one water closet, one lavatory	4,850	2,150	7,000
1400	Two water closets, two lavatories	8,950	4,100	13,050
1500				
1600	For each additional water closet over 2, add	2,800	950	3,750
1700	For each additional lavatory over 2, add	1,050	1,025	2,075
1800				
1900				
2400	NOTE: PLUMBING APPROXIMATIONS			
2500	WATER CONTROL: water meter, backflow preventer,			
2600	Shock absorbers, vacuum breakers, mixer....10 to 15% of fixtures			
2700	PIPE AND FITTINGS: .... 30 to 60% of fixtures			
2800				
2900	Cut & patch to match existing construction, add, minimum	2%	3%	
3000	Maximum	5%	9%	
3100	Dust protection, add, minimum	1%	2%	
3200	Maximum	4%	11%	
3300	Equipment usage curtailment, add, minimum	1%	1%	
3400	Maximum	3%	10%	
3500	Material handling & storage limitation, add, minimum	1%	1%	

647

D20 Plumbing

D2010 Plumbing Fixtures

D2010 956	Plumbing - Public Restroom	COST EACH		
		MAT.	INST.	TOTAL
3600	Maximum	6%	7%	
3700	Protection of existing work, add, minimum	2%	2%	
3800	Maximum	5%	7%	
3900	Shift work requirements, add, minimum		5%	
4000	Maximum		30%	

For customer support on your Commercial Renovation Cost Data, call 877.742.4719.

D2010 Plumbing Fixtures

This data illustrates and describes a men's public restroom system including a water closet, urinal, lavatory, accessories and service piping. Lines within System Components give the unit price and total price on a cost each basis for this system. Prices for men's public restroom systems are on Line Items D2010 957 1700 thru 2200. Both material quantities and labor costs have been adjusted for the system listed.

Factors: To adjust for job conditions other than normal working situations use Lines D2010 957 2900 thru 4000.

Example: You are to install the system and match existing construction. Go to Line D2010 957 3000 and apply these percentages to the appropriate MAT. and INST. costs.

System Components	QUANTITY	UNIT	COST EACH MAT.	COST EACH INST.	COST EACH TOTAL
Public men's restroom incl. water closet, urinal, lavatory, accessories,					
And necessary service piping to install this system in one wall.					
Water closet, wall mounted, one piece	1.000	Ea.	1,050	229	1,279
Rough-in waste & vent for water closet	1.000	Set	1,075	520	1,595
Urinal, wall hung	1.000	Ea.	305	445	750
Rough-in waste & vent for urinal	1.000	Set	650	470	1,120
Lavatory, 20″ x 18″, P.E. cast iron with accessories	1.000	Ea.	291	166	457
Rough-in waste & vent for lavatory	1.000	Set	480	800	1,280
Partition, painted mtl., between walls, floor mntd.	1.000	Ea.	420	211	631
For handicap unit, add	1.000	Ea.	410		410
Grab bars	1.000	Ea.	81	63	144
Urinal screen, powder coated steel, wall mounted	1.000	Ea.	187	127	314
Mirror, 18″ x 24″ with stainless steel shelf	1.000	Ea.	218	31.50	249.50
Soap dispenser, surface mounted, liquid	1.000	Ea.	55	31.50	86.50
Toilet tissue dispenser, surface mounted, stainless steel	1.000	Ea.	21	21	42
Towel dispenser, surface mounted, stainless steel	1.000	Ea.	52	39.50	91.50
TOTAL			**5,295**	**3,154.50**	**8,449.50**

D2010 957		Plumbing - Public Restroom	COST EACH MAT.	COST EACH INST.	COST EACH TOTAL
1600					
1700		Public men's restroom, one water closet, one urinal, one lavatory	5,300	3,150	8,450
1800		Two water closets, two urinals, two lavatories	10,500	6,150	16,650
1900					
2000		For each additional water closet over 2, add	2,800	950	3,750
2100		For each additional urinal over 2, add	1,150	1,050	2,200
2200		For each additional lavatory over 2, add	1,050	1,025	2,075
2300					
2400	NOTE: PLUMBING APPROXIMATIONS				
2500		WATER CONTROL: water meter, backflow preventer,			
2600		Shock absorbers, vacuum breakers, mixer....10 to 15% of fixtures			
2700		PIPE AND FITTINGS: .... 30 to 60% of fixtures			
2800					
2900		Cut & patch to match existing construction, add, minimum	2%	3%	
3000		Maximum	5%	9%	
3100		Dust protection, add, minimum	1%	2%	
3200		Maximum	4%	11%	
3300		Equipment usage curtailment, add, minimum	1%	1%	

D2010 Plumbing Fixtures

D2010 957	Plumbing - Public Restroom	COST EACH		
		MAT.	INST.	TOTAL
3400	Maximum	3%	10%	
3500	Material handling, & storage limitation, add, minimum	1%	1%	
3600	Maximum	6%	7%	
3700	Protection of existing work, add, minimum	2%	2%	
3800	Maximum	5%	7%	
3900	Shift work requirements, add, minimum		5%	
4000	Maximum		30%	

D2010 Plumbing Fixtures

This data illustrates and describes a two fixture lavatory system including a water closet, lavatory, accessories and all service piping. Lines within System Components give the unit price and total price on a cost each basis for this system. Prices for two fixture lavatory systems are on Line Items D2010 958 1800 and 1900. Both material quantities and labor costs have been adjusted for the system listed.

Factors: To adjust for job conditions other than normal working situations use Lines D2010 958 2900 thru 4000.

Example: You are to install the system while controlling dust in the work area. Go to Line D2010 958 3200 and apply these percentages to the appropriate MAT. and INST. costs.

System Components	QUANTITY	UNIT	COST EACH MAT.	COST EACH INST.	COST EACH TOTAL
Two fixture bathroom incl. water closet, lavatory, accessories and Necessary service piping to install this system in 2 walls.					
Water closet, floor mounted, 2 piece, close coupled	1.000	Ea.	253	251	504
Rough in waste & vent for water closet	1.000	Set	360	435	795
Lavatory, 20" x 18", P.E. cast iron with accessories	1.000	Ea.	291	166	457
Rough in waste & vent for lavatory	1.000	Set	480	800	1,280
Additional service piping					
1/2" copper pipe with sweat solder joints	10.000	L.F.	40.50	91	131.50
2" black schedule 40 steel pipe with threaded couplings	12.000	L.F.	97.80	252	349.80
4" cast iron soil pipe with lead and oakum joints	7.000	L.F.	150.50	168	318.50
Accessories					
Toilet tissue dispenser, chrome, single roll	1.000	Ea.	21	21	42
18" long stainless steel towel bar	1.000	Ea.	45.50	27.50	73
Medicine cabinet with mirror, 20" x 16", unlighted	1.000	Ea.	109	45	154
TOTAL			1,848.30	2,256.50	4,104.80

D2010 958	Plumbing - Two Fixture Bathroom	COST EACH MAT.	COST EACH INST.	COST EACH TOTAL
1800	Water closet, lavavatory & accessories, in 2 walls with all service piping	1,850	2,250	4,100
1900	Above system installed in 1 wall	1,725	1,950	3,675
2000				
2400	NOTE: PLUMBING APPROXIMATIONS			
2500	WATER CONTROL: water meter, backflow preventer,			
2600	Shock absorbers, vacuum breakers, mixer....10 to 15% of fixtures			
2700	PIPE AND FITTINGS: .... 30 to 60% of fixtures			
2800				
2900	Cut & patch to match existing construction, add, minimum	2%	3%	
3000	Maximum	5%	9%	
3100	Dust protection, add, minimum	1%	2%	
3200	Maximum	4%	11%	
3300	Equipment usage curtailment, add, minimum	1%	1%	
3400	Maximum	3%	10%	
3500	Material handling & storage limitation, add, minimum	1%	1%	
3600	Maximum	6%	7%	
3700	Protection of existing work, add, minimum	2%	2%	
3800	Maximum	5%	7%	
3900	Shift work requirements, add, minimum		5%	
4000	Maximum		30%	

D2010 Plumbing Fixtures

This data illustrates and describes a three fixture bathroom system including a water closet, tub, lavatory, accessories and service piping. Lines within System Components give the unit price and total price on a cost each basis for this system. Prices for a three fixture bathroom system is on Line Item D2010 959 1700. Both material quantities and labor costs have been adjusted for the system listed.

Factors: To adjust for job conditions other than normal working situations use Lines D2010 959 2900 thru 4000.

Example: You are to install the system and protect all existing work. Go to Line D2010 959 3800 and apply these percentages to the appropriate MAT. and INST. costs.

System Components	QUANTITY	UNIT	COST EACH		
			MAT.	INST.	TOTAL
Three fixture bathroom incl. water closet, bathtub, lavatory, accessories, And necessary service piping to install this system in 1 wall.					
Water closet, tank vit china flr mtd close cpld, 2 pc. w/seat supply & stop	1.000	Ea.	253	251	504
For rough in, supply, waste and vent	.930	Set	334.80	404.55	739.35
Bath tub, P.E., cast iron, 5' long, with accessories	1.000	Ea.	1,250	300	1,550
Rough-in, supply, waste and vent, for all above tubs, add	.950	Set	370.50	608	978.50
Lavatory, 20" x 18" P.E. cast iron with accessories	1.000	Ea.	291	166	457
Rough-in, supply, waste and vent for above lavatories	1.000	Set	480	800	1,280
Accessories					
Toilet tissue dispenser, surface mounted, SS, single roll	1.000	Ea.	21	21	42
Towel bar, stainless steel, 18" long	2.000	Ea.	91	55	146
Medicine cabinets with mirror, sst. frame, 16" x 22", unlighted	1.000	Ea.	109	45	154
TOTAL			3,200.30	2,650.55	5,850.85

D2010 959	Plumbing - Three Fixture Bathroom	COST EACH		
		MAT.	INST.	TOTAL
1600				
1700	Water closet, lavatory, bathtub, in 1 wall with all service piping	3,200	2,650	5,850
2400	NOTE: PLUMBING APPROXIMATIONS			
2500	WATER CONTROL: water meter, backflow preventer,			
2600	Shock absorbers, vacuum breakers, mixer....10 to 15% of fixtures			
2700	PIPE AND FITTINGS: .... 30 to 60% of fixtures			
2800				
2900	Cut & patch to match existing construction, add, minimum	2%	3%	
3000	Maximum	5%	9%	
3100	Dust protection, add, minimum	1%	2%	
3200	Maximum	4%	11%	
3300	Equipment usage curtailment, add, minimum	1%	1%	
3400	Maximum	3%	10%	
3500	Material handling & storage limitation, add, minimum	1%	1%	
3600	Maximum	6%	7%	
3700	Protection of existing work, add, minimum	2%	2%	
3800	Maximum	5%	7%	
3900	Shift work requirements, add, minimum		5%	
4000	Maximum		30%	

D2010 Plumbing Fixtures

This data illustrates and describes a three fixture bathroom system including a water closet, tub, lavatory, accessories, and service piping. Lines within System Components give the unit price and total price on a cost each basis for this system. Prices for a three fixture bathroom system are on Line Item D2010 960 1900 thru 2000. Both material quantities and labor costs have been adjusted for the system listed.

Factors: To adjust for job conditions other than normal working situations use Lines D2010 960 2900 thru 4000.

Example: You are to install the system and protect the surrounding area from dust. Go to Line D2010 960 3100 and apply these percentages to the appropriate MAT. and INST. costs.

System Components	QUANTITY	UNIT	COST EACH MAT.	COST EACH INST.	COST EACH TOTAL
Three fixture bathroom incl. water closet, bathtub, lavatory, accessories, And necessary service piping to install this system in 2 walls.					
Water closet, floor mounted, 2 piece, close coupled	1.000	Ea.	253	251	504
Rough-in waste & vent for water closet	1.000	Set	360	435	795
Bathtub, P.E. cast iron, 5' long with accessories	1.000	Ea.	1,250	300	1,550
Rough-in waste & vent for bathtub	1.000	Set	390	640	1,030
Lavatory, 20" x 18" P.E. cast iron with accessories	1.000	Ea.	291	166	457
Rough-in waste & vent for lavatory	1.000	Set	480	800	1,280
Additional service piping					
1-1/4" copper DWV type tubing, sweat solder joints	6.000	L.F.	60.60	73.80	134.40
2" black schedule 40 steel pipe with threaded couplings	12.000	L.F.	97.80	252	349.80
Accessories					
Toilet tissue dispenser, chrome, single roll	1.000	Ea.	21	21	42
18" long stainless steel towel bar	2.000	Ea.	91	55	146
Medicine cabinet with mirror, 20" x 16", unlighted	1.000	Ea.	109	45	154
TOTAL			3,403.40	3,038.80	6,442.20

D2010 960	Plumbing - Three Fixture Bathroom	COST EACH MAT.	COST EACH INST.	COST EACH TOTAL
1900	Water closet, lavatory, bathtub, in 2 walls with all service piping	3,400	3,050	6,450
2000	Above system with corner contour tub, P.E. cast iron	5,075	3,050	8,125
2300				
2400	NOTE: PLUMBING APPROXIMATIONS			
2500	WATER CONTROL: water meter, backflow preventer,			
2600	Shock absorbers, vacuum breakers, mixer....10 to 15% of fixtures			
2700	PIPE AND FITTINGS: .... 30 to 60% of fixtures			
2800				
2900	Cut & patch to match existing construction, add, minimum	2%	3%	
3000	Maximum	5%	9%	
3100	Dust protection, add, minimum	1%	2%	
3200	Maximum	4%	11%	
3300	Equipment usage curtailment, add, minimum	1%	1%	
3400	Maximum	3%	10%	
3500	Material handling & storage limitation, add, minimum	1%	1%	
3600	Maximum	6%	7%	
3700	Protection of existing work, add, minimum	2%	2%	
3800	Maximum	5%	7%	
3900	Shift work requirements, add, minimum		5%	
4000	Maximum		30%	

D2010 Plumbing Fixtures

This data illustrates and describes a three fixture bathroom system including a water closet, shower, lavatory, accessories, and service piping, Lines within System Components give the unit price and total price on a cost each basis for this system. Prices for three fixture bathroom systems are on Line Items D2010 961 2100 thru 2200. Both material quantities and labor costs have been adjusted for the system listed.

Factors: To adjust for job conditions other than normal working situations use Lines D2010 961 2900 thru 4000.

Example: You are to install the system and protect existing construction. Go to Line D2010 961 3700 and apply these percentages to the appropriate MAT. and INST. costs.

System Components	QUANTITY	UNIT	COST EACH		
			MAT.	INST.	TOTAL
Three fixture bathroom incl. water closet, shower, lavatory, accessories, And necessary service piping to install this system in 2 walls.					
Water closet, floor mounted, 2 piece, close coupled	1.000	Ea.	253	251	504
Rough-in waste & vent for water closet	1.000	Set	360	435	795
32" shower, enameled stall, molded stone receptor	1.000	Ea.	1,275	266	1,541
Rough-in waste & vent for shower	1.000	Set	380	650	1,030
Lavatory, 20" x 18" P.E. cast iron with accessories	1.000	Ea.	291	166	457
Rough-in waste & vent for lavatory	1.000	Set	480	800	1,280
Additional service piping					
1-1/4" Copper DWV type tubing, sweat solder joints	4.000	L.F.	40.40	49.20	89.60
2" black schedule 40 steel pipe with threaded couplings	6.000	L.F.	48.90	126	174.90
4" cast iron soil with lead and oakum joints	7.000	L.F.	150.50	168	318.50
Accessories					
Toilet tissue dispenser, chrome, single roll	1.000	Ea.	21	21	42
18" long stainless steel towel bar	2.000	Ea.	91	55	146
Medicine cabinet with mirror, 20" x 16", unlighted	1.000	Ea.	109	45	154
TOTAL			3,499.80	3,032.20	6,532

D2010 961	Plumbing - Three Fixture Bathroom	COST EACH		
		MAT.	INST.	TOTAL
2100	Water closet, 32" shower, lavavatory, in 2 walls with all service piping	3,500	3,025	6,525
2200	Above system with 36" corner angle shower	4,125	3,050	7,175
2300				
2400	NOTE: PLUMBING APPROXIMATIONS			
2500	WATER CONTROL: water meter, backflow preventer,			
2600	Shock absorbers, vacuum breakers, mixer....10 to 15% of fixtures			
2700	PIPE AND FITTINGS: .... 30 to 60% of fixtures			
2800				
2900	Cut & patch to match existing construction, add, minimum	2%	3%	
3000	Maximum	5%	9%	
3100	Dust protection, add, minimum	1%	2%	
3200	Maximum	4%	11%	
3300	Equipment usage curtailment, add, minimum	1%	1%	
3400	Maximum	3%	10%	
3500	Material handling & storage limitation, add, minimum	1%	1%	
3600	Maximum	6%	7%	
3700	Protection of existing work, add, minimum	2%	2%	
3800	Maximum	5%	7%	
3900	Shift work requirements, add, minimum		5%	
4000	Maximum		30%	

D2010 Plumbing Fixtures

This data illustrates and describes a four fixture bathroom system including a water closet, shower, bathtub, lavatory, accessories, and service piping. Lines within System Components give the unit price and total price on a cost each basis for this system. Prices for four fixture bathroom systems are on Line Items D2010 962 1800 thru 1900. Both material quantities and labor costs have been adjusted for the system listed.

Factors: To adjust for job conditions other than normal working situations use Lines D2010 962 2900 thru 4000.

Example: You are to install the system during weekends and evenings. Go to Line D2010 962 4000 and apply these percentages to the appropriate MAT. and INST. costs.

System Components	QUANTITY	UNIT	COST EACH		
			MAT.	INST.	TOTAL
Four fixture bathroom incl. water closet, shower, bathtub, lavatory					
Accessories and necessary service piping to install this system in 2 walls.					
Water closet, floor mounted, 2 piece, close coupled	1.000	Ea.	253	251	504
Rough-in waste & vent for water closet	1.000	Set	360	435	795
32" shower, enameled steel stall, molded stone receptor	1.000	Ea.	1,275	266	1,541
Rough-in waste & vent for shower	1.000	Set	380	650	1,030
Bathtub, P.E. cast iron, 5' long with accessories	1.000	Ea.	1,250	300	1,550
Rough-in waste & vent for bathtub	1.000	Set	390	640	1,030
Lavatory, 20" x 18" P.E. cast iron with accessories	1.000	Ea.	291	166	457
Rough-in waste & vent for lavatory	1.000	Set	480	800	1,280
Accessories					
Toilet tissue dispenser, chrome, single roll	1.000	Ea.	21	21	42
18" long stainless steel towel bar	2.000	Ea.	91	55	146
Medicine cabinet with mirror, 20" x 16" unlighted	1.000	Ea.	109	45	154
TOTAL			4,900	3,629	8,529

D2010 962	Plumbing - Four Fixture Bathroom	COST EACH		
		MAT.	INST.	TOTAL
1800	Water closet, 32" shower, bathtub, lavavatory, in 2 walls	4,900	3,625	8,525
1900	Above system with 36" corner angle shower, plumbing in 3 walls	5,675	3,975	9,650
2000				
2300				
2400	NOTE: PLUMBING APPROXIMATIONS			
2500	WATER CONTROL: water meter, backflow preventer,			
2600	Shock absorbers, vacuum breakers, mixer....10 to 15% of fixtures			
2700	PIPE AND FITTINGS: .... 30 to 60% of fixtures			
2800				
2900	Cut & patch to match existing construction, add, minimum	2%	3%	
3000	Maximum	5%	9%	
3100	Dust protection, add, minimum	1%	2%	
3200	Maximum	4%	11%	
3300	Equipment usage curtailment, add, minimum	1%	1%	
3400	Maximum	3%	10%	
3500	Material handling & storage limitation, add, minimum	1%	1%	
3600	Maximum	6%	7%	
3700	Protection of existing work, add, minimum	2%	2%	
3800	Maximum	5%	7%	
3900	Shift work requirements, add, minimum		5%	
4000	Maximum		30%	

D2010   Plumbing Fixtures

This data illustrates and describes a five fixture bathroom system including a water closet, shower, bathtub, lavatories, accessories, and service piping. Lines within System Components give the unit price and total price on a cost each basis for this system. Prices for five fixture bathroom systems are on Line Items D2010 963 1800 thru 1900. Both material quantities and labor costs have been adjusted for the system listed.

Factors: To adjust for job conditions other than normal working situations use Lines D2010 963 2900 thru 4000.

Example: You are to install and match any existing construction. Go to Line D2010 963 2900 and apply these percentages to the appropriate MAT. and INST. costs.

System Components	QUANTITY	UNIT	MAT.	INST.	TOTAL
Five fixture bathroom incl. water closet, shower, bathtub, 2 lavatories, Accessories and necessary service piping to install this system in 1 wall.					
Water closet, floor mounted, 2 piece, close coupled	1.000	Ea.	253	251	504
Rough-in waste & vent for water closet	1.000	Set	360	435	795
36" corner angle shower, enameled steel stall, molded stone receptor	1.000	Ea.	1,900	277	2,177
Rough-in waste & vent for shower	1.000	Set	380	650	1,030
Bathtub, P.E. cast iron, 5' long with accessories	1.000	Ea.	1,250	300	1,550
Rough-in waste & vent for bathtub	1.000	Set	390	640	1,030
Countertop, plastic laminate, with backsplash	2.000	Ea.	176	84	260
Vanity base, 2 doors, 24" wide	2.000	Ea.	345	115	460
Lavatories, 20" x 18" cabinet mntd, PECI	2.000	Ea.	690	416	1,106
Rough-in waste & vent for lavatories	1.600	Set	384	928	1,312
Accessories					
Toilet tissue dispenser, chrome, single roll	1.000	Ea.	21	21	42
18" long stainless steel towel bars	2.000	Ea.	91	55	146
Medicine cabinet with mirror, 20" x 16", unlighted	2.000	Ea.	218	90	308
TOTAL			6,458	4,262	10,720

D2010 963	Plumbing - Five Fixture Bathroom	MAT.	INST.	TOTAL
1800	Water closet, shower, bathtub, 2 lavatories, on 1 wall	6,450	4,250	10,700
1900	Above system installed in 2 walls with all necessary service piping	6,525	4,425	10,950
2000				
2100				
2200				
2300				
2400	NOTE: PLUMBING APPROXIMATIONS			
2500	WATER CONTROL: water meter, backflow preventer,			
2600	Shock absorbers, vacuum breakers, mixer....10 to 15% of fixtures			
2700	PIPE AND FITTINGS: .... 30 to 60% of fixtures			
2800				
2900	Cut & patch to match existing construction, add, minimum	2%	3%	
3000	Maximum	5%	9%	
3100	Dust protection, add, minimum	1%	2%	
3200	Maximum	4%	11%	
3300	Equipment usage curtailment, add, minimum	1%	1%	
3400	Maximum	3%	10%	
3500	Material handling & storage limitation, add, minimum	1%	1%	

D2010 Plumbing Fixtures

D2010 963	Plumbing - Five Fixture Bathroom	COST EACH		
		MAT.	INST.	TOTAL
3600	Maximum	6%	7%	
3700	Protection of existing work, add, minimum	2%	2%	
3800	Maximum	5%	7%	
3900	Shift work requirements, add, minimum		5%	
4000	Maximum		30%	

D3010 Energy Supply

This data illustrates and describes an oil fired hot water baseboard system including an oil fired boiler, fin tube radiation and all fittings and piping. Lines within System Components give the unit price and total price per square foot for this system. Prices for oil fired hot water baseboard systems are on Line Items D3010 540 1600 thru 2100. Both material quantities and labor costs have been adjusted for the system listed.

Factors: To adjust for job conditions other than normal working situations use Lines D3010 540 2900 thru 4000.

Example: You are to install the system while protecting all existing work. Go to Line D3010 540 3700 and apply these percentages to the appropriate MAT. and INST. costs.

System Components	QUANTITY	UNIT	COST PER S.F.		
			MAT.	INST.	TOTAL
Oil fired hot water baseboard system including boiler, fin tube radiation And all necessary fittings and piping.					
Area to 800 S.F.					
Boiler, cast iron, w/oil piping, 97 MBH	1.000	Ea.	2,500	1,656.25	4,156.25
Copper piping	130.000	L.F.	1,436.50	1,657.50	3,094
Fin tube radiation	68.000	L.F.	550.80	1,598	2,148.80
Circulator	1.000	Ea.	640	221	861
Oil tank	1.000	Ea.	540	273	813
Expansion tank, ASME	1.000	Ea.	815	97.50	912.50
TOTAL			6,482.30	5,503.25	11,985.55
COST PER S.F.		S.F.	8.10	6.88	14.98

D3010 540	Heating - Oil Fired Hot Water	COST PER S.F.		
		MAT.	INST.	TOTAL
1600	Cast iron boiler, area to 800 S.F.	8.09	6.88	14.97
1700	To 1000 S.F.	7.05	5.95	13
1800	To 1200 S.F.	6.10	5.35	11.45
1900	To 1600 S.F.	7.25	5.35	12.60
2000	To 2000 S.F.	6.05	4.77	10.82
2100	To 3000 S.F.	4.44	4.01	8.45
2200				
2300				
2700				
2800				
2900	Cut & patch to match existing construction, add, minimum	2%	3%	
3000	Maximum	5%	9%	
3100	Dust protection, add, minimum	1%	2%	
3200	Maximum	4%	11%	
3300	Equipment usage curtailment, add, minimum	1%	1%	
3400	Maximum	3%	10%	
3500	Material handling & storage limitation, add, minimum	1%	1%	
3600	Maximum	6%	7%	
3700	Protection of existing work, add, minimum	2%	2%	
3800	Maximum	5%	7%	
3900	Shift work requirements, add, minimum		5%	
4000	Maximum		30%	

D3010 Energy Supply

This data illustrates and describes a gas fired hot water baseboard system including a gas fired boiler, fin tube radiation, fittings and piping. Lines within System Components give the unit price and total price per square foot for this system. Prices for gas fired hot water baseboard systems are on Line Items D3010 550 1400 thru 1900. Both material quantities and labor costs have been adjusted for the system listed.

Factors: To adjust for job conditions other than normal working situations use Lines D3010 550 2900 thru 4000.

Example: You are to install the system with minimal equipment usage. Go to Line D3010 550 3400 and apply these percentages to the appropriate MAT. and INST. costs.

System Components	QUANTITY	UNIT	COST PER S.F. MAT.	COST PER S.F. INST.	COST PER S.F. TOTAL
Gas fired hot water baseboard system including boiler, fin tube radiation,					
All necessary fittings and piping.					
Area to 800 S.F.					
Cast iron boiler, insulating jacket, gas piping, 80 MBH	1.000	Ea.	2,718.75	2,468.75	5,187.50
Copper piping	130.000	L.F.	1,436.50	1,657.50	3,094
Fin tube radiation	68.000	L.F.	550.80	1,598	2,148.80
Circulator, flange connection	1.000	Ea.	640	221	861
Expansion tank, ASME	1.000	Ea.	815	97.50	912.50
TOTAL			6,161.05	6,042.75	12,203.80
COST PER S.F.		S.F.	7.70	7.55	15.25

D3010 550	Heating - Gas Fired Hot Water	COST PER S.F. MAT.	COST PER S.F. INST.	COST PER S.F. TOTAL
1400	Cast iron boiler, area to 800 S.F.	7.69	7.56	15.25
1500	To 1000 S.F.	6.45	6.65	13.10
1600	To 1200 S.F.	5.60	5.95	11.55
1700	To 1600 S.F.	6.85	5.80	12.65
1800	To 2000 S.F.	6.05	5.20	11.25
1900	To 3000 S.F.	4.68	4.55	9.23
2000				
2100				
2900	Cut & patch to match existing construction, add, minimum	2%	3%	
3000	Maximum	5%	9%	
3100	Dust protection, add, minimum	1%	2%	
3200	Maximum	4%	11%	
3300	Equipment usage curtailment, add, minimum	1%	1%	
3400	Maximum	3%	10%	
3500	Material handling & storage limitation, add, minimum	1%	1%	
3600	Maximum	6%	7%	
3700	Protection of existing work, add, minimum	2%	2%	
3800	Maximum	5%	7%	
3900	Shift work requirements, add, minimum		5%	
4000	Maximum		30%	

659

D3010 Energy Supply

This data illustrates and describes a wood burning stove system including a free standing stove, preformed hearth, masonry chimney, and necessary piping and fittings. Lines within System Components give the unit price and total price on a cost each basis for this system. Prices for wood burning stove systems are on Line Items D3010 991 1400 thru 2000. Both material quantities and labor costs have been adjusted for the system listed.

Factors: To adjust for job conditions other than normal working situations use Lines D3010 991 2900 thru 4000.

Example: You are to install the system and protect existing construction. Go to Line D3010 991 3500 and apply these percentages to the appropriate MAT., and INST. costs.

System Components	QUANTITY	UNIT	COST EACH MAT.	COST EACH INST.	COST EACH TOTAL
Cast iron, free standing, wood burning stove with preformed hearth, masonry Chimney, and all necessary piping and fittings to install in chimney.					
Cast iron wood burning stove, stove pipe	1.000	Ea.	1,425	975	2,400
Wall panel or hearth, non-combustible	1.000	Ea.	1,045	173.85	1,218.85
16" x 16" brick chimney, 8" x 8" flue	20.000	V.L.F.	687.50	1,512.50	2,200
Concrete in place, chimney foundations	.500	C.Y.	85.50	130.54	216.04
TOTAL			3,243	2,791.89	6,034.89

D3010 991	Wood Burning Stoves	COST EACH MAT.	COST EACH INST.	COST EACH TOTAL
1400	Cast iron, free standing with preformed hearth	3,250	2,800	6,050
1500				
1600	Installed in existing fireplace	1,975	1,075	3,050
1700				
1800	Installed with insulated metal chimney			
1900	System including ceiling package,			
2000	Metal chimney to 10 L.F.	3,800	1,500	5,300
2100				
2200				
2300				
2400				
2500				
2600				
2700				
2900	Dust protection, add, minimum	1%	2%	
3000	Maximum	4%	11%	
3100	Equipment usage curtailment, add, minimum	1%	1%	
3200	Maximum	3%	10%	
3300	Material handling & storage limitation, add, minimum	1%	1%	
3400	Maximum	6%	7%	
3500	Protection of existing work, add, minimum	2%	2%	
3600	Maximum	5%	7%	
3700	Shift work requirements, add, minimum		5%	
3800	Maximum		30%	
3900	Temporary shoring and bracing, add, minimum	2%	5%	
4000	Maximum	5%	12%	

D3010 Energy Supply

This data illustrates and describes masonry fireplace systems including a brick fireplace, footing, foundation, hearth, firebox, chimney and flue. Lines within System Components give the unit price and total price on a cost each basis for this system. Prices for masonry fireplace systems are on Line Items D3010 992 1200 thru 1500. Both material quantities and labor costs have been adjusted for the system listed.

Factors: To adjust for job conditions other than normal working situations use Lines D3010 992 2700 thru 4000.

Example: You are to install the system with some temporary shoring and bracing. Go to Line D3010 992 3900 and apply these percentages to the appropriate MAT. and INST. costs.

System Components	QUANTITY	UNIT	COST EACH MAT.	COST EACH INST.	COST EACH TOTAL
Brick masonry fireplace, including footing, foundation, hearth, firebox,					
Chimney and flue, chimney 12' above firebox.					
Footing, 4' x 7' x 12" thick, 3000 psi concrete	1.040	C.Y.	195.52	313.29	508.81
Foundation, 12" concrete block	180.000	S.F.	963	1,971	2,934
Fireplace, brick faced, 6'-0" wide x 5'-0" high	1.000	Ea.	630	2,750	3,380
Hearth	1.000	Ea.	233	550	783
Chimney, 20" x 20", one 12" x 12" flue, 12 V.L.F.	12.000	V.L.F.	612	960	1,572
Mantle, wood	7.000	L.F.	64.05	123.20	187.25
TOTAL			2,697.57	6,667.49	9,365.06

D3010 992	Masonry Fireplace	COST EACH MAT.	COST EACH INST.	COST EACH TOTAL
1200	Chimney, 20" x 20", one 12" x 12" flue, 12 V.L.F.	2,700	6,675	9,375
1300	18 V.L.F.	3,000	7,150	10,150
1400	24 V.L.F.	3,300	7,625	10,925
1500	Fieldstone face instead of brick	2,975	6,675	9,650
1600				
1700				
1800				
1900				
2000				
2100				
2200				
2300				
2700	Cut & patch to match existing construction, add, minimum	2%	3%	
2800	Maximum	5%	9%	
2900	Dust protection, add, minimum	1%	2%	
3000	Maximum	4%	11%	
3100	Equipment usage curtailment, add, minimum	1%	1%	
3200	Maximum	3%	10%	
3300	Material handling & storage limitation, add, minimum	1%	1%	
3400	Maximum	6%	7%	
3500	Protection of existing work, add, minimum	2%	2%	
3600	Maximum	5%	7%	
3700	Shift work requirements, add, minimum		5%	
3800	Maximum		30%	
3900	Temporary shoring and bracing, add, minimum	2%	5%	
4000	Maximum	5%	12%	

D3020 Heat Generating Systems

This data illustrates and describes an oil fired forced air system including an oil fired furnace, ductwork, registers and hookups. Lines within System Components give the unit price and total price per square foot for this system. Prices for oil fired forced air systems are on Line Items D3020 122 1600 thru 2800. Both material quantities and labor costs have been adjusted for the system listed.

Factors: To adjust for job conditions other than normal working situations use Lines D3020 122 3100 thru 4200.

Example: You are to install the system during evenings and weekends. Go to Line D3020 122 4200 and apply this percentage to the appropriate INST. cost.

System Components	QUANTITY	UNIT	COST PER S.F.		
			MAT.	INST.	TOTAL
Oil fired hot air heating system including furnace, ductwork, registers And all necessary hookups.					
Area to 800 S.F., heat only					
Furnace, oil, atomizing gun type burner, w/oil piping	1.000	Ea.	2,343.75	450	2,793.75
Oil tank, steel, 275 gallon	1.000	Ea.	540	273	813
Duct, galvanized, steel	312.000	Lb.	196.56	2,683.20	2,879.76
Insulation, blanket type, duct work	270.000	S.F.	1,347.30	3,969	5,316.30
Flexible duct, 6" diameter, insulated	100.000	L.F.	323	500	823
Registers, baseboard, gravity, 12" x 6"	8.000	Ea.	59.60	252	311.60
Return, damper, 36" x 18"	1.000	Ea.	83	72.50	155.50
TOTAL			4,893.21	8,199.70	13,092.91
COST PER S.F.		S.F.	6.12	10.25	16.37

D3020 122	Heating-Cooling, Oil, Forced Air	COST PER S.F.		
		MAT.	INST.	TOTAL
1600	Oil fired, area to 800 S.F.	6.13	10.25	16.38
1700	To 1000 S.F.	4.90	8.25	13.15
1800	To 1200 S.F.	4.22	7.15	11.37
1900	To 1600 S.F.	3.38	5.75	9.13
2000	To 2000 S.F.	4.08	7.90	11.98
2100	To 3000 S.F.	3.09	6.30	9.39
2200	For combined heating and cooling systems:			
2300	Oil fired, heating and cooling, area to 800 S.F.	8.50	11.45	19.95
2400	To 1000 S.F.	6.85	9.30	16.15
2500	To 1200 S.F.	5.85	8	13.85
2600	To 1600 S.F.	4.74	6.45	11.19
2700	To 2000 S.F.	5.20	8.50	13.70
2800	To 3000 S.F.	3.89	6.70	10.59
3100	Cut & patch to match existing construction, add, minimum	2%	3%	
3200	Maximum	5%	9%	
3300	Dust protection, add, minimum	1%	2%	
3400	Maximum	4%	11%	
3500	Equipment usage curtailment, add, minimum	1%	1%	
3600	Maximum	3%	10%	
3700	Material handling & storage limitation, add, minimum	1%	1%	
3800	Maximum	6%	7%	
3900	Protection of existing work, add, minimum	2%	2%	
4000	Maximum	5%	7%	
4100	Shift work requirements, add, minimum		5%	
4200	Maximum		30%	

D3020 Heat Generating Systems

This data illustrates and describes a gas fired forced air system including a gas fired furnace, ductwork, registers and hookups. Lines within System Components give the unit price and total price per square foot for this system. Prices for gas fired forced air systems are on Line Items D3020 124 1400 thru 2600. Both material quantities and labor costs have been adjusted for the system listed.

Factors: To adjust for job conditions other than normal working situations use Lines D3020 124 2900 thru 4000.

Example: You are to install the system with material handling and storage limitations. Go to Line D3020 124 3500 and apply these percentages to the appropriate MAT. and INST. costs.

System Components	QUANTITY	UNIT	COST PER S.F.		
			MAT.	INST.	TOTAL
Gas fired hot air heating system including furnace, ductwork, registers					
And all necessary hookups.					
Area to 800 S.F., heat only					
Furnace, gas, AGA certified, direct drive, w/gas piping, 44 MBH	1.000	Ea.	793.75	406.25	1,200
Duct, galvanized steel	312.000	Lb.	196.56	2,683.20	2,879.76
Insulation, blanket type, ductwork	270.000	S.F.	1,347.30	3,969	5,316.30
Flexible duct, 6″ diameter, insulated	100.000	L.F.	323	500	823
Registers, baseboard, gravity, 12″ x 6″	8.000	Ea.	59.60	252	311.60
Return, damper, 36″ x 18″	1.000	Ea.	83	72.50	155.50
TOTAL			2,803.21	7,882.95	10,686.16
COST PER S.F.		S.F.	3.50	9.85	13.35

D3020 124	Heating-Cooling, Gas, Forced Air	COST PER S.F.		
		MAT.	INST.	TOTAL
1400	Gas fired, area to 800 S.F.	3.50	9.84	13.34
1500	To 1000 S.F.	2.84	7.95	10.79
1600	To 1200 S.F.	2.50	6.90	9.40
1700	To 1600 S.F.	2.33	6.30	8.63
1800	To 2000 S.F.	2.63	7.75	10.38
1900	To 3000 S.F.	2.07	6.20	8.27
2000	For combined heating and cooling systems:			
2100	Gas fired, heating and cooling, area to 800 S.F.	5.85	11.10	16.95
2200	To 1000 S.F.	4.80	9	13.80
2300	To 1200 S.F.	4.13	7.75	11.88
2400	To 1600 S.F.	3.69	6.95	10.64
2500	To 2000 S.F.	3.77	8.35	12.12
2600	To 3000 S.F.	2.87	6.60	9.47
2900	Cut & patch to match existing construction, add, minimum	2%	3%	
3000	Maximum	5%	9%	
3100	Dust protection, add, minimum	1%	2%	
3200	Maximum	4%	11%	
3300	Equipment usage curtailment, add, minimum	1%	1%	
3400	Maximum	3%	10%	
3500	Material handling & storage limitation, add, minimum	1%	1%	
3600	Maximum	6%	7%	
3700	Protection of existing work, add, minimum	2%	2%	
3800	Maximum	5%	7%	
3900	Shift work requirements, add, minimum		5%	
4000	Maximum		30%	

Expansion Tank
Supply Piping
Gas Supply
Pressure Relief Valve
Return Piping
Circulating Pump

Cast Iron Boiler, Hot Water, Gas Fired

This data illustrates and describes boilers including expansion tank, circulating pump and all service piping. Lines within Systems Components give the material and installation price on a cost each basis for the components. Prices for boiler systems are on Line Items D3020 130 1010 thru D3020 138 1040. Material quantities and labor costs have been adjusted for the system listed.

Factors: To adjust for job conditions other than normal working situations use Line D3020 138 2700 thru 4000.

System Components

	QUANTITY	UNIT	COST EACH		
			MAT.	INST.	TOTAL
Cast iron boiler, including piping, chimney and accessories					
Boilers, gas fired, std controls, CI, insulated, HW, gross output 100 MBH	1.000	Ea.	2,825	2,125	4,950
Pipe, black steel, Sch 40, threaded, W/coupling & hangers, 10' OC, 3/4" dia	20.000	L.F.	82.80	272.25	355.05
Pipe, black steel, Sch 40, threaded, W/coupling & hangers, 10' OC, 1" dia.	20.000	L.F.	101.70	299.93	401.63
Elbow, 90°, black, straight, 3/4" dia.	9.000	Ea.	41.31	472.50	513.81
Elbow, 90°, black, straight, 1" dia.	6.000	Ea.	47.70	342	389.70
Tee, black, straight, 3/4" dia.	2.000	Ea.	14.60	164	178.60
Tee, black, straight, 1" dia.	2.000	Ea.	25	185	210
Tee, black, reducing, 1" dia.	2.000	Ea.	40	185	225
Pipe cap, black, 3/4" dia.	1.000	Ea.	5.20	23	28.20
Union, black with brass seat, 3/4" dia.	2.000	Ea.	41	114	155
Union, black with brass seat, 1" dia.	2.000	Ea.	53	123	176
Pipe nipples, black, 3/4" dia.	5.000	Ea.	9.20	30.25	39.45
Pipe nipples, black, 1" dia.	3.000	Ea.	7.10	20.93	28.03
Valves, bronze, gate, N.R.S., threaded, class 150, 3/4" size	2.000	Ea.	189	74	263
Valves, bronze, gate, N.R.S., threaded, class 150, 1" size	2.000	Ea.	236	78	314
Gas cock, brass, 3/4" size	1.000	Ea.	17.05	33.50	50.55
Thermometer, stem type, 9" case, 8" stem, 3/4" NPT	2.000	Ea.	478	54	532
Tank, steel, liquid expansion, ASME, painted, 15 gallon capacity	1.000	Ea.	730	80	810
Pump, circulating, bronze, flange connection, 3/4" to 1-1/2" size, 1/8 HP	1.000	Ea.	1,100	221	1,321
Vent chimney, all fuel, pressure tight, double wall, SS, 6" dia.	20.000	L.F.	1,220	430	1,650
Vent chimney, elbow, 90° fixed, 6" dia.	2.000	Ea.	910	87	997
Vent chimney, Tee, 6" dia.	2.000	Ea.	532	108	640
Vent chimney, ventilated roof thimble, 6" dia.	1.000	Ea.	278	50	328
Vent chimney, flat roof flashing, 6" dia.	1.000	Ea.	119	43.50	162.50
Vent chimney, stack cap, 6" diameter	1.000	Ea.	325	28.50	353.50
Insulation, fiberglass pipe covering, 1" wall, 1" IPS	20.000	L.F.	21.40	112	133.40
TOTAL			9,449.06	5,756.36	15,205.42

D3020 130	Boiler, Cast Iron, Hot Water, Gas	COST EACH		
		MAT.	INST.	TOTAL
1000	For alternate boilers:			
1010	Boiler, cast iron, gas, hot water, 100 MBH	9,450	5,750	15,200
1020	200 MBH	10,800	6,600	17,400
1030	320 MBH	10,900	7,200	18,100
1040	440 MBH	14,800	9,275	24,075
1050	544 MBH	25,100	14,500	39,600
1060	765 MBH	30,100	15,500	45,600
1070	1088 MBH	32,100	16,600	48,700

D3020 Heat Generating Systems

D3020 130	Boiler, Cast Iron, Hot Water, Gas	COST EACH		
		MAT.	INST.	TOTAL
1080	1530 MBH	39,100	21,900	61,000
1090	2312 MBH	46,000	25,800	71,800

D3020 134	Boiler, Cast Iron, Steam, Gas	COST EACH		
		MAT.	INST.	TOTAL
1010	Boiler, cast iron, gas, steam, 100 MBH	7,900	5,600	13,500
1020	200 MBH	17,900	10,500	28,400
1030	320 MBH	19,700	11,800	31,500
1040	544 MBH	28,700	18,400	47,100
1050	765 MBH	32,200	19,200	51,400
1060	1275 MBH	37,900	21,100	59,000
1070	2675 MBH	53,500	30,700	84,200

D3020 136	Boiler, Cast Iron, Hot Water, Gas/Oil	COST EACH		
		MAT.	INST.	TOTAL
1010	Boiler, cast iron, gas & oil, hot water, 584 MBH	30,800	16,100	46,900
1020	876 MBH	38,900	17,300	56,200
1030	1168 MBH	49,000	19,500	68,500
1040	1460 MBH	59,500	23,400	82,900
1050	2044 MBH	65,000	24,200	89,200
1060	2628 MBH	78,500	29,300	107,800

D3020 138	Boiler, Cast Iron, Steam, Gas/Oil	COST EACH		
		MAT.	INST.	TOTAL
1010	Boiler, cast iron, gas & oil, steam, 810 MBH	35,300	20,900	56,200
1020	1360 MBH	40,600	22,300	62,900
1030	2040 MBH	49,900	28,500	78,400
1040	2700 MBH	53,500	31,700	85,200
2700	Cut & patch to match existing construction, add, minimum	2%	3%	
2800	Maximum	5%	9%	
2900	Dust protection, add, minimum	1%	2%	
3000	Maximum	4%	11%	
3100	Equipment usage curtailment, add, minimum	1%	1%	
3200	Maximum	3%	10%	
3300	Material handling & storage limitation, add, minimum	1%	1%	
3400	Maximum	6%	7%	
3500	Protection of existing work, add, minimum	2%	2%	
3600	Maximum	5%	7%	
3700	Shift work requirements, add, minimum		5%	
3800	Maximum		30%	
3900	Temporary shoring and bracing, add, minimum	2%	5%	
4000	Maximum	5%	12%	

D3020 Heat Generating Systems

Flow ↓   ↑ Flow

Gate Valve →
Y Strainer →
Pressure Gauge →
Expansion Joint →

← Combination Valve
← Pressure Gauge
← Expansion Joint

Base Mounted End-Suction Pump

This data illustrates and describes end suction base mounted circulating pumps, including strainer and piping. Lines within system components give the material and installation price on a cost each basis for the components. Prices for circulating pumps are on Line items D3020 330 1010 thru 1050. Material quantities and labor costs have been adjusting for the system listed.

Factors: To adjust for job conditions other than normal working situations uses Lines D3020 330 2700 thru 4000.

System Components	QUANTITY	UNIT	COST EACH		
			MAT.	INST.	TOTAL
Circulator pump, end suction, including pipe,					
Valves, fittings and guages.					
Pump, circulating, CI, base mounted, 2-1/2" size, 3 HP, to 150 GPM	1.000	Ea.	9,025	740	9,765
Pipe, black steel, Sch. 40, on yoke & roll hangers, 10' O.C., 2-1/2" dia	12.000	L.F.	139.20	358.32	497.52
Elbow, 90°, weld joint, steel, 2-1/2" pipe size	1.000	Ea.	30.50	174	204.50
Flange, weld neck, 150 LB, 2-1/2" pipe size	8.000	Ea.	280	695.92	975.92
Valve, iron body, gate, 125 lb., N.R.S., flanged, 2-1/2" size	1.000	Ea.	840	266	1,106
Strainer, Y type, iron body, flanged, 125 lb., 2-1/2" pipe size	1.000	Ea.	123	273	396
Multipurpose valve, CI body, 2" size	1.000	Ea.	525	94.50	619.50
Expansion joint, flanged spool, 6" F to F, 2-1/2" dia	2.000	Ea.	588	220	808
T-O-L, weld joint, socket, 1/4" pipe size, nozzle	2.000	Ea.	15.60	121.56	137.16
T-O-L, weld joint, socket, 1/2" pipe size, nozzle	2.000	Ea.	15.60	126.80	142.40
Control gauges, pressure or vacuum, 3-1/2" diameter dial	2.000	Ea.	41	47	88
Pressure/temperature relief plug, 316 SS, 3/4" OD, 7-1/2" insertion	2.000	Ea.	108	47	155
Insulation, fiberglass pipe covering, 1-1/2" wall, 2-1/2" IPS	12.000	L.F.	30	82.20	112.20
Water flow	1.000	Ea.	1,325	1,100	2,425
Pump balancing	1.000	Ea.		335	335
TOTAL			13,085.90	4,681.30	17,767.20

D3020 330	Circulating Pump Systems, End Suction	COST EACH		
		MAT.	INST.	TOTAL
1000	For alternate pump sizes:			
1010	Pump, base mtd with motor, end suction, 2 1/2" size, 3 HP, to 150 GPM	13,100	4,700	17,800
1020	3" size, 5 HP, to 225 GPM	15,800	5,300	21,100
1030	4" size, 7-1/2 HP, to 350 GPM	18,600	6,300	24,900
1040	5" size, 15 HP, to 1000 GPM	27,400	9,325	36,725
1050	6" size, 25 HP, to 1550 GPM	36,300	11,300	47,600
2700	Cut & patch to match existing construction, add, minimum	2%	3%	
2800	Maximum	5%	9%	
2900	Dust protection, add, minimum	1%	2%	
3000	Maximum	4%	11%	
3100	Equipment usage curtailment, add, minimum	1%	1%	
3200	Maximum	3%	10%	
3300	Material handling & storage limitation, add, minimum	1%	1%	
3400	Maximum	6%	7%	
3500	Protection of existing work, add, minimum	2%	2%	
3600	Maximum	5%	7%	

D3020 Heat Generating Systems

D3020 330	Circulating Pump Systems, End Suction	COST EACH		
		MAT.	INST.	TOTAL
3700	Shift work requirements, add, minimum		5%	
3800	Maximum		30%	
3900	Temporary shoring and bracing, add, minimum	2%	5%	
4000	Maximum	5%	12%	

D3040 Distribution Systems

Galvanized Steel Duct

Fiberglass Duct Insulation

Flexible Fiberglass Duct

Horizontal Fan Coil Air Conditioning System

This data illustrates and describes fan coil air conditioners including duct work, duct installation, piping and diffusers. Lines within Systems Components give the material and installation price on a cost each basis for the components. Prices for fan coil A/C unit systems are on Line Items D3040 118 1010 thru D3040 126 1120. Material quantities and labor costs have been adjusted for the system listed.

Factors: To adjust for job conditions other than normal working situations use Lines D3040 126 2700 thru 4000.

System Components	QUANTITY	UNIT	COST EACH		
			MAT.	INST.	TOTAL
Fan coil A/C system, horizontal, with housing, controls, 2 pipe, 1/2 ton					
Fan coil A/C, horizontal housing, filters, chilled water, 1/2 ton cooling	1.000	Ea.	1,400	170	1,570
Pipe, black steel, Sch 40, threaded, W/cplgs & hangers, 10' OC, 3/4" dia.	20.000	L.F.	79.12	260.15	339.27
Elbow, 90°, black, straight, 3/4" dia.	6.000	Ea.	27.54	315	342.54
Tee, black, straight, 3/4" dia.	2.000	Ea.	14.60	164	178.60
Union, black with brass seat, 3/4" dia.	2.000	Ea.	41	114	155
Pipe nipples, 3/4" diam	3.000	Ea.	5.52	18.15	23.67
Valves, bronze, gate, N.R.S., threaded, class 150, 3/4" size	2.000	Ea.	189	74	263
Circuit setter, balance valve, bronze body, threaded, 3/4" pipe size	1.000	Ea.	90	38	128
Insulation, fiberglass pipe covering, 1" wall, 3/4" IPS	20.000	L.F.	19.80	107	126.80
Ductwork, 12" x 8" fabricated, galvanized steel, 12 LF	55.000	Lb.	34.65	473	507.65
Insulation, ductwork, blanket type, fiberglass, 1" thk, 1-1/2 LB density	40.000	S.F.	199.60	588	787.60
Diffusers, aluminum, OB damper, ceiling, perf, 24"x24" panel size, 6"x6"	2.000	Ea.	344	90	434
Ductwork, flexible, fiberglass fabric, insulated, 1"thk, PE jacket, 6" dia	16.000	L.F.	51.68	80	131.68
Round volume control damper 6" dia.	2.000	Ea.	55	66	121
Fan coil unit balancing	1.000	Ea.		94	94
Re-heat coil balancing	1.000	Ea.		140	140
Diffuser/register, high, balancing	2.000	Ea.		260	260
TOTAL			2,551.51	3,051.30	5,602.81

D3040 118	Fan Coil A/C Unit, Two Pipe	COST EACH		
		MAT.	INST.	TOTAL
1000	For alternate A/C units:			
1010	Fan coil A/C system, cabinet mounted, controls, 2 pipe, 1/2 Ton	970	1,400	2,370
1020	1 Ton	1,375	1,575	2,950
1030	1-1/2 Ton	1,375	1,575	2,950
1040	2 Ton	2,500	1,925	4,425
1050	3 Ton	3,175	2,000	5,175

D3040 120	Fan Coil A/C Unit, Two Pipe, Electric Heat	COST EACH		
		MAT.	INST.	TOTAL
1010	Fan coil A/C system, cabinet mntd, elect. ht, controls, 2 pipe, 1/2 Ton	1,925	1,400	3,325
1020	1 Ton	2,325	1,575	3,900
1030	1-1/2 Ton	2,725	1,575	4,300
1040	2 Ton	4,200	1,950	6,150
1050	3 Ton	6,875	2,000	8,875

D3040 Distribution Systems

D3040 122	Fan Coil A/C Unit, Four Pipe	COST EACH		
		MAT.	INST.	TOTAL
1010	Fan coil A/C system, cabinet mounted, controls, 4 pipe, 1/2 Ton	1,650	2,700	4,350
1020	1 Ton	2,125	2,875	5,000
1030	1-1/2 Ton	2,150	2,900	5,050
1040	2 Ton	2,850	3,025	5,875
1050	3 Ton	3,925	3,250	7,175

D3040 124	Fan Coil A/C, Horizontal, Duct Mount, 2 Pipe	COST EACH		
		MAT.	INST.	TOTAL
1010	Fan coil A/C system, horizontal w/housing, controls, 2 pipe, 1/2 Ton	2,550	3,050	5,600
1020	1 Ton	3,475	4,850	8,325
1030	1-1/2 Ton	4,325	6,525	10,850
1040	2 Ton	5,775	8,100	13,875
1050	3 Ton	7,325	11,200	18,525
1060	3-1/2 Ton	7,600	11,500	19,100
1070	4 Ton	8,050	13,400	21,450
1080	5 Ton	9,750	17,300	27,050
1090	6 Ton	9,950	18,400	28,350
1100	7 Ton	9,925	18,900	28,825
1110	8 Ton	9,925	19,200	29,125
1120	10 Ton	10,500	19,700	30,200

D3040 126	Fan Coil A/C, Horiz., Duct Mount, 2 Pipe, Elec. Ht.	COST EACH		
		MAT.	INST.	TOTAL
1010	Fan coil A/C system, horiz. hsng, elect. ht, ctrls, 2 pipe, 1/2 Ton	2,800	3,050	5,850
1020	1 Ton	3,850	4,850	8,700
1030	1-1/2 Ton	4,675	6,525	11,200
1040	2 Ton	6,325	8,100	14,425
1050	3 Ton	9,850	11,500	21,350
1060	3-1/2 Ton	10,900	11,500	22,400
1070	4 Ton	12,300	13,500	25,800
1080	5 Ton	15,100	17,300	32,400
1090	6 Ton	15,300	18,500	33,800
1100	7 Ton	15,700	19,200	34,900
1110	8 Ton	15,900	19,800	35,700
1120	10 Ton	16,900	18,900	35,800
2700	Cut & patch to match existing construction, add, minimum	2%	3%	
2800	Maximum	5%	9%	
2900	Dust protection, add, minimum	1%	2%	
3000	Maximum	4%	11%	
3100	Equipment usage curtailment, add, minimum	1%	1%	
3200	Maximum	3%	10%	
3300	Material handling & storage limitation, add, minimum	1%	1%	
3400	Maximum	6%	7%	
3500	Protection of existing work, add, minimum	2%	2%	
3600	Maximum	5%	7%	
3700	Shift work requirements, add, minimum		5%	
3800	Maximum		30%	
3900	Temporary shoring and bracing, add, minimum	2%	5%	
4000	Maximum	5%	12%	

D3040 Distribution Systems

This data illustrates and describes plate and shell and tube type heat exchangers including pressure gauge and exchanger control system. Lines within Systems Components give the material and installation price on a cost each basis for the components. Prices for heat exchanger systems are on Line Items D3040 610 1010 thru D3040 620 1040. Material quantities and labor costs have been adjusted for the system listed.

Factors To adjust for job conditions other then normal working situations use Lines D3040 620 2700 thru 4000.

Plate Heat Exchanger   Shell and Tube Heat Exchanger

System Components	QUANTITY	UNIT	COST EACH		
			MAT.	INST.	TOTAL
Shell and tube type heat exchanger					
with related valves and piping.					
Heat exchanger, 4 pass, 3/4" O.D. copper tubes, by steam at 10 psi, 40 GPM	1.000	Ea.	5,750	340	6,090
Pipe, black steel, Sch 40, threaded, w/couplings & hangers, 10' OC, 2" dia.	40.000	L.F.	411.58	1,060.50	1,472.08
Elbow, 90°, straight, 2" dia.	16.000	Ea.	488	1,184	1,672
Reducer, black steel, concentric, 2" dia.	2.000	Ea.	67	127	194
Valves, bronze, gate, rising stem, threaded, class 150, 2" size	6.000	Ea.	1,578	402	1,980
Valves, bronze, globe, class 150, rising stem, threaded, 2" size	2.000	Ea.	1,030	134	1,164
Strainers, Y type, bronze body, screwed, 125 lb, 2" pipe size	2.000	Ea.	306	117	423
Valve, electric motor actuated, brass, 2 way, screwed, 1-1/2" pipe size	1.000	Ea.	760	57	817
Union, black with brass seat, 2" dia.	8.000	Ea.	444	624	1,068
Tee, black, straight, 2" dia.	9.000	Ea.	337.50	1,089	1,426.50
Tee, black, reducing run and outlet, 2" dia.	4.000	Ea.	214	484	698
Pipe nipple, black, 2" dia.	21.000	Ea.	4.08	10.50	14.58
Thermometers, stem type, 9" case, 8" stem, 3/4" NPT	2.000	Ea.	478	54	532
Gauges, pressure or vacuum, 3-1/2" diameter dial	2.000	Ea.	41	47	88
Insulation, fiberglass pipe covering, 1" wall, 2" IPS	40.000	L.F.	53.60	246	299.60
Heat exchanger control system	1.000	Ea.	3,000	2,375	5,375
Coil balancing	1.000	Ea.		140	140
TOTAL			14,962.76	8,491	23,453.76

D3040 610	Heat Exchanger, Plate Type	COST EACH		
		MAT.	INST.	TOTAL
1000	For alternate heat exchangers:			
1010	Plate heat exchanger, 400 GPM	59,500	17,100	76,600
1020	800 GPM	95,000	21,800	116,800
1030	1200 GPM	142,500	29,500	172,000
1040	1800 GPM	193,000	36,800	229,800

D3040 620	Heat Exchanger, Shell & Tube	COST EACH		
		MAT.	INST.	TOTAL
1010	Shell & tube heat exchanger, 40 GPM	14,962.75	8,491	23,470
1020	96 GPM	27,600	15,000	42,600
1030	240 GPM	50,000	20,500	70,500
1040	600 GPM	98,000	28,800	126,800
2700	Cut & patch to match existing construction, add, minimum	2%	3%	
2800	Maximum	5%	9%	

D30 HVAC

D3040 Distribution Systems

D3040 620	Heat Exchanger, Shell & Tube	COST EACH		
		MAT.	INST.	TOTAL
2900	Dust protection, add, minimum	1%	2%	
3000	Maximum	4%	11%	
3100	Equipment usage curtailment, add, minimum	1%	1%	
3200	Maximum	3%	10%	
3300	Material handling & storage limitation, add, minimum	1%	1%	
3400	Maximum	6%	7%	
3500	Protection of existing work, add, minimum	2%	2%	
3600	Maximum	5%	7%	
3700	Shift work requirements, add, minimum		5%	
3800	Maximum		30%	
3900	Temporary shoring and bracing, add, minimum	2%	5%	
4000	Maximum	5%	12%	

D3050 Terminal & Package Units

Gas Unit Heater

Hydronic Unit Heater

This data illustrates and describes unit heaters including piping and vents. Lines within Systems Components give the material and installation price on a cost each basis for this system. Prices for unit heater systems are on Line items D3050 120 1010 thru D3050 140 1040. Material quantities and labor costs have been adjusted for the system listed.

Factors: To adjust for job conditions other than normal working situations use Lines D3050 140 2700 thru 4000.

System Components	QUANTITY	UNIT	COST EACH		
			MAT.	INST.	TOTAL
Unit heaters complete with piping, thermostats and chimney as required					
Space heater, propeller fan, 20 MBH output	1.000	Ea.	1,250	160	1,410
Pipe, black steel, Sch 40, threaded, w/coupling & hangers, 10' OC, 3/4" dia	20.000	L.F.	77.28	254.10	331.38
Elbow, 90°, black steel, straight, 1/2" dia.	3.000	Ea.	11.40	147	158.40
Elbow, 90°, black steel, straight, 3/4" dia.	3.000	Ea.	13.77	157.50	171.27
Tee, black steel, reducing, 3/4" dia.	1.000	Ea.	13.85	82	95.85
Union, black with brass seat, 3/4" dia.	1.000	Ea.	20.50	57	77.50
Pipe nipples, black, 1/2" dia	2.000	Ea.	6.38	23.40	29.78
Pipe nipples, black, 3/4" dia	2.000	Ea.	3.68	12.10	15.78
Pipe cap, black, 3/4" dia.	1.000	Ea.	5.20	23	28.20
Gas cock, brass, 3/4" size	1.000	Ea.	17.05	33.50	50.55
Thermostat, 1 set back, electric, timed	1.000	Ea.	41.50	94.50	136
Wiring, thermostat hook-up, 25' of #18-3	1.000	Ea.	8.80	28.50	37.30
Vent chimney, prefab metal, U.L. listed, gas, double wall, galv. st, 4" dia	12.000	L.F.	108	229.80	337.80
Vent chimney, gas, double wall, galv. steel, elbow 90°, 4" dia.	2.000	Ea.	51	77	128
Vent chimney, gas, double wall, galv. steel, Tee, 4" dia.	1.000	Ea.	36	50	86
Vent chimney, gas, double wall, galv. steel, T cap, 4" dia.	1.000	Ea.	2.41	31	33.41
Vent chimney, gas, double wall, galv. steel, roof flashing, 4" dia.	1.000	Ea.	8.70	38.50	47.20
Vent chimney, gas, double wall, galv. steel, top, 4" dia.	1.000	Ea.	17.95	29.50	47.45
TOTAL			1,693.47	1,528.40	3,221.87

D3050 120	Unit Heaters, Gas	COST EACH		
		MAT.	INST.	TOTAL
1000	For alternate unit heaters:			
1010	Space heater, suspended, gas fired, propeller fan, 20 MBH	1,693.47	1,528.40	3,210.90
1020	60 MBH	2,425	1,600	4,025
1030	100 MBH	2,775	1,675	4,450
1040	160 MBH	3,300	1,850	5,150
1050	200 MBH	4,125	2,050	6,175
1060	280 MBH	4,800	2,200	7,000
1070	320 MBH	6,350	2,475	8,825

D3050 130	Unit Heaters, Hydronic	COST EACH		
		MAT.	INST.	TOTAL
1010	Space heater, suspended, horiz. mount, HW, prop. fan, 20 MBH	2,200	1,625	3,825
1020	60 MBH	2,975	1,850	4,825
1030	100 MBH	3,575	2,025	5,600
1040	150 MBH	4,075	2,225	6,300
1050	200 MBH	4,600	2,500	7,100
1060	300 MBH	5,425	2,950	8,375

D3050 Terminal & Package Units

D3050 140	Cabinet Unit Heaters, Hydronic	COST EACH		
		MAT.	INST.	TOTAL
1010	Unit heater, cabinet type, horizontal blower, hot water, 20 MBH	2,600	1,600	4,200
1020	60 MBH	3,800	1,775	5,575
1030	100 MBH	4,450	1,950	6,400
1040	120 MBH	4,550	2,025	6,575
2700	Cut & patch to match existing construction, add, minimum	2%	3%	
2800	Maximum	5%	9%	
2900	Dust protection, add, minimum	1%	2%	
3000	Maximum	4%	11%	
3100	Equipment usage curtailment, add, minimum	1%	1%	
3200	Maximum	3%	10%	
3300	Material handling & storage limitation, add, minimum	1%	1%	
3400	Maximum	6%	7%	
3500	Protection of existing work, add, minimum	2%	2%	
3600	Maximum	5%	7%	
3700	Shift work requirements, add, minimum		5%	
3800	Maximum		30%	
3900	Temporary shoring and bracing, add, minimum	2%	5%	
4000	Maximum	5%	12%	

D3050 Terminal & Package Units

Gas
Cock

Roof
Curb

Gas Piping

This data illustrates and describes packaged air conditioners including rooftop DX unit, pipes and fittings. Lines within Systems Components give the material and installation price on a cost each basis for the components. Prices for A/C systems are on Line Items D3050 175 1010 thru D3050 180 1060. Material quantities and labor costs have been adjusted for the system listed.

Factors: To adjust for job conditions other than normal working situations use Lines D3050 180 2700 thru 400.

Rooftop Air Conditioner Unit

System Components	QUANTITY	UNIT	COST EACH		
			MAT.	INST.	TOTAL
Packaged rooftop DX unit, gas, with related piping and valves.					
Roof top A/C, curb, economizer, sgl zone, elec cool, gas ht, 5 ton, 112 MBH	1.000	Ea.	4,850	2,425	7,275
Pipe, black steel, Sch 40, threaded, w/cplgs & hangers, 10' OC, 1" dia	20.000	L.F.	99.33	292.95	392.28
Elbow, 90°, black, straight, 3/4" dia.	3.000	Ea.	13.77	157.50	171.27
Elbow, 90°, black, straight, 1" dia.	3.000	Ea.	23.85	171	194.85
Tee, black, reducing, 1" dia.	1.000	Ea.	20	92.50	112.50
Union, black with brass seat, 1" dia.	1.000	Ea.	26.50	61.50	88
Pipe nipple, black, 3/4" dia	2.000	Ea.	7.36	24.20	31.56
Pipe nipple, black, 1" dia	2.000	Ea.	4.73	13.95	18.68
Cap, black, 1" dia.	1.000	Ea.	6.30	24.50	30.80
Gas cock, brass, 1" size	1.000	Ea.	34	39	73
Control system, pneumatic, Rooftop A/C unit	1.000	Ea.	3,075	2,675	5,750
Rooftop unit heat/cool balancing	1.000	Ea.		505	505
TOTAL			8,160.84	6,482.10	14,642.94

D3050 175	Rooftop Air Conditioner, Const. Volume	COST EACH		
		MAT.	INST.	TOTAL
1000	For alternate A/C systems:			
1010	A/C, Rooftop, DX cool, gas heat, curb, economizer, fltrs, 5 Ton	8,160.84	6,482.10	14,644
1020	7-1/2 Ton	9,775	6,750	16,525
1030	12-1/2 Ton	14,700	7,425	22,125
1040	18 Ton	18,700	8,125	26,825
1050	25 Ton	38,600	9,250	47,850
1060	40 Ton	51,500	12,400	63,900

D3050 180	Rooftop Air Conditioner, Variable Air Volume	COST EACH		
		MAT.	INST.	TOTAL
1010	A/C, Rooftop, DX cool, gas heat, curb, ecmizr, fltrs, VAV, 12-1/2 Ton	17,400	7,925	25,325
1020	18 Ton	22,200	8,725	30,925
1030	25 Ton	43,400	10,000	53,400
1040	40 Ton	59,500	13,600	73,100
1050	60 Ton	84,500	18,500	103,000
1060	80 Ton	106,000	23,100	129,100
2700	Cut & patch to match existing construction, add, minimum	2%	3%	
2800	Maximum	5%	9%	
2900	Dust protection, add, minimum	1%	2%	
3000	Maximum	4%	11%	
3100	Equipment usage curtailment, add, minimum	1%	1%	
3200	Maximum	3%	10%	

D3050 Terminal & Package Units

D3050 180	Rooftop Air Conditioner, Variable Air Volume	COST EACH		
		MAT.	INST.	TOTAL
3300	Material handling & storage limitation, add, minimum	1%	1%	
3400	Maximum	6%	7%	
3500	Protection of existing work, add, minimum	2%	2%	
3600	Maximum	5%	7%	
3700	Shift work requirements, add, minimum		5%	
3800	Maximum		30%	
3900	Temporary shoring and bracing, add, minimum	2%	5%	
4000	Maximum	5%	12%	

D30 HVAC

D3050 Terminal & Package Units

This data illustrates and describes packaged electrical air conditioning units including multizone control and related piping. Lines within Systems Components give the material and installation price on a cost each basis for the components. Prices for packaged electrical A/C unit systems are on Line Items D3050 210 1010 thru D3050 225 1060. Material quantities and labor costs have been adjusted for the system listed.

Factors: To adjust for job conditions other than normal working situations use Lines D3050 225 2700 thru 4000.

Condenser Supply Water Piping

Condenser Return Water Piping

Self—Contained Air Conditioner Unit, Water Cooled

System Components	QUANTITY	UNIT	COST EACH		
			MAT.	INST.	TOTAL
Packaged water cooled electric air conditioning unit					
with related piping and valves.					
Self-contained, water cooled, elect. heat, not inc tower, 5 ton, const vol	1.000	Ea.	8,975	1,775	10,750
Pipe, black steel, Sch 40, threaded, w/cplg & hangers, 10' OC, 1-1/4" dia.	20.000	L.F.	122.55	321.43	443.98
Elbow, 90°, black, straight, 1-1/4" dia.	6.000	Ea.	78.90	363	441.90
Tee, black, straight, 1-1/4" dia.	2.000	Ea.	40	190	230
Tee, black, reducing, 1-1/4" dia.	2.000	Ea.	71	190	261
Thermometers, stem type, 9" case, 8" stem, 3/4" NPT	2.000	Ea.	478	54	532
Union, black with brass seat, 1-1/4" dia.	2.000	Ea.	76	127	203
Pipe nipple, black, 1-1/4" dia	3.000	Ea.	8.55	22.43	30.98
Valves, bronze, gate, N.R.S., threaded, class 150, 1-1/4" size	2.000	Ea.	320	98	418
Circuit setter, bal valve, bronze body, threaded, 1-1/4" pipe size	1.000	Ea.	171	50.50	221.50
Insulation, fiberglass pipe covering, 1" wall, 1-1/4" IPS	20.000	L.F.	23.20	118	141.20
Control system, pneumatic, A/C Unit with heat	1.000	Ea.	3,075	2,675	5,750
Re-heat coil balancing	1.000	Ea.		140	140
Rooftop unit heat/cool balancing	1.000	Ea.		505	505
TOTAL			13,439.20	6,629.36	20,068.56

D3050 210	AC Unit, Package, Elec. Ht., Water Cooled	COST EACH		
		MAT.	INST.	TOTAL
1000	For alternate A/C systems:			
1010	A/C, Self contained, single pkg., water cooled, elect. heat, 5 Ton	13,439.20	6,629.35	20,116.50
1020	10 Ton	20,700	8,150	28,850
1030	20 Ton	27,800	10,300	38,100
1040	30 Ton	47,400	11,300	58,700
1050	40 Ton	60,500	12,500	73,000
1060	50 Ton	79,000	14,900	93,900

D3050 215	AC Unit, Package, Elec. Ht., Water Cooled, VAV	COST EACH		
		MAT.	INST.	TOTAL
1010	A/C, Self contained, single pkg., water cooled, elect. ht, VAV, 10 Ton	23,200	8,150	31,350
1020	20 Ton	32,600	10,300	42,900
1030	30 Ton	54,000	11,300	65,300
1040	40 Ton	68,000	12,500	80,500
1050	50 Ton	90,000	14,900	104,900
1060	60 Ton	103,500	18,000	121,500

676

D30 HVAC

D3050 Terminal & Package Units

D3050 220	AC Unit, Package, Hot Water Coil, Water Cooled	COST EACH		
		MAT.	INST.	TOTAL
1010	A/C, Self contn'd, single pkg., water cool, H/W ht, const. vol, 5 Ton	12,000	9,175	21,175
1020	10 Ton	17,300	10,900	28,200
1030	20 Ton	43,100	13,600	56,700
1040	30 Ton	56,000	17,200	73,200
1050	40 Ton	69,000	21,000	90,000
1060	50 Ton	84,500	25,800	110,300

D3050 225	AC Unit, Package, HW Coil, Water Cooled, VAV	COST EACH		
		MAT.	INST.	TOTAL
1010	A/C, Self contn'd, single pkg., water cool, H/W ht, VAV, 10 Ton	19,800	10,300	30,100
1020	20 Ton	28,900	13,200	42,100
1030	30 Ton	47,200	16,500	63,700
1040	40 Ton	61,500	18,100	79,600
1050	50 Ton	80,500	21,200	101,700
1060	60 Ton	95,500	23,000	118,500
2700	Cut & patch to match existing construction, add, minimum	2%	3%	
2800	Maximum	5%	9%	
2900	Dust protection, add, minimum	1%	2%	
3000	Maximum	4%	11%	
3100	Equipment usage curtailment, add, minimum	1%	1%	
3200	Maximum	3%	10%	
3300	Material handling & storage limitation, add, minimum	1%	1%	
3400	Maximum	6%	7%	
3500	Protection of existing work, add, minimum	2%	2%	
3600	Maximum	5%	7%	
3700	Shift work requirements, add, minimum		5%	
3800	Maximum		30%	
3900	Temporary shoring and bracing, add, minimum	2%	5%	
4000	Maximum	5%	12%	

D3050 Terminal & Package Units

This data illustrates and describes thru-wall air conditioning and heat pump units. Lines within Systems Components give the material and installation price on a cost each basis for the components. Prices for thru-wall unit systems are on Line Items D3050 255 1010 thru D3050 260 1050. Material quantities and labor costs have been adjusted for the system listed.

Factors: To adjust for job conditions other than normal working conditions use Lines D3050 255 2700 thru 4000.

Thru-Wall Heat Pump   Thru-Wall Air Conditioner

System Components	QUANTITY	UNIT	COST EACH MAT.	COST EACH INST.	COST EACH TOTAL
Electric thru-wall air conditioning unit.					
A/C unit, thru-wall, electric heat., cabinet, louver, 1/2 ton	1.000	Ea.	910	227	1,137
TOTAL			910	227	1,137

D3050 255	Thru-Wall A/C Unit	MAT.	INST.	TOTAL
1000	For alternate A/C units:			
1010	A/C Unit, thru-the-wall, supplemental elect. heat, cabinet, louver, 1/2 Ton	910	227	1,137
1020	3/4 Ton	1,350	273	1,623
1030	1 Ton	1,500	340	1,840
1040	1-1/2 Ton	2,525	570	3,095
1050	2 Ton	2,550	715	3,265

D3050 260	Thru-Wall Heat Pump	MAT.	INST.	TOTAL
1010	Heat pump, thru-the-wall, cabinet, louver, 1/2 Ton	2,800	170	2,970
1020	3/4 Ton	2,950	227	3,177
1030	1 Ton	3,200	340	3,540
1040	Supplemental. elect. heat, 1-1/2 Ton	3,425	880	4,305
1050	2 Ton	3,425	910	4,335
2700	Cut & patch to match existing construction, add, minimum	2%	3%	
2800	Maximum	5%	9%	
2900	Dust protection, add, minimum	1%	2%	
3000	Maximum	4%	11%	
3100	Equipment usage curtailment, add, minimum	1%	1%	
3200	Maximum	3%	10%	
3300	Material handling & storage limitation, add, minimum	1%	1%	
3400	Maximum	6%	7%	
3500	Protection of existing work, add, minimum	2%	2%	
3600	Maximum	5%	7%	
3700	Shift work requirements, add, minimum		5%	
3800	Maximum		30%	
3900	Temporary shoring and bracing, add, minimum	2%	5%	
4000	Maximum	5%	12%	

D4010 Sprinklers

This data illustrates and describes a dry type fire sprinkler system. Lines within System Components give the unit cost of a system for a 2,000 square foot building. Lines D4010 320 1700 thru 2400 give the square foot costs for alternate systems. Both material quantities and labor costs have been adjusted for the system listed.

Factors: To adjust for conditions other than normal working conditions, use Lines D4010 320 2900 thru 4000.

System Components	QUANTITY	UNIT	COST EACH		
			MAT.	INST.	TOTAL
Pre-action fire sprinkler system, ordinary hazard, open area to 2000 S.F. On one floor.					
Air compressor	1.000	Ea.	890	555	1,445
4" OS & Y valve	1.000	Ea.	890	445	1,335
Pipe riser 4" diameter	10.000	L.F.	192.50	377.30	569.80
Sprinkler head supply piping, 1"	163.000	L.F.	354.75	1,046.25	1,401
Sprinkler head supply piping, 1-1/4"	163.000	L.F.	313.50	822.25	1,135.75
Sprinkler head supply piping, 2-1/2"	163.000	L.F.	482.60	1,007	1,489.60
Pipe fittings, 1"	8.000	Ea.	100	740	840
Pipe fittings, 1-1/4"	8.000	Ea.	160	760	920
Pipe fittings, 2-1/2"	4.000	Ea.	370	592	962
Pipe fittings, 4"	2.000	Ea.	660	660	1,320
Detectors	2.000	Ea.	1,550	90	1,640
Sprinkler heads	16.000	Ea.	183.20	720	903.20
Fire department connection	1.000	Ea.	600	260	860
TOTAL			6,746.55	8,074.80	14,821.35

D4010 320	Fire Sprinkler Systems, Dry	COST PER S.F.		
		MAT.	INST.	TOTAL
1700	Ordinary hazard, one floor, area to 2000 S.F./floor	3.40	4.04	7.44
1800	For each additional floor, add per floor	1.42	3.36	4.78
1900	Area to 3200 S.F./ floor	2.64	4	6.64
2000	For each additional floor, add per floor	1.41	3.58	4.99
2100	Area to 5000 S.F./ floor	2.98	4.07	7.05
2200	For each additional floor, add per floor	1.88	3.78	5.66
2300	Area to 8000 S.F./ floor	2.23	3.43	5.66
2400	For each additional floor, add per floor	1.45	3.24	4.69
2500				
2600				
2700				
2800				
2900	Cut & patch to match existing construction, add, minimum	2%	3%	
3000	Maximum	5%	9%	
3100	Dust protection, add, minimum	1%	2%	
3200	Maximum	4%	11%	
3300	Equipment usage curtailment, add, minimum	1%	1%	
3400	Maximum	3%	10%	

D40 Fire Protection

D4010 Sprinklers

D4010 320	Fire Sprinkler Systems, Dry	COST PER S.F.		
		MAT.	INST.	TOTAL
3500	Material handling & storage limitation, add, minimum	1%	1%	
3600	Maximum	6%	7%	
3700	Protection of existing work, add, minimum	2%	2%	
3800	Maximum	5%	7%	
3900	Shift work requirements, add, minimum		5%	
4000	Maximum		30%	

D4010 Sprinklers

This data illustrates and describes a wet type fire sprinkler system. Lines within System Components give the unit cost of a system for a 2,000 square foot building. Lines D4010 420 1900 thru 2600 give the square foot costs for alternate systems. Both material quantities and labor costs have been adjusted for the system listed.

Factors: To adjust for conditions other than normal working conditions, use Lines D4010 420 2900 thru 4000.

System Components	QUANTITY	UNIT	COST EACH MAT.	COST EACH INST.	COST EACH TOTAL
Wet pipe fire sprinkler system, ordinary hazard, open area to 2000 S.F. On one floor.					
4" OS & Y valve	1.000	Ea.	890	445	1,335
Wet pipe alarm valve	1.000	Ea.	1,975	650	2,625
Water motor alarm	1.000	Ea.	480	180	660
3" check valve	1.000	Ea.	405	435	840
Pipe riser, 4" diameter	10.000	L.F.	192.50	377.30	569.80
Water gauges and trim	1.000	Set	3,300	1,300	4,600
Electric fire horn	1.000	Ea.	67	102	169
Sprinkler head supply piping, 1" diameter	168.000	L.F.	354.75	1,046.25	1,401
Sprinkler head supply piping, 1-1/2" diameter	168.000	L.F.	354.75	913	1,267.75
Sprinkler head supply piping, 2-1/2" diameter	168.000	L.F.	482.60	1,007	1,489.60
Pipe fittings, 1"	8.000	Ea.	100	740	840
Pipe fittings, 1-1/2"	8.000	Ea.	200	816	1,016
Pipe fittings, 2-1/2"	4.000	Ea.	370	592	962
Pipe fittings, 4"	2.000	Ea.	660	660	1,320
Sprinkler heads	16.000	Ea.	183.20	720	903.20
Fire department connection	1.000	Ea.	600	260	860
TOTAL			10,614.80	10,243.55	20,858.35

D4010 420	Fire Sprinkler Systems, Wet	COST PER S.F. MAT.	COST PER S.F. INST.	COST PER S.F. TOTAL
1900	Ordinary hazard, one floor, area to 2000 S.F./floor	5.30	5.10	10.40
2000	For each additional floor, add per floor	1.46	3.44	4.90
2100	Area to 3200 S.F./floor	3.83	4.64	8.47
2200	For each additional floor, add per floor	1.41	3.58	4.99
2300	Area to 5000 S.F./ floor	3.43	4.46	7.89
2400	For each additional floor, add per floor	1.88	3.78	5.66
2500	Area to 8000 S.F./ floor	2.42	3.65	6.07
2600	For each additional floor, add per floor	1.56	3.44	5
2700				
2900	Cut & patch to match existing construction, add minimum	2%	3%	
3000	Maximum	5%	9%	
3100	Dust protection, add, minimum	1%	2%	
3200	Maximum	4%	11%	
3300	Equipment usage curtailment, add, minimum	1%	1%	
3400	Maximum	3%	10%	
3500	Material handling & storage limitation, add, minimum	1%	1%	
3600	Maximum	6%	7%	

D5010 Electrical Service/Distribution

This data illustrates and describes commercial service systems including a meter socket, service head and cable, entrance switch, steel conduit, copper wire, panel board, ground rod, wire, and conduit. Lines within System Components give the unit price and total price on a cost each basis for this system. Prices for commercial service systems are on Line Items D5010 210 1500 thru 1700. Both material quantities and labor costs have been adjusted for the system listed.

Factors: To adjust for job conditions other than normal working situations use Lines D5010 210 2900 thru 4000.

Example: You are to install the system with maximum equipment usage curtailment. Go to Line D5010 210 3400 and apply these percentages to the appropriate MAT. and INST. costs.

System Components	QUANTITY	UNIT	COST EACH MAT.	COST EACH INST.	COST EACH TOTAL
Commercial electric service including service breakers, metering 120/208 Volt, 3 phase, 4 wire, feeder, and panel board.					
100 Amp Service					
Meter socket	1.000	Ea.	47.50	213	260.50
Service head	.200	C.L.F.	28.20	136	164.20
Service entrance switch	1.000	Ea.	199	360	559
Rigid steel conduit	20.000	L.F.	109	210	319
600 volt copper wire #3	1.000	C.L.F.	116	136	252
Panel board, 24 circuits, 20 Amp breakers	1.000	Ea.	1,000	1,125	2,125
Ground rod	1.000	Ea.	22	124	146
TOTAL			1,521.70	2,304	3,825.70

D5010 210	Commercial Service - 3 Phase	COST EACH MAT.	COST EACH INST.	COST EACH TOTAL
1500	120/208 Volt, 3 phase, 4 wire service, 60 Amp	1,100	1,800	2,900
1550	100 Amp	1,525	2,300	3,825
1600	200 Amp	3,150	4,075	7,225
1700	400 Amp	6,275	5,100	11,375
1800				
1900				
2000				
2100				
2200				
2300				
2400				
2500				
2900	Cut & patch to match existing construction, add, minimum	2%	3%	
3000	Maximum	5%	9%	
3100	Dust protection, add, minimum	1%	2%	
3200	Maximum	4%	11%	
3300	Equipment usage curtailment, add, minimum	1%	1%	
3400	Maximum	3%	10%	
3500	Material handling & storage limitation, add, minimum	1%	1%	
3600	Maximum	6%	7%	
3700	Protection of existing work, add, minimum	2%	2%	
3800	Maximum	5%	7%	
3900	Shift work requirements, add, minimum		5%	
4000	Maximum		30%	

D5010 Electrical Service/Distribution

This data illustrates and describes a residential, single phase system including a weather cap, service entrance cable, meter socket, entrance switch, ground rod, ground cable, EMT, and panelboard. Lines within System Components give the unit price and total price on a cost each basis for this system. Prices for a residential, single phase system are also given. Both material quantities and labor costs have been adjusted for the system listed.

Factors: To adjust for job conditions other than normal working situations use Lines D5010 220 2900 thru 4000.

Example: You are to install the system with a minimum equipment usage curtailment. Go to Line D5010 220 3300 and apply these percentages to the appropriate MAT. and INST. costs.

System Components	QUANTITY	UNIT	COST EACH MAT.	COST EACH INST.	COST EACH TOTAL
100 Amp Service, single phase					
Weathercap	1.000	Ea.	8.90	57	65.90
Service entrance cable	.200	C.L.F.	28.20	136	164.20
Meter socket	1.000	Ea.	47.50	213	260.50
Entrance disconnect switch	1.000	Ea.	199	360	559
Ground rod, with clamp	1.000	Ea.	24.50	142	166.50
Ground cable	.100	C.L.F.	14.70	42.50	57.20
Panelboard, 12 circuit	1.000	Ea.	102	570	672
TOTAL			424.80	1,520.50	1,945.30
200 Amp Service , single phase					
Weathercap	1.000	Ea.	23	85	108
Service entrance cable	.200	C.L.F.	72	195	267
Meter socket	1.000	Ea.	101	360	461
Entrance disconnect switch	1.000	Ea.	425	525	950
Ground rod, with clamp	1.000	Ea.	42.50	155	197.50
Ground cable	.100	C.L.F.	30	23.50	53.50
3/4″ EMT	10.000	L.F.	11	52.50	63.50
Panelboard, 24 circuit	1.000	Ea.	510	945	1,455
TOTAL			1,214.50	2,341	3,555.50

D5010 220	Residential Service - Single Phase	COST EACH MAT.	COST EACH INST.	COST EACH TOTAL
2800				
2810	100 Amp Service, single phase	425	1,525	1,950
2820	200 Amp Service, single phase	1,225	2,350	3,575
2900	Cut & patch to match existing construction, add, minimum	2%	3%	
3000	Maximum	5%	9%	
3100	Dust protection, add, minimum	1%	2%	
3200	Maximum	4%	11%	
3300	Equipment usage curtailment, add, minimum	1%	1%	
3400	Maximum	3%	10%	
3500	Material handling & storage limitation, add, minimum	1%	1%	
3600	Maximum	6%	7%	
3700	Protection of existing work, add, minimum	2%	2%	
3800	Maximum	5%	7%	
3900	Shift work requirements, add, minimum		5%	
4000	Maximum		30%	

683

D5020 Lighting and Branch Wiring

D5020 180	Selective Price Sheet	COST EACH		
		MAT.	INST.	TOTAL
0100	Using non-metallic sheathed, cable, air conditioning receptacle	21.50	68	89.50
0200	Disposal wiring	26.50	75.50	102
0300	Dryer circuit	35.50	124	159.50
0400	Duplex receptacle	21.50	52.50	74
0500	Fire alarm or smoke detector	98.50	68	166.50
0600	Furnace circuit & switch	29.50	114	143.50
0700	Ground fault receptacle	59.50	85	144.50
0800	Heater circuit	21	85	106
0900	Lighting wiring	21.50	42.50	64
1000	Range circuit	80.50	170	250.50
1100	Switches single pole	16.20	42.50	58.70
1200	3-way	19.75	57	76.75
1300	Water heater circuit	28.50	136	164.50
1400	Weatherproof receptacle	171	114	285
1500	Using BX cable, air conditioning receptacle	28	82	110
1600	Disposal wiring	32.50	91	123.50
1700	Dryer circuit	42	148	190
1800	Duplex receptacle	28	63	91
1900	Fire alarm or smoke detector	98.50	68	166.50
2000	Furnace circuit & switch	39	136	175
2100	Ground fault receptacle	68.50	103	171.50
2200	Heater circuit	26.50	103	129.50
2300	Lighting wiring	26	51.50	77.50
2400	Range circuit	113	207	320
2500	Switches, single pole	23	51.50	74.50
2600	3-way	26	68	94
2700	Water heater circuit	40.50	162	202.50
2800	Weatherproof receptacle	176	136	312
2900	Using EMT conduit, air conditioning receptacle	42	102	144
3000	Disposal wiring	48	114	162
3100	Dryer circuit	54	184	238
3200	Duplex receptacle	42	78.50	120.50
3300	Fire alarm or smoke detector	117	102	219
3400	Furnace circuit & switch	49	170	219
3500	Ground fault receptacle	83.50	126	209.50
3600	Heater circuit	39.50	126	165.50
3700	Lighting wiring	38	63.50	101.50
3800	Range circuit	95.50	252	347.50
3900	Switches, single pole	37	63.50	100.50
4000	3-way	36	85	121
4100	Water heater circuit	45.50	200	245.50
4200	Weatherproof receptacle	190	170	360
4300	Using aluminum conduit, air conditioning receptacle	44.50	136	180.50
4400	Disposal wiring	51	151	202
4500	Dryer circuit	65	243	308
4600	Duplex receptacle	42.50	105	147.50
4700	Fire alarm or smoke detector	126	136	262
4800	Furnace circuit & switch	55.50	227	282.50
4900	Ground fault receptacle	93.50	170	263.50
5000	Heater circuit	44	170	214
5100	Lighting wiring	49.50	85	134.50
5200	Range circuit	113	340	453
5300	Switches, single pole	47.50	85	132.50
5400	3-way	46.50	114	160.50
5500	Water heater circuit	57	273	330
5600	Weatherproof receptacle	207	227	434
5700	Using galvanized steel conduit	48.50	145	193.50
5800	Disposal wiring	50.50	162	212.50

D5020 Lighting and Branch Wiring

D5020 180	Selective Price Sheet	COST EACH		
		MAT.	INST.	TOTAL
5900	Dryer circuit	73.50	262	335.50
6000	Duplex receptacle	48.50	112	160.50
6100	Fire alarm or smoke detector	124	145	269
6200	Furnace circuit & switch	54.50	243	297.50
6300	Ground fault receptacle	100	179	279
6400	Heater circuit	45.50	179	224.50
6500	Lighting wiring	51.50	91	142.50
6600	Range circuit	119	360	479
6700	Switches, single pole	54	91	145
6800	3-way	53.50	118	171.50
6900	Water heater circuit	60.50	284	344.50
7000	Weatherproof receptacle	207	243	450
7100				
7200				
7300				
7400				
7500				
7600				
7700				
7800				
7900				
8000				
8100				
8200				
8300				
8400				

D5020 Lighting and Branch Wiring

This data illustrates and describes fluorescent lighting systems including a fixture, lamp, outlet box and wiring. Lines within System Components give the unit price and total price on a cost each basis for this system. Prices for fluorescent lighting systems are on Line Items D5020 248 1200 thru 1500. Both material quantities and labor costs have been adjusted for the system listed.

Factors: To adjust for job conditions other than normal working situations use Lines D5020 248 2900 thru 4000.

Example: You are to install the system during evening hours. Go to Line D5020 248 3900 and apply this percentage to the appropriate INST. cost.

System Components	QUANTITY	UNIT	COST EACH MAT.	COST EACH INST.	TOTAL
Fluorescent lighting, including fixture, lamp, outlet box and wiring.					
Recessed lighting fixture, on suspended system	1.000	Ea.	66.50	145	211.50
Outlet box	1.000	Ea.	3.17	38	41.17
#12 wire	.660	C.L.F.	7.76	40.92	48.68
Conduit, EMT, 1/2" conduit	20.000	L.F.	15.40	80.20	95.60
TOTAL			92.83	304.12	396.95

D5020 248	Lighting, Fluorescent	COST EACH MAT.	COST EACH INST.	TOTAL
1200	Recessed lighting fixture, on suspended system	93	305	398
1300	Surface mounted, 2' x 4', acrylic prismatic diffuser	116	320	436
1400	Strip fixture, 8' long, two 8' lamps	94	300	394
1500	Pendant mounted, industrial, 8' long, with reflectors	135	345	480
1600				
1700				
1800				
1900				
2000				
2100				
2200				
2300				
2900	Cut & patch to match existing construction, add, minimum	2%	3%	
3000	Maximum	5%	9%	
3100	Dust protection, add, minimum	1%	2%	
3200	Maximum	4%	11%	
3300	Equipment usage curtailment, add, minimum	1%	1%	
3400	Maximum	3%	10%	
3500	Material handling & storage limitation, add, minimum	1%	1%	
3600	Maximum	6%	7%	
3700	Protection of existing work, add, minimum	2%	2%	
3800	Maximum	5%	7%	
3900	Shift work requirements, add, minimum		5%	
4000	Maximum		30%	

D5020 Lighting and Branch Wiring

This data illustrates and describes incandescent lighting systems including a fixture, lamp, outlet box, conduit and wiring. Lines within System Components give the unit price and total price on a cost each basis for this system. Prices for an alternate incandescent lighting system are also given. Both material quantities and labor costs have been adjusted for the system listed.

Factors: To adjust for conditions other than normal working situations use Lines D5020 250 2900 thru 4000.

Example: You are to install the system and cut and match existing construction. Go to Line D5020 250 3000 and apply this percentage to the appropriate INST. costs.

System Components			COST EACH		
	QUANTITY	UNIT	MAT.	INST.	TOTAL
Incandescent light fixture, including lamp, outlet box, conduit and wiring.					
Recessed wide reflector with flat glass lens	1.000	Ea.	107	102	209
Outlet box	1.000	Ea.	3.17	38	41.17
Armored cable, 3 wire	.200	C.L.F.	16.10	62	78.10
TOTAL			126.27	202	328.27
Recessed flood light fixture, including lamp, outlet box, conduit & wire					
Recessed, R-40 flood lamp with refractor	1.000	Ea.	114	85	199
150 watt R-40 flood lamp	.010	Ea.	6.80	5.25	12.05
Outlet box	1.000	Ea.	3.17	38	41.17
Outlet box cover	1.000	Ea.	1.10	10.65	11.75
Romex, 12-2 with ground	.200	C.L.F.	7.30	62	69.30
Conduit, 1/2" EMT	20.000	L.F.	15.40	80.20	95.60
TOTAL			147.77	281.10	428.87

D5020 250	Lighting, Incandescent	COST EACH		
		MAT.	INST.	TOTAL
2810	Incandescent light fixture, including lamp, outlet box, conduit and wiring.	126	202	328
2820	Recessed flood light fixture, including lamp, outlet box, conduit & wire	148	281	429
2900	Cut & patch to match existing construction, add, minimum	2%	3%	
3000	Maximum	5%	9%	
3100	Dust protection, add, minimum	1%	2%	
3200	Maximum	4%	11%	
3300	Equipment usage curtailment, add, minimum	1%	1%	
3400	Maximum	3%	10%	
3500	Material handling & storage limitation, add, minimum	1%	1%	
3600	Maximum	6%	7%	
3700	Protection of existing work, add, minimum	2%	2%	
3800	Maximum	5%	7%	
3900	Shift work requirements, add, minimum		5%	
4000	Maximum		30%	

D5020 Lighting and Branch Wiring

This data illustrates and describes high intensity lighting systems including a lamp, EMT conduit, EMT "T" fitting with cover, and wire. Lines within System Components give the unit price and total price on a cost each basis for this system. Prices for high intensity lighting systems are on Line Items D5020 252 1100 thru 1800. Both material quantities and labor costs have been adjusted for the system listed.

Factors: To adjust for job conditions other than normal working situations use Lines D5020 252 2900 thru 4000.

Example: You are to install the system and protect existing construction. Go to Line D5020 252 3700 and apply these percentages to the appropriate MAT. and INST. costs.

System Components			COST EACH		
	QUANTITY	UNIT	MAT.	INST.	TOTAL
High intensity lighting system consisting of 400 watt metal halide fixture And lamp with 1/2" EMT conduit and fittings using #12 wire, high bay.					
Single unit, 400 watt	1.000	Ea.	450	296	746
Electric metallic tubing (EMT), 1/2" diameter	30.000	Ea.	23.10	120.30	143.40
1/2" EMT "T" fitting with cover	1.000	Ea.	7.40	42.50	49.90
Wire, 600 volt, type THWN-THHN, copper, solid, #12	.600	Ea.	7.05	37.20	44.25
TOTAL			487.55	496	983.55

D5020 252	Lighting, High Intensity	COST EACH		
		MAT.	INST.	TOTAL
1100	High bay: 400 watt, metal halide fixture and lamp	490	495	985
1200	High pressure sodium and lamp	455	495	950
1400	1000 watt, metal halide	690	540	1,230
1500	High pressure sodium	640	540	1,180
1700	Low bay: 250 watt, metal halide	440	415	855
1800	150 watt, high pressure sodium	400	415	815
1900				
2000				
2100				
2200				
2300				
2400				
2500				
2600				
2700				
2900	Cut & patch to match existing construction, add, minimum	2%	3%	
3000	Maximum	5%	9%	
3100	Dust protection, add, minimum	1%	2%	
3200	Maximum	4%	11%	
3300	Equipment usage curtailment, add, minimum	1%	1%	
3400	Maximum	3%	10%	
3500	Material handling & storage limitation, add, minimum	1%	1%	
3600	Maximum	6%	7%	
3700	Protection of existing work, add, minimum	2%	2%	
3800	Maximum	5%	7%	
3900	Shift work requirements, add, minimum		5%	
4000	Maximum		30%	

D5090   Other Electrical Systems

This data illustrates and describes baseboard heat systems including a thermostat, outlet box, breaker, and feed. Lines within System Components give the unit price and total price on a cost each basis for this system. Prices for baseboard heat systems are on Line Items D5090 430 1500 thru 1800. Both material quantities and labor costs have been adjusted for the system listed.

Factors: To adjust for job conditions other than normal working situations use Lines D5090 430 2900 thru 4000.

Example: You are to install the system during evenings and weekends only. Go to Line D5090 430 4000 and apply this percentage to the appropriate INST. cost.

System Components	QUANTITY	UNIT	COST EACH		
			MAT.	INST.	TOTAL
Baseboard heat including thermostat, outlet box, breaker and feed.					
Electric baseboard heater, 4' long	1.000	Ea.	39	102	141
Thermostat, integral	1.000	Ea.	29	42.50	71.50
Romex, 12-3 with ground	.400	C.L.F.	20.40	136	156.40
Panel board breaker, 20 Amp	1.000	Ea.	90.50	85	175.50
Outlet box	1.000	Ea.	3.17	38	41.17
TOTAL			182.07	403.50	585.57

D5090 530	Heat, Baseboard	COST EACH		
		MAT.	INST.	TOTAL
1400	Heat, electric baseboard			
1500	2' long	172	385	557
1550	4' long	182	405	587
1600	6' long	199	440	639
1700	8' long	205	470	675
1800	10' long	340	510	850
1900				
2000				
2100				
2200				
2300				
2400				
2500				
2600				
2700				
2900	Cut & patch to match existing construction, add, minimum	2%	3%	
3000	Maximum	5%	9%	
3100	Dust protection, add, minimum	1%	2%	
3200	Maximum	4%	11%	
3300	Equipment usage curtailment, add, minimum	1%	1%	
3400	Maximum	3%	10%	
3500	Material handling & storage limitation, add, minimum	1%	1%	
3600	Maximum	6%	7%	
3700	Protection of existing work, add, minimum	2%	2%	
3800	Maximum	5%	7%	
3900	Shift work requirements, add, minimum		5%	
4000	Maximum		30%	

689

This data illustrates and describes kitchen systems including top and bottom cabinets, custom laminated plastic top, single bowl sink, and appliances. Lines within System Components give the unit price and total price on a cost each basis for this system. Prices for kitchen systems are on Line Items E1090 310 1400 thru 1600. Both material quantities and labor costs have been adjusted for the system listed.

Factors: To adjust for job conditions other than normal working situations use Lines E1090 310 2900 thru 4000.

Example: You are to install the system and protect the work area from dust. Go to Line E1090 310 3200 and apply these percentages to the appropriate MAT. and INST. costs.

System Components	QUANTITY	UNIT	COST EACH		
			MAT.	INST.	TOTAL
Kitchen cabinets including wall and base cabinets, custom laminated					
Plastic top,sink & appliances,no plumbing or electrical rough-in included.					
Prefinished wood cabinets, average quality, wall and base	20.000	L.F.	4,000	840	4,840
Custom laminated plastic counter top	20.000	L.F.	380	420	800
Stainless steel sink, 22" x 25"	1.000	Ea.	740	237	977
Faucet, top mount	1.000	Ea.	85	74	159
Dishwasher, built-in	1.000	Ea.	330	355	685
Compactor, built-in	1.000	Ea.	760	127	887
Range hood, vented, 30" wide	1.000	Ea.	91.50	267	358.50
TOTAL			6,386.50	2,320	8,706.50

E1090 310	Kitchens	COST EACH		
		MAT.	INST.	TOTAL
1400	Prefinished wood cabinets, average quality	6,375	2,325	8,700
1500	Prefinished wood cabinets, high quality	11,800	2,675	14,475
1600	Custom cabinets, built in place, high quality	15,500	3,075	18,575
1700				
1800				
1900				
2000				
2100				
2200	NOTE: No plumbing or electric rough-ins are included in the above			
2300	Prices, for plumbing see Division 22, for electric see Division 26.			
2400				
2500				
2600				
2700				
2900	Cut & patch to match existing construction, add, minimum	2%	3%	
3000	Maximum	5%	9%	
3100	Dust protection, add, minimum	1%	2%	
3200	Maximum	4%	11%	
3300	Equipment usage curtailment, add, minimum	1%	1%	
3400	Maximum	3%	10%	
3500	Material handling & storage limitation, add, minimum	1%	1%	
3600	Maximum	6%	7%	
3700	Protection of existing work, add, minimum	2%	2%	
3800	Maximum	5%	7%	
3900	Shift work requirements, add, minimum		5%	
4000	Maximum		30%	

E1090 Other Equipment

E1090 315	Selective Price Sheet	COST EACH		
		MAT.	INST.	TOTAL
0100	Cabinets standard wood, base, one drawer one door, 12" wide	298	51	349
0200	15" wide	310	53	363
0300	18" wide	340	54.50	394.50
0400	21" wide	350	56	406
0500	24" wide	410	57	467
0600				
0700	Two drawers two doors, 27" wide	440	57.50	497.50
0800	30" wide	480	59	539
0900	33" wide	500	60.50	560.50
1000	36" wide	520	62.50	582.50
1100	42" wide	545	64	609
1200	48" wide	585	67	652
1300	Drawer base (4 drawers), 12" wide	315	51	366
1400	15" wide	320	53	373
1500	18" wide	350	54.50	404.50
1600	24" wide	385	57	442
1700	Sink or range base, 30" wide	400	59	459
1800	33" wide	430	60.50	490.50
1900	36" wide	450	62.50	512.50
2000	42" wide	470	64	534
2100	Corner base, 36" wide	740	70.50	810.50
2200	Lazy Susan with revolving door	945	77	1,022
2300	Cabinets standard wood, wall two doors, 12" high, 30" wide	269	51	320
2400	36" wide	320	53	373
2500	15" high, 30" high	274	53	327
2600	36" wide	330	56	386
2700	24" high, 30" wide	365	54.50	419.50
2800	36" wide	405	56	461
2900	30" high, 30" wide	405	65.50	470.50
3000	36" wide	460	67.50	527.50
3100	42" wide	500	68.50	568.50
3200	48" wide	565	69	634
3300	One door, 30" high, 12" wide	258	57.50	315.50
3400	15" wide	270	59	329
3500	18" wide	297	60.50	357.50
3600	24" wide	350	62.50	412.50
3700	Corner, 30" high, 24" wide	395	70.50	465.50
3800	36" wide	480	77	557
3900	Broom, 84" high, 24" deep, 18" wide	740	127	867
4000	Oven, 84" high, 24" deep, 27" wide	1,125	158	1,283
4100	Valance board, 4' long	71.50	12.80	84.30
4200	6" long	107	19.20	126.20
4300	Counter tops, laminated plastic, stock 25" wide w/backsplash, min.	19	21	40
4400	Maximum	38.50	25.50	64
4500	Custom, 7/8" thick, no splash	33.50	21	54.50
4600	Cove splash	32.50	21	53.50
4700	1-1/4" thick, no splash	39.50	22.50	62
4800	Square splash	47.50	22.50	70
4900	Post formed	11.40	21	32.40
5000				
5100	Maple laminated 1-1/2" thick, no splash	84.50	22.50	107
5200	Square splash	100	22.50	122.50
5300				
5400				
5500	Appliances, range, free standing, minimum	515	99	614
5600	Maximum	2,075	248	2,323
5700	Built-in, minimum	990	114	1,104
5800	Maximum	1,725	635	2,360

E1090 Other Equipment

E1090 315	Selective Price Sheet	COST EACH		
		MAT.	INST.	TOTAL
5900	Counter top range 4 burner, maximum	385	114	499
6000	Maximum	1,875	227	2,102
6100	Compactor, built-in, minimum	760	127	887
6200	Maximum	1,275	211	1,486
6300	Dishwasher, built-in, minimum	330	355	685
6400	Maximum	495	710	1,205
6500	Garbage disposer, sink-pipe, minimum	102	142	244
6600	Maximum	267	142	409
6700	Range hood, 30" wide, 2 speed, minimum	91.50	267	358.50
6800	Maximum	940	445	1,385
6900	Refrigerator, no frost, 12 cu. ft.	520	99	619
7000	20 cu. ft.	615	165	780
7100	Plumb. not incl. rough-ins, sinks porc. C.I., single bowl, 21" x 24"	325	237	562
7200	21" x 30"	600	237	837
7300	Double bowl, 20" x 32"	405	277	682
7400				
7500	Stainless steel, single bowl, 19" x 18"	665	237	902
7600	22" x 25"	740	237	977
7700				
7800				
7900				
8000				
8100				
8200				
8300				
8400				

G1030 Site Earthwork

This data illustrates and describes utility trench excavation with backfill systems including a trench, backfill, excavated material, utility pipe or ductwork not included. Lines within System Components give the unit price and total price on a linear foot basis for this system. Prices for utility trench excavation with backfill systems are on Line Items G1030 810 1100 thru 2500. Both material quantities and labor costs have been adjusted for the system listed.

Factors: To adjust for job conditions other than normal working situations use Lines G1030 810 3100 thru 4000.

Example: You are to install the system and protect existing construction. Go to Line G1030 810 3600 and apply these percentages to the appropriate TOTAL cost.

System Components			COST PER L.F.		
	QUANTITY	UNIT	MAT.	INST.	TOTAL
Cont. utility trench excav. with 3/8 C.Y. wheel mtd. backhoe, Backfilling with excav. mat. and comp. in 12″ lifts. Trench is 2′ wide by 4′ dp. Cost of utility piping or ductwork is not incl. Cost based on an Excavation production rate of 150 C.Y. daily no hauling included.					
Machine excavate trench, 2′ wide by 4′ deep	.296	C.Y.	.79	2.25	3.04
Dozer backfill with excavated material	.296	C.Y.	.36	.62	.98
Compact in 12″ lifts, vibrating plate	.296	C.Y.	.09	2.49	2.58
TOTAL			1.24	5.36	6.60

G1030 810	Excavation, Utility Trench	COST PER L.F.		
		MAT.	INST.	TOTAL
1100	2′ wide x 2′ deep	.62	2.67	3.29
1200	3′ deep	.92	4.02	4.94
1250	4′ deep	1.24	5.35	6.60
1300	With sloping sides, 2′ wide x 5′ deep	3.49	15.10	18.60
1400	6′ deep	4.66	20	24.50
1500	7′ deep	5.95	26	32
1600	8′ deep	7.45	32	39.50
1700	9′ deep	9.10	39	48
1800	10′ deep	10.85	47	58
2000	For hauling excavated material up to 2 miles & backfilling w/gravel			
2100	Gravel, 2′ wide by 2′ deep, add	3.77	3.87	7.64
2200	4′ deep, add	7.55	7.75	15.30
2300	6′ deep, add	28.50	29	57.50
2400	8′ deep, add	45.50	46.50	92
2500	10′ deep, add	66	68	134
2600				
2700				
2800	For shallow, hand excavated trenches, no backfill, to 6′ dp, light soil		62	62
2900	Heavy soil		124	124
3000				
3100	Equipment usage curtailment, add, minimum	1%	1%	
3200	Maximum	3%	10%	
3300	Material handling & storage limitation, add, minimum	1%	1%	
3400	Maximum	6%	7%	
3500	Protection of existing work, add, minimum	2%	2%	
3600	Maximum	5%	7%	
3700	Shift work requirements, add, minimum		5%	
3800	Maximum		30%	
3900	Temporary shoring and bracing, add, minimum	2%	5%	
4000	Maximum	5%	12%	

G2010 Roadways

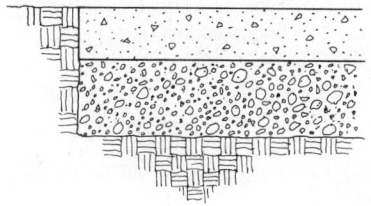

This data illustrates and describes driveway systems including concrete slab, gravel base, broom finish, compaction, and joists. Lines within System Components give the unit price and total price on a cost each basis for this system. Prices for driveway systems are on Line Items G2010 240 1200 thru 2700. Both material quantities and labor costs have been adjusted for the system listed.

Factors: To adjust for job conditions other than normal working situations use Lines G2010 240 2900 thru 4000.

Example: You are to install the system and match existing construction. Go to Line G2010 240 3000 and apply these percentages to the appropriate MAT. and INST. costs.

System Components	QUANTITY	UNIT	COST EACH MAT.	COST EACH INST.	COST EACH TOTAL
Complete driveway, 10' x 30', 4" concrete slab, 3000 psi, on 6" crushed Stone base, concrete broom finished and cured with 10' x 10' joints.					
Grade and compact subgrade	33.000	S.Y.		73.26	73.26
Place and compact 6" crushed stone base	33.000	S.Y.	171.60	50.16	221.76
Place and remove edgeforms (4 use)	80.000	L.F.	29.60	319.20	348.80
Concrete for slab, 3000 psi	4.000	C.Y.	472		472
Place concrete, 4" thick slab, direct chute	4.000	C.Y.		114.52	114.52
Screed, float and broom finish slab	4.000	C.Y.		267	267
Spray on membrane curing compound	3.000	S.F.	42	31.35	73.35
Saw cut 3" deep joints	60.000	L.F.	9	104.40	113.40
TOTAL			724.20	959.89	1,684.09

G2010 240	Driveways	COST EACH MAT.	COST EACH INST.	COST EACH TOTAL
1200	Concrete: 10' x 30' with 4" concrete on 6" crushed stone	725	960	1,685
1300	With 6" concrete	960	1,025	1,985
1400	20' x 30' with 4" concrete	1,450	1,850	3,300
1500	With 6" concrete	1,925	1,950	3,875
1600	10' wide, for each additional 10' length over 30', 4" thick, add	231	310	541
1700	6" thick, add	320	330	650
1800	20' wide, for each additional 10' length over 30', 4" thick, add	460	590	1,050
1900	6" thick, add	605	625	1,230
2000	Asphalt: 10' x 30' with 2" binder, 1" topping on 6" cr. st., sealed	625	1,100	1,725
2100	3" binder, 1" topping	760	1,125	1,885
2200	20' x 30' with 2" binder, 1" topping	1,250	1,375	2,625
2300	3" binding, 1" topping	1,525	1,400	2,925
2400	10' wide for each add'l. 10' length over 30', 2" b + 1" t, add	209	86.50	295.50
2500	3" binder, 1" topping	254	91	345
2600	20' wide for each add'l. 10' length over 30', 2" b + 1" t, add	415	174	589
2700	3" binder, 1" topping	510	183	693
2800				
2900	Cut & patch to match existing construction, add, minimum	2%	3%	
3000	Maximum	5%	9%	
3100	Dust protection, add, minimum	1%	2%	
3200	Maximum	4%	11%	
3300	Equipment usage curtailment, add, minimum	1%	1%	
3400	Maximum	3%	10%	
3500	Material handling & storage limitation, add, minimum	1%	1%	
3600	Maximum	6%	7%	
3700	Protection of existing work, add, minimum	2%	2%	
3800	Maximum	5%	7%	
3900	Shift work requirements, add, minimum		5%	
4000	Maximum		30%	

697

For customer support on your Commercial Renovation Cost Data, call 877.742.4719.

G2020 Parking Lots

This data illustrates and describes asphalt parking lot systems including asphalt binder, topping, crushed stone base, painted parking stripes and concrete parking blocks. Lines within System Components give the unit price and total price per square yard for this system. Prices for asphalt parking lot systems are on Line Items G2020 230 1400 thru 2500. Both material quantities and labor costs have been adjusted for the system listed.

Factors: To adjust for job conditions other than normal working situations use Lines G2020 230 2900 thru 4000.

Example: You are to install the system and match existing construction. Go to Line G2020 230 2900 and apply these percentages to the appropriate MAT. and INST. costs.

System Components	QUANTITY	UNIT	COST PER S.Y. MAT.	COST PER S.Y. INST.	COST PER S.Y. TOTAL
Parking lot consisting of 2″ asphalt binder and 1″ topping on 6″					
Crushed stone base with painted parking stripes and concrete parking blocks.					
Fine grade and compact subgrade	1.000	S.Y.		2.22	2.22
6″ crushed stone base, stone	.320	Ton	5.20	1.52	6.72
2″ asphalt binder	1.000	S.Y.	8.20	1.43	9.63
1″ asphalt topping	1.000	S.Y.	4.07	.94	5.01
Paint parking stripes	.500	L.F.	.09	.10	.19
6″ x 10″ x 6′ precast concrete parking blocks	.020	Ea.	.88	.42	1.30
Mobilization of equipment	.005	Ea.		4.25	4.25
TOTAL			18.44	10.88	29.32

G2020 230	Parking Lots, Asphalt	COST PER S.Y. MAT.	COST PER S.Y. INST.	COST PER S.Y. TOTAL
1400	2″ binder plus 1″ topping on 6″ crushed stone	18.45	10.90	29.35
1500	9″ crushed stone	21	11	32
1600	12″ crushed stone	23.50	11.15	34.65
1700	On bank run gravel, 6″ deep	17.60	10.20	27.80
1800	9″ deep	19.80	10.40	30.20
1900	12″ deep	22	10.60	32.60
2000	3″ binder plus 1″ topping on 6″ crushed stone	22.50	11.25	33.75
2100	9″ deep crushed stone	25	11.45	36.45
2200	12″ deep crushed stone	27.50	11.60	39.10
2300	On bank run gravel, 6″ deep	21.50	10.65	32.15
2400	9″ deep	24	10.80	34.80
2500	12″ deep	26	10.95	36.95
2600				
2700				
2900	Cut & patch to match existing construction, add, minimum	2%	3%	
3000	Maximum	5%	9%	
3100	Dust protection, add, minimum	1%	2%	
3200	Maximum	4%	11%	
3300	Equipment usage curtailment, add, minimum	1%	1%	
3400	Maximum	3%	10%	
3500	Material handling & storage limitation, add, minimum	1%	1%	
3600	Maximum	6%	7%	
3700	Protection of existing work, add, minimum	2%	2%	
3800	Maximum	5%	7%	
3900	Shift work requirements, add, minimum		5%	
4000	Maximum		30%	

G20 Site Improvements

G2020 Parking Lots

This data illustrates and describes concrete parking lot systems including a concrete slab, unreinforced, compacted gravel base, broom finished, cured, joints, painted parking stripes, and parking blocks. Lines within System Components give the unit price and total price per square yard for this system. Prices for concrete parking lot systems are on Line Items G2020 240 1500 thru 2500. All materials have been adjusted according to the system listed.

Factors: To adjust for job conditions other than normal working situations use Lines G2020 240 2900 thru 4000.

Example: You are to install the system and protect existing construction. Go to Line G2020 240 3800 and apply these percentages to the appropriate MAT. and INST. costs.

System Components			COST PER S.Y.		
	QUANTITY	UNIT	MAT.	INST.	TOTAL
Parking lot, 4″ slab, 3000 psi concrete on 6″ crushed stone base					
With 12′ x 12′ joints, painted parking strips, conc. parking blocks.					
Grade and compact subgrade	1.000	S.Y.		2.22	2.22
6″ crushed stone base, stone	.320	Ton	5.20	1.52	6.72
Place and remove edge forms, 4 use	.250	L.F.	.09	1	1.09
Concrete for slab, 3000 psi	1.000	S.F.	12.98		12.98
Place concrete, 4″ thick slab, direct chute	1.000	S.F.		3.15	3.15
Spray on membrane curing compound	1.000	S.F.	1.26	.66	1.92
Saw cut 3″ deep joints	2.250	L.F.	.34	3.92	4.26
Paint parking stripes	.500	L.F.	.09	.10	.19
6″ x 10″ x 6′ precast concrete parking blocks	.020	Ea.	.88	.42	1.30
TOTAL			20.84	12.99	33.83

G2020 240	Parking Lots, Concrete	COST PER S.Y.		
		MAT.	INST.	TOTAL
1500	4″ concrete slab, 3000 psi on 3″ crushed stone	18.25	12.95	31.20
1550	6″ crushed stone	21	13	34
1600	9″ crushed stone	23.50	13.10	36.60
1700	On bank run gravel, 6″ deep	20	12.30	32.30
1800	9″ deep	22	12.50	34.50
1900	On compacted subgrade only	15.65	11.45	27.10
2000	6″ concrete slab, 3000 psi on 3″ crushed stone	24.50	14.40	38.90
2100	6″ crushed stone	27	14.50	41.50
2200	9″ crushed stone	29.50	14.65	44.15
2300	On bank run gravel, 6″ deep	26.50	13.80	40.30
2400	9″ deep	28.50	14	42.50
2500	On compacted subgrade only	27	14.50	41.50
2600				
2700				
2900	Cut & patch to match existing construction, add, minimum	2%	3%	
3000	Maximum	5%	9%	
3100	Dust protection, add, minimum	1%	2%	
3200	Maximum	4%	11%	
3300	Equipment usage curtailment, add, minimum	1%	1%	
3400	Maximum	3%	10%	
3500	Material handling & storage limitation, add, minimum	1%	1%	
3600	Maximum	6%	7%	
3700	Protection of existing work, add, minimum	2%	2%	
3800	Maximum	5%	7%	
3900	Shift work requirements, add, minimum		5%	
4000	Maximum		30%	

This data illustrates and describes sidewalk systems including concrete, welded wire and broom finish. Lines within System Components give unit price and total price per square foot for this system. Prices for sidewalk systems are on Line Items G2030 105 1700 thru 1900. Both material quantities and labor costs have been adjusted for the system listed.

Factors: To adjust for job conditions other than normal working situations use Lines G2030 105 2900 thru 4000.

Example: You are to install the system and match existing construction. Go to Line G2030 105 2900 and apply these percentages to the appropriate MAT. and INST. costs.

System Components	QUANTITY	UNIT	COST PER S.F.		
			MAT.	INST.	TOTAL
4″ thick concrete sidewalk with welded wire fabric					
3000 psi air entrained concrete, broom finish.					
Gravel fill, 4″ deep	.012	C.Y.	.27	.10	.37
Compact fill	.012	C.Y.		.03	.03
Hand grade	1.000	S.F.		2.23	2.23
Edge form	.250	L.F.	.09	1	1.09
Welded wire fabric	.011	S.F.	.16	.43	.59
Concrete, 3000 psi air entrained	.012	C.Y.	1.42		1.42
Place concrete	.012	C.Y.		.35	.35
Broom finish	1.000	S.F.		.89	.89
TOTAL			1.94	5.03	6.97

G2030 105	Sidewalks	COST PER S.F.		
		MAT.	INST.	TOTAL
1700	Asphalt (bituminous), 2″ thick	1.15	2.87	4.02
1750	4″ thick	1.94	5.05	6.99
1800	Brick, on sand, bed, 4.5 brick per S.F.	3.14	10.95	14.09
1900	Flagstone, slate, 1″ thick, rectangular	15	11.90	26.90
2000				
2100				
2200				
2300				
2400				
2500				
2600				
2700				
2900	Cut & patch to match existing construction, add, minimum	2%	3%	
3000	Maximum	5%	9%	
3100	Dust protection, add, minimum	1%	2%	
3200	Maximum	4%	11%	
3300	Equipment usage curtailment, add, minimum	1%	1%	
3400	Maximum	3%	10%	
3500	Material handling & storage limitation, add, minimum	1%	1%	
3600	Maximum	6%	7%	
3700	Protection of existing work, add, minimum	2%	2%	
3800	Maximum	5%	7%	
3900	Shift work requirements, add, minimum		5%	
4000	Maximum		30%	

G2050 Landscaping

This data describes landscaping—lawn establishment systems including loam, lime, fertilizer, side and top mulching. Lines within system components give the unit price and total price per square yard for this system. Prices for landscaping—lawn establishment systems are on Line Items G2050 420 1000 thru 1200. Both material quantities and labor costs have been adjusted for the system listed.

Factors: To adjust for job conditions other than normal working situations use Lines G2050 420 2900 thru 4000.

Example: You are to install the system and provide dust protection. Go to Line G2050 420 3200 and apply these percentages to the appropriate MAT. and INST. costs.

System Components	QUANTITY	UNIT	COST PER S.Y. MAT.	INST.	TOTAL
Establishing lawns with loam, lime, fertilizer, seed and top mulching					
On rough graded areas.					
Furnish and place loam 4" deep	.110	C.Y.	3.71	1.01	4.72
Fine grade, lime, fertilize and seed	1.000	S.Y.	.50	3.53	4.03
Hay mulch, 1 bale/M.S.F.	1.000	S.Y.	.23	.61	.84
Rolling with hand roller	1.000	S.Y.		1.10	1.10
TOTAL			4.44	6.25	10.69

G2050 420	Landscaping - Lawn Establishment	COST PER S.Y. MAT.	INST.	TOTAL
1000	Establishing lawns with loam, lime, fertilizer, seed and hay mulch	4.44	6.25	10.69
1100	Above system with jute mesh in place of hay mulch	5.30	6.40	11.70
1200	Above system with sod in place of seed	6.25	2.19	8.44
1300				
1400				
1500				
1600				
1700				
1800				
1900				
2000				
2100				
2200				
2300				
2400				
2500				
2600				
2700				
2900	Cut & patch to match existing construction, add, minimum	2%	3%	
3000	Maximum	5%	9%	
3100	Dust protection, add, minimum	1%	2%	
3200	Maximum	4%	11%	
3300	Equipment usage curtailment, add, minimum	1%	1%	
3400	Maximum	3%	10%	
3500	Material handling & storage limitation, add, minimum	1%	1%	
3600	Maximum	6%	7%	
3700	Protection of existing work, add, minimum	2%	2%	
3800	Maximum	5%	7%	
3900	Shift work requirements, add, minimum		5%	
4000	Maximum		30%	

G2050 Landscaping

G2050 510	Selective Price Sheet	COST PER UNIT		
		MAT.	INST.	TOTAL
0100	Shrubs and trees, Evergreen, in prepared beds			
0200				
0300	Arborvitae pyramidal, 4'-5'	112	101	213
0400	Globe, 12"-15"	27	15.75	42.75
0500	Cedar, blue, 8'-10'	259	169	428
0600	Hemlock, Canadian, 2-1/2'-3'	35	42	77
0700	Juniper, andora, 18"-24"	41	18.90	59.90
0800	Wiltoni, 15"-18"	29	18.90	47.90
0900	Skyrocket, 4-1/2'-5'	119	55	174
1000	Blue pfitzer, 2'-2-1/2'	42.50	34.50	77
1100	Ketleerie, 2-1/2'-3'	58	30.50	88.50
1200	Pine, black, 2-1/2'-3'	66.50	30.50	97
1300	Mugo, 18"-24"	66.50	25	91.50
1400	White, 4'-5'	57	40.50	97.50
1500	Spruce, blue, 18"-24"	75	25	100
1600	Norway, 4'-5'	93	40.50	133.50
1700	Yew, denisforma, 12"-15"	39.50	25	64.50
1800	Capitata, 18"-24"	36.50	50.50	87
1900	Hicksi, 2'-2-1/2'	87	50.50	137.50
2000				
2100	Trees, Deciduous, in prepared beds			
2200				
2300	Beech, 5'-6'	200	60.50	260.50
2400	Dogwood, 4'-5'	140	76	216
2500	Elm, 8'-10'	275	152	427
2600	Magnolia, 4'-5'	108	152	260
2700	Maple, red, 8'-10', 1-1/2" caliper	216	305	521
2800	Oak, 2-1/2"-3" caliper	345	505	850
2900	Willow, 6'-8', 1" caliper	102	152	254
3000				
3100	Shrubs, Broadleaf Evergreen, in prepared beds			
3200				
3300	Andromeda, 15"-18", container	36.50	15.75	52.25
3400	Azalea, 15"-18", container	33	15.75	48.75
3500	Barberry, 9"-12", container	19.75	11.65	31.40
3600	Boxwood, 15"-18", B & B	48	15.75	63.75
3700	Euonymus, emerald gaiety, 12"-15", container	26.50	13.15	39.65
3800	Holly, 15"-18", B & B	40.50	15.75	56.25
3900	Mount laurel, 18"-24", B & B	79	18.90	97.90
4000	Privet, 18"-24", B & B	24	11.65	35.65
4100	Rhododendron, 18"-24", container	42	31.50	73.50
4200	Rosemary, 1 gal. container	19.85	2.52	22.37
4300	Deciduous, amalanchier, 2'-3', B & B	134	26.50	160.50
4400	Azalea, 15"-18", B & B	32.50	15.75	48.25
4500	Bayberry, 2'-3', B & B	27	26.50	53.50
4600	Cotoneaster, 15"-18", B & B	29.50	18.90	48.40
4700	Dogwood, 3'-4', B & B	36	76	112
4800	Euonymus, alatus compacta, 15"-18", container	29.50	18.90	48.40
4900	Forsythia, 2'-3', container	20.50	25	45.50
5000	Honeysuckle, 3'-4', B & B	29.50	25	54.50
5100	Hydrangea, 2'-3', B & B	33.50	26.50	60
5200	Lilac, 3'-4', B & B	31.50	76	107.50
5300	Quince, 2'-3', B & B	31.50	26.50	58
5400				
5500	Ground Cover			
5600				
5700	Plants, Pachysandra, prepared beds, per hundred	92	101	193
5800	Vinca minor or English ivy, per hundred	78.50	126	204.50

G20 Site Improvements

G2050 Landscaping

G2050 510	Selective Price Sheet	COST PER UNIT		
		MAT.	INST.	TOTAL
5900	Plant bed prep. 18" dp., mach., per square foot	2.62	.45	3.07
6000	By hand, per square foot	2.62	3.20	5.82
6100	Stone chips, Georgia marble, 50# bags, per bag	3.99	2.91	6.90
6200	Onyx gemstone, per bag	18.70	5.80	24.50
6300	Quartz, per bag	18.70	5.80	24.50
6400	Pea gravel, truck load lots, per cubic yard	30	54	84
6500	Mulch, polyethylene mulch, per square yard	.45	.50	.95
6600	Wood chips, 2" deep, per square yard	1.98	2.25	4.23
6700	Peat moss, 1" deep, per square yard	2.86	.55	3.41
6800	Erosion control per square yard, Jute mesh stapled	1.09	.76	1.85
6900	Plastic netting, stapled	.52	.61	1.13
7000	Polypropylene mesh, stapled	1.82	.61	2.43
7100	Tobacco netting, stapled	.23	.61	.84
7200	Lawns per square yard, seeding incl. fine grade, limestone, fert. and seed	.50	3.53	4.03
7300	Sodding, incl. fine grade, level ground	2.41	1.14	3.55
7500	Edging, per linear foot, Redwood, untreated, 1" x 4"	1.28	2.53	3.81
7600	2" x 4"	2.49	3.84	6.33
7700	Stl. edge strips, 1/4" x 5" inc. stakes	5.50	3.84	9.34
7800	3/16" x 4"	4.40	3.84	8.24
7900	Brick edging, set on edge	2.83	8.10	10.93
8000	Set flat	1.41	2.96	4.37
8100				
8200				
8300				
8400				

For customer support on your Commercial Renovation Cost Data, call 877.742.4719.

G3020 Sanitary Sewer

This data illustrates and describes drainage and utilities – septic system including a tank, distribution box, excavation, piping, crushed stone and backfill. Lines within System Components give the unit price and total price on a cost each basis for this system. Prices for drainage and utilities – septic systems are on Line Items G3020 610 1600 thru 2000. Both material quantities and labor costs have been adjusted for the system listed.

Factors: To adjust for job conditions other than normal working situations use Lines G3020 610 3100 thru 4000.

Example: You are to install the system with a material handling limitation. Go to Line G3020 610 3500 and apply these percentages to the appropriate MAT. and INST. costs.

System Components			COST EACH		
	QUANTITY	UNIT	MAT.	INST.	TOTAL
Septic system including tank, distribution box, excavation, piping, crushed Stone and backfill for a 1000 S.F. leaching field.					
Precast concrete septic tank, 1000 gal. capacity	1.000	Ea.	1,125	268.30	1,393.30
Concrete distribution box	1.000	Ea.	90	62	152
Sch. 40 sewer pipe, PVC, 4" diameter	25.000	L.F.	278.80	754.80	1,033.60
Sch. 40 sewer pipe fittings, PVC, 4" diameter	8.000	Ea.	118.40	648	766.40
Excavation for septic tank	120.000	C.Y.		233.28	233.28
Excavation for disposal field	120.000	C.Y.		565.60	565.60
Crushed stone backfill	76.000	C.Y.	1,558	1,020.68	2,578.68
Backfill with excavated material	26.000	C.Y.		52.78	52.78
Asphalt felt paper, 15#	6.000	Sq.	35.10	102.60	137.70
TOTAL			3,205.30	3,708.04	6,913.34

G3020 610	Septic Systems	COST EACH		
		MAT.	INST.	TOTAL
1600	1000 gal. tank with 1000 S.F. field	3,200	3,725	6,925
1700	1000 gal. tank with 2000 S.F. field	4,525	5,475	10,000
1800	With leaching pits	3,300	2,175	5,475
1900	2000 gal. tank with 2000 S.F. field	5,825	5,650	11,475
2000	With leaching pits	4,675	2,775	7,450
2100				
2200				
2300				
2400				
2500				
2600				
2700				
2800				
2900				
3100	Dust protection, add, minimum	1%	2%	
3200	Maximum	4%	11%	
3300	Equipment usage curtailment, add, minimum	1%	1%	
3400	Maximum	3%	10%	
3500	Material handling & storage limitation, add, minimum	1%	1%	
3600	Maximum	6%	7%	
3700	Protection of existing work, add, minimum	2%	2%	
3800	Maximum	5%	7%	
3900	Shift work requirements, add, minimum		5%	
4000	Maximum		30%	

G3060 Fuel Distribution

Fiberglass Underground Fuel Storage Tank, Single Wall

This data illustrates and describes fiberglass tanks including hold down slab, reinforcing steel and peastone gravel. Lines within Systems Components give the material and installation price on a cost each basis for the components. Prices for fiberglass tank systems are on Line Items G3060 310 1010 thru G3060 310 1100. Material quantities and labor costs have been adjusted for the system listed.

Factors: To adjust for job conditions other than normal working situations use Lines G3060 310 2700 thru 4000.

System Components	QUANTITY	UNIT	COST EACH		
			MAT.	INST.	TOTAL
Double wall underground fiberglass storage tanks including, excavation, hold down slab, backfill, manway extension and related piping.					
Excavating trench, no sheeting or dewatering, 6'-10 ' D, 3/4 CY hyd backhoe	18.000	Ea.		150.30	150.30
Corrosion resistance wrap & coat, small diam. pipe, 1" diam., add	120.000	Ea.	126		126
Fiberglass-reinforced elbows 90 & 45 degree, 2 inch	7.000	Ea.	332.50	619.50	952
Fiberglass-reinforced threaded adapters, 2 inch	2.000	Ea.	39.50	202	241.50
Remote tank gauging system, 30', 5" pointer travel	1.000	Ea.	5,950	380	6,330
Corrosion resistance wrap & coat, small diam. pipe. 4 " diam. add	30.000	Ea.	72.60		72.60
Pea gravel	9.000	Ea.	1,080	486	1,566
Reinforcing in hold down pad, #3 to #7	120.000	Ea.	63.60	78	141.60
Tank leak detector system, 8 channel, external monitoring	1.000	Ea.	965		965
Tubbing copper, type L, 3/4" diam.	120.000	Ea.	690	1,164	1,854
Elbow 90 degree ,wrought cu x cu 3/4 " diam.	4.000	Ea.	11.12	156	167.12
Pipe, black steel welded, sch 40, 3" diam.	30.000	Ea.	414	974.40	1,388.40
Elbow 90 degree, steel welded butt joint, 3 " diam.	3.000	Ea.	96	597.30	693.30
Fiberglass-reinforced plastic pipe, 2 inch	156.000	Ea.	1,170	778.44	1,948.44
Fiberglass-reinforced plastic pipe, 3 inch	36.000	Ea.	1,044	203.40	1,247.40
Fiberglass-reinforced elbows 90 & 45 degree, 3 inch	3.000	Ea.	450	354	804
Tank leak detection system, probes, well monitoring, liquid phase detection	2.000	Ea.	1,010		1,010
Double wall fiberglass annular space	1.000	Ea.	400		400
Flange, weld neck, 150 lb, 3 " pipe size	1.000	Ea.	38.50	99.56	138.06
Vent protector/breather, 2" diam.	1.000	Ea.	33.50	23.50	57
Corrosion resistance wrap & coat, small diam. pipe, 2" diam., add	36.000	Ea.	54.72		54.72
Concrete hold down pad, 6' x 5' x 8" thick	30.000	Ea.	93.60	50.10	143.70
600 gallon capacity, underground storage tank, double wall fiberglass	1.000	Ea.	8,300	565	8,865
Tank, for hold-downs 500-4000 gal., add	1.000	Ea.	515	180	695
Foot valve, single poppet, 3/4" diam.	1.000	Ea.	78.50	42	120.50
Fuel fill box, locking inner cover, 3" diam.	1.000	Ea.	133	151	284
Fuel oil specialties, valve,ball chk, globe type fusible, 3/4" diam.	2.000	Ea.	154	76	230
TOTAL			23,315.14	7,330.50	30,645.64

G3060 310	Fiberglass Fuel Tank, Double Wall	COST EACH		
		MAT.	INST.	TOTAL
1010	Storage tank, fuel, underground, double wall fiberglass, 600 Gal.	23,300	7,325	30,625
1020	2500 Gal.	36,900	10,100	47,000
1030	4000 Gal.	43,100	12,500	55,600
1040	6000 Gal.	50,500	14,100	64,600
1050	8000 Gal.	58,000	17,000	75,000
1060	10,000 Gal.	66,000	19,600	85,600

G3060 Fuel Distribution

G3060 310	Fiberglass Fuel Tank, Double Wall	COST EACH		
		MAT.	INST.	TOTAL
1070	15,000 Gal.	91,000	22,500	113,500
1080	20,000 Gal.	114,000	27,300	141,300
1090	25,000 Gal.	145,500	31,700	177,200
1100	30,000 Gal.	166,000	33,500	199,500
2700	Cut & patch to match existing construction, add, minimum	2%	3%	
2800	Maximum	5%	9%	
2900	Dust protection, add, minimum	1%	2%	
3000	Maximum	4%	11%	
3100	Equipment usage curtailment, add, minimum	1%	1%	
3200	Maximum	3%	10%	
3300	Material handling & storage limitation, add, minimum	1%	1%	
3400	Maximum	6%	7%	
3500	Protection of existing work, add, minimum	2%	2%	
3600	Maximum	5%	7%	
3700	Shift work requirements, add, minimum		5%	
3800	Maximum		30%	
3900	Temporary shoring and bracing, add, minimum	2%	5%	
4000	Maximum	5%	12%	

G3060 Fuel Distribution

Steel Underground Fuel Storage Tank, Double Wall

System Components	QUANTITY	UNIT	COST EACH		
			MAT.	INST.	TOTAL
Double wall underground steel storage tank including, excavation, hold down slab, backfill, manway extension and related piping.					
Fiberglass-reinforced plastic pipe, 2 inch	156.000	Ea.	1,170	778.44	1,948.44
Fiberglass-reinforced plastic pipe, 3 inch	36.000	Ea.	1,044	203.40	1,247.40
Fiberglass-reinforced elbows 90 & 45 degree, 3 inch	3.000	Ea.	450	354	804
Tank leak detection system, probes, well monitoring, liquid phase detection	2.000	Ea.	1,010		1,010
Flange, weld neck, 150 lb, 3" pipe size	1.000	Ea.	38.50	99.56	138.06
Vent protector/breather, 2" diam.	1.000	Ea.	33.50	23.50	57
Concrete hold down pad, 6' x 5' x 8" thick	30.000	Ea.	93.60	50.10	143.70
Tanks, for hold-downs, 500-2000 gal, add	1.000	Ea.	201	180	381
Double wall steel tank annular space	1.000	Ea.	400		400
Foot valve, single poppet, 3/4" diam.	1.000	Ea.	78.50	42	120.50
Fuel fill box, locking inner cover, 3" diam.	1.000	Ea.	133	151	284
Fuel oil specialties, valve, ball chk, globe type fusible, 3/4" diam.	2.000	Ea.	154	76	230
Pea gravel	9.000	Ea.	1,080	486	1,566
Reinforcing in hold down pad, #3 to #7	120.000	Ea.	63.60	78	141.60
Tank, for manways, add	1.000	Ea.	2,000		2,000
Tank leak detector system, 8 channel, external monitoring	1.000	Ea.	965		965
Tubbing copper, type L, 3/4" diam.	120.000	Ea.	690	1,164	1,854
Elbow 90 degree, wrought cu x cu 3/4" diam.	8.000	Ea.	22.24	312	334.24
Pipe, black steel welded, sch 40, 3" diam.	30.000	Ea.	414	974.40	1,388.40
Elbow 90 degree, steel welded butt joint, 3" diam.	3.000	Ea.	96	597.30	693.30
Excavating trench, no sheeting or dewatering, 6'-10' D, 3/4 CY hyd backhoe	18.000	Ea.		150.30	150.30
Fiberglass-reinforced elbows 90 & 45 degree, 2 inch	7.000	Ea.	332.50	619.50	952
Fiberglass-reinforced threaded adapters, 2 inch	2.000	Ea.	39.50	202	241.50
500 gallon capacity, underground storage tank, double wall steel	1.000	Ea.	6,150	570	6,720
Remote tank gauging system, 30', 5" pointer travel	1.000	Ea.	5,950	380	6,330
TOTAL			22,608.94	7,491.50	30,100.44

G3060 320	Steel Fuel Tank, Double Wall	COST EACH		
		MAT.	INST.	TOTAL
1010	Storage tank, fuel, underground, double wall steel, 500 Gal.	22,600	7,475	30,075
1020	2000 Gal.	30,200	9,775	39,975
1030	4000 Gal.	41,000	12,600	53,600
1040	6000 Gal.	55,500	14,000	69,500
1050	8000 Gal.	59,500	17,300	76,800
1060	10,000 Gal.	66,500	19,800	86,300

G3060 Fuel Distribution

G3060 320	Steel Fuel Tank, Double Wall	COST EACH		
		MAT.	INST.	TOTAL
1070	15,000 Gal.	75,500	23,100	98,600
1080	20,000 Gal.	87,500	27,700	115,200
1090	25,000 Gal.	136,000	31,600	167,600
1100	30,000 Gal.	155,000	33,700	188,700
1110	40,000 Gal.	195,000	42,300	237,300
1120	50,000 Gal.	236,000	50,500	286,500

G3060 325	Steel Tank, Above Ground, Single Wall	COST EACH		
		MAT.	INST.	TOTAL
1010	Storage tank, fuel, above ground, single wall steel, 550 Gal.	14,000	5,225	19,225
1020	2000 Gal.	22,000	5,350	27,350
1030	5000 Gal.	33,700	8,175	41,875
1040	10,000 Gal.	51,500	8,725	60,225
1050	15,000 Gal.	63,000	8,975	71,975
1060	20,000 Gal.	78,500	9,275	87,775
1070	25,000 Gal.	89,500	9,500	99,000
1080	30,000 Gal.	105,500	9,900	115,400

G3060 330	Steel Tank, Above Ground, Double Wall	COST EACH		
		MAT.	INST.	TOTAL
1010	Storage tank, fuel, above ground, double wall steel, 500 Gal.	14,000	5,300	19,300
1020	2000 Gal.	22,100	5,425	27,525
1030	4000 Gal.	31,100	5,525	36,625
1040	6000 Gal.	36,000	8,475	44,475
1050	8000 Gal.	43,000	8,725	51,725
1060	10,000 Gal.	46,500	8,875	55,375
1070	15,000 Gal.	64,000	9,200	73,200
1080	20,000 Gal.	71,000	9,500	80,500
1090	25,000 Gal.	83,500	9,775	93,275
1100	30,000 Gal.	90,000	10,200	100,200
2700	Cut & patch to match existing construction, add, minimum	2%	3%	
2800	Maximum	5%	9%	
2900	Dust protection, add, minimum	1%	2%	
3000	Maximum	4%	11%	
3100	Equipment usage curtailment, add, minimum	1%	1%	
3200	Maximum	3%	10%	
3300	Material handling & storage limitation, add, minimum	1%	1%	
3400	Maximum	6%	7%	
3500	Protection of existing work, add, minimum	2%	2%	
3600	Maximum	5%	7%	
3700	Shift work requirements, add, minimum		5%	
3800	Maximum		30%	
3900	Temporary shoring and bracing, add, minimum	2%	5%	
4000	Maximum	5%	12%	

G4020 Site Lighting

This data illustrates and describes light poles for parking or walkway area lighting. Included are aluminum or steel light poles, single or multiple fixture bracket arms, excavation, concrete footing, backfill and compaction. Lines within system components give the unit price and total price on a cost per each basis. Prices for systems are shown on lines G4020 210 0200 through 1440. Both material quantities and labor costs have been adjusted for the system listed.

Factors: To adjust for job conditions other than normal working situations use lines G4020 210 2700 through 3600.

Example: You are to install the system where you must be careful not to damage existing walks or parking areas. Go to line G4020 210 330 and apply these percentages to the appropriate MAT. and INST. costs.

System Components	QUANTITY	UNIT	COST EACH		
			MAT.	INST.	TOTAL
Light poles, aluminum, 20' high, 1 arm bracket					
Aluminum light pole, 20', no concrete base	1.000	Ea.	1,050	638	1,688
Bracket arm for Aluminum light pole	1.000	Ea.	135	85	220
Excavation by hand, pits to 6' deep, heavy soil or clay	2.368	C.Y.		293.63	293.63
Footing, concrete incl forms, reinforcing, spread, under 1 C.Y.	.465	C.Y.	87.42	140.08	227.50
Backfill by hand	1.903	C.Y.		85.64	85.64
Compaction vibrating plate	1.903	C.Y.		11.55	11.55
TOTAL			1,272.42	1,253.90	2,526.32

G4020 210	Light Pole (Installed)	COST EACH		
		MAT.	INST.	TOTAL
0200	Light pole, aluminum, 20' high, 1 arm bracket	1,275	1,250	2,525
0240	2 arm brackets	1,400	1,250	2,650
0280	3 arm brackets	1,550	1,300	2,850
0320	4 arm brackets	1,675	1,300	2,975
0360	30' high, 1 arm bracket	2,225	1,575	3,800
0400	2 arm brackets	2,350	1,575	3,925
0440	3 arm brackets	2,500	1,625	4,125
0480	4 arm brackets	2,625	1,625	4,250
0680	40' high, 1 arm bracket	2,725	2,125	4,850
0720	2 arm brackets	2,850	2,125	4,975
0760	3 arm brackets	2,975	2,150	5,125
0800	4 arm brackets	3,125	2,150	5,275
0840	Steel, 20' high, 1 arm bracket	1,425	1,325	2,750
0880	2 arm brackets	1,475	1,325	2,800
0920	3 arm brackets	1,450	1,375	2,825
0960	4 arm brackets	1,550	1,375	2,925
1000	30' high, 1 arm bracket	1,750	1,700	3,450
1040	2 arm brackets	1,800	1,700	3,500
1080	3 arm brackets	1,800	1,725	3,525
1120	4 arm brackets	1,900	1,725	3,625
1320	40' high, 1 arm bracket	2,300	2,300	4,600
1360	2 arm brackets	2,350	2,300	4,650
1400	3 arm brackets	2,350	2,325	4,675
1440	4 arm brackets	2,450	2,325	4,775
2700	Cut & patch to match existing construction, add, minimum	2%	3%	
2800	Maximum	5%	9%	
2900	Equipment usage curtailment, add, minimum	1%	1%	
3000	Maximum	3%	10%	

G4020 210	Light Pole (Installed)	COST EACH		
		MAT.	INST.	TOTAL
3100	Material handling & storage limitation, add, minimum	1%	1%	
3200	Maximum	6%	7%	
3300	Protection of existing work, add, minimum	2%	2%	
3400	Maximum	5%	7%	
3500	Shift work requirements, add, minimum		5%	
3600	Maximum		30%	

Reference Section

All the reference information is in one section, making it easy to find what you need to know . . . and easy to use the data set on a daily basis. This section is visually identified by a vertical black bar on the page edges.

In this Reference Section, we've included Equipment Rental Costs, a listing of rental and operating costs; Crew Listings, a full listing of all crews and equipment, and their costs; Historical Cost Indexes for cost comparisons over time; City Cost Indexes and Location Factors for adjusting costs to the region you are in; Reference Tables, where you will find explanations, estimating information and procedures, or technical data; Change Orders, information on pricing changes to contract documents; and an explanation of all the Abbreviations in the data set.

Table of Contents

Estimating Tips

- This section contains the average costs to rent and operate hundreds of pieces of construction equipment. This is useful information when estimating the time and material requirements of any particular operation in order to establish a unit or total cost. Bare equipment costs shown on a unit cost line include not only rental, but also operating costs for equipment under normal use.

Rental Costs

- Equipment rental rates are obtained from the following industry sources throughout North America: contractors, suppliers, dealers, manufacturers, and distributors.

- Rental rates vary throughout the country, with larger cities generally having lower rates. Lease plans for new equipment are available for periods in excess of six months, with a percentage of payments applying toward purchase.

- Monthly rental rates vary from 2% to 5% of the purchase price of the equipment depending on the anticipated life of the equipment and its wearing parts.

- Weekly rental rates are about 1/3 the monthly rates, and daily rental rates are about 1/3 the weekly rate.

- Rental rates can also be treated as reimbursement costs for contractor-owned equipment.

Owned equipment costs include depreciation, loan payments, interest, taxes, insurance, storage, and major repairs.

Operating Costs

- The operating costs include parts and labor for routine servicing, such as repair and replacement of pumps, filters and worn lines. Normal operating expendables, such as fuel, lubricants, tires and electricity (where applicable), are also included.

- Extraordinary operating expendables with highly variable wear patterns, such as diamond bits and blades, are excluded. These costs can be found as material costs in the Unit Price section.

- The hourly operating costs listed do not include the operator's wages.

Equipment Cost/Day

- Any power equipment required by a crew is shown in the Crew Listings with a daily cost.

- This daily cost of equipment needed by a crew includes both the rental cost and the operating cost and is based on dividing the weekly rental rate by 5 (number of working days in the week), and then adding the hourly operating cost times 8 (the number of hours in a day). This "Equipment Cost/ Day" is shown in the far right column of the Equipment Rental section.

- If equipment is needed for only one or two days, it is best to develop your own cost by including components for daily rent and hourly operating cost. This is important when the listed Crew for a task does not contain the equipment needed, such as a crane for lifting mechanical heating/cooling equipment up onto a roof.

- If the quantity of work is less than the crew's Daily Output shown for a Unit Price line item that includes a bare unit equipment cost, it is recommended to estimate one day's rental cost and operating cost for equipment shown in the Crew Listing for that line item.

Mobilization/ Demobilization

- The cost to move construction equipment from an equipment yard or rental company to the job site and back again is not included in equipment rental costs listed in the Reference Section, nor in the bare equipment cost of any unit price line item, nor in any equipment costs shown in the Crew listings.

- Mobilization (to the site) and demobilization (from the site) costs can be found in the Unit Price Section.

- If a piece of equipment is already at the job site, it is not appropriate to utilize mobilization/demobilization costs again in an estimate. ■

01 54 33 | Equipment Rental

			UNIT	HOURLY OPER. COST	RENT PER DAY	RENT PER WEEK	RENT PER MONTH	EQUIPMENT COST/DAY	
10	0010	**CONCRETE EQUIPMENT RENTAL** without operators	R015433 -10						10
	0200	Bucket, concrete lightweight, 1/2 C.Y.	Ea.	.80	23.50	70	210	20.40	
	0300	1 C.Y.		.85	28	84	252	23.60	
	0400	1-1/2 C.Y.		1.10	38.50	115	345	31.80	
	0500	2 C.Y.		1.20	45	135	405	36.60	
	0580	8 C.Y.		6.00	257	770	2,300	202	
	0600	Cart, concrete, self-propelled, operator walking, 10 C.F.		3.25	56.50	170	510	60	
	0700	Operator riding, 18 C.F.		5.35	95	285	855	99.80	
	0800	Conveyer for concrete, portable, gas, 16" wide, 26' long		12.60	123	370	1,100	174.80	
	0900	46' long		13.00	148	445	1,325	193	
	1000	56' long		13.15	157	470	1,400	199.20	
	1100	Core drill, electric, 2-1/2 H.P., 1" to 8" bit diameter		2.15	78.50	235	705	64.20	
	1150	11 H.P., 8" to 18" cores		6.00	115	345	1,025	117	
	1200	Finisher, concrete floor, gas, riding trowel, 96" wide		12.90	145	435	1,300	190.20	
	1300	Gas, walk-behind, 3 blade, 36" trowel		2.15	20	60	180	29.20	
	1400	4 blade, 48" trowel		4.25	27.50	83	249	50.60	
	1500	Float, hand-operated (Bull float) 48" wide		.08	13.35	40	120	8.65	
	1570	Curb builder, 14 H.P., gas, single screw		14.40	248	745	2,225	264.20	
	1590	Double screw		15.10	292	875	2,625	295.80	
	1600	Floor grinder, concrete and terrazzo, electric, 22" path		2.70	167	500	1,500	121.60	
	1700	Edger, concrete, electric, 7" path		1.04	51.50	155	465	39.30	
	1750	Vacuum pick-up system for floor grinders, wet/dry		1.49	81.50	245	735	60.90	
	1800	Mixer, powered, mortar and concrete, gas, 6 C.F., 18 H.P.		8.65	118	355	1,075	140.20	
	1900	10 C.F., 25 H.P.		10.85	143	430	1,300	172.80	
	2000	16 C.F.		11.15	165	495	1,475	188.20	
	2100	Concrete, stationary, tilt drum, 2 C.Y.		6.80	232	695	2,075	193.40	
	2120	Pump, concrete, truck mounted 4" line 80' boom		24.70	885	2,650	7,950	727.60	
	2140	5" line, 110' boom		32.05	1,175	3,510	10,500	958.40	
	2160	Mud jack, 50 C.F. per hr.		7.25	128	385	1,150	135	
	2180	225 C.F. per hr.		9.55	147	440	1,325	164.40	
	2190	Shotcrete pump rig, 12 C.Y./hr.		15.00	223	670	2,000	254	
	2200	35 C.Y./hr.		17.50	240	720	2,150	284	
	2600	Saw, concrete, manual, gas, 18 H.P.		6.80	45	135	405	81.40	
	2650	Self-propelled, gas, 30 H.P.		13.45	102	305	915	168.60	
	2675	V-groove crack chaser, manual, gas, 6 H.P.		2.35	17.35	52	156	29.20	
	2700	Vibrators, concrete, electric, 60 cycle, 2 H.P.		.41	8.65	26	78	8.50	
	2800	3 H.P.		.60	11.65	35	105	11.80	
	2900	Gas engine, 5 H.P.		2.05	16.35	49	147	26.20	
	3000	8 H.P.		2.80	15.35	46	138	31.60	
	3050	Vibrating screed, gas engine, 8 H.P.		2.58	71.50	215	645	63.65	
	3120	Concrete transit mixer, 6 x 4, 250 H.P., 8 C.Y., rear discharge		61.15	570	1,715	5,150	832.20	
	3200	Front discharge		72.05	700	2,100	6,300	996.40	
	3300	6 x 6, 285 H.P., 12 C.Y., rear discharge		71.10	660	1,980	5,950	964.80	
	3400	Front discharge		74.30	710	2,125	6,375	1,019	
20	0010	**EARTHWORK EQUIPMENT RENTAL** without operators	R015433 -10						20
	0040	Aggregate spreader, push type 8' to 12' wide	Ea.	3.05	25.50	76	228	39.60	
	0045	Tailgate type, 8' wide		2.90	32.50	98	294	42.80	
	0055	Earth auger, truck-mounted, for fence & sign posts, utility poles		19.10	445	1,340	4,025	420.80	
	0060	For borings and monitoring wells		46.50	690	2,070	6,200	786	
	0070	Portable, trailer mounted		3.25	33	99	297	45.80	
	0075	Truck-mounted, for caissons, water wells		98.70	2,950	8,815	26,400	2,553	
	0080	Horizontal boring machine, 12" to 36" diameter, 45 H.P.		24.40	195	585	1,750	312.20	
	0090	12" to 48" diameter, 65 H.P.		34.30	340	1,015	3,050	477.40	
	0095	Auger, for fence posts, gas engine, hand held		.60	6	18	54	8.40	
	0100	Excavator, diesel hydraulic, crawler mounted, 1/2 C.Y. cap.		24.45	415	1,250	3,750	445.60	
	0120	5/8 C.Y. capacity		32.30	550	1,655	4,975	589.40	
	0140	3/4 C.Y. capacity		35.25	620	1,860	5,575	654	
	0150	1 C.Y. capacity		49.35	695	2,090	6,275	812.80	

01 54 33 | Equipment Rental

		UNIT	HOURLY OPER. COST	RENT PER DAY	RENT PER WEEK	RENT PER MONTH	EQUIPMENT COST/DAY	
20	0200	1-1/2 C.Y. capacity	Ea.	59.40	925	2,770	8,300	1,029	**20**
	0300	2 C.Y. capacity		66.90	1,075	3,205	9,625	1,176	
	0320	2-1/2 C.Y. capacity		102.00	1,275	3,830	11,500	1,582	
	0325	3-1/2 C.Y. capacity		143.70	2,150	6,460	19,400	2,442	
	0330	4-1/2 C.Y. capacity		172.45	2,650	7,970	23,900	2,974	
	0335	6 C.Y. capacity		218.85	2,925	8,750	26,300	3,501	
	0340	7 C.Y. capacity		221.10	3,050	9,140	27,400	3,597	
	0342	Excavator attachments, bucket thumbs		3.15	245	735	2,200	172.20	
	0345	Grapples		2.70	193	580	1,750	137.60	
	0346	Hydraulic hammer for boom mounting, 4000 ft lb.		12.15	350	1,045	3,125	306.20	
	0347	5000 ft lb.		14.10	425	1,275	3,825	367.80	
	0348	8000 ft lb.		20.90	625	1,870	5,600	541.20	
	0349	12,000 ft lb.		22.90	750	2,245	6,725	632.20	
	0350	Gradall type, truck mounted, 3 ton @ 15' radius, 5/8 C.Y.		48.75	820	2,460	7,375	882	
	0370	1 C.Y. capacity		53.85	940	2,825	8,475	995.80	
	0400	Backhoe-loader, 40 to 45 H.P., 5/8 C.Y. capacity		14.85	242	725	2,175	263.80	
	0450	45 H.P. to 60 H.P., 3/4 C.Y. capacity		23.30	298	895	2,675	365.40	
	0460	80 H.P., 1-1/4 C.Y. capacity		25.90	315	945	2,825	396.20	
	0470	112 H.P., 1-1/2 C.Y. capacity		40.25	640	1,915	5,750	705	
	0482	Backhoe-loader attachment, compactor, 20,000 lb.		5.70	142	425	1,275	130.60	
	0485	Hydraulic hammer, 750 ft lb.		3.20	95	285	855	82.60	
	0486	Hydraulic hammer, 1200 ft lb.		6.20	217	650	1,950	179.60	
	0500	Brush chipper, gas engine, 6" cutter head, 35 H.P.		11.10	103	310	930	150.80	
	0550	Diesel engine, 12" cutter head, 130 H.P.		27.45	282	845	2,525	388.60	
	0600	15" cutter head, 165 H.P.		33.00	340	1,020	3,050	468	
	0750	Bucket, clamshell, general purpose, 3/8 C.Y.		1.25	38.50	115	345	33	
	0800	1/2 C.Y.		1.35	45	135	405	37.80	
	0850	3/4 C.Y.		1.50	55	165	495	45	
	0900	1 C.Y.		1.55	58.50	175	525	47.40	
	0950	1-1/2 C.Y.		2.50	80	240	720	68	
	1000	2 C.Y.		2.65	90	270	810	75.20	
	1010	Bucket, dragline, medium duty, 1/2 C.Y.		.70	23.50	70	210	19.60	
	1020	3/4 C.Y.		.75	24.50	73	219	20.60	
	1030	1 C.Y.		.75	26	78	234	21.60	
	1040	1-1/2 C.Y.		1.20	40	120	360	33.60	
	1050	2 C.Y.		1.25	45	135	405	37	
	1070	3 C.Y.		1.95	61.50	185	555	52.60	
	1200	Compactor, manually guided 2-drum vibratory smooth roller, 7.5 H.P.		7.35	210	630	1,900	184.80	
	1250	Rammer/tamper, gas, 8"		2.75	46.50	140	420	50	
	1260	15"		3.10	53.50	160	480	56.80	
	1300	Vibratory plate, gas, 18" plate, 3000 lb. blow		2.70	24.50	73	219	36.20	
	1350	21" plate, 5000 lb. blow		3.35	33	99	297	46.60	
	1370	Curb builder/extruder, 14 H.P., gas, single screw		14.40	248	745	2,225	264.20	
	1390	Double screw		15.10	292	875	2,625	295.80	
	1500	Disc harrow attachment, for tractor		.45	75	225	675	48.60	
	1810	Feller buncher, shearing & accumulating trees, 100 H.P.		47.95	755	2,270	6,800	837.60	
	1860	Grader, self-propelled, 25,000 lb.		39.40	665	1,990	5,975	713.20	
	1910	30,000 lb.		43.05	615	1,850	5,550	714.40	
	1920	40,000 lb.		64.75	1,125	3,385	10,200	1,195	
	1930	55,000 lb.		84.00	1,775	5,345	16,000	1,741	
	1950	Hammer, pavement breaker, self-propelled, diesel, 1000 to 1250 lb.		29.55	350	1,055	3,175	447.40	
	2000	1300 to 1500 lb.		44.33	705	2,110	6,325	776.65	
	2050	Pile driving hammer, steam or air, 4150 ft lb. @ 225 bpm		10.25	480	1,435	4,300	369	
	2100	8750 ft lb. @ 145 bpm		12.25	665	2,000	6,000	498	
	2150	15,000 ft lb. @ 60 bpm		13.80	800	2,405	7,225	591.40	
	2200	24,450 ft lb. @ 111 bpm		14.85	890	2,670	8,000	652.80	
	2250	Leads, 60' high for pile driving hammers up to 20,000 ft lb.		3.25	80	240	720	74	
	2300	90' high for hammers over 20,000 ft lb.		4.95	141	424	1,275	124.40	

01 54 33 | Equipment Rental

		UNIT	HOURLY OPER. COST	RENT PER DAY	RENT PER WEEK	RENT PER MONTH	EQUIPMENT COST/DAY	
20	2350	Diesel type hammer, 22,400 ft lb.	Ea.	18.00	390	1,170	3,500	378	20
	2400	41,300 ft lb.		27.45	505	1,515	4,550	522.60	
	2450	141,000 ft lb.		47.40	880	2,640	7,925	907.20	
	2500	Vib. elec. hammer/extractor, 200 kW diesel generator, 34 H.P.		54.65	645	1,930	5,800	823.20	
	2550	80 H.P.		100.35	945	2,830	8,500	1,369	
	2600	150 H.P.		190.65	1,800	5,435	16,300	2,612	
	2800	Log chipper, up to 22" diameter, 600 H.P.		63.25	640	1,920	5,750	890	
	2850	Logger, for skidding & stacking logs, 150 H.P.		52.35	830	2,490	7,475	916.80	
	2860	Mulcher, diesel powered, trailer mounted		23.95	213	640	1,925	319.60	
	2900	Rake, spring tooth, with tractor		18.08	345	1,042	3,125	353.05	
	3000	Roller, vibratory, tandem, smooth drum, 20 H.P.		8.95	152	455	1,375	162.60	
	3050	35 H.P.		11.05	252	755	2,275	239.40	
	3100	Towed type vibratory compactor, smooth drum, 50 H.P.		25.70	325	980	2,950	401.60	
	3150	Sheepsfoot, 50 H.P.		27.05	350	1,050	3,150	426.40	
	3170	Landfill compactor, 220 H.P.		88.90	1,550	4,640	13,900	1,639	
	3200	Pneumatic tire roller, 80 H.P.		16.35	360	1,085	3,250	347.80	
	3250	120 H.P.		24.55	615	1,840	5,525	564.40	
	3300	Sheepsfoot vibratory roller, 240 H.P.		67.15	1,125	3,355	10,100	1,208	
	3320	340 H.P.		97.60	1,650	4,960	14,900	1,773	
	3350	Smooth drum vibratory roller, 75 H.P.		24.75	615	1,840	5,525	566	
	3400	125 H.P.		32.10	725	2,175	6,525	691.80	
	3410	Rotary mower, brush, 60", with tractor		22.30	320	955	2,875	369.40	
	3420	Rototiller, walk-behind, gas, 5 H.P.		1.76	51.50	155	465	45.10	
	3422	8 H.P.		2.75	80	240	720	70	
	3440	Scrapers, towed type, 7 C.Y. capacity		5.65	113	340	1,025	113.20	
	3450	10 C.Y. capacity		6.50	155	465	1,400	145	
	3500	15 C.Y. capacity		6.95	180	540	1,625	163.60	
	3525	Self-propelled, single engine, 14 C.Y. capacity		130.55	1,875	5,640	16,900	2,172	
	3550	Dual engine, 21 C.Y. capacity		173.20	2,225	6,645	19,900	2,715	
	3600	31 C.Y. capacity		230.80	3,150	9,445	28,300	3,735	
	3640	44 C.Y. capacity		286.85	4,075	12,220	36,700	4,739	
	3650	Elevating type, single engine, 11 C.Y. capacity		71.80	1,000	2,995	8,975	1,173	
	3700	22 C.Y. capacity		138.90	2,400	7,185	21,600	2,548	
	3710	Screening plant 110 H.P. w/5' x 10' screen		19.62	375	1,125	3,375	381.95	
	3720	5' x 16' screen		24.98	485	1,450	4,350	489.85	
	3850	Shovel, crawler-mounted, front-loading, 7 C.Y. capacity		257.95	3,575	10,740	32,200	4,212	
	3855	12 C.Y. capacity		419.50	4,950	14,860	44,600	6,328	
	3860	Shovel/backhoe bucket, 1/2 C.Y.		2.40	66.50	200	600	59.20	
	3870	3/4 C.Y.		2.45	73.50	220	660	63.60	
	3880	1 C.Y.		2.55	83.50	250	750	70.40	
	3890	1-1/2 C.Y.		2.70	96.50	290	870	79.60	
	3910	3 C.Y.		3.05	130	390	1,175	102.40	
	3950	Stump chipper, 18" deep, 30 H.P.		6.83	198	593	1,775	173.25	
	4110	Dozer, crawler, torque converter, diesel 80 H.P.		28.80	400	1,205	3,625	471.40	
	4150	105 H.P.		35.30	535	1,605	4,825	603.40	
	4200	140 H.P.		50.95	790	2,365	7,100	880.60	
	4260	200 H.P.		76.95	1,300	3,905	11,700	1,397	
	4310	300 H.P.		100.25	1,800	5,425	16,300	1,887	
	4360	410 H.P.		132.10	2,225	6,675	20,000	2,392	
	4370	500 H.P.		170.65	3,275	9,820	29,500	3,329	
	4380	700 H.P.		280.40	5,175	15,520	46,600	5,347	
	4400	Loader, crawler, torque conv., diesel, 1-1/2 C.Y., 80 H.P.		31.25	460	1,375	4,125	525	
	4450	1-1/2 to 1-3/4 C.Y., 95 H.P.		34.40	600	1,795	5,375	634.20	
	4510	1-3/4 to 2-1/4 C.Y., 130 H.P.		53.50	895	2,680	8,050	964	
	4530	2-1/2 to 3-1/4 C.Y., 190 H.P.		65.50	1,125	3,365	10,100	1,197	
	4560	3-1/2 to 5 C.Y., 275 H.P.		87.45	1,450	4,330	13,000	1,566	
	4610	Front end loader, 4WD, articulated frame, diesel, 1 to 1-1/4 C.Y., 70 H.P.		19.75	235	705	2,125	299	
	4620	1-1/2 to 1-3/4 C.Y., 95 H.P.		24.70	300	905	2,725	378.60	

01 54 33 | Equipment Rental

		UNIT	HOURLY OPER. COST	RENT PER DAY	RENT PER WEEK	RENT PER MONTH	EQUIPMENT COST/DAY	
20								**20**
4650	1-3/4 to 2 C.Y., 130 H.P.	Ea.	29.25	365	1,095	3,275	453	
4710	2-1/2 to 3-1/2 C.Y., 145 H.P.		33.25	415	1,250	3,750	516	
4730	3 to 4-1/2 C.Y., 185 H.P.		42.60	535	1,605	4,825	661.80	
4760	5-1/4 to 5-3/4 C.Y., 270 H.P.		67.20	925	2,770	8,300	1,092	
4810	7 to 9 C.Y., 475 H.P.		113.75	1,750	5,270	15,800	1,964	
4870	9 - 11 C.Y., 620 H.P.		160.25	3,000	9,015	27,000	3,085	
4880	Skid steer loader, wheeled, 10 C.F., 30 H.P. gas		10.40	150	450	1,350	173.20	
4890	1 C.Y., 78 H.P., diesel		20.20	257	770	2,300	315.60	
4892	Skid-steer attachment, auger		.47	77.50	233	700	50.35	
4893	Backhoe		.66	110	329	985	71.10	
4894	Broom		.70	117	351	1,050	75.80	
4895	Forks		.22	37	111	335	23.95	
4896	Grapple		.53	88	264	790	57.05	
4897	Concrete hammer		1.00	167	502	1,500	108.40	
4898	Tree spade		.93	156	467	1,400	100.85	
4899	Trencher		.59	98.50	295	885	63.70	
4900	Trencher, chain, boom type, gas, operator walking, 12 H.P.		5.00	46.50	140	420	68	
4910	Operator riding, 40 H.P.		20.05	290	870	2,600	334.40	
5000	Wheel type, diesel, 4' deep, 12" wide		85.15	835	2,510	7,525	1,183	
5100	6' deep, 20" wide		91.45	1,925	5,795	17,400	1,891	
5150	Chain type, diesel, 5' deep, 8" wide		32.25	470	1,405	4,225	539	
5200	Diesel, 8' deep, 16" wide		133.50	2,975	8,960	26,900	2,860	
5202	Rock trencher, wheel type, 6" wide x 18" deep		33.65	575	1,730	5,200	615.20	
5206	Chain type, 18" wide x 7' deep		110.75	2,850	8,525	25,600	2,591	
5210	Tree spade, self-propelled		12.30	267	800	2,400	258.40	
5250	Truck, dump, 2-axle, 12 ton, 8 C.Y. payload, 220 H.P.		34.30	238	715	2,150	417.40	
5300	Three axle dump, 16 ton, 12 C.Y. payload, 400 H.P.		60.85	340	1,020	3,050	690.80	
5310	Four axle dump, 25 ton, 18 C.Y. payload, 450 H.P.		71.55	495	1,485	4,450	869.40	
5350	Dump trailer only, rear dump, 16-1/2 C.Y.		5.35	138	415	1,250	125.80	
5400	20 C.Y.		5.80	157	470	1,400	140.40	
5450	Flatbed, single axle, 1-1/2 ton rating		25.90	68.50	205	615	248.20	
5500	3 ton rating		31.05	98.50	295	885	307.40	
5550	Off highway rear dump, 25 ton capacity		73.60	1,300	3,905	11,700	1,370	
5600	35 ton capacity		82.20	1,450	4,355	13,100	1,529	
5610	50 ton capacity		102.35	1,725	5,210	15,600	1,861	
5620	65 ton capacity		105.35	1,775	5,360	16,100	1,915	
5630	100 ton capacity		150.95	2,900	8,710	26,100	2,950	
6000	Vibratory plow, 25 H.P., walking		8.50	61.50	185	555	105	
40								**40**
0010	**GENERAL EQUIPMENT RENTAL** without operators — R015433 -10							
0150	Aerial lift, scissor type, to 15' high, 1000 lb. cap., electric	Ea.	3.15	51.50	155	465	56.20	
0160	To 25' high, 2000 lb. capacity		3.60	68.50	205	615	69.80	
0170	Telescoping boom to 40' high, 500 lb. capacity, diesel		13.50	335	1,005	3,025	309	
0180	To 45' high, 500 lb. capacity		14.75	365	1,090	3,275	336	
0190	To 60' high, 600 lb. capacity		17.30	515	1,550	4,650	448.40	
0195	Air compressor, portable, 6.5 CFM, electric		.79	13	39	117	14.10	
0196	Gasoline		.70	19.65	59	177	17.40	
0200	Towed type, gas engine, 60 CFM		14.10	50	150	450	142.80	
0300	160 CFM		16.40	51.50	155	465	162.20	
0400	Diesel engine, rotary screw, 250 CFM		16.35	112	335	1,000	197.80	
0500	365 CFM		22.05	137	410	1,225	258.40	
0550	450 CFM		27.95	170	510	1,525	325.60	
0600	600 CFM		49.10	235	705	2,125	533.80	
0700	750 CFM		49.25	245	735	2,200	541	
0800	For silenced models, small sizes, add to rent		3%	5%	5%	5%		
0900	Large sizes, add to rent		5%	7%	7%	7%		
0930	Air tools, breaker, pavement, 60 lb.	Ea.	.50	10.35	31	93	10.20	
0940	80 lb.		.50	10.35	31	93	10.20	
0950	Drills, hand (jackhammer) 65 lb.		.60	17.65	53	159	15.40	

01 54 33 | Equipment Rental

		UNIT	HOURLY OPER. COST	RENT PER DAY	RENT PER WEEK	RENT PER MONTH	EQUIPMENT COST/DAY	
40	0960	Track or wagon, swing boom, 4" drifter	Ea.	62.25	910	2,735	8,200	1,045	**40**
	0970	5" drifter		73.80	1,100	3,290	9,875	1,248	
	0975	Track mounted quarry drill, 6" diameter drill		126.40	1,675	4,990	15,000	2,009	
	0980	Dust control per drill		1.04	25	75	225	23.30	
	0990	Hammer, chipping, 12 lb.		.55	26	78	234	20	
	1000	Hose, air with couplings, 50' long, 3/4" diameter		.03	4.67	14	42	3.05	
	1100	1" diameter		.03	5.65	17	51	3.65	
	1200	1-1/2" diameter		.05	8	24	72	5.20	
	1300	2" diameter		.06	10.65	32	96	6.90	
	1400	2-1/2" diameter		.11	19	57	171	12.30	
	1410	3" diameter		.13	21.50	64	192	13.85	
	1450	Drill, steel, 7/8" x 2'		.06	9.35	28	84	6.10	
	1460	7/8" x 6'		.05	8.65	26	78	5.60	
	1520	Moil points		.02	4	12	36	2.55	
	1525	Pneumatic nailer w/accessories		.55	36.50	109	325	26.20	
	1530	Sheeting driver for 60 lb. breaker		.04	6.65	20	60	4.30	
	1540	For 90 lb. breaker		.14	9	27	81	6.50	
	1550	Spade, 25 lb.		.50	7	21	63	8.20	
	1560	Tamper, single, 35 lb.		.50	33.50	100	300	24	
	1570	Triple, 140 lb.		.75	50	150	450	36	
	1580	Wrenches, impact, air powered, up to 3/4" bolt		.40	12.65	38	114	10.80	
	1590	Up to 1-1/4" bolt		.50	23.50	71	213	18.20	
	1600	Barricades, barrels, reflectorized, 1 to 99 barrels		.03	4.60	13.80	41.50	3	
	1610	100 to 200 barrels		.02	3.53	10.60	32	2.30	
	1620	Barrels with flashers, 1 to 99 barrels		.03	5.25	15.80	47.50	3.40	
	1630	100 to 200 barrels		.03	4.20	12.60	38	2.75	
	1640	Barrels with steady burn type C lights		.04	7	21	63	4.50	
	1650	Illuminated board, trailer mounted, with generator		3.50	130	390	1,175	106	
	1670	Portable barricade, stock, with flashers, 1 to 6 units		.03	5.25	15.80	47.50	3.40	
	1680	25 to 50 units		.03	4.90	14.70	44	3.20	
	1685	Butt fusion machine, wheeled, 1.5 HP electric, 2" - 8" diameter pipe		2.63	167	500	1,500	121.05	
	1690	Tracked, 20 HP diesel, 4"-12" diameter pipe		10.39	490	1,465	4,400	376.10	
	1695	83 HP diesel, 8" - 24" diameter pipe		27.35	975	2,930	8,800	804.80	
	1700	Carts, brick, hand powered, 1000 lb. capacity		.41	68.50	205	615	44.30	
	1800	Gas engine, 1500 lb., 7-1/2' lift		3.61	108	324	970	93.70	
	1822	Dehumidifier, medium, 6 lb./hr., 150 CFM		1.08	67.50	202.50	610	49.15	
	1824	Large, 18 lb./hr., 600 CFM		2.06	129	385.50	1,150	93.60	
	1830	Distributor, asphalt, trailer mounted, 2000 gal., 38 H.P. diesel		10.25	345	1,035	3,100	289	
	1840	3000 gal., 38 H.P. diesel		11.80	375	1,125	3,375	319.40	
	1850	Drill, rotary hammer, electric		1.07	26.50	80	240	24.55	
	1860	Carbide bit, 1-1/2" diameter, add to electric rotary hammer		.03	4.44	13.33	40	2.90	
	1865	Rotary, crawler, 250 H.P.		147.20	2,150	6,480	19,400	2,474	
	1870	Emulsion sprayer, 65 gal., 5 H.P. gas engine		2.79	101	303	910	82.90	
	1880	200 gal., 5 H.P. engine		8.05	168	505	1,525	165.40	
	1900	Floor auto-scrubbing machine, walk-behind, 28" path		5.13	335	1,000	3,000	241.05	
	1930	Floodlight, mercury vapor, or quartz, on tripod, 1000 watt		.44	20.50	62	186	15.90	
	1940	2000 watt		.83	41.50	124	370	31.45	
	1950	Floodlights, trailer mounted with generator, 1 - 300 watt light		3.60	71.50	215	645	71.80	
	1960	2 - 1000 watt lights		4.80	96.50	290	870	96.40	
	2000	4 - 300 watt lights		4.50	91.50	275	825	91	
	2005	Foam spray rig, incl. box trailer, compressor, generator, proportioner		33.75	495	1,485	4,450	567	
	2020	Forklift, straight mast, 12' lift, 5000 lb., 2 wheel drive, gas		26.25	207	620	1,850	334	
	2040	21' lift, 5000 lb., 4 wheel drive, diesel		20.45	250	750	2,250	313.60	
	2050	For rough terrain, 42' lift, 35' reach, 9000 lb., 110 H.P.		29.40	500	1,495	4,475	534.20	
	2060	For plant, 4 ton capacity, 80 H.P., 2 wheel drive, gas		16.05	95	285	855	185.40	
	2080	10 ton capacity, 120 H.P., 2 wheel drive, diesel		23.20	165	495	1,475	284.60	
	2100	Generator, electric, gas engine, 1.5 kW to 3 kW		3.70	11.65	35	105	36.60	
	2200	5 kW		4.80	15.35	46	138	47.60	

01 54 33 | Equipment Rental

		UNIT	HOURLY OPER. COST	RENT PER DAY	RENT PER WEEK	RENT PER MONTH	EQUIPMENT COST/DAY	
40	2300	10 kW	Ea.	9.05	38.50	115	345	95.40	40
	2400	25 kW		10.25	86.50	260	780	134	
	2500	Diesel engine, 20 kW		11.80	70	210	630	136.40	
	2600	50 kW		22.50	105	315	945	243	
	2700	100 kW		41.50	135	405	1,225	413	
	2800	250 kW		81.70	257	770	2,300	807.60	
	2850	Hammer, hydraulic, for mounting on boom, to 500 ft lb.		2.50	75	225	675	65	
	2860	1000 ft lb.		4.25	125	375	1,125	109	
	2900	Heaters, space, oil or electric, 50 MBH		1.59	7.65	23	69	17.30	
	3000	100 MBH		2.93	10.65	32	96	29.85	
	3100	300 MBH		8.54	38.50	115	345	91.30	
	3150	500 MBH		14.00	45	135	405	139	
	3200	Hose, water, suction with coupling, 20' long, 2" diameter		.02	3	9	27	1.95	
	3210	3" diameter		.03	4.33	13	39	2.85	
	3220	4" diameter		.03	5	15	45	3.25	
	3230	6" diameter		.11	17.65	53	159	11.50	
	3240	8" diameter		.32	53.50	161	485	34.75	
	3250	Discharge hose with coupling, 50' long, 2" diameter		.01	1.33	4	12	.90	
	3260	3" diameter		.01	2.33	7	21	1.50	
	3270	4" diameter		.02	3.67	11	33	2.35	
	3280	6" diameter		.06	9.35	28	84	6.10	
	3290	8" diameter		.20	33	99	297	21.40	
	3295	Insulation blower		.78	6	18	54	9.85	
	3300	Ladders, extension type, 16' to 36' long		.14	22.50	68	204	14.70	
	3400	40' to 60' long		.19	31	93	279	20.10	
	3405	Lance for cutting concrete		2.28	64	192	575	56.65	
	3407	Lawn mower, rotary, 22", 5 H.P.		1.55	39.50	118	355	36	
	3408	48" self propelled		3.36	109	327	980	92.30	
	3410	Level, electronic, automatic, with tripod and leveling rod		.91	60.50	182	545	43.70	
	3430	Laser type, for pipe and sewer line and grade		.73	48.50	145	435	34.85	
	3440	Rotating beam for interior control		.78	52	156	470	37.45	
	3460	Builder's optical transit, with tripod and rod		.09	14.65	44	132	9.50	
	3500	Light towers, towable, with diesel generator, 2000 watt		4.50	91.50	275	825	91	
	3600	4000 watt		4.80	96.50	290	870	96.40	
	3700	Mixer, powered, plaster and mortar, 6 C.F., 7 H.P.		2.70	20	60	180	33.60	
	3800	10 C.F., 9 H.P.		2.85	31.50	95	285	41.80	
	3850	Nailer, pneumatic		.55	36.50	109	325	26.20	
	3900	Paint sprayers complete, 8 CFM		1.05	70	210	630	50.40	
	4000	17 CFM		1.98	132	395	1,175	94.85	
	4020	Pavers, bituminous, rubber tires, 8' wide, 50 H.P., diesel		31.90	510	1,535	4,600	562.20	
	4030	10' wide, 150 H.P.		106.55	1,875	5,635	16,900	1,979	
	4050	Crawler, 8' wide, 100 H.P., diesel		90.15	1,900	5,670	17,000	1,855	
	4060	10' wide, 150 H.P.		113.75	2,300	6,890	20,700	2,288	
	4070	Concrete paver, 12' to 24' wide, 250 H.P.		100.45	1,550	4,685	14,100	1,741	
	4080	Placer-spreader-trimmer, 24' wide, 300 H.P.		149.30	2,675	8,010	24,000	2,796	
	4100	Pump, centrifugal gas pump, 1-1/2" diam., 65 GPM		4.05	51.50	155	465	63.40	
	4200	2" diameter, 130 GPM		5.45	56.50	170	510	77.60	
	4300	3" diameter, 250 GPM		5.75	58.50	175	525	81	
	4400	6" diameter, 1500 GPM		30.35	180	540	1,625	350.80	
	4500	Submersible electric pump, 1-1/4" diameter, 55 GPM		.40	17	51	153	13.40	
	4600	1-1/2" diameter, 83 GPM		.44	19.35	58	174	15.10	
	4700	2" diameter, 120 GPM		1.55	24.50	73	219	27	
	4800	3" diameter, 300 GPM		2.79	43.50	130	390	48.30	
	4900	4" diameter, 560 GPM		14.32	162	485	1,450	211.55	
	5000	6" diameter, 1590 GPM		21.51	217	650	1,950	302.10	
	5100	Diaphragm pump, gas, single, 1-1/2" diameter		1.08	49.50	148	445	38.25	
	5200	2" diameter		4.35	61.50	185	555	71.80	
	5300	3" diameter		4.35	65	195	585	73.80	

01 54 33	Equipment Rental	UNIT	HOURLY OPER. COST	RENT PER DAY	RENT PER WEEK	RENT PER MONTH	EQUIPMENT COST/DAY	
5400	Double, 4" diameter	Ea.	6.50	110	330	990	118	40
5450	Pressure washer 5 GPM, 3000 psi		4.95	53.50	160	480	71.60	
5460	7 GPM, 3000 psi		6.45	61.50	185	555	88.60	
5500	Trash pump, self-priming, gas, 2" diameter		4.65	22	66	198	50.40	
5600	Diesel, 4" diameter		8.95	91.50	275	825	126.60	
5650	Diesel, 6" diameter		24.25	153	460	1,375	286	
5655	Grout Pump		26.25	272	815	2,450	373	
5700	Salamanders, L.P. gas fired, 100,000 Btu		3.13	13	39	117	32.85	
5705	50,000 Btu		1.74	10.35	31	93	20.10	
5720	Sandblaster, portable, open top, 3 C.F. capacity		.60	26.50	80	240	20.80	
5730	6 C.F. capacity		.95	40	120	360	31.60	
5740	Accessories for above		.13	22	66	198	14.25	
5750	Sander, floor		.98	27.50	83	249	24.45	
5760	Edger		.53	15.65	47	141	13.65	
5800	Saw, chain, gas engine, 18" long		2.30	22	66	198	31.60	
5900	Hydraulic powered, 36" long		.75	65	195	585	45	
5950	60" long		.75	66.50	200	600	46	
6000	Masonry, table mounted, 14" diameter, 5 H.P.		1.32	56.50	170	510	44.55	
6050	Portable cut-off, 8 H.P.		2.50	33.50	100	300	40	
6100	Circular, hand held, electric, 7-1/4" diameter		.23	4.67	14	42	4.65	
6200	12" diameter		.23	8	24	72	6.65	
6250	Wall saw, w/hydraulic power, 10 H.P.		9.75	61.50	185	555	115	
6275	Shot blaster, walk-behind, 20" wide		4.80	295	885	2,650	215.40	
6280	Sidewalk broom, walk-behind		2.04	65	195	585	55.30	
6300	Steam cleaner, 100 gallons per hour		3.80	78.50	235	705	77.40	
6310	200 gallons per hour		5.40	95	285	855	100.20	
6340	Tar Kettle/Pot, 400 gallons		16.30	76.50	230	690	176.40	
6350	Torch, cutting, acetylene-oxygen, 150' hose, excludes gases		.30	15	45	135	11.40	
6360	Hourly operating cost includes tips and gas		21.00				168	
6410	Toilet, portable chemical		.13	21.50	64	192	13.85	
6420	Recycle flush type		.15	25.50	77	231	16.60	
6430	Toilet, fresh water flush, garden hose,		.19	31	93	279	20.10	
6440	Hoisted, non-flush, for high rise		.15	25	75	225	16.20	
6465	Tractor, farm with attachment		21.35	305	910	2,725	352.80	
6480	Trailers, platform, flush deck, 2 axle, 3 ton capacity		1.45	19.65	59	177	23.40	
6500	25 ton capacity		5.40	117	350	1,050	113.20	
6600	40 ton capacity		6.95	165	495	1,475	154.60	
6700	3 axle, 50 ton capacity		7.50	183	550	1,650	170	
6800	75 ton capacity		9.30	240	720	2,150	218.40	
6810	Trailer mounted cable reel for high voltage line work		5.57	265	795	2,375	203.55	
6820	Trailer mounted cable tensioning rig		11.06	525	1,580	4,750	404.50	
6830	Cable pulling rig		71.33	2,950	8,850	26,600	2,341	
6850	Portable cable/wire puller, 8000 lb max pulling capacity		3.71	167	502	1,500	130.10	
6900	Water tank trailer, engine driven discharge, 5000 gallons		6.95	143	430	1,300	141.60	
6925	10,000 gallons		9.45	198	595	1,775	194.60	
6950	Water truck, off highway, 6000 gallons		87.60	780	2,335	7,000	1,168	
7010	Tram car for high voltage line work, powered, 2 conductor		6.66	144	431	1,300	139.50	
7020	Transit (builder's level) with tripod		.09	14.65	44	132	9.50	
7030	Trench box, 3000 lb., 6' x 8'		.56	93.50	280	840	60.50	
7040	7200 lb., 6' x 20'		.75	125	375	1,125	81	
7050	8000 lb., 8' x 16'		1.08	180	540	1,625	116.65	
7060	9500 lb., 8' x 20'		1.21	201	603	1,800	130.30	
7065	11,000 lb., 8' x 24'		1.27	211	633	1,900	136.75	
7070	12,000 lb., 10' x 20'		1.50	249	748	2,250	161.60	
7100	Truck, pickup, 3/4 ton, 2 wheel drive		13.85	58.50	175	525	145.80	
7200	4 wheel drive		14.10	73.50	220	660	156.80	
7250	Crew carrier, 9 passenger		19.60	86.50	260	780	208.80	
7290	Flat bed truck, 20,000 lb. GVW		21.60	125	375	1,125	247.80	

01 54 | Construction Aids

01 54 33 | Equipment Rental

		UNIT	HOURLY OPER. COST	RENT PER DAY	RENT PER WEEK	RENT PER MONTH	EQUIPMENT COST/DAY	
40	7300	Tractor, 4 x 2, 220 H.P.	Ea.	30.10	197	590	1,775	358.80	**40**
	7410	330 H.P.		44.65	272	815	2,450	520.20	
	7500	6 x 4, 380 H.P.		51.25	315	950	2,850	600	
	7600	450 H.P.		62.10	385	1,150	3,450	726.80	
	7610	Tractor, with A frame, boom and winch, 225 H.P.		33.00	273	820	2,450	428	
	7620	Vacuum truck, hazardous material, 2500 gallons		11.85	292	875	2,625	269.80	
	7625	5,000 gallons		13.20	410	1,230	3,700	351.60	
	7650	Vacuum, HEPA, 16 gallon, wet/dry		.85	18	54	162	17.60	
	7655	55 gallon, wet/dry		.80	27	81	243	22.60	
	7660	Water tank, portable		.16	26.50	80	240	17.30	
	7690	Sewer/catch basin vacuum, 14 C.Y., 1500 gallons		17.54	615	1,850	5,550	510.30	
	7700	Welder, electric, 200 amp		3.88	16.35	49	147	40.85	
	7800	300 amp		5.76	20	60	180	58.10	
	7900	Gas engine, 200 amp		14.45	24.50	73	219	130.20	
	8000	300 amp		16.55	24.50	74	222	147.20	
	8100	Wheelbarrow, any size		.08	13	39	117	8.45	
	8200	Wrecking ball, 4000 lb.	▼	2.30	68.50	205	615	59.40	
50	0010	**HIGHWAY EQUIPMENT RENTAL** without operators R015433 -10							**50**
	0050	Asphalt batch plant, portable drum mixer, 100 ton/hr.	Ea.	80.81	1,450	4,380	13,100	1,522	
	0060	200 ton/hr.		92.38	1,550	4,655	14,000	1,670	
	0070	300 ton/hr.		109.99	1,825	5,465	16,400	1,973	
	0100	Backhoe attachment, long stick, up to 185 H.P., 10.5' long		.36	23.50	71	213	17.10	
	0140	Up to 250 H.P., 12' long		.39	25.50	77	231	18.50	
	0180	Over 250 H.P., 15' long		.54	35.50	107	320	25.70	
	0200	Special dipper arm, up to 100 H.P., 32' long		1.10	73	219	655	52.60	
	0240	Over 100 H.P., 33' long		1.37	91	273	820	65.55	
	0280	Catch basin/sewer cleaning truck, 3 ton, 9 C.Y., 1000 gal.		42.90	390	1,170	3,500	577.20	
	0300	Concrete batch plant, portable, electric, 200 C.Y./hr.		22.52	515	1,545	4,625	489.15	
	0520	Grader/dozer attachment, ripper/scarifier, rear mounted, up to 135 H.P.		2.90	58.50	175	525	58.20	
	0540	Up to 180 H.P.		3.85	88.50	265	795	83.80	
	0580	Up to 250 H.P.		5.50	142	425	1,275	129	
	0700	Pvmt. removal bucket, for hyd. excavator, up to 90 H.P.		1.85	51.50	155	465	45.80	
	0740	Up to 200 H.P.		2.05	68.50	205	615	57.40	
	0780	Over 200 H.P.		2.20	85	255	765	68.60	
	0900	Aggregate spreader, self-propelled, 187 H.P.		52.46	685	2,050	6,150	829.70	
	1000	Chemical spreader, 3 C.Y.		3.25	43.50	130	390	52	
	1900	Hammermill, traveling, 250 H.P.		70.05	2,125	6,360	19,100	1,832	
	2000	Horizontal borer, 3" diameter, 13 H.P. gas driven		6.40	55	165	495	84.20	
	2150	Horizontal directional drill, 20,000 lb. thrust, 78 H.P. diesel		31.00	680	2,045	6,125	657	
	2160	30,000 lb. thrust, 115 H.P.		38.60	1,050	3,135	9,400	935.80	
	2170	50,000 lb. thrust, 170 H.P.		55.35	1,325	4,005	12,000	1,244	
	2190	Mud trailer for HDD, 1500 gallons, 175 H.P., gas		32.25	153	460	1,375	350	
	2200	Hydromulcher, diesel, 3000 gallon, for truck mounting		23.30	247	740	2,225	334.40	
	2300	Gas, 600 gallon		8.60	98.50	295	885	127.80	
	2400	Joint & crack cleaner, walk behind, 25 H.P.		4.00	50	150	450	62	
	2500	Filler, trailer mounted, 400 gallons, 20 H.P.		9.45	212	635	1,900	202.60	
	3000	Paint striper, self-propelled, 40 gallon, 22 H.P.		7.25	157	470	1,400	152	
	3100	120 gallon, 120 H.P.		23.15	400	1,195	3,575	424.20	
	3200	Post drivers, 6" I-Beam frame, for truck mounting		19.00	385	1,155	3,475	383	
	3400	Road sweeper, self-propelled, 8' wide, 90 H.P.		39.40	585	1,755	5,275	666.20	
	3450	Road sweeper, vacuum assisted, 4 C.Y., 220 gallons		75.35	635	1,905	5,725	983.80	
	4000	Road mixer, self-propelled, 130 H.P.		47.70	785	2,350	7,050	851.60	
	4100	310 H.P.		83.25	2,125	6,395	19,200	1,945	
	4220	Cold mix paver, incl. pug mill and bitumen tank, 165 H.P.		98.45	2,400	7,190	21,600	2,226	
	4240	Pavement brush, towed		3.15	93.50	280	840	81.20	
	4250	Paver, asphalt, wheel or crawler, 130 H.P., diesel		98.15	2,375	7,125	21,400	2,210	
	4300	Paver, road widener, gas 1' to 6', 67 H.P.	▼	49.10	915	2,750	8,250	942.80	

01 54 33 | Equipment Rental

		UNIT	HOURLY OPER. COST	RENT PER DAY	RENT PER WEEK	RENT PER MONTH	EQUIPMENT COST/DAY	
50	4400	Diesel, 2' to 14', 88 H.P.	Ea.	62.75	1,075	3,205	9,625	1,143	**50**
	4600	Slipform pavers, curb and gutter, 2 track, 75 H.P.		61.85	1,000	3,035	9,100	1,102	
	4700	4 track, 165 H.P.		41.15	750	2,255	6,775	780.20	
	4800	Median barrier, 215 H.P.		66.45	1,225	3,670	11,000	1,266	
	4901	Trailer, low bed, 75 ton capacity		10.05	238	715	2,150	223.40	
	5000	Road planer, walk behind, 10" cutting width, 10 H.P.		3.70	33.50	100	300	49.60	
	5100	Self-propelled, 12" cutting width, 64 H.P.		10.50	113	340	1,025	152	
	5120	Traffic line remover, metal ball blaster, truck mounted, 115 H.P.		48.45	765	2,295	6,875	846.60	
	5140	Grinder, truck mounted, 115 H.P.		54.85	830	2,490	7,475	936.80	
	5160	Walk-behind, 11 H.P.		4.20	53.50	160	480	65.60	
	5200	Pavement profiler, 4' to 6' wide, 450 H.P.		252.75	3,400	10,165	30,500	4,055	
	5300	8' to 10' wide, 750 H.P.		395.65	4,450	13,350	40,100	5,835	
	5400	Roadway plate, steel, 1" x 8' x 20'		.08	13.65	41	123	8.85	
	5600	Stabilizer, self-propelled, 150 H.P.		48.85	625	1,870	5,600	764.80	
	5700	310 H.P.		96.75	1,775	5,295	15,900	1,833	
	5800	Striper, truck mounted, 120 gallon paint, 460 H.P.		64.90	485	1,455	4,375	810.20	
	5900	Thermal paint heating kettle, 115 gallons		8.29	25.50	77	231	81.70	
	6000	Tar kettle, 330 gallon, trailer mounted		12.54	58.50	175	525	135.30	
	7000	Tunnel locomotive, diesel, 8 to 12 ton		32.15	595	1,780	5,350	613.20	
	7005	Electric, 10 ton		26.90	680	2,035	6,100	622.20	
	7010	Muck cars, 1/2 C.Y. capacity		2.10	24.50	74	222	31.60	
	7020	1 C.Y. capacity		2.30	33	99	297	38.20	
	7030	2 C.Y. capacity		2.45	38.50	115	345	42.60	
	7040	Side dump, 2 C.Y. capacity		2.65	45	135	405	48.20	
	7050	3 C.Y. capacity		3.60	51.50	155	465	59.80	
	7060	5 C.Y. capacity		5.15	65	195	585	80.20	
	7100	Ventilating blower for tunnel, 7-1/2 H.P.		2.10	51.50	155	465	47.80	
	7110	10 H.P.		2.28	53.50	160	480	50.25	
	7120	20 H.P.		3.54	69.50	208	625	69.90	
	7140	40 H.P.		5.81	98.50	295	885	105.50	
	7160	60 H.P.		8.92	153	460	1,375	163.35	
	7175	75 H.P.		11.71	210	630	1,900	219.70	
	7180	200 H.P.		23.75	305	920	2,750	374	
	7800	Windrow loader, elevating		55.85	1,325	4,000	12,000	1,247	
60	0010	**LIFTING AND HOISTING EQUIPMENT RENTAL** without operators							**60**
	0120	Aerial lift truck, 2 person, to 80' [R015433-10]	Ea.	26.40	745	2,240	6,725	659.20	
	0140	Boom work platform, 40' snorkel		11.75	285	855	2,575	265	
	0150	Crane, flatbed mounted, 3 ton capacity [R015433-15]		16.15	198	595	1,775	248.20	
	0200	Crane, climbing, 106' jib, 6000 lb. capacity, 410 fpm [R312316-45]		39.85	1,675	5,060	15,200	1,331	
	0300	101' jib, 10,250 lb. capacity, 270 fpm		46.60	2,125	6,410	19,200	1,655	
	0500	Tower, static, 130' high, 106' jib, 6200 lb. capacity at 400 fpm		43.75	1,950	5,840	17,500	1,518	
	0600	Crawler mounted, lattice boom, 1/2 C.Y., 15 tons at 12' radius		36.96	665	2,000	6,000	695.70	
	0700	3/4 C.Y., 20 tons at 12' radius		49.28	835	2,510	7,525	896.25	
	0800	1 C.Y., 25 tons at 12' radius		65.70	1,125	3,340	10,000	1,194	
	0900	1-1/2 C.Y., 40 tons at 12' radius		65.75	1,125	3,385	10,200	1,203	
	1000	2 C.Y., 50 tons at 12' radius		69.75	1,325	3,940	11,800	1,346	
	1100	3 C.Y., 75 tons at 12' radius		74.30	1,525	4,565	13,700	1,507	
	1200	100 ton capacity, 60' boom		84.05	1,650	4,975	14,900	1,667	
	1300	165 ton capacity, 60' boom		106.80	2,050	6,170	18,500	2,088	
	1400	200 ton capacity, 70' boom		129.75	2,575	7,740	23,200	2,586	
	1500	350 ton capacity, 80' boom		181.25	3,850	11,575	34,700	3,765	
	1600	Truck mounted, lattice boom, 6 x 4, 20 tons at 10' radius		35.42	1,150	3,420	10,300	967.35	
	1700	25 tons at 10' radius		38.28	1,250	3,720	11,200	1,050	
	1800	8 x 4, 30 tons at 10' radius		41.49	1,325	3,960	11,900	1,124	
	1900	40 tons at 12' radius		44.28	1,375	4,140	12,400	1,182	
	2000	60 tons at 15' radius		49.79	1,450	4,380	13,100	1,274	
	2050	82 tons at 15' radius		55.71	1,550	4,680	14,000	1,382	
	2100	90 tons at 15' radius		62.48	1,700	5,100	15,300	1,520	

01 54 33 | Equipment Rental

		UNIT	HOURLY OPER. COST	RENT PER DAY	RENT PER WEEK	RENT PER MONTH	EQUIPMENT COST/DAY	
60	2200	115 tons at 15' radius	Ea.	70.50	1,900	5,700	17,100	1,704	60
	2300	150 tons at 18' radius		83.50	2,000	6,000	18,000	1,868	
	2350	165 tons at 18' radius		82.77	2,125	6,360	19,100	1,934	
	2400	Truck mounted, hydraulic, 12 ton capacity		41.55	545	1,630	4,900	658.40	
	2500	25 ton capacity		44.10	660	1,980	5,950	748.80	
	2550	33 ton capacity		44.65	675	2,030	6,100	763.20	
	2560	40 ton capacity		57.95	795	2,390	7,175	941.60	
	2600	55 ton capacity		74.95	870	2,615	7,850	1,123	
	2700	80 ton capacity		98.50	1,400	4,210	12,600	1,630	
	2720	100 ton capacity		92.65	1,475	4,425	13,300	1,626	
	2740	120 ton capacity		107.65	1,625	4,845	14,500	1,830	
	2760	150 ton capacity		126.15	2,100	6,320	19,000	2,273	
	2800	Self-propelled, 4 x 4, with telescoping boom, 5 ton		17.35	237	710	2,125	280.80	
	2900	12-1/2 ton capacity		31.65	375	1,130	3,400	479.20	
	3000	15 ton capacity		32.35	395	1,190	3,575	496.80	
	3050	20 ton capacity		35.10	455	1,370	4,100	554.80	
	3100	25 ton capacity		36.70	520	1,565	4,700	606.60	
	3150	40 ton capacity		45.10	585	1,755	5,275	711.80	
	3200	Derricks, guy, 20 ton capacity, 60' boom, 75' mast		23.97	415	1,238	3,725	439.35	
	3300	100' boom, 115' mast		37.86	710	2,130	6,400	728.90	
	3400	Stiffleg, 20 ton capacity, 70' boom, 37' mast		26.57	535	1,610	4,825	534.55	
	3500	100' boom, 47' mast		41.01	860	2,580	7,750	844.10	
	3550	Helicopter, small, lift to 1250 lb. maximum, w/pilot		96.80	3,325	10,000	30,000	2,774	
	3600	Hoists, chain type, overhead, manual, 3/4 ton		.15	.33	1	3	1.40	
	3900	10 ton		.75	6	18	54	9.60	
	4000	Hoist and tower, 5000 lb. cap., portable electric, 40' high		5.02	238	713	2,150	182.75	
	4100	For each added 10' section, add		.11	18.65	56	168	12.10	
	4200	Hoist and single tubular tower, 5000 lb. electric, 100' high		6.80	330	996	3,000	253.60	
	4300	For each added 6'-6" section, add		.19	32.50	97	291	20.90	
	4400	Hoist and double tubular tower, 5000 lb., 100' high		7.30	365	1,097	3,300	277.80	
	4500	For each added 6'-6" section, add		.21	35.50	107	320	23.10	
	4550	Hoist and tower, mast type, 6000 lb., 100' high		7.87	380	1,137	3,400	290.35	
	4570	For each added 10' section, add		.13	22	66	198	14.25	
	4600	Hoist and tower, personnel, electric, 2000 lb., 100' @ 125 fpm		16.61	1,000	3,030	9,100	738.90	
	4700	3000 lb., 100' @ 200 fpm		18.97	1,150	3,430	10,300	837.75	
	4800	3000 lb., 150' @ 300 fpm		21.02	1,275	3,840	11,500	936.15	
	4900	4000 lb., 100' @ 300 fpm		21.78	1,300	3,920	11,800	958.25	
	5000	6000 lb., 100' @ 275 fpm	▼	23.46	1,375	4,110	12,300	1,010	
	5100	For added heights up to 500', add	L.F.	.01	1.67	5	15	1.10	
	5200	Jacks, hydraulic, 20 ton	Ea.	.05	2	6	18	1.60	
	5500	100 ton		.40	11.65	35	105	10.20	
	6100	Jacks, hydraulic, climbing w/50' jackrods, control console, 30 ton cap.		2.05	137	410	1,225	98.40	
	6150	For each added 10' jackrod section, add		.05	3.33	10	30	2.40	
	6300	50 ton capacity		3.30	220	659	1,975	158.20	
	6350	For each added 10' jackrod section, add		.06	4	12	36	2.90	
	6500	125 ton capacity		8.60	575	1,720	5,150	412.80	
	6550	For each added 10' jackrod section, add		.59	39	117	350	28.10	
	6600	Cable jack, 10 ton capacity with 200' cable		1.72	114	343	1,025	82.35	
	6650	For each added 50' of cable, add	▼	.21	13.65	41	123	9.90	
70	0010	**WELLPOINT EQUIPMENT RENTAL** without operators							70
	0020	Based on 2 months rental							
	0100	Combination jetting & wellpoint pump, 60 H.P. diesel	Ea.	16.29	340	1,016	3,050	333.50	
	0200	High pressure gas jet pump, 200 H.P., 300 psi	"	36.68	289	868	2,600	467.05	
	0300	Discharge pipe, 8" diameter	L.F.	.01	.55	1.65	4.95	.40	
	0350	12" diameter		.01	.81	2.43	7.30	.55	
	0400	Header pipe, flows up to 150 GPM, 4" diameter		.01	.50	1.50	4.50	.40	
	0500	400 GPM, 6" diameter	▼	.01	.58	1.75	5.25	.45	

Note (row 0010): R015433 -10

01 54 33 | Equipment Rental

		UNIT	HOURLY OPER. COST	RENT PER DAY	RENT PER WEEK	RENT PER MONTH	EQUIPMENT COST/DAY	
70	0600	800 GPM, 8" diameter	L.F.	.01	.81	2.43	7.30	.55	**70**
	0700	1500 GPM, 10" diameter		.01	.85	2.56	7.70	.60	
	0800	2500 GPM, 12" diameter		.02	1.61	4.83	14.50	1.15	
	0900	4500 GPM, 16" diameter		.03	2.06	6.19	18.55	1.50	
	0950	For quick coupling aluminum and plastic pipe, add	↓	.03	2.14	6.41	19.25	1.50	
	1100	Wellpoint, 25' long, with fittings & riser pipe, 1-1/2" or 2" diameter	Ea.	.06	4.26	12.79	38.50	3.05	
	1200	Wellpoint pump, diesel powered, 4" suction, 20 H.P.		7.16	195	585	1,750	174.30	
	1300	6" suction, 30 H.P.		9.67	242	726	2,175	222.55	
	1400	8" suction, 40 H.P.		13.09	330	996	3,000	303.90	
	1500	10" suction, 75 H.P.		19.62	390	1,164	3,500	389.75	
	1600	12" suction, 100 H.P.		28.25	615	1,850	5,550	596	
	1700	12" suction, 175 H.P.	↓	41.13	685	2,050	6,150	739.05	
80	0010	**MARINE EQUIPMENT RENTAL** without operators R015433 -10							**80**
	0200	Barge, 400 Ton, 30' wide x 90' long	Ea.	17.40	1,075	3,200	9,600	779.20	
	0240	800 Ton, 45' wide x 90' long		21.05	1,300	3,880	11,600	944.40	
	2000	Tugboat, diesel, 100 H.P.		37.80	217	650	1,950	432.40	
	2040	250 H.P.		78.95	395	1,185	3,550	868.60	
	2080	380 H.P.		155.50	1,175	3,535	10,600	1,951	
	3000	Small work boat, gas, 16-foot, 50 H.P.		17.35	60	180	540	174.80	
	4000	Large, diesel, 48-foot, 200 H.P.	↓	88.50	1,250	3,735	11,200	1,455	

Crews - Renovation

Crew A-1	Hr.	Daily	Hr.	Daily	Bare Costs	Incl. O&P
1 Building Laborer	$37.90	$303.20	$61.95	$495.60	$37.90	$61.95
1 Concrete Saw, Gas Manual		81.40		89.54	10.18	11.19
8 L.H., Daily Totals		$384.60		$585.14	$48.08	$73.14

Crew A-1A	Hr.	Daily	Hr.	Daily	Bare Costs	Incl. O&P
1 Skilled Worker	$49.90	$399.20	$80.85	$646.80	$49.90	$80.85
1 Shot Blaster, 20"		215.40		236.94	26.93	29.62
8 L.H., Daily Totals		$614.60		$883.74	$76.83	$110.47

Crew A-1B	Hr.	Daily	Hr.	Daily	Bare Costs	Incl. O&P
1 Building Laborer	$37.90	$303.20	$61.95	$495.60	$37.90	$61.95
1 Concrete Saw		168.60		185.46	21.07	23.18
8 L.H., Daily Totals		$471.80		$681.06	$58.98	$85.13

Crew A-1C	Hr.	Daily	Hr.	Daily	Bare Costs	Incl. O&P
1 Building Laborer	$37.90	$303.20	$61.95	$495.60	$37.90	$61.95
1 Chain Saw, Gas, 18"		31.60		34.76	3.95	4.34
8 L.H., Daily Totals		$334.80		$530.36	$41.85	$66.30

Crew A-1D	Hr.	Daily	Hr.	Daily	Bare Costs	Incl. O&P
1 Building Laborer	$37.90	$303.20	$61.95	$495.60	$37.90	$61.95
1 Vibrating Plate, Gas, 18"		36.20		39.82	4.53	4.98
8 L.H., Daily Totals		$339.40		$535.42	$42.42	$66.93

Crew A-1E	Hr.	Daily	Hr.	Daily	Bare Costs	Incl. O&P
1 Building Laborer	$37.90	$303.20	$61.95	$495.60	$37.90	$61.95
1 Vibrating Plate, Gas, 21"		46.60		51.26	5.83	6.41
8 L.H., Daily Totals		$349.80		$546.86	$43.73	$68.36

Crew A-1F	Hr.	Daily	Hr.	Daily	Bare Costs	Incl. O&P
1 Building Laborer	$37.90	$303.20	$61.95	$495.60	$37.90	$61.95
1 Rammer/Tamper, Gas, 8"		50.00		55.00	6.25	6.88
8 L.H., Daily Totals		$353.20		$550.60	$44.15	$68.83

Crew A-1G	Hr.	Daily	Hr.	Daily	Bare Costs	Incl. O&P
1 Building Laborer	$37.90	$303.20	$61.95	$495.60	$37.90	$61.95
1 Rammer/Tamper, Gas, 15"		56.80		62.48	7.10	7.81
8 L.H., Daily Totals		$360.00		$558.08	$45.00	$69.76

Crew A-1H	Hr.	Daily	Hr.	Daily	Bare Costs	Incl. O&P
1 Building Laborer	$37.90	$303.20	$61.95	$495.60	$37.90	$61.95
1 Exterior Steam Cleaner		77.40		85.14	9.68	10.64
8 L.H., Daily Totals		$380.60		$580.74	$47.58	$72.59

Crew A-1J	Hr.	Daily	Hr.	Daily	Bare Costs	Incl. O&P
1 Building Laborer	$37.90	$303.20	$61.95	$495.60	$37.90	$61.95
1 Cultivator, Walk-Behind, 5 H.P.		45.10		49.61	5.64	6.20
8 L.H., Daily Totals		$348.30		$545.21	$43.54	$68.15

Crew A-1K	Hr.	Daily	Hr.	Daily	Bare Costs	Incl. O&P
1 Building Laborer	$37.90	$303.20	$61.95	$495.60	$37.90	$61.95
1 Cultivator, Walk-Behind, 8 H.P.		70.00		77.00	8.75	9.63
8 L.H., Daily Totals		$373.20		$572.60	$46.65	$71.58

Crew A-1M	Hr.	Daily	Hr.	Daily	Bare Costs	Incl. O&P
1 Building Laborer	$37.90	$303.20	$61.95	$495.60	$37.90	$61.95
1 Snow Blower, Walk-Behind		55.30		60.83	6.91	7.60
8 L.H., Daily Totals		$358.50		$556.43	$44.81	$69.55

Crew A-2	Hr.	Daily	Hr.	Daily	Bare Costs	Incl. O&P
2 Laborers	$37.90	$606.40	$61.95	$991.20	$39.25	$63.82
1 Truck Driver (light)	41.95	335.60	67.55	540.40		
1 Flatbed Truck, Gas, 1.5 Ton		248.20		273.02	10.34	11.38
24 L.H., Daily Totals		$1190.20		$1804.62	$49.59	$75.19

Crew A-2A	Hr.	Daily	Hr.	Daily	Bare Costs	Incl. O&P
2 Laborers	$37.90	$606.40	$61.95	$991.20	$39.25	$63.82
1 Truck Driver (light)	41.95	335.60	67.55	540.40		
1 Flatbed Truck, Gas, 1.5 Ton		248.20		273.02		
1 Concrete Saw		168.60		185.46	17.37	19.10
24 L.H., Daily Totals		$1358.80		$1990.08	$56.62	$82.92

Crew A-2B	Hr.	Daily	Hr.	Daily	Bare Costs	Incl. O&P
1 Truck Driver (light)	$41.95	$335.60	$67.55	$540.40	$41.95	$67.55
1 Flatbed Truck, Gas, 1.5 Ton		248.20		273.02	31.02	34.13
8 L.H., Daily Totals		$583.80		$813.42	$72.97	$101.68

Crew A-3A	Hr.	Daily	Hr.	Daily	Bare Costs	Incl. O&P
1 Equip. Oper. (light)	$49.15	$393.20	$77.80	$622.40	$49.15	$77.80
1 Pickup Truck, 4x4, 3/4 Ton		156.80		172.48	19.60	21.56
8 L.H., Daily Totals		$550.00		$794.88	$68.75	$99.36

Crew A-3B	Hr.	Daily	Hr.	Daily	Bare Costs	Incl. O&P
1 Equip. Oper. (medium)	$51.10	$408.80	$80.90	$647.20	$47.15	$75.22
1 Truck Driver (heavy)	43.20	345.60	69.55	556.40		
1 Dump Truck, 12 C.Y., 400 H.P.		690.80		759.88		
1 F.E. Loader, W.M., 2.5 C.Y.		516.00		567.60	75.42	82.97
16 L.H., Daily Totals		$1961.20		$2531.08	$122.58	$158.19

Crew A-3C	Hr.	Daily	Hr.	Daily	Bare Costs	Incl. O&P
1 Equip. Oper. (light)	$49.15	$393.20	$77.80	$622.40	$49.15	$77.80
1 Loader, Skid Steer, 78 H.P.		315.60		347.16	39.45	43.40
8 L.H., Daily Totals		$708.80		$969.56	$88.60	$121.19

Crew A-3D	Hr.	Daily	Hr.	Daily	Bare Costs	Incl. O&P
1 Truck Driver (light)	$41.95	$335.60	$67.55	$540.40	$41.95	$67.55
1 Pickup Truck, 4x4, 3/4 Ton		156.80		172.48		
1 Flatbed Trailer, 25 Ton		113.20		124.52	33.75	37.13
8 L.H., Daily Totals		$605.60		$837.40	$75.70	$104.68

Crew A-3E	Hr.	Daily	Hr.	Daily	Bare Costs	Incl. O&P
1 Equip. Oper. (crane)	$52.25	$418.00	$82.70	$661.60	$47.73	$76.13
1 Truck Driver (heavy)	43.20	345.60	69.55	556.40		
1 Pickup Truck, 4x4, 3/4 Ton		156.80		172.48	9.80	10.78
16 L.H., Daily Totals		$920.40		$1390.48	$57.52	$86.91

Crew A-3F	Hr.	Daily	Hr.	Daily	Bare Costs	Incl. O&P
1 Equip. Oper. (crane)	$52.25	$418.00	$82.70	$661.60	$47.73	$76.13
1 Truck Driver (heavy)	43.20	345.60	69.55	556.40		
1 Pickup Truck, 4x4, 3/4 Ton		156.80		172.48		
1 Truck Tractor, 6x4, 380 H.P.		600.00		660.00		
1 Lowbed Trailer, 75 Ton		223.40		245.74	61.26	67.39
16 L.H., Daily Totals		$1743.80		$2296.22	$108.99	$143.51

For customer support on your Commercial Renovation Cost Data, call 877.742.4719.

Crew A-3G	Hr.	Daily	Hr.	Daily	Bare Costs	Incl. O&P
1 Equip. Oper. (crane)	$52.25	$418.00	$82.70	$661.60	$47.73	$76.13
1 Truck Driver (heavy)	43.20	345.60	69.55	556.40		
1 Pickup Truck, 4x4, 3/4 Ton		156.80		172.48		
1 Truck Tractor, 6x4, 450 H.P.		726.80		799.48		
1 Lowbed Trailer, 75 Ton		223.40		245.74	69.19	76.11
16 L.H., Daily Totals		$1870.60		$2435.70	$116.91	$152.23

Crew A-3H	Hr.	Daily	Hr.	Daily	Bare Costs	Incl. O&P
1 Equip. Oper. (crane)	$52.25	$418.00	$82.70	$661.60	$52.25	$82.70
1 Hyd. Crane, 12 Ton (Daily)		877.40		965.14	109.68	120.64
8 L.H., Daily Totals		$1295.40		$1626.74	$161.93	$203.34

Crew A-3I	Hr.	Daily	Hr.	Daily	Bare Costs	Incl. O&P
1 Equip. Oper. (crane)	$52.25	$418.00	$82.70	$661.60	$52.25	$82.70
1 Hyd. Crane, 25 Ton (Daily)		1013.00		1114.30	126.63	139.29
8 L.H., Daily Totals		$1431.00		$1775.90	$178.88	$221.99

Crew A-3J	Hr.	Daily	Hr.	Daily	Bare Costs	Incl. O&P
1 Equip. Oper. (crane)	$52.25	$418.00	$82.70	$661.60	$52.25	$82.70
1 Hyd. Crane, 40 Ton (Daily)		1259.00		1384.90	157.38	173.11
8 L.H., Daily Totals		$1677.00		$2046.50	$209.63	$255.81

Crew A-3K	Hr.	Daily	Hr.	Daily	Bare Costs	Incl. O&P
1 Equip. Oper. (crane)	$52.25	$418.00	$82.70	$661.60	$49.15	$77.80
1 Equip. Oper. (oiler)	46.05	368.40	72.90	583.20		
1 Hyd. Crane, 55 Ton (Daily)		1470.00		1617.00		
1 P/U Truck, 3/4 Ton (Daily)		170.80		187.88	102.55	112.81
16 L.H., Daily Totals		$2427.20		$3049.68	$151.70	$190.60

Crew A-3L	Hr.	Daily	Hr.	Daily	Bare Costs	Incl. O&P
1 Equip. Oper. (crane)	$52.25	$418.00	$82.70	$661.60	$49.15	$77.80
1 Equip. Oper. (oiler)	46.05	368.40	72.90	583.20		
1 Hyd. Crane, 80 Ton (Daily)		2193.00		2412.30		
1 P/U Truck, 3/4 Ton (Daily)		170.80		187.88	147.74	162.51
16 L.H., Daily Totals		$3150.20		$3844.98	$196.89	$240.31

Crew A-3M	Hr.	Daily	Hr.	Daily	Bare Costs	Incl. O&P
1 Equip. Oper. (crane)	$52.25	$418.00	$82.70	$661.60	$49.15	$77.80
1 Equip. Oper. (oiler)	46.05	368.40	72.90	583.20		
1 Hyd. Crane, 100 Ton (Daily)		2216.00		2437.60		
1 P/U Truck, 3/4 Ton (Daily)		170.80		187.88	149.18	164.09
16 L.H., Daily Totals		$3173.20		$3870.28	$198.32	$241.89

Crew A-3N	Hr.	Daily	Hr.	Daily	Bare Costs	Incl. O&P
1 Equip. Oper. (crane)	$52.25	$418.00	$82.70	$661.60	$52.25	$82.70
1 Tower Cane (monthly)		1145.00		1259.50	143.13	157.44
8 L.H., Daily Totals		$1563.00		$1921.10	$195.38	$240.14

Crew A-3P	Hr.	Daily	Hr.	Daily	Bare Costs	Incl. O&P
1 Equip. Oper. (light)	$49.15	$393.20	$77.80	$622.40	$49.15	$77.80
1 A.T. Forklift, 42' lift		534.20		587.62	66.78	73.45
8 L.H., Daily Totals		$927.40		$1210.02	$115.93	$151.25

Crew A-3Q	Hr.	Daily	Hr.	Daily	Bare Costs	Incl. O&P
1 Equip. Oper. (light)	$49.15	$393.20	$77.80	$622.40	$49.15	$77.80
1 Pickup Truck, 4x4, 3/4 Ton		156.80		172.48		
1 Flatbed trailer, 3 Ton		23.40		25.74	22.52	24.78
8 L.H., Daily Totals		$573.40		$820.62	$71.67	$102.58

Crew A-4	Hr.	Daily	Hr.	Daily	Bare Costs	Incl. O&P
2 Carpenters	$48.45	$775.20	$79.15	$1266.40	$45.75	$74.38
1 Painter, Ordinary	40.35	322.80	64.85	518.80		
24 L.H., Daily Totals		$1098.00		$1785.20	$45.75	$74.38

Crew A-5	Hr.	Daily	Hr.	Daily	Bare Costs	Incl. O&P
2 Laborers	$37.90	$606.40	$61.95	$991.20	$38.35	$62.57
.25 Truck Driver (light)	41.95	83.90	67.55	135.10		
.25 Flatbed Truck, Gas, 1.5 Ton		62.05		68.25	3.45	3.79
18 L.H., Daily Totals		$752.35		$1194.56	$41.80	$66.36

Crew A-6	Hr.	Daily	Hr.	Daily	Bare Costs	Incl. O&P
1 Instrument Man	$49.90	$399.20	$80.85	$646.80	$48.30	$78.28
1 Rodman/Chainman	46.70	373.60	75.70	605.60		
1 Level, Electronic		43.70		48.07	2.73	3.00
16 L.H., Daily Totals		$816.50		$1300.47	$51.03	$81.28

Crew A-7	Hr.	Daily	Hr.	Daily	Bare Costs	Incl. O&P
1 Chief of Party	$61.65	$493.20	$96.80	$774.40	$52.75	$84.45
1 Instrument Man	49.90	399.20	80.85	646.80		
1 Rodman/Chainman	46.70	373.60	75.70	605.60		
1 Level, Electronic		43.70		48.07	1.82	2.00
24 L.H., Daily Totals		$1309.70		$2074.87	$54.57	$86.45

Crew A-8	Hr.	Daily	Hr.	Daily	Bare Costs	Incl. O&P
1 Chief of Party	$61.65	$493.20	$96.80	$774.40	$51.24	$82.26
1 Instrument Man	49.90	399.20	80.85	646.80		
2 Rodmen/Chainmen	46.70	747.20	75.70	1211.20		
1 Level, Electronic		43.70		48.07	1.37	1.50
32 L.H., Daily Totals		$1683.30		$2680.47	$52.60	$83.76

Crew A-9	Hr.	Daily	Hr.	Daily	Bare Costs	Incl. O&P
1 Asbestos Foreman	$53.90	$431.20	$86.55	$692.40	$53.46	$85.85
7 Asbestos Workers	53.40	2990.40	85.75	4802.00		
64 L.H., Daily Totals		$3421.60		$5494.40	$53.46	$85.85

Crew A-10A	Hr.	Daily	Hr.	Daily	Bare Costs	Incl. O&P
1 Asbestos Foreman	$53.90	$431.20	$86.55	$692.40	$53.57	$86.02
2 Asbestos Workers	53.40	854.40	85.75	1372.00		
24 L.H., Daily Totals		$1285.60		$2064.40	$53.57	$86.02

Crew A-10B	Hr.	Daily	Hr.	Daily	Bare Costs	Incl. O&P
1 Asbestos Foreman	$53.90	$431.20	$86.55	$692.40	$53.52	$85.95
3 Asbestos Workers	53.40	1281.60	85.75	2058.00		
32 L.H., Daily Totals		$1712.80		$2750.40	$53.52	$85.95

Crew A-10C	Hr.	Daily	Hr.	Daily	Bare Costs	Incl. O&P
3 Asbestos Workers	$53.40	$1281.60	$85.75	$2058.00	$53.40	$85.75
1 Flatbed Truck, Gas, 1.5 Ton		248.20		273.02	10.34	11.38
24 L.H., Daily Totals		$1529.80		$2331.02	$63.74	$97.13

Crew A-10D	Hr.	Daily	Hr.	Daily	Bare Costs	Incl. O&P
2 Asbestos Workers	$53.40	$854.40	$85.75	$1372.00	$51.27	$81.78
1 Equip. Oper. (crane)	52.25	418.00	82.70	661.60		
1 Equip. Oper. (oiler)	46.05	368.40	72.90	583.20		
1 Hydraulic Crane, 33 Ton		763.20		839.52	23.85	26.23
32 L.H., Daily Totals		$2404.00		$3456.32	$75.13	$108.01

For customer support on your Commercial Renovation Cost Data, call 877.742.4719.

Crews - Renovation

Crew A-11

Crew A-11	Hr.	Daily	Hr.	Daily	Bare Costs	Incl. O&P
1 Asbestos Foreman	$53.90	$431.20	$86.55	$692.40	$53.46	$85.85
7 Asbestos Workers	53.40	2990.40	85.75	4802.00		
2 Chip. Hammers, 12 Lb., Elec.		40.00		44.00	.63	.69
64 L.H., Daily Totals		$3461.60		$5538.40	$54.09	$86.54

Crew A-12

Crew A-12	Hr.	Daily	Hr.	Daily	Bare Costs	Incl. O&P
1 Asbestos Foreman	$53.90	$431.20	$86.55	$692.40	$53.46	$85.85
7 Asbestos Workers	53.40	2990.40	85.75	4802.00		
1 Trk-Mtd Vac, 14 CY, 1500 Gal.		510.30		561.33		
1 Flatbed Truck, 20,000 GVW		247.80		272.58	11.85	13.03
64 L.H., Daily Totals		$4179.70		$6328.31	$65.31	$98.88

Crew A-13

Crew A-13	Hr.	Daily	Hr.	Daily	Bare Costs	Incl. O&P
1 Equip. Oper. (light)	$49.15	$393.20	$77.80	$622.40	$49.15	$77.80
1 Trk-Mtd Vac, 14 CY, 1500 Gal.		510.30		561.33		
1 Flatbed Truck, 20,000 GVW		247.80		272.58	94.76	104.24
8 L.H., Daily Totals		$1151.30		$1456.31	$143.91	$182.04

Crew B-1

Crew B-1	Hr.	Daily	Hr.	Daily	Bare Costs	Incl. O&P
1 Labor Foreman (outside)	$39.90	$319.20	$65.20	$521.60	$38.57	$63.03
2 Laborers	37.90	606.40	61.95	991.20		
24 L.H., Daily Totals		$925.60		$1512.80	$38.57	$63.03

Crew B-1A

Crew B-1A	Hr.	Daily	Hr.	Daily	Bare Costs	Incl. O&P
1 Labor Foreman (outside)	$39.90	$319.20	$65.20	$521.60	$38.57	$63.03
2 Laborers	37.90	606.40	61.95	991.20		
2 Cutting Torches		22.80		25.08		
2 Sets of Gases		336.00		369.60	14.95	16.45
24 L.H., Daily Totals		$1284.40		$1907.48	$53.52	$79.48

Crew B-1B

Crew B-1B	Hr.	Daily	Hr.	Daily	Bare Costs	Incl. O&P
1 Labor Foreman (outside)	$39.90	$319.20	$65.20	$521.60	$41.99	$67.95
2 Laborers	37.90	606.40	61.95	991.20		
1 Equip. Oper. (crane)	52.25	418.00	82.70	661.60		
2 Cutting Torches		22.80		25.08		
2 Sets of Gases		336.00		369.60		
1 Hyd. Crane, 12 Ton		658.40		724.24	31.79	34.97
32 L.H., Daily Totals		$2360.80		$3293.32	$73.78	$102.92

Crew B-1C

Crew B-1C	Hr.	Daily	Hr.	Daily	Bare Costs	Incl. O&P
1 Labor Foreman (outside)	$39.90	$319.20	$65.20	$521.60	$38.57	$63.03
2 Laborers	37.90	606.40	61.95	991.20		
1 Aerial Lift Truck, 60' Boom		448.40		493.24	18.68	20.55
24 L.H., Daily Totals		$1374.00		$2006.04	$57.25	$83.58

Crew B-1D

Crew B-1D	Hr.	Daily	Hr.	Daily	Bare Costs	Incl. O&P
2 Laborers	$37.90	$606.40	$61.95	$991.20	$37.90	$61.95
1 Small Work Boat, Gas, 50 H.P.		174.80		192.28		
1 Pressure Washer, 7 GPM		88.60		97.46	16.46	18.11
16 L.H., Daily Totals		$869.80		$1280.94	$54.36	$80.06

Crew B-1E

Crew B-1E	Hr.	Daily	Hr.	Daily	Bare Costs	Incl. O&P
1 Labor Foreman (outside)	$39.90	$319.20	$65.20	$521.60	$38.40	$62.76
3 Laborers	37.90	909.60	61.95	1486.80		
1 Work Boat, Diesel, 200 H.P.		1455.00		1600.50		
2 Pressure Washers, 7 GPM		177.20		194.92	51.01	56.11
32 L.H., Daily Totals		$2861.00		$3803.82	$89.41	$118.87

Crew B-1F

Crew B-1F	Hr.	Daily	Hr.	Daily	Bare Costs	Incl. O&P
2 Skilled Workers	$49.90	$798.40	$80.85	$1293.60	$45.90	$74.55
1 Laborer	37.90	303.20	61.95	495.60		
1 Small Work Boat, Gas, 50 H.P.		174.80		192.28		
1 Pressure Washer, 7 GPM		88.60		97.46	10.98	12.07
24 L.H., Daily Totals		$1365.00		$2078.94	$56.88	$86.62

Crew B-1G

Crew B-1G	Hr.	Daily	Hr.	Daily	Bare Costs	Incl. O&P
2 Laborers	$37.90	$606.40	$61.95	$991.20	$37.90	$61.95
1 Small Work Boat, Gas, 50 H.P.		174.80		192.28	10.93	12.02
16 L.H., Daily Totals		$781.20		$1183.48	$48.83	$73.97

Crew B-1H

Crew B-1H	Hr.	Daily	Hr.	Daily	Bare Costs	Incl. O&P
2 Skilled Workers	$49.90	$798.40	$80.85	$1293.60	$45.90	$74.55
1 Laborer	37.90	303.20	61.95	495.60		
1 Small Work Boat, Gas, 50 H.P.		174.80		192.28	7.28	8.01
24 L.H., Daily Totals		$1276.40		$1981.48	$53.18	$82.56

Crew B-1J

Crew B-1J	Hr.	Daily	Hr.	Daily	Bare Costs	Incl. O&P
1 Labor Foreman (inside)	$38.40	$307.20	$62.75	$502.00	$38.15	$62.35
1 Laborer	37.90	303.20	61.95	495.60		
16 L.H., Daily Totals		$610.40		$997.60	$38.15	$62.35

Crew B-1K

Crew B-1K	Hr.	Daily	Hr.	Daily	Bare Costs	Incl. O&P
1 Carpenter Foreman (inside)	$48.95	$391.60	$80.00	$640.00	$48.70	$79.58
1 Carpenter	48.45	387.60	79.15	633.20		
16 L.H., Daily Totals		$779.20		$1273.20	$48.70	$79.58

Crew B-2

Crew B-2	Hr.	Daily	Hr.	Daily	Bare Costs	Incl. O&P
1 Labor Foreman (outside)	$39.90	$319.20	$65.20	$521.60	$38.30	$62.60
4 Laborers	37.90	1212.80	61.95	1982.40		
40 L.H., Daily Totals		$1532.00		$2504.00	$38.30	$62.60

Crew B-2A

Crew B-2A	Hr.	Daily	Hr.	Daily	Bare Costs	Incl. O&P
1 Labor Foreman (outside)	$39.90	$319.20	$65.20	$521.60	$38.57	$63.03
2 Laborers	37.90	606.40	61.95	991.20		
1 Aerial Lift Truck, 60' Boom		448.40		493.24	18.68	20.55
24 L.H., Daily Totals		$1374.00		$2006.04	$57.25	$83.58

Crew B-3

Crew B-3	Hr.	Daily	Hr.	Daily	Bare Costs	Incl. O&P
1 Labor Foreman (outside)	$39.90	$319.20	$65.20	$521.60	$42.20	$68.18
2 Laborers	37.90	606.40	61.95	991.20		
1 Equip. Oper. (medium)	51.10	408.80	80.90	647.20		
2 Truck Drivers (heavy)	43.20	691.20	69.55	1112.80		
1 Crawler Loader, 3 C.Y.		1197.00		1316.70		
2 Dump Trucks, 12 C.Y., 400 H.P.		1381.60		1519.76	53.72	59.09
48 L.H., Daily Totals		$4604.20		$6109.26	$95.92	$127.28

Crew B-3A

Crew B-3A	Hr.	Daily	Hr.	Daily	Bare Costs	Incl. O&P
4 Laborers	$37.90	$1212.80	$61.95	$1982.40	$40.54	$65.74
1 Equip. Oper. (medium)	51.10	408.80	80.90	647.20		
1 Hyd. Excavator, 1.5 C.Y.		1029.00		1131.90	25.73	28.30
40 L.H., Daily Totals		$2650.60		$3761.50	$66.27	$94.04

Crew B-3B

Crew B-3B	Hr.	Daily	Hr.	Daily	Bare Costs	Incl. O&P
2 Laborers	$37.90	$606.40	$61.95	$991.20	$42.52	$68.59
1 Equip. Oper. (medium)	51.10	408.80	80.90	647.20		
1 Truck Driver (heavy)	43.20	345.60	69.55	556.40		
1 Backhoe Loader, 80 H.P.		396.20		435.82		
1 Dump Truck, 12 C.Y., 400 H.P.		690.80		759.88	33.97	37.37
32 L.H., Daily Totals		$2447.80		$3390.50	$76.49	$105.95

Crew B-3C

Crew No.	Bare Costs Hr.	Bare Costs Daily	Incl. Subs O&P Hr.	Incl. Subs O&P Daily	Cost Per Labor-Hour Bare Costs	Cost Per Labor-Hour Incl. O&P
3 Laborers	$37.90	$909.60	$61.95	$1486.80	$41.20	$66.69
1 Equip. Oper. (medium)	51.10	408.80	80.90	647.20		
1 Crawler Loader, 4 C.Y.		1566.00		1722.60	48.94	53.83
32 L.H., Daily Totals		$2884.40		$3856.60	$90.14	$120.52

Crew B-4

Crew No.	Bare Costs Hr.	Bare Costs Daily	Incl. Subs O&P Hr.	Incl. Subs O&P Daily	Bare Costs	Incl. O&P
1 Labor Foreman (outside)	$39.90	$319.20	$65.20	$521.60	$39.12	$63.76
4 Laborers	37.90	1212.80	61.95	1982.40		
1 Truck Driver (heavy)	43.20	345.60	69.55	556.40		
1 Truck Tractor, 220 H.P.		358.80		394.68		
1 Flatbed Trailer, 40 Ton		154.60		170.06	10.70	11.77
48 L.H., Daily Totals		$2391.00		$3625.14	$49.81	$75.52

Crew B-5

Crew No.	Bare Costs Hr.	Bare Costs Daily	Incl. Subs O&P Hr.	Incl. Subs O&P Daily	Bare Costs	Incl. O&P
1 Labor Foreman (outside)	$39.90	$319.20	$65.20	$521.60	$41.96	$67.83
4 Laborers	37.90	1212.80	61.95	1982.40		
2 Equip. Opers. (medium)	51.10	817.60	80.90	1294.40		
1 Air Compressor, 250 cfm		197.80		217.58		
2 Breakers, Pavement, 60 lb.		20.40		22.44		
2 -50' Air Hoses, 1.5"		10.40		11.44		
1 Crawler Loader, 3 C.Y.		1197.00		1316.70	25.46	28.00
56 L.H., Daily Totals		$3775.20		$5366.56	$67.41	$95.83

Crew B-5D

Crew No.	Bare Costs Hr.	Bare Costs Daily	Incl. Subs O&P Hr.	Incl. Subs O&P Daily	Bare Costs	Incl. O&P
1 Labor Foreman (outside)	$39.90	$319.20	$65.20	$521.60	$42.11	$68.04
4 Laborers	37.90	1212.80	61.95	1982.40		
2 Equip. Opers. (medium)	51.10	817.60	80.90	1294.40		
1 Truck Driver (heavy)	43.20	345.60	69.55	556.40		
1 Air Compressor, 250 cfm		197.80		217.58		
2 Breakers, Pavement, 60 lb.		20.40		22.44		
2 -50' Air Hoses, 1.5"		10.40		11.44		
1 Crawler Loader, 3 C.Y.		1197.00		1316.70		
1 Dump Truck, 12 C.Y., 400 H.P.		690.80		759.88	33.07	36.38
64 L.H., Daily Totals		$4811.60		$6682.84	$75.18	$104.42

Crew B-6

Crew No.	Bare Costs Hr.	Bare Costs Daily	Incl. Subs O&P Hr.	Incl. Subs O&P Daily	Bare Costs	Incl. O&P
2 Laborers	$37.90	$606.40	$61.95	$991.20	$41.65	$67.23
1 Equip. Oper. (light)	49.15	393.20	77.80	622.40		
1 Backhoe Loader, 48 H.P.		365.40		401.94	15.23	16.75
24 L.H., Daily Totals		$1365.00		$2015.54	$56.88	$83.98

Crew B-6B

Crew No.	Bare Costs Hr.	Bare Costs Daily	Incl. Subs O&P Hr.	Incl. Subs O&P Daily	Bare Costs	Incl. O&P
2 Labor Foremen (outside)	$39.90	$638.40	$65.20	$1043.20	$38.57	$63.03
4 Laborers	37.90	1212.80	61.95	1982.40		
1 S.P. Crane, 4x4, 5 Ton		280.80		308.88		
1 Flatbed Truck, Gas, 1.5 Ton		248.20		273.02		
1 Butt Fusion Mach., 4"-12" diam.		376.10		413.71	18.86	20.74
48 L.H., Daily Totals		$2756.30		$4021.21	$57.42	$83.78

Crew B-6C

Crew No.	Bare Costs Hr.	Bare Costs Daily	Incl. Subs O&P Hr.	Incl. Subs O&P Daily	Bare Costs	Incl. O&P
2 Labor Foremen (outside)	$39.90	$638.40	$65.20	$1043.20	$38.57	$63.03
4 Laborers	37.90	1212.80	61.95	1982.40		
1 S.P. Crane, 4x4, 12 Ton		479.20		527.12		
1 Flatbed Truck, Gas, 3 Ton		307.40		338.14		
1 Butt Fusion Mach., 8"-24" diam.		804.80		885.28	33.15	36.47
48 L.H., Daily Totals		$3442.60		$4776.14	$71.72	$99.50

Crew B-7

Crew No.	Bare Costs Hr.	Bare Costs Daily	Incl. Subs O&P Hr.	Incl. Subs O&P Daily	Bare Costs	Incl. O&P
1 Labor Foreman (outside)	$39.90	$319.20	$65.20	$521.60	$40.43	$65.65
4 Laborers	37.90	1212.80	61.95	1982.40		
1 Equip. Oper. (medium)	51.10	408.80	80.90	647.20		
1 Brush Chipper, 12", 130 H.P.		388.60		427.46		
1 Crawler Loader, 3 C.Y.		1197.00		1316.70		
2 Chain Saws, Gas, 36" Long		90.00		99.00	34.91	38.40
48 L.H., Daily Totals		$3616.40		$4994.36	$75.34	$104.05

Crew B-7A

Crew No.	Bare Costs Hr.	Bare Costs Daily	Incl. Subs O&P Hr.	Incl. Subs O&P Daily	Bare Costs	Incl. O&P
2 Laborers	$37.90	$606.40	$61.95	$991.20	$41.65	$67.23
1 Equip. Oper. (light)	49.15	393.20	77.80	622.40		
1 Rake w/Tractor		353.05		388.36		
2 Chain Saws, Gas, 18"		63.20		69.52	17.34	19.08
24 L.H., Daily Totals		$1415.85		$2071.47	$58.99	$86.31

Crew B-7B

Crew No.	Bare Costs Hr.	Bare Costs Daily	Incl. Subs O&P Hr.	Incl. Subs O&P Daily	Bare Costs	Incl. O&P
1 Labor Foreman (outside)	$39.90	$319.20	$65.20	$521.60	$40.83	$66.21
4 Laborers	37.90	1212.80	61.95	1982.40		
1 Equip. Oper. (medium)	51.10	408.80	80.90	647.20		
1 Truck Driver (heavy)	43.20	345.60	69.55	556.40		
1 Brush Chipper, 12", 130 H.P.		388.60		427.46		
1 Crawler Loader, 3 C.Y.		1197.00		1316.70		
2 Chain Saws, Gas, 36" Long		90.00		99.00		
1 Dump Truck, 8 C.Y., 220 H.P.		417.40		459.14	37.38	41.11
56 L.H., Daily Totals		$4379.40		$6009.90	$78.20	$107.32

Crew B-7C

Crew No.	Bare Costs Hr.	Bare Costs Daily	Incl. Subs O&P Hr.	Incl. Subs O&P Daily	Bare Costs	Incl. O&P
1 Labor Foreman (outside)	$39.90	$319.20	$65.20	$521.60	$40.83	$66.21
4 Laborers	37.90	1212.80	61.95	1982.40		
1 Equip. Oper. (medium)	51.10	408.80	80.90	647.20		
1 Truck Driver (heavy)	43.20	345.60	69.55	556.40		
1 Brush Chipper, 12", 130 H.P.		388.60		427.46		
1 Crawler Loader, 3 C.Y.		1197.00		1316.70		
2 Chain Saws, Gas, 36" Long		90.00		99.00		
1 Dump Truck, 12 C.Y., 400 H.P.		690.80		759.88	42.26	46.48
56 L.H., Daily Totals		$4652.80		$6310.64	$83.09	$112.69

Crew B-8

Crew No.	Bare Costs Hr.	Bare Costs Daily	Incl. Subs O&P Hr.	Incl. Subs O&P Daily	Bare Costs	Incl. O&P
1 Labor Foreman (outside)	$39.90	$319.20	$65.20	$521.60	$43.79	$70.36
2 Laborers	37.90	606.40	61.95	991.20		
2 Equip. Opers. (medium)	51.10	817.60	80.90	1294.40		
1 Equip. Oper. (oiler)	46.05	368.40	72.90	583.20		
2 Truck Drivers (heavy)	43.20	691.20	69.55	1112.80		
1 Hyd. Crane, 25 Ton		748.80		823.68		
1 Crawler Loader, 3 C.Y.		1197.00		1316.70		
2 Dump Trucks, 12 C.Y., 400 H.P.		1381.60		1519.76	51.99	57.19
64 L.H., Daily Totals		$6130.20		$8163.34	$95.78	$127.55

Crew B-9

Crew No.	Bare Costs Hr.	Bare Costs Daily	Incl. Subs O&P Hr.	Incl. Subs O&P Daily	Bare Costs	Incl. O&P
1 Labor Foreman (outside)	$39.90	$319.20	$65.20	$521.60	$38.30	$62.60
4 Laborers	37.90	1212.80	61.95	1982.40		
1 Air Compressor, 250 cfm		197.80		217.58		
2 Breakers, Pavement, 60 lb.		20.40		22.44		
2 -50' Air Hoses, 1.5"		10.40		11.44	5.71	6.29
40 L.H., Daily Totals		$1760.60		$2755.46	$44.02	$68.89

728

Crew B-9A

Crew B-9A	Bare Costs Hr.	Daily	Incl. Subs O&P Hr.	Daily	Cost Per Labor-Hour Bare Costs	Incl. O&P
2 Laborers	$37.90	$606.40	$61.95	$991.20	$39.67	$64.48
1 Truck Driver (heavy)	43.20	345.60	69.55	556.40		
1 Water Tank Trailer, 5000 Gal.		141.60		155.76		
1 Truck Tractor, 220 H.P.		358.80		394.68		
2 -50' Discharge Hoses, 3"		3.00		3.30	20.98	23.07
24 L.H., Daily Totals		$1455.40		$2101.34	$60.64	$87.56

Crew B-9B

Crew B-9B	Bare Costs Hr.	Daily	Incl. Subs O&P Hr.	Daily	Cost Per Labor-Hour Bare Costs	Incl. O&P
2 Laborers	$37.90	$606.40	$61.95	$991.20	$39.67	$64.48
1 Truck Driver (heavy)	43.20	345.60	69.55	556.40		
2 -50' Discharge Hoses, 3"		3.00		3.30		
1 Water Tank Trailer, 5000 Gal.		141.60		155.76		
1 Truck Tractor, 220 H.P.		358.80		394.68		
1 Pressure Washer		71.60		78.76	23.96	26.35
24 L.H., Daily Totals		$1527.00		$2180.10	$63.63	$90.84

Crew B-9D

Crew B-9D	Bare Costs Hr.	Daily	Incl. Subs O&P Hr.	Daily	Cost Per Labor-Hour Bare Costs	Incl. O&P
1 Labor Foreman (outside)	$39.90	$319.20	$65.20	$521.60	$38.30	$62.60
4 Common Laborers	37.90	1212.80	61.95	1982.40		
1 Air Compressor, 250 cfm		197.80		217.58		
2 -50' Air Hoses, 1.5"		10.40		11.44		
2 Air Powered Tampers		48.00		52.80	6.41	7.05
40 L.H., Daily Totals		$1788.20		$2785.82	$44.70	$69.65

Crew B-10

Crew B-10	Bare Costs Hr.	Daily	Incl. Subs O&P Hr.	Daily	Cost Per Labor-Hour Bare Costs	Incl. O&P
1 Equip. Oper. (medium)	$51.10	$408.80	$80.90	$647.20	$46.70	$74.58
.5 Laborer	37.90	151.60	61.95	247.80		
12 L.H., Daily Totals		$560.40		$895.00	$46.70	$74.58

Crew B-10A

Crew B-10A	Bare Costs Hr.	Daily	Incl. Subs O&P Hr.	Daily	Cost Per Labor-Hour Bare Costs	Incl. O&P
1 Equip. Oper. (medium)	$51.10	$408.80	$80.90	$647.20	$46.70	$74.58
.5 Laborer	37.90	151.60	61.95	247.80		
1 Roller, 2-Drum, W.B., 7.5 H.P.		184.80		203.28	15.40	16.94
12 L.H., Daily Totals		$745.20		$1098.28	$62.10	$91.52

Crew B-10B

Crew B-10B	Bare Costs Hr.	Daily	Incl. Subs O&P Hr.	Daily	Cost Per Labor-Hour Bare Costs	Incl. O&P
1 Equip. Oper. (medium)	$51.10	$408.80	$80.90	$647.20	$46.70	$74.58
.5 Laborer	37.90	151.60	61.95	247.80		
1 Dozer, 200 H.P.		1397.00		1536.70	116.42	128.06
12 L.H., Daily Totals		$1957.40		$2431.70	$163.12	$202.64

Crew B-10C

Crew B-10C	Bare Costs Hr.	Daily	Incl. Subs O&P Hr.	Daily	Cost Per Labor-Hour Bare Costs	Incl. O&P
1 Equip. Oper. (medium)	$51.10	$408.80	$80.90	$647.20	$46.70	$74.58
.5 Laborer	37.90	151.60	61.95	247.80		
1 Dozer, 200 H.P.		1397.00		1536.70		
1 Vibratory Roller, Towed, 23 Ton		401.60		441.76	149.88	164.87
12 L.H., Daily Totals		$2359.00		$2873.46	$196.58	$239.46

Crew B-10D

Crew B-10D	Bare Costs Hr.	Daily	Incl. Subs O&P Hr.	Daily	Cost Per Labor-Hour Bare Costs	Incl. O&P
1 Equip. Oper. (medium)	$51.10	$408.80	$80.90	$647.20	$46.70	$74.58
.5 Laborer	37.90	151.60	61.95	247.80		
1 Dozer, 200 H.P.		1397.00		1536.70		
1 Sheepsft. Roller, Towed		426.40		469.04	151.95	167.15
12 L.H., Daily Totals		$2383.80		$2900.74	$198.65	$241.73

Crew B-10E

Crew B-10E	Bare Costs Hr.	Daily	Incl. Subs O&P Hr.	Daily	Cost Per Labor-Hour Bare Costs	Incl. O&P
1 Equip. Oper. (medium)	$51.10	$408.80	$80.90	$647.20	$46.70	$74.58
.5 Laborer	37.90	151.60	61.95	247.80		
1 Tandem Roller, 5 Ton		162.60		178.86	13.55	14.90
12 L.H., Daily Totals		$723.00		$1073.86	$60.25	$89.49

Crew B-10F

Crew B-10F	Bare Costs Hr.	Daily	Incl. Subs O&P Hr.	Daily	Cost Per Labor-Hour Bare Costs	Incl. O&P
1 Equip. Oper. (medium)	$51.10	$408.80	$80.90	$647.20	$46.70	$74.58
.5 Laborer	37.90	151.60	61.95	247.80		
1 Tandem Roller, 10 Ton		239.40		263.34	19.95	21.95
12 L.H., Daily Totals		$799.80		$1158.34	$66.65	$96.53

Crew B-10G

Crew B-10G	Bare Costs Hr.	Daily	Incl. Subs O&P Hr.	Daily	Cost Per Labor-Hour Bare Costs	Incl. O&P
1 Equip. Oper. (medium)	$51.10	$408.80	$80.90	$647.20	$46.70	$74.58
.5 Laborer	37.90	151.60	61.95	247.80		
1 Sheepsfoot Roller, 240 H.P.		1208.00		1328.80	100.67	110.73
12 L.H., Daily Totals		$1768.40		$2223.80	$147.37	$185.32

Crew B-10H

Crew B-10H	Bare Costs Hr.	Daily	Incl. Subs O&P Hr.	Daily	Cost Per Labor-Hour Bare Costs	Incl. O&P
1 Equip. Oper. (medium)	$51.10	$408.80	$80.90	$647.20	$46.70	$74.58
.5 Laborer	37.90	151.60	61.95	247.80		
1 Diaphragm Water Pump, 2"		71.80		78.98		
1 -20' Suction Hose, 2"		1.95		2.15		
2 -50' Discharge Hoses, 2"		1.80		1.98	6.30	6.93
12 L.H., Daily Totals		$635.95		$978.11	$53.00	$81.51

Crew B-10I

Crew B-10I	Bare Costs Hr.	Daily	Incl. Subs O&P Hr.	Daily	Cost Per Labor-Hour Bare Costs	Incl. O&P
1 Equip. Oper. (medium)	$51.10	$408.80	$80.90	$647.20	$46.70	$74.58
.5 Laborer	37.90	151.60	61.95	247.80		
1 Diaphragm Water Pump, 4"		118.00		129.80		
1 -20' Suction Hose, 4"		3.25		3.58		
2 -50' Discharge Hoses, 4"		4.70		5.17	10.50	11.55
12 L.H., Daily Totals		$686.35		$1033.55	$57.20	$86.13

Crew B-10J

Crew B-10J	Bare Costs Hr.	Daily	Incl. Subs O&P Hr.	Daily	Cost Per Labor-Hour Bare Costs	Incl. O&P
1 Equip. Oper. (medium)	$51.10	$408.80	$80.90	$647.20	$46.70	$74.58
.5 Laborer	37.90	151.60	61.95	247.80		
1 Centrifugal Water Pump, 3"		81.00		89.10		
1 -20' Suction Hose, 3"		2.85		3.13		
2 -50' Discharge Hoses, 3"		3.00		3.30	7.24	7.96
12 L.H., Daily Totals		$647.25		$990.53	$53.94	$82.54

Crew B-10K

Crew B-10K	Bare Costs Hr.	Daily	Incl. Subs O&P Hr.	Daily	Cost Per Labor-Hour Bare Costs	Incl. O&P
1 Equip. Oper. (medium)	$51.10	$408.80	$80.90	$647.20	$46.70	$74.58
.5 Laborer	37.90	151.60	61.95	247.80		
1 Centr. Water Pump, 6"		350.80		385.88		
1 -20' Suction Hose, 6"		11.50		12.65		
2 -50' Discharge Hoses, 6"		12.20		13.42	31.21	34.33
12 L.H., Daily Totals		$934.90		$1306.95	$77.91	$108.91

Crew B-10L

Crew B-10L	Bare Costs Hr.	Daily	Incl. Subs O&P Hr.	Daily	Cost Per Labor-Hour Bare Costs	Incl. O&P
1 Equip. Oper. (medium)	$51.10	$408.80	$80.90	$647.20	$46.70	$74.58
.5 Laborer	37.90	151.60	61.95	247.80		
1 Dozer, 80 H.P.		471.40		518.54	39.28	43.21
12 L.H., Daily Totals		$1031.80		$1413.54	$85.98	$117.80

Crew B-10M

Crew B-10M	Bare Costs Hr.	Daily	Incl. Subs O&P Hr.	Daily	Cost Per Labor-Hour Bare Costs	Incl. O&P
1 Equip. Oper. (medium)	$51.10	$408.80	$80.90	$647.20	$46.70	$74.58
.5 Laborer	37.90	151.60	61.95	247.80		
1 Dozer, 300 H.P.		1887.00		2075.70	157.25	172.97
12 L.H., Daily Totals		$2447.40		$2970.70	$203.95	$247.56

Crew B-10N

Crew B-10N	Bare Costs Hr.	Daily	Incl. Subs O&P Hr.	Daily	Cost Per Labor-Hour Bare Costs	Incl. O&P
1 Equip. Oper. (medium)	$51.10	$408.80	$80.90	$647.20	$46.70	$74.58
.5 Laborer	37.90	151.60	61.95	247.80		
1 F.E. Loader, T.M., 1.5 C.Y		525.00		577.50	43.75	48.13
12 L.H., Daily Totals		$1085.40		$1472.50	$90.45	$122.71

Crews - Renovation

Crew No.	Bare Costs		Incl. Subs O&P		Cost Per Labor-Hour	

Crew B-100

Crew B-100	Hr.	Daily	Hr.	Daily	Bare Costs	Incl. O&P
1 Equip. Oper. (medium)	$51.10	$408.80	$80.90	$647.20	$46.70	$74.58
.5 Laborer	37.90	151.60	61.95	247.80		
1 F.E. Loader, T.M., 2.25 C.Y.		964.00		1060.40	80.33	88.37
12 L.H., Daily Totals		$1524.40		$1955.40	$127.03	$162.95

Crew B-10P	Hr.	Daily	Hr.	Daily	Bare Costs	Incl. O&P
1 Equip. Oper. (medium)	$51.10	$408.80	$80.90	$647.20	$46.70	$74.58
.5 Laborer	37.90	151.60	61.95	247.80		
1 Crawler Loader, 3 C.Y.		1197.00		1316.70	99.75	109.72
12 L.H., Daily Totals		$1757.40		$2211.70	$146.45	$184.31

Crew B-10Q	Hr.	Daily	Hr.	Daily	Bare Costs	Incl. O&P
1 Equip. Oper. (medium)	$51.10	$408.80	$80.90	$647.20	$46.70	$74.58
.5 Laborer	37.90	151.60	61.95	247.80		
1 Crawler Loader, 4 C.Y.		1566.00		1722.60	130.50	143.55
12 L.H., Daily Totals		$2126.40		$2617.60	$177.20	$218.13

Crew B-10R	Hr.	Daily	Hr.	Daily	Bare Costs	Incl. O&P
1 Equip. Oper. (medium)	$51.10	$408.80	$80.90	$647.20	$46.70	$74.58
.5 Laborer	37.90	151.60	61.95	247.80		
1 F.E. Loader, W.M., 1 C.Y.		299.00		328.90	24.92	27.41
12 L.H., Daily Totals		$859.40		$1223.90	$71.62	$101.99

Crew B-10S	Hr.	Daily	Hr.	Daily	Bare Costs	Incl. O&P
1 Equip. Oper. (medium)	$51.10	$408.80	$80.90	$647.20	$46.70	$74.58
.5 Laborer	37.90	151.60	61.95	247.80		
1 F.E. Loader, W.M., 1.5 C.Y.		378.60		416.46	31.55	34.70
12 L.H., Daily Totals		$939.00		$1311.46	$78.25	$109.29

Crew B-10T	Hr.	Daily	Hr.	Daily	Bare Costs	Incl. O&P
1 Equip. Oper. (medium)	$51.10	$408.80	$80.90	$647.20	$46.70	$74.58
.5 Laborer	37.90	151.60	61.95	247.80		
1 F.E. Loader, W.M., 2.5 C.Y.		516.00		567.60	43.00	47.30
12 L.H., Daily Totals		$1076.40		$1462.60	$89.70	$121.88

Crew B-10U	Hr.	Daily	Hr.	Daily	Bare Costs	Incl. O&P
1 Equip. Oper. (medium)	$51.10	$408.80	$80.90	$647.20	$46.70	$74.58
.5 Laborer	37.90	151.60	61.95	247.80		
1 F.E. Loader, W.M., 5.5 C.Y.		1092.00		1201.20	91.00	100.10
12 L.H., Daily Totals		$1652.40		$2096.20	$137.70	$174.68

Crew B-10V	Hr.	Daily	Hr.	Daily	Bare Costs	Incl. O&P
1 Equip. Oper. (medium)	$51.10	$408.80	$80.90	$647.20	$46.70	$74.58
.5 Laborer	37.90	151.60	61.95	247.80		
1 Dozer, 700 H.P.		5347.00		5881.70	445.58	490.14
12 L.H., Daily Totals		$5907.40		$6776.70	$492.28	$564.73

Crew B-10W	Hr.	Daily	Hr.	Daily	Bare Costs	Incl. O&P
1 Equip. Oper. (medium)	$51.10	$408.80	$80.90	$647.20	$46.70	$74.58
.5 Laborer	37.90	151.60	61.95	247.80		
1 Dozer, 105 H.P.		603.40		663.74	50.28	55.31
12 L.H., Daily Totals		$1163.80		$1558.74	$96.98	$129.90

Crew B-10X	Hr.	Daily	Hr.	Daily	Bare Costs	Incl. O&P
1 Equip. Oper. (medium)	$51.10	$408.80	$80.90	$647.20	$46.70	$74.58
.5 Laborer	37.90	151.60	61.95	247.80		
1 Dozer, 410 H.P.		2392.00		2631.20	199.33	219.27
12 L.H., Daily Totals		$2952.40		$3526.20	$246.03	$293.85

Crew B-10Y	Hr.	Daily	Hr.	Daily	Bare Costs	Incl. O&P
1 Equip. Oper. (medium)	$51.10	$408.80	$80.90	$647.20	$46.70	$74.58
.5 Laborer	37.90	151.60	61.95	247.80		
1 Vibr. Roller, Towed, 12 Ton		566.00		622.60	47.17	51.88
12 L.H., Daily Totals		$1126.40		$1517.60	$93.87	$126.47

Crew B-11A	Hr.	Daily	Hr.	Daily	Bare Costs	Incl. O&P
1 Equipment Oper. (med.)	$51.10	$408.80	$80.90	$647.20	$44.50	$71.42
1 Laborer	37.90	303.20	61.95	495.60		
1 Dozer, 200 H.P.		1397.00		1536.70	87.31	96.04
16 L.H., Daily Totals		$2109.00		$2679.50	$131.81	$167.47

Crew B-11B	Hr.	Daily	Hr.	Daily	Bare Costs	Incl. O&P
1 Equipment Oper. (light)	$49.15	$393.20	$77.80	$622.40	$43.52	$69.88
1 Laborer	37.90	303.20	61.95	495.60		
1 Air Powered Tamper		24.00		26.40		
1 Air Compressor, 365 cfm		258.40		284.24		
2 -50' Air Hoses, 1.5"		10.40		11.44	18.30	20.13
16 L.H., Daily Totals		$989.20		$1440.08	$61.83	$90.00

Crew B-11C	Hr.	Daily	Hr.	Daily	Bare Costs	Incl. O&P
1 Equipment Oper. (med.)	$51.10	$408.80	$80.90	$647.20	$44.50	$71.42
1 Laborer	37.90	303.20	61.95	495.60		
1 Backhoe Loader, 48 H.P.		365.40		401.94	22.84	25.12
16 L.H., Daily Totals		$1077.40		$1544.74	$67.34	$96.55

Crew B-11K	Hr.	Daily	Hr.	Daily	Bare Costs	Incl. O&P
1 Equipment Oper. (med.)	$51.10	$408.80	$80.90	$647.20	$44.50	$71.42
1 Laborer	37.90	303.20	61.95	495.60		
1 Trencher, Chain Type, 8' D		2860.00		3146.00	178.75	196.63
16 L.H., Daily Totals		$3572.00		$4288.80	$223.25	$268.05

Crew B-11L	Hr.	Daily	Hr.	Daily	Bare Costs	Incl. O&P
1 Equipment Oper. (med.)	$51.10	$408.80	$80.90	$647.20	$44.50	$71.42
1 Laborer	37.90	303.20	61.95	495.60		
1 Grader, 30,000 Lbs.		714.40		785.84	44.65	49.12
16 L.H., Daily Totals		$1426.40		$1928.64	$89.15	$120.54

Crew B-11M	Hr.	Daily	Hr.	Daily	Bare Costs	Incl. O&P
1 Equipment Oper. (med.)	$51.10	$408.80	$80.90	$647.20	$44.50	$71.42
1 Laborer	37.90	303.20	61.95	495.60		
1 Backhoe Loader, 80 H.P.		396.20		435.82	24.76	27.24
16 L.H., Daily Totals		$1108.20		$1578.62	$69.26	$98.66

Crew B-11S	Hr.	Daily	Hr.	Daily	Bare Costs	Incl. O&P
1 Equipment Operator (med.)	$51.10	$408.80	$80.90	$647.20	$46.70	$74.58
.5 Laborer	37.90	151.60	61.95	247.80		
1 Dozer, 300 H.P.		1887.00		2075.70		
1 Ripper, Beam & 1 Shank		83.80		92.18	164.23	180.66
12 L.H., Daily Totals		$2531.20		$3062.88	$210.93	$255.24

Crew B-11T	Hr.	Daily	Hr.	Daily	Bare Costs	Incl. O&P
1 Equipment Operator (med.)	$51.10	$408.80	$80.90	$647.20	$46.70	$74.58
.5 Laborer	37.90	151.60	61.95	247.80		
1 Dozer, 410 H.P.		2392.00		2631.20		
1 Ripper, Beam & 2 Shanks		129.00		141.90	210.08	231.09
12 L.H., Daily Totals		$3081.40		$3668.10	$256.78	$305.68

For customer support on your Commercial Renovation Cost Data, call 877.742.4719.

Crew No.	Bare Costs		Incl. Subs O&P		Cost Per Labor-Hour	
Crew B-11W	**Hr.**	**Daily**	**Hr.**	**Daily**	Bare Costs	Incl. O&P
1 Equipment Operator (med.)	$51.10	$408.80	$80.90	$647.20	$43.42	$69.86
1 Common Laborer	37.90	303.20	61.95	495.60		
10 Truck Drivers (heavy)	43.20	3456.00	69.55	5564.00		
1 Dozer, 200 H.P.		1397.00		1536.70		
1 Vibratory Roller, Towed, 23 Ton		401.60		441.76		
10 Dump Trucks, 8 C.Y., 220 H.P.		4174.00		4591.40	62.21	68.44
96 L.H., Daily Totals		$10140.60		$13276.66	$105.63	$138.30
Crew B-11Y	**Hr.**	**Daily**	**Hr.**	**Daily**	Bare Costs	Incl. O&P
1 Labor Foreman (outside)	$39.90	$319.20	$65.20	$521.60	$42.52	$68.63
5 Common Laborers	37.90	1516.00	61.95	2478.00		
3 Equipment Operators (med.)	51.10	1226.40	80.90	1941.60		
1 Dozer, 80 H.P.		471.40		518.54		
2 Rollers, 2-Drum, W.B., 7.5 H.P.		369.60		406.56		
4 Vibrating Plates, Gas, 21"		186.40		205.04	14.27	15.70
72 L.H., Daily Totals		$4089.00		$6071.34	$56.79	$84.32
Crew B-12A	**Hr.**	**Daily**	**Hr.**	**Daily**	Bare Costs	Incl. O&P
1 Equip. Oper. (crane)	$52.25	$418.00	$82.70	$661.60	$45.08	$72.33
1 Laborer	37.90	303.20	61.95	495.60		
1 Hyd. Excavator, 1 C.Y.		812.80		894.08	50.80	55.88
16 L.H., Daily Totals		$1534.00		$2051.28	$95.88	$128.21
Crew B-12B	**Hr.**	**Daily**	**Hr.**	**Daily**	Bare Costs	Incl. O&P
1 Equip. Oper. (crane)	$52.25	$418.00	$82.70	$661.60	$45.08	$72.33
1 Laborer	37.90	303.20	61.95	495.60		
1 Hyd. Excavator, 1.5 C.Y.		1029.00		1131.90	64.31	70.74
16 L.H., Daily Totals		$1750.20		$2289.10	$109.39	$143.07
Crew B-12C	**Hr.**	**Daily**	**Hr.**	**Daily**	Bare Costs	Incl. O&P
1 Equip. Oper. (crane)	$52.25	$418.00	$82.70	$661.60	$45.08	$72.33
1 Laborer	37.90	303.20	61.95	495.60		
1 Hyd. Excavator, 2 C.Y.		1176.00		1293.60	73.50	80.85
16 L.H., Daily Totals		$1897.20		$2450.80	$118.58	$153.18
Crew B-12D	**Hr.**	**Daily**	**Hr.**	**Daily**	Bare Costs	Incl. O&P
1 Equip. Oper. (crane)	$52.25	$418.00	$82.70	$661.60	$45.08	$72.33
1 Laborer	37.90	303.20	61.95	495.60		
1 Hyd. Excavator, 3.5 C.Y.		2442.00		2686.20	152.63	167.89
16 L.H., Daily Totals		$3163.20		$3843.40	$197.70	$240.21
Crew B-12E	**Hr.**	**Daily**	**Hr.**	**Daily**	Bare Costs	Incl. O&P
1 Equip. Oper. (crane)	$52.25	$418.00	$82.70	$661.60	$45.08	$72.33
1 Laborer	37.90	303.20	61.95	495.60		
1 Hyd. Excavator, .5 C.Y.		445.60		490.16	27.85	30.64
16 L.H., Daily Totals		$1166.80		$1647.36	$72.92	$102.96
Crew B-12F	**Hr.**	**Daily**	**Hr.**	**Daily**	Bare Costs	Incl. O&P
1 Equip. Oper. (crane)	$52.25	$418.00	$82.70	$661.60	$45.08	$72.33
1 Laborer	37.90	303.20	61.95	495.60		
1 Hyd. Excavator, .75 C.Y.		654.00		719.40	40.88	44.96
16 L.H., Daily Totals		$1375.20		$1876.60	$85.95	$117.29
Crew B-12G	**Hr.**	**Daily**	**Hr.**	**Daily**	Bare Costs	Incl. O&P
1 Equip. Oper. (crane)	$52.25	$418.00	$82.70	$661.60	$45.08	$72.33
1 Laborer	37.90	303.20	61.95	495.60		
1 Crawler Crane, 15 Ton		695.70		765.27		
1 Clamshell Bucket, .5 C.Y.		37.80		41.58	45.84	50.43
16 L.H., Daily Totals		$1454.70		$1964.05	$90.92	$122.75
Crew B-12H	**Hr.**	**Daily**	**Hr.**	**Daily**	Bare Costs	Incl. O&P
1 Equip. Oper. (crane)	$52.25	$418.00	$82.70	$661.60	$45.08	$72.33
1 Laborer	37.90	303.20	61.95	495.60		
1 Crawler Crane, 25 Ton		1194.00		1313.40		
1 Clamshell Bucket, 1 C.Y.		47.40		52.14	77.59	85.35
16 L.H., Daily Totals		$1962.60		$2522.74	$122.66	$157.67
Crew B-12I	**Hr.**	**Daily**	**Hr.**	**Daily**	Bare Costs	Incl. O&P
1 Equip. Oper. (crane)	$52.25	$418.00	$82.70	$661.60	$45.08	$72.33
1 Laborer	37.90	303.20	61.95	495.60		
1 Crawler Crane, 20 Ton		896.25		985.88		
1 Dragline Bucket, .75 C.Y.		20.60		22.66	57.30	63.03
16 L.H., Daily Totals		$1638.05		$2165.74	$102.38	$135.36
Crew B-12J	**Hr.**	**Daily**	**Hr.**	**Daily**	Bare Costs	Incl. O&P
1 Equip. Oper. (crane)	$52.25	$418.00	$82.70	$661.60	$45.08	$72.33
1 Laborer	37.90	303.20	61.95	495.60		
1 Gradall, 5/8 C.Y.		882.00		970.20	55.13	60.64
16 L.H., Daily Totals		$1603.20		$2127.40	$100.20	$132.96
Crew B-12K	**Hr.**	**Daily**	**Hr.**	**Daily**	Bare Costs	Incl. O&P
1 Equip. Oper. (crane)	$52.25	$418.00	$82.70	$661.60	$45.08	$72.33
1 Laborer	37.90	303.20	61.95	495.60		
1 Gradall, 3 Ton, 1 C.Y.		995.80		1095.38	62.24	68.46
16 L.H., Daily Totals		$1717.00		$2252.58	$107.31	$140.79
Crew B-12L	**Hr.**	**Daily**	**Hr.**	**Daily**	Bare Costs	Incl. O&P
1 Equip. Oper. (crane)	$52.25	$418.00	$82.70	$661.60	$45.08	$72.33
1 Laborer	37.90	303.20	61.95	495.60		
1 Crawler Crane, 15 Ton		695.70		765.27		
1 F.E. Attachment, .5 C.Y.		59.20		65.12	47.18	51.90
16 L.H., Daily Totals		$1476.10		$1987.59	$92.26	$124.22
Crew B-12M	**Hr.**	**Daily**	**Hr.**	**Daily**	Bare Costs	Incl. O&P
1 Equip. Oper. (crane)	$52.25	$418.00	$82.70	$661.60	$45.08	$72.33
1 Laborer	37.90	303.20	61.95	495.60		
1 Crawler Crane, 20 Ton		896.25		985.88		
1 F.E. Attachment, .75 C.Y.		63.60		69.96	59.99	65.99
16 L.H., Daily Totals		$1681.05		$2213.03	$105.07	$138.31
Crew B-12N	**Hr.**	**Daily**	**Hr.**	**Daily**	Bare Costs	Incl. O&P
1 Equip. Oper. (crane)	$52.25	$418.00	$82.70	$661.60	$45.08	$72.33
1 Laborer	37.90	303.20	61.95	495.60		
1 Crawler Crane, 25 Ton		1194.00		1313.40		
1 F.E. Attachment, 1 C.Y.		70.40		77.44	79.03	86.93
16 L.H., Daily Totals		$1985.60		$2548.04	$124.10	$159.25
Crew B-12O	**Hr.**	**Daily**	**Hr.**	**Daily**	Bare Costs	Incl. O&P
1 Equip. Oper. (crane)	$52.25	$418.00	$82.70	$661.60	$45.08	$72.33
1 Laborer	37.90	303.20	61.95	495.60		
1 Crawler Crane, 40 Ton		1203.00		1323.30		
1 F.E. Attachment, 1.5 C.Y.		79.60		87.56	80.16	88.18
16 L.H., Daily Totals		$2003.80		$2568.06	$125.24	$160.50
Crew B-12P	**Hr.**	**Daily**	**Hr.**	**Daily**	Bare Costs	Incl. O&P
1 Equip. Oper. (crane)	$52.25	$418.00	$82.70	$661.60	$45.08	$72.33
1 Laborer	37.90	303.20	61.95	495.60		
1 Crawler Crane, 40 Ton		1203.00		1323.30		
1 Dragline Bucket, 1.5 C.Y.		33.60		36.96	77.29	85.02
16 L.H., Daily Totals		$1957.80		$2517.46	$122.36	$157.34

Crew B-12Q

Crew No.	Bare Costs Hr.	Bare Costs Daily	Incl. Subs O&P Hr.	Incl. Subs O&P Daily	Cost Per Labor-Hour Bare Costs	Cost Per Labor-Hour Incl. O&P
1 Equip. Oper. (crane)	$52.25	$418.00	$82.70	$661.60	$45.08	$72.33
1 Laborer	37.90	303.20	61.95	495.60		
1 Hyd. Excavator, 5/8 C.Y.		589.40		648.34	36.84	40.52
16 L.H., Daily Totals		$1310.60		$1805.54	$81.91	$112.85

Crew B-12S

Crew No.	Bare Costs Hr.	Bare Costs Daily	Incl. Subs O&P Hr.	Incl. Subs O&P Daily	Cost Per Labor-Hour Bare Costs	Cost Per Labor-Hour Incl. O&P
1 Equip. Oper. (crane)	$52.25	$418.00	$82.70	$661.60	$45.08	$72.33
1 Laborer	37.90	303.20	61.95	495.60		
1 Hyd. Excavator, 2.5 C.Y.		1582.00		1740.20	98.88	108.76
16 L.H., Daily Totals		$2303.20		$2897.40	$143.95	$181.09

Crew B-12T

Crew No.	Bare Costs Hr.	Bare Costs Daily	Incl. Subs O&P Hr.	Incl. Subs O&P Daily	Cost Per Labor-Hour Bare Costs	Cost Per Labor-Hour Incl. O&P
1 Equip. Oper. (crane)	$52.25	$418.00	$82.70	$661.60	$45.08	$72.33
1 Laborer	37.90	303.20	61.95	495.60		
1 Crawler Crane, 75 Ton		1507.00		1657.70		
1 F.E. Attachment, 3 C.Y.		102.40		112.64	100.59	110.65
16 L.H., Daily Totals		$2330.60		$2927.54	$145.66	$182.97

Crew B-12V

Crew No.	Bare Costs Hr.	Bare Costs Daily	Incl. Subs O&P Hr.	Incl. Subs O&P Daily	Cost Per Labor-Hour Bare Costs	Cost Per Labor-Hour Incl. O&P
1 Equip. Oper. (crane)	$52.25	$418.00	$82.70	$661.60	$45.08	$72.33
1 Laborer	37.90	303.20	61.95	495.60		
1 Crawler Crane, 75 Ton		1507.00		1657.70		
1 Dragline Bucket, 3 C.Y.		52.60		57.86	97.47	107.22
16 L.H., Daily Totals		$2280.80		$2872.76	$142.55	$179.55

Crew B-12Y

Crew No.	Bare Costs Hr.	Bare Costs Daily	Incl. Subs O&P Hr.	Incl. Subs O&P Daily	Cost Per Labor-Hour Bare Costs	Cost Per Labor-Hour Incl. O&P
1 Equip. Oper. (crane)	$52.25	$418.00	$82.70	$661.60	$42.68	$68.87
2 Laborers	37.90	606.40	61.95	991.20		
1 Hyd. Excavator, 3.5 C.Y.		2442.00		2686.20	101.75	111.93
24 L.H., Daily Totals		$3466.40		$4339.00	$144.43	$180.79

Crew B-12Z

Crew No.	Bare Costs Hr.	Bare Costs Daily	Incl. Subs O&P Hr.	Incl. Subs O&P Daily	Cost Per Labor-Hour Bare Costs	Cost Per Labor-Hour Incl. O&P
1 Equip. Oper. (crane)	$52.25	$418.00	$82.70	$661.60	$42.68	$68.87
2 Laborers	37.90	606.40	61.95	991.20		
1 Hyd. Excavator, 2.5 C.Y.		1582.00		1740.20	65.92	72.51
24 L.H., Daily Totals		$2606.40		$3393.00	$108.60	$141.38

Crew B-13

Crew No.	Bare Costs Hr.	Bare Costs Daily	Incl. Subs O&P Hr.	Incl. Subs O&P Daily	Cost Per Labor-Hour Bare Costs	Cost Per Labor-Hour Incl. O&P
1 Labor Foreman (outside)	$39.90	$319.20	$65.20	$521.60	$41.40	$66.94
4 Laborers	37.90	1212.80	61.95	1982.40		
1 Equip. Oper. (crane)	52.25	418.00	82.70	661.60		
1 Equip. Oper. (oiler)	46.05	368.40	72.90	583.20		
1 Hyd. Crane, 25 Ton		748.80		823.68	13.37	14.71
56 L.H., Daily Totals		$3067.20		$4572.48	$54.77	$81.65

Crew B-13A

Crew No.	Bare Costs Hr.	Bare Costs Daily	Incl. Subs O&P Hr.	Incl. Subs O&P Daily	Cost Per Labor-Hour Bare Costs	Cost Per Labor-Hour Incl. O&P
1 Labor Foreman (outside)	$39.90	$319.20	$65.20	$521.60	$43.47	$70.00
2 Laborers	37.90	606.40	61.95	991.20		
2 Equipment Operators (med.)	51.10	817.60	80.90	1294.40		
2 Truck Drivers (heavy)	43.20	691.20	69.55	1112.80		
1 Crawler Crane, 75 Ton		1507.00		1657.70		
1 Crawler Loader, 4 C.Y.		1566.00		1722.60		
2 Dump Trucks, 8 C.Y., 220 H.P.		834.80		918.28	69.78	76.76
56 L.H., Daily Totals		$6342.20		$8218.58	$113.25	$146.76

Crew B-13B

Crew No.	Bare Costs Hr.	Bare Costs Daily	Incl. Subs O&P Hr.	Incl. Subs O&P Daily	Cost Per Labor-Hour Bare Costs	Cost Per Labor-Hour Incl. O&P
1 Labor Foreman (outside)	$39.90	$319.20	$65.20	$521.60	$41.40	$66.94
4 Laborers	37.90	1212.80	61.95	1982.40		
1 Equip. Oper. (crane)	52.25	418.00	82.70	661.60		
1 Equip. Oper. (oiler)	46.05	368.40	72.90	583.20		
1 Hyd. Crane, 55 Ton		1123.00		1235.30	20.05	22.06
56 L.H., Daily Totals		$3441.40		$4984.10	$61.45	$89.00

Crew B-13C

Crew No.	Bare Costs Hr.	Bare Costs Daily	Incl. Subs O&P Hr.	Incl. Subs O&P Daily	Cost Per Labor-Hour Bare Costs	Cost Per Labor-Hour Incl. O&P
1 Labor Foreman (outside)	$39.90	$319.20	$65.20	$521.60	$41.40	$66.94
4 Laborers	37.90	1212.80	61.95	1982.40		
1 Equip. Oper. (crane)	52.25	418.00	82.70	661.60		
1 Equip. Oper. (oiler)	46.05	368.40	72.90	583.20		
1 Crawler Crane, 100 Ton		1667.00		1833.70	29.77	32.74
56 L.H., Daily Totals		$3985.40		$5582.50	$71.17	$99.69

Crew B-13D

Crew No.	Bare Costs Hr.	Bare Costs Daily	Incl. Subs O&P Hr.	Incl. Subs O&P Daily	Cost Per Labor-Hour Bare Costs	Cost Per Labor-Hour Incl. O&P
1 Laborer	$37.90	$303.20	$61.95	$495.60	$45.08	$72.33
1 Equip. Oper. (crane)	52.25	418.00	82.70	661.60		
1 Hyd. Excavator, 1 C.Y.		812.80		894.08		
1 Trench Box		81.00		89.10	55.86	61.45
16 L.H., Daily Totals		$1615.00		$2140.38	$100.94	$133.77

Crew B-13E

Crew No.	Bare Costs Hr.	Bare Costs Daily	Incl. Subs O&P Hr.	Incl. Subs O&P Daily	Cost Per Labor-Hour Bare Costs	Cost Per Labor-Hour Incl. O&P
1 Laborer	$37.90	$303.20	$61.95	$495.60	$45.08	$72.33
1 Equip. Oper. (crane)	52.25	418.00	82.70	661.60		
1 Hyd. Excavator, 1.5 C.Y.		1029.00		1131.90		
1 Trench Box		81.00		89.10	69.38	76.31
16 L.H., Daily Totals		$1831.20		$2378.20	$114.45	$148.64

Crew B-13F

Crew No.	Bare Costs Hr.	Bare Costs Daily	Incl. Subs O&P Hr.	Incl. Subs O&P Daily	Cost Per Labor-Hour Bare Costs	Cost Per Labor-Hour Incl. O&P
1 Laborer	$37.90	$303.20	$61.95	$495.60	$45.08	$72.33
1 Equip. Oper. (crane)	52.25	418.00	82.70	661.60		
1 Hyd. Excavator, 3.5 C.Y.		2442.00		2686.20		
1 Trench Box		81.00		89.10	157.69	173.46
16 L.H., Daily Totals		$3244.20		$3932.50	$202.76	$245.78

Crew B-13G

Crew No.	Bare Costs Hr.	Bare Costs Daily	Incl. Subs O&P Hr.	Incl. Subs O&P Daily	Cost Per Labor-Hour Bare Costs	Cost Per Labor-Hour Incl. O&P
1 Laborer	$37.90	$303.20	$61.95	$495.60	$45.08	$72.33
1 Equip. Oper. (crane)	52.25	418.00	82.70	661.60		
1 Hyd. Excavator, .75 C.Y.		654.00		719.40		
1 Trench Box		81.00		89.10	45.94	50.53
16 L.H., Daily Totals		$1456.20		$1965.70	$91.01	$122.86

Crew B-13H

Crew No.	Bare Costs Hr.	Bare Costs Daily	Incl. Subs O&P Hr.	Incl. Subs O&P Daily	Cost Per Labor-Hour Bare Costs	Cost Per Labor-Hour Incl. O&P
1 Laborer	$37.90	$303.20	$61.95	$495.60	$45.08	$72.33
1 Equip. Oper. (crane)	52.25	418.00	82.70	661.60		
1 Gradall, 5/8 C.Y.		882.00		970.20		
1 Trench Box		81.00		89.10	60.19	66.21
16 L.H., Daily Totals		$1684.20		$2216.50	$105.26	$138.53

Crew B-13I

Crew No.	Bare Costs Hr.	Bare Costs Daily	Incl. Subs O&P Hr.	Incl. Subs O&P Daily	Cost Per Labor-Hour Bare Costs	Cost Per Labor-Hour Incl. O&P
1 Laborer	$37.90	$303.20	$61.95	$495.60	$45.08	$72.33
1 Equip. Oper. (crane)	52.25	418.00	82.70	661.60		
1 Gradall, 3 Ton, 1 C.Y.		995.80		1095.38		
1 Trench Box		81.00		89.10	67.30	74.03
16 L.H., Daily Totals		$1798.00		$2341.68	$112.38	$146.35

Crew No.	Bare Costs Hr.	Daily	Incl. Subs O&P Hr.	Daily	Cost Per Labor-Hour Bare Costs	Incl. O&P
Crew B-13J	**Hr.**	**Daily**	**Hr.**	**Daily**	**Bare Costs**	**Incl. O&P**
1 Laborer	$37.90	$303.20	$61.95	$495.60	$45.08	$72.33
1 Equip. Oper. (crane)	52.25	418.00	82.70	661.60		
1 Hyd. Excavator, 2.5 C.Y.		1582.00		1740.20		
1 Trench Box		81.00		89.10	103.94	114.33
16 L.H., Daily Totals		$2384.20		$2986.50	$149.01	$186.66
Crew B-13K	**Hr.**	**Daily**	**Hr.**	**Daily**	**Bare Costs**	**Incl. O&P**
2 Equip. Opers. (crane)	$52.25	$836.00	$82.70	$1323.20	$52.25	$82.70
1 Hyd. Excavator, .75 C.Y.		654.00		719.40		
1 Hyd. Hammer, 4000 ft-lb		306.20		336.82		
1 Hyd. Excavator, .75 C.Y.		654.00		719.40	100.89	110.98
16 L.H., Daily Totals		$2450.20		$3098.82	$153.14	$193.68
Crew B-13L	**Hr.**	**Daily**	**Hr.**	**Daily**	**Bare Costs**	**Incl. O&P**
2 Equip. Opers. (crane)	$52.25	$836.00	$82.70	$1323.20	$52.25	$82.70
1 Hyd. Excavator, 1.5 C.Y.		1029.00		1131.90		
1 Hyd. Hammer, 5000 ft-lb		367.80		404.58		
1 Hyd. Excavator, .75 C.Y.		654.00		719.40	128.18	140.99
16 L.H., Daily Totals		$2886.80		$3579.08	$180.43	$223.69
Crew B-13M	**Hr.**	**Daily**	**Hr.**	**Daily**	**Bare Costs**	**Incl. O&P**
2 Equip. Opers. (crane)	$52.25	$836.00	$82.70	$1323.20	$52.25	$82.70
1 Hyd. Excavator, 2.5 C.Y.		1582.00		1740.20		
1 Hyd. Hammer, 8000 ft-lb		541.20		595.32		
1 Hyd. Excavator, 1.5 C.Y.		1029.00		1131.90	197.01	216.71
16 L.H., Daily Totals		$3988.20		$4790.62	$249.26	$299.41
Crew B-13N	**Hr.**	**Daily**	**Hr.**	**Daily**	**Bare Costs**	**Incl. O&P**
2 Equip. Opers. (crane)	$52.25	$836.00	$82.70	$1323.20	$52.25	$82.70
1 Hyd. Excavator, 3.5 C.Y.		2442.00		2686.20		
1 Hyd. Hammer, 12,000 ft-lb		632.20		695.42		
1 Hyd. Excavator, 1.5 C.Y.		1029.00		1131.90	256.45	282.10
16 L.H., Daily Totals		$4939.20		$5836.72	$308.70	$364.80
Crew B-14	**Hr.**	**Daily**	**Hr.**	**Daily**	**Bare Costs**	**Incl. O&P**
1 Labor Foreman (outside)	$39.90	$319.20	$65.20	$521.60	$40.11	$65.13
4 Laborers	37.90	1212.80	61.95	1982.40		
1 Equip. Oper. (light)	49.15	393.20	77.80	622.40		
1 Backhoe Loader, 48 H.P.		365.40		401.94	7.61	8.37
48 L.H., Daily Totals		$2290.60		$3528.34	$47.72	$73.51
Crew B-14A	**Hr.**	**Daily**	**Hr.**	**Daily**	**Bare Costs**	**Incl. O&P**
1 Equip. Oper. (crane)	$52.25	$418.00	$82.70	$661.60	$47.47	$75.78
.5 Laborer	37.90	151.60	61.95	247.80		
1 Hyd. Excavator, 4.5 C.Y.		2974.00		3271.40	247.83	272.62
12 L.H., Daily Totals		$3543.60		$4180.80	$295.30	$348.40
Crew B-14B	**Hr.**	**Daily**	**Hr.**	**Daily**	**Bare Costs**	**Incl. O&P**
1 Equip. Oper. (crane)	$52.25	$418.00	$82.70	$661.60	$47.47	$75.78
.5 Laborer	37.90	151.60	61.95	247.80		
1 Hyd. Excavator, 6 C.Y.		3501.00		3851.10	291.75	320.93
12 L.H., Daily Totals		$4070.60		$4760.50	$339.22	$396.71
Crew B-14C	**Hr.**	**Daily**	**Hr.**	**Daily**	**Bare Costs**	**Incl. O&P**
1 Equip. Oper. (crane)	$52.25	$418.00	$82.70	$661.60	$47.47	$75.78
.5 Laborer	37.90	151.60	61.95	247.80		
1 Hyd. Excavator, 7 C.Y.		3597.00		3956.70	299.75	329.73
12 L.H., Daily Totals		$4166.60		$4866.10	$347.22	$405.51

Crew No.	Bare Costs Hr.	Daily	Incl. Subs O&P Hr.	Daily	Cost Per Labor-Hour Bare Costs	Incl. O&P
Crew B-14F	**Hr.**	**Daily**	**Hr.**	**Daily**	**Bare Costs**	**Incl. O&P**
1 Equip. Oper. (crane)	$52.25	$418.00	$82.70	$661.60	$47.47	$75.78
.5 Laborer	37.90	151.60	61.95	247.80		
1 Hyd. Shovel, 7 C.Y.		4212.00		4633.20	351.00	386.10
12 L.H., Daily Totals		$4781.60		$5542.60	$398.47	$461.88
Crew B-14G	**Hr.**	**Daily**	**Hr.**	**Daily**	**Bare Costs**	**Incl. O&P**
1 Equip. Oper. (crane)	$52.25	$418.00	$82.70	$661.60	$47.47	$75.78
.5 Laborer	37.90	151.60	61.95	247.80		
1 Hyd. Shovel, 12 C.Y.		6328.00		6960.80	527.33	580.07
12 L.H., Daily Totals		$6897.60		$7870.20	$574.80	$655.85
Crew B-14J	**Hr.**	**Daily**	**Hr.**	**Daily**	**Bare Costs**	**Incl. O&P**
1 Equip. Oper. (medium)	$51.10	$408.80	$80.90	$647.20	$46.70	$74.58
.5 Laborer	37.90	151.60	61.95	247.80		
1 F.E. Loader, 8 C.Y.		1964.00		2160.40	163.67	180.03
12 L.H., Daily Totals		$2524.40		$3055.40	$210.37	$254.62
Crew B-14K	**Hr.**	**Daily**	**Hr.**	**Daily**	**Bare Costs**	**Incl. O&P**
1 Equip. Oper. (medium)	$51.10	$408.80	$80.90	$647.20	$46.70	$74.58
.5 Laborer	37.90	151.60	61.95	247.80		
1 F.E. Loader, 10 C.Y.		3085.00		3393.50	257.08	282.79
12 L.H., Daily Totals		$3645.40		$4288.50	$303.78	$357.38
Crew B-15	**Hr.**	**Daily**	**Hr.**	**Daily**	**Bare Costs**	**Incl. O&P**
1 Equipment Oper. (med.)	$51.10	$408.80	$80.90	$647.20	$44.70	$71.71
.5 Laborer	37.90	151.60	61.95	247.80		
2 Truck Drivers (heavy)	43.20	691.20	69.55	1112.80		
2 Dump Trucks, 12 C.Y., 400 H.P.		1381.60		1519.76		
1 Dozer, 200 H.P.		1397.00		1536.70	99.24	109.16
28 L.H., Daily Totals		$4030.20		$5064.26	$143.94	$180.87
Crew B-16	**Hr.**	**Daily**	**Hr.**	**Daily**	**Bare Costs**	**Incl. O&P**
1 Labor Foreman (outside)	$39.90	$319.20	$65.20	$521.60	$39.73	$64.66
2 Laborers	37.90	606.40	61.95	991.20		
1 Truck Driver (heavy)	43.20	345.60	69.55	556.40		
1 Dump Truck, 12 C.Y., 400 H.P.		690.80		759.88	21.59	23.75
32 L.H., Daily Totals		$1962.00		$2829.08	$61.31	$88.41
Crew B-17	**Hr.**	**Daily**	**Hr.**	**Daily**	**Bare Costs**	**Incl. O&P**
2 Laborers	$37.90	$606.40	$61.95	$991.20	$42.04	$67.81
1 Equip. Oper. (light)	49.15	393.20	77.80	622.40		
1 Truck Driver (heavy)	43.20	345.60	69.55	556.40		
1 Backhoe Loader, 48 H.P.		365.40		401.94		
1 Dump Truck, 8 C.Y., 220 H.P.		417.40		459.14	24.46	26.91
32 L.H., Daily Totals		$2128.00		$3031.08	$66.50	$94.72
Crew B-17A	**Hr.**	**Daily**	**Hr.**	**Daily**	**Bare Costs**	**Incl. O&P**
2 Labor Foremen (outside)	$39.90	$638.40	$65.20	$1043.20	$40.90	$66.70
6 Laborers	37.90	1819.20	61.95	2973.60		
1 Skilled Worker Foreman (out)	51.90	415.20	84.10	672.80		
1 Skilled Worker	49.90	399.20	80.85	646.80		
80 L.H., Daily Totals		$3272.00		$5336.40	$40.90	$66.70
Crew B-17B	**Hr.**	**Daily**	**Hr.**	**Daily**	**Bare Costs**	**Incl. O&P**
2 Laborers	$37.90	$606.40	$61.95	$991.20	$42.04	$67.81
1 Equip. Oper. (light)	49.15	393.20	77.80	622.40		
1 Truck Driver (heavy)	43.20	345.60	69.55	556.40		
1 Backhoe Loader, 48 H.P.		365.40		401.94		
1 Dump Truck, 12 C.Y., 400 H.P.		690.80		759.88	33.01	36.31
32 L.H., Daily Totals		$2401.40		$3331.82	$75.04	$104.12

733

Crew B-18

Crew No.	Bare Costs Hr.	Daily	Incl. Subs O&P Hr.	Daily	Cost Per Labor-Hour Bare Costs	Incl. O&P
1 Labor Foreman (outside)	$39.90	$319.20	$65.20	$521.60	$38.57	$63.03
2 Laborers	37.90	606.40	61.95	991.20		
1 Vibrating Plate, Gas, 21"		46.60		51.26	1.94	2.14
24 L.H., Daily Totals		$972.20		$1564.06	$40.51	$65.17

Crew B-19

Crew No.	Bare Costs Hr.	Daily	Incl. Subs O&P Hr.	Daily	Cost Per Labor-Hour Bare Costs	Incl. O&P
1 Pile Driver Foreman (outside)	$50.40	$403.20	$82.45	$659.60	$49.32	$79.69
4 Pile Drivers	48.40	1548.80	79.20	2534.40		
2 Equip. Opers. (crane)	52.25	836.00	82.70	1323.20		
1 Equip. Oper. (oiler)	46.05	368.40	72.90	583.20		
1 Crawler Crane, 40 Ton		1203.00		1323.30		
1 Lead, 90' High		124.40		136.84		
1 Hammer, Diesel, 22k ft-lb		378.00		415.80	26.65	29.31
64 L.H., Daily Totals		$4861.80		$6976.34	$75.97	$109.01

Crew B-19A

Crew No.	Bare Costs Hr.	Daily	Incl. Subs O&P Hr.	Daily	Cost Per Labor-Hour Bare Costs	Incl. O&P
1 Pile Driver Foreman (outside)	$50.40	$403.20	$82.45	$659.60	$49.32	$79.69
4 Pile Drivers	48.40	1548.80	79.20	2534.40		
2 Equip. Opers. (crane)	52.25	836.00	82.70	1323.20		
1 Equip. Oper. (oiler)	46.05	368.40	72.90	583.20		
1 Crawler Crane, 75 Ton		1507.00		1657.70		
1 Lead, 90' high		124.40		136.84		
1 Hammer, Diesel, 41k ft-lb		522.60		574.86	33.66	37.02
64 L.H., Daily Totals		$5310.40		$7469.80	$82.98	$116.72

Crew B-19B

Crew No.	Bare Costs Hr.	Daily	Incl. Subs O&P Hr.	Daily	Cost Per Labor-Hour Bare Costs	Incl. O&P
1 Pile Driver Foreman (outside)	$50.40	$403.20	$82.45	$659.60	$49.32	$79.69
4 Pile Drivers	48.40	1548.80	79.20	2534.40		
2 Equip. Opers. (crane)	52.25	836.00	82.70	1323.20		
1 Equip. Oper. (oiler)	46.05	368.40	72.90	583.20		
1 Crawler Crane, 40 Ton		1203.00		1323.30		
1 Lead, 90' High		124.40		136.84		
1 Hammer, Diesel, 22k ft-lb		378.00		415.80		
1 Barge, 400 Ton		779.20		857.12	38.82	42.70
64 L.H., Daily Totals		$5641.00		$7833.46	$88.14	$122.40

Crew B-19C

Crew No.	Bare Costs Hr.	Daily	Incl. Subs O&P Hr.	Daily	Cost Per Labor-Hour Bare Costs	Incl. O&P
1 Pile Driver Foreman (outside)	$50.40	$403.20	$82.45	$659.60	$49.32	$79.69
4 Pile Drivers	48.40	1548.80	79.20	2534.40		
2 Equip. Opers. (crane)	52.25	836.00	82.70	1323.20		
1 Equip. Oper. (oiler)	46.05	368.40	72.90	583.20		
1 Crawler Crane, 75 Ton		1507.00		1657.70		
1 Lead, 90' High		124.40		136.84		
1 Hammer, Diesel, 41k ft-lb		522.60		574.86		
1 Barge, 400 Ton		779.20		857.12	45.83	50.41
64 L.H., Daily Totals		$6089.60		$8326.92	$95.15	$130.11

Crew B-20

Crew No.	Bare Costs Hr.	Daily	Incl. Subs O&P Hr.	Daily	Cost Per Labor-Hour Bare Costs	Incl. O&P
1 Labor Foreman (outside)	$39.90	$319.20	$65.20	$521.60	$42.57	$69.33
1 Skilled Worker	49.90	399.20	80.85	646.80		
1 Laborer	37.90	303.20	61.95	495.60		
24 L.H., Daily Totals		$1021.60		$1664.00	$42.57	$69.33

Crew B-20A

Crew No.	Bare Costs Hr.	Daily	Incl. Subs O&P Hr.	Daily	Cost Per Labor-Hour Bare Costs	Incl. O&P
1 Labor Foreman (outside)	$39.90	$319.20	$65.20	$521.60	$46.09	$73.31
1 Laborer	37.90	303.20	61.95	495.60		
1 Plumber	59.20	473.60	92.30	738.40		
1 Plumber Apprentice	47.35	378.80	73.80	590.40		
32 L.H., Daily Totals		$1474.80		$2346.00	$46.09	$73.31

Crew B-21

Crew No.	Bare Costs Hr.	Daily	Incl. Subs O&P Hr.	Daily	Cost Per Labor-Hour Bare Costs	Incl. O&P
1 Labor Foreman (outside)	$39.90	$319.20	$65.20	$521.60	$43.95	$71.24
1 Skilled Worker	49.90	399.20	80.85	646.80		
1 Laborer	37.90	303.20	61.95	495.60		
.5 Equip. Oper. (crane)	52.25	209.00	82.70	330.80		
.5 S.P. Crane, 4x4, 5 Ton		140.40		154.44	5.01	5.52
28 L.H., Daily Totals		$1371.00		$2149.24	$48.96	$76.76

Crew B-21A

Crew No.	Bare Costs Hr.	Daily	Incl. Subs O&P Hr.	Daily	Cost Per Labor-Hour Bare Costs	Incl. O&P
1 Labor Foreman (outside)	$39.90	$319.20	$65.20	$521.60	$47.32	$75.19
1 Laborer	37.90	303.20	61.95	495.60		
1 Plumber	59.20	473.60	92.30	738.40		
1 Plumber Apprentice	47.35	378.80	73.80	590.40		
1 Equip. Oper. (crane)	52.25	418.00	82.70	661.60		
1 S.P. Crane, 4x4, 12 Ton		479.20		527.12	11.98	13.18
40 L.H., Daily Totals		$2372.00		$3534.72	$59.30	$88.37

Crew B-21B

Crew No.	Bare Costs Hr.	Daily	Incl. Subs O&P Hr.	Daily	Cost Per Labor-Hour Bare Costs	Incl. O&P
1 Labor Foreman (outside)	$39.90	$319.20	$65.20	$521.60	$41.17	$66.75
3 Laborers	37.90	909.60	61.95	1486.80		
1 Equip. Oper. (crane)	52.25	418.00	82.70	661.60		
1 Hyd. Crane, 12 Ton		658.40		724.24	16.46	18.11
40 L.H., Daily Totals		$2305.20		$3394.24	$57.63	$84.86

Crew B-21C

Crew No.	Bare Costs Hr.	Daily	Incl. Subs O&P Hr.	Daily	Cost Per Labor-Hour Bare Costs	Incl. O&P
1 Labor Foreman (outside)	$39.90	$319.20	$65.20	$521.60	$41.40	$66.94
4 Laborers	37.90	1212.80	61.95	1982.40		
1 Equip. Oper. (crane)	52.25	418.00	82.70	661.60		
1 Equip. Oper. (oiler)	46.05	368.40	72.90	583.20		
2 Cutting Torches		22.80		25.08		
2 Sets of Gases		336.00		369.60		
1 Lattice Boom Crane, 90 Ton		1520.00		1672.00	33.55	36.91
56 L.H., Daily Totals		$4197.20		$5815.48	$74.95	$103.85

Crew B-22

Crew No.	Bare Costs Hr.	Daily	Incl. Subs O&P Hr.	Daily	Cost Per Labor-Hour Bare Costs	Incl. O&P
1 Labor Foreman (outside)	$39.90	$319.20	$65.20	$521.60	$44.50	$72.01
1 Skilled Worker	49.90	399.20	80.85	646.80		
1 Laborer	37.90	303.20	61.95	495.60		
.75 Equip. Oper. (crane)	52.25	313.50	82.70	496.20		
.75 S.P. Crane, 4x4, 5 Ton		210.60		231.66	7.02	7.72
30 L.H., Daily Totals		$1545.70		$2391.86	$51.52	$79.73

Crew B-22A

Crew No.	Bare Costs Hr.	Daily	Incl. Subs O&P Hr.	Daily	Cost Per Labor-Hour Bare Costs	Incl. O&P
1 Labor Foreman (outside)	$39.90	$319.20	$65.20	$521.60	$43.57	$70.53
1 Skilled Worker	49.90	399.20	80.85	646.80		
2 Laborers	37.90	606.40	61.95	991.20		
1 Equipment Operator, Crane	52.25	418.00	82.70	661.60		
1 S.P. Crane, 4x4, 5 Ton		280.80		308.88		
1 Butt Fusion Mach., 4"-12" diam.		376.10		413.71	16.42	18.06
40 L.H., Daily Totals		$2399.70		$3543.79	$59.99	$88.59

Crew B-22B

Crew No.	Bare Costs Hr.	Daily	Incl. Subs O&P Hr.	Daily	Cost Per Labor-Hour Bare Costs	Incl. O&P
1 Labor Foreman (outside)	$39.90	$319.20	$65.20	$521.60	$43.57	$70.53
1 Skilled Worker	49.90	399.20	80.85	646.80		
2 Laborers	37.90	606.40	61.95	991.20		
1 Equip. Oper. (crane)	52.25	418.00	82.70	661.60		
1 S.P. Crane, 4x4, 5 Ton		280.80		308.88		
1 Butt Fusion Mach., 8"-24" diam.		804.80		885.28	27.14	29.85
40 L.H., Daily Totals		$2828.40		$4015.36	$70.71	$100.38

Crew No.	Bare Costs		Incl. Subs O&P		Cost Per Labor-Hour	
Crew B-22C	Hr.	Daily	Hr.	Daily	Bare Costs	Incl. O&P
1 Skilled Worker	$49.90	$399.20	$80.85	$646.80	$43.90	$71.40
1 Laborer	37.90	303.20	61.95	495.60		
1 Butt Fusion Mach., 2"-8" diam.		121.05		133.16	7.57	8.32
16 L.H., Daily Totals		$823.45		$1275.56	$51.47	$79.72
Crew B-23	Hr.	Daily	Hr.	Daily	Bare Costs	Incl. O&P
1 Labor Foreman (outside)	$39.90	$319.20	$65.20	$521.60	$38.30	$62.60
4 Laborers	37.90	1212.80	61.95	1982.40		
1 Drill Rig, Truck-Mounted		2553.00		2808.30		
1 Flatbed Truck, Gas, 3 Ton		307.40		338.14	71.51	78.66
40 L.H., Daily Totals		$4392.40		$5650.44	$109.81	$141.26
Crew B-23A	Hr.	Daily	Hr.	Daily	Bare Costs	Incl. O&P
1 Labor Foreman (outside)	$39.90	$319.20	$65.20	$521.60	$42.97	$69.35
1 Laborer	37.90	303.20	61.95	495.60		
1 Equip. Oper. (medium)	51.10	408.80	80.90	647.20		
1 Drill Rig, Truck-Mounted		2553.00		2808.30		
1 Pickup Truck, 3/4 Ton		145.80		160.38	112.45	123.69
24 L.H., Daily Totals		$3730.00		$4633.08	$155.42	$193.04
Crew B-23B	Hr.	Daily	Hr.	Daily	Bare Costs	Incl. O&P
1 Labor Foreman (outside)	$39.90	$319.20	$65.20	$521.60	$42.97	$69.35
1 Laborer	37.90	303.20	61.95	495.60		
1 Equip. Oper. (medium)	51.10	408.80	80.90	647.20		
1 Drill Rig, Truck-Mounted		2553.00		2808.30		
1 Pickup Truck, 3/4 Ton		145.80		160.38		
1 Centr. Water Pump, 6"		350.80		385.88	127.07	139.77
24 L.H., Daily Totals		$4080.80		$5018.96	$170.03	$209.12
Crew B-24	Hr.	Daily	Hr.	Daily	Bare Costs	Incl. O&P
1 Cement Finisher	$45.65	$365.20	$72.25	$578.00	$44.00	$71.12
1 Laborer	37.90	303.20	61.95	495.60		
1 Carpenter	48.45	387.60	79.15	633.20		
24 L.H., Daily Totals		$1056.00		$1706.80	$44.00	$71.12
Crew B-25	Hr.	Daily	Hr.	Daily	Bare Costs	Incl. O&P
1 Labor Foreman (outside)	$39.90	$319.20	$65.20	$521.60	$41.68	$67.41
7 Laborers	37.90	2122.40	61.95	3469.20		
3 Equip. Opers. (medium)	51.10	1226.40	80.90	1941.60		
1 Asphalt Paver, 130 H.P.		2210.00		2431.00		
1 Tandem Roller, 10 Ton		239.40		263.34		
1 Roller, Pneum. Whl., 12 Ton		347.80		382.58	31.79	34.97
88 L.H., Daily Totals		$6465.20		$9009.32	$73.47	$102.38
Crew B-25B	Hr.	Daily	Hr.	Daily	Bare Costs	Incl. O&P
1 Labor Foreman (outside)	$39.90	$319.20	$65.20	$521.60	$42.47	$68.54
7 Laborers	37.90	2122.40	61.95	3469.20		
4 Equip. Opers. (medium)	51.10	1635.20	80.90	2588.80		
1 Asphalt Paver, 130 H.P.		2210.00		2431.00		
2 Tandem Rollers, 10 Ton		478.80		526.68		
1 Roller, Pneum. Whl., 12 Ton		347.80		382.58	31.63	34.79
96 L.H., Daily Totals		$7113.40		$9919.86	$74.10	$103.33
Crew B-25C	Hr.	Daily	Hr.	Daily	Bare Costs	Incl. O&P
1 Labor Foreman (outside)	$39.90	$319.20	$65.20	$521.60	$42.63	$68.81
3 Laborers	37.90	909.60	61.95	1486.80		
2 Equip. Opers. (medium)	51.10	817.60	80.90	1294.40		
1 Asphalt Paver, 130 H.P.		2210.00		2431.00		
1 Tandem Roller, 10 Ton		239.40		263.34	51.03	56.13
48 L.H., Daily Totals		$4495.80		$5997.14	$93.66	$124.94

Crew No.	Bare Costs		Incl. Subs O&P		Cost Per Labor-Hour	
Crew B-25D	Hr.	Daily	Hr.	Daily	Bare Costs	Incl. O&P
1 Labor Foreman (outside)	$39.90	$319.20	$65.20	$521.60	$42.81	$69.06
3 Laborers	37.90	909.60	61.95	1486.80		
2.125 Equip. Opers. (medium)	51.10	868.70	80.90	1375.30		
.125 Truck Driver (heavy)	43.20	43.20	69.55	69.55		
.125 Truck Tractor, 6x4, 380 H.P.		75.00		82.50		
.125 Dist. Tanker, 3000 Gallon		39.92		43.92		
1 Asphalt Paver, 130 H.P.		2210.00		2431.00		
1 Tandem Roller, 10 Ton		239.40		263.34	51.29	56.42
50 L.H., Daily Totals		$4705.02		$6274.01	$94.10	$125.48
Crew B-25E	Hr.	Daily	Hr.	Daily	Bare Costs	Incl. O&P
1 Labor Foreman (outside)	$39.90	$319.20	$65.20	$521.60	$42.98	$69.30
3 Laborers	37.90	909.60	61.95	1486.80		
2.250 Equip. Opers. (medium)	51.10	919.80	80.90	1456.20		
.25 Truck Driver (heavy)	43.20	86.40	69.55	139.10		
.25 Truck Tractor, 6x4, 380 H.P.		150.00		165.00		
.25 Dist. Tanker, 3000 Gallon		79.85		87.83		
1 Asphalt Paver, 130 H.P.		2210.00		2431.00		
1 Tandem Roller, 10 Ton		239.40		263.34	51.52	56.68
52 L.H., Daily Totals		$4914.25		$6550.88	$94.50	$125.98
Crew B-26	Hr.	Daily	Hr.	Daily	Bare Costs	Incl. O&P
1 Labor Foreman (outside)	$39.90	$319.20	$65.20	$521.60	$42.56	$68.80
6 Laborers	37.90	1819.20	61.95	2973.60		
2 Equip. Opers. (medium)	51.10	817.60	80.90	1294.40		
1 Rodman (reinf.)	53.00	424.00	85.80	686.40		
1 Cement Finisher	45.65	365.20	72.25	578.00		
1 Grader, 30,000 Lbs.		714.40		785.84		
1 Paving Mach. & Equip.		2796.00		3075.60	39.89	43.88
88 L.H., Daily Totals		$7255.60		$9915.44	$82.45	$112.68
Crew B-26A	Hr.	Daily	Hr.	Daily	Bare Costs	Incl. O&P
1 Labor Foreman (outside)	$39.90	$319.20	$65.20	$521.60	$42.56	$68.80
6 Laborers	37.90	1819.20	61.95	2973.60		
2 Equip. Opers. (medium)	51.10	817.60	80.90	1294.40		
1 Rodman (reinf.)	53.00	424.00	85.80	686.40		
1 Cement Finisher	45.65	365.20	72.25	578.00		
1 Grader, 30,000 Lbs.		714.40		785.84		
1 Paving Mach. & Equip.		2796.00		3075.60		
1 Concrete Saw		168.60		185.46	41.81	45.99
88 L.H., Daily Totals		$7424.20		$10100.90	$84.37	$114.78
Crew B-26B	Hr.	Daily	Hr.	Daily	Bare Costs	Incl. O&P
1 Labor Foreman (outside)	$39.90	$319.20	$65.20	$521.60	$43.27	$69.80
6 Laborers	37.90	1819.20	61.95	2973.60		
3 Equip. Opers. (medium)	51.10	1226.40	80.90	1941.60		
1 Rodman (reinf.)	53.00	424.00	85.80	686.40		
1 Cement Finisher	45.65	365.20	72.25	578.00		
1 Grader, 30,000 Lbs.		714.40		785.84		
1 Paving Mach. & Equip.		2796.00		3075.60		
1 Concrete Pump, 110' Boom		958.40		1054.24	46.55	51.20
96 L.H., Daily Totals		$8622.80		$11616.88	$89.82	$121.01
Crew B-26C	Hr.	Daily	Hr.	Daily	Bare Costs	Incl. O&P
1 Labor Foreman (outside)	$39.90	$319.20	$65.20	$521.60	$41.70	$67.58
6 Laborers	37.90	1819.20	61.95	2973.60		
1 Equip. Oper. (medium)	51.10	408.80	80.90	647.20		
1 Rodman (reinf.)	53.00	424.00	85.80	686.40		
1 Cement Finisher	45.65	365.20	72.25	578.00		
1 Paving Mach. & Equip.		2796.00		3075.60		
1 Concrete Saw		168.60		185.46	37.06	40.76
80 L.H., Daily Totals		$6301.00		$8667.86	$78.76	$108.35

735

Crew No.	Bare Costs Hr.	Daily	Incl. Subs O&P Hr.	Daily	Cost Per Labor-Hour Bare Costs	Incl. O&P
Crew B-27	Hr.	Daily	Hr.	Daily	Bare Costs	Incl. O&P
1 Labor Foreman (outside)	$39.90	$319.20	$65.20	$521.60	$38.40	$62.76
3 Laborers	37.90	909.60	61.95	1486.80		
1 Berm Machine		295.80		325.38	9.24	10.17
32 L.H., Daily Totals		$1524.60		$2333.78	$47.64	$72.93
Crew B-28	Hr.	Daily	Hr.	Daily	Bare Costs	Incl. O&P
2 Carpenters	$48.45	$775.20	$79.15	$1266.40	$44.93	$73.42
1 Laborer	37.90	303.20	61.95	495.60		
24 L.H., Daily Totals		$1078.40		$1762.00	$44.93	$73.42
Crew B-29	Hr.	Daily	Hr.	Daily	Bare Costs	Incl. O&P
1 Labor Foreman (outside)	$39.90	$319.20	$65.20	$521.60	$41.40	$66.94
4 Laborers	37.90	1212.80	61.95	1982.40		
1 Equip. Oper. (crane)	52.25	418.00	82.70	661.60		
1 Equip. Oper. (oiler)	46.05	368.40	72.90	583.20		
1 Gradall, 5/8 C.Y.		882.00		970.20	15.75	17.32
56 L.H., Daily Totals		$3200.40		$4719.00	$57.15	$84.27
Crew B-30	Hr.	Daily	Hr.	Daily	Bare Costs	Incl. O&P
1 Equip. Oper. (medium)	$51.10	$408.80	$80.90	$647.20	$45.83	$73.33
2 Truck Drivers (heavy)	43.20	691.20	69.55	1112.80		
1 Hyd. Excavator, 1.5 C.Y.		1029.00		1131.90		
2 Dump Trucks, 12 C.Y., 400 H.P.		1381.60		1519.76	100.44	110.49
24 L.H., Daily Totals		$3510.60		$4411.66	$146.28	$183.82
Crew B-31	Hr.	Daily	Hr.	Daily	Bare Costs	Incl. O&P
1 Labor Foreman (outside)	$39.90	$319.20	$65.20	$521.60	$40.41	$66.04
3 Laborers	37.90	909.60	61.95	1486.80		
1 Carpenter	48.45	387.60	79.15	633.20		
1 Air Compressor, 250 cfm		197.80		217.58		
1 Sheeting Driver		6.50		7.15		
2 -50' Air Hoses, 1.5"		10.40		11.44	5.37	5.90
40 L.H., Daily Totals		$1831.10		$2877.77	$45.78	$71.94
Crew B-32	Hr.	Daily	Hr.	Daily	Bare Costs	Incl. O&P
1 Laborer	$37.90	$303.20	$61.95	$495.60	$47.80	$76.16
3 Equip. Opers. (medium)	51.10	1226.40	80.90	1941.60		
1 Grader, 30,000 Lbs.		714.40		785.84		
1 Tandem Roller, 10 Ton		239.40		263.34		
1 Dozer, 200 H.P.		1397.00		1536.70	73.46	80.81
32 L.H., Daily Totals		$3880.40		$5023.08	$121.26	$156.97
Crew B-32A	Hr.	Daily	Hr.	Daily	Bare Costs	Incl. O&P
1 Laborer	$37.90	$303.20	$61.95	$495.60	$46.70	$74.58
2 Equip. Opers. (medium)	51.10	817.60	80.90	1294.40		
1 Grader, 30,000 Lbs.		714.40		785.84		
1 Roller, Vibratory, 25 Ton		691.80		760.98	58.59	64.45
24 L.H., Daily Totals		$2527.00		$3336.82	$105.29	$139.03
Crew B-32B	Hr.	Daily	Hr.	Daily	Bare Costs	Incl. O&P
1 Laborer	$37.90	$303.20	$61.95	$495.60	$46.70	$74.58
2 Equip. Opers. (medium)	51.10	817.60	80.90	1294.40		
1 Dozer, 200 H.P.		1397.00		1536.70		
1 Roller, Vibratory, 25 Ton		691.80		760.98	87.03	95.74
24 L.H., Daily Totals		$3209.60		$4087.68	$133.73	$170.32

Crew No.	Bare Costs Hr.	Daily	Incl. Subs O&P Hr.	Daily	Cost Per Labor-Hour Bare Costs	Incl. O&P
Crew B-32C	Hr.	Daily	Hr.	Daily	Bare Costs	Incl. O&P
1 Labor Foreman (outside)	$39.90	$319.20	$65.20	$521.60	$44.83	$71.97
2 Laborers	37.90	606.40	61.95	991.20		
3 Equip. Opers. (medium)	51.10	1226.40	80.90	1941.60		
1 Grader, 30,000 Lbs.		714.40		785.84		
1 Tandem Roller, 10 Ton		239.40		263.34		
1 Dozer, 200 H.P.		1397.00		1536.70	48.98	53.87
48 L.H., Daily Totals		$4502.80		$6040.28	$93.81	$125.84
Crew B-33A	Hr.	Daily	Hr.	Daily	Bare Costs	Incl. O&P
1 Equip. Oper. (medium)	$51.10	$408.80	$80.90	$647.20	$47.33	$75.49
.5 Laborer	37.90	151.60	61.95	247.80		
.25 Equip. Oper. (medium)	51.10	102.20	80.90	161.80		
1 Scraper, Towed, 7 C.Y.		113.20		124.52		
1.25 Dozers, 300 H.P.		2358.75		2594.63	176.57	194.22
14 L.H., Daily Totals		$3134.55		$3775.95	$223.90	$269.71
Crew B-33B	Hr.	Daily	Hr.	Daily	Bare Costs	Incl. O&P
1 Equip. Oper. (medium)	$51.10	$408.80	$80.90	$647.20	$47.33	$75.49
.5 Laborer	37.90	151.60	61.95	247.80		
.25 Equip. Oper. (medium)	51.10	102.20	80.90	161.80		
1 Scraper, Towed, 10 C.Y.		145.00		159.50		
1.25 Dozers, 300 H.P.		2358.75		2594.63	178.84	196.72
14 L.H., Daily Totals		$3166.35		$3810.93	$226.17	$272.21
Crew B-33C	Hr.	Daily	Hr.	Daily	Bare Costs	Incl. O&P
1 Equip. Oper. (medium)	$51.10	$408.80	$80.90	$647.20	$47.33	$75.49
.5 Laborer	37.90	151.60	61.95	247.80		
.25 Equip. Oper. (medium)	51.10	102.20	80.90	161.80		
1 Scraper, Towed, 15 C.Y.		163.60		179.96		
1.25 Dozers, 300 H.P.		2358.75		2594.63	180.17	198.18
14 L.H., Daily Totals		$3184.95		$3831.39	$227.50	$273.67
Crew B-33D	Hr.	Daily	Hr.	Daily	Bare Costs	Incl. O&P
1 Equip. Oper. (medium)	$51.10	$408.80	$80.90	$647.20	$47.33	$75.49
.5 Laborer	37.90	151.60	61.95	247.80		
.25 Equip. Oper. (medium)	51.10	102.20	80.90	161.80		
1 S.P. Scraper, 14 C.Y.		2172.00		2389.20		
.25 Dozer, 300 H.P.		471.75		518.92	188.84	207.72
14 L.H., Daily Totals		$3306.35		$3964.93	$236.17	$283.21
Crew B-33E	Hr.	Daily	Hr.	Daily	Bare Costs	Incl. O&P
1 Equip. Oper. (medium)	$51.10	$408.80	$80.90	$647.20	$47.33	$75.49
.5 Laborer	37.90	151.60	61.95	247.80		
.25 Equip. Oper. (medium)	51.10	102.20	80.90	161.80		
1 S.P. Scraper, 21 C.Y.		2715.00		2986.50		
.25 Dozer, 300 H.P.		471.75		518.92	227.63	250.39
14 L.H., Daily Totals		$3849.35		$4562.23	$274.95	$325.87
Crew B-33F	Hr.	Daily	Hr.	Daily	Bare Costs	Incl. O&P
1 Equip. Oper. (medium)	$51.10	$408.80	$80.90	$647.20	$47.33	$75.49
.5 Laborer	37.90	151.60	61.95	247.80		
.25 Equip. Oper. (medium)	51.10	102.20	80.90	161.80		
1 Elev. Scraper, 11 C.Y.		1173.00		1290.30		
.25 Dozer, 300 H.P.		471.75		518.92	117.48	129.23
14 L.H., Daily Totals		$2307.35		$2866.03	$164.81	$204.72

For customer support on your Commercial Renovation Cost Data, call 877.742.4719.

Crew No.	Bare Costs Hr.	Bare Costs Daily	Incl. Subs O&P Hr.	Incl. Subs O&P Daily	Cost Per Labor-Hour Bare Costs	Cost Per Labor-Hour Incl. O&P
Crew B-33G	Hr.	Daily	Hr.	Daily	Bare Costs	Incl. O&P
1 Equip. Oper. (medium)	$51.10	$408.80	$80.90	$647.20	$47.33	$75.49
.5 Laborer	37.90	151.60	61.95	247.80		
.25 Equip. Oper. (medium)	51.10	102.20	80.90	161.80		
1 Elev. Scraper, 22 C.Y.		2548.00		2802.80		
.25 Dozer, 300 H.P.		471.75		518.92	215.70	237.27
14 L.H., Daily Totals		$3682.35		$4378.52	$263.02	$312.75
Crew B-33K	Hr.	Daily	Hr.	Daily	Bare Costs	Incl. O&P
1 Equipment Operator (med.)	$51.10	$408.80	$80.90	$647.20	$47.33	$75.49
.25 Equipment Operator (med.)	51.10	102.20	80.90	161.80		
.5 Laborer	37.90	151.60	61.95	247.80		
1 S.P. Scraper, 31 C.Y.		3735.00		4108.50		
.25 Dozer, 410 H.P.		598.00		657.80	309.50	340.45
14 L.H., Daily Totals		$4995.60		$5823.10	$356.83	$415.94
Crew B-34A	Hr.	Daily	Hr.	Daily	Bare Costs	Incl. O&P
1 Truck Driver (heavy)	$43.20	$345.60	$69.55	$556.40	$43.20	$69.55
1 Dump Truck, 8 C.Y., 220 H.P.		417.40		459.14	52.17	57.39
8 L.H., Daily Totals		$763.00		$1015.54	$95.38	$126.94
Crew B-34B	Hr.	Daily	Hr.	Daily	Bare Costs	Incl. O&P
1 Truck Driver (heavy)	$43.20	$345.60	$69.55	$556.40	$43.20	$69.55
1 Dump Truck, 12 C.Y., 400 H.P.		690.80		759.88	86.35	94.98
8 L.H., Daily Totals		$1036.40		$1316.28	$129.55	$164.54
Crew B-34C	Hr.	Daily	Hr.	Daily	Bare Costs	Incl. O&P
1 Truck Driver (heavy)	$43.20	$345.60	$69.55	$556.40	$43.20	$69.55
1 Truck Tractor, 6x4, 380 H.P.		600.00		660.00		
1 Dump Trailer, 16.5 C.Y.		125.80		138.38	90.72	99.80
8 L.H., Daily Totals		$1071.40		$1354.78	$133.93	$169.35
Crew B-34D	Hr.	Daily	Hr.	Daily	Bare Costs	Incl. O&P
1 Truck Driver (heavy)	$43.20	$345.60	$69.55	$556.40	$43.20	$69.55
1 Truck Tractor, 6x4, 380 H.P.		600.00		660.00		
1 Dump Trailer, 20 C.Y.		140.40		154.44	92.55	101.81
8 L.H., Daily Totals		$1086.00		$1370.84	$135.75	$171.35
Crew B-34E	Hr.	Daily	Hr.	Daily	Bare Costs	Incl. O&P
1 Truck Driver (heavy)	$43.20	$345.60	$69.55	$556.40	$43.20	$69.55
1 Dump Truck, Off Hwy., 25 Ton		1370.00		1507.00	171.25	188.38
8 L.H., Daily Totals		$1715.60		$2063.40	$214.45	$257.93
Crew B-34F	Hr.	Daily	Hr.	Daily	Bare Costs	Incl. O&P
1 Truck Driver (heavy)	$43.20	$345.60	$69.55	$556.40	$43.20	$69.55
1 Dump Truck, Off Hwy., 35 Ton		1529.00		1681.90	191.13	210.24
8 L.H., Daily Totals		$1874.60		$2238.30	$234.32	$279.79
Crew B-34G	Hr.	Daily	Hr.	Daily	Bare Costs	Incl. O&P
1 Truck Driver (heavy)	$43.20	$345.60	$69.55	$556.40	$43.20	$69.55
1 Dump Truck, Off Hwy., 50 Ton		1861.00		2047.10	232.63	255.89
8 L.H., Daily Totals		$2206.60		$2603.50	$275.82	$325.44
Crew B-34H	Hr.	Daily	Hr.	Daily	Bare Costs	Incl. O&P
1 Truck Driver (heavy)	$43.20	$345.60	$69.55	$556.40	$43.20	$69.55
1 Dump Truck, Off Hwy., 65 Ton		1915.00		2106.50	239.38	263.31
8 L.H., Daily Totals		$2260.60		$2662.90	$282.57	$332.86

Crew No.	Bare Costs Hr.	Bare Costs Daily	Incl. Subs O&P Hr.	Incl. Subs O&P Daily	Cost Per Labor-Hour Bare Costs	Cost Per Labor-Hour Incl. O&P
Crew B-34I	Hr.	Daily	Hr.	Daily	Bare Costs	Incl. O&P
1 Truck Driver (heavy)	$43.20	$345.60	$69.55	$556.40	$43.20	$69.55
1 Dump Truck, 18 C.Y., 450 H.P.		869.40		956.34	108.68	119.54
8 L.H., Daily Totals		$1215.00		$1512.74	$151.88	$189.09
Crew B-34J	Hr.	Daily	Hr.	Daily	Bare Costs	Incl. O&P
1 Truck Driver (heavy)	$43.20	$345.60	$69.55	$556.40	$43.20	$69.55
1 Dump Truck, Off Hwy., 100 Ton		2950.00		3245.00	368.75	405.63
8 L.H., Daily Totals		$3295.60		$3801.40	$411.95	$475.18
Crew B-34K	Hr.	Daily	Hr.	Daily	Bare Costs	Incl. O&P
1 Truck Driver (heavy)	$43.20	$345.60	$69.55	$556.40	$43.20	$69.55
1 Truck Tractor, 6x4, 450 H.P.		726.80		799.48		
1 Lowbed Trailer, 75 Ton		223.40		245.74	118.78	130.65
8 L.H., Daily Totals		$1295.80		$1601.62	$161.97	$200.20
Crew B-34L	Hr.	Daily	Hr.	Daily	Bare Costs	Incl. O&P
1 Equip. Oper. (light)	$49.15	$393.20	$77.80	$622.40	$49.15	$77.80
1 Flatbed Truck, Gas, 1.5 Ton		248.20		273.02	31.02	34.13
8 L.H., Daily Totals		$641.40		$895.42	$80.17	$111.93
Crew B-34M	Hr.	Daily	Hr.	Daily	Bare Costs	Incl. O&P
1 Equip. Oper. (light)	$49.15	$393.20	$77.80	$622.40	$49.15	$77.80
1 Flatbed Truck, Gas, 3 Ton		307.40		338.14	38.42	42.27
8 L.H., Daily Totals		$700.60		$960.54	$87.58	$120.07
Crew B-34N	Hr.	Daily	Hr.	Daily	Bare Costs	Incl. O&P
1 Truck Driver (heavy)	$43.20	$345.60	$69.55	$556.40	$47.15	$75.22
1 Equip. Oper. (medium)	51.10	408.80	80.90	647.20		
1 Truck Tractor, 6x4, 380 H.P.		600.00		660.00		
1 Flatbed Trailer, 40 Ton		154.60		170.06	47.16	51.88
16 L.H., Daily Totals		$1509.00		$2033.66	$94.31	$127.10
Crew B-34P	Hr.	Daily	Hr.	Daily	Bare Costs	Incl. O&P
1 Pipe Fitter	$60.70	$485.60	$94.65	$757.20	$51.25	$81.03
1 Truck Driver (light)	41.95	335.60	67.55	540.40		
1 Equip. Oper. (medium)	51.10	408.80	80.90	647.20		
1 Flatbed Truck, Gas, 3 Ton		307.40		338.14		
1 Backhoe Loader, 48 H.P.		365.40		401.94	28.03	30.84
24 L.H., Daily Totals		$1902.80		$2684.88	$79.28	$111.87
Crew B-34Q	Hr.	Daily	Hr.	Daily	Bare Costs	Incl. O&P
1 Pipe Fitter	$60.70	$485.60	$94.65	$757.20	$51.63	$81.63
1 Truck Driver (light)	41.95	335.60	67.55	540.40		
1 Equip. Oper. (crane)	52.25	418.00	82.70	661.60		
1 Flatbed Trailer, 25 Ton		113.20		124.52		
1 Dump Truck, 8 C.Y., 220 H.P.		417.40		459.14		
1 Hyd. Crane, 25 Ton		748.80		823.68	53.31	58.64
24 L.H., Daily Totals		$2518.60		$3366.54	$104.94	$140.27
Crew B-34R	Hr.	Daily	Hr.	Daily	Bare Costs	Incl. O&P
1 Pipe Fitter	$60.70	$485.60	$94.65	$757.20	$51.63	$81.63
1 Truck Driver (light)	41.95	335.60	67.55	540.40		
1 Equip. Oper. (crane)	52.25	418.00	82.70	661.60		
1 Flatbed Trailer, 25 Ton		113.20		124.52		
1 Dump Truck, 8 C.Y., 220 H.P.		417.40		459.14		
1 Hyd. Crane, 25 Ton		748.80		823.68		
1 Hyd. Excavator, 1 C.Y.		812.80		894.08	87.17	95.89
24 L.H., Daily Totals		$3331.40		$4260.62	$138.81	$177.53

Crew No.	Bare Costs Hr.	Daily	Incl. Subs O&P Hr.	Daily	Cost Per Labor-Hour Bare Costs	Incl. O&P
Crew B-34S	Hr.	Daily	Hr.	Daily	Bare Costs	Incl. O&P
2 Pipe Fitters	$60.70	$971.20	$94.65	$1514.40	$54.21	$85.39
1 Truck Driver (heavy)	43.20	345.60	69.55	556.40		
1 Equip. Oper. (crane)	52.25	418.00	82.70	661.60		
1 Flatbed Trailer, 40 Ton		154.60		170.06		
1 Truck Tractor, 6x4, 380 H.P.		600.00		660.00		
1 Hyd. Crane, 80 Ton		1630.00		1793.00		
1 Hyd. Excavator, 2 C.Y.		1176.00		1293.60	111.27	122.40
32 L.H., Daily Totals		$5295.40		$6649.06	$165.48	$207.78
Crew B-34T	Hr.	Daily	Hr.	Daily	Bare Costs	Incl. O&P
2 Pipe Fitters	$60.70	$971.20	$94.65	$1514.40	$54.21	$85.39
1 Truck Driver (heavy)	43.20	345.60	69.55	556.40		
1 Equip. Oper. (crane)	52.25	418.00	82.70	661.60		
1 Flatbed Trailer, 40 Ton		154.60		170.06		
1 Truck Tractor, 6x4, 380 H.P.		600.00		660.00		
1 Hyd. Crane, 80 Ton		1630.00		1793.00	74.52	81.97
32 L.H., Daily Totals		$4119.40		$5355.46	$128.73	$167.36
Crew B-34U	Hr.	Daily	Hr.	Daily	Bare Costs	Incl. O&P
1 Truck Driver (heavy)	$43.20	$345.60	$69.55	$556.40	$46.17	$73.67
1 Equip. Oper. (light)	49.15	393.20	77.80	622.40		
1 Truck Tractor, 220 H.P.		358.80		394.68		
1 Flatbed Trailer, 25 Ton		113.20		124.52	29.50	32.45
16 L.H., Daily Totals		$1210.80		$1698.00	$75.67	$106.13
Crew B-34V	Hr.	Daily	Hr.	Daily	Bare Costs	Incl. O&P
1 Truck Driver (heavy)	$43.20	$345.60	$69.55	$556.40	$48.20	$76.68
1 Equip. Oper. (crane)	52.25	418.00	82.70	661.60		
1 Equip. Oper. (light)	49.15	393.20	77.80	622.40		
1 Truck Tractor, 6x4, 450 H.P.		726.80		799.48		
1 Equipment Trailer, 50 Ton		170.00		187.00		
1 Pickup Truck, 4x4, 3/4 Ton		156.80		172.48	43.90	48.29
24 L.H., Daily Totals		$2210.40		$2999.36	$92.10	$124.97
Crew B-34W	Hr.	Daily	Hr.	Daily	Bare Costs	Incl. O&P
5 Truck Drivers (heavy)	$43.20	$1728.00	$69.55	$2782.00	$45.66	$73.13
2 Equip. Opers. (crane)	52.25	836.00	82.70	1323.20		
1 Equip. Oper. (mechanic)	52.50	420.00	83.10	664.80		
1 Laborer	37.90	303.20	61.95	495.60		
4 Truck Tractors, 6x4, 380 H.P.		2400.00		2640.00		
2 Equipment Trailer, 50 Ton		340.00		374.00		
2 Flatbed Trailers, 40 Ton		309.20		340.12		
1 Pickup Truck, 4x4, 3/4 Ton		156.80		172.48		
1 S.P. Crane, 4x4, 20 Ton		554.80		610.28	52.23	57.46
72 L.H., Daily Totals		$7048.00		$9402.48	$97.89	$130.59
Crew B-35	Hr.	Daily	Hr.	Daily	Bare Costs	Incl. O&P
1 Labor Foreman (outside)	$39.90	$319.20	$65.20	$521.60	$47.53	$75.98
1 Skilled Worker	49.90	399.20	80.85	646.80		
1 Welder (plumber)	59.20	473.60	92.30	738.40		
1 Laborer	37.90	303.20	61.95	495.60		
1 Equip. Oper. (crane)	52.25	418.00	82.70	661.60		
1 Equip. Oper. (oiler)	46.05	368.40	72.90	583.20		
1 Welder, Electric, 300 amp		58.10		63.91		
1 Hyd. Excavator, .75 C.Y.		654.00		719.40	14.84	16.32
48 L.H., Daily Totals		$2993.70		$4430.51	$62.37	$92.30

Crew No.	Bare Costs Hr.	Daily	Incl. Subs O&P Hr.	Daily	Cost Per Labor-Hour Bare Costs	Incl. O&P
Crew B-35A	Hr.	Daily	Hr.	Daily	Bare Costs	Incl. O&P
1 Labor Foreman (outside)	$39.90	$319.20	$65.20	$521.60	$46.16	$73.98
2 Laborers	37.90	606.40	61.95	991.20		
1 Skilled Worker	49.90	399.20	80.85	646.80		
1 Welder (plumber)	59.20	473.60	92.30	738.40		
1 Equip. Oper. (crane)	52.25	418.00	82.70	661.60		
1 Equip. Oper. (oiler)	46.05	368.40	72.90	583.20		
1 Welder, Gas Engine, 300 amp		147.20		161.92		
1 Crawler Crane, 75 Ton		1507.00		1657.70	29.54	32.49
56 L.H., Daily Totals		$4239.00		$5962.42	$75.70	$106.47
Crew B-36	Hr.	Daily	Hr.	Daily	Bare Costs	Incl. O&P
1 Labor Foreman (outside)	$39.90	$319.20	$65.20	$521.60	$43.58	$70.18
2 Laborers	37.90	606.40	61.95	991.20		
2 Equip. Opers. (medium)	51.10	817.60	80.90	1294.40		
1 Dozer, 200 H.P.		1397.00		1536.70		
1 Aggregate Spreader		39.60		43.56		
1 Tandem Roller, 10 Ton		239.40		263.34	41.90	46.09
40 L.H., Daily Totals		$3419.20		$4650.80	$85.48	$116.27
Crew B-36A	Hr.	Daily	Hr.	Daily	Bare Costs	Incl. O&P
1 Labor Foreman (outside)	$39.90	$319.20	$65.20	$521.60	$45.73	$73.24
2 Laborers	37.90	606.40	61.95	991.20		
4 Equip. Opers. (medium)	51.10	1635.20	80.90	2588.80		
1 Dozer, 200 H.P.		1397.00		1536.70		
1 Aggregate Spreader		39.60		43.56		
1 Tandem Roller, 10 Ton		239.40		263.34		
1 Roller, Pneum. Whl., 12 Ton		347.80		382.58	36.14	39.75
56 L.H., Daily Totals		$4584.60		$6327.78	$81.87	$113.00
Crew B-36B	Hr.	Daily	Hr.	Daily	Bare Costs	Incl. O&P
1 Labor Foreman (outside)	$39.90	$319.20	$65.20	$521.60	$45.41	$72.78
2 Laborers	37.90	606.40	61.95	991.20		
4 Equip. Opers. (medium)	51.10	1635.20	80.90	2588.80		
1 Truck Driver (heavy)	43.20	345.60	69.55	556.40		
1 Grader, 30,000 Lbs.		714.40		785.84		
1 F.E. Loader, Crl, 1.5 C.Y.		634.20		697.62		
1 Dozer, 300 H.P.		1887.00		2075.70		
1 Roller, Vibratory, 25 Ton		691.80		760.98		
1 Truck Tractor, 6x4, 450 H.P.		726.80		799.48		
1 Water Tank Trailer, 5000 Gal.		141.60		155.76	74.93	82.43
64 L.H., Daily Totals		$7702.20		$9933.38	$120.35	$155.21
Crew B-36C	Hr.	Daily	Hr.	Daily	Bare Costs	Incl. O&P
1 Labor Foreman (outside)	$39.90	$319.20	$65.20	$521.60	$47.28	$75.49
3 Equip. Opers. (medium)	51.10	1226.40	80.90	1941.60		
1 Truck Driver (heavy)	43.20	345.60	69.55	556.40		
1 Grader, 30,000 Lbs.		714.40		785.84		
1 Dozer, 300 H.P.		1887.00		2075.70		
1 Roller, Vibratory, 25 Ton		691.80		760.98		
1 Truck Tractor, 6x4, 450 H.P.		726.80		799.48		
1 Water Tank Trailer, 5000 Gal.		141.60		155.76	104.04	114.44
40 L.H., Daily Totals		$6052.80		$7597.36	$151.32	$189.93
Crew B-37	Hr.	Daily	Hr.	Daily	Bare Costs	Incl. O&P
1 Labor Foreman (outside)	$39.90	$319.20	$65.20	$521.60	$40.11	$65.13
4 Laborers	37.90	1212.80	61.95	1982.40		
1 Equip. Oper. (light)	49.15	393.20	77.80	622.40		
1 Tandem Roller, 5 Ton		162.60		178.86	3.39	3.73
48 L.H., Daily Totals		$2087.80		$3305.26	$43.50	$68.86

For customer support on your Commercial Renovation Cost Data, call 877.742.4719.

Crew No.	Bare Costs		Incl. Subs O&P		Cost Per Labor-Hour	
Crew B-37A	Hr.	Daily	Hr.	Daily	Bare Costs	Incl. O&P
2 Laborers	$37.90	$606.40	$61.95	$991.20	$39.25	$63.82
1 Truck Driver (light)	41.95	335.60	67.55	540.40		
1 Flatbed Truck, Gas, 1.5 Ton		248.20		273.02		
1 Tar Kettle, T.M.		135.30		148.83	15.98	17.58
24 L.H., Daily Totals		$1325.50		$1953.45	$55.23	$81.39
Crew B-37B	Hr.	Daily	Hr.	Daily	Bare Costs	Incl. O&P
3 Laborers	$37.90	$909.60	$61.95	$1486.80	$38.91	$63.35
1 Truck Driver (light)	41.95	335.60	67.55	540.40		
1 Flatbed Truck, Gas, 1.5 Ton		248.20		273.02		
1 Tar Kettle, T.M.		135.30		148.83	11.98	13.18
32 L.H., Daily Totals		$1628.70		$2449.05	$50.90	$76.53
Crew B-37C	Hr.	Daily	Hr.	Daily	Bare Costs	Incl. O&P
2 Laborers	$37.90	$606.40	$61.95	$991.20	$39.92	$64.75
2 Truck Drivers (light)	41.95	671.20	67.55	1080.80		
2 Flatbed Trucks, Gas, 1.5 Ton		496.40		546.04		
1 Tar Kettle, T.M.		135.30		148.83	19.74	21.71
32 L.H., Daily Totals		$1909.30		$2766.87	$59.67	$86.46
Crew B-37D	Hr.	Daily	Hr.	Daily	Bare Costs	Incl. O&P
1 Laborer	$37.90	$303.20	$61.95	$495.60	$39.92	$64.75
1 Truck Driver (light)	41.95	335.60	67.55	540.40		
1 Pickup Truck, 3/4 Ton		145.80		160.38	9.11	10.02
16 L.H., Daily Totals		$784.60		$1196.38	$49.04	$74.77
Crew B-37E	Hr.	Daily	Hr.	Daily	Bare Costs	Incl. O&P
3 Laborers	$37.90	$909.60	$61.95	$1486.80	$42.55	$68.52
1 Equip. Oper. (light)	49.15	393.20	77.80	622.40		
1 Equip. Oper. (medium)	51.10	408.80	80.90	647.20		
2 Truck Drivers (light)	41.95	671.20	67.55	1080.80		
4 Barrels w/ Flasher		13.60		14.96		
1 Concrete Saw		168.60		185.46		
1 Rotary Hammer Drill		24.55		27.00		
1 Hammer Drill Bit		2.90		3.19		
1 Loader, Skid Steer, 30 H.P.		173.20		190.52		
1 Conc. Hammer Attach.		108.40		119.24		
1 Vibrating Plate, Gas, 18"		36.20		39.82		
2 Flatbed Trucks, Gas, 1.5 Ton		496.40		546.04	18.28	20.11
56 L.H., Daily Totals		$3406.65		$4963.44	$60.83	$88.63
Crew B-37F	Hr.	Daily	Hr.	Daily	Bare Costs	Incl. O&P
3 Laborers	$37.90	$909.60	$61.95	$1486.80	$38.91	$63.35
1 Truck Driver (light)	41.95	335.60	67.55	540.40		
4 Barrels w/ Flasher		13.60		14.96		
1 Concrete Mixer, 10 C.F.		172.80		190.08		
1 Air Compressor, 60 cfm		142.80		157.08		
1 -50' Air Hose, 3/4"		3.05		3.36		
1 Spade (Chipper)		8.20		9.02		
1 Flatbed Truck, Gas, 1.5 Ton		248.20		273.02	18.40	20.23
32 L.H., Daily Totals		$1833.85		$2674.72	$57.31	$83.58
Crew B-37G	Hr.	Daily	Hr.	Daily	Bare Costs	Incl. O&P
1 Labor Foreman (outside)	$39.90	$319.20	$65.20	$521.60	$40.11	$65.13
4 Laborers	37.90	1212.80	61.95	1982.40		
1 Equip. Oper. (light)	49.15	393.20	77.80	622.40		
1 Berm Machine		295.80		325.38		
1 Tandem Roller, 5 Ton		162.60		178.86	9.55	10.51
48 L.H., Daily Totals		$2383.60		$3630.64	$49.66	$75.64

Crew No.	Bare Costs		Incl. Subs O&P		Cost Per Labor-Hour	
Crew B-37H	Hr.	Daily	Hr.	Daily	Bare Costs	Incl. O&P
1 Labor Foreman (outside)	$39.90	$319.20	$65.20	$521.60	$40.11	$65.13
4 Laborers	37.90	1212.80	61.95	1982.40		
1 Equip. Oper. (light)	49.15	393.20	77.80	622.40		
1 Tandem Roller, 5 Ton		162.60		178.86		
1 Flatbed Trucks, Gas, 1.5 Ton		248.20		273.02		
1 Tar Kettle, T.M.		135.30		148.83	11.38	12.51
48 L.H., Daily Totals		$2471.30		$3727.11	$51.49	$77.65
Crew B-37I	Hr.	Daily	Hr.	Daily	Bare Costs	Incl. O&P
3 Laborers	$37.90	$909.60	$61.95	$1486.80	$42.55	$68.52
1 Equip. Oper. (light)	49.15	393.20	77.80	622.40		
1 Equip. Oper. (medium)	51.10	408.80	80.90	647.20		
2 Truck Drivers (light)	41.95	671.20	67.55	1080.80		
4 Barrels w/ Flasher		13.60		14.96		
1 Concrete Saw		168.60		185.46		
1 Rotary Hammer Drill		24.55		27.00		
1 Hammer Drill Bit		2.90		3.19		
1 Air Compressor, 60 cfm		142.80		157.08		
1 -50' Air Hose, 3/4"		3.05		3.36		
1 Spade (Chipper)		8.20		9.02		
1 Loader, Skid Steer, 30 H.P.		173.20		190.52		
1 Conc. Hammer Attach.		108.40		119.24		
1 Concrete Mixer, 10 C.F.		172.80		190.08		
1 Vibrating Plate, Gas, 18"		36.20		39.82		
2 Flatbed Trucks, Gas, 1.5 Ton		496.40		546.04	24.12	26.53
56 L.H., Daily Totals		$3733.50		$5322.97	$66.67	$95.05
Crew B-37J	Hr.	Daily	Hr.	Daily	Bare Costs	Incl. O&P
1 Labor Foreman (outside)	$39.90	$319.20	$65.20	$521.60	$40.11	$65.13
4 Laborers	37.90	1212.80	61.95	1982.40		
1 Equip. Oper. (light)	49.15	393.20	77.80	622.40		
1 Air Compressor, 60 cfm		142.80		157.08		
1 -50' Air Hose, 3/4"		3.05		3.36		
2 Concrete Mixers, 10 C.F.		345.60		380.16		
2 Flatbed Trucks, Gas, 1.5 Ton		496.40		546.04		
1 Shot Blaster, 20"		215.40		236.94	25.07	27.57
48 L.H., Daily Totals		$3128.45		$4449.98	$65.18	$92.71
Crew B-37K	Hr.	Daily	Hr.	Daily	Bare Costs	Incl. O&P
1 Labor Foreman (outside)	$39.90	$319.20	$65.20	$521.60	$40.11	$65.13
4 Laborers	37.90	1212.80	61.95	1982.40		
1 Equip. Oper. (light)	49.15	393.20	77.80	622.40		
1 Air Compressor, 60 cfm		142.80		157.08		
1 -50' Air Hose, 3/4"		3.05		3.36		
2 Flatbed Trucks, Gas, 1.5 Ton		496.40		546.04		
1 Shot Blaster, 20"		215.40		236.94	17.87	19.65
48 L.H., Daily Totals		$2782.85		$4069.82	$57.98	$84.79
Crew B-38	Hr.	Daily	Hr.	Daily	Bare Costs	Incl. O&P
1 Labor Foreman (outside)	$39.90	$319.20	$65.20	$521.60	$43.19	$69.56
2 Laborers	37.90	606.40	61.95	991.20		
1 Equip. Oper. (light)	49.15	393.20	77.80	622.40		
1 Equip. Oper. (medium)	51.10	408.80	80.90	647.20		
1 Backhoe Loader, 48 H.P.		365.40		401.94		
1 Hyd. Hammer, (1200 lb.)		179.60		197.56		
1 F.E. Loader, W.M., 4 C.Y.		661.80		727.98		
1 Pvmt. Rem. Bucket		57.40		63.14	31.61	34.77
40 L.H., Daily Totals		$2991.80		$4173.02	$74.80	$104.33

739

Crews - Renovation

Crew B-39	Hr.	Daily	Hr.	Daily	Bare Costs	Incl. O&P
1 Labor Foreman (outside)	$39.90	$319.20	$65.20	$521.60	$40.11	$65.13
4 Laborers	37.90	1212.80	61.95	1982.40		
1 Equip. Oper. (light)	49.15	393.20	77.80	622.40		
1 Air Compressor, 250 cfm		197.80		217.58		
2 Breakers, Pavement, 60 lb.		20.40		22.44		
2 -50' Air Hoses, 1.5"		10.40		11.44	4.76	5.24
48 L.H., Daily Totals		$2153.80		$3377.86	$44.87	$70.37

Crew B-40	Hr.	Daily	Hr.	Daily	Bare Costs	Incl. O&P
1 Pile Driver Foreman (outside)	$50.40	$403.20	$82.45	$659.60	$49.32	$79.69
4 Pile Drivers	48.40	1548.80	79.20	2534.40		
2 Equip. Opers. (crane)	52.25	836.00	82.70	1323.20		
1 Equip. Oper. (oiler)	46.05	368.40	72.90	583.20		
1 Crawler Crane, 40 Ton		1203.00		1323.30		
1 Vibratory Hammer & Gen.		2612.00		2873.20	59.61	65.57
64 L.H., Daily Totals		$6971.40		$9296.90	$108.93	$145.26

Crew B-40B	Hr.	Daily	Hr.	Daily	Bare Costs	Incl. O&P
1 Labor Foreman (outside)	$39.90	$319.20	$65.20	$521.60	$41.98	$67.78
3 Laborers	37.90	909.60	61.95	1486.80		
1 Equip. Oper. (crane)	52.25	418.00	82.70	661.60		
1 Equip. Oper. (oiler)	46.05	368.40	72.90	583.20		
1 Lattice Boom Crane, 40 Ton		1182.00		1300.20	24.63	27.09
48 L.H., Daily Totals		$3197.20		$4553.40	$66.61	$94.86

Crew B-41	Hr.	Daily	Hr.	Daily	Bare Costs	Incl. O&P
1 Labor Foreman (outside)	$39.90	$319.20	$65.20	$521.60	$39.29	$63.98
4 Laborers	37.90	1212.80	61.95	1982.40		
.25 Equip. Oper. (crane)	52.25	104.50	82.70	165.40		
.25 Equip. Oper. (oiler)	46.05	92.10	72.90	145.80		
.25 Crawler Crane, 40 Ton		300.75		330.82	6.84	7.52
44 L.H., Daily Totals		$2029.35		$3146.03	$46.12	$71.50

Crew B-42	Hr.	Daily	Hr.	Daily	Bare Costs	Incl. O&P
1 Labor Foreman (outside)	$39.90	$319.20	$65.20	$521.60	$42.88	$70.31
4 Laborers	37.90	1212.80	61.95	1982.40		
1 Equip. Oper. (crane)	52.25	418.00	82.70	661.60		
1 Equip. Oper. (oiler)	46.05	368.40	72.90	583.20		
1 Welder	53.20	425.60	93.85	750.80		
1 Hyd. Crane, 25 Ton		748.80		823.68		
1 Welder, Gas Engine, 300 amp		147.20		161.92		
1 Horz. Boring Csg. Mch.		477.40		525.14	21.46	23.61
64 L.H., Daily Totals		$4117.40		$6010.34	$64.33	$93.91

Crew B-43	Hr.	Daily	Hr.	Daily	Bare Costs	Incl. O&P
1 Labor Foreman (outside)	$39.90	$319.20	$65.20	$521.60	$41.98	$67.78
3 Laborers	37.90	909.60	61.95	1486.80		
1 Equip. Oper. (crane)	52.25	418.00	82.70	661.60		
1 Equip. Oper. (oiler)	46.05	368.40	72.90	583.20		
1 Drill Rig, Truck-Mounted		2553.00		2808.30	53.19	58.51
48 L.H., Daily Totals		$4568.20		$6061.50	$95.17	$126.28

Crew B-44	Hr.	Daily	Hr.	Daily	Bare Costs	Incl. O&P
1 Pile Driver Foreman (outside)	$50.40	$403.20	$82.45	$659.60	$48.30	$78.33
4 Pile Drivers	48.40	1548.80	79.20	2534.40		
2 Equip. Opers. (crane)	52.25	836.00	82.70	1323.20		
1 Laborer	37.90	303.20	61.95	495.60		
1 Crawler Crane, 40 Ton		1203.00		1323.30		
1 Lead, 60' High		74.00		81.40		
1 Hammer, Diesel, 15K ft.-lbs.		591.40		650.54	29.19	32.11
64 L.H., Daily Totals		$4959.60		$7068.04	$77.49	$110.44

Crew B-45	Hr.	Daily	Hr.	Daily	Bare Costs	Incl. O&P
1 Equip. Oper. (medium)	$51.10	$408.80	$80.90	$647.20	$47.15	$75.22
1 Truck Driver (heavy)	43.20	345.60	69.55	556.40		
1 Dist. Tanker, 3000 Gallon		319.40		351.34		
1 Truck Tractor, 6x4, 380 H.P.		600.00		660.00	57.46	63.21
16 L.H., Daily Totals		$1673.80		$2214.94	$104.61	$138.43

Crew B-46	Hr.	Daily	Hr.	Daily	Bare Costs	Incl. O&P
1 Pile Driver Foreman (outside)	$50.40	$403.20	$82.45	$659.60	$43.48	$71.12
2 Pile Drivers	48.40	774.40	79.20	1267.20		
3 Laborers	37.90	909.60	61.95	1486.80		
1 Chain Saw, Gas, 36" Long		45.00		49.50	.94	1.03
48 L.H., Daily Totals		$2132.20		$3463.10	$44.42	$72.15

Crew B-47	Hr.	Daily	Hr.	Daily	Bare Costs	Incl. O&P
1 Blast Foreman (outside)	$39.90	$319.20	$65.20	$521.60	$42.32	$68.32
1 Driller	37.90	303.20	61.95	495.60		
1 Equip. Oper. (light)	49.15	393.20	77.80	622.40		
1 Air Track Drill, 4"		1045.00		1149.50		
1 Air Compressor, 600 cfm		533.80		587.18		
2 -50' Air Hoses, 3"		27.70		30.47	66.94	73.63
24 L.H., Daily Totals		$2622.10		$3406.75	$109.25	$141.95

Crew B-47A	Hr.	Daily	Hr.	Daily	Bare Costs	Incl. O&P
1 Drilling Foreman (outside)	$39.90	$319.20	$65.20	$521.60	$46.07	$73.60
1 Equip. Oper. (heavy)	52.25	418.00	82.70	661.60		
1 Equip. Oper. (oiler)	46.05	368.40	72.90	583.20		
1 Air Track Drill, 5"		1248.00		1372.80	52.00	57.20
24 L.H., Daily Totals		$2353.60		$3139.20	$98.07	$130.80

Crew B-47C	Hr.	Daily	Hr.	Daily	Bare Costs	Incl. O&P
1 Laborer	$37.90	$303.20	$61.95	$495.60	$43.52	$69.88
1 Equip. Oper. (light)	49.15	393.20	77.80	622.40		
1 Air Compressor, 750 cfm		541.00		595.10		
2 -50' Air Hoses, 3"		27.70		30.47		
1 Air Track Drill, 4"		1045.00		1149.50	100.86	110.94
16 L.H., Daily Totals		$2310.10		$2893.07	$144.38	$180.82

Crew B-47E	Hr.	Daily	Hr.	Daily	Bare Costs	Incl. O&P
1 Labor Foreman (outside)	$39.90	$319.20	$65.20	$521.60	$38.40	$62.76
3 Laborers	37.90	909.60	61.95	1486.80		
1 Flatbed Truck, Gas, 3 Ton		307.40		338.14	9.61	10.57
32 L.H., Daily Totals		$1536.20		$2346.54	$48.01	$73.33

Crew B-47G	Hr.	Daily	Hr.	Daily	Bare Costs	Incl. O&P
1 Labor Foreman (outside)	$39.90	$319.20	$65.20	$521.60	$41.21	$66.72
2 Laborers	37.90	606.40	61.95	991.20		
1 Equip. Oper. (light)	49.15	393.20	77.80	622.40		
1 Air Track Drill, 4"		1045.00		1149.50		
1 Air Compressor, 600 cfm		533.80		587.18		
2 -50' Air Hoses, 3"		27.70		30.47		
1 Gunite Pump Rig		373.00		410.30	61.86	68.05
32 L.H., Daily Totals		$3298.30		$4312.65	$103.07	$134.77

Crew B-47H	Hr.	Daily	Hr.	Daily	Bare Costs	Incl. O&P
1 Skilled Worker Foreman (out)	$51.90	$415.20	$84.10	$672.80	$50.40	$81.66
3 Skilled Workers	49.90	1197.60	80.85	1940.40		
1 Flatbed Truck, Gas, 3 Ton		307.40		338.14	9.61	10.57
32 L.H., Daily Totals		$1920.20		$2951.34	$60.01	$92.23

For customer support on your Commercial Renovation Cost Data, call 877.742.4719.

Crew No.	Bare Costs		Incl. Subs O&P		Cost Per Labor-Hour	
Crew B-48	Hr.	Daily	Hr.	Daily	Bare Costs	Incl. O&P
1 Labor Foreman (outside)	$39.90	$319.20	$65.20	$521.60	$43.01	$69.21
3 Laborers	37.90	909.60	61.95	1486.80		
1 Equip. Oper. (crane)	52.25	418.00	82.70	661.60		
1 Equip. Oper. (oiler)	46.05	368.40	72.90	583.20		
1 Equip. Oper. (light)	49.15	393.20	77.80	622.40		
1 Centr. Water Pump, 6"		350.80		385.88		
1 -20' Suction Hose, 6"		11.50		12.65		
1 -50' Discharge Hose, 6"		6.10		6.71		
1 Drill Rig, Truck-Mounted		2553.00		2808.30	52.17	57.38
56 L.H., Daily Totals		$5329.80		$7089.14	$95.17	$126.59
Crew B-49	Hr.	Daily	Hr.	Daily	Bare Costs	Incl. O&P
1 Labor Foreman (outside)	$39.90	$319.20	$65.20	$521.60	$45.10	$72.59
3 Laborers	37.90	909.60	61.95	1486.80		
2 Equip. Opers. (crane)	52.25	836.00	82.70	1323.20		
2 Equip. Opers. (oilers)	46.05	736.80	72.90	1166.40		
1 Equip. Oper. (light)	49.15	393.20	77.80	622.40		
2 Pile Drivers	48.40	774.40	79.20	1267.20		
1 Hyd. Crane, 25 Ton		748.80		823.68		
1 Centr. Water Pump, 6"		350.80		385.88		
1 -20' Suction Hose, 6"		11.50		12.65		
1 -50' Discharge Hose, 6"		6.10		6.71		
1 Drill Rig, Truck-Mounted		2553.00		2808.30	41.71	45.88
88 L.H., Daily Totals		$7639.40		$10424.82	$86.81	$118.46
Crew B-50	Hr.	Daily	Hr.	Daily	Bare Costs	Incl. O&P
2 Pile Driver Foremen (outside)	$50.40	$806.40	$82.45	$1319.20	$46.82	$76.02
6 Pile Drivers	48.40	2323.20	79.20	3801.60		
2 Equip. Opers. (crane)	52.25	836.00	82.70	1323.20		
1 Equip. Oper. (oiler)	46.05	368.40	72.90	583.20		
3 Laborers	37.90	909.60	61.95	1486.80		
1 Crawler Crane, 40 Ton		1203.00		1323.30		
1 Lead, 60' High		74.00		81.40		
1 Hammer, Diesel, 15K ft.-lbs.		591.40		650.54		
1 Air Compressor, 600 cfm		533.80		587.18		
2 -50' Air Hoses, 3"		27.70		30.47		
1 Chain Saw, Gas, 36" Long		45.00		49.50	22.10	24.31
112 L.H., Daily Totals		$7718.50		$11236.39	$68.92	$100.33
Crew B-51	Hr.	Daily	Hr.	Daily	Bare Costs	Incl. O&P
1 Labor Foreman (outside)	$39.90	$319.20	$65.20	$521.60	$38.91	$63.42
4 Laborers	37.90	1212.80	61.95	1982.40		
1 Truck Driver (light)	41.95	335.60	67.55	540.40		
1 Flatbed Truck, Gas, 1.5 Ton		248.20		273.02	5.17	5.69
48 L.H., Daily Totals		$2115.80		$3317.42	$44.08	$69.11
Crew B-52	Hr.	Daily	Hr.	Daily	Bare Costs	Incl. O&P
1 Carpenter Foreman (outside)	$50.45	$403.60	$82.45	$659.60	$44.33	$71.86
1 Carpenter	48.45	387.60	79.15	633.20		
3 Laborers	37.90	909.60	61.95	1486.80		
1 Cement Finisher	45.65	365.20	72.25	578.00		
.5 Rodman (reinf.)	53.00	212.00	85.80	343.20		
.5 Equip. Oper. (medium)	51.10	204.40	80.90	323.60		
.5 Crawler Loader, 3 C.Y.		598.50		658.35	10.69	11.76
56 L.H., Daily Totals		$3080.90		$4682.75	$55.02	$83.62
Crew B-53	Hr.	Daily	Hr.	Daily	Bare Costs	Incl. O&P
1 Equip. Oper. (light)	$49.15	$393.20	$77.80	$622.40	$49.15	$77.80
1 Trencher, Chain, 12 H.P.		68.00		74.80	8.50	9.35
8 L.H., Daily Totals		$461.20		$697.20	$57.65	$87.15

Crew No.	Bare Costs		Incl. Subs O&P		Cost Per Labor-Hour	
Crew B-54	Hr.	Daily	Hr.	Daily	Bare Costs	Incl. O&P
1 Equip. Oper. (light)	$49.15	$393.20	$77.80	$622.40	$49.15	$77.80
1 Trencher, Chain, 40 H.P.		334.40		367.84	41.80	45.98
8 L.H., Daily Totals		$727.60		$990.24	$90.95	$123.78
Crew B-54A	Hr.	Daily	Hr.	Daily	Bare Costs	Incl. O&P
.17 Labor Foreman (outside)	$39.90	$54.26	$65.20	$88.67	$49.47	$78.62
1 Equipment Operator (med.)	51.10	408.80	80.90	647.20		
1 Wheel Trencher, 67 H.P.		1183.00		1301.30	126.39	139.03
9.36 L.H., Daily Totals		$1646.06		$2037.17	$175.86	$217.65
Crew B-54B	Hr.	Daily	Hr.	Daily	Bare Costs	Incl. O&P
.25 Labor Foreman (outside)	$39.90	$79.80	$65.20	$130.40	$48.86	$77.76
1 Equipment Operator (med.)	51.10	408.80	80.90	647.20		
1 Wheel Trencher, 150 H.P.		1891.00		2080.10	189.10	208.01
10 L.H., Daily Totals		$2379.60		$2857.70	$237.96	$285.77
Crew B-54D	Hr.	Daily	Hr.	Daily	Bare Costs	Incl. O&P
1 Laborer	$37.90	$303.20	$61.95	$495.60	$44.50	$71.42
1 Equipment Operator (med.)	51.10	408.80	80.90	647.20		
1 Rock Trencher, 6" Width		615.20		676.72	38.45	42.30
16 L.H., Daily Totals		$1327.20		$1819.52	$82.95	$113.72
Crew B-54E	Hr.	Daily	Hr.	Daily	Bare Costs	Incl. O&P
1 Laborer	$37.90	$303.20	$61.95	$495.60	$44.50	$71.42
1 Equipment Operator (med.)	51.10	408.80	80.90	647.20		
1 Rock Trencher, 18" Width		2591.00		2850.10	161.94	178.13
16 L.H., Daily Totals		$3303.00		$3992.90	$206.44	$249.56
Crew B-55	Hr.	Daily	Hr.	Daily	Bare Costs	Incl. O&P
2 Laborers	$37.90	$606.40	$61.95	$991.20	$39.25	$63.82
1 Truck Driver (light)	41.95	335.60	67.55	540.40		
1 Truck-Mounted Earth Auger		786.00		864.60		
1 Flatbed Truck, Gas, 3 Ton		307.40		338.14	45.56	50.11
24 L.H., Daily Totals		$2035.40		$2734.34	$84.81	$113.93
Crew B-56	Hr.	Daily	Hr.	Daily	Bare Costs	Incl. O&P
1 Laborer	$37.90	$303.20	$61.95	$495.60	$43.52	$69.88
1 Equip. Oper. (light)	49.15	393.20	77.80	622.40		
1 Air Track Drill, 4"		1045.00		1149.50		
1 Air Compressor, 600 cfm		533.80		587.18		
1 -50' Air Hose, 3"		13.85		15.23	99.54	109.49
16 L.H., Daily Totals		$2289.05		$2869.92	$143.07	$179.37
Crew B-57	Hr.	Daily	Hr.	Daily	Bare Costs	Incl. O&P
1 Labor Foreman (outside)	$39.90	$319.20	$65.20	$521.60	$43.86	$70.42
2 Laborers	37.90	606.40	61.95	991.20		
1 Equip. Oper. (crane)	52.25	418.00	82.70	661.60		
1 Equip. Oper. (light)	49.15	393.20	77.80	622.40		
1 Equip. Oper. (oiler)	46.05	368.40	72.90	583.20		
1 Crawler Crane, 25 Ton		1194.00		1313.40		
1 Clamshell Bucket, 1 C.Y.		47.40		52.14		
1 Centr. Water Pump, 6"		350.80		385.88		
1 -20' Suction Hose, 6"		11.50		12.65		
20 -50' Discharge Hoses, 6"		122.00		134.20	35.95	39.55
48 L.H., Daily Totals		$3830.90		$5278.27	$79.81	$109.96

Crews - Renovation

Crew No.	Bare Costs Hr.	Daily	Incl. Subs O&P Hr.	Daily	Cost Per Labor-Hour Bare Costs	Incl. O&P
Crew B-58	Hr.	Daily	Hr.	Daily	Bare Costs	Incl. O&P
2 Laborers	$37.90	$606.40	$61.95	$991.20	$41.65	$67.23
1 Equip. Oper. (light)	49.15	393.20	77.80	622.40		
1 Backhoe Loader, 48 H.P.		365.40		401.94		
1 Small Helicopter, w/ Pilot		2774.00		3051.40	130.81	143.89
24 L.H., Daily Totals		$4139.00		$5066.94	$172.46	$211.12

Crew B-59	Hr.	Daily	Hr.	Daily	Bare Costs	Incl. O&P
1 Truck Driver (heavy)	$43.20	$345.60	$69.55	$556.40	$43.20	$69.55
1 Truck Tractor, 220 H.P.		358.80		394.68		
1 Water Tank Trailer, 5000 Gal.		141.60		155.76	62.55	68.81
8 L.H., Daily Totals		$846.00		$1106.84	$105.75	$138.35

Crew B-59A	Hr.	Daily	Hr.	Daily	Bare Costs	Incl. O&P
2 Laborers	$37.90	$606.40	$61.95	$991.20	$39.67	$64.48
1 Truck Driver (heavy)	43.20	345.60	69.55	556.40		
1 Water Tank Trailer, 5000 Gal.		141.60		155.76		
1 Truck Tractor, 220 H.P.		358.80		394.68	20.85	22.93
24 L.H., Daily Totals		$1452.40		$2098.04	$60.52	$87.42

Crew B-60	Hr.	Daily	Hr.	Daily	Bare Costs	Incl. O&P
1 Labor Foreman (outside)	$39.90	$319.20	$65.20	$521.60	$44.61	$71.47
2 Laborers	37.90	606.40	61.95	991.20		
1 Equip. Oper. (crane)	52.25	418.00	82.70	661.60		
2 Equip. Opers. (light)	49.15	786.40	77.80	1244.80		
1 Equip. Oper. (oiler)	46.05	368.40	72.90	583.20		
1 Crawler Crane, 40 Ton		1203.00		1323.30		
1 Lead, 60' High		74.00		81.40		
1 Hammer, Diesel, 15K ft.-lbs.		591.40		650.54		
1 Backhoe Loader, 48 H.P.		365.40		401.94	39.89	43.88
56 L.H., Daily Totals		$4732.20		$6459.58	$84.50	$115.35

Crew B-61	Hr.	Daily	Hr.	Daily	Bare Costs	Incl. O&P
1 Labor Foreman (outside)	$39.90	$319.20	$65.20	$521.60	$40.55	$65.77
3 Laborers	37.90	909.60	61.95	1486.80		
1 Equip. Oper. (light)	49.15	393.20	77.80	622.40		
1 Cement Mixer, 2 C.Y.		193.40		212.74		
1 Air Compressor, 160 cfm		162.20		178.42	8.89	9.78
40 L.H., Daily Totals		$1977.60		$3021.96	$49.44	$75.55

Crew B-62	Hr.	Daily	Hr.	Daily	Bare Costs	Incl. O&P
2 Laborers	$37.90	$606.40	$61.95	$991.20	$41.65	$67.23
1 Equip. Oper. (light)	49.15	393.20	77.80	622.40		
1 Loader, Skid Steer, 30 H.P.		173.20		190.52	7.22	7.94
24 L.H., Daily Totals		$1172.80		$1804.12	$48.87	$75.17

Crew B-62A	Hr.	Daily	Hr.	Daily	Bare Costs	Incl. O&P
2 Laborers	$37.90	$606.40	$61.95	$991.20	$41.65	$67.23
1 Equip. Oper. (light)	49.15	393.20	77.80	622.40		
1 Loader, Skid Steer, 30 H.P.		173.20		190.52		
1 Trencher Attachment		63.70		70.07	9.87	10.86
24 L.H., Daily Totals		$1236.50		$1874.19	$51.52	$78.09

Crew B-63	Hr.	Daily	Hr.	Daily	Bare Costs	Incl. O&P
4 Laborers	$37.90	$1212.80	$61.95	$1982.40	$40.15	$65.12
1 Equip. Oper. (light)	49.15	393.20	77.80	622.40		
1 Loader, Skid Steer, 30 H.P.		173.20		190.52	4.33	4.76
40 L.H., Daily Totals		$1779.20		$2795.32	$44.48	$69.88

Crew B-63B	Hr.	Daily	Hr.	Daily	Bare Costs	Incl. O&P
1 Labor Foreman (inside)	$38.40	$307.20	$62.75	$502.00	$40.84	$66.11
2 Laborers	37.90	606.40	61.95	991.20		
1 Equip. Oper. (light)	49.15	393.20	77.80	622.40		
1 Loader, Skid Steer, 78 H.P.		315.60		347.16	9.86	10.85
32 L.H., Daily Totals		$1622.40		$2462.76	$50.70	$76.96

Crew B-64	Hr.	Daily	Hr.	Daily	Bare Costs	Incl. O&P
1 Laborer	$37.90	$303.20	$61.95	$495.60	$39.92	$64.75
1 Truck Driver (light)	41.95	335.60	67.55	540.40		
1 Power Mulcher (small)		150.80		165.88		
1 Flatbed Truck, Gas, 1.5 Ton		248.20		273.02	24.94	27.43
16 L.H., Daily Totals		$1037.80		$1474.90	$64.86	$92.18

Crew B-65	Hr.	Daily	Hr.	Daily	Bare Costs	Incl. O&P
1 Laborer	$37.90	$303.20	$61.95	$495.60	$39.92	$64.75
1 Truck Driver (light)	41.95	335.60	67.55	540.40		
1 Power Mulcher (Large)		319.60		351.56		
1 Flatbed Truck, Gas, 1.5 Ton		248.20		273.02	35.49	39.04
16 L.H., Daily Totals		$1206.60		$1660.58	$75.41	$103.79

Crew B-66	Hr.	Daily	Hr.	Daily	Bare Costs	Incl. O&P
1 Equip. Oper. (light)	$49.15	$393.20	$77.80	$622.40	$49.15	$77.80
1 Loader-Backhoe, 40 H.P.		263.80		290.18	32.98	36.27
8 L.H., Daily Totals		$657.00		$912.58	$82.13	$114.07

Crew B-67	Hr.	Daily	Hr.	Daily	Bare Costs	Incl. O&P
1 Millwright	$51.10	$408.80	$80.00	$640.00	$50.13	$78.90
1 Equip. Oper. (light)	49.15	393.20	77.80	622.40		
1 Forklift, R/T, 4,000 Lb.		313.60		344.96	19.60	21.56
16 L.H., Daily Totals		$1115.60		$1607.36	$69.72	$100.46

Crew B-67B	Hr.	Daily	Hr.	Daily	Bare Costs	Incl. O&P
1 Millwright Foreman (inside)	$51.60	$412.80	$80.80	$646.40	$51.35	$80.40
1 Millwright	51.10	408.80	80.00	640.00		
16 L.H., Daily Totals		$821.60		$1286.40	$51.35	$80.40

Crew B-68	Hr.	Daily	Hr.	Daily	Bare Costs	Incl. O&P
2 Millwrights	$51.10	$817.60	$80.00	$1280.00	$50.45	$79.27
1 Equip. Oper. (light)	49.15	393.20	77.80	622.40		
1 Forklift, R/T, 4,000 Lb.		313.60		344.96	13.07	14.37
24 L.H., Daily Totals		$1524.40		$2247.36	$63.52	$93.64

Crew B-68A	Hr.	Daily	Hr.	Daily	Bare Costs	Incl. O&P
1 Millwright Foreman (inside)	$51.60	$412.80	$80.80	$646.40	$51.27	$80.27
2 Millwrights	51.10	817.60	80.00	1280.00		
1 Forklift, 8,000 Lb.		185.40		203.94	7.72	8.50
24 L.H., Daily Totals		$1415.80		$2130.34	$58.99	$88.76

Crew B-68B	Hr.	Daily	Hr.	Daily	Bare Costs	Incl. O&P
1 Millwright Foreman (inside)	$51.60	$412.80	$80.80	$646.40	$54.63	$85.11
2 Millwrights	51.10	817.60	80.00	1280.00		
2 Electricians	55.10	881.60	85.20	1363.20		
2 Plumbers	59.20	947.20	92.30	1476.80		
1 Forklift, 5,000 Lb.		334.00		367.40	5.96	6.56
56 L.H., Daily Totals		$3393.20		$5133.80	$60.59	$91.67

For customer support on your Commercial Renovation Cost Data, call 877.742.4719.

Crews - Renovation

Crew No.	Hr.	Daily	Hr.	Daily	Bare Costs	Incl. O&P
Crew B-68C	Hr.	Daily	Hr.	Daily	Bare Costs	Incl. O&P
1 Millwright Foreman (inside)	$51.60	$412.80	$80.80	$646.40	$54.25	$84.58
1 Millwright	51.10	408.80	80.00	640.00		
1 Electrician	55.10	440.80	85.20	681.60		
1 Plumber	59.20	473.60	92.30	738.40		
1 Forklift, 5,000 Lb.		334.00		367.40	10.44	11.48
32 L.H., Daily Totals		$2070.00		$3073.80	$64.69	$96.06
Crew B-68D	Hr.	Daily	Hr.	Daily	Bare Costs	Incl. O&P
1 Labor Foreman (inside)	$38.40	$307.20	$62.75	$502.00	$41.82	$67.50
1 Laborer	37.90	303.20	61.95	495.60		
1 Equip. Oper. (light)	49.15	393.20	77.80	622.40		
1 Forklift, 5,000 Lb.		334.00		367.40	13.92	15.31
24 L.H., Daily Totals		$1337.60		$1987.40	$55.73	$82.81
Crew B-68E	Hr.	Daily	Hr.	Daily	Bare Costs	Incl. O&P
1 Struc. Steel Foreman (inside)	$53.70	$429.60	$94.75	$758.00	$53.30	$94.03
3 Struc. Steel Workers	53.20	1276.80	93.85	2252.40		
1 Welder	53.20	425.60	93.85	750.80		
1 Forklift, 8,000 Lb.		185.40		203.94	4.63	5.10
40 L.H., Daily Totals		$2317.40		$3965.14	$57.94	$99.13
Crew B-68F	Hr.	Daily	Hr.	Daily	Bare Costs	Incl. O&P
1 Skilled Worker Foreman (out)	$51.90	$415.20	$84.10	$672.80	$50.57	$81.93
2 Skilled Workers	49.90	798.40	80.85	1293.60		
1 Forklift, 5,000 Lb.		334.00		367.40	13.92	15.31
24 L.H., Daily Totals		$1547.60		$2333.80	$64.48	$97.24
Crew B-68G	Hr.	Daily	Hr.	Daily	Bare Costs	Incl. O&P
2 Structural Steel Workers	$53.20	$851.20	$93.85	$1501.60	$53.20	$93.85
1 Forklift, 5,000 Lb.		334.00		367.40	20.88	22.96
16 L.H., Daily Totals		$1185.20		$1869.00	$74.08	$116.81
Crew B-69	Hr.	Daily	Hr.	Daily	Bare Costs	Incl. O&P
1 Labor Foreman (outside)	$39.90	$319.20	$65.20	$521.60	$41.98	$67.78
3 Laborers	37.90	909.60	61.95	1486.80		
1 Equip. Oper. (crane)	52.25	418.00	82.70	661.60		
1 Equip. Oper. (oiler)	46.05	368.40	72.90	583.20		
1 Hyd. Crane, 80 Ton		1630.00		1793.00	33.96	37.35
48 L.H., Daily Totals		$3645.20		$5046.20	$75.94	$105.13
Crew B-69A	Hr.	Daily	Hr.	Daily	Bare Costs	Incl. O&P
1 Labor Foreman (outside)	$39.90	$319.20	$65.20	$521.60	$41.73	$67.37
3 Laborers	37.90	909.60	61.95	1486.80		
1 Equip. Oper. (medium)	51.10	408.80	80.90	647.20		
1 Concrete Finisher	45.65	365.20	72.25	578.00		
1 Curb/Gutter Paver, 2-Track		1102.00		1212.20	22.96	25.25
48 L.H., Daily Totals		$3104.80		$4445.80	$64.68	$92.62
Crew B-69B	Hr.	Daily	Hr.	Daily	Bare Costs	Incl. O&P
1 Labor Foreman (outside)	$39.90	$319.20	$65.20	$521.60	$41.73	$67.37
3 Laborers	37.90	909.60	61.95	1486.80		
1 Equip. Oper. (medium)	51.10	408.80	80.90	647.20		
1 Cement Finisher	45.65	365.20	72.25	578.00		
1 Curb/Gutter Paver, 4-Track		780.20		858.22	16.25	17.88
48 L.H., Daily Totals		$2783.00		$4091.82	$57.98	$85.25

Crew No.	Hr.	Daily	Hr.	Daily	Bare Costs	Incl. O&P
Crew B-70	Hr.	Daily	Hr.	Daily	Bare Costs	Incl. O&P
1 Labor Foreman (outside)	$39.90	$319.20	$65.20	$521.60	$43.84	$70.54
3 Laborers	37.90	909.60	61.95	1486.80		
3 Equip. Opers. (medium)	51.10	1226.40	80.90	1941.60		
1 Grader, 30,000 Lbs.		714.40		785.84		
1 Ripper, Beam & 1 Shank		83.80		92.18		
1 Road Sweeper, S.P., 8' wide		666.20		732.82		
1 F.E. Loader, W.M., 1.5 C.Y.		378.60		416.46	32.91	36.20
56 L.H., Daily Totals		$4298.20		$5977.30	$76.75	$106.74
Crew B-71	Hr.	Daily	Hr.	Daily	Bare Costs	Incl. O&P
1 Labor Foreman (outside)	$39.90	$319.20	$65.20	$521.60	$43.84	$70.54
3 Laborers	37.90	909.60	61.95	1486.80		
3 Equip. Opers. (medium)	51.10	1226.40	80.90	1941.60		
1 Pvmt. Profiler, 750 H.P.		5835.00		6418.50		
1 Road Sweeper, S.P., 8' wide		666.20		732.82		
1 F.E. Loader, W.M., 1.5 C.Y.		378.60		416.46	122.85	135.14
56 L.H., Daily Totals		$9335.00		$11517.78	$166.70	$205.67
Crew B-72	Hr.	Daily	Hr.	Daily	Bare Costs	Incl. O&P
1 Labor Foreman (outside)	$39.90	$319.20	$65.20	$521.60	$44.75	$71.83
3 Laborers	37.90	909.60	61.95	1486.80		
4 Equip. Opers. (medium)	51.10	1635.20	80.90	2588.80		
1 Pvmt. Profiler, 750 H.P.		5835.00		6418.50		
1 Hammermill, 250 H.P.		1832.00		2015.20		
1 Windrow Loader		1247.00		1371.70		
1 Mix Paver 165 H.P.		2226.00		2448.60		
1 Roller, Pneum. Whl., 12 Ton		347.80		382.58	179.50	197.45
64 L.H., Daily Totals		$14351.80		$17233.78	$224.25	$269.28
Crew B-73	Hr.	Daily	Hr.	Daily	Bare Costs	Incl. O&P
1 Labor Foreman (outside)	$39.90	$319.20	$65.20	$521.60	$46.40	$74.20
2 Laborers	37.90	606.40	61.95	991.20		
5 Equip. Opers. (medium)	51.10	2044.00	80.90	3236.00		
1 Road Mixer, 310 H.P.		1945.00		2139.50		
1 Tandem Roller, 10 Ton		239.40		263.34		
1 Hammermill, 250 H.P.		1832.00		2015.20		
1 Grader, 30,000 Lbs.		714.40		785.84		
.5 F.E. Loader, W.M., 1.5 C.Y.		189.30		208.23		
.5 Truck Tractor, 220 H.P.		179.40		197.34		
.5 Water Tank Trailer, 5000 Gal.		70.80		77.88	80.79	88.86
64 L.H., Daily Totals		$8139.90		$10436.13	$127.19	$163.06
Crew B-74	Hr.	Daily	Hr.	Daily	Bare Costs	Incl. O&P
1 Labor Foreman (outside)	$39.90	$319.20	$65.20	$521.60	$46.08	$73.73
1 Laborer	37.90	303.20	61.95	495.60		
4 Equip. Opers. (medium)	51.10	1635.20	80.90	2588.80		
2 Truck Drivers (heavy)	43.20	691.20	69.55	1112.80		
1 Grader, 30,000 Lbs.		714.40		785.84		
1 Ripper, Beam & 1 Shank		83.80		92.18		
2 Stabilizers, 310 H.P.		3666.00		4032.60		
1 Flatbed Truck, Gas, 3 Ton		307.40		338.14		
1 Chem. Spreader, Towed		52.00		57.20		
1 Roller, Vibratory, 25 Ton		691.80		760.98		
1 Water Tank Trailer, 5000 Gal.		141.60		155.76		
1 Truck Tractor, 220 H.P.		358.80		394.68	94.00	103.40
64 L.H., Daily Totals		$8964.60		$11336.18	$140.07	$177.13

For customer support on your Commercial Renovation Cost Data, call 877.742.4719.

Crew B-75

	Bare Costs Hr.	Daily	Incl. Subs O&P Hr.	Daily	Cost Per LH Bare Costs	Incl. O&P
1 Labor Foreman (outside)	$39.90	$319.20	$65.20	$521.60	$46.49	$74.33
1 Laborer	37.90	303.20	61.95	495.60		
4 Equip. Opers. (medium)	51.10	1635.20	80.90	2588.80		
1 Truck Driver (heavy)	43.20	345.60	69.55	556.40		
1 Grader, 30,000 Lbs.		714.40		785.84		
1 Ripper, Beam & 1 Shank		83.80		92.18		
2 Stabilizers, 310 H.P.		3666.00		4032.60		
1 Dist. Tanker, 3000 Gallon		319.40		351.34		
1 Truck Tractor, 6x4, 380 H.P.		600.00		660.00		
1 Roller, Vibratory, 25 Ton		691.80		760.98	108.49	119.34
56 L.H., Daily Totals		$8678.60		$10845.34	$154.97	$193.67

Crew B-76

	Bare Costs Hr.	Daily	Incl. Subs O&P Hr.	Daily	Cost Per LH Bare Costs	Incl. O&P
1 Dock Builder Foreman (outside)	$50.40	$403.20	$82.45	$659.60	$49.22	$79.64
5 Dock Builders	48.40	1936.00	79.20	3168.00		
2 Equip. Opers. (crane)	52.25	836.00	82.70	1323.20		
1 Equip. Oper. (oiler)	46.05	368.40	72.90	583.20		
1 Crawler Crane, 50 Ton		1346.00		1480.60		
1 Barge, 400 Ton		779.20		857.12		
1 Hammer, Diesel, 15K ft.-lbs.		591.40		650.54		
1 Lead, 60' High		74.00		81.40		
1 Air Compressor, 600 cfm		533.80		587.18		
2 -50' Air Hoses, 3"		27.70		30.47	46.56	51.21
72 L.H., Daily Totals		$6895.70		$9421.31	$95.77	$130.85

Crew B-76A

	Bare Costs Hr.	Daily	Incl. Subs O&P Hr.	Daily	Cost Per LH Bare Costs	Incl. O&P
1 Labor Foreman (outside)	$39.90	$319.20	$65.20	$521.60	$40.96	$66.32
5 Laborers	37.90	1516.00	61.95	2478.00		
1 Equip. Oper. (crane)	52.25	418.00	82.70	661.60		
1 Equip. Oper. (oiler)	46.05	368.40	72.90	583.20		
1 Crawler Crane, 50 Ton		1346.00		1480.60		
1 Barge, 400 Ton		779.20		857.12	33.21	36.53
64 L.H., Daily Totals		$4746.80		$6582.12	$74.17	$102.85

Crew B-77

	Bare Costs Hr.	Daily	Incl. Subs O&P Hr.	Daily	Cost Per LH Bare Costs	Incl. O&P
1 Labor Foreman (outside)	$39.90	$319.20	$65.20	$521.60	$39.11	$63.72
3 Laborers	37.90	909.60	61.95	1486.80		
1 Truck Driver (light)	41.95	335.60	67.55	540.40		
1 Crack Cleaner, 25 H.P.		62.00		68.20		
1 Crack Filler, Trailer Mtd.		202.60		222.86		
1 Flatbed Truck, Gas, 3 Ton		307.40		338.14	14.30	15.73
40 L.H., Daily Totals		$2136.40		$3178.00	$53.41	$79.45

Crew B-78

	Bare Costs Hr.	Daily	Incl. Subs O&P Hr.	Daily	Cost Per LH Bare Costs	Incl. O&P
1 Labor Foreman (outside)	$39.90	$319.20	$65.20	$521.60	$38.91	$63.42
4 Laborers	37.90	1212.80	61.95	1982.40		
1 Truck Driver (light)	41.95	335.60	67.55	540.40		
1 Paint Striper, S.P., 40 Gallon		152.00		167.20		
1 Flatbed Truck, Gas, 3 Ton		307.40		338.14		
1 Pickup Truck, 3/4 Ton		145.80		160.38	12.61	13.87
48 L.H., Daily Totals		$2472.80		$3710.12	$51.52	$77.29

Crew B-78A

	Bare Costs Hr.	Daily	Incl. Subs O&P Hr.	Daily	Cost Per LH Bare Costs	Incl. O&P
1 Equip. Oper. (light)	$49.15	$393.20	$77.80	$622.40	$49.15	$77.80
1 Line Rem. (Metal Balls) 115 H.P.		846.60		931.26	105.83	116.41
8 L.H., Daily Totals		$1239.80		$1553.66	$154.97	$194.21

Crew B-78B

	Bare Costs Hr.	Daily	Incl. Subs O&P Hr.	Daily	Cost Per LH Bare Costs	Incl. O&P
2 Laborers	$37.90	$606.40	$61.95	$991.20	$39.15	$63.71
.25 Equip. Oper. (light)	49.15	98.30	77.80	155.60		
1 Pickup Truck, 3/4 Ton		145.80		160.38		
1 Line Rem.,11 H.P.,Walk Behind		65.60		72.16		
.25 Road Sweeper, S.P., 8' wide		166.55		183.21	21.00	23.10
18 L.H., Daily Totals		$1082.65		$1562.55	$60.15	$86.81

Crew B-78C

	Bare Costs Hr.	Daily	Incl. Subs O&P Hr.	Daily	Cost Per LH Bare Costs	Incl. O&P
1 Labor Foreman (outside)	$39.90	$319.20	$65.20	$521.60	$38.91	$63.42
4 Laborers	37.90	1212.80	61.95	1982.40		
1 Truck Driver (light)	41.95	335.60	67.55	540.40		
1 Paint Striper, T.M., 120 Gal.		810.20		891.22		
1 Flatbed Truck, Gas, 3 Ton		307.40		338.14		
1 Pickup Truck, 3/4 Ton		145.80		160.38	26.32	28.95
48 L.H., Daily Totals		$3131.00		$4434.14	$65.23	$92.38

Crew B-78D

	Bare Costs Hr.	Daily	Incl. Subs O&P Hr.	Daily	Cost Per LH Bare Costs	Incl. O&P
2 Labor Foremen (outside)	$39.90	$638.40	$65.20	$1043.20	$38.70	$63.16
7 Laborers	37.90	2122.40	61.95	3469.20		
1 Truck Driver (light)	41.95	335.60	67.55	540.40		
1 Paint Striper, T.M., 120 Gal.		810.20		891.22		
1 Flatbed Truck, Gas, 3 Ton		307.40		338.14		
3 Pickup Trucks, 3/4 Ton		437.40		481.14		
1 Air Compressor, 60 cfm		142.80		157.08		
1 -50' Air Hose, 3/4"		3.05		3.36		
3 Breakers, Pavement, 60 lb.		10.20		11.22	21.39	23.53
80 L.H., Daily Totals		$4807.45		$6934.95	$60.09	$86.69

Crew B-78E

	Bare Costs Hr.	Daily	Incl. Subs O&P Hr.	Daily	Cost Per LH Bare Costs	Incl. O&P
2 Labor Foremen (outside)	$39.90	$638.40	$65.20	$1043.20	$38.57	$62.96
9 Laborers	37.90	2728.80	61.95	4460.40		
1 Truck Driver (light)	41.95	335.60	67.55	540.40		
1 Paint Striper, T.M., 120 Gal.		810.20		891.22		
1 Flatbed Truck, Gas, 3 Ton		307.40		338.14		
4 Pickup Trucks, 3/4 Ton		583.20		641.52		
2 Air Compressors, 60 cfm		285.60		314.16		
2 -50' Air Hoses, 3/4"		6.10		6.71		
2 Breakers, Pavement, 60 lb.		20.40		22.44	20.97	23.06
96 L.H., Daily Totals		$5715.70		$8258.19	$59.54	$86.02

Crew B-78F

	Bare Costs Hr.	Daily	Incl. Subs O&P Hr.	Daily	Cost Per LH Bare Costs	Incl. O&P
2 Labor Foremen (outside)	$39.90	$638.40	$65.20	$1043.20	$38.48	$62.81
11 Laborers	37.90	3335.20	61.95	5451.60		
1 Truck Driver (light)	41.95	335.60	67.55	540.40		
1 Paint Striper, T.M., 120 Gal.		810.20		891.22		
1 Flatbed Truck, Gas, 3 Ton		307.40		338.14		
7 Pickup Trucks, 3/4 Ton		1020.60		1122.66		
3 Air Compressor, 60 cfm		428.40		471.24		
3 -50' Air Hose, 3/4"		9.15		10.07		
3 Breakers, Pavement, 60 lb.		30.60		33.66	23.27	25.60
112 L.H., Daily Totals		$6915.55		$9902.18	$61.75	$88.41

Crew B-79

	Bare Costs Hr.	Daily	Incl. Subs O&P Hr.	Daily	Cost Per LH Bare Costs	Incl. O&P
1 Labor Foreman (outside)	$39.90	$319.20	$65.20	$521.60	$39.11	$63.72
3 Laborers	37.90	909.60	61.95	1486.80		
1 Truck Driver (light)	41.95	335.60	67.55	540.40		
1 Paint Striper, T.M., 120 Gal.		810.20		891.22		
1 Heating Kettle, 115 Gallon		81.70		89.87		
1 Flatbed Truck, Gas, 3 Ton		307.40		338.14		
2 Pickup Trucks, 3/4 Ton		291.60		320.76	37.27	41.00
40 L.H., Daily Totals		$3055.30		$4188.79	$76.38	$104.72

Crew B-79A

	Hr.	Daily	Hr.	Daily	Bare Costs	Incl. O&P
1.5 Equip. Oper. (light)	$49.15	$589.80	$77.80	$933.60	$49.15	$77.80
.5 Line Remov. (Grinder) 115 H.P.		468.40		515.24		
1 Line Rem. (Metal Balls) 115 H.P.		846.60		931.26	109.58	120.54
12 L.H., Daily Totals		$1904.80		$2380.10	$158.73	$198.34

Crew B-79B

	Hr.	Daily	Hr.	Daily	Bare Costs	Incl. O&P
1 Laborer	$37.90	$303.20	$61.95	$495.60	$37.90	$61.95
1 Set of Gases		168.00		184.80	21.00	23.10
8 L.H., Daily Totals		$471.20		$680.40	$58.90	$85.05

Crew B-79C

	Hr.	Daily	Hr.	Daily	Bare Costs	Incl. O&P
1 Labor Foreman (outside)	$39.90	$319.20	$65.20	$521.60	$38.76	$63.21
5 Laborers	37.90	1516.00	61.95	2478.00		
1 Truck Driver (light)	41.95	335.60	67.55	540.40		
1 Paint Striper, T.M., 120 Gal.		810.20		891.22		
1 Heating Kettle, 115 Gallon		81.70		89.87		
1 Flatbed Truck, Gas, 3 Ton		307.40		338.14		
3 Pickup Trucks, 3/4 Ton		437.40		481.14		
1 Air Compressor, 60 cfm		142.80		157.08		
1 -50' Air Hose, 3/4"		3.05		3.36		
1 Breakers, Pavement, 60 lb.		10.20		11.22	32.01	35.21
56 L.H., Daily Totals		$3963.55		$5512.02	$70.78	$98.43

Crew B-79D

	Hr.	Daily	Hr.	Daily	Bare Costs	Incl. O&P
2 Labor Foremen (outside)	$39.90	$638.40	$65.20	$1043.20	$38.91	$63.46
5 Laborers	37.90	1516.00	61.95	2478.00		
1 Truck Driver (light)	41.95	335.60	67.55	540.40		
1 Paint Striper, T.M., 120 Gal.		810.20		891.22		
1 Heating Kettle, 115 Gallon		81.70		89.87		
1 Flatbed Truck, Gas, 3 Ton		307.40		338.14		
4 Pickup Trucks, 3/4 Ton		583.20		641.52		
1 Air Compressor, 60 cfm		142.80		157.08		
1 -50' Air Hose, 3/4"		3.05		3.36		
1 Breakers, Pavement, 60 lb.		10.20		11.22	30.29	33.32
64 L.H., Daily Totals		$4428.55		$6194.01	$69.20	$96.78

Crew B-79E

	Hr.	Daily	Hr.	Daily	Bare Costs	Incl. O&P
2 Labor Foremen (outside)	$39.90	$638.40	$65.20	$1043.20	$38.70	$63.16
7 Laborers	37.90	2122.40	61.95	3469.20		
1 Truck Driver (light)	41.95	335.60	67.55	540.40		
1 Paint Striper, T.M., 120 Gal.		810.20		891.22		
1 Heating Kettle, 115 Gallon		81.70		89.87		
1 Flatbed Truck, Gas, 3 Ton		307.40		338.14		
5 Pickup Trucks, 3/4 Ton		729.00		801.90		
2 Air Compressors, 60 cfm		285.60		314.16		
2 -50' Air Hoses, 3/4"		6.10		6.71		
2 Breakers, Pavement, 60 lb.		20.40		22.44	28.00	30.81
80 L.H., Daily Totals		$5336.80		$7517.24	$66.71	$93.97

Crew B-80

	Hr.	Daily	Hr.	Daily	Bare Costs	Incl. O&P
1 Labor Foreman (outside)	$39.90	$319.20	$65.20	$521.60	$42.23	$68.13
1 Laborer	37.90	303.20	61.95	495.60		
1 Truck Driver (light)	41.95	335.60	67.55	540.40		
1 Equip. Oper. (light)	49.15	393.20	77.80	622.40		
1 Flatbed Truck, Gas, 3 Ton		307.40		338.14		
1 Earth Auger, Truck-Mtd.		420.80		462.88	22.76	25.03
32 L.H., Daily Totals		$2079.40		$2981.02	$64.98	$93.16

Crew B-80A

	Hr.	Daily	Hr.	Daily	Bare Costs	Incl. O&P
3 Laborers	$37.90	$909.60	$61.95	$1486.80	$37.90	$61.95
1 Flatbed Truck, Gas, 3 Ton		307.40		338.14	12.81	14.09
24 L.H., Daily Totals		$1217.00		$1824.94	$50.71	$76.04

Crew B-80B

	Hr.	Daily	Hr.	Daily	Bare Costs	Incl. O&P
3 Laborers	$37.90	$909.60	$61.95	$1486.80	$40.71	$65.91
1 Equip. Oper. (light)	49.15	393.20	77.80	622.40		
1 Crane, Flatbed Mounted, 3 Ton		248.20		273.02	7.76	8.53
32 L.H., Daily Totals		$1551.00		$2382.22	$48.47	$74.44

Crew B-80C

	Hr.	Daily	Hr.	Daily	Bare Costs	Incl. O&P
2 Laborers	$37.90	$606.40	$61.95	$991.20	$39.25	$63.82
1 Truck Driver (light)	41.95	335.60	67.55	540.40		
1 Flatbed Truck, Gas, 1.5 Ton		248.20		273.02		
1 Manual Fence Post Auger, Gas		8.40		9.24	10.69	11.76
24 L.H., Daily Totals		$1198.60		$1813.86	$49.94	$75.58

Crew B-81

	Hr.	Daily	Hr.	Daily	Bare Costs	Incl. O&P
1 Laborer	$37.90	$303.20	$61.95	$495.60	$44.07	$70.80
1 Equip. Oper. (medium)	51.10	408.80	80.90	647.20		
1 Truck Driver (heavy)	43.20	345.60	69.55	556.40		
1 Hydromulcher, T.M., 3000 Gal.		334.40		367.84		
1 Truck Tractor, 220 H.P.		358.80		394.68	28.88	31.77
24 L.H., Daily Totals		$1750.80		$2461.72	$72.95	$102.57

Crew B-81A

	Hr.	Daily	Hr.	Daily	Bare Costs	Incl. O&P
1 Laborer	$37.90	$303.20	$61.95	$495.60	$39.92	$64.75
1 Truck Driver (light)	41.95	335.60	67.55	540.40		
1 Hydromulcher, T.M., 600 Gal.		127.80		140.58		
1 Flatbed Truck, Gas, 3 Ton		307.40		338.14	27.20	29.92
16 L.H., Daily Totals		$1074.00		$1514.72	$67.13	$94.67

Crew B-82

	Hr.	Daily	Hr.	Daily	Bare Costs	Incl. O&P
1 Laborer	$37.90	$303.20	$61.95	$495.60	$43.52	$69.88
1 Equip. Oper. (light)	49.15	393.20	77.80	622.40		
1 Horiz. Borer, 6 H.P.		84.20		92.62	5.26	5.79
16 L.H., Daily Totals		$780.60		$1210.62	$48.79	$75.66

Crew B-82A

	Hr.	Daily	Hr.	Daily	Bare Costs	Incl. O&P
2 Laborers	$37.90	$606.40	$61.95	$991.20	$43.52	$69.88
2 Equip. Opers. (light)	49.15	786.40	77.80	1244.80		
2 Dump Trucks, 8 C.Y., 220 H.P.		834.80		918.28		
1 Flatbed Trailer, 25 Ton		113.20		124.52		
1 Horiz. Dir. Drill, 20k lb. Thrust		657.00		722.70		
1 Mud Trailer for HDD, 1500 Gal.		350.00		385.00		
1 Pickup Truck, 4x4, 3/4 Ton		156.80		172.48		
1 Flatbed trailer, 3 Ton		23.40		25.74		
1 Loader, Skid Steer, 78 H.P.		315.60		347.16	76.59	84.25
32 L.H., Daily Totals		$3843.60		$4931.88	$120.11	$154.12

Crew B-82B

	Hr.	Daily	Hr.	Daily	Bare Costs	Incl. O&P
2 Laborers	$37.90	$606.40	$61.95	$991.20	$43.52	$69.88
2 Equip. Opers. (light)	49.15	786.40	77.80	1244.80		
2 Dump Trucks, 8 C.Y., 220 H.P.		834.80		918.28		
1 Flatbed Trailer, 25 Ton		113.20		124.52		
1 Horiz. Dir. Drill, 30k lb. Thrust		935.80		1029.38		
1 Mud Trailer for HDD, 1500 Gal.		350.00		385.00		
1 Pickup Truck, 4x4, 3/4 Ton		156.80		172.48		
1 Flatbed trailer, 3 Ton		23.40		25.74		
1 Loader, Skid Steer, 78 H.P.		315.60		347.16	85.30	93.83
32 L.H., Daily Totals		$4122.40		$5238.56	$128.82	$163.71

Crew No.	Bare Costs Hr.	Bare Costs Daily	Incl. Subs O&P Hr.	Incl. Subs O&P Daily	Cost Per Labor-Hour Bare Costs	Cost Per Labor-Hour Incl. O&P
Crew B-82C	Hr.	Daily	Hr.	Daily	Bare Costs	Incl. O&P
2 Laborers	$37.90	$606.40	$61.95	$991.20	$43.52	$69.88
2 Equip. Opers. (light)	49.15	786.40	77.80	1244.80		
2 Dump Trucks, 8 C.Y., 220 H.P.		834.80		918.28		
1 Flatbed Trailer, 25 Ton		113.20		124.52		
1 Horiz. Dir. Drill, 50k lb. Thrust		1244.00		1368.40		
1 Mud Trailer for HDD, 1500 Gal.		350.00		385.00		
1 Pickup Truck, 4x4, 3/4 Ton		156.80		172.48		
1 Flatbed trailer, 3 Ton		23.40		25.74		
1 Loader, Skid Steer, 78 H.P.		315.60		347.16	94.93	104.42
32 L.H., Daily Totals		$4430.60		$5577.58	$138.46	$174.30
Crew B-82D	Hr.	Daily	Hr.	Daily	Bare Costs	Incl. O&P
1 Equip. Oper. (light)	$49.15	$393.20	$77.80	$622.40	$49.15	$77.80
1 Mud Trailer for HDD, 1500 Gal.		350.00		385.00	43.75	48.13
8 L.H., Daily Totals		$743.20		$1007.40	$92.90	$125.93
Crew B-83	Hr.	Daily	Hr.	Daily	Bare Costs	Incl. O&P
1 Tugboat Captain	$51.10	$408.80	$80.90	$647.20	$44.50	$71.42
1 Tugboat Hand	37.90	303.20	61.95	495.60		
1 Tugboat, 250 H.P.		868.60		955.46	54.29	59.72
16 L.H., Daily Totals		$1580.60		$2098.26	$98.79	$131.14
Crew B-84	Hr.	Daily	Hr.	Daily	Bare Costs	Incl. O&P
1 Equip. Oper. (medium)	$51.10	$408.80	$80.90	$647.20	$51.10	$80.90
1 Rotary Mower/Tractor		369.40		406.34	46.17	50.79
8 L.H., Daily Totals		$778.20		$1053.54	$97.28	$131.69
Crew B-85	Hr.	Daily	Hr.	Daily	Bare Costs	Incl. O&P
3 Laborers	$37.90	$909.60	$61.95	$1486.80	$41.60	$67.26
1 Equip. Oper. (medium)	51.10	408.80	80.90	647.20		
1 Truck Driver (heavy)	43.20	345.60	69.55	556.40		
1 Aerial Lift Truck, 80'		659.20		725.12		
1 Brush Chipper, 12", 130 H.P.		388.60		427.46		
1 Pruning Saw, Rotary		6.65		7.32	26.36	29.00
40 L.H., Daily Totals		$2718.45		$3850.30	$67.96	$96.26
Crew B-86	Hr.	Daily	Hr.	Daily	Bare Costs	Incl. O&P
1 Equip. Oper. (medium)	$51.10	$408.80	$80.90	$647.20	$51.10	$80.90
1 Stump Chipper, S.P.		173.25		190.57	21.66	23.82
8 L.H., Daily Totals		$582.05		$837.77	$72.76	$104.72
Crew B-86A	Hr.	Daily	Hr.	Daily	Bare Costs	Incl. O&P
1 Equip. Oper. (medium)	$51.10	$408.80	$80.90	$647.20	$51.10	$80.90
1 Grader, 30,000 Lbs.		714.40		785.84	89.30	98.23
8 L.H., Daily Totals		$1123.20		$1433.04	$140.40	$179.13
Crew B-86B	Hr.	Daily	Hr.	Daily	Bare Costs	Incl. O&P
1 Equip. Oper. (medium)	$51.10	$408.80	$80.90	$647.20	$51.10	$80.90
1 Dozer, 200 H.P.		1397.00		1536.70	174.63	192.09
8 L.H., Daily Totals		$1805.80		$2183.90	$225.72	$272.99
Crew B-87	Hr.	Daily	Hr.	Daily	Bare Costs	Incl. O&P
1 Laborer	$37.90	$303.20	$61.95	$495.60	$48.46	$77.11
4 Equip. Oper. (medium)	51.10	1635.20	80.90	2588.80		
2 Feller Bunchers, 100 H.P.		1675.20		1842.72		
1 Log Chipper, 22" Tree		890.00		979.00		
1 Dozer, 105 H.P.		603.40		663.74		
1 Chain Saw, Gas, 36" Long		45.00		49.50	80.34	88.37
40 L.H., Daily Totals		$5152.00		$6619.36	$128.80	$165.48

Crew No.	Bare Costs Hr.	Bare Costs Daily	Incl. Subs O&P Hr.	Incl. Subs O&P Daily	Cost Per Labor-Hour Bare Costs	Cost Per Labor-Hour Incl. O&P
Crew B-88	Hr.	Daily	Hr.	Daily	Bare Costs	Incl. O&P
1 Laborer	$37.90	$303.20	$61.95	$495.60	$49.21	$78.19
6 Equip. Oper. (medium)	51.10	2452.80	80.90	3883.20		
2 Feller Bunchers, 100 H.P.		1675.20		1842.72		
1 Log Chipper, 22" Tree		890.00		979.00		
2 Log Skidders, 50 H.P.		1833.60		2016.96		
1 Dozer, 105 H.P.		603.40		663.74		
1 Chain Saw, Gas, 36" Long		45.00		49.50	90.13	99.14
56 L.H., Daily Totals		$7803.20		$9930.72	$139.34	$177.33
Crew B-89	Hr.	Daily	Hr.	Daily	Bare Costs	Incl. O&P
1 Equip. Oper. (light)	$49.15	$393.20	$77.80	$622.40	$45.55	$72.67
1 Truck Driver (light)	41.95	335.60	67.55	540.40		
1 Flatbed Truck, Gas, 3 Ton		307.40		338.14		
1 Concrete Saw		168.60		185.46		
1 Water Tank, 65 Gal.		17.30		19.03	30.83	33.91
16 L.H., Daily Totals		$1222.10		$1705.43	$76.38	$106.59
Crew B-89A	Hr.	Daily	Hr.	Daily	Bare Costs	Incl. O&P
1 Skilled Worker	$49.90	$399.20	$80.85	$646.80	$43.90	$71.40
1 Laborer	37.90	303.20	61.95	495.60		
1 Core Drill (Large)		117.00		128.70	7.31	8.04
16 L.H., Daily Totals		$819.40		$1271.10	$51.21	$79.44
Crew B-89B	Hr.	Daily	Hr.	Daily	Bare Costs	Incl. O&P
1 Equip. Oper. (light)	$49.15	$393.20	$77.80	$622.40	$45.55	$72.67
1 Truck Driver (light)	41.95	335.60	67.55	540.40		
1 Wall Saw, Hydraulic, 10 H.P.		115.00		126.50		
1 Generator, Diesel, 100 kW		413.00		454.30		
1 Water Tank, 65 Gal.		17.30		19.03		
1 Flatbed Truck, Gas, 3 Ton		307.40		338.14	53.29	58.62
16 L.H., Daily Totals		$1581.50		$2100.77	$98.84	$131.30
Crew B-90	Hr.	Daily	Hr.	Daily	Bare Costs	Incl. O&P
1 Labor Foreman (outside)	$39.90	$319.20	$65.20	$521.60	$42.29	$68.22
3 Laborers	37.90	909.60	61.95	1486.80		
2 Equip. Opers. (light)	49.15	786.40	77.80	1244.80		
2 Truck Drivers (heavy)	43.20	691.20	69.55	1112.80		
1 Road Mixer, 310 H.P.		1945.00		2139.50		
1 Dist. Truck, 2000 Gal.		289.00		317.90	34.91	38.40
64 L.H., Daily Totals		$4940.40		$6823.40	$77.19	$106.62
Crew B-90A	Hr.	Daily	Hr.	Daily	Bare Costs	Incl. O&P
1 Labor Foreman (outside)	$39.90	$319.20	$65.20	$521.60	$45.73	$73.24
2 Laborers	37.90	606.40	61.95	991.20		
4 Equip. Opers. (medium)	51.10	1635.20	80.90	2588.80		
2 Graders, 30,000 Lbs.		1428.80		1571.68		
1 Tandem Roller, 10 Ton		239.40		263.34		
1 Roller, Pneum. Whl., 12 Ton		347.80		382.58	36.00	39.60
56 L.H., Daily Totals		$4576.80		$6319.20	$81.73	$112.84
Crew B-90B	Hr.	Daily	Hr.	Daily	Bare Costs	Incl. O&P
1 Labor Foreman (outside)	$39.90	$319.20	$65.20	$521.60	$44.83	$71.97
2 Laborers	37.90	606.40	61.95	991.20		
3 Equip. Opers. (medium)	51.10	1226.40	80.90	1941.60		
1 Roller, Pneum. Whl., 12 Ton		347.80		382.58		
1 Road Mixer, 310 H.P.		1945.00		2139.50	47.77	52.54
48 L.H., Daily Totals		$4444.80		$5976.48	$92.60	$124.51

Crew B-90C

Crew B-90C	Hr.	Daily	Hr.	Daily	Bare Costs	Incl. O&P
1 Labor Foreman (outside)	$39.90	$319.20	$65.20	$521.60	$43.13	$69.49
4 Laborers	37.90	1212.80	61.95	1982.40		
3 Equip. Opers. (medium)	51.10	1226.40	80.90	1941.60		
3 Truck Drivers (heavy)	43.20	1036.80	69.55	1669.20		
3 Road Mixers, 310 H.P.		5835.00		6418.50	66.31	72.94
88 L.H., Daily Totals		$9630.20		$12533.30	$109.43	$142.42

Crew B-90D

Crew B-90D	Hr.	Daily	Hr.	Daily	Bare Costs	Incl. O&P
1 Labor Foreman (outside)	$39.90	$319.20	$65.20	$521.60	$42.32	$68.33
6 Laborers	37.90	1819.20	61.95	2973.60		
3 Equip. Opers. (medium)	51.10	1226.40	80.90	1941.60		
3 Truck Drivers (heavy)	43.20	1036.80	69.55	1669.20		
3 Road Mixers, 310 H.P.		5835.00		6418.50	56.11	61.72
104 L.H., Daily Totals		$10236.60		$13524.50	$98.43	$130.04

Crew B-90E

Crew B-90E	Hr.	Daily	Hr.	Daily	Bare Costs	Incl. O&P
1 Labor Foreman (outside)	$39.90	$319.20	$65.20	$521.60	$43.11	$69.47
4 Laborers	37.90	1212.80	61.95	1982.40		
3 Equip. Opers. (medium)	51.10	1226.40	80.90	1941.60		
1 Truck Driver (heavy)	43.20	345.60	69.55	556.40		
1 Road Mixers, 310 H.P.		1945.00		2139.50	27.01	29.72
72 L.H., Daily Totals		$5049.00		$7141.50	$70.13	$99.19

Crew B-91

Crew B-91	Hr.	Daily	Hr.	Daily	Bare Costs	Incl. O&P
1 Labor Foreman (outside)	$39.90	$319.20	$65.20	$521.60	$45.41	$72.78
2 Laborers	37.90	606.40	61.95	991.20		
4 Equip. Opers. (medium)	51.10	1635.20	80.90	2588.80		
1 Truck Driver (heavy)	43.20	345.60	69.55	556.40		
1 Dist. Tanker, 3000 Gallon		319.40		351.34		
1 Truck Tractor, 6x4, 380 H.P.		600.00		660.00		
1 Aggreg. Spreader, S.P.		829.70		912.67		
1 Roller, Pneum. Whl., 12 Ton		347.80		382.58		
1 Tandem Roller, 10 Ton		239.40		263.34	36.50	40.16
64 L.H., Daily Totals		$5242.70		$7227.93	$81.92	$112.94

Crew B-91B

Crew B-91B	Hr.	Daily	Hr.	Daily	Bare Costs	Incl. O&P
1 Laborer	$37.90	$303.20	$61.95	$495.60	$44.50	$71.42
1 Equipment Oper. (med.)	51.10	408.80	80.90	647.20		
1 Road Sweeper, Vac. Assist.		983.80		1082.18	61.49	67.64
16 L.H., Daily Totals		$1695.80		$2224.98	$105.99	$139.06

Crew B-91C

Crew B-91C	Hr.	Daily	Hr.	Daily	Bare Costs	Incl. O&P
1 Laborer	$37.90	$303.20	$61.95	$495.60	$39.92	$64.75
1 Truck Driver (light)	41.95	335.60	67.55	540.40		
1 Catch Basin Cleaning Truck		577.20		634.92	36.08	39.68
16 L.H., Daily Totals		$1216.00		$1670.92	$76.00	$104.43

Crew B-91D

Crew B-91D	Hr.	Daily	Hr.	Daily	Bare Costs	Incl. O&P
1 Labor Foreman (outside)	$39.90	$319.20	$65.20	$521.60	$43.95	$70.66
5 Laborers	37.90	1516.00	61.95	2478.00		
5 Equip. Opers. (medium)	51.10	2044.00	80.90	3236.00		
2 Truck Drivers (heavy)	43.20	691.20	69.55	1112.80		
1 Aggreg. Spreader, S.P.		829.70		912.67		
2 Trucks, Tractor, 6x4, 380 H.P.		1200.00		1320.00		
2 Dist. Tankers, 3000 Gallon		638.80		702.68		
2 Pavement Brushes, Towed		162.40		178.64		
2 Rollers, Pneum. Whl., 12 Ton		695.60		765.16	33.91	37.30
104 L.H., Daily Totals		$8096.90		$11227.55	$77.85	$107.96

Crew B-92

Crew B-92	Hr.	Daily	Hr.	Daily	Bare Costs	Incl. O&P
1 Labor Foreman (outside)	$39.90	$319.20	$65.20	$521.60	$38.40	$62.76
3 Laborers	37.90	909.60	61.95	1486.80		
1 Crack Cleaner, 25 H.P.		62.00		68.20		
1 Air Compressor, 60 cfm		142.80		157.08		
1 Tar Kettle, T.M.		135.30		148.83		
1 Flatbed Truck, Gas, 3 Ton		307.40		338.14	20.23	22.26
32 L.H., Daily Totals		$1876.30		$2720.65	$58.63	$85.02

Crew B-93

Crew B-93	Hr.	Daily	Hr.	Daily	Bare Costs	Incl. O&P
1 Equip. Oper. (medium)	$51.10	$408.80	$80.90	$647.20	$51.10	$80.90
1 Feller Buncher, 100 H.P.		837.60		921.36	104.70	115.17
8 L.H., Daily Totals		$1246.40		$1568.56	$155.80	$196.07

Crew B-94A

Crew B-94A	Hr.	Daily	Hr.	Daily	Bare Costs	Incl. O&P
1 Laborer	$37.90	$303.20	$61.95	$495.60	$37.90	$61.95
1 Diaphragm Water Pump, 2"		71.80		78.98		
1 -20' Suction Hose, 2"		1.95		2.15		
2 -50' Discharge Hoses, 2"		1.80		1.98	9.44	10.39
8 L.H., Daily Totals		$378.75		$578.71	$47.34	$72.34

Crew B-94B

Crew B-94B	Hr.	Daily	Hr.	Daily	Bare Costs	Incl. O&P
1 Laborer	$37.90	$303.20	$61.95	$495.60	$37.90	$61.95
1 Diaphragm Water Pump, 4"		118.00		129.80		
1 -20' Suction Hose, 4"		3.25		3.58		
2 -50' Discharge Hoses, 4"		4.70		5.17	15.74	17.32
8 L.H., Daily Totals		$429.15		$634.14	$53.64	$79.27

Crew B-94C

Crew B-94C	Hr.	Daily	Hr.	Daily	Bare Costs	Incl. O&P
1 Laborer	$37.90	$303.20	$61.95	$495.60	$37.90	$61.95
1 Centrifugal Water Pump, 3"		81.00		89.10		
1 -20' Suction Hose, 3"		2.85		3.13		
2 -50' Discharge Hoses, 3"		3.00		3.30	10.86	11.94
8 L.H., Daily Totals		$390.05		$591.13	$48.76	$73.89

Crew B-94D

Crew B-94D	Hr.	Daily	Hr.	Daily	Bare Costs	Incl. O&P
1 Laborer	$37.90	$303.20	$61.95	$495.60	$37.90	$61.95
1 Centr. Water Pump, 6"		350.80		385.88		
1 -20' Suction Hose, 6"		11.50		12.65		
2 -50' Discharge Hoses, 6"		12.20		13.42	46.81	51.49
8 L.H., Daily Totals		$677.70		$907.55	$84.71	$113.44

Crew C-1

Crew C-1	Hr.	Daily	Hr.	Daily	Bare Costs	Incl. O&P
3 Carpenters	$48.45	$1162.80	$79.15	$1899.60	$45.81	$74.85
1 Laborer	37.90	303.20	61.95	495.60		
32 L.H., Daily Totals		$1466.00		$2395.20	$45.81	$74.85

Crew C-2

Crew C-2	Hr.	Daily	Hr.	Daily	Bare Costs	Incl. O&P
1 Carpenter Foreman (outside)	$50.45	$403.60	$82.45	$659.60	$47.02	$76.83
4 Carpenters	48.45	1550.40	79.15	2532.80		
1 Laborer	37.90	303.20	61.95	495.60		
48 L.H., Daily Totals		$2257.20		$3688.00	$47.02	$76.83

Crew C-2A

Crew C-2A	Hr.	Daily	Hr.	Daily	Bare Costs	Incl. O&P
1 Carpenter Foreman (outside)	$50.45	$403.60	$82.45	$659.60	$46.56	$75.68
3 Carpenters	48.45	1162.80	79.15	1899.60		
1 Cement Finisher	45.65	365.20	72.25	578.00		
1 Laborer	37.90	303.20	61.95	495.60		
48 L.H., Daily Totals		$2234.80		$3632.80	$46.56	$75.68

Crew No.	Bare Costs		Incl. Subs O&P		Cost Per Labor-Hour	

Crew C-3	Hr.	Daily	Hr.	Daily	Bare Costs	Incl. O&P
1 Rodman Foreman (outside)	$55.00	$440.00	$89.05	$712.40	$48.99	$79.24
4 Rodmen (reinf.)	53.00	1696.00	85.80	2745.60		
1 Equip. Oper. (light)	49.15	393.20	77.80	622.40		
2 Laborers	37.90	606.40	61.95	991.20		
3 Stressing Equipment		30.60		33.66		
.5 Grouting Equipment		82.20		90.42	1.76	1.94
64 L.H., Daily Totals		$3248.40		$5195.68	$50.76	$81.18

Crew C-4	Hr.	Daily	Hr.	Daily	Bare Costs	Incl. O&P
1 Rodman Foreman (outside)	$55.00	$440.00	$89.05	$712.40	$53.50	$86.61
3 Rodmen (reinf.)	53.00	1272.00	85.80	2059.20		
3 Stressing Equipment		30.60		33.66	.96	1.05
32 L.H., Daily Totals		$1742.60		$2805.26	$54.46	$87.66

Crew C-4A	Hr.	Daily	Hr.	Daily	Bare Costs	Incl. O&P
2 Rodmen (reinf.)	$53.00	$848.00	$85.80	$1372.80	$53.00	$85.80
4 Stressing Equipment		40.80		44.88	2.55	2.81
16 L.H., Daily Totals		$888.80		$1417.68	$55.55	$88.61

Crew C-5	Hr.	Daily	Hr.	Daily	Bare Costs	Incl. O&P
1 Rodman Foreman (outside)	$55.00	$440.00	$89.05	$712.40	$52.19	$83.98
4 Rodmen (reinf.)	53.00	1696.00	85.80	2745.60		
1 Equip. Oper. (crane)	52.25	418.00	82.70	661.60		
1 Equip. Oper. (oiler)	46.05	368.40	72.90	583.20		
1 Hyd. Crane, 25 Ton		748.80		823.68	13.37	14.71
56 L.H., Daily Totals		$3671.20		$5526.48	$65.56	$98.69

Crew C-6	Hr.	Daily	Hr.	Daily	Bare Costs	Incl. O&P
1 Labor Foreman (outside)	$39.90	$319.20	$65.20	$521.60	$39.52	$64.21
4 Laborers	37.90	1212.80	61.95	1982.40		
1 Cement Finisher	45.65	365.20	72.25	578.00		
2 Gas Engine Vibrators		63.20		69.52	1.32	1.45
48 L.H., Daily Totals		$1960.40		$3151.52	$40.84	$65.66

Crew C-7	Hr.	Daily	Hr.	Daily	Bare Costs	Incl. O&P
1 Labor Foreman (outside)	$39.90	$319.20	$65.20	$521.60	$41.36	$66.78
5 Laborers	37.90	1516.00	61.95	2478.00		
1 Cement Finisher	45.65	365.20	72.25	578.00		
1 Equip. Oper. (medium)	51.10	408.80	80.90	647.20		
1 Equip. Oper. (oiler)	46.05	368.40	72.90	583.20		
2 Gas Engine Vibrators		63.20		69.52		
1 Concrete Bucket, 1 C.Y.		23.60		25.96		
1 Hyd. Crane, 55 Ton		1123.00		1235.30	16.80	18.48
72 L.H., Daily Totals		$4187.40		$6138.78	$58.16	$85.26

Crew C-8	Hr.	Daily	Hr.	Daily	Bare Costs	Incl. O&P
1 Labor Foreman (outside)	$39.90	$319.20	$65.20	$521.60	$42.29	$68.06
3 Laborers	37.90	909.60	61.95	1486.80		
2 Cement Finishers	45.65	730.40	72.25	1156.00		
1 Equip. Oper. (medium)	51.10	408.80	80.90	647.20		
1 Concrete Pump (Small)		727.60		800.36	12.99	14.29
56 L.H., Daily Totals		$3095.60		$4611.96	$55.28	$82.36

Crew C-8A	Hr.	Daily	Hr.	Daily	Bare Costs	Incl. O&P
1 Labor Foreman (outside)	$39.90	$319.20	$65.20	$521.60	$40.82	$65.92
3 Laborers	37.90	909.60	61.95	1486.80		
2 Cement Finishers	45.65	730.40	72.25	1156.00		
48 L.H., Daily Totals		$1959.20		$3164.40	$40.82	$65.92

Crew No.	Bare Costs		Incl. Subs O&P		Cost Per Labor-Hour	

Crew C-8B	Hr.	Daily	Hr.	Daily	Bare Costs	Incl. O&P
1 Labor Foreman (outside)	$39.90	$319.20	$65.20	$521.60	$40.94	$66.39
3 Laborers	37.90	909.60	61.95	1486.80		
1 Equip. Oper. (medium)	51.10	408.80	80.90	647.20		
1 Vibrating Power Screed		63.65		70.02		
1 Roller, Vibratory, 25 Ton		691.80		760.98		
1 Dozer, 200 H.P.		1397.00		1536.70	53.81	59.19
40 L.H., Daily Totals		$3790.05		$5023.30	$94.75	$125.58

Crew C-8C	Hr.	Daily	Hr.	Daily	Bare Costs	Incl. O&P
1 Labor Foreman (outside)	$39.90	$319.20	$65.20	$521.60	$41.73	$67.37
3 Laborers	37.90	909.60	61.95	1486.80		
1 Cement Finisher	45.65	365.20	72.25	578.00		
1 Equip. Oper. (medium)	51.10	408.80	80.90	647.20		
1 Shotcrete Rig, 12 C.Y./hr		254.00		279.40		
1 Air Compressor, 160 cfm		162.20		178.42		
4 -50' Air Hoses, 1"		14.60		16.06		
4 -50' Air Hoses, 2"		27.60		30.36	9.55	10.51
48 L.H., Daily Totals		$2461.20		$3737.84	$51.27	$77.87

Crew C-8D	Hr.	Daily	Hr.	Daily	Bare Costs	Incl. O&P
1 Labor Foreman (outside)	$39.90	$319.20	$65.20	$521.60	$43.15	$69.30
1 Laborer	37.90	303.20	61.95	495.60		
1 Cement Finisher	45.65	365.20	72.25	578.00		
1 Equipment Oper. (light)	49.15	393.20	77.80	622.40		
1 Air Compressor, 250 cfm		197.80		217.58		
2 -50' Air Hoses, 1"		7.30		8.03	6.41	7.05
32 L.H., Daily Totals		$1585.90		$2443.21	$49.56	$76.35

Crew C-8E	Hr.	Daily	Hr.	Daily	Bare Costs	Incl. O&P
1 Labor Foreman (outside)	$39.90	$319.20	$65.20	$521.60	$41.40	$66.85
3 Laborers	37.90	909.60	61.95	1486.80		
1 Cement Finisher	45.65	365.20	72.25	578.00		
1 Equipment Oper. (light)	49.15	393.20	77.80	622.40		
1 Shotcrete Rig, 35 C.Y./hr		284.00		312.40		
1 Air Compressor, 250 cfm		197.80		217.58		
4 -50' Air Hoses, 1"		14.60		16.06		
4 -50' Air Hoses, 2"		27.60		30.36	10.92	12.01
48 L.H., Daily Totals		$2511.20		$3785.20	$52.32	$78.86

Crew C-10	Hr.	Daily	Hr.	Daily	Bare Costs	Incl. O&P
1 Laborer	$37.90	$303.20	$61.95	$495.60	$43.07	$68.82
2 Cement Finishers	45.65	730.40	72.25	1156.00		
24 L.H., Daily Totals		$1033.60		$1651.60	$43.07	$68.82

Crew C-10D	Hr.	Daily	Hr.	Daily	Bare Costs	Incl. O&P
3 Laborers	$37.90	$909.60	$61.95	$1486.80	$41.00	$66.07
2 Cement Finishers	45.65	730.40	72.25	1156.00		
1 Concrete Mixer, 10 C.F.		172.80		190.08		
2 Trowels, 48" Walk-Behind		101.20		111.32	6.85	7.54
40 L.H., Daily Totals		$1914.00		$2944.20	$47.85	$73.61

Crew C-10C	Hr.	Daily	Hr.	Daily	Bare Costs	Incl. O&P
1 Laborer	$37.90	$303.20	$61.95	$495.60	$43.07	$68.82
2 Cement Finishers	45.65	730.40	72.25	1156.00		
1 Trowel, 48" Walk-Behind		50.60		55.66	2.11	2.32
24 L.H., Daily Totals		$1084.20		$1707.26	$45.17	$71.14

Crews - Renovation

Crew No.	Hr.	Daily	Hr.	Daily	Bare Costs	Incl. O&P
Crew C-10D						
1 Laborer	$37.90	$303.20	$61.95	$495.60	$43.07	$68.82
2 Cement Finishers	45.65	730.40	72.25	1156.00		
1 Vibrating Power Screed		63.65		70.02		
1 Trowel, 48" Walk-Behind		50.60		55.66	4.76	5.24
24 L.H., Daily Totals		$1147.85		$1777.28	$47.83	$74.05
Crew C-10E						
1 Laborer	$37.90	$303.20	$61.95	$495.60	$43.07	$68.82
2 Cement Finishers	45.65	730.40	72.25	1156.00		
1 Vibrating Power Screed		63.65		70.02		
1 Cement Trowel, 96" Ride-On		190.20		209.22	10.58	11.63
24 L.H., Daily Totals		$1287.45		$1930.84	$53.64	$80.45
Crew C-10F						
1 Laborer	$37.90	$303.20	$61.95	$495.60	$43.07	$68.82
2 Cement Finishers	45.65	730.40	72.25	1156.00		
1 Aerial Lift Truck, 60' Boom		448.40		493.24	18.68	20.55
24 L.H., Daily Totals		$1482.00		$2144.84	$61.75	$89.37
Crew C-11						
1 Struc. Steel Foreman (outside)	$55.20	$441.60	$97.35	$778.80	$52.52	$90.67
6 Struc. Steel Workers	53.20	2553.60	93.85	4504.80		
1 Equip. Oper. (crane)	52.25	418.00	82.70	661.60		
1 Equip. Oper. (oiler)	46.05	368.40	72.90	583.20		
1 Lattice Boom Crane, 150 Ton		1868.00		2054.80	25.94	28.54
72 L.H., Daily Totals		$5649.60		$8583.20	$78.47	$119.21
Crew C-12						
1 Carpenter Foreman (outside)	$50.45	$403.60	$82.45	$659.60	$47.66	$77.42
3 Carpenters	48.45	1162.80	79.15	1899.60		
1 Laborer	37.90	303.20	61.95	495.60		
1 Equip. Oper. (crane)	52.25	418.00	82.70	661.60		
1 Hyd. Crane, 12 Ton		658.40		724.24	13.72	15.09
48 L.H., Daily Totals		$2946.00		$4440.64	$61.38	$92.51
Crew C-13						
1 Struc. Steel Worker	$53.20	$425.60	$93.85	$750.80	$51.62	$88.95
1 Welder	53.20	425.60	93.85	750.80		
1 Carpenter	48.45	387.60	79.15	633.20		
1 Welder, Gas Engine, 300 amp		147.20		161.92	6.13	6.75
24 L.H., Daily Totals		$1386.00		$2296.72	$57.75	$95.70
Crew C-14						
1 Carpenter Foreman (outside)	$50.45	$403.60	$82.45	$659.60	$46.99	$76.07
5 Carpenters	48.45	1938.00	79.15	3166.00		
4 Laborers	37.90	1212.80	61.95	1982.40		
4 Rodmen (reinf.)	53.00	1696.00	85.80	2745.60		
2 Cement Finishers	45.65	730.40	72.25	1156.00		
1 Equip. Oper. (crane)	52.25	418.00	82.70	661.60		
1 Equip. Oper. (oiler)	46.05	368.40	72.90	583.20		
1 Hyd. Crane, 80 Ton		1630.00		1793.00	11.32	12.45
144 L.H., Daily Totals		$8397.20		$12747.40	$58.31	$88.52

Crew No.	Hr.	Daily	Hr.	Daily	Bare Costs	Incl. O&P
Crew C-14A						
1 Carpenter Foreman (outside)	$50.45	$403.60	$82.45	$659.60	$48.41	$78.76
16 Carpenters	48.45	6201.60	79.15	10131.20		
4 Rodmen (reinf.)	53.00	1696.00	85.80	2745.60		
2 Laborers	37.90	606.40	61.95	991.20		
1 Cement Finisher	45.65	365.20	72.25	578.00		
1 Equip. Oper. (medium)	51.10	408.80	80.90	647.20		
1 Gas Engine Vibrator		31.60		34.76		
1 Concrete Pump (Small)		727.60		800.36	3.80	4.18
200 L.H., Daily Totals		$10440.80		$16587.92	$52.20	$82.94
Crew C-14B						
1 Carpenter Foreman (outside)	$50.45	$403.60	$82.45	$659.60	$48.30	$78.51
16 Carpenters	48.45	6201.60	79.15	10131.20		
4 Rodmen (reinf.)	53.00	1696.00	85.80	2745.60		
2 Laborers	37.90	606.40	61.95	991.20		
2 Cement Finishers	45.65	730.40	72.25	1156.00		
1 Equip. Oper. (medium)	51.10	408.80	80.90	647.20		
1 Gas Engine Vibrator		31.60		34.76		
1 Concrete Pump (Small)		727.60		800.36	3.65	4.01
208 L.H., Daily Totals		$10806.00		$17165.92	$51.95	$82.53
Crew C-14C						
1 Carpenter Foreman (outside)	$50.45	$403.60	$82.45	$659.60	$46.03	$74.93
6 Carpenters	48.45	2325.60	79.15	3799.20		
2 Rodmen (reinf.)	53.00	848.00	85.80	1372.80		
4 Laborers	37.90	1212.80	61.95	1982.40		
1 Cement Finisher	45.65	365.20	72.25	578.00		
1 Gas Engine Vibrator		31.60		34.76	.28	.31
112 L.H., Daily Totals		$5186.80		$8426.76	$46.31	$75.24
Crew C-14D						
1 Carpenter Foreman (outside)	$50.45	$403.60	$82.45	$659.60	$48.04	$78.23
18 Carpenters	48.45	6976.80	79.15	11397.60		
2 Rodmen (reinf.)	53.00	848.00	85.80	1372.80		
2 Laborers	37.90	606.40	61.95	991.20		
1 Cement Finisher	45.65	365.20	72.25	578.00		
1 Equip. Oper. (medium)	51.10	408.80	80.90	647.20		
1 Gas Engine Vibrator		31.60		34.76		
1 Concrete Pump (Small)		727.60		800.36	3.80	4.18
200 L.H., Daily Totals		$10368.00		$16481.52	$51.84	$82.41
Crew C-14E						
1 Carpenter Foreman (outside)	$50.45	$403.60	$82.45	$659.60	$47.15	$76.55
2 Carpenters	48.45	775.20	79.15	1266.40		
4 Rodmen (reinf.)	53.00	1696.00	85.80	2745.60		
3 Laborers	37.90	909.60	61.95	1486.80		
1 Cement Finisher	45.65	365.20	72.25	578.00		
1 Gas Engine Vibrator		31.60		34.76	.36	.40
88 L.H., Daily Totals		$4181.20		$6771.16	$47.51	$76.94
Crew C-14F						
1 Labor Foreman (outside)	$39.90	$319.20	$65.20	$521.60	$43.29	$69.18
2 Laborers	37.90	606.40	61.95	991.20		
6 Cement Finishers	45.65	2191.20	72.25	3468.00		
1 Gas Engine Vibrator		31.60		34.76	.44	.48
72 L.H., Daily Totals		$3148.40		$5015.56	$43.73	$69.66

Crew No.	Bare Costs		Incl. Subs O&P		Cost Per Labor-Hour	

Crew C-14G

	Hr.	Daily	Hr.	Daily	Bare Costs	Incl. O&P
1 Labor Foreman (outside)	$39.90	$319.20	$65.20	$521.60	$42.61	$68.30
2 Laborers	37.90	606.40	61.95	991.20		
4 Cement Finishers	45.65	1460.80	72.25	2312.00		
1 Gas Engine Vibrator		31.60		34.76	.56	.62
56 L.H., Daily Totals		$2418.00		$3859.56	$43.18	$68.92

Crew C-14H

	Hr.	Daily	Hr.	Daily	Bare Costs	Incl. O&P
1 Carpenter Foreman (outside)	$50.45	$403.60	$82.45	$659.60	$47.32	$76.79
2 Carpenters	48.45	775.20	79.15	1266.40		
1 Rodman (reinf.)	53.00	424.00	85.80	686.40		
1 Laborer	37.90	303.20	61.95	495.60		
1 Cement Finisher	45.65	365.20	72.25	578.00		
1 Gas Engine Vibrator		31.60		34.76	.66	.72
48 L.H., Daily Totals		$2302.80		$3720.76	$47.98	$77.52

Crew C-14L

	Hr.	Daily	Hr.	Daily	Bare Costs	Incl. O&P
1 Carpenter Foreman (outside)	$50.45	$403.60	$82.45	$659.60	$44.87	$73.12
6 Carpenters	48.45	2325.60	79.15	3799.20		
4 Laborers	37.90	1212.80	61.95	1982.40		
1 Cement Finisher	45.65	365.20	72.25	578.00		
1 Gas Engine Vibrator		31.60		34.76	.33	.36
96 L.H., Daily Totals		$4338.80		$7053.96	$45.20	$73.48

Crew C-14M

	Hr.	Daily	Hr.	Daily	Bare Costs	Incl. O&P
1 Carpenter Foreman (outside)	$50.45	$403.60	$82.45	$659.60	$46.61	$75.45
2 Carpenters	48.45	775.20	79.15	1266.40		
1 Rodman (reinf.)	53.00	424.00	85.80	686.40		
2 Laborers	37.90	606.40	61.95	991.20		
1 Cement Finisher	45.65	365.20	72.25	578.00		
1 Equip. Oper. (medium)	51.10	408.80	80.90	647.20		
1 Gas Engine Vibrator		31.60		34.76		
1 Concrete Pump (Small)		727.60		800.36	11.86	13.05
64 L.H., Daily Totals		$3742.40		$5663.92	$58.48	$88.50

Crew C-15

	Hr.	Daily	Hr.	Daily	Bare Costs	Incl. O&P
1 Carpenter Foreman (outside)	$50.45	$403.60	$82.45	$659.60	$45.04	$72.99
2 Carpenters	48.45	775.20	79.15	1266.40		
3 Laborers	37.90	909.60	61.95	1486.80		
2 Cement Finishers	45.65	730.40	72.25	1156.00		
1 Rodman (reinf.)	53.00	424.00	85.80	686.40		
72 L.H., Daily Totals		$3242.80		$5255.20	$45.04	$72.99

Crew C-16

	Hr.	Daily	Hr.	Daily	Bare Costs	Incl. O&P
1 Labor Foreman (outside)	$39.90	$319.20	$65.20	$521.60	$42.29	$68.06
3 Laborers	37.90	909.60	61.95	1400.00		
2 Cement Finishers	45.65	730.40	72.25	1156.00		
1 Equip. Oper. (medium)	51.10	408.80	80.90	647.20		
1 Gunite Pump Rig		373.00		410.30		
2 -50' Air Hoses, 3/4"		6.10		6.71		
2 -50' Air Hoses, 2"		13.80		15.18	7.02	7.72
56 L.H., Daily Totals		$2760.90		$4243.79	$49.30	$75.78

Crew C-16A

	Hr.	Daily	Hr.	Daily	Bare Costs	Incl. O&P
1 Laborer	$37.90	$303.20	$61.95	$495.60	$45.08	$71.84
2 Cement Finishers	45.65	730.40	72.25	1156.00		
1 Equip. Oper. (medium)	51.10	408.80	80.90	647.20		
1 Gunite Pump Rig		373.00		410.30		
2 -50' Air Hoses, 3/4"		6.10		6.71		
2 -50' Air Hoses, 2"		13.80		15.18		
1 Aerial Lift Truck, 60' Boom		448.40		493.24	26.29	28.92
32 L.H., Daily Totals		$2283.70		$3224.23	$71.37	$100.76

Crew C-17

	Hr.	Daily	Hr.	Daily	Bare Costs	Incl. O&P
2 Skilled Worker Foremen (out)	$51.90	$830.40	$84.10	$1345.60	$50.30	$81.50
8 Skilled Workers	49.90	3193.60	80.85	5174.40		
80 L.H., Daily Totals		$4024.00		$6520.00	$50.30	$81.50

Crew C-17A

	Hr.	Daily	Hr.	Daily	Bare Costs	Incl. O&P
2 Skilled Worker Foremen (out)	$51.90	$830.40	$84.10	$1345.60	$50.32	$81.51
8 Skilled Workers	49.90	3193.60	80.85	5174.40		
.125 Equip. Oper. (crane)	52.25	52.25	82.70	82.70		
.125 Hyd. Crane, 80 Ton		203.75		224.13	2.52	2.77
81 L.H., Daily Totals		$4280.00		$6826.82	$52.84	$84.28

Crew C-17B

	Hr.	Daily	Hr.	Daily	Bare Costs	Incl. O&P
2 Skilled Worker Foremen (out)	$51.90	$830.40	$84.10	$1345.60	$50.35	$81.53
8 Skilled Workers	49.90	3193.60	80.85	5174.40		
.25 Equip. Oper. (crane)	52.25	104.50	82.70	165.40		
.25 Hyd. Crane, 80 Ton		407.50		448.25		
.25 Trowel, 48" Walk-Behind		12.65		13.91	5.12	5.64
82 L.H., Daily Totals		$4548.65		$7147.56	$55.47	$87.17

Crew C-17C

	Hr.	Daily	Hr.	Daily	Bare Costs	Incl. O&P
2 Skilled Worker Foremen (out)	$51.90	$830.40	$84.10	$1345.60	$50.37	$81.54
8 Skilled Workers	49.90	3193.60	80.85	5174.40		
.375 Equip. Oper. (crane)	52.25	156.75	82.70	248.10		
.375 Hyd. Crane, 80 Ton		611.25		672.38	7.36	8.10
83 L.H., Daily Totals		$4792.00		$7440.48	$57.73	$89.64

Crew C-17D

	Hr.	Daily	Hr.	Daily	Bare Costs	Incl. O&P
2 Skilled Worker Foremen (out)	$51.90	$830.40	$84.10	$1345.60	$50.39	$81.56
8 Skilled Workers	49.90	3193.60	80.85	5174.40		
.5 Equip. Oper. (crane)	52.25	209.00	82.70	330.80		
.5 Hyd. Crane, 80 Ton		815.00		896.50	9.70	10.67
84 L.H., Daily Totals		$5048.00		$7747.30	$60.10	$92.23

Crew C-17E

	Hr.	Daily	Hr.	Daily	Bare Costs	Incl. O&P
2 Skilled Worker Foremen (out)	$51.90	$830.40	$84.10	$1345.60	$50.30	$81.50
8 Skilled Workers	49.90	3193.60	80.85	5174.40		
1 Hyd. Jack with Rods		98.40		108.24	1.23	1.35
80 L.H., Daily Totals		$4122.40		$6628.24	$51.53	$82.85

Crew C-18

	Hr.	Daily	Hr.	Daily	Bare Costs	Incl. O&P
.125 Labor Foreman (outside)	$39.90	$39.90	$65.20	$65.20	$38.12	$62.31
1 Laborer	37.90	303.20	61.95	495.60		
1 Concrete Cart, 10 C.F.		60.00		66.00	6.67	7.33
9 L.H., Daily Totals		$403.10		$626.80	$44.79	$69.64

Crew C-19

	Hr.	Daily	Hr.	Daily	Bare Costs	Incl. O&P
.125 Labor Foreman (outside)	$39.90	$39.90	$65.20	$65.20	$38.12	$62.31
1 Laborer	37.90	303.20	61.95	495.60		
1 Concrete Cart, 18 C.F.		99.80		109.78	11.09	12.20
9 L.H., Daily Totals		$442.90		$670.58	$49.21	$74.51

Crew C-20

	Hr.	Daily	Hr.	Daily	Bare Costs	Incl. O&P
1 Labor Foreman (outside)	$39.90	$319.20	$65.20	$521.60	$40.77	$66.01
5 Laborers	37.90	1516.00	61.95	2478.00		
1 Cement Finisher	45.65	365.20	72.25	578.00		
1 Equip. Oper. (medium)	51.10	408.80	80.90	647.20		
2 Gas Engine Vibrators		63.20		69.52		
1 Concrete Pump (Small)		727.60		800.36	12.36	13.59
64 L.H., Daily Totals		$3400.00		$5094.68	$53.13	$79.60

Crew No.		Bare Costs		Incl. Subs O&P		Cost Per Labor-Hour	
Crew C-21	Hr.	Daily	Hr.	Daily	Bare Costs	Incl. O&P	
1 Labor Foreman (outside)	$39.90	$319.20	$65.20	$521.60	$40.77	$66.01
5 Laborers	37.90	1516.00	61.95	2478.00		
1 Cement Finisher	45.65	365.20	72.25	578.00		
1 Equip. Oper. (medium)	51.10	408.80	80.90	647.20		
2 Gas Engine Vibrators		63.20		69.52		
1 Concrete Conveyer		199.20		219.12	4.10	4.51
64 L.H., Daily Totals		$2871.60		$4513.44	$44.87	$70.52

Crew No.		Bare Costs		Incl. Subs O&P		Cost Per Labor-Hour	
Crew C-22	Hr.	Daily	Hr.	Daily	Bare Costs	Incl. O&P	
1 Rodman Foreman (outside)	$55.00	$440.00	$89.05	$712.40	$53.20	$86.04
4 Rodmen (reinf.)	53.00	1696.00	85.80	2745.60		
.125 Equip. Oper. (crane)	52.25	52.25	82.70	82.70		
.125 Equip. Oper. (oiler)	46.05	46.05	72.90	72.90		
.125 Hyd. Crane, 25 Ton		93.60		102.96	2.23	2.45
42 L.H., Daily Totals		$2327.90		$3716.56	$55.43	$88.49

Crew No.		Bare Costs		Incl. Subs O&P		Cost Per Labor-Hour	
Crew C-23	Hr.	Daily	Hr.	Daily	Bare Costs	Incl. O&P	
2 Skilled Worker Foremen (out)	$51.90	$830.40	$84.10	$1345.60	$50.15	$80.89
6 Skilled Workers	49.90	2395.20	80.85	3880.80		
1 Equip. Oper. (crane)	52.25	418.00	82.70	661.60		
1 Equip. Oper. (oiler)	46.05	368.40	72.90	583.20		
1 Lattice Boom Crane, 90 Ton		1520.00		1672.00	19.00	20.90
80 L.H., Daily Totals		$5532.00		$8143.20	$69.15	$101.79

Crew No.		Bare Costs		Incl. Subs O&P		Cost Per Labor-Hour	
Crew C-24	Hr.	Daily	Hr.	Daily	Bare Costs	Incl. O&P	
2 Skilled Worker Foremen (out)	$51.90	$830.40	$84.10	$1345.60	$50.15	$80.89
6 Skilled Workers	49.90	2395.20	80.85	3880.80		
1 Equip. Oper. (crane)	52.25	418.00	82.70	661.60		
1 Equip. Oper. (oiler)	46.05	368.40	72.90	583.20		
1 Lattice Boom Crane, 150 Ton		1868.00		2054.80	23.35	25.68
80 L.H., Daily Totals		$5880.00		$8526.00	$73.50	$106.58

Crew No.		Bare Costs		Incl. Subs O&P		Cost Per Labor-Hour	
Crew C-25	Hr.	Daily	Hr.	Daily	Bare Costs	Incl. O&P	
2 Rodmen (reinf.)	$53.00	$848.00	$85.80	$1372.80	$42.17	$71.17
2 Rodmen Helpers	31.35	501.60	56.55	904.80		
32 L.H., Daily Totals		$1349.60		$2277.60	$42.17	$71.17

Crew No.		Bare Costs		Incl. Subs O&P		Cost Per Labor-Hour	
Crew C-27	Hr.	Daily	Hr.	Daily	Bare Costs	Incl. O&P	
2 Cement Finishers	$45.65	$730.40	$72.25	$1156.00	$45.65	$72.25
1 Concrete Saw		168.60		185.46	10.54	11.59
16 L.H., Daily Totals		$899.00		$1341.46	$56.19	$83.84

Crew No.		Bare Costs		Incl. Subs O&P		Cost Per Labor-Hour	
Crew C-28	Hr.	Daily	Hr.	Daily	Bare Costs	Incl. O&P	
1 Cement Finisher	$45.65	$365.20	$72.25	$578.00	$45.65	$72.25
1 Portable Air Compressor, Gas		17.40		19.14	2.17	2.39
8 L.H., Daily Totals		$382.60		$597.14	$47.83	$74.64

Crew No.		Bare Costs		Incl. Subs O&P		Cost Per Labor-Hour	
Crew C-29	Hr.	Daily	Hr.	Daily	Bare Costs	Incl. O&P	
1 Laborer	$37.90	$303.20	$61.95	$495.60	$37.90	$61.95
1 Pressure Washer		71.60		78.76	8.95	9.85
8 L.H., Daily Totals		$374.80		$574.36	$46.85	$71.80

Crew No.		Bare Costs		Incl. Subs O&P		Cost Per Labor-Hour	
Crew C-30	Hr.	Daily	Hr.	Daily	Bare Costs	Incl. O&P	
1 Laborer	$37.90	$303.20	$61.95	$495.60	$37.90	$61.95
1 Concrete Mixer, 10 C.F.		172.80		190.08	21.60	23.76
8 L.H., Daily Totals		$476.00		$685.68	$59.50	$85.71

Crew No.		Bare Costs		Incl. Subs O&P		Cost Per Labor-Hour	
Crew C-31	Hr.	Daily	Hr.	Daily	Bare Costs	Incl. O&P	
1 Cement Finisher	$45.65	$365.20	$72.25	$578.00	$45.65	$72.25
1 Grout Pump		373.00		410.30	46.63	51.29
8 L.H., Daily Totals		$738.20		$988.30	$92.28	$123.54

Crew No.		Bare Costs		Incl. Subs O&P		Cost Per Labor-Hour	
Crew C-32	Hr.	Daily	Hr.	Daily	Bare Costs	Incl. O&P	
1 Cement Finisher	$45.65	$365.20	$72.25	$578.00	$41.77	$67.10
1 Laborer	37.90	303.20	61.95	495.60		
1 Crack Chaser Saw, Gas, 6 H.P.		29.20		32.12		
1 Vacuum Pick-Up System		60.90		66.99	5.63	6.19
16 L.H., Daily Totals		$758.50		$1172.71	$47.41	$73.29

Crew No.		Bare Costs		Incl. Subs O&P		Cost Per Labor-Hour	
Crew D-1	Hr.	Daily	Hr.	Daily	Bare Costs	Incl. O&P	
1 Bricklayer	$46.25	$370.00	$75.25	$602.00	$42.13	$68.55
1 Bricklayer Helper	38.00	304.00	61.85	494.80		
16 L.H., Daily Totals		$674.00		$1096.80	$42.13	$68.55

Crew No.		Bare Costs		Incl. Subs O&P		Cost Per Labor-Hour	
Crew D-2	Hr.	Daily	Hr.	Daily	Bare Costs	Incl. O&P	
3 Bricklayers	$46.25	$1110.00	$75.25	$1806.00	$43.45	$70.73
2 Bricklayer Helpers	38.00	608.00	61.85	989.60		
.5 Carpenter	48.45	193.80	79.15	316.60		
44 L.H., Daily Totals		$1911.80		$3112.20	$43.45	$70.73

Crew No.		Bare Costs		Incl. Subs O&P		Cost Per Labor-Hour	
Crew D-3	Hr.	Daily	Hr.	Daily	Bare Costs	Incl. O&P	
3 Bricklayers	$46.25	$1110.00	$75.25	$1806.00	$43.21	$70.33
2 Bricklayer Helpers	38.00	608.00	61.85	989.60		
.25 Carpenter	48.45	96.90	79.15	158.30		
42 L.H., Daily Totals		$1814.90		$2953.90	$43.21	$70.33

Crew No.		Bare Costs		Incl. Subs O&P		Cost Per Labor-Hour	
Crew D-4	Hr.	Daily	Hr.	Daily	Bare Costs	Incl. O&P	
1 Bricklayer	$46.25	$370.00	$75.25	$602.00	$42.85	$69.19
2 Bricklayer Helpers	38.00	608.00	61.85	989.60		
1 Equip. Oper. (light)	49.15	393.20	77.80	622.40		
1 Grout Pump, 50 C.F./hr.		135.00		148.50	4.22	4.64
32 L.H., Daily Totals		$1506.20		$2362.50	$47.07	$73.83

Crew No.		Bare Costs		Incl. Subs O&P		Cost Per Labor-Hour	
Crew D-5	Hr.	Daily	Hr.	Daily	Bare Costs	Incl. O&P	
1 Bricklayer	46.25	370.00	75.25	602.00	46.25	75.25
8 L.H., Daily Totals		$370.00		$602.00	$46.25	$75.25

Crew No.		Bare Costs		Incl. Subs O&P		Cost Per Labor-Hour	
Crew D-6	Hr.	Daily	Hr.	Daily	Bare Costs	Incl. O&P	
3 Bricklayers	$46.25	$1110.00	$75.25	$1806.00	$42.38	$68.97
3 Bricklayer Helpers	38.00	912.00	61.85	1484.40		
.25 Carpenter	48.45	96.90	79.15	158.30		
50 L.H., Daily Totals		$2118.90		$3448.70	$42.38	$68.97

Crew No.		Bare Costs		Incl. Subs O&P		Cost Per Labor-Hour	
Crew D-7	Hr.	Daily	Hr.	Daily	Bare Costs	Incl. O&P	
1 Tile Layer	$44.50	$356.00	$70.20	$561.60	$39.65	$62.55
1 Tile Layer Helper	34.80	278.40	54.90	439.20		
16 L.H., Daily Totals		$634.40		$1000.80	$39.65	$62.55

Crew No.		Bare Costs		Incl. Subs O&P		Cost Per Labor-Hour	
Crew D-8	Hr.	Daily	Hr.	Daily	Bare Costs	Incl. O&P	
3 Bricklayers	$46.25	$1110.00	$75.25	$1806.00	$42.95	$69.89
2 Bricklayer Helpers	38.00	608.00	61.85	989.60		
40 L.H., Daily Totals		$1718.00		$2795.60	$42.95	$69.89

Crew No.		Bare Costs		Incl. Subs O&P		Cost Per Labor-Hour	
Crew D-9	Hr.	Daily	Hr.	Daily	Bare Costs	Incl. O&P	
3 Bricklayers	$46.25	$1110.00	$75.25	$1806.00	$42.13	$68.55
3 Bricklayer Helpers	38.00	912.00	61.85	1484.40		
48 L.H., Daily Totals		$2022.00		$3290.40	$42.13	$68.55

For customer support on your Commercial Renovation Cost Data, call 877.742.4719.

751

Crew D-10

Crew No.	Hr.	Daily	Hr.	Daily	Bare Costs	Incl. O&P
1 Bricklayer Foreman (outside)	$48.25	$386.00	$78.50	$628.00	$46.19	$74.58
1 Bricklayer	46.25	370.00	75.25	602.00		
1 Bricklayer Helper	38.00	304.00	61.85	494.80		
1 Equip. Oper. (crane)	52.25	418.00	82.70	661.60		
1 S.P. Crane, 4x4, 12 Ton		479.20		527.12	14.98	16.47
32 L.H., Daily Totals		$1957.20		$2913.52	$61.16	$91.05

Crew D-11

Crew No.	Hr.	Daily	Hr.	Daily	Bare Costs	Incl. O&P
1 Bricklayer Foreman (outside)	$48.25	$386.00	$78.50	$628.00	$44.17	$71.87
1 Bricklayer	46.25	370.00	75.25	602.00		
1 Bricklayer Helper	38.00	304.00	61.85	494.80		
24 L.H., Daily Totals		$1060.00		$1724.80	$44.17	$71.87

Crew D-12

Crew No.	Hr.	Daily	Hr.	Daily	Bare Costs	Incl. O&P
1 Bricklayer Foreman (outside)	$48.25	$386.00	$78.50	$628.00	$42.63	$69.36
1 Bricklayer	46.25	370.00	75.25	602.00		
2 Bricklayer Helpers	38.00	608.00	61.85	989.60		
32 L.H., Daily Totals		$1364.00		$2219.60	$42.63	$69.36

Crew D-13

Crew No.	Hr.	Daily	Hr.	Daily	Bare Costs	Incl. O&P
1 Bricklayer Foreman (outside)	$48.25	$386.00	$78.50	$628.00	$45.20	$73.22
1 Bricklayer	46.25	370.00	75.25	602.00		
2 Bricklayer Helpers	38.00	608.00	61.85	989.60		
1 Carpenter	48.45	387.60	79.15	633.20		
1 Equip. Oper. (crane)	52.25	418.00	82.70	661.60		
1 S.P. Crane, 4x4, 12 Ton		479.20		527.12	9.98	10.98
48 L.H., Daily Totals		$2648.80		$4041.52	$55.18	$84.20

Crew E-1

Crew No.	Hr.	Daily	Hr.	Daily	Bare Costs	Incl. O&P
1 Welder Foreman (outside)	$55.20	$441.60	$97.35	$778.80	$52.52	$89.67
1 Welder	53.20	425.60	93.85	750.80		
1 Equip. Oper. (light)	49.15	393.20	77.80	622.40		
1 Welder, Gas Engine, 300 amp		147.20		161.92	6.13	6.75
24 L.H., Daily Totals		$1407.60		$2313.92	$58.65	$96.41

Crew E-2

Crew No.	Hr.	Daily	Hr.	Daily	Bare Costs	Incl. O&P
1 Struc. Steel Foreman (outside)	$55.20	$441.60	$97.35	$778.80	$52.33	$89.76
4 Struc. Steel Workers	53.20	1702.40	93.85	3003.20		
1 Equip. Oper. (crane)	52.25	418.00	82.70	661.60		
1 Equip. Oper. (oiler)	46.05	368.40	72.90	583.20		
1 Lattice Boom Crane, 90 Ton		1520.00		1672.00	27.14	29.86
56 L.H., Daily Totals		$4450.40		$6698.80	$79.47	$119.62

Crew E-3

Crew No.	Hr.	Daily	Hr.	Daily	Bare Costs	Incl. O&P
1 Struc. Steel Foreman (outside)	$55.20	$441.60	$97.35	$778.80	$53.87	$95.02
1 Struc. Steel Worker	53.20	425.60	93.85	750.80		
1 Welder	53.20	425.60	93.85	750.80		
1 Welder, Gas Engine, 300 amp		147.20		161.92	6.13	6.75
24 L.H., Daily Totals		$1440.00		$2442.32	$60.00	$101.76

Crew E-3A

Crew No.	Hr.	Daily	Hr.	Daily	Bare Costs	Incl. O&P
1 Struc. Steel Foreman (outside)	$55.20	$441.60	$97.35	$778.80	$53.87	$95.02
1 Struc. Steel Worker	53.20	425.60	93.85	750.80		
1 Welder	53.20	425.60	93.85	750.80		
1 Welder, Gas Engine, 300 amp		147.20		161.92		
1 Aerial Lift Truck, 40' Boom		309.00		339.90	19.01	20.91
24 L.H., Daily Totals		$1749.00		$2782.22	$72.88	$115.93

Crew E-4

Crew No.	Hr.	Daily	Hr.	Daily	Bare Costs	Incl. O&P
1 Struc. Steel Foreman (outside)	$55.20	$441.60	$97.35	$778.80	$53.70	$94.72
3 Struc. Steel Workers	53.20	1276.80	93.85	2252.40		
1 Welder, Gas Engine, 300 amp		147.20		161.92	4.60	5.06
32 L.H., Daily Totals		$1865.60		$3193.12	$58.30	$99.78

Crew E-5

Crew No.	Hr.	Daily	Hr.	Daily	Bare Costs	Incl. O&P
2 Struc. Steel Foremen (outside)	$55.20	$883.20	$97.35	$1557.60	$52.79	$91.34
5 Struc. Steel Workers	53.20	2128.00	93.85	3754.00		
1 Equip. Oper. (crane)	52.25	418.00	82.70	661.60		
1 Welder	53.20	425.60	93.85	750.80		
1 Equip. Oper. (oiler)	46.05	368.40	72.90	583.20		
1 Lattice Boom Crane, 90 Ton		1520.00		1672.00		
1 Welder, Gas Engine, 300 amp		147.20		161.92	20.84	22.92
80 L.H., Daily Totals		$5890.40		$9141.12	$73.63	$114.26

Crew E-6

Crew No.	Hr.	Daily	Hr.	Daily	Bare Costs	Incl. O&P
3 Struc. Steel Foremen (outside)	$55.20	$1324.80	$97.35	$2336.40	$52.82	$91.50
9 Struc. Steel Workers	53.20	3830.40	93.85	6757.20		
1 Equip. Oper. (crane)	52.25	418.00	82.70	661.60		
1 Welder	53.20	425.60	93.85	750.80		
1 Equip. Oper. (oiler)	46.05	368.40	72.90	583.20		
1 Equip. Oper. (light)	49.15	393.20	77.80	622.40		
1 Lattice Boom Crane, 90 Ton		1520.00		1672.00		
1 Welder, Gas Engine, 300 amp		147.20		161.92		
1 Air Compressor, 160 cfm		162.20		178.42		
2 Impact Wrenches		36.40		40.04	14.58	16.03
128 L.H., Daily Totals		$8626.20		$13763.98	$67.39	$107.53

Crew E-7

Crew No.	Hr.	Daily	Hr.	Daily	Bare Costs	Incl. O&P
1 Struc. Steel Foreman (outside)	$55.20	$441.60	$97.35	$778.80	$52.79	$91.34
4 Struc. Steel Workers	53.20	1702.40	93.85	3003.20		
1 Equip. Oper. (crane)	52.25	418.00	82.70	661.60		
1 Equip. Oper. (oiler)	46.05	368.40	72.90	583.20		
1 Welder Foreman (outside)	55.20	441.60	97.35	778.80		
2 Welders	53.20	851.20	93.85	1501.60		
1 Lattice Boom Crane, 90 Ton		1520.00		1672.00		
2 Welders, Gas Engine, 300 amp		294.40		323.84	22.68	24.95
80 L.H., Daily Totals		$6037.60		$9303.04	$75.47	$116.29

Crew E-8

Crew No.	Hr.	Daily	Hr.	Daily	Bare Costs	Incl. O&P
1 Struc. Steel Foreman (outside)	$55.20	$441.60	$97.35	$778.80	$52.57	$90.68
4 Struc. Steel Workers	53.20	1702.40	93.85	3003.20		
1 Welder Foreman (outside)	55.20	441.60	97.35	778.80		
4 Welders	53.20	1702.40	93.85	3003.20		
1 Equip. Oper. (crane)	52.25	418.00	82.70	661.60		
1 Equip. Oper. (oiler)	46.05	368.40	72.90	583.20		
1 Equip. Oper. (light)	49.15	393.20	77.80	622.40		
1 Lattice Boom Crane, 90 Ton		1520.00		1672.00		
4 Welders, Gas Engine, 300 amp		588.80		647.68	20.28	22.30
104 L.H., Daily Totals		$7576.40		$11750.88	$72.85	$112.99

Crew E-9

Crew No.	Hr.	Daily	Hr.	Daily	Bare Costs	Incl. O&P
2 Struc. Steel Foremen (outside)	$55.20	$883.20	$97.35	$1557.60	$52.82	$91.50
5 Struc. Steel Workers	53.20	2128.00	93.85	3754.00		
1 Welder Foreman (outside)	55.20	441.60	97.35	778.80		
5 Welders	53.20	2128.00	93.85	3754.00		
1 Equip. Oper. (crane)	52.25	418.00	82.70	661.60		
1 Equip. Oper. (oiler)	46.05	368.40	72.90	583.20		
1 Equip. Oper. (light)	49.15	393.20	77.80	622.40		
1 Lattice Boom Crane, 90 Ton		1520.00		1672.00		
5 Welders, Gas Engine, 300 amp		736.00		809.60	17.63	19.39
128 L.H., Daily Totals		$9016.40		$14193.20	$70.44	$110.88

Crews - Renovation

Crew No.	Bare Costs Hr.	Daily	Incl. Subs O&P Hr.	Daily	Cost Per Labor-Hour Bare Costs	Incl. O&P
Crew E-10						
1 Welder Foreman (outside)	$55.20	$441.60	$97.35	$778.80	$54.20	$95.60
1 Welder	53.20	425.60	93.85	750.80		
1 Welder, Gas Engine, 300 amp		147.20		161.92		
1 Flatbed Truck, Gas, 3 Ton		307.40		338.14	28.41	31.25
16 L.H., Daily Totals		$1321.80		$2029.66	$82.61	$126.85

Crew No.	Bare Costs Hr.	Daily	Incl. Subs O&P Hr.	Daily	Cost Per Labor-Hour Bare Costs	Incl. O&P
Crew E-11						
2 Painters, Struc. Steel	$41.30	$660.80	$75.05	$1200.80	$42.41	$72.46
1 Building Laborer	37.90	303.20	61.95	495.60		
1 Equip. Oper. (light)	49.15	393.20	77.80	622.40		
1 Air Compressor, 250 cfm		197.80		217.58		
1 Sandblaster, Portable, 3 C.F.		20.80		22.88		
1 Set Sand Blasting Accessories		14.25		15.68	7.28	8.00
32 L.H., Daily Totals		$1590.05		$2574.93	$49.69	$80.47

Crew No.	Bare Costs Hr.	Daily	Incl. Subs O&P Hr.	Daily	Cost Per Labor-Hour Bare Costs	Incl. O&P
Crew E-11A						
2 Painters, Struc. Steel	$41.30	$660.80	$75.05	$1200.80	$42.41	$72.46
1 Building Laborer	37.90	303.20	61.95	495.60		
1 Equip. Oper. (light)	49.15	393.20	77.80	622.40		
1 Air Compressor, 250 cfm		197.80		217.58		
1 Sandblaster, Portable, 3 C.F.		20.80		22.88		
1 Set Sand Blasting Accessories		14.25		15.68		
1 Aerial Lift Truck, 60' Boom		448.40		493.24	21.29	23.42
32 L.H., Daily Totals		$2038.45		$3068.18	$63.70	$95.88

Crew No.	Bare Costs Hr.	Daily	Incl. Subs O&P Hr.	Daily	Cost Per Labor-Hour Bare Costs	Incl. O&P
Crew E-11B						
2 Painters, Struc. Steel	$41.30	$660.80	$75.05	$1200.80	$40.17	$70.68
1 Building Laborer	37.90	303.20	61.95	495.60		
2 Paint Sprayers, 8 C.F.M.		100.80		110.88		
1 Aerial Lift Truck, 60' Boom		448.40		493.24	22.88	25.17
24 L.H., Daily Totals		$1513.20		$2300.52	$63.05	$95.86

Crew No.	Bare Costs Hr.	Daily	Incl. Subs O&P Hr.	Daily	Cost Per Labor-Hour Bare Costs	Incl. O&P
Crew E-12						
1 Welder Foreman (outside)	$55.20	$441.60	$97.35	$778.80	$52.17	$87.58
1 Equip. Oper. (light)	49.15	393.20	77.80	622.40		
1 Welder, Gas Engine, 300 amp		147.20		161.92	9.20	10.12
16 L.H., Daily Totals		$982.00		$1563.12	$61.38	$97.69

Crew No.	Bare Costs Hr.	Daily	Incl. Subs O&P Hr.	Daily	Cost Per Labor-Hour Bare Costs	Incl. O&P
Crew E-13						
1 Welder Foreman (outside)	$55.20	$441.60	$97.35	$778.80	$53.18	$90.83
.5 Equip. Oper. (light)	49.15	196.60	77.80	311.20		
1 Welder, Gas Engine, 300 amp		147.20		161.92	12.27	13.49
12 L.H., Daily Totals		$785.40		$1251.92	$65.45	$104.33

Crew No.	Bare Costs Hr.	Daily	Incl. Subs O&P Hr.	Daily	Cost Per Labor-Hour Bare Costs	Incl. O&P
Crew E-14						
1 Welder Foreman (outside)	$55.20	$441.60	$97.35	$778.80	$55.20	$97.35
1 Welder, Gas Engine, 300 amp		147.20		161.92	18.40	20.24
8 L.H., Daily Totals		$588.80		$940.72	$73.60	$117.59

Crew No.	Bare Costs Hr.	Daily	Incl. Subs O&P Hr.	Daily	Cost Per Labor-Hour Bare Costs	Incl. O&P
Crew E-16						
1 Welder Foreman (outside)	$55.20	$441.60	$97.35	$778.80	$54.20	$95.60
1 Welder	53.20	425.60	93.85	750.80		
1 Welder, Gas Engine, 300 amp		147.20		161.92	9.20	10.12
16 L.H., Daily Totals		$1014.40		$1691.52	$63.40	$105.72

Crew No.	Bare Costs Hr.	Daily	Incl. Subs O&P Hr.	Daily	Cost Per Labor-Hour Bare Costs	Incl. O&P
Crew E-17						
1 Struc. Steel Foreman (outside)	$55.20	$441.60	$97.35	$778.80	$54.20	$95.60
1 Structural Steel Worker	53.20	425.60	93.85	750.80		
16 L.H., Daily Totals		$867.20		$1529.60	$54.20	$95.60

Crew No.	Bare Costs Hr.	Daily	Incl. Subs O&P Hr.	Daily	Cost Per Labor-Hour Bare Costs	Incl. O&P
Crew E-18						
1 Struc. Steel Foreman (outside)	$55.20	$441.60	$97.35	$778.80	$53.18	$91.96
3 Structural Steel Workers	53.20	1276.80	93.85	2252.40		
1 Equipment Operator (med.)	51.10	408.80	80.90	647.20		
1 Lattice Boom Crane, 20 Ton		967.35		1064.09	24.18	26.60
40 L.H., Daily Totals		$3094.55		$4742.48	$77.36	$118.56

Crew No.	Bare Costs Hr.	Daily	Incl. Subs O&P Hr.	Daily	Cost Per Labor-Hour Bare Costs	Incl. O&P
Crew E-19						
1 Struc. Steel Foreman (outside)	$55.20	$441.60	$97.35	$778.80	$52.52	$89.67
1 Structural Steel Worker	53.20	425.60	93.85	750.80		
1 Equip. Oper. (light)	49.15	393.20	77.80	622.40		
1 Lattice Boom Crane, 20 Ton		967.35		1064.09	40.31	44.34
24 L.H., Daily Totals		$2227.75		$3216.09	$92.82	$134.00

Crew No.	Bare Costs Hr.	Daily	Incl. Subs O&P Hr.	Daily	Cost Per Labor-Hour Bare Costs	Incl. O&P
Crew E-20						
1 Struc. Steel Foreman (outside)	$55.20	$441.60	$97.35	$778.80	$52.44	$90.28
5 Structural Steel Workers	53.20	2128.00	93.85	3754.00		
1 Equip. Oper. (crane)	52.25	418.00	82.70	661.60		
1 Equip. Oper. (oiler)	46.05	368.40	72.90	583.20		
1 Lattice Boom Crane, 40 Ton		1182.00		1300.20	18.47	20.32
64 L.H., Daily Totals		$4538.00		$7077.80	$70.91	$110.59

Crew No.	Bare Costs Hr.	Daily	Incl. Subs O&P Hr.	Daily	Cost Per Labor-Hour Bare Costs	Incl. O&P
Crew E-22						
1 Skilled Worker Foreman (out)	$51.90	$415.20	$84.10	$672.80	$50.57	$81.93
2 Skilled Workers	49.90	798.40	80.85	1293.60		
24 L.H., Daily Totals		$1213.60		$1966.40	$50.57	$81.93

Crew No.	Bare Costs Hr.	Daily	Incl. Subs O&P Hr.	Daily	Cost Per Labor-Hour Bare Costs	Incl. O&P
Crew E-24						
3 Structural Steel Workers	$53.20	$1276.80	$93.85	$2252.40	$52.67	$90.61
1 Equipment Operator (med.)	51.10	408.80	80.90	647.20		
1 Hyd. Crane, 25 Ton		748.80		823.68	23.40	25.74
32 L.H., Daily Totals		$2434.40		$3723.28	$76.08	$116.35

Crew No.	Bare Costs Hr.	Daily	Incl. Subs O&P Hr.	Daily	Cost Per Labor-Hour Bare Costs	Incl. O&P
Crew E-25						
1 Welder Foreman (outside)	$55.20	$441.60	$97.35	$778.80	$55.20	$97.35
1 Cutting Torch		11.40		12.54	1.43	1.57
8 L.H., Daily Totals		$453.00		$791.34	$56.63	$98.92

Crew No.	Bare Costs Hr.	Daily	Incl. Subs O&P Hr.	Daily	Cost Per Labor-Hour Bare Costs	Incl. O&P
Crew E-26						
1 Struc. Steel Foreman (outside)	$55.20	$441.60	$97.35	$778.80	$54.34	$94.12
1 Struc. Steel Worker	53.20	425.60	93.85	750.80		
1 Welder	53.20	425.60	93.85	750.80		
.25 Electrician	55.10	110.20	85.20	170.40		
.25 Plumber	59.20	118.40	92.30	184.60		
1 Welder, Gas Engine, 300 amp		147.20		161.92	5.26	5.78
28 L.H., Daily Totals		$1668.60		$2797.32	$59.59	$99.90

Crew No.	Bare Costs Hr.	Daily	Incl. Subs O&P Hr.	Daily	Cost Per Labor-Hour Bare Costs	Incl. O&P
Crew F-3						
4 Carpenters	$48.45	$1550.40	$79.15	$2532.80	$49.21	$79.86
1 Equip. Oper. (crane)	52.25	418.00	82.70	661.60		
1 Hyd. Crane, 12 Ton		658.40		724.24	16.46	18.11
40 L.H., Daily Totals		$2626.80		$3918.64	$65.67	$97.97

Crew No.	Bare Costs Hr.	Daily	Incl. Subs O&P Hr.	Daily	Cost Per Labor-Hour Bare Costs	Incl. O&P
Crew F-4						
4 Carpenters	$48.45	$1550.40	$79.15	$2532.80	$48.68	$78.70
1 Equip. Oper. (crane)	52.25	418.00	82.70	661.60		
1 Equip. Oper. (oiler)	46.05	368.40	72.90	583.20		
1 Hyd. Crane, 55 Ton		1123.00		1235.30	23.40	25.74
48 L.H., Daily Totals		$3459.80		$5012.90	$72.08	$104.44

Crew F-5

	Bare Costs Hr.	Daily	Incl. Subs O&P Hr.	Daily	Cost Per Labor-Hour Bare Costs	Incl. O&P
1 Carpenter Foreman (outside)	$50.45	$403.60	$82.45	$659.60	$48.95	$79.97
3 Carpenters	48.45	1162.80	79.15	1899.60		
32 L.H., Daily Totals		$1566.40		$2559.20	$48.95	$79.97

Crew F-6

	Bare Costs Hr.	Daily	Incl. Subs O&P Hr.	Daily	Cost Per Labor-Hour Bare Costs	Incl. O&P
2 Carpenters	$48.45	$775.20	$79.15	$1266.40	$44.99	$72.98
2 Building Laborers	37.90	606.40	61.95	991.20		
1 Equip. Oper. (crane)	52.25	418.00	82.70	661.60		
1 Hyd. Crane, 12 Ton		658.40		724.24	16.46	18.11
40 L.H., Daily Totals		$2458.00		$3643.44	$61.45	$91.09

Crew F-7

	Bare Costs Hr.	Daily	Incl. Subs O&P Hr.	Daily	Cost Per Labor-Hour Bare Costs	Incl. O&P
2 Carpenters	$48.45	$775.20	$79.15	$1266.40	$43.17	$70.55
2 Building Laborers	37.90	606.40	61.95	991.20		
32 L.H., Daily Totals		$1381.60		$2257.60	$43.17	$70.55

Crew G-1

	Bare Costs Hr.	Daily	Incl. Subs O&P Hr.	Daily	Cost Per Labor-Hour Bare Costs	Incl. O&P
1 Roofer Foreman (outside)	$43.70	$349.60	$78.85	$630.80	$39.03	$70.42
4 Roofers Composition	41.70	1334.40	75.25	2408.00		
2 Roofer Helpers	31.35	501.60	56.55	904.80		
1 Application Equipment		193.00		212.30		
1 Tar Kettle/Pot		176.40		194.04		
1 Crew Truck		208.80		229.68	10.32	11.36
56 L.H., Daily Totals		$2763.80		$4579.62	$49.35	$81.78

Crew G-2

	Bare Costs Hr.	Daily	Incl. Subs O&P Hr.	Daily	Cost Per Labor-Hour Bare Costs	Incl. O&P
1 Plasterer	$44.90	$359.20	$71.60	$572.80	$40.28	$64.75
1 Plasterer Helper	38.05	304.40	60.70	485.60		
1 Building Laborer	37.90	303.20	61.95	495.60		
1 Grout Pump, 50 C.F./hr.		135.00		148.50	5.63	6.19
24 L.H., Daily Totals		$1101.80		$1702.50	$45.91	$70.94

Crew G-2A

	Bare Costs Hr.	Daily	Incl. Subs O&P Hr.	Daily	Cost Per Labor-Hour Bare Costs	Incl. O&P
1 Roofer Composition	$41.70	$333.60	$75.25	$602.00	$36.98	$64.58
1 Roofer Helper	31.35	250.80	56.55	452.40		
1 Building Laborer	37.90	303.20	61.95	495.60		
1 Foam Spray Rig, Trailer-Mtd.		567.00		623.70		
1 Pickup Truck, 3/4 Ton		145.80		160.38	29.70	32.67
24 L.H., Daily Totals		$1600.40		$2334.08	$66.68	$97.25

Crew G-3

	Bare Costs Hr.	Daily	Incl. Subs O&P Hr.	Daily	Cost Per Labor-Hour Bare Costs	Incl. O&P
2 Sheet Metal Workers	$57.25	$916.00	$90.35	$1445.60	$47.58	$76.15
2 Building Laborers	37.90	606.40	61.95	991.20		
32 L.H., Daily Totals		$1522.40		$2436.80	$47.58	$76.15

Crew G-4

	Bare Costs Hr.	Daily	Incl. Subs O&P Hr.	Daily	Cost Per Labor-Hour Bare Costs	Incl. O&P
1 Labor Foreman (outside)	$39.90	$319.20	$65.20	$521.60	$38.57	$63.03
2 Building Laborers	37.90	606.40	61.95	991.20		
1 Flatbed Truck, Gas, 1.5 Ton		248.20		273.02		
1 Air Compressor, 160 cfm		162.20		178.42	17.10	18.81
24 L.H., Daily Totals		$1336.00		$1964.24	$55.67	$81.84

Crew G-5

	Bare Costs Hr.	Daily	Incl. Subs O&P Hr.	Daily	Cost Per Labor-Hour Bare Costs	Incl. O&P
1 Roofer Foreman (outside)	$43.70	$349.60	$78.85	$630.80	$37.96	$68.49
2 Roofers Composition	41.70	667.20	75.25	1204.00		
2 Roofer Helpers	31.35	501.60	56.55	904.80		
1 Application Equipment		193.00		212.30	4.83	5.31
40 L.H., Daily Totals		$1711.40		$2951.90	$42.78	$73.80

Crew G-6A

	Bare Costs Hr.	Daily	Incl. Subs O&P Hr.	Daily	Cost Per Labor-Hour Bare Costs	Incl. O&P
2 Roofers Composition	$41.70	$667.20	$75.25	$1204.00	$41.70	$75.25
1 Small Compressor, Electric		14.10		15.51		
2 Pneumatic Nailers		52.40		57.64	4.16	4.57
16 L.H., Daily Totals		$733.70		$1277.15	$45.86	$79.82

Crew G-7

	Bare Costs Hr.	Daily	Incl. Subs O&P Hr.	Daily	Cost Per Labor-Hour Bare Costs	Incl. O&P
1 Carpenter	$48.45	$387.60	$79.15	$633.20	$48.45	$79.15
1 Small Compressor, Electric		14.10		15.51		
1 Pneumatic Nailer		26.20		28.82	5.04	5.54
8 L.H., Daily Totals		$427.90		$677.53	$53.49	$84.69

Crew H-1

	Bare Costs Hr.	Daily	Incl. Subs O&P Hr.	Daily	Cost Per Labor-Hour Bare Costs	Incl. O&P
2 Glaziers	$46.70	$747.20	$75.70	$1211.20	$49.95	$84.78
2 Struc. Steel Workers	53.20	851.20	93.85	1501.60		
32 L.H., Daily Totals		$1598.40		$2712.80	$49.95	$84.78

Crew H-2

	Bare Costs Hr.	Daily	Incl. Subs O&P Hr.	Daily	Cost Per Labor-Hour Bare Costs	Incl. O&P
2 Glaziers	$46.70	$747.20	$75.70	$1211.20	$43.77	$71.12
1 Building Laborer	37.90	303.20	61.95	495.60		
24 L.H., Daily Totals		$1050.40		$1706.80	$43.77	$71.12

Crew H-3

	Bare Costs Hr.	Daily	Incl. Subs O&P Hr.	Daily	Cost Per Labor-Hour Bare Costs	Incl. O&P
1 Glazier	$46.70	$373.60	$75.70	$605.60	$41.35	$67.53
1 Helper	36.00	288.00	59.35	474.80		
16 L.H., Daily Totals		$661.60		$1080.40	$41.35	$67.53

Crew H-4

	Bare Costs Hr.	Daily	Incl. Subs O&P Hr.	Daily	Cost Per Labor-Hour Bare Costs	Incl. O&P
1 Carpenter	$48.45	$387.60	$79.15	$633.20	$44.80	$72.44
1 Carpenter Helper	36.00	288.00	59.35	474.80		
.5 Electrician	55.10	220.40	85.20	340.80		
20 L.H., Daily Totals		$896.00		$1448.80	$44.80	$72.44

Crew J-1

	Bare Costs Hr.	Daily	Incl. Subs O&P Hr.	Daily	Cost Per Labor-Hour Bare Costs	Incl. O&P
3 Plasterers	$44.90	$1077.60	$71.60	$1718.40	$42.16	$67.24
2 Plasterer Helpers	38.05	608.80	60.70	971.20		
1 Mixing Machine, 6 C.F.		140.20		154.22	3.50	3.86
40 L.H., Daily Totals		$1826.60		$2843.82	$45.66	$71.10

Crew J-2

	Bare Costs Hr.	Daily	Incl. Subs O&P Hr.	Daily	Cost Per Labor-Hour Bare Costs	Incl. O&P
3 Plasterers	$44.90	$1077.60	$71.60	$1718.40	$43.02	$68.46
2 Plasterer Helpers	38.05	608.80	60.70	971.20		
1 Lather	47.30	378.40	74.55	596.40		
1 Mixing Machine, 6 C.F.		140.20		154.22	2.92	3.21
48 L.H., Daily Totals		$2205.00		$3440.22	$45.94	$71.67

Crew J-3

	Bare Costs Hr.	Daily	Incl. Subs O&P Hr.	Daily	Cost Per Labor-Hour Bare Costs	Incl. O&P
1 Terrazzo Worker	$44.60	$356.80	$70.40	$563.20	$41.00	$64.70
1 Terrazzo Helper	37.40	299.20	59.00	472.00		
1 Floor Grinder, 22" Path		121.60		133.76		
1 Terrazzo Mixer		188.20		207.02	19.36	21.30
16 L.H., Daily Totals		$965.80		$1375.98	$60.36	$86.00

Crew J-4

	Bare Costs Hr.	Daily	Incl. Subs O&P Hr.	Daily	Cost Per Labor-Hour Bare Costs	Incl. O&P
2 Cement Finishers	$45.65	$730.40	$72.25	$1156.00	$43.07	$68.82
1 Laborer	37.90	303.20	61.95	495.60		
1 Floor Grinder, 22" Path		121.60		133.76		
1 Floor Edger, 7" Path		39.30		43.23		
1 Vacuum Pick-Up System		60.90		66.99	9.24	10.17
24 L.H., Daily Totals		$1255.40		$1895.58	$52.31	$78.98

Crew J-4A

Crew No.	Bare Costs Hr.	Bare Costs Daily	Incl. Subs O&P Hr.	Incl. Subs O&P Daily	Cost Per Labor-Hour Bare Costs	Cost Per Labor-Hour Incl. O&P
2 Cement Finishers	$45.65	$730.40	$72.25	$1156.00	$41.77	$67.10
2 Laborers	37.90	606.40	61.95	991.20		
1 Floor Grinder, 22" Path		121.60		133.76		
1 Floor Edger, 7" Path		39.30		43.23		
1 Vacuum Pick-Up System		60.90		66.99		
1 Floor Auto Scrubber		241.05		265.15	14.46	15.91
32 L.H., Daily Totals		$1799.65		$2656.34	$56.24	$83.01

Crew J-4B

Crew No.	Bare Costs Hr.	Bare Costs Daily	Incl. Subs O&P Hr.	Incl. Subs O&P Daily	Cost Per Labor-Hour Bare Costs	Cost Per Labor-Hour Incl. O&P
1 Laborer	$37.90	$303.20	$61.95	$495.60	$37.90	$61.95
1 Floor Auto Scrubber		241.05		265.15	30.13	33.14
8 L.H., Daily Totals		$544.25		$760.76	$68.03	$95.09

Crew J-6

Crew No.	Bare Costs Hr.	Bare Costs Daily	Incl. Subs O&P Hr.	Incl. Subs O&P Daily	Cost Per Labor-Hour Bare Costs	Cost Per Labor-Hour Incl. O&P
2 Painters	$40.35	$645.60	$64.85	$1037.60	$41.94	$67.36
1 Building Laborer	37.90	303.20	61.95	495.60		
1 Equip. Oper. (light)	49.15	393.20	77.80	622.40		
1 Air Compressor, 250 cfm		197.80		217.58		
1 Sandblaster, Portable, 3 C.F.		20.80		22.88		
1 Set Sand Blasting Accessories		14.25		15.68	7.28	8.00
32 L.H., Daily Totals		$1574.85		$2411.74	$49.21	$75.37

Crew J-7

Crew No.	Bare Costs Hr.	Bare Costs Daily	Incl. Subs O&P Hr.	Incl. Subs O&P Daily	Cost Per Labor-Hour Bare Costs	Cost Per Labor-Hour Incl. O&P
2 Painters	$40.35	$645.60	$64.85	$1037.60	$40.35	$64.85
1 Floor Belt Sander		24.45		26.90		
1 Floor Sanding Edger		13.65		15.02	2.38	2.62
16 L.H., Daily Totals		$683.70		$1079.51	$42.73	$67.47

Crew K-1

Crew No.	Bare Costs Hr.	Bare Costs Daily	Incl. Subs O&P Hr.	Incl. Subs O&P Daily	Cost Per Labor-Hour Bare Costs	Cost Per Labor-Hour Incl. O&P
1 Carpenter	$48.45	$387.60	$79.15	$633.20	$45.20	$73.35
1 Truck Driver (light)	41.95	335.60	67.55	540.40		
1 Flatbed Truck, Gas, 3 Ton		307.40		338.14	19.21	21.13
16 L.H., Daily Totals		$1030.60		$1511.74	$64.41	$94.48

Crew K-2

Crew No.	Bare Costs Hr.	Bare Costs Daily	Incl. Subs O&P Hr.	Incl. Subs O&P Daily	Cost Per Labor-Hour Bare Costs	Cost Per Labor-Hour Incl. O&P
1 Struc. Steel Foreman (outside)	$55.20	$441.60	$97.35	$778.80	$50.12	$86.25
1 Struc. Steel Worker	53.20	425.60	93.85	750.80		
1 Truck Driver (light)	41.95	335.60	67.55	540.40		
1 Flatbed Truck, Gas, 3 Ton		307.40		338.14	12.81	14.09
24 L.H., Daily Totals		$1510.20		$2408.14	$62.92	$100.34

Crew L-1

Crew No.	Bare Costs Hr.	Bare Costs Daily	Incl. Subs O&P Hr.	Incl. Subs O&P Daily	Cost Per Labor-Hour Bare Costs	Cost Per Labor-Hour Incl. O&P
1 Electrician	$55.10	$440.80	$85.20	$681.60	$57.15	$88.75
1 Plumber	59.20	473.60	92.30	738.40		
16 L.H., Daily Totals		$914.40		$1420.00	$57.15	$88.75

Crew L-2

Crew No.	Bare Costs Hr.	Bare Costs Daily	Incl. Subs O&P Hr.	Incl. Subs O&P Daily	Cost Per Labor-Hour Bare Costs	Cost Per Labor-Hour Incl. O&P
1 Carpenter	$48.45	$387.60	$79.15	$633.20	$42.23	$69.25
1 Carpenter Helper	36.00	288.00	59.35	474.80		
16 L.H., Daily Totals		$675.60		$1108.00	$42.23	$69.25

Crew L-3

Crew No.	Bare Costs Hr.	Bare Costs Daily	Incl. Subs O&P Hr.	Incl. Subs O&P Daily	Cost Per Labor-Hour Bare Costs	Cost Per Labor-Hour Incl. O&P
1 Carpenter	$48.45	$387.60	$79.15	$633.20	$52.31	$83.46
.5 Electrician	55.10	220.40	85.20	340.80		
.5 Sheet Metal Worker	57.25	229.00	90.35	361.40		
16 L.H., Daily Totals		$837.00		$1335.40	$52.31	$83.46

Crew L-3A

Crew No.	Bare Costs Hr.	Bare Costs Daily	Incl. Subs O&P Hr.	Incl. Subs O&P Daily	Cost Per Labor-Hour Bare Costs	Cost Per Labor-Hour Incl. O&P
1 Carpenter Foreman (outside)	$50.45	$403.60	$82.45	$659.60	$52.72	$85.08
.5 Sheet Metal Worker	57.25	229.00	90.35	361.40		
12 L.H., Daily Totals		$632.60		$1021.00	$52.72	$85.08

Crew L-4

Crew No.	Bare Costs Hr.	Bare Costs Daily	Incl. Subs O&P Hr.	Incl. Subs O&P Daily	Cost Per Labor-Hour Bare Costs	Cost Per Labor-Hour Incl. O&P
2 Skilled Workers	$49.90	$798.40	$80.85	$1293.60	$45.27	$73.68
1 Helper	36.00	288.00	59.35	474.80		
24 L.H., Daily Totals		$1086.40		$1768.40	$45.27	$73.68

Crew L-5

Crew No.	Bare Costs Hr.	Bare Costs Daily	Incl. Subs O&P Hr.	Incl. Subs O&P Daily	Cost Per Labor-Hour Bare Costs	Cost Per Labor-Hour Incl. O&P
1 Struc. Steel Foreman (outside)	$55.20	$441.60	$97.35	$778.80	$53.35	$92.76
5 Struc. Steel Workers	53.20	2128.00	93.85	3754.00		
1 Equip. Oper. (crane)	52.25	418.00	82.70	661.60		
1 Hyd. Crane, 25 Ton		748.80		823.68	13.37	14.71
56 L.H., Daily Totals		$3736.40		$6018.08	$66.72	$107.47

Crew L-5A

Crew No.	Bare Costs Hr.	Bare Costs Daily	Incl. Subs O&P Hr.	Incl. Subs O&P Daily	Cost Per Labor-Hour Bare Costs	Cost Per Labor-Hour Incl. O&P
1 Struc. Steel Foreman (outside)	$55.20	$441.60	$97.35	$778.80	$53.46	$91.94
2 Structural Steel Workers	53.20	851.20	93.85	1501.60		
1 Equip. Oper. (crane)	52.25	418.00	82.70	661.60		
1 S.P. Crane, 4x4, 25 Ton		606.60		667.26	18.96	20.85
32 L.H., Daily Totals		$2317.40		$3609.26	$72.42	$112.79

Crew L-5B

Crew No.	Bare Costs Hr.	Bare Costs Daily	Incl. Subs O&P Hr.	Incl. Subs O&P Daily	Cost Per Labor-Hour Bare Costs	Cost Per Labor-Hour Incl. O&P
1 Struc. Steel Foreman (outside)	$55.20	$441.60	$97.35	$778.80	$54.61	$88.93
2 Structural Steel Workers	53.20	851.20	93.85	1501.60		
2 Electricians	55.10	881.60	85.20	1363.20		
2 Steamfitters/Pipefitters	60.70	971.20	94.65	1514.40		
1 Equip. Oper. (crane)	52.25	418.00	82.70	661.60		
1 Equip. Oper. (oiler)	46.05	368.40	72.90	583.20		
1 Hyd. Crane, 80 Ton		1630.00		1793.00	22.64	24.90
72 L.H., Daily Totals		$5562.00		$8195.80	$77.25	$113.83

Crew L-6

Crew No.	Bare Costs Hr.	Bare Costs Daily	Incl. Subs O&P Hr.	Incl. Subs O&P Daily	Cost Per Labor-Hour Bare Costs	Cost Per Labor-Hour Incl. O&P
1 Plumber	$59.20	$473.60	$92.30	$738.40	$57.83	$89.93
.5 Electrician	55.10	220.40	85.20	340.80		
12 L.H., Daily Totals		$694.00		$1079.20	$57.83	$89.93

Crew L-7

Crew No.	Bare Costs Hr.	Bare Costs Daily	Incl. Subs O&P Hr.	Incl. Subs O&P Daily	Cost Per Labor-Hour Bare Costs	Cost Per Labor-Hour Incl. O&P
2 Carpenters	$48.45	$775.20	$79.15	$1266.40	$46.39	$75.10
1 Building Laborer	37.90	303.20	61.95	495.60		
.5 Electrician	55.10	220.40	85.20	340.80		
28 L.H., Daily Totals		$1298.80		$2102.80	$46.39	$75.10

Crew L-8

Crew No.	Bare Costs Hr.	Bare Costs Daily	Incl. Subs O&P Hr.	Incl. Subs O&P Daily	Cost Per Labor-Hour Bare Costs	Cost Per Labor-Hour Incl. O&P
2 Carpenters	$48.45	$775.20	$79.15	$1266.40	$50.60	$81.78
.5 Plumber	59.20	236.80	92.30	369.20		
20 L.H., Daily Totals		$1012.00		$1635.60	$50.60	$81.78

Crew L-9

Crew No.	Bare Costs Hr.	Bare Costs Daily	Incl. Subs O&P Hr.	Incl. Subs O&P Daily	Cost Per Labor-Hour Bare Costs	Cost Per Labor-Hour Incl. O&P
1 Labor Foreman (inside)	$38.40	$307.20	$62.75	$502.00	$43.32	$71.80
2 Building Laborers	37.90	606.40	61.95	991.20		
1 Struc. Steel Worker	53.20	425.60	93.85	750.80		
.5 Electrician	55.10	220.40	85.20	340.80		
36 L.H., Daily Totals		$1559.60		$2584.80	$43.32	$71.80

For customer support on your Commercial Renovation Cost Data, call 877.742.4719.

755

Crew No.	Bare Costs Hr.	Bare Costs Daily	Incl. Subs O&P Hr.	Incl. Subs O&P Daily	Cost Per Labor-Hour Bare Costs	Cost Per Labor-Hour Incl. O&P
Crew L-10	Hr.	Daily	Hr.	Daily	Bare Costs	Incl. O&P
1 Struc. Steel Foreman (outside)	$55.20	$441.60	$97.35	$778.80	$53.55	$91.30
1 Structural Steel Worker	53.20	425.60	93.85	750.80		
1 Equip. Oper. (crane)	52.25	418.00	82.70	661.60		
1 Hyd. Crane, 12 Ton		658.40		724.24	27.43	30.18
24 L.H., Daily Totals		$1943.60		$2915.44	$80.98	$121.48
Crew L-11	Hr.	Daily	Hr.	Daily	Bare Costs	Incl. O&P
2 Wreckers	$37.90	$606.40	$65.15	$1042.40	$44.30	$72.70
1 Equip. Oper. (crane)	52.25	418.00	82.70	661.60		
1 Equip. Oper. (light)	49.15	393.20	77.80	622.40		
1 Hyd. Excavator, 2.5 C.Y.		1582.00		1740.20		
1 Loader, Skid Steer, 78 H.P.		315.60		347.16	59.30	65.23
32 L.H., Daily Totals		$3315.20		$4413.76	$103.60	$137.93
Crew M-1	Hr.	Daily	Hr.	Daily	Bare Costs	Incl. O&P
3 Elevator Constructors	$79.10	$1898.40	$121.95	$2926.80	$75.15	$115.86
1 Elevator Apprentice	63.30	506.40	97.60	780.80		
5 Hand Tools		48.00		52.80	1.50	1.65
32 L.H., Daily Totals		$2452.80		$3760.40	$76.65	$117.51
Crew M-3	Hr.	Daily	Hr.	Daily	Bare Costs	Incl. O&P
1 Electrician Foreman (outside)	$57.10	$456.80	$88.30	$706.40	$58.86	$91.77
1 Common Laborer	37.90	303.20	61.95	495.60		
.25 Equipment Operator (med.)	51.10	102.20	80.90	161.80		
1 Elevator Constructor	79.10	632.80	121.95	975.60		
1 Elevator Apprentice	63.30	506.40	97.60	780.80		
.25 S.P. Crane, 4x4, 20 Ton		138.70		152.57	4.08	4.49
34 L.H., Daily Totals		$2140.10		$3272.77	$62.94	$96.26
Crew M-4	Hr.	Daily	Hr.	Daily	Bare Costs	Incl. O&P
1 Electrician Foreman (outside)	$57.10	$456.80	$88.30	$706.40	$58.22	$90.82
1 Common Laborer	37.90	303.20	61.95	495.60		
.25 Equipment Operator, Crane	52.25	104.50	82.70	165.40		
.25 Equip. Oper. (oiler)	46.05	92.10	72.90	145.80		
1 Elevator Constructor	79.10	632.80	121.95	975.60		
1 Elevator Apprentice	63.30	506.40	97.60	780.80		
.25 S.P. Crane, 4x4, 40 Ton		177.95		195.75	4.94	5.44
36 L.H., Daily Totals		$2273.75		$3465.34	$63.16	$96.26
Crew Q-1	Hr.	Daily	Hr.	Daily	Bare Costs	Incl. O&P
1 Plumber	$59.20	$473.60	$92.30	$738.40	$53.27	$83.05
1 Plumber Apprentice	47.35	378.80	73.80	590.40		
16 L.H., Daily Totals		$852.40		$1328.80	$53.27	$83.05
Crew Q-1A	Hr.	Daily	Hr.	Daily	Bare Costs	Incl. O&P
.25 Plumber Foreman (outside)	$61.20	$122.40	$95.40	$190.80	$59.60	$92.92
1 Plumber	59.20	473.60	92.30	738.40		
10 L.H., Daily Totals		$596.00		$929.20	$59.60	$92.92
Crew Q-1C	Hr.	Daily	Hr.	Daily	Bare Costs	Incl. O&P
1 Plumber	$59.20	$473.60	$92.30	$738.40	$52.55	$82.33
1 Plumber Apprentice	47.35	378.80	73.80	590.40		
1 Equip. Oper. (medium)	51.10	408.80	80.90	647.20		
1 Trencher, Chain Type, 8' D		2860.00		3146.00	119.17	131.08
24 L.H., Daily Totals		$4121.20		$5122.00	$171.72	$213.42
Crew Q-2	Hr.	Daily	Hr.	Daily	Bare Costs	Incl. O&P
2 Plumbers	$59.20	$947.20	$92.30	$1476.80	$55.25	$86.13
1 Plumber Apprentice	47.35	378.80	73.80	590.40		
24 L.H., Daily Totals		$1326.00		$2067.20	$55.25	$86.13

Crew No.	Bare Costs Hr.	Bare Costs Daily	Incl. Subs O&P Hr.	Incl. Subs O&P Daily	Cost Per Labor-Hour Bare Costs	Cost Per Labor-Hour Incl. O&P
Crew Q-3	Hr.	Daily	Hr.	Daily	Bare Costs	Incl. O&P
1 Plumber Foreman (inside)	$59.70	$477.60	$93.05	$744.40	$56.36	$87.86
2 Plumbers	59.20	947.20	92.30	1476.80		
1 Plumber Apprentice	47.35	378.80	73.80	590.40		
32 L.H., Daily Totals		$1803.60		$2811.60	$56.36	$87.86
Crew Q-4	Hr.	Daily	Hr.	Daily	Bare Costs	Incl. O&P
1 Plumber Foreman (inside)	$59.70	$477.60	$93.05	$744.40	$56.36	$87.86
1 Plumber	59.20	473.60	92.30	738.40		
1 Welder (plumber)	59.20	473.60	92.30	738.40		
1 Plumber Apprentice	47.35	378.80	73.80	590.40		
1 Welder, Electric, 300 amp		58.10		63.91	1.82	2.00
32 L.H., Daily Totals		$1861.70		$2875.51	$58.18	$89.86
Crew Q-5	Hr.	Daily	Hr.	Daily	Bare Costs	Incl. O&P
1 Steamfitter	$60.70	$485.60	$94.65	$757.20	$54.63	$85.17
1 Steamfitter Apprentice	48.55	388.40	75.70	605.60		
16 L.H., Daily Totals		$874.00		$1362.80	$54.63	$85.17
Crew Q-6	Hr.	Daily	Hr.	Daily	Bare Costs	Incl. O&P
2 Steamfitters	$60.70	$971.20	$94.65	$1514.40	$56.65	$88.33
1 Steamfitter Apprentice	48.55	388.40	75.70	605.60		
24 L.H., Daily Totals		$1359.60		$2120.00	$56.65	$88.33
Crew Q-7	Hr.	Daily	Hr.	Daily	Bare Costs	Incl. O&P
1 Steamfitter Foreman (inside)	$61.20	$489.60	$95.40	$763.20	$57.79	$90.10
2 Steamfitters	60.70	971.20	94.65	1514.40		
1 Steamfitter Apprentice	48.55	388.40	75.70	605.60		
32 L.H., Daily Totals		$1849.20		$2883.20	$57.79	$90.10
Crew Q-8	Hr.	Daily	Hr.	Daily	Bare Costs	Incl. O&P
1 Steamfitter Foreman (inside)	$61.20	$489.60	$95.40	$763.20	$57.79	$90.10
1 Steamfitter	60.70	485.60	94.65	757.20		
1 Welder (steamfitter)	60.70	485.60	94.65	757.20		
1 Steamfitter Apprentice	48.55	388.40	75.70	605.60		
1 Welder, Electric, 300 amp		58.10		63.91	1.82	2.00
32 L.H., Daily Totals		$1907.30		$2947.11	$59.60	$92.10
Crew Q-9	Hr.	Daily	Hr.	Daily	Bare Costs	Incl. O&P
1 Sheet Metal Worker	$57.25	$458.00	$90.35	$722.80	$51.52	$81.30
1 Sheet Metal Apprentice	45.80	366.40	72.25	578.00		
16 L.H., Daily Totals		$824.40		$1300.80	$51.52	$81.30
Crew Q-10	Hr.	Daily	Hr.	Daily	Bare Costs	Incl. O&P
2 Sheet Metal Workers	$57.25	$916.00	$90.35	$1445.60	$53.43	$84.32
1 Sheet Metal Apprentice	45.80	366.40	72.25	578.00		
24 L.H., Daily Totals		$1282.40		$2023.60	$53.43	$84.32
Crew Q-11	Hr.	Daily	Hr.	Daily	Bare Costs	Incl. O&P
1 Sheet Metal Foreman (inside)	$57.75	$462.00	$91.15	$729.20	$54.51	$86.03
2 Sheet Metal Workers	57.25	916.00	90.35	1445.60		
1 Sheet Metal Apprentice	45.80	366.40	72.25	578.00		
32 L.H., Daily Totals		$1744.40		$2752.80	$54.51	$86.03
Crew Q-12	Hr.	Daily	Hr.	Daily	Bare Costs	Incl. O&P
1 Sprinkler Installer	$57.60	$460.80	$90.20	$721.60	$51.85	$81.20
1 Sprinkler Apprentice	46.10	368.80	72.20	577.60		
16 L.H., Daily Totals		$829.60		$1299.20	$51.85	$81.20

Crew Q-13	Hr.	Daily	Hr.	Daily	Bare Costs	Incl. O&P
1 Sprinkler Foreman (inside)	$58.10	$464.80	$91.00	$728.00	$54.85	$85.90
2 Sprinkler Installers	57.60	921.60	90.20	1443.20		
1 Sprinkler Apprentice	46.10	368.80	72.20	577.60		
32 L.H., Daily Totals		$1755.20		$2748.80	$54.85	$85.90

Crew Q-14	Hr.	Daily	Hr.	Daily	Bare Costs	Incl. O&P
1 Asbestos Worker	$53.40	$427.20	$85.75	$686.00	$48.05	$77.17
1 Asbestos Apprentice	42.70	341.60	68.60	548.80		
16 L.H., Daily Totals		$768.80		$1234.80	$48.05	$77.17

Crew Q-15	Hr.	Daily	Hr.	Daily	Bare Costs	Incl. O&P
1 Plumber	$59.20	$473.60	$92.30	$738.40	$53.27	$83.05
1 Plumber Apprentice	47.35	378.80	73.80	590.40		
1 Welder, Electric, 300 amp		58.10		63.91	3.63	3.99
16 L.H., Daily Totals		$910.50		$1392.71	$56.91	$87.04

Crew Q-16	Hr.	Daily	Hr.	Daily	Bare Costs	Incl. O&P
2 Plumbers	$59.20	$947.20	$92.30	$1476.80	$55.25	$86.13
1 Plumber Apprentice	47.35	378.80	73.80	590.40		
1 Welder, Electric, 300 amp		58.10		63.91	2.42	2.66
24 L.H., Daily Totals		$1384.10		$2131.11	$57.67	$88.80

Crew Q-17	Hr.	Daily	Hr.	Daily	Bare Costs	Incl. O&P
1 Steamfitter	$60.70	$485.60	$94.65	$757.20	$54.63	$85.17
1 Steamfitter Apprentice	48.55	388.40	75.70	605.60		
1 Welder, Electric, 300 amp		58.10		63.91	3.63	3.99
16 L.H., Daily Totals		$932.10		$1426.71	$58.26	$89.17

Crew Q-17A	Hr.	Daily	Hr.	Daily	Bare Costs	Incl. O&P
1 Steamfitter	$60.70	$485.60	$94.65	$757.20	$53.83	$84.35
1 Steamfitter Apprentice	48.55	388.40	75.70	605.60		
1 Equip. Oper. (crane)	52.25	418.00	82.70	661.60		
1 Hyd. Crane, 12 Ton		658.40		724.24		
1 Welder, Electric, 300 amp		58.10		63.91	29.85	32.84
24 L.H., Daily Totals		$2008.50		$2812.55	$83.69	$117.19

Crew Q-18	Hr.	Daily	Hr.	Daily	Bare Costs	Incl. O&P
2 Steamfitters	$60.70	$971.20	$94.65	$1514.40	$56.65	$88.33
1 Steamfitter Apprentice	48.55	388.40	75.70	605.60		
1 Welder, Electric, 300 amp		58.10		63.91	2.42	2.66
24 L.H., Daily Totals		$1417.70		$2183.91	$59.07	$91.00

Crew Q-19	Hr.	Daily	Hr.	Daily	Bare Costs	Incl. O&P
1 Steamfitter	$60.70	$485.60	$94.65	$757.20	$54.78	$85.18
1 Steamfitter Apprentice	48.55	388.40	75.70	605.60		
1 Electrician	55.10	440.80	85.20	681.60		
24 L.H., Daily Totals		$1314.80		$2044.40	$54.78	$85.18

Crew Q-20	Hr.	Daily	Hr.	Daily	Bare Costs	Incl. O&P
1 Sheet Metal Worker	$57.25	$458.00	$90.35	$722.80	$52.24	$82.08
1 Sheet Metal Apprentice	45.80	366.40	72.25	578.00		
.5 Electrician	55.10	220.40	85.20	340.80		
20 L.H., Daily Totals		$1044.80		$1641.60	$52.24	$82.08

Crew Q-21	Hr.	Daily	Hr.	Daily	Bare Costs	Incl. O&P
2 Steamfitters	$60.70	$971.20	$94.65	$1514.40	$56.26	$87.55
1 Steamfitter Apprentice	48.55	388.40	75.70	605.60		
1 Electrician	55.10	440.80	85.20	681.60		
32 L.H., Daily Totals		$1800.40		$2801.60	$56.26	$87.55

Crew Q-22	Hr.	Daily	Hr.	Daily	Bare Costs	Incl. O&P
1 Plumber	$59.20	$473.60	$92.30	$738.40	$53.27	$83.05
1 Plumber Apprentice	47.35	378.80	73.80	590.40		
1 Hyd. Crane, 12 Ton		658.40		724.24	41.15	45.27
16 L.H., Daily Totals		$1510.80		$2053.04	$94.42	$128.32

Crew Q-22A	Hr.	Daily	Hr.	Daily	Bare Costs	Incl. O&P
1 Plumber	$59.20	$473.60	$92.30	$738.40	$49.17	$77.69
1 Plumber Apprentice	47.35	378.80	73.80	590.40		
1 Laborer	37.90	303.20	61.95	495.60		
1 Equip. Oper. (crane)	52.25	418.00	82.70	661.60		
1 Hyd. Crane, 12 Ton		658.40		724.24	20.57	22.63
32 L.H., Daily Totals		$2232.00		$3210.24	$69.75	$100.32

Crew Q-23	Hr.	Daily	Hr.	Daily	Bare Costs	Incl. O&P
1 Plumber Foreman (outside)	$61.20	$489.60	$95.40	$763.20	$57.17	$89.53
1 Plumber	59.20	473.60	92.30	738.40		
1 Equip. Oper. (medium)	51.10	408.80	80.90	647.20		
1 Lattice Boom Crane, 20 Ton		967.35		1064.09	40.31	44.34
24 L.H., Daily Totals		$2339.35		$3212.89	$97.47	$133.87

Crew R-1	Hr.	Daily	Hr.	Daily	Bare Costs	Incl. O&P
1 Electrician Foreman	$55.60	$444.80	$85.95	$687.60	$51.52	$79.66
3 Electricians	55.10	1322.40	85.20	2044.80		
2 Electrician Apprentices	44.10	705.60	68.20	1091.20		
48 L.H., Daily Totals		$2472.80		$3823.60	$51.52	$79.66

Crew R-1A	Hr.	Daily	Hr.	Daily	Bare Costs	Incl. O&P
1 Electrician	$55.10	$440.80	$85.20	$681.60	$49.60	$76.70
1 Electrician Apprentice	44.10	352.80	68.20	545.60		
16 L.H., Daily Totals		$793.60		$1227.20	$49.60	$76.70

Crew R-1B	Hr.	Daily	Hr.	Daily	Bare Costs	Incl. O&P
1 Electrician	$55.10	$440.80	$85.20	$681.60	$47.77	$73.87
2 Electrician Apprentices	44.10	705.60	68.20	1091.20		
24 L.H., Daily Totals		$1146.40		$1772.80	$47.77	$73.87

Crew R-1C	Hr.	Daily	Hr.	Daily	Bare Costs	Incl. O&P
2 Electricians	$55.10	$881.60	$85.20	$1363.20	$49.60	$76.70
2 Electrician Apprentices	44.10	705.60	68.20	1091.20		
1 Portable cable puller, 8000 lb.		130.10		143.11	4.07	4.47
32 L.H., Daily Totals		$1717.30		$2597.51	$53.67	$81.17

Crew R-2	Hr.	Daily	Hr.	Daily	Bare Costs	Incl. O&P
1 Electrician Foreman	$55.60	$444.80	$85.95	$687.60	$51.62	$80.09
3 Electricians	55.10	1322.40	85.20	2044.80		
2 Electrician Apprentices	44.10	705.60	68.20	1091.20		
1 Equip. Oper. (crane)	52.25	418.00	82.70	661.60		
1 S.P. Crane, 4x4, 5 Ton		280.80		308.88	5.01	5.52
56 L.H., Daily Totals		$3171.60		$4794.08	$56.64	$85.61

Crew R-3	Hr.	Daily	Hr.	Daily	Bare Costs	Incl. O&P
1 Electrician Foreman	$55.60	$444.80	$85.95	$687.60	$54.73	$85.00
1 Electrician	55.10	440.80	85.20	681.60		
.5 Equip. Oper. (crane)	52.25	209.00	82.70	330.80		
.5 S.P. Crane, 4x4, 5 Ton		140.40		154.44	7.02	7.72
20 L.H., Daily Totals		$1235.00		$1854.44	$61.75	$92.72

For customer support on your Commercial Renovation Cost Data, call 877.742.4719.

757

Crews - Renovation

Crew No.	Bare Costs Hr.	Daily	Incl. Subs O&P Hr.	Daily	Cost Per Labor-Hour Bare Costs	Incl. O&P
Crew R-4						
1 Struc. Steel Foreman (outside)	$55.20	$441.60	$97.35	$778.80	$53.98	$92.82
3 Struc. Steel Workers	53.20	1276.80	93.85	2252.40		
1 Electrician	55.10	440.80	85.20	681.60		
1 Welder, Gas Engine, 300 amp		147.20		161.92	3.68	4.05
40 L.H., Daily Totals		$2306.40		$3874.72	$57.66	$96.87
Crew R-5						
1 Electrician Foreman	$55.60	$444.80	$85.95	$687.60	$48.20	$75.87
4 Electrician Linemen	55.10	1763.20	85.20	2726.40		
2 Electrician Operators	55.10	881.60	85.20	1363.20		
4 Electrician Groundmen	36.00	1152.00	59.35	1899.20		
1 Crew Truck		208.80		229.68		
1 Flatbed Truck, 20,000 GVW		247.80		272.58		
1 Pickup Truck, 3/4 Ton		145.80		160.38		
.2 Hyd. Crane, 55 Ton		224.60		247.06		
.2 Hyd. Crane, 12 Ton		131.68		144.85		
.2 Earth Auger, Truck-Mtd.		84.16		92.58		
1 Tractor w/Winch		428.00		470.80	16.71	18.39
88 L.H., Daily Totals		$5712.44		$8294.32	$64.91	$94.25
Crew R-6						
1 Electrician Foreman	$55.60	$444.80	$85.95	$687.60	$48.20	$75.87
4 Electrician Linemen	55.10	1763.20	85.20	2726.40		
2 Electrician Operators	55.10	881.60	85.20	1363.20		
4 Electrician Groundmen	36.00	1152.00	59.35	1899.20		
1 Crew Truck		208.80		229.68		
1 Flatbed Truck, 20,000 GVW		247.80		272.58		
1 Pickup Truck, 3/4 Ton		145.80		160.38		
.2 Hyd. Crane, 55 Ton		224.60		247.06		
.2 Hyd. Crane, 12 Ton		131.68		144.85		
.2 Earth Auger, Truck-Mtd.		84.16		92.58		
1 Tractor w/Winch		428.00		470.80		
3 Cable Trailers		610.65		671.72		
.5 Tensioning Rig		202.25		222.47		
.5 Cable Pulling Rig		1170.50		1287.55	39.25	43.18
88 L.H., Daily Totals		$7695.84		$10476.06	$87.45	$119.05
Crew R-7						
1 Electrician Foreman	$55.60	$444.80	$85.95	$687.60	$39.27	$63.78
5 Electrician Groundmen	36.00	1440.00	59.35	2374.00		
1 Crew Truck		208.80		229.68	4.35	4.79
48 L.H., Daily Totals		$2093.60		$3291.28	$43.62	$68.57
Crew R-8						
1 Electrician Foreman	$55.60	$444.80	$85.95	$687.60	$48.82	$76.71
3 Electrician Linemen	55.10	1322.40	85.20	2044.80		
2 Electrician Groundmen	36.00	576.00	59.35	949.60		
1 Pickup Truck, 3/4 Ton		145.80		160.38		
1 Crew Truck		208.80		229.68	7.39	8.13
48 L.H., Daily Totals		$2697.80		$4072.06	$56.20	$84.83
Crew R-9						
1 Electrician Foreman	$55.60	$444.80	$85.95	$687.60	$45.61	$72.37
1 Electrician Lineman	55.10	440.80	85.20	681.60		
2 Electrician Operators	55.10	881.60	85.20	1363.20		
4 Electrician Groundmen	36.00	1152.00	59.35	1899.20		
1 Pickup Truck, 3/4 Ton		145.80		160.38		
1 Crew Truck		208.80		229.68	5.54	6.09
64 L.H., Daily Totals		$3273.80		$5021.66	$51.15	$78.46

Crew No.	Bare Costs Hr.	Daily	Incl. Subs O&P Hr.	Daily	Cost Per Labor-Hour Bare Costs	Incl. O&P
Crew R-10						
1 Electrician Foreman	$55.60	$444.80	$85.95	$687.60	$52.00	$81.02
4 Electrician Linemen	55.10	1763.20	85.20	2726.40		
1 Electrician Groundman	36.00	288.00	59.35	474.80		
1 Crew Truck		208.80		229.68		
3 Tram Cars		418.50		460.35	13.07	14.38
48 L.H., Daily Totals		$3123.30		$4578.83	$65.07	$95.39
Crew R-11						
1 Electrician Foreman	$55.60	$444.80	$85.95	$687.60	$52.31	$81.63
4 Electricians	55.10	1763.20	85.20	2726.40		
1 Equip. Oper. (crane)	52.25	418.00	82.70	661.60		
1 Common Laborer	37.90	303.20	61.95	495.60		
1 Crew Truck		208.80		229.68		
1 Hyd. Crane, 12 Ton		658.40		724.24	15.49	17.03
56 L.H., Daily Totals		$3796.40		$5525.12	$67.79	$98.66
Crew R-12						
1 Carpenter Foreman (inside)	$48.95	$391.60	$80.00	$640.00	$45.33	$74.47
4 Carpenters	48.45	1550.40	79.15	2532.80		
4 Common Laborers	37.90	1212.80	61.95	1982.40		
1 Equip. Oper. (medium)	51.10	408.80	80.90	647.20		
1 Steel Worker	53.20	425.60	93.85	750.80		
1 Dozer, 200 H.P.		1397.00		1536.70		
1 Pickup Truck, 3/4 Ton		145.80		160.38	17.53	19.29
88 L.H., Daily Totals		$5532.00		$8250.28	$62.86	$93.75
Crew R-13						
1 Electrician Foreman	$55.60	$444.80	$85.95	$687.60	$53.34	$82.88
3 Electricians	55.10	1322.40	85.20	2044.80		
.25 Equip. Oper. (crane)	52.25	104.50	82.70	165.40		
1 Equipment Oiler	46.05	368.40	72.90	583.20		
.25 Hydraulic Crane, 33 Ton		190.80		209.88	4.54	5.00
42 L.H., Daily Totals		$2430.90		$3690.88	$57.88	$87.88
Crew R-15						
1 Electrician Foreman	$55.60	$444.80	$85.95	$687.60	$54.19	$84.09
4 Electricians	55.10	1763.20	85.20	2726.40		
1 Equipment Oper. (light)	49.15	393.20	77.80	622.40		
1 Aerial Lift Truck, 40' Boom		309.00		339.90	6.44	7.08
48 L.H., Daily Totals		$2910.20		$4376.30	$60.63	$91.17
Crew R-18						
.25 Electrician Foreman	$55.60	$111.20	$85.95	$171.90	$48.37	$74.80
1 Electrician	55.10	440.80	85.20	681.60		
2 Electrician Apprentices	44.10	705.60	68.20	1091.20		
26 L.H., Daily Totals		$1257.60		$1944.70	$48.37	$74.80
Crew R-19						
.5 Electrician Foreman	$55.60	$222.40	$85.95	$343.80	$55.20	$85.35
2 Electricians	55.10	881.60	85.20	1363.20		
20 L.H., Daily Totals		$1104.00		$1707.00	$55.20	$85.35
Crew R-21						
1 Electrician Foreman	$55.60	$444.80	$85.95	$687.60	$55.12	$85.28
3 Electricians	55.10	1322.40	85.20	2044.80		
.1 Equip. Oper. (medium)	51.10	40.88	80.90	64.72		
.1 S.P. Crane, 4x4, 25 Ton		60.66		66.73	1.85	2.03
32.8 L.H., Daily Totals		$1868.74		$2863.85	$56.97	$87.31

758

Crew No.	Bare Costs			Incl. Subs O & P		Cost Per Labor-Hour	
						Bare Costs	Incl. O&P
Crew R-22	Hr.	Daily	Hr.	Daily		Bare Costs	Incl. O&P
.66 Electrician Foreman	$55.60	$293.57	$85.95	$453.82		$50.45	$78.01
2 Electricians	55.10	881.60	85.20	1363.20			
2 Electrician Apprentices	44.10	705.60	68.20	1091.20			
37.28 L.H., Daily Totals		$1880.77		$2908.22		$50.45	$78.01
Crew R-30	Hr.	Daily	Hr.	Daily		Bare Costs	Incl. O&P
.25 Electrician Foreman (outside)	$57.10	$114.20	$88.30	$176.60		$44.67	$71.13
1 Electrician	55.10	440.80	85.20	681.60			
2 Laborers, (Semi-Skilled)	37.90	606.40	61.95	991.20			
26 L.H., Daily Totals		$1161.40		$1849.40		$44.67	$71.13
Crew R-31	Hr.	Daily	Hr.	Daily		Bare Costs	Incl. O&P
1 Electrician	$55.10	$440.80	$85.20	$681.60		$55.10	$85.20
1 Core Drill, Electric, 2.5 H.P.		64.20		70.62		8.03	8.83
8 L.H., Daily Totals		$505.00		$752.22		$63.13	$94.03

For customer support on your Commercial Renovation Cost Data, call 877.742.4719.

759

Historical Cost Indexes

The table below lists both the RSMeans® historical cost index based on Jan. 1, 1993 = 100 as well as the computed value of an index based on Jan. 1, 2016 costs. Since the Jan. 1, 2016 figure is estimated, space is left to write in the actual index figures as they become available through the quarterly *RSMeans Construction Cost Indexes*.

To compute the actual index based on Jan. 1, 2016 = 100, divide the historical cost index for a particular year by the actual Jan. 1, 2016 construction cost index. Space has been left to advance the index figures as the year progresses.

Year	Historical Cost Index Jan. 1, 1993 = 100		Current Index Based on Jan. 1, 2016 = 100		Year	Historical Cost Index Jan. 1, 1993 = 100	Current Index Based on Jan. 1, 2016 = 100		Year	Historical Cost Index Jan. 1, 1993 = 100	Current Index Based on Jan. 1, 2016 = 100	
	Est.	Actual	Est.	Actual		Actual	Est.	Actual		Actual	Est.	Actual
Oct 2016*					July 2001	125.1	60.4		July 1983	80.2	38.7	
July 2016*					2000	120.9	58.3		1982	76.1	36.8	
April 2016*					1999	117.6	56.8		1981	70.0	33.8	
Jan 2016*	207.2		100.0	100.0	1998	115.1	55.6		1980	62.9	30.4	
July 2015		206.2			1997	112.8	54.4		1979	57.8	27.9	
2014		204.9	98.9		1996	110.2	53.2		1978	53.5	25.8	
2013		201.2	97.1		1995	107.6	51.9		1977	49.5	23.9	
2012		194.6	93.9		1994	104.4	50.4		1976	46.9	22.6	
2011		191.2	92.3		1993	101.7	49.1		1975	44.8	21.6	
2010		183.5	88.6		1992	99.4	48.0		1974	41.4	20.0	
2009		180.1	86.9		1991	96.8	46.7		1973	37.7	18.2	
2008		180.4	87.1		1990	94.3	45.5		1972	34.8	16.8	
2007		169.4	81.8		1989	92.1	44.5		1971	32.1	15.5	
2006		162.0	78.2		1988	89.9	43.4		1970	28.7	13.9	
2005		151.6	73.2		1987	87.7	42.3		1969	26.9	13.0	
2004		143.7	69.4		1986	84.2	40.7		1968	24.9	12.0	
2003		132.0	63.7		1985	82.6	39.9		1967	23.5	11.3	
2002		128.7	62.1		1984	82.0	39.6		1966	22.7	11.0	

Adjustments to Costs

The "Historical Cost Index" can be used to convert national average building costs at a particular time to the approximate building costs for some other time.

Example:

Estimate and compare construction costs for different years in the same city.

To estimate the national average construction cost of a building in 1970, knowing that it cost $900,000 in 2016:

INDEX in 1970 = 28.7

INDEX in 2016 = 207.2

Note: The city cost indexes for Canada can be used to convert U.S. national averages to local costs in Canadian dollars.

Example:

To estimate and compare the cost of a building in Toronto, ON in 2016 with the known cost of $600,000 (US$) in New York, NY in 2016:

INDEX Toronto = 109.9

INDEX New York = 131.1

$$\frac{\text{INDEX Toronto}}{\text{INDEX New York}} \times \text{Cost New York} = \text{Cost Toronto}$$

$$\frac{109.9}{131.1} \times \$600,000 = .841 \times \$600,000 = \$502,975$$

The construction cost of the building in Toronto is $502,975 (CN$).

Time Adjustment Using the Historical Cost Indexes:

$$\frac{\text{Index for Year A}}{\text{Index for Year B}} \times \text{Cost in Year B} = \text{Cost in Year A}$$

$$\frac{\text{INDEX 1970}}{\text{INDEX 2016}} \times \text{Cost 2016} = \text{Cost 1970}$$

$$\frac{28.7}{207.2} \times \$900,000 = .139 \times \$900,000 = \$124,662$$

The construction cost of the building in 1970 was $124,662.

*Historical Cost Index updates and other resources are provided on the following website:
http://info.thegordiangroup.com/RSMeans.html

How to Use the City Cost Indexes

What you should know before you begin

RSMeans City Cost Indexes (CCI) are an extremely useful tool for when you want to compare costs from city to city and region to region.

This publication contains average construction cost indexes for 731 U.S. and Canadian cities covering over 930 three-digit zip code locations, as listed directly under each city.

Keep in mind that a City Cost Index number is a percentage ratio of a specific city's cost to the national average cost of the same item at a stated time period.

In other words, these index figures represent relative construction factors (or, if you prefer, multipliers) for material and installation costs, as well as the weighted average for Total In Place costs for each CSI MasterFormat division. Installation costs include both labor and equipment rental costs. When estimating equipment rental rates only for a specific location, use 01 54 33 EQUIPMENT RENTAL COSTS in the Reference Section.

The 30 City Average Index is the average of 30 major U.S. cities and serves as a national average.

Index figures for both material and installation are based on the 30 major city average of 100 and represent the cost relationship as of July 1, 2015. The index for each division is computed from representative material and labor quantities for that division. The weighted average for each city is a weighted total of the components listed above it. It does not include relative productivity between trades or cities.

As changes occur in local material prices, labor rates, and equipment rental rates (including fuel costs), the impact of these changes should be accurately measured by the change in the City Cost Index for each particular city (as compared to the 30 city average).

Therefore, if you know (or have estimated) building costs in one city today, you can easily convert those costs to expected building costs in another city.

In addition, by using the Historical Cost Index, you can easily convert national average building costs at a particular time to the approximate building costs for some other time. The City Cost Indexes can then be applied to calculate the costs for a particular city.

Quick calculations

Location Adjustment Using the City Cost Indexes:

$$\frac{\text{Index for City A}}{\text{Index for City B}} \times \text{Cost in City B} = \text{Cost in City A}$$

Time Adjustment for the National Average Using the Historical Cost Index:

$$\frac{\text{Index for Year A}}{\text{Index for Year B}} \times \text{Cost in Year B} = \text{Cost in Year A}$$

Adjustment from the National Average:

$$\frac{\text{Index for City A}}{100} \times \text{National Average Cost} = \text{Cost in City A}$$

Since each of the other RSMeans data sets contain many different items, any *one* item multiplied by the particular city index may give incorrect results. However, the larger the number of items compiled, the closer the results should be to actual costs for that particular city.

The City Cost Indexes for Canadian cities are calculated using Canadian material and equipment prices and labor rates in Canadian dollars. Therefore, indexes for Canadian cities can be used to convert U.S. national average prices to local costs in Canadian dollars.

How to use this section

1. Compare costs from city to city.

In using the RSMeans Indexes, remember that an index number is not a fixed number but a ratio: It's a percentage ratio of a building component's cost at any stated time to the national average cost of that same component at the same time period. Put in the form of an equation:

$$\frac{\text{Specific City Cost}}{\text{National Average Cost}} \times 100 = \text{City Index Number}$$

Therefore, when making cost comparisons between cities, do not subtract one city's index number from the index number of another city and read the result as a percentage difference. Instead, divide one city's index number by that of the other city. The resulting number may then be used as a multiplier to calculate cost differences from city to city.

The formula used to find cost differences between cities for the purpose of comparison is as follows:

$$\frac{\text{City A Index}}{\text{City B Index}} \times \text{City B Cost (Known)} = \text{City A Cost (Unknown)}$$

In addition, you can use RSMeans CCI to calculate and compare costs division by division between cities using the same basic formula. (Just be sure that you're comparing similar divisions.)

2. Compare a specific city's construction costs with the national average.

When you're studying construction location feasibility, it's advisable to compare a prospective project's cost index with an index of the national average cost.

For example, divide the weighted average index of construction costs of a specific city by that of the 30 City Average, which = 100.

$$\frac{\text{City Index}}{100} = \% \text{ of National Average}$$

As a result, you get a ratio that indicates the relative cost of construction in that city in comparison with the national average.

3. Convert U.S. national average to actual costs in Canadian City.

$$\frac{\text{Index for Canadian City}}{100} \times \text{National Average Cost} = \text{Cost in Canadian City in \$ CAN}$$

4. Adjust construction cost data based on a national average.

When you use a source of construction cost data which is based on a national average (such as RSMeans cost data), it is necessary to adjust those costs to a specific location.

$$\frac{\text{City Index}}{100} \times \frac{\text{Cost Based on}}{\text{National Average Costs}} = \frac{\text{City Cost}}{\text{(Unknown)}}$$

5. When applying the City Cost Indexes to demolition projects, use the appropriate division installation index.
For example, for removal of existing doors and windows, use the Division 8 (Openings) index.

What you might like to know about how we developed the Indexes

The information presented in the CCI is organized according to the Construction Specifications Institute (CSI) MasterFormat 2014 classification system.

To create a reliable index, RSMeans researched the building type most often constructed in the United States and Canada. Because it was concluded that no one type of building completely represented the building construction industry, nine different types of buildings were combined to create a composite model.

The exact material, labor, and equipment quantities are based on detailed analyses of these nine building types, and then each quantity is weighted in proportion to expected usage. These various material items, labor hours, and equipment rental rates are thus combined to form a composite building representing as closely as possible the actual usage of materials, labor, and equipment in the North American building construction industry.

The following structures were chosen to make up that composite model:

1. Factory, 1 story
2. Office, 2–4 stories
3. Store, Retail
4. Town Hall, 2–3 stories
5. High School, 2–3 stories
6. Hospital, 4–8 stories
7. Garage, Parking
8. Apartment, 1–3 stories
9. Hotel/Motel, 2–3 stories

For the purposes of ensuring the timeliness of the data, the components of the index for the composite model have been streamlined. They currently consist of:

- specific quantities of 66 commonly used construction materials;
- specific labor-hours for 21 building construction trades; and
- specific days of equipment rental for 6 types of construction equipment (normally used to install the 66 material items by the 21 trades.) Fuel costs and routine maintenance costs are included in the equipment cost.

A sophisticated computer program handles the updating of all costs for each city on a quarterly basis. Material and equipment price quotations are gathered quarterly from cities in the United States and Canada. These prices and the latest negotiated labor wage rates for 21 different building trades are used to compile the quarterly update of the City Cost Index.

The 30 major U.S. cities used to calculate the national average are:

Atlanta, GA	Memphis, TN
Baltimore, MD	Milwaukee, WI
Boston, MA	Minneapolis, MN
Buffalo, NY	Nashville, TN
Chicago, IL	New Orleans, LA
Cincinnati, OH	New York, NY
Cleveland, OH	Philadelphia, PA
Columbus, OH	Phoenix, AZ
Dallas, TX	Pittsburgh, PA
Denver, CO	St. Louis, MO
Detroit, MI	San Antonio, TX
Houston, TX	San Diego, CA
Indianapolis, IN	San Francisco, CA
Kansas City, MO	Seattle, WA
Los Angeles, CA	Washington, DC

What the CCI does not indicate

The weighted average for each city is a total of the divisional components weighted to reflect typical usage. It does not include the productivity variations between trades or cities.

In addition, the CCI does not take into consideration factors such as the following:

- managerial efficiency
- competitive conditions
- automation
- restrictive union practices
- unique local requirements
- regional variations due to specific building codes

DIVISION		UNITED STATES 30 CITY AVERAGE			ALABAMA														
					ANNISTON 362			BIRMINGHAM 350-352			BUTLER 369			DECATUR 356			DOTHAN 363		
		MAT.	INST.	TOTAL	MAT.	INST.	TOTAL	MAT.	INST.	TOTAL	MAT.	INST.	TOTAL	MAT.	INST.	TOTAL	MAT.	INST.	TOTAL
015433	CONTRACTOR EQUIPMENT		100.0	100.0		101.9	101.9		102.0	102.0		99.3	99.3		101.9	101.9		99.3	99.3
0241, 31 - 34	SITE & INFRASTRUCTURE, DEMOLITION	100.0	100.0	100.0	89.0	92.6	91.6	94.6	92.9	93.4	103.2	88.6	92.9	87.5	92.5	91.1	100.8	88.6	92.2
0310	Concrete Forming & Accessories	100.0	100.0	100.0	90.2	78.0	79.6	94.5	78.3	80.5	86.5	77.6	78.8	97.0	72.2	75.5	95.6	75.9	78.5
0320	Concrete Reinforcing	100.0	100.0	100.0	87.9	88.3	88.1	97.0	88.3	92.5	92.9	88.2	90.5	90.7	81.5	85.9	92.9	88.3	90.5
0330	Cast-in-Place Concrete	100.0	100.0	100.0	100.5	73.7	89.9	102.9	74.4	91.7	105.1	73.3	88.2	101.1	73.0	90.0	98.0	73.2	88.2
03	CONCRETE	100.0	100.0	100.0	104.1	79.4	92.1	101.8	79.8	91.1	105.1	79.0	92.4	100.0	75.3	88.0	104.3	78.2	91.6
04	MASONRY	100.0	100.0	100.0	99.7	77.7	86.2	96.3	79.1	85.7	104.9	74.7	86.3	94.3	76.6	83.4	106.4	76.7	88.1
05	METALS	100.0	100.0	100.0	95.1	96.0	95.4	95.3	95.9	95.5	94.1	94.1	94.1	97.2	93.4	96.0	94.1	94.0	94.1
06	WOOD, PLASTICS & COMPOSITES	100.0	100.0	100.0	91.9	78.9	84.7	98.0	78.9	87.4	86.6	78.9	82.3	105.1	71.8	86.6	98.5	76.8	86.4
07	THERMAL & MOISTURE PROTECTION	100.0	100.0	100.0	97.2	85.3	92.2	102.2	84.5	94.7	97.2	83.2	91.4	100.6	83.0	93.2	97.2	84.6	91.9
08	OPENINGS	100.0	100.0	100.0	98.4	77.1	93.5	107.4	78.6	100.7	98.5	75.5	93.1	110.2	71.7	101.3	98.5	74.4	92.9
0920	Plaster & Gypsum Board	100.0	100.0	100.0	83.5	78.7	80.3	89.8	78.7	82.4	80.8	78.7	79.4	90.8	71.5	77.9	90.5	76.6	81.2
0950, 0980	Ceilings & Acoustic Treatment	100.0	100.0	100.0	77.8	78.7	78.4	85.5	78.7	81.0	77.8	78.7	78.4	83.0	71.5	75.3	77.8	76.6	77.0
0960	Flooring	100.0	100.0	100.0	92.2	76.7	87.8	98.7	76.7	92.3	97.4	76.7	91.5	95.7	76.7	90.2	103.2	76.7	95.6
0970, 0990	Wall Finishes & Painting/Coating	100.0	100.0	100.0	90.4	69.1	78.0	88.1	55.9	69.4	90.4	69.1	78.0	84.3	63.7	72.3	90.4	69.1	78.0
09	FINISHES	100.0	100.0	100.0	85.2	76.8	80.6	91.9	75.7	82.9	88.4	76.6	81.9	88.5	71.8	79.2	91.3	74.1	81.8
COVERS	DIVS. 10 - 14, 25, 28, 41, 43, 44, 46	100.0	100.0	100.0	100.0	87.3	97.2	100.0	89.9	97.8	100.0	86.9	97.2	100.0	86.2	97.0	100.0	86.7	97.1
21, 22, 23	FIRE SUPPRESSION, PLUMBING & HVAC	100.0	100.0	100.0	100.9	69.8	87.7	100.0	67.3	86.2	98.1	70.7	86.5	100.0	69.2	87.0	98.1	69.2	85.9
26, 27, 3370	ELECTRICAL, COMMUNICATIONS & UTIL.	100.0	100.0	100.0	92.2	62.5	77.2	99.1	63.4	81.1	93.9	62.5	78.0	94.9	67.0	80.8	92.9	62.7	77.7
MF2014	WEIGHTED AVERAGE	100.0	100.0	100.0	97.3	77.5	88.6	99.3	77.3	89.6	97.6	76.7	88.4	98.7	76.0	88.7	97.7	76.1	88.2

DIVISION		ALABAMA																	
		EVERGREEN 364			GADSDEN 359			HUNTSVILLE 357 - 358			JASPER 355			MOBILE 365 - 366			MONTGOMERY 360 - 361		
		MAT.	INST.	TOTAL	MAT.	INST.	TOTAL	MAT.	INST.	TOTAL	MAT.	INST.	TOTAL	MAT.	INST.	TOTAL	MAT.	INST.	TOTAL
015433	CONTRACTOR EQUIPMENT		99.3	99.3		101.9	101.9		101.9	101.9		101.9	101.9		99.3	99.3		99.3	99.3
0241, 31 - 34	SITE & INFRASTRUCTURE, DEMOLITION	103.7	88.5	93.0	93.6	92.6	92.9	87.2	92.5	90.9	93.3	92.6	92.8	95.8	89.8	91.6	93.9	88.4	90.0
0310	Concrete Forming & Accessories	83.1	75.7	76.7	88.9	51.2	56.3	97.0	71.7	75.1	94.3	76.7	79.0	94.6	78.9	81.1	93.7	77.9	80.0
0320	Concrete Reinforcing	93.0	81.4	87.0	96.2	88.2	92.0	90.7	85.0	87.8	90.7	87.0	88.8	90.7	88.2	89.4	98.9	88.2	93.4
0330	Cast-in-Place Concrete	98.0	73.2	88.2	101.1	73.4	90.2	98.4	72.2	88.1	112.0	73.7	96.9	102.7	73.4	91.1	99.3	73.7	89.2
03	CONCRETE	105.4	77.0	91.6	104.8	67.3	86.6	98.7	75.5	87.4	108.5	78.6	94.0	100.6	79.8	90.5	100.3	79.3	90.1
04	MASONRY	104.9	76.7	87.5	92.7	77.7	83.5	95.7	75.1	83.0	90.2	59.0	71.0	103.1	74.7	85.6	100.1	75.6	85.0
05	METALS	94.1	93.9	94.0	95.1	95.3	95.2	97.2	94.7	96.4	95.1	92.8	94.3	96.1	96.7	96.3	95.3	95.8	95.5
06	WOOD, PLASTICS & COMPOSITES	83.1	75.8	79.6	95.7	43.7	66.7	105.1	71.8	86.6	102.2	78.9	89.2	97.2	80.5	87.9	95.1	78.9	86.1
07	THERMAL & MOISTURE PROTECTION	97.2	81.9	90.8	100.7	80.5	92.2	100.5	82.4	92.9	100.7	74.4	89.7	96.9	83.4	91.2	95.6	83.7	90.6
08	OPENINGS	98.5	74.4	92.9	106.5	52.6	94.0	109.9	72.5	101.2	106.5	73.6	98.8	101.5	78.0	96.0	101.4	77.1	95.7
0920	Plaster & Gypsum Board	80.2	76.6	77.8	83.5	42.5	56.1	90.8	71.5	77.9	87.5	78.7	81.6	87.6	80.4	82.8	87.0	78.7	81.5
0950, 0980	Ceilings & Acoustic Treatment	77.8	76.6	77.0	80.5	42.5	55.3	84.8	71.5	76.0	80.5	78.7	79.3	83.0	80.4	81.2	85.3	78.7	80.9
0960	Flooring	95.1	76.7	89.8	91.5	76.7	87.2	95.7	76.7	90.2	93.8	38.7	78.0	102.7	73.4	94.3	99.5	73.4	92.0
0970, 0990	Wall Finishes & Painting/Coating	90.4	69.1	78.0	84.3	71.7	77.0	84.3	69.1	75.5	84.3	38.6	57.7	93.7	69.1	79.4	90.1	71.7	79.4
09	FINISHES	87.7	75.3	80.9	86.1	56.1	69.5	88.9	72.0	79.6	87.3	66.6	75.9	91.4	76.9	83.4	91.2	76.5	83.0
COVERS	DIVS. 10 - 14, 25, 28, 41, 43, 44, 46	100.0	86.7	97.1	100.0	85.5	96.9	100.0	87.8	97.4	100.0	49.8	89.1	100.0	90.1	97.9	100.0	89.4	97.7
21, 22, 23	FIRE SUPPRESSION, PLUMBING & HVAC	98.1	58.2	81.2	102.1	48.5	79.4	100.0	65.6	85.5	102.1	66.6	87.1	99.9	69.2	86.9	100.0	68.8	86.8
26, 27, 3370	ELECTRICAL, COMMUNICATIONS & UTIL.	91.6	62.5	76.9	94.9	63.4	79.0	95.8	67.0	81.3	94.6	63.4	78.8	94.6	59.1	76.6	95.1	64.6	79.6
MF2014	WEIGHTED AVERAGE	97.3	73.6	86.9	98.8	67.1	84.8	98.7	75.3	88.4	99.2	71.8	87.1	98.3	76.6	88.7	97.9	77.0	88.7

DIVISION		ALABAMA									ALASKA								
		PHENIX CITY 368			SELMA 367			TUSCALOOSA 354			ANCHORAGE 995 - 996			FAIRBANKS 997			JUNEAU 998		
		MAT.	INST.	TOTAL	MAT.	INST.	TOTAL	MAT.	INST.	TOTAL	MAT.	INST.	TOTAL	MAT.	INST.	TOTAL	MAT.	INST.	TOTAL
015433	CONTRACTOR EQUIPMENT		99.3	99.3		99.3	99.3		101.9	101.9		114.7	114.7		114.7	114.7		114.7	114.7
0241, 31 - 34	SITE & INFRASTRUCTURE, DEMOLITION	107.6	88.4	94.0	100.7	88.4	92.0	87.8	92.6	91.2	129.5	130.0	129.9	123.3	130.0	128.1	139.9	130.0	132.9
0310	Concrete Forming & Accessories	90.2	76.3	78.2	87.7	77.3	78.7	96.9	78.0	80.6	119.9	119.5	119.2	128.7	119.5	120.7	130.4	119.2	120.7
0320	Concrete Reinforcing	92.9	81.5	87.0	92.9	88.2	90.5	90.7	88.3	89.5	148.9	116.0	131.8	149.8	116.0	132.3	145.4	116.0	130.2
0330	Cast-in-Place Concrete	98.0	73.6	88.4	98.0	73.4	88.3	102.6	73.7	91.2	120.0	117.7	119.1	123.9	118.4	121.7	129.7	117.7	125.0
03	CONCRETE	108.4	77.4	93.3	103.8	79.0	91.7	100.7	79.4	90.4	130.5	117.3	124.1	120.9	117.7	119.3	134.1	117.3	125.9
04	MASONRY	104.9	77.7	88.1	108.4	77.7	89.5	94.6	77.7	84.2	187.9	125.8	149.7	197.2	126.5	153.6	182.4	125.8	147.6
05	METALS	94.1	93.5	93.9	94.1	95.4	94.5	96.4	96.1	96.3	129.0	104.9	121.4	133.5	104.9	124.4	132.1	104.9	123.4
06	WOOD, PLASTICS & COMPOSITES	91.7	76.8	83.4	88.6	78.9	83.2	105.1	78.9	90.5	117.7	117.7	117.7	129.1	117.7	123.1	126.2	117.7	121.5
07	THERMAL & MOISTURE PROTECTION	97.6	84.0	91.9	97.1	84.2	91.7	100.6	84.2	93.7	156.3	118.3	140.4	171.8	117.9	149.3	177.7	118.3	152.9
08	OPENINGS	98.4	74.4	92.8	98.4	77.1	93.5	109.9	78.6	102.6	130.0	115.8	126.7	138.1	116.1	133.0	130.1	115.8	126.8
0920	Plaster & Gypsum Board	84.8	76.6	79.3	82.8	78.7	80.1	90.8	78.7	82.7	137.8	117.9	124.5	164.8	117.9	133.5	151.3	117.9	129.0
0950, 0980	Ceilings & Acoustic Treatment	77.8	76.6	77.0	77.8	78.7	78.4	84.8	78.7	80.8	121.0	117.9	119.0	124.7	117.9	120.2	133.8	117.9	123.3
0960	Flooring	99.6	73.4	92.0	97.9	28.7	78.0	95.7	76.7	90.2	131.4	128.1	130.4	125.6	128.1	126.3	131.5	128.1	130.5
0970, 0990	Wall Finishes & Painting/Coating	90.4	71.7	79.5	90.4	69.1	78.0	84.3	71.7	77.0	110.5	116.5	114.0	108.6	121.3	116.0	108.2	116.5	113.0
09	FINISHES	90.1	75.2	81.9	88.6	67.5	76.9	88.9	77.1	82.4	130.8	120.8	125.3	131.2	121.5	125.8	134.6	120.8	127.0
COVERS	DIVS. 10 - 14, 25, 28, 41, 43, 44, 46	100.0	89.2	97.7	100.0	87.3	97.2	100.0	89.4	97.7	100.0	112.5	102.7	100.0	112.5	102.7	100.0	112.3	102.7
21, 22, 23	FIRE SUPPRESSION, PLUMBING & HVAC	98.1	68.7	85.7	98.1	65.4	84.3	100.0	71.3	87.9	100.5	105.8	102.7	100.6	109.7	104.4	100.0	105.8	102.8
26, 27, 3370	ELECTRICAL, COMMUNICATIONS & UTIL.	93.4	70.7	81.9	92.7	62.5	77.4	95.4	63.4	79.2	107.1	117.3	112.3	117.5	117.3	117.4	109.5	117.3	113.5
MF2014	WEIGHTED AVERAGE	98.2	77.2	89.0	97.4	74.9	87.5	98.7	78.1	89.6	121.7	115.8	119.1	124.2	116.9	120.9	123.9	115.8	120.3

For customer support on your Commercial Renovation Cost Data, call 877.742.4719.

763

Block 1

DIVISION		ALASKA KETCHIKAN 999 MAT.	INST.	TOTAL	CHAMBERS 865 MAT.	INST.	TOTAL	FLAGSTAFF 860 MAT.	INST.	TOTAL	GLOBE 855 MAT.	INST.	TOTAL	KINGMAN 864 MAT.	INST.	TOTAL	MESA/TEMPE 852 MAT.	INST.	TOTAL
015433	CONTRACTOR EQUIPMENT		114.7	114.7		91.0	91.0		91.0	91.0		92.0	92.0		91.0	91.0		92.0	92.0
0241, 31 - 34	SITE & INFRASTRUCTURE, DEMOLITION	181.7	130.0	145.2	72.3	95.6	88.8	91.5	95.7	94.5	106.0	96.4	99.2	72.2	95.7	88.8	96.0	96.5	96.4
0310	Concrete Forming & Accessories	120.2	119.2	119.3	99.1	56.7	62.4	104.8	68.7	73.6	96.9	56.8	62.2	97.3	63.3	67.9	100.1	67.5	71.9
0320	Concrete Reinforcing	113.6	116.0	114.8	103.0	79.1	90.6	102.8	79.1	90.5	112.2	79.1	95.0	103.1	79.1	90.7	112.9	79.1	95.4
0330	Cast-in-Place Concrete	250.7	117.7	198.2	87.9	72.0	81.6	87.9	72.1	81.7	93.7	71.6	85.0	87.6	72.2	81.5	94.5	71.8	85.5
03	CONCRETE	205.3	117.3	162.5	98.1	66.3	82.6	119.4	71.7	96.2	115.7	66.2	91.6	97.8	69.3	83.9	106.3	71.1	89.2
04	MASONRY	205.4	125.8	156.4	98.7	63.3	76.9	98.9	63.4	77.0	116.6	63.2	83.7	98.7	63.4	76.9	116.8	63.3	83.8
05	METALS	133.7	104.9	124.5	101.1	72.1	91.9	101.6	72.2	92.2	99.0	72.8	90.7	101.8	72.2	92.4	99.3	73.0	90.9
06	WOOD, PLASTICS & COMPOSITES	120.7	117.7	119.0	97.5	53.1	72.7	103.4	69.4	84.4	90.0	53.2	69.5	92.5	62.0	75.5	93.3	67.5	79.0
07	THERMAL & MOISTURE PROTECTION	176.5	118.3	152.2	94.6	63.7	81.6	96.3	67.7	84.3	99.8	64.2	84.9	94.5	64.2	81.9	99.1	64.2	84.5
08	OPENINGS	133.6	115.8	129.4	105.1	58.6	94.3	105.3	71.5	97.4	93.3	58.7	85.2	105.3	65.7	96.1	93.3	68.8	87.6
0920	Plaster & Gypsum Board	153.5	117.9	129.7	92.2	51.9	65.2	95.3	68.6	77.5	95.2	51.9	66.2	84.7	61.0	68.9	97.2	66.6	76.7
0950, 0980	Ceilings & Acoustic Treatment	118.2	117.9	118.0	99.4	51.9	67.9	100.3	68.6	79.3	93.5	51.9	65.9	100.3	61.0	74.3	93.5	66.6	75.7
0960	Flooring	125.6	128.1	126.3	96.2	61.2	86.1	98.7	34.5	80.2	103.3	61.2	91.2	94.9	66.2	86.6	105.0	47.6	88.5
0970, 0990	Wall Finishes & Painting/Coating	108.6	116.5	113.2	89.5	57.5	70.9	89.5	57.5	70.9	94.1	57.5	72.8	89.5	57.5	70.9	94.1	57.5	72.8
09	FINISHES	132.2	120.8	125.9	94.5	56.2	73.3	97.7	61.0	77.4	98.8	56.4	75.3	93.2	62.4	76.2	98.5	62.7	78.7
COVERS	DIVS. 10 - 14, 25, 28, 41, 43, 44, 46	100.0	112.3	102.7	100.0	82.1	96.1	100.0	83.8	96.5	100.0	82.3	96.2	100.0	83.0	96.3	100.0	83.9	96.5
21, 22, 23	FIRE SUPPRESSION, PLUMBING & HVAC	98.6	105.8	101.6	97.8	79.0	89.9	100.2	79.1	91.3	96.7	79.1	89.3	97.8	79.0	89.9	100.1	79.1	91.2
26, 27, 3370	ELECTRICAL, COMMUNICATIONS & UTIL.	117.4	117.3	117.3	103.0	90.0	96.4	102.0	61.7	81.6	100.2	61.7	80.7	103.0	61.7	82.1	97.1	61.7	79.2
MF2014	WEIGHTED AVERAGE	134.9	115.8	126.5	98.9	73.5	87.7	102.7	71.9	89.2	100.9	69.8	87.2	98.9	71.3	86.7	100.1	72.0	87.7

Block 2

DIVISION		ARIZONA PHOENIX 850,853 MAT.	INST.	TOTAL	PRESCOTT 863 MAT.	INST.	TOTAL	SHOW LOW 859 MAT.	INST.	TOTAL	TUCSON 856 - 857 MAT.	INST.	TOTAL	ARKANSAS BATESVILLE 725 MAT.	INST.	TOTAL	CAMDEN 717 MAT.	INST.	TOTAL
015433	CONTRACTOR EQUIPMENT		92.6	92.6		91.0	91.0		92.0	92.0		92.0	92.0		91.9	91.9		91.9	91.9
0241, 31 - 34	SITE & INFRASTRUCTURE, DEMOLITION	96.5	96.7	96.7	79.1	95.5	90.7	108.3	96.4	99.9	91.5	96.5	95.0	76.0	88.0	84.5	78.4	87.9	85.1
0310	Concrete Forming & Accessories	101.0	67.7	72.2	100.9	51.1	57.8	104.3	67.5	72.5	100.5	67.3	71.8	84.7	46.4	51.6	82.6	35.0	41.4
0320	Concrete Reinforcing	111.1	79.1	94.6	102.8	79.0	90.5	112.9	79.1	95.4	93.4	79.1	86.0	97.6	67.0	81.7	103.9	66.8	84.7
0330	Cast-in-Place Concrete	94.5	71.9	85.6	87.9	71.9	81.6	93.7	71.7	85.0	97.2	71.7	87.2	74.3	75.3	74.7	79.4	75.2	77.7
03	CONCRETE	105.9	71.2	89.0	103.9	63.7	84.4	118.2	71.0	95.3	104.6	71.0	88.2	82.3	60.9	71.9	86.1	55.6	71.3
04	MASONRY	102.9	63.3	78.5	98.8	63.3	77.0	116.7	63.2	83.7	100.9	63.3	77.7	97.8	45.3	65.5	115.2	42.1	70.1
05	METALS	100.7	73.8	92.2	101.6	71.7	92.1	98.8	72.9	90.5	100.0	72.8	91.4	99.3	68.7	89.6	103.6	68.1	92.3
06	WOOD, PLASTICS & COMPOSITES	94.3	67.5	79.4	98.8	45.7	69.3	97.8	67.5	81.0	93.6	67.5	79.1	96.6	44.1	67.4	96.1	28.8	58.6
07	THERMAL & MOISTURE PROTECTION	98.7	66.0	85.0	95.0	62.8	81.6	100.0	71.4	88.0	100.2	63.5	84.8	98.8	52.8	79.5	95.6	50.3	76.7
08	OPENINGS	95.1	70.5	89.4	105.3	54.6	93.5	92.5	66.5	86.5	89.9	70.5	85.4	104.6	44.9	90.7	107.6	39.4	91.8
0920	Plaster & Gypsum Board	99.2	66.6	77.4	92.3	44.3	60.3	99.1	66.6	77.4	102.0	66.6	78.3	82.2	42.8	55.9	82.6	27.1	45.5
0950, 0980	Ceilings & Acoustic Treatment	100.9	66.6	78.2	98.6	44.3	62.7	93.5	66.6	75.7	94.3	66.6	75.9	78.6	42.8	54.9	79.6	27.1	44.8
0960	Flooring	105.3	66.2	94.1	97.2	61.2	86.8	106.9	43.3	88.6	96.1	42.5	80.7	92.5	56.8	82.2	103.0	38.1	84.3
0970, 0990	Wall Finishes & Painting/Coating	94.1	61.8	75.3	89.5	57.5	70.9	94.1	57.5	72.8	95.1	57.5	73.2	91.6	41.2	62.3	96.4	50.4	69.6
09	FINISHES	100.6	66.3	81.6	95.1	51.9	71.2	100.6	61.9	79.2	96.7	61.2	77.1	82.4	46.7	62.6	86.5	35.2	58.1
COVERS	DIVS. 10 - 14, 25, 28, 41, 43, 44, 46	100.0	83.9	96.5	100.0	81.2	95.9	100.0	83.9	96.5	100.0	83.9	96.5	100.0	68.0	93.1	100.0	66.3	92.7
21, 22, 23	FIRE SUPPRESSION, PLUMBING & HVAC	100.0	79.2	91.2	100.2	79.0	91.3	96.7	79.1	89.3	100.1	74.4	89.2	96.7	53.6	78.5	96.7	58.1	80.4
26, 27, 3370	ELECTRICAL, COMMUNICATIONS & UTIL.	102.5	64.1	83.1	101.7	61.7	81.5	97.5	61.7	79.4	99.3	61.7	80.3	98.9	63.0	80.8	96.5	61.7	78.9
MF2014	WEIGHTED AVERAGE	100.5	73.0	88.4	100.4	68.4	86.3	101.1	71.9	88.3	98.9	70.8	86.5	95.3	58.3	79.0	97.6	56.0	79.3

Block 3

DIVISION		ARKANSAS FAYETTEVILLE 727 MAT.	INST.	TOTAL	FORT SMITH 729 MAT.	INST.	TOTAL	HARRISON 726 MAT.	INST.	TOTAL	HOT SPRINGS 719 MAT.	INST.	TOTAL	JONESBORO 724 MAT.	INST.	TOTAL	LITTLE ROCK 720 - 722 MAT.	INST.	TOTAL
015433	CONTRACTOR EQUIPMENT		91.9	91.9		91.9	91.9		91.9	91.9		91.9	91.9		114.2	114.2		91.9	91.9
0241, 31 - 34	SITE & INFRASTRUCTURE, DEMOLITION	75.4	89.7	85.5	80.7	89.8	87.1	81.2	88.1	86.0	81.8	89.2	87.0	101.8	107.2	105.6	90.2	90.7	90.6
0310	Concrete Forming & Accessories	80.3	39.3	44.8	99.4	60.3	65.6	89.2	46.4	52.2	80.1	39.2	44.7	88.3	49.0	54.3	94.1	61.2	65.6
0320	Concrete Reinforcing	97.6	70.7	83.7	98.8	71.5	84.6	97.2	67.1	81.6	101.9	66.9	83.8	94.5	75.9	84.9	104.7	68.6	86.0
0330	Cast-in-Place Concrete	74.3	75.3	74.7	84.9	75.7	81.3	82.4	75.2	79.6	81.2	75.1	78.8	80.9	76.9	79.3	82.3	76.9	80.2
03	CONCRETE	82.0	58.3	70.5	89.9	68.0	79.3	89.4	60.8	75.5	89.5	57.7	74.0	86.8	65.0	76.2	91.0	68.5	80.0
04	MASONRY	88.8	45.3	62.0	96.2	52.4	69.2	98.2	46.9	66.6	86.4	43.5	59.9	90.2	44.7	62.1	94.8	71.2	80.2
05	METALS	99.3	70.0	90.0	101.6	71.5	92.1	100.4	68.6	90.3	103.5	68.1	92.3	95.9	85.4	92.6	101.8	74.9	93.2
06	WOOD, PLASTICS & COMPOSITES	93.1	34.4	60.4	113.4	61.8	84.7	102.5	44.1	70.0	93.4	34.6	60.7	100.8	46.8	70.8	100.9	61.8	79.1
07	THERMAL & MOISTURE PROTECTION	99.6	51.7	79.6	99.9	56.5	81.8	99.1	53.2	79.9	95.9	51.3	77.2	104.0	56.7	84.2	94.7	61.3	80.7
08	OPENINGS	104.6	43.9	90.4	106.7	59.3	94.6	105.4	46.1	91.6	107.6	44.2	92.9	110.6	54.0	97.4	102.5	60.0	92.6
0920	Plaster & Gypsum Board	81.6	32.9	49.0	88.2	61.1	70.1	87.3	42.8	57.5	81.2	33.1	49.1	96.0	45.3	62.1	93.1	61.1	71.7
0950, 0980	Ceilings & Acoustic Treatment	78.6	32.9	48.3	80.3	61.1	67.6	80.3	42.8	55.5	79.6	33.1	48.8	82.5	45.3	57.9	83.4	61.1	68.6
0960	Flooring	89.4	56.8	80.0	99.4	57.3	87.3	94.8	56.8	83.9	101.8	56.8	88.8	66.4	51.4	62.1	97.5	75.2	91.0
0970, 0990	Wall Finishes & Painting/Coating	91.6	34.1	58.2	91.6	53.8	69.6	91.6	41.2	62.3	96.4	55.2	72.4	80.9	48.4	62.0	92.7	54.8	70.7
09	FINISHES	81.4	40.0	58.5	85.9	59.1	71.1	84.5	46.5	63.5	86.3	42.4	62.0	80.0	48.0	62.3	88.3	62.8	74.2
COVERS	DIVS. 10 - 14, 25, 28, 41, 43, 44, 46	100.0	48.4	88.8	100.0	80.7	95.8	100.0	71.2	93.8	100.0	39.1	86.8	100.0	69.2	93.3	100.0	80.8	95.8
21, 22, 23	FIRE SUPPRESSION, PLUMBING & HVAC	96.7	51.4	77.6	100.1	51.5	79.5	96.7	51.4	77.5	96.7	53.3	78.4	100.4	53.5	80.6	99.9	53.9	80.4
26, 27, 3370	ELECTRICAL, COMMUNICATIONS & UTIL.	92.9	59.1	75.8	96.3	59.2	77.5	97.5	59.1	78.1	98.5	69.7	83.9	102.1	63.0	82.3	103.0	71.1	86.8
MF2014	WEIGHTED AVERAGE	94.1	55.6	77.1	97.8	62.3	82.2	96.6	57.6	79.4	96.9	56.9	79.3	97.4	62.6	82.1	98.2	67.4	84.6

City Cost Indexes

ARKANSAS / CALIFORNIA

DIVISION		PINE BLUFF 716 MAT.	INST.	TOTAL	RUSSELLVILLE 728 MAT.	INST.	TOTAL	TEXARKANA 718 MAT.	INST.	TOTAL	WEST MEMPHIS 723 MAT.	INST.	TOTAL	ALHAMBRA 917-918 MAT.	INST.	TOTAL	ANAHEIM 928 MAT.	INST.	TOTAL
015433	CONTRACTOR EQUIPMENT		91.9	91.9		91.9	91.9		93.0	93.0		114.2	114.2		95.5	95.5		100.4	100.4
0241, 31 - 34	SITE & INFRASTRUCTURE, DEMOLITION	83.9	90.8	88.8	77.3	88.1	85.0	95.3	91.6	92.6	109.5	107.1	107.8	101.6	106.9	105.3	100.9	107.8	105.8
0310	Concrete Forming & Accessories	79.7	60.9	63.4	85.4	56.0	60.0	85.8	44.1	49.7	93.9	49.1	55.1	115.7	120.5	119.8	103.6	120.9	118.5
0320	Concrete Reinforcing	103.8	67.9	85.2	98.2	66.9	82.0	103.4	67.0	84.5	94.5	72.1	82.9	106.5	115.2	111.0	96.3	115.1	106.0
0330	Cast-in-Place Concrete	81.2	76.9	79.5	77.8	75.2	76.8	88.6	75.4	83.3	84.8	76.9	81.7	89.4	118.0	100.7	90.9	121.8	103.1
03	CONCRETE	90.3	68.2	79.6	85.4	65.1	75.5	88.4	59.8	74.5	93.9	64.4	79.6	99.4	117.6	108.2	102.0	119.1	110.4
04	MASONRY	122.1	58.5	82.9	95.5	44.6	64.2	101.5	44.7	66.5	78.5	44.6	57.6	114.4	120.0	117.9	81.6	117.9	104.0
05	METALS	104.3	74.2	94.8	99.3	68.5	89.5	96.4	68.8	87.6	95.0	85.5	92.0	89.3	99.5	92.5	108.0	101.1	105.8
06	WOOD, PLASTICS & COMPOSITES	93.0	61.8	75.6	97.9	57.2	75.2	100.7	40.7	67.3	106.7	46.8	73.4	95.5	118.3	108.2	97.5	118.6	109.3
07	THERMAL & MOISTURE PROTECTION	96.0	58.0	80.1	99.8	54.0	80.6	96.6	52.2	78.1	104.5	58.4	85.2	95.5	116.1	104.1	101.8	119.9	109.4
08	OPENINGS	108.9	57.5	96.9	104.6	54.2	92.8	114.3	45.2	98.2	108.0	54.1	95.4	90.9	117.3	97.0	108.1	117.9	110.4
0920	Plaster & Gypsum Board	80.9	61.1	67.6	82.2	56.3	64.9	84.0	39.3	54.1	98.4	45.3	62.9	99.2	119.0	112.4	104.5	119.1	114.2
0950, 0980	Ceilings & Acoustic Treatment	79.6	61.1	67.3	78.6	56.3	63.8	82.9	39.3	54.0	80.7	45.3	57.2	101.9	119.0	113.2	105.6	119.1	114.5
0960	Flooring	101.4	57.3	88.8	92.0	56.8	81.9	103.8	49.0	88.0	68.9	51.4	63.9	97.9	107.7	100.7	101.5	108.2	103.4
0970, 0990	Wall Finishes & Painting/Coating	96.4	50.4	69.6	91.6	35.7	59.1	96.4	30.3	57.9	80.9	50.4	63.1	110.9	111.5	111.2	92.9	109.4	102.5
09	FINISHES	86.2	59.1	71.2	82.5	53.9	66.7	88.7	42.2	62.9	81.4	48.4	63.1	101.4	116.9	110.0	99.8	117.2	109.4
COVERS	DIVS. 10 - 14, 25, 28, 41, 43, 44, 46	100.0	80.8	95.8	100.0	69.5	93.4	100.0	65.0	92.4	100.0	53.0	89.8	100.0	110.6	102.3	100.0	111.3	102.5
21, 22, 23	FIRE SUPPRESSION, PLUMBING & HVAC	100.2	53.8	80.6	96.7	53.2	78.3	100.2	59.5	83.0	97.0	66.7	84.2	96.7	115.1	104.5	100.0	119.4	108.2
26, 27, 3370	ELECTRICAL, COMMUNICATIONS & UTIL.	96.7	69.7	83.1	96.3	59.1	77.5	98.4	61.0	79.5	103.7	65.2	84.2	123.0	120.3	121.6	90.6	105.3	98.0
MF2014	WEIGHTED AVERAGE	99.6	65.2	84.5	95.3	59.7	79.7	98.5	58.7	80.9	97.0	65.3	83.0	99.5	114.8	106.2	100.6	114.0	106.5

CALIFORNIA

DIVISION		BAKERSFIELD 932-933 MAT.	INST.	TOTAL	BERKELEY 947 MAT.	INST.	TOTAL	EUREKA 955 MAT.	INST.	TOTAL	FRESNO 936-938 MAT.	INST.	TOTAL	INGLEWOOD 903-905 MAT.	INST.	TOTAL	LONG BEACH 906-908 MAT.	INST.	TOTAL
015433	CONTRACTOR EQUIPMENT		98.4	98.4		101.8	101.8		98.2	98.2		98.4	98.4		97.2	97.2		97.2	97.2
0241, 31 - 34	SITE & INFRASTRUCTURE, DEMOLITION	99.7	105.6	103.9	106.9	108.0	107.7	112.5	104.7	107.0	103.8	105.3	104.9	92.8	103.6	100.5	99.9	103.6	102.5
0310	Concrete Forming & Accessories	101.9	122.9	120.1	114.3	146.6	142.3	112.4	132.4	129.7	101.0	132.0	127.8	105.6	120.9	118.8	100.4	120.9	118.1
0320	Concrete Reinforcing	99.5	115.1	107.6	95.6	116.6	106.5	105.0	116.3	110.9	83.1	116.0	100.1	97.8	115.3	106.9	97.0	115.3	106.5
0330	Cast-in-Place Concrete	89.3	120.9	101.7	120.4	126.4	122.8	98.6	123.0	108.2	96.9	122.6	107.1	84.6	120.8	98.9	96.2	120.8	105.9
03	CONCRETE	95.5	119.7	107.2	108.8	132.3	120.2	113.0	124.9	118.8	99.3	124.5	111.6	92.1	118.8	105.1	101.5	118.8	109.9
04	MASONRY	98.1	117.6	110.1	117.0	137.6	129.7	109.3	136.6	126.1	102.8	131.5	120.5	79.4	120.2	104.5	88.4	120.2	108.0
05	METALS	108.1	100.5	105.7	102.7	102.9	102.8	107.8	102.8	106.2	108.4	101.3	106.1	99.2	101.2	99.8	99.0	101.2	99.7
06	WOOD, PLASTICS & COMPOSITES	90.4	121.8	107.9	108.0	150.5	131.7	111.3	133.6	123.7	100.4	133.6	118.8	95.6	118.7	108.5	89.7	118.7	105.8
07	THERMAL & MOISTURE PROTECTION	96.6	114.9	104.3	100.7	137.5	116.1	105.7	133.3	117.2	88.4	118.6	101.1	101.9	117.3	108.3	102.2	117.8	108.7
08	OPENINGS	98.1	116.9	102.5	93.0	138.4	103.5	107.3	122.9	110.9	100.2	123.4	105.6	89.4	117.5	96.0	89.4	117.5	95.9
0920	Plaster & Gypsum Board	96.5	122.1	113.6	104.1	151.3	135.6	109.9	134.3	126.2	95.3	134.3	121.3	99.8	119.0	112.6	96.1	119.0	111.4
0950, 0980	Ceilings & Acoustic Treatment	94.5	122.1	112.8	104.0	151.3	135.3	110.6	134.3	126.3	91.1	134.3	119.7	101.2	119.0	113.0	101.2	119.0	113.0
0960	Flooring	102.4	108.2	104.1	115.3	126.5	118.5	105.4	77.1	97.3	108.5	102.0	106.6	104.7	107.7	105.6	101.9	107.7	103.6
0970, 0990	Wall Finishes & Painting/Coating	92.3	98.1	95.7	111.7	144.3	130.7	94.4	120.0	109.3	107.6	109.8	108.9	110.4	111.5	111.0	110.4	111.5	111.0
09	FINISHES	96.1	117.7	108.1	106.4	144.1	127.3	105.4	123.1	115.1	98.3	126.2	113.7	103.1	117.3	110.9	102.1	117.3	110.5
COVERS	DIVS. 10 - 14, 25, 28, 41, 43, 44, 46	100.0	109.3	102.0	100.0	125.8	105.6	100.0	122.9	105.0	100.0	122.8	104.9	100.0	111.6	102.5	100.0	111.6	102.5
21, 22, 23	FIRE SUPPRESSION, PLUMBING & HVAC	100.2	109.9	104.3	96.7	149.6	119.0	96.6	111.8	103.0	100.2	116.3	107.0	96.3	115.2	104.3	96.3	115.2	104.3
26, 27, 3370	ELECTRICAL, COMMUNICATIONS & UTIL.	103.4	102.8	103.1	104.5	143.9	124.4	96.7	115.8	106.4	94.3	100.7	97.5	99.5	120.3	110.0	99.3	120.3	109.9
MF2014	WEIGHTED AVERAGE	100.4	111.4	105.2	102.0	135.3	116.7	103.9	118.3	110.2	100.5	116.5	107.5	96.0	115.0	104.4	97.5	115.0	105.2

CALIFORNIA

DIVISION		LOS ANGELES 900-902 MAT.	INST.	TOTAL	MARYSVILLE 959 MAT.	INST.	TOTAL	MODESTO 953 MAT.	INST.	TOTAL	MOJAVE 935 MAT.	INST.	TOTAL	OAKLAND 946 MAT.	INST.	TOTAL	OXNARD 930 MAT.	INST.	TOTAL
015433	CONTRACTOR EQUIPMENT		100.3	100.3		98.2	98.2		98.2	98.2		98.4	98.4		101.8	101.8		97.3	97.3
0241, 31 - 34	SITE & INFRASTRUCTURE, DEMOLITION	100.4	107.0	105.0	108.8	104.7	105.9	103.5	104.9	104.5	96.9	105.6	103.0	112.2	108.0	109.3	104.1	103.6	103.8
0310	Concrete Forming & Accessories	102.5	121.2	118.7	102.3	132.0	128.0	98.7	132.1	127.6	112.7	122.8	121.5	103.0	146.6	140.7	104.2	121.0	118.7
0320	Concrete Reinforcing	98.6	115.5	107.3	105.0	115.7	110.6	108.8	115.7	112.4	100.1	115.1	107.9	97.8	116.6	107.5	98.3	115.0	107.0
0330	Cast-in-Place Concrete	91.3	121.4	103.1	110.2	122.7	115.1	98.6	122.8	108.2	83.8	120.9	98.4	114.2	126.4	119.0	97.6	121.1	106.9
03	CONCRETE	97.4	119.2	108.0	114.1	124.5	119.1	104.8	124.6	114.4	91.3	119.6	105.1	108.6	132.3	120.1	99.8	118.9	109.1
04	MASONRY	95.1	123.0	112.3	110.3	126.8	120.4	106.1	126.8	118.8	100.8	117.4	111.0	123.1	137.6	132.0	103.9	115.6	111.1
05	METALS	105.9	102.4	104.8	107.3	101.4	105.4	104.1	101.6	103.3	105.5	100.4	103.9	98.5	102.8	99.9	103.4	100.7	102.5
06	WOOD, PLASTICS & COMPOSITES	95.2	118.8	108.4	98.8	133.6	118.1	94.7	133.6	116.3	99.6	121.8	112.0	95.8	150.5	126.2	94.9	118.7	108.2
07	THERMAL & MOISTURE PROTECTION	101.3	119.5	108.9	105.2	120.7	111.7	104.8	121.2	111.7	94.0	110.7	101.0	98.9	137.5	115.1	97.0	118.0	105.9
08	OPENINGS	96.0	118.2	101.2	106.5	122.9	110.3	105.4	124.1	109.7	94.7	116.9	99.9	93.0	138.4	103.6	97.5	118.0	102.3
0920	Plaster & Gypsum Board	99.6	119.1	112.6	102.5	134.3	123.7	104.8	134.3	124.5	104.7	122.1	116.4	98.7	151.3	133.8	99.1	119.1	112.4
0950, 0980	Ceilings & Acoustic Treatment	111.2	119.1	116.4	109.8	134.3	126.0	105.6	134.3	124.6	92.5	122.1	112.1	106.6	151.3	136.2	94.8	119.1	110.9
0960	Flooring	102.3	108.2	104.0	101.2	113.0	104.6	101.5	110.1	104.0	109.2	105.9	108.3	108.9	126.5	114.0	101.1	108.2	103.2
0970, 0990	Wall Finishes & Painting/Coating	109.3	111.5	110.6	94.4	120.0	109.3	94.4	119.9	109.3	92.0	130.7	114.5	111.7	144.3	130.7	92.0	104.3	99.1
09	FINISHES	104.6	117.5	111.7	102.3	129.1	117.1	101.4	128.6	116.4	98.5	120.8	110.8	104.7	144.1	126.5	96.0	116.7	107.5
COVERS	DIVS. 10 - 14, 25, 28, 41, 43, 44, 46	100.0	111.8	102.5	100.0	122.8	104.9	100.0	122.8	104.9	100.0	107.0	101.5	100.0	125.7	105.6	100.0	111.5	102.5
21, 22, 23	FIRE SUPPRESSION, PLUMBING & HVAC	100.0	119.5	108.2	96.6	110.0	102.3	100.0	115.0	106.3	96.7	113.9	104.0	100.1	149.6	121.0	100.1	119.4	108.3
26, 27, 3370	ELECTRICAL, COMMUNICATIONS & UTIL.	98.0	123.5	110.9	93.8	106.5	100.3	96.0	103.9	100.0	92.7	99.1	96.0	103.8	138.2	121.2	98.1	106.5	102.3
MF2014	WEIGHTED AVERAGE	100.2	117.2	107.7	103.1	115.9	108.8	102.0	116.6	108.4	97.4	111.9	103.8	102.1	134.5	116.4	99.9	113.5	105.9

CALIFORNIA

DIVISION		PALM SPRINGS 922			PALO ALTO 943			PASADENA 910 - 912			REDDING 960			RICHMOND 948			RIVERSIDE 925		
		MAT.	INST.	TOTAL	MAT.	INST.	TOTAL	MAT.	INST.	TOTAL	MAT.	INST.	TOTAL	MAT.	INST.	TOTAL	MAT.	INST.	TOTAL
015433	CONTRACTOR EQUIPMENT		99.2	99.2		101.8	101.8		95.5	95.5		98.2	98.2		101.8	101.8		99.2	99.2
0241, 31 - 34	SITE & INFRASTRUCTURE, DEMOLITION	92.2	105.8	101.8	103.2	108.0	106.6	98.3	106.9	104.4	134.7	104.7	113.5	111.4	108.0	109.0	99.4	105.8	103.9
0310	Concrete Forming & Accessories	100.2	120.8	118.0	101.1	142.5	136.9	104.3	120.5	118.3	106.4	140.5	135.9	117.2	146.4	142.4	104.0	120.9	118.6
0320	Concrete Reinforcing	110.5	115.1	112.9	95.6	116.4	106.3	104.1	115.2	111.5	132.3	115.7	123.7	95.6	116.3	106.3	107.3	115.1	111.3
0330	Cast-in-Place Concrete	86.6	121.8	100.5	101.9	126.3	111.5	84.9	118.0	97.9	119.6	122.8	120.8	117.1	126.3	120.7	94.2	121.8	105.1
03	CONCRETE	96.6	119.1	107.5	97.9	130.4	113.7	94.9	117.6	105.9	128.2	128.3	128.2	110.6	132.1	121.1	102.8	119.1	110.7
04	MASONRY	79.8	117.6	103.1	97.9	133.8	120.0	98.2	120.0	111.7	142.0	126.8	132.6	116.8	132.2	126.3	80.8	117.6	103.5
05	METALS	108.7	101.0	106.3	96.1	102.3	98.1	89.3	99.5	92.6	113.3	101.5	109.5	96.1	102.2	98.1	108.1	101.1	105.9
06	WOOD, PLASTICS & COMPOSITES	92.5	118.6	107.0	93.2	145.3	122.2	82.3	118.3	102.3	107.4	144.9	128.3	111.8	150.5	133.3	97.5	118.6	109.3
07	THERMAL & MOISTURE PROTECTION	101.5	117.4	108.1	98.5	134.4	113.5	95.2	116.1	104.0	126.9	122.5	125.1	99.1	135.5	114.3	102.0	119.8	109.5
08	OPENINGS	103.8	117.9	107.1	93.0	133.8	102.5	90.9	117.3	97.0	121.0	129.0	122.8	93.0	137.1	103.3	106.7	117.9	109.3
0920	Plaster & Gypsum Board	99.7	119.0	112.6	97.0	146.0	129.7	92.6	119.0	110.2	103.6	146.0	131.9	106.0	151.3	136.2	103.7	119.1	114.0
0950, 0980	Ceilings & Acoustic Treatment	102.3	119.0	113.4	104.8	146.0	132.1	101.9	119.0	113.2	132.6	146.0	141.5	104.8	151.3	135.6	109.7	119.1	115.9
0960	Flooring	103.8	118.4	108.0	107.6	126.5	113.0	92.2	107.7	96.7	101.1	113.0	104.5	117.6	126.5	120.1	105.3	108.2	106.2
0970, 0990	Wall Finishes & Painting/Coating	91.4	111.3	102.9	111.7	144.3	130.7	110.9	111.5	111.2	104.0	120.0	113.3	111.7	144.3	130.7	91.4	109.4	101.8
09	FINISHES	98.4	118.9	109.7	103.1	141.1	124.1	98.5	116.9	108.7	109.5	135.9	124.1	108.0	144.1	128.0	101.4	117.2	110.1
COVERS	DIVS. 10 - 14, 25, 28, 41, 43, 44, 46	100.0	111.3	102.5	100.0	125.2	105.5	100.0	110.6	102.3	100.0	124.1	105.2	100.0	125.7	105.6	100.0	111.3	102.5
21, 22, 23	FIRE SUPPRESSION, PLUMBING & HVAC	96.6	114.6	104.2	96.7	144.1	116.8	96.7	115.1	104.5	100.4	110.0	104.5	96.7	146.5	117.7	100.0	119.4	108.2
26, 27, 3370	ELECTRICAL, COMMUNICATIONS & UTIL.	93.3	104.6	99.0	103.7	149.1	126.6	119.8	120.3	120.1	99.7	106.5	103.1	104.2	131.7	118.1	90.4	105.6	98.1
MF2014	WEIGHTED AVERAGE	98.6	112.9	104.9	98.1	133.4	113.7	97.5	114.8	105.1	112.3	117.8	114.7	101.3	132.2	114.9	100.5	113.9	106.4

CALIFORNIA

DIVISION		SACRAMENTO 942,956 - 958			SALINAS 939			SAN BERNARDINO 923 - 924			SAN DIEGO 919 - 921			SAN FRANCISCO 940 - 941			SAN JOSE 951		
		MAT.	INST.	TOTAL	MAT.	INST.	TOTAL	MAT.	INST.	TOTAL	MAT.	INST.	TOTAL	MAT.	INST.	TOTAL	MAT.	INST.	TOTAL
015433	CONTRACTOR EQUIPMENT		101.0	101.0		98.4	98.4		99.2	99.2		95.7	95.7		111.0	111.0		99.4	99.4
0241, 31 - 34	SITE & INFRASTRUCTURE, DEMOLITION	92.0	114.1	107.6	118.2	105.5	109.2	78.4	105.8	97.7	105.2	100.6	101.9	114.4	114.0	114.1	135.3	99.8	110.2
0310	Concrete Forming & Accessories	101.4	134.8	130.3	107.4	135.1	131.4	107.9	120.6	118.9	105.3	113.9	112.8	102.7	147.6	141.6	104.8	146.4	140.8
0320	Concrete Reinforcing	90.5	116.0	103.7	98.8	116.1	107.8	107.3	115.2	111.4	104.0	115.1	109.7	111.4	117.0	114.3	95.9	116.4	106.5
0330	Cast-in-Place Concrete	96.6	123.9	107.4	96.4	123.1	107.0	65.1	121.7	87.5	93.0	107.5	98.7	117.1	127.8	121.3	114.6	125.5	118.9
03	CONCRETE	98.8	125.9	111.9	109.5	126.1	117.5	76.5	119.0	97.2	100.1	111.2	105.5	111.9	133.8	122.5	111.1	132.3	121.4
04	MASONRY	98.9	131.0	118.7	100.9	131.7	119.9	86.5	119.6	106.9	90.8	116.4	106.6	123.6	140.8	134.2	136.8	133.8	134.9
05	METALS	94.4	97.6	95.4	108.3	102.6	106.5	108.1	100.9	105.8	105.9	100.7	104.3	103.7	110.6	105.9	102.3	107.7	104.1
06	WOOD, PLASTICS & COMPOSITES	90.1	136.7	116.1	99.3	136.5	120.0	101.1	118.6	110.8	95.8	110.8	104.1	95.8	150.6	126.3	106.9	150.2	131.0
07	THERMAL & MOISTURE PROTECTION	107.3	124.3	114.4	94.6	127.7	108.5	100.7	118.1	108.0	100.3	106.0	102.7	100.5	140.9	117.4	101.1	137.5	116.3
08	OPENINGS	106.0	125.9	110.7	98.7	130.8	106.2	103.9	117.9	107.2	98.1	112.2	101.4	97.2	138.5	106.8	96.4	138.3	106.1
0920	Plaster & Gypsum Board	94.3	137.3	123.0	100.1	137.3	124.9	105.5	119.0	114.5	99.6	111.0	107.2	100.6	151.3	134.5	100.5	151.3	134.4
0950, 0980	Ceilings & Acoustic Treatment	104.8	137.3	126.3	92.5	137.3	122.1	105.6	119.0	114.5	109.3	111.0	110.4	113.0	151.3	138.3	104.8	151.3	135.6
0960	Flooring	107.5	113.0	109.1	103.2	119.4	107.8	107.3	105.9	106.9	100.4	109.8	103.1	108.9	126.5	114.0	94.8	126.5	103.9
0970, 0990	Wall Finishes & Painting/Coating	108.6	132.3	122.4	99.3	144.3	122.8	91.4	100.4	96.7	107.7	114.0	111.2	111.7	158.2	138.8	94.8	144.3	123.6
09	FINISHES	101.8	132.4	118.7	98.0	134.5	118.2	99.8	116.0	108.8	104.2	112.9	109.0	106.5	145.8	128.2	100.1	143.9	124.4
COVERS	DIVS. 10 - 14, 25, 28, 41, 43, 44, 46	100.0	123.8	105.2	100.0	123.2	105.0	100.0	108.6	101.9	100.0	109.8	102.1	100.0	126.3	105.7	100.0	125.2	105.5
21, 22, 23	FIRE SUPPRESSION, PLUMBING & HVAC	99.9	120.1	108.4	96.7	121.3	107.1	96.6	114.3	104.1	100.0	118.6	107.8	100.0	178.0	133.0	100.0	149.8	121.1
26, 27, 3370	ELECTRICAL, COMMUNICATIONS & UTIL.	99.4	109.3	104.4	93.7	120.4	107.2	93.3	102.7	98.0	105.8	98.8	102.3	103.8	163.2	133.8	98.6	160.2	129.8
MF2014	WEIGHTED AVERAGE	99.6	120.1	108.6	101.0	122.3	110.4	96.4	112.3	103.4	101.4	109.8	105.1	104.1	146.0	122.5	103.9	136.9	118.4

CALIFORNIA

DIVISION		SAN LUIS OBISPO 934			SAN MATEO 944			SAN RAFAEL 949			SANTA ANA 926 - 927			SANTA BARBARA 931			SANTA CRUZ 950		
		MAT.	INST.	TOTAL	MAT.	INST.	TOTAL	MAT.	INST.	TOTAL	MAT.	INST.	TOTAL	MAT.	INST.	TOTAL	MAT.	INST.	TOTAL
015433	CONTRACTOR EQUIPMENT		98.4	98.4		101.8	101.8		101.4	101.4		99.2	99.2		98.4	98.4		99.4	99.4
0241, 31 - 34	SITE & INFRASTRUCTURE, DEMOLITION	109.8	105.6	106.8	109.2	108.0	108.4	104.1	113.7	110.8	90.6	105.6	101.3	104.0	105.6	105.1	135.0	99.7	110.0
0310	Concrete Forming & Accessories	114.4	120.8	119.9	107.2	146.4	141.1	112.2	146.4	141.8	108.0	120.6	118.9	105.0	121.0	118.8	104.8	135.2	131.1
0320	Concrete Reinforcing	100.1	115.1	107.9	95.6	116.6	106.4	96.2	116.5	106.8	111.0	115.1	113.1	98.3	115.1	107.0	118.6	116.1	117.3
0330	Cast-in-Place Concrete	103.6	121.0	110.5	113.4	126.3	118.5	131.9	125.3	129.3	83.2	121.7	98.4	97.3	121.0	106.6	113.8	125.0	118.2
03	CONCRETE	107.6	118.8	113.0	107.1	132.2	119.3	128.7	131.7	130.1	94.1	119.0	106.2	99.6	118.9	109.0	114.1	127.0	120.4
04	MASONRY	102.4	117.4	111.6	116.5	136.7	128.9	96.8	136.7	121.4	76.4	117.9	102.0	101.1	117.4	111.1	140.6	131.9	135.3
05	METALS	106.3	100.7	104.5	96.0	102.6	98.1	97.3	99.6	98.0	108.2	100.9	105.9	103.9	100.8	102.9	109.8	106.4	108.7
06	WOOD, PLASTICS & COMPOSITES	101.7	118.7	111.2	100.9	150.5	128.5	98.8	150.2	127.4	103.0	118.6	111.7	94.9	118.7	108.2	106.9	136.6	123.4
07	THERMAL & MOISTURE PROTECTION	94.8	117.3	104.2	98.9	135.5	114.2	103.1	135.6	116.7	101.8	117.5	108.4	94.3	118.4	104.3	101.0	129.7	113.0
08	OPENINGS	96.8	115.3	101.1	93.0	137.1	103.2	103.6	136.5	111.3	103.1	117.5	106.6	98.5	118.0	103.1	97.6	130.8	105.3
0920	Plaster & Gypsum Board	105.7	119.0	114.6	102.0	151.3	134.9	103.7	151.3	135.5	106.8	119.0	114.9	99.1	119.1	112.4	109.4	137.3	128.0
0950, 0980	Ceilings & Acoustic Treatment	92.5	119.0	110.0	104.8	151.3	135.6	111.4	151.3	137.8	105.6	119.0	114.5	94.8	119.1	110.9	107.3	137.3	127.1
0960	Flooring	110.2	108.2	109.6	111.2	126.5	115.6	122.3	116.4	120.6	107.8	105.9	107.3	102.3	108.2	104.0	98.9	119.4	104.8
0970, 0990	Wall Finishes & Painting/Coating	92.0	104.3	99.1	111.7	144.3	130.7	107.7	143.9	128.8	91.4	106.8	100.3	92.0	104.3	99.1	94.9	144.3	123.6
09	FINISHES	99.9	116.6	109.1	105.3	144.1	126.8	108.5	142.2	127.1	101.3	116.3	109.6	96.5	116.7	107.7	103.2	134.6	120.6
COVERS	DIVS. 10 - 14, 25, 28, 41, 43, 44, 46	100.0	118.6	104.0	100.0	125.8	105.6	100.0	125.1	105.4	100.0	111.3	102.5	100.0	111.5	102.5	100.0	123.5	105.1
21, 22, 23	FIRE SUPPRESSION, PLUMBING & HVAC	96.7	116.0	104.9	96.7	121.4	107.1	96.7	167.9	126.8	96.6	114.3	104.1	100.1	119.4	108.3	100.0	121.4	109.0
26, 27, 3370	ELECTRICAL, COMMUNICATIONS & UTIL.	92.7	105.5	99.2	103.7	146.9	125.5	100.6	119.0	109.9	93.3	106.8	100.1	91.8	109.7	100.8	97.9	120.4	109.3
MF2014	WEIGHTED AVERAGE	100.2	113.0	105.9	100.5	134.3	115.4	103.2	135.4	117.4	98.3	112.8	104.7	99.2	114.3	105.9	105.9	122.5	113.2

City Cost Indexes

CALIFORNIA / COLORADO

DIVISION		SANTA ROSA 954 MAT.	INST.	TOTAL	STOCKTON 952 MAT.	INST.	TOTAL	SUSANVILLE 961 MAT.	INST.	TOTAL	VALLEJO 945 MAT.	INST.	TOTAL	VAN NUYS 913 - 916 MAT.	INST.	TOTAL	ALAMOSA 811 MAT.	INST.	TOTAL
015433	CONTRACTOR EQUIPMENT		98.7	98.7		98.2	98.2		98.2	98.2		101.4	101.4		95.5	95.5		92.0	92.0
0241, 31 - 34	SITE & INFRASTRUCTURE, DEMOLITION	104.9	104.8	104.8	103.2	104.9	104.4	142.7	104.7	115.9	92.5	113.9	107.6	116.2	106.9	109.6	141.6	87.8	103.6
0310	Concrete Forming & Accessories	101.2	146.3	140.2	102.7	134.3	130.0	107.6	132.4	129.1	102.3	145.9	140.0	111.1	120.5	119.2	105.0	65.8	71.1
0320	Concrete Reinforcing	106.0	116.8	111.6	108.8	115.7	112.4	132.3	115.7	123.7	97.4	116.5	107.3	107.4	115.2	111.5	111.5	75.6	92.9
0330	Cast-in-Place Concrete	108.2	124.4	114.6	96.1	122.8	106.6	108.8	122.8	114.3	105.1	125.1	113.0	89.5	118.0	100.7	98.0	77.1	89.7
03	CONCRETE	114.0	131.7	122.6	103.8	125.5	114.4	130.7	124.7	127.8	104.0	131.4	117.3	108.7	117.6	113.0	118.1	71.9	95.6
04	MASONRY	107.6	140.8	128.0	106.0	126.8	118.8	140.1	126.8	131.9	71.6	136.7	111.7	114.4	120.0	117.9	134.3	71.0	95.3
05	METALS	108.5	104.7	107.3	104.3	101.6	103.4	111.9	101.5	108.6	97.2	99.1	97.8	88.4	99.5	92.0	103.4	79.7	95.9
06	WOOD, PLASTICS & COMPOSITES	93.9	150.0	125.1	100.0	136.5	120.3	109.1	134.0	123.0	87.9	150.2	122.6	90.6	118.3	106.0	94.4	65.2	78.1
07	THERMAL & MOISTURE PROTECTION	102.2	137.0	116.8	104.9	120.9	111.6	128.6	121.1	125.5	100.9	135.9	115.5	96.1	116.1	104.5	104.8	77.6	93.4
08	OPENINGS	104.8	138.2	112.5	105.4	124.4	109.8	123.0	123.1	123.0	105.4	137.0	112.7	90.7	117.3	96.9	95.7	72.4	90.3
0920	Plaster & Gypsum Board	101.5	151.3	134.7	104.8	137.3	126.5	104.4	134.7	124.7	98.4	151.3	133.7	97.3	119.0	111.8	81.1	64.0	69.7
0950, 0980	Ceilings & Acoustic Treatment	105.6	151.3	135.8	113.0	137.3	129.1	125.3	134.7	131.5	113.2	151.3	138.4	99.4	119.0	112.4	96.1	64.0	74.9
0960	Flooring	104.3	112.0	106.5	101.5	110.1	104.0	101.5	117.4	106.1	116.4	126.5	119.3	95.2	107.7	98.8	115.9	52.7	97.7
0970, 0990	Wall Finishes & Painting/Coating	91.4	143.9	122.0	94.4	119.9	109.3	104.0	120.0	113.3	108.7	143.9	129.2	110.9	111.5	111.2	103.9	55.3	75.7
09	FINISHES	100.4	141.3	123.0	103.0	130.3	118.1	109.3	130.1	120.8	105.2	143.9	126.6	100.8	116.9	109.7	103.5	62.1	80.6
COVERS	DIVS. 10 - 14, 25, 28, 41, 43, 44, 46	100.0	124.5	105.3	100.0	123.2	105.0	100.0	122.9	105.0	100.0	125.1	105.4	100.0	110.6	102.3	100.0	88.0	97.4
21, 22, 23	FIRE SUPPRESSION, PLUMBING & HVAC	96.6	171.7	128.8	100.0	115.0	106.3	97.0	110.0	102.5	100.0	133.6	114.2	96.7	115.1	104.5	96.6	71.0	85.8
26, 27, 3370	ELECTRICAL, COMMUNICATIONS & UTIL.	93.6	119.1	106.5	96.0	110.2	103.2	100.0	114.9	107.5	96.9	124.3	110.8	119.8	120.3	120.1	98.2	72.4	85.1
MF2014	WEIGHTED AVERAGE	102.6	136.3	117.4	102.1	117.9	109.0	112.0	117.2	114.3	99.1	128.9	112.2	100.4	114.8	106.7	104.3	73.1	90.5

COLORADO

DIVISION		BOULDER 803 MAT.	INST.	TOTAL	COLORADO SPRINGS 808 - 809 MAT.	INST.	TOTAL	DENVER 800 - 802 MAT.	INST.	TOTAL	DURANGO 813 MAT.	INST.	TOTAL	FORT COLLINS 805 MAT.	INST.	TOTAL	FORT MORGAN 807 MAT.	INST.	TOTAL
015433	CONTRACTOR EQUIPMENT		94.4	94.4		92.2	92.2		97.2	97.2		92.0	92.0		94.4	94.4		94.4	94.4
0241, 31 - 34	SITE & INFRASTRUCTURE, DEMOLITION	95.3	96.3	96.0	97.2	91.7	93.3	96.3	101.0	99.6	134.7	87.8	101.5	107.5	95.4	98.9	97.7	93.7	94.9
0310	Concrete Forming & Accessories	104.3	77.8	81.4	94.6	66.1	69.9	101.2	69.1	73.4	111.6	65.9	72.1	101.8	72.6	76.5	104.7	72.8	77.1
0320	Concrete Reinforcing	96.9	75.8	85.9	96.1	79.5	87.5	96.1	79.5	87.5	111.5	75.7	92.9	97.0	75.7	86.0	97.1	75.6	86.0
0330	Cast-in-Place Concrete	105.5	78.1	94.7	108.4	78.0	96.4	102.8	78.1	92.9	112.6	77.1	98.6	111.9	76.6	102.4	103.5	76.7	92.9
03	CONCRETE	107.0	77.8	92.8	110.5	73.2	92.4	104.5	74.6	90.0	120.1	72.0	96.7	118.1	74.8	97.1	105.3	74.9	90.5
04	MASONRY	94.6	72.6	81.1	95.5	72.4	81.2	97.4	72.9	82.3	120.7	71.7	90.5	112.7	74.7	89.3	109.5	72.0	86.4
05	METALS	98.9	82.5	93.7	102.1	84.7	96.5	104.6	84.6	98.2	103.4	79.9	95.9	100.2	79.4	93.5	98.7	79.3	92.5
06	WOOD, PLASTICS & COMPOSITES	104.7	80.1	91.0	93.9	64.3	77.4	102.3	67.8	83.1	103.8	65.2	82.3	102.0	74.5	86.7	104.7	74.5	87.9
07	THERMAL & MOISTURE PROTECTION	96.9	81.1	90.3	97.8	79.4	90.1	96.2	81.6	90.1	104.8	77.8	93.5	97.3	71.0	86.3	96.9	80.3	89.9
08	OPENINGS	96.3	80.5	92.6	100.4	72.9	94.0	100.9	74.9	94.9	102.8	72.4	95.7	96.3	77.5	91.9	96.2	77.5	91.9
0920	Plaster & Gypsum Board	117.7	79.8	92.4	100.4	63.4	75.7	113.0	67.2	82.4	94.4	64.0	74.1	111.4	74.1	86.5	117.7	74.1	88.6
0950, 0980	Ceilings & Acoustic Treatment	86.8	79.8	82.2	94.2	63.4	73.8	97.1	67.2	77.3	96.1	64.0	74.9	86.8	74.1	78.4	86.8	74.1	78.4
0960	Flooring	113.7	81.8	104.5	104.7	76.7	96.7	109.9	81.8	101.8	121.3	52.7	101.6	109.8	52.7	93.4	114.1	75.5	103.0
0970, 0990	Wall Finishes & Painting/Coating	102.2	66.6	81.5	101.9	55.9	75.2	102.2	69.1	83.0	103.9	55.3	75.7	102.2	40.2	66.1	102.2	53.6	73.9
09	FINISHES	105.9	77.3	90.0	102.3	65.7	82.1	106.2	70.4	86.4	106.2	61.4	81.4	104.3	65.2	82.7	105.9	71.2	86.7
COVERS	DIVS. 10 - 14, 25, 28, 41, 43, 44, 46	100.0	88.9	97.6	100.0	87.3	97.2	100.0	87.6	97.3	100.0	88.0	97.4	100.0	88.1	97.4	100.0	88.1	97.4
21, 22, 23	FIRE SUPPRESSION, PLUMBING & HVAC	96.7	76.5	88.2	100.2	83.8	93.3	100.0	82.5	92.6	96.6	81.9	90.4	100.1	76.1	89.9	96.7	75.0	87.5
26, 27, 3370	ELECTRICAL, COMMUNICATIONS & UTIL.	96.2	83.9	90.0	99.4	81.8	90.5	101.0	83.9	92.3	97.6	68.9	83.1	96.2	83.8	89.9	96.5	83.8	90.1
MF2014	WEIGHTED AVERAGE	99.1	80.3	90.8	101.3	78.7	91.3	101.6	80.4	92.3	104.7	74.9	91.6	102.4	77.6	91.5	99.7	78.0	90.1

COLORADO

DIVISION		GLENWOOD SPRINGS 816 MAT.	INST.	TOTAL	GOLDEN 804 MAT.	INST.	TOTAL	GRAND JUNCTION 815 MAT.	INST.	TOTAL	GREELEY 806 MAT.	INST.	TOTAL	MONTROSE 814 MAT.	INST.	TOTAL	PUEBLO 810 MAT.	INST.	TOTAL
015433	CONTRACTOR EQUIPMENT		95.3	95.3		94.4	94.4		95.3	95.3		94.4	94.4		93.6	93.6		92.0	92.0
0241, 31 - 34	SITE & INFRASTRUCTURE, DEMOLITION	151.0	96.2	112.2	108.2	95.5	99.2	134.3	95.7	107.0	94.2	95.3	95.0	144.3	91.7	107.1	125.9	89.2	99.9
0310	Concrete Forming & Accessories	101.8	73.2	77.0	97.1	72.8	76.1	110.2	72.7	77.7	99.7	76.2	79.4	101.2	73.2	77.0	107.3	66.1	71.7
0320	Concrete Reinforcing	110.4	75.7	92.4	97.1	75.6	86.0	110.8	75.5	92.5	96.9	74.3	85.2	110.3	75.7	92.4	106.7	79.5	92.6
0330	Cast-in-Place Concrete	97.9	76.5	89.5	103.6	76.4	92.9	108.4	75.5	95.4	99.5	57.0	82.8	97.9	76.6	89.5	97.2	78.2	89.7
03	CONCRETE	123.6	75.1	100.0	116.3	74.8	96.1	116.4	74.4	96.0	101.9	69.5	86.2	113.9	75.1	95.0	106.1	73.3	90.2
04	MASONRY	105.7	71.9	84.9	112.2	72.3	87.6	141.9	71.1	98.3	106.5	48.1	70.5	113.3	71.7	87.7	101.9	72.3	83.6
05	METALS	103.0	80.5	95.9	98.8	79.3	92.6	104.7	79.0	96.5	100.1	76.6	92.6	102.1	79.9	95.1	106.3	85.5	99.7
06	WOOD, PLASTICS & COMPOSITES	89.8	74.6	81.4	96.6	74.5	84.3	101.6	74.6	86.6	99.1	80.1	88.5	90.9	74.8	81.9	97.2	64.6	79.0
07	THERMAL & MOISTURE PROTECTION	104.7	78.8	93.9	97.8	73.3	87.5	103.8	66.8	88.3	96.8	64.1	83.1	104.9	78.8	94.0	103.3	78.0	92.7
08	OPENINGS	101.8	77.6	96.2	96.3	77.5	91.9	102.5	77.6	96.7	96.2	80.4	92.6	102.9	77.6	97.0	97.5	73.1	91.8
0920	Plaster & Gypsum Board	123.8	74.1	90.6	109.1	74.1	85.7	136.8	74.1	94.9	109.8	79.8	89.8	80.4	74.1	76.2	85.5	63.4	70.8
0950, 0980	Ceilings & Acoustic Treatment	95.3	74.1	81.3	86.8	74.1	78.4	95.3	74.1	81.3	86.8	79.8	82.2	96.1	74.1	81.5	103.6	63.4	77.0
0960	Flooring	115.0	70.6	102.3	107.4	52.7	91.6	120.6	52.7	101.1	108.6	52.7	92.5	118.6	70.6	104.8	117.2	81.8	107.0
0970, 0990	Wall Finishes & Painting/Coating	103.9	66.5	82.2	102.2	66.6	81.5	103.9	66.6	82.2	102.2	26.5	58.2	103.9	55.3	75.7	103.9	51.8	73.6
09	FINISHES	109.7	71.9	88.8	104.0	68.9	84.6	111.3	68.3	87.5	103.0	67.1	83.1	104.2	70.7	85.7	104.3	66.5	83.4
COVERS	DIVS. 10 - 14, 25, 28, 41, 43, 44, 46	100.0	88.4	97.5	100.0	88.3	97.5	100.0	88.5	97.5	100.0	88.7	97.6	100.0	88.7	97.6	100.0	88.0	97.4
21, 22, 23	FIRE SUPPRESSION, PLUMBING & HVAC	96.6	75.2	87.5	96.7	75.9	87.9	99.9	86.6	94.3	100.1	74.9	89.4	96.6	86.9	92.5	99.9	71.1	87.8
26, 27, 3370	ELECTRICAL, COMMUNICATIONS & UTIL.	95.1	68.9	81.8	96.5	83.8	90.1	97.3	53.4	75.1	96.2	83.8	89.9	97.3	55.9	76.4	98.2	72.5	85.2
MF2014	WEIGHTED AVERAGE	104.5	76.4	92.1	101.2	77.8	90.9	106.5	75.6	92.9	99.7	74.0	88.4	103.4	76.5	91.6	102.4	74.7	90.2

For customer support on your Commercial Renovation Cost Data, call 877.742.4719.

767

City Cost Indexes

Table 1

DIVISION		COLORADO SALIDA 812			CONNECTICUT BRIDGEPORT 066			BRISTOL 060			HARTFORD 061			MERIDEN 064			NEW BRITAIN 060		
		MAT.	INST.	TOTAL	MAT.	INST.	TOTAL	MAT.	INST.	TOTAL	MAT.	INST.	TOTAL	MAT.	INST.	TOTAL	MAT.	INST.	TOTAL
015433	CONTRACTOR EQUIPMENT		93.6	93.6		98.1	98.1		98.1	98.1		98.1	98.1		98.6	98.6		98.1	98.1
0241, 31 - 34	SITE & INFRASTRUCTURE, DEMOLITION	134.4	91.7	104.2	108.6	103.1	104.7	107.8	103.1	104.5	103.0	103.1	103.1	105.6	103.8	104.4	108.0	103.1	104.5
0310	Concrete Forming & Accessories	110.1	73.0	78.0	102.8	124.8	121.5	102.8	124.7	121.7	100.4	124.7	121.4	102.4	124.7	121.7	103.2	124.6	121.8
0320	Concrete Reinforcing	110.1	75.7	92.2	105.7	129.4	117.9	105.7	129.4	117.9	101.3	129.4	115.8	105.7	129.4	117.9	105.7	129.3	117.9
0330	Cast-in-Place Concrete	112.2	76.5	98.1	97.9	130.2	110.6	91.8	130.2	106.9	96.1	130.2	109.5	88.3	130.2	104.8	93.2	130.1	107.8
03	CONCRETE	114.7	75.0	95.4	102.0	127.1	114.2	98.9	127.1	112.6	99.3	127.1	112.8	97.2	127.1	111.7	99.7	127.1	113.0
04	MASONRY	143.0	71.7	99.1	120.3	139.8	132.3	110.7	139.8	128.6	118.7	139.8	131.7	110.3	139.8	128.5	113.2	139.8	129.6
05	METALS	101.8	79.7	94.8	95.2	125.6	104.9	95.2	125.5	104.8	99.7	125.5	107.9	92.1	125.5	102.7	91.3	125.4	102.1
06	WOOD, PLASTICS & COMPOSITES	98.1	74.8	85.1	104.0	122.4	114.2	104.0	122.4	114.2	89.9	122.4	108.0	104.0	122.4	114.2	104.0	122.4	114.2
07	THERMAL & MOISTURE PROTECTION	103.8	78.8	93.4	98.4	130.8	112.0	98.5	129.4	111.5	103.9	129.4	114.6	98.5	129.4	111.4	98.7	126.6	110.3
08	OPENINGS	95.8	77.6	91.6	97.8	132.5	105.8	97.8	132.5	105.8	97.3	132.5	105.4	100.1	132.5	107.6	97.8	132.5	105.8
0920	Plaster & Gypsum Board	80.8	74.1	76.3	100.8	122.5	115.3	100.8	122.5	115.3	96.6	122.5	113.9	102.0	122.5	115.7	100.8	122.5	115.3
0950, 0980	Ceilings & Acoustic Treatment	96.1	74.1	81.5	86.2	122.5	110.2	86.2	122.5	110.2	88.5	122.5	111.0	89.3	122.5	111.3	86.2	122.5	110.2
0960	Flooring	124.2	70.6	108.8	98.2	133.3	108.3	98.2	129.6	107.2	96.5	129.6	106.0	98.2	129.6	107.2	98.2	129.6	107.2
0970, 0990	Wall Finishes & Painting/Coating	103.9	55.3	75.7	93.0	126.0	112.2	93.0	126.0	112.2	93.9	126.0	112.5	93.0	126.0	112.2	93.0	126.0	112.2
09	FINISHES	104.8	70.7	85.9	94.6	125.8	111.9	94.7	125.2	111.6	93.4	125.2	111.0	95.5	125.2	112.0	94.7	125.2	111.6
COVERS	DIVS. 10 - 14, 25, 28, 41, 43, 44, 46	100.0	88.8	97.6	100.0	113.0	102.8	100.0	113.0	102.8	100.0	113.0	102.8	100.0	113.0	102.8	100.0	113.0	102.8
21, 22, 23	FIRE SUPPRESSION, PLUMBING & HVAC	96.6	71.0	85.8	99.9	118.8	107.9	99.9	118.8	107.9	99.9	118.8	107.9	96.6	118.8	105.9	99.9	118.7	107.9
26, 27, 3370	ELECTRICAL, COMMUNICATIONS & UTIL.	97.5	72.4	84.8	95.9	111.4	103.7	95.9	113.4	104.8	95.4	113.8	104.7	95.8	112.8	104.4	95.9	113.4	104.8
MF2014	WEIGHTED AVERAGE	103.9	75.4	91.3	99.5	122.0	109.4	98.7	122.2	109.6	99.2	122.2	109.6	97.5	122.2	108.4	98.3	122.1	108.8

Table 2

DIVISION		CONNECTICUT NEW HAVEN 065			NEW LONDON 063			NORWALK 068			STAMFORD 069			WATERBURY 067			WILLIMANTIC 062		
		MAT.	INST.	TOTAL	MAT.	INST.	TOTAL	MAT.	INST.	TOTAL	MAT.	INST.	TOTAL	MAT.	INST.	TOTAL	MAT.	INST.	TOTAL
015433	CONTRACTOR EQUIPMENT		98.6	98.6		98.6	98.6		98.1	98.1		98.1	98.1		98.1	98.1		98.1	98.1
0241, 31 - 34	SITE & INFRASTRUCTURE, DEMOLITION	107.8	103.8	105.0	100.2	103.8	102.8	108.4	103.1	104.7	109.0	103.1	104.9	108.4	103.1	104.7	108.4	103.1	104.6
0310	Concrete Forming & Accessories	102.5	124.7	121.7	102.4	124.8	121.7	102.8	125.2	122.1	102.7	125.2	122.1	102.8	124.7	121.7	102.7	124.7	121.8
0320	Concrete Reinforcing	105.7	129.4	117.9	82.8	129.4	106.9	105.7	129.5	118.0	105.7	129.5	118.0	105.7	129.3	117.9	105.7	129.4	117.9
0330	Cast-in-Place Concrete	94.8	131.0	109.1	80.9	131.0	100.6	96.3	131.6	110.2	97.9	131.6	111.2	97.9	130.1	110.6	91.5	130.2	106.7
03	CONCRETE	113.8	127.4	120.4	87.5	127.4	106.9	101.2	127.8	114.1	102.0	127.8	114.5	102.0	127.1	114.2	98.8	127.1	112.6
04	MASONRY	111.0	139.8	128.7	109.4	139.8	128.1	110.4	141.3	129.4	111.2	141.3	129.7	111.2	139.8	128.8	110.6	139.8	128.6
05	METALS	91.5	125.5	102.3	91.2	125.5	102.1	95.2	126.1	105.0	95.2	126.1	105.0	95.2	125.5	104.8	94.9	125.5	104.7
06	WOOD, PLASTICS & COMPOSITES	104.0	122.4	114.2	104.0	122.4	114.2	104.0	122.4	114.2	104.0	122.4	114.2	104.0	122.4	114.2	104.0	122.4	114.2
07	THERMAL & MOISTURE PROTECTION	98.6	128.8	111.2	98.4	129.6	111.5	98.6	131.4	112.3	98.5	131.4	112.3	98.5	128.6	111.1	98.7	129.6	111.6
08	OPENINGS	97.8	132.5	105.8	100.2	132.5	107.7	97.8	132.5	105.8	97.8	132.5	105.8	97.8	132.5	105.8	100.2	132.5	107.7
0920	Plaster & Gypsum Board	100.8	122.5	115.3	100.8	122.5	115.3	100.8	122.5	115.3	100.8	122.5	115.3	100.8	122.5	115.3	100.8	122.5	115.3
0950, 0980	Ceilings & Acoustic Treatment	86.2	122.5	110.2	84.3	122.5	109.6	86.2	122.5	110.2	86.2	122.5	110.2	86.2	122.5	110.2	84.3	122.5	109.6
0960	Flooring	98.2	133.3	108.3	98.2	133.3	108.3	98.2	129.6	107.2	98.2	129.6	107.2	98.2	129.6	107.2	98.2	133.3	108.3
0970, 0990	Wall Finishes & Painting/Coating	93.0	126.0	112.2	93.0	130.3	114.7	93.0	126.0	112.2	93.0	126.0	112.2	93.0	126.0	112.2	93.0	126.0	112.2
09	FINISHES	94.7	125.8	111.9	93.8	126.3	111.8	94.7	125.2	111.6	94.8	125.2	111.6	94.6	125.2	111.5	94.4	125.8	111.8
COVERS	DIVS. 10 - 14, 25, 28, 41, 43, 44, 46	100.0	113.0	102.8	100.0	113.0	102.8	100.0	113.2	102.9	100.0	113.2	102.9	100.0	113.0	102.8	100.0	113.0	102.8
21, 22, 23	FIRE SUPPRESSION, PLUMBING & HVAC	99.9	118.8	107.9	96.6	118.8	106.0	99.9	118.8	107.9	99.9	118.8	107.9	99.9	118.7	107.9	99.9	118.8	107.9
26, 27, 3370	ELECTRICAL, COMMUNICATIONS & UTIL.	95.8	112.8	104.4	93.0	113.4	103.4	95.9	165.9	131.3	95.9	165.9	131.3	95.4	111.4	103.5	95.9	113.4	104.8
MF2014	WEIGHTED AVERAGE	99.8	122.3	109.7	95.6	122.4	107.4	99.0	129.7	112.5	99.1	129.7	112.6	99.0	121.9	109.1	98.9	122.3	109.2

Table 3

DIVISION		D.C. WASHINGTON 200 - 205			DELAWARE DOVER 199			NEWARK 197			WILMINGTON 198			FLORIDA DAYTONA BEACH 321			FORT LAUDERDALE 333		
		MAT.	INST.	TOTAL	MAT.	INST.	TOTAL	MAT.	INST.	TOTAL	MAT.	INST.	TOTAL	MAT.	INST.	TOTAL	MAT.	INST.	TOTAL
015433	CONTRACTOR EQUIPMENT		108.3	108.3		117.5	117.5		117.5	117.5		117.7	117.7		99.3	99.3		92.1	92.1
0241, 31 - 34	SITE & INFRASTRUCTURE, DEMOLITION	102.8	97.6	99.1	104.3	111.7	109.6	103.3	111.8	109.3	103.5	112.2	109.6	103.1	89.0	93.7	95.1	77.3	82.5
0310	Concrete Forming & Accessories	99.3	93.4	94.2	98.6	101.4	101.0	99.8	101.4	101.2	99.0	101.4	101.1	101.0	70.2	74.4	97.5	66.7	70.8
0320	Concrete Reinforcing	101.8	93.0	97.3	101.0	101.8	101.4	97.6	101.8	99.8	101.0	101.8	101.4	94.4	74.4	84.0	90.8	68.5	79.2
0330	Cast-in-Place Concrete	112.7	87.2	102.6	98.9	103.8	100.8	86.0	103.8	93.0	93.7	103.8	97.7	90.1	72.8	83.2	95.4	70.3	85.5
03	CONCRETE	106.6	92.2	99.6	101.4	103.3	102.3	94.0	103.3	98.5	98.9	103.3	101.0	94.3	73.1	84.0	97.5	69.6	83.9
04	MASONRY	96.6	85.3	89.6	113.3	100.7	105.5	108.0	100.7	103.5	112.4	100.7	105.2	96.1	65.5	77.2	102.0	69.9	82.2
05	METALS	101.2	109.0	103.7	99.5	117.0	105.1	101.6	117.1	106.5	99.5	117.1	105.1	98.9	92.0	96.7	97.4	89.2	94.8
06	WOOD, PLASTICS & COMPOSITES	95.2	96.6	96.0	92.1	100.2	96.6	100.1	100.2	100.2	87.4	100.2	94.5	100.8	69.9	83.6	87.9	67.4	76.5
07	THERMAL & MOISTURE PROTECTION	104.4	89.0	97.9	105.6	114.3	109.2	107.9	114.3	110.6	104.6	114.3	108.6	98.0	73.8	87.9	104.4	71.1	90.4
08	OPENINGS	98.7	98.5	98.7	89.1	109.4	93.9	93.1	109.4	96.9	87.0	109.4	92.2	95.7	69.9	89.7	95.4	67.0	88.8
0920	Plaster & Gypsum Board	103.2	96.5	98.7	102.8	100.2	101.1	104.6	100.2	101.6	102.0	100.2	100.8	85.8	69.4	74.9	95.8	66.8	76.5
0950, 0980	Ceilings & Acoustic Treatment	104.3	96.5	99.1	93.0	100.2	97.7	90.5	100.2	96.9	92.4	100.2	97.6	83.3	69.4	74.1	88.7	66.8	74.2
0960	Flooring	94.5	93.7	94.2	95.2	108.6	99.0	92.2	108.6	96.9	94.7	108.6	98.7	108.7	63.8	95.8	104.6	77.3	96.8
0970, 0990	Wall Finishes & Painting/Coating	102.9	80.9	90.1	90.7	109.5	101.7	89.3	109.5	101.1	85.4	109.5	99.5	97.4	70.6	81.8	89.2	70.6	78.4
09	FINISHES	96.2	92.6	94.2	94.1	102.9	99.0	93.1	102.9	98.5	93.9	102.9	98.9	94.4	68.4	80.0	94.7	68.7	80.3
COVERS	DIVS. 10 - 14, 25, 28, 41, 43, 44, 46	100.0	101.0	100.2	100.0	101.2	100.3	100.0	101.2	100.3	100.0	101.2	100.3	100.0	86.6	97.1	100.0	86.6	97.1
21, 22, 23	FIRE SUPPRESSION, PLUMBING & HVAC	100.0	92.6	96.9	100.0	116.9	107.2	100.3	116.9	107.3	100.2	116.9	107.3	100.0	78.0	90.7	100.0	62.4	84.1
26, 27, 3370	ELECTRICAL, COMMUNICATIONS & UTIL.	100.5	105.0	102.8	97.1	111.1	104.1	98.5	111.1	104.8	100.2	111.1	105.7	94.2	68.1	81.0	94.8	72.2	83.3
MF2014	WEIGHTED AVERAGE	100.6	95.9	98.5	99.0	109.4	103.6	98.9	109.4	103.5	98.7	109.4	103.4	97.5	75.5	87.8	97.8	71.2	86.1

768

City Cost Indexes

FLORIDA

DIVISION		FORT MYERS 339,341			GAINESVILLE 326,344			JACKSONVILLE 320,322			LAKELAND 338			MELBOURNE 329			MIAMI 330 - 332,340		
		MAT.	INST.	TOTAL	MAT.	INST.	TOTAL	MAT.	INST.	TOTAL	MAT.	INST.	TOTAL	MAT.	INST.	TOTAL	MAT.	INST.	TOTAL
015433	CONTRACTOR EQUIPMENT		99.3	99.3		99.3	99.3		99.3	99.3		99.3	99.3		99.3	99.3		92.1	92.1
0241, 31 - 34	SITE & INFRASTRUCTURE, DEMOLITION	106.0	89.3	94.2	110.9	89.6	95.8	103.1	89.6	93.6	107.9	89.6	95.0	109.9	89.9	95.8	96.4	77.8	83.3
0310	Concrete Forming & Accessories	93.4	66.2	69.9	95.8	60.5	65.3	100.8	60.5	65.9	89.8	68.8	71.6	97.1	71.0	74.5	102.2	69.8	74.2
0320	Concrete Reinforcing	91.8	67.8	79.4	100.2	67.9	83.4	94.4	67.8	80.7	94.1	89.3	91.6	95.5	75.5	85.1	97.5	68.2	82.3
0330	Cast-in-Place Concrete	99.5	70.1	87.9	103.4	70.7	90.5	91.0	70.6	82.9	101.8	72.0	90.0	108.6	73.8	94.9	92.1	75.3	85.5
03	CONCRETE	98.2	69.2	84.1	105.7	66.9	86.8	94.7	66.9	81.2	100.0	74.9	87.8	106.1	74.0	90.5	97.4	72.6	85.3
04	MASONRY	95.2	58.6	72.7	110.2	60.9	79.8	95.8	60.9	74.3	112.5	75.4	89.7	93.8	67.4	77.5	102.5	74.7	85.4
05	METALS	99.5	88.0	95.8	97.8	89.1	95.1	97.5	89.0	94.8	99.4	97.4	98.8	107.1	92.4	102.4	97.9	88.7	95.0
06	WOOD, PLASTICS & COMPOSITES	84.7	67.4	75.1	94.3	59.0	74.7	100.8	59.0	77.5	80.1	68.9	73.8	96.1	69.9	81.5	93.7	67.4	79.0
07	THERMAL & MOISTURE PROTECTION	104.2	68.1	89.1	98.3	68.2	85.7	98.2	68.2	85.7	104.1	73.1	91.1	98.4	72.3	87.5	104.0	73.8	91.4
08	OPENINGS	96.7	67.0	89.7	95.3	62.4	87.7	95.7	62.4	87.9	96.6	72.9	91.1	94.9	70.2	89.2	98.9	67.0	91.5
0920	Plaster & Gypsum Board	91.9	66.8	75.2	82.3	58.3	66.2	85.8	58.3	67.4	88.9	68.4	75.2	82.3	69.4	73.7	93.8	66.8	75.8
0950, 0980	Ceilings & Acoustic Treatment	84.4	66.8	72.8	77.8	58.3	64.9	83.3	58.3	66.7	84.4	68.4	73.8	82.4	68.4	73.8	95.0	66.8	75.7
0960	Flooring	101.4	53.0	87.5	105.8	70.8	95.7	108.7	62.0	95.2	99.1	54.3	86.2	106.0	63.8	93.9	106.5	70.2	96.0
0970, 0990	Wall Finishes & Painting/Coating	93.3	70.6	80.1	97.4	70.6	81.8	97.4	70.6	81.8	93.3	70.6	80.1	97.4	93.2	95.0	86.6	70.6	77.3
09	FINISHES	94.2	63.8	77.4	92.7	62.5	76.0	94.5	60.8	75.8	93.2	65.7	78.0	93.6	71.3	81.2	96.5	69.5	81.6
COVERS	DIVS. 10 - 14, 25, 28, 41, 43, 44, 46	100.0	81.6	96.0	100.0	84.3	96.6	100.0	82.4	96.2	100.0	82.7	96.3	100.0	87.2	97.2	100.0	89.6	97.8
21, 22, 23	FIRE SUPPRESSION, PLUMBING & HVAC	98.2	62.0	82.9	98.7	65.5	84.7	100.0	65.5	85.4	98.2	63.5	83.5	100.0	80.4	91.7	100.0	60.0	83.1
26, 27, 3370	ELECTRICAL, COMMUNICATIONS & UTIL.	96.6	63.0	79.6	94.5	60.9	77.5	93.9	68.7	81.2	95.0	60.4	77.5	95.2	67.3	81.1	98.3	75.3	86.6
MF2014	WEIGHTED AVERAGE	98.1	68.8	85.1	99.0	68.8	85.7	97.3	69.6	85.1	98.8	72.7	87.3	100.1	76.6	89.8	98.9	72.3	87.2

FLORIDA

DIVISION		ORLANDO 327 - 328,347			PANAMA CITY 324			PENSACOLA 325			SARASOTA 342			ST. PETERSBURG 337			TALLAHASSEE 323		
		MAT.	INST.	TOTAL	MAT.	INST.	TOTAL	MAT.	INST.	TOTAL	MAT.	INST.	TOTAL	MAT.	INST.	TOTAL	MAT.	INST.	TOTAL
015433	CONTRACTOR EQUIPMENT		99.3	99.3		99.3	99.3		99.3	99.3		99.3	99.3		99.3	99.3		99.3	99.3
0241, 31 - 34	SITE & INFRASTRUCTURE, DEMOLITION	102.4	89.8	93.5	114.4	89.1	96.5	114.7	89.5	96.9	113.8	89.6	96.7	109.5	89.4	95.3	100.6	89.2	92.5
0310	Concrete Forming & Accessories	104.1	69.9	74.6	100.1	48.9	55.8	98.0	76.3	79.3	97.6	68.8	72.7	97.0	66.1	70.3	102.7	58.1	64.1
0320	Concrete Reinforcing	100.2	75.9	87.6	98.6	81.8	89.9	101.1	82.2	91.3	92.6	89.3	90.9	94.1	89.0	91.5	99.8	67.7	83.2
0330	Cast-in-Place Concrete	100.9	72.6	89.7	95.6	67.1	84.4	118.1	73.0	100.3	106.8	72.0	93.1	102.9	67.9	89.1	99.2	67.3	86.6
03	CONCRETE	101.1	73.3	87.6	103.8	63.1	84.0	114.1	77.4	96.2	102.7	75.0	89.2	101.7	72.4	87.4	101.0	64.6	83.3
04	MASONRY	99.2	65.5	78.4	100.7	47.1	67.6	121.9	64.4	86.4	98.2	75.4	84.2	154.9	62.9	98.2	99.3	53.6	71.1
05	METALS	97.9	92.3	96.1	98.7	93.1	96.9	99.8	94.2	98.0	100.3	97.5	99.4	100.3	97.7	99.5	95.5	88.7	93.3
06	WOOD, PLASTICS & COMPOSITES	93.5	69.9	80.4	99.6	46.9	70.3	97.4	78.4	86.8	99.2	68.9	82.3	89.3	68.9	77.9	100.6	59.0	77.4
07	THERMAL & MOISTURE PROTECTION	98.9	73.8	88.4	98.5	61.1	82.9	98.4	72.5	87.6	99.8	73.1	88.6	104.3	68.4	89.3	99.8	65.0	85.3
08	OPENINGS	101.9	70.3	94.6	93.7	59.4	85.7	93.7	76.6	89.7	99.2	72.9	93.1	95.4	72.9	90.1	100.1	62.4	91.3
0920	Plaster & Gypsum Board	89.4	69.4	76.0	85.0	45.9	58.8	88.0	78.3	81.5	91.5	68.4	76.1	94.3	68.4	77.0	98.0	58.3	71.5
0950, 0980	Ceilings & Acoustic Treatment	90.4	69.4	76.5	82.4	45.9	58.2	82.4	78.3	79.7	85.3	68.4	74.1	86.3	68.4	74.4	90.6	58.3	69.2
0960	Flooring	105.1	63.8	93.2	108.2	41.9	89.1	103.5	60.5	91.1	110.1	55.6	94.4	103.4	68.2	93.3	106.9	59.6	93.3
0970, 0990	Wall Finishes & Painting/Coating	96.9	70.6	81.6	97.4	70.6	81.8	97.4	70.6	81.8	98.3	70.6	82.2	93.3	73.4	81.7	92.1	70.6	79.6
09	FINISHES	96.5	68.4	81.0	95.2	48.2	69.2	93.9	73.1	82.4	98.7	65.9	80.6	95.8	66.8	79.8	97.0	58.8	75.9
COVERS	DIVS. 10 - 14, 25, 28, 41, 43, 44, 46	100.0	86.6	97.1	100.0	75.1	94.6	100.0	81.9	96.1	100.0	82.7	96.3	100.0	80.1	95.7	100.0	78.3	95.3
21, 22, 23	FIRE SUPPRESSION, PLUMBING & HVAC	100.0	61.2	83.6	100.0	52.9	80.1	100.0	58.1	82.3	99.9	63.3	84.4	100.0	59.5	82.9	100.1	56.3	81.6
26, 27, 3370	ELECTRICAL, COMMUNICATIONS & UTIL.	98.6	67.8	83.0	93.0	60.6	76.6	96.5	55.4	75.7	96.4	68.7	82.4	95.0	60.4	77.5	97.9	60.9	79.1
MF2014	WEIGHTED AVERAGE	99.5	72.0	87.4	98.8	62.0	82.6	101.4	71.0	88.0	100.0	73.8	88.5	101.8	70.2	87.9	98.9	65.0	84.0

FLORIDA / GEORGIA

DIVISION		TAMPA 335 - 336,346			WEST PALM BEACH 334,349			ALBANY 317,398			ATHENS 306			ATLANTA 300 - 303,399			AUGUSTA 308 - 309		
		MAT.	INST.	TOTAL	MAT.	INST.	TOTAL	MAT.	INST.	TOTAL	MAT.	INST.	TOTAL	MAT.	INST.	TOTAL	MAT.	INST.	TOTAL
015433	CONTRACTOR EQUIPMENT		99.3	99.3		92.1	92.1		93.1	93.1		94.5	94.5		95.0	95.0		94.5	94.5
0241, 31 - 34	SITE & INFRASTRUCTURE, DEMOLITION	110.0	89.6	95.6	92.0	77.3	81.6	100.3	78.6	84.9	102.0	94.2	96.5	98.7	95.8	96.7	95.5	94.4	94.7
0310	Concrete Forming & Accessories	100.0	68.8	73.0	101.0	66.5	71.1	94.4	42.5	49.5	94.5	45.7	52.3	98.6	81.2	83.6	95.8	77.4	79.9
0320	Concrete Reinforcing	90.8	89.3	90.0	93.4	67.0	79.7	90.3	79.4	84.6	95.4	77.1	85.9	94.7	80.3	87.2	95.8	73.4	84.2
0330	Cast-in-Place Concrete	100.5	72.0	89.3	90.7	70.2	82.6	91.2	70.4	83.0	108.1	71.5	93.7	108.1	71.9	93.8	102.1	71.7	90.1
03	CONCRETE	100.2	75.0	87.9	94.1	69.2	82.0	94.9	60.6	78.2	102.7	61.2	82.5	100.4	77.8	89.4	95.7	74.8	85.5
04	MASONRY	103.0	75.4	86.0	101.5	61.7	77.0	98.8	51.6	69.7	80.0	55.1	64.6	93.6	66.4	76.8	93.9	54.4	69.5
05	METALS	99.4	97.4	98.7	96.5	88.1	93.8	98.4	87.4	95.0	94.0	73.7	87.5	94.9	76.2	89.0	93.7	73.1	87.1
06	WOOD, PLASTICS & COMPOSITES	93.3	68.9	79.7	92.8	67.4	78.6	88.8	34.8	58.7	98.0	39.5	65.4	102.3	86.2	93.3	99.6	82.4	90.0
07	THERMAL & MOISTURE PROTECTION	104.6	73.1	91.4	104.1	69.6	89.7	98.6	59.7	82.3	95.3	55.0	78.5	95.1	70.7	84.9	94.9	66.0	82.8
08	OPENINGS	96.6	72.9	91.1	94.9	66.7	88.4	90.1	49.5	80.6	88.7	51.6	80.0	93.8	77.6	90.0	88.7	74.3	85.4
0920	Plaster & Gypsum Board	96.5	68.4	77.7	99.9	66.8	77.8	96.6	33.4	54.4	94.9	38.1	57.0	97.3	86.1	89.8	96.3	82.2	86.9
0950, 0980	Ceilings & Acoustic Treatment	88.7	68.4	75.3	84.4	66.8	72.8	82.2	33.4	49.9	96.1	38.1	57.7	96.1	86.1	89.5	97.0	82.2	87.2
0960	Flooring	104.6	54.3	90.2	106.6	63.9	94.3	107.1	45.9	89.5	98.6	51.8	85.1	100.2	73.2	92.4	98.8	50.5	84.9
0970, 0990	Wall Finishes & Painting/Coating	93.3	70.6	80.1	89.2	70.6	78.4	90.1	68.3	77.4	96.8	67.8	79.9	96.8	68.3	80.2	96.8	68.3	80.2
09	FINISHES	97.1	65.7	79.7	94.6	66.0	78.8	96.8	43.2	67.1	99.2	47.1	70.4	99.6	79.3	88.4	99.0	72.5	84.3
COVERS	DIVS. 10 - 14, 25, 28, 41, 43, 44, 46	100.0	84.8	96.7	100.0	86.6	97.1	100.0	80.8	95.8	100.0	81.2	95.9	100.0	86.8	97.1	100.0	86.0	97.0
21, 22, 23	FIRE SUPPRESSION, PLUMBING & HVAC	100.0	63.5	84.6	98.2	61.9	82.9	100.0	69.3	87.0	96.7	69.2	85.1	100.0	73.6	88.9	100.1	61.4	83.7
26, 27, 3370	ELECTRICAL, COMMUNICATIONS & UTIL.	94.7	60.4	77.4	95.8	73.3	84.4	93.3	60.4	76.6	100.9	69.0	84.8	100.2	70.9	85.4	101.5	65.4	83.2
MF2014	WEIGHTED AVERAGE	99.3	72.8	87.6	96.9	69.9	85.0	96.9	63.1	82.0	96.3	65.2	82.6	98.1	76.5	88.6	96.8	69.9	85.0

GEORGIA

DIVISION		COLUMBUS 318-319			DALTON 307			GAINESVILLE 305			MACON 310-312			SAVANNAH 313-314			STATESBORO 304		
		MAT.	INST.	TOTAL	MAT.	INST.	TOTAL	MAT.	INST.	TOTAL	MAT.	INST.	TOTAL	MAT.	INST.	TOTAL	MAT.	INST.	TOTAL
015433	CONTRACTOR EQUIPMENT		93.1	93.1		108.2	108.2		94.5	94.5		103.3	103.3		94.0	94.0		95.3	95.3
0241, 31 - 34	SITE & INFRASTRUCTURE, DEMOLITION	100.2	78.4	84.8	102.0	99.5	100.2	102.0	94.1	96.4	101.9	93.6	96.0	100.4	79.7	85.8	103.0	78.4	85.6
0310	Concrete Forming & Accessories	94.4	49.9	55.9	87.1	49.4	54.5	98.3	43.0	50.5	93.9	48.0	54.2	95.6	59.2	64.1	81.6	54.1	57.8
0320	Concrete Reinforcing	90.3	80.0	85.0	94.9	73.2	83.6	95.2	77.1	85.8	91.5	79.4	85.3	97.5	73.9	85.3	94.5	41.5	67.0
0330	Cast-in-Place Concrete	90.9	70.2	82.7	105.0	70.5	91.4	113.6	71.3	96.9	89.6	71.5	82.5	94.7	70.4	85.1	107.9	70.1	93.0
03	CONCRETE	94.7	64.0	79.8	101.6	62.6	82.7	104.7	59.8	82.9	94.3	63.5	79.3	97.9	67.1	82.9	102.0	58.9	81.0
04	MASONRY	98.9	62.9	76.7	81.1	46.1	59.5	88.6	55.5	68.2	112.0	51.6	74.8	95.6	62.6	75.3	82.9	49.8	62.5
05	METALS	98.0	87.7	94.7	94.9	85.1	91.8	93.2	73.1	86.8	93.8	86.9	91.6	94.9	86.4	92.2	98.4	75.7	91.2
06	WOOD, PLASTICS & COMPOSITES	88.8	44.8	64.3	80.6	45.8	61.2	102.1	36.6	65.6	95.9	42.3	66.1	95.4	57.3	74.2	73.8	52.0	61.7
07	THERMAL & MOISTURE PROTECTION	98.5	63.6	83.9	97.3	60.0	81.7	95.3	62.2	81.4	97.0	62.3	82.5	96.3	64.0	82.8	95.9	60.5	81.1
08	OPENINGS	90.1	55.0	81.9	89.5	54.4	81.3	88.7	50.0	79.7	89.8	53.7	81.4	97.7	60.6	89.1	90.4	50.7	81.1
0920	Plaster & Gypsum Board	96.6	43.7	61.2	83.6	44.6	57.6	97.3	35.1	55.7	100.7	41.1	60.9	94.8	56.5	69.2	82.9	51.0	61.6
0950, 0980	Ceilings & Acoustic Treatment	82.2	43.7	56.7	108.1	44.6	66.1	96.1	35.1	55.7	77.5	41.1	53.4	89.7	56.5	67.7	104.5	51.0	69.1
0960	Flooring	107.1	45.9	89.5	99.6	45.1	83.9	100.2	44.8	84.2	84.3	45.9	73.2	105.0	64.6	93.4	117.3	43.1	96.0
0970, 0990	Wall Finishes & Painting/Coating	90.1	68.3	77.4	87.5	68.3	76.3	96.8	68.3	80.2	92.2	68.3	78.3	88.8	68.3	76.9	94.5	68.3	79.3
09	FINISHES	96.7	49.3	70.5	108.3	48.9	75.5	99.8	43.4	68.6	85.4	47.7	64.6	96.7	60.3	76.6	111.5	52.9	79.1
COVERS	DIVS. 10 - 14, 25, 28, 41, 43, 44, 46	100.0	82.0	96.1	100.0	27.7	84.3	100.0	38.7	86.7	100.0	81.8	96.1	100.0	82.9	96.3	100.0	45.2	88.1
21, 22, 23	FIRE SUPPRESSION, PLUMBING & HVAC	100.0	64.1	84.8	96.7	61.7	81.9	96.7	69.6	85.3	100.0	70.3	87.5	100.1	64.0	84.8	97.3	66.8	84.4
26, 27, 3370	ELECTRICAL, COMMUNICATIONS & UTIL.	93.4	69.6	81.4	109.4	67.2	88.1	100.9	70.7	85.6	92.5	63.0	77.6	97.2	64.2	80.5	101.0	56.9	78.7
MF2014	WEIGHTED AVERAGE	96.8	66.1	83.3	98.0	63.0	82.6	96.9	63.6	82.2	95.8	66.2	82.7	97.8	67.6	84.5	98.2	61.0	81.8

DIVISION		GEORGIA VALDOSTA 316			GEORGIA WAYCROSS 315			HAWAII HILO 967			HAWAII HONOLULU 968			HAWAII STATES & POSS., GUAM 969			IDAHO BOISE 836 - 837		
		MAT.	INST.	TOTAL	MAT.	INST.	TOTAL	MAT.	INST.	TOTAL	MAT.	INST.	TOTAL	MAT.	INST.	TOTAL	MAT.	INST.	TOTAL
015433	CONTRACTOR EQUIPMENT		93.1	93.1		93.1	93.1		98.9	98.9		98.9	98.9		165.6	165.6		96.7	96.7
0241, 31 - 34	SITE & INFRASTRUCTURE, DEMOLITION	110.0	78.5	87.8	106.7	77.3	85.9	156.6	105.9	120.8	166.0	105.9	123.5	201.7	103.2	132.0	88.3	95.3	93.3
0310	Concrete Forming & Accessories	84.3	43.3	48.9	86.3	67.4	70.0	109.6	135.3	131.8	122.2	135.3	133.5	112.1	62.1	68.9	100.9	75.6	79.0
0320	Concrete Reinforcing	92.3	75.6	83.6	92.3	71.9	81.8	137.7	116.2	126.6	160.1	116.2	137.4	241.9	29.9	132.1	105.3	79.9	92.1
0330	Cast-in-Place Concrete	89.3	70.3	81.8	100.9	70.0	88.7	201.0	126.7	171.7	161.7	126.7	147.9	174.3	106.8	147.6	88.7	88.4	88.6
03	CONCRETE	100.1	60.3	80.8	103.2	70.1	87.1	165.3	127.6	146.9	156.0	127.6	142.2	167.5	71.9	121.1	102.7	80.9	92.1
04	MASONRY	104.6	50.4	71.2	105.4	49.7	71.1	162.6	129.3	142.0	160.0	129.3	141.1	229.1	43.4	114.6	130.2	87.3	103.8
05	METALS	97.6	85.9	93.9	96.6	80.3	91.4	120.1	105.4	115.5	134.0	105.4	125.0	154.9	76.7	130.0	107.6	80.4	98.9
06	WOOD, PLASTICS & COMPOSITES	76.6	36.3	54.2	78.3	70.0	73.6	111.1	138.5	126.3	127.4	138.5	133.6	123.4	65.5	91.1	89.6	72.5	80.1
07	THERMAL & MOISTURE PROTECTION	98.8	58.6	82.0	98.6	62.0	83.3	127.2	123.6	125.7	144.7	123.6	135.9	149.2	65.2	114.0	94.7	80.2	88.6
08	OPENINGS	86.5	49.6	77.9	86.8	64.6	81.6	120.2	131.1	122.7	133.9	131.1	133.2	127.0	52.7	109.7	96.4	66.2	89.4
0920	Plaster & Gypsum Board	89.0	35.0	52.9	89.0	69.5	76.0	106.8	139.3	128.5	149.8	139.3	142.8	214.0	53.6	106.8	95.6	71.7	79.6
0950, 0980	Ceilings & Acoustic Treatment	79.8	35.0	50.1	77.9	69.5	72.4	122.9	139.3	133.7	132.3	139.3	136.9	232.9	53.6	114.2	96.2	71.7	80.0
0960	Flooring	100.2	45.9	84.6	101.5	28.7	80.6	118.5	134.5	123.1	135.8	134.5	135.4	139.7	44.6	112.3	99.1	80.5	93.7
0970, 0990	Wall Finishes & Painting/Coating	90.1	68.3	77.4	90.1	68.3	77.4	99.0	143.8	125.1	107.2	143.8	128.5	104.0	35.9	64.4	94.0	32.6	58.3
09	FINISHES	94.0	44.1	66.4	93.6	60.7	75.4	114.0	137.4	126.9	129.4	137.4	133.8	187.5	57.5	115.6	96.4	72.0	82.9
COVERS	DIVS. 10 - 14, 25, 28, 41, 43, 44, 46	100.0	80.6	95.8	100.0	53.4	89.9	100.0	116.5	103.6	100.0	116.5	103.6	100.0	74.0	94.4	100.0	87.5	97.3
21, 22, 23	FIRE SUPPRESSION, PLUMBING & HVAC	100.0	69.4	87.1	97.9	64.1	83.6	100.4	107.4	103.3	100.5	107.4	103.4	103.3	37.1	75.3	100.0	72.7	88.5
26, 27, 3370	ELECTRICAL, COMMUNICATIONS & UTIL.	91.7	55.8	73.6	95.5	56.9	76.0	106.2	120.2	113.3	107.7	120.2	114.0	149.9	40.9	94.8	97.6	70.8	84.0
MF2014	WEIGHTED AVERAGE	97.0	62.3	81.7	97.1	64.4	82.7	120.2	119.7	120.0	125.0	119.7	122.7	143.0	57.6	105.4	101.5	77.8	91.1

DIVISION		IDAHO COEUR D'ALENE 838			IDAHO IDAHO FALLS 834			IDAHO LEWISTON 835			IDAHO POCATELLO 832			IDAHO TWIN FALLS 833			ILLINOIS BLOOMINGTON 617		
		MAT.	INST.	TOTAL	MAT.	INST.	TOTAL	MAT.	INST.	TOTAL	MAT.	INST.	TOTAL	MAT.	INST.	TOTAL	MAT.	INST.	TOTAL
015433	CONTRACTOR EQUIPMENT		91.5	91.5		96.7	96.7		91.5	91.5		96.7	96.7		96.7	96.7		102.9	102.9
0241, 31 - 34	SITE & INFRASTRUCTURE, DEMOLITION	87.0	90.9	89.7	85.9	96.0	92.3	94.0	91.7	92.4	89.1	95.3	93.5	96.0	96.1	96.1	96.6	99.5	98.7
0310	Concrete Forming & Accessories	111.9	79.3	83.7	94.6	78.8	80.9	117.2	80.5	85.5	101.1	75.4	78.9	102.2	53.1	59.7	84.8	120.0	115.2
0320	Concrete Reinforcing	112.6	95.5	103.7	107.2	79.4	92.8	112.6	95.8	103.9	105.7	79.6	92.2	107.5	79.0	92.7	95.8	109.6	103.0
0330	Cast-in-Place Concrete	95.9	86.1	92.0	84.4	73.1	80.0	99.6	86.9	94.6	91.2	88.3	90.1	93.5	62.7	81.4	98.6	114.4	104.8
03	CONCRETE	109.3	84.7	97.3	94.3	77.0	85.9	113.0	85.5	99.7	101.7	80.7	91.5	109.9	62.0	86.6	99.4	116.1	107.5
04	MASONRY	131.7	83.5	102.0	125.1	80.7	97.7	132.1	86.8	104.2	127.5	82.9	100.0	130.3	80.7	99.7	119.3	119.5	119.4
05	METALS	101.1	86.2	96.4	115.9	78.4	104.0	100.5	87.4	96.4	116.0	79.9	104.5	116.0	78.2	104.0	95.3	113.8	101.2
06	WOOD, PLASTICS & COMPOSITES	93.3	78.3	84.9	83.7	80.1	81.7	98.7	78.3	87.3	89.6	72.5	80.1	90.7	45.6	65.6	83.5	119.3	103.4
07	THERMAL & MOISTURE PROTECTION	148.2	82.1	120.5	94.1	70.1	84.1	148.4	86.1	122.3	94.6	72.5	85.3	95.3	72.6	85.8	98.0	112.5	104.1
08	OPENINGS	113.8	70.9	103.8	99.4	68.3	92.1	107.3	78.5	100.6	97.1	65.5	89.8	100.2	47.8	88.0	97.7	117.6	102.3
0920	Plaster & Gypsum Board	172.6	77.7	109.2	80.6	79.5	79.9	174.0	77.7	109.7	82.5	71.7	75.3	84.2	44.0	57.4	92.2	119.9	110.7
0950, 0980	Ceilings & Acoustic Treatment	127.5	77.7	94.5	97.0	79.5	85.4	127.5	77.7	94.5	103.6	71.7	82.5	99.4	44.0	62.7	87.0	119.9	108.8
0960	Flooring	140.8	81.3	123.7	98.8	41.4	82.3	144.3	92.2	129.3	102.5	80.5	96.2	103.8	41.4	85.9	91.9	114.1	98.2
0970, 0990	Wall Finishes & Painting/Coating	112.8	74.1	90.3	94.0	40.3	62.8	112.8	74.1	90.3	93.9	40.3	62.7	94.0	37.3	61.0	84.4	107.9	98.1
09	FINISHES	161.9	79.3	116.2	93.9	69.1	80.2	163.3	81.6	118.1	97.2	72.8	83.7	97.6	48.2	70.3	91.4	118.1	106.2
COVERS	DIVS. 10 - 14, 25, 28, 41, 43, 44, 46	100.0	86.9	97.2	100.0	46.7	88.4	100.0	87.1	97.2	100.0	87.5	97.3	100.0	42.8	87.6	100.0	105.6	101.2
21, 22, 23	FIRE SUPPRESSION, PLUMBING & HVAC	99.6	81.9	92.1	101.0	70.9	88.2	100.9	85.3	94.3	99.9	72.7	88.4	99.9	68.5	86.7	96.5	107.7	101.2
26, 27, 3370	ELECTRICAL, COMMUNICATIONS & UTIL.	90.9	77.0	83.9	90.8	70.5	80.5	89.0	81.9	85.4	95.5	70.5	82.8	92.1	59.7	75.7	96.3	93.9	95.1
MF2014	WEIGHTED AVERAGE	109.3	82.2	97.4	101.1	74.3	89.3	109.4	85.0	98.6	102.5	77.2	91.3	103.8	66.3	87.3	97.7	110.0	103.1

ILLINOIS

DIVISION		CARBONDALE 629 MAT.	INST.	TOTAL	CENTRALIA 628 MAT.	INST.	TOTAL	CHAMPAIGN 618-619 MAT.	INST.	TOTAL	CHICAGO 606-608 MAT.	INST.	TOTAL	DECATUR 625 MAT.	INST.	TOTAL	EAST ST. LOUIS 620-622 MAT.	INST.	TOTAL
015433	CONTRACTOR EQUIPMENT		110.4	110.4		110.4	110.4		103.8	103.8		98.3	98.3		103.8	103.8		110.4	110.4
0241, 31 - 34	SITE & INFRASTRUCTURE, DEMOLITION	97.6	100.7	99.8	97.9	101.4	100.4	105.8	100.4	102.0	101.2	102.7	102.2	93.9	100.5	98.6	100.0	101.4	101.0
0310	Concrete Forming & Accessories	92.1	105.7	103.9	93.8	113.3	110.6	91.4	116.9	113.5	98.9	156.0	148.3	94.3	116.5	113.5	89.5	113.6	110.3
0320	Concrete Reinforcing	92.5	110.3	101.7	92.5	110.6	101.8	95.8	105.2	100.7	99.2	164.2	132.9	90.6	101.7	96.3	92.4	116.0	101.8
0330	Cast-in-Place Concrete	90.9	101.3	95.0	91.4	119.0	102.2	114.3	111.8	113.3	105.0	152.2	123.6	99.3	113.2	104.8	92.9	116.1	102.1
03	CONCRETE	88.5	105.9	97.0	89.0	115.4	101.8	112.4	112.9	112.6	100.3	154.8	126.8	99.7	112.6	106.0	90.0	114.6	102.0
04	MASONRY	83.3	108.1	98.6	83.4	116.2	103.6	144.0	120.1	129.3	98.9	160.9	137.1	78.8	118.2	103.1	83.6	116.2	103.7
05	METALS	93.3	120.1	101.8	93.4	121.5	102.3	95.3	109.1	99.7	95.8	135.5	108.4	96.9	108.3	100.6	94.5	121.0	102.9
06	WOOD, PLASTICS & COMPOSITES	90.7	102.5	97.3	93.1	110.9	103.0	90.4	115.7	104.5	98.3	154.3	129.5	91.4	115.7	104.9	88.0	110.9	100.8
07	THERMAL & MOISTURE PROTECTION	97.8	99.9	98.6	97.8	110.1	103.0	98.7	113.7	105.0	100.0	144.5	118.6	103.8	111.2	106.9	97.8	108.2	102.2
08	OPENINGS	91.6	110.9	96.1	91.6	115.5	97.1	98.3	112.0	101.5	112.0	159.2	123.0	103.2	111.0	105.0	91.6	114.9	97.1
0920	Plaster & Gypsum Board	99.9	102.6	101.7	100.9	111.2	107.8	94.5	116.1	108.9	94.5	155.3	135.4	102.5	116.1	111.6	98.9	111.2	107.1
0950, 0980	Ceilings & Acoustic Treatment	87.2	102.6	97.4	87.2	111.2	103.1	87.0	116.1	106.3	96.8	155.7	135.8	93.8	116.1	108.6	87.2	111.2	103.1
0960	Flooring	122.1	115.5	120.2	123.0	111.3	119.6	95.2	104.4	97.8	96.4	148.7	111.4	109.2	116.8	111.4	121.0	111.3	118.2
0970, 0990	Wall Finishes & Painting/Coating	98.4	102.1	100.5	98.4	109.0	104.6	84.4	112.4	100.7	82.2	146.4	119.6	89.7	107.2	99.9	98.4	109.0	104.6
09	FINISHES	100.0	106.0	103.3	100.5	112.5	107.1	93.4	115.1	105.4	95.7	154.5	128.2	99.9	116.4	109.0	99.7	112.6	106.8
COVERS	DIVS. 10 - 14, 25, 28, 41, 43, 44, 46	100.0	99.7	99.9	100.0	101.7	100.4	100.0	105.0	101.1	100.0	124.7	105.4	100.0	104.6	101.0	100.0	101.7	100.4
21, 22, 23	FIRE SUPPRESSION, PLUMBING & HVAC	96.5	104.5	99.9	96.5	94.2	95.6	96.5	106.8	100.9	100.0	134.8	114.7	99.9	98.1	99.2	99.9	98.5	99.3
26, 27, 3370	ELECTRICAL, COMMUNICATIONS & UTIL.	97.2	106.9	102.1	98.5	106.9	102.8	99.4	94.2	96.8	96.5	128.1	112.5	101.0	89.8	95.3	98.1	102.7	100.4
MF2014	WEIGHTED AVERAGE	94.6	106.7	99.9	94.9	108.3	100.8	101.2	108.4	104.4	99.9	140.4	117.8	98.8	105.8	101.9	95.9	108.4	101.4

ILLINOIS

DIVISION		EFFINGHAM 624 MAT.	INST.	TOTAL	GALESBURG 614 MAT.	INST.	TOTAL	JOLIET 604 MAT.	INST.	TOTAL	KANKAKEE 609 MAT.	INST.	TOTAL	LA SALLE 613 MAT.	INST.	TOTAL	NORTH SUBURBAN 600-603 MAT.	INST.	TOTAL
015433	CONTRACTOR EQUIPMENT		103.8	103.8		102.9	102.9		96.2	96.2		96.2	96.2		102.9	102.9		96.2	96.2
0241, 31 - 34	SITE & INFRASTRUCTURE, DEMOLITION	97.9	100.2	99.5	99.1	99.5	99.4	101.2	101.6	101.5	95.1	101.2	99.4	98.5	100.3	99.8	100.5	101.9	101.4
0310	Concrete Forming & Accessories	99.0	115.1	113.0	91.3	118.2	114.6	100.8	158.1	150.4	94.0	142.7	136.1	105.5	123.7	121.3	100.0	154.9	147.5
0320	Concrete Reinforcing	93.5	101.7	97.7	95.3	112.2	104.1	99.2	148.5	124.8	100.1	145.7	123.7	95.3	143.0	120.1	99.2	154.8	128.0
0330	Cast-in-Place Concrete	98.9	108.8	102.8	101.5	108.1	104.1	105.0	145.5	121.0	97.9	131.4	111.1	101.4	121.4	109.3	105.0	148.5	122.2
03	CONCRETE	100.6	110.4	105.4	102.4	113.7	107.9	100.4	150.5	124.8	94.5	138.3	115.8	103.3	126.5	114.6	100.4	151.3	125.1
04	MASONRY	87.7	111.6	102.4	119.5	118.9	119.1	102.3	152.6	133.3	98.4	140.6	124.4	119.5	124.6	122.6	98.9	149.7	130.2
05	METALS	94.1	105.8	97.8	95.3	114.1	101.3	93.8	127.1	104.4	93.8	124.4	103.5	95.4	130.8	106.7	94.9	129.1	105.8
06	WOOD, PLASTICS & COMPOSITES	93.7	115.7	106.0	90.2	117.3	105.3	99.7	159.1	132.8	91.9	142.0	119.8	105.5	122.0	114.7	98.3	154.3	129.5
07	THERMAL & MOISTURE PROTECTION	103.3	106.1	104.5	98.1	110.1	103.1	99.8	142.6	117.7	99.0	134.8	114.0	98.3	119.9	107.4	100.3	140.1	116.9
08	OPENINGS	97.5	110.4	100.5	97.7	112.0	101.0	109.2	157.7	120.5	101.2	147.6	112.0	97.7	127.9	104.7	109.3	156.6	120.3
0920	Plaster & Gypsum Board	102.3	116.1	111.5	94.5	117.7	110.0	91.5	160.7	137.8	88.5	143.1	125.0	101.1	122.6	115.5	94.5	155.7	135.4
0950, 0980	Ceilings & Acoustic Treatment	87.2	116.1	106.3	87.0	117.7	107.3	96.8	160.7	139.1	96.8	143.1	127.5	87.0	122.6	110.6	96.8	155.7	135.8
0960	Flooring	110.3	115.5	111.8	95.0	114.1	100.5	96.0	159.0	114.1	92.8	126.7	102.6	102.0	117.8	106.6	96.4	134.9	107.4
0970, 0990	Wall Finishes & Painting/Coating	89.7	106.6	99.6	84.4	97.0	91.7	80.7	152.6	122.5	80.7	109.9	96.5	84.4	107.9	98.1	82.2	152.5	123.1
09	FINISHES	98.8	115.5	108.1	92.8	116.0	105.6	95.1	158.9	130.4	93.4	134.8	116.3	95.8	120.2	109.3	95.7	152.1	126.9
COVERS	DIVS. 10 - 14, 25, 28, 41, 43, 44, 46	100.0	79.7	95.6	100.0	103.2	100.7	100.0	124.1	105.2	100.0	118.2	103.9	100.0	103.8	100.8	100.0	121.6	104.7
21, 22, 23	FIRE SUPPRESSION, PLUMBING & HVAC	96.6	101.7	98.7	96.5	104.9	100.1	100.0	133.5	114.2	96.7	130.9	111.1	96.5	122.5	107.5	99.9	131.8	113.4
26, 27, 3370	ELECTRICAL, COMMUNICATIONS & UTIL.	99.0	106.8	102.9	97.2	87.4	92.2	95.8	133.3	114.7	91.2	136.1	113.9	94.5	136.1	115.5	95.6	127.5	111.7
MF2014	WEIGHTED AVERAGE	97.3	106.6	101.4	98.4	107.5	102.4	99.4	139.0	116.8	96.0	131.6	111.7	98.6	123.4	109.5	99.4	136.8	115.9

ILLINOIS

DIVISION		PEORIA 615-616 MAT.	INST.	TOTAL	QUINCY 623 MAT.	INST.	TOTAL	ROCK ISLAND 612 MAT.	INST.	TOTAL	ROCKFORD 610-611 MAT.	INST.	TOTAL	SOUTH SUBURBAN 605 MAT.	INST.	TOTAL	SPRINGFIELD 626-627 MAT.	INST.	TOTAL
015433	CONTRACTOR EQUIPMENT		102.9	102.9		103.8	103.8		102.9	102.9		102.9	102.9		96.2	96.2		103.8	103.8
0241, 31 - 34	SITE & INFRASTRUCTURE, DEMOLITION	99.5	99.5	99.5	96.8	100.1	99.1	97.2	98.5	98.1	99.0	100.7	100.2	100.5	101.9	101.4	98.4	100.5	99.9
0310	Concrete Forming & Accessories	94.3	118.8	115.5	96.8	113.6	111.3	92.9	104.0	102.5	98.8	127.6	123.7	100.0	154.9	147.5	95.1	116.3	113.4
0320	Concrete Reinforcing	92.9	112.3	102.9	93.1	104.1	98.8	95.3	106.0	100.9	87.8	135.3	112.4	99.2	154.8	128.0	93.7	103.9	99.0
0330	Cast-in-Place Concrete	98.5	116.8	105.7	99.1	102.6	100.5	99.3	100.7	99.9	100.8	125.1	110.4	105.0	148.5	122.2	94.1	108.1	99.6
03	CONCRETE	99.5	116.9	108.0	100.2	108.2	104.1	100.3	103.6	101.9	100.2	128.0	113.7	100.4	151.3	125.1	98.1	111.2	104.5
04	MASONRY	118.9	118.8	118.9	111.7	104.8	107.4	119.3	100.5	107.7	92.5	134.5	118.4	98.9	149.7	130.2	89.3	118.5	107.3
05	METALS	98.1	114.7	103.4	94.1	108.2	98.6	95.4	109.9	100.0	98.1	127.0	107.3	94.9	129.1	105.8	94.5	109.3	99.2
06	WOOD, PLASTICS & COMPOSITES	98.0	117.3	108.7	91.3	115.7	104.9	91.8	103.9	98.5	98.0	124.0	112.5	98.3	154.3	129.5	92.3	115.7	105.3
07	THERMAL & MOISTURE PROTECTION	98.9	112.0	104.4	103.3	107.0	104.8	98.1	99.4	98.6	101.4	127.3	112.2	100.3	140.1	116.9	106.0	110.9	108.1
08	OPENINGS	104.0	117.2	107.1	98.3	111.8	101.5	97.7	103.2	99.0	104.0	129.0	109.8	109.3	156.6	120.3	102.3	111.7	104.5
0920	Plaster & Gypsum Board	98.0	117.7	111.2	100.9	116.1	111.1	94.5	104.0	100.8	98.0	124.7	115.8	94.5	155.7	135.4	101.2	116.1	111.2
0950, 0980	Ceilings & Acoustic Treatment	91.9	117.7	109.0	87.2	116.1	106.3	87.0	104.0	98.2	91.9	124.7	113.6	96.8	155.7	135.8	96.9	116.1	109.6
0960	Flooring	98.8	114.1	103.2	109.2	103.9	107.6	96.3	101.5	97.8	98.8	117.8	104.3	96.4	134.9	107.4	113.6	103.6	110.7
0970, 0990	Wall Finishes & Painting/Coating	84.4	107.9	98.1	89.7	108.9	100.9	84.4	95.1	90.6	84.4	121.0	105.7	82.2	152.5	123.1	88.5	109.4	100.6
09	FINISHES	95.5	117.2	107.5	98.2	112.5	106.2	93.1	102.8	98.4	95.5	125.0	111.8	95.7	152.1	126.9	102.5	113.9	108.8
COVERS	DIVS. 10 - 14, 25, 28, 41, 43, 44, 46	100.0	105.8	101.3	100.0	81.5	96.0	100.0	99.0	99.8	100.0	113.5	102.9	100.0	121.5	104.7	100.0	104.4	101.0
21, 22, 23	FIRE SUPPRESSION, PLUMBING & HVAC	99.9	104.4	101.8	96.6	100.5	98.2	96.5	99.4	97.7	100.0	115.4	106.5	99.9	131.8	113.4	99.9	103.7	101.5
26, 27, 3370	ELECTRICAL, COMMUNICATIONS & UTIL.	98.2	97.6	97.9	96.5	81.6	89.0	89.9	99.0	94.5	98.5	132.9	115.9	95.6	128.9	112.4	103.6	92.0	97.7
MF2014	WEIGHTED AVERAGE	100.3	109.8	104.5	98.1	101.9	99.8	97.4	101.6	99.2	99.3	123.5	109.9	99.4	137.0	115.9	99.3	106.9	102.5

City Cost Indexes

INDIANA

DIVISION		ANDERSON 460			BLOOMINGTON 474			COLUMBUS 472			EVANSVILLE 476 - 477			FORT WAYNE 467 - 468			GARY 463 - 464		
		MAT.	INST.	TOTAL	MAT.	INST.	TOTAL	MAT.	INST.	TOTAL	MAT.	INST.	TOTAL	MAT.	INST.	TOTAL	MAT.	INST.	TOTAL
015433	CONTRACTOR EQUIPMENT		95.2	95.2		83.7	83.7		83.7	83.7		113.5	113.5		95.2	95.2		95.2	95.2
0241, 31 - 34	SITE & INFRASTRUCTURE, DEMOLITION	99.2	93.1	94.9	87.2	91.9	90.5	83.7	91.8	89.4	92.6	121.5	113.0	100.1	93.0	95.1	99.9	96.6	97.6
0310	Concrete Forming & Accessories	97.2	81.2	83.3	100.4	79.9	82.6	94.5	78.0	80.3	93.9	81.5	83.2	95.4	75.1	77.8	97.3	116.0	113.5
0320	Concrete Reinforcing	97.5	81.9	89.4	89.4	83.8	86.5	89.8	85.1	87.4	97.9	78.1	87.6	97.5	75.1	85.9	97.5	112.1	105.0
0330	Cast-in-Place Concrete	105.8	79.5	95.4	99.5	77.0	90.6	99.0	77.5	90.5	95.1	87.4	92.0	112.4	76.9	98.4	110.5	113.2	111.6
03	CONCRETE	99.5	81.3	90.6	103.3	79.4	91.7	102.5	79.0	91.1	103.5	83.1	93.6	102.6	76.4	89.9	101.8	114.1	107.8
04	MASONRY	91.5	81.4	85.3	95.9	76.1	83.7	95.7	76.8	84.0	91.1	84.6	87.1	95.1	80.4	86.0	92.9	115.5	106.8
05	METALS	94.5	90.0	93.1	96.8	77.4	90.7	96.9	77.7	90.8	90.4	84.8	88.6	94.5	86.1	91.8	94.5	109.2	99.2
06	WOOD, PLASTICS & COMPOSITES	100.0	81.1	89.5	110.4	80.1	93.5	105.3	77.4	89.8	91.4	80.2	85.2	99.8	73.7	85.3	97.3	114.9	107.1
07	THERMAL & MOISTURE PROTECTION	108.8	75.6	94.9	97.1	76.8	88.6	96.7	80.1	89.7	101.4	82.3	93.4	108.6	73.1	93.8	107.3	107.7	107.5
08	OPENINGS	98.5	81.0	94.4	101.7	81.0	96.9	97.8	79.9	93.6	95.6	79.6	91.9	98.5	73.2	92.6	98.5	120.1	103.5
0920	Plaster & Gypsum Board	105.8	80.8	89.1	98.2	80.2	86.2	95.5	77.5	83.5	93.9	79.1	84.0	105.0	73.2	83.7	98.7	115.5	109.9
0950, 0980	Ceilings & Acoustic Treatment	90.3	80.8	84.0	79.8	80.2	80.1	79.8	77.5	78.2	83.5	79.1	80.6	90.3	73.2	79.0	90.3	115.5	107.0
0960	Flooring	96.6	83.9	93.0	102.8	72.1	94.0	97.4	83.9	93.5	97.2	78.8	91.9	96.6	79.1	91.6	96.6	118.6	102.9
0970, 0990	Wall Finishes & Painting/Coating	95.4	71.0	81.2	89.9	82.4	85.5	89.9	84.0	86.5	95.3	87.9	91.0	95.4	73.9	82.9	95.4	126.0	113.2
09	FINISHES	94.5	80.6	86.8	93.3	78.7	85.2	91.3	79.6	84.8	91.9	81.9	86.4	94.3	75.7	84.0	93.4	117.7	106.9
COVERS	DIVS. 10 - 14, 25, 28, 41, 43, 44, 46	100.0	92.3	98.3	100.0	88.5	97.5	100.0	88.2	97.4	100.0	96.2	99.2	100.0	90.3	97.9	100.0	106.3	101.4
21, 22, 23	FIRE SUPPRESSION, PLUMBING & HVAC	99.9	78.5	90.9	99.6	78.8	90.8	96.3	78.1	88.6	99.9	80.6	91.7	99.9	72.6	88.4	99.9	106.1	102.5
26, 27, 3370	ELECTRICAL, COMMUNICATIONS & UTIL.	88.3	89.2	88.8	99.0	88.3	93.6	98.3	89.2	93.7	95.5	87.6	91.5	88.9	78.6	83.7	99.1	106.8	103.0
MF2014	WEIGHTED AVERAGE	97.1	83.6	91.1	98.8	81.2	91.0	97.1	81.3	90.1	96.6	86.6	92.2	97.6	78.6	89.3	98.4	110.0	103.5

INDIANA

DIVISION		INDIANAPOLIS 461 - 462			KOKOMO 469			LAFAYETTE 479			LAWRENCEBURG 470			MUNCIE 473			NEW ALBANY 471		
		MAT.	INST.	TOTAL	MAT.	INST.	TOTAL	MAT.	INST.	TOTAL	MAT.	INST.	TOTAL	MAT.	INST.	TOTAL	MAT.	INST.	TOTAL
015433	CONTRACTOR EQUIPMENT		90.4	90.4		95.2	95.2		83.7	83.7		102.4	102.4		93.7	93.7		92.2	92.2
0241, 31 - 34	SITE & INFRASTRUCTURE, DEMOLITION	99.0	96.2	97.0	95.5	92.9	93.6	84.6	91.8	89.7	82.5	107.3	100.0	87.0	92.3	90.8	79.4	93.8	89.6
0310	Concrete Forming & Accessories	97.6	85.1	86.8	100.5	76.8	80.0	92.1	80.6	82.2	90.9	76.3	78.3	92.0	80.7	82.3	89.1	71.2	73.6
0320	Concrete Reinforcing	97.7	85.6	91.4	88.2	85.3	86.7	89.4	84.1	86.6	88.7	73.4	80.8	98.9	81.9	90.1	90.0	76.2	82.9
0330	Cast-in-Place Concrete	99.8	86.1	94.4	104.7	81.4	95.5	99.6	82.0	92.6	93.1	74.7	85.8	104.5	82.6	95.9	96.1	72.7	86.9
03	CONCRETE	99.6	85.3	92.6	96.3	80.5	88.6	102.8	81.4	92.4	95.7	75.8	86.0	101.8	82.2	92.2	101.1	73.1	87.5
04	MASONRY	92.8	81.8	86.0	91.1	79.1	83.7	101.3	81.6	89.2	79.6	76.6	77.8	97.8	81.5	87.7	86.6	67.9	75.1
05	METALS	95.0	81.4	90.7	91.1	90.5	90.9	95.3	77.6	89.6	92.1	84.3	89.6	98.7	90.0	95.9	93.9	80.4	89.6
06	WOOD, PLASTICS & COMPOSITES	99.8	85.3	91.8	103.3	75.5	87.8	102.5	80.3	90.1	89.7	76.4	82.3	104.1	80.7	91.1	91.7	71.7	80.6
07	THERMAL & MOISTURE PROTECTION	102.2	80.2	93.0	108.4	75.9	94.8	96.7	79.6	89.5	102.4	76.1	91.4	99.7	77.0	90.2	88.6	67.8	79.9
08	OPENINGS	104.4	84.3	99.7	93.6	78.7	90.1	96.2	81.1	92.7	97.6	75.7	92.5	94.9	80.8	91.6	95.1	74.1	90.2
0920	Plaster & Gypsum Board	95.7	85.0	88.5	110.6	75.0	86.8	92.9	80.4	84.6	72.8	76.3	75.2	93.9	80.8	85.2	91.5	71.3	78.0
0950, 0980	Ceilings & Acoustic Treatment	94.4	85.0	88.2	90.3	75.0	80.2	75.7	80.4	78.8	87.6	76.3	80.1	79.8	80.8	80.5	83.5	71.3	75.4
0960	Flooring	97.0	83.9	93.2	101.0	91.4	98.2	96.3	85.3	93.1	71.8	83.9	75.3	96.3	83.9	92.7	94.6	60.4	84.8
0970, 0990	Wall Finishes & Painting/Coating	93.9	84.0	88.1	95.4	89.1	91.7	89.9	85.8	87.5	90.7	75.5	81.9	89.9	71.0	78.9	95.3	82.7	88.0
09	FINISHES	93.8	84.9	88.9	96.3	80.5	87.6	89.7	82.0	85.4	82.0	78.1	79.8	90.4	80.3	84.8	91.1	70.6	79.7
COVERS	DIVS. 10 - 14, 25, 28, 41, 43, 44, 46	100.0	93.6	98.6	100.0	88.9	97.6	100.0	91.3	98.1	100.0	42.7	87.6	100.0	91.2	98.1	100.0	41.6	87.4
21, 22, 23	FIRE SUPPRESSION, PLUMBING & HVAC	99.8	81.7	92.1	96.5	78.2	88.8	96.3	79.0	89.0	97.3	74.8	87.8	99.6	78.4	90.7	96.5	74.4	87.2
26, 27, 3370	ELECTRICAL, COMMUNICATIONS & UTIL.	102.2	89.2	95.6	92.5	78.1	85.2	97.8	82.7	90.2	93.6	75.4	84.4	91.6	87.8	89.7	94.1	75.2	84.5
MF2014	WEIGHTED AVERAGE	99.0	85.2	93.0	95.3	81.5	89.2	96.8	81.9	90.2	93.8	78.2	86.9	97.3	83.4	91.2	94.9	74.1	85.7

INDIANA / IOWA

DIVISION		SOUTH BEND 465 - 466			TERRE HAUTE 478			WASHINGTON 475			BURLINGTON 526			CARROLL 514			CEDAR RAPIDS 522 - 524		
		MAT.	INST.	TOTAL	MAT.	INST.	TOTAL	MAT.	INST.	TOTAL	MAT.	INST.	TOTAL	MAT.	INST.	TOTAL	MAT.	INST.	TOTAL
015433	CONTRACTOR EQUIPMENT		105.1	105.1		113.5	113.5		113.5	113.5		100.0	100.0		100.0	100.0		96.6	96.6
0241, 31 - 34	SITE & INFRASTRUCTURE, DEMOLITION	99.6	93.2	95.1	94.5	121.7	113.7	94.2	118.9	111.6	100.0	97.1	97.9	88.6	97.0	94.5	101.4	95.6	97.3
0310	Concrete Forming & Accessories	99.7	79.7	82.4	94.8	82.9	84.5	95.6	79.9	82.0	95.2	76.8	79.3	82.6	48.8	53.4	101.1	80.0	82.9
0320	Concrete Reinforcing	97.6	79.8	88.4	97.9	81.9	89.6	90.5	48.2	68.6	93.8	85.9	89.7	94.5	81.4	87.7	94.5	83.7	88.9
0330	Cast-in-Place Concrete	100.7	77.6	91.6	92.0	83.1	88.5	100.1	84.8	94.1	109.6	54.3	87.7	109.5	58.6	89.4	109.8	79.6	97.9
03	CONCRETE	96.6	80.4	88.7	106.5	83.0	95.0	112.3	75.7	94.5	101.5	71.8	87.1	100.3	59.7	80.5	101.6	81.2	91.7
04	MASONRY	100.5	78.5	87.0	99.3	80.7	87.8	91.3	82.2	85.7	106.4	64.6	80.6	108.4	73.7	87.1	112.8	77.7	91.1
05	METALS	94.7	102.4	97.1	91.1	86.7	89.7	85.7	68.2	80.1	85.5	95.0	88.5	85.5	90.2	87.0	87.8	92.4	89.3
06	WOOD, PLASTICS & COMPOSITES	95.5	78.9	86.2	93.5	83.0	87.7	93.9	79.9	86.1	92.2	78.4	84.5	79.0	43.3	59.1	98.9	79.8	88.2
07	THERMAL & MOISTURE PROTECTION	102.1	82.5	93.9	101.5	79.8	92.4	101.6	81.0	93.0	104.8	73.4	91.7	105.1	65.2	88.4	105.8	80.0	95.0
08	OPENINGS	96.2	78.7	92.2	96.1	82.1	92.8	92.9	67.7	87.0	98.7	71.9	92.5	103.5	52.8	91.7	103.9	79.9	98.4
0920	Plaster & Gypsum Board	96.8	78.5	84.6	93.9	82.1	86.0	94.0	78.9	83.9	104.0	77.8	86.5	99.3	41.8	60.9	108.5	79.5	89.2
0950, 0980	Ceilings & Acoustic Treatment	93.8	78.5	83.7	83.5	82.1	82.6	78.5	78.9	78.8	95.8	77.8	83.9	95.8	41.8	60.0	98.3	79.5	85.9
0960	Flooring	96.3	87.6	93.8	97.2	81.7	92.7	98.1	73.5	91.0	103.9	36.3	84.5	97.4	31.9	78.6	119.0	81.1	108.1
0970, 0990	Wall Finishes & Painting/Coating	90.4	87.1	88.5	95.3	83.1	88.2	95.3	87.9	91.0	99.3	88.6	93.1	99.3	78.7	87.3	100.6	73.4	84.8
09	FINISHES	95.3	81.8	87.9	91.9	82.8	86.9	91.4	80.3	85.3	100.1	70.5	83.8	96.0	46.9	68.8	105.8	79.5	91.2
COVERS	DIVS. 10 - 14, 25, 28, 41, 43, 44, 46	100.0	93.4	98.6	100.0	94.1	98.7	100.0	96.2	99.2	100.0	88.7	97.6	100.0	63.1	92.0	100.0	91.4	98.1
21, 22, 23	FIRE SUPPRESSION, PLUMBING & HVAC	99.9	76.8	90.1	99.9	79.2	91.2	96.5	79.5	89.4	96.7	76.0	87.9	96.7	74.1	87.1	100.1	79.7	91.4
26, 27, 3370	ELECTRICAL, COMMUNICATIONS & UTIL.	98.5	88.1	93.2	93.9	90.2	92.0	94.3	87.6	90.9	101.7	73.0	87.2	102.4	80.7	91.4	99.3	80.0	89.5
MF2014	WEIGHTED AVERAGE	97.8	84.1	91.8	97.4	86.6	92.6	95.7	82.6	89.9	97.5	76.9	88.4	97.4	71.0	85.8	99.9	82.6	92.3

City Cost Indexes

IOWA

DIVISION		COUNCIL BLUFFS 515 MAT.	INST.	TOTAL	CRESTON 508 MAT.	INST.	TOTAL	DAVENPORT 527 - 528 MAT.	INST.	TOTAL	DECORAH 521 MAT.	INST.	TOTAL	DES MOINES 500 - 503,509 MAT.	INST.	TOTAL	DUBUQUE 520 MAT.	INST.	TOTAL
015433	CONTRACTOR EQUIPMENT		95.9	95.9		100.0	100.0		100.0	100.0		100.0	100.0		101.7	101.7		95.4	95.4
0241, 31 - 34	SITE & INFRASTRUCTURE, DEMOLITION	105.1	91.7	95.6	94.5	96.1	95.6	100.3	99.2	99.5	98.6	96.1	96.8	100.9	99.5	99.9	99.2	92.9	94.7
0310	Concrete Forming & Accessories	82.0	71.1	72.6	79.1	64.6	66.6	100.6	91.0	92.3	92.7	45.1	51.5	96.9	79.4	81.7	83.4	75.1	76.2
0320	Concrete Reinforcing	96.4	78.1	86.9	89.9	81.6	85.6	94.5	101.6	98.2	93.8	76.1	84.6	94.3	85.7	89.8	93.2	83.1	88.0
0330	Cast-in-Place Concrete	114.2	79.5	100.5	112.6	63.3	93.2	105.8	91.7	100.2	106.6	79.3	95.8	95.7	80.2	89.6	107.6	97.4	103.6
03	CONCRETE	104.0	76.1	90.4	101.8	68.4	85.6	99.6	93.8	96.8	99.4	64.0	82.2	97.2	81.6	89.6	98.4	84.9	91.8
04	MASONRY	113.9	74.7	89.7	109.1	82.5	92.7	109.3	86.4	95.2	129.7	77.3	97.4	100.9	81.0	88.7	113.8	73.5	89.0
05	METALS	92.6	89.7	91.7	88.6	91.4	89.5	87.8	105.1	93.3	85.6	87.7	86.3	94.3	95.1	94.6	86.5	92.3	88.3
06	WOOD, PLASTICS & COMPOSITES	77.8	70.7	73.9	73.1	59.5	65.5	98.9	90.6	94.3	89.2	35.7	59.4	89.4	78.4	83.2	79.4	73.6	76.2
07	THERMAL & MOISTURE PROTECTION	105.2	68.4	89.8	106.1	76.5	93.7	105.2	89.2	98.5	105.0	59.7	86.0	97.4	78.5	89.5	105.4	78.8	94.3
08	OPENINGS	103.0	75.5	96.6	111.0	71.0	101.7	103.9	94.8	101.8	101.9	47.2	89.2	99.4	84.5	96.0	103.0	79.4	97.5
0920	Plaster & Gypsum Board	99.3	70.2	79.8	93.8	58.4	70.2	108.5	90.4	96.4	102.6	34.0	56.8	89.4	77.8	81.7	99.3	73.1	81.8
0950, 0980	Ceilings & Acoustic Treatment	95.8	70.2	78.8	87.2	58.4	68.2	98.3	90.4	93.1	95.8	34.0	54.9	89.7	77.8	81.8	95.8	73.1	80.8
0960	Flooring	95.9	84.1	92.5	86.9	81.7	85.4	106.6	93.2	102.7	103.3	77.8	95.9	93.4	86.1	91.3	109.0	74.0	98.9
0970, 0990	Wall Finishes & Painting/Coating	95.2	65.8	78.1	89.5	80.3	84.1	99.3	95.5	97.1	99.3	32.9	60.7	85.2	88.6	87.2	99.8	89.7	93.9
09	FINISHES	96.6	72.7	83.4	88.6	68.5	77.5	102.1	91.7	96.3	99.7	47.8	71.0	91.3	81.4	85.8	100.7	75.7	86.9
COVERS	DIVS. 10 - 14, 25, 28, 41, 43, 44, 46	100.0	89.0	97.6	100.0	67.4	92.9	100.0	94.4	98.8	100.0	80.5	95.8	100.0	91.8	98.2	100.0	90.1	97.9
21, 22, 23	FIRE SUPPRESSION, PLUMBING & HVAC	100.1	77.2	90.4	96.5	78.9	89.1	100.1	91.9	96.6	96.7	73.4	86.8	99.7	79.0	91.0	100.1	74.5	89.3
26, 27, 3370	ELECTRICAL, COMMUNICATIONS & UTIL.	104.5	80.9	92.6	94.1	80.7	87.3	97.3	92.4	94.8	99.3	44.0	71.4	104.8	80.7	92.6	103.1	79.1	91.0
MF2014	WEIGHTED AVERAGE	100.6	79.0	91.1	97.5	78.3	89.1	99.0	93.6	96.6	98.5	66.6	84.4	98.4	83.9	92.0	99.0	80.6	90.9

IOWA

DIVISION		FORT DODGE 505 MAT.	INST.	TOTAL	MASON CITY 504 MAT.	INST.	TOTAL	OTTUMWA 525 MAT.	INST.	TOTAL	SHENANDOAH 516 MAT.	INST.	TOTAL	SIBLEY 512 MAT.	INST.	TOTAL	SIOUX CITY 510 - 511 MAT.	INST.	TOTAL
015433	CONTRACTOR EQUIPMENT		100.0	100.0		100.0	100.0		95.4	95.4		95.9	95.9		100.0	100.0		100.0	100.0
0241, 31 - 34	SITE & INFRASTRUCTURE, DEMOLITION	102.8	94.9	97.2	102.9	96.0	98.0	99.6	91.0	93.5	103.5	90.8	94.5	110.2	94.8	99.3	112.0	95.5	100.3
0310	Concrete Forming & Accessories	79.7	43.7	48.6	83.9	44.4	49.7	90.6	72.2	74.7	83.7	55.1	59.0	84.1	36.7	43.1	101.1	64.7	69.6
0320	Concrete Reinforcing	89.9	66.4	77.8	89.8	81.0	85.2	93.8	86.3	89.9	96.4	67.6	81.5	96.4	65.2	80.2	94.5	77.1	85.5
0330	Cast-in-Place Concrete	105.7	43.2	81.1	105.7	56.7	86.4	110.3	66.9	93.2	110.4	58.9	90.1	108.2	43.8	82.8	108.9	54.6	87.5
03	CONCRETE	97.0	49.4	73.9	97.3	57.0	77.7	101.0	73.9	87.9	101.3	60.1	81.2	100.0	46.2	74.0	101.0	64.6	83.3
04	MASONRY	107.9	37.7	64.7	122.2	70.3	90.2	110.1	58.4	78.2	113.5	74.7	89.6	133.7	38.0	74.7	106.4	56.1	75.4
05	METALS	88.7	81.7	86.5	88.8	89.2	88.9	85.4	92.9	87.8	91.6	84.0	89.2	85.7	80.6	84.1	87.8	87.8	87.8
06	WOOD, PLASTICS & COMPOSITES	73.5	43.6	56.8	77.4	35.7	54.2	86.5	78.1	81.8	79.4	50.9	63.6	80.2	34.5	54.8	98.9	64.3	79.6
07	THERMAL & MOISTURE PROTECTION	105.4	56.6	85.0	104.9	66.4	88.8	105.6	67.6	89.7	104.4	64.6	87.7	104.8	47.0	80.6	105.2	63.7	87.8
08	OPENINGS	104.5	48.0	91.4	96.1	48.6	85.1	103.4	75.8	97.0	93.9	53.1	84.4	99.9	42.7	86.6	103.9	68.4	95.7
0920	Plaster & Gypsum Board	93.8	42.0	59.2	93.8	34.0	53.8	100.3	77.8	85.3	99.3	49.8	66.3	99.3	32.8	54.9	108.5	63.4	78.4
0950, 0980	Ceilings & Acoustic Treatment	87.2	42.0	57.3	87.2	34.0	52.0	95.8	77.8	83.9	95.8	49.8	65.4	95.8	32.8	54.1	106.6	54.7	91.7
0960	Flooring	88.3	45.3	75.9	90.5	68.3	84.1	112.4	70.7	100.4	96.7	32.9	78.4	98.4	32.7	79.5	106.6	54.7	91.7
0970, 0990	Wall Finishes & Painting/Coating	89.5	61.8	73.4	89.5	30.1	54.9	99.8	83.4	90.3	95.2	65.8	78.1	99.3	61.8	77.5	99.3	65.3	79.5
09	FINISHES	90.5	44.7	65.1	91.1	45.3	65.8	102.0	73.5	86.2	96.7	50.7	71.3	99.3	36.9	64.8	103.6	62.5	80.9
COVERS	DIVS. 10 - 14, 25, 28, 41, 43, 44, 46	100.0	79.3	95.5	100.0	84.1	96.5	100.0	84.0	96.5	100.0	64.1	92.2	100.0	78.4	95.3	100.0	87.8	97.4
21, 22, 23	FIRE SUPPRESSION, PLUMBING & HVAC	96.5	64.8	83.1	96.5	62.3	82.1	96.7	70.2	85.5	96.7	87.4	92.8	96.7	67.6	84.4	100.1	75.0	89.5
26, 27, 3370	ELECTRICAL, COMMUNICATIONS & UTIL.	100.2	41.6	70.5	99.3	55.2	77.0	101.5	71.5	86.3	99.3	80.9	90.0	99.3	41.1	69.9	99.3	72.4	85.7
MF2014	WEIGHTED AVERAGE	97.2	57.6	79.8	97.0	64.4	82.6	98.3	74.7	87.9	97.8	73.5	87.1	98.7	56.0	79.9	99.6	72.3	87.6

IOWA / KANSAS

DIVISION		SPENCER 513 MAT.	INST.	TOTAL	WATERLOO 506 - 507 MAT.	INST.	TOTAL	BELLEVILLE 669 MAT.	INST.	TOTAL	COLBY 677 MAT.	INST.	TOTAL	DODGE CITY 678 MAT.	INST.	TOTAL	EMPORIA 668 MAT.	INST.	TOTAL
015433	CONTRACTOR EQUIPMENT		100.0	100.0		100.0	100.0		104.8	104.8		104.8	104.8		104.8	104.8		102.9	102.9
0241, 31 - 34	SITE & INFRASTRUCTURE, DEMOLITION	110.3	94.8	99.3	108.1	95.8	99.4	107.7	96.4	99.7	112.6	96.5	101.3	115.2	96.3	101.8	99.7	93.6	95.4
0310	Concrete Forming & Accessories	90.5	36.7	44.0	95.5	54.1	59.7	98.2	53.2	59.3	103.9	57.6	63.8	96.8	57.2	62.6	88.8	64.2	67.6
0320	Concrete Reinforcing	96.4	66.4	80.8	90.5	83.0	86.6	104.0	55.2	78.7	103.3	55.3	78.4	100.7	55.1	77.1	102.6	55.8	78.3
0330	Cast-in-Place Concrete	108.2	43.8	82.8	113.2	76.4	98.7	118.7	55.7	93.9	120.3	67.6	99.5	122.4	55.6	96.1	114.8	53.7	90.7
03	CONCRETE	100.7	46.4	74.3	103.2	68.4	86.3	119.8	55.9	88.8	119.4	62.0	91.5	120.8	57.6	90.1	111.8	60.3	86.8
04	MASONRY	133.7	38.0	74.7	108.7	73.6	87.1	97.7	59.8	74.3	109.3	61.2	79.7	119.2	59.1	82.2	103.8	69.5	82.7
05	METALS	85.7	81.0	84.2	91.1	91.6	91.2	94.6	78.8	89.6	95.0	79.4	90.1	96.5	78.1	90.6	94.3	80.4	89.9
06	WOOD, PLASTICS & COMPOSITES	86.3	34.5	57.5	91.0	46.3	66.1	97.0	50.3	71.0	105.4	55.8	77.8	97.1	55.8	74.1	88.0	64.3	74.8
07	THERMAL & MOISTURE PROTECTION	105.7	47.0	81.2	105.1	71.3	91.0	96.9	60.3	81.6	98.9	61.7	83.3	98.8	60.0	82.6	95.1	75.6	87.0
08	OPENINGS	112.1	43.1	96.0	96.5	62.3	88.5	99.4	47.1	87.2	108.9	50.1	95.2	108.8	55.8	96.5	97.2	63.2	89.3
0920	Plaster & Gypsum Board	100.3	32.8	55.2	102.7	44.9	64.1	95.5	48.9	64.3	102.0	54.6	70.3	96.4	54.6	68.5	92.8	63.3	73.1
0950, 0980	Ceilings & Acoustic Treatment	95.8	32.8	54.1	89.7	44.9	60.0	83.0	48.9	60.4	80.6	54.6	63.4	80.6	54.6	63.4	83.0	63.3	69.9
0960	Flooring	101.5	32.7	81.7	96.3	74.0	89.9	99.9	37.7	82.0	96.5	57.7	85.4	92.7	37.7	76.8	94.7	57.7	84.1
0970, 0990	Wall Finishes & Painting/Coating	99.3	61.8	77.5	89.5	77.8	82.7	85.5	38.9	58.4	93.8	38.9	61.9	93.8	38.9	61.9	85.5	38.9	58.4
09	FINISHES	100.4	36.9	65.2	94.9	58.2	74.6	93.3	48.1	68.3	93.1	55.4	72.3	91.3	51.4	69.2	90.3	60.5	73.8
COVERS	DIVS. 10 - 14, 25, 28, 41, 43, 44, 46	100.0	78.4	95.3	100.0	87.5	97.3	100.0	39.8	86.9	100.0	40.4	87.1	100.0	40.4	87.1	100.0	39.5	86.9
21, 22, 23	FIRE SUPPRESSION, PLUMBING & HVAC	96.7	67.7	84.4	99.9	79.0	91.1	96.5	79.6	89.4	96.6	69.4	85.1	99.9	69.3	87.0	96.5	73.3	86.7
26, 27, 3370	ELECTRICAL, COMMUNICATIONS & UTIL.	101.0	41.1	70.7	95.9	62.9	79.3	100.8	67.2	83.8	100.1	78.0	88.9	97.3	73.2	85.1	98.2	72.2	85.0
MF2014	WEIGHTED AVERAGE	100.5	56.0	80.9	98.4	73.7	87.5	100.1	66.5	85.3	101.8	68.0	86.9	103.0	66.1	86.8	98.3	70.1	85.9

City Cost Indexes

KANSAS

		FORT SCOTT 667			HAYS 676			HUTCHINSON 675			INDEPENDENCE 673			KANSAS CITY 660 - 662			LIBERAL 679		
DIVISION		MAT.	INST.	TOTAL	MAT.	INST.	TOTAL	MAT.	INST.	TOTAL	MAT.	INST.	TOTAL	MAT.	INST.	TOTAL	MAT.	INST.	TOTAL
015433	CONTRACTOR EQUIPMENT		103.8	103.8		104.8	104.8		104.8	104.8		104.8	104.8		101.1	101.1		104.8	104.8
0241, 31 - 34	SITE & INFRASTRUCTURE, DEMOLITION	96.5	94.4	95.0	118.2	96.6	102.9	96.1	96.4	96.3	116.4	96.4	102.3	90.9	94.2	93.2	117.7	96.4	102.6
0310	Concrete Forming & Accessories	106.7	80.6	84.1	101.3	57.7	63.6	91.2	57.2	61.8	112.9	69.1	75.0	103.0	96.5	97.4	97.2	57.1	62.6
0320	Concrete Reinforcing	101.9	92.3	96.9	100.7	55.3	77.2	100.7	55.2	77.1	100.1	62.0	80.4	98.8	98.6	98.7	102.1	55.2	77.8
0330	Cast-in-Place Concrete	106.4	53.7	85.6	94.9	67.7	84.2	87.9	53.1	74.2	123.0	53.4	95.5	91.2	97.3	93.6	94.9	53.1	78.4
03	CONCRETE	106.9	74.3	91.1	110.2	62.0	86.8	91.9	56.8	74.8	122.1	63.5	93.6	98.7	97.5	98.1	112.3	56.8	85.3
04	MASONRY	105.1	60.1	77.4	118.8	61.2	83.3	109.3	59.7	78.7	106.2	65.6	81.2	106.5	100.2	102.6	116.4	59.7	81.7
05	METALS	94.3	93.1	93.9	94.6	79.7	89.8	94.4	78.1	89.2	94.3	81.5	90.2	102.7	101.5	102.3	94.9	78.1	89.5
06	WOOD, PLASTICS & COMPOSITES	107.5	88.9	97.1	102.2	55.8	76.4	91.9	55.8	71.8	116.1	71.1	91.1	103.3	96.9	99.7	97.7	55.8	74.4
07	THERMAL & MOISTURE PROTECTION	96.1	73.3	86.5	99.2	61.2	83.3	97.7	60.2	82.0	98.9	75.4	89.1	95.9	99.1	97.2	99.3	60.2	83.0
08	OPENINGS	97.2	85.7	94.5	108.8	55.8	96.4	108.7	55.8	96.4	106.3	65.7	96.9	98.5	90.1	96.5	108.8	55.8	96.5
0920	Plaster & Gypsum Board	97.8	88.5	91.6	99.4	54.6	69.4	95.4	54.6	68.1	109.0	70.3	83.1	91.4	96.8	95.0	97.0	54.6	68.7
0950, 0980	Ceilings & Acoustic Treatment	83.0	88.5	86.7	80.6	54.6	63.4	80.6	54.6	63.4	80.6	70.3	73.8	83.0	96.8	92.1	80.6	54.6	63.4
0960	Flooring	110.5	37.9	89.6	95.3	61.7	85.6	89.6	37.7	74.7	101.0	37.7	82.8	88.7	101.0	92.2	92.9	37.7	77.0
0970, 0990	Wall Finishes & Painting/Coating	87.2	38.9	59.1	93.8	38.9	61.9	93.8	38.9	61.9	93.8	38.9	61.9	92.7	66.7	78.2	93.8	38.9	61.9
09	FINISHES	96.0	69.8	81.5	92.9	56.1	72.5	88.6	51.4	68.0	95.6	60.5	76.2	90.6	93.1	92.0	92.1	51.4	69.6
COVERS	DIVS. 10 - 14, 25, 28, 41, 43, 44, 46	100.0	45.5	88.2	100.0	40.4	87.1	100.0	40.4	87.1	100.0	42.1	87.5	100.0	78.9	95.4	100.0	40.4	87.1
21, 22, 23	FIRE SUPPRESSION, PLUMBING & HVAC	96.5	67.5	84.3	96.6	69.4	85.1	96.6	69.3	85.1	96.6	70.4	85.5	99.8	95.4	98.0	96.6	67.4	84.3
26, 27, 3370	ELECTRICAL, COMMUNICATIONS & UTIL.	97.5	72.2	84.7	99.1	67.2	83.0	94.8	67.2	80.8	96.7	72.2	84.3	102.9	96.4	99.6	97.3	73.2	85.1
MF2014	WEIGHTED AVERAGE	98.3	73.7	87.5	101.2	66.9	86.1	97.1	65.2	83.1	101.6	70.2	87.8	99.6	95.9	98.0	101.1	65.6	85.5

KANSAS / KENTUCKY

		SALINA 674			TOPEKA 664 - 666			WICHITA 670 - 672			ASHLAND 411 - 412			BOWLING GREEN 421 - 422			CAMPTON 413 - 414		
DIVISION		MAT.	INST.	TOTAL	MAT.	INST.	TOTAL	MAT.	INST.	TOTAL	MAT.	INST.	TOTAL	MAT.	INST.	TOTAL	MAT.	INST.	TOTAL
015433	CONTRACTOR EQUIPMENT		104.8	104.8		102.9	102.9		104.8	104.8		97.2	97.2		92.2	92.2		98.8	98.8
0241, 31 - 34	SITE & INFRASTRUCTURE, DEMOLITION	105.2	96.7	99.2	94.5	92.8	93.3	101.3	95.1	96.9	114.7	83.9	92.9	79.7	94.2	90.0	88.4	95.7	93.6
0310	Concrete Forming & Accessories	93.1	54.4	59.7	100.5	47.3	54.5	99.4	50.1	56.8	87.2	101.7	99.7	85.3	84.5	84.6	88.8	80.4	81.6
0320	Concrete Reinforcing	100.1	75.4	87.3	98.3	100.6	99.5	98.2	77.1	87.3	91.4	111.6	101.9	88.7	90.5	89.6	89.6	100.6	95.3
0330	Cast-in-Place Concrete	106.3	53.9	85.6	95.7	54.9	79.6	99.4	51.6	80.6	87.5	102.9	93.6	86.8	94.2	89.7	96.8	72.3	87.2
03	CONCRETE	106.8	59.6	83.9	99.9	61.4	81.2	101.6	57.1	80.0	96.9	104.5	100.6	95.2	88.9	92.2	99.0	81.6	90.5
04	MASONRY	135.4	61.2	89.7	99.3	60.6	75.5	105.4	49.8	71.1	98.3	108.5	104.6	100.6	84.0	90.3	97.0	68.7	79.5
05	METALS	96.3	87.6	93.5	98.6	98.5	98.6	98.6	85.2	94.3	93.1	113.0	99.4	94.6	87.1	92.2	93.9	91.1	93.0
06	WOOD, PLASTICS & COMPOSITES	93.5	50.7	69.7	100.4	43.8	68.9	101.4	49.5	72.5	74.4	99.9	88.6	86.7	85.3	85.9	85.2	86.0	85.6
07	THERMAL & MOISTURE PROTECTION	98.3	61.5	82.9	99.7	70.4	87.4	97.9	54.8	79.8	92.8	94.3	93.4	88.4	81.5	85.5	101.8	76.2	91.1
08	OPENINGS	108.8	51.1	95.3	104.1	63.0	94.5	111.9	57.7	99.3	94.4	96.8	94.9	95.1	79.7	91.5	96.5	85.1	93.8
0920	Plaster & Gypsum Board	95.4	49.4	64.6	94.8	42.2	59.7	94.1	48.1	63.3	63.3	100.0	87.8	87.9	85.2	86.1	87.9	85.1	86.0
0950, 0980	Ceilings & Acoustic Treatment	80.6	49.4	59.9	87.7	42.2	57.6	84.5	48.1	60.4	78.9	100.0	92.9	83.5	85.2	84.6	83.5	85.1	84.5
0960	Flooring	91.1	61.7	82.6	100.3	76.9	93.6	98.0	58.2	86.6	77.6	104.3	85.2	92.3	83.2	89.7	94.4	39.6	78.7
0970, 0990	Wall Finishes & Painting/Coating	93.8	38.9	61.9	87.1	76.2	80.7	92.8	71.9	80.7	97.3	104.6	101.5	95.3	72.7	82.2	95.3	65.4	77.9
09	FINISHES	89.9	53.2	69.6	94.5	53.9	72.1	93.5	53.1	71.2	79.5	102.7	92.4	89.7	83.5	86.3	90.5	71.8	80.2
COVERS	DIVS. 10 - 14, 25, 28, 41, 43, 44, 46	100.0	86.7	97.1	100.0	42.6	87.6	100.0	83.6	96.4	100.0	91.8	98.2	100.0	60.9	91.5	100.0	54.0	90.0
21, 22, 23	FIRE SUPPRESSION, PLUMBING & HVAC	99.9	69.4	87.0	99.9	71.5	87.9	99.8	64.9	85.0	96.3	93.7	95.2	99.9	83.0	92.8	96.6	78.9	89.1
26, 27, 3370	ELECTRICAL, COMMUNICATIONS & UTIL.	97.1	73.3	85.0	101.7	74.1	87.7	100.7	73.2	86.8	92.0	99.8	96.0	94.4	80.9	87.6	92.0	65.6	78.7
MF2014	WEIGHTED AVERAGE	101.8	68.9	87.3	99.8	69.8	86.6	101.0	66.1	85.6	94.5	99.8	96.8	95.7	84.1	90.6	95.6	77.6	87.7

KENTUCKY

		CORBIN 407 - 409			COVINGTON 410			ELIZABETHTOWN 427			FRANKFORT 406			HAZARD 417 - 418			HENDERSON 424		
DIVISION		MAT.	INST.	TOTAL	MAT.	INST.	TOTAL	MAT.	INST.	TOTAL	MAT.	INST.	TOTAL	MAT.	INST.	TOTAL	MAT.	INST.	TOTAL
015433	CONTRACTOR EQUIPMENT		98.8	98.8		102.4	102.4		92.2	92.2		98.8	98.8		98.8	98.8		113.5	113.5
0241, 31 - 34	SITE & INFRASTRUCTURE, DEMOLITION	88.0	95.5	93.3	84.0	109.4	102.0	74.2	94.3	88.4	85.6	96.8	93.5	86.2	97.1	93.9	82.5	121.7	110.2
0310	Concrete Forming & Accessories	83.8	72.1	73.7	84.4	84.1	84.2	80.2	80.7	80.6	93.3	70.5	73.6	85.4	81.4	82.0	92.0	81.3	82.7
0320	Concrete Reinforcing	90.7	69.2	79.5	88.2	91.8	90.1	89.2	93.3	91.3	95.2	95.4	95.3	90.0	110.4	100.6	88.8	89.4	89.1
0330	Cast-in-Place Concrete	88.2	57.6	76.2	92.7	96.5	94.2	78.5	72.4	76.1	85.2	68.9	78.8	93.1	81.3	88.5	76.7	90.1	82.0
03	CONCRETE	92.0	67.1	79.9	97.3	90.3	93.9	87.1	80.4	83.8	92.5	75.0	84.0	95.7	86.9	91.4	92.2	85.9	89.2
04	MASONRY	95.0	61.6	74.5	112.3	101.1	105.4	83.8	76.9	79.6	90.4	80.8	84.5	95.6	71.2	80.6	104.4	94.0	98.0
05	METALS	89.0	78.4	85.6	92.0	97.0	93.6	93.8	88.2	92.0	91.4	89.6	90.8	93.9	95.2	94.3	85.4	88.5	86.5
06	WOOD, PLASTICS & COMPOSITES	72.8	77.7	75.5	83.2	75.2	78.8	82.0	82.4	82.2	87.6	66.3	75.7	82.3	86.0	84.3	89.1	78.4	83.1
07	THERMAL & MOISTURE PROTECTION	106.4	72.9	92.4	102.6	87.2	96.2	88.1	77.6	83.7	104.8	79.6	94.3	101.7	78.1	91.8	101.0	85.5	94.5
08	OPENINGS	92.9	66.1	86.7	98.5	82.9	94.9	95.1	83.5	92.4	97.1	76.2	92.2	96.8	78.6	92.6	93.3	80.1	90.2
0920	Plaster & Gypsum Board	95.3	76.6	82.8	70.6	75.1	73.6	87.2	82.3	83.9	95.7	64.9	75.1	87.2	85.1	85.8	90.4	77.3	81.6
0950, 0980	Ceilings & Acoustic Treatment	78.5	76.6	77.2	86.8	75.1	79.0	83.5	82.3	82.7	86.8	64.9	72.3	83.5	85.1	84.5	78.5	77.3	77.7
0960	Flooring	92.4	39.6	77.2	69.2	86.0	74.1	89.5	82.7	87.6	100.2	52.3	86.4	92.6	40.3	77.6	96.2	83.5	92.5
0970, 0990	Wall Finishes & Painting/Coating	96.1	49.4	68.9	90.7	83.2	86.3	95.3	78.7	85.6	96.9	82.8	88.7	95.3	65.4	77.9	95.3	96.1	95.7
09	FINISHES	88.6	64.3	75.2	80.9	83.4	82.3	88.4	80.6	84.1	93.9	67.5	79.3	89.7	72.6	80.3	89.6	83.1	86.0
COVERS	DIVS. 10 - 14, 25, 28, 41, 43, 44, 46	100.0	47.3	88.6	100.0	99.2	99.8	100.0	82.2	96.1	100.0	64.0	92.2	100.0	54.8	90.2	100.0	64.0	92.2
21, 22, 23	FIRE SUPPRESSION, PLUMBING & HVAC	96.6	72.0	86.2	97.3	92.1	95.1	96.7	80.6	89.9	100.0	83.2	92.9	96.6	82.5	90.6	96.7	79.1	89.3
26, 27, 3370	ELECTRICAL, COMMUNICATIONS & UTIL.	90.6	79.5	85.0	95.5	80.1	87.7	91.8	80.9	86.3	99.2	82.4	90.7	92.0	57.9	74.8	93.9	80.8	87.2
MF2014	WEIGHTED AVERAGE	93.3	71.9	83.9	95.8	91.3	93.8	92.5	82.2	87.9	96.1	80.2	89.1	95.1	78.6	87.8	93.5	86.3	90.3

KENTUCKY

DIVISION		LEXINGTON 403-405			LOUISVILLE 400-402			OWENSBORO 423			PADUCAH 420			PIKEVILLE 415-416			SOMERSET 425-426		
		MAT.	INST.	TOTAL	MAT.	INST.	TOTAL	MAT.	INST.	TOTAL	MAT.	INST.	TOTAL	MAT.	INST.	TOTAL	MAT.	INST.	TOTAL
015433	CONTRACTOR EQUIPMENT		98.8	98.8		92.2	92.2		113.5	113.5		113.5	113.5		97.2	97.2		98.8	98.8
0241, 31 - 34	SITE & INFRASTRUCTURE, DEMOLITION	90.1	98.1	95.8	85.3	94.6	91.9	92.6	122.1	113.5	85.3	121.1	110.6	126.0	82.9	95.5	79.1	96.3	91.3
0310	Concrete Forming & Accessories	95.2	71.1	74.4	94.2	84.0	85.4	90.4	84.4	85.2	88.3	82.7	83.5	96.1	86.8	88.1	86.4	79.4	80.4
0320	Concrete Reinforcing	99.5	95.6	97.5	95.7	95.6	95.7	88.8	91.5	90.2	89.4	89.1	89.2	91.9	111.4	102.0	89.2	95.4	92.4
0330	Cast-in-Place Concrete	90.3	91.4	90.7	89.4	74.7	83.6	89.5	90.2	89.8	81.8	86.0	83.5	96.2	97.7	96.8	76.7	97.6	85.0
03	CONCRETE	95.1	82.9	89.2	94.7	83.1	89.1	104.4	87.8	96.3	96.9	85.1	91.2	110.7	96.0	103.6	82.2	88.7	85.4
04	MASONRY	93.2	80.8	85.5	90.6	79.6	83.8	96.3	88.4	91.4	99.1	81.9	88.5	95.3	99.9	98.1	90.4	77.2	82.3
05	METALS	91.4	90.6	91.1	92.3	89.4	91.3	86.9	89.9	87.8	83.9	87.6	85.1	93.0	112.0	99.0	93.8	88.8	92.2
06	WOOD, PLASTICS & COMPOSITES	87.0	66.3	75.5	87.3	85.3	86.2	87.1	82.5	84.5	84.8	82.3	83.4	83.3	85.3	84.4	82.8	80.1	81.3
07	THERMAL & MOISTURE PROTECTION	106.6	80.5	95.7	103.2	79.4	93.2	101.5	85.9	95.0	101.1	85.9	94.8	93.5	89.0	91.6	101.0	79.9	92.2
08	OPENINGS	93.2	72.8	88.4	87.4	85.6	87.0	93.3	85.0	91.4	92.6	79.6	89.5	95.0	86.2	93.0	95.9	80.0	92.2
0920	Plaster & Gypsum Board	104.9	64.9	78.2	98.0	85.2	89.5	89.0	81.6	84.1	88.1	81.3	83.5	65.6	85.1	78.6	87.2	79.1	81.8
0950, 0980	Ceilings & Acoustic Treatment	82.7	64.9	70.9	87.1	85.2	85.9	78.5	81.6	80.5	78.5	81.3	80.3	78.9	85.1	83.0	83.5	79.1	80.6
0960	Flooring	97.6	64.4	88.1	94.8	82.7	91.3	95.5	83.2	92.0	94.3	57.1	83.6	82.0	104.3	88.4	92.9	39.6	77.6
0970, 0990	Wall Finishes & Painting/Coating	96.1	81.0	87.3	94.9	78.7	85.5	95.3	99.4	97.7	95.3	85.0	89.3	97.3	81.2	88.0	95.3	78.7	85.6
09	FINISHES	92.4	69.5	79.8	92.4	83.4	87.4	89.7	85.4	87.3	88.9	78.3	83.0	82.0	89.4	86.1	89.1	71.9	79.6
COVERS	DIVS. 10 - 14, 25, 28, 41, 43, 44, 46	100.0	96.4	99.2	100.0	95.7	99.1	100.0	103.5	100.8	100.0	60.9	91.5	100.0	54.5	90.1	100.0	57.7	90.8
21, 22, 23	FIRE SUPPRESSION, PLUMBING & HVAC	100.0	81.2	92.0	100.0	81.8	92.3	99.9	78.2	90.7	96.7	83.4	91.1	96.3	88.9	93.2	96.7	80.3	89.8
26, 27, 3370	ELECTRICAL, COMMUNICATIONS & UTIL.	93.0	82.4	87.7	97.8	82.4	90.0	93.9	82.8	88.3	96.1	83.6	89.8	94.8	72.9	83.7	92.4	82.4	87.3
MF2014	WEIGHTED AVERAGE	95.5	82.2	89.7	95.1	84.3	90.4	95.8	88.0	92.4	93.7	85.5	90.1	96.8	89.3	93.5	93.0	81.7	88.0

LOUISIANA

DIVISION		ALEXANDRIA 713-714			BATON ROUGE 707-708			HAMMOND 704			LAFAYETTE 705			LAKE CHARLES 706			MONROE 712		
		MAT.	INST.	TOTAL	MAT.	INST.	TOTAL	MAT.	INST.	TOTAL	MAT.	INST.	TOTAL	MAT.	INST.	TOTAL	MAT.	INST.	TOTAL
015433	CONTRACTOR EQUIPMENT		93.0	93.0		90.7	90.7		91.2	91.2		91.2	91.2		90.7	90.7		93.0	93.0
0241, 31 - 34	SITE & INFRASTRUCTURE, DEMOLITION	101.7	92.9	95.5	103.2	91.3	94.8	100.0	89.6	92.6	101.1	91.9	94.6	101.8	91.3	94.4	101.7	92.3	95.1
0310	Concrete Forming & Accessories	81.7	46.7	51.4	95.0	67.6	71.3	78.1	54.4	57.6	95.9	58.6	63.6	96.7	67.0	71.0	81.2	43.3	48.4
0320	Concrete Reinforcing	105.4	71.4	87.8	98.5	75.3	86.5	97.9	58.7	77.6	99.4	75.4	86.9	99.4	58.4	78.1	104.2	71.1	87.0
0330	Cast-in-Place Concrete	92.5	68.7	83.1	99.1	78.7	91.1	94.2	62.6	81.7	93.7	63.7	81.9	98.6	70.3	87.5	92.5	63.7	81.2
03	CONCRETE	93.9	59.9	77.4	98.6	73.3	86.3	93.0	58.8	76.4	94.1	64.2	79.6	96.6	67.1	82.2	93.7	56.7	75.7
04	MASONRY	119.9	63.8	85.3	92.6	65.1	75.7	93.7	50.0	66.8	93.7	53.4	68.9	93.0	66.4	76.6	114.1	47.8	73.2
05	METALS	95.1	79.3	90.1	99.3	80.0	93.1	91.1	71.9	85.0	90.4	80.2	87.1	90.4	72.6	84.7	95.0	78.8	89.9
06	WOOD, PLASTICS & COMPOSITES	96.1	40.4	65.1	102.6	66.7	82.4	88.4	57.2	71.0	109.3	60.8	82.3	107.5	66.7	84.8	95.5	40.9	65.1
07	THERMAL & MOISTURE PROTECTION	97.1	67.0	84.5	96.4	71.0	85.8	98.0	61.9	82.9	98.5	65.2	84.6	97.6	71.0	86.5	97.1	61.0	82.0
08	OPENINGS	116.0	48.4	100.3	105.7	64.9	96.2	97.7	56.6	88.2	101.3	60.3	91.8	101.3	59.3	91.5	116.0	51.3	100.9
0920	Plaster & Gypsum Board	81.4	39.0	53.1	96.9	66.1	76.3	96.0	56.3	69.5	104.2	60.0	74.7	104.2	66.1	78.7	81.1	39.5	53.3
0950, 0980	Ceilings & Acoustic Treatment	80.4	39.0	53.0	98.6	66.1	77.1	96.3	56.3	68.6	94.6	60.0	71.7	95.5	66.1	76.0	80.4	39.5	53.3
0960	Flooring	101.8	65.3	91.3	101.6	63.0	90.5	96.0	61.4	86.1	105.8	65.3	94.2	105.8	70.1	95.6	101.3	54.9	87.9
0970, 0990	Wall Finishes & Painting/Coating	96.4	64.4	77.8	94.7	46.3	66.6	99.8	48.3	69.9	99.8	57.9	75.5	99.8	50.0	70.8	96.4	48.1	68.3
09	FINISHES	87.7	50.3	67.0	94.8	64.2	77.9	94.2	55.0	72.5	97.9	59.2	76.5	98.2	65.8	80.3	87.5	44.3	63.6
COVERS	DIVS. 10 - 14, 25, 28, 41, 43, 44, 46	100.0	79.1	95.5	100.0	84.9	96.7	100.0	45.3	88.1	100.0	80.0	95.7	100.0	84.7	96.7	100.0	75.8	94.8
21, 22, 23	FIRE SUPPRESSION, PLUMBING & HVAC	100.2	59.7	83.1	100.0	64.8	85.2	96.7	61.5	81.8	100.1	62.7	84.3	100.1	68.3	86.7	100.2	54.2	80.7
26, 27, 3370	ELECTRICAL, COMMUNICATIONS & UTIL.	95.4	57.5	76.2	99.4	61.1	80.0	93.8	56.8	75.1	94.8	66.9	80.7	94.4	66.8	80.5	97.1	59.2	77.9
MF2014	WEIGHTED AVERAGE	99.7	63.3	83.7	99.5	69.8	86.4	95.2	61.0	80.1	97.0	66.6	83.6	97.2	69.9	85.2	99.6	59.3	81.8

LOUISIANA / MAINE

DIVISION		NEW ORLEANS 700-701			SHREVEPORT 710-711			THIBODAUX 703			AUGUSTA 043			BANGOR 044			BATH 045		
		MAT.	INST.	TOTAL	MAT.	INST.	TOTAL	MAT.	INST.	TOTAL	MAT.	INST.	TOTAL	MAT.	INST.	TOTAL	MAT.	INST.	TOTAL
015433	CONTRACTOR EQUIPMENT		91.5	91.5		93.0	93.0		91.2	91.2		98.1	98.1		98.1	98.1		98.1	98.1
0241, 31 - 34	SITE & INFRASTRUCTURE, DEMOLITION	99.3	92.9	94.8	104.3	92.3	95.8	102.3	89.9	93.5	91.3	98.6	96.5	94.5	98.8	97.5	92.3	98.6	96.8
0310	Concrete Forming & Accessories	96.4	67.7	71.6	96.1	43.8	50.9	90.1	64.4	67.8	101.8	93.2	94.3	96.6	94.2	94.5	92.0	93.3	93.1
0320	Concrete Reinforcing	99.5	59.2	78.6	106.0	60.0	82.2	97.9	58.5	77.5	103.0	102.8	102.9	92.7	104.1	98.6	91.7	104.0	98.1
0330	Cast-in-Place Concrete	93.8	69.5	84.2	95.5	63.9	83.1	101.2	51.1	81.4	95.8	108.1	100.7	75.6	118.2	92.4	75.6	108.2	88.5
03	CONCRETE	95.4	67.3	81.7	97.6	54.9	76.8	98.2	59.4	79.3	101.5	99.4	100.5	91.9	103.6	97.6	91.9	99.8	95.7
04	MASONRY	94.2	61.4	74.0	104.8	50.3	71.2	116.4	48.6	74.8	115.6	63.3	83.4	124.8	103.4	111.6	132.5	85.7	103.7
05	METALS	102.1	73.4	93.0	99.2	74.2	91.2	91.2	71.9	85.0	100.2	85.7	95.6	90.4	87.5	89.5	88.8	87.4	88.4
06	WOOD, PLASTICS & COMPOSITES	103.2	69.1	84.2	104.0	41.4	69.2	96.1	69.1	81.1	95.7	101.7	99.1	98.3	93.1	95.4	92.3	101.7	97.5
07	THERMAL & MOISTURE PROTECTION	97.1	69.7	85.6	95.2	61.8	81.2	97.8	63.3	83.4	99.5	70.6	87.4	97.9	83.3	91.8	97.6	77.2	89.2
08	OPENINGS	101.0	66.7	93.1	112.9	45.8	97.3	102.2	63.2	93.1	107.3	91.9	103.7	101.0	87.2	97.8	100.9	91.9	98.8
0920	Plaster & Gypsum Board	98.4	68.5	78.5	90.9	40.0	56.9	97.6	68.5	78.2	104.7	101.3	102.4	102.7	92.4	95.9	97.8	101.3	100.1
0950, 0980	Ceilings & Acoustic Treatment	95.2	68.5	77.6	84.3	40.0	55.0	96.3	68.5	77.9	92.7	101.3	98.4	79.9	92.4	88.2	79.0	101.3	93.7
0960	Flooring	107.5	65.3	95.4	105.9	58.9	92.4	102.8	42.4	85.4	97.5	53.1	84.7	94.6	109.9	99.0	92.6	53.1	81.2
0970, 0990	Wall Finishes & Painting/Coating	102.8	68.7	83.0	92.7	44.6	64.7	101.1	49.6	71.1	95.7	63.3	76.9	94.7	44.2	65.3	94.7	44.2	65.3
09	FINISHES	100.3	67.3	82.0	92.3	45.1	66.2	96.7	59.0	75.9	95.0	83.3	88.5	91.8	92.1	92.0	90.3	81.2	85.3
COVERS	DIVS. 10 - 14, 25, 28, 41, 43, 44, 46	100.0	83.6	96.4	100.0	79.0	95.4	100.0	81.4	96.0	100.0	97.6	99.5	100.0	100.2	100.2	100.0	97.6	99.5
21, 22, 23	FIRE SUPPRESSION, PLUMBING & HVAC	100.0	66.2	85.7	99.9	58.4	82.4	96.7	62.3	82.2	99.9	63.6	84.6	100.1	73.5	88.9	96.7	63.6	82.8
26, 27, 3370	ELECTRICAL, COMMUNICATIONS & UTIL.	96.5	74.6	85.4	102.5	66.1	84.1	92.6	73.7	83.0	101.0	79.5	90.1	99.5	76.1	87.7	97.6	79.5	88.5
MF2014	WEIGHTED AVERAGE	99.2	70.7	86.6	100.9	60.7	83.2	97.6	65.6	83.5	101.2	81.0	92.3	97.9	88.8	93.9	96.9	83.3	90.9

For customer support on your Commercial Renovation Cost Data, call 877.742.4719.

775

City Cost Indexes

MAINE

DIVISION		HOULTON 047			KITTERY 039			LEWISTON 042			MACHIAS 046			PORTLAND 040 - 041			ROCKLAND 048		
		MAT.	INST.	TOTAL	MAT.	INST.	TOTAL	MAT.	INST.	TOTAL	MAT.	INST.	TOTAL	MAT.	INST.	TOTAL	MAT.	INST.	TOTAL
015433	CONTRACTOR EQUIPMENT		98.1	98.1		98.1	98.1		98.1	98.1		98.1	98.1		98.1	98.1		98.1	98.1
0241, 31 - 34	SITE & INFRASTRUCTURE, DEMOLITION	94.3	99.7	98.1	86.4	98.7	95.1	92.0	98.8	96.8	93.6	99.7	97.9	91.6	98.8	96.7	89.9	99.7	96.8
0310	Concrete Forming & Accessories	100.7	99.9	100.0	91.3	94.3	93.9	102.7	94.2	95.4	97.5	99.8	99.5	102.5	94.2	95.4	98.6	99.8	99.7
0320	Concrete Reinforcing	92.7	102.6	97.8	88.7	104.1	96.7	113.9	104.1	108.9	92.7	102.6	97.8	103.0	104.1	103.6	92.7	102.6	97.8
0330	Cast-in-Place Concrete	75.6	117.9	92.3	73.4	109.0	87.5	77.1	118.3	93.4	75.6	117.9	92.3	91.4	118.3	102.0	77.2	117.9	93.2
03	CONCRETE	93.0	105.7	99.2	86.4	100.5	93.2	92.5	103.6	97.9	92.4	105.7	98.9	98.5	103.6	101.0	89.8	105.7	97.5
04	MASONRY	106.3	75.9	87.5	121.1	86.7	99.9	106.9	103.4	104.7	106.3	75.9	87.5	121.9	103.4	110.5	99.7	75.9	85.0
05	METALS	89.0	85.7	88.0	85.2	88.1	86.1	94.0	87.6	91.9	89.1	85.7	88.0	100.6	87.6	96.4	88.9	85.7	87.9
06	WOOD, PLASTICS & COMPOSITES	102.3	101.7	102.0	93.5	101.7	98.1	104.3	93.1	98.1	99.2	101.7	100.6	98.0	93.1	95.3	100.1	101.7	101.0
07	THERMAL & MOISTURE PROTECTION	98.0	78.2	89.7	101.5	76.5	91.1	97.7	83.3	91.7	97.9	77.2	89.3	103.4	83.3	95.0	97.6	77.2	89.1
08	OPENINGS	101.0	91.9	98.9	100.2	95.7	99.1	104.3	87.2	100.4	101.0	91.9	98.9	106.7	87.2	102.1	101.0	91.9	98.9
0920	Plaster & Gypsum Board	105.0	101.3	102.5	97.1	101.3	99.9	108.0	92.4	97.6	103.4	101.3	102.0	105.2	92.4	96.7	103.4	101.3	102.0
0950, 0980	Ceilings & Acoustic Treatment	79.0	101.3	93.7	89.6	101.3	97.3	88.1	92.4	91.0	79.0	101.3	93.7	90.9	92.4	91.9	79.0	101.3	93.7
0960	Flooring	96.0	49.7	82.7	94.3	55.4	83.1	97.8	109.9	101.2	95.0	49.7	82.0	99.5	109.9	102.5	95.5	49.7	82.3
0970, 0990	Wall Finishes & Painting/Coating	94.7	137.1	119.4	88.6	115.5	104.2	94.7	44.2	65.3	94.7	137.1	119.4	97.9	44.2	66.7	94.7	137.1	119.4
09	FINISHES	92.4	95.0	93.8	91.8	89.5	90.6	94.9	92.1	93.4	91.8	95.0	93.6	96.2	92.1	93.9	91.6	95.0	93.5
COVERS	DIVS. 10 - 14, 25, 28, 41, 43, 44, 46	100.0	103.4	100.7	100.0	97.9	99.5	100.0	109.2	102.0	100.0	103.4	100.7	100.0	109.2	102.0	100.0	103.4	100.7
21, 22, 23	FIRE SUPPRESSION, PLUMBING & HVAC	96.7	72.5	86.5	96.7	75.9	87.9	100.1	73.5	88.9	96.7	72.5	86.5	99.9	73.5	88.8	96.7	72.5	86.5
26, 27, 3370	ELECTRICAL, COMMUNICATIONS & UTIL.	101.4	79.5	90.3	92.3	79.5	85.8	101.5	80.4	90.8	101.4	79.5	90.3	103.5	80.4	91.8	101.4	79.5	90.3
MF2014	WEIGHTED AVERAGE	96.5	86.9	92.3	94.5	87.4	91.4	98.5	89.4	94.5	96.3	86.9	92.2	101.6	89.4	96.3	95.6	86.9	91.7

MAINE / MARYLAND

DIVISION		MAINE WATERVILLE 049			ANNAPOLIS 214			BALTIMORE 210 - 212			COLLEGE PARK 207 - 208			CUMBERLAND 215			EASTON 216		
		MAT.	INST.	TOTAL	MAT.	INST.	TOTAL	MAT.	INST.	TOTAL	MAT.	INST.	TOTAL	MAT.	INST.	TOTAL	MAT.	INST.	TOTAL
015433	CONTRACTOR EQUIPMENT		98.1	98.1		102.8	102.8		106.6	106.6		108.2	108.2		102.8	102.8		102.8	102.8
0241, 31 - 34	SITE & INFRASTRUCTURE, DEMOLITION	94.1	98.6	97.3	104.6	94.2	97.2	101.9	98.5	99.5	99.7	97.1	97.9	95.5	94.5	94.8	102.9	91.1	94.6
0310	Concrete Forming & Accessories	91.5	93.2	92.9	100.1	72.2	76.0	100.3	73.2	76.9	85.9	67.8	70.2	92.8	81.7	83.2	90.6	71.2	73.8
0320	Concrete Reinforcing	92.7	102.8	98.0	101.2	83.3	92.0	102.0	83.4	92.3	101.8	80.6	90.8	90.6	76.7	83.4	89.9	80.1	84.8
0330	Cast-in-Place Concrete	75.6	108.1	88.4	111.7	75.7	97.5	112.3	76.5	98.2	113.8	75.6	98.7	94.4	85.6	90.9	104.8	47.6	82.2
03	CONCRETE	93.4	99.4	96.3	105.1	76.6	91.2	107.0	77.4	92.6	106.4	74.3	90.8	92.6	83.0	87.9	100.7	65.9	83.8
04	MASONRY	117.7	63.3	84.2	106.3	72.7	85.6	102.9	72.7	84.3	110.1	71.7	86.4	102.5	90.0	94.8	117.0	42.7	71.2
05	METALS	89.0	85.7	88.0	99.7	95.8	98.4	100.0	96.4	98.9	87.9	99.6	91.6	96.3	94.3	95.7	96.6	88.6	94.0
06	WOOD, PLASTICS & COMPOSITES	91.8	101.7	97.3	97.0	72.9	83.6	107.3	74.2	88.9	79.9	67.6	73.0	94.7	80.2	86.6	92.6	78.5	84.8
07	THERMAL & MOISTURE PROTECTION	98.0	70.6	86.5	105.5	80.3	94.9	104.0	80.6	94.2	104.6	78.5	93.7	102.9	80.5	93.5	103.0	58.8	84.5
08	OPENINGS	101.0	91.9	98.9	103.7	78.9	98.0	97.6	80.6	93.6	93.2	73.7	88.7	100.8	81.7	96.3	99.0	71.5	92.6
0920	Plaster & Gypsum Board	97.8	101.3	100.1	101.6	72.4	82.1	104.9	73.6	84.0	94.6	66.7	75.9	104.9	79.9	88.2	104.9	78.2	87.0
0950, 0980	Ceilings & Acoustic Treatment	79.0	101.3	93.7	93.1	72.4	79.4	103.9	73.6	83.8	96.0	66.7	76.6	98.3	79.9	86.1	98.3	78.2	85.0
0960	Flooring	92.2	53.1	81.0	92.3	79.4	88.6	94.3	79.4	90.0	87.7	78.3	85.0	88.8	97.2	91.2	87.9	49.7	76.9
0970, 0990	Wall Finishes & Painting/Coating	94.7	63.3	76.5	86.9	78.0	81.7	95.2	78.0	85.2	102.9	78.0	88.4	92.8	95.6	94.4	92.8	78.0	84.2
09	FINISHES	90.4	83.3	86.5	92.1	73.4	81.8	99.8	74.2	85.7	91.1	69.8	79.3	95.1	85.9	90.0	95.2	69.5	81.0
COVERS	DIVS. 10 - 14, 25, 28, 41, 43, 44, 46	100.0	97.6	99.5	100.0	86.5	97.1	100.0	87.0	97.2	100.0	84.5	96.6	100.0	91.0	98.0	100.0	73.7	94.3
21, 22, 23	FIRE SUPPRESSION, PLUMBING & HVAC	96.7	63.6	82.7	100.0	81.8	92.3	100.0	81.8	92.3	96.7	83.4	91.1	96.5	74.2	87.1	96.5	77.2	88.3
26, 27, 3370	ELECTRICAL, COMMUNICATIONS & UTIL.	101.4	79.5	90.3	101.0	91.8	96.4	99.7	91.8	95.7	100.1	102.2	101.1	98.1	82.1	90.0	97.7	66.7	82.0
MF2014	WEIGHTED AVERAGE	96.8	81.0	89.8	101.0	82.7	92.9	100.0	83.4	93.2	97.1	83.8	91.2	97.3	84.1	91.5	98.9	71.2	86.7

MARYLAND / MASSACHUSETTS

DIVISION		ELKTON 219			HAGERSTOWN 217			SALISBURY 218			SILVER SPRING 209			WALDORF 206			BOSTON 020 - 022, 024		
		MAT.	INST.	TOTAL	MAT.	INST.	TOTAL	MAT.	INST.	TOTAL	MAT.	INST.	TOTAL	MAT.	INST.	TOTAL	MAT.	INST.	TOTAL
015433	CONTRACTOR EQUIPMENT		102.8	102.8		102.8	102.8		102.8	102.8		100.1	100.1		100.1	100.1		104.3	104.3
0241, 31 - 34	SITE & INFRASTRUCTURE, DEMOLITION	89.4	92.0	91.2	93.7	95.0	94.6	102.8	91.2	94.6	88.9	89.2	89.2	95.2	88.9	90.8	98.4	106.8	104.4
0310	Concrete Forming & Accessories	96.8	88.9	90.0	91.7	77.1	79.1	105.4	50.3	57.8	94.6	68.5	72.0	102.1	66.8	71.6	103.5	145.6	139.9
0320	Concrete Reinforcing	89.9	104.1	97.3	90.6	76.7	83.4	89.9	62.8	75.9	100.5	80.5	90.2	101.2	80.5	90.5	102.8	152.6	128.6
0330	Cast-in-Place Concrete	84.8	75.1	81.0	89.9	85.6	88.2	104.8	46.2	81.7	116.5	78.4	101.5	130.5	75.7	108.9	99.8	146.2	118.1
03	CONCRETE	85.3	87.7	86.5	89.0	80.9	85.1	101.7	52.9	78.0	104.8	75.5	90.5	115.5	73.8	95.2	101.2	145.9	122.9
04	MASONRY	101.4	60.1	75.9	108.5	90.0	97.1	116.9	47.0	73.8	109.3	76.7	89.2	94.0	71.7	80.2	107.9	165.2	143.2
05	METALS	96.6	101.7	98.2	96.4	94.2	95.7	96.6	82.6	92.1	92.1	95.7	93.3	92.1	95.9	93.3	99.7	130.0	109.3
06	WOOD, PLASTICS & COMPOSITES	99.9	97.0	98.3	93.8	74.1	82.8	110.7	53.2	78.7	86.6	66.8	75.6	94.0	66.8	78.8	106.4	145.5	128.1
07	THERMAL & MOISTURE PROTECTION	102.6	75.1	91.1	102.6	78.7	92.6	103.3	63.0	86.4	107.8	85.5	98.5	108.3	83.4	97.9	97.7	150.5	119.8
08	OPENINGS	99.0	86.1	96.0	99.0	75.2	93.4	99.2	60.5	90.2	85.4	72.8	82.4	86.0	73.3	83.0	99.8	147.8	111.0
0920	Plaster & Gypsum Board	108.2	97.1	100.8	104.9	73.6	84.0	115.5	52.1	73.2	100.5	66.7	77.9	103.5	66.7	78.9	104.6	146.3	132.4
0950, 0980	Ceilings & Acoustic Treatment	98.3	97.1	97.5	99.2	73.6	82.2	98.3	52.1	67.7	104.3	66.7	79.4	104.3	66.7	79.4	98.1	146.3	130.0
0960	Flooring	90.4	56.2	80.6	88.4	97.2	90.9	94.5	62.2	85.2	93.5	78.3	89.2	97.0	78.3	91.7	95.7	178.6	119.5
0970, 0990	Wall Finishes & Painting/Coating	92.8	77.6	84.0	92.8	78.0	84.2	92.8	77.6	84.0	110.9	78.0	91.7	110.9	78.0	91.7	91.3	166.2	134.9
09	FINISHES	95.6	82.8	88.5	94.9	80.3	86.9	98.7	54.7	74.4	92.4	69.7	79.8	94.2	69.1	80.3	99.6	154.6	130.0
COVERS	DIVS. 10 - 14, 25, 28, 41, 43, 44, 46	100.0	60.2	91.4	100.0	90.3	97.9	100.0	37.6	86.5	100.0	84.3	96.6	100.0	82.6	96.2	100.0	119.8	104.3
21, 22, 23	FIRE SUPPRESSION, PLUMBING & HVAC	96.5	81.5	90.2	99.9	85.9	94.0	96.5	59.3	80.8	96.7	85.5	92.0	96.7	83.0	91.0	99.9	131.8	113.4
26, 27, 3370	ELECTRICAL, COMMUNICATIONS & UTIL.	99.3	92.3	95.8	97.9	82.1	89.9	96.6	66.6	81.4	97.5	102.2	99.9	95.2	102.2	98.7	102.1	134.3	118.4
MF2014	WEIGHTED AVERAGE	96.3	84.2	91.0	97.7	85.2	92.2	99.4	61.8	82.8	96.4	84.0	90.9	97.1	82.6	90.7	100.6	139.1	117.5

City Cost Indexes

MASSACHUSETTS

DIVISION		BROCKTON 023 MAT.	INST.	TOTAL	BUZZARDS BAY 025 MAT.	INST.	TOTAL	FALL RIVER 027 MAT.	INST.	TOTAL	FITCHBURG 014 MAT.	INST.	TOTAL	FRAMINGHAM 017 MAT.	INST.	TOTAL	GREENFIELD 013 MAT.	INST.	TOTAL
015433	CONTRACTOR EQUIPMENT		100.3	100.3		100.3	100.3		101.2	101.2		98.1	98.1		99.6	99.6		98.1	98.1
0241, 31 - 34	SITE & INFRASTRUCTURE, DEMOLITION	94.8	103.2	100.8	85.1	103.1	97.9	93.8	103.3	100.5	86.7	103.1	98.3	83.4	102.9	97.2	90.4	101.7	98.4
0310	Concrete Forming & Accessories	101.1	134.6	130.1	98.6	134.6	129.7	101.1	134.8	130.3	95.5	127.0	122.8	103.5	134.9	130.7	93.7	110.0	107.8
0320	Concrete Reinforcing	101.8	152.0	127.8	81.7	158.3	121.3	101.8	158.3	131.1	85.7	151.7	119.9	85.7	152.5	120.3	89.2	125.4	108.0
0330	Cast-in-Place Concrete	95.2	142.8	114.0	79.1	142.8	104.2	92.1	143.3	112.3	80.7	142.7	105.2	80.8	146.3	106.6	83.0	124.6	99.4
03	CONCRETE	99.2	139.6	118.8	83.9	140.7	111.5	97.7	141.0	118.7	83.7	135.9	109.1	86.4	141.0	113.0	87.2	117.2	101.8
04	MASONRY	105.3	151.7	133.9	98.0	159.2	135.7	106.1	159.2	138.8	106.8	158.3	138.5	113.8	159.3	141.9	111.7	130.1	123.1
05	METALS	95.5	126.6	105.4	90.1	129.3	102.6	95.5	129.7	106.4	92.4	123.7	102.3	92.4	127.3	103.5	94.7	108.9	99.2
06	WOOD, PLASTICS & COMPOSITES	103.4	133.8	120.3	100.5	133.8	119.0	103.4	134.0	120.5	99.4	124.0	113.1	105.7	133.5	121.2	97.3	107.1	102.8
07	THERMAL & MOISTURE PROTECTION	98.3	143.8	117.4	97.5	143.6	116.8	98.2	143.0	116.9	99.0	137.8	115.2	99.1	146.5	118.9	99.0	116.4	106.3
08	OPENINGS	100.9	138.5	109.7	96.8	136.4	106.0	100.9	136.6	109.2	102.7	133.0	109.8	93.1	138.4	103.7	103.0	113.6	105.5
0920	Plaster & Gypsum Board	90.7	134.2	119.8	87.1	134.2	118.6	90.7	134.2	119.8	99.3	124.2	115.9	101.9	134.2	123.5	99.7	106.9	104.5
0950, 0980	Ceilings & Acoustic Treatment	96.8	134.2	121.6	83.7	134.2	117.2	96.8	134.2	121.6	85.3	124.2	111.1	85.3	134.2	117.7	92.8	106.9	102.1
0960	Flooring	95.3	178.6	119.3	93.1	178.6	117.7	94.3	178.6	118.5	94.8	178.6	118.9	96.7	178.6	120.2	93.9	139.3	106.9
0970, 0990	Wall Finishes & Painting/Coating	87.8	155.2	127.0	87.8	155.2	127.0	87.8	155.2	127.0	89.2	155.2	127.6	90.1	155.2	128.0	89.2	126.6	111.0
09	FINISHES	94.4	145.4	122.6	89.7	145.4	120.6	94.1	145.6	122.6	90.8	139.6	117.8	91.6	145.3	121.3	92.6	116.9	106.0
COVERS	DIVS. 10 - 14, 25, 28, 41, 43, 44, 46	100.0	117.1	103.7	100.0	117.1	103.7	100.0	117.7	103.8	100.0	105.9	101.3	100.0	116.7	103.6	100.0	105.9	101.3
21, 22, 23	FIRE SUPPRESSION, PLUMBING & HVAC	100.1	110.9	104.7	96.7	110.0	102.3	100.1	111.0	104.7	97.1	112.4	103.6	97.1	124.7	108.8	97.1	101.8	99.1
26, 27, 3370	ELECTRICAL, COMMUNICATIONS & UTIL.	98.4	99.3	98.9	95.8	99.3	97.6	98.3	99.3	98.8	98.2	105.5	101.9	95.1	127.9	111.7	98.2	99.0	98.6
MF2014	WEIGHTED AVERAGE	98.8	125.1	110.4	93.7	125.9	107.9	98.6	126.3	110.8	93.6	124.5	108.3	94.9	132.9	111.6	96.9	110.0	102.6

MASSACHUSETTS

DIVISION		HYANNIS 026 MAT.	INST.	TOTAL	LAWRENCE 019 MAT.	INST.	TOTAL	LOWELL 018 MAT.	INST.	TOTAL	NEW BEDFORD 027 MAT.	INST.	TOTAL	PITTSFIELD 012 MAT.	INST.	TOTAL	SPRINGFIELD 010 - 011 MAT.	INST.	TOTAL
015433	CONTRACTOR EQUIPMENT		100.3	100.3		100.3	100.3		98.1	98.1		101.2	101.2		98.1	98.1		98.1	98.1
0241, 31 - 34	SITE & INFRASTRUCTURE, DEMOLITION	91.2	103.1	99.6	95.6	103.2	101.0	94.6	103.2	100.7	92.3	103.3	100.1	95.6	101.6	99.8	95.0	101.8	99.8
0310	Concrete Forming & Accessories	92.7	134.6	128.9	104.8	135.1	131.0	101.0	135.2	130.6	101.1	134.8	130.3	101.0	109.1	108.0	101.3	110.3	109.1
0320	Concrete Reinforcing	81.7	158.3	121.3	106.0	148.7	128.1	106.9	148.7	128.5	101.8	158.3	131.1	88.5	120.1	104.9	106.9	125.4	116.5
0330	Cast-in-Place Concrete	86.8	142.8	108.9	93.4	143.0	113.0	84.9	143.0	107.9	81.1	143.3	105.7	92.6	123.1	104.7	88.4	125.1	102.9
03	CONCRETE	90.1	140.7	114.7	101.0	139.3	119.6	92.6	139.1	115.2	92.3	141.0	116.0	93.7	115.3	104.2	94.3	117.5	105.9
04	MASONRY	104.3	159.2	138.1	119.6	159.9	144.5	105.8	159.2	138.7	104.2	159.2	138.1	106.4	132.5	122.5	106.1	131.0	121.4
05	METALS	91.6	129.3	103.6	95.1	126.1	105.0	95.1	123.4	104.1	95.5	129.7	106.4	94.9	107.1	98.8	98.1	108.9	101.5
06	WOOD, PLASTICS & COMPOSITES	93.8	133.8	116.0	106.0	133.8	121.5	105.5	133.8	121.2	103.4	134.0	120.5	105.5	107.1	106.4	105.5	107.1	106.4
07	THERMAL & MOISTURE PROTECTION	97.8	143.6	117.0	99.6	145.7	118.9	99.3	146.6	119.1	98.2	143.0	116.9	99.4	116.6	106.6	99.3	116.8	106.6
08	OPENINGS	97.4	136.4	106.5	97.1	137.5	106.5	104.2	137.5	111.9	100.9	136.6	109.2	104.2	112.2	106.0	104.2	113.6	106.4
0920	Plaster & Gypsum Board	82.5	134.2	117.1	104.4	134.2	124.3	104.4	134.2	124.3	90.7	134.2	119.8	104.4	106.9	106.0	104.4	106.9	106.0
0950, 0980	Ceilings & Acoustic Treatment	90.2	134.2	119.3	94.6	134.2	120.9	94.6	134.2	120.9	96.8	134.2	121.6	94.6	106.9	102.7	94.6	106.9	102.7
0960	Flooring	90.4	178.6	115.7	97.2	178.6	120.6	97.2	178.6	120.6	94.3	178.6	118.5	97.7	148.1	112.2	96.5	139.3	108.8
0970, 0990	Wall Finishes & Painting/Coating	87.8	155.2	127.0	89.3	155.2	127.6	89.2	155.2	127.6	87.8	155.2	127.0	89.2	126.6	111.0	90.5	126.6	111.5
09	FINISHES	90.0	145.4	120.7	94.9	145.4	122.8	94.8	145.4	122.8	94.0	145.6	122.5	95.0	118.0	107.7	94.7	117.1	107.1
COVERS	DIVS. 10 - 14, 25, 28, 41, 43, 44, 46	100.0	117.1	103.7	100.0	117.3	103.7	100.0	117.3	103.7	100.0	117.7	103.8	100.0	105.0	101.1	100.0	106.2	101.3
21, 22, 23	FIRE SUPPRESSION, PLUMBING & HVAC	100.1	110.0	104.3	100.1	124.4	110.3	100.1	124.4	110.3	100.1	111.0	104.7	100.1	101.0	100.5	100.1	102.3	101.0
26, 27, 3370	ELECTRICAL, COMMUNICATIONS & UTIL.	96.2	99.3	97.8	97.3	127.9	112.8	97.8	127.9	113.0	99.1	99.3	99.2	97.8	99.0	98.4	97.8	99.0	98.4
MF2014	WEIGHTED AVERAGE	95.9	125.9	109.1	99.2	132.6	113.9	98.4	132.3	113.3	97.9	126.3	110.4	98.5	109.7	103.4	99.1	110.3	104.0

MASSACHUSETTS / MICHIGAN

DIVISION		WORCESTER 015 - 016 MAT.	INST.	TOTAL	ANN ARBOR 481 MAT.	INST.	TOTAL	BATTLE CREEK 490 MAT.	INST.	TOTAL	BAY CITY 487 MAT.	INST.	TOTAL	DEARBORN 481 MAT.	INST.	TOTAL	DETROIT 482 MAT.	INST.	TOTAL
015433	CONTRACTOR EQUIPMENT		98.1	98.1		109.7	109.7		99.2	99.2		109.7	109.7		109.7	109.7		95.4	95.4
0241, 31 - 34	SITE & INFRASTRUCTURE, DEMOLITION	94.9	103.1	100.7	82.3	97.3	92.9	94.7	85.1	87.9	73.7	96.2	89.6	82.1	97.4	92.9	95.3	99.0	97.9
0310	Concrete Forming & Accessories	101.6	127.1	123.6	97.0	113.3	111.1	97.8	84.2	86.0	97.1	84.3	86.1	96.9	114.1	111.8	100.7	114.1	112.3
0320	Concrete Reinforcing	106.9	151.7	130.1	96.8	112.5	104.9	94.9	91.8	93.3	96.8	111.6	104.5	96.8	112.7	105.0	98.7	112.6	105.9
0330	Cast-in-Place Concrete	87.9	142.7	109.5	87.2	105.3	94.3	92.8	97.4	94.6	83.4	88.4	85.4	85.2	103.0	92.2	99.0	103.0	100.6
03	CONCRETE	94.1	135.9	114.4	91.2	110.9	100.8	96.2	89.5	93.0	89.4	92.0	90.7	90.3	110.5	100.1	97.6	109.2	103.2
04	MASONRY	105.6	158.3	138.1	99.9	109.7	105.9	103.8	88.2	94.2	99.4	87.5	92.1	99.7	113.7	108.3	92.0	113.7	105.4
05	METALS	98.1	123.7	106.3	96.0	119.3	103.4	99.3	85.1	94.8	96.7	115.7	102.7	96.1	119.7	103.6	96.4	100.0	97.6
06	WOOD, PLASTICS & COMPOSITES	105.9	124.0	116.0	89.4	115.0	103.6	91.5	82.4	86.5	99.4	83.3	86.0	89.4	115.0	103.6	96.2	115.0	106.7
07	THERMAL & MOISTURE PROTECTION	99.3	138.5	115.7	102.8	107.3	104.7	97.6	83.3	91.6	100.5	87.0	94.9	101.2	113.0	106.2	99.9	113.0	105.3
08	OPENINGS	104.2	133.0	110.9	99.4	110.5	102.0	95.1	78.8	91.3	99.4	91.0	97.4	99.4	111.0	102.1	100.5	111.0	103.0
0920	Plaster & Gypsum Board	104.4	124.2	117.6	101.8	114.9	110.6	93.9	78.3	83.5	101.8	82.4	88.8	101.8	114.9	110.6	100.9	114.9	110.3
0950, 0980	Ceilings & Acoustic Treatment	94.6	124.2	114.2	91.2	114.9	106.9	85.8	78.3	80.9	92.1	82.4	85.7	91.2	114.9	106.9	101.2	114.9	110.3
0960	Flooring	97.2	172.8	118.9	96.6	114.5	101.8	97.6	88.3	95.0	96.6	78.4	91.4	96.3	112.3	100.9	98.0	112.3	102.1
0970, 0990	Wall Finishes & Painting/Coating	89.2	155.2	127.6	85.5	98.7	93.2	92.5	81.9	86.4	85.5	84.2	84.7	85.5	104.1	96.3	88.6	104.1	97.6
09	FINISHES	94.8	138.5	119.0	93.2	112.6	104.0	92.2	84.4	87.8	93.0	82.5	87.2	93.1	113.2	104.2	97.2	113.2	106.1
COVERS	DIVS. 10 - 14, 25, 28, 41, 43, 44, 46	100.0	110.2	102.2	100.0	105.8	101.3	100.0	93.8	98.6	100.0	91.0	98.1	100.0	106.3	101.4	100.0	106.3	101.4
21, 22, 23	FIRE SUPPRESSION, PLUMBING & HVAC	100.1	112.5	105.3	100.1	98.2	99.3	100.2	87.2	94.7	100.1	82.8	92.8	100.1	108.1	103.5	99.9	109.1	103.8
26, 27, 3370	ELECTRICAL, COMMUNICATIONS & UTIL.	97.8	105.5	101.7	96.9	108.0	102.5	93.1	83.1	88.0	96.0	89.6	92.8	96.9	106.2	101.6	97.8	106.1	102.0
MF2014	WEIGHTED AVERAGE	99.0	124.5	110.2	97.0	107.4	101.6	97.5	86.0	92.4	96.4	90.3	93.7	96.8	109.9	102.6	98.2	108.2	102.6

MICHIGAN

DIVISION		FLINT 484 - 485			GAYLORD 497			GRAND RAPIDS 493,495			IRON MOUNTAIN 498 - 499			JACKSON 492			KALAMAZOO 491		
		MAT.	INST.	TOTAL	MAT.	INST.	TOTAL	MAT.	INST.	TOTAL	MAT.	INST.	TOTAL	MAT.	INST.	TOTAL	MAT.	INST.	TOTAL
015433	CONTRACTOR EQUIPMENT		109.7	109.7		104.1	104.1		99.2	99.2		92.3	92.3		104.1	104.1		99.2	99.2
0241, 31 - 34	SITE & INFRASTRUCTURE, DEMOLITION	71.7	96.4	89.1	89.5	82.6	84.6	93.4	85.1	87.5	97.7	91.6	93.4	111.8	84.8	92.7	95.0	85.1	88.0
0310	Concrete Forming & Accessories	99.8	89.1	90.5	96.0	76.7	79.3	97.2	83.6	85.4	87.7	82.7	83.4	92.8	85.7	86.7	97.8	83.9	85.8
0320	Concrete Reinforcing	96.8	112.0	104.7	88.5	117.3	103.4	99.6	91.7	95.5	88.3	98.8	93.7	86.0	111.8	99.4	94.9	92.3	93.5
0330	Cast-in-Place Concrete	87.7	92.2	89.5	92.6	85.4	89.7	97.3	96.6	97.0	109.6	84.4	99.6	92.4	91.1	91.9	94.6	97.3	95.7
03	CONCRETE	91.7	95.5	93.5	92.7	88.4	90.6	99.9	89.0	94.6	102.2	86.5	94.5	87.4	93.4	90.3	99.7	89.4	94.7
04	MASONRY	99.9	97.0	98.1	114.8	79.9	93.3	98.0	87.6	91.6	99.5	88.9	93.0	93.4	91.9	92.5	102.2	90.1	94.7
05	METALS	96.1	116.6	102.6	100.6	113.8	104.8	96.3	84.7	92.6	99.9	92.1	97.5	100.8	114.4	105.1	99.3	85.1	94.8
06	WOOD, PLASTICS & COMPOSITES	92.9	87.4	89.9	84.9	75.7	79.8	94.9	82.0	87.7	80.4	82.2	81.4	83.6	83.5	83.6	91.5	82.4	86.5
07	THERMAL & MOISTURE PROTECTION	100.5	94.2	97.9	96.1	73.4	86.6	99.2	75.3	89.2	99.8	80.3	91.6	95.4	91.0	93.6	97.6	83.5	91.7
08	OPENINGS	99.4	94.0	98.1	95.4	75.9	90.9	103.0	83.0	98.3	102.7	75.5	96.4	94.4	91.3	93.7	95.1	78.9	91.3
0920	Plaster & Gypsum Board	104.1	86.6	92.4	93.7	73.9	80.5	95.5	77.9	83.8	52.2	82.2	72.3	92.0	81.9	85.3	93.9	78.3	83.5
0950, 0980	Ceilings & Acoustic Treatment	91.2	86.6	88.2	84.9	73.9	77.6	94.6	77.9	83.6	83.9	82.2	82.8	84.9	81.9	82.9	85.8	78.3	80.9
0960	Flooring	96.6	92.0	95.3	90.5	88.8	90.0	97.3	82.4	93.0	111.2	95.3	106.6	89.1	79.5	86.4	97.6	79.5	92.4
0970, 0990	Wall Finishes & Painting/Coating	85.5	85.4	85.5	88.8	84.2	86.1	93.6	80.8	86.2	107.6	73.6	87.8	88.8	96.8	93.4	92.5	81.9	86.4
09	FINISHES	92.8	88.6	90.5	91.6	78.2	84.2	95.6	82.9	88.6	92.8	83.9	87.9	92.5	84.9	88.3	92.2	82.4	86.8
COVERS	DIVS. 10 - 14, 25, 28, 41, 43, 44, 46	100.0	92.6	98.4	100.0	86.5	97.1	100.0	93.9	98.7	100.0	84.8	96.7	100.0	101.4	100.3	100.0	93.7	98.6
21, 22, 23	FIRE SUPPRESSION, PLUMBING & HVAC	100.1	88.7	95.3	97.0	78.6	89.2	100.1	82.4	92.6	96.9	84.6	91.7	97.0	87.8	93.1	100.2	80.7	91.9
26, 27, 3370	ELECTRICAL, COMMUNICATIONS & UTIL.	96.9	99.0	97.9	91.1	78.5	84.8	97.7	78.5	88.0	96.9	83.6	90.2	94.8	108.0	101.4	93.0	78.2	85.5
MF2014	WEIGHTED AVERAGE	96.7	95.6	96.2	96.7	83.6	90.9	98.8	83.9	92.3	98.7	85.8	93.0	96.0	94.2	95.2	97.8	83.9	91.7

MICHIGAN / MINNESOTA

DIVISION		LANSING 488 - 489			MUSKEGON 494			ROYAL OAK 480,483			SAGINAW 486			TRAVERSE CITY 496			BEMIDJI 566		
		MAT.	INST.	TOTAL	MAT.	INST.	TOTAL	MAT.	INST.	TOTAL	MAT.	INST.	TOTAL	MAT.	INST.	TOTAL	MAT.	INST.	TOTAL
015433	CONTRACTOR EQUIPMENT		109.7	109.7		99.2	99.2		92.5	92.5		109.7	109.7		92.3	92.3		98.8	98.8
0241, 31 - 34	SITE & INFRASTRUCTURE, DEMOLITION	94.7	96.4	95.9	92.6	85.0	87.2	86.5	96.6	93.7	74.7	96.1	89.9	83.4	91.1	88.9	96.7	98.2	97.8
0310	Concrete Forming & Accessories	95.5	89.6	90.4	98.2	81.6	83.8	92.6	113.6	110.7	97.0	86.7	88.1	87.7	77.4	78.8	86.5	87.6	87.5
0320	Concrete Reinforcing	98.4	111.8	105.4	95.6	91.9	93.7	87.7	118.5	103.7	96.8	111.6	104.5	89.6	91.5	90.6	96.5	112.3	104.7
0330	Cast-in-Place Concrete	97.5	91.0	94.9	92.5	93.0	92.7	76.3	103.6	87.1	86.1	88.3	87.0	85.5	80.2	83.4	104.0	102.1	103.2
03	CONCRETE	97.3	95.2	96.3	94.5	86.9	90.8	78.7	110.0	93.9	90.7	93.0	91.9	84.0	81.4	82.7	96.6	98.4	97.5
04	MASONRY	101.3	95.5	97.7	100.8	88.1	92.9	93.6	111.5	104.6	101.3	87.5	92.8	97.6	84.6	89.6	106.4	108.9	107.9
05	METALS	96.0	116.0	102.3	96.9	84.8	93.1	99.1	96.6	98.4	96.1	115.5	102.3	99.9	89.3	96.5	88.6	123.6	99.8
06	WOOD, PLASTICS & COMPOSITES	90.0	88.3	89.1	88.3	80.5	83.9	85.2	115.0	101.8	86.0	86.8	86.5	80.4	76.8	78.4	69.3	82.6	76.7
07	THERMAL & MOISTURE PROTECTION	99.5	88.0	94.7	96.5	77.2	88.4	99.1	108.8	103.1	101.4	87.5	95.6	98.8	72.0	87.6	106.7	101.4	104.5
08	OPENINGS	101.0	93.8	99.3	94.3	81.9	91.4	99.2	110.3	101.8	97.3	93.0	96.3	102.7	71.4	95.4	104.4	109.7	105.6
0920	Plaster & Gypsum Board	94.3	87.5	89.8	74.9	76.3	75.9	99.4	114.9	109.8	101.8	86.0	91.2	52.2	76.6	68.5	103.7	82.4	89.5
0950, 0980	Ceilings & Acoustic Treatment	90.6	87.5	88.5	86.6	76.3	79.8	90.5	114.9	106.7	91.2	86.0	87.8	83.9	76.6	79.1	123.1	82.4	96.2
0960	Flooring	102.2	82.4	96.5	96.3	82.4	92.3	93.3	109.5	98.0	96.6	78.4	91.4	111.2	91.2	105.4	100.0	120.2	105.8
0970, 0990	Wall Finishes & Painting/Coating	93.8	85.8	89.1	90.7	80.8	84.9	86.8	96.8	92.6	85.5	84.2	84.7	107.6	44.7	71.0	95.2	95.9	95.6
09	FINISHES	96.2	87.5	91.4	88.6	80.7	84.2	91.9	111.7	102.9	92.9	84.6	88.3	91.7	76.5	83.3	104.3	93.4	98.2
COVERS	DIVS. 10 - 14, 25, 28, 41, 43, 44, 46	100.0	101.0	100.2	100.0	93.0	98.5	100.0	100.2	100.0	100.0	91.4	98.1	100.0	83.5	96.4	100.0	96.3	99.2
21, 22, 23	FIRE SUPPRESSION, PLUMBING & HVAC	100.0	88.5	95.1	99.9	81.3	92.1	96.9	106.0	100.8	100.1	82.3	92.5	96.9	78.2	89.0	96.9	82.6	90.8
26, 27, 3370	ELECTRICAL, COMMUNICATIONS & UTIL.	99.8	92.6	96.2	93.3	78.5	85.8	98.9	103.6	101.3	94.9	89.8	92.3	92.6	78.5	85.5	105.2	105.7	105.5
MF2014	WEIGHTED AVERAGE	98.7	94.4	96.8	96.3	83.2	90.5	95.1	106.1	99.9	96.3	90.8	93.9	95.6	80.9	89.1	98.6	99.1	98.8

MINNESOTA

DIVISION		BRAINERD 564			DETROIT LAKES 565			DULUTH 556 - 558			MANKATO 560			MINNEAPOLIS 553 - 555			ROCHESTER 559		
		MAT.	INST.	TOTAL	MAT.	INST.	TOTAL	MAT.	INST.	TOTAL	MAT.	INST.	TOTAL	MAT.	INST.	TOTAL	MAT.	INST.	TOTAL
015433	CONTRACTOR EQUIPMENT		101.5	101.5		98.8	98.8		101.4	101.4		101.5	101.5		105.0	105.0		101.4	101.4
0241, 31 - 34	SITE & INFRASTRUCTURE, DEMOLITION	98.3	103.1	101.7	94.9	98.5	97.5	99.5	102.8	101.8	95.0	103.8	101.2	99.0	108.2	105.5	99.1	102.2	101.3
0310	Concrete Forming & Accessories	87.7	89.4	89.2	83.4	89.4	88.6	98.1	112.2	110.3	95.9	102.7	101.8	100.2	128.5	124.7	100.7	108.3	107.3
0320	Concrete Reinforcing	95.3	112.1	104.0	96.5	112.3	104.7	95.5	112.6	104.3	95.2	119.4	107.7	94.2	119.9	107.5	93.0	119.6	106.8
0330	Cast-in-Place Concrete	113.0	105.6	110.1	101.0	105.0	102.6	110.4	107.8	109.4	104.2	107.0	105.3	101.3	119.0	108.3	107.3	102.5	105.4
03	CONCRETE	100.7	100.4	100.5	94.1	100.2	97.1	105.6	111.5	108.4	96.1	108.1	102.0	99.4	124.0	111.3	100.6	109.3	104.8
04	MASONRY	132.6	111.5	119.6	132.1	114.5	121.2	108.3	121.8	116.7	119.9	112.6	115.4	109.9	131.5	123.2	106.6	117.2	113.1
05	METALS	89.7	123.1	100.3	88.6	123.4	99.7	94.9	125.7	104.7	89.5	127.3	101.5	95.8	130.8	106.9	95.0	129.3	105.9
06	WOOD, PLASTICS & COMPOSITES	86.4	82.4	84.2	66.4	82.6	75.4	89.2	116.0	101.1	95.7	100.6	98.4	98.1	126.6	114.0	98.1	106.5	102.8
07	THERMAL & MOISTURE PROTECTION	105.2	108.1	106.4	106.5	110.3	108.1	103.6	115.5	108.6	100.7	103.6	102.0	102.6	129.7	113.9	107.4	104.1	106.0
08	OPENINGS	90.3	109.6	94.8	104.4	109.7	105.6	111.2	118.3	112.8	95.2	117.8	100.5	106.2	136.3	113.2	106.4	125.3	110.8
0920	Plaster & Gypsum Board	89.4	82.4	84.8	102.7	82.4	89.2	93.0	111.3	105.2	94.1	101.1	98.8	92.1	127.7	115.9	100.4	107.1	104.9
0950, 0980	Ceilings & Acoustic Treatment	57.8	82.4	74.1	123.1	82.4	96.2	99.7	111.3	107.4	57.8	101.1	86.5	101.4	127.7	118.8	95.8	107.1	103.3
0960	Flooring	98.8	120.2	105.0	98.7	120.2	104.9	99.8	123.4	106.6	101.0	125.5	108.0	99.0	120.2	105.1	100.4	121.1	106.3
0970, 0990	Wall Finishes & Painting/Coating	89.3	95.9	93.1	95.2	95.9	95.6	89.3	110.3	101.5	101.0	105.1	103.4	106.1	131.3	120.8	90.5	104.9	98.9
09	FINISHES	87.1	94.6	91.2	103.6	94.7	98.7	96.8	114.4	106.5	89.1	107.1	99.0	100.3	128.0	115.6	97.0	110.4	104.4
COVERS	DIVS. 10 - 14, 25, 28, 41, 43, 44, 46	100.0	97.6	99.5	100.0	98.0	99.6	100.0	96.7	99.3	100.0	99.3	99.9	100.0	110.0	102.2	100.0	103.6	100.8
21, 22, 23	FIRE SUPPRESSION, PLUMBING & HVAC	96.1	85.3	91.5	96.9	85.2	91.9	100.0	103.5	101.3	96.1	87.6	92.5	99.8	107.2	102.9	99.8	96.4	98.4
26, 27, 3370	ELECTRICAL, COMMUNICATIONS & UTIL.	102.8	102.4	102.6	105.0	70.6	87.6	100.9	102.4	101.6	109.3	88.5	98.8	102.6	113.2	108.0	101.3	88.5	94.9
MF2014	WEIGHTED AVERAGE	97.3	100.5	98.7	99.4	96.1	98.0	101.3	110.5	105.3	97.5	102.7	99.8	100.7	119.9	109.1	100.3	106.3	103.0

MINNESOTA / MISSISSIPPI

DIVISION		SAINT PAUL 550 - 551			ST. CLOUD 563			THIEF RIVER FALLS 567			WILLMAR 562			WINDOM 561			BILOXI 395		
		MAT.	INST.	TOTAL	MAT.	INST.	TOTAL	MAT.	INST.	TOTAL	MAT.	INST.	TOTAL	MAT.	INST.	TOTAL	MAT.	INST.	TOTAL
015433	CONTRACTOR EQUIPMENT		101.4	101.4		101.5	101.5		98.8	98.8		101.5	101.5		101.5	101.5		99.9	99.9
0241, 31 - 34	SITE & INFRASTRUCTURE, DEMOLITION	95.7	103.5	101.2	93.5	105.2	101.8	95.6	98.2	97.4	92.9	102.8	99.9	86.7	101.8	97.4	104.1	88.2	92.9
0310	Concrete Forming & Accessories	97.8	129.3	125.0	84.8	116.5	112.3	87.2	86.8	86.8	84.6	88.9	88.3	89.2	83.3	84.1	94.3	52.2	57.9
0320	Concrete Reinforcing	96.3	119.9	108.5	95.4	119.4	107.8	96.9	112.2	104.8	95.0	119.1	107.5	95.0	118.6	107.2	89.3	58.0	73.1
0330	Cast-in-Place Concrete	115.4	118.4	116.6	99.8	114.3	105.5	103.1	87.2	96.8	101.3	84.3	94.6	87.7	90.3	88.7	115.4	62.4	94.5
03	CONCRETE	108.2	124.1	116.0	91.9	116.8	104.0	95.3	92.9	94.1	91.9	94.3	93.0	82.7	93.6	88.0	106.6	58.6	83.3
04	MASONRY	110.1	131.5	123.3	115.8	118.0	117.2	106.4	108.9	107.9	120.1	111.7	114.9	131.8	93.5	108.2	104.3	53.6	73.1
05	METALS	94.9	130.7	106.3	90.3	128.2	102.4	88.8	122.8	99.6	89.5	126.3	101.2	89.4	124.6	100.6	91.9	85.7	89.9
06	WOOD, PLASTICS & COMPOSITES	94.9	127.6	113.1	83.8	114.5	100.9	70.2	82.6	77.1	83.5	82.6	83.0	87.7	80.0	83.4	100.2	54.2	74.6
07	THERMAL & MOISTURE PROTECTION	102.6	129.4	113.8	105.5	116.0	109.8	107.6	98.3	103.7	105.2	106.0	105.5	105.2	90.4	99.0	97.2	60.9	82.0
08	OPENINGS	103.9	136.9	111.6	95.7	129.7	103.6	104.4	109.7	105.6	92.5	91.5	92.3	96.4	89.7	94.8	97.8	56.0	88.1
0920	Plaster & Gypsum Board	96.5	128.8	118.1	89.4	115.4	106.8	103.4	82.4	89.4	89.4	82.7	84.9	89.4	80.0	83.1	96.6	53.3	67.7
0950, 0980	Ceilings & Acoustic Treatment	98.3	128.8	118.5	57.8	115.4	95.9	123.1	82.4	96.2	57.8	82.7	74.2	57.8	80.0	72.5	87.2	53.3	64.8
0960	Flooring	96.0	120.2	103.0	95.3	120.2	102.4	99.7	120.2	105.6	96.8	116.1	102.3	99.6	116.1	104.3	98.2	69.2	89.9
0970, 0990	Wall Finishes & Painting/Coating	88.5	126.1	110.4	101.0	131.3	118.7	95.2	95.9	95.6	95.2	95.9	95.6	95.2	105.1	101.0	85.2	43.4	60.9
09	FINISHES	95.0	128.0	113.3	86.6	119.0	104.5	104.1	93.4	98.2	86.6	94.0	90.7	86.9	91.2	89.3	91.3	54.1	70.7
COVERS	DIVS. 10 - 14, 25, 28, 41, 43, 44, 46	100.0	109.9	102.1	100.0	101.2	100.3	100.0	96.2	99.2	100.0	97.6	99.5	100.0	91.1	98.1	100.0	69.1	93.3
21, 22, 23	FIRE SUPPRESSION, PLUMBING & HVAC	99.7	113.6	105.6	99.4	103.5	101.2	96.9	82.3	90.7	96.1	95.8	96.0	96.1	83.3	90.7	100.0	49.7	78.7
26, 27, 3370	ELECTRICAL, COMMUNICATIONS & UTIL.	101.4	113.2	107.4	102.8	113.2	108.1	102.3	70.6	86.3	102.8	84.6	93.6	109.3	88.5	98.8	100.4	56.5	78.2
MF2014	WEIGHTED AVERAGE	100.7	120.9	109.6	96.8	114.0	104.4	98.2	93.3	96.0	95.7	98.8	97.1	96.2	93.5	95.0	98.8	60.5	81.9

MISSISSIPPI

DIVISION		CLARKSDALE 386			COLUMBUS 397			GREENVILLE 387			GREENWOOD 389			JACKSON 390 - 392			LAUREL 394		
		MAT.	INST.	TOTAL	MAT.	INST.	TOTAL	MAT.	INST.	TOTAL	MAT.	INST.	TOTAL	MAT.	INST.	TOTAL	MAT.	INST.	TOTAL
015433	CONTRACTOR EQUIPMENT		99.9	99.9		99.9	99.9		99.9	99.9		99.9	99.9		99.9	99.9		99.9	99.9
0241, 31 - 34	SITE & INFRASTRUCTURE, DEMOLITION	97.7	87.6	90.6	102.7	87.8	92.2	103.4	88.7	93.0	100.5	87.3	91.2	99.2	88.7	91.8	108.3	87.7	93.7
0310	Concrete Forming & Accessories	86.5	48.4	53.5	83.1	49.7	54.2	83.1	67.0	69.1	96.1	47.9	54.4	92.6	67.5	70.9	83.2	51.3	55.6
0320	Concrete Reinforcing	99.8	76.7	87.9	95.7	42.0	67.9	100.4	60.2	79.5	99.8	51.9	75.0	99.3	56.9	77.3	96.3	38.6	66.4
0330	Cast-in-Place Concrete	105.8	53.1	85.0	117.7	57.8	94.1	109.0	59.4	89.4	113.8	52.6	89.6	100.5	69.2	88.2	115.1	54.5	91.2
03	CONCRETE	101.1	57.1	79.7	108.2	52.9	81.3	106.9	64.6	86.4	107.9	52.0	80.7	100.8	67.5	84.7	111.1	51.9	82.3
04	MASONRY	98.7	44.8	65.5	132.8	51.4	82.6	147.4	65.7	97.0	99.4	44.7	65.7	112.3	65.7	83.6	128.4	47.9	78.8
05	METALS	90.9	90.5	90.8	88.9	76.1	84.8	91.9	86.4	90.2	90.9	76.8	86.4	98.5	85.6	94.4	89.0	75.1	84.6
06	WOOD, PLASTICS & COMPOSITES	85.5	49.5	65.4	86.1	50.4	66.2	82.2	68.5	74.6	98.6	49.5	71.2	95.8	68.5	80.6	87.2	52.7	68.0
07	THERMAL & MOISTURE PROTECTION	97.4	54.2	79.3	97.1	50.4	77.6	97.7	66.5	84.7	97.8	54.2	79.5	95.8	67.9	84.1	97.3	52.1	78.4
08	OPENINGS	96.4	57.7	87.4	97.5	48.0	86.0	96.1	66.5	88.8	96.4	49.7	85.4	100.7	64.1	92.2	94.5	48.8	83.8
0920	Plaster & Gypsum Board	84.0	48.5	60.3	87.2	49.4	62.0	83.7	68.0	73.2	94.9	48.5	63.9	90.3	68.0	75.4	87.2	51.8	63.6
0950, 0980	Ceilings & Acoustic Treatment	82.9	48.5	60.1	82.0	49.4	60.4	85.6	68.0	74.0	82.9	48.5	60.1	89.6	68.0	75.3	82.0	51.8	62.0
0960	Flooring	103.8	50.8	88.6	91.7	57.0	81.7	102.0	50.8	87.3	110.1	50.8	93.0	96.2	69.2	88.4	90.3	50.8	78.9
0970, 0990	Wall Finishes & Painting/Coating	93.9	46.1	66.1	85.2	53.0	66.5	93.9	49.8	68.2	93.9	46.1	66.1	82.5	49.8	63.5	85.2	60.9	71.1
09	FINISHES	92.6	48.4	68.2	87.0	51.7	67.4	93.2	62.6	76.3	96.4	48.4	69.9	91.0	66.3	77.3	87.1	52.7	68.1
COVERS	DIVS. 10 - 14, 25, 28, 41, 43, 44, 46	100.0	54.1	90.1	100.0	55.2	90.3	100.0	74.6	94.5	100.0	54.1	90.1	100.0	74.6	94.5	100.0	36.3	86.2
21, 22, 23	FIRE SUPPRESSION, PLUMBING & HVAC	98.5	41.2	74.3	98.1	37.1	72.3	100.0	51.0	79.3	98.5	38.9	73.3	100.0	62.8	84.3	98.2	45.3	75.8
26, 27, 3370	ELECTRICAL, COMMUNICATIONS & UTIL.	96.3	46.0	70.9	97.9	48.7	73.0	96.3	58.4	77.1	96.3	43.1	69.4	101.8	58.4	79.9	99.4	57.0	78.0
MF2014	WEIGHTED AVERAGE	96.6	55.3	78.4	98.6	53.4	78.7	100.3	65.1	84.8	97.9	52.0	77.7	99.8	68.4	86.0	98.8	55.4	79.7

MISSISSIPPI / MISSOURI

DIVISION		MCCOMB 396			MERIDIAN 393			TUPELO 388			BOWLING GREEN 633			CAPE GIRARDEAU 637			CHILLICOTHE 646		
		MAT.	INST.	TOTAL	MAT.	INST.	TOTAL	MAT.	INST.	TOTAL	MAT.	INST.	TOTAL	MAT.	INST.	TOTAL	MAT.	INST.	TOTAL
015433	CONTRACTOR EQUIPMENT		99.9	99.9		99.9	99.9		99.9	99.9		106.9	106.9		106.9	106.9		101.8	101.8
0241, 31 - 34	SITE & INFRASTRUCTURE, DEMOLITION	95.5	87.4	89.8	99.6	88.5	91.8	95.2	87.5	89.7	92.9	94.0	93.7	94.4	93.8	94.0	106.8	92.5	96.6
0310	Concrete Forming & Accessories	83.1	49.6	54.1	80.4	66.6	68.5	83.6	50.2	54.7	92.9	92.0	92.1	85.7	82.7	83.1	86.1	91.3	90.6
0320	Concrete Reinforcing	96.8	40.5	67.7	95.7	56.9	75.6	97.7	52.0	74.0	102.7	101.8	102.2	104.0	87.7	95.5	98.8	105.4	102.2
0330	Cast-in-Place Concrete	102.2	52.4	82.5	109.3	67.7	92.9	105.8	74.0	93.2	95.5	101.7	98.0	94.5	87.0	91.6	93.1	86.1	90.3
03	CONCRETE	97.0	50.6	74.5	101.8	66.6	84.7	100.6	60.2	81.0	99.1	98.3	98.7	92.7	86.6	92.6	102.7	92.8	97.9
04	MASONRY	134.2	44.2	78.7	104.0	62.8	78.6	135.2	47.7	81.2	115.2	99.5	105.5	112.2	79.7	92.1	99.2	99.6	99.4
05	METALS	89.2	73.7	84.2	90.0	85.8	88.7	90.8	77.3	86.5	94.8	116.0	101.5	95.9	108.6	99.9	89.9	108.9	95.9
06	WOOD, PLASTICS & COMPOSITES	86.1	52.2	67.2	83.6	68.5	75.2	82.8	50.4	64.7	89.7	92.1	91.1	82.7	82.4	82.5	91.5	90.2	90.7
07	THERMAL & MOISTURE PROTECTION	96.7	52.9	78.3	96.8	66.7	84.2	97.4	57.9	80.9	100.0	99.8	99.9	99.9	85.4	93.9	92.4	96.7	94.2
08	OPENINGS	97.5	49.5	86.3	97.2	65.5	89.8	96.3	49.1	85.3	100.0	99.7	99.9	100.0	80.4	95.4	95.2	94.2	95.0
0920	Plaster & Gypsum Board	87.2	51.2	63.2	87.2	68.0	74.4	83.7	49.4	60.8	90.7	92.2	91.7	89.7	82.2	84.7	102.5	89.7	94.0
0950, 0980	Ceilings & Acoustic Treatment	82.0	51.2	61.6	83.9	68.0	73.4	82.9	49.4	60.7	96.4	92.2	93.6	96.4	82.2	87.0	95.7	89.7	91.7
0960	Flooring	91.7	50.8	79.9	90.2	69.2	84.2	102.2	50.8	87.5	97.0	96.0	96.7	93.3	82.7	90.3	97.1	101.5	98.3
0970, 0990	Wall Finishes & Painting/Coating	85.2	42.3	60.3	85.2	76.2	79.9	93.9	58.7	73.4	94.8	102.8	99.4	94.8	73.2	82.3	91.6	106.3	100.2
09	FINISHES	86.3	49.5	65.9	86.6	68.4	76.5	92.1	51.1	69.4	97.7	93.7	95.5	96.4	80.8	87.8	101.3	94.2	97.4
COVERS	DIVS. 10 - 14, 25, 28, 41, 43, 44, 46	100.0	57.4	90.8	100.0	73.7	94.3	100.0	55.2	90.3	100.0	86.3	97.0	100.0	90.5	97.9	100.0	86.6	97.1
21, 22, 23	FIRE SUPPRESSION, PLUMBING & HVAC	98.1	34.2	71.1	100.0	61.4	83.7	98.6	42.8	75.0	96.5	95.2	96.0	99.9	95.9	98.2	96.7	98.5	97.5
26, 27, 3370	ELECTRICAL, COMMUNICATIONS & UTIL.	96.4	49.1	72.5	99.4	57.4	78.2	96.0	48.9	72.2	104.8	82.5	93.5	104.7	110.1	107.4	95.1	109.4	102.3
MF2014	WEIGHTED AVERAGE	97.1	51.5	77.0	97.1	67.9	84.2	98.2	55.6	79.4	99.1	96.0	97.7	99.6	92.8	96.6	96.8	98.6	97.6

City Cost Indexes

MISSOURI

DIVISION		COLUMBIA 652			FLAT RIVER 636			HANNIBAL 634			HARRISONVILLE 647			JEFFERSON CITY 650 - 651			JOPLIN 648		
		MAT.	INST.	TOTAL	MAT.	INST.	TOTAL	MAT.	INST.	TOTAL	MAT.	INST.	TOTAL	MAT.	INST.	TOTAL	MAT.	INST.	TOTAL
015433	CONTRACTOR EQUIPMENT		110.4	110.4		106.9	106.9		106.9	106.9		101.7	101.7		110.4	110.4		105.6	105.6
0241, 31 - 34	SITE & INFRASTRUCTURE, DEMOLITION	99.8	98.7	99.0	95.5	94.0	94.4	90.5	94.0	93.0	97.7	93.5	94.7	99.6	98.6	98.9	107.4	97.8	100.6
0310	Concrete Forming & Accessories	84.6	83.3	83.4	99.0	86.5	88.2	91.1	76.0	78.0	83.2	96.6	94.8	95.3	80.2	82.2	97.6	73.6	76.8
0320	Concrete Reinforcing	96.6	120.3	108.9	104.0	106.5	105.3	102.1	101.7	101.9	98.4	117.0	108.1	104.3	112.9	108.8	102.0	97.4	99.6
0330	Cast-in-Place Concrete	90.4	87.6	89.3	98.6	93.9	96.8	90.4	101.5	94.8	95.4	103.8	98.7	95.9	86.7	92.3	100.8	77.8	91.8
03	CONCRETE	91.5	93.1	92.3	102.3	94.1	98.3	95.2	91.1	93.2	98.6	103.4	100.9	98.6	90.0	94.4	101.6	80.5	91.3
04	MASONRY	153.6	90.5	114.7	112.6	81.1	93.2	106.7	99.5	102.3	93.9	105.6	101.2	108.7	90.5	97.5	93.0	85.7	88.5
05	METALS	97.6	123.4	105.8	94.7	116.9	101.7	94.8	115.6	101.4	90.3	114.4	97.9	97.2	120.4	104.6	92.8	101.1	95.5
06	WOOD, PLASTICS & COMPOSITES	83.8	80.0	81.7	97.8	85.8	91.1	87.9	71.0	78.5	88.0	94.3	91.5	95.1	76.0	84.4	103.3	71.7	85.7
07	THERMAL & MOISTURE PROTECTION	94.6	88.3	92.0	100.2	94.4	97.8	95.8		98.2	91.6	104.3	97.0	100.9	87.8	95.4	91.7	79.4	86.5
08	OPENINGS	101.7	95.1	100.2	99.9	97.7	99.4	100.0	82.1	95.8	95.2	101.7	96.7	98.0	91.0	96.3	96.3	76.3	91.7
0920	Plaster & Gypsum Board	93.4	79.5	84.1	96.7	85.6	89.3	90.1	70.5	77.0	98.2	93.9	95.3	97.1	75.3	82.6	108.7	70.7	83.3
0950, 0980	Ceilings & Acoustic Treatment	89.7	79.5	82.9	96.4	85.6	89.3	96.4	70.5	79.2	95.7	93.9	94.5	91.9	75.3	80.9	96.6	70.7	79.4
0960	Flooring	97.6	94.6	96.7	100.4	82.7	95.3	96.2	96.0	96.2	92.6	101.5	95.1	104.5	94.6	101.7	123.5	72.0	108.7
0970, 0990	Wall Finishes & Painting/Coating	92.5	82.1	86.4	94.8	78.4	85.2	94.8	102.8	99.4	96.1	106.3	102.0	89.9	82.1	85.3	91.2	74.9	81.7
09	FINISHES	91.8	83.8	87.4	99.7	83.9	91.0	97.1	80.5	87.9	98.9	97.9	98.3	95.7	81.4	87.8	108.4	73.4	89.0
COVERS	DIVS. 10 - 14, 25, 28, 41, 43, 44, 46	100.0	96.6	99.3	100.0	91.3	98.1	100.0	84.0	96.5	100.0	88.6	97.5	100.0	96.2	99.2	100.0	85.7	96.9
21, 22, 23	FIRE SUPPRESSION, PLUMBING & HVAC	99.8	98.9	99.4	96.5	96.8	96.7	96.5	97.3	96.9	96.7	99.0	97.6	99.8	98.9	99.4	100.1	72.2	88.5
26, 27, 3370	ELECTRICAL, COMMUNICATIONS & UTIL.	99.0	119.2	109.2	109.2	100.8	104.9	103.5	82.5	92.9	100.9	98.8	99.8	104.3	119.2	111.9	93.2	76.9	85.0
MF2014	WEIGHTED AVERAGE	100.3	99.7	100.0	100.0	95.1	97.9	97.9	92.5	95.6	96.3	101.0	98.4	99.6	98.4	99.1	98.2	81.0	90.7

MISSOURI

DIVISION		KANSAS CITY 640 - 641			KIRKSVILLE 635			POPLAR BLUFF 639			ROLLA 654 - 655			SEDALIA 653			SIKESTON 638		
		MAT.	INST.	TOTAL	MAT.	INST.	TOTAL	MAT.	INST.	TOTAL	MAT.	INST.	TOTAL	MAT.	INST.	TOTAL	MAT.	INST.	TOTAL
015433	CONTRACTOR EQUIPMENT		103.1	103.1		97.6	97.6		100.2	100.2		110.4	110.4		101.1	101.1		100.2	100.2
0241, 31 - 34	SITE & INFRASTRUCTURE, DEMOLITION	99.8	96.1	97.2	94.7	89.1	90.8	81.3	93.3	89.8	98.7	98.5	98.6	97.5	93.8	94.9	84.8	93.4	90.9
0310	Concrete Forming & Accessories	96.7	103.3	102.4	83.7	79.0	79.6	84.0	82.2	82.5	92.0	95.6	95.1	89.7	79.3	80.7	85.0	77.4	78.5
0320	Concrete Reinforcing	96.9	120.9	109.3	102.9	97.7	100.2	100.6	79.4	92.2	97.1	94.1	95.5	95.6	116.5	106.4	105.4	87.6	96.2
0330	Cast-in-Place Concrete	93.8	107.1	99.0	98.6	93.9	96.7	75.9	87.4	80.4	92.5	98.3	94.8	96.7	93.7	95.5	81.0	85.5	82.8
03	CONCRETE	97.9	108.2	102.9	114.6	88.5	101.9	88.4	84.3	86.5	93.4	97.4	95.3	108.4	92.1	100.5	92.4	83.1	87.9
04	MASONRY	96.2	107.7	103.3	119.1	89.6	100.9	110.5	79.7	91.5	125.4	97.4	108.1	132.1	87.9	104.8	110.2	79.7	91.4
05	METALS	100.2	116.6	105.4	94.4	104.3	97.6	95.0	95.6	95.2	97.0	112.7	102.0	95.9	112.8	101.3	95.3	99.5	96.7
06	WOOD, PLASTICS & COMPOSITES	102.5	102.4	102.4	76.1	75.8	75.9	75.3	82.4	79.3	91.5	96.8	94.5	84.7	76.0	79.9	76.8	75.6	76.2
07	THERMAL & MOISTURE PROTECTION	91.8	105.9	97.7	106.4	94.0	101.2	104.9	85.2	96.7	94.8	96.2	95.4	100.8	93.8	97.9	105.1	85.5	96.9
08	OPENINGS	102.5	107.2	103.6	105.5	83.3	100.3	106.4	78.2	99.9	101.7	89.7	98.9	107.1	87.9	102.6	106.4	76.7	99.5
0920	Plaster & Gypsum Board	106.3	102.2	103.6	85.3	75.3	78.7	85.7	82.2	83.3	95.7	96.7	96.4	89.0	75.3	79.9	87.4	75.2	79.2
0950, 0980	Ceilings & Acoustic Treatment	104.8	102.2	103.1	94.7	75.3	81.9	96.4	82.2	87.0	89.7	96.7	94.3	89.7	75.3	80.2	96.4	75.2	82.4
0960	Flooring	99.6	104.7	101.1	75.4	94.8	80.9	88.4	82.7	86.7	101.3	94.8	99.4	78.3	94.8	83.1	88.9	82.7	87.1
0970, 0990	Wall Finishes & Painting/Coating	96.1	112.2	105.5	90.5	82.1	85.6	89.7	73.2	80.1	92.5	100.7	97.2	92.5	106.3	100.5	89.7	73.2	80.1
09	FINISHES	104.1	104.2	104.2	95.9	80.9	87.6	95.9	81.0	87.6	93.4	96.1	94.9	90.4	83.4	86.5	96.6	76.9	85.7
COVERS	DIVS. 10 - 14, 25, 28, 41, 43, 44, 46	100.0	100.5	100.1	100.0	84.5	96.6	100.0	90.5	97.9	100.0	87.9	97.4	100.0	89.3	97.7	100.0	89.8	97.8
21, 22, 23	FIRE SUPPRESSION, PLUMBING & HVAC	100.0	101.8	100.7	96.6	96.8	96.7	96.6	95.0	95.9	96.4	98.3	97.2	96.4	95.6	96.1	96.6	95.1	96.0
26, 27, 3370	ELECTRICAL, COMMUNICATIONS & UTIL.	102.5	101.6	102.0	103.7	81.9	92.7	104.0	100.7	102.3	97.5	85.8	91.6	98.6	98.8	98.7	103.1	110.0	106.6
MF2014	WEIGHTED AVERAGE	100.2	104.8	102.2	101.5	89.7	96.3	97.9	89.7	94.3	98.3	96.7	97.6	100.7	93.9	97.7	98.4	90.5	94.9

MISSOURI / MONTANA

DIVISION		SPRINGFIELD 656 - 658			ST. JOSEPH 644 - 645			ST. LOUIS 630 - 631			BILLINGS 590 - 591			BUTTE 597			GREAT FALLS 594		
		MAT.	INST.	TOTAL	MAT.	INST.	TOTAL	MAT.	INST.	TOTAL	MAT.	INST.	TOTAL	MAT.	INST.	TOTAL	MAT.	INST.	TOTAL
015433	CONTRACTOR EQUIPMENT		103.8	103.8		101.8	101.8		107.8	107.8		99.0	99.0		98.8	98.8		98.8	98.8
0241, 31 - 34	SITE & INFRASTRUCTURE, DEMOLITION	100.7	96.1	97.5	101.4	91.6	94.4	94.8	97.2	96.5	97.0	96.9	97.0	103.2	96.7	98.6	107.0	96.8	99.8
0310	Concrete Forming & Accessories	98.1	78.1	80.8	96.5	91.2	92.0	99.1	102.7	102.3	97.2	70.7	74.3	84.5	70.3	72.2	97.1	70.6	74.2
0320	Concrete Reinforcing	93.0	119.8	106.9	95.7	116.8	106.6	95.1	113.6	104.7	101.4	82.6	91.7	110.0	82.6	95.8	101.5	82.7	91.7
0330	Cast-in-Place Concrete	98.2	78.6	90.5	93.7	103.6	97.6	94.5	103.8	98.2	124.9	73.3	104.6	137.5	68.0	110.1	145.3	67.1	114.5
03	CONCRETE	104.2	86.9	95.8	97.6	100.8	99.2	97.8	106.1	101.8	107.4	74.7	91.5	111.8	72.7	92.8	117.1	72.6	95.5
04	MASONRY	101.2	80.9	88.6	96.0	97.4	96.8	92.9	109.9	103.4	137.3	76.2	99.6	133.0	78.0	99.1	137.6	77.6	100.6
05	METALS	102.2	109.5	104.5	96.5	113.7	102.0	100.6	121.7	107.3	106.7	91.6	101.9	100.8	91.3	97.8	104.2	91.6	100.2
06	WOOD, PLASTICS & COMPOSITES	92.6	76.1	83.4	103.3	89.9	95.8	97.6	101.5	99.8	90.9	69.1	78.7	78.7	69.1	73.3	92.0	69.1	79.2
07	THERMAL & MOISTURE PROTECTION	99.0	81.0	91.5	92.2	97.4	94.4	100.0	105.9	102.5	110.9	71.7	94.5	110.7	69.8	93.6	111.3	71.5	94.6
08	OPENINGS	109.5	83.3	103.4	100.6	99.4	100.3	100.1	109.4	102.3	102.0	71.2	94.9	100.2	68.8	92.9	103.3	69.9	95.5
0920	Plaster & Gypsum Board	96.7	75.4	82.5	110.5	89.4	96.4	97.7	101.8	100.5	108.0	68.5	81.6	107.0	68.5	81.3	116.7	68.5	84.5
0950, 0980	Ceilings & Acoustic Treatment	89.7	75.4	80.2	104.0	89.4	94.3	102.1	101.8	101.9	87.2	68.5	74.8	93.0	68.5	76.8	94.6	68.5	77.3
0960	Flooring	99.9	73.3	92.3	102.7	101.5	102.3	100.1	96.0	99.0	105.0	74.3	96.1	102.2	71.3	93.3	109.9	83.4	102.3
0970, 0990	Wall Finishes & Painting/Coating	86.8	74.9	79.9	91.6	83.9	87.1	94.8	108.7	102.9	98.4	96.6	97.4	97.2	51.2	70.5	97.2	86.6	91.0
09	FINISHES	95.7	77.0	85.3	105.1	91.8	97.8	101.0	101.7	101.4	98.1	73.5	84.5	98.1	68.0	81.5	102.4	74.3	86.8
COVERS	DIVS. 10 - 14, 25, 28, 41, 43, 44, 46	100.0	94.5	98.8	100.0	97.3	99.4	100.0	101.0	100.2	100.0	93.7	98.6	100.0	93.7	98.6	100.0	93.7	98.6
21, 22, 23	FIRE SUPPRESSION, PLUMBING & HVAC	99.9	74.0	88.9	100.1	88.1	95.1	99.9	105.5	102.3	100.0	75.4	89.6	100.1	74.9	89.4	100.1	69.3	87.1
26, 27, 3370	ELECTRICAL, COMMUNICATIONS & UTIL.	102.9	69.1	85.9	100.9	100.8	100.9	107.2	90.6	98.8	97.9	76.6	87.1	104.4	72.6	88.3	97.3	72.6	84.8
MF2014	WEIGHTED AVERAGE	101.8	83.3	93.7	99.4	96.7	98.2	100.2	104.3	102.0	103.7	78.8	92.7	103.7	77.1	92.0	105.2	76.8	92.7

For customer support on your Commercial Renovation Cost Data, call 877.742.4719.

City Cost Indexes

MONTANA

DIVISION		HAVRE 595 MAT.	INST.	TOTAL	HELENA 596 MAT.	INST.	TOTAL	KALISPELL 599 MAT.	INST.	TOTAL	MILES CITY 593 MAT.	INST.	TOTAL	MISSOULA 598 MAT.	INST.	TOTAL	WOLF POINT 592 MAT.	INST.	TOTAL
015433	CONTRACTOR EQUIPMENT		98.8	98.8		98.8	98.8		98.8	98.8		98.8	98.8		98.8	98.8		98.8	98.8
0241, 31 - 34	SITE & INFRASTRUCTURE, DEMOLITION	110.8	96.3	100.5	97.1	96.3	96.5	93.1	96.5	95.5	99.5	96.0	97.0	85.5	96.3	93.1	117.3	96.1	102.3
0310	Concrete Forming & Accessories	77.8	67.8	69.1	100.3	63.3	68.3	87.7	70.3	72.6	95.7	66.4	70.4	87.7	70.1	72.5	88.4	66.7	69.6
0320	Concrete Reinforcing	110.9	82.5	96.2	116.9	82.5	99.1	113.0	82.5	97.2	110.5	82.6	96.0	111.9	82.5	96.7	112.1	82.6	96.8
0330	Cast-in-Place Concrete	148.1	59.3	113.1	107.4	61.8	89.4	119.4	61.6	96.6	130.7	59.2	102.5	101.4	63.8	86.5	146.4	59.6	112.2
03	CONCRETE	120.3	68.6	95.2	103.3	67.4	85.9	100.8	70.5	86.1	108.5	68.0	88.8	88.7	71.2	80.2	123.7	68.2	96.7
04	MASONRY	134.0	72.0	95.8	127.8	77.8	97.0	131.8	78.4	98.9	139.5	67.3	95.0	160.3	78.0	109.6	140.8	68.2	96.0
05	METALS	97.0	90.5	95.0	102.9	89.7	98.7	96.9	90.8	95.0	96.2	90.8	94.5	97.4	90.8	95.3	96.3	90.8	94.6
06	WOOD, PLASTICS & COMPOSITES	71.0	69.1	69.9	92.5	60.1	74.5	81.7	69.1	74.6	89.3	69.1	78.0	81.7	69.1	74.6	81.6	69.1	74.6
07	THERMAL & MOISTURE PROTECTION	111.1	67.1	92.7	105.2	69.6	90.3	110.3	78.7	97.1	110.7	64.7	91.4	109.9	77.7	96.5	111.6	65.1	92.1
08	OPENINGS	100.7	67.0	92.9	102.8	63.7	93.7	100.7	67.1	92.9	100.2	67.0	92.5	100.2	66.8	92.4	100.2	66.8	92.5
0920	Plaster & Gypsum Board	102.7	68.5	79.9	105.9	59.3	74.8	107.0	68.5	81.3	115.9	68.5	84.2	107.0	68.5	81.3	111.0	68.5	82.6
0950, 0980	Ceilings & Acoustic Treatment	93.0	68.5	76.8	95.4	59.3	71.5	93.0	68.5	76.8	91.3	68.5	76.2	93.0	68.5	76.8	91.3	68.5	76.2
0960	Flooring	99.4	71.3	91.3	104.6	56.8	90.9	104.1	71.3	94.7	109.7	67.5	97.6	104.1	71.3	94.7	105.7	67.5	94.8
0970, 0990	Wall Finishes & Painting/Coating	97.2	48.7	69.0	93.2	47.2	66.4	97.2	47.2	68.1	97.2	47.2	68.1	97.2	62.0	76.7	97.2	48.7	69.0
09	FINISHES	97.3	66.3	80.1	99.3	59.8	77.5	98.1	67.7	81.3	101.0	64.4	80.8	97.5	69.2	81.8	100.6	64.7	80.7
COVERS	DIVS. 10 - 14, 25, 28, 41, 43, 44, 46	100.0	68.9	93.3	100.0	92.8	98.4	100.0	93.8	98.7	100.0	90.4	97.9	100.0	93.7	98.6	100.0	90.6	98.0
21, 22, 23	FIRE SUPPRESSION, PLUMBING & HVAC	96.7	66.6	84.0	99.9	69.3	87.0	96.7	68.7	84.9	96.7	68.9	84.9	100.1	68.5	86.7	96.7	69.2	85.1
26, 27, 3370	ELECTRICAL, COMMUNICATIONS & UTIL.	97.3	72.7	84.8	103.5	72.6	87.9	101.3	72.0	86.5	97.3	77.9	87.5	102.3	72.0	87.0	97.3	77.9	87.5
MF2014	WEIGHTED AVERAGE	102.7	73.0	89.6	102.8	73.6	90.0	100.4	75.5	89.5	101.6	74.0	89.4	101.1	75.7	89.9	103.8	74.2	90.8

NEBRASKA

DIVISION		ALLIANCE 693 MAT.	INST.	TOTAL	COLUMBUS 686 MAT.	INST.	TOTAL	GRAND ISLAND 688 MAT.	INST.	TOTAL	HASTINGS 689 MAT.	INST.	TOTAL	LINCOLN 683 - 685 MAT.	INST.	TOTAL	MCCOOK 690 MAT.	INST.	TOTAL
015433	CONTRACTOR EQUIPMENT		97.7	97.7		102.9	102.9		102.9	102.9		102.9	102.9		102.9	102.9		102.9	102.9
0241, 31 - 34	SITE & INFRASTRUCTURE, DEMOLITION	99.7	100.0	99.9	100.2	93.2	95.3	105.0	94.0	97.3	103.7	93.4	96.4	91.0	94.1	93.2	103.5	93.2	96.2
0310	Concrete Forming & Accessories	87.9	54.8	59.2	96.3	65.6	69.7	95.8	70.4	73.9	99.2	69.6	73.6	96.3	72.4	75.6	93.6	54.4	59.7
0320	Concrete Reinforcing	110.2	84.2	96.7	95.7	75.7	85.3	95.1	75.6	85.0	95.1	76.2	85.3	94.2	75.4	84.5	103.1	75.3	88.7
0330	Cast-in-Place Concrete	108.9	78.7	97.0	112.0	81.8	100.1	118.6	66.4	98.0	118.6	62.5	96.4	93.1	73.2	85.3	117.6	58.5	94.3
03	CONCRETE	123.3	68.9	96.8	108.2	73.9	91.5	113.3	70.9	92.7	113.5	69.3	92.0	97.6	74.1	86.2	112.0	60.9	87.2
04	MASONRY	114.5	74.7	90.0	123.8	78.4	95.8	116.1	75.7	91.2	125.7	87.6	102.2	103.8	72.4	84.4	110.1	73.6	87.6
05	METALS	98.9	77.8	92.2	90.7	85.4	89.0	92.5	87.0	90.7	93.2	86.4	91.0	94.4	86.9	92.1	94.0	84.2	90.9
06	WOOD, PLASTICS & COMPOSITES	84.3	50.3	65.4	96.4	64.4	78.5	95.7	69.7	81.2	99.2	69.7	82.8	98.3	72.2	83.8	92.6	50.4	69.1
07	THERMAL & MOISTURE PROTECTION	103.9	64.8	87.5	101.1	68.9	87.6	101.3	76.8	91.0	101.3	77.8	91.5	99.8	76.2	89.9	98.5	65.4	84.7
08	OPENINGS	97.3	56.9	87.9	97.8	64.2	90.0	97.8	67.7	90.8	97.8	67.1	90.7	107.0	65.5	97.3	97.8	54.9	87.8
0920	Plaster & Gypsum Board	82.4	49.0	60.1	95.5	63.4	74.0	94.8	68.9	77.5	96.5	68.9	78.1	105.9	71.5	82.9	94.5	49.0	64.1
0950, 0980	Ceilings & Acoustic Treatment	91.8	49.0	63.5	81.9	63.4	69.6	81.9	68.9	73.3	81.9	68.9	73.3	92.8	71.5	78.7	87.0	49.0	61.8
0960	Flooring	97.4	78.9	92.1	91.5	96.5	92.9	91.2	103.1	94.6	92.6	96.5	93.7	106.0	91.5	101.9	96.1	78.9	91.1
0970, 0990	Wall Finishes & Painting/Coating	150.0	54.8	94.6	73.2	63.7	67.7	73.2	67.6	70.0	73.2	63.7	67.7	90.6	82.5	85.9	84.4	47.7	63.1
09	FINISHES	96.3	57.7	74.9	89.3	69.8	78.5	89.4	74.8	81.3	90.0	73.0	80.6	98.6	77.0	86.7	93.8	56.8	73.3
COVERS	DIVS. 10 - 14, 25, 28, 41, 43, 44, 46	100.0	63.3	92.0	100.0	85.5	96.9	100.0	88.8	97.6	100.0	86.1	97.0	100.0	89.1	97.6	100.0	62.5	91.9
21, 22, 23	FIRE SUPPRESSION, PLUMBING & HVAC	96.7	69.7	85.3	96.6	75.1	87.5	100.0	77.3	90.4	96.6	75.2	87.6	99.9	77.3	90.3	96.5	76.1	87.9
26, 27, 3370	ELECTRICAL, COMMUNICATIONS & UTIL.	94.5	68.8	81.5	93.6	81.7	87.6	92.3	66.5	79.3	91.7	82.9	87.3	103.9	66.5	85.0	96.4	68.9	82.5
MF2014	WEIGHTED AVERAGE	101.2	70.5	87.7	98.1	77.5	89.0	99.3	76.6	89.3	99.1	78.9	90.2	99.8	76.9	89.7	99.0	70.5	86.4

NEBRASKA / NEVADA

DIVISION		NORFOLK 687 MAT.	INST.	TOTAL	NORTH PLATTE 691 MAT.	INST.	TOTAL	OMAHA 680 - 681 MAT.	INST.	TOTAL	VALENTINE 692 MAT.	INST.	TOTAL	CARSON CITY 897 MAT.	INST.	TOTAL	ELKO 898 MAT.	INST.	TOTAL
015433	CONTRACTOR EQUIPMENT		93.2	93.2		102.9	102.9		93.2	93.2		96.8	96.8		96.7	96.7		96.7	96.7
0241, 31 - 34	SITE & INFRASTRUCTURE, DEMOLITION	83.7	92.5	89.9	104.8	93.2	96.6	90.3	93.3	92.4	88.0	98.5	95.4	87.4	97.9	94.8	70.2	96.5	88.8
0310	Concrete Forming & Accessories	82.1	69.0	70.7	96.1	69.3	72.9	95.2	72.3	75.4	84.1	54.1	58.2	105.6	81.9	85.1	112.1	78.7	83.2
0320	Concrete Reinforcing	95.8	65.7	80.2	102.5	75.5	88.5	94.8	75.8	85.0	103.1	65.4	83.6	108.4	115.9	112.3	116.6	85.2	100.3
0330	Cast-in-Place Concrete	112.7	61.5	92.5	117.6	63.5	96.3	93.6	82.0	89.0	103.8	56.4	85.1	92.7	85.2	89.8	89.9	78.7	85.5
03	CONCRETE	106.6	66.2	87.0	112.1	69.3	91.3	97.9	76.5	87.5	109.4	57.8	84.3	102.4	89.6	96.2	99.1	80.3	90.0
04	MASONRY	130.6	78.7	98.6	97.5	74.5	83.3	105.5	79.6	89.5	109.7	73.6	87.5	126.7	74.9	94.8	126.1	71.5	92.5
05	METALS	94.0	73.2	87.4	93.3	84.9	90.6	94.4	78.8	89.4	104.8	73.1	94.7	102.8	100.8	102.2	106.1	87.0	100.0
06	WOOD, PLASTICS & COMPOSITES	80.2	69.2	74.1	94.5	69.7	80.7	93.6	71.7	81.4	79.1	49.9	62.8	88.6	80.4	84.0	99.5	79.1	88.1
07	THERMAL & MOISTURE PROTECTION	100.8	72.1	88.8	98.5	77.9	89.9	96.0	81.3	89.8	99.2	65.5	85.1	104.6	82.3	95.3	101.0	73.9	89.7
08	OPENINGS	99.4	64.5	91.3	97.1	67.1	90.1	104.6	71.2	96.8	99.6	53.9	89.0	98.4	101.3	99.1	100.5	78.8	95.4
0920	Plaster & Gypsum Board	95.6	68.9	77.8	94.5	68.9	77.4	99.5	71.5	80.8	96.3	49.0	64.7	101.4	79.9	87.0	108.2	78.4	88.3
0950, 0980	Ceilings & Acoustic Treatment	95.4	68.9	77.8	87.0	68.9	75.0	93.9	71.5	79.0	104.6	49.0	67.8	91.1	79.9	83.7	93.4	78.4	83.5
0960	Flooring	113.7	96.5	108.7	97.2	96.5	97.0	104.4	88.2	99.7	124.3	76.7	110.6	102.1	74.4	94.1	106.8	49.1	90.2
0970, 0990	Wall Finishes & Painting/Coating	128.4	63.7	90.8	84.4	63.7	72.4	104.5	67.2	82.8	149.5	66.1	101.0	92.7	81.0	85.9	94.0	81.0	86.5
09	FINISHES	106.8	73.0	88.1	94.1	73.0	82.5	101.1	74.3	86.3	115.6	58.0	83.7	96.9	79.5	87.3	97.7	73.4	84.2
COVERS	DIVS. 10 - 14, 25, 28, 41, 43, 44, 46	100.0	84.7	96.7	100.0	64.7	92.4	100.0	88.0	97.4	100.0	60.6	91.5	100.0	86.4	97.1	100.0	61.0	91.5
21, 22, 23	FIRE SUPPRESSION, PLUMBING & HVAC	96.3	76.0	87.7	99.9	75.1	89.4	99.8	78.6	90.8	96.2	76.0	87.7	100.0	75.2	89.5	98.4	78.5	90.0
26, 27, 3370	ELECTRICAL, COMMUNICATIONS & UTIL.	92.6	81.7	87.1	94.8	76.5	85.5	99.0	81.7	90.2	92.1	85.9	89.0	104.3	97.8	101.0	102.0	95.9	98.9
MF2014	WEIGHTED AVERAGE	99.6	76.0	89.2	98.9	75.9	88.7	99.1	79.5	90.5	101.3	71.8	88.3	101.8	86.7	95.1	101.1	81.4	92.4

For customer support on your Commercial Renovation Cost Data, call 877.742.4719.

781

NEVADA / NEW HAMPSHIRE

	DIVISION	ELY 893 MAT.	INST.	TOTAL	LAS VEGAS 889 - 891 MAT.	INST.	TOTAL	RENO 894 - 895 MAT.	INST.	TOTAL	CHARLESTON 036 MAT.	INST.	TOTAL	CLAREMONT 037 MAT.	INST.	TOTAL	CONCORD 032 - 033 MAT.	INST.	TOTAL
015433	CONTRACTOR EQUIPMENT		96.7	96.7		96.7	96.7		96.7	96.7		98.1	98.1		98.1	98.1		98.1	98.1
0241, 31 - 34	SITE & INFRASTRUCTURE, DEMOLITION	76.0	97.9	91.5	79.4	100.0	94.0	75.7	97.9	91.4	88.2	97.1	94.5	82.0	97.1	92.7	95.6	101.4	99.7
0310	Concrete Forming & Accessories	104.9	102.7	103.0	105.5	108.7	108.2	100.8	81.6	84.2	88.7	79.6	80.9	94.7	79.7	81.7	99.3	92.8	93.7
0320	Concrete Reinforcing	115.2	110.3	112.7	106.3	122.1	114.5	108.9	120.8	115.1	88.7	93.6	91.2	88.7	93.6	91.2	103.5	94.2	98.7
0330	Cast-in-Place Concrete	96.5	103.9	99.4	93.4	110.5	100.1	102.0	85.1	95.3	87.9	104.4	94.4	80.7	104.4	90.1	106.0	116.2	110.2
03	CONCRETE	107.7	104.3	106.0	102.7	111.4	106.9	106.9	90.3	98.9	95.3	90.8	93.1	87.6	90.8	89.2	104.1	101.0	102.6
04	MASONRY	131.9	102.0	113.5	118.6	104.9	110.1	125.2	74.9	94.2	101.5	84.2	90.8	101.5	84.2	90.9	118.5	103.6	109.3
05	METALS	106.1	101.4	104.6	114.3	108.3	112.4	107.7	102.4	106.0	91.8	90.0	91.3	91.8	90.0	91.3	98.1	93.8	96.8
06	WOOD, PLASTICS & COMPOSITES	90.5	103.3	97.6	88.2	106.2	98.2	84.6	80.4	82.3	91.9	87.1	89.2	97.8	87.1	91.8	92.4	90.4	91.3
07	THERMAL & MOISTURE PROTECTION	101.5	93.1	98.0	114.9	101.8	109.4	101.0	82.3	93.2	101.1	78.5	91.6	100.9	78.5	91.5	105.6	101.8	104.0
08	OPENINGS	100.4	96.0	99.4	99.4	116.7	103.4	98.3	88.9	96.1	100.1	74.4	94.1	101.3	74.4	95.0	101.0	92.4	99.0
0920	Plaster & Gypsum Board	103.2	103.3	103.3	98.1	106.3	104.5	93.1	79.9	84.3	97.1	86.2	89.8	98.4	86.2	90.3	101.0	89.7	93.4
0950, 0980	Ceilings & Acoustic Treatment	93.4	103.3	100.0	100.8	106.3	104.5	98.3	79.9	86.1	89.6	86.2	87.4	89.6	86.2	87.4	96.1	89.7	91.8
0960	Flooring	104.1	55.0	90.0	95.4	106.9	98.7	100.7	74.4	93.1	92.7	30.8	74.9	95.3	30.8	76.8	98.5	112.4	102.5
0970, 0990	Wall Finishes & Painting/Coating	94.0	121.0	109.7	96.6	121.7	111.2	94.0	81.0	86.5	88.6	45.7	63.6	88.6	46.0	63.8	89.2	116.5	105.1
09	FINISHES	96.8	96.1	96.4	95.2	109.4	103.1	95.2	79.5	86.5	90.4	67.7	77.8	90.9	67.7	78.1	93.3	98.0	95.9
COVERS	DIVS. 10 - 14, 25, 28, 41, 43, 44, 46	100.0	61.7	91.7	100.0	102.2	100.5	100.0	86.4	97.1	100.0	89.5	97.7	100.0	89.5	97.7	100.0	106.1	101.3
21, 22, 23	FIRE SUPPRESSION, PLUMBING & HVAC	98.4	103.3	100.5	100.2	107.7	103.4	100.1	75.2	89.5	96.7	39.2	72.4	96.7	39.3	72.4	99.9	83.6	93.0
26, 27, 3370	ELECTRICAL, COMMUNICATIONS & UTIL.	102.2	108.2	105.3	106.3	120.7	113.6	102.7	97.8	100.2	93.5	52.5	72.7	93.5	52.5	72.7	94.3	82.9	88.5
MF2014	WEIGHTED AVERAGE	102.4	100.6	101.6	103.6	109.4	106.1	102.2	86.5	95.3	95.7	70.3	84.5	94.9	70.4	84.1	100.0	93.8	97.3

NEW HAMPSHIRE / NEW JERSEY

	DIVISION	KEENE 034 MAT.	INST.	TOTAL	LITTLETON 035 MAT.	INST.	TOTAL	MANCHESTER 031 MAT.	INST.	TOTAL	NASHUA 030 MAT.	INST.	TOTAL	PORTSMOUTH 038 MAT.	INST.	TOTAL	ATLANTIC CITY 082,084 MAT.	INST.	TOTAL
015433	CONTRACTOR EQUIPMENT		98.1	98.1		98.1	98.1		98.1	98.1		98.1	98.1		98.1	98.1		96.0	96.0
0241, 31 - 34	SITE & INFRASTRUCTURE, DEMOLITION	95.9	97.4	97.0	82.1	97.4	92.9	93.6	101.4	99.1	97.5	101.4	100.2	90.8	101.4	98.3	95.1	102.9	100.6
0310	Concrete Forming & Accessories	93.3	82.1	83.6	106.4	81.8	85.1	99.2	92.8	93.7	101.8	92.8	94.0	90.2	92.7	92.4	115.1	129.3	127.4
0320	Concrete Reinforcing	88.7	93.8	91.3	89.5	93.7	91.7	103.6	94.2	98.7	110.9	94.2	102.2	88.7	94.2	91.5	77.6	118.4	98.7
0330	Cast-in-Place Concrete	88.3	106.9	95.6	79.2	106.1	89.8	101.1	116.6	107.2	83.6	116.6	96.6	79.3	116.6	94.0	83.3	133.0	102.9
03	CONCRETE	94.9	92.7	93.9	87.1	92.4	89.7	101.7	101.0	101.4	95.3	101.0	98.0	87.2	100.9	93.9	92.1	127.2	109.2
04	MASONRY	104.7	87.8	94.3	114.3	87.8	98.0	112.0	103.6	106.9	106.5	103.6	104.7	101.8	103.6	102.9	118.3	138.7	130.8
05	METALS	92.5	91.2	92.1	92.6	90.9	92.0	100.1	93.8	98.1	98.4	93.8	96.9	94.1	94.0	94.1	93.7	104.5	97.2
06	WOOD, PLASTICS & COMPOSITES	96.1	87.1	91.1	107.1	87.1	95.9	94.6	90.4	92.3	105.6	90.4	97.1	93.1	90.4	91.6	119.4	128.3	124.3
07	THERMAL & MOISTURE PROTECTION	101.6	80.0	92.6	101.0	80.0	92.2	105.0	101.8	103.7	101.9	101.8	101.9	101.6	114.2	106.9	102.9	128.2	113.5
08	OPENINGS	98.7	90.2	96.7	102.3	86.7	98.6	102.8	92.4	100.3	103.5	92.4	100.9	103.9	82.5	98.9	99.1	124.1	104.9
0920	Plaster & Gypsum Board	97.4	86.2	89.9	111.3	86.2	94.6	99.6	89.7	93.0	107.2	89.7	95.5	97.1	89.7	92.1	105.3	128.6	120.9
0950, 0980	Ceilings & Acoustic Treatment	89.6	86.2	87.4	89.6	86.2	87.4	97.0	89.7	92.2	101.2	89.7	93.6	90.5	89.7	90.0	78.5	128.6	111.7
0960	Flooring	94.9	50.6	82.2	106.4	30.8	84.7	92.5	112.4	98.2	99.0	112.4	102.8	92.8	112.4	98.4	101.4	153.7	116.4
0970, 0990	Wall Finishes & Painting/Coating	88.6	116.5	104.8	88.6	59.9	71.9	91.4	116.5	106.0	88.6	116.5	104.8	88.6	116.5	104.8	83.4	130.2	110.6
09	FINISHES	92.4	79.9	85.5	96.1	70.1	81.7	93.4	94.0	96.0	97.5	98.0	97.8	91.4	98.0	95.1	92.2	134.8	115.7
COVERS	DIVS. 10 - 14, 25, 28, 41, 43, 44, 46	100.0	90.7	98.0	100.0	95.7	99.1	100.0	106.1	101.3	100.0	106.1	101.3	100.0	106.1	101.3	100.0	112.5	102.7
21, 22, 23	FIRE SUPPRESSION, PLUMBING & HVAC	96.7	43.0	74.0	96.7	63.2	82.6	99.9	83.6	93.0	100.1	83.6	93.1	100.1	83.6	93.1	99.7	124.1	110.0
26, 27, 3370	ELECTRICAL, COMMUNICATIONS & UTIL.	93.5	62.7	77.9	94.4	55.5	74.7	94.3	81.9	88.0	95.6	81.9	88.7	93.9	82.9	88.4	89.4	137.0	113.5
MF2014	WEIGHTED AVERAGE	96.2	75.6	87.1	96.3	77.6	88.0	99.9	93.6	97.1	99.3	93.6	96.8	96.5	93.7	95.3	97.2	125.3	109.6

NEW JERSEY

	DIVISION	CAMDEN 081 MAT.	INST.	TOTAL	DOVER 078 MAT.	INST.	TOTAL	ELIZABETH 072 MAT.	INST.	TOTAL	HACKENSACK 076 MAT.	INST.	TOTAL	JERSEY CITY 073 MAT.	INST.	TOTAL	LONG BRANCH 077 MAT.	INST.	TOTAL
015433	CONTRACTOR EQUIPMENT		96.0	96.0		98.1	98.1		98.1	98.1		98.1	98.1		96.0	96.0		95.6	95.6
0241, 31 - 34	SITE & INFRASTRUCTURE, DEMOLITION	96.4	103.3	101.3	98.7	104.0	102.4	102.6	104.0	103.6	99.6	104.0	102.7	90.3	103.9	99.9	94.2	103.5	100.7
0310	Concrete Forming & Accessories	105.5	125.5	122.8	101.7	133.2	129.0	114.9	133.2	130.8	101.7	133.2	128.9	106.4	133.3	129.7	106.9	124.4	122.0
0320	Concrete Reinforcing	101.8	120.2	111.3	80.8	130.0	106.3	80.8	130.0	106.3	80.8	130.0	106.3	104.8	130.0	117.9	80.8	129.9	106.2
0330	Cast-in-Place Concrete	80.8	133.0	101.4	93.2	127.4	106.7	80.1	131.2	100.2	91.1	131.1	106.9	72.7	127.5	94.3	80.8	132.3	101.1
03	CONCRETE	92.8	125.9	108.9	95.6	129.5	112.1	91.7	130.8	110.7	93.9	130.7	111.8	89.4	129.4	108.8	93.3	127.0	109.7
04	MASONRY	107.4	138.7	126.7	99.5	137.5	122.9	117.4	137.5	129.7	103.7	137.5	124.5	94.2	137.5	120.8	109.2	132.4	123.5
05	METALS	99.6	105.4	101.5	91.7	113.6	98.7	93.2	113.7	99.7	91.7	113.6	98.7	97.7	111.1	101.9	91.8	110.7	97.8
06	WOOD, PLASTICS & COMPOSITES	107.6	122.9	116.1	101.0	132.9	118.8	116.7	132.9	125.7	101.0	132.9	118.8	101.9	132.9	119.2	103.0	122.9	114.1
07	THERMAL & MOISTURE PROTECTION	102.7	129.5	113.9	102.3	132.1	114.8	102.5	132.6	115.1	102.1	126.4	112.3	101.8	132.1	114.5	102.0	121.4	110.1
08	OPENINGS	101.5	121.6	106.2	105.1	129.2	110.7	103.5	129.2	109.5	102.9	129.2	109.0	101.1	129.2	107.6	97.2	123.7	103.4
0920	Plaster & Gypsum Board	100.5	123.1	115.6	98.9	133.3	121.9	106.9	133.3	124.6	98.9	133.3	121.9	102.4	133.3	123.1	100.9	123.1	115.7
0950, 0980	Ceilings & Acoustic Treatment	87.8	123.1	111.2	77.5	133.3	114.5	79.4	133.3	115.1	77.5	133.3	114.5	88.4	133.3	118.2	77.5	123.1	107.7
0960	Flooring	97.1	153.7	113.4	88.8	176.0	113.9	94.3	176.0	117.8	88.8	176.0	113.9	89.9	176.0	114.7	90.1	171.2	113.5
0970, 0990	Wall Finishes & Painting/Coating	83.4	130.2	110.6	84.5	132.5	112.4	84.5	132.5	112.4	84.5	132.5	112.4	84.6	132.5	112.5	84.6	130.2	111.1
09	FINISHES	92.1	131.6	113.9	88.0	140.5	117.0	91.6	140.5	118.6	87.9	140.5	117.0	91.0	141.2	118.8	88.9	133.6	113.6
COVERS	DIVS. 10 - 14, 25, 28, 41, 43, 44, 46	100.0	111.9	102.6	100.0	116.9	103.7	100.0	116.9	103.7	100.0	116.9	103.7	100.0	116.9	103.7	100.0	110.5	102.3
21, 22, 23	FIRE SUPPRESSION, PLUMBING & HVAC	100.0	124.2	110.2	99.7	129.0	112.1	99.9	126.5	111.2	99.7	129.0	112.1	99.9	129.1	112.2	99.7	128.1	111.7
26, 27, 3370	ELECTRICAL, COMMUNICATIONS & UTIL.	93.5	137.0	115.5	90.0	140.1	115.3	90.5	140.1	115.6	90.0	139.1	114.8	94.1	139.0	116.8	89.7	129.8	110.0
MF2014	WEIGHTED AVERAGE	98.4	124.7	110.0	96.7	129.2	111.0	97.8	128.8	111.5	96.5	129.0	110.8	96.7	128.9	110.9	96.0	124.7	108.6

City Cost Indexes

NEW JERSEY

DIVISION		NEW BRUNSWICK 088-089			NEWARK 070-071			PATERSON 074-075			POINT PLEASANT 087			SUMMIT 079			TRENTON 085-086		
		MAT.	INST.	TOTAL	MAT.	INST.	TOTAL	MAT.	INST.	TOTAL	MAT.	INST.	TOTAL	MAT.	INST.	TOTAL	MAT.	INST.	TOTAL
015433	CONTRACTOR EQUIPMENT		95.6	95.6		98.1	98.1		98.1	98.1		95.6	95.6		98.1	98.1		95.6	95.6
0241, 31 - 34	SITE & INFRASTRUCTURE, DEMOLITION	107.7	103.7	104.9	103.6	104.0	103.9	101.5	104.0	103.3	109.3	103.5	105.2	100.2	104.0	102.9	94.2	103.4	100.7
0310	Concrete Forming & Accessories	108.9	133.3	130.0	102.4	133.5	129.3	103.9	133.4	129.4	102.8	124.2	121.3	104.7	133.2	129.4	104.5	131.5	127.9
0320	Concrete Reinforcing	78.5	129.9	105.1	103.1	130.1	117.1	104.8	130.0	117.9	78.5	129.8	105.1	80.8	130.0	106.3	103.0	102.8	102.9
0330	Cast-in-Place Concrete	103.0	133.0	114.8	102.8	131.3	114.0	92.7	131.2	107.9	103.0	130.8	114.0	77.4	131.2	98.6	101.3	130.8	113.2
03	CONCRETE	109.3	131.2	120.0	102.7	130.9	116.4	99.0	130.9	114.5	108.9	126.4	117.4	88.7	130.8	109.2	102.1	124.7	113.1
04	MASONRY	116.0	136.1	128.4	109.5	137.5	126.7	100.1	137.5	123.1	103.5	131.4	120.7	102.6	137.5	124.1	113.4	132.4	125.1
05	METALS	93.8	110.9	99.2	99.4	113.9	104.0	93.4	113.8	99.9	93.8	110.4	99.1	91.7	113.7	98.7	99.4	99.9	99.5
06	WOOD, PLASTICS & COMPOSITES	112.6	132.8	123.9	95.4	132.9	116.3	103.7	132.9	120.0	104.9	122.9	114.9	104.9	132.9	120.5	99.7	132.8	118.1
07	THERMAL & MOISTURE PROTECTION	103.1	131.2	114.9	104.1	132.6	116.0	102.4	126.4	112.4	103.2	121.2	110.7	102.7	132.6	115.2	103.6	126.6	113.2
08	OPENINGS	94.0	129.2	102.2	104.7	129.2	110.4	108.0	129.2	113.0	95.9	125.2	102.7	109.1	129.2	113.8	104.3	122.1	108.4
0920	Plaster & Gypsum Board	102.4	133.3	123.1	99.1	133.3	122.0	102.4	133.3	123.1	97.4	123.1	114.6	100.9	133.3	122.6	97.6	133.3	121.5
0950, 0980	Ceilings & Acoustic Treatment	78.5	133.3	114.8	91.1	133.3	119.1	88.4	133.3	118.2	78.5	123.1	108.0	77.5	133.3	114.5	89.4	133.3	118.5
0960	Flooring	98.7	176.0	120.9	90.7	176.0	115.2	89.9	176.0	114.7	95.9	153.7	112.5	90.2	176.0	114.9	97.4	172.2	118.9
0970, 0990	Wall Finishes & Painting/Coating	83.4	132.5	112.0	85.8	132.5	113.0	84.5	132.5	112.4	83.4	130.2	110.6	84.5	132.5	112.4	87.6	130.2	112.4
09	FINISHES	92.1	141.2	119.2	90.0	141.8	118.6	91.1	141.8	119.1	90.6	130.6	112.7	89.0	140.5	117.5	92.0	139.7	118.4
COVERS	DIVS. 10 - 14, 25, 28, 41, 43, 44, 46	100.0	116.8	103.6	100.0	116.9	103.7	100.0	116.9	103.7	100.0	105.7	101.2	100.0	116.9	103.7	100.0	111.7	102.5
21, 22, 23	FIRE SUPPRESSION, PLUMBING & HVAC	99.7	130.0	112.5	100.0	128.4	112.0	99.9	129.1	112.3	99.7	127.9	111.7	99.7	126.5	111.0	100.0	127.9	111.8
26, 27, 3370	ELECTRICAL, COMMUNICATIONS & UTIL.	90.0	135.8	113.1	97.1	139.1	118.3	94.1	140.1	117.4	89.4	129.8	109.8	90.5	140.1	115.6	96.6	134.4	115.7
MF2014	WEIGHTED AVERAGE	98.8	128.7	112.0	100.3	129.3	113.1	98.5	129.4	112.1	98.2	124.0	109.6	96.7	128.8	110.9	100.2	124.9	111.1

NEW JERSEY / NEW MEXICO

DIVISION		NEW JERSEY — VINELAND 080,083			ALBUQUERQUE 870-872			CARRIZOZO 883			CLOVIS 881			FARMINGTON 874			GALLUP 873		
		MAT.	INST.	TOTAL	MAT.	INST.	TOTAL	MAT.	INST.	TOTAL	MAT.	INST.	TOTAL	MAT.	INST.	TOTAL	MAT.	INST.	TOTAL
015433	CONTRACTOR EQUIPMENT		96.0	96.0		109.4	109.4		109.4	109.4		109.4	109.4		109.4	109.4		109.4	109.4
0241, 31 - 34	SITE & INFRASTRUCTURE, DEMOLITION	99.3	103.1	102.0	91.0	102.3	99.0	110.8	102.3	104.8	98.0	102.3	101.0	97.9	102.3	101.0	107.2	102.3	103.8
0310	Concrete Forming & Accessories	99.8	127.5	123.8	100.9	63.1	68.2	99.1	63.1	67.9	99.1	63.0	67.8	100.9	63.1	68.2	100.9	63.1	68.2
0320	Concrete Reinforcing	77.6	114.8	96.9	110.0	70.3	89.4	113.5	70.3	91.1	114.8	70.3	91.7	120.3	70.3	94.4	115.1	70.3	91.9
0330	Cast-in-Place Concrete	89.8	131.0	106.1	92.1	70.1	83.4	92.8	70.1	83.8	92.7	70.1	83.8	93.0	70.1	84.0	87.5	70.1	80.7
03	CONCRETE	96.8	125.1	110.6	102.4	68.0	85.7	122.2	68.0	95.9	109.7	67.9	89.4	106.4	68.0	87.7	113.5	68.0	91.4
04	MASONRY	104.8	132.4	121.8	111.1	61.2	80.4	111.1	61.2	80.4	111.2	61.2	80.4	120.7	61.2	84.0	105.6	61.2	78.2
05	METALS	93.7	103.3	96.7	108.2	87.8	101.7	104.7	87.8	99.3	104.4	87.6	99.1	105.8	87.8	100.1	104.9	87.8	99.5
06	WOOD, PLASTICS & COMPOSITES	101.7	127.6	116.1	93.4	63.8	76.9	89.6	63.8	75.2	89.6	63.8	75.2	93.5	63.8	76.9	93.5	63.8	76.9
07	THERMAL & MOISTURE PROTECTION	102.7	125.9	112.4	98.6	70.9	87.0	100.2	70.9	87.9	99.0	70.9	87.2	98.8	70.9	87.1	99.9	70.9	87.7
08	OPENINGS	95.4	123.0	101.9	99.3	64.5	91.2	96.0	64.5	88.6	96.1	64.5	88.8	101.8	64.5	93.1	101.9	64.5	93.2
0920	Plaster & Gypsum Board	95.7	127.9	117.2	107.2	62.4	77.3	80.8	62.4	68.5	80.8	62.4	68.5	99.3	62.4	74.7	99.3	62.4	74.7
0950, 0980	Ceilings & Acoustic Treatment	78.5	127.9	111.2	92.5	62.4	72.6	96.1	62.4	73.8	96.1	62.4	73.8	90.4	62.4	71.9	90.4	62.4	71.9
0960	Flooring	94.9	153.7	111.8	98.4	66.6	89.3	104.1	66.6	93.3	104.1	66.6	93.3	100.2	66.6	90.5	100.2	66.6	90.5
0970, 0990	Wall Finishes & Painting/Coating	83.4	130.2	110.6	97.6	52.7	71.5	94.0	52.7	70.0	94.0	52.7	70.0	92.1	52.7	69.2	92.1	52.7	69.2
09	FINISHES	89.3	133.5	113.7	95.8	62.4	77.3	97.5	62.4	78.1	96.2	62.4	77.5	94.6	62.4	76.8	95.9	62.4	77.4
COVERS	DIVS. 10 - 14, 25, 28, 41, 43, 44, 46	100.0	111.2	102.4	100.0	81.7	96.0	100.0	81.7	96.0	100.0	81.7	96.0	100.0	81.7	96.0	100.0	81.7	96.0
21, 22, 23	FIRE SUPPRESSION, PLUMBING & HVAC	99.7	122.3	109.3	100.2	68.4	86.8	98.0	68.4	85.5	98.0	68.1	85.4	100.1	68.4	86.7	98.0	68.4	85.5
26, 27, 3370	ELECTRICAL, COMMUNICATIONS & UTIL.	89.4	137.0	113.5	89.2	71.3	80.2	92.5	71.3	81.8	90.2	71.3	80.6	87.1	71.3	79.1	86.5	71.3	78.8
MF2014	WEIGHTED AVERAGE	96.4	123.6	108.4	100.3	72.1	87.9	102.1	72.1	88.9	99.9	72.0	87.6	101.0	72.1	88.2	100.7	72.1	88.1

NEW MEXICO

DIVISION		LAS CRUCES 880			LAS VEGAS 877			ROSWELL 882			SANTA FE 875			SOCORRO 878			TRUTH/CONSEQUENCES 879		
		MAT.	INST.	TOTAL	MAT.	INST.	TOTAL	MAT.	INST.	TOTAL	MAT.	INST.	TOTAL	MAT.	INST.	TOTAL	MAT.	INST.	TOTAL
015433	CONTRACTOR EQUIPMENT		84.6	84.6		109.4	109.4		109.4	109.4		109.4	109.4		109.4	109.4		84.6	84.6
0241, 31 - 34	SITE & INFRASTRUCTURE, DEMOLITION	98.3	81.2	86.2	97.2	102.3	100.8	100.3	102.3	101.7	100.2	102.3	101.7	93.3	102.3	99.7	113.8	81.2	90.8
0310	Concrete Forming & Accessories	95.6	61.9	66.5	100.9	63.1	68.2	99.1	63.1	67.9	99.9	63.1	68.1	100.9	63.1	68.2	98.6	61.9	66.9
0320	Concrete Reinforcing	111.1	70.1	89.9	117.1	70.3	92.8	114.8	70.3	91.7	109.3	70.3	89.1	119.3	70.3	93.9	113.3	70.2	91.0
0330	Cast-in-Place Concrete	87.6	62.4	77.6	90.4	70.1	82.4	92.7	70.1	83.8	95.9	70.1	85.7	88.6	70.1	81.3	97.1	62.4	83.4
03	CONCRETE	86.9	64.5	76.0	103.7	68.0	86.4	110.6	68.0	89.9	103.8	68.0	86.4	102.6	68.0	85.8	95.4	64.5	80.4
04	MASONRY	106.5	60.9	78.4	105.9	61.2	78.3	122.7	61.2	84.8	107.9	61.2	79.1	105.7	61.2	78.3	102.8	60.9	77.0
05	METALS	103.2	80.6	96.0	104.6	87.8	99.3	105.6	87.8	100.0	101.8	87.8	97.3	104.9	87.8	99.5	104.5	80.7	96.9
06	WOOD, PLASTICS & COMPOSITES	79.4	62.6	70.1	93.5	63.8	76.9	89.6	63.8	75.2	96.4	63.8	78.2	93.5	63.8	76.9	85.1	62.7	72.6
07	THERMAL & MOISTURE PROTECTION	86.3	66.1	77.8	98.4	70.9	86.9	99.1	70.9	87.3	100.8	70.9	88.3	98.3	70.9	86.8	87.0	66.1	78.2
08	OPENINGS	91.8	63.9	85.3	98.0	64.5	90.2	95.9	64.5	88.6	100.8	64.5	92.4	97.9	64.5	90.1	91.6	63.9	85.2
0920	Plaster & Gypsum Board	79.3	62.4	68.1	99.3	62.4	74.7	80.8	62.4	68.5	110.3	62.4	78.3	99.3	62.4	74.7	100.6	62.4	75.1
0950, 0980	Ceilings & Acoustic Treatment	82.3	62.4	69.2	90.4	62.4	71.9	96.1	62.4	73.8	90.8	62.4	72.0	90.4	62.4	71.9	80.5	62.4	68.5
0960	Flooring	136.5	66.6	116.4	100.2	66.6	90.5	104.1	66.6	93.3	106.7	66.6	95.2	100.2	66.6	90.5	132.5	66.6	113.5
0970, 0990	Wall Finishes & Painting/Coating	83.0	52.7	65.4	92.1	52.7	69.2	94.0	52.7	70.0	96.4	52.7	71.0	92.1	52.7	69.2	84.3	52.7	65.9
09	FINISHES	106.1	61.6	81.5	94.5	62.4	76.7	96.3	62.4	77.6	99.3	62.4	78.9	94.4	62.4	76.7	107.4	61.6	82.0
COVERS	DIVS. 10 - 14, 25, 28, 41, 43, 44, 46	100.0	79.0	95.5	100.0	81.7	96.0	100.0	81.7	96.0	100.0	81.7	96.0	100.0	81.7	96.0	100.0	79.0	95.5
21, 22, 23	FIRE SUPPRESSION, PLUMBING & HVAC	100.5	68.1	86.8	98.0	68.4	85.5	99.9	68.4	86.6	100.1	68.4	86.7	98.0	68.4	85.5	98.1	68.1	85.4
26, 27, 3370	ELECTRICAL, COMMUNICATIONS & UTIL.	91.9	66.3	78.9	88.7	71.3	79.9	91.5	71.3	81.3	99.4	71.3	85.2	86.9	71.3	79.0	90.2	71.3	80.6
MF2014	WEIGHTED AVERAGE	97.5	68.1	84.6	99.0	72.1	87.1	101.4	72.1	88.5	101.1	72.1	88.3	98.6	72.1	86.9	98.3	68.8	85.3

For customer support on your Commercial Renovation Cost Data, call 877.742.4719.

783

NEW MEXICO / NEW YORK

DIVISION		TUCUMCARI 884 MAT.	INST.	TOTAL	ALBANY 120-122 MAT.	INST.	TOTAL	BINGHAMTON 137-139 MAT.	INST.	TOTAL	BRONX 104 MAT.	INST.	TOTAL	BROOKLYN 112 MAT.	INST.	TOTAL	BUFFALO 140-142 MAT.	INST.	TOTAL
015433	CONTRACTOR EQUIPMENT		109.4	109.4		112.6	112.6		116.0	116.0		106.3	106.3		110.9	110.9		96.0	96.0
0241, 31-34	SITE & INFRASTRUCTURE, DEMOLITION	97.6	102.3	100.9	84.9	103.5	98.1	96.1	92.4	93.5	98.5	113.0	108.8	120.4	124.1	123.1	97.4	97.7	97.6
0310	Concrete Forming & Accessories	99.1	63.0	67.8	100.9	100.2	100.3	101.8	88.4	90.2	97.9	171.2	161.3	108.2	178.3	168.8	102.7	116.4	114.5
0320	Concrete Reinforcing	112.5	70.3	90.6	102.4	101.9	102.1	96.9	97.4	97.2	98.1	212.2	157.2	98.3	207.9	155.1	102.4	101.1	101.7
0330	Cast-in-Place Concrete	92.7	70.1	83.8	80.4	110.3	92.2	106.0	127.1	114.3	95.7	172.9	126.1	108.1	170.9	132.9	103.7	120.1	110.2
03	CONCRETE	108.9	67.9	89.0	92.8	104.6	98.6	97.4	104.9	101.1	94.9	177.4	135.0	108.7	179.1	142.9	100.7	113.9	107.1
04	MASONRY	123.2	61.2	85.0	101.2	112.3	108.0	111.1	102.0	105.5	90.6	175.4	142.8	121.7	175.3	154.8	112.8	122.3	118.6
05	METALS	104.4	87.6	99.1	100.1	111.2	103.6	93.8	119.5	102.0	90.5	152.8	110.3	101.0	150.8	116.8	100.7	95.0	98.8
06	WOOD, PLASTICS & COMPOSITES	89.6	63.8	75.2	98.1	97.6	97.8	105.6	84.6	93.9	93.3	169.2	135.6	107.3	179.0	147.2	99.5	116.4	108.9
07	THERMAL & MOISTURE PROTECTION	99.0	70.9	87.2	110.4	105.2	108.2	109.4	96.7	104.1	107.3	163.2	130.7	109.9	162.9	132.1	102.0	111.6	106.0
08	OPENINGS	95.9	64.5	88.6	100.5	93.4	98.9	92.9	85.2	91.1	90.7	172.2	109.7	90.3	176.5	110.4	97.3	104.0	98.9
0920	Plaster & Gypsum Board	80.8	62.4	68.5	102.2	97.3	99.0	108.2	83.8	91.9	98.3	170.8	146.8	104.3	181.2	155.7	105.3	116.6	112.9
0950, 0980	Ceilings & Acoustic Treatment	96.1	62.4	73.8	90.5	97.3	95.0	91.0	83.8	86.3	84.0	170.8	141.5	87.0	181.2	149.4	101.6	116.6	111.5
0960	Flooring	104.1	66.6	93.3	97.2	109.2	100.7	103.9	102.8	103.6	97.5	189.4	124.0	110.9	189.4	133.4	96.9	119.6	103.4
0970, 0990	Wall Finishes & Painting/Coating	94.0	52.7	70.0	94.2	97.4	96.1	88.7	100.3	95.4	109.5	163.3	140.8	116.9	163.3	143.9	101.5	115.6	109.7
09	FINISHES	96.2	62.4	77.5	94.0	101.2	98.0	93.9	91.3	92.5	94.0	173.4	138.0	106.6	179.2	146.7	99.8	117.8	109.8
COVERS	DIVS. 10-14, 25, 28, 41, 43, 44, 46	100.0	81.7	96.0	100.0	99.1	99.8	100.0	97.0	99.3	100.0	132.5	107.0	100.0	132.5	107.1	100.0	106.0	101.3
21, 22, 23	FIRE SUPPRESSION, PLUMBING & HVAC	98.0	68.1	85.4	100.0	104.9	102.1	100.7	91.7	96.9	100.2	166.1	128.0	99.7	165.9	127.7	100.0	97.7	99.0
26, 27, 3370	ELECTRICAL, COMMUNICATIONS & UTIL.	92.5	71.3	81.8	96.0	103.3	99.7	98.2	101.5	99.9	91.5	186.8	139.6	97.9	186.8	142.8	98.7	101.9	100.3
MF2014	WEIGHTED AVERAGE	100.6	72.0	88.0	98.3	104.6	101.1	98.2	98.4	98.3	95.2	166.0	126.4	102.3	167.9	131.2	100.3	106.4	103.0

NEW YORK

DIVISION		ELMIRA 148-149 MAT.	INST.	TOTAL	FAR ROCKAWAY 116 MAT.	INST.	TOTAL	FLUSHING 113 MAT.	INST.	TOTAL	GLENS FALLS 128 MAT.	INST.	TOTAL	HICKSVILLE 115,117,118 MAT.	INST.	TOTAL	JAMAICA 114 MAT.	INST.	TOTAL
015433	CONTRACTOR EQUIPMENT		118.9	118.9		110.9	110.9		110.9	110.9		112.6	112.6		110.9	110.9		110.9	110.9
0241, 31-34	SITE & INFRASTRUCTURE, DEMOLITION	99.7	93.4	95.2	123.6	124.1	124.0	123.6	124.1	124.0	75.0	103.2	94.9	113.3	122.9	120.1	117.9	124.1	122.3
0310	Concrete Forming & Accessories	83.5	93.4	92.1	94.3	170.7	160.4	98.2	170.7	161.0	86.6	93.3	92.4	90.8	152.3	144.0	98.2	170.7	161.0
0320	Concrete Reinforcing	101.3	94.7	97.9	98.3	207.9	155.1	100.0	207.9	155.9	98.2	93.4	95.7	98.3	207.7	155.0	98.3	207.9	155.1
0330	Cast-in-Place Concrete	94.9	104.4	98.6	116.9	170.9	138.2	116.9	170.9	138.2	77.7	107.6	89.5	99.3	165.2	125.3	108.1	170.9	132.9
03	CONCRETE	90.4	99.2	94.7	114.9	175.8	144.5	115.4	175.8	144.7	84.4	99.1	91.6	100.5	165.4	132.0	108.0	175.8	141.0
04	MASONRY	110.0	104.9	106.9	126.2	175.3	156.5	120.1	175.3	154.1	102.8	107.7	105.8	115.7	166.4	147.0	124.0	175.3	155.6
05	METALS	94.9	120.9	103.2	101.0	150.8	116.9	101.0	150.8	116.9	94.0	108.2	98.5	102.5	147.6	116.8	101.0	150.8	116.9
06	WOOD, PLASTICS & COMPOSITES	80.7	92.2	87.1	91.0	168.9	134.3	95.5	168.9	136.3	88.7	90.7	89.8	87.6	149.4	122.0	95.5	168.9	136.3
07	THERMAL & MOISTURE PROTECTION	104.8	94.7	100.6	109.8	161.9	131.6	109.8	161.9	131.6	103.5	101.9	102.8	109.4	154.9	128.4	109.6	161.9	131.5
08	OPENINGS	99.3	88.7	96.8	89.0	171.0	108.1	89.0	171.0	108.1	93.7	87.1	92.2	89.4	160.4	105.9	89.0	171.0	108.1
0920	Plaster & Gypsum Board	100.0	91.8	94.5	93.1	170.8	145.0	95.5	170.8	145.8	94.3	90.3	91.6	92.8	150.8	131.5	95.5	170.8	145.8
0950, 0980	Ceilings & Acoustic Treatment	96.0	91.8	93.2	76.3	170.8	138.9	76.3	170.8	138.9	80.4	90.3	87.0	75.3	150.8	125.3	76.3	170.8	138.9
0960	Flooring	90.7	102.8	94.2	105.5	189.4	129.7	107.1	189.4	130.8	85.5	110.0	92.5	104.4	177.9	125.6	107.1	189.4	130.8
0970, 0990	Wall Finishes & Painting/Coating	95.8	91.7	93.5	116.9	163.3	143.9	116.9	163.3	143.9	91.7	97.4	95.0	116.9	163.3	143.9	116.9	163.3	143.9
09	FINISHES	94.1	94.9	94.5	101.5	173.2	141.1	102.3	173.2	141.5	85.2	96.1	91.2	100.0	157.3	131.7	101.8	173.2	141.3
COVERS	DIVS. 10-14, 25, 28, 41, 43, 44, 46	100.0	98.4	99.7	100.0	131.6	106.8	100.0	131.6	106.8	100.0	96.4	99.2	100.0	126.4	105.7	100.0	131.6	106.8
21, 22, 23	FIRE SUPPRESSION, PLUMBING & HVAC	96.9	91.1	94.5	96.3	165.9	125.7	96.3	165.9	125.7	96.8	101.0	98.6	99.7	155.2	123.2	96.3	165.9	125.7
26, 27, 3370	ELECTRICAL, COMMUNICATIONS & UTIL.	96.3	98.8	97.6	104.5	186.8	146.1	104.5	186.8	146.1	91.5	103.3	97.5	97.4	146.6	122.2	96.4	186.8	142.1
MF2014	WEIGHTED AVERAGE	96.9	98.3	97.6	102.5	166.2	130.6	102.4	166.2	130.5	93.3	101.1	96.7	100.3	152.7	123.4	100.7	166.2	129.6

NEW YORK

DIVISION		JAMESTOWN 147 MAT.	INST.	TOTAL	KINGSTON 124 MAT.	INST.	TOTAL	LONG ISLAND CITY 111 MAT.	INST.	TOTAL	MONTICELLO 127 MAT.	INST.	TOTAL	MOUNT VERNON 105 MAT.	INST.	TOTAL	NEW ROCHELLE 108 MAT.	INST.	TOTAL
015433	CONTRACTOR EQUIPMENT		92.8	92.8		110.9	110.9		110.9	110.9		110.9	110.9		106.3	106.3		106.3	106.3
0241, 31-34	SITE & INFRASTRUCTURE, DEMOLITION	101.1	93.6	95.8	145.7	118.8	126.7	121.4	124.1	123.3	140.5	118.6	125.1	103.8	110.5	108.5	103.4	110.5	108.4
0310	Concrete Forming & Accessories	83.5	87.3	86.8	88.1	101.5	99.7	102.7	170.7	161.6	95.6	100.3	99.7	88.2	137.0	130.4	103.1	136.9	132.4
0320	Concrete Reinforcing	101.4	97.9	99.6	98.6	139.2	119.7	98.3	207.9	155.1	97.8	139.2	119.3	97.1	207.7	154.4	97.1	207.7	154.4
0330	Cast-in-Place Concrete	98.4	103.1	100.3	104.9	136.6	117.4	111.6	170.9	135.0	98.2	134.8	112.7	106.8	147.0	122.7	106.8	147.0	122.7
03	CONCRETE	93.3	94.5	93.9	105.8	120.2	112.8	111.2	175.8	142.6	100.8	119.1	109.7	104.4	152.2	127.6	103.9	152.2	127.4
04	MASONRY	119.6	102.5	109.1	118.2	143.9	134.0	118.4	175.3	153.5	110.7	140.7	129.2	96.0	154.9	132.3	96.0	154.9	132.3
05	METALS	92.5	91.2	92.1	101.7	116.4	106.3	101.0	150.8	116.8	101.6	116.3	106.3	90.3	144.0	107.4	90.6	144.0	107.6
06	WOOD, PLASTICS & COMPOSITES	79.3	84.2	82.0	90.0	93.1	91.7	101.5	168.9	139.0	97.2	93.1	94.9	84.0	131.8	110.6	100.5	131.8	117.9
07	THERMAL & MOISTURE PROTECTION	104.2	93.4	99.7	128.4	135.8	131.5	109.7	161.9	131.6	128.0	134.4	130.7	108.2	147.2	124.5	108.2	147.2	124.6
08	OPENINGS	99.1	85.9	96.1	95.8	114.6	100.2	89.0	171.0	108.1	91.3	114.6	96.7	90.7	150.8	104.7	90.8	150.8	104.7
0920	Plaster & Gypsum Board	87.0	83.6	84.7	96.7	93.0	94.2	100.1	170.8	147.3	97.4	93.0	94.4	93.0	132.3	119.2	106.2	132.3	123.6
0950, 0980	Ceilings & Acoustic Treatment	92.7	83.6	86.7	72.1	93.0	85.9	76.3	170.8	138.9	72.1	93.0	85.9	82.4	132.3	115.4	82.4	132.3	115.4
0960	Flooring	93.3	102.8	96.0	102.7	69.7	93.2	108.8	189.4	132.0	105.2	69.7	95.0	88.9	179.6	115.0	97.0	179.6	120.7
0970, 0990	Wall Finishes & Painting/Coating	97.1	97.1	97.1	121.6	116.8	118.8	116.9	163.3	143.9	121.6	116.8	118.8	107.9	163.3	140.1	107.9	163.3	140.1
09	FINISHES	92.4	90.4	91.3	99.7	94.3	96.7	103.2	173.2	141.9	100.1	93.5	96.4	90.8	146.4	121.6	95.0	146.4	123.4
COVERS	DIVS. 10-14, 25, 28, 41, 43, 44, 46	100.0	99.7	99.5	100.0	108.8	101.9	100.0	131.6	106.8	100.0	107.8	101.7	100.0	123.0	105.0	100.0	119.8	104.3
21, 22, 23	FIRE SUPPRESSION, PLUMBING & HVAC	96.7	88.8	93.4	96.7	121.0	107.0	99.7	165.9	127.7	96.7	115.6	104.7	96.9	136.4	113.6	96.9	136.4	113.6
26, 27, 3370	ELECTRICAL, COMMUNICATIONS & UTIL.	95.3	96.6	95.9	92.8	110.2	101.6	96.9	186.8	142.3	92.8	110.2	101.6	89.9	154.0	122.3	89.9	154.0	122.3
MF2014	WEIGHTED AVERAGE	97.0	93.0	95.3	101.8	117.1	108.6	101.9	166.2	130.2	100.3	115.3	106.9	95.5	143.1	116.5	95.9	143.0	116.7

City Cost Indexes

NEW YORK

	DIVISION	NEW YORK 100-102			NIAGARA FALLS 143			PLATTSBURGH 129			POUGHKEEPSIE 125-126			QUEENS 110			RIVERHEAD 119		
		MAT.	INST.	TOTAL	MAT.	INST.	TOTAL	MAT.	INST.	TOTAL	MAT.	INST.	TOTAL	MAT.	INST.	TOTAL	MAT.	INST.	TOTAL
015433	CONTRACTOR EQUIPMENT		106.9	106.9		92.8	92.8		96.6	96.6		110.9	110.9		110.9	110.9		110.9	110.9
0241, 31 - 34	SITE & INFRASTRUCTURE, DEMOLITION	106.0	114.0	111.7	103.3	95.1	97.5	110.7	100.9	103.7	141.6	119.5	126.0	116.7	124.1	122.0	114.6	122.9	120.4
0310	Concrete Forming & Accessories	101.9	183.8	172.7	83.5	115.2	110.9	92.2	92.9	92.8	88.1	160.5	150.7	91.0	170.7	160.0	95.4	151.9	144.2
0320	Concrete Reinforcing	103.8	212.3	160.0	100.1	101.5	100.8	102.3	101.5	101.9	98.6	139.5	119.8	100.0	207.9	155.9	100.2	207.7	155.9
0330	Cast-in-Place Concrete	107.6	176.1	134.6	101.8	121.5	109.5	95.1	103.2	98.3	101.6	139.5	116.5	102.7	170.9	129.6	101.0	165.0	126.2
03	CONCRETE	105.1	184.1	143.5	95.5	113.8	104.4	97.9	97.5	97.7	103.1	147.7	124.8	103.8	175.8	138.8	101.1	165.1	132.2
04	MASONRY	100.4	179.8	149.3	127.8	130.5	129.4	96.5	101.1	99.4	110.3	147.0	132.9	112.5	175.3	151.2	121.8	166.4	149.3
05	METALS	102.4	152.8	118.4	95.0	93.1	94.4	97.9	89.6	95.2	101.7	119.2	107.3	101.0	150.8	116.8	103.0	147.1	117.0
06	WOOD, PLASTICS & COMPOSITES	97.3	184.0	145.5	79.3	110.8	96.8	95.3	90.3	92.5	90.0	168.9	133.9	87.7	168.9	132.9	92.6	149.4	124.2
07	THERMAL & MOISTURE PROTECTION	107.3	166.8	132.2	104.3	114.2	108.4	122.3	99.5	112.8	128.3	145.1	135.3	109.4	161.9	131.4	110.4	154.9	129.0
08	OPENINGS	96.7	180.2	116.1	99.2	101.1	99.6	101.6	88.2	98.5	95.8	155.0	109.6	89.0	171.0	108.1	89.4	160.4	105.9
0920	Plaster & Gypsum Board	104.4	186.0	158.9	87.0	110.9	103.0	113.9	89.4	97.6	96.7	170.8	146.2	92.8	170.8	144.9	94.0	150.8	131.9
0950, 0980	Ceilings & Acoustic Treatment	99.7	186.0	156.8	92.7	110.9	104.8	97.9	89.4	92.3	72.1	170.8	137.5	76.3	170.8	138.9	76.2	150.8	125.6
0960	Flooring	98.9	189.4	124.9	93.3	119.6	100.9	108.3	109.3	108.6	102.7	169.5	121.9	104.4	189.4	128.9	105.5	137.1	114.6
0970, 0990	Wall Finishes & Painting/Coating	109.5	159.4	138.5	97.1	115.1	107.6	117.6	94.8	104.3	121.6	117.2	119.0	116.9	163.3	143.9	116.9	163.3	143.9
09	FINISHES	99.2	182.9	145.5	92.6	116.5	105.8	97.7	95.5	96.5	99.5	160.1	133.0	100.4	173.2	140.6	100.7	149.2	127.5
COVERS	DIVS. 10 - 14, 25, 28, 41, 43, 44, 46	100.0	135.6	107.7	100.0	108.1	101.8	100.0	97.3	99.4	100.0	120.4	104.4	100.0	131.6	106.8	100.0	126.4	105.7
21, 22, 23	FIRE SUPPRESSION, PLUMBING & HVAC	100.1	168.3	128.9	96.7	101.1	98.6	96.7	101.1	98.6	96.7	125.2	108.8	99.7	165.9	127.7	100.0	152.0	122.0
26, 27, 3370	ELECTRICAL, COMMUNICATIONS & UTIL.	97.6	186.8	142.7	94.0	101.3	97.7	90.0	92.3	91.2	92.8	119.0	106.0	97.4	186.8	142.6	98.8	146.6	123.0
MF2014	WEIGHTED AVERAGE	100.7	169.8	131.1	98.0	107.2	102.1	98.4	96.7	97.7	101.0	135.2	116.1	100.3	166.2	129.3	101.1	150.9	123.0

NEW YORK

	DIVISION	ROCHESTER 144-146			SCHENECTADY 123			STATEN ISLAND 103			SUFFERN 109			SYRACUSE 130-132			UTICA 133-135		
		MAT.	INST.	TOTAL	MAT.	INST.	TOTAL	MAT.	INST.	TOTAL	MAT.	INST.	TOTAL	MAT.	INST.	TOTAL	MAT.	INST.	TOTAL
015433	CONTRACTOR EQUIPMENT		117.6	117.6		112.6	112.6		106.3	106.3		106.3	106.3		112.6	112.6		112.6	112.6
0241, 31 - 34	SITE & INFRASTRUCTURE, DEMOLITION	90.7	109.1	103.7	85.6	103.5	98.3	107.9	113.0	111.5	100.6	108.6	106.3	94.9	102.7	100.4	73.6	101.2	93.1
0310	Concrete Forming & Accessories	98.4	98.7	98.6	103.9	100.2	100.7	87.8	178.7	166.5	96.4	140.5	134.5	100.8	91.6	92.9	102.1	87.6	89.6
0320	Concrete Reinforcing	102.8	94.7	98.6	97.2	101.9	99.6	98.1	208.0	155.0	97.1	143.0	120.9	97.9	97.5	97.7	97.9	104.7	101.4
0330	Cast-in-Place Concrete	96.2	105.4	99.8	93.7	110.3	100.3	106.9	172.9	132.9	103.6	139.5	117.8	98.4	104.3	100.7	90.0	103.7	95.4
03	CONCRETE	99.3	101.1	100.2	98.4	104.6	101.4	106.0	180.1	142.0	101.0	139.4	119.7	99.9	97.9	98.9	97.7	97.2	97.4
04	MASONRY	109.1	107.8	108.3	99.3	112.3	107.3	102.7	175.4	147.5	95.8	149.9	129.2	103.2	106.9	105.5	94.3	103.5	100.0
05	METALS	99.8	109.2	102.8	98.3	111.2	102.4	88.7	152.8	109.1	88.7	120.7	98.8	97.2	107.7	100.5	95.0	110.0	99.8
06	WOOD, PLASTICS & COMPOSITES	95.0	97.5	96.4	107.9	97.6	102.1	82.6	179.4	136.5	93.1	141.7	120.2	101.9	88.8	94.6	101.9	84.6	92.3
07	THERMAL & MOISTURE PROTECTION	104.5	101.7	103.3	105.0	105.2	105.1	107.6	164.3	131.3	108.1	143.4	122.9	104.1	98.2	101.6	92.2	97.4	94.4
08	OPENINGS	108.0	92.3	104.4	99.5	93.4	98.1	90.7	177.7	110.9	90.8	141.9	102.7	95.1	86.2	93.0	98.0	85.0	95.0
0920	Plaster & Gypsum Board	100.2	97.4	98.3	103.7	97.3	99.5	93.0	181.2	151.9	97.6	142.5	127.6	99.5	88.3	92.0	99.5	84.0	89.1
0950, 0980	Ceilings & Acoustic Treatment	100.3	97.4	98.4	86.9	97.3	93.8	84.0	181.2	148.4	82.4	142.5	122.2	91.0	88.3	89.2	91.0	84.0	86.3
0960	Flooring	92.7	111.7	98.2	93.0	109.2	97.7	92.6	189.4	120.4	92.7	177.9	117.2	92.6	98.5	94.3	90.4	98.3	92.6
0970, 0990	Wall Finishes & Painting/Coating	92.2	102.9	98.4	91.7	97.4	95.0	109.5	163.3	140.8	107.9	131.5	121.6	93.2	102.7	98.8	86.5	102.7	95.9
09	FINISHES	98.0	101.7	100.0	90.9	101.2	96.6	92.6	179.5	140.7	92.3	143.6	120.7	92.8	93.6	93.3	91.0	90.5	90.7
COVERS	DIVS. 10 - 14, 25, 28, 41, 43, 44, 46	100.0	100.2	100.1	100.0	99.1	99.8	100.0	133.6	107.3	100.0	122.7	104.9	100.0	97.6	99.5	100.0	95.9	99.1
21, 22, 23	FIRE SUPPRESSION, PLUMBING & HVAC	100.0	90.1	95.9	100.2	104.5	102.0	100.2	166.1	128.0	96.9	128.6	110.3	100.3	92.0	96.8	100.3	93.2	97.3
26, 27, 3370	ELECTRICAL, COMMUNICATIONS & UTIL.	99.3	97.7	98.5	95.5	103.3	99.5	91.5	186.8	139.6	96.0	122.6	109.4	98.3	101.4	99.8	96.6	101.4	99.0
MF2014	WEIGHTED AVERAGE	100.8	100.0	100.5	98.1	104.5	101.0	96.8	167.5	128.0	95.5	131.9	111.5	98.5	98.2	98.3	96.5	97.5	97.0

NEW YORK / NORTH CAROLINA

	DIVISION	WATERTOWN 136			WHITE PLAINS 106			YONKERS 107			ASHEVILLE 287-288			CHARLOTTE 281-282			DURHAM 277		
		MAT.	INST.	TOTAL	MAT.	INST.	TOTAL	MAT.	INST.	TOTAL	MAT.	INST.	TOTAL	MAT.	INST.	TOTAL	MAT.	INST.	TOTAL
015433	CONTRACTOR EQUIPMENT		112.6	112.6		106.3	106.3		106.3	106.3		98.9	98.9		98.9	98.9		104.4	104.4
0241, 31 - 34	SITE & INFRASTRUCTURE, DEMOLITION	81.2	102.7	96.5	98.3	110.5	106.9	105.5	110.5	109.0	99.6	78.9	85.0	101.5	79.1	85.7	99.6	88.1	91.4
0310	Concrete Forming & Accessories	86.8	94.6	93.5	101.4	137.0	132.2	101.6	144.6	138.8	93.8	60.1	64.7	96.4	61.2	65.7	97.1	59.8	64.8
0320	Concrete Reinforcing	98.5	97.5	98.0	97.1	207.7	154.4	101.0	207.7	156.3	99.1	62.6	80.2	103.6	58.1	80.0	101.4	57.5	78.7
0330	Cast-in-Place Concrete	104.8	107.4	105.8	94.9	147.0	115.5	106.1	147.1	122.3	113.7	65.1	94.6	116.1	66.9	96.7	91.0	64.8	80.7
03	CONCRETE	110.2	100.3	105.4	94.5	152.2	122.6	104.0	155.7	129.1	109.1	63.8	87.1	108.8	64.0	87.0	97.7	62.6	80.6
04	MASONRY	95.5	109.2	103.9	95.2	154.9	132.0	100.0	154.9	133.8	93.4	47.8	65.3	100.1	55.5	72.6	89.2	40.8	59.3
05	METALS	95.1	107.7	99.1	90.2	144.0	107.3	98.7	144.3	113.2	94.0	85.3	91.3	94.9	83.6	91.3	109.7	82.8	101.1
06	WOOD, PLASTICS & COMPOSITES	84.1	92.1	88.5	98.2	131.8	116.9	98.1	141.7	122.4	93.9	64.6	77.6	92.4	64.6	76.9	94.5	64.6	77.8
07	THERMAL & MOISTURE PROTECTION	92.5	100.5	95.8	107.9	147.2	124.4	108.3	148.8	125.2	103.2	53.7	82.5	98.1	56.4	80.6	104.6	53.2	83.1
08	OPENINGS	98.0	88.9	95.9	90.8	150.8	104.7	94.0	156.3	108.5	96.2	64.2	88.8	97.1	63.0	89.2	107.1	63.0	96.8
0920	Plaster & Gypsum Board	90.2	91.7	91.2	100.6	132.3	121.8	104.0	142.5	129.7	97.8	63.3	74.8	97.2	63.3	74.6	97.5	63.3	74.7
0950, 0980	Ceilings & Acoustic Treatment	91.0	91.7	91.5	82.4	132.3	115.4	98.1	142.5	127.5	85.0	63.3	70.7	87.3	63.3	71.4	86.6	63.3	71.2
0960	Flooring	83.4	98.3	87.7	95.2	179.6	119.5	94.7	189.4	121.9	96.7	47.2	82.5	96.4	41.7	80.7	104.1	46.4	87.5
0970, 0990	Wall Finishes & Painting/Coating	86.5	99.3	93.9	107.9	163.3	140.1	107.9	163.3	140.1	102.1	62.0	78.8	102.3	62.0	78.9	96.0	62.0	76.2
09	FINISHES	88.2	95.5	92.3	93.1	146.4	122.6	97.4	154.2	128.8	92.8	58.3	73.7	92.3	58.1	73.3	94.1	58.1	74.2
COVERS	DIVS. 10 - 14, 25, 28, 41, 43, 44, 46	100.0	98.5	99.7	100.0	123.0	105.0	100.0	125.4	105.5	100.0	81.0	95.9	100.0	82.1	96.1	100.0	80.8	95.8
21, 22, 23	FIRE SUPPRESSION, PLUMBING & HVAC	100.3	85.8	94.2	100.3	136.4	115.6	100.3	136.5	115.6	100.5	55.6	81.5	99.9	57.1	81.8	100.6	55.2	81.4
26, 27, 3370	ELECTRICAL, COMMUNICATIONS & UTIL.	98.2	92.4	95.3	89.9	154.0	122.3	96.1	163.1	129.9	100.5	57.5	78.7	99.5	60.5	79.8	93.9	56.3	74.9
MF2014	WEIGHTED AVERAGE	98.0	96.7	97.4	95.2	143.1	116.3	99.4	146.3	120.1	99.0	62.4	82.9	99.1	63.8	83.6	100.7	61.7	83.5

For customer support on your Commercial Renovation Cost Data, call 877.742.4719.

785

NORTH CAROLINA

DIVISION		ELIZABETH CITY 279			FAYETTEVILLE 283			GASTONIA 280			GREENSBORO 270,272 - 274			HICKORY 286			KINSTON 285		
		MAT.	INST.	TOTAL	MAT.	INST.	TOTAL	MAT.	INST.	TOTAL	MAT.	INST.	TOTAL	MAT.	INST.	TOTAL	MAT.	INST.	TOTAL
015433	CONTRACTOR EQUIPMENT		108.7	108.7		104.4	104.4		98.9	98.9		104.4	104.4		104.4	104.4		104.4	104.4
0241, 31 - 34	SITE & INFRASTRUCTURE, DEMOLITION	103.8	89.7	93.8	99.0	88.0	91.2	99.5	78.9	85.0	99.5	88.1	91.4	98.5	87.8	91.0	97.6	88.7	91.3
0310	Concrete Forming & Accessories	83.4	42.3	47.9	93.4	59.5	64.1	100.4	60.0	65.5	96.9	59.8	64.8	90.2	59.6	63.8	86.6	63.5	66.6
0320	Concrete Reinforcing	99.3	45.5	71.4	103.0	57.5	79.4	99.5	58.0	78.0	100.3	57.5	78.1	99.1	57.6	77.6	98.6	45.4	71.0
0330	Cast-in-Place Concrete	91.2	66.5	81.4	119.3	64.4	97.6	111.2	65.2	93.0	90.4	64.9	80.3	113.7	64.7	94.4	109.8	71.1	94.5
03	CONCRETE	97.9	53.1	76.1	111.3	62.3	87.5	107.5	62.9	85.8	97.2	62.6	80.4	108.9	62.4	86.3	105.5	64.1	85.4
04	MASONRY	101.6	47.8	68.5	96.7	39.3	61.3	97.8	50.6	68.7	84.8	41.2	57.9	82.2	43.5	58.3	89.1	60.5	71.5
05	METALS	96.9	78.8	91.1	112.9	82.6	103.2	94.8	83.4	91.2	102.8	82.6	96.4	94.1	81.4	90.1	93.0	77.7	88.2
06	WOOD, PLASTICS & COMPOSITES	79.7	42.1	58.8	93.2	64.6	77.2	101.9	64.6	81.1	94.3	64.6	77.7	88.9	64.6	75.4	85.9	64.6	74.0
07	THERMAL & MOISTURE PROTECTION	103.9	50.7	81.6	102.7	51.5	81.3	103.5	54.2	82.8	104.4	51.3	82.2	103.5	52.3	82.1	103.4	57.6	84.2
08	OPENINGS	103.5	47.4	90.5	96.3	63.0	88.6	100.2	63.0	91.5	107.1	63.0	96.8	96.2	62.5	88.4	96.3	59.7	87.8
0920	Plaster & Gypsum Board	90.7	39.6	56.6	101.9	63.3	76.1	103.7	63.3	76.7	99.0	63.3	75.2	97.8	63.3	74.8	97.4	63.3	74.6
0950, 0980	Ceilings & Acoustic Treatment	86.6	39.6	55.5	85.8	63.3	70.9	87.5	63.3	71.5	86.6	63.3	71.2	85.0	63.3	70.7	87.5	63.3	71.5
0960	Flooring	95.9	22.3	74.8	96.9	41.8	81.1	100.2	41.0	83.2	104.1	38.2	85.1	96.6	31.6	77.9	93.8	22.0	73.1
0970, 0990	Wall Finishes & Painting/Coating	96.0	62.0	76.2	102.1	62.0	78.8	102.1	61.5	78.5	96.0	62.0	76.2	102.1	62.0	78.8	102.1	62.0	78.8
09	FINISHES	91.1	39.7	62.7	93.5	57.0	73.3	95.1	57.1	74.1	94.4	56.6	73.5	92.9	54.9	71.9	92.4	56.1	72.3
COVERS	DIVS. 10 - 14, 25, 28, 41, 43, 44, 46	100.0	84.7	96.7	100.0	80.6	95.8	100.0	81.1	95.9	100.0	81.0	95.9	100.0	81.0	95.9	100.0	84.7	96.7
21, 22, 23	FIRE SUPPRESSION, PLUMBING & HVAC	97.1	54.3	79.0	100.2	54.9	81.1	100.5	55.5	81.5	100.5	55.3	81.4	97.1	55.4	79.5	97.1	60.7	81.7
26, 27, 3370	ELECTRICAL, COMMUNICATIONS & UTIL.	93.7	34.3	63.7	99.5	51.3	75.1	99.9	59.1	79.3	93.1	57.5	75.1	98.1	59.1	78.4	97.9	45.8	71.5
MF2014	WEIGHTED AVERAGE	97.8	54.5	78.7	102.2	60.6	83.9	99.8	62.4	83.3	99.2	61.7	82.7	97.3	61.8	81.7	97.0	63.0	82.0

NORTH CAROLINA / NORTH DAKOTA

DIVISION		MURPHY 289			RALEIGH 275 - 276			ROCKY MOUNT 278			WILMINGTON 284			WINSTON-SALEM 271			BISMARCK 585		
		MAT.	INST.	TOTAL	MAT.	INST.	TOTAL	MAT.	INST.	TOTAL	MAT.	INST.	TOTAL	MAT.	INST.	TOTAL	MAT.	INST.	TOTAL
015433	CONTRACTOR EQUIPMENT		98.9	98.9		104.4	104.4		104.4	104.4		98.9	98.9		104.4	104.4		98.8	98.8
0241, 31 - 34	SITE & INFRASTRUCTURE, DEMOLITION	100.8	78.7	85.2	99.1	88.1	91.3	102.0	88.7	92.6	100.9	79.1	85.5	99.8	88.1	91.5	100.2	97.6	98.4
0310	Concrete Forming & Accessories	101.0	39.1	47.4	95.0	60.0	64.7	89.4	63.6	67.1	95.3	61.1	65.7	98.8	60.0	65.2	107.8	41.1	50.1
0320	Concrete Reinforcing	98.6	45.0	70.8	102.4	57.5	79.2	99.3	57.3	77.6	99.8	57.5	77.9	100.3	57.5	78.1	98.1	96.2	97.1
0330	Cast-in-Place Concrete	117.6	63.8	96.4	93.7	65.3	82.5	89.2	71.1	82.1	113.3	66.8	95.0	92.6	65.2	81.8	96.3	70.8	86.3
03	CONCRETE	112.4	50.6	82.4	97.5	62.8	80.7	98.9	66.3	83.0	109.0	63.8	87.1	98.4	62.8	81.1	98.7	62.9	81.3
04	MASONRY	85.3	41.1	58.1	87.6	42.0	59.5	78.5	49.6	60.7	82.8	44.6	59.2	85.0	41.5	58.2	116.5	66.9	85.9
05	METALS	92.0	77.0	87.2	95.7	82.7	91.5	96.1	81.1	91.3	93.7	82.7	90.2	100.2	82.7	94.7	96.6	92.0	95.1
06	WOOD, PLASTICS & COMPOSITES	102.5	38.3	66.8	89.1	64.6	75.5	86.2	64.6	74.1	95.8	64.6	78.4	94.3	64.6	77.7	98.9	33.4	62.5
07	THERMAL & MOISTURE PROTECTION	103.4	48.3	80.4	98.9	53.3	79.8	104.3	55.3	83.8	103.2	52.7	82.1	104.4	51.7	82.4	110.5	73.2	94.9
08	OPENINGS	96.2	45.3	84.3	103.6	63.0	94.1	102.7	62.5	93.3	96.3	63.0	88.6	107.1	63.0	96.8	111.3	49.8	97.0
0920	Plaster & Gypsum Board	103.1	36.3	58.5	91.9	63.3	72.8	92.7	63.3	73.1	99.7	63.3	75.4	99.0	63.3	75.2	94.7	31.9	52.7
0950, 0980	Ceilings & Acoustic Treatment	85.0	36.3	52.8	87.3	63.3	71.4	84.9	63.3	70.6	85.8	63.3	70.9	86.6	63.3	71.2	112.3	31.9	59.1
0960	Flooring	100.5	23.1	78.3	101.4	43.9	84.8	99.6	22.0	77.3	97.5	48.3	83.4	104.1	46.4	87.5	94.3	50.6	81.8
0970, 0990	Wall Finishes & Painting/Coating	102.1	62.0	78.8	94.4	62.0	75.6	96.0	62.0	76.2	102.1	62.0	78.8	96.0	62.0	76.2	93.5	63.3	75.9
09	FINISHES	94.8	37.5	63.1	93.3	57.8	73.7	92.2	56.1	72.3	93.4	59.4	74.6	94.4	58.2	74.4	99.0	43.7	68.4
COVERS	DIVS. 10 - 14, 25, 28, 41, 43, 44, 46	100.0	77.4	95.1	100.0	81.1	95.9	100.0	84.7	96.7	100.0	82.1	96.1	100.0	81.1	95.9	100.0	80.6	95.8
21, 22, 23	FIRE SUPPRESSION, PLUMBING & HVAC	97.1	54.1	78.9	100.1	55.6	81.3	97.1	61.1	81.9	100.5	57.1	82.1	100.5	55.5	81.5	99.9	71.7	88.0
26, 27, 3370	ELECTRICAL, COMMUNICATIONS & UTIL.	101.4	29.1	64.9	95.7	40.1	67.6	95.5	39.4	67.1	100.7	51.3	75.7	93.1	58.0	75.3	104.6	76.2	90.3
MF2014	WEIGHTED AVERAGE	98.2	51.0	77.4	97.7	59.7	81.0	96.8	61.8	81.4	98.5	61.5	82.2	98.9	62.1	82.7	102.1	70.0	87.9

NORTH DAKOTA

DIVISION		DEVILS LAKE 583			DICKINSON 586			FARGO 580 - 581			GRAND FORKS 582			JAMESTOWN 584			MINOT 587		
		MAT.	INST.	TOTAL	MAT.	INST.	TOTAL	MAT.	INST.	TOTAL	MAT.	INST.	TOTAL	MAT.	INST.	TOTAL	MAT.	INST.	TOTAL
015433	CONTRACTOR EQUIPMENT		98.8	98.8		98.8	98.8		98.8	98.8		98.8	98.8		98.8	98.8		98.8	98.8
0241, 31 - 34	SITE & INFRASTRUCTURE, DEMOLITION	106.4	95.2	98.5	114.7	95.2	100.9	99.6	97.9	98.4	110.4	95.2	99.6	105.4	95.2	98.2	107.9	98.3	101.1
0310	Concrete Forming & Accessories	101.9	36.4	45.3	91.5	36.4	43.8	98.2	48.8	55.5	95.2	36.4	44.3	93.0	36.7	44.3	91.2	70.4	73.2
0320	Concrete Reinforcing	100.5	96.4	98.4	101.4	96.1	98.7	100.9	96.9	98.9	98.9	96.4	97.7	101.0	96.8	98.8	102.4	96.3	99.2
0330	Cast-in-Place Concrete	118.3	67.4	98.2	107.0	67.4	91.4	99.3	87.5	94.6	107.0	67.4	91.4	116.8	67.5	97.4	107.0	77.0	95.2
03	CONCRETE	106.7	59.7	83.9	105.8	59.6	83.3	100.0	72.1	86.4	103.1	59.7	82.0	105.3	59.9	83.2	101.8	78.1	90.3
04	MASONRY	128.1	66.6	90.2	129.9	61.2	87.6	120.0	83.7	98.0	121.5	66.6	87.7	142.1	73.6	99.9	120.5	79.6	95.3
05	METALS	92.3	91.7	92.1	92.2	91.3	91.9	94.2	93.3	93.9	92.2	91.7	92.1	92.2	92.5	92.3	92.5	92.1	92.4
06	WOOD, PLASTICS & COMPOSITES	95.5	30.8	59.5	83.2	30.8	54.0	91.8	41.2	63.6	87.5	30.8	55.9	85.1	30.8	54.9	82.9	67.7	74.4
07	THERMAL & MOISTURE PROTECTION	109.5	72.0	93.8	110.1	70.4	93.5	107.4	79.6	95.8	109.7	72.0	93.9	109.4	74.2	94.7	109.5	81.9	97.9
08	OPENINGS	105.8	48.4	92.4	105.8	48.4	92.4	103.3	54.0	91.8	104.4	48.4	91.3	105.8	48.4	92.4	104.5	68.5	96.1
0920	Plaster & Gypsum Board	117.1	29.2	58.3	107.4	29.2	55.2	96.8	39.8	58.7	108.8	29.2	55.6	107.4	29.2	55.5	107.4	67.1	80.5
0950, 0980	Ceilings & Acoustic Treatment	111.2	29.2	56.9	111.2	29.2	56.9	105.7	39.8	62.1	111.2	29.2	56.9	111.2	29.2	56.9	111.2	67.1	82.0
0960	Flooring	106.3	50.6	90.3	98.9	50.6	85.0	103.5	50.6	88.3	101.1	50.6	86.6	99.8	50.6	85.6	98.7	50.6	84.8
0970, 0990	Wall Finishes & Painting/Coating	94.1	63.3	76.2	94.1	63.3	76.2	89.6	74.2	80.6	94.1	74.2	82.5	94.1	63.3	76.2	94.1	63.3	76.2
09	FINISHES	105.6	40.8	69.7	102.9	40.8	68.5	100.8	50.6	73.0	103.2	41.9	69.3	102.3	40.8	68.3	102.0	66.8	82.6
COVERS	DIVS. 10 - 14, 25, 28, 41, 43, 44, 46	100.0	46.6	88.4	100.0	46.7	88.5	100.0	82.9	96.3	100.0	46.7	88.4	100.0	78.3	95.3	100.0	88.1	97.4
21, 22, 23	FIRE SUPPRESSION, PLUMBING & HVAC	96.8	73.8	87.1	96.8	66.0	83.8	100.0	79.0	91.1	100.2	64.7	85.2	96.8	64.7	83.3	100.2	65.7	85.6
26, 27, 3370	ELECTRICAL, COMMUNICATIONS & UTIL.	100.6	51.1	75.5	109.5	76.2	92.7	104.5	69.2	86.7	104.2	67.0	85.4	100.6	51.1	75.5	107.5	77.2	92.2
MF2014	WEIGHTED AVERAGE	101.7	64.8	85.5	102.6	65.9	86.4	101.1	75.0	89.6	101.9	65.1	85.7	101.9	64.7	85.5	101.9	76.9	90.9

NORTH DAKOTA / OHIO

DIVISION		WILLISTON 588 MAT.	INST.	TOTAL	AKRON 442-443 MAT.	INST.	TOTAL	ATHENS 457 MAT.	INST.	TOTAL	CANTON 446-447 MAT.	INST.	TOTAL	CHILLICOTHE 456 MAT.	INST.	TOTAL	CINCINNATI 451-452 MAT.	INST.	TOTAL
015433	CONTRACTOR EQUIPMENT		98.8	98.8		91.7	91.7		87.2	87.2		91.7	91.7		97.1	97.1		96.9	96.9
0241, 31 - 34	SITE & INFRASTRUCTURE, DEMOLITION	108.3	95.2	99.0	99.0	99.1	99.0	116.1	89.5	97.3	99.1	98.9	99.0	101.0	99.8	100.2	97.8	99.6	99.1
0310	Concrete Forming & Accessories	97.0	36.4	44.6	100.9	93.6	94.6	94.9	98.4	98.0	100.9	83.5	85.8	97.3	90.9	91.8	99.3	78.2	81.0
0320	Concrete Reinforcing	103.3	96.2	99.6	100.2	92.6	96.3	94.1	87.3	90.6	100.2	76.8	88.1	91.0	80.3	85.5	96.4	79.8	87.8
0330	Cast-in-Place Concrete	107.0	67.4	91.4	91.2	98.1	94.0	111.4	96.3	105.4	92.1	95.3	93.4	101.2	98.4	100.1	93.2	91.0	92.3
03	CONCRETE	103.2	59.6	82.0	94.1	94.1	94.1	107.8	94.7	101.4	94.5	85.7	90.3	101.9	91.4	96.8	96.3	83.1	89.9
04	MASONRY	114.4	62.7	82.5	100.3	96.8	98.1	86.4	92.0	89.8	101.0	85.6	91.5	94.0	101.1	98.4	93.6	84.6	88.0
05	METALS	92.4	91.4	92.1	99.2	80.8	93.3	103.6	78.1	95.5	99.2	74.4	91.3	95.6	84.8	92.1	97.9	83.9	93.4
06	WOOD, PLASTICS & COMPOSITES	88.9	30.8	56.6	101.5	92.5	96.5	87.5	103.3	96.3	101.9	81.8	90.7	100.8	87.7	93.5	103.4	75.9	88.0
07	THERMAL & MOISTURE PROTECTION	109.7	70.8	93.4	111.7	95.8	105.0	101.9	96.9	99.8	112.8	91.0	103.7	103.9	96.9	101.0	101.7	83.5	94.1
08	OPENINGS	105.9	48.4	92.5	104.9	92.5	102.0	95.7	91.9	94.8	98.9	77.0	93.8	87.0	81.9	85.8	94.5	76.8	90.4
0920	Plaster & Gypsum Board	108.8	29.2	55.6	97.2	92.1	93.8	98.8	103.1	101.7	98.2	81.1	86.8	101.8	87.7	92.4	103.7	75.5	84.0
0950, 0980	Ceilings & Acoustic Treatment	111.2	29.2	56.9	94.4	92.1	92.9	105.9	103.1	104.1	94.4	81.1	85.6	100.6	87.7	92.1	101.4	75.5	84.3
0960	Flooring	102.0	50.6	87.2	90.9	90.9	90.9	124.0	95.1	115.7	91.1	78.5	87.5	101.9	93.9	99.6	103.1	88.1	98.8
0970, 0990	Wall Finishes & Painting/Coating	94.1	63.3	76.2	97.5	106.3	102.6	115.2	99.1	105.8	97.5	83.3	89.2	112.3	94.3	101.8	112.3	85.8	96.9
09	FINISHES	103.4	40.8	68.7	95.8	94.2	94.9	105.0	99.6	102.0	96.0	81.9	88.2	101.7	91.7	96.2	102.2	80.2	90.0
COVERS	DIVS. 10 - 14, 25, 28, 41, 43, 44, 46	100.0	46.7	88.5	100.0	98.3	99.6	100.0	53.2	89.9	100.0	95.5	99.0	100.0	93.3	98.5	100.0	89.2	97.7
21, 22, 23	FIRE SUPPRESSION, PLUMBING & HVAC	96.8	66.0	83.8	100.1	95.6	98.2	96.7	52.0	77.8	100.1	98.7	99.5	97.3	94.7	96.2	100.0	83.1	92.8
26, 27, 3370	ELECTRICAL, COMMUNICATIONS & UTIL.	104.6	77.2	90.7	101.4	93.5	97.4	98.0	97.4	97.7	100.7	94.9	97.7	97.4	84.9	91.1	96.5	80.2	88.3
MF2014	WEIGHTED AVERAGE	101.0	66.2	85.7	99.9	93.9	97.3	100.2	83.1	92.6	99.3	89.3	94.9	97.2	92.0	94.9	98.2	83.7	91.8

OHIO

DIVISION		CLEVELAND 441 MAT.	INST.	TOTAL	COLUMBUS 430-432 MAT.	INST.	TOTAL	DAYTON 453-454 MAT.	INST.	TOTAL	HAMILTON 450 MAT.	INST.	TOTAL	LIMA 458 MAT.	INST.	TOTAL	LORAIN 440 MAT.	INST.	TOTAL
015433	CONTRACTOR EQUIPMENT		92.2	92.2		91.9	91.9		91.2	91.2		97.1	97.1		90.1	90.1		91.7	91.7
0241, 31 - 34	SITE & INFRASTRUCTURE, DEMOLITION	98.9	99.6	99.4	99.0	95.5	96.5	96.6	99.0	98.3	96.5	99.2	98.4	109.4	89.6	95.4	98.4	100.3	99.7
0310	Concrete Forming & Accessories	100.9	98.4	98.7	98.7	81.9	84.2	99.3	77.1	80.1	99.3	78.3	81.1	94.8	85.9	87.1	101.0	85.8	87.8
0320	Concrete Reinforcing	100.7	93.0	96.7	101.2	83.1	91.9	96.4	82.1	89.0	96.4	79.8	87.8	94.1	82.2	87.9	100.2	92.9	96.4
0330	Cast-in-Place Concrete	89.5	106.7	96.3	93.0	89.5	91.6	86.8	86.3	86.6	92.9	91.2	92.3	102.4	95.8	99.8	86.8	104.3	93.7
03	CONCRETE	93.3	99.4	96.3	96.8	84.5	90.8	93.1	81.0	87.2	96.1	83.2	89.8	100.4	88.3	94.5	92.0	92.8	92.4
04	MASONRY	105.4	105.2	105.3	98.9	92.3	94.8	93.1	84.5	87.8	93.4	85.1	88.3	116.8	86.0	98.5	96.5	104.2	101.2
05	METALS	101.0	84.1	95.6	97.9	80.5	92.4	97.1	77.6	90.9	97.2	84.0	93.0	103.6	80.5	96.3	99.9	82.3	94.3
06	WOOD, PLASTICS & COMPOSITES	100.6	95.8	97.9	95.2	79.7	86.5	104.7	74.4	87.8	103.4	75.9	88.0	87.4	85.3	86.2	101.5	79.9	89.5
07	THERMAL & MOISTURE PROTECTION	110.4	107.9	109.3	97.7	93.2	95.8	107.3	86.2	98.4	104.0	83.7	95.5	101.4	93.3	98.0	112.8	102.1	108.3
08	OPENINGS	95.4	94.3	95.1	96.7	78.1	92.4	93.8	76.5	89.8	91.6	76.8	88.1	95.7	79.0	91.8	98.9	85.6	95.8
0920	Plaster & Gypsum Board	96.5	95.5	95.8	99.3	79.0	85.7	103.7	74.0	83.8	103.7	75.5	84.9	98.8	84.7	89.4	97.2	79.2	85.1
0950, 0980	Ceilings & Acoustic Treatment	92.8	95.5	94.6	96.2	79.0	84.8	102.3	74.0	83.6	101.4	75.5	84.3	105.0	84.7	91.6	94.4	79.2	84.3
0960	Flooring	90.7	107.1	95.4	93.0	86.8	91.2	105.6	98.7	103.6	103.1	88.1	98.4	122.8	88.6	113.0	91.1	107.1	95.7
0970, 0990	Wall Finishes & Painting/Coating	97.5	106.3	102.6	101.7	94.3	97.4	112.3	85.0	96.4	112.3	84.7	96.3	115.2	82.8	96.3	97.5	106.3	102.6
09	FINISHES	95.4	100.4	98.1	95.9	83.4	89.0	103.1	81.3	91.1	102.1	80.2	90.0	104.0	85.9	94.0	95.8	90.7	93.0
COVERS	DIVS. 10 - 14, 25, 28, 41, 43, 44, 46	100.0	102.2	100.5	100.0	92.8	98.4	100.0	89.0	97.6	100.0	89.4	97.7	100.0	94.6	98.8	100.0	100.1	100.0
21, 22, 23	FIRE SUPPRESSION, PLUMBING & HVAC	100.0	99.4	99.7	100.0	90.4	95.9	100.9	86.1	94.6	100.6	83.2	93.3	96.7	87.9	93.0	100.1	91.1	96.3
26, 27, 3370	ELECTRICAL, COMMUNICATIONS & UTIL.	101.0	104.4	102.7	98.2	87.1	92.6	95.2	82.3	88.7	95.5	84.8	90.1	98.3	83.3	90.7	100.8	87.3	94.0
MF2014	WEIGHTED AVERAGE	99.2	99.4	99.3	98.2	87.4	93.5	97.9	83.9	91.7	97.8	84.4	91.9	100.5	86.3	94.3	98.9	92.2	96.0

OHIO

DIVISION		MANSFIELD 448-449 MAT.	INST.	TOTAL	MARION 433 MAT.	INST.	TOTAL	SPRINGFIELD 455 MAT.	INST.	TOTAL	STEUBENVILLE 439 MAT.	INST.	TOTAL	TOLEDO 434-436 MAT.	INST.	TOTAL	YOUNGSTOWN 444-445 MAT.	INST.	TOTAL
015433	CONTRACTOR EQUIPMENT		91.7	91.7		91.4	91.4		91.2	91.2		95.9	95.9		94.7	94.7		91.7	91.7
0241, 31 - 34	SITE & INFRASTRUCTURE, DEMOLITION	95.0	98.9	97.7	94.7	94.3	94.4	96.9	97.6	97.4	138.5	104.3	114.3	98.2	96.3	96.8	98.9	99.4	99.3
0310	Concrete Forming & Accessories	90.4	82.7	83.8	95.1	80.7	82.6	99.3	81.5	83.9	96.6	88.6	89.7	98.7	96.4	96.7	100.9	86.5	88.4
0320	Concrete Reinforcing	91.2	77.2	83.9	93.3	83.1	88.0	96.4	82.0	89.0	90.9	86.9	88.8	101.2	84.5	92.6	100.2	86.9	93.3
0330	Cast-in-Place Concrete	84.5	90.9	87.0	84.9	89.4	86.7	89.1	86.2	87.9	92.2	93.8	92.8	93.0	101.1	96.2	90.3	97.1	93.0
03	CONCRETE	86.8	84.0	85.4	88.8	83.8	86.4	94.2	82.9	88.7	92.8	89.3	91.1	96.6	95.4	96.1	93.7	89.6	91.7
04	MASONRY	99.4	96.6	97.7	100.2	95.1	97.0	93.3	84.5	87.8	86.2	93.8	90.9	107.0	100.9	103.3	100.7	93.3	96.1
05	METALS	100.1	74.9	92.1	97.0	78.7	91.1	97.1	77.3	90.8	93.5	79.0	88.9	97.7	85.3	93.7	99.3	78.4	92.6
06	WOOD, PLASTICS & COMPOSITES	89.4	79.9	84.1	91.3	79.1	84.5	106.2	80.7	92.0	86.7	87.5	87.1	95.2	101.3	100.1	101.5	85.1	92.4
07	THERMAL & MOISTURE PROTECTION	112.2	94.2	104.7	97.3	98.5	97.8	107.1	86.6	96.9	109.1	95.0	103.2	99.1	101.3	100.1	113.0	93.2	104.7
08	OPENINGS	99.4	75.2	93.8	91.5	75.7	87.8	92.1	77.0	88.6	91.6	83.0	89.6	94.1	90.1	93.2	98.9	84.6	95.6
0920	Plaster & Gypsum Board	90.8	79.2	83.0	97.3	78.4	84.7	103.7	80.5	88.2	95.7	86.6	89.6	99.3	95.7	96.9	97.2	84.5	88.7
0950, 0980	Ceilings & Acoustic Treatment	95.2	79.2	84.6	96.2	78.4	84.4	102.3	80.5	87.9	93.3	86.6	88.8	96.2	95.7	95.9	94.4	84.5	87.9
0960	Flooring	86.0	103.4	91.0	91.5	103.4	94.9	105.6	98.7	103.6	118.2	96.5	112.0	91.9	105.1	95.7	91.1	90.3	90.9
0970, 0990	Wall Finishes & Painting/Coating	97.5	90.0	93.1	101.7	47.5	70.2	112.3	85.0	96.4	114.1	103.1	107.7	101.7	102.0	101.9	97.5	94.1	95.5
09	FINISHES	93.3	86.9	89.8	94.8	81.4	87.4	103.1	85.1	93.1	111.1	91.3	100.1	95.5	98.6	97.2	95.9	87.5	91.3
COVERS	DIVS. 10 - 14, 25, 28, 41, 43, 44, 46	100.0	95.9	99.1	100.0	53.2	89.9	100.0	89.7	97.8	100.0	93.0	98.5	100.0	99.6	99.9	100.0	96.1	99.2
21, 22, 23	FIRE SUPPRESSION, PLUMBING & HVAC	96.7	87.9	93.0	96.7	90.1	93.9	100.9	80.8	92.4	97.1	91.1	94.5	100.0	99.9	99.9	100.1	86.1	94.2
26, 27, 3370	ELECTRICAL, COMMUNICATIONS & UTIL.	98.5	77.8	88.0	92.9	77.8	85.2	95.2	87.1	91.1	88.2	113.3	100.9	98.3	103.9	101.1	100.8	85.3	93.0
MF2014	WEIGHTED AVERAGE	97.1	86.2	92.3	95.1	84.6	90.5	97.8	84.1	91.8	96.7	93.9	95.5	98.3	97.6	98.0	99.2	88.1	94.4

DIVISION		OHIO			OKLAHOMA														
		ZANESVILLE 437 - 438			ARDMORE 734			CLINTON 736			DURANT 747			ENID 737			GUYMON 739		
		MAT.	INST.	TOTAL	MAT.	INST.	TOTAL	MAT.	INST.	TOTAL	MAT.	INST.	TOTAL	MAT.	INST.	TOTAL	MAT.	INST.	TOTAL
015433	CONTRACTOR EQUIPMENT		91.4	91.4		84.3	84.3		83.4	83.4		83.4	83.4		83.4	83.4		83.4	83.4
0241, 31 - 34	SITE & INFRASTRUCTURE, DEMOLITION	97.7	96.0	96.5	98.3	96.3	96.9	99.9	94.8	96.2	96.4	94.2	94.8	101.5	94.8	96.7	104.2	94.3	97.2
0310	Concrete Forming & Accessories	92.3	81.4	82.9	93.3	42.5	49.3	91.8	48.5	54.4	85.5	46.3	51.6	95.5	36.9	44.8	98.9	49.2	55.9
0320	Concrete Reinforcing	92.8	86.4	89.5	97.1	79.6	89.0	97.6	79.7	88.3	100.6	79.3	89.6	97.0	79.7	88.1	97.6	80.5	88.8
0330	Cast-in-Place Concrete	89.5	89.5	89.5	101.2	48.1	80.3	97.9	49.8	78.9	94.5	47.9	76.1	97.9	60.1	83.0	97.9	90.0	94.8
03	CONCRETE	92.5	84.8	88.8	97.1	51.7	75.0	96.4	55.0	76.3	92.8	53.3	73.6	96.9	53.3	75.7	99.7	69.0	84.8
04	MASONRY	97.8	87.8	91.6	101.1	71.2	82.7	126.0	71.2	92.2	93.0	70.5	79.1	107.0	71.2	84.9	103.5	91.4	96.0
05	METALS	98.4	80.3	92.6	100.9	69.5	90.9	101.0	69.5	91.0	92.6	69.4	85.3	102.5	70.1	92.2	101.6	71.3	91.9
06	WOOD, PLASTICS & COMPOSITES	87.0	79.1	82.6	100.9	36.7	65.1	99.9	44.8	69.2	92.7	41.6	64.3	103.3	28.4	61.6	106.7	44.5	72.1
07	THERMAL & MOISTURE PROTECTION	97.4	93.3	95.7	108.8	64.8	90.4	108.9	65.6	90.8	99.8	64.5	85.1	109.0	65.3	90.7	109.3	76.7	95.7
08	OPENINGS	91.5	78.6	88.5	107.9	50.2	94.5	107.9	54.6	95.5	100.5	52.8	89.4	109.1	45.7	94.4	108.1	54.4	95.6
0920	Plaster & Gypsum Board	93.3	78.4	83.4	92.9	35.3	54.5	92.6	43.7	59.9	78.3	40.5	53.0	93.3	26.8	48.9	93.3	43.4	60.0
0950, 0980	Ceilings & Acoustic Treatment	96.2	78.4	84.4	81.2	35.3	50.8	81.2	43.7	56.4	78.7	40.5	53.4	81.2	26.8	45.2	81.2	43.4	56.2
0960	Flooring	89.4	86.8	88.7	97.4	41.5	81.3	96.1	39.5	79.8	98.5	59.5	87.3	98.1	81.5	93.3	99.8	23.4	77.9
0970, 0990	Wall Finishes & Painting/Coating	101.7	94.3	97.4	87.0	62.5	72.7	87.0	62.5	72.7	95.4	62.5	76.3	87.0	62.5	72.7	87.0	62.5	72.7
09	FINISHES	93.8	83.0	87.9	87.2	42.0	62.2	86.9	46.5	64.6	86.5	48.6	65.5	87.7	45.1	64.1	88.6	45.8	64.9
COVERS	DIVS. 10 - 14, 25, 28, 41, 43, 44, 46	100.0	90.2	97.9	100.0	81.0	95.9	100.0	81.9	96.1	100.0	81.5	96.0	100.0	80.1	95.7	100.0	81.8	96.1
21, 22, 23	FIRE SUPPRESSION, PLUMBING & HVAC	96.7	90.2	93.9	96.9	69.2	85.2	96.9	69.2	85.2	96.9	69.2	85.2	100.2	69.2	87.1	96.9	68.7	85.0
26, 27, 3370	ELECTRICAL, COMMUNICATIONS & UTIL.	93.1	82.7	87.8	93.5	70.1	81.7	94.4	70.1	82.1	97.4	70.1	83.6	94.4	70.1	82.2	96.0	61.0	78.3
MF2014	WEIGHTED AVERAGE	95.6	86.3	91.5	98.6	64.9	83.7	99.8	66.2	85.0	95.5	65.9	82.5	100.2	65.2	84.8	99.7	69.1	86.2

DIVISION		OKLAHOMA																	
		LAWTON 735			MCALESTER 745			MIAMI 743			MUSKOGEE 744			OKLAHOMA CITY 730 - 731			PONCA CITY 746		
		MAT.	INST.	TOTAL	MAT.	INST.	TOTAL	MAT.	INST.	TOTAL	MAT.	INST.	TOTAL	MAT.	INST.	TOTAL	MAT.	INST.	TOTAL
015433	CONTRACTOR EQUIPMENT		84.3	84.3		83.4	83.4		93.0	93.0		93.0	93.0		84.6	84.6		83.4	83.4
0241, 31 - 34	SITE & INFRASTRUCTURE, DEMOLITION	97.4	96.3	96.7	89.7	94.2	92.9	91.0	91.3	91.2	91.3	91.1	91.2	95.4	96.9	96.5	96.9	94.4	95.1
0310	Concrete Forming & Accessories	98.9	46.9	53.9	83.5	45.8	50.9	97.0	67.7	71.6	101.6	36.9	45.6	96.4	60.6	65.5	92.3	48.8	54.7
0320	Concrete Reinforcing	97.3	79.7	88.2	100.2	79.3	89.4	98.7	79.7	88.8	99.6	78.5	88.7	107.5	79.8	93.2	99.6	79.7	89.3
0330	Cast-in-Place Concrete	94.6	60.1	81.0	82.9	50.1	70.0	86.8	52.7	73.4	87.9	52.3	73.8	94.7	75.1	87.0	97.1	90.0	94.3
03	CONCRETE	93.2	57.8	76.0	83.4	53.8	69.0	88.0	65.4	77.0	89.8	51.3	71.1	96.6	69.0	83.2	95.0	68.7	82.2
04	MASONRY	103.4	71.2	83.5	110.9	71.1	86.4	95.6	71.2	80.7	113.0	55.7	77.7	105.9	71.2	84.5	88.5	71.2	77.8
05	METALS	106.4	70.2	94.9	92.6	68.8	85.0	92.5	83.1	89.5	94.0	80.8	89.8	99.2	70.6	90.1	92.5	70.1	85.4
06	WOOD, PLASTICS & COMPOSITES	105.8	41.8	70.1	90.3	41.6	63.2	105.2	70.6	85.9	109.6	29.7	65.1	95.5	59.8	75.6	100.8	44.5	69.5
07	THERMAL & MOISTURE PROTECTION	108.8	66.7	91.2	99.5	65.1	85.1	99.9	68.8	86.9	99.9	60.3	83.4	99.2	70.9	87.4	100.0	71.5	88.1
08	OPENINGS	111.1	53.0	97.5	100.5	52.8	89.4	100.5	68.7	93.1	101.7	46.3	88.8	107.7	62.8	97.3	100.5	54.5	89.8
0920	Plaster & Gypsum Board	95.0	40.6	58.7	77.3	40.5	52.7	84.0	70.1	74.7	85.9	28.0	47.3	100.4	59.1	72.8	83.0	43.4	56.5
0950, 0980	Ceilings & Acoustic Treatment	88.6	40.6	56.8	78.7	40.5	53.4	78.7	70.1	73.0	87.0	28.0	48.0	90.9	59.1	69.9	78.7	43.4	55.3
0960	Flooring	100.2	81.5	94.8	97.2	39.5	80.6	105.3	59.5	92.2	108.1	38.2	88.0	96.9	81.5	92.5	102.0	39.5	84.0
0970, 0990	Wall Finishes & Painting/Coating	87.0	62.5	72.7	95.4	62.5	76.3	95.4	62.5	76.3	95.4	62.5	76.3	88.2	62.5	73.3	95.4	62.5	76.3
09	FINISHES	89.7	53.0	69.4	85.5	44.6	62.9	88.7	65.9	76.1	91.7	37.8	61.9	91.3	63.8	76.1	88.4	46.4	65.1
COVERS	DIVS. 10 - 14, 25, 28, 41, 43, 44, 46	100.0	81.5	96.0	100.0	81.5	96.0	100.0	85.2	96.8	100.0	80.6	95.8	100.0	83.5	96.4	100.0	81.8	96.1
21, 22, 23	FIRE SUPPRESSION, PLUMBING & HVAC	100.2	69.2	87.1	96.9	65.2	83.5	96.9	65.4	83.6	100.2	65.3	85.5	100.0	69.3	87.0	96.9	65.4	83.6
26, 27, 3370	ELECTRICAL, COMMUNICATIONS & UTIL.	96.0	70.4	83.0	95.9	75.2	85.4	97.2	75.3	86.1	95.5	75.3	85.3	100.9	70.1	85.3	95.5	71.8	83.5
MF2014	WEIGHTED AVERAGE	100.7	67.5	86.1	94.8	65.3	81.8	95.2	72.1	85.0	97.5	62.9	82.3	99.8	71.3	87.3	95.5	67.6	83.2

DIVISION		OKLAHOMA												OREGON					
		POTEAU 749			SHAWNEE 748			TULSA 740 - 741			WOODWARD 738			BEND 977			EUGENE 974		
		MAT.	INST.	TOTAL	MAT.	INST.	TOTAL	MAT.	INST.	TOTAL	MAT.	INST.	TOTAL	MAT.	INST.	TOTAL	MAT.	INST.	TOTAL
015433	CONTRACTOR EQUIPMENT		91.9	91.9		83.4	83.4		93.0	93.0		83.4	83.4		98.4	98.4		98.4	98.4
0241, 31 - 34	SITE & INFRASTRUCTURE, DEMOLITION	77.7	89.2	85.9	100.1	94.4	96.1	97.7	91.3	93.2	100.2	94.8	96.4	109.2	102.6	104.5	98.6	102.6	101.4
0310	Concrete Forming & Accessories	90.2	43.3	49.6	85.4	46.1	51.4	101.7	41.5	49.6	92.0	49.1	54.9	108.4	99.7	100.9	104.9	99.6	100.3
0320	Concrete Reinforcing	100.7	79.7	89.8	99.6	79.3	89.1	99.8	79.7	89.4	97.0	79.8	88.1	100.8	98.5	99.6	105.1	98.5	101.7
0330	Cast-in-Place Concrete	86.8	50.2	72.4	100.2	48.1	79.6	95.7	49.6	77.5	97.9	90.2	94.8	102.9	102.2	102.6	99.6	102.2	100.6
03	CONCRETE	90.2	53.7	72.4	96.7	53.3	75.6	95.2	52.6	74.5	96.7	68.9	83.2	109.8	100.0	105.0	101.1	100.0	100.5
04	MASONRY	96.2	71.2	80.8	112.3	71.2	86.9	96.8	71.2	81.0	96.1	71.2	80.8	110.2	103.2	105.9	107.1	103.2	104.7
05	METALS	92.6	82.8	89.5	92.5	69.3	85.1	97.2	82.8	92.6	101.1	70.4	91.3	93.3	95.6	94.0	94.0	95.5	94.5
06	WOOD, PLASTICS & COMPOSITES	97.5	37.9	64.3	92.6	41.6	64.2	108.8	35.5	68.0	100.1	44.5	69.1	98.3	99.5	99.0	94.4	99.5	97.2
07	THERMAL & MOISTURE PROTECTION	100.0	65.1	85.4	100.0	65.4	85.5	100.0	66.8	86.1	109.0	71.5	93.3	109.3	97.8	104.5	108.4	96.2	103.3
08	OPENINGS	100.5	50.9	88.9	100.5	52.8	89.4	103.5	49.6	90.9	107.9	54.5	95.5	100.2	102.5	100.7	100.5	102.5	100.9
0920	Plaster & Gypsum Board	81.3	36.5	51.4	78.3	40.5	53.0	85.9	34.0	51.2	92.6	43.4	59.7	112.3	99.3	103.6	110.7	99.3	103.1
0950, 0980	Ceilings & Acoustic Treatment	78.7	36.5	50.8	78.7	40.5	53.4	87.0	34.0	51.9	81.2	43.4	56.2	85.2	99.3	94.5	86.2	99.3	94.8
0960	Flooring	101.2	59.5	89.2	98.5	31.5	79.2	106.9	55.7	92.1	96.1	41.5	80.4	107.0	101.2	105.4	105.2	101.2	104.1
0970, 0990	Wall Finishes & Painting/Coating	95.4	62.5	76.3	95.4	62.5	76.3	95.4	62.5	76.3	87.0	62.5	72.7	94.9	75.8	83.8	94.9	75.8	83.8
09	FINISHES	86.3	46.4	64.2	86.8	43.3	62.7	91.5	44.2	65.3	87.0	46.7	64.7	100.6	97.4	98.9	98.9	97.4	98.1
COVERS	DIVS. 10 - 14, 25, 28, 41, 43, 44, 46	100.0	81.4	96.0	100.0	81.5	96.0	100.0	81.2	95.9	100.0	81.8	96.1	100.0	102.6	100.6	100.0	102.6	100.6
21, 22, 23	FIRE SUPPRESSION, PLUMBING & HVAC	96.9	65.4	83.6	96.9	69.2	85.2	100.2	65.2	85.4	96.9	69.3	85.2	96.6	99.9	98.0	99.9	99.8	99.9
26, 27, 3370	ELECTRICAL, COMMUNICATIONS & UTIL.	95.6	75.3	85.3	97.4	70.1	83.6	97.4	64.1	80.5	95.8	70.1	82.8	99.7	95.4	97.5	98.4	95.4	96.9
MF2014	WEIGHTED AVERAGE	94.7	66.4	82.2	97.0	65.3	83.0	98.3	64.5	83.4	98.6	68.4	85.3	100.3	99.2	99.8	99.5	99.2	99.3

OREGON

DIVISION		KLAMATH FALLS 976 MAT.	INST.	TOTAL	MEDFORD 975 MAT.	INST.	TOTAL	PENDLETON 978 MAT.	INST.	TOTAL	PORTLAND 970-972 MAT.	INST.	TOTAL	SALEM 973 MAT.	INST.	TOTAL	VALE 979 MAT.	INST.	TOTAL
015433	CONTRACTOR EQUIPMENT		98.4	98.4		98.4	98.4		95.8	95.8		98.4	98.4		98.4	98.4		95.8	95.8
0241, 31 - 34	SITE & INFRASTRUCTURE, DEMOLITION	113.5	102.5	105.7	106.7	102.5	103.8	104.3	95.7	98.2	100.9	102.6	102.1	93.0	102.6	99.8	91.2	95.6	94.3
0310	Concrete Forming & Accessories	101.3	99.4	99.7	100.3	99.4	99.5	101.7	99.8	100.0	106.1	99.7	100.6	105.2	99.7	100.4	108.2	98.8	100.1
0320	Concrete Reinforcing	100.8	98.5	99.6	102.5	98.5	100.4	100.3	98.6	99.4	105.9	98.6	102.1	113.5	98.5	105.8	97.9	98.4	98.1
0330	Cast-in-Place Concrete	102.9	105.1	103.8	102.9	102.1	102.6	103.6	103.5	103.6	102.3	102.2	102.3	94.1	102.2	97.3	81.6	106.2	91.3
03	CONCRETE	112.6	100.9	106.9	107.2	99.8	103.6	93.4	100.5	96.9	102.6	100.1	101.4	98.7	100.0	99.4	78.4	100.9	89.4
04	MASONRY	125.3	103.2	111.7	104.1	103.2	103.5	114.9	103.3	107.7	109.0	103.2	105.4	114.8	103.2	107.6	113.1	103.3	107.1
05	METALS	93.3	95.3	94.0	93.6	95.3	94.1	100.0	96.1	98.7	95.1	95.6	95.2	100.7	95.6	99.1	99.9	95.0	98.3
06	WOOD, PLASTICS & COMPOSITES	89.5	99.5	95.1	88.5	99.5	94.6	91.3	99.6	95.9	95.2	99.5	97.6	91.4	99.5	95.9	99.5	99.6	99.6
07	THERMAL & MOISTURE PROTECTION	109.5	94.0	103.0	109.1	94.0	102.8	102.4	97.4	100.4	108.4	97.8	103.9	105.1	97.8	102.0	101.8	94.6	98.8
08	OPENINGS	100.2	102.5	100.7	103.3	102.5	103.1	96.9	102.5	98.2	98.2	102.5	99.2	100.1	102.5	100.6	96.8	94.0	96.2
0920	Plaster & Gypsum Board	107.4	99.3	102.0	106.8	99.3	101.8	93.8	99.3	97.5	110.2	99.3	102.9	106.7	99.3	101.7	99.8	99.3	99.4
0950, 0980	Ceilings & Acoustic Treatment	92.7	99.3	97.0	99.1	99.3	99.2	60.3	99.3	86.1	88.0	99.3	95.5	93.1	99.3	97.2	60.3	99.3	86.1
0960	Flooring	103.8	101.2	103.1	103.2	101.2	102.6	72.2	101.2	80.6	102.8	101.2	102.4	103.2	101.2	102.6	74.4	101.2	82.1
0970, 0990	Wall Finishes & Painting/Coating	94.9	70.4	80.6	94.9	70.4	80.6	87.3	72.8	78.9	94.8	72.8	82.0	93.3	75.8	83.1	87.3	75.8	80.6
09	FINISHES	101.1	96.9	98.7	101.4	96.9	98.9	71.4	97.2	85.7	98.6	97.1	97.8	97.4	97.4	97.4	71.9	97.5	86.1
COVERS	DIVS. 10 - 14, 25, 28, 41, 43, 44, 46	100.0	102.6	100.6	100.0	102.6	100.6	100.0	88.4	97.5	100.0	102.7	100.6	100.0	102.6	100.6	100.0	102.8	100.6
21, 22, 23	FIRE SUPPRESSION, PLUMBING & HVAC	96.6	99.8	98.0	99.9	99.8	99.9	98.5	112.1	104.3	99.9	99.9	99.9	100.0	99.9	99.9	98.5	72.4	87.5
26, 27, 3370	ELECTRICAL, COMMUNICATIONS & UTIL.	98.5	80.1	89.2	101.8	80.1	90.8	91.4	98.9	95.2	98.6	109.2	103.9	105.2	95.4	100.2	91.4	70.6	80.9
MF2014	WEIGHTED AVERAGE	101.3	97.1	99.4	101.0	96.9	99.2	96.2	101.4	98.5	99.7	101.1	100.3	100.9	99.2	100.2	94.1	89.1	91.9

PENNSYLVANIA

DIVISION		ALLENTOWN 181 MAT.	INST.	TOTAL	ALTOONA 166 MAT.	INST.	TOTAL	BEDFORD 155 MAT.	INST.	TOTAL	BRADFORD 167 MAT.	INST.	TOTAL	BUTLER 160 MAT.	INST.	TOTAL	CHAMBERSBURG 172 MAT.	INST.	TOTAL
015433	CONTRACTOR EQUIPMENT		112.6	112.6		112.6	112.6		111.0	111.0		112.6	112.6		112.6	112.6		111.7	111.7
0241, 31 - 34	SITE & INFRASTRUCTURE, DEMOLITION	93.6	101.9	99.5	96.7	101.4	100.1	105.3	99.2	101.0	92.5	100.5	98.1	88.0	103.0	98.6	89.7	99.0	96.3
0310	Concrete Forming & Accessories	100.2	113.7	111.8	85.1	79.8	80.5	83.8	82.7	82.9	87.2	83.8	84.3	86.6	95.9	94.6	87.9	82.1	82.9
0320	Concrete Reinforcing	97.9	108.9	103.6	94.9	98.1	96.6	94.9	98.0	96.5	96.9	97.9	97.4	95.6	110.7	103.4	92.2	107.6	100.2
0330	Cast-in-Place Concrete	89.2	105.9	95.8	99.4	82.7	92.8	117.7	87.5	105.8	94.9	89.2	92.7	87.8	99.1	92.3	96.7	70.2	86.3
03	CONCRETE	94.4	111.0	102.5	90.6	85.8	88.3	106.0	88.7	97.6	96.4	89.8	93.2	82.5	101.0	91.5	103.3	84.5	94.2
04	MASONRY	98.4	103.6	101.6	101.8	78.2	87.2	105.1	87.2	94.1	99.0	87.5	91.9	104.0	99.5	101.2	106.9	89.3	96.0
05	METALS	97.1	125.7	106.2	91.4	116.9	99.5	99.7	114.6	104.4	95.0	116.5	101.8	91.1	124.2	101.6	93.3	121.4	102.2
06	WOOD, PLASTICS & COMPOSITES	101.3	115.9	109.4	80.2	82.3	81.4	87.2	82.2	84.4	86.6	82.3	84.2	81.6	94.9	89.0	87.1	82.3	84.4
07	THERMAL & MOISTURE PROTECTION	104.0	117.8	109.8	102.8	90.9	97.9	101.9	93.5	98.4	104.0	94.5	100.0	102.5	100.2	101.6	96.1	80.1	89.4
08	OPENINGS	95.1	114.4	99.6	88.7	89.7	89.0	94.9	89.7	93.7	95.1	89.7	93.9	88.7	104.0	92.3	94.6	84.3	92.2
0920	Plaster & Gypsum Board	97.5	116.2	110.0	89.2	81.6	84.1	94.4	81.6	85.9	90.5	81.6	84.6	89.2	94.5	92.8	101.6	81.6	88.3
0950, 0980	Ceilings & Acoustic Treatment	82.8	116.2	104.9	86.9	81.6	83.4	98.6	81.6	87.4	86.2	81.6	83.2	87.8	94.5	92.3	83.2	81.6	82.1
0960	Flooring	92.6	97.4	94.0	86.3	100.0	90.3	91.6	103.0	94.9	86.8	103.0	91.5	87.2	104.2	92.1	94.9	44.4	80.3
0970, 0990	Wall Finishes & Painting/Coating	93.2	110.5	103.3	89.1	112.0	102.4	98.0	112.0	106.1	93.2	112.0	104.1	89.1	112.0	102.4	91.7	112.0	103.5
09	FINISHES	90.7	110.6	101.7	88.8	86.0	87.3	96.3	87.4	91.4	88.8	88.1	88.4	88.8	97.4	93.6	90.3	77.4	83.2
COVERS	DIVS. 10 - 14, 25, 28, 41, 43, 44, 46	100.0	105.8	101.3	100.0	95.4	99.0	100.0	98.5	99.7	100.0	99.4	99.9	100.0	102.6	100.6	100.0	96.8	99.3
21, 22, 23	FIRE SUPPRESSION, PLUMBING & HVAC	100.3	115.8	106.9	99.7	86.8	94.3	96.7	89.4	93.6	97.0	96.2	96.6	96.4	96.2	96.3	96.8	91.8	94.7
26, 27, 3370	ELECTRICAL, COMMUNICATIONS & UTIL.	97.6	100.0	98.8	88.8	111.3	100.2	94.0	111.2	102.7	91.9	111.2	101.7	89.4	111.3	100.4	88.2	89.9	89.1
MF2014	WEIGHTED AVERAGE	97.3	110.6	103.2	94.1	93.5	93.9	98.7	95.3	97.2	95.6	97.4	96.4	92.3	103.2	97.1	95.8	91.2	93.8

PENNSYLVANIA

DIVISION		DOYLESTOWN 189 MAT.	INST.	TOTAL	DUBOIS 158 MAT.	INST.	TOTAL	ERIE 164-165 MAT.	INST.	TOTAL	GREENSBURG 156 MAT.	INST.	TOTAL	HARRISBURG 170-171 MAT.	INST.	TOTAL	HAZLETON 182 MAT.	INST.	TOTAL
015433	CONTRACTOR EQUIPMENT		92.2	92.2		111.0	111.0		112.6	112.6		111.0	111.0		111.7	111.7		112.6	112.6
0241, 31 - 34	SITE & INFRASTRUCTURE, DEMOLITION	106.8	87.4	93.1	110.1	99.8	102.8	93.6	102.2	99.7	101.5	102.1	102.0	89.7	100.8	97.5	86.9	101.8	97.5
0310	Concrete Forming & Accessories	84.1	130.5	124.3	83.3	83.9	83.8	99.6	87.5	89.2	90.1	95.9	95.1	97.6	88.2	89.4	81.7	89.8	88.7
0320	Concrete Reinforcing	94.7	138.0	117.2	94.3	98.2	96.3	96.9	99.7	98.3	94.3	110.7	102.8	101.1	107.9	104.6	95.0	108.7	102.1
0330	Cast-in-Place Concrete	84.3	130.9	102.7	113.5	95.0	106.2	97.7	90.6	94.9	109.2	98.9	105.1	95.1	98.6	96.5	84.3	100.8	90.8
03	CONCRETE	89.9	131.6	110.2	107.0	92.1	99.8	89.7	92.3	91.0	100.7	100.9	100.8	97.8	96.9	97.4	87.1	98.5	92.7
04	MASONRY	101.8	130.4	119.4	105.4	91.1	96.6	90.4	93.0	92.0	115.3	99.6	105.6	105.3	91.3	96.7	111.6	99.1	103.9
05	METALS	94.8	127.1	105.1	99.7	122.0	106.8	91.6	117.2	99.8	99.6	123.2	107.1	99.7	123.7	107.3	96.8	122.9	105.1
06	WOOD, PLASTICS & COMPOSITES	82.2	131.5	109.7	86.2	82.2	84.0	97.4	85.6	90.9	93.5	94.8	94.2	93.6	87.1	90.0	80.8	87.1	84.3
07	THERMAL & MOISTURE PROTECTION	101.5	132.1	114.3	102.2	95.9	99.6	103.3	94.8	99.8	101.8	100.2	101.1	98.8	109.0	103.0	103.5	109.2	105.9
08	OPENINGS	97.4	139.2	107.1	94.9	93.4	94.5	88.9	90.2	89.2	94.8	103.9	97.0	99.2	93.5	97.9	95.6	92.3	94.8
0920	Plaster & Gypsum Board	87.8	132.3	117.5	93.4	81.6	85.5	97.5	85.1	89.2	95.5	94.6	94.9	105.3	86.6	92.8	88.2	86.6	87.1
0950, 0980	Ceilings & Acoustic Treatment	82.0	132.3	115.3	98.6	81.6	87.4	82.8	85.1	84.3	97.8	94.6	95.7	90.8	86.6	88.0	83.7	86.6	85.6
0960	Flooring	77.0	138.7	94.7	91.4	103.0	94.7	92.9	103.0	95.8	95.5	104.2	98.0	100.1	92.1	97.8	84.0	96.1	87.5
0970, 0990	Wall Finishes & Painting/Coating	92.8	151.1	126.7	98.0	112.0	106.1	98.2	97.5	97.8	98.0	106.6	103.0	92.2	99.2	96.3	93.2	115.7	106.3
09	FINISHES	82.2	132.5	110.0	96.5	89.0	92.3	91.7	90.7	91.1	97.1	97.4	97.3	94.9	89.2	91.7	86.6	91.3	89.2
COVERS	DIVS. 10 - 14, 25, 28, 41, 43, 44, 46	100.0	118.5	104.0	100.0	99.1	99.8	100.0	100.7	100.1	100.0	102.5	100.5	100.0	98.0	99.6	100.0	101.4	100.3
21, 22, 23	FIRE SUPPRESSION, PLUMBING & HVAC	96.4	138.1	111.5	96.7	90.3	94.0	99.7	97.6	98.8	96.7	95.1	96.1	99.9	99.2	99.6	96.8	92.8	95.1
26, 27, 3370	ELECTRICAL, COMMUNICATIONS & UTIL.	91.4	129.7	110.7	94.5	111.3	103.0	90.4	99.2	94.9	94.5	111.3	103.0	94.6	89.9	92.2	92.6	97.5	95.1
MF2014	WEIGHTED AVERAGE	94.8	127.4	109.1	99.1	97.5	98.4	94.0	97.5	95.5	98.7	102.8	100.5	98.5	96.5	97.6	95.2	100.3	97.5

For customer support on your Commercial Renovation Cost Data, call 877.742.4719.

789

City Cost Indexes

PENNSYLVANIA

DIVISION		INDIANA 157			JOHNSTOWN 159			KITTANNING 162			LANCASTER 175 - 176			LEHIGH VALLEY 180			MONTROSE 188		
		MAT.	INST.	TOTAL	MAT.	INST.	TOTAL	MAT.	INST.	TOTAL	MAT.	INST.	TOTAL	MAT.	INST.	TOTAL	MAT.	INST.	TOTAL
015433	CONTRACTOR EQUIPMENT		111.0	111.0		111.0	111.0		112.6	112.6		111.7	111.7		112.6	112.6		112.6	112.6
0241, 31 - 34	SITE & INFRASTRUCTURE, DEMOLITION	99.5	100.1	99.9	105.9	101.3	102.6	90.6	102.7	99.1	81.8	100.9	95.3	90.8	101.3	98.2	89.5	100.9	97.6
0310	Concrete Forming & Accessories	84.4	93.1	91.9	83.3	84.2	84.1	86.6	95.5	94.3	90.0	89.1	89.2	93.6	109.5	107.3	82.7	87.5	86.9
0320	Concrete Reinforcing	93.6	110.7	102.4	94.9	110.4	103.0	95.6	110.8	103.4	91.9	107.9	100.2	95.0	106.8	101.1	99.6	107.9	103.9
0330	Cast-in-Place Concrete	107.1	97.7	103.4	118.7	89.1	107.1	91.2	98.2	94.0	82.3	99.7	89.2	91.2	103.5	96.1	89.4	92.2	90.5
03	CONCRETE	98.2	99.2	98.7	106.8	92.2	99.7	85.1	100.5	92.6	90.7	97.7	94.1	93.6	107.9	100.5	92.2	94.5	93.3
04	MASONRY	101.6	100.0	100.6	102.5	90.3	95.0	106.8	100.0	102.6	113.4	93.8	101.3	98.4	100.6	99.7	98.4	97.3	97.7
05	METALS	99.7	122.4	106.9	99.7	121.5	106.6	91.2	123.8	101.6	93.3	123.7	102.9	96.7	122.5	104.9	95.1	123.2	104.0
06	WOOD, PLASTICS & COMPOSITES	88.0	92.2	90.3	86.2	82.2	84.0	81.6	94.9	89.0	89.8	87.1	88.3	93.1	111.9	103.6	81.6	85.4	83.7
07	THERMAL & MOISTURE PROTECTION	101.7	100.1	101.0	101.9	94.4	98.8	102.6	100.4	101.7	95.6	96.2	95.9	103.9	100.5	102.5	103.6	90.6	98.2
08	OPENINGS	94.9	98.8	95.8	94.9	93.4	94.5	88.7	104.0	92.3	94.6	97.9	95.3	95.6	111.7	99.3	92.3	91.3	92.1
0920	Plaster & Gypsum Board	94.7	91.9	92.8	93.2	81.6	85.5	89.2	94.5	92.8	103.6	86.6	92.2	91.2	112.1	105.1	88.8	84.8	86.2
0950, 0980	Ceilings & Acoustic Treatment	98.6	91.9	94.2	97.8	81.6	87.1	87.8	94.5	92.3	83.2	86.6	85.4	83.7	112.1	102.5	86.2	84.8	85.3
0960	Flooring	92.4	103.0	95.4	91.4	103.0	94.7	87.2	103.0	91.7	95.9	103.0	97.9	89.5	92.1	90.3	84.7	103.0	90.0
0970, 0990	Wall Finishes & Painting/Coating	98.0	112.0	106.1	98.0	112.0	106.1	89.1	112.0	102.4	91.7	99.2	96.1	93.2	66.4	77.7	93.2	115.7	106.3
09	FINISHES	96.2	95.5	95.8	95.9	89.4	92.3	88.9	97.0	93.4	90.3	91.6	91.0	89.0	101.1	95.7	87.7	92.5	90.4
COVERS	DIVS. 10 - 14, 25, 28, 41, 43, 44, 46	100.0	101.8	100.4	100.0	99.2	99.8	100.0	102.2	100.5	100.0	98.8	99.7	100.0	103.1	100.7	100.0	100.6	100.1
21, 22, 23	FIRE SUPPRESSION, PLUMBING & HVAC	96.7	92.8	95.1	96.7	93.0	95.2	96.4	96.5	96.4	96.8	94.2	95.7	97.0	110.8	102.8	97.0	98.1	97.5
26, 27, 3370	ELECTRICAL, COMMUNICATIONS & UTIL.	94.5	111.3	103.0	94.5	111.2	103.0	88.8	111.3	100.2	89.5	97.1	93.3	92.6	142.3	117.8	91.9	97.4	94.7
MF2014	WEIGHTED AVERAGE	97.6	101.4	99.2	98.7	98.1	98.5	92.8	103.1	97.3	94.6	98.3	96.2	95.7	112.3	103.0	94.6	98.7	96.4

PENNSYLVANIA

DIVISION		NEW CASTLE 161			NORRISTOWN 194			OIL CITY 163			PHILADELPHIA 190 - 191			PITTSBURGH 150 - 152			POTTSVILLE 179		
		MAT.	INST.	TOTAL	MAT.	INST.	TOTAL	MAT.	INST.	TOTAL	MAT.	INST.	TOTAL	MAT.	INST.	TOTAL	MAT.	INST.	TOTAL
015433	CONTRACTOR EQUIPMENT		112.6	112.6		96.7	96.7		112.6	112.6		96.1	96.1		112.1	112.1		111.7	111.7
0241, 31 - 34	SITE & INFRASTRUCTURE, DEMOLITION	88.4	103.0	98.7	96.3	99.0	98.2	87.0	100.8	96.8	99.4	98.6	98.9	105.1	104.1	104.4	84.7	100.5	95.9
0310	Concrete Forming & Accessories	86.6	97.7	96.2	85.3	131.5	125.3	86.6	92.5	91.8	101.8	139.1	134.1	97.5	98.3	98.2	81.5	89.1	88.1
0320	Concrete Reinforcing	94.4	94.1	94.2	96.2	138.2	118.0	95.6	93.9	94.7	101.7	138.2	120.6	95.4	110.7	103.3	91.1	107.9	99.8
0330	Cast-in-Place Concrete	88.6	96.9	91.9	81.2	129.2	100.1	86.1	94.7	89.5	90.1	130.8	106.2	113.5	99.0	107.8	87.6	100.2	92.6
03	CONCRETE	82.9	98.0	90.2	88.6	131.5	109.4	81.3	94.8	87.9	96.1	135.5	115.3	104.2	102.0	103.1	94.5	97.9	96.1
04	MASONRY	103.5	99.6	101.1	115.4	130.9	125.0	103.0	95.8	98.6	100.3	132.2	120.0	96.7	99.6	98.5	106.2	93.8	98.6
05	METALS	91.2	116.5	99.3	97.4	128.2	107.2	91.2	114.5	98.6	101.2	129.0	110.0	101.2	123.6	108.3	93.5	122.9	102.8
06	WOOD, PLASTICS & COMPOSITES	81.6	97.6	90.5	84.7	131.4	110.7	81.6	92.3	87.5	102.3	140.5	123.6	102.7	97.6	99.9	79.8	87.1	83.9
07	THERMAL & MOISTURE PROTECTION	102.5	97.5	100.4	107.5	134.7	118.9	102.4	98.4	100.7	102.3	137.2	116.9	102.0	101.6	101.9	95.7	108.8	101.2
08	OPENINGS	88.7	97.7	90.8	88.3	139.2	100.2	88.7	86.9	88.3	101.0	145.6	111.4	98.6	105.4	100.2	94.6	91.5	93.9
0920	Plaster & Gypsum Board	89.2	97.4	94.7	91.1	132.3	118.6	89.2	91.9	91.0	103.3	141.6	128.9	99.8	97.4	98.2	98.0	86.6	90.4
0950, 0980	Ceilings & Acoustic Treatment	87.8	97.4	94.2	85.5	132.3	116.5	87.8	91.9	90.5	94.8	141.6	125.8	98.6	97.4	97.8	83.2	86.6	85.4
0960	Flooring	87.2	103.0	91.7	87.2	138.7	102.0	87.2	103.0	91.7	97.4	138.7	109.3	99.5	104.2	100.8	91.8	103.0	95.0
0970, 0990	Wall Finishes & Painting/Coating	89.1	112.0	102.4	90.5	151.1	125.8	89.1	112.0	102.4	92.8	154.7	128.8	98.0	114.7	107.7	91.7	115.7	105.7
09	FINISHES	88.8	97.4	95.0	88.2	134.4	113.7	88.7	95.0	92.2	98.4	140.9	121.9	99.3	100.5	99.9	88.5	93.4	91.2
COVERS	DIVS. 10 - 14, 25, 28, 41, 43, 44, 46	100.0	102.9	100.6	100.0	118.4	104.0	100.0	101.9	100.4	100.0	120.4	104.4	100.0	102.8	100.6	100.0	100.4	100.1
21, 22, 23	FIRE SUPPRESSION, PLUMBING & HVAC	96.4	93.2	95.1	96.8	132.5	111.9	96.4	96.1	96.3	100.3	133.2	114.2	100.1	99.7	99.9	96.8	98.3	97.4
26, 27, 3370	ELECTRICAL, COMMUNICATIONS & UTIL.	89.4	102.8	96.2	93.4	147.2	120.6	91.0	111.2	101.2	96.2	156.2	126.5	96.5	111.3	104.0	87.8	94.8	91.3
MF2014	WEIGHTED AVERAGE	92.4	100.3	95.9	95.4	131.2	111.2	92.3	99.7	95.6	99.5	134.7	115.0	100.2	104.6	102.1	94.4	99.1	96.5

PENNSYLVANIA

DIVISION		READING 195 - 196			SCRANTON 184 - 185			STATE COLLEGE 168			STROUDSBURG 183			SUNBURY 178			UNIONTOWN 154		
		MAT.	INST.	TOTAL	MAT.	INST.	TOTAL	MAT.	INST.	TOTAL	MAT.	INST.	TOTAL	MAT.	INST.	TOTAL	MAT.	INST.	TOTAL
015433	CONTRACTOR EQUIPMENT		117.5	117.5		112.6	112.6		111.7	111.7		112.6	112.6		112.6	112.6		111.0	111.0
0241, 31 - 34	SITE & INFRASTRUCTURE, DEMOLITION	100.8	111.6	108.4	94.1	102.1	99.8	84.5	100.0	95.5	88.5	100.9	97.2	96.8	101.7	100.3	100.1	102.1	101.6
0310	Concrete Forming & Accessories	101.7	90.2	91.7	100.4	88.6	90.2	85.2	79.5	80.3	88.1	89.4	89.3	93.9	87.9	88.7	77.6	95.9	93.3
0320	Concrete Reinforcing	97.6	106.4	102.1	97.9	106.0	102.1	96.2	98.2	97.2	98.2	108.4	103.5	93.7	107.8	101.0	94.3	110.7	102.8
0330	Cast-in-Place Concrete	72.8	100.0	83.6	93.1	101.2	96.3	89.8	74.7	83.9	87.8	74.9	82.7	95.8	94.8	95.4	107.1	98.8	103.8
03	CONCRETE	87.7	98.0	92.7	96.4	97.6	97.0	96.3	83.0	89.9	91.0	89.5	90.3	98.3	95.5	97.0	97.8	100.8	99.3
04	MASONRY	104.1	95.7	98.9	98.8	99.2	99.1	104.4	79.5	89.0	96.3	102.4	100.1	106.8	91.4	97.3	116.3	99.5	106.0
05	METALS	97.8	124.0	106.1	99.3	122.0	106.5	94.8	116.7	101.7	96.8	122.1	104.9	93.2	123.3	102.8	99.4	123.1	106.9
06	WOOD, PLASTICS & COMPOSITES	102.7	86.9	93.9	101.3	85.4	92.4	88.5	82.3	85.0	87.5	85.4	86.3	87.6	87.1	87.3	80.6	94.8	88.5
07	THERMAL & MOISTURE PROTECTION	108.0	112.0	109.7	103.9	97.9	101.4	103.2	88.2	96.9	103.7	84.7	95.8	96.8	103.7	99.7	101.6	100.2	101.0
08	OPENINGS	92.8	97.8	94.0	95.1	90.4	94.0	92.9	89.7	92.1	95.6	84.9	93.1	94.7	92.0	94.1	94.8	103.9	96.9
0920	Plaster & Gypsum Board	102.0	86.6	91.7	99.5	84.8	89.7	91.0	81.6	84.7	89.8	84.8	86.5	97.2	86.6	90.1	91.5	94.5	93.6
0950, 0980	Ceilings & Acoustic Treatment	78.8	86.6	84.0	91.0	84.8	86.9	82.9	81.6	82.0	82.0	84.8	83.9	79.8	86.6	84.3	97.8	94.5	95.7
0960	Flooring	92.2	138.7	105.6	92.6	96.1	93.6	90.0	103.0	93.7	87.3	97.4	90.2	92.6	103.0	95.6	88.4	104.2	92.9
0970, 0990	Wall Finishes & Painting/Coating	89.3	109.7	101.1	93.2	115.7	106.3	93.2	112.0	104.1	93.2	115.7	106.3	91.7	115.7	105.7	98.0	112.0	106.1
09	FINISHES	90.1	99.9	95.5	92.8	91.8	92.2	88.1	86.3	87.1	87.6	91.7	89.9	89.4	92.8	91.3	94.3	97.3	96.0
COVERS	DIVS. 10 - 14, 25, 28, 41, 43, 44, 46	100.0	100.2	100.0	100.0	101.2	100.3	100.0	93.6	98.6	100.0	103.0	100.7	100.0	99.6	99.9	100.0	102.5	100.5
21, 22, 23	FIRE SUPPRESSION, PLUMBING & HVAC	100.3	111.1	104.8	100.3	99.8	100.1	97.0	86.7	92.6	97.0	102.1	99.2	96.8	92.8	95.1	96.7	95.1	96.1
26, 27, 3370	ELECTRICAL, COMMUNICATIONS & UTIL.	99.3	97.1	98.2	97.6	97.5	97.5	91.1	111.2	101.3	92.6	142.5	117.8	88.1	94.8	91.5	91.9	111.3	101.7
MF2014	WEIGHTED AVERAGE	97.1	104.5	100.4	98.1	99.8	98.8	95.1	93.0	94.2	95.1	105.0	99.4	95.3	97.3	96.2	97.7	102.8	99.9

City Cost Indexes

PENNSYLVANIA

DIVISION		WASHINGTON 153 MAT.	INST.	TOTAL	WELLSBORO 169 MAT.	INST.	TOTAL	WESTCHESTER 193 MAT.	INST.	TOTAL	WILKES-BARRE 186-187 MAT.	INST.	TOTAL	WILLIAMSPORT 177 MAT.	INST.	TOTAL	YORK 173-174 MAT.	INST.	TOTAL
015433	CONTRACTOR EQUIPMENT		111.0	111.0		112.6	112.6		96.7	96.7		112.6	112.6		112.6	112.6		111.7	111.7
0241, 31 - 34	SITE & INFRASTRUCTURE, DEMOLITION	100.2	102.1	101.6	96.0	100.7	99.3	102.2	96.4	98.1	86.5	102.1	97.6	87.9	100.7	97.0	85.5	100.9	96.4
0310	Concrete Forming & Accessories	84.5	95.9	94.4	86.6	86.2	86.2	92.2	131.1	125.9	90.7	90.3	90.3	90.5	88.3	88.6	84.9	89.1	88.5
0320	Concrete Reinforcing	94.3	110.8	102.8	96.2	107.8	102.2	95.3	138.0	117.4	96.9	108.7	103.0	93.0	108.4	101.0	93.7	107.9	101.1
0330	Cast-in-Place Concrete	107.1	98.9	103.9	94.1	90.1	92.5	90.0	131.7	106.4	84.3	101.4	91.0	81.1	92.6	85.6	88.1	99.7	92.7
03	CONCRETE	98.3	100.9	99.6	98.7	93.1	96.0	96.3	132.1	113.7	88.0	98.9	93.3	85.8	95.1	90.3	95.7	97.7	96.7
04	MASONRY	101.9	101.2	101.4	105.3	91.9	97.0	109.7	133.3	124.2	112.0	99.2	104.1	97.9	94.0	95.5	108.0	93.8	99.3
05	METALS	99.3	123.4	107.0	94.9	122.9	103.8	97.4	126.9	106.8	95.0	123.8	104.2	93.3	123.9	103.0	94.9	123.7	104.1
06	WOOD, PLASTICS & COMPOSITES	88.0	94.8	91.8	86.1	85.6	85.8	91.6	131.4	113.8	89.9	87.1	88.3	84.3	87.1	85.9	83.3	87.1	85.4
07	THERMAL & MOISTURE PROTECTION	101.7	100.6	101.2	104.2	92.0	99.1	107.9	131.3	117.7	103.5	109.2	105.9	96.3	106.2	100.4	95.8	110.0	101.7
08	OPENINGS	94.8	103.9	96.9	95.1	91.5	94.3	88.3	131.2	98.3	92.3	98.6	93.8	94.7	90.8	93.8	94.6	93.5	94.3
0920	Plaster & Gypsum Board	94.5	94.5	94.5	89.7	85.1	86.6	92.4	132.3	119.0	90.8	86.6	88.0	98.0	86.6	90.4	98.6	86.6	90.6
0950, 0980	Ceilings & Acoustic Treatment	97.8	94.5	95.7	82.9	85.1	84.3	85.5	132.3	116.5	86.2	86.6	86.4	83.2	86.5	85.4	82.2	86.6	85.1
0960	Flooring	92.4	104.2	95.8	86.4	103.0	91.2	90.3	138.7	104.2	88.2	96.1	90.5	91.4	93.4	92.0	93.2	103.0	96.0
0970, 0990	Wall Finishes & Painting/Coating	98.0	112.0	106.1	93.2	115.7	106.3	90.5	151.1	125.8	93.2	115.7	106.3	91.7	115.7	105.7	91.7	99.2	96.1
09	FINISHES	96.0	97.6	96.9	88.2	90.3	89.3	89.8	133.0	113.7	88.8	92.8	91.0	89.1	91.4	90.3	88.8	91.6	90.4
COVERS	DIVS. 10 - 14, 25, 28, 41, 43, 44, 46	100.0	102.5	100.5	100.0	99.4	99.9	100.0	118.9	104.1	100.0	101.4	100.3	100.0	98.6	99.7	100.0	98.8	99.7
21, 22, 23	FIRE SUPPRESSION, PLUMBING & HVAC	96.7	97.8	97.2	97.0	92.7	95.2	96.8	133.5	112.3	97.0	99.9	98.2	96.8	93.7	95.5	100.1	94.2	97.6
26, 27, 3370	ELECTRICAL, COMMUNICATIONS & UTIL.	94.0	111.3	102.7	91.9	97.4	94.7	93.3	114.0	103.7	92.6	97.5	95.1	88.5	85.1	86.8	89.5	84.6	87.0
MF2014	WEIGHTED AVERAGE	97.5	103.6	100.2	96.2	96.6	96.3	96.3	126.3	109.5	94.9	101.0	97.6	93.2	96.2	94.5	95.9	96.8	96.3

PUERTO RICO / RHODE ISLAND / SOUTH CAROLINA

DIVISION		PUERTO RICO SAN JUAN 009 MAT.	INST.	TOTAL	NEWPORT 028 MAT.	INST.	TOTAL	PROVIDENCE 029 MAT.	INST.	TOTAL	AIKEN 298 MAT.	INST.	TOTAL	BEAUFORT 299 MAT.	INST.	TOTAL	CHARLESTON 294 MAT.	INST.	TOTAL
015433	CONTRACTOR EQUIPMENT		87.3	87.3		100.0	100.0		100.0	100.0		104.0	104.0		104.0	104.0		104.0	104.0
0241, 31 - 34	SITE & INFRASTRUCTURE, DEMOLITION	134.0	89.2	102.3	89.5	102.5	98.7	92.2	102.5	99.5	121.0	89.3	98.6	116.3	88.7	96.8	101.4	89.0	92.6
0310	Concrete Forming & Accessories	94.1	17.2	27.6	101.0	120.9	118.3	99.0	120.9	118.0	95.1	63.8	68.0	94.2	40.5	47.7	93.3	63.4	67.5
0320	Concrete Reinforcing	192.3	12.5	99.2	101.8	147.6	125.5	102.6	147.6	125.9	98.1	64.4	80.7	97.1	25.7	60.2	97.0	65.3	80.6
0330	Cast-in-Place Concrete	97.6	29.2	70.6	77.4	122.2	95.1	97.2	122.2	107.1	82.3	68.2	76.7	82.3	67.6	76.5	96.5	68.1	85.3
03	CONCRETE	105.5	21.4	64.6	90.5	125.7	107.6	100.4	125.7	112.7	108.3	66.7	88.1	105.4	49.2	78.1	99.5	66.7	83.6
04	MASONRY	93.8	16.8	46.3	99.3	133.2	120.2	110.9	133.2	124.6	80.2	56.5	65.6	94.7	39.5	60.7	96.0	58.9	73.1
05	METALS	116.1	34.2	90.1	95.5	123.6	104.4	99.7	123.6	107.3	95.8	85.2	92.5	95.9	73.1	88.6	97.8	85.7	93.9
06	WOOD, PLASTICS & COMPOSITES	101.8	16.7	54.4	103.4	119.3	112.2	97.3	119.3	109.5	96.3	66.3	79.6	94.9	36.5	62.4	93.8	66.1	78.4
07	THERMAL & MOISTURE PROTECTION	128.3	20.7	83.3	98.1	120.0	107.2	102.1	120.0	109.6	100.5	63.2	84.9	100.3	49.6	79.1	99.5	61.7	83.7
08	OPENINGS	155.2	14.8	122.5	100.9	127.1	107.0	105.1	127.1	110.2	101.1	63.5	92.4	101.2	40.9	87.1	105.3	63.9	95.7
0920	Plaster & Gypsum Board	156.0	14.4	61.4	89.9	119.3	109.6	97.1	119.3	111.9	97.2	65.1	75.8	100.1	34.5	56.3	101.7	65.0	77.1
0950, 0980	Ceilings & Acoustic Treatment	218.5	14.4	83.4	87.0	119.3	108.4	87.2	119.3	108.5	88.3	65.1	73.0	90.8	34.5	53.5	90.8	65.0	73.7
0960	Flooring	220.6	15.9	161.8	94.3	136.8	106.5	93.6	136.8	106.1	102.7	65.8	92.1	104.3	49.2	88.4	103.9	65.4	92.8
0970, 0990	Wall Finishes & Painting/Coating	191.2	16.8	89.7	87.8	136.7	116.2	86.3	136.7	115.6	92.0	62.7	74.9	92.0	34.7	58.7	92.0	62.7	74.9
09	FINISHES	202.3	17.0	99.8	91.8	125.7	110.6	90.1	125.7	109.8	95.9	64.1	78.3	96.9	40.2	65.5	95.1	64.1	77.9
COVERS	DIVS. 10 - 14, 25, 28, 41, 43, 44, 46	100.0	16.7	81.9	100.0	108.1	101.8	100.0	108.1	101.8	100.0	72.1	94.0	100.0	76.6	94.9	100.0	72.1	94.0
21, 22, 23	FIRE SUPPRESSION, PLUMBING & HVAC	103.8	13.8	65.7	100.1	110.1	104.3	100.0	110.1	104.3	97.1	57.8	80.5	97.1	57.4	80.3	100.5	57.3	82.2
26, 27, 3370	ELECTRICAL, COMMUNICATIONS & UTIL.	120.4	18.1	68.7	99.1	99.6	99.3	96.6	99.6	98.1	98.9	65.4	81.9	102.6	33.4	67.6	100.9	59.2	79.8
MF2014	WEIGHTED AVERAGE	122.0	24.5	79.0	97.2	116.8	105.8	99.8	116.8	107.3	98.9	66.8	84.7	99.6	52.5	78.8	99.8	66.1	84.9

SOUTH CAROLINA / SOUTH DAKOTA

DIVISION		COLUMBIA 290-292 MAT.	INST.	TOTAL	FLORENCE 295 MAT.	INST.	TOTAL	GREENVILLE 296 MAT.	INST.	TOTAL	ROCK HILL 297 MAT.	INST.	TOTAL	SPARTANBURG 293 MAT.	INST.	TOTAL	ABERDEEN 574 MAT.	INST.	TOTAL
015433	CONTRACTOR EQUIPMENT		104.0	104.0		104.0	104.0		104.0	104.0		104.0	104.0		104.0	104.0		98.8	98.8
0241, 31 - 34	SITE & INFRASTRUCTURE, DEMOLITION	100.1	89.0	92.3	110.6	89.0	95.3	105.8	88.4	93.5	103.5	88.2	92.7	105.6	88.4	93.4	101.8	94.2	96.4
0310	Concrete Forming & Accessories	91.4	71.8	74.4	81.6	71.5	72.8	92.7	71.2	74.1	91.0	41.1	47.9	95.5	71.4	74.6	96.0	36.1	44.2
0320	Concrete Reinforcing	102.4	65.4	83.2	96.6	65.3	80.4	96.5	44.9	69.8	97.3	43.8	69.6	96.5	59.2	77.2	104.7	37.7	70.0
0330	Cast-in-Place Concrete	97.6	68.2	86.0	82.2	68.1	76.6	82.2	67.9	76.6	82.2	68.2	76.7	82.2	68.0	76.6	109.7	43.8	83.7
03	CONCRETE	99.2	70.5	85.2	99.4	70.3	85.3	97.9	66.4	82.6	95.7	52.8	74.8	98.1	69.2	84.0	105.8	40.6	74.1
04	MASONRY	91.2	58.9	71.3	80.2	58.9	67.1	77.7	58.9	66.1	102.2	41.6	64.8	80.2	58.9	67.1	121.0	55.1	80.4
05	METALS	95.0	86.0	92.2	96.7	85.7	93.2	96.7	78.4	90.8	95.9	75.1	89.3	96.7	85.3	93.1	92.8	65.3	84.0
06	WOOD, PLASTICS & COMPOSITES	91.6	77.1	83.5	81.4	77.1	79.0	93.5	77.1	84.3	92.1	37.6	61.7	97.3	77.1	86.0	96.1	34.7	61.9
07	THERMAL & MOISTURE PROTECTION	95.7	62.9	82.0	99.7	62.9	84.3	99.7	62.9	84.3	99.5	49.9	78.8	99.7	62.9	84.3	100.6	49.8	79.4
08	OPENINGS	101.9	69.8	94.4	101.2	69.8	93.9	101.1	65.1	92.8	101.2	43.5	87.8	101.1	69.6	93.8	101.7	35.6	86.3
0920	Plaster & Gypsum Board	95.4	76.2	82.6	91.5	76.2	81.3	95.9	76.2	82.7	95.6	35.6	55.5	98.6	76.2	83.6	107.1	33.1	57.7
0950, 0980	Ceilings & Acoustic Treatment	92.3	76.2	81.6	89.1	76.2	80.6	88.3	76.2	80.3	88.3	35.6	53.4	88.3	76.2	80.3	93.0	33.1	53.4
0960	Flooring	98.0	65.4	88.6	95.0	41.9	79.8	101.5	54.9	88.1	100.4	42.7	83.8	102.9	54.9	89.1	102.7	48.4	87.1
0970, 0990	Wall Finishes & Painting/Coating	89.4	62.7	73.9	92.0	62.7	74.9	92.0	62.7	74.9	92.0	62.7	74.9	92.0	62.7	74.9	94.2	37.1	61.0
09	FINISHES	93.1	70.6	80.6	91.8	66.6	77.9	93.8	68.8	80.0	93.2	42.4	65.1	94.6	68.8	80.3	98.1	37.6	64.7
COVERS	DIVS. 10 - 14, 25, 28, 41, 43, 44, 46	100.0	73.3	94.2	100.0	73.3	94.2	100.0	73.3	94.2	100.0	68.9	93.3	100.0	73.3	94.2	100.0	39.3	86.8
21, 22, 23	FIRE SUPPRESSION, PLUMBING & HVAC	99.9	57.8	82.1	100.5	57.4	82.2	100.5	57.1	82.1	97.1	57.0	80.2	100.5	57.3	82.2	100.0	43.4	76.1
26, 27, 3370	ELECTRICAL, COMMUNICATIONS & UTIL.	100.6	61.8	81.0	98.8	61.8	80.1	101.0	58.4	79.4	101.0	56.2	78.4	101.0	58.4	79.4	103.5	76.4	89.8
MF2014	WEIGHTED AVERAGE	98.2	68.3	85.0	98.0	67.7	84.7	98.1	66.0	83.9	98.0	56.5	79.7	98.3	67.3	84.6	101.0	53.7	80.1

For customer support on your Commercial Renovation Cost Data, call 877.742.4719.

791

SOUTH DAKOTA

DIVISION		MITCHELL 573			MOBRIDGE 576			PIERRE 575			RAPID CITY 577			SIOUX FALLS 570 - 571			WATERTOWN 572		
		MAT.	INST.	TOTAL	MAT.	INST.	TOTAL	MAT.	INST.	TOTAL	MAT.	INST.	TOTAL	MAT.	INST.	TOTAL	MAT.	INST.	TOTAL
015433	CONTRACTOR EQUIPMENT		98.8	98.8		98.8	98.8		98.8	98.8		98.8	98.8		99.8	99.8		98.8	98.8
0241, 31 - 34	SITE & INFRASTRUCTURE, DEMOLITION	98.5	94.2	95.4	98.4	94.2	95.4	101.4	94.4	96.5	100.1	94.3	96.0	93.0	96.1	95.2	98.3	94.2	95.4
0310	Concrete Forming & Accessories	95.1	36.0	44.0	85.7	35.9	42.6	98.4	38.4	46.5	103.9	35.7	44.9	100.9	45.8	53.3	82.3	35.7	42.0
0320	Concrete Reinforcing	104.1	45.8	73.9	106.8	37.7	71.0	105.5	70.3	87.3	97.9	70.5	83.7	105.5	70.5	87.4	101.2	37.8	68.4
0330	Cast-in-Place Concrete	106.6	41.4	80.9	106.6	43.6	81.8	100.6	42.8	77.8	105.8	41.9	80.6	91.0	44.9	72.8	106.6	44.9	82.3
03	CONCRETE	103.4	41.1	73.1	103.2	40.4	72.7	101.3	47.4	75.1	102.7	46.0	75.1	96.7	51.5	74.7	102.1	40.8	72.3
04	MASONRY	108.2	53.0	74.2	117.3	55.1	79.0	121.5	55.8	81.0	117.5	56.1	79.6	106.3	57.7	76.3	146.6	57.2	91.5
05	METALS	91.8	65.0	83.3	91.8	64.8	83.3	95.8	79.8	90.7	94.5	80.4	90.0	94.2	80.2	89.8	91.8	65.3	83.4
06	WOOD, PLASTICS & COMPOSITES	95.0	35.2	61.7	84.1	34.9	56.7	100.7	35.4	64.4	99.9	32.1	62.2	95.7	45.1	67.5	80.4	34.7	55.0
07	THERMAL & MOISTURE PROTECTION	100.4	47.7	78.3	100.5	49.8	79.3	102.3	49.2	80.1	101.1	49.4	79.4	98.9	52.1	79.3	100.2	50.1	79.3
08	OPENINGS	100.7	36.1	85.7	103.5	35.2	87.6	106.1	45.3	92.0	105.8	43.5	91.3	106.6	50.6	93.5	100.7	35.3	85.5
0920	Plaster & Gypsum Board	105.5	33.7	57.5	99.8	33.4	55.4	99.1	33.9	55.6	106.7	30.5	55.8	97.3	43.9	61.6	97.5	33.1	54.5
0950, 0980	Ceilings & Acoustic Treatment	88.8	33.7	52.3	93.0	33.4	53.5	93.8	33.9	54.1	95.4	30.5	52.4	88.8	43.9	59.1	88.8	33.1	52.0
0960	Flooring	102.2	48.4	86.7	97.5	51.3	84.2	101.5	34.5	82.2	102.0	75.6	94.4	97.9	75.2	91.3	95.9	48.4	82.3
0970, 0990	Wall Finishes & Painting/Coating	94.2	40.6	63.0	94.2	41.7	63.6	95.4	44.6	65.9	94.2	44.6	65.3	94.6	44.6	65.5	94.2	37.1	61.0
09	FINISHES	96.7	38.3	64.4	95.4	38.8	64.1	97.8	36.8	64.0	98.3	42.5	67.5	94.6	50.6	70.3	93.7	37.6	62.7
COVERS	DIVS. 10 - 14, 25, 28, 41, 43, 44, 46	100.0	36.3	86.2	100.0	43.8	87.8	100.0	74.5	94.5	100.0	73.5	94.3	100.0	75.6	94.7	100.0	39.2	86.8
21, 22, 23	FIRE SUPPRESSION, PLUMBING & HVAC	96.7	39.2	72.4	96.7	39.4	72.5	99.9	68.2	86.5	100.0	67.3	86.2	99.9	42.7	75.7	96.7	42.2	73.6
26, 27, 3370	ELECTRICAL, COMMUNICATIONS & UTIL.	101.8	43.6	72.4	103.5	44.5	73.6	110.2	51.4	80.5	99.9	51.4	75.4	99.3	76.7	87.9	100.9	48.9	74.6
MF2014	WEIGHTED AVERAGE	98.6	48.1	76.4	99.4	48.7	77.0	102.2	59.3	83.3	100.8	59.6	82.7	99.0	60.5	82.0	99.9	49.9	77.9

TENNESSEE

DIVISION		CHATTANOOGA 373 - 374			COLUMBIA 384			COOKEVILLE 385			JACKSON 383			JOHNSON CITY 376			KNOXVILLE 377 - 379		
		MAT.	INST.	TOTAL	MAT.	INST.	TOTAL	MAT.	INST.	TOTAL	MAT.	INST.	TOTAL	MAT.	INST.	TOTAL	MAT.	INST.	TOTAL
015433	CONTRACTOR EQUIPMENT		104.7	104.7		99.3	99.3		99.3	99.3		105.8	105.8		98.3	98.3		98.3	98.3
0241, 31 - 34	SITE & INFRASTRUCTURE, DEMOLITION	103.1	98.1	99.5	88.5	88.5	88.5	94.0	85.7	88.1	97.1	98.6	98.1	110.0	85.1	92.4	89.7	87.8	88.4
0310	Concrete Forming & Accessories	97.8	57.8	63.2	83.4	61.0	64.0	83.6	34.7	41.3	90.6	44.6	50.8	83.9	37.8	44.0	96.1	61.4	66.1
0320	Concrete Reinforcing	89.2	66.6	77.5	87.5	62.4	74.5	87.5	62.5	74.5	87.5	69.2	78.0	89.8	61.5	75.1	89.2	61.6	74.9
0330	Cast-in-Place Concrete	101.5	63.2	86.4	94.3	51.3	77.4	106.9	43.1	81.8	104.4	72.5	91.8	81.6	59.5	72.9	95.4	66.3	83.9
03	CONCRETE	100.1	63.0	82.1	99.3	59.7	80.1	110.1	45.2	78.5	100.4	60.8	81.1	109.5	51.8	81.5	97.4	64.8	81.5
04	MASONRY	108.0	52.2	73.6	124.3	54.4	81.2	118.8	43.6	72.4	124.5	47.0	76.7	123.4	44.6	74.8	84.8	54.3	66.0
05	METALS	95.7	89.9	94.3	91.9	89.1	91.0	92.0	88.5	90.9	94.2	91.8	93.4	93.0	86.8	91.0	96.2	87.8	93.5
06	WOOD, PLASTICS & COMPOSITES	106.2	59.1	80.0	72.8	62.2	66.9	73.0	32.5	50.4	89.0	43.8	63.8	78.4	35.6	54.6	92.4	60.9	74.9
07	THERMAL & MOISTURE PROTECTION	100.5	61.7	84.3	95.2	63.3	81.9	95.7	53.2	77.9	97.6	59.8	81.8	95.3	53.6	77.9	93.4	63.7	81.0
08	OPENINGS	99.2	58.0	89.6	92.2	56.8	83.9	92.2	41.5	80.4	99.2	52.4	88.3	95.6	44.3	83.6	93.3	58.2	85.1
0920	Plaster & Gypsum Board	80.5	58.4	65.7	81.8	61.5	68.3	81.8	31.0	47.9	83.8	42.6	56.3	99.2	34.3	55.8	106.4	60.3	75.6
0950, 0980	Ceilings & Acoustic Treatment	99.9	58.4	72.4	77.9	61.5	67.1	77.9	31.0	46.8	86.1	42.6	57.3	96.3	34.3	55.2	97.2	60.3	72.7
0960	Flooring	101.2	57.3	88.6	86.7	22.0	68.1	86.8	56.3	78.0	86.2	39.2	72.7	95.4	39.1	79.2	100.9	51.7	86.8
0970, 0990	Wall Finishes & Painting/Coating	98.2	63.4	78.0	84.5	63.4	72.2	84.5	63.4	72.2	86.1	63.4	72.9	95.6	63.4	76.9	95.6	63.4	76.9
09	FINISHES	96.8	58.2	75.4	88.2	54.1	69.3	88.7	40.2	61.8	88.5	43.9	63.9	100.1	39.2	66.4	94.1	59.4	74.9
COVERS	DIVS. 10 - 14, 25, 28, 41, 43, 44, 46	100.0	41.1	87.2	100.0	54.4	90.1	100.0	39.7	86.9	100.0	49.6	89.1	100.0	76.0	94.8	100.0	82.5	96.2
21, 22, 23	FIRE SUPPRESSION, PLUMBING & HVAC	100.2	62.0	84.1	97.9	79.5	90.1	97.9	74.0	87.8	100.2	68.1	86.6	99.9	58.5	82.4	99.9	67.8	86.3
26, 27, 3370	ELECTRICAL, COMMUNICATIONS & UTIL.	103.1	71.2	87.0	93.6	57.0	75.1	95.1	62.6	78.7	99.5	58.1	78.6	94.1	46.0	69.8	99.7	57.0	78.1
MF2014	WEIGHTED AVERAGE	99.9	66.6	85.2	96.2	67.2	83.4	97.5	60.0	81.0	99.0	63.5	83.4	99.9	56.4	80.7	96.5	66.8	83.5

TENNESSEE / TEXAS

DIVISION		TENNESSEE MCKENZIE 382			MEMPHIS 375, 380 - 381			NASHVILLE 370 - 372			TEXAS ABILENE 795 - 796			AMARILLO 790 - 791			AUSTIN 786 - 787		
		MAT.	INST.	TOTAL	MAT.	INST.	TOTAL	MAT.	INST.	TOTAL	MAT.	INST.	TOTAL	MAT.	INST.	TOTAL	MAT.	INST.	TOTAL
015433	CONTRACTOR EQUIPMENT		99.3	99.3		102.9	102.9		104.9	104.9		93.0	93.0		93.0	93.0		92.1	92.1
0241, 31 - 34	SITE & INFRASTRUCTURE, DEMOLITION	93.7	85.9	88.2	93.0	94.6	94.1	97.5	97.4	97.4	101.9	93.0	95.6	97.8	92.0	93.7	100.8	91.0	93.9
0310	Concrete Forming & Accessories	91.6	37.1	44.4	97.4	66.2	70.4	96.9	65.8	70.0	99.6	64.0	68.8	97.5	55.6	61.3	98.2	56.9	62.5
0320	Concrete Reinforcing	87.7	69.0	78.0	97.1	69.4	82.7	94.1	66.9	80.0	102.1	51.9	76.1	111.9	50.4	80.0	100.8	51.2	75.1
0330	Cast-in-Place Concrete	104.6	56.8	85.8	92.7	77.6	86.8	91.9	66.1	81.7	96.5	64.8	84.0	88.8	64.7	79.3	90.1	66.7	80.9
03	CONCRETE	108.6	52.1	81.1	98.7	72.2	85.8	95.7	67.7	82.1	94.7	62.7	79.2	94.3	58.6	77.0	93.2	59.9	77.0
04	MASONRY	123.2	46.0	75.6	99.4	56.2	72.8	91.2	57.7	70.6	101.2	63.4	77.9	104.2	57.3	75.3	107.4	57.1	76.4
05	METALS	92.0	91.2	91.7	97.9	93.1	96.4	96.5	91.9	95.0	102.4	70.3	92.2	97.3	69.1	88.4	100.5	68.0	90.2
06	WOOD, PLASTICS & COMPOSITES	81.8	35.0	55.7	98.3	67.6	81.2	102.4	66.3	82.3	104.8	66.3	83.4	103.2	55.4	76.5	95.9	57.0	74.2
07	THERMAL & MOISTURE PROTECTION	95.7	47.8	75.7	98.2	67.4	85.3	95.0	65.5	82.6	100.2	67.0	86.3	96.3	63.0	82.4	99.9	64.4	85.1
08	OPENINGS	92.2	45.1	81.2	99.2	56.8	91.8	98.9	67.5	91.6	105.6	61.6	95.4	109.0	55.3	96.5	109.8	54.5	96.9
0920	Plaster & Gypsum Board	84.8	33.6	50.6	93.8	67.0	75.9	92.7	65.4	74.7	86.3	65.7	72.5	96.6	54.4	68.4	87.6	56.1	66.6
0950, 0980	Ceilings & Acoustic Treatment	77.9	33.6	48.6	96.2	67.0	76.9	97.1	65.7	76.3	87.8	65.7	73.2	94.3	54.4	67.9	84.1	56.1	65.6
0960	Flooring	89.8	37.8	74.9	99.7	55.0	86.8	98.9	59.8	87.7	104.5	73.6	95.6	103.1	63.0	91.6	99.7	63.0	89.2
0970, 0990	Wall Finishes & Painting/Coating	84.5	49.6	64.2	91.4	63.4	75.1	97.4	63.4	77.6	92.9	56.1	71.5	85.3	56.1	68.3	91.6	48.8	66.7
09	FINISHES	89.9	37.1	60.7	95.0	63.4	77.5	100.1	64.0	80.1	89.9	65.2	76.3	93.0	56.6	72.9	90.0	56.8	71.6
COVERS	DIVS. 10 - 14, 25, 28, 41, 43, 44, 46	100.0	26.5	84.1	100.0	82.5	96.2	100.0	83.0	96.3	100.0	82.4	96.2	100.0	79.1	95.5	100.0	81.1	95.9
21, 22, 23	FIRE SUPPRESSION, PLUMBING & HVAC	97.9	68.3	85.4	100.0	73.4	88.8	99.9	83.2	92.9	100.2	49.9	78.9	100.0	54.3	80.7	99.9	59.5	82.8
26, 27, 3370	ELECTRICAL, COMMUNICATIONS & UTIL.	94.9	62.4	78.5	101.5	64.8	82.9	100.1	64.3	82.0	101.7	48.3	74.7	101.4	63.7	82.3	100.7	63.4	81.9
MF2014	WEIGHTED AVERAGE	97.7	59.4	80.8	98.9	72.5	87.3	98.2	74.1	87.6	100.0	62.3	83.4	99.6	62.3	83.1	100.0	63.4	83.9

TEXAS

DIVISION		BEAUMONT 776 - 777			BROWNWOOD 768			BRYAN 778			CHILDRESS 792			CORPUS CHRISTI 783 - 784			DALLAS 752 - 753		
		MAT.	INST.	TOTAL	MAT.	INST.	TOTAL	MAT.	INST.	TOTAL	MAT.	INST.	TOTAL	MAT.	INST.	TOTAL	MAT.	INST.	TOTAL
015433	CONTRACTOR EQUIPMENT		97.5	97.5		93.0	93.0		97.5	97.5		93.0	93.0		101.2	101.2		104.0	104.0
0241, 31 - 34	SITE & INFRASTRUCTURE, DEMOLITION	87.8	97.4	94.6	106.4	93.0	96.9	79.3	97.5	92.1	113.2	91.4	97.8	145.7	87.1	104.3	107.1	93.1	97.2
0310	Concrete Forming & Accessories	105.1	60.4	66.4	101.6	59.0	64.8	83.2	64.9	67.4	97.8	63.8	68.4	102.7	57.3	63.4	98.5	64.5	69.1
0320	Concrete Reinforcing	99.1	64.5	81.2	99.3	51.5	74.5	101.6	51.3	75.5	102.2	51.5	76.0	87.8	47.9	67.1	100.2	53.7	76.1
0330	Cast-in-Place Concrete	94.4	65.9	83.1	105.4	64.7	89.4	75.6	64.5	71.2	99.0	64.7	85.5	116.4	65.6	96.4	112.6	66.6	94.5
03	CONCRETE	97.0	63.8	80.9	102.4	60.4	82.0	81.4	62.9	72.4	103.1	62.5	83.4	102.2	60.0	81.7	103.9	64.6	84.8
04	MASONRY	104.3	64.6	79.8	134.7	56.5	86.5	143.7	59.5	91.8	105.3	56.5	75.2	92.0	56.4	70.1	97.7	56.5	72.3
05	METALS	97.1	76.1	90.4	98.9	69.6	89.6	96.8	70.5	88.4	99.9	69.6	90.3	96.7	80.0	91.4	101.2	82.9	95.4
06	WOOD, PLASTICS & COMPOSITES	119.2	59.9	86.2	103.9	59.9	79.4	83.0	66.3	73.7	104.2	66.3	83.1	120.9	57.5	85.6	100.3	66.6	81.6
07	THERMAL & MOISTURE PROTECTION	102.1	67.6	87.7	95.7	64.5	82.6	94.1	66.9	82.7	100.8	64.1	85.4	105.1	65.9	88.7	92.2	66.1	81.3
08	OPENINGS	101.7	61.0	92.2	101.0	58.1	91.0	103.0	61.1	93.3	100.6	61.6	91.6	110.6	53.8	97.3	96.4	62.3	88.4
0920	Plaster & Gypsum Board	103.9	59.1	74.0	86.2	59.1	68.1	92.6	65.7	74.6	85.9	65.7	72.4	97.4	56.5	70.1	93.2	65.7	74.8
0950, 0980	Ceilings & Acoustic Treatment	100.7	59.1	73.1	75.3	59.1	64.6	93.8	65.7	75.2	86.2	65.7	72.6	86.2	56.5	66.5	90.8	65.7	74.2
0960	Flooring	117.8	71.8	104.6	87.9	63.0	80.8	87.8	63.0	80.7	102.7	63.0	91.3	115.0	63.0	100.1	99.0	63.0	88.6
0970, 0990	Wall Finishes & Painting/Coating	93.9	59.4	73.8	92.9	56.1	71.5	91.5	62.2	74.5	92.9	56.1	71.5	111.7	48.5	75.1	105.2	56.1	76.6
09	FINISHES	96.9	62.0	77.6	82.4	59.3	69.6	84.5	64.4	73.4	90.2	63.1	75.2	102.6	57.2	77.5	96.3	63.4	78.1
COVERS	DIVS. 10 - 14, 25, 28, 41, 43, 44, 46	100.0	83.6	96.5	100.0	79.6	95.6	100.0	83.0	96.3	100.0	80.3	95.7	100.0	82.9	96.3	100.0	83.2	96.4
21, 22, 23	FIRE SUPPRESSION, PLUMBING & HVAC	100.2	63.5	84.7	96.8	49.5	76.8	96.8	65.3	83.5	96.8	54.3	78.9	100.2	58.3	82.5	99.9	62.7	84.2
26, 27, 3370	ELECTRICAL, COMMUNICATIONS & UTIL.	96.6	70.1	83.2	93.4	52.6	72.8	94.8	69.7	82.1	101.7	63.7	82.5	97.7	57.2	77.2	94.4	65.5	79.8
MF2014	WEIGHTED AVERAGE	98.9	68.9	85.7	99.1	60.6	82.1	96.4	68.4	84.1	99.7	64.0	84.0	102.0	63.2	84.9	99.2	67.9	85.4

TEXAS

DIVISION		DEL RIO 788			DENTON 762			EASTLAND 764			EL PASO 798 - 799,885			FORT WORTH 760 - 761			GALVESTON 775		
		MAT.	INST.	TOTAL	MAT.	INST.	TOTAL	MAT.	INST.	TOTAL	MAT.	INST.	TOTAL	MAT.	INST.	TOTAL	MAT.	INST.	TOTAL
015433	CONTRACTOR EQUIPMENT		92.1	92.1		100.7	100.7		93.0	93.0		93.1	93.1		93.0	93.0		108.6	108.6
0241, 31 - 34	SITE & INFRASTRUCTURE, DEMOLITION	126.0	91.0	101.2	105.9	84.6	90.8	109.4	91.0	96.4	102.9	91.0	94.5	101.4	93.0	95.5	103.4	95.7	98.0
0310	Concrete Forming & Accessories	99.0	57.0	62.7	109.1	64.0	70.1	102.4	63.7	68.9	98.3	61.9	66.8	100.0	64.1	69.0	92.7	62.3	66.4
0320	Concrete Reinforcing	88.4	47.9	67.4	100.7	51.6	75.3	99.5	51.1	74.4	106.7	51.5	78.1	106.8	53.6	79.2	101.1	64.3	82.0
0330	Cast-in-Place Concrete	125.3	64.3	101.2	81.5	65.9	75.4	111.4	64.7	93.0	90.3	64.9	80.3	97.0	64.9	84.3	100.4	65.6	86.7
03	CONCRETE	122.5	58.5	91.4	81.2	63.9	72.8	107.2	62.3	85.4	94.4	61.6	78.4	97.8	63.1	80.9	98.6	65.5	82.5
04	MASONRY	107.8	56.3	76.1	144.4	56.6	90.3	102.1	65.3	79.4	95.4	58.6	72.7	98.9	56.5	72.8	100.3	59.5	75.2
05	METALS	96.4	66.5	86.9	98.5	83.1	93.6	98.7	69.8	89.5	97.6	67.9	88.2	100.5	70.7	91.0	98.3	90.0	95.7
06	WOOD, PLASTICS & COMPOSITES	99.9	57.4	76.2	115.7	66.5	88.3	111.0	66.3	86.1	89.3	63.6	75.0	99.1	66.3	80.8	101.2	62.7	79.8
07	THERMAL & MOISTURE PROTECTION	101.9	64.9	86.4	93.7	66.1	82.2	96.1	67.4	84.1	98.7	64.1	84.2	92.3	65.4	81.0	93.2	67.0	82.2
08	OPENINGS	105.7	53.7	93.6	119.2	61.7	105.8	72.5	61.4	69.9	99.2	56.5	89.3	104.3	62.1	94.4	107.3	62.5	96.9
0920	Plaster & Gypsum Board	93.1	56.5	68.7	90.4	65.7	73.9	86.2	65.7	72.5	93.4	62.9	73.0	90.3	65.7	73.9	98.7	61.9	74.1
0950, 0980	Ceilings & Acoustic Treatment	82.7	56.5	65.4	80.3	65.7	70.6	75.3	65.7	69.0	89.4	62.9	71.8	85.1	65.7	72.3	97.1	61.9	73.8
0960	Flooring	99.2	63.0	88.8	82.9	63.0	77.2	111.4	63.0	97.5	101.0	63.0	90.1	108.2	63.0	95.2	103.1	63.0	91.5
0970, 0990	Wall Finishes & Painting/Coating	99.1	48.8	69.8	102.9	56.1	75.6	94.3	56.1	72.1	94.6	53.7	70.8	92.0	56.1	71.1	102.7	60.0	77.9
09	FINISHES	94.6	57.0	73.8	81.0	63.3	71.2	89.7	63.1	75.0	91.7	61.6	75.1	92.0	63.1	76.0	94.0	62.0	76.3
COVERS	DIVS. 10 - 14, 25, 28, 41, 43, 44, 46	100.0	78.8	95.4	100.0	80.7	95.8	100.0	80.3	95.7	100.0	81.5	96.0	100.0	82.5	96.2	100.0	84.2	96.6
21, 22, 23	FIRE SUPPRESSION, PLUMBING & HVAC	96.7	56.7	79.8	96.8	60.2	81.3	96.8	49.4	76.8	99.9	49.9	78.8	99.9	59.2	82.7	96.7	65.5	83.5
26, 27, 3370	ELECTRICAL, COMMUNICATIONS & UTIL.	99.8	68.3	83.9	95.6	65.5	80.4	93.3	65.5	79.2	103.7	56.9	80.1	94.7	65.5	79.9	96.3	57.2	76.5
MF2014	WEIGHTED AVERAGE	102.5	63.0	85.1	99.2	66.5	84.8	95.6	64.1	81.8	98.3	61.6	82.2	98.8	65.7	84.2	98.7	68.5	85.4

TEXAS

DIVISION		GIDDINGS 789			GREENVILLE 754			HOUSTON 770 - 772			HUNTSVILLE 773			LAREDO 780			LONGVIEW 756		
		MAT.	INST.	TOTAL	MAT.	INST.	TOTAL	MAT.	INST.	TOTAL	MAT.	INST.	TOTAL	MAT.	INST.	TOTAL	MAT.	INST.	TOTAL
015433	CONTRACTOR EQUIPMENT		92.1	92.1		101.4	101.4		108.5	108.5		97.5	97.5		92.1	92.1		94.3	94.3
0241, 31 - 34	SITE & INFRASTRUCTURE, DEMOLITION	110.8	91.0	96.8	100.4	87.5	91.3	102.2	97.9	99.2	94.5	97.4	96.6	104.3	91.0	94.9	99.8	94.5	96.0
0310	Concrete Forming & Accessories	96.2	57.1	62.4	89.5	64.1	67.5	94.8	62.3	66.7	90.5	60.0	64.2	99.1	57.1	62.8	85.4	63.9	66.8
0320	Concrete Reinforcing	89.0	45.5	66.4	100.7	51.6	75.3	100.8	54.3	76.7	101.9	51.3	75.7	88.4	47.9	67.4	99.6	51.6	74.7
0330	Cast-in-Place Concrete	106.1	64.3	89.7	102.6	65.9	88.1	97.3	65.6	84.8	103.9	65.7	88.9	89.6	64.3	79.7	119.2	64.7	97.7
03	CONCRETE	99.1	58.2	79.2	95.9	63.8	80.3	96.2	63.6	80.4	104.8	61.2	83.6	93.1	58.6	76.3	112.4	62.4	88.1
04	MASONRY	116.2	57.1	79.8	146.8	56.5	91.1	100.1	62.1	76.7	142.3	59.5	91.2	100.8	57.1	73.9	143.4	57.2	90.3
05	METALS	95.9	65.7	86.3	98.6	81.7	93.2	101.2	86.2	96.4	96.7	70.4	88.3	99.2	66.7	88.8	91.8	68.7	84.5
06	WOOD, PLASTICS & COMPOSITES	99.0	57.4	75.8	90.4	66.6	77.1	103.6	62.7	80.8	91.8	59.9	74.0	99.9	57.4	76.2	84.3	66.4	74.4
07	THERMAL & MOISTURE PROTECTION	102.5	65.0	86.8	92.2	65.4	81.0	92.8	67.4	82.2	95.1	66.2	83.1	100.8	64.5	85.6	93.8	64.8	81.7
08	OPENINGS	104.8	53.0	92.7	94.6	61.7	87.0	110.4	59.9	98.6	103.0	57.6	92.5	105.8	53.7	93.7	85.6	61.6	80.0
0920	Plaster & Gypsum Board	92.1	56.5	68.3	87.2	65.7	72.8	100.9	61.9	74.8	96.3	59.1	71.4	93.8	56.5	68.9	85.6	65.7	72.3
0950, 0980	Ceilings & Acoustic Treatment	82.7	56.5	65.4	86.7	65.7	72.8	102.0	61.9	75.4	93.8	59.1	70.8	86.0	56.5	66.5	82.6	65.7	71.4
0960	Flooring	98.6	63.0	88.3	93.4	63.0	84.6	104.6	63.0	92.6	92.5	63.0	84.0	99.0	63.0	88.6	99.5	63.0	89.0
0970, 0990	Wall Finishes & Painting/Coating	99.1	48.8	69.8	105.2	56.1	76.6	102.7	62.2	79.2	91.5	59.4	72.8	99.1	48.8	69.8	95.6	56.1	72.6
09	FINISHES	93.2	57.0	73.2	92.4	63.3	76.3	101.3	62.2	79.7	87.3	60.2	72.3	93.7	57.0	73.4	97.2	63.2	78.4
COVERS	DIVS. 10 - 14, 25, 28, 41, 43, 44, 46	100.0	79.1	95.5	100.0	80.9	95.9	100.0	84.3	96.6	100.0	79.1	95.5	100.0	80.8	95.8	100.0	80.5	95.8
21, 22, 23	FIRE SUPPRESSION, PLUMBING & HVAC	96.8	64.0	82.9	96.7	61.9	82.0	100.0	65.4	85.4	96.8	65.3	83.5	100.1	63.5	84.6	96.6	59.9	81.1
26, 27, 3370	ELECTRICAL, COMMUNICATIONS & UTIL.	96.8	63.4	79.9	91.5	65.5	78.4	98.1	69.7	83.8	94.8	67.5	81.0	100.0	60.5	80.0	91.8	55.2	73.3
MF2014	WEIGHTED AVERAGE	99.3	63.8	83.7	98.4	67.0	84.6	100.7	69.8	87.1	99.7	66.9	85.3	99.4	63.5	83.6	98.4	64.3	83.4

For customer support on your Commercial Renovation Cost Data, call 877.742.4719.

793

TEXAS

DIVISION		LUBBOCK 793-794			LUFKIN 759			MCALLEN 785			MCKINNEY 750			MIDLAND 797			ODESSA 797		
		MAT.	INST.	TOTAL	MAT.	INST.	TOTAL	MAT.	INST.	TOTAL	MAT.	INST.	TOTAL	MAT.	INST.	TOTAL	MAT.	INST.	TOTAL
015433	CONTRACTOR EQUIPMENT		103.6	103.6		94.3	94.3		101.3	101.3		101.4	101.4		103.6	103.6		93.0	93.0
0241, 31 - 34	SITE & INFRASTRUCTURE, DEMOLITION	127.3	89.9	100.8	94.8	96.0	95.7	150.4	87.1	105.7	96.8	87.5	90.3	130.6	89.8	101.8	102.1	91.4	94.5
0310	Concrete Forming & Accessories	98.2	55.7	61.5	88.7	60.4	64.2	103.8	56.8	63.1	88.5	64.1	67.4	102.2	63.8	69.0	99.5	63.7	68.6
0320	Concrete Reinforcing	103.3	52.0	76.7	101.4	64.2	82.1	87.9	47.8	67.2	100.7	50.7	74.8	104.3	51.6	77.0	102.1	51.6	75.9
0330	Cast-in-Place Concrete	96.7	65.9	84.5	106.7	65.6	90.5	126.4	65.3	102.3	96.2	65.8	84.2	102.8	65.9	88.2	96.5	64.7	84.0
03	CONCRETE	93.5	60.2	77.3	104.0	63.5	84.3	110.2	59.6	85.6	91.0	63.6	77.7	98.0	63.8	81.4	94.7	62.5	79.0
04	MASONRY	100.5	63.8	77.9	112.1	59.4	79.6	107.7	56.2	76.0	157.0	56.5	95.1	117.8	57.3	80.5	101.2	57.3	74.1
05	METALS	106.0	83.5	98.9	98.7	72.6	90.4	96.4	79.6	91.0	98.5	81.2	93.0	104.3	83.0	97.5	101.8	69.6	91.5
06	WOOD, PLASTICS & COMPOSITES	103.9	55.5	77.0	92.3	60.0	74.3	119.5	57.5	85.0	89.1	66.6	76.6	108.5	66.5	85.1	104.8	66.3	83.4
07	THERMAL & MOISTURE PROTECTION	89.6	65.6	79.6	93.6	65.1	81.7	105.4	59.6	86.3	92.1	65.4	80.9	89.9	65.0	79.5	100.2	65.4	85.6
08	OPENINGS	113.4	55.6	100.0	65.8	61.4	64.7	109.5	53.8	96.5	94.6	61.4	86.9	114.9	61.7	102.5	105.6	61.6	95.4
0920	Plaster & Gypsum Board	86.5	54.4	65.0	85.1	59.1	67.7	98.2	56.5	70.3	86.6	65.7	72.6	88.1	65.7	73.1	86.3	65.7	72.5
0950, 0980	Ceilings & Acoustic Treatment	88.6	54.4	66.0	77.6	59.1	65.4	86.0	56.5	66.5	86.7	65.7	72.8	87.0	65.7	72.9	87.8	65.7	73.2
0960	Flooring	97.0	63.0	87.3	132.8	63.0	112.7	114.3	61.3	99.1	92.9	63.0	84.3	98.5	63.0	88.3	104.5	63.0	92.6
0970, 0990	Wall Finishes & Painting/Coating	103.4	56.1	75.9	95.6	56.1	72.6	111.7	48.8	75.1	105.2	56.1	76.6	103.4	56.1	75.9	92.9	56.1	71.5
09	FINISHES	92.1	56.7	72.5	105.6	60.4	80.6	102.9	56.8	77.4	92.0	63.3	76.1	92.8	63.3	76.4	89.9	63.1	75.1
COVERS	DIVS. 10 - 14, 25, 28, 41, 43, 44, 46	100.0	81.5	96.0	100.0	79.4	95.5	100.0	81.0	95.9	100.0	80.9	95.9	100.0	81.7	96.0	100.0	81.3	96.0
21, 22, 23	FIRE SUPPRESSION, PLUMBING & HVAC	99.7	50.2	78.8	96.6	64.6	83.1	96.8	58.2	80.5	96.7	61.8	81.9	96.3	50.0	76.7	100.2	49.9	78.9
26, 27, 3370	ELECTRICAL, COMMUNICATIONS & UTIL.	100.5	63.7	81.9	92.9	70.0	81.4	97.5	36.1	66.4	91.6	65.5	78.4	100.4	63.7	81.9	101.8	63.7	82.5
MF2014	WEIGHTED AVERAGE	101.5	63.6	84.8	95.6	67.7	83.3	102.9	59.9	83.9	98.2	66.9	84.4	102.1	64.6	85.6	99.9	63.2	83.7

TEXAS

DIVISION		PALESTINE 758			SAN ANGELO 769			SAN ANTONIO 781 - 782			TEMPLE 765			TEXARKANA 755			TYLER 757		
		MAT.	INST.	TOTAL	MAT.	INST.	TOTAL	MAT.	INST.	TOTAL	MAT.	INST.	TOTAL	MAT.	INST.	TOTAL	MAT.	INST.	TOTAL
015433	CONTRACTOR EQUIPMENT		94.3	94.3		93.0	93.0		94.7	94.7		93.0	93.0		94.3	94.3		94.3	94.3
0241, 31 - 34	SITE & INFRASTRUCTURE, DEMOLITION	100.4	94.8	96.4	102.5	93.1	95.9	103.8	94.3	97.1	90.3	92.5	91.8	89.1	94.1	92.7	99.2	94.6	95.9
0310	Concrete Forming & Accessories	79.5	63.9	66.0	101.9	57.9	63.9	99.2	57.3	62.9	105.5	57.0	63.6	96.0	57.2	62.5	90.6	64.0	67.6
0320	Concrete Reinforcing	98.9	51.6	74.4	99.1	51.6	74.5	93.2	51.2	71.5	99.3	48.6	73.0	98.7	53.6	75.3	99.6	53.6	75.8
0330	Cast-in-Place Concrete	97.5	64.7	84.6	99.5	65.9	86.2	94.8	65.0	83.1	81.5	64.2	74.6	98.3	64.7	85.0	117.1	64.7	96.4
03	CONCRETE	104.5	62.5	84.0	97.8	60.3	79.6	94.2	59.5	77.3	83.9	58.6	71.6	96.6	59.8	78.7	111.8	62.9	88.0
04	MASONRY	106.3	57.2	76.0	131.0	59.3	86.8	97.7	57.1	72.7	143.5	57.1	90.2	160.8	57.2	97.0	151.9	57.2	93.5
05	METALS	98.5	68.7	89.0	99.1	69.7	89.7	101.2	68.2	90.7	98.8	67.0	88.7	91.7	69.5	84.6	98.3	69.6	89.2
06	WOOD, PLASTICS & COMPOSITES	82.8	66.4	73.7	104.2	57.4	78.2	98.4	57.5	75.6	113.8	57.4	82.4	96.1	57.5	74.6	93.9	66.4	78.6
07	THERMAL & MOISTURE PROTECTION	94.1	64.8	81.8	95.6	65.6	83.0	98.1	65.6	84.5	95.2	64.4	82.3	93.4	63.9	81.0	93.9	64.8	81.7
08	OPENINGS	65.7	61.6	64.8	101.0	56.7	90.7	104.7	54.6	93.0	69.0	55.9	65.9	85.6	57.3	79.0	65.7	62.1	64.8
0920	Plaster & Gypsum Board	82.1	65.7	71.1	86.2	56.5	66.4	94.8	56.5	69.2	86.2	56.5	66.4	89.9	56.5	67.6	85.1	65.7	72.1
0950, 0980	Ceilings & Acoustic Treatment	77.6	65.7	69.7	75.3	56.5	62.9	94.3	56.5	69.3	75.3	56.5	62.9	82.6	56.5	65.3	77.6	65.7	69.7
0960	Flooring	122.1	63.0	105.1	88.1	63.0	80.9	106.4	63.0	93.9	113.3	63.0	98.8	109.3	63.0	96.0	135.4	63.0	114.6
0970, 0990	Wall Finishes & Painting/Coating	95.6	56.1	72.6	92.9	56.1	71.5	104.6	48.8	72.1	94.3	48.8	67.8	95.6	56.1	72.6	95.6	56.1	72.6
09	FINISHES	102.6	63.2	80.8	82.1	58.3	69.0	101.3	57.1	76.8	89.0	57.0	71.3	100.1	57.9	76.8	106.7	63.2	82.7
COVERS	DIVS. 10 - 14, 25, 28, 41, 43, 44, 46	100.0	80.5	95.8	100.0	81.8	96.1	100.0	81.0	95.9	100.0	79.4	95.5	100.0	81.6	96.0	100.0	82.6	96.2
21, 22, 23	FIRE SUPPRESSION, PLUMBING & HVAC	96.6	60.4	81.3	96.8	50.9	77.4	100.1	63.4	84.6	96.8	54.1	78.8	96.6	58.7	80.6	96.6	62.6	82.2
26, 27, 3370	ELECTRICAL, COMMUNICATIONS & UTIL.	89.6	55.2	72.2	97.0	44.5	70.4	102.0	60.5	81.0	94.4	59.3	76.7	92.8	61.1	76.8	91.8	55.7	73.5
MF2014	WEIGHTED AVERAGE	94.8	64.4	81.4	98.7	59.9	81.6	100.3	64.1	84.4	94.1	61.6	79.8	97.6	63.6	82.6	98.4	65.2	83.8

TEXAS / UTAH

DIVISION		VICTORIA 779			WACO 766 - 767			WAXAHACHIE 751			WHARTON 774			WICHITA FALLS 763			LOGAN 843		
		MAT.	INST.	TOTAL	MAT.	INST.	TOTAL	MAT.	INST.	TOTAL	MAT.	INST.	TOTAL	MAT.	INST.	TOTAL	MAT.	INST.	TOTAL
015433	CONTRACTOR EQUIPMENT		107.2	107.2		93.0	93.0		101.4	101.4		108.6	108.6		93.0	93.0		96.0	96.0
0241, 31 - 34	SITE & INFRASTRUCTURE, DEMOLITION	108.3	93.2	97.6	99.2	93.0	94.8	98.5	87.5	90.7	113.3	95.6	100.8	100.0	93.0	95.1	100.4	94.6	96.3
0310	Concrete Forming & Accessories	92.6	58.2	62.8	103.9	63.9	69.3	88.5	63.9	67.2	87.3	60.3	63.9	103.9	64.1	69.5	104.7	60.7	66.6
0320	Concrete Reinforcing	97.0	48.0	71.6	99.0	51.3	74.3	100.7	50.7	74.8	101.0	51.3	75.2	99.0	51.1	74.1	105.7	80.4	92.6
0330	Cast-in-Place Concrete	112.7	66.9	94.6	88.2	67.1	79.8	101.5	65.8	87.4	115.8	66.8	96.5	94.2	64.9	82.6	84.6	72.5	79.8
03	CONCRETE	106.3	61.1	84.3	90.5	63.3	77.3	94.9	63.5	79.6	110.4	62.6	87.2	93.5	62.6	78.5	110.5	68.9	90.3
04	MASONRY	117.8	59.6	81.9	100.3	57.2	73.8	147.3	56.5	91.3	101.6	59.5	75.7	100.8	63.4	77.8	112.9	63.0	82.1
05	METALS	96.9	83.4	92.6	101.2	69.6	91.1	98.6	81.1	93.0	98.3	84.8	94.0	101.1	70.4	91.4	107.5	79.2	98.5
06	WOOD, PLASTICS & COMPOSITES	105.0	57.5	78.6	112.0	66.3	86.6	89.1	66.6	76.6	94.8	60.0	75.4	112.0	66.3	86.6	82.1	58.1	68.8
07	THERMAL & MOISTURE PROTECTION	97.1	66.7	84.4	96.0	65.3	83.2	92.2	65.3	80.9	93.6	67.0	82.5	96.0	67.0	83.9	97.2	66.2	84.2
08	OPENINGS	107.1	55.4	95.1	81.1	61.5	76.5	94.6	61.4	86.9	107.3	57.7	95.8	81.1	61.6	76.5	91.8	60.4	84.5
0920	Plaster & Gypsum Board	95.9	56.5	69.6	86.8	65.7	72.7	87.0	65.7	72.8	94.7	59.1	70.9	86.8	65.7	72.7	81.0	56.9	64.9
0950, 0980	Ceilings & Acoustic Treatment	97.9	56.5	70.5	77.8	65.7	69.8	88.3	65.7	73.4	97.1	59.1	71.9	77.8	65.7	69.8	97.0	56.9	70.4
0960	Flooring	101.2	63.0	90.2	112.5	63.0	98.3	92.9	63.0	84.3	100.0	76.9	93.4	113.2	73.6	101.8	104.1	63.1	92.3
0970, 0990	Wall Finishes & Painting/Coating	102.9	59.4	77.6	94.3	48.8	67.8	105.2	56.1	76.6	102.7	59.4	77.5	96.4	56.1	72.9	94.0	62.6	75.7
09	FINISHES	91.7	58.8	73.5	89.9	62.4	74.7	92.6	63.3	76.4	93.3	62.7	76.3	90.3	65.2	76.4	97.0	60.3	76.7
COVERS	DIVS. 10 - 14, 25, 28, 41, 43, 44, 46	100.0	79.1	95.5	100.0	82.4	96.2	100.0	80.9	95.9	100.0	79.4	95.5	100.0	80.4	95.7	100.0	86.0	97.0
21, 22, 23	FIRE SUPPRESSION, PLUMBING & HVAC	96.8	65.3	83.5	100.2	54.2	80.7	96.7	55.0	79.1	96.7	65.2	83.4	100.2	53.7	80.5	99.9	69.0	86.9
26, 27, 3370	ELECTRICAL, COMMUNICATIONS & UTIL.	100.8	57.2	78.7	97.2	59.3	78.0	91.6	65.5	78.4	99.7	67.5	83.5	98.8	60.0	79.2	96.7	75.7	86.1
MF2014	WEIGHTED AVERAGE	100.7	66.1	85.5	95.9	63.7	81.7	98.3	65.4	83.8	100.6	68.7	86.5	96.5	64.6	82.5	101.3	71.2	88.0

For customer support on your Commercial Renovation Cost Data, call 877.742.4719.

UTAH / VERMONT

DIVISION		OGDEN 842,844 MAT.	INST.	TOTAL	PRICE 845 MAT.	INST.	TOTAL	PROVO 846-847 MAT.	INST.	TOTAL	SALT LAKE CITY 840-841 MAT.	INST.	TOTAL	BELLOWS FALLS 051 MAT.	INST.	TOTAL	BENNINGTON 052 MAT.	INST.	TOTAL
015433	CONTRACTOR EQUIPMENT		96.0	96.0		95.0	95.0		95.0	95.0		96.0	96.0		98.1	98.1		98.1	98.1
0241, 31 - 34	SITE & INFRASTRUCTURE, DEMOLITION	88.1	94.6	92.7	97.7	92.6	94.1	96.4	93.0	94.0	87.7	94.6	92.5	90.3	97.9	95.7	89.6	97.9	95.5
0310	Concrete Forming & Accessories	104.7	60.7	66.6	107.2	47.6	55.7	106.4	60.7	66.9	107.1	60.7	67.0	97.3	83.2	85.1	94.9	103.7	102.5
0320	Concrete Reinforcing	105.3	80.4	92.4	113.3	80.3	96.2	114.3	80.4	96.8	107.7	80.4	93.6	84.8	84.8	84.8	84.8	84.7	84.8
0330	Cast-in-Place Concrete	85.9	72.5	80.6	84.7	60.4	75.1	84.7	72.5	79.9	93.8	72.5	85.4	85.2	114.0	96.6	85.2	114.0	96.4
03	CONCRETE	99.3	68.9	84.5	111.9	58.9	86.1	110.2	68.9	90.1	119.8	68.9	95.0	90.2	94.0	92.0	90.0	103.1	96.4
04	MASONRY	106.9	63.0	79.8	118.5	61.0	83.0	118.7	63.0	84.3	121.3	63.0	85.3	110.3	94.7	100.7	121.4	94.7	104.9
05	METALS	108.0	79.2	98.8	104.6	77.7	96.0	105.6	79.2	97.2	111.5	79.2	101.3	91.9	87.2	90.4	91.8	87.1	90.3
06	WOOD, PLASTICS & COMPOSITES	82.1	58.1	68.8	85.2	42.7	61.6	83.5	58.1	69.4	83.9	58.1	69.6	109.3	80.6	93.3	106.4	108.6	107.6
07	THERMAL & MOISTURE PROTECTION	96.1	66.2	83.6	99.0	61.0	83.1	99.0	66.2	85.2	103.4	66.2	87.9	97.0	81.0	90.3	97.0	79.4	89.6
08	OPENINGS	91.8	60.4	84.5	95.8	48.2	84.7	95.8	60.4	87.6	93.7	60.4	85.9	103.6	85.0	99.2	103.6	100.3	102.8
0920	Plaster & Gypsum Board	81.0	56.9	64.9	83.6	41.1	55.2	81.3	56.9	65.0	94.0	56.9	69.2	98.3	79.6	85.8	96.7	108.4	104.5
0950, 0980	Ceilings & Acoustic Treatment	97.0	56.9	70.4	97.0	41.1	60.0	97.0	56.9	70.4	89.6	56.9	67.9	86.1	79.6	81.8	86.1	108.4	100.8
0960	Flooring	101.8	63.1	90.7	105.3	47.2	88.6	104.9	63.1	92.9	106.2	63.1	93.8	100.9	100.6	100.8	100.0	100.6	100.2
0970, 0990	Wall Finishes & Painting/Coating	94.0	62.6	75.7	94.0	37.5	61.2	94.0	62.6	75.7	97.2	62.6	77.1	88.7	111.0	101.7	88.7	111.0	101.7
09	FINISHES	95.1	60.3	75.9	98.2	44.8	68.7	97.7	60.3	77.0	97.4	60.3	76.9	92.3	88.6	90.3	91.8	105.3	99.3
COVERS	DIVS. 10 - 14, 25, 28, 41, 43, 44, 46	100.0	86.0	97.0	100.0	58.7	91.1	100.0	86.0	97.0	100.0	86.0	97.0	100.0	94.8	98.9	100.0	97.7	99.5
21, 22, 23	FIRE SUPPRESSION, PLUMBING & HVAC	99.9	69.0	86.9	98.2	67.6	85.3	99.9	69.0	86.9	100.1	69.0	87.0	97.0	95.4	96.3	97.0	95.4	96.3
26, 27, 3370	ELECTRICAL, COMMUNICATIONS & UTIL.	97.0	75.7	86.2	102.1	75.7	88.7	97.4	75.7	86.5	99.4	75.7	87.4	100.0	88.7	94.3	100.0	60.7	80.1
MF2014	WEIGHTED AVERAGE	99.3	71.2	86.9	102.0	65.3	85.8	101.8	71.1	88.2	103.8	71.2	89.4	96.9	91.7	94.6	97.3	92.3	95.1

VERMONT

DIVISION		BRATTLEBORO 053 MAT.	INST.	TOTAL	BURLINGTON 054 MAT.	INST.	TOTAL	GUILDHALL 059 MAT.	INST.	TOTAL	MONTPELIER 056 MAT.	INST.	TOTAL	RUTLAND 057 MAT.	INST.	TOTAL	ST. JOHNSBURY 058 MAT.	INST.	TOTAL
015433	CONTRACTOR EQUIPMENT		98.1	98.1		98.1	98.1		98.1	98.1		98.1	98.1		98.1	98.1		98.1	98.1
0241, 31 - 34	SITE & INFRASTRUCTURE, DEMOLITION	91.2	97.9	95.9	96.1	97.8	97.3	89.3	96.5	94.4	92.1	97.5	95.9	93.8	97.6	96.5	89.4	96.5	94.4
0310	Concrete Forming & Accessories	97.6	83.1	85.0	98.6	85.2	87.0	95.3	76.1	78.7	97.0	81.8	83.8	97.9	85.1	86.9	93.7	76.1	78.5
0320	Concrete Reinforcing	83.9	84.7	84.4	104.7	84.8	94.4	85.6	84.6	85.1	103.7	84.8	93.9	106.0	84.8	95.1	83.9	84.6	84.3
0330	Cast-in-Place Concrete	87.9	114.0	98.2	101.7	112.9	106.1	82.6	104.4	91.2	101.6	112.9	106.0	83.5	112.9	95.1	82.6	104.4	91.2
03	CONCRETE	92.2	93.9	93.0	101.1	94.5	97.9	87.7	87.5	87.6	100.8	92.9	97.0	93.2	94.5	93.8	87.4	87.5	87.4
04	MASONRY	120.8	94.7	104.7	127.2	93.3	106.3	121.2	77.8	94.4	121.9	93.3	104.2	100.1	93.3	95.9	151.0	77.8	105.9
05	METALS	91.8	87.1	90.3	99.7	86.4	95.5	91.9	86.2	90.1	97.8	86.4	94.1	97.9	86.4	94.2	91.9	86.2	90.1
06	WOOD, PLASTICS & COMPOSITES	109.6	80.6	93.5	103.6	85.2	93.3	105.9	80.6	91.8	96.8	80.6	87.8	109.8	85.2	96.1	101.0	80.6	89.6
07	THERMAL & MOISTURE PROTECTION	97.1	76.5	88.5	104.3	78.8	93.7	96.9	69.3	85.3	104.5	78.3	93.6	97.2	78.8	89.5	96.8	69.3	85.3
08	OPENINGS	103.6	85.0	99.2	110.2	83.5	104.0	103.6	81.0	98.3	110.5	81.1	103.7	106.9	83.5	101.4	103.6	81.0	98.3
0920	Plaster & Gypsum Board	98.3	79.6	85.8	101.5	84.3	90.0	104.6	79.6	87.9	100.9	79.6	86.7	98.7	84.3	89.1	106.0	79.6	88.4
0950, 0980	Ceilings & Acoustic Treatment	86.1	79.6	81.8	92.9	84.3	87.2	86.1	79.6	81.8	91.3	79.6	83.6	89.6	84.3	86.1	86.1	79.6	81.8
0960	Flooring	101.1	100.6	100.9	100.7	100.6	100.7	104.1	100.6	103.1	104.7	100.6	103.5	100.9	100.6	100.8	107.8	100.6	105.7
0970, 0990	Wall Finishes & Painting/Coating	88.7	111.0	101.7	93.7	90.4	91.8	88.7	90.4	89.7	92.7	90.4	91.3	88.7	90.4	89.7	88.7	90.4	89.7
09	FINISHES	92.5	88.6	90.3	94.7	88.7	91.4	94.1	82.3	87.5	95.3	86.0	90.2	93.3	88.7	90.7	95.4	82.3	88.1
COVERS	DIVS. 10 - 14, 25, 28, 41, 43, 44, 46	100.0	94.8	98.9	100.0	94.7	98.8	100.0	89.1	97.6	100.0	94.2	98.7	100.0	94.7	98.8	100.0	89.1	97.6
21, 22, 23	FIRE SUPPRESSION, PLUMBING & HVAC	97.0	95.4	96.3	100.1	71.1	87.9	97.0	63.3	82.7	96.7	71.1	85.9	100.3	71.1	88.0	97.0	63.3	82.7
26, 27, 3370	ELECTRICAL, COMMUNICATIONS & UTIL.	100.0	88.7	94.3	100.4	60.7	80.3	100.0	60.7	80.1	98.6	60.7	79.4	100.0	60.7	80.1	100.0	60.7	80.1
MF2014	WEIGHTED AVERAGE	97.7	91.6	95.0	102.2	82.5	93.5	97.2	76.8	88.2	100.6	81.8	92.3	99.0	82.5	91.7	98.7	76.8	89.1

VERMONT / VIRGINIA

DIVISION		WHITE RIVER JCT. 050 MAT.	INST.	TOTAL	ALEXANDRIA 223 MAT.	INST.	TOTAL	ARLINGTON 222 MAT.	INST.	TOTAL	BRISTOL 242 MAT.	INST.	TOTAL	CHARLOTTESVILLE 229 MAT.	INST.	TOTAL	CULPEPER 227 MAT.	INST.	TOTAL
015433	CONTRACTOR EQUIPMENT		98.1	98.1		105.8	105.8		104.4	104.4		104.4	104.4		108.8	108.8		104.4	104.4
0241, 31 - 34	SITE & INFRASTRUCTURE, DEMOLITION	93.8	96.6	95.8	114.5	92.5	98.9	124.8	90.2	100.3	109.2	88.3	94.4	113.7	90.4	97.2	112.0	90.0	96.5
0310	Concrete Forming & Accessories	92.2	76.7	78.8	91.4	86.9	87.5	90.5	87.1	87.5	86.6	41.3	47.4	85.0	48.3	53.2	82.1	86.5	85.9
0320	Concrete Reinforcing	84.8	84.6	84.7	86.0	89.0	87.5	97.1	89.0	92.9	97.0	70.4	83.2	96.4	74.0	84.8	97.1	88.9	92.8
0330	Cast-in-Place Concrete	87.9	105.2	94.8	104.7	81.1	95.4	101.9	81.2	93.7	101.4	47.9	80.3	105.5	56.1	86.0	104.3	70.2	90.9
03	CONCRETE	93.9	88.0	91.0	103.8	86.0	95.2	108.6	86.3	97.7	104.7	51.0	78.6	105.4	57.6	82.1	102.3	82.2	92.5
04	MASONRY	135.5	79.3	100.9	92.7	77.5	83.4	108.4	78.0	89.6	97.3	51.2	68.9	122.2	55.1	80.8	109.0	77.9	89.8
05	METALS	91.9	86.2	90.1	101.7	97.8	100.5	100.3	99.6	100.1	99.1	88.0	95.6	99.4	92.2	97.1	99.5	97.9	99.0
06	WOOD, PLASTICS & COMPOSITES	103.3	80.6	90.7	95.9	89.9	92.6	92.6	89.9	91.1	85.5	38.6	59.4	84.1	44.1	61.8	82.9	89.9	86.8
07	THERMAL & MOISTURE PROTECTION	97.2	70.0	85.8	102.4	84.3	94.9	104.4	84.9	96.3	104.0	59.3	85.3	103.6	67.3	88.4	103.8	84.9	95.9
08	OPENINGS	103.6	81.0	98.3	102.0	85.2	98.1	100.0	85.9	96.7	103.3	45.8	89.9	101.3	51.0	89.6	101.7	85.9	98.0
0920	Plaster & Gypsum Board	95.4	79.6	84.8	102.2	89.4	93.7	98.8	89.4	92.5	95.0	36.7	56.0	95.0	41.6	59.3	95.2	89.4	91.3
0950, 0980	Ceilings & Acoustic Treatment	86.1	79.6	81.8	91.5	89.4	90.1	89.9	89.4	89.6	89.0	36.7	54.4	89.0	41.6	57.6	89.9	89.4	89.6
0960	Flooring	98.8	100.6	99.3	100.7	83.1	95.7	99.1	83.1	94.5	95.4	63.9	86.4	94.0	63.9	85.3	94.0	83.1	90.8
0970, 0990	Wall Finishes & Painting/Coating	88.7	90.4	89.7	111.5	80.3	93.4	111.5	80.3	93.4	98.4	54.3	72.7	98.4	64.0	78.4	111.5	64.0	83.9
09	FINISHES	91.6	82.6	86.7	97.1	86.0	91.0	97.0	85.9	90.8	93.5	45.8	67.1	93.1	51.2	69.9	93.8	84.0	88.4
COVERS	DIVS. 10 - 14, 25, 28, 41, 43, 44, 46	100.0	89.6	97.7	100.0	90.9	98.0	100.0	88.6	97.5	100.0	65.9	92.6	100.0	85.3	96.8	100.0	90.6	98.0
21, 22, 23	FIRE SUPPRESSION, PLUMBING & HVAC	97.0	64.1	83.1	100.4	87.4	94.9	100.4	87.6	95.0	97.0	50.0	77.1	97.0	69.9	85.6	97.0	73.7	87.1
26, 27, 3370	ELECTRICAL, COMMUNICATIONS & UTIL.	100.0	60.7	80.1	97.6	101.6	99.6	95.4	101.6	98.5	97.3	35.1	65.9	97.2	73.1	85.0	99.6	101.6	100.6
MF2014	WEIGHTED AVERAGE	98.5	77.3	89.2	100.6	89.4	95.6	101.5	89.4	96.2	99.4	54.8	79.7	100.5	67.9	86.1	99.9	85.5	93.6

For customer support on your Commercial Renovation Cost Data, call 877.742.4719.

795

VIRGINIA

DIVISION		FAIRFAX 220-221 MAT.	INST.	TOTAL	FARMVILLE 239 MAT.	INST.	TOTAL	FREDERICKSBURG 224-225 MAT.	INST.	TOTAL	GRUNDY 246 MAT.	INST.	TOTAL	HARRISONBURG 228 MAT.	INST.	TOTAL	LYNCHBURG 245 MAT.	INST.	TOTAL
015433	CONTRACTOR EQUIPMENT		104.4	104.4		108.8	108.8		104.4	104.4		104.4	104.4		104.4	104.4		104.4	104.4
0241, 31 - 34	SITE & INFRASTRUCTURE, DEMOLITION	123.4	90.1	99.9	110.7	89.8	95.9	111.6	90.0	96.3	107.0	87.0	92.9	120.3	88.5	97.9	107.9	88.7	94.3
0310	Concrete Forming & Accessories	85.0	86.2	86.0	94.8	42.4	49.5	85.0	85.9	85.7	89.6	35.3	42.7	81.0	41.4	46.7	86.6	59.5	63.2
0320	Concrete Reinforcing	97.1	89.0	92.9	93.6	49.3	70.7	97.8	88.9	93.2	95.7	49.8	71.9	97.1	64.1	80.0	96.4	70.5	83.2
0330	Cast-in-Place Concrete	101.9	93.4	98.5	103.6	52.9	83.6	103.4	69.0	89.8	101.4	46.8	79.9	101.9	57.3	84.3	101.4	61.1	85.5
03	CONCRETE	108.2	90.0	99.3	102.1	49.2	76.4	102.1	81.5	92.1	103.4	43.7	74.4	105.9	53.0	80.2	103.3	63.6	84.0
04	MASONRY	108.2	75.7	88.2	107.4	44.6	68.7	108.1	75.6	88.1	97.7	47.1	66.5	106.0	50.1	71.5	113.6	55.0	77.5
05	METALS	99.5	99.7	99.6	97.5	78.6	91.5	99.5	98.9	99.3	99.1	71.7	90.4	99.4	85.2	94.9	99.3	88.7	95.9
06	WOOD, PLASTICS & COMPOSITES	85.5	89.9	88.0	93.8	42.6	65.3	85.5	89.9	88.0	88.2	33.9	57.9	81.9	37.7	57.3	85.5	60.5	71.6
07	THERMAL & MOISTURE PROTECTION	104.3	75.6	92.3	102.0	48.2	79.5	103.8	83.8	95.4	104.0	44.3	79.0	104.2	62.6	86.8	103.8	65.9	87.9
08	OPENINGS	100.0	85.9	96.7	103.5	41.8	89.1	101.3	85.9	97.7	103.3	33.4	87.0	101.7	48.0	89.2	101.6	57.8	91.4
0920	Plaster & Gypsum Board	95.2	89.4	91.3	102.7	40.0	60.8	95.2	89.4	91.3	95.0	31.8	52.8	95.0	35.7	55.4	95.0	59.2	71.1
0950, 0980	Ceilings & Acoustic Treatment	89.9	89.4	89.6	86.8	40.0	55.8	89.9	89.4	89.6	89.0	31.8	51.1	89.0	35.7	53.7	89.0	59.2	69.3
0960	Flooring	95.8	83.1	92.1	100.4	57.9	88.2	95.8	83.1	92.1	96.8	32.4	78.3	93.7	71.1	87.2	95.4	63.9	86.4
0970, 0990	Wall Finishes & Painting/Coating	111.5	80.3	93.4	98.1	38.8	63.6	111.5	64.0	83.9	98.4	35.1	61.6	111.5	54.3	78.2	98.4	54.3	72.7
09	FINISHES	95.5	85.3	89.8	95.2	44.7	67.3	94.3	83.4	88.3	93.7	34.1	60.7	94.2	46.5	67.8	93.3	59.0	74.3
COVERS	DIVS. 10 - 14, 25, 28, 41, 43, 44, 46	100.0	89.4	97.8	100.0	44.4	88.0	100.0	85.9	96.9	100.0	42.4	87.5	100.0	65.8	92.6	100.0	69.7	93.4
21, 22, 23	FIRE SUPPRESSION, PLUMBING & HVAC	97.0	86.5	92.6	97.1	42.9	74.2	97.0	86.1	92.4	97.0	63.5	82.8	97.0	66.3	84.0	97.0	69.3	85.3
26, 27, 3370	ELECTRICAL, COMMUNICATIONS & UTIL.	98.3	101.6	100.0	91.8	47.9	69.6	95.5	101.6	98.6	97.3	44.3	70.5	97.4	96.9	97.2	98.3	52.3	75.0
MF2014	WEIGHTED AVERAGE	100.7	89.2	95.6	99.0	52.1	78.3	99.4	87.7	94.2	99.2	52.6	78.7	100.2	66.9	85.5	99.9	66.2	85.1

VIRGINIA

DIVISION		NEWPORT NEWS 236 MAT.	INST.	TOTAL	NORFOLK 233-235 MAT.	INST.	TOTAL	PETERSBURG 238 MAT.	INST.	TOTAL	PORTSMOUTH 237 MAT.	INST.	TOTAL	PULASKI 243 MAT.	INST.	TOTAL	RICHMOND 230-232 MAT.	INST.	TOTAL
015433	CONTRACTOR EQUIPMENT		108.8	108.8		109.4	109.4		108.8	108.8		108.7	108.7		104.4	104.4		108.8	108.8
0241, 31 - 34	SITE & INFRASTRUCTURE, DEMOLITION	109.6	91.0	96.5	108.0	92.0	96.7	113.3	91.2	97.7	108.2	90.9	95.9	106.2	87.7	93.2	104.1	91.2	95.0
0310	Concrete Forming & Accessories	94.1	62.4	66.7	94.4	62.5	66.8	88.1	55.8	60.2	84.4	51.6	56.1	89.6	37.6	44.6	93.3	55.8	60.9
0320	Concrete Reinforcing	93.4	71.2	81.9	102.2	71.2	86.1	93.0	74.2	83.3	93.0	71.2	81.7	95.7	63.9	79.2	102.5	74.2	87.9
0330	Cast-in-Place Concrete	100.7	65.4	86.8	103.1	65.4	88.3	106.9	53.1	85.7	99.7	65.3	86.2	101.4	47.6	80.2	97.2	53.1	79.8
03	CONCRETE	99.3	66.5	83.4	100.4	66.6	84.0	104.7	60.0	83.0	98.1	61.7	80.4	103.4	47.9	76.4	97.5	60.0	79.3
04	MASONRY	101.5	52.5	71.3	106.5	52.5	73.2	116.3	52.4	76.9	107.8	53.7	74.5	92.9	48.0	65.2	105.6	52.4	72.8
05	METALS	99.9	90.8	97.0	101.5	90.9	98.2	97.6	92.7	96.1	98.8	90.9	96.3	99.2	83.9	94.3	101.6	92.7	98.8
06	WOOD, PLASTICS & COMPOSITES	92.8	64.0	76.8	89.8	64.0	75.4	85.2	56.6	69.3	82.0	49.7	64.0	88.2	35.5	58.9	89.4	56.6	71.1
07	THERMAL & MOISTURE PROTECTION	102.0	67.0	87.4	98.9	67.0	85.6	102.0	67.0	87.3	102.0	65.2	86.6	104.0	49.9	81.3	100.6	67.0	86.5
08	OPENINGS	103.8	60.4	93.7	100.5	60.4	91.2	103.1	57.8	92.6	103.9	52.6	91.9	103.3	41.0	88.8	102.4	57.8	92.1
0920	Plaster & Gypsum Board	103.4	62.1	75.8	97.8	62.1	73.9	96.9	54.5	68.5	97.1	47.4	63.9	95.0	33.5	53.9	96.8	54.5	68.5
0950, 0980	Ceilings & Acoustic Treatment	90.1	62.1	71.5	91.6	62.1	72.1	87.6	54.5	65.7	90.1	47.4	61.8	89.0	33.5	52.3	90.7	54.5	66.7
0960	Flooring	100.4	63.9	89.9	97.8	63.9	88.1	96.0	68.4	88.1	92.8	63.9	84.5	96.8	63.9	87.3	98.2	68.4	89.6
0970, 0990	Wall Finishes & Painting/Coating	98.1	42.0	65.5	97.7	64.0	78.1	98.1	64.0	78.3	98.1	64.0	78.3	98.4	54.3	72.7	96.1	64.0	77.5
09	FINISHES	95.8	60.4	76.3	94.3	62.8	76.9	93.5	58.2	74.0	92.7	53.8	71.2	93.7	42.5	65.4	95.0	58.2	74.6
COVERS	DIVS. 10 - 14, 25, 28, 41, 43, 44, 46	100.0	87.2	97.2	100.0	87.2	97.2	100.0	84.9	96.7	100.0	70.1	93.5	100.0	42.7	87.6	100.0	84.9	96.7
21, 22, 23	FIRE SUPPRESSION, PLUMBING & HVAC	100.5	63.9	85.0	100.2	64.3	85.0	97.1	67.1	84.4	100.5	64.1	85.1	97.0	63.9	83.0	99.9	67.1	86.0
26, 27, 3370	ELECTRICAL, COMMUNICATIONS & UTIL.	94.3	65.3	79.6	97.7	61.9	79.6	94.4	73.1	83.7	92.8	61.9	77.2	97.3	54.0	75.4	97.1	73.1	85.0
MF2014	WEIGHTED AVERAGE	99.8	68.3	85.9	100.1	68.3	86.1	99.8	68.7	86.1	99.3	65.4	84.4	99.0	57.5	80.7	99.8	68.7	86.1

VIRGINIA / WASHINGTON

DIVISION		ROANOKE 240-241 MAT.	INST.	TOTAL	STAUNTON 244 MAT.	INST.	TOTAL	WINCHESTER 226 MAT.	INST.	TOTAL	CLARKSTON 994 MAT.	INST.	TOTAL	EVERETT 982 MAT.	INST.	TOTAL	OLYMPIA 985 MAT.	INST.	TOTAL
015433	CONTRACTOR EQUIPMENT		104.4	104.4		108.8	108.8		104.4	104.4		90.9	90.9		102.2	102.2		102.2	102.2
0241, 31 - 34	SITE & INFRASTRUCTURE, DEMOLITION	106.8	88.7	94.0	110.1	89.9	95.8	119.0	90.1	96.6	103.2	89.7	93.6	91.8	110.4	104.9	90.9	110.4	104.7
0310	Concrete Forming & Accessories	95.8	59.8	64.7	89.3	48.6	54.1	83.4	68.9	70.8	110.3	67.0	72.9	113.0	97.0	99.8	102.0	97.9	98.4
0320	Concrete Reinforcing	96.7	70.6	83.2	96.4	48.4	71.5	96.4	88.9	92.5	110.5	86.4	98.0	108.9	97.7	103.1	114.3	97.8	105.7
0330	Cast-in-Place Concrete	115.2	61.2	93.9	105.5	50.9	83.9	101.9	70.3	89.4	93.0	89.5	91.6	101.0	105.6	102.8	88.8	105.7	95.5
03	CONCRETE	107.8	63.8	86.4	104.7	51.2	78.7	105.4	74.3	90.3	104.6	78.5	91.9	96.1	100.0	98.0	91.9	100.1	95.9
04	MASONRY	98.9	57.8	73.6	108.6	51.1	73.2	103.6	78.4	88.0	106.9	84.8	93.3	111.8	100.4	104.8	114.1	101.9	106.6
05	METALS	101.5	89.1	97.6	99.4	82.0	93.8	99.5	98.4	99.2	100.3	81.7	94.4	101.6	91.0	98.2	102.2	91.1	98.7
06	WOOD, PLASTICS & COMPOSITES	96.8	60.5	76.6	88.2	48.2	65.9	84.1	65.8	73.9	108.4	61.7	82.4	109.9	96.6	102.5	92.8	96.6	94.9
07	THERMAL & MOISTURE PROTECTION	103.6	69.4	89.3	103.6	50.8	81.5	104.3	80.7	94.4	150.2	81.8	121.6	108.7	100.7	105.4	104.8	100.6	103.0
08	OPENINGS	102.0	57.8	91.7	101.6	46.7	88.9	103.3	72.2	96.1	121.1	66.3	108.4	109.7	97.4	106.8	112.3	97.4	108.8
0920	Plaster & Gypsum Board	102.2	59.2	73.5	95.0	45.8	62.1	95.2	64.6	74.8	146.8	60.4	89.0	110.5	96.5	101.2	103.3	96.5	98.8
0950, 0980	Ceilings & Acoustic Treatment	91.5	59.2	70.1	89.0	45.8	60.4	89.9	64.6	73.2	102.5	60.4	74.6	102.0	96.5	98.4	97.8	96.5	97.0
0960	Flooring	100.7	63.9	90.2	96.4	38.0	79.6	95.2	83.1	91.7	92.9	77.0	88.3	109.8	109.5	109.7	97.9	114.6	102.7
0970, 0990	Wall Finishes & Painting/Coating	98.4	54.3	72.7	98.4	34.5	61.2	111.5	87.2	97.3	81.3	73.9	77.0	93.3	91.5	92.2	84.2	91.5	88.4
09	FINISHES	96.0	59.8	76.0	93.6	45.3	66.8	94.8	72.1	82.2	109.8	68.1	86.8	105.2	99.0	101.8	97.0	100.0	98.7
COVERS	DIVS. 10 - 14, 25, 28, 41, 43, 44, 46	100.0	69.7	93.4	100.0	69.0	93.3	100.0	87.9	97.4	100.0	74.2	94.4	100.0	97.6	99.5	100.0	99.9	100.0
21, 22, 23	FIRE SUPPRESSION, PLUMBING & HVAC	100.4	67.4	86.4	97.0	58.2	80.6	97.0	87.6	93.0	97.3	83.9	91.6	100.2	98.1	99.3	99.9	98.2	99.2
26, 27, 3370	ELECTRICAL, COMMUNICATIONS & UTIL.	97.3	54.0	75.4	96.2	76.7	86.4	95.9	103.9	99.9	87.1	96.7	92.0	101.8	99.7	100.7	99.5	100.2	99.9
MF2014	WEIGHTED AVERAGE	101.0	66.6	85.8	99.7	61.7	83.0	100.0	85.2	93.5	103.8	81.9	94.1	102.2	99.0	100.8	101.0	99.7	100.4

WASHINGTON

DIVISION		RICHLAND 993			SEATTLE 980 - 981,987			SPOKANE 990 - 992			TACOMA 983 - 984			VANCOUVER 986			WENATCHEE 988		
		MAT.	INST.	TOTAL	MAT.	INST.	TOTAL	MAT.	INST.	TOTAL	MAT.	INST.	TOTAL	MAT.	INST.	TOTAL	MAT.	INST.	TOTAL
015433	CONTRACTOR EQUIPMENT		90.9	90.9		102.3	102.3		90.9	90.9		102.2	102.2		97.8	97.8		102.2	102.2
0241, 31 - 34	SITE & INFRASTRUCTURE, DEMOLITION	105.6	90.5	94.9	96.0	108.6	104.9	104.9	90.4	94.7	94.9	110.4	105.9	104.6	97.8	99.8	103.9	109.0	107.5
0310	Concrete Forming & Accessories	110.4	78.2	82.6	104.4	98.1	98.9	115.5	77.7	82.8	103.4	97.8	98.6	103.8	91.6	93.2	105.2	75.7	79.6
0320	Concrete Reinforcing	105.9	86.5	95.9	114.3	97.8	105.7	106.6	86.4	96.2	107.7	97.8	102.6	108.6	97.4	102.8	108.5	86.7	97.2
0330	Cast-in-Place Concrete	93.2	85.2	90.0	103.8	105.8	104.6	96.8	85.0	92.2	103.9	105.7	104.6	115.9	98.6	109.1	106.0	77.0	94.6
03	CONCRETE	104.2	82.2	93.5	99.8	100.3	100.1	106.4	81.9	94.5	97.8	100.1	98.9	107.6	94.9	101.4	105.1	78.4	92.1
04	MASONRY	108.5	85.3	94.2	109.4	101.9	104.8	109.0	85.3	94.4	110.9	101.9	105.4	110.3	95.5	101.2	114.0	91.0	99.9
05	METALS	100.7	84.7	95.6	101.1	92.8	98.5	103.3	84.2	97.2	103.2	91.1	99.4	100.8	91.6	97.9	100.8	85.1	95.8
06	WOOD, PLASTICS & COMPOSITES	108.5	75.4	90.1	102.5	96.6	99.2	117.0	75.4	93.9	99.7	96.6	98.0	91.2	91.1	91.1	101.2	73.2	85.6
07	THERMAL & MOISTURE PROTECTION	151.4	81.5	122.2	106.6	101.1	104.3	147.7	83.2	120.7	108.5	100.6	105.2	108.6	93.9	102.4	109.1	87.8	100.2
08	OPENINGS	119.0	72.7	108.2	107.2	97.4	104.9	119.6	70.7	108.2	110.5	97.4	107.4	105.7	92.2	102.6	110.0	72.3	101.2
0920	Plaster & Gypsum Board	146.8	74.5	98.5	107.0	96.5	100.0	137.1	74.5	95.3	111.2	96.5	101.4	109.3	91.1	97.1	113.8	72.4	86.1
0950, 0980	Ceilings & Acoustic Treatment	109.0	74.5	86.2	114.9	96.5	102.7	104.4	74.5	84.6	105.3	96.5	99.5	102.4	91.1	94.9	97.7	72.4	80.9
0960	Flooring	93.3	84.7	90.8	104.2	102.6	103.8	92.2	94.8	92.9	102.2	112.2	105.1	107.9	102.1	106.2	105.3	60.2	92.4
0970, 0990	Wall Finishes & Painting/Coating	81.3	73.9	77.0	106.4	91.5	97.7	81.4	70.9	75.3	93.3	91.5	92.2	95.4	71.7	81.6	93.3	72.6	81.2
09	FINISHES	111.6	77.9	93.0	107.2	97.7	101.9	108.9	79.6	92.7	103.9	99.6	101.5	102.0	91.3	96.1	104.4	71.3	86.1
COVERS	DIVS. 10 - 14, 25, 28, 41, 43, 44, 46	100.0	85.2	96.8	100.0	99.9	100.0	100.0	85.2	96.8	100.0	99.9	100.0	100.0	96.4	99.2	100.0	74.9	94.6
21, 22, 23	FIRE SUPPRESSION, PLUMBING & HVAC	100.7	111.7	105.4	100.1	114.2	106.0	100.6	84.0	93.6	100.2	98.2	99.4	100.3	92.1	96.8	96.9	85.9	92.2
26, 27, 3370	ELECTRICAL, COMMUNICATIONS & UTIL.	84.9	96.8	90.9	101.8	107.1	104.5	83.7	80.1	81.9	101.6	100.2	100.9	106.4	100.2	103.3	102.4	97.7	100.0
MF2014	WEIGHTED AVERAGE	104.4	90.7	98.4	102.3	103.8	103.0	104.8	82.6	95.0	102.6	99.6	101.3	103.4	94.4	99.4	102.8	85.9	95.3

WASHINGTON / WEST VIRGINIA

DIVISION		WASHINGTON YAKIMA 989			BECKLEY 258 - 259			BLUEFIELD 247 - 248			BUCKHANNON 262			CHARLESTON 250 - 253			CLARKSBURG 263 - 264		
		MAT.	INST.	TOTAL	MAT.	INST.	TOTAL	MAT.	INST.	TOTAL	MAT.	INST.	TOTAL	MAT.	INST.	TOTAL	MAT.	INST.	TOTAL
015433	CONTRACTOR EQUIPMENT		102.2	102.2		104.4	104.4		104.4	104.4		104.4	104.4		104.4	104.4		104.4	104.4
0241, 31 - 34	SITE & INFRASTRUCTURE, DEMOLITION	97.3	109.8	106.1	100.5	92.1	94.6	100.9	92.1	94.7	107.2	92.0	96.5	99.9	93.2	95.2	107.8	92.0	96.7
0310	Concrete Forming & Accessories	103.8	93.4	94.8	84.7	91.1	90.2	86.5	90.9	90.3	85.9	91.3	90.6	96.9	92.1	92.8	83.4	91.3	90.2
0320	Concrete Reinforcing	108.2	86.5	97.0	93.5	89.7	91.5	95.3	84.9	89.9	96.0	84.7	90.1	101.2	89.8	95.3	96.0	85.2	90.4
0330	Cast-in-Place Concrete	110.9	84.8	100.6	99.5	101.3	100.2	99.2	101.2	100.0	98.9	99.2	99.0	96.2	130.8	109.9	108.4	97.4	104.1
03	CONCRETE	102.6	88.9	95.9	100.0	94.9	97.5	99.1	93.4	96.6	102.2	93.4	97.9	98.8	105.4	102.0	106.3	92.9	99.8
04	MASONRY	103.9	84.0	91.6	96.2	96.5	96.4	94.7	96.5	95.8	106.5	90.9	96.9	92.8	98.0	96.0	110.7	90.9	98.5
05	METALS	101.5	85.3	96.3	101.5	100.3	101.1	99.4	98.5	99.1	99.5	98.4	99.2	100.1	100.8	100.3	99.5	98.5	99.2
06	WOOD, PLASTICS & COMPOSITES	100.1	96.6	98.2	88.2	90.4	89.4	87.4	90.4	89.0	86.7	91.2	89.2	97.0	90.4	93.3	83.5	91.2	87.8
07	THERMAL & MOISTURE PROTECTION	108.6	83.8	98.2	103.2	89.0	97.3	103.7	89.0	97.5	104.0	88.6	97.5	100.4	89.3	95.8	103.9	88.3	97.4
08	OPENINGS	110.0	78.6	102.7	102.2	83.9	98.0	103.7	82.8	98.8	103.7	83.3	99.0	100.8	83.9	96.9	103.7	83.4	99.0
0920	Plaster & Gypsum Board	110.8	96.5	101.3	94.6	89.9	91.4	94.2	89.9	91.3	94.6	90.7	92.0	98.0	89.9	92.6	92.6	90.7	91.4
0950, 0980	Ceilings & Acoustic Treatment	99.5	96.5	97.5	83.3	89.9	87.7	87.4	89.9	89.0	89.0	90.7	90.2	91.5	89.9	90.4	89.0	90.7	90.2
0960	Flooring	103.2	61.9	91.3	98.6	113.4	102.9	93.1	113.4	99.0	92.9	104.7	96.3	103.0	113.4	106.0	91.8	104.7	95.5
0970, 0990	Wall Finishes & Painting/Coating	93.3	70.9	80.2	89.4	94.7	92.5	98.4	94.7	96.2	98.4	94.3	96.0	88.5	100.8	95.7	98.4	94.3	96.0
09	FINISHES	103.0	85.3	93.2	91.8	95.7	93.9	91.7	95.7	93.9	92.6	94.2	93.5	96.1	96.7	96.4	91.9	94.2	93.2
COVERS	DIVS. 10 - 14, 25, 28, 41, 43, 44, 46	100.0	97.1	99.4	100.0	101.2	100.3	100.0	101.2	100.3	100.0	100.4	100.1	100.0	101.1	100.2	100.0	100.4	100.1
21, 22, 23	FIRE SUPPRESSION, PLUMBING & HVAC	100.2	111.1	104.9	97.2	98.3	97.7	97.0	91.8	94.8	97.0	97.9	97.4	100.2	99.3	99.8	97.0	97.9	97.4
26, 27, 3370	ELECTRICAL, COMMUNICATIONS & UTIL.	104.3	96.8	100.5	98.1	90.4	94.2	96.3	92.2	94.2	97.6	97.2	97.4	103.2	92.6	97.8	97.6	97.2	97.4
MF2014	WEIGHTED AVERAGE	102.8	94.9	99.3	98.8	95.1	97.2	98.2	93.6	96.2	99.6	94.7	97.4	99.7	97.4	98.7	100.2	94.7	97.7

WEST VIRGINIA

DIVISION		GASSAWAY 266			HUNTINGTON 255 - 257			LEWISBURG 249			MARTINSBURG 254			MORGANTOWN 265			PARKERSBURG 261		
		MAT.	INST.	TOTAL	MAT.	INST.	TOTAL	MAT.	INST.	TOTAL	MAT.	INST.	TOTAL	MAT.	INST.	TOTAL	MAT.	INST.	TOTAL
015433	CONTRACTOR EQUIPMENT		104.4	104.4		104.4	104.4		104.4	104.4		104.4	104.4		104.4	104.4		104.4	104.4
0241, 31 - 34	SITE & INFRASTRUCTURE, DEMOLITION	104.6	92.1	95.7	105.1	93.3	96.8	116.8	92.1	99.3	104.3	92.5	96.0	101.9	92.7	95.4	110.1	93.2	98.1
0310	Concrete Forming & Accessories	85.4	89.0	88.5	96.4	92.9	93.3	83.7	90.4	89.5	84.8	82.0	82.3	83.7	91.4	90.4	88.0	88.3	88.3
0320	Concrete Reinforcing	96.0	90.7	93.2	94.9	95.2	95.1	96.0	84.7	90.1	93.5	90.5	91.9	96.0	85.2	90.4	95.3	84.7	89.8
0330	Cast-in-Place Concrete	103.5	129.3	113.7	108.5	102.5	106.1	99.3	101.0	99.9	104.2	95.9	100.9	98.9	99.2	99.0	101.1	96.0	99.1
03	CONCRETE	102.5	103.6	103.1	105.4	97.1	101.4	109.0	93.6	101.5	103.8	89.0	96.6	98.8	93.6	96.3	104.2	91.0	97.8
04	MASONRY	111.4	92.5	99.8	94.6	101.9	99.1	97.1	93.6	94.9	97.5	90.6	93.3	128.8	90.9	105.5	84.0	93.3	89.7
05	METALS	99.5	100.7	99.9	103.9	102.6	103.5	99.4	97.8	98.9	101.9	96.3	100.1	99.6	98.6	99.3	100.2	98.4	99.6
06	WOOD, PLASTICS & COMPOSITES	85.9	87.7	86.9	98.8	90.6	94.2	83.8	90.4	87.5	88.2	80.2	83.8	83.8	91.2	87.9	86.7	86.1	86.4
07	THERMAL & MOISTURE PROTECTION	103.7	88.3	97.3	103.4	90.3	97.9	104.6	88.2	97.7	103.5	85.2	95.8	103.8	88.6	97.4	103.6	89.0	97.5
08	OPENINGS	101.7	82.7	97.3	101.5	85.2	97.7	103.7	82.8	98.8	104.3	72.4	96.9	105.0	83.4	100.0	102.6	80.5	97.4
0920	Plaster & Gypsum Board	94.0	87.1	89.4	102.1	90.1	94.1	92.6	89.9	90.8	95.2	79.4	84.7	92.6	90.7	91.4	90.5	85.5	88.6
0950, 0980	Ceilings & Acoustic Treatment	89.0	87.1	87.8	85.8	90.1	88.7	89.0	89.9	89.6	85.8	79.4	81.6	89.0	90.7	90.2	89.0	85.5	86.7
0960	Flooring	92.6	113.4	98.6	106.5	117.1	109.6	91.9	113.4	98.1	98.6	113.4	102.9	91.9	104.7	95.6	95.9	113.4	100.9
0970, 0990	Wall Finishes & Painting/Coating	98.4	100.8	99.8	89.4	94.7	92.5	98.4	75.4	85.0	89.4	41.4	61.5	98.4	94.3	96.0	98.4	94.7	96.2
09	FINISHES	92.1	94.5	93.5	95.7	97.4	96.7	93.0	93.6	93.4	92.6	82.9	87.3	91.5	94.2	93.0	93.6	93.4	93.5
COVERS	DIVS. 10 - 14, 25, 28, 41, 43, 44, 46	100.0	100.0	100.0	100.0	101.8	100.4	100.0	54.7	90.2	100.0	95.3	99.0	100.0	98.6	99.7	100.0	100.5	100.1
21, 22, 23	FIRE SUPPRESSION, PLUMBING & HVAC	97.0	91.4	94.6	100.6	100.3	100.5	97.0	98.1	97.5	97.2	85.5	92.3	97.0	97.9	97.4	100.3	90.8	96.3
26, 27, 3370	ELECTRICAL, COMMUNICATIONS & UTIL.	97.6	92.6	95.1	102.1	96.3	99.2	93.8	92.2	93.0	104.3	82.1	93.1	97.8	97.2	97.5	97.7	93.9	95.8
MF2014	WEIGHTED AVERAGE	99.5	94.4	97.3	101.4	97.8	99.8	99.8	92.8	96.7	100.4	87.0	94.5	100.2	94.8	97.8	99.6	92.5	96.5

797

For customer support on your Commercial Renovation Cost Data, call 877.742.4719.

WEST VIRGINIA / WISCONSIN

DIVISION		PETERSBURG 268 MAT.	INST.	TOTAL	ROMNEY 267 MAT.	INST.	TOTAL	WHEELING 260 MAT.	INST.	TOTAL	BELOIT 535 MAT.	INST.	TOTAL	EAU CLAIRE 547 MAT.	INST.	TOTAL	GREEN BAY 541-543 MAT.	INST.	TOTAL
015433	CONTRACTOR EQUIPMENT		104.4	104.4		104.4	104.4		104.4	104.4		100.9	100.9		101.1	101.1		98.8	98.8
0241, 31 - 34	SITE & INFRASTRUCTURE, DEMOLITION	101.1	92.6	95.1	104.0	92.7	96.0	110.9	92.8	98.1	96.7	106.4	103.5	97.6	103.9	102.1	101.0	100.1	100.4
0310	Concrete Forming & Accessories	86.9	82.3	83.0	82.9	83.0	83.0	89.6	91.4	91.2	99.1	97.5	97.8	97.4	98.5	98.4	106.3	101.1	101.8
0320	Concrete Reinforcing	95.3	81.7	88.3	96.0	90.6	93.2	94.7	85.2	89.8	91.7	114.8	103.7	93.2	98.2	95.8	91.4	91.9	91.7
0330	Cast-in-Place Concrete	98.9	96.4	97.9	103.5	96.7	100.8	101.1	94.2	98.4	102.5	99.4	101.3	102.0	98.0	100.5	105.5	109.6	107.1
03	CONCRETE	99.0	87.9	94.0	102.4	89.9	96.3	104.3	91.9	98.2	98.0	101.5	99.7	98.5	98.5	98.5	101.7	102.4	102.0
04	MASONRY	101.4	90.9	94.9	97.4	91.6	93.8	109.7	91.5	98.5	104.2	105.4	105.0	98.0	102.8	101.0	130.9	118.9	123.5
05	METALS	99.6	96.5	98.6	99.7	100.1	99.8	100.4	98.4	99.7	93.7	106.8	97.9	91.4	101.4	94.6	93.7	99.1	95.4
06	WOOD, PLASTICS & COMPOSITES	87.6	80.2	83.5	83.0	80.2	81.5	88.2	91.2	89.9	98.1	95.9	96.9	101.7	98.0	99.6	107.0	98.0	102.0
07	THERMAL & MOISTURE PROTECTION	103.8	81.3	94.4	103.9	86.2	96.5	104.0	88.0	97.3	102.6	94.6	99.2	104.2	99.9	102.4	106.2	91.4	100.0
08	OPENINGS	105.1	76.6	98.4	105.0	78.6	98.8	103.5	83.4	98.8	101.0	106.3	102.3	107.7	94.5	104.7	103.8	99.6	102.8
0920	Plaster & Gypsum Board	94.6	79.4	84.5	92.3	79.4	83.7	95.0	90.7	92.1	93.1	96.2	95.2	104.9	98.3	100.5	101.6	98.3	99.4
0950, 0980	Ceilings & Acoustic Treatment	89.0	79.4	82.7	89.0	79.4	82.7	89.0	90.7	90.2	81.8	96.2	91.3	89.2	98.3	95.2	82.8	98.3	93.0
0960	Flooring	93.7	104.7	96.3	91.7	106.0	95.8	96.8	104.7	99.1	98.0	116.6	103.3	88.9	109.7	94.9	106.5	126.8	112.3
0970, 0990	Wall Finishes & Painting/Coating	98.4	94.3	96.0	98.4	94.3	96.0	98.4	94.3	96.0	88.6	103.4	101.3	88.6	86.5	87.4	98.5	83.2	89.6
09	FINISHES	92.3	87.6	89.7	91.6	87.9	89.6	93.9	94.3	94.1	94.5	101.3	98.3	93.1	99.6	96.7	97.7	104.5	101.5
COVERS	DIVS. 10 - 14, 25, 28, 41, 43, 44, 46	100.0	61.9	91.7	100.0	99.1	99.8	100.0	93.1	98.5	100.0	97.2	99.4	100.0	97.6	99.5	100.0	100.1	100.0
21, 22, 23	FIRE SUPPRESSION, PLUMBING & HVAC	97.0	91.4	94.6	97.0	91.2	94.5	100.4	98.2	99.5	100.0	97.5	99.0	100.0	86.7	94.4	100.2	88.5	95.3
26, 27, 3370	ELECTRICAL, COMMUNICATIONS & UTIL.	100.7	82.1	91.3	100.0	82.1	91.0	95.1	97.2	96.1	99.5	87.3	93.3	104.9	86.8	95.7	99.4	81.8	90.5
MF2014	WEIGHTED AVERAGE	99.3	87.7	94.2	99.4	89.8	95.1	100.8	94.5	98.0	98.6	99.8	99.1	99.2	95.5	97.6	101.2	97.5	99.5

WISCONSIN

DIVISION		KENOSHA 531 MAT.	INST.	TOTAL	LA CROSSE 546 MAT.	INST.	TOTAL	LANCASTER 538 MAT.	INST.	TOTAL	MADISON 537 MAT.	INST.	TOTAL	MILWAUKEE 530,532 MAT.	INST.	TOTAL	NEW RICHMOND 540 MAT.	INST.	TOTAL
015433	CONTRACTOR EQUIPMENT		98.8	98.8		101.1	101.1		100.9	100.9		100.9	100.9		88.1	88.1		101.5	101.5
0241, 31 - 34	SITE & INFRASTRUCTURE, DEMOLITION	102.2	103.3	103.0	91.3	103.7	100.1	95.9	106.4	103.3	94.8	106.4	103.0	93.6	96.9	95.9	95.7	104.1	101.6
0310	Concrete Forming & Accessories	107.5	98.3	99.5	84.0	98.3	96.4	98.5	96.2	96.5	102.7	98.8	99.4	101.9	119.4	117.0	92.6	98.2	97.4
0320	Concrete Reinforcing	91.6	96.9	94.3	92.9	90.6	91.7	93.0	90.4	91.7	96.0	90.6	93.2	94.8	97.1	95.9	90.5	97.9	94.3
0330	Cast-in-Place Concrete	111.5	110.1	110.9	91.8	94.8	93.0	101.9	99.8	101.1	92.8	101.9	96.4	97.5	111.1	102.9	106.2	81.4	96.4
03	CONCRETE	102.9	102.1	102.5	89.6	95.9	92.7	97.7	96.5	97.1	95.5	98.4	96.9	96.2	111.4	103.6	96.5	92.6	94.6
04	MASONRY	101.5	117.1	111.1	97.1	105.5	102.3	104.3	103.8	104.0	107.1	104.8	105.7	108.7	119.5	115.3	125.5	101.8	110.9
05	METALS	94.6	100.6	96.5	91.3	98.3	93.5	91.2	95.5	92.6	94.6	96.8	95.3	96.0	92.8	95.0	91.5	99.4	94.0
06	WOOD, PLASTICS & COMPOSITES	102.7	93.2	97.4	87.0	98.0	93.2	97.4	95.9	96.6	105.6	97.8	96.8	103.2	120.8	113.0	90.8	99.0	95.4
07	THERMAL & MOISTURE PROTECTION	102.9	106.9	104.6	103.6	102.0	102.9	102.4	81.9	93.8	100.4	102.8	101.4	102.0	111.7	106.0	105.5	90.8	99.3
08	OPENINGS	95.3	98.2	96.0	107.7	89.3	103.4	96.8	83.3	93.6	107.1	99.1	105.2	98.9	113.3	102.3	92.6	94.7	93.1
0920	Plaster & Gypsum Board	84.9	93.4	90.5	99.7	98.3	98.8	92.4	96.2	94.9	103.5	98.1	99.9	97.7	121.6	113.7	90.2	99.5	96.4
0950, 0980	Ceilings & Acoustic Treatment	81.8	93.4	89.5	88.4	98.3	94.9	78.5	96.2	90.2	86.9	98.1	94.3	89.5	121.6	110.7	56.9	99.5	85.1
0960	Flooring	115.8	108.3	113.7	82.3	109.7	90.1	97.5	106.3	100.0	99.4	116.9	104.4	98.6	111.1	102.2	98.9	109.7	102.0
0970, 0990	Wall Finishes & Painting/Coating	109.1	117.9	114.2	88.6	77.9	82.4	98.4	103.4	101.3	94.4	103.4	99.6	103.6	120.4	113.4	101.0	86.5	92.5
09	FINISHES	99.7	101.7	100.8	89.7	98.6	94.7	93.5	99.6	96.9	95.1	102.6	99.3	99.1	119.0	110.1	87.8	98.4	93.7
COVERS	DIVS. 10 - 14, 25, 28, 41, 43, 44, 46	100.0	99.7	99.9	100.0	97.6	99.5	100.0	90.7	98.0	100.0	100.0	100.0	100.0	103.3	100.7	100.0	93.5	98.6
21, 22, 23	FIRE SUPPRESSION, PLUMBING & HVAC	100.2	96.6	98.7	100.0	86.6	94.3	96.7	86.5	92.4	99.9	94.5	97.6	99.9	102.4	100.9	96.1	86.2	91.9
26, 27, 3370	ELECTRICAL, COMMUNICATIONS & UTIL.	99.9	99.2	99.5	105.1	86.8	95.9	99.3	86.8	93.0	101.1	96.3	98.7	98.6	101.0	99.8	103.0	86.8	94.8
MF2014	WEIGHTED AVERAGE	99.2	101.7	100.3	97.6	94.8	96.4	96.8	93.7	95.4	99.3	99.1	99.2	98.9	106.9	102.4	97.1	93.8	95.6

WISCONSIN

DIVISION		OSHKOSH 549 MAT.	INST.	TOTAL	PORTAGE 539 MAT.	INST.	TOTAL	RACINE 534 MAT.	INST.	TOTAL	RHINELANDER 545 MAT.	INST.	TOTAL	SUPERIOR 548 MAT.	INST.	TOTAL	WAUSAU 544 MAT.	INST.	TOTAL
015433	CONTRACTOR EQUIPMENT		98.8	98.8		100.9	100.9		100.9	100.9		98.8	98.8		101.5	101.5		98.8	98.8
0241, 31 - 34	SITE & INFRASTRUCTURE, DEMOLITION	92.6	99.4	97.4	86.7	106.2	100.5	96.5	107.2	104.1	104.9	99.4	101.0	92.3	104.0	100.6	88.4	101.1	97.4
0310	Concrete Forming & Accessories	88.9	95.5	94.6	90.7	96.5	95.7	99.5	98.3	98.5	86.4	95.6	94.4	90.6	89.6	89.7	88.4	100.7	99.0
0320	Concrete Reinforcing	91.5	91.4	91.4	93.1	90.5	91.8	93.0	90.1	90.9	91.7	90.1	90.9	90.5	96.2	93.4	91.7	90.4	91.0
0330	Cast-in-Place Concrete	97.9	98.1	98.0	87.5	102.0	93.2	100.5	109.8	104.2	110.9	99.7	106.5	100.1	102.2	100.9	91.2	103.1	95.9
03	CONCRETE	91.4	95.8	93.6	85.6	97.4	91.3	97.0	102.1	99.5	103.5	96.2	100.0	90.9	95.5	93.2	86.4	99.7	92.9
04	MASONRY	112.2	102.5	106.2	103.2	105.4	104.6	104.2	117.1	112.1	131.3	102.5	113.6	124.7	106.7	113.6	111.6	111.1	111.3
05	METALS	91.7	96.7	93.3	91.9	95.9	93.2	95.4	100.7	97.0	91.6	97.1	93.3	92.4	100.7	95.1	91.5	97.7	93.4
06	WOOD, PLASTICS & COMPOSITES	86.9	96.0	92.0	87.9	95.9	92.4	98.4	93.2	95.5	84.6	96.0	90.9	89.2	87.0	88.0	86.4	98.0	92.9
07	THERMAL & MOISTURE PROTECTION	105.3	84.1	96.4	101.9	91.4	97.5	102.6	107.6	104.7	106.2	84.1	96.9	105.1	91.1	99.3	105.2	86.6	97.4
08	OPENINGS	99.9	93.5	98.4	96.9	98.1	97.2	101.0	98.2	100.4	99.9	84.6	96.4	92.1	90.0	91.6	100.1	94.1	98.7
0920	Plaster & Gypsum Board	89.0	96.2	93.8	86.0	96.2	92.8	93.1	93.4	93.3	89.0	96.2	93.8	90.1	87.2	88.1	89.0	98.3	95.2
0950, 0980	Ceilings & Acoustic Treatment	82.8	96.2	91.6	81.0	96.2	91.0	81.8	93.4	89.5	82.8	96.2	91.6	57.8	87.2	77.2	82.8	98.3	93.0
0960	Flooring	96.8	109.7	100.5	93.6	116.6	100.2	98.0	108.3	100.9	96.1	112.7	100.9	100.1	122.6	106.6	96.7	112.7	101.3
0970, 0990	Wall Finishes & Painting/Coating	95.2	101.4	98.8	98.4	103.4	101.3	98.3	117.9	109.7	95.2	60.4	74.9	89.3	102.8	97.2	95.2	83.2	88.2
09	FINISHES	92.2	98.1	95.4	91.4	100.3	96.3	94.5	101.7	98.5	93.0	94.7	94.0	87.1	96.7	92.4	91.9	101.7	97.3
COVERS	DIVS. 10 - 14, 25, 28, 41, 43, 44, 46	100.0	62.7	91.9	100.0	61.5	91.7	100.0	99.7	99.9	100.0	92.5	98.4	100.0	91.5	98.1	100.0	100.1	100.0
21, 22, 23	FIRE SUPPRESSION, PLUMBING & HVAC	96.9	84.5	91.7	96.7	94.4	95.7	100.0	96.6	98.6	96.9	85.8	92.2	96.1	94.4	95.3	96.9	89.7	93.9
26, 27, 3370	ELECTRICAL, COMMUNICATIONS & UTIL.	103.5	79.8	91.6	102.9	91.4	97.1	99.3	98.8	99.1	102.9	81.2	91.9	107.8	100.2	104.0	104.7	83.9	94.2
MF2014	WEIGHTED AVERAGE	97.2	91.1	94.5	95.3	96.4	95.8	98.7	102.0	100.2	99.8	91.8	96.3	96.8	97.7	97.2	96.5	96.1	96.3

WYOMING

DIVISION		CASPER 826 MAT.	INST.	TOTAL	CHEYENNE 820 MAT.	INST.	TOTAL	NEWCASTLE 827 MAT.	INST.	TOTAL	RAWLINS 823 MAT.	INST.	TOTAL	RIVERTON 825 MAT.	INST.	TOTAL	ROCK SPRINGS 829 - 831 MAT.	INST.	TOTAL
015433	CONTRACTOR EQUIPMENT		96.7	96.7		96.7	96.7		96.7	96.7		96.7	96.7		96.7	96.7		96.7	96.7
0241, 31 - 34	SITE & INFRASTRUCTURE, DEMOLITION	99.9	93.2	95.1	95.1	93.2	93.7	87.5	93.1	91.5	101.8	93.1	95.7	95.0	93.2	93.7	91.8	92.8	92.5
0310	Concrete Forming & Accessories	103.0	48.8	56.1	104.6	60.1	66.1	95.2	72.1	75.2	99.5	72.4	76.0	94.1	60.1	64.7	101.9	60.5	66.1
0320	Concrete Reinforcing	106.5	70.2	87.7	102.0	70.2	85.5	109.7	70.4	89.3	109.4	70.4	89.2	110.4	70.2	89.5	110.4	69.8	89.4
0330	Cast-in-Place Concrete	94.5	59.8	80.8	89.3	59.9	77.7	90.2	58.9	77.9	90.3	66.6	80.9	90.2	60.1	78.3	90.2	58.9	77.8
03	CONCRETE	102.1	57.2	80.3	101.1	62.4	82.3	101.4	67.4	84.9	116.0	70.1	93.7	110.2	62.4	87.0	101.9	62.1	82.5
04	MASONRY	113.5	52.3	75.8	111.0	52.3	74.8	107.5	58.4	77.3	107.5	58.4	77.3	107.6	52.3	73.5	173.9	50.1	97.6
05	METALS	104.0	71.6	93.7	106.5	71.7	95.4	102.5	71.6	92.7	102.5	71.9	92.8	102.6	72.0	92.9	103.4	71.2	93.1
06	WOOD, PLASTICS & COMPOSITES	91.5	48.0	67.3	89.4	63.3	74.8	80.4	78.3	79.2	84.0	78.3	80.8	79.4	63.3	70.4	88.8	64.6	75.3
07	THERMAL & MOISTURE PROTECTION	108.9	52.7	85.4	103.1	54.3	82.7	104.8	57.0	84.8	106.3	69.3	90.8	105.7	55.1	84.5	104.9	58.5	85.5
08	OPENINGS	106.9	51.4	94.0	106.1	59.7	95.3	110.5	67.4	100.5	110.2	67.4	100.2	110.4	59.7	98.6	110.9	59.9	99.1
0920	Plaster & Gypsum Board	98.1	46.5	63.6	90.2	62.2	71.5	86.7	77.7	80.7	87.1	77.7	80.8	86.7	62.2	70.3	98.3	63.5	75.1
0950, 0980	Ceilings & Acoustic Treatment	98.8	46.5	64.2	92.6	62.2	72.5	94.5	77.7	83.3	94.5	77.7	83.3	94.5	62.2	73.1	94.5	63.5	74.0
0960	Flooring	116.1	67.2	102.0	114.0	67.2	100.5	107.3	49.7	90.7	110.2	49.7	92.8	106.8	67.2	95.4	112.6	49.7	94.5
0970, 0990	Wall Finishes & Painting/Coating	90.7	38.5	60.4	96.4	38.5	62.7	93.0	57.4	72.3	93.0	57.4	72.3	93.0	38.5	61.3	93.0	55.4	71.1
09	FINISHES	103.0	50.3	73.9	100.2	59.4	77.6	95.2	67.5	79.9	97.6	67.5	80.9	96.0	59.4	75.7	98.5	58.1	76.1
COVERS	DIVS. 10 - 14, 25, 28, 41, 43, 44, 46	100.0	85.3	96.8	100.0	87.0	97.2	100.0	89.5	97.7	100.0	89.4	97.7	100.0	81.0	95.9	100.0	78.7	95.4
21, 22, 23	FIRE SUPPRESSION, PLUMBING & HVAC	99.9	68.1	86.5	99.9	68.1	86.5	98.1	69.3	85.9	98.1	69.3	86.0	98.1	68.1	85.4	99.9	67.3	86.1
26, 27, 3370	ELECTRICAL, COMMUNICATIONS & UTIL.	98.2	64.4	81.1	100.0	65.9	82.8	98.9	85.4	92.1	98.9	71.2	84.9	98.9	65.9	82.2	97.2	85.4	91.2
MF2014	WEIGHTED AVERAGE	102.5	63.8	85.5	102.3	66.5	86.5	100.8	72.4	88.3	103.1	71.2	89.0	102.1	66.4	86.4	104.9	68.4	88.8

WYOMING / CANADA

DIVISION		SHERIDAN 828 MAT.	INST.	TOTAL	WHEATLAND 822 MAT.	INST.	TOTAL	WORLAND 824 MAT.	INST.	TOTAL	YELLOWSTONE NAT'L PA 821 MAT.	INST.	TOTAL	BARRIE, ONTARIO MAT.	INST.	TOTAL	BATHURST, NEW BRUNSWICK MAT.	INST.	TOTAL
015433	CONTRACTOR EQUIPMENT		96.7	96.7		96.7	96.7		96.7	96.7		96.7	96.7		101.8	101.8		101.6	101.6
0241, 31 - 34	SITE & INFRASTRUCTURE, DEMOLITION	95.3	93.2	93.8	92.3	92.9	92.7	89.7	92.8	91.9	89.8	93.1	92.1	119.8	99.9	105.7	105.0	95.8	98.5
0310	Concrete Forming & Accessories	102.2	60.1	65.8	97.2	46.7	53.5	97.3	60.3	65.3	97.3	61.9	66.7	125.0	86.9	92.0	103.7	61.8	67.4
0320	Concrete Reinforcing	110.4	70.2	89.5	109.7	70.3	89.3	110.4	69.7	89.3	112.3	70.4	90.6	175.6	82.8	127.5	136.4	55.9	94.7
0330	Cast-in-Place Concrete	93.3	59.9	80.1	94.1	56.6	79.3	90.2	58.8	77.8	90.2	61.1	78.7	163.9	88.0	134.0	130.5	61.0	103.1
03	CONCRETE	110.1	62.4	86.9	106.5	55.2	81.6	101.6	62.0	82.4	101.9	63.6	83.3	150.4	86.8	119.5	123.6	61.2	93.3
04	MASONRY	107.8	52.3	73.6	107.9	54.4	74.9	107.6	42.3	67.3	107.6	58.7	77.5	173.7	97.6	126.8	161.5	64.7	101.8
05	METALS	106.4	71.7	95.4	102.4	71.3	92.5	102.6	70.6	92.5	103.3	71.6	93.2	109.3	89.7	103.0	109.9	71.6	97.7
06	WOOD, PLASTICS & COMPOSITES	90.7	63.3	75.4	82.0	46.3	62.1	82.1	64.6	72.3	82.1	64.6	72.3	118.8	85.2	100.1	100.0	61.8	78.7
07	THERMAL & MOISTURE PROTECTION	106.0	54.3	84.4	105.1	52.2	82.9	104.9	53.8	83.5	104.3	57.5	84.7	113.6	87.2	102.6	109.4	60.9	89.1
08	OPENINGS	111.1	59.7	99.2	109.0	50.7	95.5	110.8	59.9	98.9	103.3	59.9	93.2	94.7	84.0	92.2	89.3	54.3	81.2
0920	Plaster & Gypsum Board	111.2	62.2	78.4	86.7	44.7	58.7	86.7	63.5	71.2	86.9	63.5	71.3	148.8	84.7	106.0	121.2	60.6	80.7
0950, 0980	Ceilings & Acoustic Treatment	97.3	62.2	74.0	94.5	44.7	61.5	94.5	63.5	74.0	95.3	63.5	74.3	92.2	84.7	87.3	108.0	60.6	76.6
0960	Flooring	111.4	67.2	98.7	108.9	48.8	91.6	108.9	49.7	91.9	108.9	49.7	91.9	126.6	95.0	117.5	107.8	45.3	89.9
0970, 0990	Wall Finishes & Painting/Coating	95.2	38.5	62.3	93.0	45.0	65.1	93.0	55.4	71.1	93.0	55.4	71.1	107.4	89.9	97.2	113.2	51.6	77.4
09	FINISHES	103.3	59.4	79.0	96.0	45.9	68.3	95.7	58.1	74.9	95.9	59.2	75.6	114.4	88.9	100.3	109.7	57.9	81.1
COVERS	DIVS. 10 - 14, 25, 28, 41, 43, 44, 46	100.0	87.0	97.2	100.0	77.4	95.1	100.0	79.2	95.5	100.0	80.5	95.8	139.2	72.3	124.7	131.1	63.4	116.5
21, 22, 23	FIRE SUPPRESSION, PLUMBING & HVAC	98.1	68.1	85.4	98.1	67.2	85.1	98.1	67.2	85.1	98.1	69.3	85.9	103.5	98.2	101.3	103.6	69.0	89.0
26, 27, 3370	ELECTRICAL, COMMUNICATIONS & UTIL.	101.2	65.9	83.4	98.9	85.4	92.1	98.9	85.4	92.1	97.8	96.0	96.9	114.6	89.8	102.0	109.3	61.4	85.1
MF2014	WEIGHTED AVERAGE	103.7	66.5	87.3	101.4	65.5	85.6	101.0	67.4	86.2	100.2	71.5	87.6	118.3	91.7	106.6	111.8	66.3	91.8

CANADA

DIVISION		BRANDON, MANITOBA MAT.	INST.	TOTAL	BRANTFORD, ONTARIO MAT.	INST.	TOTAL	BRIDGEWATER, NOVA SCOTIA MAT.	INST.	TOTAL	CALGARY, ALBERTA MAT.	INST.	TOTAL	CAP-DE-LA-MADELEINE, QUEBEC MAT.	INST.	TOTAL	CHARLESBOURG, QUEBEC MAT.	INST.	TOTAL
015433	CONTRACTOR EQUIPMENT		104.1	104.1		101.8	101.8		101.4	101.4		106.9	106.9		102.2	102.2		102.2	102.2
0241, 31 - 34	SITE & INFRASTRUCTURE, DEMOLITION	138.3	98.8	110.4	119.2	100.2	105.8	103.9	97.5	99.4	123.0	104.8	110.2	99.4	99.1	99.2	99.4	99.1	99.2
0310	Concrete Forming & Accessories	146.7	70.4	80.7	124.1	94.1	98.2	96.7	69.5	73.2	123.4	93.3	97.4	130.0	85.4	91.4	130.0	85.4	91.4
0320	Concrete Reinforcing	173.8	52.6	111.0	160.7	81.6	119.7	136.7	46.2	89.8	128.3	73.5	99.9	136.7	77.9	106.3	136.7	77.9	106.3
0330	Cast-in-Place Concrete	131.5	75.8	109.5	150.0	109.2	133.9	154.1	71.3	121.5	177.2	104.2	148.4	121.1	96.7	111.4	121.1	96.7	111.4
03	CONCRETE	144.3	69.7	108.0	139.3	97.0	118.7	137.4	66.6	103.0	151.7	93.5	123.4	122.0	88.1	105.5	122.0	88.1	105.5
04	MASONRY	242.4	67.3	134.5	169.6	102.1	128.0	165.0	71.6	107.5	197.5	87.1	129.5	165.5	86.7	116.9	165.5	86.7	116.9
05	METALS	125.4	76.7	109.9	110.2	90.3	103.9	109.3	74.1	98.2	140.6	89.1	124.2	108.6	86.9	101.7	108.6	86.9	101.7
06	WOOD, PLASTICS & COMPOSITES	151.0	71.0	106.5	121.6	92.5	105.4	92.3	68.6	79.1	98.5	94.0	96.0	132.5	85.0	106.0	132.5	85.0	106.0
07	THERMAL & MOISTURE PROTECTION	141.4	70.3	111.7	117.1	92.6	106.9	111.8	68.5	93.7	122.5	90.9	109.3	110.6	87.9	101.1	110.6	87.9	101.1
08	OPENINGS	105.5	62.9	95.6	93.6	90.0	92.8	87.5	63.0	81.8	87.7	84.1	86.9	94.8	78.7	91.1	94.8	78.7	91.1
0920	Plaster & Gypsum Board	111.0	69.7	83.4	117.6	92.2	100.7	119.7	67.6	84.8	123.5	93.3	103.3	142.0	84.4	103.5	142.0	84.4	103.5
0950, 0980	Ceilings & Acoustic Treatment	108.2	69.7	82.7	98.9	92.2	94.5	98.9	67.6	78.2	141.8	93.3	109.7	98.9	84.4	89.3	98.9	84.4	89.3
0960	Flooring	143.1	67.5	121.4	122.5	95.0	114.6	103.5	64.6	92.3	129.7	91.2	118.6	122.5	95.7	114.8	122.5	95.7	114.8
0970, 0990	Wall Finishes & Painting/Coating	120.2	58.7	84.4	112.3	98.7	104.3	112.3	64.1	84.2	125.5	112.4	117.9	112.3	91.7	100.3	112.3	91.7	100.3
09	FINISHES	123.4	69.0	93.3	112.3	95.1	102.8	106.3	68.2	85.2	130.2	95.3	110.9	115.1	88.2	100.2	115.1	88.2	100.2
COVERS	DIVS. 10 - 14, 25, 28, 41, 43, 44, 46	131.1	65.6	116.9	131.1	74.4	118.8	131.1	64.6	116.7	131.1	94.9	123.3	131.1	82.8	120.7	131.1	82.8	120.7
21, 22, 23	FIRE SUPPRESSION, PLUMBING & HVAC	103.7	83.2	95.0	103.6	101.2	102.6	103.6	83.5	95.1	105.5	90.0	98.9	103.9	90.0	98.1	103.9	90.0	98.1
26, 27, 3370	ELECTRICAL, COMMUNICATIONS & UTIL.	116.2	69.1	92.4	108.9	89.3	99.0	113.6	63.9	88.5	103.7	100.6	102.1	108.4	72.1	90.0	108.4	72.1	90.0
MF2014	WEIGHTED AVERAGE	126.4	74.8	103.7	115.5	95.5	106.7	113.4	73.6	95.9	123.8	93.4	110.4	112.8	86.4	101.1	112.8	86.4	101.1

For customer support on your Commercial Renovation Cost Data, call 877.742.4719.

799

DIVISION		CANADA																
		CHARLOTTETOWN, PRINCE EDWARD ISLAND			CHICOUTIMI, QUEBEC			CORNER BROOK, NEWFOUNDLAND			CORNWALL, ONTARIO			DALHOUSIE, NEW BRUNSWICK			DARTMOUTH, NOVA SCOTIA		
		MAT.	INST.	TOTAL	MAT.	INST.	TOTAL	MAT.	INST.	TOTAL	MAT.	INST.	TOTAL	MAT.	INST.	TOTAL	MAT.	INST.	TOTAL
015433	CONTRACTOR EQUIPMENT		101.3	101.3		102.2	102.2		102.5	102.5		101.8	101.8		101.6	101.6		101.4	101.4
0241, 31-34	SITE & INFRASTRUCTURE, DEMOLITION	125.1	95.0	103.8	102.0	98.4	99.5	143.8	96.2	110.2	117.2	99.7	104.8	102.3	95.8	97.7	129.5	97.5	106.9
0310	Concrete Forming & Accessories	115.8	56.5	64.5	132.5	90.5	96.2	123.6	59.8	68.4	121.9	87.0	91.7	102.6	62.0	67.5	113.2	69.5	75.4
0320	Concrete Reinforcing	143.0	46.2	92.9	102.9	94.8	98.7	159.4	48.0	101.7	160.7	81.3	119.6	138.2	55.9	95.6	166.6	46.2	104.2
0330	Cast-in-Place Concrete	157.7	60.3	119.3	122.9	98.8	113.4	153.0	69.6	120.1	134.9	99.5	121.0	127.2	61.1	101.1	147.7	71.3	117.6
03	CONCRETE	143.3	56.8	101.3	113.9	94.1	104.3	182.4	62.0	123.9	131.7	90.4	111.7	129.7	61.3	96.4	161.7	66.6	115.4
04	MASONRY	179.0	62.3	107.1	163.6	94.6	121.1	238.0	63.9	130.7	168.4	93.6	122.3	162.9	64.7	102.3	253.5	71.6	141.4
05	METALS	134.3	67.8	113.2	109.7	91.6	103.9	125.8	72.8	108.9	110.2	89.0	103.5	106.6	71.8	95.5	125.9	74.1	109.5
06	WOOD, PLASTICS & COMPOSITES	96.0	55.8	73.6	133.0	90.5	109.3	129.6	58.7	90.1	120.0	86.0	101.1	99.3	61.8	78.4	117.9	68.6	90.5
07	THERMAL & MOISTURE PROTECTION	127.9	59.2	99.2	109.8	94.2	103.3	146.4	60.5	110.5	117.0	87.3	104.6	113.7	60.9	91.6	143.4	68.5	112.0
08	OPENINGS	85.7	48.5	77.0	93.5	78.2	90.0	112.1	55.4	98.9	94.8	83.7	92.3	90.4	54.3	82.0	95.6	63.0	88.0
0920	Plaster & Gypsum Board	118.5	54.4	75.7	139.5	90.0	106.4	141.5	57.3	85.3	170.7	85.6	113.8	129.8	60.6	83.6	137.5	67.6	90.8
0950, 0980	Ceilings & Acoustic Treatment	121.7	54.4	77.2	107.1	90.0	95.8	109.0	57.3	74.8	101.4	85.6	90.9	102.1	60.6	74.6	115.6	67.6	83.8
0960	Flooring	116.4	60.5	100.3	124.9	95.7	116.5	123.5	55.0	103.8	122.5	93.7	114.2	109.5	69.3	98.0	118.0	64.6	102.7
0970, 0990	Wall Finishes & Painting/Coating	124.1	43.0	76.9	113.2	106.4	109.2	120.1	61.8	86.2	112.3	91.9	100.4	115.5	51.6	78.3	120.1	61.4	87.5
09	FINISHES	121.5	56.1	85.3	117.0	93.8	104.1	122.6	59.1	87.5	120.4	89.0	103.0	111.6	62.6	84.5	120.4	68.2	91.6
COVERS	DIVS. 10-14, 25, 28, 41, 43, 44, 46	131.1	62.9	116.4	131.1	85.8	121.3	131.1	63.6	116.5	131.1	71.9	118.3	131.1	63.4	116.5	131.1	64.6	116.7
21, 22, 23	FIRE SUPPRESSION, PLUMBING & HVAC	104.1	63.1	86.8	103.6	93.5	99.3	103.7	71.0	89.9	103.9	99.1	101.9	103.6	69.0	89.0	103.7	83.5	95.1
26, 27, 3370	ELECTRICAL, COMMUNICATIONS & UTIL.	107.8	51.9	79.5	106.5	86.1	96.2	113.8	57.8	85.5	109.5	90.2	99.8	111.4	58.0	84.4	118.3	63.9	90.8
MF2014	WEIGHTED AVERAGE	120.4	61.9	94.6	111.7	92.0	103.0	131.2	66.6	102.7	115.4	92.0	105.1	112.6	66.5	92.3	127.5	73.6	103.7

DIVISION		CANADA																
		EDMONTON, ALBERTA			FORT MCMURRAY, ALBERTA			FREDERICTON, NEW BRUNSWICK			GATINEAU, QUEBEC			GRANBY, QUEBEC			HALIFAX, NOVA SCOTIA		
		MAT.	INST.	TOTAL	MAT.	INST.	TOTAL	MAT.	INST.	TOTAL	MAT.	INST.	TOTAL	MAT.	INST.	TOTAL	MAT.	INST.	TOTAL
015433	CONTRACTOR EQUIPMENT		106.9	106.9		104.2	104.2		101.6	101.6		102.2	102.2		102.2	102.2		101.4	101.4
0241, 31-34	SITE & INFRASTRUCTURE, DEMOLITION	139.3	104.8	114.9	124.4	101.0	107.8	106.8	95.9	99.1	99.2	99.1	99.1	99.7	99.1	99.3	108.2	97.7	100.7
0310	Concrete Forming & Accessories	119.2	93.3	96.8	121.9	89.8	94.1	122.5	62.3	70.4	130.0	85.3	91.3	130.0	85.2	91.3	113.2	76.6	81.5
0320	Concrete Reinforcing	128.2	73.5	99.9	148.7	73.5	109.7	133.9	56.0	93.6	144.6	77.9	110.1	144.6	77.9	110.0	142.3	61.9	100.7
0330	Cast-in-Place Concrete	183.2	104.2	152.0	199.2	104.1	161.7	124.5	61.1	99.5	119.4	96.6	110.4	123.4	96.6	112.8	128.0	80.9	109.4
03	CONCRETE	154.4	93.5	124.8	161.5	91.9	127.7	126.2	61.5	94.7	122.3	88.0	105.7	124.3	88.0	106.6	127.8	75.9	102.6
04	MASONRY	190.2	87.1	126.7	210.0	88.7	135.3	187.2	66.2	112.6	165.4	86.7	116.9	165.7	86.7	117.0	184.5	84.3	122.8
05	METALS	140.5	89.1	124.7	136.9	88.7	121.5	130.4	72.3	111.9	108.6	86.8	101.7	108.6	86.8	101.6	132.4	81.4	116.2
06	WOOD, PLASTICS & COMPOSITES	98.3	94.0	95.9	115.9	88.8	100.8	110.8	61.8	83.5	132.5	85.0	106.0	132.5	85.0	106.0	96.3	75.2	84.5
07	THERMAL & MOISTURE PROTECTION	125.4	90.9	111.0	125.7	90.0	110.7	123.6	61.9	97.8	110.6	87.9	101.1	110.6	86.4	100.5	120.6	77.7	102.7
08	OPENINGS	85.5	84.1	85.2	94.8	81.2	91.7	89.5	53.2	81.0	94.8	74.1	90.0	94.8	74.1	90.0	93.9	70.3	88.4
0920	Plaster & Gypsum Board	116.1	93.3	100.8	116.3	88.0	97.4	119.7	60.6	80.2	115.9	84.4	94.8	115.9	84.4	94.8	114.9	74.3	87.8
0950, 0980	Ceilings & Acoustic Treatment	134.6	93.3	107.2	106.3	88.0	94.2	121.9	60.6	81.3	98.9	84.4	89.3	98.9	84.4	89.3	120.2	74.3	89.9
0960	Flooring	125.6	91.2	115.7	122.5	91.2	113.5	125.2	72.4	110.0	122.5	95.7	114.8	122.5	95.7	114.8	106.8	87.0	101.1
0970, 0990	Wall Finishes & Painting/Coating	116.7	112.4	114.2	112.4	95.5	102.5	121.3	65.8	89.0	112.3	91.7	100.3	112.3	91.7	100.3	115.1	87.4	99.0
09	FINISHES	125.5	95.3	108.8	115.0	90.7	101.6	123.3	64.7	90.9	111.3	88.2	98.5	111.3	88.2	98.5	115.1	79.9	95.6
COVERS	DIVS. 10-14, 25, 28, 41, 43, 44, 46	131.1	94.9	123.3	131.1	94.6	123.2	131.1	63.4	116.5	131.1	82.8	120.7	131.1	82.8	120.7	131.1	67.8	117.4
21, 22, 23	FIRE SUPPRESSION, PLUMBING & HVAC	105.4	90.0	98.9	104.1	97.9	101.5	104.0	78.5	93.2	103.9	90.0	98.1	103.6	90.0	97.8	104.9	76.8	93.0
26, 27, 3370	ELECTRICAL, COMMUNICATIONS & UTIL.	106.4	100.6	103.5	103.7	84.8	94.2	112.8	76.5	94.4	108.4	72.1	90.0	108.9	72.1	90.3	105.2	80.6	92.8
MF2014	WEIGHTED AVERAGE	123.9	93.4	110.5	124.5	91.7	110.0	118.8	71.5	97.9	112.5	86.1	100.9	112.7	86.1	101.0	118.2	79.9	101.3

DIVISION		CANADA																
		HAMILTON, ONTARIO			HULL, QUEBEC			JOLIETTE, QUEBEC			KAMLOOPS, BRITISH COLUMBIA			KINGSTON, ONTARIO			KITCHENER, ONTARIO		
		MAT.	INST.	TOTAL	MAT.	INST.	TOTAL	MAT.	INST.	TOTAL	MAT.	INST.	TOTAL	MAT.	INST.	TOTAL	MAT.	INST.	TOTAL
015433	CONTRACTOR EQUIPMENT		108.6	108.6		102.2	102.2		102.2	102.2		105.5	105.5		104.1	104.1		104.0	104.0
0241, 31-34	SITE & INFRASTRUCTURE, DEMOLITION	112.3	112.4	112.4	99.2	99.1	99.1	99.9	99.1	99.3	121.4	102.6	108.1	117.2	103.6	107.6	99.4	104.5	103.0
0310	Concrete Forming & Accessories	120.6	92.0	95.9	130.0	85.3	91.3	130.0	85.4	91.4	122.4	88.2	92.8	122.0	87.1	91.8	112.8	84.8	88.6
0320	Concrete Reinforcing	137.3	91.0	113.3	144.6	77.9	110.1	136.7	77.9	106.3	107.2	75.9	91.0	160.7	81.3	119.6	96.8	90.9	93.7
0330	Cast-in-Place Concrete	135.9	100.0	121.7	119.4	96.6	110.4	124.4	96.7	113.5	107.7	100.5	104.8	134.9	99.5	121.0	126.0	92.6	112.9
03	CONCRETE	126.7	94.6	111.1	122.3	88.0	105.7	123.6	88.1	106.3	130.5	90.3	110.9	133.6	90.4	112.6	108.4	88.8	98.9
04	MASONRY	178.7	100.7	130.7	165.4	86.7	116.9	165.8	86.7	117.0	172.6	94.5	124.5	175.3	93.6	125.0	148.4	97.2	116.8
05	METALS	126.6	92.1	115.6	108.6	86.8	101.7	108.6	86.9	101.7	110.8	86.9	103.2	111.8	89.0	104.5	117.4	91.8	109.2
06	WOOD, PLASTICS & COMPOSITES	97.8	90.8	93.9	132.5	85.0	106.0	132.5	85.0	106.0	104.3	86.4	94.3	120.0	86.2	101.2	109.2	82.6	94.4
07	THERMAL & MOISTURE PROTECTION	118.6	93.5	108.1	110.6	87.9	101.1	110.6	87.9	101.1	127.0	85.0	109.4	117.0	88.4	105.0	111.2	90.8	102.7
08	OPENINGS	90.4	89.5	90.2	94.8	74.1	90.0	94.8	78.7	91.1	91.5	83.2	89.6	94.8	83.4	92.2	86.4	83.3	85.7
0920	Plaster & Gypsum Board	131.9	90.4	104.2	115.9	84.4	94.8	142.0	84.4	103.5	103.5	85.5	91.5	173.8	85.7	114.9	106.5	82.0	90.2
0950, 0980	Ceilings & Acoustic Treatment	127.6	90.4	103.0	98.9	84.4	89.3	98.9	84.4	89.3	98.9	85.5	90.0	114.6	85.7	95.5	101.7	82.0	88.7
0960	Flooring	120.0	100.7	114.5	122.5	95.7	114.8	122.5	95.7	114.8	121.1	53.5	101.7	122.5	93.7	114.2	109.6	100.7	107.1
0970, 0990	Wall Finishes & Painting/Coating	109.4	90.0	99.9	112.3	91.7	100.3	112.3	91.7	100.3	112.3	84.9	96.4	112.3	84.9	95.8	106.2	92.6	96.6
09	FINISHES	122.0	94.7	106.9	111.3	88.2	98.5	115.1	88.2	100.2	111.7	81.8	95.2	123.7	88.3	104.1	104.8	88.1	95.6
COVERS	DIVS. 10-14, 25, 28, 41, 43, 44, 46	131.1	98.2	124.0	131.1	82.8	120.7	131.1	82.8	120.7	131.1	92.9	122.9	131.1	71.9	118.3	131.1	96.4	123.6
21, 22, 23	FIRE SUPPRESSION, PLUMBING & HVAC	105.7	88.2	98.3	103.6	90.0	97.8	103.6	90.0	97.9	103.6	93.4	99.3	103.9	99.2	101.9	104.7	86.7	97.1
26, 27, 3370	ELECTRICAL, COMMUNICATIONS & UTIL.	101.9	99.7	100.8	110.0	72.1	90.8	108.9	72.1	90.3	111.9	81.3	96.4	109.5	88.9	99.1	106.3	97.3	101.7
MF2014	WEIGHTED AVERAGE	116.9	95.6	107.5	112.6	86.1	100.9	113.0	86.4	101.3	115.0	89.3	103.7	116.5	92.1	105.7	109.9	91.8	101.9

For customer support on your Commercial Renovation Cost Data, call 877.742.4719.

CANADA

DIVISION		LAVAL, QUEBEC			LETHBRIDGE, ALBERTA			LLOYDMINSTER, ALBERTA			LONDON, ONTARIO			MEDICINE HAT, ALBERTA			MONCTON, NEW BRUNSWICK		
		MAT.	INST.	TOTAL	MAT.	INST.	TOTAL	MAT.	INST.	TOTAL	MAT.	INST.	TOTAL	MAT.	INST.	TOTAL	MAT.	INST.	TOTAL
015433	CONTRACTOR EQUIPMENT		102.2	102.2		104.2	104.2		104.2	104.2		104.2	104.2		104.2	104.2		101.6	101.6
0241, 31 - 34	SITE & INFRASTRUCTURE, DEMOLITION	99.7	99.1	99.3	116.6	101.6	106.0	116.4	101.0	105.5	108.2	104.6	105.7	115.2	101.0	105.2	104.3	96.2	98.6
0310	Concrete Forming & Accessories	130.2	85.3	91.4	123.4	89.8	94.4	121.5	80.3	85.9	122.1	86.1	90.9	123.3	80.3	86.1	103.7	64.4	69.7
0320	Concrete Reinforcing	144.6	77.9	110.1	148.7	73.5	109.7	148.7	73.5	109.7	136.3	90.9	112.8	148.7	73.4	109.7	136.4	60.9	97.3
0330	Cast-in-Place Concrete	123.4	96.6	112.8	149.4	104.2	131.6	138.6	100.5	123.6	139.6	97.6	123.0	138.6	100.5	123.6	125.7	79.2	107.3
03	CONCRETE	124.3	88.0	106.7	137.2	91.9	115.2	131.8	86.4	109.7	128.5	91.0	110.3	131.9	86.4	109.8	121.2	69.8	96.2
04	MASONRY	165.7	86.7	117.0	183.5	88.7	125.1	165.0	82.1	113.9	178.4	98.3	129.0	165.0	82.1	113.9	161.1	66.3	102.7
05	METALS	108.7	86.8	101.7	130.7	88.8	117.4	111.0	88.6	103.9	126.6	91.4	115.4	111.0	88.5	103.8	109.9	81.1	100.7
06	WOOD, PLASTICS & COMPOSITES	132.6	85.0	106.1	119.3	89.8	102.3	115.9	79.3	95.5	107.2	83.7	94.1	119.3	79.3	97.0	100.0	63.8	79.8
07	THERMAL & MOISTURE PROTECTION	111.1	87.9	101.4	122.9	90.0	109.1	119.5	85.7	105.3	120.2	91.6	108.3	125.9	85.7	109.1	113.7	65.0	93.3
08	OPENINGS	94.8	74.1	90.0	94.8	81.2	91.7	94.8	76.0	90.5	84.0	84.6	84.1	94.8	76.0	90.5	89.3	60.9	82.7
0920	Plaster & Gypsum Board	116.2	84.4	94.9	107.7	88.0	94.5	103.6	78.3	86.7	138.0	83.2	101.4	105.9	78.3	87.4	121.2	62.6	82.0
0950, 0980	Ceilings & Acoustic Treatment	98.9	84.4	89.3	106.3	88.0	94.2	98.9	78.3	85.2	125.9	83.2	97.6	98.9	78.3	85.2	108.0	62.6	77.9
0960	Flooring	122.5	95.7	114.8	122.5	91.2	113.5	122.5	91.2	113.5	118.7	100.7	113.5	122.5	91.2	113.5	107.8	71.0	97.3
0970, 0990	Wall Finishes & Painting/Coating	112.3	91.7	100.3	112.2	104.3	107.6	112.4	81.2	94.3	114.9	99.4	105.9	112.2	81.2	94.2	113.2	52.9	78.1
09	FINISHES	111.3	88.2	98.6	112.9	91.7	101.1	110.8	82.0	94.8	123.2	89.8	104.7	110.9	82.0	94.9	109.7	64.4	84.7
COVERS	DIVS. 10 - 14, 25, 28, 41, 43, 44, 46	131.1	82.8	120.7	131.1	94.6	123.2	131.1	91.4	122.5	131.1	96.9	123.7	131.1	91.4	122.5	131.1	65.1	116.8
21, 22, 23	FIRE SUPPRESSION, PLUMBING & HVAC	104.7	90.0	98.5	103.9	94.6	100.0	103.9	94.6	100.0	105.6	85.7	97.2	103.6	91.3	98.4	103.6	71.4	90.0
26, 27, 3370	ELECTRICAL, COMMUNICATIONS & UTIL.	107.0	72.1	89.3	104.9	84.8	94.8	102.9	84.8	93.8	100.2	96.8	98.5	102.9	84.8	93.8	113.0	63.3	87.9
MF2014	WEIGHTED AVERAGE	112.8	86.1	101.1	119.1	91.2	106.8	113.9	87.9	102.5	116.3	92.2	105.7	114.1	87.2	102.2	112.0	70.7	93.8

CANADA

DIVISION		MONTREAL, QUEBEC			MOOSE JAW, SASKATCHEWAN			NEW GLASGOW, NOVA SCOTIA			NEWCASTLE, NEW BRUNSWICK			NORTH BAY, ONTARIO			OSHAWA, ONTARIO		
		MAT.	INST.	TOTAL	MAT.	INST.	TOTAL	MAT.	INST.	TOTAL	MAT.	INST.	TOTAL	MAT.	INST.	TOTAL	MAT.	INST.	TOTAL
015433	CONTRACTOR EQUIPMENT		103.8	103.8		100.3	100.3		101.4	101.4		101.6	101.6		101.8	101.8		104.0	104.0
0241, 31 - 34	SITE & INFRASTRUCTURE, DEMOLITION	111.1	98.7	102.4	116.7	95.3	101.6	123.3	97.5	105.0	105.0	95.8	98.5	138.8	99.3	110.9	110.7	103.8	105.8
0310	Concrete Forming & Accessories	118.4	90.8	94.6	106.5	58.7	65.2	113.2	69.5	75.4	103.7	62.0	67.6	147.6	84.5	93.0	118.3	89.2	93.1
0320	Concrete Reinforcing	131.7	94.9	112.6	104.9	61.2	82.2	159.4	46.2	100.8	136.4	55.9	94.7	188.3	80.9	132.6	153.3	86.4	118.7
0330	Cast-in-Place Concrete	150.2	100.1	130.4	134.5	70.2	109.1	147.7	71.3	117.6	130.5	61.1	103.1	142.3	86.1	120.1	145.7	88.7	123.2
03	CONCRETE	132.8	94.7	114.3	116.3	63.9	90.8	160.6	66.6	114.9	123.6	61.3	93.3	162.1	84.7	124.5	132.1	88.7	111.0
04	MASONRY	168.7	94.6	123.1	163.6	64.1	102.3	237.5	71.6	135.3	161.5	64.7	101.8	244.1	89.5	148.8	151.4	96.3	117.4
05	METALS	130.5	92.1	118.3	107.5	74.0	96.9	123.8	74.1	107.9	109.9	71.8	97.8	124.5	88.7	113.1	108.5	92.1	103.3
06	WOOD, PLASTICS & COMPOSITES	102.9	90.7	96.1	101.7	57.1	76.9	117.9	68.6	90.5	100.0	61.8	78.7	153.9	84.7	115.3	115.8	88.0	100.3
07	THERMAL & MOISTURE PROTECTION	119.9	94.7	109.4	110.0	62.8	90.2	143.4	68.5	112.0	113.7	60.9	91.6	150.4	83.7	122.5	112.1	87.9	102.0
08	OPENINGS	89.6	80.3	87.5	90.7	55.1	82.4	95.6	63.0	88.0	89.3	54.3	81.2	103.9	81.3	98.6	91.9	87.8	91.0
0920	Plaster & Gypsum Board	120.3	90.0	100.0	100.5	55.8	70.6	135.7	67.6	90.2	121.2	60.6	80.7	130.3	84.1	99.5	109.7	87.5	94.9
0950, 0980	Ceilings & Acoustic Treatment	130.9	90.0	103.8	98.9	55.8	70.4	108.2	67.6	81.3	108.0	60.6	76.6	108.2	84.1	92.3	98.4	87.5	91.2
0960	Flooring	122.9	98.0	115.7	111.9	60.8	97.2	118.0	64.6	102.7	107.8	69.3	96.8	143.1	93.7	128.9	112.8	103.3	110.0
0970, 0990	Wall Finishes & Painting/Coating	110.8	106.4	108.2	112.3	66.8	85.8	120.1	64.1	87.5	113.2	51.6	77.4	120.1	91.2	103.3	106.2	106.7	106.5
09	FINISHES	120.5	94.4	106.0	107.0	59.6	80.8	118.5	68.2	90.7	109.7	62.6	83.7	125.9	87.3	104.5	106.1	93.5	99.2
COVERS	DIVS. 10 - 14, 25, 28, 41, 43, 44, 46	131.1	86.4	121.4	131.1	62.7	116.3	131.1	64.6	116.7	131.1	63.4	116.5	131.1	70.7	118.0	131.1	97.3	123.8
21, 22, 23	FIRE SUPPRESSION, PLUMBING & HVAC	105.5	93.6	100.4	103.9	75.3	91.8	103.7	83.5	95.1	103.6	69.0	89.0	103.7	97.1	100.9	104.7	102.6	103.8
26, 27, 3370	ELECTRICAL, COMMUNICATIONS & UTIL.	105.7	86.1	95.8	110.9	62.3	86.3	114.1	63.9	88.7	108.7	61.4	84.8	114.4	90.1	102.1	107.4	92.6	99.9
MF2014	WEIGHTED AVERAGE	117.9	92.4	106.7	111.2	68.5	92.4	125.5	73.6	102.6	111.9	67.0	92.1	128.6	89.8	111.5	112.5	95.3	104.9

CANADA

DIVISION		OTTAWA, ONTARIO			OWEN SOUND, ONTARIO			PETERBOROUGH, ONTARIO			PORTAGE LA PRAIRIE, MANITOBA			PRINCE ALBERT, SASKATCHEWAN			PRINCE GEORGE, BRITISH COLUMBIA		
		MAT.	INST.	TOTAL	MAT.	INST.	TOTAL	MAT.	INST.	TOTAL	MAT.	INST.	TOTAL	MAT.	INST.	TOTAL	MAT.	INST.	TOTAL
015433	CONTRACTOR EQUIPMENT		104.0	104.0		101.8	101.8		101.8	101.8		104.1	104.1		100.3	100.3		105.5	105.5
0241, 31 - 34	SITE & INFRASTRUCTURE, DEMOLITION	105.9	104.3	104.8	119.8	99.7	105.6	119.2	99.6	105.4	117.9	98.7	104.4	112.1	95.5	100.4	125.0	102.6	109.2
0310	Concrete Forming & Accessories	122.1	88.9	93.4	125.0	83.1	88.7	124.1	85.6	90.8	123.5	69.9	77.1	106.5	58.5	65.0	111.9	83.1	87.0
0320	Concrete Reinforcing	129.0	90.9	109.3	175.6	82.8	127.5	160.7	81.3	119.6	148.7	52.6	98.9	109.6	61.1	84.5	107.2	75.9	91.0
0330	Cast-in-Place Concrete	133.4	98.9	119.8	163.9	82.1	131.7	150.0	87.7	125.4	138.6	75.2	113.6	121.9	70.1	101.5	134.9	100.5	121.4
03	CONCRETE	124.4	92.7	109.0	150.4	83.1	117.7	139.3	85.8	113.3	125.6	69.3	98.2	110.8	63.8	87.9	143.1	88.0	116.3
04	MASONRY	169.5	98.4	125.7	173.7	95.2	125.3	169.6	96.2	124.4	168.1	66.3	105.3	162.7	64.1	101.9	174.8	94.5	125.3
05	METALS	127.9	91.3	116.3	109.3	89.5	103.0	110.2	89.2	103.5	111.0	76.7	100.1	110.8	87.0	103.2	110.8	87.0	103.2
06	WOOD, PLASTICS & COMPOSITES	108.2	87.6	96.7	118.8	81.5	98.0	121.6	83.4	100.3	119.3	71.0	92.4	101.7	57.1	76.9	104.3	79.4	90.4
07	THERMAL & MOISTURE PROTECTION	126.1	91.7	111.7	113.6	84.5	101.4	117.1	89.2	105.4	110.4	69.8	93.4	109.9	61.7	89.7	120.9	84.2	105.6
08	OPENINGS	94.8	86.6	92.9	94.7	80.7	91.4	93.6	83.0	91.2	94.8	62.9	87.4	89.6	55.1	81.5	91.5	79.4	88.7
0920	Plaster & Gypsum Board	147.1	87.2	107.0	148.8	80.9	103.4	117.6	82.8	94.4	105.4	69.7	81.6	100.5	55.8	70.6	103.5	78.3	86.6
0950, 0980	Ceilings & Acoustic Treatment	129.0	87.2	101.3	92.2	80.9	84.7	98.9	82.8	88.3	98.9	69.7	79.6	98.9	55.8	70.4	98.9	78.3	85.2
0960	Flooring	111.4	96.0	107.0	126.6	95.0	117.5	122.5	93.7	114.2	122.5	67.5	106.7	111.9	60.8	97.2	117.7	73.3	104.9
0970, 0990	Wall Finishes & Painting/Coating	115.3	94.2	103.0	107.4	89.9	97.2	112.3	93.5	101.3	112.4	58.7	81.2	112.3	57.0	80.1	112.3	84.0	95.8
09	FINISHES	124.1	90.7	105.7	114.4	86.1	98.8	112.3	87.9	98.9	112.3	66.8	87.6	107.0	58.6	80.2	110.7	81.0	94.2
COVERS	DIVS. 10 - 14, 25, 28, 41, 43, 44, 46	131.1	95.1	123.3	139.2	71.1	124.5	131.1	72.1	118.4	131.1	65.2	116.9	131.1	62.7	116.3	131.1	92.1	122.7
21, 22, 23	FIRE SUPPRESSION, PLUMBING & HVAC	105.7	86.2	97.5	103.5	97.0	100.8	103.6	100.6	102.3	103.6	82.6	94.7	103.9	67.8	88.7	103.6	93.4	99.3
26, 27, 3370	ELECTRICAL, COMMUNICATIONS & UTIL.	105.1	98.0	101.5	115.9	88.9	102.2	108.9	89.8	99.2	110.6	59.6	84.8	110.9	62.3	86.3	109.0	81.3	95.0
MF2014	WEIGHTED AVERAGE	117.6	92.9	106.7	118.4	89.9	105.8	115.5	91.7	105.0	113.9	73.1	95.9	110.3	66.8	91.1	116.0	88.6	104.0

For customer support on your Commercial Renovation Cost Data, call 877.742.4719.

801

CANADA

DIVISION		QUEBEC CITY, QUEBEC			RED DEER, ALBERTA			REGINA, SASKATCHEWAN			RIMOUSKI, QUEBEC			ROUYN-NORANDA, QUEBEC			SAINT-HYACINTHE, QUEBEC		
		MAT.	INST.	TOTAL	MAT.	INST.	TOTAL	MAT.	INST.	TOTAL	MAT.	INST.	TOTAL	MAT.	INST.	TOTAL	MAT.	INST.	TOTAL
015433	CONTRACTOR EQUIPMENT		104.2	104.2		104.2	104.2		100.3	100.3		102.2	102.2		102.2	102.2		102.2	102.2
0241, 31 - 34	SITE & INFRASTRUCTURE, DEMOLITION	113.3	98.8	103.0	115.2	101.0	105.2	121.5	97.5	104.5	99.7	98.4	98.8	99.2	99.1	99.1	99.7	99.1	99.3
0310	Concrete Forming & Accessories	120.1	91.1	95.0	138.6	80.3	88.2	118.3	88.7	92.7	130.0	90.5	95.9	130.0	85.3	91.3	130.0	85.3	91.3
0320	Concrete Reinforcing	138.1	94.9	115.7	148.7	73.4	109.7	133.4	83.9	107.8	103.1	94.8	98.8	144.6	77.9	110.1	144.6	77.9	110.1
0330	Cast-in-Place Concrete	138.0	100.5	123.2	138.6	100.5	123.6	150.3	94.9	128.5	125.6	98.8	115.0	119.4	96.6	110.4	123.4	96.6	112.8
03	CONCRETE	127.9	95.0	111.9	132.9	86.4	110.3	138.9	89.9	115.1	119.3	94.1	107.1	122.3	88.0	105.7	124.3	88.0	106.7
04	MASONRY	166.4	94.7	122.2	165.0	82.1	113.9	183.8	96.2	129.6	165.2	94.6	121.7	165.4	86.7	116.9	165.6	86.7	117.0
05	METALS	130.5	92.4	118.4	111.0	88.5	103.8	140.1	86.6	123.1	108.1	91.6	102.9	108.6	86.8	101.7	108.6	86.8	101.7
06	WOOD, PLASTICS & COMPOSITES	108.1	90.8	98.5	119.3	79.3	97.0	104.7	88.8	95.9	132.5	90.5	109.1	132.5	85.0	106.0	132.5	85.0	106.0
07	THERMAL & MOISTURE PROTECTION	120.5	94.9	109.8	135.9	85.7	114.9	126.1	84.1	108.5	110.6	94.2	103.7	110.6	87.9	101.1	110.9	87.9	101.3
08	OPENINGS	92.6	88.0	91.5	94.8	76.0	90.5	91.3	78.7	88.3	94.4	78.2	90.6	94.8	74.1	90.0	94.8	74.1	90.0
0920	Plaster & Gypsum Board	120.7	90.0	100.2	105.9	78.3	87.4	132.2	88.4	103.0	141.8	90.0	107.2	115.7	84.4	94.8	115.7	84.4	94.8
0950, 0980	Ceilings & Acoustic Treatment	132.6	90.0	104.4	98.9	78.3	85.2	137.7	88.4	105.1	98.1	90.0	92.7	98.1	84.4	89.0	98.1	84.4	89.0
0960	Flooring	122.7	98.0	115.6	124.9	91.2	115.2	121.9	104.9	117.0	122.5	95.7	114.8	122.5	95.7	114.8	122.5	95.7	114.8
0970, 0990	Wall Finishes & Painting/Coating	126.8	106.4	114.9	112.2	81.2	94.2	118.6	92.9	103.6	112.3	106.4	108.8	112.3	91.7	100.3	112.3	91.7	100.3
09	FINISHES	122.3	94.4	106.9	111.7	82.0	95.2	130.4	93.0	109.7	114.9	93.8	103.2	111.1	88.2	98.4	111.1	88.2	98.4
COVERS	DIVS. 10 - 14, 25, 28, 41, 43, 44, 46	131.1	86.5	121.5	131.1	91.4	122.5	131.1	71.6	118.2	131.1	85.8	121.3	131.1	82.8	120.7	131.1	82.8	120.7
21, 22, 23	FIRE SUPPRESSION, PLUMBING & HVAC	105.3	93.6	100.4	103.6	91.3	98.4	105.0	94.0	96.1	103.6	93.5	99.3	103.6	90.0	97.8	100.3	90.0	95.9
26, 27, 3370	ELECTRICAL, COMMUNICATIONS & UTIL.	106.1	86.1	96.0	102.9	84.8	93.8	112.7	91.7	102.1	108.9	86.1	97.4	108.9	72.1	90.3	109.5	72.1	90.6
MF2014	WEIGHTED AVERAGE	117.9	92.8	106.8	114.6	87.2	102.5	123.0	89.0	108.0	112.3	92.0	103.4	112.5	86.1	100.9	112.0	86.1	100.6

CANADA

DIVISION		SAINT JOHN, NEW BRUNSWICK			SARNIA, ONTARIO			SASKATOON, SASKATCHEWAN			SAULT STE. MARIE, ONTARIO			SHERBROOKE, QUEBEC			SOREL, QUEBEC		
		MAT.	INST.	TOTAL	MAT.	INST.	TOTAL	MAT.	INST.	TOTAL	MAT.	INST.	TOTAL	MAT.	INST.	TOTAL	MAT.	INST.	TOTAL
015433	CONTRACTOR EQUIPMENT		101.6	101.6		101.8	101.8		100.3	100.3		101.8	101.8		102.2	102.2		102.2	102.2
0241, 31 - 34	SITE & INFRASTRUCTURE, DEMOLITION	105.1	97.3	99.6	117.7	99.7	105.0	112.8	97.5	102.0	107.5	99.3	101.7	99.7	99.1	99.3	99.9	99.1	99.3
0310	Concrete Forming & Accessories	123.9	65.2	73.1	122.8	92.5	96.6	106.3	88.8	91.1	112.1	85.4	89.0	130.0	85.3	91.3	130.0	85.4	91.4
0320	Concrete Reinforcing	136.4	61.3	97.5	114.3	82.6	97.9	110.8	83.9	96.9	103.2	81.2	91.8	144.6	77.9	110.1	136.7	77.9	106.3
0330	Cast-in-Place Concrete	128.6	80.2	109.5	138.4	101.0	123.7	130.6	94.9	116.5	124.4	86.3	109.3	123.4	96.6	112.8	124.4	96.7	113.5
03	CONCRETE	124.1	70.6	98.1	126.7	93.6	110.6	117.3	90.0	104.0	111.4	85.2	98.7	124.3	88.0	106.7	123.6	88.1	106.3
04	MASONRY	182.3	73.4	115.2	181.0	99.0	130.5	171.8	96.2	125.2	166.2	94.0	121.7	165.7	86.7	117.0	165.8	86.7	117.0
05	METALS	109.8	82.2	101.0	110.1	89.5	103.6	106.9	86.6	100.4	109.4	89.7	103.1	108.6	86.8	101.7	108.6	86.9	101.7
06	WOOD, PLASTICS & COMPOSITES	122.2	63.1	89.3	120.5	91.5	104.3	99.4	88.8	93.5	108.3	85.3	95.5	132.5	85.0	106.0	132.5	85.0	106.0
07	THERMAL & MOISTURE PROTECTION	114.0	69.4	95.3	117.2	92.6	106.9	110.1	84.5	99.3	116.0	86.5	103.6	110.6	87.9	101.1	110.6	87.9	101.1
08	OPENINGS	89.3	59.5	82.3	96.0	86.8	93.8	90.2	78.7	87.5	87.8	83.1	86.7	94.8	74.1	90.0	94.8	78.7	91.1
0920	Plaster & Gypsum Board	133.9	62.0	85.9	136.7	91.2	106.3	115.0	88.4	97.2	108.8	84.8	92.8	115.7	84.4	94.8	141.8	84.4	103.4
0950, 0980	Ceilings & Acoustic Treatment	112.9	62.0	79.2	103.0	91.2	95.2	119.5	88.4	98.9	98.9	84.8	89.6	98.1	84.4	89.0	98.1	84.4	89.0
0960	Flooring	120.3	71.0	106.2	122.5	102.4	116.7	114.4	104.9	111.7	114.9	99.3	110.4	122.5	95.7	114.8	122.5	95.7	114.8
0970, 0990	Wall Finishes & Painting/Coating	113.2	80.2	94.0	112.3	105.7	108.4	115.5	92.9	102.3	112.3	97.9	103.9	112.3	91.7	100.3	112.3	91.7	100.3
09	FINISHES	116.4	67.3	89.3	116.0	96.1	105.0	115.4	93.0	103.0	108.0	89.4	97.7	111.1	88.2	98.4	114.9	88.2	100.1
COVERS	DIVS. 10 - 14, 25, 28, 41, 43, 44, 46	131.1	65.6	116.9	131.1	73.5	118.7	131.1	72.0	118.3	131.1	95.8	123.5	131.1	82.8	120.7	131.1	82.8	120.7
21, 22, 23	FIRE SUPPRESSION, PLUMBING & HVAC	103.6	80.9	94.0	103.6	106.6	104.9	104.9	84.0	96.0	103.6	94.6	99.8	103.9	90.0	98.1	103.6	90.0	97.9
26, 27, 3370	ELECTRICAL, COMMUNICATIONS & UTIL.	115.6	90.2	102.7	111.4	92.0	101.6	110.1	91.7	100.8	110.2	90.1	100.1	108.9	72.1	90.3	108.9	72.1	90.3
MF2014	WEIGHTED AVERAGE	114.4	77.8	98.3	115.3	96.1	106.9	112.3	89.0	102.0	110.7	91.1	102.1	112.8	86.1	101.1	113.0	86.4	101.2

CANADA

DIVISION		ST. CATHARINES, ONTARIO			SAINT-JEROME, QUEBEC			ST. JOHN'S, NEWFOUNDLAND			SUDBURY, ONTARIO			SUMMERSIDE, PRINCE EDWARD ISLAND			SYDNEY, NOVA SCOTIA		
		MAT.	INST.	TOTAL	MAT.	INST.	TOTAL	MAT.	INST.	TOTAL	MAT.	INST.	TOTAL	MAT.	INST.	TOTAL	MAT.	INST.	TOTAL
015433	CONTRACTOR EQUIPMENT		101.8	101.8		102.2	102.2		102.5	102.5		101.8	101.8		101.3	101.3		101.4	101.4
0241, 31 - 34	SITE & INFRASTRUCTURE, DEMOLITION	99.7	100.9	100.6	99.2	99.1	99.1	122.6	99.0	105.9	99.8	100.5	100.3	134.1	95.0	106.5	118.8	97.5	103.7
0310	Concrete Forming & Accessories	110.9	91.6	94.2	130.0	85.3	91.3	125.6	82.3	88.1	106.7	86.5	89.3	113.5	56.6	64.2	113.2	69.5	75.4
0320	Concrete Reinforcing	97.6	90.9	94.2	144.6	77.9	110.1	159.3	74.6	115.4	98.4	90.3	94.2	157.2	46.2	99.7	159.4	46.2	100.8
0330	Cast-in-Place Concrete	120.4	99.6	112.2	119.4	96.6	110.4	158.3	95.2	133.4	121.4	96.0	111.4	140.2	60.3	108.7	114.0	71.3	97.1
03	CONCRETE	105.6	94.2	100.1	122.3	88.0	105.7	147.2	85.5	117.2	105.9	90.6	98.5	170.1	56.8	115.1	144.0	66.6	106.4
04	MASONRY	148.0	100.3	118.6	165.4	86.7	116.9	189.1	92.0	129.3	148.0	96.1	116.0	236.7	62.3	129.2	234.8	71.6	134.2
05	METALS	107.6	91.8	102.6	108.6	86.8	101.7	138.5	83.7	121.1	107.6	91.2	102.4	123.6	67.8	105.9	123.6	74.1	107.9
06	WOOD, PLASTICS & COMPOSITES	106.9	90.7	97.9	132.5	85.0	106.0	109.2	80.4	92.9	103.1	85.3	93.2	118.2	55.8	83.5	117.9	68.6	90.5
07	THERMAL & MOISTURE PROTECTION	111.2	94.3	104.2	110.6	87.9	101.1	137.4	89.4	117.3	110.7	89.6	101.9	142.6	59.9	108.0	143.4	68.5	112.0
08	OPENINGS	85.9	88.1	86.4	94.8	74.1	90.0	90.6	72.4	86.4	86.6	83.5	85.9	107.9	48.5	94.0	95.6	63.0	88.0
0920	Plaster & Gypsum Board	97.9	90.4	92.9	115.7	84.4	94.8	133.7	79.3	97.3	99.8	84.8	89.8	136.0	54.4	81.5	135.7	67.6	90.2
0950, 0980	Ceilings & Acoustic Treatment	98.4	90.4	93.1	98.1	84.4	89.0	119.4	79.3	92.8	93.5	84.8	87.7	108.2	54.4	72.6	108.2	67.6	81.3
0960	Flooring	108.4	97.3	105.2	122.5	95.7	114.8	119.8	57.0	101.7	106.4	99.3	104.3	118.1	60.5	101.5	118.0	64.6	102.7
0970, 0990	Wall Finishes & Painting/Coating	106.2	116.0	111.9	112.3	91.7	100.3	117.5	95.9	104.9	106.2	93.6	98.9	120.1	43.0	75.2	120.1	64.1	87.5
09	FINISHES	102.5	95.4	98.6	111.1	88.2	98.4	125.8	79.0	99.9	101.1	89.4	94.6	119.6	56.1	84.5	118.5	68.2	90.7
COVERS	DIVS. 10 - 14, 25, 28, 41, 43, 44, 46	131.1	74.4	118.8	131.1	82.8	120.7	131.1	70.4	118.0	131.1	96.6	123.7	131.1	62.9	116.4	131.1	64.6	116.7
21, 22, 23	FIRE SUPPRESSION, PLUMBING & HVAC	104.7	86.9	97.2	103.6	90.0	97.8	105.2	81.2	95.0	103.5	87.6	96.8	103.7	63.1	86.5	103.7	83.5	95.1
26, 27, 3370	ELECTRICAL, COMMUNICATIONS & UTIL.	107.7	98.1	102.8	109.5	72.1	90.6	111.9	82.5	97.0	108.2	98.4	103.2	113.3	51.9	82.2	114.1	63.9	88.7
MF2014	WEIGHTED AVERAGE	107.8	93.3	101.5	112.5	86.1	100.9	123.9	83.9	106.3	107.6	92.1	100.8	128.1	62.0	99.0	123.4	73.6	101.4

City Cost Indexes

CANADA

DIVISION		THUNDER BAY, ONTARIO			TIMMINS, ONTARIO			TORONTO, ONTARIO			TROIS-RIVIERES, QUEBEC			TRURO, NOVA SCOTIA			VANCOUVER, BRITISH COLUMBIA		
		MAT.	INST.	TOTAL	MAT.	INST.	TOTAL	MAT.	INST.	TOTAL	MAT.	INST.	TOTAL	MAT.	INST.	TOTAL	MAT.	INST.	TOTAL
015433	CONTRACTOR EQUIPMENT		101.8	101.8		101.8	101.8		104.2	104.2		102.2	102.2		101.4	101.4		112.6	112.6
0241, 31 - 34	SITE & INFRASTRUCTURE, DEMOLITION	104.6	100.8	101.9	119.2	99.3	105.1	128.3	105.3	112.1	119.2	99.1	105.0	104.2	97.5	99.4	114.4	107.3	109.4
0310	Concrete Forming & Accessories	118.3	89.6	93.5	124.1	84.5	89.8	119.1	99.2	101.9	155.3	85.4	94.9	96.7	69.5	73.2	120.1	88.9	93.1
0320	Concrete Reinforcing	87.3	89.8	88.6	160.7	80.9	119.4	137.5	95.7	115.9	159.4	77.9	117.2	136.7	46.2	89.8	124.2	81.1	101.9
0330	Cast-in-Place Concrete	132.5	98.7	119.1	150.0	86.1	124.8	124.2	111.3	119.1	118.0	96.7	109.6	155.8	71.3	122.4	130.7	98.7	118.1
03	CONCRETE	113.4	92.8	103.4	139.3	84.7	112.7	120.9	102.5	112.0	146.5	88.1	118.1	138.2	66.6	103.4	129.2	91.1	110.7
04	MASONRY	148.6	100.1	118.8	169.6	89.5	120.2	177.0	113.3	137.7	239.2	86.7	145.2	165.2	71.6	107.5	166.0	93.2	121.1
05	METALS	107.4	90.8	102.1	110.2	88.7	103.4	128.5	94.4	117.7	122.8	86.9	111.4	109.3	74.2	98.2	130.8	91.3	118.2
06	WOOD, PLASTICS & COMPOSITES	115.8	87.8	100.2	121.6	84.7	101.0	104.2	96.7	100.0	168.0	85.0	121.8	92.3	68.6	79.1	103.9	88.5	95.3
07	THERMAL & MOISTURE PROTECTION	111.5	91.8	103.3	117.1	83.7	103.2	127.9	101.6	116.9	141.6	87.9	119.2	111.8	68.5	93.7	127.7	86.7	110.5
08	OPENINGS	85.1	85.9	85.3	93.6	81.3	90.8	88.8	95.7	90.4	105.5	78.7	99.2	87.5	63.0	81.8	87.3	85.3	86.8
0920	Plaster & Gypsum Board	123.8	87.4	99.5	117.6	84.1	95.3	122.9	96.5	105.2	162.7	84.4	110.4	119.5	67.6	84.8	112.8	87.5	95.9
0950, 0980	Ceilings & Acoustic Treatment	93.5	87.4	89.4	98.9	84.1	89.1	120.6	96.5	104.6	107.3	84.4	92.1	98.9	67.6	78.2	129.6	87.5	101.7
0960	Flooring	112.8	104.3	110.3	122.5	93.7	114.2	114.5	106.7	112.3	143.1	95.7	129.5	103.5	64.6	92.3	129.8	94.5	119.6
0970, 0990	Wall Finishes & Painting/Coating	106.2	94.6	99.4	112.3	91.2	100.0	107.3	106.7	107.0	120.1	91.7	103.6	112.3	64.1	84.2	118.4	98.0	106.6
09	FINISHES	106.8	92.8	99.1	112.3	87.3	98.4	117.1	101.4	108.4	129.6	88.2	106.7	106.3	68.2	85.2	124.4	90.9	105.9
COVERS	DIVS. 10 - 14, 25, 28, 41, 43, 44, 46	131.1	74.4	118.9	131.1	70.7	118.0	131.1	101.0	124.6	131.1	82.8	120.7	131.1	64.6	116.7	131.1	94.5	123.2
21, 22, 23	FIRE SUPPRESSION, PLUMBING & HVAC	104.7	87.0	97.2	103.6	97.1	100.8	105.3	96.8	101.7	103.7	90.0	97.9	103.6	83.5	95.1	105.5	82.0	95.6
26, 27, 3370	ELECTRICAL, COMMUNICATIONS & UTIL.	106.3	96.3	101.2	110.2	90.1	100.1	105.1	100.3	102.7	114.4	72.1	93.0	108.7	63.9	86.0	103.8	82.6	93.1
MF2014	WEIGHTED AVERAGE	109.1	92.3	101.7	115.6	89.8	104.2	116.9	100.9	109.9	126.1	86.4	108.6	113.0	73.6	95.6	117.7	89.2	105.1

CANADA

DIVISION		VICTORIA, BRITISH COLUMBIA			WHITEHORSE, YUKON			WINDSOR, ONTARIO			WINNIPEG, MANITOBA			YARMOUTH, NOVA SCOTIA			YELLOWKNIFE, NWT		
		MAT.	INST.	TOTAL	MAT.	INST.	TOTAL	MAT.	INST.	TOTAL	MAT.	INST.	TOTAL	MAT.	INST.	TOTAL	MAT.	INST.	TOTAL
015433	CONTRACTOR EQUIPMENT		108.7	108.7		101.6	101.6		101.8	101.8		106.2	106.2		101.4	101.4		101.4	101.4
0241, 31 - 34	SITE & INFRASTRUCTURE, DEMOLITION	123.8	106.7	111.7	142.7	96.0	109.7	95.8	100.8	99.4	114.7	100.4	104.6	123.1	97.5	105.0	159.6	100.3	117.7
0310	Concrete Forming & Accessories	111.3	88.2	91.3	131.5	57.8	67.7	118.3	88.0	92.1	126.3	67.3	75.2	113.2	69.5	75.4	135.0	77.9	85.6
0320	Concrete Reinforcing	108.4	81.0	94.2	156.7	59.3	106.3	95.7	90.8	93.2	132.4	57.2	93.5	159.4	46.2	100.8	141.6	61.9	100.4
0330	Cast-in-Place Concrete	135.3	98.0	120.6	186.2	69.4	140.1	123.3	100.9	114.5	146.8	74.1	118.2	146.2	71.3	116.6	197.0	87.1	153.6
03	CONCRETE	147.5	90.3	119.7	173.7	62.9	119.8	107.3	93.0	100.3	137.6	68.4	103.9	159.8	66.6	114.5	177.0	78.5	129.1
04	MASONRY	174.8	93.2	124.5	273.0	62.5	143.3	148.1	99.5	118.1	177.0	70.5	111.4	237.4	71.6	135.2	269.2	74.5	149.2
05	METALS	109.1	86.7	102.0	140.1	73.6	118.9	107.5	91.3	102.3	149.0	75.1	125.5	123.6	74.1	107.9	138.0	78.5	119.1
06	WOOD, PLASTICS & COMPOSITES	103.3	88.4	95.0	120.8	56.3	84.9	115.8	86.0	99.2	112.1	67.8	87.4	117.9	68.6	90.5	125.9	79.1	99.9
07	THERMAL & MOISTURE PROTECTION	120.1	85.8	105.8	145.5	60.8	110.1	111.3	91.8	103.1	125.9	70.4	102.7	143.4	68.5	112.0	155.9	76.7	122.8
08	OPENINGS	91.7	81.2	89.3	110.5	53.9	97.3	84.9	85.4	85.0	85.1	60.7	79.5	95.6	63.0	88.0	102.3	66.9	94.1
0920	Plaster & Gypsum Board	108.3	87.5	94.4	154.7	54.9	88.0	110.7	85.6	93.9	119.2	66.3	83.8	135.7	67.6	90.2	155.3	78.4	103.9
0950, 0980	Ceilings & Acoustic Treatment	101.3	87.5	92.2	152.3	54.9	87.8	93.5	85.6	88.2	139.9	66.3	91.2	108.2	67.6	81.3	138.8	78.4	98.8
0960	Flooring	120.2	73.3	106.7	140.7	59.0	117.2	112.8	101.5	109.5	119.4	72.7	105.9	118.0	64.6	102.7	138.5	89.0	124.2
0970, 0990	Wall Finishes & Painting/Coating	115.5	95.6	103.9	130.7	56.0	87.3	106.2	95.5	100.0	117.3	56.5	81.9	120.1	64.1	87.5	128.8	80.3	100.6
09	FINISHES	113.7	87.0	99.0	147.7	57.6	97.8	104.6	91.2	97.2	126.0	67.5	93.7	118.5	68.2	90.7	142.2	79.6	107.6
COVERS	DIVS. 10 - 14, 25, 28, 41, 43, 44, 46	131.1	69.7	117.8	131.1	62.0	116.2	131.1	73.9	118.7	131.1	66.8	117.2	131.1	64.6	116.7	131.1	65.1	116.8
21, 22, 23	FIRE SUPPRESSION, PLUMBING & HVAC	103.6	80.1	93.7	104.2	74.0	91.4	104.7	87.0	97.2	105.4	66.2	88.8	103.7	83.5	95.1	104.3	92.9	99.5
26, 27, 3370	ELECTRICAL, COMMUNICATIONS & UTIL.	110.3	82.1	96.0	124.4	61.8	92.8	110.4	98.2	104.3	109.9	67.5	88.5	114.1	63.9	88.7	126.9	84.6	105.5
MF2014	WEIGHTED AVERAGE	116.6	86.7	103.4	137.1	67.5	106.4	108.3	92.3	101.3	122.6	70.8	99.8	125.4	73.6	102.6	136.7	82.9	113.0

Costs shown in RSMeans cost data publications are based on national averages for materials and installation. To adjust these costs to a specific location, simply multiply the base cost by the factor and divide by 100 for that city. The data is arranged alphabetically by state and postal zip code numbers. For a city not listed, use the factor for a nearby city with similar economic characteristics.

STATE/ZIP	CITY	MAT.	INST.	TOTAL
ALABAMA				
350-352	Birmingham	99.3	77.3	89.6
354	Tuscaloosa	98.7	78.1	89.6
355	Jasper	99.2	71.8	87.1
356	Decatur	98.7	76.0	88.7
357-358	Huntsville	98.7	75.3	88.4
359	Gadsden	98.8	67.1	84.8
360-361	Montgomery	97.9	77.0	88.7
362	Anniston	97.3	77.5	88.6
363	Dothan	97.7	76.1	88.2
364	Evergreen	97.3	73.6	86.9
365-366	Mobile	98.3	76.6	88.7
367	Selma	97.4	74.9	87.5
368	Phenix City	98.2	77.2	89.0
369	Butler	97.6	76.7	88.4
ALASKA				
995-996	Anchorage	121.7	115.8	119.1
997	Fairbanks	124.2	116.9	120.9
998	Juneau	123.9	115.8	120.3
999	Ketchikan	134.9	115.8	126.5
ARIZONA				
850,853	Phoenix	100.5	73.0	88.4
851,852	Mesa/Tempe	100.1	72.0	87.7
855	Globe	100.9	69.8	87.2
856-857	Tucson	98.9	70.8	86.5
859	Show Low	101.1	71.9	88.3
860	Flagstaff	102.7	71.9	89.2
863	Prescott	100.4	68.4	86.3
864	Kingman	98.9	71.3	86.7
865	Chambers	98.9	73.5	87.7
ARKANSAS				
716	Pine Bluff	99.6	65.2	84.5
717	Camden	97.6	56.0	79.3
718	Texarkana	98.5	58.7	80.9
719	Hot Springs	96.9	56.9	79.3
720-722	Little Rock	98.2	67.4	84.6
723	West Memphis	97.0	65.3	83.0
724	Jonesboro	97.4	62.6	82.1
725	Batesville	95.3	58.3	79.0
726	Harrison	96.6	57.6	79.4
727	Fayetteville	94.1	55.6	77.1
728	Russellville	95.3	59.7	79.7
729	Fort Smith	97.8	62.3	82.2
CALIFORNIA				
900-902	Los Angeles	100.2	117.2	107.7
903-905	Inglewood	96.0	115.0	104.4
906-908	Long Beach	97.5	115.0	105.2
910-912	Pasadena	97.5	114.8	105.1
913-916	Van Nuys	100.4	114.8	106.7
917-918	Alhambra	99.5	114.8	106.2
919-921	San Diego	101.4	109.8	105.1
922	Palm Springs	98.6	112.9	104.9
923-924	San Bernardino	96.4	112.3	103.4
925	Riverside	100.5	113.9	106.4
926-927	Santa Ana	98.3	112.0	104.7
928	Anaheim	100.6	114.0	106.5
930	Oxnard	99.9	113.5	105.9
931	Santa Barbara	99.2	114.3	105.9
932-933	Bakersfield	100.4	111.4	105.2
934	San Luis Obispo	100.2	113.0	105.9
935	Mojave	97.4	111.9	103.8
936-938	Fresno	100.5	116.5	107.5
939	Salinas	101.0	122.3	110.4
940-941	San Francisco	104.1	146.0	122.5
942,956-958	Sacramento	99.6	120.1	108.6
943	Palo Alto	98.1	133.4	113.7
944	San Mateo	100.5	134.3	115.4
945	Vallejo	99.1	128.9	112.2
946	Oakland	102.1	134.5	116.4
947	Berkeley	102.0	135.3	116.7
948	Richmond	101.3	132.2	114.9
949	San Rafael	103.2	135.4	117.4
950	Santa Cruz	105.9	122.5	113.2

STATE/ZIP	CITY	MAT.	INST.	TOTAL
CALIFORNIA (CONT'D)				
951	San Jose	103.9	136.9	118.4
952	Stockton	102.1	117.9	109.0
953	Modesto	102.0	116.6	108.4
954	Santa Rosa	102.6	136.3	117.4
955	Eureka	103.9	118.3	110.2
959	Marysville	103.1	115.9	108.8
960	Redding	112.3	117.8	114.7
961	Susanville	112.0	117.2	114.3
COLORADO				
800-802	Denver	101.6	80.4	92.3
803	Boulder	99.1	80.3	90.8
804	Golden	101.2	77.8	90.9
805	Fort Collins	102.4	77.6	91.5
806	Greeley	99.7	74.0	88.4
807	Fort Morgan	99.7	78.0	90.1
808-809	Colorado Springs	101.3	78.7	91.3
810	Pueblo	102.4	74.7	90.2
811	Alamosa	104.3	73.1	90.5
812	Salida	103.9	75.4	91.3
813	Durango	104.7	74.9	91.6
814	Montrose	103.4	76.5	91.6
815	Grand Junction	106.5	75.6	92.9
816	Glenwood Springs	104.5	76.4	92.1
CONNECTICUT				
060	New Britain	98.3	122.1	108.8
061	Hartford	99.6	122.2	109.6
062	Willimantic	98.9	122.3	109.2
063	New London	95.6	122.4	107.4
064	Meriden	97.5	122.2	108.4
065	New Haven	99.8	122.3	109.7
066	Bridgeport	99.5	122.0	109.4
067	Waterbury	99.0	121.9	109.1
068	Norwalk	99.0	129.7	112.5
069	Stamford	99.1	129.7	112.6
D.C.				
200-205	Washington	100.6	95.9	98.5
DELAWARE				
197	Newark	98.9	109.4	103.5
198	Wilmington	98.7	109.4	103.4
199	Dover	99.0	109.4	103.6
FLORIDA				
320,322	Jacksonville	97.3	69.6	85.1
321	Daytona Beach	97.5	75.5	87.8
323	Tallahassee	98.9	65.0	84.0
324	Panama City	98.8	62.0	82.6
325	Pensacola	101.4	71.0	88.0
326,344	Gainesville	99.0	68.8	85.7
327-328,347	Orlando	99.5	72.0	87.4
329	Melbourne	100.1	76.6	89.8
330-332,340	Miami	98.9	72.3	87.2
333	Fort Lauderdale	97.8	71.2	86.1
334,349	West Palm Beach	96.9	69.9	85.0
335-336,346	Tampa	99.3	72.8	87.6
337	St. Petersburg	101.8	70.2	87.9
338	Lakeland	98.8	72.7	87.3
339,341	Fort Myers	98.1	68.8	85.1
342	Sarasota	100.0	73.8	88.5
GEORGIA				
300-303,399	Atlanta	98.1	76.5	88.6
304	Statesboro	98.2	61.0	81.8
305	Gainesville	96.9	63.6	82.2
306	Athens	96.3	65.2	82.6
307	Dalton	98.0	63.0	82.6
308-309	Augusta	96.8	69.9	85.0
310-312	Macon	95.8	66.2	82.7
313-314	Savannah	97.8	67.6	84.5
315	Waycross	97.1	64.4	82.7
316	Valdosta	97.0	62.3	81.7
317,398	Albany	96.9	63.1	82.0
318-319	Columbus	96.8	66.1	83.3

Location Factors - Commercial

STATE/ZIP	CITY	MAT.	INST.	TOTAL
HAWAII				
967	Hilo	120.2	119.7	120.0
968	Honolulu	125.0	119.7	122.7
STATES & POSS.				
969	Guam	143.0	57.6	105.4
IDAHO				
832	Pocatello	102.5	77.2	91.3
833	Twin Falls	103.8	66.3	87.3
834	Idaho Falls	101.1	74.3	89.3
835	Lewiston	109.4	85.0	98.6
836-837	Boise	101.5	77.8	91.1
838	Coeur d'Alene	109.3	82.2	97.4
ILLINOIS				
600-603	North Suburban	99.4	136.8	115.9
604	Joliet	99.4	139.0	116.8
605	South Suburban	99.4	137.0	115.9
606-608	Chicago	99.9	140.4	117.8
609	Kankakee	96.0	131.6	111.7
610-611	Rockford	99.3	123.5	109.9
612	Rock Island	97.4	101.6	99.2
613	La Salle	98.6	123.4	109.5
614	Galesburg	98.4	107.5	102.4
615-616	Peoria	100.3	109.8	104.5
617	Bloomington	97.7	110.0	103.1
618-619	Champaign	101.2	108.4	104.4
620-622	East St. Louis	95.9	108.4	101.4
623	Quincy	98.1	101.9	99.8
624	Effingham	97.3	106.6	101.4
625	Decatur	98.8	105.8	101.9
626-627	Springfield	99.3	106.9	102.6
628	Centralia	94.9	108.3	100.8
629	Carbondale	94.6	106.7	99.9
INDIANA				
460	Anderson	97.1	83.6	91.1
461-462	Indianapolis	99.0	85.2	93.0
463-464	Gary	98.4	110.0	103.5
465-466	South Bend	97.8	84.1	91.8
467-468	Fort Wayne	97.6	78.6	89.3
469	Kokomo	95.3	81.5	89.2
470	Lawrenceburg	93.8	78.2	86.9
471	New Albany	94.9	74.1	85.7
472	Columbus	97.1	81.3	90.1
473	Muncie	97.3	83.4	91.2
474	Bloomington	98.8	81.2	91.0
475	Washington	95.7	82.6	89.9
476-477	Evansville	96.6	86.6	92.2
478	Terre Haute	97.4	86.6	92.6
479	Lafayette	96.8	81.9	90.2
IOWA				
500-503,509	Des Moines	98.4	83.9	92.0
504	Mason City	97.0	64.4	82.6
505	Fort Dodge	97.2	57.6	79.8
506-507	Waterloo	98.4	73.7	87.5
508	Creston	97.5	78.3	89.1
510-511	Sioux City	99.6	72.3	87.6
512	Sibley	98.7	56.0	79.9
513	Spencer	100.5	56.0	80.9
514	Carroll	97.4	71.0	85.8
515	Council Bluffs	100.6	79.0	91.1
516	Shenandoah	97.8	73.5	87.1
520	Dubuque	99.0	80.6	90.9
521	Decorah	98.5	66.6	84.4
522-524	Cedar Rapids	99.9	82.6	92.3
525	Ottumwa	98.3	74.7	87.9
526	Burlington	97.5	76.9	88.4
527-528	Davenport	99.0	93.6	96.6
KANSAS				
660-662	Kansas City	99.6	95.9	98.0
664-666	Topeka	99.8	69.8	86.6
667	Fort Scott	98.3	73.7	87.5
668	Emporia	98.3	70.1	85.9
669	Belleville	100.1	66.5	85.3
670-672	Wichita	101.0	66.1	85.6
673	Independence	101.6	70.2	87.8
674	Salina	101.8	68.9	87.3
675	Hutchinson	97.1	65.2	83.1
676	Hays	101.2	66.9	86.1
677	Colby	101.8	68.0	86.9

STATE/ZIP	CITY	MAT.	INST.	TOTAL
KANSAS (CONT'D)				
678	Dodge City	103.0	66.1	86.8
679	Liberal	101.1	65.6	85.5
KENTUCKY				
400-402	Louisville	95.1	84.3	90.4
403-405	Lexington	95.5	82.2	89.7
406	Frankfort	96.1	80.2	89.1
407-409	Corbin	93.3	71.9	83.9
410	Covington	95.8	91.3	93.8
411-412	Ashland	94.5	99.8	96.8
413-414	Campton	95.6	77.6	87.7
415-416	Pikeville	96.8	89.3	93.5
417-418	Hazard	95.1	78.6	87.8
420	Paducah	93.7	85.5	90.1
421-422	Bowling Green	95.7	84.1	90.6
423	Owensboro	95.8	88.0	92.4
424	Henderson	93.5	86.3	90.3
425-426	Somerset	93.0	81.7	88.0
427	Elizabethtown	92.5	82.2	87.9
LOUISIANA				
700-701	New Orleans	99.2	70.7	86.6
703	Thibodaux	97.6	65.6	83.5
704	Hammond	95.2	61.0	80.1
705	Lafayette	97.0	66.6	83.6
706	Lake Charles	97.2	69.9	85.2
707-708	Baton Rouge	99.5	69.8	86.4
710-711	Shreveport	100.9	60.7	83.2
712	Monroe	99.6	59.3	81.8
713-714	Alexandria	99.7	63.3	83.7
MAINE				
039	Kittery	94.5	87.4	91.4
040-041	Portland	101.6	89.4	96.3
042	Lewiston	98.5	89.4	94.5
043	Augusta	101.2	81.0	92.3
044	Bangor	97.9	88.8	93.9
045	Bath	96.9	83.3	90.9
046	Machias	96.3	86.9	92.2
047	Houlton	96.5	86.9	92.3
048	Rockland	95.6	86.9	91.7
049	Waterville	96.8	81.0	89.8
MARYLAND				
206	Waldorf	97.1	82.6	90.7
207-208	College Park	97.1	83.8	91.2
209	Silver Spring	96.4	84.0	90.9
210-212	Baltimore	100.9	83.4	93.2
214	Annapolis	101.0	82.7	92.9
215	Cumberland	97.3	84.1	91.5
216	Easton	98.9	71.2	86.7
217	Hagerstown	97.7	85.2	92.2
218	Salisbury	99.4	61.8	82.8
219	Elkton	96.3	84.2	91.0
MASSACHUSETTS				
010-011	Springfield	99.1	110.3	104.0
012	Pittsfield	98.5	109.7	103.4
013	Greenfield	96.9	110.0	102.6
014	Fitchburg	95.6	124.5	108.3
015-016	Worcester	99.0	124.5	110.2
017	Framingham	94.9	132.9	111.6
018	Lowell	98.4	132.3	113.3
019	Lawrence	99.2	132.6	113.9
020-022, 024	Boston	100.6	139.1	117.5
023	Brockton	98.8	125.1	110.4
025	Buzzards Bay	93.7	125.9	107.9
026	Hyannis	95.9	125.9	109.1
027	New Bedford	97.9	126.3	110.4
MICHIGAN				
480,483	Royal Oak	95.1	106.1	99.9
481	Ann Arbor	97.0	107.4	101.6
482	Detroit	98.2	108.2	102.6
484-485	Flint	96.7	95.6	96.2
486	Saginaw	96.3	90.8	93.9
487	Bay City	96.4	90.3	93.7
488-489	Lansing	98.7	94.4	96.8
490	Battle Creek	97.5	86.0	92.4
491	Kalamazoo	97.8	83.9	91.7
492	Jackson	96.0	94.2	95.2
493,495	Grand Rapids	98.8	83.9	92.3
494	Muskegon	96.3	83.2	90.5

STATE/ZIP	CITY	MAT.	INST.	TOTAL
MICHIGAN (CONT'D)				
496	Traverse City	95.6	80.9	89.1
497	Gaylord	96.7	83.6	90.9
498-499	Iron Mountain	98.7	85.8	93.0
MINNESOTA				
550-551	Saint Paul	100.7	120.9	109.6
553-555	Minneapolis	100.7	119.9	109.1
556-558	Duluth	101.3	110.5	105.3
559	Rochester	100.3	106.3	103.0
560	Mankato	97.5	102.7	99.8
561	Windom	96.2	93.5	95.0
562	Willmar	95.7	98.8	97.1
563	St. Cloud	96.8	114.0	104.4
564	Brainerd	97.3	100.5	98.7
565	Detroit Lakes	99.4	96.1	98.0
566	Bemidji	98.6	99.1	98.8
567	Thief River Falls	98.2	93.3	96.0
MISSISSIPPI				
386	Clarksdale	96.6	55.3	78.4
387	Greenville	100.3	65.1	84.8
388	Tupelo	98.2	55.6	79.4
389	Greenwood	97.9	52.0	77.7
390-392	Jackson	99.8	68.4	86.0
393	Meridian	97.1	67.9	84.2
394	Laurel	98.8	55.4	79.7
395	Biloxi	98.8	60.5	81.9
396	McComb	97.1	51.5	77.0
397	Columbus	98.6	53.4	78.7
MISSOURI				
630-631	St. Louis	100.2	104.3	102.0
633	Bowling Green	99.1	96.0	97.7
634	Hannibal	97.9	92.5	95.6
635	Kirksville	101.5	89.7	96.3
636	Flat River	100.0	95.1	97.9
637	Cape Girardeau	99.6	92.8	96.6
638	Sikeston	98.4	90.5	94.9
639	Poplar Bluff	97.9	89.7	94.3
640-641	Kansas City	100.2	104.8	102.2
644-645	St. Joseph	99.4	96.7	98.2
646	Chillicothe	96.8	98.6	97.6
647	Harrisonville	96.3	101.0	98.4
648	Joplin	98.2	81.0	90.7
650-651	Jefferson City	99.6	98.4	99.1
652	Columbia	100.3	99.7	100.0
653	Sedalia	100.7	93.9	97.7
654-655	Rolla	98.3	96.7	97.6
656-658	Springfield	101.8	83.3	93.7
MONTANA				
590-591	Billings	103.7	78.8	92.7
592	Wolf Point	103.8	74.2	90.8
593	Miles City	101.6	74.0	89.4
594	Great Falls	105.2	76.8	92.7
595	Havre	102.7	73.0	89.6
596	Helena	102.8	73.6	90.0
597	Butte	103.7	77.1	92.0
598	Missoula	101.1	75.7	89.9
599	Kalispell	100.4	75.5	89.5
NEBRASKA				
680-681	Omaha	99.1	79.5	90.5
683-685	Lincoln	99.8	76.9	89.7
686	Columbus	98.1	77.5	89.0
687	Norfolk	99.6	76.0	89.2
688	Grand Island	99.3	76.6	89.3
689	Hastings	99.1	78.9	90.2
690	McCook	99.0	70.5	86.4
691	North Platte	98.9	75.9	88.7
692	Valentine	101.3	71.8	88.3
693	Alliance	101.2	70.5	87.7
NEVADA				
889-891	Las Vegas	103.6	109.4	106.1
893	Ely	102.4	100.6	101.6
894-895	Reno	102.2	86.5	95.3
897	Carson City	101.8	86.7	95.1
898	Elko	101.1	81.4	92.4
NEW HAMPSHIRE				
030	Nashua	99.3	93.6	96.8
031	Manchester	99.9	93.6	97.1

STATE/ZIP	CITY	MAT.	INST.	TOTAL
NEW HAMPSHIRE (CONT'D)				
032-033	Concord	100.0	93.8	97.3
034	Keene	96.2	75.6	87.1
035	Littleton	96.3	77.6	88.0
036	Charleston	95.7	70.3	84.5
037	Claremont	94.9	70.4	84.1
038	Portsmouth	96.5	93.7	95.3
NEW JERSEY				
070-071	Newark	100.3	129.3	113.1
072	Elizabeth	97.8	128.8	111.5
073	Jersey City	96.7	128.9	110.9
074-075	Paterson	98.5	129.4	112.1
076	Hackensack	96.5	129.0	110.8
077	Long Branch	96.0	124.7	108.6
078	Dover	96.7	129.2	111.0
079	Summit	96.7	128.8	110.9
080,083	Vineland	96.4	123.6	108.4
081	Camden	98.4	124.7	110.0
082,084	Atlantic City	97.2	125.3	109.6
085-086	Trenton	100.2	124.9	111.1
087	Point Pleasant	98.2	124.0	109.6
088-089	New Brunswick	98.8	128.7	112.0
NEW MEXICO				
870-872	Albuquerque	100.3	72.1	87.9
873	Gallup	100.7	72.1	88.1
874	Farmington	101.0	72.1	88.2
875	Santa Fe	101.1	72.1	88.3
877	Las Vegas	99.0	72.1	87.1
878	Socorro	98.6	72.1	86.9
879	Truth/Consequences	98.3	68.8	85.3
880	Las Cruces	97.5	68.1	84.6
881	Clovis	99.9	72.0	87.6
882	Roswell	101.4	72.1	88.5
883	Carrizozo	102.1	72.1	88.9
884	Tucumcari	100.6	72.0	88.0
NEW YORK				
100-102	New York	100.7	169.8	131.1
103	Staten Island	96.8	167.5	128.0
104	Bronx	95.2	166.0	126.4
105	Mount Vernon	95.5	143.1	116.5
106	White Plains	95.2	143.1	116.3
107	Yonkers	99.4	146.3	120.1
108	New Rochelle	95.9	143.0	116.7
109	Suffern	95.5	131.9	111.5
110	Queens	100.3	166.2	129.3
111	Long Island City	101.9	166.2	130.2
112	Brooklyn	102.3	167.9	131.2
113	Flushing	102.4	166.2	130.5
114	Jamaica	100.7	166.2	129.6
115,117,118	Hicksville	100.3	152.7	123.4
116	Far Rockaway	102.5	166.2	130.6
119	Riverhead	101.1	150.9	123.0
120-122	Albany	98.3	104.6	101.1
123	Schenectady	98.1	104.5	101.0
124	Kingston	101.8	117.1	108.6
125-126	Poughkeepsie	101.0	135.2	116.1
127	Monticello	100.3	115.3	106.9
128	Glens Falls	93.3	101.1	96.7
129	Plattsburgh	98.4	96.7	97.7
130-132	Syracuse	98.5	98.2	98.3
133-135	Utica	96.5	97.5	97.0
136	Watertown	98.0	96.7	97.4
137-139	Binghamton	98.2	98.4	98.3
140-142	Buffalo	100.3	106.4	103.0
143	Niagara Falls	98.0	107.2	102.1
144-146	Rochester	100.8	100.0	100.5
147	Jamestown	97.0	93.0	95.3
148-149	Elmira	96.9	98.3	97.6
NORTH CAROLINA				
270,272-274	Greensboro	99.2	61.7	82.7
271	Winston-Salem	98.9	62.1	82.7
275-276	Raleigh	97.7	59.7	81.0
277	Durham	100.7	61.7	83.5
278	Rocky Mount	96.8	61.8	81.4
279	Elizabeth City	97.8	54.5	78.7
280	Gastonia	99.8	62.4	83.3
281-282	Charlotte	99.1	63.8	83.6
283	Fayetteville	102.2	60.6	83.9
284	Wilmington	98.5	61.5	82.2
285	Kinston	97.0	63.0	82.0

Location Factors - Commercial

STATE/ZIP	CITY	MAT.	INST.	TOTAL	STATE/ZIP	CITY	MAT.	INST.	TOTAL
NORTH CAROLINA (CONT'D)					**PENNSYLVANIA (CONT'D)**				
286	Hickory	97.3	61.8	81.7	177	Williamsport	93.2	96.2	94.5
287-288	Asheville	99.0	62.4	82.9	178	Sunbury	95.3	97.3	96.2
289	Murphy	98.2	51.0	77.4	179	Pottsville	94.4	99.1	96.5
					180	Lehigh Valley	95.7	112.3	103.0
NORTH DAKOTA					181	Allentown	97.3	110.6	103.2
580-581	Fargo	101.1	75.0	89.6	182	Hazleton	95.2	100.3	97.5
582	Grand Forks	101.9	65.1	85.7	183	Stroudsburg	95.1	105.0	99.4
583	Devils Lake	101.7	64.8	85.5	184-185	Scranton	98.1	99.8	98.8
584	Jamestown	101.9	64.7	85.5	186-187	Wilkes-Barre	94.9	101.0	97.6
585	Bismarck	102.1	70.0	87.9	188	Montrose	94.6	98.7	96.4
586	Dickinson	102.6	65.9	86.4	189	Doylestown	94.8	127.4	109.1
587	Minot	101.9	76.9	90.9	190-191	Philadelphia	99.5	134.7	115.0
588	Williston	101.0	66.2	85.7	193	Westchester	96.3	126.3	109.5
					194	Norristown	95.4	131.2	111.2
OHIO					195-196	Reading	97.1	104.5	100.4
430-432	Columbus	98.2	87.4	93.5					
433	Marion	95.1	84.6	90.5	**PUERTO RICO**				
434-436	Toledo	98.3	97.6	98.0	009	San Juan	122.0	24.5	79.0
437-438	Zanesville	95.6	86.3	91.5					
439	Steubenville	96.7	93.9	95.5	**RHODE ISLAND**				
440	Lorain	98.9	92.2	96.0	028	Newport	97.2	116.8	105.8
441	Cleveland	99.2	99.4	99.3	029	Providence	99.8	116.8	107.3
442-443	Akron	99.9	93.9	97.3					
444-445	Youngstown	99.2	88.1	94.4	**SOUTH CAROLINA**				
446-447	Canton	99.3	89.3	94.9	290-292	Columbia	98.2	68.3	85.0
448-449	Mansfield	97.1	86.2	92.3	293	Spartanburg	98.3	67.3	84.6
450	Hamilton	97.8	84.4	91.9	294	Charleston	99.8	66.1	84.9
451-452	Cincinnati	98.2	83.7	91.8	295	Florence	98.0	67.7	84.7
453-454	Dayton	97.9	83.9	91.7	296	Greenville	98.1	66.0	83.9
455	Springfield	97.8	84.1	91.8	297	Rock Hill	98.0	56.5	79.7
456	Chillicothe	97.2	92.0	94.9	298	Aiken	98.9	66.8	84.7
457	Athens	100.2	83.1	92.6	299	Beaufort	99.6	52.5	78.8
458	Lima	100.5	86.3	94.3					
					SOUTH DAKOTA				
OKLAHOMA					570-571	Sioux Falls	99.0	60.5	82.0
730-731	Oklahoma City	99.8	71.3	87.3	572	Watertown	99.9	49.9	77.9
734	Ardmore	98.6	64.9	83.7	573	Mitchell	98.6	48.1	76.4
735	Lawton	100.7	67.5	86.1	574	Aberdeen	101.0	53.7	80.1
736	Clinton	99.8	66.2	85.0	575	Pierre	102.2	59.3	83.3
737	Enid	100.2	65.2	84.8	576	Mobridge	99.4	48.7	77.0
738	Woodward	98.6	68.4	85.3	577	Rapid City	100.8	59.6	82.7
739	Guymon	99.7	69.1	86.2					
740-741	Tulsa	98.3	64.5	83.4	**TENNESSEE**				
743	Miami	95.2	72.1	85.0	370-372	Nashville	98.2	74.1	87.6
744	Muskogee	97.5	62.9	82.3	373-374	Chattanooga	99.9	66.6	85.2
745	McAlester	94.8	65.3	81.8	375,380-381	Memphis	98.9	72.5	87.3
746	Ponca City	95.5	67.6	83.2	376	Johnson City	99.9	56.4	80.7
747	Durant	95.5	65.9	82.5	377-379	Knoxville	96.5	66.8	83.5
748	Shawnee	97.0	65.3	83.0	382	McKenzie	97.7	59.4	80.8
749	Poteau	94.7	66.4	82.2	383	Jackson	99.0	63.5	83.4
					384	Columbia	96.2	67.2	83.4
OREGON					385	Cookeville	97.5	60.0	81.0
970-972	Portland	99.7	101.1	100.3					
973	Salem	100.9	99.2	100.2	**TEXAS**				
974	Eugene	99.5	99.2	99.3	750	McKinney	98.2	66.9	84.4
975	Medford	101.0	96.9	99.2	751	Waxahackie	98.3	65.4	83.8
976	Klamath Falls	101.3	97.1	99.4	752-753	Dallas	99.2	67.9	85.4
977	Bend	100.3	99.2	99.8	754	Greenville	98.4	67.0	84.6
978	Pendleton	96.2	101.4	98.5	755	Texarkana	97.6	63.6	82.6
979	Vale	94.1	89.1	91.9	756	Longview	98.4	64.3	83.4
					757	Tyler	98.4	65.2	83.8
PENNSYLVANIA					758	Palestine	94.8	64.4	81.4
150-152	Pittsburgh	100.2	104.6	102.1	759	Lufkin	95.6	67.7	83.3
153	Washington	97.5	103.6	100.2	760-761	Fort Worth	98.8	65.7	84.2
154	Uniontown	97.7	102.8	99.9	762	Denton	99.2	66.5	84.8
155	Bedford	98.7	95.3	97.2	763	Wichita Falls	96.5	64.6	82.5
156	Greensburg	98.7	102.8	100.5	764	Eastland	95.6	64.1	81.8
157	Indiana	97.6	101.4	99.2	765	Temple	94.1	61.6	79.8
158	Dubois	99.1	97.5	98.4	766-767	Waco	95.9	63.7	81.7
159	Johnstown	98.7	98.1	98.5	768	Brownwood	99.1	60.6	82.1
160	Butler	92.3	103.2	97.1	769	San Angelo	98.7	59.9	81.6
161	New Castle	92.4	100.3	95.9	770-772	Houston	100.7	69.8	87.1
162	Kittanning	92.8	103.1	97.3	773	Huntsville	99.7	66.9	85.3
163	Oil City	92.3	99.7	95.6	774	Wharton	100.6	68.7	86.5
164-165	Erie	94.0	97.5	95.5	775	Galveston	98.7	68.5	85.4
166	Altoona	94.1	93.5	93.9	776-777	Beaumont	98.9	68.9	85.7
167	Bradford	95.6	97.4	96.4	778	Bryan	96.4	68.4	84.1
168	State College	95.1	93.0	94.2	779	Victoria	100.7	66.1	85.5
169	Wellsboro	96.2	96.6	96.3	780	Laredo	99.4	63.5	83.6
170-171	Harrisburg	98.5	96.5	97.6	781-782	San Antonio	100.3	64.1	84.4
172	Chambersburg	95.8	91.2	93.8	783-784	Corpus Christi	102.0	63.2	84.9
173-174	York	95.9	96.8	96.3	785	McAllen	102.9	59.9	83.9
175-176	Lancaster	94.6	98.3	96.2	786-787	Austin	100.0	63.4	83.9

807

Location Factors - Commercial

STATE/ZIP	CITY	MAT.	INST.	TOTAL	STATE/ZIP	CITY	MAT.	INST.	TOTAL
TEXAS (CONT'D)					**WISCONSIN (CONT'D)**				
788	Del Rio	102.5	63.0	85.1	538	Lancaster	96.8	93.7	95.4
789	Giddings	99.3	63.8	83.7	539	Portage	95.3	96.4	95.8
790-791	Amarillo	99.6	62.3	83.1	540	New Richmond	97.1	93.8	95.6
792	Childress	99.7	64.0	84.0	541-543	Green Bay	101.2	97.5	99.5
793-794	Lubbock	101.5	63.6	84.8	544	Wausau	96.5	96.1	96.3
795-796	Abilene	100.0	62.3	83.4	545	Rhinelander	99.8	91.8	96.3
797	Midland	102.1	64.6	85.6	546	La Crosse	97.6	94.8	96.4
798-799,885	El Paso	98.3	61.6	82.2	547	Eau Claire	99.2	95.5	97.6
					548	Superior	96.8	97.7	97.2
UTAH					549	Oshkosh	97.2	91.1	94.5
840-841	Salt Lake City	103.8	71.2	89.4					
842,844	Ogden	99.3	71.2	86.9	**WYOMING**				
843	Logan	101.3	71.2	88.0	820	Cheyenne	102.3	66.5	86.5
845	Price	102.0	65.3	85.8	821	Yellowstone Nat'l Park	100.2	71.5	87.6
846-847	Provo	101.8	71.1	88.2	822	Wheatland	101.4	65.5	85.6
					823	Rawlins	103.1	71.2	89.0
VERMONT					824	Worland	101.0	67.4	86.2
050	White River Jct.	98.5	77.3	89.2	825	Riverton	102.1	66.4	86.4
051	Bellows Falls	96.9	91.7	94.6	826	Casper	102.5	63.8	85.5
052	Bennington	97.3	92.3	95.1	827	Newcastle	100.8	72.4	88.3
053	Brattleboro	97.7	91.6	95.0	828	Sheridan	103.7	66.5	87.3
054	Burlington	102.2	82.5	93.5	829-831	Rock Springs	104.9	68.4	88.8
056	Montpelier	100.6	81.8	92.3					
057	Rutland	99.0	82.5	91.7	**CANADIAN FACTORS (reflect Canadian currency)**				
058	St. Johnsbury	98.7	76.8	89.1	**ALBERTA**				
059	Guildhall	97.2	76.8	88.2		Calgary	123.8	93.4	110.4
						Edmonton	123.9	93.4	110.5
VIRGINIA						Fort McMurray	124.5	91.7	110.0
220-221	Fairfax	100.7	89.2	95.6		Lethbridge	119.1	91.2	106.8
222	Arlington	101.5	89.4	96.2		Lloydminster	113.9	87.9	102.5
223	Alexandria	100.6	89.4	95.6		Medicine Hat	114.1	87.2	102.2
224-225	Fredericksburg	99.4	87.7	94.2		Red Deer	114.6	87.2	102.5
226	Winchester	100.0	85.2	93.5					
227	Culpeper	99.9	85.5	93.6	**BRITISH COLUMBIA**				
228	Harrisonburg	100.2	66.9	85.5		Kamloops	115.0	89.3	103.7
229	Charlottesville	100.5	67.9	86.1		Prince George	116.0	88.6	104.0
230-232	Richmond	99.8	68.7	86.1		Vancouver	117.7	89.2	105.1
233-235	Norfolk	100.1	68.3	86.1		Victoria	116.6	86.7	103.4
236	Newport News	99.8	68.3	85.9					
237	Portsmouth	99.3	65.4	84.4	**MANITOBA**				
238	Petersburg	99.8	68.7	86.1		Brandon	126.4	74.8	103.7
239	Farmville	99.0	52.1	78.3		Portage la Prairie	113.9	73.1	95.9
240-241	Roanoke	101.0	66.6	85.8		Winnipeg	122.6	70.8	99.8
242	Bristol	99.4	54.8	79.7					
243	Pulaski	99.0	57.5	80.7	**NEW BRUNSWICK**				
244	Staunton	99.7	61.7	83.0		Bathurst	111.8	66.3	91.8
245	Lynchburg	99.9	66.2	85.1		Dalhousie	112.6	66.5	92.3
246	Grundy	99.2	52.6	78.7		Fredericton	118.8	71.5	97.9
						Moncton	112.0	70.7	93.8
WASHINGTON						Newcastle	111.9	67.0	92.1
980-981,987	Seattle	102.3	103.8	103.0		St. John	114.4	77.8	98.3
982	Everett	102.2	99.0	100.8					
983-984	Tacoma	102.6	99.6	101.3	**NEWFOUNDLAND**				
985	Olympia	101.0	99.7	100.4		Corner Brook	131.2	66.6	102.7
986	Vancouver	103.4	94.4	99.4		St. John's	123.9	83.9	106.3
988	Wenatchee	102.8	85.9	95.3					
989	Yakima	102.8	94.9	99.3	**NORTHWEST TERRITORIES**				
990-992	Spokane	104.8	82.6	95.0		Yellowknife	136.7	82.9	113.0
993	Richland	104.4	90.7	98.4					
994	Clarkston	103.8	81.9	94.1	**NOVA SCOTIA**				
						Bridgewater	113.4	73.6	95.9
WEST VIRGINIA						Dartmouth	127.5	73.6	103.7
247-248	Bluefield	98.2	93.6	96.2		Halifax	118.2	79.9	101.3
249	Lewisburg	99.8	92.8	96.7		New Glasgow	125.5	73.6	102.6
250-253	Charleston	99.7	97.4	98.7		Sydney	123.4	73.6	101.4
254	Martinsburg	100.4	87.0	94.5		Truro	113.0	73.6	95.6
255-257	Huntington	101.4	97.8	99.8		Yarmouth	125.4	73.6	102.6
258-259	Beckley	98.8	95.1	97.2					
260	Wheeling	100.8	94.5	98.0	**ONTARIO**				
261	Parkersburg	99.6	92.5	96.5		Barrie	118.3	91.7	106.6
262	Buckhannon	99.6	94.7	97.4		Brantford	115.5	95.5	106.7
263-264	Clarksburg	100.2	94.7	97.7		Cornwall	115.4	92.0	105.1
265	Morgantown	100.2	94.8	97.8		Hamilton	116.9	95.6	107.5
266	Gassaway	99.5	94.4	97.3		Kingston	116.5	92.1	105.7
267	Romney	99.4	89.8	95.1		Kitchener	109.9	91.8	101.9
268	Petersburg	99.3	87.7	94.2		London	116.3	92.2	105.7
						North Bay	128.6	89.8	111.5
WISCONSIN						Oshawa	112.5	95.3	104.9
530,532	Milwaukee	98.9	106.9	102.4		Ottawa	117.6	92.9	106.7
531	Kenosha	99.2	101.7	100.3		Owen Sound	118.4	89.9	105.8
534	Racine	98.7	102.0	100.2		Peterborough	115.5	91.7	105.0
535	Beloit	98.6	99.8	99.1		Sarnia	115.3	96.1	106.9
537	Madison	99.3	99.1	99.2					

For customer support on your Commercial Renovation Cost Data, call 877.742.4719.

Location Factors - Commercial

STATE/ZIP	CITY	MAT.	INST.	TOTAL
ONTARIO (CONT'D)				
	Sault Ste. Marie	110.7	91.1	102.1
	St. Catharines	107.8	93.3	101.5
	Sudbury	107.6	92.1	100.8
	Thunder Bay	109.1	92.3	101.7
	Timmins	115.6	89.8	104.2
	Toronto	116.9	100.9	109.9
	Windsor	108.3	92.3	101.3
PRINCE EDWARD ISLAND				
	Charlottetown	120.4	61.9	94.6
	Summerside	128.1	62.0	99.0
QUEBEC				
	Cap-de-la-Madeleine	112.8	86.4	101.1
	Charlesbourg	112.8	86.4	101.1
	Chicoutimi	111.7	92.0	103.0
	Gatineau	112.5	86.1	100.9
	Granby	112.7	86.1	101.0
	Hull	112.6	86.1	100.9
	Joliette	113.0	86.4	101.3
	Laval	112.8	86.1	101.1
	Montreal	117.9	92.4	106.7
	Quebec City	117.9	92.8	106.8
	Rimouski	112.3	92.0	103.4
	Rouyn-Noranda	112.5	86.1	100.9
	Saint-Hyacinthe	112.0	86.1	100.6
	Sherbrooke	112.8	86.1	101.1
	Sorel	113.0	86.4	101.2
	Saint-Jerome	112.5	86.1	100.9
	Trois-Rivieres	126.1	86.4	108.6
SASKATCHEWAN				
	Moose Jaw	111.2	68.5	92.4
	Prince Albert	110.3	66.8	91.1
	Regina	123.0	89.0	108.0
	Saskatoon	112.3	89.0	102.0
YUKON				
	Whitehorse	137.1	67.5	106.4

R011105-05   Tips for Accurate Estimating

1. Use pre-printed or columnar forms for orderly sequence of dimensions and locations and for recording telephone quotations.

2. Use only the front side of each paper or form except for certain pre-printed summary forms.

3. Be consistent in listing dimensions: For example, length x width x height. This helps in rechecking to ensure that, the total length of partitions is appropriate for the building area.

4. Use printed (rather than measured) dimensions where given.

5. Add up multiple printed dimensions for a single entry where possible.

6. Measure all other dimensions carefully.

7. Use each set of dimensions to calculate multiple related quantities.

8. Convert foot and inch measurements to decimal feet when listing. Memorize decimal equivalents to .01 parts of a foot (1/8″ equals approximately .01′).

9. Do not "round off" quantities until the final summary.

10. Mark drawings with different colors as items are taken off.

11. Keep similar items together, different items separate.

12. Identify location and drawing numbers to aid in future checking for completeness.

13. Measure or list everything on the drawings or mentioned in the specifications.

14. It may be necessary to list items not called for to make the job complete.

15. Be alert for: Notes on plans such as N.T.S. (not to scale); changes in scale throughout the drawings; reduced size drawings; discrepancies between the specifications and the drawings.

16. Develop a consistent pattern of performing an estimate. For example:
    a. Start the quantity takeoff at the lower floor and move to the next higher floor.
    b. Proceed from the main section of the building to the wings.
    c. Proceed from south to north or vice versa, clockwise or counterclockwise.
    d. Take off floor plan quantities first, elevations next, then detail drawings.

17. List all gross dimensions that can be either used again for different quantities, or used as a rough check of other quantities for verification (exterior perimeter, gross floor area, individual floor areas, etc.).

18. Utilize design symmetry or repetition (repetitive floors, repetitive wings, symmetrical design around a center line, similar room layouts, etc.). Note: Extreme caution is needed here so as not to omit or duplicate an area.

19. Do not convert units until the final total is obtained. For instance, when estimating concrete work, keep all units to the nearest cubic foot, then summarize and convert to cubic yards.

20. When figuring alternatives, it is best to total all items involved in the basic system, then total all items involved in the alternates. Therefore you work with positive numbers in all cases. When adds and deducts are used, it is often confusing whether to add or subtract a portion of an item; especially on a complicated or involved alternate.

R011105-50   Metric Conversion Factors

Description: This table is primarily for converting customary U.S. units in the left hand column to SI metric units in the right hand column. In addition, conversion factors for some commonly encountered Canadian and non-SI metric units are included.

	If You Know		Multiply By		To Find
Length	Inches	x	25.4[a]	=	Millimeters
	Feet	x	0.3048[a]	=	Meters
	Yards	x	0.9144[a]	=	Meters
	Miles (statute)	x	1.609	=	Kilometers
Area	Square inches	x	645.2	=	Square millimeters
	Square feet	x	0.0929	=	Square meters
	Square yards	x	0.8361	=	Square meters
Volume (Capacity)	Cubic inches	x	16,387	=	Cubic millimeters
	Cubic feet	x	0.02832	=	Cubic meters
	Cubic yards	x	0.7646	=	Cubic meters
	Gallons (U.S. liquids)[b]	x	0.003785	=	Cubic meters[c]
	Gallons (Canadian liquid)[b]	x	0.004546	=	Cubic meters[c]
	Ounces (U.S. liquid)[b]	x	29.57	=	Milliliters[c, d]
	Quarts (U.S. liquid)[b]	x	0.9464	=	Liters[c, d]
	Gallons (U.S. liquid)[b]	x	3.785	=	Liters[c, d]
Force	Kilograms force[d]	x	9.807	=	Newtons
	Pounds force	x	4.448	=	Newtons
	Pounds force	x	0.4536	=	Kilograms force[d]
	Kips	x	4448	=	Newtons
	Kips	x	453.6	=	Kilograms force[d]
Pressure, Stress, Strength (Force per unit area)	Kilograms force per square centimeter[d]	x	0.09807	=	Megapascals
	Pounds force per square inch (psi)	x	0.006895	=	Megapascals
	Kips per square inch	x	6.895	=	Megapascals
	Pounds force per square inch (psi)	x	0.07031	=	Kilograms force per square centimeter[d]
	Pounds force per square foot	x	47.88	=	Pascals
	Pounds force per square foot	x	4.882	=	Kilograms force per square meter[d]
Flow	Cubic feet per minute	x	0.4719	=	Liters per second
	Gallons per minute	x	0.0631	=	Liters per second
	Gallons per hour	x	1.05	=	Milliliters per second
Bending Moment Or Torque	Inch-pounds force	x	0.01152	=	Meter-kilograms force[d]
	Inch-pounds force	x	0.1130	=	Newton-meters
	Foot-pounds force	x	0.1383	=	Meter-kilograms force[d]
	Foot-pounds force	x	1.356	=	Newton-meters
	Meter-kilograms force[d]	x	9.807	=	Newton-meters
Mass	Ounces (avoirdupois)	x	28.35	=	Grams
	Pounds (avoirdupois)	x	0.4536	=	Kilograms
	Tons (metric)	x	1000	=	Kilograms
	Tons, short (2000 pounds)	x	907.2	=	Kilograms
	Tons, short (2000 pounds)	x	0.9072	=	Megagrams[e]
Mass per Unit Volume	Pounds mass per cubic foot	x	16.02	=	Kilograms per cubic meter
	Pounds mass per cubic yard	x	0.5933	=	Kilograms per cubic meter
	Pounds mass per gallon (U.S. liquid)[b]	x	119.8	=	Kilograms per cubic meter
	Pounds mass per gallon (Canadian liquid)[b]	x	99.78	=	Kilograms per cubic meter
Temperature	Degrees Fahrenheit	(F-32)/1.8		=	Degrees Celsius
	Degrees Fahrenheit	(F+459.67)/1.8		=	Degrees Kelvin
	Degrees Celsius	C+273.15		=	Degrees Kelvin

[a] The factor given is exact
[b] One U.S. gallon = 0.8327 Canadian gallon
[c] 1 liter = 1000 milliliters = 1000 cubic centimeters
   1 cubic decimeter = 0.001 cubic meter
[d] Metric but not SI unit
[e] Called "tonne" in England and "metric ton" in other metric countries

811

R011105-60   Weights and Measures

Measures of Length
1 Mile = 1760 Yards = 5280 Feet
1 Yard = 3 Feet = 36 inches
1 Foot = 12 Inches
1 Mil = 0.001 Inch
1 Fathom = 2 Yards = 6 Feet
1 Rod = 5.5 Yards = 16.5 Feet
1 Hand = 4 Inches
1 Span = 9 Inches
1 Micro-inch = One Millionth Inch or 0.000001 Inch
1 Micron = One Millionth Meter + 0.00003937 Inch

Surveyor's Measure
1 Mile = 8 Furlongs = 80 Chains
1 Furlong = 10 Chains = 220 Yards
1 Chain = 4 Rods = 22 Yards = 66 Feet = 100 Links
1 Link = 7.92 Inches

Square Measure
1 Square Mile = 640 Acres = 6400 Square Chains
1 Acre = 10 Square Chains = 4840 Square Yards =
  43,560 Sq. Ft.
1 Square Chain = 16 Square Rods = 484 Square Yards =
  4356 Sq. Ft.
1 Square Rod = 30.25 Square Yards = 272.25 Square Feet = 625 Square
  Lines
1 Square Yard = 9 Square Feet
1 Square Foot = 144 Square Inches
An Acre equals a Square 208.7 Feet per Side

Cubic Measure
1 Cubic Yard = 27 Cubic Feet
1 Cubic Foot = 1728 Cubic Inches
1 Cord of Wood = 4 x 4 x 8 Feet = 128 Cubic Feet
1 Perch of Masonry = 16½ x 1½ x 1 Foot = 24.75 Cubic Feet

Avoirdupois or Commercial Weight
1 Gross or Long Ton = 2240 Pounds
1 Net or Short Ton = 2000 Pounds
1 Pound = 16 Ounces = 7000 Grains
1 Ounce = 16 Drachms = 437.5 Grains
1 Stone = 14 Pounds

Power
1 British Thermal Unit per Hour = 0.2931 Watts
1 Ton (Refrigeration) = 3.517 Kilowatts
1 Horsepower (Boiler) = 9.81 Kilowatts
1 Horsepower (550 ft-lb/s) = 0.746 Kilowatts

Shipping Measure
For Measuring Internal Capacity of a Vessel:
  1 Register Ton = 100 Cubic Feet

For Measurement of Cargo:
  Approximately 40 Cubic Feet of Merchandise is considered a Shipping
  Ton, unless that bulk would weigh more than 2000 Pounds, in which case
  Freight Charge may be based upon weight.

40 Cubic Feet = 32.143 U.S. Bushels = 31.16 Imp. Bushels

Liquid Measure
1 Imperial Gallon = 1.2009 U.S. Gallon = 277.42 Cu. In.
1 Cubic Foot = 7.48 U.S. Gallons

R011110-10   Architectural Fees

Tabulated below are typical percentage fees by project size, for good professional architectural service. Fees may vary from those listed depending upon degree of design difficulty and economic conditions in any particular area.

Rates can be interpolated horizontally and vertically. Various portions of the same project requiring different rates should be adjusted proportionately. For alterations, add 50% to the fee for the first $500,000 of project cost and add 25% to the fee for project cost over $500,000.

Architectural fees tabulated below include Structural, Mechanical and Electrical Engineering Fees. They do not include the fees for special consultants such as kitchen planning, security, acoustical, interior design, etc.

Civil Engineering fees are included in the Architectural fee for project sites requiring minimal design such as city sites. However, separate Civil Engineering fees must be added when utility connections require design, drainage calculations are needed, stepped foundations are required, or provisions are required to protect adjacent wetlands.

Building Types	Total Project Size in Thousands of Dollars						
	100	250	500	1,000	5,000	10,000	50,000
Factories, garages, warehouses, repetitive housing	9.0%	8.0%	7.0%	6.2%	5.3%	4.9%	4.5%
Apartments, banks, schools, libraries, offices, municipal buildings	12.2	12.3	9.2	8.0	7.0	6.6	6.2
Churches, hospitals, homes, laboratories, museums, research	15.0	13.6	12.7	11.9	9.5	8.8	8.0
Memorials, monumental work, decorative furnishings	—	16.0	14.5	13.1	10.0	9.0	8.3

R011110-30 Engineering Fees

Typical **Structural Engineering Fees** based on type of construction and total project size. These fees are included in Architectural Fees.

Type of Construction	Total Project Size (in thousands of dollars)			
	$500	$500-$1,000	$1,000-$5,000	Over $5000
Industrial buildings, factories & warehouses	Technical payroll times 2.0 to 2.5	1.60%	1.25%	1.00%
Hotels, apartments, offices, dormitories, hospitals, public buildings, food stores		2.00%	1.70%	1.20%
Museums, banks, churches and cathedrals		2.00%	1.75%	1.25%
Thin shells, prestressed concrete, earthquake resistive		2.00%	1.75%	1.50%
Parking ramps, auditoriums, stadiums, convention halls, hangars & boiler houses		2.50%	2.00%	1.75%
Special buildings, major alterations, underpinning & future expansion		Add to above 0.5%	Add to above 0.5%	Add to above 0.5%

For complex reinforced concrete or unusually complicated structures, add 20% to 50%.

Typical **Mechanical and Electrical Engineering Fees** are based on the size of the subcontract. The fee structure for both is shown below. These fees are included in Architectural Fees.

Type of Construction	Subcontract Size							
	$25,000	$50,000	$100,000	$225,000	$350,000	$500,000	$750,000	$1,000,000
Simple structures	6.4%	5.7%	4.8%	4.5%	4.4%	4.3%	4.2%	4.1%
Intermediate structures	8.0	7.3	6.5	5.6	5.1	5.0	4.9	4.8
Complex structures	10.1	9.0	9.0	8.0	7.5	7.5	7.0	7.0

For renovations, add 15% to 25% to applicable fee.

R012153-10 Repair and Remodeling

Cost figures are based on new construction utilizing the most cost-effective combination of labor, equipment and material with the work scheduled in proper sequence to allow the various trades to accomplish their work in an efficient manner.

The costs for repair and remodeling work must be modified due to the following factors that may be present in any given repair and remodeling project.

1. Equipment usage curtailment due to the physical limitations of the project, with only hand-operated equipment being used.

2. Increased requirement for shoring and bracing to hold up the building while structural changes are being made and to allow for temporary storage of construction materials on above-grade floors.

3. Material handling becomes more costly due to having to move within the confines of an enclosed building. For multi-story construction, low capacity elevators and stairwells may be the only access to the upper floors.

4. Large amount of cutting and patching and attempting to match the existing construction is required. It is often more economical to remove entire walls rather than create many new door and window openings. This sort of trade-off has to be carefully analyzed.

5. Cost of protection of completed work is increased since the usual sequence of construction usually cannot be accomplished.

6. Economies of scale usually associated with new construction may not be present. If small quantities of components must be custom fabricated due to job requirements, unit costs will naturally increase. Also, if only small work areas are available at a given time, job scheduling between trades becomes difficult and subcontractor quotations may reflect the excessive start-up and shut-down phases of the job.

7. Work may have to be done on other than normal shifts and may have to be done around an existing production facility which has to stay in production during the course of the repair and remodeling.

8. Dust and noise protection of adjoining non-construction areas can involve substantial special protection and alter usual construction methods.

9. Job may be delayed due to unexpected conditions discovered during demolition or removal. These delays ultimately increase construction costs.

10. Piping and ductwork runs may not be as simple as for new construction. Wiring may have to be snaked through walls and floors.

11. Matching "existing construction" may be impossible because materials may no longer be manufactured. Substitutions may be expensive.

12. Weather protection of existing structure requires additional temporary structures to protect building at openings.

13. On small projects, because of local conditions, it may be necessary to pay a tradesman for a minimum of four hours for a task that is completed in one hour.

All of the above areas can contribute to increased costs for a repair and remodeling project. Each of the above factors should be considered in the planning, bidding and construction stage in order to minimize the increased costs associated with repair and remodeling jobs.

813

Reference Tables

R012909-80   Sales Tax by State

State sales tax on materials is tabulated below (5 states have no sales tax). Many states allow local jurisdictions, such as a county or city, to levy additional sales tax.

Some projects may be sales tax exempt, particularly those constructed with public funds.

State	Tax (%)	State	Tax (%)	State	Tax (%)	State	Tax (%)
Alabama	4	Illinois	6.25	Montana	0	Rhode Island	7
Alaska	0	Indiana	7	Nebraska	5.5	South Carolina	6
Arizona	5.6	Iowa	6	Nevada	6.85	South Dakota	4
Arkansas	6	Kansas	6.15	New Hampshire	0	Tennessee	7
California	7.5	Kentucky	6	New Jersey	7	Texas	6.25
Colorado	2.9	Louisiana	4	New Mexico	5.125	Utah	5.95
Connecticut	6.35	Maine	5.5	New York	4	Vermont	6
Delaware	0	Maryland	6	North Carolina	4.75	Virginia	5.3
District of Columbia	5.75	Massachusetts	6.25	North Dakota	5	Washington	6.5
Florida	6	Michigan	6	Ohio	5.75	West Virginia	6
Georgia	4	Minnesota	6.875	Oklahoma	4.5	Wisconsin	5
Hawaii	4	Mississippi	7	Oregon	0	Wyoming	4
Idaho	6	Missouri	4.225	Pennsylvania	6	Average	5.08%

Sales Tax by Province (Canada)

GST - a value-added tax, which the government imposes on most goods and services provided in or imported into Canada. PST - a retail sales tax, which three of the provinces impose on the prices of most goods and some services. QST - a value-added tax, similar to the federal GST, which Quebec imposes. HST - Five provinces have combined their retail sales taxes with the federal GST into one harmonized tax.

Province	PST (%)	QST (%)	GST(%)	HST(%)
Alberta	0	0	5	0
British Columbia	7	0	5	0
Manitoba	8	0	5	0
New Brunswick	0	0	0	13
Newfoundland	0	0	0	13
Northwest Territories	0	0	5	0
Nova Scotia	0	0	0	15
Ontario	0	0	0	13
Prince Edward Island	0	0	0	14
Quebec	0	9.975	5	0
Saskatchewan	5	0	5	0
Yukon	0	0	5	0

R012909-85   Unemployment Taxes and Social Security Taxes

State unemployment tax rates vary not only from state to state, but also with the experience rating of the contractor. The federal unemployment tax rate is 6.0% of the first $7,000 of wages. This is reduced by a credit of up to 5.4% for timely payment to the state. The minimum federal unemployment tax is 0.6% after all credits.

Social security (FICA) for 2016 is estimated at time of publication to be 7.65% of wages up to $118,500.

R013113-40   Builder's Risk Insurance

Builder's risk insurance is insurance on a building during construction. Premiums are paid by the owner or the contractor. Blasting, collapse and underground insurance would raise total insurance costs above those listed. Floater policy for materials delivered to the job runs $.75 to $1.25 per $100 value. Contractor equipment insurance runs $.50 to $1.50 per $100 value. Insurance for miscellaneous tools to $1,500 value runs from $3.00 to $7.50 per $100 value.

Tabulated below are New England Builder's Risk insurance rates in dollars per $100 value for a $1,000 deductible. For a $25,000 deductible, rates can be reduced 13% to 34%. On contracts over $1,000,000, rates may be lower than those tabulated. Policies are written annually for the total completed value in place. For "all risk" insurance (excluding flood, earthquake and certain other perils) add $.025 to total rates below.

Coverage	Frame Construction (Class 1)			Brick Construction (Class 4)			Fire Resistive (Class 6)		
	Range		Average	Range		Average	Range		Average
Fire Insurance	$.350 to	$ .850	$.600	$.158 to	$.189	$.174	$.052 to	$.080	$.070
Extended Coverage	.115 to	.200	.158	.080 to	.105	.101	.081 to	.105	.100
Vandalism	.012 to	.016	.014	.008 to	.011	.011	.008 to	.011	.010
Total Annual Rate	$.477 to	$1.066	$.772	$.246 to	$.305	$.286	$.141 to	$.196	$.180

R013113-50   General Contractor's Overhead

There are two distinct types of overhead on a construction project: project overhead and main office overhead. Project overhead includes those costs at a construction site not directly associated with the installation of construction materials. Examples of project overhead costs include the following:

1. Superintendent
2. Construction office and storage trailers
3. Temporary sanitary facilities
4. Temporary utilities
5. Security fencing
6. Photographs
7. Clean up
8. Performance and payment bonds

The above Project Overhead items are also referred to as General Requirements and therefore are estimated in Division 1. Division 1 is the first division listed in the CSI MasterFormat but it is usually the last division estimated. The sum of the costs in Divisions 1 through 49 is referred to as the sum of the direct costs.

All construction projects also include indirect costs. The primary components of indirect costs are the contractor's main office overhead and profit. The amount of the main office overhead expense varies depending on the following:

1. Owner's compensation
2. Project managers and estimator's wages
3. Clerical support wages
4. Office rent and utilities
5. Corporate legal and accounting costs
6. Advertising
7. Automobile expenses
8. Association dues
9. Travel and entertainment expenses

These costs are usually calculated as a percentage of annual sales volume. This percentage can range from 35% for a small contractor doing less than $500,000 to 5% for a large contractor with sales in excess of $100 million.

R013113-60 Workers' Compensation Insurance Rates by Trade

The table below tabulates the national averages for workers' compensation insurance rates by trade and type of building. The average "Insurance Rate" is multiplied by the "% of Building Cost" for each trade. This produces the "Workers' Compensation" cost by % of total labor cost, to be added for each trade by building type to determine the weighted average workers' compensation rate for the building types analyzed.

Trade	Insurance Rate (% Labor Cost) Range	Average	% of Building Cost Office Bldgs.	Schools & Apts.	Mfg.	Workers' Compensation Office Bldgs.	Schools & Apts.	Mfg.
Excavation, Grading, etc.	3.1 % to 21.5%	9.3%	4.8%	4.9%	4.5%	0.45%	0.46%	0.42%
Piles & Foundations	6.4 to 32.8	14.6	7.1	5.2	8.7	1.04	0.76	1.27
Concrete	4.3 to 32.5	12.6	5.0	14.8	3.7	0.63	1.86	0.47
Masonry	4.6 to 36.8	13.7	6.9	7.5	1.9	0.95	1.03	0.26
Structural Steel	6.5 to 95.8	27.5	10.7	3.9	17.6	2.94	1.07	4.84
Miscellaneous & Ornamental Metals	4.1 to 28.9	11.7	2.8	4.0	3.6	0.33	0.47	0.42
Carpentry & Millwork	5.0 to 42.8	14.4	3.7	4.0	0.5	0.53	0.58	0.07
Metal or Composition Siding	6.5 to 75.9	18.7	2.3	0.3	4.3	0.43	0.06	0.80
Roofing	6.5 to 107.1	31.4	2.3	2.6	3.1	0.72	0.82	0.97
Doors & Hardware	4.1 to 42.8	11.4	0.9	1.4	0.4	0.10	0.16	0.05
Sash & Glazing	4.8 to 28.0	13.1	3.5	4.0	1.0	0.46	0.52	0.13
Lath & Plaster	2.8 to 41.3	10.5	3.3	6.9	0.8	0.35	0.72	0.08
Tile, Marble & Floors	2.9 to 22.4	8.8	2.6	3.0	0.5	0.23	0.26	0.04
Acoustical Ceilings	3.1 to 33.8	8.6	2.4	0.2	0.3	0.21	0.02	0.03
Painting	4.1 to 34.3	11.7	1.5	1.6	1.6	0.18	0.19	0.19
Interior Partitions	5.0 to 42.8	14.4	3.9	4.3	4.4	0.56	0.62	0.63
Miscellaneous Items	2.5 to 105.9	12.6	5.2	3.7	9.7	0.65	0.47	1.22
Elevators	1.4 to 13.6	5.2	2.1	1.1	2.2	0.11	0.06	0.11
Sprinklers	2.8 to 25.9	7.6	0.5	—	2.0	0.04	—	0.15
Plumbing	1.9 to 19.5	6.9	4.9	7.2	5.2	0.34	0.50	0.36
Heat., Vent., Air Conditioning	3.8 to 18.4	8.8	13.5	11.0	12.9	1.19	0.97	1.14
Electrical	2.3 to 14.4	5.6	10.1	8.4	11.1	0.57	0.47	0.62
Total	1.4 % to 107.1%	—	100.0%	100.0%	100.0%	13.01%	12.07%	14.27%
		Overall Weighted Average	13.12%					

Workers' Compensation Insurance Rates by States

The table below lists the weighted average Workers' Compensation base rate for each state with a factor comparing this with the national average of 13.0%.

State	Weighted Average	Factor	State	Weighted Average	Factor	State	Weighted Average	Factor
Alabama	19.5%	150	Kentucky	14.3%	110	North Dakota	9.2%	71
Alaska	13.7	105	Louisiana	18.6	143	Ohio	8.2	63
Arizona	12.6	97	Maine	12.2	94	Oklahoma	12.5	96
Arkansas	8.3	64	Maryland	14.1	108	Oregon	11.8	91
California	25.8	198	Massachusetts	11.6	89	Pennsylvania	17.4	134
Colorado	7.2	55	Michigan	16.3	125	Rhode Island	12.1	93
Connecticut	23.7	182	Minnesota	22.6	174	South Carolina	18.9	145
Delaware	11.9	92	Mississippi	13.2	102	South Dakota	14.9	115
District of Columbia	11.0	85	Missouri	14.8	114	Tennessee	12.0	92
Florida	10.7	82	Montana	8.5	65	Texas	9.1	70
Georgia	28.1	216	Nebraska	17.7	136	Utah	8.8	68
Hawaii	8.1	62	Nevada	8.7	67	Vermont	13.1	101
Idaho	10.3	79	New Hampshire	20.0	154	Virginia	9.5	73
Illinois	27.1	208	New Jersey	14.4	111	Washington	10.5	81
Indiana	5.5	42	New Mexico	14.4	111	West Virginia	8.0	62
Iowa	14.6	112	New York	17.2	132	Wisconsin	13.1	101
Kansas	8.4	65	North Carolina	16.7	128	Wyoming	6.1	47
			Weighted Average for U.S. is	13.7% of payroll = 100%				

The weighted average skilled worker rate for 35 trades is 13.0%. For bidding purposes, apply the full value of Workers' Compensation directly to total labor costs, or if labor is 38%, materials 42% and overhead and profit 20% of total cost, carry 38/80 x 13.0% = 6.2% of cost (before overhead and profit)

into overhead. Rates vary not only from state to state but also with the experience rating of the contractor.

Rates are the most current available at the time of publication.

R013113-80   Performance Bond

This table shows the cost of a Performance Bond for a construction job scheduled to be completed in 12 months. Add 1% of the premium cost per month for jobs requiring more than 12 months to complete. The rates are "standard" rates offered to contractors that the bonding company considers financially sound and capable of doing the work. Preferred rates are offered by some bonding companies based upon financial strength of the contractor. Actual rates vary from contractor to contractor and from bonding company to bonding company. Contractors should prequalify through a bonding agency before submitting a bid on a contract that requires a bond.

Contract Amount	Building Construction Class B Projects			Highways & Bridges					
				Class A New Construction			Class A-1 Highway Resurfacing		
First $ 100,000 bid	\$25.00 per M			\$15.00 per M			\$9.40 per M		
Next 400,000 bid	$ 2,500	plus	\$15.00 per M	$ 1,500	plus	\$10.00 per M	$ 940	plus	\$7.20 per M
Next 2,000,000 bid	8,500	plus	10.00 per M	5,500	plus	7.00 per M	3,820	plus	5.00 per M
Next 2,500,000 bid	28,500	plus	7.50 per M	19,500	plus	5.50 per M	15,820	plus	4.50 per M
Next 2,500,000 bid	47,250	plus	7.00 per M	33,250	plus	5.00 per M	28,320	plus	4.50 per M
Over 7,500,000 bid	64,750	plus	6.00 per M	45,750	plus	4.50 per M	39,570	plus	4.00 per M

General Requirements   R0154 Construction Aids

R015423-10   Steel Tubular Scaffolding

On new construction, tubular scaffolding is efficient up to 60' high or five stories. Above this it is usually better to use a hung scaffolding if construction permits. Swing scaffolding operations may interfere with tenants. In this case, the tubular is more practical at all heights.

In repairing or cleaning the front of an existing building the cost of tubular scaffolding per S.F. of building front increases as the height increases above the first tier. The first tier cost is relatively high due to leveling and alignment.

The minimum efficient crew for erecting and dismantling is three workers. They can set up and remove 18 frame sections per day up to 5 stories high. For 6 to 12 stories high, a crew of four is most efficient. Use two or more on top and two on the bottom for handing up or hoisting. They can also set up and remove 18 frame sections per day. At 7' horizontal spacing, this will run about 800 S.F. per day of erecting and dismantling. Time for placing and removing planks must be added to the above. A crew of three can place and remove 72 planks per day up to 5 stories. For over 5 stories, a crew of four can place and remove 80 planks per day.

The table below shows the number of pieces required to erect tubular steel scaffolding for 1000 S.F. of building frontage. This area is made up of a scaffolding system that is 12 frames (11 bays) long by 2 frames high.

For jobs under twenty-five frames, add 50% to rental cost. Rental rates will be lower for jobs over three months duration. Large quantities for long periods can reduce rental rates by 20%.

Description of Component	Number of Pieces for 1000 S.F. of Building Front	Unit
5' Wide Standard Frame, 6'-4" High	24	Ea.
Leveling Jack & Plate	24	
Cross Brace	44	
Side Arm Bracket, 21"	12	
Guardrail Post	12	
Guardrail, 7' section	22	
Stairway Section	2	
Stairway Starter Bar	1	
Stairway Inside Handrail	2	
Stairway Outside Handrail	2	
Walk-Thru Frame Guardrail	2	

Scaffolding is often used as falsework over 15' high during construction of cast-in-place concrete beams and slabs. Two foot wide scaffolding is generally used for heavy beam construction. The span between frames depends upon the load to be carried with a maximum span of 5'.

Heavy duty shoring frames with a capacity of 10,000#/leg can be spaced up to 10' O.C. depending upon form support design and loading.

Scaffolding used as horizontal shoring requires less than half the material required with conventional shoring.

On new construction, erection is done by carpenters.

Rolling towers supporting horizontal shores can reduce labor and speed the job. For maintenance work, catwalks with spans up to 70' can be supported by the rolling towers.

817

R015423-20   Pump Staging

Pump staging is generally not available for rent. The table below shows the number of pieces required to erect pump staging for 2400 S.F. of building frontage. This area is made up of a pump jack system that is 3 poles (2 bays) wide by 2 poles high.

Item	Number of Pieces for 2400 S.F. of Building Front	Unit
Aluminum pole section, 24' long	6	Ea.
Aluminum splice joint, 6' long	3	
Aluminum foldable brace	3	
Aluminum pump jack	3	
Aluminum support for workbench/back safety rail	3	
Aluminum scaffold plank/workbench, 14" wide x 24' long	4	
Safety net, 22' long	2	
Aluminum plank end safety rail	2	↓

The cost in place for this 2400 S.F. will depend on how many uses are realized during the life of the equipment.

R015433-10   Contractor Equipment

Rental Rates shown elsewhere in the data set pertain to late model high quality machines in excellent working condition, rented from equipment dealers. Rental rates from contractors may be substantially lower than the rental rates from equipment dealers depending upon economic conditions; for older, less productive machines, reduce rates by a maximum of 15%. Any overtime must be added to the base rates. For shift work, rates are lower. Usual rule of thumb is 150% of one shift rate for two shifts; 200% for three shifts.

For periods of less than one week, operated equipment is usually more economical to rent than renting bare equipment and hiring an operator.

Costs to move equipment to a job site (mobilization) or from a job site (demobilization) are not included in rental rates, nor in any Equipment costs on any Unit Price line items or crew listings. These costs can be found elsewhere. If a piece of equipment is already at a job site, it is not appropriate to utilize mob/demob costs in an estimate again.

Rental rates vary throughout the country with larger cities generally having lower rates. Lease plans for new equipment are available for periods in excess of six months with a percentage of payments applying toward purchase.

Rental rates can also be treated as reimbursement costs for contractor-owned equipment. Owned equipment costs include depreciation, loan payments, interest, taxes, insurance, storage, and major repairs.

Monthly rental rates vary from 2% to 5% of the cost of the equipment depending on the anticipated life of the equipment and its wearing parts. Weekly rates are about 1/3 the monthly rates and daily rental rates are about 1/3 the weekly rates.

The hourly operating costs for each piece of equipment include costs to the user such as fuel, oil, lubrication, normal expendables for the equipment, and a percentage of the mechanic's wages chargeable to maintenance. The hourly operating costs listed do not include the operator's wages.

The daily cost for equipment used in the standard crews is figured by dividing the weekly rate by five, then adding eight times the hourly operating cost to give the total daily equipment cost, not including the operator. This figure is in the right hand column of the Equipment listings under Equipment Cost/Day.

Pile Driving rates shown for the pile hammer and extractor do not include leads, cranes, boilers or compressors. Vibratory pile driving requires an added field specialist during set-up and pile driving operation for the electric model. The hydraulic model requires a field specialist for set-up only. Up to 125 reuses of sheet piling are possible using vibratory drivers. For normal conditions, crane capacity for hammer type and size is as follows.

Crane Capacity	Hammer Type and Size		
	Air or Steam	Diesel	Vibratory
25 ton	to 8,750 ft.-lb.		70 H.P.
40 ton	15,000 ft.-lb.	to 32,000 ft.-lb.	170 H.P.
60 ton	25,000 ft.-lb.		300 H.P.
100 ton		112,000 ft.-lb.	

Cranes should be specified for the job by size, building and site characteristics, availability, performance characteristics, and duration of time required.

Backhoes & Shovels rent for about the same as equivalent size cranes but maintenance and operating expenses are higher. The crane operator's rate must be adjusted for high boom heights. Average adjustments: for 150' boom add 2% per hour; over 185', add 4% per hour; over 210', add 6% per hour; over 250', add 8% per hour and over 295', add 12% per hour.

Tower Cranes of the climbing or static type have jibs from 50' to 200' and capacities at maximum reach range from 4,000 to 14,000 pounds. Lifting capacities increase up to maximum load as the hook radius decreases.

Typical rental rates, based on purchase price, are about 2% to 3% per month.

Erection and dismantling run between 500 and 2000 labor hours. Climbing operation takes 10 labor hours per 20' climb. Crane dead time is about 5 hours per 40' climb. If crane is bolted to side of the building add cost of ties and extra mast sections. Climbing cranes have from 80' to 180' of mast while static cranes have 80' to 800' of mast.

Truck Cranes can be converted to tower cranes by using tower attachments. Mast heights over 400' have been used.

A single 100' high material **Hoist and Tower** can be erected and dismantled in about 400 labor hours; a double 100' high hoist and tower in about 600 labor hours. Erection times for additional heights are 3 and 4 labor hours

per vertical foot respectively up to 150', and 4 to 5 labor hours per vertical foot over 150' high. A 40' high portable Buck hoist takes about 160 labor hours to erect and dismantle. Additional heights take 2 labor hours per vertical foot to 80' and 3 labor hours per vertical foot for the next 100'. Most material hoists do not meet local code requirements for carrying personnel.

A 150' high **Personnel Hoist** requires about 500 to 800 labor hours to erect and dismantle. Budget erection time at 5 labor hours per vertical foot for all trades. Local code requirements or labor scarcity requiring overtime can add up to 50% to any of the above erection costs.

Earthmoving Equipment: The selection of earthmoving equipment depends upon the type and quantity of material, moisture content, haul distance, haul road, time available, and equipment available. Short haul cut and fill operations may require dozers only, while another operation may require excavators, a fleet of trucks, and spreading and compaction equipment. Stockpiled material and granular material are easily excavated with front end loaders. Scrapers are most economically used with hauls between 300' and 1-1/2 miles if adequate haul roads can be maintained. Shovels are often used for blasted rock and any material where a vertical face of 8' or more can be excavated. Special conditions may dictate the use of draglines, clamshells, or backhoes. Spreading and compaction equipment must be matched to the soil characteristics, the compaction required and the rate the fill is being supplied.

819

R015436-50   Mobilization

Costs to move rented construction equipment to a job site from an equipment dealer's or contractor's yard (mobilization) or off the job site (demobilization) are not included in the rental or operating rates, nor in the equipment cost on a unit price line or in a crew listing. These costs can be found consolidated in the Mobilization section of the data and elsewhere in particular site work sections. If a piece of equipment is already on the job site, it is not appropriate to include mob/demob costs in a new estimate that requires use of that equipment. The following table identifies approximate sizes of rented construction equipment that would be hauled on a towed trailer. Because this listing is not all-encompassing, the user can infer as to what size trailer might be required for a piece of equipment not listed.

3-ton Trailer	20-ton Trailer	40-ton Trailer	50-ton Trailer
20 H.P. Excavator	110 H.P. Excavator	200 H.P. Excavator	270 H.P. Excavator
50 H.P. Skid Steer	165 H.P. Dozer	300 H.P. Dozer	Small Crawler Crane
35 H.P. Roller	150 H.P. Roller	400 H.P. Scraper	500 H.P. Scraper
40 H.P. Trencher	Backhoe	450 H.P. Art. Dump Truck	500 H.P. Art. Dump Truck

R024119-10 Demolition Defined

Whole Building Demolition - Demolition of the whole building with no concern for any particular building element, component, or material type being demolished. This type of demolition is accomplished with large pieces of construction equipment that break up the structure, load it into trucks and haul it to a disposal site, but disposal or dump fees are not included. Demolition of below-grade foundation elements, such as footings, foundation walls, grade beams, slabs on grade, etc., is not included. Certain mechanical equipment containing flammable liquids or ozone-depleting refrigerants, electric lighting elements, communication equipment components, and other building elements may contain hazardous waste, and must be removed, either selectively or carefully, as hazardous waste before the building can be demolished.

Foundation Demolition - Demolition of below-grade foundation footings, foundation walls, grade beams, and slabs on grade. This type of demolition is accomplished by hand or pneumatic hand tools, and does not include saw cutting, or handling, loading, hauling, or disposal of the debris.

Gutting - Removal of building interior finishes and electrical/mechanical systems down to the load-bearing and sub-floor elements of the rough building frame, with no concern for any particular building element, component, or material type being demolished. This type of demolition is accomplished by hand or pneumatic hand tools, and includes loading into trucks, but not hauling, disposal or dump fees, scaffolding, or shoring. Certain mechanical equipment containing flammable liquids or ozone-depleting refrigerants, electric lighting elements, communication equipment components, and other building elements may contain hazardous waste, and must be removed, either selectively or carefully, as hazardous waste, before the building is gutted.

Selective Demolition - Demolition of a selected building element, component, or finish, with some concern for surrounding or adjacent elements, components, or finishes (see the first Subdivision (s) at the beginning of appropriate Divisions). This type of demolition is accomplished by hand or pneumatic hand tools, and does not include handling, loading, storing, hauling, or disposal of the debris, scaffolding, or shoring. "Gutting" methods may be used in order to save time, but damage that is caused to surrounding or adjacent elements, components, or finishes may have to be repaired at a later time.

Careful Removal - Removal of a piece of service equipment, building element or component, or material type, with great concern for both the removed item and surrounding or adjacent elements, components or finishes. The purpose of careful removal may be to protect the removed item for later re-use, preserve a higher salvage value of the removed item, or replace an item while taking care to protect surrounding or adjacent elements, components, connections, or finishes from cosmetic and/or structural damage. An approximation of the time required to perform this type of removal is 1/3 to 1/2 the time it would take to install a new item of like kind (see Reference Number R220105-10). This type of removal is accomplished by hand or pneumatic hand tools, and does not include loading, hauling, or storing the removed item, scaffolding, shoring, or lifting equipment.

Cutout Demolition - Demolition of a small quantity of floor, wall, roof, or other assembly, with concern for the appearance and structural integrity of the surrounding materials. This type of demolition is accomplished by hand or pneumatic hand tools, and does not include saw cutting, handling, loading, hauling, or disposal of debris, scaffolding, or shoring.

Rubbish Handling - Work activities that involve handling, loading or hauling of debris. Generally, the cost of rubbish handling must be added to the cost of all types of demolition, with the exception of whole building demolition.

Minor Site Demolition - Demolition of site elements outside the footprint of a building. This type of demolition is accomplished by hand or pneumatic hand tools, or with larger pieces of construction equipment, and may include loading a removed item onto a truck (check the Crew for equipment used). It does not include saw cutting, hauling or disposal of debris, and, sometimes, handling or loading.

R024119-20 Dumpsters

Dumpster rental costs on construction sites are presented in two ways.

The cost per week rental includes the delivery of the dumpster; its pulling or emptying once per week, and its final removal. The assumption is made that the dumpster contractor could choose to empty a dumpster by simply bringing in an empty unit and removing the full one. These costs also include the disposal of the materials in the dumpster.

The Alternate Pricing can be used when actual planned conditions are not approximated by the weekly numbers. For example, these lines can be used when a dumpster is needed for 4 weeks and will need to be emptied 2 or 3 times per week. Conversely the Alternate Pricing lines can be used when a dumpster will be rented for several weeks or months but needs to be emptied only a few times over this period.

R024119-30 Rubbish Handling Chutes

To correctly estimate the cost of rubbish handling chute systems, the individual components must be priced separately. First choose the size of the system; a 30-inch diameter chute is quite common, but the sizes range from 18 to 36 inches in diameter. The 30-inch chute comes in a standard weight and two thinner weights. The thinner weight chutes are sometimes chosen for cost savings, but they are more easily damaged.

There are several types of major chute pieces that make up the chute system. The first component to consider is the top chute section (top intake hopper) where the material is dropped into the chute at the highest point. After determining the top chute, the intermediate chute pieces called the regular chute sections are priced. Next, the number of chute control door sections (intermediate intake hoppers) must be determined. In the more complex systems, a chute control door section is provided at each floor level. The last major component to consider is bolt down frames; these are usually provided at every other floor level.

There are a number of accessories to consider for safe operation and control. There are covers for the top chute and the chute door sections. The top chute can have a trough that allows for better loading of the chute. For the safest operation, a chute warning light system can be added that will warn the other chute intake locations not to load while another is being used. There are dust control devices that spray a water mist to keep down the dust as the debris is loaded into a Dumpster. There are special breakaway cords that are used to prevent damage to the chute if the dumpster is removed without disconnecting from the chute. There are chute liners that can be installed to protect the chute structure from physical damage from rough abrasive materials. Warning signs can be posted at each floor level that is provided with a chute control door section.

In summary, a complete rubbish handling chute system will include one top section, several intermediate regular sections, several intermediate control door (intake hopper) sections and bolt down frames at every other floor level starting with the top floor. If so desired, the system can also include covers and a light warning system for a safer operation. The bottom of the chute should always be above the Dumpster and should be tied off with a breakaway cord to the Dumpster.

821

Existing Conditions — R0265 Underground Storage Tank Removal

R026510-20   Underground Storage Tank Removal

Underground storage tank removal can be divided into two categories: non-leaking and leaking. Prior to removing an underground storage tank, tests should be made, with the proper authorities present, to determine whether a tank has been leaking or the surrounding soil has been contaminated.

To safely remove liquid underground storage tanks:
1. Excavate to the top of the tank.
2. Disconnect all piping.
3. Open all tank vents and access ports.
4. Remove all liquids and/or sludge.
5. Purge the tank with an inert gas.
6. Provide access to the inside of the tank and clean out the interior using proper personal protective equipment (PPE).
7. Excavate soil surrounding the tank using proper PPE for on-site personnel.
8. Pull and properly dispose of the tank.
9. Clean up the site of all contaminated material.
10. Install new tanks or close the excavation.

Existing Conditions — R0282 Asbestos Remediation

R028213-20   Asbestos Removal Process

Asbestos removal is accomplished by a specialty contractor who understands the federal and state regulations regarding the handling and disposal of the material. The process of asbestos removal is divided into many individual steps. An accurate estimate can be calculated only after all the steps have been priced.

The steps are generally as follows:
1. Obtain an asbestos abatement plan from an industrial hygienist.
2. Monitor the air quality in and around the removal area and along the path of travel between the removal area and transport area. This establishes the background contamination.
3. Construct a two part decontamination chamber at entrance to removal area.
4. Install a HEPA filter to create a negative pressure in the removal area.
5. Install wall, floor and ceiling protection as required by the plan, usually 2 layers of fireproof 6 mil polyethylene.
6. Industrial hygienist visually inspects work area to verify compliance with plan.
7. Provide temporary supports for conduit and piping affected by the removal process.
8. Proceed with asbestos removal and bagging process. Monitor air quality as described in Step #2. Discontinue operations when contaminate levels exceed applicable standards.
9. Document the legal disposal of materials in accordance with EPA standards.
10. Thoroughly clean removal area including all ledges, crevices and surfaces.
11. Post abatement inspection by industrial hygienist to verify plan compliance.
12. Provide a certificate from a licensed industrial hygienist attesting that contaminate levels are within acceptable standards before returning area to regular use.

822

For customer support on your Commercial Renovation Cost Data, call 877.742.4719.

R028319-60   Lead Paint Remediation Methods

Lead paint remediation can be accomplished by the following methods.
1. Abrasive blast
2. Chemical stripping
3. Power tool cleaning with vacuum collection system
4. Encapsulation
5. Remove and replace
6. Enclosure

Each of these methods has strengths and weaknesses depending on the specific circumstances of the project. The following is an overview of each method.

1. **Abrasive blasting** is usually accomplished with sand or recyclable metallic blast. Before work can begin, the area must be contained to ensure the blast material with lead does not escape to the atmosphere. The use of vacuum blast greatly reduces the containment requirements. Lead abatement equipment that may be associated with this work includes a negative air machine. In addition, it is necessary to have an industrial hygienist monitor the project on a continual basis. When the work is complete, the spent blast sand with lead must be disposed of as a hazardous material. If metallic shot was used, the lead is separated from the shot and disposed of as hazardous material. Worker protection includes disposable clothing and respiratory protection.

2. **Chemical stripping** requires strong chemicals to be applied to the surface to remove the lead paint. Before the work can begin, the area under/adjacent to the work area must be covered to catch the chemical and removed lead. After the chemical is applied to the painted surface it is usually covered with paper. The chemical is left in place for the specified period, then the paper with lead paint is pulled or scraped off. The process may require several chemical applications. The paper with chemicals and lead paint adhered to it, plus the containment and loose scrapings collected by a HEPA (High Efficiency Particulate Air Filter) vac, must be disposed of as hazardous material. The chemical stripping process usually requires a neutralizing agent and several wash downs after the paint is removed. Worker protection includes a neoprene or other compatible protective clothing and respiratory protection with face shield. An industrial hygienist is required intermittently during the process.

3. **Power tool cleaning** is accomplished using shrouded needle blasting guns. The shrouding with different end configurations is held up against the surface to be cleaned. The area is blasted with hardened needles and the shroud captures the lead with a HEPA vac and deposits it in a holding tank. An industrial hygienist monitors the project. Protective clothing and a respirator are required until air samples prove otherwise. When the work is complete the lead must be disposed of as hazardous material.

4. **Encapsulation** is a method that leaves the well bonded lead paint in place after the peeling paint has been removed. Before the work can begin, the area under/adjacent to the work must be covered to catch the scrapings. The scraped surface is then washed with a detergent and rinsed. The prepared surface is covered with approximately 10 mils of paint. A reinforcing fabric can also be embedded in the paint covering. The scraped paint and containment must be disposed of as hazardous material. Workers must wear protective clothing and respirators.

5. **Removing and replacing** are effective ways to remove lead paint from windows, gypsum walls and concrete masonry surfaces. The painted materials are removed and new materials are installed. Workers should wear a respirator and tyvek suit. The demolished materials must be disposed of as hazardous waste if it fails the TCLP (Toxicity Characteristic Leachate Process) test.

6. **Enclosure** is the process that permanently seals lead painted materials in place. This process has many applications such as covering lead painted drywall with new drywall, covering exterior construction with tyvek paper then residing, or covering lead painted structural members with aluminum or plastic. The seams on all enclosing materials must be securely sealed. An industrial hygienist monitors the project, and protective clothing and a respirator are required until air samples prove otherwise.

All the processes require clearance monitoring and wipe testing as required by the hygienist.

R031113-10   Wall Form Materials

Aluminum Forms

Approximate weight is 3 lbs. per S.F.C.A. Standard widths are available from 4″ to 36″ with 36″ most common. Standard lengths of 2′, 4′, 6′ to 8′ are available. Forms are lightweight and fewer ties are needed with the wider widths. The form face is either smooth or textured.

Metal Framed Plywood Forms

Manufacturers claim over 75 reuses of plywood and over 300 reuses of steel frames. Many specials such as corners, fillers, pilasters, etc. are available. Monthly rental is generally about 15% of purchase price for first month and 9% per month thereafter with 90% of rental applied to purchase for the first month and decreasing percentages thereafter. Aluminum framed forms cost 25% to 30% more than steel framed.

After the first month, extra days may be prorated from the monthly charge. Rental rates do not include ties, accessories, cleaning, loss of hardware or freight in and out. Approximate weight is 5 lbs. per S.F. for steel; 3 lbs. per S.F. for aluminum.

Forms can be rented with option to buy.

Plywood Forms, Job Fabricated

There are two types of plywood used for concrete forms.
1. Exterior plyform which is completely waterproof. This is face oiled to facilitate stripping. Ten reuses can be expected with this type with 25 reuses possible.
2. An overlaid type consists of a resin fiber fused to exterior plyform. No oiling is required except to facilitate cleaning. This is available in both high density (HDO) and medium density overlaid (MDO). Using HDO, 50 reuses can be expected with 200 possible.

Plyform is available in 5/8″ and 3/4″ thickness. High density overlaid is available in 3/8″, 1/2″, 5/8″ and 3/4″ thickness.

5/8″ thick is sufficient for most building forms, while 3/4″ is best on heavy construction.

Plywood Forms, Modular, Prefabricated

There are many plywood forming systems without frames. Most of these are manufactured from 1-1/8″ (HDO) plywood and have some hardware attached. These are used principally for foundation walls 8′ or less high. With care and maintenance, 100 reuses can be attained with decreasing quality of surface finish.

Steel Forms

Approximate weight is 6-1/2 lbs. per S.F.C.A. including accessories. Standard widths are available from 2″ to 24″, with 24″ most common. Standard lengths are from 2′ to 8′, with 4′ the most common. Forms are easily ganged into modular units.

Forms are usually leased for 15% of the purchase price per month prorated daily over 30 days.

Rental may be applied to sale price, and usually rental forms are bought. With careful handling and cleaning 200 to 400 reuses are possible.

Straight wall gang forms up to 12′ x 20′ or 8′ x 30′ can be fabricated. These crane handled forms usually lease for approx. 9% per month.

Individual job analysis is available from the manufacturer at no charge.

R031113-40   Forms for Reinforced Concrete

Design Economy

Avoid many sizes in proportioning beams and columns.

From story to story avoid changing column dimensions. Gain strength by adding steel or using a richer mix. If a change in size of column is necessary, vary one dimension only to minimize form alterations. Keep beams and columns the same width.

From floor to floor in a multi-story building vary beam depth, not width, as that will leave the slab panel form unchanged. It is cheaper to vary the strength of a beam from floor to floor by means of a steel area than by 2″ changes in either width or depth.

Cost Factors

Material includes the cost of lumber, cost of rent for metal pans or forms if used, nails, form ties, form oil, bolts and accessories.

Labor includes the cost of carpenters to make up, erect, remove and repair, plus common labor to clean and move. Having carpenters remove forms minimizes repairs.

Improper alignment and condition of forms will increase finishing cost. When forms are heavily oiled, concrete surfaces must be neutralized before finishing. Special curing compounds will cause spillages to spall off in first frost. Gang forming methods will reduce costs on large projects.

Materials Used

Boards are seldom used unless their architectural finish is required. Generally, steel, fiberglass and plywood are used for contact surfaces. Labor on plywood is 10% less than with boards. The plywood is backed up with

2 x 4's at 12″ to 32″ O.C. Walers are generally 2 - 2 x 4's. Column forms are held together with steel yokes or bands. Shoring is with adjustable shoring or scaffolding for high ceilings.

Reuse

Floor and column forms can be reused four or possibly five times without excessive repair. Remember to allow for 10% waste on each reuse.

When modular sized wall forms are made, up to twenty uses can be expected with exterior plyform.

When forms are reused, the cost to erect, strip, clean and move will not be affected. 10% replacement of lumber should be included and about one hour of carpenter time for repairs on each reuse per 100 S.F.

The reuse cost for certain accessory items normally rented on a monthly basis will be lower than the cost for the first use.

After the fifth use, new material required plus time needed for repair prevent the form cost from dropping further; it may go up. Much depends on care in stripping, the number of special bays, changes in beam or column sizes and other factors.

Costs for multiple use of formwork may be developed as follows:

2 Uses	3 Uses	4 Uses
$\dfrac{\text{(1st Use + Reuse)}}{2} = \text{avg. cost/2 uses}$	$\dfrac{\text{(1st Use + 2 Reuses)}}{3} = \text{avg. cost/3 uses}$	$\dfrac{\text{(1st use + 3 Reuses)}}{4} = \text{avg. cost/4 uses}$

Reference Tables

R031113-60 Formwork Labor-Hours

Item	Unit	Hours Required Fabricate	Hours Required Erect & Strip	Hours Required Clean & Move	Total Hours 1 Use	Multiple Use 2 Use	Multiple Use 3 Use	Multiple Use 4 Use
Beam and Girder, interior beams, 12″ wide	100 S.F.	6.4	8.3	1.3	16.0	13.3	12.4	12.0
Hung from steel beams		5.8	7.7	1.3	14.8	12.4	11.6	11.2
Beam sides only, 36″ high		5.8	7.2	1.3	14.3	11.9	11.1	10.7
Beam bottoms only, 24″ wide		6.6	13.0	1.3	20.9	18.1	17.2	16.7
Box out for openings		9.9	10.0	1.1	21.0	16.6	15.1	14.3
Buttress forms, to 8′ high		6.0	6.5	1.2	13.7	11.2	10.4	10.0
Centering, steel, 3/4″ rib lath			1.0		1.0			
3/8″ rib lath or slab form			0.9		0.9			
Chamfer strip or keyway	100 L.F.		1.5		1.5	1.5	1.5	1.5
Columns, fiber tube 8″ diameter			20.6		20.6			
12″			21.3		21.3			
16″			22.9		22.9			
20″			23.7		23.7			
24″			24.6		24.6			
30″			25.6		25.6			
Columns, round steel, 12″ diameter			22.0		22.0	22.0	22.0	22.0
16″			25.6		25.6	25.6	25.6	25.6
20″			30.5		30.5	30.5	30.5	30.5
24″			37.7		37.7	37.7	37.7	37.7
Columns, plywood 8″ x 8″	100 S.F.	7.0	11.0	1.2	19.2	16.2	15.2	14.7
12″ x 12″		6.0	10.5	1.2	17.7	15.2	14.4	14.0
16″ x 16″		5.9	10.0	1.2	17.1	14.7	13.8	13.4
24″ x 24″		5.8	9.8	1.2	16.8	14.4	13.6	13.2
Columns, steel framed plywood 8″ x 8″			10.0	1.0	11.0	11.0	11.0	11.0
12″ x 12″			9.3	1.0	10.3	10.3	10.3	10.3
16″ x 16″			8.5	1.0	9.5	9.5	9.5	9.5
24″ x 24″			7.8	1.0	8.8	8.8	8.8	8.8
Drop head forms, plywood		9.0	12.5	1.5	23.0	19.0	17.7	17.0
Coping forms		8.5	15.0	1.5	25.0	21.3	20.0	19.4
Culvert, box			14.5	4.3	18.8	18.8	18.8	18.8
Curb forms, 6″ to 12″ high, on grade		5.0	8.5	1.2	14.7	12.7	12.1	11.7
On elevated slabs		6.0	10.8	1.2	18.0	15.5	14.7	14.3
Edge forms to 6″ high, on grade	100 L.F.	2.0	3.5	0.6	6.1	5.6	5.4	5.3
7″ to 12″ high	100 S.F.	2.5	5.0	1.0	8.5	7.8	7.5	7.4
Equipment foundations		10.0	18.0	2.0	30.0	25.5	24.0	23.3
Flat slabs, including drops		3.5	6.0	1.2	10.7	9.5	9.0	8.8
Hung from steel		3.0	5.5	1.2	9.7	8.7	8.4	8.2
Closed deck for domes		3.0	5.8	1.2	10.0	9.0	8.7	8.5
Open deck for pans		2.2	5.3	1.0	8.5	7.9	7.7	7.6
Footings, continuous, 12″ high		3.5	3.5	1.5	8.5	7.3	6.8	6.6
Spread, 12″ high		4.7	4.2	1.6	10.5	8.7	8.0	7.7
Pile caps, square or rectangular		4.5	5.0	1.5	11.0	9.3	8.7	8.4
Grade beams, 24″ deep		2.5	5.3	1.2	9.0	8.3	8.0	7.9
Lintel or Sill forms		8.0	17.0	2.0	27.0	23.5	22.3	21.8
Spandrel beams, 12″ wide		9.0	11.2	1.3	21.5	17.5	16.2	15.5
Stairs			25.0	4.0	29.0	29.0	29.0	29.0
Trench forms in floor		4.5	14.0	1.5	20.0	18.3	17.7	17.4
Walls, Plywood, at grade, to 8′ high		5.0	6.5	1.5	13.0	11.0	9.7	9.5
8′ to 16′		7.5	8.0	1.5	17.0	13.8	12.7	12.1
16′ to 20′		9.0	10.0	1.5	20.5	16.5	15.2	14.5
Foundation walls, to 8′ high		4.5	6.5	1.0	12.0	10.3	9.7	9.4
8′ to 16′ high		5.5	7.5	1.0	14.0	11.8	11.0	10.6
Retaining wall to 12′ high, battered		6.0	8.5	1.5	16.0	13.5	12.7	12.3
Radial walls to 12′ high, smooth		8.0	9.5	2.0	19.5	16.0	14.8	14.3
2′ chords		7.0	8.0	1.5	16.5	13.5	12.5	12.0
Prefabricated modular, to 8′ high		—	4.3	1.0	5.3	5.3	5.3	5.3
Steel, to 8′ high		—	6.8	1.2	8.0	8.0	8.0	8.0
8′ to 16′ high		—	9.1	1.5	10.6	10.3	10.2	10.2
Steel framed plywood to 8′ high		—	6.8	1.2	8.0	7.5	7.3	7.2
8′ to 16′ high		—	9.3	1.2	10.5	9.5	9.2	9.0

825

For customer support on your Commercial Renovation Cost Data, call 877.742.4719.

R032110-10   Reinforcing Steel Weights and Measures

Bar Designation No.**	Nominal Weight Lb./Ft.	U.S. Customary Units			SI Units			
		Nominal Dimensions*			Nominal Dimensions*			
		Diameter in.	Cross Sectional Area, in.^{2}	Perimeter in.	Nominal Weight kg/m	Diameter mm	Cross Sectional Area, cm^2	Perimeter mm
3	.376	.375	.11	1.178	.560	9.52	.71	29.9
4	.668	.500	.20	1.571	.994	12.70	1.29	39.9
5	1.043	.625	.31	1.963	1.552	15.88	2.00	49.9
6	1.502	.750	.44	2.356	2.235	19.05	2.84	59.8
7	2.044	.875	.60	2.749	3.042	22.22	3.87	69.8
8	2.670	1.000	.79	3.142	3.973	25.40	5.10	79.8
9	3.400	1.128	1.00	3.544	5.059	28.65	6.45	90.0
10	4.303	1.270	1.27	3.990	6.403	32.26	8.19	101.4
11	5.313	1.410	1.56	4.430	7.906	35.81	10.06	112.5
14	7.650	1.693	2.25	5.320	11.384	43.00	14.52	135.1
18	13.600	2.257	4.00	7.090	20.238	57.33	25.81	180.1

* The nominal dimensions of a deformed bar are equivalent to those of a plain round bar having the same weight per foot as the deformed bar.
** Bar numbers are based on the number of eighths of an inch included in the nominal diameter of the bars.

R032110-80   Shop-Fabricated Reinforcing Steel

The material prices for reinforcing, shown in the unit cost sections of the data set, are for 50 tons or more of shop-fabricated reinforcing steel and include:
1. Mill base price of reinforcing steel
2. Mill grade/size/length extras
3. Mill delivery to the fabrication shop
4. Shop storage and handling
5. Shop drafting/detailing
6. Shop shearing and bending
7. Shop listing
8. Shop delivery to the job site

Both material and installation costs can be considerably higher for small jobs consisting primarily of smaller bars, while material costs may be slightly lower for larger jobs.

R032205-30   Common Stock Styles of Welded Wire Fabric

This table provides some of the basic specifications, sizes, and weights of welded wire fabric used for reinforcing concrete.

New Designation		Old Designation		Steel Area per Foot				Approximate Weight per 100 S.F.	
Spacing — Cross Sectional Area (in.) — (Sq. in. 100)		Spacing — Wire Gauge (in.) — (AS & W)		Longitudinal		Transverse			
				in.	cm	in.	cm	lbs	kg
Rolls	6 x 6 — W1.4 x W1.4	6 x 6 — 10 x 10		.028	.071	.028	.071	21	9.53
	6 x 6 — W2.0 x W2.0	6 x 6 — 8 x 8	1	.040	.102	.040	.102	29	13.15
	6 x 6 — W2.9 x W2.9	6 x 6 — 6 x 6		.058	.147	.058	.147	42	19.05
	6 x 6 — W4.0 x W4.0	6 x 6 — 4 x 4		.080	.203	.080	.203	58	26.91
	4 x 4 — W1.4 x W1.4	4 x 4 — 10 x 10		.042	.107	.042	.107	31	14.06
	4 x 4 — W2.0 x W2.0	4 x 4 — 8 x 8	1	.060	.152	.060	.152	43	19.50
	4 x 4 — W2.9 x W2.9	4 x 4 — 6 x 6		.087	.227	.087	.227	62	28.12
	4 x 4 — W4.0 x W4.0	4 x 4 — 4 x 4		.120	.305	.120	.305	85	38.56
Sheets	6 x 6 — W2.9 x W2.9	6 x 6 — 6 x 6		.058	.147	.058	.147	42	19.05
	6 x 6 — W4.0 x W4.0	6 x 6 — 4 x 4		.080	.203	.080	.203	58	26.31
	6 x 6 — W5.5 x W5.5	6 x 6 — 2 x 2	2	.110	.279	.110	.279	80	36.29
	4 x 4 — W1.4 x W1.4	4 x 4 — 4 x 4		.120	.305	.120	.305	85	38.56

NOTES: 1. Exact W—number size for 8 gauge is W2.1
2. Exact W—number size for 2 gauge is W5.4

The above table was compiled with the following excerpts from the WRI Manual of Standard Practices, 7th Edition, Copyright 2006. Reproduced with permission of the Wire Reinforcement Institute, Inc.:
1. Chapter 3, page 7, Table 1 Common Styles of Metric Wire Reinforcement (WWR) With Equivalent US Customary Units
2. Chapter 6, page 19, Table 5 Customary Units
3. Chapter 6, Page 23, Table 7 Customary Units (in.) Welded Plain Wire Reinforcement
4. Chapter 6, Page 25 Table 8 Wire Size Comparison
5. Chapter 9, Page 30, Table 9 Weight of Longitudinal Wires Weight (Mass) Estimating Tables
6. Chapter 9, Page 31, Table 9M Weight of Longitudinal Wires Weight (Mass) Estimating Tables
7. Chapter 9, Page 32, Table 10 Weight of Transverse Wires Based on 62" lengths of transverse wire (60" width plus 1" overhand each side)
8. Chapter 9, Page 33, Table 10M Weight of Transverse Wires

827

R033105-10   Proportionate Quantities

The tables below show both quantities per S.F. of floor areas as well as form and reinforcing quantities per C.Y. Unusual structural requirements would increase the ratios below. High strength reinforcing would reduce the steel weights. Figures are for 3000 psi concrete and 60,000 psi reinforcing unless specified otherwise.

Type of Construction	Live Load	Span	Per S.F. of Floor Area				Per C.Y. of Concrete		
			Concrete	Forms	Reinf.	Pans	Forms	Reinf.	Pans
Flat Plate	50 psf	15 Ft.	.46 C.F.	1.06 S.F.	1.71 lb.		62 S.F.	101 lb.	
		20	.63	1.02	2.40		44	104	
		25	.79	1.02	3.03		35	104	
	100	15	.46	1.04	2.14		61	126	
		20	.71	1.02	2.72		39	104	
		25	.83	1.01	3.47		33	113	
Flat Plate (waffle construction) 20" domes	50	20	.43	1.00	2.10	.84 S.F.	63	135	53 S.F.
		25	.52	1.00	2.90	.89	52	150	46
		30	.64	1.00	3.70	.87	42	155	37
	100	20	.51	1.00	2.30	.84	53	125	45
		25	.64	1.00	3.20	.83	42	135	35
		30	.76	1.00	4.40	.81	36	160	29
Waffle Construction 30" domes	50	25	.69	1.06	1.83	.68	42	72	40
		30	.74	1.06	2.39	.69	39	87	39
		35	.86	1.05	2.71	.69	33	85	39
		40	.78	1.00	4.80	.68	35	165	40
Flat Slab (two way with drop panels)	50	20	.62	1.03	2.34		45	102	
		25	.77	1.03	2.99		36	105	
		30	.95	1.03	4.09		29	116	
	100	20	.64	1.03	2.83		43	119	
		25	.79	1.03	3.88		35	133	
		30	.96	1.03	4.66		29	131	
	200	20	.73	1.03	3.03		38	112	
		25	.86	1.03	4.23		32	133	
		30	1.06	1.03	5.30		26	135	
One Way Joists 20" Pans	50	15	.36	1.04	1.40	.93	78	105	70
		20	.42	1.05	1.80	.94	67	120	60
		25	.47	1.05	2.60	.94	60	150	54
	100	15	.38	1.07	1.90	.93	77	140	66
		20	.44	1.08	2.40	.94	67	150	58
		25	.52	1.07	3.50	.94	55	185	49
One Way Joists 8" x 16" filler blocks	50	15	.34	1.06	1.80	.81 Ea.	84	145	64 Ea.
		20	.40	1.08	2.20	.82	73	145	55
		25	.46	1.07	3.20	.83	63	190	49
	100	15	.39	1.07	1.90	.81	74	130	56
		20	.46	1.09	2.80	.82	64	160	48
		25	.53	1.10	3.60	.83	56	190	42
One Way Beam & Slab	50	15	.42	1.30	1.73		84	111	
		20	.51	1.28	2.61		68	138	
		25	.64	1.25	2.78		53	117	
	100	15	.42	1.30	1.90		84	122	
		20	.54	1.35	2.69		68	154	
		25	.69	1.37	3.93		54	145	
	200	15	.44	1.31	2.24		80	137	
		20	.58	1.40	3.30		65	163	
		25	.69	1.42	4.89		53	183	
Two Way Beam & Slab	100	15	.47	1.20	2.26		69	130	
		20	.63	1.29	3.06		55	131	
		25	.83	1.33	3.79		43	123	
	200	15	.49	1.25	2.70		41	149	
		20	.66	1.32	4.04		54	165	
		25	.88	1.32	6.08		41	187	

Reference Tables

R033105-10 Proportionate Quantities (cont.)

4000 psi Concrete and 60,000 psi Reinforcing—Form and Reinforcing Quantities per C.Y.					
Item	**Size**	**Forms**	**Reinforcing**	**Minimum**	**Maximum**
Columns (square tied)	10" x 10"	130 S.F.C.A.	#5 to #11	220 lbs.	875 lbs.
	12" x 12"	108	#6 to #14	200	955
	14" x 14"	92	#7 to #14	190	900
	16" x 16"	81	#6 to #14	187	1082
	18" x 18"	72	#6 to #14	170	906
	20" x 20"	65	#7 to #18	150	1080
	22" x 22"	59	#8 to #18	153	902
	24" x 24"	54	#8 to #18	164	884
	26" x 26"	50	#9 to #18	169	994
	28" x 28"	46	#9 to #18	147	864
	30" x 30"	43	#10 to #18	146	983
	32" x 32"	40	#10 to #18	175	866
	34" x 34"	38	#10 to #18	157	772
	36" x 36"	36	#10 to #18	175	852
	38" x 38"	34	#10 to #18	158	765
	40" x 40"	32	#10 to #18	143	692

Item	**Size**	**Form**	**Spiral**	**Reinforcing**	**Minimum**	**Maximum**
Columns (spirally reinforced)	12" diameter	34.5 L.F.	190 lbs.	#4 to #11	165 lbs.	1505 lb.
		34.5	190	#14 & #18	—	1100
	14"	25	170	#4 to #11	150	970
		25	170	#14 & #18	800	1000
	16"	19	160	#4 to #11	160	950
		19	160	#14 & #18	605	1080
	18"	15	150	#4 to #11	160	915
		15	150	#14 & #18	480	1075
	20"	12	130	#4 to #11	155	865
		12	130	#14 & #18	385	1020
	22"	10	125	#4 to #11	165	775
		10	125	#14 & #18	320	995
	24"	9	120	#4 to #11	195	800
		9	120	#14 & #18	290	1150
	26"	7.3	100	#4 to #11	200	729
		7.3	100	#14 & #18	235	1035
	28"	6.3	95	#4 to #11	175	700
		6.3	95	#14 & #18	200	1075
	30"	5.5	90	#4 to #11	180	670
		5.5	90	#14 & #18	175	1015
	32"	4.8	85	#4 to #11	185	615
		4.8	85	#14 & #18	155	955
	34"	4.3	80	#4 to #11	180	600
		4.3	80	#14 & #18	170	855
	36"	3.8	75	#4 to #11	165	570
		3.8	75	#14 & #18	155	865
	40"	3.0	70	#4 to #11	165	500
		3.0	70	#14 & #18	145	765

R033105-10   Proportionate Quantities (cont.)

3000 psi Concrete and 60,000 psi Reinforcing—Form and Reinforcing Quantities per C.Y.

Item	Type	Loading	Height	C.Y./L.F.	Forms/C.Y.	Reinf./C.Y.
Retaining Walls	Cantilever	Level Backfill	4 Ft.	0.2 C.Y.	49 S.F.	35 lbs.
			8	0.5	42	45
			12	0.8	35	70
			16	1.1	32	85
			20	1.6	28	105
		Highway Surcharge	4	0.3	41	35
			8	0.5	36	55
			12	0.8	33	90
			16	1.2	30	120
			20	1.7	27	155
		Railroad Surcharge	4	0.4	28	45
			8	0.8	25	65
			12	1.3	22	90
			16	1.9	20	100
			20	2.6	18	120
	Gravity, with Vertical Face	Level Backfill	4	0.4	37	None
			7	0.6	27	↓
			10	1.2	20	
		Sloping Backfill	4	0.3	31	
			7	0.8	21	
			10	1.6	15	

Live Load in Kips per Linear Foot

	Span	Under 1 Kip		2 to 3 Kips		4 to 5 Kips		6 to 7 Kips	
		Forms	Reinf.	Forms	Reinf.	Forms	Reinf.	Forms	Reinf.
Beams	10 Ft.	—	—	90 S.F.	170 #	85 S.F.	175 #	75 S.F.	185 #
	16	130 S.F.	165 #	85	180	75	180	65	225
	20	110	170	75	185	62	200	51	200
	26	90	170	65	215	62	215	—	—
	30	85	175	60	200	—	—	—	—

Item	Size	Type	Forms per C.Y.	Reinforcing per C.Y.
Spread Footings	Under 1 C.Y.	1,000 psf soil	24 S.F.	44 lbs.
		5,000	24	42
		10,000	24	52
	1 C.Y. to 5 C.Y.	1,000	14	49
		5,000	14	50
		10,000	14	50
	Over 5 C.Y.	1,000	9	54
		5,000	9	52
		10,000	9	56
Pile Caps (30 Ton Concrete Piles)	Under 5 C.Y.	shallow caps	20	65
		medium	20	50
		deep	20	40
	5 C.Y. to 10 C.Y.	shallow	14	55
		medium	15	45
		deep	15	40
	10 C.Y. to 20 C.Y.	shallow	11	60
		medium	11	45
		deep	12	35
	Over 20 C.Y.	shallow	9	60
		medium	9	45
		deep	10	40

R033105-10   Proportionate Quantities (cont.)

Table: 3000 psi Concrete and 60,000 psi Reinforcing — Form and Reinforcing Quantities per C.Y.

Item	Size	Pile Spacing	50 T Pile	100 T Pile	50 T Pile	100 T Pile
Pile Caps (Steel H Piles)	Under 5 C.Y.	24" O.C.	24 S.F.	24 S.F.	75 lbs.	90 lbs.
		30"	25	25	80	100
		36"	24	24	80	110
	5 C.Y. to 10 C.Y.	24"	15	15	80	110
		30"	15	15	85	110
		36"	15	15	75	90
	Over 10 C.Y.	24"	13	13	85	90
		30"	11	11	85	95
		36"	10	10	85	90

Basement Walls	Height	8" Thick Forms	Reinf.	10" Thick Forms	Reinf.	12" Thick Forms	Reinf.	15" Thick Forms	Reinf.
	7 Ft.	81 S.F.	44 lbs.	65 S.F.	45 lbs.	54 S.F.	44 lbs.	41 S.F.	43 lbs.
	8		44		45		44		43
	9		46		45		44		43
	10		57		45		44		43
	12		83		50		52		43
	14		116		65		64		51
	16				86		90		65
	18						106		70

R033105-20   Materials for One C.Y. of Concrete

This is an approximate method of figuring quantities of cement, sand and coarse aggregate for a field mix with waste allowance included.

With crushed gravel as coarse aggregate, to determine barrels of cement required, divide 10 by total mix; that is, for 1:2:4 mix, 10 divided by 7 = 1-3/7 barrels. If the coarse aggregate is crushed stone, use 10-1/2 instead of 10 as given for gravel.

To determine tons of sand required, multiply barrels of cement by parts of sand and then by 0.2; that is, for the 1:2:4 mix, as above, 1-3/7 x 2 x .2 = .57 tons.

Tons of crushed gravel are in the same ratio to tons of sand as parts in the mix, or 4/2 x .57 = 1.14 tons.

1 bag cement = 94#	1 C.Y. sand or crushed gravel = 2700#	1 C.Y. crushed stone = 2575#
4 bags = 1 barrel	1 ton sand or crushed gravel = 20 C.F.	1 ton crushed stone = 21 C.F.

Average carload of cement is 692 bags; of sand or gravel is 56 tons.

Do not stack stored cement over 10 bags high.

R033105-70   Placing Ready-Mixed Concrete

For ground pours allow for 5% waste when figuring quantities.

Prices in the front of the data set assume normal deliveries. If deliveries are made before 8 A.M. or after 5 P.M. or on Saturday afternoons add 30%. Negotiated discounts for large volumes are not included in prices in front of the data set.

For the lower floors without truck access, concrete may be wheeled in rubber-tired buggies, conveyer handled, crane handled or pumped. Pumping is economical if there is top steel. Conveyers are more efficient for thick slabs.

At higher floors the rubber-tired buggies may be hoisted by a hoisting tower and wheeled to the location. Placement by a conveyer is limited to three floors and is best for high-volume pours. Pumped concrete is best when the building has no crane access. Concrete may be pumped directly as high as thirty-six stories using special pumping techniques. Normal maximum height is about fifteen stories.

The best pumping aggregate is screened and graded bank gravel rather than crushed stone.

Pumping downward is more difficult than pumping upward. The horizontal distance from pump to pour may increase preparation time prior to pour. Placing by cranes, either mobile, climbing or tower types, continues as the most efficient method for high-rise concrete buildings.

831

R033105-85   Lift Slabs

The cost advantage of the lift slab method is due to placing all concrete, reinforcing steel, inserts and electrical conduit at ground level and in reduction of formwork. Minimum economical project size is about 30,000 S.F. Slabs may be tilted for parking garage ramps.

It is now used in all types of buildings and has gone up to 22 stories high in apartment buildings. Current trend is to use post-tensioned flat plate slabs with spans from 22′ to 35′. Cylindrical void forms are used when deep slabs are required. One pound of prestressing steel is about equal to seven pounds of conventional reinforcing.

To be considered cured for stressing and lifting, a slab must have attained 75% of design strength. Seven days are usually sufficient with four to five days possible if high early strength cement is used. Slabs can be stacked using two coats of a non-bonding agent to insure that slabs do not stick to each other. Lifting is done by companies specializing in this work. Lift rate is 5′ to 15′ per hour with an average of 10′ per hour. Total areas up to 33,000 S.F. have been lifted at one time. 24 to 36 jacking columns are common. Most economical bay sizes are 24′ to 28′ with four to fourteen stories most efficient. Continuous design reduces reinforcing steel cost. Use of post-tensioned slabs allows larger bay sizes.

R033543-10   Polished Concrete Floors

A polished concrete floor has a glossy mirror-like appearance and is created by grinding the concrete floor with finer and finer diamond grits, similar to sanding wood, until the desired level of reflective clarity and sheen are achieved. The technical term for this type of polished concrete is bonded abrasive polished concrete. The basic piece of equipment used in the polishing process is a walk-behind planetary grinder for working large floor areas. This grinder drives diamond-impregnated abrasive discs, which progress from coarse- to fine-grit discs.

The process begins with the use of very coarse diamond segments or discs bonded in a metallic matrix. These segments are coarse enough to allow the removal of pits, blemishes, stains, and light coatings from the floor surface in preparation for final smoothing. The condition of the original concrete surface will dictate the grit coarseness of the initial grinding step which will generally end up being a three- to four-step process using ever finer grits. The purpose of this initial grinding step is to remove surface coatings and blemishes and to cut down into the cream for very fine aggregate exposure, or deeper into the fine aggregate layer just below the cream layer, or even deeper into the coarse aggregate layer. These initial grinding steps will progress up to the 100/120 grit. If wet grinding is done, a waste slurry is produced that must be removed between grit changes and disposed of properly. If dry grinding is done, a high performance vacuum will pick up the dust during grinding and collect it in bags which must be disposed of properly.

The process continues with honing the floor in a series of steps that progresses from 100-grit to 400-grit diamond abrasive discs embedded in a plastic or resin matrix. At some point during, or just prior to, the honing step, one or two coats of stain or dye can be sprayed onto the surface to give color to the concrete, and two coats of densifier/hardener must be applied to the floor surface and allowed to dry. This sprayed-on densifier/hardener will penetrate about 1/8″ into the concrete to make the surface harder, denser and more abrasion-resistant.

The process ends with polishing the floor surface in a series of steps that progresses from resin-impregnated 800-grit (medium polish) to 1500-grit (high polish) to 3000-grit (very high polish), depending on the desired level of reflective clarity and sheen.

The Concrete Polishing Association of America (CPAA) has defined the flooring options available when processing concrete to a desired finish. The first category is aggregate exposure, the grinding of a concrete surface with bonded abrasives, in as many abrasive grits necessary, to achieve one of the following classes:

©Concrete Polishing Association of America "Glossary."

A. Cream – very little surface cut depth; little aggregate exposure
B. Fine aggregate (salt and pepper) – surface cut depth of 1/16″; fine aggregate exposure with little or no medium aggregate exposure at random locations
C. Medium aggregate – surface cut depth of 1/8″; medium aggregate exposure with little or no large aggregate exposure at random locations
D. Large aggregate – surface cut depth of 1/4″; large aggregate exposure with little or no fine aggregate exposure at random locations

The second CPAA defined category is reflective clarity and sheen, the polishing of a concrete surface with the minimum number of bonded abrasives as indicated to achieve one of the following levels:

1. Ground – flat appearance with none to very slight diffused reflection; none to very low reflective sheen; using a minimum total of 4 grit levels up 100-grit
2. Honed – matte appearance with or without slight diffused reflection; low to medium reflective sheen; using a minimum total of 5 grit levels up to 400-grit
3. Semi-polished – objects being reflected are not quite sharp and crisp but can be easily identified; medium to high reflective sheen; using a minimum total of 6 grit levels up to 800-grit
4. Highly-polished – objects being reflected are sharp and crisp as would be seen in a mirror-like reflection; high to highest reflective sheen; using a minimum total of up to 8 grit levels up to 1500-grit or 3000-grit

The CPAA defines reflective clarity as the degree of sharpness and crispness of the reflection of overhead objects when viewed 5′ above and perpendicular to the floor surface. Reflective sheen is the degree of gloss reflected from a surface when viewed at least 20′ from and at an angle to the floor surface. These terms are relatively subjective. The final outcome depends on the internal makeup and surface condition of the original concrete floor, the experience of the floor polishing crew, and the expectations of the owner. Before the grinding, honing, and polishing work commences on the main floor area, it might be beneficial to do a mock-up panel in the same floor but in an out of the way place to demonstrate the sequence of steps with increasingly fine abrasive grits and to demonstrate the final reflective clarity and reflective sheen. This mock-up panel will be within the area of, and part of, the final work.

R034105-30   Prestressed Precast Concrete Structural Units

Type	Location	Depth	Span in Ft.		Live Load Lb. per S.F.
Double Tee — 8' to 10'	Floor	28" to 34"	60 to 80		50 to 80
	Roof	12" to 24"	30 to 50		40
	Wall	Width 8'	Up to 55' high		Wind
Multiple Tee — 8'	Roof	8" to 12"	15 to 40		40
	Floor	8" to 12"	15 to 30		100
Plank	Roof		Roof	Floor	40 for Roof
		4"	13	12	
	or	6"	22	18	
or		8"	26	25	
		10"	33	29	100 for Floor
	Floor	12"	42	32	
Single Tee — 8' to 10'	Roof	28" 32" 36" 48"	40 80 100 120		40
AASHO Girder	Bridges	Type 4 5 6	100 110 125		Highway
Box Beam — 4'	Bridges	15" 27" 33"	40 to 100		Highway

The majority of precast projects today utilize double tees rather than single tees because of speed and ease of installation. As a result casting beds at manufacturing plants are normally formed for double tees. Single tee projects will therefore require an initial set up charge to be spread over the individual single tee costs.

For floors, a 2" to 3" topping is field cast over the shapes. For roofs, insulating concrete or rigid insulation is placed over the shapes.

Member lengths up to 40' are standard haul, 40' to 60' require special permits and lengths over 60' must be escorted. Excessive width and/or length can add up to 100% on hauling costs.

Large heavy members may require two cranes for lifting which would increase erection costs by about 45%. An eight man crew can install 12 to 20 double tees, or 45 to 70 quad tees or planks per day.

Grouting of connections must also be included.

Several system buildings utilizing precast members are available. Heights can go up to 22 stories for apartment buildings. The optimum design ratio is 3 S.F. of surface to 1 S.F. of floor area.

833

Reference Tables

Concrete   R0352 Lightweight Concrete Roof Insulation

R035216-10   Lightweight Concrete

Lightweight aggregate concrete is usually purchased ready mixed, but it can also be field mixed.

Vermiculite or Perlite comes in bags of 4 C.F. under various trade names. The weight is about 8 lbs. per C.F. For insulating roof fill use 1:6 mix. For a structural deck use 1:4 mix over gypsum boards, steeltex, steel centering, etc., supported by closely spaced joists or bulb trees. For structural slabs use 1:3:2 vermiculite sand concrete over steeltex, metal lath, steel centering, etc., on joists spaced 2'-0" O.C. for maximum L.L. of 80 P.S.F. Use same mix for slab base fill over steel flooring or regular reinforced concrete slab when tile, terrazzo or other finish is to be laid over.

For slabs on grade use 1:3:2 mix when tile, etc., finish is to be laid over. If radiant heating units are installed use a 1:6 mix for a base. After coils are in place, cover with a regular granolithic finish (mix 1:3:2) to a minimum depth of 1-1/2" over top of units.

Reinforce all slabs with 6 x 6 or 10 x 10 welded wire mesh.

Masonry   R0421 Clay Unit Masonry

R042110-20   Common and Face Brick

Common building brick manufactured according to ASTM C62 and facing brick manufactured according to ASTM C216 are the two standard bricks available for general building use.

Building brick is made in three grades: SW, where high resistance to damage caused by cyclic freezing is required; MW, where moderate resistance to cyclic freezing is needed; and NW, where little resistance to cyclic freezing is needed. Facing brick is made in only the two grades SW and MW. Additionally, facing brick is available in three types: FBS, for general use; FBX, for general use where a higher degree of precision and lower permissible variation in size than FBS are needed; and FBA, for general use to produce characteristic architectural effects resulting from non-uniformity in size and texture of the units.

In figuring the material cost of brickwork, an allowance of 25% mortar waste and 3% brick breakage was included. If bricks are delivered palletized with 280 to 300 per pallet, or packaged, allow only 1-1/2% for breakage. Packaged or palletized delivery is practical when a job is big enough to have a crane or other equipment available to handle a package of brick. This is so on all industrial work but not always true on small commercial buildings.

The use of buff and gray face is increasing, and there is a continuing trend to the Norman, Roman, Jumbo and SCR brick.

Common red clay brick for backup is not used that often. Concrete block is the most usual backup material with occasional use of sand lime or cement brick. Building brick is commonly used in solid walls for strength and as a fire stop.

Brick panels built on the ground and then crane erected to the upper floors have proven to be economical. This allows the work to be done under cover and without scaffolding.

R042110-50   Brick, Block & Mortar Quantities

Running Bond							For Other Bonds Standard Size Add to S.F. Quantities in Table to Left		
Number of Brick per S.F. of Wall - Single Wythe with 3/8" Joints					C.F. of Mortar per M Bricks, Waste Included				
Type Brick	Nominal Size (incl. mortar) L H W			Modular Coursing	Number of Brick per S.F.	3/8" Joint	1/2" Joint	Bond Type	Description	Factor
Standard	8 x 2-2/3 x 4			3C=8"	6.75	8.1	10.3	Common	full header every fifth course	+20%
Economy	8 x 4 x 4			1C=4"	4.50	9.1	11.6		full header every sixth course	+16.7%
Engineer	8 x 3-1/5 x 4			5C=16"	5.63	8.5	10.8	English	full header every second course	+50%
Fire	9 x 2-1/2 x 4-1/2			2C=5"	6.40	550 # Fireclay	—	Flemish	alternate headers every course	+33.3%
Jumbo	12 x 4 x 6 or 8			1C=4"	3.00	22.5	29.2		every sixth course	+5.6%
Norman	12 x 2-2/3 x 4			3C=8"	4.50	11.2	14.3	Header = W x H exposed		+100%
Norwegian	12 x 3-1/5 x 4			5C=16"	3.75	11.7	14.9	Rowlock = H x W exposed		+100%
Roman	12 x 2 x 4			2C=4"	6.00	10.7	13.7	Rowlock stretcher = L x W exposed		+33.3%
SCR	12 x 2-2/3 x 6			3C=8"	4.50	21.8	28.0	Soldier = H x L exposed		—
Utility	12 x 4 x 4			1C=4"	3.00	12.3	15.7	Sailor = W x L exposed		-33.3%

Concrete Blocks Nominal Size		Approximate Weight per S.F.		Blocks per 100 S.F.	Mortar per M block, waste included	
		Standard	Lightweight		Partitions	Back up
2" x 8" x 16"		20 PSF	15 PSF	113	27 C.F.	36 C.F.
4"		30	20		41	51
6"		42	30		56	66
8"		55	38		72	82
10"		70	47		87	97
12"		85	55		102	112

Brick & Mortar Quantities
©Brick Industry Association. 2009 Feb. Technical Notes on
Brick Construction 10:
   Dimensioning and Estimating Brick Masonry. Reston (VA): BIA. Table 1
   Modular Brick Sizes and Table 4 Quantity Estimates for Brick Masonry.

R050521-20   Welded Structural Steel

Usual weight reductions with welded design run 10% to 20% compared with bolted or riveted connections. This amounts to about the same total cost compared with bolted structures since field welding is more expensive than bolts. For normal spans of 18' to 24' figure 6 to 7 connections per ton.

Trusses — For welded trusses add 4% to weight of main members for connections. Up to 15% less steel can be expected in a welded truss compared to one that is shop bolted. Cost of erection is the same whether shop bolted or welded.

General — Typical electrodes for structural steel welding are E6010, E6011, E60T and E70T. Typical buildings vary between 2# to 8# of weld rod per

ton of steel. Buildings utilizing continuous design require about three times as much welding as conventional welded structures. In estimating field erection by welding, it is best to use the average linear feet of weld per ton to arrive at the welding cost per ton. The type, size and position of the weld will have a direct bearing on the cost per linear foot. A typical field welder will deposit 1.8# to 2# of weld rod per hour manually. Using semiautomatic methods can increase production by as much as 50% to 75%.

R051223-10   Structural Steel

The bare material prices for structural steel, shown in the unit cost sections of the data set, are for 100 tons of shop-fabricated structural steel and include:
1. Mill base price of structural steel
2. Mill scrap/grade/size/length extras
3. Mill delivery to a metals service center (warehouse)
4. Service center storage and handling
5. Service center delivery to a fabrication shop
6. Shop storage and handling
7. Shop drafting/detailing
8. Shop fabrication

9. Shop coat of primer paint
10. Shop listing
11. Shop delivery to the job site

In unit cost sections of the data set that contain items for field fabrication of steel components, the bare material cost of steel includes:
1. Mill base price of structural steel
2. Mill scrap/grade/size/length extras
3. Mill delivery to a metals service center (warehouse)
4. Service center storage and handling
5. Service center delivery to the job site

R051223-15   Structural Steel Estimating for Repair and Remodeling Projects

The correct approach to estimating structural steel is dependent upon the amount of steel required for the particular project. If the project is a sizable addition to a building, the data can be used directly from the data set. This is not the case however if the project requires a small amount of steel, for instance to reinforce existing roof or floor structural systems. To better understand this, please refer to the unit price line for a W16x31 beam with bolted connections. Assume your project requires the reinforcement of the structural members of the roof system of a 3 story building required for the installation a new piece of HVAC equipment. The project will need 4 pieces of W16x31 that are 30' long. After pricing this 120 L.F. job using the unit prices for a W16x31, an analysis will reveal that the price is wholly inadequate.

The first problem is apparent if you examine the amount of steel that can be installed per day. The unit price line indicates 900 linear feet per day can be installed by a 5 man crew with an 90 ton crane. This productivity is correct for new construction but certainly is not for repair and remodeling work. Installation of new structural steel considers that each member is installed from the foundation to the roof of the structure in a planned and systematic manner with a crane having unrestricted access to all parts of the project. Additionally each connection is planned and detailed with full field access for fit-up and final bolting. The erection is planned and progresses such that interferences and conflicts with other structural members are minimized, if not completely eliminated. All of these assumptions are clearly not the case with a repair and remodeling job, and a significant decrease in the stated productivity will be observed.

A crane will certainly be needed to lift the members into the general area of the project but in most cases will not be able to place the beams into

their final position. An opening in the existing roof may not be large enough to permit the beams to pass through and it may be necessary to bring them into the building through existing windows or doors. Moving the beams to the actual area where they will be installed may involve hand labor and the use of dollies. Finally, hoists and/or jacks may be needed for final positioning.

The connection of new members to existing can often be accomplished by field bolting with accurate field measurements and good planning but in many cases access to both sides of existing members is not possible and field welding becomes the only alternative. In addition to the cost of the actual welding, protection of existing finishes, systems and structure and fire protection must be considered.

New beams can never be installed tight to the existing decks or floors which they must support and the use of shims and tack welding becomes necessary. Additionally, further planning and cost are involved in assuring that existing loads are minimized during the installation and shimming process.

It is apparent that installation of structural steel as part of a repair and remodeling project involves more than simply installing the members and estimating the cost in the same manner as new construction. The best procedure for estimating the total cost are adequate planning and coordination of each process and activity that will be needed. Unit costs for the materials, labor and equipment can then be attached to each needed activity and a final, complete price can be determined.

R053100-10   Decking Descriptions

General - All Deck Products

A steel deck is made by cold forming structural grade sheet steel into a repeating pattern of parallel ribs. The strength and stiffness of the panels are the result of the ribs and the material properties of the steel. Deck lengths can be varied to suit job conditions, but because of shipping considerations, are usually less than 40 feet. Standard deck width varies with the product used but full sheets are usually 12″, 18″, 24″, 30″, or 36″. The deck is typically furnished in a standard width with the ends cut square. Any cutting for width, such as at openings or for angular fit, is done at the job site.

The deck is typically attached to the building frame with arc puddle welds, self-drilling screws, or powder or pneumatically driven pins. Sheet to sheet fastening is done with screws, button punching (crimping), or welds.

Composite Floor Deck

After installation and adequate fastening, a floor deck serves several purposes. It (a) acts as a working platform, (b) stabilizes the frame, (c) serves as a concrete form for the slab, and (d) reinforces the slab to carry the design loads applied during the life of the building. Composite decks are distinguished by the presence of shear connector devices as part of the deck. These devices are designed to mechanically lock the concrete and deck together so that the concrete and the deck work together to carry subsequent floor loads. These shear connector devices can be rolled-in embossments, lugs, holes, or wires welded to the panels. The deck profile can also be used to interlock concrete and steel.

Composite deck finishes are either galvanized (zinc coated) or phosphatized/painted. Galvanized deck has a zinc coating on both the top and bottom surfaces. The phosphatized/painted deck has a bare (phosphatized) top surface that will come into contact with the concrete. This bare top surface can be expected to develop rust before the concrete is placed. The bottom side of the deck has a primer coat of paint.

A composite floor deck is normally installed so the panel ends do not overlap on the supporting beams. Shear lugs or panel profile shapes often prevent a tight metal to metal fit if the panel ends overlap; the air gap caused by overlapping will prevent proper fusion with the structural steel supports when the panel end laps are shear stud welded.

Adequate end bearing of the deck must be obtained as shown on the drawings. If bearing is actually less in the field than shown on the drawings, further investigation is required.

Roof Deck

A roof deck is not designed to act compositely with other materials. A roof deck acts alone in transferring horizontal and vertical loads into the building frame. Roof deck rib openings are usually narrower than floor deck rib openings. This provides adequate support of the rigid thermal insulation board.

A roof deck is typically installed to endlap approximately 2″ over supports. However, it can be butted (or lapped more than 2″) to solve field fit problems. Since designers frequently use the installed deck system as part of the horizontal bracing system (the deck as a diaphragm), any fastening substitution or change should be approved by the designer. Continuous perimeter support of the deck is necessary to limit edge deflection in the finished roof and may be required for diaphragm shear transfer.

Standard roof deck finishes are galvanized or primer painted. The standard factory applied paint for roof decks is a primer paint and is not intended to weather for extended periods of time. Field painting or touching up of abrasions and deterioration of the primer coat or other protective finishes is the responsibility of the contractor.

Cellular Deck

A cellular deck is made by attaching a bottom steel sheet to a roof deck or composite floor deck panel. A cellular deck can be used in the same manner as a floor deck. Electrical, telephone, and data wires are easily run through the chase created between the deck panel and the bottom sheet.

When used as part of the electrical distribution system, the cellular deck must be installed so that the ribs line up and create a smooth cell transition at abutting ends. The joint that occurs at butting cell ends must be taped or otherwise sealed to prevent wet concrete from seeping into the cell. Cell interiors must be free of welding burrs, or other sharp intrusions, to prevent damage to wires.

When used as a roof deck, the bottom flat plate is usually left exposed to view. Care must be maintained during erection to keep good alignment and prevent damage.

A cellular deck is sometimes used with the flat plate on the top side to provide a flat working surface. Installation of the deck for this purpose requires special methods for attachment to the frame because the flat plate, now on the top, can prevent direct access to the deck material that is bearing on the structural steel. It may be advisable to treat the flat top surface to prevent slipping.

A cellular deck is always furnished galvanized or painted over galvanized.

Form Deck

A form deck can be any floor or roof deck product used as a concrete form. Connections to the frame are by the same methods used to anchor floor and roof decks. Welding washers are recommended when welding a deck that is less than 20 gauge thickness.

A form deck is furnished galvanized, prime painted, or uncoated. Galvanized deck must be used for those roof deck systems where a form deck is used to carry a lightweight insulating concrete fill.

R061110-30   Lumber Product Material Prices

The price of forest products fluctuates widely from location to location and from season to season depending upon economic conditions. The bare material prices in the unit cost sections of the data set show the National Average material prices in effect Jan. 1 of this data year. It must be noted that lumber prices in general may change significantly during the year.

Availability of certain items depends upon geographic location and must be checked prior to firm-price bidding.

Wood, Plastics & Comp. | R0616 Sheathing

R061636-20   Plywood

There are two types of plywood used in construction: interior, which is moisture-resistant but not waterproofed, and exterior, which is waterproofed.

The grade of the exterior surface of the plywood sheets is designated by the first letter: A, for smooth surface with patches allowed; B, for solid surface with patches and plugs allowed; C, which may be surface plugged or may have knot holes up to 1″ wide; and D, which is used only for interior type plywood and may have knot holes up to 2-1/2″ wide. "Structural Grade" is specifically designed for engineered applications such as box beams. All CC & DD grades have roof and floor spans marked on them.

Underlayment-grade plywood runs from 1/4″ to 1-1/4″ thick. Thicknesses 5/8″ and over have optional tongue and groove joints which eliminate the need for blocking the edges. Underlayment 19/32″ and over may be referred to as Sturd-i-Floor.

The price of plywood can fluctuate widely due to geographic and economic conditions.

Typical uses for various plywood grades are as follows:

AA-AD Interior — cupboards, shelving, paneling, furniture

BB Plyform — concrete form plywood

CDX — wall and roof sheathing

Structural — box beams, girders, stressed skin panels

AA-AC Exterior — fences, signs, siding, soffits, etc.

Underlayment — base for resilient floor coverings

Overlaid HDO — high density for concrete forms & highway signs

Overlaid MDO — medium density for painting, siding, soffits & signs

303 Siding — exterior siding, textured, striated, embossed, etc.

Thermal & Moist. Protec. | R0751 Built-Up Bituminous Roofing

R075113-20   Built-Up Roofing

Asphalt is available in kegs of 100 lbs. each; coal tar pitch in 560 lb. kegs. Prepared roofing felts are available in a wide range of sizes, weights and characteristics. However, the most commonly used are #15 (432 S.F. per roll, 13 lbs. per square) and #30 (216 S.F. per roll, 27 lbs. per square).

Inter-ply bitumen varies from 24 lbs. per sq. (asphalt) to 30 lbs. per sq. (coal tar) per ply, MF4@ 25%. Flood coat bitumen also varies from 60 lbs. per sq. (asphalt) to 75 lbs. per sq. (coal tar), MF4@ 25%. Expendable equipment (mops, brooms, screeds, etc.) runs about 16% of the bitumen cost. For new, inexperienced crews this factor may be much higher.

A rigid insulation board is typically applied in two layers. The first is mechanically attached to nailable decks or spot or solid mopped to non-nailable decks; the second layer is then spot or solid mopped to the first layer. Membrane application follows the insulation, except in protected membrane roofs, where the membrane goes down first and the insulation on top, followed with ballast (stone or concrete pavers). Insulation and related labor costs are NOT included in prices for built-up roofing.

837

Reference Tables

Thermal & Moist. Protec. **R0752 Modified Bituminous Membrane Roofing**

R075213-30 Modified Bitumen Roofing

The cost of modified bitumen roofing is highly dependent on the type of installation that is planned. Installation is based on the type of modifier used in the bitumen. The two most popular modifiers are atactic polypropylene (APP) and styrene butadiene styrene (SBS). The modifiers are added to heated bitumen during the manufacturing process to change its characteristics. A polyethylene, polyester or fiberglass reinforcing sheet is then sandwiched between layers of this bitumen. When completed, the result is a pre-assembled, built-up roof that has increased elasticity and weatherability. Some manufacturers include a surfacing material such as ceramic or mineral granules, metal particles or sand.

The preferred method of adhering SBS-modified bitumen roofing to the substrate is with hot-mopped asphalt (much the same as built-up roofing). This installation method requires a tar kettle/pot to heat the asphalt, as well as the labor, tools and equipment necessary to distribute and spread the hot asphalt.

The alternative method for applying APP and SBS modified bitumen is as follows. A skilled installer uses a torch to melt a small pool of bitumen off the membrane. This pool must form across the entire roll for proper adhesion. The installer must unroll the roofing at a pace slow enough to melt the bitumen, but fast enough to prevent damage to the rest of the membrane.

Modified bitumen roofing provides the advantages of both built-up and single-ply roofing. Labor costs are reduced over those of built-up roofing because only a single ply is necessary. The elasticity of single-ply roofing is attained with the reinforcing sheet and polymer modifiers. Modifieds have some self-healing characteristics and because of their multi-layer construction, they offer the reliability and safety of built-up roofing.

Thermal & Moist. Protec. **R0784 Firestopping**

R078413-30 Firestopping

Firestopping is the sealing of structural, mechanical, electrical and other penetrations through fire-rated assemblies. The basic components of firestop systems are safing insulation and firestop sealant on both sides of wall penetrations and the top side of floor penetrations.

Pipe penetrations are assumed to be through concrete, grout, or joint compound and can be sleeved or unsleeved. Costs for the penetrations and sleeves are not included. An annular space of 1" is assumed. Escutcheons are not included.

A metallic pipe is assumed to be copper, aluminum, cast iron or similar metallic material. An insulated metallic pipe is assumed to be covered with a thermal insulating jacket of varying thickness and materials.

A non-metallic pipe is assumed to be PVC, CPVC, FR Polypropylene or similar plastic piping material. Intumescent firestop sealants or wrap strips are included. Collars on both sides of wall penetrations and a sheet metal plate on the underside of floor penetrations are included.

Ductwork is assumed to be sheet metal, stainless steel or similar metallic material. Duct penetrations are assumed to be through concrete, grout or joint compound. Costs for penetrations and sleeves are not included. An annular space of 1/2" is assumed.

Multi-trade openings include costs for sheet metal forms, firestop mortar, wrap strips, collars and sealants as necessary.

Structural penetrations joints are assumed to be 1/2" or less. CMU walls are assumed to be within 1-1/2" of the metal deck. Drywall walls are assumed to be tight to the underside of metal decking.

Metal panel, glass or curtain wall systems include a spandrel area of 5' filled with mineral wool foil-faced insulation. Fasteners and stiffeners are included.

Openings   R0813 Metal Doors

R081313-20   Steel Door Selection Guide

Standard steel doors are classified into four levels, as recommended by the Steel Door Institute in the chart below. Each of the four levels offers a range of construction models and designs to meet architectural requirements for preference and appearance, including full flush, seamless, and stile & rail. Recommended minimum gauge requirements are also included.

For complete standard steel door construction specifications and available sizes, refer to the Steel Door Institute Technical Data Series, ANSI A250.8-98 (SDI-100), and ANSI A250.4-94 Test Procedure and Acceptance Criteria for Physical Endurance of Steel Door and Hardware Reinforcements.

Level		Model	Construction	For Full Flush or Seamless		
				Min. Gauge	Thickness (in)	Thickness (mm)
I	Standard Duty	1	Full Flush			
		2	Seamless	20	0.032	0.8
II	Heavy Duty	1	Full Flush			
		2	Seamless	18	0.042	1.0
III	Extra Heavy Duty	1	Full Flush			
		2	Seamless			
		3	*Stile & Rail	16	0.053	1.3
IV	Maximum Duty	1	Full Flush			
		2	Seamless	14	0.067	1.6

*Stiles & rails are 16 gauge; flush panels, when specified, are 18 gauge

Openings   R0851 Metal Windows

R085123-10   Steel Sash

An ironworker crew will erect 25 S.F. or 1.3 sash unit per hour, whichever is less.

A mechanic will point 30 L.F. per hour.

A painter will paint 90 S.F. per coat per hour.

A Glazier production depends on light size.

Allow 1 lb. special steel sash putty per 16" x 20" light.

Openings   R0853 Plastic Windows

R085313-20   Replacement Windows

Replacement windows are typically measured per United Inch.

United Inches are calculated by rounding the width and height of the window opening up to the nearest inch, then adding the two figures.

The labor cost for replacement windows includes removal of sash, existing sash balance or weights, parting bead where necessary and installation of new window.

Debris hauling and dump fees are not included.

R087110-10 Hardware Finishes

This table describes hardware finishes used throughout the industry. It also shows the base metal and the respective symbols in the three predominate systems of identification. Many of these are used in pricing descriptions in Division Eight.

US″	BMHA*	CDN	Base	Description
US P	600	CP	Steel	Primed for Painting
US 1B	601	C1B	Steel	Bright Black Japanned
US 2C	602	C2C	Steel	Zinc Plated
US 2G	603	C2G	Steel	Zinc Plated
US 3	605	C3	Brass	Bright Brass, Clear Coated
US 4	606	C4	Brass	Satin Brass, Clear Coated
US 5	609	C5	Brass	Satin Brass, Blackened, Satin Relieved, Clear Coated
US 7	610	C7	Brass	Satin Brass, Blackened, Bright Relieved, Clear Coated
US 9	611	C9	Bronze	Bright Bronze, Clear Coated
US 10	612	C10	Bronze	Satin Bronze, Clear Coated
US 10A	641	C10A	Steel	Antiqued Bronze, Oiled and Lacquered
US 10B	613	C10B	Bronze	Antiqued Bronze, Oiled
US 11	616	C11	Bronze	Satin Bronze, Blackened, Satin Relieved, Clear Coated
US 14	618	C14	Brass/Bronze	Bright Nickel Plated, Clear Coated
US 15	619	C15	Brass/Bronze	Satin Nickel, Clear Coated
US 15A	620	C15A	Brass/Bronze	Satin Nickel Plated, Blackened, Satin Relieved, Clear Coated
US 17A	621	C17A	Brass/Bronze	Nickel Plated, Blackened, Relieved, Clear Coated
US 19	622	C19	Brass/Bronze	Flat Black Coated
US 20	623	C20	Brass/Bronze	Statuary Bronze, Light
US 20A	624	C20A	Brass/Bronze	Statuary Bronze, Dark
US 26	625	C26	Brass/Bronze	Bright Chromium
US 26D	626	C26D	Brass/Bronze	Satin Chromium
US 20	627	C27	Aluminum	Satin Aluminum Clear
US 28	628	C28	Aluminum	Anodized Dull Aluminum
US 32	629	C32	Stainless Steel	Bright Stainless Steel
US 32D	630	C32D	Stainless Steel	Stainless Steel
US 3	632	C3	Steel	Bright Brass Plated, Clear Coated
US 4	633	C4	Steel	Satin Brass, Clear Coated
US 7	636	C7	Steel	Satin Brass Plated, Blackened, Bright Relieved, Clear Coated
US 9	637	C9	Steel	Bright Bronze Plated, Clear Coated
US 5	638	C5	Steel	Satin Brass Plated, Blackened, Bright Relieved, Clear Coated
US 10	639	C10	Steel	Satin Bronze Plated, Clear Coated
US 10B	640	C10B	Steel	Antique Bronze, Oiled
US 10A	641	C10A	Steel	Antiqued Bronze, Oiled and Lacquered
US 11	643	C11	Steel	Satin Bronze Plated, Blackened, Bright Relieved, Clear Coated
US 14	645	C14	Steel	Bright Nickel Plated, Clear Coated
US 15	646	C15	Steel	Satin Nickel Plated, Clear Coated
US 15A	647	C15A	Steel	Nickel Plated, Blackened, Bright Relieved, Clear Coated
US 17A	648	C17A	Steel	Nickel Plated, Blackened, Relieved, Clear Coated
US 20	649	C20	Steel	Statuary Bronze, Light
US 20A	650	C20A	Steel	Statuary Bronze, Dark
US 26	651	C26	Steel	Bright Chromium Plated
US 26D	652	C26D	Steel	Satin Chromium Plated

* - BMHA Builders Hardware Manufacturing Association
″ - US Equivalent
^ - Canadian Equivalent
Japanning is imitating Asian lacquer work

R088110-10   Glazing Productivity

Some glass sizes are estimated by the "united inch" (height + width). The table below shows the number of lights glazed in an eight-hour period by the crew size indicated, for glass up to 1/4" thick. Square or nearly square lights are more economical on a S.F. basis. Long slender lights will have a high S.F. installation cost. For insulated glass reduce production by 33%. For 1/2" float glass reduce production by 50%. Production time for glazing with two glaziers per day averages: 1/4" float glass 120 S.F.; 1/2" float glass 55 S.F.; 1/2" insulated glass 95 S.F.; 3/4" insulated glass 75 S.F.

Glazing Method	United Inches per Light							
	40"	60"	80"	100"	135"	165"	200"	240"
Number of Men in Crew	1	1	1	1	2	3	3	4
Industrial sash, putty	60	45	24	15	18	—	—	—
With stops, putty bed	50	36	21	12	16	8	4	3
Wood stops, rubber	40	27	15	9	11	6	3	2
Metal stops, rubber	30	24	14	9	9	6	3	2
Structural glass	10	7	4	3	—	—	—	—
Corrugated glass	12	9	7	4	4	4	3	—
Storefronts	16	15	13	11	7	6	4	4
Skylights, putty glass	60	36	21	12	16	—	—	—
Thiokol set	15	15	11	9	9	6	3	2
Vinyl set, snap on	18	18	13	12	12	7	5	4
Maximum area per light	2.8 S.F.	6.3 S.F.	11.1 S.F.	17.4 S.F.	31.6 S.F.	47 S.F.	69 S.F.	100 S.F.

R092000-50   Lath, Plaster and Gypsum Board

Gypsum board lath is available in 3/8″ thick x 16″ wide x 4′ long sheets as a base material for multi-layer plaster applications. It is also available as a base for either multi-layer or veneer plaster applications in 1/2″ and 5/8″ thick–4′ wide x 8′, 10′ or 12′ long sheets. Fasteners are screws or blued ring shank nails for wood framing and screws for metal framing.

Metal lath is available in diamond mesh patterns with flat or self-furring profiles. Paper backing is available for applications where excessive plaster waste needs to be avoided. A slotted mesh ribbed lath should be used in areas where the span between structural supports is greater than normal. Most metal lath comes in 27″ x 96″ sheets. Diamond mesh weighs 1.75, 2.5 or 3.4 pounds per square yard, slotted mesh lath weighs 2.75 or 3.4 pounds per square yard. Metal lath can be nailed, screwed or tied in place.

Many **accessories** are available. Corner beads, flat reinforcing strips, casing beads, control and expansion joints, furring brackets and channels are some examples. Note that accessories are not included in plaster or stucco line items.

Plaster is defined as a material or combination of materials that when mixed with a suitable amount of water, forms a plastic mass or paste. When applied to a surface, the paste adheres to it and subsequently hardens, preserving in a rigid state the form or texture imposed during the period of elasticity.

Gypsum plaster is made from ground calcined gypsum. It is mixed with aggregates and water for use as a base coat plaster.

Vermiculite plaster is a fire-retardant plaster covering used on steel beams, concrete slabs and other heavy construction materials. Vermiculite is a group name for certain clay minerals, hydrous silicates or aluminum, magnesium and iron that have been expanded by heat.

Perlite plaster is a plaster using perlite as an aggregate instead of sand. Perlite is a volcanic glass that has been expanded by heat.

Gauging plaster is a mix of gypsum plaster and lime putty that when applied produces a quick drying finish coat.

Veneer plaster is a one or two component gypsum plaster used as a thin finish coat over special gypsum board.

Keenes cement is a white cementitious material manufactured from gypsum that has been burned at a high temperature and ground to a fine powder. Alum is added to accelerate the set. The resulting plaster is hard and strong and accepts and maintains a high polish, hence it is used as a finishing plaster.

Stucco is a Portland cement based plaster used primarily as an exterior finish.

Plaster is used on both interior and exterior surfaces. Generally it is applied in multiple-coat systems. A three-coat system uses the terms scratch, brown and finish to identify each coat. A two-coat system uses base and finish to describe each coat. Each type of plaster and application system has attributes that are chosen by the designer to best fit the intended use.

Gypsum Plaster Quantities for 100 S.Y.	2 Coat, 5/8″ Thick		3 Coat, 3/4″ Thick		
	Base	Finish	Scratch	Brown	Finish
	1:3 Mix	2:1 Mix	1:2 Mix	1:3 Mix	2:1 Mix
Gypsum plaster	1,300 lb.		1,350 lb.	650 lb.	
Sand	1.75 C.Y.		1.85 C.Y.	1.35 C.Y.	
Finish hydrated lime		340 lb.			340 lb.
Gauging plaster		170 lb.			170 lb.

Vermiculite or Perlite Plaster Quantities for 100 S.Y.	2 Coat, 5/8″ Thick		3 Coat, 3/4″ Thick		
	Base	Finish	Scratch	Brown	Finish
Gypsum plaster	1,250 lb.		1,450 lb.	800 lb.	
Vermiculite or perlite	7.8 bags		8.0 bags	3.3 bags	
Finish hydrated lime		340 lb.			340 lb.
Gauging plaster		170 lb.			170 lb.

Stucco–Three-Coat System Quantities for 100 S.Y.	On Wood Frame	On Masonry
Portland cement	29 bags	21 bags
Sand	2.6 C.Y.	2.0 C.Y.
Hydrated lime	180 lb.	120 lb.

R092910-10 Levels of Gypsum Board Finish

In the past, contract documents often used phrases such as "industry standard" and "workmanlike finish" to specify the expected quality of gypsum board wall and ceiling installations. The vagueness of these descriptions led to unacceptable work and disputes.

In order to resolve this problem, four major trade associations concerned with the manufacture, erection, finish and decoration of gypsum board wall and ceiling systems developed an industry-wide *Recommended Levels of Gypsum Board Finish.*

The finish of gypsum board walls and ceilings for specific final decoration is dependent on a number of factors. A primary consideration is the location of the surface and the degree of decorative treatment desired. Painted and unpainted surfaces in warehouses and other areas where appearance is normally not critical may simply require the taping of wallboard joints and 'spotting' of fastener heads. Blemish-free, smooth, monolithic surfaces often intended for painted and decorated walls and ceilings in habitated structures, ranging from single-family dwellings through monumental buildings, require additional finishing prior to the application of the final decoration.

Other factors to be considered in determining the level of finish of the gypsum board surface are (1) the type of angle of surface illumination (both natural and artificial lighting), and (2) the paint and method of application or the type and finish of wallcovering specified as the final decoration. Critical lighting conditions, gloss paints, and thin wall coverings require a higher level of gypsum board finish than heavily textured surfaces which are subsequently painted or surfaces which are to be decorated with heavy grade wall coverings.

The following descriptions were developed by the Association of the Wall and Ceiling Industries-International (AWCI), Ceiling & Interior Systems Construction Association (CISCA), Gypsum Association (GA), and Painting and Decorating Contractors of America (PDCA) as a guide.

Level 0: Used in temporary construction or wherever the final decoration has not been determined. Unfinished. No taping, finishing or corner beads are required. Also could be used where non-predecorated panels will be used in demountable-type partitions that are to be painted as a final finish.

Level 1: Frequently used in plenum areas above ceilings, in attics, in areas where the assembly would generally be concealed, or in building service corridors and other areas not normally open to public view. Some degree of sound and smoke control is provided; in some geographic areas, this level is referred to as "fire-taping," although this level of finish does not typically meet fire-resistant assembly requirements. Where a fire resistance rating is required for the gypsum board assembly, details of construction should be in accordance with reports of fire tests of assemblies that have met the requirements of the fire rating acceptable.

All joints and interior angles shall have tape embedded in joint compound. Accessories are optional at specifier discretion in corridors and other areas with pedestrian traffic. Tape and fastener heads need not be covered with joint compound. Surface shall be free of excess joint compound. Tool marks and ridges are acceptable.

Level 2: It may be specified for standard gypsum board surfaces in garages, warehouse storage, or other similar areas where surface appearance is not of primary importance.

All joints and interior angles shall have tape embedded in joint compound and shall be immediately wiped with a joint knife or trowel, leaving a thin coating of joint compound over all joints and interior angles. Fastener heads and accessories shall be covered with a coat of joint compound. Surface shall be free of excess joint compound. Tool marks and ridges are acceptable.

Level 3: Typically used in areas receiving heavy texture (spray or hand applied) finishes before final painting, or where commercial-grade (heavy duty) wall coverings are to be applied as the final decoration. This level of finish should not be used where smooth painted surfaces or where lighter weight wall coverings are specified. The prepared surface shall be coated with a drywall primer prior to the application of final finishes.

All joints and interior angles shall have tape embedded in joint compound and shall be immediately wiped with a joint knife or trowel, leaving a thin coating of joint compound over all joints and interior angles. One additional coat of joint compound shall be applied over all joints and interior angles. Fastener heads and accessories shall be covered with two separate coats of joint compound. All joint compounds shall be smooth and free of tool marks and ridges. The prepared surface shall be covered with a drywall primer prior to the application of the final decoration.

Level 4: This level should be used where residential grade (light duty) wall coverings, flat paints, or light textures are to be applied. The prepared surface shall be coated with a drywall primer prior to the application of final finishes. Release agents for wall coverings are specifically formulated to minimize damage if coverings are subsequently removed.

The weight, texture, and sheen level of the wall covering material selected should be taken into consideration when specifying wall coverings over this level of drywall treatment. Joints and fasteners must be sufficiently concealed if the wall covering material is lightweight, contains limited pattern, has a glossy finish, or has any combination of these features. In critical lighting areas, flat paints applied over light textures tend to reduce joint photographing. Gloss, semi-gloss, and enamel paints are not recommended over this level of finish.

All joints and interior angles shall have tape embedded in joint compound and shall be immediately wiped with a joint knife or trowel, leaving a thin coating of joint compound over all joints and interior angles. In addition, two separate coats of joint compound shall be applied over all flat joints and one separate coat of joint compound applied over interior angles. Fastener heads and accessories shall be covered with three separate coats of joint compound. All joint compounds shall be smooth and free of tool marks and ridges. The prepared surface shall be covered with a drywall primer like Sheetrock first coat prior to the application of the final decoration.

Level 5: The highest quality finish is the most effective method to provide a uniform surface and minimize the possibility of joint photographing and of fasteners showing through the final decoration. This level of finish is required where gloss, semi-gloss, or enamel is specified; when flat joints are specified over an untextured surface; or where critical lighting conditions occur. The prepared surface shall be coated with a drywall primer prior to the application of the final decoration.

All joints and interior angles shall have tape embedded in joint compound and be immediately wiped with a joint knife or trowel, leaving a thin coating of joint compound over all joints and interior angles. Two separate coats of joint compound shall be applied over all flat joints and one separate coat of joint compound applied over interior angles. Fastener heads and accessories shall be covered with three separate coats of joint compound.

A thin skim coat of joint compound shall be trowel applied to the entire surface. Excess compound is immediately troweled off, leaving a film or skim coating of compound completely covering the paper. As an alternative to a skim coat, a material manufactured especially for this purpose may be applied such as Sheetrock Tuff-Hide primer surfacer. The surface must be smooth and free of tool marks and ridges. The prepared surface shall be covered with a drywall primer prior to the application of the final decoration.

Reference Tables

R142000-10   Freight Elevators

Capacities run from 2,000 lbs. to over 100,000 lbs. with 3,000 lbs. to 10,000 lbs. most common. Travel speeds are generally lower and the control is less intricate than on passenger elevators. Costs in the unit price sections are for hydraulic and geared elevators.

R142000-20   Elevator Selective Costs   See R142000-40 for cost development.

A. Base Unit	Passenger		Freight		Hospital	
	Hydraulic	Electric	Hydraulic	Electric	Hydraulic	Electric
Capacity	1,500 lb.	2,000 lb.	2,000 lb.	4,000 lb.	4,000 lb.	4,000 lb.
Speed	100 F.P.M.	200 F.P.M.	100 F.P.M.	200 F.P.M.	100 F.P.M.	200 F.P.M.
#Stops/Travel Ft.	2/12	4/40	2/20	4/40	2/20	4/40
Push Button Oper.	Yes	Yes	Yes	Yes	Yes	Yes
Telephone Box & Wire	"	"	"	"	"	"
Emergency Lighting	"	"	No	No	"	"
Cab	Plastic Lam. Walls	Plastic Lam. Walls	Painted Steel	Painted Steel	Plastic Lam. Walls	Plastic Lam. Walls
Cove Lighting	Yes	Yes	No	No	Yes	Yes
Floor	V.C.T.	V.C.T.	Wood w/Safety Treads	Wood w/Safety Treads	V.C.T.	V.C.T.
Doors, & Speedside Slide	Yes	Yes	Yes	Yes	Yes	Yes
Gates, Manual	No	No	No	No	No	No
Signals, Lighted Buttons	Car and Hall	Car and Hall	Car and Hall	Car and Hall	Car and Hall	Car and Hall
O.H. Geared Machine	N.A.	Yes	N.A.	Yes	N.A.	Yes
Variable Voltage Contr.	"	"	N.A.	"	"	"
Emergency Alarm	Yes	"	Yes	"	Yes	"
Class "A" Loading	N.A.	N.A.	"	"	N.A.	N.A.

R142000-30   Passenger Elevators

Electric elevators are used generally but hydraulic elevators can be used for lifts up to 70' and where large capacities are required. Hydraulic speeds are limited to 200 F.P.M. but cars are self leveling at the stops. On low rises, hydraulic installation runs about 15% less than standard electric types but on higher rises this installation cost advantage is reduced. Maintenance of hydraulic elevators is about the same as the electric type but the underground portion is not included in the maintenance contract.

In electric elevators there are several control systems available, the choice of which will be based upon elevator use, size, speed and cost criteria. The two types of drives are geared for low speeds and gearless for 450 F.P.M. and over.

The tables on the preceding pages illustrate typical installed costs of the various types of elevators available.

R142000-40   Elevator Cost Development

To price a new car or truck from the factory, you must start with the manufacturer's basic model, then add or exchange optional equipment and features. The same is true for pricing elevators.

Requirement: One-passenger elevator, five-story hydraulic, 2,500 lb. capacity, 12' floor to floor, speed 150 F.P.M., emergency power switching and maintenance contract.

Example:

Description	Adjustment
A. Base Elevator: Hydraulic Passenger, 1500 lb. Capacity, 100 fpm, 2 Stops, Standard Finish	1 Ea.
B. Capacity Adjustment (2,500 lb.)	1 Ea.
C. Excess Travel Adjustment: 48' Total Travel (4 x 12') minus 12' Base Unit Travel =	36 V.L.F.
D. Stops Adjustment: 5 Total Stops minus 2 Stops (Base Unit) =	3 Stops
E. Speed Adjustment (150 F.P.M.)	1 Ea.
F. Options:	
1. Intercom Service	1 Ea.
2. Emergency Power Switching, Automatic	1 Ea.
3. Stainless Steel Entrance Doors	5 Ea.
4. Maintenance Contract (12 Months)	1 Ea.
5. Position Indicator for main floor level (none indicated in Base Unit)	1 Ea.

R220102-20   Labor Adjustment Factors

Labor Adjustment Factors are provided for Divisions 21, 22, and 23 to assist the mechanical estimator account for the various complexities and special conditions of any particular project. While a single percentage has been entered on each line of Division 22 01 02.20, it should be understood that these are just suggested midpoints of ranges of values commonly used by mechanical estimators. They may be increased or decreased depending on the severity of the special conditions.

The group for "existing occupied buildings" has been the subject of requests for explanation. Actually there are two stages to this group: buildings that are existing and "finished" but unoccupied, and those that also are occupied. Buildings that are "finished" may result in higher labor costs due to the

workers having to be more careful not to damage finished walls, ceilings, floors, etc. and may necessitate special protective coverings and barriers. Also corridor bends and doorways may not accommodate long pieces of pipe or larger pieces of equipment. Work above an already hung ceiling can be very time consuming. The addition of occupants may force the work to be done on premium time (nights and/or weekends), eliminate the possible use of some preferred tools such as pneumatic drivers, powder charged drivers, etc. The estimator should evaluate the access to the work area and just how the work is going to be accomplished to arrive at an increase in labor costs over "normal" new construction productivity.

R220105-10   Demolition (Selective vs. Removal for Replacement)

Demolition can be divided into two basic categories.

One type of demolition involves the removal of material with no concern for its replacement. The labor-hours to estimate this work are found under "Selective Demolition" in the Fire Protection, Plumbing and HVAC Divisions. It is selective in that individual items or all the material installed as a system or trade grouping such as plumbing or heating systems are removed. This may be accomplished by the easiest way possible, such as sawing, torch cutting, or sledge hammering as well as simple unbolting.

The second type of demolition is the removal of some items for repair or replacement. This removal may involve careful draining, opening of unions,

disconnecting and tagging electrical connections, capping pipes/ducts to prevent entry of debris or leakage of the material contained as well as transporting the item away from its in-place location to a truck/dumpster. An approximation of the time required to accomplish this type of demolition is to use half of the time indicated as necessary to install a new unit. For example: installation of a new pump might be listed as requiring 6 labor-hours so if we had to estimate the removal of the old pump we would allow an additional 3 hours for a total of 9 hours. That is, the complete replacement of a defective pump with a new pump would be estimated to take 9 labor-hours.

R221113-50   Pipe Material Considerations

1. Malleable fittings should be used for gas service.
2. Malleable fittings are used where there are stresses/strains due to expansion and vibration.
3. Cast fittings may be broken as an aid to disassembling heating lines frozen by long use, temperature and minerals.
4. A cast iron pipe is extensively used for underground and submerged service.
5. Type M (light wall) copper tubing is available in hard temper only and is used for nonpressure and less severe applications than K and L.

6. Type L (medium wall) copper tubing, available hard or soft for interior service.
7. Type K (heavy wall) copper tubing, available in hard or soft temper for use where conditions are severe. For underground and interior service.
8. Hard drawn tubing requires fewer hangers or supports but should not be bent. Silver brazed fittings are recommended, but soft solder is normally used.
9. Type DMV (very light wall) copper tubing designed for drainage, waste and vent plus other non-critical pressure services.

Domestic/Imported Pipe and Fittings Costs

The prices shown in this publication for steel/cast iron pipe and steel, cast iron, and malleable iron fittings are based on domestic production sold at the normal trade discounts. The above listed items of foreign manufacture may be available at prices 1/3 to 1/2 of those shown. Some imported items after minor machining or finishing operations are being sold as domestic to further complicate the system.

Caution: Most pipe prices in this data set also include a coupling and pipe hangers which for the larger sizes can add significantly to the per foot cost and should be taken into account when comparing "book cost" with the quoted supplier's cost.

845

For customer support on your Commercial Renovation Cost Data, call 877.742.4719.

R260105-30 Electrical Demolition (Removal for Replacement)

The purpose of this reference number is to provide a guide to users for electrical "removal for replacement" by applying the rule of thumb: 1/3 of new installation time (typical range from 20% to 50%) for removal. Remember to use reasonable judgment when applying the suggested percentage factor. For example:

Contractors have been requested to remove an existing fluorescent lighting fixture and replace with a new fixture utilizing energy saver lamps and electronic ballast:

In order to fully understand the extent of the project, contractors should visit the job site and estimate the time to perform the renovation work in accordance with applicable national, state and local regulation and codes.

The contractor may need to add extra labor hours to his estimate if he discovers unknown concealed conditions such as: contaminated asbestos ceiling, broken acoustical ceiling tile and need to repair, patch and touch-up paint all the damaged or disturbed areas, tasks normally assigned to general contractors. In addition, the owner could request that the contractors salvage the materials removed and turn over the materials to the owner or dispose of the materials to a reclamation station. The normal removal item is 0.5 labor hour for a lighting fixture and 1.5 labor-hours for new installation time. Revise the estimate times from 2 labor-hours work up to a minimum 4 labor-hours work for just fluorescent lighting fixture.

For removal of large concentrations of lighting fixtures in the same area, apply an "economy of scale" to reduce estimating labor hours.

R260533-22 Conductors in Conduit

The table below lists the maximum number of conductors for various sized conduit using THW, TW or THWN insulations.

Copper Wire Size	1/2" TW	1/2" THW	1/2" THWN	3/4" TW	3/4" THW	3/4" THWN	1" TW	1" THW	1" THWN	1-1/4" TW	1-1/4" THW	1-1/4" THWN	1-1/2" TW	1-1/2" THW	1-1/2" THWN	2" TW	2" THW	2" THWN	2-1/2" TW	2-1/2" THW	2-1/2" THWN	3" THW	3" THWN	3-1/2" THW	3-1/2" THWN	4" THW	4" THWN
#14	9	6	13	15	10	24	25	16	39	44	29	69	60	40	94	99	65	154	142	93		143		192			
#12	7	4	10	12	8	18	19	13	29	35	24	51	47	32	70	78	53	114	111	76	164	117		157			
#10	5	4	6	9	6	11	15	11	18	26	19	32	36	26	44	60	43	73	85	61	104	95	160	127		163	
#8	2	1	3	4	3	5	7	5	9	12	10	16	17	13	22	28	22	36	40	32	51	49	79	66	106	85	136
#6		1	1		2	4		4	6		7	11		10	15		16	26		23	37	36	57	48	76	62	98
#4		1	1		1	2		3	4		5	7		7	9		12	16		17	22	27	35	36	47	47	60
#3		1	1		1	1		2	3		4	6		6	8		10	13		15	19	23	29	31	39	40	51
#2		1	1		1	1		2	3		4	5		5	7		9	11		13	16	20	25	27	33	34	43
#1					1	1		1	1		3	3		4	5		6	8		9	12	14	18	19	25	25	32
1/0					1	1		1	1		2	3		3	4		5	7		8	10	12	15	16	21	21	27
2/0					1	1		1	1		1	2		3	3		5	6		7	8	10	13	14	17	18	22
3/0					1	1		1	1		1	1		2	3		4	5		6	7	9	11	12	14	15	18
4/0						1		1	1		1	1		1	2		3	4		5	6	7	9	10	12	13	15
250 kcmil								1	1		1	1		1	1		2	3		4	4	6	7	8	10	10	12
300								1	1		1	1		1	1		2	3		3	4	5	6	7	8	9	11
350									1		1	1		1	1		1	2		3	3	4	5	6	7	8	9
400											1	1		1	1		1	1		2	3	4	5	5	6	7	8
500											1	1		1	1		1	1		1	2	3	4	4	5	6	7
600												1		1	1		1	1		1	1	3	3	4	4	5	5
700														1	1		1	1		1	1	2	3	3	4	4	5
750														1	1		1	1		1	1	2	2	3	3	4	4

Reference Tables

R312316-40 Excavating

The selection of equipment used for structural excavation and bulk excavation or for grading is determined by the following factors.

1. Quantity of material
2. Type of material
3. Depth or height of cut
4. Length of haul
5. Condition of haul road
6. Accessibility of site
7. Moisture content and dewatering requirements
8. Availability of excavating and hauling equipment

Some additional costs must be allowed for hand trimming the sides and bottom of concrete pours and other excavation below the general excavation.

Number of B.C.Y. per truck = 1.5 C.Y. bucket x 8 passes = 12 loose C.Y.

$$= 12 \text{ x } \frac{100}{118} = 10.2 \text{ B.C.Y. per truck}$$

Truck Haul Cycle:

Load truck, 8 passes	=	4 minutes
Haul distance, 1 mile	=	9 minutes
Dump time	=	2 minutes
Return, 1 mile	=	7 minutes
Spot under machine	=	1 minute
		23 minute cycle

Add the mobilization and demobilization costs to the total excavation costs. When equipment is rented for more than three days, there is often no mobilization charge by the equipment dealer. On larger jobs outside of urban areas, scrapers can move earth economically provided a dump site or fill area and adequate haul roads are available. Excavation within sheeting bracing or cofferdam bracing is usually done with a clamshell and production

When planning excavation and fill, the following should also be considered.

1. Swell factor
2. Compaction factor
3. Moisture content
4. Density requirements

A typical example for scheduling and estimating the cost of excavation of a 15′ deep basement on a dry site when the material must be hauled off the site is outlined below.

Assumptions:

1. Swell factor, 18%
2. No mobilization or demobilization
3. Allowance included for idle time and moving on job
4. No dewatering, sheeting, or bracing
5. No truck spotter or hand trimming

Fleet Haul Production per day in B.C.Y.

$$4 \text{ trucks x } \frac{50 \text{ min. hour}}{23 \text{ min. haul cycle}} \text{ x 8 hrs. x 10.2 B.C.Y.}$$

$$= 4 \text{ x } 2.2 \text{ x } 8 \text{ x } 10.2 = 718 \text{ B.C.Y./day}$$

is low, since the clamshell may have to be guided by hand between the bracing. When excavating or filling an area enclosed with a wellpoint system, add 10% to 15% to the cost to allow for restricted access. When estimating earth excavation quantities for structures, allow work space outside the building footprint for construction of the foundation and a slope of 1:1 unless sheeting is used.

R312316-45   Excavating Equipment

The table below lists theoretical hourly production in C.Y./hr. bank measure for some typical excavation equipment. Figures assume 50 minute hours, 83% job efficiency, 100% operator efficiency, 90° swing and properly sized hauling units, which must be modified for adverse digging and loading conditions. Actual production costs in the front of the data set average about 50% of the theoretical values listed here.

Equipment	Soil Type	B.C.Y. Weight	% Swell	1 C.Y.	1-1/2 C.Y.	2 C.Y.	2-1/2 C.Y.	3 C.Y.	3-1/2 C.Y.	4 C.Y.
Hydraulic Excavator	Moist loam, sandy clay	3400 lb.	40%	165	195	200	275	330	385	440
"Backhoe"	Sand and gravel	3100	18	140	170	225	240	285	330	380
15' Deep Cut	Common earth	2800	30	150	180	230	250	300	350	400
	Clay, hard, dense	3000	33	120	140	190	200	240	260	320
	Moist loam, sandy clay	3400	40	170 (6.0)	245 (7.0)	295 (7.8)	335 (8.4)	385 (8.8)	435 (9.1)	475 (9.4)
Power Shovel	Sand and gravel	3100	18	165 (6.0)	225 (7.0)	275 (7.8)	325 (8.4)	375 (8.8)	420 (9.1)	460 (9.4)
Optimum Cut (Ft.)	Common earth	2800	30	145 (7.8)	200 (9.2)	250 (10.2)	295 (11.2)	335 (12.1)	375 (13.0)	425 (13.8)
	Clay, hard, dense	3000	33	120 (9.0)	175 (10.7)	220 (12.2)	255 (13.3)	300 (14.2)	335 (15.1)	375 (16.0)
	Moist loam, sandy clay	3400	40	130 (6.6)	180 (7.4)	220 (8.0)	250 (8.5)	290 (9.0)	325 (9.5)	385 (10.0)
Drag Line	Sand and gravel	3100	18	130 (6.6)	175 (7.4)	210 (8.0)	245 (8.5)	280 (9.0)	315 (9.5)	375 (10.0)
Optimum Cut (Ft.)	Common earth	2800	30	110 (8.0)	160 (9.0)	190 (9.9)	220 (10.5)	250 (11.0)	280 (11.5)	310 (12.0)
	Clay, hard, dense	3000	33	90 (9.3)	130 (10.7)	160 (11.8)	190 (12.3)	225 (12.8)	250 (13.3)	280 (12.0)

				Wheel Loaders				Track Loaders		
				3 C.Y.	4 C.Y.	6 C.Y.	8 C.Y.	2-1/4 C.Y.	3 C.Y.	4 C.Y.
	Moist loam, sandy clay	3400	40	260	340	510	690	135	180	250
	Sand and gravel	3100	18	245	320	480	650	130	170	235
Loading Tractors	Common earth	2800	30	230	300	460	620	120	155	220
	Clay, hard, dense	3000	33	200	270	415	560	110	145	200
	Rock, well-blasted	4000	50	180	245	380	520	100	130	180

R312323-30   Compacting Backfill

Compaction of fill in embankments, around structures, in trenches, and under slabs is important to control settlement. Factors affecting compaction are:

1. Soil gradation
2. Moisture content
3. Equipment used
4. Depth of fill per lift
5. Density required

Example:

Compact granular fill around a building foundation using a 21" wide x 24" vibratory plate in 8" lifts. Operator moves at 50 F.P.M. working a 50 minute hour to develop 95% Modified Proctor Density with 4 passes.

Production Rate:

$$\frac{1.75' \text{ plate width x 50 F.P.M. x 50 min./hr. x .67' lift}}{27 \text{ C.F. per C.Y.}} = 108.5 \text{ C.Y./hr.}$$

Production Rate for 4 Passes:

$$\frac{108.5 \text{ C.Y.}}{4 \text{ passes}} = 27.125 \text{ C.Y./hr. x 8 hrs.} = 217 \text{ C.Y./day}$$

Earthwork — R3141 Shoring

R314116-40   Wood Sheet Piling

Wood sheet piling may be used for depths to 20' where there is no ground water. If moderate ground water is encountered Tongue & Groove sheeting will help to keep it out. When considerable ground water is present, steel sheeting must be used.

For estimating purposes on trench excavation, sizes are as follows:

Depth	Sheeting	Wales	Braces	B.F. per S.F.
To 8'	3 x 12's	6 x 8's, 2 line	6 x 8's, @ 10'	4.0 @ 8'
8' x 12'	3 x 12's	10 x 10's, 2 line	10 x 10's, @ 9'	5.0 average
12' to 20'	3 x 12's	12 x 12's, 3 line	12 x 12's, @ 8'	7.0 average

Sheeting to be toed in at least 2' depending upon soil conditions. A five person crew with an air compressor and sheeting driver can drive and brace 440 SF/day at 8' deep, 360 SF/day at 12' deep, and 320 SF/day at 16' deep.

For normal soils, piling can be pulled in 1/3 the time to install. Pulling difficulty increases with the time in the ground. Production can be increased by high pressure jetting.

Earthwork — R3163 Drilled Caissons

R316326-60   Caissons

The three principal types of caissons are:

(1) Belled Caissons, which except for shallow depths and poor soil conditions, are generally recommended. They provide more bearing than shaft area. Because of its conical shape, no horizontal reinforcement of the bell is required.

(2) Straight Shaft Caissons are used where relatively light loads are to be supported by caissons that rest on high value bearing strata. While the shaft is larger in diameter than for belled types this is more than offset by the savings in time and labor.

(3) Keyed Caissons are used when extremely heavy loads are to be carried. A keyed or socketed caisson transfers its load into rock by a combination of end-bearing and shear reinforcing of the shaft. The most economical shaft often consists of a steel casing, a steel wide flange core and concrete. Allowable compressive stresses of .225 f'c for concrete, 16,000 psi for the wide flange core, and 9,000 psi for the steel casing are commonly used. The usual range of shaft diameter is 18" to 84". The number of sizes specified for any one project should be limited due to the problems of casing and auger storage. When hand work is to be performed, shaft diameters should not be less than 32". When inspection of borings is required a minimum shaft diameter of 30" is recommended. Concrete caissons are intended to be poured against earth excavation so permanent forms, which add to cost, should not be used if the excavation is clean and the earth is sufficiently impervious to prevent excessive loss of concrete.

Soil Conditions for Belling		
Good	Requires Handwork	Not Recommended
Clay	Hard Shale	Silt
Sandy Clay	Limestone	Sand
Silty Clay	Sandstone	Gravel
Clayey Silt	Weathered Mica	Igneous Rock
Hard-pan		
Soft Shale		
Decomposed Rock		

Exterior Improvements — R3292 Turf & Grasses

R329219-50   Seeding

The type of grass is determined by light, shade and moisture content of soil plus intended use. Fertilizer should be disked 4" before seeding. For steep slopes disk five tons of mulch and lay two tons of hay or straw on surface per acre after seeding. Surface mulch can be staked, lightly disked or tar emulsion sprayed. Material for mulch can be wood chips, peat moss, partially rotted hay or straw, wood fibers and sprayed emulsions. Hemp seed blankets with fertilizer are also available. For spring seeding, watering is necessary. Late fall seeding may have to be reseeded in the spring. Hydraulic seeding, power mulching, and aerial seeding can be used on large areas.

849

R331113-80   Piping Designations

There are several systems currently in use to describe pipe and fittings. The following paragraphs will help to identify and clarify classifications of piping systems used for water distribution.

Piping may be classified by schedule. Piping schedules include 5S, 10S, 10, 20, 30, Standard, 40, 60, Extra Strong, 80, 100, 120, 140, 160 and Double Extra Strong. These schedules are dependent upon the pipe wall thickness. The wall thickness of a particular schedule may vary with pipe size.

Ductile iron pipe for water distribution is classified by Pressure Classes such as Class 150, 200, 250, 300 and 350. These classes are actually the rated water working pressure of the pipe in pounds per square inch (psi). The pipe in these pressure classes is designed to withstand the rated water working pressure plus a surge allowance of 100 psi.

The American Water Works Association (AWWA) provides standards for various types of **plastic pipe.** C-900 is the specification for polyvinyl chloride (PVC) piping used for water distribution in sizes ranging from 4″ through 12″. C-901 is the specification for polyethylene (PE) pressure pipe, tubing and fittings used for water distribution in sizes ranging from 1/2″ through 3″. C-905 is the specification for PVC piping sizes 14″ and greater.

PVC pressure-rated pipe is identified using the standard dimensional ratio (SDR) method. This method is defined by the American Society for Testing and Materials (ASTM) Standard D 2241. This pipe is available in SDR numbers 64, 41, 32.5, 26, 21, 17, and 13.5. A pipe with an SDR of 64 will have the thinnest wall while a pipe with an SDR of 13.5 will have the thickest wall. When the pressure rating (PR) of a pipe is given in psi, it is based on a line supplying water at 73 degrees F.

The National Sanitation Foundation (NSF) seal of approval is applied to products that can be used with potable water. These products have been tested to ANSI/NSF Standard 14.

Valves and strainers are classified by American National Standards Institute (ANSI) Classes. These Classes are 125, 150, 200, 250, 300, 400, 600, 900, 1500 and 2500. Within each class there is an operating pressure range dependent upon temperature. Design parameters should be compared to the appropriate material dependent, pressure-temperature rating chart for accurate valve selection.

Change Orders

Change Order Considerations

A change order is a written document, usually prepared by the design professional, and signed by the owner, the architect/engineer, and the contractor. A change order states the agreement of the parties to: an addition, deletion, or revision in the work; an adjustment in the contract sum, if any; or an adjustment in the contract time, if any. Change orders, or "extras" in the construction process occur after execution of the construction contract and impact architects/ engineers, contractors, and owners.

Change orders that are properly recognized and managed can ensure orderly, professional, and profitable progress for all who are involved in the project. There are many causes for change orders and change order requests. In all cases, change orders or change order requests should be addressed promptly and in a precise and prescribed manner. The following paragraphs include information regarding change order pricing and procedures.

The Causes of Change Orders

Reasons for issuing change orders include:

- Unforeseen field conditions that require a change in the work
- Correction of design discrepancies, errors, or omissions in the contract documents
- Owner-requested changes, either by design criteria, scope of work, or project objectives
- Completion date changes for reasons unrelated to the construction process
- Changes in building code interpretations, or other public authority requirements that require a change in the work
- Changes in availability of existing or new materials and products

Procedures

Properly written contract documents must include the correct change order procedures for all parties—owners, design professionals and contractors—to follow in order to avoid costly delays and litigation.

Being "in the right" is not always a sufficient or acceptable defense. The contract provisions requiring notification and documentation must be adhered to within a defined or reasonable time frame.

The appropriate method of handling change orders is by a written proposal and acceptance by all parties involved. Prior to starting work on a project, all parties should identify their

authorized agents who may sign and accept change orders, as well as any limits placed on their authority.

Time may be a critical factor when the need for a change arises. For such cases, the contractor might be directed to proceed on a "time and materials" basis, rather than wait for all paperwork to be processed—a delay that could impede progress. In this situation, the contractor must still follow the prescribed change order procedures including, but not limited to, notification and documentation.

Lack of documentation can be very costly, especially if legal judgments are to be made, and if certain field personnel are no longer available. For time and material change orders, the contractor should keep accurate daily records of all labor and material allocated to the change.

Owners or awarding authorities who do considerable and continual building construction (such as the federal government) realize the inevitability of change orders for numerous reasons, both predictable and unpredictable. As a result, the federal government, the American Institute of Architects (AIA), the Engineers Joint Contract Documents Committee (EJCDC) and other contractor, legal, and technical organizations have developed standards and procedures to be followed by all parties to achieve contract continuance and timely completion, while being financially fair to all concerned.

Pricing Change Orders

When pricing change orders, regardless of their cause, the most significant factor is when the change occurs. The need for a change may be perceived in the field or requested by the architect/engineer *before* any of the actual installation has begun, or may evolve or appear *during* construction when the item of work in question is partially installed. In the latter cases, the original sequence of construction is disrupted, along with all contiguous and supporting systems. Change orders cause the greatest impact when they occur *after* the installation has been completed and must be uncovered, or even replaced. Post-completion changes may be caused by necessary design changes, product failure, or changes in the owner's requirements that are not discovered until the building or the systems begin to function.

Specified procedures of notification and record keeping must be adhered to and enforced regardless of the stage of construction: *before, during,* or *after* installation. Some bidding documents anticipate change orders by requiring that unit prices including overhead and profit percentages—for additional as well as deductible changes—be listed. Generally these unit prices do not fully take into account the ripple effect, or impact on other trades, and should be used for general guidance only.

When pricing change orders, it is important to classify the time frame in which the change occurs. There are two basic time frames for change orders: *pre-installation change orders,* which occur before the start of construction, and *post-installation change orders,* which involve reworking after the original installation. Change orders that occur between these stages may be priced according to the extent of work completed using a combination of techniques developed for pricing *pre-* and *post-installation* changes.

Factors To Consider When Pricing Change Orders

As an estimator begins to prepare a change order, the following questions should be reviewed to determine their impact on the final price.

General

- *Is the change order work* pre-installation *or* post-installation?

  Change order work costs vary according to how much of the installation has been completed. Once workers have the project scoped in their minds, even though they have not started, it can be difficult to refocus. Consequently they may spend more than the normal amount of time understanding the change. Also, modifications to work in place, such as trimming or refitting, usually take more time than was initially estimated. The greater the amount of work in place, the more reluctant workers are to change it. Psychologically they may resent the change and as a result the rework takes longer than normal. Post-installation change order estimates must include demolition of existing work as required to accomplish the change. If the work is performed at a later time, additional obstacles, such as building finishes, may be present which must be protected. Regardless of whether the change occurs

For customer support on your Commercial Renovation Cost Data, call 877.742.4719.

851

pre-installation or post-installation, attempt to isolate the identifiable factors and price them separately. For example, add shipping costs that may be required pre-installation or any demolition required post-installation. Then analyze the potential impact on productivity of psychological and/or learning curve factors and adjust the output rates accordingly. One approach is to break down the typical workday into segments and quantify the impact on each segment.

Change Order Installation Efficiency

The labor-hours expressed (for new construction) are based on average installation time, using an efficiency level. For change order situations, adjustments to this efficiency level should reflect the daily labor-hour allocation for that particular occurrence.

- *Will the change substantially delay the original completion date?*

A significant change in the project may cause the original completion date to be extended. The extended schedule may subject the contractor to new wage rates dictated by relevant labor contracts. Project supervision and other project overhead must also be extended beyond the original completion date. The schedule extension may also put installation into a new weather season. For example, underground piping scheduled for October installation was delayed until January. As a result, frost penetrated the trench area, thereby changing the degree of difficulty of the task. Changes and delays may have a ripple effect throughout the project. This effect must be analyzed and negotiated with the owner.

- *What is the net effect of a deduct change order?*

In most cases, change orders resulting in a deduction or credit reflect only bare costs. The contractor may retain the overhead and profit based on the original bid.

Materials

- *Will you have to pay more or less for the new material, required by the change order, than you paid for the original purchase?*

The same material prices or discounts will usually apply to materials purchased for change orders as new construction. In some

instances, however, the contractor may forfeit the advantages of competitive pricing for change orders. Consider the following example:

A contractor purchased over $20,000 worth of fan coil units for an installation, and obtained the maximum discount. Some time later it was determined the project required an additional matching unit. The contractor has to purchase this unit from the original supplier to ensure a match. The supplier at this time may not discount the unit because of the small quantity, and the fact that he is no longer in a competitive situation. The impact of quantity on purchase can add between 0% and 25% to material prices and/or subcontractor quotes.

- *If materials have been ordered or delivered to the job site, will they be subject to a cancellation charge or restocking fee?*

Check with the supplier to determine if ordered materials are subject to a cancellation charge. Delivered materials not used as result of a change order may be subject to a restocking fee if returned to the supplier. Common restocking charges run between 20% and 40%. Also, delivery charges to return the goods to the supplier must be added.

Labor

- *How efficient is the existing crew at the actual installation?*

Is the same crew that performed the initial work going to do the change order? Possibly the change consists of the installation of a unit identical to one already installed; therefore, the change should take less time. Be sure to consider this potential productivity increase and modify the productivity rates accordingly.

- *If the crew size is increased, what impact will that have on supervision requirements?*

Under most bargaining agreements or management practices, there is a point at which a working foreman is replaced by a nonworking foreman. This replacement increases project overhead by adding a nonproductive worker. If additional workers are added to accelerate the project or to perform changes while maintaining the schedule, be sure to add additional supervision time if warranted. Calculate the

hours involved and the additional cost directly if possible.

- *What are the other impacts of increased crew size?*

The larger the crew, the greater the potential for productivity to decrease. Some of the factors that cause this productivity loss are: overcrowding (producing restrictive conditions in the working space), and possibly a shortage of any special tools and equipment required. Such factors affect not only the crew working on the elements directly involved in the change order, but other crews whose movement may also be hampered. As the crew increases, check its basic composition for changes by the addition or deletion of apprentices or nonworking foreman, and quantify the potential effects of equipment shortages or other logistical factors.

- *As new crews, unfamiliar with the project, are brought onto the site, how long will it take them to become oriented to the project requirements?*

The orientation time for a new crew to become 100% effective varies with the site and type of project. Orientation is easiest at a new construction site, and most difficult at existing, very restrictive renovation sites. The type of work also affects orientation time. When all elements of the work are exposed, such as concrete or masonry work, orientation is decreased. When the work is concealed or less visible, such as existing electrical systems, orientation takes longer. Usually orientation can be accomplished in one day or less. Costs for added orientation should be itemized and added to the total estimated cost.

- *How much actual production can be gained by working overtime?*

Short term overtime can be used effectively to accomplish more work in a day. However, as overtime is scheduled to run beyond several weeks, studies have shown marked decreases in output. The following chart shows the effect of long term overtime on worker efficiency. If the anticipated change requires extended overtime to keep the job on schedule, these factors can be used as a guide to predict the impact on time and cost. Add project overhead, particularly supervision, that may also be incurred.

Days per Week	Hours per Day	Production Efficiency					Payroll Cost Factors	
		1st Week	2nd Week	3rd Week	4th Week	Average 4 Weeks	@ 1-1/2 Times	@ 2 Times
5	8	100%	100%	100%	100%	100%	100%	100%
	9	100	100	95	90	96.25	105.6	111.1
	10	100	95	90	85	91.25	110.0	120.0
	11	95	90	75	65	81.25	113.6	127.3
	12	90	85	70	60	76.25	116.7	133.3
6	8	100	100	95	90	96.25	108.3	116.7
	9	100	95	90	85	92.50	113.0	125.9
	10	95	90	85	80	87.50	116.7	133.3
	11	95	85	70	65	78.75	119.7	139.4
	12	90	80	65	60	73.75	122.2	144.4
7	8	100	95	85	75	88.75	114.3	128.6
	9	95	90	80	70	83.75	118.3	136.5
	10	90	85	75	65	78.75	121.4	142.9
	11	85	80	65	60	72.50	124.0	148.1
	12	85	75	60	55	68.75	126.2	152.4

Effects of Overtime

Caution: Under many labor agreements, Sundays and holidays are paid at a higher premium than the normal overtime rate.

The use of long-term overtime is counterproductive on almost any construction job; that is, the longer the period of overtime, the lower the actual production rate. Numerous studies have been conducted, and while they have resulted in slightly different numbers, all reach the same conclusion. The figure above tabulates the effects of overtime work on efficiency.

As illustrated, there can be a difference between the *actual* payroll cost per hour and the *effective* cost per hour for overtime work. This is due to the reduced production efficiency with the increase in weekly hours beyond 40. This difference between actual and effective cost results from overtime work over a prolonged period. Short-term overtime work does not result in as great a reduction in efficiency and, in such cases, effective cost may not vary significantly from the actual payroll cost. As the total hours per week are increased on a regular basis, more time is lost due to fatigue, lowered morale, and an increased accident rate.

As an example, assume a project where workers are working 6 days a week, 10 hours per day. From the figure above (based on productivity studies), the average effective productive hours over a 4-week period are:

$$0.875 \times 60 = 52.5$$

Depending upon the locale and day of week, overtime hours may be paid at time and a half or double time. For time and a half, the overall (average) *actual* payroll cost (including regular and overtime hours) is determined as follows:

$$\frac{40 \text{ reg. hrs.} + (20 \text{ overtime hrs.} \times 1.5)}{60 \text{ hrs.}} = 1.167$$

Based on 60 hours, the payroll cost per hour will be 116.7% of the normal rate at 40 hours per week. However, because the effective production (efficiency) for 60 hours is reduced to the equivalent of 52.5 hours, the effective cost of overtime is calculated as follows:

For time and a half:

$$\frac{40 \text{ reg. hrs.} + (20 \text{ overtime hrs.} \times 1.5)}{52.5 \text{ hrs.}} = 1.33$$

Installed cost will be 133% of the normal rate (for labor).

Thus, when figuring overtime, the actual cost per unit of work will be higher than the apparent overtime payroll dollar increase, due to the reduced productivity of the longer workweek. These efficiency calculations are true only for those cost factors determined by hours worked. Costs that are applied weekly or monthly, such as equipment rentals, will not be similarly affected.

Equipment

- *What equipment is required to complete the change order?*

Change orders may require extending the rental period of equipment already on the job site, or the addition of special equipment brought in to accomplish the change work. In either case, the additional rental charges and operator labor charges must be added.

Summary

The preceding considerations and others you deem appropriate should be analyzed and applied to a change order estimate. The impact of each should be quantified and listed on the estimate to form an audit trail.

Change orders that are properly identified, documented, and managed help to ensure the orderly, professional and profitable progress of the work. They also minimize potential claims or disputes at the end of the project.

For customer support on your Commercial Renovation Cost Data, call 877.742.4719.

853

A	Area Square Feet; Ampere	Brk., brk	Brick	Csc	Cosecant
AAFES	Army and Air Force Exchange Service	brkt	Bracket	C.S.F.	Hundred Square Feet
		Brs.	Brass	CSI	Construction Specifications Institute
ABS	Acrylonitrile Butadiene Stryrene; Asbestos Bonded Steel	Brz.	Bronze		
		Bsn.	Basin	CT	Current Transformer
A.C., AC	Alternating Current; Air-Conditioning; Asbestos Cement; Plywood Grade A & C	Btr.	Better	CTS	Copper Tube Size
		BTU	British Thermal Unit	Cu	Copper, Cubic
		BTUH	BTU per Hour	Cu. Ft.	Cubic Foot
		Bu.	Bushels	cw	Continuous Wave
ACI	American Concrete Institute	BUR	Built-up Roofing	C.W.	Cool White; Cold Water
ACR	Air Conditioning Refrigeration	BX	Interlocked Armored Cable	Cwt.	100 Pounds
ADA	Americans with Disabilities Act	°C	Degree Centigrade	C.W.X.	Cool White Deluxe
AD	Plywood, Grade A & D	c	Conductivity, Copper Sweat	C.Y.	Cubic Yard (27 cubic feet)
Addit.	Additional	C	Hundred; Centigrade	C.Y./Hr.	Cubic Yard per Hour
Adj.	Adjustable	C/C	Center to Center, Cedar on Cedar	Cyl.	Cylinder
af	Audio-frequency	C-C	Center to Center	d	Penny (nail size)
AFUE	Annual Fuel Utilization Efficiency	Cab	Cabinet	D	Deep; Depth; Discharge
AGA	American Gas Association	Cair.	Air Tool Laborer	Dis., Disch.	Discharge
Agg.	Aggregate	Cal.	Caliper	Db	Decibel
A.H., Ah	Ampere Hours	Calc	Calculated	Dbl.	Double
A hr.	Ampere-hour	Cap.	Capacity	DC	Direct Current
A.H.U., AHU	Air Handling Unit	Carp.	Carpenter	DDC	Direct Digital Control
A.I.A.	American Institute of Architects	C.B.	Circuit Breaker	Demob.	Demobilization
AIC	Ampere Interrupting Capacity	C.C.A.	Chromate Copper Arsenate	d.f.t.	Dry Film Thickness
Allow.	Allowance	C.C.F.	Hundred Cubic Feet	d.f.u.	Drainage Fixture Units
alt., alt	Alternate	cd	Candela	D.H.	Double Hung
Alum.	Aluminum	cd/sf	Candela per Square Foot	DHW	Domestic Hot Water
a.m.	Ante Meridiem	CD	Grade of Plywood Face & Back	DI	Ductile Iron
Amp.	Ampere	CDX	Plywood, Grade C & D, exterior glue	Diag.	Diagonal
Anod.	Anodized			Diam., Dia	Diameter
ANSI	American National Standards Institute	Cefi.	Cement Finisher	Distrib.	Distribution
		Cem.	Cement	Div.	Division
APA	American Plywood Association	CF	Hundred Feet	Dk.	Deck
Approx.	Approximate	C.F.	Cubic Feet	D.L.	Dead Load; Diesel
Apt.	Apartment	CFM	Cubic Feet per Minute	DLH	Deep Long Span Bar Joist
Asb.	Asbestos	CFRP	Carbon Fiber Reinforced Plastic	dlx	Deluxe
A.S.B.C.	American Standard Building Code	c.g.	Center of Gravity	Do.	Ditto
Asbe.	Asbestos Worker	CHW	Chilled Water; Commercial Hot Water	DOP	Dioctyl Phthalate Penetration Test (Air Filters)
ASCE	American Society of Civil Engineers				
A.S.H.R.A.E.	American Society of Heating, Refrig. & AC Engineers	C.I., CI	Cast Iron	Dp., dp	Depth
		C.I.P., CIP	Cast in Place	D.P.S.T.	Double Pole, Single Throw
ASME	American Society of Mechanical Engineers	Circ.	Circuit	Dr.	Drive
		C.L.	Carload Lot	DR	Dimension Ratio
ASTM	American Society for Testing and Materials	CL	Chain Link	Drink.	Drinking
		Clab.	Common Laborer	D.S.	Double Strength
Attchmt.	Attachment	Clam	Common Maintenance Laborer	D.S.A.	Double Strength A Grade
Avg., Ave.	Average	C.L.F.	Hundred Linear Feet	D.S.B.	Double Strength B Grade
AWG	American Wire Gauge	CLF	Current Limiting Fuse	Dty.	Duty
AWWA	American Water Works Assoc.	CLP	Cross Linked Polyethylene	DWV	Drain Waste Vent
Bbl.	Barrel	cm	Centimeter	DX	Deluxe White, Direct Expansion
B&B, BB	Grade B and Better; Balled & Burlapped	CMP	Corr. Metal Pipe	dyn	Dyne
		CMU	Concrete Masonry Unit	e	Eccentricity
B&S	Bell and Spigot	CN	Change Notice	E	Equipment Only; East; Emissivity
B.&W.	Black and White	Col.	Column	Ea.	Each
b.c.c.	Body-centered Cubic	CO₂	Carbon Dioxide	EB	Encased Burial
B.C.Y.	Bank Cubic Yards	Comb.	Combination	Econ.	Economy
BE	Bevel End	comm.	Commercial, Communication	E.C.Y	Embankment Cubic Yards
B.F.	Board Feet	Compr.	Compressor	EDP	Electronic Data Processing
Bg. cem.	Bag of Cement	Conc.	Concrete	EIFS	Exterior Insulation Finish System
BHP	Boiler Horsepower; Brake Horsepower	Cont., cont	Continuous; Continued, Container	E.D.R.	Equiv. Direct Radiation
		Corr.	Corrugated	Eq.	Equation
B.I.	Black Iron	Cos	Cosine	EL	Elevation
bidir.	bidirectional	Cot	Cotangent	Elec.	Electrician; Electrical
Bit., Bitum.	Bituminous	Cov.	Cover	Elev.	Elevator; Elevating
Bit., Conc.	Bituminous Concrete	C/P	Cedar on Paneling	EMT	Electrical Metallic Conduit; Thin Wall Conduit
Bk.	Backed	CPA	Control Point Adjustment		
Bkrs.	Breakers	Cplg.	Coupling	Eng.	Engine, Engineered
Bldg., bldg	Building	CPM	Critical Path Method	EPDM	Ethylene Propylene Diene Monomer
Blk.	Block	CPVC	Chlorinated Polyvinyl Chloride		
Bm.	Beam	C.Pr.	Hundred Pair	EPS	Expanded Polystyrene
Boil.	Boilermaker	CRC	Cold Rolled Channel	Eqhv.	Equip. Oper., Heavy
bpm	Blows per Minute	Creos.	Creosote	Eqlt.	Equip. Oper., Light
BR	Bedroom	Crpt.	Carpet & Linoleum Layer	Eqmd.	Equip. Oper., Medium
Brg., brng.	Bearing	CRT	Cathode-ray Tube	Eqmm.	Equip. Oper., Master Mechanic
Brhe.	Bricklayer Helper	CS	Carbon Steel, Constant Shear Bar Joist	Eqol.	Equip. Oper., Oilers
Bric.	Bricklayer			Equip.	Equipment
				ERW	Electric Resistance Welded

Abbreviations

E.S.	Energy Saver	H	High Henry	Lath.	Lather
Est.	Estimated	HC	High Capacity	Lav.	Lavatory
esu	Electrostatic Units	H.D., HD	Heavy Duty; High Density	lb.; #	Pound
E.W.	Each Way	H.D.O.	High Density Overlaid	L.B., LB	Load Bearing; L Conduit Body
EWT	Entering Water Temperature	HDPE	High Density Polyethylene Plastic	L. & E.	Labor & Equipment
Excav.	Excavation	Hdr.	Header	lb./hr.	Pounds per Hour
excl	Excluding	Hdwe.	Hardware	lb./L.F.	Pounds per Linear Foot
Exp., exp	Expansion, Exposure	H.I.D., HID	High Intensity Discharge	lbf/sq.in.	Pound-force per Square Inch
Ext., ext	Exterior; Extension	Help.	Helper Average	L.C.L.	Less than Carload Lot
Extru.	Extrusion	HEPA	High Efficiency Particulate Air	L.C.Y.	Loose Cubic Yard
f.	Fiber Stress		Filter	Ld.	Load
F	Fahrenheit; Female; Fill	Hg	Mercury	LE	Lead Equivalent
Fab., fab	Fabricated; Fabric	HIC	High Interrupting Capacity	LED	Light Emitting Diode
FBGS	Fiberglass	HM	Hollow Metal	L.F.	Linear Foot
F.C.	Footcandles	HMWPE	High Molecular Weight	L.F. Nose	Linear Foot of Stair Nosing
f.c.c.	Face-centered Cubic		Polyethylene	L.F. Rsr	Linear Foot of Stair Riser
f'c.	Compressive Stress in Concrete;	HO	High Output	Lg.	Long; Length; Large
	Extreme Compressive Stress	Horiz.	Horizontal	L & H	Light and Heat
F.E.	Front End	H.P., HP	Horsepower; High Pressure	LH	Long Span Bar Joist
FEP	Fluorinated Ethylene Propylene	H.P.F.	High Power Factor	L.H.	Labor Hours
	(Teflon)	Hr.	Hour	L.L., LL	Live Load
F.G.	Flat Grain	Hrs./Day	Hours per Day	L.L.D.	Lamp Lumen Depreciation
F.H.A.	Federal Housing Administration	HSC	High Short Circuit	lm	Lumen
Fig.	Figure	Ht.	Height	lm/sf	Lumen per Square Foot
Fin.	Finished	Htg.	Heating	lm/W	Lumen per Watt
FIPS	Female Iron Pipe Size	Htrs.	Heaters	LOA	Length Over All
Fixt.	Fixture	HVAC	Heating, Ventilation & Air-	log	Logarithm
FJP	Finger jointed and primed		Conditioning	L-O-L	Lateralolet
Fl. Oz.	Fluid Ounces	Hvy.	Heavy	long.	Longitude
Flr.	Floor	HW	Hot Water	L.P., LP	Liquefied Petroleum; Low Pressure
FM	Frequency Modulation;	Hyd.; Hydr.	Hydraulic	L.P.F.	Low Power Factor
	Factory Mutual	Hz	Hertz (cycles)	LR	Long Radius
Fmg.	Framing	I.	Moment of Inertia	L.S.	Lump Sum
FM/UL	Factory Mutual/Underwriters Labs	IBC	International Building Code	Lt.	Light
Fdn.	Foundation	I.C.	Interrupting Capacity	Lt. Ga.	Light Gauge
FNPT	Female National Pipe Thread	ID	Inside Diameter	L.T.L.	Less than Truckload Lot
Fori.	Foreman, Inside	I.D.	Inside Dimension; Identification	Lt. Wt.	Lightweight
Foro.	Foreman, Outside	I.F.	Inside Frosted	L.V.	Low Voltage
Fount.	Fountain	I.M.C.	Intermediate Metal Conduit	M	Thousand; Material; Male;
fpm	Feet per Minute	In.	Inch		Light Wall Copper Tubing
FPT	Female Pipe Thread	Incan.	Incandescent	M^{2}CA	Meters Squared Contact Area
Fr	Frame	Incl.	Included; Including	m/hr.; M.H.	Man-hour
F.R.	Fire Rating	Int.	Interior	mA	Milliampere
FRK	Foil Reinforced Kraft	Inst.	Installation	Mach.	Machine
FSK	Foil/Scrim/Kraft	Insul., insul	Insulation/Insulated	Mag. Str.	Magnetic Starter
FRP	Fiberglass Reinforced Plastic	I.P.	Iron Pipe	Maint.	Maintenance
FS	Forged Steel	I.P.S., IPS	Iron Pipe Size	Marb.	Marble Setter
FSC	Cast Body; Cast Switch Box	IPT	Iron Pipe Threaded	Mat; Mat'l.	Material
Ft., ft	Foot; Feet	I.W.	Indirect Waste	Max.	Maximum
Ftng.	Fitting	J	Joule	MBF	Thousand Board Feet
Ftg.	Footing	J.I.C.	Joint Industrial Council	MBH	Thousand BTU's per hr.
Ft lb.	Foot Pound	K	Thousand; Thousand Pounds;	MC	Metal Clad Cable
Furn.	Furniture		Heavy Wall Copper Tubing, Kelvin	MCC	Motor Control Center
FVNR	Full Voltage Non-Reversing	K.A.H.	Thousand Amp. Hours	M.C.F.	Thousand Cubic Feet
FVR	Full Voltage Reversing	kcmil	Thousand Circular Mils	MCFM	Thousand Cubic Feet per Minute
FXM	Female by Male	KD	Knock Down	M.C.M.	Thousand Circular Mils
Fy.	Minimum Yield Stress of Steel	K.D.A.T.	Kiln Dried After Treatment	MCP	Motor Circuit Protector
g	Gram	kg	Kilogram	MD	Medium Duty
G	Gauss	kG	Kilogauss	MDF	Medium-density fibreboard
Ga.	Gauge	kgf	Kilogram Force	M.D.O.	Medium Density Overlaid
Gal., gal.	Gallon	kHz	Kilohertz	Med.	Medium
Galv., galv	Galvanized	Kip	1000 Pounds	MF	Thousand Feet
GC/MS	Gas Chromatograph/Mass	KJ	Kilojoule	M.F.B.M.	Thousand Feet Board Measure
	Spectrometer	K.L.	Effective Length Factor	Mfg.	Manufacturing
Gen.	General	K.L.F.	Kips per Linear Foot	Mfrs.	Manufacturers
GFI	Ground Fault Interrupter	Km	Kilometer	mg	Milligram
GFRC	Glass Fiber Reinforced Concrete	KO	Knock Out	MGD	Million Gallons per Day
Glaz.	Glazier	K.S.F.	Kips per Square Foot	MGPH	Million Gallons per Hour
GPD	Gallons per Day	K.S.I.	Kips per Square Inch	MH, M.H.	Manhole; Metal Halide; Man-Hour
gpf	Gallon per Flush	kV	Kilovolt	MHz	Megahertz
GPH	Gallons per Hour	kVA	Kilovolt Ampere	Mi.	Mile
gpm, GPM	Gallons per Minute	kVAR	Kilovar (Reactance)	MI	Malleable Iron; Mineral Insulated
GR	Grade	KW	Kilowatt	MIPS	Male Iron Pipe Size
Gran.	Granular	KWh	Kilowatt-hour	mj	Mechanical Joint
Grnd.	Ground	L	Labor Only; Length; Long;	m	Meter
GVW	Gross Vehicle Weight		Medium Wall Copper Tubing	mm	Millimeter
GWB	Gypsum Wall Board	Lab.	Labor	Mill.	Millwright
		lat	Latitude	Min., min.	Minimum, Minute

855

Misc.	Miscellaneous	PDCA	Painting and Decorating	SC	Screw Cover
ml	Milliliter, Mainline		Contractors of America	SCFM	Standard Cubic Feet per Minute
M.L.F.	Thousand Linear Feet	P.E., PE	Professional Engineer;	Scaf.	Scaffold
Mo.	Month		Porcelain Enamel;	Sch., Sched.	Schedule
Mobil.	Mobilization		Polyethylene; Plain End	S.C.R.	Modular Brick
Mog.	Mogul Base	P.E.C.I.	Porcelain Enamel on Cast Iron	S.D.	Sound Deadening
MPH	Miles per Hour	Perf.	Perforated	SDR	Standard Dimension Ratio
MPT	Male Pipe Thread	PEX	Cross Linked Polyethylene	S.E.	Surfaced Edge
MRGWB	Moisture Resistant Gypsum	Ph.	Phase	Sel.	Select
	Wallboard	P.I.	Pressure Injected	SER, SEU	Service Entrance Cable
MRT	Mile Round Trip	Pile.	Pile Driver	S.F.	Square Foot
ms	Millisecond	Pkg.	Package	S.F.C.A.	Square Foot Contact Area
M.S.F.	Thousand Square Feet	Pl.	Plate	S.F. Flr.	Square Foot of Floor
Mstz.	Mosaic & Terrazzo Worker	Plah.	Plasterer Helper	S.F.G.	Square Foot of Ground
M.S.Y.	Thousand Square Yards	Plas.	Plasterer	S.F. Hor.	Square Foot Horizontal
Mtd., mtd., mtd	Mounted	plf	Pounds Per Linear Foot	SFR	Square Feet of Radiation
Mthe.	Mosaic & Terrazzo Helper	Pluh.	Plumber Helper	S.F. Shlf.	Square Foot of Shelf
Mtng.	Mounting	Plum.	Plumber	S4S	Surface 4 Sides
Mult.	Multi; Multiply	Ply.	Plywood	Shee.	Sheet Metal Worker
M.V.A.	Million Volt Amperes	p.m.	Post Meridiem	Sin.	Sine
M.V.A.R.	Million Volt Amperes Reactance	Pntd.	Painted	Skwk.	Skilled Worker
MV	Megavolt	Pord.	Painter, Ordinary	SL	Saran Lined
MW	Megawatt	pp	Pages	S.L.	Slimline
MXM	Male by Male	PP, PPL	Polypropylene	Sldr.	Solder
MYD	Thousand Yards	P.P.M.	Parts per Million	SLH	Super Long Span Bar Joist
N	Natural; North	Pr.	Pair	S.N.	Solid Neutral
nA	Nanoampere	P.E.S.B.	Pre-engineered Steel Building	SO	Stranded with oil resistant inside
NA	Not Available; Not Applicable	Prefab.	Prefabricated		insulation
N.B.C.	National Building Code	Prefin.	Prefinished	S-O-L	Socketolet
NC	Normally Closed	Prop.	Propelled	sp	Standpipe
NEMA	National Electrical Manufacturers	PSF, psf	Pounds per Square Foot	S.P.	Static Pressure; Single Pole; Self-
	Assoc.	PSI, psi	Pounds per Square Inch		Propelled
NEHB	Bolted Circuit Breaker to 600V.	PSIG	Pounds per Square Inch Gauge	Spri.	Sprinkler Installer
NFPA	National Fire Protection Association	PSP	Plastic Sewer Pipe	spwg	Static Pressure Water Gauge
NLB	Non-Load-Bearing	Pspr.	Painter, Spray	S.P.D.T.	Single Pole, Double Throw
NM	Non-Metallic Cable	Psst.	Painter, Structural Steel	SPF	Spruce Pine Fir; Sprayed
nm	Nanometer	P.T.	Potential Transformer		Polyurethane Foam
No.	Number	P. & T.	Pressure & Temperature	S.P.S.T.	Single Pole, Single Throw
NO	Normally Open	Ptd.	Painted	SPT	Standard Pipe Thread
N.O.C.	Not Otherwise Classified	Ptns.	Partitions	Sq.	Square; 100 Square Feet
Nose.	Nosing	Pu	Ultimate Load	Sq. Hd.	Square Head
NPT	National Pipe Thread	PVC	Polyvinyl Chloride	Sq. In.	Square Inch
NQOD	Combination Plug-on/Bolt on	Pvmt.	Pavement	S.S.	Single Strength; Stainless Steel
	Circuit Breaker to 240V.	PRV	Pressure Relief Valve	S.S.B.	Single Strength B Grade
N.R.C., NRC	Noise Reduction Coefficient/	Pwr.	Power	sst, ss	Stainless Steel
	Nuclear Regulator Commission	Q	Quantity Heat Flow	Sswk.	Structural Steel Worker
N.R.S.	Non Rising Stem	Qt.	Quart	Sswl.	Structural Steel Welder
ns	Nanosecond	Quan., Qty.	Quantity	St.; Stl.	Steel
nW	Nanowatt	Q.C.	Quick Coupling	STC	Sound Transmission Coefficient
OB	Opposing Blade	r	Radius of Gyration	Std.	Standard
OC	On Center	R	Resistance	Stg.	Staging
OD	Outside Diameter	R.C.P.	Reinforced Concrete Pipe	STK	Select Tight Knot
O.D.	Outside Dimension	Rect.	Rectangle	STP	Standard Temperature & Pressure
ODS	Overhead Distribution System	recpt.	Receptacle	Stpi.	Steamfitter, Pipefitter
O.G.	Ogee	Reg.	Regular	Str.	Strength; Starter; Straight
O.H.	Overhead	Reinf.	Reinforced	Strd.	Stranded
O&P	Overhead and Profit	Req'd.	Required	Struct.	Structural
Oper.	Operator	Res.	Resistant	Sty.	Story
Opng.	Opening	Resi.	Residential	Subj.	Subject
Orna.	Ornamental	RF	Radio Frequency	Subs.	Subcontractors
OSB	Oriented Strand Board	RFID	Radio-frequency Identification	Surf.	Surface
OS&Y	Outside Screw and Yoke	Rgh.	Rough	Sw.	Switch
OSHA	Occupational Safety and Health	RGS	Rigid Galvanized Steel	Swbd.	Switchboard
	Act	RHW	Rubber, Heat & Water Resistant;	S.Y.	Square Yard
Ovhd.	Overhead		Residential Hot Water	Syn.	Synthetic
OWG	Oil, Water or Gas	rms	Root Mean Square	S.Y.P.	Southern Yellow Pine
Oz.	Ounce	Rnd.	Round	Sys.	System
P.	Pole; Applied Load; Projection	Rodm.	Rodman	t.	Thickness
p.	Page	Rofc.	Roofer, Composition	T	Temperature; Ton
Pape.	Paperhanger	Rofp.	Roofer, Precast	Tan	Tangent
P.A.P.R.	Powered Air Purifying Respirator	Rohe.	Roofer Helpers (Composition)	T.C.	Terra Cotta
PAR	Parabolic Reflector	Rots.	Roofer, Tile & Slate	T & C	Threaded and Coupled
P.B., PB	Push Button	R.O.W.	Right of Way	T.D.	Temperature Difference
Pc., Pcs.	Piece, Pieces	RPM	Revolutions per Minute	TDD	Telecommunications Device for
P.C.	Portland Cement; Power Connector	R.S.	Rapid Start		the Deaf
P.C.F.	Pounds per Cubic Foot	Rsr	Riser	T.E.M.	Transmission Electron Microscopy
PCM	Phase Contrast Microscopy	RT	Round Trip	temp	Temperature, Tempered, Temporary
		S.	Suction; Single Entrance; South	TFFN	Nylon Jacketed Wire
		SBS	Styrene Butadiere Styrene		

856

Abbreviations

TFE	Tetrafluoroethylene (Teflon)	U.L., UL	Underwriters Laboratory	w/	With
T. & G.	Tongue & Groove;	Uld.	Unloading	W.C., WC	Water Column; Water Closet
	Tar & Gravel	Unfin.	Unfinished	W.F.	Wide Flange
Th., Thk.	Thick	UPS	Uninterruptible Power Supply	W.G.	Water Gauge
Thn.	Thin	URD	Underground Residential	Wldg.	Welding
Thrded	Threaded		Distribution	W. Mile	Wire Mile
Tilf.	Tile Layer, Floor	US	United States	W-O-L	Weldolet
Tilh.	Tile Layer, Helper	USGBC	U.S. Green Building Council	W.R.	Water Resistant
THHN	Nylon Jacketed Wire	USP	United States Primed	Wrck.	Wrecker
THW.	Insulated Strand Wire	UTMCD	Uniform Traffic Manual For Control	WSFU	Water Supply Fixture Unit
THWN	Nylon Jacketed Wire		Devices	W.S.P.	Water, Steam, Petroleum
T.L., TL	Truckload	UTP	Unshielded Twisted Pair	WT., Wt.	Weight
T.M.	Track Mounted	V	Volt	WWF	Welded Wire Fabric
Tot.	Total	VA	Volt Amperes	XFER	Transfer
T-O-L	Threadolet	VAT	Vinyl Asbestos Tile	XFMR	Transformer
tmpd	Tempered	V.C.T.	Vinyl Composition Tile	XHD	Extra Heavy Duty
TPO	Thermoplastic Polyolefin	VAV	Variable Air Volume	XHHW	Cross-Linked Polyethylene Wire
T.S.	Trigger Start	VC	Veneer Core	XLPE	Insulation
Tr.	Trade	VDC	Volts Direct Current	XLP	Cross-linked Polyethylene
Transf.	Transformer	Vent.	Ventilation	Xport	Transport
Trhv.	Truck Driver, Heavy	Vert.	Vertical	Y	Wye
Trlr	Trailer	V.F.	Vinyl Faced	yd	Yard
Trlt.	Truck Driver, Light	V.G.	Vertical Grain	yr	Year
TTY	Teletypewriter	VHF	Very High Frequency	Δ	Delta
TV	Television	VHO	Very High Output	%	Percent
T.W.	Thermoplastic Water Resistant	Vib.	Vibrating	~	Approximately
	Wire	VLF	Vertical Linear Foot	Ø	Phase; diameter
UCI	Uniform Construction Index	VOC	Volatile Organic Compound	@	At
UF	Underground Feeder	Vol.	Volume	#	Pound; Number
UGND	Underground Feeder	VRP	Vinyl Reinforced Polyester	<	Less Than
UHF	Ultra High Frequency	W	Wire; Watt; Wide; West	>	Greater Than
U.I.	United Inch			Z	Zone

For customer support on your Commercial Renovation Cost Data, call 877.742.4719.

861

Index

863

865

Index

867

Index

Index

Index

873

875

Index

878

For customer support on your Commercial Renovation Cost Data, call 877.742.4719.

882

For customer support on your Commercial Renovation Cost Data, call 877.742.4719.

Division Notes

	CREW	DAILY OUTPUT	LABOR-HOURS	UNIT	BARE COSTS				TOTAL INCL O&P
					MAT.	LABOR	EQUIP.	TOTAL	

Division Notes

	CREW	DAILY OUTPUT	LABOR-HOURS	UNIT	BARE COSTS				TOTAL INCL O&P
					MAT.	LABOR	EQUIP.	TOTAL	

Division Notes

	CREW	DAILY OUTPUT	LABOR-HOURS	UNIT	BARE COSTS				TOTAL INCL O&P
					MAT.	LABOR	EQUIP.	TOTAL	
	CREW	DAILY OUTPUT	LABOR-HOURS	UNIT	MAT.	LABOR	EQUIP.	TOTAL	TOTAL INCL O&P

Division Notes

	CREW	DAILY OUTPUT	LABOR-HOURS	UNIT	BARE COSTS				TOTAL INCL O&P
					MAT.	LABOR	EQUIP.	TOTAL	

Division Notes

	CREW	DAILY OUTPUT	LABOR-HOURS	UNIT	BARE COSTS				TOTAL INCL O&P
					MAT.	LABOR	EQUIP.	TOTAL	

RSMeans—a tradition of excellence in construction cost information and services since 1942

For more information visit the RSMeans website at www.RSMeans.com   Unit prices according to the latest MasterFormat

Cost Data Selection Guide

The following table provides definitive information on the content of each cost data publication. The number of lines of data provided in each unit price or assemblies division, as well as the number of crews, is listed for each data set. The presence of other elements such as reference tables, square foot models, equipment rental costs, historical cost indexes, and city cost indexes, is also indicated. You can use the table to help select the RSMeans data set that has the quantity and type of information you most need in your work.

Unit Cost Divisions	Building Construction	Mechanical	Electrical	Commercial Renovation	Square Foot	Site Work Landsc.	Green Building	Interior	Concrete Masonry	Open Shop	Heavy Construction	Light Commercial	Facilities Construction	Plumbing	Residential
1	562	383	400	504		498	205	302	445	561	496	245	1040	393	177
2	776	279	85	732		991	206	399	213	775	733	480	1220	286	273
3	1691	340	230	1084		1483	986	354	2037	1691	1693	482	1791	316	389
4	960	21	0	920		725	180	615	1158	928	615	533	1175	0	445
5	1893	158	155	1090		844	1799	1098	720	1893	1037	979	1909	204	746
6	2453	18	18	2111		110	589	1528	281	2449	123	2141	2125	22	2661
7	1594	215	128	1633		580	763	531	523	1591	26	1328	1695	227	1048
8	2111	81	44	2645		257	1136	1757	105	2096	0	2219	2890	0	1534
9	2004	86	45	1828		310	451	2088	391	1944	15	1668	2252	54	1437
10	1049	17	10	645		216	27	864	157	1049	29	539	1141	233	224
11	1096	206	166	551		129	54	933	28	1071	0	231	1116	169	110
12	530	0	2	285		205	133	1533	14	500	0	259	1554	23	217
13	744	149	158	253		367	122	252	78	720	244	108	759	115	103
14	273	36	0	221		0	0	257	0	273	0	12	293	16	6
21	91	0	16	37		0	0	250	0	91	0	73	535	540	220
22	1168	7548	160	1201		1572	1066	848	20	1157	1681	874	7450	9370	718
23	1178	6975	580	929		157	903	778	38	1171	110	879	5213	1922	469
26	1380	458	10150	1033		793	630	1144	55	1372	562	1310	10099	399	619
27	72	0	339	34		13	0	71	0	72	39	52	321	0	22
28	96	58	136	71		0	21	78	0	99	0	40	151	44	25
31	1510	733	610	806		3263	288	7	1217	1455	3277	604	1569	660	613
32	813	49	8	881		4409	353	405	291	784	1826	421	1687	133	468
33	529	1076	538	252		2173	41	0	236	522	2157	128	1697	1283	154
34	107	0	47	4		190	0	0	31	62	214	0	136	0	0
35	18	0	0	0		327	0	0	0	18	442	0	84	0	0
41	60	0	0	33		8	0	22	0	61	31	0	68	14	0
44	75	79	0	0		0	0	0	0	0	0	0	75	75	0
46	23	16	0	0		274	261	0	0	23	264	0	33	33	0
48	12	0	25	0		0	25	0	0	12	17	12	12	0	12
Totals	24868	18981	14050	19783		19894	10239	16114	8038	24440	15631	15617	50090	16531	12690

Assem Div	Building Construction	Mechanical	Electrical	Commercial Renovation	Square Foot	Site Work Landscape	Assemblies	Green Building	Interior	Concrete Masonry	Heavy Construction	Light Commercial	Facilities Construction	Plumbing	Asm Div	Residential
A		15	0	188	150	577	598	0	0	536	571	154	24	0	1	378
B		0	0	848	2506	0	5666	56	329	1975	368	2098	174	0	2	211
C		0	0	647	928	0	1309	0	1628	146	0	818	249	0	3	588
D		1067	941	712	1859	72	2538	330	825	0	0	1345	1105	1088	4	851
E		0	0	86	261	0	301	0	5	0	0	258	5	0	5	390
F		0	0	0	114	0	114	0	0	0	0	114	3	0	6	357
G		527	447	318	312	3364	792	0	0	534	1349	205	293	677	7	307
															8	760
															9	80
															10	0
															11	0
															12	0
Totals		1609	1388	2799	6130	4013	11318	386	2787	3191	2288	4992	1853	1765		3922

Reference Section	Building Construction Costs	Mechanical	Electrical	Commercial Renovation	Square Foot	Site Work Landscape	Assem.	Green Building	Interior	Concrete Masonry	Open Shop	Heavy Construction	Light Commercial	Facilities Construction	Plumbing	Resi.
Reference Tables	yes	yes	yes	yes	no	yes	yes	yes	yes	yes	yes	yes	yes	yes	yes	yes
Models					111			25					50			28
Crews	575	575	575	553		575		575	575	575	551	575	551	553	575	551
Equipment Rental Costs	yes	yes	yes	yes		yes		yes	yes	yes	yes	yes	yes	yes	yes	yes
Historical Cost Indexes	yes	yes	yes	yes	yes	yes	yes	yes	yes	yes	yes	yes	yes	yes	yes	no
City Cost Indexes	yes	yes	yes	yes	yes	yes	yes	yes	yes	yes	yes	yes	yes	yes	yes	yes

893

Online   Book   CD   eBook   Online add-on available for additional fee

For more information visit our website at www.RSMeans.com

(Left margin, vertical text): **Unit prices according to the latest MasterFormat**

(Left margin, vertical text): **Annual Cost Guides**

RSMeans Building Construction Cost Data 2016

Offers you unchallenged unit price reliability in an easy-to-use format. Whether used for verifying complete, finished estimates or for periodic checks, it supplies more cost facts better and faster than any comparable source. More than 24,700 unit prices have been updated for 2016. The City Cost Indexes and Location Factors cover more than 930 areas for indexing to any project location in North America. Order and get *RSMeans Quarterly Update Service* FREE.

RSMeans Green Building Cost Data 2016

Estimate, plan, and budget the costs of green building for both new commercial construction and renovation work with this sixth edition of *RSMeans Green Building Cost Data*. More than 10,000 unit costs for a wide array of green building products plus assemblies costs. Easily identified cross references to LEED and Green Globes building rating systems criteria.

RSMeans Mechanical Cost Data 2016

Total unit and systems price guidance for mechanical construction—materials, parts, fittings, and complete labor cost information. Includes prices for piping, heating, air conditioning, ventilation, and all related construction.

Plus new 2016 unit costs for:

- Thousands of installed HVAC/controls, sub-assemblies and assemblies
- "On-site" Location Factors for more than 930 cities and towns in the U.S. and Canada
- Crews, labor, and equipment

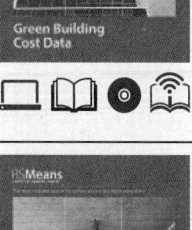

RSMeans Facilities Construction Cost Data 2016

For the maintenance and construction of commercial, industrial, municipal, and institutional properties. Costs are shown for new and remodeling construction and are broken down into materials, labor, equipment, and overhead and profit. Special emphasis is given to sections on mechanical, electrical, furnishings, site work, building maintenance, finish work, and demolition.

More than 49,800 unit costs, plus assemblies costs and a comprehensive Reference Section are included.

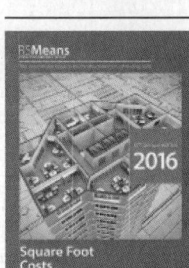

RSMeans Square Foot Costs 2016

Accurate and Easy-to-Use

- Updated price information based on nationwide figures from suppliers, estimators, labor experts, and contractors.
- Green building models
- Realistic graphics, offering true-to-life illustrations of building projects
- Extensive information on using square foot cost data, including sample estimates and alternate pricing methods

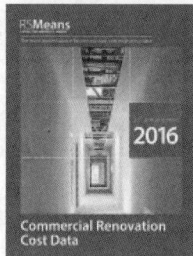

RSMeans Commercial Renovation Cost Data 2016

Commercial/Multi-family Residential

Use this valuable tool to estimate commercial and multi-family residential renovation and remodeling.

Includes: Updated costs for hundreds of unique methods, materials, and conditions that only come up in repair and remodeling, PLUS:

- Unit costs for more than 19,600 construction components
- Installed costs for more than 2,700 assemblies
- More than 930 "on-site" localization factors for the U.S. and Canada

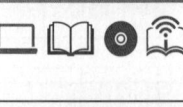

RSMeans Electrical Cost Data 2016

Pricing information for every part of electrical cost planning. More than 13,800 unit and systems costs with design tables; clear specifications and drawings; engineering guides; illustrated estimating procedures; complete labor-hour and materials costs for better scheduling and procurement; and the latest electrical products and construction methods.

- A variety of special electrical systems, including cathodic protection
- Costs for maintenance, demolition, HVAC/mechanical, specialties, equipment, and more

RSMeans Electrical Change Order Cost Data 2016

RSMeans Electrical Change Order Cost Data provides you with electrical unit prices exclusively for pricing change orders. Analyze and check your own change order estimates against those prepared when using RSMeans cost data. It also covers productivity analysis and change order cost justifications. With useful information for calculating the effects of change orders and dealing with their administration.

RSMeans Assemblies Cost Data 2016

RSMeans Assemblies Cost Data takes the guesswork out of preliminary or conceptual estimates. Now you don't have to try to calculate the assembled cost by working up individual component costs. We've done all the work for you.

Presents detailed illustrations, descriptions, specifications, and costs for every conceivable building assembly—over 350 types in all—arranged in the easy-to-use UNIFORMAT II system. Each illustrated "assembled" cost includes a complete grouping of materials and associated installation costs, including the installing contractor's overhead and profit.

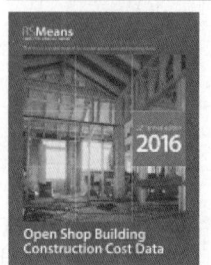

RSMeans Open Shop Building Construction Cost Data 2016

The latest costs for accurate budgeting and estimating of new commercial and residential construction, renovation work, change orders, and cost engineering.

RSMeans Open Shop "BCCD" will assist you to:

- Develop benchmark prices for change orders.
- Plug gaps in preliminary estimates and budgets.
- Estimate complex projects.
- Substantiate invoices on contracts.
- Price ADA-related renovations.

For more information visit our website at www.RSMeans.com

RSMeans Residential Cost Data 2016

Contains square foot costs for 28 basic home models with the look of today, plus hundreds of custom additions and modifications you can quote right off the page. Includes more than 3,900 costs for 89 residential systems. Complete with blank estimating forms, sample estimates, and step-by-step instructions.

Contains line items for cultured stone and brick, PVC trim, lumber, and TPO roofing.

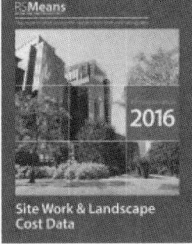

RSMeans Site Work & Landscape Cost Data 2016

Includes unit and assemblies costs for earthwork, sewerage, piped utilities, site improvements, drainage, paving, trees and shrubs, street openings/repairs, underground tanks, and more. Contains more than 60 types of assemblies costs for accurate conceptual estimates.

Includes:

- Estimating for infrastructure improvements
- Environmentally-oriented construction
- ADA-mandated handicapped access
- Hazardous waste line items

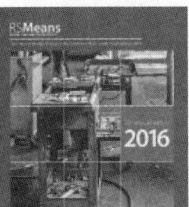

RSMeans Facilities Maintenance & Repair Cost Data 2016

RSMeans Facilities Maintenance & Repair Cost Data gives you a complete system to manage and plan your facility repair and maintenance costs and budget efficiently. Guidelines for auditing a facility and developing an annual maintenance plan. Budgeting is included, along with reference tables on cost and management, and information on frequency and productivity of maintenance operations.

The only nationally recognized source of maintenance and repair costs. Developed in cooperation with the Civil Engineering Research Laboratory (CERL) of the Army Corps of Engineers.

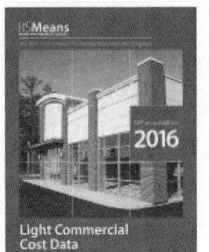

RSMeans Light Commercial Cost Data 2016

Specifically addresses the light commercial market, which is a specialized niche in the construction industry. Aids you, the owner/designer/contractor, in preparing all types of estimates—from budgets to detailed bids. Includes new advances in methods and materials.

The Assemblies section allows you to evaluate alternatives in the early stages of design/planning.

More than 15,500 unit costs ensure that you have the prices you need, when you need them.

RSMeans Concrete & Masonry Cost Data 2016

Provides you with cost facts for virtually all concrete/masonry estimating needs, from complicated formwork to various sizes and face finishes of brick and block—all in great detail. The comprehensive Unit Price Section contains more than 8,000 selected entries. Also contains an Assemblies [Cost] Section, and a detailed Reference Section that supplements the cost data.

RSMeans Labor Rates for the Construction Industry 2016

Complete information for estimating labor costs, making comparisons, and negotiating wage rates by trade for more than 300 U.S. and Canadian cities. With 46 construction trades in each city, and historical wage rates included for comparison. Each city chart lists the county and is alphabetically arranged with handy visual flip tabs for quick reference.

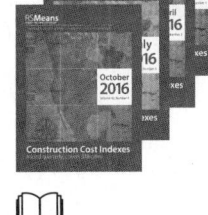

RSMeans Construction Cost Indexes 2016

What materials and labor costs will change unexpectedly this year? By how much?

- Breakdowns for 318 major cities
- National averages for 30 key cities
- Expanded five major city indexes
- Historical construction cost indexes

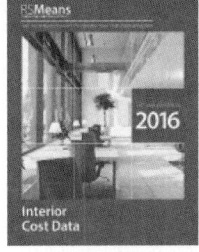

RSMeans Interior Cost Data 2016

Provides you with prices and guidance needed to make accurate interior work estimates. Contains costs on materials, equipment, hardware, custom installations, furnishings, and labor for new and remodel commercial and industrial interior construction, including updated information on office furnishings, as well as reference information.

RSMeans Heavy Construction Cost Data 2016

A comprehensive guide to heavy construction costs. Includes costs for highly specialized projects such as tunnels, dams, highways, airports, and waterways. Information on labor rates, equipment, and materials costs is included. Features unit price costs, systems costs, and numerous reference tables for costs and design.

RSMeans Plumbing Cost Data 2016

Comprehensive unit prices and assemblies for plumbing, irrigation systems, commercial and residential fire protection, point-of-use water heaters, and the latest approved materials. This publication and its companion, *RSMeans Mechanical Cost Data*, provide full-range cost estimating coverage for all the mechanical trades.

Contains updated costs for potable water, radiant heat systems, and high efficiency fixtures.

Annual Cost Guides

Unit prices according to the latest MasterFormat

895

RSMeans Online

Competitive Cost Estimates Made Easy

RSMeans Online is a web-based service that provides accurate and up-to-date cost information to help you build competitive estimates or budgets in less time.

Quick, intuitive, easy to use, and automatically updated, RSMeans Online gives you instant access to hundreds of thousands of material, labor, and equipment costs from RSMeans' comprehensive database, delivering the information you need to build budgets and competitive estimates every time.

With RSMeans Online, you can perform quick searches to locate specific costs and adjust costs to reflect prices in your geographic area. Tag and store your favorites for fast access to your frequently used line items and assemblies and clone estimates to save time. System notifications will alert you as updated data becomes available. RSMeans Online is automatically updated throughout the year.

RSMeans Online's visual, interactive estimating features help you create, manage, save and share estimates with ease! It provides increased flexibility with customizable advanced reports. Easily edit custom report templates and import your company logo onto your estimates.

	Standard	Advanced	Professional
Unit Prices	✓	✓	✓
Assemblies	X	✓	✓
Sq Ft Models	X	X	✓
Editable Sq Foot Models	X	X	✓
Editable Assembly Components	X	✓	✓
Custom Cost Data	X	✓	✓
User Defined Components	X	✓	✓
Advanced Reporting & Customization	X	✓	✓
Union Labor Type	✓	✓	✓
Open Shop*	Add on for additional fee	Add on for additional fee	✓
History (3-year lookback)**	Add on for additional fee	Add on for additional fee	Add on for additional fee
Full History (2007)	Add on for additional fee	Add on for additional fee	Add on for additional fee

*Open shop included in Complete Library & Bundles
** 3-year history lookback included with complete library

For more information visit our website at www.RSMeans.com

Estimate with Precision

Find everything you need to develop complete, accurate estimates.
- Verified costs for construction materials
- Equipment rental costs
- Crew sizing, labor hours and labor rates
- Localized costs for U.S. and Canada

Save Time & Increase Efficiency

Make cost estimating and calculating faster and easier than ever with secure, online estimating tools.
- Quickly locate costs in the searchable database
- Create estimates in minutes with RSMeans cost lines
- Tag and store favorites for fast access to frequently used items

Improve Planning & Decision-Making

Back your estimates with complete, accurate and up-to-date cost data for informed business decisions.
- Verify construction costs from third parties
- Check validity of subcontractor proposals
- Evaluate material and assembly alternatives

Increase Profits

Use RSMeans Online to estimate projects quickly and accurately, so you can gain an edge over your competition.
- Create accurate and competitive bids
- Minimize the risk of cost overruns
- Reduce variability
- Gain control over costs

30 Day Free Trial

Register for a free trial at www.RSMeansOnline.com
- Access to unit prices, building assemblies and facilities repair and remodeling items covering every category of construction
- Powerful tools to customize, save and share cost lists, estimates and reports with ease
- Access to square foot models for quick conceptual estimates
- Allows up to 10 users to evaluate the full range of collaborative functionality used items.

2016 RSMeans Seminar Schedule

Note: call for exact dates and details.

Location	Dates	Location	Dates
Seattle, WA	January and August	Bethesda, MD	June
Dallas/Ft. Worth, TX	January	El Segundo, CA	August
Austin, TX	February	Jacksonville, FL	September
Anchorage, AK	March and September	Dallas, TX	September
Las Vegas, NV	March	Philadelphia, PA	October
New Orleans, LA	March	Houston, TX	October
Washington, DC	April and September	Salt Lake City, UT	November
Phoenix, AZ	April	Baltimore, MD	November
Kansas City, MO	April	Orlando, FL	November
Toronto	May	San Diego, CA	December
Denver, CO	May	San Antonio, TX	December
San Francisco, CA	June	Raleigh, NC	December

📞 877-620-6245

Beginning early 2016, RSMeans will provide a suite of online, self-paced training offerings. Check our website www.RSMeans.com for more information.

Professional Development

eLearning Training Sessions

Learn how to use *RSMeans Online®* or the *RSMeans CostWorks®* CD from the convenience of your home or office. Our eLearning training sessions let you join a training conference call and share the instructors' desktops, so you can view the presentation and step-by-step instructions on your own computer screen. The live webinars are held from 9 a.m. to 4 p.m. or 11 a.m. to 6 p.m. eastern standard time, with a one-hour break for lunch.

For these sessions, you must have a computer with high speed Internet access and a compatible Web browser. Learn more at www.RSMeansOnline.com or call for a schedule: 781-422-5115. Webinars are generally held on selected Wednesdays each month.

RSMeans Online	RSMeans CostWorks CD
$299 per person	$299 per person

Seminar Schedule and Professional Development

Visit RSMeans.com/Online for more details on data titles in online format.

For more information visit our website at www.RSMeans.com

Professional Development

RSMeans Online Training

Construction estimating is vital to the decision-making process at each state of every project. RSMeansOnline works the way you do. It's systematic, flexible and intuitive. In this one day class you will see how you can estimate any phase of any project faster and better.

Some of what you'll learn:
- Customizing RSMeansOnline
- Making the most of RSMeans "Circle Reference" numbers
- How to integrate your cost data
- Generate reports, exporting estimates to MS Excel, sharing, collaborating and more

Also available: RSMeans Online training webinar

Maintenance & Repair Estimating for Facilities

This *two-day* course teaches attendees how to plan, budget, and estimate the cost of ongoing and preventive maintenance and repair for existing buildings and grounds.

Some of what you'll learn:
- The most financially favorable maintenance, repair, and replacement scheduling and estimating
- Auditing and value engineering facilities
- Preventive planning and facilities upgrading
- Determining both in-house and contract-out service costs
- Annual, asset-protecting M&R plan

Who should attend: facility managers, maintenance supervisors, buildings and grounds superintendents, plant managers, planners, estimators, and others involved in facilities planning and budgeting.

Facilities Construction Estimating

In this *two-day* course, professionals working in facilities management can get help with their daily challenges to establish budgets for all phases of a project.

Some of what you'll learn:
- Determining the full scope of a project
- Identifying the scope of risks and opportunities
- Creative solutions to estimating issues
- Organizing estimates for presentation and discussion
- Special techniques for repair/remodel and maintenance projects
- Negotiating project change orders

Who should attend: facility managers, engineers, contractors, facility tradespeople, planners, and project managers.

Practical Project Management for Construction Professionals

In this *two-day* course, acquire the essential knowledge and develop the skills to effectively and efficiently execute the day-to-day responsibilities of the construction project manager.

Some of what you'll learn:
- General conditions of the construction contract
- Contract modifications: change orders and construction change directives
- Negotiations with subcontractors and vendors
- Effective writing: notification and communications
- Dispute resolution: claims and liens

Who should attend: architects, engineers, owners' representatives, and project managers.

Construction Cost Estimating: Concepts and Practice

This introductory course to improve estimating skills and effectiveness starts with the details of interpreting bid documents and ends with the summary of the estimate and bid submission.

Some of what you'll learn:
- Using the plans and specifications for creating estimates
- The takeoff process—deriving all tasks with correct quantities
- Developing pricing using various sources; how subcontractor pricing fits in
- Summarizing the estimate to arrive at the final number
- Formulas for area and cubic measure, adding waste and adjusting productivity to specific projects
- Evaluating subcontractors' proposals and prices
- Adding insurance and bonds
- Understanding how labor costs are calculated
- Submitting bids and proposals

Who should attend: project managers, architects, engineers, owners' representatives, contractors, and anyone who's responsible for budgeting or estimating construction projects.

Mechanical & Electrical Estimating

This *two-day* course teaches attendees how to prepare more accurate and complete mechanical/electrical estimates, avoid the pitfalls of omission and double-counting, and understand the composition and rationale within the RSMeans mechanical/electrical database.

Some of what you'll learn:
- The unique way mechanical and electrical systems are interrelated
- M&E estimates—conceptual, planning, budgeting, and bidding stages
- Order of magnitude, square foot, assemblies, and unit price estimating
- Comparative cost analysis of equipment and design alternatives

Who should attend: architects, engineers, facilities managers, mechanical and electrical contractors, and others who need a highly reliable method for developing, understanding, and evaluating mechanical and electrical contracts.

Unit Price Estimating

This interactive *two-day* seminar teaches attendees how to interpret project information and process it into final, detailed estimates with the greatest accuracy level.

The most important credential an estimator can take to the job is the ability to visualize construction and estimate accurately.

Some of what you'll learn:
- Interpreting the design in terms of cost
- The most detailed, time-tested methodology for accurate pricing
- Key cost drivers—material, labor, equipment, staging, and subcontracts
- Understanding direct and indirect costs for accurate job cost accounting and change order management

Who should attend: corporate and government estimators and purchasers, architects, engineers, and others who need to produce accurate project estimates.

RSMeans CostWorks CD Training

This *one-day* course helps users become more familiar with the functionality of the *RSMeans CostWorks* program. Each menu, icon, screen, and function found in the program is explained in depth. Time is devoted to hands-on estimating exercises.

Some of what you'll learn:
- Searching the database using all navigation methods
- Exporting RSMeans data to your preferred spreadsheet format
- Viewing crews, assembly components, and much more
- Automatically regionalizing the database

This training session requires you to bring a laptop computer to class.

When you register for this course you will receive an outline for your laptop requirements.

Also offering web training for the RSMeans CostWorks CD!

Facilities Estimating Using the RSMeans CostWorks CD

Combines hands-on skill building with best estimating practices and real-life problems. Brings you up-to-date with key concepts and provides tips, pointers, and guidelines to save time and avoid cost oversights and errors.

Some of what you'll learn:
- Estimating process concepts
- Customizing and adapting RSMeans cost data
- Establishing scope of work to account for all known variables
- Budget estimating: when, why, and how
- Site visits: what to look for—what you can't afford to overlook
- How to estimate repair and remodeling variables

This training session requires you to bring a laptop computer to class.

Who should attend: facility managers, architects, engineers, contractors, facility tradespeople, planners, project managers and anyone involved with JOC, SABRE, or IDIQ.

Conceptual Estimating Using the RSMeans CostWorks CD

This *two-day* class uses the leading industry data and a powerful software package to develop highly accurate conceptual estimates for your construction projects. All attendees must bring a laptop computer loaded with the current year *Square Foot Costs* and the *Assemblies Cost Data* CostWorks titles.

Some of what you'll learn:
- Introduction to conceptual estimating
- Types of conceptual estimates
- Helpful hints
- Order of magnitude estimating
- Square foot estimating
- Assemblies estimating

Who should attend: architects, engineers, contractors, construction estimators, owners' representatives, and anyone looking for an electronic method for performing square foot estimating.

Assessing Scope of Work for Facility Construction Estimating

This *two-day* practical training program addresses the vital importance of understanding the scope of projects in order to produce accurate cost estimates for facility repair and remodeling.

Some of what you'll learn:
- Discussions of site visits, plans/specs, record drawings of facilities, and site-specific lists
- Review of CSI divisions, including means, methods, materials, and the challenges of scoping each topic
- Exercises in scope identification and scope writing for accurate estimating of projects
- Hands-on exercises that require scope, take-off, and pricing

Who should attend: corporate and government estimators, planners, facility managers, and others who need to produce accurate project estimates.

Unit Price Estimating Using the RSMeans CostWorks CD

Step-by-step instructions and practice problems to identify and track key cost drivers—material, labor, equipment, staging, and subcontractors—for each specific task. Learn the most detailed, time-tested methodology for accurately "pricing" these variables, their impact on each other and on total cost.

Some of what you'll learn:
- Unit price cost estimating
- Order of magnitude, square foot, and assemblies estimating
- Quantity takeoff
- Direct and indirect construction costs
- Development of contractors' bill rates
- How to use *RSMeans Building Construction Cost Data*

This training session requires you to bring a laptop computer to class.

Who should attend: architects, engineers, corporate and government estimators, facility managers, and government procurement staff.

Professional Development

Registration Information

Register early and save up to $100!
Register 30 days before the start date of a seminar and save $100 off your total fee. Note: This discount can be applied only once per order. It cannot be applied to team discount registrations or any other special offer.

How to register
Register by phone today! The RSMeans toll-free number for making reservations is 781-422-5115.

Two-day seminar registration fee - $935. One-day RSMeans CostWorks® or RSMeans Online training registration fee - $375.
To register by mail, complete the registration form and return, with your full fee, to: RSMeans Seminars, 1099 Hingham Street, Suite 201, Rockland, MA 02370.

One-day Construction Cost Estimating - $575.

Government pricing
All federal government employees save off the regular seminar price. Other promotional discounts cannot be combined with the government discount.

Team discount program
For over five attendee registrations. Call for pricing: 781-422-5115

Multiple course discounts
When signing up for two or more courses, call for pricing.

Refund policy
Cancellations will be accepted up to ten business days prior to the seminar start. There are no refunds for cancellations received later than ten working days prior to the first day of the seminar. A $150 processing fee will be applied for all cancellations. Written notice of the cancellation is required. Substitutions can be made at any time before the session starts. No-shows are subject to the full seminar fee.

AACE approved courses
Many seminars described and offered here have been approved for 14 hours (1.4 recertification credits) of credit by the AACE International Certification Board toward meeting the continuing education requirements for recertification as a Certified Cost Engineer/Certified Cost Consultant.

AIA Continuing Education
We are registered with the AIA Continuing Education System (AIA/CES) and are committed to developing quality learning activities in accordance with the CES criteria. Many seminars meet the AIA/CES criteria for Quality Level 2. AIA members may receive 14 learning units (LUs) for each two-day RSMeans course.

Daily course schedule
The first day of each seminar session begins at 8:30 a.m. and ends at 4:30 p.m. The second day begins at 8:00 a.m. and ends at 4:00 p.m. Participants are urged to bring a hand-held calculator since many actual problems will be worked out in each session.

Continental breakfast
Your registration includes the cost of a continental breakfast and a morning and afternoon refreshment break. These informal segments allow you to discuss topics of mutual interest with other seminar attendees. (You are free to make your own lunch and dinner arrangements.)

Hotel/transportation arrangements
RSMeans arranges to hold a block of rooms at most host hotels. To take advantage of special group rates when making your reservation, be sure to mention that you are attending the RSMeans seminar. You are, of course, free to stay at the lodging place of your choice. (Hotel reservations and transportation arrangements should be made directly by seminar attendees.)

Important
Class sizes are limited, so please register as soon as possible.

Note: Pricing subject to change.

Registration Form

ADDS-1000

Call 781-422-5115 to register or fax this form to 800-632-6732. Visit our website: www.RSMeans.com

Please register the following people for the RSMeans construction seminars as shown here. We understand that we must make our own hotel reservations if overnight stays are necessary.

❑ Full payment of $_____ enclosed.

❑ Bill me.

Please print name of registrant(s).

(To appear on certificate of completion)

_____
_____
_____
_____

P.O. #:_____
GOVERNMENT AGENCIES MUST SUPPLY PURCHASE ORDER NUMBER OR TRAINING FORM.

Firm name_____

Address_____

City/State/Zip_____

Telephone no._____ Fax no._____

E-mail address_____

Charge registration(s) to: ❑ MasterCard ❑ VISA ❑ American Express

Account no._____ Exp. date_____

Cardholder's signature_____

Seminar name_____

Seminar City_____

Please mail check to: 1099 Hingham Street, Suite 201, Rockland, MA 02370 USA

Professional Development